2023 IEEE 36th International Conference on Micro Electro Mechanical Systems (MEMS 2023)

Munich, Germany
15-19 January 2023

Pages 1-601

IEEE Catalog Number: CFP23MEM-POD
ISBN: 978-1-6654-9309-3

**Copyright © 2023 by the Institute of Electrical and Electronics Engineers, Inc.
All Rights Reserved**

Copyright and Reprint Permissions: Abstracting is permitted with credit to the source. Libraries are permitted to photocopy beyond the limit of U.S. copyright law for private use of patrons those articles in this volume that carry a code at the bottom of the first page, provided the per-copy fee indicated in the code is paid through Copyright Clearance Center, 222 Rosewood Drive, Danvers, MA 01923.

For other copying, reprint or republication permission, write to IEEE Copyrights Manager, IEEE Service Center, 445 Hoes Lane, Piscataway, NJ 08854. All rights reserved.

*** This is a print representation of what appears in the IEEE Digital Library. Some format issues inherent in the e-media version may also appear in this print version.

IEEE Catalog Number: CFP23MEM-POD
ISBN (Print-On-Demand):   978-1-6654-9309-3
ISBN (Online):   978-1-6654-9308-6
ISSN: 1084-6999

Additional Copies of This Publication Are Available From:

Curran Associates, Inc
57 Morehouse Lane
Red Hook, NY 12571 USA
Phone: (845) 758-0400
Fax: (845) 758-2633
E-mail: curran@proceedings.com
Web: www.proceedings.com

2023 IEEE 36th International Conference on Micro Electro Mechanical Systems (MEMS 2023)

Munich, Germany
15-19 January 2023

Pages 1-601

IEEE Catalog Number: CFP23MEM-POD
ISBN: 978-1-6654-9309-3

TABLE OF CONTENTS

Monday, 16 January
All times are Central European Time (CET).

Welcome Address

08:00   **MEMS 2023 Conference Chairs**
Núria Barniol, *Universitat Autonoma de Barcelona, SPAIN*
Franz Lärmer, *Robert Bosch GmbH, GERMANY*

• IEEE Fellows Recognition in the Field of MEMS/NEMS
• IEEE Electron Devices Society Robert Bosch
Micro and Nano Electro Mechanical Systems Award

08:35   **IEEE Electron Devices Society Robert Bosch**
Micro and Nano Electro Mechanical Systems Award Recipient
John H. (Hal) Jerman will accept on behalf of the Gas Chromatograph on a Chip Project.

Plenary Presentation I

08:50   **FROM ETCH TO EDGE AI:**
OPENING NEW HORIZONS WITH SMART SENSOR TECHNOLOGIES 1
Stefan Finkbeiner
Bosch Sensortec GmbH, GERMANY

Session I - Novel MEMS/NEMS Devices for Computing/Imaging

09:35   **SUB-300 MILLIVOLT OPERATION IN NONVOLATILE 300 NM X 100 NM**
PHASE CHANGE NANOELECTROMECHANICAL SWITCH 2
Mohammad Ayaz Masud and Gianluca Piazza
Carnegie Mellon University, USA

09:50   **A FAST AND ENERGY-EFFICIENT NANOELECTROMECHANICAL**
NON-VOLATILE MEMORY FOR IN-MEMORY COMPUTING 5
Yong-Bok Lee[1], Min-Ho Gang[2], Pan-Kyu Choi[1], Su-Hyun Kim[1],
Tae-Soo Kim[1], So-Young Lee[1] and Jun-Bo Yoon[1]
[1]Korea Advanced Institute of Science and Technology (KAIST), KOREA and
[2]National NanoFab Center (NNFC), KOREA

10:05   **TOWARDS ULTRA-HIGH SPATIAL RESOLUTION SENSING OF GHZ**
ULTRASOUND USING STRAIN MODULATION OF FIELD EFFECT TRANSISTORS 9
Rohan Sanghvi[1], Justin Kuo[2], Adarsh Ravi[1], and Amit Lal[1]
[1]Cornell University, USA and [2]Geegah Inc., USA

10:20   **A TACTILE SENSOR ARRAY WITH A MONOLITHICALLY**
INTEGRATED NEURAL NETWORK FOR EDGE COMPUTATION 13
Tengteng Lei, Yushen Hu, and Man Wong
Hong Kong University of Science and Technology, HONG KONG

10:35   **Break & Exhibit Inspection**

Session II - BioMEMS I

11:05   **EVALUATION OF LOCAL AND INTERNAL ELASTICITY OF HYDROGEL MATERIALS BY USING LIGHT-DRIVEN GEL ACTUATOR** 17
Hibiki Nakajima[1], Yuha Koike[1], Yoshiyuki Yokoyama[2], Masaya Hagiwara[3], and Takeshi Hayakawa[1]
[1]Chuo University, JAPAN, [2]Toyama Industrial Technology Research and Development Center, JAPAN, and [3]RIKEN, JAPAN

11:20   **3D PRINTED MINIATURIZED SOFT MICROSWIMMER FOR MULTIMODAL 3D AIR-LIQUID NAVIGATION AND MANIPULATION** 21
Dominique Decanini[1], Abdelmounaim Harouri[1], Ayako Mizushima[2], Beomjoon Kim[2], Yoshio Mita[2], and Gilgueng Hwang[1,2]
[1]Paris-Saclay University, FRANCE and [2]University of Tokyo, JAPAN

11:35   **SELF-DRIVEN CAPILLARIC VISCOMETER FOR DIRECT OR CASCADED BAR GRAPH READ-OUT OF RELATIVE SAMPLE VISCOSITY** 25
Daniel Mak[1], R. Claude Meffan[1,2], Julian Menges[1], Fabian Dolamore[1], Conan Fee[1], Renwick C.J. Dobson[1], and Volker Nock[1]
[1]University of Canterbury, NEW ZEALAND and [2]Kyoto University, JAPAN

11:50   **A FLEXIBLE BIOSENSING PLATFORM FOR HIGH-THROUGHPUT MEASUREMENT OF CARDIOMYOCYTE CONTRACTILITY** 29
Wenkun Dou[1], Jason Maynes[2], and Yu Sun[1]
[1]University of Toronto, CANADA and [2]Hospital for Sick Children, CANADA

12:05   **FLEXIBLE BI-DIRECTIONAL BRAIN COMPUTER INTERFACE FOR CONTROLLING TURNING BEHAVIOR OF MICE** 33
Yifei Ye[1], Ye Tian[1,2], Han Wang[1], Qian Cheng[1], Kuikui Zhang[1], Xueying Wang[1,2], Cunkai Zhou[1], Chengjian Xu[1], Xiaoling Wei[1,2], Zhitao Zhou[1,2], Tiger H. Tao[1,2,3,4,5,6], and Liuyang Sun[1,2]
[1]Chinese Academy of Sciences, CHINA, [2]University of Chinese Academy of Sciences, CHINA, [3]ShanghaiTech University, CHINA, [4]Neuroxess Co., Ltd. (Jiangxi), CHINA, [5]Guangdong Institute of Intelligence Science and Technology, CHINA, and [6]Tianqiao and Chrissy Chen Institute for Translational Research, CHINA

12:20   **Lunch & Exhibit Inspection**

Session III - MEMS Inertial Sensors and Power MEMS

13:45   **HIGH SENSITIVITY MEMS Z-AXIS ACCELEROMETER WITH IN-PLANE DIFFERENTIAL READOUT** 37
Valentina Zega[1], Gabriele Gattere[2], Manuel Riani[2], Francesco Rizzini[2], and Attilio Frangi[1]
[1]Politecnico di Milano, ITALY and [2]STMicroelectronics, ITALY

14:00   **TWO-AXIS ELECTROMAGNETIC SCANNER INTEGRATED WITH AN ELECTROSTATIC XY-STAGE POSITIONER** 41
Yuki Okamoto, Hironao Okada, and Masaaki Ichiki
National Institute of Advanced Industrial Science and Technology (AIST), JAPAN

14:15   **MEMS SHOCK ABSORBERS INTEGRATED WITH AL_2O_3-REINFORCED, MECHANICALLY RESILIENT NANOTUBE ARRAYS** 45
Hojoon Lee[1], Eunhwan Jo[1], Jae-Ik Lee[2], and Jongbaeg Kim[1]
[1]Yonsei University, KOREA and [2]Harvard Medical School, USA

14:40 **HIGH-INDUCTANCE-DENSITY MEMS 3D-SOLENOID TRANSFORMERS WITH INSERTED THIN-FILM FERRITE MAGNETIC CORE FOR ON-CHIP INTEGRATED DC-DC POWER CONVERSIONS** 49
Changnan Chen[1,2], Pichao Pan[1,2], Jiebin Gu[1,2], and Xinxin Li[1,2]
[1]Chinese Academy of Sciences, CHINA and [2]University of Chinese Academy of Sciences, CHINA

Poster/Oral Session I

14:45 **Poster/Oral Session I**
Poster presentations are listed by topic category with their assigned number starting on Page 13.

16:15 **Break & Exhibit Inspection**

MEMS Community Announcement

16:45 Clark T.-C. Nguyen, *University of California, Berkeley, USA*

Session IV - BioMEMS II

16:50 **MICRON-SIZED PARYLENE-IN-OIL WATER PROTECTION LAYER** 53
Kuang-Ming Shang[1], Haixu Shen[1], Ningxuan Dai[1], David Kong[1,2], Tzung Hsiai[3], and Yu-Chong Tai[1]
*[1]California Institute of Technology, USA, [2]Harvard University, USA, and
[3]University of California, Los Angeles, USA*

17:05 **A PIPETTE TIP INTEGRATED WITH A CAPACITIVE MICROSENSOR FABRICATED BY COMBINED 3D PRINTING AND MEMS PROCESS FOR CELL DETECTION AND TRANSPORTATION** 57
Satoshi Amaya, Hirotaka Sugiura, Bilal Turan, Shingo Kaneko, and Fumihito Arai
University of Tokyo, JAPAN

17:20 **FOLDABLE POLYMER STENT INTEGRATED WITH WIRELESS PRESSURE SENSOR FOR BLOOD PRESSURE MONITORING** 61
Nomin-Erdene Oyunbaatar and Dong-Weon Lee
Chonnam National University, KOREA

17:35 **A DYNAMIC MICROARRAY DEVICE FOR SELECTIVE PAIRING AND ELECTROFUSION OF LIPOSOMES** 65
Sho Takamori[1], Hisatoshi Mimura[1], Toshihisa Osaki[1], and Shoji Takeuchi[1,2]
[1]Kanagawa Institute of Industrial Science and Technology, JAPAN and [2]University of Tokyo, JAPAN

17:50 **REAL-TIME FUNCTIONAL BRAIN MAPPING BASED ON HIGH-CHANNEL-COUNT, ULTRA-CONFORMAL NEURAL INTERFACE** 67
*Xiner Wang[1,2], Zhaohan Chen[3], Jizhi Liang[1,2], Xiaoling Wei[1,2], Liuyang Sun[1,2],
Meng Li[1,2], Zhitao Zhou[1,2], and Tiger H. Tao[1,2,4,5,6]*
*[1]Chinese Academy of Sciences, CHINA, [2]University of Chinese Academy of Science, CHINA,
[3]Shanghai Normal University, CHINA, [4]Neuroxess Co., Ltd. (Jiangxi), CHINA,
[5]Guangdong Institute of Intelligence Science and Technology, CHINA, and
[6]Tianqiao and Chrissy Chen Institute for Translational Research, CHINA*

18:05 **Adjourn for the day**

Tuesday, 17 January

All times are Central European Time (CET).

Plenary Presentation II

08:30 ACOUSTOFLUIDICS: MERGING ACOUSTICS AND
FLUID MECHANICS FOR BIOMEDICAL APPLICATIONS 71
Tony Jun Huang
Duke University, USA

Session V - New Materials, Fabrication, and Packaging

09:15 SILICON CARBIDE REINFORCED VERTICALLY ALIGNED CARBON
NANOTUBE COMPOSITE FOR HARSH ENVIRONMENT MEMS 72
Jiarui Mo, Shreyas Shankar, Guoqi Zhang, and Sten Vollebregt
Delft University of Techonology, NETHERLANDS

09:30 A RELIABLE RELEASE METHOD FOR A BACK-END-OF-LINE NEMS SWITCH OF
A MONOLITHIC THREE-DIMENSIONAL INTEGRATED CMOS-NEMS CIRCUIT 76
Tae-Soo Kim, Yong-Bok Lee, So-Young Lee, Seung-Jun Lee, and Jun-Bo Yoon
Korea Advanced Institute of Science and Technology (KAIST), KOREA

09:45 INCREASE OF EXPANSION RATE AND DIRECTION CONTROL
OF MICROGEL ACTUATORS FOR SINGLE CELL MANIPULATIONS 80
Kyoka Nakano[1], Hiroki Wada[1], Yoshiyuki Yokoyama[2], and Takeshi Hayakawa[1]
[1]Chuo University, JAPAN and [2]Toyama Industrial Technology Research and Development Center, JAPAN

10:00 GENERALIZED-ACCUMULATED-TEMPERATURE PARAMETER FOR
CHARACTERISTIC PREDICTION OF METAL-BASED MEMS CANTILEVER 84
Yulong Zhang[1], Jianwen Sun[1], Huiliang Liu[2], and Zewen Liu[1]
[1]Tsinghua University, CHINA and [2]China Academy of Space Technology, CHINA

10:15 **Break and Exhibit Inspection**

Session VI - Micro- and Nanofluidics and Medical Applications

10:45 MEMS-BASED WATER COLLECTION CONDENSATION PARTICLE COUNTER (WCCPC)
OPTIMIZED FOR MULTI-POINT MONITORING OF AIRBORNE NANOPARTICLES 88
Seong-Jae Yoo and Yong-Jun Kim
Yonsei University, KOREA

11:00 RECONSTITUTING FUNDAMENTALS OF BACTERIA
MEDIATED CANCER
THERAPY ON A CHIP 91
Wonjun Lee[1], Jiin Park[2], Dongil Kang[3], and Seungbeum Suh[4]
[1]Seoul National University, KOREA, [2]Ewha Womans University, KOREA,
[3]Hanyang University, KOREA, and [4]Korea Institute of Science and Technology (KIST), KOREA

11:15 3D SPATIAL FOCAL CONTROL BY ARRAYED OPTOFLUIDIC PRISMS 95
Cheng-Hsun Lee, Yeonwoo Lee, and Sung-Yong Park
San Diego State University, USA

11:30 HIGH-SPEED AND PINPOINT LIQUID EXCHANGE ON MICROFLUIDIC CHIP
USING 3D PRINTED DOUBLE-BARRELED MICROPROBE WITH DUAL PUMPS 99
Xu Du[1], Shingo Kaneko[2], Hisataka Maruyama[1], Hirotaka Sugiura[2], and Fumihito Arai[1,2]
[1]Nagoya University, JAPAN and [2]University of Tokyo, JAPAN

11:45 DESIGN OF A DNA SYNTHESIS CHIP FOR DATA STORAGE WITH ULTRA-HIGH THROUGHPUT AND DENSITY FEATURING LARGE-SCALE INTEGRATED CIRCUITS AND MICROFLUIDIC CONFINEMENT 103
Ning Wang[1,2,3], Shijia Yang[1,3], Dayin Wang[1,2,3], Zhen Cao[4], Yuan Luo[1,3], and Jianlong Zhao[1,3]
[1]*Chinese Academy of Sciences, CHINA,* [2]*ShanghaiTech University, CHINA,*
[3]*University of Chinese Academy of Sciences, CHINA, and* [4]*Zhejiang University, CHINA*

MEMS 2024 Announcement

12:00 **MEMS 2024 Conference Chairs**
Wen Li, *Michigan State University, USA*
Dana Weinstein, *Purdue University, USA*

12:15 **Lunch & Exhibit Inspection**

Session VII - MEMS Fluidic Sensors

13:15 A REAL-TIME WIRELESS CALORIMETRIC FLOW SENSOR SYSTEM WITH A WIDE LINEAR RANGE FOR LOW-COST RESPIRATORY MONITORING 107
Lifeng Huang[1], Izhar[2,4], Xiaoyong Zhou[3], Mingdong Fang[3], Siwei Huang[1], Yi-Kuen Lee[2], Xiaofang Pan[1], and Wei Xu[1]
[1]*Shenzhen University, CHINA,* [2]*Hong Kong University of Science and Technology, CHINA,*
[3]*Mindray Medical International Limited, CHINA, and* [4]*University of Pennsylvania, USA*

13:30 ADVANCED THERMOPHYSICAL PROPERTIES MEASUREMENTS USING HEATER-INTEGRATED FLUIDIC RESONATORS 111
Juhee Ko, Bong Jae Lee, and Jungchul Lee
Korea Advanced Institute of Science and Technology (KAIST), KOREA

13:45 A MINIATURIZED TRANSIT-TIME ULTRASONIC FLOWMETER USING PMUTS FOR LOW-FLOW MEASUREMENT IN SMALL-DIAMETER CHANNELS 115
Yunfei Gao[1,2], Zhipeng Wu[2], Minkan Chen[2], and Liang Lou[1,2]
[1]*Shanghai University, CHINA and* [2]*Shanghai Industrial μ Technology Research Institute, CHINA*

14:00 MEMS DIFFERENTIAL THERMOPILES FOR HIGHLY-SENSITIVE HYDROGEN GAS DETECTION 119
Haozhi Zhang[1,2], Hao Jia[1,2], Ming Li[1,2], Pengcheng Xu[1,2], and Xinxin Li[1,2]
[1]*Chinese Academy of Sciences, CHINA and* [2]*University of Chinese Academy of Sciences, CHINA*

Poster/Oral Session II

14:15 **Poster/Oral Session II**
Poster presentations are listed by topic category with their assigned number starting on Page 13.

15:45 **Break & Exhibit Inspection**

Session VIII - Sonics & Ultrasonics MEMS

16:15 DOMAIN/BOUNDARY VARIATION IN CANTILEVER ARRAY FOR BANDWIDTH ENHANCEMENT OF PZT MEMS MICROSPEAKER 123
Shu-Wei Chang[1], Ting-Chou Wei[1], Sung-Cheng Lo[2], and Weileun Fang[1]
[1]*National Tsing Hua University, TAIWAN and* [2]*Transducer Star Technology Inc., TAIWAN*

16:30   **ON THE DESIGN OF PIEZOELECTRIC MEMS MICROSPEAKER WITH HIGH FIDELITY AND WIDE BANDWIDTH** 127
Ting-Chou Wei, Zih-Song Hu, Shu-Wei Chang, and Weileun Fang
National Tsing Hua University, TAIWAN

16:45   **HIGH-PERFORMANCE WAFER-SCALE TRANSFER-FREE GRAPHENE MICROPHONES** 131
Roberto Pezone, Gabriele Baglioni, Pasqualina M. Sarro, Peter G. Steeneken, and Sten Vollebregt
Delft University of Technology, NETHERLANDS

17:00   **HIGH-SPL AND LOW-DRIVING-VOLTAGE PMUTS BY SPUTTERED POTASSIUM SODIUM NIOBATE** 135
Fan Xia[1,2], Yande Peng[1,2], Sedat Pala[1,2], Ryuichi Arakawa[1,3], Wei Yue[1,2], Pei-Chi Tsao[2], Chun-Ming Chen[2], Hanxiao Liu[1,2], Megan Teng[2], Jong Ha Park[1,2], and Liwei Lin[1,2]
[1]Berkeley Sensor and Actuator Center, USA, [2]University of California, Berkeley, USA, and [3]NGK Spark Plug Co., JAPAN

17:15   **EPITAXIAL $P_B(Z_R,T_I)O_3$-BASED PIEZOELECTRIC MICROMACHINED ULTRASONIC TRANSDUCER FABRICATED ON SILICON-ON-NOTHING (SON) STRUCTURE** 139
Takuma Sekiguchi[1], Shinya Yoshida[2], Yoshiaki Kanamori[1], and Shuji Tanaka[1]
[1]Tohoku University, JAPAN and [2]Shibaura Institute of Technology, JAPAN

17:30   **Adjourn for the day**

19:00
- 22:00   **Banquet at the Löwenbräu Keller**

Wednesday, 18 January

All times are Central European Time (CET).

Plenary Presentation III

08:30   **LEVERAGING SEMICONDUCTOR ECOSYSTEMS TO MEMS** 143
Weileun Fang, Sheng-Shian Li, and Ming-Huang Li
National Tsing Hua University, TAIWAN

Session IX – Optomechanics & Photonics Integration

09:15   **PROGRAMMABLE SILICON NITRIDE PHOTONIC INTEGRATED CIRCUITS** 149
Hao Tian[1], Alaina G. Attanasio[1], Anat Siddharth[2], Andrey Voloshin[2], Viacheslav Snigirev[2],
Grigory Lihachev[2], Andrea Bancora[2], Vladimir Shadymov[2], Rui N. Wang[2], Johann Riemensberger[2],
Tobias J. Kippenberg[2], and Sunil A. Bhave[1]
[1]Purdue University, USA and [2]Swiss Federal Institute of Technology Lausanne (EPFL), SWITZERLAND

09:30   **MULTIFREQUENCY NANOMECHANICAL MASS SPECTROMETER PROTOTYPE FOR
MEASURING VIRAL PARTICLES USING OPTOMECHANICAL DISK RESONATORS** 153
Oscar Malvar[1], Eduardo Gil-Santos[1], Jose J. Ruz[1], Elena Sentre-Arribas[1], Adrián Sanz-Jiménez[1],
Priscila M. Kosaka[1], Sergio García-López[1], Álvaro San Paulo[1], Samantha Sbarra[2], Louis Waquier[2],
Ivan Favero[2], Maurits van der Heiden[3], Robert K. Altmann[3], Dimitris Papanastasiou[4],
Diamantis Kounadis[4], Ilias Panagiotopoulos[4], Jesús Mingorance[5], María Rodríguez-Tejedor[5],
Rafael Delgado[6], Montserrat Calleja[1], and Javier Tamayo[1]
*[1]Instituto de Micro y Nanotechnologis, IMN-CSIC, CSIC (CEI UAM+CSIC), SPAIN, [2]Université Paris Cité,
FRANCE, [3]The Netherland Organization for Applied Scientific Research (TNO), NETHERLANDS,
[4]Fasmatech Science and Technology, GREECE, [5]Hospital Universitario La Paz, SPAIN, and
[6]Hospital Universitario 12 de Octubre, SPAIN*

09:45   **A MICROFABRICATED DIAMOND QUANTUM MAGNETOMETER
WITH PICOTESLA SCALE SENSITIVITY** 157
Fei Xie[1,2], Qihui Liu[1,2], Yuqiang Hu[3,4], Lingyun Li[1,2], Zhichao Chen[1,2], Jin Zhang[1], Yonggui Zhang[1,2],
Yuyao Zhang[3,4], Yang Wang[1,2], Jiangong Cheng[1,2], Hao Chen[1,2], and Zhenyu Wu[1,2,3,4]
*[1]Chinese Academy of Sciences, CHINA, [2]University of Chinese Academy of Sciences, CHINA,
[3]Shanghai University, CHINA, and [4]Shanghai Industrial µTechnology Research Institute, CHINA*

10:00   **Break & Exhibit Inspection**

Session X – RF MEMS Filters & Resonators (5G & 6G)

10:30   **A NON-VOLATILE THRESHOLD SENSING SYSTEM USING A FERROELECTRIC
$HF_{0.5}ZR_{0.5}O_2$ DEVICE AND A $LiNbO_3$ MICROACOUSTIC RESONATOR** 161
Onurcan Kaya, Luca Colombo, Benyamin Davaji, and Cristian Cassella
Northeastern University, USA

10:45   **RESONANT CONFINERS FOR ACOUSTIC LOSS MITIGATION
IN BULK ACOUSTIC WAVE RESONATORS** 165
Jeronimo Segovia-Fernandez and Ernest T.-T. Yen
Texas Instruments, Kilby Labs, USA

11:00   **HIGH-CRYSTALLINITY 30% SCALN ENABLING HIGH FIGURE OF MERIT
X-BAND MICROACOUSTIC RESONATORS FOR MID-BAND 6G** 169
Gabriel Giribaldi, Pietro Simeoni, Luca Colombo, and Matteo Rinaldi
Northeastern University, USA

11:15   **FERRITE-ROD ANTENNA DRIVEN WIRELESS RESOSWITCH RECEIVER** 173
Kevin H. Zheng, Qiutong Jin, and Clark T.-C. Nguyen
University of California, Berkeley, USA

11:30   **ULTRA-WIDEBAND MEMS FILTERS USING LOCALIZED THINNED**
128° Y-CUT THIN-FILM LITHIUM NIOBATE 177
Jinbo Wu[1,2,3], Shibin Zhang[1], Pengcheng Zheng[1,2], Liping Zhang[1,2], Hulin Yao[1,2],
Xiaoli Fang[1,2], Xuedi Tian[1,2], Xiaomeng Zhao[1], Tao Wu[3], and Xin Ou[1,2]
[1]Shanghai Institute of Microsystem and Information Technology, CHINA,
[2]University of Chinese Academy of Sciences, CHINA, and [3]ShanghaiTech University, CHINA

11:45   **Lunch & Exhibit Inspection**

Session XIa - MEMS/NEMS Resonators & Non-Linear Dynamics

13:00   **ATTRACTOR EXCHANGER FOR OPEN-LOOP OPERATION OF MICROMECHANICAL**
NONLINEAR RESONATORS USING GAP-SPACING CONTINUATION 181
Chun-Pu Tsai and Wei-Chang Li
National Taiwan University, TAIWAN

13:15   **A CMOS-MEMS ULTRASENSITIVE THERMOMETER USING**
INTERNAL RESONANCE INDUCED FREQUENCY COMBS 185
Ting-Yi Chen, Chun-Pu Tsai, and Wei-Chang Li
National Taiwan University, TAIWAN

13:30   **ATOMICALLY THIN NEMS FREQUENCY COMB WITH BOTH FREQUENCY TUNABILITY**
AND RECONFIGURABLE VIA SIMULTANEOUS 1:2 AND 1:3 MODE COUPLING 189
Bo Xu, Jiankai Zhu, Chenyin Jiao, Jianglong Chen, and Zenghui Wang
University of Electronic Science and Technology of China, CHINA

13:45   **INSTRUMENTAL ANALYSIS OF ADVANCED CATALYSTS**
BASED ON RESONANT MICROCANTILEVERS 193
Xinyu Li[1,2], Pengcheng Xu[1,2], Ying Chen[1], Haitao Yu[1], and Xinxin Li[1,2]
[1]Chinese Academy of Sciences, CHINA and [2]University of Chinese Academy of Sciences, CHINA

Session XIb - BioSensors I

13:00   **A MULTIPLEXED BIOAFFINITY BIOSENSING PATCH FOR**
POINT-OF-CARE CHRONIC ULCER MONITORING 197
Md Sharifuzzaman, Dongkyun Kim, Md Selim Reza, SeongHoon Jeong,
Hye Su Song, Md Abu Zahed, and Jae Yeong Park
Kwangwoon University, KOREA

13:15   **3-DOF BIOHYBRID ACTUATOR WITH MULTIPLE SKELETAL MUSCLE TISSUES** 201
Xinzhu Ren, Yuya Morimoto, and Shoji Takeuchi
University of Tokyo, JAPAN

13:30   **A LOW NOISE MICROELECTRODE ARRAY FOR SPECIFIC**
CELL ACTIVITY MODULATION FROM CELL TO TISSUE 205
Bohan Zhang[1,2], Huiran Yang[2], Xiner Wang[2,3], Ziyi Zhu[2,3], Zongxing He[1], Wanqi Jiang[2,3], Chen Tao[1,2],
Dujuan Zou[2,3], Meng Li[2,3], Zhitao Zhou[2,3], Liuyang Sun[2,3], Tiger H. Tao[1,2,3,4,5,6], and Xiaoling Wei[2,3]
*[1]ShanghaiTech University, CHINA, [2]Chinese Academy of Sciences, CHINA, [3]University of Chinese Academy
of Sciences, CHINA, [4]Neuroxess Co., Ltd. (Jiangxi), CHINA, [5]Guangdong Institute of Intelligence Science
and Technology, CHINA, and [6]Tianqiao and Chrissy Chen Institute for Translational Research, CHINA*

13:45   **BIONIC MECHANICAL HAND INTEGRATED WITH ARTIFICIAL OLFACTORY SENSOR ARRAY FOR ENHANCED OBJECT RECOGNITION** 209

Jiachuang Wang[1,2], Xiaohui Li[1,2], MengWei Liu[1,2], Pingping Zhang[3], Tiger H. Tao[1,2,4], and Nan Qin[1,2]
[1]Chinese Academy of Sciences, CHINA, [2]University of Chinese Academy of Sciences, CHINA,
[3]Suzhou Huiwen Nanotechnology Co., Ltd., CHINA, and [4]Neuroxess Co., Ltd. (Jiangxi), CHINA

Poster/Oral Session III

14:00   **Poster/Oral Session III**
Poster presentations are listed by topic category with their assigned number starting on Page 13.

15:30   **Break & Exhibit Inspection**

Session XIIa - Force & Displacement/ Tactile Sensors & Human-Machine

16:00   **HIGH RESOLUTION TACTILE SENSOR FOR MEASUREMENT OF A COMPLICATED TACTILE FEELING OF "*SHITTORI*" WITH MOISTNESS** 213

Genki Yamada, Yuto Morita, Kyohei Terao, Fusao Shimokawa, and Hidekuni Takao
Kagawa University, JAPAN

16:15   **PYRAMIDAL STRUCTURED MXENE/ECOFLEX COMPOSITE-BASED TOROIDAL TRIBOELECTRIC SELF-POWERED SENSOR FOR HUMAN-MACHINE INTERFACE** 217

Shipeng Zhang, Sm Sohel Rana, Gagan Bahad Pradhan,
Trilochan Bhatta, Seonghoon Jeong, and Jae Yeong Park
Kangwoon University, KOREA

16:30   **LIG-BASED TRIAXIAL TACTILE SENSOR UTILIZING ROTATIONAL ERECTION SYSTEM** 221

Rihachiro Nakashima[1], Nagi Nakamura[2], Tomohiko G. Sano[1], Eiji Iwase[2], and Hidetoshi Takahashi[1]
[1]Keio University, JAPAN and [2]Waseda University, JAPAN

16:45   **A STRETCHABLE STRAIN-INSENSITIVE SMART GLOVE FOR SIMULTANEOUS DETECTION OF PRESSURE AND TEMPERATURE** 225

Sudeep Sharma, Gagan Bahadur Pradhan, Seonghoon Jeong, and Jae Yeong Park
Kwangwoon University, KOREA

17:00   **A GESTURE RECOGNITION GLOVE ASSEMBLED WITH NANOFOREST-INTEGRATED INFRARED THERMOPILES** 229

Mao Li[1,2], Meng Shi[1,2], Guidong Chen[1,2], Na Zhou[1,2], Haiyang Mao[1,2], and Chengjun Huang[1,2]
[1]Chinese Academy of Sciences, CHINA and [2]University of Chinese Academy of Sciences, CHINA

Session XIIb - BioSensors II

16:00   **ONE PUSH MEMBRANE FORMATION FOR ITERATIVE MEASUREMENT OF ION CHANNEL ACTIVITY ON ARRAYED CHIP** 233

Hisatoshi Mimura[1], Toshihisa Osaki[1,2], Sho Takamori[1], Kenji Nakao[2], and Shoji Takeuchi[1,3]
[1]Kanagawa Institute of Industrial Science and Technology (KISTEC), JAPAN,
[2]Maqsys Inc., JAPAN, and [3]University of Tokyo, JAPAN

16:15   **AN IMPLANTABLE DIFFERENTIAL SENSOR WITH PASSIVE WIRELESS INTERROGATION FOR IN-SITU EARLY DETECTION OF PERIPROSTHETIC JOINT INFECTION** 235

Jiaxin Jiang, Cole Napier, Chandrashekhar Choudhary, H. Claude Sagi,
Chia-Ying Lin, Michael T. Archdeacon, and Tao Li
University of Cincinnati, USA

16:30   **MICROMACHINED PIEZOELECTRIC FILM-BASED FLEXIBLE ELECTRONICS WITH INTEGRATION OF FILM-SELF TEMPERATURE-DETECTING BREATH SENSOR AND ACETONE GAS SENSOR** 239
Hung-Yu Yeh and Guo-Hua Feng
National Tsing Hua University, TAIWAN

16:45   **FLEXIBLE TACTILE SENSING ARRAY WITH HIGH SPACIAL DENSITY BASED ON PARYLENE MEMS TECHNIQUE** 243
Meixuan Zhang[1], Zetian Wang[1], Han Xu[2], Lang Chen[1], Yufeng Jin[2,3], and Wei Wang[1,3,4]
[1]Peking University, CHINA, [2]Peking University Shenzhen Graduate School, CHINA, [3]National Key Lab of Micro/Nano Fabrication Technology, CHINA, and [4]Beijing Advanced Innovation Center for Integrated Circuits, CHINA

17:00   **SILK-ENABLED FOLDABLE AND CONFORMAL NEURAL INTERFACE WITH IN-PLANE SHIELDING FOR HIGH-QUALITY ELECTROPHYSIOLOGICAL RECORDINGS** 247
Jizhi Liang[1,2], Zhaohan Chen[1,3], Xiner Wang[1,2], Feihong Xu[1,2], Xiaoling Wei[1,2], Liuyang Sun[1,2], Meng Li[1,2,] Tiger H. Tao[1,2,4,5,6,7], and Zhitao Zhou[1,2]
[1]Chinese Academy of Sciences, CHINA, [2]University of Chinese Academy of Sciences, CHINA, [3]Shanghai Normal University, CHINA, [4]ShanghaiTech University, CHINA, [5]Neuroxess Co., Ltd. (Jiangxi), CHINA, [6]Guangdong Institute of Intelligence Science and Technology, CHINA and [7]Tianqiao and Chrissy Chen Institute for Translational Research, CHINA

17:15   **Adjourn for the day**

Thursday, 19 January

All times are Central European Time (CET).

Plenary Presentation IV

08:30   **MATERIALS ENGINEERING FOR CHEMICAL SENSING ENHANCEMENT** 251
Navpreet Kaur, Dario Zappa, and Elisabetta Comini
University of Brescia, ITALY

Session XIII - Gas & Flow Sensors

09:15   **ON-DEMAND PREPARATION OF GAS-SENSING MATERIALS GUIDED BY RESONANT CANTILEVER-BASED THERMOGRAVIMETRIC ANALYSIS** 255
Yufan Zhou[1,2], Ming Li[1,2], Ying Chen[1,2], Xinyu Li[1,2], Pengcheng Xu[1,2], and Xinyu Li[1,2]
[1]Chinese Academy of Sciences, CHINA and [2]University of Chinese Academy of Sciences, CHINA

09:30   **AN INTELLIGENT GAS ANALYSIS SYSTEM CONSISTING OF SENSORS AND A NEURAL NETWORK IMPLEMENTED USING THIN-FILM TRANSISTORS** 259
Zong Liu[1,2], Yushen Hu[1,2], Gabriel E. Carranza[1], Fei Wang[2], and Man Wong[1]
[1]Hong Kong University of Science and Technology, HONG KONG and
[2]Southern University of Science and Technology, CHINA

09:45   **SINGLE-LAYER-ELECTRODE TEMPERATURE-MODULATED SNO$_2$ GAS SENSOR CELL WITH LOW POWER CONSUMPTION FOR DISCRIMINATION OF FOOD ODORS** 263
Chong Xing, Ruichen Liu, Yan Zhang, Dongcheng Xie, Yudong Wang, Yuan Huang, Muhammad Mustafa, Haochen Zhang, Zhongyu Shi, Lei Xu, and Feng Wu
University of Science and Technology of China, CHINA

10:00   **A PERFORMANCE ENHANCED THERMAL FLOW SENSOR WITH NOVEL DUAL-HEATER STRUCTURE USING CMOS COMPATIBLE FABRICATION PROCESS** 267
Zhongyi Liu[1], Ruoqin Wang[2], Gai Yang[1], Xinyuan Zhang[1], Rui Jiao[2], Xuejiao Li[1], Jiali Qi[3], Hongyu Yu[2], Huikai Xie[1,4], and Xiaoyi Wang[1,4]
[1]Beijing Institute of Technology, CHINA, [2]Hong Kong University of Science and Technology, HONG KONG, [3]Hangzhou Dianzi University, CHINA, and
[4]BIT Chongqing Institute of Microelectronics and Microsystems, CHINA

Session XIV - New Fabrication Techniques

10:45   **LOCAL METAL DEPOSITION ON HYDROGELS USING MICRO-PLASMA-BUBBLES** 271
Haruna Takahashi, Yu Yamashita, Naotomo Tottori, Shinya Sakuma, and Yoko Yamanishi
Kyushu University, JAPAN

11:00   **FOLDING METHOD OF KIRIGAMI STRUCTURE WITH FOLDING LINES** 275
Nagi Nakamura and Eiji Iwase
Waseda University, JAPAN

11:15   **BUBBLE-ASSISTED RE-FORMATION OF INDIVIDUAL LIPID BILAYERS IN ARRAYED DEVICE** 279
Izumi Hashimoto[1,2], Toshihisa Osaki[2], Hisatoshi Mimura[2], Sho Takamori[2], Norihisa Miki[1,2], and Shoji Takeuchi[2,3]
[1]Keio University, JAPAN, [2]Kanagawa Institute of Industrial Science and Technology, JAPAN, and
[3]University of Tokyo, JAPAN

11:30 **LARGE-SCALE ARRAYS OF TUNABLE MONOLAYER MoS$_2$ NANOELECTROMECHANICAL RESONATORS** **281**
Zuheng Liu[1], Luming Wang[3], Pengcheng Zhang[1], Maosong Xie[1], Yueyang Jia[1], Ying Chen[4], Hao Jia[4], Zenghui Wang[3], and Rui Yang[1,2]
[1]*University of Michigan – Shanghai Jiao Tong University Joint Institute, Shanghai Jiao Tong University, CHINA,* [2]*Shanghai Jiao Tong University, CHINA,* [3]*University of Electronic Science and Technology of China, CHINA, and* [4]*Chinese Academy of Sciences, CHINA*

Awards Ceremony

11:45 **Awards Ceremony**

11:55 **Final Remarks**

12:00 **Conference Ajourns**

POSTER PRESENTATIONS
All times are Central European Time (CET).

M - Monday, 16 January - 13:45 - 15:45
T - Tuesday, 17 January - 13:30 - 15:30
W - Wednesday, 18 January - 13:30 - 15:30

Classification Chart
(last character of poster number)

a - Bio and Medical MEMS
b - Emerging Technologies and New Opportunities for MEMS/NEMS
c - Industry MEMS and Advancing MEMS for Products and Sustainability
d - Materials, Fabrication and Packaging for Generic MEMS and NEMS
e - MEMS Actuators and PowerMEMS
f - MEMS Physical and Chemical Sensors
g - Micro- and Nanofluidics
h - Optical, RF and Electromagnetics for MEMS/NEMS
i - Open Posters

a - Bio and Medical MEMS
Biosensors and Bioreactors

M01-a   ANTIFOULING FOR ELECTROCHEMICALLY BIOSENSING IN BODY FLUIDS 285
Wenzheng He[1], Changdong Zhou[2], Yang Lin[2], Yuxin Tian[2], Liying Liu[2],
Qifu Zhang[2], Xiongying Ye[1], and Tianhong Cui[3]
[1]Tsinghua University, CHINA, [2]Jilin Cancer Hospital, CHINA, and [3]University of Minnesota, USA

T01-a   ELECTRO-MAGNETIC SENSOR MEDIATED BY MAGNETIC BIOMOLECULES 289
Qian Cheng[1,2], Yuqing Ge[1], Hongju Mao[1,2], Lin Zhou[1], and Jianlong Zhao[1,2]
[1]Chinese Academy of Science, CHINA and [2]University of Chinese Academy of Sciences, CHINA

W01-a   GAS-FLOW DEVICE FOR EFFECTIVE DISSOLUTION OF GAS-PHASE
ODORANTS UTILIZED FOR BIOHYBRID SENSORS 293
Takuma Nakane[1,2], Toshihisa Osaki[2], Hisatoshi Mimura[2], Sho Takamori[2],
Norihisa Miki[1,2], and Shoji Takeuchi[2,3]
*[1]Keio University, JAPAN, [2]Kanagawa Institute of Industrial Science and Technology, JAPAN, and
[3]University of Tokyo, JAPAN*

M02-a   MULTIPLE WELLS ON A CMOS-MEA FOR CELL-BASED
BIOHYBRID ODORANT SENSORS 295
Yujia Lian, Haruka Oda, Minghao Nie, and Shoji Takeuchi
University of Tokyo, JAPAN

T02-a THE INTEGRATED RGO/PEDOT: PSS-MODIFIED ULTRAFLEXIBLE
MICROELECTRODES TOWARDS LONG-TERM NEUROPHYSIOLOGICAL
SIGNALING AND DOPAMINE SENSITIVE DETECTION 297
Xueying Wang[1,2], Huiran Yang[1], Bohan Zhang[1,3], Meng Li[1,2], Liuyang Sun[1,2],
Zhitao Zhou[1,2], Tiger H. Tao[1,2,3,4,5,6], and Xiaoling Wei[1,2]
[1]Chinese Academy of Sciences, CHINA, [2]University of Chinese Academy of Sciences, CHINA, [3]Shanghai Tech
University, CHINA, [4]Neuroxess Co., Ltd. (Jiangxi), CHINA, [5]Guangdong Institute of Intelligence Science and
Technology, CHINA, and [6]Tianqiao and Chrissy Chen Institute for Translational Research, CHINA

a - Bio and Medical MEMS
Devices & Systems for Cellular and Molecular Studies

W02-a COMPARISON OF SELECTIVE FILTRATION OF ON-CHIP GLOMERULUS
COMPRISED OF ORGANOID-DERIVED AND IMMORTALIZED PODOCYTES 301
Ayumu Tabuchi[1], Kensuke Yabuuchi[2,3], Yoshiki Sahara[2], Minoru Takasato[2,4],
Kazuya Fujimoto[1], and Ryuji Yokokawa[1]
[1]Kyoto university, JAPAN, [2]RIKEN, JAPAN, and [3]Osaka University, JAPAN

M03-a CONTROLLING FIRING POINT OF MICROFIBER-SHAPED HIPSC-DERIVED
CARDIAC TISSUE WITH LOCALIZED ELECTRICAL STIMULATION DEVICE 305
Akari Masuda[1], Shun Itai[1], Yuta Kurashina[2], Shugo Tohyama[1], and Hiroaki Onoe[1]
[1]Keio University, JAPAN and [2]Tokyo University of Agriculture and Technology, JAPAN

T03-a DEVELOPMENTAL PHASES OF ON-CHIP VASCULOGENESIS
CLASSIFIED USING A DEEP LEARNING VISUAL MODEL 309
Taiga Irisa, Hang Zhou, Kazuya Fujimoto, and Ryuji Yokokawa
Kyoto University, JAPAN

W03-a HAND-DRIVEN DEVICE FOR PREPARATION
OF LINEARLY ALIGNED HYDROGEL SHEETS 313
Aoi Kato[1,2], Haruka Oda[3], Sho Takamori[2], Hisatoshi Mimura[2],
Toshihisa Osaki[2], Norihisa Miki[1,2], and Shoji Takeuchi[2,3]
[1]Keio University, JAPAN, [2]Kanagawa Institute of Industrial Science and Technology, JAPAN, and
[3]University of Tokyo, JAPAN

M04-a MICROFABRICATION AND CHARACTERIZATION OF MICRO-
STEREOLITHOGRAPHICALLY 3D PRINTED, AND DOUBLE METALLIZED
BIOPLATES WITH 3D MICROELECTRODE ARRAYS FOR
IN-VITRO ANALYSIS OF CARDIAC ORGANOIDS 315
Jorge Manrique Castro, Isaac Johnson, and Swaminathan Rajaraman
University of Central Florida, USA

T04-a OIL-SEALED RGD-MODIFIED HYDROGEL MICROWELL ARRAY WITH SIZE-
SELECTIVE PERMEATION FOR ANALYSIS ON EXOSOMES FROM SINGLE CELLS 319
Chisaki Yamagata[1], Shun Itai[1], Yuta Kurashina[2], Makoto Asai[1], Ayuko Hoshino[3], and Hiroaki Onoe[1]
[1]Keio University, JAPAN, [2]Tokyo University of Agriculture and Technology, JAPAN, and
[3]Tokyo Institute of Technology, JAPAN

W04-a PICKING SINGEL CELLS FROM 10 ML SAMPLE BASED ON A
MICROFILTRATION- LIFT COMBINATION PLATFORM 323
Qingmei Xu[1], Yuntong Wang[2,3], Xiao Ma[4], Hang Li[5], Ying Xue[5],
Yi Zhang[1], Songtao Dou[1], Huan Wang[2], Bei Li[2,5], and Wei Wang[1,6,7]
[1]Peking University, CHINA, [2]Chinese Academy of Sciences, CHINA, [3]University of Chinese Academy of
Sciences, CHINA, [4]Hangzhou Branemagic Medical Technology Co. Ltd., CHINA, [5]Hooke Laboratory,
CHINA, [6]National Key Lab of Micro/Nano Fabrication Technology, CHINA, and [7]Beijing Advanced
Innovation Center for Integrated Circuits, Beijing, CHINA

a - Bio and Medical MEMS
Flexible and Wearable Devices and Systems

M05-a A TRANSFER METHOD FOR EMBEDDING CONDUCTIVE FILLERS ON THE SURFACE OF MULTI-SCALE STRUCTURES FOR 3D FLEXIBLE CONDUCTORS 327
Dongwoo Yoo, Sangmok Kim, Jeonghyeon Hwang, and Joonwon Kim
Pohang University of Science and Technology (POSTECH), KOREA

T05-a FABRICATION OF HIGH FREQUENCY 2D FLEXIBLE PMUT ARRAY 331
Sanjog V. Joshi, Sina Sadeghpour, and Michael Kraft
KU Leuven, BELGIUM

W05-a FLEXIBLE SILK-BASED GRAPHENE BIOELECTRONICS FOR WEARABLE MULTIMODAL PHYSIOLOGICAL MONITORING 335
Sajjad Mirbakht[1], Ata Golparvar[1,2], Muhammad Umar[1], and Murat Kaya Yapici[1,3]
[1]Sabanci University, TURKEY, [2]École Polytechnique Fédérale de Lausanne (EPFL), SWITZERLAND, and [3]University of Washington, USA

M06-a HIGHLY ACCURATE MEASUREMENT OF CONTACT RESISTANCE BETWEEN GALINSTAN AND COPPER USING TRANSFER LENGTH METHOD 339
Takashi Sato and Eiji Iwase
Waseda University, JAPAN

T06-a MACHINE LEARNING ENABLED HIND FOOT DEFORMITY DETECTION USING INDIVIDUALLY ADDRESSABLE HYBRID PRESSURE SENSOR MATRIX 343
Nadeem Tariq Beigh, Faizan Beigh, Sourav Naval, Dibyajyoti Mukherjee, and Dhiman Mallick
Indian Institute of Technology, Delhi, INDIA

W06-a MULTI-MODE E-SKIN INTEGRATING CAPACITIVE-PIEZOELECTRIC SENSORS FOR STATIC-DYNAMIC MECHANORESPONSE WITH WIDE SENSING RANGE 347
Mujeeb Yousuf[1], Sushil Kumar[1], Dhairya Singh Arya[2],
Manu Garg[1], Khanjhan Joshi[1], and Pushpapraj Singh[1]
*[1]Indian Institute of Technology, Delhi, INDIA and
[2]CSIR-Central Scientific Instruments Organisation (CSIO), INDIA*

M07-a NON-INVASIVE INSTANT MEASUREMENT OF ARTERIAL STIFFNESS BASED ON HIGH-DENSITY FLEXIBLE SENSOR ARRAY 351
Fang Wang[1,2], Heng Yang[1,2], Ke Sun[1], Yi Sun[1], and Xinxin Li[1,2]
[1]Chinese Academy of Sciences, CHINA and [2]University of Chinese Academy of Sciences, CHINA

T07-a SUPPRESSION OF BIOELECTRICAL NOISE SIGNALS IN MOTION STATE BY LOW-COST MICROPILLAR HYDROGEL ELECTRODE 355
Gencai Shen, Nan Zhao, Chunpeng Jiang, Zhuangzhuang Wang, and Jingquan Liu
Shanghai Jiao Tong University, CHINA

W07-a ULTRA-THIN MEMS PACKAGING BASED ON AUXETIC STRETCHABLE STRUCTURES FOR APPLICATIONS IN WEARABLE ELECTRONICS 359
Daniel Zymelka, Toshihiro Takeshita, Yusuke Takei, and Takeshi Kobayashi
National Institute of Advanced Industrial Science and Technology, JAPAN

M08-a ULTRALOW POWER FLEXIBLE OCULAR MICROSYSTEM FOR VERGENCE AND DISTANCE SENSING BASED ON PASSIVE DIFFERENTIAL MAGNETOMETRY 362
Adwait Deshpande, Mohit U. Karkhanis, Chayanjit Ghosh, Hanseup Kim, and Carlos H. Mastrangelo
University of Utah, USA

a - Bio and Medical MEMS
Manufacturing for Bio- & Medical MEMS

T08-a   **ELECTROHYDRODYNAMIC NEBULISER (eNEB) FOR DIRECT PULMONARY DRUG DELIVERY APPLICATION** 366
Trung-Hieu Vu[1], Luan Ngoc Mai[2,3], Tuan-Hung Nguyen[1], Dang Tran[1], Tuan-Khoa Nguyen[1], Thanh Nguyen[4], Jarred Fastier-Woollel[1,5], Canh-Dung Tran[4], Toan Dinh[4], Hong-Quan Nguyen[1], Dzung Viet Dao[1], and Van Thanh Dau[1]
[1]Griffith University, AUSTRALIA, [2]Ho Chi Minh City University of Technology (HCMUT), VIETNAM
[3]Vietnam National University, VIETNAM, [4]University of Southern Queensland, AUSTRALIA, and
[5]University of Tokyo, JAPAN

W08-a   **FLEXIBLE POLYMER OPTICAL WAVEGUIDES FOR INTEGRATED OPTOGENETIC BRAIN IMPLANTS** 370
Julian A. Singer[1], Till Stramm[2], Jens Fasel[2], Oliver Schween[2], Anton Gelaeschus[1], Andreas Bahr[1,3], and Matthias Kuhl[4]
[1]Hamburg University of Technology, GERMANY, [2]TU Dortmund University, GERMANY,
[3]University of Kiel, GERMANY, and [4]University of Freiburg, GERMANY

M09-a   **HIGHLY REPRODUCIBLE TISSUE POSITIONING WITH TAPERED PILLAR DESIGN IN ENGINEERED HEART TISSUE PLATFORMS** 374
Milica Dostanic[1,2], Laura M. Windt[2], Maury Wiendels[2], Berend J. van Meer[2], Christine L. Mummery[2,3], Pasqualina M. Sarro[1], and Massimo Mastrangeli[1]
[1]Delft University of Technology, NETHERLANDS, [2]Leiden University Medical Center, NETHERLANDS, and
[3]University of Twente, NETHERLANDS

T09-a   **IN VITRO ASSEMBLY OF MUSCLE RINGS AND BIOPRINTED HYDROGEL FOR BRANCHING TUBULAR TISSUE CONSTRUCTS** 378
Tomohiro Morita, Byeongwook Jo, Minghao Nie, and Shoji Takeuchi
University of Tokyo, JAPAN

W09-a   **MICROELECTRODES FABRICATED BY VACUUM FILLING WITH LOW MELTING-POINT ALLOY FOR MUSCLE TISSUE STIMULATION** 381
Tingyu Li, Minghao Nie, Yuya Morimoto, and Shoji Takeuchi
University of Tokyo, JAPAN

M10-a   **OPTOELECTRONIC INTEGRATED ULTRAMICROELECTRODE FOR OPTICAL STIMULATION AND ELECTRICAL RECORDING OF SINGLE-CELL** 384
Qingda Xu, Ye Xi, Zhiyuan Du, Longchun Wang, Tao Ruan, Mengfei Xu, Jiawei Cao, Bin Yang, and Jingquan Liu
Shanghai Jiao Tong University, CHINA

T10-a   **THERMOFORMING OF PARYLENE C TO FORM HELICAL STRUCTURES** 388
Brianna L. Thielen and Ellis Meng
University of Southern California, Los Angeles, USA

a - Bio and Medical MEMS
Materials for Bio- and Medical MEMS

W10-a   **FABRICATION OF BIODEGRADABLE SOFT TISSUE-MIMICKED MICROELECTRODE ARRAYS FOR IMPLANTED NEURAL INTERFACING** 392
Wei-Chen Huang[1], Wan-Lou Lei[1], and Chih-Wei Peng[2]
[1]National Yang Ming Chiao Tung University, TAIWAN and [2]Taipei Medical University, TAIWAN

a - Bio and Medical MEMS
Medical Microsystems

M11-a AN OPTIMIZATION OF PERFORATION DESIGN ON A PIEZOELECTRIC-BASED
SMART STENT FOR BLOOD PRESSURE MONITORING AND LOW-FREQUENCY
VIBRATIONAL ENERGY HARVESTING 396
Jun Ying Tan[1], Sayemul Islam[2], Yuankai Li[3], Albert Kim[2], and Jungkwun "JK" Kim[1]
[1]University of North Texas, USA, [2]University of South Florida, USA, and [3]Kansas State University, USA

W11-a DEVELOPMENT OF AN ELECTRICAL-STIMULATION-INDUCED MECHANOMYOGRAM
PROBE FOR MUSCLE CONTRACTION CHARACTERISTICS EVALUATION 400
Yusuke Takei, Toshihiro Takeshita, Daniel Zymelka, and Takeshi Kobayashi
National Institute of Advanced Industrial Science and Technology (AIST), JAPAN

M12-a DUAL-FREQUENCY PIEZOELECTRIC MICROMACHINED ULTRASONIC
TRANSDUCERS FOR FUNDAMENTAL AND HARMONIC IMAGING 402
Yanfen Zhai, Waleed Maqsood, Zhou Da, Nikolai Andrianov,
Yucheng Zhang, Mohssen Moridi, and Lixiang Wu
Silicon Austria Labs GmbH, AUSTRIA

T12-a FRACTAL MICROELECTRODES INTEGRATED WITH THE
CATHETER FOR LOW-VOLTAGE PULSED FIELD ABLATION 405
Mengfei Xu[1], Mu Qin[2], Ziliang Song[3], Wen Hong[1], Qingda Xu[1], Jiawei Cao[1],
Kejun Tu[1], Longchun Wang[1], Bin Yang[1], and Jingquan Liu[1]
[1]Shanghai Jiao Tong University, CHINA,
[2]Shanghai Chest Hospital Affiliated to Shanghai Jiao Tong University, CHINA, and
[3]Shanghai General Hospital Shanghai Jiao Tong University School of Medicine, CHINA

W12-a HIERARCHICAL BONDING YIELD TEST STRUCTURE FOR FLEXIBLE HIGH
CHANNEL-COUNT NEURAL PROBES INTERFACING ASIC CHIPS 409
Marie C. Odenthal, Victor Claar, Oliver Paul, and Patrick Ruther
University of Freiburg, GERMANY

M13-a MICROWAVE-INDUCED THERMOACOUSTIC
IMAGING USING ALUMINUM NITRIDE PMUT 413
Yiwei Wang[1], Lejia Zhang[1], Junxiang Cai[1,2,3], Baosheng Wang[1,2,3],
Yuandong Alex Gu[5], Liang Lou[5], Xiong Wang[1,2,3,4], and Tao Wu[1,2,3,4]
*[1]ShanghaiTech University, CHINA and [2]Chinese Academy of Sciences, CHINA, [3]University of Chinese
Academy of Sciences, CHINA, [4]Shanghai Engineering Research Center of Energy Efficient and Custom AI
IC, CHINA, and [5]Shanghai Industrial μTechnology Research Institute, CHINA*

T13-a NEEDLE-FREE DRUG INJECTION USING A SHOCK WAVE FOCUSING SYSTEM WITH
THE FUNCTION OF REAL-TIME MICROBUBBLE-BASED DISTANCE SENSING 417
Yibo Ma, Wenjing Huang, Keita Ichikawa, and Yoko Yamanishi
Kyushu University, JAPAN

W13-a NEW WAFER-LEVEL FABRICATION OF ULTRATHIN SILICON
INSERTION SHUTTLES FOR FLEXIBLE NEURAL IMPLANTS 421
Kirti Sharma[1], Christian Boehler[1], Maria Asplund[1,2], Oliver Paul[1], and Patrick Ruther[1]
[1]University of Freiburg, GERMANY and [2]Chalmers University of Technology, SWEDEN

M14-a REAL-TIME DYNAMIC LACTATE DETECTION IN A PIPELINE USING A
MICROSENSING NEEDLE FOR ICU PATIENT MONITORING APPLICATION 425
Yuan-Sin Tang[1], Tung-Lin Yang[2], Yu-Ting Cheng[1], Hsiao-En Tsai[2,3], and Yih-Shurng Chen[3,4]
*[1]National Yang Ming Chiao Tung University, TAIWAN, [2]National Taiwan Hospital HsinChu Branch,
TAIWAN, [3]National Taiwan University College of Medicine Graduate Institute of Clinical Medicine,
TAIWAN, and [4]National Taiwan University Hospital, TAIWAN*

T14-a THREE-DIMENSIONAL FLEXIBLE NEURAL OPTO-ELECTRONIC
ARRAY WITH SILK-BASED SHUTTLE-FREE IMPLANTATION 429
Chi Gu[2,3], Huiran Yang[2], Bohan Zhang[2,4], Zhitao Zhou[2], Liuyang Sun[2,3],
Meng Li[2,3], Xiaoling Wei[2,3], and Tiger H. Tao[1,2,3,4,5,6]
*[1]Guangdong Institute of Intelligence Science and Technology, CHINA, [2]Chinese Academy of Sciences,
CHINA, [3]University of Chinese Academy of Sciences, CHINA, [4]ShanghaiTech University, CHINA, [5]Neuroxess
Co., Ltd. (Jiangxi), CHINA, and [6]Tianqiao and Chrissy Chen Institute for Translational Research, CHINA*

a - Bio and Medical MEMS
MEMS & BioMEMS for Fighting COVID-19 & Future Pandemic

W14-a A MICROFLUIDIC BIOSENSOR FOR RAPID DETECTION OF COVID-19 433
Sura A. Muhsin[1], Ying He[1], Muthana Al-Amidie[1], Karen Sergovia[1], Amjed Abdullah[1],
Yang Wang[1], Omar Alkorjia[1], Robert A. Hulsey[2], Gary L. Hunter[2], Zeynep Erdal[2],
Ryan J. Pletka[2], George S. Hyleme[2], Xiu-Feng Wan[1,2], and Mahmoud Almasri[1]
[1]University of Missouri, USA and [2]Black and Veatch, USA

M15-a A LOOP-MEDIATED ISOTHERMAL AMPLIFICATION (LAMP)-BASED POINT-OF-CARE
SYSTEM FOR RAPID ON-SITE CLINICAL DETECTION OF SARS-COV-2 VIRUSES 437
Trieu Nguyen[1], Aaydha Chidambara Vinayaka[1], Van Ngoc Huynh[1], Quyen Than Linh[1],
Sune Zoëga Andreasen[1], Mohsen Golabi[1], Dang Duong Bang[1], Jens Kjølseth Møller[2], and Anders Wolff[1]
[1]Technical University of Denmark, DENMARK and [2]University Hospital of Southern Denmark, DENMARK

a - Bio and Medical MEMS
MEMS & BioMEMS for Healthcare and Public Health

T15-a A SOLAR-DRIVEN WEARABLE MULTIPLEXED BIO-SENSING SYSTEM
FOR NONINVASIVE HEALTHCARE MONITORING IN SWEAT 440
Jujhar Singh, Bianca Ning, Paul Lee, and Lin Liu
Seattle Pacific University, USA

W15-a HIGH-THROUGHPUT MASS MEASUREMENT OF SINGLE BACTERIAL
CELLS BY SILICON NITRIDE MEMBRANE RESONATORS 444
Adrián Sanz-Jiménez[1], Oscar Malvar[1], Jose J. Ruz[1], Sergio García-López[1], Priscila M. Kosaka[1],
Eduardo Gil-Santos[1], Álvaro Cano[1], Dimitris Papanastasiou[2], Diamantis Kounadis[2], Elias Panagiotopoulos[2],
Jesús Mingorance[3], María Rodríguez-Tejedor[3], Álvaro San Paulo[1], Montserrat Calleja[1], and Javier Tamayo[1]
*[1]Instituto de Micro y Nanotecnología, SPAIN, [2]Fasmatech Science & Technology, Lefkippos TESPA,
Demokritos NCSR, Patriarchou Gregoriou & Neapoleos, GREECE, and
[3]Hospital Universitario La Paz, Madrid, SPAIN*

M16-a MICROFABRICATED ISOTHERMAL EG-FET SENSOR FOR LAMP MEDIATED
CRISPR/CAS12A DETECTION OF HEPATITIS C VIRUS 448
Hsin-Ying Ho, Wei-Sin Kao, Piyush Deval, Ling-Shan Yu, and Che-Hsin Lin
National Sun Yat-sen Universit, TAIWAN

T16-a SMART ELECTRODE ARRAY FOR COCHLEAR IMPLANTS 452
Ahmad Itawi[1], Sofiane Ghenna[1], Guillaume Tourrel[2], Sébastien Grondel[1], Cedric Plesse[3], Tran Minh Giao
Nguyen[3], Frédéric Vidal[3], Yinoussa Adagolodjo[4], Lingxiao Xun[4], Gang Zheng[4], Alexandre Kruszewski[4],
Christian Duriez[4], and Eric Cattan[1]
*[1]University Polytechnique Hauts-de-France, FRANCE, [2]Oticon Medical, FRANCE,
[3]CY Cergy Paris Université, FRANCE, and [4]University of Lille, FRANCE*

a - Bio and Medical MEMS
Tissue Engineering

W16-a   A THREE-DIMENSIONAL ARTIFICIAL INTESTINAL
TISSUE WITH A CRYPT-LIKE INNER SURFACE 456
Shuma Tanaka[1], Shun Itai[2], and Hiroaki Onoe[1]
[1]Keio University, JAPAN and [2]Tohoku University, JAPAN

M17-a   TISSUE-ENGINEERED PENNATE MUSCLES ON A CHIP 460
Motoki Ito, Yuya Morimoto, and Shoji Takeuchi
University of Tokyo, JAPAN

T17-a   WEIGHT TRAINING DEVICE TO PROMOTE MATURATION
IN SKELETAL MUSCLE TISSUES 463
Kentaro Motoi, Byeongwook Jo, Yuya Morimoto, and Shoji Takeuchi
University of Tokyo, JAPAN

a - Bio and Medical MEMS
Other Bio and Medical MEMS

W17-a   MICROSYSTEM VIBRATING MESH ATOMIZER WITH INTEGRATED
MICROHEATER FOR HIGH VISCOSITY LIQUID AEROSOL GENERATION 467
Pallavi Sharma, Irma Rocio Vazquez, and Nathan Jackson
University of New Mexico, USA

M18-a   SCALABLE MODULAR MEASUREMENT SYSTEM FOR CONTINUOUS
BLOOD MONITORING WITH PIEZOELECTRIC MEMS RESONATORS 471
Michael Schneider[1], Bernhard Kößl[1], Suresh Alasatri[1], Ingrid A.M. Magnet[2], and Ulrich Schmid[1]
[1]TU Wien, AUSTRIA and [2]Medical University of Vienna, AUSTRIA

T18-a   SILICON COMPATIBLE PROCESS TO INTEGRATE IMPEDANCE
CYTOMETRY WITH MECHANICAL CHARACTERIZATION 475
Quentin Rezard[1], Faruk Azam Shaik[1,2], Jean Claude Gerbedoen[1,2], Fabrizio Cleri[1], Dominique Collard[1,2],
Chann Lagadec[1], and Mehmet C. Tarhan[1,2]
[1]University of Lille, FRANCE and [2]University of Tokyo, Lille, FRANCE

W18-a   SORTING OF EXTRACELLULAR VESICLES BY USING OPTICALLY-INDUCED
DIELECTROPHORESIS ON AN INTEGRATED MICROFLUIDIC CHIP 479
Wei-Jen Soong, Chih-Hung Wang, Yi-Sin Chen, Chihchen Chen, and Gwo-Bin Lee
National Tsing Hua University, TAIWAN

b - Emerging Technologies and New Opportunities for MEMS/NEMS
Internet of Things (IoT) with MEMS/NEMS

M19-b   A REPROGRAMMABLE MEM SWITCH UTILIZING
CONTROLLED CONTACT WELDING 483
Tsegereda K. Esatu, Hei Kam, Lars P. Tatum, Xiaoer Hu, Urmita Sikder,
Sergio Almeida, Junqiao Wu, and Tsu-Jae King Liu
University of California, Berkeley, USA

T19-b   MICROMECHANICAL RSSI BASED ON FORCE INTERACTION DERIVED TAPPING
BANDWIDTH VARIATION IN VIBRO-IMPACT RESONATORS 487
Yi-Hsuan Huang, Hong-Sen Zheng, Chun-Pu Tsai, and Wei-Chang Li
National Taiwan University, TAIWAN

W19-b WAKE-UP IOT WIRELESS SENSING NODE BASED ON A LOW-G
THRESHOLD MEMS INERTIAL SWITCH WITH RELIABLE CONTACTS 491
Sagnik Ghosh[1], Duan Jian Goh[1], Yul Koh[1], Jaibir Sharma[1], Wei Da Toh[1],
Weiguo Chen[1], Yao Zhang[1], Eldwin Ng[1], Amit Lal[2], and Joshua E.-Y. Lee[1]
[1]Agency for Science, Technology and Research (A*STAR), SINGAPORE and [2]Cornell University, USA

b - Emerging Technologies and New Opportunities for MEMS/NEMS
Machine Learning (ML) & Artificial Intelligence (AI)
Enhanced MEMS/NEMS Design, Manufacturing, and Applications

M20-b ARTIFICIAL INTELLIGENCE (AI)-ENHANCED E-SKIN WITH
ARTIFICIAL SYNAPSE SENSORY OUTPUT FOR HUMANOID
ROBOTIC FINGER OF MULTIMODAL PERCEPTION 495
Xinge Guo[1,2] and Chengkuo Lee[1]
[1]National University of Singapore, SINGAPORE and
[2]Agency for Science, Technology and Research (A*STAR), SINGAPORE

T20-b MULTI-MEMS DIFFERENTIAL PRESSURE SENSOR ELEMENTS-BASED
AIRFLOW SENSOR WITH NEURAL NETWORK MODEL 499
Kotaro Haneda, Kenei Matsudaira, and Hidetoshi Takahashi
Keio University, JAPAN

W20-b TRIAL-AND-ERROR LEARNING FOR MEMS STRUCTURAL DESIGN
ENABLED BY DEEP REINFORCEMENT LEARNING 503
Fanping Sui[1], Wei Yue[1], Ziqi Zhang[2], Ruiqi Guo[1], and Liwei Lin[1,2]
[1]University of California, Berkeley, USA and [2]TSinghua University, CHINA

b - Emerging Technologies and New Opportunities for MEMS/NEMS
New Computing Devices and Systems with MEMS/NEMS

M21-b FULLY MICROELECTROMECHANICAL NON-VOLATILE MEMORY CELL 507
Elliott Worsey, Mukesh K. Kulsreshath, Qi Tang, and Dinesh Pamunuwa
University of Bristol, UK

T21-b NONVOLATILE STATE CONFIGURATION OF NANO-WATT PARAMETRIC ISING SPINS
THROUGH FERROELECTRIC HAFNIUM ZIRCONIUM OXIDE MEMS VARACTORS 511
Nicolas Casilli[1], Onurcan Kaya[1], Tahmid Kaisar[2], Benyamin Davaji[1],
Philip X.-L. Feng[2], and Cristian Cassella[1]
[1]Northeastern University, USA and [2]University of Florida, USA

W21-b PHYSICAL RESERVOIR COMPUTING USING NONLINEAR MEMS
RESONATOR HAVING HIGH MEMORY CAPACITY AT "EDGE OF CHAOS" 515
Hiroki Takemura, Takahiro Mizumoto, Amit Banerjee, Jun Hirotani, and Toshiyuki Tsuchiya
Kyoto University, JAPAN

M22-b PROGRAMMABLE FERROELECTRIC HZO NEMS MECHANICAL
MULTIPLIER FOR IN-MEMORY COMPUTING 519
Shubham Jadhav, Ved Gund, and Amit Lal
Cornell University, USA

T22-b STORING MEMS INTERFACES WITHOUT ELECTRICAL
AUXILIARY ENERGY FOR LONG-TIME MONITORING 522
Martin Hoffmann[1], Philip Schmitt[1], Steffen Wittemeier[3], Falk Schaller[2], Alexey Shaporin[3],
Chris Stöckel[2,3], Volker Geneiß[3], Roman Forke[3], Christian Hedayat[3], Ulrich Hilleringmann[4],
Harald Kuhn[2,3], and Sven Zimmermann[2,3]
*[1]Ruhr-Universität Bochum, GERMANY, [2]Chemnitz University of Technology, GERMANY, [3]Fraunhofer
Institute for Electronic Nano Systems ENAS, GERMANY, and [4]University of Paderborn, GERMANY*

b - Emerging Technologies and New Opportunities for MEMS/NEMS
Nonlinear Dynamics in MEMS/NEMS

W22-b A NEW FINDING ON NONLINEAR DAMPING AND STIFFNESS
OF FLEXURAL MODE CAPACITIVE MEMS RESONATORS 526
Hung-Yu Chen, Ming-Huang Li, and Sheng-Shian Li
National Tsing Hua University, TAIWAN

M23-b EXPLOITING PARAMETRIC INSTABILITY IN BISTABLE MEMS ACTUATORS 530
Daniel Platz, Johannes Fabian, Elisabeth Samm, Mahdi Mortada, Michael Schneider, and Ulrich Schmid
TU Wien, AUSTRIA

T23-b FIRST PROTOTYPE OF POLYMER MICROMACHINED
FLAPPING WING NANO AIR VEHICLE 534
Rashmikant, Ryotaro Suetsugu, Minato Onishi, and Daisuke Ishihara
Kyushu Institute of Technology, JAPAN

W23-b ITERATIVE LEARNING CONTROL FOR QUASI-STATIC
MEMS MIRROR WITH SWITCHING OPERATION 538
Matthias Macho[1], Han Woong Yoo[1], Richard Schroedter[2], and Georg Schitter[1]
[1]TU Wien, AUSTRIA and [2]TU Dresden, GERMANY

b - Emerging Technologies and New Opportunities for MEMS/NEMS
Quantum Devices and Systems with MEMS/NEMS

M24-b *Mz* ATOMIC MAGNETOMETER USING A 3D MEMS GLASS
ALKALI VAPOR CELL WITH VERTICAL SIDEWALLS 542
Jin Zhang, Jianfeng Zhang, Wenqi Li, Ziji Wang, and Jintang Shang
Southeast University, CHINA

T24-b ON-CHIP HEATING NOISE SUPPRESSION OF 3D CHIP-SCALE ATOMIC
MAGNETOMETER USING SINGLE-LAYER SHIFTED HEATER 546
Ziji Wang, Junming Wu, Jin Zhang, and Jintang Shang
Southeast University, CHINA

c - Industry MEMS and Advancing MEMS for Products and Sustainability
Barriers to Commercialization & Research Needs for Future Products

W24-c LABOR-SAVING PLATFORM FOR CHARACTERIZATION OF MEMBRANE
PROTEINS BY AUTOMATED MONITORING AND DATA REPORTING 550
Kazuto Ogishi[1], Toshihisa Osaki[2], Yuya Morimoto[1], and Shoji Takeuchi[1,2]
[1]University of Tokyo, JAPAN and [2]Kanagawa Institute of Industrial Science and Technology, JAPAN

c - Industry MEMS and Advancing MEMS for Products and Sustainability
MEMS Packaging Techniques

M25-c MODELLING IMPACT OF VISCOELASTIC PROPERTIES OF DIE-ATTACH
MATERIAL ON THE BIAS RESPONSE OF RESONANT INERTIAL SENSORS 554
Theo Miani[1], Lokesh Gurung[1], Guillermo Sobreviela-Falces[1], Douglas Young[1], Colin Baker[1], and Ashwin A. Seshia[2]
[1]Silicon Microgravity Ltd., UK and [2]University of Cambridge, UK

c - Industry MEMS and Advancing MEMS for Products and Sustainability
MEMS/NEMS - CMOS Integration

T25-c CMOS-EMBEDDED 3D MICRO/NANOFLUIDICS EMPLOYING
TOP-DOWN BEOL SINGLE-STEP WET-ETCHING TECHNIQUE 558
Wei-Yang Weng, Hung-Yu Hou, Yueh-Jung Chao, Shwu-Jen Liaw, and Jun-Chau Chien
National Taiwan University, TAIWAN

W25-c IMPLEMENTATION OF A MONOLITHIC SOC ENVIRONMENTAL
SENSING HUB USING CMOS-MEMS TECHNIQUE 562
Ya-Chu Lee[1], Tung-Lin Chien[1], Chi-Te Fang[1], Yuanyuan Huang[1], Wei-Lun Sung[2],
Yen-Chang Chu[2], Rongshun Chen[1], and Weileun Fang[1]
[1]National Tsing Hua University, TAIWAN and [2]PixArt Imaging Inc., TAIWAN

M26-c MONOLITHICALLY AND VERTICALLY INTEGRATED ENVIRONMENTAL
SENSING HUB WITH NOVEL AIR-BASED HUMIDITY SENSOR DESIGN 566
Tung-Lin Chien, Yuanyuan Huang, Fuchi Shih, and Weileun Fang
National Tsing Hua University, TAIWAN

c - Industry MEMS and Advancing MEMS for Products and Sustainability
New MEMS System Design and Integration Approaches

T26-c A SELF-CORRECTED, SELF-CLEANED MEMS AND SUITABLE FOR
ADVANCED FOUNDRY MULTI-PROJECT WAFER (MPW) 570
Sushil Kumar, Dhairya Singh Arya, Manu Garg, and Pushpapraj Singh
Indian Institute of Technology, New-Delhi, INDIA

W26-c MONOLITHIC INTEGRATION OF HUMIDITY/FLOW/TEMPERATURE SENSORS AS
ENVIRONMENT SENSING HUB FOR APPARENT-TEMPERATURE DETECTION 574
Yu-Hsuan Li, Tung-Lin Chien, Fuchi Shih, Yuanyuan Huang, and Weileun Fang
National Tsing Hua University, TAIWAN

M27-c PIEZORESISTIVE PRESSURE SENSOR WITH MONOLITHICALLY
INTEGRATED AMPLIFIER BASED ON METAL-OXIDE TRANSISTORS 578
Runxiao Shi[1], Dequan Lin[1], Kevin Chau[1,2], and Man Wang[1]
*[1]Hong Kong University of Science and Technology, HONG KONG and
[2]Chinese Academy of Sciences, CHINA*

d - Materials, Fabrication and Packaging for Generic MEMS and NEMS
Advancement in Conventional Materials for MEMS & NEMS

T27-d A PERFORMANCE ENHANCEMENT METHOD FOR THERMOPILE
SENSORS USING A CHIP PROBE TEST SYSTEM 582
Meng Shi[1,2], Mao Li[1,2], Yue Ni[3], Chenchen Zhang[1], Na Zhou[1,2], Haiyang Mao[1,2], and Chengjun Huang[1,2]
*[1]Chinese Academy of Sciences, CHINA, [2]University of Chinese Academy of Sciences, CHINA, and
[3]Jiangsu Hinovaic Technologies Co., Ltd, CHINA*

W27-d CHARACTERIZING INDUCTIVELY-COUPLED-PLASMA ETCHING OF SINGLE
CRYSTALLINE LITHIUM TANTALATE FOR MICRO-ACOUSTIC APPLICATIONS 586
Yasaman Majd, Jorge Manrique Castro, Hakhamanesh Mansoorzare, and Reza Abdolvand
University of Central Florida, USA

M28-d ROBUST POLYCRYSTALLINE 3C-SIC-ON-SI HETEROSTRUCTURES
WITH LOW CTE MISMATCH UP TO 900 °C FOR MEMS 590
Philipp Moll, Georg Pfusterschmied, and Ulrich Schmid
TU Wien, AUSTRIA

d - Materials, Fabrication and Packaging for Generic MEMS and NEMS
Digital Micromanufacturing

T28-d A 3D PRINTED FUNCTIONAL MEMS ACCELEROMETER 594
Simone Pagliano[1], David E. Marschner[1], Damien Maillard[2], Nils Ehrmann[3],
Göran Stemme[1], Stefan Braun[3], Luis Guillermo Villanueva[2], and Frank Niklaus[1]
*[1]KTH Royal Institute of Technology, SWEDEN, [2]École Polytechnique Fédérale de Lausanne (EPFL),
SWITZERLAND, and [3]Hochschule Kaiserslautern, GERMANY*

W28-d A FULLY 3D PRINTED METHOD FOR MONOLITHIC INTEGRATION
OF AN ACCELEROMETER AND A FORCE SENSOR 598
Guandong Liu[1,2], Changhai Wang[1], Kexin Wang[1], Zhili Jia[3], Ruiqi Luo[2], and Wei Ma[2,4]
*[1]Heriot-Watt University, UK, [2]Zhejiang Lab, CHINA, [3]National Institute of Metrology, CHINA, and
[4]Zhejiang University, CHINA*

d - Materials, Fabrication and Packaging for Generic MEMS and NEMS
Generic MEMS & NEMS Manufacturing Techniques

M29-d CHARACTERIZATION OF VAPOR HF SACRIFICIAL ETCHING THROUGH
SUBMICRON RELASE HOLES FOR WAFER-LEVEL VACUUM PACKAGING
BASED ON SILICON MIGRATION SEAL 602
Tianjiao Gong[1], Yukio Suzuki[1], Muhammad J. Khan[1], Karla Hiller[2], and Shuji Tanaka[1]
[1]Tohoku University, JAPAN and [2]Fraunhofer Institute for Electronic Nano Systems, GERMANY

T29-d DAMAGE PROFILE MODELING AND EXPERIMENT OF SILICON CARBIDE SUBSTRATES
IN MICRO-NANO STRUCTURE FABRICATED BY HELIUM FOCUSED ION BEAM 606
Shupeng Gao, Xi Chen, Qianhuang Chen, Qi Li, and Yan Xing
Southeast University, CHINA

W29-d LIQUID-IMMERSION INCLINED-ROTATED UV LITHOGRAPHY FOR
MICRO SUCTION CUP ARRAY 610
Gakuto Kagawa and Hidetoshi Takahashi
Keio University, JAPAN

d - Materials, Fabrication and Packaging for Generic MEMS and NEMS
New & Emerging Materials for MEMS/NEMS

M30-d PARAMETRIC AMPLIFICATION AND PHONONIC FREQUENCY COMB
GENERATION IN MoS_2 NANOELECTROMECHANICAL RESONATORS 613
S M Enamul Hoque Yousuf[1], Yunong Wang[1], Jaesung Lee[1], Steven W. Shaw[2,3], and Philip X.-L. Feng[1]
[1]University of Florida, USA, [2]Florida Institute of Technology, USA, and [3]Michigan State University, USA

T30-d PARYLENE-N AS A HIGH TEMPERATURE THIN FILM PIEZOELECTRIC MATERIAL 617
Nathan Jackson and Deepak Kunwar
University of New Mexico, USA

W30-d SILICON CARBIDE-ON-INSULATOR THERMAL-PIEZORESISTIVE
RESONATOR FOR HARSH ENVIRONMENT APPLICATION 621
Baoyun Sun[1,2], Jiarui Mo[1], Hemin Zhang[3], Henk W. van Zeijl[1], Willem D. van Driel[1], and Guoqi Zhang[1]
*[1]Delft University of Technology, NETHERLANDS, [2]China University of Petroleum, CHINA, and
[3]KU Leuven, BELGIUM*

M31-d SPIN COATING OF HIGHLY ALIGNED AGCN MICROWIRES
EPITAXIALLY GROWN ON 2D MATERIALS 625
Jimin Ham, Jaemook Lim, Joowon Lim, Gunyoung Jang, Sueng Yoon Lee,
Dohyun Lim, Sukjoon Hong, and Won Chul Lee
Hanyang Universit, Ansan, KOREA

T31-d SUSPENDED TWO-DIMENSIONAL MATERIAL MEMBRANES FOR SENSOR
APPLICATIONS FABRICATED WITH A HIGH-YIELD TRANSFER PROCESS 627
Sebastian Lukas[1], Ines Kraiem[1,2], Maximilian Prechtl[3], Oliver Hartwig[3], Annika Grundmann[1], Holger
Kalisch[1], Satender Kataria[1], Michael Heuken[1,4], Andrei Vescan[1], Georg S. Duesberg[3], and Max C. Lemme[1,2]
[1]*RWTH Aachen University, GERMANY,* [2]*AMO GmbH, GERMANY,*
[3]*University of the Bundeswehr Munich, GERMANY, and* [4]*AIXTRON SE, GERMANY*

W31-d TCF-IMPROVED SH_0 MODE ACOUSTIC RESONATORS BASED
ON 30°YX-LINBO$_3$/SIO$_2$ MEMBRANE 631
Shuxian Wu[1], Zonglin Wu[1], Hangyu Qian[1], Feihong Bao[1],
Gongbin Tang[2], Feng Xu[1], and Jie Zou[1]
[1]*Fudan University, CHINA and* [2]*Shandong University, CHINA*

M32-d WAFER SCALE MULTILAYER GRAPHENE BASED BRAIN PROBES BY
SPIN-SPRAYING METHODS FOR MAGNETIC RESONANCE IMAGING 635
Kejun Tu, Zhejun Guo, Mengfei Xu, Bin Yang, and Jingquan Liu
Shanghai Jiao Tong University, CHINA

d - Materials, Fabrication and Packaging for Generic MEMS and NEMS
New Fabrication Processes for Making MEMS/NEMS

T32-d 3D SELF-ALIGNED FABRICATION OF SUSPENDED NANOWIRES
BY CRYSTALLOGRAPHIC NANOLITHOGRAPHY 639
Erwin J.W. Berenschot, Yasser Pordeli, Lucas J. Kooijman, Yves L. Janssens,
Roald M. Tiggelaar, and Niels R. Tas
University of Twente, NETHERLANDS

W32-d A SIMPLE PROCESS FOR THE FABRICATION OF PARALLEL-PLATE ELECTROSTATIC
MEMS RESONATORS BY GOLD THERMOCOMPRESSION BONDING 643
Dolores Manrique Juarez[1], Fabrice Mathieu[1], Guillaume Libaude[1], David Bourrier[1],
Samuel Charlot[1], Laurent Mazenq[1], Véronique Conédéra[1], Ludovic Salvagnac[1],
Isabelle Dufour[2], Liviu Nicu[1], and Thierry Leïchlé[1,3]
[1]*LAAS-CNRS, FRANCE,* [2]*Université de Bordeaux, IMS UMR-CNRS, FRANCE, and* [3]*Georgia Tech, USA*

M33-d ELECTROMECHANICALLY STABLE INTERCONNECTION BETWEEN
LIG AND THICK DAM-SHAPED METALLIC ELECTRODE VIA STORED
AG MICROPARTICLE SOLUTION 647
Saeyoung Park, Yoo-Kyum Shin, and Min-Ho Seo
Pusan National University, KOREA

T33-d FREE-STANDING MEMBRANES WITH SELF-ASSEMBLED
NANOPORE ARRAYS FOR TEM OBSERVATION OF LIQUID SAMPLES 651
Joowon Lim[1], Jimin Ham[1], Sungho Jeon[1], Yuna Bae[2,3], Minho Kang[2,3],
Sueng Yoon Lee[1], Jungwon Park[2,3], and Won Chul Lee[1]
[1]*Hanyang University, KOREA,* [2]*Seoul National University, KOREA, and*
[3]*Institute of Basic Science (IBS), KOREA*

W33-d NONPLANAR NANOFABRICATION VIA INTERFACE ENGINEERING 653
Sarah O. Spector, Peter F. Satterthwaite, and Farnaz Niroui
Massachusetts Institute of Technology, USA

M34-d WAFER-LEVEL FABRICATION OF CONFORMAL SUB 10-NM NANOGAPS 657
Sayali Tope, Seungbeom Noh, and Hanseup Kim
University of Utah, USA

d - Materials, Fabrication and Packaging for Generic MEMS and NEMS
Packaging & Assembly

T34-d MEMS RESONATOR VACUUM-SEALED BY SILICON
MIGRATION AND HYDROGEN OUTDIFFUSION 661
Muhammad Jehanzeb Khan, Yukio Suzuki, Tianjiao Gong, Takashiro Tsukamoto, and Shuji Tanaka
Tohoku University, JAPAN

W34-d MEMS THIN-FILM VACUUM PACKAGE UTILIZING GLOW DISCHARGE GETTER 665
Vikram Maharshi[1], Manjeet Kumar[1], Ajay Agarwal[2], and Bhaskar Mitra[1]
[1]Indian Institute of Technology, Delhi, INDIA and [2]Indian Institute of Technology, Jodhpur, INDIA

e – MEMS Actuators and PowerMEMS
Actuator Components & Systems

M35-e LNOI THIN-FILM DUAL-AXIS RESONANT MICRO-MIRROR
WITH E16 TORSIONAL ACTUATION 669
Yaoqing Lu[1,2,3], Kangfu Liu[1,2,3], Yuxi Wang[1,2,3], Ran Nie[1], and Tao Wu[1,2,3,4]
[1]ShanghaiTech University, CHINA, [2]Chinese Academy of Sciences, CHINA,
[3]University of Chinese Academy of Sciences, CHINA, and
[4]Shanghai Engineering Research Center of Energy Efficient and Custom AI IC, CHINA

T35-e A PIEZOELECTRIC MEMS SPEAKER WITH STRETCHABLE FILM SEALING 673
Linbing Xu, Mingchao Sun, Menglun Zhang, Chengze Liu, Xiaopeng Yang, and Wei Pang
Tianjin University, CHINA

W35-e BROADBAND MEMS SPEAKER BY SINGLE-WAY MULTI-RESONANCE
ARRAY WITH ACOUSTIC DAMPING TUNING: A PROOF OF CONCEPT 677
Mingchao Sun, Menglun Zhang, Chengze Liu, and Wei Pang
Tianjin University, CHINA

M36-e IONIC LIQUID ELECTROSPRAY THRUSTER WITH TWO-STAGE
ELECTRODES ON GLASS SUBSTRATE 681
Akane Nishimura[1], Yoshinori Takao[2], Toshiyuki Tsuchiya[1], and Yoshinori Takao[2]
[1]Kyoto University, JAPAN and [2]Yokohama National University, JAPAN

W36-e MONOLITHIC INTEGRATION OF PZT ACTUATION UNITS OF VARIOUS ACTIVATED
RESONANCES FOR FULL-RANGE MEMS SPEAKER ARRAY 685
Hsu-Hsiang Cheng[1], Sung-Cheng Lo[1], Yu-Chen Chen[1], Ming-Ching Cheng[1],
Ting-Chou Wei[1], Mingching Wu[2], and Weileun Fang[1]
[1]National Tsing Hua University, TAIWAN and [2]CoretronicMEMS Co., Ltd., TAIWAN

M37-e PULL-IN VOLTAGE REDUCTION IN ELECTROSTATIC AIRGAP ACTUATOR
USING 12 NM-ULTRATHIN INTERNAL DIELECTRIC TRANSDUCTION 689
Satish K. Verma and Bhaskar Mitra
Indian Institute of Technology, New Delhi, INDIA

e – MEMS Actuators and PowerMEMS
Energy Harvesting Materials, Structures, and Transducers

T37-e A REVERSE ELECTROWETTING-ON-DIELECTRIC (REWOD) ENERGY
HARVESTER USING NONWETTING GALLIUM COATED ELECTRODE
AND ULTRATHIN GALLIUM OXIDE SHELL AS DIELECTRIC LAYER 693
Jinwon Jeong, Bokyung Suh, and Jeong Bong (JB) Lee
University of Texas at Dallas, USA

W37-e ASYMMETRIC QUAD LEG ORTHOPLANAR SPRING FOR
WIDEBAND PIEZOELECTRIC MICRO ENERGY HARVESTING 697
Ali Mohammadi, Shamin Sadrafshari, Alborz Shokrani, and Chris R. Bowen
University of Bath, UK

M38-e EVALUATION OF THERMOELECTRIC PROPERTIES OF
MONOLITHICALLY-INTEGRATED CORE-SHELL Si NANOWIRE BRIDGES 701
Akio Uesugi, Shusuke Nishiyori, Koji Sugano, and Yoshitada Isono
Kobe University, JAPAN

T38-e GLAZE TILE-INSPIRED LIQUID-SOLID POWER GENERATOR
FOR CONTINUOUS WATER FLOW ENERGY HARVESTING 705
Dezhi Nie[1], Boming Lyu[1], Yongbo Hu[1], Jian Zhang[1], Yongqing Fu[2], Honglong Chang[1], and Kai Tao[1]
[1]Northwestern Polytechnical University, CHINA and [2]Northumbria University, UK

W38-e MEMS CANTILEVERED ENERGY HARVESTER WITH TAPERED
THICKNESS FOR STRESS CONTROL 709
Takahito Yokota, Kensuke Kanda, Takayuki Fujita, and Kazusuke Maenaka
University of Hyogo, JAPAN

M39-e TAPERED HELMHOLTZ RESONATOR WIND ENERGY
HARVESTER DRIVEN BY AEROACOUSTICS 712
Chen Hua, Liyun Zhen, Jingquan Liu, and Bin Yang
Shanghai Jiao Tong University, CHINA

e – MEMS Actuators and PowerMEMS
Manufacturing for Actuators & Power MEMS

T39-e ANDROMEDA: A FLEXIBLE MEMS TECHNOLOGY PLATFORM FOR A
VARIETY OF PIEZOELECTRICALLY ACTUACTED MICROMIRRORS 716
Irene Martini, Anna Alessandri, Marta Carminati, Roberto Carminati, Paolo Ferrarini, Daniela A.L. Gatti,
Riccardo Gianola, Borka Lazarova, Carla M. Lazzari, Andrea Nomellini, Laura Oggioni, Claudia Pedrini,
Carlo L. Prelini, Riccardo Tacchini, and Michele Vimercati
STMicroelectronis, ITALY

W39-e DESIGN OF BUTTERFLY PLATE PIEZOELECTRIC ACTUATOR
WITH DUAL DRIVING ELECTRODES FOR MEMS MICRO-MIRROR 720
Si-Han Chen[1], Shih-Chi Liu[1], Hao-Chien Cheng[1], Hung-Yu Lin[1],
Kai-Chih Liang[2], Mingching Wu[2], and Weileun Fang[1]
[1]National Tsing Hua University, TAIWAN and [2]Coretronic MEMS Corporation, TAIWAN

M40-e FULLY-FLEXIBLE MICRO-SCALE ACTUATOR ARRAY
WITH THE LIQUID-GAS PHASE CHANGE MATERIALS 724
Sangjun Sim, Kyubin Bae, and Jongbaeg Kim
School of Mechanical Engineering, Yonsei University, KOREA

e – MEMS Actuators and PowerMEMS
Power MEMS Components & Systems

T40-e A NOVEL COMB DESIGN FOR ENHANCED POWER AND
BANDWIDTH IN ELECTROSTATIC MEMS ENERGY CONVERTORS 728
Jinglun Li[1], Habilou Ouro-Koura[1], Hannah Arnow[1], Arian Nowbahari[2], Mathew Galarza[1], Meg Obispo[1],
Xing Tong[1], Mehdi Azadmehr[2], Mona M. Hella[1], John A. Tichy[1], and Diana-Andra Borca-Tasciuc[1]
[1]Rensselaer Polytechnic Institute, USA and [2]University of South-Eastern Norway, NORWAY

e – MEMS Actuators and PowerMEMS
Self-Powered Devices and Microsystems

W40-e   A HYBRID NANOGENERTOR-DRIVEN SELF-POWERED
WEARABLE PERSPIRATION MONITORING SYSTEM 732
Md Abu Zahed, S M Sohel Rana, Md Sharifuzzaman, Seonghoon Jeong,
Gagan Bahadur Pradhan, Hye Su Song, and Jae Yeong Park
Kwangwoon University, KOREA

M41-e   A MONOLITHIC INTEGRATED AND TRANSPARENT MICROSYSTEM
CONSTRUCTED BY USING AMORPHOUS INGAZNO FILM 736
Bin Jia, Chao Zhang, and Xiaodong Huang
Southeast University, CHINA

T41-e   FLOWING WATER ENABLES STEERABLE CHARGE
DISTRIBUTION ON ELECTRET SURFACE 740
Boming Lyu[1], Jian Zhang[1], Yunjia Li[2], Yongqing Fu[3], Honglong Chang[1], Weizheng Yuan[1], and Kai Tao[1]
[1]*Northwestern Polytechnical University, CHINA*, [2]*Xi'an Jiaotong University, CHINA, and*
[3]*University of Northumbria, UK*

W41-e   SELF-POWERED FLEXIBLE PIEZOELECTRET
ARRAY FOR WEARABLE APPLICATIONS 744
Hao Yang[1,2], Rui M.R. Pinto[1], Pedro González[1], Alar Ainla[1],
Mohammadmahdi Faraji[1], and K.B. Vinayakumar[1]
[1]*International Iberian Nanotechnology Laboratory, PORTUGAL and* [2]*Xi'an Jiaotong University, CHINA*

f - MEMS Physical and Chemical Sensors
Fluidic Sensors

M42-f   A BULK-TYPE PRESSURE SENSOR WITH FULL-BRIDGE
IMPLEMENTATION ENABLED BY STRESS-MODIFYING TRENCHES 748
Dequan Lin[1], Man Wong[1], and Kevin Chau[1,2]
[1]*Hong Kong University of Science and Technology, CHINA*, [2]*Chinese Academy of Science, CHINA*

T42-f   A CMOS COMPATIBLE MICRO PIRANI GAUGE WITH STRUCTURE
OPTIMIZATION FOR PERFORMANCE ENHANCEMENT 752
Rui Jiao[1], Gai Yang[2], Ruoqin Wang[1], Yue Tang[2], Zhongyi Liu[2],
Huikai Xie[2,3], Hongyu Yu[1], and Xiaoyi Wang[2,3]
[1]*Hong Kong University of Science and Technology, HONG KONG*, [2]*Beijing Institute of Technology, CHINA,*
and [3]*BIT Chongqing Institute of Microelectronics and Microsystems, CHINA*

W42-f   A THERMAL AIRFLOW SENSOR BASED ON MN-CO-NI-O THIN FILM 756
Jie Wang, Yunfei Liu, Zhezheng Zhu, Chengchen Gao, Zhenchuan Yang, and Yilong Hao
Peking University, CHINA

M43-f   HIGHLY SENSITIVE WAVE HEIGHT SENSOR WITH MEMS
PIEZORESISTIVE CANTILEVER AND WATERPROOF MEMBRANE 760
Takuto Hirayama and Hidetoshi Takahashi
Keio University, JAPAN

T43-f   MEMS CAPACITANCE DIAPHRAGM GAUGE WITH
TWO SEALED REFERENCE CAVITIES 763
Xiaodong Han[1,2], Jingzhen Li[3], Gang Li[4], and Yongjian Feng[1]
[1]*Xiamen University, CHINA*, [2]*University of Twente, NETHERLANDS,*
[3]*Beijing University of Technology, CHINA, and* [4]*Lanzhou Institute of Physics, CHINA*

W43-f   TOWARDS A GAS INDEPENDENT THERMAL FLOW METER 767
Shirin Azadi Keari[1], Remco J. Wiegerink[1], Remco G.P. Sanders[1], and Joost C. Lotters[1,2]
[1]*University of Twente, NETHERLANDS and* [2]*Bronkhorst High-Tech BV, NETHERLANDS*

f - MEMS Physical and Chemical Sensors
Force & Displacement Sensors

M44-f   AN INTEGRATED MEMS DEVICE FOR *IN-SITU* FOUR-PROBE
ELECTRO-MECHANICAL CHARACTERIZATION OF PT NANOBEAM 771
Yuheng Huang, Meng Nie, Binghui Li, Kuibo Yin, and Litao Sun
Southeast University, CHINA

T44-f   FINGERLIKE TACTILE TEXTURE INTEGRATED SENSOR WITH
COLD AND WARM SENSATIONS OF SUB-MM SPATIAL RESOLUTION 775
Nachi Mise, Mitsuki Kozasa, Kyohei Terao, Fusao Shimokawa, and Hidekuni Takao
Kagawa University, JAPAN

W44-f   MODIFIED BEAM STRUCTURES FOR IMPROVED RESONANT SENSING 779
Erfan Ghaderi and Behraad Bahreyni
Simon Fraser University, CANADA

M45-f   OCCLUSAL PAPER-BASED FLEXIBLE PRESSURE SENSOR FOR
IN SITU MEASURING ORAL OCCLUSAL FORCE 783
Wenduo Wang, Xin Zhang, Ning Zhao, Jingquan Liu, and Bin Yang
Shanghai Jiao Tong University, CHINA

T45-f   SUCTION CUP ARRAY WORKING ALSO AS TACTILE SENSOR
TO DETECT CUPS DEFORMATION USING KCF AND CNN 787
Toshihiro Shiratori, Jinya Sakamoto, Yuki Kumokita, Masato Suzuki,
Tomokazu Takahashi, and Seiji Aoyagi
Kansai University, JAPAN

W45-f   VERTICAL INTEGRATION OF FORCE TRANSMISSION STRUCTURE ON CAPACITIVE
CMOS-MEMS TACTILE FORCE SENSOR FOR SENSITIVITY IMPROVEMENT 791
Yuanyuan Huang, Yen-Lin Chen, Shihwei Lin, Fuchi Shih, Zihsong Hu, and Weileun Fang
National Tsing Hua University, TAIWAN

f - MEMS Physical and Chemical Sensors
Gas & Chemical Sensors

M46-f   1-OCTADECANETHIOL SAM ON CMOS-MEMS GOLD PLATED
RESONATOR VIA DIP-CAST FOR VOCs SENSING 795
Rafel Perelló-Roig[1,2], Jaume Verd[1,2], Sebastià Bota[1,2],
Bartomeu Soberats[1], Antonio Costa[1], and Jaume Segura[1,2]
[1]*University of the Balearic Islands, SPAIN and* [2]*Health Research Institute of the Balearic Islands, SPAIN*

T46-f   APPLICATION OF DEEP LEARNING NETWORK FOR HUMIDITY
COMPENSATION OF SEMICONDUCTOR METAL OXIDE GAS SENSORS 799
Mingu Kang, Incheol Cho, and Inkyu Park
Korea Advanced Institute of Science and Technology (KAIST), KOREA

W46-f   DEVELOPMENT OF MONOLITHIC MICRO-LED GAS SENSOR BASED
E-NOSE SYSTEM FOR REAL-TIME, SELECTIVE GAS PREDICTION 803
Kichul Lee, Mingu Kang, and Inkyu Park
Korea Advanced Institute of Science and Technology (KAIST), KOREA

M47-f   ELECTRONIC-NOSE: AN ARRAY OF 16 MOS-GAS SENSORS INTEGRATED
WITH TEMPERATURE AND MOISTURE SENSING CAPABILITIES 807
Xiawei Yue[1,2], Shuai Wei[1,2], Pingping Zhang[3], Zhitao Zhou[1], Tiger Tao[1,2,4,5,6], and Nan Qin[1]
[1]*Chinese Academy of Sciences (CAS), CHINA,* [2]*University of Chinese Academy of Sciences, CHINA,*
[3]*Suzhou Huiwen Nanotechnology Co. Ltd., CHINA,* [4]*ShanghaiTech University, CHINA,* [5]*Shanghai Research
Center for Brain Science and Brain-Inspired Intelligence, CHINA and* [6]*Neuroxess Co., Ltd. (Jiangxi), CHINA*

T47-f   ENHANCEMENT OF SENSITIVITY IN PHOTONIC CRYSTAL BASED
CHEMICAL SENSOR USING CHEMO-MECHANICAL BILAYER EFFECT 811
Seyeon Lee[1], Naik T. Banabathi[1], Dongwon Kang[3], Sookyung Kang[2],
Kyungsuk Cho[2], Jungwook Kim[1], and Jungyul Park[1]
[1]Sogang University, KOREA, [2]Iwha University, KOREA, and [3]University of California, Los Angeles, USA

W47-f   METAL ION RECOGNITION SENSOR BASED ON RESISTANCE SWITCHING EFFECT 814
Tian Kang, Yusa Chen, Guanzhou Lin, Shengxiao Jin,
Liye Li, Hongshun Sun, Senyong Hu, and Wengang Wu
Peking University, CHINA

M48-f   MULTI-HOTSPOT MID-IR NANOANTENNAS WITH MATCHED LOSS
AND HIGH-INTENSITY NEAR-FIELD FOR SUB-PPM-LEVEL GAS DETECTION 818
Hong Zhou, Zhihao Ren, Cheng Xu, Liangge Xu, Xinge Guo, and Chengkuo Lee
National University of Singapore, SINGAPORE

T48-f   PALLADIUM BASED MEMS HYDROGEN SENSORS 822
Max Hoffmann[1], Marion Wienecke[1], Maren Lengert[2], Michael H. Weidner[2], and Jan Heeg[2]
*[1]Hochschule Wismar, Institut für Oberflächen- und Dünnschichttechnik, GERMANY and
[2]Materion GmbH, GERMANY*

W48-f   SELECTIVE DISCRIMINATION OF PPB-LEVEL VOCS USING MOS GAS
SENSOR IN PULSE-HEATING MODE WITH THE MODIFIED HILL'S MODEL 826
Gaoqiang Niu, Yi Zhuang, Yushen Hu, Zong Liu, and Fei Wang
Southern University of Science and Technology, CHINA

M49-f   THERMAL CONDUCTIVITY DETECTOR (TCD)-TYPE GAS SENSOR BASED
ON THE SUSPENDED 1D NANOHEATER FOR IOT APPLICATIONS 830
Wootaek Cho, Jong-Hyun Kwak, Taejung Kim, and Heungjoo Shin
Ulsan National Institute of Science and Technology (UNIST), KOREA

f - MEMS Physical and Chemical Sensors
Inertial Sensors

T49-f   120 PPM QUALITY FACTOR THERMAL STABILITY FROM -40°C TO +60°C OF A
DUAL-AXIS MEMS GYROSCOPE BASED ON JOULE EFFECT DYNAMIC CONTROL 833
Jian Cui[1,2] and Qiancheng Zhao[1,2]
[1]Peking University, CHINA and [2]Beijing Advanced Innovation Center for Integrated Circuits, CHINA

W49-f   A FORCE-BANLANCE CAPACITIVE MEMS GRAVIMETER
WITH SUPERIOR RESPONSE TIME, SELF-NOISE AND DRIFT 837
Le Gao[1], Fangzheng Li[1], Jian Zhang[1], Bingyang Cai[1], Wenjie Wu[1], and Liangcheng Tu[2]
[1]Huazhong University of Science and Technology, CHINA and [2]Sun Yat-sen University, CHINA

M50-f   A MEMS-BASED GRAVIMETER FOR SIMULTANEOUS VERTICAL
AND HORIZONTAL EARTH TIDES MEASUREMENTS 841
Lujia Yang[1], Xiaochao Xu[1], Qian Wang[1], Ji'ao Tian[1], Yanyan Fang[1],
Chun Zhao[1], Wenjie Wu[1], Fangjing Hu[1], and Liangcheng Tu[1,2]
[1]Huazhong University of Science and Technology, CHINA and [2]Sun Yat-sen University, CHINA

T50-f   A NOVEL MULTIPLE FOLDED BEAM DISK RESONATOR FOR MAXIMIZING
THE THERMOELASTIC QUALITY FACTOR 845
Xiaopeng Sun[1], Xin Zhou[1], Lei Yu[2], Kaixuan He[2], Xuezhong Wu[1], and Dingbang Xiao[1]
*[1]National University of Defense Technology, CHINA and
[2]East China Institute of Photo-Electronic IC, CHINA*

W50-f A TIME-SERIES CONFIGURATION METHOD OF MODE REVERSAL MEMS GYROSCOPES UNDER DIFFERENT TEMPERATURE-VARYING CONDITIONS 849
Liangqian Chen, Tongqiao Miao, Qingsong Li, Peng Wang, Junjian Li, Xuezhong Wu, Dingbang Xiao, and Xiang Xi
National University of Defense Technology, CHINA

M51-f ACOUSTICALLY ISOLATED MEMS BAW GYROSCOPES 853
Diego Emilio Serrano, Amir Rahafrooz, Duane Younkin, Kieran Nunan, Mitul Dalal, Sagnik Pal, and Ijaz Jafri
Panasonic Device Solutions Laboratory of Massachusetts, USA

T51-f ACTIVE QUALITY FACTOR STABILIZATION OF MEMS RESONATOR UTILIZING ELECTRICAL DISSIPATION REGULATION 857
Yang Zhao, Qin Shi, Guoming Xia, and Anping Qiu
Nanjing University of Science and Technology, CHINA

W51-f DEMONSTRATION OF GYRO-LESS NORTH FINDING USING A T-SHAPED MEMS DIFFERENTIAL RESONANT ACCELEROMETER 861
Kei Masunishi, Etsuji Ogawa, Daiki Ono, Fumito Miyazaki, Hiroki Hiraga, Kengo Uchida, Jumpei Ogawa, Hideaki Murase, and Yasushi Tomizawa
Toshiba Corporation, JAPAN

M52-f ENHANCED STIFFNESS SENSITIVITY IN A MODE LOCALIZED SENSOR USING INTERNAL RESONANCE ACTUATION 865
Jianlin Chen[1], Hemin Zhang[2], Takashiro Tsukamoto[1], Michael Kraft[2], and Shuji Tanaka[1]
[1]Tohoku University, JAPAN and [2]KU Leuven, BELGIUM

T52-f MODELING STRESS EFFECTS ON FREQUENCIES OF A MEMS RING GYROSCOPE 869
Mehran Hosseini-Pishrobat, Baha Erim Uzunoglu, and Erdinc Tatar
Bilkent University, TURKEY

W52-f RATE INTEGRATING GYROSCOPE TUNED BY FOCUS ION BEAM TRIMMING AND INDEPENDENT CW/CCW MODES CONTROL 873
Jianlin Chen[1], Takashiro Tsukamoto[1], Giacomo Langfelder[2], and Shuji Tanaka[1]
[1]Tohoku University, JAPAN and [2]Politecnico di Milano, ITALY

M53-f TEMPERATURE DEPENDENCE OF QUALITY FACTORS AT HIGH FREQUENCIES IN MEMS GYROSCOPES 877
Daniel Schiwietz[1,2], Eva M. Weig[2], and Peter Degenfeld-Schonburg[1]
[1]Robert Bosch GmbH, GERMANY and [2]Technical University of Munich, GERMANY

f - MEMS Physical and Chemical Sensors
Manufacturing Techniques for Physical Sensors

T53-f 0.5MM×0.5MM 150KPA-MEASURE-RANGE HIGH-TEMPERATURE PRESSURE SENSOR WITH HIGH-PERFORMANCE AND LOW FABRICATION-COST 881
Peng Li[1,2], Wei Li[1], Changnan Chen[1,3], Ke Sun[1], Min Liu[1], Sheng Wu[1], Pichao Pan[1,3], Jiachou Wang[1,3], and Xinxin Li[1,2,3]
[1]Chinese Academy of Sciences, CHINA, [2]Fudan University, CHINA, and [3]University of Chinese Academy of Sciences, CHINA

W53-f AUTOMATIC PICO LASER TRIMMING SYSTEM FOR SILICON MEMS RESONANT DEVICES BASED ON IMAGE RECOGNITION 885
Yuxian Liu[1], Qiancheng Zhao[1,2], Dacheng Zhang[1], and Jian Cui[1,2]
[1]Peking University, CHINA and [2]Beijing Advanced Innovation Center for Integrated Circuits, CHINA

M54-f MICROMACHINING FUSED SILICA MICRO SHELL RESONATOR WITH QUARTZ GLASS MOLD BY THERMAL REFLOW 889
Zhaoxi Su, Bin Luo, Qiankai Tang, Linqian Zhu, and Jintang Shang
Southeast University, CHINA

T54-f WAFER-LEVEL PATTERNING OF TIN OXIDE
NANOSHEETS FOR MEMS GAS SENSORS 893
Mingjie Li, Wenxin Luo, Xiaojiang Liu, Gaoqiang Niu, and Fei Wang
Southern University of Science and Technology, CHINA

f - MEMS Physical and Chemical Sensors
Materials for Physical Sensors

W54-f AIR DAMPING EFFECTS ON DIFFERENT MODES OF AlN-on-Si
MICROELECTROMECHANICAL RESONATORS 897
Yuncong Liu[1], S M Enamul Hoque Yousuf[1], Afzaal Qamar[2], Mina Rais-Zadeh[2,3], and Philip X.-L. Feng[1]
[1]University of Florida, USA, [2]University of Michigan, USA, and [3]California Institute of Technology, USA

M55-f A NOVEL PIEZORESISTIVE PRESSURE SENSOR
BASED ON CR-DOPED V_2O_3 THIN FILM 901
Michiel Gidts, Wei-Fan Hsu, María Recaman Payo, Shashwat Kushwaha, Chen Wang,
Frederik Ceyssens, Dominiek Reynaerts, Jean-Pierre Locquet, and Michael Kraft
KU Leuven, BELGIUM

f - MEMS Physical and Chemical Sensors
Metrology and Measurement Techniques for MEMS/NEMS Sensors

T55-f A NOVEL FEEDTHROUGH CANCELLATION TECHNIQUE FOR
PIEZOELECTRIC MEMS RESONANT SENSORS IN IONIC LIQUID MEDIUM 905
Cheng-Yen Wu, Zhong-Wei Lin, and Sheng-Shian Li
National Tsing Hua University, TAIWAN

W55-f CHARACTERIZATION OF PACKAGING STRESS WITH
A CAPACITIVE STRESS SENSOR ARRAY 909
Tolga Veske[1], Derin Erkan[1], and Erdinc Tatar[1,2]
[1]Bilkent University, TURKEY and [2]National Nanotechnology Research Center (UNAM), TURKEY

M56-f MILLISECOND-LEVEL PULSE-HEATING SENSING SYSTEM
FOR MEMS-BASED GAS SENSORS 913
Yi Zhuang, Gaoqiang Niu, Lang Wu, and Fei Wang
Southern University of Science and Technology, CHINA

T56-f MULTIPLE PARAMETER DECOUPLING USING A SINGLE
RESONANT MEMS SENSOR VIA BLUE SIDEBAND EXCITATION 917
Jingqian Xi[1], Lei Xu[1], Yuan Wang[2], Fangjing Hu[1], Chengxin Li[4], Linlin Wang[4],
Huafeng Liu[1], Chen Wang[4], Michael Kraft[4], and Chun Zhao[3]
*[1]Huazhong University of Science and Technology, CHINA, [2]University of Macau, CHINA,
[3]University of York, UK, and [4]University Leuven, BELGIUM*

f - MEMS Physical and Chemical Sensors
Nanoscale Physical Sensors

W56-f DIAMOND NANOWIRES ARRAY PREPARED BY ANNEALING NANO-CRYSTALLINE
DIAMOND IN AIR AND ITS APPLICATION IN FIELD EMISSION 921
Yang Wang, Chen Lin, and Jinwen Zhang
Peking University, CHINA

M57-f QUANTIFIED STRESS RELAXATION IN CARBON NANOTUBE RESONATORS 925
Morten Vollmann, Cosmin Roman, Miroslav Haluska, and Christofer Hierold
ETH Zürich, SWITZERLAND

T57-f   **SELF-REFERENCED TEMPERATURE SENSORS BASED ON CASCADED SILICON RING RESONATOR** 929
Xiantao Zhu, Minmin You, Zude Lin, Bin Yang, and Jingquan Liu
Shanghai Jiao Tong University, CHINA

f - MEMS Physical and Chemical Sensors
Sonic & Ultrasonic MEMS Transducers

W57-f   **A 0.35 mm^2 SYSTEM ON CHIP LEVEL DETECTOR BASED ON AN ANNULAR PMUT-ON-CMOS ARRAY** 933
Eyglis Ledesma, Iván Zamora, Francesc Torres, Arantxa Uranga, and Núria Barniol
Universitat Autònoma de Barcelona, SPAIN

M58-f   **AN ALSCN PMUT-ON-CMOS SENSOR FOR MONITORING FLUIDS' DENSITY, VISCOSITY, SOUND VELOCITY, AND COMPRESSIBILITY** 937
Eyglis Ledesma, Iván Zamora, Jesús Yanez, Arantxa Uranga, and Núria Barniol
Universitat Autònoma de Barcelona, SPAIN

T58-f   **AUTO-POSITIONING AND HAPTIC STIMULATIONS VIA A 35 MM SQUARE PMUT ARRAY** 941
Wei Yue[1], Yande Peng[1], Hanxiao Liu[1], Fan Xia[1], Fanping Sui[1], Seiji Umezawa[2],
Shinsuke Ikeuchi[2], Yasuhiro Aida[2], and Liwei Lin[1]
[1]University of California, Berkeley, USA and [2]Murata Manufacturing Co., Ltd., JAPAN

W58-f   **BODY FORCE BASED DROPLET EJECTION BY GHZ ACOUSTIC MICRO-TRANSDUCER** 945
Haitao Zhang, Yangchao Zhou, Menglun Zhang, Wenlan Guo, Chen Sun, Xuexin Duan, and Wei Pang
Tianjin University, CHINA

M59-f   **BONE CONDUCTION PICKUP BASED ON PIEZOELECTRIC MICROMACHINED ULTRASONIC TRANSDUCERS** 949
Chongbin Liu[1], Xiangyang Wang[1], Yong Xie[2], and Guoqiang Wu[1]
[1]Wuhan University, CHINA and [2]Xidian University, CHINA

T59-f   **BREAKING THE DEAD ZONE LIMITATION OF PMUTS BASED ON A PHASE SHIFT OF DRIVING WAVEFORM WITH WINDOW FUNCTION** 953
Chun-You Liu, Chin-Yu Chang, and Sheng-Shian Li
National Tsing Hua University, TAIWAN

W59-f   **DRONE-MOUNTED LOW-FREQUENCY PMUTS FOR > 6-METER RANGEFINDER IN AIR** 957
Hanxiao Liu[1], Yande Peng[1], Wei Yue[1], Seiji Umezawa[2], Shinsuke Ikeuchi[2], Yasuhiro Aida[2],
Chunming Chen[1], Peggy Tsao[1], and Liwei Lin[1]
[1]University of California, Berkeley, USA and [2]Murata Manufacturing Co., Ltd., JAPAN

M60-f   **MASS PRODUCED MICROMACHINED ULTRASONIC TIME-OF-FLIGHT SENSORS OPERATING IN DIFFERENT FREQUENCY BANDS** 961
Richard J. Przybyla[1], Stefon E. Shelton[1], Cathy Lee[1], Ben Eovino[1], Quy Chau[1],
Mitchell H. Kline[1], Oleg I. Izyumin[1], and David A. Horsley[1,2]
[1]TDK Invensense, USA and [2]University of California, Davis, USA

T60-f   **MEMS FIRST-ORDER BESSEL BEAM ACOUSTIC TRANSDUCER FOR PARTICLE TRAPPING AND CONTROLLABLE ROTATING** 965
Jiaqi Li[1], Zhenhuan Sun[1], Yuyu Jia[1], Teng Li[1], Haojian Lu[2], Lurui Zhao[3], Hai Liu[3], and Song Liu[1]
*[1]ShanghaiTech University, CHINA, [2]Zhejiang University, CHINA, and
[3]University of Southern California, Los Angeles, USA*

W60-f NON-INVASIVE CAROTID ARTERY MONITORING BY USING
ALUMINUM NITRIDE PMUT CLOSE-PACKED ARRAYS 969
Sheng Wu[1,2,3], Kangfu Liu[2], Shuai Shao[2], Wei Li[1,3], Ying Chen[1,3], Tao Wu[2], and Xinxin Li[1,3]
[1]Chinese Academy of Sciences, CHINA, [2]ShanghaiTech University, CHINA, and
[3]University of Chinese Academy of Sciences, CHINA

M61-f NON-LINEAR BEHAVIORAL MODELING OF CAPACITIVE MEMS MICROPHONES 973
Sebastian Anzinger[1,2], Hutomo Suryo Wasisto[1], Abhiraj Basavanna[1], and Alfons Dehé[2,3]
[1]Infineon Technologies AG, GERMANY, [2]University of Freiburg, GERMANY, and
[2]Hahn-Schickard-Gesellschaft, GERMANY

T61-f VORTEX-BEAM ACOUSTIC TRANSDUCER FOR UNDERWATER PROPULSION 977
Jaehoon Lee, Kianoush Sadeghian Esfahani, and Eun S. Kim
University of Southern California, USA

W61-f WIDEBAND AND HIGHLY SENSITIVE MICROMACHINED PZT FILM-BASED
ULTRASONIC MICROPHONE WITH PARYLENE FILM AND FLEXIBLE
HELMHOLTZ RESONATOR ENHANCEMENT 981
Chung-Hao Huang and Guo-Hua Feng
National Tsing Hua University, TAIWAN

f - MEMS Physical and Chemical Sensors
Other Physical Sensors

M62-f HALBACH-ARRAY MAGNETIC COIL ARRANGEMENT ON CMOS
CHIP FOR SENSITIVITY ENHANCEMENT OF INDUCTIVE TACTILE SENSOR 985
Tien Chou, Zih-Song Hu, and Weileun Fang
National Tsing Hua University, TAIWAN

T62-f *ON-MEMS-CHIP* COMPACT TEMPERATURE SENSOR FOR
LARGE-VOLUME, LOW-COST SENSOR CALIBRATION 989
Paolo Frigerio[1], Andrea Fagnani[1], Valentina Zega[1], Gabriele Gattere[2],
Attilio Frangi[1], and Giacomo Langfelder[1]
[1]Politecnico di Milano, ITALY and [2]STMicroelectronics, ITALY

W62-f PARTICULATE MATTER SENSOR BASED ON TWO STAGE CASCADE
VIRTUAL IMPACTORS AND THERMOPHORETIC MICROHEATERS 993
Kwang-Wook Choi[1], Ilhwan Kim[1], Seokwhan Chung[1], Gi-Bong Sung[2], and Se-Jin Yook[2]
[1]Samsung Advanced Institute of Technology, KOREA and [2]Hanyang University, KOREA

g – Micro- and Nanofluidics
Biological and Medical Microfluidics and Nanofluidics

M63-g A MICROFLUIDIC OXYGEN GRADIENT GENERATOR FOR
THE STUDY OF AEROTROPISM IN HYPHAE OF OOMYCETES 997
Ayelen Tayagui[1,2], Yiling Sun[1,2], Ashley Garrill[1], and Volker Nock[1,2]
[1]University of Canterbury, NEW ZEALAND and
[2]MacDiarmid Institute for Advanced Materials and Nanotechnology, NEW ZEALAND

T63-g A PAPER-BASED DUAL APTAMER ASSAY ON AN INTEGRATED
MICROFLUIDIC SYSTEM FOR DETECTION OF HNP 1 AS A
BIOMARKER FOR PERIPROSTHETIC JOINT INFECTIONS 1001
Rishabh Gandotra[1], Feng-Chih Kuo[2], Mel S. Lee[3], and Gwo-Bin Lee[1]
[1]National Tsing Hua University, TAIWAN, [2]Kaohsiung Chang Gung Memorial Hospital, TAIWAN, and
[3]Paochien Hospital, TAIWAN

W63-g AN INTEGRATED MICROFLUIDIC PLATFORM FOR TUMOR CELL SEPARATION AND FLUORESCENCE IN SITU HYBRIDIZATION AT SINGLE CELL LEVEL 1005
Shihui Qiu[1,2], Na Li[1,2], Zhenhua Wu[1,2], Jianlong Zhao[1,2], and Hongju Mao[1,2]
[1]Chinese Academy of Science, CHINA and [2]University of Chinese Academy of Sciences, CHINA

M64-g CHARACTERIZATION OF OOCYTE HARDENING USING A MICROFLUIDIC ASPIRATION-ASSISTED ELECTRICAL IMPEDANCE SPECTROSCOPY SYSTEM 1009
Yuan Cao, Julia Floehr, and Uwe Schnakenberg
RWTH Aachen University, GERMANY

T64-g DOUBLE PULSE IRRADIATION OF FS LASER FOR ENHANCING THE PERFORMANCE OF PRECISE LASER SORTING METHOD 1013
Ryota Kiya[1], Yoshinaga Rintaro[1], Yo Tanaka[2], Yaxiaer Yalikun[1,2], and Yoichiroh Hosokawa[1]
[1]Nara Institute of Science and Technology, JAPAN and
[2]Institute of Physical and Chemical Research (RIKEN), JAPAN

W64-g DROPLET BASED HIGH THROUGHPUT SINGLE-SPERM CRYOPRESERVATION PLATFORM 1017
Na Li[1,2], Shihui Qiu[1,3], Zhenhua Wu[1,3], and Hongju Mao[1,3]
[1]Chinese Academy of Sciences, CHINA, [2]ShanghaiTech University, CHINA, and
[3]University of Chinese Academy of Sciences, CHINA

M65-g DUAL ION-SELECTIVE MEMBRANE DEPOSITED ION-SENSITIVE FIELD-EFFECT TRANSISTOR (DISM-ISFET) INTEGRATING WHOLE BLOOD PROCESSING MICROCHAMBER FOR IN SITU BLOOD ION TESTING 1021
Xiao-Wen Chen, Syuan-Rong Huang, and Nien-Tsu Huang
National Taiwan University, TAIWAN

---

g – Micro- and Nanofluidics
Generic Microfluidics & Nanofluidics

W65-g STRONG MICROSTREAMING FROM A PINNED OSCILLATING MEMBRANE AND APPLICATION TO GAS EXCHANGE 1025
Anthony L. Mercader and Sung Kwon Cho
University of Pittsburgh, USA

M66-g TUNABLE NANOPORE-INTEGRATED MICRO-/NANOFLUIDIC PLATFORM FOR ION TRANSPORT CONTROL IN THE PRESENCE OF CONCENTRATION AND TEMPERATURE GRADIENTS 1029
Dongwoo Seo[1], Dongjun Kim[1], Jongwan Lee[1], Cong Wang[2], Jungyul Park[2], and Taesung Kim[1]
[1]Ulsan National Institute of Science and Technology (UNIST), KOREA and [2]Sogang University, KOREA

---

g – Micro- and Nanofluidics
Integrated/Embedded Microfluidics and Nanofluidic Systems & Platforms

W66-g QUANTITATIVE ASSESSMENT OF CAPTURED MAGNETIC NANOPARTICLES USING SELF-POWERED MAGNETOELECTRIC PLATFORM FOR BIOLOGICAL APPLICATIONS 1033
Pankaj Pathak, Vinit K. Yadav, Samaresh Das, and Dhiman Mallick
Indian Institute Of Technology Delhi, INDIA

M67-g REAL-TIME OPERATION OF MICROCANTILEVER-BASED IN-PLANE RESONATORS PARTIALLY IMMERSED IN A MICROFLUIDIC SAMPLER 1037
Jiushuai Xu, Entian Cao, Michael Fahrbach, Vladislav Agluschewitsch, Andreas Waag, and Erwin Peiner
Technische Universität Braunschweig, GERMANY

T67-g SUSPENDED NANOCHANNEL RESONATORS MADE BY
NANOIMPRINT AND GAS PHASE DEPOSITION 1041
Manuel Müller[1], Jeremy Teuber[1], Rukan Nasri[1], Francesc Torres Canals[2], Núria Barniol[2],
Jordi Llobet Sixto[3], Xavier Borrise[3], Francesc Perez-Murano[3], and Irene Fernandez-Cuesta[1]
[1]University of Hamburg, GERMANY, [2]Universitat Autónoma de Barcelona, SPAIN, and
[3]IMB-CNM CSIC, SPAIN

g – Micro- and Nanofluidics
Manufacturing for Micro- and Nanofluidics

W67-g DEVELOPING AN EXTREMELY HIGH FLOW RATE PNEUMATIC PERISTALTIC
MICROPUMP FOR BLOOD PLASMA SEPARATION WITH INERTIAL PARTICLE
FOCUSING TECHNIQUE FROM FINGERTIP BLOOD WITH LANCETS 1045
Tuan N.A. Vo[1,2,3], Pin-Chuan Chen[1], and Pai-Shan Chen[4]
[1]National Taiwan University of Science and Technology, TAIWAN, [2]Ho Chi Minh City University of
Technology (HCMUT), VIETNAM, [3]Vietnam National University, VIETNAM, and [4]National Taiwan
University, TAIWAN

M68-g DIRECT PATTERNING ON POROUS SURFACE USING DROP IMPACT PRINTING 1049
Bheema Sankar Reddy[1], Chandantaru Dey Modak[1,2], Rutvik Lathia[1],
Bhawana Agarwal[1,3], Ebinesh Abraham R[1], and Prosenjit Sen[1]
[1]Indian Institute of Science, Bangalore, INDIA, [2]CNRS - ESPCI PSL, FRANCE, and
[3]Johns Hopkins University, USA

T68-g MANUFACTURING 3D-PRINTED PAPER MICROFLUIDICS INTEGRATED
WITH IONIZATION MASS-SPECTROMETRY FOR ILLICIT DRUGS
ANALYSIS AND ON-CHIP CHROMATOGRAPHY 1052
Muhammad Faizul Zaki[1], Pin-Chuan Chen[1], Yi-Xin Wu[2], and Pai-Shan Chen[2]
[1]National Taiwan University of Science and Technology, TAIWAN and
[2]National Taiwan University, TAIWAN

g – Micro- and Nanofluidics
Materials for Micro & Microfluidics

W68-g DETECTION LIMITS IN NANOMECHANICAL MASS FLOW SENSING
FOR NANOFLUIDICS WITH NANOWIRE OPEN CHANNELS 1056
Javier E. Escobar, Juan Molina, Eduardo Gil-Santos, José J. Ruz, Óscar Malvar,
Priscila M. Kosaka, Javier Tamayo, Álvaro San Paulo, and Montserrat Calleja
Instituto de Micro y Nanotecnología, IMN-CNM (CSIC), SPAIN

g – Micro- and Nanofluidics
Modeling of Micro & Nanofluidics

M69-g CONTROLLING PARTICLE AGGREGATION AND SEPARATION
IN LIQUID ON MEMBRANE RESONATORS 1060
Haoran Zhang[1,2], Hao Jia[1,2], and Xinxin Li[1,2]
[1]Chinese Academy of Sciences, CHINA and [2]University of Chinese Academy of Sciences, CHINA

T69-g DEVELOPMENT OF BOAT MODEL POWERED BY
ELECTRO-HYDRODYNAMIC PROPULSION SYSTEM 1064
Luan Ngoc Mai[1,2], Tuan-Khoa Nguyen[3], Trung Hieu Vu[3], Thien Xuan Dinh[4], Canh-Dung Tran[5], Hoang-
Phuong Phan[6], Toan Dinh[5], Thanh Nguyen[5], Nam-Trung Nguyen[3], Dzung Viet Dao[3], and Van Thanh Dau[3]
[1]Ho Chi Minh City University of Technology, VIETNAM, [2]Vietnam National University Ho Chi Minh City,
VIETNAM, [3]Griffith University, AUSTRALIA, [4]Explosion Research Institute Inc., JAPAN, [5]University of
Southern Queensland, AUSTRALIA, and [6]University of New South Wales, AUSTRALIA

W69-g HEMODYNAMIC ANALYSIS OF CARDIOMEMS:
ADVERSE HEMODYNAMIC EFFECTS 1068
Zhenhao Liu[1], Jiangli Han[2], and Xing Chen[1]
[1]Beihang University, CHINA and [2]Peking University Third Hospital, CHINA

M70-g MODAL QUALITY FACTOR INVERSION OF NON-SLENDER
MEMS RESONATORS BETWEEN GASES AND LIQUIDS 1072
Andre L. Gesing, Thomas Tran, Daniel Platz, and Ulrich Schmid
TU Wien, AUSTRIA

g – Micro- and Nanofluidics
Other Micro- and Nanofluidics

T70-g CLASSIFYING CELL CYCLE BY ELECTRICAL
PROPERTIES USING MACHINE LEARNING 1076
Jian Wei and Xiaoxing Xing
Beijing University of Chemical Technology, CHINA

W70-g HIGH-THROUGHPUT SPHERICAL SUPRAPARTICLE SELF-ASSEMBLY
BY ENHANCED EVAPORATION OF COLLOIDAL WATER DROPLETS
THROUGH THIN FILM OF WATER-SOLUBLE OIL 1080
Wonhyung Lee, Joowon Rhee, and Joonwon Kim
Pohang University of Science and Technology (POSTECH), KOREA

M71-g IN-ICE POLYMERIZATION FOR FUNCTIONAL HYDROGEL MICROBEAD
WITH FLASH FREEZING CENTRIFUGAL MICROFLUIDIC DEVICE 1084
Tomomi Murayama[1], Koki Yoshida[1], Yuta Kurashina[2], and Hiroaki Onoe[1]
[1]Keio University, JAPAN and [2]Tokyo University of Agriculture and Technology, JAPAN

T71-g TEMPERATURE-RESPONSIVE MICROCAPSULES MANUFACTURED BY PROMOTING
CONTROLLED CLOAKING WITH THE HELP OF MICRO/NANOPARTICLES 1088
Rutvik Lathia[1], Bheema Sankar Reddy[1], Chandantaru Dey Modak[1,2],
Satchit Nagpal[1,3], and Prosenjit Sen[1]
[1]Indian Institute of Science, INDIA, [2]CNRS - ESPCI PSL, FRANCE, and [3]Texas A&M University, USA

W71-g WATER VITRIFICATION IN A MICROCHANNEL AT LOW COOLING RATE 1091
Ayane Sato, Tomohiro Hayashi, and Tadashi Ishida
Tokyo Institute of Technology, JAPAN

h - Optical, RF and Electromagnetics for MEMS/NEMS
Electrical Field and Magnetic Field Sensors and Transducers

M72-h A HIGHLY SENSITIVE 3-AXIS MICRO SEARCH-COIL MAGNETOMETER
ENABLED BY HIGH DENSITY THROUGH-SILICON-VIA PROCESS 1095
Hadi Tavakkoli, Mingzheng Duan, Longheng Qi, Izhar, Xu Zhao, and Yi-Kuen Lee
Hong Kong University of Science and Technology, HONG KONG

T72-h FULLY INTEGRATED BACK-BIASED 3D HALL SENSOR WITH
WAFER-LEVEL INTEGRATED PERMANENT MICROMAGNETS 1099
Björn Gojdka[1], Daniel Cichon[2], Markus Stahl-Offergeld[2], Dominik Schröder[3],
Niels Clausen[1], Christian Hedayat[3], Hans-Peter Hohe[2], and Thomas Lisec[1]
*[1]Fraunhofer Institute for Silicon Technology ISIT, GERMANY, [2]Fraunhofer Institute for Integrated Circuits
IIS, GERMANY, and [3]Fraunhofer Institute for Electronic Nano Systems ENAS, GERMANY*

h - Optical, RF and Electromagnetics for MEMS/NEMS
Free Space Optical Components & Systems

W72-h A LARGE-STROKE TIP-TILT-PISTON MICROMIRROR WITH
ELECTROMAGNETIC ACTUATORS BASED ON METALLIC GLASS 1103
Chuan-Hui Ou, Nguyen V. Toan, and Takahito Ono
Tohoku University, JAPAN

M73-h ARBITRARY SHAPED BACKSIDE REINFORCEMENT FOR
TWO DIMENSIONAL RESONANT MICROMIRRORS 1107
Takashi Sasaki, Adrien Piot, Anton Lagosh, Clement Fleury, Markus Bainschab, Yanfen Zhai,
Marcus Baumgart, Sara Guerreiro, Dominik Holzmann, Aleš Travnik, and Mohssen Moridi
Silicon Austria Labs, AUSTRIA

T73-h HIGH TRANSMITTANCE METASURFACE HOLOGRAMS USING SILICON NITRIDE 1111
Masakazu Yamaguchi, Hiroki Saito, Satoshi Ikezawa, and Kentaro Iwami
Tokyo University of Agriculture and Technology, JAPAN

W73-h MULTIFUNCTIONAL OPTICAL METASURFACE FOR ANOMALOUS REFLECTION,
STRUCTURAL COLOR, AND SURFACE LATTICE RESONANCE 1115
Liye Li[1], Hongshun Sun[1], Yifan Ouyang[1], Shengxiao Jin[1], Tian Kang[1], Zhimei Qi[2], and Wengang Wu[1]
[1]Peking University, CHINA and [2]Chinese Academy of Science, CHINA

M74-h NOVEL WAVEFRONT-SPLITTING INTERFEROMETER FOR ULTRA-COMPACT
BROADBAND FT-IR SPECTROSCOPY EXTENDING TO VISIBLE RANGE 1119
Bassem Mortada[1], Yasser M. Sabry[1,2], Bassam Saadany[1], Tarik Bourouina[3], and Diaa Khalil[2]
[1]Si-Ware Systems, EGYPT, [2]Ain Shams University, EGYPT, and [3]Université Gustave Eiffel, FRANCE

T74-h PIEZOELECTRICALLY ACTUATED MICROMIRROR WITH
DYNAMIC DEFORMATION COMPENSATION MECHANISM 1123
Takashi Sasaki, Adrien Piot, Jaka Pribošek, Anton Lagosh, Clement Fleury, Markus Bainschab, Yanfen Zhai,
Marcus Baumgart, Sara Guerreiro, Dominik Holzmann, Aleš Travnik, and Mohssen Moridi
Silicon Austria Labs, AUSTRIA

W74-h RESONANT d_{33} MODE PZT MEMS MIRROR EXCITED WITH
DIRECTIONAL INTERDIGITATED ELECTRODES 1127
Pooja Thakkar, Anton Lagosh, Takashi Sasaki, Markus Bainschab, and Jaka Pribošek
Silicon Austria Labs GmbH, AUSTRIA

M75-h RESONANT PIEZOELECTRIC VARIFOCAL MIRROR WITH ON-CHIP INTEGRATED
DIFFRACTIVE OPTICS FOR INCREASED FREQUENCY RESPONSE 1131
Jaka Pribošek, Anton Lagosh, Pooja Thakkar, Takashi Sasaki, and Markus Bainschab
Silicon Austria Labs, AUSTRIA

T75-h UNIQUE DISPERSION RELATION FOR PLASMONIC
PHOTODETECTORS WITH SUBMICRON GRATING 1135
Yuki Kaneda[1], Masaaki Oshita[1], Utana Yamaoka[1], Shiro Saito[2], and Tetsuo Kan[1]
[1]University of Electro-Communications, JAPAN and [2]IMRA JAPAN Co., LTD., JAPAN

h - Optical, RF and Electromagnetics for MEMS/NEMS
Infrared (IR) Sensors and Imaging Systems

W75-h INTEGRATION OF A HIGH TEMPERATURE TRANSITION METAL OXIDE
NTC THIN FILM IN A MICROBOLOMETER FOR LWIR DETECTION 1139
Sarah Risquez[1], Sebastian Redolfi[2], Clement Fleury[1], Matthias Wulf[2], Ali Roshanghias[1],
Adrien Piot[1], Jeremy Streque[1], Kerstin Schmoltner[2], Thang Duy Dao[1], Markus Puff[2], and Mohssen Moridi[1]
[1]Silicon Austria Labs GmbH, AUSTRIA and [2]TDK Electronics GmbH & Co OG, AUSTRIA

M76-h   **PERIODIC CAVITIES ON THE IR-ABSORBER FOR RESPONSIVITY ENHANCEMENT OF CMOS-MEMS THERMOELECTRIC IR SENSOR** 1143
Yung-Chen Li, Tien Chou, Pen-Sheng Lin, Yu-Cheng Huang, Fuchi Shih,
You-An Lin, Da-Jen Yen, Mei-Feng Lai, and Weileun Fang
National Tsing Hua University, TAIWAN

T76-h   **ULTRA-LARGE PIXEL ARRAY PHOTOTHERMAL TRANSDUCER AND ITS THERMAL PERFORMANCE PREDICTION STRATEGY** 1147
Defang Li[1,3], Jinying Zhang[1,2], Jiushuai Xu[3], Erwin Peiner[3], Zhuo Li[1,2],
Xin Wang[1], Suhui Yang[1], and Yanze Gao[1]
[1]Beijing Institute of Technology, CHINA, [2]Yangtze Delta Region Academy of Beijing Institute of Technology, CHINA, and [3]Technische Universität Braunschweig, GERMANY

h - Optical, RF and Electromagnetics for MEMS/NEMS
MEMS for Timing & Frequency Control

W76-h   **A CMOS-MEMS BEAM RESONATOR WITH $Q > 10,000$** 1151
Ting-Yi Chen and Wei-Chang Li
National Taiwan University, TAIWAN

M77-h   **GENERIC TEMPERATURE COMPENSATION SCHEME FOR CMOS-MEMS RESONATORS BASED ON ARC-BEAM DERIVED ELECTRICAL STIFFNESS FREQUENCY PULLING** 1155
I-Chieh Hsieh, Hong-Sen Zheng, Chun-Pu Tsai, Ting-Yi Chen, and Wei-Chang Li
National Taiwan University, TAIWAN

T77-h   **HIGH-Q AND LOW-MOTIONAL IMPEDANCE PIEZOELECTRIC MEMS RESONATOR THROUGH MECHANICAL MODE COUPLING** 1159
Linhai Huang[1], Zhihong Feng[1], Yuhao Xiao[2], Fengpei Sun[1], and Jinghui Xu[1]
[1]Huawei Technologies Company Ltd., CHINA and [2]Wuhan University, CHINA

h - Optical, RF and Electromagnetics for MEMS/NEMS
Photonic Components & Systems

W77-h   **CROSSTALK-FREE LARGE APERTURE 2D GIMBAL MICROMIRROR** 1163
Behrad Ghazinouri and Siyuan He
Toronto Metropolitan University, CANADA

M78-h   **INVERSE INTERFERENCE EFFECT-ENHANCED ULTRASENSITIVE SENSING VIA MID-IR NANOANTENNAS** 1167
Hong Zhou, Dongxiao Li, Xinge Guo, Zhihao Ren, and Chengkuo Lee
National University of Singapore, SINGAPORE

T78-h   **TWISTED AND CONTACTED AU MICRO-RODS 3D CHIRAL METAMATERIALS WITH CIRCULAR DICHROISM VIA AN ABSORPTIVE ROUTE IN LONG-WAVELENGTH INFRARED** 1171
Gaku Furusawa[1], Natsuki Kanda[2], Ryusuke Matsunaga[2], and Tetsuo Kan[1]
[1]University of Electro-Communications, JAPAN and [2]University of Tokyo, JAPAN

h - Optical, RF and Electromagnetics for MEMS/NEMS
RF MEMS Components & Systems

W78-h 3D HYBRID ACOUSTIC RESONATOR WITH COUPLED FREQUENCY RESPONSES
OF SURFACE ACOUSTIC WAVE AND BULK ACOUSTIC WAVE 1175
Liping Zhang[1,2], Shibin Zhang[1], Jinbo Wu[1,2], Pengcheng Zheng[1,2], Hulin Yao[1,2],
Yang Chen[1,2], Kai Huang[1,2], Xiaomeng Zhao[1], Min Zhou[1], and Xin Ou[1,2]
[1]Chinese Academy of Sciences, CHINA and [2]University of Chinese Academy of Sciences, CHINA

M79-h A C/K$_U$ DUAL-BAND RECONFIGURABLE BAW FILTER
USING POLARIZATION TUNING IN LAYERED SCALN 1179
Dicheng Mo, Shaurya Dabas, Sushant Rassay, and Roozbeh Tabrizian
University of Florida, USA

T79-h ACOUSTOELECTRIC-DRIVEN FREQUENCY MIXING IN
MICROMACHINED LITHIUM NIOBATE ON SILICON WAVEGUIDES 1183
Hakhamanesh Mansoorzare and Reza Abdolvand
University of Central Florida, USA

W79-h EFFECT OF SCANDIUM COMPOSITION ON THE PHONON SCATTERING LIFETIME
OF ALUMINUM SCANDIUM NITRIDE ACOUSTIC WAVE RESONATORS 1186
Yue Zheng[1], Mingyo Park[1], Chao Yuan[2], and Azadeh Ansari[1]
[1]Georgia Institute of Technology, USA and [2]Wuhan University, CHINA

M80-h LITHIUM NIOBATE THIN FILM BASED A$_1$ MODE RESONATORS WITH FREQUENCY
UP TO 16 GHZ AND ELECTROMECHANICAL COUPLING FACTOR NEAR 35% 1190
Rongxuan Su[1], Zhenyi Yu[1], Sulei Fu[1], Huiping Xu[1], Shuai Zhang[1], Peisen Liu[1],
Yu Guo[2], Cheng Song[1], Fei Zeng[1], and Feng Pan[1]
[1]Tsinghua University, CHINA and [2]Jiangnan University, CHINA

T80-h SUB-3 DB INSERTION LOSS BROADBAND ACOUSTIC DELAY LINES AND
HIGH FOM RESONATORS IN LINBO$_3$/SIO$_2$/SI FUNCTIONAL SUBSTRATE 1194
Chun-Chen Yeh, Chia-Hsien Tsai, Guan-Lin Wu, Tzu-Hsuan Hsu, and Ming-Huang Li
National Tsing Hua University, TAIWAN

W80-h SUPPRESSION OF SPURIOUS MODES IN ALUMINUM NITRIDE S$_1$ LAMB WAVE
RESONATORS USING A MECHANICAL SOFT-CONTACT SCHEME 1198
Shao-Siang Tung[1], Tzu-Hsuan Hsu[1], Yens Ho[2], Yung-Hsiang Chen[2],
Yelehanka R. Pradeep[3], Rakesh Chand[3], and Ming-Huang Li[1]
*[1]National Tsing Hua University, TAIWAN, [2]Vanguard International Semiconductor Corporation, TAIWAN,
and [3]Vanguard International Semiconductor Corporation Singapore PTE. Ltd., SINGAPORE*

h - Optical, RF and Electromagnetics for MEMS/NEMS
THz MEMS Components & Systems

M81-h TERAHERTZ REFLECTIVE METALENS FOR ARBITRARY
OFF-AXIS FOCUSING WITH LARGE DEPTH OF FOCUS 1202
Jiahao Miao, Yi Liu, Cong Lin, Zhanxuan Zhou, and Xiaomei Yu
Peking University, CHINA

h - Optical, RF and Electromagnetics for MEMS/NEMS
Other Electromagnetic MEMS/NEMS

T81-h TOWARDS A BETTER CMOS-MEMS RESOSWITCH USING
ELECTROLESS PLATING FOR CONTACT ENGINEERING 1206
Ting-Jui Liou, Chun-Pu Tsai, Ting-Yi Chen, and Wei-Chang Li
National Taiwan University, TAIWAN

i - Open Posters

W81-i **A MEMS-CMOS INFRA-RED MICROSYSTEM WITH IN-SENSOR MACHINE LEARNING CAPABILITIES**
Marco Castellano, Ugo Garozzo, Luca Gandolfi, Davide Ruggiero, and Giuseppe Bruno
STMicroelectronics, ITALY

M82-i **A NOVEL BAROMETRIC PRESSURE SENSOR WITH A CAPACITVE TRANSDUCER AND IMPROVED PERFORMANCE**
Thomas Friedrich[1], Volkmar Senz[1], and Ferenc Lukacs[2]
[1]Robert Bosch GmbH, GERMANY and [2]Robert Bosch Kft., HUNGARY

T82-i **A NOVEL CLASS OF MOTION SENSORS FEATURED WITH AN ELECTRIC POTENTIAL SENSING CHANNEL**
Enrico R. Alessi, Fabio Passaniti, and Emanuele Lavelli
STMicroelectronics, ITALY

W82-i **A STABLE MIR PHOTODETECTOR BASED ON 2D PTSI/P-SI NANOHOLE ARRAYS**
Ashenafi A. Elyas, Masahiko Shiraishi, and Tetsuo Kan
University of Electro-communications, JAPAN

M83-i **AN EQUIVALENT CIRCUIT MODEL FOR THE PHASE GRADIENT METASURFACE ANALYSIS IN VISIBLE BAND**
Liye Li[1], Senyong Hu[1], Yifan Ouyang[1], Yusa Chen[1], Meizhang Wu[2], and Wengang Wu[1]
[1]Peking University, CHINA and [2]University of Science and Technology Beijing, CHINA

T83-i **DETECTION OF MASS AND MATERIAL NATURE OF MICROPARTICLES BY A PIEZOELCTRIC MEMS**
Francesco Foncellino and Luigi Barretta
STMicroelectronics, ITALY

W83-i **ELECTRO-OPTICAL TESTING SOLUTION FOR TMOS MEMS SENSOR SENSITIVITY ASSESSMENT AT WAFER LEVEL**
Roberta Carbone, Dario Premi, and Marco Rossi
STMicroelectronics, ITALY

T84-i **HIGH PERFORMANCE SPUTTERED PZT PMUTS OPERATING IN THE ULTRASOUND IMAGING RANGE REPRODUCIBLE AT WAFER-SCALE**
Jihang Liu[1], David Sze Wai Choong[1], Duan Jian Goh[1], Merugu Srinivas[1], Qing Xin Zhang[1], Steven Lee Hou Jang[1], Huamao Lin[1], Fabio Quaglia[3], Domenico Giusti[3], Laura Castoldi[3], Claudia Pedrini[3], Luca Barabani[3], Annachiara Esposito[3], Luigi Barretta[3], Rossana Scaldaferri[3], Alberto Leotti[2], Adriyan Hidayat Mohamed Hamsah[3], Peter Chang Hyun Kee[1], and Lee En-Yuan Joshua[1]
[1]Institute of Microelectronics, SINGAPORE, [2]ST Microelectronics, SINGAPORE, and [3]ST Microelectronics, ITALY

W84-i **PIEZOELECTRIC ACTUATOR INTRODUCTION FOR ACCURATE POSITIONING READ/WRITE ELEMENT IN HARD DISK DRIVE (HDD)**
Domenico Giusti and Marco Ferrera
STMicroelectronics, ITALY

M85-i **PIEZOELECTRIC MEMS FOR MICROPARTICLES DETECTION: ALTERNATIVE READOUT FOR MASS DETECTION**
Luigi Barretta and Francesco Foncellino
STMicroelectronics, ITALY

T85-i **SIDE WALL DETECTION TYPE SPR SENSOR WITH GOLD GRATING ON GLASS**
Masaaki Oshita, Shinichi Suzuki, Kazuto Masamoto, and Tetsuo Kan
University of Electro-Communications, JAPAN

W85-i **SPUTTERED PZT AIR-COUPLED PMUTS WITH WIDE BANDWIDTH AND LONG DETECTION RANGE FOR RANGING APPLICATIONS**
Mantalena Sarafianou[1], David Sze Wai Choong[1], Duan Jian Goh[1], Jihang Liu[1], Joshua En-Yuan Lee[1], Srinivas Merugu[1], Qing Xin Zhang[1], Peter Hyun Kee Chang[1], Fabio Quaglia[2], Domenico Giusti[2], Laura Castoldi[2], Filippo D'Ercoli[2], Riccardo Tacchini[2], Alberto Leotti[3], and Dao Hao Sim[3]
[1]Institute of Microelectronics, SINGAPORE, [2]ST Microelectronics, ITALY, and [3]ST Microelectronics, SINGAPORE

M86-i **THERMOELECTRIC MIROPHONE**
Akash Gupta[1], Dr. Achim Bittner[1], Prof. Dr.-Ing, and Alfons Dehe[1,2]
[1]Hahn-Schickard Institute for Applied Research e.V., GERMANY and [2]University of Freiburg, GERMANY

T86-i **ULTRA-PRECISE DEPOSITION: DIGITAL MICROMANUFACTURING FOR ADVANCED PACKAGING**
Lukasz Witczak, Jolanta Gadzalinska, Iwona Gradzka-Kurzaj, Mateusz Lysien, Ludovic Schneider, Aneta Wiatrowska, Karolina Fiaczyk, Piotr Kowalczewski, Lukasz Kosior, and Filip Granek
XTPL SA, POLAND

W86-i **WAFER-LEVEL DEFECT CHARACTERIZATION AND POLARITY-DEPENDENT RESISTANCE DEGRADATION OF SPUTTERED SODIUM POTASSIUM NIOBATE THIN FILMS**
Kuan-Ting Ho[1], Daniel Monteiro Diniz Reis[1], and Karla Hiller[2]
[1]Robert Bosch GmbH, GERMANY and [2]Technical University Chemnitz, GERMANY

WELCOME

Welcome to the 36th IEEE International Conference on Micro Electro Mechanical Systems (MEMS 2023) in Munich, Germany!

The IEEE MEMS Conference series originated in 1987, and has been known as the IEEE International Conference on Micro Electro Mechanical Systems since 1999. Over the last decade, the MEMS community has experienced immense progress in the science and technology of miniaturization, as well as increasing technical maturity and commercialization of ever smarter products encompassing embedded artificial intelligence and wireless connectivity to the Internet-of-Things. Since 2020, the conference is sponsored by the IEEE MEMS Technical Community, a new body within the IEEE dedicated to supporting and developing our MEMS community.

This Conference brings together annually the international MEMS community consisting of top players in academia and industry by providing them with the latest results on every aspect of MEMS. The organizers this year made a special effort to attract technical contributions from industry, also showcased in the Industry Workshop. We hope that you will enjoy the presentations by our accomplished invited plenary speakers and stringently selected contributed oral and poster presentations.

We would like to express our sincerest gratitude to all the authors who submitted their abstracts. Their high-quality work serves as the foundation for the success of this conference. A total of 314 papers out of 636 submitted abstracts were carefully selected by 47 experts comprising the Technical Program Committee (TPC) using a well-established double-blind review process that ensures scientific quality as the sole selection criterion. The TPC comprises academic and industrial members, with equal representation from three regional divisions: the Americas, Europe & Africa, and Asia & Oceania. To allow for more focused and careful deliberation, the actual abstract review process divided the overall TPC into eight sub-committees, each with six members to evaluate and rate each abstract. We are grateful to all TPC members who volunteered their valuable time, including participation in a two-day virtual meeting last October, for paper selection.

The conference arranges presentation of accepted papers in a mixed single/parallel session format with 4 invited plenary, 70 oral, 238 poster presentations, and 15 open poster presentations. In addition, the TPC collectively nominated, based on quality, abstract submissions as finalists for the Outstanding Student Paper Awards and images for the Art in Microtechnology prize. These awards aim to recognize excellence amongst work presented by students and will be announced in a special ceremony to conclude the conference late Thursday morning.

We gratefully acknowledge the industrial support groups, exhibitors, and benefactors for their contributions to this conference. The dedicated and relentless effort of Ms. Sara Stearns and her team at PMMI in managing this conference is highly appreciated.

In closing, we hope you enjoy the networking, technical presentations, exhibition booths, and events of the 2023 IEEE International MEMS Conference this week in Munich!

Nuria Barniol
Universita Autonoma di Barcelona, Spain

Franz Laermer
Robert Bosch GmbH, Germany

BENEFACTORS

We gratefully acknowledge, at the time of printing, the financial contributions from the following:

CONFERENCE SPONSORS

BENEFACTOR

Bavarian Ministry of Economic Affairs, Regional Development and Energy

Technische Universität München
Chair for Biomolecular Nanotechnology (DIETZ LAB)

DIAMOND BENEFACTORS

Robert Bosch GmbH
STMicroelectronics
SUSS MicroTec Solutions GmbH & Co. KG
TDK

GOLD BENEFACTORS

STMicroelectronics

SILVER BENEFACTOR

Plasma-Therm, LLC

BRONZE BENEFACTOR

NXP Semiconductors

OUTSTANDING STUDENT POSTER PRESENTATION AWARD SPONSOR

MDPI - sensors

OUTSTANDING STUDENT ORAL PRESENTATION AWARD SPONSOR

Microsystems & Nanoengineering/Springer Nature

ART IN MICROTECHNOLOGY AWARD SPONSOR

MDPI - micromachines

BENEFACTORS (continued)

LANYARD BENEFACTOR

Lyncee Tec

EXHIBITORS

Coventor, SARL
Heidelberg Instruments Mikrotechnik GmbH
IEEE MEMS Technical Community
i-ROM GmbH
idonus sarl
JX Nippon Mining & Metals Europe GmbH
KLA Corporation
Lyncee Tec
memsstar ltd.
Microqubic AG
NAGASE (EUROPA) GmbH
NexGen Wafer Systems GmbH
Nextron Corporation
Plasma-Therm, LLC
Polytec GmbH
Polyteknik AS
Purdue University
Robert Bosch GmbH
scia Systems GmbH
Silicon Austria Labs
SmarAct GmbH & Co. KG
STMicroelectronics
Sumitomo Precision Products Co., Ltd.
SUSS MicroTec Solutions GmbH & Co. KG
TDK
Tousimis
Zurich Instruments

MEDIA BENEFACTORS

MDPI - machines
Microtech Ventures, Inc.
MDPI - Sensors Journal

CONFERENCE OFFICIALS

Conference Chairs

Núria Barniol .... Universidad Autonoma Barcelona, SPAIN
Franz Lärmer .... Bosch, GmbH, GERMANY

International Steering Committee

Zhihong Li (Chair) .... Peking University, CHINA
Shuji Tanaka (Chair) .... Tohoku University, JAPAN
Núria Barniol .... Universidad Autonoma Barcelona, SPAIN
Karen Cheung .... University of British Columbia, CANADA
Philip Feng .... University of Florida, USA
David Horsley .... University of California, Davis, USA
Hyunjoo Jenny Lee .... KAIST, KOREA
Sheng-Shian Li .... National Tsing Hua University, TAIWAN
Wen Li .... Michigan State University, USA
Zhihong Li .... Peking University, CHINA
Franz Lärmer .... Bosch, GmbH, GERMANY
Niclas Roxhed .... KTH Royal Institute of Technology, SWEDEN
Shuji Tanaka .... Tohoku University, JAPAN
Dana Weinstein .... Purdue University, USA
Haixia "Alice" Zhang .... Peking University, CHINA

Executive Technical Program Committee

Taeko Ando .... Ritsumeikan University, JAPAN
Cédric Ayela .... CNRS, FRANCE
Nuria Barniol .... Universitat Autonoma de Barcelona, SPAIN
Sarah Bedair .... US Army Research Laboratory, USA
Cristian Cassella .... Northeastern University, USA
Eric Chiou .... University of California, Los Angeles, USA
Caroline Coutier .... CEA-Leti, FRANCE
Philip Feng .... University of Florida, USA
Irene Fernandez-Cuesta .... University of Hamburg, GERMANY
Frank Goldschmidtboeing .... University Freiburg, GERMANY
Songbin Gong .... University of Illinois, USA
Nicole Hashemi .... Iowa State University, USA
Fatimah Ibrahim .... University Malaya, MALAYSIA
Nathan Jackson .... University of New Mexico, USA
Michael Kraft .... KU Leuven, BELGIUM
Franz Laermer .... Robert Bosch GmbH, GERMANY
Hyunjoo Jenny Lee .... KAIST, KOREA (ROK)
Thierry Leichle .... CNRS, USA
Sheng-Shian Li .... National Tsing Hua University, TAIWAN
Wen Li .... Michigan State University, USA
Zhihong Li .... Peking University, CHINA
Andreu Llobera .... Silicon Austria Labs, AUSTRIA
Ruochen Lu .... University of Texas, Austin, USA
Michel Maharbiz .... UC Berkeley / iota Biosciences, USA
Jianmin Miao .... Shanghai Jiaotong University, CHINA
Hiroshi Miyajima .... Sumitomo Precision Products Co., Ltd., JAPAN
Mina Rais-Zadeh .... NASA JPL, USA

Executive Technical Program Committee (continued)

Niclas Roxhed ....KTH Royal Institute of Technology, SWEDEN
Jose-Luis Sanchez-Rojas .... Universidad de Castilla - La Mancha, SPAIN
Hamed Sattari .... CSEM, SWITZERLAND
Takaaki Suzuki .... Gunma University, JAPAN
Massood Tabib-Azar .... University of Utah, USA
Shuji Tanaka .... Tohoku University, JAPAN
Sindy Tang .... Stanford University, USA
Jaume Verd .... University of the Balearic Islands, SPAIN
Guillermo Villanueva .... EPFL, SWITZERLAND
Sandra Vos .... NXP, USA
Fei Wang .... SUSTech, CHINA
Dana Weinstein .... Purdue University, USA
Anders Wolff .... Technical University of Denmark DTU, DENMARK
Guoqiang Wu .... Wuhan University, CHINA
Yao-Joe Yang .... National Taiwan University, TAIWAN
Shinya Yoshida .... Shibaura Institute of Technology, JAPAN
Roland Zengerle .... Hahn-Schickard & University of Freiburg, GERMANY
Siyang Zheng .... Carnegie Mellon University, USA
Rong Zhu .... Tsinghua University, CHINA
Yao Zhu .... Institute of Microelectronics, A*STAR, SINGAPORE

FROM ETCH TO EDGE AI:
OPENING NEW HORIZONS WITH
SMART SENSOR TECHNOLOGIES

Stefan Finkbeiner
Bosch Sensortec GmbH, GERMANY

ABSTRACT

Almost no state-of-the-art consumer electronics device exists that does not contain at least one MEMS sensor. From smartphones over tablets to smartwatches and hearables, MEMS sensors have taken over important key functions in these devices.

MEMS sensors themselves are undergoing continuous development processes to increase performance, functionality, reduce power consumption and shrink size.

In parallel, more and more new sensor use cases and applications are being developed with MEMS sensors. Either by intelligently combining them with other measurement technologies or by applying leading-edge technologies like artificial intelligence and machine learning.

In his talk, Dr. Stefan Finkbeiner, CEO at Bosch Sensortec will introduce latest solution examples, explaining how new technological developments enrich sensor related use cases for Consumer Electronics without forgetting the roots for smart sensing technology: MEMS technologies.

SUB-300 MILLIVOLT OPERATION IN NONVOLATILE 300 NM X 100 NM PHASE CHANGE NANOELECTROMECHANICAL SWITCH

Mohammad Ayaz Masud and Gianluca Piazza
Carnegie Mellon University, Pittsburgh, USA

ABSTRACT

This paper reports the design, fabrication, and characterization of the Fin Phase Change Nanoelectromechanical Relay (FinPCNR), a truly nanoscale switch. We harness the nonvolatile volume transformation of GeTe, a widely used phase change material, to drive the laterally actuating relay. The PCNR is turned ON by applying a sub-300 mV short actuation pulse (< 500 ns) to a fin-shaped heater (l: 300 nm, w: 100 nm, h> 100 nm). Novel nanofabrication techniques are developed to self-align multiple functional vertical layers on the heater sidewall. In the OFF state, we achieve near zero leakage (23 fA) by maintaining a sub-5 nm airgap between the metal channel and the drain/source electrodes. This device is an ideal candidate for high density, ultra low-leakage memory circuitry as it combines the nonvolatility of GeTe and the high ON-OFF current ratio of NanoElectroMechanical (NEM) relays.

KEYWORDS

Phase Change Materials, NEM Relay, GeTe, Emerging Memory, Nonvolatility.

INTRODUCTION

Complementary metal oxide semiconductor (CMOS) technology has been at the core of all computing devices for the last five decades. Aggressive scaling of CMOS devices has led to the high efficiency, speed and performance of both processing and memory units in today's computers. As current CMOS technology is projected to reach its scaling limit by 2028, further scaling will be enabled by 3D fabrication. This, however, will not improve the energy efficiency [1]. Novel applications, such as, Internet-of-Things (IoT), look-up memory, wearable devices, and neuromorphic computing, calls for research and development of ultra-low-power nanoscale devices. Researchers are exploring several emerging technologies that can augment or replace CMOS in these energy-constrained applications. All these emerging memory devices, such as, resistive memory (ReRAM), phase change memory (PCRAM), magnetoresistive memory (MRAM), and ferroelectric field effect transistors (FeFET) demonstrate some degree of non-volatility, a property which enables data retention in the absence of a power supply [2]. However, most of these devices suffer from their limited contrast in resistance between the ON and OFF states (ON-OFF ratio). High OFF current leads to high leakage, resulting in greater static power consumption [3].

The Phase Change Nanoelectromechanical Relay (PCNR) is one such emerging technology, which is very suitable for the aforementioned applications as it combines the nonvolatility of phase change materials (PCM) and the high ON-OFF current ratio (10^8) of nanoelectromechanical (NEM) relay architectures. Different from all previously demonstrated relays, the PCNR is highly scalable [4]. PCNR utilizes the 10% reversible volume expansion of

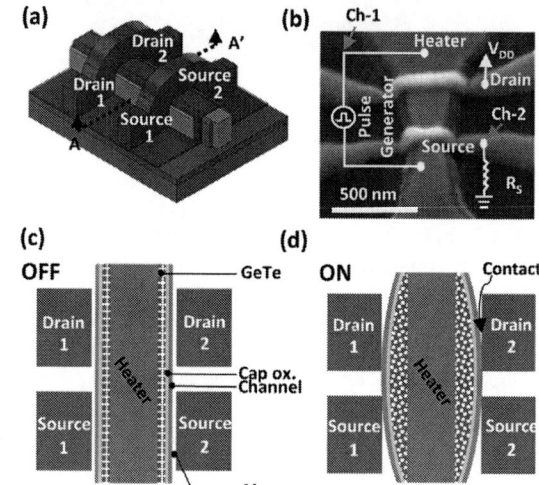

Figure 1:(a) Schematic model of a 4-terminal FinPCNR device. (b)SEM image of a fabricated device with schematic test setup. (c-d) A-A' horizontal cross-section of the device in the as fabricated OFF and ON states, respectively. The lateral actuator connects Drain 1 (2) and Source 1(2) through the left (right) side channel. The electrodes of each side are connected, as shown in (a), to reduce contact resistance.

GeTe to open and close a pair of metal contacts [5]. Microscale vertically actuating PCNR devices have been reported to have very low actuation voltage (1.1 V) and very high ON-OFF current ratio (10^8) due to its 20 nm airgap [6]. Here we present an advanced scaled down design of PCNR – the FinPCNR. This fin-shaped laterally actuating relay turns on with a sub-300 mV actuation pulse. Sub-5 nm airgap successfully maintains the desired low leakage during operation. While the prototype FinPCNR already occupies an area (300 nm x 100 nm) smaller than any other NEM relay design, there is plenty of room left for further scaling as this architecture does not require any anchored flexure.

OPERATING PRINCIPLE

Fig. 1 shows a schematic diagram of the FinPCNR, illustrating its (a) 3D structure, (b) SEM image with a schematic test setup and horizontal cross-sections in the (c) OFF and (d) ON states. Voltage is applied across the heater using a pulse generator. Heater voltage and channel current are recorded using the ch-1 and ch-2 oscilloscope probes, respectively. The FinPCNR comprises of two main components- the contact pair and the lateral actuator. The contact pair is formed by a pair of Pt drain and source electrodes, and a Pt channel placed on the sidewall of the actuator. The metal channel bridges the drain and source electrodes when the relay is in the ON state. The actuator consists of a tungsten heater, a sidewall layer of GeTe, and a thin layer of Al_2O_3 to encapsulate GeTe during phase

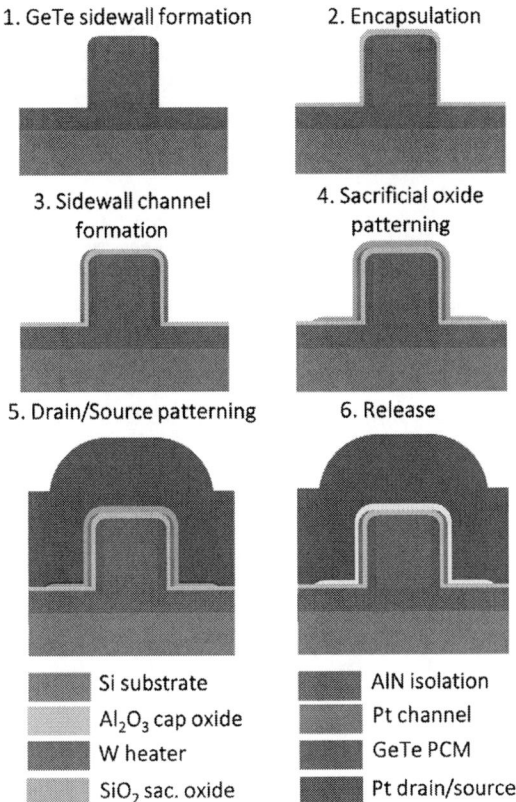

1. GeTe sidewall formation
2. Encapsulation
3. Sidewall channel formation
4. Sacrificial oxide patterning
5. Drain/Source patterning
6. Release

■ Si substrate
■ Al₂O₃ cap oxide
■ W heater
■ SiO₂ sac. oxide

■ AlN isolation
■ Pt channel
■ GeTe PCM
■ Pt drain/source

Figure 2: Sidewall formation steps around the heater mandrel. Preliminary steps of AlN deposition and heater mandrel formation are not shown.

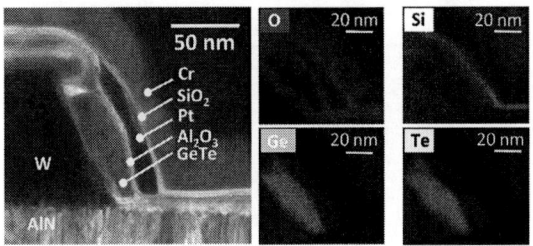

Figure 3:Cross-sectional TEM image of a device before release. EDX element map from STEM-HAADF shows the regions with GeTe and sacrificial SiO2.

channel from the drain and source, once again creating an airgap and turning the relay OFF. The switch ON and OFF pulses can also be called actuation and recrystallization pulse, respectively.

FABRICATION PROCESS

The fabrication process involves multiple self-aligned sidewall formation steps around the heater mandrel. These self-aligned processes prevent misalignment from conventional patterning with electron-beam lithography (EBL). At first, 100 nm AlN is deposited on a Si substrate. This layer provides electrical insulation but good thermal conductivity between the heater and the substrate. Next, 100 nm tungsten is deposited and patterned with SF_6 plasma in a reactive ion etching (RIE) tool to produce sub-100 nm wide heaters. The high aspect ratio heater acts as

change. Transformation of GeTe between the crystalline (smaller) and amorphous (larger) states results in lateral expansion and contraction of the actuator. We use GeTe as the PCM as its volume changes by 10% during phase transformation. It is also capable of retaining its state at higher temperature than other widely used phase change materials, such as $Ge_2Sb_2Te_5$ (GST) [7].

As fabricated, GeTe is in the low volume crystalline state. The thickness of the sacrificial material determines the size of the gap after release. The gap is sized to ensure that the expected volume change in GeTe layer is sufficient to make contact between the electrodes and the channel. We apply an electric pulse on the heater to raise the temperature of GeTe over its melting point (T_{melt}^{GeTe}=1000K). Molten GeTe is quickly quenched into the amorphous state. The expanded layer of GeTe presses the metallic channel into the drain and source, forming a contacting bridge between the two electrodes. A similar process returns the FinPCNR to the OFF state. This increases the PCM temperature above its glass transition point (500 K for GeTe) and rapidly crystallizes any amorphous portion of the film. Unlike the switching ON process, the PCM does not need to be melted this time. As a result, the relay can be switched back at a lower voltage. However, an input pulse with a slow trailing edge is preferred for RESET as it ensures complete recrystallization of the PCM. At the end of this pulse, the crystalline PCM contracts and separates the metallic

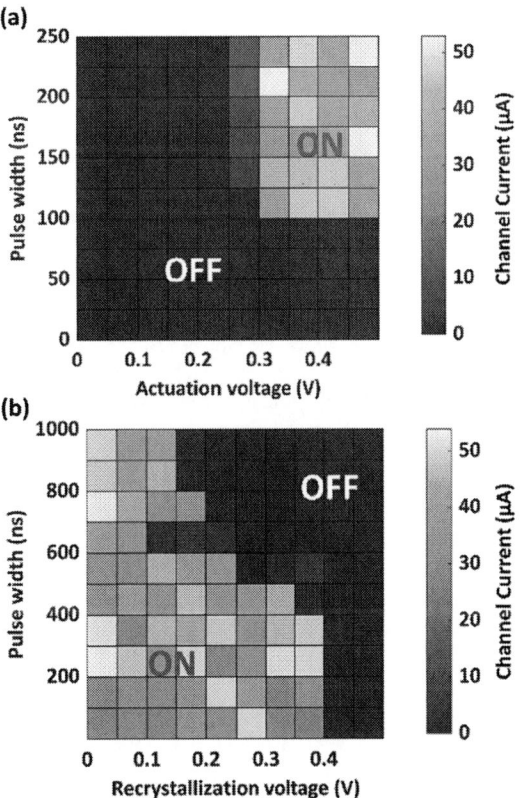

Figure 4: Channel current measured after applying a wide range of pulse amplitude and duration for (a) actuation and (b) recrystallization.

978-1-6654-9309-3/23 $31.00 © 2023 IEEE

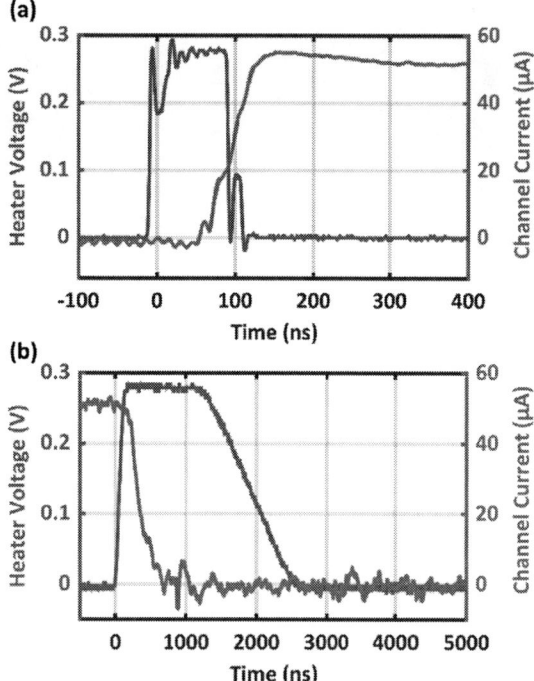

Figure 5: Heater voltage and resulting change in channel current during (a) actuation and (b) recrystallization.

the mandrel for the following self-aligned sidewall patterning steps. In the next step, 30 nm GeTe (PCM) is deposited in a co-sputtering tool that ensures semi-conformal deposition with the help of a rotating chuck. The substrate is heated at 220^0C to crystallize the PCM film. Then it is etched with Ar^+ plasma in an inductively coupled plasma (ICP) RIE tool. The etch process is precisely timed so that only the sidewall PCM survives. We use cross-sectional TEM as well as electrical testing to determine the exact sidewall etch time. Next, a layer of 5 nm alumina is deposited in the ALD tool as an encapsulation layer. This alumina layer is crucial as it contains the molten GeTe during phase transformation. This deposition is done at a high temperature (250^0C) to reduce the number of pinholes in the alumina film. Next, 10 nm ALD Pt is deposited as the channel material. This layer is etched using Ar^+ plasma in an ion-milling tool for effective sidewall metallization [8]. Next, 3~5 nm sacrificial oxide is deposited to separate the actuator stack from the following layer of electrodes (drain and source). Finally, 100 nm thick metal (Pt with a Cr adhesion layer) is patterned to form the electrodes. Fig. 3 illustrates all layers before the release step in a cross-sectional TEM image. EDX results clearly demonstrate a thin SiO_2 layer between the metal contacts. As a final step, the sacrificial oxide is selectively etched using vapor HF to release the device and form the desired airgap.

DEVICE CHARACTERIZATION

A wide range of input voltages and pulse duration (width) are tested to find the most suitable combination of amplitude and duration for both actuation and recrystallization voltages, as depicted in Fig. 4. Actuation pulses have fast rising and falling time (2 ns) to facilitate

the melt-quench cycle. Recrystallization pulses have slow falling edge (up to 1 µs) to ensure that complete recrystallization is achieved at every cycle. The OFF state channel current is measured (23 fA) with a B1500 parameter analyzer.

Fig. 5 shows the dynamic change in channel current in response to applied pulses. The RC delay is caused by the current limiting resistor connected to the source electrode. 100 ns wide 290 mV actuation pulse switches the relay ON. A recrystallization pulse of the same amplitude but longer duration, with 100 ns rising and 1 µs falling edge is applied to turn it OFF. It is evident that the contacts are opened within the first 100 ns of the onset of the pulse.

Operating voltage in the reported device dimensions is already compatible with CMOS power supplies. Faster switching and further voltage scaling should be expected from smaller device dimensions.

CONCLUSIONS

This paper reports the first demonstration of a highly scalable nonvolatile NEM relay with <5 nm air-gap. A very low actuation voltage makes it an ideal candidate for high-density nonvolatile memory operation. This vertical structure of Fin PCNR paves the way for further scaling and the development of more reliable emerging non-volatile and low-leakage memory architectures.

REFERENCES

[1]S. Barraud *et al.*, "Vertically stacked-NanoWires MOSFETs in a replacement metal gate process with inner spacer and SiGe source/drain," in *2016 IEEE International Electron Devices Meeting (IEDM)*, San Francisco, CA, USA, Dec. 2016, p. 17.6.1-17.6.4.
[2]S. Hong, O. Auciello, and D. Wouters, Eds., *Emerging Non-Volatile Memories*. Boston, MA: Springer US, 2014.
[3]Nam Sung Kim *et al.*, "Leakage current: Moore's law meets static power," *Computer*, vol. 36, no. 12, pp. 68–75, Dec. 2003.
[4]J. T. Best, M. A. Masud, M. P. de Boer, and G. Piazza, "Phase Change NEMS Relay," in *2019 IEEE International Electron Devices Meeting (IEDM)*, San Francisco, CA, USA, Dec. 2019, p. 34.1.1-34.1.4.
[5]J. Best and G. Piazza, "High Work Density Gete Mechanical Phase Change Actuator," in *2019 IEEE 32nd International Conference on Micro Electro Mechanical Systems (MEMS)*, Seoul, Korea (South), Jan. 2019.
[6]J. T. Best, M. A. Masud, M. P. Boer, and G. Piazza, "Phase Change Nanoelectromechanical Relay for Nonvolatile Low Leakage Switching," *Adv. Electron. Mater.*, p. 2200085, Apr. 2022.
[7]A. Fantini *et al.*, "Comparative assessment of GST and GeTe materials for application to embedded phase-change memory devices," presented at the 2009 IEEE International Memory Workshop, 2009, pp. 1–2.
[8]Y.-C. Lin, V. P. J. Chung, S. Santhanam, T. Mukherjee, and G. K. Fedder, "Sidewall Metallization on CMOS MEMS by Platinum ALD Patterning," *J. Microelectromechanical Syst.*, vol. 29, no. 5, Oct. 2020.

CONTACT

*MA Masud, tel: +14126526524; mmasud@cmu.edu

A FAST AND ENERGY-EFFICIENT NANOELECTROMECHANICAL NON-VOLATILE MEMORY FOR IN-MEMORY COMPUTING

Yong-Bok Lee[1], Min-Ho Gang[2], Pan-Kyu Choi[1], Su-Hyun Kim[1], Tae-Soo Kim[1],
So-Young Lee[1], and Jun-Bo Yoon[1]

[1]School of Electrical Engineering, Korea Advanced Institute of Science and Technology (KAIST),
Daejeon, Republic of Korea
[2]National NanoFab Center (NNFC), Daejeon, Republic of Korea

ABSTRACT

This paper reports a fast, energy-efficient, and complementary metal–oxide–semiconductor (CMOS) compatible nanoelectromechanical non-volatile memory (NEM-NVM) for in-memory computing. To achieve a fast speed and ultra-low energy consumption, we introduced an out-of-plane electrode configuration that enables efficient scaling of the actuation air gap and stiffness, the most important parameters of the operating speed and energy. We also utilized well-established CMOS manufacturing techniques to reliably fabricate the NEM-NVM. The fabricated NEM-NVM has a fast programming speed (< 100 ns) and ultra-low programming energy (< 10 fJ/bit). Furthermore, for the first time, we demonstrated that our NEM-NVM can perform logical operations such as AND, XOR, and NAND logic gates by exploiting the operation mechanisms and non-volatility of the NEM-NVM.

KEYWORDS

Nanoelectromechanical (NEM) non-volatile memory, Fast speed, Energy-efficiency, CMOS-compatible, In-memory computing.

INTRODUCTION

With the explosive growth of data-centric applications, the von Neumann architecture, in which computation and storage are physically separated, faces a significant performance bottleneck associated with latency and energy [1]. Recently, an in-memory computing architecture, in which computational tasks are processed in the memory itself, has been proposed to overcome the performance bottleneck [2, 3]. The new architecture can dramatically improve the latency and energy issues by eliminating the data shuttling between the processing unit and memory unit. Implementing in-memory computing typically requires non-volatile memory with a fast speed (< 100 ns), ultra-low energy consumption (< 10 fJ/bit), and compatibility with complementary metal–oxide–semiconductor (CMOS) fabrication processes [4]. However, it is still challenging to satisfy such requirements because emerging memory devices, such as magnetic random-access memory (MRAM) and resistive random-access memory (RRAM), have a slow speed and high energy consumption, and their materials are not compatible with CMOS processes [5]. Recently, a nanoelectromechanical non-volatile memory (NEM-NVM) has drawn considerable attention as a promising candidate due to its ultra-low static power consumption and CMOS compatibility [6-8]. However, a conventional NEM-NVM has an in-plane electrode configuration that limits the efficient scaling of the actuation air gap and stiffness which are the most important parameters of the operating speed and dynamic energy [9, 10]. Moreover, in-memory computing using the NEM-NVM has never been demonstrated before. In this work, we report a fast and energy-efficient NEM-NVM fabricated with CMOS-compatible processes. To achieve both a fast speed and ultra-low programming energy, an out-of-plane electrode configuration was introduced. As a result, the fabricated NEM-NVM shows a fast programming speed (< 100 ns) and ultra-low programming energy (< 10 fJ/bit). Moreover, we demonstrated that the fabricated NEM-NVM can perform in-memory computing such as AND, XOR, and NAND logic gates through interaction between various combinations of applied voltage (input) and resulting contact resistance (output).

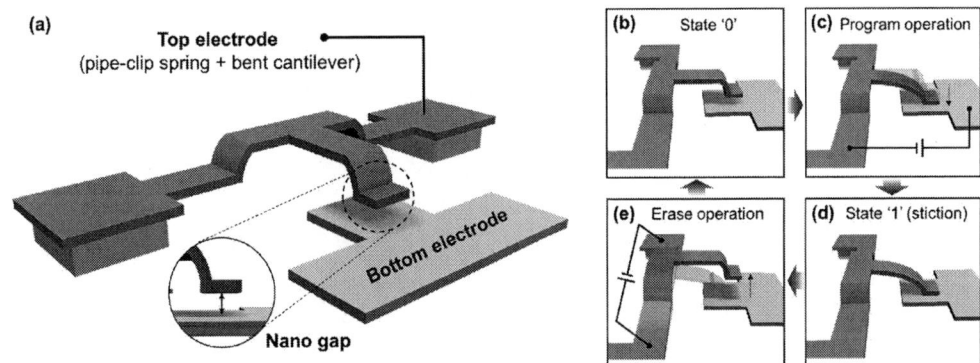

Figure 1:(a) Schematic of the proposed NEM-NVM with an out-of-plane electrode configuration. (b) Initial state of the NEM-NVM (state '0'). (c) Electrostatic program operation. (d) Programmed state of the NEM-NVM (state '1'). (e) Electrothermal erase operation.

CONCEPT

Figure 1 shows the structure and overall operating mechanisms of the NEM-NVM with an out-of-plane electrode configuration. The NEM-NVM consists of top and bottom electrodes, and the two electrodes are physically separated with a nanoscale airgap in the initial state (Figure 1(b)). Therefore, an ultra-low off current and static power consumption can be achieved in the initial state (state '0'). When an electrostatic force is applied between the top and bottom electrodes for program operation, the bent cantilever is deflected in the vertical direction (Figure 1(c)) and the deflected bent cantilever is maintained by the adhesion force even if the applied voltage is switched off (Figure 1(d)). In other words, non-volatility is achieved by intentionally using the stiction phenomenon, which is one of the major failures of N/MEMS devices. Figure 1(e) shows the electrothermal erase operation of the NEM-NVM. The adhered bent cantilever is recovered to its initial state by the thermal expansion force generated from the pipe-clip spring during the joule heating process.

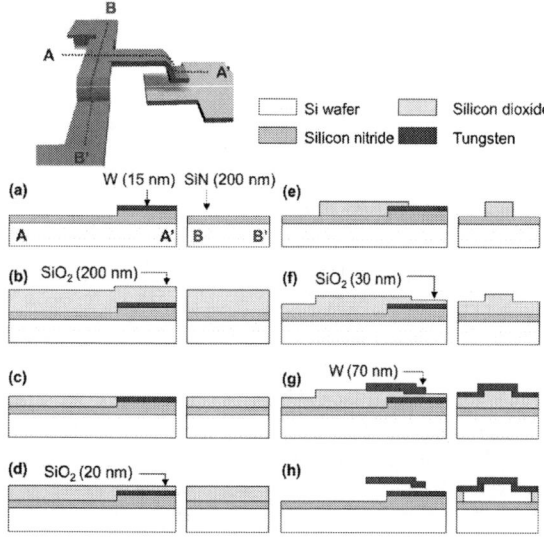

Figure 2: CMOS-compatible fabrication process. (a) Silicon nitride deposition (200 nm thick) and bottom electrode formation (15 nm thick). (b) 1st silicon dioxide deposition (200 nm thick). (c) Chemical-mechanical polishing process. (d) 2nd silicon dioxide deposition (20 nm thick). (e) Silicon dioxide patterning. (f) 3rd silicon dioxide deposition (30 nm thick). (g) Top electrode formation (70 nm thick). (h) BOE wet etching and CPD process.

FABRICATION

We fabricated the memory devices only with CMOS-compatible materials and equipment, potentially enabling monolithic integration with conventional CMOS circuits. In addition, the possibility of mass production was demonstrated by patterning the NEM-NVM through a KrF scanner system. Figure 2 shows the overall fabrication processes of the proposed NEM-NVM. First, silicon nitride layer was deposited on an 8-inch silicon wafer for electrical insulation using low pressure chemical vapor deposition.

Then, molybdenum was deposited using a sputter system and patterned to create the bottom electrode (Figure 2(a)). Next, to make the surface flat, a thick layer of silicon dioxide was deposited using plasma enhanced chemical vapor deposition (PECVD), and then, chemical mechanical polishing was performed (Figure 2(b-c)). These steps were essential for the accurate fabrication of the designed mechanical structure. Then, an atomic layer deposition process was used to deposit a silicon dioxide layer with an exact thickness, and it was patterned to define the pipe-clip structure. To achieve a high speed and low programming energy, a thin silicon dioxide layer, which serves as sacrificial layer between the top and bottom electrodes, was deposited using the PECVD. Molybdenum was deposited using sputter system and patterned to create the top electrode. As a final step, sacrificial silicon dioxide layer was removed using a buffered oxide etchant, and then we conducted critical point drying process to reliably dry the NEM-NVM without initial stiction. Figure 3 (a-b) show the optical images of the fabricated NEM-NVM on an 8-inch wafer and diced chip. The successfully fabricated NEM-NVM was confirmed by a top-view scanning electron microscope (SEM) image (Figure 3(c)) and cross-sectional transmission electron microscope (TEM) images of the pipe-clip spring (Figure 3(d)) and bent cantilever (Figure 3(e)) of the NEM-NVM. Because we utilized the well-established CMOS manufacturing processes, scalability, high-density, and delicate fabrication were achieved.

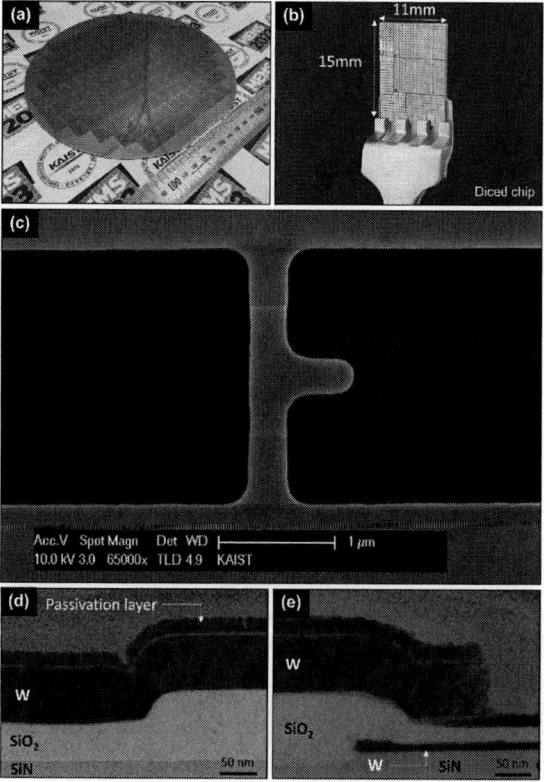

Figure 3: Fabrication results. (a-b) Optical image of the fabricated 8-inch wafer and diced chip, respectively. (c) Top-view SEM image of the fabricated NEM-NVM. (d-e) Cross-sectional TEM images of the pipe-clip spring (left) and bent cantilever (right) of the NEM-NVM, respectively.

EXPERIMENTAL RESULTS

Electrical measurements were performed to verify the operating characteristics of the fabricated NEM-NVM. Figure 4 shows the measured I-V curve of the program operation with a 100 nA current compliance. Before the turn-on voltage was applied, an ultra-low leakage current was observed due to the physical gap between the top and bottom electrodes. And then, the NEM-NVM was mechanically pulled down at 9.9 V, exhibiting an abrupt switching characteristic. The programming energy was less than 10 fJ/bit because the device has a very small air gap and stiffness due to the out-of-plane electrode configuration. The operation speed is also an important parameter of the non-volatile memory for in-memory computing. Thus, the programming speed of the NEM-NVM was measured through a customized voltage divider circuit with a load resistor shown in Figure 5 (a). Figure 5 (b) shows the transient response and programming speed ($<$ 100 ns) of the NEM-NVM, which is faster than those of previous NEM memory devices. For the erase operation, a voltage pulse of 0.9V for a time duration of 400 ns was applied to the programmed device. Figure 6 shows the contact resistance of the NEM-NEM in the programmed state and erased state, respectively. This result indicates that the fabricated NEM-NVM could operate successfully according to the proposed operating methods and has a high on-off ratio ($> 10^3$).

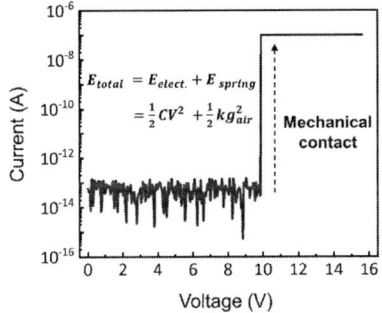

Figure 4: Current-voltage (I-V) curve of the fabricated NEM-NVM. The programming voltage was 9.9 V

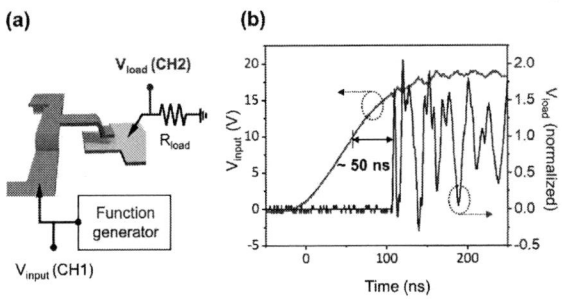

Figure 5: (a) Customized voltage divider circuit with a load resistor to measure the programming speed of the NEM-NVM. (b) Transient response of the NEM-NVM according to the applied voltage. The applied voltage reached about 10 V and then was divided by the load resistor after 50ns, which is the programming speed.

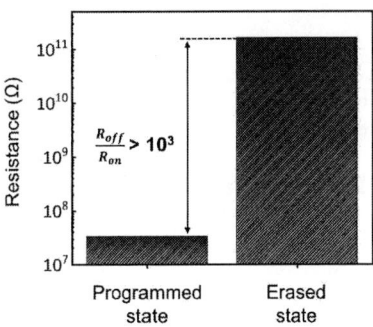

Figure 6: Comparison of the contact resistance between the programmed state and erased state.

Finally, various logical operations such as AND, XOR, and NAND logic gates were demonstrated through the interaction between the various combinations of applied voltage (input) and contact resistance (output). The logical operations are performed in two sequential steps. The first step is to initialize the contact resistance of the NEM-NVM to the 'H' or 'L' state ('H' represents a high contact resistance, and 'L' represents a low contact resistance). As a second step, when the various combinations of voltage are applied to the NEMS memory, the initialized contact resistance is changed and new information is stored in the memory itself, which is a widely used method in non-volatile memory for in-memory computing [11, 12]. Figure 7(a) shows the schematic illustration, truth table, and experimental results of the AND logical operation. The two input voltages were prepared as V_{top} ($= V_{program}/2$) and V_{bot} ($= -V_{program}/2$), respectively, and the contact resistance was initialized to the 'H' state. In the truth table, '1' means that the prepared input voltage is applied to the electrode, and '0' means that ground voltage is applied to the electrode. The AND gate was performed as follows: If both of the input voltages are '0', or one of the input voltages is '1', the voltage drop between the top and bottom electrode is smaller than $V_{program}$, and the program operation does not occur. Therefore, the output contact resistance maintains the 'H' state. However, if both of the input voltages are all '1', the voltage drop between the top and bottom electrodes satisfies the $V_{program}$, and the program operation occurs. Therefore, the output contact resistance is changed to the 'L' state. As shown in the measured results of the AND logical operation using the fabricated NEM-NVM, only when both of the two input voltages were '1', the contact resistance was changed to the 'L' state. Next, Figure 7(b) shows the schematic illustration, truth table, and experimental results of the XOR logical operation. The two input voltages were prepared as $V_{program}$ and the contact resistance was initialized to the 'H' state. In this case, when one of the input voltages was '1', the voltage difference between the electrodes satisfied the $V_{program}$, and the program operation occurred. Thus, the output contact resistance was changed to the 'L' state. Finally, Figure 7(c) shows the schematic illustration, truth table, and experimental results of the NAND logical operation. Unlike the previous logical operations, the NAND operation uses the erase mechanism of the NEM-

NVM. Therefore, two input voltages were prepared as $V_{top\_1}(= V_{erase}/2)$ and $V_{top\_2}(= -V_{erase}/2)$, respectively. The contact resistance was initialized to the 'L' state. In this case, only when both of the input voltages were '1', the voltage difference between the top electrodes satisfied the V_{erase}, and the erase operation occurred. Therefore, the output the contact resistance changed to the 'H' state. The experimental results indicate that various in-memory computing such as AND, XOR, and NAND gates can be implemented in the single NEM-NVM.

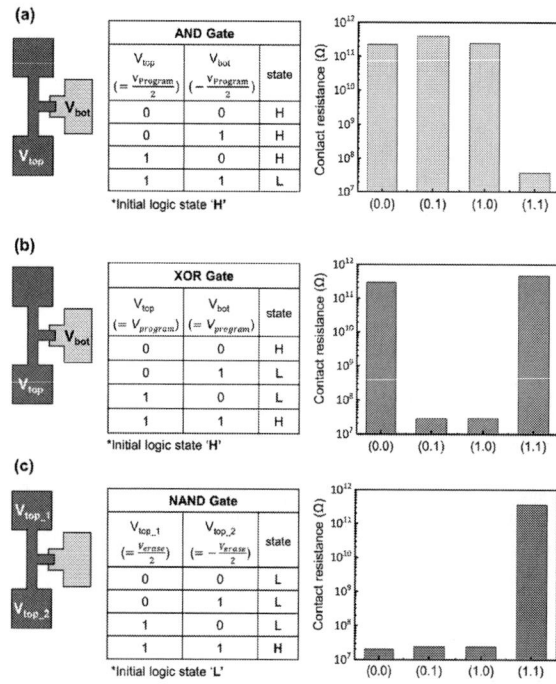

Figure 7: Schematic illustration, truth table and measured results of the (a) AND gate, (b) XOR gate, and (c) NAND gate. Logical operations in the NEM-NVM are enabled through the interaction between the applied voltage (input) and resulting contact resistance (output). 'H' represents a high contact resistance, and 'L' represents a low contact resistance.

CONCLUSION

In summary, we have demonstrated a fast, energy-efficient, and CMOS compatible NEM-NVM. Using well-established CMOS manufacturing techniques, the NEM-NVM was successfully fabricated on an 8-inch wafer. The fabricated NEM-NVM exhibited a fast programming speed (< 100 ns) and low programming energy (< 10 fJ/bit), which are the lowest values compared to those of other state-of-the-art technology. Furthermore, we showed that various logical operations such as AND, XOR, and NAND gates can be implemented through the NEM-NVM for the first time. Our approach will provide new inspiration for the realization of MEMS based non-von Neumann computing.

ACKNOWLEDGEMENTS

This work was supported by National Research and Development Program through the National Research Foundation of Korea (NRF) funded by Ministry of Science and ICT (2020M3F3A2A01082600). This work was also supported by Samsung Electronics. The EDA tool was supported by the IC Design Education Center (IDEC).

REFERENCES

[1] M. A. Zidan, J. P. Strachan, W. D. Lu, "The future of electronics based on memristive systems", *Nat. Electron.*, vol. 1, pp. 22-29, 2018.

[2] A. Sebastian, M. Le Gallo, R. Khaddam-Aljameh, E. Eleftheriou, "Memory devices and applications for in-memory computing", *Nat. Nanotechnol.*, vol. 15, pp. 529-544, 2020.

[3] Q. Xia, J. J. Yang, "Memristive crossbar arrays for brain-inspired computing", *Nature materials*, vol. 18, pp. 309-323, 2019.

[4] D. Ielmini, H.-S. P. Wong, "In-memory computing with resistive switching devices", *Nat. Electron.*, vol. 1, pp. 333-343, 2018.

[5] H.-S. P. Wong, S. Salahuddin, "Memory leads the way to better computing", *Nat. Nanotechnol.*, vol. 10, pp. 191-194, 2015.

[6] J. Zhang, Y. Deng, X. Hu, J. P. Nshimiyimana, S. Liu, X. Chi, P. Wu, F. Dong, P. Chen, W. Chu, "Nanogap-Engineerable Electromechanical System for Ultralow Power Memory", *Advanced Science*, vol. 5, pp. 1700588, 2018.

[7] C. Kim, R. Marsland, R. H. Blick, "The nanomechanical bit", *Small*, vol. 16, pp. 2001580, 2020.

[8] O. Y. Loh, H. D. Espinosa, "Nanoelectromechanical contact switches", *Nat. Nanotechnol.*, vol. 7, pp. 283-295, 2012.

[9] U. Sikder, K. Horace-Herron, T.-T. Yen, G. Usai, L. Hutin, V. Stojanović, T.-J. K. Liu, "Toward Monolithically Integrated Hybrid CMOS-NEM Circuits", *IEEE Trans. Electron. Devices*, vol. 68, pp. 6430-6436, 2021.

[10] S. Rana, J. Mouro, S. J. Bleiker, J. D. Reynolds, H. M. Chong, F. Niklaus, D. Pamunuwa, "Nanoelectromechanical relay without pull-in instability for high-temperature non-volatile memory", *Nat. Commun.*, vol. 11, pp. 1-10, 2020.

[11] B. C. Jang, Y. Nam, B. J. Koo, J. Choi, S. G. Im, S. H. K. Park, S. Y. Choi, "Memristive Logic-in-Memory Integrated Circuits for Energy-Efficient Flexible Electronics", *Advanced Functional Materials*, vol. 28, pp. 1704725, 2018.

[12] B. C. Jang, S. Y. Yang, H. Seong, S. K. Kim, J. Choi, S. G. Im, S.-Y. Choi, "Zero-static-power nonvolatile logic-in-memory circuits for flexible electronics", *Nano Research*, vol. 10, pp. 2459-2470, 2017.

CONTACT

Y.-B. Lee, tel: +82-42-350-5476;
bok6155@kaist.ac.kr

978-1-6654-9309-3/23 $31.00 © 2023 IEEE

TOWARDS ULTRA-HIGH SPATIAL RESOLUTION SENSING OF GHZ ULTRASOUND USING STRAIN MODULATION OF FIELD EFFECT TRANSISTORS

Rohan Sanghvi[1], Justin Kuo[2], Adarsh Ravi[1] and Amit Lal[1]
[1]SonicMEMS Laboratory, Cornell University, USA and
[2]Geegah Inc., USA

ABSTRACT

This work demonstrates the detection of GHz ultrasonic (U/S) pulses using a 130nm technology-based PFET transistor, monolithically integrated with a GHz Fresnel focusing ultrasonic transducer. The focusing transducer enables the concentration of the ultrasonic strain at the PFET location resulting in observable modulation of the transistor current due to piezoresistivity in the transistor conduction channel. 50ns wide U/S pulses with 1-2 GHz frequencies were successfully detected using the PFET transistor. The outcome of measuring strain with nm-scale transistors indicates the potential application of 3D arrays of CMOS transistors for sub-μm spatial resolution and high bandwidth active sensing of high-frequency ultrasonic pulses and waves.

KEYWORDS

GHz imaging, ultrasonic, Fresnel focusing transducer, CMOS, strain modulation, piezoresistance

INTRODUCTION

The development of high-frequency resonant devices has played a crucial role in the advancement of critical fields such as telecommunication [1], [2], sub-surface microscopy [3], acoustic characterization of biological specimen [4] and fingerprint sensing [5], [6]. The ability to measure displacements and vibration mode shapes of high-frequency resonators and oscillators is critical for extracting important device parameters. These parameters include the quality factor, electromechanical coupling, acoustic velocities, and dissipation characteristics, which can further inform design and performance optimization [7].

The predominant methods for measuring sub-nm vibrations at frequencies over 1 GHz are based either on optical interferometry or piezoelectric elements as receive-mode transducer pixels. Optical methods including continuous-wave, pump-probe, and stroboscopic interferometry [7]–[11] have been employed to investigate vibrations of micro and nano-scale oscillators and resonators. However, these are massively complex systems that use expensive and delicate optical components, require stringent calibration, and are difficult to miniaturize for on-chip integration with MEMS and NEMS systems. Although piezoelectric detection methods have been miniaturized, the spatial resolution has been limited to 10s of microns owing to microfabrication challenges and reduced transducer capacitance exacerbating detection challenges.

Therefore, to overcome these key challenges, a CMOS transistor as an imaging nano-pixel for sensing high-frequency ultrasound by leveraging the strain modulation of the piezoresistive channel silicon is presented here. To demonstrate this, a thin film Aluminum Nitride (AlN) transducer was designed and integrated with CMOS for launching short beam-formed high-frequency ultrasonic pulses into a silicon substrate. Fresnel zone plate architecture is used to achieve localized high-intensity strains focused onto a PFET. The strain wave interacts with the PFET and causes small signal modulation in channel conductance, which is intercepted to provide magnitude, frequency, and phase information about the strain wave. Experimental characterization of the PFET for a range of parameters, such as RF drive voltage, drive frequency, and gate bias voltage, has been conducted. The results are validated using theoretical models and FEA simulations, hence establishing the potential for application of 3D arrays of transistors for sub-μm resolution, high bandwidth imaging of high frequency (> 1GHz) ultrasonic waves.

PRINCIPLE OF OPERATION

The device consists of a 2μm thin-film Aluminum Nitride (AlN) piezoelectric layer fabricated on a 669μm thick silicon substrate with integrated 130nm CMOS transistors. The bulk acoustic wave (BAW) mode resonance frequency is 1.77-1.87 GHz. The top electrodes of the transducer are patterned as a Fresnel zone plate (figure 1a). The radii of the rings are chosen such that the 1.77 GHz acoustic pulse focuses at a distance of 1338μm. This distance corresponds to the total transit length from the transmitter to the bottom reflector and back to the transmitting plane. The strain wave is focused right below the center of the concentric rings (figure 1b). The FWHM of the ultrasonic beam at the focus can be approximated as

$$d \approx 1.22\,(r_N - r_{N-1}) \qquad (1)$$

where $r_N - r_{N-1}$ is the width of the outermost ring of the Fresnel zone plates [12]. Therefore, a significant share of the acoustic energy transmitted by the transducer can be expected to be focused within a 15μm diameter region. Focusing effect of the Fresnel transducer was simulated using COMSOL. Figure 2 shows the results for a 50ns 1.77 GHz pulse with an input stress amplitude of 2kPa. A time snapshot of the longitudinal stress at the axis (σ_{zz}) at 182ns is presented in figure 2a. As the pulse travels through the silicon and arrives back to the transmit side (at 0μm), the focusing increases the longitudinal stress amplitude to 18kPa at the transistor location. Moreover, the effects of the Fresnel focusing can be observed in figure 2b as dense red and blue zones, corresponding to tensile and compressive longitudinal stresses respectively. This highlights the localization of strain in a 15μm diameter focus and facilitates characterizing the interaction of the acoustic pulse with a single transistor.

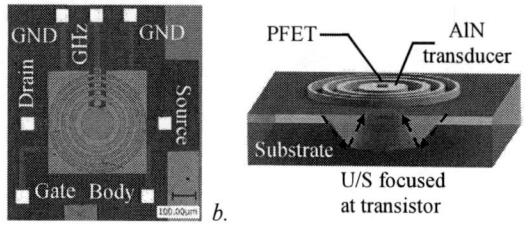

Figure 1: a. Microscope image of the AlN Fresnel transducer device with PFET fabricated at the focal point (not visible in this view); b. Cross-section of device operation – Fresnel AlN transducer (green), resonates in BAW mode at 1.77 GHz, transmitting a high-intensity focused strain wave (purple) onto the PFET, thereby inducing piezoresistance modulation in the PFET source-to-drain channel

A 5μm wide PFET, based on 130nm technology, fabricated at the focal point (W/L=5000/130). As the GHz strain wave passes through the DC-biased PFET, it causes a transient change in the mobility of the transistor channel due to the piezoresistive properties of the channel silicon. The modulation in mobility ($\Delta\mu$) depends linearly on the applied stress (σ_{Si}):

$$\frac{\Delta\mu}{\mu} = [\pi_{Si}][\sigma_{Si}] \qquad (2)$$

where π_{Si} is the piezoresistance coefficient of silicon. This transient change in mobility induces a small signal modulation in the drain current (ΔI_D), and, consequently, in the drain voltage (ΔV_D) [13].

$$\Delta V_D = \Delta I_D R_S = I_{D,DC}\left(\frac{\Delta\mu}{\mu}\right) R_D \qquad (3)$$

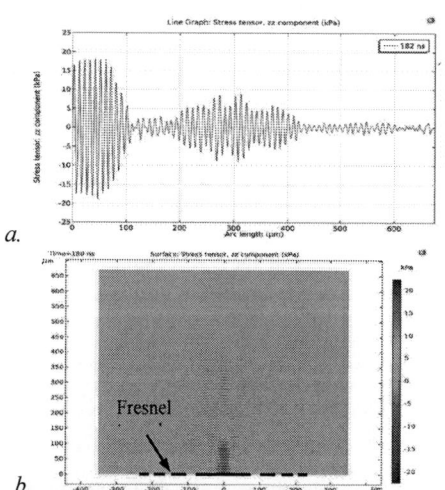

Figure 2: COMSOL simulation of Fresnel focusing for 2kPa drive stress shows ~15μm wide focal region with ~18kPa stress amplitude

where $I_{D,DC}$ is the DC drain current, and R_D is the termination resistor at the drain. ΔV_D is amplified and measured using a high-frequency oscilloscope.

EXPERIMENTAL SETUP

A schematic of the experimental setup for actuating the ultrasonic pulses and detecting strain modulation in transistor channel is shown in figure 3. An RF generator is used to produce a continuous wave RF signal at BAW mode resonance frequency of 1.77 GHz. The continuous wave is converted to 50ns pulses by routing through a high-speed single pull double throw (SPDT) RF switch controlled by a pulse generator. The input pulse from the RF-switch passes through a high-frequency power amplifier which drives the thin film AlN transducer. After completing a return trip through the substrate, this acoustic pulse is captured as the ultrasonic first echo using a high frequency oscilloscope. The PFET is DC biased using the circuit shown in figure 4. An off-chip source measure unit (SMU) was used to supply the 1.2V rail voltage and the bias voltage (V_g). The PFET drain is terminated with a 50Ω resistor, and the small signal modulation at the drain is amplified using a GHz amplifier with a gain of 36 dB. The GHz RF and PFET biasing modules were controlled using a custom python program.

The strain modulation in the PFET is investigated by sweeping the RF drive voltage (V_{pp} ultrasound drive), the bias voltage (V_g), and the RF signal frequency. As the transducer is pulsed, waveform data of the first ultrasonic echo and the PFET drain voltage is recorded. Figure 5 presents an example of the recorded waveforms from the AlN transducer (red) and the PFET (green) for 35mV$_{pp}$ U/S drive and 0.1V V_g. Further, the peak amplitude of either waveform is extracted and analyzed.

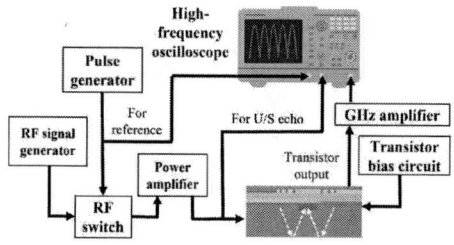

Figure 3: Testing setup for pulse-echo characterization

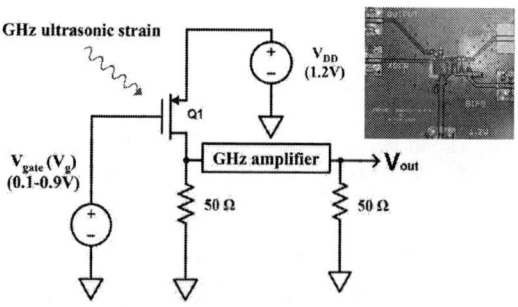

Figure 4: PFET biasing and stain modulation detection circuit; (inset) picture of PCB setup with connections to transducer and transistor gate, source and drain

978-1-6654-9309-3/23 $31.00 © 2023 IEEE

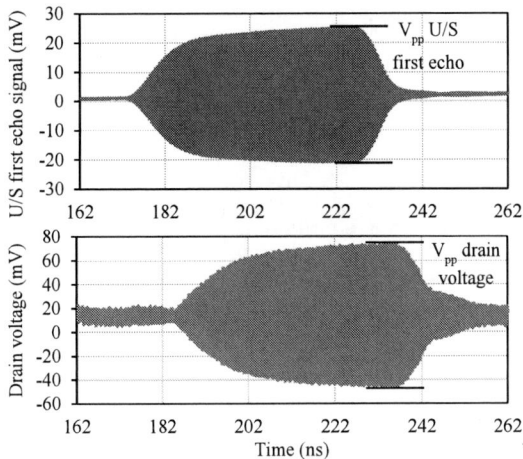

Figure 5: Example waveforms of first ultrasonic echo from the transducer (red) and corresponding PFET drain modulation (green) due to pulsed drive of the transducer

RESULTS AND DISCUSSION

The performance of the transducer and the imaging PFET for a range of V_{pp} U/S drive voltages (0-50mV$_{pp}$) and V_g (0.1 - 0.9V), is presented in figure 6. The ultrasonic first echo amplitude (V_{pp} U/S first echo) is found to increase linearly from 0-35mV$_{pp}$ for V_{pp} U/S drive voltages up to 20mV$_{pp}$ until the power amplifier starts saturating (figure 6a). As expected, change in V_g has minimal effect on V_{pp} U/S first echo, which is directly related to the induced stress. Furthermore, the output amplitude of the PFET small signal modulation (V_{pp} drain voltage (amp.)) also shows a similar trend, as it increases from 0-180mV$_{pp}$ for the chosen range of U/S drive voltages (figure 6b). However, this modulation is observed to be dependent on V_g. As V_g is decreased from 0.9V - 0.1V, the transistor moves from cut-off to saturation, thereby increasing $I_{D,DC}$. Since V_{pp} drain voltage (amp.) is directly proportional to $I_{D,DC}$, the modulation owing to the same stress is observed to be higher for lower gate voltages. The V_{pp} U/S first echo and V_{pp} drain voltage (amp.) data in figures 6a and 6b is correlated in figure 6c to demonstrate this relationship. V_{pp} drain voltage (amp.) shows a linear relationship with V_{pp} U/S first echo. Additionally, the slopes of the traces in figure 6c, which can be understood as the sensitivity of detection, increase in the same ratio as $I_{D, DC}$ in agreement with theory.

The influence of temperature variations on the PFET imaging performance due to potential ultrasonic heating was analyzed by recording the sensitivity for different RF drives. The overlapping traces in the figure 7 plot shows that the sensitivity primarily depends on V_g and, consequently the drain current and has minimal deviation as drive voltage is increased from 10-30mV$_{pp}$. This result agrees with the theoretical model discussed previously and serves as confirmation to exclude heating effects.

Figure 8 presents the frequency response of the transducer (red) and transistor (green) signals as the drive frequency is swept from 1.7 to 1.9GHz. The transistor response peaks at 1.77GHz due to the coincidence of Fresnel focusing on the transistor.

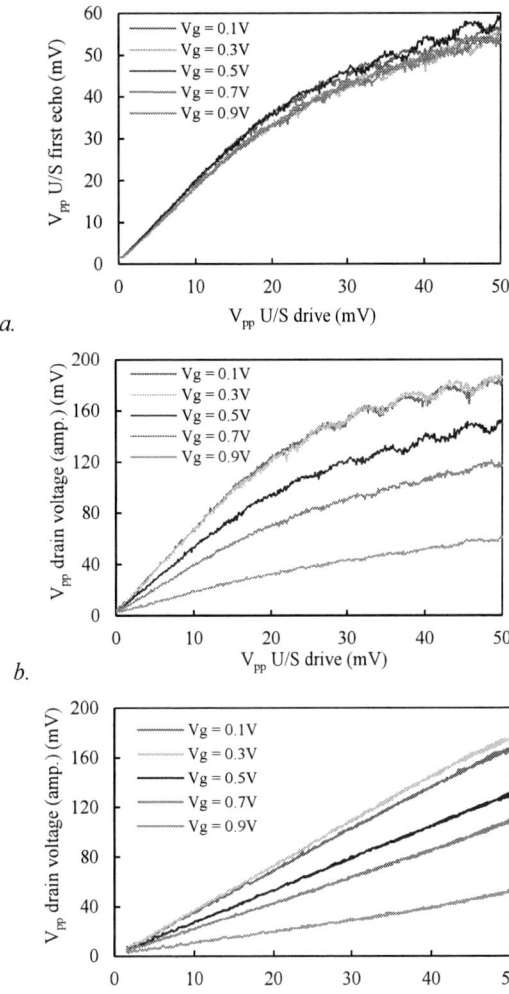

a.

b.

c.

Figure 6: a. Variation in first ultrasonic echo (V_{pp} U/S first echo) amplitude with RF drive voltage (V_{pp} U/S drive) for various gate voltages (V_g) ; b. Variation in drain voltage small signal modulation amplitude (V_{pp} drain voltage (amplified)) with RF drive voltage for various gate voltages; c. V_{pp} drain voltage (amplified) vs. V_{pp} U/S first echo for various gate voltages

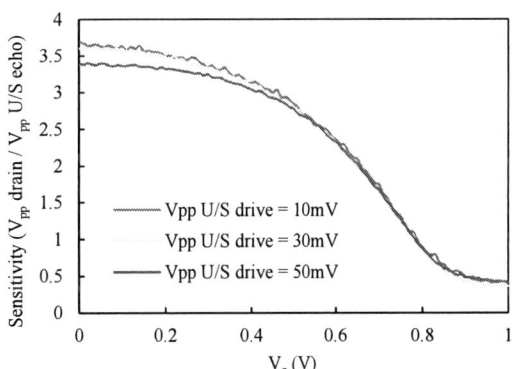

Figure 7: Variation of sensitivity with gate voltage (V_g) for different RF drive amplitudes

Moreover, peaks spaced at every 6MHz are observed in the transistor output, which can be attributed to the interference of acoustic pulses from the multiple annular transducers in the Fresnel configuration. This phenomenon was modelled by calculating the normalized axial field due each ring using the expression [14]

$$|p(z,f)| =$$

$$\left| j \sin\left[\frac{k}{2}\left(\sqrt{r_a{}^2 + z^2} - z\right)\right] e^{-j\frac{k}{2}\left(\sqrt{r_a{}^2+z^2}+z\right)} \sin\left(k\sqrt{r_a{}^2 + z^2}\right) \right|$$

$$where \; k \; = \; \frac{2\pi c}{f} \; and \; r_a = \frac{r_o + r_i}{2} \tag{4}$$

with c being the speed of sound in silicon, f being the operation frequency, z being the focal length, and r_o and r_i being the rings' inner and outer radii. The theoretical response of the transistor due to the strain wave (blue) was calculated by obtaining the magnitude of the total pressure field from all rings and using it to calculate the transistor conductance modulation. The close agreement of the theoretical curve with the experimental data confirms the strain modulation effect in the transistor.

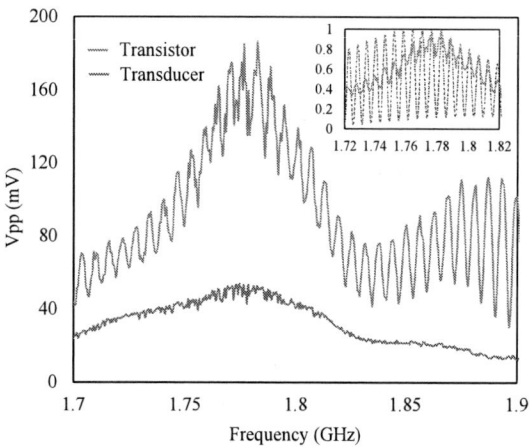

Figure 8: Frequency response of the transducer (red) and transistor (green); inset: theoretical (blue) vs. experimental (green) curves for normalized drain voltage modulation amplitude vs. frequency, showing interferences at 6 MHz intervals.

CONCLUSION

The application of a monolithically integrated PFET for imaging the magnitude, frequency, and phase of GHz pulses and waves is successfully demonstrated. 50ns wide ultrasonic pulses with 1.77 GHz frequency were generated using a Fresnel-type AlN transducer and focused on the transistor. The strain wave was sensed as a small signal modulation in drain current facilitated by the piezoresistance of the channel silicon. The results of this study substantiate the application of 2D and 3D arrays of transistors for high bandwidth, sub-μm spatial resolution imaging of high-frequency strain waves, and subsequently, displacements and mode shapes of GHz resonant devices.

REFERENCES

[1] A. Gao, K. Liu, J. Liang, and T. Wu, "AlN MEMS filters with extremely high bandwidth widening capability," *Microsyst Nanoeng*, vol. 6, no. 1, p. 74, 2020, doi: 10.1038/s41378-020-00183-5.

[2] C. Zuo, N. Sinha, and G. Piazza, "Very high frequency channel-select MEMS filters based on self-coupled piezoelectric AlN contour-mode resonators," Sens Actuators A Phys, vol. 160, no. 1, pp. 132–140, 2010, doi: https://doi.org/10.1016/j.sna.2010.04.011.

[3] A. Baskota, J. Kuo, and A. Lal, "Gigahertz Ultrasonic Multi-Imaging of Soil Temperature, Morphology, Moisture, and Nematodes," in 2022 IEEE 35th International Conference on Micro Electro Mechanical Systems Conference (MEMS), 2022, pp. 519–522. doi: 10.1109/MEMS51670.2022.9699813.

[4] P. S. Balasubramanian, A. Singh, C. Xu, and A. Lal, "GHz Ultrasonic Chip-Scale Device Induces Ion Channel Stimulation in Human Neural Cells," Sci Rep, vol. 10, no. 1, p. 3075, 2020, doi: 10.1038/s41598-020-58133-0.

[5] J. C. Kuo, J. T. Hoople, M. Abdelmejeed, M. Abdel-moneum, and A. Lal, "64-Pixel solid state CMOS compatible ultrasonic fingerprint reader," in 2017 IEEE 30th International Conference on Micro Electro Mechanical Systems (MEMS), 2017, pp. 9–12. doi: 10.1109/MEMSYS.2017.7863326.

[6] J. Hoople, J. Kuo, S. Ardanuç, and A. Lal, "Chip-scale reconfigurable phased-array sonic communication," in *2014 IEEE International Ultrasonics Symposium*, 2014, pp. 479–482. doi: 10.1109/ULTSYM.2014.0119.

[7] L. Shao *et al.*, "Femtometer-amplitude imaging of coherent super high frequency vibrations in micromechanical resonators," *Nat Commun*, vol. 13, no. 1, pp. 1–9, 2022, doi: 10.1038/s41467-022-28223-w.

[8] J. v Knuuttila, P. T. Tikka, and M. M. Salomaa, "Scanning Michelson interferometer for imaging surface acoustic wave fields.," *Opt Lett*, vol. 25, no. 9, pp. 613–615, May 2000, doi: 10.1364/ol.25.000613.

[9] K. Kokkonen and M. Kaivola, "Scanning heterodyne laser interferometer for phase-sensitive absolute-amplitude measurements of surface vibrations," *Appl Phys Lett*, vol. 92, no. 6, p. 63502, Feb. 2008, doi: 10.1063/1.2840183.

[10] K. Misiakos *et al.*, "Broad-band Mach-Zehnder interferometers as high performance refractive index sensors: Theory and monolithic implementation," *Opt Express*, vol. 22, no. 8, pp. 8856–8870, 2014, doi: 10.1364/OE.22.008856.

[11] Z. Shen, X. Han, C.-L. Zou, and H. X. Tang, "Phase sensitive imaging of 10 GHz vibrations in an AlN microdisk resonator," *Review of Scientific Instruments*, vol. 88, no. 12, p. 123709, Dec. 2017, doi: 10.1063/1.4995008.

[12] Z. Zhang *et al.*, "Hybrid-level Fresnel zone plate for diffraction efficiency enhancement," *Opt Express*, vol. 25, no. 26, pp. 33676–33687, 2017, doi: 10.1364/OE.25.033676.

[13] R. Marathe, W. Wang, and D. Weinstein, "Si-based unreleased hybrid MEMS-CMOS resonators in 32nm technology," in *2012 IEEE 25th International Conference on Micro Electro Mechanical Systems (MEMS)*, 2012, pp. 729–732. doi: 10.1109/MEMSYS.2012.6170289.

[14] C. H. Sherman and J. L. Butler, "Acoustic Radiation from Transducers BT - Transducers and Arrays for Underwater Sound," C. H. Sherman and J. L. Butler, Eds. New York, NY: Springer New York, 2007, pp. 438–466. doi: 10.1007/978-0-387-33139-3_10.

CONTACT

*A. Lal, amit.lal@cornell.edu

A TACTILE SENSOR ARRAY WITH A MONOLITHICALLY INTEGRATED NEURAL NETWORK FOR EDGE COMPUTATION

Tengteng Lei, Yushen Hu, and Man Wong

The Hong Kong University of Science and Technology, Hong Kong, CHINA

ABSTRACT

Described presently is a monolithically integrated "intelligent" sensor system consisting of a piezoelectric tactile sensor array and an artificial neural network (ANN). Low-temperature processed, dual-gate thin-film transistors (TFTs) based on semiconducting indium-tin-zinc oxide are deployed in the former for signal acquisition and in the latter for executing the "multiply-accumulate" operation. With the ANN appropriately trained, the system is capable of both "imaging" and "recognizing" a distributed force load applied on the sensor array using 3D-printed stamps. The proposed integration scheme can be deployed to construct intelligent sensor systems for applications other than tactile sensing.

KEYWORDS

Tactile sensor array, artificial neural network, edge computation, dual-gate, metal-oxide, thin-film transistors.

INTRODUCTION

Tactile sensors are deployed in systems for health monitoring, robotic contact sensing, and prosthetic control [1], etc. A host of studies have been carried out to implement large-area and high-resolution tactile sensors based on different transduction mechanisms [2]-[4]. Due to its spontaneous self-biasing, long durability, and good malleability, piezoelectric polyvinylidene fluoride (PVDF) has been applied to construct arrays for the tactile sensing of textures and pulse waves [5].

However, the tactile perception of a biological somatosensory system (Fig. 1a) is a comprehensive activity of sensing, learning, and interpretation [6] that cannot be accomplished by the mere signal-acquisition function of a conventional physical sensor. Consequently, "perception" in the latter is supported by a separate signal-processing system. Software-based artificial neural networks (ANNs) are being intensely investigated for learning and interpretation, but their performance is limited by the von Neumann architecture of the typical hardware processors on which the ANNs are implemented.

Recent advances in neuromorphic devices enabling computation in memory have accelerated the integration of sensors with ANNs for more effective signal interpretation and edge computation. Though attempts have been made to combine a single sensory node and an artificial synapse [7]-[8], it is still far from the scale required for accomplishing more complex neuromorphic computing tasks, such as pattern recognition. Though a sensor array combined with an ANN has been reported for tactile perceptual learning, the optical fibers required by the photo-memristors [9] or the ionic cables between the tactile sensors and the synaptic transistors [10]-[11] deteriorate the reliability of the systems.

Recently, a tactile sensor array [12] based on PVDF has been reported. Dual-gate (DG) thin-film transistors

(TFTs) have been deployed in the array to couple the PVDF sensor output signals to an in-pixel amplifier. The same technology has been deployed to construct neuromorphic logic gates [13] and ANN for pattern recognition [14]. Presently described is a biomimetic sensor system (Fig. 1b) consisting of a PVDF tactile sensor array monolithically integrated with an ANN. Processed at a temperature below 300 °C, DG TFTs based on semiconducting indium-tin-zinc oxide (ITZO) are deployed in the sensor array for signal acquisition and in the ANN for neuromorphic computation. With the ANN appropriately trained, the system is capable of both "imaging" and "recognizing" a distributed force load applied using 3D-printed stamps on the sensor array. Clearly, the TFT technology can be deployed to construct intelligent sensor systems for applications other than tactile sensing.

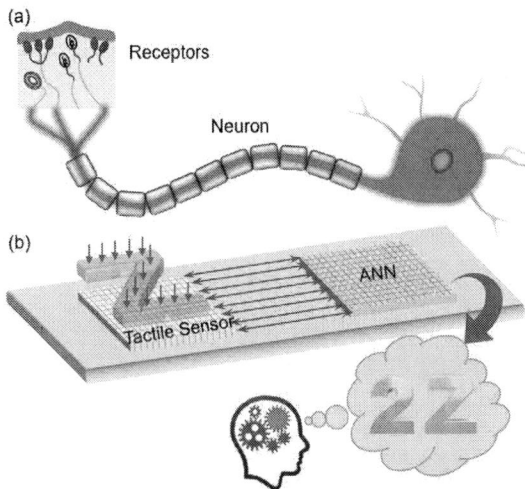

Figure 1. Schematic diagram of (a) a biological neuron and (b) conceptual design of an intelligent tactile sensing system.

EXPERIMENTAL

Fabrication Process

Starting with a glass substrate covered with a 300-nm-thick low-temperature oxide (LTO) buffer layer, 100-nm-thick molybdenum (Mo) is sputtered and patterned as the bottom gate electrode (BG) before the plasma-enhanced chemical vapor deposition (PECVD) of a bottom gate-insulator (GI) stack consisting of 75-nm-thick silicon oxide (SiO_x) on 50-nm-thick silicon nitride. Following a 2-hr thermal treatment at 300 °C in oxygen (O_2), a 20-nm-thick ITZO active layer is sputtered in an atmosphere of 60% argon and 40% O_2. After patterning the ITZO layer to form an active island (AC), a 100-nm-thick SiO_x is deposited as the top GI. Following another thermal treatment at 300 °C for 6 hrs in O_2, a 150-nm-thick Mo is sputter-deposited and patterned as a top-gate electrode (TG) before covering a 200-nm-thick SiO_x. This is followed by an additional annealing process at 250 °C for 8 hrs in O_2. Source/drain (S/D) activation in an O_2 plasma is carried out before the opening of contact holes to access the TG, AC, and BG. A stack of 200-nm-thick aluminum (Al) on 150-nm-thick Mo is deposited before its patterning to form the S/D electrodes.

After the TFT array processing, a PVDF film is fixed atop the sensor array region using a UV-curable glue. Shown in Fig. 2a is a photograph of a monolithically integrated system. Shown in Fig. 2b is the layout of the system of an 8×8 tactile sensor (left) and two 8×8 ANNs (right).

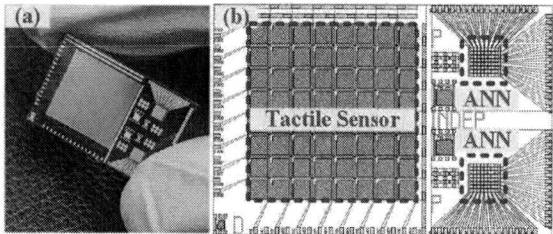

Figure 2. (a) Die photograph and (b) layout of the monolithically integrated 8×8 tactile sensor array (left) and 8×8 ANNs (right).

Sensor Nodes and Artificial Synapses

The tactile sensor and the ANN are composed of respective arrays of sensor nodes and artificial synapses. A sensor node (Fig. 3a) contains three transistors (TFTs M1, M2, and M3) and a piezoelectric transducer (PVDF). Regulated by the switch TFT M3, tactile signals as the output of the in-pixel amplifier consisting of TFTs M1 and M2 are sampled. The desired operating point of the amplifier is set by the bias voltage applied on the BG of the DG TFT M1. A synapse (Fig. 3b) contains three transistors (TFTs M4, M5, and M6) and two capacitors (C1 and C2). The input signal (I_{ij}) on the BG and the weight signal (W_{ij}) on the TG of the DG TFT M6 are applied respectively through the addressing TFTs M4 and M5. These signals are stored on C1 and C2 when M4 and M5 are turned off. The DG TFT M6 emulates the activities across a synapse, with its BG acting as the pre-synapse and its channel as the post-synapse.

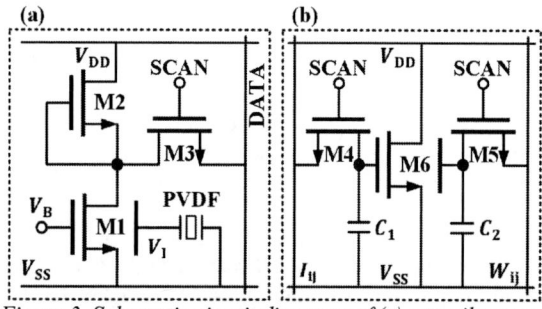

Figure 3. Schematic circuit diagrams of (a) a tactile sensor node and (b) a synapse.

Respectively shown in Figs. 4a and 4b are the photographs of a sensor node with an area of 1250×1250 μm^2 and a collection of synapses, each with an area of 206×206 μm^2.

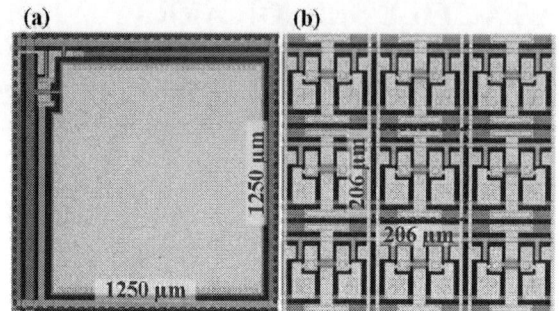

Figure 4. photographs of (a) a tactile sensor node and (b) a portion of an ANN with a synapse highlighted with a dotted frame.

ANN Training

As a demonstration, the integrated system is trained to distinguish two tactile images of 3D printed stamps of "2" and "Z" applied on a sensor array. Based on the Mini-Batch Gradient Descent Algorithm (MBGDA) [15], a pair of "2" and "Z" is randomly selected to form a mini-batch. Eight mini-batches containing a total of sixteen tactile images for "2" and "Z" are packaged as the training data sets.

Illustrated in Fig. 5 is the structure and the training procedure of the hardware ANN for a mini-batch input. Tactile signals (I_{ij}) obtained by the sensor array and the weight signals (W_{ij}) are fed into an 8×8 ANN by active-matrix addressing. A deeper network is implemented by reusing the first row of the 8×8 ANN as a second 8×1 layer. The eight sigmoid-activated outputs of the first layer are fed as inputs to the second layer. The activated output of the latter layer is the final ANN output V_O.

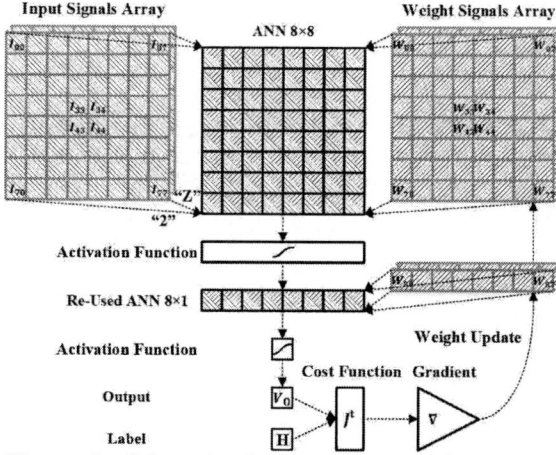

Figure 5. Schematic diagram showing the training procedure of an 8×8×1 ANN for a mini-batch input.

A cost function (J) is computed by the mean squared errors between output voltages (V_O) and the "label" values. The respective labels of patterns "2" and "Z" are "5 V (H)" and "0 V (L)". The partial derivative (∇) of J with respect to W_{ij} denotes the gradient. With an appropriately selected learning rate (α), W_{ij} can be iteratively updated according to

$$W_{ij}^{t+1} = W_{ij}^{t} - \alpha \times \nabla_{W_{ij}} J^{t}(W_{ij}^{t}, Z_t), \quad (1)$$

978-1-6654-9309-3/23 $31.00 © 2023 IEEE   14

where Z is the input tactile signals of a mini-batch, t $(0, ...,7)$ denotes the index of the mini-batches, and the respective i $(0, ...,8)$ and j $(0, ...,7)$ represents the number of rows and columns.

Shown in Fig. 6 is the training procedure of the hardware ANN for eight different mini-batches inputs in an epoch. During training Epoch n, each mini-batch input provides a cost function (J^t), and hence the updated gradient array. Iterating through the eight mini-batches, thus all 16 tactile inputs, the cost function (J^n) for training epoch n can be obtained. The convergence of training can be ascertained when J^n converges to a small steady-state value.

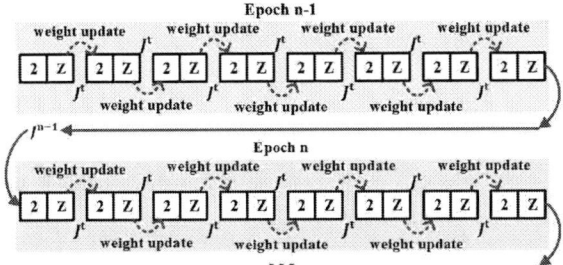

Figure 6. Schematic diagram showing the training procedure of an 8×8×1 ANN for eight different mini-batch inputs in Epoch n.

RESULTS AND DISCUSSION
Dataset

Used for training the integrated ANN, a number "2" and a letter "Z" stamp with a similar shape were placed on the 8×8 sensor array, the active area of which is 1 cm². Sixteen tactile images acquired with different positions are deployed as the training data sets (Fig. 7).

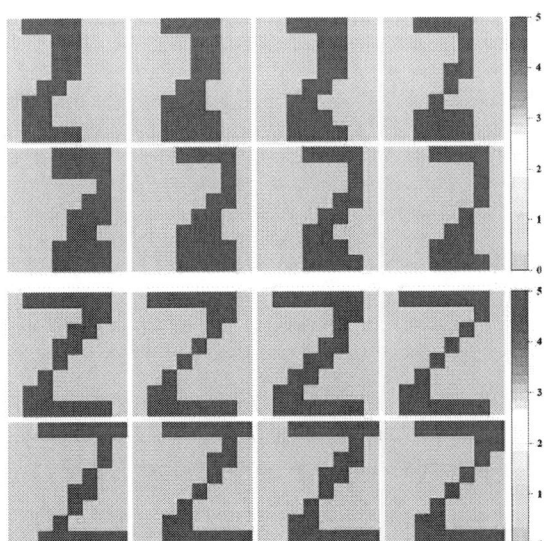

Figure 7. Tactile images of 16 different "2" and "Z" as training data sets.

In addition to the training data sets, another eight tactile signals of "2" and "Z", each with four patterns, are utilized as the test data set (Fig. 8).

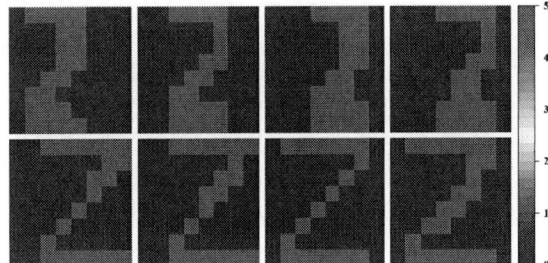

Figure 8. Tactile images of 8 different "2" and "Z" as test data sets.

Cost Function

The cost function, defined as the mean squared errors between predictions and labels, is an effective evaluation of the training process. For the MBGDA, the cost function J^t is the mean squared error for each mini-batch input, specifically a "2" and a "Z". The cost function J^n reveals the mean squared errors of overall sixteen tactile signals. Due to the miniaturization of the batch size, the training process is accelerated and the details of training effects for each mini-batch can be observed. After the initialization of epoch 0, J^t (Fig. 9a) and J^n (Fig. 9b) exhibit a gradual downward trend for the remaining 11 epochs. However, J^t in last few epochs shows a significant fluctuation, some of which are increased while others are smaller. Overfitting of the ANN model occurred on some datasets, resulting in a slight polarization effect of training. Defining both the noise high margin and noise low margin of 1.5 V with respect to the labels, the accuracy of the correct classification for all training datasets and test datasets is therefore decided as follows:

$$\text{Accuracy} = \text{Correct Predictions/Total.} \qquad (2)$$

As shown in the red dash curve in Fig. 9b, the recognition accuracy increased with the training epochs and eventually saturates at 96%.

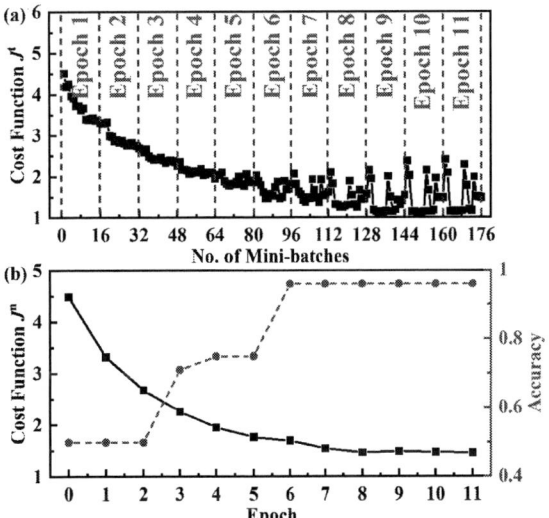

Figure 9. (a) Cost function J^t of each mini-batch vs the number of mini-batches; (b) Cost function J^n of all training sets (black curve) and accuracy (red curve) vs training epochs.

Classification

Shown in Fig. 10 is the evolution of the output voltage V_O *vs.* training epochs for both training and test datasets. In the very beginning, there is no distinct difference in output voltage corresponding to 24 tactile input signals. After 11 epochs of training, the two patterns are properly resolved except for the first number "2". Due to the row-dimensionality reduction [16] of the first ANN layer, the weight parameters are decreased, leading to a faster training process but a slightly reduced accuracy.

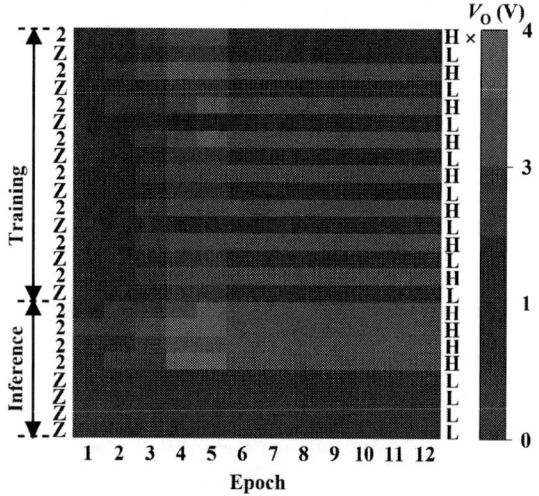

Figure 10. The evolution of the output voltage V_O vs. training epochs for all training and inference sets.

CONCLUSION

An active-matrix tactile sensor array integrated with an artificial neural network for edge computation is demonstrated. Dual-gate thin-film transistors are deployed in the sensor nodes for signal acquisition and in the artificial neurons for neuromorphic computation. Mimicking the behavior of tactile perception learning of human beings, amplified tactile signals acquired by the sensor array are subsequently utilized for training and inference by the ANN. Based on the architecture of row-dimensionality reduction and the training method of MBGDA, pattern recognition results show an accuracy of 96%. The integrated system offers good potential for the realization of intelligent neuromorphic sensor systems.

ACKNOWLEDGMENT

This work was supported by The State Key Laboratory on Advanced Displays and Optoelectronics Technologies under Grant ITC-PSKL12EG02, Grant RGC 16215720, GDNSF-Joint Funds GDST22EG04 and in part by the Science and Technology Program of Shenzhen under Grant JCYJ20200109140601691.

REFERENCES

[1] L. E. Osbron *et al.*, "Prosthesis with neuromorphic multilayered e-dermis perceives touch and pain," *Sci. Robot.*, vol. 3, no. 19, 2018.

[2] J. Weichart *et al.*, "Tactile sensing with scalable capacitive sensor arrays on flexible substrates," *J. MEMS*, vol. 30, no. 6, pp. 915-929, 2021.

[3] S. Wang *et al.*, "Skin electronics from scalable fabrication of an intrinsically stretchable transistor array," *Nature*, vol. 555, pp. 83-88, 2018.

[4] Z. Chen *et al.*, "Flexible piezoelectric-induced pressure sensors for static measurements based on nanowires/graphene heterostructures," *ACS Nano*, vol. 11, no. 5, pp. 4507-4513, 2017.

[5] W. Lin *et al.*, "Skin-inspired piezoelectric tactile sensor array with crosstalk-free row + column electrodes for spatiotemporally distinguishing diverse stimuli," *Adv. Sci*, 8, 2002817, 2021.

[6] J. C. Yeo *et al.*, "Wearable mechanotransduced tactile sensor for haptic perception," *Adv. Mater. Technol.*, vol. 2, no. 6, 1700006, 2017.

[7] Y. Kim *et al.*, "A bioinspired flexible organic artificial afferent nerve," *Sci.*, vol. 360, no. 6392, pp. 998-1003, 2018.

[8] C. Zhang *et al.*, "Oxide synaptic transistors coupled with triboelectric nanogenerators for bio-inspired tactile sensing application," *IEEE Electron Device Lett.*, vol. 41, no. 4, pp. 617-620, 2020.

[9] H. Tan *et al.*, "Tactile sensory coding and learning with bio-inspired optoelectronic spiking afferent nerves," *Nat. Commun.*, 11, 1369, 2020.

[10] C. Wan *et al.*, "An artificial sensory neuron with tactile perceptual learning," *Adv. Mater.*, vol. 30, no. 30, 1801291, 2018.

[11] X. Wu *et al.*, "Artificial multisensory integration nervous system with haptic and iconic perception behaviors," *Nano Energy*, vol. 85, 106000, 2021.

[12] T. Lei *et al.*, "Active-matrix tactile sensor array based on the monolithic integration of PVDF and dual-gate transistors," *MEMS*, pp. 71-74, 2022.

[13] Y. Hu *et al.*, "Neuromorphic implemented of logic functions based on parallel dual-gate thin-film transistors," *IEEE Electron Device Lett.*, vol. 43, no. 5, pp. 741-744, 2022.

[14] Y. Hu *et al.*, "An artificial neural network implemented using parallel dual-gate thin-film transistors," *IEEE Trans. Electron Devices*, vol. 69, no. 10, pp. 5574-5579, 2022.

[15] A. Mustapha *et al.*, "An overview of gradient descent algorithm optimization in machine learning: application in the ophthalmology field," *SADASC 2020*, CCIS 1207, pp. 349-359, 2020.

[16] H. Tan *et al.*, "Bioinspired multisensory neural network with cross-modal integration and recognition," *Nat. Commun.*, 12, 1120, 2021.

CONTACT

*T. Lei, tel: +852-5565-7461; tlei@connect.ust.hk

EVALUATION OF LOCAL AND INTERNAL ELASTICITY OF HYDROGEL MATERIALS BY USING LIGHT-DRIVEN GEL ACTUATOR

Hibiki Nakajima[1], Yuha Koike[1], Yoshiyuki Yokoyama[2],
Masaya Hagiwara[3], and Takeshi Hayakawa[1]
[1]Department of precision engineering, Chuo University, JAPAN,
[2]Toyama Industrial Technology Research and Development Center, Toyama, JAPAN, and
[3]RIKEN, Saitama, JAPAN

ABSTRACT

We propose a method for evaluating local and internal elasticity of hydrogel materials. In this method, we deform inside of a sample by using a light-driven gel actuator and evaluate internal elasticity from strain of the material. Since the actuator can be driven by irradiating light through a microscope without wiring, this method enables us to evaluate local and internal elasticity which is difficult to be measured with conventional methods. In this study, we compared strains of two samples with different elasticities. It was observed that strain of a softer sample was larger than that of a harder sample. Furthermore, we can estimate the harder sample is approximately 1.3 times harder than the softer sample. Therefore, we succeeded in evaluating local and internal elasticities of samples with the proposed method.

KEYWORDS

Gel actuator, Light drive, PNIPAAm, Elasticity evaluation

INTRODUCTION

Living cells reside in a natural hydrogel called extracellular matrix (ECM) in *in-vivo* environment. Substantial researches in two decades have established that ECM not only provides mechanical support for cells, but also affects several fundamental cellular processes, including differentiation, migration, growth, and adhesion [1-5]. These fundamental cellular processes are important keys on pattern formation, wound healing, and tumor invasion in detail.

Since elasticity or stiffness of ECM is one of the important factors to decide cellular behaviors, physical interactions between cells and surrounding ECM have discussed by using hydrogel materials with tunable elastic modulus. Also, recent works have revealed that cellular movement alters mechanical property of environment surrounding cell. Moreover, this alteration guides cellular movement [6]. Therefore, to understand physical interactions between cells and surrounding environment, it is necessary to evaluate elasticity of ECM. In other words, evaluation of local and internal elasticity of hydrogels as materials of ECM are highly demanded.

Conventionally, to evaluate mechanical property in microscale, several methods such as atomic force microscopy (AFM) [7, 8], magnetic twisting cytometry (MTC) [9, 10], and magnetic resonance elastography (MRE) [11] have been proposed. First, AFM is the most popular method to evaluate local elasticity. However, a probe of AFM can access and measure only a surface of samples, and it is difficult for AFM to evaluate internal elasticity of a sample. Second, MTC evaluates an elasticity by measuring displacements of magnetic beads adhered on a sample surface when applying twisting magnetic field. However, MTC is not suitable for local and internal evaluation because it is difficult to apply magnetic field locally. MRE can acquire distributions of elastic modulus in biological tissues by imaging propagating elastic waves. Although MRE based techniques are suitable for evaluation of internal elasticity, it is difficult to achieve evaluation of local elasticity with high spatial resolutions. MRE has spatial resolutions as 100-200 μm at best [11]. As explained as above, it is difficult for conventional methods to evaluate local and internal mechanical property of ECM with high special resolution as cellular sizes.

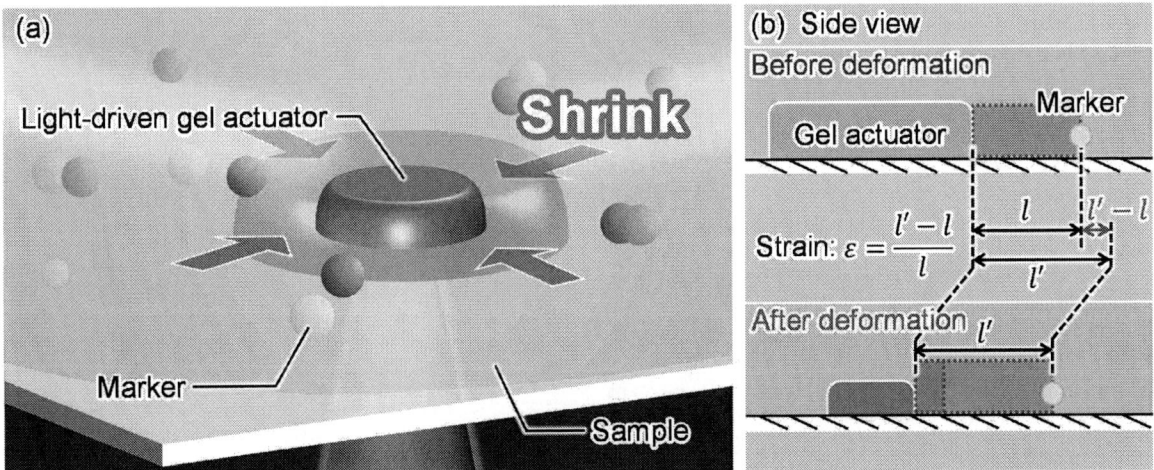

Figure 1: Concept of evaluation method of local and internal elasticity (a)Internal deformation of a sample by using the light-driven gel actuator (b)Schematic diagram of measuring a strain of a sample

Here, to realize this internal and local evaluation of elasticity, we propose an evaluation method by using a microactuator which can be driven locally without wiring. In this study, we demonstrate evaluation of elasticity of collagen, which is a hydrogel often used as ECM, by using proposed method.

CONCEPT

We propose a method to evaluate local and internal elasticity by using the light-driven gel actuator, as shown in Fig. 1 (a). The actuator is made of a temperature-responsive gel which shrinks at high temperature and swells at low temperature [12]. This actuator is patterned on a substrate and can be driven by irradiating light to control its temperature. We put hydrogel sample around the light-driven gel actuator patterned on a glass substrate. Then, we irradiate the actuator with light and shrink it to deform the hydrogel sample. Thus, we can evaluate local and internal elasticity of the hydrogel by measuring strain of the hydrogel sample calculated from the deformation of a sample (Fig. 1 (b)). Therefore, proposed method enables us to evaluate local and internal elasticity of samples which is difficult with conventional methods. In this study, we evaluate elasticities of two samples with different elasticities by comparing strains.

EXPERIMENTS

Fabrication and drive of the gel actuator

Light-driven gel actuator is made of Poly (N-isopropylacrylamide) (PNIPAAm) which is temperature-responsive gel. PNIPAAm swells at lower temperature than the Lower Critical Solution Temperature (LCST) (≈ 32 °C) and shrinks at higher temperature than LCST (Fig. 2). To fabricate gel actuator made of PNIPAAm in microscale, we use photo-processable PNIPAAm (Bioresist, Nissan Chemical Corporation, Tokyo, JAPAN) [13]. Also, to control temperature of a gel actuator by irradiating light, we mix carbon nanotubes (eDIPS INK EC-DH, MEIJO NANO CARBON Co., Ltd., Aichi, Japan, density: 0.2 wt%, Solution type: EtOH) as a light absorber into PNIPAAm. Since carbon nanotubes convert light into heat, we can drive the actuator without wiring by irradiating an infrared (IR) laser (SP-020P-A-HS-S, SPI Laser, Southampton, United Kingdom, wavelength: 1064nm, maximum power: 20W).

Observation of sample's deformation

In this study, we evaluate elasticity of collagen (Cellmatrix® Type I-A, Nitta Gelatin, Osaka, Japan) as samples because collagen is commonly used as an ECM material for cell culture. Then, we prepare collagen samples with different elasticities by adding cross-linker genipin (G0458, Tokyo Chemical Industry Co., Ltd., Tokyo, Japan) to a collagen [14].

To visualize internal deformation of collagens, we mix marker microbeads (4010A, Thermo Fisher Scientific, Massachusetts, U.S., diameter: 1.0 μm) into collagens. Then, we observe the actuator and microbeads with camera (BFS-U3-32S4, Teledyne FLIR LLC, Oregon, U.S.) through objective lens (LUCPLFLN40X, Olympus, Tokyo, JAPAN).

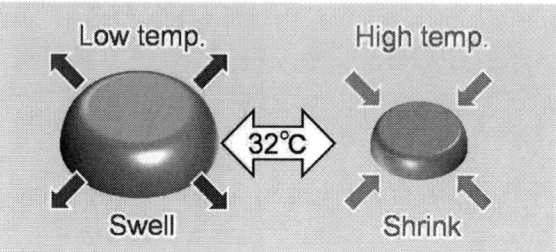

Figure 2: Characteristics of temperature-responsive gel

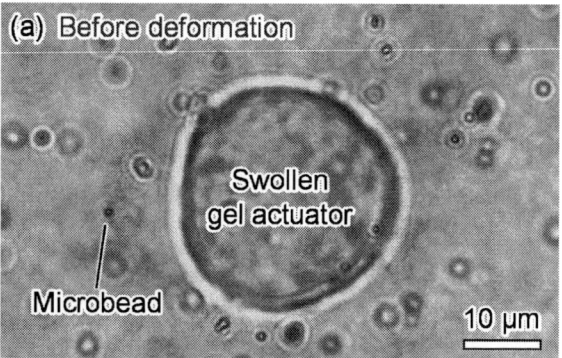

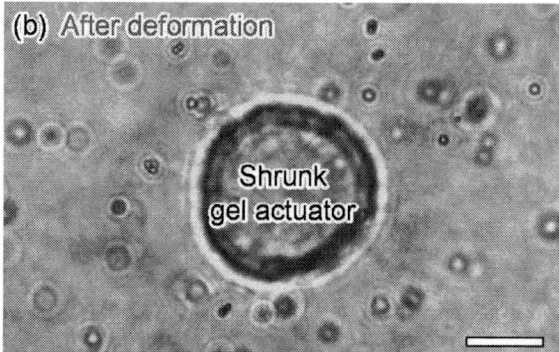

Figure 3: Observation results of internal deformation of collagen with marker microbeads. (a) Before deformation (b) After deformation

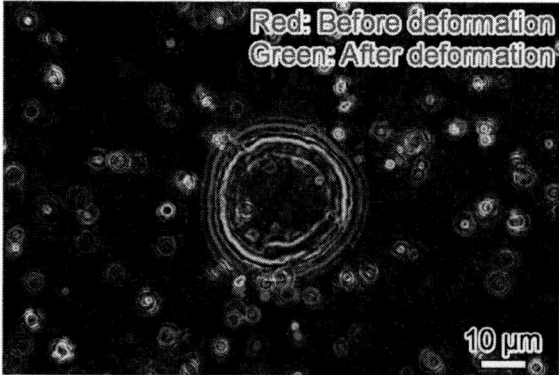

Figure 4: Overlaid image of before and after deformation. Each color indicates results of edge detection by visual analysis.

978-1-6654-9309-3/23 $31.00 © 2023 IEEE   18

RESULTS

Observation of internal deformations of samples

We observed internal deformations of collagens. First, we fabricated disc shaped light-driven gel actuators on a glass substrate. Diameters of the actuators were approximately 30 μm and thickness were approximately 5 μm in shrunk state. Second, we put a collagen around the patterned gel actuators. Third, we irradiate the actuator with laser and shrink it to deform the hydrogel sample.

Figure 3 (a) shows that the gel actuator was swollen in collagen before the laser irradiation, and Fig. 3 (b) shows the gel actuator was shrunk by laser irradiation. We successfully deform the collagen with proposed method by using the light-driven gel actuator. Figure 4 shows an overlaid image of before and after deformation, red color image shows edge of the gel actuator before the deformation of sample and green color image shows that of after the deformation [15]. As shown in Fig. 4, we succeeded in observing displacements of microbeads according to the shrinkage of the gel actuator.

Evaluating elasticity from strain

We performed proof of concept of the proposed method by comparing strains of two samples with different elasticities. To evaluate local and internal elasticity of samples, we calculated strains of samples by using measured position of a microbead, as shown in Fig. 5. Figure 5 shows that we set the radial axis with the center of the gel actuator as the origin. In this geometrical configuration, strain ε can be calculated by

$$\varepsilon = \frac{l' - l}{l} = \frac{(r_i' - r_0') - (r_i - r_0)}{r_i - r_0} \quad (1)$$

where l is length between the microbead and surface of the gel actuator, r_i is coordinate of each microbead, r_0 is coordinate of surface of the gel actuator, and letters without prime indicate before deformation and letters with prime indicate after deformation.

We deformed two samples with different elasticities by using the light-driven gel actuator, and measured strains of each sample. One sample was normal collagen, and the other sample was harder collagen cross-linked by adding genipin (50 mM). In two samples, we calculated the strain from the displacements of several beads that were in the same direction seen from the actuator and at different distances from center of the actuator. Figure 6 shows results of calculated strain. As shown in Fig. 6. It was observed that the strains of the normal collagen were larger than those of the harder collagen. This result was consistent with general tendency that soft sample shows larger strain than hard sample. Furthermore, we estimated that the Young's modulus of the harder collagen was approximately 1.3 times higher than that of the softer collagen based on Hooke's law. Therefore, we succeeded in quantitatively evaluating local and internal elasticity from strain.

CONCLUSION

In this study, we demonstrate proof of concept of the proposed evaluation method of local and internal elasticity of hydrogels. We confirmed that the light-driven gel actuator can deform collagens and its deformation can be

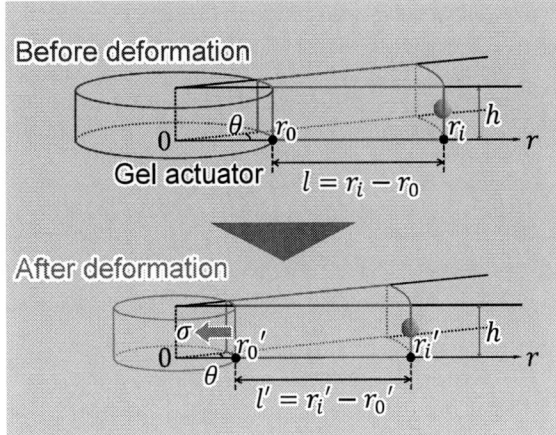

Figure 5: Schematic diagram of calculation of strain

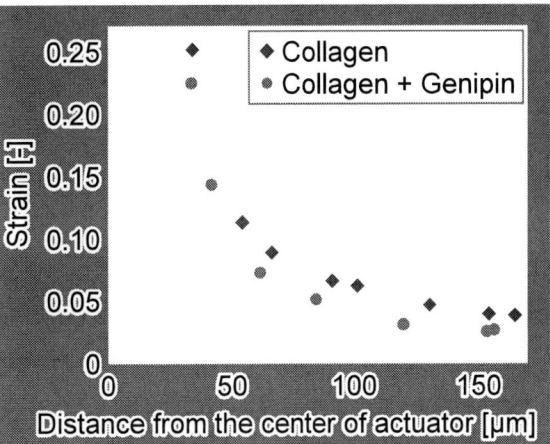

Figure 6: Result of measured strains with different elasticities

measured by observing marker microbeads. Also, we can evaluate local and internal elasticity of collagens by comparing strains of each collagen with different elasticities. In the future, we will measure the actuator force and quantitatively evaluate elastic modulus of actual cell culture enviromnemts.

ACKNOWLEDGEMENTS

This research was supported by Chuo University Personal Research Grant and JSPS KAKENHI grant Number 21K18703.

Photo-processable PNIPAAm (Bioresist) is provided by Nissan Chemical Corporation.

REFERENCES

[1] D. E. Discher, P. Janmey, Y. L. Wang, "Tissue cells feel and respond to the stiffness of their substrate", *Science*, 310.5751, pp.1139-1143, 2005.

[2] O. Chaudhuri, J. Cooper-White, P. A. Janmey, D. J. Mooney, V. B. Shenoy, "Effects of extracellular matrix viscoelasticity on cellular behaviour", *Nature*, 584(7822), pp.535-546, 2020.

[3] C. M. Lo, H. B. Wang, M. Dembo, Y. L. Wang, "Cell movement is guided by the rigidity of the substrate",

Biophysical journal, 79.1, pp.144-152, 2000.

[4] A. J. Engler, S. Sen, H. L. Sweeney, D. E. Discher, "Matrix elasticity directs stem cell lineage specification", *Cell*, 126.4, pp.677-689, 2006.

[5] A. Sparavigna, "Role of the extracellular matrix in skin aging and dedicated treatment-State of the art" *Plastic and Aesthetic Research*, 7, 14, 2020.

[6] M. Hagiwara, H. Maruyama, M. Akiyama, I. Koh, F. Arai, "Weakening of resistance force by cell-ECM interactions regulate cell migration directionality and pattern formation", *Communications biology*, 4.1, pp.1-13, 2021.

[7] Y. M. Efremov, T. Okajima, A. Raman, "Measuring viscoelasticity of soft biological samples using atomic force microscopy", *Soft matter*, 16.1, pp.64-81, 2020.

[8] P. D. Garcia, C. R. Guerrero, R. Garcia, "Nanorheology of living cells measured by AFM-based force–distance curves", *Nanoscale*, 12.16, pp.9133-9143, 2020.

[9] P. H. Wu, D. R. B. Aroush, A. Asnacios, W. C. Chen, M. E. Dokukin, B. L. Doss, P. Durand-Smet, A. Ekpenyong, J. Guck, N. V. Guz, P. A. Janmey, J. S. H. Lee, N. M. Moore, A. Ott, Y. C. Poh, R. Ros, M. Sander, I. Sokolov, J. R. Staunton, N. Wang, G. Whyte, D. Wirtz, "A comparison of methods to assess cell mechanical properties", *Nature Methods*, 15, pp.491-498, 2018.

[10] H. Li, J. M. Mattson, Y. Zhang, "Integrating structural heterogeneity, fiber orientation, and recruitment in multiscale ECM mechanics". *Journal of the mechanical behavior of biomedical materials*, 92, pp.1-10, 2019.

[11] K. Zhang, M. Zhu, E. Thomas, S. Hopyan, Y. Sun, "Existing and potential applications of elastography for measuring the viscoelasticity of biological tissues in vivo", *Frontiers in Physics*, 9, 670571, 2021.

[12] Y. Koike, Y. Yokoyama, T. Hayakawa, "Light-driven hydrogel microactuators for on-chip cell manipulations", *Frontiers in Mechanical Engineering*, 6: 2, 2020.

[13] Y. Yokoyama, M. Umezaki, T. Kishioka, E. Tamiya, Y. Takamura, "Micro-and nano-fabrication of stimulus-responsive polymer using nanoimprint lithography", *Photopolymer Science and Technology*, vol. 24, pp. 63-70, 2011.

[14] H. G. Sundararaghavan, G. A. Monteiro, N. A. Lapin, Y. J. Chabal, J. R. Miksan, D. I. Shreiber, "Genipin-induced changes in collagen gels: Correlation of mechanical properties to fluorescence", *Journal of Biomedical Materials Research Part A: An Official Journal of The Society for Biomaterials, The Japanese Society for Biomaterials, and The Australian Society for Biomaterials and the Korean Society for Biomaterials*, 87.2, pp.308-320, 2008.

[15] M. D. Abramoff, P. J. Magalhaes, S. J. Ram, "Image Processing with ImageJ", *Biophotonics International*, volume 11, issue 7, pp. 36-42, 2004.

CONTACT

*T. Hayakawa, tel: +81-3-3817-1834; e-mail: hayaka-t@mech.chuo-u.ac.jp

3D PRINTED MINIATURIZED SOFT MICROSWIMMER FOR MULTIMODAL 3D AIR-LIQUID NAVIGATION AND MANIPULATION

Dominique Decanini,[1] Abdelmounaim Harouri[1], Ayako Mizushima[2], Beomjoon Kim[3,4], Yoshio Mita[2,4] and Gilgueng Hwang[1,2,3,4]*

[1]C2N/CNRS (UMR9001), Paris-Saclay University, FRANCE and
[2]Dept. of Electrical Engineering and Information Systems, The University of Tokyo, JAPAN and
[3]Institute of Industrial Science, The University of Tokyo, JAPAN and
[4]LIMMS/CNRS-IIS (IRL2820), The University of Tokyo, JAPAN

ABSTRACT

Major challenges of microscale swimmers for biomedical applications could be listed as the 3D motion for air-liquid navigation, selective motion control of multiple swimmers, visual servo automation control and micromanipulation functionality. Here we report a miniaturized soft microswimmer for serving as a wireless microrobotic manipulator in air-liquid environments. We fabricated these micrometric soft swimmers by two-photon 3D nanoprinting of Polydimethylsiloxane (PDMS) and metallization of ferromagnetic nickel layer for magnetic propulsion. We demonstrated for the first time 3D air-liquid navigation and selective motion control of multiple soft swimmers thanks to the multimodal propulsion with magnetic and buoyancy force. We further reduced stick-slip phenomenon by applying high frequency vibration by piezo disc actuator. This allowed real-time visual tracking of the soft microswimmers and also the micromanipulation of microfibers. We believe that this 3D printed and miniaturized mobile agents should enlarge the application areas from *in-vitro* microfluidics to bio/chemical assay.

KEYWORDS

3D printing, Two-photon nanolithography, Soft microswimmer, Magnetic Actuation, Microfluidics

INTRODUCTION

Microfluidics is an essential platform for lab-on-a-chip based point-of-care diagnostics or organ-on-a-chip [1]. There is huge need for micromanipulation of biological or chemical samples in confined microfluidic environment [2]. In the recent decade, wirelessly propelled mobile microrobotic swimmers have drawn much interest due to the small footprint with a high accessibility to confined microfluidic environments [3,4]. More recently soft artificial swimmers have shown their great promise due to their excellent compatibility to biological environments which have more complex mechanics and dynamics [5]. Although they are promising, we now face more complex manufacturing challenges than ever before. Manufacturing of such soft 3D microstructures often requires other fabrication methods such as soft lithography based on mold-casting process than the standard cleanroom microfabrication processes. This adds more fabrication steps to the cleanroom microelectromechanical system (MEMS) process.

Recently 3D printing has been widely used for mostly the rapid-prototyping. In this context, 3D printing by direct laser writing could be a promising solution for printing 3D micro/nanostructures. 3D printing of microscale mechanical transducers could accelerate and facilitate the manufacturing process. Although 3D printing was attempted for microscale actuators [6], they are still very limited in the manufacturable scale, material and morphology. Alternatively, the cleanroom micro/nanofabrication technologies have been used to fabricate microscale soft actuators [7]. This is a promising alternative but the fabricated microstructures based on this method can swim in only liquid environment which limits their potential applications. Millimeter scale soft robots have been developed by on-demand magnetization assisted printing [8] and this has further been extended to be a better fit for liquid/air operation [9]. This work is to further miniaturize the millimeter scale soft swimmers [9] in microscale and also to extend the propulsion in 3D for air-liquid navigation by multimodal propulsion combining the magnetic and buoyancy force due to gravity compensation by bubble trap. In addition, this can separate the motions of

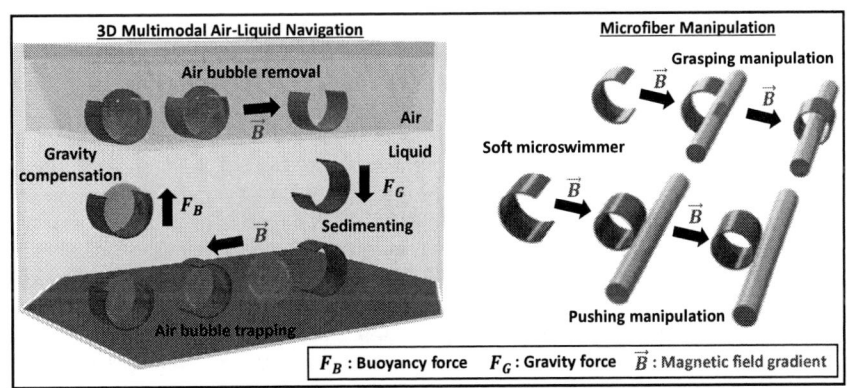

Figure 1: Schematic concept of multimodal soft microswimmer: (left) 3D motion by bubble trap (right) 2D manipulation of microfibers.

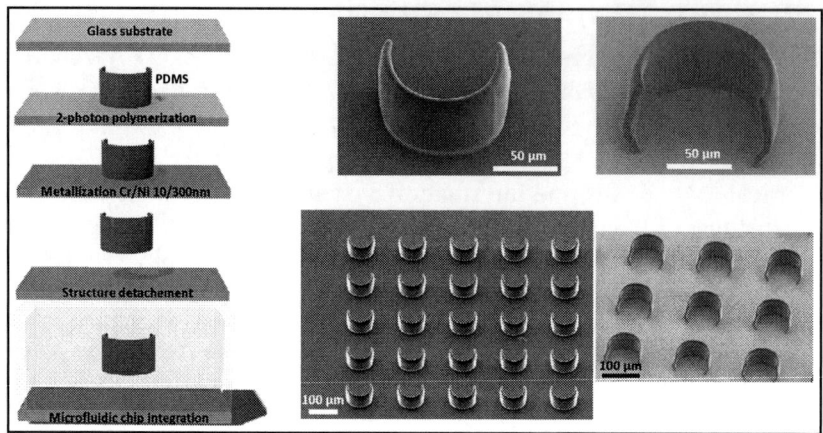

Figure 2: Fabrication process (left): 1) 3D printing PDMS by two-photon polymerization. 2) Metallization of ferromagnetic nickel layer by sputtering. 3) Structure detachment by micromanipulation. 4) Microfluidic chip integration. (right) SEM photograph of the fabricated soft microswimmers.

multiple soft microswimmers thus selective motion control is possible (Figure 1). The proposed 3D multimodal air-liquid navigation is achieved by combining the propulsion by magnetic field gradient and the gravity compensation from air bubble trapping as shown in the Figure 1. The soft and deformable mechanics should also facilitate robust 2D manipulation of microfiber structures by pushing or pulling.

FABRICATION

Microfabrication of 3D soft microswimmers

We chose 3D printing method to fabricate the proposed soft microswimmers. Figure 1 shows the design and fabrication of the proposed soft microswimmer. In this work, we miniaturized the millimeter scale soft swimmers [9] to microscale by two-photon 3D nanoprinting (GT+, Nanoscribe GmbH) of PDMS and magnetic layer metallizations (*Cr/Ni* 10/300 nm). Photosensitive PDMS (IP-PDMS, Nanoscribe) deposited on top of a glass substrate coated with indium-tin-oxide (ITO) layer. After developing the lithography, the sample was rinsed with isopropyl alcohol and carefully dried. Then additional metal layers (*Cr/Ni* 10/300 nm) were sputtered. Finally, each soft microswimmer was detached by micromanipulation with a nanoprobe. The detached soft microswimmer was placed onto a glass substrate then a PDMS fluidic chamber is bounded by oxygen plasmas treatment on both the glass and the PDMS chamber. This allows to fill liquid and make the swimmer experiments.

Magnetic propulsion system

A microscale soft swimmer was developed to be able to move by external magnetic field gradient so to trap a micro air bubble in liquid by the curved morphology (Figure 1). For the magnetic propulsion, single electromagnetic coil mounted upward direction on the robotic arm manipulator for generating and controlling magnetic field gradient pulling the soft microswimmers downward direction. Magnetic field can switch on/off and/or the intensity of magnetic field is controlled either by the applied current to the coils and also by the distance between the soft microswimmer and the coil. The magnetic

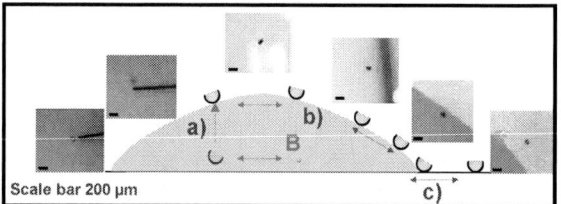

Figure 3: Air-liquid 3D motion of soft microswimmer by bubble trap gravity compensation and droplet surface motion.

field trapped soft microswimmers can move horizontally by moving the coil by the robotic arm. A similar configuration of this magnetic field generator is described in more detail elsewhere [4].

RESULT

Major results obtained by the proposed soft microswimmers are the multimodal 3D air-liquid navigation and the selective motion control of multiple soft microswimmers. These are important characteristics for them to be applied for micromanipulation in wet or dry environments.

Multimodal 3D navigation

First for the 3D motions, instead of adding upward pulling electromagnetic coil, we used air bubble trap for gravity compensation. This allows to secure an optical path and objective lens for upright microscope and also the microfluidic circuit tube connections. In addition, this avoids additional consumption of energy and also heat dissipation. Instead, we propose to use buoyancy force of the soft microswimmer by trapping an air bubble. Air bubbles are injected by pushing air via a syringe and a tube to fluidic chamber. Thanks to the increased buoyancy force by trapped air bubble, the soft microswimmer is able to compensate the gravity without magnetic field gradient pulling upward so it can come out of the liquid surface. Then the trapped air bubble is removed at the air-liquid interface thus the buoyancy force of soft microswimmer decreases. Soft microswimmer should then sediment by also applying a downward direction magnetic field

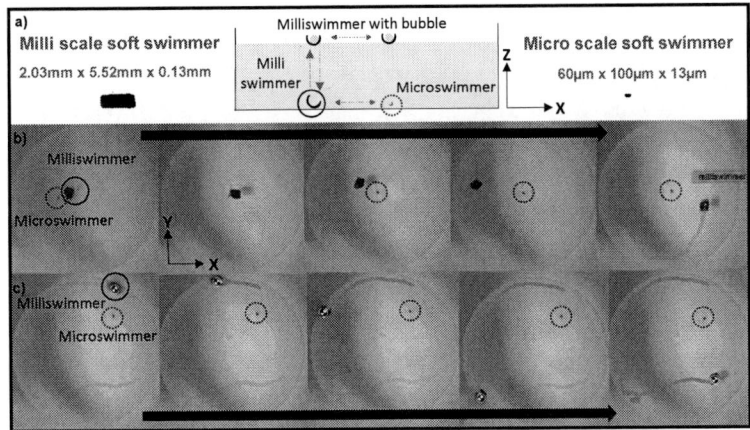

Figure 4: Multiswimmer motion control selectivity. The floating milliswimmer with bubble trap and the sinking microswimmer are selectively controlled by external magnetic field gradient.

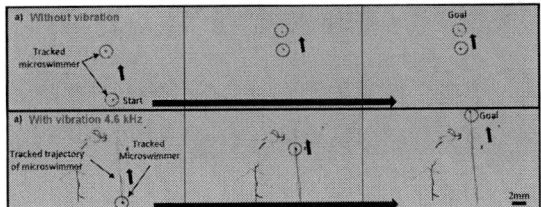

Figure 5: (a) Stick-slip motion without vibration, (b) Enhancement of motion control and visual tracking by adding high frequency vibration (4.6 kHz).

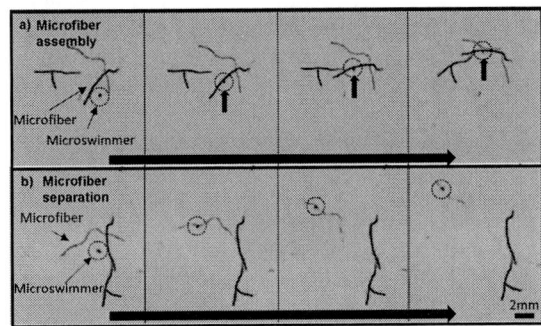

Figure 6: Manipulation of microfiber by soft microswimmer (a) assembly and (b) separation.

gradient. This multimodal 3D motion by buoyancy and magnetic force is an energy efficient 3D motion control strategy for microscopic soft swimmers. Figure 3 shows the air-liquid 3D navigation by the trapped air bubble of soft microswimmer propelled by magnetic field gradient. Thanks to the compensated gravity by the trapped air bubble and the soft deformable morphology, they can navigate in 3D and they can also move in/out of a droplet. This is one of the most requiring features for microswimmers to be able to navigate in between air-liquid environments. This seamless air-liquid navigation must be an important asset towards their real-life biological or environmental applications which require seamless handlings of liquid or solid samples.

Selective motion control of multiple swimmers

Second for the selective motion control of multiple swimmers, the selectivity of each swimmer is essential. Typically, the selectivity can be attained by differentiating the magnetization of each swimmer [10]. Compared to the horizontally placed electromagnetic coils, the electromagnetic field gradient generated by the upward direction coil placed on the bottom of the fluidic chamber in this work could counterbalance the surface tension at the water surface. In this way, multiple soft microswimmers can be separately controlled by global magnetic field gradient. A better selectivity can be achieved by differentiating the size of the soft microswimmers. Applying two swimmers (mm scale and micro scale) could separate the motions of the two swimmers (one floating and the other sinking) (Figure 4). Therefore, manufacturing scalability is highly demanding. In this work, we miniaturized the conventional millimetric soft swimmers

[9] to micro scale soft swimmers by nanoscale 3D printing and consecutive ferromagnetic metal layer deposition by sputtering. As shown in the Figure 4, we fabricated a milli scale soft swimmer and a micro scale soft swimmer. The size of the micro scale soft swimmer is miniaturized by 55 times of the milli scale one. Both swimmers are sedimented inside a fluidic chamber filled with water and the milliswimmer traps a bubble so to float to water/air surface. Adjusting the magnetic field intensity of bottom coil and its distance to the fluidic chamber, only the floating milliswimmer can move by moving the coil but the microswimmer can stay static on the bottom substrate. This is due to the weak magnetization of the small volume microswimmer compared to the milliswimmer and also due to the substrate friction. Thanks to the proposed concept of the multimodal swimmers, we were able to demonstrate the selective motion control of multiple swimmers with using a single electromagnetic coil.

Micromanipulation

In addition to the 3D navigation, the soft microswimmers are also able to manipulate microstructures. In this work, we aim to further demonstrate their micromanipulation capabilities. However, the bottom pulling magnetic force has a disadvantage of increasing surface stiction. While this was useful for the selective motion control of multiple swimmers in the prior section, the motion of soft microswimmer becomes to be the stick-slip phenomenon. This is problematic for both the stable micromanipulation and also the stable visual tracking which are essential to

Table 1. Comparison table of soft swimmers.

	[6]	[7]	[8]	[9]	This work
Size	**~ 100 μm**	~ 5 mm	~ 5 mm	~ 5 mm	**~ 100 μm**
Environment	Liquid	Liquid	Liquid or Air	**Liquid & Air**	**Liquid & Air**
Motion	**3D**	2D	2D	2D	**3D**
Propulsion	Single (Magnetic)	Single (Magnetic)	Single (Magnetic)	Single (Magnetic)	**Multi (Magnetic+Buoyancy)**
Selective motion	N	N	N	N	**Y**

their applications and automation control. To avoid the stick-slip phenomenon, we applied high frequency vibrations to the fluidic chamber containing the soft microswimmers by a piezoelectric disc actuator placed on the bottom of the fluidic chamber. The soft microswimmers under piezoelectric disc vibrations in 4.6 kHz allowed to reduce the typical stick-slip phenomenon of the soft microswimmers propelled by magnetic field gradient pulling. This stabilizes the motion control and enables the real-time visual tracking of soft microswimmers. We demonstrated the closed-loop motion control as shown in the Figure 5.

For the micromanipulation, we used microfibers as the target to manipulate. To manipulate microstructures on the substrate, soft microswimmers need to be sufficiently stiff in their mechanics and also the surface friction needs to be sufficiently high to be able to generate enough manipulation force while pushing. The magnetic force applied to the bottom direction allows the soft swimmer to deform to make itself stiffer. Moreover, this increases the friction between the swimmer and the surface which is necessary for the manipulation of very large footprint (millimeter-long) microfibers or other microstructures. Finally, this vibration-induced motion control and the soft deformable microswimmer by magnetic field gradient allowed successful manipulation of mm-long microfibers (Figure 6).

CONCLUSIONS

In this work, we proposed a magnetically actuated micrometric soft swimmers fabricated by nanoscale 3D printing. We achieved the manufacturing scalability of the soft microswimmers and their multimodal 3D air-liquid navigations by combining magnetic and buoyancy force. Thanks to these features, we demonstrated the selective motion control of multiple swimmers and this is one most challenging but highly demanding. Furthermore, the proposed soft microswimmers showed their possible visual tracking of motion by avoiding stick-slip phenomenon. This shows a great promise for the closed-loop automation control. Finally, we demonstrated micromanipulation of microfiber structures. The achieved results of the miniaturization in sub-millimeter scale, the motion compatibility in air-liquid environments, the multimodal 3D navigation, the selective motion control of multiple swimmers and the micromanipulation capability. These are indispensable features for soft microswimmers to get a step closer to their biological or environmental applications.

ACKNOWLEDGEMENTS

This work was partially supported by the JSPS Core-to-Core Program (JPJSCCA20190006), MEXT ARIM JPMXP1222UT1006, JST CREST JPMJCR20T2 and JSPS KAKENHI KIBAN C (21K03931).

REFERENCES

[1] D. Huh, B. D. Matthews, A. Mammotto, M. M. Zavala, H. Y. Hsin, D. E. Ingber, "Reconstituting Organ-Level Lung Functions on a Chip", *Science*, vol. 328, pp. 1662-1668, 2010.

[2] T. Kawahara, M. Sugita, M. Hagiwara, F. Arai, H. Kawano, I. Shihira-Ishikawa, A. Miyawaki, "On-chip microrobot for investigating the response of aquatic microorganisms to mechanical stimulation", *Lab. Chip.*, vol. 13, pp. 1070-1078, 2013.

[3] A. Barbot, D. Decanini, G. Hwang, "On-chip Microfluidic Multimodal Swimmer toward 3D Navigation", *Sci. Rep.*, vol. 6, pp. 19041, 2016.

[4] G. Hwang, I. A. Ivan, J. Agnus, H. Salmon, S. Alvo, N. Chaillet, S. Regnier, A. Haghiri-Gosnet, "Mobile Microrobotic Manipulator in Microfluidics", *Sens. Act. A: Phys.*, vol. 215, pp. 56-64, 2014.

[5] C. Hu, S. Pane, B. J. Nelson, "Soft Micro- and Nanorobotics", *Annu. Rev. Con. Rob. Auto. Sys.*, vol. 1, pp. 53-75, 2018.

[6] M. Tyagi, G. M. Spinks, E. W.H. Jager, "3D Printing Microactuators for Soft Microrobots", *Soft Rob.*, vol. 8, pp. 19-27, 2021.

[7] J. Cui, T.-Y. Huang, Z. Luo, P. Testa, H. Gu, X.-Z. Chen, B. J. Nelson, L. J. Heyderman, "Nanomagnetic encoding of shape-morphing micromachines", *Nature*, vol. 575, pp. 164-168, 2019.

[8] T. Xu, J. Zhang, M. Salehizadeh, O. Onaizah, E. Diller, "Millimeter-scale flexible robots with programmable three-dimensional magnetization and motions", *Sci. Rob.*, vol. 4, pp. eaav4494, 2019.

[9] G. Hwang, B. Kim, A. Toyokura, A. Higo, Y. Mita, "Miniaturized Soft Transformable Swimmer for Environmentally Friendly and Sustainable Fluidic Carrier", in *Proc. DTIP 2021*, pp. 1-4.

[10] E. Diller, J. Giltinan, M. Sitti, "Independent control of multiple magnetic microrobots in three dimensions", *Intl. J. Rob. Res.*, vol. 32, pp. 614-631, 2013.

CONTACT

*G. Hwang, tel: +33-6-7575-4072;
gilgueng.hwang@c2n.upsaclay.fr

SELF-DRIVEN CAPILLARIC VISCOMETER FOR DIRECT OR CASCADED BAR GRAPH READ-OUT OF RELATIVE SAMPLE VISCOSITY

*Daniel Mak[1], R. Claude Meffan[1,2], Julian Menges[1], Fabian Dolamore[1], Conan Fee[1], Renwick C.J. Dobson[1] and Volker Nock[*1]*

[1]Biomolecular Interaction Centre, MacDiarmid Institute for Advanced Materials and Nanotechnology, University of Canterbury, Christchurch, NZ and
[2] Department of Micro-Engineering, Kyoto University, Kyoto, Japan

ABSTRACT

The viscosity of samples is a key property, where differences can indicate variation or abnormalities. A biological example is that high blood viscosity is a predictor of cardiovascular events. However, viscosity is difficult to measure in a simple and affordable way, such as for Point-of-Care applications. Thus, we developed a simple to use platform that eliminates external effects like temperature by measuring the sample viscosity relative to a control. The platform employs a novel capillaric driven circuit, including the recently developed capillaric field effect transistors and a single integrated capillaric pump, to apply an identical pressure to parallel sample and control channels simultaneously. This induces flow in the parallel channels relative to the viscosity, which presents as a difference in the distance travelled along the bar graph read-out section of the device.

Using poly-ethylene glycol solutions as exemplars, we demonstrate that the device can visualize relative viscosity of colored samples and control consistently within 2%. Additionally, the design of the capillaric circuitry inherently accounts for viscosity variation due to temperature, improving the applicability of the device to less-controlled environments. Finally, we demonstrate how the design can be cascaded to read-out translucent samples *via* colored indicator liquid and how, based on blood mimicking liquids, it can be used to compare blood viscosity.

KEYWORDS

Capillaric Circuits, Co-flow Viscometer, Capillaric Field Effect Transistor, Point-of-care Testing.

INTRODUCTION

Capillary viscometers constitute important analytical tools, providing insight into fluid properties and sample composition in an easy-to-use format. Measurements with these devices are typically cheaper and faster than with cone and plate viscometers, all while requiring smaller sample volumes [1]. However, this comes with the trade-off of reduced accuracy and precision. Viscosity of a sample and changes thereof, can be a key indicator for assays in the monitoring of industrial, chemical, biological, and medical samples [2]. Most implementations of microfluidic viscometers to date either use external pumps [1,2] or, in case of capillaric circuits, paper as a substrate [3] to actuate flow, limiting their uptake and precision. The current work differs from previous work, in that it combines self-driven capillary circuits [4] and capillaric Field Effect Transistors (cFETs) [5-8] to implement a bar-graph type flow stream viscometer.

We have previously used the cFETs to demonstrate feedback-controlled binary switching [5], metering and logic operations [6], and field effect transistor-like control of capillaric flow [7,8]. In this paper we utilize the cFET valve, and related control structures based on it, to realize a fully automated capillaric viscometer capable of analyzing both colored and translucent samples. Read-out is implemented as a bar-graph type format, where the distance a sample flows in a channel, relative to a control liquid in a parallel channel, relates to the viscosity (Fig. 1). For potential applications, this read-out can, for example, be related to an analyte concentration via the observed change in viscosity read off using a visual guide on the device.

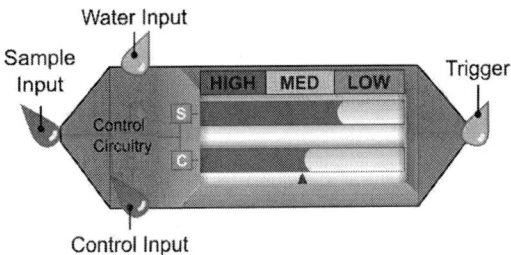

Figure 1: Concept of operation of the bar-graph viscometer. Relative viscosities of a sample and control are reported as two bar lengths for visual read-out. In the example shown, sample liquid viscosity would be 30% less than the control liquid.

EXPERIMENTAL

Device design

The viscometer uses a single integrated capillary pump to apply homogenous capillary pressure to multiple viscous samples simultaneously [9]. This allows an advantage of microfluidic rheometers, the insensitivity to interface conditions, to be preserved whilst still using capillary forces for operation. A capillaric circuit diagram is shown in Fig. 2(a). This includes preprogrammed metered reservoir charging and sealing blocks for the sample and control liquid intakes (Fig. 2(b)). The use of separate water and trigger inlets allowed for prefilling of the measurement circuitry followed by timed initiation. To measure translucent samples, additional intake flow resistance blocks were added between the transducing resistance and a connecting valve to be filled with colored liquid [9]. Liquid from these dye reservoirs fills the measurement channels, rather than from the sample and control reservoirs.

Chip Fabrication

Test devices used in this work were fabricated from cross-linked poly(methyl methacrylate) (PMMA; 4.5 mm general purpose acrylic; PSP Plastics) as previously described [5,8]. In brief, channel milling was performed using a Mini-Mill/GX micro milling machine running an NSK-3000 Spindle (Minitech Machinery Corporation), with a minimum addressable step size of 1 μm. Machining tools were purchased from Performance Micro Tool in diameters of 3.175 mm (SR-4-1250-S), 250 μm (250M2X750S) and 100 μm (100M2X300S) for the square heads and 200 μm (TR-2-0080-BN) for ball nose. Design files and milling parameters (G-code) were prepared using CAD software (Autodesk Fusion 360© 2020 Autodesk, V2.0.10356).

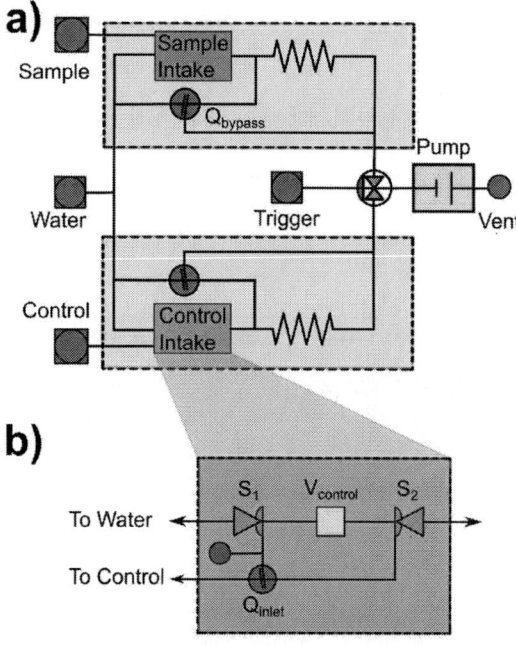

Figure 2: Capillaric circuit diagram for the co-flow viscometer. (a) Functional components of the viscometer illustrating water supply, sample, control and trigger inlets, a single capillary pump, and cFETs Q_{bypass} sealing sample and control chamber by-pass lines. (b) Diagram of the Control *intake block, which takes liquid from the control inlet and fills reservoir $V_{control}$ situated between trigger valves S_1 and S_2, providing metering in the process. This block is mirrored for the* Sample *intake.*

Each sample was fabricated by an initial face cut (3.175 mm cutter) to level out the surface, followed by milling of each channel. The surface was then polished using acrylic polish (aluminium oxide-based CRC, code 9230), followed by ultra-sonication for 1 min in ~5% (v/v) aqueous isopropyl alcohol solution, washing with acetone, isopropyl alcohol and water and blow drying with nitrogen. To close microscopic cracks that occurred during the milling process, the surface was coated with high molecular weight PMMA solution (average M_w = 996,000, 2.5% in xylene; Sigma Aldrich). Any remaining solvent

was removed by drying samples at 90°C for 5 min on a hotplate and keeping the hot sample under vacuum for at least 1 min. Finally, samples were plasma-treated ten times for 1 min, each time at 25 W, pulsed mode (ratio 50) using O_2 gas (3 sccm; Tergeo Plasma Cleaner, Pie Scientific). A thin (2–3 mm) polydimethylsiloxane (PDMS) slab was prepared by mixing 10:1 w:w base:curing agent and curing at 80°C for 2 h and used as a cover for the chip (Sylgard 184, Electropar).

Device Testing

For testing, liquids were loaded into the inlets of the device using a pipette. An over-head camera (Canon EOS 760D, Canon Macro lens EF 100 mm 1:2.8 USM, 1080p, 25 fps) filmed the uptake of liquids into the device [8]. Where visualization was required, various dyes were added to color the liquids (Tartrazine, Ponceau-4R, Brilliant Blue, Sigma Aldrich) [5]. A range of solutions with well-controlled fluid viscosities were prepared from stock concentrations of 2 kDa and 4 kDa molecular weight polyethylene glycol (PEG) (stock concentration 50% with water, Sigma Aldrich). The concentrations tested were (2 kDa(1-x) : 4 kDa(x)) for x∈{1.0, 0.75, 0.5, 0.25, 0.0}. These volumetric mixtures created a viscosity range, spanning between that of 2 kDa (~31 mPa.s at 20°C, 50 s⁻¹) and 4 kDa (~108 mPa.s at 20°C, 50 s⁻¹) PEG. The control liquid was stock 2 kDa PEG solution (stock concentration 50% with water, Sigma Aldrich). For reference values, the PEG mixtures were measured with a conventional cone-plate rheometer (MCR 302, Anton-Paar) over a temperature range of 15-45°C at a shear rate of 50 s⁻¹.

To measure the performance as a function of temperature, the device was mounted on a testing platform integrating two Peltier modules [8]. Temperature was controlled by the input voltage to the Peltier modules and the heat-sink was immersed in an ice-bath to maintain a constant temperature on the cold side. For device tests at elevated or lowered temperatures, the devices were preheated to within ±1°C of the target temperature and maintained within this range throughout. Video data from the camera was converted uncompressed into raw .avi format using ffmpeg (v4.3.1) and analyzed using ImageJ (FIJI) [10]. Image processing involved background subtraction, contrast adjustment, binary color thresholding and average brightness extraction. The data extracted was then processed and plotted using Python (v3.8.3.33) [11].

RESULTS

Basic user operation of the co-stream differential viscometer is illustrated in Fig. 3, together with resulting actions on a PMMA chip. For testing, devices were initially filled from the water inlet using yellow dye for visualization (Fig. 3(1)). This filled each parallel branch of the device up to the point where parallel channels met the outlet distribution channel. At this point, flow into the outlet channel was prevented by two-level trigger valves (2LTVs). Intake volumes for sample and control remained unfilled due to 2LTVs either side. The bypass channels are automatically sealed via the Q_{bypass} cFETs prior to the sample loading phase. Blue coloured sample and control liquids were pipetted into their respective inlets with each liquid filling their corresponding intake chambers (Fig.

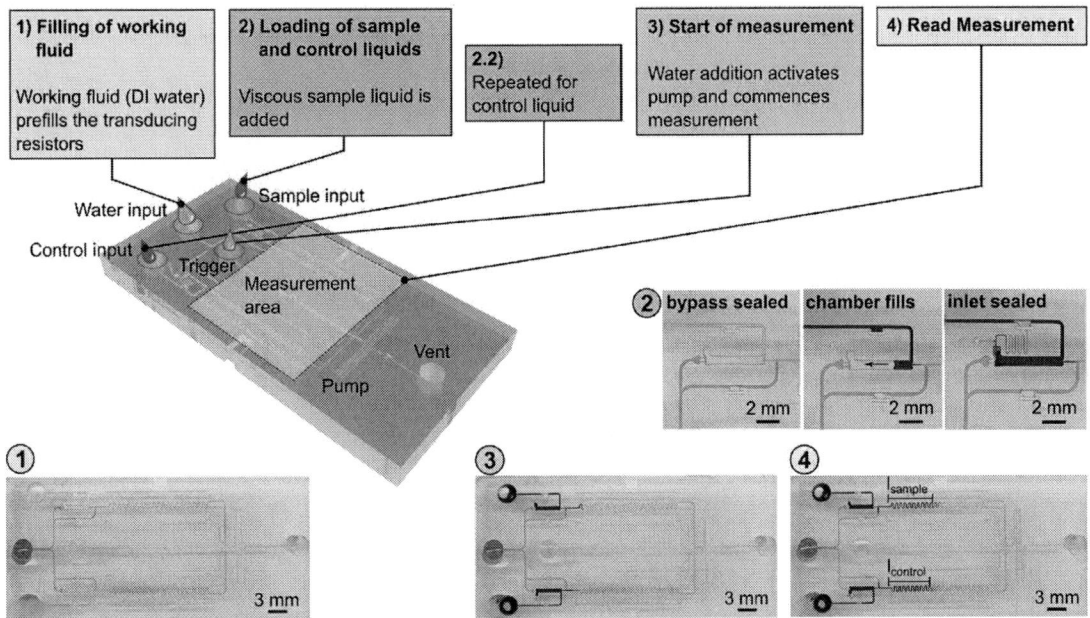

Figure 3: User input and resulting on-chip actions (colored for visualization) for basic operation of the viscometer. First, working fluid is added to the water inlet (1). Under capillary self-pumping, liquid flows through the bypasses channel, and fills the transducing flow resistors up to the release junction. Viscous sample and control materials are then added to their respective inlets (2). The user triggers a measurement by adding liquid to the trigger inlet (3). Once fluids have displaced from the sample and control chambers through the flow resistors, result can be read from the colored resistor lengths (4).

3(2)). Once the chambers had filled, the connections to the inlets self-sealed via cFETs Q_{inlet} and $Q_{control}$. Due to the self-triggering cFET circuitry, where the trigger channel is automatically actuated upon filling of the device, each cFET closed to its predetermined bubble volume [5,8]. A measurement was started by addition of water to the trigger inlet, which activated the 2LTVs linking each branch to the outlet and created a fluidic connection to the capillary pump (Fig. 3(3)). Finally, results of a measurement could be read-out by comparing the distances the sample and control liquids had travelled along their respective channels once the flow had stopped (Fig. 3(4)).

Direct comparison of measurement lengths yielded the anticipated equal and opposite linear trend between sample and control which results from one fluid flowing faster than the other. Variation in the total filling percentage between comparable experiments was attributed to the sealing time of the self-sealing connecting valve. Recording the ratio of the measurement lengths instead, the deviation between experiments was significantly reduced to a standard deviation of less than 2% (Fig. 4(a)). In addition to reducing the sensitivity to pump displacement, the ratio method has other advantages. In this mode, the viscometer is mostly temperature independent (Fig. 4(b)), despite the absolute viscosity more than halving over the tested temperature range. In all of these experiments, the pumping time was less than 2 min, and as fast as 30 s in some cases, demonstrating the platforms suitability for in-field testing.

For translucent samples, additional intake blocks are added between the sample/control intake blocks and their respective measurement areas. Filling of the device proceeds as per the original design, with the addition of a colored dye to both new intake blocks during step 2 (see https://youtu.be/W7tbpLeFUJQ). As the capillary pressure is provided by the single capillary pump and the flow to each side comes from the same water inlet, the new intake blocks have no effect on the distance travelled by either side. However, rather than the colorless control or sample entering the measurement area, the colored dye fills this area (equal to the amount of liquid that flows through the sample and control intakes respectively) and can be read out and related to viscosity as in the previous design. In one test, the dye on the sample side (4 kDa PEG) flowed 0.98 cm, while the dye on the control side (2 kDa PEG) flowed 1.45 cm. This shows that the device does transduce a difference in viscosity between the sample and control material. However, the ratio of these flow distances is 0.675, which is considerably different to the expected ratio of 0.33 for the viscous materials. This difference is caused by the increased base resistance of the device due to the added read-out sub circuitry. This increased base resistance reduces the sensitivity.

We also tested the viscometers using blood-mimicking test liquids (Joninn TL3 and TL4, Jonsman Innovation ApS). These replicate the density, viscosity, surface tension, and non-Newtonian properties of whole blood at 45% and 60% hematocrit values respectively. Initially, these were tested in the original direct-readout devices, however, the light red color made it difficult to define the distance travelled in the measurement area. Thus, the cascade viscometer was used, which provided an improved read out.

978-1-6654-9309-3/23 $31.00 © 2023 IEEE

CONCLUSION

The results show that capillaric microfluidic devices can be used for comparative rheology measurements. This allows the design of cheap microfluidic devices for in-field rheology testing such as blood viscometry. A benefit of the design is it also minimizes the influence of temperature on measurements, widening the applicability of the devices.

We developed both a direct read-out and separated, cascade read-out capillary microfluidic viscometer which simplifies testing. The use of capillary action and circuitry allows for simplicity, ease of use, and low cost, making the devices ideal for point-of-care applications.

The device builds on the cFET we developed previously, providing a use-case where the cFET is integral in controlling the metering of sample and control liquids and closing off bypass or triggering channels.

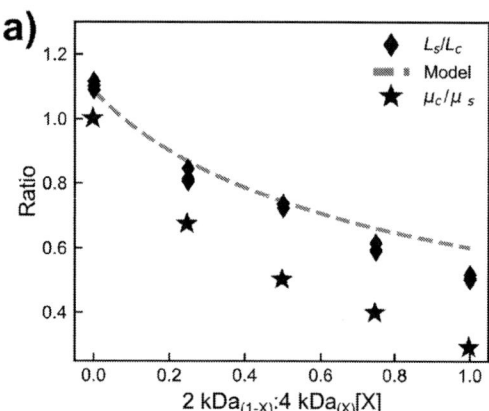

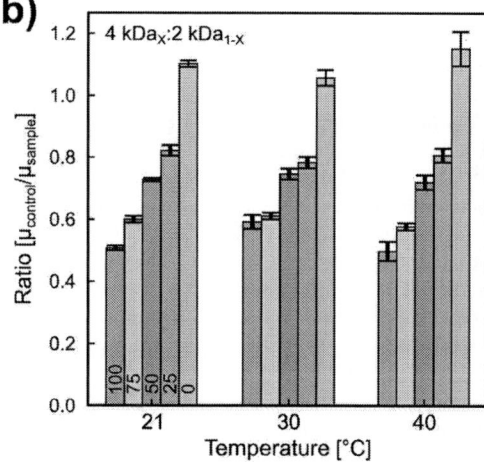

Figure 4: Characterization of the viscometer using volumetric mixes of 2 kDa and 4 kDa PEG. (a) Plot of length ratio L_S/L_C of the two bars compared to modelled flow, and cone-plate rheometer results μ_C/μ_S. Note the inversion of length (S/C) to viscosity (C/S). (b) Temperature rejection capability of the viscometer, indicating measurement consistency in presence of temperature-induced viscosity changes.

ACKNOWLEDGEMENTS

The authors would like to thank Linda Chen and Gary Turner of the University of Canterbury Nanofabrication Laboratory for technical support. Funding was provided by MBIE Grant UOCX1706. C.M. acknowledges the Japan Society for the Promotion of Science and V.N. Rutherford Discovery Fellowship RDF-19-UOC-019 for additional funding.

REFERENCES

[1] Y. J. Kang and S. Yang, "Integrated microfluidic viscometer equipped with fluid temperature controller for measurement of viscosity in complex fluids," *Microfluid. Nanofluid.*, vol. 14, no. 3, pp. 657-668, 2013.

[2] H. Hong, J. M. Song, and E. Yeom, "3D printed microfluidic viscometer based on the co-flowing stream," *Biomicrofluidics*, vol. 13, no. 1, p. 014104, 2019.

[3] J. Xu, X. Hu, H. Khan, M. Tian, and L. Yang, "Converting solution viscosity to distance-readout on paper substrates based on enzyme-mediated alginate hydrogelation: Quantitative determination of organophosphorus pesticides," *Anal. Chim. Acta*, vol. 1071, pp. 1-7, 2019.

[4] A. Olanrewaju, M. Beaugrand, M. Yafia, and D. Juncker, "Capillary microfluidics in microchannels: from microfluidic networks to capillaric circuits," *Lab Chip*, vol. 18, no. 16, pp. 2323-2347, 2018.

[5] J. Menges, C. Meffan, F. Dolamore, C. Fee, R. Dobson, and V. Nock, "New flow control systems in capillarics: off-valves," *Lab Chip*, vol. 21, no. 1, pp. 205–214, 2021.

[6] C. Meffan, J. Menges, F. Dolamore, C. Fee, R. Dobson, V. Nock, "Transistor off-Valve Based Feedback, Metering and Logic Operations in Capillary Microfluidics," in *Proc. of IEEE MEMS`21 Conference*, Virtual, Jan. 2021, pp. 218–221.

[7] C. Meffan et al., "Field Effect Transistor-Like Control of Capillaric Flow Using Off-Valves," in *Proc. of IEEE MEMS'22 Conference*, Tokyo, Japan, Jan. 2022, pp. 263-266.

[8] C. Meffan et al., "Capillaric field effect transistors," *Microsys. Nanoeng.*, vol. 8, no. 1, p. 33, 2022.

[9] C. Meffan, J. Menges, F. Dolamore, D. Mak, C. Fee, V. Nock, R. Dobson, "A versatile capillaric microfluidics viscometer platform for bar-graph type point-of-care diagnostics," *ChemRxiv*, 2021

[10] J. Schindelin et al., "Fiji: an open-source platform for biological-image analysis," *Nat. Meth.*, vol. 9, no. 7, pp. 676-682, 2012.

[11] G. Van Rossum and F. Drake, "Python 3 Reference Manual CreateSpace," *Scotts Val. CA*, 2009.

CONTACT

*V. Nock; tel: +64 3 3694303;
volker.nock@canterbury.ac.nz

A FLEXIBLE BIOSENSING PLATFORM FOR HIGH-THROUGHPUT MEASUREMENT OF CARDIOMYOCYTE CONTRACTILITY

Wenkun Dou[1], Jason Maynes[2], and Yu Sun[1]*

[1]Department of Mechanical and Industrial Engineering, University of Toronto, Toronto, Canada
[2]Program in Molecular Medicine, Hospital for Sick Children, Toronto, Canada

ABSTRACT

Contractile force generated by cardiomyocyte beatings is critical for pumping oxygenated blood supply from the heart to other organs. Measuring cardiomyocyte contractility is critical in exploring cardiac disease mechanisms and quantifying drug efficacy. This paper reports a novel biosensing platform that is integrated with an ultrathin membrane array, flexible carbon black (CB)-PDMS strain sensors, and carbon fiber electrodes for continuous and high-throughput measurement of contractility, beating rate, and beating rhythm in a monolayer of human induced pluripotent stem cell-derived cardiomyocytes (hiPSC-CMs). The flexible biosensing array has been utilized to conduct high-throughput measurement of cardiomyocyte contractile function in responses to cardiac drug candidates.

KEYWORDS

hiPSC-CMs, Microdevice array, Contractility, Drug testing

INTRODUCTION

Heart disease represents a major threat to human health, leading to ~30% of human deaths annually, and causing significantly diminished life quality of patients [1][2]. Developing *in vitro* cardiac models and biosensing platforms to accelerate cardiac drug discoveries would be beneficial for the treatment of heart diseases. Traditional *in vitro* drug screening assays commonly relied on cardiomyocytes from neonatal mouse hearts or standard cell lines, for example, HL-1 and H9c2 cell lines [3]. However, those models encountered problems such as nonhuman proteome and differences in beating physiologies, resulting in poor translations to human physiology, low drug assessment accuracy, and failures in late clinical trials [4]. Recent breakthrough in induced stem cell technology has been applied to generate stem cell-induced cardiomyocyte with a human proteome for the establishment of *in vitro* cardiac models [5].

Cardiomyocyte contraction is one of the key cardiac functional properties. Patients with either congenital or acquired heart diseases commonly have abnormal contraction symptoms, such as weak contraction force, heart fibrillation, and arrhythmic beatings. To this end, developing biosensing technologies to quantify contractility changes of *in vitro* cardiac models is essential for analyzing cardiac physiology, investigating cardiac disease mechanisms, and testing the therapeutic efficacy or potential cardiotoxicity of drug candidates [6][7]. To date, several biosensing techniques/platforms have been developed for the measurement of cardiac contraction from single cardiomyocytes, 2D monolayers, and 3D *in vitro* cardiac tissues [8][9]. For example, video-based beating analysis [10], atomic force microscopy (AFM) [11],

impedance measurement, and optical tracking of cantilever deflections [12]. Furthermore, recent studies have reported dintegrating piezoresistive strain sensors into flexible structures to realize continuous electrical readout of cardiomyocyte contractility[13][14]. However, current techniques challenges exist in how to increase sensor sensitivity to accurately capture weak mechanical contraction signals generated by *in vitro* cardiac models and how to increase platform capacity for high-throughput drug testing.

Here, we proposed a new flexible biosensing platform in a 24-well plate format to realize continuous and high-throughput measurement of contractile function of *in vitro* hiPSC-CM monolayers. Ultrathin and suspended membranes integrated with flexible CB-PDMS strain sensors were developed for cell culture and in-situ cardiomyocyte contractility measurement. Platform efficacy was validated by assessing dynamic contractility changes in response to different concentrations of cardiac drugs.

METHODS AND MATERIALS

Device fabrication

The flexible biosensing platform was fabricated in a 24-well plate format. The PDMS base structure was formed in an aluminum mold, peeled off, treated by plasma, and bonded onto a glass substrate. Then, the bottom PDMS membrane layer (20:1 curing ratio) was transferred onto the PDMS base structure. An air-spray gun and a shadow mask were applied to deposit the CB-PDMS stain sensor in a defined pattern. The spray printing method enabled fabricating ultrathin, porous, and serpentine-shaped flexible strain sensors with high sensitivity. The microgrooved top insulation PDMS layer was spin-coated onto a silanized PDMS stamp and bonded onto the bottom membrane layer. Two parallel carbon fiber wires (1k, Toray) were attached around the suspended membrane and connected to a stimulator (Master-8, AMPI) for the electrical pacing of hiPSC-CM monolayers. Customized glass cylinders (ID 8 mm, OD 10 mm, and height 10 mm) were bonded onto the top membrane layer to hold cell culture media.

hiPSC-CM culture and drug testing

hiPSC-CM (iCell2, Cellular Dynamics International) monolayers were cultured on the suspended membrane surfaces based on previously published protocol [14]. After fabrication, the flexible biosensing platforms were sterilized by gamma irradiation. Prior to cell culture, the PDMS membrane surface was activated by oxygen plasma for 1 min. Then, each culture chamber was immediately coated with extracellular matrix mixture solution and incubated at 37°C overnight. Cells were thawed and dispersed in plating medium and seeded at a density of 1.5 $\times 10^5$ cells/cm^2 to form a cell monolayer. Cell monolayers were cultured at 37 °C/5% CO_2 for 10 days. Culture medium was replaced every second day.

Before drug testing, the culture media was replaced with 250 µl fresh maintenance media and incubated for 2 hours to avoid the influence of nutrients and temperature on beating behaviors. Dose-dependent responses were recorded by adding 2.5 µl (1:100 volume ratio) of drug stock solutions subsequently every 10 mins. Drug testing was repeated in three independent wells.

RESULTS AND DISCUSSION
Platform design and working principle

The device's compositions are schematically shown in Fig. 1, consisting of a bottom glass slide, a PDMS base structure, a suspended membrane array, a microgrooved top surface, CB-PDMS flexible strain sensors for contractility measurement, carbon fiber electrodes for electrical pacing, and glass cylinders for cell culture. The thin membrane is 25 µm in total thickness. Circular microgrooves are 5 µm in depth and 20 µm in thickness, integrated onto the membrane surface to induce cell alignment in circumferential directions. Two carbon fiber electrodes were distributed around the suspended membrane to conduct electrical pacing of cardiomyocyte monolayers.

After seeding, cardiomyocytes attached onto the membrane surface on day 1. With the increase of culture days, hiPSC-CM formed intracellular connections, initiated synchronous beating and generated compressive stress on the top membrane surface. Synchronous cell contraction caused deflections of each ultrathin, suspended membrane, and induced electrical resistance changes of the embedded CB-PDMS strain sensor. The magnitude of contractile stress correlates with the membrane deflection magnitude and the relative resistance change of the embedded strain sensor.

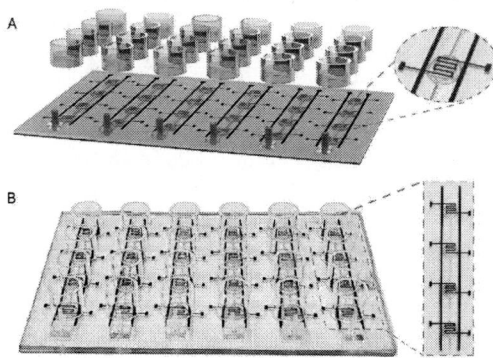

Figure 1: (A) Schematic diagram and (B) image of the carbon-based biosensing array for high-throughput cardiomyocyte contractility measurement. Each element consists of a bottom glass slide, a PDMS base structure, a suspended membrane array, CB-PDMS flexible strain sensors, carbon fiber electrodes for electrical pacing and, glass cylinders for cell culture.

Characterization and calibration of CB-PDMS strain sensors

CB-PDMS composite (25% CB ratio) was utilized as the strain sensing material. The spherical shape of zero-

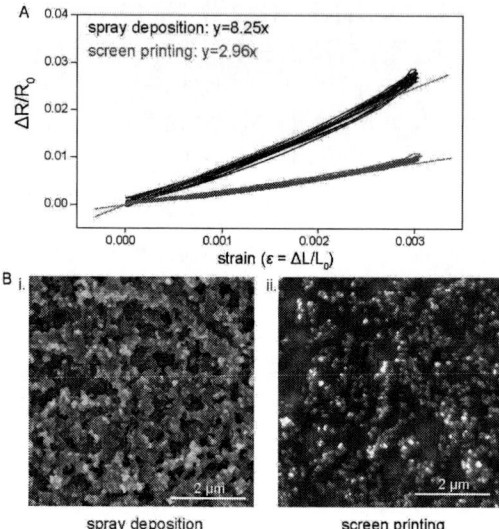

Figure 2: Calibration of CB-PDMS Strains sensors. (A) Uniaxial tensile testing was conducted to measure the gauge factor of the strain sensor fabricated by the spray deposition method and the screen-printing method. (B) SEM images showing the microstructure and surface morphologies of CB-PDMS (25% CB ratio) composites fabricated by spray deposition and screen printing.

dimensional CB particles could form point contact between nanoparticles, which could be easily disrupted under low strain magnitude. To evaluate strain sensing performance, tensile testing with a maximum strain magnitude of 0.3% was conducted to mimic the strain magnitude produced by cardiomyocyte monolayers [14]. The corresponding resistance changes were simultaneously recorded by a digital multimeter. Results showed that the CB-PDMS strain sensors displayed a highly linear relationship between the applied strain magnitude and the relative resistance change with limited hysteresis (Fig. 2A). In addition, the flexible sensor fabricated by the spray deposition method showed a higher gauge factor value (8.25) than that fabricated by the screen-printing method (2.96). The improvement of sensitivity is attributed to the unique coarse morphology and porous microstructure generated by direct spray deposition, in contrast to the solid CB-PDMS material fabricated by the screening printing method (SEM images in Fig. 2B).

Cell morphology

Both the bright field image and confocal image in Fig. 3 showed that cardiomyocytes generated a confluent cell monolayer on the top membrane surface. Cell distributions, sarcomere structures, and intracellular connections between adjacent iPSC-CMs were analyzed by immunostaining of cell nuclei, α-actinin, and gap junction protein (connexin-43). As shown in the figure, cardiomyocytes cultured on the microgrooved membrane surface exhibited morphological alignment along the circular microgrooves. Cardiomyocyte contraction is produced by the dynamic sliding between actin and myosin filaments in each sarcomere unit. The α-actinin protein could anchor actin filaments to the z-lines [15]. α-actinin immunostaining results showed the striated sarcomere structures for contractile force generation. In addition, the clear expression of connexin-43 protein on cell boundaries validated that hiPSC-CM in a cell monolayer could establish effective intracellular communications for synchronous cell beatings.

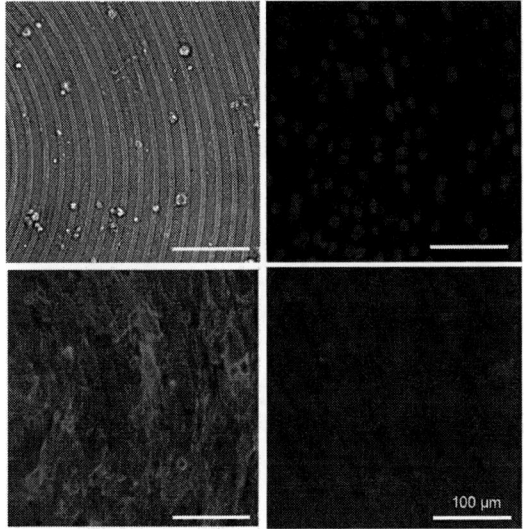

Figure 3: Bright-field image and confocal images of aligned hiPSC-CMs cultured on the microgroove membrane surface. Cardiomyocytes generated morphological alignment along circular microgrooves. Cells were stained with α-actinin (green), connexin-43 (red), and nuclei (DAPI, blue). Scale bar: 100 μm.

Drug testing

Platform validation was conducted by monitoring the cardiomyocyte contractility changes under drug administration. Isoproterenol can activate the β-adrenergic pathway, accelerate calcium influx through L-type calcium channel and further increase calcium release from the sarcoplasmic reticulum [16]. Here, isoproterenol (I6504, Sigma) was tested on iPSC-CM monolayers after day 5 when the periodic and stable signals could be generated by synchronous cell beatings. Eight isoproterenol concentrations were tested in our experiments. Fig. 4A demonstrated representative contractility signal curves before and after isoproterenol administration. As shown in Fig. 4B and C, both contractility and beating rate increased with the increase of isoproterenol concentration in a dose-dependent manner. The EC_{50} value for contractility is 348 nM, and for beating rate is 225 nM. Results confirmed that our flexible biosensing array is capable of measuring drug effects on cardiomyocyte contraction.

CONCLUSION

This paper reported a new flexible biosensing platform that enables in-situ and continuous measurement of contractile properties of hiPSC-CM monolayers. Ultrathin membrane array embedded with flexible CB-PDMS strain sensors was utilized for high-sensitive contractility measurement. The synchronous beating of cardiomyocytes generated sufficient contractile force to induce deflections of ultrathin-suspended membranes and deformations of the embedded CB-PDMS strain sensors for contractility

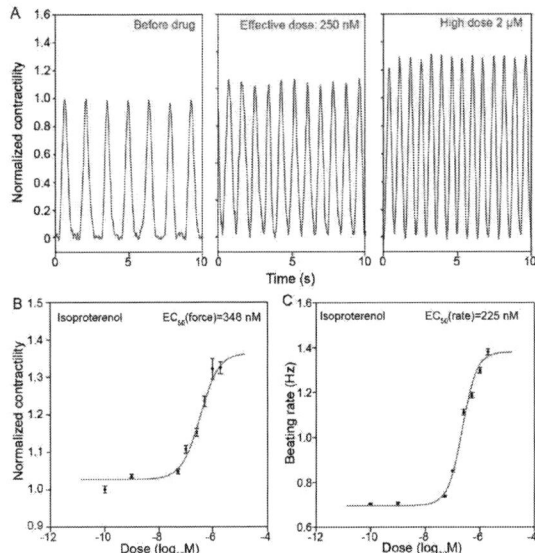

Figure 4: Measurement of CMs contractility for cardiac drug testing. (A) Representative contractility signals that show contractility changes before and after isoproterenol administration. Isoproterenol doses were increased sequentially, and the dose-dependent responses on (B) contractility and (C) beating rate were recorded.

measurement. Experimental data showed that the hiPSC-CM monolayers displayed cell alignment along circular microgrooves. Platform validation was conducted by applying isoproterenol with various concentrations. The platform precisely reflected the contractility and beating rate changes in response to isoproterenol in a dose-dependent manner. Our flexible biosensing platform realized comprehensive analysis of cardiomyocyte contractile function for high-throughput drug testing and therapeutic discoveries.

ACKNOWLEDGEMENTS

This work was supported by Canadian Institutes of Health Research (CIHR). W.D. acknowledges a Restracomp Fellowship from The Hospital for Sick Children. J.T.M. acknowledges the Wasser Family and the SickKids Foundation for chair funds. Y.S. also acknowledges the Canada Research Chairs Program (CRC Tier I) for financial support.

REFERENCES

[1]   H. J. Aparicio *et al.*, *Heart Disease and Stroke Statistics-2021 Update A Report from the American Heart Association.* 2021.

[2]   H. Savoji *et al.*, "Cardiovascular disease models: A game changing paradigm in drug discovery and screening," *Biomaterials*, vol. 198, no. May 2018, pp. 3–26, 2019.

[3]   A. Diokmetzidou *et al.*, *Strategies to Study Desmin in Cardiac Muscle and Culture Systems*, 1st ed., vol. 568. Elsevier Inc., 2016.

[4]   R. Breckenridge, "Heart failure and mouse models," *DMM Dis. Model. Mech.*, vol. 3, no. 3–4,

pp. 138–143, 2010.

[5] A. S. T. Smith, J. Macadangdang, W. Leung, M. A. Laflamme, and D. H. Kim, "Human iPSC-derived cardiomyocytes and tissue engineering strategies for disease modeling and drug screening," *Biotechnol. Adv.*, vol. 35, no. 1, pp. 77–94, 2017.

[6] W. Dou *et al.*, "Microengineered platforms for characterizing the contractile function of in vitro cardiac models," *Microsystems Nanoeng.*, vol. 8, no. 1, 2022.

[7] M. C. Ribeiro *et al.*, "Functional maturation of human pluripotent stem cell derived cardiomyocytes invitro - Correlation between contraction force andelectrophysiology," *Biomaterials*, vol. 51, pp. 138–150, 2015.

[8] Y. Dai, N. E. Oyunbaatar, B. K. Lee, E. S. Kim, and D. W. Lee, "Spiral-shaped SU-8 cantilevers for monitoring mechanical response of cardiomyocytes treated with cardiac drugs," *Sensors Actuators, B Chem.*, 2017.

[9] Y. S. Zhang *et al.*, "Bioprinting 3D microfibrous scaffolds for engineering endothelialized myocardium and heart-on-a-chip," *Biomaterials*, vol. 110, pp. 45–59, 2016.

[10] K. J. Hansen, J. T. Favreau, J. R. Gershlak, M. A. Laflamme, D. R. Albrecht, and G. R. Gaudette, "Optical Method to Quantify Mechanical Contraction and Calcium Transients of Human Pluripotent Stem Cell-Derived Cardiomyocytes," *Tissue Eng. - Part C Methods*, vol. 23, no. 8, pp. 445–454, 2017.

[11] J. Liu, N. Sun, M. A. Bruce, J. C. Wu, and M. J. Butte, "Atomic force mechanobiology of pluripotent stem cell-derived cardiomyocytes," *PLoS One*, vol. 7, no. 5, 2012.

[12] N. E. Oyunbaatar, A. Shanmugasundaram, Y. J. Jeong, B. K. Lee, E. S. Kim, and D. W. Lee, "Micro-patterned SU-8 cantilever integrated with metal electrode for enhanced electromechanical stimulation of cardiac cells," *Colloids Surfaces B Biointerfaces*, vol. 186, no. May 2019, p. 110682, 2020.

[13] J. U. Lind *et al.*, "Instrumented cardiac microphysiological devices via multimaterial three-dimensional printing," *Nat. Mater.*, vol. 16, no. 3, pp. 303–308, 2017.

[14] W. Dou *et al.*, "A Carbon-Based Biosensing Platform for Simultaneously Measuring the Contraction and Electrophysiology of iPSC-Cardiomyocyte Monolayers," *ACS Nano*, vol. 16, no. 7, pp. 11278–11290, 2022.

[15] J. G. Jacot *et al.*, "Cardiac myocyte force development during differentiation and maturation," *Ann. N. Y. Acad. Sci.*, vol. 1188, pp. 121–127, 2010.

[16] A. Lymperopoulos, G. Rengo, and W. J. Koch, "Adrenergic nervous system in heart failure: Pathophysiology and therapy," *Circ. Res.*, vol. 113, no. 6, pp. 739–753, 2013.

CONTACT
*Yu Sun, yu.sun@utoronto.ca

FLEXIBLE BI-DIRECTIONAL BRAIN COMPUTER INTERFACE FOR CONTROLLING TURNING BEHAVIOR OF MICE

Yifei Ye[1], Ye Tian[2,3], Han Wang[2], Qian Cheng[1], Kuikui Zhang[2], Xueying Wang[1,3], Cunkai Zhou[2], Chengjian Xu[2], Xiaoling Wei[1,3], Zhitao Zhou[1,3], Tiger H. Tao[1,2,3,4,5,6,7,8,9,*], and Liuyang Sun[1,2,3,*]

[1] State Key Laboratory of Transducer Technology, Shanghai Institute of Microsystem and Information Technology, Chinese Academy of Sciences, Shanghai, China

[2] 2020 X-Lab, Shanghai Institute of Microsystem and Information Technology, Chinese Academy of Sciences, Shanghai, China

[3] School of Graduate Study, University of Chinese Academy of Sciences, Beijing, China

[4] Center of Materials Science and Optoelectronics Engineering, University of Chinese Academy of Sciences, Beijing, China

[5] School of Physical Science and Technology, ShanghaiTech University, Shanghai, China

[6] Center for Excellence in Brain Science and Intelligence Technology, Chinese Academy of Sciences, Shanghai, China

[7] Neuroxess Co., Ltd. (Jiangxi), Nanchang, Jiangxi, China

[8] Guangdong Institute of Intelligence Science and Technology, Hengqin, Zhuhai, Guangdong, China
and

[9] Tianqiao and Chrissy Chen Institute for Translational Research, Shanghai, China

ABSTRACT

We developed a bi-directional brain computer interface enabling both electrical stimulation and recording through the same flexible implantable probe. An electrical stimulation threshold as small as 5 µA was demonstrated by effectively controlling the turning behavior of mice with the probe implanted in the motor cortex area. More importantly, both the neurons and electrodes remained healthy upon electrical stimulation as the spike waveform recorded from the same neuron remained unchanged by the same electrode before and after stimulation, showing the stability of the proposed system for neural stimulation and recording.

KEYWORDS

Brain computer interface, Flexible probe, Turning behavior, Neuromodulation, Motor control

INTRODUCTION

Brain computer interface (BCI) is a pathway for communication between the brain and the outside environment by recording and modulating neural activity. Generally, BCI systems could be divided into non-invasive systems and invasive systems. In early researches, non-invasive BCI systems with electroencephalograms (EEGs) are dominant, primarily affording control of computer cursors or other devices [1]. However, the non-invasive systems encountered obstacles to achieving more complex neural communications due to limited signal quality and transfer rate. On the contrary, the invasive BCIs implanted into brain tissue could directly communicate with neurons, which could realize high-quality neural signal acquisition, moreover, localized neural modulation.

Most invasive BCI systems focus on neural signal acquisition for improved performance on the control of external devices or sensory recognition [2, 3]. Recently, people have paid more attention to neural modulation by applying certain stimuli to neurons for applications such as

tactile perception and visual reconstruction [4, 5]. However, current electrical stimulation mainly relies on rigid electrodes, which has limitations on brain tissue damage, inflammatory response, and scar formation. Since good mechanical matching between flexible electrodes and brain has attracted widespread attention in long-term neural recording, whether flexible electrodes can achieve effective neural stimulation remains to be explored [6].

In this work, we proposed a bi-directional BCI system with a flexible neural probe as a key component, capable of both neural recording and stimulation (Fig. 1). Flexible electrodes were utilized to apply electrical stimulation to the secondary motor cortex (M2) of mice, which could successfully control the turning behavior of mice and stably record neural signals before and after stimulation.

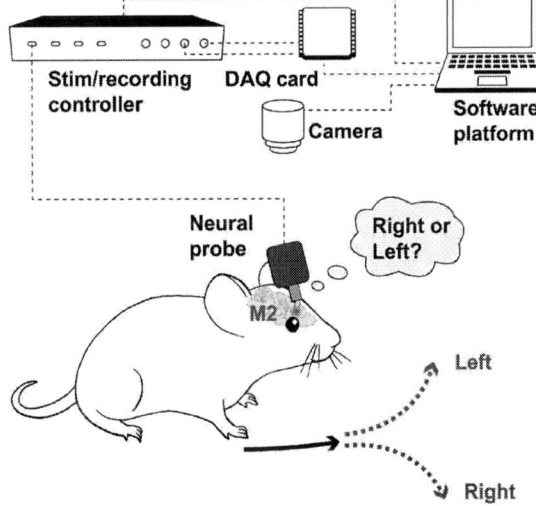

Figure 1: Schematic of bi-directional flexible BCI system for the control of mouse turning behavior.

MATERIALS AND METHODS
Device design and fabrication

We designed a 32-channel electrode array distributed on two shanks of a flexible neural probe (Fig. 2a, Fig. 2b). The neural probe is composed of the top and bottom encapsulation layers and the middle layer of interconnectors and electrodes. The fabrication process began with deposition of a sacrificial layer on a silicon substrate. A layer of Cr/Ni with a thickness of 10/150 nm was deposited by e-beam evaporation. After a polyimide (PI) bottom encapsulation layer with a thickness of 2.5 microns was cured, Ti/Au/Ti (5/100/5 nm) was evaporated with following lift-off to get interconnector and electrode structure. Then Ti/Ni/Au (5/150/50 nm) was deposited and patterned as backend pads by lift-off. After that, another PI layer was cured as top encapsulation. RIE was adopted to etch PI for the device profile and expose electrodes and pads, and 100 nm of aluminum is sputtered and patterned by lithography and wet etching as a hard mask. After PI etching, the wafer was wet etched to wipe off aluminum mask. Then, the conducting polymer PEDOT: PSS was coated on Au electrodes by electroplating technique to reduce the impedance. Finally, the device was released in Nickel etchant to obtain the flexible probe (Fig. 2c).

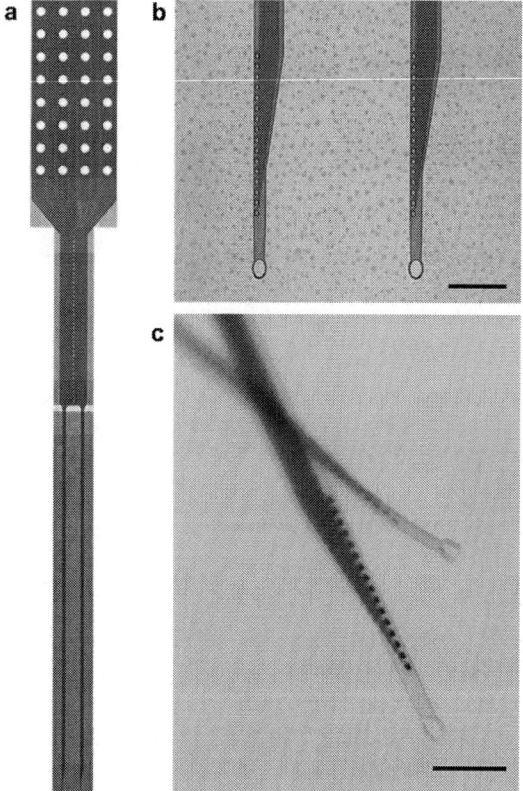

Figure2: (a) Micrograph of the neural probe with two shanks. (b) Electrode arrays on two shanks of the neural probe. Scale bar: 200 μm. (c) The released shanks show great flexibility in PBS. Scale bar: 200 μm.

Device assembly and implantation

As the flexible probe was too soft to penetrate the brain tissue, the tip-etched tungsten wires were utilized to assist probe implantation. Firstly, the front of the tungsten wires was etched in NaOH solution with a positive voltage of 8V applied until a T-shaped tip was formed. Then, two flexible shanks with holes in the front were sleeved into the tip-etched tungsten wires and temporally bonded by dissolvable PEG (Fig. 3a, Fig. 3b). In operation, the shanks were implanted into the secondary motor cortex (M2: Anteroposterior (AP) +1.5 mm, mediolateral (ML) ±0.75 mm, dorsoventral (DV) −0.75mm) of left and right brain (Fig. 3c). After implantation, the tungsten wires were removed, and the device was fixed by dental cement (Fig. 3d).

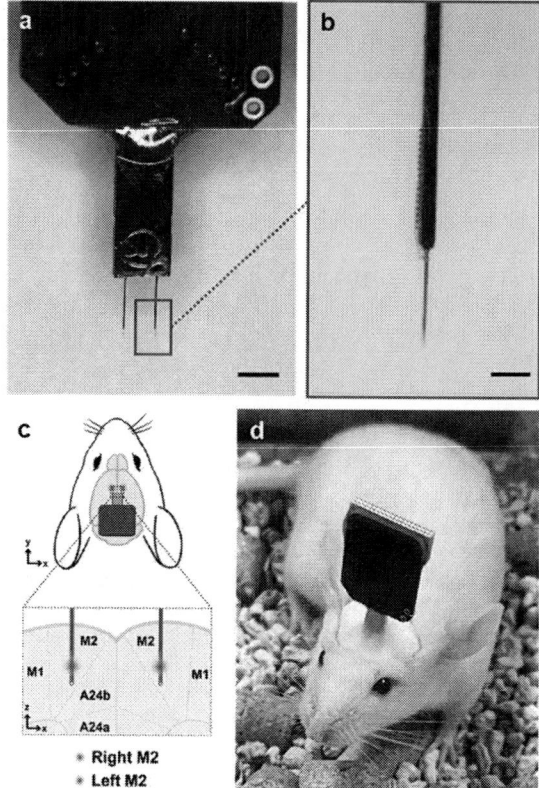

Figure 3: (a) Two flexible shanks were assembled to two tip-etched tungsten wires. Scale bar: 2 mm. (b)The details of tip-etched tungsten wire assembled with the shank. Scale bar: 200 μm. (c) Schematic of the implantation location of the neural probe. Two shanks of the neural probe were implanted in the right and left M2 respectively for electrical stimulation and recording. (d) The mouse with neural probe implanted in both right and left M2.

Bi-directional BCI system

The proposed bi-directional BCI system contains a flexible probe, a stim/recording controller, a DAQ card, a camera to record mice behavior, and a software platform (Fig. 1). The developed software platform outputs instructions to the stim/recording controller through DAQ card to realize synchronous triggering of video and neural recording, as well as specified electrical stimulation instructions. The stimulation waveform consists of a 200 μs, 5 μA (1 nC/phase) cathodic phase, a 100 μs interphase period, and a 400 μs, 2.5 μA anodic phase with a frequency of 100 Hz. Compared to the conventional rigid electrodes (~hundreds of μA), the stimulation current for effective motor control of the flexible electrodes decreased by two orders of magnitude [7].

RESULTS AND DISCUSSION
Stimulation-induced turning behavior

Fig. 4a shows the mouse trajectory when electrical stimulation was applied to the left and right M2 respectively. The black hollow circle marks the mouse's starting point, and the red and blue lines draw the mouse's movement during 10 s electrical stimulation. The stimulation of left or right M2 cortex regions results in significant differences in mouse movements. The mouse circled to the left during the right M2 stimulation, on the contrary, to the right during the left M2 stimulation.

Quantitatively, we tracked mouse head rotation before, during, and after stimulation, and calculated angular displacement, where positive indicates counter-clockwise rotation (left) and negative indicates clockwise rotation (right) (Fig. 4b). Before stimulation, the mouse was stabilized. Once stimulation was on, the mouse rapidly rotated unilaterally. The rotation direction in multiple tests was specifically determined by the stimulated cortex area, reflecting the accuracy and repeatability of electrical stimulation. When the stimulation was off, the mouse tended to return to quiescence.

Meanwhile, the mean angular speed before, during, and after stimulation was calculated (Fig. 4c). The electrical stimulation can make stationary mice become active, and perform noticeable turning behavior. The results indicate that this flexible BCI system could afford accurate control of mouse right or left turning behavior by stimulating left or right M2 cortex regions.

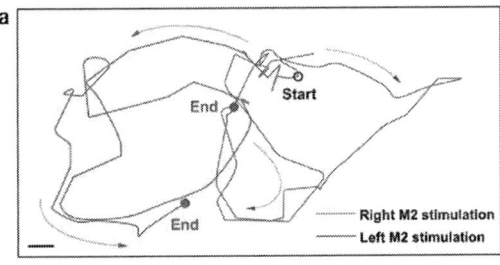

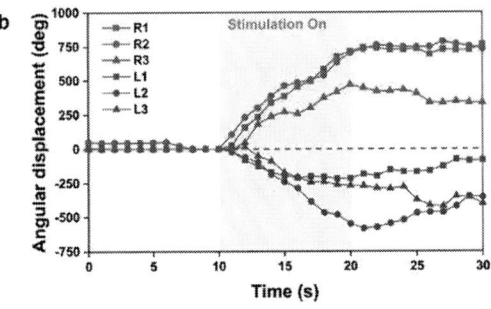

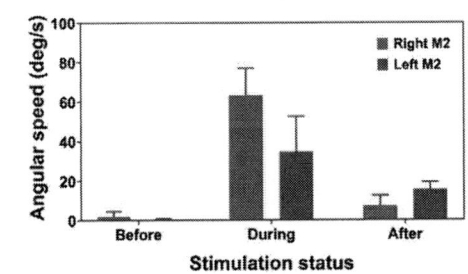

Figure4: (a) Mouse trajectory when stimulating right or left M2. Scale bar: 1cm. (b) Angular displacement of the mouse before (0-10 s), during (10-20 s), and after (20-30 s) stimulation of right M2 (R1, R2, R3) or left M2 (L1, L2, L3). (c) Angular speed before, during, and after stimulation of the right and left M2.

Neural recording before and after stimulation

Fig. 5a shows the recorded neural activities before and after stimulation. Obvious spikes could be recorded both before and after stimulation. To examine the recording stability after stimulation, we extracted the neuronal activity and the mean spike waveform of a single neuron (neuron 1) from traces recorded by electrode Ch. 1. The firing rate of this neuron was significantly increased after stimulation, while the mean spike waveform was nearly unchanged, showing the stability of electrical recording after stimulation (Fig. 5b). In addition, we compared the electrode impedance before and after electrical stimulation. The impedance of most electrodes remained stable after stimulation (Fig. 5c). Therefore, the experimental results reflected that the flexible electrodes showed strong stability for both electrical stimulation and recording.

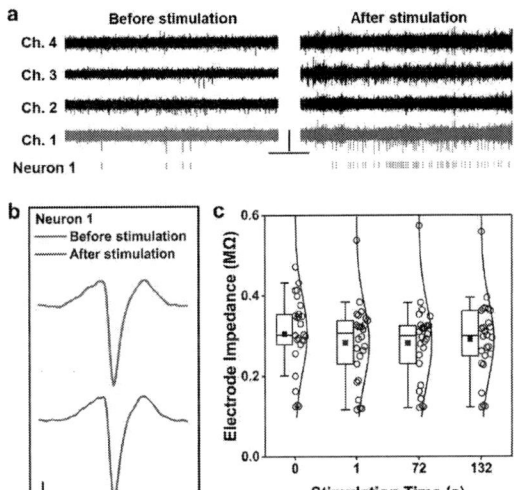

Figure 5: (a) Typical neural activities recorded by four electrodes (Ch. 1-Ch. 4) and the sorted action potential raster plot of neuron 1 recorded by Ch. 1 before and after stimulation. Scale bar: 1s (horizontal), 100 µV(vertical). (b) Mean spike waveform of neuron 1 before and after stimulation. Scale bar: 0.5 ms (horizontal), 10 µV (vertical). (c) Electrode impedance before and after stimulation.

CONCLUSIONS

In this paper, we proposed a flexible bi-directional brain computer interface that afforded electrical stimulation and neural recording. The minimized brain tissue damage led to less glial scar after implantation thus better electrical contact between electrodes and neurons. Therefore, we successfully achieved an electrical stimulation threshold as small as 5 µA, demonstrated by effective control of the mouse turning behavior. Furthermore, we showed that both the neurons and electrodes remained healthy upon the electrical stimulation as the spike waveform recorded from the same neuron remained unchanged by the same electrode before and after stimulation. This bi-directional BCI system provides a strong platform for exploring neural networks as well as the treatment of neural disorders based on electrical stimulation.

ACKNOWLEDGEMENTS

This work was partially supported by the National Key R & D Program of China (Grant Nos. 2021ZD0201600, 2019YFA0905200, 2021YFC2501500, 2021YFF1200700, 2022ZD0209300, 2022ZD0212300), National Natural Science Foundation of China (Grant No. 61974154), Key Research Program of Frontier Sciences, CAS (Grant No. ZDBS-LY-JSC024), Shanghai Pilot Program for Basic Research-Chinese Academy of Science, Shanghai Branch (Grant No. JCYJ-SHFY-2022-01), Shanghai Municipal Science and Technology Major Project (Grant No. 2021SHZDZX), CAS Pioneer Hundred Talents Program, Shanghai Pujiang Program (Grant Nos. 21PJ1415100, 19PJ1410900), the Science and Technology Commission Foundation of Shanghai (No. 21JM0010200), Shanghai Rising-Star Program (Grant No. 22QA1410900), the Innovative Research Team of High-level Local Universities in Shanghai, the Jiangxi Province 03 Special Project and 5G Project (Grant No. 20212ABC03W07), Fund for Central Government in Guidance of Local Science and Technology Development (Grant No. 20201ZDE04013), Special Fund for Science and Technology Innovation Strategy of Guangdong Province (Grant Nos. 2021B0909060002, 2021B0909050004).

REFERENCES

[1] M. A. Lebedev and M. A. Nicolelis, "Brain-machine interfaces: past, present and future," *Trends Neurosci,* vol. 29, no. 9, pp. 536-46, 2006.

[2] M. M. Shanechi, "Brain-machine interfaces from motor to mood," *Nat Neurosci,* vol. 22, no. 10, pp. 1554-1564, 2019.

[3] F. R. Willett, D. T. Avansino, L. R. Hochberg, J. M. Henderson, and K. V. Shenoy, "High-performance brain-to-text communication via handwriting," *Nature,* vol. 593, no. 7858, pp. 249-254, 2021.

[4] S. N. Flesher *et al.*, "A brain-computer interface that evokes tactile sensations improves robotic arm control," *Science,* vol. 372, no. 6544, pp. 831-836, 2021.

[5] X. Chen, F. Wang, E. Fernandez, and P. R. Roelfsema, "Shape perception via a high-channel-count neuroprosthesis in monkey visual cortex," *Science,* vol. 370, no. 6521, pp. 1191-1196, 2020.

[6] C. Gu, J. Jiang, T. H. Tao, X. Wei, and L. Sun, "Long-term flexible penetrating neural interfaces: materials, structures, and implantation," *Science China Information Sciences,* vol. 64, no. 12, 2021.

[7] B. Koo *et al.*, "Manipulation of Rat Movement via Nigrostriatal Stimulation Controlled by Human Visually Evoked Potentials," *Sci Rep,* vol. 7, no. 1, p. 2340, 2017.

CONTACT

*Tiger H. Tao, tel: +86-21-62511070; tiger@mail.sim.ac.cn
*Liuyang Sun, tel: +86-19821080802; liuyang.sun@mail.sim.ac.cn

HIGH SENSITIVITY MEMS Z-AXIS ACCELEROMETER WITH IN-PLANE DIFFERENTIAL READOUT

Valentina Zega[1], Gabriele Gattere[2], Manuel Riani[2], Francesco Rizzini[2], and Attilio Frangi[1]
[1]Civil and Environmental Engineering Department, Politecnico di Milano, ITALY and
[2]STMicroelectronics, ITALY

ABSTRACT

This paper reports the innovative design and a preliminary experimental characterization of a miniaturized *z*-axis accelerometer. For the first time in a MEMS accelerometer, a motion conversion mechanism is implemented to allow an electrostatic readout based on in-plane comb fingers, thus overcoming the main limitations of out-of-plane parallel plate detection, e.g. nonlinearities, trade-off between sensitivity and full-scale range, pull-in, etc. The first prototype fabricated by exploiting the features of the Thelma-Double fabrication process shows an experimental sensitivity of 12.9 fF/g which agrees well with numerical predictions computed by considering nominal geometric dimension of the sensor.

KEYWORDS

MEMS, *z*-axis accelerometer, motion conversion, high-sensitivity, differential readout, comb fingers.

INTRODUCTION

Since the early 1990s, a large variety of *z*-axis electrostatic accelerometers have been proposed [1]-[2]. They mainly belong to two families exploiting either translational or rotational motion depending on the possibility to manufacture top and bottom electrodes through the different techniques adopted for fabrication. Due to the out-of-plane nature of such sensors, parallel-plate based differential readout has been usually implemented with consequent compromise between sensitivity, linearity and full-scale [3].

Only few designs able to implement a readout system based on capacitors with varying overlap have been proposed so far. In [4], for example, a force rebalance readout is proposed by combining lateral comb fingers electrodes (used for sensing) with a parallel plate capacitor located beneath the proof mass (used to electrostatically pull-down the mass in closed-loop operation). In [5]-[6] *z*-axis accelerometers with asymmetric vertical comb fingers are presented as valid solution to overcome limitations of parallel-plates based readout schemes. In [7], regular comb fingers are instead employed to measure the out-of-plane displacement of the proof mass induced by external acceleration in a non-differential way.

However, innovative, compact and high-performance solutions compatible with standard MEMS fabrication processes are still required to allow an efficient and differential readout of out-of-plane accelerations. To this purpose, the idea to design proper suspension springs able to convert an out-of-plane (in-plane) force into an in-plane (out-of-plane) displacement developed in the last years for MEMS actuators seems very promising.

The first out-of-plane (in-plane) to in-plane (out-of-plane) motion-conversion mechanism in MEMS actuators has been introduced by Ando et al. [8]-[9]. They exploited slanted cross-section beams to obtain an out-of-plane motion as consequence of an imposed in-plane force. Thanks to the non-symmetric cross-section, indeed, this spring provides intrinsic coupling between out-of-plane and in-plane movements [10]. The main drawback of their solution is related to the complex fabrication process it requires. Slanted cross-sections can indeed be realized in MEMS through anisotropic wet etching that is however not suitable for rectangular cross-sectioned comb fingers needed for actuation. To solve this problem, Hotzen et al. [11]-[12] proposed a different motion conversion mechanism based on ladder-shaped suspension springs which are more compatible with MEMS mass-production despite preserving the ability to generate out-of-plane motion as consequence of an applied in-plane force.

Here we start from the idea presented in [11]-[12] and we propose an innovative design for the suspension springs of a *z*-axis accelerometer able to convert the out-of-plane translation of the proof mass induced by the inertial force in an in-plane motion of the external frames where readout comb fingers are located.

The resulting design represents the first prototype of a new generation of high sensitivity *z*-axis MEMS accelerometer with in-plane differential readout fully compatible with a commercial fabrication process, i.e. Thelma-Double by STMicroelectronics [13]-[14].

MECHANICAL DESIGN

In Fig. 1a, a schematic and simplified view of the proposed suspension spring is reported for the sake of clarity. From the structural perspective, it consists of two elongated springs fabricated on the first (EPI 1) and second (EPI 2) polysilicon layers, respectively and connected through auxiliary blocks built with full silicon thickness (EPI 1 + EPI 2). Thanks to the non-symmetric cross-section, a force applied along the *z*-direction on one side of the spring, will cause a displacement along both the *z*- and the *y*- directions [10]. If a proof mass is then connected on one side of such springs and an external frame properly constrained to move only along the *y*-axis direction, is attached on the other side, a *z*-axis accelerometer with in-plane differential readout can be obtained.

In Fig. 1b, a 3D-view of the proof-mass suspension spring here proposed is reported together with the definition of the main geometric quantities whose dimensions are collected in Table 1. With respect to the simplified scheme shown in Fig. 1a, auxiliary masses (colored in grey in Fig. 1b) can be seen. They only guarantee fabricability of the structure through the Thelma-Double process without playing any structural role.

The proposed *z*-axis accelerometer is schematically shown in Fig. 2a. It consists in a rectangular proof mass

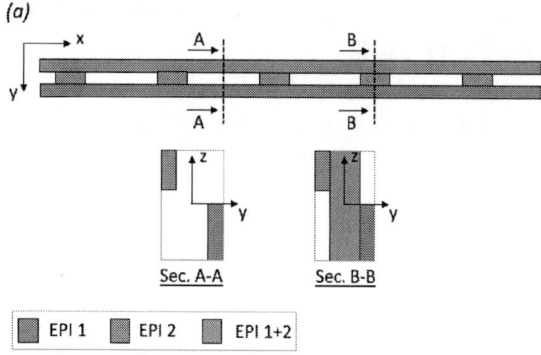

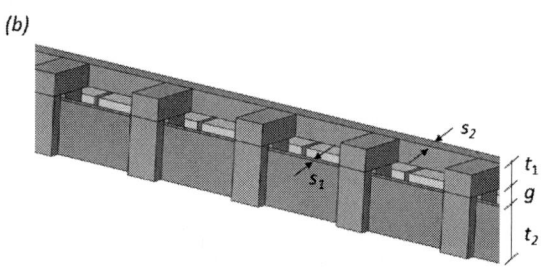

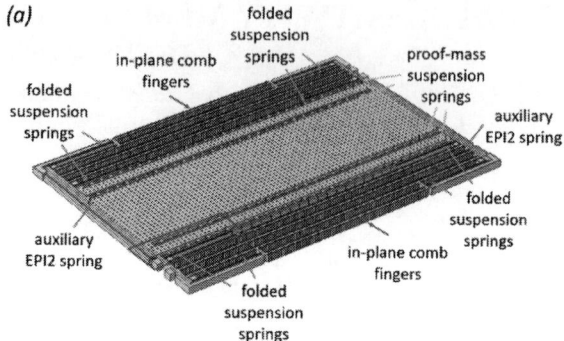

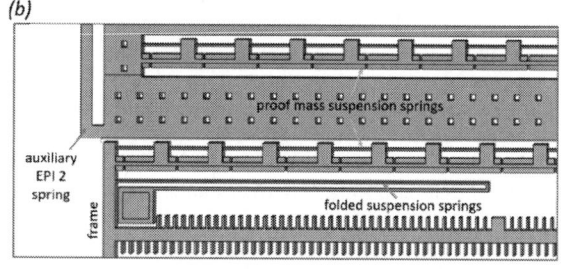

Figure 1: (a) Schematic in-plane view of the proposed suspension spring with non-symmetric cross-sections. It consists of two elongated springs fabricated on the first (EPI 1) and second (EPI 2) polysilicon layers, respectively and connected through auxiliary blocks built with full silicon thickness (EPI1 + EPI2). (b) 3D view of the spring configuration implemented in the proposed z-axis accelerometer.

Figure 2: (a) Schematic view of the proposed z-axis accelerometer. (b) Close-up view of the bottom-left corner of the proposed z-axis accelerometer.

Table 1: Geometric nominal dimensions of the proposed non-symmetric cross sectioned z-axis accelerometer proof mass suspension springs.

s_1	1.7 μm	t_1	20 μm
s_2	3.0 μm	t_2	8.2 μm
l_1	1440 μm	g	1.8 μm
l_2	1440 μm		

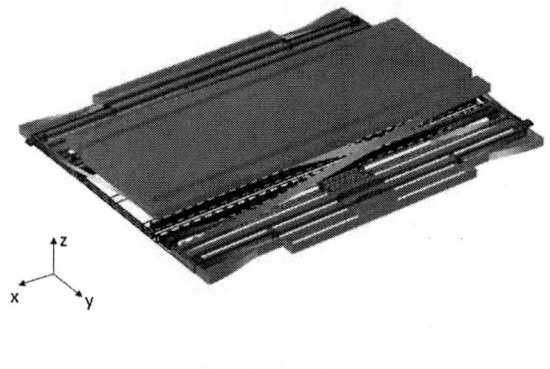

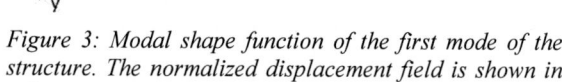

Figure 3: Modal shape function of the first mode of the structure. The normalized displacement field is shown in color.

connected through four non-symmetric cross-sectioned springs (Fig. 1) to two external frames suspended from the substrate through standard rectangular cross-sectioned folded springs. The proof mass, the two auxiliary frames and the folded suspension springs are fabricated such as to have a EPI1+EPI2 out-of-plane thickness. Comb fingers electrodes are located inside the two external frames for the in-plane readout of the z-axis external acceleration. Two additional suspension springs made in EPI2 are finally connected to the proof mass to avoid unwanted spurious torsional modes at low frequencies while not penalizing the desired movement of the structure.

In Fig. 2b, a close-up view of the bottom-left corner of the proposed accelerometer is shown for the sake of clarity. Proof mass and external frames suspensions springs are shown together with a portion of the auxiliary EPI 2 spring. Stators allowing the differential in-plane readout of the z-axis acceleration are not reported in Fig. 2b for simplicity, while holes in the proof mass are necessary to guarantee the structure release through the adopted MEMS

fabrication process.

The first mode of the structure is computed in COMSOL Multiphysics through a modal analysis and reported in Fig. 3: it consists in a simultaneous translation of the proof mass in the z-direction and of the two external frames in the y-direction. The natural frequency of the first mode of the proposed structure computed numerically by considering nominal geometric dimensions is 2100 Hz as reported in Table 2.

When an external acceleration acts on the accelerometer, the proof mass translates in the z-direction according to the first mode of the device shown in Fig. 3, thanks to the inertial force. The nearly 1:1 motion conversion ratio between the z-axis movement of the proof mass and the *y*-axis displacement of the external frames is achieved through a proper optimization of the non-symmetrical cross-sections of the proof mass suspension springs. The optimal design of the folded suspension springs guarantees instead the pure translational motion along the *y*-axis direction of the external frames and thus a differential readout through comb fingers electrodes located in them.

By considering nominal dimensions of the proposed MEMS accelerometer, a sensitivity of 14.4 fF/g is numerically estimated. The full-scale range of the proposed accelerometer in the present realization is limited to 17g by the gap between the proof mass and the substrate, i.e. 1.8 μm, and can in principle be improved by releasing such constraint during fabrication.

Finally, from a set of nonlinear static analyses performed in COMSOL Multiphysics, we numerically demonstrate a nonlinearity below 1%, i.e. 0.37%, for external accelerations in the range 0-50g. Note that numerically estimated nonlinearities take into account both geometric contributions deriving from large displacements of both the proof mass and the auxiliary frames and electrostatic contributions coming from spurious unwanted out-of-plane displacements of comb fingers during the regular functioning of the proposed accelerometer. Out-of-plane displacements of the external frames induced by the non-symmetric cross-sectioned springs' response are minimized by design, but are in principle different from zero and must be taken into account in the nonlinearity estimation.

EXPERIMENTAL RESULTS

The *z*-axis accelerometer has been fabricated through the Thelma-Double fabrication process of STMicroelectronics [13]-[14] in polysilicon (*E*= 160GPa, v=0.23, ρ = 2330 Kg/m^3) and its Scanning Electron Microscope (SEM) image is reported in Fig. 4a. The *z*-axis accelerometer footprint is 1524 μm x 980 μm.

The sensor is then wire-bonded on a ceramic carrier and mounted on a Printed Circuit Board (PCB) which has the only function of redirecting the electrical signals from the MEMS transducer to capacitance meter.

The PCB containing the accelerometer under study is then mounted on a mechanical support able to rotate from the +1g to the -1g condition as shown in Fig. 4b.

In Fig. 5 the differential capacitance variation between the movable mass and two sense stator electrodes is reported for different levels of external accelerations in the range -1g - +1g. By fitting the curve, a sensitivity of 12.9fF/g is obtained, which well agrees with numerical predictions summarized in Table 2.

Table 2: Numerical predictions computed in COMSOL Multiphysics by considering nominal geometric dimensions of the accelerometer and experimental results.

	Numerical Predictions	Experimental Measurements
Natural Frequency [Hz]	2100	2300
Sensitivity [fF/g]	14.4	12.9
Linearity [@50g]	0.37%	N.A.

Figure 4: (a) SEM image of the z-axis accelerometer fabricated through the Thelma-Double process of STMicroelectronics. (b) Experimental set-up: the MEMS is wire-bonded to a ceramic carrier, mounted on a PCB and then on a mechanical support able to rotate from the -1g condition to the +1g condition.

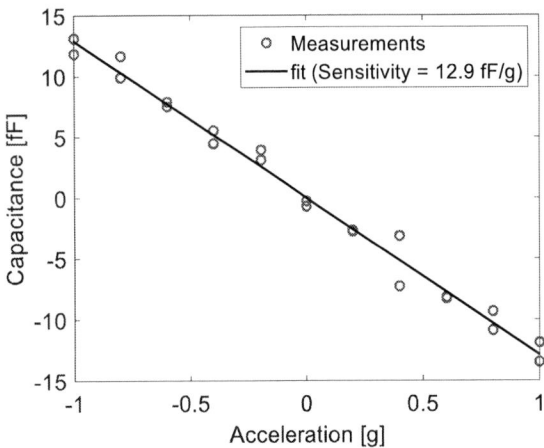

Figure 5: Experimental differential capacitance variation measured on the z-axis accelerometer here proposed when an external acceleration in the range -1g - +1g is applied by rotating the mechanical support to which the PCB is mounted. Linear fitting employed to determine the sensor sensitivity.

978-1-6654-9309-3/23 $31.00 © 2023 IEEE

A frequency response measurement, not reported here for the sake of brevity, has been also performed by applying a bias voltage on the MEMS z-axis accelerometer here proposed, an alternate current signal on one set of electrodes and performing the readout on the other set of electrodes originally designed for the differential readout of the z-axis acceleration. The experimental resonant frequency of the first mode (Fig. 3) of the fabricated device is equal to 2300 Hz. It is slightly higher than the one numerically predicted in Table 2, thus suggesting the presence of fabrication imperfections, i.e. over etch and pre-stresses, not correctly catched by the actual model based on nominal geometric dimensions of the mechanical structure.

CONCLUSIONS

A novel high-sensitivity z-axis MEMS accelerometer with in-plane differential readout fully compatible with a commercial fabrication process, i.e. Thelma-Double by STMicroelectronics, has been designed, fabricated and preliminary tested.

It shows an experimental sensitivity of 12.9 fF/g which is in good agreement with theoretical estimation computed by considering nominal geometric dimensions of the accelerometer. The small discrepancy between numerical predictions and experiments can be ascribed to fabrication imperfections, i.e. over etch and pre-stresses, not taken into account in the model.

A very promising 0.37% of nonlinearity in the acceleration range 0-50g has been also estimated through numerical analyses and will be experimentally validated in future works.

The proposed accelerometer, thanks to the exploitation of the motion conversion capability of non-symmetric cross-sectioned beams opens the way to a new class of z-axis MEMS high-sensitivity accelerometer with in-plane differential readout.

REFERENCES

[1] V. Kempe, *Inertial MEMS: Principles and practice*, Cambridge: Cambridge University Press, 2011.

[2] A. Corigliano, R. Ardito, C. Comi, A. Frangi, A. Ghisi, S. Mariani, *Mechanics of Microsystems*, Wiley, 2018.

[3] G. Langfelder, M. Bestetti, M. Gadola, "Silicon MEMS inertial sensors evolution over a quarter century", *J. Micromech. Microeng.*, vol. 31, pp. 084002, 2021.

[4] Y. Terzioglu, T. Kose, K. Azgin, T. Akin, "A simple out-of-plane capacitive MEMS accelerometer utilizing lateral and vertical electrodes for differential sensing", in *Proc. IEEE SENSORS 2015,* Busan, South Korea, 1-4 November, 2015, pp. 1-3.

[5] T. Tsuchiya, H. Funabashi, "A Z-axis differential capacitive SOl accelerometer with vertical comb electrodes", in *Proc. IEEE MEMS 2004*, Maastricht, Netherlands, 25-29 January, 2004, pp. 524- 527.

[6] J. Wang, Z. Yang, G. Yan, "A Silicon-an-Glass Z-axis Accelerometer with Vertical Sensing Comb Capacitors", in *Proc. 2012 7th IEEE International Conference on Nano/Micro Engineered and Molecular Systems (NEMS)*, Kyoto, Japan, 5-8 March, 2012, pp. 583-586.

[7] A. Sharaf, S. Sedky, "Design and simulation of a high-performance z-axis SOI-MEMS accelerometer", *Microsyst. Technol.*, vol. 19, pp. 1153–1163, 2013.

[8] Y. Ando, T. Ikehara, S. Matsumoto, "Design, fabrication and testing of new comb actuators realizing three-dimensional continuous motions", *Sensors and Actuators A: Physical*, vol. 97-98, pp. 579-586, 2002.

[9] Y. Ando, "Fabrication and testing of three-dimensional stages providing displacement of up to 8um", in *Proc. Solid-State Sensors, Actuators and Microsystems IEEE Transducers 2005*, Seoul, Korea, 5-9 June, 2005.

[10] J. M. Gere, S. Timoshenko, *Mechanics of materials*, 3rd ed. Boston: PWS-KENT Pub. Co., 1990.

[11] I. Hotzen, O. Ternyak, S. Shmulevich, D. Elata, "Mass-fabrication compatible mechanism for converting in-plane to out-of-plane motion", in *Proc. IEEE MEMS 2015*, Estoril, Portugal, 18-22 January, 2015, pp. 897-900.

[12] I. Hotzen, O. Ternyak, S. Shmulevich, D. Elata, "Selective Stiffening for Producing Motion Conversion Mechanisms", *Procedia Eng.*, vol. 87, pp. 1589-1592, 2014.

[13] F. Vercesi, L. Corso, G. Allegato, G. Gattere, L. Guerinoni, C. Valzasina, A. Novellini, A. Alessandri, I. Gelmi, "Thelma-Double: a new technology platform for manufacturing of high-performance MEMS inertial sensors", in *Proc. IEEE MEMS 2022*, Tokyo, Japan, 9-13 January, 2022, pp. 778-781.

[14] G. Gattere, F. Rizzini, L. Guerinoni, L. Falorni, C. Valzasina, F. Vercesi, L. Corso, G. Allegato, "High Performance Inertial MEMS IMU with ThELMA-Double Technology" in *Proc. IEEE INERTIAL 2022*, Avignon, France, 8-11 May 2022.

CONTACT

*V. Zega, tel: +39 02-23994315; valentina.zega@polimi.it

TWO-AXIS ELECTROMAGNETIC SCANNER INTEGRATED WITH AN ELECTROSTATIC XY-STAGE POSITIONER

Yuki Okamoto[1], Hironao Okada[1], Masaaki Ichiki[1]
[1]National Institute of Advanced Industrial Science and Technology (AIST), JAPAN

ABSTRACT

We developed a two-axis resonant electromagnetic scanner monolithically integrated with an in-plane electrostatic XY stage. The XY stage is composed of two-axis comb-drive electrostatic actuators, and the XY stage can control the scanner's position. The scanner is placed at the center of the XY stage's moving part and rotates in two-axis using electromagnetic force. As the electromagnetic actuators in the scanner and the comb-drive electrostatic actuators in the XY stage are independent, the proposed device can stabilize the scanning point dynamically.

KEYWORDS

MEMS Scanner, stabilizer, positioner, comb-drive actuator, electromagnetic actuator.

INTRODUCTION

Microelectromechanical systems (MEMS)-based scanners have become essential for laser scanning because of their compact sizes and fast responses [1]. By utilizing these advantages, MEMS scanners have been integrated into various applications, such as small projectors (known as pico projectors) and light detection and ranging (LiDAR) [1]. Especially in LiDAR applications, a high-performance MEMS scanner is required [1, 2] for long-distance measurement in autonomous cars, autonomous mobile robots (AMR), and drones.

In long-distance measurements under moving conditions, even small vibration of the laser-scanning device results in large errors at distant measurement points, as shown in Fig. 1. Therefore, anti-vibration treatment needs to be applied. In previous studies, the scanning angle and focal length were actively controlled by using integrated sensors and actuators, such as piezoresistive sensors on hinges [3] or integrated piezoelectric actuators on the mirror [4]. However, these studies cannot compensate for in-plane vibrations, shown in Fig. 1(a). To overcome the in-plane vibrations, a bulky gimbal stabilizer is typically used. However, the bulky gimbal stabilizer requires moving the entire LiDAR enclosure. As a result, using a gimbal stabilizer leads to an increase in the size of the LiDAR. On the other hand, as the available space in an AMR or drone is limited, reducing the space for sensor equipment is required. Therefore, integrating a stabilizer into a laser scanning component is required.

For MEMS optical imaging systems (OIS), previous studies on XY-stages use out-of-plane-working [5] or in-plane-working actuators [6-9]. Although out-of-plane actuation in OIS for CMOS image sensors is acceptable [5], it causes undesired scanner rotation. Besides, previous MEMS XY-stages, including in-plane-type XY-stages [6-8], require to be stacked on the laser scanning device. However, the stacking process degrades the working responses by increasing the mass on the XY-stage and attenuating the amount of light by passing the light on a

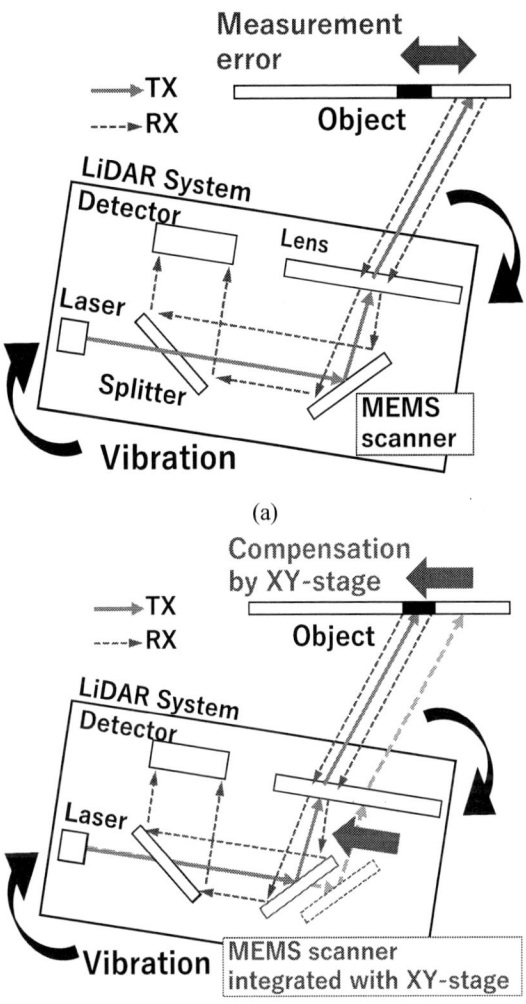

(a)

(b)

Figure 1: Concept of stabilization using integrated XY-stage. (a) LiDAR using a conventional MEMS scanner and (b) LiDAR using the proposed MEMS scanner with XY-stage.

lens [9]. Moreover, fabrication cost increases as the assembling process of the XY-stage and the laser scanning device is complicated.

In this paper, we proposed a two-axis resonant electromagnetic scanner monolithically integrated with an in-plane electrostatic XY stage. The XY stage is composed of two-axis comb-drive electrostatic actuators and two thick frames, and the XY stage can control the scanner's position. As the XY stage is driven only by in-plane electrostatic actuators and the thick frames avoid unexpected deformation of the mirror part, the work of the stage does not inhibit laser scanning. Therefore, the

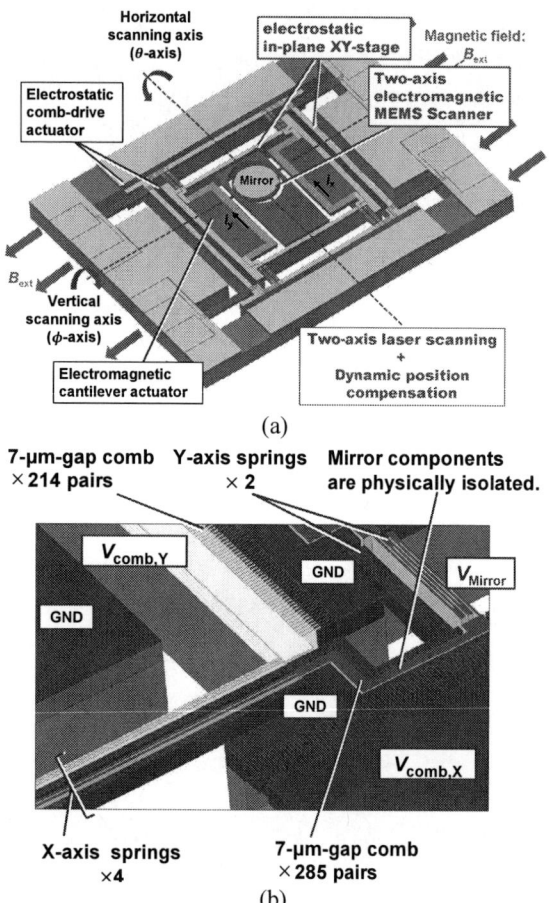

(a)

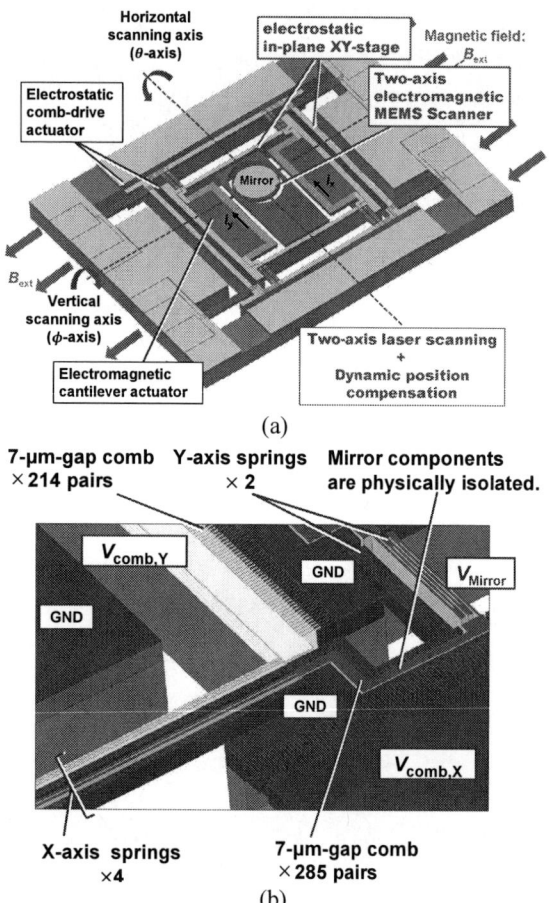

(b)

Figure 2: (a) Schematic of the proposed two-axis scanner's monolithically integrated design with a comb-drive electrostatic XY-stage. (b) Close-up view of the XY-stage's frame. The scanner and XY-stage parts are physically isolated and supported by the substrate layer.

proposed device can stabilize the scanning point dynamically. Furthermore, as the XY-stage part is monolithically integrated, it does not reduce the light amount.

METHODOLOGY

Figure 2(a) shows the proposed two-axis scanner's monolithically integrated design with a comb-drive electrostatic XY-stage. The scanning mirror rotates in two axes driven by the electromagnetic actuators. In the scanner part, we applied an asymmetric structure to realize two-axis scanning using only a one-axis lateral magnetic field to reduce the external magnets [10]. The scanner comprises a silicon-on-insulator (SOI) device layer and is placed at the center of the XY stage. The XY stage is composed of comb drive actuators, springs, and two movable thick frames. The frames in the XY-stage parts are composed of an SOI device, BOX, and substrate layers to prevent unintended out-of-plane actuation caused by the in-plane actuation. The device is 1.8×1.1 cm², and the diameter of the mirror part is 1.6 mm. Figure 2(b) shows the close-up view of the

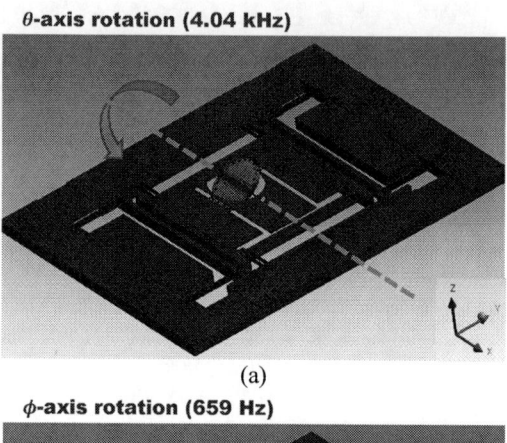

(a)

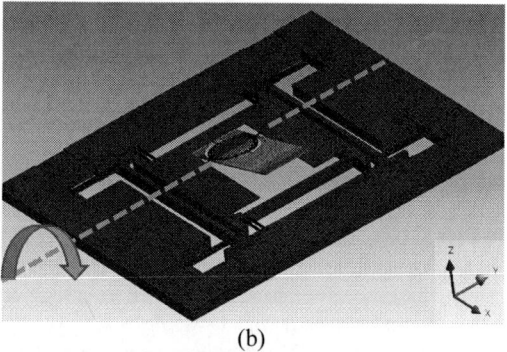

(b)

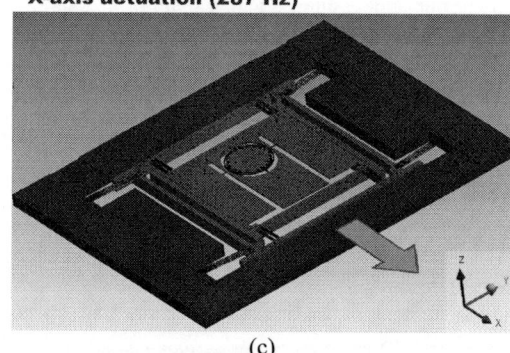

(c)

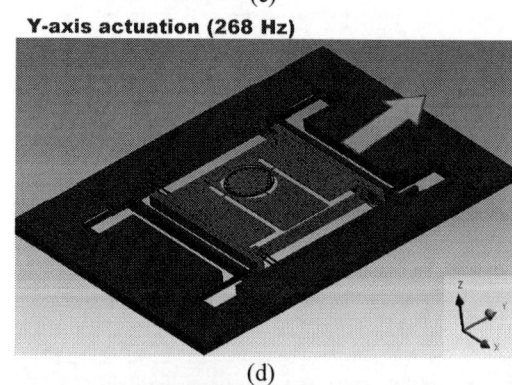

(d)

Figure 3: FEM simulation results of various resonant modes of the proposed device at (a) θ-axis rotation resonant mode, (b) φ-axis rotation mode, (c) X-axis actuation mode, and (d) Y-axis actuation mode. Out-of-plane rotation is driven by electromagnetic actuators, and in-place actuation is driven by electrostatic actuators.

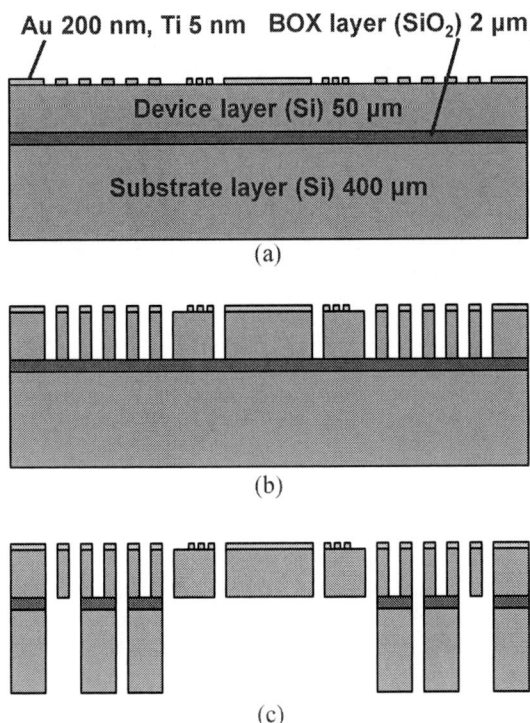

Au 200 nm, Ti 5 nm BOX layer (SiO$_2$) 2 µm

Device layer (Si) 50 µm

Substrate layer (Si) 400 µm

(a)

(b)

(c)

Figure 4: Process flow of the proposed device. (a) The titanium and gold electrode layers were patterned. (b) The Topside DRIE of the device layer was performed. (c) Backside DRIE of the backside substrate was performed, and then the moving structures were released by etching the BOX layer using CHF$_3$ plasma.

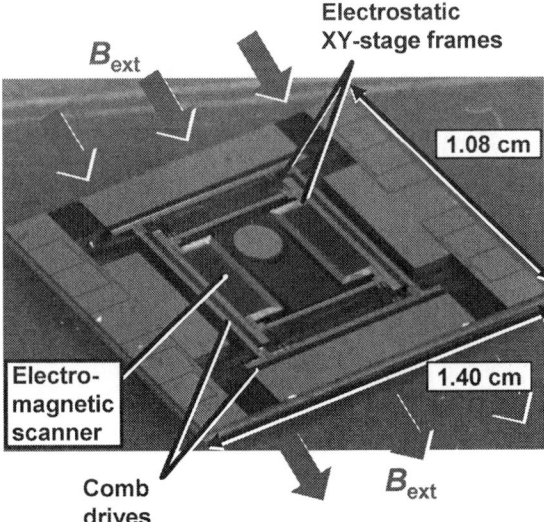

Figure 5: Image of the fabricated device.

Figure 6: PCB-mounted device with external permanent magnets. Owing to the asymmetric frame structure of the MEMS scanner part, two-axis rotation is obtained only using one pair of magnets.

the proposed device derived from the finite-element method (FEM) simulation using ANSYS Mechanical. Two resonant modes exist in the out-of-plane rotating modes, and the two resonant modes in the in-plane actuation modes exist. In the ϕ-axis rotation mode, we reported slow and fast resonant modes previously [10]. In this study, we chose slow-resonant mode. The resonant frequency of the θ-axis and ϕ-axis rotation modes are 4.04 kHz and 659 Hz, respectively, as shown in Figs 3(a) and 3(b). On the other hand, in the in-plane actuation modes, the X-axis and Y-axis actuation modes were observed at 287 Hz and 268 Hz, respectively, as shown in Fig. 3(c) and 3(d).

FABRICATION

Figure 4 shows the process flow of the proposed device. We used a 50 µm-2 µm-400 µm SOI wafer as a substrate. The device layer of the SOI wafer is highly doped to increase the conductivity. First, 20-nm-thick titanium and 200-nm-thick gold were sputtered on the silicon. The gold and titanium layers were etched using an Ar ion milling machine, as shown in Fig.4(a). Subsequently, the 50 µm-thick device silicon layer was etched by deep reactive ion etching (DRIE), as shown in Fig. 4(b). Next, the backside is etched by DRIE. Finally, the moving structures were released by etching the BOX layer using CHF$_3$ plasma, as shown in Fig. 4(c).

RESULTS

Figure 6 shows the fabricated device mounted on a printed circuit board (PCB). As shown in Fig. 6, we put two external permanent magnets on both sides of the device to generate B_{ext}.

Figure 7 shows the frequency responses of the in-plane X-axis and Y-axis actuation measured by the micro system analyzer (MSA-500, Polytec, Inc.). We applied a 30-V and 12-V sinusoidal wave to the X-axis and Y-axis axis electrostatic comb-drive actuator in this measurement. As shown in Fig. 7, the resonant frequencies of the X-axis and Y-axis in-plane actuation mode frequencies were 225 Hz

XY-stage's frame. The scanner and XY-stage parts are physically isolated and supported by the substrate layer. Owing to the structure, we used the device silicon layer and Au/Ti electrode layer for both the scanner and XY-stage parts. Besides, these driving signals can be applied independently owing to the physical-isolated structures.

Figure 3 shows the various resonant mode shapes of

978-1-6654-9309-3/23 $31.00 © 2023 IEEE   43

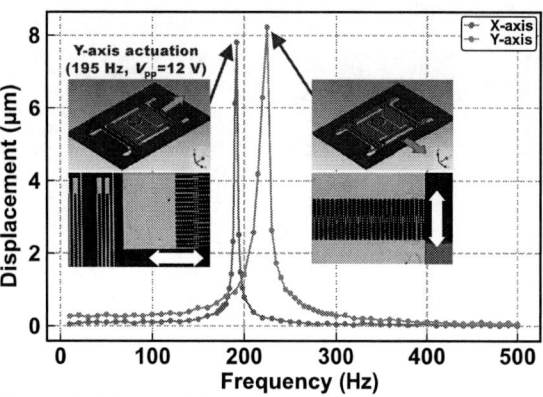

Figure 7: Measured frequency response of the X-axis and Y-axis actuation driven by the comb-drive electrostatic actuators.

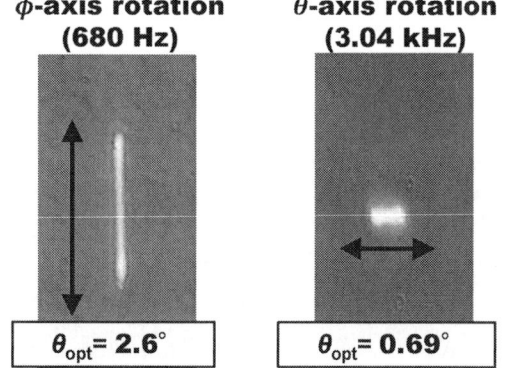

Figure 8: Scanned laser pattern using the θ-axis and φ-axis rotation mode. We applied 21 V to each actuator.

and 195 Hz, respectively.

Figure 8 shows the scanned laser pattern. The voltage to each axis was applied independently using the two electromagnetic cantilever actuators in the scanner part. We applied 21 V to each actuator. The optical scanning angle was 2.6° for the 680 Hz φ-axis scan and 0.69° for the 3.04 kHz θ-axis scan. These resonant frequencies were higher than those of the in-plane actuation frequencies. Therefore, the rotation of the scanner part and the actuation of the XY-stage do not affect each other.

CONCLUSIONS

We proposed a two-axis resonant electromagnetic scanner monolithically integrated with an in-plane electrostatic XY stage. The XY stage is composed of two-axis comb-drive electrostatic actuators and can control the scanner's position. The scanner is placed at the center of the XY stage's moving part and rotates in two-axis using electromagnetic force. As the electromagnetic actuators in the scanner and the comb-drive electrostatic actuators in the XY stage are independent, the proposed device can stabilize the scanning point dynamically. We successfully demonstrated out-of-plane biaxial scanning and in-plane XY actuation using the proposed device with a 1.6-mm mirror. The optical scanning angle was 2.6° for the 680 Hz φ-axis scan and 0.69° for the 3.04kHz θ-axis scan. The results indicate that the proposed method enables us to realize a small anti-vibration laser scanning system.

ACKNOWLEDGEMENTS

This research was supported by Advanced Research Infrastructure for Materials and Nanotechnology in Japan (ARIM) of MEXT and The University of Tokyo (Proposal No. JPMXP1222UT1029). The photomask was fabricated using System Design Lab. (d.lab) 8-inch EB writer F5112+VD01 donated by Advantest Corporation. This project is jointly supported by JSPS KAKENHI (No. 20K22422 and No. 21K14219).

REFERENCES

[1] S. T. S. Holmström, U. Baran, and H. Urey, "MEMS Laser Scanners: A Review," *J. Microelectromech. Syst.*, vol. 23, no. 2, pp. 259–275, Apr. 2014.

[2] Y. Okamoto *et al.*, "High-uniformity centimeter-wide Si etching method for MEMS devices with large opening elements," *Jpn. J. Appl. Phys.*, vol. 57, no. 4S, p. 04FC03, Mar. 2018.

[3] A. Vergara, T. Tsukamoto, W. Fang, and S. Tanaka, "Feedback Controlled Pzt Micromirror with Integrated Buried Piezoresistors," in *MEMS 2022*, Jan. 2022, pp. 243–246.

[4] S. Inagaki, Y. Okamoto, A. Higo, and Y. Mita, "High-Resolution Piezoelectric Mems Scanner Fully Integrated With Focus-Tuning and Driving Actuators," in *Transducers 2019 - EUROSENSORS XXXIII*, Jun. 2019, pp. 474–477.

[5] H. Wang, S. Yamada, and S. Tanaka, "Moving coil type electromagnetic microactuator using metal/silicon driving springs and parylene connecting beams for pure in-plane large motion in three axes," *Sens. Actuators A Phys.*, vol. 342, p. 113606, Aug. 2022.

[6] M. Rakotondrabe, A. G. Fowler, and S. O. R.Moheimani, "Control of a Novel 2-DoF MEMS Nanopositioner With Electrothermal Actuation and Sensing," *IEEE Trans. Control Syst. Technol.*, vol. 22, no. 4, pp. 1486–1497, Jul. 2014.

[7] L. Sun, J. Wang, W. Rong, X. Li, and H. Bao, "A silicon integrated micro nano-positioning XY-stage for nano-manipulation," *J. Micromech. Microeng.*, vol. 18, no. 12, p. 125004, Oct. 2008.

[8] K. Laszczyk *et al.*, "A two directional electrostatic comb-drive X–Y micro-stage for MOEMS applications," *Sens. Actuators A Phys.*, vol. 163, no. 1, pp. 255–265, Sep. 2010.

[9] K. Takahashi, H. N. Kwon, K. Saruta, M. Mita, H. Fujita, and H. Toshiyoshi, "A two-dimensional *f*-θ micro optical lens scanner with electrostatic comb-drive XY-stage," *IEICE Electronics Express*, vol. 2, no. 21, pp. 542–547, 2005.

[10] Y. Okamoto, T. -V. Nguyen, H. Okada, and M. Ichiki, "Via-Less Two-Axis Electromagnetic Scanner Using An Asymmetric Frame On A One-Axis Lateral Magnetic Field," in *MEMS 2022*, pp. 967-970.

CONTACT

*Y. Okamoto, tel: +81-29-861-2601;
yuki-okamoto@aist.go.jp

MEMS SHOCK ABSORBERS INTEGRATED WITH Al₂O₃-REINFORCED, MECHANICALLY RESILIENT NANOTUBE ARRAYS

Hojoon Lee[1†], Eunhwan Jo[1†], Jae-Ik Lee[2], and Jongbaeg Kim[1]*

[1]School of Mechanical Engineering, Yonsei University, Seoul, Republic of Korea and
[2]Massachusetts General Hospital, Harvard Medical School, Boston, Massachusetts, USA

ABSTRACT

This paper presents MEMS-integrated nanotube arrays as an in-plane shock absorber to ensure device reliability to external mechanical shocks. The geometrically structured nanotube arrays, consisting of vertically-aligned carbon nanotubes (CNTs) coated with alumina (Al_2O_3) layer, are directly integrated onto designated areas, allowing them to be compatible with conventional batch-fabrication processes for MEMS. The deformable carbon nanotube arrays with an exceptionally high aspect ratio dissipate the abruptly applied kinetic energy, while the reinforcing Al_2O_3 layer provides outstanding mechanical resilience to withstand extreme deformations. Therefore, in the shock-drop test, specimens with the nanotube arrays showed approximately 115 % and 80 % higher survival rate (with an average fracture acceleration of 10,000 g) compared to control groups (without the nanotubes) with a hard stop and compliant spring-based shock absorbers, respectively.

KEYWORDS

Shock absorber, alumina, carbon nanotube, mechanical shock

INTRODUCTION

One of the major issues hindering the commercialization of MEMS is its low reliability in shock environments. Mechanical shock induces excessive stresses on beams suspending moving parts in MEMS, leading to the fracture and malfunction of the MEMS devices [1]. Changing the dimension or materials to simply increase the beam's stiffness degrades the device performances, such as sensitivity, detection limits, and operable ranges. Numerous designs and materials for MEMS shock absorbers have been proposed as separate components that do not affect the existing device [2-5]. However, difficulties lay in their practical application to MEMS devices due to their insufficient shock reliability and/or their complicated and serially processed manufacturing methods. This work introduces MEMS-integrable and vertically-aligned nanotube arrays for in-plane shock absorbers to achieve high shock reliability. The compressible and mechanically resilient nanotube arrays [6], as a cushioning structure of shock absorbers, successfully prevent the fractures of the micro-structures.

CONCEPT

Figure 1(a) shows schematic diagrams of the specimen consisting of a proof mass, a micro-beam, and shock absorbers. As the specimen is exposed to mechanical shock, relative motion is induced between the proof mass and the shock absorbers. As the proof mass comes in contact with the compliant spring in the shock absorber, the

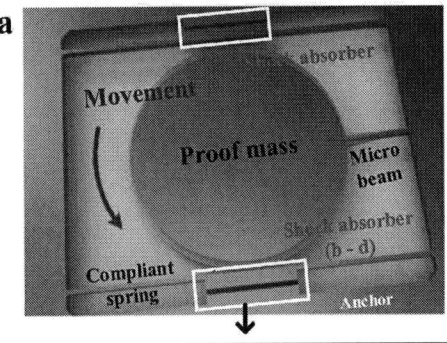

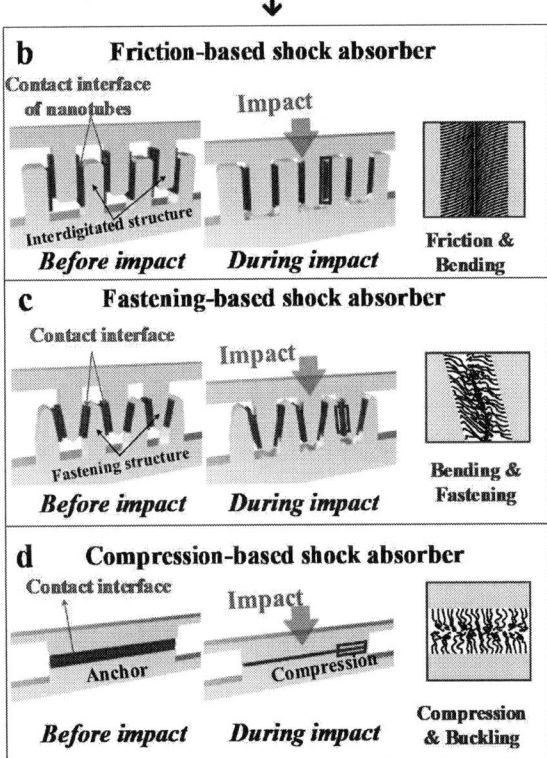

Figure 1: Schematic of the shock absorbers with nanotube arrays and mechanisms for shock energy dissipation. (a) Schematic diagram of the specimen with the shock absorbers located at both ends up to where the proof mass can move. The shock absorber starts to absorb the kinetic energy from the proof mass as soon as the shock absorber comes in contact with the proof mass. (b) The friction-based shock absorber dissipating the kinetic energy through friction between the nanotube arrays and bending deformation of the nanotube arrays. (c) The fastening-based shock absorber dissipating energy through the bending deformation of the nanotube arrays and the fastening mechanism. (d) The compression-based shock absorber dissipating energy through the buckling deformation of the nanotube arrays.

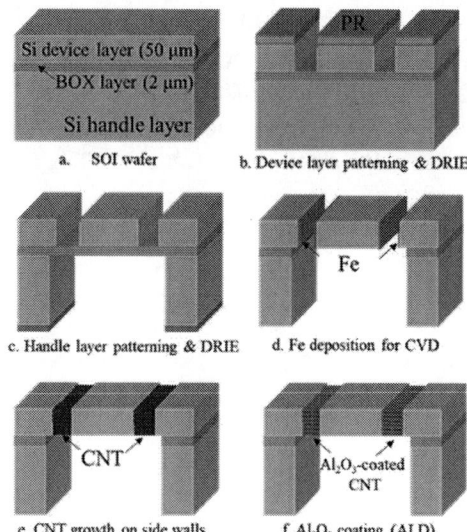

Figure 2: Fabrication process of the device. (a) A 4-inch silicon-on-insulator (SOI) wafer with a device layer of 50 µm is prepared. (b) The device layer of the SOI wafer is patterned using photolithography and deep reactive ion etching (DRIE) for the micro-structures. (c) The handle layer is also patterned through photolithography and DRIE. The structure is then released by etching the buried oxide (BOX) layer through hydrogen fluoride (HF) etching. (d) 3 nm of iron (Fe) catalyst is deposited selectively using sputtering through a shadow mask, fabricated from the same SOI wafer. The Fe catalyst on the top surface is then removed via reactive ion etching (RIE). (e) CNTs are synthesized only on the sidewalls of the micro-structure through chemical vapor deposition (CVD). (f) Al_2O_3 is then coated using atomic layer deposition (ALD).

mechanical shock begins to be absorbed. During the impact, the nanotube arrays are deformed, increasing the impact time and dissipating the kinetic energy. To utilize both the compression and bending of the nanotubes, we introduce three types of silicon structures for the shock absorbers: friction-based shock absorber (Figure 1(b)), fastening-based shock absorber (Figure 1(c)), and compression-based shock absorber (Figure 1(d)).

FABRICATION

Figure 2 shows the fabrication process of the device including the integration of the CNTs coated with an Al_2O_3 layer. The device is fabricated on a 4-inch silicon-on-insulator (SOI) wafer with a 50 µm-thick device layer. The device layer and the handle layer of the SOI wafer are patterned using photolithography followed by deep reactive ion etching (DRIE). Then, hydrogen fluoride (HF) solution is used for releasing the patterned micro-structure by etching a 2 µm-thick buried oxide (BOX) layer. After releasing the micro-structure, iron (Fe) catalyst of 3 nm-thickness for CNT growth is deposited only on the designated areas by sputtering through a shadow mask, which was fabricated on the same SOI wafer. After the growth of the CNT arrays through chemical vapor deposition (CVD), the Al_2O_3 layer is coated by atomic

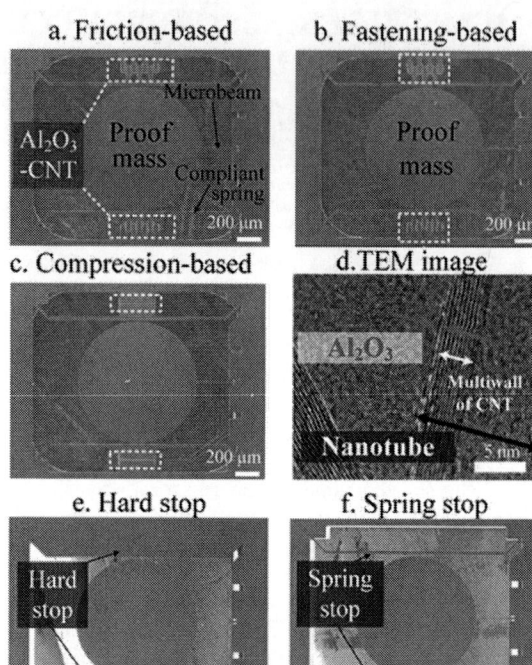

Figure 3: SEM and TEM images of the fabricated specimens. (a) The friction-based shock absorber. (b) The Fastening-based shock absorber. (c) The compression-based shock absorber. (d) A TEM image of Al_2O_3-coated CNT. (e, f) SEM images of the control devices including the hard stop (e) and the spring stop (f).

layer deposition (ALD) to form the Al_2O_3-reinforced, mechanically resilient nanotube arrays.

RESULTS & DISCUSSION

Figure 3(a-c) shows scanning electron microscopy (SEM) images of the fabricated shock absorbers with the integrated nanotube arrays. The multi-walled CNTs are directly synthesized on the shock absorber structures and uniformly coated by a thin Al_2O_3 layer to form the nanotube arrays. As shown in the TEM image of Figure 3(d), multi-walled CNTs are uniformly coated by a thin Al_2O_3 layer through ALD. Along with the fiction-based, fastening-based, and compression-based shock absorbers shown in Figure 3(a-c), respectively, hard stop and spring stop shock absorbers are also fabricated for the control devices. The hard stop shock absorber shown in Figure 3(e), is designed to physically restrict the movement of the proof mass to prevent failure of the device [7]. Another control device, shown in Figure 3(f), is the spring stop shock absorber which is designed to increase the contact time of the proof mass and the silicon (Si) structure to reduce the impact force and prevent failure [5].

Figure 4 shows enlarged SEM images of the linear contact interface of the aligned nanotube arrays. As shown in Figure 4(a), the friction-based shock absorber has linear contact interfaces of nanotube arrays. These interfaces induce friction as the nanotube arrays move in parallel but opposite directions while absorbing the mechanical shock. The fastening-based shock absorber, shown in Figure 4(b),

978-1-6654-9309-3/23 $31.00 © 2023 IEEE 46

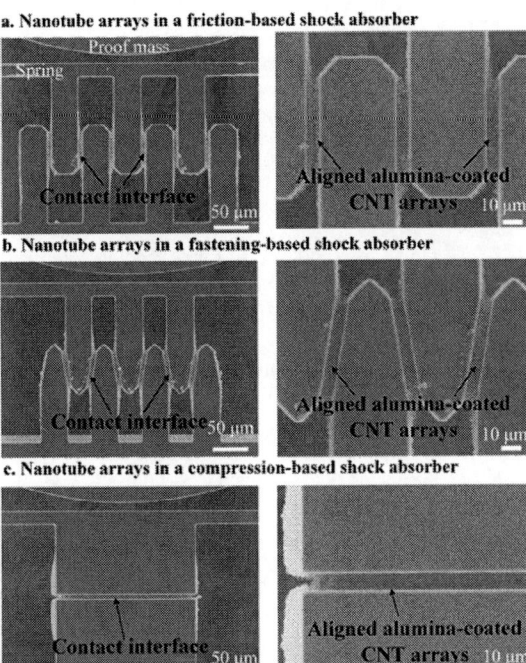

a. Nanotube arrays in a friction-based shock absorber

Proof mass
Spring
Contact interface
50 μm
Aligned alumina-coated CNT arrays
10 μm

b. Nanotube arrays in a fastening-based shock absorber

Contact interface
50 μm
Aligned alumina-coated CNT arrays
10 μm

c. Nanotube arrays in a compression-based shock absorber

Contact interface
50 μm
Aligned alumina-coated CNT arrays
10 μm

Figure 4: SEM images of the contact interfaces of the vertically-aligned, MEMS-integrated nanotube arrays. The contact interfaces of the friction-based shock absorber (a), fastening-based shock absorber (b), and compression-based shock absorber (c) are shown with their enlarged images.

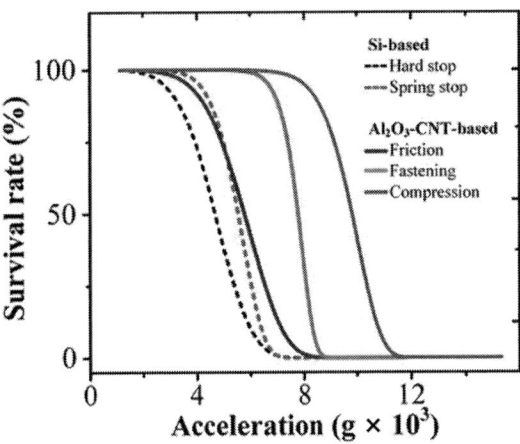

Figure 5: The survival rate of the devices plotted through their measured acceleration. Seven devices were tested for each of the designs and the devices were considered to have survived if the proof mass was not removed. The maximum acceleration that each device had survived was recorded and fit to Weibull survival distribution.

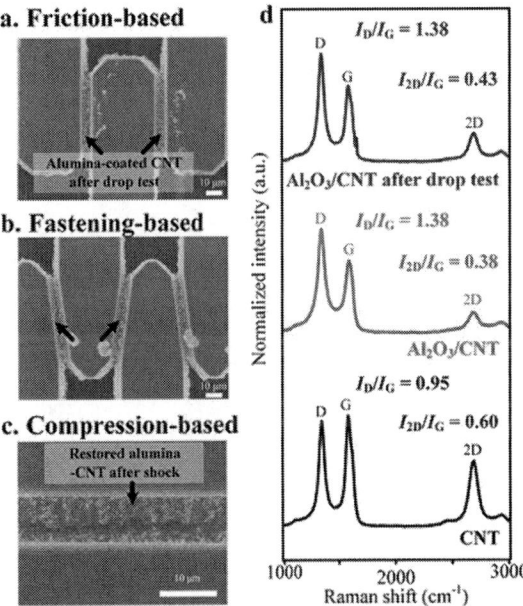

Figure 6: SEM images of the (a) friction-based shock absorber, (b) fastening-based shock absorber, and (c) compression-based shock absorber after the drop test experiment. (d) Raman spectra of Al_2O_3-CNT array after undergoing excessive deformation with the Raman spectra of a bare CNT array and Al_2O_3-CNT array.

has interfaces of nanotubes that generate not only friction but also compression between the nanotubes. As the structure absorbs the shock, the nanotubes are bent and the interdigitated Si structure is fastened to dissipate the kinetic energy of the proof mass. Figure 4(c) shows the compression-based shock absorber with the enlarged image of the contact interface with the Al_2O_3-reinforced nanotube arrays. As the mechanical shock is applied to the compression-based shock absorber, the aligned nanotube arrays are compressed and buckled to dissipate the energy.

Survival rates of the shock absorbers were measured and fit to Weibull survival distribution (Figure 5). Seven specimens for each design of shock absorbers, including the control devices, were fabricated and tested. The shock absorbers were attached to a moving carriage on which a commercial accelerometer was fixed. The moving carriage was released and dropped from specified heights along a guide rail. As the carriage hits the bottom mechanical shock is generated and the acceleration is constantly measured through the commercial accelerometer. A total of 35 specimens were tested and the devices were considered to have failed if the proof mass was removed. The maximum acceleration experienced by the shock absorber before failure was measured for each specimen and the data were fit to Weibull survival distribution (Figure 5) for predicting the continuous probability distribution of the specimens [8].

As a result, the compression-based shock absorber failed at an average acceleration of 10,000 g, which is 115 % and 80 % improvement compared with that of the hard stop (average 4,600 g) and the spring stop (average 5,500 g),

respectively. The friction-based shock absorber and the fastening-based shock absorber had an average failure acceleration of 5,700 g and 7,800 g, respectively.

After the excessive mechanical shock, the optical and material analysis confirms no significant characteristic change in Al_2O_3-CNT arrays. Figure 6(a-c) shows the SEM images of the friction-based, fastening-based, and compression-based shock absorbers. As can be seen in the SEM images, the nanotube arrays in all three designs have

978-1-6654-9309-3/23 $31.00 © 2023 IEEE

no noticeable damage. Raman spectra of the Al_2O_3-reinforced nanotubes before and after the drop test are shown in Figure 6(d) along with the Raman spectra of CNTs. The intensity ratio between the disorder-induced D band and the tangential G band (I_D/I_G) indicates the defects in the CNTs [9]. Due to the addition of the Al_2O_3 layer, the ratio (I_D/I_G) increased from 0.95 of the bare CNT to 1.38 of the Al_2O_3-reinforced nanotubes which indicates the CNTs are coated with Al_2O_3 [10]. The intensity ratio (I_D/I_G) of the reinforced nanotubes did not change after the drop test experiment which confirms that the nanotubes have not been damaged after the drop test.

CONCLUSION

We utilized Al_2O_3-reinforced CNT arrays as shock-absorbing nanostructure to dissipate the kinetic energy from mechanical shocks; these mechanically resilient nanotube arrays increased the absorbing capability of MEMS shock absorbers. When the abrupt shock causes a collision between the compliant spring and the proof mass, the deformable and resilient Al_2O_3-CNT nanotube arrays, integrated into the compliant spring, increase the contact time of the collision. As a result, the specimens with the nanotube arrays exhibited up to 115% increment in the average failure acceleration values compared to that with the hard stop devices. Also, we confirmed that there is no significant change in the material characteristics of the reinforced nanotube arrays via SEM and Raman spectroscopy.

ACKNOWLEDGEMENT

This work was supported by the National Research Foundation of Korea (NRF) grant funded by the Korea government (MSIT). (No. 2021R1A2B5B03002850).

REFERENCES

[1] V. T. Srikar and S. D. Senturia, "The Reliability of Microelectromechanical Systems (MEMS) in Shock Environments", *J. Microelectromech. Syst.*, vol. 11, pp. 206-214, 2002.

[2] K. Xu, F. Jiang, W. Zhang, and Y. Hao, "Micromachined Integrated Self-Adaptive Nonlinear Stops for Mechanical Shock Protection of MEMS", *J. Micromech. Microeng.*, vol. 28, p. 064006, 2018.

[3] J.-I. Lee, D. Kwon, and J. Kim, "Shock Protection Based on Confined Self-Adjusting Carbon Nanotube Arrays". in *Digest Tech. Papers Transducers'15 Conference*, Anchorage, June 21-25, 2015, pp. 287-290.

[4] A. K. Delahunty and W. T. Pike, "Metal-Armouring for Shock Protection of MEMS", *Sensors Actuators, A Phys.*, vol. 215, pp. 36–43, 2014.

[5] S. W. Yoon, S. Lee, N. C. Perkins, and K. Najafi, "Shock-Protection Improvement Using Integrated Novel Shock-Protection Technologies", *J. Microelectromech. Syst.*, vol. 20, pp. 1016-1031, 2011.

[6] S. J. Park, J. Shin, D. J. Magagnosc, S. Kim, C, Cao, K. T. Turner, P. K. Purohit. D. S. Gianola, and A. J. Hart, "Strong, Ultralight Nanofoams with Extreme Recovery and Dissipation by Manipulation of Internal Adhesive Contacts", *ACS Nano*, vol. 14, pp. 8383–8391, 2020.

[7] R. Abdolvand, B. V. Amini, and F. Ayazi, "Sub-Micro-Gravity In-Plane Accelerometers with Reduced Capacitive Gaps and Extra Seismic Mass", *J. Microelectromech. Syst.*, vol. 16, pp. 1036-1043, 2007.

[8] D. G. Kleinbaum and M. Klein, *Survival Analysis: a Self-Learning Text*, New York: Springer, 2012.

[9] E. Jo, Y. Lee, Y. Jung, S. Kim, Y. Kang, M. Seo, J. Yoon, and J. Kim, "Integration of Gold Nanoparticle–Carbon Nanotube Composite for Enhanced Contact Lifetime of Microelectromechanical Switches with Very Low Contact Resistance", *ACS Appl. Mater. Interfaces*, vol. 13, pp. 16959-16967, 2021.

[10] C. Silvestri, M. Riccio, R. H. Poelma, A. Jovic, B. Morana, S. Vollebregt, A. Irace, G. Q. Zhang, and P. M. Sarro, "Effects of Conformal Nanoscale Coatings on Thermal Performance of Vertically Aligned Carbon Nanotubes", *Small*, vol. 14, p. 1800614, 2018.

CONTACT

* J. Kim, tel: +82-2-2123-2812; kimjb@yonsei.ac.kr

† Equally contributed authors

HIGH-INDUCTANCE-DENSITY MEMS 3D-SOLENOID TRANSFORMERS WITH INSERTED THIN-FILM FERRITE MAGNETIC CORE FOR ON-CHIP INTEGRATED DC-DC POWER CONVERSIONS

Changnan Chen[1,2], Pichao Pan[1,2], Jiebin Gu[1], and Xinxin Li[1,2]

[1]State Key Laboratory of Transducer Technology, Shanghai Institute of Microsystem and Information Technology, Chinese Academy of Sciences, Shanghai 200050, CHINA

[2]University of Chinese Academy of Sciences, Beijing 100049, CHINA

ABSTRACT

This paper presents high-inductance-density MEMS 3D-solenoid magnetic core transformers that are wafer-level fabricated by a novel thick-metal micro-casting technique. With the embedded thin-film high-permeability ferrite core, an over 7 times boost in inductance (from 43.3 nH to 334.2 nH) and 25% increment in coupling coefficient (from 0.76 to 0.96) are achieved. For the batch-fabricated transformer in a tiny footprint of 6 mm², an over 1.6 A saturated current and one of the best-reported power transfer efficiency of 89.5% at a lower frequency of 10 MHz are demonstrated. The proposed magnetic core transformers are ideal for high-efficiency integrated DC-DC power conversion applications.

KEYWORDS

Power MEMS, Micro integrated transformers, PwrSoC, Isolation technology, MEMS-casting, Ferrite magnetic core, DC-DC power conversions

INTRODUCTION

With the rapid development of electric vehicles (EVs) and hybrid EVs, there is an urgent need to miniaturize and fully integrate the motor drive system to achieve higher power density and energy efficiency [1]. The gate drivers (GDs) powered by integrated transformer-based DC-DC converters are the interfaces between the controller and the high-voltage power stage, which control the energy flow in the motors by driving the IGBT or SiC switching devices at high frequency. High-power transfer efficiency of the micro transformers in isolated DC-DC converters for GDs is mandatory, which limits the effectiveness of the fully integrated systems [2]. Unfortunately, current chip-based transformers are structurally constrained by the low coupling coefficient and the poor quality factor, which largely affect the efficiency [3]. Moreover, it is difficult to obtain sufficient magnetic flux by the laminated magnetic film to boost the inductance value. Low inductance density will lead to a larger device footprint or a relatively higher working frequency, neither is favorable for the power system integration and efficiency improvements. In addition, the micro-transformers proposed recently are fabricated based on the standard electroplating process [4], which is time-consuming, since photolithography and electroplating need to be repeated to grow the 3D coil structures in both the horizontal and vertical directions in sequence, and for the thick metal constructions, up to 10 hours of electroplating is always needed. Therefore, a novel structure design and batch fabrication process for the magnetic core transformer with a highly improved electromagnetic performance is proposed.

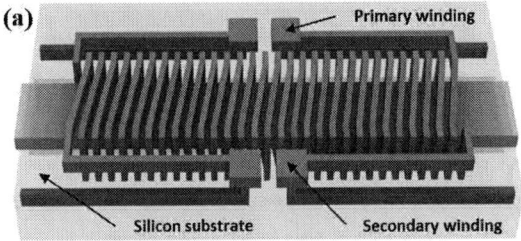

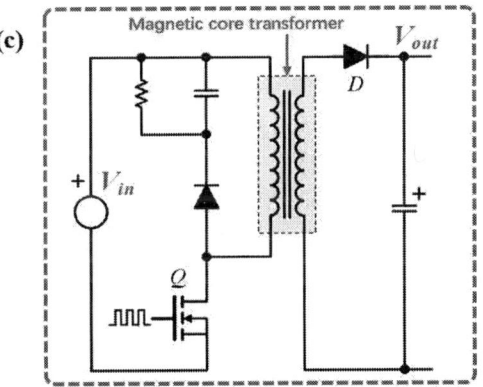

Figure 1: (a) Schematic of the proposed ferrite film inserted 3D-solenoid magnetic core transformer. (b) Cross-sectional view showing the details of the etched cavity and the magnetic core. (c) Diagram of an isolated DC-DC converter for powering GDs.

DESIGN AND FABRICATION

In this study, we develop a high-inductance-density MEMS magnetic core transformer that consists of interleaved 3D-solenoidal windings with ultra-low dc resistance and an inserted thin-film ferrite magnetic core. The interleaved solenoids have the advantages of a high coupling coefficient and low leakage inductance compared to the planar coil structures. Traditionally, it is challenging to interconnect the top and bottom metal layers through an electroplating process to form solenoidal structures. However, by using the MEMS micro-casting technique, the metallization for the interleaved 3D-solenoids can be achieved through the low-melting-point Zn-Al alloy injecting and re-filling in the pre-etched high-resistance silicon wafers in few minutes [5]. The proposed magnetic core transformer is schematically in Fig. 1, the windings can fully exploit the flux enhancement from the high-permeability ferrite to enable an increase in inductance value and coupling coefficient. Additionally, the unclosed thin-film magnetic core design effectively avoids core saturation under high current load conditions.

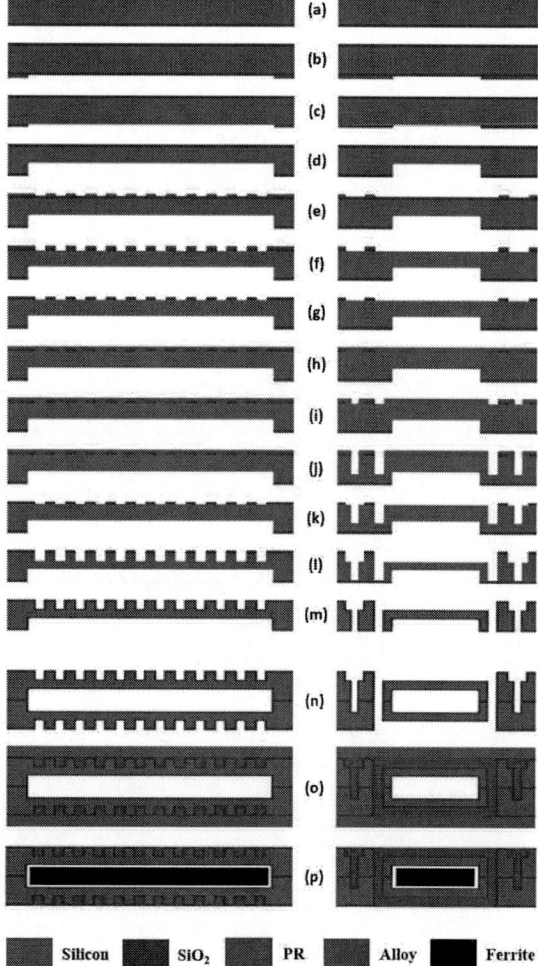

Figure 2: Fabrication steps of the 3D-solenoid magnetic core transformer. (a-d) Formation of the magnetic core inserting cavity on the backside of the structural wafer by deep-RIE. (e-l) Forming different depths of grooves and vias by multiple deep-RIE. (m) Thermal oxidation to insulate the wafer surface. (n) Aligning and stacking the structural wafers and tooling wafers to form the casting mold. (o-p) Molten alloy injecting and re-filling. And a high permeability thin-film ferrite magnetic core is then inserted into the pre-etched cavity.

The fabrication steps are shown in Fig. 2, including the structural wafer MEMS micromachining process and the Zn-Al alloy micro-casting process.

(a) 2 μm-thick SiO_2 is grown on high resistance silicon wafer by thermal oxidation and is later patterned to form etching windows.

(b-d) Formation of the magnetic core inserting cavity on the backside of the structural wafer by deep-RIE.

(e-g) The SiO_2 on the frontside is patterned to form the etching windows for the grooves and vias.

(h-l) Different depths of grooves and vias in the structural wafer are formed by multiple deep-RIE.

(m) Thermal oxidation is performed to form a 2 μm oxide layer on the surface of the silicon wafer as well as the sidewall of the grooves and vias.

(n) Aligning and stacking the structural wafers and tooling wafers to form the casting mold.

(o-p) After the molten alloy injecting and re-filling, complete thick metal 3D-solenoid coil structures are formed in the high-resistance silicon wafer. And a high permeability thin-film ferrite magnetic core is then inserted into the pre-etched cavity.

RESULTS AND DISCUSSION

The optical image of the fabricated 3D-solenoid magnetic core transformer is shown in Fig. 3 (a). With a tiny footprint, the transformer chip can be easily placed on a pen tip. Shown in Fig. 3 (b)-(c) are the X-ray perspective side-view as well as the top view of the chip. Two coupled Zn-Al alloy 3D-solenoids both have 15 turns, the metal width is 60 μm and the space between the coils is 70 μm.

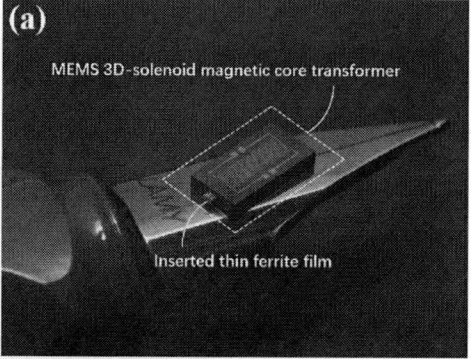

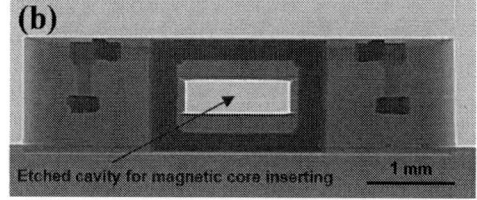

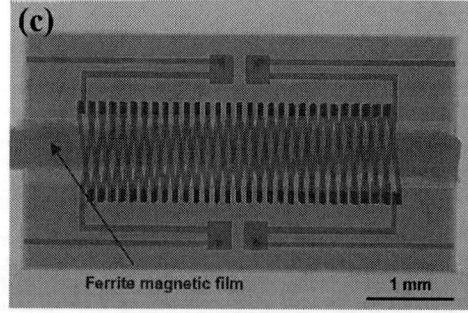

Figure 3: (a) Optical image of the fabricated 3D-solenoid magnetic core transformer placed on a pen tip. (b) X-ray perspective side-view as well as the (c) top view of the chip.

A 150 μm thick thin film ferrite magnetic core is inserted into the pre-etched cavity to enhance the magnetic flux generated by the tight-coupled 3D-solenoids. Dual Ground-Signal (GS-SG) testing pads are placed on the side of the solenoids for two-port S-parameters measurement.

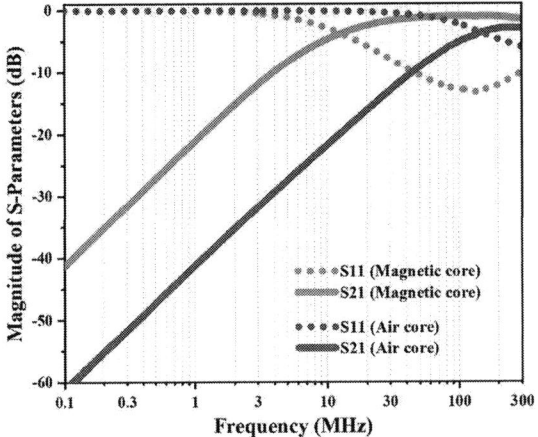

Figure 4: The comparisons of the measured S-parameters between the magnetic core transformer and the air core one.

Shown in Fig. 4 are the measured S-parameters with both the magnetic core transformer and the air core transformer. The measured primary coil inductance value, secondary coil inductance value and coupling coefficient are shown in Fig. 5. As indicated in Fig. 5, with the inserted ferrite magnetic core, an almost 7 times boost in the inductance value is achieved where the inductance is effectively increased from 43.3 nH to 334.2 nH at 10 MHz. Moreover, benefiting from the inserted magnetic core, the coupling coefficient between the primary and the secondary coil obtained a nearly 25% increment from 0.76 to 0.96 at 10 MHz, indicating that the leakage inductance is well controlled. Additionally, with a tiny footprint of 6 mm^2, the inductance density of the magnetic core transformer device reaches 55.7 nH/mm^2, which is several times higher than the other devices (see Table.1).

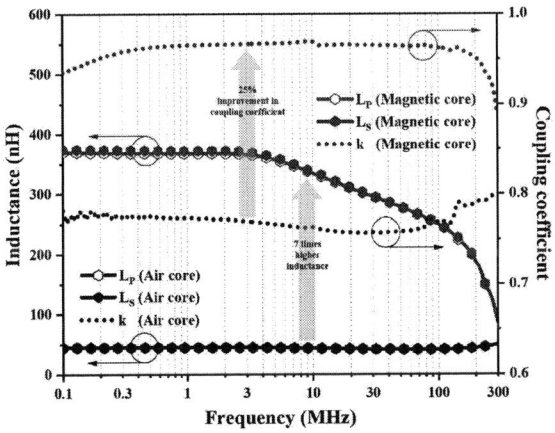

Figure 5: Measured inductance values and the coupling coefficient of the 3D-solenoidal windings in the magnetic core transformer compared with the air core one.

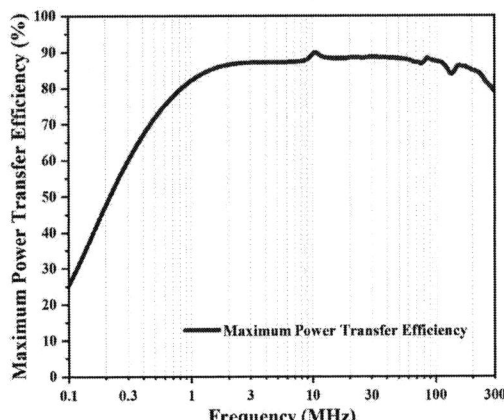

Figure 6: The extracted maximum power transfer efficiency of the 3D-solenoid magnetic core transformer from the measurement.

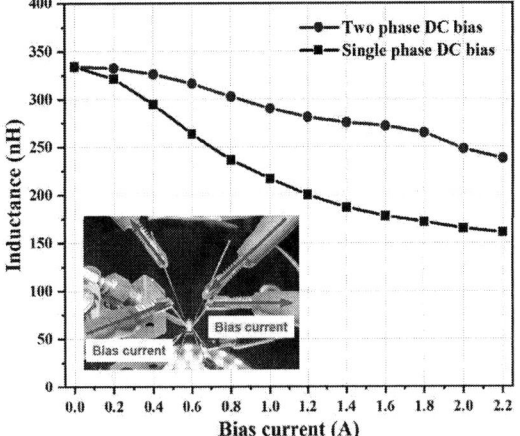

Figure 7: Measured inductance of the magnetic core transformer at 10 MHz with DC bias loading at primary and/or secondary windings.

As shown in Fig. 6, the extracted maximum power transfer efficiency of our magnetic core transformers from the measurements exceeds 89% at a lower frequency of 10 MHz. And in Fig. 7, under the two-phase DC bias loading condition, the measured saturation current exceeds 1.6 A when the inductance drops by 20% from the initial value. Table 1 compares the performance of our device with prior works [3,4,6,7]. In addition to the superior inductance density, our device exhibits both a higher coupling coefficient and power transfer efficiency even at a lower frequency of 10 MHz, which is of great advantage in miniaturized and fully integrated DC-DC converters.

Table 1. The performance comparisons of the proposed 3D-solenoid magnetic core transformer in this study with other related works.

	Chen [3]	Mundotiya [4]	Meyer [6]	Wang [7]	**This Work**
Magnetic Core	No	Yes	No	Yes	**Yes**
Frequency (MHz)	170	70	125	20	**10**

Inductance (nH)	16	47	46	80	**334.2**
Inductance Density (nH/mm²)	8	10.44	20	22.85	**55.7**
DC Resistance (Ω)	1.6	0.5	2	0.34	**0.45**
L/R (nH/Ω)	10	94	23	235	**742**
Coupling Coefficient k	0.87	0.65	0.63	N/A	**0.96**
Power Transfer Efficiency η_{max}	70%	N/A	78%	37%	**89.5%**

CONCLUSIONS

High-inductance-density MEMS 3D-solenoid transformers with inserted thin-film ferrite magnetic core have been wafer-level batch fabricated through MEMS and molten Zn-Al alloy micro-casting technique. The interleaved 3D-solenoidal windings are with ultra-low dc resistance of 0.45 Ω. With the embedded thin-film high-permeability ferrite core, the 2mm×3mm transformer chip demonstrates a superior inductance density of over 55.7 nH/mm², and a high coupling coefficient of 0.96 is achieved. The unclosed thin-film magnetic core design effectively avoids core saturation and leads to a high saturation current of over 1.6 A. At a lower frequency of 10 MHz, one of the best-reported transformer efficiency of 89.5% is measured. The developed MEMS magnetic core micro transformers are of great potential to be integrated into the future isolated DC-DC converters and other PwrSoC applications.

ACKNOWLEDGEMENTS

The authors would like the thank the financial support by National Key Research and Development Program of the Ministry of Science and Technology of China (2016YFA0200803), Innovation Team and Talents Cultivation Program of National Administration of Traditional Chinese Medicine (No: ZYYCXTD-D-202002, No: ZYYCXTD-D-202003), and the National Science Foundation of China Projects (61834007, 62074151, 61674160, 61527818).

REFERENCES

[1] I. Husain et al., "Electric Drive Technology Trends, Challenges, and Opportunities for Future Electric Vehicles," *Proceedings of the IEEE*, vol. 109, no. 6, pp. 1039-1059, 2021.

[2] S. Ma, T. Zhao, and B. Chen, "4A isolated half-bridge gate driver with 4.5V to 18V output drive voltage," in *Conference Proceedings - IEEE Applied Power Electronics Conference and Exposition - APEC*, ed, 2014, pp. 1490-1493.

[3] B. Chen, "Fully integrated isolated DC-DC converter using micro-transformers," in *Conference Proceedings - IEEE Applied Power Electronics Conference and Exposition - APEC*, ed, 2008, pp. 335-338.

[4] B. M. Mundotiya, D. Dinulovic, L. Rissing, and M. C. Wurz, "Fabrication and characterization of a Ni-Fe-W core microtransformer for high-Frequency power applications," *Sensors and Actuators, A: Physical*, Article vol. 267, pp. 42-47, 2017.

[5] J. Gu, B. Liu, H. Yang, and X. Li, "A metal micro-casting method for through-silicon Via(TSV) fabrication," in *2017 IEEE Electron Devices Technology and Manufacturing Conference (EDTM)*, 2017, pp. 211-212.

[6] C. D. Meyer, S. S. Bedair, B. C. Morgan, and D. P. Arnold, "High-Inductance-Density, Air-Core, Power Inductors, and Transformers Designed for Operation at 100–500 MHz," *IEEE Transactions on Magnetics*, vol. 46, no. 6, pp. 2236-2239, 2010.

[7] N. Wang et al., "High efficiency Si integrated micro-transformers using stacked copper windings for power conversion applications," in *2012 Twenty-Seventh Annual IEEE Applied Power Electronics Conference and Exposition (APEC)*, 2012, pp. 411-416.

CONTACT

*Xinxin Li, Tel: +86-21-62131794; xxli@mail.sim.ac.cn

MICRON-SIZED PARYLENE-IN-OIL WATER PROTECTION LAYER

Kuang-Ming Shang[1], Haixu Shen[1], Ningxuan Dai[1], David Kong[1,2], Tzung Hsiai[3], and Yu-Chong Tai[1]
[1]Department of Medical and Electrical Engineering, Caltech, Pasadena, CA, USA,
[2]Department of Electrical Engineering, Harvard University, Cambridge, MA, USA, and
[3]Department of Bioengineering, UCLA, Los Angeles, CA, USA

ABSTRACT

This work studies the parylene-in-oil composite layer as a water protection layer for biomedical applications where adhesion is critical. Compared to solid parylene, the parylene-in-oil film is thermodynamically favorable for preventing water condensation at the interface, subsequent parylene delamination, and underneath material corrosion. To study the effects of silicone oil thicknesses on the composite film, we fabricate test structures and measure both the dry and wet adhesion using a 180° peeling test and an accelerated soak-to-delamination test in saline at 67°C, respectively. Results show that when the oil thickness of the film is between 1-2 μm, the wet adhesion increases a few folds while the dry adhesion remains adequate for implantable medical device applications.

KEYWORDS

Parylene, Parylene-in-oil, Silicone oil, Adhesion, Corrosion, Medical devices, Implantable devices

INTRODUCTION

Parylene C (PAC) is a biocompatible polymer used substantially in medical implants as mechanical substrates and insulating layers. Due to its conformal nature, PAC has become a popular material for neural stimulating and recording devices requiring flexibility, lightweight, and tissue-mimicking form factors. Although PAC exhibits many good properties, its chemical inertness results in poor adhesion between PAC and underlying materials for biomedical applications that require robustness and long-term reliability [1].

Several approaches have been investigated to enhance PAC adhesion, including mechanically roughening the surfaces with plasma [2] and chemically treating the surfaces with adhesion promotors (e.g., A-174 silane) [3]. It has also been shown that a stringent cleaning of the underlying surfaces (e.g., Buffered HF dipping) and subsequent low-temperature annealing in a vacuum chamber could enhance the PAC adhesion strength by minimizing the voids at the interfaces [4-6].

However, most existing methods only address PAC adhesion under dry conditions. Under wet conditions, which are the clinical reality, long-term PAC adhesion remains an engineering challenge because PAC as a polymer is subject to water vapor permeation, which breaks the adhesion and causes PAC delamination because water condensation occurs at the interfaces [7]. One major drawback of PAC delamination is that the underlying materials are no longer being insulated, meaning that if there are metallic materials in the implant, they are subject to corrosion and malfunction. On the other hand, instead of using a polymer as a coating film to prevent water condensation and subsequent corrosion, a widely applied method for anti-corrosion in the metal industry is to use lubricant oil as a temporary protective layer. Although the oil needs to be reapplied frequently, it can act as a water vapor boundary to prolong the lifetime of metal.

We combine the strategies above and propose a new parylene-in-oil composite approach to prevent interface water condensation and later delamination for medical device applications. Within the parylene-in-oil layer, the hydrophobic and biocompatible silicone oil prevents water condensation. The stalactite-like parylene structures in and through the oil secure the oil from leaching out and provide adhesion to the substrate.

FABRICATION PROCESS OF THE PARYLENE-IN-OIL COMPOSITE LAYER

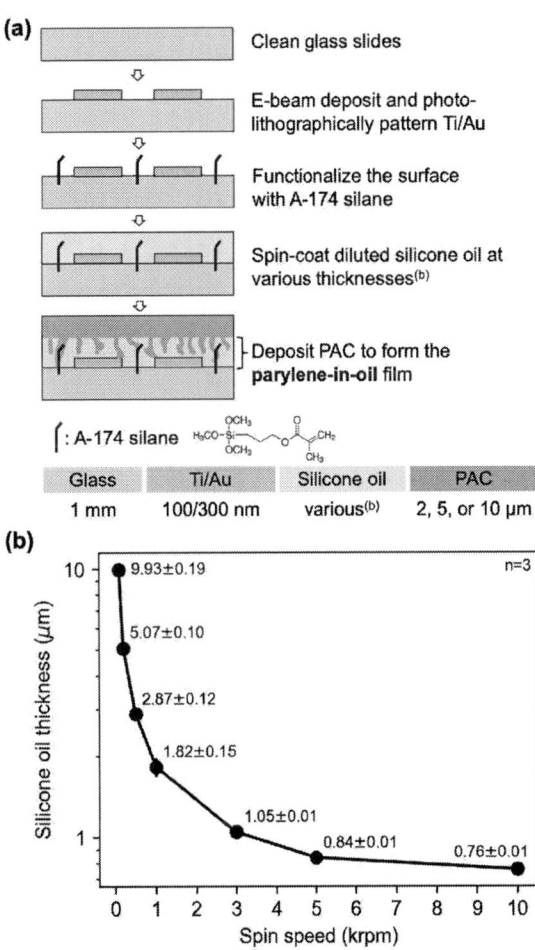

Figure 1: (a) Schematic of the fabrication process of parylene-in-oil film structure on a glass-based circuit board. (b) The characteristics of silicone oil thickness versus spin speed with heptane-diluted 100k cSt silicone oil (10% (w/v)).

We fabricate the parylene-in-oil composite layer on a glass-based circuit board test structure using standard microfabrication methods (Fig. 1a). First, a 100-nm thick titanium and a 300-nm thick gold films are e-beam evaporated onto a cleaned glass substrate, followed by a photolithographic method to create desired patterns. Then, the surfaces are treated with an oxygen plasma for descumming (200 mTorr, 50W, 1 min) before the A-174 surface silanization to promote the adhesion between PAC and glass. The surface silanization is achieved by submerging the substrates into a 0.5% (v/v) A-174 solution using a 1:1 DI water/IPA solvent. In order to achieve a thin oil coating, 100k cSt silicone oil is heptane-diluted to 10% (w/v) and spin-coated onto the surface. A gentle bake is followed to remove the heptane residual. High viscosity silicone oil is preferred because of its mechanical stability and biocompatibility. Oil thickness is measured by the weight difference before and after oil spin-coating using an analytical balance (XSR105DU, Mettler-Toledo, Columbus, USA) and calculated based on the surface area and 100k cSt silicone oil density (0.971 g/mL). By adjusting the spin speed, various thicknesses as small as 760 nm are achieved (Fig. 1b). Finally, a 2-10 μm PAC layer is deposited onto the silicone oil to form the parylene-in-oil composite layer for the following peeling or soak-to-delamination tests. Notice that only the soak-to-delamination test required metallization.

CROSS-SECTIONAL MORPHOLOGIES OF THE PARYLENE-IN-OIL COMPOSITE LAYER

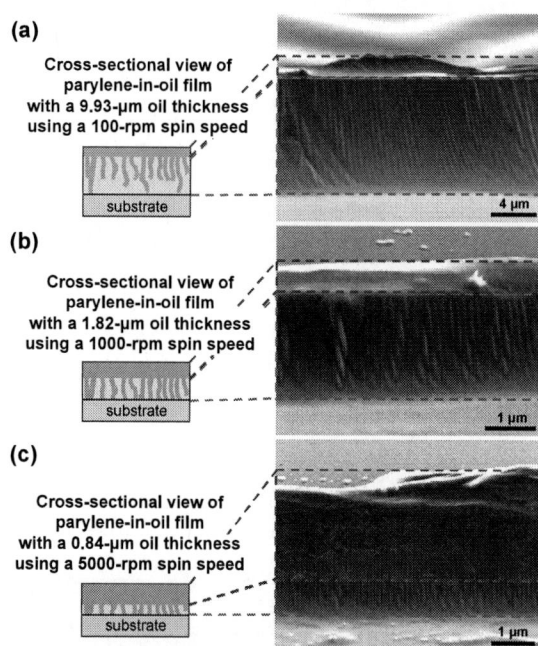

Figure 2: SEM images of parylene-in-oil film with different silicone oil spin rates and thicknesses: (a) 100 rpm / 9.93 μm, (b), 1000 rpm / 1.82 μm, and (c) 5000 rpm / 0.84 μm. In the cross-sectional views, blue-dotted boxes and red-dotted boxes indicate solid parylene and parylene-in-oil mixed structures, respectively.

The silicone oil thickness affects the cross-sectional morphologies of parylene-in-oil composites. To assess the micron-sized structures, we prepare the films with different oil thicknesses of 9.93, 1.82, or 0.84 μm using a spin speed of 100, 1000, or 5000 rpm, respectively. After a 2-μm PAC deposition, films are diced in liquid nitrogen to preserve the structural integrity and imaged under SEM (Fig. 2). All three cases comprise a solid PAC layer on top and a parylene-in-oil mixed layer on the bottom. The top PAC layer automatically provides excellent mechanical sealing for the underneath stalactite-like parylene-in-oil structures, which encapsulates the silicone oil for water protection.

The bottom parylene-in-oil layers exhibit slightly different morphologies as the oil thickness varies. The structures are shown in the red-dotted area in Fig. 2, where the brighter part indicates the PAC, and the darker part indicates the silicone oil. When the silicone oil is thin, the volume ratio of oil and PAC is relatively uniform across the composite film (Fig. 2b, 2c). However, when the silicone oil becomes thicker, fewer PAC monomers are able to reach to, polymerize at, and form bonding with the bottom substrate. This phenomenon is clearly observed in Fig. 2a, where the silicone oil is the primary composition near the underlying substrate. Although the ability to prevent water permeation is superior when silicone oil is thicker, less PAC bonding to the substrate will weaken the adhesion strength under the thicker oil condition. The dichotomy motivates us to investigate further the optimal oil thickness of the composite parylene-in-oil layer for both adhesion strength and water protection.

DRY ADHESION STRENGTH OF THE PARYLENE-IN-OIL COMPOSITE LAYER

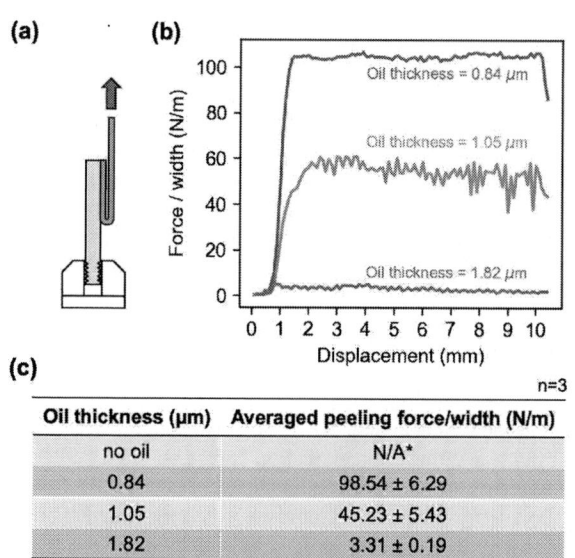

Oil thickness (μm)	Averaged peeling force/width (N/m)
no oil	N/A*
0.84	98.54 ± 6.29
1.05	45.23 ± 5.43
1.82	3.31 ± 0.19

* Cannot peel off the film without breaking the film

Figure 3: (a) Schematic of 180° peeling test to characterize the adhesion of parylene-in-oil films on glass slides. Films with a width of 2 inches are used. (b) Representative peeling force profiles of 180° peeling tests of parylene-in-oil films with different oil thicknesses. (c) Summary of the peeling test results.

To study the effects of silicone oil thickness on dry adhesion strength of the parylene-in-oil layer, different films are prepared with the oil thickness of 1.82, 1.05, or 0.84 μm using a spin speed of 1000, 3000, or 5000 rpm, respectively. A film without silicone oil is also prepared as a control. After the oil is spun on the A-174-treated glass substrate, a 10-μm thick PAC is deposited onto the oil to form the composite layer. Subsequently, the films with a 2-inch width are peeled at the 180° configuration using a motorized test stand and a force gauge (ESM303 and M5-10, MARK-10, Copiague, USA) at the speed of 5 mm/min (Fig. 3a).

Under different conditions of oil thickness, both the representative peeling force profiles in Fig. 3b and the averaged peeling forces in Fig. 3c demonstrate that the measured force decreases as the oil thickness increases, suggesting that fewer parylene can bond to the glass substrate when the oil film is thick.

WET SOAK-TO-DELAMINATE TEST OF THE PARYLENE-IN-OIL COMPOSITE LAYER

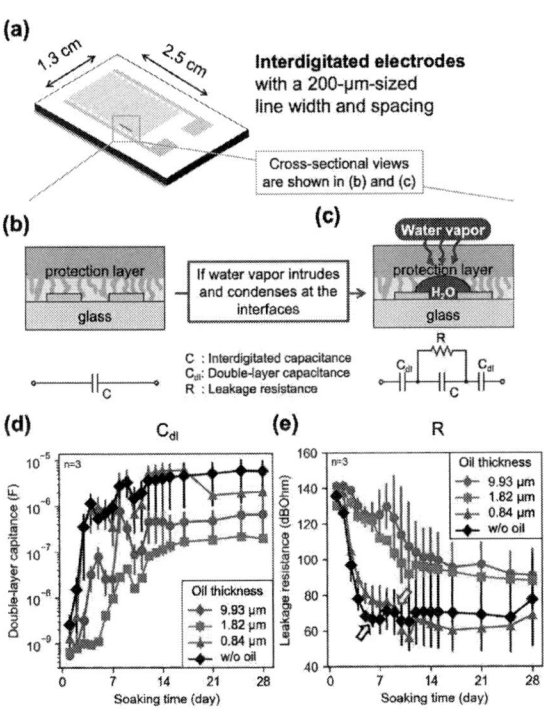

Figure 4: (a) Schematic of a pair of microfabricated interdigitated gold electrodes on a glass slide coated with parylene-in-oil film for a 67° C accelerated soak-to-delamination test in saline (0.9 % (w/v) NaCl) to characterize the wet adhesion of different insulating layers using EIS. Schematic and equivalent circuit for interdigitated electrodes and protection layer with (b) an excellent and (c) a poor wet adhesion. Poor wet adhesion will lead to water condensation in between the protection layer and electrodes. (d) Double-layer capacitance (C_{dl}) and (e) leakage resistance (R) are extracted from the EIS using an equivalent circuit in (c). Arrows indicate the device failure time.

Under wet environments, polymer coatings, including PAC, are subject to water permeation and film delamination. To quantify the proposed parylene-in-oil layer's water protection effect, we design the interdigitated gold electrodes as test structures (Fig. 4a) with parylene-in-oil layer and with PAC layer on top to compare. Both test structures are submerged into saline at an elevated temperature (67 °C) to mimic the failure mode in bodily fluids at an accelerated rate. Electrochemical impedance spectroscopy (EIS) between two electrodes is conducted to monitor the film integrity for four weeks. If the film remains intact, only the capacitive component will appear for the EIS between two electrodes (Fig. 4b). However, if water vapor permeates the film and condenses at the interfaces, the electrical impedance will change rapidly because of the additional metal-liquid interfaces and the conducting path that bridges two electrodes (Fig. 4c). As a result, the EIS between two electrodes will exhibit double-layer capacitances (Fig. 4d) and a leakage resistance (Fig. 4e) as extra components. Here, we define the failure time when the leakage resistance drops to a saturated low value. Results show that the resistance quickly drops with a failure time of 5 days when there is no oil, and it drops at a slower rate with a failure time of 9 days when there is a 0.84-μm thick silicone oil. For oil thicker than 1.82 μm, the resistances had not dropped to the baseline within a four-week monitoring time, demonstrating that the parylene-in-oil composite layer prevents water condensation and prolongs the device lifetime by at least a few folds.

TRADE-OFF BETWEEN DRY AND WET CONDITIONS OF THE PARYLENE-IN-OIL COMPOSITE LAYER

Fig. 5 illustrates the trade-off between the adhesion force and water condensation rate using the parylene-in-oil composite film. For the failure time at a physiological temperature of 37°C, we extrapolate the failure time at an 8-fold difference from results at 67°C in Fig. 4e using the Arrhenius equation:

$$FT \propto e^{\frac{E_a}{kT}} \qquad (1)$$

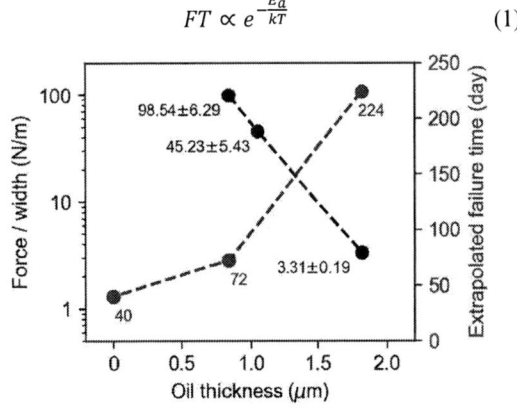

Figure 5: A summary of dry and wet adhesion strength of parylene-in-oil film w.r.t different oil thickness. Peeling force per width on the left y-axis is captured from experimental results shown in Fig. 3c. Failure time on the right y-axis is extrapolated from Fig. 4e using the Arrhenius equation to a physiological body temperature (i.e., 8-fold).

where FT is the failure time, E_a is the failure mode activation energy, T is the absolute temperature, and k is the Boltzmann constant. In summary, a thinner oil is beneficial for strong adhesion strength, while a thicker oil is favorable for high water resistance. Depending on applications, one can choose the optimal oil thickness for desirable properties. Nevertheless, an oil thickness of 1-2 µm provides both proper adhesion and water protection.

DISCUSSION

This work aims to propose a novel method of protecting implantable medical devices using only soft materials. Although hard materials like metal or glass are currently used as encapsulating cases for medical device protection, it has been clinically shown that hard materials exacerbate immune responses and trigger severe foreign body reactions [8]. Therefore, this type of implantable device could gradually lose its function due to the fibrosis, which encapsulate interfaces between the device and the target tissue.

Instead of using hard materials to block the water vapor permeation as a protection mechanism, we aim to prevent water nucleation and condensation at the device interface, which is the critical step for corrosion to occur. Here, we utilize the hydrophobic nature of biocompatible silicone oil to form a parylene-in-oil protective layer, in which the water vapor is thermodynamically unfavorable to form droplets at the interfaces because silicone oil is presented.

The PAC deposition process onto a silicone oil provides an automatic seal for the oil. Although a similar process has been previously shown [9, 10], we propose to further thin down the oil to a micron-sized thickness so that PAC is allowed to penetrate the thin oil to form bonding with the underlying substrates. As such, we have successfully demonstrated the water protection effect of this parylene-in-oil layer on a hard-to-protect hydrophilic glass substrate, which is usually prone to water condensation. Although there is a trade-off between water protection and adhesion, our results suggest that using A-174 as an adhesion promotor still provides adequate mechanical strength. Going forward, to be able to widely apply the proposed method to medical devices that possess various kinds of substrates, including noble metals or PAC itself, further investigation is needed for the optimal combination of oil viscosity, oil thickness, types of adhesion promotor, PAC deposition rate, and post-process thermal treatments.

CONCLUSIONS

We propose and fabricate a parylene-in-oil composite layer as a method to prevent water condensation and protect implantable medical devices. We then characterize the thickness effects of silicone oil in parylene-in-oil layer on the cross-sectional morphologies, the dry adhesion strength, and the wet soak-to-delamination failure time. We conclude that using the parylene-in-oil layer with an oil thickness between 1-2 µm is most suitable for medical implants that require anti-corrosion and proper adhesion.

ACKNOWLEDGEMENTS

This project is funded by NIH R01HL149808. The authors would also like to thank Mr. Vince Wu and Dr. Chi Ma on the preparation and imaging techniques of SEM samples. The authors would then like to thank Mr. Suhash Aravindan and Mr. Trevor Roper for the insightful discussion regarding this work.

REFERENCES

[1] J. Ordonez, M. Schuettler, C. Boehler, T. Boretius, and T. Stieglitz, "Thin films and microelectrode arrays for neuroprosthetics," *MRS Bulletin*, vol. 37, no. 6, pp. 590–598, Jun. 2012.

[2] C. D. Lee and E. Meng, "Mechanical Properties of Thin-Film Parylene–Metal–Parylene Devices," *Frontiers in Mechanical Engineering*, vol. 1, 2015.

[3] R. Huang and Y. C. Tai, "Parylene to silicon adhesion enhancement," *TRANSDUCERS 2009 - 2009 International Solid-State Sensors, Actuators and Microsystems Conference*, Jun. 2009, pp. 1027–1030.

[4] E. M. Schmidt, J. S. Mcintosh, and M. J. Bak, "Long-term implants of Parylene-C coated microelectrodes," *Med. Biol. Eng. Comput.*, vol. 26, no. 1, pp. 96–101, Jan. 1988.

[5] E. M. Davis, N. M. Benetatos, W. F. Regnault, K. I. Winey, and Y. A. Elabd, "The influence of thermal history on structure and water transport in Parylene C coatings," *Polymer*, vol. 52, no. 23, pp. 5378–5386, Oct. 2011.

[6] W. Li, D. C. Rodger, E. Meng, J. D. Weiland, M. S. Humayun, and Y.-C. Tai, "Wafer-Level Parylene Packaging With Integrated RF Electronics for Wireless Retinal Prostheses," *Journal of Microelectromechanical Systems*, vol. 19, no. 4, pp. 735–742, Aug. 2010.

[7] J. Ortigoza-Diaz, K. Scholten, and E. Meng, "Characterization and Modification of Adhesion in Dry and Wet Environments in Thin-Film Parylene Systems," *Journal of Microelectromechanical Systems*, vol. 27, no. 5, pp. 874–885, Oct. 2018.

[8] A. Carnicer-Lombarte, S.-T. Chen, G. G. Malliaras, and D. G. Barone, "Foreign Body Reaction to Implanted Biomaterials and Its Impact in Nerve Neuroprosthetics," *Frontiers in Bioengineering and Biotechnology*, vol. 9, 2021.

[9] N. Binh-Khiem, K. Matsumoto, and I. Shimoyama, "Porous Parylene and effects of liquid on Parylene films deposited on liquid," *2011 IEEE 24th International Conference on Micro Electro Mechanical Systems*, Jan. 2011, pp. 111–114.

[10] A. Shapero and Y.-C. Tai, "Parylene-oil-encapsulated low-drift implantable pressure sensors," *2018 IEEE Micro Electro Mechanical Systems (MEMS)*, Jan. 2018, pp. 47–50.

CONTACT

*K.M. Shang; kshang@caltech.edu

A PIPETTE TIP INTEGRATED WITH A CAPACITIVE MICROSENSOR FABRICATED BY COMBINED 3D PRINTING AND MEMS PROCESS FOR CELL DETECTION AND TRANSPORTATION

Satoshi Amaya[1], Hirotaka Sugiura[1], Bilal Turan[1], Shingo Kaneko[1], and Fumihito Arai[1]
[1]The University of Tokyo, JAPAN

ABSTRACT

This paper reports on the detection and manipulation of cells using a pipette tip with a capacitive microsensor fabricated by combined 3D printing and MEMS processes. This pipette tip with microsensor was fabricated by 3D printing directly onto a MEMS device. First, a pair of comb electrode capacitive sensor was fabricated using the traditional MEMS process. Then, the pipette tip structure was 3D-printed directly on the sensor substrate by projection micro stereolithography. This process is possible to form nm- to cm-scale structures integrally with high precision and three-dimensionally. The fabricated device successfully detected cell suction and ejection. In addition, cell transfer was performed to show the possibility of applying the device to cell manipulation by a robot.

KEYWORDS

Capacitive microsensor, 3D printing, Cell Detection

INTRODUCTION

In order to reduce the workload of scientists in their daily routines, laboratory automation has been developed to automate experimental operations. In addition to simple experimental operations, technologies have been developed to automatically optimize complicated tasks and experimental conditions using AI robots that combine artificial intelligence (AI) and robot technology. These technologies are expected not only to reduce the workload of scientists but also to reduce human error and accelerate scientific research. In the biomedical field, cell manipulation requires many routine tasks, and automated robotic systems for cell manipulation are being developed to reduce the workload of operators [1, 2]. A commercial pipette tip is widely employed for automated cell manipulation systems due to its cost and compatibility with current cell manipulation procedures. However, stable operation of automated systems needs sensors and imaging systems to detect errors during cell manipulation. For this reason, a micropipette with an integrated sensor fabricated with MEMS technology was proposed as a tool for a fully automated cell-picking system [3].

On the other hand, in MEMS fabrication technology, the application of 3D printing has been attracting attention due to the improvement of modeling accuracy [4, 5]. 3D printing can fabricate complex three-dimensional structures that are difficult to achieve with machining and MEMS process and is used in a variety of research fields. Among them, 3D printers using photo-curable resins have been widely adopted. For example, the two-photon polymerization method enables nanometer-scale processing and has been applied to the fabrication of comb-type actuators, microrobots, and micro three-dimensional structures combined with electroforming technology [6-8]. Furthermore, processing by projection micro stereolithography can fabricate micro- to centimeter-sized structures, and there have been proposals for the direct fabrication of microstructures in fluidic devices [9, 10]. However, 3D printing has not been combined as a part of the MEMS process, and there are no reports of its application as a tool for robotic systems.

Therefore, we propose a versatile and applicable process including a 3D printing process in the general MEMS process. And we report on the fabrication of a pipette tip integrated with a capacitive microsensor for Xenopus oocyte suction and detection. Then, a cell transportation demonstration using the fabricated device will be described.

CONCEPT

In this study, we focused on an experiment using Xenopus oocytes, which is one of the experiments in the field of drug discovery that require a high skill level in cell manipulation. Cells are supposed to be transported one by one. Figure 1 shows a schematic diagram of the pipette tip integrated with a capacitive microsensor. The device consists of a substrate with a capacitance sensor at the tip and a 3D-printed pipette tip (Figure 1(a)). The device sucks

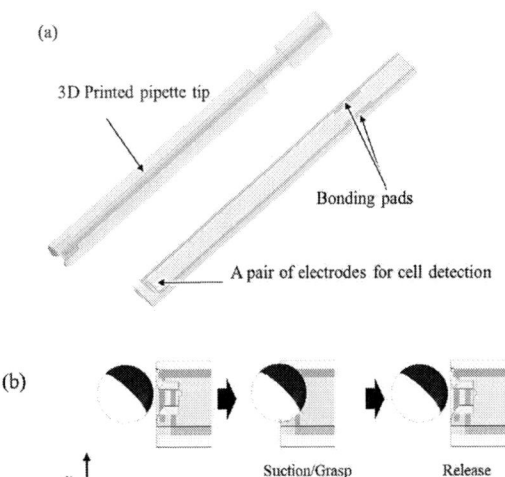

Figure 1: Illustration of the structure of pipette tip integrated with a capacitive microsensor and detection principle. (a) Disassembled image of the device, (b) illustration of the ideal capacitance signal obtained from the sensor.

and ejects a cell through a pumping system. The device is designed to detect the suction/grasp state of a cell from the change in capacitance (Figure 1(b)). When a cell is sucked and grasped, the cell is placed on the electrodes. At that time, the capacitance changes. After the cell manipulation experiment, the capacitance returns to its original value by ejecting the cell.

DEVICE DESIGN

Xenopus oocytes with a diameter of about 1 mm have been used as a robust model for heterologous expression [11, 12]. We designed a device that can detect/transport cells one by one, in accordance with a typical heterologous expression experimental system. Therefore, the sensor electrodes and pipette tip shape were designed to ensure that when a cell is sucked, it is placed on the sensor to be detected and transported. A pair of coplanar interdigit electrode capacitors, which are suitable for microfabrication, is employed as a capacitive microsensor. According to Figure 2(a), the electric flux lines generated by the electrode pair have a penetration depth T, which is the maximum height measured perpendicular to the electrode surface. Here, consider a microchannel of height h filled with fluid; if T < h, the capacitance is not affected by the top wall of the microchannel. Therefore, it should be designed to be lower than the top wall of the pipette tip (Figure 2(a)). The effective width of the electrode to be activated is defined by Equation 1 [13, 14]

$$W_{eff} = a\left(\sqrt{1+\left(\frac{h}{a}\right)^2} - 1\right) \quad (1)$$

where w_{eff} is the effective width of the electrode and a is the gap between the electrodes. For these reasons, the electrode width and gap of the capacitance sensor were set to 500 μm and 10 μm, respectively (Figure 2(c)), considering the device dimension and fabrication process. The external dimension of the device is shown in Figure 2(b). For trapping and transporting each Xenopus oocyte, a cavity the same size as the cell is designed at the tip of the device. A channel with an inner diameter of 0.5 mm is designed to prevent a cell from passing through the cavity.

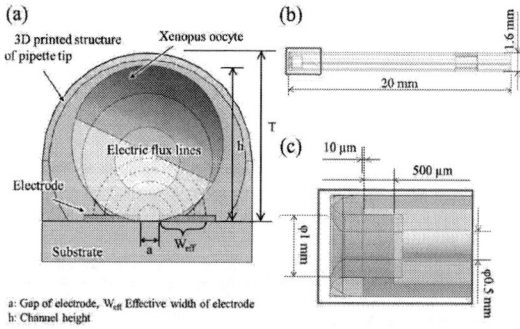

Figure 2: Structure of a pipette tip integrated with a capacitive microsensor. (a) Schematic image of the front view of a pipette tip and electric flux line, (b) external image of pipette tip structure, (c) enlarged image of device tip.

FABRICATION

The fabrication process of the pipette tip integrated with a capacitive microsensor is shown in Figure 3. And the fabricated device is shown in Figure 4. This process was constructed by including a 3D printing process in the traditional MEMS process. First, the glass substrate was cut by a dicing machine to align the 3D printing process with the outline of the substrate. Next, a pair of electrodes were fabricated as capacitance sensors on the glass substrate using a traditional MEMS process. The electrode

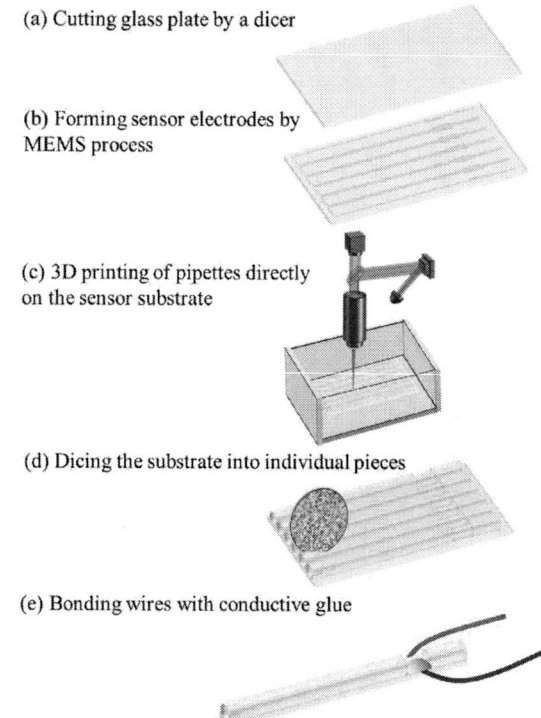

Figure 3: Fabrication process which combines 3D printing and MEMS process.

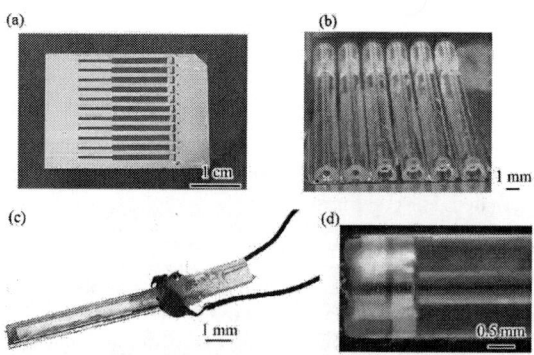

Figure 4: Fabricated pipette tip with integrated with a capacitive microsensor. (a) A photograph of the sensor electrodes made by MEMS process, (b) a photograph of the pipette tip structures, (c) a photograph of the fabricated device with wiring, (d) a microscopic image of the electrode aligned at the pipette tip.

materials were Cr 10 nm thick and Au 100 nm thick (Figure 4(a)). The glass substrate with electrodes was aligned and placed on a 3D printer (S130, Boston Micro Fabrication), and the pipette tip shape was formed directly on the substrate (Figure 4(b)). After that, the substrate on which the pipette tips were formed was diced into pieces. Finally, wires were bonded to the bonding pad with conductive adhesive, and a silicone rubber tube was attached for connecting to the pump system (Figure 4(c)). A microscopic image of the tip of the fabricated device showed that the capacitance electrode was accurately aligned with the device structure (Figure 4(d)). This result shows that 3D printing can be combined as part of the MEMS process. This fabrication process employs 3D printing, which makes it possible to fabricate many types of devices simultaneously by changing the shape of each device to be fabricated.

EXPERIMENTS AND RESULTS

A schematic diagram of the experimental system is shown in Figure 5. The fabricated device was connected to a piezo pump via a fixture jig and attached to a micromanipulator (Quick Pro, Micro Support Co., Ltd.) for transfer operations. The cells were placed in a buffer solution to soak a cell loading device. The cell loading device is a fluidic device that provides a continuous supply of cells and is connected via tubing to a cell storage tank. An LCR meter (ZM2372, NF Co., Ltd.) was used to measure capacitance for cell detection. The measurement system is a device designed for voltage clamp experiments.

Evaluation of cell detection

As a cell detection experiment, the capacitance change was measured when a cell was sucked and ejected using a fabricated pipette tip with a microsensor. The cell could be sucked and ejected after being trapped in the cavity of the pipette tip by driving the piezo pump (Figure. 6(a)). Figure 6(b) shows one of the results of capacitance change during suction and ejection of the cell measured by the fabricated device. The red line in the measurement results represents a moving average. The capacitance increased around 8 nF before and after the cell was sucked out using the device. The capacitance decreased around 5 nF by ejecting the cell. This difference in capacitance change in suction and ejection seems to have occurred by the remaining adhesive material of the cell surface on the electrodes.

Cell transportation experiment

Cell transportation operation was demonstrated using the fabricated pipette tip with a microsensor and cell detection/transportation system. In this experiment, the device tip was operated manually while checking the top view by microscope and oblique image by another camera. The experimental procedure involved first sucking and grasping the cell in the buffer solution on the cell loading device. Then, the cell was picked up from the loading device in this state. Finally, the cell was transported to the measurement system and put in a predetermined position (Figure 7). It was shown that the fabricated device and experimental system enabled the transportation of cell without breakage.

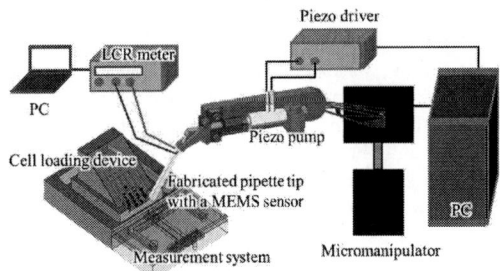

Figure 5: Schematic diagram of cell detection and transportation experiment system. This experimental system is designed for continuous measurement of cellular characteristics.

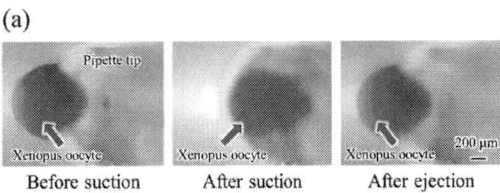

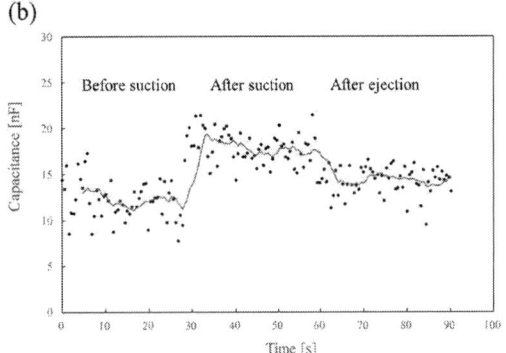

Figure 6: Result of suction in a buffer solution of oocyte using the fabricated device. (a) Photos of device tip in suction and ejection of a cell, (b) The capacitance measurement result of the state of before and after sucking cell, the red line is moving average.

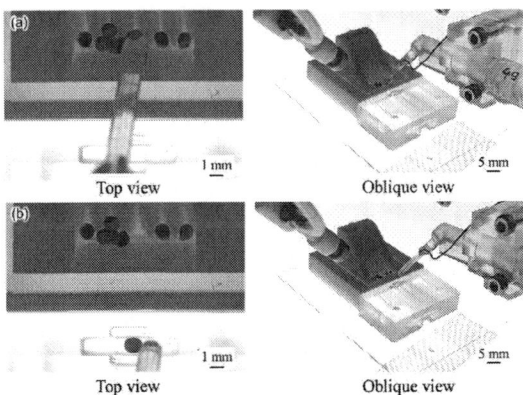

Figure 7: Photos of cell transportation operation experiment. (a) Cell pickup on cell loading device, (b) After cell transfer and put on to measurement system.

CONCLUSION

We have constructed a fabrication process that combined 3D printing and a traditional MEMS process. The constructed process is expected to enable high-precision and three-dimensional integrated formation of nm- to cm-scale structures and is expected to have a variety of applications. In addition, we have fabricated a pipette tip with a capacitive microsensor by the constructed process. The fabricated device could detect the cell suction state. Furthermore, a cell transportation experiment using the fabricated device was demonstrated to show its feasibility for robotic cell manipulation.

ACKNOWLEDGEMENTS

This work was supported by JST Moonshot R&D – MILLENNIA Program Grant Number JPMJMS2033-08. And a part of this work was supported by "Advanced Research Infrastructure for Materials and Nanotechnology in Japan (ARIM)" of the Ministry of Education, Culture, Sports, Science and Technology (MEXT). Proposal Number JPMXP1222UT1033.

REFERENCES

[1] Ochiai K, Motozawa N, Terada M, et al., "A Variable Scheduling Maintenance Culture Platform for Mammalian Cells", SLAS TECHNOLOGY: Translating Life Sciences Innovation. 2021;26(2):209-217.

[2] Burger, B., Maffettone, P.M., Gusev, V.V. et al., "A mobile robotic chemist", Nature 583, 237–241 (2020).

[3] K. Tashiro, T. Masuda and F. Arai, "Single-cell picking by integrated micropipette having sensor for passage detection of cell," 2014 International Symposium on Micro-NanoMechatronics and Human Science (MHS), 2014.

[4] Chee Meng Benjamin Ho, Sum Huan Ng, King Ho Holden Li, Yong-Jin Yoon, "3D printed microfluidics for biological applications", Lab Chip, 2015,15, 3627-3637.

[5] Takahiro Tamura, Takaaki Suzuki, "Seamless fabrication technique for micro to millimeter structures by combining 3D printing and photolithography", 2019 Jpn. J. Appl. Phys. 58 SDDL10.

[6] https://www.nanoscribe.de/en/media-press/press-releases/chip-3d-microprinting-photonic-and-mems-systems/

[7] Toshiro Yamanaka, Fumihito Arai, "Self-Propelled Swimming Microrobot Using Electroosmotic Propulsion and Biofuel Cell", IEEE Robotics and Automation Letters, VOL. 3, NO. 3, JULY 2018, pp.1787-1792,

[8] Williams, G., Hunt, M., Boehm, B. et al. Two-photon lithography for 3D magnetic nanostructure fabrication. Nano Res. 11, 845–854 (2018).

[9] Ogishi, Kazuto and Osaki, Toshihisa and Morimoto, Yuya and Takeuchi, Shoji,"3D printed microfluidic devices for lipid bilayer recordings", Lab Chip, 2022,22, 890-898.

[10] Liu, J., Li, Z., Ding, Y., Chen, A., Liang, B., Yang, J., Cheng, J.-C., Christensen, J., Twisting Linear to Orbital Angular Momentum in an Ultrasonic Motor. Adv. Mater. 2022, 34, 2201575.

[11] Dumont, J.N., "Oogenesis in Xenopus laevis (Daudin). I. Stages of oocyte development in laboratory maintained animals". J. Morphol., 136: 153-179. 1972.

[12] Barbara Sottocornola, Sabina Visconti, Sara Orsi, Sabrina Gazzarrini, Sonia Giacometti, Claudio Olivari, Lorenzo Camoni, Patrizia Aducci, Mauro Marra, Alessandra Abenavoli, Gerhard Thiel, Anna Moroni,"The Potassium Channel KAT1 Is Activated by Plant and Animal 14-3-3 Proteins*", Journal of Biological Chemistry, Volume 281, Issue 47, 2006, Pages 35735-35741.

[13] J. Z. Chen, A. a Darhuber, S. M. Troian, and S. Wagner, "Capacitive sensing of droplets for microfluidic devices based on thermocapillary actuation.," Lab Chip, 2004, vol. 4, no. 5, pp. 473–80.

[14] Caglar Elbuken, Tomasz Glawdel, Danny Chan, Carolyn L. Ren,"Detection of microdroplet size and speed using capacitive sensors, Sensors and Actuators A: Physical", Volume 171, Issue 2, 2011, Pages 55-62.

CONTACT

*Satoshi Amaya, amaya@mesl.t.u-tokyo.ac.jp

FOLDABLE POLYMER STENT INTEGRATED WITH WIRELESS PRESSURE SENSOR FOR BLOOD PRESSURE MONITORING

Nomin-Erdene Oyunbaatar[1] and Dong-Weon Lee[1,2,3]*

[1]Department of Mechanical Engineering, Chonnam National University, Gwangju, SOUTH KOREA
[2]Center for Next-generation Sensor Research and development, Chonnam National University, Gwangju, SOUTH KOREA
[3]Advanced Medical Device Research Center for Cardiovascular Disease, Chonnam National University, Gwangju, SOUTH KOREA

ABSTRACT

Cardiovascular disease (CVD) is the leading cause of death in worldwide. One of the main diseases is atherosclerosis, in which plaques build up in the walls of the arteries, causing them to narrow gradually. This narrowing effect of the arteries blocks blood flow and interrupts oxygen supply. One of the most common treatments utilized in the narrowed vessel is the insertion of a stent into the blood vessel. However, despite various stents decreasing the risk of recurrence, in-stent restenosis keeps the possibility of late thrombosis and subsequent heart attack. Therefore, to reveal restenosis at an early stage, non-invasive and accessible methods are required. The proposed new fabrication method for integrating wireless pressure sensor and foldable polymer stent is based on MEMS processes. Photosensitive polymer SU-8 was utilized as a body of mesh stent structure and LC wireless pressure sensor. Wireless pressure sensor connection and working principle was evaluated based on inductance and capacitance coupling circuit.

KEYWORDS

Polymer stent, wireless sensor, LC type pressure sensor

INTRODUCTION

Cardiovascular disease (CVD) is one of the leading factors of fatalities in the world. In 2019, over 850,000 deaths were attributed to CVD in the United States and 17.9 million worldwide. Heart attacks and strokes are the cause of nearly 85% of all deaths worldwide [1]. Atherosclerosis is a major CVD causing heart attacks by the buildup of plaques (fats, cholesterol, calcium etc.) in and on the arterial walls that causes them to thicken. The continuing thickening or narrowing of the arteries can impede the blood supply to and from the heart, eventually leading to a heart attack. In such a situation, an angioplasty followed by stenting is a common treatment option to open the blood vessels and resume the blood flow. A stent is a small and expandable mesh tube placed inside the constricted artery to keep it open for a long term. The stent usually made from metal has been in use over the last 20 years [3-6]. A limitation of these stents is that they cannot prevent the further accumulation of plaque, that can cause inflammation and the resultant re-narrowing of the stent portion, known as in-stent restenosis. Over time, this situation can cause an abrupt thrombotic occlusion of the artery (stent thrombosis) leading to a myocardial infarction or death. Drug eluting stents (DES) are more commonly used now, as the drug coated on the stent is released gradually [7-10]. DES can prevent the occurrence of a thrombus and lower the plaque build-up. However, it cannot fully stop this continues growing tissue, resulting in in-stent restenosis, and the patients may suffer from late thrombosis in the long run leading to a heart attack. Nearly 50% of the stented patient can suffer from this restenosis [11].

To prevent restenosis at the early stage, it is desirable to monitor its onset by the use of "intelligent/smart" stents that can sense the pressure change and transmit the information wirelessly in real time. Tiny pressure sensors made by the micromachining process can be integrated with the metallic stent and implanted into the artery, where the stent acts as an antenna for wireless communication. This integration was first attempted by Takahata et al. with a silicon based capacitive pressure sensor and dual inductor stentenna [12]. Park et al. then demonstrated the use of biocompatible polymer stents with the incorporation of miniature MEMS based wireless pressure sensors and data sensing [13]. The proposed approaches offered increasing the inductance value and quality factors as well as promoting the sensing distance. The use of MEMS-based pressure sensor in this application was also introduced by Chen et al. with its integration with an antenna stent, thereby enhanced the electromechanical performance in-vivo [14]. Their team further developed a new chip made of pure steel (316LSS) with an encompassing pressure sensor that was connected by laser micro-welding [15]. In addition, the use of 3D printing technology is advanced way of fabricating biodegradable stent [16], however the combination of the pressure sensor and stent still questionable, and the integration process was proceeding with the help of the various biocompatible glue. Till now, the integration of the sensor with the stent was achieved by mechanical assembly like bonding, gluing and conventional soldering, which is low-throughput and time-consuming process. More reliable integration techniques were needed to mass-produce a combination of stents and pressure sensors.

In this work, we developed a new process method to fabricate integrated stents and pressure sensors using the standard MEMS fabrication process. Both the stent and the pressure sensor were patterned using biocompatible photosensitive SU-8 polymer on a silicon wafer. The pressure sensor is combined with an inductor coil built on the top of the stent strut, which increases the inductor size and improves the output performance of the wireless sensor. The proposed smart stent is designed to follow standard catheter insertion procedures with required mechanical strength and compact footprints. The sensor main concept is shown in Figure 1.

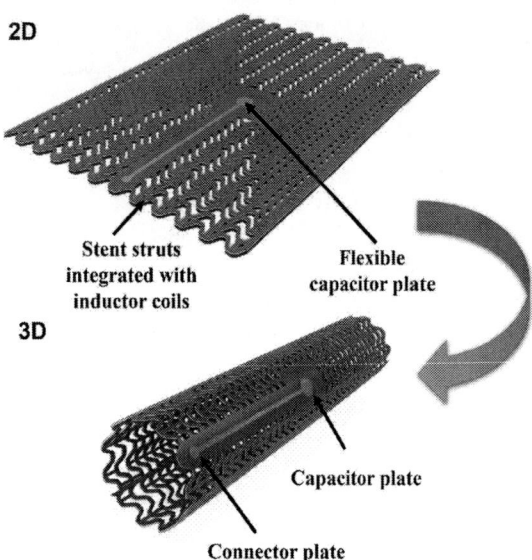

2D

Stent struts
integrated with
inductor coils

Flexible
capacitor plate

3D

Capacitor plate

Connector plate

Figure 1. Concept of the polymer stent integrated with LC type pressure sensor for sensing signal wirelessly to convert portable device. Self-foldable effect from 2D to 3D helical structure under elevating temperature.

FABRICATION OF SENSOR AND MEASUREMETNT METHODS
Design and fabrication of stent

Wireless sensor fabrication begins with a thin layer of silicon dioxide on two 4-inch silicon wafers, and the top and bottom layers fabricated on each wafer were manufactured in the form of a pressure sensor through thermal compression bonding. Final the fabrication process of the smart stent was completed by removing sacrificial layer made by silicon dioxide.

The detailed process flow is as follows. First, a SU-8 photoresist with a thickness of 10 μm was coated on a 300 nm pre-deposited layer of silicon dioxide. After depositing Ti/Au metal by sputtering, a metal layer was patterned using a wet etching technique and a bottom capacitor plate was prepared. After that, the lower capacitor plate was protected with a SU-8 thin film with a thickness of 2 μm. For the formation of the air cavity between the upper capacitor layer and the lower capacitor layer, a PermiNex 1010 adhesive layer was patterned to a thickness of 10 μm on the base layer. The process flow of the top layer is illustrated in Figure 2(a).

An oxide-based sacrificial layer for the bottom part was prepared using the same method as the upper layer. In order to ensure electrical insulation, a negative photoresist having a thickness of about 2 μm was coated on the silicon substrate. To form pair of inductor coil made of Cu on the stent strut structure, a Ti/Cu metal layer was deposited on the upper part of the insulating layer and then a thick photoresist was masked for the electroplating process. The thickness of the photoresist mold pattern was 20 μm for ensuring to form pair of copper inductor coil with a thickness of about 10 μm. A gap distance between each inductor pattern was fixed at 20 μm. After the

electroplating process, a thick SU-8 layer was coated on the inductor coil and patterned to complete the lower layer of the pressure sensor. A schematic illustration of the fabrication process flow of bottom layer was shown in Figure 2(b). These processes were separately performed on each silicon wafer for the upper layer and the lower layer, and then thermally bonded together by using a hot press technique (~10 kN at 150°C for 30 seconds). The smart stent was rolled up from 2D to 3D using an annealing process at a temperature of 250°C for 3 minutes after the sacrificial layer was removed to release the stent structure. The diameter of the smart stent produced through the additional thermal treatment was about 3 μm.

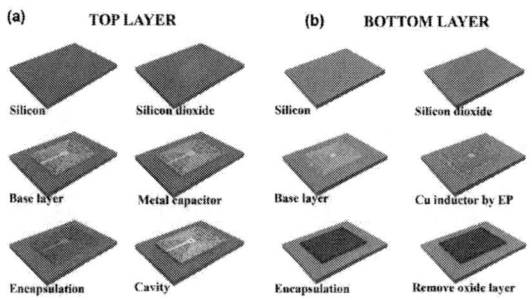

Figure 2. Fabrication process flow of the LC pressure integrated with polymer stent composed of silicon substrate; (a) schematic of top layer fabrication; (b) schematic of bottom layer fabrication.

Measurement system and theoretical description

The LC type pressure sensor response was detected by the electromagnetic coupling (wirelessly coupling) effect between the wireless sensor and the external loop antenna. Firstly, the inductance and the capacitance of the LC pressure sensor was measured by using a LCR meter (E4980AL, Keysight, USA). After that, the external antenna was connected to a portable network analyzer (Field Fox RF Analyzer N19913B, Keysight, USA) to record the change of the electrical resonance of the LC circuit as a function of the capacitance change. In order to characterize the output signal of the fabricated wireless sensor, both of theoretical and experimental methods were carried out. The initial resonance frequency, inductance and capacitance are calculated by following equations 1, 2 and 3, respectively.

The resonance frequency was calculated using the following equation:

$$f_0 = \frac{1}{2\pi\sqrt{LC}} \qquad (1)$$

where f_0 is the resonance frequency (Hz), L is the inductance, and C is the capacitance.

The inductance of the planar square inductor plate was calculated using the following equation:

$$L = K_1 \mu_0 \frac{n^2 d_{avg}}{1 + K_2 \rho} \qquad (2)$$

where L is the inductance of the coil (Henry), n is the number of turns, d_{avg} is average diameter of inductor coils, K_1 and K_2 are the layout constants (2.34 and 2.75, respectively) for the square electrodes, and ρ is the filling

ratio of the coil, which represents the hollow of the spiral coils.
Capacitance was calculated using the following equation:

$$C = \frac{\varepsilon_r \varepsilon_0 A}{d} \qquad (3)$$

where ε_r is the relative permittivity, ε_0 is the permittivity of air, A is the overlap of the electrode area, and d is the spacing between the capacitor plates.

RESULT AND DISCUSSION

The fabrication process was performed based on MEMS techniques that made it possible to fabricate very small as well as mass manufacturing of smart sensors. Figure 3(a) illustrates an optical image before releasing 2D stent integrated with LC type pressure sensor from the silicon substrate with a sacrificial layer made of silicon dioxide. The dimensions of the smart stent were 12×18 mm in width and length, respectively. Figure 3(b) illustrates an optical image of the 2D stent after removing the silicon dioxide sacrificial layer by the BHF solution and then releasing the 2D stent on the silicon substrate.

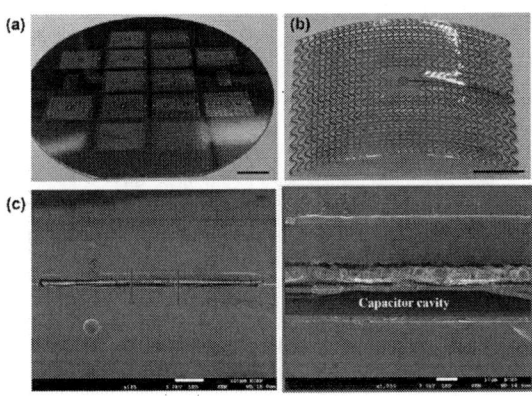

Figure 3. (a) Optical image of the fabricated smart stent integrated with the LC pressure sensor before releasing from silicon wafer; (b) 2D smart stent after releasing from silicon substrate by removing silicon dioxide layer; (c) side view of FE-SEM image of the capacitor, and magnified image for cavity between top and bottom capacitor plate.

Two capacitor plates were positioned at the center of the 2D stent structure. One capacitor located at the top was designed and manufactured as a thin layer to enable a change in capacitance. Theoretically, changes in the resonant frequency occur only due to changes in capacitance caused by deformation of the top thin film. The initial cavity of the parallel capacitor was fixed at 10 μm to ensure the deformation of the flexible capacitor plate (top plate) while different pressure ranges were applied. A cross-sectional view of the capacitor plate is shown in Figure 3(c). In order to induce a folding effect from a 2D to a 3D structure, high temperatures were applied in different ramping ranges. The heat treatment and duration were optimized according to the thickness of the polymer stent and its mechanical properties. The proposed thickness of the stent was about 40 μm and the optimized temperature and duration were 250°C for 4 minutes, respectively. The

optical image of the rolled stent is shown in Figure 4 (inset). For the characterization of the fabricated wireless sensor, an external antenna was connected to the VNA, and the S_{11} parameter for two types of smart stents (before and after the rolling) was recorded separately. As shown in Figure 4, the difference in resonant frequency before and after rolling the smart stents are 33 MHz and 45 MHz, respectively. The increase in the resonance frequency is predicted to be caused by the bending effect of the inductor. The calculated resonance frequency was 35.5 MHz for the polymer stent with a built-in wireless pressure sensor, and the experimentally obtained values were 33 MHz, respectively. The difference between the theoretical resonance frequency and the experimental resonance frequency was less than 5%, showing an acceptable range for further characterization.

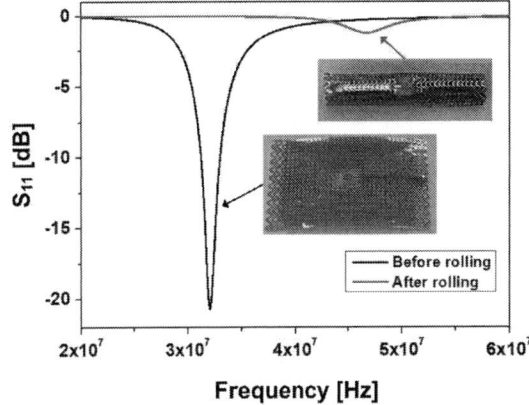

Figure 4. Measured resonance frequency of before and after rolled smart stent integrated with LC type wireless pressure sensor.

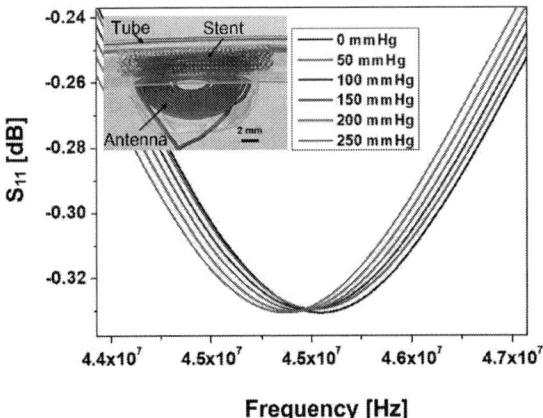

Figure 5. The resonace frequency shift of rolled smart stent integrated with LC type wireless pressure sensor as function of different applied pressure.

In order to confirm the detection response of the smart stent according to applied pressures, the wireless sensor was placed in a 4 mm tube and the pressure was consistently increased up to 250 mmHg by 50 mmHg. As the applied pressure increased, the resonance frequency moved to the left, and the sensitivity of the obtained sensor was 1.28 kHz/mmHg. It was found that the working distance was greatly improved as the entire stent structure was employed as an inductor, differentiating from the smart stents proposed in the previous study.

CONCLUSION

This paper demonstrated smart stent integrated with LC type wireless pressure sensor. The proposed fabrication technique allows for solving the challenges of integrating stent and wireless LC pressure sensors. The unique design enables mass production based on MEMS techniques in 2D planners. After that, it exhibits a folding effect to make a 3D structure without changing the performance of the smart stent by using heat treatment that induces permanent deformation of the polymer stents. The sensitivity of the proposed sensor was evaluated in a various pressure range, and the improved sensitivity of the smart sensor was experimentally confirmed through various experiments.

ACKNOWLEDGEMENTS

This study was supported by a National Research Foundation of Korea (NRF) grant funded by the Korean government (MSIT) (2020R1A5A8018367) and Basic Science Research program through National Research Foundation of Korea (NRF) funded by the ministry of Education (2022R1I1A1A01072651).

REFERENCES

[1] C.W. Tsao, A.W. Aday, Z.I. Almarzooq, A. Alonso, A.Z. Beaton, M.S. Bittencourt, A.K. Boehme, A.E. Buxton, A.P. Carson, Y. Commodore-Mensah, and M.S. Elkind, "Heart disease and stroke statistics—2022 update: a report from the American Heart Association," Circulation, vol. 145, no. 8, pp. e153-e639, 2022.

[2] J. Canonge, J. Jayet, F. Heim, N. Chakfé, M. Coggia, M, R. Coscas, and F. Cochennec, "Comprehensive review of physician modified aortic stent grafts: technical and clinical outcomes," European Journal of Vascular and Endovascular Surgery, vol. 61, no. 4, pp. 560-569, 2021.

[3] K. Takahata, Y.B. Gianchandani, and K.D. Wise, "Micromachined antenna stents and cuffs for monitoring intraluminal pressure and flow," Journal of Microelectromechanical Systems, vol. 15, no. 5, pp. 1289-1298, 2006.

[4] K. Yamaji, K. Inoue, T. Nakahashi, M. Noguchi, T. Domei, M. Hyodo, Y. Soga, S. Shirai, K. Ando, K. Kondo, and K. Sakai, "Bare metal stent thrombosis and in-stent neoatherosclerosis," Circulation: Cardiovascular Interventions, vol. 5, no. 1, pp. 47-54, 2012.

[5] T. Sawas, S. Al Halabi, M.A. Parsi, and J.J. Vargo, "Self-expandable metal stents versus plastic stents for malignant biliary obstruction: a meta-analysis," Gastrointestinal Endoscopy, vol. 82, no. 2, pp. 256-267, 2015.

[6] Y. Yamada, T. Sasaki, T. Takeda, T. Mie, T. Furukawa, A. Kasuga, M. Matsuyama, M. Ozaka, Y. Igarashi, and N. Sasahira, "A novel laser-cut fully covered metal stent with anti-reflux valve in patients with malignant distal biliary obstruction refractory to conventional covered metal stent," Journal of Hepato-Biliary-Pancreatic Sciences, vol. 28, no. 7, pp.563-571, 2021.

[7] T.F. Lüscher, J. Steffel, F.R. Eberli, M. Joner, G. Nakazawa, F.C. Tanner, R. Virmani, "Drug-eluting stent and coronary thrombosis: biological mechanisms and clinical implications," Circulation, vol. 115, no. 8, pp. 1051-1058, 2007.

[8] R. Wessely, "New drug-eluting stent concepts," Nature Reviews Cardiology, vol 7, no. 4, pp. 194-203, 2010.

[9] R. Virmani, A. Farb, G. Guagliumi, and F.D. Kolodgie, "Drug-eluting stents: caution and concerns for long-term outcome," Coronary artery disease, vol. 15, no. 6, pp. 313-318, 2004.

[10] S. Kuramitsu, S. Sonoda, K. Ando, H. Otake, M. Natsuaki, R. Anai, Y. Honda, K. Kadota, Y. Kobayashi, and T. Kimura, "Drug-eluting stent thrombosis: Current and future perspectives," Cardiovascular intervention and therapeutics, vol. 36, no. 2, pp. 158-168, 2021.

[11] Z. Feng, G. Duan, P. Zhang, L. Chen, Y. Xu, B. Hong, W. Zhao, J. Liu, and Q. Huang, "Enterprise stent for the treatment of symptomatic intracranial atherosclerotic stenosis: an initial experience of 44 patients," BMC neurology, vol. 15, no. 1, pp. 1-7, 2015.

[12] K. Takahata, Y.B. Gianchandani, and K.D. Wise, "Micromachined antenna stents and cuffs for monitoring intraluminal pressure and flow," Journal of Microelectromechanical Systems, vol. 15, no. 5, pp. 1289-1298, 2006.

[13] J. Park, J.K. Kim, D.S. Kim, A. Shanmugasundaram, S.A. Park, S. Kang, S.H. Kim, M.H. Jeong, and D.W. Lee, "Wireless pressure sensor integrated with a 3D printed polymer stent for smart health monitoring," Sensors and Actuators B: Chemical, vol. 280, pp. 201-209, 2019.

[14] X. Chen, B. Assadsangabi, Y. Hsiang, and K. Takahata, "Enabling angioplasty-ready "Smart" Stents to detect in-stent restenosis and occlusion," Advanced Science, vol. 5, no. 5, p. 1700560, 2018.

[15] X. Chen, D. Brox, B. Assadsangabi, M.S.M. Ali, and K. Takahata, "A stainless-steel-based implantable pressure sensor chip and its integration by microwelding," Sensors and Actuators A: Physical, vol. 257, pp. 134-144, 2017.

[16] J. Park, J.K. Kim, S.J. Patil, J.K. Park, S. Park, D.W. Lee, "A wireless pressure sensor integrated with a biodegradable polymer stent for biomedical applications," Sensors, vol 16, pp. 809, 2016.

CONTACT

*D.W. Lee, tel: +82-062-530-1684; mems@jnu.ac.kr

A DYNAMIC MICROARRAY DEVICE FOR SELECTIVE PAIRING AND ELECTROFUSION OF LIPOSOMES

Sho Takamori[1], Hisatoshi Mimura[1], Toshihisa Osaki[1], and Shoji Takeuchi[1,2]
[1]Kanagawa Institute of Industrial Science and Technology, JAPAN and
[2]The University of Tokyo, JAPAN

ABSTRACT

This paper reports a microfluidic device for selective pairing and electrofusion of cell-sized liposomes. The device contains two dynamic microarrays systems arranged in a planar symmetric orientation and connected at trapping sites for capturing liposomes. In addition, electrodes made of a low melting point alloy are installed for the electrofusion of trapped liposomes. This design enables the selective pairing of target liposomes and their electrofusion. As a demonstration, we pair two species of liposomes and implement their electrofusion. We believe our device is useful for selectively fusing specific liposomes/cells with a specific size range dispersed in polydisperse bulk samples.

KEYWORDS

Giant Liposomes, Selective Pairing, Selective Electrofusion, Micro Electrodes

INTRODUCTION

Electrofusion of lipid membranes is performed with various aims in research including intra-cellular cargo delivery by liposome-cell fusion [1], nuclear transfer by cell fusion [2], zygote production by gamete fusion [3]. As the lipid membrane is formed by the self-assembly of amphiphilic lipid molecules *via* hydrophobic interactions, membrane fusion also occurs by the self-assembly of disrupted membranes. In a typical setting of electrofusion procedure, alternating electric field (AC) is applied for making chain-like aggregates of vesiculated membranes (liposomes) and electric pulses are further imposed for making pores on the membranes and inducing their reformation into hybrid membranes [4].

Standard electrofusion method mediating AC-induced non-selective formation of target chains accompanies both wanted and unwanted fusion in the single run and exhibits a low fusion selectivity [1]. To improve the selectivity, DNA-mediated selective electrofusion (DASE) has been proposed by utilizing the specificity of Watson-Crick base pairing in liposome-cell adhesion [1]. Although the DASE significantly improved the selectivity of fusion, it relies on adhesion of targets in a polydisperse bulk mixture and unpreferable adhesions (*e.g.*, with small liposomes, dead cells and debris) occupies a large amount of the product.

For the one-to-one selective fusion of vesiculated membranes with a designated size, Tan *et al.*'s microfluidic array system for dynamic target trapping [5] could be used (Fig. 1A). Sugahara *et al.* [6] introduced arrayed trapping sites and microelectrodes in the microarray system for the pairing and electrofusion of liposomes in a specific size range (Fig. 1B). However, this device lacks a mechanism for selective pairings. In another study by Teshima *et al.* [7], the authors duplicated the dynamic microarray system

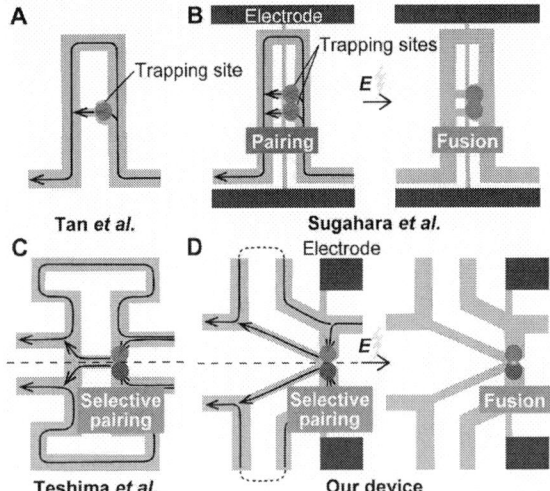

Figure 1: Comparisons of previously reported dynamic microarray devices and a device proposed in this study. (A) Tan et al.'s microarray for dynamic trap. (B) Sugahara et al.'s device with microelectrodes. (C) Teshima et al.'s dual dynamic microarray device. (D) Our device.

in a planar symmetric orientation connecting both at trapping sites. This design enabled the authors to selectively capture two different target objects (stained hydrogels) in a specific size range at the trapping site (Fig. 1C).

In this study we introduce a device for selective pair formation and electrofusion of target liposomes by taking advantages of both Sugahara *et al.*'s microelectrodes for electrofusion and Teshima *et al.*'s selective pairing system using planar symmetric dual dynamic microarrays (Fig. 1D). We fabricate a device and demonstrate the selective pairing and electrofusion of cell-sized liposomes.

MATERIALS AND METHODS

Fabrication of Microfluidic Device

The device was fabricated by standard soft lithography (Fig. 2). In brief, a SU-8 mold was prepared and microchannels were transferred to PDMS. The PDMS was bonded to a glass. Fusible alloy (T_m: 47 °C) was loaded into electrode channels by vacuum filling [6].

Formation of Giant Liposomes

Giant liposomes encapsulating 500 mM sucrose with either 5 μM AF488-dex_{10k} or 5 μM TexRed-dex_{10k} were formed by a standard electroformation protocol [1] using DOPC as a sole lipid component. Electroformed liposomes were dispersed into electrofusion buffer (300 mM glucose, 100 mM NaCl, 0.1 mM $MgCl_2$, 0.1 mM $CaCl_2$).

IEEE MEMS 2023, Munich, GERMANY
15 - 19 January 2023

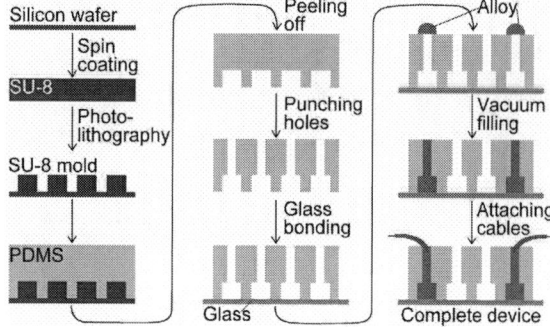

Figure 2: Fabrication steps of the proposed device.

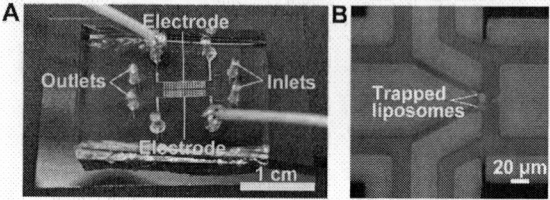

Figure 3: (A) Fabricated device. (B) Trapping of two cell-sized liposomes at the trapping sites in the device.

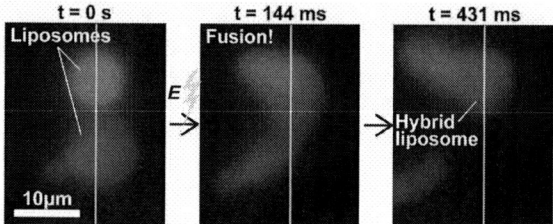

Figure 4: Electrofusion of cell-sized liposomes at the trapping site.

Microscopy

Microscopy was performed on an inverted fluorescence microscope (IX71, Olympus) equipped with a mercury lamp, color CMOS camera (DP74, Olympus) and 60× NA1.45 objective (Plan Apo N, Olympus).

Electrofusion experiment

The liposomes were loaded into the device by pumping at 1 µL min^{-1} (11 plus, Harvard Apparatus). After liposomes were captured at the trapping sites, the flow rate was reduced to 0.3 µL min^{-1}, AC (2.9 MHz, 10 V$_{RMS}$) was applied for 10 s and an electric pulse (50 µs width, 300 V) was applied using a voltage supply (LF101 Electro Cell Fusion Generator, Nepagene, Japan).

RESULTS AND DISCUSSION

Trapping of Cell-Sized Liposomes at Trapping Sites

Fig. 3 shows the fabricated device and cell-sized liposomes trapped in the trapping sites. Here, liposomes in magenta are overlayed on a green image of the device to highlight the trapping of the cell-sized liposomes at the designated trapping sites (Fig. 3B).

Electrofusion of Cell-Sized Liposomes

Cell-sized liposomes encapsulating AF488-dex$_{10k}$ (green) and TexRed-dex$_{0k}$ (red) were paired at a trapping site and electrofused (Fig. 4). At $t = 0$ s, the liposomes are tightly paired by the applied AC (Fig.4 left). Right after $t = 0$ s, an electric pulse (50 µs width, 1,500 V mm^{-1}) is imposed. At $t = 144$ ms, the paired liposomes are in the middle of fusion (Fig. 4 middle). At $t = 433$ ms, the fusion is completed (Fig. 4 right).

CONCLUSION

In summary, we fabricated a microfluidic device equipped with i) dual dynamic microarray systems for the selective pairing of liposomes and ii) two microelectrodes made of fusible alloy for the electrofusion of paired liposomes (Fig. 3A). We captured cell-sized liposomes at the trapping site in the fabricated device (Fig. 3B). We performed the electrofusion of fluorescently labelled two cell-sized liposomes (Fig. 4). The micrographs indicate the complete content mixing throughout the observed fusion event. We believe our microfluidic approach is advantageous for the selective fusion of liposomes/cells in polydisperse bulk samples.

ACKNOWLEDGEMENTS

This work is supported by JST CREST (JPMJCR18S5) and JSPS KAKENHI (JP22K15080).

REFERENCES

[1] S. Takamori, P. Cicuta, S. Takeuchi and L. Di Michele, "DNA-assisted selective electrofusion (DASE) of *Escherichia coli* and giant lipid vesicles", *Nanoscale*, 14, 14255-14267, 2022.

[2] A. Ogura, K. Inoue, K. Takano, T. Wakayama, and R. Yanagimachi, "Birth of mice after nuclear transfer by electrofusion using tail tip cells." *Molecular Reproduction and Development: Incorporating Gamete Research*, 57(1), pp.55-59, 2000.

[3] E. Toda, N. Koiso, A. Takebayashi, M. Ichikawa, T. Kiba, K. Osakabe, Y. Osakabe, H. Sakakibara, N. Kato and T. Okamoto, "An efficient DNA-and selectable-marker-free genome-editing system using zygotes in rice." *Nature plants*, 5(4), pp.363-368, 2019.

[4] K.A. Riske, N. Bezlyepkina, R. Lipowsky and R. Dimova, "Electrofusion of model lipid membranes viewed with high temporal resolution", *Biophysical Reviews and Letters*, 1(04), pp.387-400, 2006.

[5] W.H. Tan and S. Takeuchi, "A trap-and-release integrated microfluidic system for dynamic microarray applications", *Proceedings of the national academy of sciences*, 104(4), pp.1146-1151, 2007.

[6] K. Sugahara, Y. Morimoto, S. Takamori and S. Takeuchi, "A dynamic microarray device for pairing and electrofusion of giant unilamellar vesicles", *Sensors and Actuators B: Chemical*, 311, p.127922, 2020.

[7] T. Teshima, H. Ishihara, K. Iwai, A. Adachi and S. Takeuchi, "A dynamic microarray device for paired bead-based analysis", *Lab on a Chip*, 10(18), pp.2443-2448, 2010.

CONTACT

*S. Takamori, tel: +81-44-8192037; takamori@iis.u-tokyo.ac.jp

REAL-TIME FUNCTIONAL BRAIN MAPPING BASED ON HIGH-CHANNEL-COUNT, ULTRA-CONFORMAL NEURAL INTERFACE

Xiner Wang[1,3], Zhaohan Chen[4], Jizhi Liang[1,3], Xiaoling Wei[2,3], Liuyang Sun[1,3], Meng Li[2,3], Zhitao Zhou[2,3] and Tiger H. Tao[1,2,3,5,6,7,8,9]

[1]2020 X-Lab, Shanghai Institute of Microsystem and Information Technology, Chinese Academy of Sciences, Shanghai, CHINA

[2]State Key Laboratory of Transducer Technology, Shanghai Institute of Microsystem and Information Technology, Chinese Academy of Sciences, Shanghai, CHINA

[3]School of Graduate Study, University of Chinese Academy of Sciences, Beijing, CHINA

[4]College of Life Sciences, Shanghai Normal University, Shanghai, CHINA

[5]Center of Materials Science and Optoelectronics Engineering, University of Chinese Academy of Sciences, Beijing, CHINA

[6]Center for Excellence in Brain Science and Intelligence Technology, Chinese Academy of Sciences, Shanghai, CHINA

[7]Neuroxess Co., Ltd. (Jiangxi), Nanchang, Jiangxi, CHINA

[8]Guangdong Institute of Intelligence Science and Technology, Hengqin, Zhuhai, Guangdong, CHINA and

[9]Tianqiao and Chrissy Chen Institute for Translational Research, Shanghai, CHINA

ABSTRACT

We report a real-time functional brain mapping technique based on high-channel-count, ultra-conformal electrocorticographic (ECoG) electrodes. The MEMS-based 64-channel ECoG electrodes can well conform to the curvilinear surface of the cortex, thus showing good neural signal recording quality. This neural interface was applied to Labrador dogs for delineating cortical areas related to eyelid, nose, and limb motor function. The result was in strong congruence to that derived by electrical cortical stimulation (ECS) mapping, which is the gold standard in mapping the eloquent cortex. Our method can provide an instant functional map, which promises a great potential of this technique in the real-time passive functional mapping of the eloquent cortex.

KEYWORDS

Passive functional brain mapping, ECoG, Neural interface, Ultra-conformal contact.

INTRODUCTION

Functional mapping of the eloquent cortex is a common clinical practice in invasive brain surgery for localization and preservation of the eloquent issue, thus optimizing neurosurgical results and minimizing postoperative morbidity. Brain mapping can also deepen our understanding of the brain's functional organization. The most traditional functional mapping technique, called electrical cortical stimulation (ECS) [1], requires the application of electrical current to the exposed cortex tissue to induce an apparent phenomenon. Although it remains the clinical gold standard, ECS is very time-consuming, carries the risk of provoking after-discharges or seizures, and cannot easily be used in pediatric patients.

With the advances in neural recording and clinical techniques, passive brain mapping modalities have emerged, which avoid the need to apply electrical stimulation. Positron emission tomography (PET) [2] and functional magnetic resonance imaging (fMRI) [3] are both examples of mapping techniques for passive functional localization. Regional cerebral blood flow (rCBF) and blood oxygen level dependent (BOLD) signals are continuously monitored respectively as patients perform behavioral tasks. However, there still exist some limitations. PET and fMRI require expensive medical equipment and lack sufficient sample rate and spatial resolution for spectral analysis and functional mapping.

Hence, patients who require functional localization of the eloquent cortex would be greatly benefited from a mapping technique that addresses the shortcomings of the existing mapping methodology. A real-time, reliable, safe, and low-cost mapping method is still less explored. Recently, several studies have shown that ECoG signals reflect task-related modulations as the experimental subjects perform tasks [4-6], which can provide a new technique (i.e., ECoG mapping) with all those desirable characteristics. This method is also particularly attractive because the ECoG electrodes are usually involved in invasive brain surgery.

In this work, we present a real-time passive functional mapping technique based on high-channel-count, ultra-conformal ECoG electrodes, fabricated by a standard microelectromechanical system (MEMS) process. That could offer significant promise in their application in ECoG-based brain mapping.

DESIGN

To realize real-time, reliable ECoG signals recording and mapping, the neural interface is designed following two main goals. One is to cover the desired area of the cerebral cortex with high-density recording sites. The high-spatial-resolution recording is of great importance to achieve accurate functional localization. The other is to maintain conformal contact with the curvilinear surface of the brain, to obtain stable and high-fidelity ECoG signals during intraoperative recording.

In our 64/529-channel ECoG electrodes, electrode sites were distributed in an area of 3.5 cm×3.5 cm, which exhibits high electrode density to resolve high spatio-temporal ECoG signals from the surface of the cortex. To improve the conformability to the pleated brain cortex, the thickness of the top and bottom polyimide (PI) encapsulation layers is reduced, which gives rise to a total thickness of 7 μm. A specialized mesh design is also adopted to facilitate conformal contact to the soft and curved brain tissue.

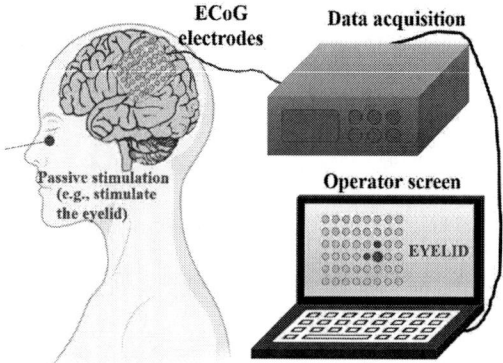

Figure 1: Equipment setup and interface of the ECoG-based real-time passive functional mapping system to the patient.

To explicitly demonstrate real-time passive functional mapping of the eloquent cortex using the ultra-conformal ECoG electrodes, **Figure 1** shows the equipment setup and interface of the ECoG-mapping system, including the high-channel-count, ultra-conformal ECoG electrodes, data acquisition equipment, and the operator screen. Our 64/529-channel ECoG electrodes are placed on the cortex of the subject, then record ECoG signals while the subject is exposed to passive stimulation. The data acquisition equipment receives the signals, performs data preprocessing including amplification, filtering, and digitalization, then transmits these processed data to the computer, thereby effectively reducing noise interference. The computer is used not only to further process and analyze the ECoG signals but also to present the final mapping results in a topographical form in real time on the operator screen. Such a setup is clinically verified in Huashan Hospital.

FABRICATION

The high-channel-count, ultra-conformal ECoG electrodes are fabricated with a sandwich structure using a similar process in our previous work [7]. As shown in **Figure 2**, flexible polyimide (PI) was spin-coated on the silicon wafer and patterned by photography to form the bottom insulating layer. The Cr/Au metal layer was then patterned by photography, evaporation, and lift-off process to define the recording sites and interconnectors. Spin-coating formed a layer of PI as the encapsulation layer. Sputtering the aluminum as the hard mask followed by UV photolithography, aluminum corrosion, and oxygen plasma dry etching exposed the recording sites and defined a pattern of the mesh structure. Immersion in

hydrofluoric acid to release the electrode device completed the fabrication. Bonding the back-end pads to external connectors yielded points for connections to data acquisition equipment.

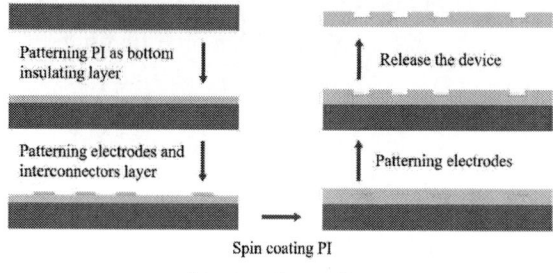

Figure 2: The flow chart of the process to fabricate the high-channel-count, ultra-conformal ECoG electrodes.

EXPERIMENTS RESULTS

Characterization

After the high-channel-count ECoG electrodes were fabricated, a few tests were carried out to characterize the mechanical and electrical performance of our ultra-conformal neural interface.

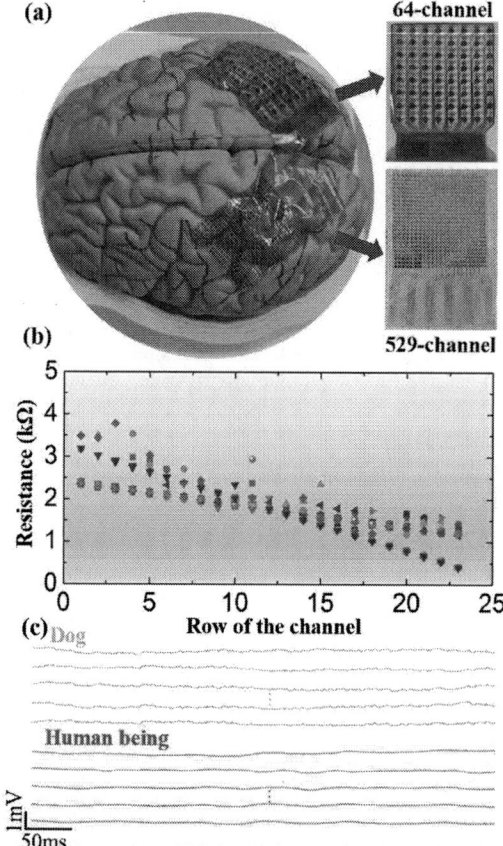

Figure 3: (a) The paragraphs of 64-channel and 529-channel ECoG electrodes, both can achieve excellent conformal coverage over the highly convoluted brain surface. (b) Channel resistance of a 529-channel ECoG device. (c) The ECoG signals acquired from Labrador dogs and human beings under anesthesia.

Achieving conformability is a challenging but crucial task given the highly convoluted mechanical properties of the brain surface. As shown in **Figure 3a**, our 64 and 529-channel ECoG electrodes can both achieve excellent conformal coverage over the curved brain surface. Electrochemical impedance was measured by immersing our electrodes in the phosphate buffer saline (PBS) solution. **Figure 3b** shows the resistance of a 529-channel ECoG device. The resulting impedances of the recording sites with a diameter of 800 μm are below 5 kΩ at 1 kHz. The great conformal contact and the large recording area significantly lower the contact impedance and, ultimately, improve signal-to-noise ratios (SNRs) in the electrophysiological recording. The signals shown in **Figure 3c** were acquired from Labrador dogs and human beings under anesthesia. The low-noise signals without artifacts confirmed the recording capabilities of our ECoG electrodes.

In vivo application

To validate the feasibility and reliability of our ECoG-based mapping system, we conducted a neural recording and mapping experiment on a Labrador dog model. **Figure 4** shows the mapping results of sensory and motor cortical areas. Our 64-channel ECoG electrodes were placed under the dura mater of the anesthetic Labrador dog (**Figure 4a**). When mechanical stimuli were applied to the eyelid of the dog, the ECoG amplitudes of some specific channels typically increased in the delta (1–4 Hz) band, as shown in **Figure 4b**. Furtherly, the first responsive channels located in the sensory area, then spread to the motor area in a short time (typically several hundred milliseconds), which suggested a functional linkage between sensory and motor areas (**Figure 4c**).

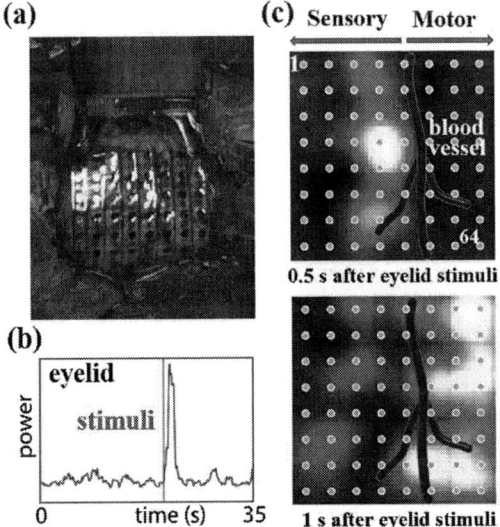

Figure 4: (a) Paragraph of the ECoG electrodes placed under the dura mater of a Labrador dog for functional mapping of sensory and motor cortical areas. (b) Time domain response of the ECoG signals to the mechanical stimuli in the delta (1–4 Hz) band. (c) The spread of task-related EEG signals between sensory and motor areas.

CONCLUSION

In summary, we provide a novel, real-time functional brain mapping method, using the high-channel-count, ultra-conformal ECoG electrodes. Device thinness and compliance were improved to realize conformal contact with the highly convoluted cortex surface and acquire effective neural signals. In addition, the flexible ECoG electrodes can identify and map cortical areas related to motor function with high spatial resolution. Our ECoG-based mapping technique can localize stimuli-related cortical areas, which will open up prospects in the field of intraoperative passive functional brain mapping in clinical practice.

ACKNOWLEDGEMENTS

Xiner Wang and Zhaohan Chen contributed equally to this work. The work was partially supported by the Key-Area Research and Development Program of Guangdong Province (Grant No. 2021B0909060002), National Key R & D Program of China (Grant Nos. 2019YFA0905200, 2021ZD0201600, 2021YFC250150, 2021YFF1200700, 2022ZD0209300, 2022ZD0212300), National Natural Science Foundation of China (Grant No. 61974154), Key Research Program of Frontier Sciences, CAS (Grant No. ZDBS-LY-JSC024), Shanghai Pilot Program for Basic Research–Chinese Academy of Science, Shanghai Branch (Grant No. JCYJ-SHFY-2022-01), Shanghai Municipal Science and Technology Major Project (Grant No. 2021SHZDZX), CAS Pioneer Hundred Talents Program, Shanghai Pujiang Program (Grant Nos. 19PJ1410900, 21PJ1415100), the Science and Technology Commission Foundation of Shanghai (No. 21JM0010200), Shanghai Rising-Star Program (Grant No. 22QA1410900), the Innovative Research Team of High-level Local Universities in Shanghai, the Jiangxi Province 03 Special Project and 5G Project (Grant No. 20212ABC03W07), Fund for Central Government in Guidance of Local Science and Technology Development (Grant No. 20201ZDE04013).

REFERENCES

[1] G. A. Ojemann, "Brain Organization for language from the perspective of electrical stimulation mapping," Behavioral and Brain Sciences, vol. 6, no. 2, pp. 189–206, 1983.

[2] P. T. Meyer, "Preoperative motor system brain mapping using positron emission tomography and statistical parametric mapping: Hints on cortical reorganisation," *Journal of Neurology, Neurosurgery & Psychiatry*, vol. 74, no. 4, pp. 471–478, 2003.

[3] A. Chakraborty and A. W. McEvoy, "Presurgical functional mapping with functional MRI," *Current Opinion in Neurology*, vol. 21, no. 4, pp. 446–451, 2008.

[4] R. Arya, P. S. Horn, and N. E. Crone, "ECoG high-gamma modulation versus electrical stimulation for presurgical language mapping," Epilepsy & Behavior, vol. 79, pp. 26–33, 2018.

[5] A. B. Martin, X. Yang, Y. B. Saalmann, L. Wang, A. Shestyuk, J. J. Lin, J. Parvizi, R. T. Knight, and S. Kastner, "Temporal Dynamics and response

modulation across the human visual system in a spatial attention task: An ECOG study," The Journal of Neuroscience, vol. 39, no. 2, pp. 333–352, 2018.

[6] R. A. Domingo, T. Vivas-Buitrago, G. De Biase, E. H. Middlebrooks, P. S. Bechtle, D. S. Sabsevitz, A. Quiñones-Hinojosa, and W. O. Tatum, "Intraoperative seizure detection during active resection of glioblastoma through a novel hollow circular electrocorticography array," Operative Neurosurgery, vol. 21, no. 2, 2021.

[7] F. Zheng, S. Zhang, Y. Zhou, and T. H. Tao, "A silk-enabled conformal brain electrode for recording and disease treatment," 2019 IEEE 32nd International Conference on Micro Electro Mechanical Systems (MEMS), 2019.

CONTACT

*Tiger H. Tao, tel: +86-21-62511070;
tiger@mail.sim.ac.cn
*Zhitao Zhou, tel: +86-21-62511070;
ztzhou@mail.sim.ac.cn

ACOUSTOFLUIDICS: MERGING ACOUSTICS AND FLUID MECHANICS FOR BIOMEDICAL APPLICATIONS

Tony Jun Huang

William Bevan Distinguished Professor of Mechanical Engineering and Materials Science

Pratt School of Engineering, Duke University

Group Website: https://acoustofluidics.pratt.duke.edu/

ABSTRACT

The use of sound has a long history in medicine. Dating back to 350 BC, the ancient Greek physician Hippocrates, regarded as "the father of medicine", devised a diagnostic method for detecting fluid in the lungs by shaking patients by their shoulders and listening to the resulting sounds emanating from their chest. As acoustic technology has advanced, so too has our ability to "listen" to the body and better understand underlying pathologies. The 18th century invention of the stethoscope allowed doctors to gauge the health of the heart; the 20th century invention of ultrasound imaging revolutionized the field of biomedical imaging and enabled doctors to diagnose a range of conditions in the fields of obstetrics, emergency medicine, cardiology, and pulmonology. In the last decade, a new frontier in biomedical acoustic technologies has emerged, termed acoustofluidics, which joins cutting-edge innovations in acoustics with micro- and nano-scale fluid mechanics. Advances in acoustofluidics have enabled unprecedented abilities in the early detection of cancer, the non-invasive monitoring of prenatal health, the diagnoses of traumatic brain injury and neurodegenerative diseases, and have also been applied to develop improved therapeutic approaches for transfusions and immunotherapies. In this talk, I summarize our lab's recent progress in this exciting field and highlight the versatility of acoustofluidic tools for biomedical applications through many unique examples, ranging from the development of high-purity, high-yield methods for the separation of circulating biomarkers such as exosomes and circulating tumor cells, to highly precise, biocompatible platforms for manipulating cells and studying cell-cell communication, to high-throughput therapeutic approaches for platelet isolation and enrichment, to strategies for high-resolution 3D bioprinting, to programable, contact-free technologies for digital fluid manipulation. These acoustofluidic devices can precisely manipulate objects across 7 orders of magnitude (from a few nanometers to a few centimeters). Thanks to these favorable attributes (e.g., versatility, precision, and biocompatibility), acoustofluidic devices harbor enormous potential in becoming a leading technology for a broad range of applications, playing a critical role for translating innovations in technology into advances in biology and medicine.

SILICON CARBIDE REINFORCED VERTICALLY ALIGNED CARBON NANOTUBE COMPOSITE FOR HARSH ENVIRONMENT MEMS

Jiarui Mo, Shreyas Shankar, Guoqi Zhang, and Sten Vollebregt

Laboratory of Electronic Components, Technology and Material (ECTM), Department of Microelectronics, Delft University of Technology, Delft, The Netherlands

ABSTRACT

Fabricating high-aspect-ratio (HAR) structures with silicon carbide (SiC) is a challenging task. This paper presents a silicon carbide (SiC) reinforced vertically aligned carbon nanotubes (VACNT) composite as a promising candidate to fabricate HAR MEMS devices for harsh environment applications. The use of a VACNT array allows the fast realization of HAR structures as a template for MEMS fabrication. The template can later be easily filled by amorphous-SiC due to the porous nature of the VACNT forest. The SiC-CNT nanocomposite has electrical properties dominated by VACNT arrays and mechanical stability dominated by the a-SiC. Based on this concept, a thermal actuator is fabricated and proven to function up to 450°C for the first time.

KEYWORDS

SiC-CNT composite, HAR structures, harsh environment, thermal actuator

INTRODUCTION

With the advancement of the micro-/nano-fabrication technology, M-/NEMS devices play a significant role in various fields. For instance, many silicon (Si) MEMS devices have been applied to automotive products. In recent years, MEMS devices have been increasingly needed in high-temperature environments, for instance, oil drilling, combustion monitoring, and aerospace applications. However, Si-based MEMS can hardly meet the requirement as silicon's electrical and mechanical properties experience degradations at elevated temperatures.

SiC, a well-known wide bandgap semiconductor, has numerous advantages over Si. SiC has higher thermal conductivity, high critical E-field, high chemical inertness, and high Young's modulus. Thanks to the wide bandgap, SiC-based active devices maintain electrical behaviors above 300°C, which is the temperature limit of Si-based active devices. Most importantly, SiC is not plastically deformable even at elevated temperatures above 500°C. These properties are favourable for designing robust MEMS devices for harsh environments. Researchers have put much effort into SiC MEMS sensor development in recent years. These sensors include but are not limited to pressure sensors [1, 2, 3, 4], motion sensors [5, 6, 7], and resonators [8, 9, 10, 11, 12]. Currently, SiC MEMS development is limited by insufficient micro-machining techniques. Due to the mechanical and chemical stability of the SiC, traditional etching methods are not efficient. Typically, the SiC etching rate by dry etching is below 1 μm/min. In 2021, Erbacher et al. reported the highest etching rate (4 μm/min) so far by reactive ion etching (RIE) with $SF_6/O_2/He$ gas mixture [13]. However, this etching rate is still relatively low compared to Si bulk etching method, and it requires a particular inductively coupled plasma (ICP) etcher with high power density. In addition to the slow etch rate, rough etching surface, non-vertical sidewalls, and micro-trenching effect are often observed in the literature [14, 15, 16].

The CNT forrest in combination with various filler materials is an attractive material for MEMS applications [17,18]. In this work, we propose using SiC-CNT composite as an alternative for the fast fabrication of HAR structure. VACNT arrays can be grown at a high growthrate (a few tens of micrometers per minute) while maintaining a vertical sidewall, and HAR structures can be obtained easily with the array. After the growth, the porous HAR template can be filled by LPCVD SiC; therefore, the mechanical property of the array is reinforced by SiC [19]. To demonstrate the potential of this composite thermal actuators were fabricated and tested up to 450°C.

SIC-CNT COMPOSITE
Fabrication of the composite

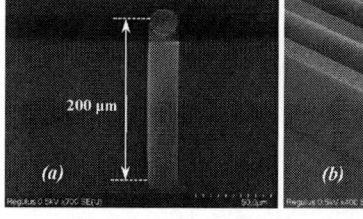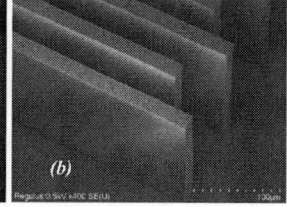

Figure 1: As-deposited HAR VACNT (a) pillars; (b) blocks from a 30° tilted view.

The fabrication of the composite consists of only a one-mask step, i.e. the patterning of the catalyst. First, an oxide layer is thermally grown, acting as a diffusion barrier from the metal catalyst to the Si substrate [17]. Then Al_2O_3 (20 nm) and Fe (2 nm) layers are evaporated by a lift-off process. The CNT growth is performed with an Aixtron Blackmagic CVD system. The precursor gases are C_2H_2/H_2 (50/700 sccm), reacting at 80 mbar and 600°C. After the CNT growth, the sample is loaded into an LPCVD furnace for SiC deposition. The SiC deposition is done with

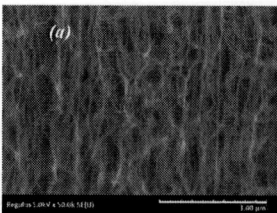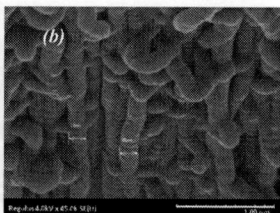

Figure 2: CNT array (a) before and (b) after being coated by a-SiC.

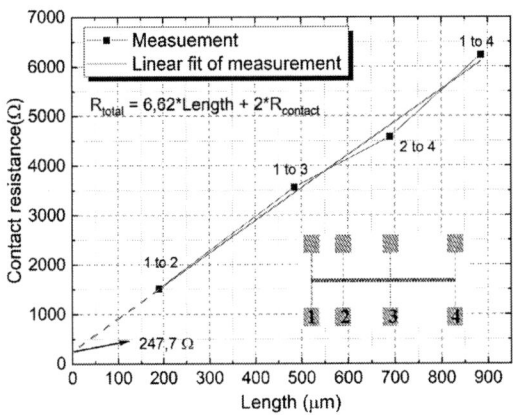

Figure 3: Measured resistances from different segments of the TLM structure. The intersection between the curve and y-axis is the value of the contact resistance.

SiH_2Cl_2 and C_2H_2 diluted in H_2 at 760°C at 1 mbar.

Figure 1 shows as-deposited CNT pillars and blocks with a maximum aspect ratio of 10. Figure 2a and Figure 2b provide close-up views of the CNT array surface before and after the SiC deposition. As can be observed, the CNTs are densified by the SiC and are not only weakly bonded by van der Waals force.

Electrical properties of SiC-CNT composite

In many MEMS applications, the construction material is required to be conductive. Examples are capacitive accelerometers, thermal actuators, and resonators. However, the filler material, i.e. amorphous SiC, is non-conductive. Therefore, it is essential to investigate the bulk resistivity of the composite. In addition, as the device needs to interface with the instrument/readout, the contact resistance from composite to metal is also characterized. For this purpose, resistors with different dimensions and test structures for the transmission line method (TLM) were fabricated to evaluate the bulk resistivity of the SiC-CNT composite and the contact resistance. To measure the test structure, a metal layer was sputtered and patterned as electrodes before fabricating SiC-CNT. Titanium nitride (TiN) was used because it provides suffiecient thermal budget for the subsequent process.

The resistivity of the composite is extracted and summarized in Table 1. The resistivity of the measured samples ranges from 6.84×10^{-3} to 7.09×10^{-3} $\Omega \cdot m$. The resistivity of the intrinsic VACNT is 2.23×10^{-3} $\Omega \cdot m$, which shows better conductivity than the SiC-CNT composite. Nevertheless, the conductivity of the composite still shows an improvement of a few orders of magnitude compared to a-SiC ($\sim M\Omega \cdot cm$). Hence, it implies that the

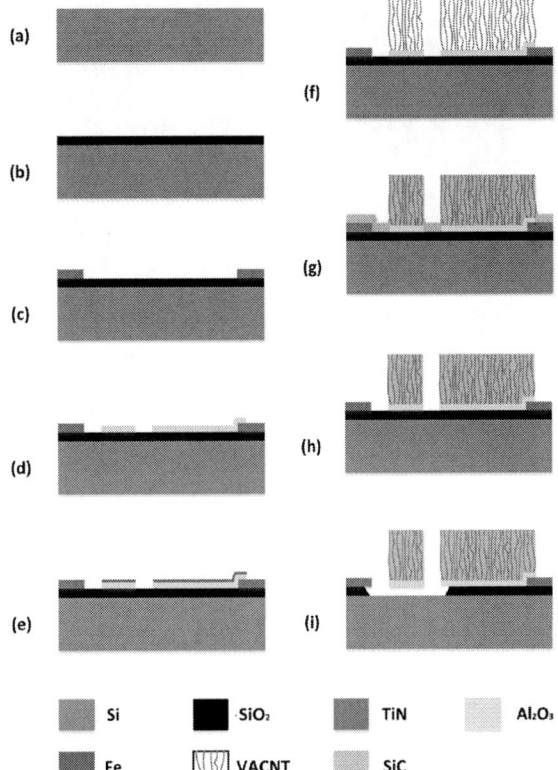

Fig. 4. Process overview of the SiC-CNT composite. (a) Si substrate is used as starting material; (b) SiO_2 sacrificial layer is deposited; (c) TiN is sputtered as electrodes; (d)(e) Al_2O_3/Fe bilayer is evaporated, and the patterning of the bilayer is done by lift-off process; (f) CNT arrays are grown on the area defined by Fe; (g) The CNT array is infiltrated by a-SiC; (h) Floor layer removal by dry etching; (i) Vapor HF to release the structure.

conductive paths between individual CNTs remain and dominate the electrical behaviour of the composite.

The TLM measurement result is given in Figure 3. The intersection between the linear fit and y-axis is the resistance between a 20×10 μm^2 TiN/SiC-CNT interface. The resulting specific contact resistivity is 2.48×10^{-4} $\Omega \cdot cm^2$.

SIC-CNT THERMAL ACTUATOR

A thermal actuator is fabricated and characterized as a demonstrator of using the SiC-CNT composite as an emerging material for MEMS applications.

Fabrication of the thermal actuator

The process flow of the thermal actuator is illustrated in Figure 4. Since the thermal actuator requires a suspended

Table 1. Resistance measurement of SiC-coated CNTs resistors(* This sample is not coated with a-SiC)

Sample (#)	CNT Height (μm)	Width (μm)	Resistance (Ω)	Resistivity (Ω·m)
1	10	10	20540	0.00684
2	10	20	10580	0.00705
3	10	40	5321	0.00709
4	20	10	10610	0.00707
5*	33	10	2030	0.00223

978-1-6654-9309-3/23 $31.00 © 2023 IEEE

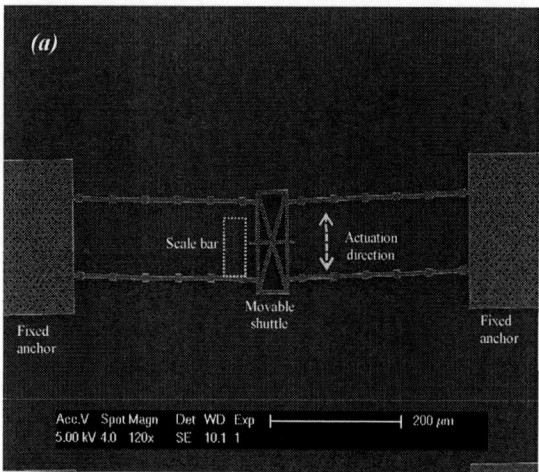

Figure 5: (a) A complete view of the thermal actuator based on SiC-CNT composite; (b) A zoom-in view of the suspended shuttle.

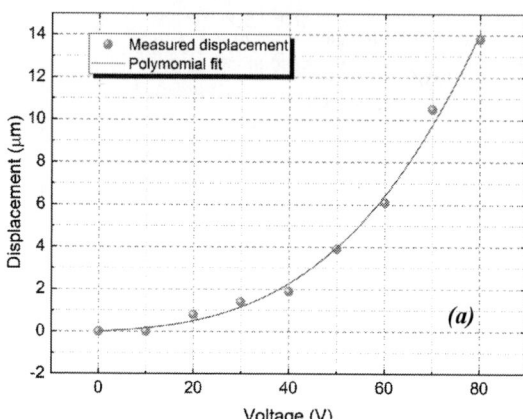

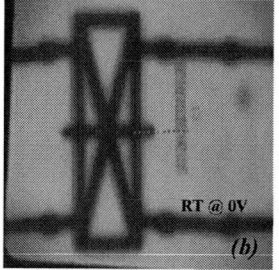
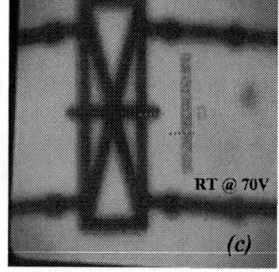

Figure 6: (a) The displacement of the movable shuttle vs. V_{bias}; (b) The actuator at rest position at 25°C (0 V); (c) The device is actuated by 70 V bias at 25°C

structure for actuation, SiC-CNT composite fabrication process is incorporated with a surface micro-machining process. The process starts with a 100 mm Si wafer (Figure 4a). First, a 3 µm thermal oxide is grown on the substrate as the sacrificial layer (Figure 4b). Then, Ti/TiN is sputtered and patterned, acting as electrodes for the device (Figure 4c). Figure 4d to 4g describe the fabrication of SiC-CNT composite, which is the same process as mentioned earlier. Before releasing the SiC-CNT structure, the SiC floor layer needs to be removed by dry etching. This is not only to make sure that the sacrificial layer can be exposed to the etchant, but also to ensure proper probing of the metal pad during measurement. The final release is done by vapour HF so that the risk of stiction can be eliminated.

The final device is demonstrated in Figure 5. The actuator consists of two fixed anchors, one movable shuttle, and suspended beams. The beams are designed to be pre-bended to a certain direction, which determines the actuation direction. From Figure 5b, it can be seen that the movable shuttle is suspended without sagging after being released.

Device characterization
The actuator is first tested at room temperature. The

device is biased at different voltages (V_{bias}), from 0 to 80 V, with a step of 10 V. The Joule heat generated on the device will induce thermal expansion, resulting in a displacement towards the pre-bended direction. The movement of the movable shuttle is read by a scale bar fabricated alongside the shuttle. The displacement as a function of the bias voltage is plotted in Figure 6a. The V_{bias} and displacement relationship aligned well with a third-order polynomial fit. The movement can hardly be seen until V_{bias} reaches 20 V. The maximum displacement is around 13.8 µm, obtained at 80 V. It is worth mentioning that the device breaks down when V_{bias} is approximately 90 V due to the excessive Joule heating.

To verify the high-temperature stability of the composite and the thermal actuator, we measured the device at elevated temperatures, i.e. at 150°C, 300°C, and 450°C. The biasing voltage was set to 70 V to avoid the breakdown of the device due to the external heat source in combination with self-heating. In fact, the breakdown voltage decreases to around 80 V when the ambient temperature rises to 450°C. Figure 7a to 7c demonstrate the thermally-induced displacement, i.e. 9.4 µm (150°C), 11.0 µm (300°C), and 9.9 µm (450°C). The displacement did not show a significant change compared to actuation at room temperature (10.5 µm). Figure 7d shows that the rest position at 450°C is almost the same as that at room temperature; therefore, it can be concluded that the external heat source does not contribute much to the thermal actuation temperature. The reason could be that the SiC-CNT composite has a similar coefficient of thermal expansion (CTE) so that the device expand to a similar extent as the Si substrate.

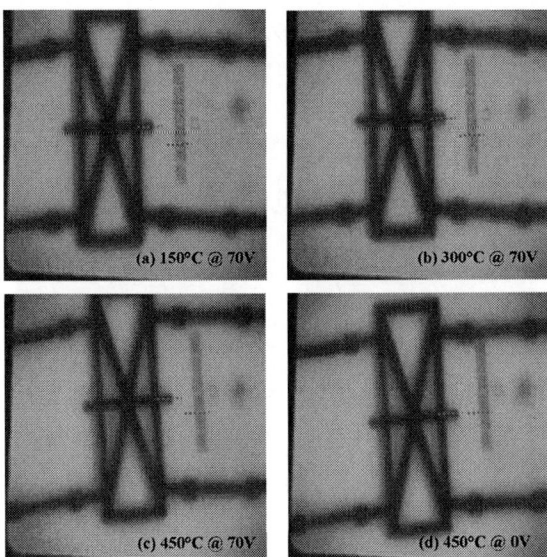

Figure 7: The thermal actuator biased at 70 V at (a) 150°C, (b) 300°C, and (c) 450°C; (d) The shuttle returned to rest position, showing no plastic deformation.

The high-temperature measurement result shows that the thermal actuator can work properly at 450°C , and the SiC-CNT composite showed stable mechanical properties at the same temperature. It needs to be mentioned that the device can possibly work at a higher temperature as the measurement setup was limited to 450°C.

CONCLUSION

In this paper, we presented a generic fabrication technique for harsh-environment applications. The method uses the SiC reinforced VACNT array, enabling fast fabrication of SiC-based HAR structures. It is proven that the electrical behavior of the SiC-CNT composite is dominated by CNT.

The fabrication process of the composite can be incorporated into standard process techniques, such as the surface micromachining process. To illustrate this, a thermal actuator was fabricated and characterized. The device showed stable operation at 450°C successfully, and the composite can potentially withstand an even higher temperature without degradation. The result shows that the SiC-CNT composite can be used as an alternative for SiC for harsh environment MEMS design.

REFERENCES

[1] L. M. Middelburg et al., "Toward a Self-sensing Piezoresistive Pressure Sensor for All-SiC Monolithic Integration", *IEEE Sensors Journal,* vol. 20, pp. 11265-11274, 2020.

[2] J. Mo et al., "Surface-micromachined Silicon Carbide Pirani Gauges for Harsh Environments", *IEEE Sensors Journal,* vol. 21, pp. 1350-1358, 2020.

[3] H-P. Phan et al., "Highly sensitive pressure sensors employing 3C-SiC nanowires fabricated on a free standing structure", *Materials & Design,* vol. 156, pp. 16-21, 2018.

[4] R. S. Okojie et al., "4H-SiC Piezoresistive Pressure Sensors at 800°C with Observed Sensitivity Recovery", *IEEE Electron Device Letters,* vol. 36, pp. 16-21, 2018.

[5] S. Rajgopal et al., "A Silicon Carbide Accelerometer for Extreme Environment Applications", *Materials Science Forum,* vol. 600, pp. 859-862, 2009.

[6] A. Alfaifi et al., "Optimization of In-plane SiC Capacitive Accelerometer Design Parameters", *in 2016 14th IEEE International New Circuits and Systems Conference (NEWCAS),* Vancouver, June 26-29, 2016, pp. 1-4.

[7] A. R. Atwell et al., "Simulation, Fabrication, and Testing of Bulk Micromachined 6H-SiC High-g Piezoresistive Accelerometer", *Sensors and Actuators A: Physical,* vol. 104, pp. 11-18, 2003.

[8] K. Adachi et al., "Single-crytalline 4H-SiC Micro Cantilevers with a High Quality Factor", *Sensors and Actuators A: Physical,* vol. 197, pp. 122-125, 2013.

[9] L. Jiang et al., "SiC Cantilever Resonators with Electrothermal Actuation", *Sensors and Actuators A: Physical,* vol. 128, pp. 376-386, 2006.

[10] L. Belsito et al., "SiC Cantilever Resonators with Electrothermal Actuation", *Journal of Microeletromechanical Systems,* vol. 29, pp. 117-128, 2019.

[11] F. Nabki et al., "Low-stress CMOS-compatible Silicon Carbide Surface-micromachining technology – Part II: Beam Resonators for MEMS above IC", *Journal of Microeletromechanical Systems,* vol. 20, pp. 730-744, 2011.

[12] R. G. Azevedo et al., "A SiC MEMS Resonator Strain Sensor for Harsh Environment Applications", *IEEE Sensors Journal,* vol. 7, pp. 568-576, 2007.

[13] K. Erbacher et al., "Investigation of Deep Dry Etching of 4H SiC Material for MEMS Applications Using DOE Modelling", *in 2021 IEEE 34th International Conference on Micro Electro Mechanical Systems (MEMS),* Gainesville, January 25-29, 2021, pp. 634-637.

[14] K. Dowling et al., "Profile evolution of high aspect ratio silicon carbide trenches by inductive coupled plasma etching", *Journal of Microeletromechanical Systems,* vol. 26, pp. 135-142, 2017.

[15] L. E. Lunaet et al., "Dry Etching of High Aspect Ratio of 4H-SiC Microstructures", *ECS Journal of Solid State Science and Technology,* vol. 6, pp. 207, 2017.

[16] R. Liu et al., "A Dry Etching Method for 4H-SiC via Using Photoresist Mask", *Journal of Crystal Growth,* vol. 531, pp. 125351, 2020.

[17] L. Ci et al., "Continuous Carbon Nanotube Reinforced Composites", *Nano Letters,* vol. 8, pp. 2762-2766, 2008.

[18] D. N. Hutchison et al., "Carbon Nanotubes as a Framework for High-Aspect-Ratio MEMS Fabrication", *Jounal of Microelectromechnical System,* vol. 19, pp. 75-82, 2009.

[19] R. Poelma et al., "Tailoring the Mechnical Properties of High-Aspect-Ratio Carbon Nanotube Arrays using Amorphous Silicon Carbide Coatings", *Advanced Functional Materials,* vol. 24, pp. 5737-5744, 2014.

CONTACT

*J.Mo, tel: +31 630794557; J.Mo@tudelft.nl.

A RELIABLE RELEASE METHOD FOR A BACK-END-OF-LINE NEMS SWITCH OF A MONOLITHIC THREE-DIMENSIONAL INTEGRATED CMOS-NEMS CIRCUIT

Tae-Soo Kim[†], Yong-Bok Lee[†], So-Young Lee, Seung-Jun Lee, and Jun-Bo Yoon

School of Electrical Engineering, Korea Advanced Institute of Science and Technology (KAIST),
KOREA
[†]These authors contributed equally to this work.

ABSTRACT

This paper reports a reliable release method for an aluminum (Al) back-end-of-line (BEOL) NEMS switch of a monolithic three-dimensional (M3D) integrated CMOS-NEMS circuit. We examined the causes of Al damage during the etching process of sacrificial silicon dioxide (SiO_2) when using a hydrofluoric acid (HF solution) and determined that water (H_2O) in the HF solution is the main cause of the Al damage. Therefore, by adding sulfuric acid (H_2SO_4) with strong hygroscopicity to the HF solution, we developed a reliable release method that achieves superb selectivity ($> 10^5$) of SiO_2 over Al. Using the proposed method, we successfully released Al BEOL NEMS switches and demonstrated an M3D CMOS-NEMS circuit with a demultiplexer (DEMUX) function.

KEYWORDS

Reliable Release Method, Monolithic Three-Dimensional Integrated CMOS-NEMS Circuit, Aluminum Back-End-Of-Line NEMS Switch

INTRODUCTION

A monolithic three-dimensional integrated (M3D) CMOS-NEMS circuit [1-5], which is composed of a front-end-of-line (FEOL) circuit and back-end-of-line (BEOL) NEMS switches, has attracted attention due to its reduced path delay and low power consumption compared to those of standard CMOS circuits [5, 6]. In order for the BEOL NEMS switch composed of an aluminum (Al) interconnect to operate mechanically, it is necessary to etch sacrificial silicon dioxide (SiO_2) surrounding the Al BEOL NEMS switch. In general, a hydrofluoric acid (HF solution) is used to etch the sacrificial SiO_2, but fluoride ions (F^-) [7] contained in the HF solution also etch Al [8]. Therefore, dry etch methods including the use of vapor HF [1], carbon

tetrafluoride (CF_4) plasma [2], and sulfur hexafluoride (SF_6) plasma [2] have been employed to etch the sacrificial SiO_2. However, these methods leave undesired residue on the surface of the BEOL NEMS switch after the release process [1, 2, 9, 10]. This residue can increase the contact resistance between the electrodes, resulting in reduced reliability. It is therefore still challenging to reliably release the Al BEOL NEMS switch without any damage or undesired residue during the etching process of the sacrificial SiO_2. Therefore, in this work, we introduce a reliable release method with a high selectivity ($> 10^5$) of SiO_2 over Al using a mixture of HF and sulfuric acid (H_2SO_4), and demonstrate a M3D CMOS-NEMS circuit by reliably releasing Al BEOL NEMS switches using a mixture of HF and H_2SO_4.

CONCEPT

The HF solution etches both SiO_2 and Al. The SiO_2 is etched while reacting with HF to form SiF_4, and Al is etched by reacting with F^- [10] formed by the reaction of water (H_2O) with HF in the HF solution [9].

$$Si\,O_2 + 4HF \rightleftarrows Si\,F_4 + 2H_2O \qquad (1)$$

$$H_2O + 2HF \rightleftarrows H^+ + F^- + (H_3O^+ \cdot F^-) \qquad (2)$$

$$Al^{3+} + 3F^- \rightleftarrows AlF_3 \qquad (3)$$

In the reaction of HF and H_2O, the equilibrium constant (K) representing the ratio of the product to the reactant is so small, 6.7×10^{-4}, that most of the reactants HF and H_2O remain, and only a very small amount of the product F^- is formed. Although the amount of F^- formed in the HF solution is very small, the Al is etched in the solution because the reactivity between Al and F^- is very strong ($K = 5 \times 10^{16}$).

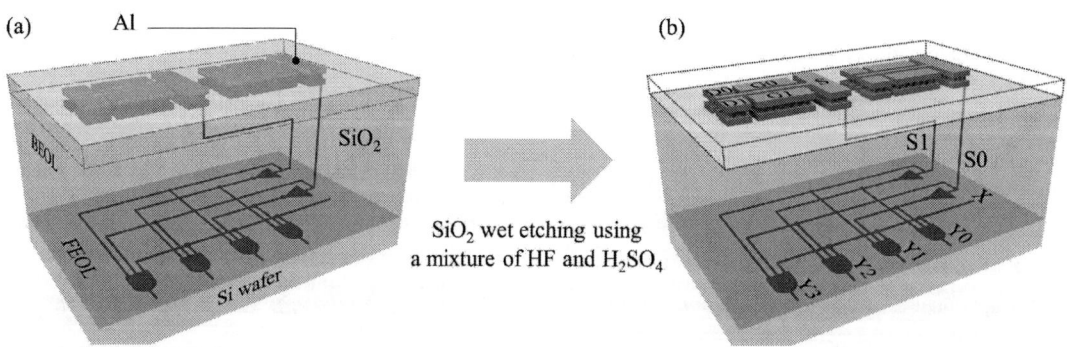

Figure 1: Schematic diagram of the M3D CMOS-NEMS DEMUX (a) before and (b) after the release step.

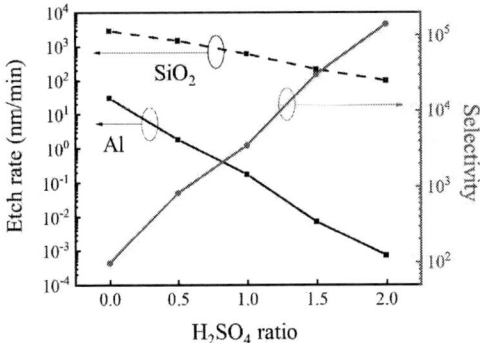

Figure 2: Etch rate and selectivity of SiO_2 (PECVD) and Al (sputtered) according to the ratio of H_2SO_4 in the mixture. 'H_2SO_4 ratio' refers to the volume ratio of the 98 % H_2SO_4 solution to the 48 % HF solution.

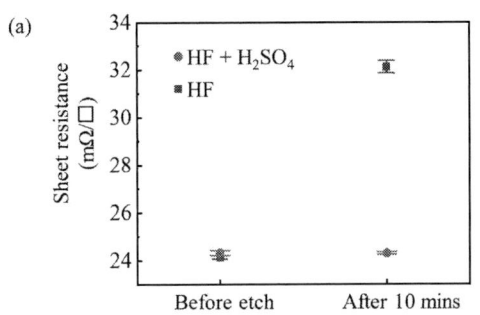

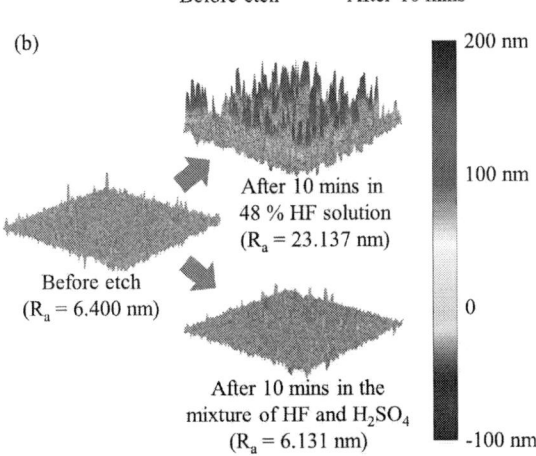

Figure 3: (a) Changes in the sheet resistance of Al after 10 minutes in the 48 % HF solution and in the mixture of HF and H_2SO_4. (b) Atomic force microscope (AFM) measurement results. Changes in the surface roughness (R_a) of Al.

We noted that the formation of F^- that etches Al originates from H_2O in the HF solution. In other words, reducing H_2O in the HF solution can suppress the Al etching [11]. Therefore, we mixed HF with H_2SO_4, which is highly hygroscopic and reacts strongly with H_2O. As a result, in the mixture of HF and H_2SO_4, there is almost no H_2O required to form F^-, thus suppressing the formation of F^-.

$$H_2SO_4 + H_2O \rightleftarrows HSO_4^- + H_3O^+ \qquad (4)$$

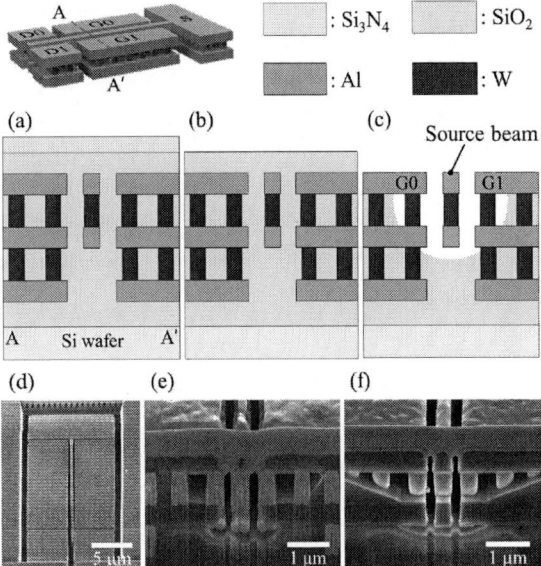

Figure 4: Fabrication process of the Al BEOL NEMS switch. (a) Before release step of the device manufactured by commercial standard IC process. (b) Etching passivation Si_3N_4 layer using CF_4 RIE. (c) Sacrificial SiO_2 wet etching using a mixture of HF and H_2SO_4. (d) SEM image of the device after release step. SEM image of an FIB cross-section of the device released with (e) the 48 % HF solution and with (f) the mixture of HF and H_2SO_4.

EXPERIMENTAL RESULTS

We designed an experiment to investigate whether a mixture of HF and H_2SO_4 would not etch Al. Figure 2 shows the experimental results of the etch rate of SiO_2 and Al according to the composition ratio of a 48 % HF solution and a 98 % H_2SO_4 solution. The etch selectivity of SiO_2 over Al increased as the proportion of H_2SO_4 in the mixture increased. This indicates that the hygroscopicity of H_2SO_4 inhibits the formation of F^- and increases the selectivity. In particular, when enough H_2SO_4 was mixed to react with all H_2O molecules present in the 48 % HF solution (when the volume ratio of the 98 % H_2SO_4 solution to the 48 % HF solution was greater than 2), the etch rate of Al was less than 7×10^{-4} nm/min and the Al hardly reacted.

The resistivity and surface roughness of a material are important factors directly related to the performance of a NEMS switch. Therefore, the sheet resistance and surface roughness of the Al were compared before and after exposure to the mixture. The sheet resistance of the Al film increased by 33.1 % after exposure to the 48 % HF solution for 10 minutes, but increased only by 0.25 % when exposed to the mixture of HF and H_2SO_4 (Fig. 3(a)). The surface roughness (R_a) of the Al film after being exposed to the HF solution for 10 minutes was increased by 262 % and became very rough, but when exposed to the mixture of HF and H_2SO_4 for 10 minutes, the surface roughness was decreased by only 4.20 % and the change was insignificant (Fig. 3(b)).

978-1-6654-9309-3/23 $31.00 © 2023 IEEE

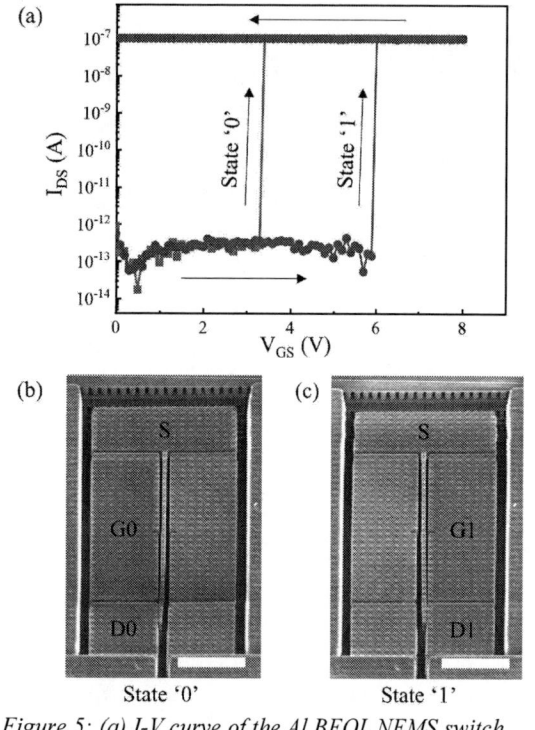

Figure 5: (a) I-V curve of the Al BEOL NEMS switch. SEM image of the Al BEOL NEMS switch programmed to (b) state '0' and (c) state '1' (Scale bars: 5 μm).

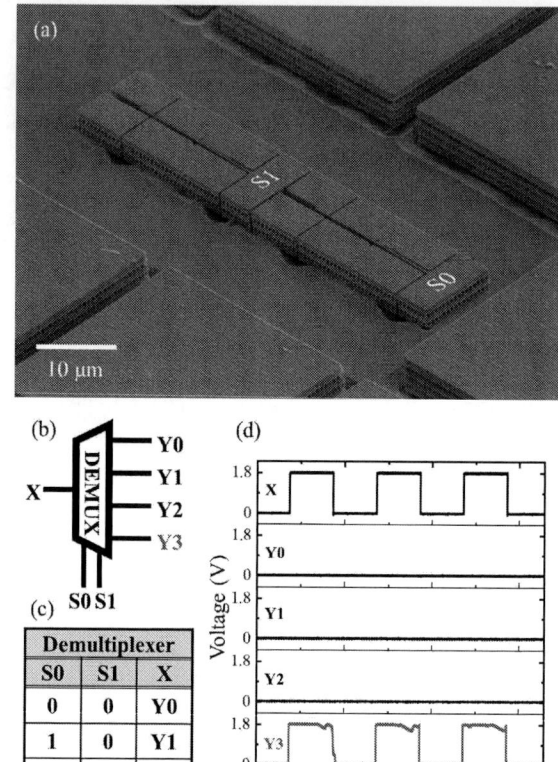

Figure 6: SEM image of the M3D CMOS-NEMS DEMUX released with the mixture of HF and H_2SO_4. (b) Circuit diagram of the M3D CMOS-NEMS DEMUX. (c) Truth table of the DEMUX. (d) Operation results of the M3D CMOS-NEMS DEMUX when both Al BEOL NEMS switches were programmed to state '1'.

FABRICATION

Through the experimental results, we confirmed that the mixture of HF and H_2SO_4 is suitable for releasing the sacrificial SiO_2 of the Al BEOL NEMS switches because it hardly etches Al and preserves the physical properties of the Al such as the sheet resistivity and surface roughness.

We fabricated a monolithic three-dimensional integrated (M3D) CMOS-NEMS circuit through a commercial standard integrated circuit (IC) process. In the initial state, the cross-sectional image of the Al BEOL NEMS switch is as shown in Fig. 4(a). Before etching the sacrificial SiO_2, the top passivation layer, silicon nitride (Si_3N_4), was etched with CF_4 reactive ion etching (RIE) (Fig. 4(b)). The sacrificial SiO_2 was then etched using the proposed mixture of HF and H_2SO_4 (Fig. 4(c)). After etching, the mixture was removed by rinsing with isopropyl alcohol, not water, in order not to damage the device by reacting the mixture with water. Finally, critical point drying (CPD) was used to dry the Al BEOL NEMS switches without initial stiction.

Figure 4(d) shows the released Al BEOL NEMS switch using the mixture of HF and H_2SO_4. We compared cross-sectional SEM images of the Al BEOL NEMS switches released with the 48 % HF solution and the mixture of HF and H_2SO_4, respectively. When the 48 % HF solution was used, Al was etched and the structure of the device was damaged (Fig. 4(e)). On the other hand, when the mixture of HF and H_2SO_4 was used, the device was successfully released without any damage (Fig. 4(f)).

DEMONSTRATIONS

The Al BEOL NEMS switches were designed to be non-volatile through the stiction phenomenon, and thus could store the selection information of the DEMUX.

Figure 5(a) shows the current-voltage (I-V) characteristics and non-volatility of the Al BEOL NEMS switch. When the Gate '0' (G0) voltage reached 3.4 V in the initial state of the Al BEOL NEMS switch, the source beam was deflected to Drain '0' (D0), which is the programmed state '0' of the Al BEOL NEMS switch. Even when the G0 voltage returned to 0, the source beam maintained its adhered state, showing the non-volatility of the Al BEOL NEMS switch. When the voltage of Gate '1' (G1) reached 6.0 V in state '0', the source beam was deflected to Drain '1' (D1) and switched to state '1'. When switching from state '0' to '1', the switching voltage (6.0 V) was increased from the first programming voltage (3.4 V). This was because the actuation airgap between the G1 electrode and the source beam was increased, and a stronger electrostatic force was required to break the adhesion between the source beam and D0. The state '1' also shows the non-volatility of the device that maintained its adhered state even when the voltage of G1 was switched off. Figures 5(b) and (c) present SEM images of programmed state '0' and '1', respectively.

Finally, the M3D CMOS-NEMS circuit with a DEMUX function was demonstrated. Figure 6(a) shows a SEM image of the fabricated CMOS-NEMS circuit corresponding to the schematic diagram shown in Figure 1. Two Al BEOL NEMS switches were M3D integrated on the FEOL circuit, and the selection information of DEMUX was stored in the Al BEOL NEMS switches. A simplified circuit diagram and truth table of the M3D CMOS-NEMS DEMUX are presented in Figs. 6(b-c), respectively. Figure 6(d) shows the operation results of the M3D CMOS-NEMS circuit. When both Al BEOL NEMS switches were programmed to state '1', the input signal 'X' was routed only to output 'Y3' among 4 outputs. This result indicates that the Al BEOL NEMS switches were monolithically well fabricated on a CMOS circuit and it could successfully replace SRAM with low energy consumption.

CONCLUSIONS

In this work, we demonstrated a reliable release method for Al BEOL NEMS switches of a M3D CMOS-NEMS circuit. To reliably release the Al BEOL NEMS switches, we utilized the strong hygroscopicity of H_2SO_4 to remove the H_2O in a HF solution, thereby suppressing the formation of F^- that etches Al. The release method using the mixture of HF and H_2SO_4 achieved superb selectivity of SiO_2 over Al greater than 10^5. The method also preserved the physical properties of Al such as the sheet resistance and surface roughness. Using a mixture of HF and H_2SO_4, we released the Al BEOL NEMS switch without any damage or residue, and confirmed reliable operation of the Al BEOL NEMS switch. Finally, operation of the M3D CMOS-NEMS circuit with the DEMUX function was successfully demonstrated. Furthermore, we anticipate that the release method using HF and H_2SO_4 can reliably release various M/NEMS structures made of Al fabricated by etching sacrificial SiO_2 and will not be limited to the release of M3D CMOS-NEMS circuits.

ACKNOWLEDGEMENTS

This work was supported by the National Research and Development Program through the National Research Foundation of Korea (NRF) by Ministry of Science and ICT (2020M3F3A2A01082600). This work was also supported by Samsung Electronics. The chip fabrication and EDA tool were supported by the IC Design Education Center (IDEC), Korea.

REFERENCES

[1] H. S. Kwon, "Island-style monolithic three-dimensional CMOS-nanoelectromechanical logic circuits.", *IEEE Electron Device Lett*, vol. 41, no. 8, pp. 1257-1260, 2020.

[2] U. Sikder, "Toward Monolithically Integrated Hybrid CMOS-NEM Circuits.", *IEEE Trans Electron Devices*, vol. 68, no.12, pp. 6430-6436, 2021.

[3] J. S. Lee, "In-Memory Nearest Neighbor Search With Nanoelectromechanical Ternary Content-Addressable Memory.", *IEEE Electron Device Lett*, vol. 43, no. 1, pp. 154-157, 2021.

[4] Back, G., "Tri-state nanoelectromechanical memory switches for the implementation of a high-impedance state.", *IEEE Access*, vol. 8, pp. 202006-202012, 2020.

[5] Kim, Y. J., "Nonvolatile nanoelectromechanical memory switches for low-power and high-speed field-programmable gate arrays.", *IEEE Trans Electron Devices*, vol. 62, no. 2, pp. 673-679, 2014.

[6] Qin, T., "Performance analysis of nanoelectromechanical relay-based field-programmable gate arrays.", *IEEE Access*, vol. 6, pp. 15997-16009, 2018.

[7] Harris, D. C., *Quantitative Chemical Analysis*, Macmillan, 2010.

[8] Xue, T., "Effect of fluoride ions on the corrosion of aluminium in sulphuric acid and zinc electrolyte.", *J Appl Electrochem*, vol. 21, no. 3, pp. 238-246, 1991.

[9] Xu, S., "Fluorocarbon polymer formation, characterization, and reduction in polycrystalline–silicon etching with CF 4-added plasma.", *J Vac Sci Technol A*, vol. 19, no. 3, pp. 871-877, 2001.

[10] Choi, W. Y., "Three-dimensional integration of complementary metal-oxide-semiconductor-nanoelectromechanical hybrid reconfigurable circuits.", *IEEE Electron Device Lett*, vol. 36, no. 9, pp. 887-889, 2015.

[11] I. Gablech, "Infinite selectivity of wet SiO_2 etching in respect to Al.", *Micromachines*, vol. 11, no.4, pp. 365, 2020.

CONTACT

†T.-S. Kim, tel: +82-42-350-5476; retaesu@kaist.ac.kr
†Y.-B. Lee, tel: +82-42-350-5476; bok6155@kaist.ac.kr
†Equally contributed authors

INCREASE OF EXPANSION RATE AND DIRECTION CONTROL OF MICROGEL ACTUATORS FOR SINGLE CELL MANIPULATIONS

Kyoka Nakano[1], Hiroki Wada[1], Yoshiyuki Yokoyama[2] and Takeshi Hayakawa[1]
[1]Department of Precision Mechanics, Chuo University, JAPAN and
[2]Toyama Industrial Technology Research and Development Center, Toyama, JAPAN

ABSTRACT

We propose a partial-constraint and side constraint fabrication process of gel actuators to increase expansion rates and control direction of actuation. We patterned dextran as a sacrificial layer to realize partially peel off gel actuator from a substrate. Furthermore, we change the constraint location from bottom to side of gel actuator by peeling off from a substrate. By decreasing constraint area and changing constraint location, we succeeded in increasing volume expansion rates of gel actuators from 1.7 to 5.0 and length expansion rates of gel actuators from 1.1 to 1.9. Furthermore, by changing constraint location, we succeeded in controlling directions of actuations of gel actuators. Finally, we demonstrate a mechanical stimulation device to stimulate living microorganisms by using the proposed method.

KEYWORDS

Gel actuator, Cell manipulation, On-chip actuator, PNIPAAm, Lift-off process

INTRODUCTION

Conventional cell assays are conducted for bulk cell samples that can acquire only an averaged data. It is difficult to obtain differences of individual cells with conventional bulk cell assays. Therefore, single cell analysis that can reveal cellular heterogeneity are attracting large attention in various fields, such as cell biology and medicine [1]. To realize single cell analysis, single cell manipulations are essential techniques. Conventionally, mechanical micromanipulators are widely used for single cell manipulations, such as a trap, transport, rotation, enucleation, or injection of single cells. However, micromanipulator requires mature skills for an operator and it leads to low-throughput and low repeatable process. Thus, single cell manipulations using a microfluidic chip, that are called on-chip cell manipulations, are potential candidate to realize high-throughput and highly repeatable single cell manipulations. On-chip cell manipulation s realize low-cost and high-throughput single cell manipulation and analysis [2]. Generally, on-chip manipulations are classified into contact and non-contact approaches. Non-contact manipulations realize cell manipulations without physical contact to target cells, by using external field such as optical [3], magnetic [4], electric [5] or acoustic fields[6]. These methods are commonly low invasive to cells because they avoid direct contact. However, manipulation forces of these methods are lower than contact manipulations. Therefore, applications of non-contact manipulations are limited. On the other hand, contact manipulations can generate larger force than non-contact manipulations. Conventional contact manipulations are realized by using microtools [7] or microstructures [8] actuated by on-chip actuators. These contact manipulations are used cell stiffness measurement, mechanical cell stimulation and so on [9]. Conventionally, hard materials such as Si or metal are widely used for on-chip actuators to realize strong force and high positioning accuracy. However, these materials are much harder than cells. Thus, this manipulation can damage target cells cells.

Under these circumstances, on-chip soft actuators are proposed for single cell manipulations. They are suitable for low-invasive manipulations because their softness is similar to living cells. Poly (N-isopropylacrylamide: PNIPAAm) are widely used for material of on-chip soft actuators [10-12]. PNIPAAm is thermoresponsive hydrogel which swells at low temperature (32°C ≤) by absorbing surrounding water. Conversely, PNIPAAm shrinks at high temperature (32°C ≥) by exhaling retained water. Thus, we can use this volume change as an actuator because this volume change occurs reversible. Previously, we use PNIPAAm by patterning it on a glass substrate. However, in this fabrication method, volume expansion rates of gel actuators are small as approximately 2 times because they are constrained to the substrate as shown in Fig.1 (a). Furthermore, they can only isotropically swell and direction of actuation cannot be controlled.

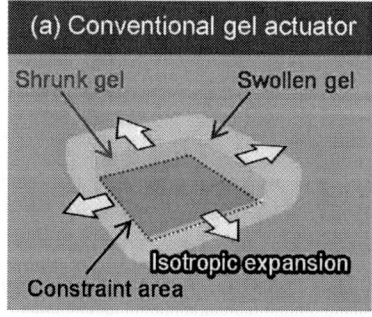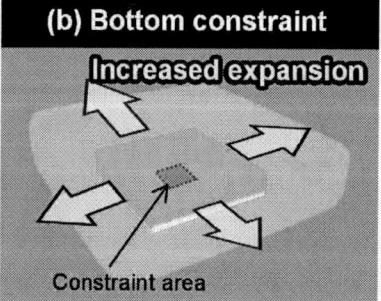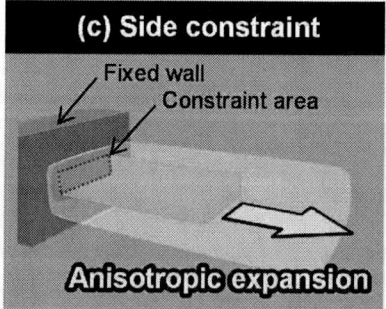

Figure 1: Concept of partially constraint gel actuator on bottom and side
(a) Conventional method (b) partial bottom constraint (c) partial side constraint

In this study, we propose a method that can enlarge deformation of an on-chip soft actuator and control direction of actuation of the actuators.

METHODS

In this study, we propose a partial-constraint fabrication process of gel actuators to realize higher expansion rate and control of direction of actuation. We use Poly (N-isopropylacrylamide) (PNIPAAm) (Bioresist, Nissan Chemical Corporation, Tokyo, JAPAN). By decreasing constraint area at a bottom of a gel actuator, expansion rate becomes higher, as shown in Fig.1 (b). Furthermore, by changing constraint location from bottom to side of a gel actuator, the gel actuator anisotropically expands to an opposite direction of the constraint area, as shown in Fig.1 (c). To realize these constraints, we use dextran as a sacrificial layer because it is dissolved to water but not dissolved to developers of photoresists. For partial constraints at bottoms of gel actuators, we first spincorted dextran solution of 30wt% concentration, as shown in Fig.2 (a) (ii). After that, we patterned the dextran by using lift-off process as shown in Fig.2 (a). Then, we patterned gel actuators over the sacrificial layer, and we can partially constrain bottom of the actuators. For side constraints, we first fabricate SU-8 3025 (Tokyo Ohka Kogyo Corporation, Tokyo, JAPAN) wall and gel actuators on a dextran layer, as shown in Fig.2 (b). The SU-8 wall works as fixed wall to hold gel actuators on its side. Then we realize side constrained gel actuators by dissolving the dextran.

EXPERIMENTS

In this study, we evaluate volume expansion rates and length expansion rate of gel actuators with different constraint areas and locations. We measure volume expansion rate by using confocal laser scanning microscopy (CLSM). We mix fluorescein in gel actuator for 3D observation by using CLSM. We observe 3D structures of shrunk and swollen gel actuators and measure each volume and length from 3D observation results by using ImageJ [13]. The volume of shrunk gel actuator is V_{shrink} and the volume of swollen gel actuator is V_{swell}. We calculate volume expansion rate ε_v of gel actuators by using following equation.

$$\varepsilon_v = \frac{V_{swell}}{V_{shrink}} \quad (1)$$

Next, we measure length expansion rate of gel actuators with different constraint areas and locations. We observe shrunk and swollen state of gel actuators by using bright field images of an inverted microscope. We measure sizes of the actuators in horizontal direction from observation results by using ImageJ [13]. Lengths of shrunk gel actuators are L_{shrink} and lengths of swollen gel actuators are L_{swell}. We calculate length expansion rate ε_L of gel actuators by using the following equation.

$$\varepsilon_L = \frac{L_{swell}}{L_{shrink}} \quad (2)$$

RESULTS AND DISCUSSION
Evaluation of volume and length expansion rates

First, we evaluated volume expansion rates and directions of actuation of gel actuators with different

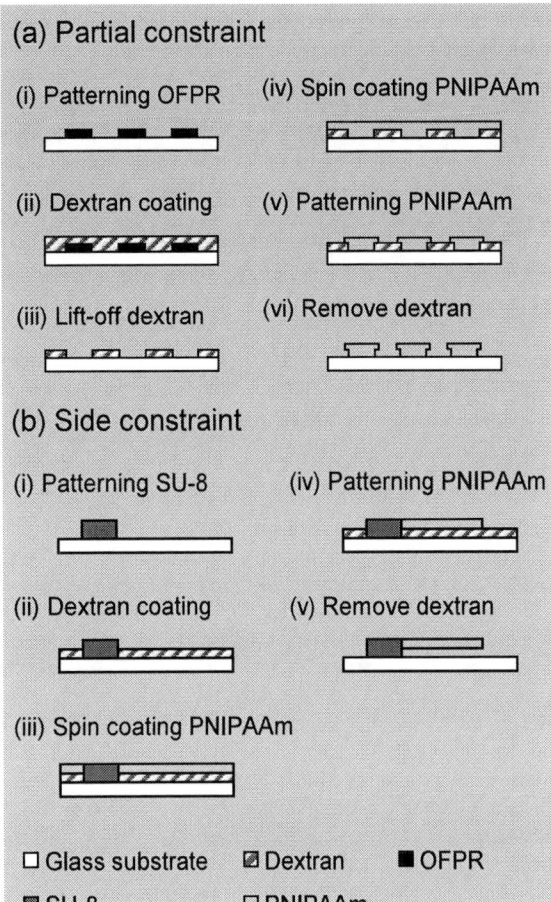

Figure 2: Partial-constraint fabrication process
(a) Partial-constraint process at bottom
(b) Side constraint process

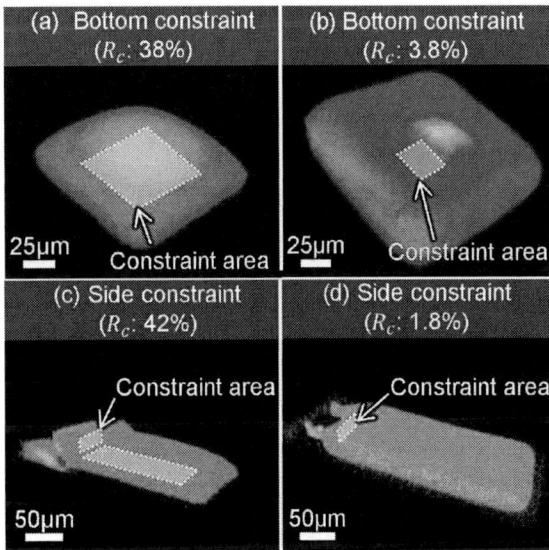

Figure 3: CLMS image of swollen gels
with difference constraint areas and locations
(a) Bottom constraint (b) Partial-constraint at bottom
(c) Side and bottom constraint (d) Side constraint

constraint areas and locations. Figure 3 shows 3D images of swollen gel actuators acquired by using CLSM with different constraint areas and locations. We evaluated volume expansion rates according to ratio of constraint area/ total surface area: R_c. From Fig. 3, gel actuators with lower R_c had higher expansion rates. Furthermore, a gel actuator with side constraint deforms in horizontal direction much larger than actuators with bottom constraints. These results are summarized in Fig.4. By changing R_c from 1.7 % to 45 %, it is possible to change volume expansion rates from approximately 1.7 to 5.0, as shown in Fig.4 (a). Furthermore, it is also possible to increase length expansion rates in horizontal direction from 1.1 to 1.9, as shown in Fig.4 (b). Therefore, we succeeded in changing volume expansion rates and direction of actuation of gel actuators with different constraint areas and locations.

Demonstration of cell manipulation

Finally, we demonstrate mechanical stimulation device using side-constraint on-chip gel actuator to realize mechanical stimulation for living microorganisms. Mechanical cell stimulation is important cell manipulation to analyze mechanotransduction or motility of living cells or microorganisms. We fabricated mechanical stimulation device for living microorganisms. The fabricated device is shown in Fig.5. We fabricated gel actuators with side constraint and attached a SU-8 needle on the actuator for the stimulation. We irradiate infrared (IR) laser to drive gel actuators. We patterned light absorber under the gel actuator. When we irradiate IR laser at light absorber, we can increase temperature and shrink gel actuators [14]. After that, by stopping the laser and swelling the actuator, we can swell the gel actuator and SU-8 needle can push a target cell. By using this device, we demonstrate mechanical stimulation of *Euglena* that is a kind of microalgae. As shown in Fig. 5 (a) and (b), we succeeded in stimulating single *Euglena*.

CONCLUSION

In this study, we propose partial-constraint process of gel actuators to realize higher expansion rate and control of direction of actuation. To realize this process, we use dextran 30wt % solution as a sacrificial layer and we fabricate partial-constraint and side constraint gel actuator by lift-off dextran. We decrease constraint area and change constraint location of gel actuator by partial-constraint process. By decreasing and changing the constraint area of gel actuators, we succeeded in realizing large deformation. The volume expansion changes from approximately 1.7 to 5.0 and length expansion rate enlarge from approximately 1.1 to 1.9 by using this proposal fabrication process. Finally, we demonstrate mechanical cell stimulation as an application by using proposed fabrication process, and we succeeded in stimulating *Euglena*. We will apply the proposed process to various cell manipulations in the future.

ACKNOWLEDGEMENTS

This research is supported by Chuo University Personal Research Grant and JSPS KAKENHI grant Number 21K18703. Photo-prosessable PNIPAAm

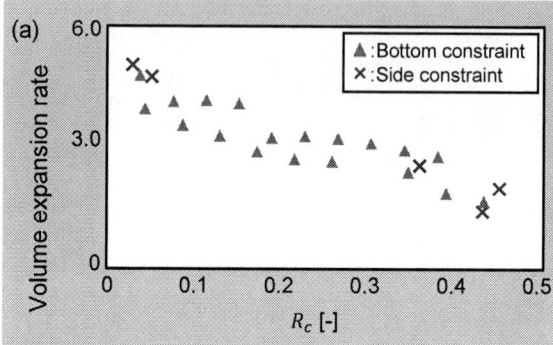

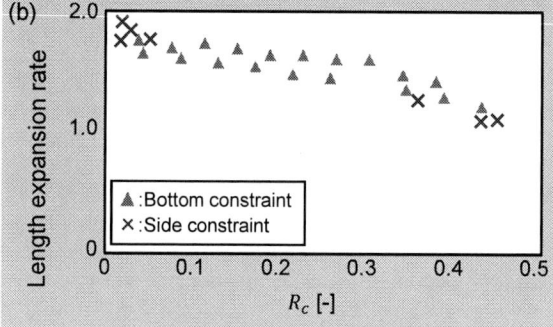

Figure 4: Results of (a) Volume expansion rate and (b) length expansion rate with various R_c values.

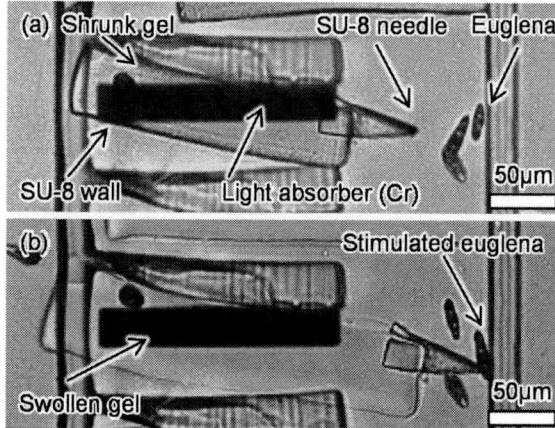

Figure 5: Demonstration of mechanical stimulation of microorganisms by using a side constraint gel actuator (a)Before stimulation, (b) During stimulation

(Bioresist) is provided by Nissan Chemical Corporation.

REFERENCES

[1] Wang, D., Bodovitz, S., "Single cell analysis: the new frontier in 'omics'". *Trends in biotechnology*, 28.6, pp.281-290, 2010.

[2] Yeo, L. Y., Chang, H. C., Chan, P. P., Friend, J. R. (2011). "Microfluidic devices for bioapplications", *small*, 7.1, pp.12-48, 2011

[3] Chiou, P.Y., Ohta, A.T., Wu, M.C., "Massively parallel manipulation of single cells and microparticles using optical images", *Nature*, 436.7049, pp.370–372, 2005.

[4] Liu, W., Dechev, N., Foulds, I. G., Burke, R., Parameswaran, A., Park, E. J., "A novel permalloy based magnetic single cell micro array", *Lab on a Chip*, 9.16, pp.2381-2390, 2009.

[5] C. Wu, R. Chen, Y. Liu, Z. Yu, Y. Jiang, X. Cheng, "A planar dielectrophoresis-based chip for high-throughput cell pairing," *Lab on a Chip*, vol. 17, no. 23, pp. 4008–4014, 2017.

[6] Ding, X., Lin, S. C. S., Lapsley, M. I., Li, S., Guo, X., Chan, C. Y., Huang, T. J., "Standing surface acoustic wave (SSAW) based multichannel cell sorting", *Lab on a Chip*, 12.21, 4228-4231, 2012.

[7] Kawahara, T., Sugita, M., Hagiwara, M., Arai, F., Kawano, H., Shihira-Ishikawa, I., Miyawaki, A., "On-chip microrobot for investigating the response of aquatic microorganisms to mechanical stimulation." *Lab on a Chip*, 13.6, pp.1070-1078, 2013.

[8] Tran, Q. D., Kong, T. F., Hu, D., Lam, R. H. "Deterministic sequential isolation of floating cancer cells under continuous flow", *Lab on a Chip*, 16.15, pp.2813-2819, 2016.

[9] Sugiura, H., Sakuma, S., Kaneko, M., Arai, F, "On-chip method to measure mechanical characteristics of a single cell by using moiré fringe", *Micromachines*, 6.6, pp.660-673, 2015

[10] Arai, F., Ng, C., Maruyama, H., Ichikawa, A., El-Shimy, H., Fukuda, T., "On chip single-cell separation and immobilization using optical tweezers and thermosensitive hydrogel", *Lab on a Chip*, 5.12, pp.1399-1403, 2005.

[11] Ito, K., Sakuma, S., Yokoyama, Y., Arai, F., "On-chip gel-valve using photoprocessable thermoresponsive gel", *ROBOMECH Journal*, 1.1, pp.1-8, 2014.

[12] Hayakawa, T., Sakuma, S., Fukuhara, T., Yokoyama, Y., Arai, F., "A single cell extraction chip using vibration-induced whirling flow and a thermo-responsive gel pattern", *Micromachines*, 5.3, pp.681-696, 2014.

[13] Schneider, C. A., Rasband, W. S., Eliceiri, K. W., "NIH Image to ImageJ: 25 years of image analysis", *Nature methods*, 9.7, pp.671-675, 2012.

[14] Koike, Y., Yokoyama, Y., Hayakawa, T., "Light-driven hydrogel microactuators for on-chip cell manipulations", *Frontiers in Mechanical Engineering*, 6, 2, 2020.

CONTACT

*T. Hayakawa, tel: +81-3-3817-1834; email: hayaka-t@mech.chuo-u.ac.jp

GENERALIZED-ACCUMULATED-TEMPERATURE PARAMETER FOR CHARACTERISTIC PREDICTION OF METAL-BASED MEMS CANTILEVER

Yulong Zhang[1], Jianwen Sun[1], Huiliang Liu[2], and Zewen Liu[1]
[1]School of Integrated Circuits, Tsinghua University, CHINA and
[2]Institute of Telecommunication and Navigation Satellites, China Academy of Space Technology, CHINA

ABSTRACT

The Generalized-Accumulated-Temperature (GAT) is proposed for characteristic prediction of metal-based MEMS cantilever in this work for the first time. GAT is a thermal footprint parameter for thermo-sensitive devices with both temperature and time considered. RF MEMS switches are employed as a typical alloy-cantilever-based devices to investigate the effect of GAT. The GAT parameter is used in thermal budget design and device lifetime prediction of RF MEMS switch, which shows better fitting performance than traditional Larson-Miller (LM) method. Moreover, the GAT parameter can not only help metal-cantilever-based MEMS device realization, will also be used as an important parameter in other thermo-sensitive MEMS devices.

KEYWORDS

MEMS, Metal Cantilever, Lifetime, Generalized-Accumulated-Temperature (GAT) Parameter.

INTRODUCTION

Cantilever, as a typical structure, is widely used in Micro-Electro-Mechanical Systems (MEMS) devices, sensors and actuators included. However, some cantilevers, especially metal-based cantilevers, are sensitive with thermal processes, owing to thermal mismatch[1-2], creep[3-4], recrystallization[5] and so on, which causes a limitation for device reliability and lifetime. Previously, researchers[6] have reported a Larson-Miller (LM) method to evaluate the influence of temperatures on cantilevers. The LM method is suitable for constant temperature analysis, but not suitable for variable temperature processes. Actually, the MEMS devices need to experience complex thermal process in fabrication processes and application with variable temperatures, which results in need of better characteristic prediction method for these MEMS devices.

Accumulated temperature (AT) is an important parameter in agricultural meteorology and climatology [7-8], which shows out the influence of ambient temperature on plants and crops. The crops will not bloom, fruit and mature until the accumulated temperature reaches certain values [9]. It is a way to include temperature and time into one dimension for quantitative evaluation of growth speed of plants. Usually the index of accumulated temperature data is used to create models of crop growth.

In this work, we introduce a novel Generalized-Accumulated-Temperature (GAT) parameter into MEMS field to characterize the influence of complex thermal processes, in which temperature and time are both taken into consideration. Alloy-cantilever-based RF MEMS switches are designed, fabricated and tested to study the GAT parameter. After that, the GAT parameter is used in characteristic prediction of metal-based cantilever, including thermal budget design and device lifetime prediction.

GENERALIZED-ACCUMULATED-TEMPERATURE IN MEMS CANTILEVER

Figure 1 shows the schematic diagram of the GAT-caused deformation of the MEMS cantilever. There is a temperature-induced deformation (Δz_{tip}) on the MEMS cantilever after the thermal process. When temperature is lower than third of melting temperature (T_m), the main influence factor is creep of the cantilever material. The deformation of the cantilever is a function of temperature (T) and time (t), which can be illustrated by GAT parameter, as expressed in (1).

$$\Delta z_{tip} = f(GAT) = G(T,t) \approx \int A(T)e^{-1/T} dt \qquad (1)$$

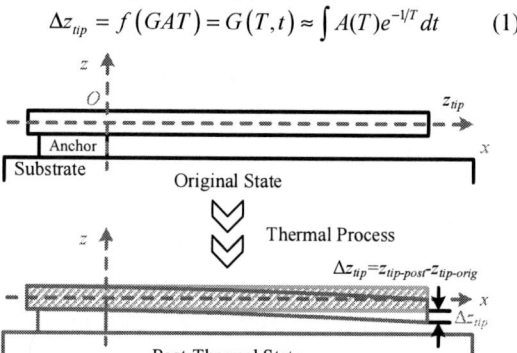

Figure 1: Schematic diagram of the GAT-caused deformation of the cantilever.

FABRICATION OF CANTILEVER-BASED MEMS DEVICES

A simplified gold-alloy-cantilever-based RF MEMS switch is employed to study and verify the GAT method. Figure 2 illustrates a fabrication process at room-temperature, which is used to manufacture the RF MEMS switch, without any extra thermal effect before test strictly. The switches are fabricated on quartz substrate.

The detailed fabrication processes are listed as follows:

(a) First, a 500 nm-thick SiO_2 layer is formed as a buffer layer. A 300 nm-thick Si_3N_4 layer is patterned by low pressure chemical vapor deposition (LPCVD) to define the protection layer (Fig.2a).

(b) Second, a 20/200 nm-thick Cr/Au seed layer is sputtered and a 2 μm-thick gold layer is electroplated as co-planar waveguide (CPW) transmission line (TML) (Fig. 2b).

(c) Then, the Cr/Au seed layer is patterned as contacts (Fig. 2c).

(d) After that, a 2 μm-thick SiO₂ layer is deposited as sacrificial layer (Fig. 2d).

(e) Then, the sacrificial layer is patterned to form contact dimples and anchor (Fig. 2e).

(f) The 2 μm-thick gold-alloy cantilever is electroplated (Fig. 2f).

(g) Finally, the sacrificial layer is released by buffered hydrofluoric acid. The cross-section schematic view of the device is shown in Fig. 2g.

The scanning electron microscope (SEM) photos of the device are shown in Figure 3, with the whole view (Fig. 3a), cantilever view (Fig. 3b) and contact view (Fig. 3c). A red line is labeled in Fig. 3b, which is the observation line by 3D optical microscope (OM) in the followed measurement section.

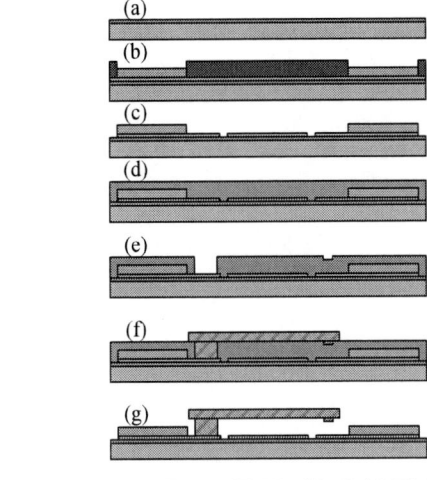

□ Quartz ▨ Si₃N₄ ▨ SiO₂ ■ PR ▨ Gold ▨ Gold Alloy

Figure 2: *Fabrication process of the gold-alloy-cantilever-based RF MEMS switch: (a) Si₃N₄ deposition by LPCVD; (b) Cr/Au seed layer deposition and CPW electroplating; (c) Seed layer patterning; (d) SiO₂ sacrificial layer deposition; (e) Contact dimple and anchor patterning; (f) Gold alloy cantilever electroplating; (g) Sacrificial layer releasing.*

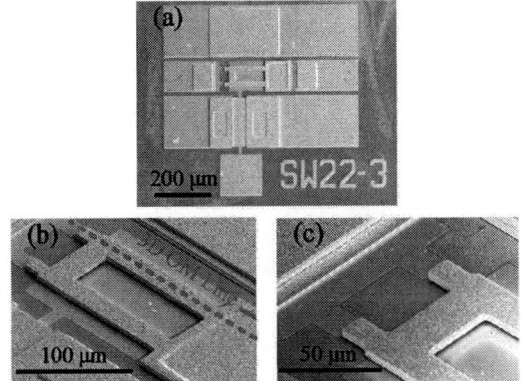

Figure 3: *SEM photos of the device under test with (a) Whole view; (b) Cantilever view; and (c) Contact view.*

MEASUREMENT AND ANALYSIS

The devices are tested by an *in-situ* co-testing method using 3D-OM and RF probe station, in which RF part is used for long-term test, and OM part is used to calculate and calibrate the z_{tip} from S-parameter [5].

The effect of the test method are shown in Figure 4, with the equivalent circuit of the device. The capacitance (C) at off-state is calculated by S-parameters from 50 MHz to 20 GHz. The z_{tip} is tested by 3D-OM. The relationship between C and z_{tip} can be expressed by (2).

$$C = C_0 + \alpha \times e^{\beta \times z_{tip}} \qquad (2)$$

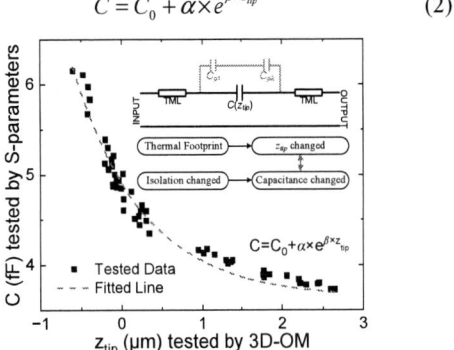

Figure 4: *In-situ co-testing method, with equivalent circuit, and relationship between capacitance and z_{tip}.*

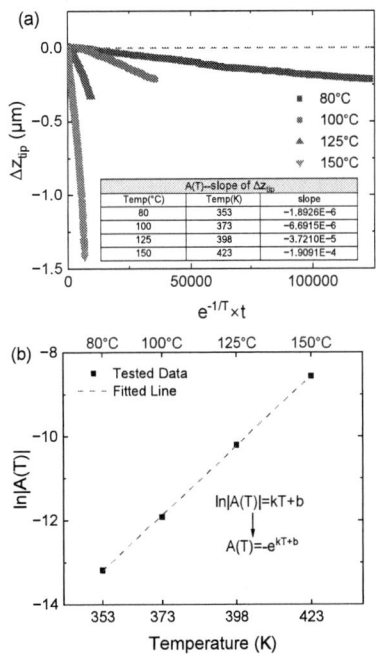

Figure 5: *Tested deformations under different temperature: (a) Relationship between exp(-1/T) × t and Δz_{tip}; (b) Relationship between ln|A(T)| and temperature (T).*

The tested deformations (Δz_{tip}) under different temperatures are plotted in Figure 5. Figure 5a shows the relationship between exp(-1/T)×t and Δz_{tip}, and the slopes of Δz_{tip} ($A(T)$) are listed in Fig. 5a. Figure 5b shows the linear relationship between ln|A(T)| and temperature. $A(T)$ can be expressed by (3), in which k and b are slope and intercept, respectively.

978-1-6654-9309-3/23 $31.00 © 2023 IEEE

$$A(T) = -e^{kT+b} \qquad (3)$$

Based on Fig. 5, the function $f(GAT)$ can be expressed by a formula as (4), which is suitable for calculating Δz_{tip} during complex thermal processes.

$$\Delta z_{tip} \approx \int A(T)e^{-1/T}\,dt = -\int e^{kT+b-1/T}\,dt \qquad (4)$$

CHARACTERISTIC PREDICTION

Thermal processes, for example annealing, packaging, and high-temperature storage, are important influencing factors for metal-based cantilevers, which could result in failure of the cantilever. With the help of GAT parameter, responses of the cantilever can be predicted.

Thermal Budget Design

Thermal budget design is a typical application of the GAT parameter, by which the warping of the cantilever can be controlled.

The gold-alloy cantilevers in devices show different stresses under different electroplating current densities. The controllable residual stress of the gold-alloy cantilever is shown in Figure 6, by which we can control $z_{tip-orig}$ by modifying the electroplating current density.

Effect of thermal budget design at 0.75 ampere per square decimeter (ASD) are shown in Figure 7. Figure 7a illustrates the curling-up profile of the cantilever before thermal process. After a room temperature (RT)~200°C~RT thermal process, the original curly cantilever shows a flat profile, as illustrated in Fig. 7b. The z_{tip} and Δz_{tip} before and after the thermal process are listed in Table 1.

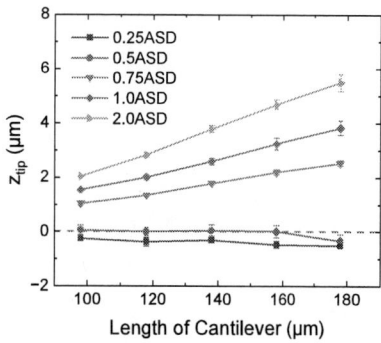

Figure 6: *Residual-stress-caused z_{tip} of the gold-alloy cantilever under different electroplating current density.*

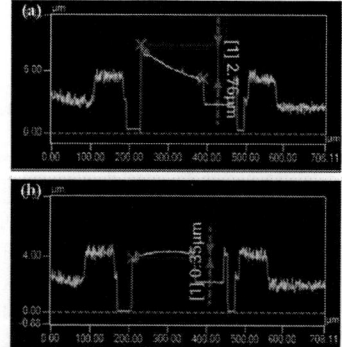

Figure 7: *Profile of the gold-alloy cantilever at 0.75 ASD electroplating current density: (a) Before thermal process; (b) After thermal process.*

Table 1. *$z_{tip-orig}$ $z_{tip-post}$ and Δz_{tip} before and after the thermal process.*

Before Thermal Process	After Thermal Process	Δz_{tip}
$z_{tip-orig}$	$z_{tip-post}$	
2.76μm	-0.35μm	-3.11μm

Lifetime Prediction Method

On the basis of the GAT parameter, lifetime prediction under high-temperature storage can be conducted accurately.

Figure 8a illustrates the GAT-based lifetime prediction method. The Δz_{tip} shows out linear relationship with GAT parameter. The correlation coefficient of the linear fitting is 0.9786. Figure 8b shows the conventional LM method. The Δz_{tip} shows out exponential relationship with LM parameter. The correlation coefficient of the exponential fitting is 0.9746.

Both the predicted lifetimes of the device by the two methods are about 25 days at 50°C. But, the plotted line by GAT ($R^2=0.9786$) shows more accurate fitting performance than LM ($R^2=0.9746$). The GAT lifetime prediction method is more suitable for high-temperature storage than LM.

Moreover, because of the integral form in (3), the GAT method can be used in complex variable thermal process, rather than constant temperature process in LM method.

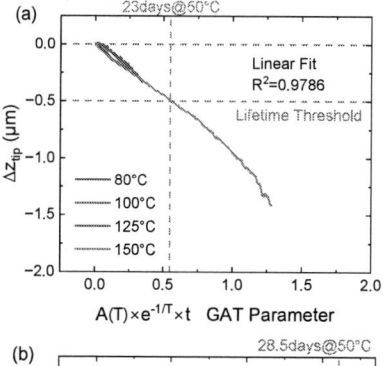

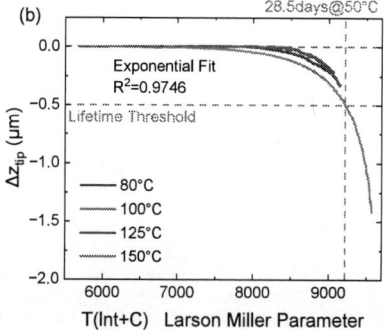

Figure 8: *Comparison of lifetime prediction by GAT method and LM method: (a) GAT method; (b) LM method; in which the plotted line of GAT method ($R^2=0.9786$) shows more accurate fitting performance than LM method ($R^2=0.9746$).*

CONCLUSION

The Generalized-Accumulated-Temperature (GAT) is proposed for MEMS cantilever applications in this work for the first time, which is a thermal footprint parameter for temperature sensitive devices. The GAT method is used in thermal budget design and device lifetime prediction. This work shows three distinctive advancements: (a) first GAT concept introduction into MEMS field to evaluate the influence of different thermal process, for MEMS cantilevers especially; (b) first combination and verification of GAT-based co-design of controllable residual stress and thermal budget for electroplated metal cantilevers; (c) introduction of the GAT-based lifetime prediction method, with its more accurate fitting performance than traditional Larson-Miller (LM) method. These advancements above are studied and verified on basis of simplified gold-alloy-cantilever-based RF MEMS switches. As such, the GAT can not only help metal-cantilever-based MEMS device realization, will also be widely used as a generic concept in other thermal sensitive MEMS devices.

ACKNOWLEDGEMENTS

The authors gratefully acknowledge Mr. Linsong Li and Ms. Wang Zhou from SiMEMS and Si-Era for their help on device fabrication. The research is supported by National Key R&D Program of China No. 2018YFB2002801.

REFERENCES

[1] T.J. Kang, J.G. Kim, J.H. Kim, K.C. Hwang, B.W. Lee, C.W. Baek, Y.K. Kim, D. Kwon, H.Y. Lee, Y.H. Kim, "Deformation characteristics of electroplated MEMS cantilever beams released by plasma ashing", *Sens. Actuat. A Phys.*, vol. 148, pp. 407-415, 2008.

[2] M. Teranishi, C. Chen, T.M. Chang, T. Konishi, D. Yamane, K. Machida, K. Masu, M. Sone, "Enhancement in structure stability of gold micro-cantilever by constrained fixed-end in MEMS devices", *Microelectron. Eng.*, vol. 187–188, pp. 105–109, 2018.

[3] M. Barbato, A. Cester, G. Meneghesso, "Viscoelasticity Recovery Mechanism in Radio Frequency Microelectromechanical Switches", *IEEE T. Electron. Dev.*, vol. 63, pp. 3620–3626, 2016.

[4] H. Hsu, M. Koslowski, D. Peroulis, "An Experimental and Theoretical Investigation of Creep in Ultrafine Crystalline Nickel RF-MEMS Devices", *IEEE T. Microw. Theory*, vol. 59, pp. 2655–2664, 2011.

[5] Y. Zhang, J. Sun, H. Liu, Z. Liu, "Modeling and Measurement of Thermal–Mechanical-Stress-Creep Effect for RF MEMS Switch Up to 200 °C", *Micromachines*, vol. 13, 166, 2022.

[6] T. Moran, C. Keimel and T. Miller, "", in *Proc. EuMIC 2016*, London, UK, October 3–4 2016, pp. 440-443.

[7] A. de Candolle, *Géographie Botanique Raisonné.* vol. 2, Paris: V. Masson, 1855.

[8] Accumulated Temperature, [Online]. Available: https://glossary.ametsoc.org/wiki/Accumulated_temperature

[9] E. Bunting, "Accumulated temperature and maize development", England. *The Journal of Agricultural Science*, vol. 87, pp. 577-583, 1976.

CONTACT

*Zewen Liu, tel: +86-010-62789241; liuzw@tsinghua.edu.cn

MEMS-BASED WATER COLLECTION CONDENSATION PARTICLE COUNTER (WCCPC) OPTIMIZED FOR MULTI-POINT MONITORING OF AIRBORNE NANOPARTICLES

Seong-Jae Yoo[1], and Yong-Jun Kim[1]
[1]School of Mechanical Engineering, Yonsei University, Korea

ABSTRACT

Although monitoring of airborne nanoparticles is important due to their fatal health risk, personal devices for this purpose have not been developed yet. For accurate and sensitive monitoring of nanoparticle in personal environments, we developed a compact and low-cost MEMS-based WCCPC. Our system was able to accurately detect test nanoparticles through quantitative experiments (within 7.57% difference), and showed performance comparable to that of the reference CPC for one week in an external environment.

KEYWORDS

Water collection condensation particle counter, Airborne nanoparticle, Multi-point monitoring.

INTRODUCTION

According to recent studies, airborne particulate matter (PM) is the primary cause of respiratory and heart diseases. [1] Nanoparticles (NPs; particles less than 100 nm in diameter) account for the majority of PM number concentration. [2] Notably, NPs are more likely to infiltrate into the alveoli without being filtered by the bronchi. Infiltrated NPs are viewed as more important from a health standpoint because they are more toxic than micro-sized particles. [3] Urban NPs are predominantly generated from fossil fuel combustion. In addition, human exposure to NPs has increased in daily life owing to the use of nanomaterials in the medical, beauty, food, and packaging industries, etc. Accordingly, global projects, such as EXPOSoMICS, are conducting research on the effects of NP exposure on respiratory conditions. [4]

CPC is used as a reference instrument because it is high accurate and reliable in the counting of nanoparticles (It used the nucleation light scattering technique to count individual nanoparticles). However, commercial CPCs are high-cost/oversized instruments that is mainly used for nanoparticle measurements by being fixed in a station [3]. In addition, most CPCs used an alcohol-based working fluid vapor that is harmful to the human body for the condensation growth of nanoparticles and the working fluid requires continuous replenishment.

To overcome the limitations, our research them developed a new MEMS-based WCCPC that used water as the working fluid and collects used water from the atmosphere. Our system was able to collect remaining vapor after condensation growth using the moderated water condensation method. Our system has been processed to characterization in the laboratory and long-term testing in an outdoor environment, and through this, the stable operation of our system and high accuracy with reference equipment has been demonstrated.

SYSTEM OPRTATION

Our system grows nanoparticles to the micro-sized droplet by water vapor onto them and then count the grown droplets by the light-scattering method. Our system consists of three key components: reservoir, nanoparticles growth section and miniature optical particle counter (OPC) (Fig.1). Working fluid was transported from the reservoir to the growth section channel wall by the capillary action of the SUEX dry film-based hydrophilic micropillar array wick. As the entrained air sample enters the conditioner, it reaches the conditioner wall temperature (10°C) and 100% RH. The initiator (40°C) creates supersaturated air using the mass diffusion rate of water vapor and the thermal diffusion rate of air. Since the mass diffusivity of water vapor (0.28cm2·s-1) is higher than the air thermal diffusivity (0.19cm2·s-1), water vapor reaches the center of the sample stream at a faster than the temperature. At this point, the air sample becomes supersaturated, nanoparticles acted as condensed nuclei to grow into micro-sized droplets.

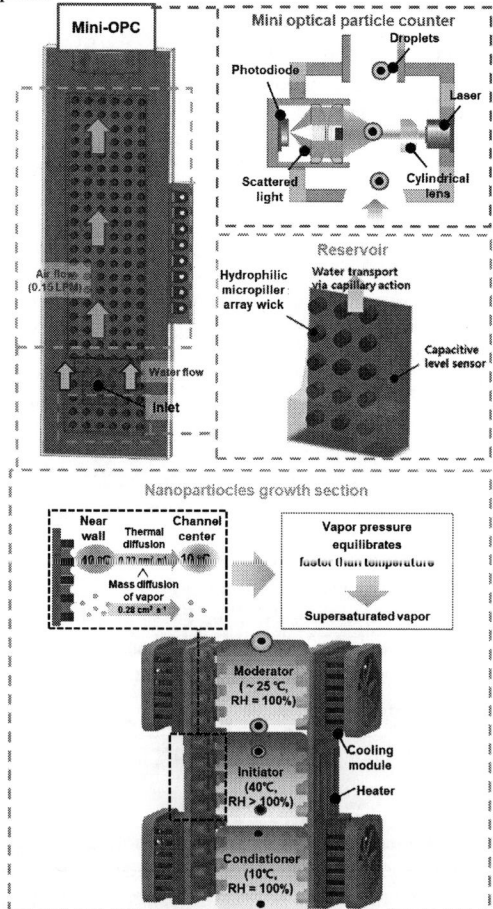

Figure 1: Operating principle of our system.

The air sample containing the micro-sized droplets reaches the moderator, which re-cools the air sample to a lower temperature. The cooled air sample is saturated and the water vapor remaining after the micro-sized droplet condenses on the channel walls. The condensed water is transported to the condenser for reuse. The grown micro-sized droplets are counted by mini OPC.

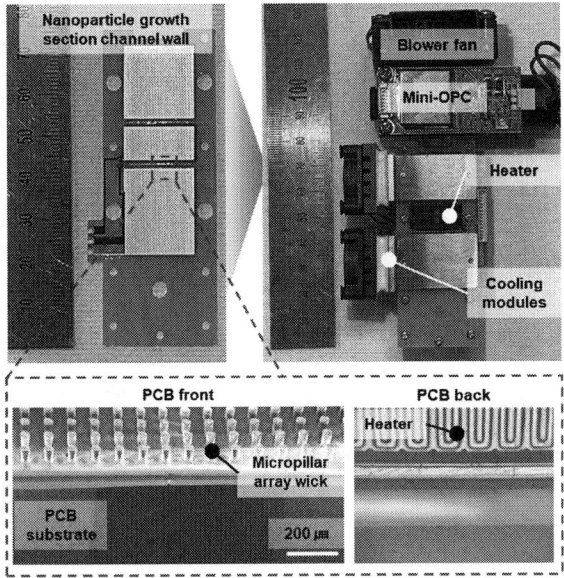

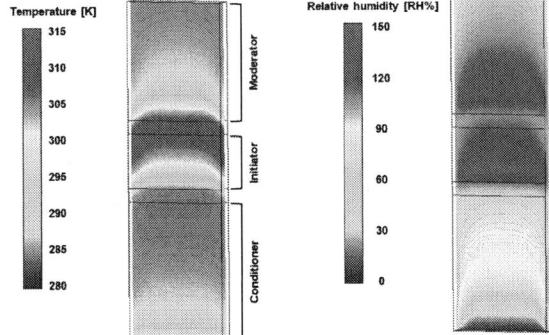

Figure 2: Our system fabricated with two printed circuit boards and a 3D printed channel.

RESULTS

The temperature, relative humidity profile inside nanoparticle growth channel were calculated using computational fluid dynamics (CFD) flow simulation (Fig. 3). As a result of the simulation, it was confirmed that the air introduced into our system reached the designed temperature and relative humidity.

[Simulation figure placed here in original flow — see below]

Figure 3: Simulation results of our system

Our system was evaluated using NaCl and Ag2O nanoparticles (5~100nm) and compared with the reference instrument (Aerosol electrometer 3068b/CPC 3776, TSI Inc. USA). Our system was able to measure particles with a size of 6 nm, and the 50% counting efficiency, which is the main performance indicator of CPC, was 9.4 and 23.7nm respectively (Fig. 4).

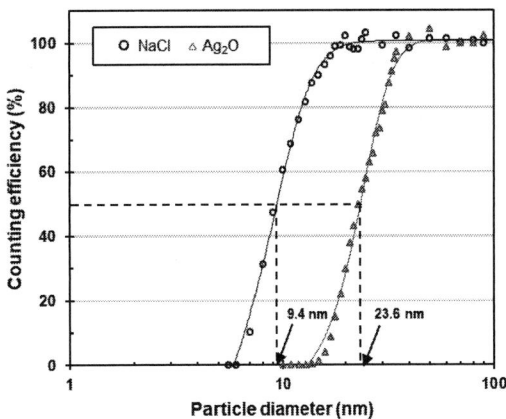

Figure 4: Comparison of counting efficiency between our system and reference aerosol instrument (aerosol electrometer).

Fig. 5 is the result of measuring the nanoparticle measurement accuracy of our system in the low and high number concentration range (0~270000N/cm^3). Our system showed a difference of 7.08% from the reference CPC in the range of 0~20000N/cm^3, and when there were more than 20000N/cm^3 particles, a difference of 7.57% was showed when simple calibration was performed.

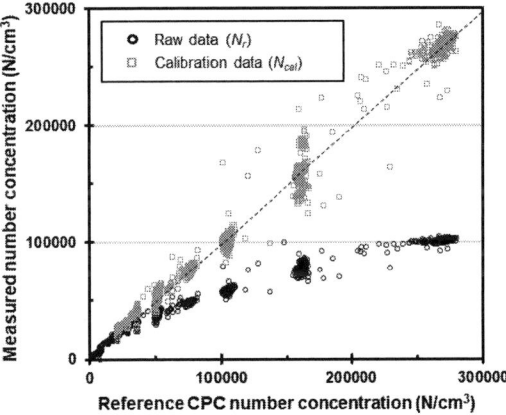

Figure 5: Comparison of nanoparticle measurement accuracy between our system and reference CPC in the low and high number concentration.

When compared with the reference CPC for one week in an external environment, a correlation coefficient of 0.91 was shown, and stable operation was possible (Fig. 6). The above results indicate that our system is not only inexpensive/compact, but also shows the possibility of stable and high-accuracy measurement of nanoparticles.

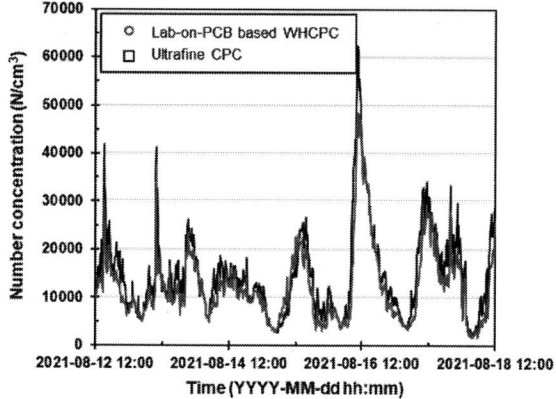

Figure 6: Comparative evaluation of our system and reference CPC in an outdoor environment for one week.

ACKNOWLEDGEMENTS

This work was supported by the National Research Foundation of Korea(NRF) grant funded by the Korea government(MSIT). (No, 2022M3C1B6090739).

REFERENCES

[1] Neuberger, Manfred, et al. "Acute effects of particulate matter on respiratory diseases, symptoms and functions: epidemiological results of the Austrian Project on Health Effects of Particulate Matter (AUPHEP)." Atmospheric Environment 38.24 (2004): 3971-3981.

[2] Watson, John G., et al. "Variations of nanoparticle concentrations at the Fresno Supersite." Science of the Total Environment 358.1-3 (2006): 178-187.

[3] Schmid, Otmar, and Tobias Stoeger. "Surface area is the biologically most effective dose metric for acute nanoparticle toxicity in the lung." Journal of Aerosol Science 99 (2016): 133-143.

[4] Viana, Mar, et al. "Workplace exposure and release of ultrafine particles during atmospheric plasma spraying in the ceramic industry." Science of the Total Environment 599 (2017): 2065-2073.

CONTACT

* Yong-Jun Kim, tel: +82-2-2123-7212;
  yjk@yonsei.ac.kr

RECONSTITUTING FUNDAMENTALS OF
BACTERIA MEDIATED CANCER THERAPY ON A CHIP

Wonjun lee[1†], Jiin Park[2†], Dongil Kang[3] and Seungbeum Suh[4]*

[1]Department of Mechanical Engineering, Seoul National University, Republic of Korea
[2]Department of Life Science, Ewha Womans University, Republic of Korea
[3]Department of Biomedical Engineering, Hanyang University, Republic of Korea
[4]Center for Healthcare Robotics, Korea Institute of Science & Technology, Republic of Korea

ABSTRACT

This paper reports a microfluidic platform tailored for emulating fundamentals of bacteria-mediated cancer therapy (BMCT). Specifically, we attempted to replicate the basic features of the bacteria-colonized tumor microenvironment and the accompanying immune response in a single well of the device. The effects of bacterial infection or stimulation on tumor spheroids were observed to experimentally examine bacteria's innate antitumor cytotoxicity and potential as a therapeutic agent carrier *in vitro*. Consequently, we observed enhanced dead signals as well as decreased stemness of tumor spheroids as a result of bacterial cytotoxicity and IFN-β secretion, respectively. Furthermore, enzyme-linked immunosorbent assay results from cell culture media indicative of co-cultured macrophage's M1 polarization demonstrated bacteria's potential for immune-activating adjuvant. Our device, with a user-friendly platform leveraging spontaneous capillary, will offer a chance to scrutinize BMCT at cellular and tissue levels.

KEYWORDS

Organ-on-a-chip, Bacteria mediated cancer therapy, Tumor microenvironment

INTRODUCTION

In recent years, technical advances in the biomedical area have dramatically augmented our understanding of oncology, and this body of knowledge has been successfully translated into novel treatments or strategies for cancer. However, the hypoxic environment of the tumor microenvironment (TME) still remains to be a major hurdle for researchers since it not only restricts therapeutic efficacy but also directly promotes the establishment of a tumor immunosuppressive milieu [1]. The workaround adopted by some researchers was to utilize hypoxia-responsive nanoparticles to achieve a preferential engagement in hypoxic tumor tissues, but the inactive nature of nanoparticles resulted in unsatisfactory outcomes [2]. In this context, the hypoxia-targeting capacity of bacteria has drawn attention to using bacteria as an effective therapeutic agent in cancer therapy. In addition to hypoxia targeting ability, their immunogenicity produced a positive antitumor immune response, further breaking new ground for bacteria-based tumor immunotherapy [3]. The complexity of bacteria as living organisms was ambivalent to researchers. Achievement of unprecedented antitumor action was possible with fine-tuning of many functionalities, though challenging.

However, due to the intricacy of the triangular interactions between bacteria, tumors, and the immune system, contemporary works on BMCT are mostly academic rather than practical and limited to engineered model organisms. Their method of examining resected tumors or mice inoculated with bacteria and cancer precludes the analysis of intrinsic features of individual cells underlying the phenomenon. Furthermore, given the substantial disparities in fundamental physiology between humans and engineered model organisms, it is evident that such restrictions might lead to translational failure and impede future "bench-to-bedside" BMCT deployment. Therefore, a more human-centric *in vitro* solution to support current *in vivo* models for their use as "pre-preclinical" assessments of BMCT is in demand.

To this end, we propose a novel microfluidic platform optimized using a rapid prototyping technique to recapitulate the fundamentals of BMCT *in vitro*. Here, we aimed to emulate the basics of 1) a bacteria-colonized tumor microenvironment and 2) the corresponding immune response in a single well of the microfluidic device. Macrophages and bacterial-infected tumor spheroids were selectively patterned into defined regions of the device by spontaneous capillary flow under hydrophilic conditions. Increased pro-inflammatory cytokine production from macrophages indicated the implementation of the bacterial inflammation concept as patterned cells proliferated and generated a corresponding 3D tissue environment. As a disclaimer, let us emphasize that even though the assays we've done here were straightforward, our methodology is easily extensible for advanced systems that are more sophisticated.

MATERIALS AND METHODS

Platform design and fabrication

The PMMA chip body of the microfluidic device was fabricated by a commercial $CO2$ laser cutter (Epilog Mini 24, Epilog). The inlets and open-channel reservoirs were created with vector cut, whereas the microchannels were created by raster engraving mode. According to the desired microchannel dimensions, the laser's speed and power were mutually adjusted (Figure 1B). Solidworks (Dassault System, USA) was used to transfer drawings to the laser cutter. 3M™ 9795R advanced polyolefin diagnostic microfluidic medical tape (thickness: 50 μm) as the substrate for the chip was then subsequently bonded to the UV sterilized PMMA body part to complete the device.

Bacterial strains and materials

The microbial species used are *Escherichia coli* (*E. coli*), *Salmonella Typhimurium* (*S. typhimurium*), and IFN-β secreting *S. typhimurium* (kindly provided by Dr. Heung Jin Jeon, Chungbuk National University). The microbial species were cultured overnight at 37°C in a Luria-Bertani (LB) broth under shaking conditions. 100 μL overnight cultures were inoculated in 10 mL fresh LB broth and incubated at 37 °C and 100 rpm in the shaking incubator. Once the bacteria were confirmed to be in the log phase, cells were collected by centrifugation at 6000 rpm for 5 min. After discarding the supernatant, the bacterial pellet was suspended in 1x phosphate-buffered saline (PBS; Gibco), and the concentrations were properly adjusted according to the experimental requirements.

Cell cultivation and spheroid preparation

Human umbilical endothelial cells (HUVECs; Cefobio, Korea) were cultured in endothelial growth medium 2 (EGM-2; Lonza) and used in experiments between passages 4-5. NIH-3T3 (KCLB, Korea) cell line and macrophages RAW 264.7 (KCLB, Korea) were cultured in Dulbecco's modified Eagle's medium (DMEM; Hyclone) supplemented with 10% fetal bovine serum (FBS; HyClone, USA) and 1% penicillin-streptomycin (PS; Gibco, USA). The murine breast cancer cell line 4T1 (ATCC, USA) was cultured in Roswell Park Memorial Institute 1640 (RPMI; Hyclone) supplemented with 10% FBS and 1% PS. All cells were maintained in a humidified incubator at 37°C and 5% CO2.

Cell suspensions were harvested in a 1% agarose (Promega, USA) pre-coated 96-well plate with a U-shaped bottom well (SPL Lifescience, Korea) to prepare spheroids for the experiments. Co-culture tumor spheroids were initiated by mixing 4T1s and NIH-3T3s (at a ratio of 2:1; total cell was 15,000). After pre-culturing in a wellplate for 4 days, spheroids were infected with different bacterial agents at different multiplicity of infection (MOI). In the case of IFN-β secreting *S. typhimurium*, L-arabinose dissolved in the medium was added at 0.01% (w/v) into wells for protein induction. After infection, spheroids were washed with 1× PBS gently and treated with 100 μg/mL gentamicin for 2 h to kill extracellular bacteria.

Hydrogel and cell patterning

Prior to loading, the surface of the microfluidic device was hydrophilized with plasma surface treatment at 70W for 3 min (Femto Science, Korea). The following steps were taken to pattern the cells in the chip: the central channel was patterned with 9 μL of cellular (RAW 264.7, final concentration 1×10^6 mL^{-1}) Matrigel solution (final concentration 2.5 mgmL^{-1}; Corning, USA). Infected spheroid and control spheroid were encapsulated into cellular Matrigel solution (NIH-3T3, final concentration 1×10^6 mL^{-1} & HUVECs, final concentration 3×10^6 mL^{-1}) and were then introduced to the upper side channel and lower side channel, respectively. After waiting 45 min for the gels to polymerize, EGM-2 was added to the reservoirs, and the chips were kept in an incubator.

RESULTS AND DISCUSSION

Design and function of Microfluidic device

The microfluidic device used in this study consists of single PMMA body housing the microfluidic patterning geometries and the media reservoir, with an adhesive

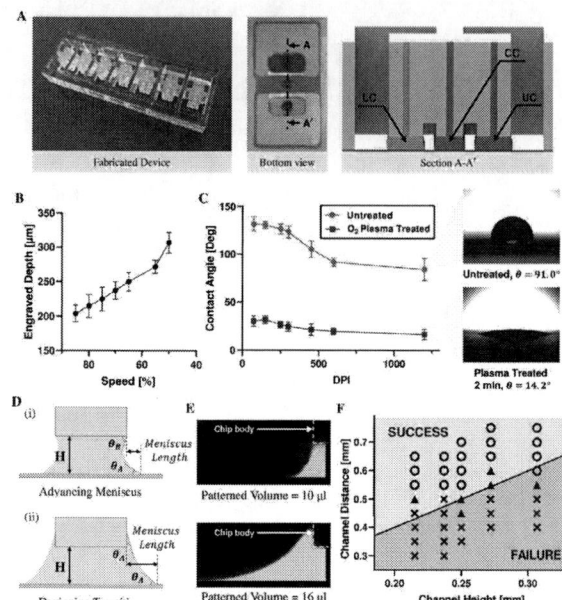

Figure 1: A) Representative images and conceptual rendering of the fabricated microfluidic device. B) Laser engraved depth of PMMA with different laser speeds where 100% corresponds to the actual speed of 85mm/s [4]. C) The results of contact angle measurements before and after plasma treatment on engraved PMMA with different DPI. D) Schematic illustration of two possible capillary liquid configurations. E) Images of the cross-sections of the devices filled with a FITC solution under the noted volume. F) Design criteria for optimal fluid patterning between distinct channels in terms of volume and distance. Experimental results were derived through three consecutive successes, three consecutive failures, or the remaining cases.

bonded to a polycarbonate (PC) film substrate. Each unit well consists of 3 microchannels that are disposed in parallel and partitioned along perpendicular distances (Figure 1A). Working as a pathway of nutrition and oxygen, all three channels are directly exposed to the media chamber for cell feeding. The cell patterning region contains a single hole in the center, designed to capture a large spheroid. Liquid can be patterned in the designated structure by directly sticking a pipette into the injection hole. Compared to equivalent conventional soft lithographic PDMS-based platforms, our laser-cutter-manufactured platform is capable of much higher throughput and flexibility in terms of operation and fabrication [4]. Specifically, including device fabrication, substrate bonding, and plasma treatment, preparing a final device for the experiment took less than 10 minutes. In contrast to soft lithography, which takes multiple steps, this rapid prototyping requires only one: transferring a prepared file to the laser cutter and then waiting for the output. This simplicity has the consequence of increasing design flexibility, which adds to the convenience of optimization. The primary issue with PDMS-based devices, small molecule adsorption, is also resolved by the material choice of a PMMA body and a silicon-based substrate [5]. The

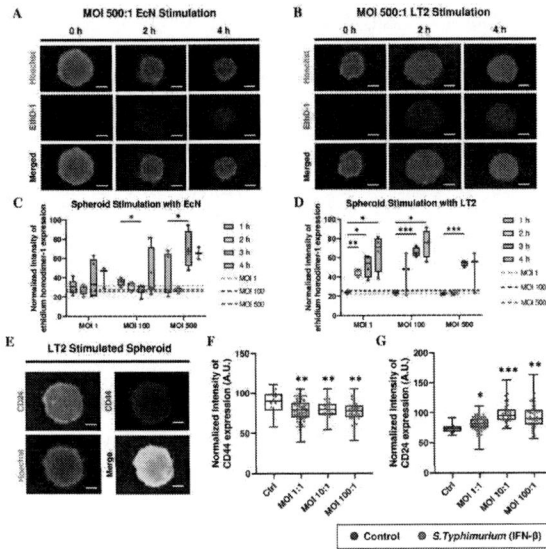

Figure 2: A-B) Representative images of co-culture tumor spheroid (4T1 and NIH3T3) stained with ethidium homodimer-1 (red; dead cells) and Hoechst (blue; nuclei) in the presence of bacterial inflammation of Escherichia coli (E.coli) and S.typhimurium, respectively. Scale bars, 100 μm. C-D) Box plot of normalized fluorescent intensity of ethidium homodimer-1 from the E.coli and S.typhimurium inflammation model, respectively. E) Representative images of tumor spheroids stained with Hoechst (blue), CD24 (green), and CD44 (red) from the S.typhimurium (IFN-β) inflammation model. Scale bars, 100 μm. F-G) Normalized intensities of CD24 and CD44 expression in tumor spheroids from the S.typhimurium (IFN-beta) inflammation model after 18 h of infection under different MOI. $*p < 0.05$, $**p < 0.005$, and $***p < 0.0005$ in two-tailed unpaired t-tests.

aforementioned traits all together made it easier to optimize the microfluidic device and deploy a 3D culture for emulating a bacteria-colonized TME.

Selective Microfluidic Patterning

The designed microfluidic device utilizes air plasma-induced hydrophilic surface modification to facilitate spontaneous capillary flow patterning (SCP) of droplets [6] (Figure 1-C). In essence, liquid confined in a slit of hydrophilic surface does not maintain a right-angled air-liquid contact, which might result in the failure of hydrogel patterning. The capillary liquid underneath the body will initially form a concave interface when loaded along the injection port and advance toward the structure's perimeter while maintaining θ_A as a contact angle with the bottom substrate. Since the bottom substrate is large enough, the menisci meet it at Young's contact angle θ_A, but there are two alternatives for each menisci's top portion: (1) They may be pinned at the edges, forming an angle of θ_B w.r.t. the horizontal, and (2) the menisci may be unpinned and placed outside the open ends, with the contact angle θ_A wetting sidewall of the rail [7] (Figure 1D). The air-liquid meniscus length on the bottom substrate must be meticulously managed under these circumstances in order

for the liquid in each channel to be precisely patterned without interfering with one another, making instance (1) more conducive to patterning [8]. Accordingly, the proper liquid volume was estimated under the assumption that the menisci were circumferential and that their surfaces were entirely hydrophilic ($\theta_A = 0$, $\theta_B = \pi/2$). Half of the distance between the two channels must be less than the meniscus length for the channel to be selectively patterned, and meniscus length L can be formulated as follows:

$$\mathbf{L} = \mathbf{H_0} tan(\frac{\theta_B}{2} - \frac{\theta_A}{2}) < \mathbf{H_0} \qquad (1)$$

where θ_A, θ_B, and H_0 are the bottom contact angle, body contact angle, and channel height, respectively. A parametric study was conducted to validate this theoretical analysis. Different channel heights and spacing were used during the experiment. A designated volume of PBS was injected into each channel, which was then examined to see if any coloring mixture had happened. The correctness of the theoretical analysis was validated in the majority of experimental groups and corroborated the design rule for successful patterning (Figure 1F). We hypothesize that the laser cutter's instability is to blame for minor failures that occur close to the boundary.

Bacterial infection on tumor spheroid

One of the primary advantages of using bacteria for

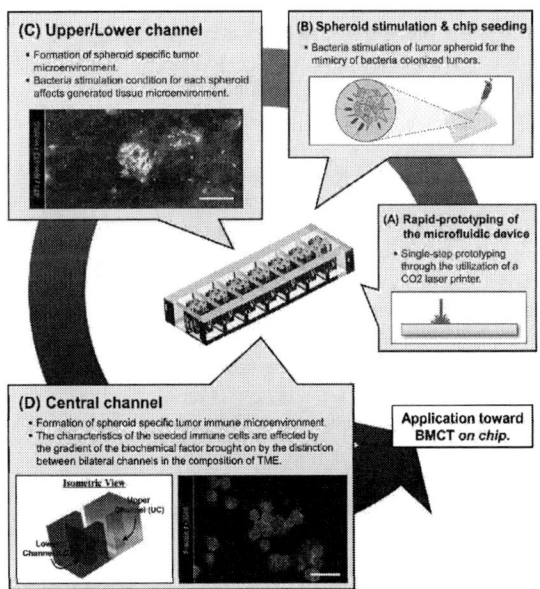

Figure 3: The overall experiment schematic for on-chip generation of bacteria-colonized TME. A) The fabricated device was cut and assembled in minutes through rapid prototyping. B) A conceptual illustration of the spheroid infection process. C) Representative confocal image of generated TME stimulated with S.typhimurium. Scale bar = 500 μm. D) Computational simulation showing the effect of the source configured in the upper channel to the lower channel through concentration gradients. Representative confocal image of RAW264.7 showing pro-inflammatory polarization after 4 days of coculture. Scale bar = 25 μm.

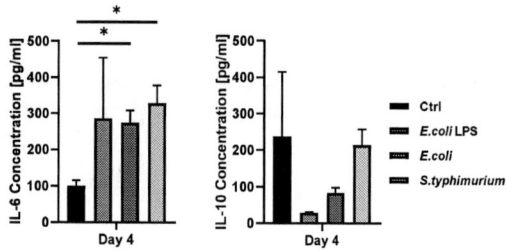

Figure 4: ELISA results from cell culture media after 4 days of co-culture. The secretion of pro-inflammatory cytokines was greater in RAW264.7 cells cocultured with tumor spheroids infected by bacterial agents (E.coli, S.typhimurium, E.coli lipopolysaccharide). Bars indicate mean ±SEM. *p < 0.05, **p < 0.005, and ***p < 0.0005 in two-tailed unpaired t-tests.

cancer treatment is their ability to target tumors and colonize them preferentially. In this respect, the intrinsic antitumor cytotoxicity of bacteria produced synergistic outcomes with endeavors to design bacteria as a producer of nanomaterials or immunotherapeutic molecules by genetic engineering, thus forging new ground for the cancer immunotherapy paradigm. To verify this notion, the antitumor cytotoxicity of bacteria was first observed with tumor spheroid infection (Figure 2A). The proportion of dead cells grew with time in both *E.coli* and *S.typhimurium* infections, and the trend was clearly more prominent in the more virulent *S.typhimurium* infection (Figure 2D). Furthermore, when *S.typhimurium* was genetically engineered to secrete IFN-β, it not only reduced tumor survival rate but also successfully repressed cancer stem cell properties (Figure 2F). This observation is consistent with previous studies demonstrating positive outcomes for the therapeutic usage of IFN-β in treating drug-resistant, highly aggressive triple-negative breast cancer [9]. The biological properties of the spheroid prepared through this method mirror the chemical properties of the used bacterial stimulant, i.e., toxicity, strain characteristic, and designed secreting factor. When transferred to the chip, these infected spheroids retain the key features of bacteria-cancer interaction and thus contribute to the generation of bacteria-colonized TME *in vitro*.

Reconstituting bacteria colonized TME

Figure 3 depicts a schematic representation of the experimental procedure using our apparatus. As a proof of concept, RAW 264.7 cells were seeded in the CC with *S. typhimurium*-infected tumor spheroid (6 h, MOI 100:1) in the UC and control spheroid without any infection in the LC. Seeded cells proliferated and created a corresponding 3D tissue environment after 4 days of co-culture, contributing to the development of multifunctional TME on-chip (Figure 3C).

Enzyme-linked immunosorbent assay (ELISA) was conducted to confirm whether the concept of the bacterial inflammation model was successfully implemented in the device. Likewise, RAW264.7 cells were cocultured with tumor spheroids infected by different bacterial agents: *E.coli* (6 h, MOI 100:1), *S.typhimurium* (6 h, MOI 100:1),

and *E.coli* lipopolysaccharide (6 h, 20 μg/mL). As a result, we were able to establish that the bacterial treatment of tumor spheroids elicited an activated immune response in the microfluidic device since RAW 264.7 cells secreted more pro-inflammatory cytokines when co-cultured with bacterial-infected tumor spheroids (Figure 4). Promoted pro-inflammatory M1 polarization of macrophages indicates that the designed device can emulate the basic immunostimulatory mechanism of BMCT.

CONCLUSION

In this research, we presented a microfluidic platform that can mimic the fundamental pathophysiology of BMCT. The design rules for precise liquid patterning and reproduced salient features of bacteria colonized TME utilizing bacterial-infected tumor spheroid ensured successful *in vitro* reconstruction of tumor microbe microenvironment. The proposed methodology is of generic impact and easily transposable to more advanced systems with different cell culture approach. We envision that our strategy will expand existing research on BMCT that has been restricted to the traditional mouse model, thereby bridging laboratory and clinical practice.

REFERENCES

[1] M. Binnewies, et al. "Understanding the tumor immune microenvironment (TIME) for effective therapy", *Nature medicine*, vol. 24, pp. 541-550, 2018.

[2] X. Huang, et al. "Hypoxia-tropic protein nanocages for modulation of tumor-and chemotherapy-associated hypoxia.", *Acs nano*, vol. 13, pp. 236-247, 2018.

[3] X. Huang, et al. "Bacteria-based cancer immunotherapy", *Advanced Science*, vol. 8, 2003572, 2021.

[4] A. E Ongaroa, N. Howartha, V. La Carrubbac, M. Kersaudy-Kerhoas, "Rapid prototyping for micro-ngineering and microfluidic applications", *Advances in Manufacturing Technology XXXII*, vol. 8, pp. 107-112, 2018.

[5] M. W Toepke, D. J Beebe, "PDMS absorption of small molecules and consequences in microfluidic applications", *Lab on a Chip*, vol. 6, pp. 1484-1486, 2006.

[6] S. B Berry, et al. "Upgrading well plates using open microfluidic patterning", *Lab on a Chip*, vol. 17, pp. 4253-4264, 2017.

[7] A. Malijevsky, A. O Parry, "Edge contact angle, ` capillary condensation, and meniscus depinning", *Physical Review Letters*, vol. 127, 115703, 2021.

[8] S. Kim, et al. "Anchor-impact: A standardized microfluidic platform for high-throughput antiangiogenic drug screening", *Biotechnology and Bioengineering*, vol. 118, pp. 2524-2535, 2021.

[9] M. R Doherty, et al. "Interferon-beta represses cancer stem cell properties in triple-negative breast cancer", *Proceedings of the National Academy of Sciences*, vol. 114, pp. 13792-13797, 2017.

CONTACT

*S. Suh. Public, tel: +82-10-9427-8195; keenhurt81@kist.re.kr

3D SPATIAL FOCAL CONTROL BY ARRAYED OPTOFLUIDIC PRISMS

Cheng-Hsun Lee, Yeonwoo Lee, and Sung-Yong Park*

Department of Mechanical Engineering, San Diego State University, San Diego, CA, USA

ABSTRACT

A new 3D focal control system composed of arrayed optofluidic prisms is presented. Through dynamic control of the fluid-fluid interface via electrowetting, incoming rays are spatially steered to achieve 3D focal control. Analytical study identifies the prism angle required to obtain a focal point at $P_{focal} = (f_x, f_y, f_z)$ located in 3D space. Experimentally, an arrayed system has demonstrated its 3D focal tunability along $0 \leq f_x \leq 30$, $0 \leq f_y \leq 30$, and $500 \leq f_z \leq \infty$ in millimeters. This new lens capability for 3D focal control can be potentially used for tracking eye movement for smart displays, or solar tracking for smart compact concentrated photovoltaic systems.

KEYWORDS

Electrowetting, liquid prism, optofluidics, focal length, arrayed system

INTRODUCTION

Conventional solid-type optical devices typically require bulky and complex moving parts for rapid and spatial beam steering. These additional mechanical components make the systems bulky, costly, and complex, and lack their agility [1]. To address the issue, recent studies have demonstrated rapid and wide-angle beam steering using an electrowetting-based microfluidic technology [2, 3]. Electrowetting has been widely developed as an effective means for small-scale liquid handling due to the dominance of surface tension forces over body forces in micro/meso scales [4, 5]. It enables to control the shape and position of a liquid interface for effective light control through the applications of bias voltages instead of mechanical inputs. A liquid lens is one common example of the optofluidic devices controlled by the electrowetting principle [6, 7]. Smooth surface of a liquid drop allows to function it as an optical lens where electrowetting actively controls its curvature for focal tuning of a lens. Another type of optofluidic device relying on electrowetting is the liquid prism for beam steering [8, 9]. A prism is filled with two immiscible liquids with different refractive indices. The electrowetting effect controls the prism apex angle which in turn steers beam pathway. However, all previous works [2, 3, 6-8] was limited to linear tuning focal control.

This study presents a new 3D focal control system composed of $n \times n$ arrayed optofluidic prisms. Dynamic control of the fluidic interface via electrowetting enables convergence of incoming light rays on a focal point located in 3D space. An advantage of such an optofluidic system is that light control can be achieved instantaneously without the need for bulky and complicated mechanical moving parts. Analytical study predicts the prism angle required to achieve 3D focal tunability of the optofluidic system. Experimental studies have further demonstrated 3D focal tunability of the system by simultaneously performing light focusing along lateral, longitudinal, and axial directions as

much as $\leq f_x \leq 30$, $0 \leq f_y \leq 30$, and $500 \leq f_z \leq \infty$ in millimeters. Such 3D focal tunability opens up a door to the development of tunable, compact optical devices for crucial applications such as tracking eye movement for smart displays or solar tracking for smart compact concentrated photovoltaic systems.

AN ARRAYED OPTOFLUIDIC PRISM SYSTEM FOR 3D FOCAL CONTROL

Figure 1(a) shows a schematic of the optofluidic beam steering system that enables spatial focal control of incoming rays in 3D space. It consists of $n \times n$ arrayed liquid prisms filled with two immiscible liquids such as water and oil. A curved meniscus is formed at the interface with an initial contact angle of θ_0 on the sidewall surface of the prism. When bias voltages are applied to the prism's sidewalls, the electrowetting effect is induced to modify the surface tension force and results in the contact angle changes. Such contact angle modifications enable to have a straight profile of the interface with an apex angle (φ) of the prism (Figure 1b). With the two liquids whose refractive indices are different each other, incoming light can be manipulated for 3D beam steering without any mechanical moving components like gimbaled mirrors. This beam steering performance has been demonstrated in previous studies using an electrowetting-driven liquid prism [2, 3, 6-8]. However, all previous demonstrations were limited to linear optical tuning. To achieve 3D focal control, our study herein proposes an $n \times n$ arrayed form of a single prism module. Each prism of the arrayed optofluidic system is individually controlled to provide its own prism angle (φ) using the electrowetting principle.

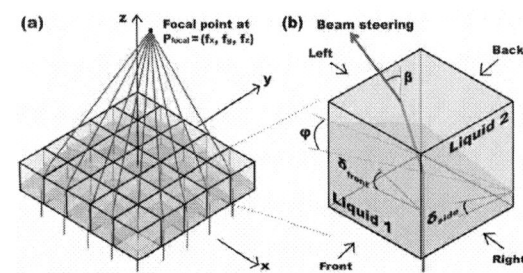

***Figure 1:** **(a) An n × n arrayed form of the optofluidic prism system for 3D focal control and (b) an enlarged image of a single prism module filled with two immiscible liquids.** A liquid prism is controlled via electrowetting to have a straight profile of the fluidic interface with an apex angle (φ) of the prism. Due to the refractive index differences ($n_1 \neq n_2 \neq n_{air}$) of each medium, an incoming light beam is dynamically manipulated at the interface for 3D beam steering. By symmetrically or asymmetrically controlling individual prisms in the array system, light focusing can be achieved in 3D free space, truly offering a new lens capability for 3D focal control without the need of bulky and complex mechanical moving components.*

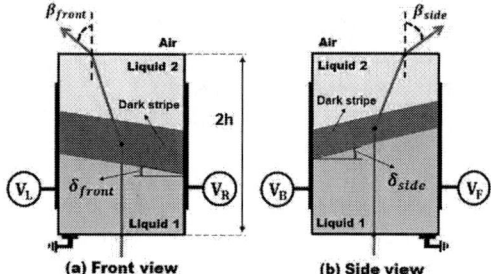

Figure 2: (a) Front and (b) side views of a single prism at an arbitrary location of P (x, y, 0). *When bias voltages (V_L, V_R, V_F, and V_B) are applied to four sidewalls of the prism, the fluidic interface is controlled via electrowetting to have the straight profile with the prism angle (φ). The tilted angles (δ_{front} and δ_{side}) of a dark stripe simply observed in the front and side views can be used to confirm the prism angle (φ) controlled in 3D space.*

Consequently, incoming light can be spatially steered for light focusing on a focal point that is located at P_{focal} (f_x, f_y, f_z) in 3D space. This 3D optical tunability enables to dynamically modulate the lens' power that couldn't be implemented by solid-type optics without bulky and complex mechanical moving components.

PRISM ANGLE CONTROL IN 3D

Performance analysis: To control the fluidic interface in 3D via electrowetting, bias voltages (V_F, V_B, V_L, and V_R) are separately applied at four sidewalls of the prism, respectively. The surface tension modification results in the contact angle changes on each sidewall of the prism. The resultant contact angles (θ_F, θ_B, θ_L, and θ_R) modulated from an initial angle θ_0 on four sidewalls are mathematically estimated as using the Young-Lippmann equation [2, 3]:

$$\cos\theta_F = \cos\theta_0 + \frac{1}{2\gamma}cV_F^2, \quad \cos\theta_B = \cos\theta_0 + \frac{1}{2\gamma}cV_B^2$$
$$\cos\theta_L = \cos\theta_0 + \frac{1}{2\gamma}cV_L^2, \quad \cos\theta_R = \cos\theta_0 + \frac{1}{2\gamma}cV_R^2 \quad (1)$$

where c is the specific capacitance of a dielectric layer and γ is the interfacial tension between two liquids. An incoming ray perpendicularly projected from the bottom can be refracted as much as the prism angle (φ) at the interfaces due to the refractive index difference ($n_1 \neq n_2 \neq n_{air}$) between two media. Using the geometrical relations and Snell's law at the interfaces, the beam steering angle β of the ray as it exits the prism can be also estimated as:

$$\beta = \sin^{-1}\left[n_1\sin\varphi\cos\varphi - n_2\sin\varphi\sqrt{1-\left(\frac{n_1}{n_2}\sin\varphi\right)^2}\right] \quad (2)$$

which is expressed as a function of the refractive indices (n_1 and n_2) of the two liquids and the prism angle (φ).

Since the fluidic interface is spatially modulated to have a straight profile, it is difficult to recognize the prism

angle (φ) controlled in 3D space. To simply estimate φ, we have instead used two observable parameters shown in the front and side views. Figure. 2 shows schematic illustrations visualized on the front and side views of the prism positioned at an arbitrary location of P_C ($x, y, 0$) when the fluidic interface is spatially modulated with the prism angle (φ). The prism has its height of $2h$. Due to less light transmission, the 3D manipulated interface can be visualized as a dark stripe in the front and side views. This dark stripe is characterized with the tilted angles (δ_{front} and δ_{side}) simply observed from experiments, as presented in Fig. 2(a) and (b). To have a straight profile (i.e., $\theta_L + \theta_R = 180°$ and $\theta_F + \theta_B = 180°$) of the interface with the prism angle (φ) in 3D, the contact angles are modulated in such a manner that fulfils the following relations:

$$\delta_{front} = \theta_L - 90° \quad \text{when} \quad \theta_L \geq 90° \geq \theta_R$$
$$\delta_{side} = \theta_F - 90° \quad \text{when} \quad \theta_F \geq 90° \geq \theta_B \quad (3)$$

These two observable parameters (δ_{front} and δ_{side}) obtained from experiments can be expressed as a function of φ and ω using the geometrical relations:

$$\tan\delta_{front} = \tan\varphi\cos\omega$$
$$\tan\delta_{side} = \tan\varphi\sin\omega \quad (4)$$

where ω indicates the azimuth angle of the ray's pathway from an x axis. With the given information on the prism location at P_C ($x, y, 0$) and the desired focal point P_{focal} (f_x, f_y, f_z) to be achieved by a liquid prism, the value of ω can be estimated as:

$$\omega = \tan^{-1}\left(\frac{f_y - y}{f_x - x}\right) \quad (5)$$

Thus, Eq. (4) informs us that the two observable parameters, δ_{front} and δ_{side}, can be simply used to ensure the magnitude of the prism angle (φ) achieved. When combing Eq. (1), (3), and (4), one can know the bias voltages required to obtain the prism angle (φ) in 3D. With the given values of ω and φ, we can further express the direction of the fluidic interface in a unit vector ($\hat{n}$) normal to the surface of the interface as:

$$\hat{n} = \begin{bmatrix} -\sin\varphi\cos\omega \\ -\sin\varphi\sin\omega \\ \cos\varphi \end{bmatrix} \quad (6)$$

Experimental demonstrations: When electric potentials ($V_F = 107$ V, $V_B = 51$ V, $V_L = 106$ V, and $V_R = 51$ V) were applied to four sidewalls of the prism, the fluidic interface was modulated to have a straight profile with the prism angle (φ) in 3D. Fig. 3(a) and Fig.3(b) shows the front and side views of the prism experimentally obtained, where the interface was visualized as a dark stripe that is characterized with a non-zero tilted angle of δ_{front} = -7° in the front view, while δ_{side} = 7° in the side view, respectively. For an illustration purpose, let's assume that a prism is located on a 45° diagonal direction and desired to have its focal point along a z-axis for axial focal control

978-1-6654-9309-3/23 $31.00 © 2023 IEEE

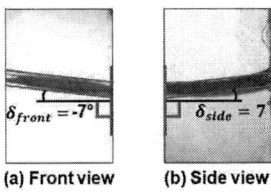

(a) Front view   **(b) Side view**

Figure 3: Experimental demonstrations of the fluid-fluid interface control in 3D via electrowetting. Control of the fluidic interface in 3D is visualized as a dark stripe with the tilted angles ($\delta_{front} = \delta_{side} = 7°$) shown in (a) the front and (b) side views, respectively. These observation parameters can be simply used to confirm the prism angle at $\varphi = 10°$ achieved.

(i.e., $\omega = 45°$, $f_x = f_y = 0$, and $f_z \neq 0$). Simply using these two observation parameters of tilted angles (δ_{front} and δ_{side}) and the information on the prism location and desired focal point, experimental results in Fig. 3 confirmed that the fluid-fluid interface was successfully modulated via electrowetting to have a straight profile with the prism angle at $\varphi = -10°$ and the unit normal vector $\hat{n} = [0.1228,\ 0.1228,\ 0.9848]$ to indicate the direction of the fluidic interface.

AXIAL FOCAL CONTROL

Performance analysis: This section discusses about the prism angle requirement to achieve axial focal control along the central lens axis (i.e., the z axis). Through symmetrical modulation of individual prisms, all incoming rays can be steered to be converged at a single point of P_{focal} $(0, 0, f_z)$ on the z axis. A key consideration is to find f_z as a function of φ and ω for the prism located at P $(r, \theta, 0)$, where $r = \sqrt{x^2 + y^2}$ is a radial distance of P_C from an origin and $\theta = \tan^{-1}(y/x)$ is an azimuth angle from an x axis. For axial focal control, $f_x = f_y = 0$. Thus, it is given that $\omega = \theta$. Then, f_z can be written as:

$$f_z = (r - a)\cot\beta + h \qquad (7)$$

where a is denoted as the distance between the intersection point of the ray on the top surface of the prism and the vertical line extended from the prism center, which can be estimated as:

$$a = h\tan\left\{\sin^{-1}\left(\frac{\sin\beta}{n_2}\right)\right\} \qquad (8)$$

By substituting Eq. (2) and (8) into Eq. (7), the focal point f_z on a z axis can be plotted as a function of φ for various prism locations of r along a radial direction, which is shown in Fig. 4. For this analytical study, a liquid prism is assumed to be filled with binary liquids of 1-bromonaphthalene (1-BN) as a high-refractive-index oil ($n_1 = 1.65$) and water ($n_2 = 1.33$), and to have its height $2h = 25$ mm, which will be the same conditions of our succeeding experimental tests. Fig. 4 provides critical information on the prism angle φ required for the prism positioned at P_C $(r, \theta, 0)$ to achieve the ray focusing on a

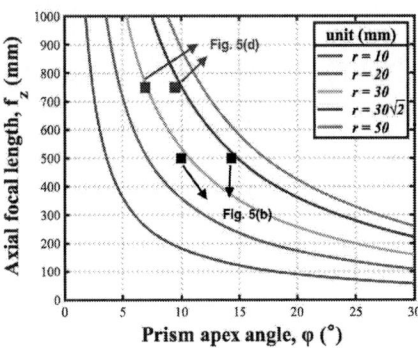

Figure 4: Axial focal tunability of the arrayed optofluidic system. An axial focal length f_z on a z axis is presented as a function of the prism angle φ for various prism distances of r in a radial direction from an origin. At higher prism angles, more beam steering and thus shorter focal lengths can be achieved, resulting in higher lens power. Four data points added in the graph indicate the axial focal length f_z obtained from the experimental tests shown in Fig. 5 (b) and (c).

particular focal length f_z on a z axis for axial focal control. In general, a large apex angle is required for larger beam steering of incoming light, which leads to high lens power with a short focal length. Similarly, for the prisms with large r values (i.e., further from the origin), a high apex angle is required to compensate for their distance from the central axis.

For any other focal length f_z desired to be achieved, the required prism angles may be similarly determined for any prism location. For comparison with our analytical studies, four experimental data points are added to the graph for $r = 30$ and $30\sqrt{2}$ mm to demonstrate the system's capability for axial focal control.

Experimental demonstrations: Figure 5 shows experimental results to demonstrate the lens' capability for axial focal control along a z axis. With the applications of bias voltages to the sidewalls of the three prisms, they were initially modulated to have all prism angles at $\varphi_1 = \varphi_2 = \varphi_3 = 0°$ and the normal directions to the fluid interfaces as $\hat{n}_1 = \hat{n}_2 = \hat{n}_3 = [0,\ 0,\ 1]$. As a result, incoming laser beams pass through the three prisms in parallel with no light refraction at the interfaces. The top view as shown of Figure 5(a) indicates that the laser spots are located at the same positions as the prisms' locations at $\varphi_1 = \varphi_2 = \varphi_3 = 0°$. This experimental result indicates that a focal point is said to be at infinity, i.e., $f_z = \infty$. These three prisms were further controlled to achieve axial focal control. With the applications of the bias voltages, the three prisms were symmetrically controlled to have the prism angels at $\varphi_1 = 10°$, $\varphi_2 = 10°$, and $\varphi_3 = 14°$ where the incoming laser beams were steered to be focused at P_{focal} $(f_x, f_y, f_z) = (0, 0, 500$ mm) for axial focal control on a z axis. Figure 5(b) shows the top view of the grid paper located at $z = 50$ cm where the three laser beams were focused on the origin. Figure 5(c) shows the convergence of three laser beams on the origin located at $z = 75$ cm to represent lower lens' power performance, when the prisms were controlled to have $\varphi_1 = 10°$, $\varphi_2 = 10°$, and $\varphi_3 = 14°$.

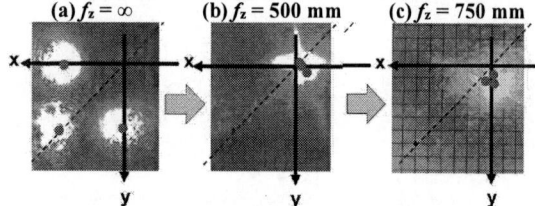

(a) $f_z = \infty$ **(b)** $f_z = 500$ mm **(c)** $f_z = 750$ mm

Figure 5: Experimental demonstrations of the focal control along an axial (z) direction by symmetrical prism control. (a) and (c) Three prisms are manipulated to have all prism angles at $\varphi = 0°$. Then, the laser beams pass through each prism in parallel without beam steering, resulting an axial focal length located at $f_z = \infty$ where the laser spots initially located at P_1 (30, 0), P_2 (0, 30), and P_3 (30 $\sqrt{2}$, 30 $\sqrt{2}$) in millimeters. After symmetrically controlling the prism angles, three laser beams are spontaneously steered to focus on the origin that is located at two different x-y planes of (b) $f_z = 50$ cm and (d) $f_z = 75$ cm. The test results demonstrated the lens capability for an axial focal control in z direction.

3D FOCAL CONTROL

Figure 6 presents the focal tunability along both lateral and longitudinal directions by asymmetrically controlling the three prisms. With the prism angles of $\varphi_1 = 6°$, $\varphi_2 = 9°$, and $\varphi_3 = 8°$ controlled via electrowetting, the arrayed system was able to show 3D focal tunability for which the focal point can be located at P_{focal} (f_x, f_y, f_z) = (20, 10, 500) in millimeters (Figure 6a). Figure 6(b) shows an experimental result of the focal point located at P_{focal} (f_x, f_y, f_z) = (30, 30, 500) in millimeters by controlling the prism angles of $\varphi_1 = 10°$, $\varphi_2 = 10°$, and $\varphi_3 = 0°$. Figure 6(c) presents 3D focal tunability to manipulate the focal point at P_{focal} = (30, 0, 500) in millimeters when $\varphi_1 = 0°$, $\varphi_2 = 14°$, and $\varphi_3 = 10°$. Experimental results presented in Figure 6 have successfully demonstrated 3D focal tunability of the arrayed optofluidic system by asymmetrical control of individual prisms.

CONCLUSION

To address the issue of conventional solid-type optical devices that typically require bulky and complex moving parts for rapid and spatial beam steering, recent studies have demonstrated wide-angle beam steering using optofluidic prism devices driven by electrowetting. However, the lens performance was limited to linear tuning, failing to fully exhibit the focal tunability in 3D space. This paper presents a new lens capability for 3D focal control, which couldn't be achieved by solid-type devices without bulky and complex mechanical moving parts. An optofluidic system consists of $n \times n$ arrayed liquid prisms filled with two immiscible liquids such as water and oil. Dynamic control of the fluidic interface via electrowetting enables convergence of incoming light rays on a focal point located in 3D space. We have analytically studied to predict the prism angle required to achieve the desired focal location in 3D. Experimental studies have further demonstrated 3D focal tunability of the system by simultaneously performing light focusing along lateral,

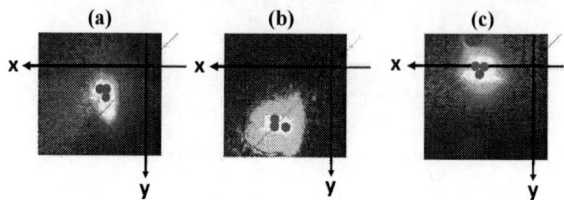

(a) **(b)** **(c)**

Figure 6: Experimental demonstrations of the 3D focal control along both lateral and longitudinal directions at $f_z = 500$ mm through asymmetrical prism control. Light focusing of the three laser beams are variously tuned to have the focal points located at (a) $f_x = 20$ and $f_y = 10$, (b) $f_x = f_y = 30$, and (c) $f_x = 30$ and $f_y = 0$ in millimeters, respectively.

longitudinal, and axial directions as much as $\leq f_x \leq 30$, $0 \leq f_y \leq 30$, and $500 \leq f_z \leq \infty$ in millimeters. This new lens capability for 3D focal control can be potentially used for smart tracking systems such as eye tracking for smart displays, or solar tracking.

ACKNOWLEDGEMENTS

This work was partially supported by the NSF CAREER Award (ECCS - 2046134), USA.

REFERENCES

[1] E. Arbabi, A. Arbabi, S. M. Kamali, Y. Horie, M. Faraji-Dana, and A. Faraon, "MEMS-tunable dielectric metasurface lens," *Nature Communications,* vol. 9, no. 1, p. 812, 2018.

[2] C. Clement, S. K. Thio, and S.-Y. Park, "An optofluidic tunable Fresnel lens for spatial focal control based on electrowetting-on-dielectric (EWOD)," *Sensors and Actuators B: Chemical,* vol. 240, pp. 909-915, 2017.

[3] C. E. Clement and S.-Y. Park, "High-performance beam steering using electrowetting-driven liquid prism fabricated by a simple dip-coating method," *Applied Physics Letters,* vol. 108, p. 191601, 2016.

[4] A. R. Wheeler, "Putting Electrowetting to Work," *Science,* vol. 322, pp. 539 - 540, 2008.

[5] S. K. Thio and S.-Y. Park, "A review of optoelectrowetting (OEW): from fundamentals to lab-on-a-smartphone (LOS) applications to environmental sensors," *Lab on a Chip,* vol. 22, no. 21, pp. 3987-4006, 2022.

[6] B. W. Hendriks, S. Kuiper, M. J. VAN As, C. A. Renders, and T. W. Tukker, "Electrowetting-based variable-focus lens for miniature systems," *Optical Review,* vol. 12, no. 3, pp. 255-259, 2005.

[7] S. Kuiper and B. H. W. Hendriks, "Variable-focus liquid lens for miniature cameras," *Applied Physics Letters,* vol. 85, p. 1128, 2004.

[8] S. K. Thio, D. Jiang, and S.-Y. Park, "Electrowetting-driven solar indoor lighting (e-SIL): an optofluidic approach towards sustainable buildings," *Lab on a Chip,* vol. 18, pp. 1725-1735, 2018.

[9] S. K. Thio, S. W. Bae, and S.-Y. Park, "Plasmonic nanoparticle-enhanced optoelectrowetting (OEW) for effective light-driven droplet manipulation," *Sensors and Actuators B: Chemical,* vol. 308, p. 127704, 2020.

CONTACT

*S.-Y. Park, tel: +1-619-594-2319; spark10@sdsu.edu

HIGH-SPEED AND PINPOINT LIQUID EXCHANGE ON MICROFLUIDIC CHIP USING 3D PRINTED DOUBLE-BARRELED MICROPROBE WITH DUAL PUMPS

Xu Du[1], Shingo Kaneko[2], Hisataka Maruyama[1], Hirotaka Sugiura[2], and Fumihito Arai[1,2]

[1]Department of Micro-Nano Mechanical Science and Engineering, Nagoya University, JAPAN and
[2]Department of Mechanical Engineering, The University of Tokyo, JAPAN

ABSTRACT

We propose a method to exchange the ultra-small volume of liquid on the microfluidic chip in high-speed by using a 3D printed double-barreled microprobe with dual pumps. The probe tip is designed double-barreled and assembled with two piezoelectric stacks to realize the simultaneous liquid injection and suction. We evaluated the performances of the developed liquid exchange system. The results showed that high-speed and pinpoint liquid exchange can be achieved within 10 ms. The highest flow speed is approximately 18 mm/s. The sensitivity and resolution are 22 pL/v and 0.54 pL respectively. This system will contribute to revealing the detailed cell deformation process under rapid osmolarity change, and characterizing the physiological function of mechanosensitive channels without destroying cells.

KEYWORDS

Liquid exchange, Microfluidic chip, 3D printed double-barreled microprobe, Dual pumps.

INTRODUCTION

Changing the extracellular environment is an important process in cell measurement, which can simultaneously expose cells in stimulation and detection [1-4]. For example, the mechanosensitive (MS) channels are one kind of the ion channels of *Synechocystis*, which can sense the cell membrane tension and release cytoplasmic solutes into the extracellular environment to protect the cell from decreases of osmotic concentration. We can reveal the functional mechanism of MS channels by measuring the mechanical properties of *Synechocystis* under alterable osmotic concentrations [5].

In our previous research, extracellular solutions were changed by moving the liquid-liquid interface of laminar flow in a closed microfluidic chip to measure the dynamic deformation of cells under alterable osmotic concentrations. However, it is difficult to control the position and pressure of liquid exchange in a closed microchannel. In addition, over 0.3 seconds of liquid-exchange time is too long to observe cell deformation accurately, some research exposed that the response time of *Synechocystis* cells to osmolarity change is approximately 30 ms [6].

We propose a liquid exchange system, which can easily be integrated with an opened chip. The system assembled two piezo pumps and 3D printed probe compactly to reduce the response time. Two barrels in the 3D printed probe inject and suck liquid respectively, preventing diffusion before the 3D printed probe connects the original liquid. An external manipulator is used to position the probe tip near the upper surface of an open chip.

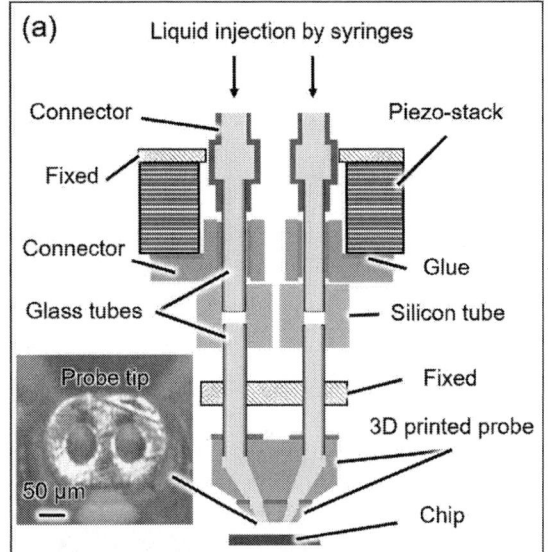

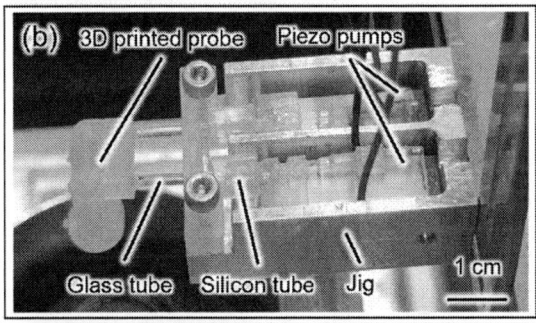

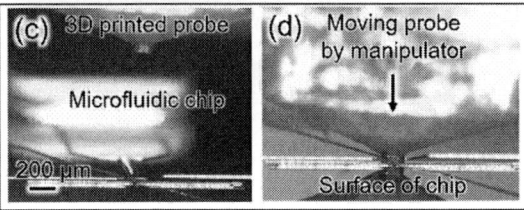

Figure 1: (a) Schematic of the proposed 3D printed probe with dual pumps and image of the probe tip. (b) image of chip and 3D printed probe. (c) Configuration of the chip and the tip of the 3D printed probe.

MATERIAL AND METHOD

We fabricated the probe by 3D printing since it's difficult to make a double-barreled structure by traditional glass capillary. Owing to the size limitation of the high-precision 3D printer, the 3D printed probe is divided into two parts for printing. Two piezo-actuators were assembled with silicon tubes and glass tubes to form pumps that provide each barrel with individual and picoliter-level volume control (Fig. 1 (a) and (b)). The compact structure design shortens the liquid-exchange time. Two chip probes were integrated with the chip, to measure the reactive force and deformation of cells [5]. The synchronous injection and suction from the outlet hole (diameter 45 μm, Fig. 1 (a)) limit the liquid-exchange area which decreases the disturbance to the chip probes. The external manipulator can move the 3D printed probe to a specified position on the chip surface (Fig.1 (c) and (d)). After positioning the probe tip, the pumps were excited to inject and suck the liquid respectively and synchronously, the silicon tube will be compressed/elongated, and it will apply positive/negative pressure on the outlet holes of the probe tip, the original solution around the hole will be replaced with the injected solution.

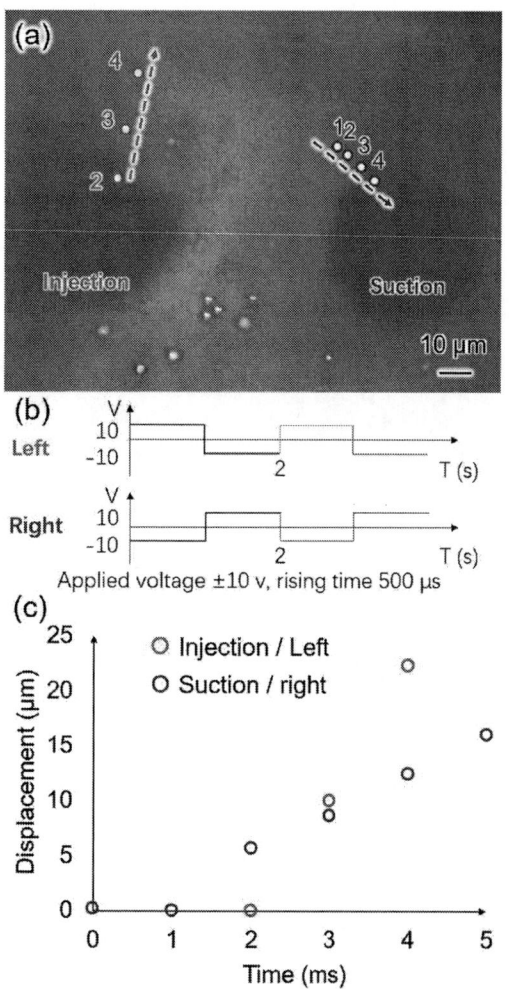

Figure 2: (a) Microscopic image of the probe tip and the movement route of beads, the number on the figure indicates the time or frame when the beads move to the position; (b) Conceptual diagram of applied anti-phase voltage by signal generator in each pump (±10 v, rising time 500 μs); (c) Displacement of target microbeads when the liquid exchange happened.

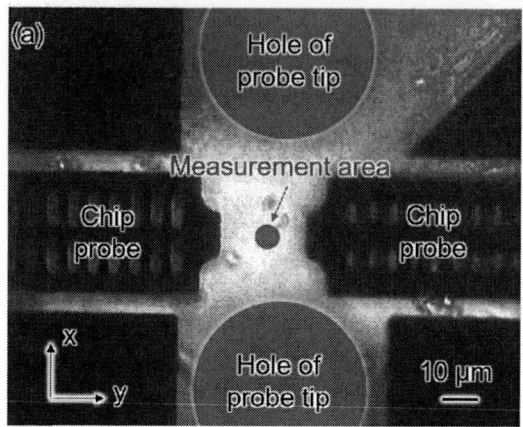

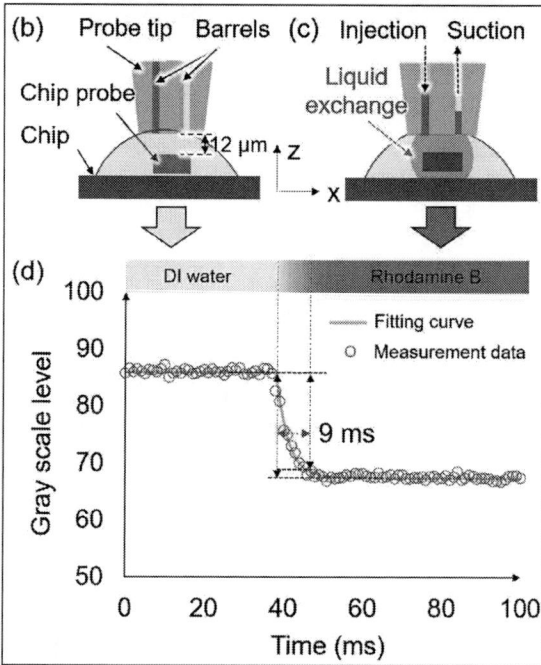

Figure 3: (a) Microscopic image of the position of the probe tip, chip probes, and measurement area; (b) Conceptual diagram of the probe tip and chip; (c) Conceptual diagram of liquid exchange on the measurement area; (d) Experimental data for the evaluation of liquid exchange time.

EXPERIMENTS AND RESULTS
Synchronization of two pumps

The driving signal of two piezo-pumps were synchronized by signal generator (±10 v, rising time 500 μs). The displacement of piezo-actuators deformed silicon tubes, which changed the volume of solution in silicon tubes. Hence, solutions were injected and sucked from two holes of the 3D printed probe synchronously, and the solution around the measurement area was replaced. To confirm the synchronization of injection and suction, we added polystyrene beads in the original solution and measured the displacement of beads when liquid exchange happened. High-speed camera (1000 fps) was used to record the displacement of beads. As shown in the figure 2 (c), the time difference between the initial movement of the injected (left) and sucked (right) beads was only 1 ms. Given that the limitation of frame rate of high-speed camera, the error of synchronization was approximately 2 ms.

Liquid exchange time

The DI water and DI water mixed with 10g/L Rhodamine B solution were injected into the barrels respectively. The holes of the probe tip were positioned at the upper of the chip probes (The distance between the probe tip and the upper surface of chip probes is approximately 12 μm (Fig 3)). To evaluate the liquid exchange time, we defined a certain time when gray scale level change reached 90% of its stable state as the liquid exchange time, which is approximately 9 ms (Fig.3 (d)).

Flow speed

To evaluate the effect of diffusion and estimate disturbance to trapped cells, we added polystyrene beads in the original solution and measured the total displacement and flow speed of beads (Fig. 4). The movement route of the microbead when liquid exchange happened was shown in the Fig. 4 (a). The highest flow speed was approximately 18 mm/s (Fig. 4 (b)). Although the flow speed was higher, no obvious disturbance could be observed on chip probes.

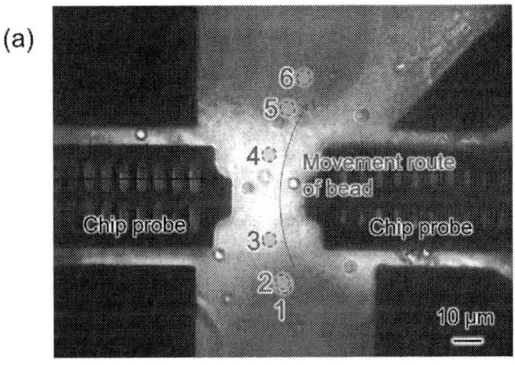

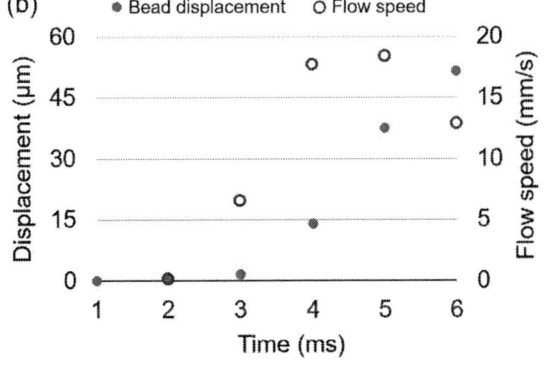

Figure 4: (a) Movement route of bead on the microfluidic chip, the orange points are the position of the bead in consecutive frames; (b) Displacement and flow speed of target microbead when the liquid exchange happened. The highest flow speed was approximately 18 mm/s

Sensitivity and resolution

The sensitivity and resolution of controlled volume were evaluated by measuring the volume change corresponding to the displacement of the beads under different voltages (Fig. 5). Given that the sensitivity and resolution of the liquid exchange system are only related to the size of the silicon tube and the excited voltage of pumps, but not to the structure of probe tip. To observe the displacement of beads into the probe tip, we used a glass probe instead of the 3D printed probe [7] (Fig. 5 (a)). The sensitivity of the proposed liquid exchange system was calculated by the slope of fitting curve in Fig. 5 (c), which was approximately 22 pL/v. The resolution is determined by the sensitivity and the stability of input voltage. The standard deviation of the input voltage was measured as 25 mV. Hence, the theoretical resolution of volume control was 0.54 pL.

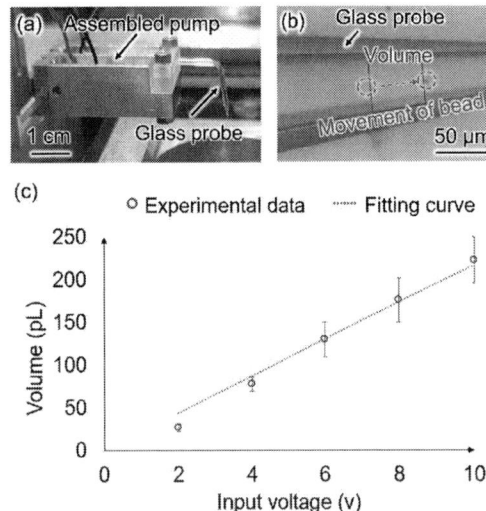

Figure 5: (a) Image of assembled pump and glass probe; (b) Microscopic image of glass probe and the volume change corresponding to the displacement of the beads; (c) Experiment data of volume control.

CONCLUSION

In this study, we proposed a method to exchange liquid on the microfluidic chip with high-speed and small volume by using a 3D printed probe with dual pumps. The system assembled the two piezo pumps and 3D printed probe compactly to reduce the response time. Two barrels in the 3D printed probe injected and sucked liquid respectively and synchronously, preventing diffusion before the 3D printed probe connects the original liquid. The error of synchronization of injection and suction was lower than 1 ms. Using this proposed 3D printed probe with pumps system, we obtained high speed liquid exchange time of approximately 9 ms and high resolution of 0.54 pL respectively. The highest flow speed was approximately 18 mm/s, no obvious disturbance was observed on chip probes of the microfluidic chip. This system provides a novel method to exchange the solutions on chip using 3D printed probe, which will contribute to accomplishing high-speed liquid exchange and low-disturbance force detection.

ACKNOWLEDGEMENTS

This work was partially supported by Grants-in-Aid for Scientific Research from the Ministry of Education, Culture, Sports, Science and Technology (18H03762 and 21H04543 to F. A. and N. U.). We thank the Chinese Scholarship Council for supporting the author Xu Du in his PhD study.

REFERENCES

[1] S. Takayama, E. Ostuni, P. LeDuc, K. Naruse, D. E. Ingber and G. M. Whitesides, "Subcellular positioning of small molecules", Nature, vol. 411, pp. 1016-1016, 2001.

[2] M. D. Tarn, M. J. Lopez-Martinez, N. Pamme, "On-chip processing of particles and cells via multilaminar flow streams", Anal. Bioanal. Chem., vol. 406, pp. 139-161, 2014.

[3] S. E. Ong, S. Zhang, H. Du and Y. Fu, "Fundamental principles and applications of microfluidic systems", Front. Biosci, vol. 13, pp. 2757–2773, 2008.

[4] A. Ainla, G. D. Jeffries, R. Brune, O. Orwar and A. Jesorka, "A multifunctional pipette", Lab Chip, vol. 12, pp. 1255-1261, 2012.

[5] D. Chang, S. Sakuma, K. Kera, N. Uozumi, F. Arai. "Measurement of the mechanical properties of single Synechocystis sp. strain PCC6803 cells in different osmotic concentrations using a robot-integrated microfluidic chip", Lab Chip, vol. 18, pp. 1241-1249, 2018.

[6] K. Nanatani, T. Shijuku, M. Akai, Y. Yukutake, M. Yasui, S. Hamamoto, K. Onai, M. Morishita, M. Ishiura and N. Uozumi, "Characterization of the role of a mechanosensitive channel in osmotic down shock adaptation in Synechocystis sp PCC 6803", Channels, vol. 7, pp. 238–242, 2013.

[7] Y. Kasai, S. Sakuma and F. Arai, "Isolation of single motile cells using a high-speed picoliter pipette", Microfluid Nanofluidics, vol. 23 pp. 1-9, 2019.

CONTACT

*X. Du, TEL: +81-52-789-5220;
duxu@biorobotics.mech.nagoya-u.ac.jp

DESIGN OF A DNA SYNTHESIS CHIP FOR DATA STORAGE WITH ULTRA-HIGH THROUGHPUT AND DENSITY FEATURING LARGE-SCALE INTEGRATED CIRCUITS AND MICROFLUIDIC CONFINEMENT

Ning Wang[1,2,3], Shijia Yang[1,3], Dayin Wang[1,2,3], Zhen Cao[4], Yuan Luo[1,3]*, and Jianlong Zhao[1,3]

[1]State Key Laboratory of Transducer Technology, Shanghai Institute of Microsystem and Information Technology, Chinese Academy of Sciences, P. R. of China

[2]ShanghaiTech University, P. R. of China

[3]Center of Materials Science and Optoelectronics Engineering, University of Chinese Academy of Sciences, P. R. of China

[4]College of Information Science and Electronic Engineering, Zhejiang University, P. R. of China

ABSTRACT

This paper proposes a novel DNA synthesis chip design targeting data storage application, achieving ultra-high throughput(10^7) and unit density($10^7/cm^2$) via the synergetic function of a dynamic random access memory (DRAM)-like integrated circuit (IC) and microfluidic confinement architecture. Through theoretical and simulation study, we verify individual addressability to 10^7 synthesis units by maintaining stable voltage output. We also demonstrate effective proton confinement through microfluidic structures at $10^7/cm^2$ unit density.

KEYWORDS

DNA Data Storage, DNA Synthesis, Microfluidic, DRAM

INTRODUCTION

With the rapid development of information technology, the amount of information generated by human society has grown exponentially. According to the forecast of the International Data Corporation (IDC), the total amount of global data will reach an astonishing level of 175ZB by 2025[1, 2]. The current storage media will not be able to meet the rapidly growing data storage demands. Therefore, it is imperative to pursue next-generation storage technology.

DNA is considered a promising alternative data storage medium thanks to its ultra-high data density and storage longevity[3]. The DNA data storage process consists of four steps, which are data encoding, DNA synthesis, DNA sequencing, and data decoding. In particular, DNA synthesis and DNA sequencing are the core foundation of DNA data storage, corresponding to the writing and reading of information respectively. From the further comparison between them, we can see that DNA synthesis has become the major bottleneck due to its insufficient throughput and high cost[4, 5]. Therefore, achieving high-throughput and high-density DNA synthesis is of utmost importance for the practical application and the further development of DNA data storage.

The state-of-art synthesis platforms generally report achievable throughput at $10^4 \sim 10^6$ and unit density at $10^4 \sim 10^5/cm^2$[6, 7]. The current high-throughput synthesis methods are all based on the column-type oligonucleotide chemical synthesis method that is divided into four steps: deprotection, coupling, capping, and oxidation, with the only modification on the deprotection step to adapt to their platform[8]. In particular, the inkjet printing technology, using printed Trichloroacetic acid (TCA) droplets to complete deprotection, is hindered by its mechanical positioning accuracy to further improve unit density. The photochemical synthesis method achieves photo-deprotection via Digital Micromirror Arrays (DMA)-assisted light direction, yet inevitably suffers from optical interference and complexity of optoelectronic components resulting in limited unit density. Electrochemical-based method that deprotects by electrochemical reaction could be the most promising candidate[7], yet cross-talk issues caused by proton diffusion stymies further improvement on throughput and density.

In this work, we propose a DRAM-like IC with a simple two-component unit that drastically increases system density and throughput, together with a novel microfluidic approach to restrict proton diffusion in electrochemical synthesis. Our design enables individual addressability and voltage control for synthesis unit density up to $10^7/cm^2$, and throughput at 10^7.

EXPERIMENTAL

The overall device design

A schematic illustration of the overall design of the device is presented in Fig. 1. Our DNA synthesis chip comprises an underlying IC with exposed electrodes and the microfluidic architecture located on-top (Fig. 1a). Four solutions for the electrochemical synthesis of phosphoramidite-oligonucleotides are sequentially circulated through the microfluidic architecture. The microfluidic chambers serve as the synthesis reaction sites and are configured to align with the electrodes. The synthesis reaction can be turned on/off by simply controlling the voltage supply on electrodes, triggering proton generation[7].

When the electrolyte solution containing the nucleotide monomers is injected into the microfluidic device and fills the chamber, voltage is applied to the selected electrodes until enough protons are generated to complete the deprotection of oligonucleotides. Diffusion of the excess protons generated therein is restricted by a physical barrier between the chambers to avoid cross-talk between sites. New nucleotide monomers will couple quickly to the deprotected oligonucleotides. The solution in the chamber is then replaced, and capping and oxidation reactions occur on the sites to complete one round of

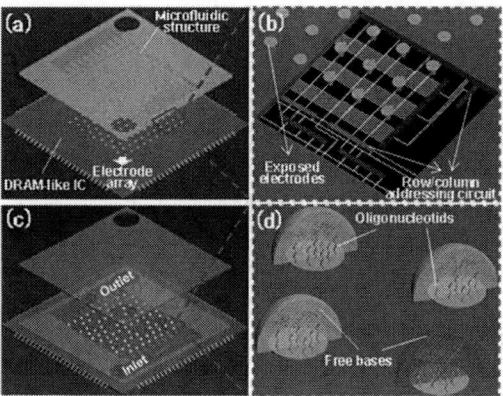

Figure 1: (a) & (c) Schematic illustration of the overall design of DNA synthesis chip with enlarged view showing (b) the electrode-IC connection and (d) individual synthesis unit. red symbolizes energized electrode in (c-d).

synthesis. After that, the IC will select the next batch of electrodes to apply voltage for a new round of synthesis reactions. In this manner, our design can implement individual addressability to each synthesis unit.

IC design

The IC (Fig.2a) features a DRAM-like architecture consisting of row/column addressing circuits, and a two-dimension electrode array ($3163 \times 3163 = 10^7$). The individual unit within the array is a single transistor connected to a capacitor (Fig.2b), which occupies a size of 3.2×3.2 μm². We further designed the row/column addressing circuit to perform the required logic for individual addressability to each of the 10^7 units. Each output of the row addressing circuit is connected to the transistor gate, and the transistor drain connects with the column addressing circuit output. The row addressing is achieved using a 12-3163 decoder, hence the numbers of input ports and output ports are 12 and 3163 respectively. For column addressing, we assign 133 input ports in total, with the first 128 as free input and the last 5 to act as 5-25 decoder (since [3163/128]=25). This design enables us to select 128 columns simultaneously. Similar to DRAM, although our circuit can only apply voltage to 128 units at any given instance, the capacitor enables the voltage to drop at a sufficiently slow rate during the off period. The voltage will be replenished after the row/column addressing cycles through all 10^7 units.

The design process of the integrated circuit is as follows: 1) Firstly, according to the requirements of the addressing function, we wrote the Verilog code and performed pre-simulation to verify the correctness of the function. 2) Next, we compile the Verilog code to generate netlist and check the logic equivalence between netlist and the code. 3) Then, we wrote the Cadence Encounter Script to generate the Layout using the netlist. 4) Finally, after passing DRC and LVS checking, we extract the layout to perform post-simulation.

Microfluidic design

For the microfluidic design, we first adopted a simple microchamber array interconnected through microchannels. The repeat unit in the array consists of a square chamber and four inclined channels connecting

adjacent

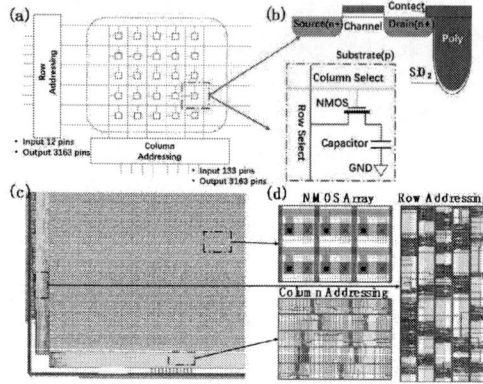

Figure 2: (a) The overall architecture of the DRAM-like Integrated Circuit Design: row/column addressing logic circuit and a two-dimensional array with 10^7 units, each consisting of (b) a transistor connected to a capacitor. (c) The layout of IC design with enlarged view in (d) showing detailed components in row/column addressing logic circuit and the two-dimensional array

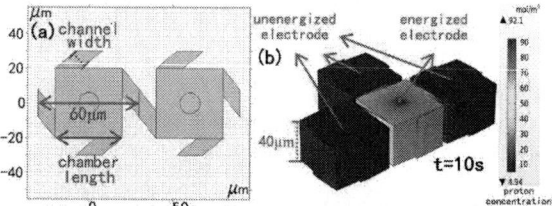

Figure 3: (a) Top view of the micro-chamber/channel structure. (b) Simulation schematic of microchamber/channel structure.

chambers diagonally, and the size of the unit is 60×60 μm² (Fig. 3a). The square chamber and channel in the unit are 40 μm high, and the 20 μm diameter circular electrode is located at the bottom center of the chamber(Fig.3b).

Alternatively, we adapted the microfluidic design of static droplet formation[9], as shown in Fig. 4. The main structure of the design is an array of capillary valve units, in which the adjacent units are symmetric about the x-axis. The capillary valve unit consists of three parts: channel, chamber, and capillary valve, with an overall size of 3.16×3.16 μm². Therefore the density of cells in the array can reach 10^7/cm².

The channel in the unit is bent at 90 degrees, two chambers are connected at each end, and the bend is connected to the capillary valve. The neck where the chamber is connected to the channel is contracted, making the neck wider than the channel and narrower than the chamber. This is designed to ensure that the chamber can be filled effectively when the liquid flows through it, and when liquid flows away, the liquid in the chamber can be cut off and retained to form stable droplets.

The capillary valve connects the chamber to the bend of the channel and is 1/9 of the width of the chamber. It is due to the capillary valve that the liquid is subjected to strong resistance after filling the chamber and therefore flows away from the channel. After finalizing the microfluidic design, we conducted corresponding simulations with Comsol Multiphysics software packages.

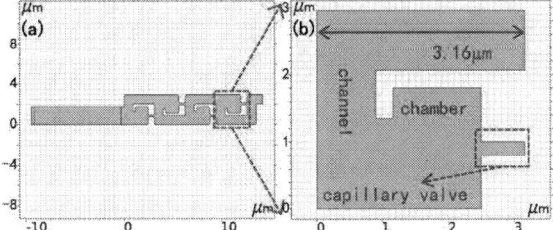

Figure 4: (a) The adapted microfluidic structure for static droplet formation, with the enlarged view (b) on each unit showing the main flow channel, micro-capillary structure, and the reaction site, the total size is $3.16 \times 3.16 \mu m^2$, equivalence of $10^7/cm^2$.

RESULTS AND DISCUSSION

Integrated circuit simulation

Firstly, we seek to verify the addressing function of the integrated circuit. Based on the row/column addressing logic described above, for instance, we assigned the input of row addressing circuit to 0b001010011010=666. Then, we set the 44th of the first 128 inputs of the column addressing to high (5V) and the last 5 inputs to 0b01011=11, which selected the 44x25+11=1111th output port. By observing the post-simulation waveform (Fig.5a), the (666, 1111) unit is charged (red), while the voltages of other cells are close to zero (grey), which confirms the addressing function is working properly.

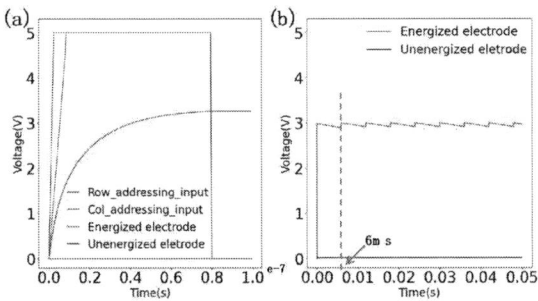

Figure 5: The post-simulation waveform: (a) with row/column addressing (green and blue) input set at 5V, energized electrode (red) is charged to 3V in 80 ns and unenergized electrodes (black) are kept at 0V. (b) The cycling charge-discharge motion of the electrode maintains a relatively stable voltage output

Moreover, the post-simulation waveform also shows the charging behavior of this unit, which rises to the specified voltage in ~80 ns. Therefore, it takes $80ns \times 10^7/128 = 6ms$ for 10^7 units to complete one round of consecutive charging. According to our simulation, after 6ms, the first transistor discharges only by <5% before it gets recharged. The time scale (6ms) is negligible compared to the typical synthesis reaction (seconds). Hence, such cycling charge-discharge motion is eventually able to output a relatively stable voltage (Fig. 5b).

Microfluidic simulation

For the microchamber/channel structure, we simulate an array with four cells. This is because in each round of oligonucleotide synthesis, only 1/4 of the electrodes on the chip will be energized to generate protons for deprotection reactions. Although these energized electrodes are not uniformly distributed on the chip, we choose a more ideal case to simulate.

In the four-cell array used for the simulation, we set the electrode of the center cell to be energized. That is, we assign a flux of $6.37 \times 10^{-3} mol/(m^2 \cdot s)$ on the electrode surface in the simulation to represent the proton production rate[10]. Since the diffusion coefficient of the protons (hydrogen ions) we use in this study is not found in the literature, we choose a value that is in the same order of magnitude as other small molecules: 2×10^{-5} cm²/s[11]. Finally, we fix the dimensions of the unit and electrode, only scan through the channel width and chamber length in an appropriate range, and observe the distribution of proton concentration in the simulated array after 10 s.

The simulation shows, when the microchamber length and microchannel width are 55μm and 1μm respectively, the proton concentrations in the energized/unenergized chambers differ by >10 times (Fig.6, dashed line), effectively preventing interference from proton diffusion. However, when downsizing to density of $10^7/cm^2$ ($3.2 \times 3.2 \mu m^2$), it shows poor performance in diffusion restriction.

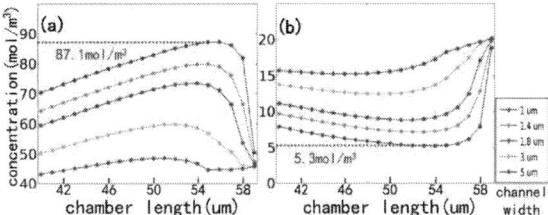

Figure 6: Proton concentration in the micro-chamber with (a) energized and (b) unenergized electrode plotted against varying chamber length and channel width. The combination of 55um length and 1um width.

For the microfluidic design of static droplet formation, we designed a structure with four capillary valve units in series with fluid injection channels, as shown in Figure 7. According to the actual injection situation, we set up a section of the aqueous phase in the injection channel, while the rest of the channel is filled with a non-aqueous phase. At the inlet on the left side of the structure, we set the injection rate to 3 mm/s. This is to ensure that the solution in the entire structure can be replaced within 10s when the sequence contains 3163 capillary valve units with a density of $10^7/cm^2$. The simulation shows that the aqueous phase is restricted to form isolated droplets in each site in front of the capillary due to an increase in flow resistance, thus completely avoiding the proton diffusion between the reaction sites. Therefore, we successfully reduce the size of the reaction sites to ~$3.2 \times 3.2 \mu m^2$, an equivalent of unit density at $10^7/cm^2$

978-1-6654-9309-3/23 $31.00 © 2023 IEEE

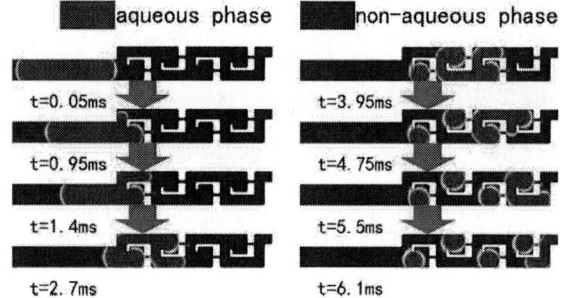

Figure 7: Simulation of two-phase flow: the aqueous phase flows in the capillary array driven by the non-aqueous phase, forming independent droplets in the reaction site.

CONCLUSION

In summary, we demonstrate a novel DNA synthesis chip design for data storage application with DRAM-like IC and microfluidic confinement. We show from multiple simulation analyses that the microfluidic structure is capable of achieving the functions of liquid exchange and confinement. Combined with the addressable and controllable DRAM-like circuit, our design is set to fulfill the goal of high-throughput DNA synthesis. Our DNA synthesis chip is projected to gain >10 times improvement in both throughput and unit density compared with current platforms.

ACKNOWLEDGEMENTS

This work is supported by the National Key R&D Program of China (2021YFF1200300).

REFERENCES

[1]   Y. Dong, F. Sun, Z. Ping, Q. Ouyang, and L. Qian, "DNA storage: research landscape and future prospects," *Natl Sci Rev,* vol. 7, no. 6, pp. 1092-1107, Jun 2020.

[2]   D. R.-J. G.-J. Rydning, J. Reinsel, and J. Gantz, "The digitization of the world from edge to core," *Framingham: International Data Corporation,* vol. 16, 2018.

[3]   L. Ceze, J. Nivala, and K. Strauss, "Molecular digital data storage using DNA," *Nature Reviews Genetics,* vol. 20, no. 8, pp. 456-466, 2019.

[4]   D. D. S. Alliance, "Alliance, DNA Data Storage, Preserving Our Digital Legacy: an Introduction To Dna Data Storage, tech. rep. June," tech. rep. June2021.

[5]   A. Extance, "How DNA could store all the world's data," *Nature,* vol. 537, no. 7618, 2016.

[6]   P. L. Antkowiak *et al.*, "Low cost DNA data storage using photolithographic synthesis and advanced information reconstruction and error correction," *Nat Commun,* vol. 11, no. 1, p. 5345, Oct 22 2020.

[7]   B. H. Nguyen *et al.*, "Scaling DNA data storage with nanoscale electrode wells," vol. 7, no. 48, p. eabi6714, 2021.

[8]   S. Kosuri and G. M. Church, "Large-scale de novo DNA synthesis: technologies and applications," *Nature Methods,* vol. 11, no. 5, pp. 499-507, 2014.

[9]   H. Boukellal, Š. Selimović, Y. Jia, G. Cristobal, and S. Fraden, "Simple, robust storage of drops and fluids in a microfluidic device," *Lab on a Chip,* vol. 9, no. 2, pp. 331-338, 2009.

[10] K. Maurer *et al.*, "Electrochemically generated acid and its containment to 100 micron reaction areas for the production of DNA microarrays," *PLoS One,* vol. 1, no. 1, p. e34, 2006.

[11] M. H. Abraham and W. E. J. J. o. S. C. Acree, "Limiting diffusion coefficients for ions and nonelectrolytes in solvents water, methanol, ethanol, propan-1-ol, butan-1-ol, octan-1-ol, propanone and acetonitrile at 298 K, analyzed using Abraham descriptors," vol. 48, no. 5, pp. 748-757, 2019.

CONTACT

*Y. Luo, E-mail: yuanluo@mail.sim.ac.cn

A REAL-TIME WIRELESS CALORIMETRIC FLOW SENSOR SYSTEM WITH A WIDE LINEAR RANGE FOR LOW-COST RESPIRATORY MONITORING

Lifeng Huang[1], Izhar[2,4], Xiaoyong Zhou[3], Mingdong Fang[3], Siwei Huang[1], Yi-Kuen Lee[2], Xiaofang Pan[1], and Wei Xu[1]

[1]Shenzhen University, CHINA
[2]Hong Kong University of Science and Technology, Hong Kong, CHINA
[3]Mindray Medical International Limited, CHINA and
[4]University of Pennsylvania, USA

ABSTRACT

In this paper, a real-time wireless respiratory monitoring system using a dual pair of detectors-based thermal flow (D^2TF) sensor with a wide linear range is reported. The D^2TF sensor structure and packaging are optimized to achieve a linear range of -70 SLM ~ 70 SLM (-4 m/s ~ 4 m/s) and a sensitivity of 9.18 mV/SLM. Moreover, the developed wireless sensor system shows a system response time of less than 23 ms with a data sample rate of 2 kHz, which can capture sharp signal changes in respiration. Therefore, the proposed respiratory monitoring system can significantly reduce the complexity of calibration and simplify signal processing, which is promising for low-cost intelligent medical care applications.

KEYWORDS

Thermal flow sensor, respiratory monitoring, wide linear range, wireless sensing system

INTRODUCTION

Respiratory monitoring is vital for the prevention and treatment of many diseases, such as Asthma, Apnea syndrome, and chronic obstructive pulmonary disease [1, 2]. In some reported works, piezoelectric sensors, temperature sensors, and humidity sensors were used to monitor respiration. However, these sensors are difficult to detect respiratory flow rate and cannot obtain critical respiratory characteristics such as tidal volume (TV), which limits their broad use in precise medical applications [3, 4]. On the contrary, flow sensors fit well into the sensing component of respiratory monitoring systems due to their ability to reveal more details of respiration. In particular, micro thermal flow sensors have no moving parts and are compatible with the CMOS fabrication process, which can significantly reduce sensor cost and footprint while achieving high sensitivity [5].

In recent years, many micro thermal flow sensors for respiratory monitoring have been reported [6-8]. However, these flow sensors generally cannot achieve a linear response with good accuracy over a wide flow range. Meanwhile, the nonlinear sensor response requires extensive calibration efforts and time. Therefore, a low-cost respiratory monitoring system with a wide linear range that can satisfy daily respiratory monitoring is necessary.

In this paper, we designed a CMOS-compatible D^2TF sensor with a wide linear range based on the theory of Peclet number-based linear criterion [9]. Besides, a wireless sensing platform was built to validate the respiratory monitoring capability of the D^2TF sensor.

CONCEPT AND SYSTEM

As shown in Fig. 1(a), the real-time wireless respiratory monitoring system includes the packaged D^2TF sensor, interface circuit, monitor, power supply, and breathing mask. The D^2TF sensor is first embedded in a PCB and packaged into a 20mm diameter 3D-printed channel, which is then connected to a breathing mask. Fig. 1(b) shows the schematic structure of the silicon-based D^2TF sensor. The polysilicon-based microheater (R_h) is placed in the middle of a 4 μm thick suspended thin film, with two pairs of Poly-Si detectors placed symmetrically upstream ($R_{u1,2}$) and downstream ($R_{d1,2}$). In the absence of gas flow, the temperature profile is distributed symmetrically on both sides of the microheater, so the temperature difference between the upstream and downstream detectors is 0, i.e., $\Delta T = T_d - T_u = 0$. When there is a gas flow, a temperature difference related to the fluid flow U will be generated, i.e., $\Delta T = T_d - T_u \neq 0$. Thus, the respiratory flow rate can be detected.

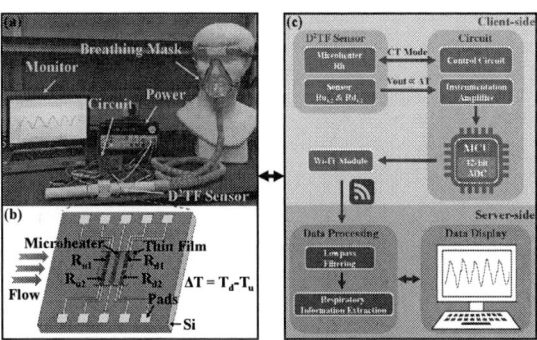

Figure 1. (a) Wireless flow sensor system for respiratory monitoring, (b) schematic of the D^2TF sensor with two pairs of upstream ($R_{u1,2}$) and downstream ($R_{d1,2}$) detectors, (c) the signal processing from the MEMS sensor to the cloud server.

The signal processing from the MEMS sensor to the cloud server is shown in Fig. 1(c). The temperature difference sensed by the detectors is firstly converted into an analog voltage signal through a Wheatstone bridge, and then the amplified output voltage (V_{out}) is encoded by the microcontroller unit (MCU) through an integrated 12-bit analog-to-digital converter (ADC), thus the flow rate measurement and acquisition are realized. Later, a WIFI module is used to upload the digital flow rate data to the cloud server every 0.5 ms, resulting in a sample rate of 2 kHz, which is fast enough for respiratory monitoring applications.

To suppress the high-frequency noise, the respiratory signal is filtered by a software-based 22 Hz low-pass filter before the final respiratory information is extracted. At the same time, the monitor will display respiratory patterns in real-time, so that abnormal respiration can be identified in time. In this way, the real-time wireless respiratory monitoring system is successfully implemented.

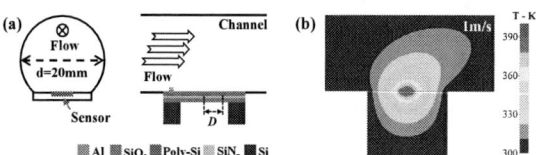

Figure 2. (a) Cross-sectional view of the packaged D²TF sensor in a 20 mm diameter tube, (b) CFD simulation shows a distorted temperature profile around R_h under an input flow velocity of 1 m/s.

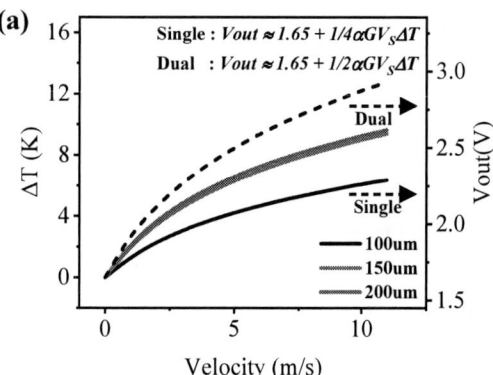

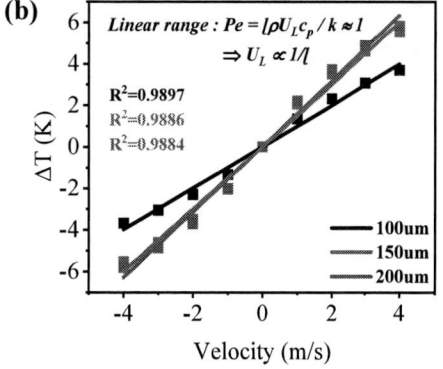

Figure 3. The effect of distance D between microheater and detectors on the (a) sensitivity and (b) the linearity of the D²TF sensor. Note, the D is chosen as 100 μm with good linearity, while two pairs of detectors are designed to remedy the sensitivity loss.

SENSOR THEORY

The D²TF is fabricated by using a CMOS-compatible process [10] and then placed at the bottom center of a 20 mm diameter tube, so the respiratory resistance can be reduced with a sufficiently large cross-sectional area, as

shown in Fig. 2(a). From the CFD simulation, Fig. 2(b) shows the distorted temperature distribution around R_h at an average input flow of 1 m/s, confirming that the sensor works well for small gas flow. On the other hand, the nonlinear response of the conventional calorimetric flow sensor requires someone to spend a lot of time and effort in the calibration, so achieving a linear sensor response over a wide flow range would be of great benefit. According to [9], the linearity criteria based on the Peclet number ($Pe \approx 1$) can be used to optimize the linearity of the calorimetric flow sensor, and it can be inferred that the linearity of the flow sensor is inversely proportional to the distance D between the microheater and detectors, as illustrated in (1).

$$U_L \propto D^{-1} \qquad (1)$$

where U_L is the linear range.

Fig. 3 shows the simulated effect of D on the sensitivity and the linearity of the D²TF sensor, indicating that a small D provides better linearity, but lower sensitivity. Therefore, a D of 100 μm was chosen to obtain good linearity within -4 m/s to 4 m/s, and two pairs of detectors were designed to double the sensitivity. Moreover, by setting the offset voltage of 1.65 V, the doubled and amplified output voltage of the developed D²TF sensor is given by:

$$V_{out} \approx 1.65 + \frac{1}{2}\alpha G V_s \Delta T \qquad (2)$$

where α is the TCR of the Poly-Si resistor, G is the gain of the instrumentation amplifier, and V_s is the supply voltage of the Wheatstone bridge. Eventually, the D²TF sensor achieves good linearity while maintaining high sensitivity.

EXPERIMENT AND DISCUSSION

Fig. 4 shows the calibration setup for the fabricated and packaged D²TF sensor. A commercial flow sensor (SFM3000, Sensirion, Switzerland) was adopted as the reference flow meter. The microheater is configured in constant temperature (CT) mode with an operating temperature of 125 °C and a consumed power of 20 mW, while the power consumption of detectors is negligible due to their high resistance and low supply voltage $V_s = 0.73$ V.

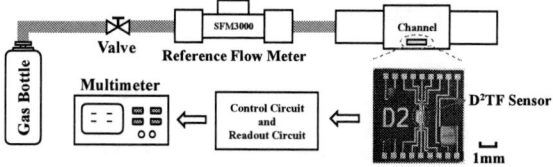

Figure 4. Calibration of the fabricated and packaged D²TF sensor.

The measured sensor output as a function of input nitrogen gas flow from -200 SLM to 200 SLM is plotted in Fig. 5, which demonstrates that the packaged D²TF sensor is capable of detecting bidirectional gas flow over a wide range. Moreover, the D²TF sensor shows a prominent sensitivity of 9.18 mV/SLM within a wide linear range of -

70 SLM to 70 SLM, which can satisfy the requirements of daily respiratory monitoring. Therefore, the developed D^2TF sensor is able to help us reduce the time-consuming calibration work and realize low-cost respiratory monitoring.

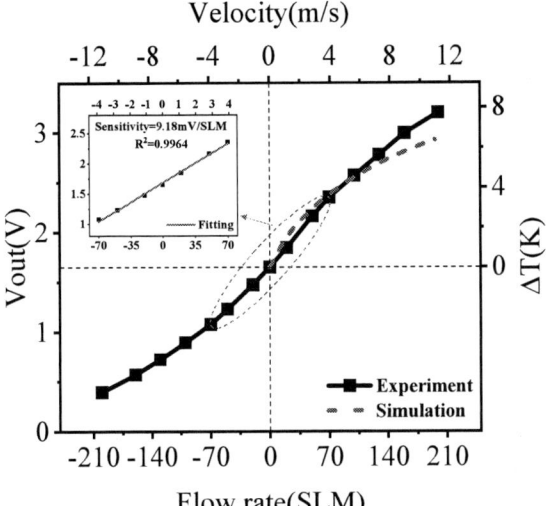

Figure 5. The D^2TF sensor shows a prominent sensitivity of 9.18 mV/SLM in a linear range of -70 SLM ~ 70 SLM (-4 m/s ~ 4 m/s), based on the tested nitrogen gas flow from -200 SLM to 200 SLM. Note such a wide linear range helps us reduce the time-consuming calibration work for sensors applied in daily respiratory monitoring.

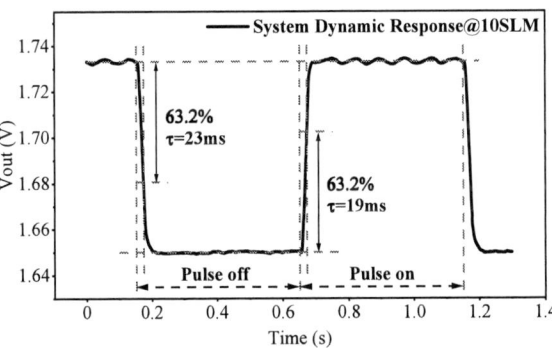

Figure 6. The measured response time of the respiratory monitoring system under a nitrogen gas flow of 10 SLM, the microheater is powered by a pulse signal and the time constant of the wireless sensor system is less than 23 ms.

The response time of the respiratory monitoring system can indicate whether it is fast enough to capture rapidly changing respiratory signals. When the microheater is powered by a pulse signal, the response time of the flow sensor can be defined as the time required for the output signal to reach 63.2% of the amplitude from the beginning. Herein, a 1 Hz pulse signal with a peak-to-peak voltage of 5 V is applied to the microheater to initiate such a purpose. As shown in Fig. 6, the data recorded by the cloud server shows that the response time of the respiratory monitoring system is less than 23 ms when the input gas flow rate is 10 SLM. Accordingly, the system bandwidth is calculated to be around 6.9 Hz, which is much higher than the normal

respiratory rate of about 0.3 Hz. Evidently, the flow sensor system has ample ability to monitor rapid changes in respiration.

A simulated respiratory experiment was performed to validate the ability of the respiratory monitoring system in the detection of abnormal respiratory patterns. As shown in Fig. 7(a), the wirelessly recorded flow rates of a subject with the simulated symptoms of apnea, polypnea, and hypopnea are recognized together with the normal respiratory pattern by the respiratory monitoring system in the linear range. In the beginning, the subject breathes smoothly, and the change in the flow rate is relatively stable and also normal. The subject then stopped breathing for more than 10 s to simulate the symptom of apnea, at which point the flow rate approached 0 SLM, as indicated in the gray region. The light green area indicates an increase in respiratory rate and tidal volume (TV) in the presence of polypnea. The simulated hypopnea pattern is also shown in the dusty blue region, indicating that the respiratory monitoring system has sufficient sensitivity to detect shallow breathing.

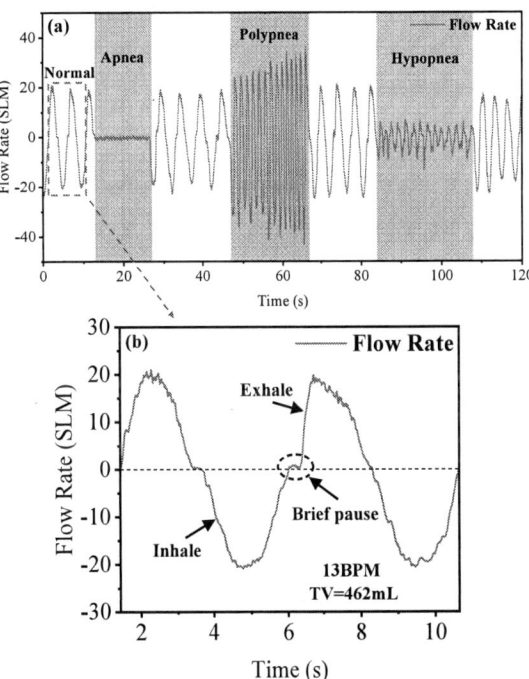

Figure 7. (a) Wirelessly recorded flow rates of a subject with the simulated abnormal respiration, the situation of Apnea, Polypnea, and Hypopnea diseases are successfully captured in the linear range of the D^2TF sensor, thus the reduced calibration worked is feasible. (b) Details of normal respiratory flow rate signals that enlarged from (a).

Fig. 7(b) shows the details of normal respiratory flow rate signals in the dotted box of Fig. 7(a), where the peak-to-peak flow rate is approximately 40 SLM, the respiratory rate is 13 BPM, and the tidal volume is 462 mL. Furthermore, the respiratory monitoring system can successfully capture small changes, such as brief pauses during the respiratory transition. The rapid expiratory flow rate pattern of the subject got coughing is shown in Fig. 8, which demonstrates the rapid response capability of the developed respiratory monitoring system. Nevertheless, the monitored peak expiratory flow (PEF) of 114 SLM exceeds the linear range, which further guides us to optimize the sensor system to expand the linear range in the future.

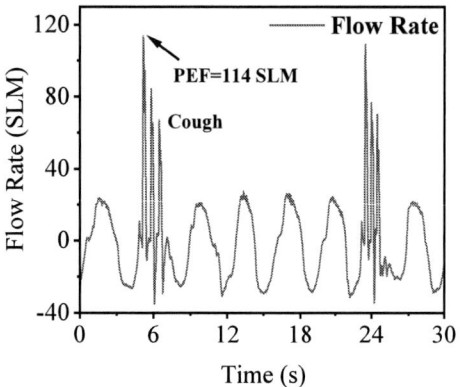

Figure 8. The respiratory flow pattern of a subject got coughing shows the flow rate exceeds the linear range, which guides us to redesign the sensor system with an extended linear range.

Table 1. Performance comparison between the reported respiratory monitoring system and this work.

Reference	Jiang[11]	Nguyen[12]	Wang[13]	This work
Power, mW	60	40#	6.5	20
Response Time, ms	193a	1600a	6a	23b
Flow Range, m/s	0 ~ 4	N/A	-8.6 ~ 8.6	-11 ~ 11
Linear Range, m/s	0 ~ 1#	N/A	-1.3 ~ 1.3#	-4 ~ 4
Sensitivity, (m/s)-1	-10%#	340mV	830mV	164mV
Working Principle	Hot-film		Calorimetric	

a response time of the sensor, b response time of the wireless sensor system.
Power, Linear Range, and Sensitivity are calculated from the available data.

CONCLUSION

In summary, a real-time wireless respiratory monitoring system with a wide linear range was successfully developed using a dual pair of detectors-based thermal flow (D²TF) sensor. The respiratory monitoring system achieved a sensitivity of 9.18 mV/SLM over a wide linear range of -70 SLM ~ 70 SLM (-4 m/s ~ 4 m/s), indicating that the presented wireless sensor system can reduce the time-consuming calibration work for the sensor applied in daily respiratory monitoring (Table 1). Furthermore, with a system response time of less than 23 ms, the sensor system is capable of capturing sharp signal changes in respiration. As a result, the proposed respiratory monitoring system has the potential to be used in low-cost intelligent medical care applications.

ACKNOWLEDGEMENTS

This work was supported by the National Natural Science Foundation of China (52105582). The Natural Science Foundation of Guangdong Province (2022A1515010894), Shenzhen Science and Technology Program (JCYJ20210324095210030, JCYJ20220818095810023), and Shenzhen-Hong Kong-Macau S&T Program (Category C: SGDX20210823103200004).

REFERENCES

[1] Z. Cao, R. Zhu, and R. Y. Que, "A Wireless Portable System with Microsensors for Monitoring Respiratory Diseases," *IEEE Transactions on Biomedical Engineering*, vol. 59, no. 11, pp. 3110-3116, 2012.

[2] E. Vanegas, R. Igual, and I. Plaza, "Sensing systems for respiration monitoring: A technical systematic review," *Sensors*, vol. 20, no. 18, pp. 5446, 2020.

[3] I. Mahbub et al., ''A low-power wireless piezoelectric sensor-based respiration monitoring system realized in CMOS process,'' *IEEE Sensors Journal*, vol. 17, no. 6, pp. 1858-1864, Mar. 2017.

[4] T. Dinh et al., "Stretchable respiration sensors: Advanced designs and multifunctional platforms for wearable physiological monitoring," *Biosensors and Bioelectronics*, vol. 166, pp. 112460, 2020.

[5] W. Xu et al., "A wireless dual-mode micro thermal flow sensor system with extended flow range by using CMOS-MEMS process," *Proc. of IEEE MEMS Conf.*, Belfast, UK, Jan. 2018, pp. 824-827.

[6] T. Dinh et al., "Solvent-free fabrication of biodegradable hot-film flow sensor for noninvasive respiratory monitoring," *Journal of Physics D: Applied Physics*, vol. 50, no. 21, pp. 215401, 2017.

[7] Y. Liu et al., "Epidermal electronics for respiration monitoring via thermo-sensitive measuring," *Materials Today Physics*, vol. 13, pp. 100199, 2020.

[8] P. Jiang, S. Zhao, and R. Zhu, "Smart sensing strip using monolithically integrated flexible flow sensor for noninvasively monitoring respiratory flow," *Sensors*, vol. 15, no. 12, pp. 31738-31750, 2015.

[9] B. Li, W. Xu, M. Paszkiewicz, Z. Li, R. Wang, and Y. K. Lee, "Theoretical and Experimental Study of Peclet Number Effect on the Linearity of Thermoresistive Micro Calorimetric Flow Sensors," *Proc. of APCOT Conf.*, Hong Kong, 2018.

[10] Izhar, W. Xu, L. J. Yang and Y. K. Lee, "CMOS Compatible MEMS Air Velocity Sensor with Improved Sensitivity and Linearity for Human Thermal Comfort Sensing Applications," *IEEE Sensors Journal*, vol. 21, no. 21, pp. 23872-23879, 2021.

[11] T. Jiang et al., "Wearable breath monitoring via a hot-film/calorimetric airflow sensing system," *Biosensors and Bioelectronics*, vol. 163, pp. 112288, 2020.

[12] T. Nguyen et al., "A wearable, bending-insensitive respiration sensor using highly oriented carbon nanotube film," *IEEE Sensors Journal*, vol. 21, no. 6, pp. 7308-7315, 2021.

[13] X. Wang, Z. Ke, G. Liao, X. Pan, Y. Yang, and W. Xu, "A fast-response breathing monitoring system for human respiration disease detection," *IEEE Sensors Journal*, vol. 22, no. 11, pp. 10411-10419, 2022.

CONTACT

*Wei Xu, tel: +86-755 2653-4853; weixu@szu.edu.cn

ADVANCED THERMOPHYSICAL PROPERTIES MEASUREMENTS USING HEATER-INTEGRATED FLUIDIC RESONATORS

Juhee Ko[1], Bong Jae Lee[1], and Jungchul Lee[1]
[1]Korea Advanced Institute of Science and Technology, SOUTH KOREA

ABSTRACT

Measurement of thermophysical properties of liquids on microscale has been developed with increasing interest in thermal management or energy storage/transport systems. Thermal conductivity, heat capacity, and density are required to predict heat transfer performance, however, simultaneous measurement of thermophysical properties of small-volume liquids has been rarely studied. Recently, we have proposed new metrology for the three intrinsic properties simultaneously by heater-integrated fluidic resonators (HFRs) in an atmospheric pressure environment, which consist of a microchannel, a resistive heater/thermometer, and a mechanical resonator. Thermal conductivity is measured from a temperature response by using a resistive thermometer upon heating, and the specific heat capacity is obtained from the volumetric heat capacity along with the density by the resonance densitometer. In this paper, we show improvement in thermophysical properties measurement performance with HFRs by switching the environment around the sensor to the vacuum. It is validated by numerical analysis that the dynamic range of thermal conductivity is expanded by ~50 times and the sensitivity of heat capacity is increased by ~2 times. This improves the resolution of thermal conductivity and heat capacity with the same measurement resolution in vacuum environment.

KEYWORDS

Channel resonator, Simultaneous measurements, Heat capacity, Thermal conductivity.

INTRODUCTION

Thermophysical properties of liquids are important for a number of applications, including thermal management of electronic devices [1] and energy storage systems [2], and photothermal therapy using light waves [3]. With the advantages of fast response time and small sample consumption, there have been continuous efforts for thermophysical properties measurement with microfluidic channel integrated sensors [4]–[6]. In the case of thermal conductivity sensors, the representative methods are heat flow measurement of the sample placed between a heater and a heat sink [7], measuring the resistance of the heater inside the flowing fluid [8], [9], or the 3-omega method, measuring the amplitude of the temperature oscillation at 3 times of heating frequency [10], [11]. By using a microchamber system with a heater, energy consumption analysis at a fixed temperature is representative of heat capacity sensors [12]. However, studies with micro-fabricated sensors have not been able to simultaneously extract different thermal properties using an identical sample. In addition, the mass or volume of the sample is not quantitatively measured, and thus has lower accuracy compared to conventional bulky measuring devices due to its size effects.

Meanwhile, fluidic resonators with a hollow channel inside the solid structure have been used for buoyant mass sensing of cells, micro-/nanoparticles, or liquids [13] with high accuracy. While most of the previous researches focused on biological matters thus have been operated at or near room temperature [14], [15], only a few studies have utilized photothermal temperature modulation [16] or the global heating of the entire system [17] to measure heat capacity or temperature-dependent density of liquid samples, respectively. However, their sensitivity, response time, and operating range of temperature have been limited due to increased thermal mass and mismatched thermal expansion of different packaging materials. The integration of on-chip heaters towards fast and quantitative heating has not been well studied. For these reasons, we have recently introduced a fast, precise, and highly sensitive thermophysical properties measurement technique using heater-integrated fluidic resonators (HFRs) [18]. Thermal conductivity (k), specific heat capacity (c_p), and density (ρ) are simultaneously measured by resistive heating/thermometry and resonant densitometry.

To take one step further, it is assumed that if the convection heat loss is decreased in vacuum compared to the atmospheric pressure environment, the difference in the thermophysical properties of the fluid in the channel will be easier to be investigated by temperature measurements. In this paper, we introduce a sensing performance enhancement method by vacuum environment around the HFRs (Figure 1). We utilized numerical analysis of heat transfer along the HFR to validate the idea and demonstrated the vacuum operation experimentally.

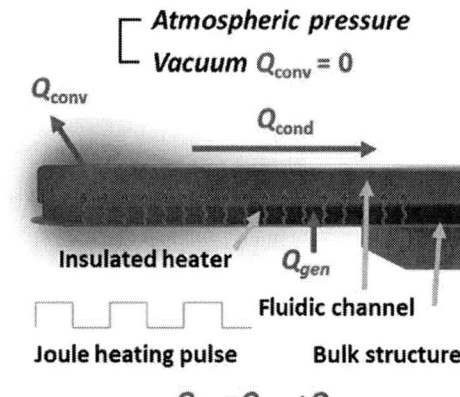

Figure 1: Schematic diagram of heat transfer system in a heater-integrated fluidic resonator (HFR) with a joule heating pulse in atmospheric pressure and vacuum environment.

METHODS

Thermophysical Properties Measurement

The transient temperature response of HFR is dependent on the thermophysical properties of liquid inside the microfluidic channel (Figure 2a). With the same heating power, the amount of steady-state temperature difference (ΔT) is inversely related to thermal conductivity (k), and the characteristic time to reach the steady-state value (time constant, τ) is inversely related to thermal diffusivity ($k/\rho c_p$). Under a square pulsed power input by joule heating, the temperature response is monitored by resistive thermometry, where ΔT can be represented by a steady-state resistance difference (ΔR) and τ can be extracted from transient resistance change in time. A volumetric heat capacity (ρc_p) is decoupled from $\rho c_p/k$ by using τ and ΔR together. A specific heat capacity (c_p) is exclusively extracted by dividing the volumetric heat capacity (ρc_p) by density (ρ), obtained from resonance frequency shift (Δf) as shown in Figure 2b. Physical relationships between these three intrinsic properties (k, c_p, and ρ) and measurands (ΔR, τ, and Δf) are obtained by dimensional analysis of the governing heat equation, $k \propto 1/\Delta R$, $\rho c_p \propto \tau/\Delta R$, and $c_p \propto \tau/\Delta R/|\Delta f|$ as described in [18].

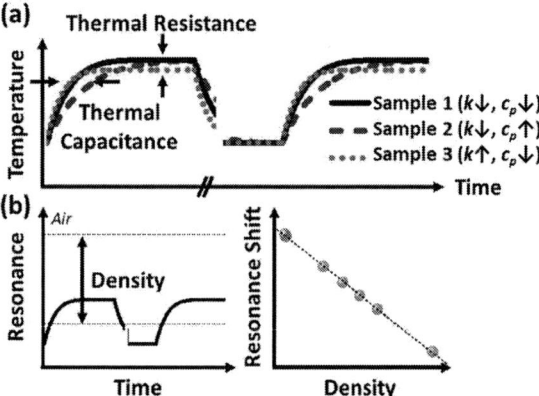

Figure 2: Conceptual illustration of (a) temperature measurements dependent on thermophysical properties of three samples with different thermal conductivity and specific heat capacity. Higher thermal conductivity shows a higher thermal loss, thus lower temperature increase with the fixed power input. Higher specific heat capacity shows a slower response due to increased thermal mass. Conceptual illustrations of (b) real-time resonance frequency measurements, which represent the density of the various liquid samples inside HFR.

Experimental System

HFR has a suspended fluidic channel of 200-μm length, 50-μm width, and 3-μm thickness. Figure 3a shows the optical microscope of HFR from the top view and the scanning electron microscope of FIB-milling-processed HFR to clearly show the cross-section of the suspended fluidic channel. In order to experimentally implement thermophysical measurement in a vacuum, a custom vacuum chamber that can control the environment around the sensor is configured. The chamber is designed with fluid ports to transport fluid samples into the sensor, electrical connections for joule heating and resistance measurements, and a transparent glass window that allows laser access to measure the resonant frequency. The space in the chamber with a volume of ~0.2 mL is connected by a vacuum pump and maintained 1.2E-4 mbar to reduce the convective heat transfer coefficient around the HFRs.

To measure the resistance changing in real-time simultaneously with joule heating pulses, the voltage and current are recorded using a voltage divider circuit (Figure 3b). The laser, which is focused on and reflected from the free end of the cantilever, is monitored by a photodetector to measure the resonant frequency, by converting a cantilever vibration to a sinusoidal voltage signal. A piezo actuator is placed outside the vacuum chamber for cantilever excitation. A phase-locked loop in a lock-in amplifier is used to record resonance frequency in real time. Specific information for the data acquisition has been explained in the supplementary information of our previous paper [18].

Numerical Analysis

Finite element method (FEM) analysis is performed to investigate the effect of the vacuum environment on the measurement of thermophysical properties. The boundary conditions are set for the bottom surface of the bulk part of the sensor to be maintained at room temperature, and for the top surface of the bulk and the entire cantilever to have convective heat transfer. The heat transfer coefficient ($h = 1000$ W/m²-K) is set by comparing the experimental results of temperature mapping with Raman thermometry. It corresponds to the coefficient at the characteristic dimension of 10-μm from the scaling analysis of a long horizontal cylinder [19]. To assume the vacuum environment around the sensor, the heat transfer coefficient is set as zero ($h = 0$ W/m²-K). Transient response of temperature is calculated in the time range of 40-ms with the step of 0.5-ms.

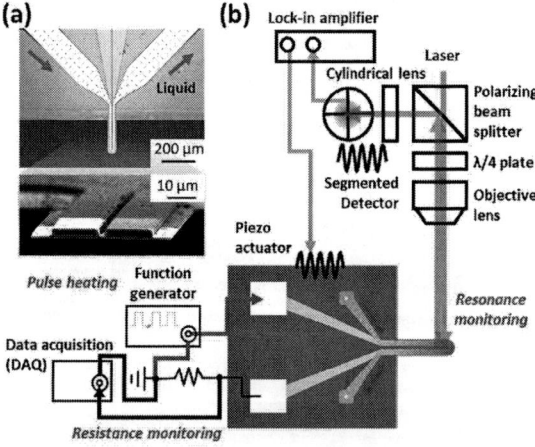

Figure 3: (a) Optical micrograph of HFR and scanning electron micrograph after focused ion beam milling (FIB) of the cross-section of HFR. (b) A schematic illustration of measurement setup with resonance monitoring, pulse heating, and resistance monitoring.

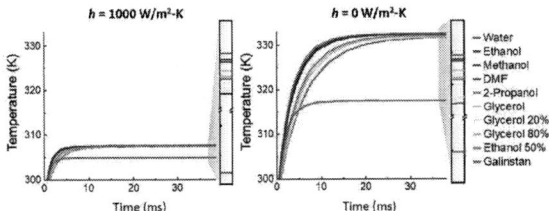

Figure 4: Comparison of temperature response in atmospheric and vacuum environment by using FEM during a 3-mW pulse heating of HFRs. Average temperature along the fluidic channel is used.

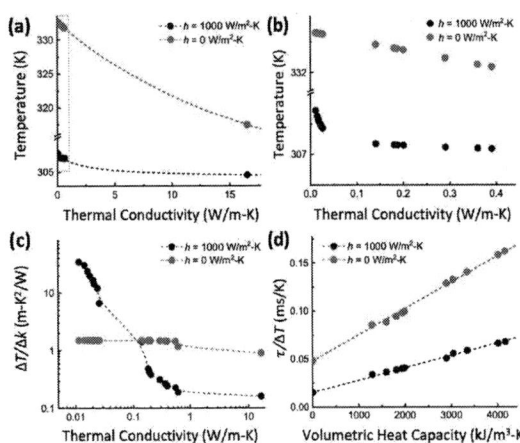

Figure 5: Steady-state temperature and time constants of HFRs filled with various thermophysical properties. (a) Steady-state temperature as a function of thermal conductivity and (b) its zoomed-in view. (c) Comparison of slopes in the piecewise linear function of ΔT as a function of thermal conductivity ($\Delta T/\Delta k$). (d) The ratio of τ to ΔT as a function of volumetric heat capacity.

RESULTS AND DISCUSSION

Temperature Response with Numerical Analysis

The temperature response of HFR filled with fluids of various thermophysical properties (Table 1) is compared between the atmospheric pressure and the vacuum environment as shown in Figure 4. In a vacuum environment, the temperature increase is higher (~3.5 times) than in the atmospheric pressure environment, owing to the convective loss around the cantilever. Temperature change in the vacuum environment also shows a slower (~2.4 times) response than in the atmospheric pressure environment.

Figure 5 shows the results of temperature response versus thermal conductivity and volumetric heat capacity. In the case of thermal conductivity (Figures 5a and 5b), the linearity increases in vacuum environment, which means enhanced sensitivity in the high thermal conductivity range. The linear dynamic range is enhanced about 50 times in the vacuum environment. Interestingly, in the case of low thermal conductivity (under 0.07 W/m-K), which is in the range of gases in general, it shows that there was a difference of more than 10 times in sensitivity as shown in Figure 5c. As shown in Figure 5d, the sensitivity of volumetric heat capacity measurement is also increased by about 2 times.

Table 1: Thermophysical properties of various liquids at 300 K used in the numerical analysis [20].

	Density [kg/m³]	Thermal conductivity [W/m-K]	Heat capacity [J/kg-K]
DI water	997	0.608	4181
Ethanol	795	0.177	2280
Methanol	786	0.2	2534
DMF	944	0.180	2050
2-propanol	781	0.137	2040
Glycerol	1259	0.292	2414
Glycerol 20%	1049	0.546	3829
Glycerol 80%	1207	0.356	2768
Ethanol 50%	896	0.394	3227
Galinstan	6440	16.5	200

Experimental Results

The measurement result of the resistance change as a function of temperature is shown in Figure 6. Considering

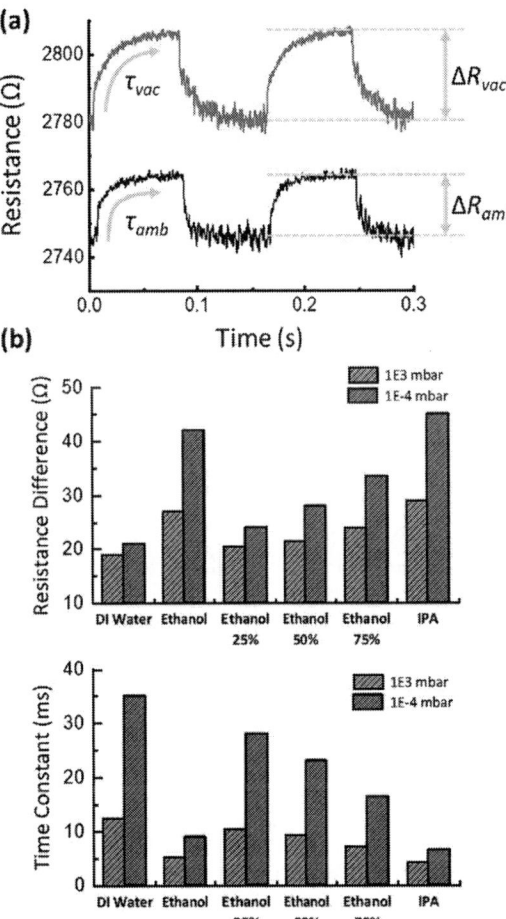

Figure 6: (a) Resistance change of HFR filled with ethanol 50% during repeated heating pulses. (b) Resistance differences and time constants extracted from the transient changes of the resistance.

978-1-6654-9309-3/23 $31.00 © 2023 IEEE

the time constants in numerical analysis, pulse heating with 80-ms width is applied. In vacuum, as the convective heat loss decreases, the resistance differences increase. The higher the concentration of water, which has high thermal conductivity compared to ethanol, the smaller the difference in resistance between the atmospheric pressure and the vacuum environments. In addition, the time constants representing the rate of temperature response increase, which is related to the smaller thermal diffusivity. The differences in time constants increase with the concentration of water, which has a higher heat capacity than ethanol.

ACKNOWLEDGEMENTS

This research was supported by National Research Foundation of Korea (NRF) grants funded by the Korean government (Ministry of Science and ICT) (NRF-2020R1A2C3004885 and NRF-2020R1A4A2002728). This work was also supported by the Technology Innovation Program (00144157, Development of Heterogeneous Multi-Sensor Micro-System Platform) funded By the Ministry of Trade, Industry & Energy (MOTIE, Korea).

REFERENCES

[1] M. Ghanbarpour, E. Bitaraf Haghigi, R. Khodabandeh, "Thermal Properties and Rheological Behavior of Water Based Al_2O_3 Nanofluid as a Heat Transfer Fluid", *Exp. Therm. Fluid. Sci.*, vol. 53, pp. 227–235, 2014.

[2] D. Shin and D. Banerjee, "Enhanced Thermal Properties of SiO_2 Nanocomposite for Solar Thermal Energy Storage Applications", *Int. J. Heat Mass Transf.*, vol. 84, pp. 898–902, 2015.

[3] Y. Ren, Y. Yan, H. Qi, "Photothermal Conversion and Transfer in Photothermal Therapy: From Macroscale to Nanoscale", *Adv. Colloid. Interface Sci.*, vol. 308, p. 102753, 2022.

[4] S. R. Choi and D. Kim, "Real-time Thermal Characterization of 12 nl Fluid Samples in a Microchannel", *Rev. Sci. Instrum.*, vol. 79, no. 6, p. 064901, 2008.

[5] B. K. Park, N. Yi, J. Park, D. Kim, "Note: Development of a Microfabricated Sensor to Measure Thermal Conductivity of Picoliter Scale Liquid Samples", *Rev. Sci. Instrum.*, vol. 83, no. 10, p. 106102, 2012.

[6] G. Paul, M. Chopkar, I. Manna, P. K. Das, "Techniques for Measuring the Thermal Conductivity of Nanofluids: A Review", *Renew. Sustain. Energy Rev.*, vol. 14, no. 7, pp. 1913–1924, 2010.

[7] R. Shrestha, K. M. Lee, W. S. Chang, D. S. Kim, G. H. Rhee, T. Y. Choi, "Steady Heat Conduction-based Thermal Conductivity Measurement of Single Walled Carbon Nanotubes Thin Film Using a Micropipette Thermal Sensor", *Rev. Sci. Instrum.*, vol. 84, no. 3, p. 034901, 2013.

[8] B. Zhao, F. Feng, B. Tian, Z. Yu, X. Li, "Micro Thermal Conductivity Detector Based on SOI Substrate with Low Detection Limit", *Sens. Actuators B, Chem.*, vol. 308, p. 127682, 2020.

[9] A. Mahdavifar, M. Navaei, P. J. Hesketh, M. Findlay, J. R. Stetter, G. W. Hunter, "Transient Thermal Response of Micro-thermal Conductivity Detector (µTCD) for the Identification of Gas Mixtures: An Ultra-fast and Low Power Method", *Microsyst. Nanoeng.*, vol. 1, p. 15025, 2015.

[10] S. Gauthier, A. Giani, P. Combette, "Gas Thermal Conductivity Measurement Using the Three-omega Method", *Sens. Actuators A, Phys.*, vol. 195, pp. 50–55, 2013.

[11] D. G. Cahill, "Thermal Conductivity Measurement from 30 to 750 K: the 3ω Method", *Rev. Sci. Instrum.*, vol. 61, no. 2, pp. 802–808, 1990.

[12] W. Lee, W. Fon, B. W. Axelrod, M. L. Roukes, "High-Sensitivity Microfluidic Calorimeters for Biological and Chemical Applications", *Proc. Natl. Acad. Sci.*, vol. 106, no. 36, pp. 15225–15230, 2009.

[13] T. P. Burg *et al.*, "Weighing of Biomolecules, Single Cells and Single Nanoparticles in Fluid", *Nature*, vol. 446, no. 7139, pp. 1066–1069, 2007.

[14] J. Ko, D. Lee, B. J. Lee, S. K. Kauh, J. Lee, "Micropipette Resonator Enabling Targeted Aspiration and Mass Measurement of Single Particles and Cells", *ACS Sens.*, vol. 4, no. 12, pp. 3275–3282, 2019.

[15] J. Ko, J. Jeong, S. Son, J. Lee, "Cellular and Biomolecular Detection Based on Suspended Microchannel Resonators", *Biomed. Eng. Lett.*, vol. 11, no. 4, pp. 367–382, 2021.

[16] M. F. Khan, N. Miriyala, J. Lee, M. Hassanpourfard, A. Kumar, T. Thundat, "Heat Capacity Measurements of Sub-nanoliter Volumes of Liquids Using Bimaterial Microchannel Cantilevers", *Appl. Phys. Lett.*, vol. 108, no. 21, p. 211906, 2016.

[17] M. Yun, I. Lee, S. Jeon, J. Lee, "Facile Phase Transition Measurements for Nanogram Level Liquid Samples Using Suspended Microchannel Resonators", *IEEE Sens. J.*, vol. 14, no. 3, pp. 781–785, 2014.

[18] J. Ko, F. Khan, Y. Nam, B. J. Lee, J. Lee, "Nanomechanical Sensing Using Heater-integrated Fluidic Resonators", *Nano Lett.*, vol. 22, no. 19, pp. 7768–7775, 2022.

[19] J. Peirs, D. Reynaerts, H. van Brussel, "Scale Effects and Thermal Considerations for Micro-actuators", in *Proceedings. 1998 IEEE International Conference on Robotics and Automation*, 1998, pp. 1516–1521.

[20] E. W. Lemmon, H. B. Ian, M. L. Huber, M. O. McLinden, "Thermophysical Properties of Fluid Systems." *NIST Chemistry Webbook*, vol. 69, p. 20899, 2022.

CONTACT

*J. Lee, tel: +82-042-350-3212; jungchullee@kaist.ac.kr

A MINIATURIZED TRANSIT-TIME ULTRASONIC FLOWMETER USING PMUTS FOR LOW-FLOW MEASUREMENT IN SMALL-DIAMETER CHANNELS

Yunfei Gao[1,2], Zhipeng Wu[2], Minkan Chen[2], and Liang Lou[1,2*]*
[1]School of Microelectronics, Shanghai University, China
[2]Shanghai Industrial μTechnology Research Institute, China

ABSTRACT

In this paper, we report a miniaturized transit-time ultrasonic flowmeter (TTUF) for low-flow measurement in small-diameter channels. The proposed TTUF contains two pieces of ScAlN-based piezoelectric micromachined ultrasonic transducers (PMUTs) and a π-type channel. The PMUTs contain 169 (13 × 13) square cells with the dimension of 2.8 × 2.8 mm^2, showing good transmitting sensitivity and receiving sensitivity at frequency of 1MHz in water. Experiment results show that the developed TTUF provides a wide range of flow measurements from 2 to 300 L/h with 1 % repeatability in the channel of 4 mm diameter, which is smaller than most of the reported ones.

KEYWORDS

Aluminum nitride (AlN), PMUTs, Transit-time ultrasonic flowmeter (TTUF), Small-diameter

INTRODUCTION

As new-technology flowmeters, transit-time ultrasonic flowmeter (TTUF) is considered one of the most widely used devices for flow measurements [1], [2]. So far, its application in small-diameter channels is still limited [3]. One of the reasons for the limitation is that the subtle differences in transit-time are difficult to detect. Although the π-type configuration is an effective way to increase transit-time in small-diameter channels, the inner diameter of π-type flowmeter still can't reach below 8 mm due to the limitation of the oversized bulk ultrasonic transducer [4], [5].

Recently, micromechanical ultrasonic sensors (MUTs) based on microelectromechanical systems (MEMS) technology offer great superiority in terms of miniaturization and integration[6]. The MUTs family can be simply classified as capacitive micromachined ultrasonic transducers (CMUTs) and piezoelectric micromachined ultrasonic transducers (PMUTs) [7]. Compared with CMUTs, PMUTs offer the advantages of no bias voltage, relatively linear behavior, and low power consumption[8]. Among the piezoelectric materials commonly used in PMUTs, the lead-free AlN is a promising candidate for low-cost highly integrated PMUT device, since AlN is compatible with standard complementary metal oxide semiconductor (CMOS) fabrication processes [9]. Thus, AlN-based PMUTs have been widely applied as acoustic transceiver in flow measurement [10], [11]. Ultrasonic liquid flow meters based on AlN PMUTs have been reported [12], [13], Zhu *et al.* successfully realize the flow measurement with measuring channel diameter of 8 mm. The size of the PMUTs in their work is 3.2mm × 3.2 mm^2, and the operating frequency of the system is around 2.5MHz. To achieve flow measurement in smaller-diameter channels with high precision, PMUTs used in the flowmeter need further miniaturization and improved performance.

In this paper, we report a miniaturized TTUF using ScAlN PMUTs which is fully integrated in commercial hardware system. The PMUTs are fabricated based on Cavity Silicon-on-Insulator (CSOI) platform. The fabricated PMUTs have a size of 2.8 × 2.8 mm^2, which operate well at 1 MHz in water. The experiments verify that the proposed TTUF can realize high-precision monitoring of flow in the channel of 4 mm diameter, which is half of the smallest channel diameter reported so far.

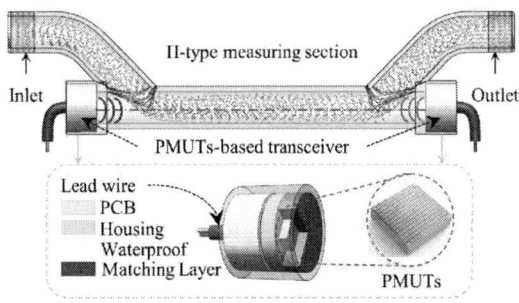

Figure 1: Illustration of a miniaturized ultrasonic flowmeter based on PMUTs.

THEORY AND MODEL

Figure 1 illustrates the schematic diagram of the developed miniaturized TTUF. The flowmeter consists of two PMUTs-based transceivers and a measuring channel. PMUTs are mounted on a small PCB using wire bonding. To compensate for the acoustic impedance mismatch, the front side of the PMUT is covered by an acoustic matching layer. Here, polyurethane (PU) is chosen as the matching layer for the transceiver because its acoustic impedance is close to the that of water (1.42 MRayl).

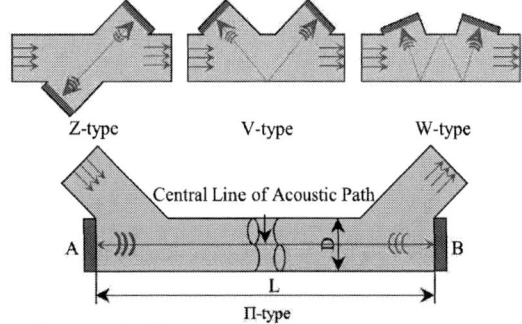

Figure 2: Working principle of the TTUF based on common and π-type mounting configuration.

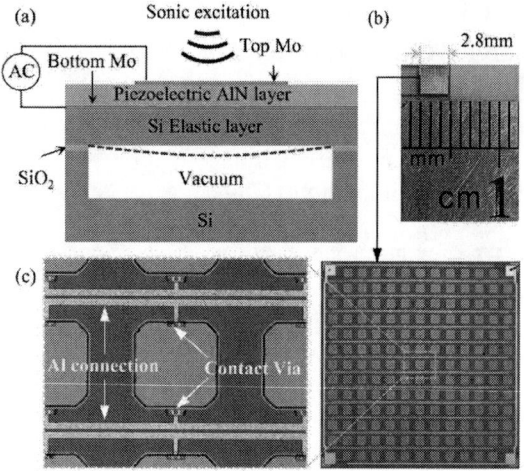

Figure 3: (a) Schematic of the PMUT cross-section. (b) Photograph of a fabricated device. (c) Close-view of the PMUTs array.

As shown in Figure 2, the axial interrogation type structural configuration like π-type is especially useful for accurate measurement of small-diameter channels rather than zigzag-type acoustic transmission path such as "Z-type" or "V-type" and "W-type". The operation of π-type TTUF by means of transit-time method is based on the measurement of the difference of time-of-flight (dToF). There are two PMUTs-based transceivers mounted on both sides of the measuring section, which act alternately as transmitter and receiver. When the flowmeter starts working, the transducer A emits an ultrasonic signal and transducer B receives. Then the process is repeated in the opposite direction against the flow. The transit time of the ultrasonic signals propagating with the flow (T_1) and against the flow (T_2) can be expressed as:

$$T_1 = \frac{L}{c - \overline{v}}, \qquad (1)$$

$$T_2 = \frac{L}{c + \overline{v}}, \qquad (2)$$

where L is the length of acoustic path, c is the speed of sound in the controlled media and $\overline{v}$ is the flow average velocity of the ultrasonic wave propagation path, the difference of transit-time (ΔT) can be calculated through (1) and (2) as

$$\Delta T = T_1 - T_2 = \frac{2L\overline{v}}{c^2 - \overline{v}^2}. \qquad (3)$$

Since $c^2 \gg \overline{v}^2$ in the liquid, a rational approximation for ΔT is

$$\Delta T \approx \frac{2L\overline{v}}{c^2}. \qquad (4)$$

The flow rate Q can be obtained from (3) as

$$Q \approx v_a \frac{\pi d^2}{4} = k_c \overline{v} \frac{\pi d^2}{4} = k_c \frac{\pi}{8} \frac{c^2 d^2}{L} \Delta T, \qquad (5)$$

where d is the pipe inner diameter, v_a is average flow velocity on the cross-section of a fully developed flow inside a straight pipe section, k_c is the correction factor, which is ratio of the v_a to the $\overline{v}$.

As a miniaturized alternative to bulk piezoelectric transducer, the PMUTs operates in a flexure mode. The

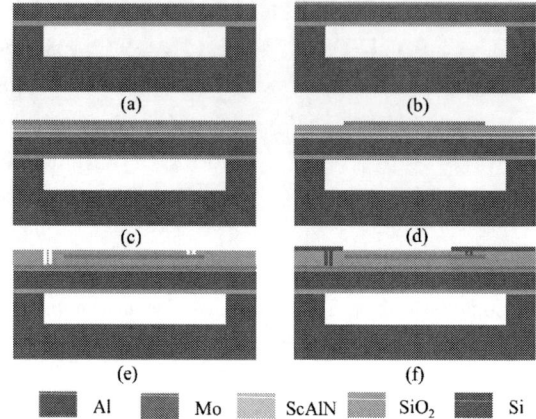

Figure 4: Fabrication process flow of the ScAlN-based PMUT. (a) customizing a cavity-SOI substrate, (b) depositing AlN seed layer, (c) multiple layers sputtering, (d) etching top electrodes, (e) depositing and etching a SiO₂ insulation layer via holes, and (f) Deposition and patterning of Al metal layer for electrical connections and to form bonding pads.

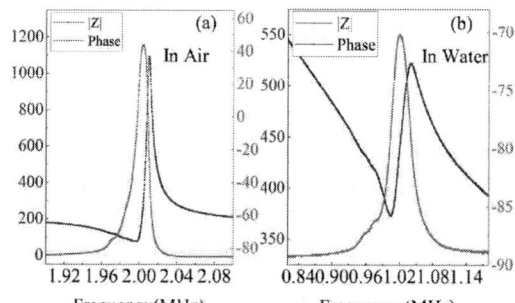

Figure 5: The impedance-frequency spectrum of the PMUTs in (a) air and (b) DI-water.

PMUTs are comprised of a thin film ScAlN piezoelectric layer sandwiched between two Molybdenum (Mo) electrodes and a Silicon (Si) passive layer, as shown in Figure 3(a). Due to the piezoelectric properties of ScAlN, this structure converts electrical potential to mechanical vibrations and vice versa. The dimension of the fabricated PMUT array is about 2.8 × 2.8 mm², as shown in Figure 3(b). In this paper, the PMUTs are designed to be rectangular to achieve a higher fill factor [14], the side of the square cavity is approximately 180 μm, as shown in Figure 3(c).

FABRICATION AND CHARACTERIZATION

The fabrication process of the PMUTs array starts from a cavity silicon-on-insulator (CSOI) wafer (Figure 4a). Prior to the deposition of Mo/ScAlN/Mo stack, a 20 nm AlN seeding layer is deposited (Figure 4b). Next, 0.2 μm Mo/1 μm ScAlN/0.2 μm Mo stack on the ScAlN seeding layer (Figure 4c). The top Mo layer is formed by plasma etching (Figure 4d). An oxide layer is deposited to form the isolation layer, and followed by the etching of SiO₂ and ScAlN to pattern the via opening for the top and bottom electrodes (Figure 4e). Deposition and patterning of aluminum (Al) leads and bonding pads (Figure 4f).

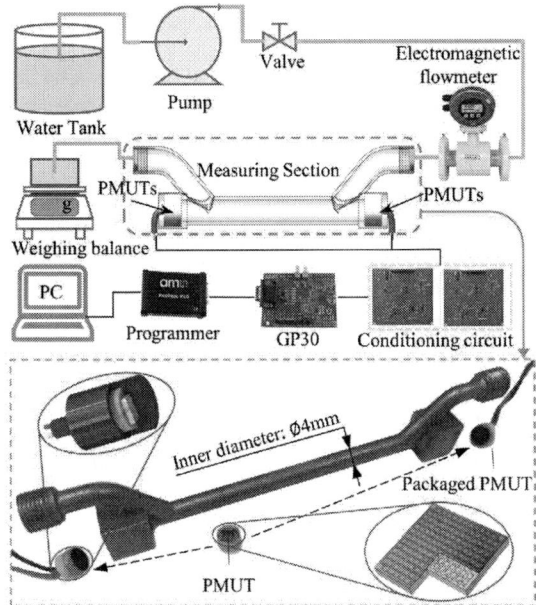

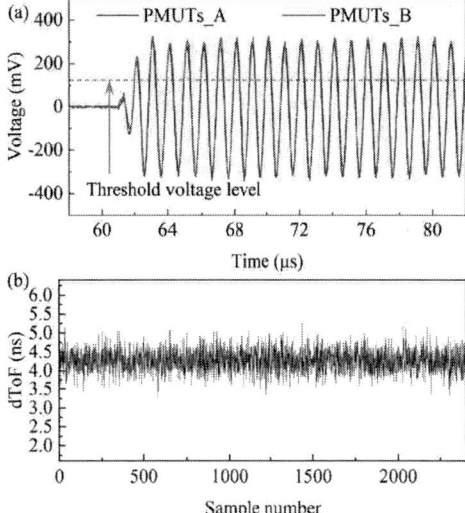

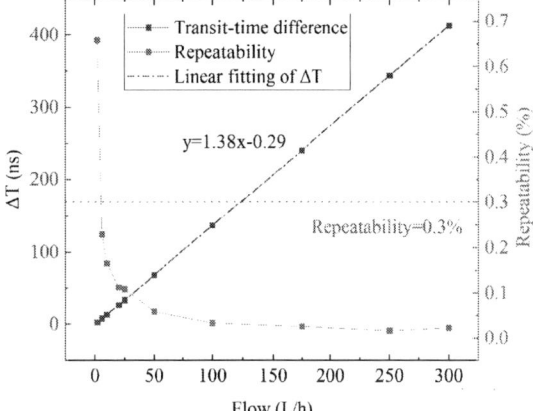

Figure 7: (a) Upstream and downstream waveform signals at zero-flow condition. (b) Transit-time difference data collected in zero-flow condition.

Figure 6: Experimental setup for the testing of the performance of the proposed flowmeter at different flow rates.

As shown in Figure 5(a), the impedance of the PMUTs is measured in air. According to the impedance-frequency spectrum, the resonant frequency of the PMUTs is 1.996 MHz and the anti-resonant frequency is 2.01 MHz. As shown in Figure 5(b), the PMUTs also perform well in DI-water. The resonant frequency drops to 1.04 MHz due to the added mass effect, which is very close to 1 MHz.

EXPERIMENTAL METHOD AND RESULTS

In order to test the performance of the fabricated small-diameter TTUF, a calibration experiment is conducted in the standard liquid flow calibration system. Optical images of the developed small-diameter TTUF with hardware for signal processing are shown in Figure 6. The diameter of the fabricated TTUF is 4 mm. In this paper, the time of ultrasound propagation upstream and downstream are measured by a time-to-digital converter (TDC) chip. Whether the TDC chip can detect the time difference correctly depends on the quality of the signal. An amplitude of 20 V is generally required to excite PMUTs to realize higher acoustic pressure output. Hence, the signal processing unit includes circuits for filtering and amplifying the received signals. The integrated circuit is connected to the PC via PicoProg device, which acts as a USB-to-SPI converter. The transit-time measured data are processed through computer and converted into the corresponding flow value, which are displayed via the graphical user interface (GUI).

Figure 7(a) shows the received signals from both transceivers under zero-flow conditions after amplification and filtering. The maximum amplitude of the two received signals is approximately 300 mV. Since the two received signals have approximately the same amplitude, the same threshold voltage level can be set to detect the first wave. In order to reduce the effects of noise and obtain the correct transit time. Here, the threshold voltage level is set to the average of the first and second amplitudes.

Figure 8: Measurement repeatability and the difference of transit-time with respect to varying flow rates.

Prior to performing dynamic flow measurements, we conduct flow measurement experiments at zero flow conditions. Figure 7(b) shows the difference in transit time (dToF) data collected over a 5-minute period. Theoretically, when the flow rate is zero, the transit times of ultrasonic waves propagating against and with the flow are equal. However, offsets often occur in practical measurements, which will result in significant errors in the measured flow data. Offset errors are mainly caused by non-reciprocity in the circuit, temperature variations, etc. To compensate for the effects of reciprocity, the actual measured flow value is subtracted from the average of the measured dToF data under zero flow conditions.

In order to verify the performance of the developed flowmeter, flow experiments are conducted according to the standard verification regulations [15]. Each experiment is conducted three times under ten standard flow rates, the experimental results corresponding to preset flow points are illustrated in Figure 8. After calibrating the flowmeter, the maximum repeatability is less than 1% within the range of 2 ~ 300 L/h. And it decreases as the flow rate increases, showing the challenges in achieving high accuracy of measurement of very low flow rates. The sensitivity of the flow meter is 1.38 ns/(L/h). In the future, the PMUTs could be integrated with application-specific integrated circuits (ASICs) to achieve further miniaturization and low power consumption.

CONCLUSION

In this paper, a π-type TTUF based on ScAlN PMUTs is proposed to measure the liquid flow rate. The PMUT array contains 169 (13 × 13) elements with the size of 2.8 mm × 2.8 mm^2. Due to the small size of PMUTs, the packaged transceiver can be properly installed on the flowmeter with 4 mm diameter channel. Experiment results verify that the proposed TTUF has excellent performance. The developed TTUF is ideal for flow measurement, especially for small diameter applications.

ACKNOWLEDGEMENTS

The authors would like to acknowledge Yuandong (Alex) Gu, Songsong Zhang and all the other members of the Shanghai Industrial µTechnology Research Institute (SITRI) for their help and support.

REFERENCES

[1] U. Salmaz, M. A. H. Ahsan, and T. Islam, "High-Precision Capacitive Sensors for Intravenous Fluid Monitoring in Hospitals," *IEEE Transactions on Instrumentation and Measurement*, vol. 70, pp. 1–9, 2021, doi: 10.1109/TIM.2021.3102681.

[2] T. Kim, J. Kim, and X. Jiang, "Transit Time Difference Flowmeter for Intravenous Flow Rate Measurement Using 1–3 Piezoelectric Composite Transducers," *IEEE Sensors Journal*, vol. 17, no. 17, pp. 5741–5748, Sep. 2017, doi: 10.1109/JSEN.2017.2727340.

[3] L. C. Lynnworth and Y. Liu, "Ultrasonic flowmeters: Half-century progress report, 1955–2005," *Ultrasonics*, vol. 44, pp. e1371–e1378, Dec. 2006, doi: 10.1016/j.ultras.2006.05.046.

[4] Y. Yu and G. Zong, "Note: Ultrasonic liquid flow meter for small pipes," *Review of Scientific Instruments*, vol. 83, no. 2, p. 026107, Feb. 2012, doi: 10.1063/1.3687780.

[5] Y. Chen, Y. Chen, S. Hu, and Z. Ni, "Continuous ultrasonic flow measurement for aerospace small pipelines," *Ultrasonics*, vol. 109, p. 106260, Jan. 2021, doi: 10.1016/j.ultras.2020.106260.

[6] F. Akasheh, T. Myers, J. D. Fraser, S. Bose, and A. Bandyopadhyay, "Development of piezoelectric micromachined ultrasonic transducers," *Sensors and Actuators A: Physical*, vol. 111, no. 2, pp. 275–287, Mar. 2004, doi: 10.1016/j.sna.2003.11.022.

[7] X. Le, Q. Shi, P. Vachon, E. J. Ng, and C. Lee, "Piezoelectric MEMS—evolution from sensing technology to diversified applications in the 5G/Internet of Things (IoT) era," *J. Micromech. Microeng.*, vol. 32, no. 1, p. 014005, Dec. 2021, doi: 10.1088/1361-6439/ac3ab9.

[8] X. Le, Q. Shi, P. Vachon, E. J. Ng, and C. Lee, "Piezoelectric MEMS—evolution from sensing technology to diversified applications in the 5G/Internet of Things (IoT) era," *J. Micromech.*

Microeng., vol. 32, no. 1, p. 014005, 2021, doi: 10.1088/1361-6439/ac3ab9.

[9] Z. Tong *et al.*, "An Ultrasonic Proximity Sensing Skin for Robot Safety Control by Using Piezoelectric Micromachined Ultrasonic Transducers (PMUTs)," *IEEE Sensors Journal*, pp. 1–1, 2021, doi: 10.1109/JSEN.2021.3068487.

[10] B. E. Eovino, Y. Liang, S. Akhbari, and L. Lin, "A single-chip flow sensor based on bimorph PMUTs with differential readout capabilities," in *2018 IEEE Micro Electro Mechanical Systems (MEMS)*, 2018, pp. 1084–1087. doi: 10.1109/MEMSYS.2018.8346748.

[11] X. Chen, C. Liu, D. Yang, X. Liu, L. Hu, and J. Xie, "Highly Accurate Airflow Volumetric Flowmeters via pMUTs Arrays Based on Transit Time," *Journal of Microelectromechanical Systems*, vol. 28, no. 4, pp. 707–716, Aug. 2019, doi: 10.1109/JMEMS.2019.2916987.

[12] K. Zhu *et al.*, "An ultrasonic flowmeter for liquid flow measurement in small pipes using AlN piezoelectric micromachined ultrasonic transducer arrays," *J. Micromech. Microeng.*, vol. 30, no. 12, p. 125010, Nov. 2020, doi: 10.1088/1361-6439/abc100.

[13] K. Zhu, X. Chen, M. Qu, D. Yang, L. Hu, and J. Xie, "Non-Contact Ultrasonic Flow Measurement for Small Pipes Based on AlN Piezoelectric Micromachined Ultrasonic Transducer Arrays," *Journal of Microelectromechanical Systems*, vol. 30, no. 3, pp. 480–487, Jun. 2021, doi: 10.1109/JMEMS.2021.3066408.

[14] T. Wang and C. Lee, "Zero-Bending Piezoelectric Micromachined Ultrasonic Transducer (pMUT) With Enhanced Transmitting Performance," *Journal of Microelectromechanical Systems*, vol. 24, no. 6, pp. 2083–2091, 2015, doi: 10.1109/JMEMS.2015.2472958.

[15] L. Li *et al.*, "Experimental and Numerical Analysis of a Novel Flow Conditioner for Accuracy Improvement of Ultrasonic Gas Flowmeters," *IEEE Sensors Journal*, vol. 22, no. 5, pp. 4197–4206, Mar. 2022, doi: 10.1109/JSEN.2022.3145668.

CONTACT

*Liang Lou; tel: +86-131-2237-0109;
liang.lou@sitrigroup.com
*Minkan Chen; tel: +86-188-1781-8597;
vincent.chen@sitrigroup.com;

MEMS DIFFERENTIAL THERMOPILES FOR HIGH-SENSITIVITY HYDROGEN GAS DETECTION

Haozhi Zhang[1,2], Hao Jia[1,2], Ming Li[1,2], Pengcheng Xu[1,2] and Xinxin Li[1,2]

[1]State Key Lab of Transducer Technology, Shanghai Institute of Microsystem and Information Technology, Chinese Academy of Sciences, Shanghai 200050, CHINA and
[2]University of Chinese Academy of Sciences, Beijing 100049, CHINA

ABSTRACT

We report a MEMS differential thermopile sensor for high-sensitivity hydrogen (H_2) gas detection. By exploiting the single-sided microhole inter-etch and sealing (MIS) process, pairs of MEMS single-crystalline silicon thermopiles are batch fabricated into $2 \times 1 mm^2$ dies. Each suspended thermopile is integrated with high-density thermocouples, allowing us to achieve a temperature sensitivity of ~27.8mV/°C. Such design ensures the detection of tiny temperature changes (down to ~1mK) caused by catalytic combustion of H_2-air mixture at 120°C when the sensing thermopile is loaded with platinum nanoparticles-decorated aluminum oxide (Pt NPs@Al_2O_3) nanosheets. The sensors exhibit a linear response to H_2 concentration from 1.0% down to 5ppm with a response time of ~2.5s and a recovery time of ~1.7s (in 1% H_2), which are more sensitive than the previously reported MEMS thermopile H_2 sensors. The proposed sensors are also selective against other combustible gases and remain stable after 30 days. Our highly-sensitive, cost-effective MEMS differential thermopile sensors hold promise for H_2 leakage detection in industrial applications.

KEYWORDS

Hydrogen sensor, MEMS thermopile, MIS process, High Sensitivity

INTRODUCTION

As a promising alternative to fossil fuels, hydrogen (H_2) plays a pivotal role nowadays in the energy transition toward a green and sustainable global economy. Hydrogen has been utilized in a wide range of applications, such as the aerospace and automotive industries. However, its explosive nature (4% in ambient) demands sensitive, fast H_2 gas detection techniques to ensure safe storage, transportation, and use in the industry [1]. Moreover, increasing interest has been paid to miniature, cost-effective MEMS gas sensors with low power consumption and scalable fabrication [2-4]. To date, electrochemical and semiconductor H_2 sensors based on H_2 adsorption and diffusion processes have been intensively studied. Despite their good detection limits (~0.1-10ppm), electrochemical [5] and semiconductor [6] H_2 sensors often show a linear detection range ~1-2 orders of magnitude and response and recovery time typically from tens to hundreds of seconds [7-8]. On the other hand, MEMS H_2 sensors based on catalytic combustion process (*e.g.*, thermopile-type) attract increasing attention by offering a large linear detection range and fast response and recovery (second-level) while still showing limited sensitivity (~10-100ppm) [9-10].

To further push for trace detection and early warning of H_2 leakage, endeavors need to be continuously made to achieve high sensitivity down to the ppm level. Currently, studies have been mostly focused on improving the catalyst efficiency, and a detection limit down to ~10 ppm has been reported [11]. Efforts on sensor design and fabrication have been scant. For thermopile-type sensors, the number of thermocouples and the Seebeck coefficient of the thermoelectric material directly affect the capability of detecting tiny temperature changes due to the catalytic combustion of H_2 on the thermopiles. Therefore, it is necessary to innovate and optimize the structural design and fabrication process of MEMS differential thermopile-based H_2 sensors.

In this work, we design and fabricate MEMS differential thermopile H_2 sensors using the MIS process on (111) single-crystalline silicon wafers. We characterize the temperature sensitivity and hydrogen sensing performance, showing high-sensitivity good detection limit, selectivity, and stability.

DEVICE DESIGN

The sensor is composed of a pair of temperature-controlled MEMS thermopiles, in differential form, as illustrated in Figure 1. The left thermopile is a sensing thermopile that will be loaded with a catalyst and used for selective detection of the gas. To eliminate the environmental interferences such as gas thermal conductivity and flow rate, the thermopile on the right was designed as the reference.

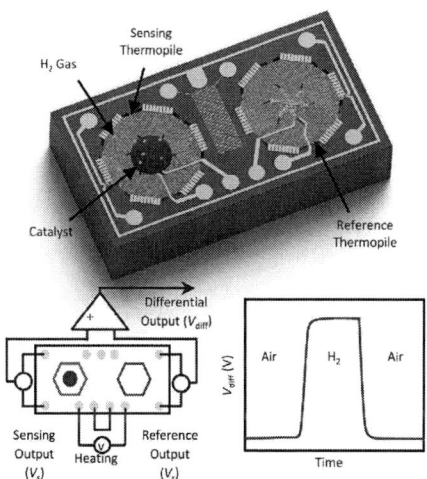

Figure 1: Concept of differential thermopile H_2 sensor. A pair of MEMS thermopiles with the sensing thermopile covered by Pt NPs@Al_2O_3 catalyst output a differential voltage signal in response to the temperature change during catalytic combustion of H_2-air over Pt.

A single thermopile has a diameter of 640um and

consists of 54 pairs of single-crystalline silicon thermocouples in series, which are suspended below the silicon nitride (SiN_x) membrane and thermally isolated from the silicon substrate by an insulating cavity. Due to the high Seebeck coefficient [12] and high density of single-crystalline silicon thermocouples, we can increase the temperature sensitivity of the thermopiles by several times [11].

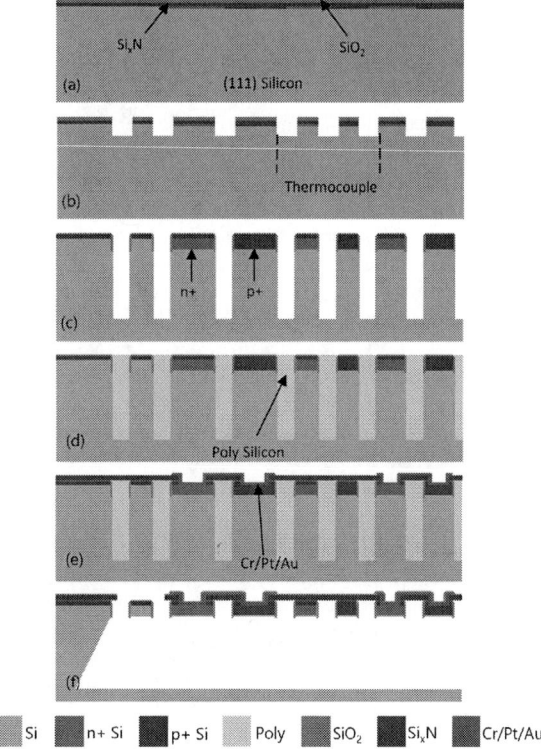

Figure 2: Fabrication process of the differential thermopile H_2 sensor. (a) Ion implantation and SiN_x deposition. (b) Thermopile patterning. (c) DRIE etching. (d) Poly-silicon deposition. (e) Electrodes patterning. (f) TMAH wet etching.

As Pt $NPs@Al_2O_3$ is loaded on the sensing thermopile and both sensing and reference thermopiles are heated at ~120°C in H_2/air mixture, a differential voltage output (V_{diff}) between sensing and reference thermopiles is read by the circuitry, which reveals the exothermic process of H_2 catalytic combustion on Pt catalyst.

DEVICE FABRICATION

The devices are batched fabricated on a 4-inch <111> wafer by exploiting the microhole inter-etch and sealing (MIS) process, providing good uniformity and low cost (Fig. 2). The fabrication starts with a single-side-polished <111> wafer. Boron ion implantation and phosphorus ion implantation are first applied to the thermocouple region, and then low-stress SiN_x/SiO_2 films are deposited by low-pressure chemical vapor deposition (LPCVD) (Fig. 2a). Then, thermocouples are patterned by etching the SiN_x, SiO_2, and single-crystalline silicon sequentially (Fig. 2b). A LPCVD SiO_2 layer is deposited and anisotropically

etched by RIE to form the sidewall structure. Then, deep trenches with a depth of 40μm are etched to define the depth of the thermally isolated cavity (Fig. 2c). Then, polysilicon is deposited with LPCVD to fill the deep trenches and then surface polished using chemical mechanical polishing (CMP) (Fig. 2d). A 1μm thick low-stress SiN_x layer is deposited as a supporting film. Then, the contact holes of the thermocouples are patterned by etching SiN_x. The interconnection between thermocouples is performed by sputtering and patterning Cr/Pt/Au metal films with thicknesses of 40 nm/100 nm/3000 nm, respectively (Fig. 2e). Then, A SiO_2 layer is deposited using plasma-enhanced chemical vapor deposition (PECVD) as the insulating layer. Finally, silicon is etched in the 25% tetramethylammonium hydroxide (TMAH) at 80°C to release the whole device (Fig. 2f).

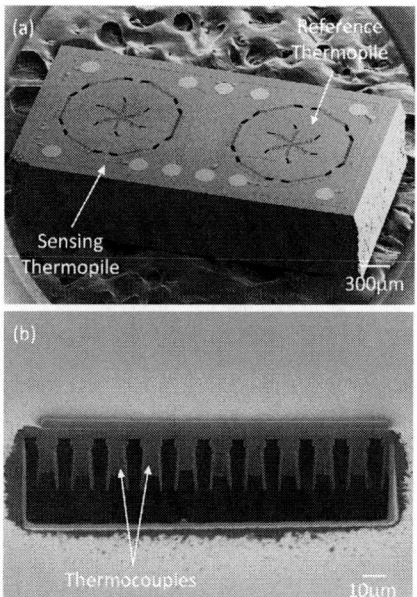

Figure 3: Fabricated differential thermopile H_2 sensor. (a) SEM image of a typical sensor, consisting of a sensing thermopile and reference thermopile. (b) Cross-sectional view of the single-crystalline Si thermocouples underneath the supporting SiN_x membrane after FIB cut.

RESULTS AND DISCUSSION
Characterization of MEMS Thermopile H_2 sensors

The fabricated differential thermopile H_2 sensor is characterized using scanning electron microscopy (SEM), as illustrated in Figure 3a. The size of a sensor die is 1mm×2mm. The FIB image provides a cross-sectional of the single-crystalline Si thermocouples, which are densely suspended under the insulating membrane, as shown in Figure 3b. The optical image of a catalyst-loaded H_2 sensor is shown in Figure 4a, where the Pt $NPs@Al_2O_3$ catalyst is uniformly loaded on the suspended membrane of the sensing thermopile on the left. In contrast, the reference thermopile on the right side is kept blank (Fig. 4a). The $NPs@Al_2O_3$ catalyst is characterized by the transmission electron microscopy (TEM) image. As shown in 4b-c, Pt nanoparticles are seen uniformly grown on the Al_2O_3 nanosheets, with a diameter of 5-10 nm.

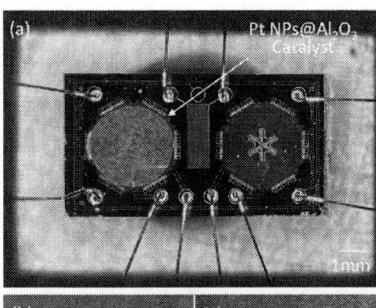

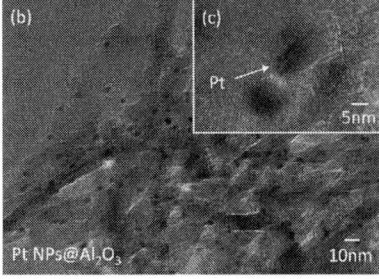

Figure 4: Fabricated differential thermopile H_2 sensor. (a) Optical image of the sensor with Pt NPs@Al$_2$O$_3$ catalyst loaded on the sensing thermopile. (b)&(c) TEM image of Pt NPs synthesized on Al$_2$O$_3$ nanosheets.

Calibration of Temperature Response

We first calibrate the temperature response of the MEMS differential thermopiles by using the integrated heating resistors on the device, which heat the sensing region (at the device center). The temperature is measured non-contact using an infrared thermal imager with a spatial resolution of 20μm. The relationship between the heating voltage and the averaged temperature within the sensing region is shown in Fig. 5a, which shows reasonable agreement with the finite element simulation results. Then, the temperature response is obtained by detecting the output voltage of a single thermopile during the heating process, as shown in Fig. 5b, which indicates a temperature sensitivity of ~27.8mV/°C.

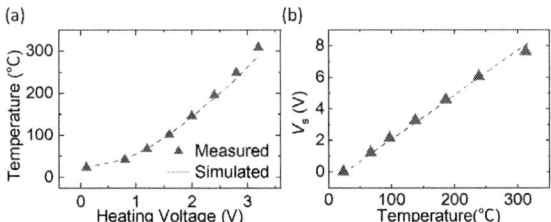

Figure 5: Temperature response of the MEMS differential thermopile H_2 sensor. (a) Simulated vs. measured sensor temperature averaged within the heated region (red). (b) Calibration of temperature sensitivity of a typical sensing thermopile, which is ~27.8mV/°C.

H₂ Sensing Performance

We test our MEMS differential thermopiles for H_2 detection in air. The sensor is loaded in a chamber with an H_2 flow rate of 200 sccm. The High-End-MEMS intelligent gas distribution system controls the gas flow, with a high-precision mass flow meter.

To optimize the working temperature of the H_2 sensors,

we switch the supplied gases to the sensors between air and 1% hydrogen-air mixture, and the differential outputs (V_{diff}) are recorded at working temperatures from 40°C to 240°C, as shown in Fig. 6a. We determine that a temperature of 120°C can balance the device sensitivity and power consumption. The sensor power consumption is thus ~40mW at 120°C (Fig. 6b).

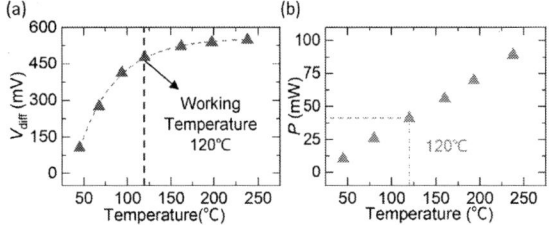

Figure 6: Sensor characterization. (a) The differential output of a typical sensor in 1% H_2/air mixture at different working temperatures, a temperature of 120°C is chosen with balanced differential output and power consumption. (b) Power consumption of ~40mW measured at 120°C.

The sensors are then tested in a series of H_2 concentrations, as shown in Fig. 7a. Figure 7b illustrates the amplitude of V_{diff} to H_2 over the concentration range of 5ppm - 1% and the temperature rise corresponding to V_{diff}, which is 441mV in 1% H_2 and 22uV in 5 ppm H_2. It can be inferred that the 22uV output is caused by a 0.78mK temperature rise, validating the high temperature resolution, hence the good detection limit of the differential thermopiles for H_2 sensing.

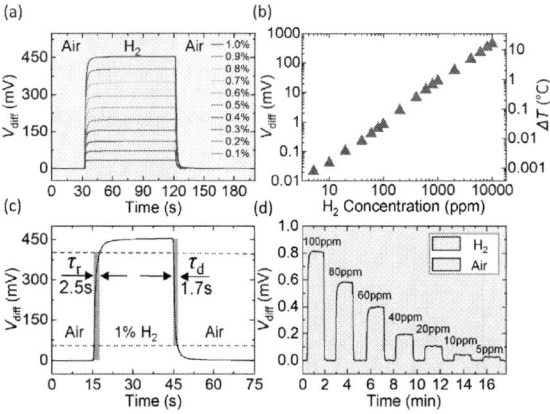

Figure 7: Sensor response to the H_2 concentration gradient. (a) Differential output in 0.1% - 1% H_2/air mixtures. (b) Typical linear response to H_2 concentrations from 10ppm to 10000ppm. (c) Sensor response and recovery time (t$_{90}$) in 1% H_2/air mixture. (d) Sensor differential output under a continuous H_2 concentration gradient from 100ppm down to 5ppm.

The response and recovery time (t$_{90}$) exposed to 1% hydrogen can be calculated from V_{diff} of the sensor, as shown in Fig. 7c, which are 2.5s and 1.7s, respectively. Figure 7d shows zoom-in V_{diff} curves when detecting low concentrations of hydrogen from 100 ppm to 5 ppm, demonstrating that the sensor can easily detect 5 ppm H_2.

Meanwhile, we evaluate the selectivity and stability of our MEMS differential thermopile H_2 sensor. A sensor is tested in different atmospheres of the combustible gases CO (1%), CH_4 (1%), C_2H_6 (1%), and the common VOCs toluene (C_7H_8, 1%). The results are compared with the sensor response in 0.1% H_2, as shown in Fig. 8a, showing good selectivity of our MEMS differential thermopile H_2 sensors. Besides, we record V_{diff} of the same sensor to 1% H_2 gas every week for 28 days, as shown in Fig. 8b. It can be inferred that the V_{diff} is not degraded in the ambient environment and remains stable with a fluctuation within $\pm2.5\%$.

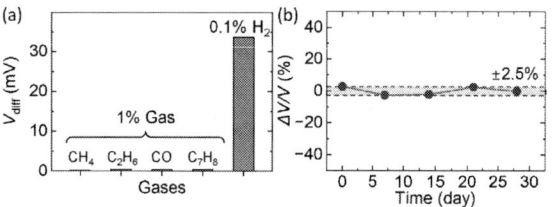

Figure 8: Sensor selectivity and stability. (a) Sensor differential output to 0.1% H_2 in air vs. 1% CH_4, 1% C_2H_6, 1% CO and 1% C_7H_8 in air, showing good selectivity. (b) The differential output of the same sensor in 28 days, showing a small variation of only $\pm2.5\%$.

CONCLUSION

In summary, we have designed and fabricated a MEMS differential thermopile sensor for high-performance hydrogen gas detection. The sensor consists of two identical temperature-controlled thermopiles, and each thermopile is integrated with 54 pairs of single crystal thermocouples and achieves a high temperature sensitivity of 27.8mV/°C. Such design ensures the detection of tiny temperature changes caused by catalytic combustion of H_2-air mixture at 120°C when the sensing thermopile is loaded with Pt NPs@Al_2O_3 nanosheets. The MEMS differential thermopile sensors exhibit a linear response to H_2 concentration from 1.0% down to 5 ppm (across >3 orders of magnitude) with a response time of 2.5s and a recovery time of 1.7s (in 1% H_2), which are more sensitive than the reported MEMS thermopile H_2 sensors, and show much wider linear range and faster response than electrochemical, and semiconductor H_2 sensors. Moreover, the sensors have good H_2 selectivity and stability. Our high-sensitivity MEMS differential thermopile sensors hold promise for H_2 leakage detection and early warning in industrial applications.

ACKNOWLEDGEMENTS

The authors acknowledge financial support from the National Key R&D Program of China (2021YFB3200800), National Natural Science Foundation of China (61974155, 61831021, 62104241, 62271473), Science and Technology Innovation Plan of Shanghai (19510744600), Scientific Instrument Project of the Chinese Academy of Sciences (YJKYYQ20210024), Shanghai Pujiang Program (20PJ1415600), Innovation Team and Talents Cultivation Program of National Administration of Traditional Chinese Medicine (ZYYCXTD-D-202002, ZYYCXTD-D-202003).

REFERENCES

[1] I. Darmadi, F. A. A. Nugroho, and C. Langhammer, "High-performance nanostructured palladium-based hydrogen sensors—current limitations and strategies for their mitigation", *ACS Sens.*, vol. 5, no. 11, pp. 3306–3327, Nov. 2020.

[2] H. Jia, P. Xu, and X. Li, "Integrated resonant micro/nano gravimetric sensors for bio/chemical detection in air and liquid", *Micromachines*, vol. 12, no. 6, Art. no. 6, Jun. 2021.

[3] F. Yao, P. Xu, H. Jia, X. Li, H. Yu, and X. Li, "Thermogravimetric analysis on a resonant microcantilever", *Anal. Chem.*, vol. 94, no. 26, pp. 9380–9388, Jul. 2022.

[4] Y. Chen, P. Xu, X. Li, Y. Ren, and Y. Deng, "High-performance H_2 sensors with selectively hydrophobic micro-plate for self-aligned upload of Pd nanodots modified mesoporous In_2O_3 sensing-material", *Sens. Actuators B Chem.*, vol. 267, pp. 83–92, Aug. 2018.

[5] Z. Zhi, W. Gao, J. Yang, C. Geng, B. Yang, S. Fan, *et al.*, "Amperometric hydrogen gas sensor based on Pt/C/Nafion electrode and ionic electrolyte", *Sens. Actuators B Chem.*, vol. 367, p. 132137, Sep. 2022.

[6] X. Wang, M. Li, P. Xu, Y. Chen, H. Yu, and X. Li, "In Situ TEM technique revealing the deactivation mechanism of bimetallic Pd–Ag nanoparticles in hydrogen sensors", *Nano Lett.*, vol. 22, no. 7, pp. 3157–3164, Apr. 2022.

[7] J. Moon, H. Hedman, M. Kemell, A. Tuominen, and R. Punkkinen, "Hydrogen sensor of Pd-decorated tubular TiO_2 layer prepared by anodization with patterned electrodes on SiO_2/Si substrate", *Sens. Actuators B Chem.*, vol. 222, pp. 190–197, Jan. 2016.

[8] J. Hu, T. Zhang, Y. Chen, P. Xu, D. Zheng, and X. Li, "Area-selective, *in-situ* growth of pd-modified zno nanowires on mems hydrogen sensors", *Nanomaterials*, vol. 12, no. 6, Art. no. 6, Jan. 2022.

[9] M. Nishibori, W. Shin, N. Izu, T. Itoh, I. Matsubara, S. Yasuda, and S. Ohtani, "Robust hydrogen detection system with a thermoelectric hydrogen sensor for hydrogen station application", *Int. J. Hydrog. Energy*, vol. 34, no. 6, pp. 2834–2841, Mar. 2009.

[10] E. Brauns, E. Morsbach, S. Kunz, M. Bäumer, and W. Lang, "A fast and sensitive catalytic gas sensors for hydrogen detection based on stabilized nanoparticles as catalytic layer", *Sens. Actuators B Chem.*, vol. 193, pp. 895–903, Mar. 2014.

[11] A. S. Pranti, D. Loof, S. Kunz, V. Zielasek, M. Bäumer, and W. Lang, "Characterization of a highly sensitive and selective hydrogen gas sensor employing Pt nanoparticle network catalysts based on different bifunctional ligands", *Sens. Actuators B Chem.*, vol. 322, p. 128619, Nov. 2020.

[12] W. Li, Z. Ni, J. Wang, and X. Li, "A front-side microfabricated tiny-size thermopile infrared detector with high sensitivity and fast response", *IEEE Trans. Electron Devices*, vol. 66, no. 5, pp. 2230–2237, May 2019.

CONTACT

*H. Jia; hao.jia@mail.sim.ac.cn
*X.X. Li; xxli@mail.sim.ac.cn

DOMAIN/BOUNDARY VARIATION IN CANTILEVER ARRAY FOR BANDWIDTH ENHANCEMENT OF PZT MEMS MICROSPEAKER

Shu-Wei Chang[1], Ting-Chou Wei[2], Sung-Cheng Lo[3], Weileun Fang[1,2]

[1] Inst. of NanoEng. and Microsyst., National Tsing Hua University, Hsinchu City, Taiwan
[2] Dept. of Power Mech. Eng., National Tsing Hua University, Hsinchu City, Taiwan
[3] Transducer Star Technology Inc., Hsinchu City, Taiwan

ABSTRACT

This study demonstrates a novel piezoelectric micro-electromechanical system (MEMS) microspeaker with wide bandwidth at high range for in-ear applications. The microspeaker is composed of eight high-frequency range triangular cantilevers anchored on a square frame. Merits of the design to modulate resonant frequencies of cantilevers are, (1) domain variation: change the dimensions of cantilevers inside the supporting frame, (2) boundary variation: change the boundaries of cantilevers outside the support frame. Thus, multiple sound pressure level (SPL) poles and wide bandwidth are achieved. Measurements in the standard ear simulator (driven at 0.707Vrms input signal with out-of-phase driving method) demonstrate the proposed design increases the bandwidth to 3.7~14.0kHz when SPL is 90dB, as summarized in Table1.

KEYWORDS

MEMS, microspeaker, piezoelectric, acoustics, cantilever array, out-of-phase driving, wide bandwidth

INTRODUCTION

In recent years, the use of MEMS acoustic transducers in the consumer electronics market has been increasing, where the true wireless bluetooth earphones is the most prominent device. The speakers of traditional earphones consist of dynamic speaker or balanced armatures. Though actuated by the Lorentz force with a higher driving amplitude, the assembly accuracy and miniaturization are critical challenges for these speakers [1]. In this regard, the MEMS microspeakers fabricated by using the wafer level micromachining process could offer the advantages of size reduction and consistency.

MEMS microspeakers with different designs attract attentions recently to achieve the high SPL and wide bandwidth requirement of the audio market. Driving methods of MEMS microspeaker can be categorized as electrostatic, electromagnetic and piezoelectric actuation. Electrostatic microspeakers utilize parallel electrode plates to generate sound pressure by electrostatic force [2]. However, insufficient SPL and high driving voltage owing to limited displacement between tiny gaps of electrodes are the concern. Electromagnetic microspeakers adapted from conventional dynamic drivers can be realized by MEMS technology [3]. Although outstanding performance has been demonstrated, the addition assembly process required for magnetic components is the concern. Piezoelectric microspeakers attract attention due to their large dynamic displacement at low driving voltage. As piezoelectric thin films become available in foundries, the piezoelectric microspeakers are considered as promising components for the related applications.

The goal of micropspeaker design is generally SPL and bandwidth enhancement. Various structures are reported to strengthen the SPL with higher displacement under small diaphragm area [4]. However, the sound pressure is only enhanced near resonances or SPL poles, so that the bandwidth is not improved. Hence, different driving methods are further presented to expand the bandwidth [5-7]. For example, the two-way microspeakers leverage both woofers and tweeters to enhance bandwidth, yet the relatively low driving efficiency is a concern [6]. The concept to have multiple structures with out-of-phase driving could enhance bandwidth within limited diaphragm area [5]. Therefore, this study presents novel structure designs with out-of-phase driving method [5] to expand the bandwidth while reaching required SPL with low driving voltage.

DESIGN CONCEPT

This study utilizes unimorph piezoelectric structure to generate sound pressure by inverse piezoelectric effect. Fig.1a depicts the proposed $4\times4mm^2$ microspeaker chip composed of eight triangular cantilevers with $8mm^2$ equivalent diaphragm area. The cantilevers are anchored to the central rigid frame with 100μm in width. The cantilevers inside the frame are designed to have different dimensions (domain variation) to modulate their resonant frequencies. The out-of-plane bending stiffness of the triangular cantilever varies with its length. By varying the crossover points of the four tips, cantilevers of four

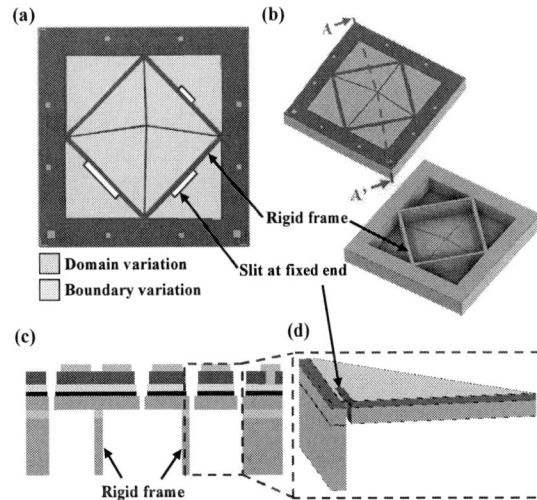

Figure 1: *Proposed microspeaker design concept, (a) bandwidth enhancement design including domain variation and boundary variation, (b) Frontside & backside 3D schematic, (c) AA' cross-section view of the device, (d) 3D cross-section of boundary variation units.*

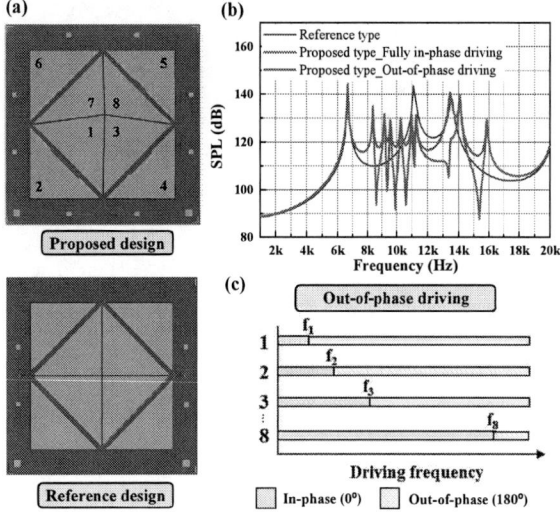

Figure 2: *Driving method and simulation results, (a) reference design and unit sequence, (b) frequency response simulation results, (c) out-of-phase driving methods.*

different bending stiffnesses are achieved, leading to diverse resonant frequencies. Moreover, the triangular cantilevers outside the frame are designed to have different fixed boundaries (boundary variation) to modulate their resonant frequencies. The slit at the fixed-end of the cantilever is exploited to reduce the boundary constraint of the structure. Thus, the boundary condition as well as the stiffness of the cantilever varies with the length of the slit. The stiffness of the cantilever decreases with the length of the slit. This study designs slits of three different lengths ($300\mu m$, $600\mu m$, $900\mu m$) to offer diverse resonant frequencies of cantilevers. Fig.1c-d illustrates cross-sections of chip to show cantilevers, rigid frames, and the slit to modulate boundary conditions of cantilever.

Fig.2a present the comparison between proposed and reference designs. The SPL poles in a frequency response is contributed by mode shapes of structures. The performance of the proposed design is expected to generate more SPL poles, broadening wider bandwidth than the reference design, which is composed of eight identical cantilevers within the same footprint. Simulations in Fig.2b show the frequency responses of SPL for different designs and driving methods. As the gray line indicated in figure, the proposed design driven by "fully in-phase driving" exhibits ten different SPL poles including eight cantilevers and two Helmholtz resonances in the pressure field simulation. The simulations indicate the SPL may drop between poles in such driving approach. It is well-known that the vibration of structure will be out-of-phase as the driving frequency exceeds its resonant frequency. Thus, the out-of-phase vibration between cantilevers may cause the drop of SPL. The concept of "out-of-phase driving" [5] is employed to prevent the SPL cancellation when the excitation frequency exceeding the 1st resonant frequency of each cantilever (Fig.2c). When a cantilever is driven right after the resonant frequency, the phase of the driving signal will be reversed to avoid the out-of-phase response. Simulations in Fig.2b demonstrate that the proposed design with out-of-phase driving (red-line) increases the SPL in

the high frequency range. The resonant frequencies of domain and boundary variation cantilevers then be modulated with out-of-phase driving to reach a decent SPL. In other words, for a higher SPL, the proposed design could increase the bandwidth at the high frequency range.

FABRICATION

Fig.3 presents the fabrication processes including five masks. In Fig.3a, the ZrO_2 (adhesion layer), Pt (bottom electrode layer), and PZT layers were deposited onto the 4 inches SOI wafer with $5\mu m$ thick device layer, and then the PZT was wet etched (in $HF:HCl:H_2O$ etchant) to define the ground and structures. In Fig.3b, the Cr/Au (top electrode layer) were then deposited by evaporation and patterned by lift-off to define the electrical routings. In Fig.3c, the frontside layers including bottom electrode and the Si device layer were patterned to define domains and boundaries of microspeaker array. The Pt and ZrO_2 were first dry etched by high density plasma reactive ion etching system (HDP-RIE). After that the Si device layer was also etched by HDP-RIE to define the structures and boundaries. Note that the $3\mu m$ wide slits were designed to ensure lower acoustic short. In Fig.3d, the back chamber and the rigid frame were patterned by DRIE from the backside of Si wafer. Finally, the buried oxide was removed by RIE to release the cantilever array.

Micrograph in Fig.4a shows the typical fabricated microspeaker, noting that cantilevers bent by residual stresses will cause dark color on their tips. The domain variation and boundary variation units are respectively fabricated inside and outside the rigid frame. The zoom-in micrograph in Fig.4b displays gaps and domain variations for four cantilever units. Fig.4c exhibits the slit for boundary variation. The focused ion beam (FIB) micrograph in Fig.4d further shows the layer stacking of the structures.

MEASUREMENT

Fig.5a shows the microspeaker wire-bonded on a customized PCB with a $3 \times 3mm^2$ acoustic hole at the center as the device-under-test (DUT). Fig.5b demonstrates the DUT is mounted with a 3D-printed acrylic adapter to match the size of the ear simulator. The DUT is then fixed onto the standard coupler (G.R.A.S. RA0401) of the ear simulator system (G.R.A.S. 43AG-6) in an anechoic box, as shown in Fig.5c. When the driving signal is transmitted

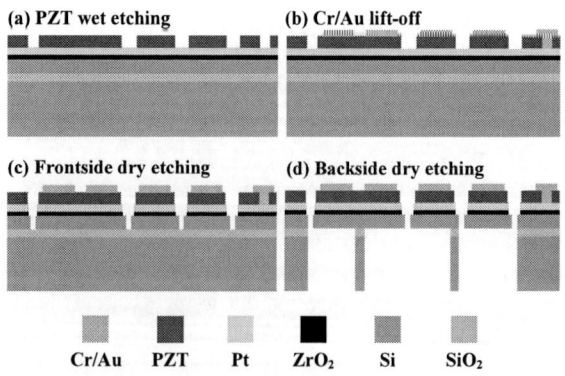

Figure 3: *Fabrication process flow.*

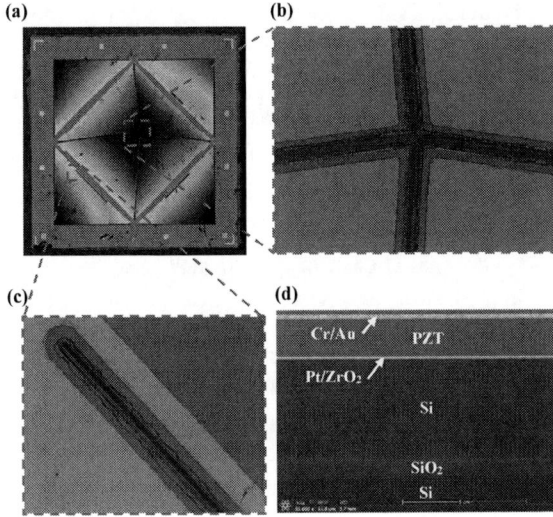

Figure 4: Fabrication result, (a) proposed design, (b) domain variation, (c) boundary variation, (d) layer stacking image of electrical pad by FIB.

from the pulse spectrum analyzer (B&K) to the DUT, the sound pressure will be generated, and then pass through the ear simulator to reach the standard pressure field microphone (G.R.A.S. 46BE). The acoustic excitation is then transformed to electrical signal back to the analyzer, so that the frequency response is recorded and presented by the computer.

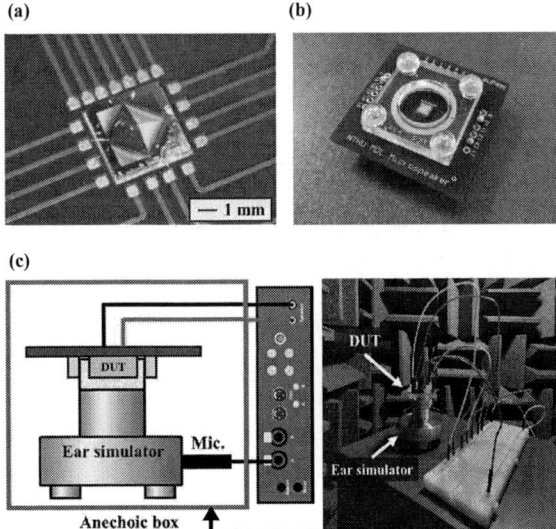

Figure 5: The measurement setup, (a) proposed design after wire bonding on PCB, (b) DUT, (c) schematic and real picture of measurement setup.

Measurements in Fig.6a show frequency responses of proposed/reference designs driven with $0.707V_{rms}$. As depicted by the blue line, the reference design with "fully in-phase driving" exhibits maximum SPL 111 dB at its resonant frequency (10kHz). Meanwhile, it demonstrates ±3dB flat response within the frequency range of 160Hz~1.4kHz at the sound level of 73dB. Below 160Hz,

SPL drops significantly due to acoustic short which was induced by the structure deformation resulted from the residual stress. As depicted by the gray line, the proposed design with "fully in-phase driving" exhibits maximum SPL 110dB at 12kHz. In general, most of the cantilevers present 105dB~110dB at their SPL poles, showing high SPL performance from cantilever array. Nevertheless, due to fabrication defects, the SPL and frequency (15.6kHz) of one cantilever did not fall into the design range. It demonstrates ±3dB flat response within the frequency range of 93Hz~1.3kHz at the sound level of 79dB, which is higher than that of the reference design due to larger average displacement of diaphragm area.

This study focuses on the bandwidth in which SPL>90dB. Fig.6b depicts the zoom-in of high frequency range. The proposed design with "fully in-phase driving" shows the SPL cancellation between two adjacent poles which resulting from the phase inverse response of structure after resonant, as predicted in Fig.2b. Thus, only some small and discrete bandwidths could meet the required SPL. The red line with boosted SPL represents the responses of proposed design with out-of-phase driving. The SPL between any two adjacent poles is increased by up to 30dB, showing outstanding sound pressure repairing ability. In comparison, the bandwidths of 90dB SPL are: 3.7~14.0kHz (proposed design) and 5.3~12.6kHz (reference design). Moreover, the proposed design presents even higher SPL (>100dB) within the frequency range of 5.2~12.6kHz. Thus, the present design could contribute high SPL at high frequency range. As summarized in Table1, the bandwidth of microspeaker with limited diaphragm area is expanded by the proposed structure design with the out-of-phase driving method.

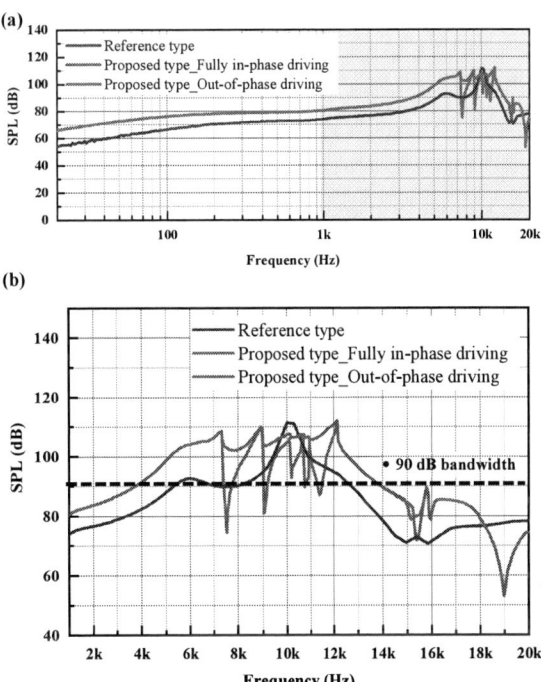

Figure 6: Measurement result, (a)the frequency response, (b) the response at 1kHz~20kHz frequency range.

Table 1: Comparison of piezoelectric MEMS speakers

Design	Proposed	Reference	[5]	[7]
Area (mm^2)	8.1 mm^2	8.1 mm^2	4 mm^2	2.5 mm^2
Signal (V$_{rms}$)	0.707			
Measurement	G.R.A.S. RA0401 ear simulator			
BW (kHz) @SPL>90dB	3.7~14.0	5.3~12.6	1.5~2.5	17.0~18.0

CONCLUSION

This study presents a high SPL and wide bandwidth piezoelectric microspeaker at high frequency range. The microspeaker consists of eight triangular cantilever diaphragms, grouped by four domain variation units and four boundary variation units inside and outside the rigid frame, respectively. Hence, the resonant frequencies of the eight units can be modulated and multiple SPL poles can be achieved. SPL cancellation occurs between any two poles due to vibration phase inverse. Thus, this study further employs the out-of-phase driving to enhance SPL and expand the bandwidth. The microspeaker was implemented using the processes on the SOI wafer with 5μm device layer and 2μm PZT thin film. Measurements indicate the final bandwidth of SPL>90dB driven with 0.707V$_{rms}$ by out of phase driving is 3.7~14.0kHz in the ear simulator, which presents outstanding performance with small size (4×4×0.5mm^3), low power, and wide bandwidth at high frequency range.

ACKNOWLEDGEMENT

This project was supported by the National Science and Technology Council under grant number, NSTC 111-2221-E-007-070-MY3, NSTC 111-2218-E-007-014-MBK, NSTC 111-2923 - E - 007 - 014 - MY2, NSTC 110 - 2926 - I - 007 - 506 -.The author would like to thank the Center for Nanotechnology, Materials Science, and Microsystems of National Tsing Hua University (CNMM), Taiwan Semiconductor Research Institute, Taiwan (TSRI), Nano Facility Center of National Yang Ming Chiao Tung University, Taiwan (NFC), and Nano-Electro-Mechanical-System Research Center of National Taiwan University, Taiwan (NEMS) for providing the fabrication facilities and IC manufacturing.

REFERENCE

[1] F. Stoppel, A. Männchen, F. Niekiel, D. Beer, T. Giese, and B. Wagner, "New integrated full-range MEMS speaker for in-ear applications," *IEEE MEMS Conmference*, (2018), pp. 1068-1071.

[2] G. De Pasquale, L. Rufer, S. Basrour, A. Somà, "Modeling and validation of acoustic performances of micro-acoustic sources for hearing applications," *Sensors and Actuators A,* 274 (2016), pp. 614-628.

[3] I. Shahosseini, E. Lefeuvre, J. Moulin, E. Martincic, M. Woytasik, and G. Lemarquand, "Optimization and microfabrication of high performance silicon-based MEMS microspeaker," *IEEE Sensors Journal*, 13, (2013), pp. 273-284.

[4] H. H. Cheng, S. C. Lo, Z. R. Huang, Y. J. Wang, M. Wu, and W. Fang, "On the design of piezoelectric MEMS microspeaker for the sound pressure level enhancement," *Sensors and Actuators A*, 306, (2020), 111960.

[5] Y. J. Wang, S. C. Lo, M. L. Hsieh, S. D. Wang, Y. C. Chen, M. Wu, W. Fang, "Multi-way in-phase/out-of-phase driving cantilever array for performance enhancement of PZT MEMS microspeaker," *IEEE MEMS Conference*, (2021), pp. 83-84.

[6] F. Stoppel, C. Eisermann, S. Gu-Stoppel, D. Kaden, T. Giese, and B. Wagner, "Novel membrane-less two-way MEMS loudspeaker based on piezoelectric dual-concentric actuators," *Transducers*, (2017), pp. 2047-2050.

[7] Y. T. Lin, S. C. Lo, and W. Fang, "Two-way piezoelectric MEMS microspeaker with novel structure and electrode design for bandwidth enhancement," *Transducers*, (2021), pp. 230-233.

CONTACT

*W.Fang, Tel: +886-3-5742923; fang@pme.nthu.edu.tw

ON THE DESIGN OF PIEZOELECTRIC MEMS MICROSPEAKER WITH HIGH FIDELITY AND WIDE BANDWIDTH

Ting-Chou Wei[1], Zih-Song Hu[2], Shu-Wei Chang[2], and Weileun Fang[1,2]
[1]Dept. of Power Mech. Eng., National Tsing Hua University, Hsinchu City, TAIWAN
[2]Inst. of NanoEng. and Microsyst., National Tsing Hua University, Hsinchu City, TAIWAN

ABSTRACT

This study presents a novel piezoelectric MEMS microspeaker consisted of the cantilever-plate actuator, central-diaphragm, and connecting meandering-springs. Features of the design: (1) cantilever-plate actuator: supply large actuation force with simple electrical routing, (2) central-diaphragm: simple Si layer structure to provide relatively flat piston motion, (3) connecting meandering-spring: transmit actuation force to central-diaphragm and reduce the influence of thin film residual stresses. Thus, large average central-diaphragm out-of-plane displacement is achieved, and the piston mode of microspeaker excited by cantilever-plate actuators could ensure good THD (Total-Harmonic-Distortion i.e. high fidelity) response. Measurements show the design has a maximum SPL (Sound-Pressure-Level) of 123 dB and full-range (20 Hz~20 kHz) 80dB bandwidth with overall THD below 7 % under 0.707 V_{rms}+10 V_{DC} driving voltage in standard ear simulator.

KEYWORDS

Piezoelectric MEMS Microspeaker, PZT, acoustic devices, SPL, bandwidth, high fidelity

INTRODUCTION

Microspeaker plays a significant role on audio experience enhancement, especially for in-ear applications. The Dynamic Drivers (DD) and Balanced Armatures (BA) are two key components in modern acoustic market. The DD is features in low cost because of its mature fabrication process. Wide bandwidth and stable response in audio band (20~20 kHz) is implemented by Lorentz force driven. However, joule heat issue provided by voice coil results in poor power consumption, and its bulky membrane size also limits its potential of miniaturization. The BA, also driven by Lorentz force, is designed in relatively small size and high stiffness, which features in better Treble (high frequency) band performance. Yet, narrow bandwidth and assemble requirement show its downsides.

Piezoelectric MEMS microspeaker is now a suitable alternative of BA that enables low power consumption and maintains good performance in the meantime, due to its outstanding output efficiency and small size. In addition, batch fabrication of devices can also be simplified by standard MEMS process. In general, flat response with sufficient SPL and low THD are major design concerns for microspeakers. To attain sufficient output, the PZT (Lead Zirconate Titanate) will be a good candidate aside from other typical piezoelectric materials like AlN, ZnO, $TiBaO_3$ due to its dramatically higher d_{31} coefficient [1], which will directly affect the output displacement as well as sound pressure level of devices.

Recent micro speakers are roughly sorted into fully-clamped diaphragm [2-4], cantilever [5-7], and spring-diaphragm [8-9] designs. The fully-clamped diaphragm eliminates air leakage issue and in-band(< 20 kHz) higher modes because of high stiffness and no trench involved. However, the hardening effect posed by thin film residual stresses, which frequently occurred in fabrication processes, will downgrade its performance. Cantilever structures can reach the largest out-of-plane displacement with simple electrical routing, yet suffer from air leakage problem caused by pre-deformation after fabrication. The spring-diaphragm structure is normally operated in piston mode. As in [10], the up-down piston vibration of diaphragm with lower dynamic deformation could offer better THD performance, yet complicated electrical-routing and out-of-phase driving are required. Hence, this study presents a piston mode microspeaker with simple electrical routing and single driving signal, which provides responses with good THD and wide bandwidth.

DESIGN CONCEPT

Fig.1 shows the proposed microspeaker (chip: 3×3 mm^2, microspeaker: 4 mm^2). The central-diaphragm connects to four cantilever-plate actuators through meandering springs (Fig.1a-b). The in-phase periodic bending excitation of cantilever-plate actuators could induce the piston (up-down) vibration of the central-diaphragm and meandering-springs (Fig.1c) to generate acoustic pressure. Moreover, flexible meandering-springs, despite the hardening effect from pre-stress, are exploited to release residual stresses of the films on actuator. To ensure the flatness of central diaphragm, all redundant layers included metals and PZT will be removed to reduce pre-deformation. Besides, in order to strike a balance

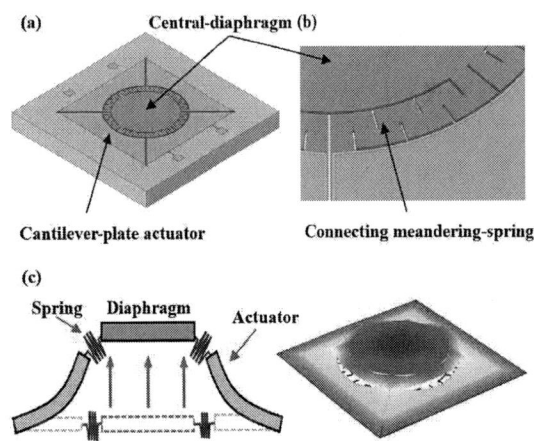

Figure 1: Design concept, (a) 3D schematic (b) zoom-in of components (c) working principle of piston mode excitation.

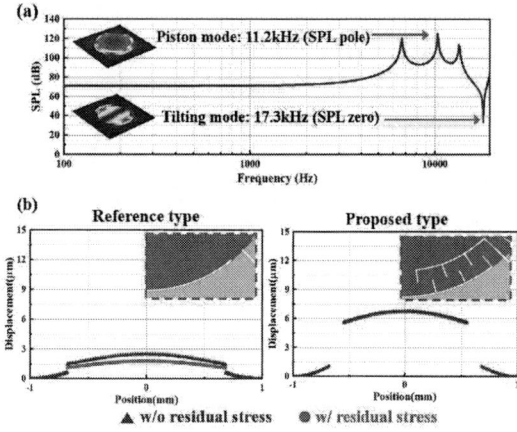

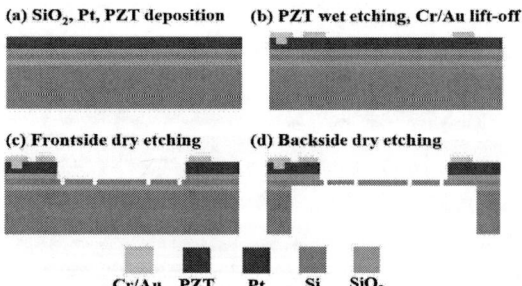

Figure 2: Simulation results, (a) frequency response (b) out-of-plane displacement of reference and proposed type at 0.7 V_{rms}, 10 kHz.

between fabrication yield and acoustic short circuit, gap size of 5 μm is chosen for the proposed microspeaker.

The simulation of frequency response within 20 kHz in pressure field of the proposed design is indicated in Fig.2a. Note that the 1st and 3rd SPL peaks are due to Helmholtz resonance from ear simulator, while the piston mode (the 1st structure mode) at 11.2 kHz is the 2nd SPL peak. On the other hand, the SPL zero at 17 kHz is resulted from the tilting mode of spring-diaphragm structure. The amplitude of central-diaphragm piston motion is also predicted by simulation, as shown in Fig.2b. The results depict that the proposed design with meandering-springs can not only enhance the central-diaphragm displacement, but also minimize the influence of residual stress.

FABRICATION

Fig.3 illustrates the process steps. In Fig.3a, SiO$_2$ and bottom electrode (Pt) were deposited by PECVD and sputtering respectively. Then, PZT film was deposited by sol-gel process on an SOI wafer with 5μm device layer. After that, as shown in Fig.3b, PZT wet etching were utilized to define bottom electrode pads. Subsequently, Cr/Au thin film were deposited by E-gun and patterned by lift-off as top-electrode. In Fig.3c, after removal of PZT, Pt, and SiO$_2$ layers by ICP-RIE, Si device layer was etched by DRIE to define the spring-diaphragm structure. Finally, the diaphragm of microspeaker was suspended by backside Si DRIE and buried oxide RIE (Fig.3d).

Figure 3: Fabrication process flow.

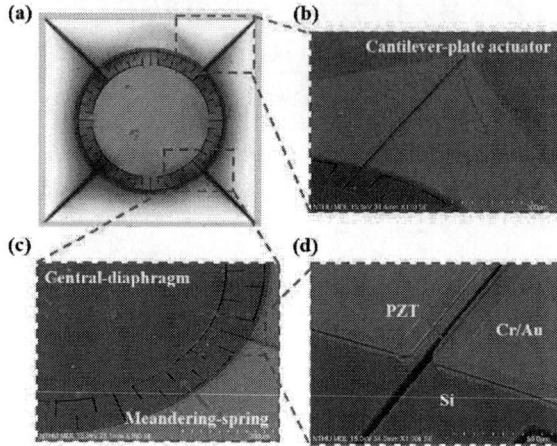

Figure 4: Fabrication results, (a) frontside overview, SEM images of (b) cantilever-plate actuator (c) meandering-spring and central-diaphragm, (d) stacked layers.

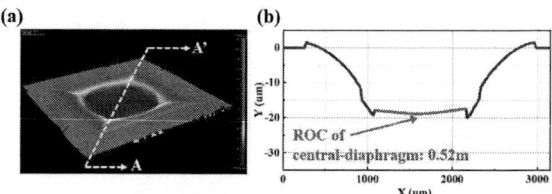

Figure 5: Static measurement, (a) pre-deformation of the device (b) AA' cross section view.

Figs.4a-c reveal typical fabricated microspeaker chip and the zoom-in of actuator and meandering-spring. Fig.4d shows Si layer for spring/diaphragm, and stacked films for actuator. In addition, Fig.5 demonstrates the pre-deformation (before driving) of microspeaker measured by optical interferometer. About 20 μm pre-deformation of actuators due to compressive residual stress was observed. The 0.52 m ROC (Radius of Curvature) indicates the good flatness of central diaphragm.

MEASUREMENT

The measurement setup for evaluation of acoustic characteristics was illustrated in Fig.6. Fig.6a shows the chip wire-bonded on PCB as the DUT (Device-Under-Test). As depicted in Figs.6b-c, the DUT was fixed on the standard ear-simulator (G.R.A.S. RA0401) inside the anechoic box, and then the input signal was provided by the B&K audio analyzer. When the microspeaker was driven, the output sound pressure was received by standard microphone (G.R.A.S. 40AG) in the ear simulator. At last, the signal was recorded through the power module and the audio analyzer.

Figs.7-10 indicate the measurement results of acoustic performance. Fig.7 demonstrates the various frequency responses of unipolar AC driving at different DC voltage (driving at 0.7 V$_{rms}$). The application of DC voltage main is to avoid reversal of polarization. After applying DC voltage, the responses are significantly improved by more than 20 dB SPL. In addition, the SPL will grow gradually

978-1-6654-9309-3/23 $31.00 © 2023 IEEE   128

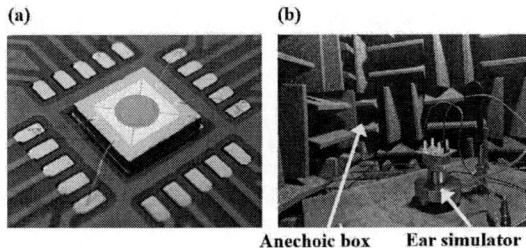

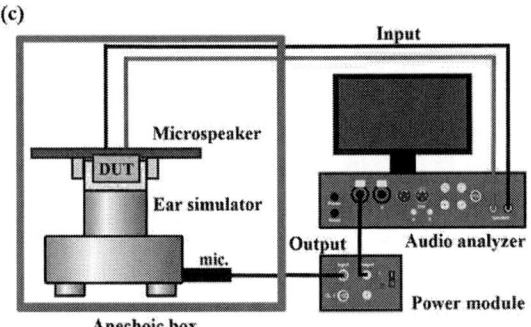

Anechoic box   Ear simulator

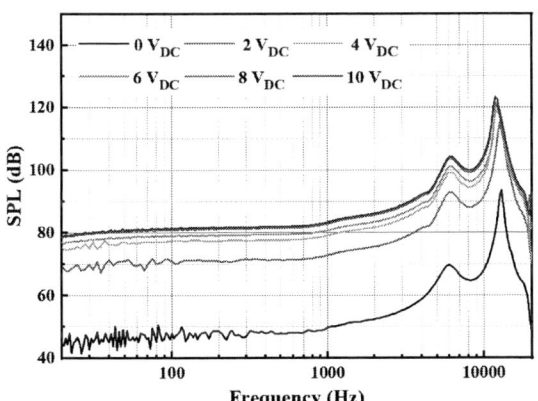

Figure 6: Measurement setup, (a) wire-bonded device on PCB (b) acoustic measurement environment (c) schematic of acoustic measurement system.

along with increasing DC voltage and saturate at around 8~10 V_{DC}. These results indicate the employment of DC voltage can not only prevent the concern of polarization reversal, but also boosts the d_{31} coefficient of PZT film used in this study. Measurements in Fig.8 reveal the responses with different AC signal at 10 V_{DC}. From 20 Hz~20 kHz, SPL is over 80 dB when unipolar AC driving above 1.4 V_{rms} (10 V_{DC}), and even higher than 93 dB at 3.5 V_{rms} (10 V_{DC}). The THD performance under 0.7 V_{rms} and 10 V_{DC} is depicted in Fig.9. With the overall SPL above 80 dB, the proposed design can achieve a THD below 7 %. Moreover, exclude the THD peaks caused by acoustic short circuit and subharmonics, the THD at other bands are even below 1 %, which again demonstrates the favorable sound quality of piston motion. Fig.10 presents the linearity of measured responses. Good linearity of 0.977 and 0.987 respectively at 1 kHz and 10 kHz shows the high stability of response at both low and high frequency range.

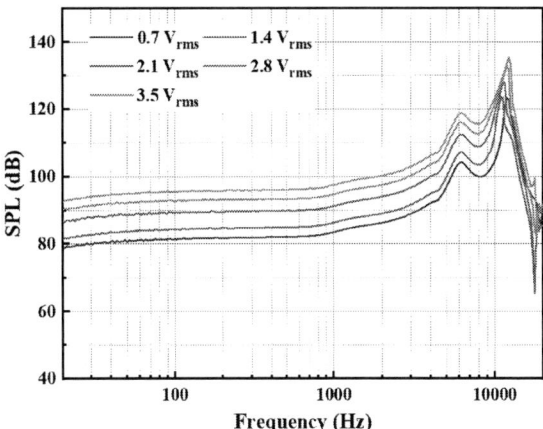

Figure 8: Acoustic performance measurement under 10 V_{DC} with different AC signal

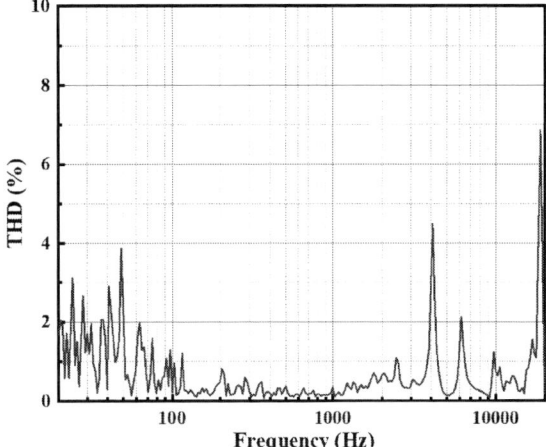

Figure 9: THD response at 0.7 V_{rms} +10 V_{DC}

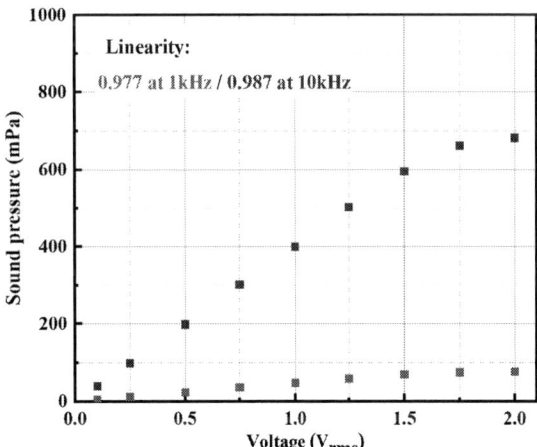

Figure 10: Linearity of sound pressure and driving signal at 1 kHz and 10 kHz

CONCLUSION

This study proposed a novel piezoelectric MEMS microspeaker integrating cantilever and spring-diaphragm structure. The cantilever actuators benefit the simple electrical routings while the spring-diaphragm structure

Figure 7: Acoustic performance measurement under 0.7 V_{rms} with different DC bias

has an upside in terms of good sound quality because of its piston motion. The proposed microspeaker is implemented by standard MEMS fabrication process on an SOI wafer. By applying only 0.7 V_{rms} with 10 V_{DC} bias, the proposed design with the small membrane merely 4mm^2 is capable of producing SPL over 80 dB (max value of 123 dB at 13 kHz) with THD below 7 % in full audio band (20 Hz~20 kHz) in standard ear-simulator measurement system. In this work, the acoustic characteristics under various driving conditions were investigated as well. Additional application of DC voltage can boost overall SPL remarkably without sacrificing the performance with regard to THD. The raise of SPL due to d_{31} coefficient improvement of PZT will saturate at 8~10 V_{DC}. On the other hand, increasing AC voltage will deteriorate THD although enhancement of SPL can be achieved.

Table 1 summarizes the measurement results and the comparison with the existing designs. It can be seen that the proposed design has superior performance over the small-active area designs [3,4]. Moreover, under only a quarter of the diaphragm area in [2], comparable accomplishments are attained in this study.

ACKNOWLEDGEMENT

This project was supported by the National Science and Technology Council under grant number NSTC 111-2221-E-007-069-MY3, NSTC 111-2923-E-007-014 -MY2, NSTC 110-2926-I-007 -506 -, and NSTC 110-2218-E-007-032-. The author would like to thank the Center for Nanotechnology, Materials Science, and Microsystems of National Tsing Hua University (CNMM), Taiwan Semiconductor Research Institute, Taiwan (TSRI), Nano Facility Center of National Yang Ming Chiao Tung University, Taiwan (NFC), and Nano-Electro-Mechanical-System Research Center of National Taiwan University,

Taiwan (NEMS) for providing the fabrication facilities and IC manufacturing.

REFERENCES

[1] A. Fawzy and M. Zhang, "Piezoelectric Thin Film Materials for Acoustic MEMS Devices," *IEEE ACCS& PEIT*, Hurgada, Egypt, 2019, pp. 82-86.

[2] H. Wang, P. X. -L. Feng and H. Xie, "A Dual-Electrode MEMS Speaker Based on Ceramic PZT with Improved Sound Pressure Level by Phase Tuning," *IEEE MEMS*, Gainesville, FL, 2021, pp. 701-704.

[3] S.H. Yi and E. S. Kim, "Piezoelectric microspeaker with compressive nitride diaphragm," *IEEE MEMS*, Las Vegas, NV, 2002, pp. 260-263.

[4] R. Gao, X. Chu, Y. Huan, Y. Sun, J. Liu, X. Wang, and L. Li. "A study on (K, Na) NbO3 based multilayer piezoelectric ceramics micro speaker," *Smart materials and structures* 23, no. 10, pp.105018, 2014.

[5] S. S. Lee, R. P. Ried and R. M. White, "Piezoelectric cantilever microphone and microspeaker," *Journal of Microelectromechanical Systems*, vol. 5, no. 4, 1996, pp. 238-242.

[6] F. Stoppel, C. Eisermann, S. Gu-Stoppel, D. Kaden, T. Giese and B. Wagner, "Novel membrane-less two-way MEMS loudspeaker based on piezoelectric dual-concentric actuators," *Transducers*, Kaohsiung, Taiwan, 2017, pp. 2047-2050.

[7] F. Stoppel, A. Männchen, F. Niekiel, D. Beer, T. Giese and B. Wagner, "New integrated full-range MEMS speaker for in-ear applications," *IEEE MEMS*, Belfast, UK, 2018, pp. 1068-1071.

[8] H.H. Cheng, S.C. Lo, Z.R. Huang, Y.J. Wang, M. Wu, and W. Fang. "On the design of piezoelectric MEMS microspeaker for the sound pressure level enhancement," *Sensors and Actuators A: Physical* 306, pp. 111960, 2020.

[9] Y.T. Lin, S.C. Lo and W. Fang, "Two-Way Piezoelectric MEMS Microspeaker with Novel Structure and Electrode Design for Bandwidth Enhancement," *Transducers*, Orlando, FL, 2021, pp. 230-233.

[10] I. Shahosseini, E. Lefeuvre, J. Moulin, E. Martincic, M. Woytasi and G. Lemarquand, "Optimization and Microfabrication of High Performance Silicon-Based MEMS Microspeaker," *IEEE Sensors Journal*, vol. 13, no. 1, 2012, pp. 273-284.

CONTACT

*W. Fang, Tel: +886-3-5742923; fang@pme.nthu.edu.tw

Table 1: Comparison of piezoelectric MEMS speaker

Type	This study	F. Stoppel [2]	H. H. Cheng [3]	Y. T. Lin [4]
Area (mm^2)	4	16	2.5	2.5
Signal (V_{rms})	0.7(10V_{dc})	0.7	0.7	0.7
Bandwidth SPL> 80dB	20Hz~ 20kHz	20Hz~ 20kHz	1.5kHz~ 3kHz	16kHz~ 19kHz
Bandwidth THD < 10%	20Hz~ 20kHz	20Hz~ 20kHz	100Hz~ 3kHz	N/A

HIGH-PERFORMANCE WAFER-SCALE TRANSFER-FREE GRAPHENE MICROPHONES

Roberto Pezone[1], Gabriele Baglioni[2], Pasqualina M. Sarro[1], Peter G. Steeneken[2,3], and Sten Vollebregt[1]

[1]Laboratory of Electronic Components, Technology and Materials (ECTM), Department of Microelectronics, Delft University of Technology, The Netherlands

[2]Kavli Institute of Nanoscience, Department of Quantum Nanoscience, Delft University of Technology, The Netherlands

[3]Department of Precision and Microsystems Engineering (PME), Delft University of Technology, The Netherlands

ABSTRACT

A repeatable method to fabricate multi-layer graphene (ML-gr) membranes of $2r = 85 - 155$ μm (t < 10 nm) with a 100% yield on 100 mm wafers is demonstrated. These membranes show higher sensitivity than a commercial MEMS-Mic combined with an area reduction of 10x. The process overcomes one of the main limitations when integrating graphene diaphragms in microphones due to the absence of automatic transfer methods on non-planarized target substrates. This method aims to overcome this limitation by combining a full-dry release of Chemical Vapor Deposition (CVD) graphene by Deep Reactive Ion Etching (DRIE) and vapor HF (VHF).

KEYWORDS

Graphene, Microphone, Membrane, Wafer-Scale, MEMS, Transfer-Free.

INTRODUCTION

Condenser MEMS-Mic lack a clear direction for miniaturization and high performance due to the Si-metal diaphragm's physical properties, i.e., high-tension. The microphone sensitivity S (eq.1) is determined by the membrane displacement per pressure load (C_m) towards the counter electrode at distance x_0 and the constant bias voltage (V_b) that maintains a fixed charge on the electrodes [1].

$$S = \frac{V_b C_m}{x_0} \qquad (1)$$

To achieve a more significant sensitivity without incurring high power consumption, due to large voltage bias, or microfabrication issues, such as membrane stiction caused by tiny gaps, a higher mechanical compliance (C_m) is needed. Because C_m is inversely proportional to the membrane tension and thickness, graphene, due to its high aspect-ratio, low-tension, and capacity to conduct electrical current, is the perfect candidate for enabling larger deflection under sound pressure for a smaller membrane size. In the last five years, advanced methods to fabricate such large membranes made by CVD graphene, primarily via polymer-based transfer, have been investigated [2 - 10]. However, despite the very high aspect ratios and high crystallinity of the demonstrated free-standing membranes (single-layer and few-layer graphene) in the mentioned works do not provide a clear route toward industrialization of the devices. This is because the transfer-based methods employed are not easily scalable toward high-volume wafer-level fabrication. The main reason is addressed to the absence of commercial equipment for transferring graphene over large cavities. Furthermore, current transfer methods often suffer from low yield, polymer contamination, cracks, and folding, leading to adhesion issues, especially for non-planarized target substrates. This work proposes an alternative approach in order to overcome these limitations with a wafer-scale transfer-less method where multi-layered graphene (ML-gr) drums are grown and released on the same substrate. More than 50 membranes with $2r = 85 - 300$ μm have been measured, showing a great tension uniformity. Peak mechanical compliances of ≈ 92 nm Pa^{-1} and ≈ 9 nm Pa^{-1} for the proposed membranes of $2r = 300$ μm and $2r = 85 - 155$ μm are obtained. A 100% yield of functional devices with diameters between $2r = 85 - 155$ μm is shown, differently than the 18% yield of the $2r = 300$ μm membranes.

EXPERIMENTAL SECTION

Process-flow

A 100 mm Si p-type wafer is thermally oxidized, forming 1 μm of SiO_2. LPCVD silicon-rich low-stress silicon nitride (110 nm, SiH_2Cl_2 315 sccm/NH_3 85 sccm) at 850 °C is first deposited and then patterned on the top-side and entirely removed on the back-side. PECVD TEOS-based 5 μm-thick silicon oxide is deposited on the back-side to be used as an etching mask for the bulk Si DRIE. A thin film of 50 nm molybdenum is sputtered at 50°C and etched by dry-etching with Cl and O_2 chemistry. The photoresist is stripped by O_2 plasma followed by rinsing in N-methyl-2-pyrrolidone (NMP) and DI-water (Fig. 1 (1)). Graphene is synthesized at 935 °C with an AIXTRON Black Magic at 25 mbar and H_2/CH_4 gas sources (Fig. 1 (2)). Next, Cr/Au (20/200 nm) is evaporated by ion-beam evaporation and patterned using a lift-off technique with NMP at 65 °C (Fig. 1 (3)). Mo is wet-etched with H_2O_2 and rinsed with DI-water (Fig. 1 (4)). The Bosch process is used to etch the bulk Si on the back-side with the graphene side facing the chuck, avoiding SF_6 exposure (Fig. 1 (5)). Finally, VHF etch is performed at 45 °C with 100% anhydrous HF, N_2, and EtOH in a commercially available Primaxx μEtch system, at 125 Torr, on diced chips of 10 mm x 10 mm (Fig. 1 (6)). All the residuals originating from the LPCVD SiN_x and VHF reaction are removed with a post-bake at $T > 110$ °C in vacuum.

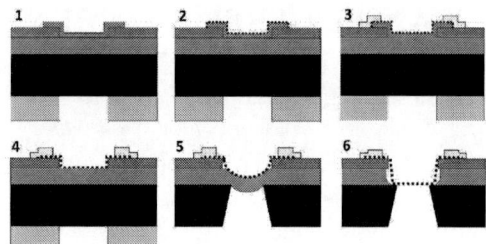

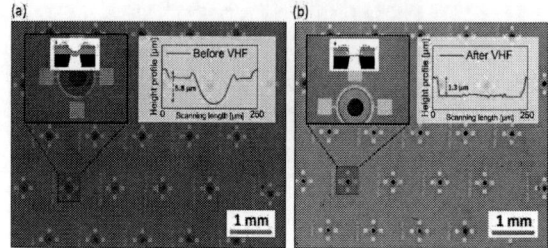

Figure 1: Main fabrication steps. (1) Patterning of the 110 nm SiN$_x$ layer (top-side) and 5 µm PECVD TEOS layer (back-side). Sputtering and dry-etching of 50 nm Mo. (2) Graphene CVD at 935 °C and 25 mbar. (3) Cr/Au (20/200 nm) evaporation and patterning by lift-off, which are used as additional anchoring and electrical contact. (4) Mo wet-etching in H$_2$O$_2$. (5) Backside DRIE of bulk Si. (6) Vapor HF of thermal SiO$_2$ (950 nm).

Acoustic measurement setup

A reference microphone (Sonarworks XREF20) measures the input sound pressure from the speaker, and it is placed close to the sample to have equidistance from the loudspeaker, avoiding significant phase differences. The signal captured by the reference microphone, the mechanical frequency response of the graphene membrane detected by the Polytec vibrometer at the center of the membrane, and the output signal of the speaker are monitored by a Moku:Lab (Liquid Instruments) hardware platform. The mechanical compliance is obtained from the ratio of the two signals received by the Moku:Lab after correction and calibration steps of the corresponding sensitivities of the vibrometer controller and reference microphone. Acoustic actuation at a sound pressure level of 1 Pa ($\approx$ 94 dB SPL) is used to test the fabricated membranes [11].

RESULTS AND DISCUSSION
Membrane fabrication

The fabricated heterostructures composed of SiO$_2$/ML-gr (before VHF) and the released ML-gr (after VHF) show different topography under optical inspection by a 3D laser scanning confocal microscope. A maximum downward out-of-plane deflection ranging from h_0 = 3.05 – 6.08 µm for membranes with $2r$ = 85 – 155 µm and 10 – 11.4 µm for $2r$ = 300 – 350 µm are measured. A buckling state addresses these deformations due to different thermal expansion coefficients between the SiO$_2$ layer and Si substrate. After the VHF release of the SiO$_2$ layer, the ML-gr recovers its original flat shape in the suspended region, showing nano-wrinkling due to imprinting left by the Mo topography on which graphene was grown (Fig. 2b). In addition, after the release, due to the thin SiO$_2$ film, the graphene attaches to the Si substrate, probably because of sagging or stiction during the final release. With a laser 3D microscope, a step height of $\approx$ 1.3 µm, equal to the sum of the thickness of the SiO$_2$, SiN$_x$, and the electrode, is measured, confirming its collapse on the silicon underneath (Fig. 2b). All Membranes with diameters of $2r$ = 85 – 155 µm show a 100% yield with all 132 suspended devices surviving after the vapor HF

Figure 2: Chip-view with topographic analysis along each membrane before and after SiO$_2$ etching by using 3D laser scanning confocal microscope. (a) Suspended SiO$_2$/ML-gr heterostructure with step height measurement of one single drum showing 1st mode buckling deformation. (b) Released ML-gr (after VHF) where buckling has disappeared showing a flat suspended membrane. These results are extendable to all devices present in the chips with only some small peak height variations due to membrane size differences.

release. Large SiO$_2$/ML-gr heterostructures with diameters of 300–350 µm show a 37% yield on 117 fabricated drums after the DRIE. After SiO$_2$ etching, the same drums decrease their yield from 37% to 18%. Larger membranes have lower yield because of higher buckling modes that cause significant distortions, cracks, and high deformations in the oxide, compared to the first modes that have been found for $2r$ = 85–155 µm membranes. This can be improved by tuning the stress of the SiO$_2$ film or compensating the compressive stress with the tensile SiN$_x$ frames [12].

Graphene characterization

A Horiba HR800 Raman spectrometer equipped with a 514.5 nm Ar$^+$ laser maintained at 5mW is used to inspect the crystallinity of the 25 test drums before and after the VHF step. A 100× objective with a numerical aperture of 0.9 is used with a spot size of about 696 nm. A large defectivity is highlighted for the final ML-gr drums, with I_D/I_G ranging from $0.52 - 0.88$ after normalization with respect to the G-band of the measured data. The high D band is related to any kind of defect that distorts the graphene lattice, like edges, wrinkles, Stone–Wales defects, and vacancies. In addition, the D intensity shows the invasiveness of the process on the graphene compared to previous works where lower defect intensity was reported for the same material [13]. This increased defectivity source could be attributed to the lift-off step where NMP or the short ultrasonic bath has probably negatively affected the quality of the material. Finally, it can be seen that the Raman 2D-peak ratios are representative of multi-layer graphene, as shown in Fig. 3a, where $I_{2D}/I_G < 1$ are measured [14]. The difference in I_D/I_G, I_{2D}/I_G is due to thickness variation and variations in the number of defects. A graphene thickness of $\approx 7 \pm 2$ nm is measured with an atomic force microscope (AFM) from Cypher Asylum Research in semi-contact mode (Fig. 3b). The AFM thickness measurements are made on graphene which is processed with all the reported steps except the VHF. Since the SiN$_x$ cannot be marked as flat reference point due to a partial over-etching during Mo

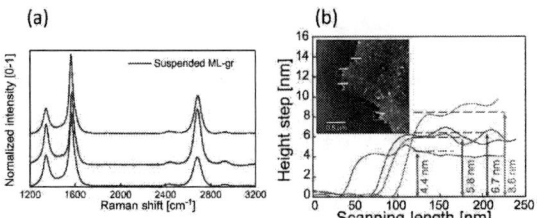

Figure 3: (a) Raman spectrum from 1200 to 3200 cm^{-1} of three different ML-gr drums in the suspended region. The difference in I_D/I_G, I_{2D}/I_G is due to thickness variation and variations in the number of defects.(b) Thickness measurement of the ML-gr after transfer on bare Si/SiO_2 substrate by AFM in semi contact mode.

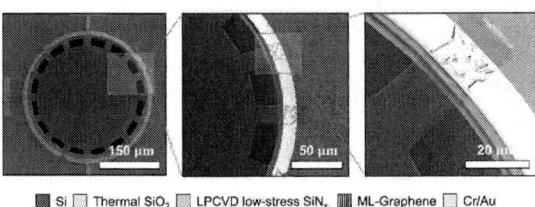

■ Si □ Thermal SiO_2 □ LPCVD low-stress SiN_x ▥ ML-Graphene □ Cr/Au

Figure 4: Scanning electron microscope (SEM) pictures made by a Hitachi Regulus 8230 of suspended 300 μm trampoline ML-gr.

patterning, it is transferred in DI-water on a clean thermally oxidized silicon chip. Fig. 4 shows a suspended membrane that is patterned in a trampoline geometry with a diameter of 300 μm where the Cr–Au/SiN_x/Si interface acts as a clamping support for the suspended graphene.

Resonance frequency and mechanical compliance

Dynamic visualization of the proposed membranes is managed by a digital holographic microscope (LynceeTec) equipped with a laser wavelength of 666 nm and a 10x objective lens in stroboscopic mode. The interference between the reflected laser beam from the sample and the reference path provides the intensity and phase of each pixel, defining the 3D topography. At its resonance frequency, mechanical motion is exhibited because of the optical phase shifts that result from the oscillating membrane. Mounting 1 x 1 cm chips on a piezo-shaker controlled by a sine wave with 0.5V and a frequency range $f_0 = 50 - 350$ kHz in a vacuum chamber at 10^{-4} kPa to reduce any air damping effects, the first mode of the resonance frequencies are visualized. More than ten membranes have been inspected, and they show resonance frequencies over $\approx 244 - 318$ kHz for diameters of $2r = 120 - 155$ μm as shown in Fig. 5a. Using a graphene thickness of 7 nm, a density of 2260 kg/m^3, a diameter range of $120 - 155$ μm, the experimental results fit the analytical values in a pre-tension range of $n_0 = 0.03 - 0.05$ N/m. The analytical values are calculated with Eq. 2 [15],

$$f_0 = \frac{2.405}{2\pi R} \sqrt{\frac{n_0}{\rho t}} \qquad (2)$$

where n_0 is the pre-tension (N/m) of the graphene, ρ the mass density, and t is the thickness of the graphene. Here,

the bending rigidity influence is omitted since it is estimated to be small. The membrane displacement z caused by sound actuation is measured by a single-point laser Doppler vibrometer (LDV, OFV-5000 Polytec GmbH) at the drum's center, where the most significant deflection is expected. The obtained results are referred to a sound pressure of 1 Pa (94 dB). More than 50 devices with diameters ranging from $85 - 300$ μm are inspected, showing a mechanical compliance range of $\approx 3 - 92$ nm Pa^{-1} at 1 kHz as shown in Fig. 5b. This specific frequency is considered to be the middle of the audio band where the human ear has the highest sensitivity. Then, the analytical results are calculated and compared with the experimental ones according to Eq. 3 where the displacement z is quadratically related to the membrane size [15]. More importantly, the cubic terms in z are neglected due to the linear regime in this relation. The proposed drums are compared with the experimental results in Fig. 5b. Within the $2r = 85 - 155$ μm, the quadratic dependency is found with a pre-tension fit n_0 of 0.2 N/m, unlike the larger membranes with 2r = 300 μm that show n_0 of $0.07 - 0.1$ N/m depending on their clamping geometry.

$$\Delta P = \frac{4 n_0 z}{R^2} \qquad (3)$$

Spring structures show lower pre-tension of ≈ 0.03 N/m compared the fully-clamped (≈ 0.1 N/m) for same membrane diameter of 300 μm. This also translates in a 1.4 times higher mechanical compliance of the spring structure than the fully-clamped. The quadratic dependency is not shown in the entire diameter range of $85 - 300$ μm, and the reason might be addressed to a more profound sagging of larger membranes compare to the smaller one. When comparing the pre-tension extracted from Eq. 2 - 3, we note that different values of n_0 are obtained as in Fig 5a, b. These differences might be caused by uncertainty in the mass and thickness that affect Eq. 2, gas damping and permeation effects at 1 kHz that affect Eq. 3, and differences in the deflected mode shapes from theory.

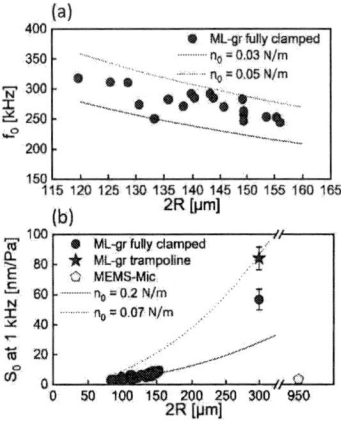

Figure 5: Resonance frequency and mechanical compliance measurements. (a) Resonance frequency for ML-gr with 120–155 um. (b) Comparison of the mechanical compliance of the ML-gr membranes with a commercial MEMS microphone MP23DB01HP, MP34DT04 STMicroelectronics (yellow pentagon).

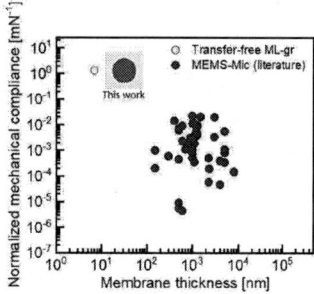

Figure 6: Benchmark of mechanical compliance normalized by the membrane area of proposed ML-gr and MEMS-Si based microphone state-of-the-art.

Additional investigation is needed to quantitatively account for these differences. These compliances exceed that of a larger membrane of $2r = 950$ μm used in commercially available MEMS Si microphones. Peak compliance of 92 nm Pa^{-1} is achieved with a $2r = 300$ μm ML-gr trampoline. With this, we demonstrate the potential of graphene for future miniaturized and high-performance microphones as compared to traditional MEMS-Mic (Fig. 6), with a scalable fabrication method suitable for uniform device performance and allowing more accurate, repeatable statistical studies of free-standing graphene-based devices [15].

CONCLUSIONS

This work demonstrates a repeatable and scalable process flow to efficiently fabricate wafer-scale multi-layer graphene drums with diameters from $2r = 85$ to 300 μm. A 100% yield was demonstrated for membranes with $2r = 85 - 155$ μm. These drums can operate as microphones with mechanical compliances up to 92 nm Pa^{-1} that exceeds that of a larger membrane of $2r = 950$ μm used in commercially available MEMS Si microphones that generally have few nm Pa^{-1}. The results presented here show the great potential of these devices for next-generation, high-volume, wafer-scale graphene microphone technologies once developed in a capacitive architecture.

ACKNOWLEDGEMENTS

The authors thank the Delft University of Technology Else Kooi Lab staff for processing support and thank Herre van der Zant for useful discussions. This project has received funding from Union's Horizon 2020 research and innovation program under Grant Agreement No. 881603 (Graphene Flagship).

REFERENCES

[1] M. Fueldner, Chapter 48 – Microphones, Handbook of Silicon Based MEMS Materials and Technologies (Third Edition), Elsevier, 2020, pp.937-948.

[2] D. Todorović, A. Matković, M. Milićević, D. Jovanović, R. Gajić, I. Salom and M. Spasenović, "Multilayer Graphene Condenser Microphone", *2D Materials*, 2, 045013, 2015.

[3] S. Woo, J. -H. Han, J. -H Lee, S. Cho, K. -W. Seong, M. Choi, J.-H. Cho, "Realization of a High Sensitivity Microphone for a Hearing Aid Using a Graphene–

PMMA Laminated Diaphragm", *ACS Appl. Mater. Interfaces*, 9, 1237– 1246, 2017.

[4] S. Wittmann, C. Glacer, S. Wagner, S. Pindl, M.C. Lemme, "Graphene Membranes for Hall Sensors and Microphones Integrated with CMOS-Compatible Processes", *ACS Applied Nano Materials*, 2, 5079– 5085, 2019.

[5] S. Wagner, C. Weisenstein, A. Smith, M. Östling, S. Kataria, M. Lemme, "Graphene Transfer Methods for the Fabrication of Membrane-Based NEMS Devices", *Microelectron. Eng.*, 159, 108– 113, 2016.

[6] C. –K. Lee, Y. Hwangbo, S. –M. Kim, S. –K. Lee, S. –M. Lee, S. –S. Kim, K. –S. Kim, H. –J. Lee, B. –I. Choi, C. –K. Song, J. –H. Ahn and J. –H. Kim, "Monatomic Chemical-Vapor-Deposited Graphene Membranes Bridge a Half-Millimeter-Scale Gap", *ACS Nano*, 8, 2336– 2344, 2014.

[7] S. A. Akbari, V. Ghafarinia, T. Larsen, M. M. Parmar and L. G. Villanueva, "Large Suspended Monolayer and Bilayer Graphene Membranes with Diameter up to 750 μm", *Sci. Rep.*, 10, 6426, 2016.

[8] Y. –M. Chen, S. –M. He, C. –H. Huang, C. –C. Huang, W. –P. Shih, C. –L. Chu, J. Kong, J. Li and C. –Y. Su, "Ultra-large Suspended Graphene as a Highly Elastic Membrane for Capacitive Pressure Sensors", *Nanoscale*, 8, 3555– 3564, 2016.

[9] A. F. Carvalho, A. J. Fernandes, M. B. Hassine, P. Ferreira, E. Fortunato and F. M. Costa, "Millimeter-Sized Few-Layer Suspended Graphene Membranes", *Applied Materials Today*, 21, 100879, 2020.

[10] J. Xu, G. S. Wood, E. Mastropaolo, M. J. Newton and R. Cheung, "Realization of a Graphene/PMMA Acoustic Capacitive Sensor Released by Silicon Dioxide Sacrificial Layer", *ACS Appl. Mater. Interfaces*, 13, 38792– 38798, 2021.

[11] G. Baglioni, R. Pezone, S. Vollebregt, K. C. Zobenića, M. Spasenović, D. Todorović, H. Liu, G. Verbiest, H. S. van der Zant and P. G. Steeneken, "Optical Characterization of Ultra Sensitive Graphene Membranes for Microphone Applications", *Under review, 2022*.

[12] A. Shchepetov, M. Prunnila, F. Alzina, L. Schneider, J. Cuffe, H. Jiang, E. I. Kauppinen, C. M. S. Torres and J. Ahopelto, "Ultra-Thin Free-Standing Single Crystalline Silicon Membranes with Strain Control", *Appl. Phys. Lett.*, 102, 192108, 2013.

[13] F. Ricciardella, S. Vollebregt, T. Polichetti, M. Miscuglio, B. Alfano, M. L. Miglietta, E. Massera, G. Di Francia and P. M. Sarro, "Effects of Graphene Defects on Gas Sensing Properties towards NO2 Detection", *Nanoscale*, 9, 6085– 6093, 2017.

[14] L. Malard, M. Pimenta, G. Dresselhaus, and M. Dresselhaus, "Raman Spectroscopy in Graphene", *Phys. Rep*, 473, 51 - 87, 2009.

[15] R. Pezone, G. Baglioni, P. M. Sarro, P. G. Steeneken and S. Vollebregt, "Sensitive Transfer-Free Wafer-Scale Graphene Microphones", *ACS Appl. Mater. Interfaces*, 14 (18), 21705-21712, 2022.＝

CONTACT

*R. Pezone, public mail: r.pezone@tudelft.com

HIGH-SPL AND LOW-DRIVING-VOLTAGE PMUTS BY SPUTTERED POTASSIUM SODIUM NIOBATE

Fan Xia[1,2], Yande Peng[1,2], Sedat Pala[1,2], Ryuichi Arakawa[1,3], Wei Yue[1,2], Pei-Chi Tsao[2],
Chun-Ming Chen[2], Hanxiao Liu[1,2], Megan Teng[2], Jong Ha Park[1,2], and Liwei Lin[1,2]

[1]Berkeley Sensor and Actuator Center, Berkeley, CA, USA
[2]Department of Mechanical Engineering, University of California, Berkeley, CA, USA and
[3]NGK Spark Plug Co., JAPAN

ABSTRACT

This work presents an air-coupled piezoelectric micromachined ultrasonic transducer (pMUT) with high transmitting acoustic pressure by using sputtered potassium sodium niobate (K,Na)NbO₃ (KNN) thin film with a high piezoelectric coefficient ($e_{31} \sim$ 8-10 C/m^2) and low dielectric constant ($\epsilon_r \sim$ 260-300) for the first time. The fabricated KNN pMUT with a resonant frequency at 104.5 kHz has been tested to exhibit unprecedented results: (1) high sound pressure level (SPL) of 109 dB/V at a distance of 10 cm, which is 8 times higher than that of AlN-based pMUTs at a similar frequency; (2) low-voltage operation of only 4 volts peak-to-peak amplitude (V_{p-p}); and (3) good receiving sensitivity. As such, this work presents a new class of high-SPL and low-driving-voltage pMUTs for potential applications in various fields, including consumer electronics, such as but not limited to haptic feedback, loudspeaker, and AR/VR systems.

KEYWORDS

Ultrasound, pMUT, acoustic pressure, low driving voltage, piezoelectric, KNN.

INTRODUCTION

Ultrasonic transducers are widely used in robotics and automotives for object detection. Compared with the bulk ultrasonic transducers, the small-footprint piezoelectric micromachined ultrasonic transducers (pMUTs) have low power consumption and wide bandwidth for applications in consumer electronics and Internet of Things (IoT), such as range-finding [1], gesture recognition [2], fingerprint sensing [3], and 3D imaging [4]. However, relatively low output pressure remains a significant challenge for AlN-based pMUTs, which limits the signal transmissions in various applications. For instance, state-of-the-art AlN-based rangefinder can only reach a travel distance of 4 m with an array design [5]. For other applications where large acoustic pressure output is critical, such as mid-air haptic feedback [6], this becomes an engineering challenge.

It is well-known that the transmitting performance of pMUTs is mainly determined by the device structure and piezoelectric material property. Continuous efforts have been made in optimizing the device structures, such as to modify the boundary condition [7], construct the bimorph architecture [8] and adjust the mode shape [9]. Even so, the output pressure of pMUT is still limited by the intrinsically low piezoelectric coefficient of the material such as AlN with the e_{31} around 1 C/m^2. There are also reports on increasing the piezoelectric coefficient of materials by tuning material compositions, for example, the scandium-doped AlN [10]. Other materials, like lead zirconate titanate (PZT)-based pMUTs [11,12] have relatively higher output pressure, but their low receiving sensitivity and the lead-contents are key drawbacks. Thus, there's a strong need for better piezoelectric materials of batch fabrication process to further boost the performance of pMUTs.

Here, we report an air-coupled pMUT based on sputtered, lead-free KNN thin film with high piezoelectric coefficient ($e_{31} \sim$ 8-10 C/m^2) and low dielectric constant ($\epsilon_r \sim$ 260-300) for the first time. The excellent transmitting sensitivity (8 times higher than that of AlN-based pMUTs) and good receiving performance under a low-voltage excitation are demonstrated.

DESIGN AND FABRICATION

Principle of Design

Figure 1a shows the structure of KNN pMUTs with a circular unimorph diaphragm. The device consists of the 2-um sensor-type KNN thin film [13] as the active piezoelectric layer and a 5-um Si device layer as the elastic layer. The dual-electrode geometry has a circular-shape inner electrode (67% in radius [14]) and a ring-shape outer electrode. The differential drive can be applied, where the inner and outer electrode are excited with opposite polarity to enhance the vibration displacement and output pressure correspondingly. The simulated fundamental mode shape with clamped boundary condition is shown in **Figure 1b**.

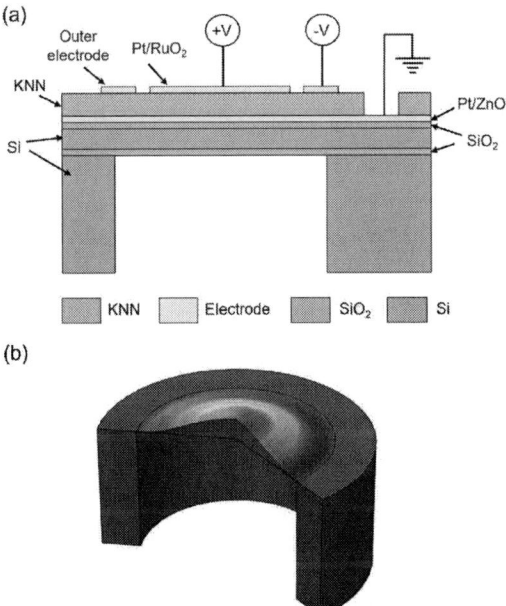

Figure 1: (a) The schematic cross-sectional view of a dual electrode, circular unimorph KNN pMUT; (b) simulated fundamental mode shape.

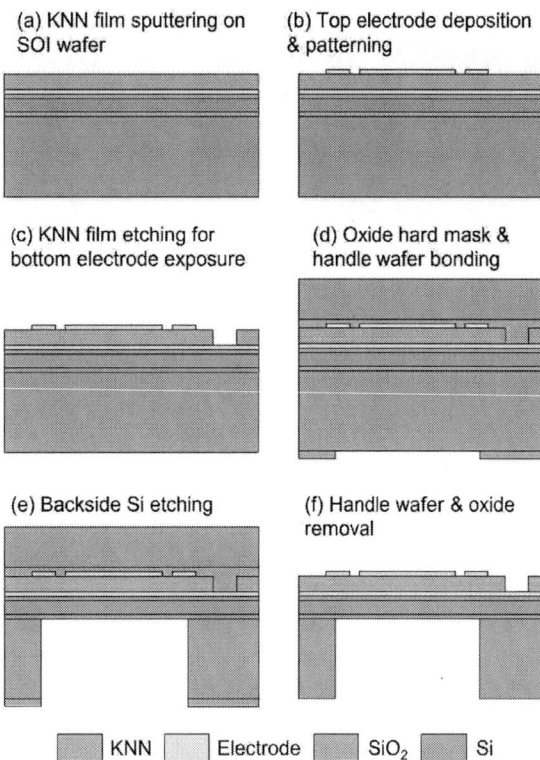

(a) KNN film sputtering on SOI wafer

(b) Top electrode deposition & patterning

(c) KNN film etching for bottom electrode exposure

(d) Oxide hard mask & handle wafer bonding

(e) Backside Si etching

(f) Handle wafer & oxide removal

KNN · Electrode · SiO₂ · Si

Figure 2: Fabrication process.

Fabrication process

The detailed fabrication process is shown in **Figure 2**. The process starts with the sputtering of 25 nm-thick ZnO adhesion layer and 200 nm-thick Pt bottom electrode on a 6-inch SOI wafer which consists of a 5 um-thick Si device layer, 1 um-thick buried silicon oxide (BOX) layer and 610 um-thick (100) Si handle substrate. Then 2 um-thick KNN thin film which serves as the active piezoelectric layer, is deposited via a radio frequency (RF) magnetron-sputtering process on top of the Pt/ZnO bottom electrode layer. After that, 10 nm-thick RuO_2 and 100 nm-thick Pt are sputtered and patterned as the inner circular and outer ring Pt/RuO_2 top electrode, where RuO_2 is used to promote the adhesion strength between Pt and KNN thin film [13]. Afterwards, the via openings to access the bottom electrode is realized by patterning the KNN film through a wet-etching process. Finally, backside silicon cavity is defined by a deep reactive-ion etching (DRIE) process where the BOX layer acts as the etching stop layer, and the device wafer is temporarily bonded to a handle wafer during the process.

RESULTS

Characterization

Figure 3-4 shows the characterization results of the fabricated KNN pMUTs. The crystal structure of the sputtered KNN film is characterized by using the X-ray diffraction to monitor the film quality, and the sharp peak in the (001) orientation confirms its perovskite structure with good crystallinity (**Figure 3**). The benign surface morphology is highlighted by the optical image of the pMUTs as well (**Figure 4a-4b**).

The cross-sectional scanning electron microscope (SEM) images, as shown in **Figure 4c-4d**, display the well-defined backside silicon cavity, the tightly stacked Pt/RuO_2/KNN/Pt/ZnO/SiO_2/Si multi-layered diaphragm structure, and further show the thickness of each layer (1.9 um-thick KNN and 5.2 um-thick Si device layer). The dense columnar structure of the KNN thin film validates the good crystal orientation which plays an important role in the piezoelectric performances.

Transmission performance

The transmission pressure of a single KNN pMUT with a radius of 500 um is measured utilizing a high-sensitivity microphone (Bruel & Kjar, Type 4138). The prototype KNN pMUT is driven by continuous differential sine-wave signals for the inner and outer electrodes. The receiving signal is collected by placing the microphone 1 cm above the pMUT and the sound pressure level (SPL) is extracted through the corresponding sensitivity. As shown in **Figure 5a**, a maximum transmission pressure of 135.7 dB SPL (121.73 Pa) is achieved at 104.5 kHz with the 4 V_{p-p} excitation, and the non-linear hardening phenomena can be observed near the peak [15].

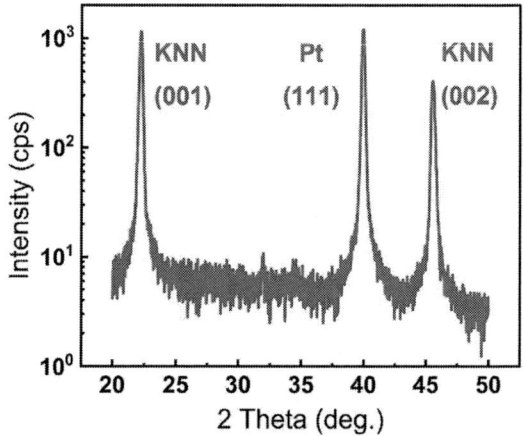

Figure 3: XRD of sputtered KNN thin film with (001) crystal orientation.

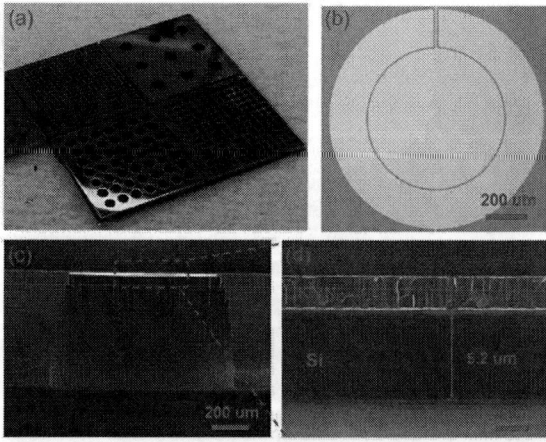

Figure 4: Characterization results. (a) Backside of KNN pMUTs with different sizes; (b) a microscope top image of prototype KNN pMUT; (c-d) cross-sectional SEM images.

978-1-6654-9309-3/23 $31.00 © 2023 IEEE

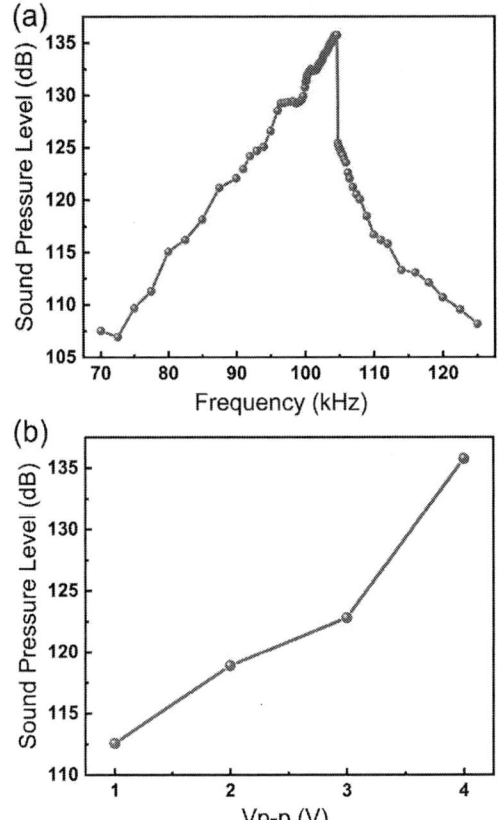

Figure 5: Transmission performance of single KNN pMUT. Sound pressure level (SPL) @ 1 cm axial distance versus (a) frequency, and (b) excitation voltage (V_{p-p}).

The influences of excitation voltage and axial distance on the transmission pressure are further evaluated at 104.5 kHz. As shown in **Figure 5b**, the SPL at 1 cm axial distance increases from 112.6 dB (8.5 Pa) to 135.8 dB (122.71 Pa) as V_{p-p} increases from 1 V to 4 V. **Figure 6** shows the relationship between the SPL and axial distance, where the SPL under a 4 V_{p-p} input gradually decreases from 150.5 dB (666.85 Pa) @ 0.1cm to 115.1 dB (11.3 Pa) @ 10cm, and 102.6 dB (2.71 Pa) @ 30 cm. Thus, the transmission sensitivity of the single KNN pMUT reaches 129.3 dB/V SPL (58.27 Pa/V) @ 1 cm, 109 dB/V SPL (5.66 Pa/V) @ 10 cm, and 96.6 dB/V SPL (1.36 Pa/V) @ 30 cm axial distance.

Finally, a summary of the transmission performance comparison with other devices in the state-of-the-art literature is shown in Table 1. The KNN pMUT stands out with its transmitting sensitivity 8~10 times higher than those of AlN pMUTs, highlighting its extraordinary advantage of high SPL under a low driving voltage.

Receiving performance

The receiving performance is evaluated by using two identical pMUTs (500 um in radius) as the transmitter (T_x) and receiver (R_x) respectively. The pMUT transmitter is excited via 7-cycle sine-wave pulse signal (104.5 kHz, 4 V_{p-p}) with the differential drive scheme. For the receiving pMUT, only the inner region is utilized. Furthermore, a

charge amplifier is applied, and the amplified receiving signals with respect to distances are measured. **Figure 7a** exhibits the excitation and receiving signal with a travel distance of 2 m. The receiving signal peak-value has a Time-of-Flight (ToF) of 6.13 ms which matches well with the real distance measured by the laser distance meter. The relationship between the receiving signal magnitude and travel distance is displayed in **Figure 7b**, and the signal can be clearly identified with the travel distance up to 6.1 m (3 m for the round trip), demonstrating good receiving performance.

CONCLUSION

This work has demonstrated the first sputtered KNN-based pMUTs. The KNN film exhibits good crystal quality in (001) orientation, which contributes to the high piezoelectric coefficient. The fabricated KNN pMUT with a resonant frequency of 104.5 kHz is used to evaluate the transmission and receiving performance. Specifically, a high SPL of 135.8 dB can be obtained from the single pMUT with axial distance of 1 cm under a low driven voltage of 4 V_{p-p}, and it remains larger than 100 dB with distance up to 40 cm. The corresponding transmission sensitivity turns out to be 8 times larger than that of the state-of-the-art AlN-based pMUTs. The good sensitivity is also demonstrated by the clear receiving signal which travels up to 6.1 m. As such, this work sheds light on high-SPL and low-driving-voltage pMUTs for potential applications, like haptic feedback, loudspeaker, and AR/VR systems.

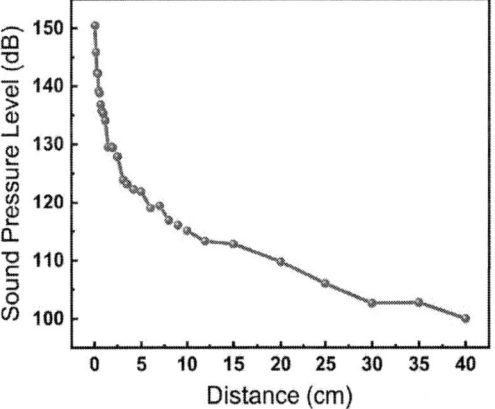

Figure 6: SPL @ resonance under the continuous sine-wave excitation (4 V_{p-p}) versus the axial distance.

Table 1: Transmission sensitivity comparison with other devices in the state-of-the-art literature.

Material	Frequency	V_{p-p}	SPL (dB/V)	Ref.
AlN	101k	20	90@10 cm	[4]
36% Sc doped AlN	60k	/	71.6@30 cm	[10]
Single-crystal PZT	46k	10	93@14.4 cm 86.3@33 cm	[11]
PZT	100k	6	93.5@3.5 cm	[12]
PMN-PZT	151k	3	94.6@6 cm; 88.6@13 cm	[16]
KNN	104.5k	4	**109@10 cm; 96.6@30 cm**	**This work**

978-1-6654-9309-3/23 $31.00 © 2023 IEEE

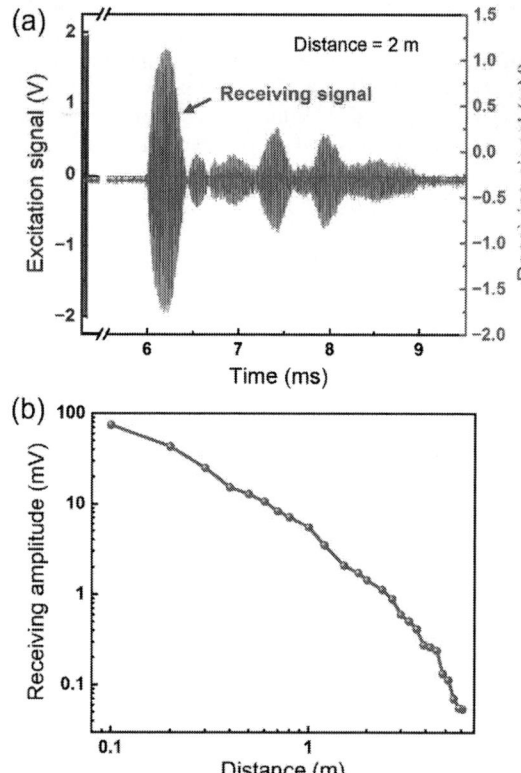

Figure 7: Receiving performance. (a) Amplified receiving signal with a travelling distance of 2 m; (b) the receiving signal magnitude vs the travel distance.

ACKNOWLEDGEMENTS

This work was supported in part by BSAC (Berkeley Sensor and Actuator Center) and the NSF grant ECCS-2128311. The authors thank SCIOCS Co. Ltd. for KNN film deposition/etching processes, UC Berkeley Marvell Nanofabrication Lab for the rest of the fabrication process.

REFERENCES

[1] R. J. Przybyla, S. E. Shelton, A. Guedes, I. I. Izyumin, M. H. Kline, D. A. Horsley, and B. E. Boser. "In-air rangefinding with an aln piezoelectric micromachined ultrasound transducer", *IEEE Sensors Journal* 11, no. 11 (2011): 2690-2697.

[2] R. J. Przybyla, H. Tang, S. E. Shelton, D. A. Horsley, and B. E. Boser. "12.1 3D ultrasonic gesture recognition", in *2014 IEEE International Solid-State Circuits Conference Digest of Technical Papers (ISSCC)*, pp. 210-211. IEEE, 2014.

[3] X. Jiang, H. Tang, Y. Lu, E. J. Ng, J. M. Tsai, B. E. Boser, and D. A. Horsley. "Ultrasonic fingerprint sensor with transmit beamforming based on a PMUT array bonded to CMOS circuitry", *IEEE Trans. Ultrason. Ferroelectr. Freq. Control*, 64, no. 9 (2017): 1401-1408.

[4] Z. Shao, Y. Peng, S. Pala, Y. Liang, and L. Lin. "3D ultrasonic object detections with> 1 meter range", in *2021 IEEE 34th International Conference on Micro Electro Mechanical Systems (MEMS)*, pp. 386-389. IEEE, 2021.

[5] Z. Shao, S. Pala, Y. Peng, and L. Lin. "Bimorph Pinned Piezoelectric Micromachined Ultrasonic Transducers for Space Imaging Applications", *J. Microelectromech. Syst.*, 30, no. 4 (2021): 650-658.

[6] S. Pala, Z. Shao, Y. Peng, and L. Lin. "Ultrasond-induced haptic sensations via PMUTS", in *2021 IEEE 34th International Conference on Micro Electro Mechanical Systems (MEMS)*, pp. 911-914. IEEE, 2021.

[7] Y. Liang, B. Eovino, and L. Lin. "Piezoelectric micromachined ultrasonic transducers with pinned boundary structure", *J. Microelectromech. Syst.*, 29, no. 4 (2020): 585-591.

[8] S. Akhbari, F. Sammoura, B. Eovino, C. Yang, and L. Lin. "Bimorph piezoelectric micromachined ultrasonic transducers", *J. Microelectromech. Syst.*, no. 2 (2016): 326-336.

[9] T. Wang, R. Sawada, and C. Lee. "A piezoelectric micromachined ultrasonic transducer using piston-like membrane motion", *IEEE Electron Device Letters*, 36, no. 9 (2015): 957-959.

[10] Y. Kusano, I. Ishii, T. Kamiya, A. Teshigahara, G. Luo, and D. A. Horsley. "High-SPL air-coupled piezoelectric micromachined ultrasonic transducers based on 36% ScAlN thin-film", *IEEE Trans. Ultrason. Ferroelectr. Freq. Control*, 66, no. 9 (2019): 1488-1496.

[11] G. Luo, Y. Kusano, M. N. Roberto, and D. A. Horsley. "High-pressure output 40 kHz air-coupled piezoelectric micromachined ultrasonic transducers", in *2019 IEEE 32nd International Conference on Micro Electro Mechanical Systems (MEMS)*, pp. 787-790. IEEE, 2019.

[12] G. Massimino, L. D'Alessandro, F. Procopio, R. Ardito, M. Ferrera, and A. Corigliano. "Air-coupled PMUT at 100 kHz with PZT active layer and residual stresses: Multiphysics model and experimental validation", in *2017 18th International Conference on Thermal, Mechanical and Multi-Physics Simulation and Experiments in Microelectronics and Microsystems (EuroSimE)*, pp. 1-4. IEEE, 2017.

[13] K. Shibata, K. Watanabe, T. Kuroda, and T. Osada. "KNN lead-free piezoelectric films grown by sputtering", *Appl. Phys. Lett.*, 121, no. 9 (2022): 092901.

[14] S. Pala, and L. Lin. "An Improved Lumped Element Model for Circular-shape pMUTs", *IEEE Open Journal of Ultrasonics, Ferroelectrics, and Frequency Control (2022)*.

[15] G. Massimino, B. Lazarova, F. Quaglia, and A. Corigliano. "Air-coupled PMUTs array with residual stresses: experimental tests in the linear and non-linear dynamic regime", *Int. J. Smart Nano Mater.*, 11, no. 4 (2020): 387-399.

[16] Z. Zhou, S. Yoshida, and S. Tanaka. "Epitaxial PMnN-PZT/Si MEMS ultrasonic rangefinder with 2 m range at 1 V drive", *Sens. Actuator A Phys.*, 266 (2017): 352-360.

CONTACT

*F. Xia, tel: +1-341-766-8361; fxia21@berkeley.edu

EPITAXIAL $P_B(Z_R,T_I)O_3$-BASED PIEZOELECTRIC MICROMACHINED ULTRASONIC TRANSDUCER FABRICATED ON SILICON-ON-NOTHING (SON) STRUCTURE

Takuma Sekiguchi[1], Shinya Yoshida[2], Yoshiaki Kanamori[1], and Shuji Tanaka[1]
[1]Tohoku University, JAPAN and
[2]Shibaura Institute of Technology, JAPAN

ABSTRACT

This study reports the fabrication method of a piezoelectric micromachine ultrasonic transducers (pMUTs) using a Silicon on Nothing (SON) structure and a monocrystalline thin film of lead zirconate titanate (PZT). For the SON structure, an array pattern of small hole was fabricated on a (100) Si substrate. Hydrogen annealing was then used to induce Si reflow. As a result, a plate-shaped cavity with a thickness of 0.5 μm was formed under a (100) Si diaphragm of 75 μm diameter. PZT was epitaxially grown on the diaphragm via buffer layers. A pMUT was fabricated and preliminarily characterized. The merits of this fabrication method include single side process, no release and sealing process and compatibility with most of major piezoelectric materials.

KEYWORDS

Piezoelectric micro ultrasonic transducers (pMUTs), Silicon on Nothing (SON), $Pb(Zr,Ti)O_3$ (PZT) monocrystalline thin films, silicon migration, hydrogen anneal

INTRODUCTION

Piezoelectric micromachined ultrasonic transducers (pMUTs) are expected to be used not only in a medical field but also in various consumer applications such as range finders and biometric identification[1, 2]. In general, pMUTs, a diaphragm structure laminated with an elastic plate and a piezoelectric transducer thin film vibrates for transmitting and receiving ultrasonic waves. Thus, the performance of the piezoelectric thin film is one of the critical factors determining that of the device. Recently, a lead zirconate titanate (PZT) monocrystalline (Mono PZT) thin films was demonstrated to have a large figure of merit (FOM) and to provide high-performance pMUTs[3, 4]. The Mono PZT thin film is epitaxially grown on a (100) Si substrate via a buffer layer by sputter deposition. If such a high-performance pMUT could be fabricated affordably, the applications would be expanded more.

However, several issues remain in the fabrication of pMUTs. A silicon-on-insulator (SOI) wafer has been often used as a base wafer. The piezoelectric thin film and electrode layers are first deposited on the topside of the wafer, and then the backside is etched by Deep Reactive Ion Etching (DRIE) to form a diaphragm. This backside process often causes dimensional and/or alignment errors. Also, a notching effect at the interface between SiO_2 layer and Si device layer may enlarge the dimensional error. In particular, the errors will be significant in case the pMUT is small. Furthermore, flipping the wafer is not preferable in the standard semiconductor process line and may raise the cost for special contamination avoidance.

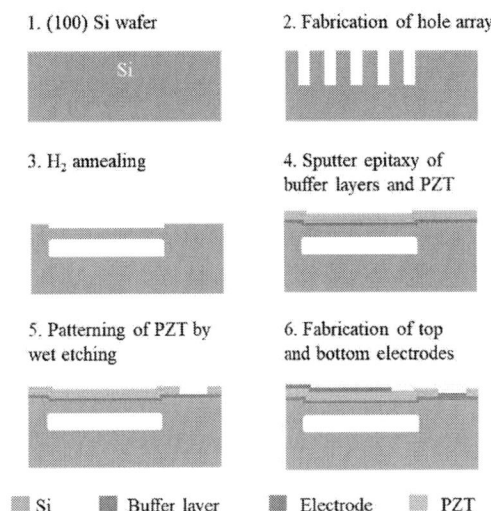

Fig. 1 Fabrication process using Silicon on Nothing (SON) structure for pMUTs with the $Pb(Zr,Ti)O_3$ (PZT) monocrystalline thin film.

The use of a cavity-SOI wafer has been proposed as an alternative[5]. This wafer is normally fabricated as follows. First, a cavity for the pMUT diaphragm is patterned on a Si wafer. Then, the wafer is oxidized thermally, and another wafer is bonded to cover the cavity. Finally, the capped wafer is thinned down to a few micrometers to form an inner cavity structure. This complex manufacturing process increases the cost. The fabrication method based on surface micromachining using a sacrificial layer is also proposed[6]. However, this method does not allow the epitaxial growth of Mono PZT on the sacrificial layer, which is normally a polycrystalline material.

Thus, we have proposed to utilize a Silicon on Nothing (SON) structure[7]. Figure 1 shows the schematic of the fabrication process developed in this study. First, regularly arranged small holes are fabricated on the surface of a (100) Si wafer. Next, hydrogen annealing is performed for inducing migration of Si atoms on the surface. As a result, a cavity structure surrounded by Si is formed, which is called as "SON" structure. Since this SON structure is formed by reflowing a (100) Si wafer as a base, its surface should be (100) Si plane. Thus, buffer layers and a Mono PZT thin film can be epitaxially grown on it. Finally, a pMUT with the Mono PZT thin film on the SON structure is completed after fabricating electrodes. This fabrication method is probably simpler and less expensive than that using a cavity-SOI wafer. The whole process is completed

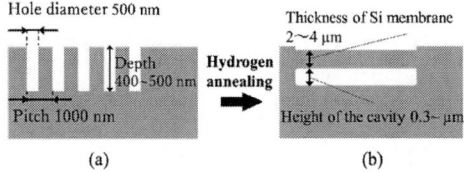

Fig. 2 (a) Design of the hole array for a target dimensions of SON structure (b) Expected dimensions of SON structure after hydrogen annealing.

from the front side without flipping the wafer. Thus, the dimensional and/or alignment errors will be reduced. An AlN-based pMUT on a SON structure was reported[8]. If the Mono PZT thin film with a much higher FOM is formed on the SOI structure, the performance of the pMUT should be enhanced. In this study, we developed the fabrication process mentioned above and characterized the fundamental properties of a prototyped Mono-PZT-based pMUT on the SON structure.

DESIGN OF PMUT AND SON STRUCTURE pMUT

Design of pMUT

In this study, the target resonant frequency of the pMUT was 4~6 MHz assuming body-coupled applications. The diameter of the pMUT diaphragm was set to 75μm. The thickness of the Mono PZT film mounted on the pMUT was about 1.5 μm. Considering the design conditions, the target thickness of the Si diaphragm in the SON structure was determined to be 2~4 μm.

Design of hole array pattern for SON structure

Considering the target dimensions of the SON structure above, the hole array formed in a Si base wafer was designed as illustrated in Fig. 2 (a). The diameter of each hole is 500 nm. The holes are arranged in a square shape with the pitch of 1000 nm. The hole depth is 4~5 μm. Based on the previous work[9], the buried cavity higher than 0.3 μm and a diaphragm of 2~3 μm thickness should be formed after hydrogen annealing, as shown in Fig. 2 (b). This dimension is suitable for the full vibration of the pMUT diaphragm for body-coupled applications.

DETAILS OF FABRICATION PROCESS

The details of the fabrication process shown in Fig. 1 are as follows. (1) A (100) Si substrate was prepared as a base substrate. (2) A positive-tone photoresist (TDMR-AR80, Tokyo Ohka Kogyo) was spin-deposited on the substrate, and a hole array pattern was formed using an *i*-line stepper (FPA-3030i5+, Canon), as shown in Fig. 3 (a). Then, the deep holes were fabricated by DRIE. As seen in Fig. 3 (b), 4-μm-deep holes were successfully formed almost as designed. (3) The Si substrate was annealed under hydrogen atmosphere at 1200°C and 50 kPa[8]. As a result, the holes were sealed, as seen in Fig. 3 (c).

(4) SrRuO$_3$(SRO)/La$_{0.5}$Sr$_{0.5}$CoO$_3$(LSCO)/CeO$_2$/yttria-stabilized zirconia (Y$_2$O$_3$: 8 mol%) buffer layers were epitaxially grown on the Si substrate by pulse laser deposition with a KrF excimer laser[4]. The thicknesses of SRO, LSCO, CeO$_2$, and YSZ layers were approximately

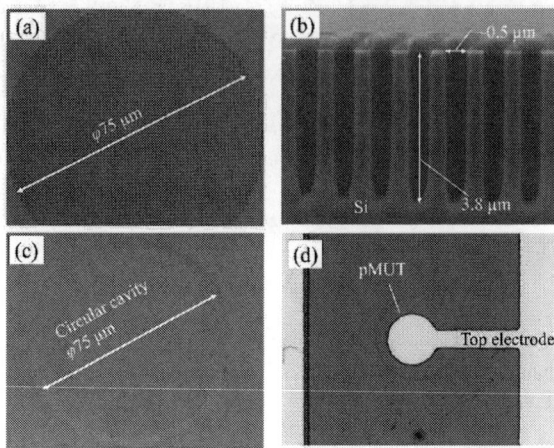

Fig. 3 (a) Circularly arranged holes of the photo-resist for the next dry etching process. (500 nm in diameter and 1 μm in pitch) (b) Cross-sectional image of the holes after dry etching. (c) Formation of circular SON structure after hydrogen annealing. (d) Prototyped pMUT with a Mono PZT thin film on SON structure.

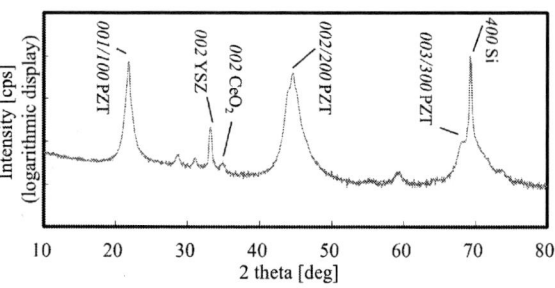

Fig. 4 Typical X-ray diffraction pattern of the Mono PZT thin film sputter-deposited on the specimen after hydrogen annealing.

200, 10, 20, and 10 nm, respectively. The SRO layer was used as a conductive bottom electrode. After that, 1.5-μm-thick Mono PZT was sputter deposited on the buffer layer at ~600°C and 0.5 Pa with a mixture of Ar and O$_2$. The sputter target was prepared by mixing PbO, PbZrO$_3$, and PbTiO$_3$ powders at a composition of Pb$_{1.1}$(Zr$_{0.5}$, Ti$_{0.5}$)O$_3$.

Figure 4 is a typical X-ray diffraction pattern of the PZT deposited on the Si substrate after the hydrogen annealing. The PZT thin film exhibited a strong (001/100) crystal orientation. No significant peak corresponding to other orientations for PZT was observed. If PZT was epitaxially grown on the buffer layers in the cube-on-cube relationship, only the (001/100) crystal orientation should be obtained. Thus, this XRD pattern supports the fact that a Mono PZT thin film was successfully deposited.

(5) The Mono PZT thin film was patterned by wet etching with an etchant (Pure Etch PT204, Hayashi Pure Chemical) to expose the SRO buffer layer. (6) A Au/Cr pattern with thicknesses of 200 nm/10 nm was fabricated as a top electrode via a lift-off process. The exposed SRO region was also metalized in this step. Figure 3 (d) shows

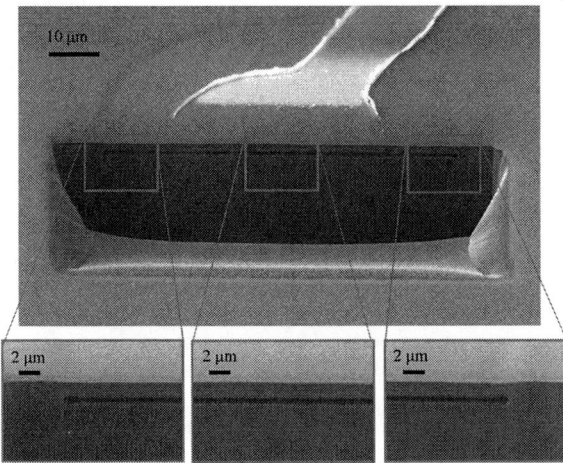

Fig. 5 Cross-sectional view of the prototyped pMUT.

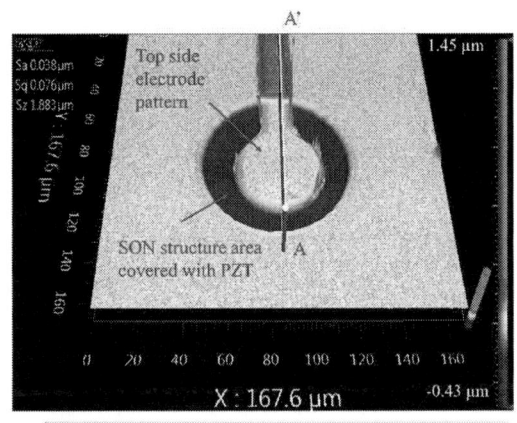

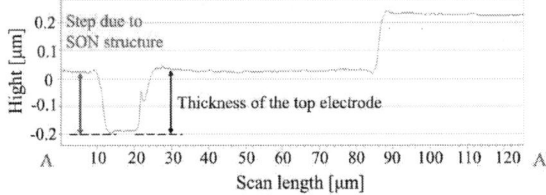

Fig. 6 Topographic image of pMUT and surface profile along A-A' line in the image.

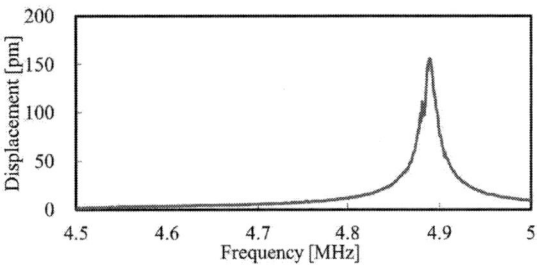

Fig. 7 Displacement-frequency response measured with a laser doppler vibrometer.

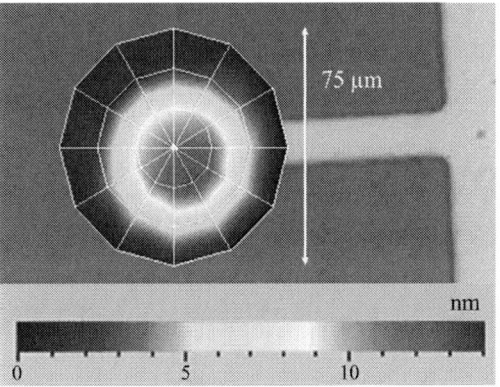

Fig. 8 Displacement map of the pMUT actuated with 250 mV_{P-P} at the resonant frequency.

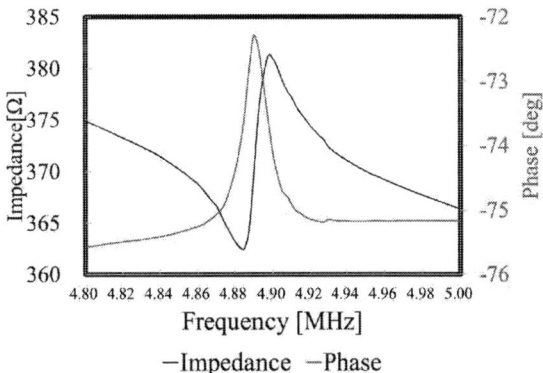

Fig. 9 Impedance and phase spectrum of the pMUT.

one of the prototyped pMUTs. The diameter of the top electrode is 45 μm.

The diaphragm of the pMUT was etched by focused ion beam, and the cross section was observed, as shown in Fig. 5. The holes were completely sealed by hydrogen annealing, and the plate-shaped cavity was formed inside the substrate as expected. The thickness of the Si diaphragm is about 2.2 μm, and the cavity height is about 500 nm. These dimensions are close to the expected ones. In addition, the thickness of the Si diaphragm looks nearly uniform at any position.

The surface properties of the fabricated pMUT were analyzed using a white light interferometer, as shown in Fig. 6. A step of about 200 nm was observed between the SON and the original Si substrate region. Such a step is always

formed in the fabrication of a SON structure. However, no disconnection of the top electrode line was observed in this study. Thus, the slight step is not a practical problem.

CHARACTERIZATION OF THE PROTOTYPED PMUT

The displacement-frequency response of the prototyped device was measured using a laser doppler vibrometer (MSA-500, Polytec) (Fig. 7). The device was actuated by applying a chirp signal of 250 mV$_{p-p}$. The measured resonant frequency and mechanical Q-factor were 4.89 MHz and 332, respectively. The measured resonant frequency almost agree with the designed one.

Figure 8 shows the map of the vibration amplitude in the pMUT at the resonant frequency. The center position of the vibration was slightly shifted from that of the diaphragm to the lower right in this picture. This may be because the topside circular electrode was misaligned by about 6.2 μm from the center of the diaphragm, as shown in Fig. 3 (d). But the vibration center does not coincide with the misaligned position of the top electrode. Thus, this misalignment is not only the cause of the shift of the vibration center. Another cause may be the non-uniformity of the thickness of the SON structure, which is difficult to be observed in scanning electron microscopy. Further investigation should be needed.

The impedance and phase spectra of the pMUT was measured by an impedance analyzer (4194A impedance/gain analyzer, Keysight Technologies), as plotted in Fig. 9. The resonant frequency was almost the same as the mechanical resonant frequency. The electromechanical coupling coefficient (k_{eff}^2) was estimated to be 0.53% from the following equation

$$k_{eff}^2 = \frac{f_a^2 - f_r^2}{f_a^2}, \qquad (1)$$

where f_a and f_r are the anti-resonance and resonance frequencies, respectively. In this prototyped pMUT, the inactive region such as electrode pads, which formed a capacitance structure with PZT, was relatively large compared to the active region in the diaphragm. This large capacitance gave the low k_{eff}^2 value. In practical, an insulator layer such as SiO$_2$ or resin can be deposited on the PZT thin film in the inactive region for reducing the capacitance. For example, if a 1.5-μm-thick polymer layer, of which relative dielectric constant assumed to be 3, was deposited as well as reference[10], the k_{eff}^2 value will be improved to approximately 4%. This value is comparable to those of pMUTs reported previously.

CONCLUSION

In order to create high-performance and affordable pMUTs with small dimensional errors, we developed the fabrication method utilizing the SON structure for the Mono-PZT-based pMUT in this study. The array pattern of the deep holes (diameter: 500 nm, pitch: 1000 nm, depth: 4 μm) in (100) Si was fabricated, and then a SON diaphragm structure was formed by hydrogen annealing at 1200°C. The thickness of the Si diaphragm was about 2.2 μm, and the cavity height was about 500 nm. Eventually, the pMUT with (100/001)-oriented Mono PZT was successfully fabricated. The fabricated pMUT had a resonant frequency of 4.89 MHz and an electromechanical coupling coefficient of 0.53%. This achievement is a great step forward creating affordable and high-performance pMUTs.

REFERENCES

[1] J. Jung, W. Lee, W. Kang, E. Shin, J. Ryu, and H. Choi, "Review of piezoelectric micromachined ultrasonic transducers and their applications," *J. Micromech. Microeng.*, vol. 27, 2017.

[2] S. Sun, J. Wang, M. Zhang, Y. Yuan, Y. Ning, D. Ma,

P. Niu, Y. Gong, X. Yang, and W. Pang, "Eye-Tracking Monitoring Based on PMUT Arrays," *J. Microelectromechanical Syst.*, vol. 31, pp. 45–53, 2022.

[3] S. Yoshida, H. Hanzawa, K. Wasa, and S. Tanaka, "Fabrication and characterization of large figure-of-merit epitaxial PMnN-PZT/Si transducer for piezoelectric MEMS sensors," *Sens. Actuators, A Phys.*, vol. 239, pp. 201–208, 2016.

[4] S. Yoshida, H. Hanzawa, K. Wasa, M. Esashi, and S. Tanaka, "Highly c-axis-oriented monocrystalline Pb(Zr, Ti)O$_3$ thin films on si wafer prepared by fast cooling immediately after sputter deposition," *IEEE Trans. Ultrason. Ferroelectr. Freq. Control*, vol. 61, no. 9, pp. 1552–1558, 2014.

[5] Q. Wang, Y. Lu, S. Mishin, Y. Oshmyansky, and D. A. Horsley, "Design, Fabrication, and Characterization of Scandium Aluminum Nitride-Based Piezoelectric Micromachined Ultrasonic Transducers," *J. Microelectromechanical Syst.*, vol. 26, no. 5, pp. 1132–1139, 2017.

[6] G. L. Luo and D. A. Horsley, "Piezoelectric Micromachined Ultrasonic Transducers with Corrugated Diaphragms Using Surface Micromachining," *TRANSDUCERS 2019*, pp. 841–844, 2019.

[7] J. Su, X. Zhang, G. Zhou, C. Xia, W. Zhou, and Q. Huang, "A review: crystalline silicon membranes over sealed cavities for pressure sensors by using silicon migration technology," *J. Semicond.*, vol. 39, no. 7, p. 071005, 2018.

[8] D. S. W. Choong D. S. W. Choong, D. S.-H. Chen, D. J. Goh, J. Liu, S. Ghosh, Y. Koh, J. Sharma, S. Merugu, F. Quaglia, M. Ferrera, A. S. Savoia, and E. J. Ng, "Silicon-On-Nothing ScAlN pMUTs," *IEEE Int. Ultrason. Symp.*, pp. 5–8, 2021.

[9] T. Sato, I. Mizushima, S. Taniguchi, K. Takenaka, S. Shimonishi, H. Hayashi, M. Hatano, K. Sugihara, and Y. Tsubashima,"Fabrication of silicon-on-nothing structure by substrate engineering using the empty-space-in-silicon formation technique," *Jpn. J. Appl. Phys.*, vol. 43, pp. 12–18, 2004.

[10] Ziyi Liu, Shinya Yoshida and Shuji Tanaka, "Fabrication and characterization of annular-shaped piezoelectric micromachined ultrasonic transducer mounted with Pb(Zr,Ti)O$_3$-based monocrystalline thin film," *J. Micromech. Microeng.* vol. 31, 125014, 2021.

ACKNOWLEDGMENT

This work was supported in part by A-STEP (Adaptable and Seamless Technology Transfer Program through Target-driven R&D).

CONTACT

*Shuji Tanaka, tel: +81 22 795 6934; mems@tohoku.ac.jp

LEVERAGING SEMICONDUCTOR ECOSYSTEMS TO MEMS

Weileun Fang,[1,2] Sheng-Shian Li[1,2], and Ming-Huang Li[1]

[1]Power Mechanical Engineering Department, National Tsing Hua University, Hsinchu, TAIWAN
[2] Nano Engineering and Microsystems Institute, National Tsing Hua University, Hsinchu, TAIWAN

ABSTRACT

Taiwan, with population of near 23 million and area of 36000km^2, is active in the semiconductor related industries and researches, especially in Hsinchu city where the National Tsing Hua University (NTHU) is located. The faculties and students of NTHU have the opportunity to frequently and closely interact with the semiconductor industries. This article would like to share the experience of NTHU MEMS group regarding how they leverage the huge semiconductor resources to promote MEMS technologies in the following four stages. First, employing the CMOS-MEMS technologies serves as the bridge to communicate with the semiconductor industries. Second, by preventing various mechanical issues from thin films, promising applications for CMOS-MEMS technologies are demonstrated. Third, the MEMS above CMOS technology established in the foundry further exhibits the win-win collaboration for MEMS and semiconductor technologies. Finally, due to the potential applications in Smart-X and Metaverse, semiconductor industries are even developing processes with new functional materials for MEMS recently. In conclusions, it is a win-win strategy between academia/research and industry/market to leverage the resources in mature semiconductor ecosystems for the development and commercialization of MEMS.

KEYWORDS

CMOS, CMOS-MEMS, semiconductor, MEMS, sensors, actuators, transducers, piezoelectric

INTRODUCTION

The Microelectromechanical Systems (MEMS) technologies have been developed for near half century. Many thanks for the efforts and contributions from numerous pioneers, researchers, engineers, etc. in both academia and industry to not only bring novel ideas but also realize useful applications. The demand for MEMS sensors has been growing from billions in 2012 to more than trillions in 2020. In the meantime, the semiconductor industry has made even greater advances in technologies and markets. Since the mature semiconductor industry dominates a tremendous amount of resources, leveraging the existing semiconductor ecosystem to develop MEMS devices and products may bring advantages. On the other hand, as indicated in the technology development roadmap of semiconductor in Fig.1 [1], MEMS is one of the important techniques to achieve diversification chips (so called the "More than Moore" solution, in the horizontal-axis) when the miniaturization of semiconductor devices reaches the physical limitation (so called the "More Moore" solution, in the vertical-axis). Thus, from this point of view, MEMS could continue as well as extend the importance and influence of semiconductor related technologies.

Taiwan, with population of near 23 million and area of 36000km^2, is active in the semiconductor related industries and researches. There are various world renowned semiconductor companies including design house, foundry, packaging and testing house, and system companies distributed in Taiwan, such as the TSMC, ASE, MediaTek, etc. Moreover, several common laboratories equipped with semiconductor fabrication and testing facilities have also been established in the universities and national laboratories. It takes only 1~2 hours to travel among these organizations through high speed rail. These organizations form the semiconductor networks/ecosystems in Taiwan, and also enable diverse ways of collaboration. Thus, it's cost-effective to leverage such semiconductor ecosystems to develop MEMS devices and products (Fig.2).

Since the National Tsing Hua University (NTHU) is adjacent to the Hsinchu Science Park of Taiwan, the faculties and students of NTHU have the opportunity to frequently and closely interact with the semiconductor industries. This article would like to share the experience of NTHU MEMS group regarding their collaboration with the mature and huge semiconductor resources and industries for the promotion of MEMS technologies. Hopefully this experience sharing could bring useful information and give inspiration to the MEMS society.

CMOS AS THE BRIDGE

The complementary metal oxide semiconductor (CMOS) processes are mature fabrication technologies to implement microelectronics components. To date, many existing CMOS processes have been established in

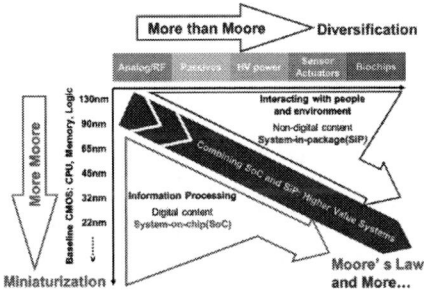

Fig.1: The semiconductor technology roadmap from ITRS [1].

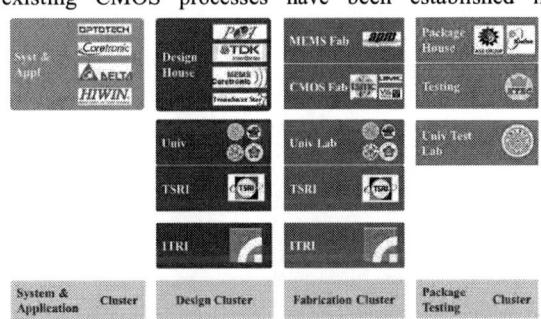

Fig.2: MEMS/semiconductor ecosystems in Taiwan networking with NTHU: examples to showcase "Leveraging semiconductor ecosystems to MEMS".

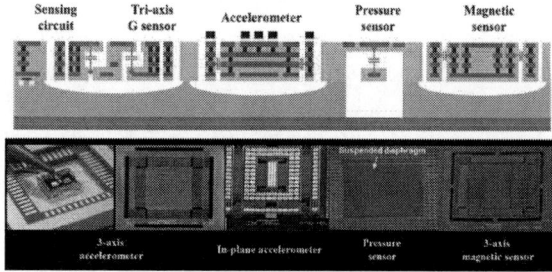

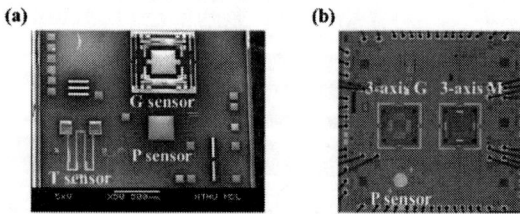

Fig.3: The CMOS-MEMS inertial sensors platform.

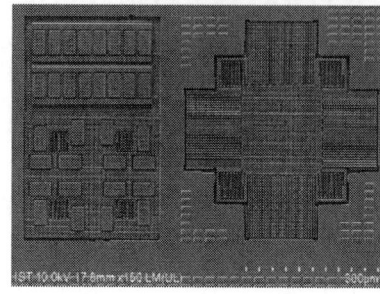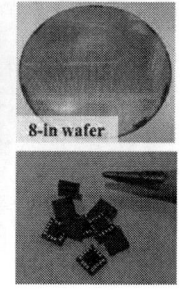

Fig.5: The CMOS-MEMS inertial sensors fabricated and packaged through the semiconductor supply chain.

commercial foundries, such as the 0.35μm 2P4M (two poly-Si and four metal layers), the 0.18μm 1P6M (one poly-Si and six metal layers), etc. processes. To easily communicate with semiconductor people, the CMOS processes would be the "bridge" for us, i.e., MEMS people, during the initial engagement. Moreover, to attract the attention of industry, inertial sensors were selected as our initial target in early 2000. Many thanks for the inspiration from Prof. Gary Fedder to demonstrate the possibility by the use of CMOS process to fabricate inertial sensors [2]. Thus, various inertial sensors have been developed and implemented through the standard CMOS platforms in our group [3-5] to show semiconductor people real devices and applications about what is MEMS.

Fig.3 shows a typical CMOS-MEMS process platform to fabricate and integrate inertial sensors [3-11]. In this platform, the TSMC helped to prepare the CMOS chip using the standard 0.35μm 2P4M process. After that, the post-CMOS release processes developed by us were used to define and suspend MEMS structures. The CMOS-MEMS inertial sensors such as the accelerometers, magnetometers, gyroscopes, and altimeters (pressure sensors) can be realized on the same MPW (multi project wafer). By following the design rules of standard CMOS process, springs, diaphragms, driving and sensing electrodes, and so on are developed to improve the performances or reduce the size of sensors [3]. Since these sensors are implemented using the same process platform, they can be easily integrated on a single chip (the system-on-chip solution) as the inertial sensing hub for different applications [7-11]. For example, the SoC tire pressure monitoring sensor (TPMS) in Fig.4a consists of pressure, acceleration, and temperature sensors [11]. Moreover, the 7-axis inertial sensing hub including accelerometer, magnetometer, and altimeter is demonstrated in Fig.4b. Our group also collaborated with design house to develop inertial sensors on 8-in CMOS wafer through their supply chain (Fig.5). These examples identify various connections (in Fig.2) to leverage semiconductor resources to MEMS.

However, the residual stresses and the mismatch of the coefficient of thermal expansion (CTE) of CMOS films are critical challenges for CMOS-MEMS inertial sensors [12-14]. The suspended flexible springs of CMOS-MEMS inertial sensors have unwanted initial deflection caused by the thin film residual stresses, and also experience the unwanted thermal deformation during operation resulted from the CTE mismatch of thin films. Moreover, the creep of the aluminum alloy metal layers is another concern for the suspended CMOS-MEMS structures [15], which may cause the shape of flexible spring for inertia sensors varying with time, thus leading to the reliability issue of CMOS-MEMS devices [14]. We know from such lesson learned, and these issues have to be addressed so that CMOS-MEMS inertial sensors can find practical applications.

APPLICATIONS FOR CMOS-MEMS

According to the residual stresses, CTE mismatch, creep, etc. of thin films, it is challenging to fabricate reliable CMOS-MEMS devices with flexible movable/deformable structures, such as the inertial sensors and microphone. In this regard, the environment sensors (such as the humidity, temperature, gas, infrared, sensors, etc.) typically with no flexible structures will be good candidates for the applications of CMOS-MEMS platforms. Prof. Henry Baltes and his group in ETH demonstrated various CMOS-MEMS environment sensors, and also successfully extended the results to establish the company Sensirion [16]. For instance, gas sensors were implemented using the BCD processes (Bipolar-CMOS-DMOS) with backside Si etching [17-18]. In this product, the MEMS structure is used to improve the thermal isolation.

Fig.6 displays the typical CMOS-MEMS process platform to fabricate and integrate environment sensors [19-24]. Similarly, the TSMC fabricates the CMOS chip

Fig.4: Two inertial sensing hubs for different applications.

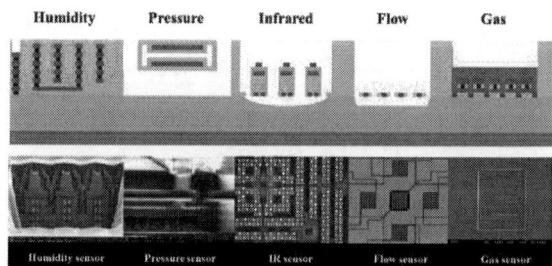

Fig.6: The CMOS-MEMS environment sensors platform.

978-1-6654-9309-3/23 $31.00 © 2023 IEEE

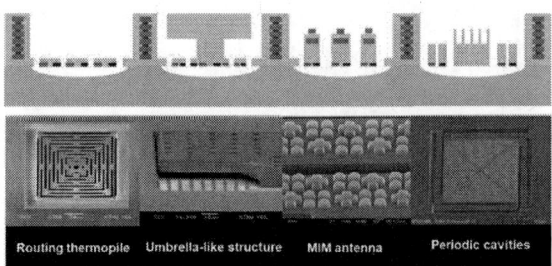

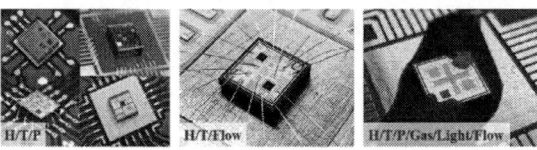

Fig.7: IR sensors of various designs are realized by the platform

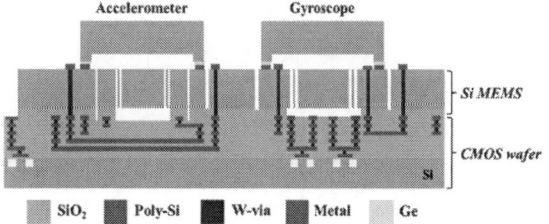

Fig.10: The MEMS above CMOS platform.

Fig.8: Environment sensing hubs for different applications.

(using the standard 0.18 μm 1P6M process) where our group develops the post-CMOS release processes to realize CMOS-MEMS environment sensors. The inherent materials in CMOS processes enable us to realize sensors with capacitive, piezoresistive, inductive, and thermoelectric sensing mechanisms. The CMOS process equipped with multiple-layer thin films offers the design flexibility for the integration of MEMS components and electrical routings. For example, in Fig.7, the mushroom, periodic-cavities, and MIM-array infrared absorbers are vertically integrated with thermoelectric sensing layers for performance enhancement [25-29]. Again, these sensors are fabricated using the same process platform, and hence various types of environment sensors are further integrated as the SoC environment sensing hub for different applications [19]. Fig.8 indicates typical environment sensing hubs with different sensors integration [30-33]. Our group also collaborated with design house to develop infrared sensors on 8-in CMOS wafer through their supply chain (Fig.9). The company further integrates the thermal image with the CCD image for COVID-19 detection and home security applications. The CMOS-MEMS platforms can also be exploited to develop tactile force sensors [34]

MEMS ABOVE CMOS

In addition to CMOS-MEMS, Mr. Steven Nasiri, the former CEO and co-founder of the InvenSense Inc., presented the architecture of MEMS above CMOS to provide another approach to monolithically integrate micro mechanical structures with sensing circuits. In this novel architecture (Fig.10), MEMS structures are defined on the device layer of the cavity SOI (silicon on insulator) and then wafer-level bonded with the CMOS substrate [35]. Thus, MEMS devices can vertically integrate with the ASIC on a single chip. Similar to the CMOS-MEMS

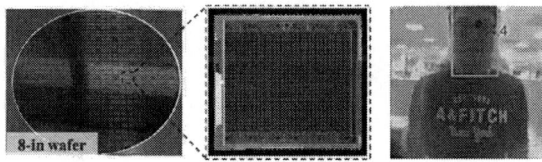

Fig.9: The CMOS-MEMS IR sensors on 8-in wafer.

technology, both of the footprint of chip and the bonding/assembly cost can be reduced. Moreover, since the MEMS structures can be fabricated using the Si device layer, the concerns about residual stresses, CTE mismatch, and reliability of the thin films for CMOS-MEMS technologies can be prevented. Mr. Nasiri further leveraged the resources of semiconductor ecosystems in Taiwan to prove the concept and commercialize motion sensors. The process line was also successfully upgraded from 6-in to 8-in wafer. As a result, the InvenSense contributed a significant amount of wafer requirements, which is an encouraging case for the engagement of MEMS and semiconductor industries.

Since 2010, Mr. Nasiri was invited to share the progress and status of InvenSense in the SEMICON Taiwan (an important international semiconductor annual event) for many years. In the meantime, Mr. Benedetto Vigna, who was the president of the Analog, MEMS, and Sensors of the ST Microelectronics (STM) then, was also invited to present the business achievement and technology roadmap about the MEMS sensors in his company. Mr. Vigna emphasized the advantages of two chip solution adopted by the STM to integrate the separate MEMS and ASIC chips. Thus, in SEMICON Taiwan, Steven and Benedetto had many inspiring debates regarding the advantages of "single chip solution" and "two chips solution". Through their discussions, we could appreciate the advantages of single chip solution such as to reduce the footprint of chip and the bonding/assembly cost, and to shorten electrical routings between sensors and circuits. However, the single chip solution also has some concerns such as the yield drop and the extra cost caused by miss-matching of chip size. Anyway, it's nice to have two options for chip integration to fulfill different applications.

The TSMC MEMS above CMOS platform can fabricate and integrate motion sensors (accelerometer and gyroscope) on a single chip to form the inertial sensing hub. In addition to that, these two motion sensors have to be sealed and operated in different pressure conditions. We collaborated with the TSMC MEMS team to develop wafer-level bonding technologies to enable various MEMS sensors respectively sealed in chambers of different pressure (ambient/vacuum) conditions [36-37], in Fig.11a. Thus, accelerometers were sealed in approximately 100mbar chamber pressure (for air damping), and gyroscopes were sealed in single-digit mbars vacuum conditions (for high quality factor). Moreover, TSMC also offered us the MEMS above CMOS platform for joint development research projects. Based on the design of various devices and test structures in Fig.11b, we could help to find potential problems or design concerns for this

978-1-6654-9309-3/23 $31.00 © 2023 IEEE

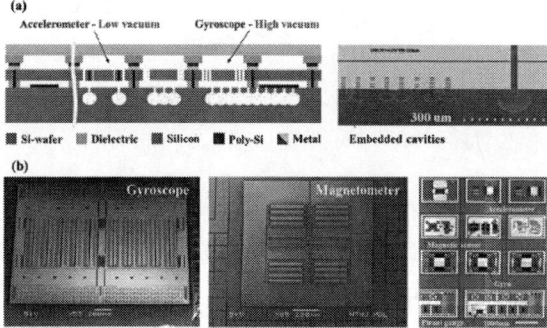

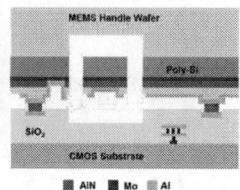

Fig.11: (a) Motion sensors sealed in different chamber pressure, (b) test devices and structures on the TSMC MEMS platform.

platform in the early stage to shorten the development time for customers. On the other hand, we could leverage the professional and advanced foundry processes to implement novel MEMS designs [38-39]. In fact, many useful and mature process modules (in foundries) to fabricate MEMS devices are not available in research institutions, for example, manufacturing the cavity SOI wafer, thinning the device layer, and preparing the TSVs (through silicon vias). The collaboration guided us towards a new research direction. Thus, we can develop more broad and challenging designs through existing semiconductor fabrication and packaging resources, and further bridging the gap between researches and commercial applications.

NEW MATERIALS NEW ERA

The MEMS pioneer, Dr. Kurt Petersen, has pointed out in Transducers 2017 and many events that micro actuators and the related functional materials are one of the important topics worthy of attention. In fact, the driving mechanisms and fabrication processes for micro actuators have been extensively investigated. However, as compared with the micro sensors, the MEMS actuators still have limited commercial applications. The micro actuators are frequently facing the challenges from operation and fabrication; for example, the electrostatic actuator has the concerns of pull-in effect and limited moving space, while the electromagnetic actuator has the concerns of magnets assembly and power consumption [40]. Moreover, it is still not straightforward to find the right products or killer applications for micro actuators. Nevertheless, the two well-known products equipped with MEMS key components, the TI DMD (digital mirror display) [41] and the HP inkjet printer head [42], are respectively enabled by the electrostatically driven micro mirror array and the thermal bubble actuator. These two products convinced us that micro actuators have potential for practical applications.

Piezoelectric transductions have a long development history and broad applications, such as quartz and *PZT* (Lead Zirconate Titanate) ceramic. However, their bulky size impedes the miniaturization and integration. Thanks to thin-film piezoelectric process through semiconductor-based wafer process, such as *ZnO*, *PZT*, and *AlN*, the micro actuators driven by the piezoelectric films exhibit the merits of low power consumption and large driving force in a more compact chip size. Thus, piezoelectric thin film

Fig.12: (a) VIS 8-in wafer AlN platform, (b) InvenSense 8-in wafer AlN+CMOS platform.

is a promising material to realize MEMS actuators. Recently, as predicted by Dr. Petersen, piezoelectric films and their applications on micro actuators are attracting significant attentions, for both academia and industry. The fabrication technologies for thin piezoelectric films have been established in foundries and equipment suppliers. In particular, *AlN* and *PZT* piezoelectric films are gradually adopted in renowned research institutes and leading companies for potential commercialization in a variety of applications [43]. For example, Leti/STM's piezoelectric technology possesses both *PZT* and *AlN* platforms with potential applications on ink jet heads, auto focus, ultrasonic devices, etc. The MEMSCAP Inc. offers MPW *AlN* on SOI piezoelectric process platform (PiezoMUMPs) while SilTerra provides *AlN* on CMOS process to enable piezoelectric MEMS and circuit integration. The TSMC also offers *PZT* and *AlN* process platforms to their customers targeted on acoustic devices.

- ● *AlN piezoelectric film*

AlN is suitable for high frequency operation thanks to its low dielectric constant and dielectric loss even though its piezoelectric coefficients (d_{33} and d_{31}) are relatively low among its thin-film *ZnO* and *PZT* counterparts. As a result, bulk acoustic wave (BAW) devices, such as FBAR (thin-film bulk acoustic resonator) and SMR (solidly mounted resonator), are dominant in RF front-end modules for wireless communications. Therefore, our group is currently working with Vanguard International Semiconductor Corp. (VIS) to develop high performance *AlN* based resonators and filters for 5G communications [44-45] (Fig.12a). In the past few years, our group also collaborated with

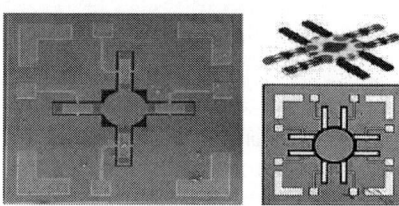

Fig.13: (a) Piezoelectric support transducer based resonator and (b) gyroscope realized by MEMSCAP PiezoMUMPs.

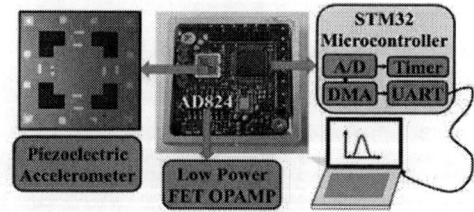

Fig.14: In-house AlN based accelerometer as a vibration sensing module for smart manufacturing.

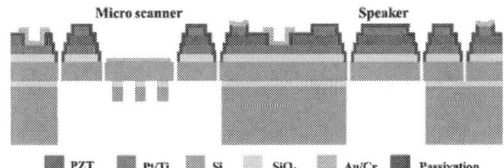

Fig.15: The foundry available piezoelectric actuator platform.

InvenSense to explore SHF (super high frequency) composite FBAR resonators implemented by the company's *AlN* MEMS-CMOS platform [46] in Fig.12b. In recent years, we further uses MEMSCAP Inc. PiezoMUMPs to deliver high-*Q* support transducer based resonators [47], low phase noise oscillators [48], channel-select filters [49], and sensors [50-51] in Fig.13. In particular, we collaborated with Delta Electronics to develop an in-house *AlN* SOI process platform to fabricate wide-bandwidth accelerometers enabling a vibration sensing module for applications on smart manufacturing [50] in Fig.14. As a result, thin-film *AlN* is expected to extensively play an important role in piezoelectric MEMS for a variety of commercial applications.

● *PZT piezoelectric film*

In general, the *PZT* film has a relatively large piezoelectric coefficient (comparing with other piezoelectric films like *AlN*, *ZnO*, etc.) which could provide higher driving force for micro actuators. Although the commercial application of *PZT* film is not as mature as that of *AlN*, many equipment suppliers and foundries are dedicated in the development of related technologies recently. In recent years, we leverage semiconductor ecosystems to access the wafer-level *PZT* processes to develop micro speakers and scanners and to explore their applications. Fig.15 shows the typical architecture of actuators consisted of *PZT* film, mechanical structure layer, and top/bottom driving electrodes. Moreover, Fig.16 shows typical fabrication results on the 6-inch wafer. The MPW wafer contains micro scanners and speakers of different designs.

In Fig.17, the micro scanners of different designs have been implemented [52-53]. The figure of merits for the design of micro scanners are the scanning frequency, the scanning angle, and the mirror size, under a given chip size. In this regard, many mechanical structures such as transmission springs, actuators, torsional springs, etc. are designed to improve the performances of scanners [54-57]. Here shows the results from our collaborations with a systems company (Coretronic Corp.) and a design house (Coretronic MEMS Corp.). Based on the specifications requested by the system company, we develop the micro scanners with design house for LiDAR and HUD (head-up display) applications. After that, the system company will

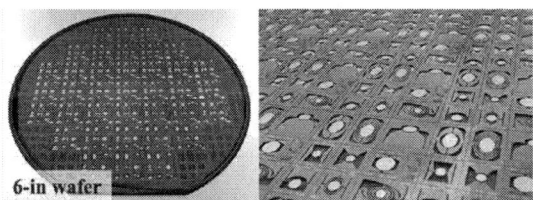

Fig.16: The PZT scanners fabricated on the 6-in MPW wafer.

Fig.17: The PZT scanners of different designs for testing.

realize the applications using micro scanners, as the LiDAR demonstrated in Fig.18. We appreciate industry partners to offer us the *PZT* platform through their suppliers. Moreover, we also collaborate with the MEMS and CMOS foundries to evaluate their *PZT* platforms. In this regard, test keys are developed to characterize the mechanical and piezoelectric properties of thin films. Reliability issues of *PZT* actuators are also monitored at various conditions.

As a second example, micro speakers of different designs shown in Fig.19a have also been developed [58-61]. The micro speaker is a promising device for in-ear applications to enhance the audio experience [62]. The *PZT* film has high electro-mechanical conversion efficiency and large driving force, and hence could offer the advantages of low power consumption and high sound pressure level (SPL) for micro speakers. In general, the major concerns for micro speakers are the flat response with sufficient SPL and low THD (total harmonic distortion). Thus, micro speakers with different passive mechanical structures (springs and diaphragms), actuators, and driving electrodes are designed to improve their performances [63-65]. Part of the designs in Fig.19b are the joint development results with *T* design house, and we also support *x* design house with testing facilities and technologies for micro speakers. Again, these examples show various connections (Fig.2) to leverage semiconductor ecosystem to MEMS development.

CONCLUSION AND FUTURE OUTLOOK

After the development of near six decades, the semiconductor industries form a complete eco-system and dominate a tremendous amount of resources. It is cost-effective to leverage the resources in mature semiconductor ecosystems for the researches, technology developments, and commercialization of MEMS devices. This article shares the experience of the diversified interaction and collaboration between the NTHU MEMS group and the semiconductor industries in Taiwan. We could contribute technical information and solutions and also serve as the think-tank to some of the companies. On the other hand, we also receive many technical supports from these companies, for example accessing advanced process platforms,

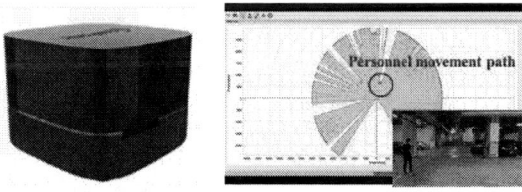

Fig.18: The LiDAR system implemented by the industry partner.

978-1-6654-9309-3/23 $31.00 © 2023 IEEE 147

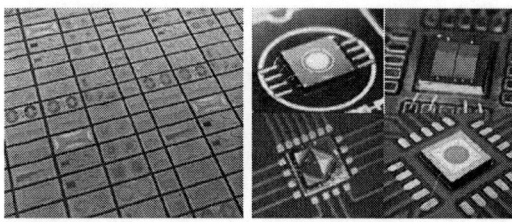

Fig.19: The PZT speakers fabricated on the 6-in MPW wafer, and the packaged samples for acoustic testing.

realizing packaging and system integration, etc. To bring more impacts and contributions to the MEMS society in Taiwan, we established the Micro Sensors and Actuators Technology Consortium (μSAT) in 2013. So far we have near 40 industry members distributed from system, design houses, foundries, and packaging houses, including the tier-one companies such as the TSMC and ASE. The μSAT has a nick name "the MEMS seven-eleven" since our service, like the convenience store, is anytime, anywhere, for anyone in MEMS area. In case you need any help regarding the MEMS academia or industry in Taiwan, please feel free to contact me at the "MEMS 7-11".

ACKNOWLEDGEMENTS

This work was supported in part by the National Science Tech. Council of Taiwan under the Grant of NSTC 111-2221-E-007-069-MY3, NSTC 111-2221-E-007-070-MY3. The author appreciates the support from industry collaborators: TSMC, apm, VIS, UMC, ASE, CoretronicMEMS, Pixart, Sensirion Taiwan, TDK-Invensense, Transducers Star, xMEMS, Delta, Coretronic; and the National Taiwan Semiconductor Research Inst. (TSRI).

REFERENCES

[1] https://irds.ieee.org/
[2] G. K. Fedder, *IEEE Sensors*, 2005.
[3] W. Fang, S.-S. Li, C.-L. Cheng, C.-I. Chang, W.-C. Chen, Y.-C. Liu, M.-H. Tsai, and C. Sun, *Transducers*, 2013.
[4] C.-M. Sun, C.-W. Wang, and W. Fang, *Sens. Act. A Phys.*, 2008.
[5] C.-I. Chang, M.-H. Tsai, Y.-C. Liu, C.-M. Sun, and W. Fang, *Transducers, 2011*.
[6] Y.-C. Liu, M.-H. Tsai, T.-L. Tang, and W. Fang, *J. Micromech. Microeng.*, 2011.
[7] C.-M. Sun, C. Wang, M.-H. Tsai, H.-S. Hsieh, and W. Fang, *J. Micromech. Microeng.*, 2009.
[8] C.-M. Sun, M.-H. Tsai, Y.-C. Liu, and W. Fang, *IEEE Trans Electron Devices*, 2010.
[9] C.-I. Chang, M.-H. Tsai, C.-M. Sun, and W. Fang, *J. Micromech. Microeng.*, 2014.
[10] M.-H. Tsai, Y.-C. Liu, C.-M. Sun, C. Wang, C.-W. Cheng, and W. Fang, *Transducers*, 2011.
[11] C.-M. Sun, M. -H. Tsai, C. Wang, Y.-C. Liu, and W. Fang, *Transducers*, 2009.
[12] C.-L. Cheng, M.-H. Tsai, W. Fang, *J. Micromech. Microeng*, 2015.
[13] H. Lakdawala and G. K. Fedder, Transducers, 1999.
[14] W. Fang, S.-S. Li, Y. Chiu, and M.-H. Li, book chapter in the "*3D and Circuit Integration of MEMS,*" 1st ed. Weinheim, Germany: Wiley, 2020.
[15] R. Modlinski, A. Witvrouw, P. Ratchev, R. Puers, J.M.J. den Toonder, and I. De Wolf, *Microelectron Eng*, 2004.
[16] A. Hierlemann, O. Brand, C. Hagleitner, and H. Baltes, *Proceedings of the IEEE*, 2003.
[17] https://www.sensirion.com
[18] D. Rüffer, F. Hoehne, and J. Bühler, *Sensors*, 2018.
[19] Y.-C. Lee, M.-L. Hsieh, P.-S. Lin, C.-H. Yang, S.-K. Yeh, T. T. Do and W. Fang, *J. Micromech. Microeng.*, 2021.
[20] W.-C. Lin, C.-L. Cheng, C.-L. Wu, W. Fang, *Transducers, 2017*.
[21] T.-L. Chien, Y.-C. Lee, C.-T. Chou, J.-Y. Lin, H.-Y. Chen, and W. Fang, *IEEE MEMS*, 2022.
[22] Y.-C. Lee, T.-L. Chien, C.-T. Fang, Y.-Y. Huang, W.-L. Sung, Y.-C. Chu, R. Chen, and W. Fang, *IEEE MEMS*, 2023.
[23] C.-H. Yang, C.-C. Chang, Y.-C. Lee, Y.-L. Chen, and W. Fang, *Transducers*, 2021.
[24] P.-S. Lin, Y. Wang, M.-C. Cheng, Y.-C. Chen, Y.-C. Huang, and W. Fang, *IEEE MEMS*, 2020.
[25] K.-C. Chang, Y.-C. Lee, C.-M. Sun, W. Fang, *IEEE MEMS*, 2017.
[26] T.-W. Shen, K.-C. Chang, C.-M. Sun, and W. Fang, *J. Micromech. Microeng.*, 2019.
[27] Y.-C. Huang, P.-S. Lin, Y.-L. Chen, C.-F. Hu, and W. Fang, *IEEE MEMS*, 2021.
[28] P.-S. Lin, T.-W. Shen, K.-C. Chan, W. Fang, *IEEE Sens. J.*, 2020.
[29] Y.-C. Li, T. Chou, P.-S. Lin, Y.-C. Huang, F. Shih, Y.-A. Lin, T.-J. Yen, M.-F. Lai, and W. Fang, *IEEE MEMS*, 2023.
[30] C.-C. Chang, P.-H. Hong, S.-K. Yeh, Y.-C. Lin, M.-F. Lai, and W. Fang, *IEEE MEMS*, 2020.
[31] Y.-C. Lin, P.-H. Hong, S.-K. Yeh, C.-C. Chang, and W. Fang, *IEEE MEMS*, 2020.
[32] Y.-C. Lin, Y.-C. Lee, C.-H. Yang, W. Fang, *IEEE Sensors*, 2021.
[33] T.-L. Chien, Y. Huang, F. Shih, and W. Fang, *IEEE MEMS*, 2023.
[34] S.-K. Yeh, M.-L. Hsieh, and W. Fang, *IEEE Sensors J.*, 2021.
[35] S. Nasiri, S. Winkler, and R. Ramadoss, book chapter in the "*MEMS Packaging,*" 1st Ed., New Jersey, the World Scientific Publishing Co., 2018.
[36] C.-W. Cheng, K.-C. Liang, C.-H. Chu, and W. Fang, *J. Micromech. Microeng.*, 2017.
[37] S.-W. Cheng, J.-C. Weng, K.-C. Liang, Y.-C. Sun, and W. Fang, *J. Micromech. Microeng.*, 2018.
[38] F.-Y. Lee, K.-C. Liang, E. Cheng, S.-S. Li, and W. Fang, *IEEE MEMS*, 2016.
[39] W. L. Sung, F. Y. Lee, C. L. Cheng, C. I. Chang, E. Cheng, and W. Fang, *IEEE MEMS*, 2016.
[40] S. T. S Holmström, U. Baran, and H. Urey, *J Microelectromech Syst.*, 2014.
[41] https://www.ti.com/dlp-chip/overview.html
[42] N. J. Nielsen et al., *Hewlett-Packard Journal*, 1985
[43] G. Pillai and S.-S. Li, *IEEE Sens. J.*, 2021.
[44] C.-Y. Chang, Y.-M. Huang, T.-H. Hsu, Y.-H. Chen, R. Chand, Y. R. Pradeep, Y. Ho, M.-H. Li, W. Fang, S.-S. Li, *EFTF/IFCS*, 2022.
[45] Y.-M. Huang, C.-Y. Chang, T.-H. Hsu, Y. Ho, Y.-H. Chen, Y. Pradeep, R. Chand, S.-S. Li, W. Fang, and M.-H. Li, *IEEE MTT-S Int. Microw. Symp. (IMS'22)*, 2022.
[46] G. Pillai, A. A. Zope, J. M.-L. Tsai, S.-S. Li, *IEEE T-UFFC*, 2017.
[47] G. Pillai, A. A. Zope, and S.-S. Li, *IEEE/ASME J. Microelectromech. Syst. (JMEMS)*, 2019.
[48] H.-T. Jen, G. Pillai, S.-I. Liu, and S.-S. Li, *IEEE T-UFFC*, 2021.
[49] G. Pillai, C.-Y. Chen, and S.-S. Li, *IEEE IFCS-EFTF*, 2019.
[50] J. Satija, P.-W. Huang, S. Singh, T. Shen, H.-Y. Chen, and S.-S. Li, *IEEE MEMS*, 2022.
[51] N. M. Nguyen, C.-Y. Chang, G. Pillai, S.-S. Li, *IEEE MEMS*, 2021.
[52] H. Cheng, S. Liu, C. Hsu, H. Lin, F. Shih, M. Wu, K. Liang, M. Lai, and W. Fang, *Sens. Actuator A Phys.*, 2022 (in press).
[53] S. Chen, S. Liu, H. Cheng, H. Lin, K. Liang, M. Wu, and W. Fang, *IEEE MEMS*, 2023.
[54] K. Meinel, C. Stoeckel, M. Melzer, S. Zimmermann, R. Forke, K. Hiller, and T. Otto, *Transducers*, 2019.
[55] U. Baran, D. Brown; S. Holmstrom, D. Balma, W. O. Davis, P. Muralt, and H. Urey, *J. Micromech. Microeng.*, 2012.
[56] S. Gu-Stoppel, T. Giese, H. Quenzer, U. Hofmann, and W. Benecke, *MikroSystemTechnik; Congress*, 2017.
[57] A. Vergara, T. Ts., W. Fang, and S. Tanaka, *J. Micromech. Microeng.*, 2020.
[58] H. H. Cheng, S. Lo, Y. Wang, Y. Chen, W. Lai, M. Hsieh, M. Wu, and W. Fang, *IEEE MEMS*, 2019.
[59] Y. J. Wang, S. C. Lo, M. L. Hsieh, S. D. Wang, Y. C. Chen, M. Wu, and W. Fang, *IEEE MEMS*, 2021.
[60] Y. T. Lin, S. C. Lo, and W. Fang, *Transducers*, 2021.
[61] H. Cheng, S. Lo, Z. Huang, Y. Wang, M. Wu, and W. Fang, *Sens. Act. A Phys.*, 2020.
[62] https://xmems.com/products/
[63] H. Cheng, S. Lo, Y. Chen, M. Cheng, T. Wei, M. Wu, and W. Fang, *IEEE MEMS*, 2023.
[64] S. Chang, T. Wei, S. Lo, and W. Fang, *IEEE MEMS*, 2023.
[65] T. Wei, Z. Hu, S. Chang, and W. Fang, *IEEE MEMS*, 2023.

CONTACT

*W. Fang, tel: +886-3-5742923; fang@pme.nthu.edu.tw

PROGRAMMABLE SILICON NITRIDE PHOTONIC INTEGRATED CIRCUITS

Hao Tian[1], Alaina G. Attanasio[1], Anat Siddharth[2], Andrey Voloshin[2], Viacheslav Snigirev[2], Grigory Lihachev[2], Andrea Bancora[2], Vladimir Shadymov[2], Rui N. Wang[2], Johann Riemensberger[2], Tobias J. Kippenberg[2], and Sunil A. Bhave[1],*

[1] OxideMEMS Lab, Purdue University, West Lafayette, IN, USA
[2] Laboratory of Photonics and Quantum Measurements, Swiss Federal Institute of Technology Lausanne (EPFL), Lausanne, Switzerland

ABSTRACT

Silicon Nitride (Si_3N_4) photonic integrated circuits (PICs) have emerged as core technology in a variety of applications ranging from LIDAR to quantum control and computing. However, the need for high-speed, low-voltage tuning and modulation has been a long standing challenge because Si_3N_4 lacks electro-optic effect. In this work, we demonstrate power efficient (nW) and fast (sub-µs) piezoelectric tuning of Si_3N_4 optical microring resonators by monolithically integrating Lead Zirconate Titanate (PZT) piezoMEMS actuators. We achieve, for the first time, the co-integration of PZT and thermal actuators, enabling the tuning of optical resonances over one free spectral range (FSR). The programmability is demonstrated by aligning two optical resonators fabricated on separate chips, which is the first step towards fiber-optic connected optomechanical sensors and computing networks. We further show acousto-optic modulation (AOM) of Si_3N_4 photonics using the High-overtone Bulk Acoustic Resonators (HBAR) excited by PZT actuators with modulation frequency up to 2 GHz.

KEYWORDS

Piezoelectric actuator; Si_3N_4 microring resonator; Stress-optic effect; Acousto-optic modulator.

INTRODUCTION

Integrated photonics has attracted much attention recently, because of its ability to miniaturize the optical technologies developed in research labs into commercialized products from atomic clocks [1], integrated semiconductor lasers [2], light detection and ranging (LiDAR) [3], to optical gyroscopes [4]. In the past decade Si_3N_4 photonic circuits have emerged as one key platform, due to its ultra-low optical loss from visible to mid-infrared, and large Kerr nonlinearity $\chi^{(3)}$ [5]. It has been successfully applied in the generation of dissipative Kerr soliton microcombs [6], supercontinuum generation [7], as well as photonic quantum computing circuits [8]. However, the lack of an electro-optic property has presented a major challenge to actively tune Si_3N_4 photonics, a feature that is highly desirable for feedback control, compensating fabrication errors, and programmable PICs for optical sensing and computing networks.

Conventionally, thermal-optical tuning with integrated micro-heater is employed, at the expense of speed (~ms) and power consumption (~mW) [9]. Although hybrid integration with electro-optical materials, such as lead zirconate titanate (PZT) [10], lithium niobate ($LiNbO_3$)

[11], and barium titanate ($BaTiO_3$) [12], has provided an alternative way, further improvements in fabrication technology are necessary to increase the optical Q and engineer the optical dispersion for frequency comb generation.

Most recently, many efforts have been devoted to the development of stress-optic tuning of Si_3N_4 by monolithically integrating piezoelectric actuators on top of the Si_3N_4 photonics. Upon applying voltage to the actuator, it will deform due to the piezoelectric effect, which generates stress around the Si_3N_4 optical waveguide and changes its refractive index via the stress-optic effect. The change of refractive index tunes the optical response of the photonic devices. As a well-known piezoelectric material, Aluminum Nitride (AlN) actuators have been applied in tuning Si_3N_4 photonics [13]. Many promising applications have been successfully demonstrated from tunable optical frequency combs [14], semiconductor lasers [15], entangled photon pairs [16], to photonic computing networks [17]. However, the achievable tuning efficiency of an optical resonance is limited to around 20MHz/V, due to the relatively small piezoelectric coefficient of AlN. This inevitably requires a large applied voltage on the order of 100 V. While CMOS circuits can achieve such high voltages on-chip, the charge-discharge pump circuits are slow, which limits the system tuning speed.

As the traditional piezoelectric ceramic, PZT has regained much attention in tuning photonic circuits due to the maturity of depositing high quality PZT thin films on photonic chips. Many pioneering works have been done in PZT tuning of photonic circuits in both released [18] and unreleased structures [19], [20]. In this work, by integrating a disk PZT actuator on an unreleased photonic chip, we demonstrate the most efficient tuning of a Si_3N_4 microring resonator to date. Due to its much larger piezoelectric coefficient, 578 MHz/V tuning is achieved, more than one order of magnitude bigger than the AlN actuator. In addition to the PZT actuator, we also co-integrate a micro-heater on the same optical resonator, which enables a much larger tuning range. Beyond the quasi-DC tuning, we realize microwave frequency AOM by exciting HBAR modes in the substrate. This combination of PZT actuator, heater, and AOM provides a comprehensive toolset to program the PICs with more degrees of freedom, making PZT a universal platform.

DEVICE DESIGN

Figure. 1(a) shows the optical image of the fabricated disk shaped PZT actuator on top of a Si_3N_4 microring resonator, which consists of a microring and a bus

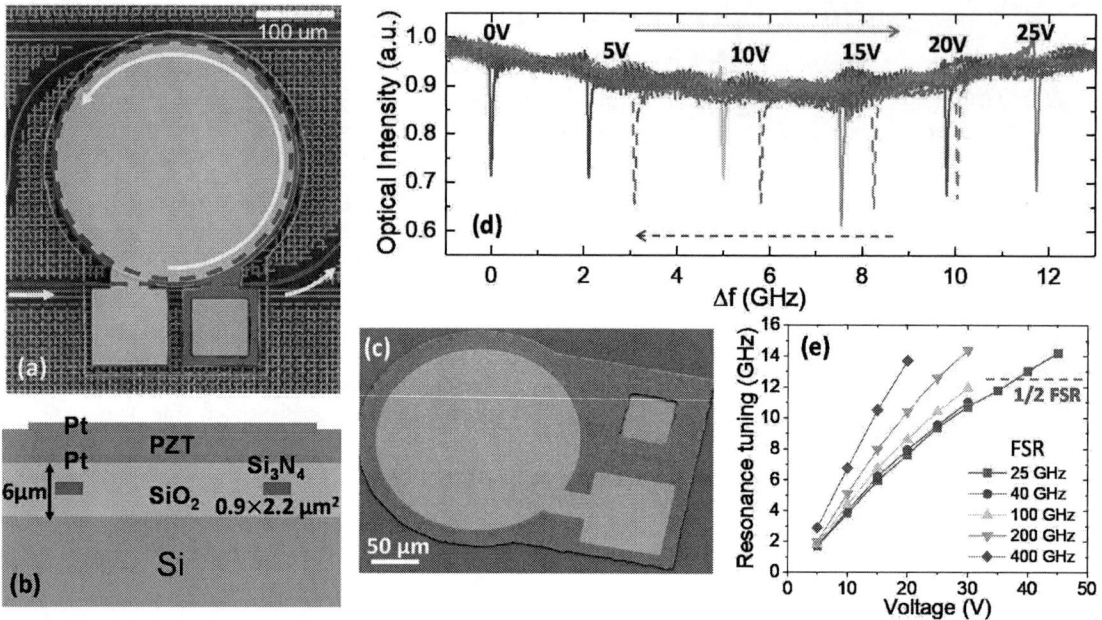

Figure 1: (a) Optical image and (b) cross-section of the PZT actuator and optical microring resonator (red lines). (c) False colored SEM of the PZT actuator. (d) Optical spectrum showing tuning of one optical resonance with increasing (solid lines) and decreasing (dashed lines) applied voltage. (d) Resonance tuning vs. Voltage for different FSR microrings.

waveguide (red lines). Light is injected and guided through the bus waveguide, which is then coupled into the microring via an evanescent field and trapped in the optical resonator at optical resonances. The resonator supports a series of equidistant optical resonances, with the spacing between them defined as the FSR, which is inversely proportional to the radius R. Devices with different radii were studied systematically. The cross-section is illustrated in Fig. 1(b). 1 μm PZT film is sandwiched between top and bottom electrodes made from 100 nm Platinum (Pt). The Si_3N_4 waveguide (0.9×2.2 μm^2) is embedded inside a 6 μm SiO_2 cladding and is 3 μm away from the bottom electrode for preserving the low loss of the optical waveguide. The whole structure sits on an unreleased high resistivity Si substrate. Optimum tuning is achieved when the waveguide is 8 μm inside the top metal. The Scanning Electron Microscopy (SEM) image of the PZT actuator is shown in Fig. 1(c), where the bottom electrode can be accessed through a via in the PZT layer.

OPTICAL RESONANCE TUNING

PZT is a ferroelectric material deposited using sol-gel technology. We apply 25 V to the actuator for at least one second to align the ferroelectric domains before device operation. We found the poling increased the tuning efficiency by a factor of two. Figure 1(d) shows the piezoelectric tuning of an optical resonance of a microring with R=115 μm (FSR=200 GHz) under different voltages. The resonance frequency is red-shifted monotonically with increasing voltages and we determine a tuning of up to 10 GHz with a voltage of less than 20V, which corresponds to a modulation efficiency of 500 MHz/V. Due to the ferroelectricity of PZT, hysteresis is observed as the voltage is gradually decreased (dashed curves). This feature can serve as an optical-readout memory element in the future. To maintain the direction of the polarization of

the film, only positive voltage is applied. The application of negative voltage will reverse the polarization which tends to align with the electric field. Hence unlike AlN [13], we find that the tuning under positive and negative voltages has the same tuning direction. Thus, in real applications, an offset voltage should be applied if bi-directional tuning is required. Moreover, a high optical Q of 3 Million is maintained, which is important for tunable frequency comb and frequency-agile laser. The leakage current is kept below 1 nA at 20 V due to the high quality of PZT film. This enables ultra-low static power consumption on the order of tens of nano-Watts. The power efficiency, defined as the power consumption per tuning frequency, is 2 pW/MHz.

The tuning of microring resonators with different radii (and FSRs) is studied in Fig. 1(e). Small radius rings have larger FSR. It can be found that the tuning efficiency increases as the device size decreases (FSR increases), since the relative change of radius $\Delta r/r$ increases as the ring deforms which contributes to the relative change of resonance frequency $\Delta f/f_0$, besides the stress-optic effect. All the devices achieve tuning bigger than 10 GHz when the voltage is larger than 30 V. However, as the voltage further increases, the tuning tends to saturate due to the saturation of the polarization of PZT at high electric fields. This leads to a nonlinear response for large signals, which can be compensated with a pre-conditioned input signal.

Interestingly, for the 25 GHz FSR device, tuning over half FSR is achieved under 40 V, corresponding to a π phase shift of light. This can be understood that, with 2π phase shift, light will shift from one resonance to the next harmonic, which is one FSR. We calculate the figure of merit (modulation length product) $V_\pi \cdot L = 22$ V·cm, where V_π is the voltage that produces π phase shift of the light. Although this value is an order of magnitude larger than that for the state of the art electro-optic tuning using

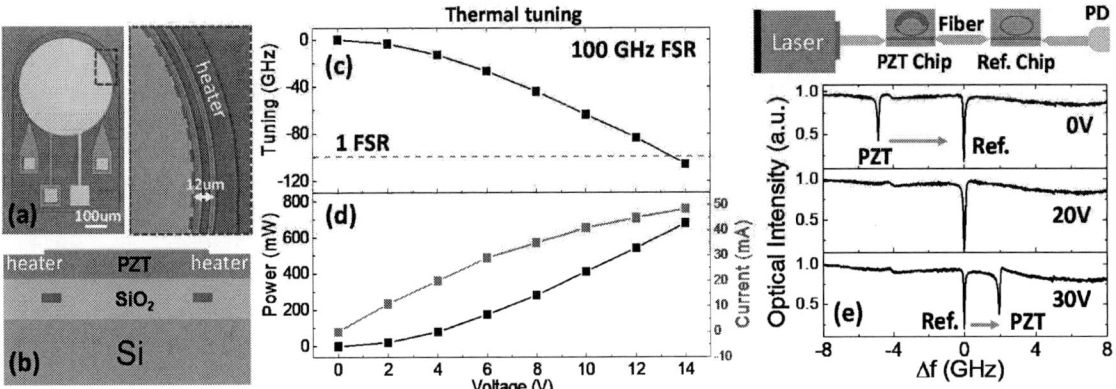

Figure 2: (a) Optical image and (b) cross-section of PZT actuator co-integrated with heater. (c) Resonance tuning via heater under different voltages, and correspondingly (d) power consumption (left Y-axis) and current (right Y-axis). (e) Demonstration of aligning one resonance to a reference (Ref.) optical resonator on a separate chip using solely PZT actuator. The resonances become aligned under 20 V with initial misalignment of 5 GHz. PD: Photodetector.

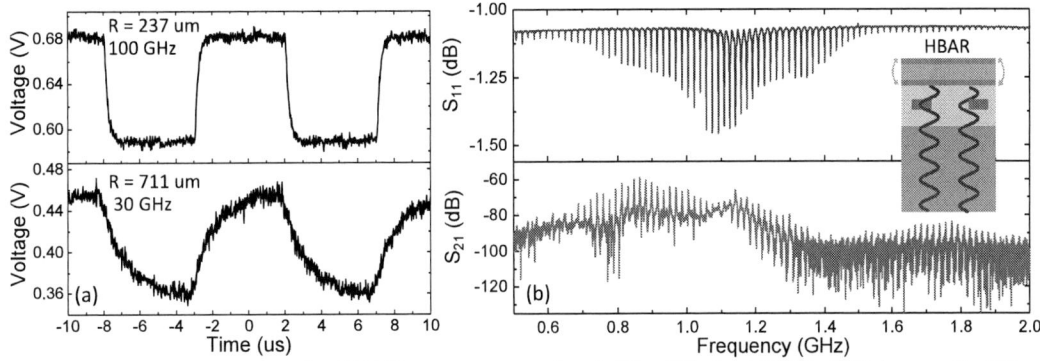

Figure 3: (a) Dynamic response under 100kHz square wave signals(V_{PP}=1V) for 100GHz FSR (upper, R=237μm) and 30GHz FSR (lower, R=711μm) devices. The laser is biased at the slope of the optical resonance, and Y-axis is the voltage output from the photodetector. (b)Electromechanical S_{11}(upper) and Optomechanical (S_{21}) responses, showing periodic (spacing of 17.6MHz) HBAR modes and efficient AOM. Inset: schematic of HBAR mode formed in the substrate.

LiNbO₃ [21], the modulation loss product $V_\pi \cdot L \cdot \alpha = 1.87$ V·dB is comparable to 1 V·dB of LiNbO₃, thanks to the ultra-low loss of the Si₃N₄ waveguide (0.085 dB/cm).

CO-INTEGRATION WITH HEATER

We further demonstrate the compatibility of piezo-electric and thermal tuning by incorporating micro-heaters into the bottom metal layer at no cost for extra masks (Fig. 2(a), (b)). However, to have efficient thermal tuning, the optical ring has to be brought close to the micro-heater such that the ring is congruent with the top metal edge. This reduces the overlap of the optical mode and stress field and lowers the stress-optic tuning by half. Thermal tuning of a device with 100 GHz FSR (R=237 μm) is shown in Fig. 2(c) where over one full FSR is achieved under 14 V, at the expense of 700 mW (Fig. 2(d)). This shows the ability of coarse tuning (long range) via heater along with fine tuning (fast speed) with piezoelectric actuators (see Fig. 2(e)).

To demonstrate programmability, the resonances of optical resonators fabricated on two separate chips are aligned through the PZT actuator (Fig. 2(e)). Initially at 0V, the spacing between them is 5 GHz. By applying 20V at the PZT actuator, there is only one resonance in the transmission spectrum showing good alignment. As we further increase the voltage, the resonance of the PZT chip

crosses over the reference chip. This demonstrates the ability to compensate resonance misalignment induced by the fabrication variation, which is required to build scalable PIC multi-chip modules (MCMs).

ACOUSTO-OPTIC MODULATION

The tuning speed is characterized by applying a 100 kHz square wave voltage to the PZT, while measuring the transmitted light intensity by biasing the laser at the slope of the optical resonance, as shown in Fig. 3(a). For a small device with R=237 μm, the PZT actuator achieves switching between two states, with the response time limited to 0.2 μs by the capacitive time constant due to the large permittivity of PZT. It becomes more obvious for a large device with R=711 μm, where it takes 1.4 μs to respond.

Beyond this quasi-DC tuning, HBAR modes can be excited at microwave frequencies through the PZT actuator, where bulk acoustic waves are confined in the acoustic Fabry-Pérot cavity formed by the top and bottom surfaces of the photonic chip. A series of mechanical resonances is formed up to 2 GHz, which are distributed evenly with spacing of 17.6 MHz (see Fig. 3(b)). Efficient acousto-optic modulation is achieved between 0.8-1 GHz. This high frequency AOM will add more functionalities to

Table 1: Comparison with state of the art piezoelectric tuning of Si_3N_4 microring resonators.

Ref.	Material	Structure	FSR (GHz)	Radius (μm)	Optical Q	Tuning (MHz/V)	Speed (μs)	Power (pW/MHz)	$V_\pi \cdot L$ (V·cm)
[13]	AlN	Unreleased	200	118	6.5M	25	0.05	20	124
[18]	PZT	Released	52	580	0.086M	3250	3.7	3	3.6
[19]	PZT	Unreleased	48	625	3.6M	200	NA	5	43
PZT w/o heater	PZT	Unreleased	400	57	3.56M	578	0.1	1.7	18
PZT w/ heater	PZT	Unreleased	100	237	2.2M	243	0.2	4	30

the platform, such as the generation of Pound-Drever-Hall error signal for locking the optical resonance to an external laser source [19].

In conclusion, we demonstrate a programmable Si_3N_4 photonic circuit that is capable of low power, fast tuning through PZT piezoMEMS actuators, and large range tuning via micro-heaters. From the comparison in Table 1, it can be seen our PZT actuator possesses both large tuning efficiency, sub-μs speed, and the highest power efficiency. Therefore, our PZT-on-SiN Photonics platform lays the foundation for Si_3N_4 PICs for sensing and communication networks.

ACKNOWLEDGEMENTS

The authors would like to thank DARPA MTO under contract No. W911NF2120248 (NINJA LASER); European Union's H2020 research and innovation program; FET Proactive Grant No. 732894 (HOT); NSF QISE-NET DMR 17-47426; SNSF Ambizione Fellowship (201923);ESA Contract No. 4000135357/21/NL/GLC/my. Photonic chips were fabricated in the EPFL Center of MicroNano-Technology (CMi), and PZT actuators were fabricated by Radiant Technologies Inc.

REFERENCES

[1] Z. L. Newman et al., "Architecture for the photonic integration of an optical atomic clock," Optica, vol. 6, no. 5, pp. 680–685, 2019.

[2] W. Jin et al., "Hertz-linewidth semiconductor lasers using CMOS-ready ultra-high-Q microresonators," Nat. Photonics, pp. 1–8, 2021.

[3] P. Trocha et al., "Ultrafast optical ranging using microresonator soliton frequency combs," Science (80-.)., vol. 359, no. 6378, pp. 887–891, 2018.

[4] S. Gundavarapu et al., "Interferometric Optical Gyroscope Based on an Integrated Si_3N_4 Low-Loss Waveguide Coil," J. Light. Technol., vol. 36, no. 4, pp. 1185–1191, 2018.

[5] D. J. Moss, R. Morandotti, A. L. Gaeta, and Lipson Michal, "New CMOS-compatible platforms based on silicon nitride and Hydex for nonlinear optics," Nat. Photonics, vol. 7, no. 8, pp. 597–607, 2013.

[6] T. J. Kippenberg, R. Holzwarth, and S. A. Diddams, "Microresonator-based optical frequency combs.," Science, vol. 332, no. 6029, pp. 555–9, 2011.

[7] A. L. Gaeta, M. Lipson, and T. J. Kippenberg, "Photonic-chip-based frequency combs," Nat. Photonics, vol. 13, no. 3, pp. 158–169, 2019.

[8] J. M. Arrazola et al., "Quantum circuits with many photons on a programmable nanophotonic chip," Nature, vol. 591, no. 7848, pp. 54–60, 2021.

[9] X. Xue et al., "Thermal tuning of Kerr frequency combs in silicon nitride microring resonators," Opt. Express, vol. 24, no. 1, p. 687, 2016.

[10] K. Alexander et al., "Nanophotonic Pockels modulators on a silicon nitride platform," Nat. Commun., vol. 9, no. 1, p. 3444, 2018.

[11] A. N. R. Ahmed, S. Shi, M. Zablocki, P. Yao, and D. W. Prather, "Tunable hybrid silicon nitride and thin-film lithium niobate electro-optic microresonator," Opt. Lett., vol. 44, no. 3, pp. 618–621, 2019.

[12] S. Abel et al., "Large Pockels effect in micro- and nanostructured barium titanate integrated on silicon," Nat. Mater., vol. 18, no. 1, pp. 42–47, 2019.

[13] H. Tian et al., "Hybrid integrated photonics using bulk acoustic resonators," Nat. Commun., vol. 11, 3073, 2020.

[14] J. Liu et al., "Monolithic piezoelectric control of soliton microcombs," Nature, vol. 583, no. 7816, pp. 385–390, 2020.

[15] G. Lihachev et al., "Low-noise frequency-agile photonic integrated lasers for coherent ranging," Nat. Commun., vol. 13, 3522, 2022.

[16] T. Brydges et al., "An Integrated Photon-Pair Source with Monolithic Piezoelectric Frequency Tunability," arXiv Prepr. arXiv2210.16387, pp. 1–9, 2022.

[17] M. Dong et al., "High-speed programmable photonic circuits in a cryogenically compatible, visible–near-infrared 200 mm CMOS architecture," Nat. Photonics, vol. 16, no. 1, pp. 59–65, 2022.

[18] W. Jin, R. G. Polcawich, P. A. Morton, and J. E. Bowers, "Piezoelectrically tuned silicon nitride ring resonator," Opt. Express, vol. 26, no. 3, pp. 3174–3187, 2018.

[19] J. Wang, K. Liu, M. W. Harrington, R. Q. Rudy, and D. J. Blumenthal, "Silicon nitride stress-optic microresonator modulator for optical control applications," Opt. Express, vol. 30, no. 18, p. 31816, 2022.

[20] N. Hosseini et al., "Stress-optic modulator in TriPleX platform using a piezoelectric lead zirconate titanate (PZT) thin film," Opt. Express, vol. 23, no. 11, pp. 14018–14026, 2015.

[21] C. Wang et al., "Integrated lithium niobate electro-optic modulators operating at CMOS-compatible voltages," Nature, vol. 562, no. 7725, pp. 101–104, 2018.

CONTACT

*Sunil A. Bhave, mobile: +1-510-390-3269; bhave@purdue.edu

MULTIFREQUENCY NANOMECHANICAL MASS SPECTROMETER PROTOTYPE FOR MEASURING VIRAL PARTICLES USING OPTOMECHANICAL DISK RESONATORS

Oscar Malvar[1], Eduardo Gil-Santos[1], Jose J. Ruz[1], Elena Sentre-Arribas[1], Adrián Sanz-Jiménez[1], Priscila M. Kosaka[1], Sergio García-López[1], Álvaro San Paulo[1], Samantha Sbarra[2], Louis Waquier[2], Ivan Favero[2], Maurits van der Heiden[3], Robert K. Altmann[3], Dimitris Papanastasiou[4], Diamantis Kounadis[4], Ilias Panagiotopoulos[4], Jesús Mingorance[5], María Rodríguez-Tejedor[5], Rafael Delgado[6], Montserrat Calleja[1] and Javier Tamayo[1]*

[1]Bionanomechanics Lab., Instituto de Micro y Nanotecnología, IMN-CSIC, CSIC (CEI UAM+CSIC), 28760, Tres Cantos, Madrid, SPAIN
[2]Matériaux et Phénomènes Quantiques, Université Paris Cité, CNRS, UMR 7162, 75013, Paris, FRANCE
[3]The Netherland Organization for Applied Scientific Research, TNO, NETHERLANDS
[4]Fasmatech Science and Technology, Athens, GREECE
[5]Hospital Universitario La Paz, Madrid, SPAIN and
[6]Hospital Universitario 12 de Octubre, Madrid, SPAIN

ABSTRACT

Nanomechanical mass spectrometry allows characterization of analytes with broad mass range, from small proteins to bacterial cells, and with unprecedented mass sensitivity. In this work, we show a novel multifrequency nanomechanical mass spectrometer prototype designed for focusing, guiding and soft-landing of nanoparticles and viral particles on a nanomechanical resonator surface placed in vacuum. The system is compatible with optomechanical disk resonators, with an integrated optomechanical transduction method, and with the laser beam deflection technique for the measurement of the vibrations of microcantilever resonators. The prototype allows the in-vacuum alignment of resonators thanks to a dedicated visualization system. Finally, in this work, we have demonstrated the detection of gold nanoparticles, polystyrene nanoparticles and phage G viruses with optomechanical disks and microcantilever resonators.

KEYWORDS

Nanomechanical mass spectrometry, optomechanical disk resonators, microcantilevers, mass sensing.

INTRODUCTION

State of the art

Conventional mass spectrometry can measure the mass to charge ratio of small analytes with unprecedented sensitivity, reaching the 18 MDa limit measuring the mass of the bacteriophage HK97 [1]. On the other hand, charge detection mass spectrometry (CDMS) has moved into the mainstream for measuring mass distributions above the MDa regime [2], [3]. However, these technologies need the ionization of the particles and its mass sensitivity degrades for larger analytes. Since the invention of the first nanomechanical mass spectrometer (NMS) system in 2009 in Caltech by Michael Lee Roukes's group [4] this promising technology has demonstrated its potential by measuring the mass of proteins [5], viruses [6] and even bacterial cells [7]. NMS allows weighing individual particles in real time with outstanding mass sensitivity and with high dynamic range. Furthermore, NMS has demonstrated the measurement of bacteria cell stiffness [8] and even the possibility to extract information of the analyte's shape [9]. Contrary to conventional mass spectrometry or CDMS, NMS doesn't need the ionization of the particles [10], therefore this technique can be used to measure the mass of particles that cannot be ionized, opening the door to new possibilities.

On the other hand, optomechanical resonators present an extraordinary displacement sensitivity, high quality factor together with small masses and high mechanical frequencies, emerging as promising sensors in the nanomechanical spectrometry field [11], [12]. Recently, optomechanical devices have been used in a nanomechanical spectrometer as a sensor element, measuring the mass of tantalum clusters [13]. Moreover, optomechanical disk resonators have proven the detection of vibration modes of a single bacterium [14] and the characterization of the mass of polystyrene nanoparticles in air [15].

Here, we present a multifrequency NMS prototype designed to measure optomechanical disk resonators in vacuum and compatible with the beam deflection technique, for the characterization of nanoparticles and viral particles.

DESCRIPTION OF THE SYSTEM

Nanomechanical mass spectrometer prototype

Figure 1a shows a picture of the multifrequency NMS prototype designed to measure the mass of individual analytes. The prototype is designed for soft-landing, guiding and focusing of the particles on a reduced area by an in-vacuum optical alignment system. The prototype comprises four stages with decreasing pressure. The first stage is at atmospheric pressure and it holds an electrospray ionization (ESI) system for the nebulization of the particles in a controlled atmosphere (Figure 1b), thus avoiding that undesirable particles present in the air can go inside the

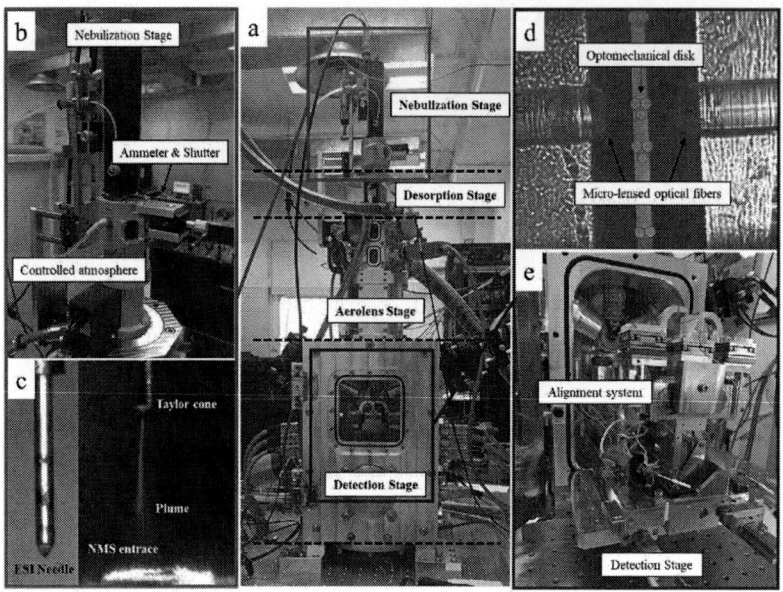

Figure 1: (a) Picture of the nanomechanical mass spectrometer prototype where the four stages with decreasing pressure are shown. (b) Nebulization stage with a controlled atmosphere that contains the ESI system. An ammeter is used to check the stability of the nebulization process and a shutter is used to close or open the system. (c) Picture of the ESI needle used in this work (left) and microscope image showing the Taylor cone and the subsequent expansion plume (right) during the nebulization process. (d) Microscope image of the optomechanical disk resonators taken with the in-vacuum visualization system. (e) Picture of the detection stage where the high precision alignment system is shown.

system and interfering the measurements. An ammeter is used to control the stabilization of the nebulization process measuring the current intensity of the ESI needle with high precision. The ESI nebulization flow rate is set to 0.2 µl/min by means of a high-pressure syringe pump controller. A sheath and a curtain gas of N_2 is used with a flow rate of 0.3 l/min between the ESI stage and the input of the spectrometer prototype. A high voltage (3-4 kV) is applied to the ESI needle in order to generate the Taylor cone (Figure 1c). A fine jet emerges from tip of the Taylor cone that expand into a plume containing microdroplets with the analytes. The microdroplets evaporate during its way to the next stage to finally have individual particles that can be detected with a nanomechanical sensor element placed in the detection stage. A heated capillary placed at the second stage is set to 200 °C in order to improve solvent evaporation. Additionally, a water-cooling system isolates this stage to the rest of the prototype in order to reduce thermal effects that can affect the mechanical stability of the system.

The third stage comprises an aerolens designed to reduce the velocity and increase the transportation efficiency of the particles [16]. The resonator element is placed in the fourth stage at 10^{-3} mbar that is designed to measure several resonance frequencies of optomechanical disk or microcantilever resonator by means of an optomechanical transduction method or beam deflection technique, respectively. A designed element focuses the particles in a reduced area of R = 150 µm, as measured for 100 nm gold nanoparticles experiments. The prototype also has a visualization/alignment system (Figure 1e) designed to perfectly align the particle beam and the resonator surface and to in-vacuum alignment of the micro-lensed optical fibers and the waveguide integrated into the

optomechanical disk resonator device (Figure 1d) [17]. An automated feedback software continuously tracks the optical transmission in order to maintain the system aligned for long periods of time. The prototype was tested with two different configurations: microcantilever configuration and optomechanical disk resonator set-up.

RESULTS
Microcantilever configuration set-up

The NMS prototype was tested with microcantilever resonators. First, we use a Si_3N_4 microcantilevers array to demonstrate the in-vacuum alignment between the particle beam and the resonator surface. Figure 2a shows a dark field microscope image of an array of 100 microcantilevers and a zoomed image (red dashed circle) showing the nebulization of 100 nm gold nanoparticles. The prototype effectively focuses these particles in a reduced area of around 150 µm of radius with high-efficiency, with a detection of around (12-15) particles per minute. In this experiment, we have tracked the four first flexural modes of a microcantilever with dimensions of 60 µm length, 10 µm width and 100 nm thickness, showed in figure 2a (blue dashed box). Figure 2b shows a real-time record of the relative frequency shifts during the nebulization experiments. After applying the inverse problem algorithm [18], we obtain the mass probability density distribution of these particles.

On the other hand, we have tested the prototype with the nebulization of phage G viral particles. Figure 3a shows the real-time record of the relative frequency shifts of the first three flexural modes during the adsorptions of phage G viral particles on a Si_3N_4 microcantilevers. The dimension of the microcantilever is 40 µm length, 6 µm width and 100 nm thickness. Figure 3b shows the mass distribution of the viral particles after applying the inverse problem algorithm [18], obtaining a mass distribution between 1 fg and 2 fg, consistent with the expected value, demonstrating the potential of the NMS for viral particles mass sensing using microcantilever resonators.

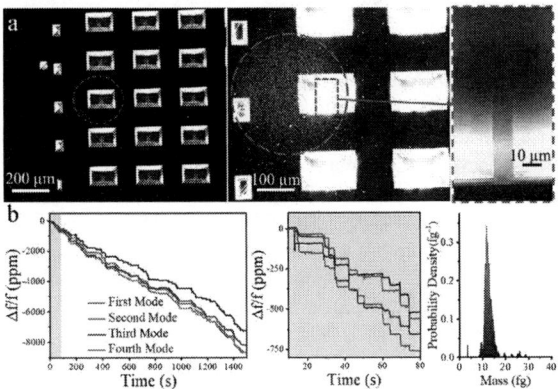

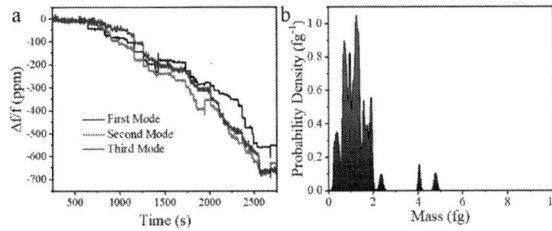

Figure 2: (a) Dark field microscope image of an array of microcantilevers, zoomed image of the nebulized region (red-dashed circle) and microcantilever covered with gold nanoparticles after the experiment (right). (b) Relative frequency shifts of the four tracked modes during the adsorption of 100 nm gold nanoparticles and zoomed area (light-blue). Probability density mass distribution (right) obtained after applying the inverse problem algorithm.

Figure 3: (a) Relative frequency shifts during the phage G experiments on a microcantilever resonator surface. (b) Mass probability density after applying the inverse problem algorithm.

Optomechanical disk resonator set-up

Finally, we have demonstrated the use of 11 μm of radius optomechanical disk resonators as sensor element in the NMS. Figure 4a shows a scanning electron microscope image of an optomechanical disk resonator used in this work. Figure 4b shows the thermomechanical spectrum of the used modes (RBM1, RBM2, M190 and M290) [15] and the simulated modes (inset) given by finite element method (FEM). We have tested the prototype with two different analytes: 140 nm polystyrene nanoparticles (PNPs) and phage G viral particles. Figure 5a shows the relative frequency shifts obtained during a nebulization experiment of 140 nm PNPs. Figure 5b shows the obtained mass distribution and the landing position in polar plot representation of two tentative events after applying the invers problem algorithm with the optomechanical disk resonators configuration set-up.

On the other hand, Figure 5c shows the real-time record of the relative frequency shifts during the phage G viral particles nebulization experiments. Figure 5d shows

two mass distributions and landing positions in polar plot representation obtained after applying the inverse problem algorithm of two tentative events. The obtained results demonstrate the detection of polystyrene and viral particles with optomechanical disk resonators.

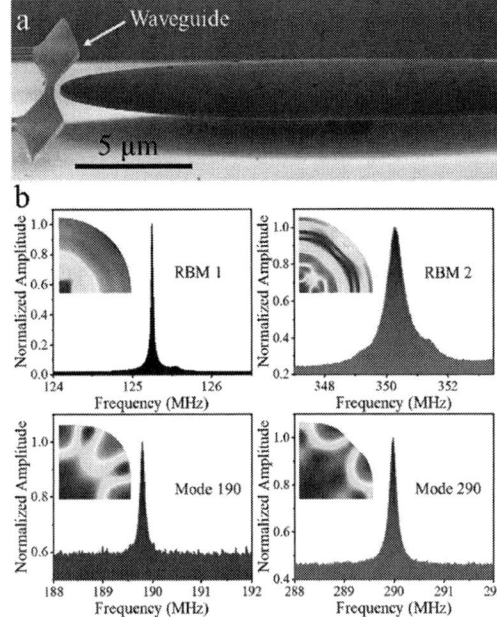

Figure 4: (a) Scanning electron microscope image of an optomechanical disk resonator of 11 μm of radius used in this work. (b) Thermomechanical frequency response of the tracked modes measured in the nanomechanical mass spectrometer prototype. Inset: FEM simulation mode shape representation.

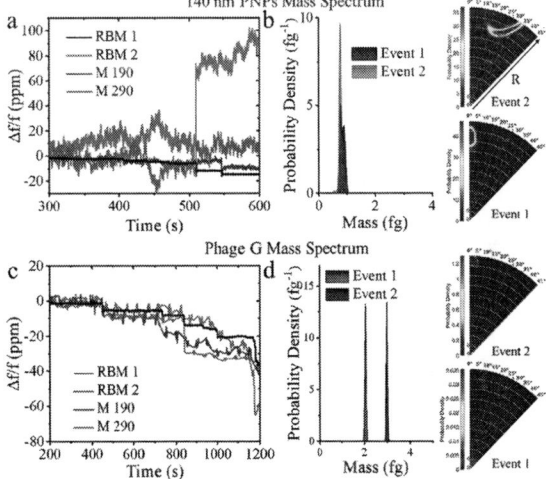

Figure 5: (a) Relative frequency shifts during the nebulization of 140 nm polystyrene nanoparticles. (b) Mass spectrum and landing position in polar plot representation obtained from two tentative events. (c) Relative frequency shift during the nebulization of phage G viral particles. (d) Mass spectrum and landing position in polar plot representation of two tentative events during the nebulization of phage G viral particles.

However, there are two important factors that need to be considered in order to extract the mass and position with more accuracy. First, we need to perfectly know the used mode shapes of the optomechanical disk resonators in order to apply the inverse problem algorithm, what is complicate due to the unknown of the pedestal geometry that modify the expected mode shapes (Figure 4b insets). Second, some of the particle adsorptions could produce a change in the optomechanical coupling or in the optical quality factor, altering the mechanical frequency shifts (like positive frequency shifts observed in Figure 5a) and interfering in the correct mass and position calculations. Despite these undesirable effects, we have obtained mass distribution values of some tentative measured events that are in good agreement with the expected masses, suggesting that optomechanical mass spectrometry can play an important role in the future, with huge applications in the environment or biomedical fields, for instance, in the detection of nanoparticles and microbiological entities.

CONCLUSION

In conclusion, we show here a novel multifrequency nanomechanical mass spectrometer prototype, compatible with microcantilever and with optomechanical disk resonators, that can measure several frequencies at the same time. The prototype is able to nebulize, guide and soft-land nanoparticles and viral particles from solvent solution to the surface of optomechanical disk and microcantilever resonators in vacuum. Finally, we have tested the prototype with gold nanoparticles, polystyrene nanoparticles and phage G viral particles obtaining promising results.

ACKNOWLEDGEMENTS

This work was supported by the European Union's Horizon 2020 Research and Innovation Program under Grant Agreement No. 731868-VIRUSCAN, by the EIC Pathfinder Innovation Council 101034583-H2020 VIRAIR and by the project PCL2021-007892 PTI Salud Global (CSIC). We acknowledge the service from the Micro and Nanofabrication Laboratory and X-SEM laboratory at IMN-CNM funded by the Comunidad de Madrid (Project S2018/NMT-4291 TEC2SPACE) and by MINECO (Project CSIC12-4E-1794 with support from FEDER, FSE). E. G. S. acknowledges financial support by the Spanish Science and Innovation Ministry through Ramón y Cajal grant RYC-2019-026626-1 and by the BBVA Foundation through "*Leonardo Grant investigadores y creadores culturales 2021*" – DISOM project.

REFERENCES

[1] J. Snijder, R. J. Rose, D. Veesler, J. E. Johnson, and A. J. R. Heck, "Studying 18 MDa Virus Assemblies with Native Mass Spectrometry," *Angewandte Chemie International Edition*, vol. 52, no. 14, pp. 4020–4023, Apr. 2013.

[2] D. Z. Keifer, E. E. Pierson, and M. F. Jarrold, "Charge detection mass spectrometry: weighing heavier things," *Analyst*, vol. 142, no. 10, pp. 1654–1671, 2017.

[3] A. R. Todd, L. F. Barnes, K. Young, A. Zlotnick, and M. F. Jarrold, "Higher Resolution Charge Detection Mass Spectrometry," *Anal. Chem.*, vol. 92, no. 16, pp. 11357–11364, Aug. 2020.

[4] A. K. Naik, M. S. Hanay, W. K. Hiebert, X. L. Feng, and M. L. Roukes, "Towards single-molecule nanomechanical mass spectrometry," *Nat. Nanotechnol.*, vol. 4, no. 7, pp. 445–450, 2009.

[5] M. S. Hanay *et al.*, "Single-protein nanomechanical mass spectrometry in real time," *Nat. Nanotechnol.*, vol. 7, no. 9, pp. 602–608, 2012.

[6] S. Dominguez-Medina *et al.*, "Neutral mass spectrometry of virus capsids above 100 megadaltons with nanomechanical resonators," *Science (1979)*, vol. 362, no. 6417, pp. 918–922, Nov. 2018.

[7] A. Sanz-Jiménez *et al.*, "High-throughput determination of dry mass of single bacterial cells by ultrathin membrane resonators," *Commun. Biol.*, vol. 5, no. 1, p. 1227, 2022.

[8] O. Malvar *et al.*, "Mass and stiffness spectrometry of nanoparticles and whole intact bacteria by multimode nanomechanical resonators," *Nat. Commun.*, vol. 7, no. 1, p. 13452, 2016.

[9] M. S. Hanay, S. I. Kelber, C. D. O'Connell, P. Mulvaney, J. E. Sader, and M. L. Roukes, "Inertial imaging with nanomechanical systems," *Nat. Nanotechnol.*, vol. 10, no. 4, pp. 339–344, 2015.

[10] E. Sage *et al.*, "Neutral particle mass spectrometry with nanomechanical systems," *Nat. Commun.*, vol. 6, Mar. 2015.

[11] M. Aspelmeyer, P. Meystre, and K. Schwab, "Quantum optomechanics," *Phys. Today*, vol. 65, no. 7, pp. 29–35, Jul. 2012.

[12] X. Liu *et al.*, "Progress of optomechanical micro/nano sensors: a review," *Int. J. Optomechatronics*, vol. 15, no. 1, pp. 120–159, Jan. 2021.

[13] M. Sansa *et al.*, "Optomechanical mass spectrometry," *Nat. Commun.*, vol. 11, no. 1, p. 3781, 2020.

[14] E. Gil-Santos *et al.*, "Optomechanical detection of vibration modes of a single bacterium," *Nat. Nanotechnol.*, vol. 15, no. 6, pp. 469–474, 2020.

[15] S. Sbarra, L. Waquier, S. Suffit, A. Lemaître, and I. Favero, "Multimode Optomechanical Weighting of a Single Nanoparticle," *Nano Lett.*, vol. 22, no. 2, pp. 710–715, Jan. 2022.

[16] D. Papanastasiou *et al.*, "Experimental and numerical investigations of under-expanded gas flows for optimal operation of a novel multipole differential ion mobility filter in the first vacuum-stage of a mass spectrometer," *Int. J. Mass Spectrom.*, vol. 465, p. 116605, Jul. 2021.

[17] E. Gil-Santos *et al.*, "High-frequency nano-optomechanical disk resonators in liquids," *Nat. Nanotechnol.*, vol. 10, no. 9, pp. 810–816, 2015.

[18] J. J. Ruz, O. Malvar, E. Gil-Santos, D. Ramos, M. Calleja, and J. Tamayo, "A review on theory and modelling of nanomechanical sensors for biological applications," *Processes*, vol. 9, no. 1, 2021.

CONTACT

*O. Malvar, tel: +34-918060700 (ext. 440858); oscar.malvar@csic.es

A MICROFABRICATED DIAMOND QUANTUM MAGNETOMETER WITH PICOTESLA SCALE SENSITIVITY

Fei Xie[1,2], Qihui Liu[1,2], Yuqiang Hu[3,4], Lingyun Li[2,5], Zhichao Chen[2,5], Jin Zhang[1], Yonggui Zhang[1,2], Yuyao Zhang[3,4], Yang Wang[1,2], Jiangong Cheng[1,2*], Hao Chen[1,2*], and Zhenyu Wu[1,2,3,4*]

[1] State Key Laboratory of Transducer Technology, Shanghai Institute of Microsystem and Information Technology, Chinese Academy of Sciences, Shanghai, China
[2] School of Graduate Study, University of Chinese Academy of Sciences, Beijing, China
[3]School of Microelectronics, Shanghai University, Shanghai, China
[4]Shanghai Industrial μTechnology Research Institute, Shanghai, China and
[5]Center for Excellence in Superconducting Electronics, Shanghai Institute of Microsystem and Information Technology, Chinese Academy of Sciences, Shanghai, China

ABSTRACT

This research reports a microfabricated diamond quantum magnetometer based on ensemble nitrogen-vacancy (NV) centers. The sensor device is fabricated with a standard MEMS process and integrated with optical fiber and photodiode (PD). The sensitivity of the sensor and signal-to-noise ratio (SNR) are determined by the detection efficiency of photons emitted from NVs in bulk diamonds. The miniaturized and portable diamond sensor shows a 132-fold enhancement of fluorescence collection efficiency over the fiber confocal system and an unshielded sensitivity of 334 pT·Hz$^{-1/2}$ under ambient conditions. The essential sensor component is $1\times1\times0.1$ cm^3 in volume. Sensors with improved integration and sensitivity are possible to be used for practical applications, such as magnetocardiography and precision current sensing.

KEYWORDS

Diamond quantum magnetometer, nitrogen-vacancy (NV) centers, microfabrication, miniaturization, integration, portability.

INTRODUCTION

As a solid-state quantum platform for measuring magnetic fields, diamond NV sensor has attracted broad interest due to a combination of high spatial resolution, sensitivity and thermostability. Miniaturization, integration, and portability are the current trends in biological and industrial fields. To the best of our knowledge, the highest sensitivity of fiber-integrated diamond magnetometers is 310 pT·Hz$^{-1/2}$ [1]. However, cubic centimeter size limits certain applications for near-field sensing. The smallest diamond magnetometer is only 1.5 mm^2 but limited by a poor magnetic sensitivity of 245,000 pT·Hz$^{-1/2}$ [2]. Based on an approach reported recently [3], this research focuses on the development of the prototype with improved sensitivity. Optimized optical architecture with an on-spot integrated photodetector greatly enhances the efficiency of fluorescence collection leading to a 5-fold sensitivity improvement compared to previous work.

DESIGN AND FABRICATION
INTEGRATED SENSOR HEAD

Regarding the geometry design of diamond quantum magnetometers, there are two considerations: efficient excitation of the NV centers and collection of the emitted fluorescence signal. Both of key factors are critical for the improvement of SNR. Several methods have been proposed to address this issue, including light-trapping diamond waveguide [4], side wall collection [5], resonant optical cavity [6], and so on. The 3D structure of the proposed compact diamond sensor device is demonstrated in Figure 1(a). It is composed of four parts: a diamond chip, a fiber coupling module (FCM), a microwave (MW) printed circuit board (PCB) and a photoelectric detection module. The diamond is encapsulated in the stacked silicon wafers and the laser is guided through the FCM, then focused on the diamond for spin initialization. The MW PCB provides a path of frequency-adjustable MWs for spin control. After passing a long pass filter, the fluorescence generated by the NV centers is collected by the PD mounted on another PCB converting to a photocurrent for signal read-out. A physical representation of our integrated diamond sensor head is shown in Figure 1(b).

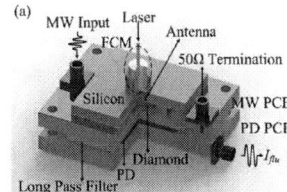 

Figure 1: (a) Schematic diagram and (b) photograph of proposed compact diamond sensor device.

FABRICATION

As the essential component of the sensing device, the diamond chip is fabricated with a standard MEMS process. Figure 2(a-d) show the fabrication steps for the diamond chip, including the preparation of three monolithic processes (fiber slot, spacer and diamond) and their order of bonding.

(a) Fiber slot process: 500-nm-thick SiO$_2$ is grown by thermal oxidation and 30/50/700-nm-thick Ti/Pt/Au are sputtered on a high resistance silicon wafer. The metal layers are patterned to form MW transmission line with the first photolithography and subsequent ion beam etching (IBE). A second-round photolithography was carried out, followed by reactive ion etching (RIE) for SiO$_2$ removal. The exposed area of silicon was engraved with deep reactive ion etching (DRIE).

(b) Spacer process: Thermal oxidation of 500-nm-thick SiO_2 is later patterned to open etching windows. After wet etching with 40% KOH at 50 °C for 48 hours, the wafer forms 54.7° deep holes. Metal films are sputtered on the back of the silicon for subsequent bonding and patterned with the second photolithography and IBE. Then, SiO_2, grown by plasma enhanced chemical vapor deposition (PECVD), and metal layers are deposited on the sloping sidewall of 54.7° to increase optical reflectivity.

(c) Diamond process: The commercially available single-crystal-diamonds are grown via CVD with the natural ^{13}C content from Element Six. The samples have a concentration of $[^{14}N] < 13$ ppm with dimensions of $3 \times 3 \times 0.5$ mm^3. Diamonds are boiled in piranha solution, followed by ultrasonic cleaning with acetone, anhydrous ethanol, and deionized water. After diamond surface plasma activation, the omega-shaped transmission line is patterned with the same process of fiber slot.

(d) Bonding process: The diamond and fiber slot are bonded to each other with a silicon holder, then bonded with spacer, both with Au-Au thermal compression bonding at 380 °C for 30 min.

Here, both SiO_2 and Pt films play a role of insulating layer to avoid Au-Si interdiffusion. After slicing, the front and rear side views of the as-fabricated diamond chip are shown in Figure 2(e) and (f), with a mass of 0.17 g and a volume of 0.1 cm^3. Note that the stick out part on both sides of fiber slot is used to lead out MW transmission line. The shear force test indicated the bonding strength reach more than 20 MPa, which is robust enough for subsequent processes.

After the microfabrication process, the diamond chip is electrically connected to the MW PCB (Rogers 4350B and FR-4) by flip-chip process. Here, a pathway of PCB-Si-Diamond-Si-PCB-50 Ω load for MW transmission is formed, where all with Au as conductive material. The FCM, mainly containing glass tubes for fixing, a fiber (Thorlabs, FG200LEA) and a gradient index lens (Thorlabs, G1P10), is fixed to the diamond chip with epoxy resin by on-line alignment technology. Afterward, a long pass filter and a PD (Hamamatsu, S12915-66R) are glued to the MW PCB with UV adhesive, respectively. PD PCB is fixed to MW PCB and welded with the PD for photocurrent output.

EXPERIMENTAL

The optical and signal path for the measurement system is presented in Figure 3(a). Light from 532-nm laser (Coherent, Verdi-V6) enters port 1 of fiber circulator (Thorlabs, WMC2L1F) via a polarization beam splitter and exits port 2 arriving in the diamond sensor through the FCM. Thanks to the long pass filter (600 nm), only the excited fluorescence (637-800 nm) is collected by the PD1 at the bottom. A fraction of the laser is picked up by the PD2 as a reference for laser intensity fluctuation suppression. By using a self-designed differential amplifier, the resulted common mode rejection voltage signal is sent to a lock-in amplifier (LIA, Zurich Instruments, MFLI) for phase-sensitive demodulation and read out by a data acquisition (DAQ) card, thus realizing optically detected magnetic resonance (ODMR) [7]. The MW signal generated from MW source is sent to MW PCB with an amplification gain of 46 dB. A permanent magnet positioned 5 cm away from the integrated diamond head produces a bias magnetic field of 12 mT for subsequent measurement.

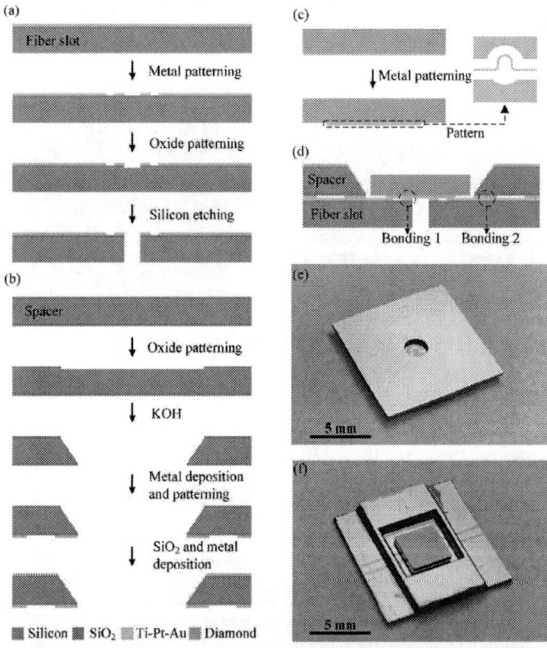

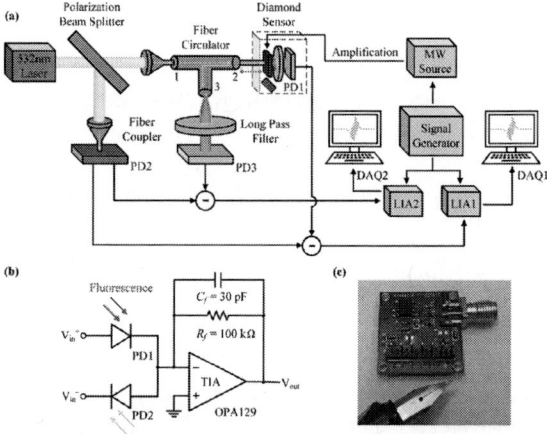

Figure 3: (a) Schematic of the optical and signal path for the measurement system. (b) The NV fluorescence signal and the reference laser signal were recorded using a differential amplifier circuit. (c) Image of self-designed differential amplifier PCB.

In order to compare the PD-integrated mode with the previously reported fiber confocal mode [3], fluorescence signal collected by PD3 is read out by a balanced detector (Thorlabs, PDB450A) with reference laser from PD2. Then,

Figure 2: Fabrication steps of the diamond chip containing (a) fiber slot, (b) spacer and (c) diamond. (d) Bonding processes: diamond and fiber slot are bonded, then bonded with spacer, both with Au–Au thermal compression bonding. Photographs of (e) the front and (f) rear side views of the as-fabricated diamond chip.

the same demodulation system is used for ODMR detection. Figure 3(b) shows the self-designed differential amplifier circuit. Both PD1 and PD2 are working in the reverse bias mode to obtain a large dark current. Feedback capacitance C_f and resistance R_f are set as 30 pF and 100 kΩ, respectively, thus determining the bandwidth of 53 kHz and the transimpedance amplification of 10^5 V/A. Figure 3(c) shows a photograph of self-designed differential amplifier PCB.

RESULTS AND DISCUSSION

A comparison of fluorescence detected with two operation modalities at various laser power is shown in Figure 4(a). The fluorescence intensity of PD-integrated mode is 132-times higher than that of fiber confocal mode. High refractive index of diamond results in inefficient fluorescence collection in all directions. For fiber confocal mode, the fluorescence emitted from the diamond top surface is picked up by the fiber, and a 20~30% insertion loss is observed for the fiber coupling system. For PD-integrated mode, the fluorescence from diamond bottom surface and four sides of the diamond is reflected to the PD at the bottom through the 54.7° sidewall. As presented in Figure 4(b), normalized ODMR spectra with ^{14}N hyperfine splitting [8] driven by a laser power of 500 mW and a MW power of -4 dBm, indicate that the ratio of the full width half maximum to the contrast of diamond sensor is comparable due to the same demodulation system. Therefore, it is verified that the sensitivity improvement is due to the enhancement of the fluorescence detection efficiency.

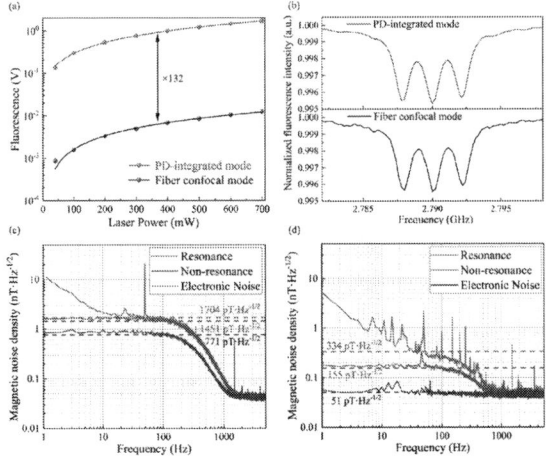

Figure 4: Comparison of results measured with two modalities: (a) NV fluorescence (converted to voltage), as a function of laser power, shows a 132-fold enhancement. (b) Normalized ODMR spectra with ^{14}N hyperfine splitting. Diamond magnetometer magnetic noise density spectrum with (c) fiber confocal mode and (d) PD-integrated mode.

Figure 4(c) and (d) show the magnetic noise density spectrum of diamond magnetometer in two modes, respectively. An optimized magnetic sensitivity of 334 pT·Hz$^{-1/2}$ for PD-integrated sensor configuration is obtained, achieving a 5-fold sensitivity improvement compared to previous sensitivity of 1,704 pT·Hz$^{-1/2}$

obtained in fiber confocal mode. In addition, low frequency noise is observed in the spectrum of PD-integrated mode due primarily to the electronic noise of the self-designed circuit which could be further optimized.

As demonstrated in Figure 5, resonance, non-resonance magnetic sensitivity and equivalent electronic noise floor vary with the laser power ranging from 100 mW to 500 mW. With the increase of laser power, the magnetic sensitivity is improved and tends to saturate. Compared with the fiber confocal mode, a much weaker laser power is required to obtain the same level of magnetic sensitivity for PD-integrated mode.

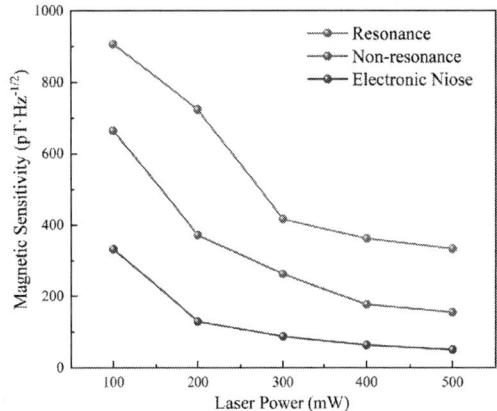

Figure 5: Resonance, non-resonance magnetic sensitivity and equivalent electronic noise floor vary with the laser power ranging from 100 mW to 500 mW.

Table 1: Comparison of state-of-art integrated diamond magnetometers.

Ref.	Volume (cm³)	Sensitivity (pT·Hz$^{-1/2}$)	Diamond Characterization
[1]	90	310	[NV] ~ 2.8 ppm, 99.995% ^{12}C enriched diamond.
[2]	1.5 mm² (CMOS level)	245,000	[NV] ~ 0.01 ppm.
[9]	539	7,000	[N] ~ 1 ppm, [NV] ~ 0.2 ppb, with the natural ^{13}C content.
[10]	2.9	344	[NV] ~ 0.4 ppm, (111)-oriented, 99.97% ^{12}C enriched diamond.
Our work	0.1(chip) 7.2(sensor head)	334	[N] ~ 13 ppm, with the natural ^{13}C content.

Finally, the performances of our device in comparison with other reports are summarized in Table 1. Our proposed compact diamond sensor device has a relatively small size and a high magnetic sensitivity. Further optimization is focused on the diamond material, geometry for extending the optical path and readout circuits. High quality diamonds have a potential for further improvements in magnetic sensitivity, such as diamonds synthesized with purified ^{12}C.

CONCLUSIONS

In summary, this paper reports an integrated and portable diamond quantum magnetometer using a standard MEMS process, achieving a resonance magnetic sensitivity of 344 pT·Hz$^{-1/2}$, a non-resonance magnetic sensitivity of 155 pT·Hz$^{-1/2}$ and an electronic noise floor of 51 pT·Hz$^{-1/2}$. The overall size of the essential component of diamond sensor device is less than $1 \times 1 \times 0.1$ cm^3 with a mass of 0.17 g. By optimizing the geometry of the sensor device, the magnetic sensitivity is improved to achieve picotesla scale due to the enhancement of fluorescence collection efficiency. The proposed microfabrication process is scalable, which offers possibilities for the fabrication of miniaturized, integrated, and portable diamond quantum magnetometers for practical applications.

ACKNOWLEDGEMENTS

The authors thank Prof. Xiaohong Ge and Shuna Wang from the Shanghai Institute of Microsystem and Information Technology, Chinese Academy of Sciences, for sample preparation and useful discussion. This work was supported by CAS Strategic Pilot Project (No. XDC07030200), R&D Program of Scientific Instruments and Equipment, Chinese Academy of Sciences (No. YJKYYQ20190026) and National Key R&D Program of China (No. 2021YFB3202500).

REFERENCES

[1] R. L. Patel, L. Q. Zhou, A. C. Frangeskou, G. A. Stimpson, B. G. Breeze, A. Nikitin, M. W. Dale, E. C. Nichols, W. Thornley, B. L. Green, M. E. Newton, A. M. Edmonds, M. L. Markham, D. J. Twitchen, G. W. Morley, "Subnanotesla magnetometry with a fiber-coupled diamond sensor", *Physical Review Applied*, vol. 14, pp. 044058, 2020.

[2] M. I. Ibrahim, C. Foy, D. R. Englund, R. Han, "High-scalability CMOS quantum magnetometer with spin-state excitation and detection of diamond color centers", in *IEEE Journal of Solid-State Circuits*, 2021, pp. 1001-1014.

[3] F. Xie, Y. Hu, L. Li, C. Wang, Q. Liu, N. Wang, L. Wang, S. Wang, J. Cheng, H. Chen, Z. Wu, "A microfabricated fiber-integrated diamond magnetometer with ensemble nitrogen-vacancy centers", *Applied Physics Letters*, vol. 120, pp. 191104, 2022.

[4] H. Clevenson, M. E. Trusheim, C. Teale, T. Schröder, D. Braje, D. Englund, "Broadband magnetometry and temperature sensing with a light-trapping diamond waveguide", *Nature Physics*, vol. 11, pp. 393-397, 2015.

[5] D. L. Sage, L. M. Pham, N. Bar-Gill, C. Belthangady, M. D. Lukin, A. Yacoby, R. L. Walsworth, "Efficient photon detection from color centers in a diamond optical waveguide", *Physical Review B*, vol. 85, pp. 121202, 2012.

[6] S. R. Nair, L. J. Rogers, X. Vidal, R. P. Roberts, H. Abe, T. Ohshima, T. Yatsui, A. D. Greentree, J. Jeske,

T. Volz, "Amplification by stimulated emission of nitrogen-vacancy centres in a diamond-loaded fibre cavity", *Nanophotonics*, vol. 9, pp. 4505–4518, 2020.

[7] J. F. Barry, M. J. Turner, J. M. Schloss, D. R. Glenn, Y. Song, M. D. Lukin, H. Park, and R. L. Walsworth, "Optical magnetic detection of single-neuron action potentials using quantum defects in diamond", *Proceedings of the National Academy of Sciences*, vol. 113, pp. 14133-14138, 2016.

[8] E. Moreva, E. Bernardi, P. Traina, G. Petrini, S. Ditalia Tchernij, J. Forneris, F. Picollo, V. Pugliese, A. Sosso, Ž. Pastuović, I. P. Degiovanni, P. Olivero, and M. Genovese, "Magnetic sensing with nitrogen-vacancy centers based on lock-in detection", in *Conference on Precision Electromagnetic Measurements (CPEM)*, 2020, pp. 1-2.

[9] J. L. Webb, J. D. Clement, L. Troise, S. Ahmadi, G. J. Johansen, A. Huck, and U. L. Andersen, "Nanotesla sensitivity magnetic field sensing using a compact diamond nitrogen-vacancy magnetometer", *Applied Physics Letters*, vol. 114, pp. 231104, 2019.

[10] F. M. Stürner, A. Brenneis, T. Buck, J. Kassel, R. Rölver, T. Fuchs, A. Savitsky, D. Suter, J. Grimmel, S. Hengesbach, M. Förtsch, K. Nakamura, H. Sumiya, S. Onoda, J. Isoya, and F. Jelezko, "Integrated and portable magnetometer based on nitrogen-vacancy ensembles in diamond", *Advanced Quantum Technologies*, vol. 4, pp. 2000111, 2021.

CONTACT

*Zhenyu Wu, tel: +86-21-69075572; zhenyu.wu@mail.sim.ac.cn

*Hao Chen, tel: +86-21-69075572; haochen@mail.sim.ac.cn

*Jiangong Cheng, tel: +86-21-62511070; jgcheng@mail.sim.ac.cn

A NON-VOLATILE THRESHOLD SENSING SYSTEM USING A FERROELECTRIC Hf$_{0.5}$Zr$_{0.5}$O$_2$ DEVICE AND A LiNbO$_3$ MICROACOUSTIC RESONATOR

Onurcan Kaya, Luca Colombo, Benyamin Davaji, and Cristian Cassella
Northeastern University, Electrical Engineering, Boston, MA, USA

ABSTRACT

This work reports a novel threshold sensor system that is able to detect and memorize the occurrence of temperature violations by relying on a 20 nm-thick ferroelectric Hafnium Zirconium Oxide (HZO) varactor and a LiNbO$_3$ shear-horizontal (SH$_0$) Lamb wave microacoustic resonator, both microfabricated in-house. The reported sensor system is driven by a continuous-wave signal at a frequency (f_{in}) slightly detuned from the LiNbO$_3$ device's resonance frequency (f_{res}~33.3 MHz). When the ambient temperature changes, the voltage at f_{in} across the varactor increases proportionally to the resonator's figure-of-merit (FoM), ultimately causing a partial ferroelectric polarization switching of the HZO varactor for a temperature exceeding a certain programmable threshold (T_{th}). Following such a switching event, the capacitance (C_T) of the HZO varactor experiences a sudden change, causing a non-volatile 0.75-1 MHz shift of the read-out resonance frequency (f_{read} ~260 MHz) that is equal to the resonance frequency of an LC-tank formed by a lumped inductor and by the series of C_T with the LiNbO$_3$ device's capacitance (C_0). The ability to generate temperature-induced non-volatile changes of f_{read} through HZO ferroelectric varactors and microacoustic resonators is demonstrated for the first time in this work, and represents the key to implement a threshold sensing functionality and to memorize the occurrence of any temperature violations.

KEYWORDS

Ferroelectricity, Temperature Threshold Sensing, Hafnium Zirconium Oxide, Lithium Niobate, Acoustic Resonators

INTRODUCTION

Improper refrigeration of food and drugs along the cold chain represents a significant problem, generating threats to human health safety and severe economic losses. Fueled by the Radio-Frequency-Identification (RFID) revolution, several temperature sensing technologies have been developed, aiming at timely identifying and permanently marking any items undergoing temperature violations [1-10]. In this regard, thanks to their superior electromechanical performance and high Temperature-Coefficient-of-Frequency (TCF), various resonant microacoustic temperature sensors have been reported [11, 12]. Such devices can sense their ambient temperature with high sensitivity, yet they cannot be used for threshold sensing. Also, they cannot keep track of any previously occurred temperature violations.

Hafnium Zirconium Oxide (Hf$_{0.5}$Zr$_{0.5}$O$_2$ or HZO) varactors have been used in resistive and capacitive memory applications [13, 14]. This paper leverages an HZO varactor as a readout element in a novel threshold

sensing system. The proposed system relies on a ferroelectric 20 nm-thick HZO varactor and a Lithium Niobate (LiNbO$_3$) microacoustic RF resonator. The LiNbO$_3$ resonator is used as a temperature-sensing element and the HZO varactor is utilized in a series LC resonant circuit as a non-volatile memory element to detect and memorize the occurrence of temperature violations. In the proposed system the LiNbO$_3$ microacoustic resonator is driven by a continuous-wave signal at a frequency (f_{in} = 33 MHz) that is very close to the device's resonance frequency (f_{res}~ 33.3 MHz), and the LC readout circuitry resonates at around 260 MHz (f_{read}). Ambient temperature change shifts the resonance frequency of the microacoustic resonator causing a passive voltage amplification across the ferroelectric varactor. When the ambient temperature change is larger than the designed threshold value, the voltage across the HZO varactor becomes large enough to initiate the ferroelectric switching, which creates a non-volatile 0.75-1 MHz shift of f_{read}.

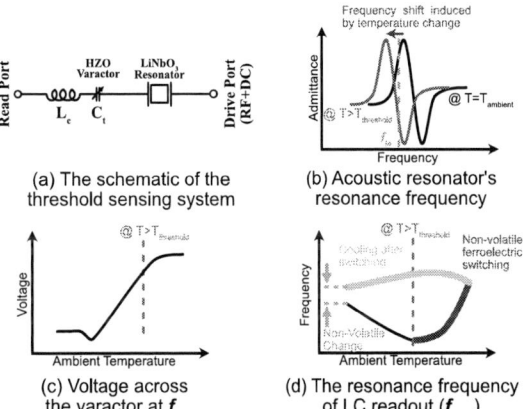

(a) The schematic of the threshold sensing system

(b) Acoustic resonator's resonance frequency

(c) Voltage across the varactor at f_{in}

(d) The resonance frequency of LC readout (f_{read})

Figure 1: (a) Schematic of the reported threshold sensing system; (b, c, and d) Summary of the working principle of our reported threshold sensing system; (b) The acoustic resonator is excited by a continuous-wave signal with a frequency (f_{in}) close to its resonance frequency. When the ambient temperature changes, the resonance frequency of the resonator shifts left, (c) Such a shift in the resonance frequency of the acoustic resonator triggers a passive voltage amplification across the varactor at f_{in}. (d) When the voltage across the HZO varactor increases the resonance frequency (f_{read}) of the LC-readout circuitry decreases until the voltage becomes large enough to initiate the ferroelectric switching of the HZO varactor (black line segment). When the ambient temperature exceeds a certain threshold temperature ($T_{threshold}$) the voltage across the varactor becomes large enough to trigger the ferroelectric switching (blue line segment). As a result, f_{read} undergoes a non-volatile change, allowing to capture and memorize the occurrence of the temperature violation.

DEVICE OPERATION

This system consists of a LiNbO₃ microacoustic resonator, a Hf₀.₅Zr₀.₅O₂ (HZO) ferroelectric varactor and an inductor (Fig. 1-a). Figs. 1-b,c,d summarize the basic operating principle of the system. The LiNbO₃ microacoustic resonator is used as a temperature sensing element and it is excited with a continuous-wave signal at a frequency (f_{in}), which is slightly detuned from the resonance frequency of the resonator (f_{res}). Meanwhile, the inductor forms a series LC resonant circuit with the HZO varactor and the static capacitance of the LiNbO₃ device, whose resonance frequency (f_{read}) is used as a readout parameter. When the ambient temperature changes, the LiNbO₃ resonator's resonance frequency shifts to the left (Fig. 1-b). This triggers a passive amplification of the varactor's voltage at f_{in} (Fig. 1-c) that induces a ferroelectric switching when the temperature exceeds a certain threshold (T_{th}) that can be programmed by setting f_{in}. As a result, f_{read} undergoes a non-volatile change, allowing one to capture and memorize the occurrence of any temperature violations (Fig. 1-d).

FABRICATION

The fabrication flow of the LiNbO₃ resonator is given in Fig. 2 (a). The process started with a 2.5 um thin film X-cut LiNbO₃ on a high resistivity silicon wafer. The thin film LiNbO₃ was bonded to the silicon wafer via Surface Activated Bonding (SAB) and thinned to the desired thickness (performed by NGK, Inc.). Then the release pits were formed by using ion milling, followed by the sputter deposition and patterning by lift-off of a 400 nm-thick AlSiCu layer to form the resonator's top electrodes.

Finally, the device was released through the XeF₂ isotropic etch.

The fabrication flow of the HZO varactor is given in Fig. 2 (b). A low-resistivity silicon wafer with a 150 nm thermal oxide was used as the varactor's substrate. 100 nm-thick Platinum (Pt) bottom electrodes were sputtered and patterned via lift-off. At this step, to minimize any fencings along the bottom electrode's edges, a bi-layer lift-off process was optimized. It is worth mentioning that the HZO device's bottom electrode was designed to be slightly wider than its top electrode to decrease the number of overlapping edges. Then, a 20 nm thick ferroelectric HZO layer and a 3 nm thick Al₂O₃ capping layer were deposited using Atomic Layer Deposition. The binary HZO layer was deposited by alternating the pulses of tetrakis (dimethylamido) hafnium (TFMAHf) and tetrakis (dimethylamino) zirconium (TDMAZr) precursors, each of which is followed by water pulses as the O₂ source. The Al₂O₃ layer was instead deposited using alternating pulses of trimethylaluminium (TMA) and water precursors. Next, vias to reach the bottom electrode were formed through a dry etch process. Finally, 150 nm thick gold top electrodes were deposited using e-beam evaporation. After the fabrication, HZO varactors were annealed using a rapid thermal processor (RTP) under vacuum in a Nitrogen (N₂) environment for 40 seconds at 400°C.

SEM images of the fabricated LiNbO₃ microacoustic resonator and HZO varactor are shown in Fig. 2 (c) and (d), respectively. As can be seen from Fig. 2 (c) the resonator consists of 7 identical parallel resonators with a pitch of 57 μm. Structural layers of the HZO varactor are highlighted in Fig. 2 (d), together with close-up views of the overlap area and of the via the region.

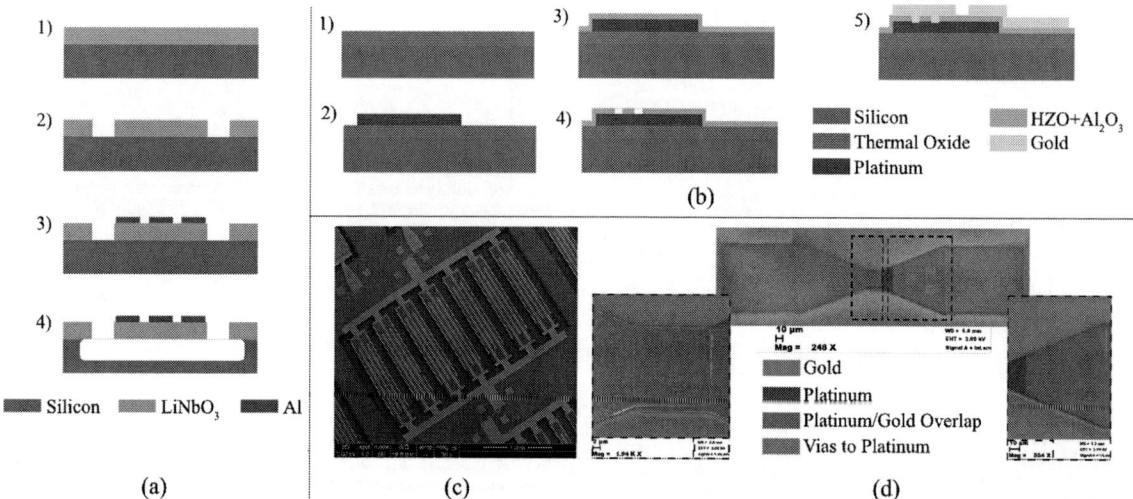

Figure 2: (a) Fabrication flow of the LiNbO₃ resonator; (1) The fabrication of the LiNbO₃ resonator relies on a 2.5 μm thin film X-cut LiNbO₃ wafer, bonded to the silicon substrate via Surface Activated Bonding and thinned to the desired thickness. (2) We started by forming the release pits through ion milling. (3) Then a 400 nm Al top electrode was sputtered and patterned through lift off. (4) Finally, the device was released through XeF₂ isotropic etch; (b) Fabrication flow of the HZO varactor; (1) A silicon wafer with a 150 nm thermal oxide was used as substrate. (2) A 100 nm thick Platinum bottom electrode was sputtered and patterned by lift-off. (3) Then, 20 nm HZO and 3 nm Al₂O₃ films were deposited using Atomic Layer Deposition. (4) Following this step, the vias were formed through the HZO film to reach the bottom Pt electrode. (5) Finally, a 150 nm top gold electrode was deposited by using e-beam evaporation. After the fabrication, the HZO varactor was annealed through a Rapid Thermal Processor in an N₂ environment for 40 seconds at 400°C; (c) SEM image of the fabricated LiNbO₃ resonator; (d) SEM image of the fabricated HZO varactor.

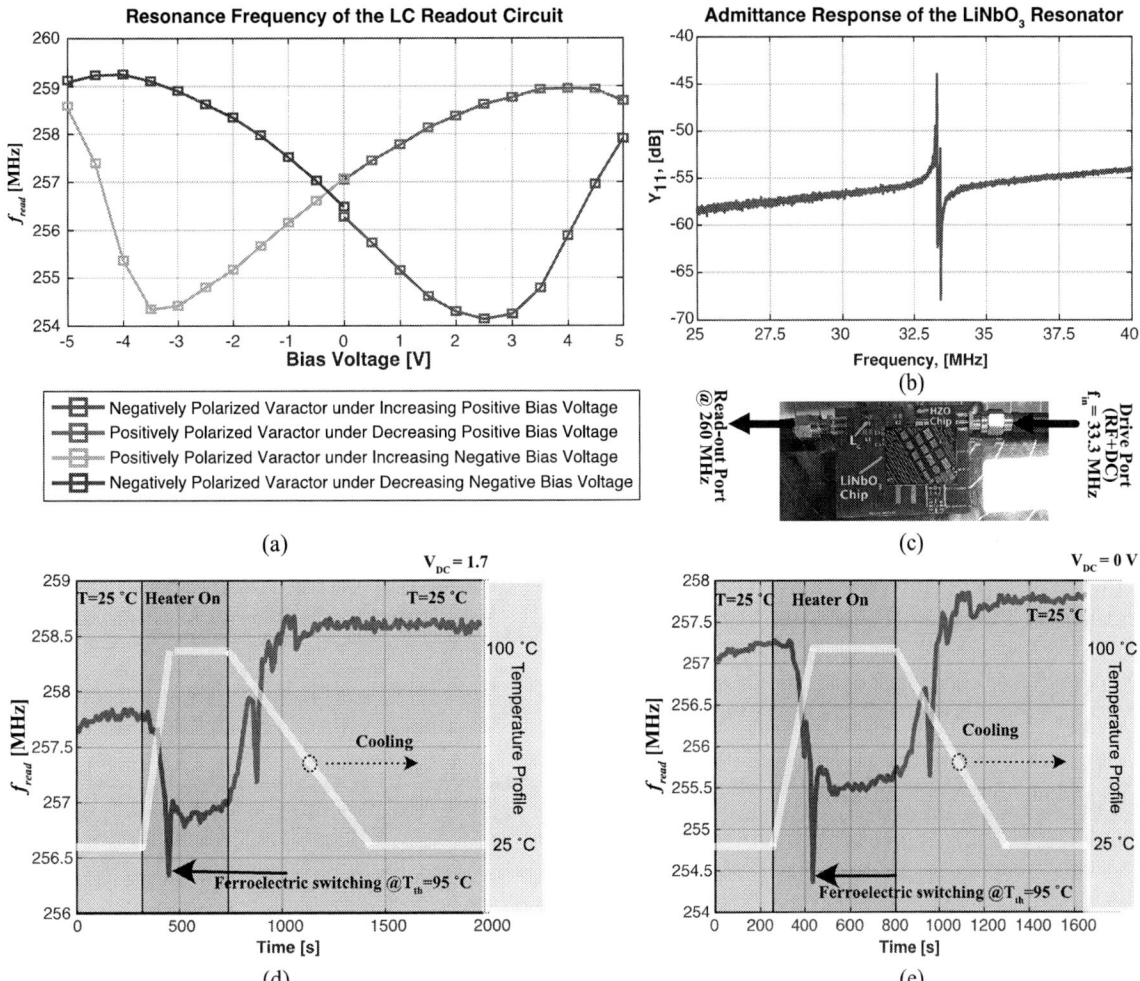

Figure 3: (a) Measured trend of f$_{read}$ vs. the applied DC voltage, extracted at ambient temperature. The varactor was initially negatively polarized, and it was subjected to a positive bias voltage (shown by the blue line segment). The resonance frequency decreased until ferroelectric switching occurs, followed by a rapid and sharp increase. Next, the bias voltage was brought back to zero (shown by the orange line segment). At that point, the varactor was positively polarized. The procedure was repeated for negative bias signals (yellow and purple curves). (b) Admittance response of the LiNbO₃ resonator. (c) Top-view picture of the assembled threshold sensing system on top of a PCB. In order to characterize its threshold sensing performance, the system was DC-biased and was excited at f$_{in}$ from one of its ports (the "Drive port") while tracking f$_{read}$ from the other port (the "Read port") by using a vector network analyzer; (d,e) Temperature response of the reported threshold sensing system under two DC-biasing voltages (1.7 V (d) and 0 V (e)). The right curve was obtained after applying a reset voltage pulse to reset the ferroelectric varactor's state. The temperature was increased up to 100 °C and later brought back to 25 °C. Evidently, a ferroelectric switching occurred producing a sudden notch when the temperature exceeded 95 °C. Moreover, as the temperature re-reached 25 °C, the f$_{read}$ value became different from the original value. Such a non-volatile shift is the key to recognize that a violation has occurred.

EXPERIMENTAL RESULTS

The two key components of the threshold sensing system, *i.e.* the LC readout circuit and the LiNbO₃ microacoustic resonator, were first tested separately at ambient temperature. Fig.1-a shows the ferroelectric behavior of the f$_{res}$ under a swept DC bias voltage at room temperature. Initially, the ferroelectric varactor was negatively polarized, and it was subjected to a positive bias voltage. The resonance firstly decreased until the varactor's ferroelectric domains start switching. The ferroelectric switching started at around 2.5 V, above

which the resonance frequency started to increase sharply (see the blue line in Fig. 1-a). At that point, most of the ferroelectric domains in the HZO varactor were positively polarized. After that, the DC bias was brought back to 0 V. The resonance frequency of the LC circuit decreased following the orange curve. Then the same procedure was repeated for the negative DC bias values. The behavior of the LC circuit's resonance frequency for negative DC bias voltages is shown by the yellow and purple curves. Fig. 2-b shows the admittance response of the LiNbO₃ resonator. As can be seen, its resonance frequency was measured around 33.3 MHz.

The threshold system was assembled on a Printed Circuit Board (PCB), as shown in Fig. 3-c. The system was DC-biased and excited from the drive port at a frequency f_{in} slightly detuned from the resonance frequency of the resonator, while tracking f_{read} from the read-port by using a network analyzer. The temperature responses of the reported threshold sensing system were obtained under two different DC-biasing voltages of 1.7 V and 0 V, and are shown in Fig. 3- d and Fig. 3-e. The system was heated up to 100 °C and then cooled down to room temperature again. During our experiment we found the passive voltage amplification across the varactor to become large enough when the temperature exceeded 95°C, after which a ferroelectric partial switching occurred, producing a sudden notch on f_{read}. Even more, when the temperature re-reached 25°C, we found f_{read} to be different from its original value. Such a non-volatile change is the key to recognizing that a temperature violation has occurred.

CONCLUSIONS

In this work, a novel temperature threshold sensing system based on a LiNbO₃ microacoustic resonator and a ferroelectric HZO varactor was reported. The system uses the LiNbO₃ resonator as a temperature sensing element and leverages the HZO varactor as a non-volatile element of an LC tank whose resonance frequency was used as a read-out parameter. The proposed system was implemented on top of a PCB and its temperature response was measured under two different DC-biasing voltages. The obtained results revealed that the combination of a ferroelectric element with a MEMS device enables the realization of sense, computation, and storage functions.

ACKNOWLEDGEMENTS

This work has been funded by the National Science Foundation CCF-FET program (Grant #2103351 & #2103091). The authors wish to thank the staff of the George J. Kostas Nanoscale Technology and Manufacturing Research Center at Northeastern University for assistance in the device fabrication.

REFERENCES

[1] G. Bruckner, J. Bardong, C. Gruber, and V. Plessky, "A Wireless, Passive ID Tag and Temperature Sensor for a Wide Range of Operation," *Procedia Engineering,* vol. 47, pp. 132-135, 2012/01/01/ 2012, doi: https://doi.org/10.1016/j.proeng.2012.09.102.

[2] H. Campanella, M. Narducci, S. Merugu, and N. Singh, "Dual MEMS Resonator Structure for Temperature Sensor Applications," *IEEE Transactions on Electron Devices,* vol. 64, no. 8, pp. 3368-3376, 2017, doi: 10.1109/TED.2017.2708129.

[3] R. Bhattacharyya, C. Floerkemeier, and S. Sarma, "RFID tag antenna based temperature sensing," in *2010 IEEE International Conference on RFID (IEEE RFID 2010),* 14-16 April 2010 2010, pp. 8-15, doi: 10.1109/RFID.2010.5467239.

[4] I. Zalbide, E. D. Entremont, A. Jiménez, H. Solar, A. Beriain, and R. Berenguer, "Battery-free wireless sensors for industrial applications based on UHF RFID technology," in *SENSORS, 2014 IEEE,* 2-5 Nov. 2014

2014, pp. 1499-1502, doi: 10.1109/ICSENS.2014.6985299.

[5] T. T. Thai *et al.,* "Design and Development of a Novel Passive Wireless Ultrasensitive RF Temperature Transducer for Remote Sensing," *IEEE Sensors Journal,* vol. 12, no. 9, pp. 2756-2766, 2012, doi: 10.1109/JSEN.2012.2201463.

[6] C. Ghouila-Houri *et al.,* "MEMS high temperature gradient sensor for skin-friction measurements in highly turbulent flows," in *2019 IEEE SENSORS,* 27-30 Oct. 2019 2019, pp. 1-4, doi: 10.1109/SENSORS43011.2019.8956802.

[7] A. Jiménez-Sáez *et al,* "Chipless Wireless High Temperature Sensing Based on a Multilayer Dielectric Resonator," in *2019 IEEE SENSORS,* 27-30 Oct. 2019 2019, pp. 1-4, doi: 10.1109/SENSORS43011.2019.8956863.

[8] H. M. E. Hussein and C. Cassella, "Giant Sensitivity through Fully-Passive and Chip-Less Parametric Temperature Sensors," in *2020 IEEE SENSORS,* 25-28 Oct. 2020 2020, pp. 1-4, doi: 10.1109/SENSORS47125.2020.9278907.

[9] H. M. E. Hussein, M. Rinaldi, M. Onabajo, and C. Cassella, "Capturing and recording cold chain temperature violations through parametric alarm-sensor tags," *Applied Physics Letters,* vol. 119, no. 1, p. 014101, 2021/07/05 2021, doi: 10.1063/5.0054022.

[10] H. M. E. Hussein, M. Rinaldi, M. Onabajo, and C. Cassella, "A chip-less and battery-less subharmonic tag for wireless sensing with parametrically enhanced sensitivity and dynamic range," *Scientific Reports,* vol. 11, no. 1, p. 3782, 2021/02/12 2021, doi: 10.1038/s41598-021-82894-x.

[11] X.-G. Tian *et al,* "High-resolution, high-linearity temperature sensor using surface acoustic wave device based on LiNbO₃/SiO₂/Si substrate," *AIP Advances,* vol. 6, no. 9, p. 095317, 2016/09/01 2016, doi: 10.1063/1.4963797.

[12] J. Zhao *et al.,* "The research of dual-mode film bulk acoustic resonator for enhancing temperature sensitivity," *Semiconductor Science and Technology,* vol. 36, no. 2, p. 025018, 2021/01/06 2021, doi: 10.1088/1361-6641/abd15c.

[13] S. Jadhav *et al,* "HZO-based FerroNEMS MAC for in-memory computing," *Applied Physics Letters,* vol. 121, no. 19, p. 193503, 2022/11/07 2022, doi: 10.1063/5.0120629.

[14] V. Gund *et al.,* "Multi-level Analog Programmable Graphene Resistive Memory with Fractional Channel Ferroelectric Switching in Hafnium Zirconium Oxide," in *2022 Joint Conference of the European Frequency and Time Forum and IEEE International Frequency Control Symposium (EFTF/IFCS),* 24-28 April 2022 2022, pp. 1-4, doi: 10.1109/EFTF/IFCS54560.2022.9850768.

CONTACT

*Onurcan Kaya, tel: +1-605-2520066; kaya.on@northeastern.edu

*Cristian Cassella, tel: +1-267-9925507; c.cassella@northeastern.edu

RESONANT CONFINERS FOR ACOUSTIC LOSS MITIGATION IN BULK ACOUSTIC WAVE RESONATORS

Jeronimo Segovia-Fernandez, and Ernest T.-T. Yen
Texas Instruments, Kilby Labs, USA

ABSTRACT

This paper presents a new method to mitigate the lateral acoustic energy leakage and improve the parallel Q factor (Q_p) of thickness-extensional BAW resonators. The proposed technique consists in surrounding the top electrode with concentric $\lambda/2$ metal rings (*a.k.a.* resonant confiners) that vibrate at the A_0-mode present in the substrate outskirts at the parallel resonance frequency (f_p). Experimental results from dual-Bragg acoustic resonators (DBARs) of different electrode shapes (circle, square and pentagon) and normalized area ratios (100%, 75%, 50% and 25%) show double-digit percentage boosts, validating our technique to improve Q_p and setting a precedent for its implementation in other types of BAW technologies.

KEYWORDS

BAW resonator, Q-factor, acoustic energy losses, antisymmetric mode, resonant confiners.

INTRODUCTION

The proliferation of new wireless communication protocols for Industry 4.0 (*e.g.*, WirelessHART), telemedicine (*e.g.*, NB-IoT), AgTech (*e.g.*, ZigBee), home automation (*e.g.*, Wi-Fi 7), and self-driving EV (*e.g.*, V2X) (among others) have urged international spectrum stakeholders to create new frequency bands and constrain the existing Tx/Rx channel requirements. For decades, RF filters and oscillators based on acoustic wave resonators have been considered as the preferred technologies to select or generate signals in the L and S bands due to their precise frequency response, small form factor, and low manufacturing cost. Compared with surface acoustic wave (SAW) resonators, bulk acoustic wave (BAW) resonators can be co-fabricated with semiconductor circuits, operate at higher resonance frequencies, and attain better Q factors, improving insertion loss and roll-off in RF filters, and phase noise and power consumption in oscillators [1]. In BAW resonators, the main energy loss mechanisms defining their parallel Q-factor (Q_p) have either an intrinsic nature and depend on volume (anelastic damping) or an extrinsic nature and depend on the area-to-perimeter ratio (acoustic leakage) [2]. The latter is especially important for higher resonance/smaller devices and has led BAW designers to find innovative methods that can efficiently contain the energy within the main resonator body.

Based on the mode of vibration, every type of BAW technology requires a specific framing technique to mitigate acoustic leakage. For example, Lamb wave resonators exhibiting a low-dispersion symmetric S_0-mode rely on reflective gratings (or Bragg reflectors) to minimize the lateral acoustic losses [3]. These reflective gratings are formed by metal strips of $\lambda/4$ linewidth and separation and take advantage of the large acoustic velocity mismatch that exists between metalized and non-metalized regions. In the case of thickness-extensional BAW resonators, the antisymmetric A_0-mode exhibited at the parallel resonance frequency (f_p) shows low acoustic velocity mismatch [4], which hinders the implementation of reflective gratings. As an alternative, film bulk acoustic resonator (FBAR) and solidly-mounted resonator (SMR) designers have opted for placing dual-step frames at the edge of the top electrode in order to improve Q_p and, simultaneously, reduce spurious modes [5], [6]. However, this technique has its own drawbacks. For example, the staircase construction requires additional metal patterning steps, which raises process cost and complexity, and the frames only work when placed within the resonator active region, which limits their size and range of energy confinement.

In this paper, we present a new technique to improve the Q_p of thickness-extensional BAW resonators by surrounding the top electrode with $\lambda/2$ metal rings that resonate at the A_0-mode and confine the lateral acoustic energy traveling into the substrate. To validate the proposed technique, the so-called resonant confiners are implemented within an emerging type of BAW resonator known as dual-Bragg acoustic resonator (DBAR) based on an unreleased AlN structure sandwiched between top and bottom Mo electrodes and acoustic Bragg mirrors (Fig. 1). The Bragg mirrors are formed by alternating $\lambda/4$ SiO_2 and TiW layers acting as low and high acoustic impedances, respectively, co-optimized to reduce transmission of both shear and longitudinal acoustic waves [7]. Simulations and experiments confirm that DBARs using resonant confiners can improve their Q_p irrespective of the electrode shape (*e.g.*, circle, square and pentagon) and active area.

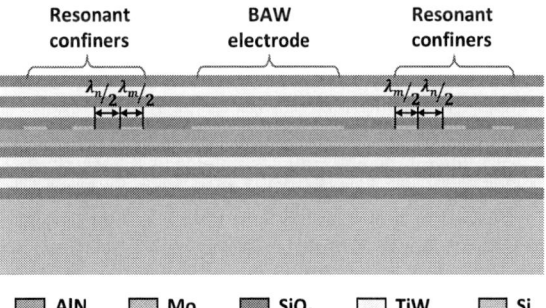

Figure 1: Cross-section of a DBAR with lateral resonant confiners for Q_p improvement.

ANTISYMMETRIC MODE ANALYSIS

To simulate the DBAR mode of vibration, we use COMSOL Multiphysics. The plate stress and strain components are computed by means of a 2D axisymmetric model in which the center of resonator is aligned with the rotation axis ($r=0$) and perfectly matched layers (PML) are added to the bottom and right sides of geometry to emulate a semi-infinite like substrate [8]. At f_p, the mode of vibration in the resonator active region, which coincides with the area covered by top electrode, exhibits a symmetric motion that makes the AlN plate to periodically expand and contract in the vertical direction (Fig. 2.a). In comparison to the "piston" mode displayed at series resonance [9], this symmetric mode shows high lateral dispersion and wave reflection at the edge of top electrode. Although some energy gets reflected, a small portion of the energy generated in the active region travels into the substrate and gets dissipated. If we amplify the vertical displacement beyond the top electrode, we find that the traveling waves exhibit an antisymmetric mode of vibration through which the AlN plate periodically moves up and down along its radius (Fig. 2.b). This antisymmetric motion shows no spatial attenuation, which proves the propagating (real) nature of the mode. To compute the acoustic wavelength in the non-metallized region (λ_n), we set a Cut Line 2D between the AlN and bottom electrode, and simulate the vertical displacement. Hence, we obtain $\lambda_n = 1.48\mu m$ (Fig. 2.c).

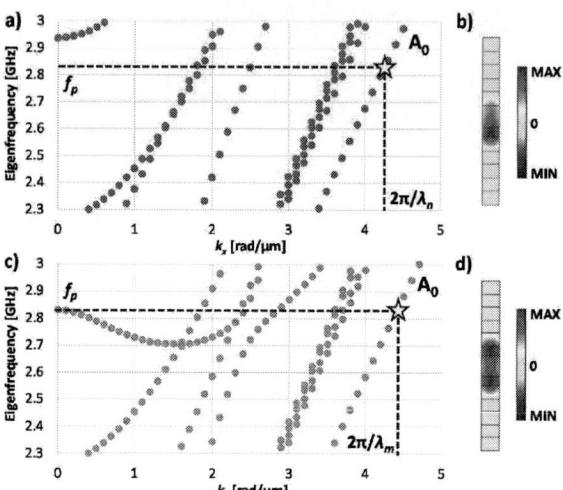

Figure 3: Dispersion characteristics of a) non-metalized and c) metalized regions. A_0-mode displayed in b) non-metalized and d) metalized regions at f_p and identification of their respective acoustic wavelengths (λ_n and λ_m).

After finding $\lambda_n/2$ and $\lambda_m/2$, we can design the resonant confiners around the DBAR and, via FEM, adjust the gap that exists between top electrode and first metal confiner. This gap is important to set the phase of acoustic waves reaching the substrate. Based on simulations, the optimum gap is the one that guarantees maximum displacement at the edges of metal rings. With the adoption of resonant confiners, the mode of vibration in the active region changes from symmetric to antisymmetric, replicating the substrate mode and mitigating lateral reflections (Fig. 4.a). Beyond the top electrode, we observe that resonant confiners vibrate at the A_0-mode, generating as a standing waveform whose amplitude decreases over distance (Fig. 4.b). This exponential decay is more evident if we set a Cut Line 2D between the AlN and bottom electrode and plot the vertical displacement (Fig. 4.c).

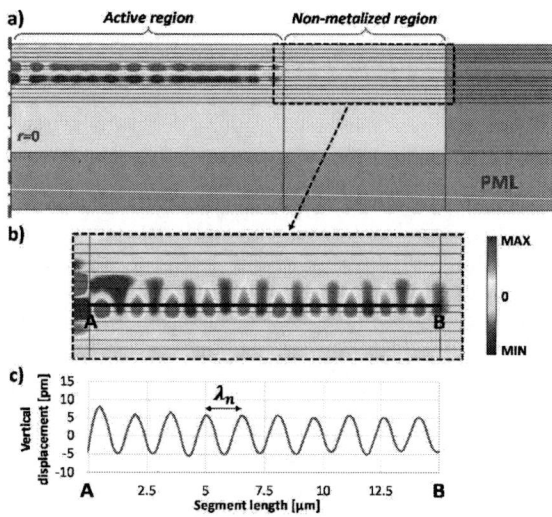

Figure 2: Axisymmetric FEM of DBAR. a) Mode shape at f_p, b) amplified mode shape outside the top electrode region, and c) vertical displacement along the substrate showing no attenuation and high leakage.

To unveil the antisymmetric mode order, we simulate the dispersion characteristics of an infinite plate that replicates the substrate cross-section by applying Floquet periodicity to the lateral edges in both electrical and mechanical domains [10]. In this way, we find that f_p and the lateral dispersion wavenumber ($2\pi/\lambda_n$) intersect at the A_0-mode (Fig. 3.a). In addition, we compare the vertical displacement displayed in the substrate (Fig. 2.b) with the mode shape generated through dispersion analysis (Fig. 3.b) and find that they are equivalent. The proposed resonant confiners technique relies on implementing an array of metal rings concentric to the DBAR active region where the width and separation are equivalent to $\lambda/2$ of the propagating A_0-mode. The metal ring separation, whose cross-section coincides with the non-metalized region, is set as $\lambda_n/2$, and the metal ring width is set as $\lambda_m/2$. To compute λ_m, we include the top electrode metal as an electrically-inactive layer in the infinite-plate substrate and re-run dispersion analysis. Hence, we find that f_p intersects with the A_0-mode at $\lambda_m = 1.43\mu m$ (Fig. 3.c).

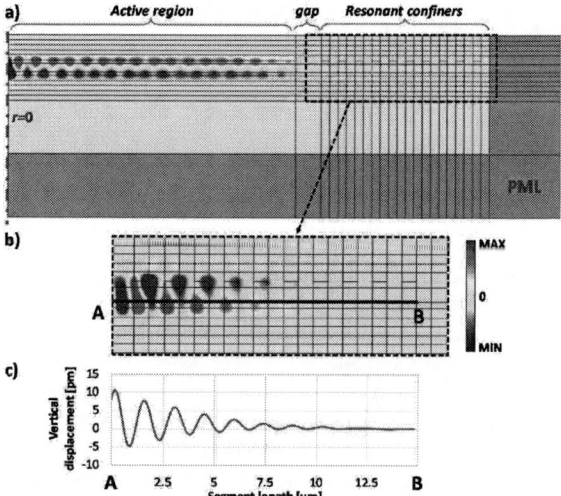

Figure 4: Axisymmetric FEM of DBAR with 10 pairs of resonant confiners. a) Mode shape at f_p, b) amplified mode shape in the confiners, and c) vertical displacement along the substrate showing high attenuation and low leakage.

ADMITTANCE SIMULATIONS

To validate the proposed energy confinement method, we simulate the admittance response and Q_p of DBARs formed by a 0.7μm-thick AlN layer sandwiched between 0.3μm-thick top and bottom Mo electrodes. These simulations are performed using the 2D axisymmetric COMSOL model previously described, in which the only source of damping is substrate leakage. Fig. 5.a shows the admittance response, zoomed in around f_p, of DBARs having non-metalized region and 4 different radii of active region: 76, 66, 54, and 38μm. From this plot, we find that Q_p, computed as the ratio $f_p / \Delta f_{-3dB}$, increases with radius, going from 1,475 (r=38μm) to 4,115 (r=76μm). In contrast, Fig. 5.b shows the admittance response of DBARs having the same radii of active region and including 10 pairs of resonant confiners. From this plot, we find that Q_p does not necessarily increase with radius and is around 2x greater than the values calculated for DBARs without resonant confiners.

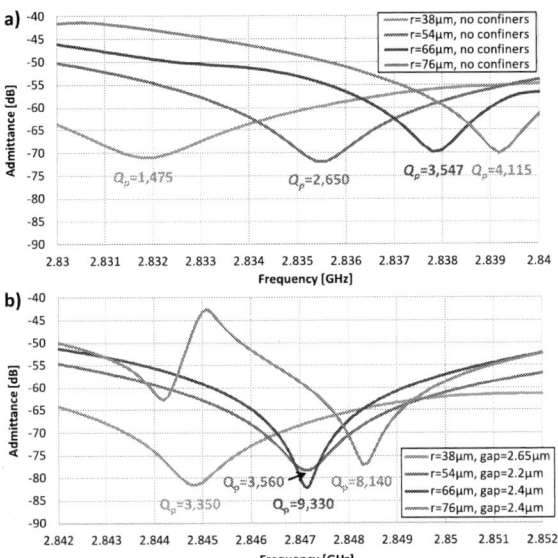

Figure 5: Frequency domain simulations of DBARs of different radius a) without and b) with resonant confiners. The gap is optimized to achieve the largest Q_P.

EXPERIMENTAL RESULTS

The effectiveness of resonant confiners to mitigate lateral acoustic leakage is also validated experimentally. In a first experiment, we built DBARs of equivalent active area and different electrode shape: circle, square, and pentagon, with and without resonant confiners (Fig. 6). At the time of this investigation, our lithographic resolution was 1μm, so we could not tape-out confiners with $\lambda_n/2$ and $\lambda_m/2$ dimensions. As an alternative, we designed metal rings having width and separation equal to λ, which also enables maximum displacement at the edges of metalized regions and, based on simulations, produces similar Q_p results. The DUTs were measured at room conditions directly from the wafer using a semi-automatic Cascade probe station and an Agilent N5230A vector network analyzer to record the S_{11}-parameter and convert it into admittance.

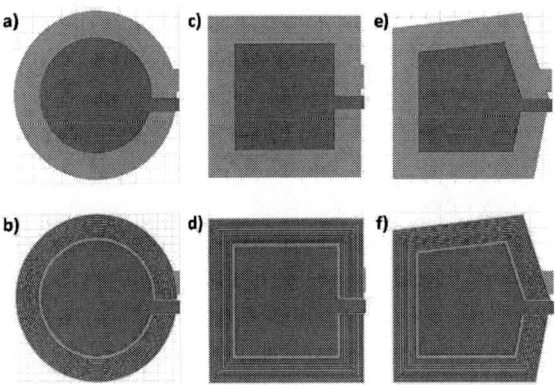

Figure 6: Layout of DBARs of equivalent active area (18,200μm²) and different electrode shapes. a), c) and e) show devices without resonant confiners, and b), d) and f) show devices with resonant confiners. Red and blue layers represent top and bottom electrodes, respectively.

In total, we measured 42 devices (14 per electrode shape). From this experiment, we find that Q_p improves 11% for the pentagon shape and 15% for the circular shape. Instead, no improvement is observed for square-shaped resonators due to a spurious mode near f_p distorting our Δf_{-3dB} measurements (Table 1). Although the improvement is not as high as our predictions, we observe a steady trend that establishes that resonant confiners can work for any geometry of active region. In our opinion, no further improvement is obtained due to three main reasons: 1) there are other energy losses, such as dielectric losses and mechanical losses of anelastic nature, dominating Q_p, 2) the top Bragg mirror shows an uneven surface over the resonant confiners that is difficult to simulate with accuracy, and 3) the thicknesses of DBAR layers are not as expected, modifying the dispersion characteristics of metallized and non-metalized regions and the targeted λ_m and λ_n values.

Table 1. Measured Q_p from DBARs of different electrode shape, with and without resonant confiners. Statistics are generated from a sample of 14 equivalent devices.

	Parallel Q					
	Average		Median		Standard deviation	
	w/o confiners	w/ confiners	w/o confiners	w/ confiners	w/o confiners	w/ confiners
Circle	1066	1224	1095.5	1220	130.41	57.65
Square	1110	1121	1111.5	1167	94.25	197.28
Pentagon	662	734	675.5	742	53.30	41.07

In a second experiment, we built pentagon-shaped DBARs of different area ratios: 100%, 75%, 50% and 25% (Fig. 7). The goal of this experiment is to evaluate the effect of resonator area on acoustic leakage and if the resonant confiners can achieve higher Q_p improvement when acoustic leakage becomes dominant. In addition, we extract the parallel resistance (R_p), which is key to improve return loss in RF filters and power consumption in oscillators, and is calculated as $k_t^2 Q_p/(2\pi f_p C_0)$, where k_t^2 and C_0 stand for electromechanical coupling and dielectric capacitance, respectively.

978-1-6654-9309-3/23 $31.00 © 2023 IEEE

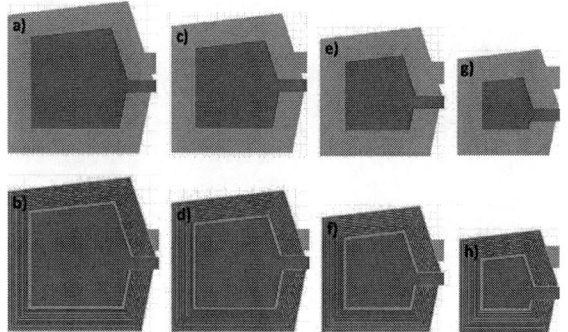

Figure 7: Layout of pentagon-shaped DBARs of normalized electrode area ratios equal to 1, 0.75, 0.5 and 0.25. a), c) e) and g) show devices without resonant confiners, and b), d), f) and h) show devices with resonant confiners. Reference area is 18,200μm².

In total, we measured 56 devices (14 per area ratio) and find that the average Q_p improves 11% when the area is 18,200μm², and 26% when the area is 4,550μm² (or one quarter of the reference area) (Fig. 8). These ratios can be compared with results from Fig. 5 since the simulated radii correspond to circular-shaped DBARs of equivalent active areas. Through this comparison, we corroborate that 1) experiments always show less improvement than simulations and 2) only experiments show a clear trend with the radius. In terms of R_p, we observe a similar trend with area, but smaller improvement ratios due to a decrease in k_t^2 (the inclusion of resonant confiners makes the resonator structure more complaint and less energy efficient). On the other hand, the Q_p-k_t^2 product is still higher, which anticipates the benefits of resonant confiners to improve the performance of BAW resonators operating at higher frequencies (*e.g.*, 5G) when small devices are needed to match RF network conditions.

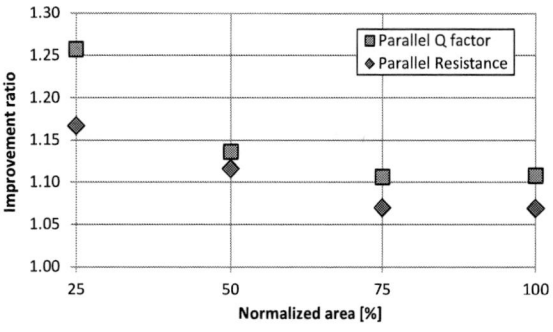

Figure 8: Measured Q_p and R_p improvement ratios due to the inclusion of resonant confiners vs. normalized electrode area.

CONCLUSION

In this paper, a new layout-based technique to enhance the Q_p and R_p of thickness-extensional BAW resonators affected by acoustic leakage has been introduced. This technique relies on surrounding the BAW top electrode with $\lambda/2$ metal rings that resonate at the A_0-mode displayed in the substrate as a way to confine the laterally-traveling acoustic energy. To optimize the width and separation of resonant confiners, infinite plate dispersion analysis is run for the non-metalized (λ_n) and metalized (λ_m) regions forming the substrate. Moreover, the gap that remains in between top electrode and first metal confiner is adjusted via FEM to ensure maximum displacement at the edges of metal rings. Simulations and experiments run on the emerging DBAR validate our technique and demonstrate that resonant confiners can be adopted for any electrode shape and attain higher improvement for smaller areas, setting a precedent for its implementation in other types of BAW technologies, such as FBAR and SMR.

REFERENCES

[1] Y. Liu, Y. Cai, Y. Zhang, A. Tovstopyat, S. Liu, and C. Sun, "Materials, Design, and Characteristics of Bulk Acoustic Wave Resonator: A Review," *Micromachines*, vol. 11, no. 7, Art. no. 7, Jul. 2020.

[2] M.-A. Dubois and C. Muller, "Thin-Film Bulk Acoustic Wave Resonators," in *MEMS-based Circuits and Systems for Wireless Communication*, C. C. Enz and A. Kaiser, Eds. Boston, MA: Springer US, 2013, pp. 3–28.

[3] V. Yantchev and I. Katardjiev, "Design and fabrication of thin film lamb wave resonators utilizing longitudinal wave and interdigital transducers," in *IEEE Ultrasonics Symposium, 2005.*, Sep. 2005, vol. 3, pp. 1580–1583.

[4] B. D. Zaitsev and S. G. Joshi, "Reflection of ultrasonic Lamb waves produced by thin conducting strips," *IEEE Trans Ultrason Ferroelectr Freq Control*, vol. 46, no. 6, pp. 1539–1544, 1999.

[5] N. Nguyen, A. Johannessen, S. Rooth, and U. Hanke, "A Design Approach for High-Q FBARs With a Dual-Step Frame," *IEEE Trans Ultrason Ferroelectr Freq Control*, vol. 65, no. 9, pp. 1717–1725, Sep. 2018.

[6] X. Li *et al.*, "Use of double-raised-border structure for quality factor enhancement of type II piston mode FBAR," in *2017 Joint Conference of the European Frequency and Time Forum and IEEE International Frequency Control Symposium (EFTF/IFCS)*, Jul. 2017, pp. 547–550.

[7] E. T.-T. Yen *et al.*, "Integrated High-frequency Reference Clock Systems Utilizing Mirror-encapsulated BAW Resonators," in *2019 IEEE International Ultrasonics Symposium (IUS)*, Oct. 2019, pp. 2174–2177.

[8] J. Segovia-Fernandez and G. Piazza, "Analytical and Numerical Methods to Model Anchor Losses in 65-MHz AlN Contour Mode Resonators," *Journal of Microelectromechanical Systems*, vol. 25, no. 3, pp. 459–468, Jun. 2016.

[9] J. Kaitila, M. Ylilammi, J. Ella, and R. Aigner, "Spurious resonance free bulk acoustic wave resonators," in *IEEE Symposium on Ultrasonics, 2003*, Oct. 2003, pp. 84-87.

[10] D. L. Jaggard and C. Elachi, "Floquet and coupled-waves analysis of higher-order Bragg coupling in a periodic medium*," *J. Opt. Soc. Am., JOSA*, vol. 66, no. 7, pp. 674–682, Jul. 1976.

CONTACT

*J. Segovia-Fernandez, jeronimo.segovia@ti.com

HIGH-CRYSTALLINITY 30% SCALN ENABLING HIGH FIGURE OF MERIT X-BAND MICROACOUSTIC RESONATORS FOR MID-BAND 6G

Gabriel Giribaldi, Pietro Simeoni, Luca Colombo, and Matteo Rinaldi
SMART Center, Northeastern University, Boston, MA, USA

ABSTRACT

This paper presents an experimental comparison of high-performance X-band piezoelectric 30% Sc-doped Aluminum Nitride (ScAlN) Cross-sectional Lamé Mode resonators (CLMRs) for two different thin-film's crystallinity levels. The presented MEMS devices stand out in terms of electromechanical coupling (k_t^2) and quality factor (Q) in this frequency range, while being fabricated with a low-complexity 3-masks micro-machining process. Nevertheless, the resonators employing a higher crystallinity film feature motional $Q \cdot k_t^2$ Figures of Merit (FoM) that are up to 5.6 times higher per same geometry, highlighting the impact of the piezoelectric layer quality in the final device performance. Moreover, motional quality factors (Q_m) larger than similar X-band AlN devices are achieved while attaining a 6-fold increase in k_t^2. This demonstrates that ScAlN can deliver resonator performance well beyond the one achievable by AlN ones and, more importantly, that the doping process does not degrade the mechanical properties of microelectromechanical devices, provided that a high-quality piezoelectric film is used.

This work features a motional $Q \cdot k_t^2$ Figure of Merit of 73 at 9.5 GHz, being, to the authors' knowledge, the highest ever shown above 8 GHz and a bank of lithographically frequency-definable resonators with record-breaking metrics covering the 8 to 11 GHz range, all fabricated on the same substrate. Such metrics have the power of enabling the synthesis of compact, wide-bandwidth, and low insertion loss passive pass-band filters for the next generation 5G and 6G cellular radios.

KEYWORDS

MEMS, microacoustic resonators, CLMR, ScAlN, 5G, 6G

INTRODUCTION

The recent developments of mobile communication are fostering an always-faster progress in virtually any human field, drastically diminishing distance between individuals and time between discoveries, applications, and new life standards. Such progress has latterly culminated in mass-deployment of the fifth-generation mobile network (5G), promising outstanding improvements in Key Parameter Indicators (KPIs) such as data-rate, spectral efficiency, and latency [1]. Nevertheless, almost only the sub-6GHz band of 5G (FR-1), has become part of our daily life, providing a marginal upgrade to the KPIs from the previous generation. The mm-wave spectrum (5G FR-2), *i.e.* the frequency range where the full 5G potential resides, is today far from full scale deployment. This is due to a lack of technological maturity in the hardware (HW) components of Radiofrequency Front-Ends (RFFEs), especially for mobile handsets, and for the need of a high density of base-stations, given the high propagation losses

of electromagnetic (EM) waves at such high carrier frequencies ($f > 24$ GHz). The emerging 7-20 GHz band, on the other hand, provides an excellent trade-off between network capacity and coverage, and is under study as a possible 5G third frequency range (FR-3). Moreover, it is foreseen to become the 6G mid-band [2], devoted to dense urban areas coverage. In that framework, the requirements on the HW components will become more stringent, given the implementation of massive MIMO antenna arrays and 400 MHz or more component carrier bandwidth [2]. In particular, the passive band-pass filters in mobile RFFEs represent a bottleneck, given the lack of miniaturized and cost-effective solutions targeting the 7-20 GHz range. The microacoustic resonator technology, state-of-the-art in this field up to the 5G sub-6 GHz band, has already been demonstrated to be able to be scaled beyond 10 GHz [3]–[6]. In [3] and [4], the state-of-the-art thin-Film Bulk Acoustic Resonator (FBAR) technology is up-scaled to the mm-wave spectrum. Nevertheless, FBARs (and their overtones) lack the possibility of lithographically tuning the resonance frequency, and in commercial products on-chip frequency diversity is achieved via complex fabrications including several mass-loading and trimming steps, notably increasing the cost per device. In [6], on the other hand, top electrode-only devices with k_t^2 up to 4.9% have been demonstrated in the 7-11 GHz range using $Sc_{0.3}Al_{0.7}N$ as active layer. Nevertheless, the quality factor results degraded with respect to similar X-band AlN devices [5], and the presented resonators feature similar motional $Q \cdot k_t^2$ FoM. Such outcome is in line with the literature [7], [8], and looks inevitable: ScAlN devices attain higher k_t^2 but lower Q, and the FoM does not change much.

In this work, we tackle this paradigm, showing resonators' performance well beyond the one attainable by AlN ones, with loaded FoM ($Q_{3dB} \cdot k_t^2$) more than quadruple on similar devices in the same frequency range

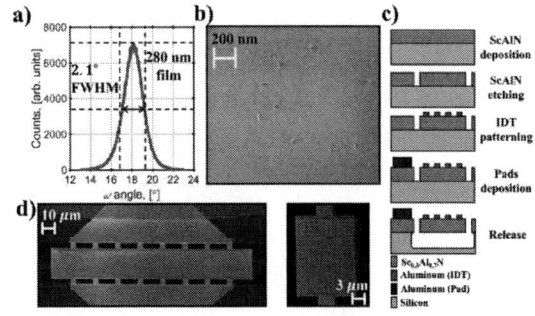

Figure 1: a) X-Ray Diffraction (XRD) Rocking Curve scan of the optimized 280 nm film employed in this work, b) top surface SEM micrograph of the film's surface, c) micro-machining process followed to fabricate the CLMRs, and d) SEM micrographs of showcase devices.

(X-band) [5], and mechanical FoM ($Q_m \cdot k_t^2$) almost 7 times larger. Furthermore, we show a higher motional quality factor than the one obtained on the AlN device of [5] and a bank of high-performance devices in the 8-11 GHz range, all fabricated on the same substrate. Such results stem from the employment of high-crystallinity piezoelectric thin-films, main players responsible of the ultimate device performance.

SCALN AND THIN-FILM SPUTTERING OPTIMIZATION

Sc-doping of AlN provides a material softening and an increase in the dielectric constant and the piezoelectric coefficients [9]. The latter translates into the possibility of achieving higher electromechanical couplings in ScAlN resonators [10], [11], which directly translates into wider filter bandwidth (BW).

The growth of high quality ScAlN is, nevertheless, not trivial and very sensitive to the deposition conditions [12], [13]. Of particular problem is the formation of Abnormally Oriented Grains (AOGs), well known to reduce the performance of microacoustic devices [14], and limiting the maximum thickness achievable by thin-films of this material.

The films employed in this work are sputtered, taking advantage of an industrial-grade Evatec® Clusterline II tool. A thorough optimization process was followed to obtain the desired crystallinity and low-presence of AOGs. The process is similar to the one of [6], utilizing reactive magnetron sputtering, a 12" compound casted $Al_{0.7}Sc_{0.3}$ target, and sole Nitrogen (*i.e.* no Argon) as carrier gas. Of particular interest is that the film is directly deposited on high-resistivity 200 mm silicon wafers, without the need for an AlN seed layer or any metal template.

As in [15], it was found that a lower process pressure ensures higher *c*-axis orientation and much lower presence of AOGs. Therefore, starting from the recipe of [6], the N_2 flow was reduced from 90 to 30 sccm. Moreover, the power applied to the casted target was reduced from 7 to 5 kW. The deposition temperature was 300°C, compatible with standard CMOS processing. Fig.1a-b shows the final Full Width-Half Maximum (*FWHM*) of the target 280 nm film, featuring a 2.1° value, and a top surface Scanning Electron Microscope (SEM) micrograph, showcasing a very low density of AOGs. By further reducing the deposition pressure, higher crystallinity values can be achieved, but the film's stress becomes more compressive, and has to be traded off with the *FWHM*. Finally, it was found that the base pressure has a strong influence on the final crystallinity and process repeatability. Repeatable results were obtained with base pressures below 5e-8 Torr, and a room-temperature deposition chamber conditioning helps reaching such low values.

DEVICE FABRICATION AND TESTING

The devices were fabricated with a standard and low-complexity 3-mask process as in [6]. The top aluminum electrode thickness was increased from 60 to 95 nm, in order to slightly reduce the value of series parasitic resistance. The micro-machining flow, together with top surface SEMs of showcase devices are shown in Fig.1c-d.

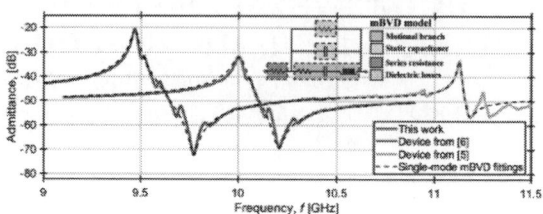

Figure 2: Comparison between the best resonator of this work with the best one from [6] (30% ScAlN) and the one from [5] (AlN). The single-mode fittings are shown. The main metrics and FoM of the resonators are shown in Table 1.

The testing was performed with a VNA (Keysight N5221A), two Cascade Ground-Signal-Ground (GSG) RF probes with 150 μm pitch, two low-loss RF cables, and a probe station in laboratory conditions. The experimental setup was calibrated via the Transmission-Reflection-Line (TRL) standard.

Table 1: Comparison of the main metrics and FoM of the devices shown in Fig.2. R_s is the mBVD series resistance and Z_0 the matching impedance.

Device	[5]	[6]	This work
Material	AlN	$Sc_{0.3}Al_{0.7}N$	$Sc_{0.3}Al_{0.7}N$
f_s **[GHz]**	11	10	9.5
$k_{t\,SM}^2$ **[%]**	1.3	4.9	8.05
Q_{3dB}	615	264	420
Q_m	800	260	905
R_s **[Ω]**	12	8	7
Z_0 **[Ω]**	290	290	215
$Q_{3dB} \cdot k_{t\,SM}^2$	8	13	33.8
$Q_{3dB} \cdot f_s$ **[THz]**	6.76	2.64	3.99
$Q_m \cdot k_{t\,SM}^2$	10.7	13	73
$Q_m \cdot f_s$ **[THz]**	8.8	2.6	8.6

EXPERIMENTAL RESULTS

Impact of Sc-doping and crystallinity on resonators' performance

In Fig.2, the admittance curves of three X-band Lateral Field Excited (LFE) CLMRs are compared. Their active layers are made of AlN (yellow curve) and of $Sc_{0.3}Al_{0.7}N$ (blue and red). The AlN device is the one reported in

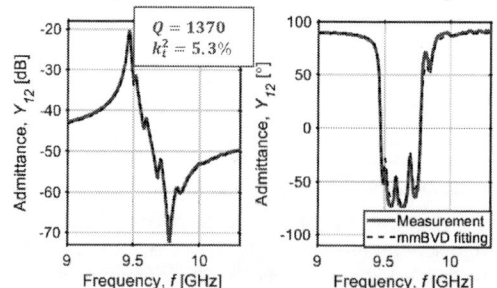

Figure 3: Multi-mBVD fitting of the best resonator from this work, featuring higher motional quality factor and lower k_t^2, but same motional FoM of the single-mode-fitted one.

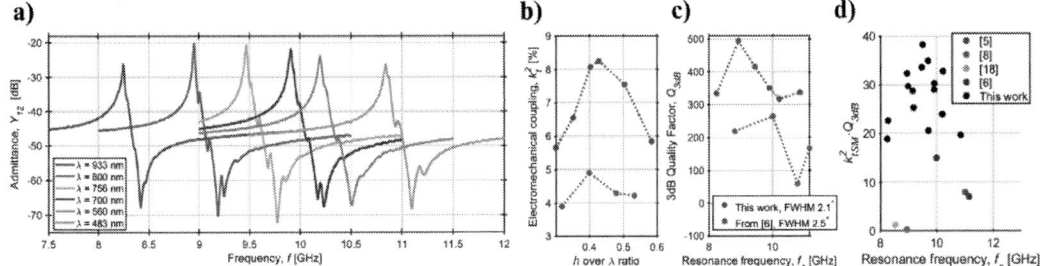

Figure 4: a) Showcase of 6 resonators from this work in the 8-11 GHz of frequency operation. Their metrics and FoM are reported in Table 2. In b-c), the k_t^2 and Q of the devices in a) and the ones of [6] above 8 GHz are compared. In d), the $Q \cdot k_t^2$ FoM of the devices in a) plus others made on the same chip is compared with fundamental-mode Lamb Wave resonators in the literature.

[5], while the ScAlN devices are the best one from [6] and the best one from this work and differ only in the crystallinity of their piezoelectric layer (*FWHM* of 2.5° and 2.1°, respectively), and in the top electrode thickness, $t_{top} = 60$ nm and $t_{top} = 95$ nm, respectively. Their metrics and FoM are reported in Table 1. The three devices are non-degenerate CLMRs, meaning CLMRs where the ratio of the thickness of the piezoelectric layer (h) and the horizontal acoustic wavelength (λ) are optimal, giving the maximum attainable electromechanical coupling [16]. Such geometrical ratio is found to be around 0.4 for LFE CLMRs [6], [11]. The AlN device has $\lambda = 830$ nm, while the ScAlN ones have $\lambda = 700$ nm and a thinner piezo. The AlN resonator has a higher resonance frequency compared to the ScAlN ones even if the horizontal acoustic wavelength is larger since AlN is stiffer than ScAlN. The difference in resonance frequency (f_s) for the ScAlN devices is explained by the different top electrode thicknesses.

The resonators were fitted with a modified Butterworth-Van Dyke (mBVD) model. Compared to AlN, the ScAlN devices feature larger single-mode k_t^2 ($k_{t\ SM}^2 = \pi^2/8(f_p^2 - f_s^2)/f_s^2$, f_p being the anti-resonance frequency), 3.7 and 6.2 times higher, respectively. The $k_{t\ SM}^2$ value of the higher crystallinity ScAlN is in line with Finite Element Modeling (FEM) simulations, experimentally demonstrating the impact of the *FWHM* on the final resonator performance. Nevertheless, also the 3dB quality factor is higher per lower *FWHM*, and the reason is attributed to the much lower density of AOGs. By utilizing the model of [17], the series resistance was calculated for the ScAlN devices and added to the contact resistance, therefore also the motional quality factors of the devices could be extracted, featuring 260 and 905, respectively. The contact resistance was extracted as in [17]. A recorded motional $Q \cdot k_t^2$ FoM equal to 73 is the largest ever reported above 8 GHz. The computed motional quality factors demonstrate with no doubt the potentialities of ScAlN in delivering resonators with higher performance compared to AlN.

Multi-mBVD fitting

As can be observed in Fig.2, the devices suffer from the presence of transversal spurious modes. Such spurious responses interfere with the extraction of the motional parameters of the main mode. In fact, a single-mode fitting tends to overestimate the k_t^2 and underestimate the Q. To

have a more accurate metrics extraction, a multi-mBVD (mmBVD) fitting can be used. Such fitting process is not trivial given the non-unicity of the solution. Fig.3 shows the same resonator of Fig.2, but with a multi-mBVD fitting, also including the phase response. Moreover, the $Q_{m_{mmBVD}} \cdot f_s$ product FoM of 13.02 THz is, to the author's knowledge, the highest ever reported when employing the microacoustic technology in a fundamental mode (no overtones). When using a multi-mBVD fitting, the final value of the electromechanical coupling results to be smaller than the predicted one from FEM. Therefore, more room for improvement of the present results exists.

On-chip multifrequency

Harnessing CLMR's lithographic tunability of the resonance frequency, devices in the 8-11 GHz were fabricated on the same chip. The admittance curves are shown in Fig.4a, and their record-breaking metrics and FoM reported in Table 2. The devices are all matched close to 200 Ω, therefore not too far from 50 Ω. For simplicity, and to be able to make fair comparisons, the 3dB quality factor and the single mode k_t^2 are reported. In Fig.4b-c, the $k_{t\ SM}^2$ vs. h/λ and Q_{3dB} vs. resonance frequency are compared for the devices of [6] and the ones of Fig.4a, once more highlighting how vital is the role played by the film's crystallinity in setting the final devices' performance.

Finally, in Fig.4d, the same devices' metrics, together with the ones of other resonators fabricated in the same chip with the same process flow, are compared with the X-band fundamental mode Lamb-Wave resonators present in the literature [5], [6], [8], [18], clearly standing out.

Table 2: Main metrics of the resonators of Fig.4a.

Res. λ [μm]	f_s [GHz]	$k_{t\ SM}^2$ [%]	Q_{3dB}	$Q_{3dB} \cdot k_{t\ SM}^2$
933	8.24	5.65	334	18.9
800	8.95	6.54	494	32.32
700	9.47	8.05	420	33.81
650	9.9	8.25	350	28.94
560	10.2	7.54	316	23.9
483	10.85	5.84	337	20

CONCLUSIONS

In this work, we presented resonators in the X-band with record-breaking performance. In particular, the highest $Q_{3dB} \cdot k_{t\ SM}^2$, $Q_m \cdot k_{t\ SM}^2$, and $Q_{m_{mmBVD}} \cdot f_s$ FoM above 8 GHz are attained when employing the microacoustic technology in a fundamental mode. More importantly, the present paper, following up on the preliminary results of [6], irrefutably proves that the dogma of ScAlN devices having lower quality factors with respect to AlN ones is wrong. Sc-doped AlN allows, therefore, the fabrication of resonators with undisputed performance in the X-band and enable the synthesis of wide-BW and compact passive ladder filters for future 5G and mid-band 6G applications.

REFERENCES

[1] "IEEE 5G and beyond technology roadmap white paper," 2017.

[2] H. Holma, H. Viswanathan, and P. Mogensen, "Extreme massive MIMO for macro cell capacity boost in 5G-Advanced and 6G." Nokia Bell Labs.

[3] M. Hara et al., "Super-High-Frequency Band Filters Configured with Air-Gap-Type Thin-Film Bulk Acoustic Resonators," *Jpn J Appl Phys*, vol. 49, no. 7S, p. 07HD13, Jul. 2010, doi: 10.1143/JJAP.49.07HD13.

[4] Z. Schaffer, P. Simeoni, and G. Piazza, "33 GHz Overmoded Bulk Acoustic Resonator," *IEEE Microwave and Wireless Components Letters*, vol. 32, no. 6, pp. 656–659, 2022, doi: 10.1109/LMWC.2022.3166682.

[5] M. Assylbekova, G. Chen, G. Michetti, M. Pirro, L. Colombo, and M. Rinaldi, "11 GHz Lateral-Field-Excited Aluminum Nitride Cross-Sectional Lamé Mode Resonator," in *IFCS-ISAF 2020 - Joint Conference of the IEEE International Frequency Control Symposium and IEEE International Symposium on Applications of Ferroelectrics, Proceedings*, 2020. doi: 10.1109/IFCS-ISAF41089.2020.9234874.

[6] G. Giribaldi et al., "X-Band Multi-Frequency 30% Compound SCALN Microacoustic Resonators and Filters for 5G-Advanced and 6G Applications," in *2022 Joint Conference of the European Frequency and Time Forum and IEEE International Frequency Control Symposium (EFTF/IFCS)*, 2022, pp. 1–4. doi: 10.1109/EFTF/IFCS54560.2022.9850563.

[7] R. H. Olsson, Z. Tang, and M. D'Agati, "Doping of Aluminum Nitride and the Impact on Thin Film Piezoelectric and Ferroelectric Device Performance," in *Proceedings of the Custom Integrated Circuits Conference*, 2020, vol. 2020-March. doi: 10.1109/CICC48029.2020.9075911.

[8] M. Park, Z. Hao, R. Dargis, A. Clark, and A. Ansari, "Epitaxial Aluminum Scandium Nitride Super High Frequency Acoustic Resonators," *Journal of Microelectromechanical Systems*, vol. 29, no. 4, 2020, doi: 10.1109/JMEMS.2020.3001233.

[9] M. A. Caro et al., "Piezoelectric coefficients and spontaneous polarization of ScAlN," *Journal of Physics: Condensed Matter*, vol. 27, no. 24, May 2015.

[10] G. Esteves et al., "Al0.68Sc0.32N Lamb wave resonators with electromechanical coupling coefficients near 10.28%," *Appl Phys Lett*, vol. 118, no. 17, p. 171902, 2021, doi: 10.1063/5.0047647.

[11] G. Giribaldi, L. Colombo, F. Bersano, C. Cassella, and M. Rinaldi, "Investigation on the Impact of Scandium-doping on the kt2 of ScxAl1-xN Cross-sectional Lamé Mode Resonators," in *IEEE International Ultrasonics Symposium, IUS*, 2020, vol. 2020-Septe. doi: 10.1109/IUS46767.2020.9251829.

[12] M. Pirro, B. Herrera, M. Assylbekova, G. Giribaldi, L. Colombo, and M. Rinaldi, "Characterization of Dielectric and Piezoelectric Properties of Ferroelectric AlScN Thin Films," in *Proceedings of the IEEE International Conference on Micro Electro Mechanical Systems (MEMS)*, 2021, vol. 2021-Janua. doi: 10.1109/MEMS51782.2021.9375427.

[13] S. Yasuoka et al., "Effects of deposition conditions on the ferroelectric properties of (Al1−xScx)N thin films," *J Appl Phys*, vol. 128, no. 11, p. 114103, 2020, doi: 10.1063/5.0015281.

[14] C. Liu, B. Chen, M. Li, Y. Zhu, and N. Wang, "Evaluation of the Impact of Abnormally Orientated Grains on the Performance of ScAlN-based Laterally Coupled Alternating Thickness (LCAT) Mode Resonators and Lamb Wave Mode Resonators," in *2020 IEEE International Ultrasonics Symposium (IUS)*, 2020, pp. 1–3. doi: 10.1109/IUS46767.2020.9251507.

[15] R. Beaucejour, V. Roebisch, A. Kochhar, C. G. Moe, M. D. Hodge, and R. H. Olsson, "Controlling Residual Stress and Suppression of Anomalous Grains in Aluminum Scandium Nitride Films Grown Directly on Silicon," *Journal of Microelectromechanical Systems*, vol. 31, no. 4, pp. 604–611, 2022, doi: 10.1109/JMEMS.2022.3167430.

[16] C. Cassella, Y. Hui, Z. Qian, G. Hummel, and M. Rinaldi, "Aluminum Nitride Cross-Sectional Lamé Mode Resonators," *Journal of Microelectromechanical Systems*, vol. 25, no. 2, 2016, doi: 10.1109/JMEMS.2015.2512379.

[17] L. Colombo, A. Kochhar, G. Vidal-Álvarez, P. Simeoni, U. Soysal, and G. Piazza, "Sub-GHz X-Cut Lithium Niobate S_0 Mode MEMS Resonators," *Journal of Microelectromechanical Systems*, pp. 1–13, 2022, doi: 10.1109/JMEMS.2022.3204449.

[18] M. Rinaldi, C. Zuniga, and G. Piazza, "5-10 GHz AlN Contour-Mode Nanoelectromechanical Resonators," in *2009 IEEE 22nd International Conference on Micro Electro Mechanical Systems*, 2009, pp. 916–919. doi: 10.1109/MEMSYS.2009.4805533.

CONTACT

*G. Giribaldi, giribaldi.g@northeastern.edu

FERRITE-ROD ANTENNA DRIVEN WIRELESS RESOSWITCH RECEIVER

Kevin H. Zheng[1], Qiutong Jin[1], and Clark T.-C. Nguyen[1]
[1]University of California, Berkeley

ABSTRACT

A micromechanical resoswitch receiver coupled to a 23.7-kHz wireless channel via a ferrite-rod antenna successfully demonstrates short-range wireless reception with resoswitch sensitivity better than –62 dBm at an on-off keying (OOK) bit rate of 0.9 kbit/s. The demonstrated performance confirms models that predict the maximum bit rate of such a receiver, revealing an ability to adapt to bit rates as high as 8 kbit/s sans antenna that surprisingly are not limited mechanically, but then reduce upon introduction of finite electrical bandwidth antennas. Indeed, the demonstrated wireless resoswitch receiver displays a bit rate-to-sensitivity trade-off governed by the finite rise time of the signal received by the antenna. Meanwhile, the receiver consumes no power while waiting and listening for valid inputs. Unlike an RFID tag, this resoswitch receiver rejects strong interferers via the much higher Q of its mechanical resonance, allowing it to serve applications that must operate in congested wireless environments.

KEYWORDS

Resoswitch, wireless, receiver, antenna, bit rate, low power, resonator, micromechanical, quality factor.

INTRODUCTION

By harnessing resonance vibration to reduce the power required to close a mechanical switch, micromechanical resonant switches ("resoswitches") [1] [2] provide mechanical means to filter, amplify, and demodulate communication signals, making possible transistor-less mechanical communication receivers capable of 1-kbit/s bit rates and –60-dBm sensitivities while consuming no power when waiting and listening for valid input codes [3]. These attributes enable continuous monitoring without any need for sleep/wake cycling to save power, making resoswitches particularly attractive for always-on monitoring applications. A recent advancement that employs pre-energization to increase the bit rate capacity of resoswitches to 8 kbit/s [4] now provides an opportunity to expand the application range from low bit-rate tagging and label updates to higher-speed voice communications. In fact, theory predicts that pre-energization allows a resoswitch to adapt to any needed bit rate if not constrained by other system limitations, such as noise or finite system bandwidth.

To investigate the effect of finite system bandwidth on practical wireless resoswitch receivers, this work evaluates wireless reception using a resoswitch fed by a ferrite-rod antenna. Ferrite-rod antennas are particularly suited for the resoswitch devices in this work because they can be tuned to the 20-kHz range and support a relatively high output impedance in the >10 kΩ range. While pre-energizing theory suggests that resoswitch device Q poses no limit on the maximum bit rate of a resoswitch receiver [4], this work shows that bandwidth limitations imposed by the Q of the antenna matching network can still limit the bit rate. Ultimately, harnessing the high bit rate supportable by a resoswitch requires proper co-design with the antenna.

RECEIVER STRUCTURE & OPERATION

Fig. 1 summarizes the wireless demonstration set-up employing ferrite-rod antennas and a resoswitch configured as a wireless receiver. The resoswitch consists of a movable gold shuttle suspended 2.5 μm above a silicon substrate by folded-beam springs, flanked on opposite sides by two comb-drive transducer electrode pairs with interdigitated fingers and two sharp impactor protrusions separated 1 μm from spring-softened impact output electrodes. The ferrite-rod receive antenna connects to the left transducer pair, which serves as the input. The right transducer pair is unused but, if desired, can be used to boost the input electromechanical coupling when a differential input signal is available, e.g., as supplied by a balun [1].

Proper resoswitch operation requires application of a dc bias V_{DD} to its conductive shuttle, in this case 5V, that serves to amplify the force induced by a voltage input signal applied to its input electrode, e.g., by the antenna in Fig. 1. Note that merely applying V_{DD} to the resoswitch shuttle

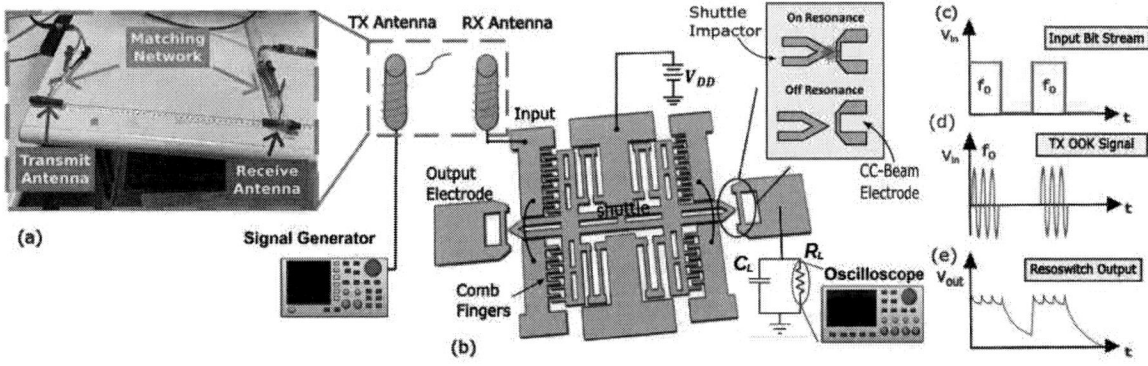

Fig. 1: Wireless communication demonstration set-up employing (a) ferrite-rod antennas and (b) an all-mechanical resoswitch to receive and demodulate the (d) OOK-modulated data bit stream of (c) into the matching resoswitch output bit stream of (e).

draws no dc current and consumes no power, since when at rest, the resoswitch shuttle is effectively one plate of a charged capacitor. In this state, the resoswitch is effectively listening for valid inputs while consuming no power. It consumes power only upon reception of a valid input that it must demodulate and amplify.

In this wireless demonstration, data in the form of a bit stream as shown in Fig. 1(c) is modulated onto a carrier frequency before sending it through the communication channel. Here, on-off keying (OOK) serves as a convenient modulation scheme, where at the transmitter, each '1' maps to a sinusoid at the resonance frequency of the resoswitch, and each '0' maps to 0V. Thus, when sending a '0' bit, no signal is transmitted nor received, the shuttle remains stationary, and any charge on the load capacitor C_L drains to ground via a large bleed resistor R_L, which in the demonstration setup is the 1 MΩ oscilloscope input impedance. Since the output across the load capacitor measures 0V, this indicates successful transmission of the '0' bit. On the other hand, when sending a '1' bit, the transmit antenna outputs a sinusoid at the resoswitch resonance frequency. Once received by the receive antenna, the signal drives the resoswitch shuttle to vibrate at resonance and periodically impact its output electrode. The metal-to-metal contact of each impact transfers charge from V_{DD} to C_L, charging the output to V_{DD}, which constitutes successful transmission and reception of a '1' bit.

Note that only input signals with frequencies in the passband of the resoswitch can excite it to impact, since only these signals benefit from amplification by the high Q of the mechanical device. Stray off-resonance signals, e.g., from other wireless devices, will not generate enough amplitude for impact. In this way, the resoswitch rejects out-of-band interferers as a good wireless receiver should.

ANTENNA-RESOSWITCH CO-DESIGN

Feeding a resoswitch receiver efficiently using a ferrite-rod antenna poses several challenges. One challenge is impedance matching, where the very high impedance of a comb-driven resonator practically precludes matching; another is the small, tuned bandwidth of the ferrite-rod antenna and its matching network.

Ferrite-Rod Antenna

Each ferrite-rod antenna in Fig. 1 comprises a magnetic rod wrapped in winds of wire, which allows it to be electrically small. In such a small loop antenna, the current around the perimeter of the loop is in phase, since the total conductor length is a small fraction of the signal wavelength, λ. The ferrite core increases the effective area of the antenna by concentrating magnetic field lines. Increasing the loop area and the number of turns raises the voltage between the two antenna terminals for a fixed field strength at the cost of increasing the physical size of the antenna.

Electrically, a ferrite-rod antenna simply behaves like an inductor with reactance $X_L = 2\pi f L$, where f is frequency and L is the inductance of the coil. While there is also radiation resistance R_R and additional loss resistance R_L in series with the inductor, they are small in magnitude compared to the reactance [5]. Ferrite-rod antennas are inefficient due to their small size (compared to λ) and low

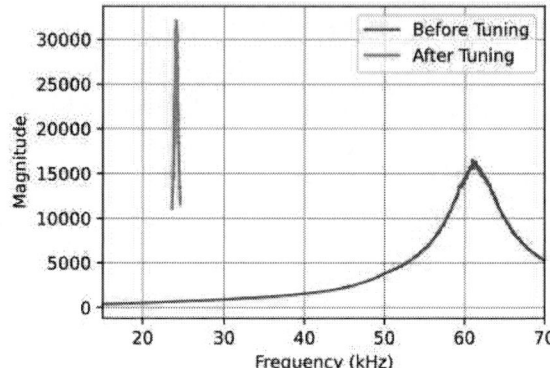

Fig. 2: Z parameter magnitude, derived from measured S parameters, of the ferrite-rod antenna before and after tuning.

radiation resistance. As a result, practical ferrite-rod antennas are typically tuned with a parallel capacitor across the loop terminals. This forms a parallel-resonant LC circuit that boosts the output voltage Q_{ant} times greater. While Q_{ant} boosts the voltage, it also limits the bandwidth to $f_{BW} = f_o/Q_{ant}$, where f_o is the center frequency.

The antennas demonstrated in this work are commercial ferrite-rod antennas pre-tuned for cross-continent domestic time broadcast reception near 60-kHz. Both S_{11} and LCR meter measurements indicate coil inductances in the 10-15 mH range. Tuning this antenna to the 23.7-kHz measured resonance frequency of the resoswitch requires an external 6.8 nF tuning capacitor connected in parallel. Fig. 2 plots the real part of the measured S_{11} converted to Z parameters. Before tuning, the antenna is resonant near 60 kHz, but only has about 22 Ω of real resistance at 23.7 kHz. After tuning, the resonant frequency shifts down, boosting the real part of the antenna's impedance up to 31 kΩ, which helps the match to a resoswitch. Note also how the bandwidth of the antenna reduces from 5.6 kHz at 60 kHz to only 3 kHz at 23.7 kHz—a rather consequential event that (as will be seen) will reduce the maximum receiver bit rate.

Impedance Matching

The motional resistance R_x of the resoswitch is that of a comb-driven resonator [6], so is a strong function of the finger gap spacing, as well as the dc bias voltage across the drive and shuttle fingers. Device 1 in Table 1 summarizes the design of the resoswitch used herein and computes an expected R_x of 1654 MΩ, which is clearly much too high to match the 31kΩ of the receive antenna.

A more aggressive resoswitch, such as summarized in Device 2 of Table 1 which employs the lithography technology of [6] and the high Q offered by [7] could be made to match the antenna impedance. For present purposes, however, using the sensitivity formula from [1], the 1654 MΩ to 31 kΩ resoswitch-to-antenna impedance mismatch of the demonstrated device imposes 47 dB of loss between the antenna and resoswitch versus a matched condition, meaning that -18 dBm must appear on the antenna to cause impacting for a -65 dBm resoswitch sensitivity. This is still sufficient for short range applications, e.g., in RFID tags.

EXPERIMENTAL RESULTS

A 23.7-kHz gold micromechanical resoswitch was fabricated via a one-mask process similar to that of [1],

Table 1: Resoswitch Design and Performance

Parameter	This Work (Device 1)	Scaled (Device 2)
Thickness, h [μm]	1.3	1.3
Finger Gap, d [μm]	1.0	0.1
Impact Gap, d_0 [μm]	1.0	0.1
Number of Fingers, N_f	38	270
Finger Width [μm]	1	0.1
Shuttle Mass [kg]	1.55×10^{-12}	1.50×10^{-12}
Frequency [kHz]	23.712	23.712
Q_{res}	706	7,200
V_P [V]	5	5
R_x [MΩ]	1654	0.032
Sensitivity [dBm]	-65.2	-92

with key device attributes summarized in Table 1 and its SEM shown in Fig. 3.

The resoswitch was placed in a LakeShore FWPX vacuum probe station at 1 mTorr to reduce gas damping and improve Q. Input, output, and shuttle electrodes were probed within the vacuum chamber, then routed outside the chamber to the receive antenna, a Keysight DSO1024 oscilloscope, and a dc power supply, successively. A Tektronix AFG3102 function generator delivered the OOK-modulated bit stream to the ferrite-rod transmit antenna. Fig. 4 plots the power received by the resoswitch as a function of the distance between antennas when sourcing 13.6 dBm (~23 mW) from the function generator, showing an expected inverse square dependence on distance. Despite the antenna's small size, which makes it inefficient for transmit, the setup shown in Fig. 1 with 23 mW output still provides sufficient transmission to meet the −65.2 dBm sensitivity of the resoswitch receiver at up to 18 cm.

Antenna-Imposed Bit Rate Reduction

The ability of a resoswitch to adapt to its input bit rate described in [4] relies on resonance energy stored in the device during non-driven periods, i.e., during '0' inputs, where the high Q of the device allows it to continue vibrating after the input drive is gone. The more energy in the device when the drive returns, i.e., when a '1' bit arrives, the faster it can return to impacting, since it need only traverse the distance from its initial pre-energized vibration amplitude to the impact threshold.

The amount of stored energy in a resoswitch depends on the drive efficiency during the pre-energization period, which practically means it depends on the drive amplitude during a '1' input. OOK signals directly fed to a resoswitch input would appear as the orange curves in Fig. 5 (measured at the function generator output), where the gated-sinusoid envelopes have sharp edge transitions, indicating the presence of energy at higher harmonic frequencies. On the other hand, if the OOK signals go through a band-limited channel before appearing at the resoswitch receiver input, the loss of high frequency content results in the blue curves measured at the 3-kHz bandwidth receive antenna output, showing reduced amplitude and more gradual up and down transitions. These less efficient waveforms not only hurt the receive sensitivity of the resoswitch but also deliver energy more slowly, so reduce its pre-energization.

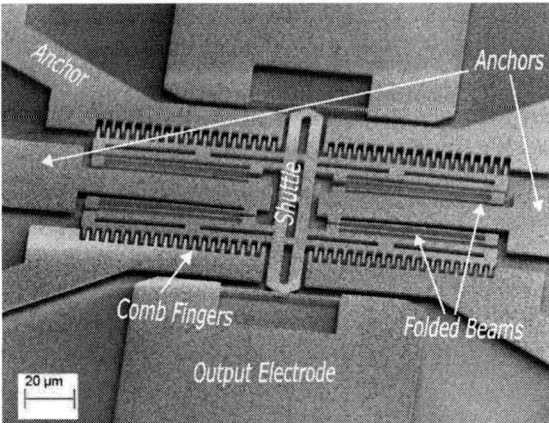

Fig. 3: SEM micrograph of the single-mask electroplated-gold resoswitch fabricated for this work.

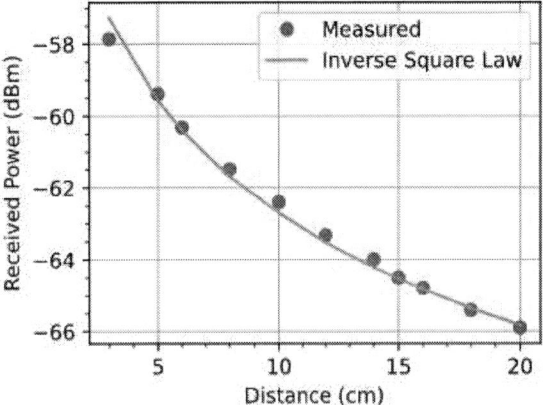

Fig. 4: Measured power delivered to the resoswitch vs. antenna separation distance falls off with an inverse square law.

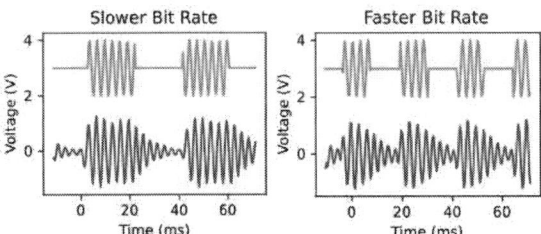

Fig. 5: Transmitted (orange) and antenna received (blue) waveforms (decimated to show sinusoid) for the cases of slow and fast bit rates.

This then compromises its ability to adapt to the incoming bit rate, thereby reducing its maximum bit rate.

To illustrate this phenomenon, Fig. 6 presents measured OOK-modulated waveforms transmitted at 13.6 dBm (green) over 15 cm and received (blue) by the respective antennas, clearly showing waveform distortion caused by finite channel bandwidth. Despite this, the resoswitch still successfully demodulates the waveform, where the output bit stream (orange) matches the input up to 900bit/s. Fig. 7 presents similar plots using a 1 kbit/s rate that ensue when first using a much wider channel bandwidth, then attempting this same bit rate via the antenna-band-limited wireless channel. Without the antennas, the bit rate adapting ability of the resoswitch can keep up with the 1 kbit/s rate and further permits a rate of 8 kbit/s [4], which is 8 times faster

978-1-6654-9309-3/23 $31.00 © 2023 IEEE

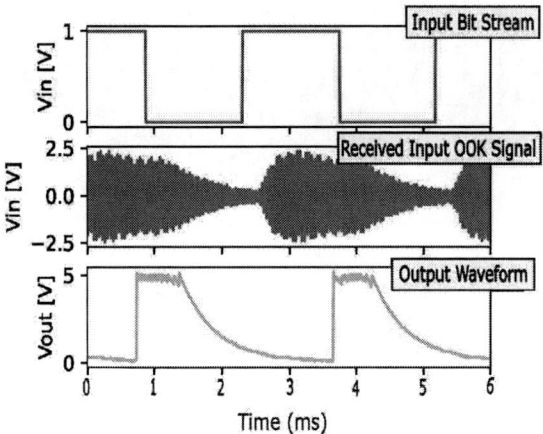

Fig. 6: Measured waveforms showing successful wireless demodulation using the antenna-fed resoswitch receiver.

than this and the highest yet demonstrated. The addition of the antennas and consequent bandlimited channel results in the last plot (green), where the resoswitch clearly can no longer track the incoming 1 kbit/s data stream, confirming the impact of the 3-kHz channel limit.

Fig. 8 finally presents a measured plot of wireless sensitivity (i.e., the minimum detectable power delivered to the resoswitch) vs. the input bit rate, showing a sensitivity of –65 dBm at 100 bit/s that worsens (due to waveform distortion) to –62.2 dBm at the maximum bit rate of 900 bits/s, both sufficient for short range sensor network applications.

CONCLUSIONS

The 900 bit/s data rate over a wireless channel via an all-mechanical receiver, despite high impedance mismatch, is an encouraging demonstration. While the observed antenna-derived bandwidth constraints dampen somewhat the excitement surrounding a resoswitch's bit rate adaptability, it is important to note that the employed ferrite-rod antennas were of the wrong frequency, so required wide tuning that ultimately limited their bandwidth. Antennas that start at the proper frequency could easily provide much more bandwidth, which would then allow the resoswitch receiver to operate at a non-limited 8 kbit/s good enough for voice communications. The use of even smaller ferrite-rode antennas at frequency co-designed with a more aggressively dimensioned resoswitch for better impedance matching (e.g., Device 2 in Table 1) is the next logical step towards longer range voice communications.

REFERENCES

[1] R. Liu, *et. al.*, "Zero quiescent power VLF mechanical communication…," in *2015 Transducers*, Jun. 2015, pp. 129–132. doi: 10.1109/TRANSDUCERS.2015.7180878.

[2] C.-P. Tsai, *et. al.*, "A 125-KHZ CMOS-MEMS Resoswitch …," in *MEMS 2020*, Jan. 2020, pp. 106–109. doi: 10.1109/MEMS46641.2020.9056120.

[3] C. T.-C. Nguyen, *et. al.*, "An ultra-low power mechanical trigger detector (invited)," Proceedings, 2017 Gov. Microcircuit Applications & Critical Technology Conf., Reno, Nevada, Mar. 2017, pp. 104–107.

[4] Q. Jin, *et. al.*, "Bit Rate-Adapting Resoswitch," Tech. Digest, 2022 IEEE IEDM, to be published.

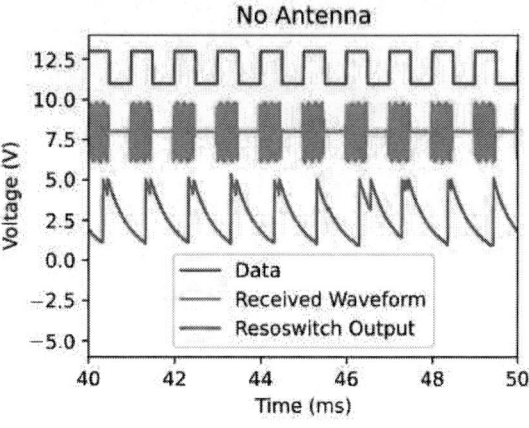

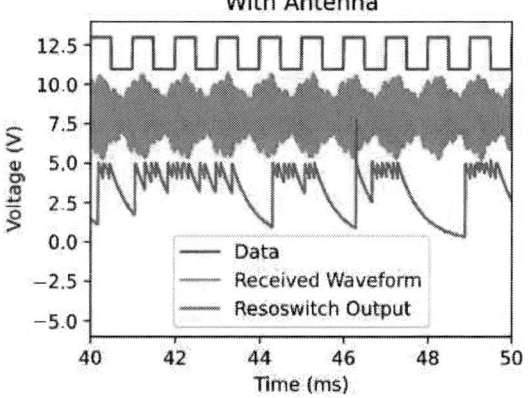

Fig. 7: Measured comparison of received waveforms (orange) and resoswitch outputs (green) at 1 kbit/sec connected directly (top) and wirelessly using an antenna (bottom).

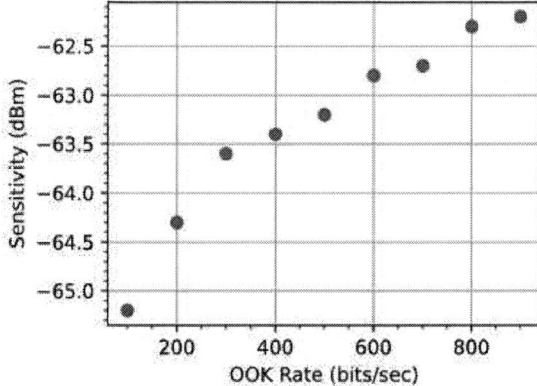

Fig. 8: Measured ferrite-rod antenna-driven resoswitch receiver sensitivity vs. transmit bit rate.

[5] W. H. Silver, *The ARRL Antenna Book For Radio Communications*, 22nd ed. Newington, CT, USA: ARRL, 2011.

[6] H. G. Barrow *et al.*, "A real-time 32.768-kHz clock oscillator using a 0.0154-mm2 micromechanical resonator frequency-setting element," in *IFCS 2012*, May 2012, pp. 1–6. doi: 10.1109/FCS.2012.6243740.

[7] A. Ozgurluk, *et. al.*, "Q-boosting of metal MEMS resonators via localized anneal-induced tensile stress," in *EFTF/IFCS 2017*, Jul. 2017, pp. 10–15. doi: 10.1109/FCS.2017.8088786.

ULTRA-WIDEBAND MEMS FILTERS USING LOCALIZED THINNED 128° Y-CUT THIN-FILM LITHIUM NIOBATE

Jinbo Wu[1,2,3,#], Shibin Zhang[1,#,], Pengcheng Zheng[1,2], Liping Zhang[1,2], Hulin Yao[1,2], Xiaoli Fang[1,2], Xuedi Tian[1,2], Xiaomeng Zhao[1], Tao Wu[3,*], Xin Ou[1,2,*]*

[1]The State Key Laboratory of Functional Materials for Informatics, Shanghai Institute of Microsystem and Information Technology, Shanghai 200050, CHINA
[2]The Center of Materials Science and Optoelectronics Engineering, University of Chinese Academy of Sciences, Beijing 100049, CHINA and
[3]ShanghaiTech University, Shanghai 201210, CHINA

ABSTRACT

This work demonstrates a first-order antisymmetric Lamb wave (A1) mode MEMS filter with a fractional bandwidth (FBW) exceeding 24%. Based on the localized thinned 128° Y-cut lithium niobate (LiNbO$_3$) films, the series and shunt A1 mode resonators exhibit excellent effective electromechanical coupling coefficients (k_{eff}^2) of 43.6% and 39.8%, respectively, which is the key to realizing ultra-wideband RF acoustic filters. The fabricated A1 mode filter with a center frequency of 1.79 GHz achieves a minimum insertion loss of 0.61 dB and a maximum 3-dB FBW of 24.6%. After further optimizations, the A1 mode filters will have great potential as broadband acoustic-only filters for 5G New Radio bands.

KEYWORDS

5G New Radio, MEMS, acoustic, resonators, filters, ultra-wideband, lithium niobate, A1 mode.

INTRODUCTION

Radio-frequency acoustic devices including surface acoustic wave (SAW) and bulk acoustic wave (BAW) resonators have been widely applied in modern wireless communication systems as filtering solutions [1]-[7]. Unlike the 4G Long Term Evolution (LTE), the 5G New Radio (NR) bands place more stringent requirements on the frequency and bandwidth of the filters. For example, the fractional bandwidth (FBW) of the N77 band (3.3-4.2 GHz, as schematically shown in Fig. 1(a)) is as high as 24%. The conventional SAW and BAW devices can hardly meet the bandwidth requirements of 5G-NR bands due to their limited electromechanical coupling coefficients.

In recent years, several creative methods have been proposed to build broadband filters [8]-[13]. BAW filters based on Sc-doped aluminum nitride (AlN) [8]-[9] instead of pure AlN can achieve broadened bandwidths, while their FBWs are still difficult to exceed 20%. SAW filters based on the heterogeneous substrates can achieve complete coverage of the N77 band [10], but the electrode width (W$_e$) is scaled to less than 0.25 μm due to the limitation of SAW velocity. The higher lithography costs and reduced power tolerance will hinder the application of the heterogeneous substrate-based SAW filters. Integration of acoustic resonators with lumped elements [11]-[13] is another broadband filtering solution, and the enlarged losses as well as the footprint (of the lumped-LC elements) are huge challenges.

In addition to the above methods, high-order plate

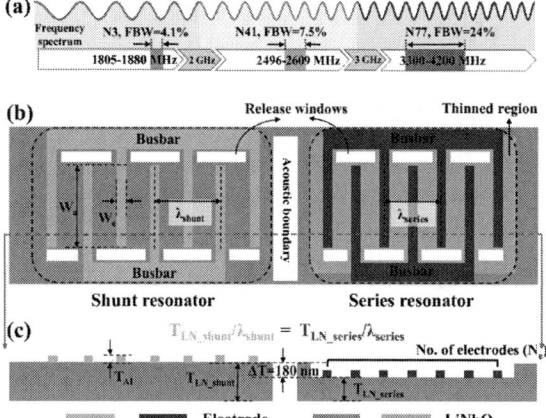

Figure 1: (a) Typical 5G-NR frequency bands. Top view and (b) cross-sectional view of the schematic of the shunt and series A1 mode resonators.

wave devices based on single-crystal piezoelectric thin films may be one of the most promising solutions to break the limitation of frequency and bandwidth of acoustic devices. Such as the first-order shear horizontal wave (SH1) and the first-order asymmetric Lamb wave (A1) resonators on the rotated Y-cut thin-film lithium niobate (LiNbO$_3$) demonstrate extremely high effective electromechanical coupling coefficients (k_{eff}^2) of larger than 40% [14]-[16]. At the same time, A1 mode filters based on the Z-cut LiNbO$_3$ thin films show FBWs exceeding 10% [17]-[18]. However, the existing high-order plate wave filters do not take full advantage of their high k_{eff}^2, and the bandwidth of the high-order plate wave filters can be further enhanced.

In this work, the A1 mode devices (without additional lumped elements) on the 128° Y-cut LiNbO$_3$ thin films were designed and fabricated. Both the shunt and series A1 mode resonators exhibit excellent k_{eff}^2 of around 40% by locally thinning LiNbO$_3$ thin films. The fabricated A1 mode filter exhibits a minimum insertion loss (IL$_{min}$) of 0.61 dB and an excellent FBW of 24.6%, showing great potential as an acoustic-only broadband filtering solution for 5G-NR bands.

DEVICES DESIGN AND SIMULATION

A1 mode devices consist of top interdigital transducers (IDTs) and single-crystal LiNbO$_3$ thin films, as shown in Figs. 1(b) and (c). Sufficient frequency offset (Δf) between the shunt and series resonators and large k_{eff}^2 is the key to

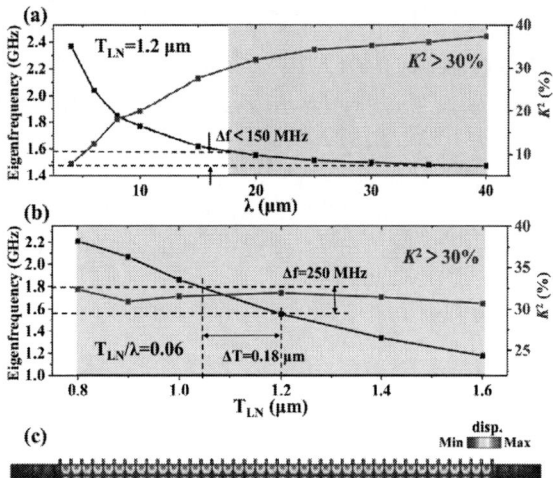

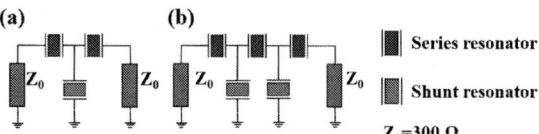

Figure 2: Simulated eigenfrequency and K^2 of the A1 mode resonators with different (a) λ and (b) T_{LN}. (c)Simulated displacement mode shape of the A1 mode.

Figure 3: The schematics of ladder filter topologies ($Z0=300\Omega$) of (a) Design I and (b) Design II.

Table I: The key parameters of the series and shunt resonators.

Parameter	T_{LN} (μm)	W_e (μm)	W_a (μm)	λ (μm)	N_e	T_{Al} (nm)
Series resonator	1.02	2	50	17	61	300
Shunt resonator	1.2	2	50	20	71	300

realizing the ultra-wideband filters. To achieve the ultra-wideband acoustic filters, A1 mode resonators based on 128° Y-cut LiNbO₃ thin films with different wavelengths (λ) and thicknesses of LiNbO₃ thin films (T_{LN}) were simulated. The corresponding eigenfrequency, intrinsic electromechanical coupling (K^2, directly related to k_{eff}^2 [19]), and displacement mode shape are shown in Fig. 2. K^2 is defined by $(v_m^2 - v_0^2)/v_0^2$, where v_m and v_0 are the phase velocities of the A1 mode in LiNbO₃ thin films with metalized and free surface boundary conditions, respectively. As shown in Fig. 2(a), when the T_{LN} is determined, K^2 decreases as λ decreases and all resonators with $K^2 > 30\%$ have Δf less than 150 MHz, severely limiting the bandwidth of the constructed filters. As shown in Fig. 2(b), when the T_{LN}/λ maintains a small value, a sufficient Δf can be achieved with $K^2 > 30\%$. This means that ultra-wideband filters can be implemented when the series and shunt resonators have different T_{LN}.

Based on the simulated results, the A1 mode filters were designed. The schematics of the ladder filter topologies are shown in Fig. 3. The key parameters of the series and shunt resonators are listed in Table I, and the parameters are marked in Fig. 1. The small duty factor

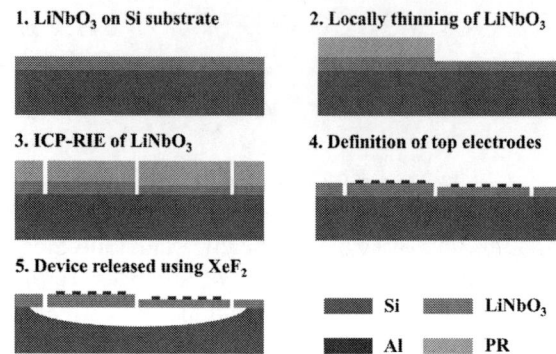

Figure 4: Fabrication process for the A1 mode filter.

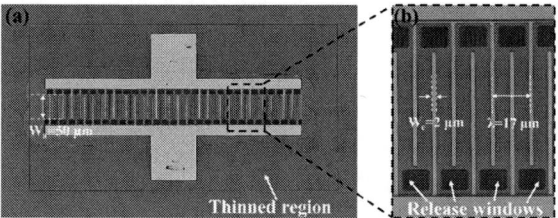

Figure 5: (a) Optical microscope image and (b) zoomed-in image of an A1 mode resonator on thinned LiNbO₃.

($2\times W_e/\lambda$) is used here to mitigate the spurious responses. The termination impedance is set to 300 Ω due to the small static capacitance of the A1 mode devices. T_{LN_series} and T_{LN_shunt} were chosen to be 1.02 μm and 1.2 μm, respectively. The difference in thickness (ΔT) is achieved by locally thinning the LiNbO₃ film in the area where the series resonator is located. Both the series and shunt resonators possess a W_e of 2 μm. The moderate W_e of A1 mode devices means low lithography costs and small ohmic losses.

FABRICATION AND MEASUREMENT

The fabrication process of A1 mode MEMS filters based on 128° Y-cut LiNbO₃ thin films is shown in Fig. 3. First, the 1.2 μm thick single-crystal LiNbO₃ thin films was transferred onto a high resistivity silicon substrate. To achieve frequency offset and high k_{eff}^2, the photoresist SPR220-7 was patterned on the surface of LiNbO₃ thin films as a mask, and then the area where series resonators were located was locally thinned to 1.08 μm using ICP-RIE. After that, the photoresist SPR220-7 was patterned again to form the release window, and the etching of LiNbO₃ thin films in this step is also achieved by ICP-RIE. Next, the 300 nm thick aluminum is defined as IDT electrodes on top of the LiNbO₃ thin films through photolithography, metal evaporation, and lift-off processes. Finally, the A1 mode devices were suspended by removing the silicon underneath the LiNbO₃ through XeF₂-based dry etching.

The optical microscope image and zoomed-in image of an A1 mode resonator (series resonator) are shown in Figs. 4(a) and (b), respectively. It can be seen from Fig. 4(a) that the series resonator is built on the locally thinned region of LiNbO₃ thin films.

The fabricated A1 mode devices were characterized using a vector network analyzer (Keysight E5071C) at

978-1-6654-9309-3/23 $31.00 © 2023 IEEE

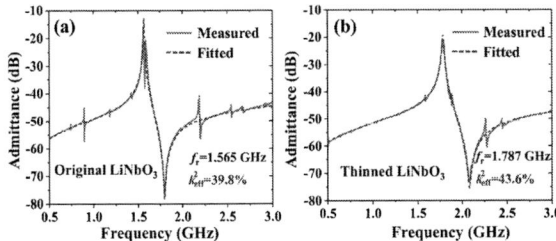

Figure 6: Measured and fitted admittance curves of A1 mode resonators on (a) original and (b) thinned LiNbO₃ thin films.

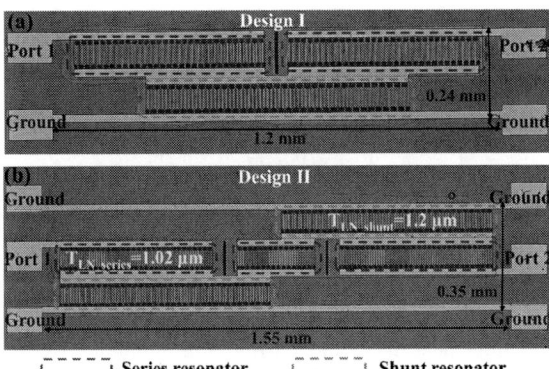

Series resonator   Shunt resonator

Figure 7: Optical microscope images of the fabricated filters: (a) Design I and (b) Design II.

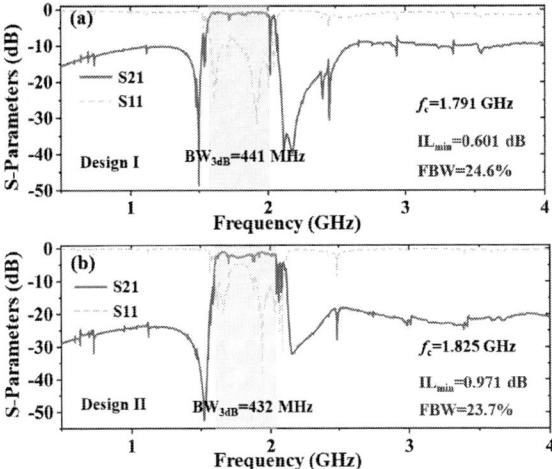

Figure 8: Measured S21 and S11 of the fabricated filters: (a) Design I and (b) Design II.

room temperature in the air. The measured and fitted admittance curves of A1 mode resonators on original and thinned LiNbO₃ thin films are shown in Figs. 6(a) and (b), respectively. The resonant frequencies (f_r) of the above fabricated resonators are 1.565 GHz and 1.787 GHz, respectively. Importantly, the k_{eff}^2 of the above fabricated resonators are 39.8% and 43.6%, respectively. The sufficient Δf and high k_{eff}^2 ensure the ultra-wideband of the filters.

The optical microscope images of the fabricated filters with different designs are shown in Fig. 7, and the overall footprints of the two filters are 0.24×1.2 mm² and 0.35×1.55 mm², respectively. The measured S-parameters of the fabricated filters are shown in Fig. 8. Design I shows a 3dB-FBW of 24.6%, a center frequency (f_c) of 1.79 GHz, and an IL_min of 0.601 dB, while design II shows a 3dB-FBW of 23.7%, an f_c of 1.83 GHz, and an IL_min of 0.971 dB. The fabricated filters show excellent performance, especially the maximum FBW is larger than 24%, which can meet the bandwidth requirements of 5G-NR bands.

However, there is still a lot of work to be done before the practical applications. Such as some in-band ripples caused by the non-uniformity of the thickness of the LiNbO₃ thin films and the imperfection of the device design, thus the device design and fabricated process need to be optimized. Besides, the operating frequency of A1 mode devices in this work is not high enough for the N77 band and other high-frequency bands. By reducing the thickness of the piezoelectric film rather than the linewidth of the lithography, the operating frequency of the A1 mode device can be increased, thus the A1 mode device can

achieve high-frequency applications under easy lithography conditions. It is worth noting that the terminal impedance of filters here is larger than 50 Ω due to the small static capacitance of the A1 mode device. Using alternative methods instead of the small duty factor to suppress spurious mode (e.g., embedded electrodes [20]), and using composite piezoelectric films [16], [21] can increase the static capacitance of the A1 mode device. The poor thermal conductivity and large temperature coefficient of frequency (TCF) can be improved by combining LiNbO₃ thin films with other materials [16], [22]. Although there are still some imperfections, after further optimizations, the A1 mode filters will have great potential as a broadband filtering solution for 5G-NR bands.

CONCLUSION

In this work, the first-order asymmetric Lamb wave (A1) mode devices were designed and fabricated based on the 128° Y-cut LiNbO₃ thin films. By locally thinning the LiNbO₃ thin films, the measured results of the series and shunt resonators show excellent effective electromechanical coupling coefficients (k_{eff}^2) of 43.6% and 39.8%, respectively. The fabricated filter shows a large 3-dB fractional bandwidth (FBW) exceeding 24%, which meets the bandwidth requirements of 5G-NR bands. After further optimizations in the device design and fabricated process, it can be expected to realize the high-performance acoustic-only MEMS filter for 5G-NR bands.

ACKNOWLEDGEMENTS

This work was supported by the National Key R&D Program of China (2022YFB3606701). The authors also appreciate the device fabrication support from the ShanghaiTech Quantum Device Laboratory (SQDL), ShanghaiTech University.

REFERENCES

[1] K. Hashimoto, M. Kadota, T. Nakao, M. Ueda, and M. Miura, "Recent development of temperature compensated SAW devices," in *Proc. IEEE Ultrasonics Symp.*(IUS), 2011, pp. 79–86, doi:

10.1109/ULTSYM.2011.0021.

[2] S. Zhang, R. Lu, H. Zhou, S. Link, Y. Yang, Z. Li, K. Huang, X. Ou, and S. Gong, "Surface acoustic wave devices using lithium niobate on silicon carbide," *IEEE Trans. Microw. Theory Techn.*, vol. 68, no. 9, pp. 3653–3666, Sep. 2020, doi: 10.1109/TMTT.2020.3006294.

[3] T. Takai, H. Iwamoto, Y. Takamine, T. Fuyutsume, T. Nakao, M. Hiramoto, T. Toi, and M. Koshino, "High-performance SAW resonator with simplified LiTaO$_3$/SiO$_2$ double layer structure on Si substrate," *IEEE Trans. Ultrason. Ferroelectr. Freq. Control*, vol. 66, no. 5, pp. 1006–1013, May 2019, doi: 10.1109/TUFFC.2019.2898046.

[4] S. Inoue and M. Solal, "Spurious Free SAW Resonators on Layered Substrate with Ultra-High Q, High Coupling and Small TCF," in *Proc. IEEE Int. Ultrason. Symp. (IUS)*, 2018, pp. 1-9, doi: 10.1109/ULTSYM.2018.8579852.

[5] V. Plessky, S. Yandrapalli, P. J. Turner, L. G. Villanueva, J. Koskela, M. Faizan, A. De Pastina, B. Garcia, J. Costa, and R. B. Hammond, "Laterally excited bulk wave resonators (XBARs) based on thin lithium niobate platelet for 5GHz and 13 GHz filters," in *IEEE MTT-S Int. Microw. Symp. Dig.*, Jun. 2019, pp. 512–515, doi: 10.1109/MWSYM.2019.8700876.

[6] R. Vetury *et al.*, "High Rejection, 160MHz Bandwidth, High Q-factor 6 GHz RF Filters for Wi-Fi 6E manufactured in a Novel BAW Process," in *Proc. IEEE Int. Ultrason. Symp. (IUS)*, 2021, pp. 1-4, doi: 10.1109/IUS52206.2021.9593678.

[7] T. Yokoyama, Y. Iwazaki, Y. Onda, Y. Sasajima, T. Nishihara and M. Ueda, "Highly piezoelectric co-doped AlN thin films for wideband FBAR applications," in *Proc. IEEE Int. Ultrason. Symp. (IUS)*, 2014, pp. 281-288, doi: 10.1109/ULTSYM.2014.0070.

[8] S. Gupta, E. Mehdizadeh, K. Cheema and J. B. Shealy, "Miniaturized Ultrawide Bandwidth WiFi 6E Diplexer Implementation Using XBAW RF Filter Technology," in *Proc. IEEE Int. Microw. Symp. (IMS)*, 2022, pp. 880-882, doi: 10.1109/IMS37962.2022.9865319.

[9] M. Akiyama, T. Kamohara, K. Kano, A. Teshigahara, Y. Takeuchi, and N. Kawahara, "Enhancement of piezoelectric response in scandium aluminum nitride alloy thin films prepared by dual reactive cosputtering," *Adv. Mater.*, vol. 21, no. 5, pp. 593–596, Feb. 2009.

[10] Su, Rongxuan, et al. "Over GHz bandwidth SAW filter based on 32° YX LN/SiO$_2$/poly-Si/Si heterostructure with multilayer electrode modulation." *Appl. Phys. Lett.*, 120.25 (2022): 253501.

[11] C. Zuo, C. He, W. Cheng and Z. Wang, "Hybrid Filter Design for 5G using IPD and Acoustic Technologies," in *Proc. IEEE Int. Ultrason. Symp. (IUS)*, 2019, pp. 269-272, doi: 10.1109/ULTSYM.2019.8925918.

[12] J. Mateu *et al.*, "Acoustic Wave Transversal Filter for 5G N77 Band," in *IEEE Trans. Microw. Theory Techn.*, vol. 69, no. 10, pp. 4476-4488, Oct. 2021, doi: 10.1109/TMTT.2021.3091766.

[13] D. Psychogiou, R. Gómez-García and D. Peroulis, "High-Q bandpass filters using hybrid acoustic-wave-lumped-element resonators (AWLRs) for UHF applications," in *Proc. IEEE Int. Microw. Symp. (IMS)*, 2015, pp. 1-4, doi: 10.1109/MWSYM.2015.7166855.

[14] M. Kadota, S. Tanaka and T. Kimura, "First shear horizontal mode plate wave in LiNbO3 showing 20 km/s phase velocity," in *Proc. IEEE Int. Ultrason. Symp. (IUS)*, 2015, pp. 1-4, doi: 10.1109/ULTSYM.2015.0458.

[15] R. Lu, Y. Yang, S. Link and S. Gong, "A1 Resonators in 128° Y-cut Lithium Niobate with Electromechanical Coupling of 46.4%," *J. Microelectromech. Syst.*, vol. 29, no. 3, pp. 313-319, June 2020, doi: 10.1109/JMEMS.2020.2982775.

[16] P. Zheng *et al.*, "Electromechanical Coupling Enhancement in A1 Mode Acoustic Resonators with Bi-Layer Structure," in *Proc. IEEE Int. Micro Electro Mech. Syst. (MEMS)*, 2022, pp. 1010-1013, doi: 10.1109/MEMS51670.2022.9699816.

[17] Y. Yang, R. Lu, L. Gao and S. Gong, "4.5 GHz Lithium Niobate MEMS Filters With 10% Fractional Bandwidth for 5G Front-Ends," *J. Microelectromech. Syst.*, vol. 28, no. 4, pp. 575-577, Aug. 2019, doi: 10.1109/JMEMS.2019.2922935.

[18] P. J. Turner *et al.*, "5 GHz band n79 wideband microacoustic filter using thin lithium niobate membrane," *Electron. Lett.*, vol. 55, no. 17, pp. 942–944, Aug. 2019.

[19] S. Gong and G. Piazza, "Design and analysis of lithium-niobate based high electromechanical coupling RF-MEMS resonators for wideband filtering," *IEEE Trans. Microw. Theory Tech.*, vol. 61, no. 1, pp. 403–414, Jan. 2013, doi: 10.1109/TMTT.2012.2228671.

[20] Y. Yang, L. Gao, R. Lu and S. Gong, "Lateral Spurious Mode Suppression in Lithium Niobate A1 Resonators," in *IEEE Trans. Ultrason. Ferroelectr. Freq. Control*, vol. 68, no. 5, pp. 1930-1937, May 2021, doi: 10.1109/TUFFC.2020.3049084.

[21] R. Lu, Y. Yang, S. Link and S. Gong, "Enabling Higher Order Lamb Wave Acoustic Devices With Complementarily Oriented Piezoelectric Thin Films," *J. Microelectromech. Syst.*, vol. 29, no. 5, pp. 1332-1346, Oct. 2020, doi: 10.1109/JMEMS.2020.3007590.

[22] A. E. Hassanien, R. Lu and S. Gong, "Near-Zero Drift and High Electromechanical Coupling Acoustic Resonators at > 3.5 GHz," in *IEEE Trans. Microw. Theory Tech.*, vol. 69, no. 8, pp. 3706-3714, Aug. 2021, doi: 10.1109/TMTT.2021.3079497.

CONTACT

*Shibin Zhang, sbzhang@mail.sim.ac.cn
*Tao Wu, wutao@shanghaitech.edu.cn
*Xin Ou, ouxin@mail.sim.ac.cn
*Corresponding authors
#Authors contributed equally to this work

ATTRACTOR EXCHANGER FOR OPEN-LOOP OPERATION OF MICROMECHANICAL NONLINEAR RESONATORS USING GAP-SPACING CONTINUATION

Chun-Pu Tsai and Wei-Chang Li

Institute of Applied Mechanics, National Taiwan University, Taipei, Taiwan

ABSTRACT

This work demonstrates for the first time a micromechanical "*attractor exchanger*" that allows one to swab operation between the attractors at the high- and low-energy branches that coexist in the hysteresis region of a nonlinear resonator. In particular, instead of using the typical frequency sweeping approach to reach the attractor at the high energy branch, gap-spacing continuation is employed here as an alternative space to transfer the stable operation from lower-energy attractor to the other with the permission of a single-frequency drive. The technique enables a new operation scheme that avoids circuits required for generating frequency sweeping signal for applications based on open-loop operation of nonlinear resonators.

KEYWORDS

Attractor, gap-spacing continuation, parameter sweeping, vibro-impact resonator.

INTRODUCTION

Recently, nonlinear phenomena and dynamical behaviors in MEMS devices have been extensively studied, the nonlinear operation become useful on improvement of signal-to-noise ratio and frequency stability of device, such as gyroscopes or oscillators, and acquisition of small signal sensing, such as AFM and mass determination [1-5]. However, the nonlinearity induced hysteresis region in the frequency response, where two branches (or, response states) occur with forward and backward frequency sweepings, creates difficulty in operating these devices within this region—the motion would always settle on the low energy state/branch if being driven with a single frequency from rest.

In fact, an experimental method called control-based continuation (CBC) has been developed to locate the unstable orbits of nonlinear device in macro scale and has been applied to approach a specific state or analyze the bifurcation of devices [6-8]. Note that the so-called orbits of a system represent the periodic solutions of this system. The technique involves simultaneous control of two parameters or even more for a nonlinear device, including, for example, the driving force, displacement, and/or phase. Note that, in order to identify the orbits of the original system, the controller must be non-invasive, meaning that the action of feedback/constrain must vanish when the system lies on the orbit (i.e. the feedback gain needs to be zero after stabilization) [6]. Therefore, the choice of the additional controlled parameter is critical in CBC. While this technique can be applied to micro scale [9], where a phase shifter is used to select the operating point of nonlinear resonators by controlling the I/O phase difference, it serves for only the nonlinear oscillator subset. For other applications, though, such as the vibro-impact

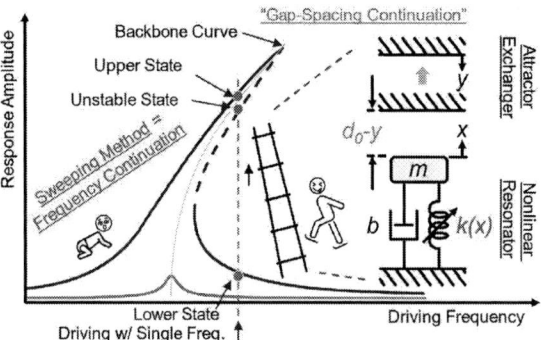

Fig. 1: Conceptual illustration of gap-spacing continuation technique and corresponding mechanical model for the attractor exchanger with a nonlinear device.

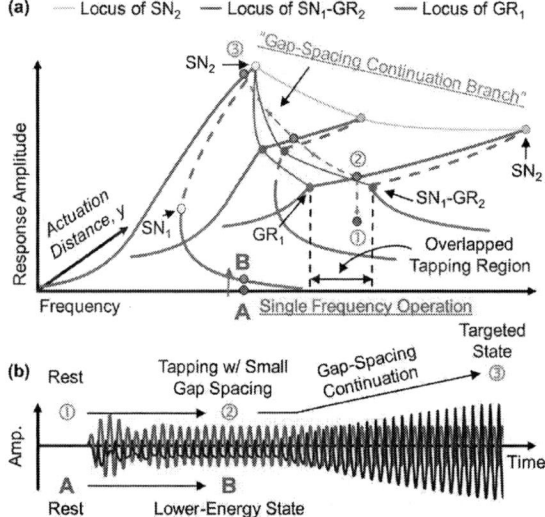

Fig. 2: (a) Illustration of gap-spacing continuation technique, showing frequency response curves of a Duffing resonator with varying gap spacing and a new smooth gap-spacing branch. (b) Simulated time-domain displacement amplitude evolvement of a Duffing oscillator during gap-spacing continuation operation, where $SN_{i=1,2}$ and $GR_{i=1,2}$ are saddle-node and grazing bifurcation, respectively, while $SN_1\text{-}GR_2$ is the label where saddle-node and grazing bifurcation are very closed.

resonator based surface condition monitoring [10], open-loop frequency sweeping is typically needed to reach certain orbits/periodic solutions. The frequency sweeping scheme entails sophisticated control circuits for frequency synthesizing and data flow, which makes it less practical towards circuit integration.

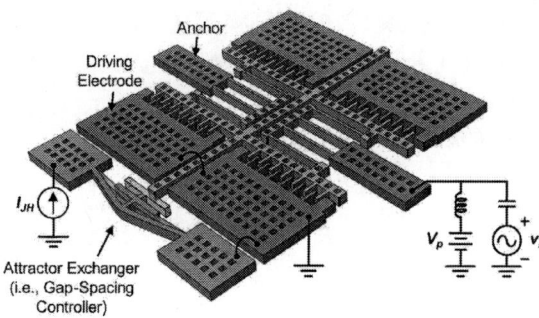

Fig. 3: (a) Schematic of the attractor exchanger with the comb-driven resonator and set-up. The attractor exchanger consists of an arc-beam thermal actuator and is driven by the joule-heating current I_{JH}.

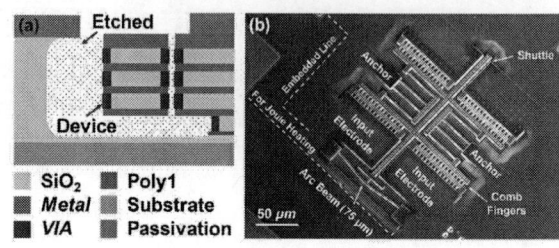

Fig. 4: (a) Cross-section after the release process and (b) the SEM photo of a released arc-beam thermal actuator on a comb-driven vibro-impact resonator. Note that both attractor exchanger and the folded-beam comb-driven vibro-impact resonator are composed of entire metal layer stacks from METAL1 to METAL4.

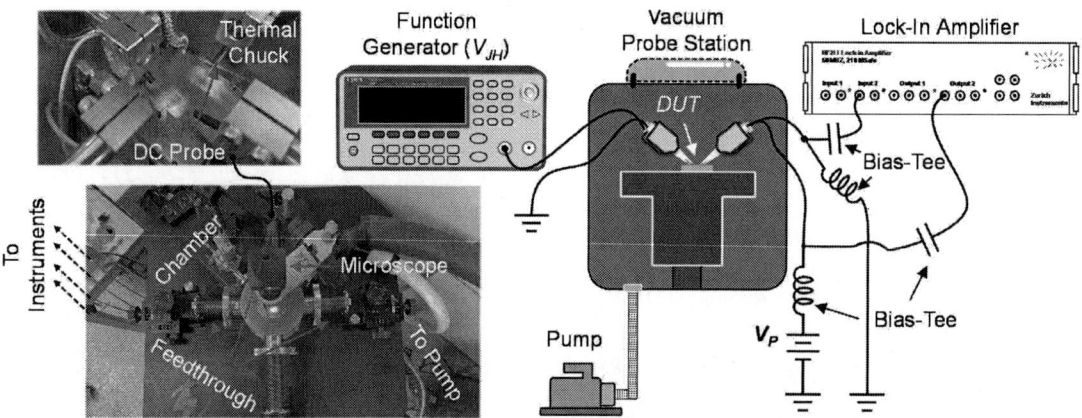

Fig. 5: Measurement set-up for frequency transmission and the functionality validation of gap-spacing continuation. The inset shows the real picture of probe station, associated feedthrough, and zoom-in view of DC probe.

ATTRACTOR EXCHANGE WITH GAP-SPACING SWEEPING TECHNIQUE

Inspired by the CBC technique and closed-loop operation, Fig. 1 illustrates the concept of the gap-spacing continuation technique. The variable gap spacing provides another parameter space to control the state of system without the need of feedback loop and frequency sweeping generators. As shown on the Fig. 1, differing from the typical frequency sweeping method as a baby slowly "crawling uphill", the gap-spacing continuation sets up an alternative branch on the gap-spacing space for one to approach, analogously "climbing a ladder" towards, the targeted high-energy attractor. The attractor exchanger initially sets the gap spacing at a smaller distance, corresponding to a wider overlapped-tapping region, where the device can be driven by a single frequency to enter the tapping-mode operation, like stepping on the first step of the gap-continuation ladder (cf. ① → ② in Fig. 2 (a)). Next, the attractor exchanger gradually enlarges the gap spacing to change the device orbit as frequency sweeping method does but along the gap-spacing space (cf. ② → ③ in Fig. 2 (a)). Since the device will tend to trace a new attractor in the vicinity of the current attractor as the gap varies, the response of the device is continuously "guided" and finally reach the high-energy attractor. Fig. 2 (b) plots the qualitative numerical simulation using MATLAB and Simulink for the time-domain displacement amplitude evolevment for ① → ② → ③ realized by the gap-spacing continuation technique compared to that of single-frequency drive on the original device set-up (A → B).

DEVICE STRUCTURE AND OPERATION

Fig. 3 presents the schematic of an arc-beam thermal actuator as the attractor exchanger for the gap-spacing controllable stopper for a CMOS-MEMS comb-driven vibro-impact resonator. The resonator is driven by an ac signal v_i with a polarization bias V_P on the driving electrode, and the attractor exchanger is deformed by applying joule heating current I_{JH}, which can smoothly control and vary the gap spacing during operation.

EXPERIMENTAL RESULTS

Post-CMOS Fabrication of Device

Similar to [11], the device is fabricated on a standard 0.35-μm CMOS process platform and released in Al-compatible HF-based Silox Vapox etchant III at room temperature for ~60 min. Fig. 4 (a) and (b) plot the cross-section schematic and the SEM photo of a released device, respectively.

Measurement Set-Up

Fig. 5 shows photos of the measurement set-up used for the device. The chip is placed in a custom-built vacuum

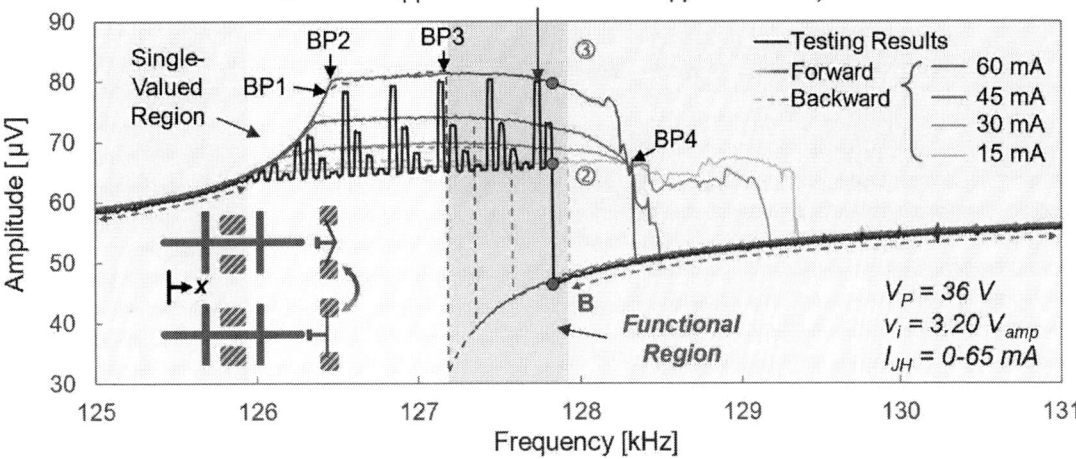

Fig. 6: Functionality validation of gap-spacing continuation technique with lock-in amplifier's sweeper function (HF2LI, Zurich Instruments). The measured frequency transmissions of a vibro-impact resonator with applied current of 0/15/30/48/65 mA are shown in gradual red lines and, solid line and dashed line correspond to the forward and backward sweeping methods. The attractor exchanger is tested with pulses of joule-heating current, showing the operating state would go from the lowest attractor to one of the rest ones, then back when each of the applied pulse is off.

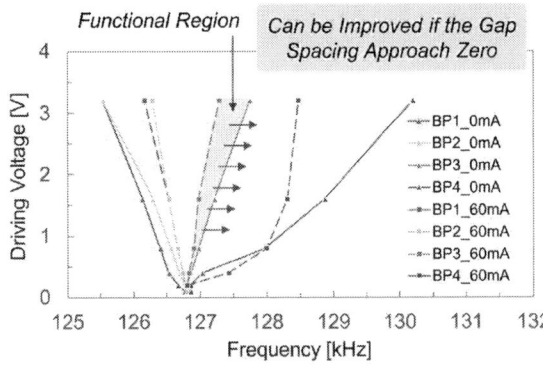

Fig. 7: Measured bifurcation set (driving voltage vs. the BP locations in frequency) of the vibro-impact resonator with and without applied current of 60 mA, showing the detailed functional region.

probe station pumped down to below 5×10^{-3} Torr. The arbitrary waveform generator (33220A, Agilent Technologies, USA) provides the DC or modulated pulse signal to the attractor exchanger, and the lock-in amplifier (HF2LI, Zurich Instruments, Switzerland) measures the frequency response with standard frequency sweeping method. Note that, due to the instrument limitation, the functionality is verified on frequency domain indirectly. In particular, 1200 steps along the interesting frequency region are chosen, and the period of modulated pulse signal is 8 sec (i.e. 1-sec hold time for each step).

Gap-Spacing Continuation Technique

Fig. 6 plots the frequency response of the vibro-impact resonator with varying joule-heating current to the attractor exchanger—a larger transmission level is associated with a larger gap-spacing. The key bifurcation points are labeled as BP1-BP4. To show the functionality of attractor exchanger, the exchanger is driven with alternating joule-heating current during frequency sweeping measurement.

The results show that the exchanger works only within the functional region and overlapped-tapping region as indicated in Fig. 6 and that the device response can swab between the low-energy attractor to the high-energy attractors. Fig. 7 further plots the driving voltage versus the BP locations known as the bifurcation set for the comb-driven vibro-impact resonator with two different applied current of 0 and 60 mA. For each attractor, the overlapped-tapping region is between BP2-BP3. The attractor exchange functional region is between the BP3 of the two states.

CONCLUSIONS

This work demonstrated attractor exchange functionality based on a simple CMOS-MEMS arc-beam thermal actuator that allows swabbing between the attractors at the high- and low-energy branches that coexist in the hysteresis region of a nonlinear resonator in open-loop operation. In addition, gap continuation is employed here as an alternative space to transfer the stable operation from one attractor to the other with the permission of a single-frequency drive. The technique enables a new operation scheme that avoids circuits required for generating frequency sweeping signal for applications based on open-loop operation of nonlinear resonators. However, in order to utilize and design this functionality on nonlinear devices, the underlying nonlinear dynamical motion should be well modeled. Work towards this goal continues.

ACKNOWLEDGEMENTS

This research was funded by the National Science and Technology Council, Taiwan (MOST-109-2628-E-002-004-MY3). The chip fabrication was supported by Taiwan Semiconductor Research Institute (TSRI) and Taiwan Semiconductor Manufacturing Company (TSMC), Hsinchu, Taiwan. The authors would like to thank the staff in NEMS Research Center at NTU for providing technical support.

REFERENCES

[1] A. Chandrashekar, P. Belardinelli, U. Staufer, and F. Alijani, "Robustness of attractors in tapping mode atomic force microscopy," *Nonlinear Dynamics,* vol. 97, no. 2, pp. 1137-1158, 2019.

[2] M. F. Daqaq, R. Masana, A. Erturk, and D. Dane Quinn, "On the role of nonlinearities in vibratory energy harvesting: a critical review and discussion," *Applied Mechanics Reviews,* vol. 66, no. 4, 2014.

[3] A. Hajjaj, N. Jaber, S. Ilyas, F. Alfosail, and M. I. Younis, "Linear and nonlinear dynamics of micro and nano-resonators: Review of recent advances," *International Journal of Non-Linear Mechanics,* vol. 119, p. 103328, 2020.

[4] S. Tiwari and R. N. Candler, "Using flexural MEMS to study and exploit nonlinearities: A review," *Journal of Micromechanics and Microengineering,* vol. 29, no. 8, p. 083002, 2019.

[5] G. Chakraborty and N. Jani, "Nonlinear dynamics of resonant microelectromechanical system (MEMS): A review," *Mechanical Sciences,* pp. 57-81, 2021.

[6] G. Abeloos, L. Renson, C. Collette, and G. Kerschen, "Stepped and swept control-based continuation using adaptive filtering," *Nonlinear Dynamics,* vol. 104, no. 4, pp. 3793-3808, 2021.

[7] E. Bureau, F. Schilder, I. Santos, J. J. Thomsen, and J. Starke, "Experimental bifurcation analysis for a driven nonlinear flexible pendulum using control-based continuation."

[8] E. Bureau, F. Schilder, I. F. Santos, J. J. Thomsen, and J. Starke, "Experimental bifurcation analysis of an impact oscillator—tuning a non-invasive control scheme," *Journal of Sound and Vibration,* vol. 332, no. 22, pp. 5883-5897, 2013.

[9] H. K. Lee, R. Melamud, S. Chandorkar, J. Salvia, S. Yoneoka, and T. W. Kenny, "Stable operation of MEMS oscillators far above the critical vibration amplitude in the nonlinear regime," *Journal of microelectromechanical systems,* vol. 20, no. 6, pp. 1228-1230, 2011.

[10] S.-C. Lu, C.-P. Tsai, Y.-C. Huang, W.-R. Du, and W.-C. Li, "Surface condition influence on the nonlinear response of MEMS CC-beam resoswitches," *IEEE Electron Device Letters,* vol. 39, no. 10, pp. 1600-1603, 2018.

[11] J.-R. Liu, S.-C. Lu, C.-P. Tsai, and W.-C. Li, "A CMOS-MEMS clamped–clamped beam displacement amplifier for resonant switch applications," *Journal of Micromechanics and Microengineering,* vol. 28, no. 6, p. 065001, 2018.

CONTACT

*W.-C. Li, Tel: +886-2-3366-5636; wcli@iam.ntu.edu.tw

A CMOS-MEMS ULTRASENSITIVE THERMOMETER USING INTERNAL RESONANCE IUDUCED FREQUENCY COMBS

Ting-Yi Chen, Chun-Pu Tsai, and Wei-Chang Li
Institute of Applied Mechanics, National Taiwan University, Taipei, Taiwan

ABSTRACT

This work demonstrates a CMOS-MEMS resonant thermometer with a temperature coefficient of frequency (TCF) as high as 59.2× compared to that of a regular resonator-based counterpart using frequency combs in an internal resonating (IR) beam resonator. Particularly, the time-varying amplitude modulated resonance signal due to energy exchange between the 1st and 3rd flexural modes in the IR resonator yields combs in the frequency domain. The comb generation occurs within a certain frequency range, and the comb spacing exhibits a certain dependency on the offset from the onset IR. By driving the resonator at a fixed frequency around its 1st mode, the offset changes and therefore the comb spacing changes in response to temperature changes, yielding a TCF up to -11,481 ppm/°C in the comb spacing shift, a 59.2× higher against the -194 ppm/°C of the resonance frequency of a reference beam resonator. The results verify the IR induced frequency combs can serve as a promising approach towards ultrasensitive sensors.

KEYWORDS

Frequency combs, internal resonance, CMOS-MEMS, temperature sensors.

INTRODUCTION

Recently, mechanical frequency combs have been extensively investigated due to their promising potentials in improving the performances of various applications [1-3]. For example, frequency combs have been used to enhance the energy extraction in energy harvesting [1], and [2] uses comb spacing as the sensing metric and shows a higher sensitivity compared to regular resonators in response to DC voltage difference. Based on these, this work utilizes IR induced frequency combs with the spacing as the output metric and uses ambient temperature change to demonstrate an ultrasensitive sensing scheme. To illustrate, Fig. 1 (a) depicts the 1st and 3rd FEA simulated mode shapes of this work, illustrating the concept of time varying energy transfer between the two modes. Fig. 1 (b) then shows the mechanism of continuous energy exchange between two correlated modes. The increasing and decreasing vibrational amplitude over time results in an amplitude modulated signal, which essentially corresponds to combs in the frequency domain. Interestingly, when the frequency response backbone changes with environmental disturbances, e.g., temperature perturbations, the comb spacing also varies. (*cf*. Fig. 1 (c)).

DEVICE STRUCTURE AND OPERATION

This work adopts the previously proposed stepped beam resonator exhibiting 1:6 IR in [3] of which the SEM photo is shown in Fig. 2. In particular, the capacitively transduced beam resonator consists of stacked aluminum

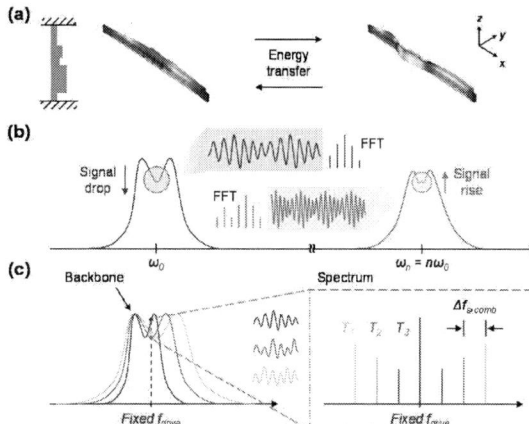

Fig. 1: Illustration of the IR system. (a) When IR is triggered, some parts of energy of the low-order mode serve as the driving source to activate the high-order mode, and vice versa, which causes energy to transfer back and forth between two correlated modes. (b) The beating envelope formed by the time-varying vibrational amplitude results in frequency combs with specific comb spacing. (c) The frequency response backbone as well as the comb spacing change with environmental disturbances.

Fig. 2: SEM photo of a CMOS-MEMS stepped beam resonator with a zoom-in to the step location.

and tungsten vias surrounded by intermediate oxide layers. To operate, a sufficiently large driving signal is required to trigger IR so that the sixth harmonic tone derived from the geometric nonlinearity matches the resonance frequency of the 3rd flexural mode.

ANALYTICAL MODELING

To model 1:6 IR with low-order mode (LM) excitation, the general equation of motion could be expressed using hexic nonlinearity as

$$\ddot{u}_1 + \omega_1^2 u_1 = -2\hat{\mu}_1 \dot{u}_1 + \alpha_2 u_1^5 u_3 + F_1 \cos(\Omega t + \tau_1)$$
$$\ddot{u}_3 + \omega_3^2 u_3 = -2\hat{\mu}_3 \dot{u}_3 + \alpha_8 u_1^6 \tag{1}$$

where ω_1 and ω_3 are natural frequencies, and u_1 and u_3

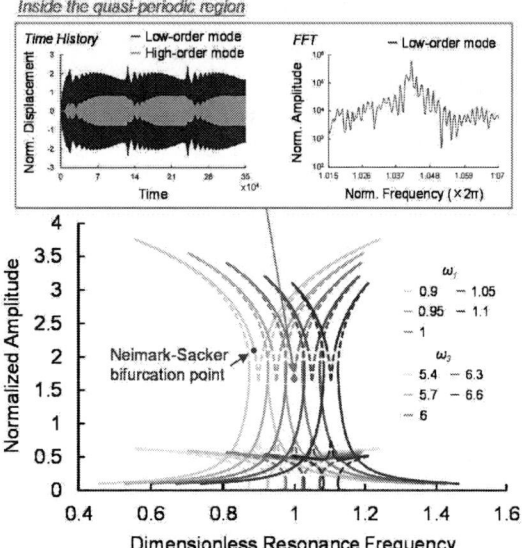

Fig. 3: Simulated M-shaped frequency responses based on the model derived in this work, where green and blue curve series represent the low-order and the high-order mode, respectively. Zooming into the center of the energy exchange region, the time- and frequency-domain results show the amplitude-modulated waveform and frequency combs, respectively.

are modal amplitudes of LM and the high-order mode (HM), respectively. F_1 is the harmonic driving force with frequency of $\Omega \approx \omega_1$. This work borrows the derivation of [4] to obtain the amplitude-frequency relationship, which, after eliminating secular terms, could be shown as

$$
\left[64\omega_1\mu_1 a_1 - \alpha_2 a_3 a_1^5 \sin(\gamma_2)\right]^2 + \left[64\omega_1 a_1\sigma_1 + \alpha_2 a_3 a_1^5 \cos(\gamma_2)\right]^2 - (32f_1)^2 = 0
$$

$$
a_1 = \left(\frac{\omega_3 a_3 \Gamma}{\alpha_8}\right)^{\frac{1}{6}}
$$

$$
\sin(\gamma_2) = -\frac{\mu_3}{\Gamma}, \qquad \cos(\gamma_2) = -\frac{(6\sigma_1 - \sigma_2)}{\Gamma}
$$

$$
\Gamma = [\mu_3^2 + (6\sigma_1 - \sigma_2)^2]^{\frac{1}{2}}
$$

where a_1 and a_3 are amplitudes of harmonic motions of the 1st and the 3rd mode, respectively. ϵ is a small dimensionless scale factor, $\sigma_1 = \frac{\Omega - \omega_1}{\epsilon^5}$ is the frequency detuning parameter used to show the proximity of the driving frequency to the resonance frequency, $\mu_1 = \frac{\hat{\mu}_1}{\epsilon^5}$, $\mu_3 = \frac{\hat{\mu}_3}{\epsilon^5}$, and $f_1 = \frac{F_1}{\epsilon^6}$. Fig. 3 shows the simulated M-shaped frequency responses with varying non-dimensional resonance frequencies, of which the solid and dashed lines represent stable and quasi-periodic solutions, respectively. At this transition point, the eigenvalues are pure complex and thus could be realized with Neimark-Sacker bifurcation [5]. Zooming in to the center of the quasi-periodic region, the simulated results show the occurrence of the energy transferring phenomenon in both time- and frequency-domains, i.e., the beating envelope and frequency combs. Furthermore, it has been demonstrated by [4] that the comb spacing would vary as the spacing

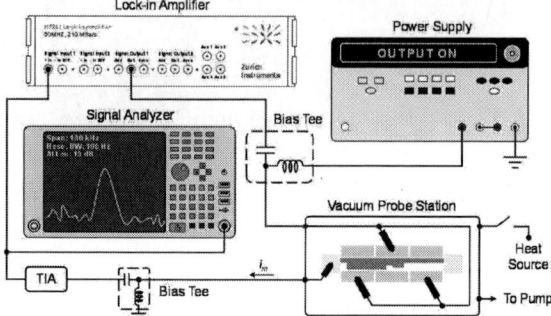

Fig. 4: The measurement set-up. A lock-in amplifier and a power supply are used for measuring frequency responses, and a signal analyzer is parallelly connected to capture frequency combs simultaneously.

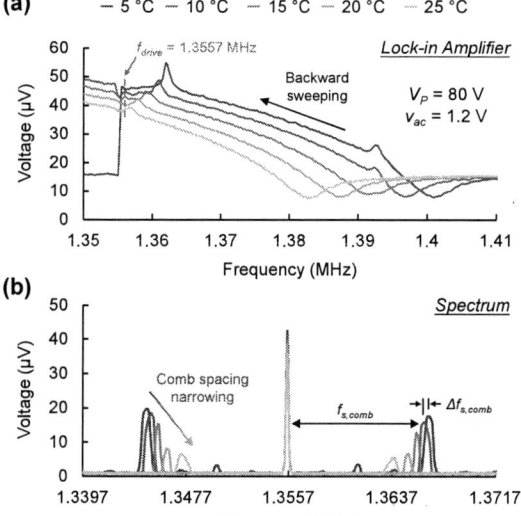

Fig. 5: (a) Measured backward frequency responses with varying temperatures from 5 °C to 25 °C at $V_P = 80$ V and $v_{ac} = 1.2$ V. To observe the frequency comb population over all temperature conditions, the driving frequency is hold at 1.3557 MHz in the IR region, where energy transfer occurs. (b) Measured frequency comb population at different temperatures, revealing that the comb spacing decreases as temperature increases.

between the onset IR and the driving point changes, and therefore in resonant type sensors, the comb spacing variation could reflect the disturbances to be sensed.

EXPERIMENTAL RESULTS

Fabrication and Measurement Set-Up

To demonstrate this, this work performs temperature sweeping to induce deviations in resonance frequencies on the 1:6 IR stepped resonator. The demonstration vehicle is fabricated using a 0.35-μm 2P4M CMOS-MEMS process platform of [6] and wet released in Silox Vapox III. Fig. 4 illustrates the measurement set-up. This work utilizes Zurich HF2LI Lock-in Amplifier (LIA) to generate an ac voltage signal v_{ac} combined with a DC-bias V_P from a power supply to drive the resonator. The motional current is then picked up by LIA and a signal analyzer through a

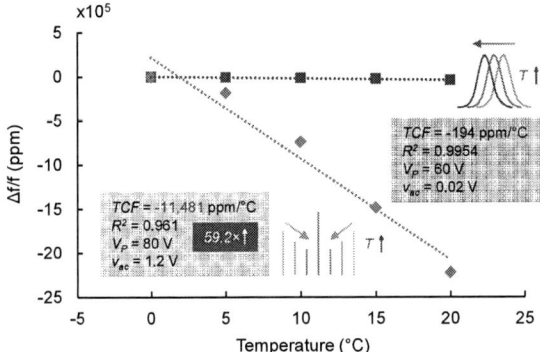

Fig. 6: Measured sensitivity compared to that using the linear sensing scheme, showing a 59.2× enhancement using the comb spacing modulation sensing scheme.

transimpedance amplifier to obtain the frequency response and comb population, respectively.

Temperature Dependency of Frequency Responses

This work adopts the backward sweeping and electrical softening to tune the 1st resonance frequency to be exactly one-sixth of the 3rd mode. Fig. 5 (a) plots the measured backward frequency responses at $V_P = 80$ V and $v_{ac} = 1.2$ V with varying temperatures from 5 °C to 25 °C, where the peaks around 1.36 MHz indicate the onset points of IR. The measurement results agree well to the simulated M-shaped frequency response even though the second rise signal, which could possibly be covered by the Duffing effect, is not captured. After passing through the onset IR points, the vibration system starts energy transfer and undergoes the quasi-periodic motion within a specific IR bandwidth that increases as temperature decreases. Therefore, by holding the driving frequency at 1.3557 MHz so that for every temperature point the motion lies in the quasi-periodic region, the comb spacing would be modulated since the distance from the onset point of IR to the driving point changes.

Temperature Response on Frequency Combs

To further observe frequency combs in the overall temperature interval, Fig. 5 (b) shows the measured combs and reveals that the comb spacing reduces as temperature decreases and therefore the TCF regarding the frequency comb sensing scheme could be expressed as

$$TCF = \frac{\Delta f_{s,comb}/f_{s,comb}}{\Delta T} \qquad (3)$$

where T is temperature, and $f_{s,comb}$ and $\Delta f_{s,comb}$ denote the original frequency comb spacing and the comb spacing difference, respectively. Fig. 6 compares the sensitivity of the comb sensing scheme to that using the typical sensing strategy, i.e., the resonance frequency drift in response to physical perturbations, examining that this work achieves TCF up to -11,481 ppm/°C and 59.2× sensitivity enhancement compared to a regular CC-beam resonator. Table 1 finally compares the work with other micromechanical resonant temperature sensors, showing that using comb spacing as a new sensing strategy could achieve an unprecedented ultrahigh sensitivity. Indeed, using frequency comb-based sensors, although still shows

Table 1: Comparison of Temperature Sensors.

Ref.	[7]	[8]	[9]	This work
Device	Double-resonator DETF	CC-beam	Strain-amplifying DETF	CC-beam
Material or Process	Si/SiO$_2$	CMOS-MEMS	Silicon-on-glass	CMOS-MEMS
Resonance Frequency	1 MHz	640 kHz	46 kHz	~1.36 MHz
Sensing Technique	Change of f_{beat}	f_r drift	f_r drift	Change of $f_{s,comb}$
TCF (ppm/°C)	-570	-4,855	604	-11,481

f_{beat}, f_r, and $f_{s,comb}$ denote the beating frequency, resonance frequency and frequency comb spacing, respectively.

a compromise of a limited operating temperature range, could already satisfy the requirement for ultrasensitive application fields.

CONCLUSION

This work demonstrates in a geometrically nonlinear CMOS-MEMS stepped beam resonator that IR induced frequency comb sensing strategy could serve as a promising approach for ultrasensitive applications. Particularly, by using temperature as a demonstration physical quantity, TCF based on the comb spacing modulation in response to environmental disturbances shows an unprecedented sensitivity up to -11,481 ppm/°C, which is a 59.2× enhancement compared to a typical sensing scheme. While limited operating conditions require more work to make frequency comb-based resonant sensors towards practical usage, the idea of adopting nonlinearity is successfully realized for designing sensors with ultrahigh sensitivity.

ACKNOWLEDGEMENT

This research was funded by National Science and Technology Council, Taiwan (MOST-109-2628-E-002-004-MY3). The chip fabrication was supported by the Taiwan Semiconductor Research Institute (TSRI) and Taiwan Semiconductor Manufacturing Company (TSMC), Hsinchu, Taiwan. The authors would like to thank the staff in NEMS Research Center at NTU for providing technical support.

REFERENCES

[1] L. Bu, E. Arroyo and A. A. Seshia, "Frequency Combs: A New Mechanism for MEMS Vibration Energy Harvesters," in *the 21st International Conference on Solid-State Sensors, Actuators and Microsystems (TRANSDUCERS)*, Online, Jun. 20-25, 2021.

[2] M. Park and A. Ansari, "Self-Sustained Dual-Mode Mechanical Frequency Comb Sensors," in *Proc. Solid-State Sensors, Actuat. Microsyst. Workshop*, Kaohsiung, Taiwan, Jun. 18-22, 2017.

[3] T.-Y. Chen, C.-P. Tsai and W.-C. Li, "1:6 Internal Resonance in CMOS-MEMS Multiple-Stepped CC-Beam Resonators," in *the 35th IEEE Int. Conf. on*

Micro Electro Mechanical Systems (*MEMS'22*), Tokyo, Japan, Jan. 9-13, 2022.

[4] A. H. Nayfeh and D. T. Mook, Nonlinear Oscillations, John Wiley & Sons, 2008.

[5] A. Keşkekler, H. Arjmandi, P. G. Steeneken and F. Alijani, "Symmetry-Breaking Induced Frequency Combs in Graphene Resonators," *Nano Lett.,* vol. 22, no. 15, pp. 6048-6054, 2022.

[6] J.-R. Liu, S.-C. Lu, C.-P. Tsai and W.-C. Li, "A CMOS-MEMS Clamped-Clamped Beam Displacement Ampifier for Resonant Switch Applications," *J. Micromech. Microeng.,* vol. 28, no. 6, p. 065001, 2018.

[7] C. M. Jha, G. Bahl, R. Melamud, S. A. Chandorkar, M. A. Hopcroft, B. Kim, M. Agarwal, J. Salvia, H. Mehta and T. W. Kenny, "CMOS-Compatible Dual-Resonator MEMS Temperature Sensor with Milli-Degree Accuracy," in *the 14th International Conference on Solid-State Sensors, Actuators, and Microsystems* (*TRANSDUCERS*), Lyon, France, Jun. 10-14, 2007.

[8] H. Göktaş, K. L. Turner and M. E. Zaghloul, "Enhancement in CMOS-MEMS Resonator for High Sensitive Temperature Sensing," *IEEE Sensors Journal,* vol. 17, no. 3, pp. 598-603, 2017.

[9] T. Kose, K. Azgin and T. Akin, "Design and Fabrication of a High Performance Resonant MEMS Temperature Sensor," *J. Micromech. Microeng.,* vol. 26, no. 4, p. 045012, 2016.

[10] D. A. Czaplewski, C. Chen, D. Lopez, O. Shoshani, A. M. Eriksson, S. Strachan and S. W. Shaw, "Bifurcation Generated Mechanical Frequency Comb," *Phys. Rev. Lett.,* vol. 121, no. 24, p. 244302, 2018.

[11] G. Gobat, L. Guillot, A. Frangi, B. Cochelin and C. Touzé, "Backbone curves, Neimark-Sacker boundaries and appearance of quasi-periodicity in nonlinear oscillators: application to 1:2 internal resonance and frequency combs in MEMS," *Meccanica,* vol. 56, pp. 1937-1969, 2021.

[12] A. Ganesan, C. Do and A. A. Seshia, "Phononic Frequency Comb via Intrinsic Three-Wave Mixing," *Phys. Rev. Lett.,* vol. 118, no. 3, p. 033903, 2017.

CONTACT
*W.-C. Li, Tel: +886-2-3366-5636; wcli@iam.ntu.edu.tw

ATOMICALLY THIN NEMS FREQUENCY COMB WITH BOTH FREQUENCY TUNABILITY AND RECONFIGURABLE VIA SIMULTANEOUS 1:2 AND 1:3 MODE COUPLING

Bo Xu, Jiankai Zhu, Chenyin Jiao, Jianglong Chen and Zenghui Wang

Institute of Fundamental and Frontier Sciences, University of Electronic Science and Technology of China, Chengdu, China

ABSTRACT

We report a frequency comb via simultaneous internal resonances (1:2 & 1:3) in a nanoelectromechanical system (NEMS) resonator based on 2D semiconductor molybdenum disulfide (MoS_2). By applying a single-tone excitation with amplitude V_{ac} beyond the threshold and carefully setting the gate voltage V_g, multiple internal resonances can be realized and equidistant spectral lines around the fundamental mode frequency are observed. The frequency spacing and the number of spectral lines can be tuned by the driving amplitude V_{ac}, and switched between different configurations by the driving frequency f_d. The combination of such resonant responses can be leveraged to realize NEMS frequency combs with high reconfigurability and fine tunability.

KEYWORDS

Molybdenum disulfide (MoS_2), internal resonance, nanoelectromechanical system (NEMS), mode coupling, frequency comb, two-dimensional (2D) materials.

INTRODUCTION

Frequency comb is a type of device that is capable of producing a series of frequency spectral lines around the central frequency with equidistant space. Since the discovery of optical frequency comb, it has been used in sensing, signal transduction and communication applications [1-2]. Similar to optical frequency combs, phononic frequency combs have also been realized in mechanical systems [3-4]. In particular, enabled by microelectromechanical system (MEMS) devices, MEMS phononic frequency combs are playing an increasingly important role in fundamental research and practical applications, and have been widely demonstrated [5-6]. To date, such devices often leverage frequency mixing of two driving tones or nonlinear responses like internal resonance [7-8].

Two-dimensional (2D) materials such as graphene and MoS_2 have enabled atomically thin NEMS resonators [9-11]. Compared with MEMS devices based on silicon, 2D NEMS resonators with far thinner thickness behave mechanically much more like membranes and thus exhibit much larger frequency tuning range. They can also efficiently exhibit multimode resonance and rich nonlinear dynamics. All these can be achieved by adjusting tension in 2D NEMS devices [12]. Among all frequency tuning methods, electrostatic force is most commonly used by applying voltage to the gate electrode underneath 2D material, which can offer a continuous frequency tuning mechanism. This makes 2D NEMS resonators promising platforms for achieving internal resonance and thus realizing highly tunable phononic frequency comb [13-14].

In this work, we fabricate an MoS_2 NEMS resonator and measure its resonance under different gate voltages. By carefully monitoring multiple modes and the frequency ratio between them at different voltages, we find unique mode coupling among these modes in which multiple internal resonances can be excited by single-tone driving. When $V_g = 2$ V, we demonstrate both 1:2 and 1:3 internal resonances in this device while achieving phononic frequency comb near the fundamental resonance at 7.7 MHz. By sweeping f_d and V_{ac}, both the frequency spacing and the number of spectral lines can be continuously tuned and discretely reconfigured, providing a new strategy for frequency control. This can enable new devices such as 2D NEMS radio-frequency (RF) filters, frequency-shift based sensing, and ultra-low power computing and memory systems [15].

RESONATOR DEVICE FARBICATION AND MEASUREMENT

The design of the 2D MoS_2 NEMS resonator involves a circular cavity in the substrate covered by one piece of MoS_2 flake. The substrate includes a 300 nm silicon oxide (SiO_2) layer on silicon substrate, and the cavity depth is about 260 nm.

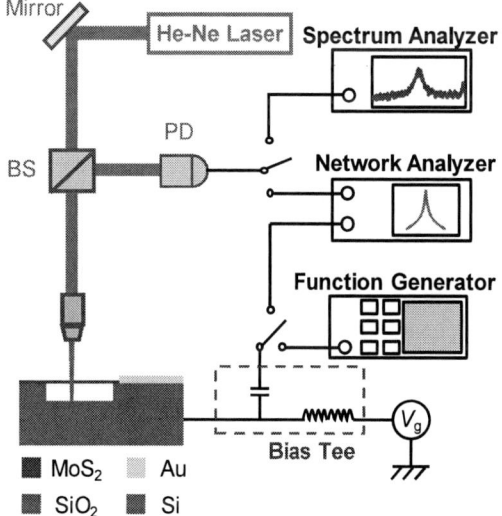

Figure 1: The customized-built measurement set-up and device illustration.

MoS_2 flake is mechanically exfoliated and then transferred onto the predefined substrate with prefabricated cavity. The fabrication process has been reported previously [16]. The resulting MoS_2 nanomechanical resonators can be excited by modulating the electrical signal on the Au electrode contacting the MoS_2 flake.

The resonant responses of MoS_2 NEMS resonator is characterized using our customized measurement set-up, as shown in figure 1. The resonant motion is interferometrically detected using a 633 nm laser, with all measurements performed in a vacuum chamber (about 10^{-3} mbar) under room temperature. Device motion modulates the reflected laser light intensity [17], and the photodetector (PD) transduces the optical signal carrying information about device motion to electrical signal, which is measured by a network analyzer or spectrum analyzer.

RESULTS AND DISCUSSION

After device fabrication (optical image shown in figure 2a inset), we perform Raman and photoluminescence (PL) characterization for the MoS_2 NEMS resonator. From the Raman spectra (figure 2a) E_{2g}^1 and A_{1g} peaks at 383.5 cm^{-1} and 408.5 cm^{-1} can be clearly observed. From the PL spectra in figure 2b, a PL peak corresponding to the indirect band transition of 1.31eV can be seen.

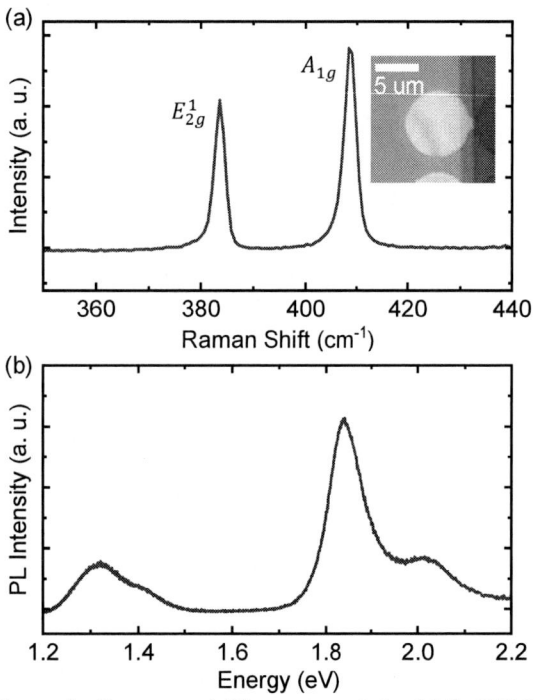

Figure 2: Raman and PL spectra of the MoS_2 NEMS resonator shown in (a) and (b), respectively. The image of device is shown in figure 2a inset.

Using our customized measurement set-up, we perform characterization for the resonant responses for this MoS_2 NEMS resonator. The DC voltage and AC driving signal are applied to the gate electrode using a Bias-Tee. The resonance motion of device is electrically excited and interferometrically detected.

We first characterize multimode resonances in the MoS_2 NEMS resonator under a small DC voltage using a network analyzer. Then, by sweeping gate voltage from -10 V to 10V and fixing AC driving amplitude V_{ac} at 50 mV,

we can tune the resonance frequency of different resonant modes simultaneously. We measure multimode resonance frequencies at every DC gate voltage. We find that at V_g =2 V, 4 V and 6 V, the frequency ratios of the $f_1 : f_4 : f_6$ are near 1: 2 : 3, as shown in figure 3.

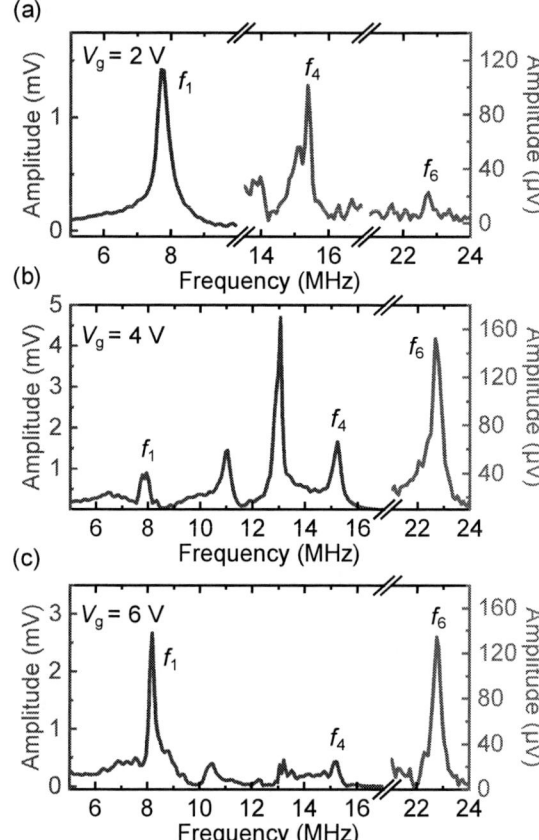

Figure 3: Multimode resonance spectra of MoS_2 NEMS resonator at V_g= 2 V, 4 V and 6 V (the AC driving amplitude V_{ac} fixed at 50 mV), shown in (a), (b) and (c), respectively.

Among shown in figure 3, when V_g= 2 V, the frequency ratio of multimode resonance $f_1 : f_4 : f_6$ are closest to 1: 2: 3 (f_1 = 7.7 MHz, f_4 = 15.4 MHz, f_6 = 22.7 MHz, also shown in the figure 4). To achieve internal resonance among these resonance modes, we apply larger driving on the device and characterize its nonlinear response. At V_g = 2 V, we increase V_{ac} from 20 mV to 1000 mV with steps of 5 mV while fixing the driving frequency at f_1, and monitor the device response using a spectrum analyzer.

The evolution of the resonant response is shown in the figure 5a. When V_{ac} is about 300 mV, a phononic frequency comb is achieved and the frequency spacing and the numbers of frequency lines can be tuned by AC driving amplitude. The frequency spacing Δf is 0, 0.46 MHz, 0.58 MHz, and 0.64 MHz, for V_{ac}=200 mV, 400 mV, 600 mV, and 800mV, respectively (the resonance spectrum are shown in figure 5b-e).

978-1-6654-9309-3/23 $31.00 © 2023 IEEE

(ZYGX2020ZB014 and ZYGX2020J029), and China Postdoctoral Science Foundation (Grant 2021M690554).

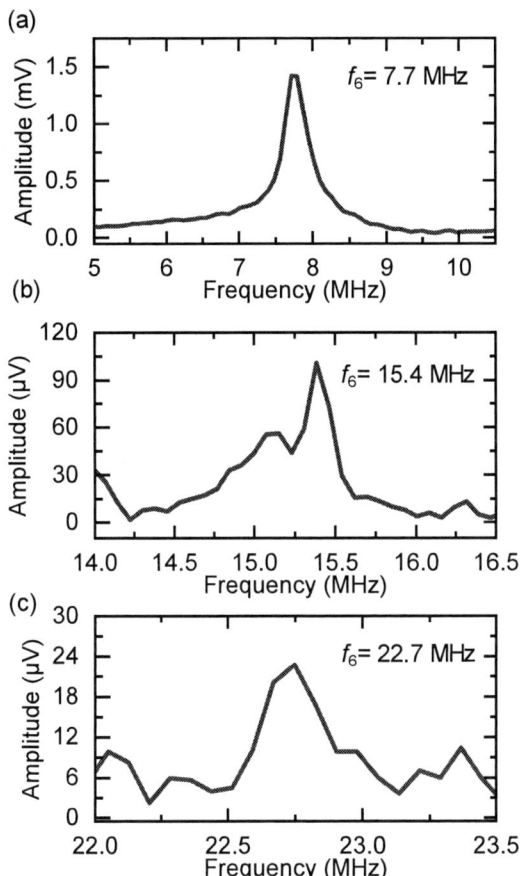

Figure 4: Individual resonance modes when $V_g = 2$ V and $V_{ac} = 50$ mV. The 1st, 4th and 6th resonance mode of the MoS_2 NEMS resonator are shown (a), (b) and (c), respectively.

On the other hand, when we fix $V_{ac} = 300$mV and sweep the driving from 7.6 MHz to 7.8 MHz, the energy transfer among these three modes can also be finely tuned. We find that the phononic frequency comb pattern can be reconfigured, i.e., the frequency spacing and the number of frequency lines are both tunable with driving frequency (figure 6a-d).

CONCLUSION

In summary, we demonstrated phononic frequency comb with 1:2 and 1:3 internal resonance in an MoS_2 NEMS resonator. By sweeping driving amplitude V_{ac} and frequency f_d, the frequency spacing and the number of frequency lines can be tuned. This suggests that atomically thin NEMS frequency comb can hold promises for future frequency control applications.

ACKNOWLEDGEMENTS

The authors gratefully acknowledge support from National Natural Science Foundation of China (Grants 62150052, U21A20459, 62104029, 62004026, 62004032, and 12104086), Sichuan Science and Technology Program (Grants 2021YJ0517 and 2021JDTD0028), Fundamental Research Funds for the Central Universities

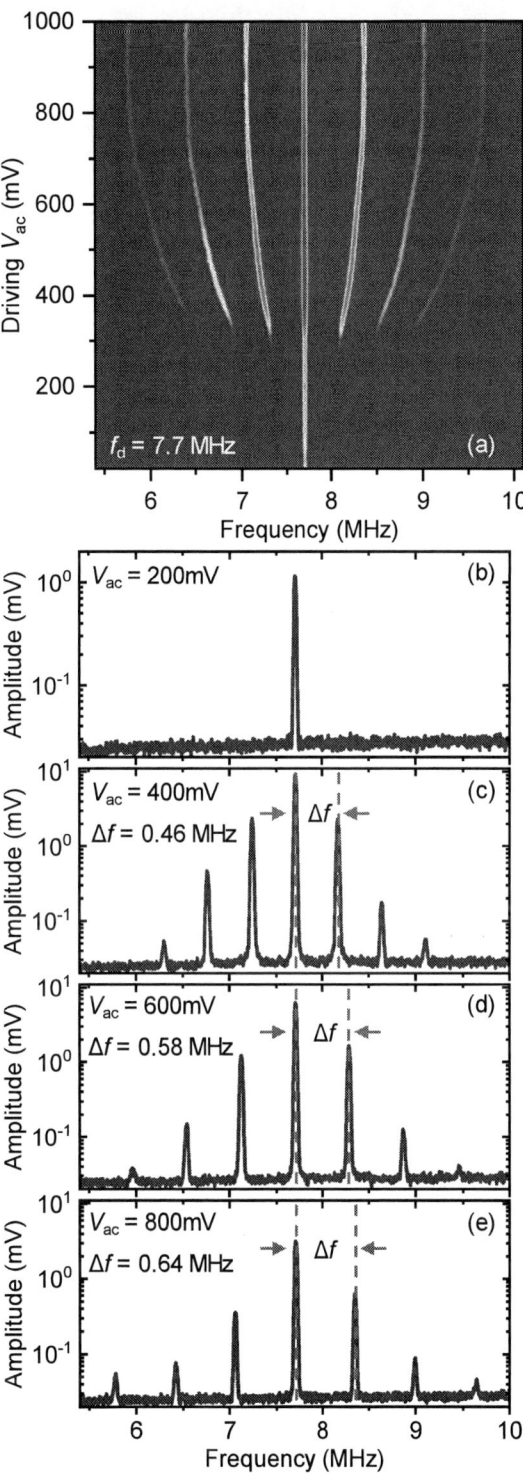

Figure 5: Generation and tuning of the frequency comb. When $V_g = 2$ V and $f_d = 7.7$ MHz, the multimode resonance spectrum evolution with V_{ac} sweeping from 20 mV to 1000 mV is shown in (a). The resonance spectrum at $V_{ac} = 200$ mV, 400mV, 600 mV and 800 mV are shown in (b-e), respectively.

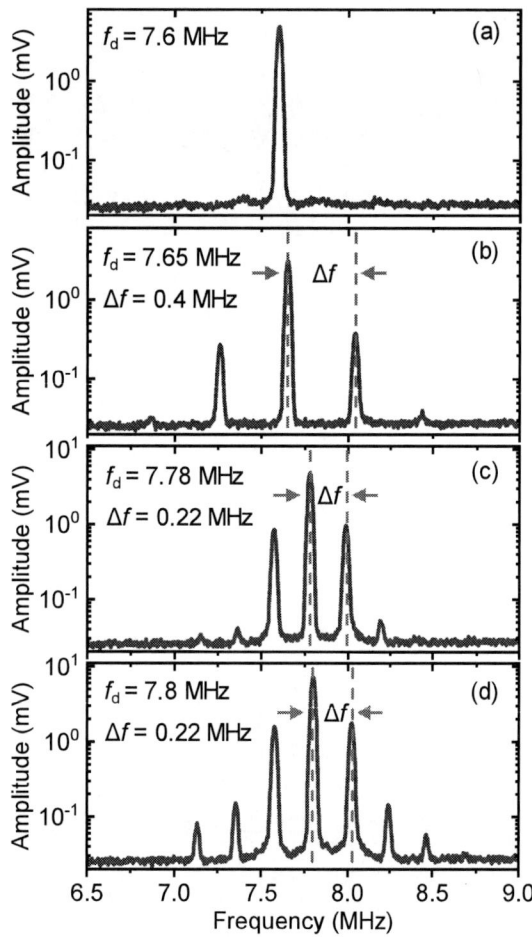

Figure 6: Tuning the frequency comb. When $V_g = 2\,V$ and $V_{ac} = 300\,mV$, the multimode resonance spectrum evolution with f_d sweeping from 7.6 MHz to 7.8 MHz are shown in (a-d).

REFERENCES

[1] T. Udem, R. Holzwarth and T. W. Hänsch, "Optical Frequency Metrology", *Nature*, vol. 416, 6877, pp. 233-237, 2002.

[2] R. J. Jones and J. C. Diels, "Stabilization of Femtosecond Lasers for Optical Frequency Metrology and Direct Optical to Radio Frequency Synthesis", *Phys. Rev. Lett.*, vol. 86, 3288, 2001.

[3] A. Ganesan, C. Do and A. Seshia, "Phononic Frequency Comb via Intrinsic Three-Wave Mixing", *Phys. Rev. Lett.*, vol. 118, 033903, 2017.

[4] A. Ganesan, C. Do and A. Seshia, "Frequency Transitions in Phononic Four-Wave Mixing", *Appl. Phys. Lett.*, vol. 111, 064101, 2017.

[5] R. Wei, J. Lee, T. Mei, Y. Xie, M. S. Islam, S. Mandal and P. X.-L. Feng, "A Self-Sustained Frequency Comb Oscillator via Tapping Mode Comb-Drive Resonator Integrated with a Feedback ASIC", in *Proc. 32th IEEE Int. Conf. on Micro Electro Mechanical Systems (MEMS 2019)*, Seoul, Jan. 27-31 2019, pp. 165-168.

[6] G. Gobat, V. Zega, P. Fedeli, L. G. Falorni, L. Guerinoni, C. Touzé and A. Frangi, "Experimental Evidence of Mechanical Frequency Comb in a Quad-Mass MEMS Structure", in *Proc. 34th IEEE International Conference on Micro Electro Mechanical Systems (MEMS 2021)*, Gainesville, Jan. 25-29, 2021, pp. 615-618.

[7] X. Wang, Q. Yang, R. Huan, Z. Shi, W. Zhu, Z. Jiang, Z. Deng and X. Wei, "Frequency Comb in 1: 3 Internal Resonance of Coupled Micromechanical Resonators", *Appl. Phys. Lett.*, vol. 120, 173506, 2022.

[8] D. A. Czaplewski, C. Chen, D. Lopez, O. Shoshani, A. M. Eriksson, S. Strachan and S. W. Shaw, "Bifurcation Generated Mechanical Frequency Comb", *Phys. Rev. Lett.*, vol. 121, 244302, 2018.

[9] J. S. Bunch, A. M. Van Der Zande, S. S. Verbridge, I. W. Frank, D. M. Tanenbaum, J. M. Parpia, H. G. Craighead and P. L. Mceuen, "Electromechanical Resonators from Graphene Sheets", *Science*, vol. 315, pp. 490-493, 2007.

[10] J. Lee, Z. Wang, K. He, J. Shan and P. X. L. Feng, "High Frequency MoS_2 Nanomechanical Resonators", *ACS Nano*, vol. 7, pp. 6086-6091, 2013.

[11] Z. Wang, B. Xu, S. Pei, J. Zhu, T. Wen, C. Jiao, J. Li, M. Zhang and J. Xia, "Recent Progress in 2D van der Waals Heterostructures: Fabrication, Properties, and Applications", *Sci. China Inf. Sci.*, vol. 65, 211401, 2022.

[12] J. Lee, Z. Wang, K. He, R. Yang, J. Shan and P. X. L. Feng, "Electrically Tunable Single-and Few-Layer MoS_2 Nanoelectromechanical Systems with Broad Dynamic Range", *Sci. Adv.*, vol. 4, eaao6653, 2018.

[13] A. Keşkekler, H. Arjmandi-Tash, P. G. Steeneken and F. Alijani, "Symmetry-Breaking-Induced Frequency Combs in Graphene Resonators", *Nano Lett.*, vol. 22, pp. 6048-6054, 2022.

[14] J. Lee, S. W. Shaw and P. X.-L. Feng, "Phononic Frequency Comb Generation via 1:1 Mode Coupling in MoS_2 2D Nanoelectromechanical Resonators", in *34th IEEE International Conference on Micro Electro Mechanical Systems (MEMS 2022)*, Tokyo, Jan. 9-13, 2022, pp. 503-506.

[15] Z. Wang, J. Fang, P. Zhang and R. Yang, "Nanomechanics: Emerging Opportunities for Future Computing", *China Inf. Sci.*, vol. 64, 206401, 2021.

[16] F. Xiao, B. Xu, Y. Liang, J. Zhu, S. Pei, T. Wen, J. Li, S. Wu, J. Xia and Z. Wang, "Frequency Tuning in Resonant Nano Electromechanical Devices Based on Anisotropic Two-Dimensional Semiconductor Rhenium Disulfide", in *Proc. 2021 Joint Conference of the European Frequency and Time Forum and IEEE International Frequency Control Symposium (EFTF/IFCS 2021)*, Gainesville, Jul. 7-17, 2021, pp1-3..

[17] J. Zhu, P. Zhang, R. Yang and Z. Wang, "Analyzing Electrostatic Modulation of Signal Transduction Efficiency in MoS_2 Nanoelectromechanical Resonators with Interferometric Readout", *Sci. China Inf. Sci.*, vol. 65, 122409, 2022.

CONTACT

* Z. Wang, email: zenghui.wang@uestc.edu.cn

INSTRUMENTAL ANALYSIS OF ADVANCED CATALYSTS BASED ON RESONANT MICROCANTILEVERS

Xinyu Li[1,2], Pengcheng Xu[1,2], Ying Chen[1], Haitao Yu[1] and Xinxin Li[1,2]

[1]State Key Lab of Transducer Technology, Shanghai Institute of Microsystem and Information Technology, Chinese Academy of Sciences, Shanghai 200050, CHINA and
[2]University of Chinese Academy of Sciences, Beijing 100049, CHINA

ABSTRACT

Advanced catalysts are highly influential in various research and application fields. The catalytic performance parameters, such as catalytic activity and kinetics, were mainly characterized using the commercial-available TPD instruments (i.e., temperature-programmed desorption), which *ex-situ* measure the probe molecules in the gas flow. In this paper, a special-designed resonant microcantilever has been utilized as the mass-weighing component of a scientific instrument for characterizing advanced catalysts. Compared to the available TPD instruments, the proposed cantilever-based TPD instrument *in-situ* records the number of desorbed probe molecules (i.e., the mass-loss Δm of the material) via the frequency-change Δf during the heating process, and the bulky furnace and detector are not required. The cantilever-TPD instrument only consumes *ng*-level samples for one-time measurement, which is 6 orders of magnitude lower than the commercial TPD instrument. With a single-time *in-situ* TPD measurement, desorption activation energy can be directly calculated. Various catalysts (e.g., ZSM-5) have been successfully characterized by using the cantilever-TPD instrument, which is expected to become the next generation of TPD scientific instrument.

KEYWORDS

Resonant microcantilever, TPD, advanced catalysts, instrument

INTRODUCTION

Characterization plays a vital role during the advanced catalyst development and optimization [1]. The frequently characterization techniques include X-ray photoelectron spectroscopic (XPS), infrared spectroscopy (IR), nuclear magnetic resonance (NMR), etc. [2, 3]. Among them, temperature-programmed desorption (TPD) provides significant fundamental information on catalysts (such as type of active sites and catalytic activity) according to the desorption temperatures and stages. TPD can be used to characterize almost all practical catalysts and provide numerous information about the surface reaction of active sites from the perspective of energy [4], which makes TPD a pivotal characterization technology for advanced catalyst research.

However, the available TPD instruments are invariably the *ex-situ* measurement method, and there has been no significant technological breakthrough in principle for decades. The instruments typically utilize a furnace to heat the sample programmatically, and detect the probe molecules in the elution gas flow by *ex-situ* detection method. 10-100 mg of the sample is generally consumed to meet the detection limit of the thermal conductivity detector (TCD) or mass spectrometer (MS) [5]. More critically, since the available instruments cannot perform accurate and quantitative characterization, it is challenging to obtain the desorption activation energy (E_d) [6]. In summary, the *ex-situ* measurement of the available TPD instruments lacks sufficient sensitivity to measure the desorbed probe molecules, severely limiting the development of advanced catalysts. Therefore, there is an urgent need to improve this measurement method of the available TPD instruments.

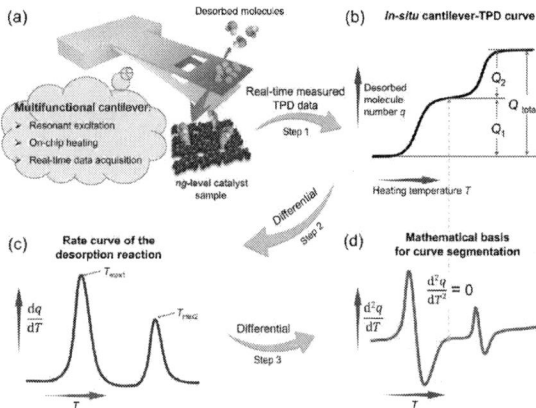

Figure 1: Concept of in-situ cantilever-TPD technology and the TPD data processing. (a) Schematically showing the core measuring tool and principle. (b) Desorbed probe molecule number q is real-time recorded as the in-situ cantilever-TPD curve. (c) and (d) are the first-order and second-order differential of the in-situ cantilever-TPD curve, which represent the desorption rate of adsorbed molecules and the mathematical basis for TPD curve segmentation, respectively.

Herein we demonstrate an *in-situ* TPD technology by developing a resonant microcantilever that integrates resonance-excitation/piezoresistive-sensing elements and a heating micro-resistor. The novel technology only requires *ng*-level samples, and the TCD or MS is no longer required since the cantilever can real-time measure the number of the desorbed molecules *in situ* where the desorption reaction occurs. As illustrated in Fig. 1(a), when the heating-induced desorption initials, the mass-loss of the sample generates a frequency-increasing signal [7]. Based on the cantilever-TPD curve in Fig. 1(b), the number Q of the desorbed probe molecules increases along with the rising temperature T. The first-order differential of the TPD curve represents the desorption rate (Fig. 1(c)), and the second-order differential gives the mathematical basis for dividing the desorption stages (Fig. 1(d)). Furthermore, the desorption kinetics can be calculated based on the Arrhenius equation using the cantilever-TPD data from a single-time measurement.

EXPERIMENTAL

Chemicals

ZSM-5 zeolite catalyst is purchased from Nankai University Catalyst Co., Ltd. Ethanol is purchased from Shanghai Lingfeng Chemical Reagent Co., Ltd. Mixture of NH_3 and Ar (10% NH_3 in Ar), Ar with the purity of 99.99% are purchased from Shanghai APK Gas Co., Ltd. Silicon-on-insulator (SOI) wafers are purchased from Soitec.

Characterization

The appearance of the microcantilever is characterized by Hitachi S4800 field emission scanning electron microscopy (FE-SEM). The transmission electron microscopy (TEM, FEI Tecnai G^2 F20) with an accelerating voltage of 200 kV is used to characterize the morphologies and microstructure of the ZSM-5 zeolites to be measured. The commercial TPD result is characterized by Micromeritics AutoChem II 2920.

Temperature distribution simulation of the on-chip heating microcantilever

The COMSOL Multiphysics software as a finite-element analysis tool is implemented to simulate the heating temperature of the Pt-made microheater. The specific strategy of the simulation is to analyze the heating temperature uniformity of the microheater on the free end of the microcantilever by applying a voltage to the microheater to generate heat. The simulation is carried out under the steady state, with the physical fields including current in the shell, heat transfer in solids and fluids, and laminar flow. And the multi-physical field coupling of electromagnetic field and non-isothermal flow are considered. The thermal conductivity and electrical conductivity of Pt are set to be 71.6 W/(m·K) and 8.9×10^6 S/m, respectively. Other material parameters are the default values in COMSOL Multiphysics software.

Fabrication process of the multifunctional microcantilever

The fabrication process is shown in Fig. 2 and described as follows. (a) Thermal oxidation of the SOI wafer with 380 μm silicon substrate, 700 nm buried oxidation layer, and 3 μm top silicon layer. 350 nm-thick SO_2 layers are formed on the front and back of the wafer. (b) Boron ion implantation process is implemented after photoresist patterning to form piezo-resistors for resonance excitation and detection. (c) The Pt-made microheater is formed by patterning photoresist, sputtering 100 nm-thick Pt layer, and lift-off technology. (d) The contact holes are opened by reactive-ion etching (RIE), followed by sputtering and patterning 330 nm-thick Cr/Au. (e) The electrical insulation layer composed of the 100 nm-thick SiO_2 film is fabricated by plasma-enhanced chemical vapor deposition (PECVD). Photolithography and wet etching process are implemented at the pad areas for further interconnection. (f) The dielectric layers and active silicon layer are removed by photolithography, RIE, and deep reactive-ion etching (DRIE), then the shapes of the

resonant microcantilever and the rectangular heat-insulated window is formed. (g) The 380 nm-thick silicon substrate beneath the cantilever is removed by backside DRIE. (h) The cantilever is released into free-standing by removing the 700 nm-thick buried oxidation layer with buffered HF etchant. When the above fabrication process is successfully completed, a laser sawing equipment (Disco DFL7341) is implemented to dice the wafer into chips. Once the chips are wired on the designed PCB, they can be ready for the *in-situ* TPD analysis.

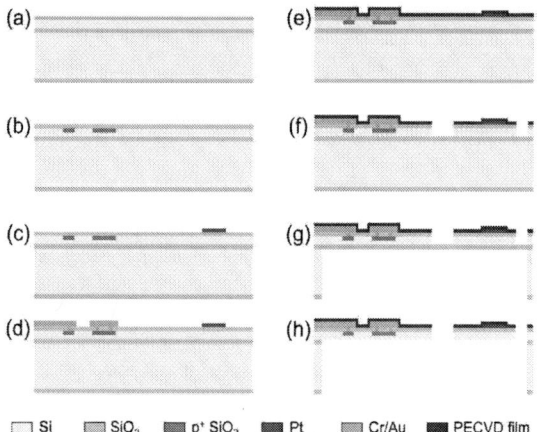

Figure 2: MEMS fabrication process flow of the cantilevers for in-situ TPD measurements.

Preparation of the *in-situ* TPD microcantilever

To load the sample to be measured onto the above fabricated silicon-based cantilever, a suspension for sample loading is first prepared by ultrasonic dispersion of the catalyst in ethanol. Typically, 2 mg of catalyst is added to 1 ml of ethanol, and ultrasound dispersion is implemented for 10 min to form a uniformly dispersed suspension. Then a commercial micromanipulator (Eppendorf, model: PatchMan NP2) is implemented to load one drop of the suspension onto the sample area of the microcantilever.

Measurement process of the cantilever-TPD

The prepared microcantilever is put into a specially designed test chamber which is connected to the controllable gas flow, operation control and data acquisition system. During the TPD measurement, the cantilever firstly pretreats the sample under the Ar atmosphere with a flow rate of 100 ml/min with the on-chip heating temperature of 100°C for 10 min to remove the impurities adsorbed on the sample. Then the probe molecules (100 ml/min) are introduced to the test chamber for the probe pre-adsorption process. After 30 min, the atmosphere is changed to Ar flow, with the sample heated at 100°C for 30 min to remove the physically adsorbed probe molecules. The sample is continually heated to the target maximum temperature at the preset heating rate (β), the frequency signal and the real-time temperature are acquired with the increase of the temperature. When a complete temperature-programmed process has been completed, the process is repeated once more under the same experimental conditions as the baseline. The final *in-*

situ TPD curve is obtained by subtracting the two curves measured by the above operation.

RESULTS AND DISCUSSION

Characterization results of the microcantilever

Fig. 3 shows the special-designed microcantilever for *in-situ* TPD measurements. The temperature distribution simulation result shown in Fig. 3(a) indicates that on the microcantilever with a heat-insulated window, the temperature in the sample region reaches above 650°C, while the temperature at the fixed end is below 100°C. Meanwhile, the heating temperature is evenly distributed in the sample loading region. Fig. 3(b) shows the linear relationship between the temperature and resistance of the microheater, which facilitates the conversion between time and temperature, thus ensuring constant heating rates. Here, the cantilever-TPD technology is used to characterize the typical catalyst of ZSM-5, as shown in Fig. 3(c). Multifunction of resonant excitation and detection, heating and temperature measurement are integrated in the micron-scale cantilever. The nanoporous structure of the ZSM-5 catalyst can be clearly observed in Fig. 3(d).

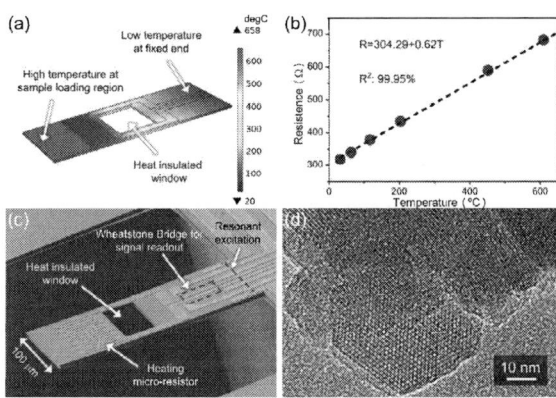

Figure 3: The special-designed cantilever for catalyst characterization. (a) COMSOL simulation showing the temperature distribution. (b) Temperature calibration according to the resistance of the microheater. (c) SEM (scanning electron microscopy) image of the fabricated cantilever. (d) TEM (transmission electron microscopy) image of the ZSM-5 catalyst.

The *in-situ* cantilever-TPD instrument

The instrument of the *in-situ* TPD measurement with a cantilever as the core is schematically shown in Fig. 4, which mainly includes a gas-flow control module, an operation control system, a data acquisition and recording workstation. The microcantilever is placed in the test chamber, which provides a stable atmosphere for the cantilever, and connects the excitation and detection signals with the cantilever. The inset of Fig. 4 shows the circuit structure of the heating and resonance control systems. The phase-locked loop (PLL) module shown in the inset is designed to acquire resonant frequency signals. It has a maximum sampling rate of 100 pts/sec, which provides sufficient data for real-time TPD measurement and desorption kinetic calculation.

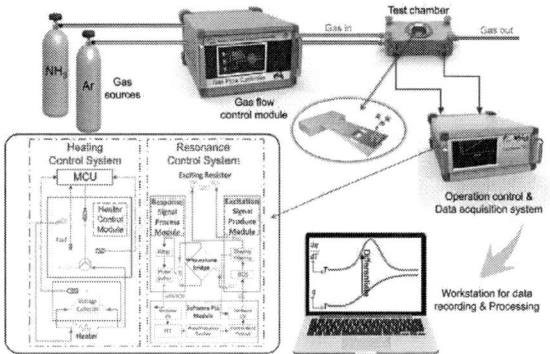

Figure 4: Schematic scheme of the instrument based on the in-situ cantilever-TPD technology. The inset shows the digital modules and circuits for the in-situ cantilever-TPD measurements.

Cantilever-TPD results of ZSM-5 zeolite

The typical acidic catalyst of the ZSM-5 zeolite is used for validation of the cantilever-TPD instrument, and NH_3 is chosen as the probe to perform TPD measurements of the ZSM-5 zeolite. When the adsorption is saturated, the temperature-programmed heating is initiated at a preset range and rate (β) under the inert atmosphere of Ar.

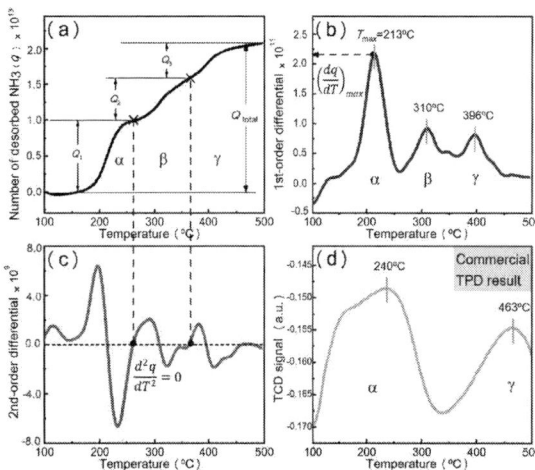

Figure 5: TPD characterization results of zeolite ZSM-5 catalyst. (a) The in-situ cantilever-TPD curve. (b) first-order and (c) second-order differential curves of the in-situ TPD curve in (a). (d) The TPD result obtained by a commercial TPD instrument. α, β, and γ stages represent weak acid, medium acid, and strong acid sites, respectively.

The cantilever-TPD curve shown in Fig. 5(a) reflects the real-time change in the number of desorbed NH_3 molecules during the heating process. The rate curve of the desorption reaction shown in Fig. 5(b) and the mathematical basis for curve segmentation shown in Fig. 5(c) are obtained by first-order and second-order differentiation of Fig. 5(a), respectively. According to the above data processing, the cantilever-TPD curve can be classified successfully into three desorption stages,

978-1-6654-9309-3/23 $31.00 © 2023 IEEE

corresponding to α (weak acid), β (medium acid), and γ (strong acid) sites. The temperature value corresponding to the maximum desorption rate at each acid region can be obtained through the desorption rate curve. The cantilever-TPD results in Fig. 5(a)-(c) indicate that there are three acid sites in ZSM-5, the peak temperatures are 213°C, 310°C, and 396°C, respectively. And the number of the desorbed NH_3 molecules at each site can be accurately obtained, which allows the calculation of the desorption kinetics based on the Arrhenius equation.

Fig. 5(d) shows the TPD curve measured with a commercial instrument, where only two desorption stages can be identified, and the specific number of desorbed molecules cannot be obtained. Compared to the commercial TPD instrument, the *in-situ* cantilever-TPD technology provides more rich information for catalyst characterization and the detailed results are summarized in Table 1 for comparison.

Table 1: Summary of the measured parameters of the ZSM-5 catalyst. T_{max}, Q, and E_d are peak temperature, number of the desorbed probe molecules, and desorption activation energy, respectively.

Types of values		Measurement methods	
		In-situ TPD	Commercial TPD
T_{max}	T_{max} (α)	213°C	240°C
	T_{max} (β)	310°C	
	T_{max} (γ)	396°C	463°C
Q	Q (α)	9.17×10^{12}	\
	Q (β)	2.87×10^{12}	\
	Q (γ)	5.98×10^{12}	\
E_d	E_d (α)	89 kJ/mol	\
	E_d (β)	181 kJ/mol	\
	E_d (γ)	239 kJ/mol	\

CONCLUSION

In summary, an *in-situ* TPD instrument based on the multifunctional microcantilever with resonance excitation and detection, heating and temperature measurement is proposed and developed for advanced catalysts analysis. In this paper, the characterization of ZSM-5 zeolite catalyst as an example shows the advantages of the *in-situ* cantilever-TPD. Compared with the commercial *ex-situ* TPD instruments, the *in-situ* TPD technology reduces the sample mass by 6 orders of magnitude to the nanogram level and eliminates the need for bulky furnaces and *ex-situ* TCD/MS detectors. More significantly, the *in-situ* TPD provides more rich information for catalyst characterization, which means the acid sites are more clearly distinguished and the desorption activation energy can be directly calculated with a single-time *in-situ* TPD measurement. This *in-situ* TPD characterization instrument is expected to be a next-generation enabling tool for high-performance catalyst development.

ACKNOWLEDGEMENTS

This work is supported by the National Natural Science Foundation of China (62227815, 61974155, 61874130, 61804156, 62104241), the National Key R&D Program of China (2021YFB3200800, 2020YFB2008603), Key Research Program of Frontier Sciences of the Chinese Academy of Sciences (QYZDJ-SSW-JSC001), Shanghai "Road and Belt" International Young Scientist Exchange Program (19510744600), Scientific Instrument Project of the Chinese Academy of Sciences (YJKYYQ20210024), Shanghai Pujiang Program (20PJ1415600). The authors also acknowledge the support of the Innovation Team and Talents Cultivation Program of the National Administration of Traditional Chinese Medicine (ZYYCXTD-D-202002, ZYYCXTD-D-202003).

REFERENCES

[1] S. Kondrat, P. Smith, P. Wells, *et al.*, "Stable Amorphous Georgeite as a Precursor to a High-Activity Catalyst", *Nature*, vol. 531, pp. 83-87, 2016.

[2] R. Borade, A. Sayari, A. Adnot, S. Kaliaguine, "Characterization of Acidity in ZSM-5 Zeolites: An X-Ray Photoelectron and IR Spectroscopy Study", *J. Phys. Chem.*, vol. 94, pp. 5989-5994, 1990.

[3] W. Zhang, S. Xu, X. Han, X. Bao, "In Situ Solid-State NMR for Heterogeneous Catalysis: A Joint Experimental and Theoretical Approach", *Chem. Soc. Rev.*, vol. 41, pp. 192-210, 2012.

[4] P. Wang, X. Li, S. Fan, X. Chen, M. Qin, *et al.*, "Impact of Oxygen Vacancy Occupancy on Piezo Catalytic Activity of $BaTiO_3$ Nanobelt", *Appl. Catal. B*, vol. 279, pp. 119340, 2020.

[5] J. L. Falconer, J. A. Schwarz, "Temperature-Programmed Desorption and Reaction: Applications to Supported Catalysts", *Catal. Rev. Sci. Eng.*, vol. 25, pp. 141-227, 1983.

[6] D. Mykhalyk, M. Petryk, M. Petryk, O. Petryk, I. Mudryk, "Mathematical Modeling of Hydrocarbons Adsorption in Nanoporous Catalyst Media using Nonlinear Langmuir's Isotherm using Activation Energy", *ACIT*, Ceske Budejovice, June 5-7, 2019, pp. 72-75.

[7] J. Hao, P. C. Xu, X. X. Li, "Integrated Resonant Micro/Nano Gravimetric Sensors for Bio/Chemical Detection in Air and Liquid", *Micromachines*, vol. 12, pp. 645, 2021.

CONTACT

*P.C. Xu, tel: +86-21-6213-1794;
  xpc@mail.sim.ac.cn
*X.X. Li, tel: +86-21-6213-1794;
  xxli@mail.sim.ac.cn

978-1-6654-9309-3/23 $31.00 © 2023 IEEE

A MULTIPLEXED BIOAFFINITY BIOSENSING PATCH FOR POINT-OF-CARE CHRONIC ULCER MONITORING

Md Sharifuzzaman, Dongkyun Kim, Md Selim Reza, SeongHoon Jeong, Hye Su Song, Md Abu Zahed, J. Y. Park*

Department of Electronic engineering, Kwangwoon University, Republic of Korea

ABSTRACT

This paper reports a flexible multiplexed biosensing patch integrated with a bioinspired microfluidic wound exudate collector for dynamic assessment of the wound's inflammatory milieu and infection status at the point-of-care. The immunosensors platform was functionalized by integrating a MXene ($Ti_3C_2T_x$)-based antifouling nanocomposite that resists biofouling while preserving electroconductivity. As a proof of concept, we successfully demonstrated three panels profiling including interleukin-6 (linear range 0-40 ng mL^{-1}), temperature (sensitivity 0.08% $^oC^{-1}$ in the physiological range of 25–50 ^oC), and pH (sensitivity of -60 mV pH^{-1} in the wound relevant pH range of 3–9) in human serum.

KEYWORDS

Multiplexed biosensing patch, Microfluidic wound exudate collector, Antifouling coating, Chronic ulcer monitoring

INTRODUCTION

Due to an aging population, chronic wounds impose a growing financial and social burden on health care systems worldwide. For instance, venous ulcers, which can affect up to 15% of adults over 70 and have recurrence rates that range from 54 to 78%, require long-term treatments to heal [1], [2]. Due to a variety of environmental and physiological conditions, chronic wounds are unable to heal naturally. When wound healing is progressing, the wound exudate fluid exhibits a dynamic mixing of cytokines, growth factors, and microbes [3].

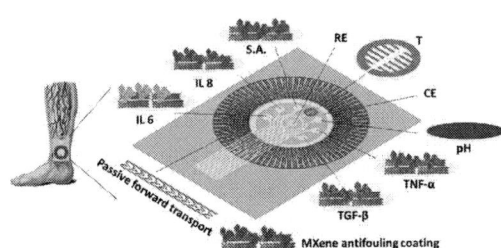

Figure 1: Schematic of a multiplexed immunosensing system for chronic ulcer monitoring. CE (counter electrode) and RE (reference electrode).

Current flexible sensors developed for wound care are only capable of monitoring a limited number of parameters, such as uric acid, oxygen, pH, or temperature [4]. Therefore, multibiomarker detection has been suggested for identifying the specific stage of wound healing. For instance, wound fluids from nonhealing ulcers contain more IL-6 from tumor necrosis factor than those from healing ulcers [5]. The pH of difficult-to-heal wounds is often alkaline, ranging from 7.15 to 8.93 [6]. Higher temperatures seem to impede the rate of wound healing. Moreover, accurate in situ biomarker detection is hard to do when using microfluidic channels to transport wound fluid to the sensing area. In comparison to earlier, single-marker sensors, this suggested platform includes a number of improvements, including (i) a microdrop functionalization, (ii) microfluidic layouts, and (iii) materials for sensing (iv) multibiomarker analysis.

For low-cost, single-use, sensitive multiplexed point-of-care diagnostics for home healthcare, affinity-based electrochemical biosensors (such as those utilizing antibodies, aptamers, and so on) present a potential substitute [7]. The concern is that, unlike optical biosensors, electrochemical biosensing relies on an electrode configuration that conforms to a closed electrical and ionic circuit, where the interrogation-reading process is coupled. By creating a three-dimensional (3D) nanocomposite matrix made of BSA interspersed with conductive nanomaterials, which maintain electron transfer to the underlying electrode while simultaneously increasing analytical performances, allowing functionalization, reducing non-specific binding, and with antibodies or other bioreceptors [8]. Crosslinking glutaraldehyde (GA) to BSA molecules, forming a porous 3D protein matrix loaded with conducting nanomaterials, and so enhancing the electrochemical performance overall [9].

Here, we present a microfluidic immunosensing device for the point-of-care quantitative assessment of a panel of biomarkers that potentially enable in situ characterization of the wound inflammation, microenvironment, and infection state. With the use of this platform, we present the chronic ulcer care patch, termed CUcare, which interacts with wounds as a bioanalytical dressing and measures a variety of healing biomarkers such as inflammatory mediators, bacterial load, and physicochemical characteristics (Figure 1). The selected biomarkers for the panel include interleukin-6 (IL-6), pH, and temperature (T), all of which are higher in wound fluids from nonhealing ulcers compared to healing ulcers.

EXPERIMENTAL

Materials and Apparatus

1-ethyl-3-(3-dimethylaminopropyl) carbodiimide hydrochloride (EDC), N-hydroxysuccinimide (NHS), Iron (III) chloride ($FeCl_3$), potassium hexacyanoferrate (III) [$K_3Fe(CN)_6$], potassium hexacyanoferrate (II) [$K_4Fe(CN)_6$], potassium chloride (KCl), hydrochloric acid (HCl), aniline, buffer solution, bovine serum albumin, 70% glutaraldehyde were purchased from Sigma. Human serum and human plasma were purchased from Sigma-Aldrich. From MELINEX, polyethylene terephthalate (PET) films were acquired. Recombinant human IL-6 and anti-IL-6 were purchased from abcam. We performed electrochemical analysis of the fabricated device at room temperature using an electrochemical workstation (CHI 660D, CH Instruments, USA). The liquid transport system was simulated using the computational fluid dynamics module of COMSOL Multiphysics version 5.3a.

MXene-based Antifouling Coating Preparation

The MXene-BSA-GA antifouling nanocomposites were produced by combining 8 mg mL^{-1} of MXene with 5 mg mL^{-1} of bovine serum albumin (BSA and IgG-Free) in PBS solution (pH 7.4) according to previously modified method [10]. To denature the protein, the solution was sonicated in a ultrasonicator for 30 minutes, followed by 5 minutes of heating at 100 °C. The final opaque black solution was centrifuged for 10 minutes at 500 rpm to remove superfluous aggregates. The supernatant solution of the semitransparent nanocomposite was then diluted with 70% glutaraldehyde for crosslinking at a ratio of 70:1.

Design of the Multiplexed Immunosensing Platform

Figure 1 depicts a chronic ulcer care (CUcare) patch placed on an open wound which includes pH, T, immunosensors, and a microfluidic wound exudate collector. The microfabrication processes for the CUcare patch are shown in Figure 2. First, a 70 μm thick PET substrate was bonded to a SiO$_2$ wafer. To increase the surface roughness of PET, the PET-mounted wafer was treated with O$_2$ plasma. The e-beam evaporator was then used to deposit Pt (thickness: 1000 Å) for the temperature sensor, which was then patterned using a lift-off technique. Following O$_2$ plasma deposition, Au (2500 Å) was then deposited and patterned as a device layer. After that, the SU-8 photoresist was patterned for electrical and Pt-based T sensor passivation. Inspired by the pitcher plant, we devised the topological surface with simple V-shaped posts arranged in parallel channels, which permits unidirectional

2100 (~150 μm). Finally, using a laser, all the electrode layers were cut into the shape of the biosensing patch and gently removed from the SiO$_2$ wafer. The working electrodes were functionalized by MXene-bovine serum albumin-glutaraldehyde (MXene-BSA-GA) antifouling coatings, antibodies (anti IL-6), antigens (IL-6), and polyaniline (PANI) for pH sensor. To determine the optimal coating method for the antifouling nanocomposite on the gold electrodes, the cleaned and plasma-treated electrodes were placed over a hot plate for three minutes to allow the temperature to reach 95°C. The 5 μL antifouling nanocomposite was then drop cast onto each chip and incubated for 3 min. The sensors were then rinsed by dipping them in PBS either immediately or after 10 minutes of cooling and then washing. Before electrochemical analysis, the unreacted glutaraldehyde groups were neutralized by exposing the chips to 0.5 M ethanolamine in PBS.

RESULTS AND DISCUSSION

Figure 3a depicts a transparent and flexible CUCare prototype for monitoring chronic ulcers. It comprised a polar array of circular-shaped working electrodes that shared a central Ag/AgCl reference electrode and a perimeter Au counter electrode to form a compact circular layout suitable for microvolume analysis. Figure 3b shows the designed and simulated optical appearance of the contact line in the direction of pinning. The cavity structure on the V-shaped post impedes the backflow significantly. The directional liquid transport system was designed to capture extra wound fluid and transfer it to the sensor within 100 seconds, ensuring accurate sensing regardless of the size or shape of the ulcer.

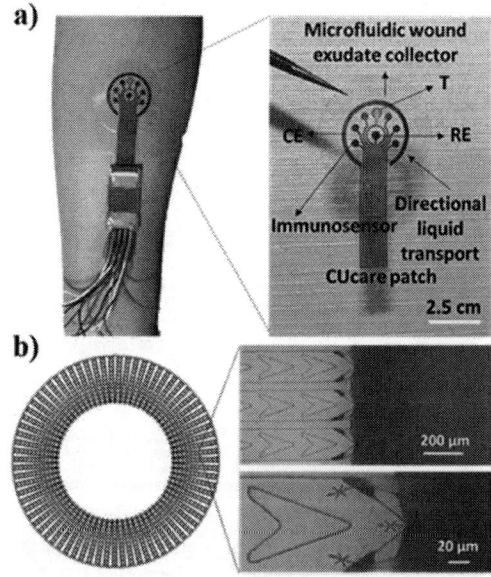

Figure 3: a) CUcare concept for envisioned monitoring of chronic ulcers. The prototype was demonstrated by applying it to a leg. b) Pinning mechanism on the bioinspired surface. The backflow is strongly thwarted by the cavity structure on the V-shaped posts.

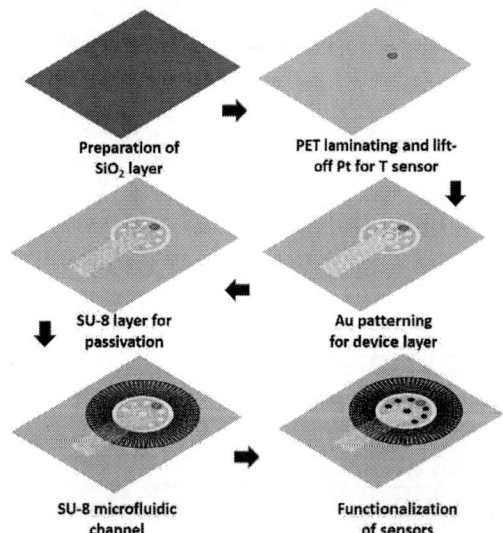

Figure 2: Microfabrication sequences of the CUcare biosensing patch.

wound exudate transport. The microfluidic wound exudate collector was engineered by spin-coating a layer of SU-8

Electrochemical-Physiological Performance Analysis of the Biosensing Patch

Using cyclic voltammetry (CV) in 5 mM $K_3Fe(CN)_6$/$K_4Fe(CN)_6$ redox probe, the sequential fabrication of different layers on the working electrode was validated. When MXene-BSA was covalently crosslinked with GA (Figure 4a), it displayed a higher redox peak than MXene-BSA and bare Au because the porous coating facilitates electroactive species movement. We assumed that the layer operates as an array of indented nanoelectrodes, with each pore functioning as a separate nanoelectrode [11]. Figure 4b depicts a FESEM picture of MXene-BSA-GA antifouling coating. This highly porous structure reflects why the MXene-BSA-GA coated electrode had a stronger surface-diffusion-controlled redox reaction.

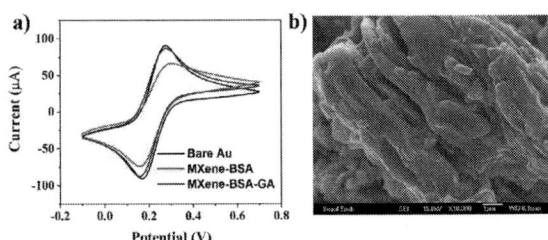

Figure 4: a) Cyclic voltammetry representing oxidation and reduction of 5 mM ferri-/ferrocyanide solution using gold electrode. b) FESEM image of MXene-BSA-GA antifouling coating showing accordion and porous morphology.

A polyaniline (PANI) polymer-based working electrode is used for pH monitoring [12]. The working electrode was functionalized by dipping it into a 0.1 m aniline aqueous solution containing 1 M HCl. Then, 50 segments of CV were performed in the potential range of (0.2-1) V at a scan rate of 0.50 Vs^{-1} to electropolymerized PANI. The pH variations were obtained by monitoring the open-circuit potential (OCP) using a pH range of 3–9 with a good sensitivity of −60 mV pH^{-1} and linearity (R^2 = 0.9997) (Figure 5a and b).

Fig. 5c shows the characterization of the Pt-based temperature sensor. It exhibits positive temperature coefficient resistance (PTCR) behavior, where the resistance rises with temperature because the metal conductor makes the atoms vibrate more and thus makes it harder for the electrons to flow [13]. The sensitivity, as measured by the relative change in resistance, is 0.08 °C^{-1} with a linearity of R^2 = 0.999. To ensure coverage of the pH and temperature fluctuations in a wound fluid environment, the working ranges of the pH sensor (pH 3 to 9) and the T sensor (30° to 45°C) were taken into consideration.

The affinity-based immunosensor was characterized using differential pulse voltammetry (DPV) to monitor variations of peak current height. With increasing target concentration, the peak current height was found to decrease. We employed serum to resemble wound exudates due to its molecular similarity to wound fluid. The immunosensor's effectiveness against the IL-6 analyte at varying concentrations (0-40 ng mL^{-1}) was examined based on concentrations found in wound fluids from chronic ulcer patients [14], revealing an excellent limit of detection (LOD) = 5 ng mL^{-1} and linearity (Figure 5d).

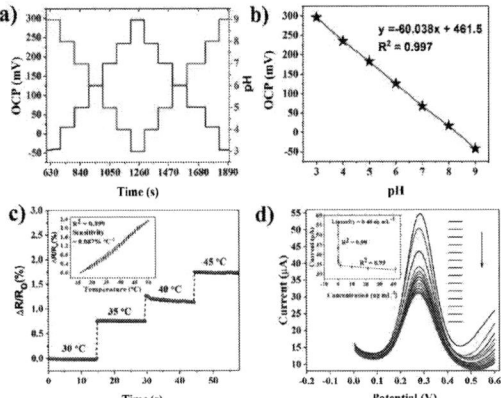

Figure 5: a) OCP of the pH sensor in real-time at various pH levels. b) The OCP of the pH sensor is calibrated against serum pH measurements. c) Temperature-dependent dynamic response of the T sensor and relative resistance changes in the temperature sensor (inset). d) Relative peak height reduction of the IL-6 sensor versus the concentration of the target in serum and linearity curve (inset).

CONCLUSION

In this paper, using cutting-edge sensor architecture and functionalization techniques, an integrated flexible microfluidic multiplexed immunosensing system was newly developed that enables simultaneous monitoring of multibiomarker profiles. The CUCare platform is the first in its class of point-of-care devices that can give precise and pertinent individualized clinical diagnostic data. In addition, the developed sensor technology permits different panels of biomarkers for a range of applications requiring multiplexed analysis, such as high-content screening and diagnostic pathology. Future work entails incorporating the biosensors into a smart dressing that meets safety, regulatory, and mass production requirements, as well as examining the use of sensor data within existing clinical workflows.

ACKNOWLEDGEMENT

This research was supported the Competency Development Program for Industry Specialists of the Korean Ministry of Trade, Industry and Energy (MOTIE), operated by the Korea Institute for Advancement of Technology (KIAT) (No. P0002397, HRD program for Industrial Convergence of Wearable Smart Devices) and by the Technology Innovation Program (RS-2022-00154983, Development of Low-Power Sensors and Self-Charging Power Sources for Self-Sustainable Wireless Sensor Platforms) funded by the Ministry of Trade, Industry & Energy (MI, Korea). The authors are grateful to the group members of the Advanced Sensor and Energy Research (ASER) Laboratory of Kwangwoon University for their valuable suggestions and support.

978-1-6654-9309-3/23 $31.00 © 2023 IEEE

REFERENCES

[1] L. Leren, E. Johansen, H. Eide, R. S. Falk, L. K. Juvet, and T. M. Ljoså, "Pain in persons with chronic venous leg ulcers: A systematic review and meta-analysis," *Int. Wound J.*, vol. 17, no. 2, pp. 466–484, Apr. 2020, doi: 10.1111/iwj.13296.

[2] M. S. Brown, B. Ashley, and A. Koh, "Wearable Technology for Chronic Wound Monitoring: Current Dressings, Advancements, and Future Prospects," *Front. Bioeng. Biotechnol.*, vol. 6, no. APR, pp. 1–21, Apr. 2018, doi: 10.3389/fbioe.2018.00047.

[3] S. L. Drinkwater, A. Smith, and K. G. Burnand, "What Can Wound Fluids Tell Us About the Venous Ulcer Microenvironment?," *Int. J. Low. Extrem. Wounds*, vol. 1, no. 3, pp. 184–190, Sep. 2002, doi: 10.1177/153473460200100307.

[4] P. Mostafalu *et al.*, "Smart Bandage for Monitoring and Treatment of Chronic Wounds," *Small*, vol. 14, no. 33, p. 1703509, Aug. 2018, doi: 10.1002/smll.201703509.

[5] L. E. Edsberg, J. T. Wyffels, M. S. Brogan, and K. M. Fries, "Analysis of the proteomic profile of chronic pressure ulcers," *Wound Repair Regen.*, vol. 20, no. 3, pp. 378–401, May 2012, doi: 10.1111/j.1524-475X.2012.00791.x.

[6] L. A. Schneider, A. Korber, S. Grabbe, and J. Dissemond, "Influence of pH on wound-healing: a new perspective for wound-therapy?," *Arch. Dermatol. Res.*, vol. 298, no. 9, pp. 413–420, Jan. 2007, doi: 10.1007/s00403-006-0713-x.

[7] M. Zourob, *Recognition receptors in biosensors.* 2010.

[8] S. Kumar, W. Ahlawat, R. Kumar, and N. Dilbaghi, "Graphene, carbon nanotubes, zinc oxide and gold as elite nanomaterials for fabrication of biosensors for healthcare," *Biosensors and Bioelectronics*. 2015, doi: 10.1016/j.bios.2015.03.062.

[9] J. Sabaté del Río, O. Y. F. Henry, P. Jolly, and D. E. Ingber, "An antifouling coating that enables affinity-based electrochemical biosensing in complex biological fluids," *Nat. Nanotechnol.*, vol. 14, no. 12, pp. 1143–1149, Dec. 2019, doi: 10.1038/s41565-019-0566-z.

[10] S. S. Timilsina, N. Durr, M. Yafia, H. Sallum, P. Jolly, and D. E. Ingber, "Ultrarapid Method for Coating Electrochemical Sensors with Antifouling Conductive Nanomaterials Enables Highly Sensitive Multiplexed Detection in Whole Blood," *Adv. Healthc. Mater.*, vol. 11, no. 8, p. 2102244, Apr. 2022, doi: 10.1002/adhm.202102244.

[11] U. Zupančič, P. Jolly, P. Estrela, D. Moschou, and D. E. Ingber, "Graphene Enabled Low-Noise Surface Chemistry for Multiplexed Sepsis Biomarker Detection in Whole Blood," *Advanced Functional Materials*. 2021, doi: 10.1002/adfm.202010638.

[12] H. Y. Y. Nyein *et al.*, "A Wearable Electrochemical Platform for Noninvasive Simultaneous Monitoring of Ca2+ and pH," *ACS Nano*, vol. 10, no. 7, pp. 7216–7224, 2016, doi: 10.1021/acsnano.6b04005.

[13] "Adv Funct Materials - 2022 - Zahed - A Nanoporous Carbon-MXene Heterostructured Nanocomposite-Based Epidermal Patch for.pdf." .

[14] R. Zillmer, H. Trøstrup, T. Karlsmark, P. Ifversen, and M. S. Ågren, "Duration of wound fluid secretion from chronic venous leg ulcers is critical for interleukin-1α, interleukin-1β, interleukin-8 levels and fibroblast activation," *Arch. Dermatol. Res.*, 2011, doi: 10.1007/s00403-011-1164-6.

CONTACT

*J.Y. Park, tel: +82-2-940-5113; jaepark@kw.ac.kr

3-DOF BIOHYBRID ACTUATOR WITH MULTIPLE SKELETAL MUSCLE TISSUES

Xinzhu Ren[1], Yuya Morimoto[1, 2], and Shoji Takeuchi[1, 2]
[1]Faculty of Engineering, the University of Tokyo, JAPAN and
[2]Graduate School of Information Science and Technology, the University of Tokyo, JAPAN

ABSTRACT

We proposed a biohybrid actuator powered by a bundle of multiple skeletal muscle tissues. This actuator can contract partially depending on the position of stimulating electrodes, which makes it possible to achieve 3-DoF (Degree of Freedom) motions with a single biohybrid actuator. The multiple muscle tissue actuator was constructed by a pair of rollable devices on the sides of tissues; the devices were flat during culture to make an equal culture condition for each muscle tissue, and were rolled up to obtain an actuator after tissue maturation. We believe that this biohybrid actuator with 3-DoF will be a useful tool in biohybrid robotics to mimic motions of living animals.

KEYWORDS

Electrical stimulation, Muscle contraction, 3D printing, Human myoblast, Hand-rolled structure.

INTRODUCTION

Muscle tissue has attracted attentions as a suitable material for actuators due to its advantages such as high energy efficiency, low noise and self-healing ability [1, 2]. Especially skeletal muscle tissue, a kind of voluntary striated muscle, is a promising actuator because of its ON/OFF controllability and high power/weight ratio that can be up to 1000 W/kg [3]. Also, skeletal muscle has voltage-dependent Ca^{2+} channels allowing it to contract synchronously with electrical stimulation [4]. With the advancement of biofabrication techniques, cultured skeletal muscle tissue can be constructed in vitro. Therefore, cultured skeletal muscle tissue has become widely used as living actuators in biohybrid robotics to achieve simple movements such as gripping [5], crawling [6], actuating in the air [7], and rotating joints [8]. However, conventional muscle-powered actuators have been limited to 1-DoF motions. In nature, we can see many complex motions in animals like movements of earthworms and elephants' trunks. To realize life-like movements in biohybrid robots, biohybrid actuators capable of multi-DoF motions are needed. In the case of earthworm , longitudinal muscles contribute to multi-DoF movements; the longitudinal muscles on the one side contract while ones on the opposite side stretch to make the body bending, and contract or stretch together to make the body moving forward [9].

In this paper, referring to the pattern of multiple muscles in the longitudinal muscles of earthworm, we propose a biohybrid actuator called multiple muscle tissue actuator. Our actuator composed of 8 separate muscle tissues which can selectively contract based on the position of the stimulating electrodes to provide 3-DoF motions (Fig. 1). When the electrodes are on the side of the actuator,

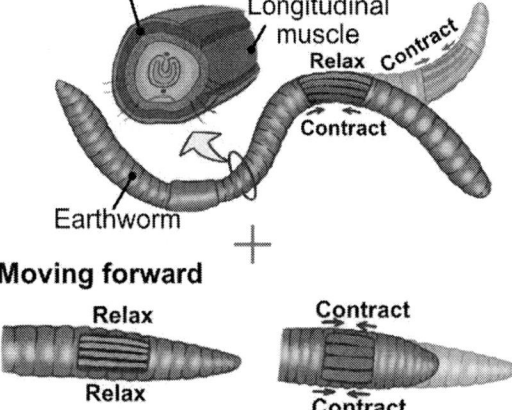

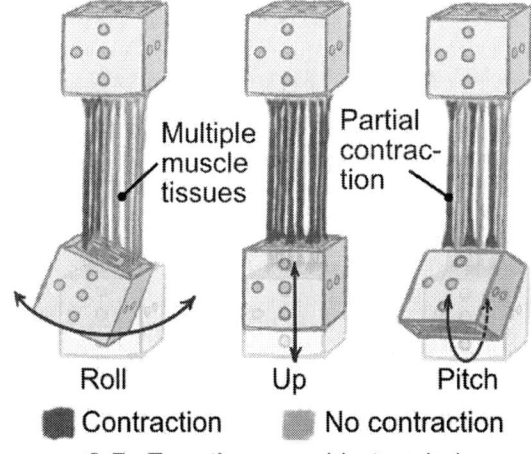

Figure 1. Conceptual illustration of biohybrid actuator with multiple skeletal muscle tissues. Conventional biohybrid actuators only provide 1-DoF motions. Proposed biohybrid actuator can produce 3-DoF motions by selective muscle contractions.

only the muscle tissues near the electrodes contract, allowing 2-dimension rotation of the actuator. All muscle tissues contract when the electrodes are in the middle, allowing the actuator to move straight.

MATERIALS AND METHODS

Cell Preparation

Human skeletal muscle myoblasts (Cell Applications, Inc) were maintained at 37°C in a 5% CO_2 atmosphere in skeletal muscle cell growth medium-2 (SkGM-2, Lonza). We used Trypsin-EDTA to peel the cells off from culture dishes at 80% confluence to prepare the single-cell suspension.

Fabrication of Devices

The multiple muscle tissue actuator consists of muscle tissues, anchors to fix the tissues, and a pair of anchor holders. The anchors were printed by a projection micro stereolithography with a 3D printer (microArch S140, BMF). We coated a 2 μm parylene and fibronectin (0.1 mg/ml in phosphate buffered saline (PBS)) on the anchors. The anchor holders were composed of parts with holes fabricated by a 3D printer (DigitalWax 028J) using resin (DM210); the parts were connected by a 5 μm thick flexible parylene sheet and the interval between two parts were fixed at 450 μm, allowing the anchor holder to be rolled into a cube-like form (Fig. 2(a)). We also used a 3D printer (Agilista, Keyence Corp.) to form a base and a stopper for tissue culture, and templates for making polydimethylsiloxane (PDMS) molds, which were used for patterning myoblast-laden hydrogel. We coated a 2 μm parylene layer on the 3D printed template and cast PDMS prepolymer containing 10 to 1 silicone elastomer and curing agent, then heat it at 75°C for 1.5 h to make it solidify. After peeled the PDMS molds from the templates, we coated them using MPC polymer solution (0.5wt% Lipidure-CM5206 in 99% ethanol) and bovine serum solution (BSA) solution (1% BSA in PBS) to prevent the hydrogel from sticking to the PDMS molds. All the devices were sterilized by UV light and O_3 gas for 30 min before being used in the experiments.

Construction of Multiple Muscle Tissue Actuator

We put the anchors in the PDMS molds after 3 times washes in PBS (Fig. 2(b-i)) [10], and cast hydrogel solution composed of myoblasts (1.5×10^7 cells/mL), Matrigel, the growth medium, fibrinogen and thrombin. After gelling the hydrogel in 30 min incubation at 37°C (Fig. 2(b-ii)), the striped structures of myoblast-laden hydrogel were transferred to the culture device, ensuring that the length of the tissue was maintained during the culture (Fig. 2(b-iii)). We cultured the striped structures for 2 weeks in the growth medium with 1% amino caproic acid (ACA) for 2 days, and in differentiation medium composed of Dulbecco's modified eagle's medium (DMEM) (low glucose) with 1% ACA, 2% horse serum (HS), 100 U/mL penicillin and 100 μg/mL streptomycin for 2 weeks (Fig. 2(b-iv)). After the formation of cultured skeletal muscle tissue, we removed the stoper and rolled up the pair of anchor holders using a rectangular glass stick (Fig. 2(b-v)). During the rolling process, we mounted two disks with square hole in the center at the ends of the rectangular glass stick (Fig. 3(a)), in order to make the process easily complete by hands without directly touching the device for decreasing the risk of contamination. Finally, we fixed the rolled anchor holders by inserting two rods into their holes to construct a multiple muscle actuator (Fig. 2(b-vi), Fig. 3(b)).

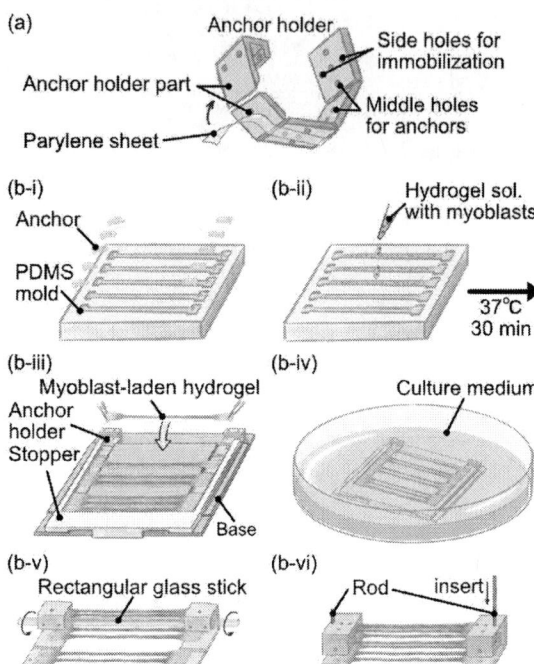

Figure 2. Fabrication process. (a) Parts assembly of an anchor holder by attaching a parylene sheet on anchor holder parts to link them. (b-i, ii) Shaping myoblast-laden hydrogel with a PDMS mold and anchors by (i) putting anchors into a mold and (ii) patterning hydrogel in the mold. (b-iii) Setting myoblast-laden hydrogel. (b-iv) Culture of myoblast-laden hydrogel. (b-v) Rolling a pair of anchor holders (b-vi) Fixing the anchor holder with rods to obtain a multiple muscle tissue actuator.

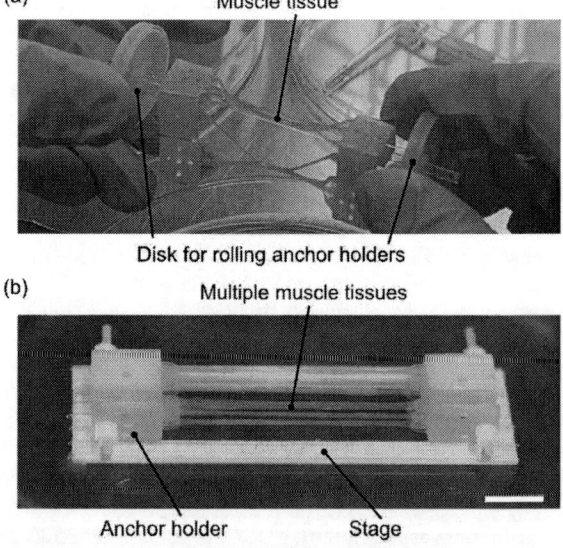

Figure 3. (i) Photo during the manual rolling process for formation of a multiple muscle tissue actuator. Two disks with square hole in the center were used to roll the stick easily. (ii) Photo of the multiple muscle tissue actuator in transparent culture medium. Scale bar: 5 mm.

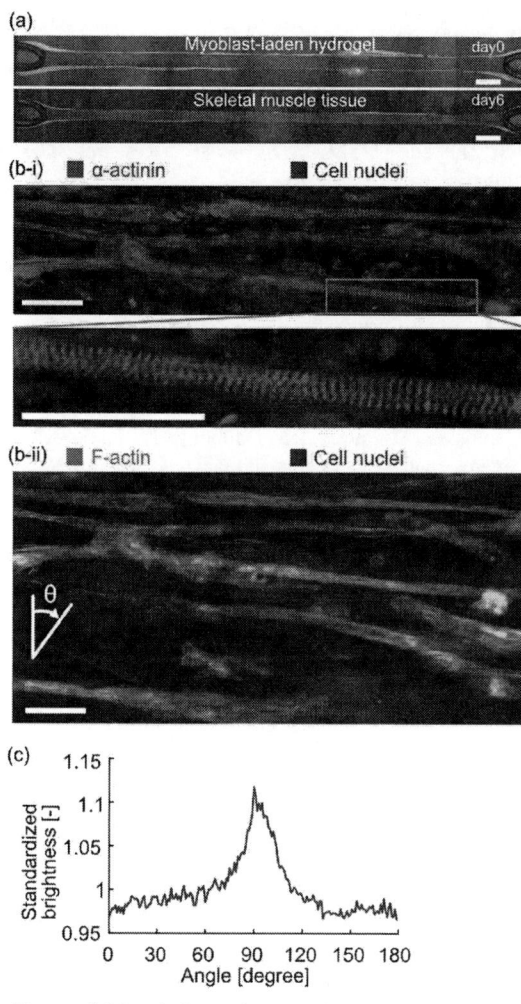

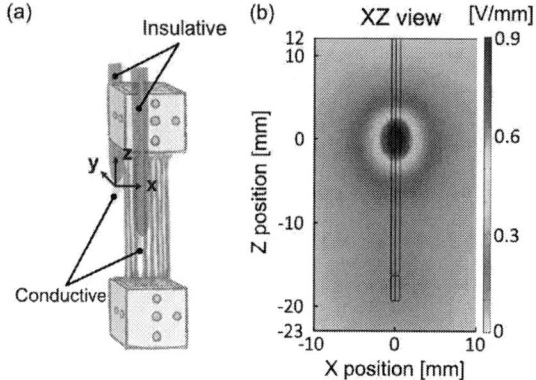

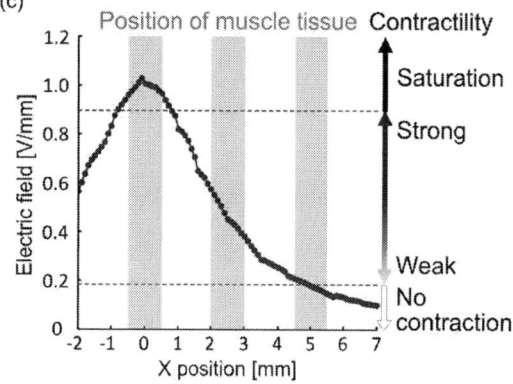

Figure 5. Electric field simulation. (a) Sketch for the condition of electrodes. (b) Color map of electric field (11 V, electrodes center distance: y; 9 mm, z; 15 mm). (c) Variation in electric field at z = 0 on the x-axis.

Figure 4. Morphological assay. (a) Formation of mouse skeletal muscle tissue. (b-i, ii) Immunostaining image of cultured human skeletal muscle tissue. Red; α-actinin. Green; F-actin. Blue; cell nuclei. (c) Directional distribution of brightness in the image (b-ii) calculated by fast Fourier transform. Scale bars: (a) 1 mm, (b-i, ii) 50 μm.

Immunostaining

We immunostained the muscle tissues to investigate its maturity. After being washed 3 times with PBS, the muscle tissues were fixed with 4% paraformaldehyde for 20 min, permeabilized with 0.3% Triton X-100 for 20 min, and blocked with 2.5% BSA overnight at 4°C. We incubated the muscle tissues with 0.1% monoclonal anti-α-actinin antibody (Sigma-Aldrich Co. LLC) and 0.2% phalloidin (Alexa Fluor 488, Thermo Fisher Scientific Inc.) overnight at 4°C, then with 0.2% secondary antibody (goat anti-mouse IgG Alexa Fluor 568, Thermo Fisher Scientific Inc.) and 0.1% DAPI (4',6-diamidino-2-phenylindole, Invitrogen) at room temperature for 2 h.

Simulation

To examine the possibility of selective muscle contraction in the multiple muscle actuator, we estimated the electric field each muscle tissue takes during the

electrical stimulation. We prepared 3D static models of electrical conductivity using the finite element method simulation software (COMSOL Multiphysics). Electrodes were placed parallel to the first (left) column of muscle tissue matrix with 15 mm vertical distance in culture medium ($60 \times 60 \times 55$ mm³). 3 mm of the tip of electrodes were conductive and the rest were insulated.

Electrical Stimulation

We generated an alternative-current square wave as electrical pulses using a function generator (Agilent Technologies Inc.) and an amplifier (AE-TDA2030, Akizuki Denshi Tsusho Co., Ltd.). We covered gold electrodes with parafilm to make them insulated, except for 3 mm conductive section at the ends of the electrodes. The condition of electrical stimulation was 11 V and 1 Hz (pulse width 2 ms, frequency 250 Hz, period 0.2 s).

RESULT AND DISCUSSION
Morphology

We confirmed that the formation of muscle tissue cultured on our device was successful, as the tissue shrunk during the formation (Fig. 4(a)). After 14 days incubation, striped patterns of α-actinin were clearly found, indicating the formation of sarcomeres (Fig. 4(b-i)). Also, the orientations of F-actin were basically aligned in the same direction (Fig. 4(b-ii, c)). These results show a high maturity of the muscle tissue.

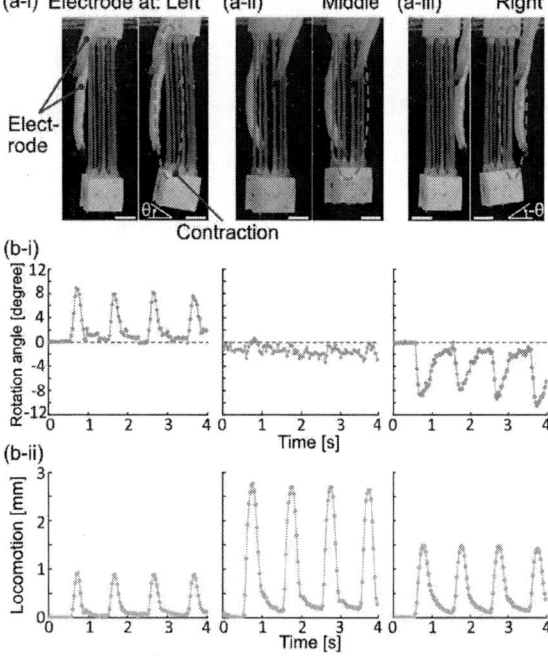

Figure 6. Performance of multiple muscle tissue actuator. (a) Photos before and during the electrical stimulation (11 V, 1 Hz). The electrodes are at (i) left, (ii) middle, (iii) right of the actuator. (b) Motion analysis with a tracker. (i) Rotation of the bottom rolled anchor holder. (ii) Locomotion of the center of bottom rolled anchor holder. Scale bars: 5 mm.

Distribution of Electric Field

We showed changes in the amplitude of electric field depending on distance from the electrode (Fig. 5(a, b)), and compared estimated muscle contractility of each column, based on the relationship between applied electric field and muscle contractility mentioned in the previous research [8]. The muscle tissue in the first (left) column can be fully stimulated, while the stimulation on the muscle tissue in the third (right) column is not enough to make the tissue contract.

Actuation Performance

As a demonstration, we evaluated the performance of the biohybrid actuator. When the electrodes were on the side of the actuator, the bottom anchor holder rotated, and when they were in the middle, the bottom anchor holder moved straight up and down (Fig. 6). Due to the isotropic arrangement of muscle tissues, the rotation suggests availability of pitch and roll. Thus, the biohybrid actuator can provide 3-DoF motions through selective muscle contractions.

CONCLUSION

We succeeded in creating a biohybrid actuator with multiple skeletal muscle tissues, which can provide 3-DoF motions. Moreover, the contractile length of the actuator was approximately 10%, close to the living muscle [11]. Thus, we believe that the multiple muscle tissue actuator will be widely used to mimic the motions of living bodies.

ACKNOWLEDGEMENTS

This work was partially supported by JST-Mirai Program JPMJMI20C1, the JSPS Grants-in-Aid for Scientific Research (KAKENHI) (Grant Number 21H00321) and UTEC-UTokyo FSI Research Grant Program.

REFERENCES

[1] M. Juhas, G. C. Engelmayr, Jr., A. N. Fontanella, G. M. Palmer, and N. Bursac, "Biomimetic engineered muscle with capacity for vascular integration and functional maturation in vivo," *Proc Natl Acad Sci U S A,* vol. 111, pp. 5508-5513, 2014.

[2] R. Raman, L. Grant, Y. Seo, C. Cvetkovic, M. Gapinske, A. Palasz, H. Dabbous, H. Kong, P. P. Pinera, and R. Bashir, "Damage, Healing, and Remodeling in Optogenetic Skeletal Muscle Bioactuators," *Adv Healthc Mater,* vol. 6, p. 1700030, 2017.

[3] L. Ricotti and A. Menciassi, "Bio-hybrid muscle cell-based actuators," *Biomed Microdevices,* vol. 14, pp. 987-998, 2012.

[4] W. Melzer, E. Rios, and M. F. Schneider, "Time course of calcium release and removal in skeletal muscle fibers," *Biophys J,* vol. 45, pp. 637-641, 1984.

[5] K. Kabumoto, T. Hoshino, Y. Akiyama, and K. Morishima, "Voluntary movement controlled by the surface EMG signal for tissue-engineered skeletal muscle on a gripping tool," *Tissue Eng Part A,* vol. 19, pp. 1695-1703, 2013.

[6] G. J. Pagan-Diaz, X. T. Zhang, L. Grant, Y. Kim, O. Aydin, C. Cvetkovic, E. Ko, E. Solomon, J. Hollis, H. Kong, T. Saif, M. Gazzola, and R. Bashir, "Simulation and Fabrication of Stronger, Larger, and Faster Walking Biohybrid Machines," *Adv Funct Mater,* vol. 28, p. 1801145, 2018.

[7] Y. Morimoto, H. Onoe, and S. Takeuchi, "Biohybrid robot with skeletal muscle tissue covered with a collagen structure for moving in air," *APL Bioeng,* vol. 4, p. 026101, 2020.

[8] Y. Morimoto, H. Onoe, and S. Takeuchi, "Biohybrid robot powered by an antagonistic pair of skeletal muscle tissues," *Sci Robot,* vol. 3, p. eaat4440, 2018.

[9] X. Zhan, J. Xu, and H. Fang, "In-plane gait planning for earthworm-like metameric robots using genetic algorithm," *Bioinspir Biomim,* vol. 15, p. 056012, 2020.

[10] B. Jo, Y. Morimoto, and S. Takeuchi, "3D-Printed Centrifugal Pump Driven by Magnetic Force in Applications for Microfluidics in Biological Analysis," *Adv Healthc Mater,* p. e2200593, 2022.

[11] D. Sangian, S. Naficy, G. M. Spinks, and B. Tondu, "The effect of geometry and material properties on the performance of a small hydraulic McKibben muscle system," *Sensor Actuat a-Phys,* vol. 234, pp. 150-157, 2015.

CONTACT

*Xinzhu Ren, tel: +81-3-5841-6488; ren@hybrid.t.u-tokyo.ac.jp

A LOW NOISE MICROELECTRODE ARRAY FOR SPECIFIC CELL ACTIVITY MODULATION FROM CELL TO TISSUE

Bohan Zhang[1,2], Huiran Yang[2], Xiner Wang[3,4], Ziyi Zhu[2,4], Zongxing He[1], Wanqi Jiang[2,4], Chen Tao[1,2], Dujuan Zou[3,4], Meng Li[2,4], Zhitao Zhou[2,4], Liuyang Sun[3,4], Tiger H.Tao[1,2,3,4,5,6,7,8,9] and Xiaoling Wei[2,4]

[1] School of Physical Science and Technology, ShanghaiTech University, Shanghai, China
[2] State Key Laboratory of Transducer Technology, Shanghai Institute of Microsystem and Information Technology, Chinese Academy of Sciences, Shanghai, China
[3]2020 X-Lab, Shanghai Institute of Microsystem and Information Technology, Chinese Academy of Sciences, Shanghai, China
[4]School of Graduate Study, University of Chinese Academy of Sciences, Beijing, China
[5]Center of Materials Science and Optoelectronics Engineering, University of Chinese Academy of Sciences, Beijing, China
[6]Center for Excellence in Brain Science and Intelligence Technology, Chinese Academy of Sciences, Shanghai, China
[7]Neuroxess Co., Ltd. (Jiangxi), Nanchang, Jiangxi, China
[8]Guangdong Institute of Intelligence Science and Technology, Hengqin, Zhuhai, Guangdong, China
[9]Tianqiao and Chrissy Chen Institute for Translational Research, Shanghai, China.

ABSTRACT

We report a microelectrode array (MEA) system that enables stimulation and in situ electrophysiological recording by integrating a 32-channel microelectrode array with highly transparency fabricated on a glass substrate. The specially designed double shielding rings allow to reduce electromagnetic shielding and reduced the noise of the MEA to only about 3 μV, which is much lower than that of commercial electrodes with around 10μV [1]. As a proof of concept, simultaneous electrical stimulation and electrophysiological recordings with signal to noise ratios up to 30 were achieved in cellular models. In contrast to existing technologies, our MEA will allow to specifically modulate the activity of neurons.

KEYWORDS

Microelectrode array, Specific modulation, Low noise, High signal to noise ratio.

INTRODUCTION

In neuroscience, it is important and difficult to match single cell activities and electrical signal specifically in vivo [2,3]. To explain the mechanism of neuronal electrical activity, researchers use a method known as extracellular electrophysiology [4,5]. As a powerful tool of extracellular electrophysiology, microelectrode arrays (MEAs) have good compatibility, which can be applied to any electrogenic tissue [6]. Microelectrodes are now widely used to detect and regulate the activity of neuronal networks. The current MEA includes two types: commercial and scientific research. Commercial electrodes tend to pursue higher channel numbers, while scientific MEAs focus more on functionality, such as lower noise and the function of the entire microelectrode array system, such as electrochemical modification and assembly.

In order to achieve functions including biocompatibility and high signal to noise ratio (SNR), various materials have been used in microelectrode manufacturing, and conductive polymers are one of the most widely explored materials [7]. To date, the researchers have modified the electrodes using plating on electrodes such as poly (3,4-ethylenedioxythiophene) - poly (phenylene sulfonate) (PEDOT: PSS) to improve various performances of electrodes for research purposes [8]. Hydrogels also show good potential for electrochemical modification devices [9]. Polymeric materials such as silk protein or PFA are also widely used as electrode modification materials for electrode fabrication and assembly to achieve better biocompatibility and ductility [10]. Polyaniline, as a black conductive polymer, can be synthesized by redox reaction, and can meet the differentiation of electrical and optical properties in cells [11]. Inspired by electrode modification and organic redox reactions, targeting assembly for cells to achieve discrimination of electrical signals among cell species is validated to be feasible [12].

FABRICATION

MEA

Most MEA electrodes use conventional opaque materials such as silicon as the substrate, which will lead to the inability to observe the morphology of cell growth during cell culture. In order to achieve optical recording, we use transparent glass as the substrate of MEA.

A silicon nitride (Si-N) layer (~50nm) was deposited for increasing the adhesion between glass and metal. Chromium/nickel/aurum (Cr/Ni/Au) (~5/100/50nm) Pads, electrodes and interconnects were patterned through UV contact lithography, electron beam evaporation, and lift-off processes. Then a Polyimide (PI) layer was coated on the wafer through spin coating and heating.

On the PI layer, an aluminum mask layer (~300nm) was sputtered and then patterned through UV contact lithography and etching processes. As the final step of the process, the PI cap with O$_2$ etching process and then etched the remaining aluminum mask.

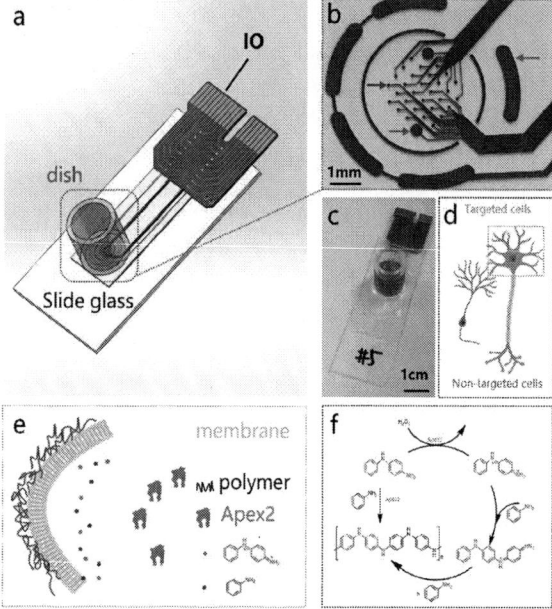

Figure 1:(a) Schematic diagram of the assembled microelectrode array system. (b) Zoom-in image of the electrode array in (a) (red dashed box). Scale bar: 1 mm. (c) Image of fabricated microelectrode array integrated neural probe. Scale bar: 1cm. (d)Different nerve cells in vitro (e) Schematic diagram of the extramembrane redox reaction. (f) Reaction of Apex2-mediated polymerization from aniline and N-phenyl-1,4-phenylenediamine.

Device Assembly

The electrodes were connected to the back-end recording system by welding to a printed circuit board (PCB). A custom-made transparent glass tube with a diameter of 1 cm was used as a Petri dish and connected to a glass plate at the bottom of the electrode by Polydimethylsiloxane (PDMS) for waterproof. The whole MEA system was immobilized on a glass slide by PDMS to achieve a favorable light transmission with operability, which made the cell growth process and optical testing feasible (Fig. 1a).

Redox reactions

The synthesis of an electroactive polymer that achieves biocompatible within living animal cells is an important part of the whole experiment. Polyaniline (PANI) and poly(3,4-ethylenedioxythiophene) (PEDOT) are appropriate because of aqueous synthesis for excellent biocompatibility [13]. We used a single-enzyme–facilitated polymerization using chemically modified monomers for which polymerization is triggered by an enzyme that can be expressed in specific cells (Fig. 1d) [14]. We used a specific intracellularly expressed enzyme facilitating this cellular catalytic redox reaction. Benefiting from the shorter mean diffusion length of radical cations in aqueous solution and the lower solubility of the reaction

generated polymers in water, the substrates would be deposited on the target cells near the membrane (Fig. 1f).

CHARACTERIZATION
Shield design

The shielding effect simulation was conducted using the COMSOL Multiphysics 5.6 software. A 2D simulation model was built with one substrate layer, one encapsulation layer and 32 round electrodes, including two electrodes with larger areas designed for electrical stimulation. The substrate and encapsulation layer are both composed of 2μm thick polyimide (PI) membrane. The round electrodes and the shielding ring are Au electrodes with 2μm thickness.

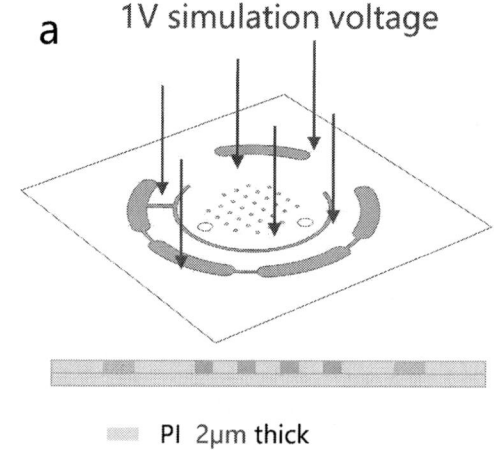

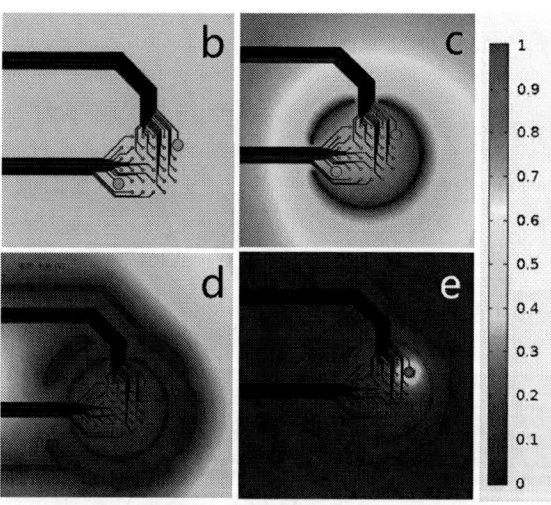

Figure 2: (a) Schematic diagram of thermal noise simulation. (b) Simulation diagram of thermal noise of unshielded ring electrode. (c) Simulation diagram of thermal noise of a single shield ring electrode. (d) Simulation diagram of thermal noise of double shielded ring electrode. (e) Simulation diagram of thermal noise effect after large electrode point current stimulation.

To compare the shielding effect under different shield ring structures, an input voltage of 1V was applied to the periphery of the above device to simulate the thermal noise. Double shield ring structure achieves excellent shielding performance, significantly better than single shield ring and unshielded structures (Fig. 2d). We simulate displays the potential heatmap under single point electrical stimulation as well (Fig. 2e). Under our electrode design, the effects of artifacts and crosstalk can be well reduced, thus improving the recording quality of neural signals. The shielding ring is directly grounded, which has the advantage of avoiding the risk of bacterial contamination from the PCB lead to the culture dish.

Image of targeted cells

We selected neurons in CA1 region of C57 strain mice of P0 as specific cells. Neurons were cultured in 90% DMEM medium and 10% FBS within 6 hours after extraction. The purpose of this step is for the initial adaptation and growth of neurons. After 6 hours, the new culture medium was replaced by 98% Neurobasal and 2% B27 supplement solution, which was beneficial to the growth of neurons and inhibition of glial cells. The composition of the culture medium remained unchanged throughout the subsequent culture process, and the medium was changed every three days. During the whole process of neuronal cell culture, double antibody is necessary to prevent bacterial infection [15]. After one week of culture, we took the bright field images of neurons (Fig. 3a).

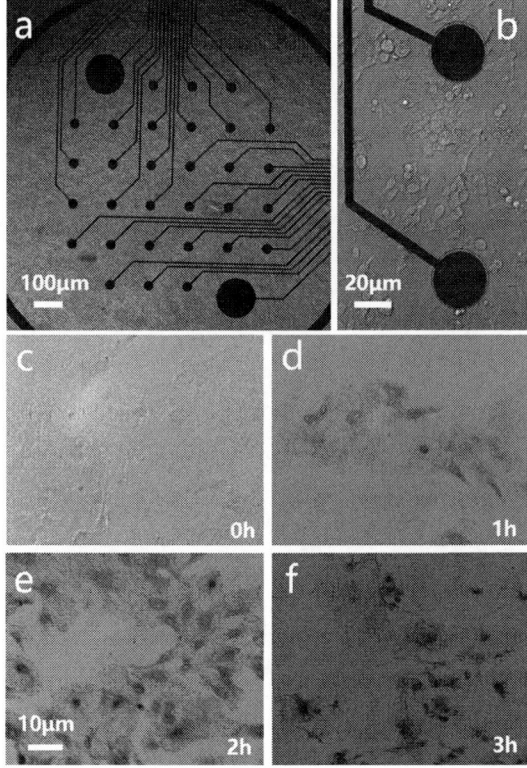

Figure 3: (a) Close-up image of cell culture well. Scale bar: 100μm. (b) Zoom-in image of the electrode array in (a). Scale bar: 30μm. (c-f) Image of cell surface topography in targeted assembly from 0 to 3 hours. Scale bar: 10μm.

Apurinic/apyrimidinic endodeoxyribonuclease2 (Apex2) protein gene was transfer into neuron cells by electroporation to specifically secrete hydrogen peroxide. When aniline is added to the stably growing neuron cells, the hydrogen peroxide secreted by the cells will catalyze the redox reaction to produce the black product polyaniline. The imaging in the bright field proved that with the increase of time, polyaniline would adhere to the surface of the cell membrane, forming a visible black film (Fig. 3c-f).

Electrical characterization

Previous experiments showed that the resistance value of PANI decreased after reaction, and the frequency of PANI release decreased after current stimulation. We air dry the cells on the electrode points, give the voltage values of two adjacent electrode points through the electrochemical workstation, measure the current, and calculate the cell resistance by voltammetry (Fig. 4a). Within 4 hours after the start of redox reaction (targeted assembly), the electrical resistance of cells decreased as expected (Fig. 4b). After the targeted assembly is completed, the cell resistance decreases by about two orders of magnitude (Fig. 4c).

When conductive black deposits accumulate on the cell surface, greater electrical stimulation becomes available. After the current stimulation of 2.5μA, the cells showed continuous discharge [16]. With the progress of self-assembly, the neural point activity of cells in local field potential will be inactivated (Fig. 4d).

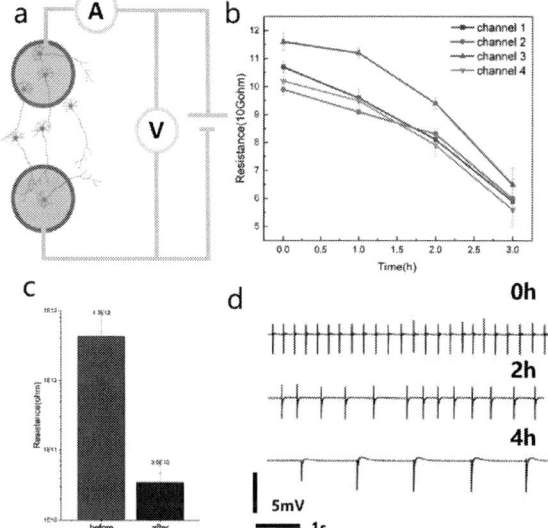

Figure 4:(a) Schematic diagram of measuring cell resistance by voltammetry. (b) Cell resistance values next to different channels within 0 to 3 hours of targeted assembly. (c) Histogram of final resistance values before and after cell assembly. (d) Local field potential recordings of targeted assembly cells from of 0 to 4 hours. Scale bar: 100μV,100ms.

CONCLUSION

To summarize, we designed and manufactured a microelectrode array for cells and tissues. The transparency of the glass bottom provides the optical imaging capability for the entire system. The conformal connection of ultrathin MEA realizes the integration of 32 channel electrophysiological recording function and high anti-interference ability to external signals. Based on the fabricated microelectrode arrays, we have conducted targeted assembly tests in cell models. In the future, this targeted assembly method and microelectrode array can be used for specific neuron regulation and high spatial and temporal resolution in situ electrophysiological recording, and provide a new platform for neural circuit tracking in neuroscience research.

ACKNOWLEDGEMENTS

This work was partially supported by Key-Area Research and Development Program of Guangdong Province（2021B0909060002）, Guangdong high level Innovation Research Institute（2021B0909050004）,the National Key R & D Program of China (Grant Nos. YFA0905200, 2021ZD0201600, 2021YFC2501500, 2021YFF1200700, 2022ZD0209300, 2022ZD0212300), National Natural Science Foundation of China (Grant No.61974154), Key Research Program of Frontier Sciences, CAS (Grant No. ZDBSLY-JSC024), Shanghai Pilot Program for Basic Research—Chinese Academy of Science, Shanghai Branch (Grant No. JCYJ-SHFY-2022-01), Shanghai Municipal Science and Technology Major Project (Grant No. SHZDZX), CAS Pioneer Hundred Talents Program, Shanghai Pujiang Program (Grant Nos.19PJ1410900, 21PJ1415100), the Science and Technology Commission Foundation of Shanghai (No. JM0010200), Shanghai Rising-Star Program (Grant No. QA1410900), the Innovative Research Team of High-level Local Universities in Shanghai, the Jiangxi Province 03 Special Project and 5G Project (Grant No. ABC03W07), Fund for Central Government in Guidance of Local Science and Technology Development (Grant No. ZDE04013).

REFERENCES

[1] Xiang, Y., Liu, H., Yang, W. et al. "A biosensing system employing nanowell microelectrode arrays to record the intracellular potential of a single cardiomyocyte", *J. Microsyst Nanoeng*, vol. 8, pp. 70 ,2022.

[2] Jacobs Joshua,Kahana Michael J, Ekstrom Arne D et al. "Brain oscillations control timing of single-neuron activity in humans", *J. J Neurosci*, vol.27,pp. 3839-3844, 2007.

[3] Baker S N,Philbin N,Spinks R et al. "Multiple single unit recording in the cortex of monkeys using independently moveable microelectrodes", *J. J Neurosci Methods*, vol. 94,pp. 5-17, 1999.

[4] Bayat F Kemal,Alp M İkbal,Bostan Sevginur et al. "An improved platform for cultured neuronal network electrophysiology: multichannel optogenetics integrated with MEAs", *J. Eur Biophys J*, vol. 51, pp. 503-514, 2022.

[5] Churchland Mark M,Yu Byron M,Sahani Maneesh et al. "Techniques for extracting single-trial activity patterns from large-scale neural recordings", *J. Curr Opin Neurobiol*, vol. 17, pp. 609-618,2007.

[6] Stett Alfred,Egert Ulrich,Guenther Elke et al. "Biological application of microelectrode arrays in drug discovery and basic research", *J. Anal Bioanal Chem*, vol. 377, pp. 486-495, 2003.

[7] Kim Raeyoung,Nam Yoonkey, "Polydopamine-doped conductive polymer microelectrodes for neural recording and stimulation", *J. J Neurosci Methods,* vol. 326, 108369, 2019.

[8] Fan, X., Nie, W., Tsai, H., Wang, N., Huang, H., Cheng, Y., Wen, R., Ma, L., Yan, F., Xia, Y., "PEDOT: PSS for Flexible and Stretchable Electronics: Modifications, Strategies, and Applications", *J. Adv. Sci*, vol. 6, 1900813, 2019.

[9] Green Rylie A,Baek Sungchul,Poole-Warren Laura A et al. "Conducting polymer-hydrogels for medical electrode applications", *J. Sci Technol Adv Mater*, vol. 11, 014107, 2010.

[10] Zhou, Y., Gu, C., Liang, J. et al. "A silk-based self-adaptive flexible opto-electro neural probe", *J. Microsyst Nanoeng*, vol. 8, pp. 118 ,2022.

[11] Niu Jia,Lunn David J,Pusuluri Anusha et al. "Engineering live cell surfaces with functional polymers via cytocompatible controlled radical polymerization", *J .Nat Chem*, vol. 9, pp. 537-545, 2017.

[12] Pan Lijia,Yu Guihua,Zhai Dongyuan et al. "Hierarchical nanostructured conducting polymer hydrogel with high electrochemical activity", *J. Proc Natl Acad Sci USA*, vol. 109, pp. 9287-9292, 2012.

[13] Dvir, T., Timko, B., Brigham, M. et al. "Nanowired three-dimensional cardiac patches", *J. Nature Nanotech*, vol. 6, pp. 720–725,2011.

[14] Liu Jia,Kim Yoon Seok,Richardson Claire E et al. "Genetically targeted chemical assembly of functional materials in living cells, tissues, and animals", *J .Science*, vol.367, pp.1372-1376, 2020.

[15] Kaech, S., Banker, G. "Culturing hippocampalneurons", *J. Nat Protoc*, vol. 1, pp. 2406-2415 ,2006.

[16] Cerea Andrea,Caprettini Valeria,Bruno Giulia et al. "Selective intracellular delivery and intracellular recordings combined in MEA biosensors", *J .Lab Chip*, vol. 18, pp. 3492-3500, 2018.

CONTACT

*Tiger H. Tao, tel: +86-21-62511070;
tiger@mail.sim.ac.cn
*X. Wei, tel: +86-21-62511070;
xlwei-jerry@mail.sim.ac.cn

BIONIC MECHANICAL HAND INTEGRATED WITH ARTIFICIAL OLFACTORY SENSOR ARRAY FOR ENHANCED OBJECT RECOGNITION

Jiachuang Wang[1,2], Xiaohui Li[1,2], MengWei Liu[1,2], Pingping Zhang[3], Tiger H. Tao[1,2,4], and Nan Qin[1,2]

[1]State Key Laboratory of Transducer Technology, Shanghai Institute of Microsystem and Information Technology, Chinese Academy of Sciences, Shanghai, CHINA
[2]University of Chinese Academy of Sciences, Beijing, CHINA
[3]Suzhou Huiwen Nanotechnology Co., Ltd, Suzhou, CHINA and
[4]Neuroxess Co., Ltd. (Jiangxi), Nanchang, CHINA

ABSTRACT

We report a bionic mechanical hand integrated with MEMS gas sensor array for improving the performance of object recognition by the fusion of the artificial olfactory and tactility. Combined with the algorithm inspired by Drosophila, the bionic mechanical hand is able to respond to various kinds of gases within 1 s and identify 9 gases with an excellent recognition accuracy of 97.8% within 5 s. Moreover, the success probability of recognizing object and grasping target for the rescue robot equipped with our bionic mechanical hand is above 90%. It suggests that this bionic mechanical hand has great potential in analyzing complex gas components, precise object search and rapid emergency rescue.

KEYWORDS

Object recognition, olfactory sensor array, bionic Drosophila algorithm, mechanical hand, MEMS.

INTRODUCTION

MEMS sensors have the detection ability beyond human beings in many environmental perception fields, which is an important basis for the feasibility of integrated intelligent multi-sensor systems [1-3]. Tactile sensor can sense the applied force and feedback the local characteristics of the contact object accurately. Olfactory sensor can respond to the ambient gas and identify objects with special smell quickly. Integrating the above two sensory information has an application value that should not be ignored.

In the emergency, analyzing environment gas comprehensively and grabbing target objects stably are of significance, which can greatly reduce the risk and difficulty of rescue [4,5]. Previous research mainly used visual images to achieve object recognition and real-time navigation tasks, and rarely consider the olfactory sensory information during overall analyzing and processing. The addition of the artificial olfactory sensing not only can help rescuer avoid hazardous gases, but also can identify objects with unique smell in the dark and narrow environment during searching and rescuing [6]. Combined with the above environmental awareness information, the mobile robot equipped can be controlled to adopt efficient rescue strategies for the target object in distress [7,8].

With the development of the miniaturization and intelligence of MEMS sensors, the combination of multisensory information and intelligent algorithm has gradually attracted extensive attention of researchers [9-12]. Here, we demonstrate the bionic mechanical hand integrated with the artificial olfactory sensor array and mounted on the mobile rescue robot. After collecting and processing the response data of multi-channel gas sensors, 9 gases and 5 common objects are identified immediately, and the bionic manipulator can be further controlled according to the feedback information from the additional pressure sensor, offering new opportunities for object recognition and decision-making.

EXPERIMENTAL METHODS

As shown in Figure 1, we integrate a multi-channel MEMS gas sensor array on the palm of bionic mechanical hand, which is fabricated by combining 8 single channel gas sensors with different sensitive materials. The enlarged SEM (scanning electron microscopy) image presents the microstructure of the single-channel MEMS gas sensor. When the olfactory sensor array works in an appropriate temperature and humidity range, the sensitive materials of each channel will adsorb specific gas molecules, thus obtaining the corresponding responses to various gases with improved accuracy.

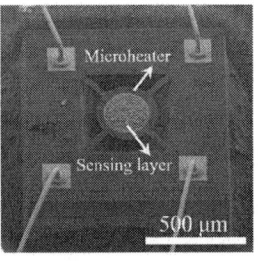

Figure 1: Optical image of the bionic mechanical hand integrated with multi-channel MEMS gas sensor array and SEM image of the single-channel MEM gas sensor.

The bionic Drosophila neural network algorithm mainly consists of three important steps, including random sparse projection, winner take all and recognition (Figure 2). Preprocess and standardization of the response data collected by artificial olfactory sensor array is a significant method to reduce interference factors in the environment. Afterwards, the bionic algorithm reduces the response data complexity by increasing the dimensions and binary values and obtains odor tags by recoding. It dramatically speeds up the operation rate and improves the success probability of objects capturing, which is critical for the recognition and decision-making strategy.

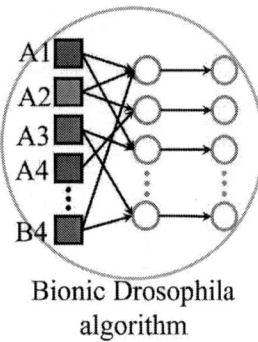

Bionic Drosophila
algorithm

Figure 2: Data process with a bionic olfactory neural network algorithm.

The bending degree of fingers of the bionic mechanical hand can be controlled by the five steering gears separately, so that we can gain at least 1024 grab gestures. Meanwhile, the tactile sensors on the fingertip feedback the condition of touching targets to achieve stable grabbing and effectively avoid damaging fragile samples. Therefore, we can judge the shape and category of the object through its grasping gesture preliminarily.

The artificial olfactory sensor array and data acquisition module are integrated on the palm of bionic hand mounted on the mobile rescue robot, which can be programmed and controlled using the development board. The whole system has the ability of automatic navigation and search, which can identify and capture target objects according to the pre-collected data and the intelligent algorithm.

EXPERIMENTAL RESULTS

The gas response experiment is carried out in a fume hood with standard operation. First, we place the multi-channel gas sensor in a transparent gas box to work for a period to ensure its stable working state. Then, the prepared test gas is pumped into the gas box along the pipeline. Then, the response of each gas sensor is collected through the data acquisition module for subsequent algorithm processing and analysis. For the safety and correctness of the experiment, the gas box must be opened for ventilation after each gas response experiment to avoid mutual interference.

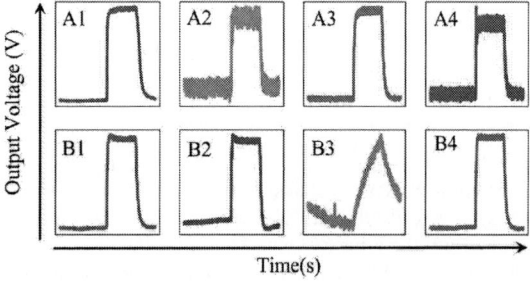

Figure 3: Response curve of the artificial olfactory sensor array toward acetone.

As shown in the Figure 3, artificial olfactory sensor

array is capable of responding to acetone within 1s during the experiment, indicating its outstanding sensitivity. The output voltage and response curve of each channel are different so that we are able to obtain the comprehensive odor data information. In the preprocess phase, we set the maximum voltage as 3.3 V and the minimum voltage as 0 V. Consequently, it is natural to implement the normalization operation.

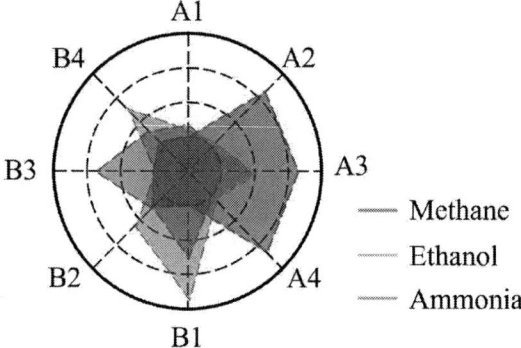

Figure 4: The olfactory mappings of three different gases including methane, ethanol and ammonia.

In order to explore the repeatability and specificity of the multi-channel sensor array, we select three different gases, including methane, ethanol and ammonia. The data obtained by repeating the same gas response experiment is close to each other, which shows excellent stability of MEMS gas sensors. Meanwhile, the olfactory mappings composed of the stable output voltage of the multi-channel sensor array adsorbing different gases present significant differences, ensuring the acquisition and recognition of odor tags (Figure 4).

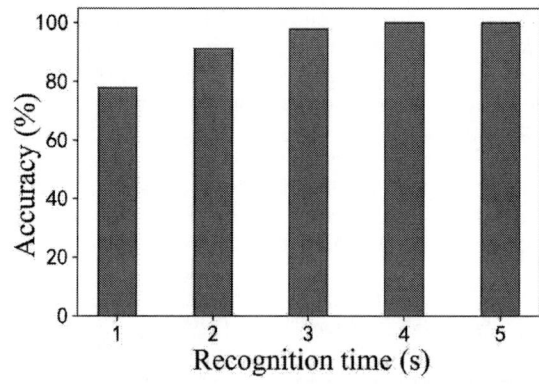

Figure 5: The change in recognition accuracy of the bionic neural network algorithm with the increase of recognition time.

The whole system identifies the environmental gas by the bionic Drosophila neural network algorithm. With the increase of recognition time, the recognition accuracy of the bionic algorithm gradually increases and reaches 97.8% at 5 s (Figure 5). Stable sensor response and high recognition accuracy are gained simultaneously within 5 s, which shows the superiority of the whole system.

Orange   Tomato   Cup

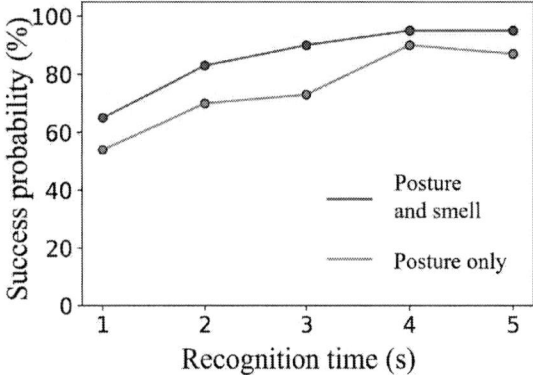

Towel   Stone

Figure 6: Target objects for building the dataset and the bionic mechanical hand integrated with the artificial olfactory sensor array mounted on the mobile robot grabing an orange.

We further demonstrate the bionic mechanical hand to grab the target objects in a tight manner and select 5 target objects to build the dataset for evaluating the detailed performance of the rescue mobile robot using bio-inspired Drosophila olfactory algorithm (Figure 6).

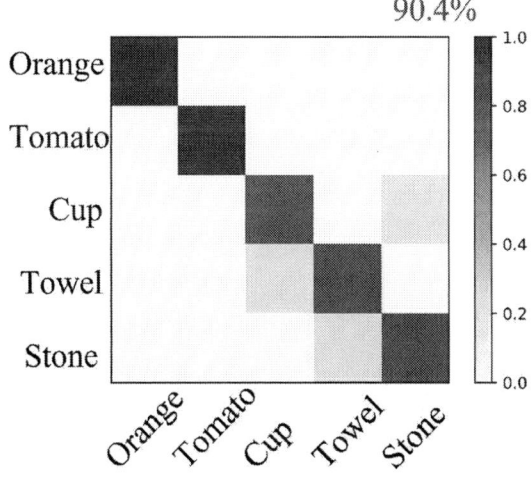

Figure 7: The confusion matrix of the five selected target objects by using the bionic olfactory neural network algorithm.

The confusion matrix of the five selected objects shows that the recognition accuracy is up to 90.4% (Figure 7). In the process of building dataset, objects with special smell, including orange and tomato, can be correctly identified. Meanwhile, the accurate recognition for other objects is partial, which means that the algorithm produces confusion. In other words, there is huge difficult for distinguishing these targets clearly only by smell.

Therefore, it is necessary to combine tactile information in further analysis.

Figure 8: The change in success probability of identifying and grabbing objects with the increase of recognition time.

The success probability of identifying and grasping objects remains more than 90% when recognition time exceeds 3 s (Figure 8). Moreover, the combination of posture and smell has a higher accuracy than single posture recognition, the strategy of fusing multi-sensor information has great prospects.

CONCLUSION

We report a bionic mechanical hand that combines artificial olfactory and tactile with an inspired Drosophila neural network algorithm to achieve efficient recognition of multiple gases and target objects without using visual information. Benefiting from the high performance of gas and pressure sensors, the mechanical hand can precisely obtain the environmental gas composition and object shape information. This work is expected to be used for identification tasks in emergency rescue and target search, providing inspiration for future integrated advanced sensors and fusion multisensory recognition algorithm strategies.

ACKNOWLEDGEMENTS

This work was supported by National Natural Science Foundation of China (grant no. 62236005), National Science and Technology Major Project from the Minister of Science and Technology of China (grant no. 2018AAA0103100), National Science Fund for Excellent Young Scholars (grant no. 61822406), Key Research Program of Frontier Sciences, CAS (grant no. ZDBS-LY-JSC024) and Fund of Youth Innovation Promotion Association CAS (grant no. 2022234).

REFERENCES

[1] M. Wang, Z. Yan, T. Wang, *et al.* Gesture recognition using a bioinspired learning architecture that integrates visual data with somatosensory data from stretchable sensors, *Chen, Nat. Electron.*, 3, pp. 563–570, 2020.

[2] C. Jirayupat, K. Nagashima, T. Hosomi, *et al.* Breath odor-based individual authentication by an artificial olfactory sensor system and machine learning, *Chem.*

Commun., 58, pp. 6465-6465, 2022.

[3] S. Dasgupta, C. F. Stevens, S. Navlakha, A neural algorithm for a fundamental computing problem, *Science*, 358, pp. 793–796, 2017.

[4] N. Imam, T. A. Cleland, Rapid online learning and robust recall in a neuromorphic olfactory circuit, *Nat. Mach. Intell.*, 2, pp. 181–191, 2020.

[5] F. Zhao, Y. Zeng, A. Guo, *et al.* A neural algorithm for Drosophila linear and nonlinear decision-making, *Sci. Rep.*, 10, 18660, 2020.

[6] P. Y. Wang, Y. Sun, R. Axel, *et al.* Evolving the olfactory system with machine learning, *Neuron*, 109, pp. 3879–3892, 2021.

[7] Z. Sun, M. Zhu, X. Shan, *et al.* Augmented tactile-perception and haptic-feedback rings as human-machine interfaces aiming for immersive interactions, *Nat. Commun.*, 13, 5224, 2022.

[8] G. Li, R. Zhu, A Multisensory Tactile System for Robotic Hands to Recognize Objects, *Adv. Mater. Technol.*, 4, 1900602, 2019.

[9] S. Sundaram, P. Kellnhofer, Y. Li, *et al.* Learning the signatures of the human grasp using a scalable tactile glove, *Nature*, 569, pp. 698-702, 2019.

[10] M. Q. Hill, C. J. Parde, C. D. Castillo, *et al.* Deep convolutional neural networks in the face of caricature, *Nature Machine Intelligence*, 1, pp. 522-529, 2019.

[11] A. Borthakur, T. A. Cleland, A spike time-dependent online learning algorithm derived from biological olfaction, *Front. Neurosci.*, 13, 656, 2019.

[12] N. Mandairon, F. Kermen, C. Charpentier, *et al.* Context-driven activation of odor representations in the absence of olfactory stimuli in the olfactory bulb and piriform cortex, *Front. Behav. Neurosci.*, 8, 138, 2014.

CONTACT

* Nan Qin, tel: +86-21-62511070;
qinnan@mail.sim.ac.cn

HIGH RESOLUTION TACTILE SENSOR FOR MEASUREMENT OF A COMPLICATED TACTILE FEELING OF *"SHITTORI"* WITH MOISTNESS

Genki Yamada, Yuto Morita, Kyohei Terao, Fusao Shimokawa, and Hidekuni Takao

Faculty of Engineering and Design, Kagawa University, Japan

ABSTRACT

In this paper, we present a high-resolution MEMS tactile sensor that can measure the *"shittori* feel", which is becoming increasingly important in tactile quality evaluation. The Japanese word *"shittori"*, refers to a complicated tactile sensation that includes moistness and smoothness, describing the pleasant tactile sensation of fine leather and well-moisturized skin. Although *shittori* is one of the most important tactile sensations in measuring high-quality tactile feelings, the principles of its quantitative measurement are not well understood. In this study, we propose a new hypothesis regarding the adhesive frictional properties of moist materials. Moreover, we have realized the first tactile sensor that can simultaneously detect moistness and smoothness to distinguish the *"shittori* feel" of moist materials.

KEYWORDS

Tactile sensors, Touch feeling, Haptics, Surface topography, Frictional forces

INTRODUCTION

Humans can recognize textures from their smoothness and softness by tracing objects with their fingertips. Textures are judged by detecting physical stimuli such as surface roughness, friction, and temperature characteristics using tactile receptors on the fingertips, integrating these characteristics, and comparing them with memory.

The *"shittori* feel" is recognized as a feeling of moistness with smoothness [1], and is one of the preferred textures of leather, hair, skin, and food surfaces. Note that moistness can be perceived in the absence of actual moisture content. Since there are no receptors that directly detect moisture content at the fingertips, it can be expected when the properties of a surface resemble those of a wet surface, the surface is perceived to be moist. Previous studies have attempted to explain this dry-wet feeling by using the mean values of the surface roughness of an object [2] or the friction coefficient and its mean deviation [3,4] as characteristic quantities. In recent report on *"shittori* feel", Nonomura et al. modeled the feeling of moistness based on the difference between the static and dynamic coefficients of friction [1]. However, it clarifies an empirical rule based on the phenomenological theory and is simply based on a rough evaluation of friction with macro-scale load cell sensors.

In this study, we propose a new hypothesis for the mechanism of moistness perception by focusing on the adhesive friction characteristics unique to moist materials, and we develop a new tactile sensor that can identify the *"shittori* feel" by simultaneously acquiring both the moistness and surface texture of a surface, including its micro-roughness and friction.

CONCEPT AND CONFIGURATION

We have discovered experimentally that adhesive friction based on surface activation energy is closely related to "moistness". On the basis of the relationship, a completely new tactile sensor for quantifying moistness can be configured. Figure 1 shows the concept and detection principle of the device for moistness sensing in this study.

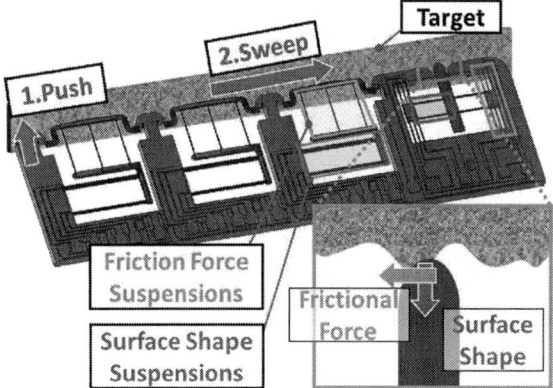

Figure 1: Concept and detection principle of the new tactile sensor for moistness sensing.

To perform a measurement, this tactile sensor is pressed against an object and then scans it laterally. As the object is being scanned, the contactor follows its surface, and both the surface micro-roughness and its related friction are precisely acquired by the piezoresistive sensor circuit integrated on the silicon beam suspension, which supports the movement of the contactor. This makes it possible to determine the important relationship between the vertical load on a surface with low roughness and the generated friction force in a single scanning measurement.

Figure 2 illustrates the two friction modes that explain the presence or absence of a "moist" feeling. The friction force $f(N)$ follows Amonton's law, and is proportional to the vertical load N. If $f(N)$ is a dominant part of the total friction force (Figure 2 upper), the surface feels dry rather than moist. "*Shittori*" is felt when the surface is sufficiently smooth [1] and, at the same time, the adhesive friction $f(A)$ is significant (Figure 2, bottom). The adhesive friction $f(A)$ is generated by the viscoelasticity or surface tension of the liquid, and it depends on the apparent contact area A.

Figure 3 illustrates the principle of simultaneously acquiring the surface roughness, vertical load, and friction coefficient using the tactile sensor with high spatial resolution. Surfaces exhibiting frictional properties that obey Amonton's law, such as dry surface, show a constant coefficient of friction for a vertical load.

In contrast, on a surface that feels moist, the friction coefficient is expected to increase in Region A, where the vertical load is small and the effect of the adhesive friction force is dominant, and to decrease as the load increases, and the effect of adhesion decreases, as in Region B.

In the configuration of the sensor shown in Figure 1, the rightmost device is a sensor with a fingerprint structure that mimics a fingertip and aims to acquire surface texture and adhesive friction characteristics based on the measurement principle of Figure 3. The other three devices are used for screening the moisturized materials based on macroscale friction at three different levels of load, as explained later.

Friction characteristics based on Amonton's Friction Laws

Vertical load N

Shear Force F

$$F = f(N)$$

Friction characteristics by viscoelasticity

$$F = f(N) + f(A)$$

Figure 2: Two friction models that explain the presence or absence of a "moist" feeling.

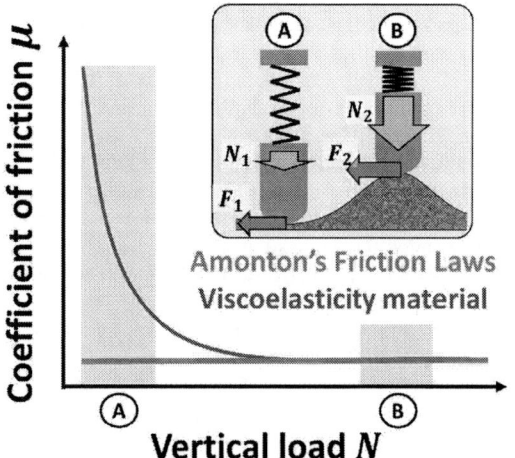

Figure 3: Two different behaviors of friction coefficient among materials with/without moistness. The behavior can be detected by using the tactile sensor with high spatial resolution in Figure 1.

FABRICATION OF SENSOR

Figure 4 shows the fabrication process of the high-resolution tactile sensor. The sensor has a p-type silicon on insulator (SOI) wafer with an active layer of 50 μm thickness, a buried oxide layer of 0.5 μm thickness and a support substrate of 300 μm thickness.

First, the SOI wafer is cleaned. Then, an oxide film is

formed over the active layer. Next, a diffusion layer for the piezoresistive circuit wiring is patterned using photolithography, and a high concentration diffusion layer is formed by a thermal phosphorus diffusion process. Piezoresistive elements are formed by implanting phosphorus ions using an ion implanter and by annealing. To form a hard mask for deep reactive-ion etching (RIE), chromium films are patterned on the front and back sides of the active layer. The first deep-RIE process forms the movable structures of the contactor and silicon beam suspension on the active layer side, and a second deep-RIE is performed to etch the handle layer under the movable structures. Finally, the chromium film and intermediate oxide film are removed with chemicals to complete the sensor.

Figure 5 shows a photograph of the sensor chip fabricated through these processes and details of the structure.

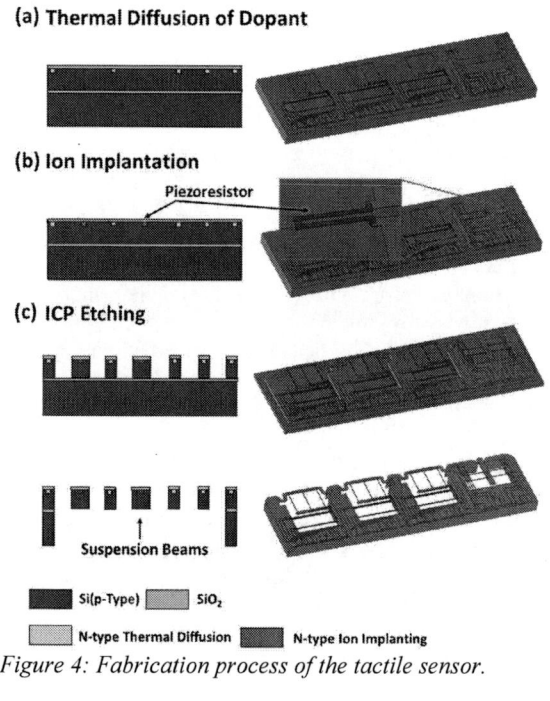

Figure 4: Fabrication process of the tactile sensor.

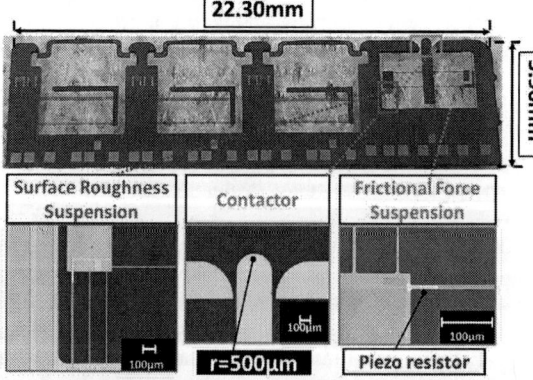

Figure 5: Photograph of tactile sensor chip for moisture perception fabricated using 50μm-thick SOI active layer.

DEVICE EVALUATION

First, the sensitivity of the fabricated sensor was valuated. The response of the sensor to a two-axis input load was evaluated using micromechanical test equipment (Femtotools FT-MTA03) with a nN-order force resolution. Since the equipment can measure the displacement and force simultaneously, the sensitivity and mechanical spring constant of the sensing unit were precisely determined. Separete evaluations were conducted on the fingerprint structure sensor and the macroscale sensor.

The response of the sensors to vertical displacement was measured by pressing the test probe against the contactor vertically. Similarly, the response to horizontal forces was measured by pressing the probe horizontally. The output responses of the sensors were highly linear with respect to the vertical displacement and horizontal load. The measurement resolutions of the fingerprint structure sensor were 1.6×10^{-1} μm for vertical displacement and 2.1×10^{-2} mN for horizontal friction. The measurement resolutions of the macro-scale sensors were 1.7 μm for vertical displacement and 1.8 μN for horizontal friction.

SENSING EXPERIMENT

Figure 6 shows the experimental setup for measuring the "*shittori* feel" with this sensor. A sample fixed to a single-axis stage was pressed against the sensor, which was moved laterally at a constant speed to measure the sample surface.

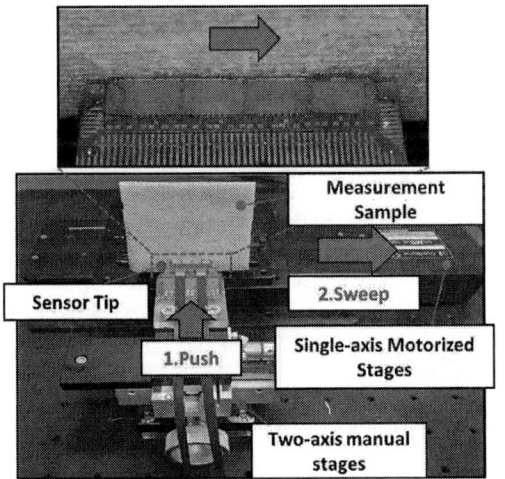

Figure 6: Experimental system for measuring "shittori feel" with tactile sensor in this study.

In the first experiment, measurements were conducted on two samples with clearly different tactile sensations: artificial leather with a moist feel and a paper with a dry feel. Measurements were taken three times at a distance of 5 mm, and the sample scanning speed of the single-axis stage was set at 100 μm/s.

To predict the material property, the sample was initially measured with the macroscale sensor, and the friction coefficients at three different loads were extracted.

The three vertical loads were set at small values because differences in friction properties are enhanced at low loads as shown in Figure 3. Figure 7 shows the coefficient of friction measured by the macro-scale sensors with different load levels. The paper sample has a similar friction coefficient regardless of the vertical load, whereas the coefficient of friction of the artificial leather sample (with a *shittori* feel) varies with the vertical load as expected.

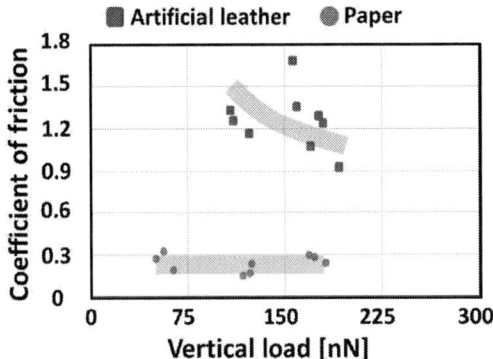

Figure 7: Relationship between average vertical load and average friction coefficient obtained from macro-scale sensors with three different load levels.

The above measurements with the macroscale sensors are screening measurement of moistness touch feeling. This result is used as a screening to determine the need for "moistness measurement by scanning in low load areas"

Having obtained an overview of the friction characteristics, the change in the friction coefficient with the vertical load was measured by applying an appropriate contact load (~5 mN) to the fingertip sensor.

Figure 8 shows the relationships between the vertical load and friction coefficient on the two samples measured by the fingerprint structure sensor. This scatter plot shows the friction coefficient obtained by dividing the friction load at each vertical load. The paper sample shows a constant coefficient of friction up to very low loads of around 0.1 mN. In contrast, the artificial leather sample shows a rapid increase in the friction coefficient below a vertical load of 1.3 mN. To quantitatively evaluate moistness, it is necessary to quantify the trend of this frictional property.

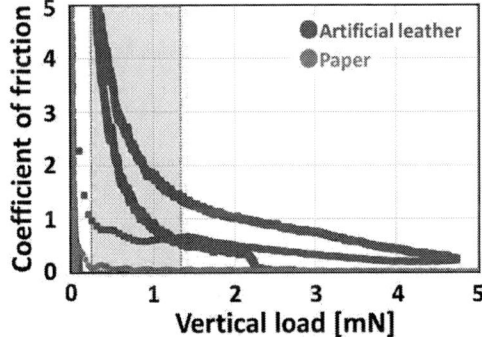

Figure 8: Relationship between vertical load and friction coefficient measured by the fingerprint structure sensor.

This detection method is applicable to other materials. In this study, we compared the friction property between dry and moist states of the same material and verified the validity of moistness detection using the direct change in moistness as a reference. Cotton and rayon fabrics were used for the comparison of wet and dry conditions. To create a "*shittori* feel", each fabric was moistened by spraying a small amount of water (0.64 mg/cm²).

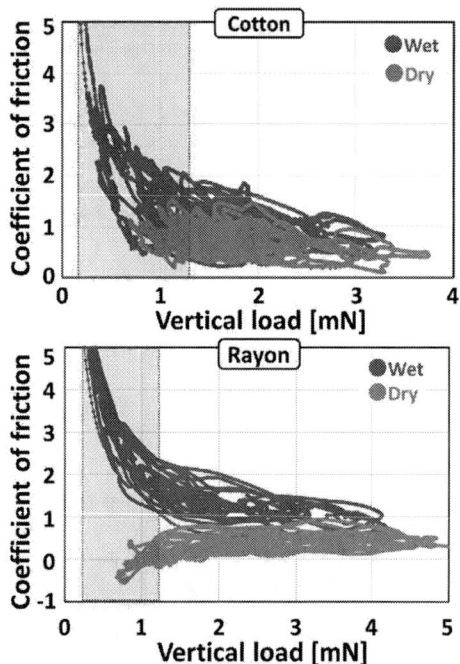

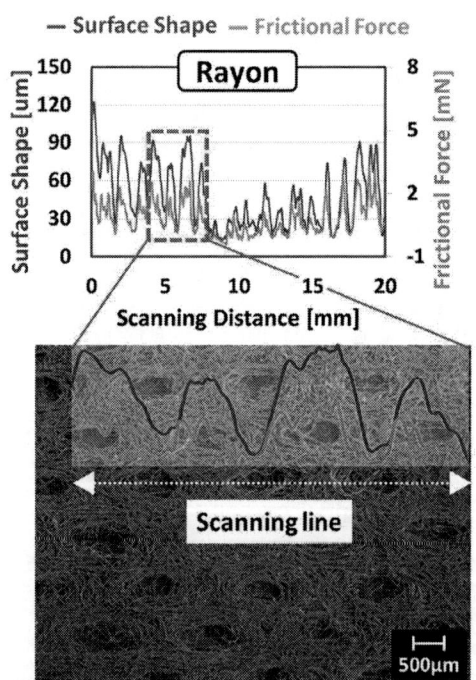

Figure 9: Comparison of friction properties of cotton and rayon fabrics under wet and dry conditions.

Figure 10: Comparison of sensor-acquired results with SEM images.

Figure 9 shows scatter plots comparing the load dependences of the friction coefficient under of cotton and rayon fabrics under "wet" and "dry". The scanning speed was 5 mm/s and the scanning distance was 20 mm. For both materials, an exponential increase in the coefficient of friction was clearly observed below a load of 1 mN for the wet samples.

These results shown in Figures 8 and 9 are consistent with our hypothesis that the characteristic frictional properties of materials with a moist feeling are due to adhesive friction caused by the viscoelasticity or surface tension of a liquid on the surface of the object.

Figure 10 is one of the original data from which the aforementioned friction characteristics were calculated. The inset shows the surface shape and friction waveform of a dry rayon surface traced by the fingertip sensor. Comparison with the SEM image of the sample shows that the periodic surface structure of the rayon fabric closely matches the waveform. In other words, the tactile sensor proposed in this study can evaluate "smoothness" at the same time as evaluating "moistness", and therefore the quantitative measurement of the "*shittori* feel" can be achieved with a single scanning measurement.

CONCLUSION

In this study, a novel high-resolution tactile sensor was developed to measure a "*shittori* feel", a Japanese term that that is becoming increasingly important in tactile quality evaluation. The tactile sensor, fabricated on the basis of a new hypothesis regarding the adhesive frictional properties of moist materials, successfully detecting the characteristic frictional properties of materials.

ACKNOWLEDGEMENTS

This research was partially supported by the JST-CREST research funding program (Grant Number JPMJCR20C2)

REFERENCES

[1] M. Egawa, R. Kuhara, Y. Nonomura, "Physical origin of a complicated tactile sensation: 'shittori feel'", Royal Society Open Science, vol.6, no.7, Article ID.190039, 2019.

[2] Y. Okajima, Y. Takeda, "Tactile dryness of building materials", Transactions of the Architectural Institute of Japan, vol. 327, pp.12-19, 1983.

[3] I. H. M. Hashim, S. Kumamoto, K. Takemura, T. Maeno, S. Okuda, Y. Mori, "Tactile evaluation feedback system for multi-layered structure inspired by human tactile perception mechanism", Sensors, vol.17, p.2601, 2017.

[4] M. Egawa, M. Oguri, T. Hirao, M. Takahashi, M. Miyakawa, "The evaluation of skin friction using a friction feel analyzer", Skin Research and Technology, vol.8, pp.41-51, 2002.

CONTACT

Prof. Hidekuni Takao,
Faculty of Engineering, Kagawa University,
2217-20, Hayashi, Takamatsu, Japan.
Tel/Fax: +81-87-864-2331;
E-mail: takao.hidekuni@kagawa-u.ac.jp

PYRAMIDAL STRUCTURED MXENE/ECOFLEX COMPOSITE-BASED TOROIDAL TRIBOELECTRIC SELF-POWERED SENSOR FOR HUMAN-MACHINE INTERFACE

Shipeng Zhang, SM Sohel Rana, Gagan Bahadur Pradhan, Trilochan Bhatta, Seonghoon Jeong, and Jae Yeong Park*

Department of Electronic Engineering, Kwangwoon University, 447-1 Wolgye-dong, Nowon-gu, Seoul 01897, Republic of Korea.

ABSTRACT

In this work, we newly developed a high-performance self-powered toroidal triboelectric sensor (STTS). The pyramidal array structure of MXene/Ecoflex nanocomposites was created by 3D printing technology to create space between the finger skin and the negative electrical layer. This avoids the need for spacers in previous work and simplifies the structure of the sensor to improve wearability.The highly negatively charged MXene/Ecoflex nanocomposite also acts as a negative friction layer and uses a flexible conductive fabric as an electrode to improve flexibility. The developed STTS features high flexibility, high triboelectric peak-to-peak voltage (19.91 V) output, high sensitivity (0.088 VkPa^{-1}), and a wide pressure detection range (0-120 kPa.) The STTS can be easily worn on the finger to accurately detect various finger movements for virtual reality gaming experiences and human-machine interface (HMI) control of appliance switches.

KEYWORDS

Toroidal TENG, self-powered sensor, human-machine interface, MXene/Ecoflex composite.

INTRODUCTION

Over the past few decades, the rapid development of human-computer interaction technology has expanded into various industry sectors such as remote control and virtual reality gaming experiences, and it has become a dynamic emerging technology in the field of human-computer interaction today. In order to make user interaction with virtual environments portable and comfortable, more compact and lightweight sensing systems with high-resolution feedback output signals are required.

To implement various feedback sensing functions, various pressure sensors and electronic skins based on capacitive [1] and piezoresistive [2] operations have been developed. However, these sensors require a driving voltage to operate, making them bulky. Triboelectric self-powered sensors can operate without any external drive voltages. The MXene/PET/ITO nanogenerator was proposed by Dong et al. and the Ecoflex/Aluminium film triboelectric sensor was reported by Zhu et al. [3]. However, the PET substrate and aluminum film electrodes limit flexibility and wearing comfort. In addition, it is difficult to obtain high output performance for triboelectric sensors based on a single material, resulting in low sensitivity. Triboelectric-based self-powered sensors can provide interactive visual and haptic experiences in simulated virtual worlds. [4-8] Pukar et al. proposed a self-powered triboelectric sensor with high sensitivity, but it was based on a complex structured two-electrode system. In order to work in triboelectric separated contact mode, spacers must be designed between the two electrodes and need to be fixed to the finger with adhesive tape, reducing the wearing convenience [7].

Considering these issues and challenges, herein, for the first time, conductive fabrics are applied as flexible electrodes on composite surfaces to fabricate wearable triboelectric self-powered sensors on fingers based on highly negatively charged MXene/Ecoflex nanocomposites. An extremely simplified structural and fabrication design is newly developed using 3D printing technology to create pyramidal array structures of MXene/Ecoflex nanocomposite creating space between the positively charged finger skin and the negatively charged layer avoiding the spacing structure in previous works. This simplified single-electrode toroidal structure does not require any fixation and can be easily worn on the finger. The prepared STTS exhibits outstanding output performance, including high peak-to-peak voltage, high sensitivity, and wide pressure detection range. Based on its outstanding performance, the developed STTS has been successfully applied to virtual reality and gaming experiences (flying car games) and human-machine interface (home appliance control). The developed STTS shows excellent performance as a self-powered motion sensor for interactive human-machine interfaces, showing great potential for future virtual reality and human-machine interaction technologies.

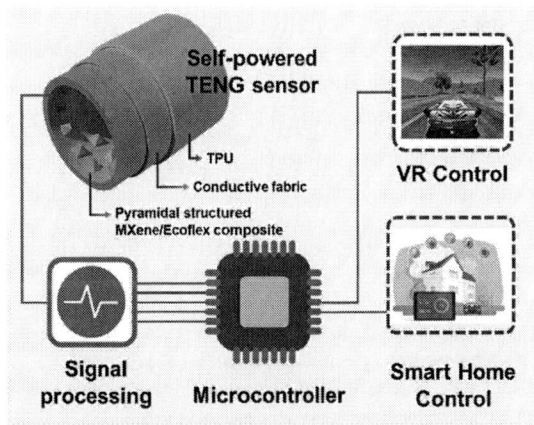

Figure 1: Conceptual diagram of proposed STTS for human-machine interface and virtual reality applications.

MATERIALS AND METHOND
Materials

Materials: MAX phase-Ti_3AlC_2 (particle sizes $\geq$ 38 μm) was purchased from Carbon-Ukraine Ltd. The hydrofluoric acid (HF) and hydrochloric acid (HCl) were purchased from Sigma-Aldrich. The Ecoflex-50 was purchased from Smooth-On Ltd. All commercially available reagents were used as received without further purification.

Synthesis of MXene

MXene was synthesized by selective etching of the aluminum layer from the MAX phase using hydrofluoric acid (HF) and hydrochloric acid (HCl). Add 6 mL deionized water in a plastic (HF-safe) bottle, begin stirring at 100 rpm at 35 °C, then add 12 mL HCl (12 M), then add 2 mL HF (48-51 wt. %) final concentration will be 5 wt%. Over the course of 5 minutes, slowly add Ti_3AlC_2 to the mixture. Increase the stirring speed to 300 rpm. Once all MAX is added, gently close the lid. Let the reaction proceed for 24 hours. Use a 50 ml centrifuge tube, pre-filled with water, to which the etching mixture is added, and centrifuge the mixture at 3500 rpm for 10 minutes. Then decant the supernatant and check the pH. Continue washing until the pH is neutral (>6). Collect the dark supernatant at each wash step until it becomes clear. After collection, the mixture was centrifuged at 10,000 rpm for 15 minutes and the final multilayer MXene was obtained by decanting clear water.

Fabrication of pyramidal structured MXene/Ecoflex nanocomposite

To make the triboelectric negative electrode material, Ecoflex-50 silicone rubber solution A and solution B were mixed in a 1:1 weight ratio. The synthesized Mxene is mixed with Ecoflex and stirred after which it is poured onto a 3D-printed mask pre-printed with a pyramidal notched structure. The top surface is then covered with conductive fabric electrodes and the MXene/Ecoflex solution is dried with the fabric at room temperature for 24 hours. Finally, the dried MXene/Ecoflex nanocomposite was separated from the 3D-printed mold. The cut size was 7.5 cm x 2 cm for optimizing the TENG material.

Fabrication of the flexible finger ring package

The encapsulation layer was printed with a 3D printer using TPU-95A flexible material and electrically connected using copper wire bonded to the fabric electrodes on the back of the TENG sensor. Finally, the MXene/Ecoflex/conductive fabric electrodes are rolled up and attached to the inner surface of the finger ring-type package.

Characterization and measurement.

The electrical output performance of the STTS was tested with an oscilloscope (wavesurfer 510). The voltage was determined using an oscilloscope and a 100 MΩ high impedance probe (P6015A). An electrometer (Keithley) and a current preamplifier (model SR570), were used to measure charge and current, respectively. In order to optimize the electrical properties of the material, an electrodynamic shaker was used as a vibration source. The frequency of the vibrations is managed by a function generator and a power amplifier.

RESULTS AND DISCUSSION

Figure 1 shows the structural composition of STTS and its application in virtual game control and home appliance switching. Figure 2 shows the step-by-step fabrication process of the STTS. MXene/Ecoflex nanocomposites are in a gel state before drying and have good flow and immersion properties. The flexible conductive fabric has a porous structure, and the MXene/Ecoflex nanocomposites in the gel state can be easily immersed into the voids of the conductive fabric to form a strong physical bond after drying. The prepared pyramidal structure of the MXene/Ecoflex nanocomposite has good flexibility and can be crumpled, rolled, and twisted arbitrarily. By combining MXene/Ecoflex nanocomposites with flexible conductive fabric electrodes, a highly flexible self-powered triboelectric sensor was fabricated.

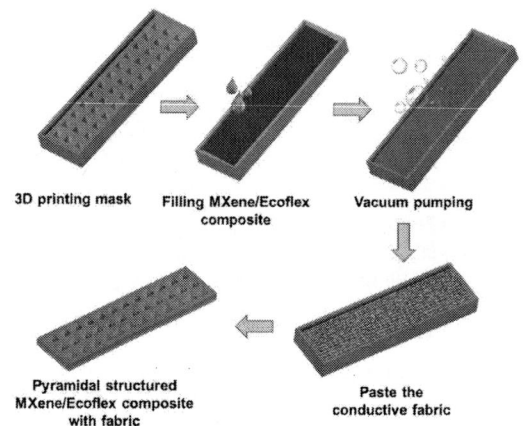

Figure 2: Fabrication sequences of self-powered toroidal triboelectric sensor (STTS).

The design of the STTS is based on the contraction and expansion mechanism of the finger muscles. As the finger begins to flex, the finger muscles expand, increasing the contact area between the skin of the finger and the MXene/Ecoflex composite layer of the pyramidal structure. Figure 3 shows the operating mechanism of STTS incorporating contact separation and electrostatic induction. The fabricated STTS was operated in contact separation mode single electrode using the finger skin as the positive and MXene/Ecoflex as the negative layer. Due to triboelectrification and electrostatic induction at the contact surface, the potential between the connected fabric electrodes and the ground will change, driving electron flow and generating triboelectric output. In equilibrium, triboelectric charges with opposite polarity are induced, resulting in no electron flow. These triboelectric charges are not compensated once the separation of MXene/Ecoflex and human skin occurs. the negative charge on the MXene/Ecoflex surface induces a positive charge on the fabric electrode, leading to a flow of free electrons from the fabric electrode to the ground. This electrostatic induction phase can produce an output voltage/current signal that finally ends when the negative triboelectric charges on the MXene/Ecoflex almost cancel out the positive charges induced on the fabric electrodes.

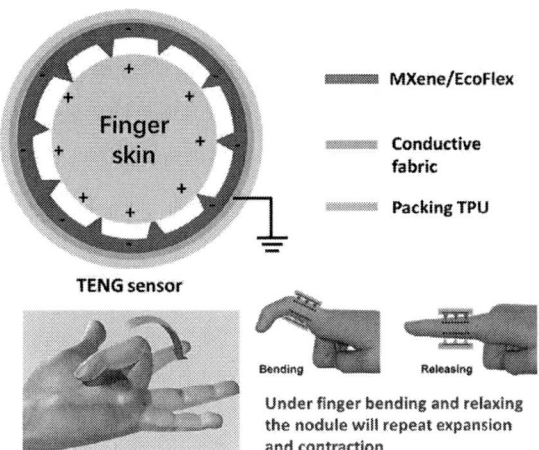

TENG sensor

MXene/EcoFlex

Conductive fabric

Packing TPU

Bending   Releasing

Under finger bending and relaxing the nodule will repeat expansion and contraction

Figure 3: Operation principle of the STTS that illustrates the charge generation and current flow in the bent and released states of the sensor device.

As the human skin approaches the MXene/Ecoflex, the induced positive charge on the fabric electrode decreases, and electrons flow from the ground to the fabric electrode until the human skin is fully in contact again. This is a typical theoretical model of the contact separation mode.

Figure 4a shows that the output voltage of the TENG increases when the MXene concentration increases. However, MXene concentration above 0.6 wt% shows

saturation behavior with a slight decrease in output voltage and current, which may be due to its accumulation [49]. The results show that a concentration of 0.5 wt% MXene produces the maximum voltage output performance.

The peak-to-peak voltage of the triboelectric voltage signal generated by the STTS is as high as 19.91 V during the application and the release of the external force when the finger is bent, as shown in Figure 4b. the performance of STTS in sensing pressure, from low to high pressures, was also investigated. To quantify this experiment, we used a pressure-controllable dynamometer as an external pressure source. The experimental results demonstrate that the fabricated STTS is highly sensitive over a wide range of pressures. The calculated sensitivities at these different pressure state: 0.088 VkPa^{-1}, 0.063 VkPa^{-1}, and 0.018 VkPa^{-1} for the low, medium, and high-pressure states, respectively, are shown in Figure 4c. As external pressure increases, pressure sensitivity gradually decreases. However, the sensitivity curves for low, medium and high pressures have a linear relationship with the applied pressure.

Figure 5a shows the system architecture diagram of STTS applied to virtual game control (Figure 5b) and home appliance switching (Figure 5c). Bending the thumb, index finger, and middle finger enables on/off control of the fan, alarm, and light, respectively. STTS also enables the control of virtual games by the human-machine interface. The thumb is set to accelerate, the index finger to turn left, and the middle finger to turn right. Figure 4a shows the triboelectric output voltage of the TENG when the concentration of MXene is increased. Figure 4b shows the peak-to-peak voltage of the TENG (19.91 V) when the finger is bent. Figure 5d shows the signal processing circuit and test setup. Table 1 shows a comparison of the STTS with previous works.

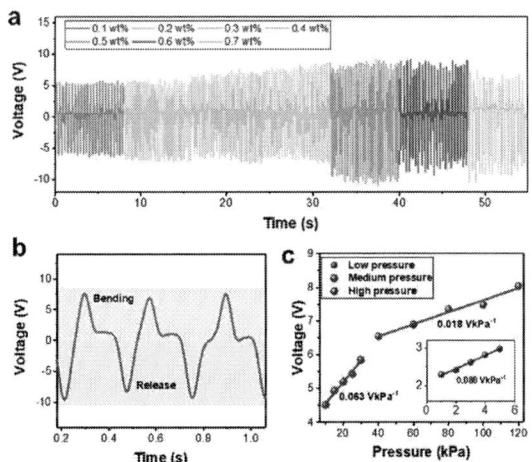

Figure 4: (a) Peak-to-peak voltage of triboelectric output of MXene/Ecofelx at different mixing conditions. (b) Nature of triboelectric voltage waveforms under the action of external pressure generated during finger bent and release. (c) Variation of triboelectric output voltage and pressure sensitivity with the different ranges of applied pressure.

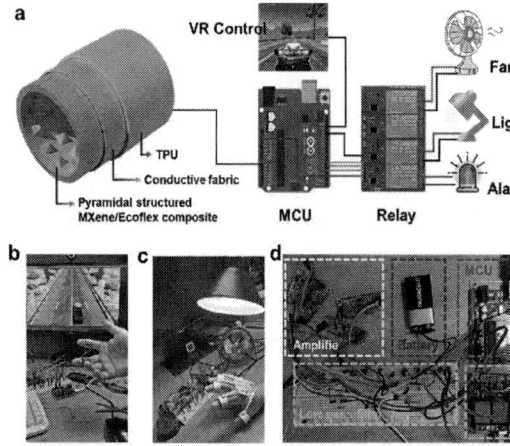

Figure 5: (a) System architecture diagram of the STTS applied to virtual reality human-machine interconnection. (b) Control the settings and schematics of home electronics. (c) Virtual reality experience to control racing games. (d) Circuit and test setup.

Table 1. Comparison of the fabricated STTS with other works.

Friction materials	Working mode	Sensitivity/ detection range	Triboelectric output	Refs.
Styrene/PDMS	Double electrode	0.048 V kPa^{-1}/0−20 kPa	17 mV	[8]
PTFE/AgNWs/ PDMS	Double electrode	127.22 mV kPa^{-1}/5−50 kPa	3.48 V	[9]
PDMS/PEP/ITO	Double electrode	NA	7 V	[10]
MXene/Ecoflex/ conductive fabric	Single electrode	0.088 V kPa^{-1}/0-120 kPa	19.91 V	This study

CONCLUSIONS

Herein, a self-powered ring-type triboelectric sensor with pyramidal structured MXene/Ecoflex nanocomposites, and highly flexible fabric electrodes was newly proposed, developed, and evaluated for virtual reality experiences and human-machine interaction scenarios. High surface charge density high triboelectric output performance and ultra-flexibility were achieved by improving the performance of MXene/Ecoflex nanocomposites. Pyramid array structures are created on MXene/Ecoflex nanocomposites by using 3D-printed templates. The developed STTS provides high triboelectric peak-to-peak voltage (19.91V) output, high sensitivity (0.088VkPa-1), and wide pressure detection range (0-120kPa) to successfully capture finger movements for human-machine interface applications in the next-generation smart and interactive products such as virtual reality and gaming experiences (flying car games) and human-machine interface (appliance control).

ACKNOWLEDGEMENTS

This research was supported by Korea Institute for Advancement of Technology (KIAT) grant (P0020967, The Competency Development Program for Industry Specialist), and by the Technology Innovation Program (RS-2022-00154983, Development of Low-Power Sensors and Self-Charging Power Sources for Self-Sustainable Wireless Sensor Platforms) funded by the Ministry of Trade, Industry & Energy (MI, Korea). The authors are grateful to the group members of the Advanced Sensor and Energy Research (ASER) Laboratory of Kwangwoon University for their valuable suggestions and support.

REFERENCES

[1] Z. Zhang, T. He, M. Zhu, Z. Sun, Q. Shi, J. Zhu, B. Dong, M. R. Yuce, C. Lee, npj Flex. Electron. 2020, 4, 1.

[2] T. Jin, Z. Sun, L. Li, Q. Zhang, M. Zhu, Z. Zhang, G. Yuan, T. Chen, Y. Tian, X. Hou, C. Lee, Nat. Commun. 2020, 11, 1.

[3] J. Si, R. Duan, M. Zhang, X. Liu, Nanomaterials 2022, 12.]

[4] B. Nie, R. Li, J. Cao, J. D. Brandt, T. Pan, *Adv. Mater.* 2015, *27*, 6055.

[5] H. Chen, L. Miao, Z. Su, Y. Song, M. Han, X. Chen, X. Cheng, D. Chen, H. Zhang, *Nano Energy* 2017, *40*, 65.

[6] M. Zhu, Z. Sun, Z. Zhang, Q. Shi, T. He, H. Liu, T. Chen, C. Lee, *Sci. Adv.* 2020, *6*, 1.

[7] P. Maharjan, T. Bhatta, M. Salauddin, M. S. Rasel, M. T. Rahman, S. M. S. Rana, J. Y. Park, *Nano Energy* 2020, *76*, 105071.

[8] H. J. Lee, K. Y. Chun, J. H. Oh, C. S. Han, *ACS Sensors* 2021, *6*, 2411.

[9] G. Yao, L. Xu, X. Cheng, Y. Li, X. Huang, W. Guo, S. Liu, Z. L. Wang, H. Wu, *Adv. Funct. Mater.* 2020, *30*, 1907312.

[10] C. M. Chiu, S. W. Chen, Y. P. Pao, M. Z. Huang, S. W. Chan, Z. H. Lin, *Sci. Technol. Adv. Mater.* 2019, *20*, 964.

CONTACT

* Jae Yeong Park , tel: +82-2-940-5113; jaepark@kw.ac.kr

LIG-BASED TRIAXIAL TACTILE SENSOR
UTILIZING ROTATIONAL ERECTION SYSTEM

Rihachiro Nakashima[1], Nagi Nakamura[2], Tomohiko G. Sano[1], Eiji Iwase[2] and Hidetoshi Takahashi[1]
[1]Keio University, Japan and [2]Waseda University, Japan

ABSTRACT

This paper reports a laser-induced graphene (LIG)-based triaxial tactile sensor utilizing rotational erection system (RES). The proposed process realizes a sensing element with a piezoresistive layer by irradiating a polyimide (PI) film with two types of lasers only once each. Furthermore, the sensing element formed with origami/kirigami method maintains a three-dimensional structure by applying a single external force. The prototype triaxial tactile sensor fabricated by the proposed fabrication process responded independently to normal and shear forces.

KEYWORDS

Triaxial tactile sensor, Laser-induced graphene, Rotational erection system

INTRODUCTION

MEMS multi-axial tactile sensors have been widely developed to evaluate human tactile sensation and safely control robots [1, 2]. One of the sensing elements of these tactile sensors is standing piezoresistive cantilevers in elastic material [3, 4]. Conventionally, Si-based cantilevers were maintained standing by magnetic forces or residual stresses. However, controlling such a small force was difficult. In addition, the ultra-thin Si cantilever was fragile for use as a tactile sensor. Against this background, PI film, a flexible and easy-to-process material, has been used for the tactile sensor with standing cantilevers [5]. The piezoresistive layer of those devices uses graphene obtained by CO_2 laser irradiation, called LIG [6]. However, the sensor design is limited by a thermal deformation process to deform the PI cantilever into a standing position.

In this paper, we propose a triaxial tactile sensor with standing cantilevers utilizing RES on a PI film. RES is a three-dimensional structure obtained by folding and cutting into a flat sheet, as shown in Fig. 1(a) [7]. When an external force is applied to the hub at the center of the RES, the three arms stand up while the hub rotates and then maintains the standing position. In other words, three standing cantilevers can be realized on PI film without using the thermal deformation process. We designed and fabricated a triaxial tactile sensor using RES and LIG. Then, we evaluated its response to normal force and two-axis shear force.

PRINCIPLE

Fig. 1(b) shows the concept of a triaxial tactile sensor using LIG. A UV laser cuts a polyimide film into a cantilever shape. A CO_2 laser forms a LIG at the base of the cantilever. Then, three LIG cantilevers are embedded in an elastic body to realize the triaxial tactile sensor. As shown in Fig. 1(c), the standing LIG cantilevers in the elastic body deform when normal or shear forces are applied to the tactile sensor surface. When a normal force

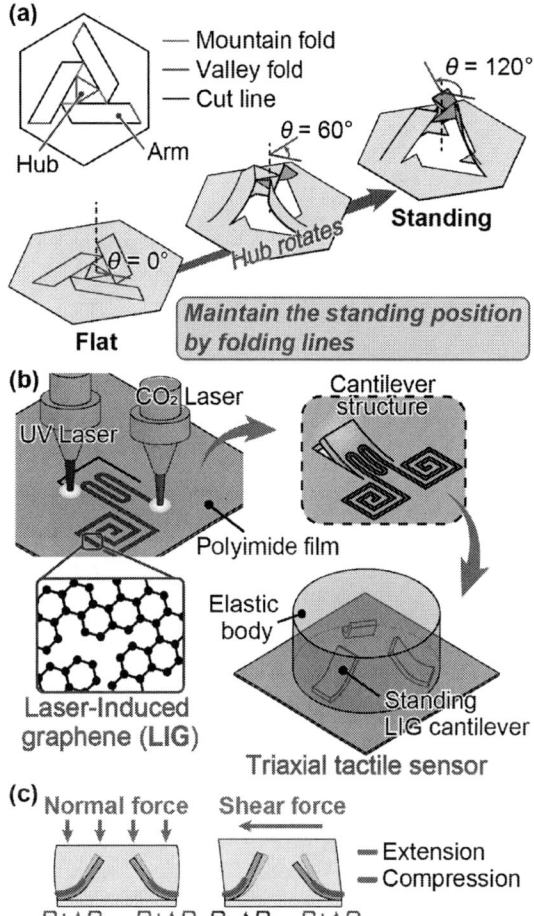

Fig. 1: (a) Design and deformation sequence of the RES. (b) Concept sketch of the LIG-based triaxial tactile sensor. (c) Schematic illustration of the cantilevers deforms when normal and shear forces are applied to the sensor surface.

is applied, the bases of all cantilevers are extended by deformation. Thus, all LIG cantilevers generate positive fractional resistance changes. On the other hand, when a shear force is applied, the cantilevers deform in different directions depending on their position. Thus, the LIG cantilevers generate positive or negative fractional resistance changes. Therefore, by arranging three LIG cantilevers in a circle at equal intervals, normal force and two-axis shear force can be measured.

DESIGN AND FABRICATION
Sensor design

Fig. 2(a-i) shows the design of the proposed sensing element before standing. The shape of the sensing element is a regular hexagon with a 36 mm diameter inscribed circle

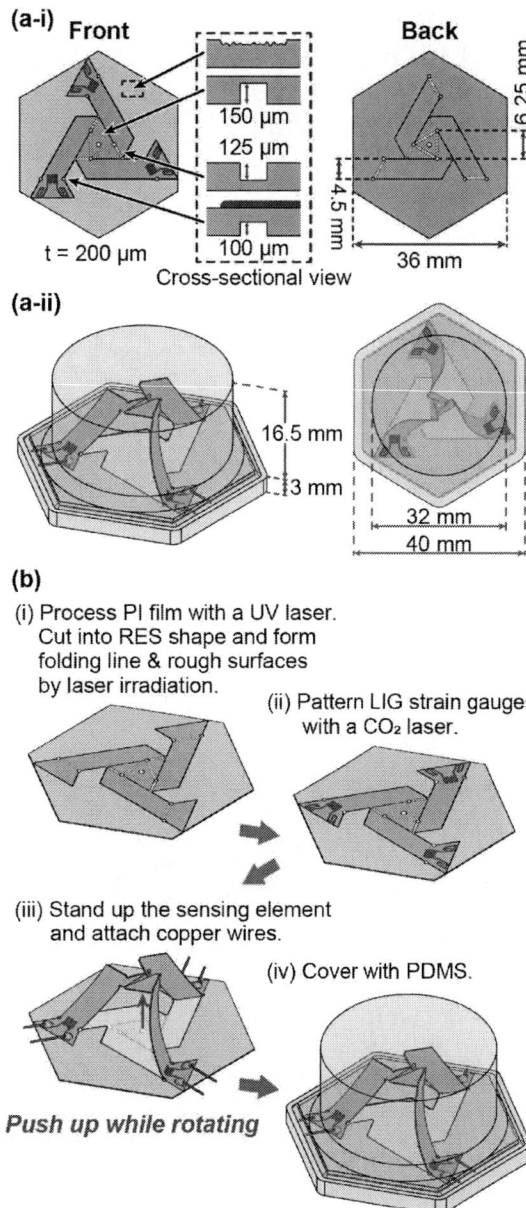

Fig. 2: (a) Design of (i) the triaxial sensing element and (ii) the proposed tactile sensor. (b) Fabrication process of the proposed tactile sensor.

and a thickness of 200 μm. The RES structure on the sensing element consists of cut lines and 0.5 mm width grooves. The grooves are formed on the mountainside of the folding line, and the depth depends on the fold's angle. Holes at both ends of the groove prevent breakage due to stress concentration. The front side of the sensing element has a rough surface except for the RES structure area. This roughening aims to improve the adhesion of the elastic body by increasing the surface area. The arms of the RES are 4.5 mm wide, and LIG strain gauges are formed at the base of the arms. Fig. 2(a-ii) shows the design of the proposed tactile sensor. The elastic body covers the sensing element with a 30 mm diameter and 16.5 mm height.

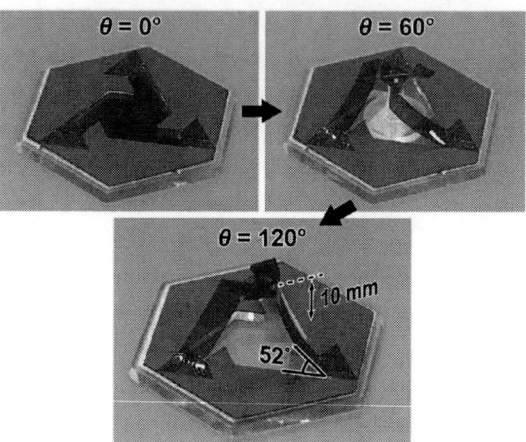

Fig. 3: Photograph of the deformation sequence of the fabricated sensing element.

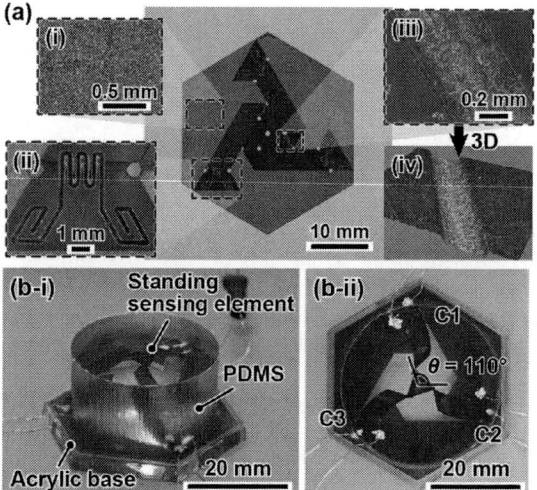

Fig. 4: (a) Photographs of the fabricated sensing element. Enlarged view of (i) the rough surface, (ii) the LIG strain gauge, and (iii) the folding line. (iv) 3D view of the folding line. (b) Photographs of the fabricated tactile sensor from (i)bird's eye view and (ii)top view.

Fabrication

Fig. 2(b) shows the fabrication process of the proposed sensor. First, a PI film was processed with a UV laser. Specifically, cutting the RES shape, forming the grooves, and surface roughening were performed simultaneously. At this time, the grooves on the back side were also formed. Next, the LIG strain gauges were patterned at the base of the RES with a CO_2 laser. Then, after standing up the sensing element, the copper wire was attached to the LIG strain gauge using a conductive paste. Finally, the deformed sensing element was glued to an acrylic base with a hole in the center and embedded in the Polydimethylsiloxane (PDMS). Thus, the fabrication process was relatively easy.

Photographs of the deformation sequence of the fabricated sensing element are shown in Fig. 3. From the photographs, it was observed that the hub stood up while rotating. After deformation, the height of the RES hub was 10 mm, and the angle of the RES arm was 52°.

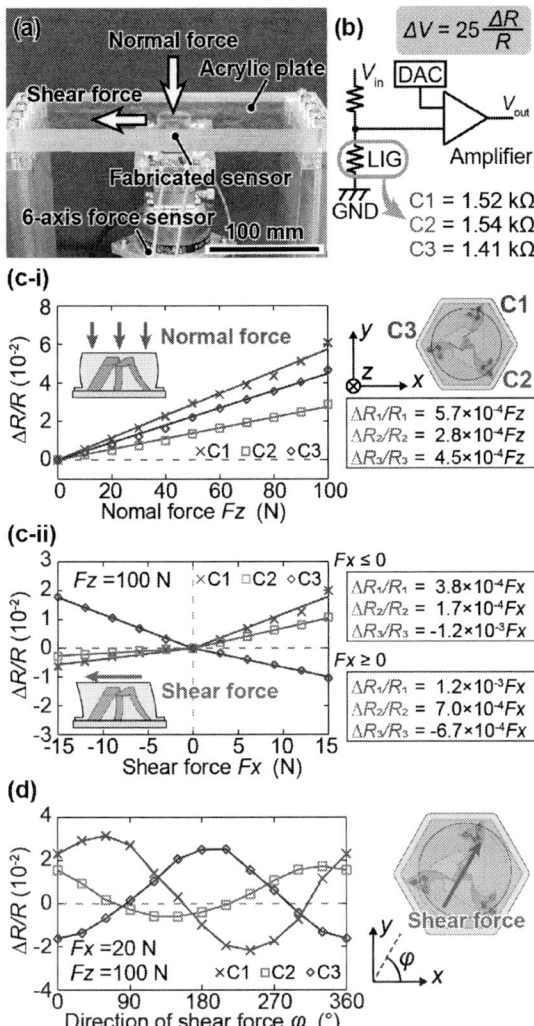

Fig. 5: (a) Photographs of the experimental setup. (b) Schematic diagram of the bridge circuit. (c) Response of the sensor to (i) normal force F_z and (ii) shear force F_x. (d) Relationship between the shear force angle and the sensor response.

The fabricated sensing element before standing up is shown in Fig. 4(a). The LIG strain gauges were formed as a zigzag line at the base of the RES arms (Fig. 4(a-ii)). Also, the rough surfaces and grooves were accurately formed (Fig. 4(a-i), (a-iii), (a-iv)). Fig. 4(b) shows photographs of the fabricated tactile sensor. There are no air bubbles inside the PDMS. The RES structure was fixed in a standing position. The LIG cantilevers with the LIG strain gauges were defined as C1, C2, and C3, respectively. The initial resistances of the LIG strain gauges were C1 = 1.52 kΩ, C2 = 1.54 kΩ, and C3 = 1.41 kΩ, respectively.

EXPERIMENT AND RESULT
Experimental setup

Fig. 5 (a) shows the photograph of the experimental setup to evaluate the fabricated tactile sensor. The fabricated sensor was placed on top of a six-axis force sensor (Minebea, OPFT-220) and pressed against a fixed acrylic plate using an $xyz\theta$ stage. The fractional resistance

change of the LIG strain gauges was converted to the voltage changes by a bridge circuit, as shown in Fig. 5(b).

Tactile force measurement

We measured the response of the fabricated sensor to normal and two-axis shear force. Fig. 5(c) shows the relationship between the applied force and the fractional resistance changes of the LIG cantilevers. All cantilevers showed a linear positive resistance change when a normal force from 0 N to 100 N was applied (Fig. 5(c-i)). The sensitivities were calculated to be $\Delta R_1/R_1 = 5.7 \times 10^{-4} \, Fz$, $\Delta R_2/R_2 = 2.8 \times 10^{-4} \, Fz$ and $\Delta R_3/R_3 = 4.5 \times 10^{-4} \, Fz$, respectively. On the other hand, when an x-axis shear force from -15 N to 15 N was applied while holding the sensor with 100 N normal force, C1 and C2 showed linearly positive resistance changes, while C3 showed linearly negative resistance changes. (Fig. 5(c-ii)).

The relationship between the direction of shear force and the fractional resistance changes is shown in Fig. 5(d). As a result, each response increased or decreased like a sinusoidal curve depending on the shear force direction. These results suggested that the sensor was able to detect the triaxial components of normal and shear forces.

CONCLUSION

We proposed the triaxial tactile sensor using RES structure and LIG cantilevers on a PI film. We were able to easily fabricate the sensing element with origami/kirigami structure and the piezoresistive layer using different laser processing. In addition, the fabricated sensing element maintained the standing position only once external force was applied. The fabricated tactile sensor showed different responses to normal and shear forces. Therefore, refining the process and design will realize a more accurate triaxial tactile sensor with the proposed method.

ACKNOWLEDGEMENT

This work was partially supported by JSPS KAKENHI Grant numbers JP22K18421 and JP21H01274, The Takano Science Foundation, and Amada Foundation.

REFERENCES

[1] L. Ascari *et al.*, "Bio-inspired grasp control in a robotic hand with massive sensorial input," *Biol Cybern*, vol. 100, no. 2, pp. 109–128, 2009.

[2] C. Chi *et al.*, "Recent progress in technologies for tactile sensors," *Sensors*, vol. 18, no. 4, 2018.

[3] K. Noda *et al.*, "Flexible tactile sensor for shear stress measurement using transferred sub-μm-thick Si piezoresistive cantilevers," *J Micromech Microeng*, vol. 22, no. 11, 2012.

[4] M. Sohgawa *et al.*, "Tactile sensor array using microcantilever with nickel-chromium alloy thin film of low temperature coefficient of resistance and its application to slippage detection," *Sens. Actuators A Phys.*, vol. 186, pp. 32–37, 2012.

[5] R. Nakashima *et al.* "Multi-axial tactile sensor using standing LIG cantilevers on polyimide film," *Proc. of MEMS2022*, pp. 688–690, 2022.

978-1-6654-9309-3/23 $31.00 © 2023 IEEE

[6] R. Ye *et al.* "Laser-Induced Graphene: From Discovery to Translation," *Adv. Mate.*, vol. 31, no. 1, 2019.

[7] T. Yoneda *et al.*, "Structure, Design, and Mechanics of a Pop-Up Origami with Cuts," *Phys Rev Appl*, vol. 17, no. 2, 2022.

CONTACT

*Rihachiro Nakashima, tel:+81-3-5450-8076; na8.3436@keio.jp

A STRETCHABLE STRAIN-INSENSITIVE SMART GLOVE FOR SIMULTANEOUS DETECTION OF PRESSURE AND TEMPERATURE

Sudeep Sharma, Gagan Bahadur Pradhan, Seonghoon Jeong, and Jae Yeong Park*

Department of Electronic Engineering, Kwangwoon University, Seoul, Republic of Korea

ABSTRACT

In this work, we newly developed a stretchable strain-insensitive smart glove, capable of simultaneously monitoring pressure and temperature. A low-cost CO_2 laser engraving and electrospinning technology were used for the first time to build a multifunctional smart glove with layer-by-layer geometry. The key elements of the e-glove fabrication are CNT-coted laser engrave graphene (CNT/LEG) and ionic nano-fibrous membrane (INM). The proposed smart glove can detect pressure through the variation of electric double-layer capacitance (EDLC) between the two CNT/LEG electrodes and INM, while the temperature is detected by the thermoresistivity of the CNT/LEG electrode. The pressure and temperature sensitivity were recorded as 0.506 kPa^{-1} (0-200 kPa) and 0.212%$°C^{-1}$ (20-70°C), respectively.

KEYWORDS

Smart glove, strain-insensitive, stretchable electronics, pressure sensor, temperature sensor

INTRODUCTION

Human skin is endowed with an extraordinary network of sensors able to detect vital information on the sense of touch [1]. The smart glove that reproduces the feeling of touch like human skin, offers a remarkable application in man-machine interfacing, skin prosthetics, and healthcare surveillance [2, 3]. However, mimicking the human sense of touch in the smart glove is a significant challenge that requires effort at the component level spanning from signal processing and communication to materials engineering [4]. Additionally, simultaneous detection and discrimination of a variety of tactile stimuli, including pressure and temperature, are critical challenges. Although high scalability, sensitivity, and biodegradability have already been achieved, the use of the complex photolithography process is the production process on a large scale [5]. Therefore, the simple manufacturing process using an inexpensive and flexible inorganic material with detection features even greater remains to be investigated.

Rigid inorganic nanomaterials are integrated into substrates that are naturally flexible to create stretchable electronics. Serpentine and open mesh connection layouts are employed to minimize the stress impact. Lately, a smart glove with the capability of simultaneous monitoring of pressure and temperature is fabricated by a photolithography process on a stretchable substrate [6]. However, there is still a significant drawback since the embedded sensing region is not stretchy, which results in a significant alteration in the performance of the sensors due to apparent cracks generated by the mechanical deformation of substrate stretching in various directions [7]. Therefore, to prevent the intrinsic coupling of mechanical deformation by substrate stretching, the sensing region, including interconnections, must be strain-insensitive to provide complete mechanical stretchability in smart gloves.

Here, we developed a stretchable strain-insensitive smart glove able to simultaneously detect and differentiate pressure and temperature stimuli in real-time. A low-cost fabrication process utilizing laser engraving technology is being explored to produce strain-insensitive patterns. A highly sensitive strain-insensitive pattern of CNT/LEG on styrene-ethylene-butylene-styrene (SEBS) stretchable substrate (CNT/LEG@SEBS) is used as an active sensing material for the temperature sensor. The unique strain-insensitive structure is tailored to stretch according to the deformation and movement of the hand without a significant change in the sensing performance. The incorporation of the CNT in LEG also acts as a bridge between the adjacent graphene sheets, which compensate for the resistance change under stretching conditions, preserving the sensitivity of the sensors. In addition, ionic liquid (IL) is mixed with the poly (vinylidene fluoride-cohexafluoropropylene) (P(VdF-HFP)) followed by the electrospinning process to produce the iontronic nanofibrous membrane (INM). The thin INM is inserted between the CNT/LEG@SEBS electrode to realize the electrical double layer-based pressure sensor. As proof of concept, we demonstrate our smart glove for simultaneously detecting pressure and temperature Additionally, the feasibility of the smart glove for human-machine interactions is also successfully demonstrated.

EXPERIMENTAL METHODS
Fabrication of CNT/LEG@SEBS

Firstly, the strain-insensitive pattern was loaded into the laser machine to transfer the designed pattern into the LEG conductive network. LEG is produced by radiating a CO_2 laser beam of 10.6 μm wavelength on a polyimide sheet with an intensity of 2.7 W and a speed of 5%. Subsequently, 3 g of SEBS powder was dissolved in 20 ml of a toluene solution and continuously stirred for 6 h until the solution becomes transparent. The transparent SEBS solution is cast into the LEG patterns and left in an ambient condition for drying. The LEG@SEBS film was peeled off from the polyimide substrate and dipped into a CNT-COOH solution for 10 min to coat a CNT layer on the LEG pattern. Finally, the CNT/LEG@SEBS film was rinsed with water to avoid unnecessary deposition outside the pattern followed by drying inside the convection oven for 2 h.

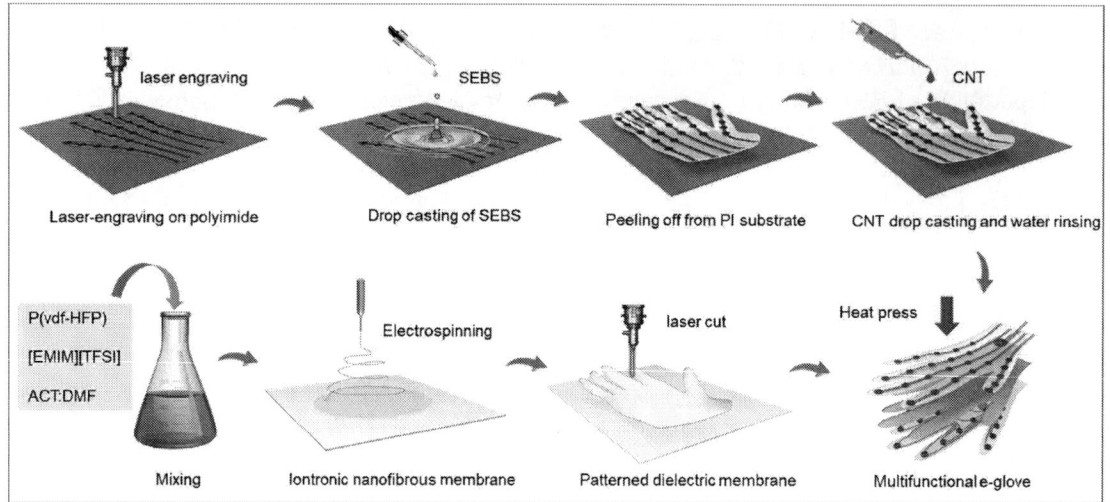

Figure 1. Schematic of the fabrication sequence describing various stages to prepare the multifunctional e-glove. The fabrication of the CNT/LEG@SEBS was based on laser engraving a PI sheet followed by the SEBS solution rod casting and CNT deposition, while the INM was prepared by blending [EMIM][TFSI] into the P(VdF-HFP) solution followed by electrospinning process.

Fabrication of INM

The electrospinning technique was used to prepare INM. In this process, 20% of P(VdF-HFP) pellets were dissolved in dimethylformamide: acetone (3:2) solvent by stirring for 1 h at a temperature of 35 °C. Subsequently, different concentration of the IL (10%, 20%, 30%, and 40%) was dissolved to the P(VdF-HFP) weight ratio and further stirred for 15 min for uniform dispersion. Then the P(VdF-HFP)/IL transparent gel was loaded into a 20ml syringe with a 21 G needle for electrospinning. The gel was injected from the needle at a continuous injection rate of $0.2 \ mLh^{-1}$ with 25 kV at various spinning times (30 min, 1 h, 1.5 h). Finally, the INM was peeled from the aluminum foil and dried inside a convection oven for 3 h at 50 °C.

Fabrication of smart glove

The boundary line of the upper and lower CNT/LEG@SEBS film with the INM was retracted by laser cutting along a palm structure. Then, the INM was inserted between the upper and lower electrode film followed by a hot press at 100 C for 10 min at 4 MPa. The 25-pressure point of the upper electrode is serially connected, while the lower electrode was parallelly connected to realize pressure sensing performance. The upper electrode patterns with pressure points were utilized as a temperature sensors. Afterward, the patterned CNT/LEG@SEBS film with the INM was infused with a commercial thermoplastic glove by a hot-press machine. Finally, the copper tape was attached for the ohmic contact.

Characterization

The electromechanical performance of the capacitive pressure sensor was characterized by using a force gauge (HF-100) with a moving fixture (JSV-H1000). The force gauge was used to provide uniform compression force to the sensor by manually quantifying speed and distance. The corresponding change in capacitance is recorded using an LCR meter (Hioki, IM 3536). For real-time demonstration, the 25 pressure sensors were connected to an 8-channel multiplexer (74HC4051) through an operational amplifier (OPA602BP). For temperature characterization, Increase and decrease in temperature on the sensor surface were measured by a commercial IR sensor while the continuous increase in temperature was provided via a hot plate. Resistance response was measured by a digital multimeter (DMM6500). For real-time monitoring, the sensor was connected to a Wheatstone bridge circuit (R1, R2, R3 = 15 kΩ). The mismatch voltage signal equivalent to the temperature stimuli was transmitted to the signal analyzer. An ARM cortex M33 microprocessor was used to coordinate all the modules and transmit the signal to the smartphone via the Bluetooth unit. For material characterization, the surface morphology of the CNT/LEG@SEBS film and INM was observed by the field effect scanning electron microscope (FESEM) machine (JSM-6700F). Molecular vibration and defects on graphite surfaces were measured by Raman spectroscopy (Renishaw, in Via Raman Microscope).

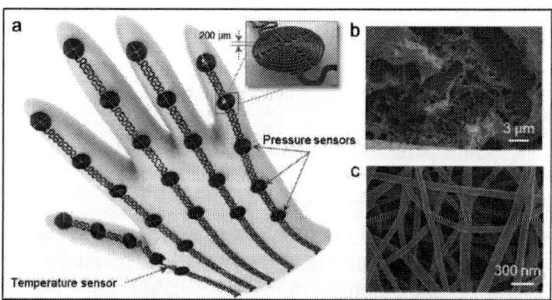

Figure 2. (a) Schematic of the smart glove with pressure and temperature sensors. The engraved pattern width is ~200 μm. (b) FESEM image of multilayered porous graphene produced by laser engraving. (c) FESEM image of INM with uniform nanofiber diameter of ~150 nm.

EXPERIMENTAL RESULTS
Structural geometry and fabrication of the smart glove

Figure 1 shows the schematic diagram of the fabrication sequence describing the smart glove preparation. The fabrication sequence consists of two steps: the preparation of CNT/LEG@SEBS and INM film. The fabrication of the CNT/LEG@SEBS includes laser engraving on a polyimide sheet followed by the casting of the SEBS solution on the laser engraved design and CNT drop casting. Additionally, the fabrication of INM involves blending IL and P(VdF-HFP) followed by the electrospinning process. The individual components were stacked together and assembled in a commercial thermoplastic glove to realize temperature and pressure sensing features. Figure 2a shows the schematic diagram of the smart glove with temperature and pressure sensing components. The sensors network was designed to be strain-insensitive in all directions, ranging from the open mesh (200 μm) to the serpentine structure (400 μm). The FESEM image of CNT/LEG in Figure 2b shows multilayered graphene flakes with porous morphology. Furthermore, the CNTs are evenly distributed, forming a bridge between the graphene sheets. Figure 2c illustrates the FESEM image of the INM showing the hierarchical nanofibrous assembly with a uniform thickness of ~150 nm.

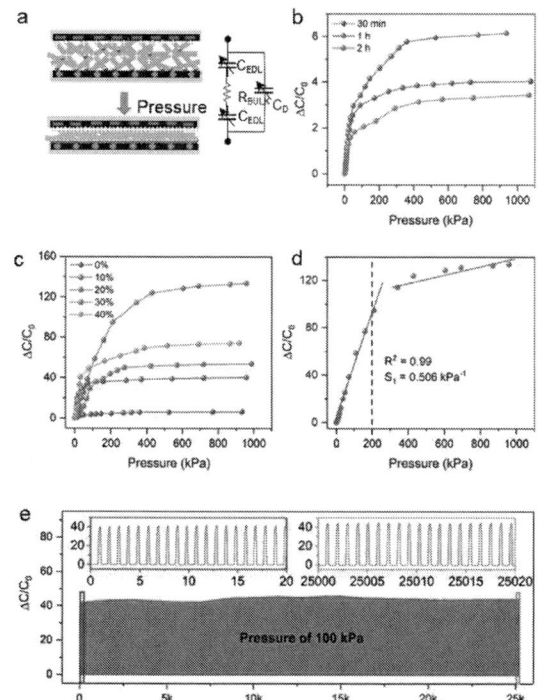

Figure 3. (a) Schematic showing the operation of the pressure sensor before and after pressure with an equivalent circuit diagram. (b) Pressure sensitivity profile with various INM thicknesses. (c) Pressure sensitivity profile with various IL concentrations. (d) Detailed investigation of the pressure sensitivity of the INM with 30% IL concentration and 1 h spinning time. (e) Cyclic stability of the sensor under 25,000 loading/unloading cycles.

Pressure sensor characterization of the smart glove

Figure 3a shows a working mechanism of the iontronic pressure sensor with the equivalent circuit diagram. The fabricated pressure sensor consists of an INM sandwiched between the CNT/LEG@SEBS electrodes. The output capacitance depends on the EDL at the interface of the CNT/LEG@SEBS electrodes and INM (C_{EDL}), resistance from the contact area between the electrode and the membrane (R_{BULK}), and the parallel plate capacitor (C_D). Under pressure, the intermediate layer (INM) deforms elastically, increasing the contact area between the electrode and the membrane, and allowing more channels for ion conduction. The phenomenon ultimately reduces the R_{BULK} and significantly increases the EDLC.

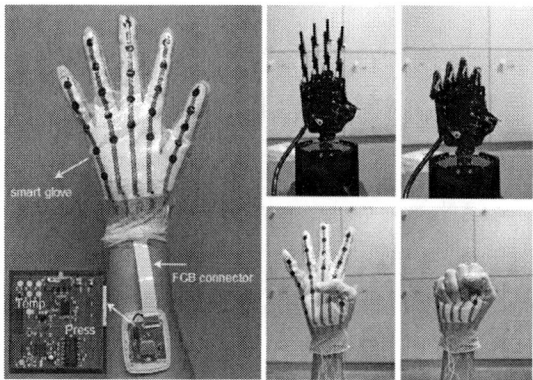

Figure 4. Demonstration of the smart glove in human-machine interfacing.

The INM is a crucial element for the pressure sensor and its thickness and ion concentration affect the pressure sensitivity. The INM with a spinning time of 1 hour shows a high-pressure sensitivity performance in relation to the other spinning time (Figure 3b). This is attributed due to the higher change in the contact area between the electrode and membrane due to the higher area fraction in larger spinning time [8]. However, the porosity of the membrane is compromised by a longer spinning time, which creates bulkiness and tenacity in the membrane, reducing compressibility [9]. The sensitivity graph in Figure 3c shows that the IL at 30% concentration has higher sensitivity and linearity due to larger $\Delta C/C_0$ compared to the other concentration. The detailed plot of 30% IL with 1 h spinning time is shown in Figure 3d, where the sensor showed a high sensitivity of 0.506 kPa^{-1} in a wide pressure range (200 kPa). The fabricated smart glove pressure sensor showed excellent durability with over 25k loading and unloading cycles at a high pressure of 100 kPa. There is no obvious change in the capacitance value in the beginning and ending cycles. Furthermore, for a real-time demonstration, the smart glove was worn by a person and integrated with a robotic hand. The finger motion was detected by the pressure sensors of the smart glove, which controls the robot hand by the human hand gesture.

Temperature sensor characterization of the smart glove

In carbon-based materials, a high temperature enables a significant number of electrons to move across neighboring conducting materials by hopping electrons as shown in Figure 4a. The higher the temperature, the probability of the electron hopping is larger with a longer range, so a lower resistance is expected [10]. In addition, the addition of the CNT particles efficiently enhances the charge-carrying properties of graphene. Thermal activation enables the electrons to transport effectively between CNTs through the contact points, providing an efficient charge transport route [11]. The phenomenon is distinctly proven in Figure 4b, where the temperature coefficient of resistance (TCR) of the CNT/LEG@SEBS shows more than four times greater than the LEG@SEBS (-0.212% °C^{-1}). Furthermore, to analyze the temperature sensing properties, the smart glove was subjected to various temperature conditions for 1 min (Figure 4c). The sensor shows a stable and distinct resistance response to the temperature. The glove was slowly touched/removed from a human body, the body temperature triggers the resistance to change proportionately as shown in Figure 4d. The sensor responded successfully to the body temperature and returned to its original baseline after being removed.

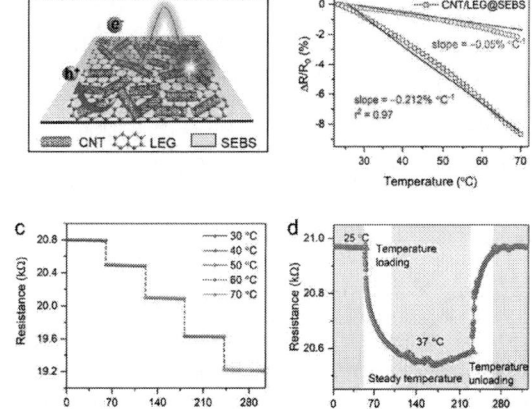

Figure 5. (a) Schematic diagram of the working mechanism of the temperature sensor. (b) Temperature coefficient of resistance (TCR) of LEG@SEBS and CNT/LEG@SEBS. (c) Sensor response with sequential loading of temperature and holding for 1 min. (d) Response of the sensor while loading and unloading onto a human chest.

CONCLUSION

In summary, the stretchable and strain-insensitive glove is successfully designed and fabricated for the simultaneous detection of pressure and temperature. The facile and cost-effective laser engraving technology is used for the fabrication of the smart glove. This approach simplifies the manufacturing process by avoiding multiple fabrication steps and waiting time. The result of this study showed that the cross-link of the CNT in the graphene flakes not only enhances the temperature sensitivity but also compensate for the resistance alteration under strain condition. In addition, high-pressure sensitivity and linearity are obtained by using INM due to its larger area

and volume fraction. Notably, these characteristics have enabled our smart glove to be used in a promising application for a prosthetic hand, humanoid robotics, and human-machine interfacing.

ACKNOWLEDGEMENTS

This research was supported by Korea Institute for Advancement of Technology (KIAT) grant (P0020967, The Competency Development Program for Industry Specialist)) and by the Technology Innovation Program (RS-2022-00154983, Development of Low-Power Sensors and Self-Charging Power Sources for Self-Sustainable Wireless Sensor Platforms) funded by the Ministry of Trade, Industry & Energy (MI, Korea). The authors are grateful to the group members of the Advanced Sensor and Energy Research (ASER) Laboratory of Kwangwoon University for their valuable suggestions and support.

REFERENCES

[1] A. Chortos, J. Liu, and Z. Bao, "Pursuing prosthetic electronic skin," *Nat. Mater.*, vol. 15, pp. 937–950, 2016.

[2] G. Gu *et al.*, "A soft neuroprosthetic hand providing simultaneous myoelectric control and tactile feedback," *Nat. Biomed. Eng.*, 2021.

[3] F. Liu *et al.*, "Neuro-inspired electronic skin for robots," *Sci. Robot.*, vol. 7, pp. 1-15, 2022.

[4] J. C. Yang *et al.*, "Electronic Skin: Recent Progress and Future Prospects for Skin-Attachable Devices for Health Monitoring, Robotics, and Prosthetics," *Adv. Mater.*, vol. 31, pp. 1–50, 2019.

[5] M. Zhu *et al.*, "Haptic-feedback smart glove as a creative human-machine interface (HMI) for virtual/augmented reality applications," *Sci. Adv.*, vol. 6, pp. 1–15, 2020.

[6] J. Kim *et al.*, "Stretchable silicon nanoribbon electronics for skin prosthesis," *Nat. Commun.*, vol. 5, 5747-5758, 2014.

[7] F. Zhang *et al.*, "Flexible and self-powered temperature-pressure dual-parameter sensors using microstructure-frame-supported organic thermoelectric materials," *Nat. Commun.*, vol. 6, pp. 1–10, 2015.

[8] V. Amoli *et al.*, "Ionic Tactile Sensors for Emerging Human-Interactive Technologies: A Review of Recent Progress," *Adv. Funct. Mater.*, vol. 30, pp. 1–32, 2020.

[9] S. Sharma *et al.*, "Hydrogen-Bond-Triggered Hybrid Nanofibrous Membrane-Based Wearable Pressure Sensor with Ultrahigh Sensitivity over a Broad Pressure Range," *ACS Nano*, vol. 15, pp. 4380–4393, 2021.

[10] D. H. Ho *et al.*, "Stretchable and Multimodal All Graphene Electronic Skin," *Adv. Mater.*, vol. 28, pp. 2601–2608, 2016.

[11] R. Wu *et al.*, "Silk Composite Electronic Textile Sensor for High Space Precision 2D Combo Temperature–Pressure Sensing," *Small*, vol. 15, pp. 1–11, 2019.

CONTACT

*J.Y. Park tel: +82-2-940-5113; jaepark@kw.ac.kr

A GESTURE RECOGNITION GLOVE ASSEMBLED WITH NANOFOREST-INTEGRATED INFRARED THERMOPILES

Mao Li[1,2], Meng Shi[1,2], Guidong Chen[1,2], Na Zhou[1,2], Haiyang Mao[1,2*], and Chengjun Huang[1,2]*
[1]Institute of Microelectronics of Chinese Academy of Sciences, Beijing 100029, CHINA and
[2]University of Chinese Academy of Sciences (UCAS), Beijing 100029, CHINA

ABSTRACT

In this work, we reported a glove assembled with nine infrared thermopiles integrated with nanoforests (NFs) for non-contact gesture recognition. Ascribed to the large infrared absorption of the NFs due to the hybrid surface plasmon resonance, the thermopiles exhibit high sensitivity, and are able to characterize the relative position of a finger to the palm and to other fingers, making discernment of different gestures possible. Compared with other non-contact gesture recognition apparatuses, such a glove is able to realize thermoelectric response and gesture recognition without thermal imaging, thus is much simpler for identification.

KEYWORDS

Gesture recognition, wearable, nanoforests, infrared thermopiles

INTRODUCTION

Human communication has evolved over a long period of time and has developed a variety of effective systems. Unlike languages that vary from country to country, easy-to-understand gestures have a wider universality for interaction. A variety of sign language recognition systems have emerged to make sign language communication easier [1-3]. Furthermore, with the popularization of virtual reality and augmented reality, human-machine interfaces based on gesture control have brought significant interest because of the diverse control and high accuracy.

With gestures playing an increasingly important role in sign language recognition and non-contact control, different gesture recognition systems have been aroused recently. Among these systems, wearable gesture apparatuses, gloves for instance, have brought new avenues for non-contact control and sign language identification. Generally, flexible contact sensors that can detect bending angles of fingers are often used for wearable gesture detection. Contact sensors sticking around the knuckles for motion sensing are prerequisite for such gloves [4-8]. However, these sensors tend to cause restrictions on finger movement and discomfort in gesture variations.

Non-contact sensors also have plenty of applications in the field of gesture recognition through excellent imaging capabilities combined with image recognition. Non-contact optical/thermal sensors applied in gesture recognition systems can characterize deformation of the hand and fingers without touching the knuckles [9-11]. However, such gesture recognition systems usually require complex computational networks and are separated from the hand, making it difficult to achieve wearable applications [12, 13]. Wearable gesture recognition systems consisting of non-contact sensors are seldomly in sight yet.

In this work, to fill the gap of non-contact sensors in wearable gesture recognition applications, nine nanoforest (NF)-integrated thermopiles are assembled onto a glove, which is further used for non-contact sensing without involving imaging. NF-integrated thermopiles are with high sensitivity because of the enhanced infrared absorption brought by NFs, which makes it possible to distinguish distance. With these NF-integrated thermopiles, relative positions of the detected fingers to the palm and to other fingers are discriminated. Consequently, the realization of gestures recognition for Chinese sign language is successfully achieved. Such a designed glove is expected to open new and promising applications in sign language identification and non-contact control.

METHOD

Schematic of the gesture recognition glove with NF-integrated thermopiles assembled on its surface is illustrated in Figure 1, and each thermopile sticked to the rubber glove will be wired to the processing system. NF-integrated thermopiles located at the positions corresponding to the belly of fingers from thumb to pinky are defined as B1-B5, while those located on inner sides of the thumb, index finger, ring finger and pinky finger are defined as S1-S4, respectively. Inset in Figure 1 demonstrates the NF-integrated thermopile, which is with excellent sensitivity and has been verified in our previous work [14]. Benefited from its superior sensitivity, the distances between heat source and thermopiles can be differed. In this work, heat emitted from the human hand is considered as a heat source. B1-B5 are used to characterize the distances (D1) between the belly of the bending fingers and the palm, meanwhile, S1-S4 are used for distinguishing the distances (D2) between the fingers with thermopiles and their adjacent fingers. By collecting D1 and D2, relative positions of the detected fingers to the palm and to other fingers can be characterized, leading to identification of the corresponding gestures.

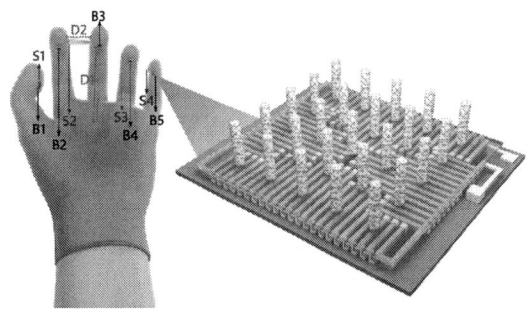

Figure 1: Schematic of the designed gesture recognition glove assembled with NF-integrated thermopiles.

A microscope image of a NF-integrated thermopile is shown in Figure 2 (a). Fabrication process of the NF-integrated thermopile has been reported in our previous work [14]. The NFs were prepared through a reactive ion etching process involving only O_2 and Ar plasma, which is simple and low-cost. Top view of a scanning electron microscope (SEM) image of NFs is shown in Figure 2 (b). The light trapping effect brought by the high aspect ratio structure and the hybrid plasmon resonances brought by the composite NFs with Ag nanoparticles result in high absorption in the infrared wavelength where the hand thermal signal radiation exists. Moreover, such a NF-integrated sensor can be easily fabricated with a conventional micromachining process, thus is suitable for mass production, which helps control the cost of the recognition glove [15].

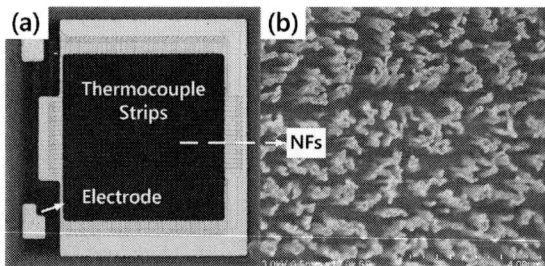

Figure 2: (a) A microscope image of the thermopile with NFs. (b) A SEM image of the NFs.

RESULTS AND DISCUSSION

Simulation has been performed to verify the principle of a thermopile to recognize distance, and the relationship between the output voltage of NF-integrated thermopiles and distance has been tested experimentally. In the simulation, an original thermopile with a heat source which was sufficiently large to cover the absorption area and had a consist temperature at 37 °C has been built for the proposed model. The surrounded air in the model was considered to be still. Figure 3 (a) exhibits the simulation results of output voltages (illustrated as red curve) and temperature differences (illustrated as black curve) of an original thermopile, under different distances from the heat source. From the curves, when distance varies, the temperature of the hot ends of thermocouple strips changes, which causes a temperature difference in thermopile. Subsequently, according to the Seebeck effect, thermocouples convert the temperature difference into output voltages. Thus, with the ideal environment in the simulation, the original thermopile is able to discriminate distance.

However, the results in the actual test are somewhat different. To describe the relative position of the detected fingers to the palm and to other fingers, D1 and D2 were defined within a certain range. A volunteer with a hand length of 19 cm (from the wrist joint to the tip of the middle finger) has carried out the test. Taking the longest middle finger for example, when the finger is straightened, the output voltage of the thermopile is 0, while when the finger is bent enough for B3 to be able to detect the thermal signal of the palm, D1 would not exceed 5 cm. Therefore,

experimentally, the output voltages of thermopiles were measured in the range of 0-5 cm from the heat source. According to the measured results in Figure 3 (b), ten original thermopiles were used to test their output voltages under different distances from a hot plate set at 37 °C. Nevertheless, the black curve of the original thermopiles shows no obvious ability to differentiate distances. Such a result is caused by the low sensitivity of the original thermopiles and the complexity of the environment with unpredictable thermal convection in the air. Contrastingly, ten NF-integrated thermopiles display much better distance differentiation ability under the same condition due to their high sensitivity which is mentioned above. Therefore, NF-integrated thermopiles are essential for distance discrimination.

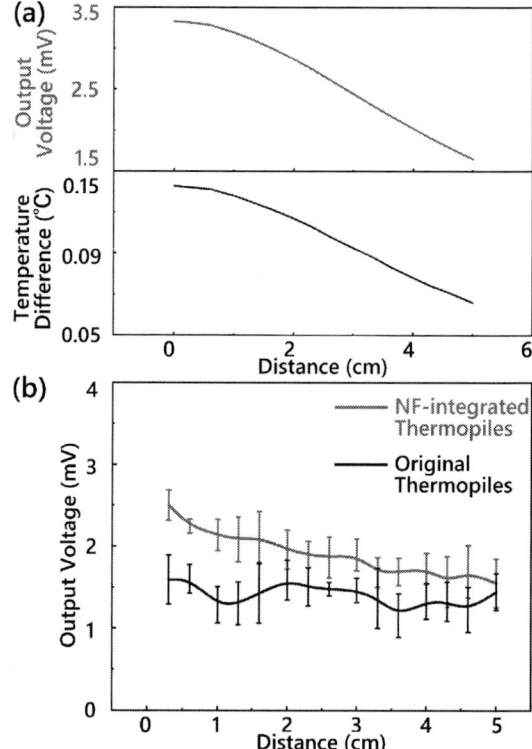

Figure 3: (a) Simulated temperature differences between hot and cold ends of thermocouple strips, and simulated output voltages of an original thermopile under different distances from the heat source at 37 °C. (b) Measured output voltages of the thermopiles under different distances from the heat source at 37 °C.

Based on the distance differentiation ability of NF-integrated thermopiles, a gesture recognition glove assembled with nine NF-integrated thermopiles was prepared, as illustrated in Figure 4. Surface Mount Technology has been applied to package the NF-integrated thermopiles, as shown in inset. Notably, the device has a surface area less than 10 mm², which has very little effect on the movements of fingers. Each thermopile has been sticked to a rubber glove at the corresponding positions of B1-B5 and S1-S4, and then wired to the system for further

978-1-6654-9309-3/23 $31.00 © 2023 IEEE   230

processing.

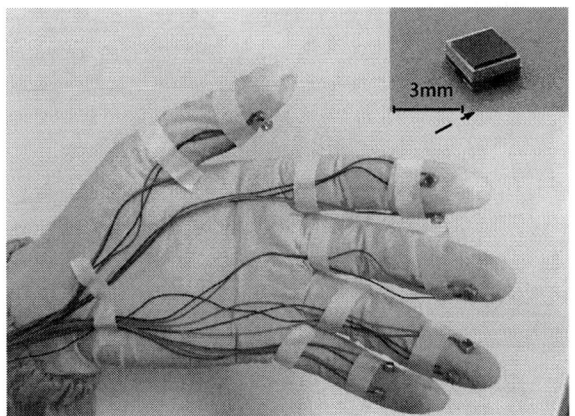

Figure 4: Photo of the gesture recognition glove assembled with nine NF-integrated thermopiles.

Figure 5 illustrates the gesture recognition system from signal acquisition to color mapping. High sensitivity of NF-integrated thermopiles has brought effective recognition of distances, but will introduce a high ambient thermal noise, which brings disturbance to the output voltages. To address this issue and convert the signal of thermopiles into digital signal, a filtering and amplification module has been applied in the system to filter the noise and amplify the output voltages of the detected finger signal. For different hand length of different people, a certain hand gesture might have different D1 and D2 for the detected fingers, resulting in fluctuations of output voltages within a fuzzy range. To solve the problem, a data analysis module capable of discriminating the range of the output signal has been used. The module enables the adjustment of the range to which the output voltages belong by inputting different hand lengths, so that the gestures of different hands can be accurately recognized.

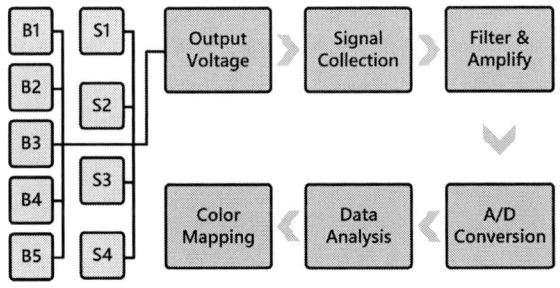

Figure 5: Schematic diagram of the gesture recognition system from signal acquisition to data processing, and then to color mapping.

The output signals of the nine-channel NF-integrated thermopiles were processed by the system and the corresponding gestures were recognized. Color mapping presenting the output voltages of the NF-integrated thermopiles of B1-B5 and S1-S4 is shown in Figure 6. Chinese sign language from 1-10 were implemented and tested. For a volunteer with a hand length of 19 cm, the

output voltages were separated into three ranges: <1 mV, 1~2 mV and >2 mV. As demonstrated in the mapping, the NF-integrated thermopiles of B1-B5 are able to distinguish most of the gestures by recognizing the distances from the finger belly to the palm of the hand. But for gestures in Figure 6 (g) and (j), S1-S4 are essential because B1, B2 and B3 detect the thermal signal of other fingers instead of the palm when the fingers are close enough such as Chinese language "7" which is performed in (g). By placing S1-S4, the relative positions between the fingers could be identified, thus providing a better identification of the gestures. As illustrated above, nine NF-integrated thermopiles enable the relative position of each finger to be described and expressed through output voltages. Combining these signals, different gestures can be recognized accurately.

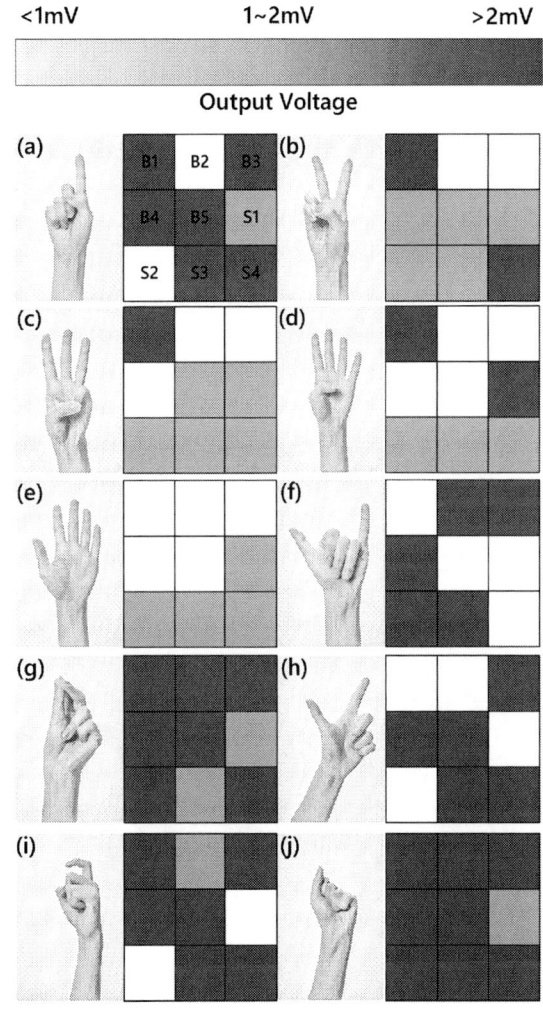

Figure 6: Color mapping of the nine NF-integrated thermopiles corresponding to the gesture language from one to ten.

CONCLUSION

In summary, a wearable gesture recognition glove assembled with non-contact NF-integrated thermopiles is proposed. Based on the high sensitivity brought by the NFs, the proposed thermopiles are able to distinguish the relative positions of the detected fingers to the palm and to other fingers, making it possible to recognize the corresponding gestures. The thermopile is with a small size, which has no effect on the flexibility of hand movements. By combining the glove with a processing system, Chinese sign language from 1 to 10 has been implemented and recognized. Such a system is able to discriminate gestures without causing restrictions in finger action. Therefore, the glove has a promising future with diverse applications in non-contact control and sign language recognition.

ACKNOWLEDGEMENTS

This work was supported by National Natural Science Foundation of China (Grant No. 62201567), Youth Innovation Promotion Association, Chinese Academy of Sciences (Grant No. 2022117), State Key Laboratory of Dynamic Test jointly built by Province and Ministry Open Fund (Grant No. 2022-SYSJJ-07), High Technology Research and Development Project of Guangdong (Grant No. 2019B010117001).

REFERENCES

[1] H. Zhang, D. Zhang, B. Zhang, D. Wang, and M. Tang, "Wearable Pressure Sensor Array with Layer-by-Layer Assembled MXene Nanosheets/Ag Nanoflowers for Motion Monitoring and Human–Machine Interfaces", *ACS Appl. Mater. Interfaces,* vol. 14, pp. 48907-48916, 2022.

[2] X. Guo, X. Lu, P. Jiang, and X. Bao, "SrTiO$_3$/CuNi-Heterostructure-Based Thermopile for Sensitive Human Radiation Detection and Noncontact Human-Machine Interaction", *Adv. Mater.*, vol. 34, 2204355, 2022.

[3] Y. Luo, Z. Wang, J. Wang, X. Xiao, Q. Li, W. Ding, and H. Y. Fu, "Triboelectric Bending Sensor Based Smart Glove towards Intuitive Multi-Dimensional Human-Machine Interfaces", *Nano Energy*, vol. 89, 106330, 2021.

[4] P. Maharjan, T. Bhatta, H. Cho, and J. Y. Park, "A Highly Sensitive Self-Powered Flex Sensor for Prosthetic Arm and Interpreting Gesticulation", in *Digest Tech. Papers IEEE MEMS 2020,* Vancouver, January 18-22, 2020, pp. 665-668.

[5] J. Zhang, Y. Song, H. Chen, X. Cheng, X. Chen, B. Meng, Q. Yuan, and H. Zhang, "Stretchable, Transparent and Wearable Sensor for Multifunctional Smart Skins", in *Digest Tech. Papers IEEE MEMS 2017*, Las Vegas, January 22-26, 2017, pp. 1025-1028.

[6] M. Zhu, Z. Sun, Z. Zhang, Q. Shi, T. Chen, H. Liu, and C. Lee, "Sensory-Glove-Based Human Machine Interface for Augmented Reality (AR) Applications", in *Digest Tech. Papers IEEE MEMS 2020*, Vancouver, January 18-22, 2020, pp. 16-19.

[7] J. Pan, Y. Li, Y. Luo, X. Zhang, X. Wang, D. L. T. Wong, C-H. Heng, C-K. Tham, and A. V-Y. Thean, "Hybrid-Flexible Bimodal Sensing Wearable Glove System for Complex Hand Gesture Recognition", *ACS Sens.*, vol. 6, pp. 4156-4166, 2021.

[8] P. C. Uzabakiriho, M. Wang, K. Wang, C. Ma, and G.

Zhao, "High-Strength and Extensible Electrospun Yarn for Wearable Electronics", *ACS Appl. Mater. Interfaces,* vol. 14, pp. 46068-46076, 2022.

[9] F. Liang, C. Cai, K. Zhang, L. Zhang, J. Li, H. Bi, P. Wu, H. Zhu, C. Wang, H. Wang, Z. Dong, C. Luo, Z. Luo, C. Shan, W. Hu, and X. Wu, "Infrared Gesture Recognition System Based on Near-Sensor Computing", *IEEE Electron Device Lett.*, vol. 42, pp. 1053-1056, 2021.

[10] Y. Sugiura, F. Nakamura, W. Kawai, T. Kikuchi and M. Sugimoto, "Behind the palm: Hand gesture recognition through measuring skin deformation on back of hand by using optical sensors", in *Digest Tech. Papers SICE 2017,* Kanazawa, September 19-22, 2017, pp. 1082-1087.

[11] N. Le Ba, S. Oh, D. Sylvester, and T. T. Kim, "A 256 Pixel, 21.6 µW Infrared Gesture Recognition Processor for Smart Devices", *Microelectronics J.*, vol. 86, pp. 49-56, 2019.

[12] J. Liu, K. Furusawa, T. Tateyama, Y. Iwamoto and Y. -W. Chen, "An Improved Hand Gesture Recognition with Two-Stage Convolution Neural Networks Using a Hand Color Image and Its Pseudo-Depth Image", in *Digest Tech. Papers IEEE ICIP 2019*, Taipei, September 22-25, 2019, pp. 375-379.

[13] T. L. Dang, S. D. Tran, T. H. Nguyen, S. Kim, and N. Monet, "An Improved Hand Gesture Recognition System Using Keypoints and Hand Bounding Boxes", *Array*, vol. 16, 100251, 2022.

[14] M. Li, M. Shi, B. Wang, C. Zhang, S. Yang, Y. Yang, N. Zhou, X. Guo, D. Chen, S. Li, H. Mao, and J. Xiong, "Quasi-Ordered Nanoforests with Hybrid Plasmon Resonances for Broadband Absorption and Photodetection", *Adv. Funct. Mater.*, vol. 31, 2102840, 2021.

[15] C. Zhang, H. Mao, M. Shi, J. Xiong, K. Long, and D. Chen, "A Fiber-Si$_3$N$_4$ Composite Nanoforest with High 7.6 to 11.6 µm Absorption for MEMS Infrared Sensors", in *Digest Tech. Papers IEEE MEMS 2020*, Vancouver, January 18-22, 2020, pp. 949-952.

CONTACT

*H.Y. Mao, tel: 86-010-82995934;
maohaiyang@ime.ac.cn
*N. Zhou, tel: 86-010-82995794;
zhouna@ime.ac.cn

ONE PUSH MEMBRANE FORMATION FOR ITERATIVE MEASUREMENT OF ION CHANNEL ACTIVITY ON ARRAYED CHIP

Hisatoshi Mimura[1], Toshihisa Osaki[1,2], Sho Takamori[1],
Kenji Nakao[2], and Shoji Takeuchi[1,3]

[1]Kanagawa Institute of Industrial Science and Technology (KISTEC),
[2]Maqsys Inc., and [3]The University of Tokyo, JAPAN

ABSTRACT

This paper reports a tracing device for the simultaneous formation of planar lipid bilayers on an arrayed double-well chip and repetitive measurements of ion channel activities. Artificial membrane systems are excellent platforms for studying ion channels at the single-molecule level. However, since single-molecule measurements can detect differences in the activity of individual ion channel molecules, a large amount of data is required to converge the measurement results. Here we develop a device with small teeth that move up and down to trace micro holes for forming membranes and detecting ion channel activities repetitively. The device is demonstrated using a 16-channel arrayed double-well chip, realizing efficient membrane formation and repetitive detection of ion channel signals per unit time.

KEYWORDS

Planar lipid bilayer, Artificial membrane, Ion channel activity measurement

INTRODUCTION

Ion channels are membrane proteins and play essential roles in the membranes of both cells and organelles. They form tiny pores in the membranes and transport specific ions selectively in the passive mode, according to the electrochemical potential gradient across the membrane. In living organisms, ion channels regulate ion transport from one compartment to another, separated by the membrane, to transmit electrical signals between and within cells. Due to their crucial role in maintaining the biological function, ion channels are considered important potential targets for drug development against various diseases such as neurological, cardiovascular, and cancer. There are at least 400 ion channels in humans, and a survey of human genes relevant to drug discovery shows that the ion channels account for 19% of important drug targets in human gene families [1].

Electrophysiological techniques can be applied for drug screening of ion channels. The artificial membrane systems, one such approach, are excellent platforms to readily detect the activity of ion channels at the single-molecule level [2,3]. Therefore, artificial membrane systems are expected to be used to screen drug candidates and characterize interactions between the candidates and ion channels. However, due to high sensitivity enough to reflect the activity of individual molecules in the measurement results, numerous measurements are necessary to converge the data. In this regard, conventional methods have limitations. To solve the problem, the measurement throughput needs to be significantly improved.

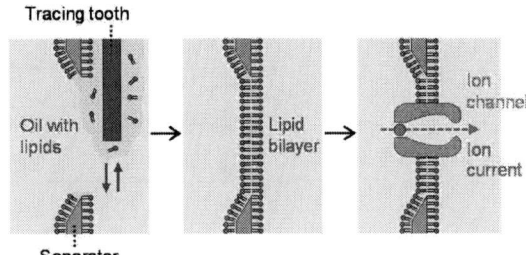

Figure 1: Conceptual diagram of membrane formation and ion channel activity detection using a tracing device.

In this work, we design and fabricate a tracing device aiming for simultaneous membrane formation on an arrayed double-well chip and repetitive measurements of ion channel activities. We apply the idea of using small plates to trace micro holes for forming membranes to the 16-ch arrayed double-well chip reported previously [2,3], verifying both effectiveness in forming membranes and measuring ion channel activities per unit time.

EXPERIMENTAL

Design and fabrication of the tracing device

Fig. 1 illustrates a conceptual diagram of how the tracing device works. The developed device consists of two parts: an arrayed double-well chip and a tracing top. The chip accommodates 16 double wells arranged in the shape of Θ. A pair of Ag/AgCl electrodes are embedded in the bottom of the wells to detect ion currents. The wells with working and reference electrodes are called R (recording) and G (ground) wells, respectively. A separator (0.3 mm thick) with a micro hole (0.7 mm diameter) is placed in the center of the double well to serve as a scaffold for forming an artificial membrane. The micro hole is tapered, and its diameter decreases from the front side (1.1 mm) to the back side (0.7 mm). The back side of the separator is placed facing the R well and fixed with an adhesive.

The tracing top has the same number of small plates (4.2 mm long, 1.5 mm wide, and 0.4 mm deep) as double wells. The small plates, which we called tracing teeth, were designed to be inserted into the position and depth of the well on the R side.

A spring-based mechanism is used to move the trace top up and down repetitively. The arrayed double-well chip has pillars that install springs and serve as guide posts for the up-and-down motion of the tracing top. By pushing the trace top with a finger, the tracing teeth descend to the micro hole of the separator, and then the teeth return thanks to the springs when the finger releases, resulting in the formation of membranes. Since the back side of the separators is flat around the micro holes, the distance between the holes and the tracing teeth is kept constant.

The arrayed double-well chip, the separators, and the tracing top were all fabricated with acrylic plates using an NC micro-milling machine. To prevent depletion of the lipid-dispersed oil in the wells during use, the tracing top was hydrophilized with oxygen plasma to reduce oil adsorption.

Repetitive measurement of ion channel activity

The measurement of ion channel activities using the tracing device is done as follows. First, an arrayed double-well chip is connected to a 16-ch patch clamp amplifier via an in-house-made mounter. Next, lipid-dispersed oil and buffer solution are dispensed to the double wells. The springs are installed to the guide posts, and then the tracing top is attached to the chip. Finally, ion current recording is started after pushing the tracing top to form membranes simultaneously on the chip, followed by adding aliquots of ion channel samples.

RESULTS AND DISCUSSION

To evaluate the performance of the tracing device, we measured the activity of porcine potassium ion channels (Fig. 2). Recordings were performed for 30 minutes using a 16-channel double-well chip. Membrane formation and ion channel activity measurement were repeated by pushing the tracing top with a finger approximately every 3 min to move the tracing teeth up and down along the separators. We examined the relationship between the distance from the tracing teeth to the separators, the membrane retention time, and the number of ion channel signals acquired (Fig. 3). The tracing devices with the tracing teeth with 0.05 mm distance to the separators showed about six times longer membrane retention time and nearly three times more signals per unit time, compared to the tracing teeth with a distance of 0.2 mm.

CONCLUSION

In this study, we developed the tracing device and verified its performance using the 16-ch arrayed double-well chip. We were able to find effective distances between the tracing teeth and the separators to obtain sufficient membrane retention time and ion channel signals per unit time. We believe the developed methodology can be applied to an arrayed chip with more double wells, dramatically increasing the throughput of measuring ion channel activities at the single-molecule level using artificial cell membranes.

ACKNOWLEDGEMENTS

This work was partly supported by JST START Grant (JPMJST1181), JSPS KAKENHI Grant Number JP17H02758 (T.O.), and NEDO Entrepreneurs Program (NEP).

REFERENCES

[1] R. Santos, et al., "A comprehensive map of molecular drug targets." *Nat. Rev. Drug Discov.*, 16, 19-34, 2017.

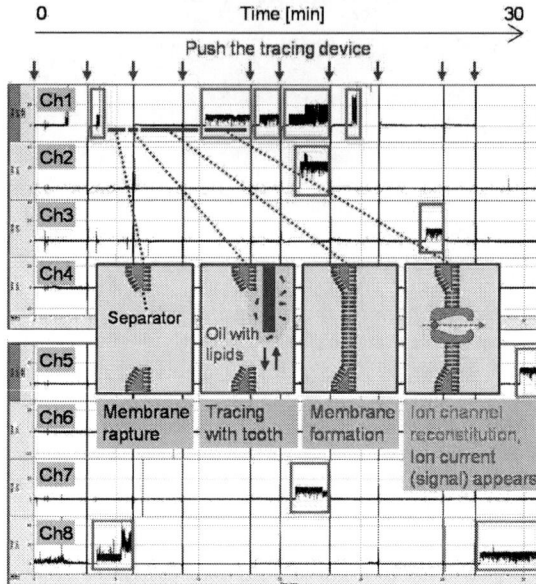

Figure 2: Representatives of ion channel currents measured simultaneously using a tracing device with a 16-ch arrayed double-well chip. Eight of the sixteen recordings are shown here. Ion channel currents are boxed in pink. The purple arrows indicate the points of the tracing top pushed, generating noises shown as vertical lines during the measurement. The cartoons illustrate the membrane situations assumed from the recorded signals.

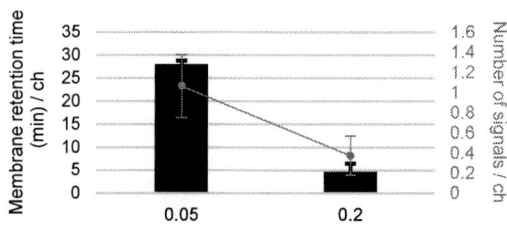

Figure 3: Comparison of the performance of tracing devices with different distances from the tracing teeth to the separators. Mean values and standard errors of 16 double wells for 30 min measurement are shown.

[2] R. Kawano, Y. Tsuji, T. Osaki, K. Kamiya, M. Hirano, T. Ide, N. Miki, S. Takeuchi, "Automated parallel recordings of topologically identified single ion channels", *Sci. Rep.*, 3, 1995, 2013.

[3] K. Kamiya, T. Osaki, K. Nakao, R. Kawano, S. Fujii, N. Misawa, M. Hayakawa, S. Takeuchi, "Electrophysiological measurement of ion channels on plasma/organelle membranes using an on-chip lipid bilayer system", *Sci. Rep.*, 8, 17498, 2018.

CONTACT

*Hisatoshi Mimura, tel: +81-44-819-2037; h-mimura@iis.u-tokyo.ac.jp

AN IMPLANTABLE DIFFERENTIAL SENSOR WITH PASSIVE WIRELESS INTERROGATION FOR IN-SITU EARLY DETECTION OF PERIPROSTHETIC JOINT INFECTION

Jiaxin Jiang[1]*, Cole Napier[1], Chandrashekhar Choudhary[1], H. Claude Sagi[2],
Chia-Ying Lin[2], Michael T. Archdeacon[2], and Tao Li[1]#

[1]Department of Electrical and Computer Engineering, University of Cincinnati, USA
[2]Department of Orthopaedic Surgery, University of Cincinnati, USA

ABSTRACT

This paper presents an implantable, passive wireless biosensor for integration with knee joint prostheses to facilitate *in situ* early detection of periprosthetic joint infection (PJI). The magnetoelastic (ME) transduction mechanism of the sensor enables wireless interrogation without the need of an antenna or battery. Antibodies are immobilized on the sensor to target specific types of bacteria of interest. A novel differential configuration of the ME sensor is introduced to distinguish the response to target bacteria from variations in properties of the surrounding fluid medium. A triangular sensor geometry is used to help enhance mass sensitivity. Differential sensor operation was successfully verified *in vitro* for bacterial detection in *E. coli* suspensions with different viscosities ranging between 1-5.9 cP, demonstrating effective elimination of the effect of the fluid medium.

KEYWORDS

Magnetoelastic sensor, implantable device, resonant, differential, bacterial detection, mass loading, viscosity

INTRODUCTION

Total knee arthroplasty (TKA), a surgical procedure to treat chronic degenerative conditions of the knee by replacement, is becoming increasingly prevalent due to an aging population; the annual number of patients receiving TKA in the US was ≈1 million in 2020 and is expected to double by 2030 [1,2]. Periprosthetic joint infection (PJI) is a devastating complication of TKA and is typically caused by bacterial contamination [3]. Despite a low but relatively stable incidence rate of ≈2%, PJI can require revision surgeries that are accompanied by substantial functional and socioeconomic burden for both the patient and society. This is further aggravated with the rapidly increasing number of TKA cases; with an average hospital cost of over $28k per PJI case and estimated 40k cases per year, the annual financial impact is estimated to exceed $1.1 billion by 2030 in the US alone [4].

No unanimously accepted approach has been established to date for the diagnosis of PJI. Abnormal concentrations of serum biomarkers [5] and reduction in viscosity of synovial fluids [6] have both been reported as criteria to evaluate the infection progression; however, the time required for these factors to reach levels sufficient for diagnosis can lead to further accumulation of pathogen and worsening infection. A wireless implantable biosensor that can enable real-time, *in situ* detection of target bacteria during early stages of colonization and infection (e.g. 48-72 hours after surgery) is thus highly desirable. Limited research has been reported in this area. Acrylamide hydrogels have been investigated for antigen sensitivity, intended to be used in a resonating LC wireless sensor for early detection of post-surgical infection [7].

Magnetoelastic (ME) sensors have been used in a wide range of applications such as the measurements of temperature [8], pressure [9], stress/strain [10], pH [11], metal ion concentration [12], cell growth [13], and concentrations of pathogen [14] and biomarkers [15]. The inherent passive wireless capability of ME sensors makes them highly desirable for implantable applications without the need for an antenna or local power source, such as their use with orthopedic implants to monitor potential structural failures [16] and with biliary stents to minotor sludge accumulation and restenosis [17].

This paper reports an implantable ME biosensor for integration with a knee arthroplasty prosthesis to facilitate *in situ* early detection of PJI [18]. Functionalization of sensor surfaces with selected antibodies allows the sensor to target specific types of bacteria of interest. A novel differential sensor configuration is introduced, which is essential to distinguish the effect of target bacteria from variations caused by surrounding fluid medium. Optimizations of sensor geometry, differential configuration parameters, and functionalization method enabled successful *in vitro* demonstration of bacteria detection while eliminating effects of medium variations.

DEVICE DESIGN

Figure 1 illustrates the concept of the ME biosensor for *in situ* early detection of PJI. The wireless biosensor functionalized with antibodies targeting specific types of bacteria is mounted in a biocompatible package for

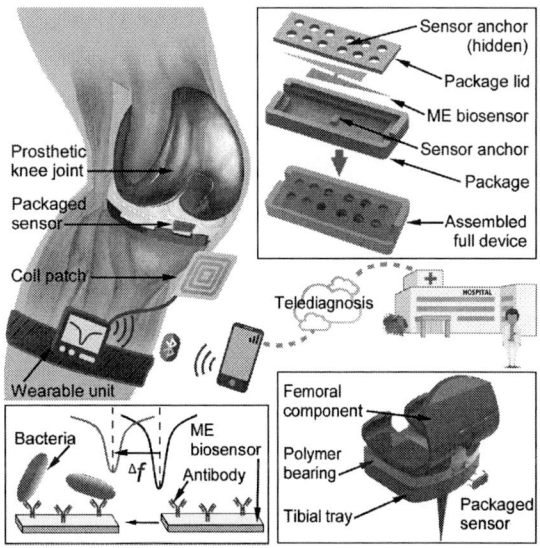

Figure 1: Device concept diagram and application scenario of wireless biosensor for early detection of PJI.

integration into a recess on a prosthetic knee joint. The package has anchors to suspend the sensor inside, preventing physical interference from tissue surrounding the implant area, while the perforations on the package lid allow exchange of fluid. An external coil patch can be attached on the skin near the ME sensor and connected to a wearable unit to wirelessly interrogate the sensor. The wearable unit can be further connected wirelessly with a smartphone, allowing remote access and telediagnosis through a cellular link.

The working principle of the ME biosensor is illustrated in Fig. 2. When the ME sensor is excited by a time-varying magnetic field generated from a transmit coil, it produces a longitudinal vibration. This generates a magnetic flux with a resonance frequency, which can vary with changes in the boundary conditions of the sensor such as the mass and fluid medium in contact with the sensor. When a small mass, Δm, is applied to the ME sensor of an initial mass M, the resonance frequency shift, Δf, can be derived from equations given in [19] as

$$\Delta f = -\frac{\Delta m}{4LM}\sqrt{\frac{E}{\rho(1-v^2)}} \qquad (1)$$

where L is the length of the sensor; E, ρ, and v are Young's modulus, density, and Poisson's ratio for the ME material, respectively. When the properties of the fluid medium change, the corresponding Δf is given as [14]

$$\Delta f = -\frac{\sqrt{\pi f_0}}{2\pi\rho d}\sqrt{\eta\rho_m} \qquad (2)$$

where f_0 is the resonance frequency in air, ρ and d are the density and thickness of the ME sensor, and ρ_m and η are the density and viscosity of the medium, respectively.

ME sensors have shown high performance as wireless immunosensors for *in vitro* applications in liquid media [12-15]. The medium properties and conditions such as viscosity, density, temperature, and pH can cause significant changes in the resonance frequency of the ME immunosensors. Therefore, these parameters are usually carefully controlled to maintain sensor performance under *in vitro* conditions. However, the properties of the surrounding body fluid, particularly its viscosity and density, can change at any time in the *in vivo* environment; the effect of such changes on the sensor resonance frequency must be eliminated to provide meaningful sensor outputs. To achieve this, a novel differential sensor consisting of two ME elements is proposed, one with and the other without functionalization. The element without

functionalization acts as a reference. When the target bacteria are present in the medium and bind to the antibodies, mass loading is applied only to the functionalized sensing element. Any common mode changes between the sensing and reference elements, such as those from changes in medium properties, are eliminated by subtracting the outputs of the two elements.

A triangular geometry, instead of the traditional rectangular geometry commonly used for ME immunosensors, is chosen to provide a higher mass sensitivity [18]. Effort has gone into the design and optimization of the differential sensor configuration. Particularly, to minimize the magnetic coupling and thus the interference between the two elements, a separation gap g and a length difference ΔL between the two elements are critical parameters of the differential design. Figure 3 shows the geometric design of the differential ME sensor and the simulation results obtained using COMSOL® Multiphysics. A design with ΔL=0.6 mm and g=1.5 mm was found to provide adequately low coupling of magnetic flux between the two elements, and was used for implementation in this work. The anchor of each element is placed near the null point of vibration of the geometry.

DEVICE FABRICATION

The ME differential sensors were fabricated from ribbons of Metglas® 2826MB ($Fe_{45}Ni_{45}Mo_7B_3$) alloy from Metglas Inc. using an in-house high-precision micro-electro-discharge machine (Smaltec® EM203 µEDM). The sensing element was coated on one side with a Cr/Au layer (40/60 nm thick) by e-beam evaporation while the reference element was protected from deposition with photoresist. The Au surface provides a critical, biocompatible layer for antibody immobilization while Cr serves as an adhesion layer.

Surface functionalization included forming a self-assembled monolayer (SAM) on the Au surface using cysteamine (CYSTE), immobilizing protein G on the SAM via the N-(3-Dimethylaminopropyl)-N'-ethylcarbodiimide (EDC) and N-hydroxysulfosuccinimide (Sulfo-NHS) protocol, incubating antibodies to bind with protein G on the sensors, and finally treating with bovine serum albumin (BSA) to block non-specific binding sites. Protein G is used as a linker molecule to achieve orthogonal antibody immobilization, allowing higher antigen capture rates and greater antibody binding density [20]. For this work, lyophilized cells of strain K12 *E. coli*, Sulfo-NHS, protein G and BSA were purchased from Sigma Aldrich; CYSTE (98%), phosphate buffered saline (PBS, pH 7.4), rabbit anti-*E. coli* polyclonal antibody, goat anti-rabbit antibody

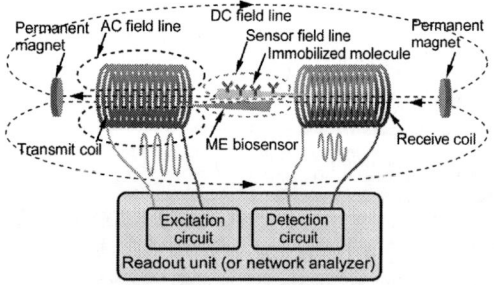

Figure 2: Schematic diagram showing wireless interrogation method for readout of the passive differential ME sensor for bacterial detection. A network analyzer was used as the readout unit for preliminary experiments.

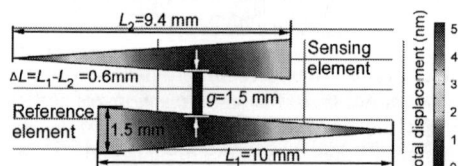

Figure 3: COMSOL® Multiphysics simulation of differential ME sensor. The selected gap and length difference between the two elements were g=1.5mm and ΔL=0.6mm, respectively. Smaller displacement observed for sensing element due to mass loading as expected.

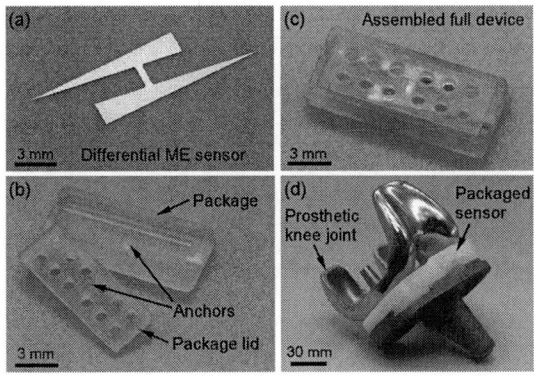

Figure 4: (a) Differential ME sensor (Metglas 2826MB); (b) package and lid with anchors (3D printed from VisiJet M3 resin); (c) fully assembled device; (d) packaged device attached to prosthetic knee joint for demonstration of an intended integration site.

conjugated with Alexa Fluo® 488, and EDC were acquired from Fisher Scientific.

The sensor functionalization procedure began with thorough cleaning of the ME sensors in an ultrasonic cleaner using acetone, isopropanol, and DI water, sequentially. The sensors were then immersed in a CYSTE solution (10 mM) for 16 h to deposit the SAM. Protein G (2 µg/mL) was activated in a solution containing EDC (0.01 mM) and Sulfo-NHS (0.02 mM) for 1 h at 37°C. After rinsing with PBS, the ME sensors were soaked in an activated protein G solution for 2 h at 37°C. Another PBS rinsing step was done to remove loosely-bonded protein G. This was followed by incubating the sensors in an activated antibody solution for 2 h at 37°C to immobilize the anti-*E. coli* antibodies. Loosely-bonded antibodies were removed by another PBS rinsing. The sensors were then treated with a 1% w/w BSA solution for 30 min, rinsed with PBS and dried under a nitrogen stream to become ready for testing.

The sensor package with the perforated lid was 3D printed from a biocompatible resin (VisiJet® M3). Two anchors in the package are used to clamp the joint area of the differential sensor and suspend it in the package for free vibration (Fig. 4). The packaged sensor can then be integrated in a recess on the prosthetic knee joint.

EXPERIMENTAL RESULTS

Imaging techniques including scanning electron microscopy (SEM), atomic force microscopy (AFM) and fluorescence microscopy were used to evaluate the performance of surface functionalization with or without the use of protein G. Fluorescence images obtained using secondary antibodies (goat anti-rabbit antibody conjugated with Alexa Fluor 488, 4 µg/mL) with the Olympus IX81 microscope verified an improved antibody coverage when protein G was used (Fig. 5g-h). SEM images showed higher antibody binding density with the use of protein G (Fig. 5e-f). AFM images showed the different surface roughness with and without protein G (Fig. 5c-d).

All sensor experiments described below were performed using a 50-turn coil of 3/4-inch diameter made from 28 AWG magnet wire. The coil was connected to a network analyzer (Keysight® E5061B) for extraction of the

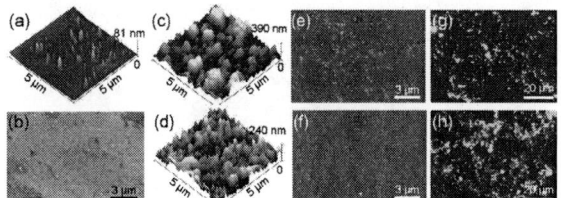

Figure 5: ME biosensor functionalization. (a) AFM and (b) SEM images of bare gold surface; (c) AFM, (e) SEM and (g) fluorescence images after direct antibody immobilization without protein G; (d) AFM, (f) SEM and (h) fluorescence images after antibody immobilization with protein G, showing improvements.

resonance frequencies of the ME sensors.

The feasibility of the differential sensor for *in vitro* bacterial detection was validated experimentally. Functionalized sensors were first tested in 1 mL PBS solution for 30 min as the control. Then 10 µL *E. coli* suspension was added to the PBS solution and the test continued for another 60 min. The concentration of *E. coli* in the solution was 5×10^6 cfu/mL as determined by plate count. The test was repeated four times using different sensors with results averaged (N=4) and shown in Fig. 6. During the control period in PBS, the responses of both sensing and reference elements were stable at about zero. After the bacteria were added, the sensing element showed increasing Δf magnitude due to *E. coli* binding, while the Δf of the reference element had a slight increase and then remained stable. The slight increase may be caused by minor change in the viscosity of the solution when the bacteria suspension was added.

The capability of the differential sensor to eliminate the effect of varying medium properties were also verified by *in vitro* tests. The sensors were first tested in PBS solution for 20 min as the control, and then sequentially in three *E. coli* suspensions with different medium for 20 min each to emulate the impact of varying medium properties. The three *E. coli* suspensions were made with PBS as well as 40% and 50% glycerol/water solutions, resulting in viscosity of 1 cp, 3.8 cp and 5.9 cp, respectively as determined by NDJ-5S rotational viscometer. As shown in Fig. 7(a), a large Δf corresponding to the change in medium properties can lead to significant error when interpreting the sensor response attributable to mass loading from bacteria binding. After differential correction with an algorithm, the corrected Δf of the reference

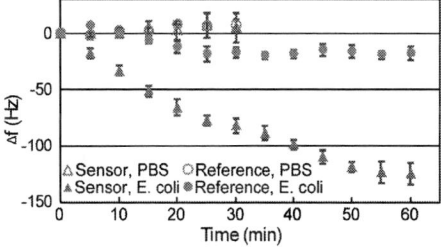

Figure 6: In vitro test results of differential sensor for E. coli detection. Data averaged from readings of 4 sensors (N=4). Error bar shows standard error. The sensing element had increasing Δf magnitude with E. coli binding, while Δf of reference remained close to 0.

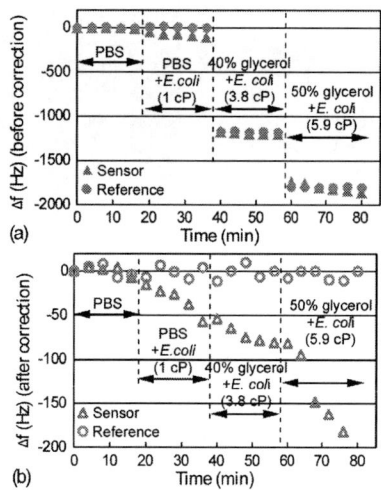

Figure 7: In vitro test results of differential sensors (g=1.5mm, ΔL=0.6mm) in E. coli suspensions with different viscosities (1-5.9 cP). The sensor was tested in PBS solution for 20 min, and then 3 types of E. coli suspensions (5×10⁶ cfu/mL) for 20 min each. (a) Raw data of measured Δf before differential correction, showing large Δf for both sensing and reference elements due to changes in the medium properties. (b) Δf after differential correction with an algorithm, showing corrected Δf of the reference element stayed close to 0 while that of the sensing element gradually increased as expected, demonstrating effective elimination of medium effect using the differential operation.

element stayed close to zero while the corrected Δf of the sensing element gradually increased in magnitude as expected. This validated the differential sensor mechanism by demonstrating effective elimination of medium effect.

CONCLUSIONS

An implantable differential biosensor with passive wireless interrogation capability has been designed to facilitate *in situ* early detection of PJI. Functionalized with selected antibodies, the sensor can target specific types of bacteria known to be present during the early stages of PJI. Improved performance of functionalization was obtained by using protein G. Wireless differential operation of the sensor for bacterial detection was successfully tested *in vitro*, in PBS and in *E.coli* suspensions with viscosity ranging between 1-5.9 cP, demonstrating effective elimination of the medium effect. Future works will focus on further performance improvements for the sensor, the package and the interrogation unit, as well as the evaluation of sensor characteristics and safety for the target *in vivo* environment.

ACKNOWLEDGMENTS

This work was supported in part by internal funding at the University of Cincinnati. Facilities used include the Mantei Center Cleanroom at the University of Cincinnati.

REFERENCES

[1] E.C. Rodríguez-Merchán and S. Oussedik, eds., *Total Knee Arthroplasty: A Comprehensive Guide*, Springer, 2015.

[2] J.A. Singh, *et. al.*, "Rates of total joint replacement in the united states: future projections to 2020-2040 using the national inpatient sample," *J. Rheumatology*, 46(9), 2019.

[3] J.L. Del Pozo and R. Patel, "Infection associated with prosthetic joints," *N. Engl. J. Med.*, 361(8), 787-94, 2009.

[4] A. Premkumar, *et. al.*, "Projected economic burden of periprosthetic joint infection of the hip and knee in the United States," *J. Anthroplasty*, 36(5), 2021.

[5] J. Parvizi, *et. al.*, "The 2018 definition of periprosthetic hip and knee infection: an evidence-based and validated criteria," *J. Arthroplasty*, 33(5), pp. 1309-14.e2, 2018.

[6] J. Fu, *et. al.*, "Synovial fluid viscosity test is promising for the diagnosis of periprosthetic joint infection," *J. Anthroplasty*, 34(6), pp. 1197-1200, 2019.

[7] E.A. Capogna, D.J. Munoz-Pinto, M.S. Hahn, and E.H. Ledet, "Identifying a crosslink density relationship of acrylamide hydrogels to be used with passive sensors for detecting periprosthetic infection," *41st Annual Northeast Biomedical Engineering Conference*, Troy, NY, 2015.

[8] C. Mungle, *et. al.*, "Magnetic field tuning of the frequency-temperature response of a magnetoelastic sensor," *Sensors and Actuators A: Physical*, 101(30), pp. 143-149, 2002.

[9] M.K. Jain and C.A. Grimes, "A wireless magnetoelastic micro-sensor array for simultaneous measurement of temperature and pressure," *IEEE Trans. Magn.*, 37, 2001.

[10] L. Liu, *et. al.*, "Stress monitoring of prestressed steel strand based on magnetoelastic effect under weak magnetic field considering material strain," *Progress in Electromagnetics Research C*, 104, pp. 157-170, 2020.

[11] K.G. Ong, E.L. Tan, C.A. Grimes, and R. Shao, "Removal of temperature and earth's field effects of a magnetoelastic pH sensor," *IEEE Sensors J.*, 8(4), pp. 341-6, 2008.

[12] X. Guo, *et. al.*, "A bovine serum albumin-coated magnetoelastic biosensor for the wireless detection of heavy metal ions," *Sensors and Actuators B: Chemical*, 256, pp. 318-324, 2018.

[13] S. Shekhar, S.S. Karipott, R.E. Guldberg, and K.G. Ong, "Magnetoelastic sensors for real-time tracking of cell growth," *Biotechnol. Bioeng.*, 118(6), pp. 2380-85, 2021.

[14] P. Pang, et. al., "Detection of Mycobacterium tuberculosis in sputum sample based on a wireless magnetoelastic-sensing device," *Talanta*, 76, pp. 360-4, 2008.

[15] R. Liu, *et. al.*, "High sensitivity detection of human serum albumin using a novel magnetoelastic immunosensor," *Journal of Materials Science*, 54, pp. 9679-88, 2019.

[16] D.E. Mouzakis, *et. al.*, "Contact-free magnetoelastic smart microsensors with stochastic noise filtering for diagnosing orthopedic implant failures," *IEEE Transactions on Industrial Electronics*, 56(4), pp. 1092-1100, 2009.

[17] S.R. Green, R.S. Kwon, G.H. Elta, and Y.B. Gianchandani, "In vivo and in situ evaluation of a wireless magnetoelastic sensor array for plastic biliary stent monitoring," *Biomed. Microdevices*, 15, pp. 509-517, 2013.

[18] J. Jiang, K. Sureshkumar, C. Choudhary, T. Kerkes, H.C. Sagi, C.-Y. Lin, M.T. Archdeacon, and T. Li, "A passive wireless differential sensor for in-situ early detection of periprosthetic joint infection," *Tech. Digest Hilton Head 2022 Workshop*, Jun. 2022, pp. 5-8.

[19] W. Shen, *et. al.*, "Design and characterization of a magnetoelastic sensor for the detection of biological agents," *J. Phys. D: Applied Physics*, 43(1), 015004, 2009.

[20] C. Menti, *et. al.*, "Influence of antibody immobilization strategies on the analytical performance of a magneto-elastic immunosensor for Staphylococcus aureus detection," *Mat. Sci. and Eng: C*, 76, pp. 1232-39, 2017.

CONTACT

*J. Jiang, tel: +1(513)208-3047; jiangj8@mail.uc.edu

#T. Li, tel: +1(513)556-3508; litao@uc.edu

MICROMACHINED PIEZOELECTRIC FILM-BASED FLEXIBLE ELECTRONICS WITH INTEGRATION OF FILM-SELF TEMPERATURE-DETECTING BREATH SENSOR AND ACETONE GAS SENSOR

Hung-Yu Yeh[1], and Guo-Hua Feng[1,2]

[1] Department of Power Mechanical Engineering, National Tsing Hua University, Hsinchu, TAIWAN
[2] Institute of Nano Engineering & MicroSystems, National Tsing Hua University, Hsinchu, TAIWAN

ABSTRACT

The state-of-the-art wearable micromachined flexible electronics consisting of the breath sensor with self-temperature monitoring and acetone gas sensor are presented (Fig. 1). The breath sensor constructed by a freestanding patterned titanium film coated with 1 µm-thick piezoelectric $BaTiO_3$ film possessing two novelties: (1) the stiffness designable triangular cantilever with uniform stress distribution when utilizing its piezoelectric property for airflow sensing. (2) The fixed film region utilizing its negative temperature coefficient (NTC) thermistor effect as a temperature sensor. We believe this simple-structured breath sensor is reported for the first time worldwide. The film-based breath sensor is integrated with the charge amplifier circuit implemented on a polyimide film as wearing flexible electronic device. This device also integrated with the acetone gas sensor, possessing high sensitivity at room temperature, based on tin dioxide (SnO_2) reduced graphene oxide (RGO) composite film.

KEYWORDS

Piezoelectric, wearable device, breath sensor, barium titanate, thermistor, acetone.

INTRODUCTION

With the continuous deterioration of air quality, the number of people suffering from respiratory diseases is growing rapidly. In order to detect the health of the respiratory system at any time, the convenient wearable breathing detection devices become more and more important. Wearable breathing detection devices could detect whether breathing is abnormal at any time and get proper treatment immediately. In addition, the advantage of low power consumption of the wearable device makes it able to work for a long time [1].

Asthma and chronic obstructive pulmonary disease (COPD) are currently the most common respiratory diseases. Many studies have pointed out that airway obstruction disease can lead to poor breathing. Respiratory rate and the tidal volume will be key indicators to determine whether people get disease [2]. It has also been confirmed that most of airway obstruction diseases will be accompanied by symptoms of airway inflammation which will also slightly affect the temperature of the air exhaled. Therefore, in addition to using the respiratory rate and tidal volume as the crucial parameters for judging respiratory diseases, it is also possible to measure changes in the temperature of the breathing air. This could help to increase the accuracy of diagnosing respiratory disease [3].

In the environment where people live, the air may consist of harmful chemical substances at any time. These harmful substances cause considerable deterioration to

health. However, in addition to being exposed to these dangerous situations without consciousness, we could use gas sensors to determine whether the air contains unsafe constituents. The gas sensors detect the content of volatile organic compounds (VOC) in the air exhaled or inhaled by human such as acetone, alcohol, ammonium and carbon monoxide. Studies have pointed out that patients suffering from lung cancer or diabetes will have a higher concentration of acetone in the exhaled air than the normal range of 0.3-0.9 ppm of the healthy people. Therefore, if the proper gas sensors are integrated into the breath sensing device, better monitoring human health condition could be achieved [4].

Here, the proposed device not only consists of the piezoelectric cantilever beams to sense breathing rate but a thermistor element located around the cantilever beams to simultaneously measure the breathing temperature. This enhances the reliability of judging breathing health. The acetone gas sensing element is also integrated so the device can analyze the breath gas composition in addition to monitoring the respiratory flow pattern. At present, several wearable devices based on different sensing technologies have been developed, but not all of them are completely suitable for human usage. In order to be worn for a long time, the wearable device should measure the signal correctly and has minimal interference with normal human breathing activity [5]. Our device design is ergonomic so the wearer can wear it for a long time without discomfort. The most important thing is that the device greatly reduces the obstruction to the airway and detects human breathing signal more effective.

DEVICE DESIGN AND FABRICATION

This wearable device integrated the breathing, temperature and acetone sensors and the charge amplifier circuit on a polyimide film. The whole electronic thin film device was flexible and easy to attach on nonplanar surfaces of targets. Figure 2 shows details of the flexible electronic device. Figure 2(a) and (b) demonstrated how the sensors and circuit were attached to the polyimide film, and show the flexibility of the functional thin film element. Figure 2(c) displayed the inner part of the complete device which assembled the functional thin film element into the 3D printed shell. It also confirmed that the flexible film element can be applied to curved surfaces of the 3D printed shell to benefit the housing fabrication for well contacting human skin. To make people wear it comfortably and firmly, we also employed the silicone material to fabricate nasal plug (Fig.2(d)). The demonstration of the subject wearing the complete device is shown in figure 2(e). The maximum size of the complete device is about 6 cm, suitable for most people usage.

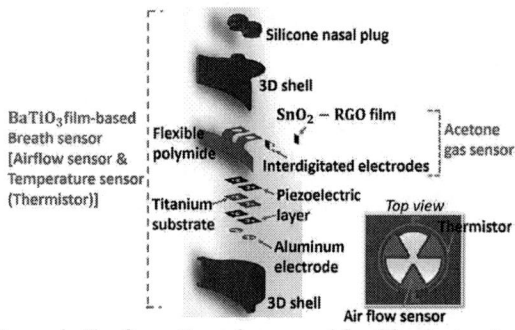

Figure 1: Configuration of proposed flexible electronics.

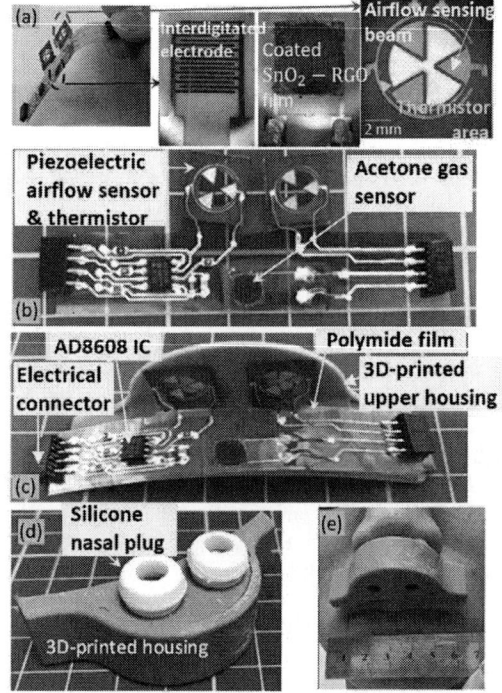

Figure 2: Fabrication results: (a) The core part (airflow, temperature, acetone gas sensors) of the developed flexible electronic device. (b) Top view of the core part. (c) The core part integrated with the 3D printed housing. (d) Finished device with nasal plugs. (e) A subject wearing the device ready for testing.

Breathing and temperature sensors design

The breathing and temperature sensing elements both utilized barium titanate as the main material. The breathing sensing element used the piezoelectric property to measure the respiratory flow signal, while the temperature sensing part employed the thermistor property of barium titanate to measure the temperature. The breathing element part was designed as a circular array of three piezoelectric cantilever beams in the airway. The beams performed bending motions to output piezoelectric signals when the breathing air passed through. Beyond previous work [6-8], the newly-designed triangular shape beams with larger area and uniform strain distribution, which could improve the signal amplitude and sensitivity [9]. The temperature sensing element was designed to surround the fixed region of the cantilever beams. This configuration not only unaffected by the beam deformation, but also instantly

detected the temperature change of the air passing through.

Barium titanate film fabrication

We started with a 10 μm-thick titanium foil as a substrate. After the photolithography process to pattern the substrate as the designed shape, we executed two times of hydrothermal processes to grow the $BaTiO_3$ layer on the surface of the patterned substrate [10]. The first hydrothermal process was to grow a layer of rutile titanium dioxide layer on the surface of titanium foil.

We immersed the titanium foil in a mixed solution of titanium trichloride and hydrochloric acid, then put it into an autoclave, slowly heated it to 180 °C. The pressure was kept about 10 kg/cm² and continued for 6 hours before taking it out. After removing the foil with rutile titanium dioxide layer, we immersed it in the second autoclave containing a mixture of barium chloride and sodium hydroxide, then slowly heated to 225 °C. The pressure is set about 26 kg/cm², maintained for 72 hours. It was followed by taking out from the autoclave to obtain the element with $BaTiO_3$ piezoelectric layer on its surface. This element with the barium titanate piezoelectric layer was attached to the glass plate, and then the designed shadow mask was covered on it. The aluminum metal electrode layer was sputtered on the surface of the piezoelectric layer.

Acetone gas sensor design and fabrication

For fabricating the acetone gas sensor, we patterned 8 pairs of interdigitated electrodes with a width and a spacing of 150 μm on the polyimide film surface. Then a layer of tin dioxide ($SnO2$)-reduced graphene oxide (RGO) hybrid composite film was covered onto the electrode region as the acetone gas sensing material.

The fabrication process of the sensing material was divided into two parts. We started with the fabrication of interdigital electrodes. First, a very thin layer of chromium is sputtered on the polyimide film as an adhesive layer, and then a layer of aluminum metal was sputtered on top of it. Finally, the metal layer was etched to form the designed interdigital electrodes. The second part was the sensing material fabrication. SnO_2-RGO powder was fabricated by adding 0.75 mg RGO and 24 mg $SnCl_4$-$5H_2O$ into 21.5 ml DI water. After the mixture blended with ultrasound assisted vibration for one hour, we poured the mixture into the autoclave and put the autoclave into the oven as well as heated it to 120 °C for 12 hours. After that, the solution was centrifuged at 3500 rpm for 15 minutes. Once the centrifugation procedure was completed, the precipitated material was washed with DI water for 10 times, and then dripped on the interdigitated electrodes with pipette and baked to dry to finish the sensing material fabrication [11].

Integrated device design

The whole flexible thin film element was based on a polyimide film. First, the amplifier circuit was printed on the polyimide film with the Voltera V-One circuit printer (Voltera Inc., Canada). The amplifier circuit here integrated with AD8606. Both the $BaTiO_3$ and the SnO_2-RGO sensors were mounted on the reserved positions of the flexible polyimide film. Finally, the processed polyimide film was attached to the 3D printed shell housing to complete the overall wearable device fabrication.

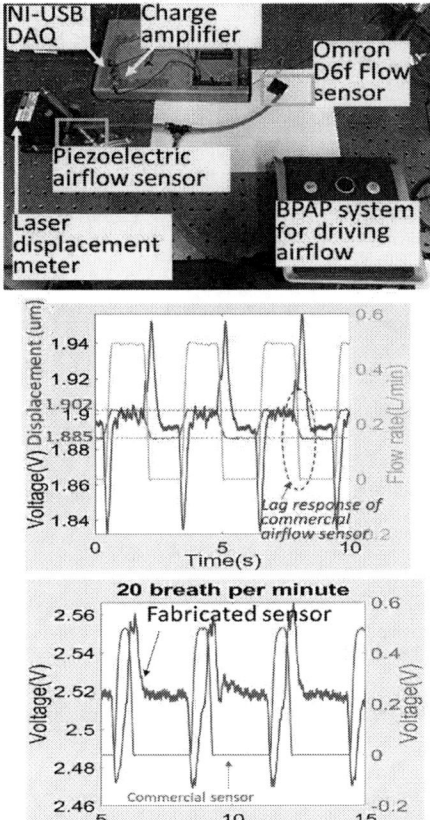

Figure 3: Experimental setup and results of fabricated airflow sensor characterized by commercial sensor and laser displacement meter.

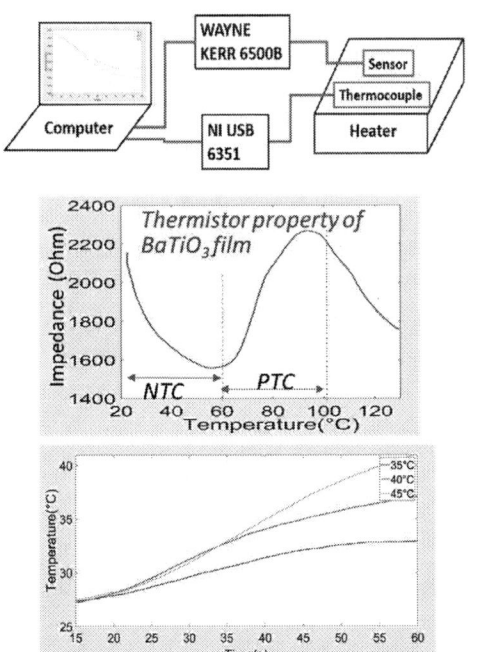

Figure 4: Experimental setup and characterization results of the fabricated barium titanate film element as a thermistor.

EXPERIMENTAL SETUP AND RESULTS

For characterizing the breathing sensing element, we used an OMRON D6f commercial flowmeter. Moreover, a laser displacement meter was applied to measure the actual movement of the cantilever beam at the front end of the breathing element when the air passed through. we use NI-USB DAQ to connect to the computer Labview program to obtain the component signals (Fig. 3). By comparing the signals from the breathing sensing element, the commercial flowmeter and the laser displacement meter at the same time, we found that the signal of commercial flowmeter was slightly behind the signal of the laser displacement meter, while the signal response of our component to the flow change is relatively in time. Further comparing the different parameters of respiratory conditions, such as different respiratory rate, rise time and pressure difference signals of our element and the commercial product, we could observe similar signal trends.

The thermistor element employed the impedance analyzer to measure the change in resistance and a thermocouple to calibrate the temperature. For temperature change within a large range, it could be seen that the element had an NTC effect in the low temperature range and a PTC effect in the high temperature range, which was in line with the properties of the material itself [12]. We also calibrated three different temperature rising conditions. The resulting trend of the resistance change of the thermistor almost matched the temperature measured from the thermocouple (Fig. 4).

The acetone gas sensor was tested by different acetone gas concentrations. In order to precisely control the testing gas environment for measuring the impedance, we built a customized gas test chamber and exploited a pump to provide the air flow pressure. The ratio of the inlet air flow of the acetone gas source to the atmospheric air flow was controlled to obtain the acetone gas concentration settings. Figure 5 shows the results that the fabricated sensor responded to acetone gas instantly. Furthermore, the acetone sensor exhibited a good sensing reliability.

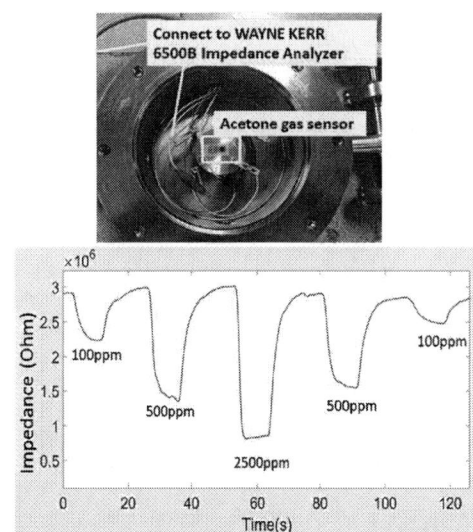

Figure 5: Experimental setup and results of the fabricated acetone gas sensor

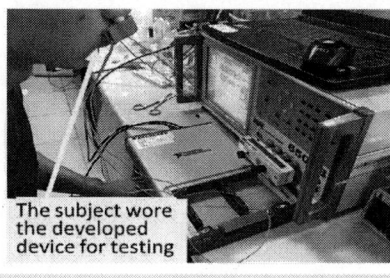

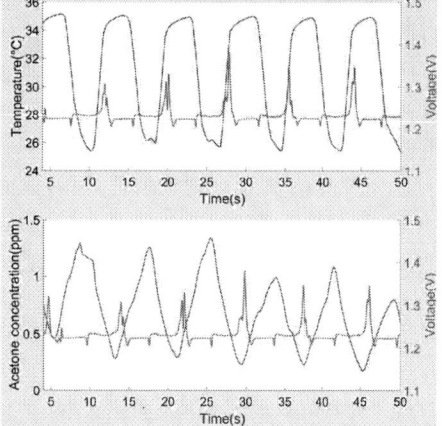

Figure 6: A subject tests the device to acquire airflow, air temp. and acetone gas responses.

We also carried out actual human wearing test (Fig. 6). The results show that the signal of temperature and acetone concentration variations matched the frequency of the breathing signal. The developed device also responded to the breathing movement of the subject and reflected the breathing signal on different breathing patterns.

CONCLUSION

The proposed micromachined flexible electronic device was successfully developed. This device integrated the piezoelectric cantilever beam breathing element, thermistor temperature sensing element, acetone sensor as well as the signal process circuit component on a single polyimide film. Through this functional flexible electronic device, we could acquire the signals easily by just connecting the reserved plug. In breathing testing, the developed element displayed the performance comparable to, or even better than, a commercial flow meter. Moreover, we not only found an excellent thermistor effect of the barium titanite film, but also verified that the constructed thermistor could perfectly respond to different temperature changes in the temperature range close to the human body. The acetone gas sensor made of SnO_2-RGO material possessed a good sensitivity and repeatability for detecting acetone gas concentration. Finally, the human wearing test was performed, the results show that the developed flexible electronic device was indeed a wearable sensing device for human application.

ACKNOWLEDGEMENTS

This work was supported by Ministry of Science and Technology in Taiwan under grant No. 111-2923-E-007 -004-MY2 and No. 109-2221-E-007-111-MY2.

REFERENCES

[1] M. Cuartero, M. Parrilla, G. A. Crespo, "Wearable potentiometric sensors for medical applications", *Sensors*, vol. 19, no. 2, pp. 363, 2019.

[2] E. Tufvesson, E. Nilsson, T. A. Popov, R. Hesselstrand, L. Bjermer, "Fractional exhaled breath temperature in patients with asthma, chronic obstructive pulmonary disease, or systemic sclerosis compared to healthy controls", *European Clinical Respiratory Journal*, vol. 7, no. 1, pp. 1747014, 2020.

[3] P. J. Barnes, B. R. Celli, "Breath acetone as a potential marker in clinical practice", *European Respiratory Journal*, vol. 33, pp. 1165-1185, 2009.

[4] Veronika Ruzsányi, Miklós Péter Kalapos, "Breath acetone as a potential marker in clinical practice", *Journal of Breath Research*, vol. 11, no. 2, pp. 024002, 2017.

[5] D. Naranjo-Hernández, A. Talaminos-Barroso, J. Reina-Tosina, L. M. Roa, G. Barbarov-Rostan, P. Cejudo-Ramos, E. Márquez-Martín, F. Ortega-Ruiz, "Smart Vest for Respiratory Rate Monitoring of COPD Patients Based on Non-Contact Capacitive Sensing", *Sensors*, vol. 18, no. 7, pp. 2144, 2019.

[6] G. H. Feng, P. C. Su, "Wearable Piezoelectric Thin-Film Based Breath Sensing Device with Highly Sensitive Ammonia Detection Ability for Examining Kidney Disease", *in Digest Tech. Papers MEMS 2020 Conference*, Jan. 25-29, 2021, pp. 25-29.

[7] G. H. Feng, P. C. Su, "Barium titanate piezoelectric-film-based beam-array airflow sensor for wearable breath-monitoring application", *Journal of Micromechanics and Microengineering*, vol. 32, no. 1, pp. 015009, 2021.

[8] G. H. Feng, P. C. Su, "Low-Power Wearable Breath Monitoring Device with humidity-and Ammonia-Sensing Functions (May 2022)", *IEEE Sensors Journal*, vol. 22, no. 19, pp. 18295-18305, 2022.

[9] B. Tian, H. Li, N. Yang, H. Liu, Y. Zhao, "A MEMS-based flow sensor with membrane cantilever beam array structure", *Proc. IEEE 12th Int. Conf. Nano/Micro Engi. Mol. Syst. (NEMS)*, pp. 185-189, 2017.

[10] G. H. Feng, K. Y. Lee, "Hydrothermally synthesized PZT film grown in highly concentrated KOH solution with large electromechanical coupling coefficient for resonator", *Royal Society open science*, vol. 4, no. 12, pp. 171363, 2017.

[11] D. Zhang, A. Liu, H. Chang, B. Xia, "Room-temperature high-performance acetone gas sensor based on hydrothermal synthesized SnO_2-reduced graphene oxide hybrid composite Wearable potentiometric sensors for medical applications", *RSC Advances*, vol. 5, pp. 3016-3022, 2015.

[12] V. Shut, S. Kostomarov, A. Gavrilov, "PTCR barium titanate ceramics obtained from oxalate-derived powders with varying crystallinity", *Journal of Materials Science*, vol. 43, pp. 5251-5257, 2008.

CONTACT

*G.-H. Feng, tel: +886-3-571-5131; ghfeng@pme.nthu.edu.tw

FLEXIBLE TACTILE SENSING ARRAY WITH HIGH SPATIAL DENSITY BASED ON PARYLENE MEMS TECHNIQUE

Meixuan Zhang[1], Zetian Wang[1], Han Xu[2], Lang Chen[1], Yufeng Jin[2,3], and Wei Wang[1,3,4]
[1]School of Integrated Circuits, Peking University, Beijing, CHINA
[2]School of Electronic and Computer Engineering, Peking University Shenzhen Graduate School, Shenzhen, CHINA
[3]National Key Lab of Micro/Nano Fabrication Technology, Beijing, CHINA and
[4]Beijing Advanced Innovation Center for Integrated Circuits, Beijing, CHINA

ABSTRACT

This paper reports a novel flexible tactile sensing array with high spatial density, large range and high linearity based on Parylene MEMS technique to mimic the human fingertips' ability of static pressure detection. The proposed array is composed of piezoresistive pressure sensors and Parylene-filled trenches in between for high-resolution sensing and flexible connection separately. Inspired by sensory receptors in human skin, we successfully fabricated a flexible tactile sensing array with high spatial density ($\sim$156 cm^{-2}), large range (0-2 MPa) and low nonlinearity (0.46% FS (full scale)). The excellent compatibility of Parylene MEMS technique to conventional microfabrication makes it possible for future microsystem to achieve both high performance and good flexibility.

KEYWORDS

Flexible tactile sensing array, Parylene-filled trench, Silicon-based flexible electronics, Parylene MEMS

INTRODUCTION

Flexible tactile sensing has shown significant potential in various applications, including electronic skins, human–machine interfaces and wearable health monitoring devices [1-4]. In order to achieve flexible tactile sensors with high performance (high sensitivity, high accuracy, low detection limit, etc.), active materials such as carbon nanotubes (CNTs)[5, 6], graphene[7], conductive polymers[8, 9], metal and semiconductor nanowires[10, 11] have been widely used. However, limited by the processing technologies and the electrical properties of these active materials, a series of problems exist in flexible tactile sensors fabricated by the above sensitive materials, such as poor uniformity, poor linearity, poor stability, large temperature drift and incompatibility with MEMS and CMOS technologies. On the one hand, behavior like poor linearity will increase the constraints in construction of signal processing circuits and restrict the integration capabilities. On the other hand, the incompatibility of the processing technologies will limit its mass production and future applications.

Here, we present a novel flexible tactile sensing array with high spatial density, large range and high linearity based on Parylene MEMS technique, as shown in Figure 1. Inspired by the high-density sensory receptors in human skin (140 cm^{-2} at fingertips)[12], silicon-based piezoresistive pressure sensor is chosen as the sensing unit to mimic the human fingertips' ability of static pressure detection, and Parylene-filled flexible trenches are

introduced between the sensing units to realize the flexibility of rigid devices (Figure 1a-b). The structure of the sensing unit is shown in Figure 1c. Piezoresistors at the edge of the diaphragm are connected by Au wires to form the Wheatstone bridge, which can convert pressure signals to the corresponding electrical signals. This CMOS compatible approach retains the superior properties of silicon-based device and exhibits promising potentials for high-performance flexible electronics technologies.

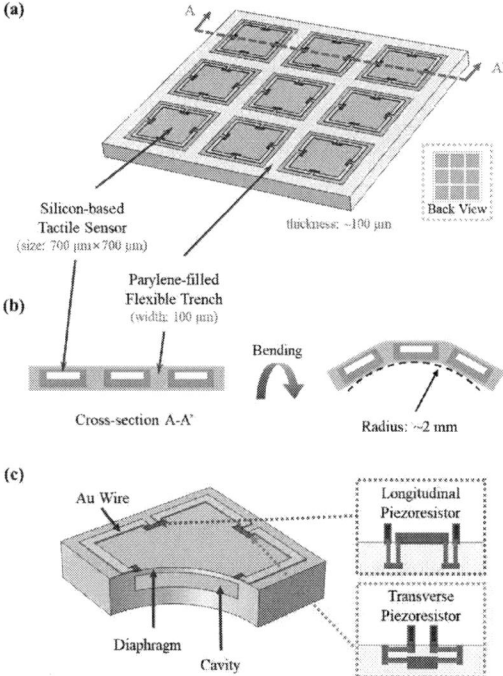

Figure 1: (a) Schematic illustration of the flexible tactile sensing array. (b) Cross-section A-A' and the bending model of the proposed array. (c) Structure of the sensing unit: piezoresistive pressure sensor.

DESIGN AND FABRICATION

Equivalent Young's Modulus Model of the Flexible Array

Parylene-filled flexible trenches are the key structure for the flexibility of the rigid silicon-based array. Therefore, the mechanical property of the proposed flexible array was studied by establishing the equivalent Young's modulus model of the Parylene/silicon composite materials. As illustrated in Figure 2a, the equivalent Young's modulus of the flexible silicon-based array can be calculated as the following Equation (1):

$$E = \left[\frac{E_{pac}E_{Si}}{E_{Si}+(E_{pac}-E_{Si})V_1} - E_{pac} \right] \cdot V_2 + E_{pac} \qquad (1)$$

where E_{pac} (2.8 GPa) and E_{Si} (170 GPa) are the Young's modulus of Parylene C and monocrystalline silicon, $V_1 = V_{Si}/V_{unit}$ and $V_2 = t_{Si}/t$ are the silicon proportion in horizontal and vertical directions.

Considering the functional density of the flexible array as well as the wiring area of the electrical interconnection between units, the value of V_1 in the layout design is generally below 85%, where the theoretical value of the equivalent Young's modulus of the flexible silicon-based array is below 10 GPa, as shown in Figure 2b. This value is equivalent to the Young's modulus of the collagen fiber in organisms[13], exhibiting strong robustness while achieving flexibility, just as the collagen fiber providing structural support for biological tissue.

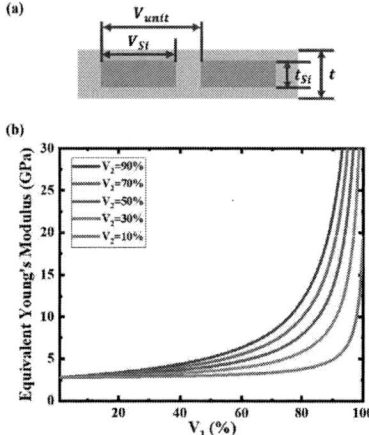

Figure 2: (a) The equivalent Young's modulus model of the proposed array. (b) Theoretical value of the equivalent Young's modulus versus silicon proportion in horizontal (V1) and vertical (V2) directions.

Fabrication Process

The fabrication process of the flexible tactile sensing array is shown in Figure 3. To obtain higher spatial density and good flexibility, the size of each sensing unit is identified as 700×700 μm^2 and the flexible trench width is 100 μm (Figure 1a-b). Instead of traditional KOH anisotropic etching to form cavity and square diaphragm of the sensor, deep reactive ion etching (DRIE) and wafer-to-wafer bonding was chosen to realize smaller chip size for higher resolution. As shown in Figure 3, DRIE was firstly used to define the cavity of sensing units. Wafer-to-wafer bonding followed by top silicon thinning was used to define the sensing diaphragm with a precise thickness. The Wheatstone bridge in each unit was formed by Boron implantation and metallization. Based on the conformal deposition of Parylene C thin film in high-aspect-ratio trenches, the flexible connection between functional units was fabricated by five steps: (1) etching micro-trenches (10 μm wide) by DRIE, (2) depositing 5 μm Parylene C to fill the micro-trenches, (3) etching the surface Parylene C by oxygen plasma, (4) etching the remnant silicon between micro-trenches, and (5) depositing Parylene C to fully fill the flexible trench. The multilayer wiring and bonding pads

were fabricated via electroplating process, Parylene deposition and selective etching. Finally, the flexible array was released by back thinning to expose the Parylene-filled trenches, and another Parylene layer was deposited to protect the backside. The whole fabrication process is MEMS and post-CMOS compatible.

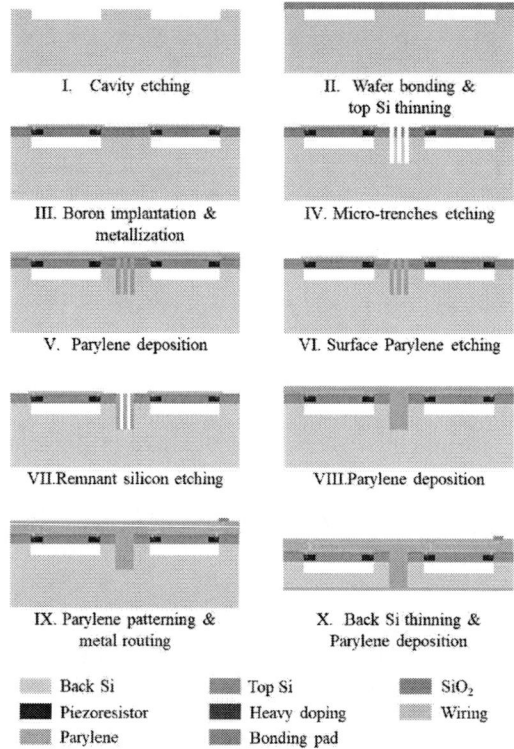

Figure 3: Fabrication process of the flexible tactile sensing array. (i-iii) Sensing units. (iv-viii) Parylene-filled trenches. (ix-x) Electrical interconnection between sensing units and flexible release of the array.

EXPERIMENTAL RESULTS
Mechanical Performance of the Flexible Array

Flexible silicon-based array with Parylene-filled trenches was fabricated to demonstrate the mechanical properties of the proposed strategy. As shown in Figure 4a, the 1×1 cm^2 array with silicon proportion of 70% has a radius of curvature of 2 mm, exhibiting excellent flexibility. Stress-strain tests were carried out to obtain the equivalent Young's modulus (Figure 4a) and elongation to break (Figure 4b) of arrays with different functional density. The measured value of the equivalent Young's modulus of the array is 9.5±0.9 GPa for silicon portion of 70% and 13.3±1.4 GPa for silicon portion of 83%, which shows high consistency with the theoretical value in Figure 2b.

Characterization of the Flexible Tactile Sensing Array

The fabricated flexible tactile sensing array is shown in Figure 5. 4-inch wafer was processed and gently released in a flexible form (Figure 5a), then the 3×3 and 5×5 flexible tactile sensing arrays (Figure 5b-c) were diced by laser cutting. Microscope image and SEM image of the electroplated copper wiring over Parylene-filled flexible trench (Figure d-e) presents the well electrically interconnected sensing units for power supply and sensing characterization. The tactile sensing array was wire bonded with flexible printed circuit board and microforce gauge was used to apply loads on the diaphragm of the tactile sensor for static pressure testing. To evaluate the influence of Parylene-filled flexible trenches fabrication process on the sensing properties of the sensors, we compared the sensitivity and nonlinearity of the tactile sensors before (w/o Parylene) and after (w/ Parylene) flexible fabrication process, as shown in Figure 6. The experimental results indicated that the sensing range was up to 2 MPa. With the Parylene-filled trench and flexible wiring over it, the sensitivity decreased from 0.162 mV/kPa to 0.113 mV/kPa, and the nonlinearity increased from 0.26% FS to 0.46% FS, still remaining high performance of the silicon-based device.

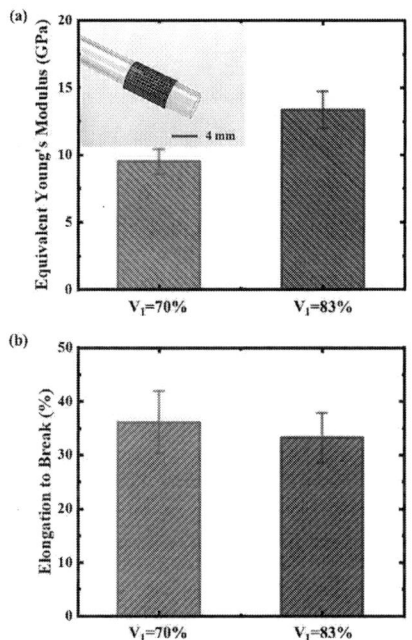

Figure 4: Experimental results of the equivalent Young's modulus and elongation to break of fabricated flexible arrays with different silicon proportion.

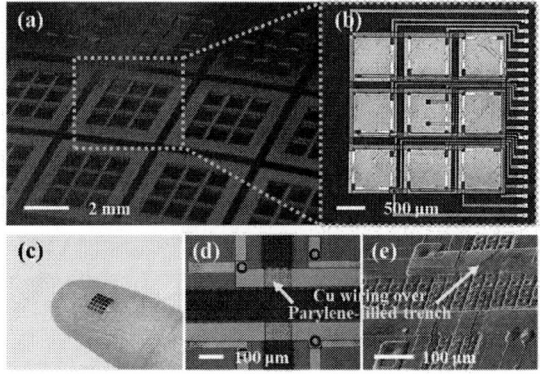

Figure 5: Images of the flexible tactile sensing array. (a) Wafer level. (b) 3×3 array. (c) 5×5 array on fingertip. (d-

e) Microscope image and SEM image of Cu wiring over Parylene-filled trench.

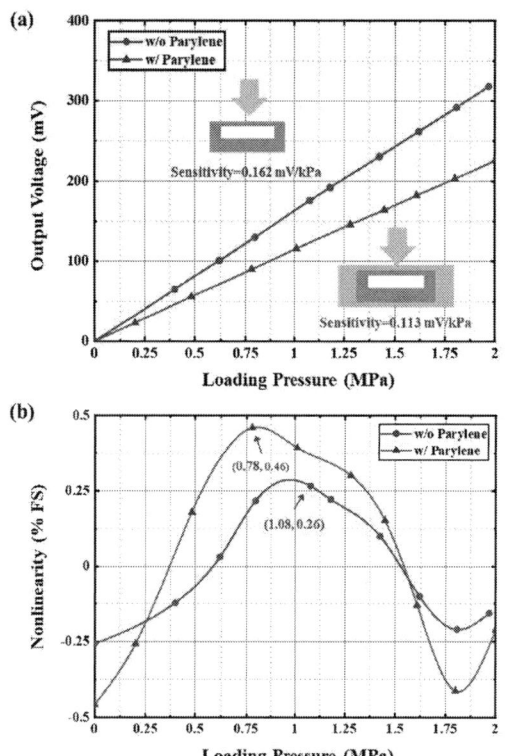

Figure 6: Characterization of the tactile sensors before (w/o Parylene) and after (w/ Parylene) flexible fabrication process. (a) Output voltage with loading pressure. (b) Nonlinearity error with loading pressure.

CONCLUSIONS

In this paper, we proposed a novel flexible tactile sensing array based on Parylene MEMS technique. The silicon-based sensing units (piezoresistive pressure sensors) and Parylene-filled trenches in between exhibit high-performance static pressure sensing and good flexibility. A theoretical model of the equivalent Young's modulus of the Parylene/silicon composite structure was established to evaluate the mechanical property of the proposed array and well verified by stress-strain tests. Based on MEMS and post-CMOS compatible fabrication process, a flexible tactile sensing array with high spatial density ($\sim$156 cm^{-2}), large range (0-2 MPa) and low nonlinearity (0.46% FS) was successfully fabricated. The promising results illustrated powerful applicability of Parylene MEMS technique in mass production of high-performance silicon-based flexible electronics.

ACKNOWLEDGEMENTS

This work is supported by the National Natural Science Foundation of China under Grant No. 62074003. The authors would like to thank the staff of National Key Laboratory of Science and Technology on Micro/Nano Fabrication for their help with the fabrication process.

REFERENCES

[1] S. Pyo, J. Lee, K. Bae *et al.*, "Recent progress in flexible tactile sensors for human-interactive systems: from sensors to advanced applications", *Adv. Mater.*, vol. 33, pp. 2005902, 2021.

[2] S. Chun, J. S. Kim, Y. Yoo *et al.*, "An artificial neural tactile sensing system", *Nat. Electron.*, vol. 4, pp. 429-438, 2021.

[3] M. L. Hammock, A. Chortos, B. C. K. Tee, J. B. H. Tok and Z. Bao, "25th anniversary article: the evolution of electronic skin (e-skin): a brief history, design considerations, and recent progress", *Adv. Mater.*, vol. 25, pp. 5997–6038, 2013.

[4] J. Li, R. Bao, J. Tao, Y. Peng and C. Pan, "Recent progress in flexible pressure sensor arrays: from design to applications", *J. Mater. Chem. C*, vol. 6, pp. 11878–11892, 2018.

[5] X. Sun, J. Sun, T. Li *et al.*, "Flexible tactile electronic skin sensor with 3D force detection based on porous CNTs/PDMS nanocomposites", *Nano-Micro Lett.*, vol. 11, pp. 1-14, 2019.

[6] X. Fu, J. Zhang, J. Xiao *et al.*, "A high-resolution, ultrabroad-range and sensitive capacitive tactile sensor based on a CNT/PDMS composite for robotic hands", *Nanoscale*, vol. 13, pp. 18780-18788, 2021.

[7] S. Chen, K. Jiang, Z. Lou *et al.*, "Recent developments in graphene-based tactile sensors and E-skins", *Adv. Mater. Technol.*, vol. 3, pp. 1700248, 2018.

[8] L. Pan, A. Chortos, G. Yu *et al.*, "An ultra-sensitive resistive pressure sensor based on hollow-sphere microstructure induced elasticity in conducting polymer film", *Nat. Commun.*, vol. 5, pp. 1-8, 2014.

[9] X. Su, X. Wu, S. Chen *et al.*, "A highly conducting polymer for self-healable, printable, and stretchable organic electrochemical transistor arrays and near hysteresis-free soft tactile sensors", *Adv. Mater.*, vol. 34, pp. 2200682, 2022.

[10] W. Wu, X. Wen, Z. L. Wang, "Taxel-addressable matrix of vertical-nanowire piezotronic transistors for active and adaptive tactile imaging", *Science*, vol. 340, pp. 952-957, 2013.

[11] W. Cheng, L. Yu, D. Kong *et al.*, "Fast-response and low-hysteresis flexible pressure sensor based on silicon nanowires", *IEEE Electron Device Lett*, vol. 39, pp. 1069-1072, 2018.

[12] A. Chortos, J. Liu, Z. Bao, "Pursuing prosthetic electronic skin" *Nature Mater.*, vol. 15, pp. 937-950, 2016.

[13] M. P. Wenger, L. Bozec, M. A. Horton, P. Mesquida, "Mechanical properties of collagen fibrils", *Biophys. J.*, vol. 93, pp. 1255-1263, 2007.

CONTACT

*W. Wang, Peking University, Beijing 100871, China; Tel: +86-10-62750175; Email: w.wang@pku.edu.

SILK-ENABLED FOLDABLE AND CONFORMAL NEURAL INTERFACE WITH IN-PLANE SHIELDING FOR HIGH-QUALITY ELECTROPHYSIOLOGICAL RECORDINGS

Jizhi Liang[1,2], Zhaohan Chen[1,3], Xiner Wang[1,2], Feihong Xu[2,4], Xiaoling Wei[2,4], Liuyang Sun[1,2,4], Meng Li[2,4], Tiger H. Tao[1,2,4,5,6,7,8,9,10], and Zhitao Zhou[2,4]

[1]2020 X-Lab, Shanghai Institute of Microsystem and Information Technology,
Chinese Academy of Sciences, Shanghai, CHINA

[2]School of Graduate Study, University of Chinese Academy of Sciences, Beijing, CHINA

[3]College of Life Sciences, Shanghai Normal University, Shanghai, CHINA

[4]State Key Laboratory of Transducer Technology, Shanghai Institute of Microsystem and
Information Technology, Chinese Academy of Sciences, Shanghai, CHINA

[5]Center of Materials Science and Optoelectronics Engineering,
University of Chinese Academy of Sciences, Beijing, CHINA

[6]School of Physical Science and Technology, ShanghaiTech University, Shanghai, CHINA

[7]Center for Excellence in Brain Science and Intelligence Technology,
Chinese Academy of Sciences, Shanghai, CHINA

[8]Neuroxess Co., Ltd. (Jiangxi), Nanchang, Jiangxi, CHINA

[9]Guangdong Institute of Intelligence Science and Technology,
Hengqin, Zhuhai, Guangdong, CHINA and

[10]Tianqiao and Chrissy Chen Institute for Translational Research, Shanghai, CHINA

ABSTRACT

This paper reports a flexible, foldable and conformal neural interface with in-plane shielding fabricated by MEMS process. Based on the tailored planar pattern derived from hemisphere, it can conformally attach to curved surfaces with featured radiuses of typical intracranial space-occupying tumors or brain nuclei. Introducing the controlled degradable silk fibroin as the temporary immobilization material, it can be temporarily fixed into ball shape to minimize implantation trauma during surgery, and conformally cover the curved surface after silk dissolved by tissue fluids, which makes it suitable for electrophysiological monitoring of deep brain regions. Additionally, coplanar metal shielding layer design reduces the interference of mechanical artifacts and power frequency noises, which significantly improves the quality of recorded signals.

KEYWORDS

Electrophysiological recordings, Conformal contact, In-plane shielding, Silk-enabled foldable

INTRODUCTION

Flexible electrodes with good mechanical match to brain tissue are often used on the cerebral cortex or in the superficial cortex layer but not in the deeper brain regions by now, although the deep brain stimulation and epilepsy studies have witnessed the widely use of intracranial electroencephalography electrodes [1-4]. As a typical example in terms of function and anatomical structure, hypothalamus is one of the most important deep brain regions (Figure 1a), helping projecting the accepted nerve impulses to other regions of the brain. However, hypothalamus slices study disrupts neural circuits inevitably, and implanting stainless steel electrodes in rodent animals to approach deep brain regions can cause penetrating injuries. Therefore, it could facilitate the acute recordings and neurological studies of periventricular and deep brain regions, if flexible electrodes can be delivered through catheter or implanted minimally invasively. Meanwhile, the electrically noisy environment of laboratory and operating room will cause challenges to electrophysiological recordings [5]. To record high-quality in vivo signals, a flexible, foldable and conformal neural interface with in-plane shielding has been developed here.

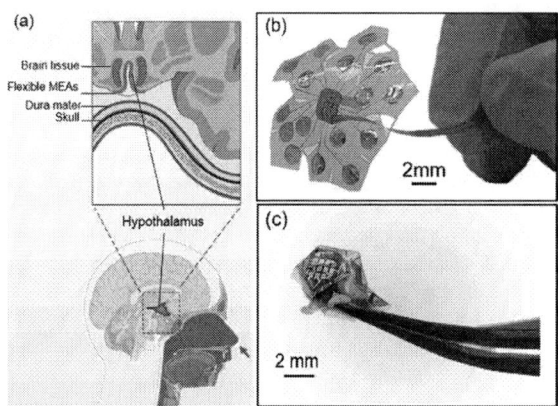

Figure 1. Flexible, foldable electrode array designed for conformal coverage over the featured curved surfaces. (a) The inspiration and conceptual graph of the electrode array applying in endoscopic skull base surgery. Electrode array is delivered into brain along the path of transsphenoidal craniopharyngioma resection and closely attached to the hypothalamus after removing the craniopharyngioma. Photographs of assembled electrodes array with unfolded form (b) and folded form (c).

DESIGN

The neural interface in the form of flexible sheet electrode is designed as the shape of the unfolding pattern of hemisphere and with the multilayer interconnection technology, making it more space-saving to form conformal contact to curved surfaces similar to those deep brain regions after tumor resection. A silk-enabled temporary folding strategy is developed and integrated with the matured transsphenoidal craniopharyngioma resection to achieve the minimally invasive implantation. A coplanar metal shielding layer besieges recording sites and interconnectors. This in-plane shielding is designed for the better trade-off between electromagnetic shielding effectiveness and neural interface conformal contact which directly affected by the device thickness [6,7].

FABRICATION

The neural interface in the form of flexible polymer substrate foldable electrode is fabricated using a standard microelectromechanical systems (MEMS) process [8,9]. However, it introduced the multilayer interconnection technology to form reliable electrical connection between the reflow soldering pads and electrode sites on the opposite face.

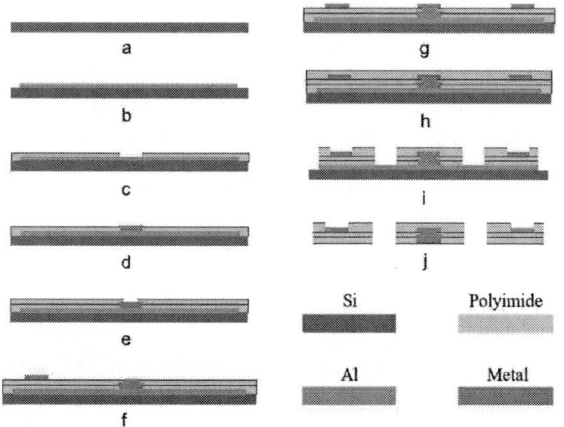

Figure 2: Fabrication flow diagram. (a) Pre-cleaned silicon wafer (b) Sputtering sacrificial layer of Al; (c, e, h) Spin coating and patterning of polyimide; (d) Deposition of soldering pads metal layer (Cr/Ni/Au); (f) Sputtering vias and coplanar metal shielding layer (Cr/Au); (g) Sputtering electrode sites and interconnectors (Cr/Au); (i) Patterning and wet etching of mesh structures and electrode sites; (j) Releasing the electrode.

As shown in Figure 2, the sacrificial layer of 1 μm aluminum is firstly sputtered on the silicon wafer, and a 3 μm polyimide substrate layer is spin-coated thereon. After photolithography patterning and wet etching of the PI substrate, the metal (Cr/Ni/Au, 100/1000/5000Å) layer of reflow soldering pads is formed by electron beam evaporation. Then, a 2 μm PI insulation interlayer is fabricated in the same manner but with small arrayed patterns on pads exposure area, followed with the sputtering of the metal (Cr/Au, 150/4000Å) to form multilayer interconnecting vias. Meanwhile, the coplanar metal (Cr/Au, 150/4000Å) shielding layer is sputtered on the PI interlayer (not for electrodes T0 without shield). Subsequently, the metal (Cr/Au, 150/4000Å) layer of electrode sites and interconnectors is formed in the besiege of shielding metal. Finally, an 8 μm PI encapsulation layer with the same tailored patterns and mesh structure as the previous two PI layer is fabricated, and the electrode contact sites is formed simultaneously. The processing is completed by etching Al sacrificial layer by buffered hydrofluoric acid.

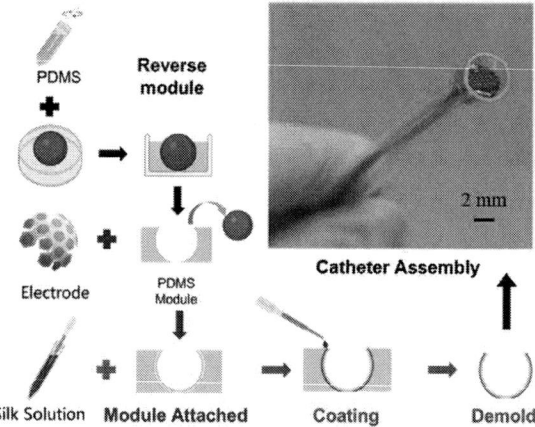

Figure 3: Assembly flow diagram of silk-enabled foldable electrode array being reshaped in a home-made PDMS mold, being temporarily immobilized by controlled degradable silk fibroin solution and finally being assembled in a typical medical drainage catheter for minimally invasive implantation.

The electrical bonding between this neural interface and flexible printed circuit board (FPC) with connecters to electrophysiological signal acquisition system is formed by low temperature solder paste reflow soldering technology. Subsequently, waterproof UV curing adhesive is used on the soldering pads area to protect the bonding.

To assemble this neural interface into a typical catheter for external ventricular drainage, we develop a silk-enable folding and temporary immobilizing strategy (as shown in Figure 3). Firstly, a hemispheric pit with the size appropriate to electrodes is formed in a home-made PDMS mold using a 3D printed reverse resin model. Then, the soldering pads area is put in the bottom of the pit cautiously, and the petal-like electrodes are naturally folded and attach onto the pit surface. After dropping a little prepared silk fibroin solution into the pit, the electrodes and PDMS mold either undergo air dry or water vapor annealing to form controlled degradable immobilization coating [10]. Finally, the neural interface and the 2.0 mm width FPC in customized length can be easily fit into the catheter.

RESULTS

The normalized distribution of principal stress on electrodes during the temporary silk immobilization process according to mechanics simulation does not deteriorate the gold metal trace on electrodes (Figure 4a), and it is further validated by electrical testing (Figure 4c). Upon adjusting the degree of crosslink in silk fixing process, the electrode array can be controlled to unfold naturally within minutes to hours (Figure 4b).

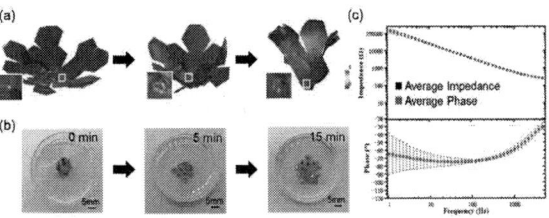

Figure 4: Simulation and in vitro characterizations. (a) Mechanics simulation with different force constraints, increasing from right to left. Inserts are the magnified view of maximum stress areas under each state. (b) The electrode is fixed into a ball shape using silk solution and can self-unfold totally within 15 min in normal saline. (c) The electrochemical impedance spectra (EIS) and phase curves of each active recording sites, which are measured in the phosphate buffer saline (PBS) under the frequency from 1 Hz to 5kHz.

In order to validate the effectiveness of in-plane shielding layer to reduce external electrical interference. We conducted in vitro PBS recording test and the noise density spectral indicates better performance in T1 electrodes with shielding design (Figure 5b), and T1 can substantially reduce 50 Hz power noise in statistics (Figure 5a). Furthermore, in vivo animal experiment was conducted on epileptic rat model, T1 electrodes recorded lower signal power in the low-frequency band and 50 Hz line noise than T0 did during the isoflurane anesthesia state (Figure 5d), and recorded typical electrophysiological signals when seizure was onset (Figure 5c).

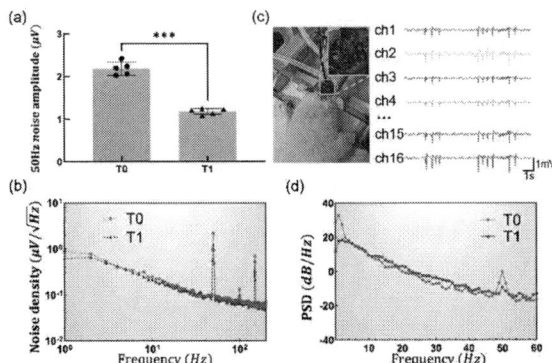

Figure 5: Electrostatic shielding characterization and high-quality electrophysiological recording in vivo. In vitro characterization of noise amplitude at 50 Hz (a) and noise density spectra (b) of electrodes T0 (without shield) and T1 (with shield). (c) In vivo shielding performance validation based on epileptic rat model, electrodes are conformally attached to the brain surface after the silk being dissolved, and typical epileptiform discharges were recorded. (d) Power spectral density comparison of T0 and T1 in anesthesia state.

CONCLUSION

The neural interface demonstrated here can fit into medical catheter used in minimally invasive surgery, conformally cover featured curved surfaces, and perform high-quality electrophysiological recordings with less external electrical interference both in vitro and in vivo. We plan to further explore the potential of this neural interface being a new electrophysiological tool for the direct and minimally invasive study of periventricular nuclei and deep brain nuclei, as well as for the clinical applications of corresponding targets.

ACKNOWLEDGEMENTS

Jizhi Liang and Zhaohan Chen contributed equally to this work. This work was partially supported by the National Key R & D Program of China (Grant Nos. 2019YFA0905200, 2021ZD0201600, 2021YFC2501500, 2021YFF1200700, 2022ZD0209300, 2022ZD0212300), Key-Area Research and Development Program of Guangdong Province (2021B0909060002), Guangdong high level Innovation Research Institute (2021B0909050004), National Natural Science Foundation of China (Grant No. 61974154), Key Research Program of Frontier Sciences, CAS (Grant No. ZDBS-LY-JSC024), Shanghai Pilot Program for Basic Research – Chinese Academy of Science, Shanghai Branch (Grant No. JCYJ-SHFY-2022-01), Shanghai Municipal Science and Technology Major Project (Grant No. 2021SHZDZX), CAS Pioneer Hundred Talents Program, Shanghai Pujiang Program (Grant Nos. 19PJ1410900, 21PJ1415100), the Science and Technology Commission Foundation of Shanghai (No. 21JM0010200), Shanghai Rising-Star Program (Grant No. 22QA1410900), the Innovative Research Team of High-level Local Universities in Shanghai, the Jiangxi Province 03 Special Project and 5G Project (Grant No. 20212ABC03W07), Fund for Central Government in Guidance of Local Science and Technology Development (Grant No. 20201ZDE04013).

REFERENCES

[1] M. R. Mercier et al., "Advances in human intracranial electroencephalography research, guidelines and good practices", NeuroImage, vol. 260, p. 119438, 2022.

[2] A. Cometa et al., "Clinical neuroscience and neurotechnology: An amazing symbiosis", iScience, vol. 25, no. 10, p. 105124, 2022.

[3] D. Wang et al., "Electrophysiological properties and seizure networks in hypothalamic hamartoma", Ann Clin Transl Neurol, vol. 7, no. 5, pp. 653–666, 2020.

[4] B. Thielen and E. Meng, "A comparison of insertion methods for surgical placement of penetrating neural interfaces", J Neural Eng, vol. 18, no. 4, Apr. 2021.

[5] A. C. Paulk et al., "Large-scale neural recordings with single neuron resolution using Neuropixels probes in human cortex", Nat Neurosci, vol. 25, no. 2, pp. 252–263, Feb. 2022.

[6] D.-H. Kim et al., "Dissolvable films of silk fibroin for ultrathin conformal bio-integrated electronics", Nature Mater, vol. 9, no. 6, pp. 511–517, 2010.

[7] B. Ji et al., "Flexible and stretchable opto-electric neural interface for low-noise electrocorticogram

recordings and neuromodulation in vivo", Biosensors and Bioelectronics, vol. 153, p. 112009, 2020.

[8] F. Zheng, S. Zhang, Y. Zhou, and T. H. Tao, "A Silk-Enabled Conformal Brain Electrode for Recording and Disease Treatment", in 2019 IEEE 32nd International Conference on Micro Electro Mechanical Systems (MEMS), Seoul, Korea (South), 2019, pp. 625–627.

[9] T. Xiao, S. Zhang, Y. Zhou, and T. H. Tao, "A MEMS-Based Flexible High-Density Brain Electrode for Multi-Modal Neural Encoding/Decoding", in 2019 IEEE 32nd International Conference on Micro Electro Mechanical Systems (MEMS), 2019, pp. 615–616.

[10] Z. Zhou et al., "The Use of Functionalized Silk Fibroin Films as a Platform for Optical Diffraction-Based Sensing Applications", Adv Mater, vol. 29, no. 15, p. 1605471, 2017.

CONTACT

* Tiger H. Tao, tel: +86-21-62511070;
tiger@mail.sim.ac.cn
* Zhitao Zhou, tel: +86-21-62511070;
ztzhou@mail.sim.ac.cn

MATERIALS ENGINEERING FOR CHEMICAL SENSING ENHANCEMENT

Navpreet Kaur, Dario Zappa, and Elisabetta Comini
SENSOR Laboratory, University of Brescia and INSTM UdR Brescia
Via D. Valotti 9, Brescia 25133, Italy

ABSTRACT

Metal oxides nanowires and novel heterostructures are synthesized using different techniques and finally integrated into gas sensing platform. In particular, nanowires were synthesized using thermal oxidation and VLS mechanism. While, heterostructures i.e. NiO/ZnO (p-n) and NiO/NiWO$_4$/WO$_3$ (p-p-n) were synthesized using VLS and VS mechanisms. Detailed investigations reveal the dependence of sensors selectivity and sensitivity on nanowires synthesis techniques. While, the superior performance of heterostructures as compared to bare nanowires presents the novel pathway to further enhance the performance of nanostructured gas sensors.

KEYWORDS

Metal oxide, nickel oxide, tungsten oxide, zinc oxide, nanowire, heterostructures, chemical/gas sensor.

INTRODUCTION

Day by day environmental monitoring is getting more and more essential due to the continuous rise in human activities, especially in the direction of industrial and technological developments. Further, the increase of atmospheric pollutants rises the demand for effective and inexpensive systems for the detection of environmentally hazardous gases. In this context, for the last two decades, one–dimensional (1D) metal oxide nanostructures have been attracting much interest in the field of gas sensors due to their remarkable physical and chemical properties, distinguishing them from bulk materials[1], [2]. Moreover, they possess a high surface-to-volume ratio, single crystallinity and better stability.

Furthermore, advances in fabrication methods have enabled the production of low-cost sensor materials with enhanced sensitivity and stability [3]. In nanostructured materials, the enhancement in the sensor response can be achieved by increasing the active surface area of the sensor. Various strategies have been used to increase the gas response and selectivity, including modulating the sensing temperature, morphological control, catalyst doping/loading, catalytic filtering of interference gases, and construing a junction between two materials [3], [4].

Herein, we are presenting the synthesis and characterization of different p and n-type metal oxide-based NWs and their novel heterostructures morphologies such as NiO [5], WO$_3$ [6], and ZnO [7] NWs, and NiO/ZnO [8], NiO/NiWO$_4$/WO$_3$ [9] for chemical sensing applications. These nanostructured materials based sensors were found to exhibit different selectivity and sensitivity at different experimental conditions, making them useful to generate array of sensors (e-nose).

EXPERIMENTAL

Evaporation-condensation technique works on the basis of two mechanisms i.e., vapor-liquid-solid (VLS) and vapor-solid (VS). The major difference between the two mechanisms is the involvement of noble metal catalysts (Au, Pt, Pd, etc) in VLS to promote the nucleation of source material on the substrate during the growth process. The nanowire growth was carried out in a lab-made tubular furnace. Different materials such as NiO and ZnO were grown using the VLS mechanism. The ultra-thin layer of Gold (Au) nano particles was used as a catalyst on the alumina substrate prior to the growth of nanowires. In particular, for the Growth of NiO nanowires, the evaporation temperature of NiO powder was set at 1400 °C and Au-deposited Al$_2$O$_3$ (2x2 mm^2 Kyocera, Japan, 99% purity) substrates were placed at a temperature of 930 °C inside an alumina tube of the tubular furnace. On the other hand, ZnO NWs were grown at a lower temperature compared to NiO. The evaporation temperature was set to 1200 °C and the substrate temperature was around 450 °C. The deposition time was set to 15 minutes. An argon gas was used as a carrier gas and its flow was set at 100 SCCM during both depositions.

Further, the same condition was used to grow the branched heterostructures of NiO/ZnO using a two-step deposition method: i) NiO NWs were grown on Au-catalyzed alumina substrate using the VLS mechanism. ii) ZnO NWs were deposited directly on top of NiO NWs using VS mechanism, resulting in branch-like nanostructures. While, in the case of NiO/NiWO$_4$/WO$_3$ heterostructures, WO$_3$ NWs were grown on top of NiO NWs at an evaporation temperature of 1100 °C, at 1 mbar pressure with the substrate at a temperature of 530 °C.

Thermal oxidation was used to grow the WO$_3$ NWs directly on the alumina substrate. The growth of nanostructures consists of two steps: the deposition of a metallic layer followed by thermal oxidation. A metallic tungsten film was deposited on 2x2 mm^2 alumina substrates (Kyocera, Japan, 99% purity) by RF magnetron sputtering (100nm thickness, 100 W argon plasma, 100 °C, 5.5×10^{-3} mbar), starting from a pure W target (>99.9% purity). The growth of the nanowires was performed inside a custom vacuum tubular furnace, where samples were oxidized at a certain temperature (600 °C) and atmosphere conditions (10 SCCM Ar flow). The pressure inside the furnace was set at 1 mbar. Prior to gas sensing measurements, samples were annealed in air at 400 °C for 24 hours, to completely oxidize the residual metallic tungsten layer.

Furthermore, for the fabrication of sensing devices, firstly a TiW adhesion layer was deposited by DC magnetron sputtering (70W argon plasma, @ 300 °C, 5.5×10^{-3} mbar), followed by the deposition of interdigitated platinum contact using the same conditions explained before. A platinum heater was deposited on the back side of the alumina substrate via the same procedure used for contact deposition. Finally, the prepared devices were mounted on TO packages using electro-soldered gold wires. The conductometric response of these sensing devices was measured towards different gas analytes and concentrations at different temperatures. The response of n-type metal oxide was calculated using the following equations [10]

For oxidizing gases,

$$\text{Response} = (R_{gas} - R_{air})/R_{air} = (G_{air} - G_{gas})/G_{gas} \quad (1)$$

and for reducing gases,

$$\text{Response} = (G_{gas} - G_{air})/G_{air} \quad (2)$$

Vice versa for p-type metal oxide sensors.

RESULTS AND DISCUSSION

The morphology of the nanostructures was investigated by FE-SEM. Figure 1a reports the morphology of NiO NWs grown by using the VLS mechanism on alumina substrates to exhibit scattered and homogenous. The nanowire diameter was found in the range of 15 to 60 nm and a length at the micrometer scale. Further Figure 1b shows WO_3 NWs morphology on alumina substrate using thermal oxidation of the W layer. The average diameter of the NWs was found in the range of 20–30 nm while the length is approximately 1–2 μm.

In Figure 1c–d the heterostructure morphology of NiO/ZnO and $NiO/NiWO_4/WO_3$ is shown, respectively. In particular, Figure 1c shows the small ZnO NWs and leaf-like structures grown epitaxially on the NiO nanowires. Similarly, Figure 1d shows the growth of small WO_3 nanowires on top of each NiO nanowire, effectively designing branch-like nanostructures. Interestingly, in this case, the WO_3 vapors first form ternary material seeds (nickel tungstate - $NiWO_4$) on NiO NWs prior to the growth of WO_3 NW, resulting in the growth of (p/p/n) $NiO/NiWO_4/WO_3$ heterostructures.

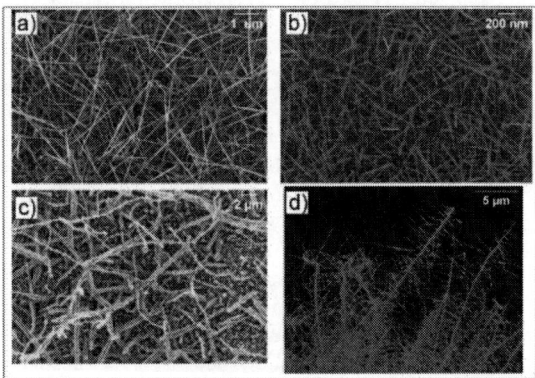

Figure 1: SEM images of a) NiO nanowires, b) WO_3 nanowires, c) NiO/ZnO nanowire-based heterostructures, d) $NiO/NiWO_4/WO_3$ heterostructures grown on alumina substrates.

Figure 2 (a, b) reports the isothermal dynamic-transient response of fabricated NiO and WO_3 NW sensors towards hydrogen (50; 200; 500 ppm) at the optimal working temperatures. Clearly, in the case of NiO, sensors show an electrical conductance decrease when exposed to a reducing gas. This is the typical behavior of a p-type metal oxide sensor under reducing compound exposure. The opposite behavior can be seen in the case of WO_3 NWs.

Moreover, both the sensing devices were tested toward different reducing and oxidizing gases in a range of temperatures from 200 to 500 °C. Figure 3 (a, b) reports the calculated response (from eq^n 1 and 2) for NiO and WO_3 NW sensors towards H_2 (50 ppm), ethanol (50 ppm), acetone (100 ppm) and NO_2 (1 ppm) at 50 % relative humidity.

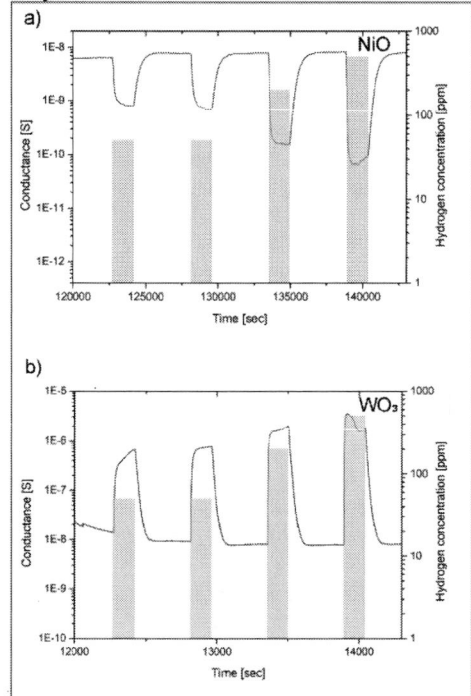

Figure 2: Dynamics response of a) NiO NWs at 300 °C, b) WO_3 NWs at 200 °C towards hydrogen [50; 50; 200; 500 ppm] at RH 50%.

It is clear that both sensors possess highly different sensitivity and selectivity when operated at different temperatures.

In particular, WO_3 exhibits a higher response towards H_2 at low temperature (200 °C) which is 1000 times higher compared to the NiO NW sensors at its optimal working temperature of 300 °C for H_2. Both the sensors do not show a high response towards VOCs (ethanol, acetone) compared to H_2. On the other hand, for oxidizing gases such as NO_2, NiO sensors show a 50 times higher response at low temperature (200 °C) in comparison to the WO_3 sensors. Moreover, WO_3 shows the highest response toward NO_2 at higher temperatures (ranging within 400 - 500 °C).

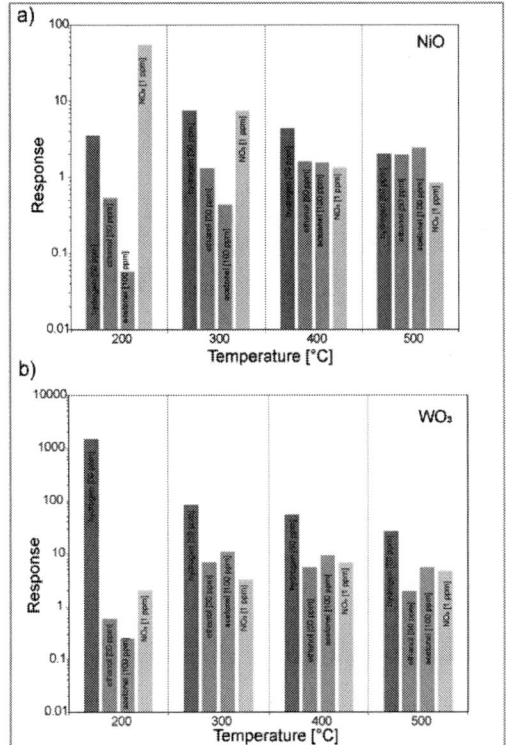

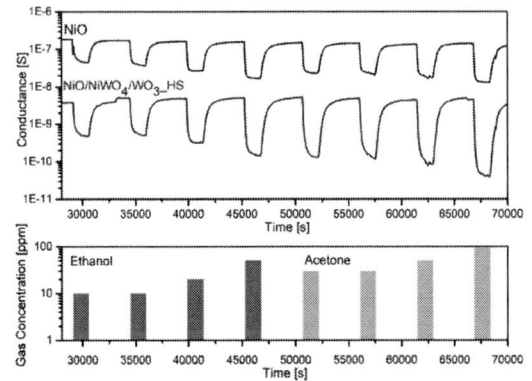

Figure 3: Temperature dependence response of a) NiO and b) WO₃ sensors for ethanol (50 ppm) and acetone (100 ppm) hydrogen(50 ppm) and NO2 (1 ppm).

Furthermore, combining two different nanostructured materials on a single sensing platform opens new possibilities to tailor the platform sensing characteristics. Figure 4 shows the isothermal dynamic-transient response of fabricated heterostructures of NiO and WO₃ on a single sensing device. In particular, these heterostructures were tested for ethanol and acetone gases analyte. The dynamic response curves in Figure 4 clearly show that the response of NiO/NiWO₄/WO₃ heterostructures is much higher compared to the bare NiO NWs.

Figure 4: Dynamics response of a) NiO NWs b) NiWO₄ HS at 400 ºC towards ethanol [10; 10; 20; 50 ppm] and acetone [30; 30; 50; 100 ppm]at RH 50%.

Figure 5 presents the response comparison of bare NiO and heterostructured sensor towards ethanol and acetone (50 ppm) at 400 °C. The heterostructure sensors exhibit a 30 times higher response towards ethanol and 40 times

higher towards acetone compared to bare NiO sensing device.

Finally, these results emphasize that employing material engineering techniques such as fabricating complex nano-structures (branch-like heterostructures) and using different growth methods can enhance the sensitivity and selectivity of sensing systems. In addition to this, surface functionalization of nanowires with self-assembled monolayer is another excellent strategy to enhance their sensing performance, especially high response and selectivity at relatively lower working temperatures as reported in [11]–[13].

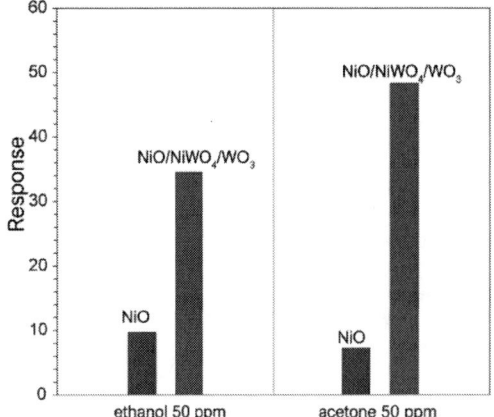

Figure 4: Response comparison of NiO NWs and NiWO₄ HS at 400 ºC ethanol and acetone of 50 ppm at RH 50 %.

CONCLUSIONS

In conclusion, metal oxides nanowires and heterostructures were proposed for gas sensing applications. The nanowires prepared using different techniques exhibit dense morphologies and are highly crystalline in nature.

At different operating temperatures, NiO and WO₃ nanowires possess different selectivity and sensitivity. In particular, WO₃ exhibits a higher response towards H_2 at a low temperature of 200 °C which is 1000 times higher compared to the NiO NW. While, NiO NWs sensors exhibits 50 times higher response at a low temperature of 200 °C in comparison to the WO₃ sensor towards oxidizing gases such as NO_2.

Furthermore, the combination of these two oxides into heterostructures i.e. NiO/NiWO₄/WO₃ (p-p-n) showed superior performances toward reducing gases as compared to bare NiO NWs.

Finally, all these nanostructured sensing materials can be potential candidates to fabricate e-noses or sensor array systems.

ACKNOWLEDGEMENTS

This work was partially funded by NATO Science for Peace and Security Programmer (SPS) under grant G5634 AMOXES – "Advanced Electro-Optical Chemical Sensors", MIUR "Smart Cities and Communities and social innovation" project titled "SWaRM Net/Smart Water Resource Management Networks, and Regione

Lombardia Call Hub Ricerca e Innovazione, within the project "MoSoRe@ Unibs—Infrastrutture e servizi per la Mobilità Sostenibile e Resiliente".

REFERENCES

[1] E. Comini *et al.*, "Metal oxide nanoscience and nanotechnology for chemical sensors," *Sensors Actuators B Chem.*, vol. 179, pp. 3–20, Mar. 2013, doi: 10.1016/j.snb.2012.10.027.

[2] E. Şennik, U. Soysal, and Z. Z. Öztürk, "Pd loaded spider-web TiO2 nanowires: Fabrication, characterization and gas sensing properties," *Sensors Actuators B Chem.*, vol. 199, pp. 424–432, 2014, doi: https://doi.org/10.1016/j.snb.2014.03.052.

[3] N. Kaur, M. Singh, and E. Comini, "One-Dimensional Nanostructured Oxide Chemoresistive Sensors," *Langmuir*, vol. 36, no. 23, pp. 6326–6344, Jun. 2020, doi: 10.1021/acs.langmuir.0c00701.

[4] D. Zappa, V. Galstyan, N. Kaur, H. M. M. Munasinghe Arachchige, O. Sisman, and E. Comini, "'Metal oxide -based heterostructures for gas sensors'- A review," *Anal. Chim. Acta*, vol. 1039, pp. 1–23, 2018, doi: https://doi.org/10.1016/j.aca.2018.09.020.

[5] N. Kaur, E. Comini, D. Zappa, N. Poli, and G. Sberveglieri, "Nickel oxide nanowires: vapor liquid solid synthesis and integration into a gas sensing device.," *Nanotechnology*, vol. 27, no. 20, p. 205701, May 2016, doi: 10.1088/0957-4484/27/20/205701.

[6] D. Zappa, A. Bertuna, E. Comini, M. Molinari, N. Poli, and G. Sberveglieri, "Tungsten oxide nanowires for chemical detection," *Anal. Methods*, vol. 7, no. 5, pp. 2203–2209, 2015, doi: 10.1039/C4AY02637C.

[7] A. Moumen, N. Kaur, N. Poli, D. Zappa, and E. Comini, "One Dimensional ZnO Nanostructures: Growth and Chemical Sensing Performances," *Nanomaterials*, vol. 10, no. 10. 2020, doi: 10.3390/nano10101940.

[8] N. Kaur *et al.*, "Branch-like NiO/ZnO heterostructures for VOC sensing," *Sensors Actuators, B Chem.*, vol. 262, pp. 477–485, 2018, doi: 10.1016/j.snb.2018.02.042.

[9] N. Kaur, D. Zappa, V.-A. Maraloiu, and E. Comini, "Novel Christmas Branched Like NiO/NiWO4/WO3 (p–p–n) Nanowire Heterostructures for Chemical Sensing," *Adv. Funct. Mater.*, vol. 31, no. 38, p. 2104416, Jul. 2021, doi: https://doi.org/10.1002/adfm.202104416.

[10] N. Kaur, D. Zappa, N. Poli, and E. Comini, "Integration of VLS-Grown WO$_3$ Nanowires into Sensing Devices for the Detection of H$_2$S and O$_3$," *ACS Omega*, vol. 4, no. 15, pp. 16336–16343, Oct. 2019, doi: 10.1021/acsomega.9b01792.

[11] M. Singh, N. Kaur, and E. Comini, "Role of Self-Assembled Monolayers in Electronic Devices," *J. Mater. Chem. C*, vol. 8, pp. 3938–3955, 2020, doi: 10.1039/D0TC00388C.

[12] M. Singh, N. Kaur, G. Drera, A. Casotto, L. S. Ermenegildo, and E. Comini, "SAM Functionalized ZnO Nanowires for Selective Acetone Detection: Optimized Surface Specific Interaction Using APTMS and GLYMO Monolayers," *Adv. Funct. Mater.*, vol. 30, no. 38, p. 2003217, Jul. 2020, doi: 10.1002/adfm.202003217.

[13] M. Singh, N. Kaur, A. Casotto, L. Sangaletti, N. Poli, and E. Comini, "Methyl (–CH$_3$)-terminated ZnO nanowires for selective acetone detection: a novel approach toward sensing performance enhancement via self-assembled monolayer," *J. Mater. Chem. A*, vol. 10, no. 6, pp. 3178–3189, 2022, doi: 10.1039/D1TA09290A.

CONTACT

*Elisabetta Comini, tel: +39 030 3715771; elisabetta.comini@unibs.it

ON-DEMAND PREPARATION OF GAS-SENSING MATERIALS GUIDED BY RESONANT CANTILEVER-BASED THERMOGRAVIMETRIC ANALYSIS

Yufan Zhou[1,2], Ming Li[1,2], Ying Chen[1,2], Xinyu Li[1,2], Pengcheng Xu[1,2], and Xinxin Li[1,2]

[1] State Key Laboratory of Transducer Technology, Shanghai Institute of Microsystem and Information Technology, Chinese Academy of Sciences, Shanghai 200050, CHINA and

[2] School of Microelectronics, University of Chinese Academy of Sciences, Beijing 100049, CHINA

ABSTRACT

In this work, the preparation conditions of Mn_3O_4 nanowires for formaldehyde (HCHO) sensing are optimized with a resonant cantilever-based thermogravimetric analysis (referred to as cantilever-TGA) technology. The cantilever-TGA technology only consumes about 20 ng of samples for one measurement, which is six orders of magnitude lower than the mainstream commercial-available TGA instruments, making it ideal for exploring optimal sample preparation conditions. Herein the cantilever-TGA technology has been successfully used to investigate the preparation conditions of Mn_3O_4 sensing material from the precursor of β-MnO_2 nanowires. With the guidance of cantilever-TGA, pure-phase and morphology well-maintained Mn_3O_4 gas-sensing materials can be obtained under H_2 atmosphere at 330°C, which is 230°C lower than N_2 atmosphere. Due to the well-maintained nanowire-like structure and high mobility, the response of the $Mn_3O_4@H_2$ material to HCHO is one-fold higher than that of the $Mn_3O_4@N_2$ material.

KEYWORDS

Resonant microcantilever, thermogravimetric analysis, calcination, nanomaterial

INTRODUCTION

Metal oxide semiconductor (MOS) materials such as Mn_3O_4 with desired nanostructures, including phase and morphology, are widely used for high-sensitive gas detection [1]. Calcination is the frequently used heat treatment route for preparing MOS sensing materials [2]. Under suitable calcination conditions, MOS materials with pure phase and desired morphology can be prepared from precursors [3]. Calcination at high temperatures bring high energy consumption, and some specific morphologies of the precursor may be destroyed [4]. The available calcination conditions are often determined by research experience [5], which is challenging to achieve comprehensive optimization in terms of structural transformation and morphological inheritance of precursors. To achieve on-demand sensing material preparation, systematic optimization of calcination conditions is necessary.

Figure 1 schematically shows the cantilever-TGA technology used to optimize the calcination conditions of Mn_3O_4-based sensing materials. With the heating procedure, the temperature of the β-MnO_2 precursor loaded on the microcantilever increases accordingly. Meanwhile, the frequency-change Δf of the microcantilever is recorded in real-time. According to the proportional relationship of

$\Delta f \propto \Delta m$, the mass-loss Δm of the precursor can be quantitatively obtained [6, 7]. With the measured cantilever-TGA curve, structural transformation can be obtained.

Figure 1: Schematic of the cantilever-TGA technology for exploring the calcination process.

EXPERIMENTIAL

Experimental Setup

Figure 2 shows the experimental setup for the TGA measurements, where the microcantilever is placed in a chamber with a controlled atmosphere.

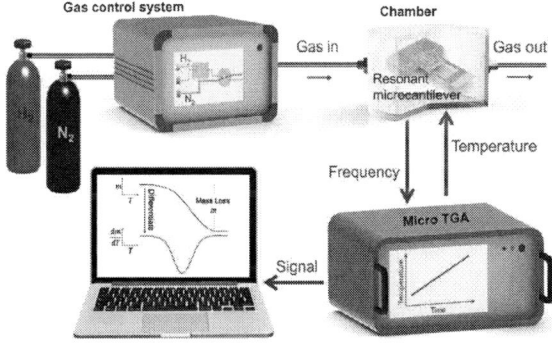

Figure 2: Schematic diagram of the experimental setup for cantilever-TGA measurements.

Preparation of MnO₂

β-MnO₂ was synthesized by hydrothermal method. 2.56 g NaMnO₄ was dissolved in 80 mL water and 4.06 g MnSO₄·5H₂O was added to the solution with stirring and stirred for a further 30 min. The mixture was transferred to a stainless steel autoclave with Teflon vessel liner. Then the mixture was heated at 433 K for 12 h. After cooling down to room temperature, the products were collected via high-speed centrifugation and washed sequentially with deionized water as well as ethanol to remove the residual chemicals. Finally, the products were dried at 353 K overnight to give a black powder of β-MnO₂ [8].

Characterization

Figure 3a shows the resonant microcantilever with the β-MnO₂ precursor loaded at its free-end. To provide desired temperatures on the material, the special-designed microcantilever integrates a micro-heater. The SEM mages in Figures 3b clearly show the one-dimensional (1D) nanostructure of the β-MnO₂ sample. The high-resolution TEM (HRTEM) images and selected-area electron diffraction (SAED) of β-MnO₂ show that the lattice fringes are 0.28 nm and 0.31 nm. That is associated with the (001) and ($1\bar{1}0$) crystal plane respectively. The XRD pattern in Figure 3d indicates that the sample used for calcination can be assigned to pure β-phase MnO₂ (JCPDS card No. 24-0735).

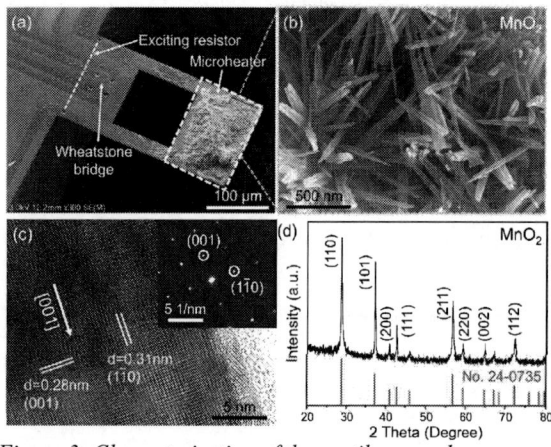

Figure 3: Characterization of the cantilever and precursor materials. (a) SEM image of the cantilever; (b) SEM image shows the nanowire shape of the precursor; (c) HRTEM image and selected area electron diffraction (SAED) of the β-MnO₂ precursor; (d) XRD pattern shows the pure β-MnO₂ phase of the precursor.

RESULTS & DISCUSSION

As shown in Figure 4a, the 1D β-MnO₂ precursor decomposed along with the temperature increase and two mass-loss stages can be observed in the obtained TGA curve. The first stage with a mass-loss of approximately 9.6% is brought by the deoxidation of β-MnO₂ to Mn₂O₃. With the increasing temperature, the produced Mn₂O₃ will continue the deoxidation process, followed by the second mass-loss stage. According to the TGA results, Mn₃O₄ can be produced with a heating temperature above 560°C under the N₂ atmosphere. In contrast, we find that if β-MnO₂

nanowires are calcined in a N₂ atmosphere mixed with 5% H₂, the calcination conditions become moderate. As shown in Figure 4b, two mass-loss can be observed in the TGA-on-chip curve obtained under a 5% H₂-contained N₂ atmosphere. Moreover, the mass loss curve is steeper than that in N₂. Due to the strong reducibility of H₂, the β-MnO₂ precursor can be reduced to Mn₃O₄ at a moderate temperature as low as about 330°C, which is more than 200°C lower than the calcination under a pure N₂ atmosphere. Obviously, the lower calcination temperature is beneficial for the 1D morphology preservation of the β-MnO₂ precursor to the Mn₃O₄ product. According to the TGA-on-chip results, a calcination temperature of 400°C is sufficient to obtain Mn₃O₄ in an H₂-contained N₂ atmosphere. When the temperature rises to 650°C, the produced Mn₃O₄ can be further reduced to MnO. The TGA-on-chip measurements indicate that the deoxidation process of the β-MnO₂ nanowires under a pure N₂ atmosphere is longer than under the H₂-contained N₂ atmosphere, which means the β-MnO₂ nanowires are deoxidated with a slow rate. Therefore, if the calcination time is insufficient, the β-MnO₂ precursor will decompose incompletely, and therefore, the product will contain impurities.

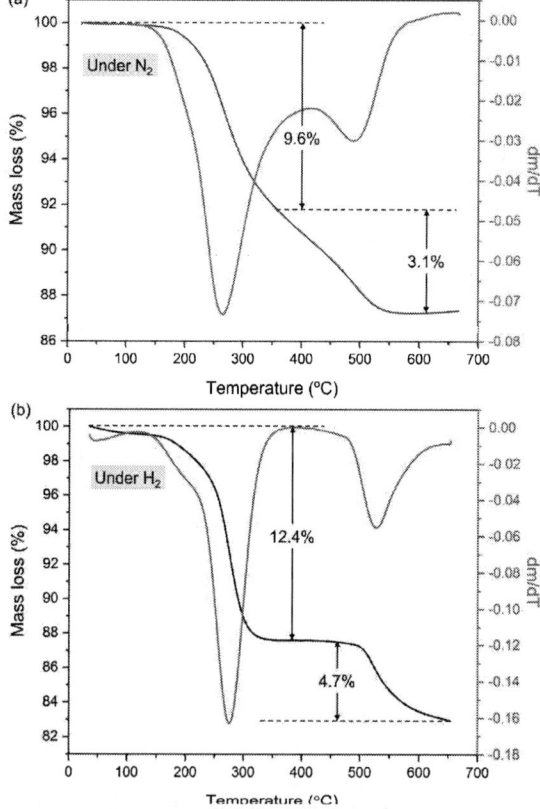

Figure 4: Cantilever-TGA curves of the β-MnO₂ precursor calcinated under two atmospheres. (a) under N₂ and (b) H₂ atmosphere. The red curve shows the decomposition rate of the sample and indicates the temperature at which the maximum rate occurs.

The *in-situ* TEM results in Figure 5a-b indicate the 1D

nanowire morphology of the β-MnO₂ has been destroyed during the calcination under the N₂ atmosphere, while the Mn₃O₄@H₂ sample well maintains the nanowire structure.

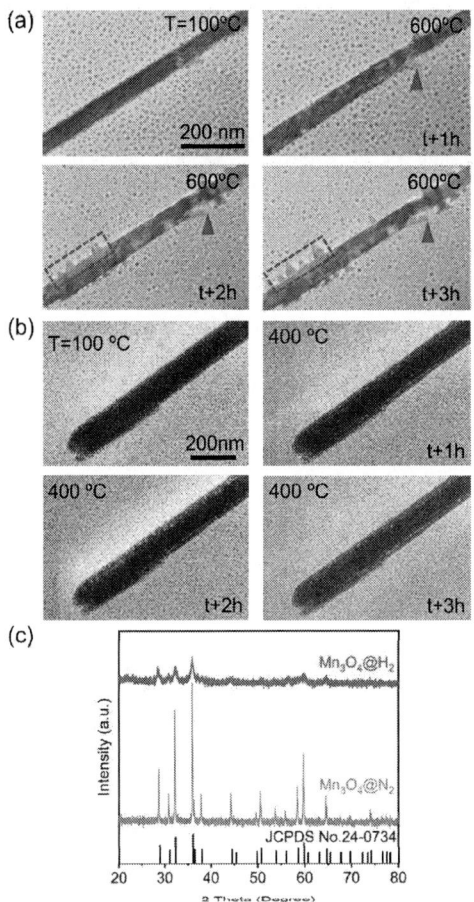

Figure 5: (a) In-situ TEM images of β-MnO₂ nanowires under N₂ atmosphere. The process of nanowire destruction is marked in red. (b) In-situ TEM images of β-MnO₂ nanowires under H₂ atmosphere. (c) XRD patterns of the two Mn₃O₄ samples.

Table 1: Physical properties of the two Mn₃O₄ samples.

	Resistivity (ohm cm)	Mobility (cm²/V·s)	Carrier Density (10¹⁶ cm⁻³)	Band gap (eV)
Mn₃O₄@N₂	105	0.944	6.27	1.83
Mn₃O₄@H₂	87	1.387	7.25	1.32

XRD characterizes the two products obtained with varying processes of calcination, and the comparison results are shown in Figure 5c. The Mn₃O₄ obtained under the pure N₂ atmosphere at 600°C has stronger peak intensity and narrower half peak width than Mn₃O₄ obtained under the H₂-contained N₂ atmosphere at 400°C. That means the average particle sizes of Mn₃O₄@N₂ are larger than those of Mn₃O₄@H₂. The Debye-Scherrer

equation determines the average crystallite size of nanomaterials as

$$D = \frac{K\lambda}{\beta \cos\theta}$$

where D is the average crystallite size; K is Scherrer constant, 0.89; λ is the X-ray wavelength,1.54Å; β is the full width of half-maximum intensity (FWHM); θ is Bragg's peak angle. The particle sizes of Mn₃O₄@H₂ and Mn₃O₄@N₂ are calculated by the Scherrer equation as 17.2 and 42.9 nm, respectively. Table 1 compares the physical properties of the two Mn₃O₄ products, indicating that the Mn₃O₄@H₂ sample has higher mobility and is more suitable for gas sensing applications.

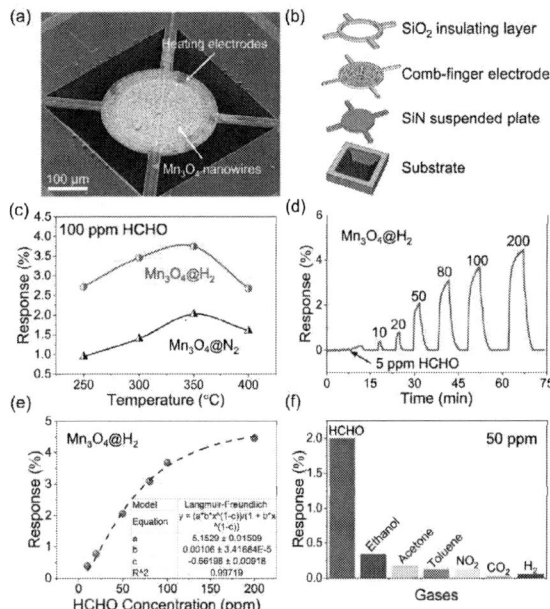

Figure 6: Sensing results of the materials. (a) SEM images of the micro-hotplate sensor; (b) Schematic diagram of the micro-hotplate sensor structure (c) Temperature-dependent responses of the two sensors to 100 ppm HCHO; (d) Sensing curves of the Mn₃O₄@H₂ materials to HCHO. (e) Langmuir-Freundlich fitting of the Mn₃O₄@H₂ material to HCHO. (f) Selectivity results of the Mn₃O₄@H₂ material.

Gas sensing properties for HCHO of the two products are measured. As shown in Figure 6a, manganese oxide nanowires are loaded onto an integrated MEMS chip with the assistance of inkjet printing. Figure 6b shows the schematic structure of the device, which consists of a substrate, a Si$_x$N$_y$ suspended plate, a comb finger electrode and a SiO₂ insulating layer, respectively [9]. According to the response results of 100 ppm HCHO vapor in Figure 6c, Mn₃O₄@H₂ shows a better sensing response than Mn₃O₄@N₂. Besides, the response signal reaches the maximum at 350°C. Figure 5d plots multi-concentration profiles of Mn₃O₄@H₂ at an optimal working temperature of 350°C. When the HCHO concentration is as low as 5 ppm, only the sensor loaded with Mn₃O₄@H₂ still has a response signal. Obviously, the LOD is lower than 5 ppm. As shown in Figure 5e, we found the relationship between the sensing response signal and the HCHO concentration range of 5-200 ppm fits well with Langmuir-Freundlich equation. To investigate the selectivity of the sensor, six kinds of common gases including ethanol, acetone, toluene, NO_2, H_2 and CO_2 are selected as interfering gases. As shown in Figure 6f, the sensor outputs a response value up to 2 for 50 ppm HCHO, while the same sensor only shows a response value under 0.5 to the interfering gases with the same concentration.

CONCLUSION

In summary, the chip-based TGA techniques were successfully used to optimize the calcination parameters of β-MnO_2 nanowires. According to our chip-based TGA results, Mn_3O_4 can be obtained by calcining β-MnO_2 precursors at 600°C under a pure N_2 atmosphere, but lowered to 400°C under 5% H_2 atmosphere. Meanwhile, through *in-situ* TEM we observe that when the temperature is kept at 600°C under pure N_2, the morphology of the nanowires changes and some nanoparticles appear around the nanowires. In contrast, the one-dimensional morphology can be calcined at 400°C under 5% H_2 for more than three hours without change. Guided by the calcination parameters revealed by TGA, the desired phase products are prepared. The calcination products are carefully characterized by SEM, XRD and Hall effect and compared their gas-sensitive performance for HCHO. The results clearly show that $Mn_3O_4@H_2$ has smaller nanoparticle sizes and greater gas sensitivity for HCHO. Our research not only provides a kind of manganese oxide catalyst for a great performance HCHO sensor but also provides a method of MEMS-based TGA to optimize calcination parameters for the on-demand preparation of materials.

ACKNOWLEDGEMENTS

The authors gratefully acknowledge financial support by the National Key R&D Program of China (2020YFB2008603, 2021YFB3201302), National Natural Science Foundation of China (61974155, 61831021, 62104241, 62271473, U21A20500), Key Research Program of Frontier Sciences of Chinese Academy of Sciences (QYZDJ-SSW-JSC001), Science and Technology Innovation Plan of Shanghai (19510744600), Scientific Instrument Project of the Chinese Academy of Sciences (YJKYYQ20210024), Shanghai Pujiang Program (20PJ1415600), Innovation Team and Talents Cultivation Program of National Administration of Traditional Chinese Medicine (ZYYCXTD-D-202002, ZYYCXTD-D-202003).

REFERENCES

[1] A. Dey, "Semiconductor Metal Oxide Gas Sensors: A Review", *Mat. Sci. Eng. B*, vol. 229, pp. 206-217, 2018.

[2] M. H. Seo, M. Yuasa, T. Kida, J. S. Huh, K. Shimanoe, N. Yamazoe, "Gas Sensing Characteristics and Porosity Control of Nanostructured Films Composed of TiO_2 Nanotubes", *Sensor. Actuat. B-Chem.*, vol. 137, pp. 513-520, 2009.

[3] S. Li, N. Hasan, H. Ma, G. Zhu, L. Pan, F. Zhang, N. Son, M. Kang, C. Liu, "Hierarchical V_2O_5/ZnV_2O_6 Nanosheets Photocatalyst for CO_2 Reduction to Solar Fuels", *Chem. Eng. J.*, vol. 430, pp. 132863, 2022.

[4] Y. Yu, J. Zhang, C. Chen, C. He, J. Miao, H. Li, J. Chen, "Effects of Calcination Temperature on Physicochemical Property and Activity of $CuSO_4/TiO_2$ Ammonia-selective Catalytic Reduction Catalysts", *J. Environ. Sci.*, vol. 91, pp. 237-245, 2020.

[5] M. Wu, W. Zhan, Y. Guo, Y. Wang, Y. Guo, X. Gong, L. Wang, G. Lu, "Solvent-free Selective Oxidation of Cyclohexane with Molecular Oxygen over Manganese Oxides: Effect of the Calcination Temperature", *Chinese J. Catal.*, vol. 37, pp. 184-192, 2016.

[6] F. Yao, P. Xu, H. Jia, X. Li, H. Yu, X. Li, "Thermogravimetric Analysis on a Resonant Microcantilever", *Anal. Chem.*, vol. 94, pp. 9380-9388, 2022.

[7] H. Jia, P. Xu, X. X. Li, "Integrated Resonant Micro/Nano Gravimetric Sensors for Bio/Chemical Detection in Air and Liquid", *Micromachines*, vol. 12, pp. 645, 2021.

[8] E. Hayashi, Y. Yamaguchi, K. Kamata, N. Tsunoda, Y. Kumagai, F. Oba, M. Hara, "Effect of MnO_2 Crystal Structure on Aerobic Oxidation of 5-Hydroxymethylfurfural to 2,5-Furandicarboxylic Acid", *J. Am. Chem. Soc.*, vol. 141, pp. 890-900, 2019.

[9] Y. Chen, P. Xu, X. Li, Y. Ren, Y. Deng, "High-performance H_2 Sensors with Selectively Hydrophobic Micro-plate for Self-aligned Upload of Pd Nanodots Modified Mesoporous In_2O_3 Sensing-material", *Sensor. Actuat. B-Chem.*, vol. 267, pp. 83-92, 2018.

CONTACT

*P.C. Xu, tel: +86-21-62131794; xpc@mail.sim.ac.cn
*X.X. Li, tel: +86-21-62131794; xxli@mail.sim.ac.cn

AN INTELLIGENT GAS ANALYSIS SYSTEM CONSISTING OF SENSORS AND A NEURAL NETWORK IMPLEMENTED USING THIN-FILM TRANSISTORS

Zong Liu[1, 2], Yushen Hu[1, 2], Gabriel E. Carranza[1], Fei Wang[2], and Man Wong[1]
[1]The Hong Kong University of Science and Technology, HONG KONG and
[2]Southern University of Science and Technology, CHINA

ABSTRACT

Described presently is an "intelligent" gas-analysis system consisting of a gas-sensor array and an artificial neural network (ANN) consisting of dual-gate thin-film transistors (TFTs) for in-system inferencing. The micro-hotplates needed for implementing gas sensors with improved specificity are constructed using a silicon-migration technology for the spontaneous creation of a suspended diaphragm. The TFTs deployed in both the array and the network sub-systems are based on semiconducting indium-gallium-zinc oxide. The two sub-systems are packaged on a printed-circuit board. The ANN infers from the raw sensor signals the identity and the concentration of the gas, thus demonstrating the feasibility of building TFT-based intelligent sensor systems.

KEYWORDS

Gas-sensor, thin-film transistor, indium-gallium-zinc oxide, artificial neural network, monolithic integration.

INTRODUCTION

Gas sensors are demanded and deployed in industrial production, scientific research, air quality monitoring, disease diagnosis, etc. The properties of a sensor can be differentiated by a variety of techniques, such as the deployment of different functional materials or operating at different temperatures, etc. Deployment of an array of differentiated sensors is an effective way of improving gas sensing specificity. With the ubiquitous deployment of sensor systems when implementing the internet-of-things technology, the resulting large volume of sensor data may incur high computation costs and suffer from latency problems with Cloud-based data processing. Consequently, "edge computation" enabled by local hardware data processing systems is desired.

Presently reported is the construction on micro-hotplates (MHPs) of a gas-sensor array addressed using a monolithically integrated active-matrix of thin-film transistors (TFTs). A silicon-migration technology (SiMiT) requiring no sacrificial layer etch [1] is deployed for the spontaneous realization of suspended diaphragms. Coated with tin (IV) oxide (SnO_2) as the functional material for gas sensing, sensors in the array are differentiated by modulating the temperature of the different MHPs.

Recently, artificial neural network (ANN) circuits deploying parallel, double-gate (DG) TFTs have been reported [2] and trained for pattern-recognition. Presently, an ANN as an edge inferencing engine and a gas-sensing MHP array are packaged together on a printed-circuit board (PCB) to realize an "intelligent" gas analysis system.

FABRICATION OF THE GAS SENSOR ARRAY AND THE ANN

Shown in Figure 1 is the fabrication process flow of the TFT-addressed, gas-sensing MHPs and the TFTs for realizing an ANN. Their high degree of similarity illustrates the inherent compatibility and potential of monolithic integration of the two sub-systems.

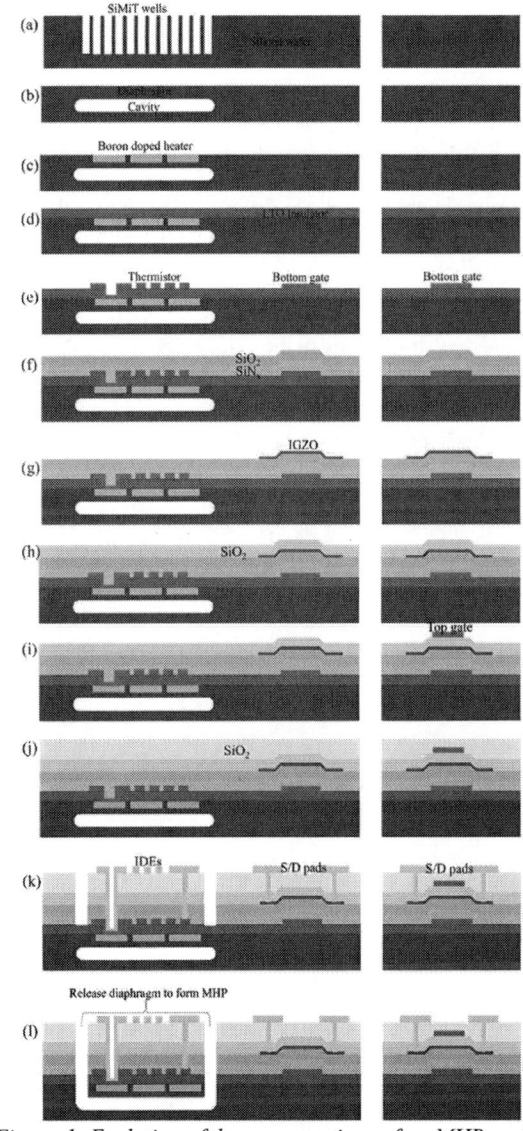

Figure 1: Evolution of the cross-sections of an MHP and a bottom-gate TFT (left) in a gas-sensor array and a parallel, dual-gate TFT (right) in an ANN.

Construction of the array starts with the definition of a SiMiT well array on a silicon (Si) substrate using deep reactive-ion etching (Fig. 1a). Annealing in a hydrogen atmosphere at a temperature of 1120 °C results in the spontaneous formation of a vacuum cavity sealed by a 2.5-μm -thick Si diaphragm (Fig. 1b). Compared to other established processes [3] of making suspended structures, SiMiT requires no sacrificial layer etch. Consequently, common process incompatibility issues inhibiting the monolithic integration of MEMS and electronic devices, such as transistors, are partially resolved.

Boron is selectively implanted to form the resistive heater on the suspended diaphragm (Fig. 1c). After the deposition of an insulating low-temperature oxide (LTO) layer (Fig. 1d), the thermistor on the diaphragm and the bottom gate electrode of an adjacent TFT are patterned from the same molybdenum (Mo) layer (Fig. 1e). After the formation of the gate dielectric layer, a 20-nm indium-gallium-zinc oxide (IGZO) as the active layer is sputtered in an atmosphere of oxygen (O_2):argon = 2:18 sccm and patterned (Fig. 1g). Following the formation of a passivation layer and the opening of the contact holes, a stack of Mo under aluminum (Al) is deposited to form both the source/drain (S/D) electrodes of the TFTs and the gas-sensing interdigitated electrodes (IDEs) (Fig. 1k). An annealing at 400 °C in O_2 for 4 h is carried out to passivate the channel of the TFT and to form its thermally induced S/D regions. Thermally insulated from the substrate and anchored using 4 tethers, a suspended MHP (Fig. 1l) is defined by dry-etching of the diaphragm.

A separate ANN made of parallel, DG, IGZO TFTs is fabricated using a similar TFT process, as shown on the right-hand-side of Figure 1. In principle, the processing of the ANN can be done simultaneously with that of the TFTs on the gas sensor chip, with the small difference of inserting (Fig. 1i) a top gate Mo electrode on a DG TFT.

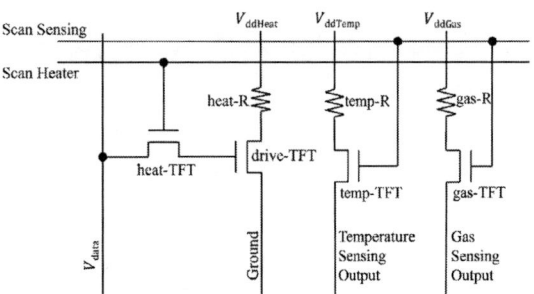

Figure 2: The circuit schematic of a unit in the gas-sensing MHP array.

Shown in Figure 2 is the schematic of the circuit in one unit of the gas-sensing MHP array. This unit is replicated to form an "$m \times n$" array. The current flowing through the resistive heater is regulated by passing the voltage V_{data} on the data line to the gate of the drive-TFT when the heat-TFT is turned on. Benefitting from the exceptionally low off-state current of an IGZO TFT, V_{data} is stored and maintained on the capacitor associated with the gate of the drive-TFT when the heat-TFT is turned off. The currents passing through the thermistor temp-R and the functionalized gas-sensing IDEs gas-R are passed to the

respective output lines when the respective temp-TFT and gas-TFT are turned on. Active-matrix addressing allows the temperature of the MHP associated with each sensor unit to be independently modulated and measured, and the sensing signal associated with gas-R to be individually monitored. The number of interconnections can be significantly reduced from $m \times n$ to $m + n$ when deploying such an addressing scheme.

Shown in Figure 3 is the schematic of a computation unit of the ANN. This unit is replicated to construct ANNs of an arbitrary size. At a given V_{ds}, the current flowing through the DG TFT is modulated by both the input and weight signals applied respectively on its bottom and top gate electrodes. From a device operation point-of-view, the weight signal modulates the threshold voltage "seen" by the input signal. Applying active-matrix addressing by controlling the signal on the scan line, the weight signal can be written in and stored on the weight storage capacitor. The units in the same column share a "Signal in" line while those in the same row share an "Output" line.

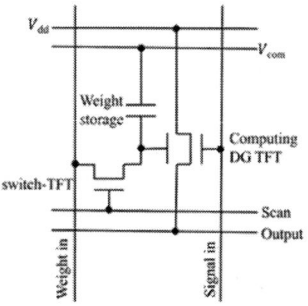

Figure 3: The circuit schematic of a unit in the ANN.

Shown in Figure 4a is an 8 × 8 ANN chip packaged on a PCB with its peripheral components consisting of a demultiplexer, multiplexers, analog/digital converter, and microcontroller. Shown in Figure 4b is a 2 × 2 gas sensor array with four MHPs.

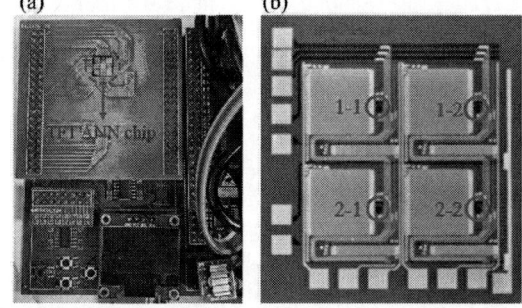

Figure 4: Photographs of (a) an ANN chip and its peripheral circuit. (b) a 2 × 2 gas sensor array with four MHPs labeled 1-1 to 2-2.

MEASUREMENT RESULTS

Shown in Figure 5 are the temperature vs. voltage thermal characteristics of an MHP. Shown in the inset is the image obtained using an infrared camera of the MHP with 10 V applied across its heater. The respective dimensions of the MHP and each of its tethers are 130 μm × 130 μm and 135 μm × 20 μm. Together with the cavity below the MHP, the long and thin tethers reduce the heat loss from the MHP to the surrounding substrate. The temperature is seen to rise with increasing voltage applied across the heater. Based on Joule's heating, the rise exhibits a quadratic rather than a linear dependence on voltage. Benefitting from the thermal insulation of the suspended MHP, a relatively low 66 mW is needed to obtain 300 °C.

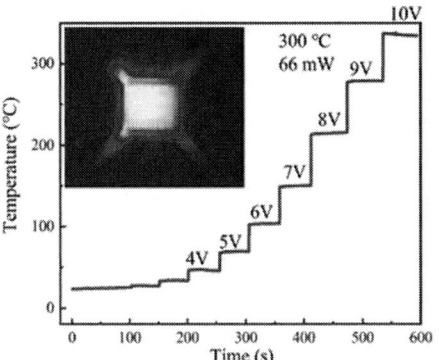

Figure 5: The thermal characteristics of an MHP. The infrared image of the MHP at 10 V shows a relatively uniform temperature distribution.

The IDEs of the 2 × 2 gas-sensor array are coated with SnO_2 as the functional material. The temperatures of MHP Units 1 to 4 are set at approximately 300, 250, 200, and 150 °C. Each gas-sensing output terminal is connected in series with a resistor to form a voltage divider and to generate an output voltage signal. In each cycle of the test, the gas being detected is sealed in a chamber to maintain the concentration for about 10 mins. While the optimum working temperature for both gases is ~300 °C, the response patterns of the two gases are clearly different (Fig. 6), particularly for the lower-temperature sensors.

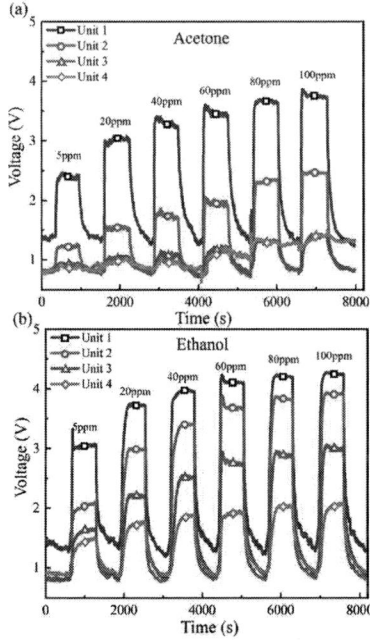

Figure 6: The measured output voltage of a sensor with SnO_2 as the gas-sensing medium when exposed to (a) acetone and (b) ethanol from 5 to 100 ppm.

A schematic of the sensor-ANN system is exhibited in Figure 7a, showing how a 3-level 4 × 4 × 1 ANN (Fig. 9b) is implemented by passing data twice through an 8 × 8 neuromorphic computation array. For each gas and from Figure 6, two sets of 30 data points covering each concentration level are randomly extracted for training and for inferencing.

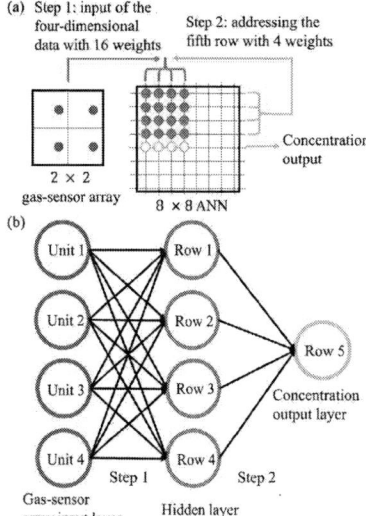

Figure 7: (a) Schematic of the training process of concentration extraction. (b) Corresponding schematic of the structure of the ANN.

Concentration extraction is performed using the analog outputs directly as the inputs of the ANN. A supervised, feedforward training algorithm is used to calculate the gradient of the mean-squared error using the difference between the cost function before and after a small change is applied to the weight signal on the top gate electrode. The gradient in turn is used to continuously adjust the weight signal. The iteration is terminated and training is complete when the cost function is below a preset value (Fig. 8). Respectively 29 and 20 epochs are executed to train the ANN for acetone and ethanol.

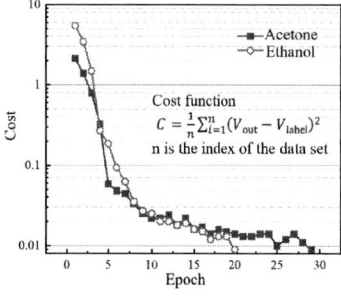

Figure 8: Evolution of the cost function during training for finding the concentration of ethanol and acetone.

The concentration extraction results of the training and inference data sets refed into the post-trained ANN can be seen in Figure 9. The output (red hollow circles) of the ANN is very close to the concentration labels (black solid squares). These results show that the concentration of gases can be successfully inferred.

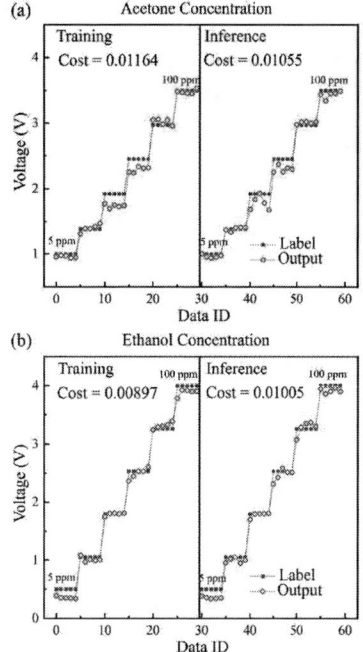

Figure 9: (a) The concentration extraction of (a) acetone after 29 epochs of training and (b) ethanol after 20 epochs of training.

For gas species identification, a $3 \times 4 \times 1$ ANN is implemented using the same 8×8 computation array. From Figure 6, two sets of 40 data points including air and each concentration level are randomly extracted for training and for inferencing. The label voltage for acetone, ethanal and air are respectively 3, 2 and 1 V. For each test, the outputs of the units are normalized w.r.t. that of Unit 3, thus generating 3 independent sensor outputs for the ANN.

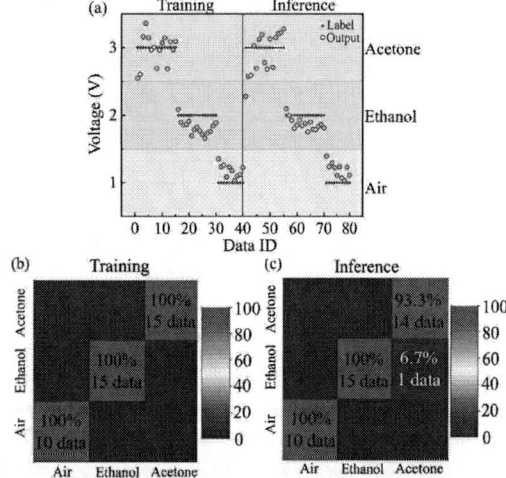

Figure 10: (a) The identification result of the post-trained ANN for the training set and the inference set. High accuracy for identifying acetone, ethanol, and air for (b) training set and (c) inference set.

Shown in Figure 10 are the results of identifying acetone, ethanol, and air. Upon completion of the training process, the ANN outputs of the different gases (red hollow circles) cluster around the their corresponding labels (black solid squares). The ANN achieves 100% correct identification for the training set as shown in Figure 10b. The trained ANN is applied to the inference set and the results are shown in Figure 10c, showing similar identification accuracies.

CONCLUSION

Based on SiMiT technology, monolithically integrated 2×2 gas sensor array incorporating TFT-based active-matrix addressing has been implemented. The temperatures of the MHPs can be individually modulated to improve sensing specificity. Combined with an ANN chip based on DG IGZO TFTs, gas concentration and species identification have been achieved. These encouraging results demonstrate the feasibility of building a monolithically integrated, intelligent gas-analysis system, not just for gas-sensing but also for other applications.

ACKNOWLEDGMENT

This work was supported in part by the Science and Technology Program of Shenzhen under Grant JCYJ20200109140601691 and in part by the Innovation and Technology Fund under Grant GHP/013/19SZ, jointly with the Science and Technology Program of Shenzhen under Grant SGDX20190918105001787. Device fabrication was carried out at the Nanosystem Fabrication Facility of The Hong Kong University of Science and Technology.

REFERENCES

[1] F. Zeng and M. Wong, "A Self-Scanned Active-Matrix Tactile Sensor Realized Using Silicon-Migration Technology," in *Journal of Microelectromechanical Systems*, vol. 24, no. 3, pp. 677-684, June 2015, doi: 10.1109/JMEMS.2014.2344025.

[2] Y. Hu, Y. Wang, T. Lei, F. Wang and M. Wong, "Neuromorphic Implementation of Logic Functions Based on Parallel Dual-Gate Thin-Film Transistors," in IEEE Electron Device Letters, vol. 43, no. 5, pp. 741-744, May 2022, doi: 10.1109/LED.2022.3164684.

[3] A. Hierlemann, O. Brand, C. Hagleitner and H. Baltes, "Microfabrication techniques for chemical/biosensors," in Proceedings of the IEEE, vol. 91, no. 6, pp. 839-863, June 2003, doi: 10.1109/JPROC.2003.813583.

CONTACT

*Zong Liu, zliucy@connect.ust.hk
*Fei Wang, wangf@sustech.edu.cn
*Man Wong, eemwong@ust.hk

SINGLE-LAYER-ELECTRODE TEMPERATURE-MODULATED SNO₂ GAS SENSOR CELL WITH LOW POWER CONSUMPTION FOR DISCRIMINATION OF FOOD ODORS

Chong Xing[1], Ruichen Liu[1], Yan Zhang[1], Dongcheng Xie[1], Yudong Wang[1], Yuan Huang[1], Muhammad Mustafa [1], Haochen Zhang[1], Zhongyu Shi[1], Lei Xu[1], and Feng Wu[2]

[1] School of Microelectronics, University of Science and Technology of China, Hefei, CHINA and
[2] School of Information Science and Technology, University of Science and Technology of China, Hefei, CHINA

ABSTRACT

This paper presents a SnO_2-based six-channel single-layer-electrode temperature-modulated metal oxide (MOX) gas sensor cell with excellent gas discrimination and ultra-low power consumption. The proposed device features horizontally arranged heater and detecting electrodes on the same layer without an isolation layer. As a result, the fabrication steps are largely reduced and the device endurance under high-temperature operation is significantly improved. More importantly, the temperature distribution in the active region of the MOX gas sensor can be effectively tuned by the strategically designed heater width on each sensing channel to achieve high-resolution gas detection. Such gas discrimination ability was further verified by detecting the food-related gases and odors from different food species.

KEYWORDS

gas sensor, temperature-modulated, process simplification, low power.

INTRODUCTION

Metal oxide (MOX) gas sensor is a competitive candidate for gas detection applications thanks to its CMOS compatibility, low cost, and high gas responsivity. Recently, the emerging requirements of MEMS-based gas sensing are greatly promoting the development of MOX gas sensors towards miniaturization, high integration density, and low power consumption [1]. In this regard, rational designs of both device structure and fabrication process of the MOX gas sensors are required to realize decent gas discrimination ability on a single device/material platform [2] [3]. In this work, we demonstrate a SnO_2-based six-channel MOX gas sensor cell with excellent gas discrimination, low power consumption, and a simple fabrication process based on MEMS fabrication techniques. The proposed device structure features the horizontally arranged heater and detection electrodes, and the strategically designed heater width on each sensing channel for temperature gradient modulation and high-resolution gas detection. Further comparative analysis of the response of the MOX gas sensors under different food-related gases was performed, showing great potential for future implementation of E-nose applications.

DESIGN AND FABRICATION PROCESS

Design of the MOX Gas Sensor Cell

In a conventional MOX gas sensor, the size of the active region is typically as large as about $500{\times}500$ μm². Therefore, a large-scale heater has to be coil-like to uniformly distribute the heat distribution at the active region. To miniaturize the lateral scale of the MOX gas sensor, the heater and detecting electrode need to be vertically stacked and isolated by a $PECVD-SiN_x$ insulation layer. Notably, the large internal strain stress in such an insulation layer is the main cause of weak mechanical strength and performance degradation of MOX gas sensors under high-temperature operation.

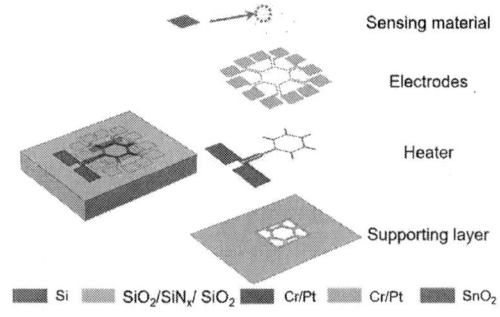

Figure 1: Schematic of the proposed six-channel single-layer-metal temperature-modulated MOX gas sensor cell.

Thanks to the advanced MEMS fabrication techniques, the active region of a MOX gas sensor can be scaled down to $10{\times}10$ μm². Hence, the heater and the detecting electrodes can be spatially arranged on the same horizontal plane, as shown in Fig. 1. This structure design can significantly reduce the steps and cost of the fabrication process and improve the device's durability to withstand a higher heating voltage and operating temperature. Furthermore, the heater in the proposed MOX gas sensor consists of six in-series micro-heaters for each sensing channel (Ch. 1-Ch. 6). Since the current flowing through each micro-heater is the same, the heat generation and temperature distribution at different sensing areas can be effectively tuned by the metal width of the micro-heater. Fig. 2 shows the optimized width for each micro-heater, which is verified by thermal simulation conducted by COMSOL Multiphysics software. As shown in Fig. 3(a) and 3(b), the uniformity of temperature on the 34-μm-width suspended bridge at the active region was verified. A temperature gradient of about 40°C between two adjacent sensing channels can be achieved, fundamentally enabling a high-gas-resolution MOX sensor based on a single device/material platform.

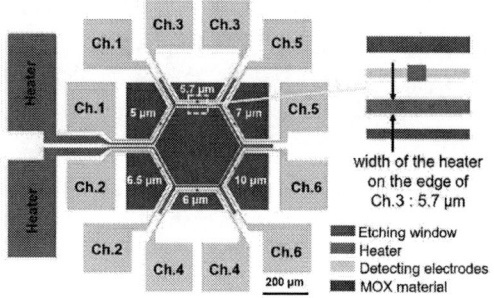

Figure 2: Layout design (the width of the heater on each edge of the gas sensor cell is marked).

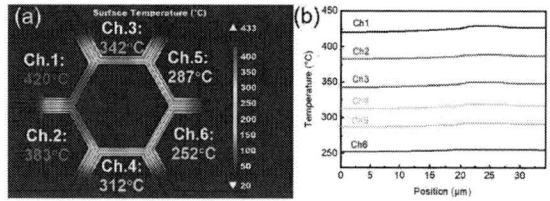

Figure 3: (a)Thermal simulation of the MOX gas sensor cell; (b)temperature distribution of the 34-μm-width suspended bridge at the active region.

Fabrication Process

The fabrication process of the proposed MOX gas sensor is schematically shown in Fig. 4, which can be concluded as followed:

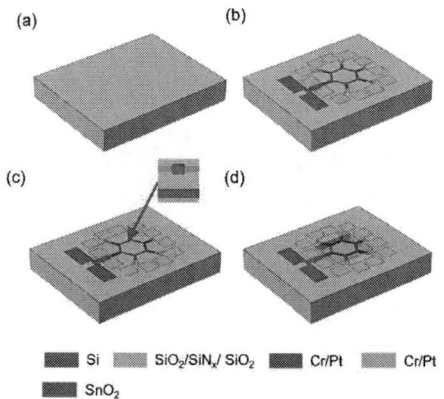

Figure 4: Schematic of the fabrication process of the proposed MOX gas sensor cell. (a) supporting layer by LPCVD; (b) metal deposition; (c) SnO_2 deposition; (d) Dry etching.

(a) A 2.6-μm-thick supporting layer consisting of $SiO_2/SiN_X/SiO_2$ (400/1200/1000 nm) was formed on a 4-inch silicon wafer by LPCVD (Low Pressure Chemical Vapor Deposition).

(b) The heater and the detecting electrodes (Cr/Pt = 10/200 nm) were simultaneously magnetron sputtered and patterned by a lift-off process.

(c) The 400-nm-thick SnO_2 as gas-sensing material

was patterned on the active region by a lift-off process.

(d) Multiple dry etching steps were carried out to form the final structure including (1) three RIE (Reactive Ion Etching) steps to remove $SiO_2/SiN_X/SiO_2$ at the etching window; (2) isotropic etching to release the silicon under the active region by SF_6-based ICP (Inductive Coupled Plasma).

RESULTS AND DISCUSSION

Surface Characterization

Fig. 5(a) shows the 45°-tilted SEM image of the gas sensor cell. It can be observed that the silicon underneath the active region was successfully removed. This structure can reduce the undesired heat dissipation paths towards the substrate, thereby increasing the utilization ratio of the generated heat from the micro-heater and realizing low power consumption. Fig. 5(b) shows the enlarged image of one of the six gas sensing channels. The energy-dispersive X-ray spectroscopy (EDS) element mapping including Sn and O is shown in Fig. 5(c)-(e), demonstrating good uniformity of the sputtered SnO_2 material.

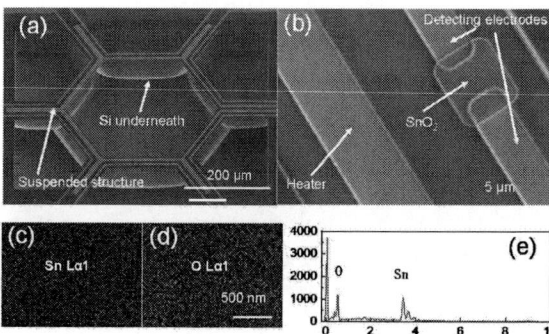

Figure 5: SEM images of (a) 6-channel gas sensor cell and (b) one sensing region of the gas sensor cell; EDS mapping images of the SnO_2 film including (c) Sn and (d) O elements; (e) EDS energy spectrum of the SnO_2 film.

Performance Evaluation of the Heater

First of all, we selected the proper heating voltage of the heater, considering that the power consumption of the micro heater accounts for the majority of the total power consumption of the MOX gas sensor. Fig. 6(a) shows the relationship between the heating power consumption and the applied heating voltage. At the heating voltage of 3.4-4.2 V used in the following tests, the power consumption is as low as 5.5 mW.

Thereafter, we employed the infrared thermal imager to check the heat distribution of the gas sensor. As shown in Fig. 6(b), the temperature is distributed as a gradient in the six active regions where the SnO2 was patterned, which is consistent with the previous simulation results.

In addition, the mechanical strength of gas sensors with micro-heater on suspended structures is another vital factor for the sake of device durability. In our proposed structure, since the PECVD-SiN_x insulation layer is eliminated, the mechanical strength of the device would be enhanced to bear a heating voltage of up to 9 V.

978-1-6654-9309-3/23 $31.00 © 2023 IEEE

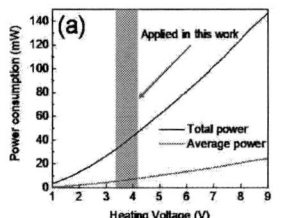
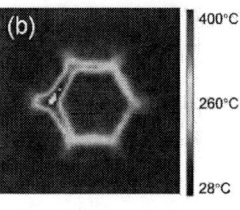

Figure 6: (a) Relationship between the power consumption of the heater and heating voltage; (b) Thermal image of the gas sensor cell taken by infrared imager at the average heating power consumption of 5.5 mW.

Setup for Gas Sensing Test

The performance of the sensor cell is characterized using the test setup shown in Fig. 7(a).

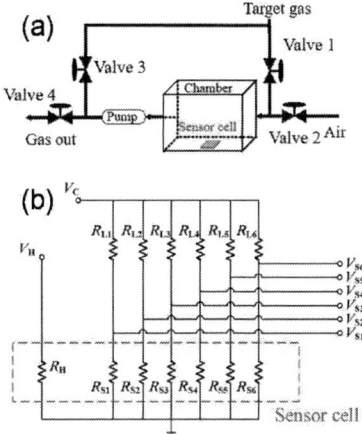

Figure 7: (a) Gas-sensing setup; (b) Gas-detecting circuit of the proposed gas sensor cell.

At the initial state before the target gas injection into the chamber, Valves 1 and 3 are closed and Valves 2 and 4 are open. At this point, the gas inside the chamber and the air in the atmosphere are circulated under the drive of the air pump. Here, we define the resistance between each pair of detecting electrodes at the initial state as the baseline resistance R_0. Thereafter, we inject the target gas into the chamber with closed Valves 2&4 and open Valves 1&3. Under different gas species and gas concentrations, the sensing resistance (R_S) of SnO_2 at each channel would be changed to different values and can be extracted by the gas detection circuit, as shown in Fig. 7(b). Here, R_{S1} to R_{S6} stand for the sensing resistance of six channels and R_{L1} to R_{L6} are the matching resistors. V_{S1} to V_{S6} are the response voltages between six pairs of detecting electrodes determined by $V_C \times R_S/(R_L+R_S)$, where V_C is the total supply voltage. R_H and V_H stand for the heater resistance and heating voltage, respectively.

Food-related Gas Detection

To evaluate the gas-sensing properties of the proposed SnO_2 gas sensor cell, six representative gases generated from the food ripening/spoiling were chosen for testing, including ethanol (C_2H_6O), formaldehyde (CH_2O), ammonia (NH_3), hydrogen sulfide (H_2S), ethyl acetate ($C_4H_8O_2$), and trimethylamine (TMA). At the optimal heating power consumption of 5.5 mW, the response behavior of the six channels to these food-related gases is shown in Fig. 8.

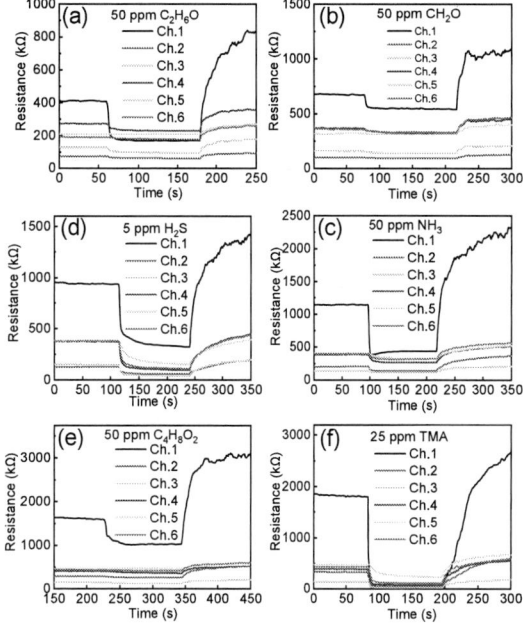

Figure 8: Response to food-related gases (@ 5.5mW) including (a) ethanol, (b) formaldehyde, (c) ammonia, (d) hydrogen sulfide, (e) ethyl acetate, and (f) trimethylamine.

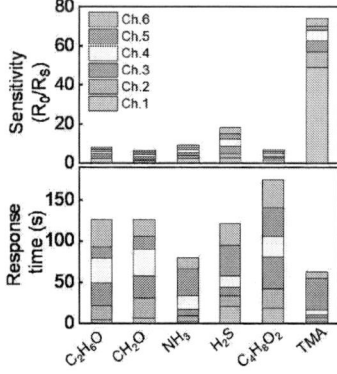

Figure 9: Stacked column chart of sensitivity and response time to food-related gases.

Based on these results, we calculated the sensitivity and response time of the device, as concluded in Fig. 9, where the sensitivity and response time are defined as R_0/R_S and the time that the relative change of resistance takes from 10% to 90%, respectively. Overall, the gas sensor cell has the highest average sensitivity and shortest average response time for TMA detection, while the opposite results for $C_4H_8O_2$ detection.

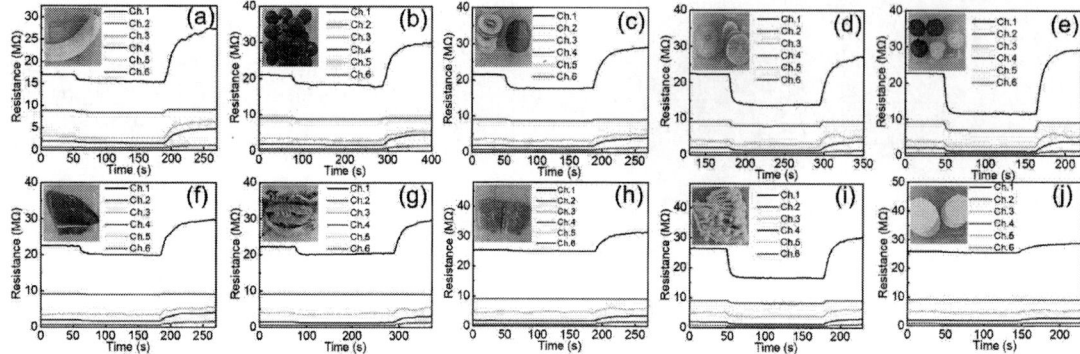

Figure 10: Smell-sensing tests of the gas sensor cell to different food species (@7.8 mW), including (a) banana, (b) grape, (c) kiwi, (d) citrus, (e) lychee, (f) beef, (g) shrimp, (h) rainbow trout, (i) beech mushroom, (j) radish.

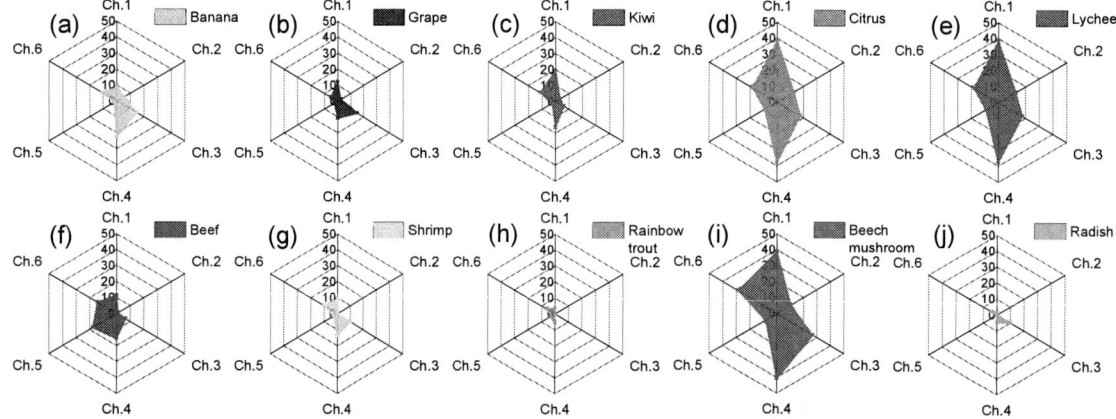

Figure 11: Radar chart of the ΔR (%) of the gas sensor cell to food odors (@7.8 mW).

Besides, the device exhibits similar sensitivity for C_2H_6O and NH_3 while the response time for NH_3 is much lower than for C_2H_6O. When checking the specific sensing properties of each channel, we can find that Ch. 1 possesses the largest sensitivity and shortest response time for all the gases except H_2S while Ch. 4 is found to have the best selection for H_2S. In addition, although the sensitivity of all channels to CH_2O is relatively low and identical, the response speed of different channels can be very distinct. These device merits of the multiple-channel temperature-modulated gas sensor cell can be helpful to distinguish food-related gases.

Food Odors Test

The food odors test was performed by the same setup except that the gas sources are from the crispers containing different food species, as shown in the insert pictures of Fig. 10. The rate of resistance change for smells from these food odors is denoted as $\Delta R=(1-R_S/R_0)\times100\%$. We plotted the radar charts (Fig. 11) as a "gas-detection report" for different food species gathered from the six channels. These different food odors can be well distinguished based on the comparison of ΔR of the six channels. The very specific figures for each food are very referential and informative for precise food discrimination.

ACKNOWLEDGEMENTS

This work was partially carried out at the USTC Center for Micro and Nanoscale Research and Fabrication and supported by the National Key Research and Development Program of China under Grant 2018YFE0194500, the National Natural Science Foundation of China under Grant 31827803, and the Innovation and Entrepreneurship fund of USTC (No. WK5290000003).

REFERENCES

[1] M. I. A. Asri, M. N. Hasan, M. R. A. Fuaad, Y. M. Yunos, and M. S. M. Ali, "MEMS Gas Sensors: A Review", *IEEE Sens. J.*, vol. 21, pp. 18381-18397, 2021.

[2] D. Xie, R. Liu, G. Adedokun, L. Xu, and F. Wu, "A Novel Low Power Hexagonal Gas Sensor Cell for Multi-Channel Gas Detection", in *2021 IEEE 34th International Conference on Micro Electro Mechanical Systems (MEMS)*, Virtual, January 25-29, 2021, pp. 430-433.

[3] F. Xue, G. Adedokun, D. Xie, R. Liu, Y. Zhang, M. Muhammad, L. Xu, and F. Wu, "A Low Power Four-Channel Metal Oxide Semiconductor Gas Sensor Array With T-Shaped Structure", *J. Microelectromech. Syst.*, vol. 31, pp. 275-282, 2022.

CONTACT

*Lei Xu, Email: okxulei@ustc.edu.cn

A PERFORMANCE ENHANCED THERMAL FLOW SENSOR WITH NOVEL DUAL-HEATER STRUCTURE USING CMOS COMPATIBLE FABRICATION PROCESS

*Zhongyi Liu[1], Ruoqin Wang[2], Gai Yang[1], Xinyuan Zhang[1], Rui Jiao[2], Xuejiao Li[1], Jiali Qi[3], Hongyu Yu[2], Huikai Xie[1, 4, *], and Xiaoyi Wang[1,4, *]*

[1]Beijing Institute of Technology, Beijing, CHINA
[2] Hong Kong University of Science and Technology, Hong Kong, China
[3]Hangzhou Dianzi University, Hangzhou, China and
[4]BIT Chongqing Institute of Microelectronics and Microsystems, Chongqing, China.

ABSTRACT

For the first time, a CMOS compatible calorimetric flow sensor with a dual-heater (DH) structure is proposed instead of using the traditional single-heater (SH) design. CFD simulation results demonstrated that the DH design can enhance the performance of the device with high sensitivity. Thereof, the DH sensor was fabricated with four suspended bridges and tested under three configuration modes: parallel-heater (PH) mode, full-bridge (FB) mode, and half-bridge (HB) mode. Experimental results demonstrated that the PH mode can achieve a high sensitivity of 403.25 mV/(m/s)/W, which is over 4 times that of the HB mode (83.7mV/(m/s)/W). This DH flow sensor under its PH mode can also provide an ultra large measurement range (±40 m/s) and fast response time (1.8 ms @16.7 m/s). The good performance of this novel flow sensor demonstrated its potential applications in respiration monitoring in medical areas and airflow control in HVAC systems for smart buildings.

KEYWORDS

CMOS compatible, flow sensor, dual-heater

INTRODUCTION

Thermal flow sensors have been widely used in our daily life including wind measurement, HVAC monitoring, and so on [1-3]. Generally, flow sensors are based on thermal [4], piezoelectric [7], piezoresistive [8], capacitive [9], or surface acoustic wave-based [10] sensing mechanisms. Among these types, thermal flow sensors have attracted the attention of many researchers for their simple structures and easy implementation. Thermal flow sensors typically are composed of heaters and temperature sensors, which can be categorized into three types: hot-wire/hot-film, calorimetric, and time-of-flight [11].

At present, most flow sensors are designed with a hot-wire or calorimetric structure with a single heater (Fig. 1(a)), making them difficult to achieve high sensitivity, large measurement range, and fast response simultaneously. For example, Djuzhev N A et al. reported a flow sensor with a response time of 24 ms, but its measurement range was only 10 m/s [2]. K. Makinwa et al. presented a flow sensor with the measurement range improved to 18 m/s using thermal sigma-delta modulation techniques, but its response time was as long as 1.59 s [3]. D. Moser et al. demonstrated a flow sensor with a measurement range of up to 25 m/s, while, its sensitivity was only 68 mV/(m/s)/W [12].

In this paper, we propose a novel calorimetric flow sensor designed with a dual-heater (DH) structure (Fig. 1(b)) to enhance all three parameters in balance.

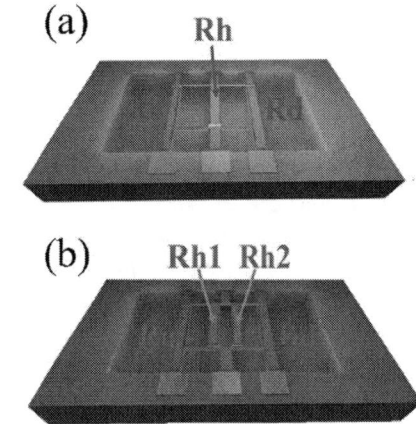

Figure 1: The schematics of a flow sensor with (a) a single-heater (SH) structure and (b) a DH structure.

CONCEPT AND FABRICATION
Methodology and Structure Design

Hot-wire flow sensors have a wide measuring range, but they cannot identify the flow direction and have relatively low sensitivity at the low-velocity range. A thermal calorimetric flow sensor with a single heater structure, as illustrated in Fig. 1(a), can be used to measure flow direction, but it is still difficult to enhance the performance indicators of measuring range, sensitivity, and response speed simultaneously.

Thus, a calorimetric flow sensor with dual heaters (DH) is proposed. The schematic of the DH flow sensor is shown in Fig. 1(b), where the four suspended bridges carry two heaters and two temperature sensors, respectively. This DH design can provide more uniform heat distribution and higher temperature difference which can enhance the sensitivity. Firstly, the CFD simulations were performed using ANSYS FLUENT (Fig. 2), revealing that the DH design can increase the sensitivity by a factor of around 3, compared to the corresponding single-heater design.

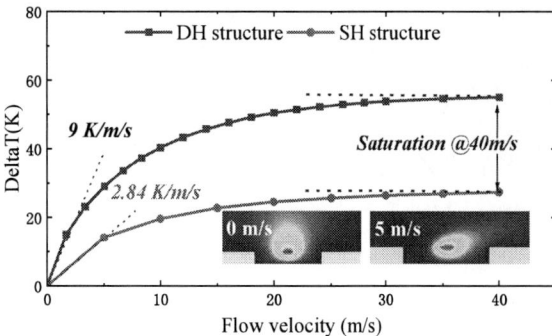

Figure 2: Performance comparison between SH and DH design by CFD simulation.

Fabrication process

Based on the design analysis and simulation verification, the proposed DH flow sensor was fabricated with a CMOS compatible fabrication process (Fig. 3). First, a 1.5μm LPCVD silicon oxide layer was deposited on the cleaned Si wafer, followed by a 1μm LPCVD nitride layer deposited to reduce the stress of the film. Then a 1μm n-type polysilicon was deposited on the nitride layer. After patterning by photolithography and RIE (reactive ion etching), the heaters and temperature sensors layers were formed, as shown in Fig. 3(d).

To protect the polysilicon from damage during the substrate silicon etching process, another layer of 1.5 μm passivation silicon oxide was deposited, as shown in Fig. 3(e). Then, vias were fabricated by an advanced oxide etcher (AOE), followed by Al deposition and patterning (Fig. 3(f)). To achieve a deep cavity and reduce the over-etching issue, a Deep Reactive Ion Etching (STS Multiplex ICP DRIE etcher) was used to make deep trenches (Fig. 3(h)) and then XeF_2 isotropic etching was used to release the microstructures (Fig. 3(i)). The dimensions of the sensor are 800 μm × 700 μm × 525 μm with a cavity height of 120 μm, and the widths of the heater and sensor bridges are all 45 μm. Fig. 4 showed an image of this fabricated flow sensor.

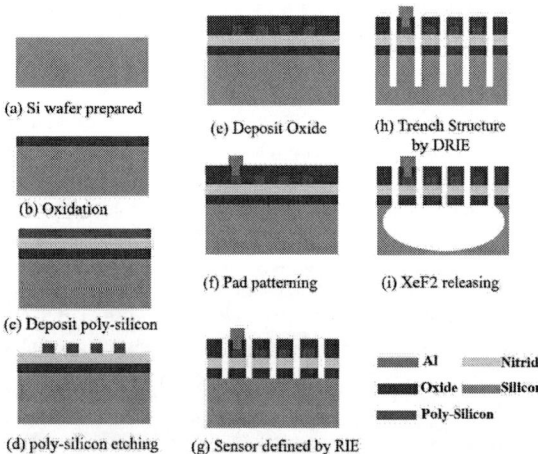

Figure 3: CMOS compatible fabrication process flow.

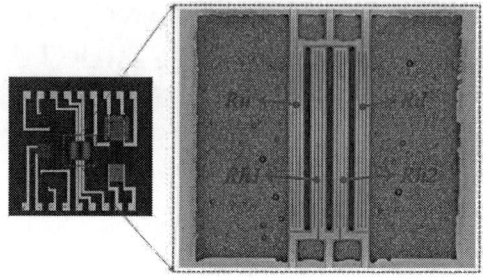

Figure 4: Micro image of the fabricated DH flow sensor.

EXPERIMENTS
Initial Characterization

Before testing the performance of the sensor, the thermal coefficient of resistance (TCR) α of the n-type polysilicon heaters and sensors was measured, which was 1.63×10^{-3}/K (Fig.5(a)). Fig. 5(b) showed that this sensor has a high power efficiency of 4.99 °C/mW.

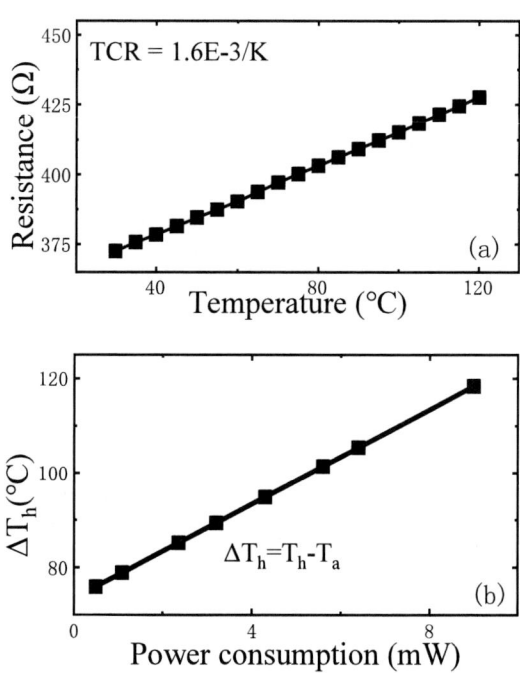

Figure 5: (a) TCR characterization; (b) heater power efficiency characterization.

Test Setup

The fabricated DH flow sensor isshown in Fig. 4. There are multiple sensors on the chip. Only the flow sensor was characterized. The closed-up picture in Fig. 4 showed the four thermal resistors of the flow sensor, which are labeled from left to right as Ru1, Rh1, Rh2, and Ru2. Three different readout circuit connections can be used, as shown in Figure 6 with the specific assignment of four thermal resistors. Under PH mode, the two heaters are connected in parallel as the heat source. Under the HB mode, the two heaters are connected on the separate arms of a Wheatstone bridge. Under the FB mode, besides the configuration with the HB mode, the two sensors are also arranged on two bridge arms. The signal from the Wheatstone bridge is amplified by a low-noise instrumentation amplifier (INA188 IDR).

The testing setup is shown in Fig. 7. The fabricated DH sensor chips were packaged with a flow channel for N_2 flow measurements. The flow channel was designed with a cross-sectional area of 10 mm × 2 mm. The nitrogen gas flow is generated by a high-pressure gas cylinder and controlled by a mass flow controller (range 0~30 SLM: HORIBA SEF-Z514). The sensor was tested with a reference sensor (Honeywell AWM5104VN) for data calibration.

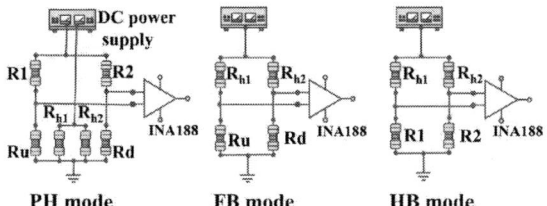

Figure 6: The device was tested under three testing modes (PH mode, FB mode, HB mode).

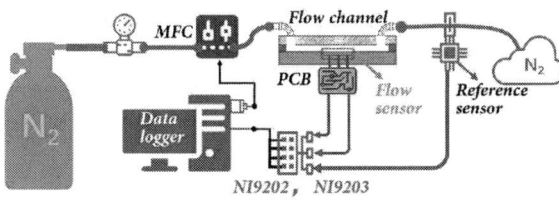

Figure 7: Schematic of the test setup.

RESULTS AND DISCUSSION

The DH flow sensor's performance was characterized by the test setup described above under all three working modes. Fig 8(a) showed the relationships between the flow rate and the output of the sensor in different operating modes. These data demonstrated that the PH mode achieved the highest sensitivity (403.25 mV/(m/s)/W) followed by the FB mode (221.93 mV/(m/s)/W) and the HB mode (83.7 mV/(m/s)/W). The sensitivity of the PH mode was more than 4 times that of the HB mode, and nearly twice that of the FB mode. The results revealed the excellent performance of the DH structure and experimentally verified the feasibility of the proposed design.

Furthermore, under the input velocity of 16.7 m/s, which was the maximum velocity that the experimental setup could provide, only the PH mode sensor did not saturate. With the assistance of CFD simulation and validation of the experimental data in the prior segment (<16.7 m/s), the expected maximum flow range of the PH mode could reach up to 40 m/s (Fig. 8(b)), which was better than most of the reported works.

The response time was characterized with a pulsed signal powered on the heaters and measured the output of the Wheatstone bridge readout circuit. This DH flow sensor

also has a fast response. Fig. 8(c) illustrated that the DH flow sensor owned a rise time of 3.5 ms and a fall time of 4.1 ms at the velocity of 8.3 m/s (a medium-velocity flow). Fig. 8(d) also revealed that the response time decreases as the flow rate increases, resulting from the enhanced micro heat transfer by forced convection. And limited to the experimental equipment (max U @16.7 m/s), the fastest response time this sensor can achieve was 1.8 ms at 16.7 m/s (Fig. 8(d)).

Finally, the stability of the device under different flow velocities was characterized. Fig. 8(e) showed a set of 30 minutes-long measurement data, indicating the maximum variation was only 0.43%. A comparison table was summarized in Table I, showing that the proposed flow sensor's normalized sensitivity and flow range are far superior to others' work in Table I while keeping the response time in par with others'.

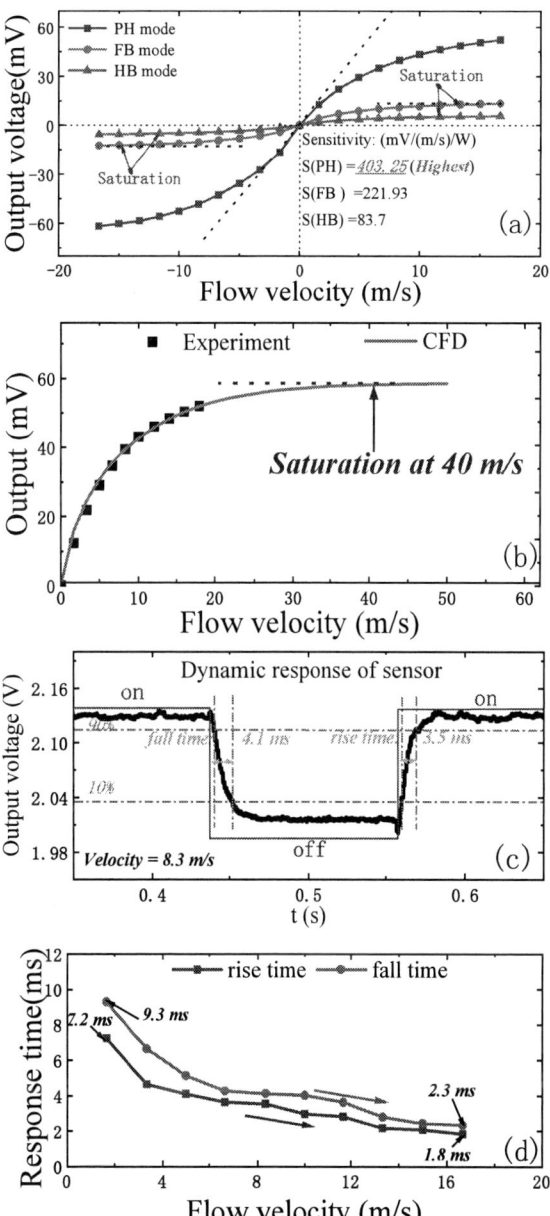

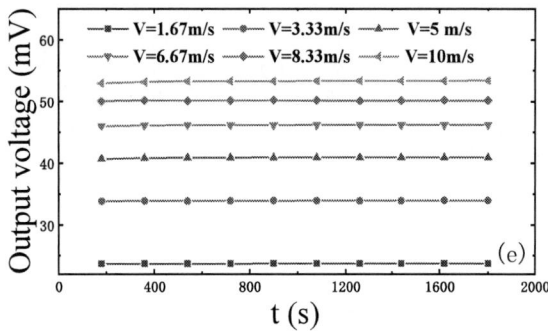

Figure 8: (a)Performance comparison between three different modes as a function of input flow; (b) measurement range determination based on experiment and CFD simulation results; (c) response time test at 8.3m/s (d) response time test(0-16.7m/s); (e) stability test under six types of flow rates.

Table I.: comparison with reported works.

Ref.	SS[a]	flow range	S* [b]mV/(m/s)/W	r[c](ms)
[1]	SH	-26~26(m/s)	112	3.6
[2]	DH	0-10m/s	N/A	24
[6]	SH	1200(sccm)	123	250
this work	DH	±40(m/s) (±4.8e+4 (sccm))	403.25	<9.3 t_{min}=1.8

[a] SS:sensor structure; [b] S*: normalized sensitivity over power; [c]:response time.

CONCLUSION

In this paper, a DH calorimetric flow sensor is proposed and fabricated using a CMOS compatible fabrication method. This sensor can be configured with three working modes (PH mode, FB mode, and HB mode). Testing results demonstrate that the PH mode achieved the highest sensitivity (403.25 mV/(m/s)/W). The maximum measurement range can reach up to 40 m/s, better than many reported works. Moreover, this sensor shows fast response time (less than 1.8 ms @16.7 m/s) and good stability. In the future, this DH sensor will be applied in the human breathing health monitoring field.

ACKNOWLEDGEMENTS

Zhongyi Liu and Ruoqin Wang contributed equally to this paper. This work was supported in part by a grant from the Beijing Institute of Technology and the General Research Fund from the Research Grants Council of Hong Kong. The authors acknowledge the technical support of the staff at BIT and HKUST NFF and MCPF. The work was supported in part by the Beijing Institute of Technology Research Fund Program for Young Scholars (XSQD-202206004), in part by the Natural Science Foundation of Chongqing (2022NSCQ-MSX5423), and in part by the Shenzhen-Hong Kong-Macau S&T Program (Category C) Grant No. SGDX20210823103200004

REFERENCES

[1] Xu W, Gao B, Ahmed M, et al. A wafer-level encapsulated CMOS MEMS thermoresistive calorimetric flow sensor with integrated packaging design[C]//2017 IEEE 30th International Conference on Micro Electro Mechanical Systems (MEMS). IEEE, 2017: 989-992.

[2] Djuzhev N A, Novikov D V, Demin G D, et al. An experimental study on MEMS-based gas flow sensor for wide range flow measurements[C]//2018 IEEE Sensors Applications Symposium (SAS). IEEE, 2018: 1-4.

[3] Makinwa K A A, Pertijs M A P, vd Meer J C, et al. Smart sensor design: The art of compensation and cancellation[C]//ESSCIRC 2007-33rd European Solid-State Circuits Conference. IEEE, 2007: 76-82.

[4] Cubukcu A S, Zernickel E, Buerklin U, et al. A 2D thermal flow sensor with sub-mW power consumption[J]. Sensors and Actuators A: Physical, 2010, 163(2): 449-456.

[5] Dong Z, Chen J, Qin Y, et al. Fabrication of a micromachined two-dimensional wind sensor by Au–Au wafer bonding technology[J]. Journal of microelectromechanical systems, 2012, 21(2): 467-475.

[6] Ke W, Liu M, Li T, et al. MEMS thermal gas flow sensor with self-test function[J]. Journal of Micromechanics and Microengineering, 2019, 29(12): 125009.

[7] H. C. Liu, S. S. Zhang, R. Kathiresan, T. Kobayashi, and C. Lee, "Development of piezoelectric microcantilever flow sensor with wind-driven energy harvesting capability," Applied Physics Letters, vol.100, no. 22, May 2012, Art. no. 223905.

[8] Q. Zhang, W. Ruan, H. Wang, Y. Zhou, Z. Wang, and L. Liu, "A self-bended piezoresistive microcantilever flow sensor for low flow rate measurement," Sensors and Actuators A: Physical, vol. 158, no. 2, pp. 273-279, 2010/03/01/ 2010

[9] Liao S H, Chen W J, Lu M S C. A CMOS MEMS capacitive flow sensor for respiratory monitoring[J]. IEEE Sensors Journal, 2013, 13(5): 1401-1402.

[10] Q. Zhang et al., "Flexible ZnO thin film acoustic wave device for gas flow rate measurement," Journal of Micromechanics Microengineering, vol. 30, no. 9, p. 095010, 2020.

[11] Kuo J T W, Yu L, Meng E. Micromachined thermal flow sensors—A review[J]. Micromachines, 2012, 3(3): 550-573.

[12] Moser D, Lenggenhager R, Wachutka G, et al. Fabrication and modelling of CMOS microbridge gas-flow sensors[J]. Sensors and Actuators B: Chemical, 1992, 6(1-3): 165-169.

[13] Qiu L, Hein S, Obermeier E, et al. Micro gas-flow sensor with integrated heat sink and flow guide[J]. Sensors and Actuators A: Physical, 1996, 54(1-3): 547-551.

CONTACT

* Xiaoyi Wang, tel: +86-13611325054
E-mail: xiaoyiwang@bit.edu.cn

*Huikai Xie, tel: +86-13611325054
E-mail: hk.xie@bit.edu.cn

LOCAL METAL DEPOSITION ON HYDROGELS USING MICRO-PLASMA-BUBBLES

Haruna Takahashi, Yu Yamashita, Naotomo Tottori, Shinya Sakuma, and Yoko Yamanishi
Kyushu University, JAPAN

ABSTRACT

We propose a local metal deposition method that enables patterning on hydrogels using micro-plasma-bubbles (MPBs). In this method, we generate MPBs in a solution containing metal ions. Local metal deposition on hydrogels has a wide range of possibilities, however, it has been a considerable challenge since appropriate methods have not been established. In this work, we reduced metal ions in an electrolyte solution and successfully deposited several kinds of metals such as copper, nickel, copper-nickel alloy, and even gold nanoparticles on hydrogels by using MPBs. These results show that the proposed method opens up new possibilities of an on-demand metallization process for hydrogels that can serve as soft actuators and electronics.

KEYWORDS

Hydrogel, Microbubble, Plasma, Metal deposition, Nanoparticle

INTRODUCTION

Hydrogel-Metal Composites

Hydrogel-metal composites (HMCs) have diverse applications especially in soft robotics, such as stretchable strain sensors for human skin [1], light-driven soft actuators using temperature-responsive hydrogels and metal nanoparticles as photothermal converters [2,3]. Conventional fabrication methods for HMCs are mixing metal or metal-compound particles into a hydrogel matrix, or deposition of liquid metals on hydrogels. However, these methods are only able to incorporate metal materials homogeneously or simply to lay them on the surface of hydrogels, and unable to locally "modify" the prefabricated hydrogel itself. If it is possible, HMCs will have no internal discontinuity between hydrogel and metal, making them mechanically preferable, as shown in Fig. 1. Therefore, local modification of hydrogels by metal deposition, not just mixing or laying metals on hydrogels, is highly demanded in terms of designing novel soft actuators and sensors.

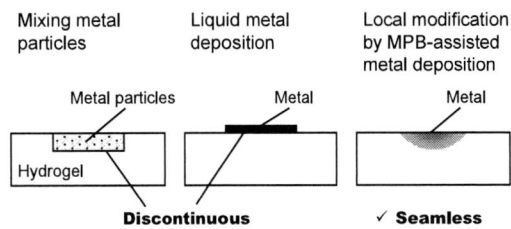

Figure 1: Fabrication of hydrogel-metal composites.

Micro-Plasma-Bubbles

To achieve such local metal deposition on hydrogels, we utilized micro-plasma-bubbles (MPBs). MPBs can locally provide reactive species generated by discharge in an electrolyte solution containing metal ions. Previously, we proposed an MPB-assisted metal deposition method [4]. In this work, we aim to locally deposit metal on hydrogels by reducing metal ions near/inside hydrogels into simple metals using MPBs.

CONCEPT

Figure 2 shows the concept of MPB-assisted metal deposition on hydrogels. To generate MPBs, we have proposed a bubble injector (BI), which consists of an electrode and insulator tube. We retained a BI above hydrogel in a solution containing metal ions, as shown in Fig. 2(a). When we applied pulsed direct current (DC) to BI, the electric field is concentrated on the tip of BI thanks to the surrounding insulator tube, generating bubble(s). If the electric field surpasses the dielectric strength of water vapor, it induces discharge and plasma. Fig. 2(b) shows that active species such as hydrated electrons e_{aq}^- and atomic hydrogen H in plasma can reduce metal ions around an MPB. Since standard reduction potentials (E^o) versus standard hydrogen electrode of hydrated electrons ($E^o = -2.87$ V) [5] and atomic hydrogen ($E^o = -2.30$ V) [5] are lower than those of reactive metal ions such as zinc ion ($E^o = -0.76$ V) [6], they can reduce a variety of metals in theory. As a result, metal ions near/inside hydrogels are reduced,

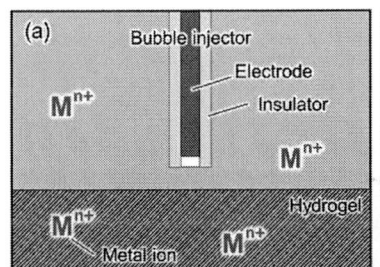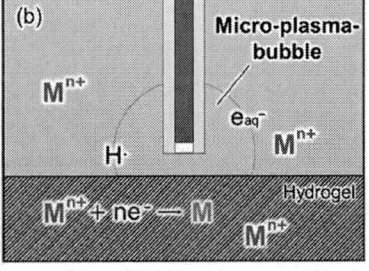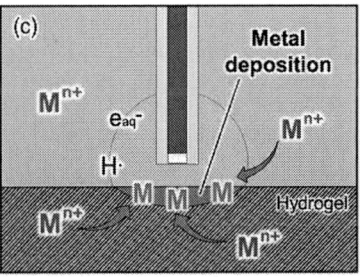

Figure 2: Concept of metallization of hydrogels. (a) Bubble injector and hydrogel substrate in metal ion solution. (b) Micro-plasma-bubble and reducing agents are generated when electric pulses are applied to the injector. Active species derived from plasma reduce metal ions. (c) Metal is deposited on hydrogel.

and form deposits as shown in Fig. 2(c). Although plenty of previous works have synthesized metal or its compound by plasma in liquid [7-9], our method is distinct in that it utilizes the bubbles as local reactive regions for metal deposition, whereas they are generally considered to be by-products.

MATERIAL AND METHODS
Hydrogels and Electrolyte Solutions

Acrylamide (AAm), N,N'-methylenebis(acrylamide) (Bis), ammonium peroxodisulfate (APS), N,N,N',N'-tetramethylethylenediamine (TEMED), copper(II) acetate monohydrate (CuAc·H_2O), nickel(II) acetate tetrahydrate (NiAc·$4H_2O$), hydrogen tetrachloroaurate(III) tetrahydrate (HAuCl$_4$·$4H_2O$), and potassium chloride (KCl) were purchased from Wako Pure Chemical Industries, Ltd..

The hydrogel substrates used for the following experiments were made of 15 wt% poly(acrylamide). AAm (1.45 g) and Bis (0.05 g) were dissolved in 10 mL of deionized water. To this solution, 50 µL of 10 wt% APS aqueous solution and 4.0 µL of TEMED were added. We poured this solution between two parallel glass plates with a 1 mm-thick rubber spacer and waited until gelation under room temperature.

For electrolyte solutions containing metal ions, we prepared 0.3 mol/L CuAc (6.7 mS/cm) and 0.3 mol/L NiAc (14.2 mS/cm) aqueous solutions. We added KCl to 0.2 mmol/L HAuCl$_4$ aqueous solution until the conductivity reaches 6.6 mS/cm. As for alloy, we mixed the same volume of CuAc and NiAc solutions prepared above. We soaked hydrogels in each solution for a sufficient time before experiments. After that, metal ions were removed by soaking hydrogels in pure water.

Bubble Injector and Experimental Setup

Figure 3 shows the fabricated BI. We inserted a tungsten wire as an active electrode in a glass capillary as an insulator. As a counter electrode, we used a stainless-steel pipe.

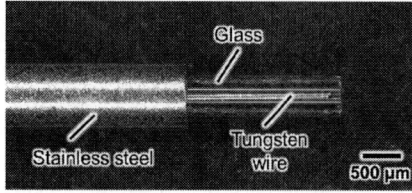

Figure 3: Bubble injector to generate MPBs.

Figure 4 shows the experimental setup for MPB-assisted metal deposition on hydrogels. We controlled the horizontal and vertical positions of a BI with a micromanipulator. After soaking the BI above hydrogels in electrolyte solutions, we applied pulsed DC with a power supply (PLT1500; BEX Co. Ltd.) to the BI. A resistor of 1 kΩ was placed between the BI and the power supply. The applied voltage was 800 V and the frequency was 50 kHz. Pulse widths/number of pulses were 10 µs/500 times for CuAc, NiAc, and their mixture solutions; 1 µs/2000 times for HAuCl$_4$ solution. We observed the deposition process with a high-speed camera (VHX; Keyence Corporation), and light source.

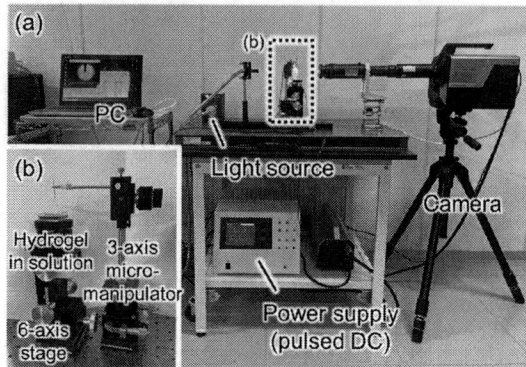

Figure 4: Experimental setup. (a) Configuration of the system for observing the MPB-assisted metal deposition process. (b) Setup around the bubble injector and hydrogel soaked in an electrolyte solution.

EXPERIMENTS AND RESULTS

First, we observed the MPB-assisted metal deposition process for hydrogels, as shown in Figure 5. It was confirmed that when the pulse width was 10 µs and the number of pulses was 500 times, the BI generated one single bubble for all pulses. Metal deposits were formed where the MPB was applied.

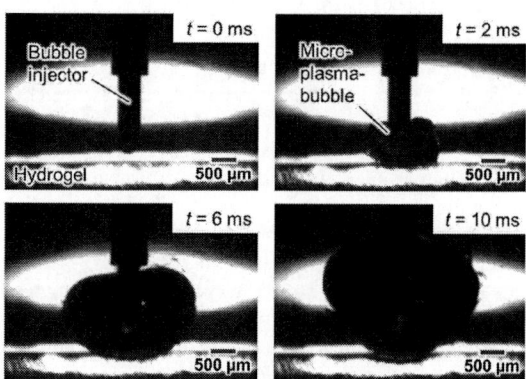

Figure 5: Image sequence of metal deposition process. The hydrogel was placed under the BI in a solution. Recording speed was 10000 fps.

Next, we conducted elemental analyses of nickel, copper, and copper-nickel alloy deposits shown in Fig. 6(a) by laser-induced breakdown spectroscopy (EA-300; Keyence Corporation). As for nickel and copper, we evaluated the element detection probabilities from 5 trials according to the distance between the BI and hydrogels, as shown in Fig. 6(b). The result indicates that both Ni and Cu were detected with high probability when the distance was under 100 µm. Based on this result, we deposited copper-nickel alloy with a distance of 100 µm. We confirmed that the alloy was successfully deposited (the atomic ratio of Cu to Ni was 54% to 46%, respectively). Additionally, we observed the cross-sectional view of the nickel deposit, as shown in Fig. 6(c). It implies that the active species derived from MPBs reduced nickel ions inside the hydrogel as well as the ones in the solution.

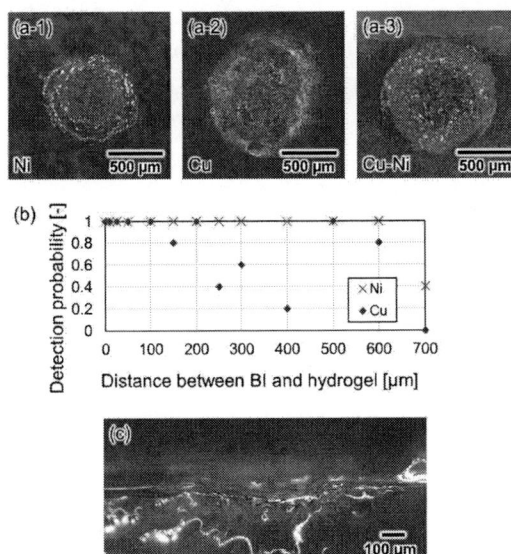

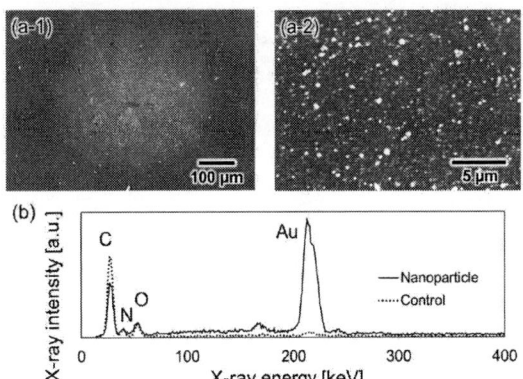

Figure 8: Deposition of Au nanoparticles on hydrogels. (a-1) SEM image of the deposition area. (a-2) SEM image of the magnified part in the deposition area. (b) Elemental analysis of (a-2) (red cross: analysis point). Dotted line represents control (hydrogel).

Figure 6: Metal deposition of (a-1) Ni, (a-2) Cu, and (a-3) Cu-Ni alloy. (b) Elemental detection probabilities change in accordance with the distance between BI and hydrogel. (c) Cross-sectional image of Ni deposit.

Demonstration of Magnetic Property

Figure 7(a) shows the patterned nickel line on the hydrogel. We drew the line by manipulating the BI by 500 μm for each deposition. It was confirmed that the adhesion of nickel was strong enough to bear the following deposition. Using the nickel-patterned hydrogel, we examined the magnetic property of the deposition. We observed that the nickel pattern was selectively attracted to the magnets with around 400 mT in water.

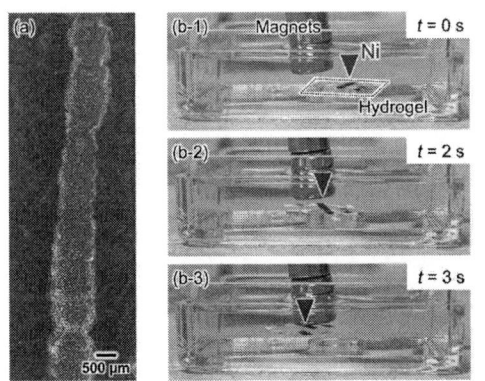

Figure 7: (a) Patterning of a nickel line on hydrogel. (b) Image sequence of a piece of hydrogel with nickel patterns being attracted to magnets in water.

Deposition of Gold Nanoparticles

We synthesized and deposited gold nanoparticles on hydrogels from $HAuCl_4$ solution using MPBs. We repeated the application of pulsed DC 20 times. SEM images and elemental analysis by energy dispersive X-ray spectroscopy demonstrate that gold nanoparticles of several hundred nanometers were synthesized and selectively deposited where the MPB was applied, as shown in Fig. 8.

DISCUSSION

According to Fig.6(a), it appears that metals are mainly deposited in the central area. We assume that this is because the plasma density is higher closer to the active electrode. The peripheral area is considered to be just surface-modified; we detected the elements of similar composition to the hydrogel substrates. It is necessary to explore the conditions for more uniform metal deposition in detail.

CONCLUSION

We demonstrated the MPB-assisted local metal deposition method for hydrogels. In this method, copper, nickel, and copper-nickel alloy were successfully deposited with the help of reductive species in the plasma. As to nickel, we confirmed the capability of patterning and magnetic property. Additionally, we realized *in-situ* synthesis and deposition of gold nanoparticles of several hundred nanometers. From these results, it is now clear that MPBs are effective in the on-demand functionalization of hydrogels for unconventional soft robotics.

ACKNOWLEDGEMENTS

This study was supported by JSPS KAKENHI Grant No. JP22H00198 and JP22K18783.

REFERENCES

[1] C. Xu, B. Ma, S. Yuan, C. Zhao, H. Liu, "High-Resolution Patterning of Liquid Metal on Hydrogel for Flexible, Stretchable, and Self-Healing Electronics", *Advanced Electronic Materials*, vol. 6, issue 1, 2020.

[2] S. Watanabe, K. Arikawa, M. Uda, S. Fujii, M. Kunitake, "Multimotion of Marangoni Propulsion Ships Controlled by Two-Wavelength Near-Infrared Light", *Langmuir*, vol. 37, issue 50, pp. 14597–14604, 2021.

[3] M. Kim, J. Choi, S. Y. Kim, "Low-intensity near-infrared light-tunable gold nanorod-incorporated hydrogel actuator system for remotely controlled human–robot interface applications", *Materials Today Chemistry*, vol. 26, 2022.

[4] Y. Yamashita, S. Sakuma, Y. Yamanishi, "On-Demand Metallization System Using Micro-Plasma Bubbles", *Micromachines*, vol. 13, issue 8, p. 1312, 2022.

[5] H. A. Schwarz, "Free Radicals Generated by Radiolysis of Aqueous Solutions", *J. Chem. Educ.*, vol. 58, No. 2, pp. 101-105, 1981.

[6] D. C. Harris, *Quantitative Chemical Analysis*, W. H. Freeman, 2010.

[7] Q. Chen, J. Li, Y. Li, "A review of plasma-liquid interactions for nanomaterial synthesis", *Journal of Physics D: Applied Physics*, vol. 48, No. 42, 2015.

[8] N. Saito, J. Hieda, O. Takai, "Synthesis process of gold nanoparticles in solution plasma", *Thin Solid Films*, vol. 518, issue 3, pp. 912-917, 2009.

[9] H. J. Kim, J. G. Shin, C. S. Park, D. S. Kum, B. J. Shin, J. Y. Kim, H. D. Park, M. Choi, H. S. Tae, "In-Liquid Plasma Process for Size- and Shape-Controlled Synthesis of Silver Nanoparticles by Controlling Gas Bubbles in Water", *Materials*, vol. 11, issue 6, p. 891, 2018.

CONTACT

*H. Takahashi, tel: +81-92-802-3229; takahashi.haruna.441@s.kyushu-u.ac.jp

FOLDING METHOD OF KIRIGAMI STRUCTURE WITH FOLDING LINES

Nagi Nakamura and Eiji Iwase

Department of Applied Mechanics and Aerospace Engineering, Waseda University, Tokyo, JAPAN

ABSTRACT

We proposed a kirigami structure with folding lines (referred to as a "kiri-origami") for flexible electronic devices and its folding methods. Kiri-origami structure has advantages in mounting rigid components and batch fabrication for the folding structure with many panels. We achieved the ideal folding-up of the kiri-origami structure by using a buffer structure and 2-axis extension in order to remove the cause of distortion and effectively utilize the tensile force for bending. We compared shape uniformity with/without the buffer structure and evaluated the panel warping under a 2-axis extension. In addition, we fabricated a kiri-origami LED display with more than 500 hinges and we show the applicability of our structure and methods for devices.

KEYWORDS

Origami, Kirigami, Flexible device, Self-folding, Batch fabrication.

INTRODUCTION

In recent years, origami and kirigami structures have been interested in as a method to realize flexible/stretchable electronic devices using non-stretchable materials. In the conventional origami structures, a thin substrate is folded up at folding lines, and in the kirigami structures, a thin substrate is deformed out-of-plane using slits and extension. Both the origami and kirigami structures realize a large stretching deformation as the whole structure by bending/torsional deformation with small strain. So the origami and kirigami structures can enable non-stretchable but electrically high-performance materials to be a stretchable substrate, and applied for flexible displays[1][2], sensors[3], generating devices[4], etc. The origami structures have flat panel regions that are advantageous for mounting rigid electronic elements[1][4], but are difficult to fold into the target shape because a compression-based deformation has many mode advantages for batch fabrication because a tension-based deformation has a few mode bifurcation[2]. The kirigami structures have, however, no flat panel regions. Therefore, we propose "kirigami structures with folding lines (kiri–origami structure)" that combines the advantages of both origami and kirigami structures. The structures are shown in Fig. 1(a). Only by macroscopic extension, the kiri-origami structure enables batch fabrication of stretchable electronic devices with rigid electric elements and large amounts of units. The kiri-origami structure is defined as a structure that has slits and folding lines and is folded up by macroscopic extensional deformation.

In this study, we propose two effective folding methods for the kiri-origami structures with mutual orthogonal slits and 2-line connections[5] as shown in Fig. 1(b). The first method is controlling the fixing condition by buffer structures and the second method is controlling the 2-axis extension. We evaluated the methods in folding ideality. First, we compared shape uniformity with/without the buffer structures. Next, we compared the deformations under 1-axis or 2-axis extension and evaluated folding ideality as the panel warpage by 3D scanning. In addition, we demonstrated a fabrication of the kiri-origami LED display and showed that our methods can be applied for fabrication of stretchable electronic devices.

FOLDING METHOD AND DESIGN

Fig. 1(b) shows the folding methods proposed in this study. Generally, kiri-origami structures have fixed part to apply tension at the edges. Contrary to the deformable part with slits, the fixed part dose not deform. So geometrical mismatch occurs as the structure is folded up. The mismatch is usually canceled by distortion of the deformable part. But the deformation of kirigami and kiri-origami structures is significantly affected by the fixed condition at the edges, and depending on the type, folding up is almost impossible. The buffer structures inserted between the fixed and deformable part expand anisotropically, thus we can apply the tensile force without distortion. In addition, the actual deformation under 1-axis

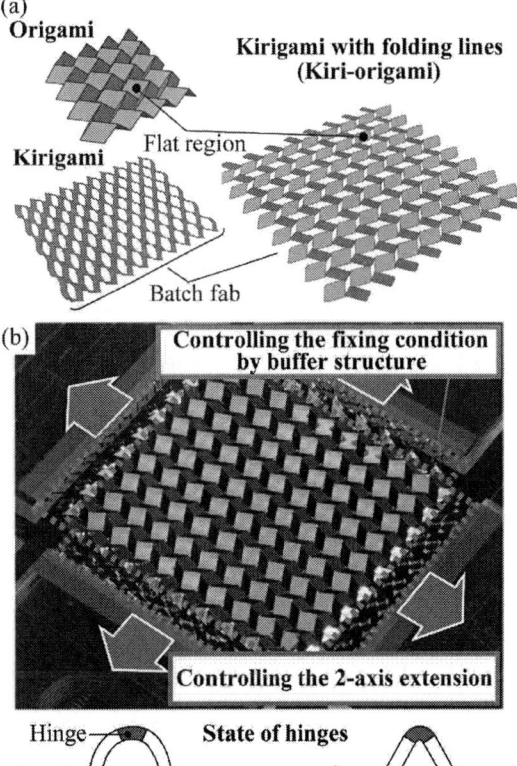

Figure 1: Concept of this study, (a) Kirigami structure with folding lines (Kiri-origami), (b) Methods of kirigami folding for ideal shapes.

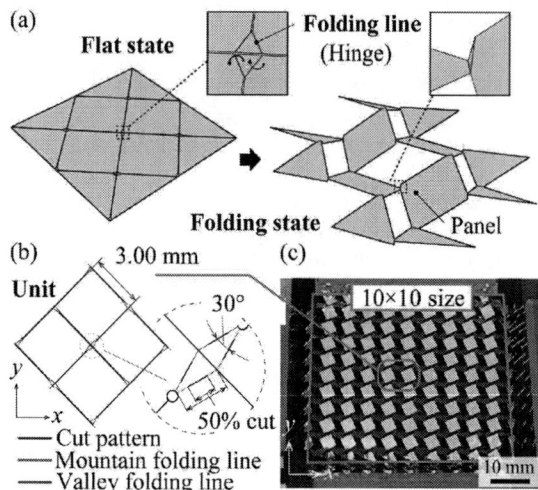

Figure 2: The kirigami structure for criterion, (a) Theoretical deformation of Mutual orthogonal cut pattern, (b) Design of the unit, (c) Kirigami array (10×10 size).

extension is different from the ideal deformation without considering the fixed condition due to warpage of the panels. The geometric model does not take equilibrium of force into account, so in reality the panels warp and make the structures balanced as different shapes. In some types of kiri-origami structures, folding up is capped due to the warpage. Thus, we propose to extend in 2-axis. The 2-axis extension enables to correct the deformation of the whole structure as the ideal shape. This method works as to control the situation of applying force to the hinge and makes it possible to continue folding-up beyond the restriction under 1-axis extension.

Next, Fig. 2(a) shows the deformation of the kiri-origami structure used in this study to evaluate the fold up methods. The structure has mutual orthogonal cut pattern which constructs square panels and has 2 folding lines between adjacent square panels. The structures of mutual orthogonal cut pattern[5]-[7] have many types of folding lines and various deformation patterns. The structure of the test sample is auxetic. Fig. 2(b) shows the design of the test sample. The minimum unit which composes the test sample is made of four square panels. The angle between the pair of folding lines is 30°, which is the principal parameter for the deformation types. Each panel is 3-mm square, and 15 % (0.45 mm) remains for connection. To keep from stress concentration every corner point of slits has a hole. The folding lines are marked by 50 % cut aimed to locally reduce the stiffness. Fig. 2(c) is entire view of the kiri-origami structure. The test sample has overlapping 10 units in each row and column. All of the fabrications are deal with by a 355 nm UV-Laser processing machine, and the test sample is made of a polyimide copper substrate (Polyimide : 25 μm, Cu : 8 μm, Toray Metaloyal®□)

EXPERIMENTS
Evaluation of the effect by the buffer structure

Fig. 3(a) shows the mechanism of the buffer structure[2]. In addition to the parallel slit pattern, the

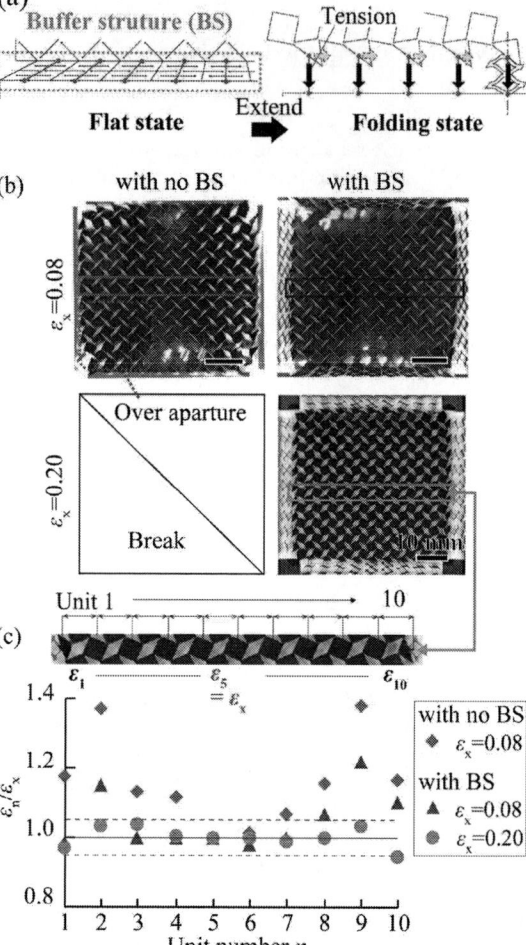

Figure 3: Effect by the buffer structure (BS), (a) Mechanism of BS for uniform deformation over the entire region, (b) Image comparison for uniformity of deformation, BS reduce dispersion of the strain distribution, (c) Deformation of units at the center row.

buffer structure has slits which divide connection in shear direction. Therefore, the rings surrounding each slit can extend freely. When tensile force is applied, the ring arrays extend like a spring and entirely deform as a buffer structure. To prevent clacking in diagonal direction of the deformable part, the length of the fixed part should be longer than initial length of the deform part. When the ring arrays become aligned tensile force is transmitted in the most ideal way. The situation is designed to be at $\varepsilon_x = 0.200$, $\varepsilon_y = 0.142$ of the deformable part in the test sample. In addition, the buffer structure should start buckling before the deform part so that the buffer structure works correctly. In the case of the test design, the pitch of the parallel lines is 0.5 mm, and the ring arrays line should have more than 2 ring by a line.

The difference in deformation uniformity for presence of the buffer structures is evaluated in Fig. 3(b). In this comparison ε_x is measured from the center unit of the test sample. Fig. 3(c) shows a graph of the normalized strain $\varepsilon_n/\varepsilon_x$ measured for each unit of the center row in the x-

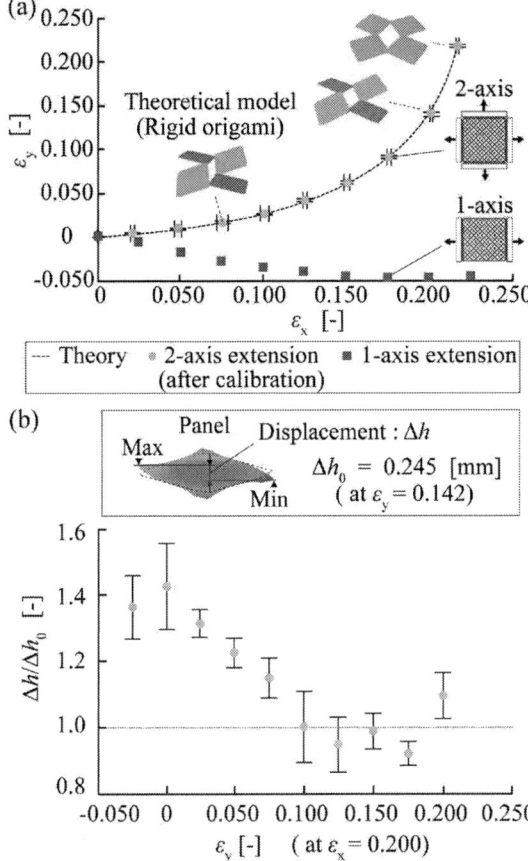

Figure 4: Utility verification of the folding method, (a) Stain correction by 2-axis extension, this kirigami structure cannot be folded theoretically by 1-axis extension, (b) Relationship between distance from the theoretical path and panel warpage (ideality of folding), the minimum locates at the vicinity of theoretical deformation.

direction. Without the buffer structure, the corner areas expanded excessively and began to fracture at the edge of the corner at $\varepsilon_x = 0.080$. Contrary to the corner, the columns near the center slightly expanded. In Fig. 2(c) The deviation of $\varepsilon_n/\varepsilon_x$ is largest in the unit next to the edge, approximately 40% than the center unit. In addition, up to three-unit interior from the edge, the deviations are around 10% so the uniform deformation area (the deviation is less than 5 %) is only two-unit size. On the other hand, with the buffer structure the test samples were able to be folded up to $\varepsilon_x = 0.200$ without cracking. At $\varepsilon_x = 0.080$, the deformation became uniformed and the maximum deviation became about 20% due to the buffer strucures. The uniform deformation area is except for the edge units and the next interior units. At $\varepsilon_x = 0.200$, which is design point, the entire structure deformed more uniformly. In Fig. 3(c) the uniform deformation area is except for the edge units. Compared with the case without the buffer structures, the introduction of the buffer structure results more uniform folding, namely ideal folding that suppresses the occurrence of panel warpage.

Evaluation of the effect by 2-axis extension

Fig. 4(a) shows the results of measuring the strain in the central units of the test samples. The sample for 2-axis extension has the buffer structure in x and y-direction and fixed onto translation stages. The sample for 1-axis has the structures in only x-direction. In addition, the theoretical model and deformation path is on Fig. 4(a). The model simulated by Rhinoceros® and Grasshopper® and called as "rigid origami". The rigid origami is constructed by rigid body panels and rotation axes. The right edge point of the theoretical path is the maximally extended state of this structure. At the beginning of extension, the orthogonal cuts begin to be opened, and the panels incline, which is similar to the deformation of parallel-cut kirigami, so the change of ε_y is low. As approach to the end, the panels fall down and rotate horizontally, and the entire deformation becomes more isotropic. At the beginning half, the one hinge located between the two square panels mainly bends, and at the ending half, the other hinge mainly bends. The 1-axis plots are the mean values of 3 samples, and the step is $\varepsilon_x = 0.025$. The 2-axis plots are the mean values of 5 samples when the stages were set according to the prior calibration. In the 1-axis result, the samples were extended along the x-axis up to $\varepsilon_x = 0.225$, but did not extend in the y-axis direction. ε_y continued to decline against the ideal deformation. By observation of the hinges, we found the hinge which folds after was not sufficiently folded up. The situation indicates that in 1-axis extension the tensile force not effectively works as bending force as the structure is folded up. On the contrary, in the 2-axis extension result all plots up to the maximally expanded state were on the theoretical path. From $\varepsilon_x = 0.000$ to $\varepsilon_x = 0.100$, the x-direction error increased, and the maximum is $\sigma_{max} = 0.004$. In contrast from $\varepsilon_x = 0.150$ to $\varepsilon_x = 0.225$ the y-direction error had been increased, and the maximum is $\sigma_{max} = 0.002$. Because of adding one more axis extension, it becomes possible to apply effective bending force for the hinges.

Next, in order to confirm that ideal folding is realized by deforming along the theoretical path, we measured panel warpage when the distance from the path changes. ε_y is changed, and ε_x is fixed. When ideal folding occurs, the hinge is only bent locally, and the panel is flattened. Since the adjacent panels fold over horizontally and form a dead angle as the structure is folded up, in this study the panel warpage was used to evaluate the folding ideality. We used the optical 3D measuring machine (Keyence VR-5000). In the analysis, we took the concentric circular area for the panel, which diameter is 1/3 of the width, as the standard plane, and defined that Δh [mm] is the displacement between the highest and lowest point and panel warpage is $\Delta h/\Delta h_0$. Fig. 4(b) shows the warpages of the panel evaluated by fixing $\varepsilon_x = 0.200$ and changing ε_y by 0.200 in Fig. 4(a). The measurement started from $\varepsilon_y = -0.025$ and continued upward. The panel warping $\Delta h/\Delta h_0$ declined from $\Delta h/\Delta h_0 = 0.143$ ($\varepsilon_y = 0.000$) to $\Delta h/\Delta h_0 = 0.95$, 0.93 ($\varepsilon_y = 0.125, 0.175$), and after decline, the rate ascended again up to $\Delta h/\Delta h_0 = 0.110$ ($\varepsilon_y = 0.200$). Overall, the warpage tends to decrease as approached to the theoretical path. The reason why the warpage does not disappear even on the theoretical path is due to the effect of the reaction moment generated by the hinge in response to the bending angle, which can be reduced by increasing the difference in

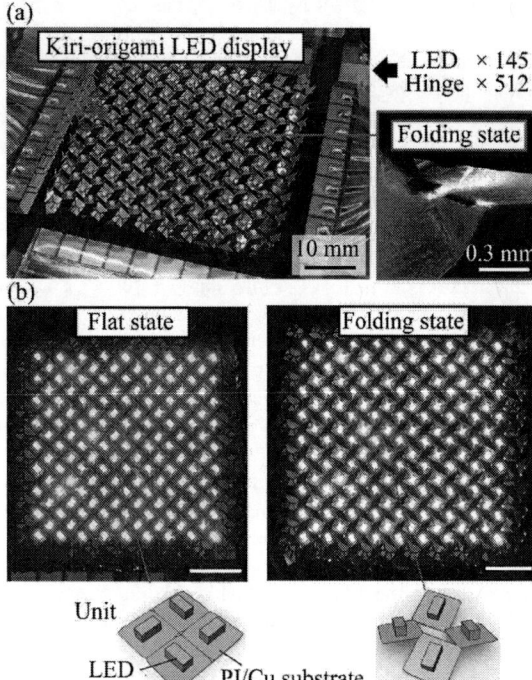

Figure 5: Demonstration of the adaptivity for electronic devices, (a) LED matrix display using the kiri-origami structure, ideal folding is made at the folding line. (b) State of the structure before/after folding and 3D shape data of hinges on the device, folded precisely regardless of electrical elements, Authentic deformation of the display.

stiffness between the hinge and the panel.

Demonstration of an electric device using the kiri-origami structure

To show the adaptivity of our folding methods to electronic devices, we demonstrated to make a LED matrix display using kiri-origami structure. We folded up the substate after mounting LEDs. Fig. 5(a) shows the overview and the hinge shape. The display has origami-like flat regions so there is less fear to be isolated using rigid LEDs than kirigami devices or elastic substate, and like kirigami, is able to fold up at once regardless of the structure has numerous units. In reality, it was succussed to fold up 512 hinges after 145 LEDs are mounted. Folding was proceeded along the theoretical path up to the point ε_x = 0.200. As the enlarged figure shows, the 50 % slit is folded up as a hinge, and the flatness of the panel is maintained. Finally, as shown in Fig. 5(b), the electrical performance is maintained before and after folding. Our folding methods can be used to fold up a kiri-origami with electric elements.

CONCLUSION

In this study, we proposed "kirigami structure with folding lines (kiri-origami structure)" and gave methods to fold up ideally. Kiri-origami structure excels in mounting rigid elements and batch fabrication due to combination of conventional kirigami and origami structure. Controlling the fixing condition by buffer structures and controlling 2-axis extension is necessary to fold up ideally. We compared the shape uniformity with/without the buffer structure and showed the buffer structures reduce the error from a maximum of 40 % to less than 5 % overall. Furthermore, we showed that 2-axis extension enables the kiri-origami structure to be folded up theoretically and that ideal folding is achieved by approaching the deformation to the theoretical path. We then demonstrated the fabrication of a kiri-origami LED display with more than 500 hinges. In conclusion, we have shown that the kiri-origami structure can realize batch fabrication only by extension to make a flexible electronic device, which has not been realized in the past.

ACKNOWLEDGEMENTS

This work was partially supported by JSPS KAKENHI Grant Number 22H04954.

REFERENCES

[1] Y. Deng, W. Liu, Y.K. Cheung, Y. Li, W. Hong, and H. Yu, "Curved Display Based on Programming Origami Tessellations" *Microsyst. Nanoeng.*, vol. 7, 101, 2021.

[2] H. Taniyama, and E. Iwase, "Design of Rigidity and Breaking Strain for a Kirigami Structure with Non-Uniform Deformed Regions", *Micromachines*, vol. 10, 6, 395, 2019.

[3] R. Sun, B. Zhang, L. Yang, W. Zhang, I. Farrow, F. Scarpa, and J. Rossiter, "Kirigami Stretchable Strain Sensors with Enhanced Piezoelectricity Induced by Topological Electrodes", *Appl. Phys. Lett.*, vol. 112, 251904, 2018.

[4] Y. Sato, S. Terashima, and E. Iwase, "Origami-type Flexible Thermoelectric Generator Fabricated by Self-folding Using Linkage Mechanism", *Proc. of the 35th IEEE Int. Conf. on MEMS (MEMS2022)*, Tokyo, January 9-13, 2022, pp. 31-34.

[5] K. Sempuku, and K. Tachi, "Self-folding Rigid Origami Based on Auxetic Kirigami", *J. Int. Assoc. Shell Spat. Struct.*, vol. 62, pp. 294-304, 2021.

[6] A. Rafsanjani, and K. Bertoldi, "Buckling-Induced Kirigami", *Phys. Rev. Lett.*, vol. 118, 8, 084301, 2017.

[7] Y. Tang, Y. Li, Y. Hong, S. Yang, and J. Yin, "Programmable Active Kirigami Metasheets with More Freedom of Actuation", *Proc. Natl. Acad. Sci. U.S.A.*, vol. 116, 52, pp. 26407-26413, 2019.

CONTACT

Nagi Nakamura, Tel: +81-3-5286-2741, E-mail: nakamura@iwaselab.amech.waseda.ac.jp.

Eiji Iwase, Tel: +81-3-5286-2741, E-mail: iwase@waseda.jp.

BUBBLE-ASSISTED RE-FORMATION OF INDIVIDUAL LIPID BILAYERS IN ARRAYED DEVICE

Izumi Hashimoto[1,2], Toshihisa Osaki[2], Hisatoshi Mimura[2], Sho Takamori[2], Norihisa Miki[1,2], Shoji Takeuchi[2,3].

[1]Department of Mechanical Engineering, Keio University, JAPAN
[2]Kanagawa Institute of Industrial Science and Technology, JAPAN
[3]Grad. School of Information Science and Technology, The University of Tokyo, JAPAN

ABSTRACT

This paper describes a method capable of individual and selective re-formation of ruptured BLMs (bilayer lipid membranes) by using air bubbles in arrayed devices. BLMs play an important role in membrane protein studies, biosensor applications, and drug screening. In such applications, data throughput is a key issue, but BLMs are easily ruptured by physical/electrical disturbance, which makes it difficult to operate a large BLM array. In this work, we propose a bubble-assisted BLM re-formation method: An aqueous phase is divided by a small air bubble covered with a lipid monolayer. After the bubble departed, the divided aqueous phases are brought into contact with the lipid monolayer, resulting in formation of BLM. This method enables selective BLM re-formation in a multi-channel device. To verify individual and selective BLM re-formation using this method, we developed a 4ch bubble-assisted device and demonstrated repetitive BLM re-formation.

KEYWORDS

Bilayer lipid membrane, Air bubble, Re-formation.

INTRODUCTION

A cell membrane is composed of a phospholipid bilayer and membrane proteins. The membrane proteins are considered as important targets of drugs as well as highly sensitive elements for biohybrid sensors (Fig. 1a). To reconstitute the membrane proteins in vitro, there have been numbers of microfluidic devices developed including double-well device with a droplet interface bilayer [1]. By using the device, BLM was formed just by contacting two

aqueous droplets. The BLM devices have been further improved for a decade and many applications are now proposed based on the BLM platforms.

Although the platforms have been well developed, BLMs are still easily collapsed by physical and electrical disturbances because of its thickness of 5 nm. Therefore, the re-formation of BLM is a main challenge to improve data throughput with a BLM array, because the manual BLM re-formation is a time-consuming and labor-intensive process. Previous works introduced microfluidic devices for simultaneous re-formation of a BLM array using a mechanical motion, but they were unsuitable for individual re-formation of a specific BLM (Fig. 1b) [1,2].

In this paper, we propose a BLM re-formation using air bubbles (Fig. 2). In this technique, an aqueous droplet, merged by BLM rupture, is divided by an air bubble, followed by re-formation of BLM by bringing the divided droplets into contact each other after departure of the bubble. Using the technique, we demonstrate individual re-formation of ruptured BLMs in an array device.

EXPERIMENTAL

We fabricated a bubble-assisted BLM re-formation device. The device was composed of a double well with a bubble inlet and a pair of through-holes for electrodes, made of acrylic plates. An acrylic separator with a 600-μm aperture was glued at the middle of the well and Ag/AgCl electrodes were embedded at the bottom of the well (Fig. 3a). The detailed design of the bubble inlet is shown in Fig. 3b. A step with a height of 0.5 mm and an angle of 90° was designed at the inlet to guide air bubbles to the aperture. In this work, 4 pairs of the double well were integrated in a

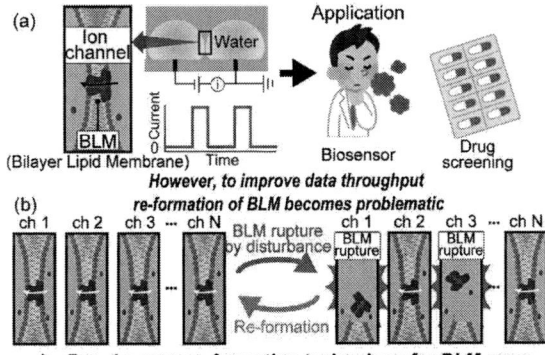

Figure 1: (a) Ion channels incorporated in a BLM are applied for highly sensitive biosensors and for drug screening. (b) Easy and reproducible BLM re-formation is necessary to improve the data throughput with a BLM array because of fragility of BLM.

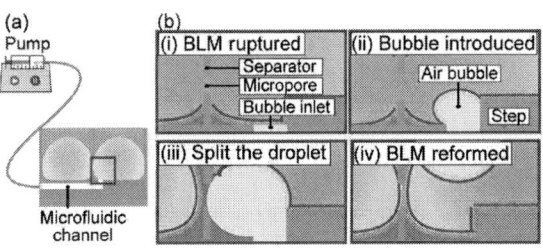

Figure 2: Schematic diagrams of the bubble-assisted re-formation of a BLM. (a) Device overview. A microchannel was integrated underneath the wells, and bubbles were generated using an external pump. (b) An enlarged cross-sectional view of the device, illustrating the reformation process of a ruptured BLM by an air bubble.

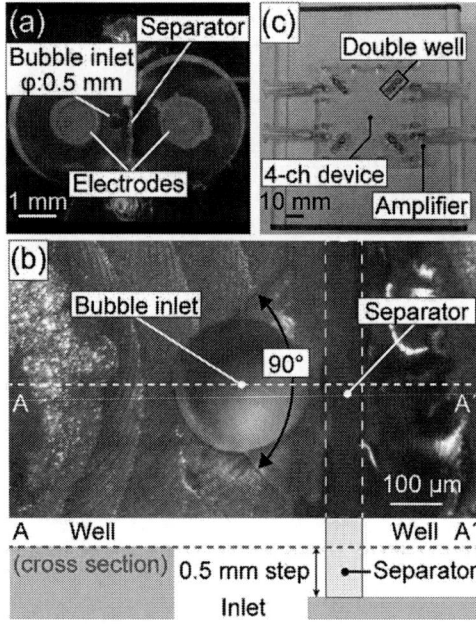

Figure 3: Photographs of the fabricated device. (a) A double well. (b) Enlarged view of the bubble inlet; 0.5-mm step was fabricated to control the direction of bubbles. (c) 4 double wells were integrated in a single device, and mounted on an in-house developed 4ch-amplifier for electrical recordings.

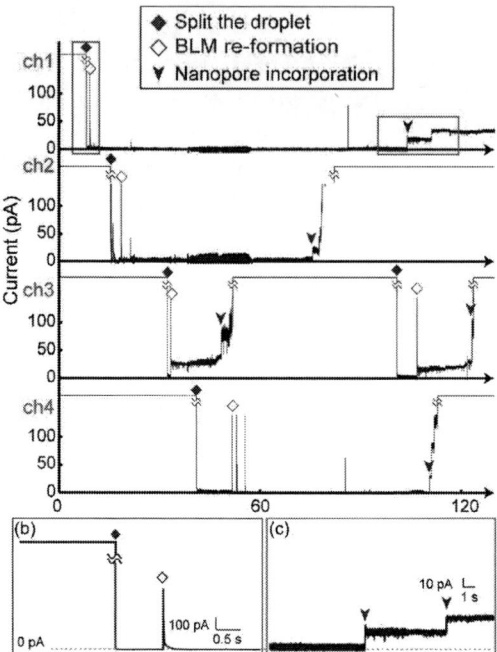

Figure 4: Parallel recording of αHL nanopore currents. (a) A 30-minute current traces using the 4ch device. (b) Enlarged view of division of merged droplets by a bubble and BLM re-formation. (c) Enlarged view of stepwise signals. Each step indicates incorporation of αHL nanopore in BLM.

single device, and mounted on an in-house developed 4ch-amplifier for electrical recordings (Fig. 3c). Air bubbles were generated using syringe pumps connected to the microchannels in the device. The flow rate was set at 100 μL/min.

In this paper, we verified individual BLM re-formation using the 4ch bubble-assisted device. First, we prepared BLMs in the device by subsequently adding 2.4 μL of DPhPC (20 mg/mL in decane) and 25 μL of aqueous solution (1 M KCl with 10 mM phosphate buffer, pH7.4), and all BLMs were ruptured once. Then, re-formation was performed each time when the BLM was ruptured. BLM formation was confirmed by using a nanopore-forming protein, α-hemolysin (αHL).

RESULTS AND DISCUSSION

We demonstrated individual re-formation of ruptured BLMs with the device. We observed 3 current patterns for the BLM re-formation process by air bubbles. The first one is the successful case: Injection of a bubble divides the merged droplet, resulting in zero current, followed by BLM re-formation and nanopore incorporation that presents step current signals. The second one is a failure of BLM re-formation where no step signal is confirmed, indicating formation of a thick oil membrane. The third one is another failure case where the current overflows after the bubble departed, considered as a lack of sufficient lipids for BLM re-formation.

Fig. 4a shows the current traces of 4 double-wells. When a BLM was ruptured and the current overflowed, a bubble was manually injected that resulted in zero current

due to separation of two wells by a bubble. Subsequently, the departure of the bubble shortly disturbed the current (Fig. 4b). BLM re-formation was confirmed by the stepwise current attributed to nanopore incorporation in the BLM (Fig. 4c). As shown in the overall traces in Fig. 4, BLMs were individually re-formed in respective wells after introduction of bubbles, without disturbing the other wells.

CONCLUSION

We proposed a BLM reformation method that utilized an air bubble. Individual and selective BLM re-formation was demonstrated with the 4ch device by presenting current traces with αHL nanopores. The method will be feasible for real-time and on-site applications and contribute to improve their data throughput.

ACKNOWLEDGEMENTS

This work was partly supported by JST START Grant Number JPMJST1811, and by the Program for Building Regional Innovation Ecosystem of MEXT, Japan.

REFERENCES

[1] M. Gotanda et al., *Sensors and Actuators B: Chemical*, 292 (2019), pp. 57-63.
[2] Juan M. Rio Martinez et al., *Small*, 11 (2015), pp. 119-125.

CONTACT

*Izumi Hashimoto, tel: +81 44-819-2037
springer0827@keio.jp

LARGE-SCALE ARRAYS OF TUNABLE MONOLAYER MoS$_2$ NANOELECTROMECHANICAL RESONATORS

*Zuheng Liu[1], Luming Wang[3], Pengcheng Zhang[1], Maosong Xie[1], Yueyang Jia[1],
Ying Chen[4], Hao Jia[4], Zenghui Wang[3,*], and Rui Yang[1,2,*]*

[1]University of Michigan – Shanghai Jiao Tong University Joint Institute,
Shanghai Jiao Tong University, Shanghai 200240, CHINA
[2]School of Electronic Information and Electrical Engineering,
Shanghai Jiao Tong University, Shanghai 200240, CHINA
[3]Institute of Fundamental and Frontier Sciences,
University of Electronic Science and Technology of China, Chengdu 610054, CHINA
[4]Shanghai Institute of Microsystem and Information Technology,
Chinese Academy of Sciences, Shanghai 200050, CHINA

ABSTRACT

We report on the first demonstration on scaled fabrication of electrically tunable monolayer molybdenum disulfide (MoS$_2$) nanoelectromechanical resonator arrays. A centimeter-sized monolayer MoS$_2$ film is grown by chemical vapor deposition (CVD), and then transferred onto the substrate with pre-patterned microtrench and electrode arrays, resulting in freely suspended nanoelectromechanical resonator arrays based on two-dimensional (2D) MoS$_2$ membranes. Raman and photoluminescence (PL) measurements confirm the monolayer thickness and uniformity of material properties. We experimentally demonstrate resonance excitation and detection, nonlinear resonances, as well as electrostatic frequency tuning, in these arrayed MoS$_2$ resonators. The resonators show resonance frequency tuning ranges ($\Delta f/f_0$) up to 21%, as well as gate tuning of quality (Q) factors ($|\Delta Q/Q_0|$) by up to 44%. These arrays of 2D nanoelectromechanical resonators can enable more complex circuits at large scale.

KEYWORDS

2D NEMS, MoS$_2$ resonators, Large-scale resonator arrays, Frequency tuning, Scalable fabrication.

INTRODUCTION

Resonant micro-/nano-electromechanical systems (MEMS/NEMS) have been widely used in sensing and radio-frequency (RF) communication applications [1–3]. As devices continue to scale down, 2D NEMS resonators have attracted tremendous interest due to their atomic scale thickness, ultra-high frequency tunability, and board dynamic range [4,5]. Device-level demonstrations of 2D NEMS resonators have been explored for sensing, computing, and RF signal processing [6–10]. Towards larger-scale 2D NEMS devices, arrays of CVD graphene NEMS resonators with contact electrodes have been demonstrated [11]. Beyond graphene, 2D semiconductors such as MoS$_2$ have also emerged as promising materials for 2D NEMS resonators due to the large and tunable bandgap, strong electromechanical coupling, and ultralow power consumption [12,13]. However, while CVD MoS$_2$ resonators arrays without electrical contacts have been demonstrated [14], and the single MoS$_2$ NEMS resonator with electrical readout has been achieved [15–17], electrically accessible NEMS resonator arrays based on 2D semiconductors have remained elusive, which limits the future mass production and integration with circuits.

Here, we demonstrate large-scale electrically tunable monolayer MoS$_2$ NEMS resonator arrays, with both resonance frequency and Q tunable by DC gate voltages. Following the fabrication of monolayer MoS$_2$ NEMS resonator arrays, we perform optical interferometry measurement and verify the consistency of the devices, and then measure the gate tuning properties. Such capability for scalable production of uniform 2D NEMS resonator arrays with electrically tunable resonance frequency and Q holds promises for all-electrical 2D NEMS voltage-controlled oscillators and RF filters, frequency-shift based sensing, and ultra-low power computing and memory systems [18].

WORKING PRINCIPLE OF THE ARRAY

Fabrication of the electrically tunable and large-scale MoS$_2$ nano-resonator arrays has been challenging due to several aspects: growing continuous 2D material film with a large area, suspending the 2D material on surface microtrenches and contacting them with metal electrodes at the same time, and increasing the density of resonators each with necessary contact electrodes, *etc*. In this work, to ensure the material property, we use chemical vapor deposition (CVD) to grow centimeter-scale continuous monolayer MoS$_2$. To increase the fabrication yield, we apply a water-assisted transfer technique, which has been previously used to transfer large-scale MoS$_2$ onto microtrenches [14]. To increase the resonator array density, we design the substrate with regularly arranged electrodes and circular microtrenches on an oxidized silicon (Si) wafer (Fig. 1). For each probing pad, four pairs of contact electrodes and thus four resonators are connected to it, with two resonators along the horizontal direction, and another two along the vertical direction. As such, the array is densely packed, and each resonator can be electrically accessed. The whole chip occupies an area of about 1 square-centimeter, and based on the electrode arrangement, there can be up to thousands of 2D NEMS resonators on the same chip, as long as the transferred 2D material can cover the whole area.

Following wire bonding to the corresponding contact pads, the neighboring resonators can be measured. After grounding the contact pad, and applying the AC excitation voltage (v_{ac}) and DC gate voltage (V_{GS}) to the back gate, the resonator can be capacitively excited, and each resonator can be measured using a custom-built interferometry system using a 633 nm wavelength laser (Fig. 1a) [13,19]. The large-scale 2D resonators with electrical contacts also hold the potential for electrical signal transduction of each resonator, making it promising for future applications.

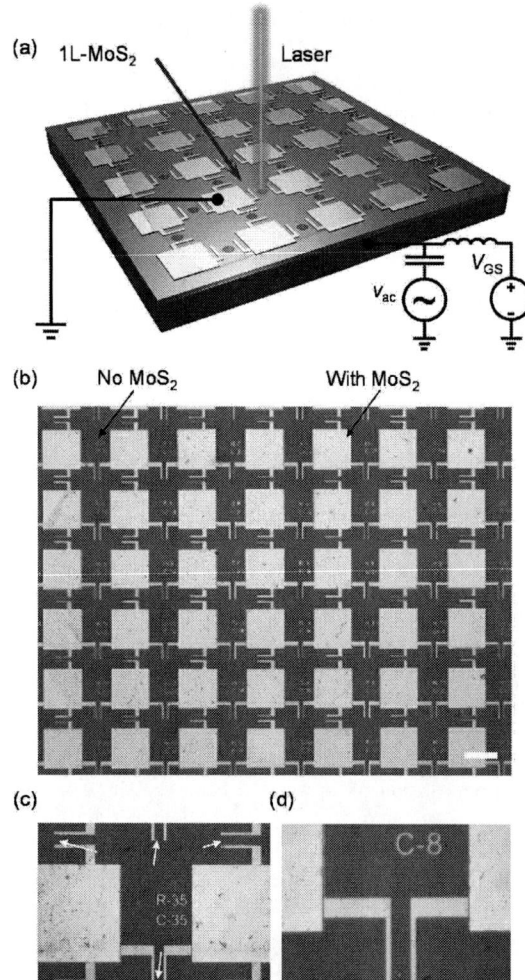

Figure 1: Large-scale arrays of 2D MoS₂ NEMS resonators. (a) 3D schematic illustration of an array of 2D MoS₂ NEMS resonators with a laser for optical interferometry motion detection, and electrodes for motion excitation and gate tuning. (b) Optical images of the fabricated monolayer (1L) MoS₂ resonator array. Scale bar: 100 μm. (c) Zoom-in optical image of a region with 6 MoS₂ resonators that are all in contact with the electrodes. Scale bar: 20 μm. (d) Zoom-in optical image of a single MoS₂ resonator with contact electrodes. Scale bar: 10 μm.

FABRICATION PROCESS

We fabricate the devices on a Si wafer with 300 nm silicon dioxide (SiO_2) on the surface (Fig. 2a). We first etch the circular-shape microtrenches with 2 μm diameters on the wafer (Fig. 2b). Then we pattern the contact electrodes by evaporating 5 nm chromium (Cr) and 35 nm gold (Au), which forms the substrate (Fig. 2c). We then perform water-assisted transfer of centimeter-scale CVD MoS₂ continuous film onto a polydimethylsiloxane (PDMS) stamp [14], followed by all-dry transfer of MoS₂ onto the

back-gated SiO_2/Si substrate with surface circular microtrenches and densely-packed local contact electrodes [20] (Fig. 2d). During the dry transfer process, we first press the PDMS with MoS₂ onto the chip gently, then heat up the substrate after they are fully in contact, and finally slowly lift up the transfer stage, so that the MoS₂ will remain suspended over the microtrenches and in contact with the electrodes. This forms the large-scale array of electrically tunable 2D MoS₂ resonators.

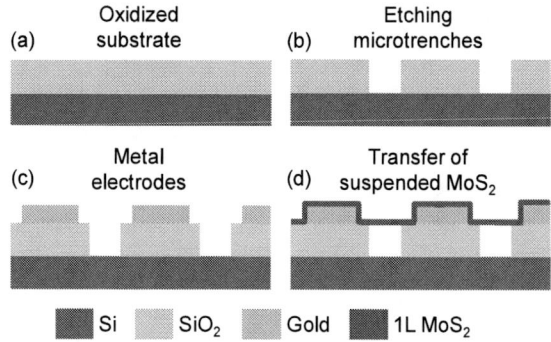

Figure 2: Cross-sectional illustration of the fabrication processes for 1L MoS₂ resonator arrays, showing (a) oxidized Si substrate, (b) microtrench etching, (c) contact electrode deposition, and (d) transfer of 1L MoS₂ film.

2D MATERIAL CHARACTERIZATION

Before measuring the resonances, we first characterize the properties of the MoS₂ material using Raman spectroscopy and photoluminescence (PL) spectra, with 532 nm excitation. Raman peak separation of ~19.5 cm⁻¹ is consistent with the distinct signatures of 1L MoS₂ (Fig. 3a). PL peak position of ~1.86 eV and large PL intensity also corresponds to 1L MoS₂ with high quality (Fig. 3b).

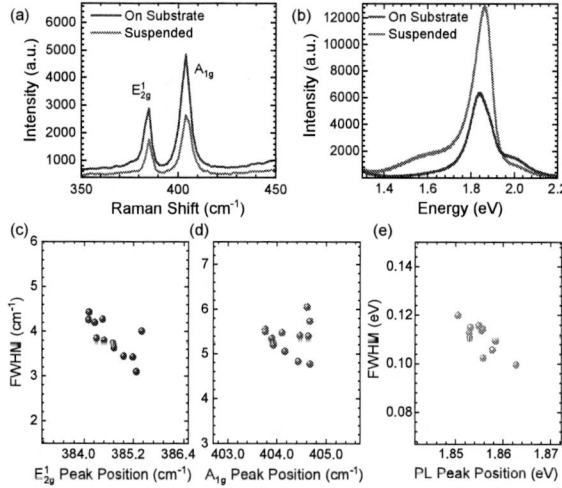

Figure 3: Measurements of the Raman and PL spectra for monolayer MoS₂ NEMS resonator arrays. (a-b) Representative (a) Raman (b) PL spectra for suspended monolayer MoS₂, with the spectra for substrate-supported monolayer MoS₂ as references. (c-e) Statistics of (c) E_{2g}^1 and (d) A_{1g} Raman modes, and (e) PL peak positions for the resonator array, with the full width at half maximum (FWHM) obtained from fitting.

For evaluating the uniformity of the MoS_2 film, we measure the Raman and PL characteristics on the array of resonators. The statistics of E_{2g}^1 and A_{1g} Raman peaks and PL peak positions, and their FWHM show good uniformity of material properties across devices (Fig. 3c–e).

RESONANCE MEASUREMENTS

To drive the device motion, V_{GS} and v_{ac} are connected through a bias-tee, thus the electrostatic force at driving frequency is proportional to the $V_{GS} \cdot v_{ac}$, and we can properly adjust the drive strength *via* controlling either the DC or AC voltage. The devices are measured at room temperature in vacuum.

Due to the extremely thin material, the resonance spectra of 1L MoS_2 resonators are typically more difficult to measure than bulk materials. We optimize the resonance signal amplitude by gradually increasing the DC voltage from the DC power supply and the AC driving voltage from the network analyzer, during the continuous frequency sweep, which allows us to record linear to Duffing nonlinear resonances [21]. We measure the resonant responses of the devices with V_{GS} fixed at 10 V and v_{ac} sweeping from 100 mV to 1000 mV. Four representative devices in Fig. 4 show clear transition from linear to nonlinear resonances as we gradually increase v_{ac}. The resonance frequencies are consistently around 46–49 MHz, and nonlinear resonances show hardening behavior.

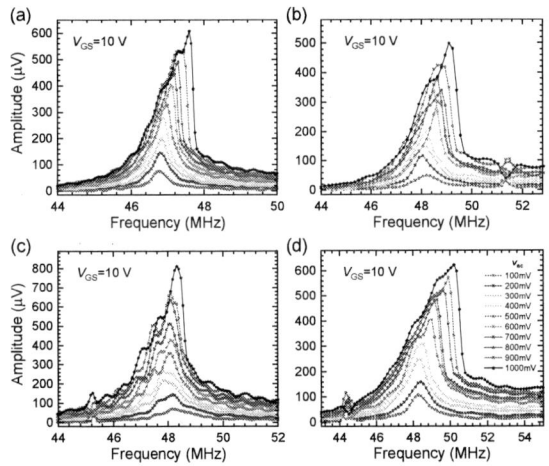

Figure 4: Measurements of the linear and nonlinear resonances for 1L MoS_2 NEMS resonators. The resonance responses with increasing v_{ac} from 100 mV to 1000 mV show Duffing nonlinearity, with a fixed V_{GS} of 10 V, for four representative resonators.

GATE TUNING PROPERTIES

The DC gate voltage applied between the MoS_2 drumheads and the back gate not only provides the electrostatic driving force for exciting the resonance, but also induces static deflection and thus straining of the membrane, leading to tuning of resonance frequency and Q. Therefore, we gradually increase the V_{GS} applied to the back gate from 0 V to 12 V with a 1 V interval, at a fixed v_{ac} of 1000 mV, and repeat the measurements for 12 MoS_2 resonators. From the recorded resonance spectra (Fig. 5), a consistent upward frequency shift with larger V_{GS} is

observed due to the increased tensile strain. Four representative resonators show consistent tuning ranges from ~36 MHz to ~41 MHz. Such consistent resonance frequency tuning for large-scale MoS_2 resonators benefits from the uniform material characteristics and efficient material transfer process with minimal contamination.

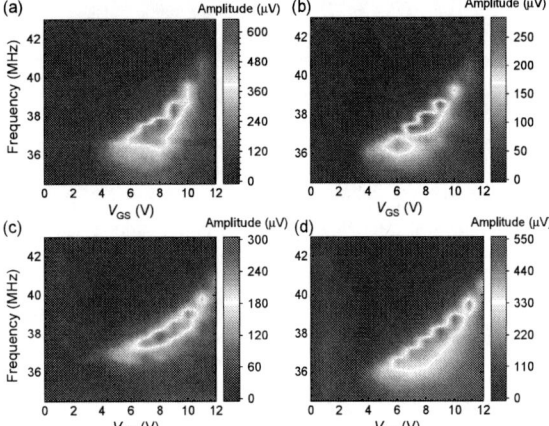

Figure 5: Measurements for the dependence of resonances on V_{GS} for 1L MoS_2 NEMS resonators. The resonance signal amplitude is shown in color scale for V_{GS} from 0 V to 12 V and the frequencies show consistent resonance tuning for four resonators, with v_{ac} fixed at 1000 mV.

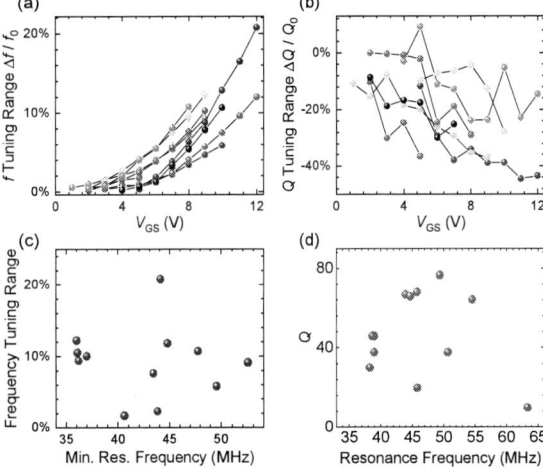

Figure 6: Summary of frequency and Q tuning by V_{GS}. (a) Frequency tuning by V_{GS} for different devices as shown by different colors. (b) Q tuning by V_{GS} for different devices as shown by different colors. (c) Summary of the largest frequency tuning range and the minimum measurable resonance frequency for each device. (d) Summary of resonance frequency and Q, measured at v_{ac}=1000 mV, and V_{GS}=7 V.

We further summarize the gate tuning characteristics for all 12 devices that we have measured (Fig. 6). We first extract their resonance frequency (f_0) and Q under each V_{GS} from fitting, and summarize the frequency tuning ranges by taking the differences between the maximum and minimum resonance frequencies (Δf), and then divide it by the minimum resonance frequency, to obtain $\Delta f/f_0$. The frequency tuning range can reach up to 21%, and the tuning

978-1-6654-9309-3/23 $31.00 © 2023 IEEE

trend of increasing frequency with a larger V_{GS} is also consistent (Fig. 6a and 6c). Similarly, we also summarize the Q tuning range by V_{GS}, which shows $|\Delta Q/Q_0|$ up to 44% (Fig. 6b and 6d). We find that Q generally decreases with a larger V_{GS} for these fully-clamped circular resonators, which is due to the combined effects from increased tensile strain and increased vibration amplitude as V_{GS} increases [22,23]. Some data points at large V_{GS} are missing, because the devices are already driven into nonlinear resonance.

CONCLUSION

In summary, we design and fabricate electrically tunable large-scale monolayer MoS_2 resonator arrays, and measure the frequency response. These devices consistently show Duffing nonlinearity under large drive voltages. For the 12 resonators measured, the resonance frequency increases and Q decreases with a larger V_{GS} quite consistently. Our work paves the way towards scalable production of electrically tunable NEMS resonator circuits based on 2D semiconductors.

ACKNOWLEDGEMENTS

The authors thank the support from National Natural Science Foundation of China (NSFC) (grants U21A20505, 62104140, 62104241), Science and Technology Commission of Shanghai Municipality (STCSM) Natural Science Project General Program (grant 21ZR1433800), Shanghai Pujiang Program(20PJ1415600), Lingang Laboratory Open Research Fund (grant LG-QS-202202-11), and Biren Technology–Shanghai Jiao Tong University Joint Laboratory Open Research Fund.

REFERENCES

[1] H. Jia, *et al.*, "Integrated Resonant Micro/Nano Gravimetric Sensors for Bio/Chemical Detection in Air and Liquid", *Micromachines*, vol. 12, 645, 2021.

[2] F. Yao, *et al.*, "Thermogravimetric Analysis on a Resonant Microcantilever", *Anal. Chem.*, vol. 94, pp. 9380-9388, 2022.

[3] X. L. Feng, *et al.*, "A Self-Sustaining Ultrahigh-Frequency Nanoelectromechanical Oscillator", *Nat. Nanotechnol.*, vol. 3, pp. 342-346, 2008.

[4] B. Xu, *et al.*, "Nanomechanical Resonators: Toward Atomic Scale", *ACS Nano*, vol. 16, pp. 15545-15585, 2022.

[5] Z. Wang, *et al.*, "Recent Progress in 2D van der Waals Heterostructures: Fabrication, Properties, and Applications", *Sci. China Inf. Sci.*, vol. 65, 211401, 2022.

[6] J. Lee, *et al.*, "Electrically Tunable Single- and Few-Layer MoS_2 Nanoelectromechanical Systems with Broad Dynamic Range", *Sci. Adv.*, vol. 4, eaao6653, 2018.

[7] C. Chen, *et al.*, "Graphene Mechanical Oscillators with Tunable Frequency", *Nat. Nanotechnol.*, vol. 8, pp. 923–927, 2013.

[8] J. Fang, *et al.*, "A Cantilever-based Resonator for Reconfigurable Nanomechanical Computing", *J. Micromech. Microeng.*, vol. 31, 124003, 2021.

[9] Z. Wang, *et al.*, "Thermal Hysteresis Controlled Reconfigurable MoS_2 Nanomechanical Resonators", *Nanoscale.*, vol. 13, pp. 18089-18095, 2021.

[10] P. Zhang, *et al.*, "Nanoelectromechanical Memories based on Nonlinear 2D MoS_2 Resonators", in *Proc. 35th IEEE Int. Conf. on Micro Electro Mechanical Systems (MEMS 2022)*, Tokyo, Jan. 9-13, 2022, pp. 208-211.

[11] A. M. van der Zande, *et al.*, "Large-Scale Arrays of Single-Layer Graphene Resonators", *Nano Lett.*, vol. 10, pp. 4869–4873, 2010.

[12] R. Yang, *et al.*, "Electromechanical Coupling and Design Considerations in Single-Layer MoS_2 Suspended-Channel Transistors and Resonators", *Nanoscale*, vol. 7, pp. 19921-19929, 2015.

[13] R. Yang, *et al.*, "Raman Spectroscopic Probe for Nonlinear MoS_2 Nano-electromechanical Resonators", *Nano Lett.*, vol. 22, pp. 5780-5787, 2022.

[14] H. Jia, *et al.*, "Large-Scale Arrays of Single- and Few-Layer MoS_2 Nanomechanical Resonators", *Nanoscale*, vol. 8, pp. 10677–10685, 2016.

[15] C. Samanta, *et al.*, "Nonlinear Mode Coupling and Internal Resonances in MoS_2 Nanoelectromechanical System", *Appl. Phys. Lett.*, vol. 107, 173110, 2015.

[16] R. Yang, *et al.*, "Local-Gate Electrical Actuation, Detection, and Tuning of Atomic-Layer MoS_2 Nanoelectromechanical Resonators", in *Proc. 30th IEEE Int. Conf. on Micro Electro Mechanical Systems (MEMS 2017)*, Las Vegas, Jan. 22-26, 2017, pp. 163-166.

[17] R. Yang, *et al.*, "All-Electrical Readout of Atomically Thin MoS_2 Nanoelectromechanical Resonators in the VHF Band", in *Proc. 29th IEEE Int. Conf. on Micro Electro Mechanical Systems (MEMS 2016)*, Shanghai, Jan. 24-28, 2016, pp. 59-62.

[18] Z. Wang, *et al.*, "Nanomechanics: Emerging Opportunities for Future Computing", *Sci. China Inf. Sci.*, vol. 64, 206401, 2021.

[19] J. Zhu, *et al.*, "Analyzing Electrostatic Modulation of Signal Transduction Efficiency in MoS_2 Nanoelectromechanical Resonators with Interferometric Readout", *Sci. China Inf. Sci.*, vol. 65, p. 122409, 2022.

[20] R. Yang, *et al.*, "Multilayer MoS_2 Transistors Enabled by a Facile Dry-transfer Technique and Thermal Annealing", *J. Vac. Sci. Technol. B.*, vol. 32, 061203, 2014.

[21] A. Eichler, *et al.*, "Nonlinear Damping in Mechanical Resonators Made from Carbon Nanotubes and Graphene", *Nat. Nanotechnol.*, vol. 6, pp. 339-342, 2011.

[22] P. Zhang, *et al.*, "Strain-Modulated Dissipation in Two-Dimensional Molybdenum Disulfide Nanoelectromechanical Resonators", *ACS Nano*, vol. 16, pp. 2261-2270, 2022.

[23] P. Zhang, *et al.*, "Strain-Modulated Equivalent Circuit Model and Dissipation Model for 2D MoS_2 NEMS Resonators", in *Proc. 34th IEEE International Conference on Micro Electro Mechanical Systems (MEMS 2021)*, Gainesville, 658-661, Jan. 25-29, 2021, pp. 658-661.

CONTACTS

*Z. Wang, email: zenghui.wang@uestc.edu.cn

*R. Yang, email: rui.yang@sjtu.edu.cn

978-1-6654-9309-3/23 $31.00 © 2023 IEEE

ANTIFOULING FOR ELECTROCHEMICALLY BIOSENSING IN BODY FLUIDS

Wenzheng He[1], Changdong Zhou[2], Yang Lin[2], Yuxin Tian[2], Liying Liu[2], Qifu Zhang[2], Xiongying Ye[1]*, and Tianhong Cui[3]*

[1] State Key Laboratory of Precision Measurement Technology and Instruments, Department of Precision Instruments, Tsinghua University, Beijing, 100084, China,
[2] Department of Urology, Jilin Cancer Hospital, Jilin Province, Changchun, 130012, China, and
[3] Department of Mechanical Engineering, University of Minnesota, 111 Church St. S.E., Minneapolis, MN 55455, USA

ABSTRACT

This paper presents a simple and robust antifouling sensor based on a nano-wrinkle electrode with bovine serum albumin for electrochemically detecting small molecules in complex body fluids. The nano-wrinkle was prepared by a one-step shrinking technique, being capable of excluding large proteins in physical structures and enhancing the response of small molecules. The prepared sensor demonstrates excellent antifouling capability under discontinuous or continuous exposure to proteins or fetal calf serum, which preserves 97% of its original signal after a 21-day exposure to fetal calf serum and enabled the quantification of dopamine in the protein-coexisted environment with a low limit of detection (LOD) (1.09 μM) and a wide detection range from 20 to 1105 μM.

KEYWORDS

Antifouling, Nano-wrinkle, Electrochemical sensor, Dopamine, Shrinking technique, Serum

INTRODUCTION

There is a growing need for electrochemical detection of small molecules (e.g. dopamine, uric acid, etc. [1]) owing to its high sensitivity, low cost, and ease of operation. One of the main challenges is that the concentration of targets in the real samples (e.g. serum) is usually far lower than the coexisting background proteins which may adhere to electrochemical sensing regions and foul the sensor, leading to surface passivation, response reduction, or even false positives, especially the label-free sensor [2]. Thus, an effective, simple, and robust antifouling strategy is highly desired to keep the sensor at high sensitivity and stability in complex body fluids.

Two methods based on nanoporous structures and poly (ethylene glycol) coating were most frequently employed for antifouling according to the mechanism of steric repulsion and strong hydration layer, respectively [3]. Though effective, the preparation of nanoporous structures relies on dangerous acid treatment and the poly (ethylene glycol) can be oxidized in the presence of oxygen, limiting their long-term applications to complex body fluids.

In this work, a simple antifouling strategy is presented with high performance, which uses the nano-wrinkle as the physical shield to prevent from the absorption of proteins and improve the response area and use a strong cationic electrolyte to fix bovine serum albumins (BSA). The nano-wrinkle was prepared and regulated by a simple shrink process without acid treatment. Besides, the modified BSA (isoelectric point: 4.7 [4]) was negatively charged in the normal human blood (pH:7.34~7.45 [5]), which can selectively attract the positively charged dopamine and repulse other background proteins by electrostatic force. These features enable stable detection of small molecules in complex biological fluids with high performance.

MATERIALS AND METHODS

Chemicals and Apparatus

The shrink films (polystyrene, KSF6-ASST) were purchased from Grafix. The BSA, 1× PBS, prostate specific antigen (PSA) and its monoclonal antibody, and fetal bovine serum (FBS) were obtained from Sangon Biotech Co., Ltd (Shanghai, China). Dopamine was from J&K Scientific Ltd, Beijing, China. Poly (diallyldimethylammonium chloride) (PDDA, 1.5 wt%) were from Sigma-Aldrich. Other chemicals were acquired from Sigma-Aldrich.

The surface topography of the prepared sensor was evaluated by a scanning electron microscope (SEM, Zeiss Merlin Compact) and analyzed by Image J (National Institutes of Health). The electrochemical tests including cyclic voltammetry (CV) and differential pulse voltammetry (DPV) were carried out by an electrochemical workstation (CHI 660E).

Sensor Fabrication

The antifouling sensor was fabricated, as shown in Figure 1(A). First, gold was sputtered on a shrink film with different thicknesses and heated at 160°C for 3 mins to create a shrink electrode with wrinkles owing to the mismatch of Young's modulus between gold and shrink polymer. Then the working electrode was pretreated by air plasma for 3 mins to eliminate the hydrophobicity and make use of the inner space of the wrinkles. Next, a 10 μL PDDA solution (positive charged) and 30 mg/mL BSA were dropped on the electrode successively and incubated for 30 mins to immobilize BSA by electrostatic force. Next, clear this electrode with ultrasonic cleaning (40 kHz, 144 W) for 1 min, immerse it in 1× PBS for 30 mins to remove the unstable BSA, and obtain the final antifouling electrochemical electrode. The prepared electrode experienced a reduction of 2.6 times in side size after heating, as shown in Figure1(B). The feature sizes of wrinkles including wavelength (λ) and amplitude (A) can be theoretically regulated by the thickness (h) of gold film, as shown in Figure 1(D) [6].

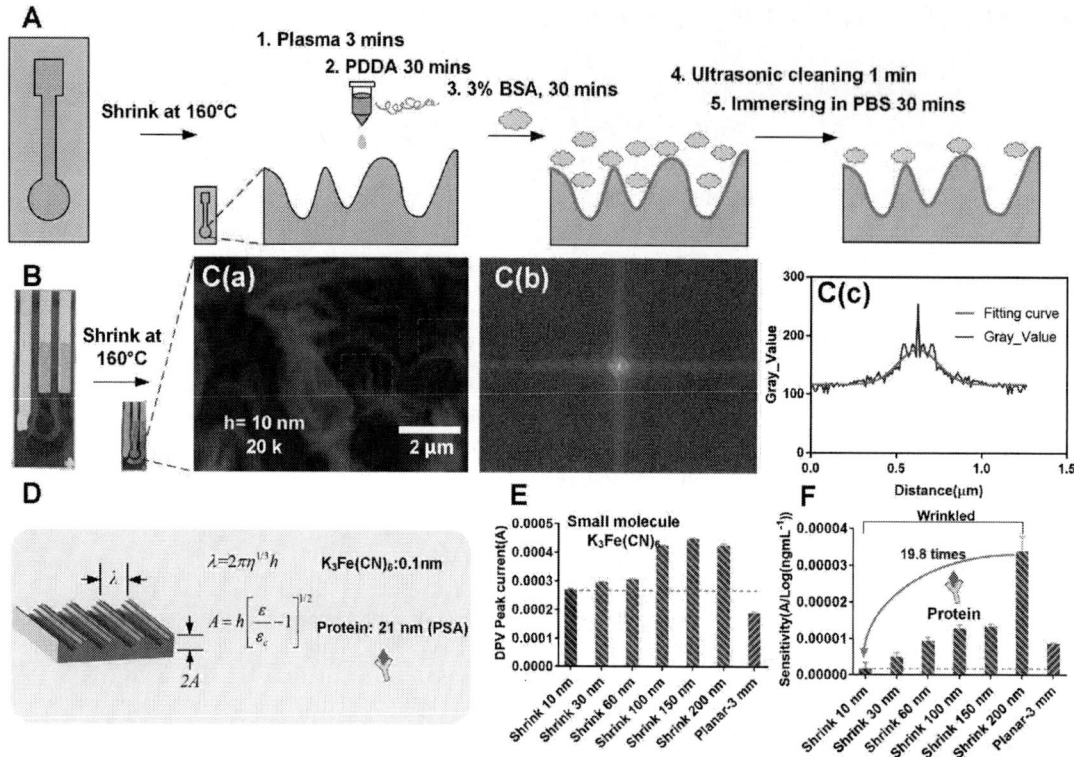

Figure 1: (A) Fabrication of antifouling electrochemical biosensor. (B) Photos of electrodes before and after shrinking process. (C) SEM figure of shrink 10 nm electrode, its fast Fourier transform (FFT) image of selected areas, and its plot profile at 0 degrees. (D) Regulation theory of wrinkles and feature sizes of biomolecules. (E) DPV peak currents of different electrodes in 5 mM $K_3Fe(CN)_6$. (F) Comparison of response sensitivity to PSA among different electrodes

RESULTS AND DISCUSSION
Shrink electrode regulation and response

Owing to the difference in physical size between small molecules and proteins, the fouling influence of proteins could be regulated by the surface topography of the sensor, especially in nano-scale. Here, the wavelength of the shrink electrode can vary from 118.8 to 2,376.6 nm and the amplitude varies from 17.1 to 343.5 nm theoretically when the thickness of the gold film was regulated from 10 nm to 200 nm. These feature sizes provide different diffusion freedom for small molecules and proteins.

To characterize the real wavelength, the selected regions of the SEM image of shrink electrodes were analyzed and measured by FFT, as shown in Figure 1(C). The real wavelength of the 10 nm shrink electrode is about 118 ± 8.25 nm close to the results above, confirming the accuracy of the theoretical calculation.

To obtain the best physical antifouling effect and larger response, the shrink electrodes with different thicknesses were tested in the 5 mM $K_3Fe(CN)_6$ (small molecule, about 0.1 nm), as shown in Figure 1(E). All of the shrink electrodes demonstrates larger DPV peak currents than the planar electrode, indicating a larger response area and higher sensitivity to small molecules. A better physical antifouling sensor means a lower sensitivity to proteins. Here, the PSA (about 21 nm) was considered the target protein for electrochemical detection with a label-free method. As shown in Figure 1(F), the 10 nm and

30 nm shrink electrodes exhibits a smaller response sensitivity, compared with the planar electrode. The 10 nm shrink electrode demonstrates the lowest sensitivity to PSA, almost one-twentieth of the 200 nm shrink electrode and one-fifth of the planar electrode because of the small size of wrinkles, indicating a potential resistivity to proteins. Consequently, the 10 nm shrink electrode is regarded as the optimal antifouling sensor for detecting small molecules owing to its larger DPV response to $K_3Fe(CN)_6$ than the planar electrode and a super weaker sensitivity to PSA.

Antifouling coating

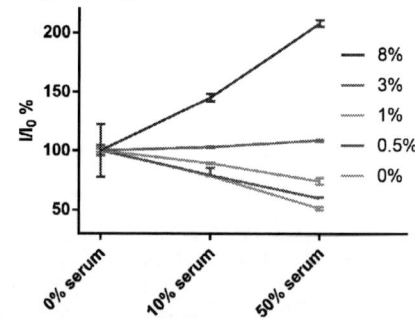

Figure 2: Peak current variation of electrodes modified with different BSA after exposure to serums.

The mechanism of antifouling coating is based on charge repulsion. Most of the common proteins are negatively charged in normal human blood [6], and albumin constitutes a massive ~60% of total protein. Besides, BSA is also negatively charged and will repulse other background proteins in human blood. Thus, BSA was utilized as the antifouling coating. Here, the concentration of BSA for coating was optimized first. As shown in Figure 2, the peak current of the sensor modified with BSA at a low concentration (0%~1%) decreases with the increase of serum, indicating a weak antifouling effect for the non-specific absorption of proteins. The peak current of the sensor at a high concentration (8%) increases with the increase of serum, because of the dissociation of unstable BSA, leading to a resistance reduction. Therefore, 3% BSA was considered the optimal coating.

To address the issue of unstable BSA, the immobilization method with different reagents and concentrations was optimized for a strong bonding, and the antifouling sensor was pretreated by ultrasonic cleaning and PBS soaking to remove the redundant BSA. As demonstrated in Figure 3 (A), the sensor immobilized BSA with different concentrations of poly(ethyleneimine) (PEI) exhibited a better antifouling effect than the sensor without immobilization reagent (Figure 2, 0%) and the sensor with 10 mg/mL PEI demonstrated the lowest peak current variation (only 5%) after exposure to 50% serum. A cationic electrolyte (PDDA) was expected to fix the BSA more strongly. As shown in Figure 3 (B), the sensors with PDDA exhibits a smaller variation than the ones with PEI and the one with 15 mg/mL PDDA presented the smallest variation (only 2%) after fouling with 50% serum. Next, ultrasonic cleaning at different powers was utilized and optimized to remove the unstable BSA. The sensor treated with ultrasonic cleaning at a low power (144 W) exhibits excellent antifouling capacity and preserves 99.2% of its original signal after fouling.

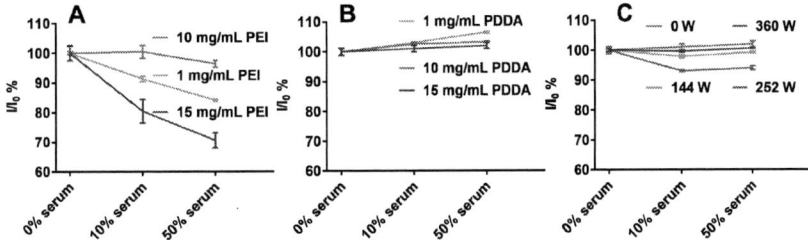

Figure 3: (A) Variation of DPV peak currents of electrode immobilized BSA with differnet (A) PEI and (B) PDDA after fouling with FBS. (C) Anti-serum fouling performance comparision of the electrode with 15 mg/mL PDDA cleaned ultrasonicly with different powers.

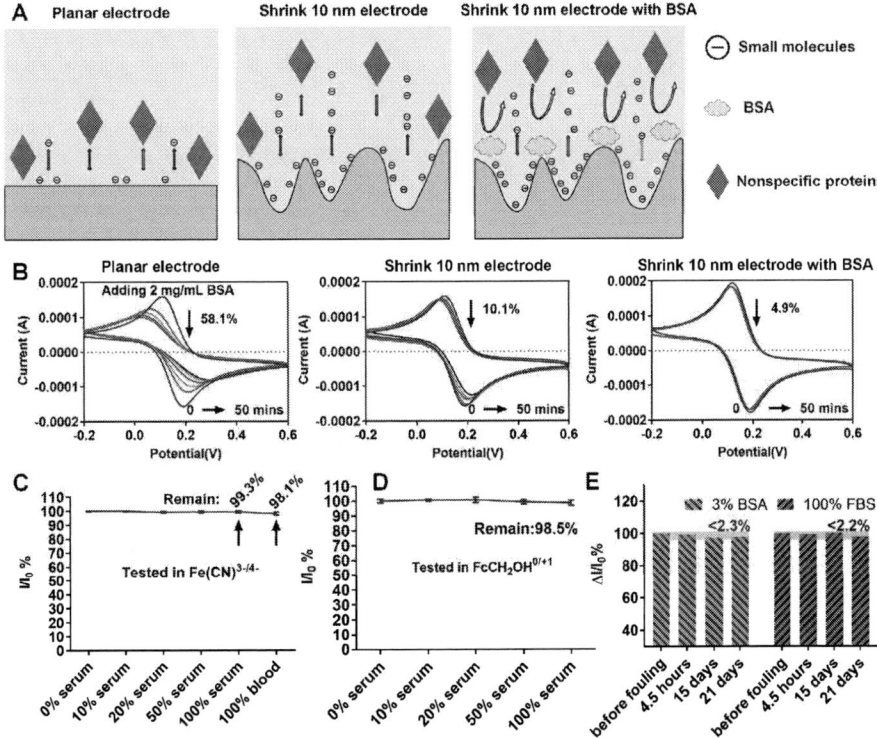

Figure 4: (A) Antifouling strategy of shrink electrode with BSA. (B)Variation of CV curve after adding 2 mg/mL BSA in 5 mM $K_3Fe(CN)_6$ solution. Variation of DPV peak currents tested in (C) $Fe(CN)^{3-/4-}$ and (D) $FcCH_2OH^{0/+1}$ after fouling in FBS or rabbit whole blood at different concentrations. (E)Variation of DPV peak currents after the long-term fouling in 3% BSA and 100% FBS.

Antifouling Performance Evaluation

Figure 4 (A) presents the antifouling mechanism. Here, the antifouling capacity of the 10 nm shrink electrode relied on the small wavelength of wrinkle which was capable of blocking many proteins. Having modified with BSA by the method above, the antifouling capacity was expected to improve further. The CV test confirmed this deduction. According to Figure 4 (B), adding 2 mg/mL BSA, the peak current of the planar electrode continuously decreased by 58.1% after 50 mins and the 10 nm shrink electrode decreased by 10.1%. Immobilized with BSA, the shrink electrode merely decreased by 4.9%, exhibiting less adsorption of proteins owing to the electrostatic repulsing.

Exposed to fetal bovine serum with different concentrations and blood for 30 mins, the 10 nm shrink electrode with BSA still preserved 98% of its original signal (Figure 4(C)), indicating the strong antifouling capacity in complex environments. When tested in different charged redox probes ($FcCH_2OH^{0/+1}$), the current remained above 98%, as shown in Figure 4(D), confirming that the antifouling capacity is not affected by test environments. Exposed to FBS and BSA for 21 days, the peak current only dropped by less than 3%, demonstrating long-term stability. This method exhibits perfect antifouling performance, comparable to or superior to PEG coating and BSA cross-linking coating reported by Sabaté et al. in Nature Nanotechnology[4].

Detection Dopamine Coexisted with Proteins

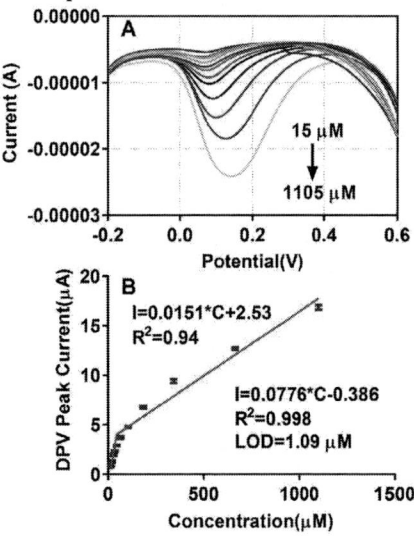

Figure 5:(A) DPV curves tested in dopamine solution from 15 to 1105 µM mixed with 2 mg/mL BSA and (B) their peak current calibration curve.

The antifouling sensor was utilized to detect dopamine, an important neurotransmitter associated with some nervous system diseases (e.g. Parkinson's [6]) because the modified BSA is negatively charged and capable of attracting dopamine [1] and eliminating the contamination from dopamine oxide. As shown in Figure 5, the peak current of the prepared sensor demonstrated a piecewise linear relationship with the increase of dopamine concentrations. Although a high concentration of BSA was mixed in the target solution, there is no obvious impact on the response, indicating the sensor having an excellent antifouling performance which is hard to achieve for the common gold electrode [1]. In the BSA-coexisting solution, the prepared sensor exhibited a low LOD (1.09 µM) and a wide detecting range (15~1105 µM), showing a competent performance with the previous graphene sensor [7].

CONCLUSION

This paper presents a wrinkle-based antifouling sensor for electrochemically detecting small molecules in complex biological fluids with easy preparation and high stability capable of preserving over 97% of its original signal after exposure to 100% serum for 21 days and keeping a good antifouling capability in a different charged environment. The prepared sensor can detect dopamine with high performances even in the protein-coexisted environment, showing its potential feasibility in clinical serum or blood diagnosis.

ACKNOWLEDGEMENTS

This work was supported by the Science and Technology Development Plan of Jilin Province (20210204108YY).

REFERENCES

[1] W. He, X. Ye, and T. Cui, "Flexible Electrochemical Sensor With Graphene and Gold Nanoparticles to Detect Dopamine and Uric Acid," *IEEE Sens. J.*, vol. 21, no. 23, pp. 26556–26565, 2021.

[2] C. Jiang, G. Wang, R. Hein, N. Liu, X. Luo, and J. J. Davis, "Antifouling Strategies for Selective in Vitro and in Vivo Sensing," *Chem. Rev.*, vol. 120, no. 8, pp. 3852–3889, 2020.

[3] P. H. Lin and B. R. Li, "Antifouling strategies in advanced electrochemical sensors and biosensors," *Analyst*, vol. 145, no. 4, pp. 1110–1120, 2020.

[4] J. Sabaté, O. Y. F. Henry, P. Jolly, and D. E. Ingber, "An antifouling coating that enables affinity-based electrochemical biosensing in complex biological fluids," *Nat. Nanotechnol.*, vol. 14, no. December, pp. 1143–1149, 2019.

[5] P. T. L. Antony, P. Sinduja, and R. Priyadharshini, "Comparison of Random Blood Sugar and pH of Blood- A Clinicopathological Study," *J. Pharm. Res. Int.*, vol. 33, pp. 302–309, 2021.

[6] X. Zhuang, P. Mazzoni, and U. J. Kang, "The role of neuroplasticity in dopaminergic therapy for Parkinson disease," *Nat. Rev. Neurol.*, vol. 9, no. 5, pp. 248–256, 2013.

[7] L. Fu *et al.*, "Defects regulating of graphene ink for electrochemical determination of ascorbic acid, dopamine and uric acid," *Talanta*, vol. 180, no. July 2017, pp. 248–253, 2018.

CONTACT

*X.Ye,tel:+86-010-62793166;
xyye@mail.tsinghua.edu.cn
*T. Cui, tel: +1-612-616-1636;
cuixx006@umn.edu

ELECTRO-MAGNETIC SENSOR MEDIATED BY MAGNETIC BIOMOLECULES

Qian Cheng[1,2], Yuqing Ge[1], Hongju Mao[1,2], Lin Zhou[1,*], Jianlong Zhao[1,2,*]

[1]State Key Laboratory of Transducer Technology, Shanghai Institute of Microsystem and Information Technology, Chinese Academy of Sciences, Shanghai 200050, China

[2]Center of Materials Science and Optoelectronics Engineering, University of Chinese Academy of Sciences, Beijing 100049, China

ABSTRACT

This report presents a magnetic-electrical sensor based on a gold side-gate electrode Graphene field effect transistor and mediated by magnetic biomolecules. The magnetic sensitive protein act as a key part in the animal's precise perception of the geomagnetic field. In our work, this magnetic biomolecule is coupled to a biocompatible interface, which is made in a sample and low-casts way. The spatial conformation of the magnetic biomolecules is altered by the magnetic field, in addition affects the scattering on electric charges while in motion. And then convert the magnetic signals to electrical signals. The sensor promises to be a new type magnetic field detecting method as it reacts rapidly to varying magnetic field.

KEYWORDS

magnetic-electrical sensor, Magnetic biomolecules, Side-gate configured, Graphene Electrolyte-gated Transistors.

INTRODUCTION

The ability of species like monarch butterflies, pigeons, and many others to migrate long distances using the geomagnetic field has left present scientists confused as to how creatures perceive the existence of the Earth's magnetic field [1, 2]. The phenomenon of biological magneto-reception is governed by anisotropic hyperfine coupling between unpaired electrons and nuclear spins, according to the widely used chemical compass (Spin-correlated radical pair, SCRP) model [3, 4]. The Cry-centered mechanism cannot effectively explain the phenomenon of biological magneto-reception, despite cryptochrome being a significant candidate for this theory [5]. MagR, a new magneto-receptor, and the mechanism of Cry4 protein-induced magnetism were both identified. A stick-like complex that is produced by MagR and Cry responds spontaneously to magnetic fields [6].

The electrochemical impedance and magnetic field effect spectrometers were ingeniously created in order to achieve the innovative transducing mechanism[7]. To evaluate the response of MagR/Cry4 complexes to changes in magnetic signals, however, requires the use of large instruments or optical systems due to the relatively low sensitivity of proteins to magnetic fields in living beings. Therefore, we use electrolyte-gated transistors (EGTs) to convert this faint biological information into an electrical signal. EGTs may convert biochemical and biological inputs into amplified electrical signals and can work steadily in aqueous conditions with a low voltage bias[8]. Graphene is also an especially attractive channel material

for EGTs biosensors due to its ambipolar field effect, amazing electrical properties, atomically thin structures, and significant electrostatic interaction with proteins. Traditional electrolyte-gated gates frequently employ the external electrode as the gate electrode, which makes it difficult to integrate, miniaturize, and really make available devices[9, 10]. To make this process simpler, we develop a piece of metal electrode that works as the gate electrode directly on the side of the source/drain electrode. The same simple fabrication processes can be used to incorporate both the source-drain electrodes and the on-chip gate electrodes. Additionally, the side gate electrode provides the same gate voltage functionality as an ordinary external gate.

Here, we firstly demonstrate natural MagR/Cry4 complex configured conformal graphene EGT bionic magnetic sensor with a completely integrated on-chip gold side gate electrode. This bionic sensor could reflect the significant variation of magnetic field based on the MagR/Cry4 complex and responds very quickly.

EXPERIMENTAL

Fabrication of graphene EGTs with side-gate

The side-gate electrode and drain/source electrodes were patterned by photolithography, metal sputtering, and lift-off process with titanium/gold (20/50nm) on a silicon oxide substrate (4 in.) at the same time. Between the source and drain electrodes, single layer graphene was transferred to electrode surface by classical wet transfer method. To prepare uniform dBSA film, BSA (Bovine Serum Albumin, Purity: >96% was obtained from Sangon Biotech (Shanghai. China)) solution with the concentration of 0.01ug/mL was deposited over the surface of graphene and heated at 80°C for 3 minutes. The dBSA functional channels of the graphene EGTs arrays were defined by photolithography and O_2 plasma were used to etch the surrounding dBSA functional graphene under vacuum. Except for the nano-dBSA/graphene film between the source and drain electrodes, other parts of that were etched. Thus, it is guaranteed that there is no physical connection between the side-gate and the source-drain. Then, Polydimethysiloxane (PDMS) was utilized as passivation layer, which limits current leakage during measurements by covering the metal electrodes and the parts of the graphene and source/drain contacts. The schematic diagram of magnetic-electrical sensor was clearly visible in figure 1(a), and an optical microscope image of it was illustrated in figure 1(b). There are three groups EGTs with distance 250, 500, 700μm from the side-gate electrode. And the conductive channel dimensions were 250μm in length and 100μm in width. And the dBSA/graphene appears light blue-green. The surface topography and the thickness of different graphene samples was evaluated by AFM with tapping mode. As depicted in figure 1(c) and (d), the graphene surface was flat at the nanoscale and has a minimum thickness of about 1nm. And the BSA forms a film with a thickness of about 12nm after heat denaturation. Raman spectra at various positions on each sample were recorded in figure 1(e). Two typical Raman characteristics can be seen in the pristine graphene: a G peak at 1582 cm-1 and a 2D peak at 2700 cm^{-1}. The existence of a D peak at 1350 cm^{-1}, which is commonly associated with the presence of defects in the graphene structure and may lead to defects scattering, is seen in the Raman spectra of graphene when it is coated with a dBSA film. as well as a D' peak at 1620 cm^{-1}, which is connected to impurities and surface charges.

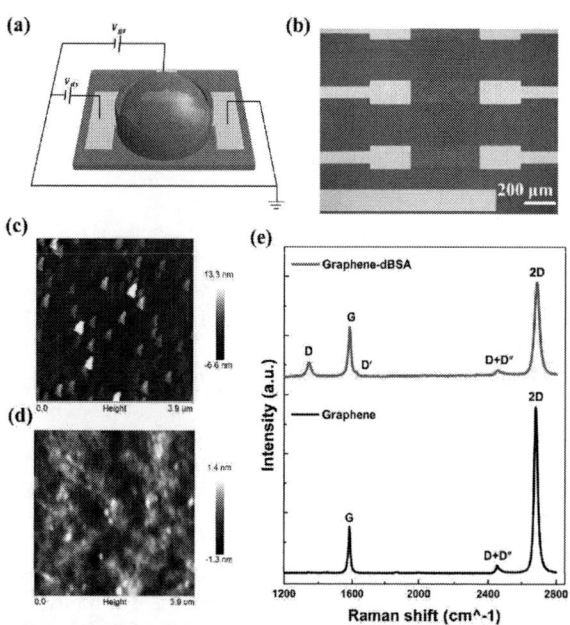

Figure 1: Device schematic and characterisation. (a) schematic diagram(b) optical picture. AFM image of dBSA films (c) and graphene (d). (e) Raman spectra.

The Magnetic biomolecule modification

In order to covalently bond the MagR/Cry4 complex, we specifically utilized the carboxyls on the nano-dBSA/graphene channel surfaces by using straightforward 1-ethyl-3-(3-(dimethylamino)-propyl) carbodiimide (EDC) and N-hydroxysuccinimide (NHS) chemistry (Darmstadt, Germany). After combining the EDC, NHS, MagR/Cry4 solution in a 1:1:1 volume ratio, the dBSA films were incubated with the combination solution at room temperature. EDC, NHS, and MagR/Cry4 were each present in quantities of 500mM, 100mM, and 0.2 mg/mL, respectively. After incubation, Phosphate Buffered Saline (PBS, Thermo Fisher Scientific, Waltham, MA) was used to extract the unconjugated and physically adsorbed protein complex and remove it. The entire incubation process lasts two hours in complete darkness.

Customized magnetic field generation system

We used an electrifield solenoid to provide a uniform magnetic field for our research. The magnetic field system is composed of a copper solenoid, a cooling water system, a magnetic shield, and a DC power source. The amount of current produced by the DC power source determined how intense the magnetic field will be. Magnetic field detection and calibration using a Gauss meter (TD8620, TunKia). Circulating water continuously cools the heat produced by the solenoid, ensuring that the magnetic field is stable and consistent. The shield ensures that the test object is really only exposed to the magnetic field of the solenoid by shielding it from interference from the environment's magnetic fields.

Electrical measurement

The output and transfer characteristics of this conformal graphene EGT were measured by the Portable NI Digital Source Measure Module (SMU PXle-4140). By combining with customized magnetic field generation system, electrical signals of this bionic sensor responding to magnetic fields were recorded by this module. Prior to the start of electrical experiment, 50μl of 1X PBS buffer was injected into the chambers. Schematic diagram of measurement principle was shown in figure 1(a).

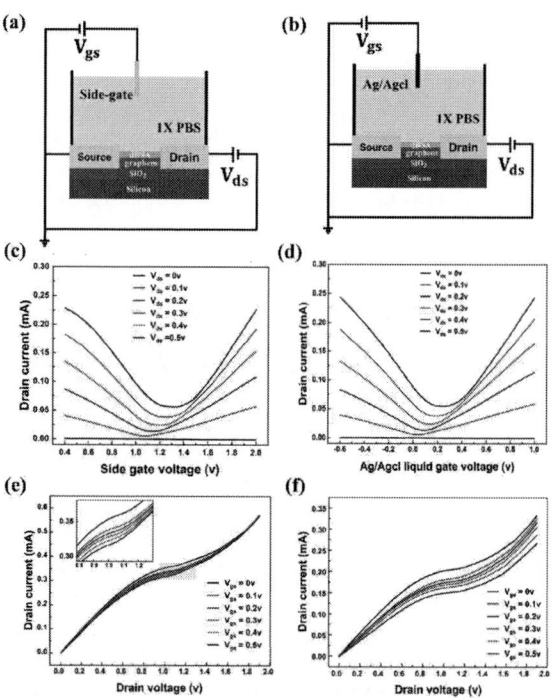

Figure 2: Comparison of the Side-gate and external Ag/Agcl electrode. Schematic diagram of the measurement setup of Side-gate electrode (a), Ag/Agcl electrode (b). (c) (e) transfer/output family curves with Side-gate voltage. (d) (f) transfer/output family curves with external AgAgcl electrode.

RESULTS

Comparison of side-gate and Ag/Agcl electrode EGTs

We evaluated how well the graphene EGTs performed

978-1-6654-9309-3/23 $31.00 © 2023 IEEE

when gate voltage was applied by both the on-chip gold side electrode and the external Ag/AgCl electrode. Figures 2(a) and 2(b) schematically show the measuring setup. Figures 2(c) and 2(d) illustrate the drain current-gate voltage (Ids-Vgs) family curves across a side-gate electrode and an Ag/AgCl electrode, respectively. The Dirac point, where the populations of electronics and holes are equal between two branches (left branch: p-branch, right branch: n-branch), is where the density of charge carriers, and subsequently the current, reaches a minimum. The relationship between the source-drain current and gate voltage displays a V-like curve because the polarity of the

majority carriers shifts during the gate voltage sweep. And, Figure. 2(e) and 2(f) shows the output curves (drain current - drain voltage, Ids-Vds) of dBSA/graphene EGTs through side-gate and Ag/AgCl electrode respectively. A linear ohmic regime is anticipated at low bias, implying excellent electrical and graphene interactions. For the graphene EGTs device, Figure 2 shows a statistical comparison of the gate biased through the Ag/AgCl electrode and the gold side gate, suggesting that the two gates configured may have equal gating functions. As a result, we can claim that the on-chip gate electrode has perfect control over the dBSA/graphene EGTs in an aqueous environment.

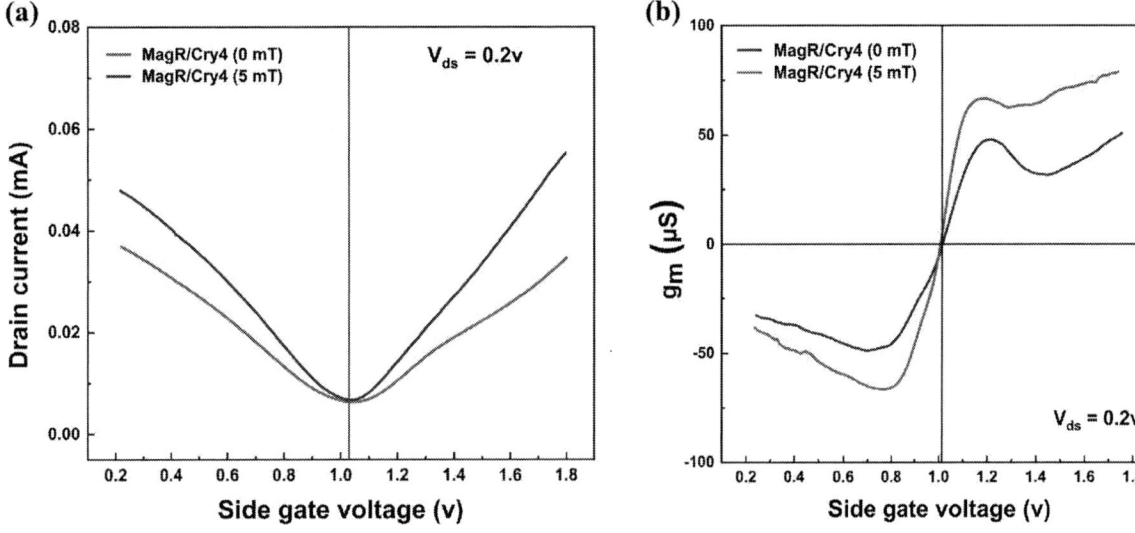

Figure 3 Variation exposed to magnetic field. (a) Transfer curve after incubation of MagR/Cry4 complex in the presenc of 0 and 5mT magnetic fields. (b) Transductance at 0 and 5mT magnetic fields.

Detection of Magnetic Field

The MagR/Cry4 complex configured graphene EGTs with the side-gate electrode were applied to detect magnetic field by the portable NI digital source measure module. By scanning the gate voltage under a constant source-drain voltage (Vds = 0.2v), the transfer characteristic curve of MagR/Cry4 modified graphene EGTs with the side-gate is obtained. The source-drain current increased substantially after exposure to a magnetic field (5mT), as can be seen in Figure 3(a), which compares the transfer curves under a 0 and 5mT magnetic field. The increase in source-drain current was caused by the MagR/Cry4 complex's response to the magnetic field. The drain current is higher under a 5mT magnetic field at the same gate voltage, and the transfer curve is steeper in the presence of the magnetic field than in the without of it. Additionally, Figure 3 depicted the transconductance of the MagR/Cry4 configured graphene EGTs at 0 and 5mT magnetic fields (b). As can be seen, the transconductance was significantly higher in the presence of a magnetic field than it was in the absence of one, suggesting that the magnetic field involves changing the spatial conformation of MagR/Cry4 complex to lessen the effect of scattering on channel elections and increase carrier mobility.

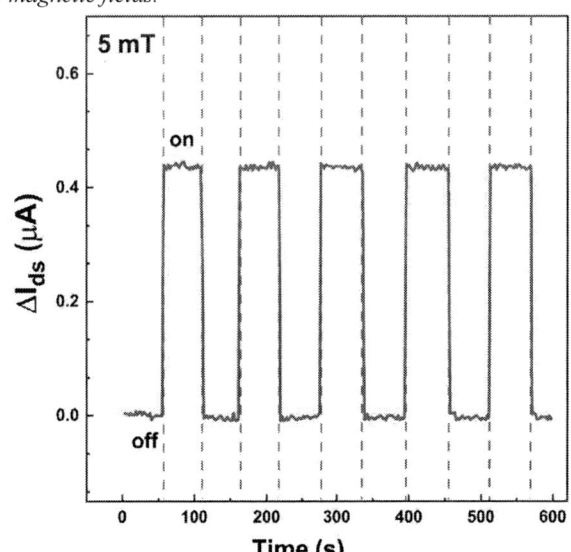

Figure 4: The net current responses to the "on-off" magnetic field with the frequency of 100 seconds.

Figure 4 depicts the net current's responses to the magnetic field's 100-second "on-off" cycle. The output current of this bionic magnetic device instantaneously stepped to a higher state value when a magnetic field was supplied, and in a similar manner, the output current stepped to a lower state value when the magnetic field was removed. And for numerous cycles after that, it keeps instantly following the magnetic field. This experimental finding revealed that the MagR/Cry4 complex based magnetic-electrical sensor has the capacity to react right away to the magnetic field.

CONCLUSION

This study marks the first to demonstrate a unique magnetic-electrical sensor based on magnetic biomolecules modified graphene EGTs with integrated side-gate electrode. On the dBSA/graphene film, the magnetic biomolecules have been successfully modified. Additionally, the external Ag/Agcl electrode's effective gating function was replaced with the on-chip side-gate electrode. The source-drain currents of these MagR/Cry4 complexes modified dBSA/graphene EGTs were employed to reflect changes in MagR/Cry4 conformation caused by magnetic fields. We anticipate that this bionic magnetic sensor will play a significant role in bionic geomagnetic monitoring or navigation systems.

ACKNOWLEDGEMENTS

This work was financially supported by National Key Research and Development Program of China (Nos. 2021YFD2000204), Science and Technology Commission of Shanghai Municipality (No. 20dz1101200, No. 21JM0010402, No. 22ZR1473500), National Natural Science Foundation of China (No. 62231025, No. 61971410 and No. 62001458), Shanghai Sailing Program (No. 20YF1457100), and China Postdoctoal Science Foundation (2020000246).

REFERENCES

[1] R. Wiltschko and W. Wiltschko, "Animal navigation: how animals use environmental factors to find their way," (in English), Eur. Phys. J.-Spec. Top., Review; Early Access p. 16, 2022.

[2] H. Mouritsen, "Long-distance navigation and magnetoreception in migratory animals," (in English), Nature, Review vol. 558, no. 7708, pp. 50-59, Jun 2018.

[3] T. Ritz, S. Adem, and K. Schulten, "A model for photoreceptor-based magnetoreception in birds," (in English), Biophys. J., Article vol. 78, no. 2, pp. 707-718, Feb 2000.

[4] T. Ritz, P. Thalau, J. B. Phillips, R. Wiltschko, and W. Wiltschko, "Resonance effects indicate a radical-pair mechanism for avian magnetic compass," Nature, vol. 429, no. 6988, pp. 177-80, 2004.

[5] H. Mouritsen and P. J. Hore, "The magnetic retina: light-dependent and trigeminal magnetoreception in migratory birds," Current Opinion in Neurobiology, vol. 22, no. 2, pp. 343-352, 2012.

[6] S. Qin, Y. Hang, C. Yang, Y. Dou, and Z. Liu, "A magnetic protein biocompass," Nature Materials, vol. 15, no. 2, pp. 217-226, 2016.

[7] L. Xue et al., "A Novel Biomimetic Magnetosensor Based on Magneto-Optically Involved Conformational Variation of MagR/Cry4 Complex," (in English), Adv. Electron. Mater., Article vol. 6, no. 4, p. 7, Apr 2020, Art no. 1901168.

[8] F. Torricelli et al., "Electrolyte-gated transistors for enhanced performance bioelectronics," (in English), Nature reviews. Methods primers, vol. 1, 2021 (Epub 2021 Oct 2021.

[9] G. Seo et al., "Rapid Detection of COVID-19 Causative Virus (SARS-CoV-2) in Human Nasopharyngeal Swab Specimens Using Field-Effect Transistor-Based Biosensor (vol 14, pg 5135, 2020)," (in English), ACS Nano, Correction vol. 14, no. 9, pp. 12257-12258, Sep 2020.

[10] L. Q. Wang et al., "Rapid and ultrasensitive electromechanical detection of ions, biomolecules and SARS-CoV-2 RNA in unamplified samples," (in English), Nat. Biomed. Eng, Article vol. 6, no. 3, pp. 276-+, Mar 2022.

CONTACT

*L. Zhou, Tel: +86-21-62511070-8706;

zhoulinzlw@mial.sim.ac.cn.

*J. Zhao, Tel: +86-21-62511070-8701;

jlzhao@mail.sim.ac.cn.

GAS-FLOW DEVICE FOR EFFECTIVE DISSOLUTION OF GAS-PHASE ODORANTS UTILIZED FOR BIOHYBRID SENSORS

Takuma Nakane[1,2], Toshihisa Osaki[2], Hisatoshi Mimura[2], Sho Takamori[2],
Norihisa Miki[1,2], and Shoji Takeuchi[2,3]

[1]Keio University, JAPAN, [2]Kanagawa Institute of Industrial Science and Technology, JAPAN, and
[3]The University of Tokyo, JAPAN

ABSTRACT

This study describes a gas-flow device that enhances the solubility of gas-phase odorants into aqueous solution for cell-based sensors. Cell-based sensors using olfactory receptors have attracted attention because of their high sensitivity and selectivity. However, most of previous studies used liquid-phase odorants as a target sample due to their low solubility of gas-phase odorants even though the applications expect the gas-phase sensing. Therefore, we focus on gas-phase odorant detection using the gas-flow device. Here, we clarify the impact of the velocity of convection inside the solution to enhancement of the solubility, based on computational simulation. Moreover, we experimentally verify efficiency of odorant dissolution in two types of devices, compared with the negative control using a sample odorant.

KEYWORDS

Biohybrid sensor, Cell-based sensor, Gas sensor, Olfactory receptor, Gas-phase odorant, Gas-flow channel

INTRODUCTION

In the environment, it is estimated that there are approximately 300,000 to 400,000 different odorants. Living organisms are able to sense the odors composed of these odorant mixtures by using a set of olfactory receptors. In recent years, the superior sensing mechanism of living organisms has attracted great attention in odorant-sensor development. The sensors that directly utilize the sensing mechanism are called biohybrid sensors. One of the sensing elements for the biohybrid sensors is cultured cells that express olfactory receptors (ORs), which is called cell-based sensors. As shown in Fig. 1, insect ORs transport calcium ions into the cells when odorants bind to the ORs. By using a fluorescent calcium indicator, binding of odorants to ORs can be converted to fluorescence [1,2]. Owing to their high sensitivity and selectivity, a number of studies have been conducted aiming for various applications such as healthcare, safety/security, and environmental monitoring [3,4]. However, most of previous studies used liquid-phase odorants as a target sample, i.e., the odorants dissolved in aqueous solution in advance due to their low solubility, even though the applications expect the gas-phase sensing.

We, therefore, focus on gas-phase odorant detection using the gas-flow device. Here, we investigate the impact of the velocity of convection inside the solution to enhance the solubility, based on computational simulation. Moreover, we experimentally verify efficiency of odorant dissolution in two types of devices, compared with the negative control simply filled with gas-phase odorants without gas flow.

EXPERIMENTAL

Computational Simulation of the Solubility of Odorants

We simulated the solubility with three conditions: two types of gas-flow channels (Fig. 2a) and a negative control that is simply exposed with gas-phase odorants without flow. To clarify the dominant factor of dissolution of gas-phase odorants into the aqueous solution, we analyzed the movement of the gas-liquid interface caused by the introduction of gas-flow and the velocity of convection inside the solution.

Fabrication of Gas-flow Device

Based on implications obtained by the simulation, we fabricated two types of the gas-flow devices, which consists of 4 layers of acrylic plates (Acrylite, Mitsubishi Chemical, Japan) by using micromilling (MM-100; Modia Systems, Japan) (Fig. 3a). The bottom layers making up a well were assembled by screws, then fixed together with the top layers of gas-flow channel with bolts. The channel design is easily exchangeable only reattaching the top 2-layer.

Quantification of Dissolved Concentration of Odorants

Gas chromatography (GC, 2010 Plus, Shimadzu, Japan) was used for the quantitative analysis of the concentration of odorants dissolved in the aqueous solution. We estimated the dissolved concentration using the calibration curve obtained by preliminary experiments. Gas-phase odorants were flowed by a gas generator (Permeator PD-1B-2, Gastec, Japan) into the device, and the gas from the outlet was sucked by fume hood (Fig. 3b).

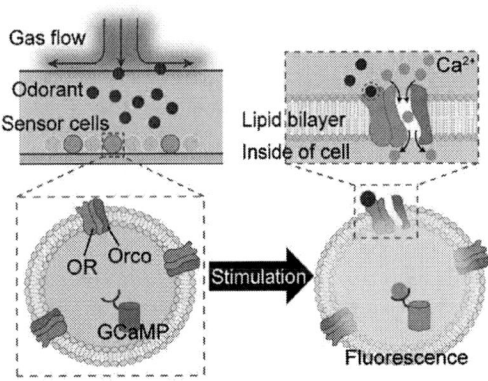

Figure 1: Schematic image of concept of this work and process for odorants detection using a sensor cell.

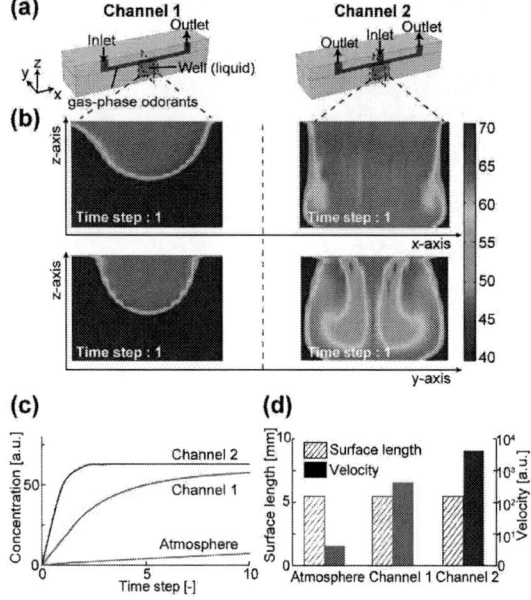

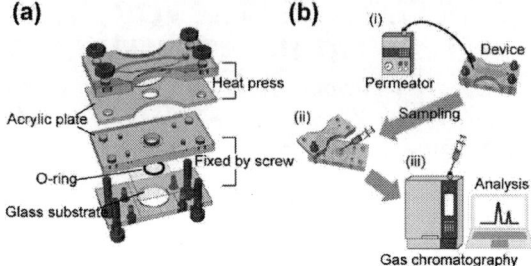

Figure 3: (a) Device structure and procedure of the device assembly. (b) Experimental set up to analyze odorant solubility using gas chromatography.

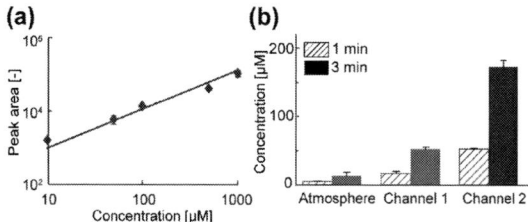

Figure 4: (a) Calibration curve of 1-octen-3-ol. (b) Dissolved concentration of 1-octen-3-ol at 1 and 3 minutes.

Figure 2: (a) Cross-sectional view of the gas flow devices. (b) Visualization of the solubility of odorants in a well. (Top : x-z plane, Bottom : y-z plane) (c) Time-course of concentration with two types of channels and the atmosphere filled with gas-phase odorants. (d) A comparison of the surface length and the velocity of convection.

RESULTS AND DISCUSSION

Computational Simulation of the Solubility of Odorants

We conducted computational simulation to analyze the solubility of gas-phase odorants in the aqueous solution. As shown in Fig. 2b-c, we observed different concentration distributions for the two channels, resulted in the difference of the solubility. To clarify what causes the difference of the solubility, we analyzed the surface length of gas-liquid interface and the velocsity of convection inside the solution. Fig. 2d shows that the velocity of convection indicates the same tendency as the solubility of each condition while the surface length is negligible regardless of conditions. These results indicate that the difference of the solubility in each condition is most probably attributed to the velocity of convection. In other words, the velocity of convection inside the solution could be a significant factor to enhance the solubility of odorants.

Quantification of Dissolved Concentration of Odorants

We fabricated two types of the gas-flow devices based on computational simulation. Subsequently, we verified the solubility of odorants using a gas chromatograph, which was calculated based on the calibration curve shown in Fig. 4a. As a result, we confirmed a significant difference between the two channel types and efficient dissolution of odorants compared with the negative control using a sample odorant (Fig. 4b).

CONCLUSION

In this work, we clarified that the convection inside the solution could be a significant factor to enhance the

solubility of odorants. Moreover, we successfully observed an effective dissolution of odorants by the devices, compared with the negative control simply filled with gas-phase odorants without flow. We believe that this study provides the design factor of effective dissolution of gas-phase odorants, and especially applied for designing cell-based sensors for gas-phase odorant detection.

ACKNOWLEDGEMENTS

This work was partly supported by JSPS KAKENHI Grant Number JP21H05013, and the Program for Building Regional Innovation Ecosystem of MEXT, Japan.

REFERENCES

[1] K. Sato, M. Pellegrino, T. Nakagawa, T. Nakagawa, L. B. Vosshall, K. Touhara, "Insect olfactory receptors are heteromeric ligand-gated ion channels," *Nature*, vol. 452, pp. 1002–1006, 2008.

[2] R. Glatz, K. Bailey-Hill, "Mimicking nature's noses: From receptor deorphaning to olfactory biosensing," *Prog. Neurobiol.*, vol. 93, pp. 270–296, 2011.

[3] M. Termtanasombat, H. Mitsuno, N. Misawa, S. Yamahira, T. Sakurai, S. Yamaguchi, T. Nagamune, R. Kanzaki, "Cell-Based Odorant Sensor Array for Odor Discrimination Based on Insect Odorant Receptors," *J. Chem. Ecol.*, vol. 42, pp. 716–724, 2016.

[4] Y. Hirata, H. Oda, T. Osaki, S. Takeuchi, "Biohybrid sensor for odor detection," *Lab on a Chip*, vol. 21, pp. 2643–2657, 2021.

CONTACT

*Takuma Nakane, tel: +81-44-819-2037; takuma0929@keio.jp

MULTIPLE WELLS ON A CMOS-MEA FOR CELL-BASED BIOHYBRID ODORANT SENSORS

Yujia Lian, Haruka Oda, Minghao Nie, and Shoji Takeuchi

Grad. School of Information Science and Technology, The University of Tokyo, JAPAN

ABSTRACT

This paper demonstrates a method to construct multiple wells on a CMOS microelectrode array (MEA). CMOS-MEA is featured with high-spatial resolution and high sensitivity and can be used to electronically detect cell signals. By adhering to a 3D-printed multiple-well frame on the CMOS-MEA using uncured PDMS glue, no leakage was found between the wells. After seeding the olfactory receptor-expressing cells on the CMOS-MEA, we detected the electrical signal triggered by odorant molecules. In addition, a separate observation of cell signaling was achieved in our proposed multiple wells. With this approach, different types of cells can be seeded on a CMOS-MEA separately for the sensing of odorant molecules.

KEYWORDS

Biohybrid odorant sensor, CMOS-MEA, neuron cell, multiple wells.

INTRODUCTION

Many animals recognize a variety of odorant molecules with high sensitivity naturally. For example, police dogs perform odorant detection tasks to identify drugs owing to their keen sense of smell. Recently, biohybrid odorant sensors have been proposed to incorporate functional biomaterials such as proteins and cells into artificial systems to mimic the olfaction of animals with high sensitivity and selectivity [1, 2]. Amongst the various types of artificial odorant sensors, cell-based odorant sensors have the potential to achieve a more stable expression and maintenance of olfactory sensing function [3-5]. Previous works on cell-based odorant sensors have mainly used the calcium-induced cellular fluorescence to detect odorants [4, 5]. However, those methods take a long reaction time for the signal to be transduced from calcium ion influx to fluorescence. When odorant molecules bind to the olfactory receptor, calcium ions influx into the intracellular, and it takes time for that ion to combinate with calcium indicator.

In this work, we proposed to use CMOS-MEA to detect the calcium ion influx electrically, therefore skipping the transduce time for calcium ion influx to fluorescence, so that it will be more suitable to be loaded onto robots for applications such as real-time odorant detection in the future. In addition, we further proposed a method to construct completely isolated multiple wells on a CMOS-MEA, which can allow the separate seeding of different types of cells onto a single chip to avoid money and time consumption (Figure. 1A-B).

EXPERIMENTAL METHODS

Fabrication of CMOS-MEA Chip with separating wells

We prepared biohybrid odorant sensors with multiple

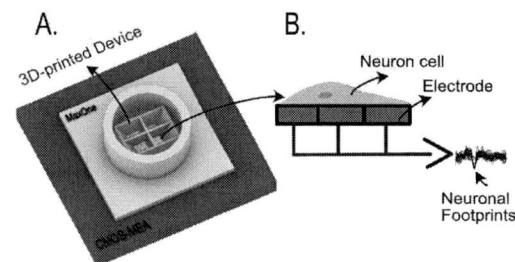

Figure 1: Conceptual illustration of the biohybrid odorant sensor. A. An illustration of the CMOS-MEA with separated wells. B. The diagram for CMOS-MEA measurement of extracellular electrical signals.

wells on a CMOS-MEA (Maxwell, Switzerland). A 3D-printed device was created using a 3D printer DigitalWax 028J (DWS, Japan). The surface of the 3D-printed device was coated with Parylene-C. A layer of uncured PDMS with a 1:10 weight ratio of silicone elastomer diluted with hexane at a volume ratio of 1:4 was applied to the bottom of the 3D-printed device [6]. The 3D-printed device coated with PDMS-Hexane was then placed in the center of the CMOS-MEA and cured at 80°C for 2 h (Figure. 2B-C).

Cell Culture and Transfection

Mouse hypothalamic GnRH neuronal (GT1-7 cells) cell lines were cultured in Dulbecco's modified Eagle's medium (DMEM) (Nakarai, Japan) with 10% (v/v) fetal bovine serum (FBS) (Biosera, France), 100 U mL-1 penicillin, and 100 μg mL-1 streptomycin (Sigma Aldrich, USA), at 37 °C under a humidified atmosphere containing 5% CO_2. The cells were passaged before they became confluent.

Then, we prepared GT1-7 cells expressing olfactory receptors. Recent studies have reported that when co-expressed with olfactory receptors, the short form of the chaperone factor RTP protein (RTP1s) that promotes olfactory receptors, Ric8b, a facilitator of GDP-GTP exchange for G protein Gαolf, and Gαolf have the highest cell surface expression [7]. To enhance the expression of olfactory receptors in GT1-7, which is known for having a high electrical activity characterized. We constructed a plasmid including RTP1s, Ric8b, and Gαolf. After 24 h of seeding GT1-7 cells in the CMOS-MEA chip directly, the plasmid including RTP1s, Ric8b, and Gαolf, and the olfactory receptor plasmid was then co-transfected into GT1-7 cells using Lipofectamine 2000 (Thermo Fisher Scientific, USA). The expression of olfactory receptors was confirmed by immunostaining, with the wild-type of GT1-7 cells as the negative control.

Odorant Assay Using the Biohybrid Odorant Sensor

The odorant assays were performed after 24 h post-transfection. Before the odorant assay, the odorant molecules were diluted in the Hanks' balanced salt solution (HBSS, Bioworld). After incubation for 10 min at 37 °C, the odorant solution (2 mM Eugenol) was added to the biohybrid odorant sensor.

To confirm the amplitude change in the GT1-7 cells expressing olfactory receptors after the addition of odorant molecules, we seeded the GT1-7 cells in the CMOS-MEA without separating wells.

To confirm the amplitude change in the GT1-7 cells expressing olfactory receptors in the CMOS-MEA with separating wells as a biohybrid odorant sensor, we seeded the GT1-7 cells after fabricating the CMOS-MEA with separating wells. And different plasmids were transfected into GT1-7 cells.

RESULTS AND DISCUSSION

Figure. 2A depicts the temporal change in amplitude of GT1-7 cells expressing the olfactory receptor on CMOS-MEA without separating wells. The amplitude rapidly decreased within 1 second after the addition of 2 mM Eugenol. This indicates that after extracellular calcium ions having an influx in GT1-7 cells, Ca^{2+}-activated Cl^- is elicited by odorant molecules. We have confirmed cell viability by Calcein-AM (Figure. 2B).

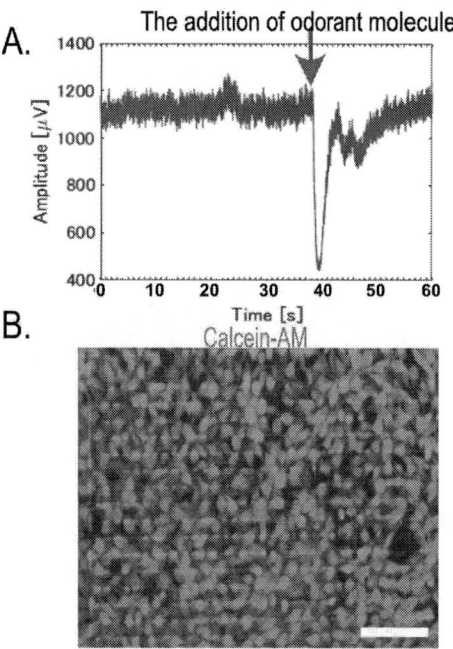

Figure 2: Electrical measurement and live-cell staining of GT1-7 cells on CMOS-MEA. A. The amplitude of all electrodes after the addition of odorant molecules in GT1-7 cells expressing olfactory receptors. B. Microscope image of calcein-AM positive (green) stained GT1-7 cells on CMOS-MEA chips. Scale bar: 200 μm

CONCLUSION

In this paper, we proposed a biohybrid odorant sensor with four wells. With this biohybrid odorant sensor, we can culture different types of cells on only one chip and perform at least four odorant assays using a single chip, which can save time and budgets. We found that GT1-7 cells expressing the olfactory receptors on CMOS-MEA have detected stimuli after adding the odorant molecules. After culturing three different types of GT1-7 cells on CMOS-MEA in separate wells, only the cells expressing olfactory receptors exhibited amplitude changes after the addition of odorant molecules.

ACKNOWLEDGEMENTS

This work was supported by Grant-in-Aid for Scientific Research (S) Grant Number JP21H05013.

REFERENCES

[1] Misawa, Nobuo, et al. "Highly sensitive and selective odorant sensor using living cells expressing insect olfactory receptors." *Proceedings of the National Academy of Sciences.,* vol.107, pp.15340-15344, 2010.

[2] Misawa, Nobuo, et al. "Construction of a biohybrid odorant sensor using biological olfactory receptors embedded into bilayer lipid membrane on a chip." *ACS sensors.,* vol.19, pp.711-716, 2019.

[3] Misawa, Nobuo, et al. "Membrane protein-based biosensors." *Journal of the Royal Society Interface.,* vol. 15, 20170952, 2018.

[4] Hirata, Yusuke, et al. "Portable biohybrid odorant sensors using cell-laden collagen micropillars." *Lab on a Chip.,* vol.19, pp.1971-1976, 2019.

[5] Oda, Haruka, et al. "Cell-based biohybrid sensor device for chemical source direction estimation." *Cyborg and Bionic Systems.,* vol. 2021, 8907148, 2021.

[6] Duru, Jens, et al. "Engineered biological neural networks on high-density CMOS microelectrode arrays." *Frontiers in neuroscience.,* vol.16, 829884, 2022.

[7] Shepard, Blythe D., et al. "A cleavable N-terminal signal peptide promotes widespread olfactory receptor surface expression in HEK293T cells." *PLoS One.,* vol.8, e68758, 2013.

CONTACT

* Yujia LIAN; phone +81-3-5841-6488; lian@hybrid.t.u-tokyo.ac.jp

THE INTEGRATED RGO/PEDOT:PSS-MODIFIED ULTRAFLEXIBLE MICROELECTRODES TOWARDS LONG-TERM NEUROPHYSIOLOGICAL SIGNALING AND DOPAMINE SENSITIVE DETECTION

Xueying Wang[1,2], Huiran Yang[1], Bohan Zhang[1,3], Meng Li[1,2], Liuyang Sun[2,4], Zhitao Zhou[1,2], Tiger H. Tao[1,2,3,4,5,6,7,8,9] and Xiaoling Wei[1,2]

[1]State Key Laboratory of Transducer Technology, Shanghai Institute of Microsystem and Information Technology, Chinese Academy of Sciences, Shanghai, China

[2]University of Chinese Academy of Sciences, Beijing, China

[3]School of Physical Science and Technology, Shanghai Tech University, Shanghai, China

[4]2020 X-Lab, Shanghai Institute of Microsystem and Information Technology, Chinese Academy of Sciences, Shanghai, China

[5]Center of Materials Science and Optoelectronics Engineering, University of Chinese Academy of Sciences, Beijing, China

[6] Center for Excellence in Brain Science and Intelligence Technology, Chinese Academy of Sciences, Shanghai, China

[7]Neuroxess Co., Ltd. (Jiangxi), Nanchang, Jiangxi, China

[8]Guangdong Institute of Intelligence Science and Technology, Hengqin, Zhuhai, Guangdong, China

[9]Tianqiao and Chrissy Chen Institute for Translational Research, Shanghai, China

ABSTRACT

We reported a flexible neural electrode that can simultaneously detect neural electrophysiological signals and dopamine concentration fluctuation, where the thickness is only 5 microns with bending stiffness ~4.7×10^{-12}N·m² that is ~3-4 orders of lower than reported [1]. The extremely thin thickness and ultralow bending stiffness cause less chemo-mechanical mismatch between the probes and the neural tissues enabling long-term stable recording. The dopamine detection electrode modified by rGO/PEDOT:PSS further improves the sensitivity of dopamine detection due to the introduction of rGO reinforcing the adsorption to dopamine, which helps to reach the sensitivity as high as 20.2 pA/μM and nanomolar detection limit with high selectivity for dopamine. The device is verified to simultaneously detect neurophysiological signals and dopamine changes *in vivo*. In vitro experiments verified its detection performance did not decrease significantly within two months (accelerated aging test), providing assistance for simultaneously detect dopamine and electrophysiological signals in vivo for a long time in neuroscience research. The proposed devices would be useful for neuroscience studies in small rodents, large animal and ultimately non-human primates.

KEYWORDS

penetrating flexible probe, neural electrophysiological signal detection, dopamine detection

INTRODUCTION

The development and application of brain microelectrodes provides the possibility for exploring brain science. At present, the brain microelectrodes mainly focus on the acquisition of neuroelectric signals, but neural science researches and neurodegenerative diseases are also closely related to the fluctuation of neurotransmitters. It is necessary to develop devices which can simultaneously detect neural electrophysiological signals and neurotransmitters precisely for a period of time. With the development of flexible electrode, it is possible to track and record single neurons *in vivo* for a long time. At present, many research groups have focused on the synchronous detection of neural electrophysiological signals and neurotransmitters such as dopamine, glutamate, etc. but most of their electrodes are made of rigid materials, such as glass carbon or silicon [1-2]. There are many researches about flexible probe for recording neural neurophysiological signals or neurotransmitters individually but few studies on flexible electrodes with integrated electrical signal detection and electrochemical detection of neurotransmitters [3-5]. Based on the fabrication of ultra-thin flexible electrode, the dopamine (DA) detection electrode site could be achieved by modifying the sensitive materials on the special electrode site of the flexible probe, so that the electrode site can turn to neurotransmitter detection channel to achieve synchronous *in vivo* recording of electrophysiological signals and neurotransmitters of the flexible probe. Moreover, the minimally invasiveness of the flexible electrodes to the brain tissue makes it promising to achieve long-term in vivo recording of both electrophysiological signals and neurotransmitters change.

FABRICATION AND MODIFICATION

Flexible PI Probe Fabrication Process

The flexible probe was fabricated by microelectro-mechanical system (MEMS) surface processing as shown in figure 1. First, the flexible electrode release area called sacrificial layer is patterned by photolithography, metal evaporation and lift-off of nickel. Second, spin coating and curing polyimide (PI) is used as the insulating layer at the bottom of the electrode. Third, the photolithography and patterned Cr/Au/Cr is used as the metal wiring to connect the electrode and bonding pad. Fourth, the evaporation Cr/Ni/Au is used to form the bonding pad. Fifth, spin coating and curing polyimide are used as the top insulating layer. Sixth, the electrode and bonding pad are exposed by etching the PI with the aluminum as hard shadow mask. Finally, the probe can be released by etching the nickel sacrificial layer.

The electrode site for recording neural neurophysiological signals or detection for DA are designed by backend connection printed circuit board (PCB). According to the distribution of DA in brain region, the DA detection site was sited at the nethermost of the shank which could arrive the caudate putamen (striatum) at the deep of approximately 3-4 mm which is rich in dopamine.

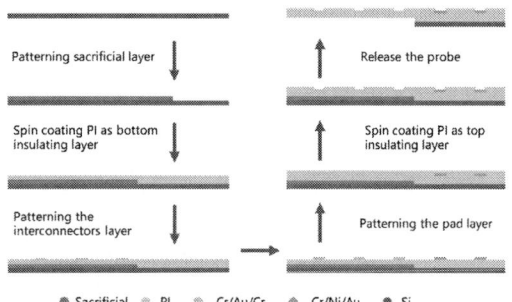

Figure 1: The process of the flexible probe's fabrication. All the metallization process done by UV photography, evaporation and lift off.

Modification of the dopamine detection electrode

In order to achieve a highly sensitive and specific response of the electrode to dopamine, rGO/PEDOT:PSS (rGO: reduced Graphene Oxide; PEDOT: poly(3,4-ethylenedioxythiophene); PSS: polystyrene sulfonate) nanomaterials were electroplated on the electrode. The electroplating solution was composed of 0.1M PSS(Sigma-Aldrich) and 2 mg/ml rGO(XFNANO) for 12 h ultrasonic dispersion, and then 0.02 M EDOT(Sigma-Aldrich) was added and ultrasonic for 30 min. Electroplating adopts cyclic voltammetry method. 0-0.95 V is applied to the electroplated electrode, and the sweep rate is 0.1 V/s. Different cycles are carried out. In this way, graphene doped PEDOT: PSS can be modified on the electrode, greatly increasing the reactive area, and the carboxyl group carried by rGO further increases the response to dopamine. As shown in figure 2, the photo of whole flexible probe contains 128 electrode with the modified electrode in red box in (a) .Figure 2(b) and (c) are the SEM photo of modified electrode. The rough surface provides activity site for DA oxidation.

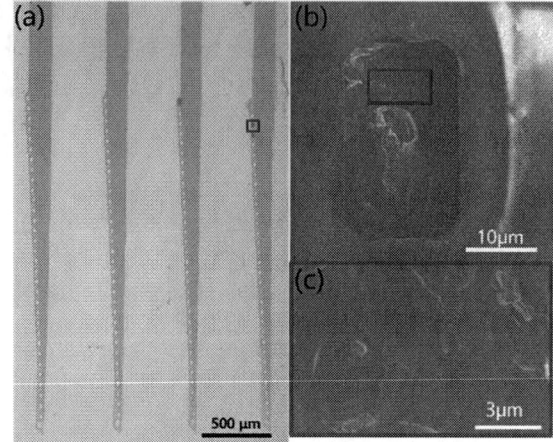

Figure 2: The photo of the rGO/PEDOT:PSS modified electrode. (a) Photo of 128-channel probe with rGO/PEDOT:PSS covering nafion film(red box); (b)SEM image of the electrode electroplated rGO/PEDOT:PSS; the enlarged

In order to achieve the specific response of the electrode to dopamine, Nafion was added to the surface of the modified electrode to isolate the influence of common neurotransmitters such as ascorbic acid (AA) and uric acid (UA) as shown in Figure 3.

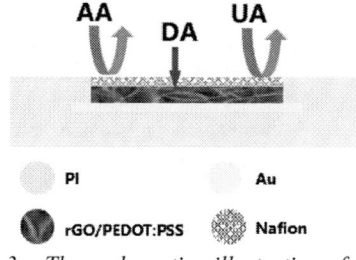

Figure 3: The schematic illustration for dopamine detection of modified electrode

EXPERIMENTAL RESULTS
Characterization in *vitro*

The detection sensitivity of the electrode to dopamine is a very important parameter, and the detection sensitivity is related to the effective surface area of the electrode. The larger the effective surface area is, the higher the sensitivity will be. In addition, the larger the electrode surface area, the smaller the impedance, so we studied the impedance under different electroplating cycles to obtain the best electroplating parameters. As shown in Figure 4, the electrode impedance decreases at 10-30 electroplating cycles and becomes stable at 30 cycles to 40cycles. Additional electroplating cycles more than 30 do not increase the surface area of the electrode. However, thicker coatings affect electron transport efficiency and increase the response time to DA.

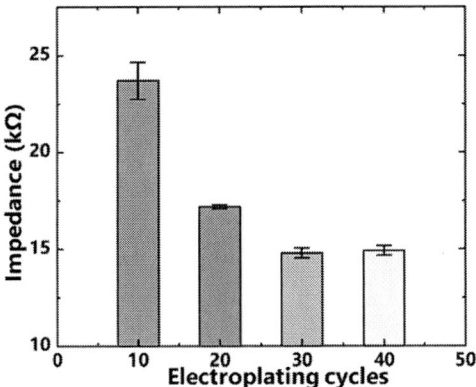

Figure 4: The impedance of modified electrodes at the frequency of 1k Hz at different electroplating cycles

Compared with PEDOT:PSS, the electrode modified with rGO/PEDOT:PSS can make the surface rougher and the impedance lower, as shown in Figure 5.

The response of rGO/PEDOT:PSS modified electrode and PEDOT:PSS modified electrode are also compared. As shown in Figure 6, the cyclic voltammograms of electrodes to 5μM dopamine under different modification conditions. The response of rGO/PEDOT:PSS modified electrode to dopamine is nearly three times larger than PEDOT:PSS modified electrode. Both of the modified electrodes are more obvious than the bare electrode. The oxidation potential of dopamine is +0.16V.

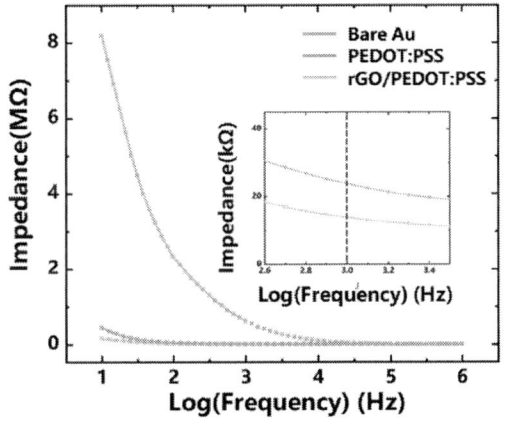

Figure 5: The impedance of bare, PEDOT:PSS and rGO/PEDOT:PSS modified microelectrode at different frequencies.

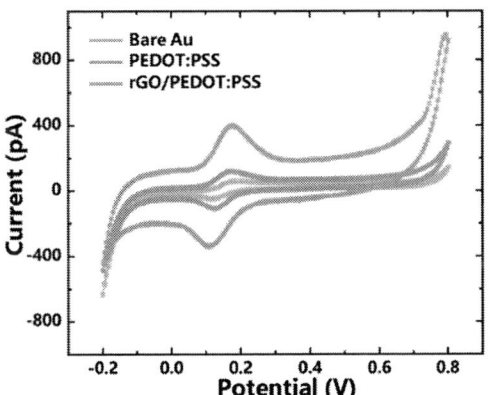

Figure 6: Cyclic voltammograms of the bare microelectrodes, PEDOT:PSS-modified microelectrodes, and rGO/PEDOT:PSS-modified microelectrodes in 5 μM DA respectively.

In addition to dopamine, AA and UA are also electroactive substances in brain tissue. In order to prevent the influence of these substances from the detection of dopamine, the surface of the electrode modified with rGO/PEDOT: PSS was added with 0.5% nafion alcohol solution and dried at room temperature. Nafion membrane is a cationic membrane, which can selectively pass through positively charged dopamine and isolate negatively charged AA and UA, so as to achieve dopamine specific detection. As shown in Figure 7, Cyclic voltammetry was performed on 25μM of dopamine and other common neurotransmitters, and the electrodes after covering Nafion showed a specific response to dopamine.

To calibrate the relationship between the dopamine concentration and the current, the oxidation potential of +0.16V was applied to the modified electrode to record the response current at different concentrations. As shown in Figure 8, the modified electrode showed a good linear relationship to dopamine from the nanomole level to tens of micromoles, with a sensitivity of 20.2pA/μM, which is sufficient to detect fluctuation of dopamine *in vivo*.

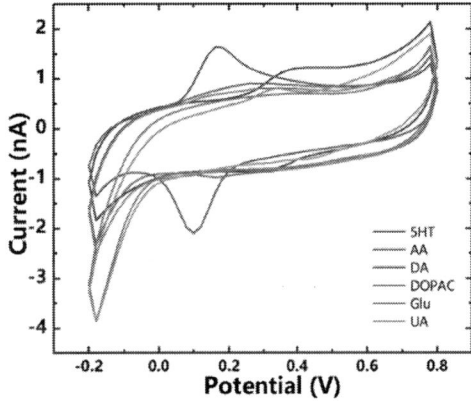

Figure 7: Cyclic voltammograms of the rGO/PEDOT:PSS and Nafion modified microeletrode to different neurotransmitters at the concentration of 25μM

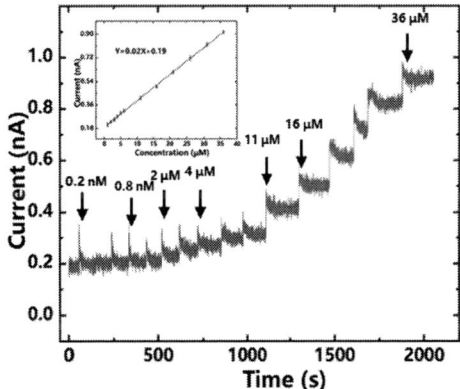

Figure 8: The Calibration curves of rGO/PEDOT:PSS modified microelectrode and linear fitting curve of the

current with concentration of dopamine (0.2 μM–36 μM).

Accelerated aging experiments were carried out to verify the long-term effectiveness of the modified electrodes. The accelerated aging of the electrode was carried out at 60℃, and the reaction rate coefficient was selected as 2, so the accelerated aging factor could be calculated as 13. The accelerated aging of 5 days could be equivalent to 65 days. The response to dopamine during the aging experiment was recorded, as shown in Figure 9. The response current decreased slightly but could remain above 90%.

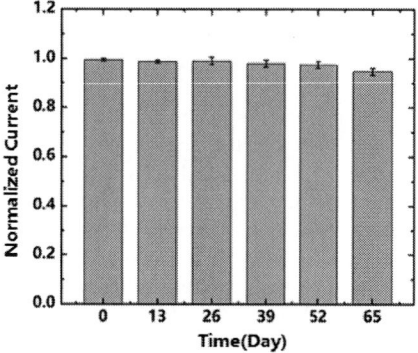

Figure 9: The normalized response for dopamine with the accelerated aging time.

Implantation of Flexible Probe

The flexible probe was implanted into mouse brain caudate putamen (AP: 0.5mm; ML: - 2 mm; DV: -3mm;) to functional verification. As shown in figure 10, there are six channels neural electrophysiological signals containing low frequency (f < 250 Hz) which is called local field potential and high frequency (f > 250 Hz) which reflects the neural spikes. the microelectrode can synchronously detect the changes of electrophysiological activity and dopamine fluctuation without obvious interference between the signals.

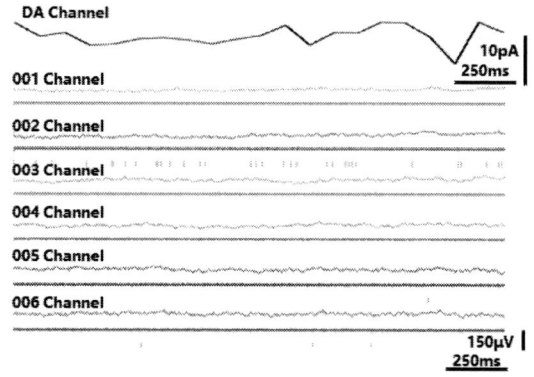

Figure 10: The simultaneously fluctuation of dopamine and neural electrophysiological signals

CONCLUSION

In this article, we designed and prepared a flexible neural electrode which could simultaneously detect neural electrophysiological signals and dopamine concentration fluctuation with the thickness of 5 microns of flexible polyimide. Profit from the extremely thin thickness, the bending stiffness of the flexible probe is lower than silicon probe which is beneficial to less chemo-mechanical mismatch between the probes and the neural tissues to the application of long-term *in vivo* research. Due to the introduction of rGO, the dopamine detection electrode modified by rGO/PEDOT:PSS further improves the sensitivity of dopamine detection reaching the sensitivity as high as 20.2 pA/μM and nanomolar detection limit with high selectivity for dopamine. With the implantation of the probe, the device is verified to simultaneously record neural electrophysiological signals and dopamine concentration fluctuation *in vivo*. According to the accelerated aging test, the detection performance did not decrease significantly within two months, providing possibility for simultaneously detect dopamine and electrophysiological signals *in vivo* for a long time in neuroscience research.

ACKNOWLEDGEMENTS

This work was partially supported by the National Key R & D Program of China (Grant Nos. 2021YFF1200700, 2019YFA0905200, 2021ZD0201600, 2021YFC2501500, 2022ZD0209300, 2022ZD0212300), National Natural Science Foundation of China (Grant No. 61974154), Key Research Program of Frontier Sciences, CAS (Grant No. ZDBS- LY-JSC024), Shanghai Pilot Program for Basic Research—Chinese Academy of Science, Shanghai Branch (Grant No. JCYJ-SHFY-2022-01), Shanghai Municipal Science and Technology Major Project (Grant No. 2021SHZDZX), CAS Pioneer Hundred Talents Program, Shanghai Pujiang Program (Grant Nos. 19PJ1410900, 21PJ1415100), the Science and Technology Commission Foundation of Shanghai (No. 21JM0010200), Shanghai Rising-Star Program (Grant No. 22QA1410900), the Innovative Research Team of High-level Local Universities in Shanghai, the Jiangxi Province 03 Special Project and 5G Project (Grant No. 20212ABC03W07), Fund for Central Government in Guidance of Local Science and Technology Development (Grant No. 20201ZDE04013), Special Fund for Science and Technology Innovation Strategy of Guangdong Province (Grant Nos. 2021B0909060002, 2021B0909050004)

REFERENCES

[1]. E. He et al. ACS Sens. 6(2021), pp.3377–3386.
[2]. E. Castagnola et al. Biosensors and Bioelectronics, 191(2022), pp.113-440.
[3]. X. L. Wei et al. Advanced Science, 5(2018), pp.1-9.
[4]. J. X. Liu et al. Nature 606(2022), pp. 94–101.
[5]. G. Wu et al. Nano Letters 22 (9)(2022), pp.3668-3677.

CONTACT

*X. Wei, tel: +86-21-62511070;
xlwei-jerry@mail.sim.ac.cn

COMPARISON OF SELECTIVE FILTRATION OF ON-CHIP GLOMERULUS COMPRISED OF ORGANOID-DERIVED AND IMMORTALIZED PODOCYTES

Ayumu Tabuchi[1], Kensuke Yabuuchi[2,3], Yoshiki Sahara[2], Minoru Takasato[2,4], Kazuya Fujimoto[1], and Ryuji Yokokawa[1]

[1]Department of Micro Engineering, Kyoto University, JAPAN,
[2]Center for Biosystems Dynamics Research, RIKEN, JAPAN,
[3]Graduate School of Medicine, Osaka University, JAPAN, and
[4] Graduate School of Biostudies, Kyoto University, JAPAN

ABSTRACT

We compared two on-chip glomerular filtration barrier models using immortalized podocytes and kidney organoid-derived podocytes from induced pluripotent stem cells (iPSCs). Podocytes and human umbilical vein endothelial cells (HUVECs) were seeded on the side wall of fibrin gel in a three-channel polydimethylsiloxane (PDMS) device. Both models showed higher filtration rate for inulin than albumin, and significantly higher selectivity was measured when organoid-derived podocytes were used. This study demonstrated the importance of using highly functional podocytes derived from iPSC organoids than conventional immortalized podocytes. The results will be the basis for the creation of glomerular chips that can mimic more biological functions in the future.

KEYWORDS

hiPSCs, kidney organoid, filtration function, microphysiological systems (MPS)

INTRODUCTION

Glomerulus is a part of nephron in kidney that is responsible for blood filtration [1]. It is carried out by glomerular filtration barrier, which consists of vascular endothelial cells, glomerular basement membrane, and podocytes. When the glomerular filtration barrier is damaged by drugs such as an anticancer agent or hyperglycemia, podocytes are damaged and the filtration function is impaired.

Therefore, the evaluation of glomerular toxicity is essential in developing a new drug. Currently, animal studies are conducted using mice and rats. However, animal studies show species differences for the efficacy of candidate drugs.

For more accurate assessment of human glomerular drug toxicity, in vitro evaluation method is necessary as an alternative to animal models. Though the on-chip glomerulus models have been reported [2-4], podocytes were derived from animals or immortalized cells [5]. In on-chip models, it is not possible to accurately evaluate human filtration function when using animal cells because of species differences. In addition, immortalized cells are not suitable because their imperfect shape and protein expression, compared with those of podocytes in living organisms, affect their filtration function in a glomerular chip. In this study, we constructed a glomerular chip using kidney organoids from hiPS cells [6] as a cell source, and

showed that podocytes derived from kidney organoid was superior in functionality to immortalized cells.

EXPERIMENTAL

Device design and fabrication

The design of the microfluidic device is shown in Figure 1. The device consists of polydimethylsiloxane (PDMS) with three channels separated by hexagonal micropillars with the inter-pillar spacing of 100 μm. A cover glass is bonded to the bottom of the channels. Channels 1 and 3 of 700 μm and channel 2 of 500 μm in width.

Negative photoresist, SU-8 3050, was patterned onto a silicon substrate with a height of 100 μm as a mold. The PDMS solution with 10:1 mixture of main agent and hardener poured into the mold and cured at 80°C for at least 12 hours. The PDMS and cover glass were treated with oxygen plasma for bonding.

Cell culture

Human umbilical vein endothelial cells (RFP-HUVEC, Angio-Proteomie), human conditionally immortalized

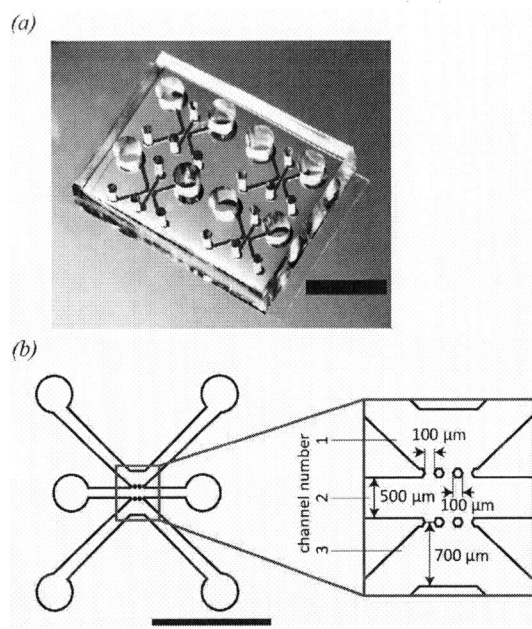

(a)

(b)

Figure 1: Device design. (a) The image of the device with blue ink poured into the channels. Scale bar: 10 mm. (b) The design of the device. Scale bar: 5 mm

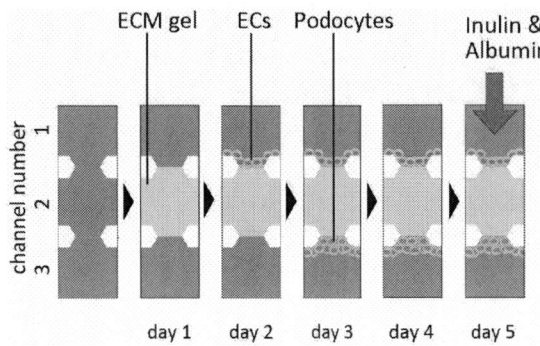

Figure 2: Schematic image of cell culture in the device.

podocytes (hciPods), and hiPS cells were used for experiments. EGM-2 (Lonza) was used for RFP-HUVECs, RPMI1640 (Fujifilm) was used for hciPod, and StemFit (Ajinomoto) was used for hiPSCs as the growth medium.

Kidney organoid culture and dissociation

Kidney organoid was generated by an established protocol [7]. HiPSCs were thawed and passaged three times, then exchanged for APEL2 medium (Stemcell Technologies) containing CHIR99021 (R&D) to start inducing differentiation into posterior primitive streak. On day 4, cells were transferred to APEL2 medium containing FGF9 (R&D) and Heparin (Sigma), and differentiation into intermediate mesoderm was induced until day 7. Cells were centrifuged at a concentration of 2.5×10^5 cells/pellet to form pellets and transferred onto a culture insert (Corning). Pellets were incubated for 1 hour in APEL2 containing CHIR99021, and then transferred to APEL2 containing FGF9 and Heparin. After incubated at the air-liquid interface until day 12, organoids were cultured with APEL2 until day 28.

Kidney organoids were dissociated using a method by Hale *et al.* [8]. Kidney organoids (day 28) were suspended with cell-dissociating enzymes (TrypLE Select, Gibco) and dissociated by pipetting every 3 min during the incubation. In kidney cells, podocytes are firmly bound to each other by foot processes, so they are not separated by enzymatic treatment and remain as aggregates of podocytes. On the other hand, other cells are separated into single cells by enzymatic treatment. The cell suspension was then filtered in sequence through 70-μm and 40-μm cell strainers to collect cell aggregates ranging in size from 40 μm to 70 μm. The 70-μm strainer was used to remove undissociated cells, and the 40-μm cell strainer separates the podocyte cell aggregates from single cells. After the dissociation process, the aggregates of these collected podocyte cells were suspended in RPMI1640 (Fujifilm) to the desired concentration level.

Maturation of immortalized podocytes

After seeding, hciPods were cultured in a 33 °C incubator for 3 to 4 days to reach about 60% confluence. The cells were then transferred to a 37°C incubator to begin maturation, and medium was changed every 2 days. After 2 weeks of incubation, the cells were used for experiments.

Construction of on-chip glomerulus filtration barrier

Fibrin gel was introduced into channel 2 of the device and incubated for 10 min. The fibrin gel was prepared by mixing fibrinogen solution (Sigma) with collagen IV (Corning), laminin 521 (Sigma), aprotinin (Sigma), and thrombin (Sigma). Coating solution for channel 1 was prepared by mixing fibronectin (Sigma) and PBS(-). Coating solution for channel 3 was prepared by mixing collagen IV (Corning), laminin521 (Sigma) in PBS(-).

After incubation overnight, HUVECs adjusted to a concentration of 1×10^7 cells/ml was introduced into channel 1 and the device was rotated 90 degrees every 5 minutes and then again every 10 minutes to ensure that the HUVECs adhered to all surfaces of the channel. Next day, the podocyte aggregate suspension obtained from the organoids or hciPods were then introduced into channel 3. Podocytes were attached to the gel by incubating the device with channel 3 side up for 1 hour. The device was then returned to its original orientation, medium was replaced, and incubated overnight.

Filtration experiments

Fluorescently labeled inulin and albumin were used in this experiment. In vivo, inulin is an indicator substance for assessing glomerular filtration capacity because it readily permeates the glomerular filtration barrier and is not reabsorbed by the proximal tubules. On the other hand, albumin is known to be present in the blood vessels and does not permeate the glomerular filtration barrier. In this experiment, fluorescent-labeled albumin (Alexa fluor 647) and fluorescent-labeled inulin (FITC) at 100 μg/ml were dissolved in 10 mg/ml bovine serum albumin solution.

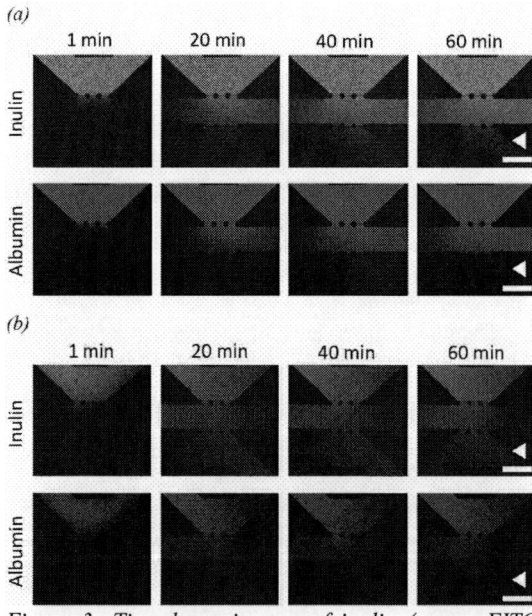

Figure 3: Time lapse images of inulin (green, FITC conjugated) and albumin (blue, Alexa fluor 647 conjugated) diffusion in the device. (a) Device with organoid derived podocytes. (b) Device with hciPods. White arrows indicate the leaked inulin/albumin in channel 3 Scale bar:500 μm.

Then, the solution was introduced into the channel 1 and

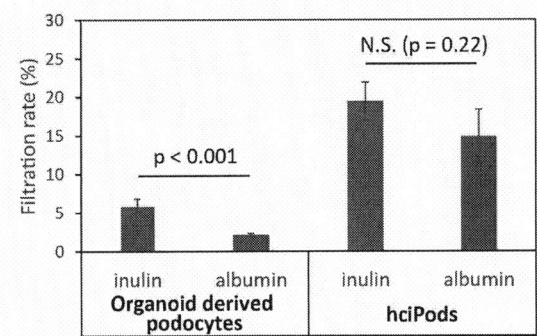

Figure 4: Graph of normalized fluorescence intensity of the channel 3 after 1 hour of filtration test. The fluorescence intensity of inulin and albumin permeating the cellular barriers of the endothelium and podocytes is shown. Values are mean ± SD.

observed to diffuse with time under a confocal microscope. Filtration rate was calculated by dividing the fluorescence intensity of channel 3 by the fluorescence intensity of channel 1.

RESULT AND DISCUSSION

In the device using organoid-derived podocytes, time-lapse images (Figure 3 (a)) show no difference in the permeation of the vascular endothelial cell layer for either inulin or albumin. On the other hand, in the permeation of the podocyte layer, inulin permeates but albumin does not. In the device using immortalized podocytes (Figure 3 (b)), the time-lapse images show no difference in the permeation of inulin and albumin in the vascular endothelial cell layer, as in the device using organoid-derived podocytes. However, inulin and albumin permeation was also observed in the podocyte layer.

The permeability of the podocyte layer is shown in the graph (Figure 4). The permeability of inulin was significantly higher than that of albumin in podocytes derived from organoids. On the other hand, in immortalized podocytes, inulin permeability was higher than that of albumin, but the difference was not significant. In conclusion, on-chip glomeruli with organoid-derived podocytes showed a significant difference in the permeability of inulin and albumin in the podocyte layer. This trend is consistent with the selective filtration of inulin in glomerular filtration in living organisms. On the other hand, this trend was not observed in devices using immortalized podocytes.

CONCLUSION

In this study, we constructed on-chip glomeruli. We compared two cell sources of podocytes in the construction: kidney organoid-derived podocytes and human immortalized podocytes. We mimicked the structure of the glomerular filtration barrier by seeding vascular endothelial cells, ECM gel, and podocytes into a three-channel device. To evaluate the functionality of the constructed filtration barrier, the permeability of fluorescently labeled inulin and albumin was quantified. The device using organoid-derived podocytes showed

selectivity for filtration of inulin, similar to that of living organisms. On the other hand, devices using immortalized podocytes showed no selectivity for inulin and albumin. These results suggest that organoid-derived podocytes are more suitable than immortalized podocytes for the construction of on-chip glomeruli. In the future, to further mimic the filtration function of human kidney, we will optimize the components of the ECM gel and evaluate the selectivity of filtration due to differences in charge as observed in glomerular filtration in living organism.

ACKNOWLEDGEMENTS

This research was supported by AMED-MPS project under Grant Number JP22be1004204 and JP17be0304205, Japan, Tateisi Science and Technology Foundation. Microfluidic devices for this work were fabricated at Kyoto University, Nanotechnology Hub, supported by "Nanotechnology Platform Program" of the Ministry of Education, Culture, Sports, Science and Technology (MEXT), Japan, Grant Number JPMXP09F19KT0107.

REFERENCES

[1] M. R. Pollak, S. E. Quaggin, M. P. Hoenig, L. D. Dworkin, "The Glomerulus: The Sphere of Influence" *Clin J Am Soc Nephrol.* vol. 9: pp. 1461–1469, 2014.

[2] Li Wang, Tingting Tao, Wentao Su, Hao Yu, Yue Yu, Jianhua Qin, "A disease model of diabetic nephropathy in a glomerulus-on-a-chip microdevice", *Lab Chip*, vol. 17, pp. 1749-1760, 2017.

[3] S. Musah, A. Mammoto, T. C. Ferrante, S. S. F. Jeanty, M. Hirano-Kobayashi, T. Mammoto, K. Roberts, S. Chung, R. Novak, M. Ingram, T. Fatanat-Didar, S. Koshy, J. C. Weaver, G. M. Church, and D. E. Ingber, "Mature-induced-pluripotent-stem-cell-derived human podocytes reconstitute kidney glomerular-capillary-wall function on a chip", *Nat Biomed Eng.* 0069, 2017.

[4] A. Petrosyan, P. Cravedi, V. Villani, A. Angeletti, J. Manrique, A. Renieri, R. E. De Filippo, L. Perin, and S. Da Sacco, "A glomerulus-on-a-chip to recapitulate the human glomerular filtration barrier", *Nat Com.* vol. 10, no. 1. 2019.

[5] M. A. Saleem, M. J. O'Hare, J. Reiser, R. J. Coward, C. D. Inward, T. Farren, C. Y. Xing, L. Ni, P. W. Mathieson, P. Mundel. "A Conditionally Immortalized Human Podocyte Cell Line Demonstrating Nephrin and Podocin Expression", *J Am Soc Nephrol* vol. 13, pp. 630–638, 2002

[6] M. Takasato, P. X. Er, H. S. Chiu, B. Maier, G. J. Baillie, C. Ferguson, R. G. Parton, E. J. Wolvetang, M. S. Roost, S. M. Chuva De Sousa Lopes, and M. H. Little, "Kidney organoids from human iPS cells contain multiple lineages and model human nephrogenesis", *Nature*, vol. 526, no. 7574. pp. 564-568, 2015.

[7] M. Takasato, P. X. Er, H. S. Chiu, M. H. Little, "Generation of kidney organoids from human pluripotent stem cells", *Nat. Protoc.*, vol. 11, no. 9, pp. 1681–1692, 2016.

[8] L. J. Hale, S. E. Howden, B. Phipson, A. Lonsdale, P. X. Er, I. Ghobrial, S. Hosawi, S. Wilson, K. T. Lawlor,

S. Khan, A. Oshlack, C. Quinlan, R. Lennon, and M. H. Little, "3D organoid-derived human glomeruli for personalised podocyte disease modelling and drug screening", *Nat Com,* vol. 9, no. 1, 2018.

CONTACT

*Ayumu TABUCHI, tel: +81-75-383-3684; tabuchi.ayumu.34u@st.kyoto-u.ac.jp

CONTROLLING FIRING POINT OF MICROFIBER-SHAPED HIPSC-DERIVED CARDIAC TISSUE WITH LOCALIZED ELECTRICAL STIMULATION DEVICE

Akari Masuda[1], Shun Itai[1], Yuta Kurashina[2], Shugo Tohyama[3], and Hiroaki Onoe[1]
[1]Faculty of Science and Technology, Keio University, Kanagawa, JAPAN,
[2]Institute of Engineering, Tokyo University of Agriculture and Technology, JAPAN
[3]School of medicine, Keio University, Tokyo, JAPAN

ABSTRACT

We propose a device to apply localized electrical stimulation to control cardiac conduction in *in vitro* cardiac tissue. Our system can initiate the generation of electrical impulses at any location by applying local electrical stimulation to the desired point of the microfiber-shaped cardiac tissue. Our system generated a strong and intense electric field between the electrodes in FEM analysis, indicating only the cardiac tissue between the electrodes was electrically stimulated. We could observe the cardiac conduction optically by visualizing the timing and location of impulses with calcium imaging at any location and time on the tissue. The system enables accurate analysis of cardiac potential and conduction velocity of the heart model, and could contribute to the elucidation of pathological dynamics such as arrhythmia, whose mechanism of occurrence still remains unknown.

KEYWORDS

iPSCs, Cardiac Tissue, Tissue Engineering, Electrical Stimulation, Microfiber, Cardiac Conduction.

INTRODUCTION

Cardiovascular diseases (CVDs) such as ischemic heart disease and dilated cardiomyopathy are the leading cause of death in the world. CVDs kills 17.9 million people each year and accounts for 32% of all deaths worldwide [1]. Novel medicine is highly demanded to improve the patients' prognosis, because effective basic treatments for CVDs have not been developed. Therefore, *in vitro* tissue models are attracting attention as a platform for the next generation of drug efficacy research. In particular, the development of models representing real diseases using patient-derived human induced pluripotent stem cells (hiPSCs) can facilitate rare drug development, especially for monogenic rare genetic diseases, by compensating for the lack of predictive *in vitro* human models for drug discovery [2].

Altered impaired cardiac conduction is closely related to the development of arrhythmias, a major cause of death in heart diseases [3]. Therefore, *in vitro* heart models with the ability to analyze cardiac conduction are quite longed as a platform for pathological and pharmacokinetic researches to develop effective treatment. Among the models of various shapes (fiber, sheet, spheroid previously reported) [4][5], the fiber-shaped models are the most similar to the muscle fiber of *in vivo*. In one of the previous researches, the cardiac conduction of the fiber-shaped tissue was evaluated by applying electrical stimulation while fixing the tissue to electrodes [5]. However, this system could not precisely reproduce the behavior of a biological heart because the tissue was applied to overall

stimulation. In case of *in vivo*, a fired impulse from a small part of the heart tissue (sinoatrial node) propagates through the gap junction [6]. Thus, local electrical stimulation is required on the heart model to mimic the *in vivo* conduction mechanism.

Here, we propose a firing point designable localized electrical stimulation device for microfiber-shaped cardiac tissue (Fig. 1). The device permits the selection of localized electrical stimulation sites, allowing stable analysis of conduction in the tissue mimicking *in vivo*.

METHODOLOGY
Fabrication of cardiac microfiber

The hiPSC-derived cardiac microfiber was fabricated with a double-coaxial laminar flow microfluidic device [7][8] (Fig. 2 (a)). Cell suspensions were prepared by suspending hiPSC-derived cardiomyocytes and human ventricular cardiac fibroblasts (NHCF-V) in collagen

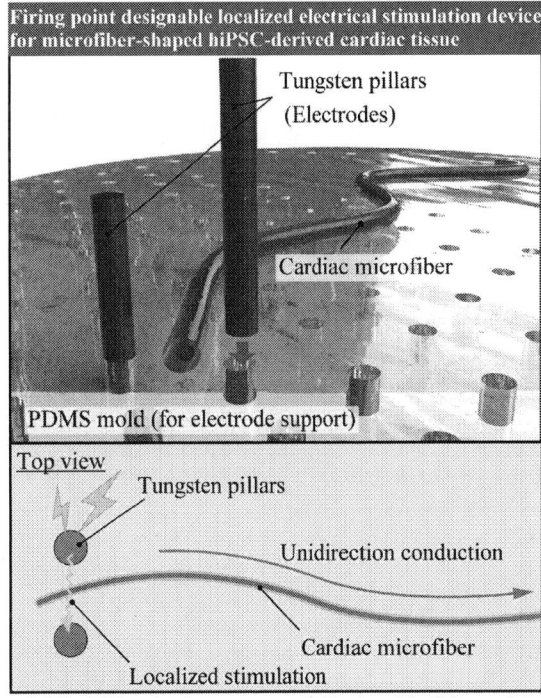

Figure 1: Concept of our localized electrical stimulation device for microfiber-shaped cardiac tissue. The firing point of cardiac microfiber would be designed by placing the pillars on both sides of the tissue.

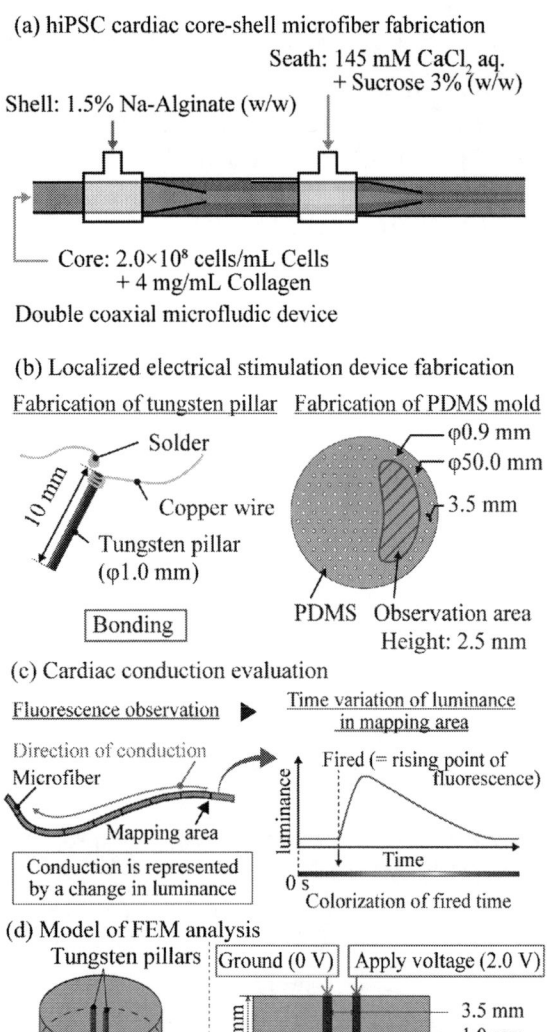

(a) hiPSC cardiac core-shell microfiber fabrication

(b) Localized electrical stimulation device fabrication

(c) Cardiac conduction evaluation

(d) Model of FEM analysis

Figure 2: Fabrication and analysis method of hiPSC-derived cardiac microfiber and localized electrical stimulation device. (a) Fabrication of coaxial shaped core-shell microfiber with a double-coaxial laminar flow microfluidic device. (b) Fabrication of localized electrical stimulation device. (c) The analytical flow of the conduction of the cardiac microfiber. (d) The dimensions of FEM model

solution. The prepared cell suspension was poured into the core of the microfluidic device and sodium alginate solution was poured outside of the microfluidic device. Sodium alginate solution was gelatinized by flowing calcium chloride from the outside of the sodium alginate solution. Finally, a microfiber with a protective shell for the cells was fabricated.

Fabrication of localized electrical stimulation device

The localized electrical stimulation device is composed of two electrodes and a PDMS mold supporting electrodes (Fig. 2 (b)). The electrode was obtained by wrapping a copper wire around a tungsten pillar and fixing

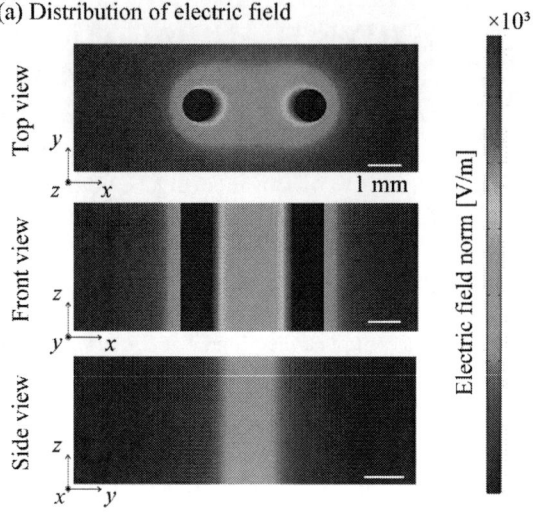

(a) Distribution of electric field

(b) Range of electrical stimulation

Graph of electrical field on front plane

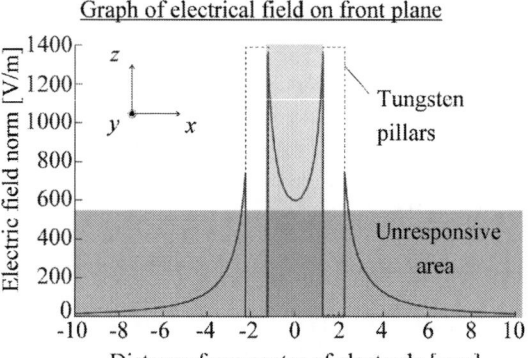

Graph of electrical field on side plane

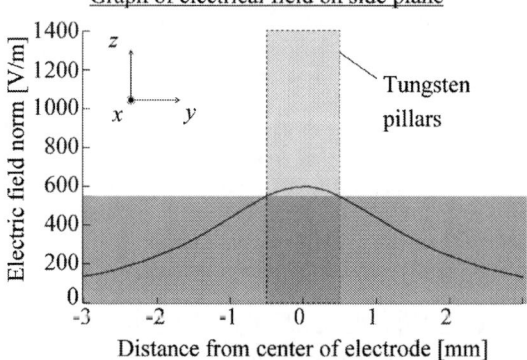

Figure 3: FEM analysis of the electric field on our device. (a) Distribution diagram of the electric field at each cross section. (b) Range of electrical stimulation applied to cardiac microfibers. Electric fields sufficient to stimulate cardiac microfiber are generated only between electrodes.

it with solder covered with a heat shrink tube. The PDMS mold consisted of a area with holes (for fixing the fiber-shaped tissue) and a area without holes (for observation) was fabricated with an aluminum mold. The PDMS pre-gel solution was heated at 75°C for 1 hour to gelatinize the

(a) Applied to Fiber A **(b) Applied to Fiber C**

Figure 4: Confirmation of selective electrical stimulation of cardiac microfibers. (a) (b) Changes in beating rate of cardiac microfiber to electrical stimulation: mean ± S.D., n = 3.

PDMS mold.

Analysis of electrical stimulation

The distribution of the electric field norm generated in our electrical stimulation device was analyzed by finite element method (COMSOL Multiphysics). A three-dimensional model of the tungsten pillar and the culture medium used in the experiment was created (Fig. 2 (c)). Two tungsten pillars were fabricated, and a voltage of 2.0 V was applied to one pillar while the other was grounded. The electric field norm distribution around the tungsten pillar was visualized in the top, front and side views. Graphs of the intensity variation of the electric field norm at the centerline of the front and side views were prepared and the intensity distribution of the electric field norm was analyzed.

Applying localized electrical stimulation

The fabricated electrical stimulation device was used to apply local electrical stimulation to cardiac microfibers. Tungsten pillars were placed on both sides of the target cardiac microfiber, and 1.5-Hz biphasic pulses were applied to the electrodes. The beating of the fibers was filmed while electrical stimulation was applied, and the beating rate of the cardiac microfibers was analyzed by motion video analysis (Keyence, motion analyzer).

Controlling the firing point

To analyze the firing point of the electrical signal, the cardiac conduction was visualized by calcium imaging. Optical mapping images were produced and compared before and after the application of electrical stimulation. Fluo-8 was added to the cardiac microfibers, and the conduction was captured by fluorescence imaging. The image of cardiac microfiber was divided at equal intervals, and the timing of maximum fluorescence intensity was mapped. The optical mapping images were compared before and after the application of electrical stimulation.

EXPERIMENTAL RESULTS
Fabricated hiPSC cardiac microfiber and localized electrical stimulation device

The cardiac microfiber had a double-coaxial structure consisting of a core layer of cardiac tissue with an outer

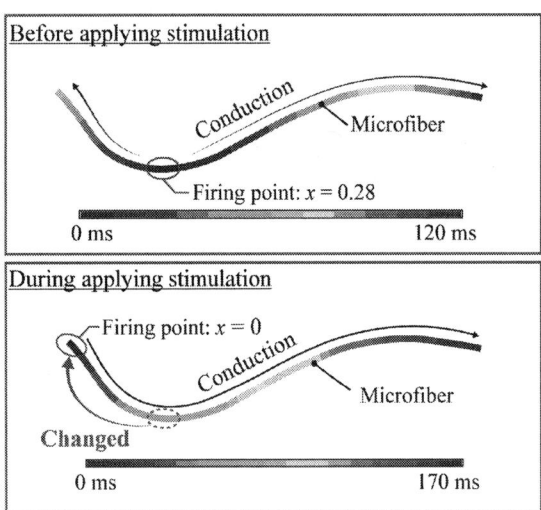

Figure 5: The optical mapping for evaluation cardiac conduction. (a) Image of calcium imaging setup. (b) Controlled firing point of cardiac microfiber with our stimulation device.

diameter of 100 μm, and a shell layer of sodium alginate with an outer diameter of 200 μm. The cardiac microfibers beated without breaking from the second day of culture. The electrical stimulation device had an outer diameter of 50 mm, and tungsten pillars could be inserted into the hole. The device could be integrated in a culture dish filled with culture medium. Therefore, our device could set electrodes at any position on the cardiac microfibers being cultured.

Analysis of electrical stimulation

The simulation results of the electric field norm distribution generated in our device is shown in Fig. 3 (a). The electric field norm distribution diagram shows that the electric field is strongly distributed around and between the electrodes. The graphs of the intensity variation of the electric field norm at the centerlines of the front and side views are shown in Fig. 3 (b). The graph shows that a high electric field norm is generated at the position of the tungsten pillar. Based on the result graph, we designed a voltage such that the electric field norm applied outside the tungsten pillar is in the unresponsive region of the cardiac tissue (<556.5 V/m). In our device, electrical stimulation was applied only to the tissue between the tungsten pillars when a voltage of 2.0 V is applied.

Applying localized electrical stimulation

The results of the analysis of the beating rate of the cardiac microfibers are shown in Fig. 4. When electrical stimulation was applied, only the fiber between the electrodes (Fig. 4 (a): Fiber A, Fig. 4 (b): Fiber C) beat at 1.5 Hz. Therefore, electrical stimulation could be applied only to the target fiber by using our device.

Controlling the firing point

Calcium imaging visualized the electrical signals conducting through the fiber as changes in fluorescent intensity. The optical mapping images before and after the application of electrical stimulation are shown in Fig. 5. From the optical mapping images, the change in the firing point of the cardiac microfiber from the center of the fiber to the end point due to local electrical stimulation was confirmed. Therefore, our system could arbitrarily design the firing point of the cardiac microfiber.

CONCLUSION

We proposed a device to apply localized electrical stimulation to control cardiac conduction in in vitro cardiac tissue. The electrical stimulation device was confirmed to be capable of applying electricity locally by FEM analysis. Electrical stimulation was used to control the beating of the tissue at any given position. Optical observation of cardiac conduction confirmed that the firing point was controlled by electrical stimulation. The results show the ability of our system for the pathology of arrhythmia and pharmacokinetic testing.

ACKNOWLEDGEMENTS

This work was partly supported by Takahashi Industrial and Economic Research Foundation, Japan.

REFERENCES

[1] R. Xue, Q. Li, Y. Geng, H. Wang, F. Wang, and S. Zhang, "Abdominal obesity and risk of CVD: A dose-response meta-analysis of thirty-one prospective studies," *Br. J. Nutr.*, vol. 126, no. 9, pp. 1420–1430, 2021, doi: 10.1017/S0007114521000064.

[2] C. Y. Huang *et al.*, "Enhancement of human iPSC-derived cardiomyocyte maturation by chemical conditioning in a 3D environment," *J. Mol. Cell. Cardiol.*, vol. 138, no. July 2019, pp. 1–11, 2020, doi: 10.1016/j.yjmcc.2019.10.001.

[3] J. H. King, C. L. H. Huang, and J. A. Fraser, "Determinants of myocardial conduction velocity: Implications for arrhythmogenesis," *Front. Physiol.*, vol. 4 JUN, no. June, pp. 1–14, 2013, doi: 10.3389/fphys.2013.00154.

[4] C. S. Ong *et al.*, "Biomaterial-Free Three-Dimensional Bioprinting of Cardiac Tissue using Human Induced Pluripotent Stem Cell Derived Cardiomyocytes," *Sci. Rep.*, vol. 7, no. 1, Dec. 2017, doi: 10.1038/s41598-017-05018-4.

[5] T. M. Spencer *et al.*, "Fibroblasts Slow Conduction Velocity in a Reconstituted Tissue Model of Fibrotic Cardiomyopathy," *ACS Biomater. Sci. Eng.*, vol. 3, no. 11, pp. 3022–3028, Nov. 2017, doi: 10.1021/acsbiomaterials.6b00576.

[6] S. Verheule, M. J. A. Van Kempen, S. Postma, M. B. Rook, and H. J. Jongsma, "Gap junctions in the rabbit sinoatrial node," *Am. J. Physiol. - Hear. Circ. Physiol.*, vol. 280, no. 5 49-5, pp. 2103–2115, 2001, doi: 10.1152/ajpheart.2001.280.5.h2103.

[7] H. Onoe *et al.*, "Metre-long cell-laden microfibres exhibit tissue morphologies and functions," *Nat. Mater. 2013 126*, vol. 12, no. 6, pp. 584–590, Mar. 2013, doi: 10.1038/nmat3606.

[8] A. Masuda *et al.*, "Fixation-Free Evaluation of Cardiac Contractile Force by Human iPSC-Derived Cardiac Core-Shell Microfiber," *IEEE Symp. Mass Storage Syst. Technol.*, vol. 2022-January, pp. 172–175, 2022, doi: 10.1109/MEMS51670.2022.9699631.

CONTACT

*A. Masuda, tel: +81-45-566-1507; scarlet.msd1260@keio.jp

DEVELOPMENTAL PHASES OF ON-CHIP VASCULOGENESIS CLASSIFIED USING A DEEP LEARNING VISUAL MODEL

Taiga Irisa, Hang Zhou, Kazuya Fujimoto, and Ryuji Yokokawa

Kyoto University, JAPAN

ABSTRACT

Vasculogenesis, the formation of de novo blood vessels, occurs in several fundamental physiological and pathological processes, including embryonic development, adult angioblast mobilization, and tumor development. Although substantial research has been performed on vasculogenesis, a comprehensive understanding of the morphological changes during the developmental process of vasculature remains elusive. Here, we report a deep learning-based visual methodology in conjugation with a vasculature-on-a-chip model that can recognize and classify the developmental phases of an on-chip vascular network. Combining the unsupervised and supervised learning strategies, three distinct time-dependent morphological phases during vasculogenesis were identified. The temporal variation trajectory of vascular morphology during entire vascularization process was revealed.

KEYWORDS

Vasculogenesis, deep learning, morphology analysis, microvasculature-on-a-chip

INTRODUCTION

The vasculature is the earliest organ to develop during embryogenesis [1]. The formation process of de novo vasculature is called vasculogenesis. In vasculogenesis, mesodermal cells differentiate into endothelial progenitor cells and those cells form a primitive vessel network that serves as the basis for the mature vascular system. During this process, cells experience morphological changes such as elongation, network formation, and lumenization (Figure 1) [2]. However, it is unclear how these cells transfer from one morphology to another. Especially, the critical time points initiating the morphological changes and relevant genomic regulators during vasculogenesis have not yet been identified [3].

To identify the key time points and transcription factors in morphological changes, a precise quantitative classification of the vascular morphology through vasculogenesis is required. Conventional methods for analyzing the morphology use geometric information of vascular networks such as numbers and length of vascular branches to compare the activity of endothelial cells in response to different biochemical and biophysical stimuli [4, 5]. These methodologies oversimplify the morphological features of the vascular network based on biasedly handcrafted algorithms, which were mostly designed for the relatively matured network. None of them was demonstrated to be capable of analyzing diverse morphological states. A comprehensive study on the dynamic morphological changes during the vasculogenic process, including sparsely scattered cells, conjunct vascular lumens, and complicated matured networks is significantly lacking. Therefore, an unbiased method for analyzing the changes in vascular morphology is highly needed.

In the past few years, deep learning models based on convolutional neural networks (CNNs) have enabled diverse computer vision tasks including image classification. CNN models can extract rough morphological and contour information automatically by filtering images, which leads to the extraction of further useful image information to classify the image types. Thus, we aimed to establish an unbiased and systematic method of analyzing the morphology of vasculogenesis using a CNN model, which can classify the distinct phases of vasculogenesis and extract the corresponding features. By introducing this method to detect the morphological changes in real-time vascular formation, it is expected that the genetic information obtained from the specific morphological phases will reveal the key transcription factors (Figure 2).

EXPERIMENTAL METHODS

Device fabrication

The microfluidic device was fabricated by assembling a 3-channel PDMS layer onto a 35-mm glass bottom dish (Figure 3a, b). The three channels patterned on PDMS were separated by micro-posts. One end of the side channel is open and connected to the glass bottom dish, which worked as a medium reservoir.

Vasculogenesis assay

GFP-labeled human umbilical vein endothelial cells

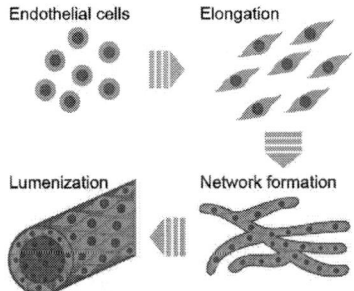

Figure 1: Morphological changes during vasculogenesis.

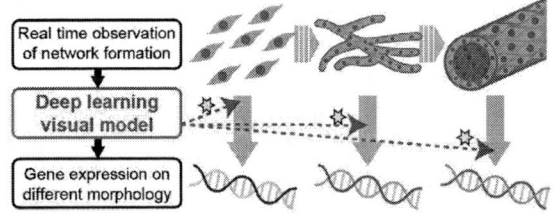

Figure 2: Proposed methodology of acquiring the genetic information from the specific morphologies.

(GFP-HUVECs, Angio Proteomie) were suspended in a fibrin-collagen gel at a concentration of 8×10^6 cells/mL. The GFP-HUVECs suspension was injected into the center channel. After gel polymerization, EGM-2 (Lonza) containing 50 ng/mL VEGF and 50 ng/mL bFGF was introduced into the side channels (Figure 3c), which are essential to promote vascular network formation. As soon as EGM-2 was introduced, the device was incubated in a stage-top incubator (Tokai hit, STXG-IX3WX-SET) mounted on the confocal microscopy (Olympus, FV-3000).

Fluorescent images of GFP-HUVECs were taken every 30 minutes for 48 hours. The EGM-2 introduction time point was defined as 0 hour of the cultured time.

Unsupervised clustering

The images from nine independent experiments were first projected using maximum intensity projections of 19 Z-sections to represent three-dimensional structures of the vascular morphologies. The publicly available VGG16 model pretrained on ImageNet datasets to classify 1000 classes was firstly used to extract the image features [6]. Since it has a very deep network structure and pretrained with many images, it was used as a general image feature extractor when the final dense layers (classifier) were removed [7]. All the preprocessed images were input into the VGG16 model without the final dense layers. Next, the output features from the VGG16 model were hierarchically clustered using Euclidean distances to process Ward's method. We set the labeling threshold as 50% of the largest Euclidean distances between the clusters. After labeling as their clusters, the data points in each cluster were plotted according to the obtained time.

CNN classifier training

A convolutional neural network (CNN)-based image classifier was constructed to classify and predict the different vasculogenic phases. The images taken at 2 to 3.5 hours, 15 to 16.5 hours, and 40 to 41.5 hours were labeled as Phase 1, Phase 2, and Phase 3, respectively. The dataset consisted of 1443 images including 939 training images, 312 validation images. All images were normalized. The training images were augmented by rotation and mirroring to increase the generalizability of the model. The model was trained for 20 epochs.

CNN classifier performance test

The performance of the trained CNN classifier was tested by 192 test images including 64 images for each phase and the prediction accuracy was calculated. A Score Class Activation Mapping (Score-CAM) was introduced to visually interpret the prediction strategy of the CNN classifier by obtaining the weights of each activation map on the predicted class [8]. The whole-time images from three experiment, namely batch 1, 2 and 3, were labeled as their obtained time point and input into the constructed CNN classifier to demonstrate a real-time detection of phase transition.

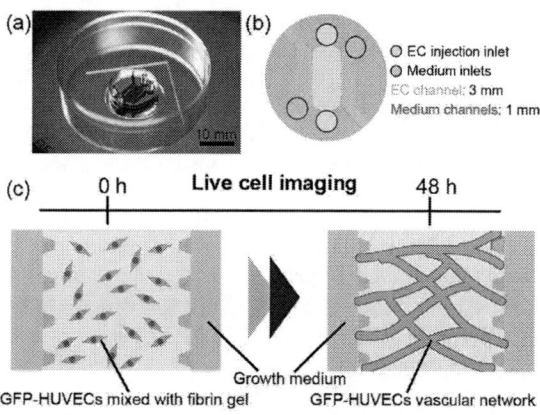

Figure 3: Experimental setup. (a) Microfluidic device. (b) Schematic diagram of the device. (c) Experimental flow.

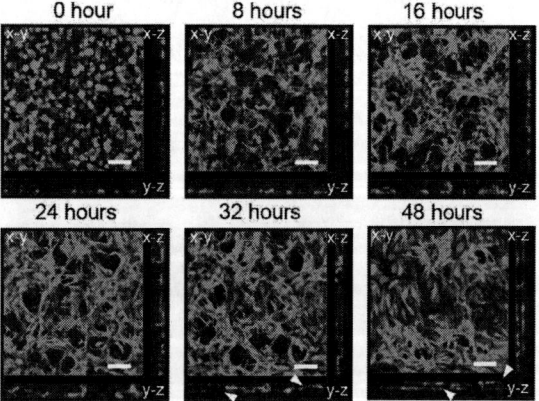

Figure 4: Time-lapse images of vasculogenic assay. Scale bar: 100 μm

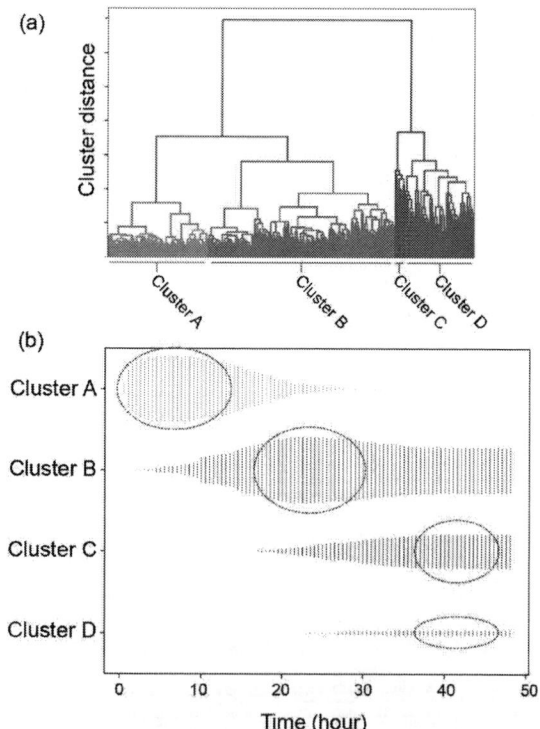

Figure 5: Image feature clustering. (a)Hierarchical clustering. (b) Cluster's time-dependent distribution.

RESULTS AND DISCUSSION
Vascular network formed within 48 hours

Initially, HUVEC shapes were rounded and distributed randomly (Figure 4, 0 hour). Their shapes started to elongate (Figure 4, 8 hours) and connected (Figure 4, 16 and 24 hours). Then, lumen structures formed in the network, which can be confirmed by the hollow circle consisted of cells in the cross section of the device (Figure 4, 32 hours white arrows). Finally, these lumens widened gradually (Figure 4, 48 hours white arrows). During in vivo vasculogenesis, it is reported that endothelial cells change their shape from rounded to elongated [9]. Next, they gather at distinct embryonic locations and form lumenless aggregates, and then they rearrange their junctional contact points to make a hollow structure to transport substances to organs [10]. Our

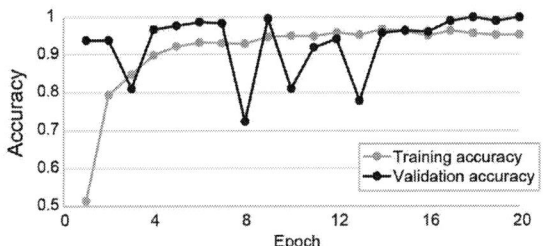

Figure 6: Training process of the CNN model.

observation results match the in vivo phenomena, so our vasculogenesis-on-a-chip mimicked well the changes in dynamic vasculogenic morphology.

Time-dependent image features

There were four major clusters in the features extracted from images in the observation for 48 hours (Figure 5a). The clusters showed the different distribution in the cultured time (Figure 5b). The distribution of data in cluster A peaked between 0 and 10 hours. Cluster B's peak was approximately between 15 and 30 hours. Clusters C and D showed the similar distributions and their peaks were in between 35 to 48 hours. Images at those time courses were easily distinguishable by their morphologies, such as scattered cells (Figure 4, 0 hour), aggregated and connected cells (Figure 4, 16 hours), and cells having wider connected regions (Figure 4, 48 hours). These results suggested that there were three major morphological phases during the 48-hour vasculogenesis in the microfluidic device. Correspondingly, we decided to classify the three phases in vasculogenesis-on-a-chip using the constructed CNN model.

CNN classifier detected morphological changes

The CNN classifier was trained to classify the three different vasculogenic phases. The accuracy of prediction to both datasets reached above 95% in eighteen learning epochs (Figure 6), which demonstrated the feasibility of

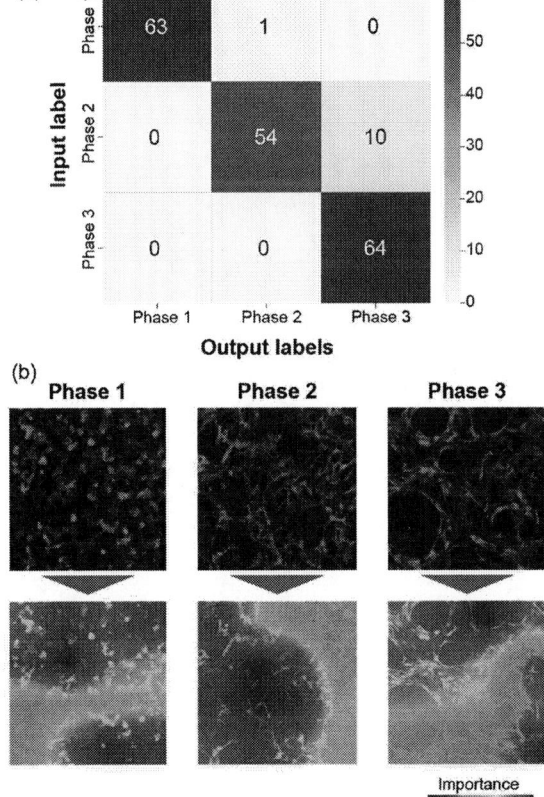

Figure 7: The CNN classifier performance test. (a) Prediction results on the test dataset. (b) Visual interpretation of the CNN classifier's prediction strategy.

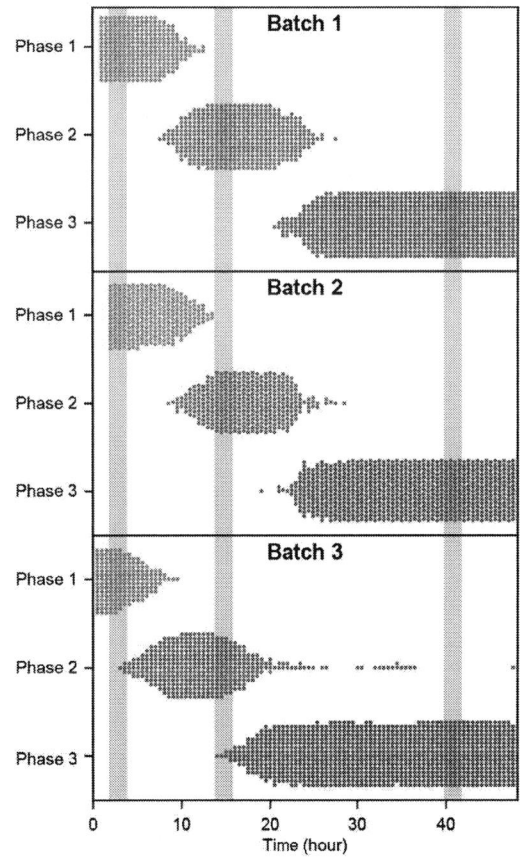

Figure 8: The Phase transition detection. Data points highlighted in blue lines are used for training, validation or test of the CNN model.

the trained CNN classifier.

The performance of the trained CNN was tested using the test dataset. Those images were predicted with 94.2% accuracy (Figure 7a). The representative pattern features in the distinct 3 phases, which were recognized by the trained CNN model, were visually highlighted by Score-CAM (Figure 7b). In Phase 1, the importance of the area where there were rounded cells scattered was high. In Phase 2, the importance of the area where cells aggregated and connected was high. In Phase 3, the importance of the blank areas and the area where cells made lumen structures was high. These results suggested that the regions where the CNN model focused on to distinguish the three phases appeared to match our description mentioned in the observation part, so we concluded that the CNN model learned the features of each vasculogenic phases well and can be used for further classification tasks.

Inputting the real-time images into the constructed CNN model showed that each phase feature had its time range and the phase transition happened gradually in all three experiments (Figure 8). These results suggested that the constructed CNN model successfully worked and detected the phase transition by predicting the labels in real time among different experiments.

Interestingly, the time points of the phase transition were different from experiment to experiment. In experiment batch 2, Phase 2 starts from 9.5 to 10.5 hours and Phase 3 starts from 22 to 23 hours, which were about 1 hour later than batch 1 (8.5 to 9 hours, 21.5 to 22 hours, respectively). In contrast, in experiment batch 3, Phase 2 starts from 3 to 3.5 hours and Phase 3 from 15 to 16 hours, which were 5 to 6 hours earlier than batch 1 and 2. These results suggested that our proposed model was able to detect the difference in the rate of vascular development, which cannot be detected by conventional quantitative methods. Using our model, it is possible to collect the cellular samples, obtain the genetic information, and analyze the link between morphology and gene expression by the quantitative morphological evaluation, independent of unreliable indexes such as cultured time.

CONCLUSION

In this study, we constructed an unbiased quantification method for classifying and detecting the distinct morphology during vasculogenesis in vitro. Firstly, we confirmed there are three timeframes of distinct vascular development within 48 hours using unsupervised clustering of image features. Then, we constructed a CNN classifier to distinguish and predict three different phases of vascular development. We confirmed the model was able to detect the phase transition in real time. Thus, this model enables us to determine the time points to collect samples to analyze gene expression level to reveal the transcription factors playing important roles in those observed morphological changes. We believe this visual model-based system will provide wide applications such as analyzing the mechanism of self-organizing organoids.

ACKNOWLEDGEMENTS

This study was partially supported by AMED under Grant Number JP21be0304205. Microfabrication was supported by Kyoto University Nano Technology Hub in "Nanotechnology Platform Project" sponsored by MEXT, Japan.

REFERENCES

[1] O. C. Velazquez, "Angiogenesis and vasculogenesis: inducing the growth of new blood vessels and wound healing by stimulation of bone marrow-derived progenitor cell mobilization and homing," *J Vasc Surg*, vol. 45 Suppl A, no. Suppl A, pp. A39–A47, 2007.

[2] C. W. Peak, L. Cross, A. Singh, and A. K. Gaharwar, "Microscale technologies for engineering complex tissue structures," in *Microscale Technologies for Cell Engineering*, Springer International Publishing, 2015, pp. 3–25.

[3] K. Tsuji-Tamura and M. Ogawa, "Morphology regulation in vascular endothelial cells," *Inflammation and Regeneration 2018 38:1*, vol. 38, no. 1, pp. 1–13, Sep. 2018.

[4] E. Zudaire, L. Gambardella, C. Kurcz, and S. Vermeren, "A Computational Tool for Quantitative Analysis of Vascular Networks," *PLoS One*, vol. 6, no. 11, p. e27385, Nov. 2011.

[5] H. Z. Sailem and A. al Haj Zen, "Morphological landscape of endothelial cell networks reveals a functional role of glutamate receptors in angiogenesis," *Scientific Reports 2020 10:1*, vol. 10, no. 1, pp. 1–14, Aug. 2020.

[6] K. Simonyan and A. Zisserman, "Very Deep Convolutional Networks for Large-Scale Image Recognition," Sep. 2014.

[7] D. Albashish, R. Al-Sayyed, A. Abdullah, M. H. Ryalat, and N. Ahmad Almansour, "Deep CNN Model based on VGG16 for Breast Cancer Classification," in *2021 International Conference on Information Technology, ICIT 2021 - Proceedings*, Jul. 2021, pp. 805–810.

[8] H. Wang *et al.*, "Score-CAM: Score-weighted visual explanations for convolutional neural networks," *IEEE Computer Society Conference on Computer Vision and Pattern Recognition Workshops*, vol. 2020-June, pp. 111–119, Jun. 2020.

[9] C. J. Drake, A. LaRue, N. Ferrara, and C. D. Little, "VEGF regulates cell behavior during vasculogenesis," *Dev Biol*, vol. 224, no. 2, pp. 178–188, Aug. 2000.

[10] K. Xu and O. Cleaver, "Tubulogenesis during blood vessel formation," *Seminars in Cell and Developmental Biology*, vol. 22, no. 9. Elsevier Ltd, pp. 993–1004, 2011.

CONTACT

*T. Irisa, tel: +81-75-383-3687; irisa.taiga.88w@st.kyoto-u.ac.jp

HAND-DRIVEN DEVICE FOR PREPARATION OF LINEARLY ALIGNED HYDROGEL SHEETS

Aoi Kato [1,2], Haruka Oda [3] Sho Takamori [2], Hisatoshi Mimura [2], Toshihisa Osaki [2], Norihisa Miki [1,2], and Shoji Takeuchi [2,3]

[1]Department of Mechanical Engineering, Keio University, JAPAN
[2]Kanagawa Institute of Industrial Science and Technology, JAPAN
[3]Graduate School of Information Science and Technology, The University of Tokyo, JAPAN

ABSTRACT

In this paper, we present a three-layered device for preparation of heterologous core layers linearly aligned in a hydrogel sheet by a simple four-step process without using pumps. Recently, cell-based sensors have been developed for detection of odorant molecules owing to the sensitivity and selectivity of olfactory receptors to chemical substances. To detect various odorant molecules with these sensors, it is necessary to array multiple types of sensor cells expressing different olfactory receptors. We therefore propose a sheet-shaped sensor chip for a sensor-cell array. Previously, hydrogel sheets with heterogeneous cell patterns were produced using microfluidic devices with syringe pump systems, causing loss of both sample volumes and time for settings. Here, we develop a device that enables preparation of hydrogel sheets with a small amount of samples and with a single manual pipette.

KEYWORDS

Cell-based sensor, Insect olfactory receptor, Surface tension, Hydrogel sheet.

INTRODUCTION

Living organisms sense environmental stimuli such as light, sound, and chemicals using various types of receptors. Among them, human beings are known to differentiate the smell of more than 300,000 to 400,000 odors with less than 400 olfactory receptors (ORs), which are therefore considered as promising sensing elements for biohybrid sensors with high sensitivity and selectivity. Cell-based odorant sensors are one of those sensors, utilizing ORs expressed on the cell (sensor cells) [1]. Similar to the living organisms, a set of various sensor cells is expected to discriminate multiple and complex mixtures of odorant molecules by their signal patterns, and possibly applied for breath and body odor diagnosis, environmental assessment, explosive detection, and so on.

In a previous study, a few types of sensor cells were directly patterned on a substrate via a biocompatible anchor [2]. For further miniaturization and mass production, we considered that a sheet-shaped sample will be beneficial for both process of sample preparation and reproduction of samples because a sheet form will be prepared by a microfluidic device and the sheet can be cut into any sizes. In previous works, microfluidic devices were developed for preparation of a hydrogel sheet encapsulating different types of cells [3,4]. Although the devices were able to precisely control the arrangement of each cell type, the use of pump systems has disadvantages in loss of sample volumes and time-consuming for setting the systems. In

this study, we propose a hand-driven device for preparation of a hydrogel sheet with encapsulating linearly aligned heterologous core-layers.

EXPERIMENTAL

Design and Fabrication of the Device

The design of the proposed device is shown in Fig. 1. The device consists of three layers made of acrylic plates with a thickness of 5 mm (Acrylite, Mitsubishi Chemical, Japan) and channel was prepared using an NC micro-machining machine (MiniMiller MM100, Modia Systems, Japan). The top layer is the inlet of the shell solution (Fig. 1a). The middle layer has the wells to fulfill the core solution. The outlet of the core solution is designed at the bottom of the wells (Fig. 1b). The bottom layer provides a shallow channel to arrange the core and shell solution by a laminar flow (Fig. 1c). At the outlet of the bottom well, gelation is conducted and a hydrogel sheet is formed.

Preparation of Hydrogel Sheet Using the Device

Hydrogel sheet was formed by the following four-step procedure, using the developed device (Fig. 2). (i) Put the middle layer on the bottom layer, and inject the colored alginate-Na solutions (core solutions) into the wells of the middle layer with a manual pipette (12 μL each). The solution is retained in the wells without leaking from the outlet at the bottom due to the effect of a capillary stop valve. (ii) Place the top layer on the middle layer. (iii)

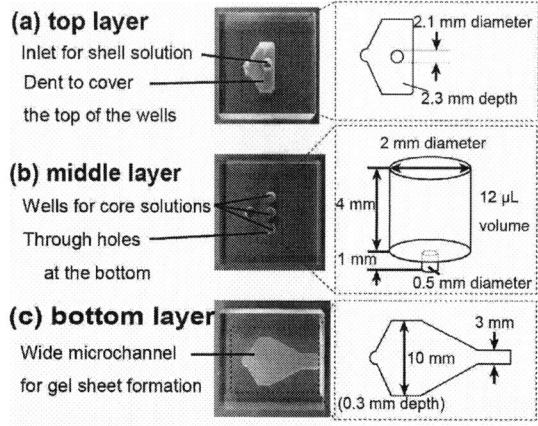

Figure 1: Device design. Details of the three components of the proposed device. The Device was fabricated with a micromilling process. (a) top layer. (b) middle layer. (c) bottom layer.

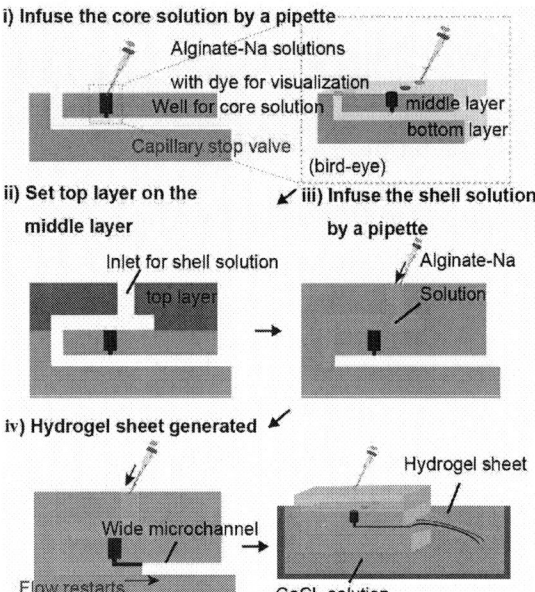

i) Infuse the core solution by a pipette
Alginate-Na solutions
with dye for visualization
Well for core solution
Capillary stop valve
middle layer
bottom layer
(bird-eye)

ii) Set top layer on the middle layer
Inlet for shell solution
top layer

iii) Infuse the shell solution by a pipette
Alginate-Na Solution

iv) Hydrogel sheet generated
Wide microchannel
Flow restarts
Hydrogel sheet
CaCl₂ solution

Figure 2: Four steps of hydrogel-sheet preparation using the device. (i) Put the middle layer on the bottom layer, and inject colored alginate-Na (core) solutions at each well by a pipette. (ii) Put the top layer on top, and (iii) inject the alginate-Na (shell) solution from the inlet. When the shell solution comes to the bottom of the well, the core solutions start to flow. (iv) A hydrogel sheet forms by dipping the device into CaCl₂ solution.

Slowly inject a colorless alginate-Na solution (shell solution) through the inlet of the top layer with a pipette (250 µL). When the shell solution arrived at the bottom of the capillary valve, the core solutions in the wells begin to flow due to the change in surface tension. The core solutions are linearly aligned by the laminar flow. (iv) At the outlet of the bottom layer, gelation is occurred by immersing in 100 mM CaCl₂ solution.

RESULTS AND DISCUSSION

By infusion of the shell solution from the inlet, the laminar flow was observed and the core solutions linearly aligned within the shell solution (Fig. 3). The gelation of the alginate solution was immediately occurred after immersion into the CaCl₂ solution. Care should be taken to avoid uneven pipetting speed, which may cause variations of the core shape in the sheet. More than 15 cm of hydrogel sheet was obtained by the single procedure using 250 µL of shell solution (Table 1). The width of the sheet was small compared to the outlet width of the device by the gelation, while the thickness of the sheet was similar to the device design. The thickness and the width of the core layer will be suitable for encapsulation of sensor cells because the size is sufficiently thin for the diffusion of nutrients. Also, the hydrogel sheet was easily cut by scissors to replicate the samples (Fig. 3c). Compared to the previous device with pump systems, the developed device uses just a manual pipette, supporting fast and easy manipulation, and consumes less core solution for production.

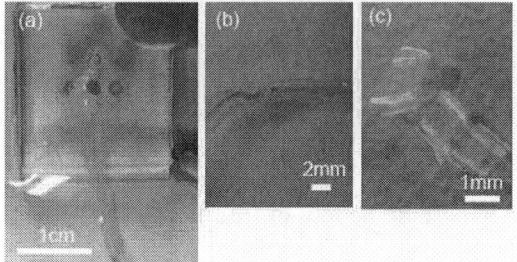

Figure 3: (a) Image of a hydrogel sheet preparation with colored core layers inside the device. (b) Alginate solution was gelated in CaCl₂ solution. (c) The gel sheet was cut by scissors for replication.

Table 1. Size of hydrogel sheet

length		176	±	19	mm
width	start	1300	±	188	µm
	middle	1159	±	226	µm
	end	1289	±	385	µm
thickness	start	258	±	127	µm
	middle	262	±	206	µm
	end	277	±	100	µm

Table 2. Width of core layers

width		
	pink	126 ± 25 µm
	blue	131 ± 49 µm
	green	123 ± 37 µm

CONCLUSION

We developed the device to generate a hydrogel sheet with linearly aligned multiple core layers, simply by using a manual pipette. Using the device, we demonstrated production of the sheet aligning three core layers. The device had another advantage in the volume of the core solution without the dead volume. Since the sheet is cut into arbitrary lengths, the products can be replicated. In the future, the device design has to be optimized for the cell-based sensor applications.

ACKNOWLEDGEMENTS

This work was partly supported by JSPS KAKENHI Grant Number JP21H05013, and the Program for Building Regional Innovation Ecosystem of MEXT, Japan.

REFERENCES

[1] Y. Hirata, H. Oda, T. Osaki, S. Takeuchi, "Biohybrid sensor for odor detection," *Lab on a Chip*, vol. 21, pp. 2643–2657, 2021.

[2] M. Termtanasombat, H. Mitsuno, N. Misawa, S. Yamahira, T. Sakurai, S. Yamaguchi, T. Nagamune, R. Kanzaki, "Cell-Based Odorant Sensor Array for Odor Discrimination Based on Insect Odorant Receptors," *J. Chem. Ecol.*, vol. 42, pp. 716–724, 2016.

[3] A. Kobayashi, K. Yamakoshi, Y.Yajima, R. Utoh, M. Yamada, M. Seki, "Preparation of stripe-patterned heterogeneous hydrogel sheets using microfluidic devices for high-density coculture of hepatocytes and fibroblasts," *J. Biosci. Bioeng.*, vol 116, pp. 761-767, 2013.

[4] L. Leng, A. McAllister, B. Zhang, M. Radisic, A. Günther, "Mosaic hydrogels: One-step formation of multiscale soft materials," vol 24, pp. 3650–3658, 2012.

CONTACT

*Aoi Kato, tel: +81-44-819-2037; aoi4012@keio.jp

MICROFABRICATION AND CHARACTERIZATION OF MICRO-STEREOLITHOGRAPHICALLY 3D PRINTED, AND DOUBLE METALLIZED BIOPLATES WITH 3D MICROELECTRODE ARRAYS FOR *IN-VITRO* ANALYSIS OF CARDIAC ORGANOIDS

Jorge Manrique Castro, Isaac Johnson, and Swaminathan Rajaraman
University of Central Florida, USA

ABSTRACT

This paper reports for the first time a bioplate microfabrication process using 3D printing technology and a two-step metallization process (ink-casting and sputtering deposition) that enables *in-vitro* studies of cardiac organoids through 3D microelectrode arrays (MEAs). The fabricated bioplate is in high-throughput (HT) platform, containing 24 wells and *9 spaghetti-shaped electrodes* per well, representing 192 recording sites and 24 stimulation sites in total. The 3D MEA was characterized structurally by optical, SEM and 3D confocal microscopy, and functionally by electrochemical impedance spectroscopy (EIS) with and without a synthetic organoid. Lastly, the bioplate is designed to interface with commercially available HT systems such as Maestro (Axion Biosystems).

KEYWORDS

Micro-stereolithography, 3D printing, Ink-casting, 3D Microelectrode Array (MEA), Electrochemical Impedance Spectroscopy (EIS), High-Throughput Bioplates, Cardiac Organoids.

INTRODUCTION

Organoids studies are the next-generation models that provides a better understanding of human physiology [1]. Several bioelectronics platforms (e.g., impedance electrodes, 2D and 3D MEAs) have been fabricated to culture and monitor brain organoids, which have been widely studied [2]. Cardiac organoids are being explored and engineered to replace dysfunctional hearts [3], heart disease analysis, and localized drug response [4].

Design and fabrication of sensing platforms that interact and procure data from cardiac organoids and spheroids are being currently in their infancy. These platforms are required to have low noise microelectrodes for high quality recordings and increased capabilities like throughput, i.e., multiple assays running in parallel in the same device. "3D rolling" MEA approach by Kalmykov et al. [5], provides a clever way to envelope a 3D biological structure like organoids by using a self-rolling array based on soft electronics. Microelectrode arrays surround the 3D sample and record the propagation of electrical signals on surface. More recently, Rogers group [6] presented a 3D Multifunctional Mesoscale Framework for neural interfacing of cortical spheroids and assembloids, providing multimodal capabilities at different geometries due to its exquisite shape control. Another recent example to characterize cardiac organoids relies on the classic Patch-Clamp technique [7] to target one specific cell on the 3D construct, and measure its electrophysiological response upon cell membrane excitation. In addition to the methods described above, Machine Learning (ML) is providing an alternative in the analysis of organoids by processing vast amounts of data including fluorescent or phase contrast images [8, 9]. Information about vascularity, size, shape, and feature counts can be measured and predicted using this approach. In a previous work of our group presented in IEEE MEMS 2022 [10], cardiac beat sensing was recorded from monolayer of confluent cells in a micro-stereolithographic 3D printed platform in a single well demonstrating that this microfabrication approach might serve as starting point to future biomedical devices development where high electrode density and defined cell organizations are required for large data procurement and analysis.

The excitement around organoids, spheroids and assembloids makes it imperative to develop new platforms that interface with such cell constructs in a bioplate configuration to enable HT assays. This is challenging since high spatial and temporal resolution are required to monitor cardiac electrophysiology. Adequate definition of size, shape, and distribution of 3D structures (i.e. nano- and microelectrodes) that interface with electrogenic entities is vital to couple electronic platforms to electrical signaling of the biological entity [11]. Such HT configurations scramble microfabrication and packaging efforts due to their awkward size. 3D printing of bioplates [12] is a

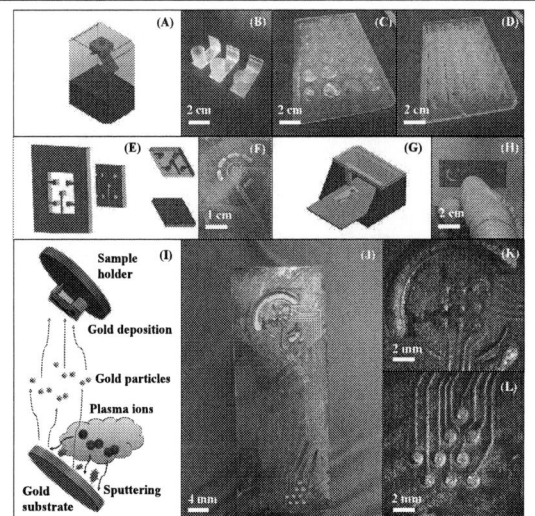

Figure 1: Microfabrication process flow schematic and optical images. (A-D) 3D printing of single well and multi-well organoid bioplates. (E) Silver ink casting onto 3D printed substrate. Ink-cast 3D MEA from the bottom view (F). (G-H) Stencil mask processing using precision cutter plotter on Kapton®. (I-L) Sputter deposition (Au) on MEA (K) and contact pads (L) areas.

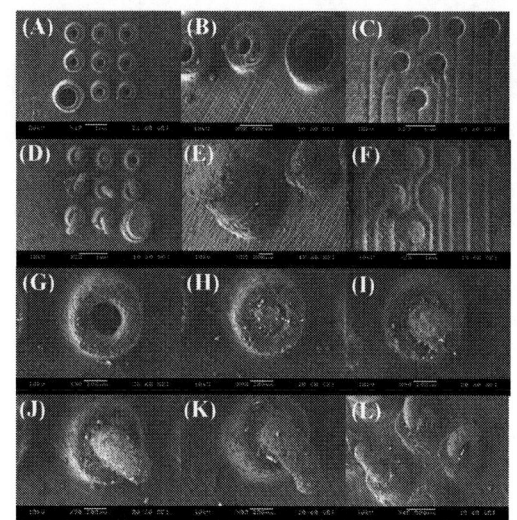

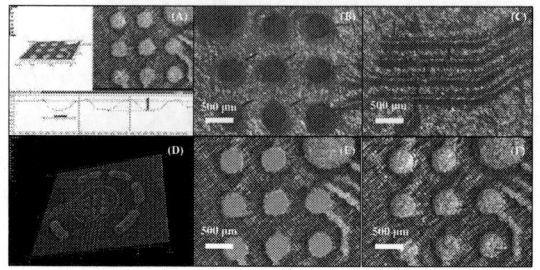

Figure 2: SEM analysis of unfilled (A-C) and filled (D-F) devices on 3D MEA and contacts pads areas. (G-K) Spaghetti-shaped microelectrode formation from empty microtower (G) to fully developed electrode (K). (L) Multiple spaghetti 3D electrodes.

Figure 3: Laser confocal characterization. (A) 3D profile of metallized MEA. (B-C) Physical characterization of microtower diameters and microchannel widths. (D) 3D surface of the MEA area from the bottom side of the bioplate. (E) Surface roughness characterization of the deposited metals (Ag+Au) on the back side of the microelectrodes. Blue circles represent the region of interest. (F) Physical characterization of microelectrode diameters after Ag+Au metallization.

flexible approach that can overcome complex, highly specialized, and multi-step microfabrication methods. In this work, a monolithically integrated, HT bioplate is fabricated with double metallized 3D microelectrodes for cardiac organoids to enhance electrical interfacing.

MATERIALS AND METHODS
Bioplate and 3D MEA Design

High-Throughput 24-well bioplates and single-well devices were designed using CAD software (Solidworks 2022, Dassault Systèmes) and micro-stereolithographically 3D printed (Form 3, Formlabs) as depicted in **Figure 1A-D**. The topside of the bioplate contains a 3x3 array of 3D electrodes inside each well to record and stimulate cardiac organoids. 3D microtowers (W:400 μm; H:500 μm) were used as vertical through vias and supporting structures for self-insulated [13] and metallized electrodes. The backside microchannels (W:150 μm; D:500 μm) served as traces connecting the microelectrodes to contact pads.

Two-step Metallization

Microelectrodes and traces metallization was a two-step process starting with Ag paste (EP3HTSMed, Masterbond) ink-casting (**Figure 1E-F**) [14], followed by sputter metallization (Au) using a pre-designed, plotter cut Kapton® stencil mask (**Figure 1G-I**). Precise cutting of Kapton® was carried out with an automatic plotter cutter (Cameo 4, Silhouette) using settings – Force:10, Speed:1, Passes:1, Acceleration:1, and Blade Depth:2. Gold was deposited using a sputter coating system (EMS150T ES, Quorum Technologies) under vacuum conditions of 2×10^{-3} mbar. This two-step metallization process allowed 150 nm Au to encapsulate Ag layer avoiding flaking of the cured paste, shorting, etc. Fully defined 3D MEA and contact pads to interface with pogo pins of Axion Maestro HT electrophysiology system are depicted in **Figure 1J-L**.

Scanning Electron Microscopy and Optical Imaging

Optical micrographs were captured (Annlov Digital Microscope) to identify fully open microtowers and microchannels, and scanning electron microscopy (SEM JSM-6480, JEOL) was used to analyze the surface structure of 3D microelectrodes and traces. Lastly, a 3D surface profiler (VK-X3000, Keyence) allowed the characterization of design to device and surface roughness.

Impedance Analysis

EIS was performed using a vector network analyzer (Bode 100, Omicron Lab) at Single-Port configuration in the 1 Hz - 10 MHz range with phosphate buffer saline

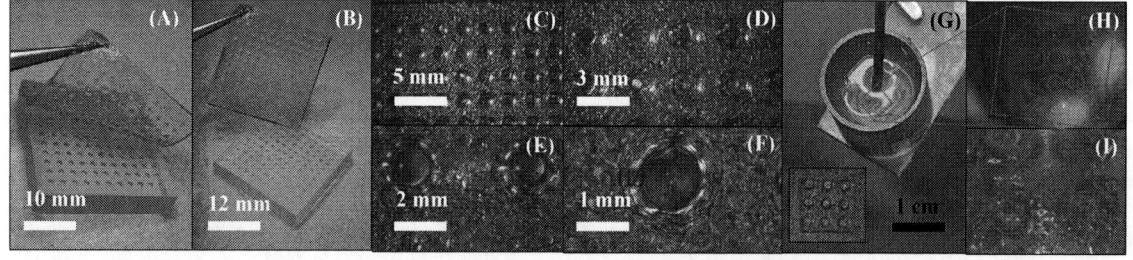

Figure 4: Synthetic organoid development and testing. (A-B) PDMS/Polystyrene-based biocompatible layer being released from 3D printed mold to mimic a network of 100 organoids (1 mm diameter each). (C-F) Close-up images of synthetic organoids functionalized with 10μm size blue polystyrene beads, which are clearly visible on the network. (G-I) Single-well electrochemical impedance setup implementing the organoid network showcased in the (G) inset. Enclosed in dotted red circles are integrated synthetic organoids atop the 3D microelectrodes.

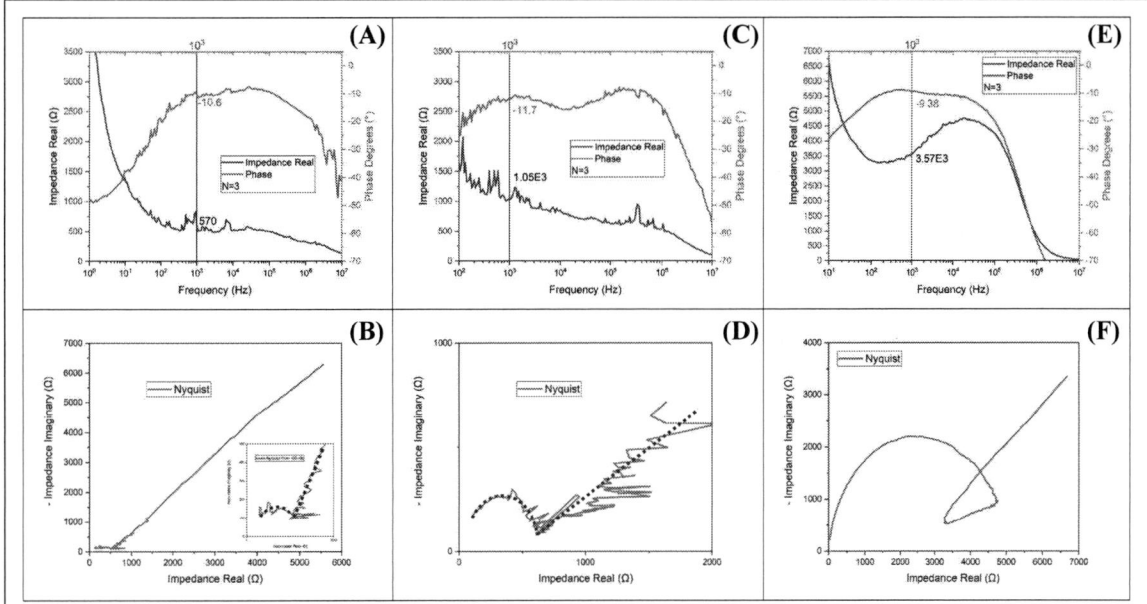

Figure 5: Electrochemical impedance response of the 3D MEA. (A-B) and (C-D) depicting Bode/Nyquist plot pairs from PBS solution/electrode interface at frequency intervals within the 1Hz-10MHz range. (E-F) pair depicts increased impedance signature @1KHz, charge exchange ($R \Leftrightarrow O + ze^-$) on the semicircle kinetic region, and diffusion behavior at higher frequencies with synthetic organoids.

(PBS) as electrolyte and synthetic organoids atop the 3D electrodes. Values of magnitude and phase at the electrophysiologically relevant 1 kHz frequency were recorded.

Synthetic Organoids

Synthetic organoids were fabricated using a Polydimethylsiloxane/Polystyrene (PDMS/PS) bio compatible composite. To create a 100 organoids network, 3 grams of PDMS (10:1 ratio) were mixed with 20 µL blue PS microbeads. The mixing was layered onto a 3D printed array of 100 organoid-like cavities and cured at 70°C for 20 mins. A square piece of the synthetic organoids network was placed atop the 3D MEA and EIS was performed to investigate the impact of these structures atop the MEAs.

RESULTS

Figure 2 depicts the conductive, fully cured, and Au coated silver paste *forming the 3D spaghetti-shaped* (due to geometry and casting force) *microelectrodes to interface with organoids*. **Figure 3** details precision definition of traces and electrodes on the bioplate before and after metallization. Surface roughness was evaluated to be *as low as 6.1μm* (**Figure 3E**) in areas where Au was deposited onto Ag. Design to device for the narrowest features in the microfabrication was *5.75% (lowest)* for the

3D printed microtowers, and *7.33% (highest)* for the microchannels as presented in **Table 1**.

EIS for the 3D microelectrodes was *570 Ω / -10.6°* and *1.05 kΩ / -11.7°* at 1 kHz frequency for spectra from 1 Hz/ 100 Hz up to 10 MHz respectively. Upon addition of the synthetic organoids network (**Figure 4G-I**), *electrical response increased to 3.57 kΩ / -9.38°* as showed in Bode plots from **Figure 5 Top** indicating feasibility for cardiac organoids monitoring. This value is similar to other reported on electroactive cell studies @ 1 kHz frequency [15]. At this frequency, barrier integrity, and permeability of the cellular construct can be further analyzed and compared to other samples on the HT bioplate in a single run. *A well-defined biological response with kinetic (semicircle) and diffusion (tilted line) components is demonstrated* on the Nyquist plots **Figure 5 Bottom** which is typical for polarizable electrodes [16].

CONCLUSIONS

This paper reports for the first time to our knowledge a HT bioplate microfabrication process using 3D printing and multi-step metallization (ink-casting and sputtering) for *in-vitro* studies of cardiac organoids through 3D MEAs. Upon addition of a synthetic organoid network to the MEA, impedance at 1 kHz increased approximately 3x from

Table 1. Surface roughness and deviation on the microfabrication process of a representative single-well device.

Samples (N)	Feature	Mean Surface Roughness (μm)	Mean Width/Diameter Design (μm)	Mean Width/Diameter Device (μm)	Error (%)	Microfabrication Process
8	Metallized microtower	6.1	400	429	7.25	Ink casting + Metal sputtering
8	Non-metallized microtower	Not applicable	400	377	5.75	3D printing
4	Microchannels	Not applicable	150	139	7.33	3D printing

conditions with no organoid. This impedance response at the interface level, can provide insights about cardiac organoids electrophysiology and its value in preclinical models. Finally, the right selection of technology (3D printing), and optimization of the microfabrication process (additive manufacturing + double metallization) can be performed by only adjusting a few parameters in a rapid prototyping environment. Such a simple hybrid microfabrication technique will enable functional, ready to use devices for high-throughput *in-vitro* assays.

ACKNOWLEDGEMENTS

The authors would like to acknowledge Materials Characterization Facility (MCF) at the University of Central Florida (UCF) for access to instruments as part of the device fabrication and characterization. The authors also would like to acknowledge Florida High Tech Corridor Match grants that provided funding for this work; and NBSS, NSTC, SGA, and CGS from UCF that provided funding for presenting this work at MEMS 2023.

REFERENCES

[1] D. Zhao, W. Lei, and S. Hu, "Cardiac organoid — a promising perspective of preclinical model," *Stem Cell Research & Therapy,* vol. 12, no. 1, p. 272, 2021/05/06 2021, doi: 10.1186/s13287-021-02340-7.

[2] K. Tasnim and J. Liu, "Emerging Bioelectronics for Brain Organoid Electrophysiology," *Journal of Molecular Biology,* vol. 434, no. 3, p. 167165, 2022/02/15/ 2022, doi: https://doi.org/10.1016/j.jmb.2021.167165.

[3] M. A. C. Williams, D. B. Mair, W. Lee, E. Lee, and D.-H. Kim, "Engineering Three-Dimensional Vascularized Cardiac Tissues," *Tissue Engineering Part B: Reviews,* vol. 28, no. 2, pp. 336-350, 2022/04/01 2021, doi: 10.1089/ten.teb.2020.0343.

[4] Y. S. Zhang *et al.*, "From cardiac tissue engineering to heart-on-a-chip: beating challenges," *Biomedical Materials,* vol. 10, no. 3, p. 034006, 2015/06/11 2015, doi: 10.1088/1748-6041/10/3/034006.

[5] A. Kalmykov *et al.*, "Organ-on-e-chip: Three-dimensional self-rolled biosensor array for electrical interrogations of human electrogenic spheroids," *Science Advances,* vol. 5, no. 8, p. eaax0729, doi: 10.1126/sciadv.aax0729.

[6] Y. Park *et al.*, "Three-dimensional, multifunctional neural interfaces for cortical spheroids and engineered assembloids," *Science Advances,* vol. 7, no. 12, p. eabf9153, doi: 10.1126/sciadv.abf9153.

[7] B. Joddar *et al.*, "Engineering approaches for cardiac organoid formation and their characterization," *Translational Research,* 2022/08/19/ 2022, doi: https://doi.org/10.1016/j.trsl.2022.08.009.

[8] N. Gritti *et al.*, "MOrgAna: accessible quantitative analysis of organoids with machine learning," *Development,* vol. 148, no. 18, 2021, doi: 10.1242/dev.199611.

[9] H. A. Strobel, A. Schultz, S. M. Moss, R. Eli, and J. B. Hoying, "Quantifying Vascular Density in Tissue Engineered Constructs Using Machine Learning," (in English), *Frontiers in Physiology,* Methods vol. 12, 2021-April-27 2021, doi: 10.3389/fphys.2021.650714.

[10] C. M. Didier, A. Kundu, J. M. Castro, C. Hart, and S. Rajaraman, "Compact Micro-Stereolithographic (μSLA) Printed, 3D Microelectrode Arrays (3D MEAS) with Monolithically Defined Positive and Negative Relief Features For in Vitro Cardiac Beat Sensing," in *2022 IEEE 35th International Conference on Micro Electro Mechanical Systems Conference (MEMS)*, 9-13 Jan. 2022 2022, pp. 325-328, doi: 10.1109/MEMS51670.2022.9699662.

[11] F. Santoro *et al.*, "Interfacing Electrogenic Cells with 3D Nanoelectrodes: Position, Shape, and Size Matter," *ACS Nano,* vol. 8, no. 7, pp. 6713-6723, 2014/07/22 2014, doi: 10.1021/nn500393p.

[12] J. M. Castro, A. Kundu, A. Rozman, and S. Rajaraman, "Investigation of the Effect of Printing Angle and Device Orientation on Micro-Stereolithographically Printed, and Self-insulated, 24-well, High-Throughput 3D Microelectrode Arrays," in *2021 IEEE Sensors*, 31 Oct.-3 Nov. 2021 2021, pp. 1-4, doi: 10.1109/SENSORS47087.2021.9639637.

[13] J. M. Castro and S. Rajaraman, "Experimental and Modeling Based Investigations of Process Parameters on a Novel, 3D Printed and Self-Insulated 24-Well, High-Throughput 3D Microelectrode Array Device for Biological Applications," *Journal of Microelectromechanical Systems,* pp. 1-14, 2022, doi: 10.1109/JMEMS.2022.3160663.

[14] A. Kundu, T. Ausaf, and S. Rajaraman, "3D Printing, Ink Casting and Micromachined Lamination (3D PICLμM): A Makerspace Approach to the Fabrication of Biological Microdevices," (in eng), *Micromachines,* vol. 9, no. 2, p. 85, 2018, doi: 10.3390/mi9020085.

[15] D. A. Soscia *et al.*, "A flexible 3-dimensional microelectrode array for in vitro brain models," *Lab on a Chip,* 10.1039/C9LC01148J vol. 20, no. 5, pp. 901-911, 2020, doi: 10.1039/C9LC01148J.

[16] S. Wang, J. Zhang, O. Gharbi, V. Vivier, M. Gao, and M. E. Orazem, "Electrochemical impedance spectroscopy," *Nature Reviews Methods Primers,* vol. 1, no. 1, p. 41, 2021/06/10 2021, doi: 10.1038/s43586-021-00039-w.

CONTACT

*Swaminathan Rajaraman, tel: +1-407-823-4339; swaminathan.rajaraman@ucf.edu

OIL-SEALED RGD-MODIFIED HYDROGEL MICROWELL ARRAY WITH SIZE-SELECTIVE PERMEATION FOR ANALYSIS ON EXOSOMES FROM SINGLE CELLS

Chisaki Yamagata[1], Shun Itai[1], Yuta Kurashina[2],
Makoto Asai[3], Ayuko Hoshino[4], and Hiroaki Onoe[1]

[1]Faculty of Science and Technology, Keio University, JAPAN,
[2]Faculty of Engineering, Tokyo University of Agriculture and Technology, JAPAN,
[3]Keio University Global Research Institute, Keio University, JAPAN and
[4]School of Life Science and Technology, Tokyo Institute of Technology, JAPAN

ABSTRACT

We propose an oil-sealed arginine-glycine-aspartate (RGD)-modified hydrogel microwell array for analyzing exosomes secreted from single cells. Our device realizes the collection of exosomes from single cells in parallel. The size-selective permeation of the RGD-modified alginate hydrogel allows both stable cell culturing of single or few cells in closed wells and the confinement of exosomes secreted in each space. The nutrients (< 20 nm) that are necessary for cell culture can pass through the hydrogel, while exosomes (30-150 nm) do not permeate the walls of microwells. The cell culture property and the permeability of our device were examined, showing the capability to collect exosomes from single cells. We believe that our device would contribute to understanding the mechanisms of various diseases.

KEYWORDS

Exosome, Extracellular vesicle, Single-cell analysis, Microwell array, Alginate, Hydrogel, Arginine-glycine-aspartate (RGD) motif

INTRODUCTION

Exosomes are the key to revealing cancer metastatic mechanisms and developing a new treatment [1,2], as tumor-derived exosomes were discovered to prepare pre-metastatic niches [3]. To understand the relationship between cancer metastasis and tumor-derived exosomes, single-cell-level exosome secretion analysis has become of great interest because each cell in a heterogeneous cell population in tumor tissues has turned out to secrete a unique distribution of various exosomes (Fig. 1 (a)).

Microfluidic technologies have been powerful tools to analyze single cells thanks to their high throughputs and high automatability [4,5]. So far, microdroplet-based single-cell analysis [6] and microwell array for single-cell [7,8] have been developed. However, previous devices have never analyzed exosomes secreted by single cells because of the difficulty in isolating the exosomes and the continuous cell culture simultaneously.

Here, we propose an oil-sealed hydrogel microwell array for analyzing exosomes secreted from single cells (Fig. 1 (b)). Our device realizes the collection of exosomes from single cells in parallel. The size-selective permeation of the RGD-modified alginate hydrogel allows both stable cell culturing of single or few cells in closed wells and the confinement of exosomes secreted in each space. The nutrients (< 20 nm) that are necessary for cell culture can pass through the hydrogel, while exosomes (30-150 nm) do not permeate the walls of microwells.

METHODOLOGY

Cell culture experiments for the material design

To design the material of the microwell array, four different types of calcium-alginate hydrogel were examined for cell culture capability (Fig. 2 (a)). We set two parameters for the experiment: the presence of RGD modification and concentrations of alginate hydrogel. First, four sheets of calcium-alginate hydrogel in different material conditions were fabricated: 2 wt% calcium-alginate, 4 wt% calcium-alginate, 2 wt% RGD-modified calcium-alginate, and 4 wt% RGD-modified calcium-alginate. Then, Human breast cancer MDA-MB-231 organotropic lines 4175 (4175 cells) were seeded above the hydrogel sheets. The concentration of seeded cells was 3.0 x 10^4 cells/mL. After 12 hours of culture, aspect ratios of seeded cells were calculated to examine the cell adhesion properties of four types of calcium-alginate hydrogel.

Comparison of cell culture media

We compared normal Dulbecco's Modified Eagle Medium (DMEM) and conditioned medium in terms of the ability to culture a small density of cells (Fig. 2 (b)). The

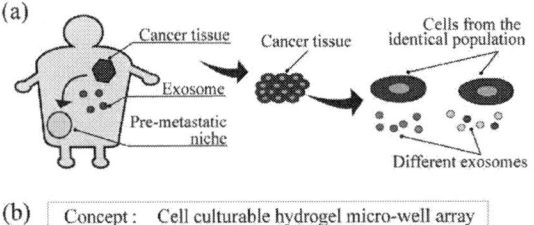

(a)

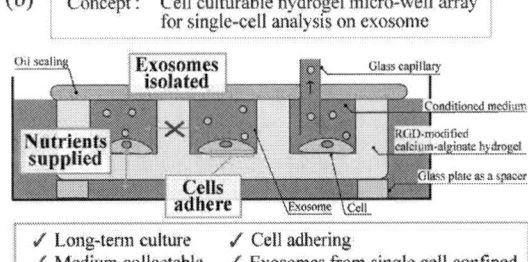

(b)

Figure 1: (a) Concept of the importance of analyze exosomes from single cells. (b) Concept of oil-sealed RGD-modified hydrogel microwell array for analyzing exosomes secreted from single cells.

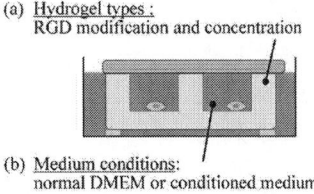

(a) Hydrogel types :
RGD modification and concentration

(b) Medium conditions:
normal DMEM or conditioned medium

Figure 2: Concept of experimental parameters for designing hydrogel microwell array. (a) Hydrogel type in terms of the presence of RGD modification and concentration of alginate. (b) Medium condition.

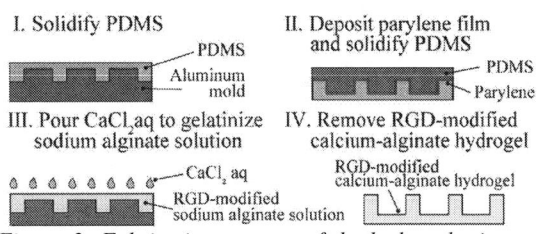

I. Solidify PDMS

II. Deposit parylene film and solidify PDMS

III. Pour CaCl₂aq to gelatinize sodium alginate solution

IV. Remove RGD-modified calcium-alginate hydrogel

Figure 3: Fabrication process of the hydrogel microwell array.

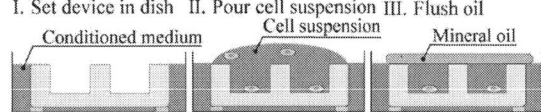

I. Set device in dish   II. Pour cell suspension   III. Flush oil

Figure 4: Method of cell culture examination.

conditioned medium was prepared by collecting DMEM from cell culture dishes where 4175 cells were cultured. First, 4175 cells were seeded in normal DMEM and in conditioned medium. Then, the cells in each medium were observed. This experiment was held in two conditions of cell density: 2.5×10^2 cells/mL and 5.0×10^2 cells/mL.

Fabrication of the microwell array

The microwell array was fabricated by repeated micromolding (Fig. 3). Firstly, the aluminum mold was prepared by machining. Secondly, poly-dimethylpolysiloxane (PDMS) was solidified on the aluminum mold to fabricate a first PDMS mold. After peeling the first PDMS mold off from the aluminum mold, a parylene layer was deposited on the first PDMS mold. Then, additional PDMS prepolymer was solidified on the first PDMS to obtain a second PDMS mold. Lastly, RGD-modified sodium alginate solution was poured into the second PDMS mold. Then, calcium chloride aqueous solution was poured to gelatinize the sodium alginate solution. The fabrication process was completed by peering off the hydrogel.

Permeation analysis

We examined the permeability of our device. We used three fluorescent substances with different diameters: fluorescein isothiocyanate (FITC)-dextran (6.6 nm), fluorescently labeled liposomes (90-120 nm), and fluorescently labeled exosomes (30-150 nm). FITC-dextran is the model of nutrients necessary for cell culture.

First, fluorescent substances were poured over the hydrogel microwell entirely. Secondly, oil was flushed right above the microwell. We took fluorescence microscopic images right after flushing the oil and two

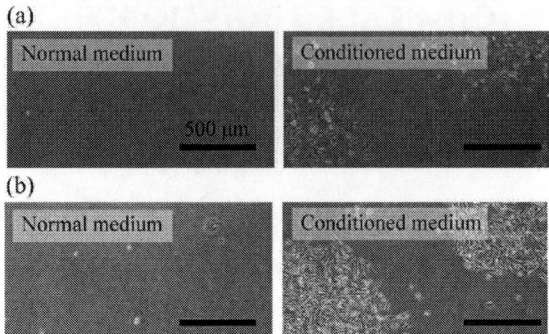

Figure 5: Results of cell culture experiments for comparing cell culture media. (a) Experiment with 2.5×10^2 cells/mL cell concentration. (b) Experiment with 5.0×10^2 cells/mL cell concentration.

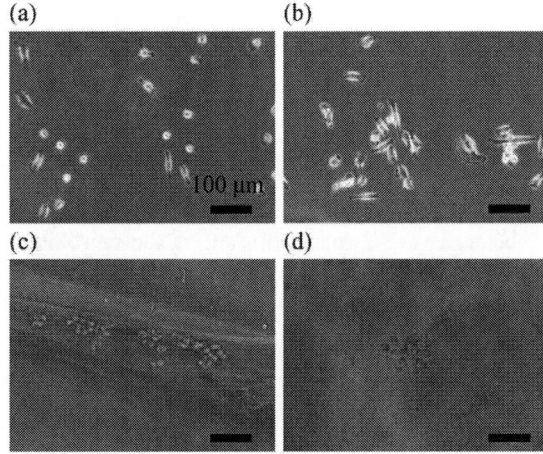

Figure 6: Results of the cell culture experiments for the material design. (a) RGD-modified calcium-alginate with 2 wt% alginate. (b) RGD-modified calcium-alginate with 4 wt% alginate. (c) Calcium-alginate with 2 wt% alginate. hours later. Then, the fluorescence intensities of the images were analyzed.

Cell culture examination

To examine the cell culture capability of our device, a small number of 4175 cells were cultured in our device (Fig. 4). First, 4175 cells were seeded in the device with conditioned medium. Then, oil was flushed over the microwell. We observed the cells with phase-contrast microscopes.

RESULTS AND DISCUSSION
Cell culture experiments for the material design

Cells on the 2 wt% calcium-alginate and 4 wt% calcium-alginate did not adhere to the hydrogel. On the other hand, cells on the 2 wt% RGD-modified calcium-alginate and 4 wt% RGD-modified calcium-alginate did adhere to the hydrogel. These results showed that the RGD modification is necessary for cell culture (Fig. 5 (a-d)). Comparing cells on the 2 wt% RGD-modified calcium-alginate and the 4 wt% RGD-modified calcium-alginate, the latter has larger aspect ratio, meaning stronger adhesion. From these experimental results, we confirmed that RGD-modified calcium-alginate with 4 wt% alginate is suited as our device.

Figure 7: Fabricated hydrogel microwell array.

Comparison of cell culture media

The cells cultured with normal DMEM did not adhere to the microwell at either of the cell concentrations. However, the cells cultured with the conditioned medium adhered to the microwell at either of the cell concentrations (Fig. 6 (a,b)). These results indicated that conditioned medium is appropriate to culture the low density of the 4175 cells. Therefore, conditioned medium was used in the cell culture experiments in our device.

Fabrication of the microwell array

RGD-modified microwell array was successfully fabricated using repeated micromolding. The shape and size were mostly completed as designed (Fig. 7).

Permeation analysis

Permeation characteristics were evaluated with FITC-dextran, fluorescent liposomes, and fluorescent labeled exosomes (Fig. 8 (a-e)). All substances were confined in microwells right after flushing oil. After two hours, liposomes and exosomes were still confined in the microwell. On the other hand, the FITC-dextran was diffused out of the microwell.

The results proved that the nutrients (< 20 nm) that are necessary for cell culture can pass through the hydrogel, while liposomes (90-120 nm) and exosomes (30-150 nm) do not permeate the walls of microwells. This difference in permeation among different substances makes it possible to achieve our aim of analyzing exosomes from single cells. In other words, the size-selective permeation of the RGD-modified alginate hydrogel allows both stable cell culturing of single or few cells in closed wells and the confinement of exosomes secreted in each space.

Cell culture examination

4157 cells were successfully adhered to and proliferated in the microwell (Fig. 9). These results showed the device's capability to culture a small number of cells in a closed space, suggesting the capability to culture single cells.

CONCLUSIONS

We proposed an RGD-modified hydrogel microwell array for analyzing exosomes secreted from single cells. The experimental results show that the size-selective permeation of the RGD-modified alginate hydrogel allows both stable cell culturing of single or few cells in closed wells and the confinement of exosomes secreted in each space. Also, we successfully cultured a small number of cells in hydrogel microwell. These results indicate that our

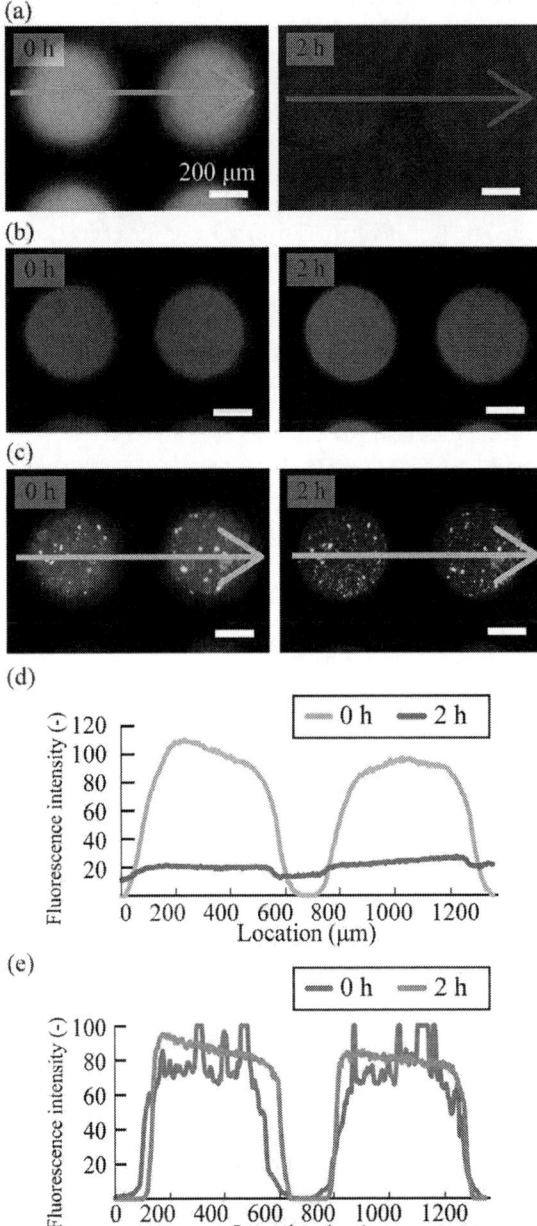

Figure 8: Results of Permeation analysis. Fluorescent images of (a) FITC-dextran, (b) fluorescent liposomes, and (c) fluorescent labeled exosomes. The fluorescence intensities of the fluorescent images of (d) FITC-dextran, and (e) fluorescent labeled exosomes.

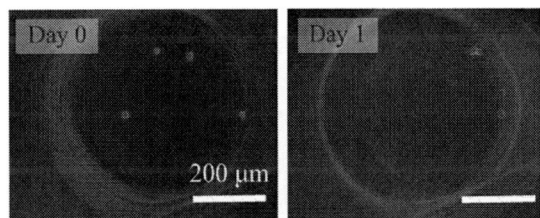

Figure 9: Results of cell culture examination in hydrogel microwell array.

device can be an effective parallel platform to collect the single cell derived exosome for analysis of exosome characteristics. We believe that our device would contribute to understanding the mechanisms of various diseases.

ACKNOWLEDGEMENTS

This work was partly supported by Research Project Keio 2040 (Creativity Initiative) at Keio University Global Research Institute, Japan.

REFERENCES

[1] W. S. Toh, *et al.,* "MSC exosome as a cell-free MSC therapy for cartilage regeneration: Implications for osteoarthritis treatment," *Semin. Cell Dev. Biol.,* vol. 67, pp. 56–64, 2017.

[2] Ladan Mashouri, *et al.,* "Exosomes: composition, biogenesis, and mechanisms in cancer metastasis and drug resistance" *Molecular cancer,* vol. 18, 75, 2019.

[3] A. Hoshino, *et al.,* "Tumour exosome integrins determine organotropic metastasis", *Nature*, vol. 527, pp. 329-335, 2015.

[4] K. Klepárník and F. Foret, "Recent advances in the development of single cell analysis-A review," *Anal. Chim. Acta,* vol. 800, pp. 12–21, 2013.

[5] T. W. Murphy, *et al.,* "Recent advances in the use of microfluidic technologies for single cell analysis," Analyst, vol. 143, *Royal Society of Chemistry*, pp. 60–80, 2018.

[6] J. Edd, *et al.,* "Controlled encapsulation of single-cells into monodisperse picolitre drops", *Lab on a Chip*, vol. 8, pp. 1262-1264, 2008.

[7] A. Revzin, *et al.,* "Development of a microfabricated cytometry platform for characterization and sorting of individual leukocytes", *Lab on a Chip*, vol. 5, pp. 30-37, 2005.

[8] D. Di Carlo, *et al.,* "Dynamic single cell culture array," *Lab on a Chip*, vol. 6, pp. 1445–1449, 2006.

CONTACT

*C.Yamagata, tel: +81-045-566-1507; chisaki.ymgt@keio.jp

PICKING SINGLE CELLS FROM 10 ML SAMPLE BASED ON A MICROFILTRATION- LIFT COMBINATION PLATFORM

Qingmei Xu[1, #], Yuntong Wang[2,3, #], Xiao Ma[4], Hang Li[5], Ying Xue[5], Yi Zhang[1], Songtao Dou[1], Huan Wang[2], Bei Li[2,5], Wei Wang[1,6,7*]*

[1] School of Integrated Circuits, Peking University, Beijing, CHINA
[2] State Key Laboratory of Applied Optics, Changchun Institute of Optics, Fine Mechanics and Physics, Chinese Academy of Sciences, Changchun, CHINA
[3] University of Chinese Academy of Sciences, Beijing, CHINA
[4] Hangzhou Branemagic Medical Technology Co. Ltd., Hangzhou, CHINA
[5] Hooke Laboratory, Changchun, CHINA
[6] National Key Lab of Micro/Nano Fabrication Technology, Beijing, CHINA
[7] Beijing Advanced Innovation Center for Integrated Circuits, Beijing, CHINA
# The two authors contributed equally to this work.

ABSTRACT

This paper reported a microfiltration-laser induced forward transfer (microfiltration-LIFT) combination platform that fulfilled an automatic single-cell picking-up capability from a large volume sample (more than 10 mL). Cells were first rapidly enriched by a gravity-driven microfiltration based on previously-reported PERFECT filters, and the captured-cells on the filter could then be accurately picked up at single-cell resolution by the LIFT system using a focused laser with a spot size of ~10 μm. This platform is a promising method to pick up single cells from large volume sample for subsequent heterogeneity molecular analyses.

KEYWORDS

Single cell picking, Filtration, Laser-induced forward transfer (LIFT)

INTRODUCTION

Circulating tumor cells (CTCs) are rare cells with extremely low abundance in peripheral blood and have shown extensive molecular heterogeneity [1]. So far, numerous CTC isolation methods, such as gradient centrifuge [2], filtration [3, 4], microfluidic technology [5, 6] and marker immunoaffinity [7, 8] have been proposed and achieved the cell capture. However, the CTC-related research was relatively stagnant at the stage of the quantitative characterization of CTCs due to the contamination introduced by non-specifically captured white blood cells (WBCs) [9]. As it is known that just enumeration of CTCs could not reflect the heterogeneity of tumor biology.

To unravel the heterogeneity of CTCs, there is an urgent need to develop single-CTC picking methods to better understand the role of individual CTC at molecular biology level. Currently, the working mechanisms of single-CTC picking include manual micro-manipulation [10], laser capture microdissection (LCM) [11, 12], mechanical cell-picker [13], and laser-induced forward transfer (LIFT) [14]. Unfortunately, the efficiency of widely used manually piking-up of cells by manipulator is low, which dramatically affects its real application [15]. LCM and mechanical cell-picker were reported a tendency to collect more than one cell per pick due to the large cutting area. In contrast, the laser-induced forward transfer (LIFT) technology could automatic picking the single cells under high resolution. Therefore, LIFT was a promising way to picking single-CTC from the pretreated samples.

In this study, we developed a novel microfiltration-laser induced forward transfer (microfiltration-LIFT) combination platform that allow for high-throughput isolation and automatic picking of single CTC from a large volume sample (more than 10 mL). The microfiltration-LIFT platform coupled the bilayer of micropore-arrayed filters (PERFECT filters) with fluorescence recognition LIFT system. Beyond enumeration, the preliminary performance of the platform showed that the target cells could be rapidly isolated and concentrated by the micropore-arrayed filters under gravity and precisely picked up with single-cell resolution using the LIFT technique in several seconds. The microfiltration-LIFT platform offered a unique pathway for efficient single-CTC picking, which lay the foundation for biological properties analyses of CTC.

MATERIALS AND METHODS

System Design

The microfiltration was composed of a contacted bilayer of micropore-arrayed filters (PERFECT filters) with two different pore diagonal diameters [4]. The upper layer with a large pore size could isolate the liquid layer of the trapped cell, eliminating the stiction of the liquid layer to the captured cells, and thus reduce the energy required for cell transfer during the LIFT, while the lower layer with a small pore size was used for size-based cell separation and enrichment. A 50 nm TiN was deposited on the surface of the lower micropore array (see Figure 1(f)), which was used for LIFT-based single-cell picking-up after enrichment (see Figure 1 (a)). The sample with spiking HepG2 cells was firstly filtrated using the contacted-bilayer of micropore-arrayed filters (see Figure 1(b-c)). After filtration, the microfiltration device was immediately bonded onto a transparent glass slide of the 3D platform in the LIFT system with the cells facing down (see Figure 1(d)). Finally, the single cell was identified and transferred by the LIFT from the microfiltration device to a culture dish.

IEEE MEMS 2023, Munich, GERMANY
15 - 19 January 2023

978-1-6654-9309-3/23 $31.00 © 2023 IEEE

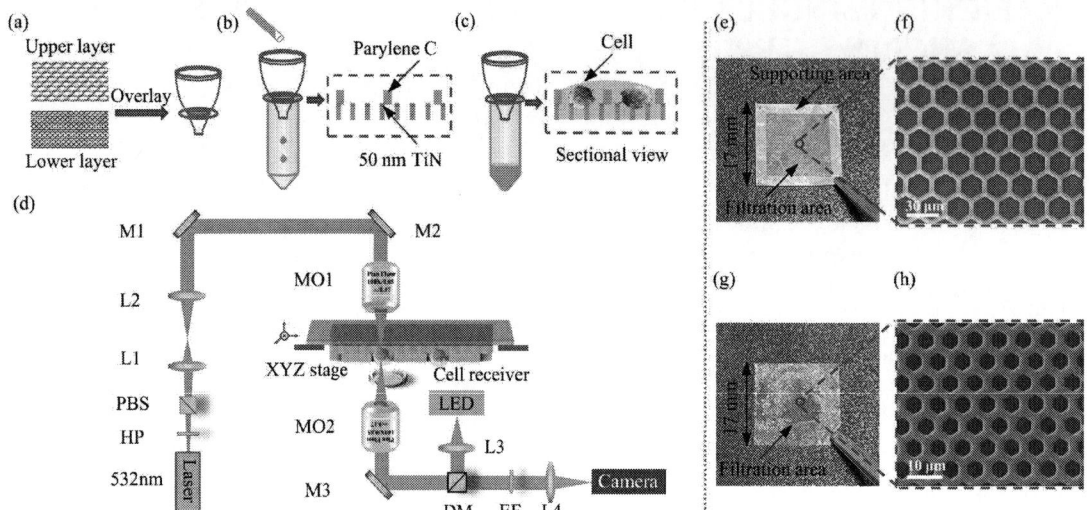

Figure 1: The schematic (a-d) of the microfiltration-laser induced forward transfer (microfiltration-LIFT) combination platform: (a) package of the contacted bilayer of micropore-arrayed device, (b-c) size-based cell separation and enrichment, (d) identification and picking-up of the single cell. L1 to L4, lenses; HP, half-waveplate; PBS, polarizing beam splitter; M1 to M3, mirrors; DM, dichroic mirror; EF, emission filter; MO1 to MO2, microscopy objectives. The photograph and SEM images of the upper(e-f) and lower(g-h) PERFECT filters.

Simulation

The temperature simulation was implemented in the Heat Transfer Module from COMSOL Multiphysics 5.6. Heat transfer in the solid and liquid domain was modeled by the laser acting on the TiN metal film caused the surrounding liquid to heat up. The following parameters were imposed in the numerical simulation: metal absorption rate, 0.508; metal thickness, 50nm; liquid layer thickness, 5μm; laser pulse time, 5ns. Laser pulse energy $E\_0=400nj$ was set for solving the power density.

Cell Preparation

Human liver cancer cell (HepG2) was used to evaluate the performance of the system. The cells were cultured in Minimum Essential Medium (MEM) (Gibco, ThermoFisher, USA) supplemented with 10% fetal bovine serum (Gibco, ThermoFisher, Australia) at 37 °C in a humidified atmosphere with 5% CO_2.

RESULTS AND DISCUSSION

We performed COMSOL simulation analysis to assess the damage to cells during the single cell picking-up process (Figure 2a). The metal film that laser spotted on was around 2700°C (Figure. 2b), while the liquid 0.6 μm away from the metal layer was less than 27°C within 100 ns (Figure 2b). The transfer time of the pulse laser (6 ns) was far less than 100 ns. The temperature variation in the entire fluid domain was shown in Figure 2c, which indicated the LIFT operation was safe to the target cell.

Figure 3 showed the whole process of microfiltration-LIFT platform for single cell picking-up. The contacted bilayer of micropore-arrayed filters was attached onto the transparent glass slide of the LIFT system after filtration (see Figure 3(a) and (j)). The target cell was identified by fluorescent staining method, as shown in Figure 3(b-c), (f-g) and (k-l). Then the target cell was instantaneously

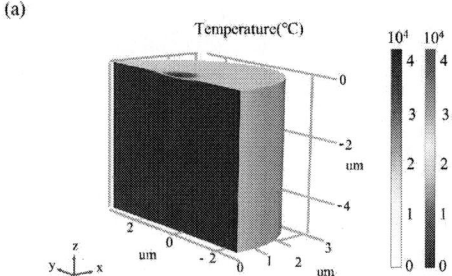

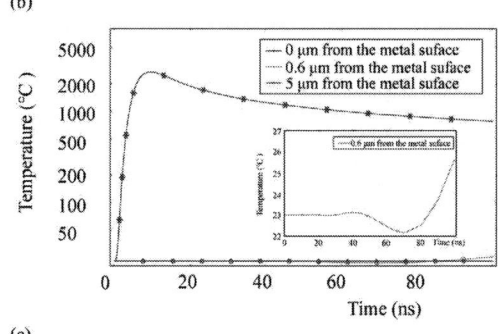

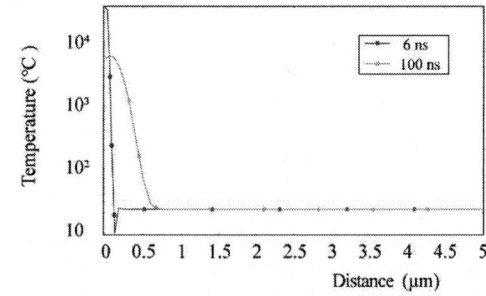

Figure 2: Finite-element analysis of the cell sorting process. The temperature field model during the sorting process (a) and the temperature fields at different locations during the sorting process change over time (b). The maximum temperature of the laser ablated metal film was less than 2700°C, while the liquid 0.6 μm away from the metal layer maintained below 27°C within 100 ns. (c) The temperature variation in the entire fluid domain at different times. Liquid domain temperatures exceeding 0.6um of the laser-ablated metal film were below 27°C.

Monolayer micropore-arrayed filters

Contacted bilayer of micropore-arrayed filters

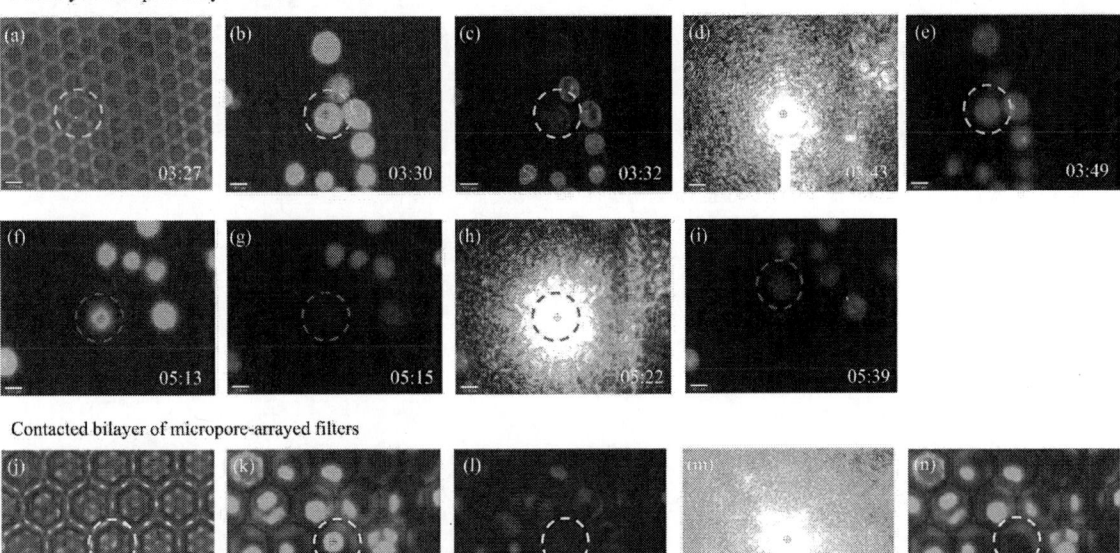

Figure 3: The dynamic process of micropore-LIFT platform for single cell picking. (a-i) The single cell picking based on monolayer micropore-arrayed filters: the targeted cell picking was failed and the cell did not move at the beginning (a-e), while the cell produced a small displacement after a period of time (f-i) due to the liquid layer was decreased over time. (j-p) The single cell picking based on contacted bilayer of micropore-arrayed filters: the targeted cell picking was achieved owing to the upper micropore array which could incise the liquid layer of the trapped cell. The picked single cell was received by the cell receiver (o-p). The cells were prestained with cell tracker green and Hoechst.

transferred off the microfiltration device at 350 nJ in 6 ns (see Figure 3(m)). Figure 3(n) showed that the target cell was picked up successfully and the cell was found on the cell receiver right under the glass slide (see Figure 3(o-p)). The contacted bilayer of micropore-arrayed filters is able to pick up single cells in 30 s using the LIFT system, while the monolayer micropore-arrayed filters could only move the cell in 6 min. As shown in Figure 4, the shape of the isolated CTC was morphologically intact.

CONCLUSIONS

In this work, we successfully developed a novel platform facilitating the single cell isolation and picking from a large volume sample by the contacted bilayer of micropore-arrayed filters and LIFT technique. By using our microfiltration-LIFT platform, the heterogeneity molecular analyses of CTC will be feasible at different cancer staging for the same patients.

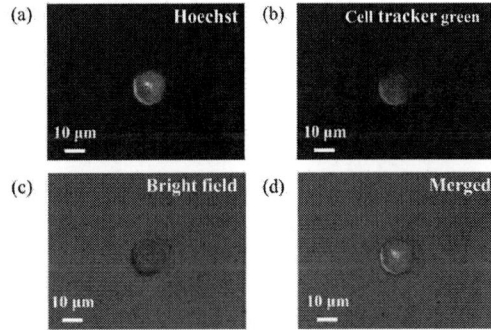

Figure 4: Typical fluorescent images of the intact single-CTC received by the cell receiver after isolation.

ACKNOWLEDGEMENTS

We are grateful for the support from the National Natural Science Foundation of China (Grant No. 82027805, 62104227) and the Strategic Priority Research Program of the Chinese Academy of Sciences (No. XDA22020403). The authors also want to thank the staffs from the National Key Laboratory of Science and Technology on Micro/Nano Fabrication for their help with the fabrication process.

REFERENCES

[1] D. Ishwar et al., "Minimally invasive detection of cancer using metabolic changes in tumor-associated natural killer cells with Oncoimmune probes," *Nature communications,* vol. 13, no. 1, pp. 1-20, 2022.

[2] Z. Shen et al., "Current detection technologies for circulating tumor cells," *Chemical Society Reviews,* vol. 46, no. 8, pp. 2038-2056, 2017.

[3] M. S. Kim et al., "SSA-MOA: a novel CTC isolation platform using selective size amplification (SSA) and a multi-obstacle architecture (MOA) filter," *Lab on a Chip,* vol. 12, no. 16, pp. 2874-2880, 2012.

[4] Y. Liu et al., "2.5-Dimensional Parylene C micropore array with a large area and a high porosity for high-throughput particle and cell separation," *Microsystems & Nanoengineering,* vol. 4, no. 1, pp. 1-12, 2018.

[5] M. He et al., "Integrated immunoisolation and protein analysis of circulating exosomes using microfluidic technology," *Lab on a Chip,* vol. 14, no. 19, pp. 3773-3780, 2014.

[6] S. Nagrath et al., "Isolation of rare circulating tumour cells in cancer patients by microchip technology," *Nature,* vol. 450, no. 7173, pp. 1235-1239, 2007.

[7] H. Im et al., "Label-free detection and molecular profiling of exosomes with a nano-plasmonic sensor," *Nature biotechnology,* vol. 32, no. 5, pp. 490-495, 2014.

[8] E. Racila et al., "Detection and characterization of carcinoma cells in the blood," *Proceedings of the National Academy of Sciences,* vol. 95, no. 8, pp. 4589-4594, 1998.

[9] D. Lin et al., "Circulating tumor cells: Biology and clinical significance," *Signal transduction and targeted therapy,* vol. 6, no. 1, pp. 1-24, 2021.

[10] J. Fröhlich et al., "New techniques for isolation of single prokaryotic cells," *FEMS microbiology reviews,* vol. 24, no. 5, pp. 567-572, 2000.

[11] M. R. Emmert-Buck et al., "Laser capture microdissection," *Science,* vol. 274, no. 5289, pp. 998-1001, 1996.

[12] V. Espina et al., "Laser capture microdissection technology," *Expert review of molecular diagnostics,* vol. 7, no. 5, pp. 647-657, 2007.

[13] M. Kamal et al., "PIC&RUN: An integrated assay for the detection and retrieval of single viable circulating tumor cells," Scientific reports, vol. 9, no. 1, pp. 1-11, 2019.

[14] N. T. Kattamis et al., "Thick film laser induced forward transfer for deposition of thermally and mechanically sensitive materials," *Applied Physics Letters,* vol. 91, no. 17, p. 171120, 2007.

[15] F. Tang et al., "mRNA-Seq whole-transcriptome analysis of a single cell," *Nature methods,* vol. 6, no. 5, pp. 377-382, 2009.

CONTACT

*B. Li, tel: +86-431-86708977; beili@ciomp.ac.cn
*W. Wang, tel: +86-10-62750175; w.wang@pku.edu.cn

A TRANSFER METHOD FOR EMBEDDING CONDUCTIVE FILLERS ON THE SURFACE OF MULTI-SCALE STRUCTURES FOR 3D FLEXIBLE CONDUCTORS

Dongwoo Yoo, Sangmok Kim, Jeonghyeon Hwang, and Joonwon Kim
Pohang University of Science and Technology (POSTECH), Republic of Korea

ABSTRACT

This paper demonstrates a transfer method for embedding conductive fillers (e.g., metal nanowires (MWs) and multi-walled carbon nanotubes (MWCNTs)) on the surface of polymers (e.g., polydimethylsiloxane (PDMS), polyurethane (PU)) with various multi-scale structures to fabricate 3D flexible conductors. Conductive fillers were embedded on the surfaces of macro- and microstructures considering the optimal surface energy conditions for transferring spray-coated conductive fillers from 3D molds to the polymer networks. This transfer method achieves highly conductive and robust 3D flexible conductors without changing the mechanical properties (e.g., elastic modulus, ductility, and brittleness), generating cracks in the conductive layer, or losing the conductive fillers under repeated pressure. Consequently, the developed conductors applied to pressure sensors exhibited sensing stability by maintaining a constant initial value.

INTRODUCTION

Various flexible pressure sensors, such as those in capacitive, piezoresistive, and triboelectric devices, require 3D flexible conductors with multi-scale structures and various shapes for improved sensor performance (e.g., response time, sensitivity, and dynamic range) [1]. Flexible conductors with high mechanical durability and conductivity are crucial for the widespread application of these devices but are challenging to realize. Previous studies have reported conductor fabrication methods by addition of conductive fillers to (or directly incorporating conductive fillers into) the polymer matrix and by spray-coating the conductive fillers directly on the surface of polymers [2, 3]. The first method facilitates the fabrication of conductive composites but limits the ductility, causing brittleness of the composites [4]. By contrast, the spray-coating technique results in high conductivity while maintaining the mechanical properties of the polymers. However, this method also has several drawbacks: 1) cracks occur in the conductive layer under repeated pressure, resulting in increased resistance, and 2) conductive fillers are loosely attached to the surface of the polymers, resulting in filler loss. Therefore, a transfer method for embedding conductive fillers on the surface of polymers would be beneficial for the fabrication of robust conductors that maintain the high conductivity and inherent mechanical properties of the polymers.

This study demonstrates a transfer method for embedding MWs on the surfaces of flexible polymers with various multi-scale structures to fabricate 3D flexible conductors. The MWs were successfully embedded on the surfaces of the macro- and microstructures under the optimal surface energy conditions for transferring the spray-coated MWs from 3D molds to the polymer networks. Compared with the addition of conductive fillers to the polymer matrix, MWs embedded on the surface of the polymer do not affect the mechanical properties (e.g., elastic modulus, ductility, and brittleness) of the polymer. In addition, compared with spray-coated MWs on the surface of polymers, the developed conductors improve durability by preventing the loss of MWs from the contact surface under repeated high pressure. Consequently, our developed conductor applied to the pressure sensor exhibits

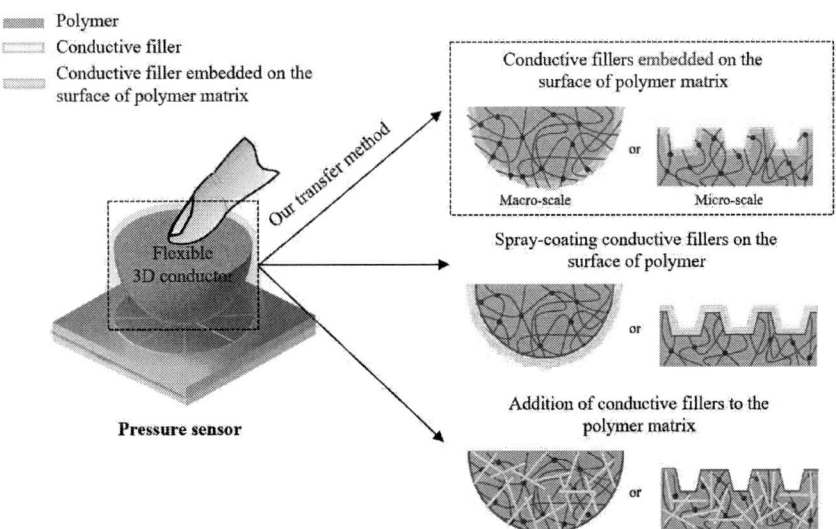

Figure 1: Design concept of the transfer method for embedding conductive fillers on the surface of multi-scale structures for 3D flexible conductors.

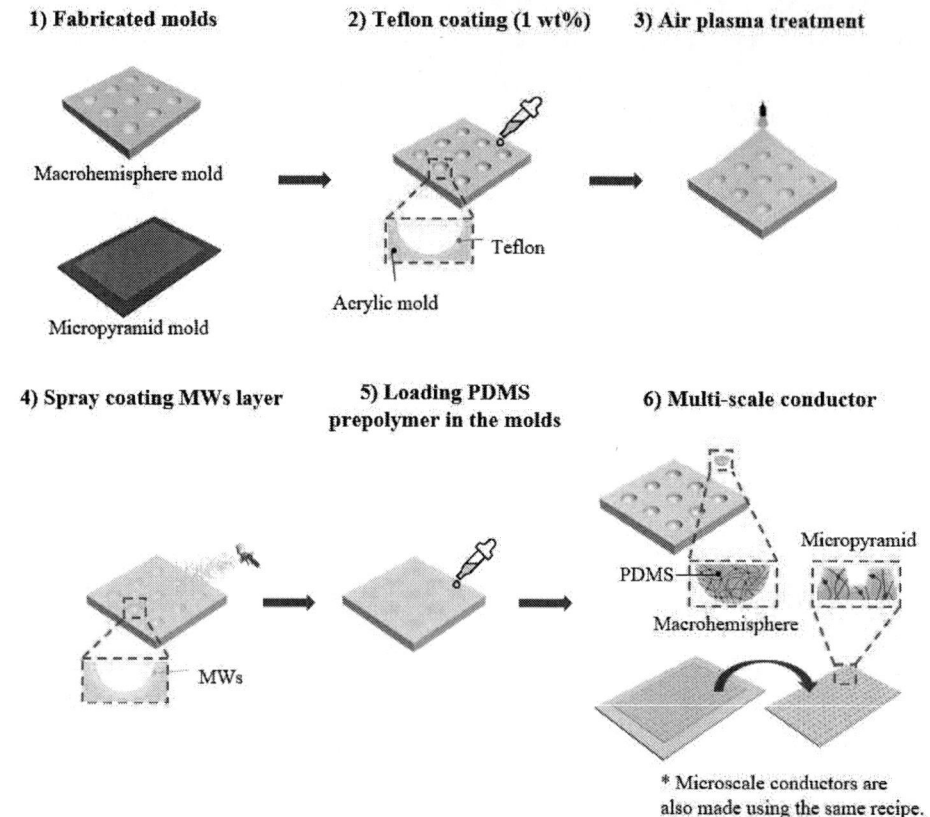

Figure 2: Fabrication process of 3D flexible conductors with surface-embedded metal nanowires (MWs).

sensing stability characteristics of the sensor by maintaining the initial value constant compared with the conductor fabricated by the spray-coating method. In addition, the sensing performance (e.g., response time, sensitivity, dynamic range, and robustness) can be improved because the conductors applied to the sensor can be fabricated in multi-scale structures and various shapes.

MATERIALS AND METHODS

Figure 1 shows the design concept of the transfer method in comparison to two previously reported 3D flexible conductor fabrication methods. Figure 2 shows a schematic of the proposed fabrication process. 3D molds with truncated pyramidal micro- and hemispheric macrostructures were fabricated for the conductors using

photolithography and milling, respectively. Before spray-coating MWs dispersed in isopropyl alcohol (IPA) on the molds, a Teflon solution (1 wt%) was poured into the molds to lower the surface energy, thereby facilitating the transfer of MWs from the molds to the polymer. The molds were then placed in an oven at 70 °C for 6 h to volatilize the solvent. To achieve homogeneous spray-coated MW layers on the hydrophobic mold surface, the molds were pretreated with air plasma at 50 W for 30 s to make them temporarily hydrophilic. After spray-coating the MWs, the prepolymer (i.e., PDMS) was poured into the molds, and the specimen was placed in a vacuum chamber to infiltrate the prepolymer into the MW networks. The curing of the prepolymer infiltrated into the MW networks was firmly

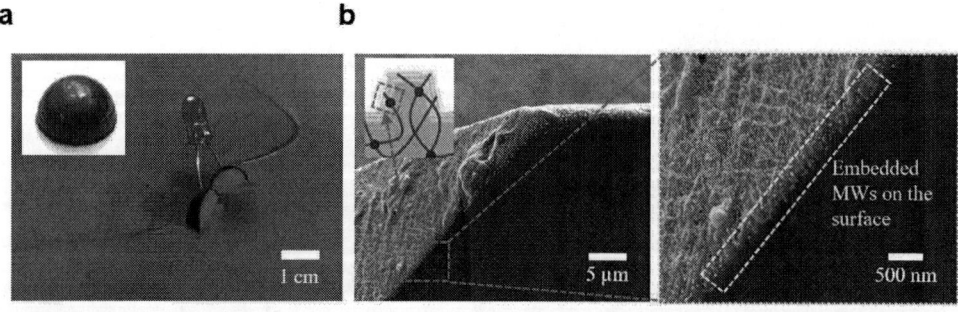

Figure 3: (a) Macrohemisphere conductor connected to an LED. (b) Cross-section of specimens fabricated with a truncated pyramidal microstructure.

maintained by the crosslinked networks of the polymer. By releasing the cured polymer from the molds, 3D flexible conductors with conductive fillers embedded on the surface of the multi-scale structures were fabricated.

RESULTS

There are several methods (e.g., the addition of conductive fillers to the polymer matrix and spray-coating conductive fillers on the surface of the polymer) for fabricating 3D flexible conductors. In the method of addition of conductive fillers into the polymer matrix, conductive composites can be easily fabricated if the percolation threshold is satisfied by adjusting the volume percentages of the prepolymer and conductive fillers [5]. However, the fabricated composites exhibited reduced ductility and brittleness. Therefore, the application of composites to pressure sensors is difficult because of the robustness and sensing stability of the sensor under repeated high pressure. In contrast, the fabrication method for spray-coating conductive fillers on the surface of polymers exhibits high conductivity while maintaining the mechanical properties. However, the sensor to which the conductor is applied causes cracks in the conductive layer or loss of conductive fillers under repeated high pressure. Therefore, the application of conductors to sensors is difficult in terms of the robustness and sensing stability of the sensor.

To apply robust and flexible conductors to pressure sensors, a method for fabricating conductors that maintains the inherent mechanical properties and high conductivity of the polymer is required. Therefore, by using the transfer method of embedding conductive fillers on the surface of polymers through optimal surface energy control of the 3D mold, conductors with the described advantages were fabricated. In addition, the developed method can be fabricated into multi-scale structures and various shapes to improve the sensor performance (e.g., response time, sensitivity, and dynamic range).

A light-emitting diode (LED) connection and scanning electron microscopy (SEM) imaging were performed on conductors to confirm whether the conductors fabricated using the proposed method had conductivity and were

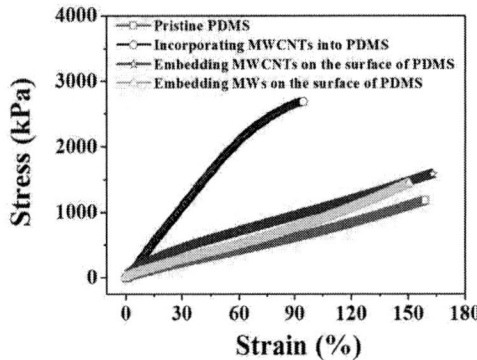

Figure 4: Stress-strain curves for comparing mechanical properties of pristine polydimethylsiloxane (PDMS) and different conductors with multi-walled carbon nanotubes (MWCNTs) incorporated into PDMS, MWCNTs embedded on the surface of PDMS, and MWs embedded on the surface of PDMS.

properly embedded with MWs on the surface of the polymers. Figure 3a shows the connection of a power source to the macrohemisphere conductor to turn on the LED. The current can flow when wires or LEDs are connected to conductors because they are embedded on the surface of the polymer. Figure 3b shows two SEM images of MWs embedded on the surface by cutting the cross-section of the micro-pyramidal conductor using a focused ion beam (FIB). As shown in the SEM images, the MWs infiltrated the polymer networks on the surface of the conductor.

The conductors fabricated by the developed method were compared with pristine PDMS and the addition of MWCNTs to PDMS to ensure that their intrinsic mechanical properties were maintained. Figure 4 shows that the addition of MWCNTs to PDMS results in a lower elongation at break compared with pristine PDMS, and conductors with conductive fillers (i.e., MWCNTs and MWs) embedded on the PDMS surface.

The conductors fabricated using the developed method were confirmed to be robust and durable when applied to the sensor under repeated high pressure. Figure 5 shows

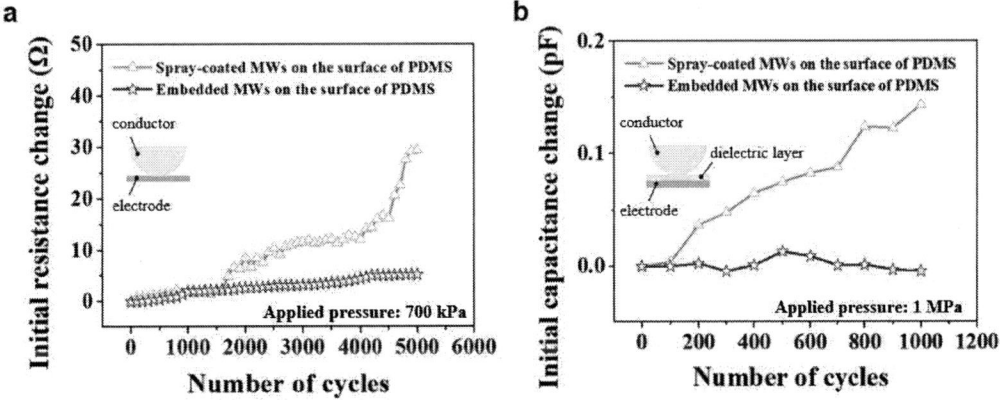

Figure 5: Durability test of conductors made of spray-coated and embedded MWs on the surface of the polymer by measuring the changes in (a) resistance and (b) capacitance at repeated high pressure.

that the developed transfer method is more robust under repeated high pressure than the spray-coating method in terms of resistivity and capacitance. Figure 5a shows that the conductors fabricated by the spray-coating method cracked in the conductive layer under repeated pressure, resulting in increased resistance. Figure 5b shows that the initial capacitance of conductors with spray-coated MWs on the PDMS surface increases, whereas the initial capacitance of conductors with embedded MWs on the PDMS surface is maintained even at repeated high pressures. The conductive fillers were loosely attached to the polymer surface of the conductors fabricated by spray coating, which resulted in the loss of the conductive fillers into the dielectric layer, thereby increasing the initial capacitance. However, when conductors are fabricated by the proposed transfer method, MWs are strongly embedded in the polymer networks and do not stick to the dielectric layer of the sensor, contributing to sensing stability.

CONCLUSION

We demonstrate a transfer method for embedding conductive fillers on the surface of flexible polymers with various macro- and microstructures to fabricate 3D flexible conductors. To achieve this, the surface energy conditions of the 3D molds were optimized to transfer spray-coated conductive fillers from the molds to the polymer networks. FIB and SEM images confirmed that the conductive fillers were successfully embedded on the surface of the polymers. In addition, the conductors maintained their inherent mechanical properties without changing the physical properties of the polymer. The pressure sensor with conductors fabricated by the developed transfer method exhibited sensing stability without resistance change and initial capacitance change under repeated high pressures. In addition, the sensing performance (e.g., response time, sensitivity, dynamic range, and robustness) can be improved because the conductors applied to the sensor can be fabricated in various multi-scale structures.

ACKNOWLEDGEMENTS

This research is supported by Rediscovery of the Past R&D Result through the Ministry of Trade, Industry and Energy (MOTIE) and the Korea Institute for Advancement of Technology (KIAT) (grant number: P0022631)

REFERENCES

[1] J. Qin, L. J. Yin, Y. N. Hao, S. L. Zhong, D. L. Zhang, K. Bi, Y. X. Zhang, Y. Zhao, Z. M. Dang, "Flexible and Stretchable Capacitive Sensors with Different Microstructures", *Adv. Mater.,* vol. 33, p. 2008267, 2021.

[2] J. Du, L. Wang, Y. Shi, F. Zhang, S. Hu, P. Liu, A. Li, J. Chen, "Optimized CNT-PDMS Flexible Composite for Attachable Health-Care Device", *Sensors,* vol. 20, p. 4523, 2020.

[3] A. Kumar, M. O. Shaikh, C. H. Chuang, "Silver Nanowire Synthesis and Strategies for Fabricating Transparent Conducting Electrodes", Nanomaterials, vol. 11, p. 693, 2021.

[4] L. J. Waldman, M. W. Keller, "Remendable conductive polyethylene composite with simultaneous restoration of electrical and mechanical behavior", *Polym. Eng. Sci.*, vol. 62, pp. 991-998, 2022.

[5] K. Chu, D. Kim, Y. Sohn, S. Lee, C. Moon, S. Park, "Electrical and Thermal Properties of Carbon-Nanotube Composite for Flexible Electric Heating-Unit Applications", *IEEE Electron Device Lett.*, vol. 34, pp. 668-670, 2013.

CONTACT

* J. Kim, tel: +82-54-279-2185; joonwon@postech.ac.kr

FABRICATION OF HIGH-FREQUENCY 2D FLEXIBLE PMUT ARRAY

Sanjog V. Joshi[1], Sina Sadeghpour[1], and Michael Kraft[1]

[1]Faculty of Electrical Engineering (ESAT-MNS), KU Leuven, Belgium

ABSTRACT

This paper reports an 8x8 2D flexible pMUT array fabrication process along with preliminary characterization results. The top-down fabrication process uses a Silicon wafer as a temporary carrier from which a flexible pMUT array is released by DRIE. The flexible pMUT membrane consists of a 1 μm PZT and 6 μm polyimide layers. A 14 μm thick polyimide sidewall with a cavity of 100 μm in diameter is defined by reactive ion etching (RIE). The released structure is bonded to PDMS for making the assembly mechanically robust. The flexible pMUTs demonstrate a 5 mm bending radius, 40% fractional bandwidth of the 4.2 MHz center frequency in-air, and 7 mm·s^{-1}·V^{-1} transmission sensitivity.

KEYWORDS

Flexible Microsystems, Wearable Medical Devices, Ultrasound Transducers, PMUT, Piezoelectric Thin Films, PZT, Piezo-MEMS

INTRODUCTION

Ultrasound assists millions of people in monitoring and diagnosing health problems where, in particular, medical imaging with rigid ultrasound probes has proven to be extremely useful [1]. Ultrasound transducers are at the heart of such systems and generate ultrasound waves typically using vibration that emits acoustic pressure waves. These vibrations are commonly produced using a piezoelectric effect in a bulk piezoelectric transducer. Here, the piezoelectric material is utilized as thick piezoelectric ceramic in d$_{33}$ thickness mode vibration [2]. Ultrasound is being extensively used in nondestructive evaluation (NDE) [3], ultrasonic actuation [4], medical imaging [5], drug delivery and other therapeutic applications [6], particle and cell manipulation [7], and object recognition [8]. Due to limited propagation distance in air, ultrasound is desirable for low-power consumer applications with spatial bandwidth sharing, and unregulated sound spectrum beyond the audible range, and the same has been witnessed in the bio-medical field.

Flexible ultrasound transducers broaden the ultrasound application areas by providing 1) conformal contact to the skin, 2) user-friendliness with an opportunity for continuous monitoring and reduced operator variability, and 3) increased angular coverage compared to rigid planar transducer probes with the possibility of power efficient mechanical focusing. Flexible ultrasound transducer devices can be built on flexible substrates, such as plastic, which allows folding, wrapping, rolling, and twisting with a negligible effect on its electronic function. The flexible device system is typically a multilayered integrated system of bendable or stretchable substrates, conducting electrodes, active materials, instrumentation, encapsulation, and packaging materials. The flexible devices are wearable, conformal, light and thin, unbreakable, and cost-effective for large-area applications. As such, flexible electronics and microsystems are

well - researched topics with various reviews summarizing the state of the art. [9-11].

The bulk piezoelectric transducer, a mature technology, dominates flexible ultrasound. However, they have several drawbacks- 1) Their performance is restricted by fabrication techniques, e.g., mechanical dicing limits the interelement pitch [12]. 2) Their bulky design results in high operating voltages. 3) There is little room for design variations as resonance frequency depends only on device thickness [2]. 4) Custom fabrication processes and integration with electronics often give poor yields. 5) Thick piezoelectric results in poor acoustic matching, and low penetration depths. Micromachined ultrasound transducer (MUT) alternatives, piezoelectric- MUT, and capacitive- MUT benefit from small form factor, reduced power consumption, batch fabrication, and the possibility for CMOS integration. In a piezoelectric micromachined ultrasound transducer (pMUT), a piezoelectric thin film in the membrane actuates the device in d$_{31}$ flexural mode [13]. On the other hand, a MUT based on the electrostatic principle is called capacitive MUT or cMUT. These transducers also work in a flexural vibration mode with their vibrations set up with electrostatic forces between electrodes [14].

However, flexible MUTs show a trade-off between performance and bendability, e.g., Silicon-based approaches provide high acoustic pressures and hence satisfactory output performance. To make a silicon-based device bendable, there are mainly three possibilities: 1) Thinning down silicon to such a thickness that it can bend easily and can be mounted on curved substrates or fixtures [14]. However, such a design is fragile. 2) Island-spring-based designs where the transducers reside on islands and spring structures render bendability [15]. This approach suffers from a limited bending radius, interconnect area, and fill factor. 3) Silicon-based transducers on islands, connected by flexible link structures made from low-modulus materials such as polyimide or PDMS [16]. Low fill factors and electronics integration in such design present challenges. On the other hand, polymer-based designs are easily bendable, but they suffer from low sensitivities as a result of the low rigidity modulus of the materials [17].

Ultrasound transducer arrays with dynamic control of each element allow beam steering and focusing. The arrays could bring benefits in the reduction of test time, better reliability, and measurement quality for various practical applications, including medical imaging. Especially, 2D arrays can provide more functionality and attractively, a 3D scan as compared with 1D arrays focused and steered only in the azimuthal plane [2]. However, there are no studies on high-frequency 2D flexible MUT arrays for imaging applications. Hence, towards developing such technology, we report fabrication process of an 8x8 2D flexible pMUT array. The polymer-based bendable array benefits from high bandwidth around the resonance frequencies of 4 - 5 MHz which is interesting for imaging applications.

TRANSDUCER DESIGN

Any MUT can be described as a membrane, cavity sidewalls, and a substrate operating in flexural mode (fig. 1-b). The pMUT membrane typically contains a patterned layer stack of passive layers, electrodes, and piezoelectric thin film. The membrane is driven by applying an excitation voltage between the top and bottom electrodes. The applied electric field forms transverse stress in the active piezoelectric layer, which causes the membrane to displace out-of-the plane, generating a pressure wave in the outer medium. The resonant frequency of the pMUT, because of the flexural (d_{31}) mode, is mainly related to the geometry, boundary conditions, and mechanical stiffness of membranes [13].

Fig. 1-a shows the cross-section illustration of a single flexible pMUT element fabricated, which consists of a 1 mm thick PDMS substrate and 14 μm thick polyimide sidewalls defining a 100 μm cavity. The membrane is made of a 7 μm thick polyimide passive layer and a 1 μm thick PZT thin film. PZT is widely used in piezo-MEMS because of its excellent piezoelectric properties [18]. PZT is sandwiched by 200 nm thick top and bottom Pt electrodes as well as 2.5 μm thick polyimide dielectric.

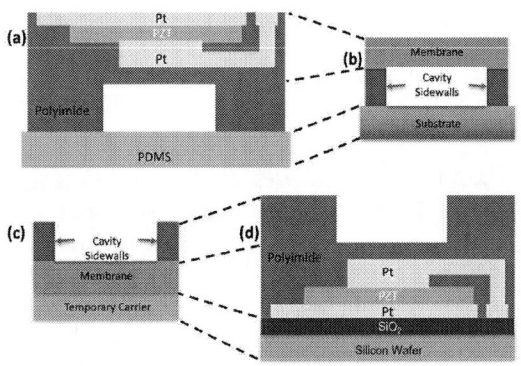

Figure 1: (a) schematic of a flexible pMUT element. (b) High level description of a MUT related to layer stacks from part a. (c) Simplified layout and (d) detailed stacks of top-down pMUT fabrication scheme (PDMS bonding not shown here). Schematics are not to scale.

FABRICATION PROCESS

A top-down approach is used for the realization of the design. Hence, the device layers are stacked in reverse order where first, membrane structures followed by cavity sidewalls are realized on a temporary carrier. Then they are bonded to a flexible substrate. Finally, the carrier is removed to complete the fabrication. The fabrication stack on a silicon wafer carrier is shown in fig. 1-d.

Optical microscopic images of each array fabrication step are provided in fig. 2. The silicon wafer is first cleaned with Piranha and BHF. This is followed by wet oxidation resulting in 500 nm oxide on both sides. Then, the bottom Pt layer is patterned acting as the seed layer for PZT. Pt also acts as the electrode pad layer- the array corners are electrode pads common to all elements whereas the other pads provide electrical access for each individual pMUT element in the array (fig. 2-a). This prepares the substrate for further steps.

After substrate preparation, sol-gel PZT deposition is followed which involves spin coating, pyrolysis, and rapid thermal annealing processing [15]. PZT is then patterned by wet etching with the s1818 photoresist mask and the two-step etching process described in [19] (fig. 2-b). Then, a 2.5 μm polyimide interlayer (acting as a dielectric and reducing parasitic capacitance and roughness across the wafer) was deposited. For this, the PI2611 solution (HD Microsystems) was spun, soft-baked on a hot plate, and cured in an oven. For patterning the polyimide, a Chromium thin film served as a hard mask (fig. 2-c). Pt interconnects from the pMUT element to individual electrode pads were then realized again with the liftoff process (fig. 2-d). After interconnects, the polyimide membrane and sidewall are realized (fig. 2-e).

Finally, the release procedure is performed which involves Deep Reactive Ion Etching (DRIE) on the stack shown in fig. 1-d to release the flexible array from the temporary silicon carrier. For this, the wafer backside is patterned with a negative photoresist. The oxide is then removed to open electrode pads (in fig. 2-f which are the same Pt electrode pads in fig. 2-a viewed from the backside). The array is then cut from the wafer and bonded to a 1 mm thick PDMS to complete the fabrication.

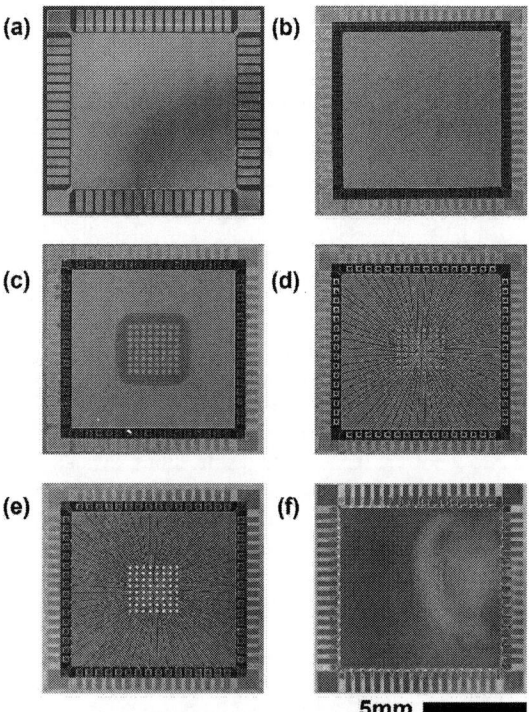

Figure 2: Optical microscope images of 2D array during fabrication- (a) patterning of the bottom Pt layer on thermal oxide. (b) Sol-gel deposited PZT is patterned in pad area. (c) Polyimide interlayer patterning. (d) Top Pt layer patterning of electrodes and interconnects for individual pMUT element. (e) Polyimide membrane layer deposition followed by polyimide sidewalls patterning. (f) After DRIE and oxide removal, exposing first Pt layer from backside.

978-1-6654-9309-3/23 $31.00 © 2023 IEEE

RESULTS AND DISCUSSION

To characterize the deposited PZT material, two measurements were performed. 1) XRD measurements after the PZT deposition step and 2) P-E hysteresis loop measurements of the PZT thin-film after top electrode deposition. These are shown in fig. 3. (111)-orientated PZT without any pyrochlore (normally around 29°) results in a satisfactory d_{31} response and transmission sensitivity.

For the P-E hysteresis loop measurement, a sawyer-tower circuit was assembled. Briefly, the PZT capacitor and the sample capacitor (C_o) of a known value were connected in series. The value of the sample capacitor (~30 times the PZT capacitance) was chosen such that almost all the applied voltage drops across the sample capacitor. From the applied voltage signal, the electric field was calculated by dividing the voltage by the PZT thickness whereas polarization charge density across the PZT capacitor was determined by dividing the charge across the sample capacitor, C_oV_o, (same as the charge across the PZT capacitor, being in series) by area of PZT capacitor. Remnant polarization and coercive field of 14 µc/cm2 and ~50 kV/cm compare well with recent reports [20].

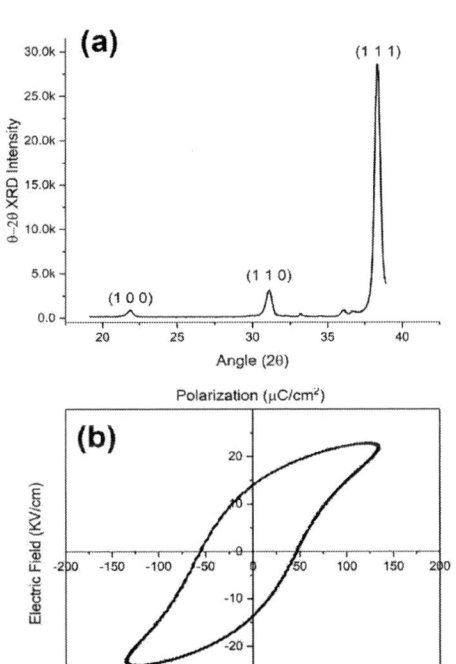

Figure 3: (a) XRD showing (111) orientation, and (b) polarization vs. electric field (P-E) hysteresis loop at 1 kHz, of deposited PZT thin film.

Images of the released device are shown in fig. 4-a. The total thickness of the device before bonding with PDMS is 25 µm. The device bending was illustrated in fig. 4-b by wrapping it around a finger. This indicates a bending radius of about 5 mm. The bendability was retained even after PDMS bonding (fig. 4-c). The high - modulus PZT and electrode layers showed no cracking or breaking. Moreover, no substrate curling was observed before and after PDMS bonding.

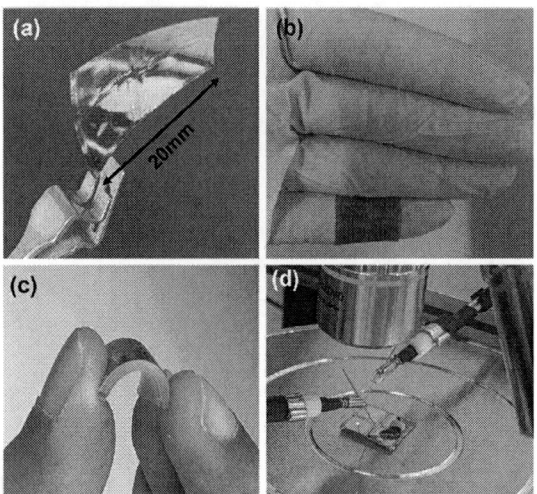

Figure 4: (a) Image of the released device. (b) Flexible array wrapped around a finger. (c) Bendability after bonding to PDMS. (d) pMUTs measured with LDV.

After the preliminary mechanical characterization of the bendability and robustness, the vibration performance was determined. With membrane-based devices such as pMUTs or cMUTs, in-air frequency response can be recorded by a Laser Doppler vibrometer (LDV). When an AC voltage is applied across the transducer electrodes (with or without DC bias), the elements are set into vibration at the application frequency. At the resonance frequency, the highest vibration amplitude is recorded for the MUT resonators. To measure such a response, the transducer is excited with a periodic chirp signal applied over a band of frequencies. In the signal, the input energy around the resonance of the transducer is very small compared to the sine wave signal at resonance. This is the reason in-air sensitivity should be characterized by a sine wave at the resonance frequency [21]. However, the periodic chirp measurement allows the measurement of the transducer bandwidth at the pMUT surface.

Fig. 4-d shows pMUT sample on LDV setup. In-air characterization (fig. 5) with an LDV system (Polytec MSA-600) shows resonance at 4.2 MHz with a -3dB bandwidth of about 1.7 MHz, i.e., 40 % fractional bandwidth, as a response to a periodic chirp. This is the highest in-air bandwidth for the reported flexible pMUTs according to our knowledge. 14 Vp-p sinus excitation at the resonance frequency resulted in 100 mm/s velocity, thus, ~7mm/s/V sensitivity for a pMUT element- 100 times that of the recent polymer-based flexible pMUT [17].

CONCLUSION

Fabrication of a PZT-based and fully addressed 2D flexible pMUT array with 64 channels is proposed in this paper, describing the design and top-down fabrication process in detail. The basic functionality of the pMUTs was confirmed by bending demonstration and in-air LDV characterization. The resonance frequencies in the order of 4-5 MHz and higher bandwidth in the frequency response reported in the paper are particularly interesting for imaging applications. The 2D array will further be characterized towards the application.

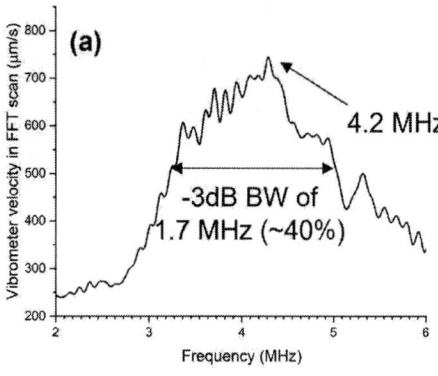

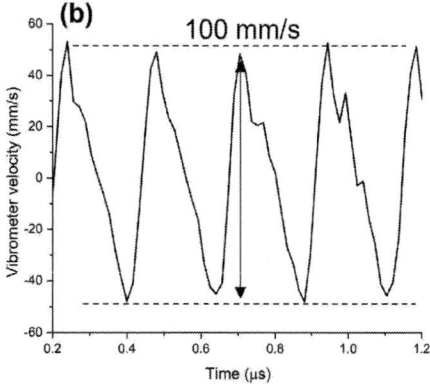

Figure 5: LDV measurements of a pMUT element-(a) FFT with periodic chirp signal. (b) Time domain measurement with 14 V_{p-p} sinusoid at resonance resulted in 100 mm/s velocity, thus, 7mm/s/V sensitivity.

ACKNOWLEDGEMENTS

This work was supported by the KU Leuven Interdisciplinary Networks (ID-N) under Grant ZKD-6578 and C3 Grant ZKD-8726.

REFERENCES

[1] Ramona Luca et al., An educational overview of ultrasound probe types and their fields of application. Archives of Acoustics, 46:3–15, 03 2021.

[2] K. K. Shung et al., Ultrasonic transducers and arrays. IEEE Engineering in Medicine and Biology Magazine, 15(6):20 30, 1996.

[3] Bruce W. Drinkwater et al., Ultrasonic arrays for non-destructive evaluation: A review. NDT & E International, 39(7):525 – 541, 2006.

[4] B. Watson et al., Piezoelectric ultrasonic micro/milli-scale actuators. Sensors and Actuators A: Physical, 152(2):219 – 233, 2009.

[5] Paul Suetens. Fundamentals of Medical Imaging. Cambridge University Press, 2 edition, 2009.

[6] William G. Pitt et al., Ultrasonic drug delivery–a general review. Expert opinion on drug delivery, 1(1):37–56, Nov 2004.

[7] Yongqiang Qiu et al., Acoustic devices for particle and cell manipulation and sensing. Sensors (Basel, Switzerland), 14(8):14806–14838, Aug 2014.

[8] P. Kleinschmidt et al., Ultrasonic remote sensors for noncontact object detection. Siemens Forschungs und Entwicklungsberichte, 10(2):110– 118, January 1981.

[9] Mingzhi Zou et al., Flexible devices: from materials, architectures to applications. Journal of Semiconductors, 39(1):011010, Jan 2018.

[10] Muhammad A. Alam et al., Flexible Electronics, pages 860–865. Springer Netherlands, Dordrecht, 2012.

[11] I-Chun Cheng et al., Overview of Flexible Electronics Technology, pages 1–28. Springer US, Boston, MA, 2009.

[12] Chen et al., in Medical Imaging: Technology and Applications pp. 253–271 (CRC Press, 2013).

[13] S Sadeghpour et al., Design and fabrication strategy for an efficient lead zirconate titanate based piezoelectric micromachined ultrasound transducer. Journal of Micromechanics and Microengineering, 29(12), 2019.

[14] K. A. Wong et al., "Curved micromachined ultrasonic transducers", Proc. IEEE Ultrason. Symp, vol. 1, pp. 572-576, 2003.

[15] S. Sadeghpour et al., Bendable piezoelectric micromachined ultrasound transducer (PMUT) arrays based on Silicon-On-Insulator (SOI) technology, Journal of Microelectromechanical Systems, Vol. 29, No. 3, pp. 378-386, 2020.

[16] Jin Hyung Lee et al., Flexible piezoelectric micromachined ultrasonic transducer (pMUT) for application in brain stimulation. Microsystem Technologies, 23(7):2321–2328, 2017.

[17] Jeong et al., Fully Flexible PMUT Based on Polymer Materials and Stress Compensation by Adaptive Frequency Driving, Journal of Microelectromechanical Systems, vol. 30, no. 1, pp. 137-143, Feb. 2021.

[18] Gabriel Smith et al., Pzt-based piezoelectric mems technology. Journal of the American Ceramic Society, 95(6), 1777–1792, 2012.

[19] Haoran Wang et al., A one-step residue-free wet etching process of ceramic PZT for piezoelectric transducers, Sensors and Actuators A, 290, 130–136, (2019).

[20] Rodríguez-Aranda et al., Ferroelectric hysteresis and improved fatigue of PZT (53/47) films fabricated by a simplified sol–gel acetic-acid route, J. Mater. Sci. Mater. Electron, 25, 4806–4813, 2014.

[21] S. Sadeghpour et al., "Novel Phased Array Piezoelectric Micromachined Ultrasound Transducers (pMUTs) for Medical Imaging," in IEEE Open Journal of Ultrasonics, Ferroelectrics, and Frequency Control, vol. 2, pp. 194-202, 2022.

CONTACT

*Sanjog V. Joshi, sanjogvilas.joshi@kuleuven.be

FLEXIBLE SILK-BASED GRAPHENE BIOELECTRONICS FOR WEARABLE MULTIMODAL PHYSIOLOGICAL MONITORING

Sajjad Mirbakht [1], Ata Golparvar [1,2], Muhammad Umar [1], and Murat Kaya Yapici [1,3,4]

[1] Faculty of Engineering and Natural Sciences, Sabanci University, 34956 Istanbul, Turkey

[2] IClab, École Polytechnique Fédérale de Lausanne (EPFL), 2002 Neuchâtel, Switzerland

[3] Sabanci University SUNUM Nanotechnology Research Center, 34956 Istanbul, Turkey

[4] Department of Electrical Engineering, University of Washington, 98195 Seattle, USA

ABSTRACT

We introduce a skin-inspired epidermal bioelectronic patch (EBP) with unprecedented integration of graphene on silk with a facile and cost-effective fabrication process, excluding complex ink formulation. We advance the state-of-the-art by directly screen-printing pristine graphene ink on "donor" substrates like polyethylene terephthalate (PET) and successfully receiving the printed graphene traces onto silk protein-based "carrier" substrates by a controlled drop-casting, curing, and transfer printing process. The presented approach retains the conductivity of the pristine graphene ink prepared by reducing graphene oxide suspension through "green" chemistry based on a water-soluble vitamin C reagent. We demonstrate that silk/graphene epidermal patches (SGEP) can conformally attach to human skin without additional adhesives or gel materials. Our study reveals a superior response of the SGEP by successfully acquiring multimodal physiological signals such as cardiac-ECG, macular-EMG, and ocular-EOG with almost perfect correlation reaching a record high of 98.48% to those recorded simultaneously with medical-grade, pre-gelled with adhesive backings, wet Ag/AgCl electrodes.

KEYWORDS

Electronic skin (e-skin), epidermal bioelectronics patch, continuous monitoring, flexible electronics, green electronics, reduced graphene oxide (rGO), silk protein, skin-like electronics.

INTRODUCTION

Epidermal bioelectronic patches (EBP) provide continuous biopotential signal monitoring by taking advantage of elastic and stretchable materials [1]. Additionally, EBP's better electrical and mechanical performances with comfortable skin attachment possibility have the potential to replace the commercial "wet" and gel-free "dry" silver/silver chloride (Ag/AgCl) electrodes in long-term wearable biopotential signal monitoring applications.

In such commercial electrodes, the performance of the "wet" electrodes degrades in time by dehydration of the gel, and red/swollen skin may develop immediately upon the removal of the electrodes by mechanical peeling due to strong adhesive backing, which also reportedly introduces skin irritation [2]. On the other hand, "dry" electrodes fail to accurately mimic skin deformations on a micro/nanometer scale, reducing the recording's signal-to-noise ratio (SNR) and leaving the system vulnerable to motion artifacts [3]. Alternatively, EBP offers a reliable platform for continuous physiological biosignal monitoring by correctly mimicking skin deformations and allowing conformal attachment by Van Der-Waal forces without extra adhesive materials, provided that their modulus and thickness match that of the human skin [4].

For the development of EBP, various materials and substrates were suggested, including polymers, textiles, and plastics [5]. Polymers are distinguished candidates for epidermal devices due to their natural elasticity and intrinsic stretchability [6]. However, complexities in the molecular engineering of synthesized polymers to obtain desired physical properties with cost-effective manufacturing are major obstacles to developing intelligent material interfaces [7]. On the other hand, the synthesis of silk, an organic biopolymer, is simple and inherently advantageous for electronic skin applications due to silk's natural self-healing, breathability, biodegradability, biocompatibility, stretchability, and low-cost production [8].

Another challenge in developing reliable skin-like patches is on identifying a suitable functional material to enable conductivity. Although metal or metallic materials such as silver and gold are widely used in flexible electronics, they require costly, specialized deposition techniques and suffer from intrinsic hardness, resulting in material delamination upon integration with ultra-soft materials under deformations [9].

On the other hand, 2D materials such as graphene are favored for functionalization in highly flexible and stretchable wearable sensory systems, including innovative textile biosensors [10]. However, for instance, the usual chemistry for the reduction of graphene oxide relies on using hydrazine or hydrazine hydrate, which is toxic [11]. Therefore, the "green" graphene oxide reduction process using l-ascorbic acid (Vitamin C), which is an aqueous and environment-friendly solution by nature, is proposed [12].

Among various approaches to printing high-quality graphene patterns, screen-printing remains the only low-cost printing technique compared to inkjet-, and spray printing that does not rely on ink formulation engineering without sacrificing homogeneity [13]. To date, only a single report exists on developing a skin-like health monitoring system by merging silk with graphene by optimizing a complex ink suspension formulation [14]. Therefore, this study advances state-of-art by introducing a simple fabrication process to exploit the inherent advantages of silk and graphene without tedious ink formulation engineering and introduces a new health monitoring platform for continuous multimodal physiological biosignal monitoring.

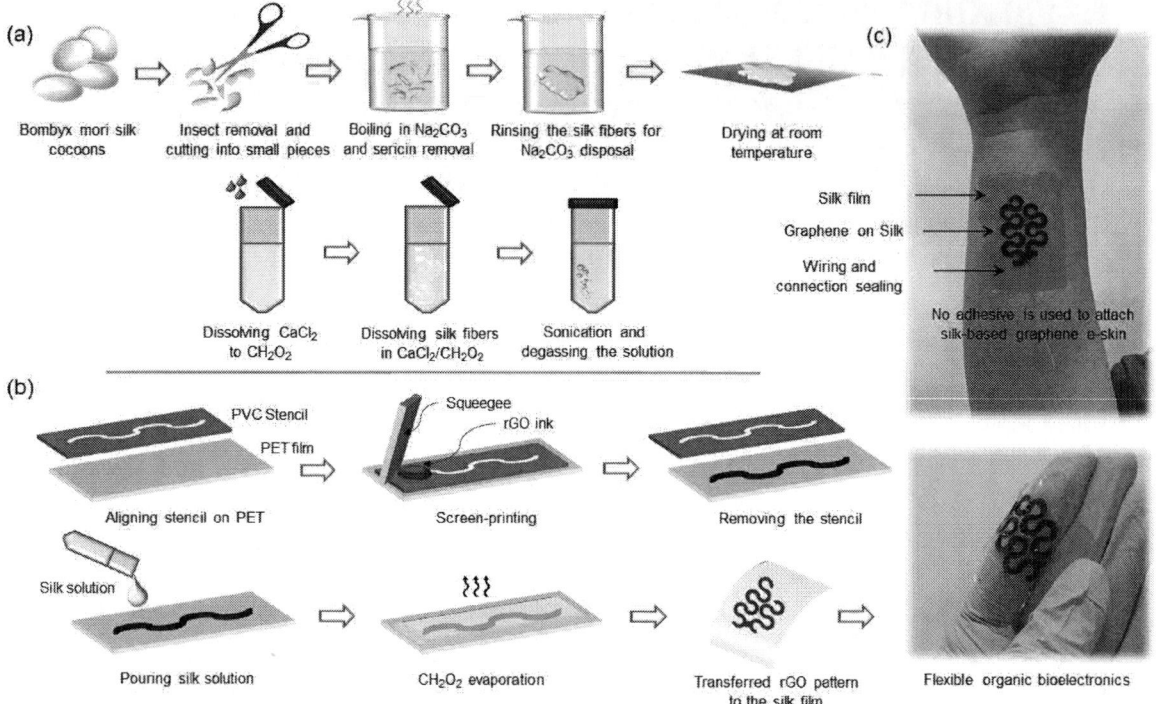

Figure 1: (a) Synthesis of silk solution (b) The screen-printing process of rGO on flexible silk film substrates (top) and transferring rGO patterns onto the flexible silk film by drop-casting and drying (bottom) (c) Images show final epidermal patch with patterned serpentine rGO traces on silk film attached to the skin without applying external adhesives for gel-free, adhesive-less, multimode biopotential signal recording.

METHODOLOGY

Figure 1a and 1b summarizes the development cycle of silk/graphene epidermal patch (SGEP). The fabrication begins with the preparation of the silk solution involving degumming the Bombyx mori cocoons (purchased from a local vendor) by boiling them for 30 minutes in 0.02 mmol/l sodium carbonate-dissolved deionized water and subsequently rinsed in freshwater three times [15]. The extracted fibers were then dissolved in formic acid solution ($CaCl_2/CH_2O_2$) in a weight ratio of 1:45 g to prepare the silk film.

For the preparation of the screen-printing graphene ink, graphene oxide (GO) solution (0.4 wt%) was mixed with l-ascorbic acid ($C_6H_8O_6$) in 114 mmol/l concentration, followed by stirring at room temperature and sonication at 50 °C for 1 h. The solution's color change from golden-yellow to black certifies the success of the reduction process.

Polyethylene terephthalate (PET) film was screen-printed and used as a donor substrate for transferring the rGO patterns to the silk film. To increase and activate the surface energy of the carrier PET film, corona treatment (BD-20AC, Electro-Technic Products, US) was introduced prior to rGO printing. For screen printing, 75 μm adhesive polyvinyl chloride (PVC, ORACAL® 641, Europe GmbH) foils were prepared as stencils with a desktop vinyl cutter (gs-24, Ronald, USA). Stencils were placed onto PET films, and rGO ink was cast and spread using a smooth glass. Next, screen-printing of the prepared rGO solution

was performed. The printed rGO-PET films were annealed for 30 minutes at 80°C, and stencils were removed after that.

Later, the silk solution was drop-casted onto a screen-printed PET substrate and left in the fume hood overnight to evaporate the solvent. Finally, the graphene on silk electrodes was separated from the PET films and was connected to the acquisition unit by thin copper wires for quick prototyping (Figure 1c).

RESULTS AND DISCUSSIONS

Sensor Characterization

Successful transformation of graphene oxide (GO) to reduced graphene oxide (rGO) was observed by Raman spectroscopy. The Raman scattering spectra were recorded in the spectral region of 100–3000 cm^{-1} with confocal micro-Raman microscopy (532 nm, Renishaw, England, UK). Crystalline graphite shows only the G-band around 1600 cm^{-1} indicative of the primary in-plane vibrational mode, while GO and rGO display another broader D-band around 1345 cm^{-1} indicating defects in crystal symmetricity (Figure 2) [16]. The intensity ratio of the D and G bands (I_D/I_G) is a metric used to signify the quality of the reduction, whereby an increase in the I_D/I_G ratio is expected upon reduction. While the I_D/I_G ratio was 0.77 for GO, it increased to 1.54, validating the success of the "green" chemical reduction process and formation of rGO. The electrode's resistance was measured continuously for 8 days after the electrode fabrication (Figure 3). We found

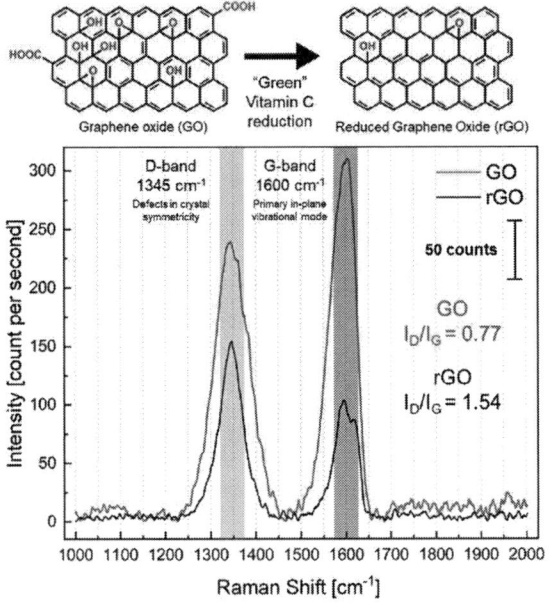

Figure 2: Increase in ratiometric Raman intensity depicts the successful conversion of GO to rGO.

that the resistance of the electrode significantly decreased from 9.96 kΩ to 4.44 kΩ in 5 days, due to the presence of vitamin C agents in the rGO ink, but the resistance is stable around 4 kΩ after day 5. Although lower resistance implies higher quality biopotential recording, in practice, we did not observe any significant change in the recorded signal quality on day one and day eight.

The skin-electrode impedance was acquired using high-impedance input instrumentation amplifiers and a Howland current pump circuit. The skin-electrode impedance of the electrodes was 50 kΩ at DC, 34 kΩ at 50 Hz, and 23 kΩ at 100 Hz, while the Ag/AgCl electrode was 110 kΩ, 47 kΩ, and 32 kΩ respectively.

Multimode Biopotential Recordings
Multiple biopotential signals were acquired from a healthy

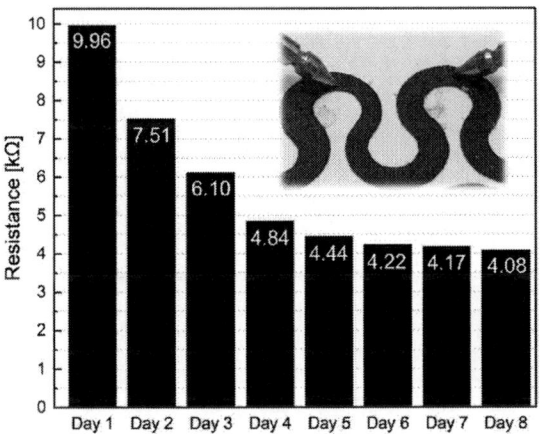

Figure 3: The resistance changes of silk/graphene epidermal patch (SGEP) in 8 days.

participant (25-year-old, BMI 24.9). In particular, our novel SGEP is used in successfully recording lead-I electrocardiograph (ECG), biceps surface electromyography (EMG), and forehead electrooculography (EOG). Both commercial and SGEP were placed on the forearms for recording ECG signals for synchronous ECG recording. Figure 4a illustrates the recorded ECG signals with correlation coefficient of 98.49% compared to those recorded simultaneously with conventional Ag/AgCl electrodes with clearly distinguishable P-QRS-T complex.

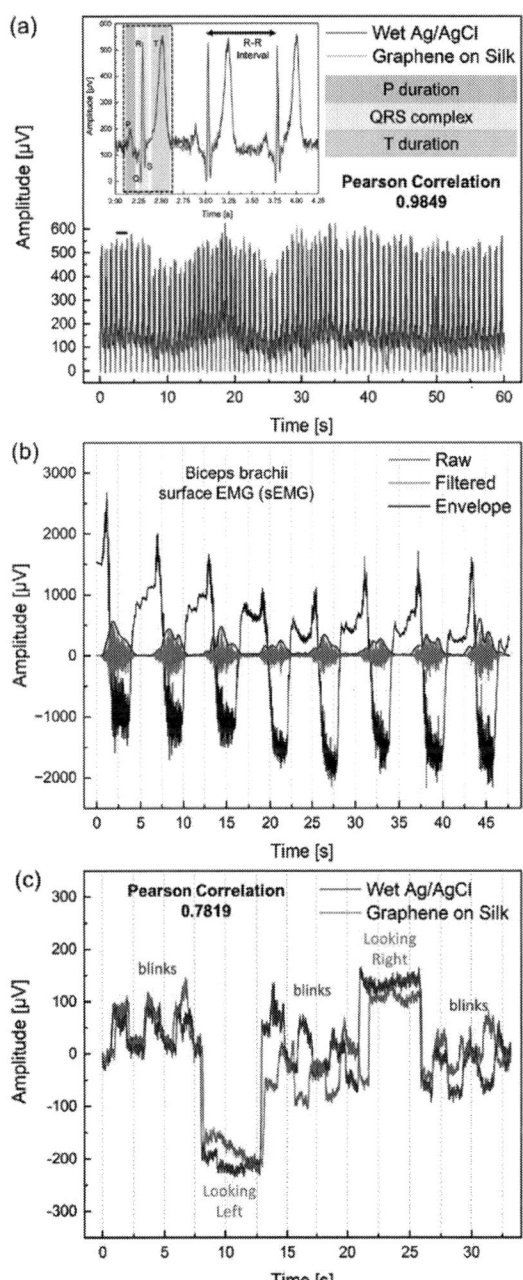

Figure 4: Multimodal (a) ECG, (b) EMG, and (c) EOG sensing with introduced silk/graphene epidermal patch.

EMG signals were recorded by placing electrodes on the bicep when a volunteer was instructed to clench a fist and then open it in intervals of 5 s, resulting in clear shifts in EMG waveform after signal processing (Figure 4b). For obtaining EOG signals, two electrodes were placed on the forehead near the temples, and the volunteer was asked to perform saccadic left and right eye movements, fixation at the center, right and left sides, and blinks (Figure 4c). After completing the experimental protocol with our developed electrodes, the same experiment was repeated with Ag/AgCl electrodes. The achieved correlation coefficient was 78.19% which is an acceptable value considering the recordings were not simultaneous.

CONCLUSION

In conclusion, we report a facile method to yield highly flexible, conformable, and adhesive-free skin-attachable epidermal electronic patches by using green and cost-effective fabrication process to record biopotential signals, including ECG, EMG, and EOG. In order to obtain this device, a low-cost screen-printed rGO was transferred onto the silk polymer to obtain a skin-like patch. We showed that the execration of silk, reducing the GO, and the screen-printing process are green cost-effective methods which lead to an epidermal patch that shows reliable performance in recording ECG, EMG, and EOG signals with 98.49% correlation in simultaneous recording of ECG signals and 78.19% in asynchronous EOG recordings with respect to commercial Ag/AgCl electrodes.

ACKNOWLEDGEMENTS

The study was conducted according to the guidelines of the declaration of Helsinki and approved by the ethics committee of Sabanci university (fens-2020-48). Professor Murat Kaya Yapici appreciates the support of the Turkish Academy of Sciences (TUBA) within the framework of the TUBA Outstanding Young Scientist Award Program (GEBIP).

REFERENCES

[1] Y. Liu et al., "Epidermal mechano-acoustic sensing electronics for cardiovascular diagnostics and human-machine interfaces," Sci Adv, vol. 2, no. 11. 2016.

[2] A. J. Golparvar and M. K. Yapici, "Toward graphene textiles in wearable eye tracking systems for human-machine interaction," Beilstein Journal of Nanotechnology, vol. 12, pp. 180–189, Feb. 2021.

[3] Y. Kim et al., "Soft Wireless Bioelectronics Designed for Real-Time, Continuous Health Monitoring of Farmworkers," Adv Healthc Mater, no. 13, 2022.

[4] D.-H. Kim et al., "Epidermal Electronics," Science (1979), vol. 333, no. 6044, pp. 838–843, Aug. 2011.

[5] W. Wu and H. Haick, "Materials and Wearable Devices for Autonomous Monitoring of Physiological Markers," Advanced Materials, no. 41Oct. 2018.

[6] S. Wang, J. Y. Oh, J. Xu, H. Tran, and Z. Bao, "Skin-Inspired Electronics: An Emerging Paradigm," Acc Chem Res, vol. 51, no. 5, pp. 1033–1045, May 2018.

[7] H. Tran, V. R. Feig, K. Liu, Y. Zheng, and Z. Bao, "Polymer Chemistries Underpinning Materials for Skin-Inspired Electronics," Macromolecules, vol. 52, no. 11, pp. 3965–3974, Jun. 2019.

[8] C. Wang, K. Xia, Y. Zhang, and D. L. Kaplan, "Silk-Based Advanced Materials for Soft Electronics," Acc Chem Res, vol. 52, no. 10, pp. 2916–2927, Oct. 2019.

[9] J. A. Fan et al., "Fractal design concepts for stretchable electronics," Nat Commun, vol. 5, no. 1, May 2014.

[10] G. Acar, O. Ozturk, A. J. Golparvar, T. A. Elboshra, K. Böhringer, and M. K. Yapici, "Wearable and Flexible Textile Electrodes for Biopotential Signal Monitoring: A review," Electronics (Basel), vol. 8, no. 5, p. 479, Apr. 2019.

[11] A. Furst, R. C. Berlo, and S. Hooton, "Hydrazine as a Reducing Agent for Organic Compounds (Catalytic Hydrazine Reductions)," Chem Rev, vol. 65, no. 1, pp. 51–68, Feb. 1965.

[12] J. Zhang, H. Yang, G. Shen, P. Cheng, J. Zhang, and S. Guo, "Reduction of graphene oxide via L-ascorbic acid," Chem. Commun., vol. 46, no. 7, pp. 1112–1114, 2010.

[13] L. Liu, Z. Shen, X. Zhang, and H. Ma, "Highly conductive graphene/carbon black screen-printing inks for flexible electronics," J Colloid Interface Sci, vol. 582, pp. 12–21, Jan. 2021.

[14] Q. Wang, S. Ling, X. Liang, H. Wang, H. Lu, and Y. Zhang, "Self-Healable Multifunctional Electronic Tattoos Based on Silk and Graphene," Adv Funct Mater, vol. 29, no. 16, p. 1808695, Apr. 2019.

[15] D. N. Rockwood, R. C. Preda, T. Yücel, X. Wang, M. L. Lovett, and D. L. Kaplan, "Materials fabrication from Bombyx mori silk fibroin," Nat Protoc, vol. 6, no. 10, pp. 1612–1631, Oct. 2011.

[16] I. Childres, L.A. Jauregui, W. Park, H. Cao, and Y.P., Chen, "Raman spectroscopy of graphene and related materials," New developments in photon and materials research, 2013.

CONTACT

*M.K. Yapici, tel: +90-216-4839553; murat.yapici@sabanciuniv.edu

HIGHLY ACCURATE MEASUREMENT OF CONTACT RESISTANCE BETWEEN GALINSTAN AND COPPER USING TRANSFER LENGTH METHOD

Takashi Sato and Eiji Iwase
Waseda University, JAPAN

ABSTRACT

We have studied a measurement configuration for highly accurate measurement of the contact resistance (R_c) between gallium-based liquid metals (LMs) and solid metals (SMs) using the Transfer Length Method (TLM). In the conventional TLM, target materials with low conductivity are used as wiring and contact to SM electrodes with high conductivity. However, since LMs have almost the same conductivity as SMs, the measurement configuration for LMs is not obvious. We compared the measurement configurations of LMs and SMs. As a result, we found that the R_c for the configuration of SM wiring and LM electrodes was six times higher than that of LM wiring and SM electrodes. Our method accelerates research on LM interfaces and provides new opportunities for the development of stretchable electronics using LMs.

KEYWORDS

Liquid metal, Galinstan, Contact resistance, Transfer Length Method, Stretchable electronics

INTRODUCTION

Recently, stretchable electronic devices using gallium-based LMs have attracted significant attention[1]-[5]. In particular, stretchable electronic devices consisting of LMs and rigid electronic components such as chip LEDs and Micro-Electro-Mechanical System (MEMS) sensors have high potential in stretchability and performance. In such devices, the R_c between the LM and SM, such as electronic components or electrodes, is critical. Since LMs have high surface tension[6], the TLM is suitable for R_c measurement. In the TLM, the measured resistance (R) is the combined resistance of the volume resistance of wiring (R_{wiring}) and R_c[7]. R_{wiring} is obtained from the relationship between the wiring length (L_{wiring}) and R, and R_c is calculated by subtracting R_{wiring} from R.

The R_c in the TLM precisely includes the volume resistance of the electrode ($R_{electrode}$). Conventionally, $R_{electrode}$ is much lower than R_c and is assumed to be negligible[7]. For LMs and SMs, however, R_c may be so low that $R_{electrode}$ cannot be neglected[8]. Therefore, we compared $R_{electrode}$ and R_c and demonstrated a connection method of terminals to measure R that does not include $R_{electrode}$. Furthermore, in TLM measurement of materials with low conductivity, such as semiconductors, graphene, and conductive adhesives, these materials are used as wiring to improve the signal-to-noise ratio[7],[9]. However, since both LMs and SMs have high conductivity (10^6 to 10^7 S/m), the measurement configuration for LMs is not obvious. Thus, we studied TLM measurement configurations of LMs and SMs. Our study accelerates research on LM interfaces and provides new opportunities for developing stretchable electronics using LMs.

METHODS

Figure 1 shows the schematic illustrations of a method to measure R_c between LMs and SMs using the TLM with high accuracy. As shown in Figure 1(a), we compared the connection of measuring terminals. In the one-sided connection, the sensing and source terminals of the four-probe method were connected to the same end of the electrodes. The measured R is

$$R = R_{wiring} + 2R_c + 2R_{electrode} \\ = \alpha L_{wiring} + 2R_c + 2R_{electrode}. \tag{1}$$

In contrast, in the two-sided connection, the sensing and source terminals were connected to both ends of the electrodes. The measured R is

$$R = R_{wiring} + 2R_c \\ = \alpha L_{wiring} + 2R_c. \tag{2}$$

We compared the one-sided and two-sided connections to evaluate the effect of $R_{electrode}$ and demonstrated measuring R that does not include $R_{electrode}$ by the two-sided connection. As shown in Figure 1(b), we further compared the TLM measurement configurations: (i) configuration with LM wiring in contact with SM electrodes and (ii) configuration with SM wiring in contact with LM electrodes. In the TLM, the transfer length (L_T) expresses a length at which current mainly flows at the interface between wiring and electrode[7]. Using the L_T, the contact resistance per effective contact area, *i.e.*, the contact resistivity (ρ_c), is expressed by the following equation (3).

$$\rho_c = R_c L_T W_{wiring} \tag{3}$$

The W_{wiring} means the width of the wiring. We compared the R_c, L_T, and ρ_c for different measurement configurations.

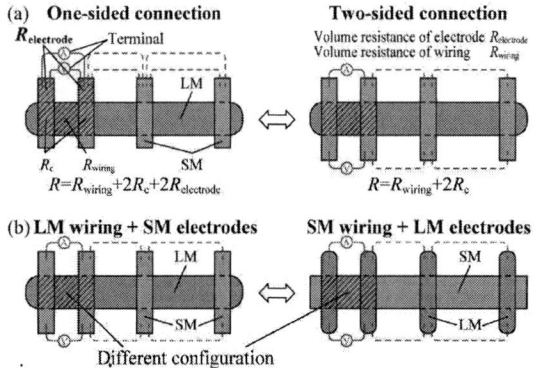

Figure 1: Schematic illustrations of measurement configuration for highly accurate measurement of contact resistance between gallium-based liquid metals (LMs) and solid metals (SMs) using Transfer Length Method (TLM).

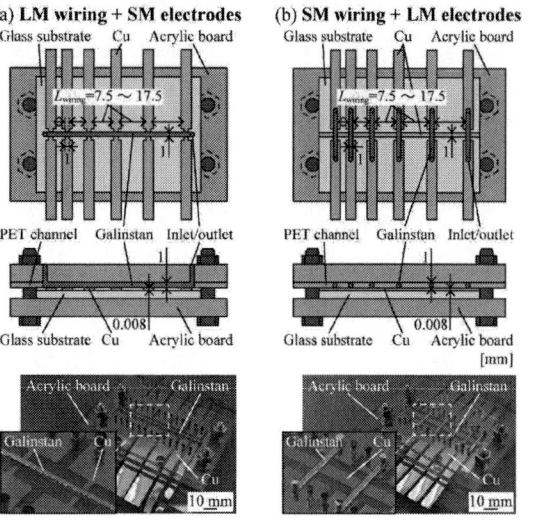

Figure 2: Schematic illustrations and photographs of TLM measurement devices with (a) LM (galinstan) wiring and SM (copper) electrodes, (b) SM wiring and LM electrodes.

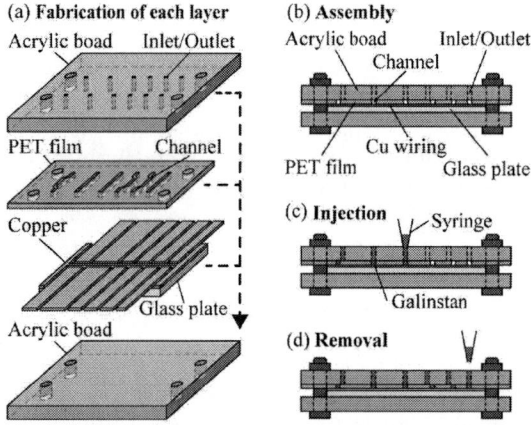

Figure 3: Schematic illustrations of fabrication procedure of TLM measurement devices.

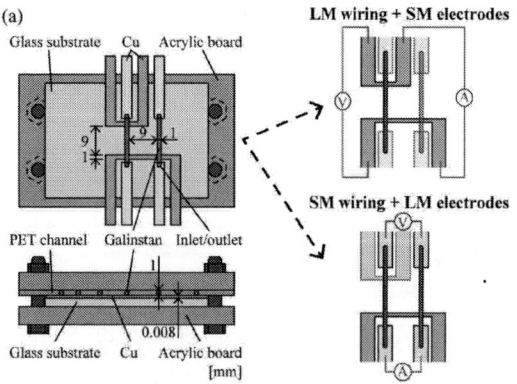

Figure 4: Schematic illustrations of L-shaped devices for measuring contact resistance at same contact point by different measurement configurations.

Figure 2 shows schematic illustrations and photographs of TLM measurement devices with LM wiring and SM electrodes (Figure 2(a)) and with SM wiring and LM electrodes (Figure 2(b)). We used galinstan (Changsha Rich Nonferrous Metals, China) as an LM material and copper substrate (Toray, Japan)) as an SM material. In the measurement device, galinstan contacted a copper substrate. The copper substrate was patterned into electrode or wiring shapes with a UV laser cutter. The galinstan filled a cutout in the PET substrate in the shape of the electrodes or wirings. Acrylic substrates sandwiched a PET substrate with cutouts filled with galinstan and a copper substrate. This structure can seal LM without changing the surface condition of the SM substrate. The copper substrate and galinstan were 0.008 mm thick and 1 mm thick, respectively. The electrodes and wiring were 1 mm wide, and L_{wiring} was from 7.5 mm to 17.5 mm.

Figure 3 shows the fabrication process of the measurement devices. As shown in Figure 3(a) and (b), two acrylic substrates, a PET substrate with cutouts, and a copper substrate were assembled and fixed. The copper substrate was treated with flux (NS30, Nihon Superior, Japan) before assembly. Then, as shown in Figures 3(c) and (d), galinstan was injected with a syringe in the cutout and sealed. The resistance of the device R was measured with an ohm meter (RM3545, Hioki, Japan) and a switch main frame (SW1002, Hioki, Japan). A device had five different L_{wiring}s, and the resistance R was measured ten times for each L_{wiring}. An approximate straight line was drawn for the measurement result (200 plots) for the four devices, and Y-intercept and its standard deviation were calculated.

Since the R_{c} of LM may differ depending on the contact procedures[8], we also studied the cause of the difference in TLM measurement results. Figure 4 shows an L-shaped device for measuring contact resistance at the same contact point with different measurement configurations. In the L-shaped device, two LM lines and two SM lines crossed each other. For the configuration with LM wiring and SM electrodes, one LM wiring and two SM electrodes were measured. For the configuration with SM wiring and LM electrodes, one SM wiring and two LM electrodes were measured. The copper substrate and galinstan were 0.008 mm thick and 1 mm thick, respectively. The electrodes and wiring were 1 mm wide, and L_{wiring} was 10 mm. The measured R is

$$
\begin{aligned}
R &= R_{\mathrm{wiring}} + 2R_{\mathrm{c}} \\
&= \alpha L_{\mathrm{wiring}} + 2R_{\mathrm{c}}.
\end{aligned} \tag{4}
$$

R_{wiring} was calculated from α obtained from Figure 4 and L_{wiring}, and R_{c} is calculated by subtracting R_{wiring} from R.

RESULTS AND DISCUSSION

Figure 5 shows the TLM measurement results for the connection of measuring terminals. Since we used the same device for one-sided and two-sided connections, the slope of the approximate straight line α of equations (1) and (2) was 0.239 mΩ/mm in both results. The measured α was almost the same (-16.7%) as α of 0.279 mΩ/mm estimated from the conductivity and dimensions of the galinstan

978-1-6654-9309-3/23 $31.00 © 2023 IEEE

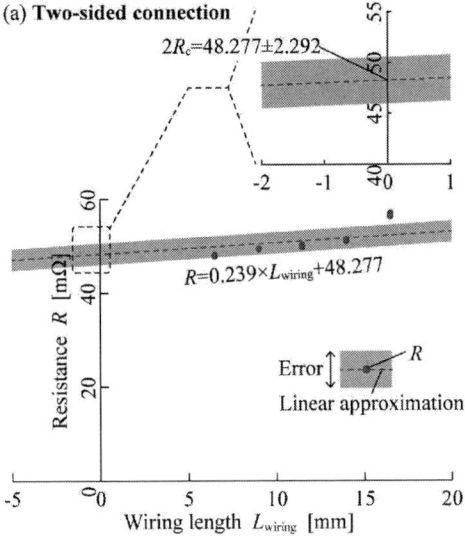

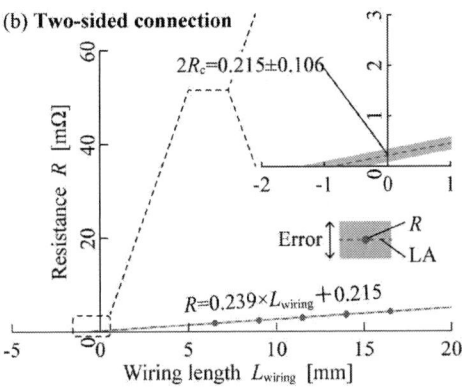

Figure 5: TLM measurement results of contact resistance on connections of measurement terminals. (N=4)

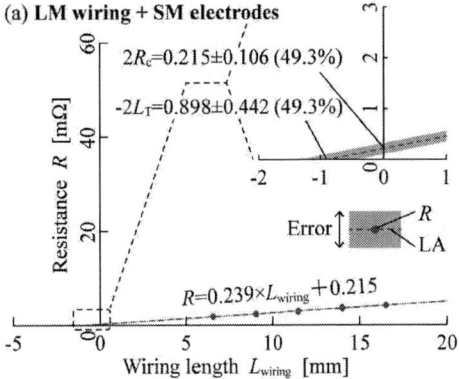

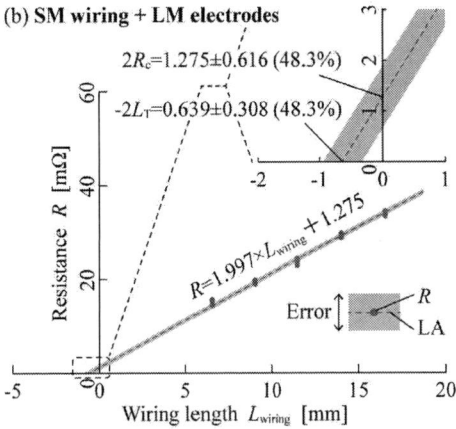

Figure 6: TLM measurement results of contact resistance on configurations with (a) LM wiring and SM electrodes, (b) SM wiring and LM electrodes. (N=4)

wiring. The Y-intercept (= $2R_{electrode} + 2R_c$) was 48.277 ± 2.292 mΩ for the one-sided connection, while the Y-intercept (= $2R_c$) was 0.215 ± 0.106 mΩ for the two-sided connection. The difference in Y-intercept between one-sided and two-sided connections was 48.062 mΩ. The measured difference was almost the same (-0.8%) as the difference of 48.444 mΩ estimated from the conductivity and dimensions of the copper electrode. Therefore, $2R_c$ and $2R_{electrode}$ were 0.215 mΩ and 48.062 mΩ, respectively. These results show that $R_{electrode}$ cannot be ignored for the TLM measurement of LMs and SMs since $R_{electrode}$ was higher than R_c by more than two orders of magnitude. Moreover, we demonstrated that R could be measured without including $R_{electrode}$ by the two-sided connection.

Figure 6 shows the TLM measurement results for the measurement configurations. The slope α for the configuration of LM wiring and SM electrodes and that of SM wiring and LM electrodes were 0.239 mΩ/mm and 1.997 mΩ/mm, respectively. The measured α for the configuration of SM wiring and LM electrodes was almost the same (-5.2%) as α of 2.100 mΩ/mm estimated from the conductivity and dimensions of the copper wiring. The Y-

intercept was 0.215 ± 0.106 mΩ (error: 49.3%) for the configuration of LM wiring and SM electrodes. In contrast, Y-intercept was 1.275 ± 0.616 mΩ (error: 48.3%) for the configuration of SM wiring and LM electrodes. These results show that R_c differed by six times for the configurations.

Figure 7(a) shows the R_c for the measurement configurations measured by L-shaped and TLM devices obtained from equation (4). For the configuration of LM wiring and SM electrodes, R_c in L-shaped devices was 0.091 ± 0.028 mΩ, almost the same (-15.0%) as in the TLM devices. For the configuration of SM wiring and LM electrodes, R_c in L-shaped devices was 0.635 ± 0.182 mΩ, almost the same (-0.5%) as in the TLM devices. These results show that the difference in R_c between configurations was due to the measurement configuration, not the fabrication process. Figures 7(b) and (c) shows the L_T and ρ_c at the same contact point by the different measurement configurations. The L_T for the configurations of LM wiring and SM electrodes and that of SM wiring and LM electrodes were 0.449 ± 0.221 mm and 0.319 ± 0.154 mm, respectively. Moreover, the ρ_c for the configurations of LM wiring and SM electrodes and that of SM wiring and LM electrodes were 0.048 ± 0.034 mΩmm^2 and $0.204 \pm$

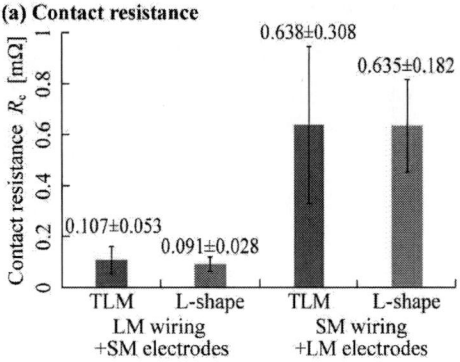

(a) Contact resistance

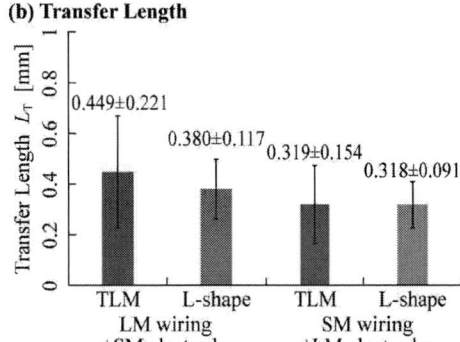

(b) Transfer Length

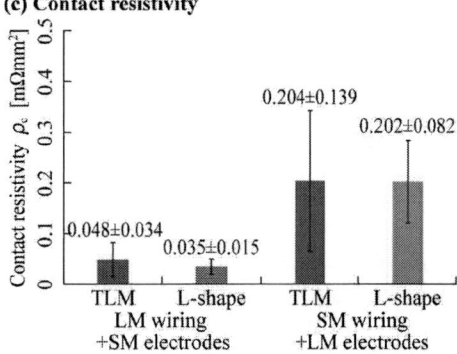

(c) Contact resistivity

Figure 7: (a) Contact resistance, (b) transfer length, (c) contact resistivity at same contact point by different measurement configurations. (N=6)

0.139 $m\Omega mm^2$, respectively. Therefore, we found that the ρ_c for the configuration of SM wiring and LM electrodes was four times higher than that of LM wiring and SM electrodes.

CONCLUSION

We have studied a measurement configuration for highly accurate measurement of the R_c between gallium-based LMs and SMs using the TLM. First, we evaluated the effect of $R_{electrode}$ on TLM measurement. From the TLM measurement results for the connection of measurement terminals, we found that $R_{electrode}$ cannot be ignored for the

TLM measurement of LMs and SMs since $R_{electrode}$ was higher than R_c by more than two orders of magnitude. We also demonstrated that R could be measured without including $R_{electrode}$ by the two-sided connection. Furthermore, we investigated the measurement configurations of LMs and SMs. From the TLM measurement results for the configurations, we found that the R_c for the configuration of SM wiring and LM electrodes was six times higher than that of LM wiring and SM electrodes. Our study accelerates research on LM interfaces and provides new opportunities for the development of stretchable electronics using LMs.

ACKNOWLEDGEMENTS

This work was partly supported by JST ACT-X JPMJAX21K6 and JSPS KAKENHI JP22J13665.

REFERENCES

[1] S. Zhu *et al.*, "Ultrastretchable fibers with metallic conductivity using a liquid metal alloy core," *Adv. Funct. Mater.*, vol. 23, no. 18, pp. 2308–2314, May 2013.

[2] Y. Gao *et al.*, "Wearable Microfluidic Diaphragm Pressure Sensor for Health and Tactile Touch Monitoring," *Adv. Mater.*, vol. 29, no. 39, 1701985, 2017.

[3] Y. R. Jeong *et al.*, "A skin-attachable, stretchable integrated system based on liquid GaInSn for wireless human motion monitoring with multi-site sensing capabilities," *NPG Asia Mater.*, vol. 9, no. 10, e443, 2017.

[4] K. B. Ozutemiz, J. Wissman, O. B. Ozdoganlar, and C. Majidi, "EGaIn–Metal Interfacing for Liquid Metal Circuitry and Microelectronics Integration," *Adv. Mater. Interfaces*, vol. 5, no. 10. 1701596, 2018.

[5] M. Zadan, M. H. Malakooti, and C. Majidi, "Soft and Stretchable Thermoelectric Generators Enabled by Liquid Metal Elastomer Composites," *ACS Appl. Mater. Interfaces*, vol. 12, no. 15, pp. 17921–17928, 2020.

[6] T. Daeneke *et al.*, "Liquid metals: Fundamentals and applications in chemistry," *Chem. Soc. Rev.*, vol. 47, no. 11, pp. 4073–4111, 2018.

[7] D. K. Schroder, *Semiconductor material and device characterization*, 3rd ed., *Wiley-Blackwell*, 2006.

[8] T. Sato, K. Yamagishi, M. Hashimoto, and E. Iwase, "Method to Reduce the Contact Resistivity between Galinstan and a Copper Electrode for Electrical Connection in Flexible Devices," *ACS Appl. Mater. Interfaces*, vol. 13, no. 15, pp. 18247–18254, 2021.

[9] M. Shaygan *et al.*, "Low resistive edge contacts to CVD-grown graphene using a CMOS compatible metal," *Ann. Phys.*, vol. 529, no. 11, 1600410, 2017.

CONTACT

T. Sato, tel: +81-3-5286-2741; machotakashi@fuji.waseda.jp
E. Iwase, tel: +81-3-5286-2741; iwase@waseda.jp

MACHINE LEARNING ENABLED HIND FOOT DEFORMITY DETECTION USING INDIVIDUALLY ADDRESSABLE HYBRID PRESSURE SENSOR MATRIX

Nadeem Tariq Beigh, Faizan Beigh, Sourav Naval, Dibyajyoti Mukherjee, and Dhiman Mallick
Department of Electrical Engineering, Indian Institute of Technology Delhi, New Delhi, INDIA

ABSTRACT

We demonstrate a scalable, cost-effective hybrid sensor based on piezoelectric-triboelectric mechanisms for pressure mapping of 36 regions of the hindfoot and precise detection of any deformity. The matrix sensor comprises of individually addressable 6×6 sensors, which prevent pressure cell crosstalk and enhance the detection limit of the sensor up to 170kPa. The sensor can be easily integrated as an in-sole device because it is constructed on a flexible substrate using photo-patternable BTO (Barium Titanate)/SU-8 nanocomposite. The developed sensor has a pressure sensitivity of 34mV/kPa for 0-5kPa and 2.7mV/kPa for 5-170kPa, respectively. Additionally, we apply a convolution neural network (CNN-2D) to extract features from the 36 hybrid signals for precise deformity detection. The developed CNN model has a model loss of 0.1%, 98% accuracy, and a cross-feature differentiation value of 2%.

KEYWORDS

Piezoelectric, Triboelectric, Pressure Sensor, Convolution Neural Network, Flexible Sensor.

INTRODUCTION

Wearable or flexible sensors have emerged as a rapidly expanding research field in recent years because of their inherent flexibility and lightness. It offers a wide range of applications, including human-activity monitoring, human-machine interface, physiological disorder diagnosis, etc., to improve the quality of life. Wearable sensors can be directly affixed to the human body and worn as accessories to precisely identify the body's motion and dynamics. Combined with data analytic algorithms, these wearable devices can provide valuable data-based insights into biomechanics to quantify health parameters for biomedical monitoring, diagnosis, and several other critical applications [1]. Primarily, wearable sensors are developed into clinical and commercial devices. Commercial wearables provide information to consumers regarding primary biomarkers such as heart rate, blood glucose, and activity tracking. Clinical wearables have been developed for healthcare professionals to monitor and analyze walking patterns and body posture and perform gait assessment tasks in different scenarios.

In the modern world, gait and postural deformities are common due to an improper lifestyle [2]. Monitoring and detecting these deformities become essential before they lead to permanent body impairment or cause a grave injury. Researchers have been working extensively on postural deformities and correction. Studies indicate that the most prominent deformities include splay foot, flat foot, unstable hind foot with protruding heels, high arches, and irregular gait that cause difficulty in walking and correct posture.

Hindfoot deformities that correlate to the plantar pressure distribution are the prime reason for knee weakness and joint pains. There have been several investigations on plantar pressure mapping of the foot. Most of these studies indicate that the hind foot generates the highest plantar pressure during walking, running, or jumping.

Several pressure sensors have been developed in the recent past to map hind foot pressure The majority of devices that have been described use a sensor with one of the following transduction mechanisms: piezoresistive [3], capacitive [4], piezoelectric [5], or triboelectric [6]. A single transduction mechanism solely relies on a specific sensing parameter (strain for piezoelectric and contact area for triboelectric etc.) to estimate the applied stimulus. Combining different transduction mechanisms in a sensor device allows a precise pressure measurement. This has motivated the researchers to switch to a hybrid device involving multiple types of transducers [7]. A large fraction of the developed hybrid topologies involves distinct generators, making the device bulky and unsuitable for wearable applications. Piezoelectric nanoparticles mixed with photoresists possess dual piezoelectric-triboelectric properties and inherent photopatternability, making them suitable for micro-patterning [8]. Albeit their salient characteristics, the works involving photopatternable nanocomposites are still minimal. It is also important to note that the amount of data collected using a single sensor is insufficient to accurately map the pressure generated by different hind foot regions. An array of pressure cells placed beneath different areas of the foot have been reported in the literature to provide a higher resolution, better sensitivity, and efficient pressure mapping to differentiate among various pressure map patterns [9].

Here, we report a novel, individually addressable 6×6 matrix-mapped pressure sensor for diagnosing foot deformities based on photopatternable BTO/SU-8 nanocomposites with dual piezoelectric/triboelectric transduction. The designed sensor has a high resolution, mapping 36 pressure points uniquely and evaluating the generated pattern, correlating with the type of deformity. This paper elaborately discusses the proposed sensor's design and the associated fabrication details. The performance is characterized by simulation analysis and backed by rigorous experimental validation. Finally, the hind foot data acquired as a set of 32 hybrid voltage signals are processed by a 2D-CNN model to differentiate between normal and abnormal hindfoot.

MATERIALS AND METHODS
Device Fabrication

The device fabrication process flow is shown in Figures 1 (a-e). Initially, a 5cm×5cm×50μm polyamide (Kapton) substrate is patterned, followed by the subsequent steps of bottom electrode deposition and etching. Each

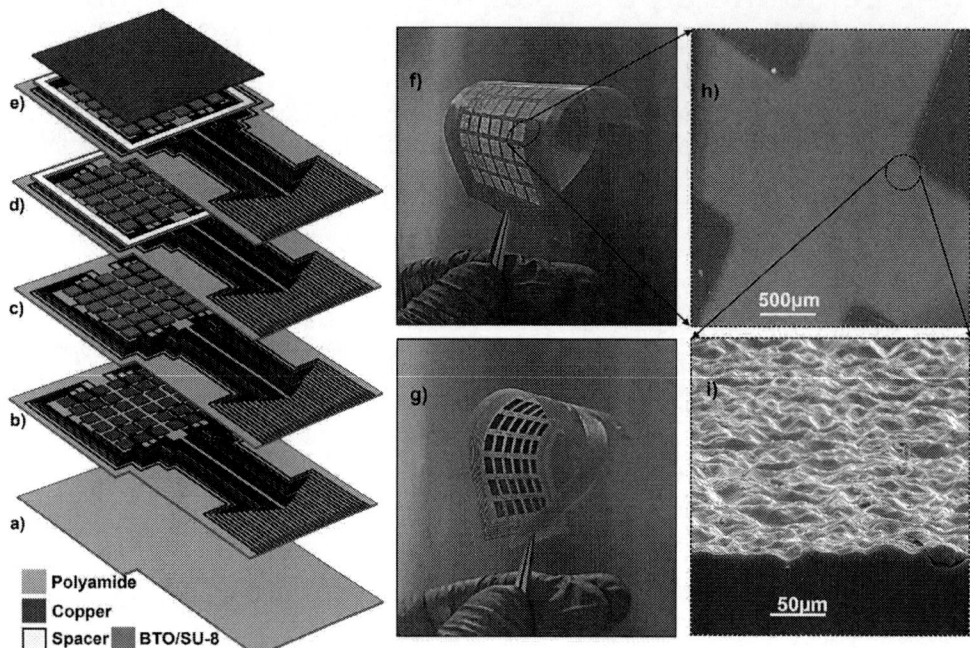

Figure 1: Device fabrication process flow: a) Kapton polyamide substrate b) Copper electrode deposition and etching c) BTO/SU-8 nanocomposite deposition and patterning d) spacer placement and e) Top electrode placement. Actual device prototype f) front and g) back view. SEM images of device h) top and i) magnified view.

electrode is 5mm×5mm and is individually addressable. Next, a 20% weight ratio BTO/SU-8 nanocomposite is spin-coated at 1000rpm for 30 seconds to achieve a film thickness of ~50 μm on the patterned substrate. The nanocomposite thin film is pre-baked at 95°C for 15 minutes and given a UV exposure dose of 250 mJ/cm². This is followed by a post exposure bake of 1 minute and 5 minutes at 65°C and 95°C, respectively. The film is dipped in SU-8 3000 developer and rigorously sonicated for 5 minutes to achieve distinct and clean patterns. Finally, the developed substrate is cleaned in IPA (Isopropyl alcohol) and dried using a nitrogen gun, followed by a moisture bake for 2 minutes at 95°C. The process parameters for thin film deposition, exposure, and development are according to [8]. The front and back view of the fabricated sensor is shown in Figure 1(f) and 1(g), respectively. The topology of the deposited/patterned nanocomposite thin film is shown by the SEM (scanning electron microscopy) image in Figure 1(h), and the corresponding magnified view is presented in Figure 1(i).

A Kapton spacer of 150 μm is placed around the pressure cells. A top contacting surface of copper tape adhered on the ITO/PET sheet is affixed over the Kapton spacer using double-sided tape. Therefore, providing the necessary gap between the contacting layers. Finally, the protruding copper contacts are attached to a 40-pin flexible printed circuit (FPC) to a 2×20 insulation-displacement connector (IDC) for interfacing with circuitry.

RESULTS AND DISCUSSIONS
Finite Element Simulation

The hybrid pressure sensor is fabricated by using a dual transduction and photo patternable 20% BTO/SU-8

nanocomposite layer. The dual transduction properties and reliability of this nanocomposite is well established in the previous works [7,8], and is further verified by the

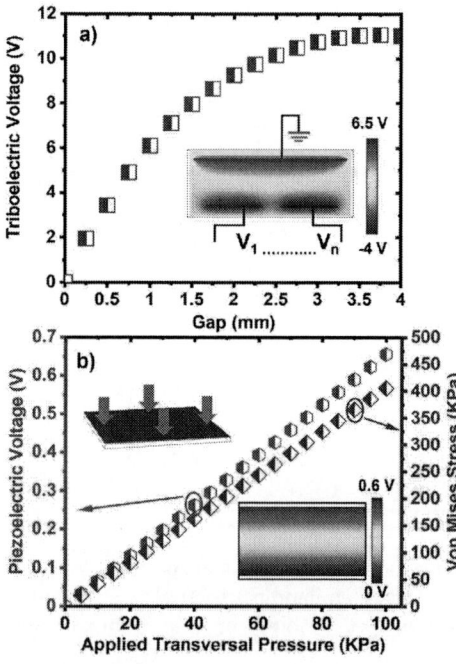

Figure 2: Simulation results a) gap varying triboelectric voltage with voltage distribution for 2mm gap shown in inset and b) pressure varying piezoelectric voltage and stress study with transversal voltage distribution for 100kPa given in inset.

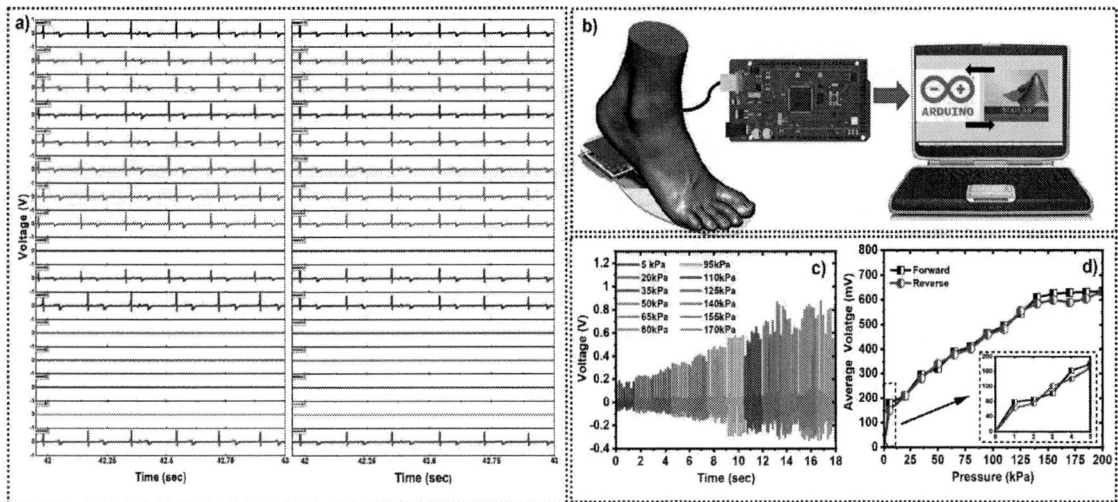

Figure 3: Experimental results a) individual pressure cell voltage responses of a normal foot b) Illustration of Data acquisition and circuit interface with GUI c) Output response for a varying pressure amplitude sweep d) Voltage-Pressure hysteresis curve of the unit pressure cell with low pressure response shown in the inset.

multiphysics finite element method (FEM) simulations. The simulation parameters are given in Table I. The variation of output voltage as a function of the gap between the triboelectric layers is shown in Figure 2(a). The resulting differential floating potential between the contacting layers gives a linear relation with the gap distance and reaches saturation at 11.8V for a gap greater than 3.5mm. The observed trend is similar to that reported in the previous works on the triboelectric sensor, wherein the triboelectric voltage saturates beyond a threshold gap due to charge equalization [11]. The variation of piezoelectric output voltage as a function of applied pressure is presented in Figure 2(b). The study is conducted over the pressure range of 0-100 kPa, which covers the hind foot's walking, running, and standing pressure. However, the unique device design allows only a single electrode to extract piezoelectric and triboelectric output, depending on the contact state. The piezoelectric output aids the triboelectric output and sustains the sensor output in the complete contact mode, which is zero for a triboelectric-only thin film.

Table 1: Simulation parameters

Parameter	Value
Cell Area	4 mm^2
Surface Charge Density	100nC/cm^2
d_{31}	10pm/V

Mechanical characterization

The mechanical characterization of the developed pressure sensor cell is performed using a mechanical pressure imparter setup (PTI-3) with a variable impacting pressure of 0-200kPa at 4Hz. The response shows a typical pressure sensor behavior with a multiple working regime, the deep linear, linear, and saturation region. The output response saturates beyond 170kPa due to charge neutralization as well as by the deformation limit of the nanocomposite (Figure 3(c)). The sensor shows two sensitivities over the linear region with a value of 34 mV/kPa for 0-5kPa and 2.7mV/kPa for 5-170kPa. The

hysteresis study is depicted in Figure 3(d), wherein the sensor demonstrates a reliable response with a negligible hysteresis over forward and backward pressure sweeps. The average deviation between respective test points is 2.7% and thereby confirms the sensor's consistency.

Hind foot deformity detection

Accurate gait analysis is vital for many systems that help clinicians diagnose and manage abnormal gait. It can also help identify falls, hind foot, and plantar pressure anomalies, which are frequent in the elderly and sports. Force plates are used as gait sensors. Force plates are lab-only, expensive, and require professional operation. Wearable sensor technology makes gait analysis practical. Sensor insoles that monitor foot pressure are utilized for gait analysis since each gait event can be defined by a pressure pattern [12].

The developed pressure sensor matrix maps the hind foot pressure via 32 hybrid outputs given to two 16:1 MUXs

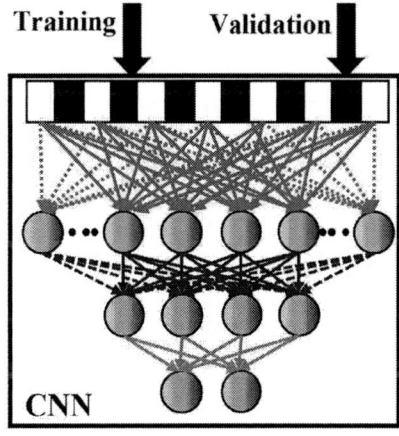

Figure 4: The developed CNN model illustration with input data set classification

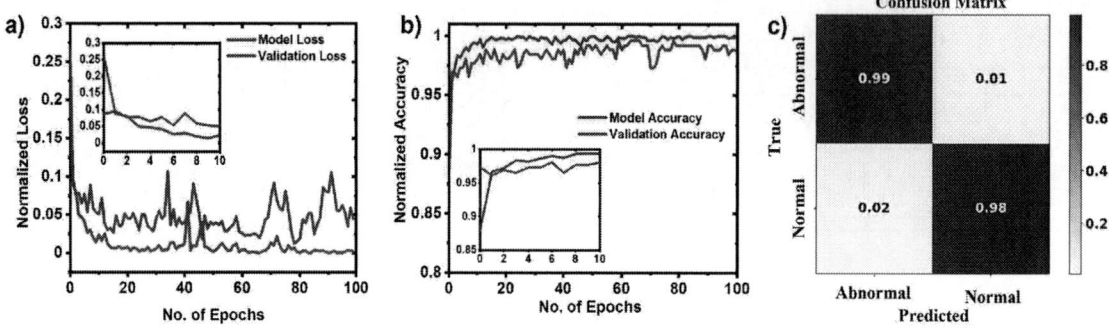

Figure 5: a) Resulting model/validation loss b) Model/validation accuracy and c) confusion matrix.

connected separately to two 10 bit-ADCs clocked by ATMEGA2560 microcontroller. The output data drives a customized 2D-CNN model (Figure 4) with an Adam optimizer, wherein the data is acquired with a specific window length of 80 and hoop size of 40. The training data is set at 20% of the total data set with a learning rate of 0.01 and the model is run for 100 epochs. The resulting loss and accuracy values for the model and validation are depicted in Figures 5 (a) and (b), respectively. The obtained 0.1% model loss indicates excellent learning efficiency. The 98.84% model accuracy, 98.01% precision, 2% false positivity, and a 0.98 F1 score given in Figure 5(c) imply that exceptional feature identification and differentiation is achieved. The model metrics demonstrate that deformities of the hind foot can be differentiated precisely and hence enable early deformity diagnosis.

CONCLUSION

A scalable, cost-effective hybrid pressure sensor based on piezoelectric-triboelectric mechanisms was demonstrated. The individually addressable 6×6 matrix sensors inhibit pressure cell crosstalk and boost the sensor's detection limit to 170kPa. The sensor's flexible substrate and photopatternable BTO/SU-8 nanocomposite make it easy to incorporate as an in-sole device. The developed sensor is 34mV/kPa for 0-5kPa and 2.7mV/kPa for 5-170kPa. We also used a convolution neural network (CNN-2D) to detect hind foot abnormalities with 0.1% model loss, 98% accuracy, 2% cross-feature differentiation, 98.01 % precision, 2% false positivity, and a 0.98 F1 score.

ACKNOWLEDGEMENTS

This work is financially supported by I-HUB Foundation for Cobotics, Technology Innovation Hub, IIT Delhi. The authors also acknowledge the cleanroom facilities at NRF and CRF, IIT Delhi.

REFERENCES

[1] X. Wang, Z. Liu, and T. Zhang, "Flexible Sensing Electronics for Wearable/Attachable Health Monitoring," Small, vol. 13, no. 25, p. 1602790, 2017,

[2] W. Tao, T. Liu, R. Zheng, and H. Feng, "Gait Analysis Using Wearable Sensors," Sensors, vol. 12, no. 2, Art. no. 2, Feb. 2012

[3] L. Pan et al., "An ultra-sensitive resistive pressure sensor based on hollow-sphere microstructure induced elasticity in conducting polymer film," Nat Commun, vol. 5, no. 1, Art. no. 1, Jan. 2014

[4] K. F. Lei, K.-F. Lee, and M.-Y. Lee, "Development of a flexible PDMS capacitive pressure sensor for plantar pressure measurement," Microelectronic Engineering, vol. 99, pp. 1–5, Nov. 2012

[5] Y. Cha, H. Kim, and D. Kim, "Flexible Piezoelectric Sensor-Based Gait Recognition," Sensors, vol. 18, no. 2, Art. no. 2, Feb. 2018,

[6] Q. Zhang et al., "Wearable Triboelectric Sensors Enabled Gait Analysis and Waist Motion Capture for IoT-Based Smart Healthcare Applications," Advanced Science, vol. 9, no. 4, p. 2103694, 2022,

[7] S. Naval, N.T. Beigh, D. Mukherjee, A. Jain and D. Mallick, "Flexible V-Shaped Piezoelectric-Triboelectric Device for Biomechanical Energy Harvesting and Sensing," Journal of Physics D: Applied Physics, 2022,

[8] N.T. Beigh, S. Singh, A. Goswami, and D. Mallick," Dual Piezoelectric/Triboelectric Behavior of BTO/SU-8 Photopatternable Nanocomposites for Highly Efficient Mechanical Energy Harvesting," Adv. Electron. Mater, 2022.

[9] R. Matsuda et al., "Highly stretchable sensing array for independent detection of pressure and strain exploiting structural and resistive control," Sci Rep, vol. 10, no. 1, p. 12666, Dec. 2020,

[10] N. T. Beigh and D. Mallick, "Low-Cost, High-Performance Piezoelectric Nanocomposite for Mechanical Energy Harvesting," IEEE Sensors Journal, vol. 21, no. 19, pp. 21268–21276, Oct. 2021,

[11] Wang, Z., Liu, W., Hu, J., He, W., Yang, H., Ling, C., Xi, Y., Wang, X., Liu, A. and Hu, C., 2020. Two voltages in contact-separation triboelectric nanogenerator: from asymmetry to symmetry for maximum output. *Nano Energy, 69*, p.104452.

[12] Díaz, S., Stephenson, J.B. and Labrador, M.A., 2019. Use of wearable sensor technology in gait, balance, and range of motion analysis. *Applied Sciences, 10*(1), p.234

CONTACT

*Nadeem Tariq Beigh, Mob.: +91-7006-503-538; Nadeem_Beigh@ee.iitd.ac.in

MULTI-MODE E-SKIN INTEGRATING CAPACITIVE-PIEZOELECTRIC SENSORS FOR STATIC-DYNAMIC MECHANORESPONSE WITH WIDE SENSING RANGE

Mujeeb Yousuf[1], Sushil Kumar[1], Dhairya Singh Arya[2], Manu Garg[1], Khanjan Joshi[1], and Pushpapraj Singh[1]*

[1]Indian Institute of Technology, Delhi and
[2] CSIR-Central Scientific Instruments Organization (CSIO), Chandigarh

ABSTRACT

Emulating human somatosensory response requires integrated (dynamic and static) mechanoresponsive capabilities that demand multimode implementation. Here, we introduce a multimode mechanoreceptor for detecting dynamic stimuli using piezoelectric sensing-mode (P-mode) and static stimuli using capacitive mode (C-mode). For P-mode, the piezoelectric layer is sandwiched between electrodes to mimic dynamic response and to respond to static stimuli μ-pyramidal shaped capacitive device is integrated. This synergetic effect allows for a broader detection range of 1 kPa to 140 kPa, which is simply missing in their standalone responses (C-mode: 1 kPa-to-25 kPa @ 30fF/kPa and P-mode: 20 kPa-to-140 kPa @ 150 mV/kPa). Our method simplifies the sensory integration for modern e-skins. A real-time human grabbing and holding activity validates its functionality for static/dynamic stimuli.

KEYWORDS

Multimode Sensor, Mechanoreceptors, μ-Pyramidal, Static, Dynamic Stimuli

INTRODUCTION

Human skin can perceive and differentiate different spatiotemporal tactile stimuli (static and dynamic pressure, vibrations) through different mechanoreceptors [1]. Slowly adapting mechanoreceptors (i.e., Merkel and Ruffini corpuscles) perceive static or sustained pressure on the skin [2]. Similarly, fast-adapting mechanoreceptors (i.e., Meissner and Pacinian corpuscles) detect dynamic pressure and vibrations. For E-skin to emulate human somatosensory response, perceiving and discriminating between static and dynamic stimuli with a wide sensing range is of paramount importance [3]. Such Multimode flexible E-skin finds tremendous potential in the field of robotics, artificial skin, human-machine interface, and wearable health monitoring devices. For example, these capabilities in E-skin assist the robots in grasping, holding, and manipulating objects [4].

Earlier efforts used either single-mode piezoresistive/ capacitive sensors for static stimuli perception or piezoelectric for dynamic stimuli detection with low sensitivity range [5-6]. Capacitive or piezoresistive mode depends on the continuous change in electrical response (current, voltage, resistance, capacitance) due to sustained pressure stimuli. The P-mode responds only to dynamically changing stimuli. To increase the sensitivity and detection range, E-skin with different novel materials and designs (μ-pillar, μ-pyramid, graded structure) are reported [7].

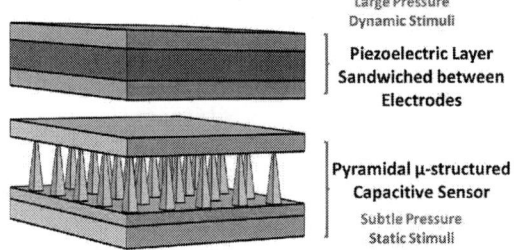

Figure 1: Conceptual Image of the multi-mode device. The device integrates the capacitive-piezoelectric sensors for the static and dynamic mechano-response. This synergetic effect allows detection of pressure stimuli over a broad sensing range (1-140 kPa).

However, using a single-mode sensor to simultaneously detect the pressure over a broad pressure range with high sensitivity is still difficult. Few works reported multimode performances but are limited to texture recognition only and suffer from the poor sensing-range (~1-60 kPa) [8].

Herein we extended the functionality by integrating a capacitive-piezoelectric device for simultaneous dynamic and static signals detection with a wide sensing range. The C-mode detects subtle and static pressure. The P-mode detects the dynamic touch with a large pressure range. The synergetic effect of P-mode and C-mode extends the sensitive range (1-140 kPa) and enables multifunctionality (static and dynamic perception) in the device. The epidermal μ-pyramidal dielectric (Polydimethylsiloxane) PDMS in the capacitive device enhances the sensitivity and enables subtle pressure detection. As a proof of concept, the multimode device is demonstrated for real-time grabbing, holding, and slipping of the object by discerning the static and dynamic signals.

DEVICE CONCEPT

We report a wide detection range multimode (P/C-mode) sensory platform by integrating epidermal μ-structured capacitive (for C-mode) and piezoelectric device (P-mode). Inspired by human skin, the device (Fig. 1) integrates the piezo-sensor along with C-mode device. Here, C-mode perceives static stimuli whereas P-mode detects the dynamic response simultaneous. In the demonstrated device, top layer piezoelectric film is responsible for dynamic and large stimuli detection, whereas the bottom μ-pyramidal shaped capacitive sensor responds to static and subtle stimuli. Improved stress-

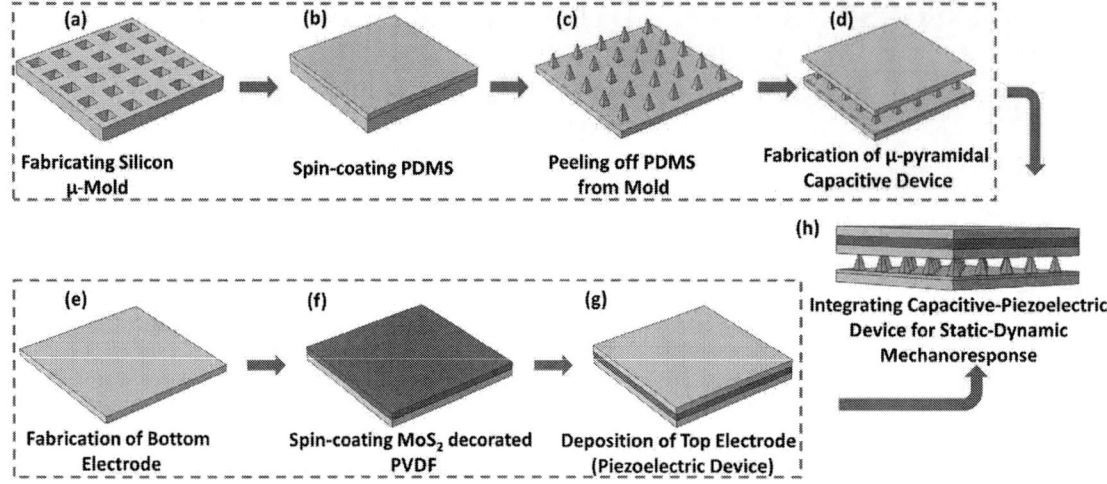

Figure 2: Schematic of the fabrication process of the multi-mode device. The fabrication process is divided into three steps viz: fabrication of capacitive sensor, fabrication of piezoelectric sensor and the integration of both devices. (a) fabrication of the silicon mold by standard lithography process followed by the wet etching of silicon using KOH (b) spinning coating of PDMS on silicon mold (c) peeling off the PDMS from silicon mold (d) sandwiching the μ-pyramidal PDMS between the electrodes (e) gold electrode deposition (using e-beam) on PET sheet for bottom electrode (f) spin coating piezoelectric layer on bottom electrode (g) top electrode fabrication (h) integrated multi-mode device.

displacement transduction of μ-pyramidal μ-structure of PDMS offers improved sensitivity [9]

EXPERIMENTAL METHODS

Fabrication and Characterizations

The multimode E-skin is integrated by vertical stacking of capacitive and piezoelectric sensors. The fabrication process is divided into three steps: fabrication of capacitive sensor, fabrication of piezoelectric sensor, and assembly of these sensors. First, the μ-pyramidal PDMS dielectric layer is fabricated using silicon molds. The silicon is lithographically patterned by opening a window of 30 μm with a pitch size of 50 μm followed by wet etching with potassium hydroxide (KOH). Then, the PDMS is spin-coated on the mold. The PDMS dielectric layer is peeled-off after curing (for an hour @ 120⁰ C). The top and bottom electrodes are formed by depositing gold on a (Polyethylene Terephthalate) PET sheet. Both electrodes are spin-coated with a very thin layer of PDMS, and after pre-curing it for 10 minutes, the μ-pyramidal layer is sandwiched between the electrodes. The thin layer of pre-cured PDMS serves as an adhesion promoter between the electrodes and the sandwiched μ-pyramidal dielectric layer. Secondly, the piezoelectric sensor is fabricated by a spin-coating piezoelectric film on a bottom electrode (fabricated on a PET sheet). The top electrode is formed by depositing the gold using an e-beam evaporator. Finally, the individual devices are assembled and encapsulated with the PDMS after taking contacts. Fig.3 (a) and (b) show the top view of the Sem Image of the μ-pyramidal PDMS layer.

WORKING MECHANISM

Mimicking the somatosensory response of the human skin, the multimode sensor E-Skin is fabricated by

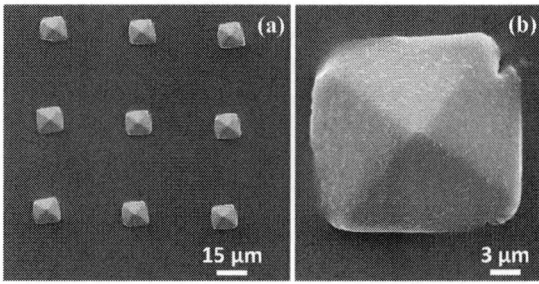

Figure 3: (a) SEM image of the μ-pyramidal PDMS (b) magnified image showing one μ-pyramid

integrating a capacitive-piezoelectric layer for a wide sensing range. The dynamic pressure stimuli on the piezoelectric layer generate the output voltage. The generated voltage is quantified by the equation given as [10]:

$$V_{oc} = g_{33} \times \sigma \times Y \times g = \frac{d_{33}}{\varepsilon_0 \varepsilon_r} \times \sigma \times Y \times g \qquad (1)$$

where g_{33} is the voltage constant, σ is the stress-induced, g is the thickness of the material, and Y is young's modulus. Therefore, the output voltage is proportional to the stress-induced, which in turn depends on the pressure applied.

The response of the capacitive sensor can be realized by the parallel plate configuration, the capacitance is given as

$$C = \frac{\varepsilon_0 \varepsilon_r A}{d} \qquad (2)$$

where $\varepsilon_0 \varepsilon_r$ is the relative permittivity, A is the overlapping area, and d is the thickness of the dielectric layer. With the

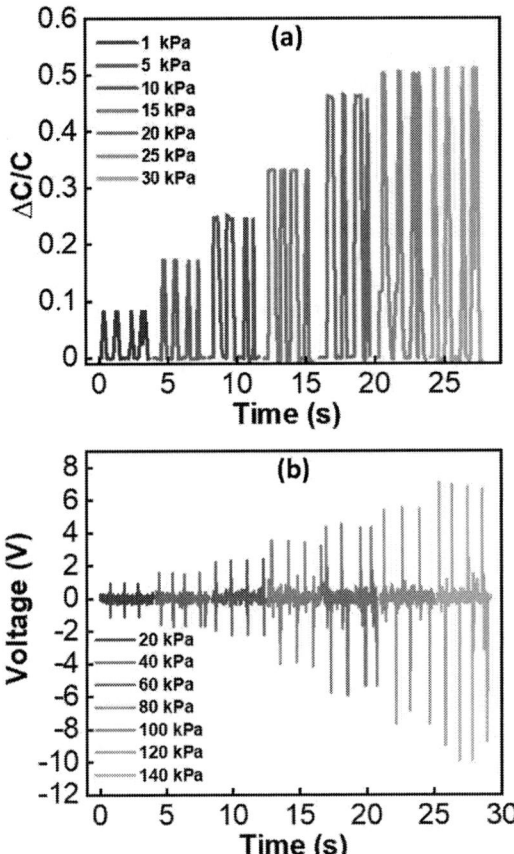

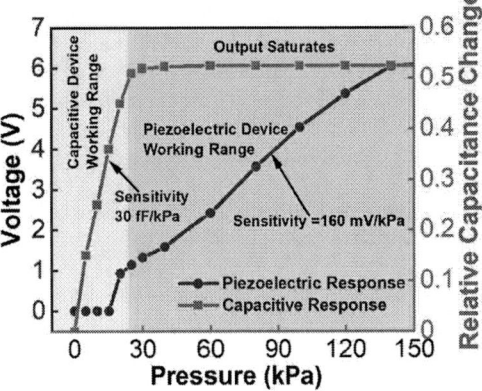

Figure 5: Sensing range of the multi-mode mechanoreceptor. The capacitive device detects minimum pressure of 1 kPa while piezoelectric device has working range of 20-140 kPa

Figure 4: Capacitive response of the multi-mode mechanoreceptor. The device can detect the minimum pressure of 1 kPa (b) Dynamic electromechanical characterization of piezoelectric device. The device has a working range of 20-140 kPa with peak-to-peak voltage generation of 16V.

increasing static pressure, the permittivity increases while the distance between the plates decreases [11]. This doubling effect leads to greater capacitive change. The capacitive sensor with a planar dielectric layer is less sensitive compared to the μ-pyramidal structure.

RESULTS AND DISCUSSIONS

The electromechanical characterization of the devices is performed using a standard push tester, and the corresponding electrical response is recorded with the Keithley Parameter analyzer (4200 SCS). Fig.4 (a) shows the response of the μ-pyramidal based capacitive device. Results verify that the sensor responds to the static load range of 1-25 kPa. C-Mode detects minimum pressure of 1 kPa and saturates after 25 kPa. In P-Mode, the device responds to a dynamic pressure range of 20 -140 kPa (Fig. 4(b)). The presence of MoS_2 in the PVDF matrix results in a higher d_{33} coefficient, resulting in a higher voltage generation (16 V at 140 kPa).

Fig. 5 describes the device working range characteristics. Notably, the synergetic effect of the multimode mechano-receptor broadens the sensing range from 1 kPa to 140 kPa. The capacitive mode responds to 25 kPa with a sensitivity of 30 kPa^{-1}. The better sensitivity

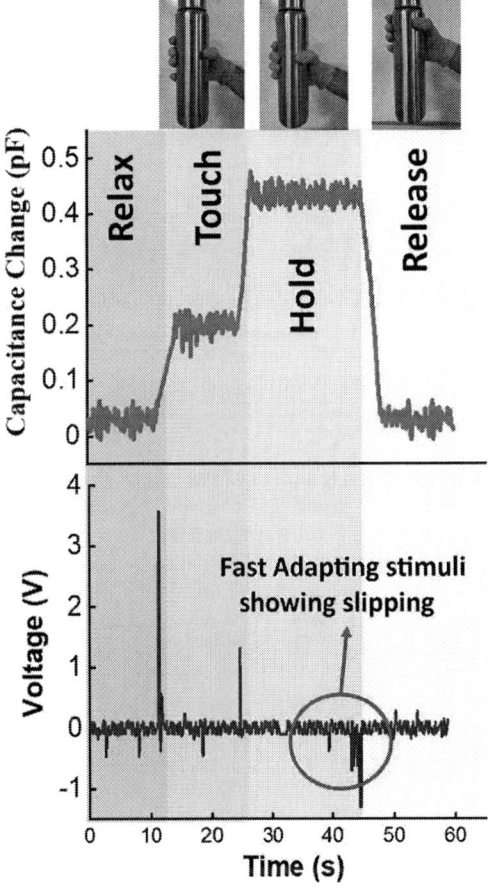

Figure 6: The voltage response of Fast adapting (Piezoelectric device) and Slow adapting (capacitive device) mechano-receptor from holding to releasing of the object

in C-mode is due to the μ-pyramidal structure. P-mode broadens the working range to 140 kPa with a sensitivity of 160 mV/kPa. Hence, the dual effect of the multimode sensor increases the working range and enables static-dynamic characteristics of the device.

As the device can successfully discriminate the static and dynamic stimuli, a real-time response of the device's mutual touch and pressure sensation is presented in Fig. 6. In a relaxed state (when fingers are not in touch with the bottle), both the devices show zero response. As the first touch is made, the instantaneous voltage (~4 V) generation occurs dynamically and is recorded by a piezoelectric device, whereas the capacitive device shows a shift in response from the base value. When a glass is grabbed firmly, the piezoelectric device initially shows a small peak (~2 V) and zero afterward. The capacitive response value increases as the glass is grabbed tightly (more pressure results in more capacitive change). Similarly, during release, multiple voltage peaks are generated within piezo-device, confirming the slipping of the bottle. The capacitive response just returns to the initial value without inferring slipping details. Therefore, real-time detailed information can be obtained for decoding complex human motions covering subtle-to-large stimuli (0-150 kPa).

CONCLUSION

In conclusion, we introduce a multimode-integrated sensory platform for simultaneously detecting static-dynamic stimuli with a wide pressure sensing range. The static and subtle stimuli are perceived by the μ-pyramidal capacitive device, while the piezoelectric device detects the dynamically changing stimuli. The synergetic effect of the multimode device allows for the detection of pressure over a range of 1-140 kPa. The C-mode detects a subtle pressure stimulus of 1-25 kPa with sensitivity of 30fF/kPa. P-Mode increases the detection range from 20-140 kPa with a sensitivity of 150 mV/kPa. Our method simplifies the sensory integration for modern E-skin. For the proof of concept, a real-time human grabbing and holding activity validates its functionality for static/dynamic stimuli.

ACKNOWLEDGEMENTS

The authors would like to thank National Research Facility (NRF) and Central Research Facility (CRF) for their support to the research and would also thank Prof. Dhiman Mallick and Nadeem Tariq Beigh from Electrical Department of IIT Delhi for providing electromechanical characterization facility.

REFERENCES

[1] A. Chortos, J. Liu, and Z. Bao, "Pursuing Prosthetic Electronic Skin," *Nature Materials*, vol. 15, no. 9, pp. 937–950, 2016.

[2] S. Maksimovic, M. Nakatani, Y. Baba, A. M. Nelson, K. L. Marshall, S. A. Wellnitz, P. Firozi, S.-H. Woo, S. Ranade, A. Patapoutian, and E. A. Lumpkin, "Epidermal merkel cells are mechanosensory cells that tune mammalian touch receptors," *Nature*, vol. 509, no. 7502, pp. 617–621, 2014

[3] A. Chortos, J. Liu, and Z. Bao, "Pursuing Prosthetic Electronic Skin," *Nature Materials*, vol. 15, no. 9, pp. 937–950, 2016.

[4] G. Li, S. Liu, Q. Mao, and R. Zhu, "Multifunctional electronic skins enable robots to safely and dexterously interact with human," *Advanced Science*, vol. 9, no. 11, p. 2104969, 2022.

[5] M. Yousuf, D. S. Arya, M. Garg, D. Varshney, and P. Singh, "Asymmetric Capacitive Pixel Array (E-cap) for mapping of epidermal vectored movement," *IEEE Journal on Flexible Electronics*, vol. 1, no. 2, pp. 134–140, 2022.

[6] P. Fang, X. Ma, X. Li, X. Qiu, R. Gerhard, X. Zhang, and G. Li, "Fabrication, structure characterization, and performance testing of Piezoelectret-film sensors for Recording body motion," *IEEE Sensors Journal*, vol. 18, no. 1, pp. 401–412, 2018.

[7] S. Sharma, A. Chhetry, D. Kim, and J. Y. Park, "Polyaniline-Nanospikes modified hybrid nanofibrous membrane based flexible piezoresistive sensor for Physiological Signal Monitoring," *2022 IEEE 35th International Conference on Micro Electromechanical Systems Conference (MEMS)*, 2022.

[8] Y. Pang, X. Xu, S. Chen, Y. Fang, X. Shi, Y. Deng, Z.-L. Wang, and C. Cao, "Skin-inspired textile-based tactile sensors enable multifunctional sensing of wearables and soft robots," *Nano Energy*, vol. 96, p. 107137, 2022.

[9] S. R. Ruth and Z. Bao, "Designing tunable capacitive pressure sensors based on material properties and microstructure geometry," *ACS Applied Materials & Interfaces*, vol. 12, no. 52, pp. 58301–58316, 2020.

[10] G. Liu, X. Chen, X. Li, C. Wang, H. Tian, X. Chen, B. Nie, and J. Shao, "Flexible, equipment-wearable piezoelectric sensor with piezoelectricity calibration enabled by in-situ temperature self-sensing," *IEEE Transactions on Industrial Electronics*, vol. 69, no. 6, pp. 6381–6390, 2022.

[11] S. R. Ruth, L. Beker, H. Tran, V. R. Feig, N. Matsuhisa, and Z. Bao, "Rational design of capacitive pressure sensors based on pyramidal microstructures for specialized monitoring of Biosignals," *Advanced Functional Materials*, vol. 30, no. 29, p. 1903100, 2019.

CONTACT
*M Yousuf, tel: +91-7006341905, immujeeb4u@gmail.com

NON-INVASIVE INSTANT MEASUREMENT OF ARTERIAL STIFFNESS BASED ON HIGH-DENSITY FLEXIBLE SENSOR ARRAY

Fang Wang[1,2], Heng Yang[1,2], Ke Sun[1], Yi Sun[1,*] and Xinxin Li[1,2,*]

[1]State Key Laboratory of Transducer Technology, Shanghai Institute of Microsystem and Information Technology, Chinese Academy of Sciences, Shanghai 200050, CHINA and
[2]University of Chinese Academy of Sciences, Beijing 100049, CHINA

ABSTRACT

This paper reports an innovative, non-invasive and instant method of arterial stiffness measurement. We delineate the tactile signals of arteries with varying stiffness using a dense sensor array, and analyze the factors affecting human tactility of stiffness and softness at fingertip with a deep learning model. Based on the most influential factor, an algorithm to instantaneously estimate the arterial stiffness is developed for the first time. The five stiffness grades ranked in order of the stiffness index (SI) may serve as an indicator of arterial aging and are expected to assist in early screening for certain cardiovascular diseases (CVD).

KEYWORDS

Arterial stiffness measurement, high-density flexible sensor array, tactile perception, deep learning model, stiffness index.

INTRODUCTION

Extremely threatening to the human health are CVD that are the leading cause of death. Arteriosclerosis is one of the most important inducements of CVD, bringing about multiple organ diseases. Early detection and diagnosis of atherosclerosis can largely achieve prevention of CVD, thus reducing the incidence of CVD [1].

Atherosclerosis index (AI) and the degree of arteriosclerosis are the main indicators of arteries stiffening [2]. However, AI is susceptible to drugs, for instance, a range of antihypertensive drugs will reduce the AI value sharply. Besides, it is not instant because blood sampling and testing is required. This detection of arteries stiffness belongs to a kind of invasive methods [3-4], which are more sophisticated. Another Invasive detection methods, such as angiography, which is an interventional detection method that injects developer into arteries. Because X-ray cannot penetrate the developer, angiography uses this feature to diagnose vascular diseases through the image displayed by the developer under X-ray. This kind of method can accurately evaluate the lumen and structure of arteries, but it will cause hurt to patients, and its operation is too complicated and expensive, requiring precise professional instruments, which is inconvenient for the instantaneous detection. Hitherto, invasive measurements of arterial stiffness require the sophisticated operations and instruments as the evaluation of arterial stiffness [5].

Non-invasive detection methods include pulse wave velocity (PWV), ankle brachial index (ABI), and pulse pressure (PP) measurement methods. These methods, which are non-invasive, have been used to evaluate arterial wall distensibility and hardness. The arterial stiffening degree is currently evaluated by the PWV indirectly [6].

However, this method has low detection accuracy, poor repeatability and is not convenient for wearable detection. Higher PWV is attributed to low elastance and less compliance. But the measurement of PWV involves multiple segments of arteries, which is inconvenient, and the exact blood traveling distance is hard to estimate.

In this paper, we report an innovative, non-invasive and instant method of arterial stiffness measurement. Our sensor array can sense the tactile details among arteries with different stiffness immediately for the first time. Furthermore, an arterial stiffness index (SI) is calculated by the characteristics of the signal enveloping curve, which is serviceable in assessing the risk of CVD.

EXPERIMENTAL METHODS

Preparation of materials

To address the problem of measurement instrument, we achieve a high-density flexible tactile sensor array by 18 ultra-small MEMS silicon pressure sensors, with a pitch of 0.65mm and total span 11.05mm [7]. The sensor are fabricated by advanced microhole inter-etch and sealing process, which can acquire real-time tactile sequence information and exhibit high output sensitivity of 0.65 mV KPa-1/3.3 V, hysteresis of 0.10% full scale (FS), low non-linearity of about ±0.10% FS, low repeatability error of 0.11% FS.

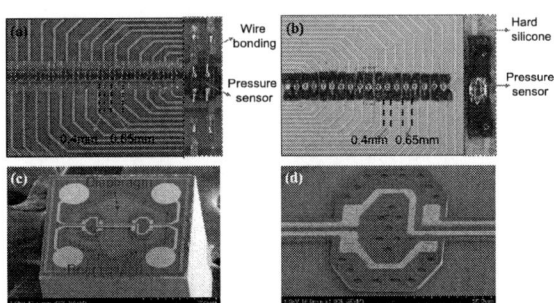

Figure 1: High-density flexible sensor array. (a) 18 ultra-small MEMS pressure sensors (0.4 mm × 0.4 mm × 0.2 mm each) are integrated on a flexible printed circuit board with a pitch of 0.65 mm, total span 11.05mm. (b) The sensors are coated with silicon. (c) Electron microscopy image. (d) Close-up view of the beam-island structure.

The arterial structure is so complex that explicitly interpreting the tactility of stiffness and softness is rather intractable. Hence, we contrive an orthogonal physical experiment that is easy to handle variables to investigate all sorts of factors related to arterial stiffness [8]. Various experimental materials are fabricated by silica gel mixed solution in order to make silicone materials with different stiffness, and then silicone tubes of various elastic levels are insert into them, such as hard-metal tube, elastic tube and so on. At last, the embedded silicone materials are left for 3-4 hours, then they are taken out for the subsequent experiments.

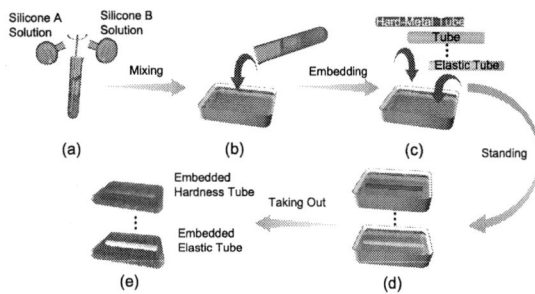

Figure 2: Manufacturing process of orthogonal physical experimental materials. Silicone materials of varying stiffness and embedded tubes of different elasticities.

Orthogonal physical experiment

Orthogonal physical experimental refers to an experimental method to explore multiple factors and levels. When there are 3 or more factors involved in the test, and there may be interactions between the factors, the test workload will be very large, or even difficult to implement.

Hence, orthogonal physical experiment can help us reduce the number of experiments and improve work efficiency by arranging experiments according to the orthogonal table. In order to simplify the complex structure of human arteries, we use the prepared silicone material to simulate the different tactile test of human arteries.

We adopt the orthogonal test method of three factors with three levels, and one factor with four levels. Among them, the factor refers to the physical quantities that affect the value of test indicators, the level refers to the state of factors in the test.

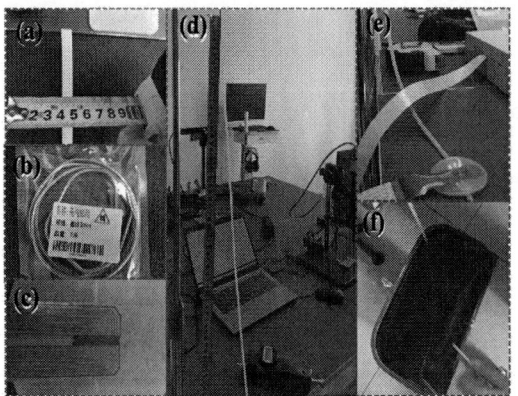

Figure 3: Experimental procedure. (a) Preparation for silicone materials. (b) Hard metal tube (c) High-density sensor array touching silicone embedded with hard metal tube. (d) Water flowing through the tubes. (e) Sensor array pressing silicone embedded with hollow elastic tube. (f) Water recycling.

There are four factors in the experiment, namely: 1. silicone material stiffness; 2. silicone material thickness; 3.

the diameters of embedded tubes; 4. the depths of embedded tubes. Additionally, these factors are respectively employed to simulate the skin softness, fatty thickness, arterial diameters, and arterial depths.

Four factors are considered in this experiment. One factor has four levels, and the other factors have three levels. In the experiment with different levels, too many testing times need to be carried out (should be used orthogonal table $L_{16}(4 * 3^3)$ in trials). To reduce the number of experiment times, we introduced the evaluation coefficient of the experiment, and adopted the pseudo level method. In the fourth level experiment, we repeated other variables except the depth from the surface, so we used the orthogonal table $L_{16}(4^4)$ to conduct experiments and only 16 groups levels are needed in total.

Hence, in the levels design, we set the stiffness grades of 0, 10 and 20 Shore Hardness (SH) for silicone materials, 7, 9 and 11mm for the thickness of silicone materials, 3, 4 and 5mm for the diameters of embedded tubes, 2, 3, 4 and 5mm for the depths of embedded tubes.

Table 1: Levels design of orthogonal physical experiments

levels	Stiffness /SH	diameter /mm	depth/ mm	thickness/ mm
1	0	3	2	7
2	10	4	3	9
3	20	5	4	11
4	20	5	5	11

Especially, water from a certain height difference flow through the various elastic tube to analogy the pressure in arteries as shown in Figure 3. When the water with a certain height goes through the elastic tube, we use our dense sensor array at the other end to collect tactile data.

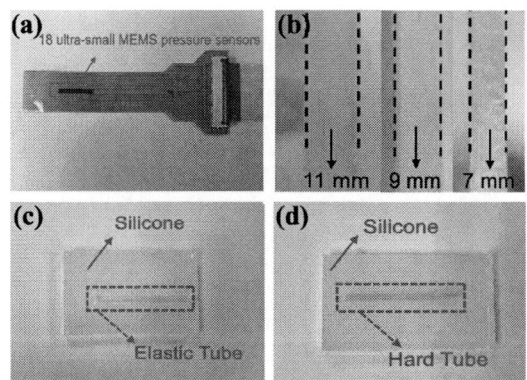

Figure 4: Experimental materials and sensor array. (a) High-density Flexible Sensor Array. (b) Silicone materials with varying thickness: 7mm; 9mm; 11mm. (c) Silicone embedded with elastic tube for blind test. (d) Silicone embedded with hard tube for blind test.

Deep learning model

For each orthogonal group we conduct 30 experiments repeatedly, and a total of 480 groups of data was collected. Furthermore, abundant tactile information characterized by envelope curves are acquired as shown in Figure 6b.

Moreover, we propose an explainable indication deep

978-1-6654-9309-3/23 $31.00 © 2023 IEEE

learning model to explore influence weight of four factors on stiffness of tactility. The model can convert the collected sensor data into a 3D spatiotemporal sequence matrix, whose features are used in two states of deep convolutional neural network: "softness" and "hardness".

In the two states, the data are categorized into two classes: "low impact class" and "high impact class", counting the inter class probability of each state, which is used to calculate correlation between factors and tactility. Then we use layer-wise relevance propagation algorithm to visualize semantic information of tactile sensation. Finally, we explore the most influential weight of the above four

factors in the form of confusion matrix in term of the stiffness of tactile. Additionally, 10 volunteers are invited to blind test various objects embedded in silicone to verify the correlation between tactility and the factors above.

In further experiments, the data collected through 10 groups of human tactile blind experiments show that most of the human tactile sensation of stiffness focuses on the "depth" information of the objects but is not sensitive to the thickness of the objects and the diameter changes of the embedded objects, which is the same as our theoretical experiments.

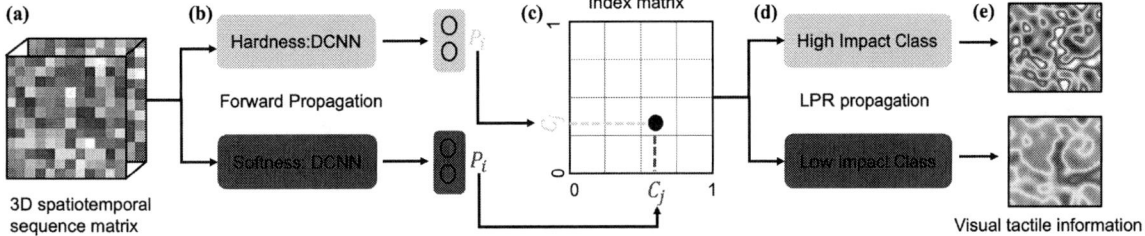

Figure 5: An explainable indication deep learning model. (a) 3D spatiotemporal sequence matrix. (b) Deep neural network forward propagation. (c) Index matrix. (d) LPR propagation. (e) Correlation of tactile information.

EXPERIMENTAL RESULTS

Similar phenomena occur impressively in our orthogonal physical experiments.

Arterial stiffness is associated with the elasticity of arterial wall and peripheral resistance. Poor elasticity is related to a tactility of "stiffness". Pressing the high-density sensor array on the human radial artery, the signal envelope curves of samples labeled as "normal" by doctors show a "concave" state, while with respect to the samples of "stiffness" a "pinnacle" shape is obtained. These characteristics remain invariable even in condition of increasing external pressures.

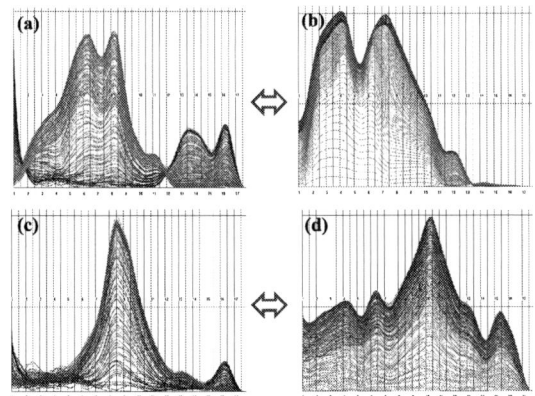

Figure 6: Pressing high-density flexible sensor array on the radial artery and silicone materials. (a) Envelope curves labeled as "normal" by doctors show a "concave" form. (b) Envelope curves appear "concave" in experiment with embedding elastic tube. (c) Envelope curves labeled as "stiff" manifest "pinnacle". (d) Envelope curves demonstrate "pinnacle" in experiment with embedded with hard tube.

In silicone materials embedded with elastic tubes (elastic tube), "concave" is found in envelope curves, while in terms of that with poor elasticity (hard metal tube), "pinnacle" appears instead. When pressing high-density flexible sensor array on the human radial artery, we find that envelope curves labeled as "normal" by doctor show a "concave" form, which is consistent with envelope curves in orthogonal experiment using elastic tube, as shown in Figure 6a and 6b. On the contrary, the envelop curves labeled as "stiff" manifest a "pinnacle" form, which is identical to the experiment using hard tube.

Further, we explore the sensor tactile information by deep learning model, and find that the models recognition of "depth" information will become more and more obvious with the increase of layers of neural network. In contrast to the difference of other factors, with the increase of the layers of neural network, the learning information of each layer does not change much, which is consistent with the learning parameters. By means of layer-wise relevance propagation algorithm, combined with blind testing, we conclude that the embedding "depth" has the most dominant impact on finger tactility for its learning-rates are up to 0.85 and 0.83 from deep learning confusion matrices, as shown in Figure 7.

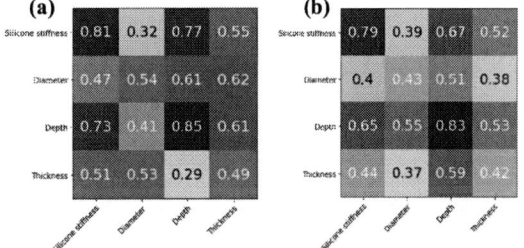

Figure 7: Tactile confusion matrices. (a) Confusion matrix in hard tube. (b) Confusion matrix in elastic tubes.

Taking the "depth" as the reference parameter, we define the stiffness index (SI) of envelope curves: ratio of "concave" height to peak height, as shown in Figure 8a.

Furthermore, the SI is divided into five grades, which visualization by the SI bar. The closer the SI gets to 0, the softer the artery is, which is consistent with the label "normal" in clinical trials; the closer the SI to 1, a "pinnacle" is more likely to be formed on the envelope curve, which coincides with the diagnosis label of arteriosclerosis as shown in Figure 8.

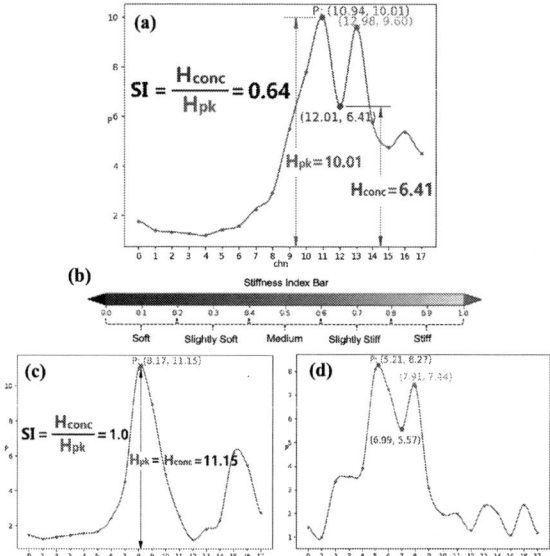

Figure 8: Calculation of stiffness index (SI) with envelope curves. (a) Definition of SI. (b) Stiffness Index Bar (The closer the SI gets to 1, the stiffer it is; the closer to 0, the softer). (c) A case of arteriosclerosis in clinical diagnosis (SI = 1) (d) Envelope curve appears "concave" in a case with label "normal", which is consistent with the observation in clinical trials.

The Root Mean Square Error between physicians' labels and our SI grades among 232 samples reaches 0.1032. As a result, the developed SI facilitates the measurement of the arteries stiffening degree to a great extent and are expected to assist in early screening for certain CVD.

CONCLUSION

In conclusion, the paper reports an innovative, non-invasive, and instant method to measure the stiffness of arterial. We adopt the high-density flexible sensor array using 18 ultra-small MEMS pressure sensors, which can delineate the tactile signals of arteries with various stiffness. Through the orthogonal physical experiment, we are able to analyze the four factors affecting human tactility of stiffness and softness at fingertip with a deep learning model. With the assistance of layer-wise relevance propagation algorithm, we draw a conclusion that the embedding "depth" has the most dominant impact on finger tactility since its learning-rates are up to 0.85 and 0.83 through deep learning algorithm.

Accordingly, based on the most influential factor, an algorithm to instantaneously estimate the arterial stiffness is developed for the first time. We define the SI of envelope curves, which can rank arterial stiffness into five grades. Moreover, we conduct an evaluation, among 232 samples, between physicians' labels and SI grades by the Root Mean Square Error, which can reach 0.1032. The SI turns out to be an indicator of arterial aging, which can assist in assessing the risk of CVD.

ACKNOWLEDGEMENTS

This work is partially supported by the R&D Plan Project in Key Fields of Guangdong Province (grant no. 2020B1111120004).

REFERENCES

[1] O. Sajdeya et al., "Triglyceride glucose index for the prediction of subclinical atherosclerosis and arterial stiffness: A meta-analysis of 37,780 individuals," *Curr. Probl. Cardiol.*, vol. 47, no. 12, p. 101390, 2022, doi: 10.1016/j.cpcardiol.2022.101390.

[2] Herbert. R. et al., "Fully implantable batteryless soft platforms with printed nanomaterial-based arterial stiffness sensors for wireless continuous monitoring of restenosis in real time", *Nano Today* 46, 101557, 2022.

[3] Y. Sun, M. Liu, Y. Xiao, and Y. Chen, "A novel molecular communication inspired detection method for the evolution of atherosclerosis," *Comput. Methods Programs Biomed.*, vol. 219, p. 106756, 2022, doi: 10.1016/j.cmpb.2022.106756.

[4] D. Chakraborty and M. Chattopadhyay, "Finite Element Method based Modeling of a Sensory System for Detection of Atherosclerosis in Human Using Electrical Impedance Tomography," *Procedia Technol.*, vol. 10, pp. 262–270, 2013, doi: 10.1016/j.protcy.2013.12.360.

[5] Z. Wang et al., "A Novel Methodology for Semi-automatic Measurement of Arterial Stiffness by Doppler Ultrasound: Clinical Feasibility and Reproducibility," Ultrasound Med. Biol., vol. 47, no. 7, pp. 1725–1736, 2021, doi: 10.1016/j.ultrasmedbio.2021.03.004.

[6] Cuomo F, Roccabianca S, Dillon-Murphy D, Xiao N, Humphrey JD, Figueroa CA. Effects of age-associated regional changes in aortic stiffness on human hemodynamics revealed by computational modeling. *PLoS One*. 2017; 12:e0173177.

[7] D. Jiao et al., "Ultra-small pressure sensors fabricated using a scar-free microhole inter-etch and sealing (MIS) process", *Journal of Micromechanics and Microengineering*, vol 30, no 6, 2020.

[8] L. Fang et al., "An orthogonal experimental design and QuEChERS based UFLC-MS/MS for multi-pesticides and human exposure risk assessment in Honeysuckle," *Ind. Crops Prod.*, vol. 164, no. March, p. 113384, 2021, doi: 10.1016/j.indcrop.2021.113384.

CONTACT

*Yi Sun, tel: +86-21-62511070; sunyi@mail.sim.ac.cn
*Xinxin Li, tel: +86-21-62511070; xxli@mail.sim.ac.cn

SUPPRESSION OF BIOELECTRICAL NOISE SIGNALS IN MOTION STATE BY LOW-COST MICROPILLAR HYDROGEL ELECTRODE

Gencai Shen[1,2], Nan Zhao[1,2], Chunpeng Jiang[1,2], Zhuangzhuang Wang[1,2], and Jingquan Liu[1]

[1] National Key Laboratory of Science and Technology on Micro/Nano Fabrication, Shanghai Jiao Tong University, 200240, Shanghai, CHINA and

[2] DCI Joint Team, Collaborative Innovation Center of IFSA, Department of Micro/Nano-electronics, Shanghai Jiao Tong University, 200240, Shanghai, CHINA.

ABSTRACT

This paper reports a micropillar hydrogel electrode to suppress bioelectrical noise signals in motion state. The hydrogel micropillar arrays with uniform spacing and same morphology are prepared through MEMS technology. The performance of micropillar hydrogel electrode is almost identical to that of the Ag/AgCl wet electrode which is regarded as "gold standard" in clinics. Moreover, compared with the dry electrode, the micropillar electrode shows lower noise signal when subject is talking or moving, making it has great potential to record bioelectrical signals in real scenes.

KEYWORDS

Bioelectrical signals, MEMS technology, hydrogel electrode, micropillar arrays.

INTRODUCTION

With the rapid development of electronic technology and people's attention to health, wearable physiological monitoring devices have received wide attention from academia and industry [1, 2]. Compared with magnetic resonance imaging (MRI) and magnetoencephalography (MEG) equipment, wearable bioelectric acquisition systems have the advantages of strong portability and low cost, having great potential for daily health monitoring. The bioelectrical signals of the human body contain a variety of important information about physiological states, which is the basic characteristic of life activities. By recording and analyzing bioelectric signals in daily life, real-time monitoring and feedback on human health can be realized. As the interface between human body and bioelectricity acquisition equipment, physiological electrode can convert ion current into electronic current. Therefore, it is necessary to prepare high performance physiological electrodes to collect bioelectric signals.

The conventional Ag/AgCl wet electrodes have been widely used for electroencephalograph (EEG) recording, which are regarded as the "gold standard" in clinics and hospitals [3, 4]. Nevertheless, the procedure for installing wet electrodes is very complex, and it is not suitable for daily EEG acquisition [5]. In recent years, portable bioelectric dry electrodes have been reported by many scholars, and these electrodes are convenient to install and do not need complicated operation. Tseghai et al. proposed a flexible textile-based dry EEG electrode which is a washable and reusable electrode with good portability [6]. Compared with the traditional Ag/AgCl electrodes, this kind of electrodes are more convenient to install and do not need to be installed by professionally trained personnel. Pei et al. reported an electrode with stable impedance performance, which does not need to clean hair and has no residue after use [7]. The impedance of the hydrogel

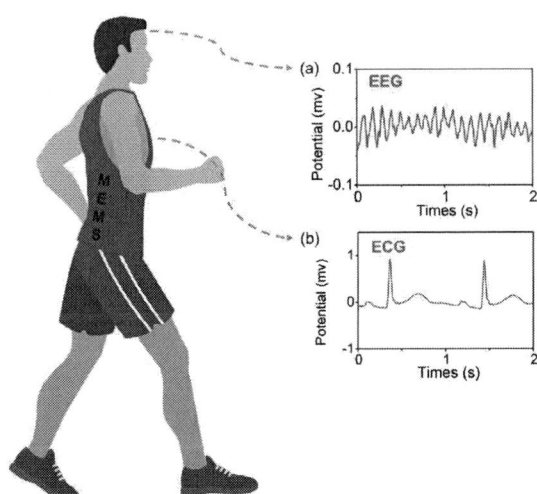

Figure 1: Bioelectrical signal collection: (a) EEG signal. (b) ECG signal.

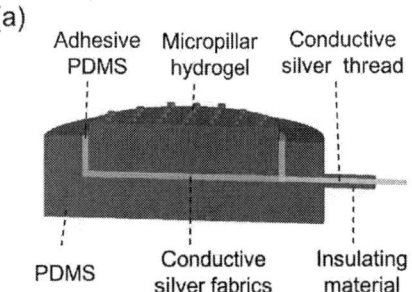

Figure 2: (a) Schematic representation of the micropillar electrodes. (b) Photo of the micropillar electrode.

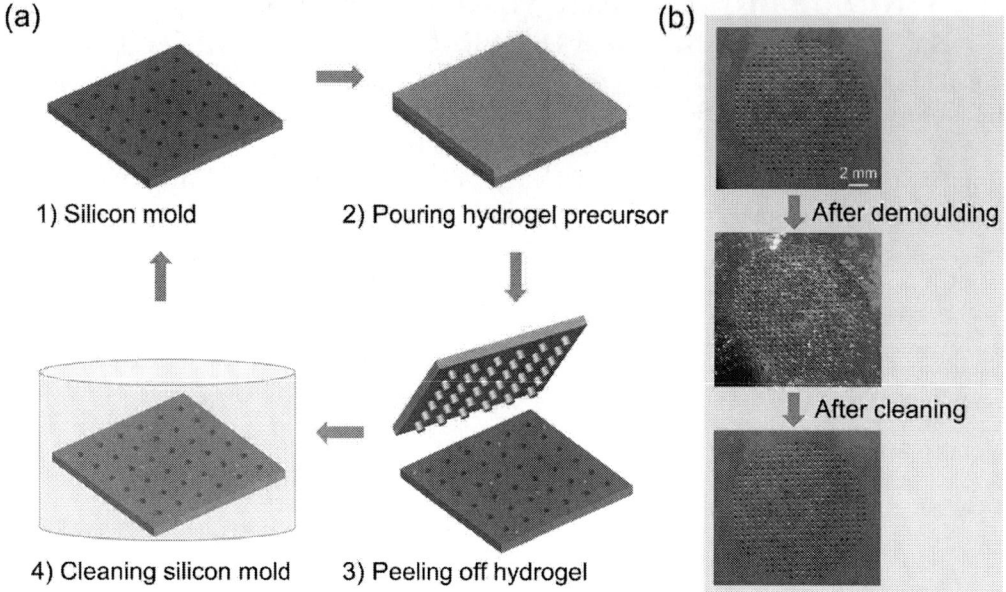

Figure 3: (a) The process of preparing the micropillar hydrogel using reusable silicon mold. (b) The photos of reusable silicon molds.

electrodes are lower than most dry electrodes, and it has great potential in brain-computer interface field.

In the process of recording physiological electrical signals, subjects are usually required to be as motionless as possible to reduce motion artifacts. However, body motion is unavoidable, especially in the case of collecting bioelectrical signals in daily life, so it is urgent to find a solution to reduce the interference of motion artifacts. It is noteworthy that the micropillar array on electrode can increase the friction between the electrode and skin, reducing the extra noise signal caused by body motion [8]. The micropillar arrays with uniform spacing and the same morphology can be fabricated by MEMS technology according to the actual requirements [9]. Whereas, it is difficult to prepare micropillars in large quantities at low

(a) Spin coating photoresist (b) Lithography process

(c) DRIE process (d) Removing photoresist

(e) Precooling silicon mold (f) Pouring prepolymer solution

(g) Micropillar hydrogel (h) Reusable silicone mold

Si   Photoresist   Hydrogel   Mobile cold plate

Figure 4: The fabrication processes of the silicon mold and the micropillar hydrogel.

cost.

In this paper, we propose a low-cost micropillar hydrogel electrode to suppress bioelectrical noise signals in a motion state. In the process of the preparing micropillar hydrogel electrodes, the clean mold can be obtained just by boiling the contaminated ones in 90°C hot water for 30 minutes, improving the manufacturing speed of micropillars and reducing the production cost. The correlation between the micropillar hydrogel electrode and the wet electrode reaches 0.93. Compared with the dry electrode, the micropillar electrodes display a low noise signal when the subject is talking or moving.

DESIGN AND FABRICATION

Figure 1 shows the application scenes of collecting electroencephalograph (EEG) and electrocardiogram (ECG) signals with micropillar hydrogel electrodes. Figure 2a and Figure 2b show the schematic diagram and photo of the electrode, respectively. The prepared physiological electrode can convert ion current into electronic current which can be collected by external equipment. The micropillar arrays on the hydrogel are designed to improve the friction between electrode and skin. In order to reduce the cost of fabricating the hydrogel micropillars, we adopted a reusable silicon mold scheme.

Figure 3a demonstrates the preparation process of hydrogel micropillars on a reusable silicon mold. In our work, the clean silicon mold can be obtained simply by boiling the contaminated ones in 90°C hot water for 30 minutes, improving the manufacturing speed of micropillars and reducing the production cost. The detailed preparation steps of the silicon mold are presented in Figure 4a-d. In brief, a 6 μm positive photoresist was spun on a silicon wafer (Figure 4a) and then solidified in a hot plate. Subsequently, the photoresist was pattern by the lithography process (Figure 4b). Figure 4c shown that the silicon wafer was prepared by the DRIE etching process.

978-1-6654-9309-3/23 $31.00 © 2023 IEEE   356

(a)

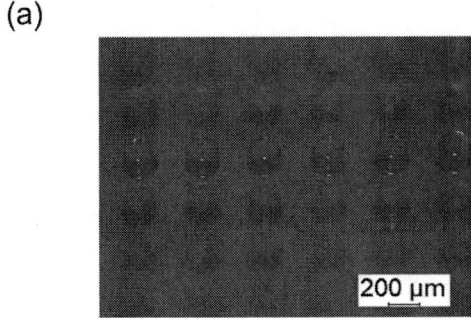

(b)

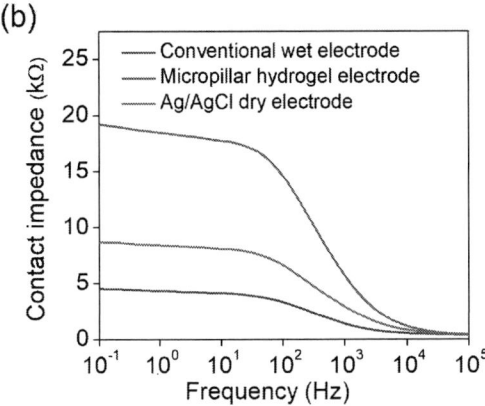

Figure 5: (a) Optical micrograph of the micropillar hydrogel. (b) Contact resistance of conventional wet electrode, micropillar electrode, and dry electrodes.

After cleaning the photoresist using acetone and ethanol, the silicon mold was obtained, as shown in Figure 4d. The preparation process of hydrogel refers to the previously reported literature [10, 11]. Before pouring the hydrogel precursor solution into the silicon mold, as shown in Figure 4e-f, a cold iron plate needs to be placed under the silicon mold to reduce its temperature, increasing the solidification time of the hydrogel. After that, it was evacuated under a vacuum environment to remove excess air bubbles in the mold, so that the hydrogel could fill the silicon mold. As shown in Figure 4g, the micropillar hydrogel could be peeled off from the silicon mold. As shown in Figure 4h and Figure 3b, the silicon mold can be reused after cleaning in hot water, reducing the production cost of the micropillar electrodes. Then, the electrodes are assembled as demonstrated in Figure 2 and then connected with the EEG acquisition equipment.

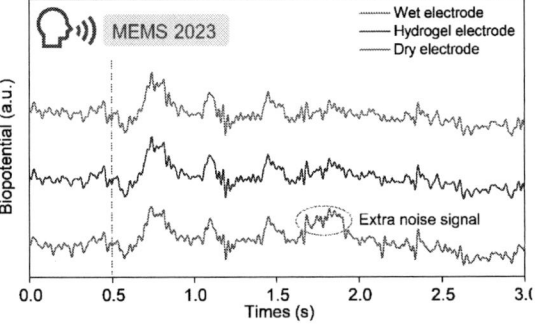

Figure 6: Bioelectric signal collected by wet electrode, micropillar electrode, and dry electrode during speech.

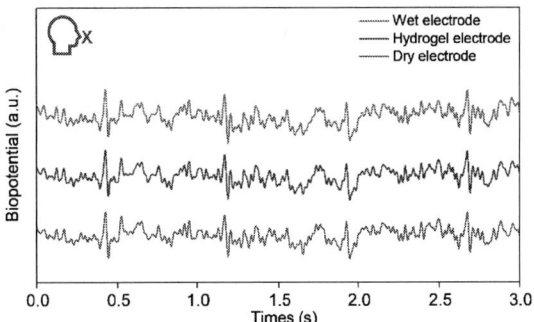

Figure 7: Bioelectric signal collected by wet electrode, micropillar electrode, and dry electrode without speech.

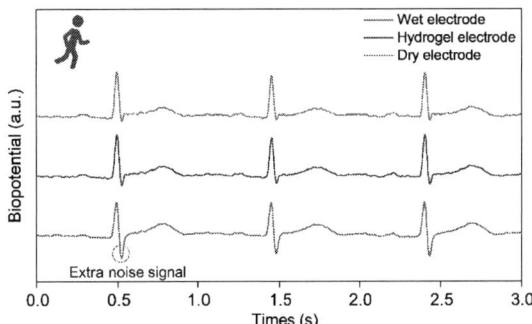

Figure 8: Bioelectric signal collected by wet electrode, micropillar electrode, and dry electrode during walking.

RESULTS AND DISCUSSION

Figure 5a demonstrates that the micropillar hydrogel arrays have uniform spacing and the same morphology, indicating that the silicon mold has been successfully prepared. To test the ability of the prepared electrodes, the micropillar electrodes were tested in comparison with standard wet electrodes and dry electrodes. And the skin-electrode contact impedance is recorded, shown in Figure 5b. The impedance of micropillar hydrogel electrode is obviously lower than that of dry electrode and slightly higher than that of standard wet electrode. Because the hydrogel electrode and the wet electrode contain electrolytes which can wet the skin stratum corneum with poor conductivity, thus reducing the contact impedance. However, for the dry electrode, due to the lack of electrolyte, the electric conductivity of skin stratum corneum can not be increased, so the measured impedance is higher.

The installation of wet electrodes requires the injection of conductive paste, which is very time consuming and requires specialized skills, limiting the application in many cases [12]. Our micropillar hydrogel electrode has the advantages of strong portability and moderate impedance. To evaluate the noise suppression effect of the micropillar hydrogel electrode, we measured the noise level of the electrode during speech. As shown in Figure 6, there is an obvious extra noise signal in the dry electrodes when subjects said "MEMS 2023", but not on the micropillar electrodes and the wet electrodes. And the correlation between micropillar electrodes and the wet electrodes reaches 0.93. As shown in Figure 7, when the subjects are silent, the performance of the three electrodes is comparable. As shown in Figure 8, the ECG measured by the dry electrode also shows more noise signals during exercise. In general, the micropillar hydrogel electrodes have similar performance to standard wet electrode, while the performance of dry electrode is slightly worse. This indicates that under the talking or moving state, the performance of micropillar electrodes is similar to that of wet electrodes, while the dry electrode shows a relatively high extra noise level.

CONCLUSION

In this paper, we present a low-cost micropillar hydrogel electrode to suppress bioelectrical noise signals in motion state. The hydrogel micropillar arrays are prepared by MEMS technology, and the inverted molds can be reused by simple processing, reducing the manufacturing cost. Compared with the dry electrode, the prepared micropillar hydrogel electrodes show the effect of suppressing the noise signal in speaking or moving state. And the correlation between micropillar electrodes and conventional Ag/AgCl wet electrodes reaches 0.93, showing that the micropillar hydrogel electrodes have great potential for convenient acquisition of EEG signals in real scenes.

ACKNOWLEDGMENTS

This work was partially supported by the Oceanic Interdisciplinary Program of Shanghai Jiao Tong University (Grant No. SL2020ZD205) and the Scientific Research Fund of Second Institute of Oceanography, MNR (Grant No. SL2020ZD205), the Strategic Priority Research Program of Chinese Academy of Sciences (Grant No. XDA 25040000), the National Key R&D Program of China (Grant No. 2020YFB1313502), the Program of Shanghai Academic/Technology Research Leader (Grant No. 18XD1401900). The authors are also grateful to the Center for Advanced Electronic Materials and Devices (AEMD) of Shanghai Jiao Tong University.

REFERENCES

[1] C. Wang, H. Wang, B. Wang et al., "On-skin paintable biogel for long-term high-fidelity electroencephalogram recording," Science Advances, vol. 8, pp. 1396, 2022.

[2] L. Feng, H. Shan, Y. Zhang et al., "An Efficient Model-Compressed EEG Net Accelerator for Generalized Brain-Computer Interfaces with Near Sensor Intelligence," IEEE Transactions on Biomedical Circuits and Systems, 2022, DOI: 10.1109/TBCAS.2022.3215962.

[3] B.-C. Kang, and T.-J. Ha, "Noninvasive electroencephalogram sensors based on all-solution-processed trapezoidal electrode array," Applied Physics Letters, vol. 120, pp. 213301, 2022.

[4] G. Li, S. Wang, M. Li et al., "Towards real-life EEG applications: Novel superporous hydrogel-based semi-dry EEG electrodes enabling automatically 'charge–discharge' electrolyte," Journal of Neural Engineering, vol. 18, pp. 046016, 2021.

[5] Pedrosa, Paulo, Fiedler et al., "In-service characterization of a polymer wick-based quasi-dry electrode for rapid pasteless electroencephalography," Biomedizinische Technik, vol. 63, pp. 349-359, 2018.

[6] G. B. Tseghai, B. Malengier, K. A. Fante et al., "Velcro Hook Electroencephalogram Textrode for Brain Activity Monitoring," 2022 IEEE International Conference on Flexible and Printable Sensors and Systems, pp. 1-4, 2022.

[7] W. H. Pei, X. T. Wu, X. Zhang et al., "A Pre-Gelled EEG Electrode and Its Application in SSVEP-Based BCI," IEEE Transactions on Neural Systems And Rehabilitation Engineering, vol. 30, pp. 843-850, 2022.

[8] X. Niu, L. Z. Wang, H. Li et al., "Fructus Xanthii-Inspired Low Dynamic Noise Dry Bioelectrodes for Surface Monitoring of ECG," Acs Applied Materials & Interfaces, vol. 14, pp. 6028-6038, 2022.

[9] Z. Chen, J. Yu, M. Xu et al., "Highly deformable and transparent triboelectric physiological sensor based on anti-freezing and antidrying ionic conductive hydrogel." 2021 IEEE 34th International Conference on Micro Electro Mechanical Systems, pp. 525-528, 2021.

[10] G. Shen, K. Gao, N. Zhao et al., "A Fully Flexible Hydrogel Electrode for Daily EEG Monitoring," IEEE Sensors Journal, vol. 22, pp. 12522-12529, 2022.

[11] Q. Wu, J. Wei, B. Xu et al., "A robust, highly stretchable supramolecular polymer conductive hydrogel with self-healability and thermo-processability," Scientific Reports, vol. 7, pp. 1-11, 2017.

[12] J. C. Liu, S. Lin, W. Z. Li et al., "Ten-Hour Stable Noninvasive Brain-Computer Interface Realized by Semidry Hydrogel-Based Electrodes," Research, vol. 2022, 2022.

CONTACT

*J.Q. Liu, tel: +86-21-34207209; jqliu@sjtu.edu.cn

ULTRA-THIN MEMS PACKAGING BASED ON AUXETIC STRETCHABLE STRUCTURES FOR APPLICATIONS IN WEARABLE ELECTRONICS

Daniel Zymelka, Toshihiro Takeshita, Yusuke Takei, and Takeshi Kobayashi
National Institute of Advanced Industrial Science and Technology

ABSTRACT

This paper reports a new packaging method based on auxetic structures, intended for highly stretchable wearable devices incorporating ultra-thin MEMS components. The auxetics are a class of mechanical metamaterials with a negative Poisson's ratio. Thus, auxetics exhibit the specific property of becoming wider when stretched and narrower when compressed. Such a feature makes auxetic structures conformable, yet flexible and stretchable. Owing to the specific features, the demonstrated integration method of ultra-thin MEMS components that is based on the use of auxetic structures can be very promising for wearable electronic devices.

KEYWORDS

Auxetics, mechanical metamaterials, MEMS, packaging, printed electronics, stretchable electronics.

INTRODUCTION

Rigid and flexible printed circuit boards have been carriers for electronic systems for decades. However, to meet growing requirements for the new trends in electronics, especially regarding large-area wearable devices, flexible and stretchable electronic systems have been widely developed in recent years [1,2]. Although there was rapid progress in organic electronics [3], there are still challenges related to stretchable hybrid electronics incorporating silicon-based components (including ultra-thin MEMS) and printed wiring. The main challenges are associated with the large stress to which the electronic components as well as interconnections between these components and the wiring system are subjected. Especially ultra-thin MEMS devices are prone to damage even under moderate mechanical stress.

To address this problem, this research focuses on the design and fabrication methods of hybrid electronics with packaging that is based on auxetic mechanical metamaterials (structures that have a negative Poisson's ratio). The auxetic structures besides becoming wider when stretched and narrower when compressed, experience local strain domains lower than the macroscopic strain of standard flexible and stretchable substrates. In this work, we use this feature to integrate the electronic devices into the low-strain domains, which isolates them from excessive strain and prevents them from being damaged.

MATERIALS AND METHODS

The preparation of devices starts with the screen printing of conductive wires using silver-based ink (Fig.1(a)). A commercially available silver ink (TAIYO INK, ELEPASTE NP1), designed for stretchable electronics, was printed onto a 100 μm thick polyurethane substrate and dried in a convection oven at 90°C for 45 min. Next, the ultra-thin PZT/silicon chips (Fig. 1b) were integrated with the printed wires. A detailed description of the fabrication process for the ultra-thin PZT/silicon chips has been already described in a previous study [4]. The silicon chips were bonded to the polyurethane substrate using a hot-melt adhesive. To provide a connection between the printed wiring system and the silicon chips, the silver ink was deposited onto the chips' terminals using a dispenser (Musashi Engineering ML-5000XII). In the next step, the entire construction was laminated with an additional polyurethane sheet (Fig. 1(c)). Both polyurethane layers were bonded using a hot-melt adhesive. After that, the laminated structures were cut into the desired form using a laser cutter (Fig. 1(d)). An example of the fabricated device is shown in Figure 1(e)

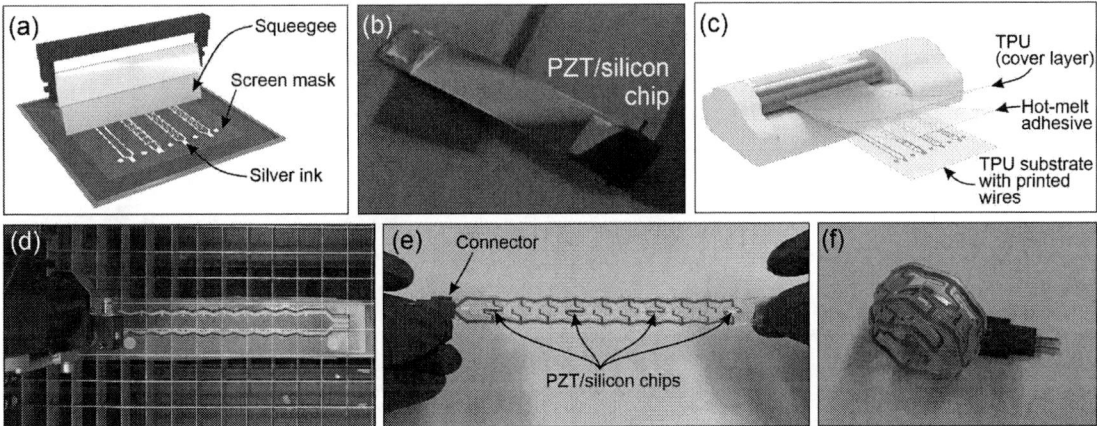

Figure 1: Fabrication process, (a) screen printing of conductive silver wires, (b) assembling of thin PZT/silicon chips, (c) lamination, (d) laser cutting, (e),(f) examples of a fabricated device with the auxetic structure.

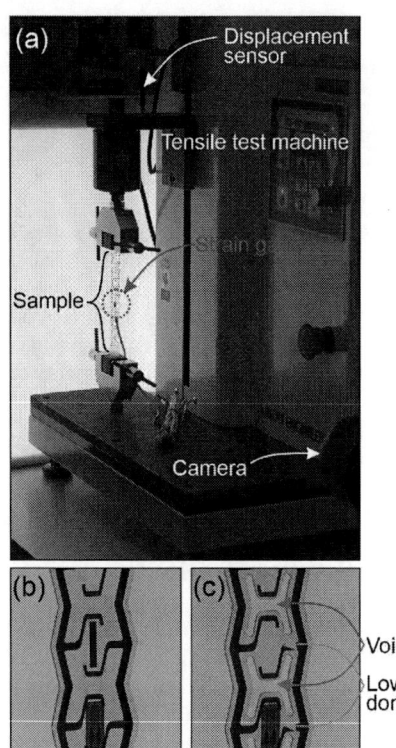

Figure 2: (a) experimental setup for mechanical analysis, (b) sample with no voids, (c) sample with auxetic structure (voids). Both samples had an integrated conventional strain gauge intended for local strain analysis.

and Figure 1(f), demonstrating its flexibility.

Such prepared devices were analyzed using a tensile test machine (Aikoh Engineering) and the results were compared with numerical simulations (Ansys). The experimental analysis on the tensile test machine was recorded using a camera (Basler acA2500-14uc), as demonstrated in Figure 2(a). Besides the silicon chips, commercially available 2 mm long strain gauges were used for the experimental analysis of local strain. Two types of samples were investigated in this study. Samples with a standard construction (there were no voids) and with the auxetic structure (with voids), as shown in Figure 2(b) and Figure 2(c), respectively. While the local strain was measured using the strain sensors, the global strain to which the entire samples were subjected was measured using a linear displacement sensor integrated with the tensile test machine (Fig. 1(a)). The output signals from the linear displacement sensor, strain sensors, and the camera were recorded simultaneously by a computer program prepared for this task (LabView, NI).

RESULTS

All experimental analysis and mechanical simulations were performed up to the same maximal strain of 20%. In the case of experimental analysis, this strain was measured

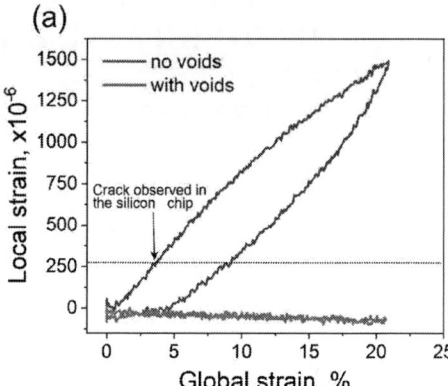

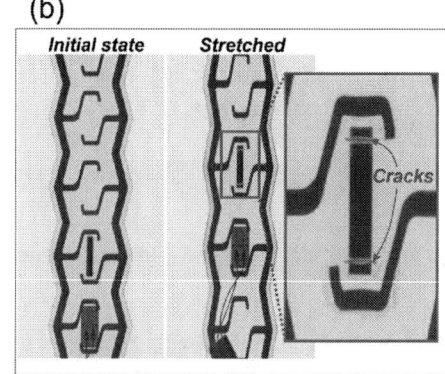

Figure 3: (a) Results of the local strain analysis showing large local strains for the sample with a standard structure, and negligible local strain variations for the sample based on the auxetic structure (with voids). (b) An example showing crack formation in a silicon chip integrated into the structure with no voids. Note that to make it easier to observe crack formation, in this case, the silicon chips were not connected to the wiring system using silver ink.

using the linear displacement sensor integrated with the tensile test machine.

Figure 3(a) demonstrates the results of the comparative analysis conducted with the standard structure (no voids) and the auxetic structure. Large local strains were measured for the sample with the standard construction. Compared to these results, it was demonstrated that the engineered voids in the auxetic structure reduce the local strain significantly. The

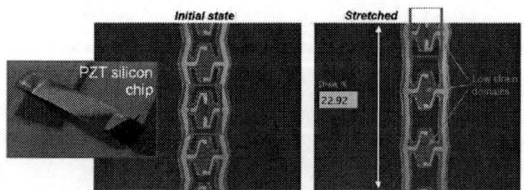

Figure 4: Sample with the auxetic structure and integrated PZT/silicon chips (shown in Figure 1e) subjected to large strain. Despite strain exceeding 20%, the integrated silicon chips were not damaged.

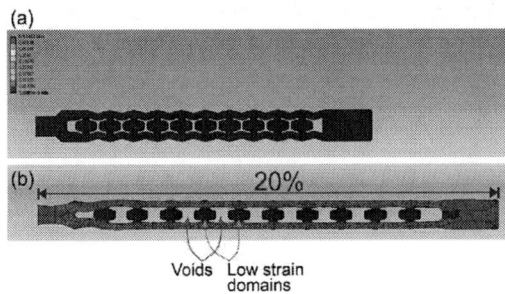

Figure 5: Results of the numerical simulations for strain analysis in samples with the auxetic structure (a) initial state (b) when stretched up to 20%.

measured strain using the small strain sensors located in the low-strain domains was almost negligible. The collected data demonstrate that the investigated auxetic structure was very effective in local strain isolation and in contrast to the standard structure, reveals to be suitable for the integration of electronic components, including ultra-thin MEMS devices.

To demonstrate the negative effect of local strain on the integrated components, in Figure 3(b) a crack formation in the PZT/silicon chip integrated into a sample with the standard construction is shown. The crack formation has been observed even under moderate strain. In contrast to these results, the same type of ultra-thin silicon chips incorporated into a sample with the auxetic structure could withstand strain exceeding 20% (Fig. 4). Despite the large strain, the silicon chips integrated into the auxetic structure were fully operational.

The conducted numerical simulations (Fig. 5) are consistent with the experimental results. Even though the entire sample was subjected to a large strain of 20%, the local domains exhibit negligible strain.

CONCLUSION

In this study, we demonstrated an original concept of hybrid electronics that is based on auxetic structures. It was demonstrated experimentally and based on numerical simulations, that thanks to the engineered voids in the auxetic structures, the fabricated devices experience local strain domains lower than the macroscopic strain of standard flexible and stretchable substrates. We used these low-strain domains to integrate the ultra-thin PZT/silicon chips. With such construction, the integrated silicon chips were isolated from large strains, which prevented them from being damaged. The results of this study can be especially promising for wearable electronics integrated into textiles. Such applications require stretchable, conformable, yet lightweight constructions. As we demonstrated in this work, it can be achieved by the implementation of auxetic structures for the packaging of electronic components.

Further work will be extended to large 2D structures and an analysis of other geometries of the auxetic mechanical metamaterials.

ACKNOWLEDGEMENTS

This work was supported by the Core Research for Evolutional Science and Technology (CREST) program of the Japan Science and Technology Agency (JST) JPMJFR206G and JSPS KAKENHI Grant JP 22K18421.

REFERENCES

[1] M. H. Behfar, B. Khorramdel, A. Korhonen, E. Jansson, A. Leinonen, M. Tuomikoski, et al., *Advanced Engineering Materials*, Vol. 23 Issue 12, 2021.

[2] N. Munzenrieder, G. Cantarella, C. Vogt, L. Petti, L. Buthe, G. A. Salvatore, et al., "Stretchable and Conformable Oxide Thin-Film Electronics", *Advanced Electronic Materials,* Vol. 1 Issue 3, 2015.

[3] M. Berggren, D. T. Simon, D. Nilsson, P. Dyreklev, P. Norberg, S. Nordlinder, et al.," Browsing the Real World using Organic Electronics, Si-Chips, and a Human Touch", *Advanced Materials*, Vol. 28 Issue 10 Pages 1911-1916, 2016.

[4] T. Takeshita, D. Zymelka, Y. Takei, N. Makimoto and T. Kobayashi, "Development of a nail-deformation haptics device fabricated adopting ultra-thin PZT-MEMS technology", *Jpn. J. Appl. Phys.*, 61 SN1024, 2022.

CONTACT

*D. Zymelka, tel: +81-29-861-8248; daniel.zymelka@aist.go.jp

ULTRALOW POWER FLEXIBLE OCULAR MICROSYSTEM FOR VERGENCE AND DISTANCE SENSING BASED ON PASSIVE DIFFERENTIAL MAGNETOMETRY

Adwait Deshpande[1#], Mohit U. Karkhanis[1#], Chayanjit Ghosh[1#], Hanseup Kim[1],*
and Carlos H. Mastrangelo[1]

[1]Department of Electrical and Computer Engineering, University of Utah, Salt Lake City, Utah, USA
[#]Equal contributing authors

ABSTRACT

We report the theory, construction, and testing of a flexible ocular, on-the-eye microsystem used for ultra-low power object distance sensing suitable for smart adaptive contact lenses. The microsystem determines object distance by vergence angle triangulation. Vergence angle is determined from passive measurements of the earth's magnetic field at each eye. Vergence measurements were performed every 5-degree interval over 35 degrees in total for each eye to accommodate the entire human visual range. Vergence angle measurements had an RMS error of 1.74 degrees and a distance ranging RMS error of 14.04 mm. The energy requirement per magnetic field measurement was estimated to be approximately 2 μJ per eye.

KEYWORDS

Object distance sensing, vergence angle triangulation, magnetometry, smart contact lens.

INTRODUCTION

As the human eye ages, the crystalline lens within the eye loses its flexibility which leads to a persistent refractive error called presbyopia. More than 1.8 billion people around the globe suffer from this refractive error [1]. Presbyopia manifests itself as an inability of the human ocular system to focus on objects at different distances, near and far. This condition leads to visual impairment, blurred vision, and reduced quality of life. To date, corrective eyewear technologies that consist of bifocal, trifocal, or progressive lens designs have been extensively used for managing and mitigating the effects of presbyopia. A fundamental drawback of such lenses is that they severely reduce the field of view and partition the user's vision into areas of different refractive indices and magnifications. Such division of the field of view has been known to cause accidents in the elderly. Additionally, users are also subject to a false perception of depth, image jumps, and the inherent optical aberrations induced during the manufacturing of such lenses [2].

A better approach for vision correction is the use of smart eyeglasses or smart contact lenses embedded with user-specific power-vs-distance models [3], [4]. The smart contacts adaptively change the optical power of the lens depending on the distance of the object. This results in sharp object images in the user's field of view. The realization of smart contacts require many innovative technologies and subsystems such flexible wiring [5], energy harvesters [6], [7], power storage [8], eye tracking sensors [9], tunable focus lenses [10], supporting circuitry,

and soft packaging integration. [11]

The practical realization of this autofocusing system requires the determination of the object distance from the user and the vergence angle between the user's eyes with a very low power consumption. In this work, we make use of MEMS magnetometers, one on each eye, to sense the geomagnetic field. The vergence angle is calculated from this differentially sensed magnetic field. We achieve vergence angle determination with an RMS error of 1.74 degrees and object distance determination with an RMS error of 14.04 mm with an ultra-compact and ultralow-power approach.

DISTANCE RANGING METHODS

Both active and passive distance ranging methods have been previously used to determine object distance. In the case of active ranging, a beam of light is emitted from the user and it is returned back after reflecting from the object. The range is determined by the reflected signal angle or time of flight. The power consumption of this method depends on the object distance. The further the object is, the more powerful beam is required. Such systems are power-hungry and may require battery packs or need to be constantly plugged in. This makes them unsuitable for mobile applications like standalone smart contact lenses where limited energy is available. Passive ranging is more energy-efficient and hence desirable [12].

The most commonly used mobile eye-tracking method utilizes camera-based image processing techniques. Here, the face is lit with multiple IR LEDs, the cameras take multiple pictures, and a powerful computer processes them to determine the gaze direction [13]–[15]. This technique, again, is power inefficient and cannot be used for contact lenses. More recently, VCSEL pairs were used for eye-tracking with an IR camera [16]. Camera-less approaches are desirable to reduce power consumption. IR LEDs and photodiodes-based camera-less eye trackers were implemented in a standalone smart eyeglasses system [17] and in a contact lens system (with photodiodes) paired with eyeglasses (with IR LEDs) [18], a system where light from VCSELs is directed with a micromirror onto the eye and detected with photodiodes is also used for gaze tracking in smart eyeglasses [19]. Purkinje reflections have also been used in literature for eye tracking [20]. Another approach involves using scleral coils, which require a headgear to be worn. This headgear has large, power-hungry, external AC-driven electromagnets. An eye-angle-dependent EMF is induced in the scleral coils, which behave like the secondary windings of a transformer [21], [22]. Though this method is highly accurate, the continuous requirement of wearing bulky headgear is undesirable and may cause the user some discomfort.

Recently, a low-power, low-profile quad scleral coil-based approach was developed for vergence angle detection [12]. In this approach, two sets of coils were placed on each eye. One set of coils generates a magnetic field, and emf is induced on the other set of coils on the other eye. The vergence angle can be determined from the amount of emf induced. This approach completely eliminates the need for any external emf-generating circuitry and hence, any bulky headgear involved. Further improvements were made to this system to drastically reduce the energy consumption to just 340 nJ per measurement [23].

In this paper we discuss a method of determination of object distance that does not require any radiating energy emissions; hence ultimately requiring less power. The method is based on the measurement of a vector field that is uniform on both eyes.

WORKING PRINCIPLE

The magnetic field generated by the quad coil setup in [23] is unnecessary if another external field exists in the space between the two eyes. Two types of such fields exist. One is the Earth's gravitational field, and the other is the magnetic field of the Earth. The gravitational field can be measured with the help of an accelerometer. However, the the downward facing gravitational field is stronger than its horizontal component which are parallel to the ground. The other field i.e., the Earth's magnetic field points downward and to the north. This field can be affected by nearby magnetic materials and can be measured using tiny magneto-resistive sensors. Such magnetic field sensors can be routinely found as a digital compass in consumer electronics.

To determine the vergence angle between the eyes, the magnetic field should roughly point along the same direction on both the eyes. The interpupillary distance (IPD) of human eyes is approximately 60 mm. It can be safely assumed that no other magnetic object will come in close proximity to the eyeballs, lower than the IPD. In such a case the magnetic field will be equal at both eyeballs. If there exists a magnetic object in the vicinity of the eyes, lesser than 60 mm in distance, it will distort the surrounding magnetic field. Also, this method of vergence triangulation does not depend on the direction of the magnetic field.

In this work, we place the magnetometers at a 45-degree angle from the pupil, on the side of the nose bridge. This angle placement is arbitrary and does not affect the field measured. The vergence triangulation approach can be understood from Figure 1. Here it can be seen that, if the vergence angle is θ_V, then the angle between the magnetometers is given by $\theta_V + \pi/2$. When the eyes focus on an object, the eyes rotate about the z-axis. However, the magnetometers are placed such that the reference z-axis is mapped as magnetometer x-axis. The vergence angle can be calculated using (1) if object distance (d) is considered much greater than the IPD, and the object distance can be triangulated using (2).

$$d = \frac{IPD}{2\tan\frac{\theta_V}{2}} \approx \frac{IPD}{\theta_V} \qquad (1)$$

$$\theta_V = -\frac{\operatorname{asin}\left((b_{yl}.b_{yr}+b_{zl}.b_{zr})\right)}{\sqrt{b_{yl}^2+b_{zl}^2}.\sqrt{b_{yr}^2+b_{zr}^2}} \qquad (2)$$

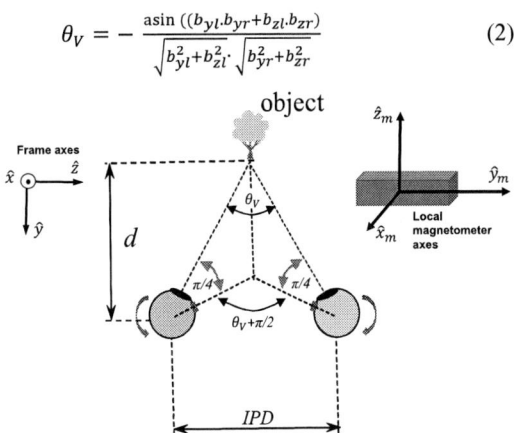

Figure 1: Vergence angle and distance ranging measurement using magnetometers placed at 45 degrees from the pupil.

EXPERIMENTAL
Magnetometer on Flexible Smart Contacts

A contact lens needs to withstand the bending it experiences while a user puts it on and takes it off. It also needs to accommodate appropriate circuit components. Also, attaching components to a curved substrate is difficult. Hence a special origami-based approach was adapted for the substrate of the smart contact lens (SCL) which involved fabricating a polyimide-based flex-PCB, populating it with components and then folding and joining it such that it resembles a dome-shaped contact lens.

A flex-PCB for assembling the magnetometer with the SCL platform was designed using EAGLE (Autodesk). The thickness of the flex-PCB was 80 µm, including a copper trace thickness of 18 µm. Supporting circuit components as shown in Figure 2 were then mounted on the PCB using reflow soldering. The magnetometer used is HMC5883L (Honeywell) and the microcontroller used is the ultralow-power, 32-bit Cortex-M4 MAX32660. The assembled, folded and joined PCB is shown in Figure 3.

A Goldberg polyhedron shell was selected as the base shape and then sliced, from top to have a flat area for the lens and from the bottom to define the SCL size, in Rhino 6. Then using simple cut and unfold origami techniques, this shell was flattened in JavaView. Next a 2D projection of this flattened shell was taken in OpenSCAD. A 4 mm aperture was added to this design, for the lens, at the center of the top sliced face. For permanent folding of the faces, small protrusions or tabs were designed in AutoCAD (Autodesk) to be added to the edge of the cut faces which sported copper pads, designed in EAGLE (Autodesk). Similar pads were also present on the opposite face sharing the edge with the faces with tabs, but with tiny 200 µm holes. Copper pads were also placed on the bottom side of this face, under the holes to allow for proper adhesion with solder. Solder was reflowed on the copper pads of the tabs. The two opposite edge sharing faces were brought together such that the tabs were directly underneath the holes. Heat was applied through the holes and the tab was soldered to the opposite face. All such tabs were joined which resulted in a curved, dome-shaped PCB in the shape of a scleral contact lens.

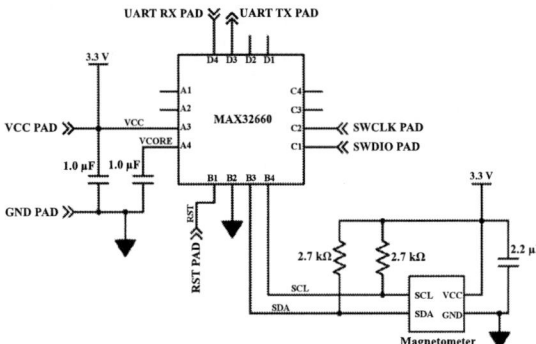

Figure 2: Circuit diagram of magnetic field measurement circuit on the smart contact lens.s

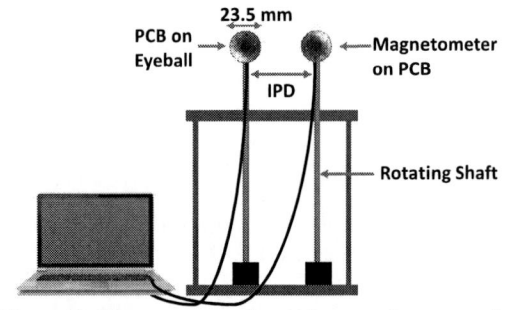

Figure 3: HMC 5883L magnetometer on flexible PCB for smart contact lens.

Data Collection

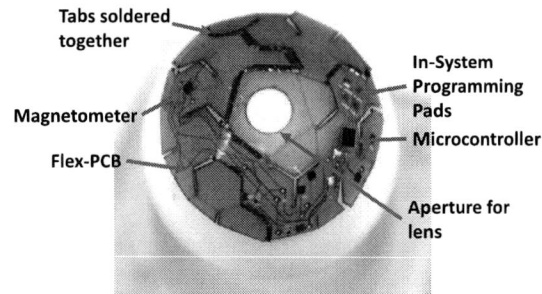

Figure 4: Rotary experimental setup for measuring magnetic field components from magnetometers on each eye.

The folded flex-PCB was mounted on rubber balls. These balls were attached to a plastic shaft, which in turn was attached to a rotating setup. Embedded C program was written in Maxim Integrated Eclipse IDE. The program was uploaded to the microcontroller through the extra pads designed on the flex-PCB for In-System Programming (ISP). The microcontroller used I2C communication to obtain magnetic field data from the magnetometer. The microcontroller was also externally wired so that data could be stored using UART on a computer for further analysis. The magnetometers need to be calibrated before use to account for hard-iron and soft-iron offsets. These offsets were added to the program uploaded to the microcontroller. The human eye can converge up to 35 degrees on each eye to focus on an object. Hence, individual eyes were rotated inwards at an interval of 5

degrees and data was recorded. Figure 4 shows the measurement setup. The measurements were repeated three times.

RESULTS AND DISCUSSION

Figure 5 shows the plot of measured vergence angles vs ideal vergence angles. It can be observed that the RMS error was 1.74 degrees. Apart from the data point at 60 degrees, most angles were within 1 degree of the ideal value. Table 1 compares the predicted object distance with the expected object distance. The RMS error was 14.04 mm. At smaller distances, the error was less than 1.5 mm. The magnetometer consumes 100 μA at 3.3V at an output data rate of 160 Hz, thus consuming approximately 2 μJ of energy per magnetic field measurement per eye. This energy consumption is magnetometer dependent and dominated by the magnetometer internal circuitry and operating modes.

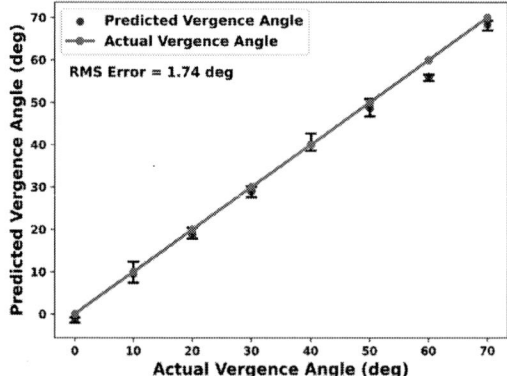

Figure 5: Predicted vergence angles calculated from magnetometers vs expected vergence angles.

Table 1. Expected vs Predicted Object Distance.

Expected Object Distance (mm)	Predicted Object Distance (mm)
342.9	377.57
170.1	181.67
111.9	116.1
82.4	82.3
64.3	66.39
51.9	56.33
42.8	44.21
	RMSE = 14.04 mm

SUMMARY

A novel approach for eye tracking was developed which measured the eye vergence angle and object distance by sensing Earth's magnetic field with the help of magnetometers. This was achieved with and RMSE of 1.74 degrees. The total energy consumption per reading per eye is 2 μJ.

REFERENCES

[1] R. R. A. Bourne *et al.*, "Trends in prevalence of blindness and distance and near vision impairment over 30 years: An analysis for the Global Burden of Disease Study," *Lancet Glob. Heal.*, vol. 9, no. 2, pp. e130–e143, Feb. 2021, doi: 10.1016/S2214-109X(20)30425-3.

[2] J. S. Wolffsohn and L. N. Davies, "Presbyopia: Effectiveness of correction strategies," *Prog. Retin. Eye Res.*, vol. 68, pp. 124–143, Jan. 2019, doi: 10.1016/J.PRETEYERES.2018.09.004.

[3] M. U. Karkhanis *et al.*, "Correcting Presbyopia with Autofocusing Liquid-Lens Eyeglasses," Jan. 2021, Accessed: Feb. 01, 2021. [Online]. Available: http://arxiv.org/abs/2101.08782.

[4] M. U. Karkhanis *et al.*, "Compact Models of Presbyopia Accommodative Errors for Wearable Adaptive-Optics Vision Correction Devices," *IEEE Access*, 2022, doi: 10.1109/ACCESS.2022.3187036.

[5] A. Deshpande *et al.*, "High-Toughness Aluminum-N-Doped Polysilicon Wiring for Flexible Electronics," Jun. 2022, doi: 10.1109/FLEPS53764.2022.9781554.

[6] E. Pourshaban *et al.*, "Flexible and Semi-Transparent Silicon Solar Cells as a Power Supply to Smart Contact Lenses," *ACS Appl. Electron. Mater.*, vol. 4, no. 8, pp. 4016–4022, Aug. 2022, doi: 10.1021/ACSAELM.2C00665/ASSET/IMAGES/LARGE/EL2C00665_0007.JPEG.

[7] E. Pourshaban *et al.*, "Eye Tear Activated Mg-Air Battery Driven by Natural Eye Blinking for Smart Contact Lenses," *Adv. Mater. Technol.*, p. 2200518, 2022, doi: 10.1002/ADMT.202200518.

[8] M. Nasreldin *et al.*, "High performance stretchable Li-ion microbattery," *Energy Storage Mater.*, vol. 33, pp. 108–115, Dec. 2020, doi: 10.1016/j.ensm.2020.07.005.

[9] C. Ghosh, A. Mastrangelo, A. Banerjee, H. Kim, and C. H. Mastrangelo, "Micropower Object Range and Bearing Sensor for Smart Contact Lenses," in *2020 IEEE Sensors*, Oct. 2020, pp. 1–4, doi: 10.1109/SENSORS47125.2020.9278622.

[10] A. Banerjee *et al.*, "Microfabricated Low-Profile Tunable LC-Refractive Fresnel (LCRF) Lens for Smart Contacts," in *Conference on Lasers and Electro-Optics, Technical Digest Series (Optica Publishing Group)*, 2022, p. AW4C.3, doi: 10.1364/CLEO_AT.2022.AW4C.3.

[11] A. Deshpande, E. Pourshaban, C. Ghosh, A. Banerjee, H. Kim, and C. Mastrangelo, "Adhesion Strength of PDMS to Polyimide Bonding with Thin-film Silicon Dioxide," Jun. 2021, doi: 10.1109/FLEPS51544.2021.9469779.

[12] C. Ghosh *et al.*, "Low-Profile Induced-Voltage Distance Ranger for Smart Contact Lenses," *IEEE Trans. Biomed. Eng.*, 2020, doi: 10.1109/TBME.2020.3040161.

[13] M. Mehrubeoglu, L. M. Pham, H. T. Le, R. Muddu, and D. Ryu, "Real-time eye tracking using a smart camera," *Proc. - Appl. Imag. Pattern Recognit. Work.*, 2011, doi: 10.1109/AIPR.2011.6176373.

[14] H.-C. Kim, J. Cha, and W. D. Lee, "Eye Detection for Gaze Tracker with Near Infrared Illuminator," in *2014 IEEE 17th International Conference on Computational Science and Engineering*, 2014, pp. 458–464, doi: 10.1109/CSE.2014.111.

[15] A. Al-Rahayfeh and M. Faezipour, "Eye tracking and head movement detection: A state-of-art survey," *IEEE J. Transl. Eng. Heal. Med.*, vol. 1, pp. 11–22, 2013, doi: 10.1109/JTEHM.2013.2289879.

[16] A. Khaldi *et al.*, "A laser emitting contact lens for eye tracking," *Sci. Rep.*, vol. 10, no. 1, p. 14804, 2020, doi: 10.1038/s41598-020-71233-1.

[17] A. S. Mastrangelo *et al.*, "A low-profile digital eye-tracking oculometer for smart eyeglasses," in *Proceedings - 2018 11th International Conference on Human System Interaction, HSI 2018*, Aug. 2018, pp. 506–512, doi: 10.1109/HSI.2018.8431368.

[18] L. Massin *et al.*, "Development of a new scleral contact lens with encapsulated photodetectors for eye tracking," *Opt. Express*, vol. 28, no. 19, pp. 28635–28647, 2020, doi: 10.1364/OE.399823.

[19] N. Sarkar, D. Stratheam, G. Lee, A. Olfat, A. Rohani, and R. R. Mansour, "A large angle, low voltage, small footprint micromirror for eye tracking and near-eye display applications," in *2015 Transducers - 2015 18th International Conference on Solid-State Sensors, Actuators and Microsystems (TRANSDUCERS)*, 2015, pp. 855–858, doi: 10.1109/TRANSDUCERS.2015.7181058.

[20] M. R. Clark, "A two-dimensional Purkinje eye tracker," *Behav. Res. Methods Instrum. 1975 72*, vol. 7, no. 2, pp. 215–219, Mar. 1975, doi: 10.3758/BF03201330.

[21] E. Whitmire *et al.*, "EyeContact: Scleral coil eye tracking for virtual reality," in *International Symposium on Wearable Computers, Digest of Papers*, Sep. 2016, vol. 12-16-September-2016, pp. 184–191, doi: 10.1145/2971763.2971771.

[22] D. A. Robinson, "A Method of Measuring Eye Movement Using a Scleral Search Coil in a Magnetic Field," *IEEE Trans. Bio-medical Electron.*, vol. 10, no. 4, pp. 137–145, 1963, doi: 10.1109/TBMEL.1963.4322822.

[23] C. Ghosh *et al.*, "A Nano-Joule Burst-Mode Eye-Gaze Angle and Object Distance Sensor for Smart Contact Lenses," *Proc. IEEE Sensors*, vol. 2021-October, 2021, doi: 10.1109/SENSORS47087.2021.9639572.

CONTACT

*A. Deshpande; adwait.p.deshpande@utah.edu

ELECTROHYDRODYNAMIC NEBULISER (eNEB) FOR DIRECT PULMONARY DRUG DELIVERY APPLICATION

Trung-Hieu Vu[1,*], Luan Ngoc Mai[2,3], Tuan-Hung Nguyen[1], Dang Tran[1], Tuan-Khoa Nguyen[4], Thanh Nguyen[5], Jarred Fastier-Woollel[1,6], Canh-Dung Tran[5], Toan Dinh[5], Hong-Quan Nguyen[1], Dzung Viet Dao[1], and Van Thanh Dau[1,7,**]

[1]School of Engineering and Built Environment, Griffith University, Queensland, AUSTRALIA
[2]Ho Chi Minh City University of Technology (HCMUT), Ho Chi Minh City, VIETNAM
[3]Vietnam National University Ho Chi Minh City, Ho Chi Minh City, VIETNAM
[4]Queensland Micro and Nano Technology Centre, Griffith University, AUSTRALIA
[5]School of Mechanical and Electrical Engineering, University of Southern Queensland, AUSTRALIA
[6]School of Engineering, The University of Tokyo, Tokyo, JAPAN
[7]Centre for Catalysis and Clean Energy, Griffith University, AUSTRALIA.

ABSTRACT

We present a new method of nebulisation based on electrohydrodynamic atomisation (EHDA) for use in pulmonary drug delivery application. Our device named electrohydrodynamic nebuliser (eNEB) is a single-step platform to directly produce and deliver micro/nano particles. We applied the working principle of EHDA but employed a sharp pin as reference electrode. As the device works, solution is atomised into plume of particles with sizes under 1 μm and the pin generates ion wind which results in not only reduction of net charge but also propulsion in the particle cloud. The device was able to fabricate particles and fibres from polyvinylidene fluoride (PVDF), at concentrations 5 wt% and 20 wt%. A 5-day growth of breast cancer cells was assayed to investigate the effectiveness for medical applications.

KEYWORDS

Ion wind, neutralisation, atomised plume, electrohydrodynamic

INTRODUCTION

Therapeutics intake, either via oral or nasal route, is a non-invasive and self-administered technique to drug delivery in patient's local respiratory systems or throughout the human body [1]. The monodisperses of 1–5 μm bronchodilator particles are optimal in terms of drug-delivery efficacy inside the human body [2]. However, other experiments also note that the range of optimal aerosol particle sizes may be much smaller than 1–5 μm, covering a range of 2.0 to 3.5 μm [3], [4]. This size range can be achieved by electrohydrodynamic atomisation, a technique capable of producing monodispersed structure suitable for drug delivery.

Electrohydrodynamic atomisation (EHDA) has been widely used to fabricate micro/nano substances such as particles or fibres [5], [6]. This method of atomisation utilises electrostatic force to disintegrate solution into mist of droplets. For a polymeric solution with a mixture of solvent and polymers, the droplets quickly turn into polymer particles whose sizes can be as small as a few nanometres. The fabricated particles are highly charged which can be collected on a substrate of opposite charge [7]. For drug substances fabricated by this method, the particles require additional treatment such as purification,

loading, etc. followed by atomisation [8], [9]. For instance, commercial devices like nebulisers, powder inhalers, mist inhalers, puffers, etc. can transfer particles to a target (such as the lung, nasal membranes, etc. The indirect administration of drug particles creates a challenge because of the discontinuation between the generation process and the particle delivery. Therefore, it is necessary to enhance the current EHDA in order to provide a direct technique that enables simultaneous particle generation and delivery [10]. One of the challenges with conventional EHDA is to discharge the highly charged generated particles before collection. To discharge the particles, it is necessary to install an extra subsystem to create a cloud of opposite charge. Charged particles generated from capillary will be discharged and become electroneutral while flowing through the counter-ion cloud. The mechanism of such neutralisation method is detailed in numerous literatures [7], [8], [11]. The method was improved by creating biologically active nano aerosol from a volatile solvent that was neutralised by a cloud of oppositely charged ions or oppositely charged twin-head droplets [12], [13].

The present work developed an eNEB as a solution to overcome the existing challenge in EHDA and offer concurrent generation and delivery of atomised particles in one step as described in Fig. 1. The device uses two electrodes connected to opposite polarity of a high voltage source (positive to a capillary nozzle and negative to the pin). As high voltage is applied, elevated electric field generates charged particles from the capillary and an oppositely counter ion wind in the pin. Due to the continuous interaction with the ion wind, atomised particles are neutralised. In addition, the momentum accumulated in the wind is used to propel droplets forward to a desired target [14]. The approach poses significant strength such as high portability, compact size, no additional sub-systems, simple to employ, etc. and the key advantage is the ability to neutralise particulates in free space [15]. This approach has great potential for practical uses such as respiratory drug delivery or cell treatment.

DEVELOPMENT OF eNEB
Working principle

The key idea of our novel approach is applying EHDA in combination with generation of ion wind for neutralization and propulsion as described in Figure 1. The capillary is a 22 Ga nozzle, and the negative electrode is a pin with a sphere radius of 35 μm located in the same plane of the capillary at a selected distance d = 5 mm from the nozzle. As high voltage is applied, electrostatic force breaks the surface tension in liquid and generates mist of fine particles [16]. On the negative side, the pin emits ion wind with two primary functions: (i) to provide negative ions to neutralise/discharge the highly positive particles, and (ii) to propel the plume of particles forward. Numerous literatures have investigated the effect of ion wind in neutralisation of using a pin or sharp electrode [17], [18]. Ion wind is also used as a sufficient source of propulsion for different purposes as seen in previous works [14], [18]–[20]. In this configuration, the success in depositing neutralised particles relies heavily on the generation of ion wind during atomising process.

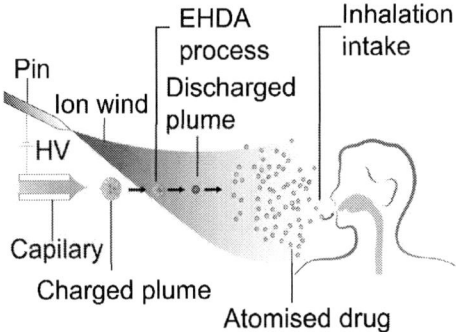

Figure 1: Atomisation process of neutralised particles for pulmonary drug delivery via EHDA

Generation of ion wind with different pins

The geometry if the pin plays an important role in the generation of ion wind. For the atomisation and delivery to happen simultaneously, the applied voltage needs to fulfill the requirements: (i) achieve onset electric field to initiate EHDA and (ii) electric field at the pin tip is sufficient to generate ion wind. The theoretical equations are as follow:

$$E_w = E_o \left(1 + 2.62 \times 10^{-2}/R_c^{\frac{1}{2}}\right) \quad (1)$$

where $E_o = 3.23 \times 10^6$ V/m [21], and R_c is spherical tip radius. This equation assesses the onset voltage at which ion wind is initiated at a pin.

$$V_o = \sqrt{\frac{\gamma R_o}{\varepsilon_o}} \, ln \frac{4d}{R_o} \quad (2)$$

where γ, ε_o, R_o and d presents, respectively, the liquid's surface tension, vacuum dielectric constant, capillary outer radius, and the electrode distance. Eq (2) outlines the value of applied voltage at which EHDA is established.

$$E_{tip} = -\frac{2V/R_c}{ln\left(\frac{4d}{R_c}\right)} \quad (3)$$

where E_{tip} is the electric field at the pin's tip, and V is the applied voltage. In practice, the value V_o in Eq (2) is used to calculate the electric field at a pin with certain spherical radius. The ion wind is generated when $E_{tip} > E_w$, or when the applied field is greater than onset field of ion wind. The significant parameter in Eq (1) and (3) is the pin spherical radius (SR).

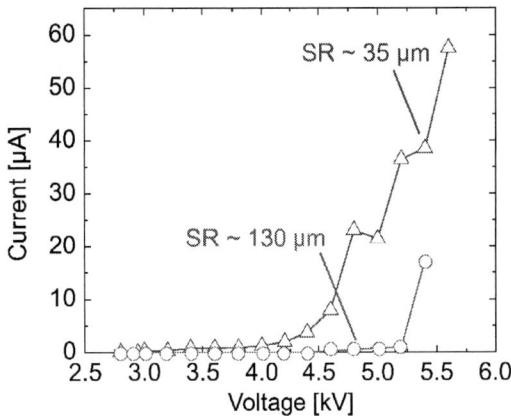

Figure 2: I-V relationship reflecting the onset voltage for simultaneous generation of ion wind and particle plume

The importance of the pin tip is investigated by corresponding the I – V curve for different spherical radii. Two pins were used to depict the initiation of ion wind. The experimented SR includes 35 and 130 μm. Experimental solution was isopropyl alcohol with the following specifications: surface tension γ ≈ 20.8 mN/m, density ρ ≈ 0.785 g/ml, viscosity μ ≈ 1.66 mPas, conductivity K ≈ 6 μS/m and relative permittivity ε ≈ 18.6. The pins were placed 5 mm from the capillary. The estimated value of V_o = 3 kV is the onset voltage for EHDA. The current jump in two pins seen in Figure 2 describes that greater radius results in higher onset voltage to ignite the EHDA process and ion wind. From calculation, 35 μm pin has $E_{tip} > E_w$ at 3 kV so the ion wind can be sufficiently generated. In the same regard, 130 μm pin requires 3.5 kV to achieve the effective ion wind, which is well beyond the onset voltage for EHDA process, thus it is not ideal for our system.

Fabrications of polymer particles

We demonstrated that the device can fabricate particles in one step. PVDF 5 wt% was used as spraying solution (PVDF powder in a mixture of solvent of DMF and acetone). The 35 μm pin is used for optimal ion wind propulsion. The collector is a piece of aluminium foil held 8 cm from the nozzle. This distance ensures the solvent can evaporate completely and does not dissolve particles as they reach collector. The device showed successful fabrication and delivery of particles at a collecting distance of 8 cm, which is 16 times greater than the electrode distance. Figure 3 indicates the collected particles and their size analysis. Overall, particles show monodispersed size and uniform distribution. The analysis shows that the size is within 1 – 5 μm for collected particles. In the following sections, size of atomised particles demonstrates that fabricated particles can even be smaller but only the larger particles can be collected due to efficiency loss in collection. The choice of a biocompatible polymer as PVDF implies the versatile use of this device in the field of drug delivery and biomedical applications. In conventional EHDA, various materials have been researched for use in medical treatment. Thus, the results suggest a highly potential approach to generating monodispersed particles by applying the eNEB.

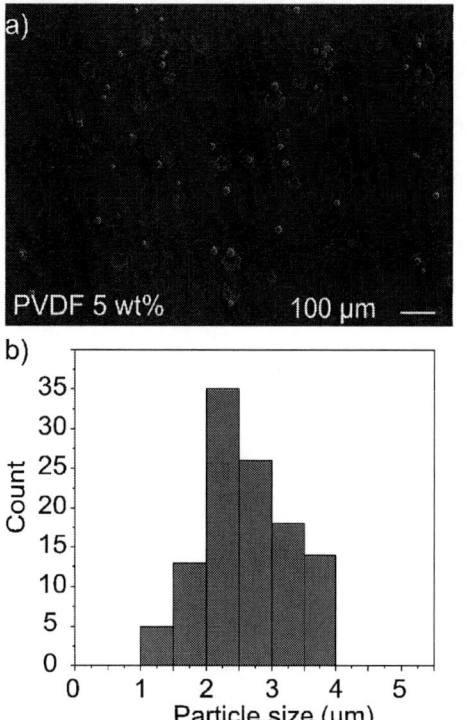

Figure 3: PVDF generated from eNEB; a) SEM images; b) size analysis of PVDF particles.

Size and charge data of particles in the atomised plume
To investigate the size and charge of fabricated particles in the generated plume, experiment was carried out and results are shown in Figure 4. For consistency across experiments, PVDF 5 wt% was selected as atomised solution. An optical particle sizer (TSI, OPS 3330) and aerosol electrometer (TSI, 3068B) were used to assess the size and charge, respectively. Both devices were put in an isolated chamber to prevent contamination from ambience (dust particulate, etc.). Both devices were placed 15 cm away from the nozzle. Therefore, during the spraying process, eNEB would generate plume of particles and get processed by the two devices. The limitation of optical sizer is that it cannot catch particles > 10 μm. For each device, data for 3 samples was collected and each sample lasted 60 seconds. In Fig.4a, the average size of particles was shown to be mostly 0.3 – 1.5 μm. However, previous section could only collect particles between 1 – 5 μm. This loss is due to highly sensitive nature of small particles to external airflow

The aerosol electrometer recorded the charge in the atomised plume as shown in Fig.4b. Overall, the generation of ion wind reduces the charge significantly with the average value of only 200 fA. This value is close to the electrostatic noise in the air [7] with several advantages. Owing to the effective mixing of the positive particles and negative ions in the wind, the eNEB was proven to effectively generate particles with low charge. In addition, the low charged particles can be easily collected on any substrate placed at a distance from the nozzle.

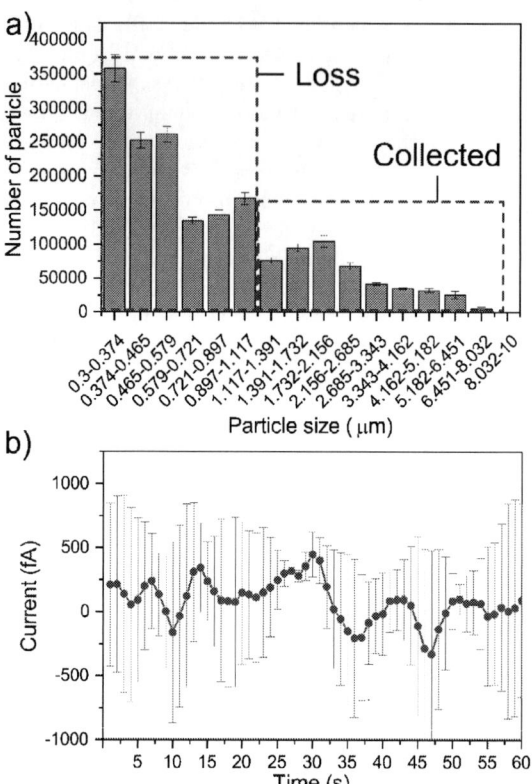

Figure 4: Analysis of generated particles; a) Size distribution of atomise particles, "collected" indicates the range of particle sized collected for SEM observation while "loss" is the particle range observed only by OPS; b) charge measured by electrometer for the atomised plume of particles

Potential in pulmonary drug delivery via proliferation
Demonstration experiment was conducted to assess the potential of the device for use in biomedical application such as a single-step method for drug atomisation and delivery.

Figure 5 summarises the spraying experiment on breast cancer cells. The experiment was conducted as follow: Petri dish containing cells were placed horizontally the eNEB. The spraying plume was controlled by adjusting the applied voltage while the flow rate is constant at 0.5 ml/hr. PVDF 5 wt% was used to spray onto cells. The parameters such as spraying period and distance can affect the final results significantly because of unevoporated solvent. PVDF is a biocompatible polymer but the solvent can eliminate cell growth. So, eNEB was used to spray particles onto cells at different conditions as shown in Fig. 5b, to determine the best conditions for spraying. As shown in Figure 5, cell samples were able to grow normally in comparision to the control sample after 5 days. Other experiments indicates that long spraying period of up to 90s and close distance of 5 cm pose significant toxicity to cell growth.

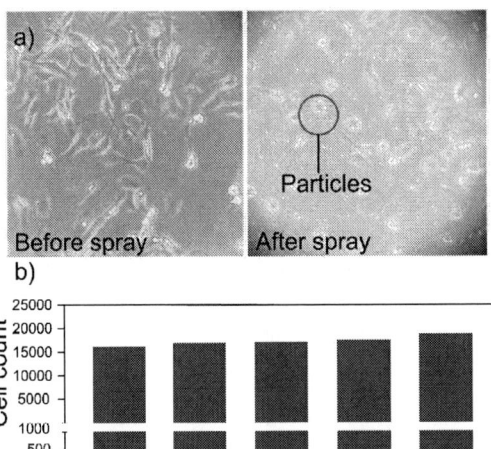

Figure 5: Experimental setup for proliferation assay; a) experiment procedure; a) microscopic images showing the particles attached to cells; b) proliferation results for various conditions of spraying duration and collecting distance after 5 days.

CONCLUSION

Herein, we present the development of a new generation of nebuliser based on electrohydrodynamic (eNEB). This device utilises a pin as a source of ions to solve the existing charge problem in EHDA. The ion wind is also the key factor to provide propulsion to the plume of atomised particles. SEM images and various data imply that the device can genearte particles at monodispersed size with uniform distribution and they contain low charge in the plume. A test with growth of breast cancer further implies that the eNEB has high potential to be the first single-step method in medical application such as pulmonary drug delivery.

REFERENCES

[1] N. Kumar, V. Gautam, V. Kumar, and P. K. Maurya, "Nanoparticle-based macromolecule drug delivery to lungs," *Target. Chronic Inflamm. Lung Dis. Using Adv. Drug Deliv. Syst.*, pp. 227–259, 2020, doi: 10.1016/B978-0-12-820658-4.00011-X.

[2] P. Demoly, P. Hagedoorn, A. H. de Boer, and H. W. Frijlink, "The clinical relevance of dry powder inhaler performance for drug delivery," *Respir. Med.*, vol. 108, no. 8, pp. 1195–1203, 2014.

[3] P. Zanen, L. T. Go, and J. W. J. Lammers, "The optimal particle size for parasympathicolytic aerosols in mild asthmatics," *Int. J. Pharm.*, vol. 114, no. 1, pp. 111–115, 1995, doi: 10.1016/0378-5173(94)00224-S.

[4] M. J. Telko and A. J. Hickey, "Dry powder inhaler formulation," *Respir. Care*, vol. 50, no. 9, pp. 1209–1227, 2005.

[5] T.-H. Vu *et al.*, "Enhanced Electrohydrodynamics for Electrospinning a Highly Sensitive Flexible Fiber-Based Piezoelectric Sensor," *ACS Appl. Electron. Mater.*, 2022, doi: 10.1021/acsaelm.2c00030.

[6] J. W. Fastier-Wooller, T. H. Vu, C.-D. Tran, T. Dinh, V. T. Dau, and D. V. Dao, "In-Situ Depostion of Pressure and Temperature Sensitive E-Skin for Robotic Applications," in *2021 21st International Conference on Solid-State Sensors, Actuators and Microsystems (Transducers)*, Jun. 2021, pp.

1267–1270, doi: 10.1109/Transducers50396.2021.9495600.

[7] V. T. Dau *et al.*, "In-air particle generation by on-chip electrohydrodynamics," *Lab Chip*, vol. 21, no. 9, pp. 1779–1787, 2021, doi: 10.1039/d0lc01247e.

[8] V. T. Dau, T. K. Nguyen, and D. V. Dao, "Charge reduced nanoparticles by sub-kHz ac electrohydrodynamic atomization toward drug delivery applications," *Appl. Phys. Lett.*, vol. 116, no. 2, 2020, doi: 10.1063/1.5133714.

[9] V. T. Dau, T. X. Dinh, C. D. Tran, T. Terebessy, T. C. Duc, and T. T. Bui, "Particle precipitation by bipolar corona discharge ion winds," *J. Aerosol Sci.*, vol. 124, no. December 2017, pp. 83–94, 2018, doi: 10.1016/j.jaerosci.2018.07.007.

[10] T.-H. Vu, H.-D. Vu, H. T. Nguyen, D. V. Dao, and V. T. Dau, "Simultaneous generation and delivery of neutral polymeric aerosol by electrohydrodynamic nebulizer," 2022.

[11] V. T. Dau, T. X. Dinh, T. Terebessy, and T. T. Bui, "Bipolar corona discharge based air flow generation with low net charge," *Sensors Actuators, A Phys.*, vol. 244, pp. 146–155, 2016, doi: 10.1016/j.sna.2016.03.028.

[12] F. Mou, C. Chen, J. Guan, D.-R. Chen, and H. Jing, "Oppositely charged twin-head electrospray: a general strategy for building Janus particles with controlled structures.," *Nanoscale*, vol. 5, no. 5, pp. 2055–64, 2013, doi: 10.1039/c2nr33523a.

[13] V. N. Morozov, "Generation of biologically active nano-aerosol by an electrospray-neutralization method," *J. Aerosol Sci.*, vol. 42, no. 5, pp. 341–354, 2011, doi: 10.1016/j.jaerosci.2011.02.008.

[14] V. T. Dau, T. X. Dinh, T. T. Bui, and T. Terebessy, "Corona anemometry using dual pin probe," *Sensors Actuators, A Phys.*, vol. 257, pp. 185–193, 2017, doi: 10.1016/j.sna.2017.02.025.

[15] V. T. Dau, T. X. Dinh, C. D. Tran, T. Terebessy, and T. T. Bui, "Dual-pin electrohydrodynamic generator driven by alternating current," *Exp. Therm. Fluid Sci.*, vol. 97, no. October 2017, pp. 290–295, 2018, doi: 10.1016/j.expthermflusci.2018.04.028.

[16] T.-H. Vu *et al.*, "Electric Field-Enhanced Electrohydrodynamic Process For Fabrication of Highly Sensitive Piezoelectric Sensor," in *2022 IEEE 35th International Conference on Micro Electro Mechanical Systems Conference (MEMS)*, Jan. 2022, pp. 337–340, doi: 10.1109/MEMS51670.2022.9699674.

[17] V. T. Dau, T. X. Dinh, T. T. Bui, and T. Terebessy, "Bipolar corona assisted jet flow for fluidic application," *Flow Meas. Instrum.*, vol. 50, pp. 252–260, 2016, doi: 10.1016/j.flowmeasinst.2016.07.005.

[18] N. T. Van *et al.*, "A circulatory ionic wind for inertial sensing application," *IEEE Electron Device Lett.*, vol. 40, no. 7, pp. 1182–1185, 2019, doi: 10.1109/LED.2019.2916478.

[19] V. T. Dau *et al.*, "Electrospray propelled by ionic wind in a bipolar system for direct delivery of charge reduced nanoparticles," *Appl. Phys. Express*, vol. 14, no. 5, p. 55001, Mar. 2021, doi: 10.35848/1882-0786/abf36b.

[20] T. X. Dinh, D. B. Lam, C. D. Tran, T. T. Bui, P. H. Pham, and V. T. Dau, "Jet flow in a circulatory miniaturized system using ion wind," *Mechatronics*, vol. 47, no. January, pp. 126–133, 2017, doi: 10.1016/j.mechatronics.2017.09.007.

[21] M. Robinson, "Movement of air in the electric wind of the corona discharge," *Trans. Am. Inst. Electr. Eng. Part I Commun. Electron.*, vol. 80, no. 2, pp. 143–150, 1961, doi: 10.1109/TCE.1961.6373091.

CONTACT

*Trung-Hieu Vu, trunghieu.vu@griffithuni.edu.au
**Van Thanh Dau, v.dau@griffith.edu.au

FLEXIBLE POLYMER OPTICAL WAVEGUIDES FOR INTEGRATED OPTOGENETIC BRAIN IMPLANTS

Julian A. Singer[1], Till Stramm[2], Jens Fasel[2], Oliver Schween[2], Anton Gelaeschus[1], Andreas Bahr[1,3], and Matthias Kuhl[4]

[1]Institute for Integrated Circuits, Hamburg University of Technology, GERMANY,
[2]Chair for High Frequency Technology, TU Dortmund University, GERMANY,
[3]Research Group of Sensor System Electronics, University of Kiel, GERMANY and
[4]Laboratory for Microelectronics, IMTEK, University of Freiburg, GERMANY

ABSTRACT

This paper presents a manufacturing process for flexible optical waveguides made of biocompatible polydimethylsiloxane (PDMS) to be attached to future optogenetic implants. Based on a complementary metal-oxide-semiconductor (CMOS) application-specific integrated circuit (ASIC), the system uses integrated single photon avalanche diodes (SPADs) as sensors for fluorescence imaging (FI) and post-CMOS micro light-emitting diodes (µLEDs) for stimulation [1]. Attaching fibers to SPADs and µLEDs to collect and distribute light from and to deeper brain layers can unfold previously unachieved medical applications. Two technologies for biocompatible waveguide manufacturing are presented: thin-film and molding. Such a combination of integrated bidirectional implantable ASICs with custom waveguide manufacturing is not limited to medical scenarios but opens up a broad field of applications.

KEYWORDS

MEMS, optogenetics, waveguide, polymer, brain implants, neural implants, integrated circuits, ASIC, biomedical, SPAD, µLED, thin-film, molding.

INTRODUCTION

The emerging healthcare topic of optogenetic implants allows targeting specific regions of the brain and monitoring its activity using FI [2]. A deeper understanding of how the brain works, the treatment of neurological diseases, and interaction with the central nervous system are becoming increasingly important. With progress in biomedical methods, new demands are also being placed on technical implementation [3]. The new field of optogenetics enables interactions with the brain by optical means using light-sensitive biochemical markers [4]. The specificity of these opsins opens up new possibilities for stimulating the brain, but in turn, requires miniaturized light sources with high precision and low power consumption. New methods to study these relationships may emerge from the steady miniaturization of integrated circuits in micro- and nanoelectronics, as their small size allows direct and continuous measurement as well as manipulation of neurons.

By using different opsins, brain activity can either be induced or suppressed. The applicable light is mainly in the visible spectrum. In [5], a bidirectional pair of opsins has recently been published that suppresses action potentials when excited with light at 490 nm wavelength and induces action potentials at 595 nm, which are the wavelengths targeted for the proposed system. Further applications do not only require stimulating, but also observing optical processes within neural tissue meanwhile. Therefore, FI can be used with the help of calcium or voltage indicators [6, 7]. By targeting those indicators with a light source, their fluorescence can be monitored using highly sensitive photodetectors to make brain structure and activity visible [7]. Combining those concepts with optogenetic tools renders closed-loop optogenetic platforms possible, which can adapt stimulation in real-time according to observed activity.

In consequence, using integrated optoelectronic devices, new fields of application in the areas of neural sensing, stimulation, inhibition, and tooling for bioelectronic medicine can be designed. CMOS-integrated SPADs can be used for calcium imaging, as their high sensitivity is well suited for the application. Simultaneously, their small size and low power consumption make them suitable for implantation. Combined with post-CMOS µLEDs on top of a CMOS ASIC, this allows for small, low-power, wireless, implantable devices with high resolution [1, 8].

However, area coverage and insertion depth are limited. ASICs with sensors and µLEDs on top can only be placed on top of the brain reaching only a limited amount of neuron populations. To target more specific regions deeper in the brain, insertion into the brain is necessary. Compared to electrical neural implants being miniaturized towards needle size to reduce tissue damage [9], optical devices can hardly be made small enough to insert them deep into the brain without damaging the tissue. Thus, it is targeted to place the electronics mounted to the skull and waveguides can be used to bring the light into and out of the brain. Thin flexible fibers with waveguides can target deep brain regions with minimal tissue damage and reduced risk of inflammation.

A variety of waveguides for medical applications, including optogenetics, have been proposed [10]. Therein, PDMS has been found a suitable material due to its optical transparency, flexibility, and biocompatibility. PDMS has been widely used in different medical applications including optogenetics, ranging from optical lenses [11] to waveguides [12]. PDMS waveguides have been manufactured in a variety of processes [12, 13]. Such waveguides need to transmit light power in the range of mW/mm² for the desired optogenetic markers [5], of which PDMS is capable due to its low losses. To attach large numbers of waveguides to an ASIC, a custom-made waveguide bundle matching the array spacing is required. This work presents manufacturing processes for such waveguide arrays.

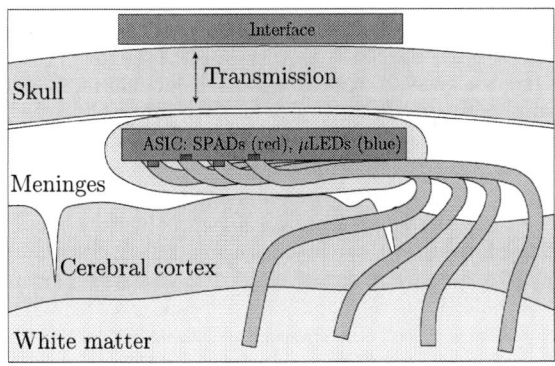

Figure 1: Concept of the wireless optogenetic implant with attached polymer waveguides to optically access regions deep in the brain.

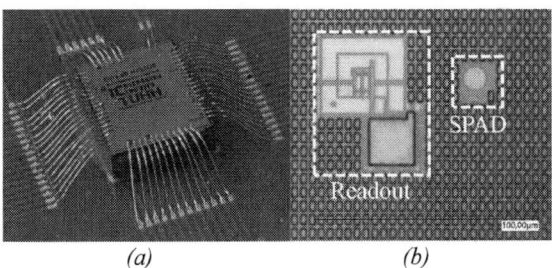

Figure 2: CMOS ASIC for optical measurements (a) and detailed view of a single SPAD with readout circuit (b), to which the waveguides are to be connected.

APPLICATION SCENARIO

Flexible waveguides are to be connected to an ASIC as shown in Figure 1, to access a greater area and deeper tissue regions than the ASIC itself could. The ASIC in development (Figure 2a) houses CMOS-integrated SPADs (Figure 2b) as detectors for light gathering and post-CMOS applied μLEDs [1]. The final system will incorporate all drive and readout circuits in CMOS as well as compression architectures [14] and wireless transmission to an interface outside of the head to reduce inflammation risk (Figure 1). As the ASIC with dimensions of several millimeters cannot be inserted into the brain matter, flexible polymer waveguides are connected to it to reach deep brain regions. Core dimensions of the waveguides greater than 50 μm are targeted to ensure multi-mode operation and match the μLEDs and SPADs with diameters of 10 to 100 μm on the developed ASIC (Figure 2) [1].

MANUFACTURING PROCESS

To manufacture the waveguides, Wacker Elastosil® RT 601 and RT 604 are chosen for core and cladding due to their processability, optical properties, and biocompatibility. Using glass as a carrier, two custom processes are developed to manufacture waveguides in the desired size with optical characteristics suitable for optogenetic applications. The first is a 15-step thin-film process with lift-off as shown in Figure 3. Therein, three SU-8 lithography steps are applied to create the PDMS waveguide using a printed film mask. In steps (1-4), the lower cladding is created with a lithography step in Kayaku

Advanced Materials SU-8 2035. Using a negative resist of the type AR-N 4340 by Allresist and another SU-8 step, the core is created in (5-9). A lift-off (10) is performed to free the core, and site and top cladding are deposited using another SU-8 lithography (11-14). As an alternative to the lift-off, reactive-ion etching has also been studied, yielding even higher precision. Though for the application, this precision is not needed, making the lift-off preferable due to its simplicity in processing. The waveguide can then be separated (15) or used as an array.

The second developed process is an optimized transfer process using micro-molding based on [13]. The twelve-step process shown in Figure 4 uses a lithography-created SU-8 mold (1-4). First, the mold is polished in step (5), before the core is deposited using a blading process in step (6) into the molds shown in Figure 5a. Using two more blading steps with a lift-off in between (7-10), the cladding is deposited before the waveguides can be separated (11-12). The lift-off process done by hand is shown in Figure 5b. For the first characterization, the separation is done by hand-cutting with a scalpel. For future applications, single waveguides or arrays can be precisely laser-cut.

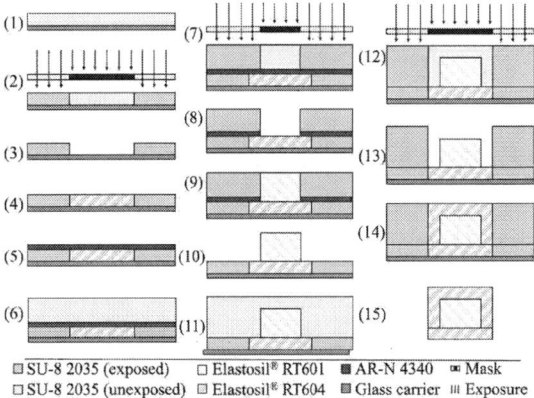

Figure 3: Thin-film process to manufacture PDMS waveguides. The lift-off executed by dissolving the AR-N 4340 in acetone is shown in step (10).

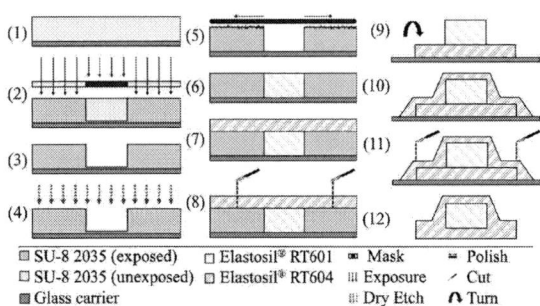

Figure 4: Molding process using a blading technique to manufacture PDMS waveguides. In step (9), the waveguide is removed from the mold and the remaining cladding is deposited.

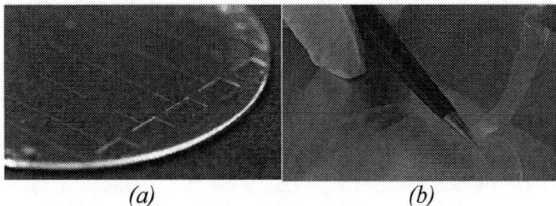

Figure 5: SU-8 mold for the blading process (a) and manual peel-off of PDMS waveguides (b).

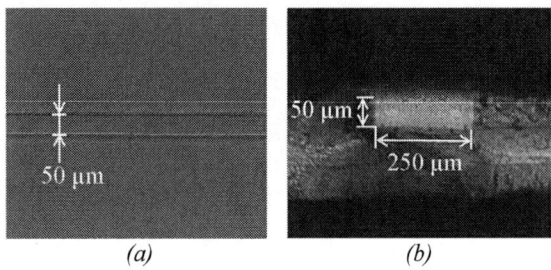

Figure 6: Top view (a) of a 50 × 50 μm² waveguide core and cross-sectional view (b) of a waveguide core with 250 × 50 μm² under the microscope.

EXPERIMENTAL RESULTS

Using the thin-film process, waveguides with core dimensions down to 10 μm can be manufactured. Conversely, the transfer process is limited to 50 μm minimum core size. It is favorable to have a core with greater dimensions than the LEDs and sensors to improve coupling and simplify alignment. Therefore, the transfer process is preferred for the given application, as it meets the requirements with fewer process steps and higher reproducibility. For other applications requiring higher precision and smaller sizes, the developed thin-film process may prove to be superior. Waveguides with two different core sizes are manufactured for the given application, with 50 × 50 μm² (Figure 6a) core cross-sectional array for smaller μLEDs and SPADs and with 250 × 50 μm² (Figure 6b) for larger commercial LEDs. The waveguides exhibit cores with sub-μm sidewall roughness, as shown in Figure 6a.

The manufactured waveguides are characterized by an optical measurement setup using a single-mode lensed glass fiber to couple light into the waveguide. As the manufactured waveguides operate in multi-mode, the coupling from the single-mode fiber to the waveguides is lossy. Therefore, a cutback measurement is performed, where the PDMS waveguide is cut by a fraction of a millimeter for every measurement. With this method, the attenuation per length of the waveguide can be determined neglecting the low coupling efficiency. The coupling can be reproduced easily since the lensed fiber with a 1.7 μm output diameter is much smaller than the 50 × 50 μm² PDMS waveguide, which is used in all following measurements.

The measurement setup is shown in Figure 7. Two different lasers at different wavelengths are used. Of these, one is in the visible range at 650 nm for proof of principle. For exact determination of the losses, a Keysight 8164B Lightwave Measurement System is used with a Keysight 81640A source module at 1550 nm. It outputs a defined 1 mW of power and is matched to the used glass fibers. From the glass fiber, the light is coupled into the PDMS waveguide core using an optical alignment table under a microscope. The light coupling out of the waveguide is focused onto a collimator using a 20× lens and fed back into a Keysight 81633A power measurement module.

Figure 8 shows the optical alignment with the fiber on the left, the PDMS waveguides on a glass substrate in the middle, and the lens on the right. Using visible red light the light guidance through the core can be observed as well as the square-shaped core focused onto the collimator by the lens (Figure 8). At 1550 nm a defined cutback is performed and the transmitted power is measured. The results shown in Figure 9 express a linear behavior as expected. Using a linear fit, the loss can be determined to be approximately 0.47 dB/mm for the waveguide. With 0.19 dB/mm being the theoretical limit for RT 601 at 1550 nm [15], the process induces acceptable losses due to geometric mismatch and sidewall roughness, and the attenuation is suitable for the desired application. At visible wavelengths used in biomedical applications, the material can be expected to have even lower attenuation since the material absorption in the range of 0.02 dB/mm is significantly lower [15].

Extrapolation of the measurements shows the coupling losses to be around 27.1 dB. Therein, roughly 6 · 0.2 dB = 1.2 dB are due to the losses in the 6 FC/APC connectors used in the setup, leaving 25.9 dB of combined losses due to single- to multi-mode coupling and lens misalignment.

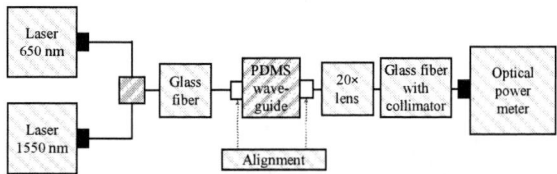

Figure 7: Measurement setup to evaluate the manufactured waveguides at different wavelengths.

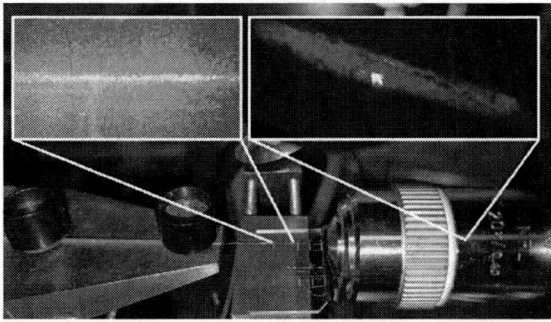

Figure 8: Single waveguide in characterization setup. Glass fibers are used to couple light into the polymer waveguide. The light guided along the core can be observed under the microscope (upper left). Behind the lens, the square-shaped core of the waveguide can be seen transmitting most of the light while the scattered light intensity is significantly lower (upper right).

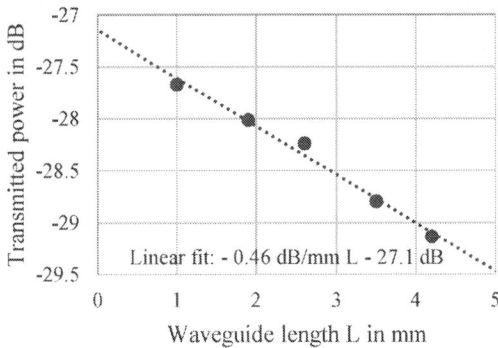

Figure 9: Cutback measurement of a manufactured 50 × 50 µm² waveguide. The total transmitted power is measured for different waveguide lengths. The loss in the waveguide is estimated using a linear fit.

CONCLUSION AND OUTLOOK

Two different ways of manufacturing biocompatible PDMS waveguides are shown in this work, namely one thin-film and one molding process. Measurements prove the waveguides' functionality with losses suitable for application in optogenetic implants. The processes can be used to manufacture waveguides for attachment to different array spacings of µLEDs or SPADs in optogenetic implants. For future applications, laser-cutting of the waveguide arrays will enhance attachment precision. In addition, custom-made waveguide bundles can be created to simultaneously attach high numbers of waveguides to large on-chip arrays using the proposed processes.

ACKNOWLEDGEMENTS

The authors gratefully thank Lukas Rennpferdt from the Institute of Microsystems Technology, Hamburg University of Technology, for his great assistance in setting up the measurement environment. Furthermore, thanks are given to the German Federal Ministry of Education and Research (BMBF) and the Hamburg University of Technology for funding the ForLab HELIOS and I³Lab HELIOS projects that have made this work possible. Also, we want to acknowledge everyone incorporated from Hamburg University of Technology and TU Dortmund University for making this joint work possible.

REFERENCES

[1] J. A. Singer, A. Geläschus, C. Schuster, H. K. Trieu, M. Kuhl, "Integrierte optoelektronische Systeme basierend auf hochempfindlichen Einzelphotonen-detektoren für optogenetische Anwendungen", in *MikroSystemTechnik Kongress 202*1, pp. 212–215.

[2] M.K. Kim et al., "Plugging Electronics Into Minds: Recent Trends and Advances in Neural Interface Microsystems", *IEEE Solid-State Circuits Magazine*, vol. 11, no. 4, pp. 29–42, 2019.

[3] K. Deisseroth, M.J. Schnitzer, "Engineering approaches to illuminating brain structure and dynamics", Neuron, 80(3), pp. 568-77, 2013.

[4] L. Fenno, O. Yizhar, K. Deisseroth, "The development and application of optogenetics", *Annual review of neuroscience*, 34, 2011.

[5] J. Vierock et al., "BiPOLES is an optogenetic tool developed for bidirectional dual-color control of neurons", *Nature Communication*, vol. 12, no. 1, p. 4527, 2021.

[6] C. Grienberger, A. Konnerth, „Imaging calcium in neurons," *Neuron*, 73(5), pp. 862-885, 2012.

[7] T. Knöpfel, C. Song, "Optical voltage imaging in neurons: moving from technology development to practical tool," *Nature Reviews Neuroscience*, pp. 1-9, 2019.

[8] S. Moazeni et al., "A Mechanically Flexible, Implantable Neural Interface for Computational Imaging and Optogenetic Stimulation Over 5.4×5.4mm2 FoV", in *IEEE Transactions on Biomedical Circuits and Systems (BioCAS)*, vol. 15, no. 6, pp. 1295-1305, Dec. 2021.

[9] B. Shui, D. De Dorigo, A. Sayed Herbawi, P. Ruther, O. Paul, Y. Manoli, M. Kuhl, "A Slim Needle Neural Probe with 160 Active Recording Sites and Selectable ADCs," in *2019 IEEE Biomedical Circuits and Systems Conference (BioCAS)*, 2019.

[10] J. Wang, J. Dong, "Optical Waveguides and Integrated Optical Devices for Medical Diagnosis, Health Monitoring and Light Therapies", *Sensors*, 20(14), pp. 3981, 2020.

[11] E. Klein, Y. Kaku, O. Paul, P. Ruther, "Flexible µLED-Based Optogenetic Tool with Integrated µ-Lens Array and Conical Concentrators Providing Light Extraction Improvements above 80%", in *2019 IEEE 32nd International Conference on Micro Electro Mechanical Systems (MEMS)*, 2019.

[12] R. Nazempour, Q. Zhang, R. Fu, X. Sheng, "Biocompatible and Implantable Optical Fibers and Waveguides for Biomedicine", *Materials*, vol. 11, no. 8, p. 1283, Jul. 2018.

[13] S. Kopetz et al. "PDMS-based optical waveguide layer for integration in electrical–optical circuit boards", *AEU-International Journal of Electronics and Communications*, 61.3, pp. 163-167, 2007.

[14] J. D. Rieseler, M. Kuhl, "A Superposition-Based Analog Data Compression Scheme for Massively-Parallel Neural Recordings", in *2017 IEEE Biomedical Circuits and Systems Conference (BioCAS)*, 2017.

[15] M. Maluck, *Replikationstechniken zur Herstellung einmodiger integriert-optischer Komponenten aus neuartigen und kommerziellen Polymeren*, PhD Thesis, TU Dortmund University, 2007.

CONTACT

*J. A. Singer, tel: +49 40 42878 2752; julian.singer@tuhh.de

HIGHLY REPRODUCIBLE TISSUE POSITIONING WITH TAPERED PILLAR DESIGN IN ENGINEERED HEART TISSUE PLATFORMS

Milica Dostanić[1,2], Laura M. Windt[2], Maury Wiendels[2], Berend J. van Meer[2],
Christine L. Mummery[2,3], Pasqualina M. Sarro[1] and Massimo Mastrangeli[1]

[1] Microelectronics, TU Delft, Delft, The Netherlands
[2] Anatomy and Embryology, LUMC, Leiden, The Netherlands
[3] Applied Stem Cell Technology, University of Twente, Enschede, The Netherlands

ABSTRACT

We present a novel design of elastic micropillars for tissue self-assembly in engineered heart tissue (EHT) platforms. The innovative tapered profile confines reproducibly the tissue position along the main micropillar axis, increasing the accuracy of tissue contraction force measurement. Polydimethylsiloxane-based pillars were designed and fabricated by wafer-level molding in an hourglass shape, with symmetric tapering producing a restriction for tissue movement in the middle of the pillars' length. Confinement efficacy of the new geometry was validated by comparing the tissue performance in straight versus tapered (75° or 80° tapering angle) micropillars. While in all three cases compact tissues formed successfully, for both tapered designs the functionality assays evidenced yield increase from 15% to 100%, higher spatial tissue confinement, and correspondingly higher accuracy and smaller dispersion in measurements of tissue contraction force.

KEYWORDS

engineered heart tissue, heart-on-chip, organ-on-chip, microfabrication

INTRODUCTION

Engineered heart tissue (EHT) models have demonstrated valuable potential to reproduce the (patho)physiology of human cardiac tissue *in vitro* [1]. They are composed of cardiomyocytes (CMs) and non-cardiomyocyte cells within an extracellular matrix (ECM), which self-assemble into tissue-like constructs around two or multiple elastic anchoring points, here called (micro)pillars. For a platform meant to culture EHTs, the design of substrate, the shape and number of elastic pillars, and their mechanical properties are crucial for tissue formation. In our previous miniature EHT platform based on polydimethylsiloxane (PDMS)[2], the cell-gel mixture self-assembled into a tissue-like construct around a pair of micropillars with uniform, rectangular cross-section within an elliptic microwell. However, tests showed that such straight pillar geometry is not optimal for long-term tissue culture, as tissues tend in time to increase the contraction force and move upwards towards the pillars' tip, eventually jumping off and detaching from the pillars in extreme cases. Variation in tissues' adherence position along the pillars hampers precise and consistent measurements of tissue contractile parameters. To address this issue, herein we introduce a tailored tapering in pillar design to constrain the tissue on a precisely defined position along the pillars' length.

DESIGN OF TAPERED PILLARS

The tapered design of pillars aims to pre-determine the tissue position over time. This increased positional control is

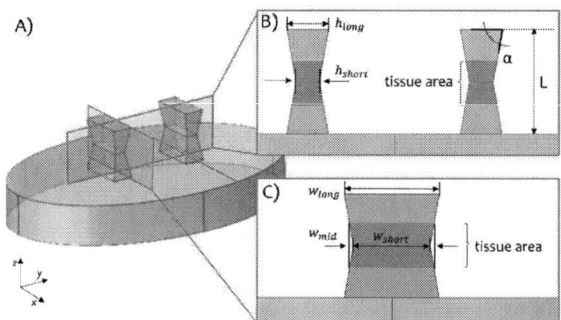

Figure 1. Design and geometrical parameters of tapered pillars implemented in Comsol Multiphysics: A) 3D geometry of tapered pillars on PDMS substrate; B) cross-section of the pillars along the y axis; C) cross-section of pillars along the x axis.

achieved by providing mechanical resistance against tissue movement outside of the intended region. Moreover, the new design aims to preserve the volume of the elliptic well (2 μL) and pillar stiffness close to that of the previously developed pillars [2]. Additionally, tapering should provide mechanical constraint to the tissue movement in the middle of the pillars' length, without affecting significantly the dynamics of tissue formation observed in the previously developed models.

To meet these requirements, an hourglass pillar profile with central symmetric tapering was introduced, whereby the pillar walls form a 75° or 80° angle α with the horizontal planes (Fig. 1). Such a profile is chosen as a compromise between sharper angles that provide stricter mechanical confinement and an angle close to 90° that enables easy removal of PDMS structures from the mould. The dimensions of the pillars for both tapering angles were determined from numerical simulations implemented in Comsol Multiphysics®. Cross-sections and all relevant dimensions of the tapered pillars are shown in Fig. 1 and the values given in Table 1. These dimensions were used to design masks for the platform microfabrication.

Table 1 Design parameters for the tapered pillars.

Symbol	Description	Value	
α [deg]	tapering angle	80	75
h_{short} [μm]	pillars' middle thickness	142	157
h_{long} [μm]	pillars' tip thickness	252	280
w_{short} [μm]	pillars' middle width	383	367
w_{long} [μm]	pillars' tip width	471	447
L [μm]	pillars' length	500	500

FABRICATION

The hourglass profile of tapered pillars was achieved by molding PDMS into a Si substrate processed on both sides by alternating anisotropic deep reactive ion etching (DRIE, Bosch process) and isotropic etching of silicon. Such etching approach creates a staircase-like profile while precisely defining the tapering angle of the etched holes and the location of their conjunction [3], [4]. Fine tuning of the tapering angle was controlled by knowing the etch rates of both isotropic and anisotropic etching processes. The tapering angle α (Fig. 1.B) can be calculated according to the following relation:

$$tg\alpha = \frac{c_B \cdot r_B(z) + t_{iso} \cdot r_{iso}^{\downarrow}(z)}{(t_{iso} - t_{C_4F_8}) \cdot r_{iso}^{\rightarrow}(z)}$$

Here c_B is the total number of Bosch cycles, t_{iso} is the duration of isotropic etch, and $t_{C_4F_8}$ is the time required to remove the residual C_4F_8 layer from the sidewalls after the Bosch process. Vertical etch rate of the Bosch process, as well as vertical and lateral isotropic etch rates, are defined as r_B, $r_{iso}^{\downarrow}$ and $r_{iso}^{\rightarrow}$, respectively. All etching parameters are highly dependent on process temperature, geometry of the masks (density of structures) and depth of etching, which requires etch rate adjustments per each etching cycle.

Fabrication of the mould for tapered pillars started from plasma-enhanced chemical vapor deposition (PE-CVD) of a 5 μm-thick oxide layer on a double-side polished, 500 μm-thick Si wafer. The oxide layer was patterned using standard photolithography and dry etching to

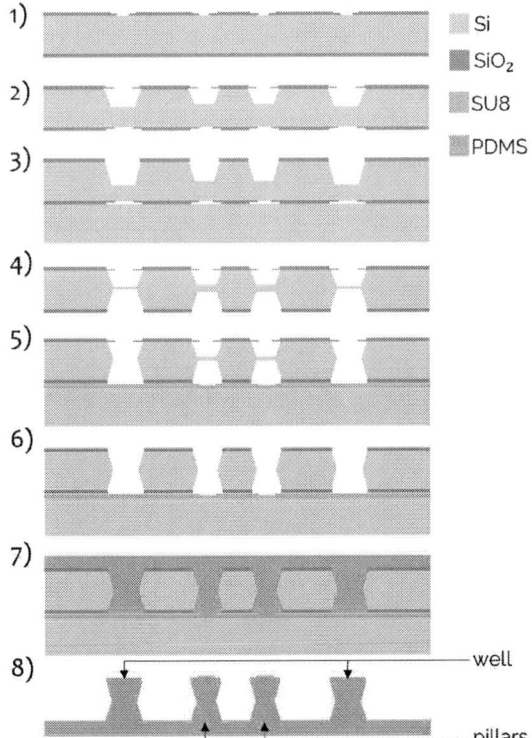

Figure 2. Sketch of the fabrication process of the PDMS-based EHT platform with tapered pillars: 1) definition of SiO₂ hard mask; 2-4) DRIE of tapered holes into the Si mold; 5) wafer bonding; 6) etching finalization of the tapered structures; 7) PDMS molding; 8) released EHT platforms with tapered pillars

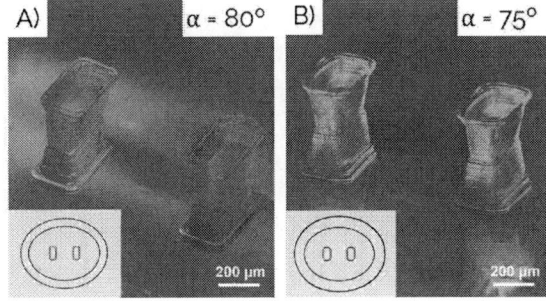

Figure 3. Optical images of PDMS tapered pillars released from the silicon mould, with two different tapering angles: A) α=75° and B) α=80°. Top view of the platforms with elliptic microwell are shown in the insets.

create a two-step hard mask for DRIE, on both wafer sides (Fig. 2.1). In the patterning, the windows in the first mask determined the narrow rectangular shapes which defined the middle part of the pillars. The larger windows of the second mask defined the top and bottom dimensions of the micropillars at the final etching stage, and fine-tuned the tapering angle.

After hard mask definition, DRIE etching followed, yielding a scalloped, staircase-like profile. First, the front side of the wafer was etched by alternating Bosch process and isotropic etching, until reaching a depth of 240 μm (Fig. 2.2). For the last 20 μm of the front-side etch, the inner part of the hard mask was removed by wet etching of SiO₂ in a buffered HF solution. Frontside DRIE was finalized by transferring the process wafer on a Si carrier wafer (Fig. 2.3). Transfer to the carrier wafer ensured the same etching conditions on both front- and backside of the process wafer, and thus symmetry of tapered shapes. After etching, the wafer was cleaned in O₂ plasma to remove polymer residues from the Bosch process.

The same etching strategy was applied to the backside, until reaching a depth of 200 μm (Fig. 2.4). However, finalization of etching required bonding the process wafer to a support wafer to enclose etched pillars and wells from one side. Direct wafer bonding was performed using an adhesive polymer as intermediate layer in between the two silicon wafers (Fig. 2.5). In this case, SU8 – an epoxy-based negative resist – was used as the bonding adhesive, spin-coated and soft-baked on the support Si wafer. The bonding was performed by applying 2.5 kN force on the wafers, which enabled wetting of the Si surface of the process wafer by the SU8 layer. In this configuration, final cross-linking and hard baking of SU8 were performed at 120 °C, to ensure strong and reliable wafer bonding. Once bonding was completed, etching could be finalized to merge the tapered holes in the middle of the wafer (Fig 2.6).

The final step of the fabrication is PDMS moulding. Prior to polymer spin-coating, the wafer surface was made hydrophobic by adsorbing a self-assembled monolayer of perfluorosilane from vapor phase. Elastomer mixed with the curing agent in 10:1 ratio was spin-coated for 30 s at 400 rpm on top of the Si mould (Fig. 2.7). After degassing and curing at 90 °C for 1 h, PDMS structures were released from the mould (Fig. 2.8). Both tapered pillar structures were optically characterized using optical profilometry and the corresponding images are shown in Fig. 3.

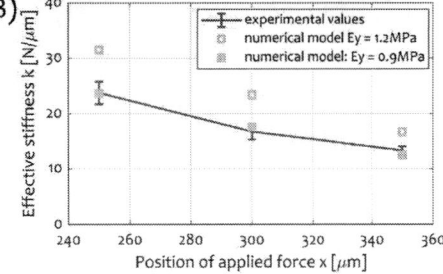

Figure 4. A) Setup for the mechanical characterization of the elastic micropillars using nanoindentation; B) Stiffness analysis for 80° tapered pillars. Data from finite element simulations (green) compared and fitted to the measured data (blue) to estimate the actual Young's modulus of PDMS.

MECHANICAL CHARACTERIZATION

A FemtoTools Nanomechanical Testing System (FT-NMT03) was used for the mechanical characterization of the tapered pillars [2]. A silicon cantilever with flat circular tip, 50 μm in diameter, was used to apply force on different positions along the pillars length. The setup is illustrated in Fig. 4A. Measurements were performed on 3 different positions on each pillar, and repeated for 3 samples of both 75° and 80° tapered pillars. The maximum applied force was 80 μN and the pillars were displaced accordingly. The corresponding displacement–force curve was obtained by measuring the pillars' displacement with a piezo scanner. The stiffness of pillars in the point of force application was obtained from the unloading curve upon pillars' return to the initial position. The measured pillar stiffness was 14 N/m and 16 N/m for 80° and 75° tapered designs, respectively, compared to 10 N/m for 90° straight pillars.

Nanoindentation measurements were used to calculate the Young's modulus (E_Y) of the PDMS and to compare their elastic response curve to the ones obtained from numerical and analytical models. Both models assumed E_Y = 1.2 MPa. The conditions from the experimental setup were recreated in Comsol to enable data comparison. The shift between measured values and simulations for E_Y = 1.2 MPa is noticeable from Fig. 4B. This shift was corrected by finding the value of E_Y that minimizes the mean square difference between the values in both curves. The new value of E_Y was found to be 900 KPa, for both 80° and 75° pillars. Comparison of measured and simulated values, before and after Young's modulus correction for the case of 80° tapered pillars is shown in Fig 4B.

INCLUSION OF CARDIAC CELLS

The assessment of the novel pillar design efficiency was performed by culturing EHTs on both tapered designs and comparing them to the tissue performance in the case of straight pillars. The main questions of the study were the efficiency of tissue confinement in tapered design and its

effect on the tissue contractile properties, yield of the experiment, and variation of the results.

For EHT formation on tapered pillars the same protocol was used as reported previously [5]. Briefly, three different cell types: cardiac fibroblast (cFB), endothelial cells (ECs) and cardiomyocytes (CMs) were derived from an hiPSC line (LUMC0020iCTRL-06). The hiPSCs were differentiated into CMs as described previously [6]. The EHTs were composed from 70% hiPSC-CMs,15% hiPSC-derived cFB and 15% ECs. For the ECM gel mixture, 41% of acid solubilized collagen I (3.3 mg/mL), 5% of DMEM (10X), 6% of NaOH, 9% of growth factor reduced Matrigel and 39% of formation medium were used. The EHTs were cultured in formation medium for 72 h combined with VEGF (50 ng/mL) and FGF (5 ng/mL). After 72 h the culture medium was switched to MBEL+ VEGF (50 ng/mL)+ FGF (5 ng/mL) and maintained in culture for 14 days. The tissues successfully formed in all three different pillar designs. Representative images of tissues formed around straight, 80° tapered and 75 ° tapered pillars are shown in Fig. 5. Brightfield images were taken live on day 7 since the beginning of the experiment with a Nikon Eclipse Ti2.

To assess tissue confinement with the tapered geometry, the position of the tissue along pillars length were measured for both designs. Optical images were taken using Keyence VHX-900F microscope at 60° angle. The imaging was performed in the end of the experiment, after fixing the tissues on the pillars. The measurements were repeated for at least three tissues of each tapered pillar design, and confinement in the middle of pillars was observed in all of them. In Fig. 6 representative examples of tissue confinement in the middle region are shown. By visual inspection it was determined that the new geometry successfully provided mechanical constraints for the tissue movement outside of the narrow area in the middle of the pillars.

The comparative study between tapered and straight pillar designs was performed based on the yield of the experiment and contractile tissue parameters. The yield of the experiment increased from 15% for the straight design to 100% for both 80° and 75° tapered pillars. All tissues from tapered designs reached the end point of the experiment, proving that spatial confinement solved the problem of tissue jumping off and enabled long-term cell culture. The duration of the experiment was 14 days.

The contractile parameters of tissues in different pillars were measured. Videos of contracting tissues were recorded on day 4, 7 and 11, during spontaneous contraction

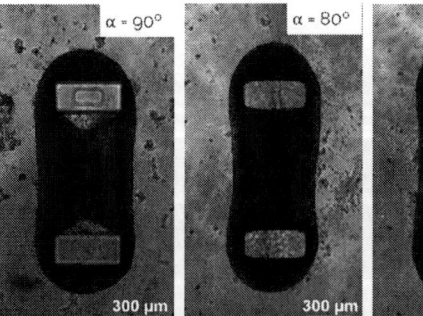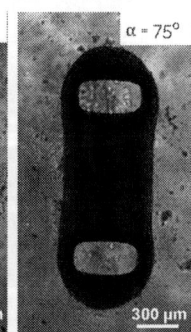

Figure 5. Brightfield imaging of EHT formation around straight pillars (i.e., α=90°) and both versions of tapered pillars (α=80° and α=75°). Tissue shown in black.

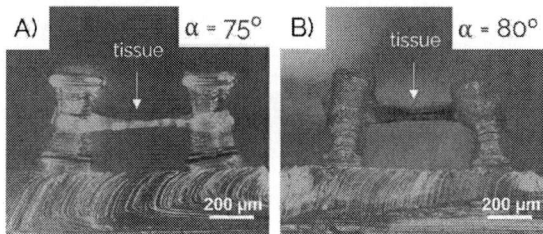

Figure 6. Example of spatial confinement of tissues in the middle of pillar length in both A) 75° and B) 80° tapered designs.

as well as during electrical pacing at 1 Hz and 2 Hz. Videos of contracting tissues of 10 s duration were recorded using an inverted optical microscope (Nikon Eclipse Ti2) with a high-speed camera. The videos were analyzed using a custom-made standalone app for tissue contractile performance assessment, by tracking the displacement of pillars tips. The results of the analysis and comparison of all three designs on day 7 are shown in Fig. 7.

From the graphs it can be noticed that the variability in the contractile behavior is reduced (by 85%) in case of both tapered pillar designs compared to the straight ones. The other parameters describing contractile kinetics, such as contraction and relaxation time to reach 10% and 90% of the upstroke and downstroke of the contraction cycle, are comparable in all three designs. It is yet to be determined to which extent the decrease in variability of contractile properties can be attributed to the new pillar design and tissue confinement or to the increase in stiffness of the tapered pillars compared to previous designs.

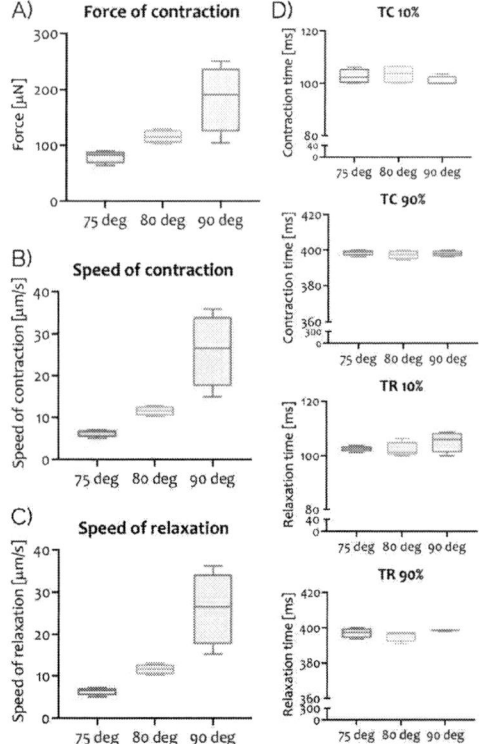

Figure 7. Comparison of contractile performance of tissues on straight and tapered pillars: (A) Contractile force of EHTs; contraction (B) and relaxation velocity (C) of EHTs over time. (D) Time of contraction (TC) and relaxation (TR) to reach 10% and 90% of the upstroke and downstroke of the contraction cycle.

CONCLUSIONS

We presented a novel design of PDMS pillars for tissue self-assembly in EHT platforms. The design is meant to increase the control over tissue position over time and consequently reproducibility of contraction force measurements. PDMS-based pillars were designed and fabricated in an hourglass shape, with symmetric tapering, producing a restriction in the middle of the pillars' length. The pillars were fabricated with two different tapering angles to test their confinement efficiency. In addition, mechanical characterization of both tapered designs was performed using a nanoindentation measurement system. This resulted in correction of the assumed Young's modulus and consequently precise measurements of force of contraction.

The efficiency of new designs in tissue confinement was tested by performing a comparative study between straight and both tapered pillar designs. It was noticed that the tissue formation was unaffected by the new pillar design. Tissue confinement was confirmed optically for both tapered designs. Consequently, the yield of experiments significantly increased, and the issue of tissues jumping off the pillars was solved for this specific EHT platform. Furthermore, the effect of the tapered design on tissue contractile properties was analyzed. A decrease in variability of the contractile force measurements was observed, while the other contractile kinetic parameters were comparable in all three designs.

ACKNOWLEDGEMENTS

The authors thank the staff at the Else Kooi Laboratory of TU Delft for their support with microfabrication. This work was supported by the Netherlands Organ-on-Chip Initiative, an NWO gravitation project funded by the Ministry of Education, Culture and Science of the government of the Netherlands (024.003.001).

REFERENCES

[1] J. M. Stein, C. L. Mummery, and M. Bellin, "Engineered models of the human heart: Directions and challenges," *Stem Cell Rep.*, vol. 16, no. 9, pp. 2049–2057, Sep. 2021.

[2] M. Dostanić *et al.*, "A Miniaturized EHT Platform for Accurate Measurements of Tissue Contractile Properties," *J. Microelectromechanical Syst.*, vol. 29, no. 5, pp. 881–887, Oct. 2020.

[3] R. Li, Y. Lamy, W. F. A. Besling, F. Roozeboom, and P. M. Sarro, "Continuous deep reactive ion etching of tapered via holes for three-dimensional integration," *J. Micromech. Microeng.*, vol. 18, no. 12, p. 125023, Nov. 2008,

[4] Z. Ren and M. E. McNie, "Inductively coupled plasma etching of tapered via in silicon for MEMS integration," *Microelectron. Eng.*, vol. 141, pp. 261–266, Jun. 2015.

[5] A. Hansen *et al.*, "Development of a drug screening platform based on engineered heart tissue," *Circ. Res.*, vol. 107, no. 1, pp. 35–44, Jul. 2010.

[6] E. Giacomelli *et al.*, "Human-iPSC-Derived Cardiac Stromal Cells Enhance Maturation in 3D Cardiac Microtissues and Reveal Non-cardiomyocyte Contributions to Heart Disease," *Cell Stem Cell*, vol. 26, no. 6, pp. 862-879.e11, Jun. 2020.

IN VITRO ASSEMBLY OF MUSCLE RINGS AND BIOPRINTED HYDROGEL FOR BRANCHING TUBULAR TISSUE CONSTRUCTS

Tomohiro Morita, Byeongwook Jo, Minghao Nie and Shoji Takeuchi
Graduate School of Information Science and Technology, The University of Tokyo, JAPAN

ABSTRACT

This paper describes a method to assemble muscle rings and a bioprinted hydrogel for fabrication of tubular tissue constructs. Muscle rings with circumferential alignment with contraction under electrical stimulation was achieved by culturing in ring-shaped PDMS mold. Bioprinted polyethylene glycol (PEG) hydrogel with 10 µm resolution enabled printing of various structures such as honeycomb, tubular and bifurcated structure. Followed by the glass capillary based transferring method, the muscle rings were effectively aligned to a PEG hydrogel with tubular structure. Our approach provides effective fabrication of circumferentially aligned tubular tissue constructs which have promising outlook in the field of regenerative medicine and drug screening models.

KEYWORDS

Polyethylene glycol, Bioprinting, Circumferential cell alignment, Electrical stimulation, Stretch stimulation, A glass capillary-based transferring method.

INTRODUCTION

Tubular tissues such as blood vessels, intestines, and bronchi have unique structures forming a circumferential aligned by muscle tissues, and this alignment functions effectively to transport and permeate oxygen and nutrients [1-2]. To replicate this tissue alignment, there have been previous studies for fabricating tubular structures such as hydrogel molding [3-4] and fiber assembly [5]. However, it is still challenging to achieve branching tubular tissue with circumferential cell alignment. The hydrogel molding method cannot effectively control the cell alignment, the fiber assembly method is difficult to construct a tissue with a branched structure.

In this work, we propose a method to fabricate tubular tissues with circumferential alignment by arranging multiple muscle rings to surround the tubular structure constructed from hydrogel. Unlike muscle fibers used in [5], each muscle ring has a pre-formed cell alignment in the circumferential direction and can be manipulated one by one. Therefore, by assembling muscle rings to the hydrogel with bifurcated structure, we believe above problems can be solved. For fabricating such a fine and complicated structured hydrogel, bioprinting is a promising technology. Photolithography is one of the bioprinting methods, curing the exposed area by the laser spot and printing with higher resolution than that of other printing methods such as drop-based bioprinting (DBB) and extrusion-based bioprinting (EBB) [6-7]. Here, we propose a method to insert muscle rings into tubular structured hydrogels printed by a commercial stereolithography bioprinter BIO NOVA X (CELLINK, Sweden). As the resolution of this bioprinter is as high as 10 µm, it is expected to print a hydrogel with a branching

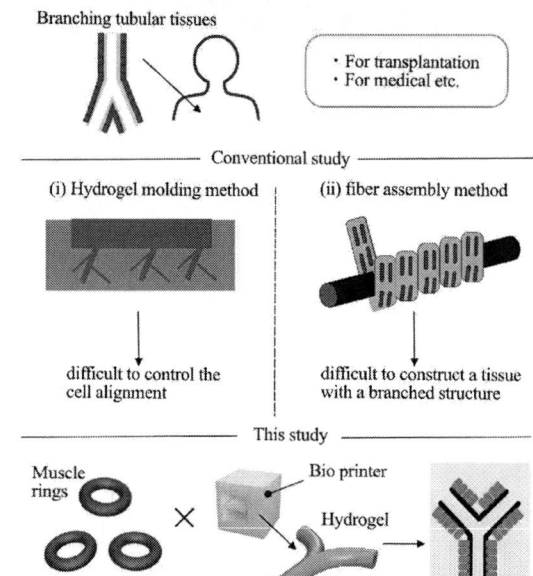

Figure 1. Conceptual illustration of this study for the fabrication of branching tubular constructs and the comparison with the previous studies

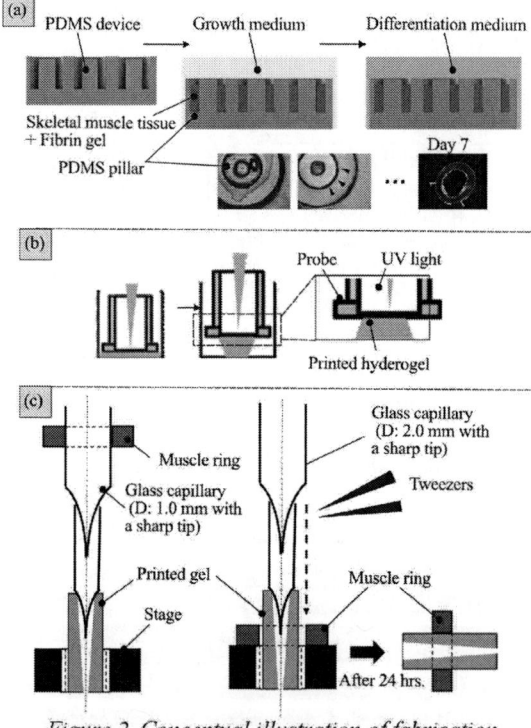

Figure 2. Conceptual illustration of fabrication method. (a) Fabrication of muscle rings (b) Process of printing hydrogel (c) Attachment of muscle rings to gel constructs

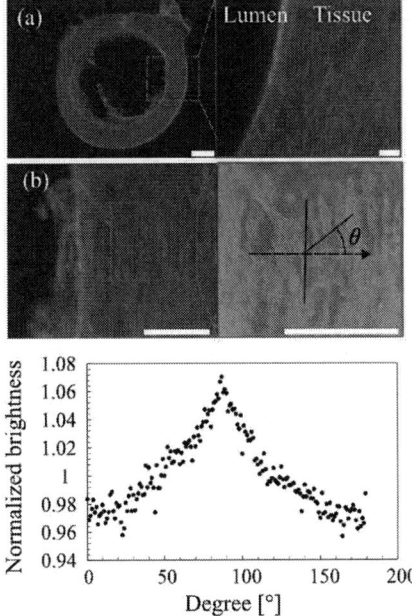

Figure 3. The morphological analysis of a muscle ring. (a) The confocal microscopic images of fabricated muscle ring after 5 days from differentiation. (b) Evaluation of the orientation angle of muscle fibers (Green: F-actin, Blue: Nuclei). (Scale bars: (a)500 μm (b)100 μm)

tubular structure and with a diameter of less than 3 mm for supporting muscle rings. In this report, we propose a method to fabricate a tubular tissue constructs by assembling ring-shaped tissues to the periphery of a bioprinted hydrogel.

FABRICATION METHODS

Skeletal muscle rings were fabricated by gelling a cell-fibrin gel solution in a PDMS mold (Fig. 2 (a)) [8]. After filling gel solution in PDMS mold, growth medium with 1 % aminocaproic acid is added to cover them. The growth medium was replaced with a differentiation medium after two days for myoblasts to differentiate into myocytes. On the fifth day of differentiation, the skeletal muscle rings were taken out and used for experiments.

As shown in Fig. 2 (b), a 10 % concentration of PEG hydrogel (Stiffness: 14 kPa) was printed by a commercial bioprinter (BIO NOVA X, CELLINK, Sweden). After printing out a hydrogel, a glass capillary with a sharpened tip was passed through the hole of the hydrogel and made the muscle ring (outer diameter: 2.7 mm, inner diameter: 2.2 mm) to reach the printed hydrogel (Fig. 2 (c)). The sharpened tip was created by using a glass capillary puller (PC-100 puller, NARISHIGE, Japan). To facilitate the insertion of the muscle ring from the glass capillary into the hydrogel, the outer diameter of the hydrogel (1.8 mm) was determined to be 200 μm smaller than that of the glass capillary (2 mm). Finally, the glass capillary was removed after the muscle ring adhered to the outer surface of the hydrogel.

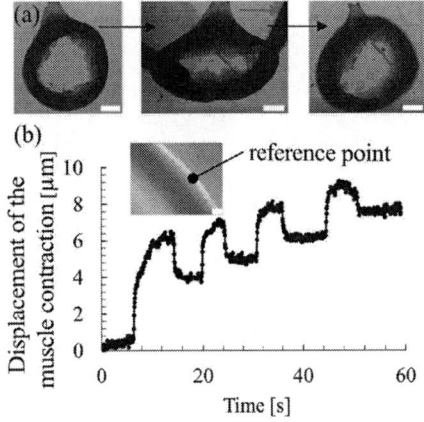

Figure 4. The deformation and the response of a muscle ring (a) The deformation by stretch stimulation. (b) The response by electrical stimulation and amount of movement of referencing point. (Scale bars: (a) 500 μm (b) inset: 1 mm)

RESULTS AND DISCUSSIONS

Morphology and function evaluation of a muscle ring

To evaluate the morphological characteristics of the fabricated muscle ring, Phalloidin and Hoechst33342 were used to stain the skeletal of the myocytes and cell nuclei, respectively. As shown in Fig. 3(a), f-actin and nuclei were aligned in the circumferential direction. Also, the degree of orientation was 89.2° by Fast Fourier Transform analysis (Fig. 3(b)). These results indicate that high alignment of the cells along circumferential direction was achieved through culturing in the PDMS molding.

Next, Fig. 4(a) shows the deformation of the muscle ring by tweezers' stretch stimulation. When a tension was applied and released by tweezers, the deformation was restored elastically. This result indicates that the muscle ring has enough strength to be manipulated without breaking. Also, electrical stimulation was applied to confirm the functional characteristics of the fabricated muscle ring. Fig. 4(b) shows the value of displacement of the muscle contraction plotted over time. Contraction of the muscle tissue was synchronized with the applied electrical pulses with 20.0 Hz of frequency.

Bioprinting of the hydrogel constructs

For validating the printing fidelity of the bioprinter, we printed the lattice and honeycomb structure with a wall width of 15 μm and 75 μm in Fig. 5(a) and (b), respectively. Both standard deviations were within 10 %, indicating that the high printability of using this bioprinter was verified.

Next, the tubular structures with different diameters in the longitudinal direction (maximum outer diameter: 1.8 mm, length: 3.8 mm) were printed. Fig. 5(c) shows an image of the printed hydrogel, and shows the inner diameters plotted with the expected values. The inner diameter had a maximum error of 10 %.

Finally, Fig. 5(d) show an external and an enlarged view of a printed hydrogel construct with a bifurcated structure (outer diameter: 1.8 mm, inner diameter: 1.2 mm), respectively. Also, the optical coherence tomography images of a hydrogel construct divided by A-A' and B-B'

978-1-6654-9309-3/23 $31.00 © 2023 IEEE

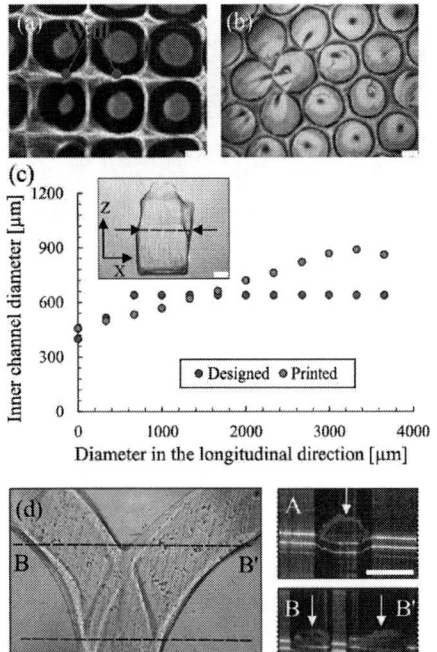

Figure 5. Fabrication of a bioprinted hydrogel component with (a, b) lattice and honeycomb structure, (c) a tubular, and (d) a bifurcated structure (Scale bars: (a) 50 μm, (b) 200 μm, (c) inset: 500 μm, (d) 1 mm, respectively)

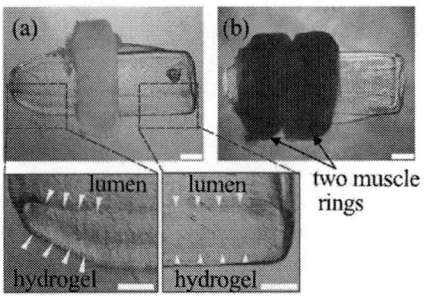

Figure 6. Fabrication of a hydrogel with a tubular structure with (a) a muscle ring (b) two muscle rings (Scale bars: 500μm)

cross-sections were obtained. A fine outer shape and tubular structures were obtained due to the high resolution of the bioprinter.

In-vitro assembly of muscle rings and bioprinted hydrogel constructs

To fabricate tubular tissue construct, we manipulated the muscle rings tissue with the glass capillary-based transferring method mentioned above. As shown in Fig. 6(a), we assembled muscle rings to the periphery of a bioprinted hydrogel successfully. Also, the second ring was able to place next to the first one. By stacking multiple muscle rings in this way, it is expected to fabricate a tissue with muscle rings oriented to the tubular structure.

CONCLUSION

In this paper, we proposed a method to assemble the muscle rings and a bioprinted hydrogel construct for tubular tissue constructs. Fabrication of muscle rings through PDMS mold showed circumferential alignment of myoblasts with functional contraction under electrical stimulation. A bioprinted PEG hydrogel with high resolution enabled fabrication of complex structure such as honeycomb, bifurcated and tubular structure. By glass capillary-based transferring method, alignment of muscle rings to the tubular hydrogel structure was confirmed. Using our proposed method, tubular tissue construct with circumferential alignment can be fabricated which have potentials in applications for regenerative medicine and drug screening models.

ACKNOWLEDGEMENTS

This work was partially supported by the JST-Mirai Program, Grant Number JPMJMI20C1, and JST SPRING, Grant Number JPMJSP2108.

REFERENCES

[1] Z. Liu, C. Wen, and S.-L. Zhang, "Evaluation of the fluid-movement contribution to the oxygen transport in tissue," *bioRxiv*, p. 2021.01.28.427630, 29-Jun-2022.

[2] K. Przewłócka, M. Folwarski, K. Kaźmierczak-Siedlecka, K. Skonieczna-Żydecka, and J. J. Kaczor, "Gut-muscle AxisExists and may affect skeletal muscle adaptation to training," *Nutrients*, vol. 12, no. 5, p. 1451, May 2020.

[3] T. Ching et al., "Biomimetic Vasculatures by bioprinted Porous Molds," *Small*, vol. 18, no. 39, p. e2203426, Sep. 2022.

[4] D. Taniguchi et al., "Scaffold-free trachea regeneration by tissue engineering with bio-bioprinting," *Interact. Cardiovasc. Thorac. Surg.*, vol. 26, no. 5, pp. 745–752, May 2018.

[5] A. M. Dostie, H. G. Lea, U. N. Lee, T. L. van Neel, E. Berthier, and A. B. Theberge, "Freestanding hydrogel lumens for modeling blood vessels and vasodilation," *SLAS Technol*, Aug. 2022.

[6] S. Itai and H. Onoe, "In Vitro Artery Model with Circumferentially Aligned & Contractible Smooth Muscle by Unfixed Molding & Screwing Fabrication," in *2022 IEEE 35th International Conference on Micro Electro Mechanical Systems Conference (MEMS)*, 2022, pp. 275–278.

[7] Y. Wu, P. Kennedy, N. Bonazza, Y. Yu, A. Dhawan, and I. Ozbolat, "Three-Dimensional Bioprinting of Articular Cartilage: A Systematic Review," *Cartilage*, vol. 12, no. 1, pp. 76–92, Jan. 2021.

[8] H. A. Strobel, E. L. Calamari, B. Alphonse, T. A. Hookway, and M. W. Rolle, "Fabrication of Custom Agarose Wells for Cell Seeding and Tissue Ring Self-assembly Using bioprinted Molds," *J. Vis. Exp.*, no. 134, Apr. 2018.

CONTACT

*T. Morita, tel: +81-90-4839-8851; tmhr-morita@hybrid.t.u-tokyo.ac.jp

MICROELECTRODES FABRICATED BY VACUUM FILLING WITH LOW MELTING-POINT ALLOY FOR MUSCLE TISSUE STIMULATION

Tingyu Li[1], Minghao Nie[1], Yuya Morimoto[1] and Shoji Takeuchi[1]

[1] Graduate School of Information Science and Technology, The University of Tokyo, Japan

ABSTRACT

In this work, we proposed a method to fabricate the low melting-point alloy microelectrodes (LMPA-MEs) for muscle tissue stimulation. Different from the common design where electrodes and anchors are separate, the LMPA-MEs are the integration of the electrodes and anchors. It can both stimulate and immobilize the muscle tissue. In this manner, the LMPA-MEs can save space for the miniaturization of the biohybrid actuators and stimulate muscle tissue under lower driving voltage (± 7.5V), compared with conventional Au electrode (± 10.5V). By utilizing the vacuum filling method, the fabrication process is simple and hands free. Therefore, we believed that the LMPA-MEs can improve the efficiency of electrical stimulation to the muscle tissue and has great potential for the compact arrangement of muscle actuators.

KEYWORDS

Low-melting-point alloy; microelectrode; vacuum filling; biohybrid actuator.

INTRODUCTION

As the integration of cultured muscle tissue and artificial supporting device, biohybrid robot actuated by muscle tissue has attracted many researchers' attention due to its flexibility, biomimetic movements, and miniaturization of devices [1]. The biohybrid robot can achieve different kinds of movements, like swimming [2], rotating [3], gripping [4] and pumping [5]. And these movements of the muscle tissue can be controlled by electricity [6], light [7], and chemicals [8]. Among these methods, electrical stimulation shows great advantages since it requires no gene modification and the response time is rapid. In the electrically driven biohybrid actuators, the metal electrodes are usually placed away from the muscle tissue to apply electrical pulses [3], as shown in Fig.1A. However, the away placement of the electrodes limited the compact arrangements of actuators, especially for the system with multiple actuators [9]. In addition, there will be an extra voltage drop across the culture medium between the electrode and the muscle tissue and cause inefficient electrical stimulation.

Here, we presented the LMPA-MEs, incorporating the MEs into the tissue anchors. The as-prepared LMPA-MEs were embedded inside the muscle tissue, so it can fix the tissue, avoiding its self-contraction, and shrink the size of the whole device. Moreover, the contact area of LMPA-MEs with muscle tissue increased, and subsequently, the efficiency of electrical stimulation to the muscle tissue improved. The design of the proposed anchor with electrodes is shown in Fig.1B. By creating the air pressure difference between the inner and outer of the device, the vacuum filling method can push the liquid metal into the hollow device [10].

FABRICATION

Device Design

Current vacuum filling method succeeded only in microchannels with dead-ends. In such device, the ME was covered inside the microchannels and cannot be exposed, therefore causing the problem of large driving voltage. To improve the efficiency of electrical stimulation to the muscle tissue, it's necessary to prepare open windows for the exposure of the MEs.

In this work, we proposed a novel vacuum filling method to fabricate the LMPA-MEs with 6 open windows. The LMPA-ME was composed of hollow pillars, liquid metal, and package. And during vacuum filling, the open windows were covered by PDMS mold to avoid the leakage of the liquid metal.

LMPA-ME Fabrication

The fabricating process of LMPA-MEs was shown in Fig. 2. Firstly, we fabricated the anchors with and without microchannels and open windows by a stereolithographic 3D printer (microArch S140, Boston micro fabrication, Inc.). And the polydimethylsiloxane (PDMS, Sylgard 184 Silicone Elastomer, Dow Corning Toray Co. Ltd.) was poured to the solid pillars. After solidification, the PDMS mold was released from the solid pillars and covered the anchor with the PDMS mold, as shown in Fig. 2A; the

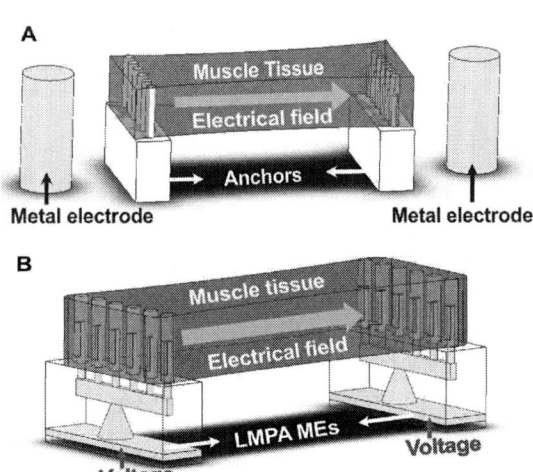

Figure 1. Conceptual illustration of microelectrodes in skeletal muscle tissue. A. Conventional metal electrode placed away from the muscle tissue. B. The microelectrodes serve as anchors for the muscle tissue and apply electrical stimulation to it. C. Cross sectional view of microelectrodes. Microelectrodes are exposed from the open window to muscle tissue.

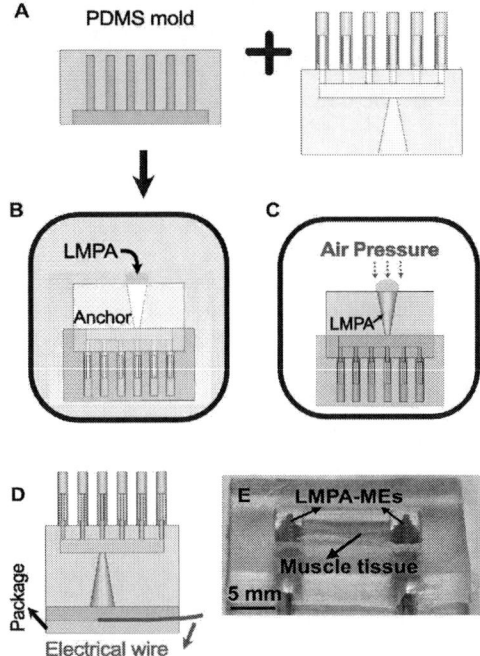

Figure 2. Fabrication flow of microelectrodes: A. Cover the anchor with PDMS mold. B. Pour LMPA to the inlet. Vacuum pumping for 1 h at 85 °C. C. Recovering standard atmospheric pressure. D. Package by glue bonding with electrical wires and 10-μm parylene coating. E. Microelectrodes are fixed by the PDMS device, and muscle tissue is anchored by the microelectrodes.

PDMS mold protected the LMPA from leaking out of the anchor in the subsequent process. Next, the LMPA was added to the inlet of the anchor in a vacuum oven under 85°C. The air inside the anchor was removed by vacuum pumping for 1 h (Fig. 2B). After recovering to standard atmospheric pressure, the LMPA was pushed into the microchannels (Fig. 2C). Then the bottom of the LMPA-MEs was connected to the electrical wire by glue bonding and coated with 10 μm parylene C (Dichloro-p-cyclophane, Specialty Coating System, Inc.) for biosafety (Fig. 2D). The PDMS mold was kept until parylene coating finished.

Muscle Tissue Culture

To construct the skeletal muscle tissue, two LMPA-MEs were immobilized by a PDMS culture device so that the distance between the pair of LMPA-MEs was 1 cm (Fig. 2E). Muscle tissue was formed by pouring C2C12 cell suspension in a mixture of collagen and Matrigel.

RESULTS AND DISCUSSION
Evaluation of Skeletal Muscle Tissue Formation

The microscopic image displays the whole cultured muscle tissue at day 5 after differentiation (Fig. 3A). The muscle tissue is 1 cm long and shrank in the middle. To further observe the distribution of cell nuclei and actin filament inside the muscle tissue, the muscle tissue was stained with Hoechst 33342 and phalloidin at day 7 after

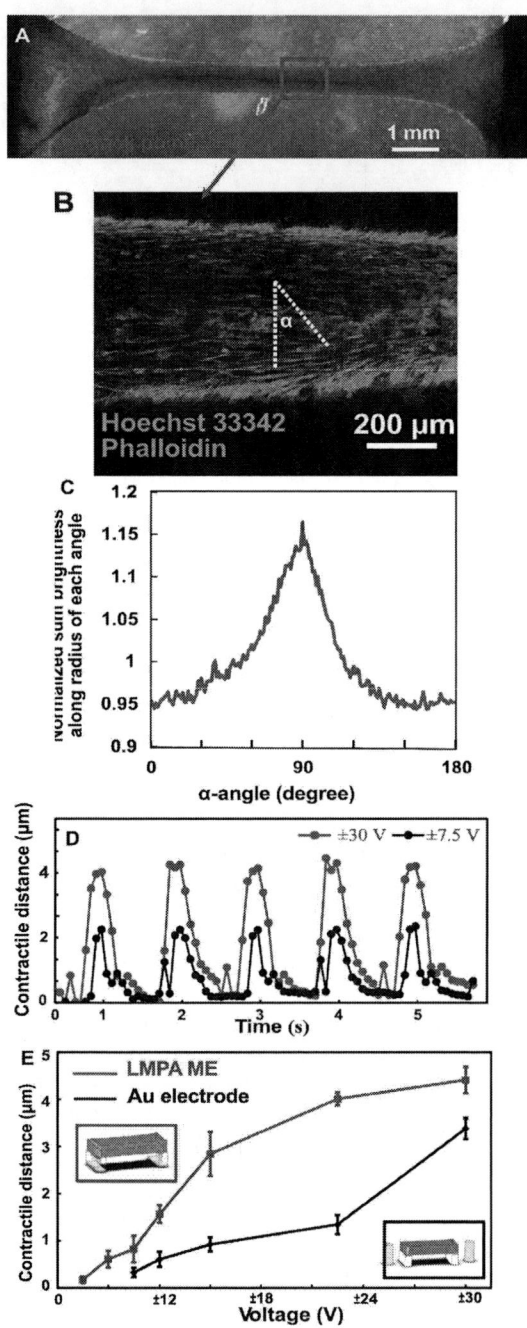

Figure 3. A. Microscopic image of muscle tissue at Day 5 after differentiation. B. Confocal microscopic image of the muscle tissue at Day 7 after differentiation. Green: F-actin. Blue: nuclei. C. Directional distribution of myotubes in the muscle tissue shown in image B. D. Contractile distance of muscle tissue under ± 30 V and ± 7.5 V by LMPA microelectrodes. Frequency, 1 Hz, duration, 10 ms. E. Average contractile distance of muscle tissue under different voltage applied by LMPA microelectrodes and Au electrodes.

differentiation (Fig. 3B). As a result of collecting the normalized sum brightness at different angles, Fig. 3C shows that the peak normalized sum brightness appeared at 90 degrees, indicating that the myotubes were highly aligned.

Electrical Stimulation

The muscle tissue was stimulated by LMPA-MEs and Au electrodes under sequentially electrical pulses (frequency, 1 Hz; duration, 10 ms), and the contractile distance at the track point, shown in Fig. 4A, was recorded. Fig. 3D shows the contractile distance of the muscle tissue under the electrical stimulation of LMPA-MEs. As the applying voltage increased from $\pm$ 7.5 V to $\pm$ 30 V, the contractile distance of the muscle tissue increased from 2 to 4 µm, showing that the contractile distance can be controlled by different applying voltage. Fig. 3E shows the average contractile distance of the muscle tissue under different voltage by LMPA-MEs and Au electrodes. For the driving voltage, LMPA-MEs possessed lower minimum driving voltage ($\pm$ 7.5 V), comparing with that of Au electrodes ($\pm$ 10.5 V). Moreover, the average contractile distance of the muscle tissue under LMPA-MEs kept larger than that of Au electrodes under the same applying voltage (Fig. 3E). Specifically, to achieve the same contractile distance of 3 µm, the applying voltage of LMPA-MEs and Au electrodes was $\pm$ 15 V and $\pm$ 28 V, respectively. These results demonstrated that the LMPA-MEs possess more efficient stimulating property to the muscle tissue and require lower power consumption to actuate biohybrid actuators.

CONCLUSION

In this work, we presented a vacuum filling method to fabricate LMPA-MEs for the electrical stimulation of biohybrid actuators in a simple and hands-free way. LMPA-MEs serve as anchors to fix the cultured muscle tissue, hence it can save space for the further miniaturization of the biohybrid robot system. And the C2C12 cell suspension with LMPA-ME anchors differentiated to muscle tissue successfully and the myotubes inside the muscle tissue aligned uniformly. These indicated that the LMPA-ME exhibited certain biosafety. In addition, in terms of electrical stimulation, the LMPA-MEs showed lower driving voltage and larger contractile distance of the muscle tissue, comparing with that of Au electrodes. These results demonstrated that LMPA-MEs can actuate the biohybrid actuator in a more efficient way and reduce the power consumption of biohybrid robot.

ACKNOWLEDGEMENTS

This work was partially supported by China Scholarship Council (CSC202106010042), National Natural Science Foundation of China (62004007), China Postdoctoral Science Foundation Grant (2021M700258), JST-Mirai Program Grant Number JPMJMI20C1, Japan, and the JSPS Grants-in-Aid for Scientific Research (KAKENHI) (Grant Number 21H01779), Japan.

REFERENCES

[1] C. Appiah, C. Arndt, K. Siemsen, A. Heitmann, A. Staubitz, and C. Selhuber-Unkel, "Living Materials Herald a New Era in Soft Robotics", *Adv. Mater.*, vol. 31, no. 36, 1807747, 2019.

[2] M. Guix, *et al.*, "Biohybrid soft robots with self-stimulating skeletons", *Sci. Robot.*, vol. 6, eabe7577, 2020.

[3] Y. Morimoto, H. Onoe, and S. Takeuchi, "Biohybrid robot powered by an antagonistic pair of skeletal muscle tissues", *Sci. Robot.*, vol. 3, eaat4440, 2018.

[4] K. Kabumoto, T. Hoshino, Y. Akiyama, and K. Morishima, "Voluntary movement controlled by the surface EMG signal for tissue-engineered skeletal muscle on a gripping tool". *Tissue Eng. Part A.,* vol. 19, pp. 1695–1703, 2013.

[5] Z. Li, *et al.*, "Biohybrid valveless pump-bot powered by engineered skeletal muscle", *Proc. Natl. Acad. Sci. U.S.A.*, vol. 116, no. 5, pp.1543-1548, 2022.

[6] Y. Morimoto, H. Onoe, and S. Takeuchi, "Biohybrid robot with skeletal muscle tissue covered with a collagen structure for moving in air," *APL Bioeng.*, vol. 4, 026101, 2020.

[7] J. Wang, Y. Wang, Y. Kim, T. Yu, and R. Bashir, "Multi-actuator light-controlled biological robots," *APL Bioeng.*, vol. 6, 036103, 2022.

[8] Z. Chen, F. Fu, Y. Yu, H. Wang, Y. Shang, and Y. Zhao, "Cardiomyocytes-Actuated Morpho Butterfly Wings," *Adv. Mater.*, vol. 31, no. 8, 1805431, 2019.

[9] P. Won, S. Ko, C. Majidi, A. Feinberg, and V. Webster-Wood, "Biohybrid Actuators for Soft Robotics: Challenges in Scaling Up", *Actuators*, vol. 9, 96, 2020.

[10] Y. Lin, O. Gordon, M. R. Khan, N. Vasquez, J. Genzer, and M. D. Dickey, "Vacuum filling of complex microchannels with liquid metal," *Lab Chip.*, vol. 17, no. 18, pp. 3043–3050, 2017.

CONTACT

*Tingyu Li, the University of Tokyo, Japan, tel: +81-03-5841-6488; E-mail: li-tingyu966@g.ecc.u-tokyo.ac.jp

OPTOELECTRONIC INTEGRATED ULTRAMICROELECTRODE FOR OPTICAL STIMULATION AND ELECTRICAL RECORDING OF SINGLE-CELL

Qingda Xu[1,2], Ye Xi[1,2], Zhiyuan Du[1,2], Longchun Wang[1,2], Tao Ruan[1,2], Mengfei Xu[1,2], Jiawei Cao[1,2], Bin Yang[1,2], and Jingquan Liu[1,2]

[1]National Key Laboratory of Science and Technology on Micro/Nano Fabrication, Shanghai Jiao Tong University, Shanghai, 200240, CHINA and

[2]DCI Joint Team, Collaborative Innovation Center of IFSA, Department of Micro/Nano Electronics, Shanghai Jiao Tong University, Shanghai, 200240, CHINA

ABSTRACT

Single-cell detection is of increasing interest with the development of micro-nano electrodes. To achieve single-cell optical stimulation and electrical recording, this paper reports an optoelectronic integrated ultramicroelectrode (OE-UME). The OE-UME is fabricated from optical fiber and quartz capillary by laser-assisted pulling method, which has the advantages of a simple process and high integration. Besides, the electrochemistry, electrical artifacts, and optical stimulation of OE-UME are tested. For the application, the OE-UME has the potential to efficiently transmit specific wavelengths of light to single-cell regions and study the relationship between intracellular optogenetics and membrane potential signals.

KEYWORDS

Ultramicroelectrode, Optogenetics, Optoelectronic Integration, Single-Cell Stimulation.

INTRODUCTION

The application of brain-computer interface devices is of great significance for neuroscience and brain-computer interaction. Today, the action potentials of brain cells and local field potentials generated by cell populations are widely used to study brain science problems such as neural circuits and pathogenesis. In addition to electrical methods, methods such as light, magnetism, and sound are also used in brain-computer interface technology.

Due to high spatial and temporal resolution and high specificity, optogenetics has great potential in neuroscience [1]. To understand the information of the cell and neural circuits more comprehensively, many works have been carried out *in vivo* experiments by the method of optoelectronic integration [2,3]. In 2016, Li et al. used optoelectronic integrated devices to reveal the relationship between preparatory activity and large-scale unilateral silencing in the mouse premotor cortex [2]. In 2021, Kim et al. reported a wirelessly rechargeable, fully implantable, soft optoelectronic system, and made a successful experiment on freely behaving rats [3]. But these studies cannot target single-cell or subcellular information.

For single-cell studies, most of the research works quantitatively study the mechanism of cell activity in ultramicrometer or nanometer space by voltage, current and electrochemical methods, such as membrane potential change, the distribution of intracellular substance concentration, and vesicle contents. In recent years, there have also been several optical works on single cells [4,5].

In 2012, Yan et al. realized the optical stimulation of intracellular compartments and delivered QDs into cells with spatial and temporal specificity [4]. However, their device can not record the electrical information and further greatly reduce the variety of substances that could be detected. In 2021, Hunt et al. achieved the optical stimulation and electrical recording of single cells [5]. Unsatisfactorily, the device has complex processes with low integration (the distance between the light outlet and the electrode >10μm). To date, compared to electrical methods, there is still a lack of optogenetic studies in single-cell applications. Therefore, the optoelectronic integrated ultramicroelectrode with a simple fabrication process and high integration is of great significance for the single-cell technique.

This paper shows a detailed fabrication process of a bimodal ultramicroelectrode. With the same ductility and malleability of optical fiber and glass capillary, the OE-UME at the cell level is fabricated based on the laser-assisted pulling method. And the emitted light from the light outlet has a concentrated optical path through the tip polishing process. Besides, benefiting from the wax package, OE-UME enables the integration of optical stimulation and electrical recording sites within a range of less than 2μm. Moreover, different from previous works, the highly integrated device provides a valuable tool for future research in cell-level optogenetics, single-cell technique, and brain science.

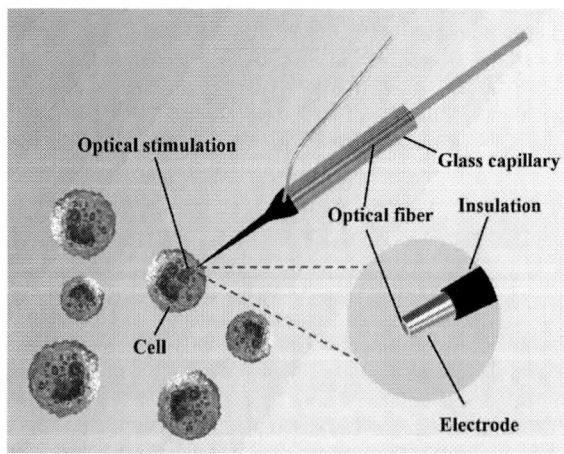

Figure 1: Conceptual diagram of optoelectronic integrated ultramicroelectrode for single-cell application.

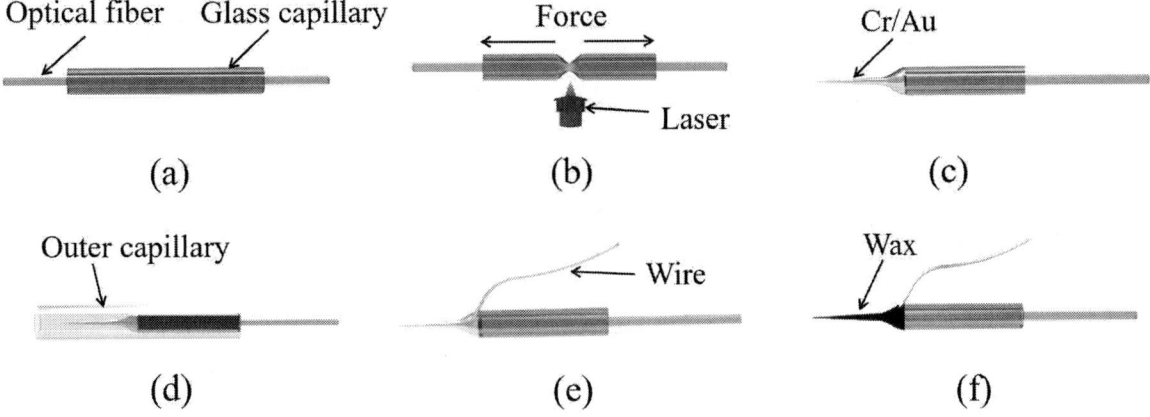

Figure 2: Fabrication process of OE-UME: (a) optical fiber insertion, (b) laser-assisted pulling, (c) sputtering, (d) tip polishing, (e) Cu wire-leading, and (f) wax package.

DESIGN AND FABRICATION

Figure 1 shows the conceptual diagram of OE-UME. The light outlet is below the electrode point, and the recording electrode is a nano-ring electrode whose exposed area is controlled by the wax package. Due to the advantage of low loss, optical fiber was chosen as the optical waveguide for optical stimulation. To ensure miniaturization and high integration, the OE-UME was made by laser-assisted pulling. And to optimize the bonding force of optical fiber, the quartz capillary whose material is the same as the optical fiber was used as the dielectric layer of the light outlet and electrical recording site. The diagram of the fabrication process flow of OE-UME is shown in Figure 2.

In Figure 2(a), the optical fiber was first pretreated with ultrasonic cleaning in acetone, ethanol, and de-ionized water and dried in a drying oven at 150°C. Then, the optical fiber was inserted into a quartz capillary (o.d. = 1.0mm, i.d. = 0.5mm). Figure 2(b) shows the step of laser-assisted pulling. The quartz capillary with optical fiber was pulled using a micropipette puller system (P2000; Sutter Instrument Co.). The pull parameters are as follows: Heat = 850, Velocity = 30, Delay = 128, Filament = 2, Pull = 200. Then, 6nm Cr and 40nm Au were sputtered onto the pulled capillary surface to form the electrode layer. And the step is shown in Figure 2(c). The Cr and Au layers were sputtered at the chamber pressure of 1Pa and the power of 100W for both 1min. It was worth noting that, in order to ensure the binding force between the electrode layer and the quartz capillary, steps (a-c) were carried out in the clean room, and ultrasonic cleaning or nitrogen blowing was carried out before sputtering.

There are two reasons for carrying out step (d): one is to ensure that the light outlet of the optical fiber at the electrode tip is in the same plane as the quartz capillary after step (b), and the other is to remove the Cr and Au sputtered to the light outlet in step (c). In step 2(d), the electrode after sputtering was inserted into another thicker borosilicate capillary (o.d. = 2.0mm, i.d. = 1.56mm). Then, the adhesive wax was melted by heating and injected into the borosilicate capillary, where it was solidified for

subsequent steps. During the wax curing process, it was necessary to ensure that the tip of the inner quartz capillary does not extend beyond the end face of the outer borosilicate capillary. And the capillary was further polished with sandpapers of different grits. The polishing effect was judged and the polishing length was controlled by observing whether there was light at the tip, which can ensure the subcellular size of the tip. After polishing, the adhesive wax was removed with acetone. Figure 2(e) shows the wire-leading process. Cu wire was fixed on the metal layer with conductive silver paste and further baked in a vacuum oven at 100°C for 1h until the paste was cured. For the next step, the electrode surface was insulated (Figure 2(f)). The Apiezon wax was first melted on a hot plate at 200°C [6]. Then the electrode was passed through the liquid Apiezon wax by a micromanipulator. Benefiting from the surface tension of the wax, the electrode site was exposed at the tip. And to minimize background noise, the Apiezon wax was used to wrap as much of the electrode front as possible.

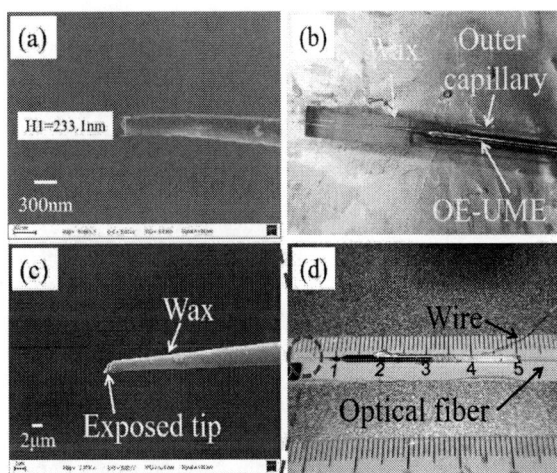

Figure 3: SEM images of OE-UME after laser pulling (a), and after wax package (c). (b) Tip polishing with wax and outer glass capillary. (d) Picture of OE-UME.

The optical and SEM images at different micromachining phases are shown in Figure 3. The diameter of the tip after pulling is less than 300nm (Figure 3(a)), besides, the tip after tip polishing is less than 2µm, and the distance between the electrode site and the light outlet is less than 2µm. (Figure 3(c)). The SEM images were taken with a high vacuum SEM (ULTRA55, Zeiss, Oberkochen, Germany). Figure 3(b) shows a picture before removing the adhesive wax and after tip polishing. Before polishing, it was necessary to ensure that the adhesive wax was filled to the electrode neck to prevent mechanical damage to the OE-UME. Figure 3(d) shows a physical view of OE-UME. And the length of OE-UME is about 5cm, excluding the optical fiber.

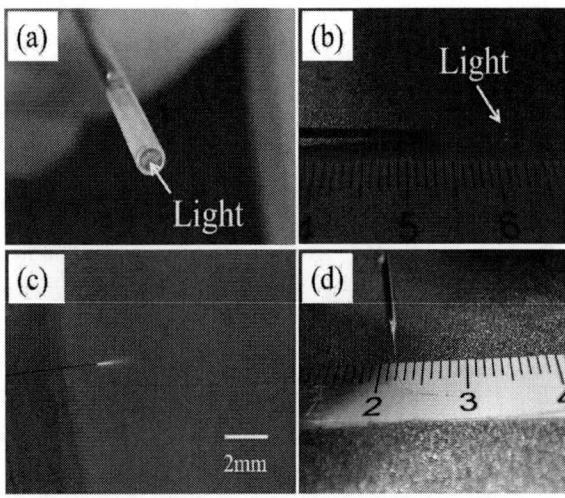

Figure 4: Optical path conduction during tip polishing (a) and after wax package (b); optical path in agar with fluorescein sodium (c); (d) bending of OE-UME.

Figure 4 shows the optical path of OE-UME in different steps and test conditions. During the tip polishing process, light spots can be seen on the borosilicate capillary section, indicating that the metal layer on the tip of OE-UME has been removed and the light outlet is in a plane with the quartz capillary port (Figure 4(a)). And Figure 4(b) shows a picture of the light spot after the wax package. The light path in 0.6% agar with fluorescein sodium is shown in Figure 4(c). It can be seen that the light path of OE-UME is concentrated, indicating that the tip polishing effect is satisfactory. Figure 4(d) shows the OE-UME bent 90°, demonstrating its robustness, which means it can meet more experimental requirements.

EXPERIMENTS AND DISCUSSION

The electrical characteristic and optical stimulation function of OE-UME was tested. Figure 5(a) shows a steady-state cyclic voltammogram curve in 5mM $[Ru(NH_3)_6]^{3+}$ solution. The test was performed on Autolab (PGSTAT204, Metrohm, Herisau, Switzerland). It can be seen that OE-UME presents a good S-shape. Then, it is well known that electrical artifact is a common problem in optoelectronic integrated devices. Especially in thin metal electrodes, intense electrical artifacts can be generated to

vitiate useful signals due to the photoelectric effect [7]. Therefore, the electrical artifact of OE-UME was measured. The signal was recorded by an RHD2000 Evaluation System (Intan Technology, LA, CA) connected to a computer. And the light source was a laser generator with a frequency of 1Hz. The experimental results are shown in Figure 5(b). It can be seen that the background noise measured in PBS is about 30µV and there are no obvious electrical artifacts when the laser is turned on. One possible reason is that the light outlet of OE-UME is below the electrode site. In this specific structure, the electrode layer absorbs so little photon energy that no electrical artifact is observed. Thus, OE-UME has a good optoelectronic integration capability and is feasible in the application of optical stimulation and electrical recording.

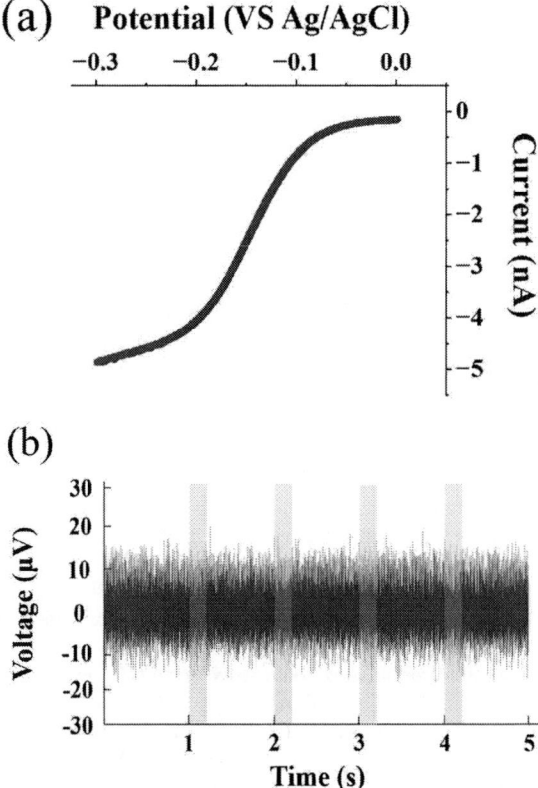

Figure 5: (a) Diagram of limited current in a solution of 5mM $[Ru(NH_3)_6]^{3+}$; (b) Test curve of background noise and light artifact. The yellow area is the period of the optical stimulation.

Then, the optical stimulation ability for single-cell was tested. Hippocampal neuronal cell line (HT-22) transfected with the green fluorescent protein (GFP) was used for the experiment and Dulbecco's modified Eagle medium (DMEM) containing 10% fetal bovine serum and 1% penicillin (100 IU/mL)-streptomycin (0.1 mg/mL) was used as cell culture. OE-UME was held on a micromanipulator. And the manipulation process under the microscope is shown in Figure 6(a). Figure 6(b) shows the position of OE-UME relative to the cell under optical stimulation. Figures 6(c) and 6(d) show fluorescence photographs of a cell under different light intensities. It can be concluded that OE-UME can successfully achieve single-cell optical stimulation, and the cell fluorescence intensity increases with the increase of luminescence intensity of OE-UME.

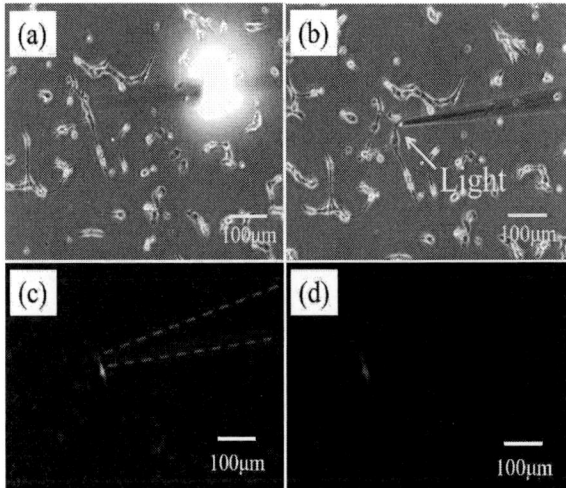

Figure 6: Micrographs of a single cell alignment process (a), and optical stimulation (b). Fluorescence photographs of a single cell under higher light intensity (c) and lower light intensity (d). (Bright white spot in Figure (a) is the liquid level of the medium, and the dashed red line in Figure (c) is the electrode trace.)

CONCLUSION

This paper reported a simple and optoelectronic integrated ultramicroelectrode based on laser-assisted pulling. OE-UME has a concentrated optical path and good robustness. Moreover, experiments show that OE-UME can achieve single-cell optical stimulation and no photoelectric artifact. All these results in this paper verify the outstanding performance of our device, implying numerous practical applications and providing many possibilities for the optogenetics development of single-cell science.

ACKNOWLEDGEMENTS

This work was partially supported by the National Key R&D Program of China under the grant（2022ZD0208601, 2020YFB1313502）, the Strategic Priority Research Program of Chinese Academy of Sciences (Grant No.XDA25040100, XDA25040200 and XDA25040300), the National Natural Science Foundation of China (No. 42127807-03), Project supported by Shanghai Municipal Science and Technology Major Project（2021SHZDZX）, Shanghai Pilot Program for Basic Research - Shanghai Jiao Tong University (No.21TQ1400203), SJTU Trans-med Award (No.2019015, 21X010301627), the Oceanic Interdisciplinary Program of Shanghai Jiao Tong University (No.SL2020ZD205, SL2020MS017, SL2103), Scientific Research Fund of Second Institute of Oceanography, MNR (No.SL2020ZD205).

The authors are also grateful to the Center for Advanced Electronic Materials and Devices (AEMD) of Shanghai Jiao Tong University.

REFERENCES

[1] S. Liu, Z. Wang, Y. Su, and et al., "A neuroanatomical basis for electroacupuncture to drive the vagal–adrenal axis", *Nature*, vol. 598, pp. 641-645, 2021.

[2] N. Li, K. Daie, K. Svoboda, and S. Druckmann, "Robust neuronal dynamics in premotor cortex during motor planning", *Nature*, vol. 537, 122, 2016.

[3] C. Y. Kim, M. J. Ku, R. Qazi, and et al., "Soft subdermal implant capable of wireless battery charging and programmable controls for applications in optogenetics", *Nat. Commun.*, 12, 535, 2021.

[4] R. Yan, J. Park, Yeonho Choi and et al., "Nanowire-based single-cell endoscopy", *Nat. Nanotechnol.*, 7, pp. 191-196, 2012.

[5] D. L. Hunt, C. Lai, R. D. Smith and et al., "Multimodal *in vivo* brain electrophysiology with integrated glass microelectrodes", *Nat. Biomed. Eng.*, 3, pp. 741-753, 2019.

[6] R. Pan, M. Xu, D. Jiang, and et al., "Nanokit for single-cell electrochemical analyses", *Proc. Natl. Acad. Sci. USA*, 113, pp. 11436-11440, 2016.

[7] L. Wang, M. Wang, F. Ge, and et al., "The use of a double-layer platinum black-conducting polymer coating for improvement of neural recording and mitigation of photoelectric artifact", *Biosens. Bioelectron.*, 145, 111661, 2019.

CONTACT

*J.Q. Liu, tel: +86-21-34207209; jqliu@sjtu.edu.cn

THERMOFORMING OF PARYLENE C TO FORM HELICAL STRUCTURES

Brianna L. Thielen and Ellis Meng

Department of Biomedical Engineering, University of Southern California, Los Angeles, CA, USA

ABSTRACT

Microfabricated, thin film, Parylene C helices down to a 0.25 mm diameter were achieved using a post-fabrication thermoforming process. This overcomes the planar configuration imposed by MEMS fabrication techniques. By utilizing a thermoplastic, Parylene C backbone, planar devices can be fixtured into a desired 3D geometry and thermoformed, permanently transforming the device into the fixtured shape. This work characterizes the thermoforming process and necessary parameters to achieve 0.25 to 2.6 mm diameter helices in bare Parylene strips and microfabricated Parylene and metal devices without insulation (Parylene) or electrical (metal) failure.

KEYWORDS

Parylene C, thermoforming, three-dimensional, fabrication.

INTRODUCTION

Microfabrication can be used to build complex MEMS devices on a flat substrate. These planar devices can be assembled into 3D structures via stacking or linkages, however biomedical applications often require more complex 3D geometries to produce seamless, intimate interfaces with anatomical features. By using flexible polymer substrates instead of a traditional silicon backbone, planar, thin film, microfabricated devices can be transformed into intricate 3D geometries through post-processing, achieving a conformation matching the targeted and surrounding tissue and thereby minimizing the body's immune response to implanted devices [1].

Several polymers are compatible with existing microfabrication techniques (the most common of which are polydimethylsiloxane (PDMS), polyimide, and Parylene C) and are significantly less stiff than silicon, resulting in flexible devices that can be transformed from a planar formation into 3D shapes. Parylene C (here on referred to as Parylene) is particularly useful due to its thermoplasticity, which allows it to be transformed permanently into complex geometries by thermoforming against a template [1]. PDMS and polyimide are thermoset polymers, so their shape cannot be easily modified; non-planar structures fabricated using these materials are achieved by attachment to a supporting structure or plastic deformation [2]–[5] (which can damage the device).

Thermoforming is performed by fixturing a Parylene film or microfabricated device into a desired shape and heating it above the glass transition temperature (approximately 60-90 °C [1], [6], [7]), which softens the amorphous regions of the polymer while the film is still in a solid state. As the film cools back to room temperature it re-hardens and more crystalline regions form, holding the Parylene permanently in the fixtured shape and increasing the stiffness and insulative properties [1].

Parylene is commonly used in microfabricated medical devices, with most devices consisting of multiple layers of Parylene and metal to form sensors, electrodes, or microfluidic channels. Several thermoformed Parylene devices have been developed to match anatomical features, to provide mechanical strain relief, or to form a mechanical anchor to interface with another device (Table 1). One of the most common and useful geometries is a helix, which can be used for strain relief, to interface with features such as nerves, muscle fibers, and blood vessels, or can be mounted on cylindrical supports (such as catheters, stents, or probes) to produce "smart" devices.

Table 1: Examples of microfabricated Parylene medical devices formed into 3D geometries via thermoforming.

Device	Curvature Shape	Curvature Diameter
This work	Helix	0.25-2.6 mm
Cable strain relief [1], [8]	Helix	1.1-5 mm
Cuff electrode [9]	Tube	3 mm
Retinal electrode array [1], [10], [11]	Sphere section	5 mm
Retinal electrode array [12]	Cylinder section	1 mm
Penetrating cortical electrode array [1], [11]	Cylinder or cone (between Parylene layers)	0.25 mm

An evaluation of the design space for thermoformed helices (and other shapes) is presented towards enabling novel device geometries. The smallest Parylene helix reported in literature is 1.1 mm in diameter [8]. The smallest thermoformed Parylene feature reported is a 0.25 mm cylinder on a Parylene-metal-Parylene electrode array, however features at this size exhibited cracking in both the Parylene and metal layers, rendering the device non-functional [1], [7]. This work describes a method for thermoforming Parylene helices down to 0.25 mm (a 4× improvement) while maintaining electrical conductivity in the metal layer of a Parylene-metal-Parylene device.

MATERIALS AND METHODS

Bare Parylene Helix Fabrication

Bare Parylene strips of varying thickness were fabricated and thermoformed into 0.25, 1.6, and 2.6 mm diameter helices. First, Parylene was deposited via chemical vapor deposition (CVD) onto a bare 4" silicon wafer. For some wafers, a second coat of Parylene was deposited using the same procedure (after breaking vacuum, removing the samples, and re-loading the samples) to produce thicker layers and/or to evaluate the effects of multi-layered Parylene. Thicknesses of 5.4 to 26.3 μm were achieved by varying the amount of dimer loaded into the machine and the number of coating runs.

After Parylene was deposited, the film was removed from the wafer by cutting around the edge of the wafer with a scalpel and peeling the film off the wafer surface. The freed film was then cut into strips (300 μm width by 20 mm length) using a cutting plotter.

Cut Parylene strips were fixtured into a helical shape by wrapping around a stainless steel mandrel and held in place using non-adhesive, 10 μm thick Teflon tape. Parylene strips with two layers were tested wrapping in both directions (i.e., one sample with the base Parylene layer towards the inside of the helix, and a second sample with the base Parylene layer towards the outside of the helix). Helix diameter (0.25, 1.6, 2.6 mm) was defined by the mandrel and helix angle (angle between the axis of the helix and the long edge of the Parylene strip, Figure 2; 15°, 30°, 45°) was imposed by using a template.

After fixturing, parts were baked at 200 °C (~0.7 °C/min ramp) for 12 hours under vacuum in a programmable oven. After cooling to room temperature, parts were removed from the oven and separated from the mandrel. Some samples were also thermoformed at 100 and 150 °C using the same method to determine the impact of lower temperature thermoforming.

The resulting Parylene helices were visually inspected using a microscope for their ability to retain the desired helical shape/size and for cracking failure in the Parylene.

Parylene-Metal-Parylene Helix Fabrication

After optimizing the thermoforming method on bare Parylene strips, devices with a functional metal layer were evaluated. Three-layer devices were fabricated, with a base Parylene layer, a metal layer patterned with electrodes on one end, bondpads on the other end, and traces connecting each electrode to a single bondpad, and a top Parylene layer with etched openings to expose the metal bondpads and electrodes. Two groups of devices were fabricated, one with asymmetric Parylene layers (layers of different thicknesses) and one with symmetric Parylene layers.

Parylene-metal-Parylene (PMP) devices were fabricated by first depositing a base Parylene layer (3.4 μm thickness for the asymmetric group, 4.4 μm for the symmetric group) via CVD onto a bare 4" silicon wafer (Figure 1A). In the symmetric group only, the base Parylene layer was annealed under vacuum with nitrogen flow at 150 °C for 4 hours (which causes minor shrinkage (<2%) of the Parylene [1]). Next, 15 nm titanium and 200 nm platinum were deposited via e-beam evaporation and patterned into electrodes, traces, and bondpads using a lift-off technique (Figure 1B). A top Parylene layer (11.5 μm for the asymmetric group, 4.7 μm for the symmetric group) was deposited via CVD. Openings in the top Parylene for exposed metal electrodes and bondpads were etched using O_2 reactive ion etching and the outer edge of the device (through both Parylene layers) was etched using the same method (Figure 1C). Devices were manually released from the wafer using water and tweezers (Figure 1D).

Released devices were thermoformed into 1.6 and 0.25 mm helices (Figure 2) at a 45° helix angle using the same thermoforming method as was used for bare Parylene strips. PMP helices were visually inspected for their ability to retain the desired helical shape/size and for cracking failure in the Parylene and electrically tested for continuity between the exposed bondpads and electrodes using an LCR meter before and after thermoforming.

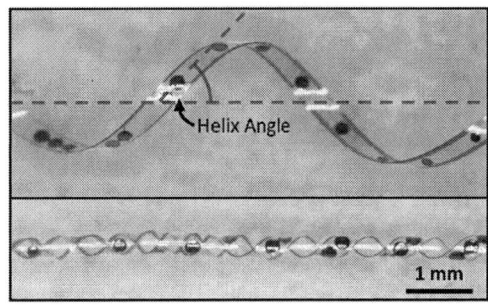

Figure 2: Photos of (top) 1.6 mm and (bottom) 0.25 mm Parylene-metal-Parylene device helices.

RESULTS AND DISCUSSION

Bare Parylene Helices

In bare Parylene strips thermoformed into 1.6 or 2.6 mm diameter helices, no cracking was observed at any thickness. In 0.25 mm helices, thinner Parylene strips (≤11.1 μm thickness) exhibited no cracking (Figure 3A) or minor cracking (cracks not spanning the full film thickness; Figure 3B). The thinnest samples (5.6 μm thick) did not exhibit any cracking at either helix angle, while thicker samples (9.9 and 11.1 μm thicknesses) yielded minor cracking in some samples. All thicker samples (≥13.3 μm thickness) exhibited full thickness cracking (Figure 3C) at all helix angles. Number of Parylene layers and uneven layer thicknesses did not impact shape retention or cracking of bare Parylene strips.

Parylene strips thermoformed at a 30° or 45° helix angle held the desired shape at all helix diameters (except for one outlier), while several parts formed at a 15° helix angle did not. Loose helix shape (Figure 3D) was not correlated to Parylene thickness or helix diameter, suggesting a fixturing issue (i.e., more difficulty wrapping Parylene strips around the mandrel at a 15° helix angle).

These results demonstrate that Parylene thickness must be considered when thermoforming to smaller helix diameters (thin Parylene is less likely to crack but is fragile and requires careful handling). When thermoforming to a larger diameter, thicker Parylene can be safely used without cracking failure. When producing helices with

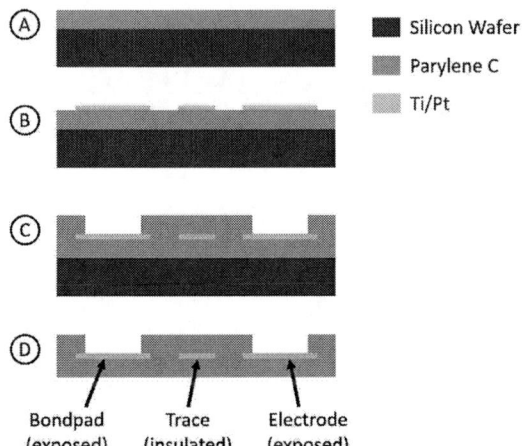

Figure 1: Fabrication process flow for Parylene-metal-Parylene devices.

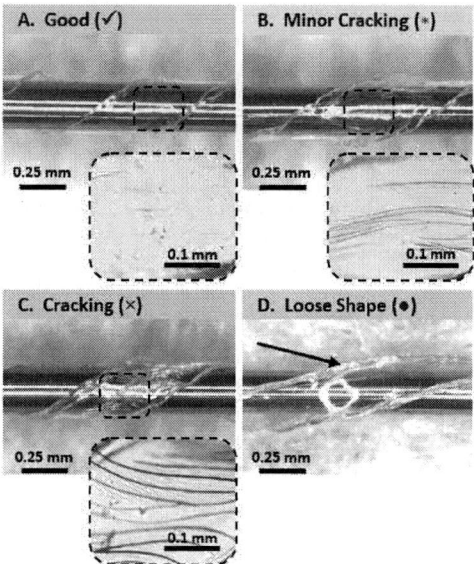

Figure 3: Examples of bare Parylene strips thermoformed into 0.25 mm helices, showing (A) a good result, (B) minor (partial-thickness) cracking, (C) cracking, and (D) loose shape, with detailed views of good and cracked Parylene.

small angles, devices should be tightly fixtured to ensure the desired shape is achieved.

Thermoforming tests were repeated on Parylene strips of 13.3 μm (single layer) and 18.5 μm (two asymmetric layers) thickness at 100 and 150 °C. Temperature did not impact the results at any helix diameter or angle. Full results for bare Parylene strip thermoforming are summarized in Table 2.

Parylene-Metal-Parylene Helices

PMP devices exhibited similar cracking results to bare Parylene strips: no cracking (Figure 4A) at 1.6 mm helix diameter, no cracking or minor cracking (Figure 4B) in thin devices (9.1 μm total thickness) at 0.25 mm helix diameter, and cracking (Figure 4C) in thick devices (14.9 μm total thickness) at 0.25 mm helix diameter. In thin/symmetric

devices when the pre-annealed base Parylene layer was oriented to the inside of the helix, most regions of the device exhibited no cracking, with minor cracking visible in some areas. Thin/symmetric devices were curled towards the base layer when released off the carrier wafer due to pre-shrinkage of the base Parylene. This natural curvature likely alleviated bending stress in the Parylene during fixturing, leading to minor cracking only in localized areas on the device.

All traces were conductive between bondpads and electrodes when thermoformed to 1.6 mm diameter. Trace continuity was maintained in thin/symmetric devices thermoformed with the pre-shrunk base layer towards the inside of the helix. All other devices had discontinuous traces after thermoforming to 0.25 mm diameter.

Thin/symmetric devices did not retain the desired shape in some areas when formed into 1.6 mm helices, likely due to the high-stress interposing metal layer (centered in the device when symmetric Parylene layers are used, or off-center when asymmetric Parylene layers are used). Thick/asymmetric devices formed to the desired shape at all diameters, and thin/symmetric devices formed to the desired shape at 0.25 mm diameter. Thicker Parylene layers can counteract the forces in the metal film more effectively (and thus are more likely to hold the desired thermoformed shape).

Thermoforming tests were repeated on thin/symmetric PMP devices (9.1 μm total thickness, formed in both directions) at 100 and 150 °C. Temperature did not impact the results at either diameter. Full results for PMP device thermoforming are summarized in Table 3.

CONCLUSION

Microfabricated Parylene devices are most commonly in a planar configuration. Biomedical devices, however, benefit from a 3D structure that can more closely interface with complex anatomical features. By utilizing thermoforming, intricate, Parylene-based, planar devices can be transformed into 3D geometries.

The thermoforming method described here produces mm- and sub mm- sized helices (0.25 to 2.6 mm diameter)

Table 2: Thermoforming results over varying thicknesses, helix angles, and helix diameters for bare Parylene strips treated at 200 °C for 12 hours.

✓ good result		* minor cracking	× cracking			● loose shape		
Parylene Thickness (μm)			**30°, 45° Helix Angle**			**15° Helix Angle**		
			Helix Diameter (mm)			**Helix Diameter (mm)**		
Inner Layer	**Outer Layer**	**Total**	**2.6**	**1.6**	**0.25**	**2.6**	**1.6**	**0.25**
5.6	n/a	5.6[1]	✓	✓	✓	✓	✓	✓
9.9	n/a	9.9[1]	✓	✓	*	●	✓	*
5.5	5.6	11.1	✓	✓	*	●	●	✓
13.3	n/a	13.3[2]	✓	✓	×	✓	✓	×
5.3	13.2	18.5[1,2]	✓	✓	×	●	●	×
13.2	5.3	18.5[1,2]	✓	✓	×	●	✓	× ●
20.9	n/a	20.9[1]	✓	✓	×	●	✓	×
5.4	20.4	25.8	✓	✓	× ●	✓	✓	× ●
20.4	5.4	25.8	✓	✓	×	✓	✓	* ●
11.9	14.4	26.3[1]	✓	✓	×	✓	✓	× ●
14.4	11.9	26.3[1]	✓	✓	×	●	●	×

[1] Only the indicated parts were tested at a 30° helix angle (all parts were tested at 15° and 45°).

[2] Indicated parts were also tested at 100 and 150 °C thermoforming temperature; temperature did not impact results.

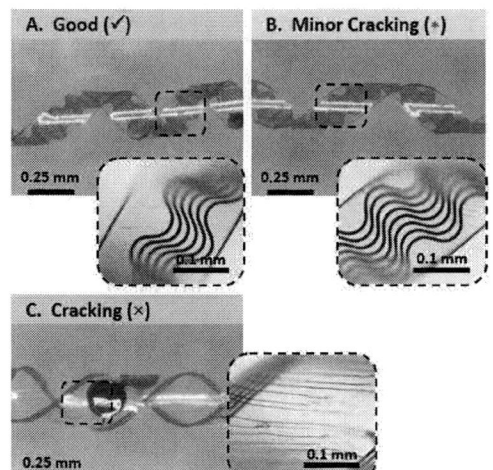

Figure 4: Examples of PMP devices thermoformed into 0.25 mm helices, showing (A) a good result, (B) minor (partial-thickness) cracking, and (C) cracking, with detailed views of good and cracked Parylene

Table 3: Thermoforming result for varying thicknesses and helix diameters for PMP devices at 45° helix angle and treated at 200 °C for 12 hours.

✓ good result			* minor cracking	
✕ cracking			● loose shape	
Parylene Thickness (µm)			**Helix Diameter (mm)**	
Inner Layer	**Outer Layer**	**Total**	**1.6**	**0.25**
4.4[1]	4.7	9.1[2]	✓ ●	✓ *[3]
4.7	4.4[1]	9.1[2]	✓ ●	* —
3.4	11.5	14.9	✓	✕ —
11.5	3.4	14.9	✓	✕ —

[1] *4.4 µm layer was deposited and annealed (150 °C, 4 hours) before adding metal and top Parylene.*

[2] *Indicated parts were also tested at 100 and 150 °C; temperature did not impact results.*

[3] *Most regions had no cracking; some minor cracking was visible in a few areas.*

from microfabricated Parylene films and devices, a 4× improvement from prior published work. When thermoforming into small helices (0.25 mm diameter), thin Parylene must be used to prevent cracking failure. Devices with interposing metal layers are sometimes unable to hold the desired shape due to high film stress in the metal, however thicker Parylene can more effectively overcome the metal stress and hold the desired shape.

More work is necessary to elucidate the relationship between Parylene thickness, film stress, and thermoforming process parameters on PMP devices. In addition, device regions with exposed metal (such as electrode sites, where Parylene is selectively removed) must be evaluated to determine the impact of large, unbalanced stress on thermoforming capability.

ACKNOWLEDGEMENTS

This work was funded by the University of Southern California under the Provost New Directions in Research and Scholarship Award.

REFERENCES

[1] B. J. Kim, B. Chen, M. Gupta, and E. Meng, "Formation of three-dimensional Parylene C structures via thermoforming," *J. Micromechanics Microengineering*, vol. 24, no. 6, p. 065003, 2014.

[2] C. Li *et al.*, "Multifunctional lab-on-a-tube (LOT) probe for simultaneous neurochemical and electrophysiological activity measurements," in *Transducers & Eurosensors XXVII: The 17th International Conference on Solid-State Sensors, Actuators and Microsystems (TRANSDUCERS & EUROSENSORS XXVII)*, 2013, pp. 880–883.

[3] C. Li, P. M. Wu, J. Han, and C. H. Ahn, "A flexible polymer tube lab-chip integrated with microsensors for smart microcatheter," *Biomed. Microdevices*, vol. 10, no. 5, pp. 671–679, 2008.

[4] C. Li and R. K. Narayan, "Development of a novel catheter for early diagnosis of bacterial meningitis caused by the ventricular drain," in *IEEE 25th International Conference on Micro Electro Mechanical Systems (MEMS)*, 2012, pp. 120–123.

[5] Z. Guo, B. Ji, M. Wang, X. Wang, B. Yang, and J. Liu, "A polyimide-based 3-D ultrathin bioelectrode with elastic sites for neural recording," *J. Microelectromechanical Syst.*, vol. 27, no. 6, pp. 1035–1040, 2018.

[6] S. Dabral *et al.*, "Stress in thermally annealed parylene films," *J. Electron. Mater.*, vol. 21, no. 10, pp. 989–994, 1992.

[7] B. J. Kim *et al.*, "3D Parylene sheath neural probe for chronic recordings," *J. Neural Eng.*, vol. 10, no. 4, p. 045002, 2013.

[8] R. Huang and Y. C. Tai, "Flexible parylene-based 3-D coiled cable," *IEEE 5th Int. Conf. Nano/Micro Eng. Mol. Syst.*, pp. 317–320, 2010.

[9] X. Kang, J. Q. Liu, H. Tian, B. Yang, Y. Nuli, and C. Yang, "Self-closed parylene cuff electrode for peripheral nerve recording," *J. Microelectromechanical Syst.*, vol. 24, no. 2, pp. 319–332, 2015.

[10] E. Yoon *et al.*, "An implantable microelectrode array for chronic in vivo epiretinal stimulation of the rat retina," *J. Micromechanics Microengineering*, vol. 30, no. 12, p. 124001, 2020.

[11] S. A. Hara, B. J. Kim, J. T. W. Kuo, C. D. Lee, E. Meng, and V. Pikov, "Long-term stability of intracortical recordings using perforated and arrayed Parylene sheath electrodes," *J. Neural Eng.*, vol. 13, no. 6, p. 066020, 2016.

[12] A. C. Johnson and K. D. Wise, "A self-curling monolithically-backed active high-density cochlear electrode array," in *IEEE 25th International Conference on Micro Electro Mechanical Systems (MEMS)*, 2012, pp. 914–917.

CONTACT

*E. Meng, tel: +1-213-7406952; ellis.meng@usc.edu

FABRICATION OF BIODEGRADABLE SOFT TISSUE-MIMICKED MICROELECTRODE ARRAYS FOR IMPLANTED NEURAL INTERFACING

Wei-Chen Huang[1], Wan-Lou Lei[1], and Chih-Wei Peng[2]
[1]National Yang Ming Chiao Tung University, Hsinchu, Taiwan
[2]Taipei Medical University, Taipei, Taiwan

ABSTRACT

This paper reports a method to fabricate biodegradable, adhesive and soft tissue-compliant microelectrode arrays (MEAs). MEAs composed of polylactic acid (PLA) biodegradable insulation layers and Pt/PLA hybrid electrodes can be fabricated in a batch by photolithography with a tunable photoresist/developer composition. The coordinated sacrificial layer dissolution and interfacial adhesion driven by an adhesive hydrogel enable directly transfer printing Pt-PLA MEAs stacks. This integrated process will allow, for first time, the batch production of implanted hydrogel MEAs composed of metal/organic composites as conformal electrodes and fully biodegradable insulation layers to show significantly lower impedance (0.1-2.5 kΩ) at 1 kHz as compared with those reported from most literatures. This device enables the success of both in-vivo site-specific neural stimulation and recording.

KEYWORDS

Neural interfaces; Microelectrode arrays; Biodegradable electronics; Hydrogel electronics; Platinum; Transfer printing; Polylactic acid (PLA)

INTRODUCTION

Neural interfaces with functions of biosignal recording and electrical stimulation enable the communication between external computers and nerve tissues, leading to significant contribution in treating peripheral nerve disorders [1]. The device performance relies on the capacity of collecting sufficient amounts of electrical charge flows bidirectionally from nerve and power circuits, which is mainly dependent on the mechanical compatibility and stable contact between the device and nerve fibers. Currently, implanted peripheral neural interfaces are generally designed in the form of thin-film and plastic-based cuff electrode arrays composed of metal and the polyimides, parylene C, SU-8, or PDMS [2-4]. Despite the device can wrap around the peripheral nerve fibers, both the mechanical strength of MEAs tracks and substrate cannot perfectly match with that of nerve tissue. In addition, as peripheral nerve tissue healing and regeneration occurs in a certain therapeutic time window, MEAs composed of nondegradable materials will become a physical barrier restricting the tissue reconstruction and have risk causing a chronic inflammatory response.

Many efforts have been made to develop implanted biodegradable electronics in recent years because surgical removal can be eliminated to prevent infections associated with foreign tissue responses [5]. Implanted neural interfaces are expected to exhibit transient performance on a time scale appropriate for tissue repair. Implantable device substrate or coating layers made of biodegradable and biocompatible polymer materials such as silk, PVA, and PLGA with suitable mechanical properties can meet the clinical needs for short-term applications [6-8]. However, integration of biodegradable polymer materials into MEA technology by photolithography faces challenges due to the complicated device fabrication process. The thermal stability, solvent compatibility, and pressure tolerance of these biodegradable polymers need to be considered in all manufacturing steps.

Polylactic acid, PLA, is a commonly known recyclable and biodegradable polymer for medical usages. Because of the biodegradable capacity, PLA has been regarded as one of the most prospecting hydrophobic polymers for the replacement of the conventional nondegradable implantable polymer including polyethylene terephthalate (PET), low-density polyethylene (LDPE) and high-density polyethylene (HDPE) [9]. Here, inspired by the hydrogel-mediated transfer printing reported in 2018 [10], the work presents a complete process of manufacturing transient soft tissue-compliant MEAs. The method reported here differs from previous work in a specific method to fabricate metal-organic hybrid electrode and biodegradable polyesters (PLA) insulation layer (Figure 1). This paper will also show a new set of experimental results using the device connected with flexible flat cables in the practical application including implanted neural stimulation and signals recording, to highlight the superiority for neural interfacing.

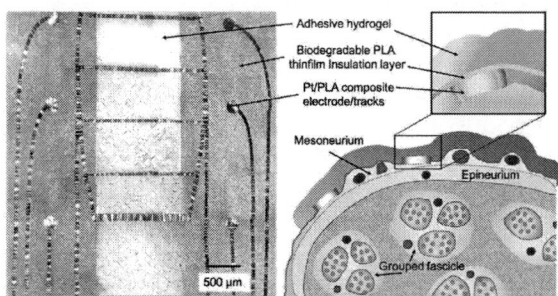

Figure 1: A real optical microscopic view of Pt/PLA MEAs on a hydrogel substrate (left) and an illustration (right) showing the side view of the device-peripheral neural interfaces.

DEVICE FABRICATION

Figure 2 is an illustration showing the fabrication process of the device. O_2 plasma treatment was conducted to increase the hydrophilicity of a 4-inch Si wafer surface. To fabricate the sacrificial layer, 10 % polyacrylic acid (PAA) solution was spin-coated on the wafer at 600 rpm for 6 s followed by 3800 rpm for 30 s. After baked at 100

°C for 1min, the sample was then immersed in 4 M Calcium chloride ($CaCl_2$) solution for 10 min to ensure completely crosslinking of calcium-crosslinked PAA (PAA-Ca) which is insoluble in water. Then the wafer was baked again at 100 °C for 1 min.

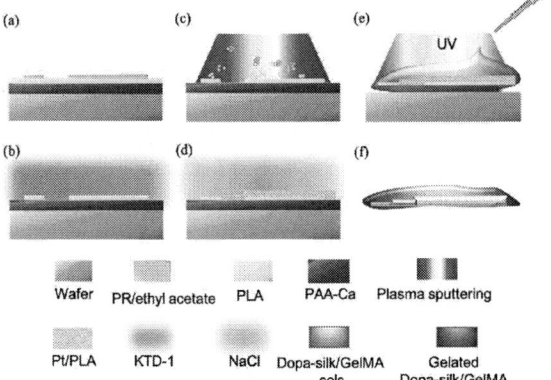

Figure 2: A schematic illustration showing the fabrication processes of hydrogel-based Pt/PLA MEAs.

An uniform PLA thin film with thickness of 267 ± 30 nm was obtained through spin-coating of 2 % PLA chloroform solution at 1500 rpm for 10 s. It is worthy to mention that the volatile PLA solution caused non-uniform thickness of the deposited film on the wafer where the film thickness got gradually reduced from the central to the edge regions. This situation limited the implement of dry etch, thus wet etch was used instead. For patterning PLA, FH6400 Photoresist (PR) was diluted by ethyl acetate at volume ratio of PR to ethyl acetate of 1:10 at 60 °C (Figure 2(a)). The diluted PR was spin-coated at 3000 rpm for 15 s, followed by baked at 90°C for 90 s. With KTD-1 as a developer, PLA was dissolved fast, which permitted simultaneous PR removal and selective etch of PLA (Figure 2(b)). In fact, KTD-1 cannot dissolve PLA but the ingredient, 2.38% of TMAH, is the catalyst allowing ethyl acetate to decompose PLA [11]. Such approach enabled directly micropatterning of PLA with a final PLA thickness of ~95 nm on PAA-Ca.

The platinum (Pt) conductive layer with thickness of ~150 nm was sputtered using Dual E-Gun Evaporation System (ULVAC EBX-10C, Japan). The photoresist AZ6112 with 1.2 µm was used for patterning Pt via photolithography. It was later etched by High Density Plasma Reactive Ion Etching System, HDP-RIE (Duratek company) (Figure 2(c)). The Cl_2/ BCl_3 gas flow rate were 30 sccm, the ICP power was 600 W, the bias was 180 W, the chamber pressure was 10 mTorr, and the etching time was around 238 s. Because the process temperature was higher than glass transition temperature of PLA (T_g = 80°C), Pt sputtering on PLA resulted in the formation Pt/PLA hybrid, which can be demonstrated in the spectrum of X-ray photoelectron spectroscopy (XPS, 1486.6 eV Al Kα, Thermo Fisher Scientific Theta Probe), where the elements of Pt, C, and O are observable (Figure 3). Such an inorganic/organic hybrid electrode with a relatively lower Young's modulus than that of pure Pt can avoid cracks produced when the device is deformed and can also enhance conformal contact to the soft nerve tissues.

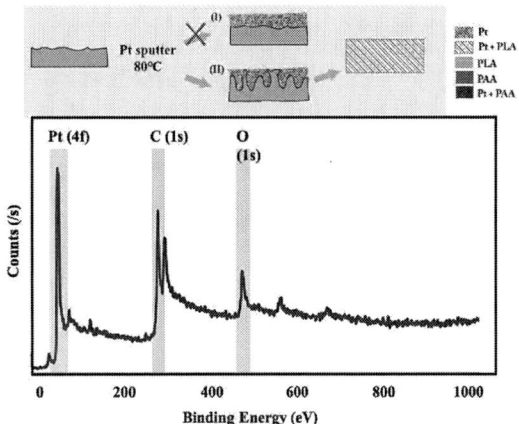

Figure 3: A real optical microscopic view of Pt/PLA MEAs on a hydrogel substrate (left) and an illustration (right) showing the side view of the device-peripheral neural interfaces.

The microscopic images of the electrode sites are shown in Figure 4(a). The electrode sites were designed with two different sizes, i.e 50 µm and 100 µm in diameter. Before transferring of MEAs to the hydrogel substrate, the sacrificial layer PAA-Ca was dissolved by the immersion of 4 M NaCl solution for 2 h at 50 °C (Figure 2(d)). After immersion for a few seconds, the changed color of the PAA-Ca indicates successful dissolution (Figure 4(a)). The mono-ion Na^+ enabled the ionic exchange with the Ca^{2+} in PAA, turning the sacrificial layer dissolvable in water. The immersion was lasted to ~2 h to confirm complete dissolution of PAA-Ca.

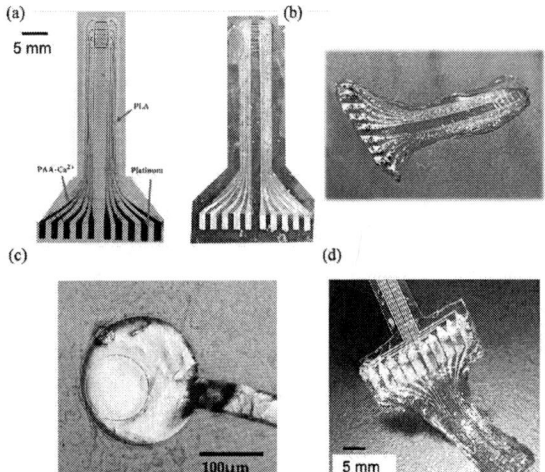

Figure 4: (a) Optical image of MEAs fabricated on PAA before (left) and after (right) immersed in NaCl solution. (b) The transfer printed MEAs on adhesive hydrogels. (c)A optical image showing the electrode site with corrugated morphology. (d) Hydrogel-based MEAs connected Ethylene-vinyl acetate (EVA)-based FFC.

The MEA substrate was fabricated by hydrogel precursors composed of dopamine-modified silk/gelatin methacryloyl (Sdopa-GelMA). GelMA is a photopolymerizable gelatin derivative that is a highly-recommended tissue engineering scaffold material and can be quickly crosslinked by UV into a stable hydrogel with tunable mechanical strength, structure and biodegradability. Here, GelMA combined with silk forms an elastic and tough hydrogel. The modification of dopamine on silk fibroins can provide strong hydrogen bonding interactions to strengthen the stickiness of the hydrogel. The hydrogel precursor solution was then poured onto the MEA patterns, followed by cured with UV radiation for fully gelation. Followed with the addition of water around the hydrogel to facilitate detachment the device, the resultant hydrogel MEA was obtained (Figure 2(e) and (f)). Figure 4(b) shows the transparent hydrogel MEAs obtained by transfer printing, where the MEA tracks show slightly corrugated without cracks on the soft hydrogel substrate (Figure 4(c)). The bond pads were connected to a customized printed ethylene-vinyl acetate (EVA)-based flexible flat circuit board (Brave C&H Supply, Taiwan) using silver paste and are encased in PDMS silicone (Sylgard® 184A, Dow corning) for protection (Figure 4(d)).

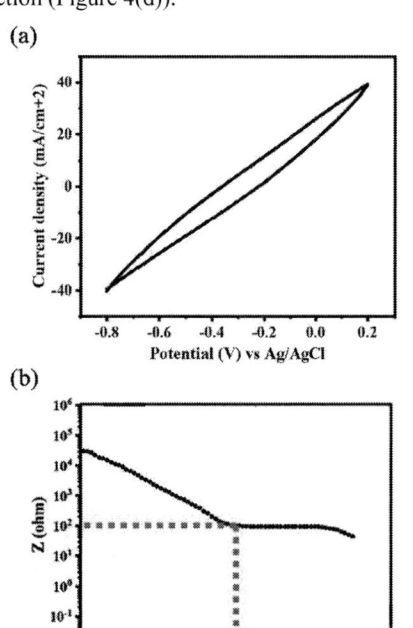

Figure 5: (a) A CV curve of the channel site with 50 μm in diameter recorded in PBS. (b) A Bode plot showing the impedance values versus frequency.

IN VITRO ELECTROCHEMICAL PERFORMANCE

The electrochemical properties of the device were determined by electrochemical impedance and charge storage capacity (CSC) using electrochemical instrument 600E Potentiostat/Galvanostat (CHI Instruments, Austin, TX, USA) with three-electrode system: The hydrogel MEAs were used as the working electrode, platinum was used as the counter electrode, and Ag/AgCl was used as working electrode in 0.1M PBS. The impedance response was determined with an amplitude sinusoid input of 100 mV at the frequency ranged from 1 Hz to 1 MHz. The average impedance was demonstrated in Bode plot. For the cyclic voltametric (CV) investigation, the scan rate was 50 mV/s and the applied potential range from -0.8 to 0.2 V and from -1.4 to 0.6 V with 20 cycles in scan number. Figure 5 shows the electrical impedance and CV measurements of the device. The calculated CSC for electrical sites with diameters of 50 μm and 100 μm are 39.27 μC/cm^2 and 140.3 μC/cm^2 in PBS solution respectively. The impedance of both electrode sites is decreased with increasing frequency as expected. The electrochemical impedance was around 2.5 kΩ and 100 Ω. at 1 kHz. As compared with those reported from most literatures [12-15], the hydrogel MEAs exhibit relatively lower impedance, which implies better signal transduction.

IN VIVO NEURAL SIGNAL RECORDING AND STIMULATION

The hydrogel MEAs were placed on the sciatic nerve of a rat for neural stimulation and recording (Figure 6(a)). The skin of the left hind paw was mechanically stimulated by using a clip with pressure intensity. Each mechanical stimulus lasted ~1 s and was repeated approximately every 10 s. The neural signal recording was lasted for 60 s, and each trial included 6 stimuli. Figure 6(b) shows the in vivo recording of the evoked action potentials in a time window of 10 s, where the highest peak potential can go up to 0.15 mV and the calculated SNR is approximately 8 dB. For neural stimulation, the amplitude of electromyographic (EMG) responses in the calf muscle evoked by MEA stimulation can be used as readouts of peripheral nerve excitability. A single biphasic stimulation using 5 mA at 0.1 Hz induces a response in the EMG by 48 ms in Figure 6(c), indicating that MEA can provide excitability of motor function. These results reveal that the hydrogel MEAs can provide conformal contact with the nerve tissues to monitor functions of sciatic nerves.

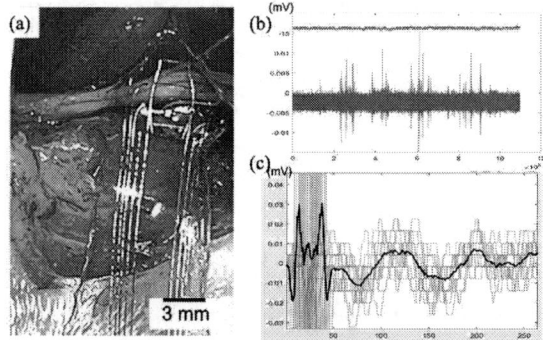

Figure 6: (a) Optical image of hydrogel MEAs laminated on a sciatic nerve. (b) In-vivo neural recording. (c) Motor evoked potentials (MEP) obtained in response to electrical stimulation using 5 mA at 0.1 Hz.

CONCLUSION

A neural implant was developed with a MEA architecture composed of a degradable PLA insulation layer, Pt/PLA metal-organic hybrid electrodes and an adhesive hydrogel substrate. The adjustment of PR composition combined with a specific PR developer overcome the limitation of using MEMS process to micropattern biodegradable polymer in a batch. With a low melting point, PLA can be involved in the metal sputtering process to form organic/inorganic thin film as a bionic electrode with improved biocompatibility. Transfer printing of MEAs on an adhesive hydrogel overcame the challenge of combining hydrophilic polymer networks and hydrophobic complex electronic microstructure. This resultant hydrogel-based device laminated on a sciatic nerve offers functions of neural signal recording and neural stimulation, showing a promising potential to improve the clinical application of peripheral neural interfaces.

ACKNOWLEDGEMENTS

This work was supported by the Ministry of Science and Technology in Taiwan (MOST) with the granted funds MOST-110-2636-E-009-012 and MOST-111-2636-E-A49-007. In addition, the authors acknowledge the National Applied Research Laboratories for XPS analysis.

REFERENCES

[1] Y. Cho, J. Park, C. Lee, S. Lee, "Recent Progress on Peripheral Neural Interface Technology Towards Bioelectronic Medicine", *Bioelectron. Med.*, vol. 6, no. 1, pp.1-10, 2020.

[2] Y. Cho, S. Park, J. Lee, K. J. Yu, "Emerging Materials and Technologies with Applications in Flexible Neural Implants: A Comprehensive Review of Current Issues with Neural Devices", *Adv. Mater.*, vol. 33, no. 47, pp. 2005786, 2021.

[3] C. E. Larson, E. Meng, "A Review for The Peripheral Nerve Interface Designer", *J. Neurosci. Methods.*, vol. 332, pp.108523, 2020.

[4] E. Redolfi Riva, S. Micera, "Progress and Challenges of Implantable Neural Interfaces Based on Nature-Derived Materials", *Biosens. Bioelectron.*, vol. 7, no. 1, pp. 1-10, 2021.

[5] A. Fanelli, D. Ghezzi, "Transient Electronics: New Opportunities for Implantable Neurotechnology", *Curr. Opin. Plant Biol.*, vol. 72, pp. 22-28, 2021.

[6] A. C. Patil, Z. Xiong, N. V. Thakor, "Toward Nontransient Silk Bioelectronics: Engineering Silk Fibroin for Bionic Links", *Small Methods.*, vol. 4, no. 10, pp. 2000274, 2020.

[7] J. Yoon, J. Han, B. Choi, Y. Lee, Y. Kim, J. Park, M. Lim, M. H. Kang, D. H. Kim, D. M. Kim, S. Kim, and S. J. Choi, "Three-dimensional Printed Poly (vinyl alcohol) Substrate with Controlled On-demand Degradation for Transient Electronics", *ACS Nano.*, vol. 12, no. 6, pp. 6006-6012, 2018.

[8] J. K. Chang, H. Fang, C. A. Bower, E. Song, X. Yu, J. A. Rogers, "Materials and Processing Approaches for Foundry-compatible Transient Electronics", *PNAS.*, vol. 114, no. 28, pp. E5522-E5529, 2017.

[9] D. H. Jiang, T. Satoh, S. H. Tung, C. C. Kuo, "Alternatives to Nondegradable Medical Plastics", *ACS Sustain. Chem. Eng.*, vol. 10, no. 15, pp. 4792-4806, 2022.

[10] W. C. Huang, X. C. Ong, I. S. Kwon, C. G. Lee, E. Haosheng, W. Gary, K, Fedder, R. A. Gaunt, C. J. Bettinger, "Ultracompliant Hydrogel-Based Neural Interfaces Fabricated by Aqueous-Phase Microtransfer Printing", *Adv. Funct. Mater.*, vol. 28, no. 29, pp. 180105, 2018.

[11] S. Xie, Z. Sun, T. Liu, J. Zhang, T. Li, X. Ouyang, X. Qiu, S. Luo, W. Fan, H. Lin, "Beyond Biodegradation: Chemical Upcycling of Poly (lactic acid) Plastic Waste to Methyl Lactate Catalyzed by Quaternary Ammonium Fluoride", *J Catal.*, vol. 402, pp. 61-71, 2021.

[12] C. Boehler, S. Carli, L. Fadiga, T. Stieglitz, M. Asplund, "Tutorial: Guidelines for Standardized Performance Tests for Electrodes Intended for Neural Interfaces and Bioelectronics", *Nat. Protoc.*, vol. 15, no. 11, pp. 3557-3578, 2020.

[13] M. Vomero, E. Castagnola, F. Ciarpella, E. Maggiolini, N. Goshi, E. Zucchini, S. Carli, L. Fadiga, S. Kassegne, D. Ricci, "Highly Stable Glassy Carbon Interfaces for Long-Term Neural Stimulation and Low-Noise Recording of Brain Activity", *Sci. Rep.*, vol. 7, no. 1, pp. 40332, 2017.

[14] C. Boehler, T. Stieglitz, M. Asplund, "Nanostructured Platinum Grass Enables Superior Impedance Reduction for Neural Microelectrodes", *Biomaterials.*, vol. 67, pp. 346-353, 2015.

[15] G. Márton, G. Orbán, M. Kiss, R. Fiáth, A. Pongrácz, I. Ulbert, "A Multimodal, SU-8 - Platinum - polyimide Microelectrode Array for Chronic In Vivo Neurophysiology", *PLoS One.*, vol. 10, no. 12, pp. e0145307, 2015.

CONTACT

*W.C. Huang, weichenh@nycu.edu.tw

AN OPTIMIZATION OF PERFORATION DESIGN ON A PIEZOELECTRIC-BASED SMART STENT FOR BLOOD PRESSURE MONITORING AND LOW-FREQUENCY VIBRATIONAL ENERGY HARVESTING

Jun Ying Tan[1], Sayemul Islam[2], Yuankai Li[3], Albert Kim[2], and Jungkwun 'JK' Kim[1]

[1]Department of Electrical Engineering, University of North Texas, Denton, TX 76207, USA.
[2]Department of Medical Engineering, the University of South Florida, Tampa, FL 33620, USA.
[3]Department of Electrical and Computer Engineering, Kansas State University, Manhattan, KS 66506, USA.

ABSTRACT

This paper presents the perforated design of a piezoelectric tube that will be employed as a self-powered Smart Stent for real-time blood pressure monitoring. The proposed Smart Stent was made of polyvinylidene fluoride (PVDF), which can harvest energy from pulse-motion low-frequency vibration such as blood flow. This study focuses on a unique pattern of the perforation added to the Smart Stent. We observed that the perforation design of Smart Stent varies its sensitivity to pressure change and produces different energy harvesting performances. The eight different perforations design of the Smart Stent were fabricated, examined, and reported their performances.

KEYWORDS

Smart Stent, piezoelectric, pressure sensor, low-frequency energy harvester, implantable medical device.

INTRODUCTION

Abdominal aortic aneurysm (AAA) accounts for approximately 60% of mortality once it ruptures [1]. A common preventative treatment, known as endovascular aneurysmal repair (EVAR), has been a popular intervention by which placing a fabric-covered stent graft at the abnormal site to redirect the blood flow away from the aortic wall [2]. The practice of EVAR has a lower 30-day postoperative mortality rate of 3.5% compared to 7.1% for conventional open surgical repair [3]. It also reduced time under general anesthesia [4], length of stay in the hospital and intensive care unit (ICU) [5], and likelihood of intraoperative blood transfusions [6], [7]. Despite the advantages, a fatal mechanical failure of the stent, known as endoleak, has remained unsolved [8], [9]. Hence, a post-EVAR surveillance is recommended. Radiography, ultrasonography, computed tomography (CT), nuclear imaging, magnetic resonance angiography (MRA), and conventional angiography are the current post-EVAR surveillance techniques [8]. However, these techniques are expensive and require trained personnel, while some might require an invasive procedure, long scanning time and could cause radiation exposure [8].

Previously, we demonstrated a Smart Stent system that can measure the blood flow information wirelessly through an ultrasonic excitation [10]. In this work, we further explore the Smart Stent as a pressure sensor and energy harvester based on a similar architecture. The Smart Stent simultaneously harvests ambient vibrational energy (i.e., blood flow) while monitoring the blood pressure at the implanted site in real time. The harvested energy can be used to power the transmission circuit and other implanted devices to eliminate the necessity for an external source.

To achieve that, we incorporate a custom-perforated, piezoelectric-based Smart Stent that can be deployed onto the stent graft, as shown in Fig. 1. The function of the perforation has been discussed in the previous study [10]. In short, it induced a negative Poisson's ratio (NPR) of the Smart Stent, which allows it to expand transversely under axial strain. However, the previous design was characterized to enhance ultrasonic powering efficiency, while the proposed Smart Stent aimed to harvest low-frequency vibrational energy from blood flow (1 to 1.67 Hz or 60 to 100 beats per min). Therefore, it is necessary to optimize the perforation design.

In this paper, the perforation design was characterized based on the size, orientation, and shape of the perforation. Eight types of perforation designs were demonstrated including seven perforated samples and one unperforated sample as a control. All Smart Stents had the same length and diameter of 70 and 14.5 mm, respectively. The Smart Stents were examined in a closed-loop circulation system and a complete set of experimental results were reported. The optimum design is recommended based on the sensitivity to pressure change and the magnitude of generated voltage.

MATERIALS AND FABRICATION

Poly(vinylidene fluoride) (PVDF) powder and N, N-Dimethylformamide (DMF) 99% were purchased from Alfa Aesar Chemicals. Polydimethylsiloxane (PDMS) elastomer was purchased from Dow Corning Corporation.

The fabrication process of the Smart Stent is depicted in Fig. 2. The PVDF film casting was based on Cardoso et al.'s work [11]. In brief, 15 wt. % of PVDF solution was prepared by dissolving the PVDF powder in DMF solution, and continuously mixing with a magnetic stirrer for 15 min at 30 °C. The substrate for film casting was a 2" × 3" glass slide, that had been cleaned with organic solvents (acetone, methanol, and isopropanol), dried with nitrogen, and

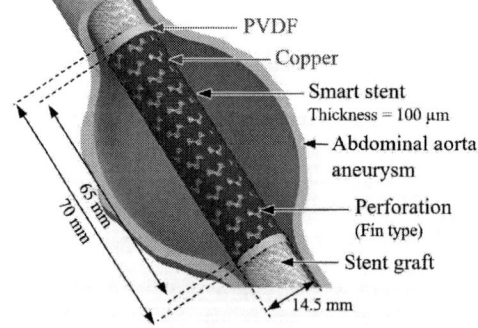

Figure 1: Conceptual drawing of a Smart Stent mounted on a stent graft in an abdominal aorta aneurysm.

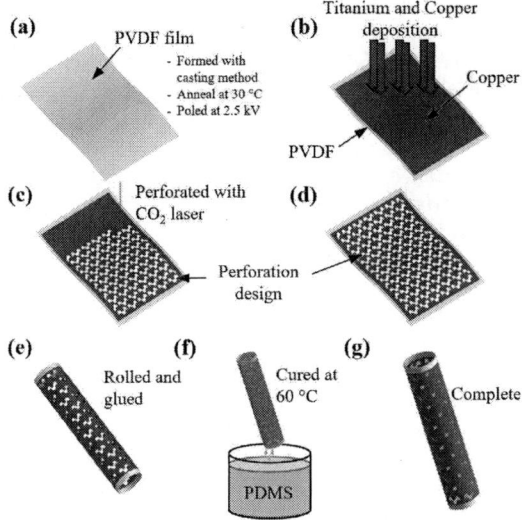

Figure 2: Fabrication process of the Smart Stent. (a) PVDF film. (b) Metallization. (c) Laser machining. (d) Perforated film. (e) Rolled to form tube. (f) PDMS passivation. (g) Smart Stent complete.

treated with oxygen plasma to remove all potential contaminants. 3 g of the PVDF solution was cast onto the glass slide and allowed to spread until uniform thickness was achieved. The sample was annealed for at least 24 h at 30 °C and 30% humidity. Once dried, the PVDF film was delaminated from the glass slide by immersing the sample in deionized (DI) water for 5 min due to hydrophobicity. For poling, the film was sandwiched between two electrodes and applied voltage of 2.5 kV for 1 h at 80 °C. A 100-μm PVDF film was obtained via this process as shown in Fig. 2(a). The film was metalized on both sides with titanium (100 nm) and copper (300 nm) using a DC sputter as depicted in Fig. 2(b). For perforation, the metalized film was laser-machined with a CO_2 laser with 22.5 W at cutting speed of 200 mm/s as illustrated in Fig. 2(c-d). The perforated film was rolled into a tube and glued together using DMF solution as demonstrated in Fig. 2(e). The sample was dip-coated into a PDMS solution for passivation, cured at 60 °C for 3 hours as shown in Fig. 2(f), and the completed Smart Stent is shown in Fig. 2(g).

PERFORATION DESIGNS AND SETUP

The perforation designs of the Smart Stent were characterized based on the pore size (defined as "big" or "small"), orientation (i.e., normal or parallel), and shape (i.e., fin or wave). The combination of these features results in eight different perforation designs, as listed in Table 1. The acronym of each design is derived from its corresponding features, for instance, BNF is short for a sample with big-sized (B), normally-oriented (N), and fin-shaped (F) perforation. Note that the BNF sample was adopted from our previous work [10]. In addition, an unperforated sample was included as a control, abbreviated as "plain."

A closed-loop circulation system was prepared to examine the Smart Stents, which consists of a DI water reservoir, two peristaltic pumps, a 3D printed abdominal aorta model, a pressure gauge, a roller clamp, and a Smart Stent as shown in Fig. 3. The abdominal aorta model was 3D printed with a flexible resin (Elastic 50A, Formlabs Inc.). A pressure gauge was installed as close to the model as possible to monitor the water pressure prior entering the model. A constant flow of DI water was pumped by peristaltic pump 1 to maintain a steady 80 mmHg pressurized environment inside the model. In the meantime, the peristaltic pump 2 was programmed to intermittently pump pulses of water with 40 mmHg, which result in a cumulative pressure of 120 mmHg inside the model. The pressure can be adjusted based on the water volume that was pumped into the model by pump 1 and 2, where pump 1 defines the diastolic pressure and pump 2 defines the systolic pressure, typically, 80/120 mmHg was used. The roller clamp at the end of the model was used to fine tune the pressure. In order to monitor the Smart Stents after being implanted into the system, wires were connected on each side of the Smart Stent to enable measuring the induced voltage via an oscilloscope (Analog Discovery 2, Digilent) and a computer. A ground bar (GND) was introduced in the system to eliminate the 60 Hz noise. Due to the limitation of the peristaltic pumps, the fastest pulse rate that can be generated by the described system was 30 pulses per minute (0.5 Hz), which is less than the typical human heart rate of 60 to 100 beats per minute (1 to 1.67 Hz). Nonetheless, this limitation does not impair the experiment in determining the optimal

Table 1. Parameters of the perforation designs

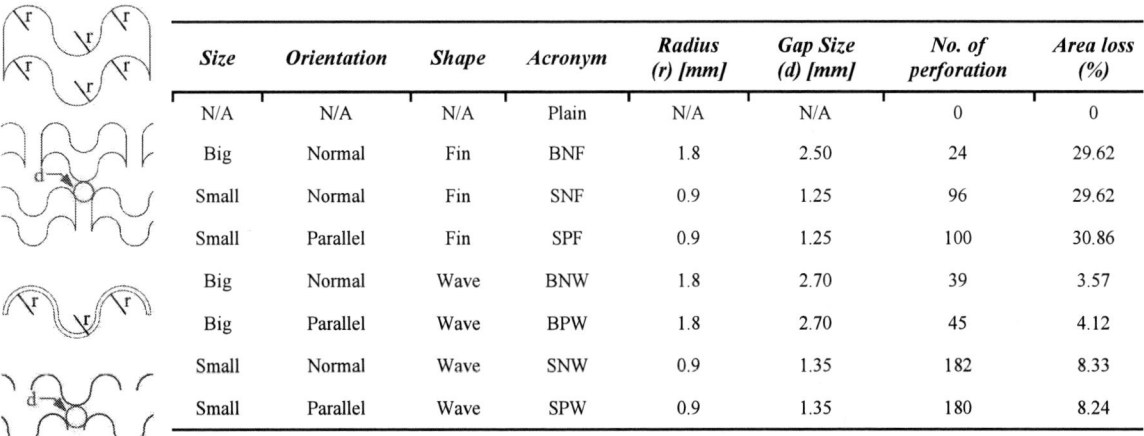

Size	Orientation	Shape	Acronym	Radius (r) [mm]	Gap Size (d) [mm]	No. of perforation	Area loss (%)
N/A	N/A	N/A	Plain	N/A	N/A	0	0
Big	Normal	Fin	BNF	1.8	2.50	24	29.62
Small	Normal	Fin	SNF	0.9	1.25	96	29.62
Small	Parallel	Fin	SPF	0.9	1.25	100	30.86
Big	Normal	Wave	BNW	1.8	2.70	39	3.57
Big	Parallel	Wave	BPW	1.8	2.70	45	4.12
Small	Normal	Wave	SNW	0.9	1.35	182	8.33
Small	Parallel	Wave	SPW	0.9	1.35	180	8.24

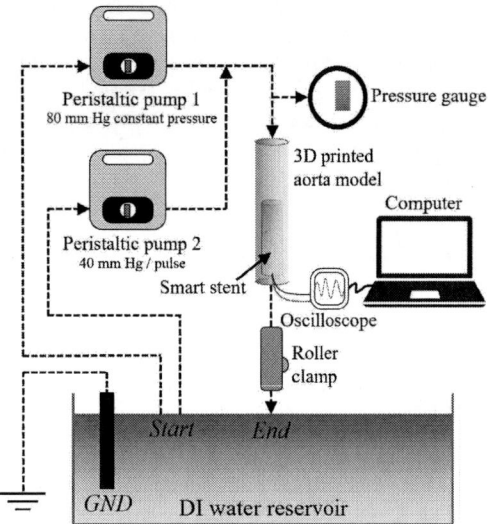

Figure 3: Experimental setup of a closed-loop circulation system for testing the Smart Stents.

perforation design on the Smart Stent as the pulse rate does not affect the generated voltage.

RESULTS AND DISCUSSION

Figure 4(a-h) present the fabrication results of eight types of Smart Stents, including plain (control), BNF, SNF, SPF, BNW, BPW, SNW, and SPW. All Smart Stents were evaluated in the described system to determine their sensitivity to pressure change and voltage generation. All samples were tested at least three times and the average values were reported with corresponding standard deviation. All samples were wired and implanted into the system, and the results were shown in Fig. 5.

Figure 5(a) shows a 30-second continuous measurement of the voltage generated by the SNF sample in three different scenarios. The first scenario was represented by the black line, showing that the Smart Stent generated a steady voltage of 110 mV under 80 mmHg while the pump 1 was turned on and pump 2 remained off. The ripples were induced by the vibration caused by pump 1. The magnitude of the ripples was measured of approximately 1 mV, which is less than 1% of the measured voltage, and thus, negligible. This result demonstrates the stability of the system and the reliability of the Smart Stents. The second scenario was presented in blue line, with pump 1 constantly on, and pump 2 on intermittently, imitating the typical blood pressure variation of 80 to 120 mmHg every pulse. In this case, the voltage increased from 110 to 180 mV at each pulse, then decreased to 110 mV as the pulse pressure released. A maximum of 2.5% variation of the peak values was measured throughout the 30-second measurement, indicating the precision of the device. The third scenario, depicted in red line, had pump 1 constantly on, while pump 2 was on intermittently with lower pulse pressure of 10 mmHg, resulting in a cumulative pressure of 90 mmHg at each pulse. The described scenario imitated the situation when an endoleak or malfunction occurred in the system, which resulted in a drop in pulse pressure. The peak voltage was reduced to roughly 140 mV in response to the

decreased pulse pressure, confirming its applicability as an early warning indication for endoleak.

Figure 5(b) shows the comparison of the Smart Stent's sensitivity with different perforation designs in response to pressure change. It should be noted that the legends were arranged in descending order in terms of the measured peak-to-peak voltage to better analyze the data. Among the seven perforated samples, all small-sized perforated samples (SPF, SPW, SNW, SNF) have demonstrated significant improvement in sensitivity to pressure change in comparison to the plain sample, whereas the big-sized perforated samples (BNW, BPW, BNF) appeared to perform worse. This result suggests that by introducing small-sized perforation on the Smart Stent, the sensitivity to pressure change can be improved up to 40%.

To evaluate the feasibility of the Smart Stents as low-frequency energy harvester, the average voltages generated by the Smart Stents were analyzed, and the result is shown in Fig. 5(c). Similarly, the legends were arranged in descending order in terms of the average voltage. Based on the results, SNW and SNF outperformed the plain sample in voltage generation, whereas the other perforated samples were inferior to the plain sample. The orientation of the perforation was critical in determining the magnitude of the generated voltage. Among the small-sized perforated samples, the normally perforated samples (SNW and SNF) have a positive impact on voltage generation, whereas the parallelly perforated samples (SPW and SPF) have a negative impact. However, of the two samples (SNW and SNF) that have positive impact, the SNW sample has received a 30% increase in voltage generation over the plain sample, while the SNF sample only received roughly 5% increment. This is due to the perforation shape, where the SNF sample has lost about 30% of its piezoelectric material, while the SNW sample only lost about 8%. In addition, the footprint's size of the wave pattern is smaller than the fin pattern, which 182 wave-patterns can fit onto the SNW sample while only 96 fin-patterns on the SNF sample. Similar results were observed in other comparison such as BNW and BNF, and SPW and SPF samples, where wave-shaped perforated samples yielded higher voltage than the fin-shaped perforated samples. In low frequency voltage generation, the wave-shaped perforation has shown superiority over fin-shaped perforation.

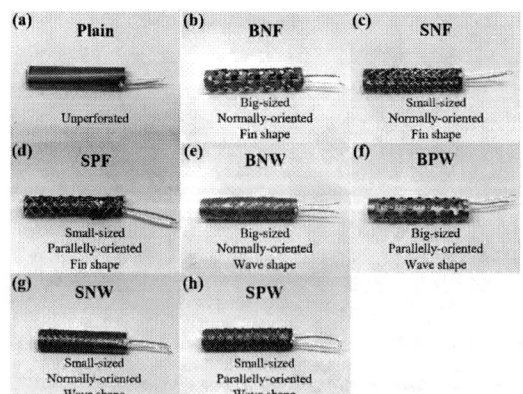

Figure 4: Fabrication results of the Smart Stents. (a) Plain. (b) BNF. (c) SNF. (d) SPF. (e) BNW. (f) BPW. (g) SNW. (h) SPW.

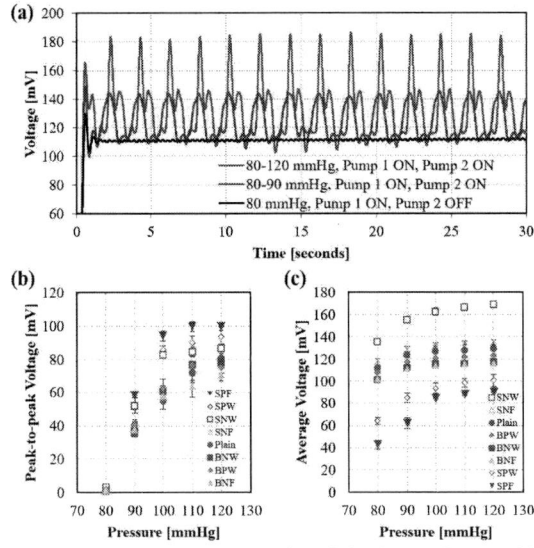

Figure 5: Experimental results of the Smart Stents. (a) 30-s continuous measurement of the SNF sample at 80, 90, and 120 mmHg cumulative pressure. (b) Sensitivity and (c) voltage generation of the Smart Stents.

CONCLUSIONS

With the perforation design as proposed, Smart Stent with 29.62% loss in piezoelectric material, such as the SNF sample, was able to generate higher voltage than the plain sample with better sensitivity to pressure change. The size of the perforation showed great importance in determining the device's sensitivity to pressure change, with small-sized perforation favored over big-sized perforation. The orientation of the perforation also highly affects the device's sensitivity and generated voltage. Parallelly-perforated samples are desirable for increasing the device sensitivity, whilst normally-perforated samples are desirable for higher voltage generation. Lastly, the shape of the perforation plays crucial role in the perforation design of the Smart Stents. Although there is no significant effect on the device sensitivity, it has considerable effect on the magnitude of the generated voltage, where wave-shaped perforation offers superior performance over fin-shaped perforation. In consideration of all the aforementioned parameters, the SNW sample has the optimum perforation design on the Smart Stent as pressure sensor and low-frequency energy harvester. Although SPF and SPW samples have offered excellent sensitivity to pressure change, their poor voltage generation hinders their practicality as Smart Stent. Considering the Smart Stent will be required to sustain other integrated circuits, especially a transmission circuit for wireless monitoring, a full-wave bridge rectifier integrated chip (IC) is needed to be integrated onto the device. While the IC typically requires at least 100 mV of forward voltage drop to operate, neither the SPF nor the SPW samples generate 100 mV or more under ideal conditions, rendering them ineffective as energy harvesters. In contrast, the SNW sample generated the highest voltage at various pressures while being moderately sensitive to pressure change, making it the optimum perforation design on the Smart Stent for pressure sensing and low-frequency, passive energy harvesting.

ACKNOWLEDGEMENTS

The research was supported by National Science Foundation (NSF) CNS 2039014, ECCS 2054567, ECCS 2029086, and ECCS 2029077.

REFERENCES

[1] Z. Yuan, Y. Lu, J. Wei, J. Wu, J. Yang, and Z. Cai, "Abdominal Aortic Aneurysm: Roles of Inflammatory Cells," *Frontiers in Immunology*, vol. 11, 2021.

[2] A. England and R. Mc Williams, "Endovascular Aortic Aneurysm Repair (EVAR)," *Ulster Med J*, vol. 82, no. 1, pp. 3–10, Jan. 2013.

[3] R. Greenhalgh, "Comparison of endovascular aneurysm repair with open repair in patients with abdominal aortic aneurysm," *The Lancet*, vol. 364, no. 9437, pp. 843–848, Sep. 2004.

[4] D. A. Bettex, M. Lachat, T. Pfammatter, D. Schmidlin, M. I. Turina, and E. R. Schmid, "To Compare General, Epidural and Local Anaesthesia for Endovascular Aneurysm Repair (EVAR)".

[5] G. Soulez *et al.*, "Pain and Quality of Life Assessment after Endovascular Versus Open Repair of Abdominal Aortic Aneurysms in Patients at Low Risk," *Journal of Vascular and Interventional Radiology*, vol. 16, no. 8, pp. 1093–1100, Aug. 2005.

[6] M. V. Raval and M. K. Eskandari, "Outcomes of elective abdominal aortic aneurysm repair among the elderly: Endovascular versus open repair," *Surgery*, vol. 151, no. 2, pp. 245–260, Feb. 2012.

[7] D. L. Davenport, S. D. O'Keeffe, D. J. Minion, E. E. Sorial, E. D. Endean, and E. S. Xenos, "Thirty-day NSQIP database outcomes of open versus endoluminal repair of ruptured abdominal aortic aneurysms," *Journal of Vascular Surgery*, vol. 51, no. 2, pp. 305-309.e1, Feb. 2010.

[8] D. Daye and T. G. Walker, "Complications of endovascular aneurysm repair of the thoracic and abdominal aorta: evaluation and management," *Cardiovasc Diagn Ther*, vol. 8, no. Suppl 1, pp. S138–S156, Apr. 2018.

[9] H. O. Kim, N. Y. Yim, J. K. Kim, Y. J. Kang, and B. C. Lee, "Endovascular Aneurysm Repair for Abdominal Aortic Aneurysm: A Comprehensive Review," *Korean J Radiol*, vol. 20, no. 8, p. 1247, 2019.

[10] S. Islam, X. Song, E. T. Choi, J. Kim, H. Liu, and A. Kim, "In Vitro Study on Smart Stent for Autonomous Post-Endovascular Aneurysm Repair Surveillance," *IEEE Access*, vol. 8, pp. 96340–96346, 2020.

[11] V. F. Cardoso, G. Minas, C. M. Costa, C. J. Tavares, and S. Lanceros-Mendez, "Micro and nanofilms of poly(vinylidene fluoride) with controlled thickness, morphology and electroactive crystalline phase for sensor and actuator applications," *Smart Materials and Structures*, vol. 20, no. 8, p. 087002, Jul. 2011.

CONTACT

*Jungkwun 'JK' Kim, tel: +1-940-369-7027; jungkwun.kim@unt.edu

DEVELOPMENT OF AN ELECTRICAL-STIMULATION-INDUCED MECHANOMYOGRAM PROBE FOR MUSCLE CONTRACTION CHARACTERISTICS EVALUATION

Yusuke Takei[1], Toshihiro Takeshita[1], Daniel Zymelka[1], and Takeshi Kobayashi[1]
[1] National Institute of Advanced Industrial Science and Technology (AIST), JAPAN

ABSTRACT

In this study, we developed a probe to evaluate the contraction characteristics of muscles. This probe measures the mechanomyogram which is induced by the electrical stimulation. The probe consists of two electrodes for electrical muscle stimulation and an ultra-thin piezoelectric membrane for measuring mechanomyogram. The device evaluates the contraction characteristics of muscles by analyzing the speed of contraction of fast and slow twitch muscles based on the mechanomyogram response to electrical stimulation. As a demonstration, the probe system successfully evaluated how the stretch characteristics of the gastrocnemius muscle improved with the physiotherapist's treatment.

KEYWORDS

Muscle, Mechanomyogram, Lead Zirconate Titanate, Ultrathin MEMS

INTRODUCTION

As life expectancy increases worldwide, it is becoming increasingly important to age in good health. Especially among the elderly, frailty, a gradual decrease in musculoskeletal mass, is a trigger for the transition to a state of long-term care required [1]. Thus, early detection of frailty is the key to healthy aging. In recent years, the muscles of the lower limbs have become stiff and contractile, and the "change of slow-twitch muscle to fast-twitch muscle" has been attracting attention as a predictor of frailty. However, the only method for evaluating muscle contraction characteristics was based on posture information obtained when the subject voluntarily assumed a pose, such as in body forward bending measurements, which made measurement reproducibility and accuracy difficult. In this study, we focused on muscle contraction induced by external electrical stimulation and fabricated a probe that can evaluate muscle contraction characteristics from mechanomyogram induced by electrical stimulation (Fig.1).

METHOD

Figure 2 shows an overview of our fabricated probe. The device consists of two conductive rubber electrodes for electrical muscle stimulation and an acoustic sensor for mechanomyogram measurement. The acoustic sensor uses an ultrathin piezoelectric element that is transfer and mounted on a film in anticipation of future miniaturization and flexibility of the device. The fabrication process of the acoustic sensor is shown in Figure 3 [2]. Figure 4 shows the measurement of electrical stimulation-induced mechanomyogram with our probe and an example of the measured data. The procedure of mechanomyogram analyzation is shown in Figure 5. First, the electrical

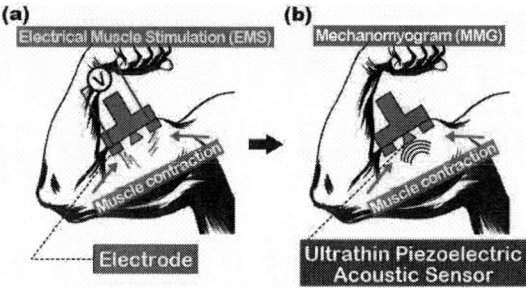

Figure 1: Concept of our research. We have developed a medical probe which can measure muscle contractility for early detection of "frailty", the key of healthy aging. (a)Muscles contracts by electrical stimulation. (b) Measure the mechanomyogram by acoustic sensor.

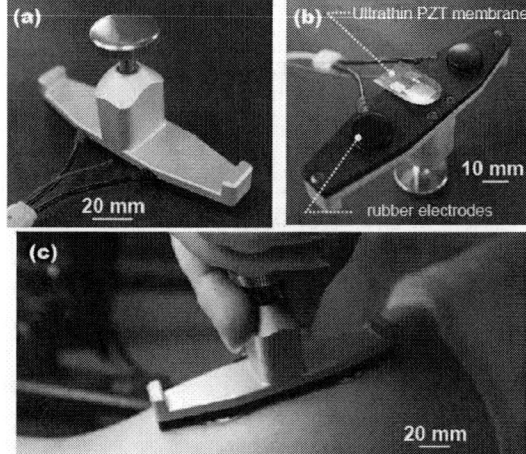

Figure 2: (a) Top view of the fabricated probe. (b) Bottom surface of the probe composed of two rubber electrodes and an acoustic sensor with an ultrathin PZT film. (c) The probe in use.

stimulation pulse and the corresponding mechanomyogram are extracted, and then noise removal and rectification processing are performed. As shown in Figure 6, the mechanomyogram waveform consists of two peaks: the peak on the left with the fastest contraction speed is the muscle sound derived from the fast twitch muscle, and the peak on the right with the relatively slow contraction speed is the muscle sound derived from the slow twitch muscle. The time from the electrical stimulus signal input of these two peaks corresponds to the muscle contraction time of the slow and fast muscles [3].

EXPERIMENT

978-1-6654-9309-3/23 $31.00 © 2023 IEEE

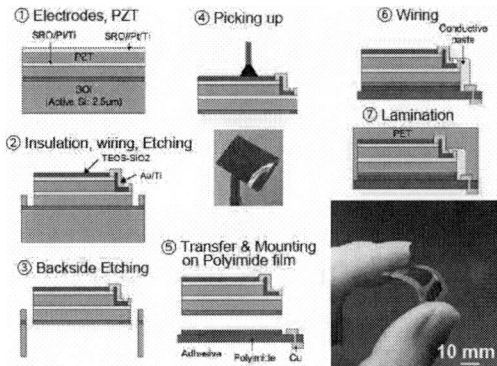

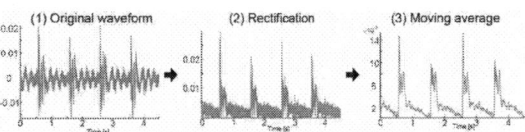

Figure 5: Processing procedure of mechanomyogram.

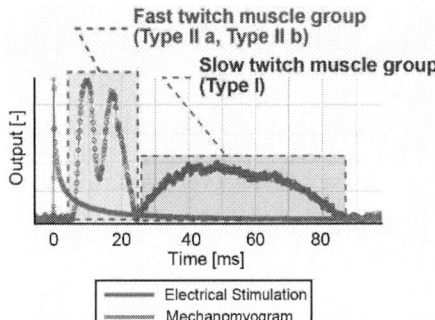

Figure 3: Fabrication process of ultrathin PZT membrane mounted on the polyimide film. The total thickness of the PZT membrane is under 10μm. So it can be bent as shown in the photograph below right.

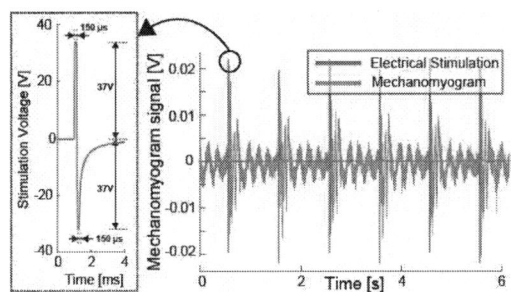

Figure 6: Evaluation of muscle contractility by mechanomyogram. Fast and slow twitch muscle groups can be observed depending on the time from the electrical stimulus input.

Figure 4: Electrical stimulation and mechanomyogram measured by the probe.

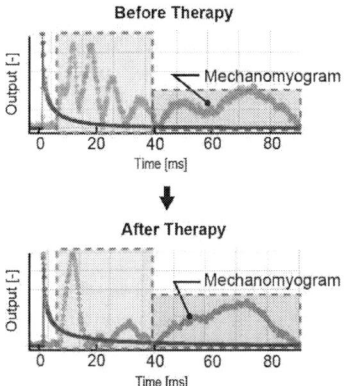

As an evaluation, we measured the effect of a physical therapist's massage on the improvement of gastrocnemius (calf muscle) elasticity to relieve ankle contractures with our fabricated probe. Figure 7 shows the measurement results. It can be seen that before the treatment, the peak of the fast twitch muscle was divided into two peaks, but immediately after the treatment, the peak of the fast twitch muscle became one. This means that the elasticity of the fast-twitch muscles was adjusted by the massage therapy, and as a result, the contraction speed of the fast-twitch muscles as a whole was aligned. In addition, 90 minutes after the end of the treatment, the peaks of the fast-twitch muscles were again divided into two, suggesting that the effect of the treatment was temporary.

Figure 7: Example of mechanomyogram evaluation on improvement of muscle contractility by physical therapy.

laminated film vibrator using an ultra-thin MEMS actuator", J. Micromech. Microeng., 32, 105001, 2022.

[3] A. Islam *et.al.,* *"Mechanomyogram for Muscle Function Assessment: A Review",* PLoS ONE, DOI: 10.1371/journal.pone.0058902, 2013.

ACKNOWLEDGEMENTS

This research is based on results obtained from a project (JPNP21004, JPNP20004) subsidized by the New Energy and Industrial Technology Development Organization (NEDO) and JSPS KAKENHI Grant Number JP 22K18421.

CONTACT

*Yusuke Takei, tel: +81-29-861-2802;
yusuke-takei@aist.go.jp

REFERENCES

[1] T. Yamada, "Mechanisms of decline in muscle quality in sarcopenia", Sarcopenia Molecular Mechanism and Treatment Strategies, pp.295-322, 2021.

[2] T. Takeshita *et.al.,* "Mechanical characteristics of

DUAL-FREQUENCY PIEZOELECTRIC MICROMACHINED ULTRASONIC TRANSDUCERS FOR FUNDAMENTAL AND HARMONIC IMAGING

Yanfen Zhai, Waleed Maqsood, Zhou Da,
Nikolai Andrianov, Yucheng Zhang, Mohssen Moridi, and Lixiang Wu*
Silicon Austria Labs GmbH, 9500 Villach, Austria

ABSTRACT

A piezoelectric micromachined ultrasonic transducer (PMUT) with single oval vibrating diaphragm and dual frequency bands is developed for fundamental and harmonic imaging. The feature of dual-frequency bands of single PMUT element is verified by ultrasonic imaging experiments. The dual-frequency PMUT doubles the fractional bandwidth and halves the device dimensions compared to state of the art, which will enable ultrasonic imaging with combined advantages of high resolution via harmonic imaging and large depth via fundamental imaging. Therefore, it holds great promise for next-generation intravascular ultrasound (IVUS).

KEYWORDS

Piezoelectric micromachined ultrasonic transducer (PMUT), dual frequency bands, harmonic imaging.

INTRODUCTION

Dual-frequency PMUTs were originally for super-harmonic imaging. [1] Recently, the dual-frequency PMUT or PMUT array was considered as a desirable option for applications such as IVUS and photoacoustic imaging. [2-4] Since two frequency bands can be utilized for fundamental and harmonic imaging, respectively, it is beneficial to combine the advantages of fundamental and harmonic images, i.e., large penetration depth and high spatial resolution. In comparison, there is always a tradeoff between the penetration depth and spatial resolution for single frequency ultrasonic transducers, which limits the clinical applications. [2]

By far, dual-frequency PMUTs have been usually implemented by either combining two ultrasonic elements with different center frequencies [3] or exploiting the first two natural vibration modes of one ultrasonic element by dual top electrodes. [1,4] For IVUS and endoscopic ultrasound with strict restriction on the form factor, the single-diaphragm dual-frequency PMUT is much preferred than the dual-frequency PMUT array with two types of diaphragms. Moreover, it also has advantages of high fill ratio and high spatial resolution.

METHOD

A new single-diaphragm dual-frequency PMUT design with much more simplified structures is proposed, which combines an oval vibration diaphragm and a single top electrode, comparing with state of the art which typically has a circular diaphragm and dual top electrodes. As shown in Figure 1, PMUTs with an oval or oval-ring top electrode were fabricated and the dimensions of a PMUT diaphragm were characterized by scanning electron microscope (SEM) with focused ion beam (FIB).

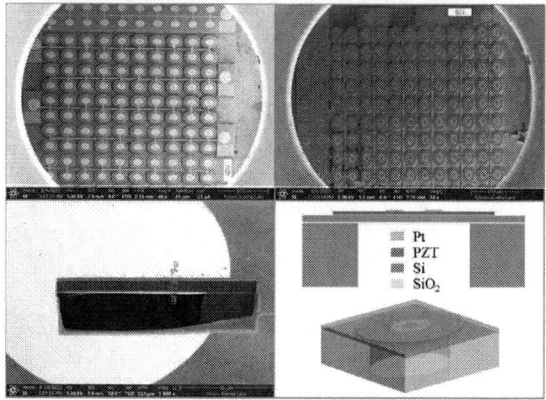

Figure 1: SEM images of (a) a PMUT sample with oval top electrode, (b) a PMUT sample with oval ring top electrode, (c) a FIB cut of PMUT diaphragm; (d) Schematic and cross-section of a PMUT with single oval ring electrode.

One sample with 70% top electrode coverage (Sample S70) and another sample with oval-ring top electrode (covering the radius from 20% to 50%, Sample S32) were selected and characterized electrically and acoustically. The electrical characterization was performed using a Keysight Impedance Analyzer E4990A with sweep frequency signal from 1MHz – 10 MHz, and an AC voltage of 500mV with 15V DC bias was applied for monitoring higher amplitudes. The acoustic characterization was conducted using the Ultrasound Measurement System (Precision Acoustic Ltd) with a 3D scanning tank. The fundamental and harmonic imaging experiment was conducted in the same 3D scanning tank with an imaging phantom (or metal pins). As shown in Figure 2a, Samples S70 and S32 were immersed in the Galden D-02 liquid, which has excellent electrical insulation properties, and measured by a needle hydrophone placed 10 mm above the PMUT sample, which was driven by a short square wave (one cycle, 8.3 MHz, 10V) generated by an ultrasound pulser (STMicroelectronics STEVAL IME-009V1).

In addition, the beam profile measurement of the linear array 10x2 with an aperture of 1.8mmx0.3mm for each sample was performed by translating the hydrophone location and mapping the pressure field, respectively. A driven signal, consisting of 1 cycle of 8.3 MHz square wave with a peak-to-peak amplitude of 20V, was generated by a STMicroelectronics STEVAL IME-009V1 Ultrasound Pulser. The pre-amplifier output waveform was averaged 32 times by the oscilloscope to reduce the noise presented on the signal. The maximum peak-to-peak amplitude of the hydrophone pre-amplifier output for all samples is used to derive the output pressure at hydrophone tip by utilising the given sensitivity of hydrophone and amplifier gain(54dB) of Sonotec variable amplifier. As shown in Figure 2c, the -6 dB focal spot was measured to be around 4 mm by 1.2 mm.

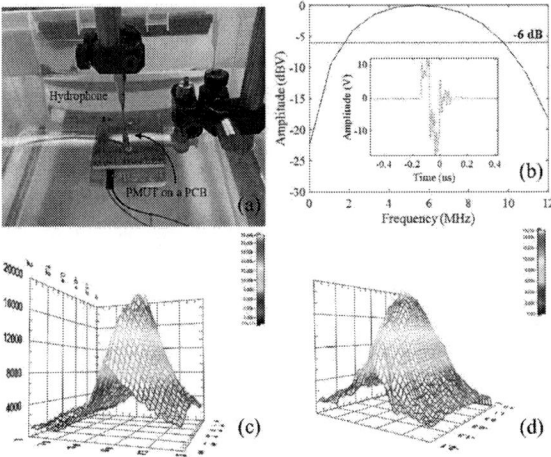

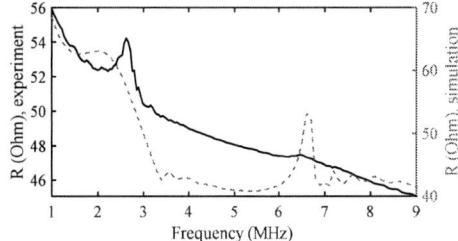

Figure 2: (a) PMUT characterization by a needle hydrophone in a 3D scanning tank, (b) the pulse signal for driving PMUTs, beam profiles of Sample S70 (e) and Sample S32 (f).

The data for image reconstruction comes from M-scan, which is to scan the object by moving the transducer in one direction with one scan per movement. As shown in Figure 5a, the scanning object is an array of metal pins, which was immersed in the Galden D-02 liquid along with the transducer. The metal pins were placed in a small angle to the liquid surface for object recognition. The scanning step is set to be 0.1 mm and 100 steps in total, and the echoes of 100 A-scans were collected to form a B-scan image. Due to the small size of the transducer, the aperture of each scan was limited, which further restricted the lateral resolution. To overcome this disadvantage, the image reconstruction is aided with Synthetic Aperture algorithm to increase the lateral resolution. The basic concept of this method is to sum up the signals that reached the same pixel from each scan, thus creating a larger equivalent aperture.

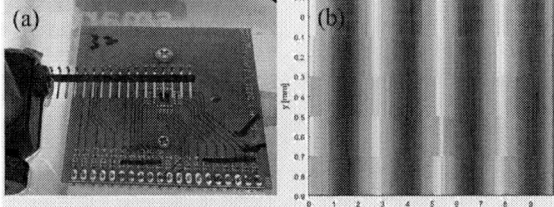

Figure 3: (a) The ultrasonic imaging experiment setup and (b) a C-Scan image of metal pins.

As shown in Figure 3b, A C-scan was also conducted to validate the performance of the transducer. The scan movement was pointed in another direction for a sweep distance of 0.1 mm, which is perpendicular to the A-scan direction, then another 100-step A-scan is performed. 10 A-scans were carried out, which resulted in 1000 scanning steps in total. To generate a C-scan image, Hilbert-transformation was conducted for each A-scan time series. The largest absolute value of each A-scan array represents the intensity of the current pixel. A top view of the metal pins was expected to be observed in a rectangular shape.

RESULTS AND DISCUSSION

Figure 4 shows the impedance comparison between

experimental results and FEM simulation results of Sample S70. The first and second resonance frequencies are 2.6MHz and 6.5 MHz, respectively.

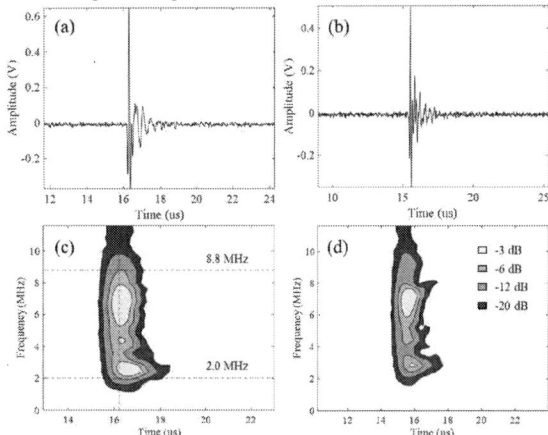

Figure 4: Comparison between the impedance measurement result (left) and the FEM simulation result (right).

Figure 5a/c shows the echo signal received by the needle hydrophone and its short-time Fourier Transform of Sample 70. The -6 dB bandwidth is 125.9% (i.e., 2.0-8.8 MHz) and the peak amplitude is 647.2 mV (after 54 dB gain). Figure 5b/d shows that the high- and low-frequency bands of Sample 32 are not connected at the -6dB level. The bandwidths are 40% (2-3 MHz) and 67% (4-8 MHz) and both peak amplitudes are 504.5 mV.

Figure 5: The ultrasonic signals transmitted by Sample S70 (a) and Sample S32 (b), and their corresponding short-time Fourier transforms (c/d).

The imaging reconstruction results are shown in Figure 6. For Sample S32, both the high-frequency (5-12 MHz) and low-frequency (0-3 MHz) B-scan images display the structures of metal pins and liquid surface with good resolution, as shown in Figure 6a and 6b. However, Figure 6c and 6d (Sample S70) show that the low-frequency image is heavily blurred while the high-frequency image has a good resolution. This indicates that Sample S32 can acquire the high- and low-frequency ultrasonic waves within its dual frequency bands simultaneously and the crosstalk between the high- and low-frequency components is very low even though they occur in the same PMUT diaphragm. In comparison, for Sample S70, the high-frequency component dominates the received echo signal, which indicates that the two-frequency bands are not balanced properly in the receive mode. In addition, the comparison between the two samples suggests that dual frequency bands of a single PMUT element can be finely tuned by optimizing the top electrode geometry.

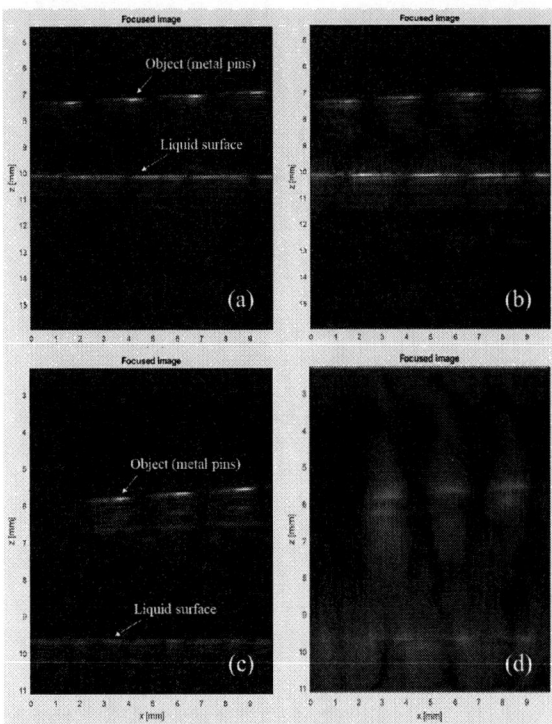

Figure 6: High-frequency (5-12MHz) and low-frequency (0-3MHz) B-Scan images with Sample 32 (a, b) and Sample 70 (c, d), C-Scan image I, and the imaging experiment setup (f).

CONCLUSION

Single-diaphragm dual-frequency PMUTs have been developed to leverage the combined advantages of fundamental and harmonic imaging. The ultrasonic imaging results of two PMUT samples suggest that the fundamental and harmonic imaging can be conducted with a single PMUT element simultaneously, which is promising for next-generation IVUS applications.

ACKNOWLEDGEMENTS

We thank Javad Abbaszadeh for designing PCBs and help on vibration measurements.

REFERENCES

[1] L. Wu, X. Chen, G. Wang, and Q. Zhou, "Dual-frequency piezoelectric micromachined ultrasonic transducers," *Applied Physics Letters*, vol. 115, no. 2, p. 023501, 2019.

[2] C. Peng, H. Wu, S. Kim, X. Dai, and X. Jiang, "Recent advances in transducers for intravascular ultrasound (IVUS) imaging," Sensors, vol. 21, no. 10, p. 3540, 2021

[3] H. Wang, H. Yang, Z. Chen, Q. Zheng, H. Jiang, P. X.-L. Feng, and H. Xie, "Development of dual-frequency PMUT arrays based on thin ceramic PZT for endoscopic photoacoustic imaging," Journal of Microelectromechanical Systems, vol. 30, no. 5, pp. 770–782, 2021.

[4] J. Cai, Y. Wang, D. Jiang, S. Zhang, Y. A. Gu, L. Lou, F. Gao, and T. Wu, "Beyond fundamental resonance mode: high-order multi-band AlN PMUT for in vivo photoacoustic imaging," Microsystems & Nanoengineering, vol. 8, no. 1, pp. 1–12, 2022.

[5] L. Wu, M. Moridi, G. Wang, and Q. Zhou, "Microfabrication and characterization of dual-frequency piezoelectric micromachined ultrasonic transducers," in *2021 IEEE International Symposium on Applications of Ferroelectrics (ISAF)*. IEEE, 2021, pp. 1–4.

CONTACT

*Lixiang Wu, Tel.: +43 664 88 39 06 10;
Email: Lixiang.Wu@silicon-austria.com

FRACTAL MICROELECTRODES INTEGRATED WITH THE CATHETER FOR LOW-VOLTAGE PULSED FIELD ABLATION

Mengfei Xu[1,2], Mu Qin[3], Ziliang Song[4], Wen Hong[1,2], Qingda Xu[1,2], Jiawei Cao[1,2], Kejun Tu[1,2], Longchun Wang[1,2], Bin Yang[1], and Jingquan Liu[1]

[1]National Key Laboratory of Science and Technology on Micro/Nano Fabrication,
Shanghai Jiao Tong University, CHINA

[2]DCI Joint Team, Collaborative Innovation Center of IFSA, Department of Micro/Nano Electronics,
Shanghai Jiao Tong University, CHINA

[3]Department of Cardiology, Shanghai Chest Hospital Affiliated to Shanghai Jiao Tong University,
CHINA and

[4]Shanghai General Hospital Shanghai Jiao Tong University School of Medicine, CHINA

ABSTRACT

This paper reports a novel flexible fractal microelectrode integrated with the catheter to achieve diagnosis and treatment for low-voltage pulsed field ablation (PFA). The fractal design increases the ratio of the edge length to the geometric surface area of the electrode, which contributes to the transfer of current from the electrode into the tissue. Besides, the fractal structure reduces the spacing between the electrodes, which can obtain a large electric field. Pt-Black coating enhances the electrical stimulation and electrocardiogram (ECG) recording capability. Foremost, it achieves *in vivo* ablation at 100 V on rat hearts. In short, the device can perform ablation procedures independently, including low-voltage PFA and ECG recording.

KEYWORDS

Atrial Fibrillation, Catheter, Fractal Microelectrode, Pulsed Field Ablation

INTRODUCTION

Atrial fibrillation is one of the most serious diseases with a prevalence of approximately 2% [1]. PFA has emerged as a new ablation modality for atrial fibrillation due to its tissue specificity and few complications. Several types of electrodes have been developed to promote PFA towards clinical application, which can be classified into two categories, namely, rigid electrodes such as Farawave [2], PVAC-GOLD catheter [3], and flexible electrodes such as soft electronic arrays [4]. Rigid electrodes have large electrode spacing and therefore require voltages with thousands of volts. Flexible electrodes are usually manufactured by microfabrication processes and have small electrode spacing, avoiding troubles caused by high voltage [5]. In the last few years, balloon catheters equipped with flexible electronics have been widely used in ablation therapy such as radiofrequency ablation [6, 7], cryoablation [8, 9], and PFA [5, 10]. These devices integrated with sensors and actuators promote the safety and efficiency of ablation procedures. However, flexible electrodes are difficult to manipulate to perform ablation at the target location. A microelectrode integrated with the catheter is necessary to ablate with high position accuracy at low voltage.

In this work, flexible microelectrodes with fractal design are proposed to be integrated with a catheter. The novel fractal design can reduce the electrode spacing, resulting in a large electric field at the same voltage. Fractal microelectrodes and the catheter are integrated to

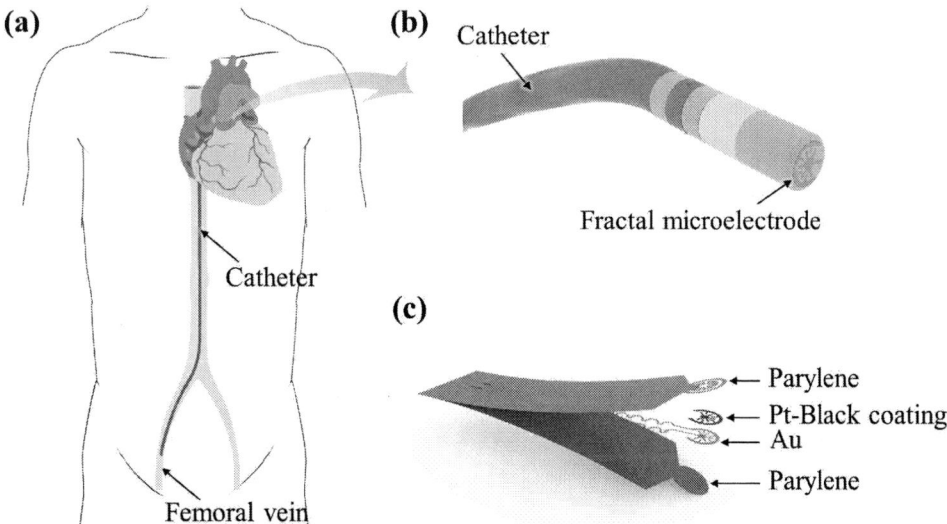

Figure 1: (a) Concept of applying (b) fractal microelectrodes integrated with the catheter for low-voltage PFA via the minimally invasive intervention. (c) Explosion schematic illustration of fractal microelectrodes.

form a smart catheter, which can record ECG. The Pt-Black coating on the electrode enhances the signal-to-noise ratio (SNR) of ECG. Also, potato ablation experiments and *in vivo* experiments confirm its ability of low-voltage PFA with location accuracy.

DESIGN AND FABRICATION

When atrial fibrillation occurs, the integration of fractal microelectrodes and the catheter allows for targeted ablation of the cardiac tissue (Figure 1a). This method has the advantages of accuracy and low damage and has the potential for minimally invasive interventional procedures. Figure 1b shows the design of the device. The electrode is integrated on the head of the catheter. As shown in Figure 1c, the electrode contains a 4-layer structure, namely a substrate layer (Parylene-C), an electrode layer (Au), an encapsulation layer (Parylene-C), and a modification layer (Pt-Black). A fractal design is innovatively introduced for the microelectrode to reduce the electrode spacing, which can obtain a large electric field. In addition, the fractal design contributes to the transfer of current from electrodes to tissues.

The microelectrodes were manufactured by conventional microfabrication processes. A 4-inch silicon wafer was selected as the substrate on which a 5 μm Parylene-C was deposited by chemical vapor deposition.

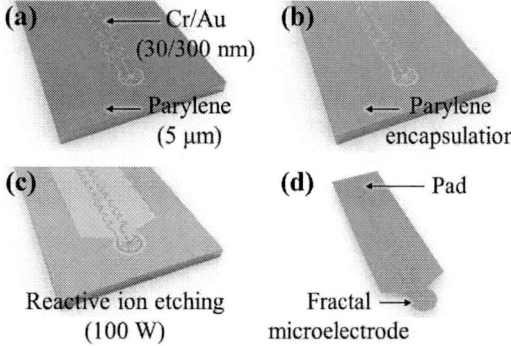

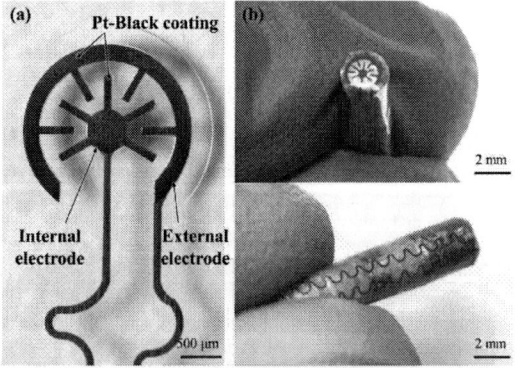

Figure 2: The microfabrication procedure. (a) Deposition of the bottom layer Parylene-C and patterning of the metal layer. (b) Deposition of the top layer Parylene-C. (c) Exposure of electrodes, pads, and the device outline. (d) Releasing electrodes.

Figure 3: Photographs of (a) fractal microelectrodes and (b) fractal microelectrodes integrated with the catheter.

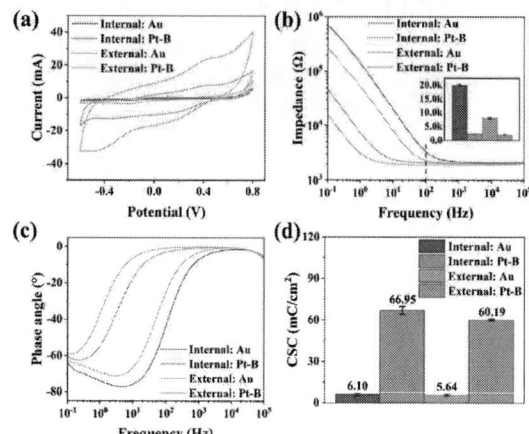

Figure 4: The electrochemical characterization of bare Au and Pt-Black (Pt-B) coated electrodes. (a) CV, (b, c) EIS and (d) CSC diagrams of internal and external fractal electrodes (n=3). The inset of (b) shows the impedance at 100 Hz.

Then, a layer of Cr/Au was sputtered as a metal layer. It is worth mentioning that pretreating Parylene-C by reactive ion etching at 100 W for 2 min is necessary to enhance the adhesion force between Parylene-C and the metal layer. Then, the metal layer was patterned by photolithography and wet etching to define the shape of microelectrodes (Figure 2a). Next, the second layer of Parylene-C (5 μm) was deposited as an encapsulation layer (Figure 2b). A 15 μm photoresist was patterned by photolithography and used as a mask to remove Parylene-C on electrodes, pads, and surrounding area by reactive ion etching (Figure 2c). Finally, the wafer was immersed in deionized water to release the microelectrodes from the substrate (Figure 2d). The electrochemical modification of Pt-Black and measurements were described in the previous work [11]. With the help of the anisotropic conductive film and hot-pressing process, the electrical connection between the pads and the enameled wire was achieved. Integration of fractal microelectrodes and the catheter was completed using the biocompatible silica gel and Kapton tape.

RESULTS AND DISCUSSION

Fractal microelectrodes include the external electrode and internal electrode with fractal design (Figure 3a). The fractal structure reduces the electrode spacing and obtains a large electric field, which is conducive to reaching the electric field threshold for the irreversible electroporation

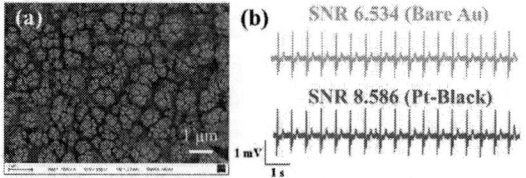

Figure 5: Enhanced ECG recording capability with Pt-Black. (a) The SEM image of Pt-Black coating on fractal microelectrodes. (b) ECG recorded by bare Au and Pt-Black coated electrodes.

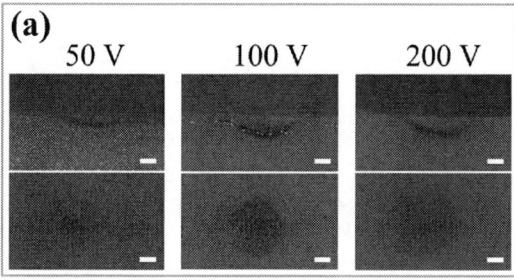

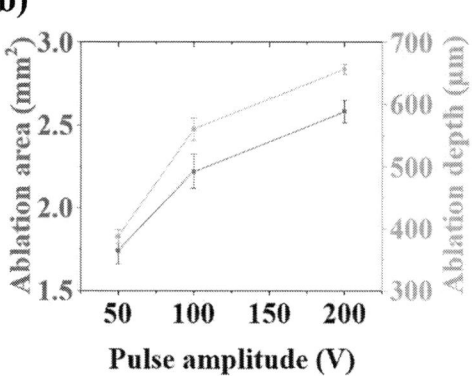

Figure 6: Ablation experiment on potato tissues. (a) Cross-section and surface photographs of potato tissues after PFA at different pulse amplitudes. Scale bar, 500 μm. (b) Ablation area and depth at different pulse amplitudes (n=3).

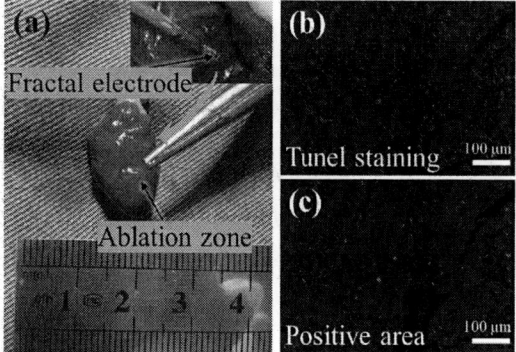

Figure 7: In vivo ablation experiment on a rat. (a) Fractal electrodes integrated on the catheter for epicardial ablation at 100 V. (b) Fluorescent photograph of the ablation area. (c) Almost all cardiomyocytes undergo apoptosis (green).

of cells. Microelectrodes are modified with Pt-Black to improve the electrochemical properties and charge storage capacity [12]. In addition, microelectrodes with a diameter of 2 mm allow for *in vivo* point-by-point ablation with high resolution, compared to commercial implantable electrodes (diameter of 31 mm) [13]. Figure 3b shows photographs of the fractal microelectrode integrated with a catheter, where the flexibility of microelectrodes ensures a tight fit between them.

Pt-Black is electroplated on microelectrodes for better electrochemical properties. The typical cyclic voltammetry (CV), electrochemical impedance spectra (EIS), and charge storage capacity (CSC) are shown in Figure 4. Obviously, Pt-Black significantly reduces impedance at the low-frequency range (0.1-1000 Hz), which can improve the SNR of ECG. The impedance for the internal and external electrodes reduces by an order of magnitude at 100 Hz. Also, Pt-Black coated microelectrodes exhibit a larger CSC than bare Au. In a word, Pt-Black coating facilitates electrical stimulation and recording during the heart surgery.

Accurate mapping of electrical activity helps focus the targeted treatment area and confirm the treatment effect before and after cardiac ablation procedures. The Pt-Black coating on microelectrodes has a uniformly distributed cauliflower-like microstructure (Figure 5a), which helps to improve the signal quality. Figure 5b illustrates the ECG signal recorded on the skin of a human chest by the smart catheter with fractal microelectrodes. The SNR is enhanced from 6.534 to 8.586 after modification of Pt-Black (SNR is calculated from the average peak of the signal and the root mean square of the noise region). By recording high-quality ECG signals, abnormal signatures of cardiac arrhythmia can be easily captured by the smart catheter.

After the targeted treatment area is identified, PFA is performed by fractal microelectrodes with a series of electrical pulses. The duration of each pulse is 100 μs, avoiding a large amount of Joule heat generation. During the ablation process, high-intensity electrical pulses are applied to targeted tissues and cell apoptosis occurs due to irreversible electroporation. For potato tissues, cells release the phenoloxidase enzyme after the cell membrane is disrupted, which promotes the oxidation of phenolic compounds to yield dark regions. Electrical pulses (number of 90, amplitude of 50 V-200 V, duration of 100 μs, frequency of 1 Hz) are applied to potato tissues, and the ablation area and maximum depth are measured after 24 hours (Figure 6a). With the increase of pulse amplitude, the ablation area and depth gradually increase, and the maximum values are 2.6 mm^2 and 660 μm at 200 V, respectively (Figure 6b). Figure 7a demonstrates the *in vivo* ablation experiment of the smart catheter on a rat heart and apoptotic cells in the ablation zone prove the success

Table 1: Comparison with previous electrodes for PFA.

References	Flexibility	Voltage	PFA
[2]	Ordinary	2000 V	*In vivo*/Porcine
[3]	Ordinary	1500 V	*In vivo*/Porcine
[14]	No	200 V	Potato tissue
[4]	Good	200 V	*In vitro*/Rabbit
This work	Excellent	100 V	*In vivo*/Rice

of low-voltage ablation (Figure 7b and Figure 7c). Table 1 shows the comparison with previous electrodes and this work demonstrates excellent performance.

CONCLUSION

In this paper, we present a fractal microelectrode and its integration with the catheter for PFA. Fractal design enhances electrode performance in terms of both electric field and current density. In addition, the device ensures ECG recording and low-voltage ablation on biological tissues. The Pt-Black coating is beneficial to improve the electrochemical properties of the electrode, thus facilitating ECG recording and ablation. Both potato experiments and *in vivo* experiments confirm the ability of low-voltage PFA. This work opens up new prospects for research on the novel smart catheter.

ACKNOWLEDGEMENTS

This work was supported by the National Key R&D Program of China under grant (2022ZD0208601), the Strategic Priority Research Program of Chinese Academy of Sciences (Grant No. XDA25040100, XDA25040200 and XDA25040300), the National Natural Science Foundation of China (No.42127807-03), Project supported by Shanghai Municipal Science and Technology Major Project (2021SHZDZX), SJTU Trans-med Award (No.2019015, 21X010301627). The authors are grateful to the Center for Advanced Electronic Materials and Devices (AEMD) of Shanghai Jiao Tong University.

REFERENCES

[1] A. Wojtaszczyk, G. Caluori, M. Pesl, K. Melajova, and Z. Starek, "Irreversible electroporation ablation for atrial fibrillation," *J Cardiovasc Electrophysiol.*, vol. 29, no. 4, pp. 643-651, Apr, 2018.

[2] J. S. Koruth, K. Kuroki, I. Kawamura, R. Brose, R. Viswanathan, E. D. Buck, E. Donskoy, P. Neuzil, S. R. Dukkipati, and V. Y. Reddy, "Pulsed Field Ablation Versus Radiofrequency Ablation: Esophageal Injury in a Novel Porcine Model," *Circ Arrhythm Electrophysiol.*, vol. 13, no. 3, pp. e008303, Mar, 2020.

[3] M. T. Stewart, D. E. Haines, D. Miklavcic, B. Kos, N. Kirchhof, N. Barka, L. Mattison, M. Martien, B. Onal, B. Howard, and A. Verma, "Safety and chronic lesion characterization of pulsed field ablation in a Porcine model," *J Cardiovasc Electrophysiol.*, vol. 32, no. 4, pp. 958-969, Apr, 2021.

[4] M. Han, L. Chen, K. Aras, C. Liang, X. Chen, H. Zhao, K. Li, N. R. Faye, B. Sun, J. H. Kim, W. Bai, Q. Yang, Y. Ma, W. Lu, E. Song, J. M. Baek, Y. Lee, C. Liu, J. B. Model, G. Yang, R. Ghaffari, Y. Huang, I. R. Efimov, and J. A. Rogers, "Catheter-integrated soft multilayer electronic arrays for multiplexed sensing and actuation during cardiac surgery," *Nat Biomed Eng.*, vol. 4, no. 10, pp. 997-1009, Oct, 2020.

[5] M. F. Xu, W. Hong, M. Qin, Z. L. Song, Y. Z. Shi, B. Yang, and J. Q. Liu, "Low-Voltage Flexible Interdigital Electrode for Pulsed Field Ablation with Effect Evaluation," *2022 IEEE 35th International Conference on Micro Electro Mechanical Systems (MEMS)*, pp. 106-109, 2022.

[6] D. H. Kim, N. Lu, R. Ghaffari, Y. S. Kim, S. P. Lee, L. Xu, J. Wu, R. H. Kim, J. Song, Z. Liu, J. Viventi, B. de Graff, B. Elolampi, M. Mansour, M. J. Slepian, S. Hwang, J. D. Moss, S. M. Won, Y. Huang, B. Litt, and J. A. Rogers, "Materials for multifunctional balloon catheters with capabilities in cardiac electrophysiological mapping and ablation therapy," *Nat Mater.*, vol. 10, no. 4, pp. 316-23, Apr, 2011.

[7] L. Klinker, S. Lee, J. Work, J. Wright, Y. Ma, L. Ptaszek, R. C. Webb, C. Liu, N. Sheth, M. Mansour, J. A. Rogers, Y. Huang, H. Chen, and R. Ghaffari, "Balloon catheters with integrated stretchable electronics for electrical stimulation, ablation and blood flow monitoring," *Extreme Mech Lett.*, vol. 3, pp. 45-54, 2015.

[8] W. Hong, M. Xu, Z. Guo, L. Wang, M. Qin, and J. Liu, "Multi-functional sensor array on the cryoablation balloon for atrial fibrillation," *Sens Actuators A Phys.*, vol. 341, 2022.

[9] Y. J. Zhan, W. Hong, W. X. Sun, and J. Q. Liu, "Flexible Multi-Positional Microsensors for Cryoablation Temperature Monitoring," *IEEE Electron Device Lett.*, vol. 40, no. 10, pp. 1674-1677, Oct, 2019.

[10] M. Xu, W. Hong, Z. Song, Z. Guo, L. Wang, M. You, Z. Lin, B. Yang, M. Qin, and J. Liu, "Multifunctional Microelectrode for Low-Voltage Pulsed Field Ablation in Atrial Fibrillation," *IEEE Electron Device Lett.*, vol. 43, no. 11, pp. 1965-1968, 2022.

[11] L. Wang, C. Ge, F. Wang, Z. Guo, W. Hong, C. Jiang, B. Ji, M. Wang, C. Li, B. Sun, and J. Liu, "Dense Packed Drivable Optrode Array for Precise Optical Stimulation and Neural Recording in Multiple-Brain Regions," *ACS Sens.*, vol. 6, no. 11, pp. 4126-4135, Nov 26, 2021.

[12] S. H. Sunwoo, S. I. Han, H. Kang, Y. S. Cho, D. Jung, C. Lim, C. Lim, M. j. Cha, S. P. Lee, T. Hyeon, and D. H. Kim, "Stretchable Low-Impedance Nanocomposite Comprised of Ag–Au Core–Shell Nanowires and Pt Black for Epicardial Recording and Stimulation," *Adv Mater Technol.*, vol. 5, no. 3, pp. 1900768, 2019.

[13] V. Y. Reddy, P. Neuzil, J. S. Koruth, J. Petru, M. Funosako, H. Cochet, L. Sediva, M. Chovanec, S. R. Dukkipati, and P. Jais, "Pulsed Field Ablation for Pulmonary Vein Isolation in Atrial Fibrillation," *J Am Coll Cardiol.*, vol. 74, no. 3, pp. 315-326, Jul 23, 2019.

[14] T. K. Nguyen, S. Yadav, T. A. Truong, M. Han, M. Barton, M. Leitch, P. Guzman, T. Dinh, A. Ashok, H. Vu, V. Dau, D. Haasmann, L. Chen, Y. Park, T. N. Do, Y. Yamauchi, J. A. Rogers, N. T. Nguyen, and H. P. Phan, "Integrated, Transparent Silicon Carbide Electronics and Sensors for Radio Frequency Biomedical Therapy," *ACS Nano*, vol. 16, no. 7, pp. 10890-10903, Jul 11, 2022.

CONTACT

*J.Q. Liu, tel: +86-21-34207209; jqliu@sjtu.edu.cn

HIERARCHICAL BONDING YIELD TEST STRUCTURE FOR FLEXIBLE HIGH CHANNEL-COUNT NEURAL PROBES INTERFACING ASIC CHIPS

Marie C. Odenthal[1,+], Victor Claar[1,+], Oliver Paul[1,2], and Patrick Ruther[1,2,]*

[1] Department of Microsystems Engineering (IMTEK), University of Freiburg, Germany and
[2] BrainLinks-BrainTools Center, University of Freiburg, Germany
[+] equal contributions

ABSTRACT

This study presents novel, hierarchical bonding yield test structures designed to establish and validate a high-density interfacing process between CMOS ASIC chips and highly flexible neural probes made of polyimide. The efficient test procedure allows to identify open circuits within the $n \times n$ bonding pad array in order to locate electrical defects between contact pairs using a minimal number of electrical measurements. Applying flip-chip bonding to interface PI-based test structures with electroplated pads and silicon dummy chips mimicking neural probes and ASIC chips, respectively, a bonding yield of 97.2 % has been achieved.

KEYWORDS

Hybrid integration, flexible neural probes, flip-chip bonding, process development

INTRODUCTION

Long-term stable neural probes interfacing brain tissue are a prerequisite for basic neuroscientific research [1, 2] and its clinical translation into a variety of neuroprostheses (Fig. 1) [3, 4]. State-of-the-art microelectromechanical system (MEMS) technologies using silicon or polymeric substrates enable a distinct miniaturization of these implants at an improved channel count. By implementing them in complementary metal-oxide-semiconductor (CMOS) technology [5], implants have been realized that provide highest channel counts at minimal electrode pitch along slender probe shanks with widths of 100 μm and below [6–8]. Due to the distinct mechanical mismatch in the elastic moduli of brain tissue and the silicon probe substrate, these tools may cause increased tissue trauma in

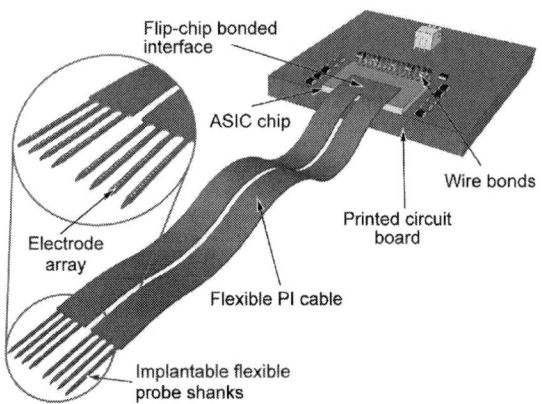

Figure 1: Schematic of a high-channel-count flexible neural probe interfacing an ASIC chip applied as neuroprosthetic device to record electrophysiological activity and electrically stimulate brain tissue.

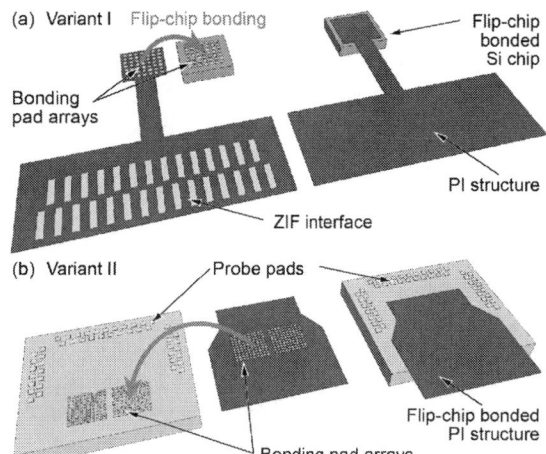

Figure 2: Bond yield test structures of Variant I and II based on dummy Si chips and PI structures mimicking the CMOS ASIC and the high channel-count flexible neural probe, respectively. Schematics are not drawn to scale; interconnecting lines are omitted.

chronic in-vivo applications and, in consequence, show a reduced long-term recording stability [9]. In contrast, neural probes realized using highly flexible polymeric substrates such as polyimide, SU-8 and parylene C are known to cause minimal tissue damage at improved long-term recording stability [10, 11]. Due to the fact that integrated circuitry at minimal line width is not available to be implemented in polymeric substrates, the overall channel count is in general limited by the achievable line width along the probe shanks. The growing demand however for an increased number of electrical recording and/or stimulation channels applicable with these flexible implants asks for the hybrid integration of neural probes and application-specific integrated circuits (ASICs) [12]. In this context, the overall channel count and the request for minimal chip dimensions asks for high-density interfaces between ASIC and the flexible neural probes.

In order to establish and validate respective assembly processes, dedicated test structures are required using cost-effective dummy structures designed to mimic highly flexible neural probes, as described e.g. in [11, 13], and the CMOS-based ASIC chips [12]. Furthermore, these test structures preferably allow determining the complete interface functionality applying a single electrical measurement.

HIERARCHICAL TEST STRUCTURE CONCEPT AND DESIGN

As schematically illustrated in Fig. 2, the bonding yield test structure concept applies silicon (Si)-based

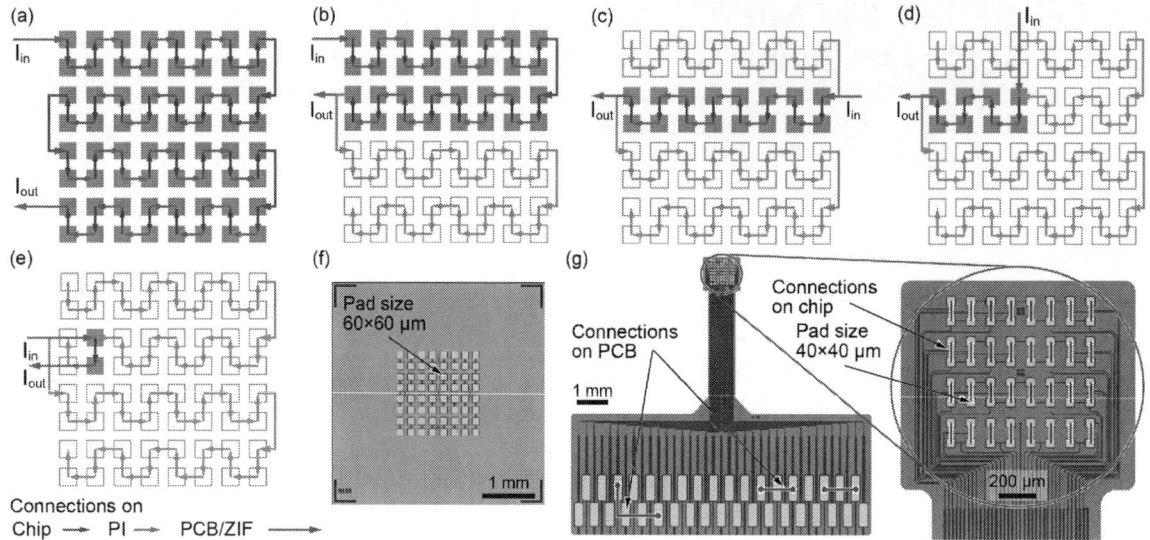

Figure 3: Extraction of bonding yield by measuring daisy chain integrity: (a) Entire array, (b-e) blocks of 4×8, 2×8, 2×4 and 2×1 bonded pads (grey squares), (f,g) mask layouts of Variant I of the (f) compact Si dummy chip and (g) PI structure with a 37-channel ZIF interface. Arrow colors indicate the shorting lines on either Si chip (blue) or PI cable (orange); green lines represent connections on a test board to the external instrumentation.

dummy chips mimicking the CMOS ASIC and highly flexible polyimide (PI) interfaces. The two components of these test structures comprise arrays of 8×8 pairwise shorted bonding pads made of aluminum (Al) and electroplated gold (Au) on the Si chip and the PI structure, respectively, with a pitch of 50, 100, 150 and 200 μm. The test structures are available in two design variants differing mainly in the type of electrical interface to the external electrical instrumentation. In case of Variant I (Fig. 2(a)), the highly compact Si chip slightly extends beyond the pad array and is bonded to the PI structure comprising respective contact pads to connect to an external instrumentation via a zero insertion force (ZIF) connector. In contrast, Variant II transfers the design complexity of the PI structure to the Si chip which is equipped with test pads along its perimeter that are accessible via prober needles or a probe card (Fig. 2(b)).

The overall test structure concept is further detailed in Fig. 3. Here, differently colored arrows indicate the current flow between the electrically shorted pairs of pads positioned on the Si chip and PI component. The paired bonding pads are arranged as a daisy-chained line of interconnections between the dummy chip and PI structure. The respective line of contacts is electrically accessible either as a whole or block-wise, i.e. through two, four, eight and 32 blocks of 4×8, 2×8, 2×4, and 2×1 bonding pads, respectively, as exemplarily illustrated in Figs. 3(a-e), with green arrows representing feed lines from the larger contacts, i.e. ZIF pads or probe pads to the blocks of bonded pads. Using the ZIF contact pads of Variant I {cf. mask layouts of dummy chip and PI structure with a pad pitch of 100 μm shown in Fig. 3(g)} and the probe pads of Variant II, the bonding yield can be broken down to individual pairs of pads between the Si-chip and PI structure by testing the entire array or different bonding pad blocks for open circuits.

TEST STRUCTURE FABRICATION

The test structures are realized using in-house MEMS processes. In case of the Si-based test chips, a 1.5-μm-thick insulation layer is realized using plasma-enhanced chemical vapor deposition (PECVD) of silicon nitride (Si_xN_y) and silicon oxide (SiO_x). The aluminum (Al)-based pads and metal tracks (thickness 500 nm; 30 nm titanium adhesion promoter) are deposited by evaporation and patterned using a lift-off process applying an image reversal photo resist. The metal structures are passivated by a second 1.5-μm-thick PECVD layer stack of Si_xN_y and SiO_x. The upper passivation layer stack is patterned using reactive ion etching (RIE) before the wafer is diced into chips of 2×2 mm². In case of the PI-based test structure components, MEMS processes similar to those applied to fabricate flexible neural probes are used [11, 13]. In short, polyimide is spin-coated onto 4-inch Si wafers to a thickness of 5 μm. Next, platinum is sputter deposited (thickness 250 μm) and patterned by lift-off. Subsequently, a second 5-μm-thick PI layer is spin-coated and patterned by RIE using a positive resist as masking layer. Finally, bonding pads and ZIF contact pads, as applied in case of Variant I, are thickened by gold (Au) electroplating before the PI structures are peeled off the fabrication wafer. We either deposited a gold thickness of ca. 3 μm partially filling the bonding pad recesses or overgrew the PI surface by 2 to 3 μm of Au which resulted in a Au thickness of 7 to 8 μm. While the thicker electroplated pads are bonded directly onto the Si dummy chips with the Al pads left blank, Au stud bumps are applied as an alternative on a subset of dummy chips. These stud bumps are realized by applying a conventional wire bonder (F&S Bondtec Semiconductor GmbH, Braunau, Austria) using a 25-μm-diameter gold wire. These stud bumps, shown as prepared in Fig. 4(a), are further flattened by compression using a flip-chip bonder (Finetech GmbH, Fineplacer®Sigma, Berlin, Germany) in combination with a bare silicon chip.

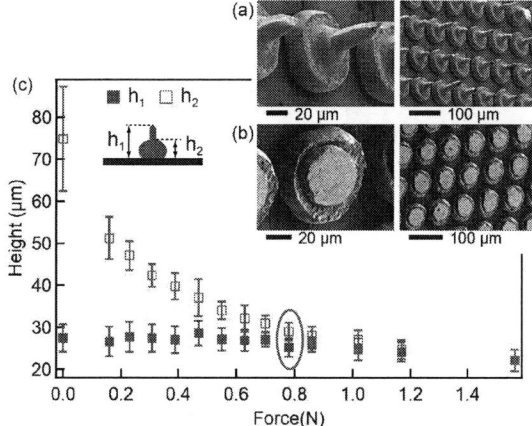

Figure 4: (a) Stud bump heights h_1 and h_2 vs. compression force. Stud bumps planarized at 0.78 N are applied for bonding tests. (b,c) Representative stud bumps (b) as deposited and (c) compressed at 1 N per bump.

A representative example of a flattened stud bump achieved after applying a vertical compression force of 1 N is shown in Fig. 4(b). The achievable stud bump height as a function of the applied force is shown in Fig. 4(c), with the respective heights h_1 and h_2 as defined in the inset.

EXPERIMENTAL RESULTS

Figure 5 shows flip-chip-bonded dummy chips and PI structures of Type I with a pad pitch of 100 μm fabricated in-house. The 10-μm-thin PI structures are equipped with electroplated Au bonding pads that were bonded either directly onto the Al pads or to stud bumps that were planarized for these bonding tests at 0.78 N/bump. The integrity of the daisy chain of bonded pads was evaluated using a multimeter (Keithley 2700) in combination with a multiplexing card (Keithley 7702) controlled by a dedicated Python program. Starting with the entire daisy-chain, smaller blocks are addressed in case that the entire array or a larger contact block shows an open circuit. Applying a bonding force of 100 N and an ultrasonic power of 1 W, bond yields of 93.7% ($n = 20$) and 97.2% (n = 70) for chips with and without stud bumps (planarized at a compression force 0.78 N/bump), respectively, were achieved.

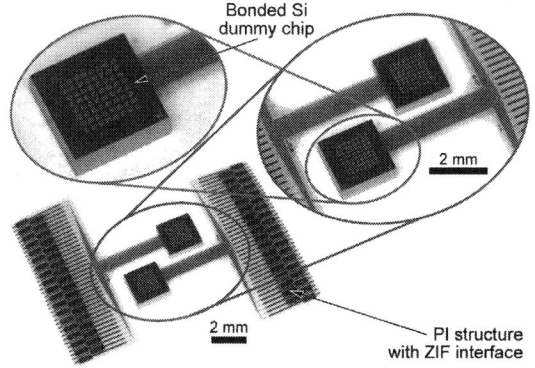

Figure 5: Optical micrographs of flip-chip bonded test structures of Type I (bonding pad pitch 100 μm).

CONCLUSION

As reported here, the innovative test structure approach for a 8×8 bonding pad array implementing a hierarchical interfacing scheme is characterized by a compact design with only 37 contact pads to the external instrumentation. It applies Si-based dummy chips and polymeric test structures mimicking a flexible neural probe made from PI. The overall concept allows to electrically characterize the flip-chip bonding process for open circuits between the dummy chips and PI test structures. The electrical interface to an external instrumentation enables to identify potential open circuits with pairs of bonding pads, as detailed by the set of differently sized bonding pad blocks. The process validation and establishment either applied electroplated Au pads in combination with bare Al pads on the CMOS dummy chips or thicker stud bumps deposited and planarized on the Si chip in combination with bonding pads on the PI structure having a typical thickness of the electroplated Au of 3 μm. Depending on the bonding pad type combination, bonding yields of 93.7% ($n = 20$) and 97.2% (n = 70) have been achieved for chips with and without stud bumps (compression force 0.78 N/bump), respectively.

ACKNOWLEDGEMENTS

This work received funding from the European Union's Horizon 2020 research and innovation program (Grant Agreement number 899287, NeuraViPeR). The authors wish to thank the team of the Cleanroom Service Center at the Department of Microsystems Engineering (IMTEK) for technical support during processing and flip-chip assembly, J. Joos from our Microsystem Materials Laboratory for establishing the stud bump process, and K. Steffen, Laboratory for Process Technology, for electroplating the PI structures.

REFERENCES

[1] B. Rubehn and T. Stieglitz, "In vitro evaluation of the long-term stability of polyimide as a material for neural implants," *Biomaterials*, vol. 31, no. 13, pp. 3449–3458, 2010.

[2] T. D. Y. Kozai *et al.*, "Chronic tissue response to carboxymethyl cellulose based dissolvable insertion needle for ultra-small neural probes," *Biomaterials*, vol. 35, no. 34, pp. 9255–9268, 2014.

[3] X. Chen, F. Wang, E. Fernandez, and P. R. Roelfsema, "Shape perception via a high-channel-count neuro-prosthesis in monkey visual cortex," *Science*, vol. 370, no. 6521, pp. 1191–1196, 2020.

[4] S. Raspopovic *et al.*, "Restoring natural sensory feedback in real-time bidirectional hand prostheses," *Sci. Transl. Med.*, vol. 6, no. 222, pp. 222ra19-222ra19, 2014.

[5] P. Ruther and O. Paul, "New approaches for CMOS-based devices for large-scale neural recording," *Curr. Opin. Neurobiol.*, vol. 32, pp. 31–37, 2015.

[6] B. C. Raducanu *et al.*, "Time multiplexed active neural probe with 1356 parallel recording sites," *Sensors*, vol. 17, no. 10, p. 2388, 2017.

[7] A. S. Herbawi *et al.*, "CMOS neural probe with 1600 close-packed recording sites and 32 analog output

channels," *J. Microelectromech. Syst.*, vol. 27, no. 6, pp. 1023-1034, 2018.

[8] N. A. Steinmetz *et al.*, "Neuropixels 2.0: A miniaturized high-density probe for stable, long-term brain recordings," *Science*, vol. 372, no. 6539, p. eabf4588, 2021.

[9] S. P. Lacour, G. Courtine, and J. Guck, "Materials and technologies for soft implantable neuroprostheses," *Nat. Rev. Mater.*, vol. 1, no. 10, p. 16063, 2016.

[10] L. Luan *et al.*, "Recent advances in electrical neural interface engineering: Minimal invasiveness, longevity, and scalability," *Neuron*, vol. 108, no. 2, pp. 302–321, 2020.

[11] P. Čvančara *et al.*, "Stability of flexible thin-film metallization stimulation electrodes: analysis of explants after first-in-human study and improvement of in vivo performance," *J. Neural Eng.*, vol. 17, no. 4, p. 046006, 2020.

[12] S.-Y. Park *et al.*, "A Miniaturized 256-channel neural recording interface with area-efficient hybrid integration of flexible probes and CMOS integrated circuits," *IEEE Trans. Biomed. Eng.*, vol. 69, no. 1, pp. 334–346, 2022.

[13] C. Boehler *et al.*, "Actively controlled release of Dexamethasone from neural microelectrodes in a chronic in vivo study," *Biomaterials*, vol. 129, pp. 176–187, 2017.

CONTACT

*P. Ruther, tel: +49-761-203-7197 ruther@imtek.uni-freiburg.de

MICROWAVE-INDUCED THERMOACOUSTIC IMAGING USING ALUMINUM NITRIDE PMUT

Yiwei Wang[1]*, Lejia Zhang[1]*, Junxiang Cai[1,2,3], Baosheng Wang[1,2,3],
Yuandong Alex Gu[5], Liang Lou[5], Xiong Wang[1,2,3,4], and Tao Wu[1,2,3,4]

[1]School of Information Science and Technology, ShanghaiTech University, China
[2]Shanghai Institute of Microsystem and Information Technology, Chinese Academy of Sciences, Shanghai, China
[3]University of Chinese Academy of Sciences, Beijing, China
[4]Shanghai Engineering Research Center of Energy Efficient and Custom AI IC, Shanghai, China and
[5]Shanghai Industrial µTechnology Research Institute, China

ABSTRACT

This paper reports the novel application of an aluminum nitride (AlN) multi-cell piezoelectric micromachined ultrasonic transducer (PMUT) device in microwave-induced thermoacoustic imaging (MITAI). PMUT is used as the ultrasonic transducer (UT) in thermoacoustic imaging for the first time. This proposed PMUT device has an improved frequency response due to the design of cells with cavities in variant diameters. The device was characterized and tested in a microwave-induced thermoacoustic reconstruction process. The utilization of PMUT device was proved to achieve significantly higher quality images from reconstruction. The results from the proposed PMUT showed a decrease of 53% in mean square error (MSE) and increased the structural similarity (SSIM) by 19.7 times. This will potentially enable smaller yet better medical MITAI equipment.

KEYWORDS

Aluminum nitride, PMUT, thermoacoustic imaging.

INTRODUCTION

Thermoacoustic imaging (TAI) and thermoacoustic tomography (TAT) have been studied [1-2] and are prosperous in medical diagnosis and treatment [3-7]. Similar to photoacoustic imaging (PAI) [8-10], when the target tissue is heated by pulsed microwave (or laser in PAI), it will expand elastically and result in ultrasound emission. As the attenuation of ultrasound is relatively low in human body and soft tissues, a sensor is able to collect thermoacoustic signals generated by the sample in any angular positions from the whole heated area. In contrast to traditional medical imaging technologies, such as ultrasound, x-ray, and MRI, thermoacoustic imaging reflects more physical and chemical information of the target tissue. There has been research focusing on non-invasive foreign body detection [11] and guided hyperthermia therapies [12] powered by MITAI, in which traditional centimeter scaled ultrasonic transducers (UTs) are commonly used. Unlike these UT devices, PMUTs enabled by MEMS fabrication technology allow miniaturized and more flexible designs. In this work, specifically, the multi-cell PMUT design extends the frequency bandwidth which is more flexible to detect variable thermoacoustic signals in different types of tissues. Moreover, it is possible to optimize the overall

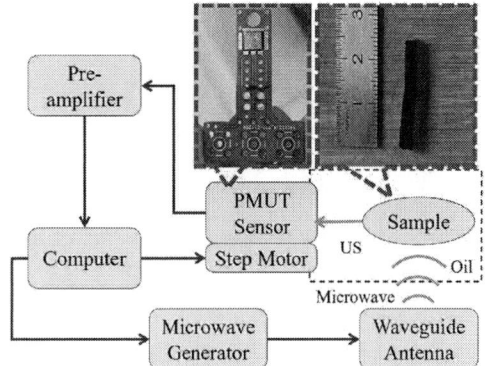

Figure 1: System setup. The scanning process is automatically controlled with a computer program.

Table 1: Microwave Generator Configurations.

Frequency	Peak Power	Pulse Width	Duty
2.45 GHz	20 kW	500 ns	0.05 %

frequency response of the device to work perfectly for specific tissues or organs. In addition, there are studies focusing on reducing the size of the medical equipment [13]. Currently, there has been tiny microwave generators [14] and small-scaled endoscopes [15-16]. With the proposed miniaturized ultrasonic transducers and methods in this work, the clinical MITAI equipment can be even smaller yet higher quality in the future. This will also enable wearable thermoacoustic monitoring devices [17].

METHODS

The Setup and Structures

The prototype is composed of an oil tank, a blood vessel sample, a microwave source and an ultrasonic transducer. Figure 1 shows the system schematics and the structure of the setup. The microwave source is positioned under a pool of oil where the sample is immersed. The microwave pulses are emitted from the bottom of the tank. An ultrasonic transducer is mounted on a 3D-printed holder which is driven and controlled by step motors. The generated thermoacoustic signal travels through the pre-amplifier unit and band-pass filter circuit, and is finally sampled by the computer. Two types of UTs were applied

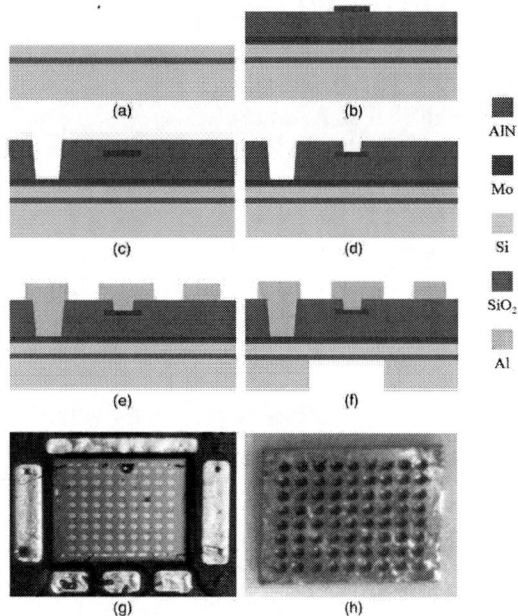

Figure 2: Fabrication process flow. (a) Starting SOI wafer. (b) Piezoelectric/Metal stack deposition and top electrode patterning. (c) SiO₂ deposition & via etch. (d) Top electrode via etching. (e) Pad pattern. (f) Backside cavity release. Images of the fabricated device, in (g) front side and (h) back side.

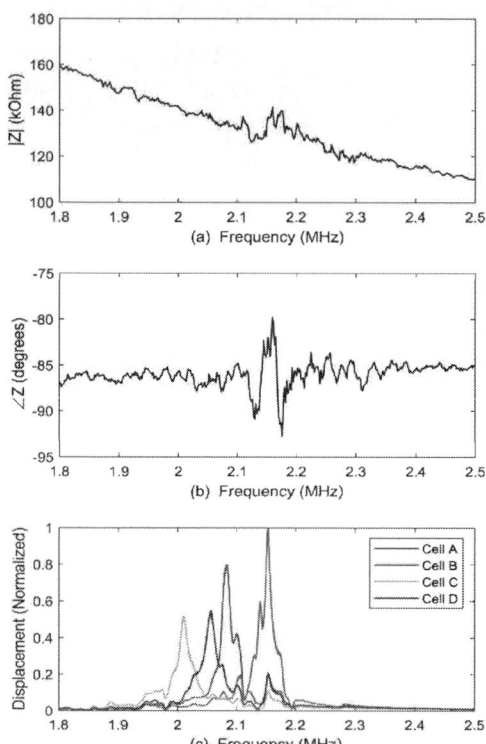

Figure 3: Frequency response in terms of admittance, (a) amplitude and (b) phase. (c) Displacements of four representative cells.

in this experiment, one is a traditional transducer and the other is our proposed PMUT device. The PMUT device contains 8 × 9 AlN membrane cells in circle shape with different diameters, resulting in different resonance frequencies. The sample object was a rectangular block with 25 mm in length and 3 mm in width.

To study the feasibility of the PMUT substitution and the performance of the device in the system, a traditional UT was also used in comparison. The experiments were performed in the same condition and configuration, except the different types of transducers.

System Configurations

The distance from the center of the sample target to the ultrasonic transducer was set to 69 mm. The transducer was rotated 360 degrees in total in the same plane where the sample was fixed on with a step of 1 degree. The signal was averaged for 300 times per measurement to improve signal-to-noise ratio (SNR). The configurations of the microwave generator are listed in Table 1.

PMUT DEVICE FABRICATION AND CHARACTERIZATION

Fabrication Process

The fabrication process of the proposed multi-cell PMUT device is schematically shown in Figure 2, which is similar to the PMUT work reported before [16]. The device was built on an SOI substrate of 5 μm silicon device layer, which was P-type doped (a). The bottom Mo electrode and 1 μm AlN were deposited. Then the top Mo electrode was deposited and patterned (b). Upon them another SiO₂ layer was deposited. Then, the hard mask SiO₂ was deposited (c)

through PECVD process and AlN layers were etched to make vias to top (c) and bottom (d) electrodes. Next, aluminum (Al) pads were deposited and patterned (e) for probing and wire bonding. Finally, backside cavities were etched using DRIE (f) to form vibrating membranes.

A close-up detail of both sides is shown in Figures 2(g) and 2(h). The cells are divided into 9 rows (vertically in Figure 2(g)), with 8 cells in each row. The cells in one row are connected in parallel through their top electrodes. And the pads of all 9 rows are electrically connected. As a result, all 72 cells in the rectangular array are connected together and used as one single-port device.

For convenience in thermoacoustic imaging experiments, the PMUT device was mounted on a customized PCB and wire bonded (shown previously in Figure 1). This made it easier to connect the detection circuit to the device through coaxial cables and fix the device on the frames with plastic latches and clamps precisely.

Characterization

The electrical and mechanical properties of the PMUT device were characterized. As is mentioned during the fabrication process, all cells on the PMUT have different resonant frequencies. Figures 3(a) and 3(b) show the frequency response of the device as a one-port unit through an impedance analyzer (Keysight E4990). The peaks in Figures 3(a) and 3(b) indicate that there is a electrical resonant frequency at 2.1 MHz, which matches the frequency of the compared ULSO transducer.

Figure 3(c) shows normalized displacement as a

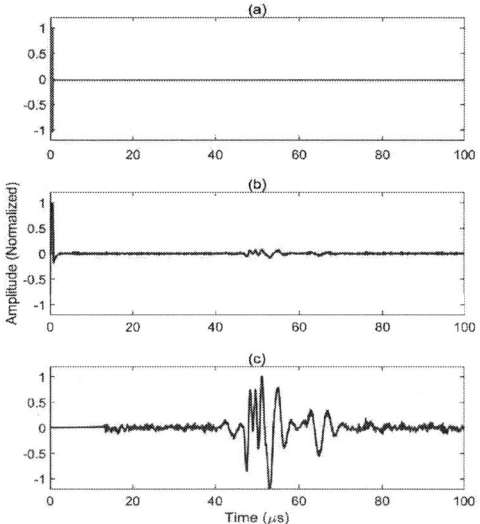

Figure 4: Time domain signals. (a) Generated microwave signal. (b) Received US signal, and (c) filtered and denoised signal.

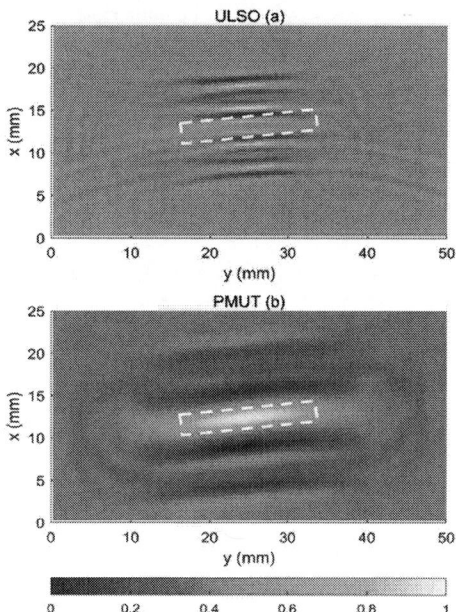

Figure 5: Comparison of imaging results. (a) Result of reconstructed image with ULSO transducer, and (b) result with the proposed multi-cell PMUT.

Table 2: Comparison of imaging quality.

Transducer	SSIM	MSE
ULSO	3e-4	0.2752
PMUT	59e-4	0.1468

function of driving frequency, of four representative cells by Polytec MSA-600. The peaks of displacement waveform represents mechanical resonant frequencies, indicating that the cell has the best energy conversion capability working at certain frequency. The passbands of the cells are interleaved, covering a larger range of mechanical vibration frequencies.

All cells have different membrane diameters, due to variance in backside cavity etching. Because it is used as a single-port device, the frequency responses of each cell can be added together as its overall frequency characteristics. All cells on the device are connected in parallel so that its overall frequency pass band will be extended. This design of cell array with variant parameter helps to achieve larger bandwidth.

RESULTS

The imaging experiment was performed with the setup and configurations mentioned in Method section. The generated microwave signal is shown in Figure 4(a). The corresponding thermoacoustic signal from the target is shown in Figure 4(b). The strong pulse at the beginning is resulted from electromagnetic coupling between the circuit on the transducer device and the initial microwave pulse. It is filtered out to obtain the real responding ultrasonic signal as is shown in Figure 4(c). When an object is heated by the microwave, it will expand due to thermal expansion and then restore to original scale due to elastic stress. As it vibrates back and forth, an ultrasound signal is emitted, carrying information about the thermal absorption properties of the target. Thus, the waveform in Figure 4(c) is composed of thermoacoustic signals at different frequencies and different delays, determined by the physical and chemical properties of the target material. Time domain signals from each angular position were collected and stored.

A MATLAB program was used to reconstruct the image of the original object from signals at all angles. The

image reconstruction algorithm we used, which is a conventional time-domain imaging algorithm, is distinguished by its high accuracy. It works by back projecting (BP) the ultrasound signal received by the transducer onto each pixel in the imaging area. The pixel values are imaged by calculating the delay value of the ultrasound signal at the distance between the transducer and the image pixel point. This algorithm can be used in any geometric mode, and it has no special requirements for the trajectory of the transducer (straight trajectories and circular trajectories are both possible). For the BP algorithm, the more points measured, the more accurate the imaging results. Tissues with different microwave absorption characters would show different signal strength. Thus, the different tissues can be segmented and visible in the resulting image. The comparison of 2-D imaging results is shown in Figure 5.

In the experiment, the traditional UT showed a larger spatial resolution, while the PMUT could detect a wider range along the long axis. The white rectangles in Figure 5 indicate the location of the target object where there is supposed to be strong signals. It can be seen from the resulting images that the traditional transducer had higher resolution along x-axis, while the PMUT imaging revealed more accurate details. As for the y-axis, the PMUT detected signals more widely in terms of angular ranges. As a result, the multi-cell PMUT device heard significantly better signals and reconstructed the shape of the original object more accurately, indicating the object's position and internal properties. The accuracy includes shape, length, volume, and others, which could be numerically evaluated here in terms of structural similarity (SSIM) indicating the similarity of the reconstructed pattern and the original pattern and mean square error (MSE) indicating the difference between them.

Table 2 shows the statistics indicating the quality of reconstruction, in terms of SSIM and MSE.

CONCLUSION

We have successfully demonstrated the application of multi-cell PMUT in MITAI and the device significantly improved the reconstruction imaging quality. This method utilized the miniaturization and broadened bandwidth of the multi-cell PMUT device, achieving easier detection and clearer imaging. The utilization of PMUT in MITAI will potentially reduce the demand of microwave power, doing less harm to the target tissues while keeping efficiency and functionalities. This is supposed to have the prospect to help build higher quality medical thermoacoustic imaging systems in the future.

ACKNOWLEDGEMENTS

This work was supported in part by the National Natural Science Foundation of China under Grant 61874073, and in part by the Lingang Laboratory under Grant LG-QS-202202-05. The work was also sponsored by Double First-Class Initiative Fund of ShanghaiTech University. The authors appreciate ShanghaiTech Quantum Device Lab (SQDL), Puyang Huang, Zhenghang Zhi, and Yuxi Wang for their kind assistance with wire bonding and tests on the device.

REFERENCES

[1] Q. Liu, X. Liang, W. Qi, Y. Gong, H. Jiang, and L. Xi, "Biomedical microwave-induced thermoacoustic imaging," *J. Innov. Opt. Health Sci.*, vol. 15, no. 04, p. 2230007, Jul. 2022.

[2] M. Xu and L. V. Wang, "Time-domain reconstruction for thermoacoustic tomography in a spherical geometry," *IEEE Trans. Med. Imaging*, vol. 21, no. 7, pp. 814–822, Jul. 2002.

[3] X. Wang, D. R. Bauer, R. Witte, and Hao Xin, "Microwave-Induced Thermoacoustic Imaging Model for Potential Breast Cancer Detection," *IEEE Trans. Biomed. Eng.*, vol. 59, no. 10, pp. 2782–2791, Oct. 2012.

[4] J. Song, Z. Zhao, J. Wang, X. Zhu, J. Wu, Z. Nie, and Q. H. Liu, "An Integrated Simulation Approach And Experimental Research on Microwave Induced Thermo-Acoustic Tomography System," *PIER*, vol. 140, pp. 385–400, 2013.

[5] X. Zhu, Z. Zhao, J. Wang, G. Chen, and Q. H. Liu, "Active Adjoint Modeling Method in Microwave Induced Thermoacoustic Tomography for Breast Tumor," *IEEE Trans. Biomed. Eng.*, vol. 61, no. 7, pp. 1957–1966, Jul. 2014.

[6] Y. Huang, M. Omar, W. Tian, H. Lopez-Schier, G. G. Westmeyer, A. Chmyrov, G. Sergiadis, and V. Ntziachristos, "Noninvasive visualization of electrical conductivity in tissues at the micrometer scale," *Sci. Adv.*, vol. 7, no. 20, p. eabd1505, May 2021.

[7] Y. Huang, S. Kellnberger, G. Sergiadis, and V. Ntziachristos, "Blood vessel imaging using radiofrequency-induced second harmonic acoustic response," *Sci Rep*, vol. 8, no. 1, p. 15522, Dec. 2018.

[8] C. Huang, K. Wang, L. Nie, L. V. Wang, and M. A. Anastasio, "Full-Wave Iterative Image Reconstruction in Photoacoustic Tomography With Acoustically Inhomogeneous Media," *IEEE Trans. Med. Imaging*, vol. 32, no. 6, pp. 1097–1110, Jun. 2013.

[9] L. V. Wang, "Tutorial on Photoacoustic Microscopy and Computed Tomography," *IEEE J. Select. Topics Quantum Electron.*, vol. 14, no. 1, pp. 171–179, 2008.

[10] J. Cai, Y. Wang, D. Jiang, S. Zhang, Y. A. Gu, L. Lou, F. Gao, and T. Wu, "Beyond fundamental resonance mode: high-order multi-band ALN PMUT for in vivo photoacoustic imaging," *Microsyst Nanoeng*, vol. 8, no. 1, p. 116, Nov. 2022.

[11] Y. Sun, C. Li, B. Wang, and X. Wang, "A Low-Cost Compressive Thermoacoustic Tomography System for Hot and Cold Foreign Bodies Detection," *IEEE Sensors Journal*, vol. 21, no. 20, pp. 23588–23596, 2021.

[12] J. Li, B. Wang, D. Zhang, C. Li, Y. Zhu, Y. Zou, B. Chen, T. Wu, and X. Wang, "A Preclinical System Prototype for Focused Microwave Breast Hyperthermia Guided by Compressive Thermoacoustic Tomography," *IEEE Trans. Biomed. Eng.*, vol. 68, no. 7, pp. 2289–2300, Jul. 2021.

[13] L. Wu, Z. Cheng, Y. Ma, Y. Li, M. Ren, D. Xing, and H. Qin, "A Handheld Microwave Thermoacoustic Imaging System With an Impedance Matching Microwave-Sono Probe for Breast Tumor Screening," *IEEE Transactions on Medical Imaging*, vol. 41, no. 5, pp. 1080–1086, 2022.

[14] L. Huang, Z. Zheng, Z. Chi, and H. Jiang, "Low-cost miniaturized microwave generator for thermoacoustic imaging," in *2020 International Conference on Microwave and Millimeter Wave Technology (ICMMT)*, 2020, pp. 1–2.

[15] J. Cai, Y. Wang, L. Lou, S. Zhang, Y. Gu, F. Gao, and T. Wu, "Photoacoustic and Ultrasound Dual-Modality Endoscopic Imaging Based on ALN Pmut Array," in *2022 IEEE 35th International Conference on Micro Electro Mechanical Systems Conference (MEMS)*, Tokyo, Japan, Jan. 2022, pp. 412–415.

[16] J. Cai, K. Liu, L. Lou, S. Zhang, Y. A. Gu, and T. Wu, "Increasing Ranging Accuracy of Aluminum Nitride Pmuts by Circuit Coupling," in *2021 IEEE 34th International Conference on Micro Electro Mechanical Systems (MEMS)*, Gainesville, FL, USA, Jan. 2021, pp. 740–743.

[17] M. Lin, H. Hu, S. Zhou, and S. Xu, "Soft wearable devices for deep-tissue sensing," *Nat Rev Mater*, vol. 7, no. 11, pp. 850–869, Mar. 2022.

CONTACT

*Yiwei Wang; wangyw@shanghaitech.edu.cn
*Lejia Zhang; zhanglj@shanghaitech.edu.cn
*Xiong Wang; wangxiong@shanghaitech.edu.cn
*Tao Wu; wutao@shanghaitech.edu.cn
The first two authors contributed equally to this work.

NEEDLE-FREE DRUG INJECTION USING A SHOCK WAVE FOCUSING SYSTEM WITH THE FUNCTION OF REAL-TIME MICROBUBBLE-BASED DISTANCE SENSING

Yibo Ma, Wenjing Huang, Keita Ichikawa and Yoko Yamanishi

Dept. of Mechanical Engineering, Faculty of Engineering, Kyushu University, Japan

ABSTRACT

In this paper, we proposed a novel needle-free injection method using electrically induced microbubbles. This microbubble needle-free injector is developed to achieve the depth-controlled drug injection. A focusing device was developed to enhance the pressure of microbubble-induced shock wave and improve the perforation ability of injector. Due to the injection depth being related to the distance between injector tip and target, we propose to control the injection depth by sensing and adjusting the distance. We developed the distance measurement by microbubble injector itself. In this work, the microbubble injector was both the actuator and sensor. This needle-free microbubble is a promising tool to achieve the injection depth-controlled in the future.

KEYWORDS

Electrically induced microbubbles; needle-free injector; shock wave focusing; drug delivery

INTRODUCTION

Drug injections and vaccinations often use metal needle injector. Metal needle injector has been used for the long term due to its easy skin penetration and controllable drug introduction. However, because of its direct needle puncture of the patient's tissue and blood vessels, infectious diseases caused using these injectors have been a problem in the world for a long time. The stinging sensation of needle insertion can cause anxiety and panic in some patients. Also, such as long-term insulin injections can lead to hardening of the skin.

In order to solve these problems, drug delivery systems were developed worldwide such as skin permeation and implantable drug delivery devices and needle-free injection systems incorporating biomimetic drug delivery methods[1]. Especially, the needle-free injector which uses high-pressure jet of water to perforate the skin and introduce drugs have been developed and are being commercially produced and used in some countries. The fluid left in the patient's tissues after the injection can cause pain and bruising[2]. Also, precise injection depth control and real-time injection distance control are yet to be settled. Therefore, we proposed a novel needle-free injection method, the microbubble injection.

In our laboratory, a device that generates the microbubbles by concentrating the high voltage electrical field in the tip of wiring has been developed (named as electrically induced microbubbles injector)[3]. When the high-voltage electric field is concentrated at the tip of the injector, the ions near the tip move at high speed and collide with each other due to the high voltage electric field. The collision of ions causes the molecules to disperse, which eventually leads to the production of tiny bubbles[3][4][5]. When the generated electric field exceeds the dielectric constant of the bubble, the plasma and discharge are generated[6]. The temperature in the tiny bubble rises rapidly and the pressure of the bubble is not released in the process[7][8][9][10]. The thermoelastic stress caused by the in this process is confined inside the tiny bubble, which leads to a maximum pressure rise inside the bubble[10]. Then the tiny bubble expands rapidly and with the rapid expansion of the tiny bubble, shock waves are generated. The microbubble injection has excellent perforation ability.

In this work, a microbubble-induced shock wave focusing injection with ability of drug introduction depth-controlled was proposed. The reagent was introduced by microbubble injection and the introduction depth was improved by shock wave focusing. Also results of micron-level distance sensing by microbubble injection were shown in this paper. In the future, the distance sensing technique will be integrated to achieve the low-cost, low-pain, introduction depth-controlled injection by developing microbubble injector.

MECHANISM AND CONCEPT

Figure 1 showed the concept of this research. The distance from injector tip to target was measured by the microbubble itself. Moved the device to the position where target was located on reflector focusing point. Microbubble was injected to perforate on target and the drug was introduced into target. When target was located on reflector focusing point, the introduction depth was maximum depth. By adjusting the distance, the introduction depth can be adjusted. In this study, we showed a novel distance sensing method with microbubble injector itself to realize micron-level distance measurement. As the microbubble expanded and touched the target, the resistance of device reduced. The longer the distance, the longer it takes for the expanding bubble to reach the target. The distance sensing was developed based on this mechanism. Shock wave was generated from microbubble and reflected by the reflector, then the focused/enhanced shock wave reached the target and perforate on target. Materials with higher acoustic impedance than saline can reflect shock waves. In this study, we chose Tin to produce the reflector because of its high acoustic impedance. For now, we had achieved the distance sensing by microbubble and reagent introduction by focused shock wave.

EXPERIMENT

Microbubble injector fabrication and distance sensing development experiment

A 100 μm diameter tungsten wire was inserted into a

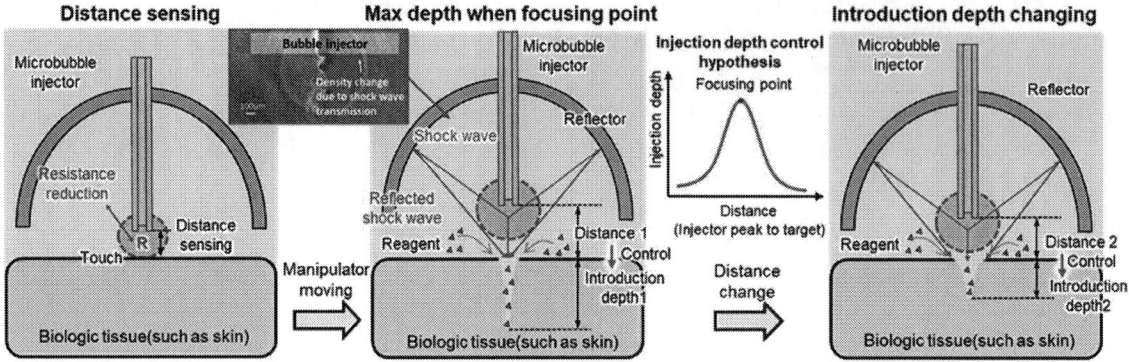

Figure 1: Concept of needle-free drug injection using a shock wave focusing system with the function of real-time microbubble-based distance sensing

150μm diameter polyimide(PI) tube. Then it was inserted into a metal injection needle. This is what we called "microbubble injector". Figure 2 (a) showed the structure of a microbubble injector. Injection needle is the counter electrode. A function generator supplied the voltage to the injector. The injector was applied with a square wave voltage of 700 v for 10 μs. The microbubble generated with the voltage application. 0.9% saline is the solution.

Figure 2 (b) showed the distance sensing method development experiment setup. The injector is in a case and the silicon wafer was the target for distance sensing. A high-speed camera was used to observe the sensing field and record the microbubble generation and collapse. The bubble injector was gripped by manipulator. Located the sensor tip at the zero distance from the target. Injected the microbubble and recorded the current of the injector and the voltage across the injector. Then moved up the injector 10 μm each time and record the current and voltage.(n =5) After data analysis. Place the microbubble at each distance and evaluate the injector's current output.

Reagent injection without/with reflector

In this experiment, we injected the fluorescent beads into the target, chicken meat with the microbubble injector. And the chicken meat was cut to observe the introduction depth of fluorescent beads under the fluorescence phase contrast microscope (Nikon TI-DH). Injection without reflector and injection with reflector were done in this experiment.

The reflector was made with Tin due to its large acoustic impedance with respect to that of saline. When a shock wave reaches an object interface that differs from the transmitting medium, reflection and transmission occur. The ratio of reflection to transmission is defined by the reflectance R, which is expressed by the following equation using the acoustic impedance of the object[11].

$$R = \frac{A_2 - A_1}{A_2 + A_1} \qquad (1)$$

Where A_1 is the acoustic impedance of the medium transmitting the shock wave and A_2 is the acoustic impedance of the material at the destination. Equation (1) shows that the larger the difference in acoustic impedance between the two media, the greater the reflectance, and the greater the fraction of the shock wave that is reflected. When A_2 is smaller than A_1, the reflectance R takes a negative value. In this case, the phase of the shock wave is reversed, and a negative pressure expansion wave is generated[12]. The positive pressure shock wave pushes the object in the direction of expansion, while the negative pressure expansion wave pulls the object in the opposite direction. In our case, the positive pressure shock wave (the focused shock wave) should be used, so the reflector material should be large acoustic impedance with respect to that of saline material. The Tin was selected due to its large acoustic impedance, low-cost and non-toxic in most situations. The reflector was manufactured by die casting. The shape of the reflector is half of an oval. The diameter of reflector was 10 mm. And a diameter of 0.5 mm hole was opened at the top of the reflector for the injector insertion.

The reagent experimental setup was shown in Figure 3

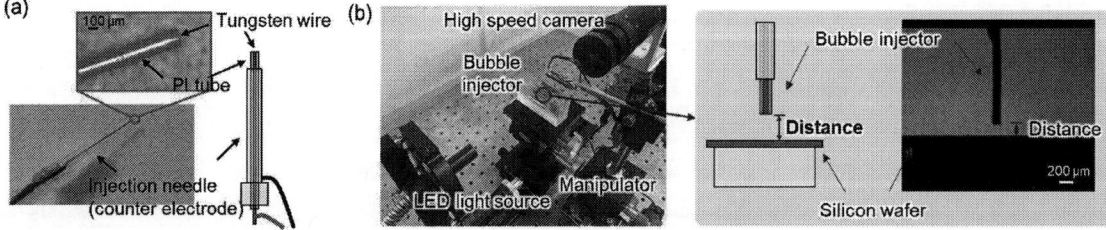

Figure 2: (a) Schematic of microbubble injector and the picture of it; (b) Distance sensing method development experiment setup

978-1-6654-9309-3/23 $31.00 © 2023 IEEE

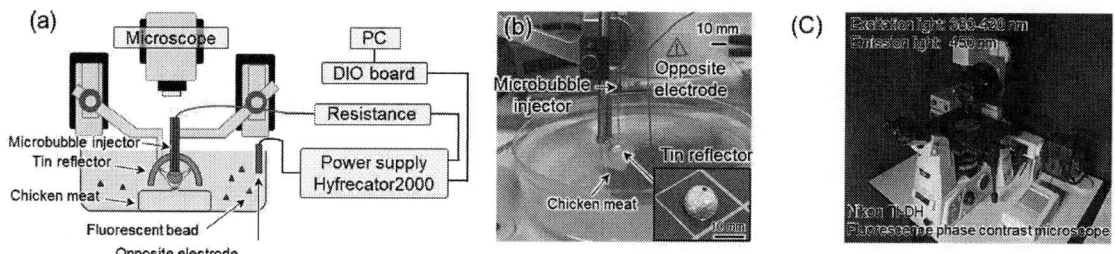

Figure 3: (a) Experiment of fluorescent beads introduction with microbubble-induced shock wave focusing device; (b) Reagent injection with microbubble injector and Tin reflector; (c) Target fluorescence observation under fluorescence phase contrast microscope

(a). The power source was High frequency power supply, which is for electric scalpels. The power source was connected to the bubble injector with a resistance. The reagent was a mixture of fluorescent beads (Fluoro-Max, Thermo scientific Co. Ltd) with a diameter of 2.1 μm in 0.9% NaCl solution. The microbubble injector tip and the chicken meat were located at the two focusing point of the oval reflector. The microbubble injector was located on Tin reflector side focusing point. Microbubbles were generated and shock wave generated from the microbubble. The fluorescent beads were introduced into the chicken meat. Then we removed the reflector and repeated the experiment. The delivery results were observed by a fluorescence phase contrast microscope. Reagent introduction and depth measurement were achieved by real-time observation of fluorescence.

RESULT AND DISCUSSION

Microbubble injector fabrication and distance sensing development experiment

The current of the injector at distance of a 50 μm was shown in Figure 4 (a). The current curve consisted of two segments, a smooth curve, and a sudden rise of the square wave. We recorded the microbubble movement in 10 μs voltage application. As showed in Figure, the current suddenly raised at 5.8 μs which was the time microbubble touching the target surface. And in other current curve at each distance, as the distance increases, the expansion microbubble would take more time to touch the target surface. Thus, the current sudden rising time was delayed. The average value of the current for 10 μs was used as a measure, and the Figure 4 (b) was made based on this. We can see from the Figure 4 (b) that the relationship between current and distance was approximately a linear relationship. Then we set the distance from 10 to 100 μm and measured the distance using the microbubble distance-current curve we obtained. We found that the obtained current data matched the curve.

Figure 4 (c) showed the mechanism of distance sensing with electrically induced microbubble sensor. The microbubble injector was located at the target surface top. Voltage was applied to the microbubble injector. Microbubble was generated and expanded. The microbubble touched the target and sensor system resistance reduced. The current of the injector was record and the distance sensing was achieved.

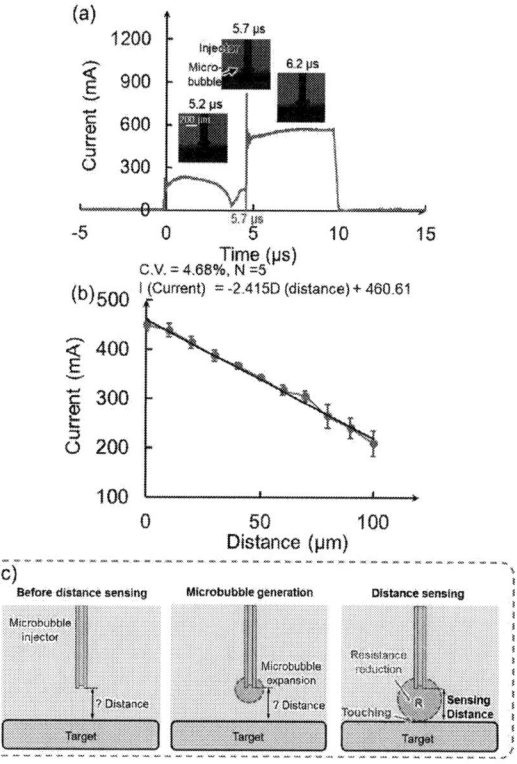

Figure 4: (a) Current of the injector at distance 50 μm; (b) Distance-current curve by microbubble distance sensor; (c) Schematic of microbubble injection distance sensing mechanism

Reagent injection without/with reflector

The Figure showed the top view and the cross section of fluorescent beads injected chicken meat sample without/with reflector. The injection without reflector was done 5 times. As shown in Figure 5 (a), the top of the injected sample was a line. And the injection depth was approximately 600 μm at each sample. As shown in Figure 5 (b), the top view of the injected area was an irregular area. And the injection depth is from 1500 μm to 2500 μm in 5 samples. The inaccurate control of the distance between the target and the microbubble injector tip and that the microbubble injector tip was not accurately located in focusing point. We thought these were the main reasons for the different depths of each injection. Also, for now, the

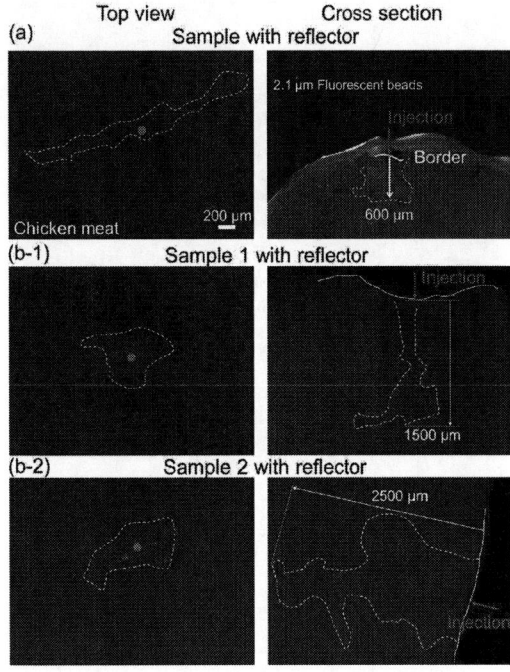

Figure 5: Top view and the cross section of fluorescent beads injected chicken meat sample (a) without reflector; (b-1) and (b-2)with reflector

Tin reflector was manufactured by die casting, which caused the imperfect shape of reflector and unsmooth surface.

CONCLUSION AND FUTURE PLANS

In this research, we proposed a needle-free drug injection system using focused shock wave which has the function of controllable introduction depth by real-time microbubble-based distance sensing. In this drug introduction depth-control injector, focused microbubble induced shock wave was applied. The distance between the target and the focal point of shock wave reflector affects the effect of the focused shock wave on the target, and thus the injection depth. We propose to control the injection depth by sensing/adjusting the distance by microbubble injector itself. We developed a novel micron-level distance sensing method based on electrically induced microbubble. For now, the microbubble distance sensor had achieved the 10-100 μm, 10 μm distance measurement accuracy. In the future, we would continue the sensing method development and integrate it as an individual device to achieve the non-contact, low-cost micron-level distance sensing which is a field yet to be developed. We also developed the reagent introduction by microbubble-induce shock wave focusing. And the introduction depth is 1500-2500 μm which meet the basic drug injection depth. But the injection depth is not stable due to the not perfect experiment setup and fabricated reflector. In the future, we shall re-design the reflector to achieve the stable drug introduction depth. After finishing the microbubble distance sensing and stable drug injection depth development, we shall integrate the two systems and achieve the drug introduction depth-control by signal microbubble injector.

ACKNOWLEDGMENT

This work was supported by JSPS KAKENHI Grant Number, JP21J13114, JP22H00198, JP22K17783 and JPMJSP2136. We would also like to thanks to Associate Professor Tokitada Hashimoto of Saga University for his knowledge of shock waves and experimental advice in conducting this research.

REFERENCES

[1] K. Oka, S. Aoyagi et al., : "Fabrication of a micro needle for a trace blood test", Sensors and Actuators A, Vol.97-98, pp.478-485 (2002)

[2] N. Petrovsky, Y. Honda-Okada et al., : "A randomized controlled study to assess the immunogenicity and tolerability of a 2012 trivalent seasonal inactivated influenza vaccine administered via a disposable syringe jet injector device versus a traditional pre-filled syringe and needle", Trials in Vaccinology, Vol.2, pp.39-44 (2013).

[3] K. Ichikawa, S. Maeda and Y. Yamanishi : "Evaluation of invasiveness by collapsing of electrically induced bubbles for a needle-free injector", Journal of Microelectromechanical systems, Vol.27, No.2, pp.305-311 (2018)

[4] O. Lesaint et al., "A comparison of negative and positive," J. Phys. D. Appl. Phys., vol. J. Phys. D (2001)

[5] J. F. Kolb, R. P. Joshi, S. Xiao, and K. H. Schoenbach, "Streamers in water and other dielectric liquids," J. Phys. D. Appl. Phys., vol. 41, no. 23 (2008)

[6] I. M. Gavrilov, V. R. Kukhta, V. V. Lopatin, and P. G. Petrov, "Dynamics of prebreakdown phenomena in a uniform field in water," IEEE Trans. Dielectr. Electr. Insul., vol. 1, no. 3, pp. 496–502 (1994)

[7] G. Paltauf and H. Schmidt-Kloiber, "Photoacoustic cavitation in spherical and cylindrical absorbers," Appl. Phys. A Mater. Sci. Process., vol. 68, no. 5, pp. 525–532 (1999)

[8] G. Paltauf and P. E. Dyer, "Photomechanical processes and effects in ablation," Chem. Rev., vol. 103, no. 2, pp. 487–518 (2003)

[9] A. Vogel and V. Venugopalan, "Mechanisms of pulsed laser ablation of biological tissues," Chem. Rev., vol. 103, no. 2, pp. 577–644 (2003)

[10] A. Vogel, J. Noack et al., : "Mechanisms of femtosecond laser nanosurgery of cell and tissues", Applied physics B, Vol.81, pp.1015-1047 (2005)

[11] K. Takayama : "Shock Wave Handbook", Springer Fairlark Tokyo (1995)

[12] T. Nakabaru, T. Hashimoto et al., : "Generating and Focusing of Underwater Expansion Wave using a Silicon Resin Reflector", Journal of Thermal Science, Vol.22, No.3, pp.209-215 (2013)

CONTACT

Yibo Ma, ma.yibo.190@s.kyushu-u.ac.jp

NEW WAFER-LEVEL FABRICATION OF ULTRATHIN SILICON INSERTION SHUTTLES FOR FLEXIBLE NEURAL IMPLANTS

Kirti Sharma[1,2], Christian Boehler[1,2], Maria Asplund[1,2,3,4], Oliver Paul[1,2], and Patrick Ruther[1,2]

[1] Department of Microsystems Engineering (IMTEK), University of Freiburg, GERMANY,
[2] BrainLinks-BrainTools Center, University of Freiburg, GERMANY,
[3] Freiburg Institute for Advanced Studies (FRIAS), University of Freiburg, GERMANY, and
[4] Department of Microtechnology and Nanoscience, Chalmers University of Technology, SWEDEN

ABSTRACT

This paper reports a novel, cost-effective process for the fabrication of ultrathin silicon (Si) shuttles applied as insertion tools for highly flexible polyimide (PI) neural implants. The process exploits the so-called etching before grinding (EBG) process established to realize Si-based neural probes of the Michigan style. In this study, EBG is combined for the first time with a subsequent deep reactive ion etch (DRIE) process applied on the wafer-level. The innovative approach allows to realize insertion shuttles with a base thickness > 50 µm using wafer grinding and to reliably thin down the slender shuttle shanks (width ≥ 35 µm) to thicknesses as small as 15 µm using DRIE. The backgrinding liquid wax applied during wafer grinding enables the safe release of the delicate shuttle structures from their carrier wafer using isopropanol. Flexible, 15-µm-thin neural probes made from PI are precisely aligned and temporarily bonded to the custom-designed insertion shuttles applying polyethylene glycol (PEG) and reliably deployed into cortical tissue.

KEYWORDS

Silicon thinning, grinding, silicon shuttles, ultrathin probes, polyimide probes, self-alignment, backgrinding liquid wax, ultrathin chips

INTRODUCTION

Neuroscientific research and its clinical translation rely on a wide variety of neural implants fabricated using microelectromechanical system (MEMS) technologies. While stiff probes based on complementary metal-oxide-semiconductor (CMOS) fabrication technology enable a pronounced increase in the number of electrodes [1, 2], flexible implants made from polymers such as parylene-C, SU-8, and polyimide (PI) are known to elicit minimal tissue response and provide an improved long-term recording stability [3-5]. However, the enhanced mechanical flexibility of these implants requires dedicated implantation strategies for their reliable and precisely targeted probe insertion into neural tissue.

Strategies applied to improve the probe shaft insertion capability include incorporating a stiff material at the probe tip [6], or integrating channels into the probe shanks temporarily filled with a material enhancing the shank stiffness during implantation [7]. Alternatively, dissolvable polyethylene glycol (PEG) mechanically braces the flexible probe shanks, exposing only a short section of the probe length during the implantation process [8]. This temporary shortening of the probe shanks increases their stiffness and buckling force for implantation, while the

PEG is gradually dissolved as the probe is advanced into the brain. Methods based on stiff insertion shuttles use microwires [4, 9] or Si combs [10, 11] to implant flexible polymeric probes to the desired depth, followed by the shuttle retraction. As illustrated in Fig. 1(a), tip-sharpened wire shuttles deliver probes by being mechanically engaged to their tips. On the other hand, Si shuttles may be used together with PEG to temporarily attach the flexible probes {Fig. 1(b)}, and rely on PEG dissolution in brain tissue in order to decouple the probes from their insertion shuttles after implantation. The preparation of microwires and in particular microwire arrays for the implantation of multi-shank flexible probes is a labor-intensive and serial process [9]. In contrast, microfabricated Si shuttles provide as a key advantage the parallel processing of hundreds of devices on a Si wafer using MEMS technologies.

Here, we report on a novel fabrication technology of ultra-thin Si-based insertion shuttles and an experimental approach to precisely align and assemble flexible neural probes onto these shuttles. In contrast to shuttles presented by Felix et al. [10] and Joo et al. [11], our new approach avoids channel structures for probe fixation and enables a further thinning of the shuttle shanks {Fig. 1(c)}. In addition, the fabrication process achieves ultra-thin Si devices without using expensive silicon-on-insulator (SOI) wafers, as applied in [10, 11]. The fabrication process, abbreviated in the following as EBAG, i.e. etching before and after grinding, applies EBG [12] in combination with a subsequent DRIE process. The EBG process, introduced by

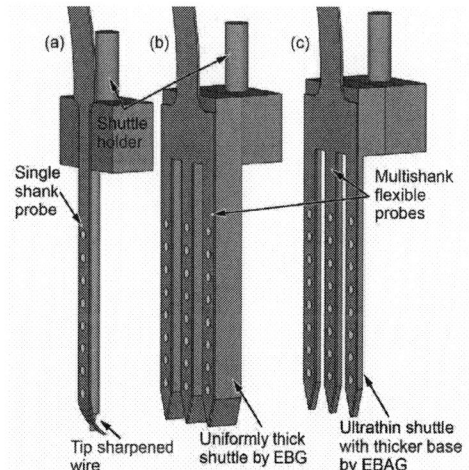

Figure 1: Insertion tool strategies for flexible probes: (a) Tip-sharpened wire engaged to a hole in the flexible probe shank, and Si-based multi-shank shuttles realized using the (b) standard EBG and (c) novel EBAG processes.

Herwik et al. as a CMOS-compatible process to realize ultrathin neural probes, enables structures of homogeneous thickness down to 25 μm defined by wafer grinding {Fig. 1(b)}. In contrast, EBAG allows to realize stepped Si structures locally thinned beyond the EBG-defined substrate thickness using DRIE. The stepped device geometry avoids the delicate handling of small and ultrathin devices representing a potential risk of fracture due to excessive mechanical loads. Here, this innovative process is applied for the fabrication of ultrathin, stepped Si insertion shuttles. The geometry of the shuttles closely matches the dimensions of the thin flexible probes, and helps to minimize the insertion trauma, which is essential for ensuring successful probe-tissue integration [13].

MATERIAL AND METHODS

Shuttle Design

The shuttles proposed in this work have a larger base of increased thickness for improved device handling, which is combined with slender, ultrathin shanks intended to minimize tissue damage during implantation. The stepped shuttle design is compared in Fig. 1(b,c) to EBG-realized shuttles of uniform thickness. The shuttle design can be customized to fit the contour of any single- or multi-shank flexible probe, with the shuttle shanks being only 20 μm wider than those of the flexible probes.

Shuttle Fabrication

The new EBAG process is summarized in Fig. 2. It applies standard, single-side polished Si wafers with a thickness of 525 μm. First, the in-plane geometry of the shuttles is defined by DRIE using a photolithographically patterned photoresist (AZ10XT, MicroChemicals GmbH, Ulm, Germany) as a masking layer, like in the EBG process {Fig. 2(i)}. In contrast to EBG, where the wafer is mounted on a grinding tape prior to rear side grinding, the shuttle wafer is mounted on a rigid carrier wafer {Fig. 2(ii)}. For this purpose, a backgrinding liquid wax (BGL7120, AI

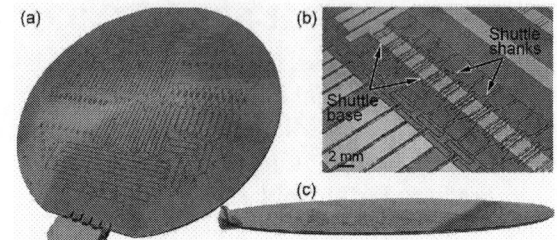

Figure 3: (a) Wafer ground to 50 μm on a carrier wafer. Devices on the wafer are separated by DRIE trenches but fixed to the carrier using BGL7120. (b) Section of a wafer after post-grinding lithography; shuttle bases are protected by photoresist while shanks are exposed, ready to be thinned by DRIE. (c) Side view of the wafer assembly shown in (a).

Technology, Inc., Princeton Junction, NJ, USA) is spin-coated onto the carrier wafer, and the DRIE-processed shuttle wafer is bonded upside down using the substrate bonder SB6 Gen2 (SÜSS MicroTec, Garching, Germany) at a temperature and pressure of 160 °C and 300 mbar, respectively. Next, the wafer stack is ground from the rear of the shuttle wafer to the desired base thickness using an automatic surface grinder (DAG810, DISCO HI-TEC Europe Gmbh, Munich, Germany) {Fig. 2(iii)}. The rear-side grinding separates the individual shuttle structures while they stay securely attached to the carrier wafer by the grinding wax. Figures. 3(a,c) show a processed 4-inch wafer, carrying shuttles and Si probes, ground to a thickness of 50 μm on top of a 525-μm-thick carrier wafer.

Subsequently, a second photolithography is performed using the photoresist AZ10XT to mask those areas on the shuttle wafer that are not to be thinned. Thereby, the base of the shuttles is covered with photoresist, while the shanks are exposed, as clearly indicated in Fig. 3(b). The shanks are selectively thinned by applying a second DRIE step using the inductively-coupled plasma (ICP) etcher STS Multiplex (Surface Technology Systems, Newport, U.K.). The processing parameters for this etching step are adapted such that the shanks are anisotropically etched perpendicularly to the top surface while minimizing any lateral etching from the sidewalls. Process gases SF_6 and O_2 (flow rates 130 and 13 sccm, respectively) are applied during the etch cycles, while C_4F_8 (flow rate 85 sccm) is applied during the passivation cycles. The durations of the individual etching and passivation cycles strongly influence the etch rate and the achievable etch profile, and are further discussed in the results section.

Finally, the photoresist is removed by a quick rinse in dimethyl sulfoxide (DMSO), followed by dry resist stripping in oxygen plasma to completely remove any resist residues. In order not to affect the BGL7120 layer, temperatures during the dry resist stripping process are limited to a maximum of 160 °C. For insertion shuttle release, the wafer stack is soaked in isopropanol overnight to dissolve the BGL7120 layer between the carrier wafer and shuttle structures.

Assembly of Flexible Probe to Insertion Shuttle

A liquid-assisted self-alignment process is used to precisely align flexible polyimide probes to their respective

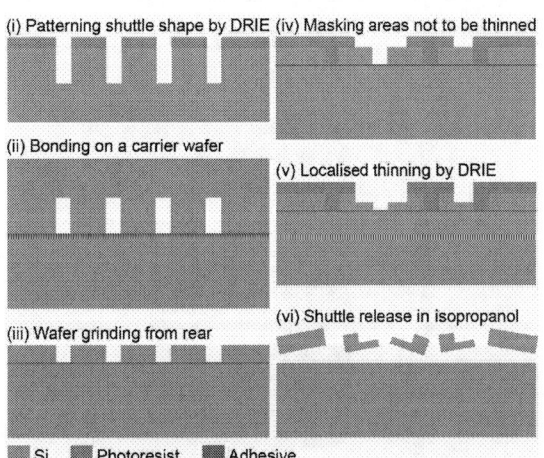

Figure 2: Process flow for the fabrication of silicon shuttles selectively thinned at the wafer level: (i) DRIE to define shuttle contour, (ii) carrier wafer mount, (iii) wafer grinding, (iv) 2nd photolithography, (v) DRIE of shuttle shanks, and (vi) device release using isopropanol.

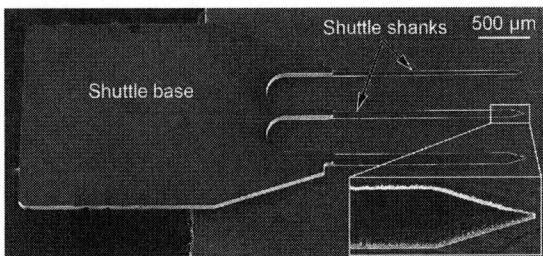

Figure 4: SEM of an EBAG-fabricated shuttle comprising three shanks 55, 90, and 125 μm in width. The shuttle base is 75 μm thick, while the shanks are selectively thinned to 25 μm.

shuttles. First, a shuttle is wetted by an aqueous ethanol solution before the PI probe is brought into contact with the shuttle. Surface tension aligns the corresponding shanks of PI probe and insertion shuttle to each other. Once the probe tips are positioned slightly behind the shuttle tips, the ethanol is allowed to evaporate, resulting in a temporary fixation of the probe on the shuttle through weak adhesive forces. Shank tips are then dipped up to a depth of 500 μm into a PEG solution (5% PEG in 50% ethanol). This forms a thin PEG layer at the probe-shuttle interface, thus holding the two devices together. The applied PEG has a molecular weight of 35 kg/mol.

RESULTS

Figure 4 presents a scanning electron micrograph (SEM) of an insertion shuttle realized by EBAG with a 70-μm-thick base and shanks thinned down to 25 μm. The three shuttle shanks have different widths of 55, 90 and 125 μm, and are designed to accommodate a flexible probe with shank widths of 35, 70, and 105 μm, respectively. In the magnified view of the shank tip in Fig. 4, one can observe chamfering at the DRIE etched top surface of the shanks. It is suspected that this mainly results from the surface damage caused by wafer grinding. Figure 5(a) shows typical grinding marks on the surface of a shuttle tip ground to 60 μm. Grinding introduces micro-defects on the top as well as on the edges of the ground surface. As the shanks are subsequently thinned during the second DRIE step, chamfering is caused by the amplification of these micro-defects, as clearly seen on the shuttle tip thinned to 25 μm in Fig. 5(b). In addition, the cross section of the thinned shanks and the extent of chamfering strongly

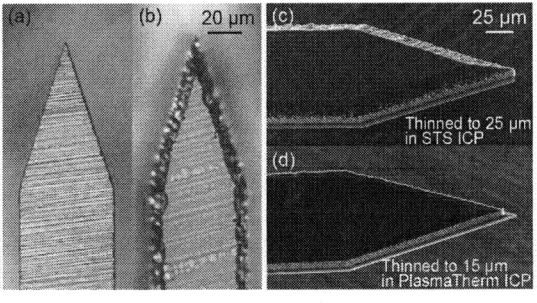

Figure 5: (a,b) Optical micrographs of shuttle shank rear side after (a) grinding to a thickness of 60 μm and (b) further thinning to 25 μm. (c,d) Comparison of two shuttle tips etched in different DRIE systems.

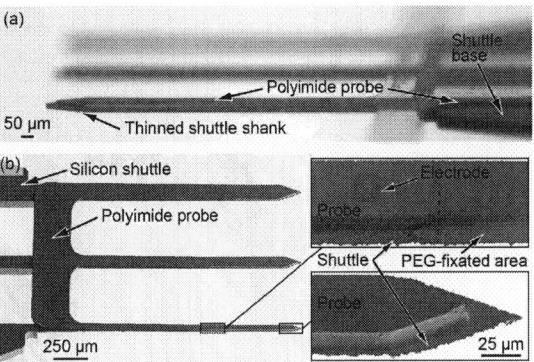

Figure 6: (a) Optical micrograph of a 15-μm-thick flexible PI probe aligned and fixed using PEG on a selectively thinned silicon shuttle. (b) SEM of a probe-shuttle assembly highlighting the accurate alignment of the PI test probe using PEG. The small amount of PEG used is clearly observable at the shuttle tips.

depend on the etch and passivation cycle durations.

Based on a detailed parameter study, etch and passivation durations of 8 s and 6 s per cycle provided the best thinning results with a mostly anisotropic etch profile. An average etch rate of 400 nm/cycle was observed over the wafer for these parameters. At a fixed passivation cycle duration of 6 s, increasing the length of etch cycles results in higher etch rates and a more isotropic etch with a reduction in the shank width due to lateral etching. On the other hand, decreasing the length of etch cycles produces shanks with lesser chamfering but leads to the formation of undesirable microneedles, i.e. Si nanograss at the probe edges. While chamfering is not an issue for wide shanks, it might compromise the stability of shanks narrower than 50 μm due to excessive material loss. However, chamfering can as well be seen as a feature as it leads to tip sharpening that lowers insertion forces, as demonstrated by Joo et al. [11].

In a second set of experiments, we performed the shank thinning using a more advanced DRIE system (Versaline ICP, Plasma-Therm, St. Petersburg, FL, USA) which provides faster switching between the etch and passivation cycles. It allows cycle durations of less than 1 s in contrast to the minimum duration of 6 s reliably achieved with the STS ICP system. Additionally, for each passivation cycle it applies two etch cycles that vary in the bias voltage, gas flows and etch pressure. Through a preliminary tuning of parameters, we could achieve anisotropic thinning of shuttles with this system, and also observed a significant reduction in chamfering. A comparison of two shuttle tips etched using the STS ICP and PlasmaTherm ICP is presented in Fig. 5(c,d). Since the observed etch rate per cycle was comparable for the two etch systems, a detailed investigation is required to conclude the exact reason for this reduction in chamfering.

A 15-μm-thick PI probe aligned and attached to 25-μm-thick shuttle shanks is demonstrated in Fig. 6(a). The precision of probe alignment and the minimal amount of PEG deposited during the fixation process are illustrated in the SEM micrographs in Fig. 6(b). Here, we applied a minimalistic probe comprised of three shanks interfaced through a broader base. The PEG application area is confined to the front section of the probe and does not cover the electrodes that are further away from the tips, as shown in Fig. 6(b). The developed probe-shuttle assembly approach using PEG as an adhesive material does not lead to any measurable increase in the overall dimensions of the implant {Fig. 6}.

The implantation and release of PI-based probes attached to custom-designed Si shuttles using PEG was further verified by reliably implanting probes in an agar gel-based brain phantom (0.65 wt%). After insertion, the probe-shuttle interface is rinsed with 0.01 M phosphate-buffered saline (PBS). The dissolution of PEG takes ~2 min, providing sufficient time for an accurate insertion, yet a short release time which is beneficial during animal surgeries. The successful implantation of PI probes into the mouse cortex using PEG-fixed ultrathin shuttles was further verified by our neuroscientific partners.

CONCLUSION

A novel and cost-effective process termed EBAG, developed for the fabrication of stepped shuttles with a thicker base and ultrathin shanks was presented. An easy, liquid-assisted self-alignment procedure is used to reliably attach 15-μm-thick flexible PI probes on custom-tailored 25-μm-thick Si shuttle shanks. The developed assembly procedure uses a minimal amount of PEG at the probe-shuttle interface and therefore maintains a small implantation footprint. This is expected to minimize insertion trauma, which is important for a successful integration of probes in brain tissue. The feasibility of inserting multi-shank PI probes PEG-fixed onto ultrathin shuttles was successfully demonstrated in the mouse cortex. Recording data and histological analysis from these implants will be presented in a future publication. Although presented here for the fabrication of Si shuttles, the EBAG process can also be used to fabricate stepped, ultrathin neural probes, as already demonstrated by Otte et al. with a similar process applied on the device level [14].

ACKNOWLEDGEMENTS

This work received funding from the European Union's Horizon 2020 research and innovation program (Grant Agreement number 899287, NeuraViPeR) and by the Federal Ministry of Economics, Science and Arts, state of Baden-Württemberg, Germany, supporting the BrainLinks-BrainTools Center within the sustainability programme for projects of the excellence initiative II (Cluster EXC 1086). The authors gratefully acknowledge technical support by the RSC team at the Department of Microsystems Engineering (IMTEK) regarding cleanroom fabrication and SEM imaging as well as Calogero Gueli, Laboratory for Biomedical Microtechnology for sharing the BGL7120 sample and processing parameters.

REFERENCES

[1] N. A. Steinmetz et al., "Neuropixels 2.0: A Miniaturized High-density Probe for Stable, Long-term Brain Recordings," *Science*, vol. 372, no. 6539, p. eabf4588, 2021.

[2] D. De Dorigo et al., "Fully Immersible Subcortical Neural Probes with Modular Architecture and a Delta-Sigma ADC Integrated Under Each Electrode for Parallel Readout of 144 Recording Sites," *IEEE J. Solid-State Circuits*, vol. 53, no. 11, pp. 3111–3125, 2018.

[3] E. Song, J. Li, S. M. Won, W. Bai, and J. A. Rogers, "Materials for Flexible Bioelectronic Systems as Chronic Neural Interfaces," *Nat. Mater.*, vol. 19, no. 6, pp. 590–603, 2020.

[4] L. Luan et al., "Ultraflexible Nanoelectronic Probes Form Reliable, Glial Scar-free Neural Integration," *Sci. Adv.*, vol. 3, no. 2, p. e1601966, 2017.

[5] M. Vomero et al., "On the Longevity of Flexible Neural Interfaces: Establishing Biostability of Polyimide-based Intracortical Implants," *Biomaterials*, vol. 281, p. 121372, 2022.

[6] K.-K. Lee et al., "Polyimide-based Intracortical Neural Implant with Improved Structural Stiffness," *J. Micromech. Microeng.*, vol. 14, no. 1, pp. 32–37, 2004.

[7] S. Takeuchi, D. Ziegler, Y. Yoshida, K. Mabuchi, and T. Suzuki, "Parylene Flexible Neural Probes Integrated with Microfluidic Channels," *Lab. Chip*, vol. 5, no. 5, pp. 519–523, 2005.

[8] X. Wang et al., "A Parylene Neural Probe Array for Multi-Region Deep Brain Recordings," *J. Microelectromech. Syst.*, vol. 29, no. 4, pp. 499–513, 2020.

[9] Z. Zhao, X. Li, F. He, X. Wei, S. Lin, and C. Xie, "Parallel, Minimally-invasive Implantation of Ultra-flexible Neural Electrode Arrays," *J. Neural Eng.*, vol. 16, no. 3, p. 035001, 2019.

[10] S. H. Felix et al., "Insertion of Flexible Neural Probes Using Rigid Stiffeners Attached with Biodissolvable Adhesive," *JoVE J. Vis. Exp.*, no. 79, p. e50609, 2013.

[11] H. R. Joo et al., "A Microfabricated, 3D-sharpened Silicon Shuttle for Insertion of Flexible Electrode Arrays Through Dura Mater into Brain," *J. Neural Eng.*, vol. 16, no. 6, p. 066021, 2019.

[12] S. Herwik, O. Paul, and P. Ruther, "Ultrathin Silicon Chips of Arbitrary Shape by Etching Before Grinding," *J. Microelectromechanical Syst.*, vol. 20, no. 4, pp. 791–793, 2011.

[13] E. Otte, A. Vlachos, and M. Asplund, "Engineering Strategies Towards Overcoming Bleeding and Glial Scar Formation around Neural Probes," *Cell Tissue Res.*, vol. 387, no. 3, pp. 461–477, 2022.

[14] E. Otte, V. Cziumplik, P. Ruther, and O. Paul, "Customized Thinning of Silicon-based Neural Probes Down to 2 μm," in *42nd Annu. Int. Conf. IEEE Eng. Med. Biol.- Proc.*, Montreal, QC, Canada, 2020, pp. 3388–3392.

CONTACT

K.Sharma, tel: +49-761-203-7196, kirti.sharma@imtek.de

REAL-TIME DYNAMIC LACTATE DETECTION IN A PIPELINE USING A MICROSENSING NEEDLE FOR ICU PATIENT MONITORING APPLICATION

Yuan-Sin Tang[1], Tung-Lin Yang[2], Yu-Ting Cheng[1], Hsiao-En Tsai[2,3], and Yih-Shurng Chen[3,4]

[1] Microsystems Integration Laboratory, National Yang Ming Chiao Tung University, TAIWAN
[2] Department of Surgery, National Taiwan Hospital HsinChu Branch,
[3] National Taiwan University College of Medicine Graduate Institute of Clinical Medicine
[4] Department of Surgery, National Taiwan University Hospital, TAIWAN

ABSTRACT

In this paper, we demonstrate a non-enzymatic microsensing needle for lactate detection in a dynamic environment mimicking the urine flowing through the pipeline of urine bag. Experimental results show the microneedle can exhibit similar linearity of 0.97, sensitivity of 5.9 x 10^3 nA/mM·mm^2 and LoD (Limit of Detection) of 0.29 mM for the lactate detection both under a static and dynamic environment where the flow rate ranges from 0, 10, 50, to 100 mL/min. It is our belief that the microsensing needle with the capability of the real-time lactate monitoring in urine bag pipeline can effectively predict and prevent postoperative patients in an intensive care unit (ICU) from sudden cardiac death by earlier diagnosis and intervention.

KEYWORDS

Non-enzymatic lactate sensor, needle-typed biosensor, Real-time detection, Dynamic environment

INTRODUCTION

Recently, real-time detection of biomarkers in human body has been crucial for clinical diagnosis to the patients possible with microcirculatory failure [1-6]. For instance, kidney tubular biomarkers, mostly elevated urinary NGAL, can be used to predict death in critically ill COVID-19 patients [1]. 4-hour plasma creatinine clearance measurements can help clinician for the early detection and intervention for treatment of acute kidney injury in critically ill patients [2]. Urinary d-lactate concentration significantly relating to neurological symptoms, encephalopathy, and short bowel syndrome and renal function in patients with diabetes can be used as an early renal damage marker [5, 6].

Previous studies have shown that lactate can also be detected from body fluid such as serum, urine, sweat etc. for monitoring the patients with high risk in myocardial injury/impaired tissue oxygenation during acute coronary syndrome, acute decompensated heart failure and shock status [7, 8]. The urinary lactate/creatinine ratio in newborn infants with asphyxia can be used for predicting the development of hypoxic–ischemic encephalopathy [7]. Initial blood lactate can be an independent outcome predictor, whose time course mirrors organ dysfunction in COVID-19 ICU patients with poor clinical outcomes [8]. Thus, there is a high demand for real-time lactate detection in clinical diagnosis.

On the other hand, while the patients with terrible conditions of the heart and lungs have cardiac surgery of extracorporeal membrane oxygenation (ECMO), immediate post-operative and delayed post-operative parameters such as lactate, pH, etc. are required to monitor patients' conditions and observe whether the patients develop other complications including renal failure, liver failure, etc., that potentially exist [9,10]. A patient with renal failure likely to require to take a urine bag for assessing renal function while he/she is on hemodialysis therapy [11]. The traditional hematology and urine test methods in hospitals take a long time to deliver patients' blood or urine sample for biochemistry analysis, thereby resulting in delayed diagnosis and treatment. These situations reveal the imminent need of new sensing technology for fulfilling the real-time detection under not only a static environment but also a dynamic environment.

Previously, we have successfully developed a fully inkjet-printed flexible lactate sensor with high anti-interference ability for clinical blood plasma tests [12,13]. In this paper, we will present a non-enzymatic microneedle based on the flexible sensor technology to facilitate the real-time lactate sensing to meet the needs. A microsensing needle is proposed to be inserted in patient's urine bag pipeline for real-time lactate monitoring as shown in Fig. 1. Via wireless data transmission, the sensing information can be transmitted to handheld devices for allowing medical doctors to do early diagnosis and intervention. The schematic of lactate microsensing needle is shown in Fig. 2 where a flexible inkjet-printed non-enzymatic electrochemical lactate or pH sensor is housed inside a 26G stainless steel microneedle with an open sensing window. Moreover, we believe the scheme can be extended for the application of the microsensing needle in the ECMO system for immediate monitoring post-operative physiological parameters in the future.

SENSING PRINCIPLE OF THE LACTATE SENSOR

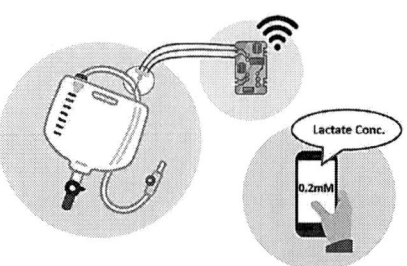

Figure 1: The scheme of real-time lactate monitoring on the urine bag for ICU application using the wireless sensing microneedle.

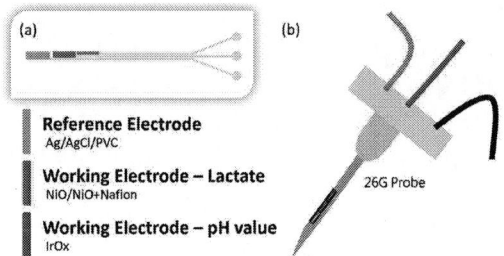

Figure 2: The schematic of lactate microsensing needle (a) A flexible inkjet-printed non-enzymatic electrochemical lactate and pH sensor. (b) A sensor is embedded in a 26G stainless steel needle with a sensing window opened by laser-cutting.

The redox reaction of the NiO$_x$ with lactate is listed as follows:

$$Ni^{2+} \rightarrow Ni^{3+} + e^- \qquad (1)$$
$$Ni^{3+} + Lactate \rightarrow Ni^{2+} + Pyruvic\ Acid \qquad (2)$$

Initially, the active Ni^{2+} ions are converted to Ni^{3+} ions during the anodic scan through the electrochemical oxidation reaction, and Ni^{3+} ions rapidly oxidize the lactate molecules to form pyruvic acid via a two-electron transfer process. Thus, via the measurement of the redox current from the sensing needle in the solutions with different lactate concentrations, a relationship of the current versus lactate concentration can be obtained [13]. All the measurements are conducted under a constant potential biased at 0.45V, which is the anodic peak of lactate found by cyclic voltammetry (CV).

FABRICATION OF THE MICROSENSING NEEDLE AND MEASUREMENT SETUP

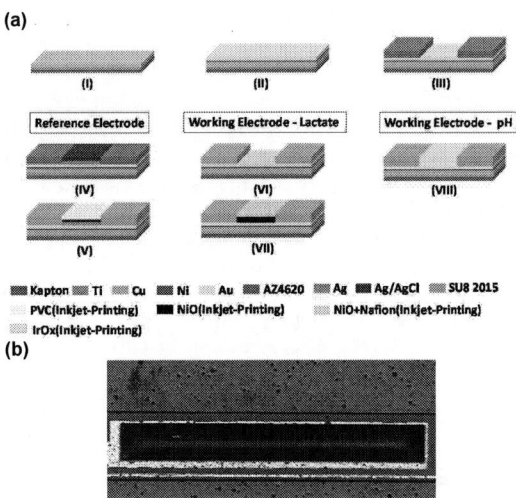

Figure 3: (a) The microsensing needle's fabrication process flow. The fabrication includes photolithography, inkjet printing, and laser cutting processes to form a sensing needle comprising three electrodes, two working electrodes (NiO for lactate and IrO$_x$ for pH detection), one Ag/AgCl$_{(s)}$ reference electrode. The detail fabrication process can be referred to [12,13]. (b) The microphotograph of the working electrode for lactate detection.

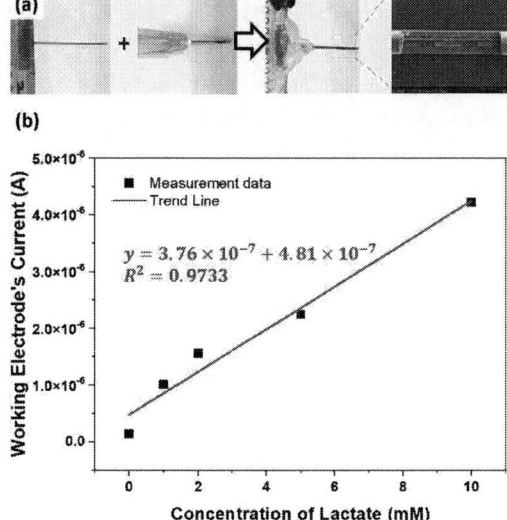

Figure 4: (a) As-fabricated needle (top), and (b) corresponding static lactate measurement, output current vs. lactate concentration, where the lactate sensing probe can exhibit a linearity of 0.97, sensitivity of 5.9 x 10^3 nA/mM·mm^2 and LoD of 0.29 mM (bottom).

Fig. 3 depicts the fabrication process of the microsensing needle. The fabrication comprises nine major process steps for the fabrication of reference and sensing electrodes and the microneedle integration. Regarding the electrode fabrication as shown in Fig. 3(a), it includes (i) sputtering a 30 nm/120 nm Ti/Cu layer on a Kapton substrate, (ii) photopatterning a 9 μm thick AZ4620 photoresist followed by electroplating and electroless plating to form Ni/Au electrodes, (iii) photopatterning another 9 μm thick AZ4620 photoresist followed by electroplating Ag to form sensing and working electrodes, (iv) selective chemical treatment to form a Ag/AgCl$_{(s)}$ reference electrode [12,13], (v) stripping AZ 4620 followed by Cu/Ti wet etching using the mixture of HAC : H$_2$O$_2$: H$_2$O = 5 : 5 : 100 and BOE respectively, (vi) photopatterning 20 μm thick SU-8 to define lactate and pH sensing electrode surface, (vii) and (viii) inkjet printing NiO/NiO-Nafion nanocomposite for the fabrication lactate sensing electrode and IrO$_x$ for the fabrication of pH sensing

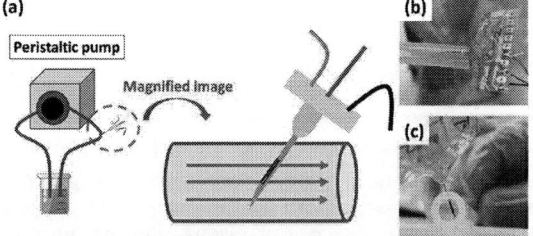

Figure 5: (a) The schematic setup for dynamic lactate measurement, (b) and (c) are the top and side views of the photographs on the sensing needle, which is inserted in the tube for lactate detection indicating the needle can provide enough mechanical strength to penetrate the tube wall.

electrode, respectively. Fig. 3(b) shows the as-printed lactate electrode. Regarding the microneedle integration, it begins with laser cutting the Kapton substrate to form a needle-typed sensor tape followed by inserting the tape in a 26G medical needle with an opening cut by laser as shown in Fig. 4(a).

To mimic the pipeline of patient's urine bag, a rubber tube connected to a peristaltic pump was used. By controlling the peristaltic pump's flow rate, we develop a constantly dynamic environment for lactate detection as shown in Fig. 5. In this experiment, we chose four kinds of flow rate i.e., 0, 10, 50, and 100 mL/min to present the dynamic environment and a microsensing needle was inserted into the tube for detecting lactate concentration. 1 mM, 2 mM, 5 mM, and 10 mM lactate solutions were formed by adding 50mM lactate solution into 0.408 mL, 0.833 mL, 2.222 mL, and 5 mL pH=7.4 PBS respectively.

(a)

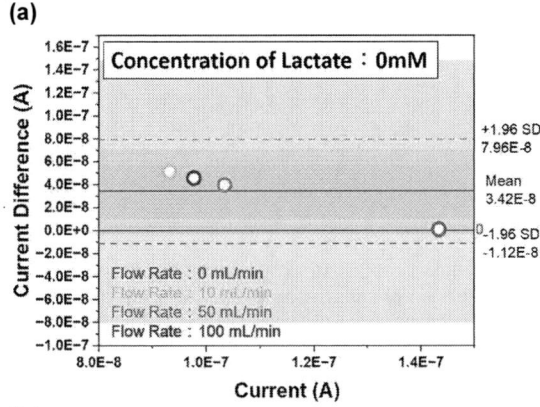

(b)

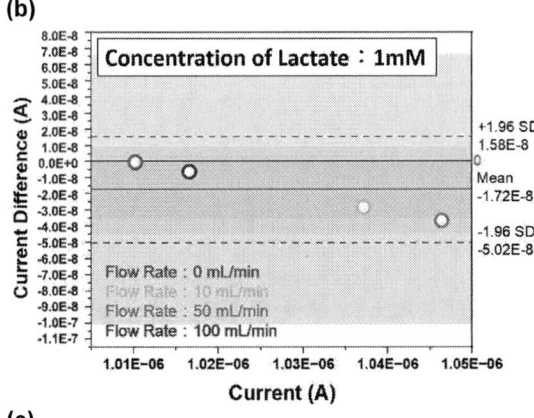

(c)

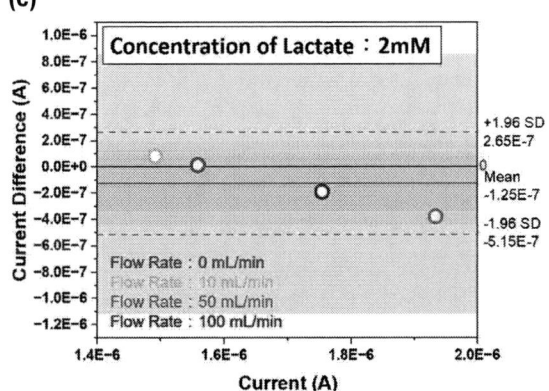

(d)

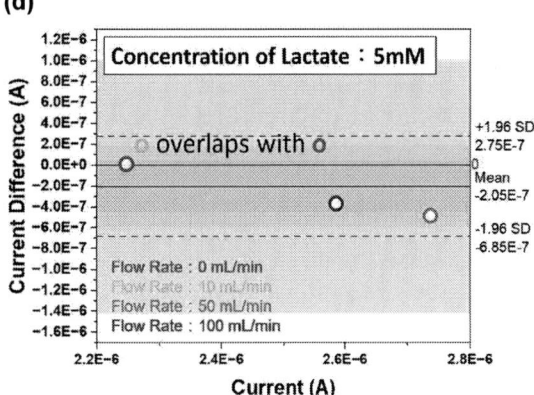

(e)

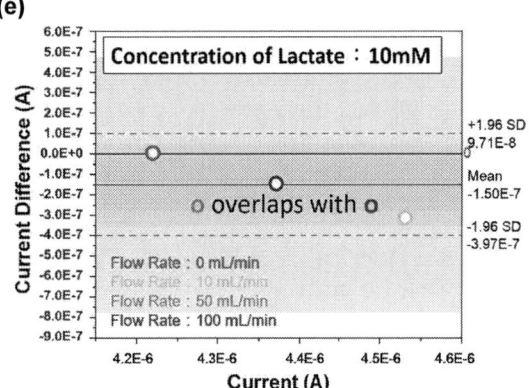

(f)

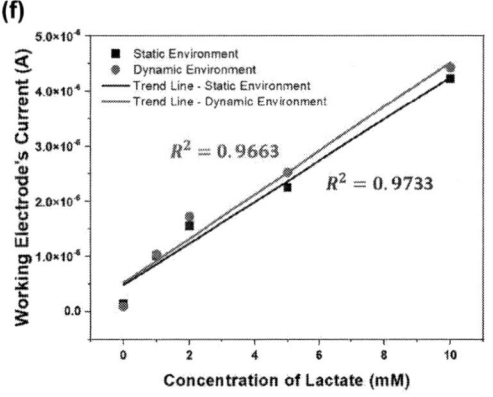

Figure 6: (a)-(e) Bland-Altman plots of the dynamic lactate measurement: detected current under different flow rate, i.e., 0, 10, 50, and 100 mL/min. in the pH=7.4 PBS solutions with different lactate concentrations. The solid blue line shows the mean of the detected current difference between each flow rate and flow rate=0 mL/min; the dashed lines show the upper (mean + 1.96 standard deviation) and lower limits (mean − 1.96 standard deviation) of the interval of agreement indicating that 95% of the measurements' differences between each flow rate and flow rate = 0 mL/min are within this range, (f) The performance of the lactate sensing needle under the static and dynamic conditions, where the linearity and sensitivity of the needle are almost the same, i.e., 0.97 and ~5.9 x 10³ nA/mM·mm².

EXPERIMENTAL RESULTS AND DISCUSSION

Fig. 6 shows the static and dynamic responses of the microsensing needle. The lactate sensing probe exhibits a linearity of 0.97, sensitivity of 5.9 x 10^3 nA/mM·mm² and limit of detection (LoD) of 0.29 mM plotted as the back line in Fig. 6(f). Fig. 6(a)-(e) are Bland-Altman plots from the dynamic lactate measurement where the tested solution flowing through the tube with different flow rate i.e., 0, 10, 50, and 100 mL/min. Besides, the solutions are prepared with five kinds of lactate concentrations from 1 to 10 mM. The value in Y-axis is the current difference between the measured value from the sensing needle inside the tube with flowing liquid and the one with static liquid. From these plots, we can find that all the differences within ±1.96 standard deviation are all smaller than 5x10^{-7} A indicating that the microsensing needle can exhibit similar sensing characteristic whether it is operated in the static or dynamic condition. Furthermore, the performance of the lactate microsensing needle has been characterized under the static and dynamic condition in Fig. 6(f). The results show almost the same sensing characteristic in terms of the linearity and sensitivity of the sensing needle, which are ~0.97 and ~5.9 x 10^3 nA/mM·mm² respectively. The similarity can be attributed to low flowing rate or small effective cross area of the needle inside the pipeline, which will not induce enough turbulence in the sensing region. Since higher flowing rate ranging from 200 to 1000 mL/min will happen in the ECMO circulation system, further investigation is still underway.

CONCLUSION

A microsensing needle based on the flexible sensor technology, photolithography and inkjet printing methods has been successfully developed to facilitate the lactate sensing. Experimental results show that this non-enzymatic lactate sensor can exhibit similar linearity of 0.97, sensitivity of 5.9 x 10^3 nA/mM·mm² and LoD of 0.29 mM for both the real-time lactate detection under a static and dynamic environment.

ACKNOWLEDGEMENT

This work was supported by Center for Neuromodulation Medical Electronics Systems" from The Featured Areas Research Center Program within the framework of the Higher Education Sprout Project by the Ministry of Education (MOE) in Taiwan and in part by the National Taiwan University Hospital Hsin-Chu Branch under the Grant Number 111-HCH001, MOST 110-2321-B-009-004, and NSTC 111-2221-E-A49-110. The graduate student scholarship and travel fund was supported by Google. The authors would like to thank the Nano Facility Center (NFC) in NYCU for the support of fabrication facility.

REFERENCES

[1] G. F. Bezerra, G. C. Meneses, P. L. Albuquerque, N. C. Lopes, R. S. Santos, J. C. da Silva, S. M. Mota, R. R Guimarães, F. R. Guimarães, Á. R Guimarães, C. M. Adamian, P. R. de Lima, I. C. Bandeira, M. M. Dantas, G. B. Junior, R. B Oriá, E. F Daher, and A. M. Martins, "Urinary tubular biomarkers as predictors of death in critically ill patients with COVID-19. Biomarkers in Medicine, vol. 6, no. 9, pp. 681-692, 2022.

[2] J. W. Pickering, C. M. Frampton, R. J. Walker, G. M. Shaw, and Z. H. Endre, "Four hour creatinine clearance is better than plasma creatinine for monitoring renal function in critically ill patients." Critical Care, vol. 16, no. 3, pp. 1-13, 2012.

[3] C. Jung and M. Kelm, "Evaluation of the microcirculation in critically ill patients." Clinical Hemorheology and Microcirculation, vol. 61, pp. 213-224, Nov. 2015.

[4] K. Werdan, B. Patel, M. Girndt, H. Ebelt, J. Schröder, and S. Nuding, "Monitoring of the kidneys, liver, and other vital organs." The ESC Textbook of Intensive and Acute Cardiovascular Care, 126, 2015.

[5] J. P. Talasniemi, S. Pennanen, H. Savolainen, L. Niskanen, and J. Liesivuori, "Analytical investigation: assay of D-lactate in diabetic plasma and urine." Clinical Biochemistry, Elsevier, vol. 41, pp. 1099-1103, June 2008.

[6] C. K. Chou, Y. T. Lee, S. M. Chen, C. W. Hsieh, T. C. Huang, Y. C. Li, and J. A. Lee, "Elevated urinary D-lactate levels in patients with diabetes and microalbuminuria." Journal of Pharmaceutical and Biomedical Analysis, vol. 116, pp. 65-70, June 2015.

[7] C. C. Huang, S. T. Wang, Y. C. Chang, K. P. Lin, and P. L. WU, "Measurement of the urinary lactate: creatinine ratio for the early identification of newborn infants at risk for hypoxic–ischemic encephalopathy: Nejm." New England Journal of Medicine, vol. 341, no. 5, pp. 328-335, July 1999.

[8] A. Thirumurugan, A. Thewles, R. D. Gilbert, S. A. Hulton, D. V. Milford, C. J. Lote, and C. M. Taylor, "Urinary L-lactate excretion is increased in renal fanconi syndrome." Nephrology Dialysis Transplantation, vol. 19, no. 7, pp. 1767–1773, July 2004.

[9] P. Ariyaratnam, L. A. McLean, A. R. J. Cale, and M. Loubani, "Extra-corporeal membrane oxygenation for the post-cardiotomy patient." Heart Failure Reviews, Springer US, vol. 19, pp. 717–725, Mar. 2014.

[10] K. Subramaniam, M. Boisen, P. R. Shah, V. Ramesh, A. Pete, "Mechanical circulatory support for cardiogenic shock." Best Practice & Research Clinical Anaesthesiology, vol. 26, pp. 131–146, Aug. 2012.

[11] J. Himmelfarb and T. A. Ikizler, "Hemodialysis" New England Journal of Medicine, vol. 363, pp. 1833–1845, 2010.

[12] K. L. Tsou, K. Y. Chen, Y. D. Chou, Y. T. Cheng, H. E. Tsai, and C. K. Lee, "Inkjet-printed, flexible, non-enzymatic lactate sensor with high sensitivity and low interference using a stacked NiOx/NiOx-Nafion nanocomposite electrode with clinical blood test verification." Talanta, vol. 249, pp. 123598, 2022.

[13] Y. S. Huang, K. Y. Chen, Y. T. Cheng, C. K. Lee, and H. E. Tsai, "An inkjet-printed flexible non-enzymatic lactate sensor for clinical blood plasma Test. " IEEE EDL, vol. 41, pp. 597-600, 2020.

CONTACT

*Yu-Ting Cheng, ytcheng@g2.nctu.edu.tw.

THREE-DIMENSIONAL FLEXIBLE NEURAL OPTO-ELECTRONIC ARRAY WITH SILK-BASED SHUTTLE-FREE IMPLANTATION

Chi Gu[2,3], Huiran Yang[2], Bohan Zhang[2,4], Zhitao Zhou[2], Liuyang Sun[3,5], Meng Li[2,3], Xiaoling Wei[2,3] and Tiger H. Tao[1,2,3,4,5,6,7,8,9]

[1]Guangdong Institute of Intelligence Science and Technology, Hengqin, Zhuhai, Guangdong, China
[2] State Key Laboratory of Transducer Technology, Shanghai Institute of Microsystem and Information Technology, Chinese Academy of Sciences, Shanghai, China
[3] University of Chinese Academy of Sciences, Beijing, China
[4] School of Physical Science and Technology, Shanghai Tech University, Shanghai, China
[5] 2020 X-Lab, Shanghai Institute of Microsystem and Information Technology, Chinese Academy of Sciences, Shanghai, China
[6] Center of Materials Science and Optoelectronics Engineering, University of Chinese Academy of Sciences, Beijing, China
[7] Center for Excellence in Brain Science and Intelligence Technology, Chinese Academy of Sciences, Shanghai, China
[8]Neuroxess Co., Ltd. (Jiangxi), Nanchang, Jiangxi, China
[9]Tianqiao and Chrissy Chen Institute for Translational Research, Shanghai, China.

ABSTRACT

We report a flexible three-dimensional (3D) opto-electronic array of 512 electrophysiological recording channels with silk-based shuttle-free implantation method. The shuttle-free implantation provides minimally invasive and high-efficiency implantation of the 3D neural electrode arrays with large coverage area and high electrode density. Therefore, we achieved a stable chronic recording of 1 month *in vivo*. We further integrate a silk fiber with highly transparency and low optical loss into our neural electrode array to provide optogenetic stimulation. We believe that this 3D flexible neural opto-electronic array will provide a high-resolution, long-term stable platform for optogenetic-related neural circuit studies.

KEYWORDS

3D Neural Opto-electronic Array, Silk fiber, Shuttle-free Implantation

INTRODUCTION

Neural interfaces build bi-directional communications between electronics and brains. Among them, electrophysiological recording, a method with a long history in neuroscience research, provides the study of the function of neural circuits in the brain. As a milestone product in this research field, Utah array, which is a widely commercialized FDA-approved neural interface, have been used in neurological disease treatment, external device control and other applications [1]. However, it is limited by the large implantation damage and low resolution. Rigid silicon-based probes cause severe nerve scarring, which in turn shows a low signal-to-noise ratio. The relative movement between nerve cells and hard probes also affects long-term stable electrophysiological signal recording. The flexible polymer electrodes fabricated by MEMS-compatible processes significantly addressed these issues [2-6], but they only distribute electrodes along the z-axis of the shank, instead of covering a large area of multiple brain regions in the x- and y-axis directions. In addition, flexible probes often require hard shuttle tools to aid implantation, which still results in a large implant footprint. This shuttle-based implantation also faces the problem of low implantation efficiency.

In addition to electrophysiological recordings, optogenetics also provides a platform for neuroscience research [7]. Through gene modification, specific activation of cells is achieved, which can realize efficient research on the working mechanism of neural circuits. In recent years, a large number of photoelectrode platforms that integrate electrophysiological recording and optogenetics have been proposed through different methods. However, in situ stimulation/recording, minimally invasive, and long-term stability remain important research topics in this field [8-16].

In this work, we combine the advantages of these two schemes to report a flexible electrode array with 3D distribution that covers a large area, and realize its efficient shuttle-free implantation. In addition, we integrated silk fibers to provide optogenetic stimulation. Compared with the previous work about silk neural opto-electronic probe which the electrodes were attached to the surface of the fiber [15-16], the opto-electronic array reported here has the electrophysiological recording sites distributed under the fiber and directly contact with the neurons stimulated by optogenetics, so as to achieve precise and high-resolution opto-electronic stimulation and recording.

ELECTRODE ARRAY

As shown in Figure 1a-b, the 512-channel electrode array is composed of four 128-channel flexible electrodes. The flexible electrodes include polyimide (PI) substrates and packaging, as well as gold electrodes and interconnects (Figure 1c) are fabricated with MEMS-compatible processes. Firstly, the nickel (Ni) sacrificial layer is graphically drawn by electron beam evaporation, UV lithography deposition. Secondly, a PI substrate is deposited by spin coating. Thirdly, the chromium/aurum (Cr/Au) interconnect layer and electrodes are deposited and patterned by UV lithography, electron beam evaporation, lift-off. The PI insulating encapsulation layer is then deposited again by spin coating. Finally, the flexible probe shape is obtained by UV lithography and etching. An optical microscope photograph of the fabricated flexible electrodes is shown in Figure 1d.

Figure 1e shows the silk coating process. The polydimethylsiloxane (PDMS) mold that matches the probe is fabricated using a MEMS-compatible process. Then we carefully align the flexible shank into the mold that has been injected with silk. After drying, the flexible probe is temporary curing. The width and thickness of the silk reinforcement layer are adjusted, therefore minimizing the implantation damage while meeting the bending stiffness required for implantation.

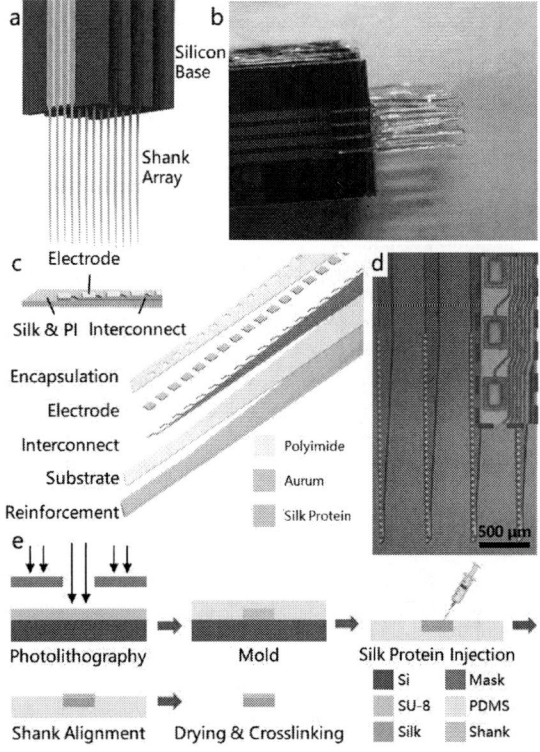

Figure 1: (a-b). Schematic and photographs of the 3D array electrodes. (c) Schematic diagram of the layered structure of flexible electrodes. (d) Optical microscope photo of the flexible electrodes. (e) Schematic diagram of the process flow of silk coating.

SHUTTLE-FREE IMPLANTATION

Figure 2a shows the stress-strain curve of silk film with different boiled time. Although the strain that the membrane can withstand before breaking is greater as the crosslinking time is shorter, different membranes still show similar low Young's modulus (~20 MPa), resulting in a smaller neuroinflammatory response after implantation. Compared to many of the available commonly used neural probes (include microwire, silicon probe, carbon fiber, etc.), the reported electrode array has both low cross-sectional area and high electrode density as shown in Figure 2b. This

allows us to achieve high-resolution distribution of electrodes with a small implant footprint. The feasibility of this silk-based shuttle-free implantation method is demonstrated by the simulation experiments of implanting agar in Figure 2c-d. Agar blocks (~1 wt%) were used to mimic brain tissue. The silk solution used to temporarily cure the flexible probe is doped with a yellow pigment for easy observation. The flexible probe after protein curing can be easily inserted into the agar block in parallel and regain flexibility after implantation.

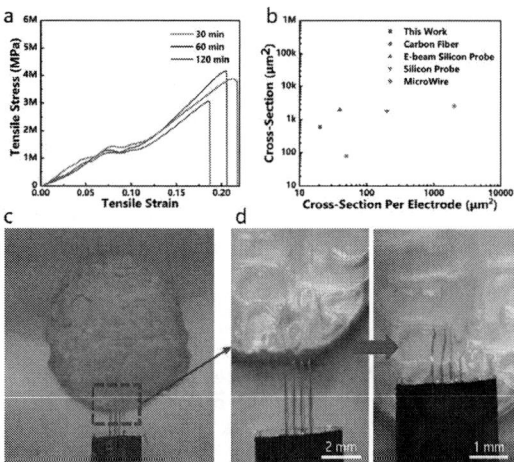

Figure 2: (a) The stress-strain curve of silk film with different boiled time. (b) Cross-section and electrode density of this work compared with previously reported neural probes. (c-d) In vitro simulation of silk-based shuttle-free implantation using an agar-brain model.

ELECTROPHYSIOLOGICAL RECORDING

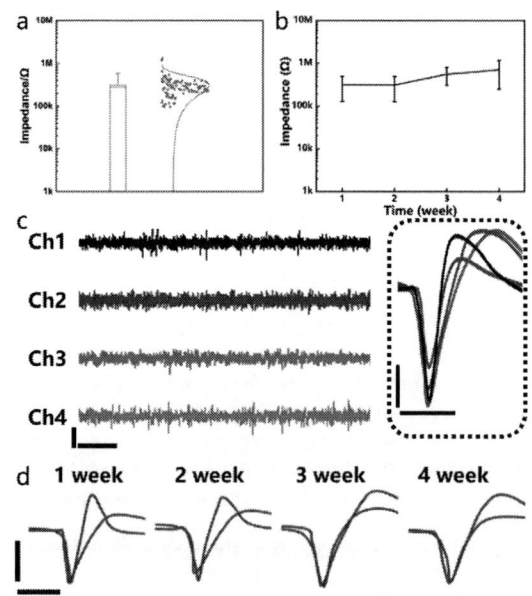

Figure 3: (a) The measured impedance of electrode array. (b) Changes in impedance within one month of implantation. (c) Electrophysiological signals and sorted spikes acquired from the 4 channels of the electrode array. Scale bar: left: 100μV, 50ms; right: 20μV, 0.4ms. (d) Spikes acquired from 2 channels of the electrode array within one month of implantation. Scale bar: 50μV, 0.5ms.

As shown in Figure 3, we verified the electrophysiological long-term recording capability of the electrodes array by in vivo experiments in mice. Before implantation, the electrode sites are electrochemically modified by PEDOT/PSS (EDOT 0.02M, PSS 0.2M, plating current 10nA, plating time 10s) to further reduce the electrode impedance and thus improve the quality of the recorded electrophysiological signal. The modified electrode sites have an average initial impedance of ~250 kΩ and remained stable for 1 month after implantation as shown in Figure 3a-b.

The left in Figure 3c shows the record trace (250Hz high-pass filter) from 4 channels of the electrodes array, and the right are the sorted single-unit from these 4 channels, which demonstrate the single-neuron level electrophysiological recording capability of the reported electrode array. To verify the good biocompatibility of the electrode array, we performed weekly electrophysiological signal recording for a period of 1 month. Figure 3d shows the sorted single-unit from 2 channels of the electrodes array. Both channels acquire a stable single-unit signal during this period. Combined with the stable electrode impedance described above during this period, they illustrate the long-term in-vivo stability of the reported electrodes array.

integration (b) of silk fiber. (c) SEM of silk fiber. (d) Optical loss of silk fibers fabricated with different concentration of silk solution. (e-f) Implantation and optogenetic stimulation capabilities verified using an agar brain model.

OPTO-ELECTRONIC INTEGRATION

The fabrication of silk fibers begins with the extraction of silk solutions based on previous research [17]. We fabricated silk fibers through a wet-spinning-like process. The silk solution (8-20 wt%) is perfused in a syringe and extruded at a uniform rate in a methanol solution. In methanol, the α-helix of silk changes rapidly to β-sheet, resulting in a crosslinking reaction to form silk fiber. The diameter of the silk fiber can be changed (100μm – 1mm) by adjusting the size of the syringe needle and the speed at which the solution is extruded. After washing in deionized water, Silk fiber is soaked in a silk solution doped with antimicrobials, thus coating the surface with an antibacterial layer to further enhance the biocompatibility of silk fiber. As shown in Figure 4b, silk fiber is aligned with a commercial fiber via a ferrule and further connected to a laser source, to integrate it with the electrode array.

Figure 4c is the scanning electron microscope (SEM) image of the silk fiber, showing its smooth surface, which leads to in its high transparency and low transmission loss. This is confirmed by the results of the transmission loss characterization of silk fiber. The lower the mass fraction of the silk solution, the lower the transmission loss of the silk fiber. In our work, silk fiber with transmission loss as low as ~2.8dB/cm was achieved (Figure 4d), which would easily meet the needs of optogenetic stimulation.

The implantation and optogenetic stimulation capabilities of the opto-electronic array were verified by the simulation experiments of implanted agar in vitro shown in Figure 4e-f. The flexible neural opto-electronic array is implanted in parallel and efficiently into agar blocks of simulated brain tissue by the shuttle-free method (Figure 4e). The agar block was doped with sodium fluorescein, and under 485 nm laser excitation, the photoexcitation range after implantation into brain tissue was simulated (Figure 4f).

CONCLUSION

To summarize, we report a flexible 3D neural opto-electronic array with silk-based shuttle-free implantation method. The assembly of multiple flexible electrode probes constructs a 3D electrode array, enabling high-resolution electrophysiological recording over a wide coverage. The Silk-based shuttle-free implantation method reduces implantation damage, improves device biocompatibility, and achieves a single-neuron level electrophysiological record of 1 month. The introduction of silk fiber further integrates the opto-stimulation function on the electrode array, providing an effective optogenetic research platform. We foresee that this 3D neural opto-electronic array will be widely used in optogenetics-related neural circuit research.

ACKNOWLEDGEMENTS

This work was partially supported by This work was partially supported by Key-Area Research and

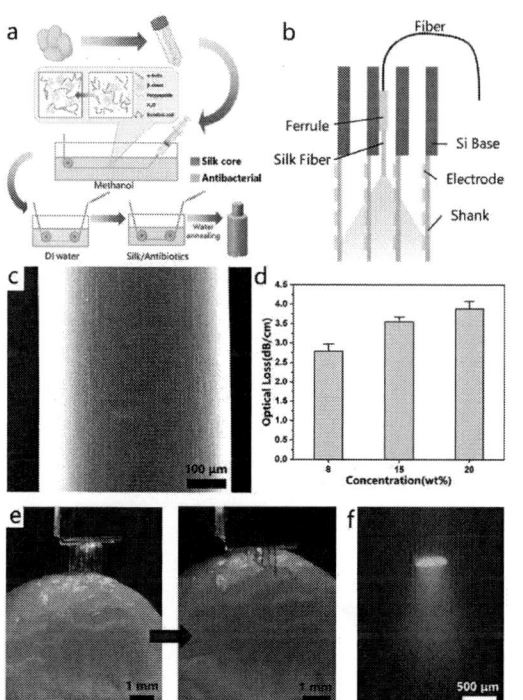

Figure 4: Schematic diagram of process flow (a) and

Development Program of Guangdong Province （2021B0909060002）, Guangdong high level Innovation Research Institute（2021B0909050004）, the National Key R & D Program of China (Grant Nos. 2019YFA0905200, 2021ZD0201600, 2021YFC2501500, 2021YFF1200700, 2022ZD0209300, 2022ZD0212300), National Natural Science Foundation of China (Grant No. 61974154), Key Research Program of Frontier Sciences, CAS (Grant No. ZDBSLY-JSC024), Shanghai Pilot Program for Basic Research—Chinese Academy of Science, Shanghai Branch (Grant No. JCYJ-SHFY-2022-01), Shanghai Municipal Science and Technology Major Project (Grant No. 2021SHZDZX), CAS Pioneer Hundred Talents Program, Shanghai Pujiang Program (Grant Nos.19PJ1410900, 21PJ1415100), the Science and Technology Commission Foundation of Shanghai (No. 21JM0010200), Shanghai Rising-Star Program (Grant No. 22QA1410900), the Innovative Research Team of High-level Local Universities in Shanghai, the Jiangxi Province 03 Special Project and 5G Project (Grant No. 20212ABC03W07), Fund for Central Government in Guidance of Local Science and Technology Development (Grant No. 20201ZDE04013).

REFERENCES

[1] E. M. Maynard, C. T. Nordhausen, R. A. Normann, Electroencephalogr. Clin. Neurophysiol., 102 (1997), pp. 228-239.

[2] L. Lan, X. Wei, Z. Zhao, et al., Sci. Adv., 3 (2017), e1601966.

[3] X. Wei, L. Lan, Z. Zhao, et al., Adv. Sci., 5 (2018), 1700625.

[4] Viveros R D, Zhou T, Hong G, et al. Advanced One- and Two-Dimensional Mesh Designs for Injectable Electronics. Nano Letters, 2019, 19(6): 4180-4187.

[5] Yang X, Zhou T, Zwang T J, et al. Bioinspired neuron-like electronics. Nature Materials, 2019, 18(5): 510-+.

[6] Xie C, Liu J, Fu T-M, et al. Three-dimensional macroporous nanoelectronic networks as minimally invasive brain probes. Nature Materials, 2015, 14(12): 1286-1292.

[7] M. Shaaya, et al., "Optogenetics: The Art of Illuminating Complex Signaling Pathways", *Physiology*, vol. 36, pp. 52-60, 2021.

[8] Rossi M A, Go V, Murphy T, et al. A wirelessly controlled implantable LED system for deep brain optogenetic stimulation[J]. Frontiers in Integrative Neuroscience, 2015, 9.

[9] Jeong J W, McCall J G, Shin G. Wireless Optofluidic Systems for Programmable In Vivo Pharmacology and Optogenetics[J]. Cell, 2015, 162: 662-674.

[10] Kim T, McCall J G, Jung Y H, et al. Injectable, cellular-scale optoelectronics with applications for wireless optogenetics[J]. Science, 2013, 340: 211-216.

[11] Canales A, Park S, Kilias A, et al. Multifunctional Fibers as Tools for Neuroscience and Neuroengineering[J]. Accounts of Chemical Research, 2018, 51: 829-838.

[12] Mohanty A, Li Q, Tadayon M A, et al. Reconfigurable nanophotonic silicon probes for sub-millisecond deep-brain optical stimulation[J]. Nature Biomedical Engineering, 2020, 4(2): 1-9.

[13] Shim E, Chen Y, Masmanidis S C, et al. Multisite silicon neural probes with integrated silicon nitride waveguides and gratings for optogenetic applications[J]. Scientific Reports, 2016, 6(1): 22693-22693.

[14] Segev E, Reimer J, Moreaux L, et al. Patterned photostimulation via visible-wavelength photonic probes for deep brain optogenetics[J]. Neurophotonics, 2016, 4(1): 011002-011002.

[15] C. Gu, H. Yang, B. Zhang, et al., in 2021 Transducers, pp. 291-294.

[16] Zhou, Y., Gu, C., Liang, J. et al. A silk-based self-adaptive flexible opto-electro neural probe. Microsyst Nanoeng 8, 118 (2022).

[17] D. N. Rockwood, et al., "Materials fabrication from Bombyx mori silk fibroin", *Nat. Protoc.*, vol. 6, no. 10, 2011.

CONTACT

*X. Wei, tel: +86-21-62511070; xlwei-jerry@mail.sim.ac.cn
*Tiger H. Tao, tel: +86-21-62511070; tiger@mail.sim.ac.cn

A MICROFLUIDIC BIOSENSOR FOR RAPID DETECTION OF COVID-19

Sura A. Muhsin[1], Ying He[2,3,4], Muthana Al-Amidie[1], Karen Sergovia[2,3,4], Amjed Abdullah[1], Yang Wang[2,3,4], Omar Alkorjia[1], Robert A. Hulsey[5], Gary L. Hunter[5], Zeynep Erdal[5], Ryan J. Pletka[5], George S. Hyleme[5], Xiu-Feng Wan[1,2,3,4], and Mahmoud Almasri[1]

[1]Department of Electrical Engineering and Computer Science, College of Engineering; [2]Center for Influenza and Emerging Infectious Diseases; [3]Department of Molecular Microbiology and Immunology, School of Medicine; [4]Bond Life Sciences Center, University of Missouri, Columbia, Missouri, USA. [5]Black and Veatch, Overland Park, Kansas.

ABSTRACT

We have designed, fabricated, and tested a MEMS-based impedance biosensor for accurate and rapid detection of severe acute respiratory syndrome coronavirus 2 (SARS-COV-2) using of clinical samples. The device consists of focusing region that concentrate low quantities of the virus present in the samples to a detectable threshold, trap region hat maximize the captured virus, and detection region to detect the virus with high selectivity and sensitivity, using an array of interdigitated electrodes (IDE) coated with a specific antibody. Changes in the impedance value due to the binding of the SARS-COV-2 antigen to the antibody will indicate positive or negative result. The device was able to detect inactivated SARS-COV-2 antigen present in phosphate buffer saline (PBS) with a concentration as low as 50 TCID50/ml in 30 minutes. In addition, the biosensor was able to detect SARS-COV-2 in clinical samples (swabs) with a sensitivity of 84 TCID50/ml, also in 30 minutes.

KEYWORDS

Microfluidic channel, MEMS, biosensor, SARS-COV-2, dielectrophoresis, microfabrication.

INTRODUCTION

COVID-19 pandemic expects to cause multiple waves and remains to become an endemic disease in the future [1]. As November 20 of 2022, it has caused > 634 million laboratories confirmed infections worldwide, of which > 6.59 million were fatal. The total estimated deaths caused by COVID-19 infection in the United States alone were > 1.07 million [2,3,4,5].

The majority of the currently FDA-approved tests to detect COVID viruses are nucleic acid- based on methods and monoclonal antibody based rapid antigen test (RAT). The nucleic acid based methods such as RT-PCR has high sensitivity and specificity but take up to six hours to obtain the results [4]. The recent Abbott Diagnostics's nucleic acid-sequence based method requires as short as 15 minutes; however, the data for sensitivity and specificity is still not publicly available [5]. Nevertheless, these methods can result in potential false negative results due to rapid mutations in viruses, especially RNA viruses such as COVID-19 [6]. RAT, particularly the strip paper, is a rapid immunoassay with the results acquired within approximately 15 minutes and has been widely for self-test. However, RAT can reliably detect viral loads in the clinical samples with 3,000 TCID50/ml [7] thus cannot detect viruses during early stage of diseases Multiple diagnosis platforms for diagnosis of COVID-19 were released in the past several months [8]. Of those methods, the antibody-based testing is proved to be rapid and reliable to identify infection. A countless number of groups have investigated various diagnostic techniques for point-of-care (POC) as well as self-detection kits. However, the sensitivity of these kits depends on period of time between infection and testing to produce and detectable response of IgM [9].

We rapidly developed a diagnostic method to detect the SARS-COV-2 virus in clinical samples. Given the limits of present testing, it is evident that a more sensitive, real-time, field deployable detection approach is urgently needed to make testing reliable and feasible, which might have a significant impact on virus surveillance.

MATERIALS AND METHODS

Biosensor Design

The device consists of fluidic microchannel, which include a region for focusing the SARS-COV-2 virus into the centerline of this region and directing them toward the detection region to obtain highly concentrated samples, a trapping region surrounding the detection electrodes, and detection region (Figure 1). The focusing region consists of two set of electrode pairs, each set uses a ramp down vertical electrode pair made of electroplated gold along with tilted thin film finger pairs with a ramp down channel that generates positive dielectrophoresis (p-DEP) forces to focus and concentrate the bacteria into the center of the microchannel, and direct them toward the sensing microchannel. This ensures detection of low concentrations of the virus. The virus detection region consists of two set interdigitated electrode (IDE) arrays surrounded by the trapping electrode pairs. The first set is used for virus detection, while the second set is used for negative control. It was not coated with antibody. The trapping electrodes is used to trap the virus on top of the detection electrode. It is made of electroplated gold with a thickness of 15 μm. The biosensor was sealed using PDMS and wire-bonded to a printed circuit board.

Device Fabrication Methodology

The biosensor was fabricated on a glass substrate using surface micromachining processing steps. Figure 2 shows a side view along with optical images of the fabricated focusing, trapping and detection electrode and the packaged device. The glass substrate was first cleaned using piranha solution for a 3 minutes. The Pirhana consists of a mixture of hydrogen peroxide (H_2O_2) and sulfuric acid (H_2SO_4) with a ratio of 1 : 3. The substrate was then washed using DI water and blown dry with

IEEE MEMS 2023, Munich, GERMANY
15 - 19 January 2023

nitrogen. The fabrication process sequence is as follow: (1) A layer of SU-8 with a thickness of 4 μm was spin-

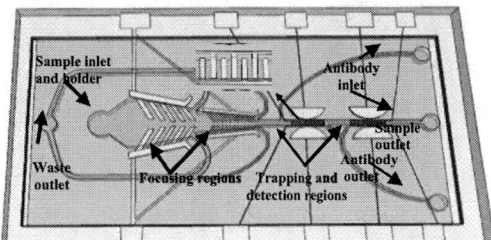

Figure 1. A schematic of the impedance-based biosensor.

coated to improve the adhesion between the SU-8 channel and glass substrate. It was first prebaked at 65 °C for1 minute, and 95 °C for 2 minutes, exposed to flood UV light, and then post baked at 65 °C for for 30 seconds, and 95 °C for 90 seconds. The substrate was then hard baked at 150 °C for 30 minutes to harden the photoresist layer. (2) Thin films made of Chromium (Cr) and gold (Au) were deposited and patterned to create the tilted thin film finger pairs, IDE arrays, electrode traces and bonding pads, and seed layer for the locations corresponding to the ramp down focusing electrodes, and the trapping electrodes. The Au was etched using a mixture of potassium iodide (KI) iodine (I_2), and DI water. (3) A photoresist mold (AZ4620) was formed, and Au with a thickness of 12 μm was electroplated to create the focusing electrodes sidewall and trapping electrodes. The photoresist mold was then washed away using Acetone, and Isopropanol (IPA), and the Cr layer was etched in chromium etchant. (4) The microchannel was then fabricated using SU-8 photoreisst (SU8-2050). The channel height was 50 μm. (5) A polydimethylsiloxane (PDMS) cover was used to seal the microchannel (Figure 2b). the PDMS mold was prepared with inlets and outlets using a 3D printed microstructure to seal the microchannel. After mixing the PDMS with a curing agent, it was poured into the mold and left to dry overnight. This PDMS slab was used to serve as top cover along with fluidic connectors. An oxygen plasma treatment was applied on the PDMS cover in order to change its surface to hydrophilic, and then SU-8 will be spin coated onto it to serve as glue. The oxygen plasma step was used to improve the adhesion of SU-8 to PDMS. The microchannel was aligned and bonded to the PDMS cover manually. The PDMS/SU-8 cover was cross-linked with SU-8 microchannel and form a strong bond (Figure 2e). The device was fixed and wire bonded to PCB for electrical connections.

Antibody and Antigen Sample Preparation

Anti-SARS-CoV-2 ferret polyclonal antibody was used as the capture antibody. IgG was purified from the polyclonal ferret anti-SARS-COV-2 serum, and the purified polyclonal antibody original concentration was determined with the working stock concentration of 1.8 mg/mL. The antibody was then mixed with a cross-linker sulfosuccinimidyl 6-[3-(2-pyridyldithio) propionamido] hexanoate (Sulfo-LC-SPDP) (sulfosuccinimidyl 6-[3-(2-pyridyldithio) propionamido] hexanoate (Sulfo-LC-

SPDP) in order to improve the adhesion of the antibody to the first IDE set without contaminating the second IDE array set that is used for negative control. The final concentration of antibody was then determined by Quant-iT Protein Assay Kit to be 1.66 mg/ml, which was then diluted and tested at various concentrations between 0.0166 - 1.66 mg/ml. To coat the antibody cross-linker, the mixure was loaded into the antibody inlet while suction was applied to the anibody outlet (All other inlets and outlets were closed). Once the detection channel was filled with the solution, the flow was stopped for 1.5 hours to allow the antibody to non-specifically adsorbed to the gold surface of the electrode. Any unbounded antibodies were washed using distilled water for 15 minutes. The anybody impedance was then measured.

The SARS-CoV-2 viruses were prepared with a final concentration of 2×10^4/well at 37 °C in 5% CO_2. The viruses were then inactivated and serially diluted into various concentrations with phosphate buffer saline

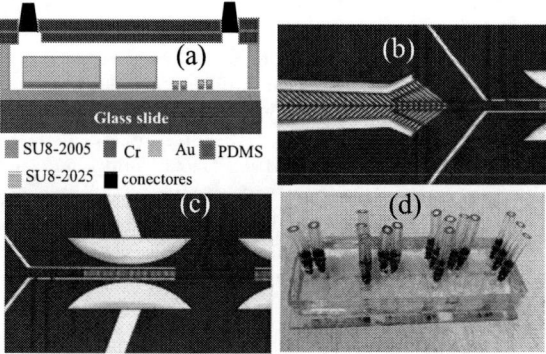

Figure 2. (a) Side-view of the fabricated biosensor, (b,c) Optical images of the focusing and detection regions. (d) completed and packaged device.

(PBS). The human nasopharyngeal or nasal swab samples were collected from COVID-19 patients in the summer of 2020, inactivated and diluted 1:10 in PBS before testing.

The virus testing sample was then loaded inrto the sample inlet and suction was applied into the sample outlet while all other inlets are closed. And thus, the sample was directed to flow toward the focusing region and then toward the detection region. Once the channel was filled with the sample, the flow was stopped for 30 minutes so that virus antigen could bind to its specific antibody. Any unbound particles or viruses were then washed away using distilled water for 15 minutes. The impedance was measured and subtracted from the antibody impedance. The difference determine the status of the sampel, i.e., positive or negative. The sensitivity was measured for various concentrations in order to determine the lowest measurable concentration (highest dilution) in known positive and negative samples.

RESULTS AND DISCUSSIONS
Focusing Effect

To demonstrate the focusing and trapping capabilities, polystyrene nanobeads with diameter < 1μm suspended in water were loadedinto the sample inlet while suction was applied into the sample outlet directing the flow toward the focusing electrode region, which concentrates the microbeads at the centerline of the microchannel. The concentrated microbeads sample was continued to flow toward the detection/trapping regions. We have determined optimum AC voltages and frequencies for the focusing and trapping electrodes, i.e., 4 Vp-p at 5 MHz, 5 Vp-p at 6 MHz, respectively. The applied voltages generated a nonuniform electric field gradient that caused the microbeads to be focused in the center of the channel and then be trapped on top of the detection electrode (Figure 3). The figure shows the microbeads after applying both the focusing and trapping effects.

Figure 3. An optical image of the focusing and detection regions showing the nanobeads after applying the focusing and trapping effects.

Sensitivity and Selectivity Testing

In the first experiment, we have tested several antibody concentrations to obtain an optimum coating concentration. Each antibody dilution was tested using propagated inactivated SARS-CoV-2 virus at original concentration (9.05×10^{10} viral RNA/ml in PBS) and a fixed antibody coating time, i.e., 1 hour. We have determined that a concentration of 2 µg/ml was the optimum one with the highest impedance value. The coating time was fixed at 1 hour in this experiment (Figure 4).

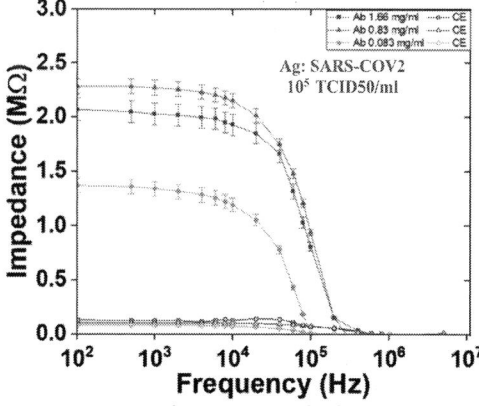

Figure 4: Testing of optimum antibody concentration.

In the second experiment, the lowest measurable concentration (LOD) was determined. An original concentration of SARS-CoV-2 virus was diluted with PBS to achieve various viral RNA concentrations from 9.05×10^{11} /ml to 9.05×10^{7} /ml. The samples were then loaded into the sample inlet while suction was applied to the sample outlet with first IDE array coated with the appropriate antibody at a concentration of 0.083 µg/ml. The results were plotted in (Figure 5) and demonstrated

that the detection electrode that has matching antibody to the antigen showed strong impedance change. The impedance change of the control electrode was very small confirming the signal was correct. The LOD was 4.52×10^7 copies/ml.

In the third experiment, the specificity of the

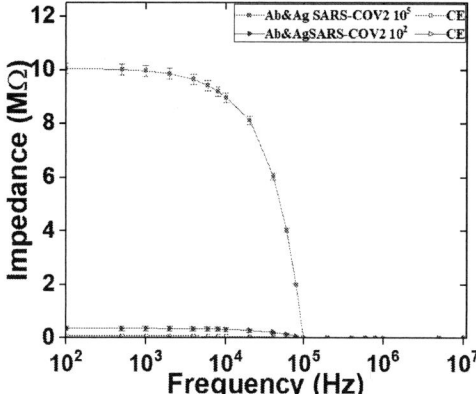

Figure 5: Testing of SARS-CoV-2 virus at two dilutions. Ab refers to antibody, Ag refers to antigen, CE refers to negative control.

biosensor was confirmed. The antibody anti-SWZ/13 influenza was diluted to 0.083 mg/ml and then immobilized on the sensing electrode while SARS-CoV-2 samples with a fixed concentration of 10^5 TCID50/ml

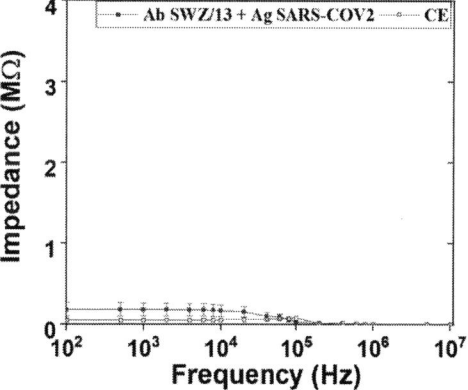

Figure 6: Testing of specificity by immobilizing antibody (Ab) SWZ/13 against SARS-CoV-2 samples.

(correlated to 9.05×10^{10} RNA copies/ml) in PBS were tested as an antigen. The measured response of the influenza antibody showed no difference in the impedance measurement values with respect to the baseline impedance (antibody value) of the detection electrode confirming the specificity of the biosensor (Figure 6).

The fourth experiment determined the selectivity of the biosensor. After coating the detection electrode with an SARS-CoV-2 antibody (0.083 mg/ml), influenza virus samples with a concentration of 10^6 TCID50/ml (correlated to 10^9 RNA copies/ml) were loaded into the biosensor sample inlet. The virus sample was delivered in the same manner. The impedance was measured again for both the detection electrode and the control electrode. The impedance change was very small indicating the biosensor is selective (Figure 7).

Two inactivated clinical human swab samples with

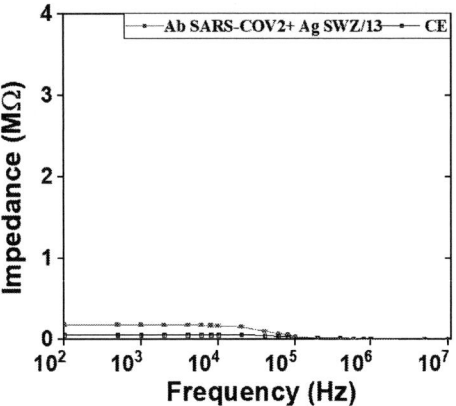

Figure 7: The selectivity was measured using influenza virus samples with a concentration of 10^6 TCID50/ml with the sensing electrode coated with SARS-CoV-2

various viral RNA levels between 3.63×10^6-5.88×10^{12} copies/ml were tested using the same procedures discussed to determine the biosensor sensitivity. Prior to testing, the samples were subjected to 10-fold dilution. The testing results showed that the 2 diluted samples with final concentrations between 2.37×10^7 and 5.88×10^{11} RNA copies /ml were tested positive. The results indicate as low as 2.37×10^7 RNA copies/ml could be detected by our biosensor device (Figure 8).

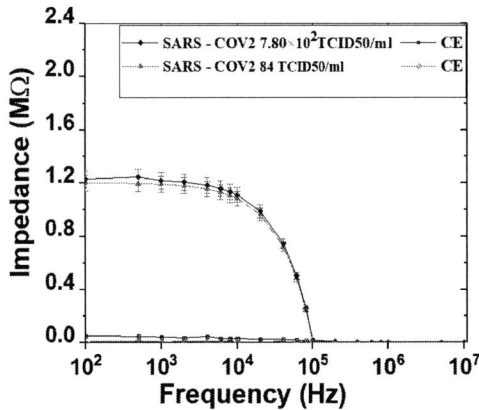

Figure 8: Testing of two inactivated clinical human samples with final titrations between $84 - 7.80 \times 10^5$ TCID50/ml after 10-fold dilution.

CONCLUSION

This paper has presented design, fabrication, and testing of a disposable impedance based biosensor for SARS-CoV-2 detection in clinical human samples for use in-field for clinical settings or in laboratories. A combination of focusing and trapping electrode pairs were used to maximize the virus concentration available for binding with the antibody to increase the sensitivity. The device sensitivity was demonstrated by testing the virus in

PBS. The biosensor was able to detect SARS-CoV-2 antigen in clinical human samples with a dilution as low as 84 TCID50/ml in 40 minutes.

ACKNOWLEDGEMENTS

This project was funded by Black and Veatch. We acknowledge Drs Xiaojiang Zhang and Minhui Guan for their technical assistance in this project, and Dr. Jane McElroy, Naser Ashiekh, and Cynthia Tang for their assistance in acquiring and de-identifying human clinical samples. We are grateful for Erin Keys, George Gering, Hannah Langbart , and Tami Hansen from Columbia City Sanitary Sewer and Storm Water Utilities of City for their assis-tance in collecting water samples used in this study. The first two authors have equal contributions in the experiment. The last two authors have equal contribution.

REFERENCES

[1] Zhou, Peng, Xing-Lou Yang, Xian-Guang Wang, Ben Hu, Lei Zhang, Wei Zhang, Hao-Rui Si et al. "A pneumonia outbreak associated with a new coronavirus of probable bat origin." nature 579, no. 7798 :270-273, 2020.

[2] https://covid19.who.int/

[3] https://covid.cdc.gov/covid-data-tracker/#datatracker-

[4] Shaffaf, Tina, and Ebrahim Ghafar-Zadeh. "COVID-19 diagnostic strategies. Part I: Nucleic acid-based technologies." Bioengineering 8, no. 4: 49, 2021.

[5] J. Liu et al., "An integrated impedance biosensor platform for detection of pathogens in poultry products," Scientific Reports vol. 8, no. 1, pp. 1–10, doi: 10.1038/s41598-018-33972-0, 2018.

[6] Tahamtan, Alireza, and Abdollah Ardebili. "Real-time RT-PCR in COVID-19 detection: issues affecting the results." Expert review of molecular diagnostics 20, no. 5 ,453-454, (2020).

[7] Deerain, Joshua M., Thomas Tran, Mitchell B. Batty, Yano Yoga, Julian Druce, Charlene Mackenzie, George Taiaroa et al. "Assessment of twenty-two SARS-CoV-2 rapid antigen tests against SARS-CoV-2: A laboratory evaluation study." medRxiv (2021).

[8] S. Stanley et al., "Limit of Detection for Rapid Antigen Testing of the SARS-CoV-2 Omicron and Delta Variants of Concern Using Live-Virus Culture," J Clin Microbiol, vol. 60, no. 5, doi: 10.1128/JCM.00140-22, 2022.

[9] M. Yüce, E. Filiztekin, and K. G. Özkaya, "COVID 19 diagnosis —A review of current methods," Biosens Bioelectron, vol. 172, p. 112752, doi: 10.1016/J.BIOS.2020.112752, 2021.

CONTACTS

M. Almasri: 1-(573) 882-0813, almasrim@missouri.edu.

A LOOP-MEDIATED ISOTHERMAL AMPLIFICATION (LAMP)-BASED POINT-OF-CARE SYSTEM FOR RAPID ON-SITE CLINICAL DETECTION OF SARS-COV-2 VIRUSES

Trieu Nguyen[1], Aaydha Chidambara Vinayaka[1], Van Ngoc Huynh[1], Quyen Than Linh[1], Sune Zoëga Andreasen[1], Mohsen Golabi[1], Dang Duong Bang[1], Jens Kjølseth Møller[2] and Anders Wolff[1]*

[1]Department of Biotechnology and Biomedicine, Technical University of Denmark, Kongens Lyngby, Denmark and

[2]Department of Clinical Microbiology, University Hospital of Southern Denmark, Vejle Hospital, Beriderbakken 4, DK-7100 Vejle, Denmark

ABSTRACT

In the ongoing COVID-19 pandemic, sensitive and rapid on-site detection of the SARS-CoV-2 coronavirus has been one of crucial objectives. A point-of-care (PoC) device called PATHPOD for quick, on-site detection of SARS-CoV-2 employing a real-time reverse-transcription loop-mediated isothermal amplification (RT-rLAMP) reaction on a polymer cartridge. The PATHPOD consists of a standalone device (weighing under 1.2 kg) and a cartridge, and can identify 10 distinct samples and 2 controls in less than 50 minutes. The PATHPOD PoC system is fabricated and clinically validated for the first time in this work

KEYWORDS

SARS-CoV-2, point-of-care, COVID-19, LAMP, limit of detection, open-source hardware, total internal reflection, TIR.

INTRODUCTION

More rapid and widespread tests is crucial to mitigate the COVID-19 pandemic and limit lockdowns[1] Current method for SARS-CoV-2 detection is real-time reverse-transcription polymerase chain reaction (RT-rPCR)[2] which can take 16 - 48 hours to have the results due to the sample transportation and laboratory works [1]. Point of care (PoC) testing, which can perform the test on-site, rapidly and cost-efficiently, is a potential candidate to overcome the drawbacks of conventional RT-rPCR[3]. Our goal is to provide on-site PoC systems that will give reliable results in much shorter time.

Loop-mediated isothermal amplification (LAMP) reaction is a nucleic acid amplification method that amplifies DNA under isothermal conditions with high specificity, sensitivity, and speed. The technique synthesizes up to 10^9 copies from target DNA in less than an hour at a constant temperature of approximately 65°C using a set of four specifically designed primers and a DNA polymerase with high strand displacement activity[4]. The reaction produces pyrophosphate as a byproduct (Figure 1a). The pyroposphate byproduct combines with magnesium ion to form the precipitate magnesium pyrophosphate, which can be used to monitor the reaction (Figure 1b).

MATERIALS AND METHODS

The PATHPOD/CORONADX system

The PATHPOD system consist of a cartridge and a stand-alone instrument.

Cartridge

The cartridges consist of an injection molded microfluidic cyclic olefin copolymer (COC) (Topas 5013L-10), a laser-cut holder made of black poly methyl methacrylate (PMMA) polymer, laser-cut double-side PSA tape for assembly, a laser-cut dust lid, and a laser-cut permanent lid (figure 2). The cartridge has 12 wells for 10 individual samples and 2 controls. The reagents are pre-loaded in the cartridge in a gelified form, and has a shelf life of up to 1 year at 5°C.

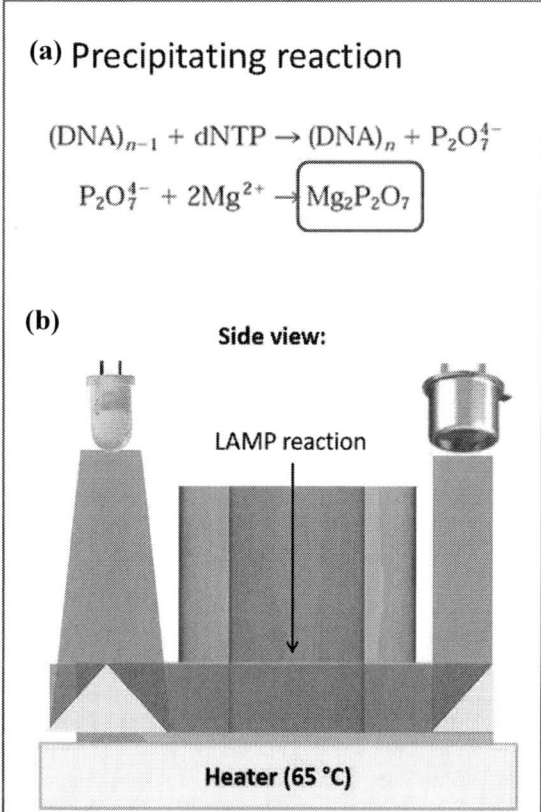

Figure 1 (a) Precipitate forming (magnesium pyrophosphate) during the amplification of DNA is detected by sending light across the reaction well and measure the absorption of light (b).

PATHPOD instrument

The PATHPOD CORONADX standalone instrument (weight 1.2 kg, Figure 3b) can heat the samples in the cartridge and keep the reaction temperature constant at 65°C ± 0.25°C by using a PID algorithm (controlled by an Arduino board). The instrument also contains the optical detection system. The principle of the optical system used to monitor the reaction are shown in (figure 1b): The incident LED light (λ= 535 nm) hits a pyramid structure in the COC (refractive index np = 1.53) and is totally internal reflected (TIR) at the COC/air interface. The reflected light continues across the reaction well to another pyramid structure where the light is TIR reflected out of the chip to a photodiode where the amount of transmitted light is measured. As the LAMP reaction progresses more and more magnesium pyrophosphate.

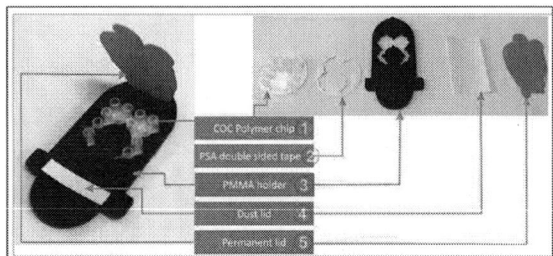

Figure 2. Components of the PATHPOD cartridge

RT-rLAMP reactions

Loading of samples in the cartridge and of cartridge in instrument is shown in figure 3a and 3b. The RT-rLAMP reactions are performed at 65°C for 50 min, and then terminated by heating to 90°C for 5 min in the PATHPOD instrument. The detection principle is light absorption as the result of precipitate forming (magnesium pyrophosphate) during the amplification of DNA (Figure 1).

EXPERIMENTAL RESULTS

LAMP primers

To design LAMP primer set to detect COVID19, genome sequences of 800 SARS-CoV-2 virus originated from EU countries, China and USA were extracted. Highly conserved regions of the Nucleosis (N) gene (involved in Viral RNA packaging of the SARS-CoV-2 viruses) were selected for designing the RT-rLAMP primer sets using Primer Explorer V5 software

Study design

Throughout the COVID-19 pandemic, routine samples were submitted to the Department of Clinical Microbiology for COVID-19 diagnoses (March 2020). Both primary healthcare facilities and hospitals provided throat swabs in EswabTM medium (Copan, Italy). The standard laboratory van service or internal hospital transport were used for the transportation.

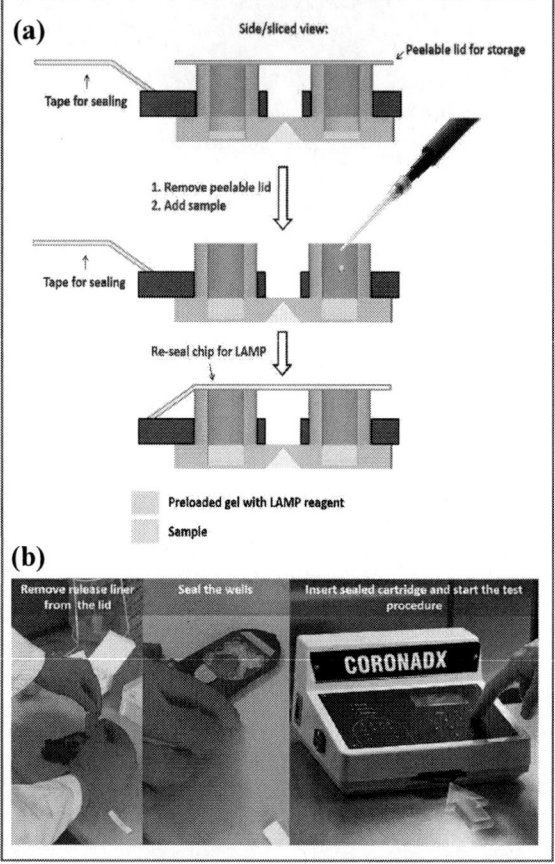

Figure 3. Protocol of loading and sealing the cartridge for running LAMP on the PATHPOD system. (a) Schematic with side-view showing the protocol for loading sample and sealing the cartridge/wells. (b) Picture series showing the sealing and starting the test on the instrument.

Detection with/without PATHPOD

Current diagnostic methods of SARS-CoV-2, such as RT-rPCR and other studies using RT-rLAMP often required the use of purified RNA samples. However, the purification method requires advanced centralized laboratory facilities, and well-trained personnel, which are costly, time-consuming, and not compatible with PoC. We developed a simple boiling method wherein clinical samples were heated to 95 °C for 5 min before added to the cartridge.

A total of 210 ESwab COVID-19 positive and negative samples that were routinely examined in March 2020 at the Department of Clinical Microbiology, Vejle Hospital are selected for the study. Two different sets of the samples that included the 110 RNA extracts typically used in the altona PCR assay, and the other set included the remaining of 100 ESwab heated at 95°C for 5 minutes (no RNA extraction/purification). The samples were then shipped frozen to DTU for analysis. Table 1 displays the result distribution.

For identification of the SARS-CoV-2 virus in the RNA extracts of 110 samples that were purified by Altona assay as a standard test in Vejle Hospital, the PATHPOD

Coronadx LAMP assay had a sensitivity of 94.0% (95% CI: 83.8 -97.9) and a specificity of 96.7% (95% CI: 88.6 - 99.1) respectively. While the sensitivity and specificity of the PATHPOD Coronadx LAMP tests were 73.4% (95% CI: 59.7 - 83.8) and 92.2% (95% CI: 81.5 - 96.9), respectively, based on direct testing without initial RNA extraction of 110 samples (Table 2).

Figure 4 displays a scatterplot of pairings of Ct values and times to detection for samples that tested positive for SARS-CoV-2. The Altona real-time PCR assay's CT-values and the PATHPOD LAMP assay's time to detection were observed to have a positive direct linear correlation both with and without the usage of purified RNA. When materials were just heated before testing, however, the PATHPOD LAMP assay's time to detection increased.

Table 1. *summary the result of testing PATHPOD with real sample from a hospital in Denmark*

Altona PCR (RNA extraction)	CRONADX (RNA extraction)	CORONADX (Heating 95°C, 5 min)	No. of samples
+	+	NA	47
+	-	NA	3
Inhibition (I)	+	NA	1
-	+	NA	1
-	-	NA	58
+	NA	+	36
+	NA	-	13
Inhibition (I)	NA	-	2
-	NA	+	4
-	NA	-	45

+ = PCR positive, - = PCR negative, I = Inhibited, NA = not applicable

Table 2. *Comparison the performance of the PATHPOD system with Altona PCR with real samples*

	ATLONA RT-PCR				Coronadx		95% confidence interval	
	POSITIVE	NEGATIVE						
Coronadx Positive	47	2	49		Sensitivity	94.0%	83.8	97.9
Coronadx Negative	3	58	61		Specificity	96.7%	88.6	99.1
TOTAL	50	60	110		NPV	95.1%	86.5	98.3
					PPV	95.9%	86.3	98.9

1 Altona inhibited sample

Series 2 (95°C, 5 minutes):	PATHPOD RT-LAMP				Coronadx		95% confidence interval	
	POSITIVE	NEGATIVE						
Coronadx Positive	36	4	40		Sensitivity	73.4%	59.7	83.8
Coronadx Negative	13	47#	60		Specificity	92.2%	81.5	96.9
	49	51	100		NPV	78.3%	66.4	86.9
					PPV	90.0%	77.0	96.0

2 Altona inhibited sample

REFERENCES

[1] Nguyen, T.; Bang, D. D.; Wolff, A. 2019 Novel Coronavirus Disease (COVID-19): Paving the Road for Rapid Detection and Point-of-Care Diagnostics. *Micromachines* **2020**. https://doi.org/10.3390/mi11030306.

[2] VM, C.; O, L.; M, K.; R, M.; A, M.; Chu, D. Detection of 2019 Novel Coronavirus (2019-NCoV). *Euro Surveill* **2020**, *0* (0).

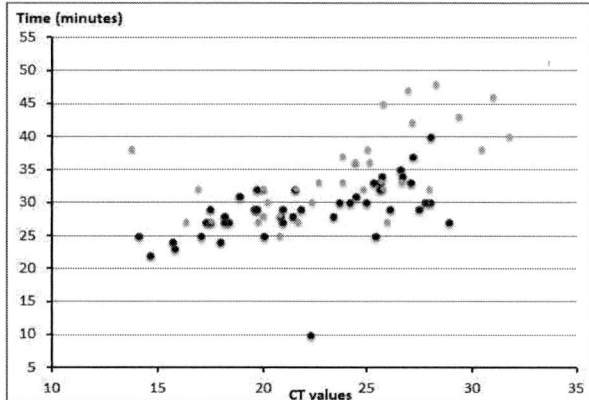

Figure 4. *Altona PCR assay (CtT values) versus PATHPOD CORONADX-LAMP assay (time to detection).*
●: *Both Assays performed with RNA extracted specimens,*
◐: *Coronadx assays performed with ESwab medium heated to 95°C for 5 minutes.*

ACKNOWLEDGEMENTS

Financial support was received from the EU-projects CORONADX, grant number 101003562 and VIVALDI, grant number 773422 as well as from COVIDTESTS, a Danish government grant.

We sincerely thank Gitte Kronborg and Professor Dr. Jan Gorm Lisby from Hvidorve Hospital in Denmark for donating the SARS-CoV-2 samples, as well as Dr. Charlotta Polacek Strandh from Statens Serum Institute in Denmark for the inactivated SARS-CoV-2 viruses.

CONTACT

Anders Wolff, tel. +4522450209, awol@dtu.dk

[3] Nguyen, T.; Chidambara, V. A.; Andreasen, S. Z.; Golabi, M.; Ngoc, H. Van; Linh, Q. T.; Bang, D. D.; Wolff, A. Point-of-Care Devices for Pathogen Detections: The Three Most Important Factors to Realize towards Commercialization. *Trends Anal. Chem.* **2020**. https://doi.org/10.1016/j.trac.2020.116004.

[4] T. Notomi *et al.*, "Loop-mediated isothermal amplification of DNA.," *Nucleic Acids Res.*, vol. 28, no. 12, p. E63, 2000, doi: 10.1093/nar/28.12.e63.

A SOLAR-DRIVEN WEARABLE MULTIPLEXED BIO-SENSING SYSTEM FOR NONINVASIVE HEALTHCARE MONITORING IN SWEAT

Jujhar Singh, Bianca Ning, Paul Lee, Lin Liu*

Department of Engineering and Computer Science,
Seattle Pacific University, Seattle, WA, 98119, USA

ABSTRACT

In this study, we developed a novel solar-powered wearable miniaturized human sweat tracker that specifically monitors key metabolic analytes (pH, glucose, and cortisol levels) and body temperature. Its wearability and self-monitoring platform allow users to track their physical and mental status. Our fully integrated system in smartwatch construction offers an accessible avenue for quantitative analysis. Furthermore, the laser printing electrode method reduces the cost of fabrication, eliminates the complexity of multiple target biomarkers screening, and enables large-scale production. It can also contribute to revolutionizing self-powered wireless sensor networks for other biomedical devices.

KEYWORDS

Electrochemical methods; Wearable and flexible sensors; Point of Care (POC); Mental health; Cortisol, Sweat; Glucose; pH monitoring

INTRODUCTION

Wearable biosensing devices provide critical insights into overall health status and quantify human performance, which is becoming widespread to empower continuous measurement of critical biomarkers in biological fluids like saliva, blood, and sweat [1,2]. Human sweat is rich in physiological and psychological information that enables non-invasive monitoring through various biomarkers. For example, cortisol is one potential biomarker for stress estimation [3-6]. The interest in another biomarker, glucose, is related to diabetes that results in abnormal glucose levels [7-10]. The device reported here differs from previous works [6, 12-14] as we fully integrated the precise glucose and cortisol monitoring by calibrating the sensor reading with the sweat pH and body temperature.

With the substantial advances in wearable biosensing systems, most bioanalytic devices are still centralized due to the demanding requirements for flexible and standalone wearable systems [1, 11]. In this work, we demonstrated an innovative design of a medical device that could sense, analyze, and track a user's metabolic levels whilst providing an in-depth examination of their physical and mental status. The device consists of three layers (Figure 1a). The top energy harvesting layer utilizes a round solar panel (Diameter: 2.2 inches) as a self-sustainable and efficient power supply, freeing the device from the need for recharging or replacing batteries. The smartwatch circuitry middle layer integrates other electronic modules onto a circular PCB board (Diameter: 1.2 inches). In the third layer, multiplexed laser-printed sensors (1.67×1.17 inches) are integrated on a flexible polyimide film substrate, facing the wrist for fluid sample access. The device can be wirelessly connected to the user-friendly smartphone tracking application to provide information regarding glucose and cortisol levels.

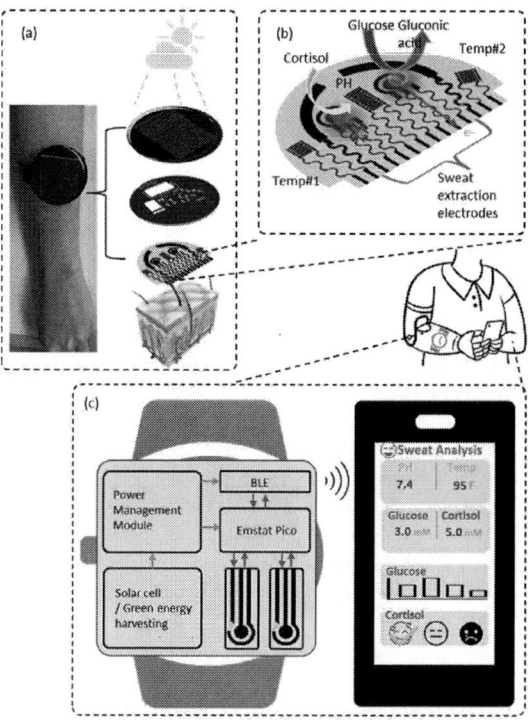

Figure 1: (a) The bio-sensing system breakdown highlighting the different layers (b) Schematic diagram of the multiplexed biosensors. (c) Schematic Illustration of tracking body temperature, cortisol, glucose, and pH in sweat using the phone app and block diagram of the biosensing system.

METHODS AND MATERIALS

Multiplexed Sensors Fabrication

The fully integrated multiplexed sensors were patterned by laser-induced graphene (LIG) on the flexible Polyimide (PI) film (Figure 1b& 2e). PI film (300HPP-ST, Kapton) was chosen as the flexible substrate for their commercial availability and ease of fabrication. LIG-based electrodes were produced by a 3D laser printer (Glowforge Pro, Glowforge).

Facile additive materials deposition facilitated multiplexed sensing in a single miniaturized piece of PI film. Two temperature sensors were deposited with Graphene Oxide (GO) (Sigma-Aldrich) (Figure 2a) [12]. The centered pH sensor was created by dropping-casting polyaniline (PANI) dispersion (regents: dodecyl benzene sulfonic acid, DBSA, and regent grade ammonium persulfate, $[NH_4]_2S_2O_8$, acetone, and PVA (polyvinyl alcohol) were purchased from Sigma-Aldrich) to form a pH-responsive film (Figure 2c) [13]. Silver/silver chloride ink (Ag/AgCl) (Ercon Inc.) was screen-printed onto the reference electrodes of both glucose and cortisol sensors. The working electrode of the glucose sensor was modified by electrodepositing the Prussian blue (PB) layer and coating a glucose oxide (GOx) selective membrane (Figure 2d). The working electrode of the cortisol sensor was modified with a molecularly imprinted polymer containing PB and Polypyrrole (Ppy) with cortisol cavities (Figure 2b). All the other chemicals for the glucose and cortisol sensors were purchased from Sigma-Aldrich. More fabrication details were discussed here [6,14].

Circular Circuit Design

The circular PCB circuit is based around two dual-channel potentiostat modules (Emstat Pico, Palmsens), which are connected to the round solar panel (5V/40mA) through power regulating circuitry (LTC3536EMSE, and LT1761ES5-3.3). A low-power Bluetooth module (nRF52832) is designed for wireless analytical data from Emstat Pico on board to a cellphone or computer (Figure 1c). One Emstat Pico module works in the Open Circuit Potentiometry (OCP) mode connecting to both temperature and pH sensors, the other Emstat Pico module operates in the Chronoamperometry (CA) mode with both cortisol and glucose sensors.

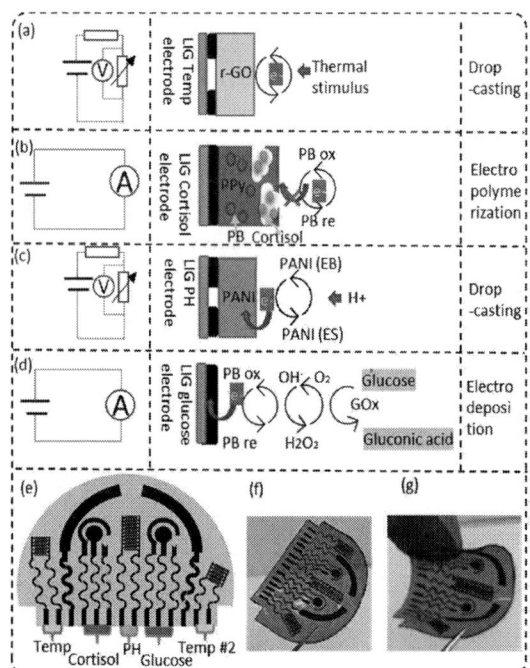

Figure 2) (a)-(d) Schematic diagrams of multiplexed LIG-electrode temperature, cortisol, pH, and glucose sensors. (e) Schematic diagram of the multiplexed biosensors. (f)& (g) Photos images of the multiplexed biosensors on PI film.

RESULTS AND DISCUSSION

The bottom LIG-based electrodes of the fully integrated multiplexed sensors were patterned by 3D laser printing on the flexible substrate PI Film (Figure 1b& 2e). To ensure the performance of the single biosensor units, channels were tested individually.

Cortisol Sensor

The Cortisol sensor developed here is a non-invasive molecularly imprinted polymer (MIP) electrochemical sensor. The sensor detects sweat cortisol via highly selective binding of cortisol to the MIP to impede the electron transfer process of the embedded Prussian blue redox probes (Figure 2b). The cortisol molecules can occupy the MIP cavities to alter the charge transfer rates of PB by its oxidation and reduction, which is shown in the following equation [15]. Inhibition of the PB transducer, leading to an amperometric response, reflects the fluctuation of the cortisol concentration. Figure 3a displays the detection of 10 nM cortisol at 0.1 V in 0.1 M Phosphate-Buffered Saline (PBS, pH =7.4).

$$PB_{re} + e^- \leftrightarrow PB_{ox} \qquad (1)$$

Glucose Sensor

The working LIG electrode of the glucose sensor was modified by a PB transducer layer, a GOx selective membrane, and a Nafion film [14]. In the presence of glucose, GOx catalyzes the following reaction:

$$glucose + O_2 \xrightarrow{GOx} H_2O_2 + gluconic\ acid \qquad (2)$$

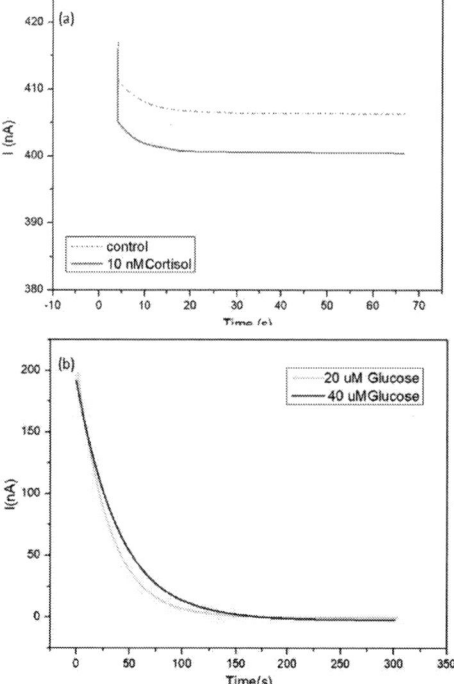

Figure 3: (a) &(b) Amperometric responses of the cortisol sensor (10 nM) and glucose sensor (20/40 µM)

The hydrogen peroxide (H_2O_2) product is then reduced by PB, generating amperometric responses with the change of glucose concentration. The reference electrode was also treated with Ag/AgCl ink. Figure 3b demonstrates the amperometric responses under 20 μM and 40 μM Glucose in 0.1 M PBS (pH=7.4) with -0.15 V. It nearly follows the inversely proportional relationship between current and time of the electrochemical sensors with microvolume solutions [16-18].

Temperature Sensor

In the temperature sensor, we fabricated a highly sensitive graphene-based temperature sensor combining the reduced graphene oxide (rGO) and LIG-based interdigital electrodes. rGO demonstrates a negative temperature coefficient (NTC) of resistance for temperature sensing, its sensing mechanism to the temperature can be explained by generating charge carriers under the thermal stimulus, this increases the carrier mobility as the temperature increases, leading to a drop in the resistance of rGO [12, 19]. The proposed rGO LIG-based sensor works in a comparable way as other NTC thermistors connecting to a 100 kΩ resistor and constant 3.3 V from the Emstat Pico as a voltage divider. The voltage reduction corresponding to the resistance drop of the rGO LIG-based sensor was recorded by open circuit potentiometry function from Emstat Pico. In Figure, 4a room temperature is responded at 0.43± 0.011 V (95%), and the output voltage varies with different subjects' body temperatures.

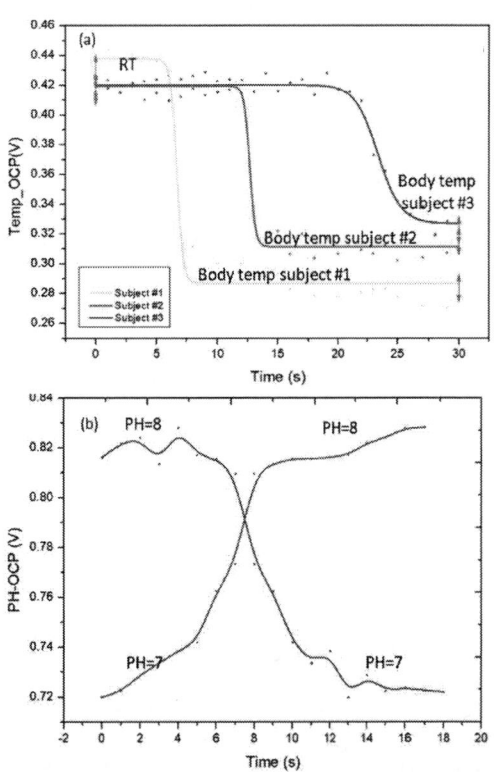

pH Sensor

The pH sensor is based on the combination of PANI membrane, interdigital electrode, and PI film substrate. PANI membrane as the pH-sensitive layer results in a corresponding rise or fall in the mobility of the ions that tunes in the sensor resistance with pH fluctuations on the surface of the PANI membrane. The mobility change is due to the reversible protonation process on the PANI membrane in the acid solution [13]. In the acid solution, the sensor is doped with H+ ions to create the emeraldine salt (ES) form of PANI. ES is highly electrically conductive leading to decreasing resistance of the sensor and a corresponding voltage change. The voltage measurement was collected by Emstat Pico as well. For each test, the sensors were immersed in the 0.1 M PBS with pH=7 or pH = 8. Figure 4b shows the response time of the pH sensors from pH 7 to pH 8 and a reverse change. The average response time is 10 seconds. This response time could vary based on the thickness of the PANI membrane deposited on the surface of LIG sensor (Figure 4b) [20].

CONCLUSION

To conclude, we developed a self-powered non-invasive sweat monitoring system that would contribute to the prevention of diabetes, stress eating, and other mental diseases caused by unhealthy lifestyles. This smartwatch-like design gives the user a lightweight and flexible health monitoring experience by providing sweat pH, glucose, cortisol levels, and body temperature data. When connected with smartphone-based algorithms, the real-time data could help to make recommendations to people on lifestyles and athletic performance or provides early warnings on diseases. The system at hand has the potential to revolutionize self-powered wireless biosensors and other biomedical devices.

Future work will be required to fully validate the multiplexed sensors in vitro and in vivo and address the reliability by comparing them to the standard test results.

ACKNOWLEDGEMENTS

The authors would like to thank Dr. Karisa Pierce's Laboratory and Digital Manufacturing Laboratory at Seattle Pacific University for providing the lab facilities.

REFERENCES

[1] T. R. Ray et al., "Bio-Integrated Wearable Systems: A Comprehensive Review," Chemical Reviews, vol. 119, no. 8, pp. 5461–5533, Jan. 2019.

[2] A. Sharma, M. Badea, S. Tiwari, and J. L. Marty, "Wearable Biosensors: An Alternative and Practical Approach in Healthcare and Disease Monitoring," Molecules, vol. 26, no. 3, p. 748, Feb. 2021.

[3] M. Pali et al., "CATCH (Cortisol Apta WATCH): 'Bio-mimic alarm' to track Anxiety, Stress, Immunity in human sweat," *Electrochimica Acta*, vol. 390, p. 138834, Sep. 2021.

[4] S. C. E. Schmidt, J.-P. Gnam, M. Kopf, T. Rathgeber, and A. Woll, "The Influence of Cortisol, Flow, and Anxiety on Performance in E-Sports: A Field Study,"

Figure 4(a) The voltage response of the temperature sensor. (b) The dynamic response of the pH sensor (a rising from pH=7 to 8 and a reverse fall)

BioMed Research International, vol. 2020, pp. 1–6, Jan. 2020.

[5] E. Epel, R. Lapidus, B. McEwen, and K. Brownell, "Stress may add bite to appetite in women: a laboratory study of stress-induced cortisol and eating behavior," *Psychoneuroendocrinology*, vol. 26, no. 1, pp. 37–49, Jan. 2001.

[6] W. Tang, L. Yin, J., and J. Wang, *Advanced Materials*, 33(18), p.2008465, 2021

[7] E. Nery, M. Kundys, P. Jelen, and M Jonsson-Niedziolka, *Analytical Chemistry*.88(23), pp. 11271–11282, 2016,

[8] Y. Yu, H. Y. Y. Nyein, W. Gao, and A. Javey, "Flexible Electrochemical Bioelectronics: The Rise of In Situ Bioanalysis," *Advanced Materials*, vol. 32, no. 15, p. 1902083, Aug. 2019.

[9] J. Kim, A. S. Campbell, and J. Wang, "Wearable non-invasive epidermal glucose sensors: A review," *Talanta*, vol. 177, pp. 163–170, Jan. 2018.

[10] H. Li et al., "Recent advances in biofluid detection with micro/nanostructured bioelectronic devices," *Nanoscale*, vol. 13, no. 6, pp. 3436–3453, Feb. 2021.

[11] R. Torre, E. Costa-Rama, H. P. A. Nouws, and C. Delerue-Matos, "Screen-Printed Electrode-Based Sensors for Food Spoilage Control: Bacteria and Biogenic Amines Detection," *Biosensors*, vol. 10, no. 10, p. 139, Sep. 2020.

[12] R. Han et al., "Facile fabrication of rGO/LIG-based temperature sensor with high sensitivity," *Materials Letters*, vol. 304, p. 130637, Dec. 2021.

[13] Y. Li, Y. Mao, C. Xiao, X. Xu, and X. Li, "Flexible pH sensor based on a conductive PANI membrane for pH monitoring," *RSC Advances*, vol. 10, no. 1, pp. 21–28, 2020.

[14] T. Chang et al., "Highly integrated watch for noninvasive continual glucose monitoring," *Microsystems & Nanoengineering*, vol. 8, no. 1, pp. 1–9, Feb. 2022.

[15] A. A. Karyakin, "Prussian Blue and Its Analogues: Electrochemistry and Analytical Applications," *Electroanalysis*, vol. 13, no. 10, pp. 813–819, Jun. 2001.

[16] D. A. Aikens, "Electrochemical methods, fundamentals and applications," *Journal of Chemical Education*, vol. 60, no. 1, p. A25, Jan. 1983.

[17] D. M. R. de Rooij, "Electrochemical Methods: Fundamentals and Applications," *Anti-Corrosion Methods and Materials*, vol. 50, no. 5, Oct. 2003.

[18] L. Chang, C. Liu, Y. He, H. Xiao, and X. Cai, "Small-volume solution current-time behavior study for application in reverse iontophoresis-based non-invasive blood glucose monitoring," *Science China Chemistry*, vol. 54, no. 1, pp. 223–230, Oct. 2010.

[19] Abid, P. Sehrawat, S. S. Islam, P. Mishra, and S. Ahmad, "Reduced graphene oxide (rGO) based wideband optical sensor and the role of Temperature, Defect States and Quantum Efficiency," *Scientific Reports*, vol. 8, Feb. 2018.

[20] R. G. Bates, M. Paabo, and R. A. Robinson, "Interpretation of pH measurements in alcohol-water solvents," *The Journal of Physical Chemistry*, vol. 67, no. 9, pp. 1833–1838, Sep. 1963.

CONTACT

*L. Liu, tel: +1- 206-281-2961; liul5@spu.edu

HIGH-THROUGHPUT MASS MEASUREMENT OF SINGLE BACTERIAL CELLS BY SILICON NITRIDE MEMBRANE RESONATORS

Adrián Sanz-Jiménez[1], Oscar Malvar[1], Jose J. Ruz[1], Sergio García-López[1], Priscila M. Kosaka[1],
Eduardo Gil-Santos[1], Álvaro Cano[1], Dimitris Papanastasiou[2], Diamantis Kounadis[2],
Elias Panagiotopoulos[2], Jesús Mingorance[3], María Rodríguez-Tejedor[3], Álvaro San Paulo[1],
Montserrat Calleja[1] and Javier Tamayo[1]

[1]Instituto de Micro y Nanotecnología, IMN-CSIC (CEI UAM+CSIC), 28760 Tres Cantos, Madrid, SPAIN,

[2]Fasmatech Science & Technology, Lefkippos TESPA, Demokritos NCSR, Patriarchou Gregoriou & Neapoleos, 15341 Athens, GREECE and

[3]Hospital Universitario La Paz, IdiPAZ, 28046 Madrid, SPAIN

ABSTRACT

We present a technological approach to precisely measure the dry mass of many individual cells of a bacteria colony. In this technique, bacteria are transported from aqueous solution into gas phase and subsequently guided to the surface of a silicon nitride membrane resonator. Abrupt downshifts in the membrane eigenfrequencies are measured upon every bacterium adhesion and are related to the dry mass of the cell by theoretical methods. We measure the dry mass of *Escherichia coli* K-12 and *Staphylococcus epidermidis* with an unprecedented throughput of 20 cells/min and with a mass resolution of ~1%. Finally, we apply the Koch & Schaechter model to assess the intrinsic sources of growth stochasticity.

KEYWORDS

Nanomechanical mass spectrometry, high-throughput and high-precision membrane resonators, mass sensing, Koch & Schaechter model, bacteria growth stochasticity.

INTRODUCTION

In spite of decades of research, how bacteria are able to maintain their size remains an open question. Cell populations are intrinsically heterogeneous due to stochastic and deterministic mechanisms [1]. The heterogeneity of cell populations plays a fundamental role in the biological function. In order to understand the mechanisms of bacteria size regulation, techniques that can measure individual cells in populations are required. The dry mass, the mass of the non-aqueous content of the cell, is a particularly valuable parameter to quantify the size regulation of bacteria. It provides information of the amount of proteins, nucleic acids, carbohydrates and lipids present in a cell [2]. Despite its importance, the precise and rapid quantification of the dry mass of single bacterial cells remains technically challenging. Techniques such as quantitative phase imaging (QPI), microchannel resonators and mass spectrometry (MS) have been developed to measure the dry mass of single cells. QPI assumes a constant value for the ratio between the refractive index and the biomolecular mass density of cells [3]. While this ratio is similar for most globular proteins, it can vary by almost 20% for carbohydrates or lipids. However, the major limitation of QPI is the optical spatial resolution and phase noise, which limits the precision for measuring the dry mass of small cells below 1 μm, such as bacteria.

Microchannel resonators measure the buoyant mass of suspended cells at sub-femtogram accuracy. The buoyant mass is the total mass of the cell minus that of the fluid displaced by the cell. Determination of the dry mass of the bacteria requires measuring the same bacterial cell in two fluids of different density, which is technically complex and limits the throughput of the assay [4]. Finally, MS approaches have measured the mass of individual biological particles, previously ionized. For instance, charge-detection MS (CDMS) simultaneously measure the charge and the mass-to-charge ratio of single ions with masses up to hundreds of MDa [5]. However, mass resolution decreases for large ions.

Nanomechanical mass spectrometry (NMS) has emerged as the only technique that can directly measure the mass of single biological entities from the nanometer to the micrometer scale, namely proteins, viruses and bacterial cells without requiring the measurement of the charge state of the ions [6]-[10]. In this technique, analytes are gently ionized from aqueous solution into gas phase by electrospray ionization (ESI) and guided to the surface of micro- and nanomechanical resonators, while retaining its original native structure as much as possible. When the analyte adheres to the resonator, abrupt downshifts in the resonance frequencies of the normal vibration modes of the resonator are measured, and can be related to the mass of the analyte. The fundamental limitation of the technique is the detection efficiency, defined as the ratio between the total nebulized particles and the particles that accrete on the resonator. This limitation comes from the trade-off between the resonator size and the minimum detectable mass. Smaller resonators are more sensitive but also dispose a smaller capture area. Here, we propose the use of ultrathin membrane resonators (*400×350×0.05 μm*) for the measurement of the dry mass of bacterial cells. These structures present a high mass resolution together with a high capture area. Additionally, we have developed an inverse theoretical algorithm to determinate the dry mass of individual bacterial cells from the resonance frequency downshifts of several vibration modes. Finally, we have obtained the dry mass distributions of two bacteria species: *Escherichia coli* K-12 and *Staphylococcus epidermidis*.

RESULTS AND DISCUSSION
Nanomechanical mass spectrometer

The nanomechanical mass spectrometer comprises four differential pressure stages (Figure 1a, left). In the first stage, a high voltage (2-3 kV) is applied to a solution with the bacterial cells to generate a Taylor cone at the electrospray ionization (ESI) source, which is at atmospheric pressure. The bacterial cells travel through the second stage at 10 mbar, where a capillary inlet operated at 150 °C promotes desolvation. Subsequently, the cells are transferred to an aerolens tube designed to laminarize the underexpanded gas flow. Numerical simulations by finite element method (FEM) show that the bacteria injected with supersonic speeds (600-700 m/s) into the bore of the aerolens are decelerated to a few tens of m/s inside the laminarized flow (Figure 1a, right). A pair of skimmer-shaped lens electrodes are used to focus the bacterial cells onto the surface of the membrane resonator, which is at the fourth pressure stage, at 0.1 mbar. The soft-landing of our instrument enables to retain the native conformation of the bacteria in the resonator surface, as confirmed by scanning electron microscope (SEM) images (Figure 1b).

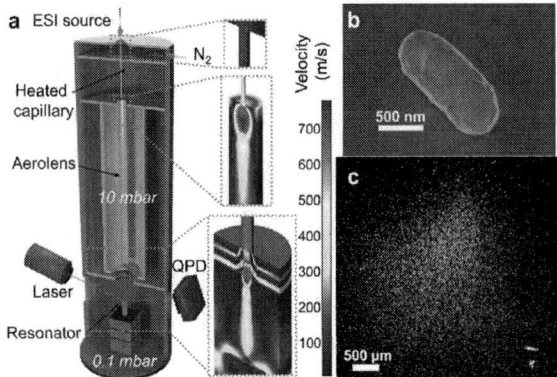

Figure 1: Nanomechanical mass spectrometer. a) Schematic of the instrument (left) and numerical simulation of the flow speed by FEM (right). The speed of the cells is very similar to the flow speed. b) Scanning electron microscopy image of an E. coli bacterium on a silicon surface. c) E. coli bacterial cells over a silicon surface placed at the resonator position.

Figure 1c shows a dark-field optical microscope image of *Escherichia coli* K-12 cells over a silicon surface placed at the resonator position. The full width at half maximum (FWHM) of the cells distribution is of about 2.2×1.7 mm for *E. coli* cells and 1.6×1.2 mm for *S. epidermidis* cells [10]. The transmission efficiency, defined as the ratio between the number of nebulized particles and the number of cells that reach the resonator stage, is of about 0.03%, which is similar to previous works [7][9]. Moreover, the capture efficiency, defined as the ratio between the effective sensing area of the resonator and the cross section area of the cells beam at the resonator position, is of about 5% [10], which represents an increase of two orders of magnitude with respect to previous NMS works in vacuum conditions [7][9].

Membrane nanomechanical resonators
The nanomechanical resonators used in this work are silicon nitride rectangular membranes with side lengths of $L_x = 400$ μm and $L_y = 350$ μm, and thickness of 50 nm (Figure 2a). The experimental vibration mode shapes can

be described by linear elasticity theory (equation 1) [10].

$$\psi_{ij}(x,y) = 2 \sin\left(\frac{i\pi}{L_x}x\right)\sin\left(\frac{j\pi}{L_y}y\right) \quad (1)$$

where *(i,j)* refers to the number of antinodes in *x* and *y* directions, respectively.

We have measured the resonance frequencies of six vibration modes simultaneously by phase-locked loop (PLL) systems and driving the resonator by a piezoelectric actuator placed underneath the membrane. In particular, these modes were (1,1), (2,1), (1,2), (3,1), (1,3) and (3,2). The spectra of these resonances are shown in Figure 2b, with frequencies of about 470.4, 714.6, 771.6, 996.6, 1104.2 and 1169.6 kHz and Q-factors of about 1500, 2700, 3000, 3000, 3500 and 4100, respectively.

The motion of the membrane resonator was measured by the laser-beam deflection technique. This technique is sensitive to the slope of the vibration modes shapes of the membrane [11]. Figure 2c show the squares of the theoretical partial slope of the beam displacement along the *x*-direction of the six vibration modes, respectively. In order to measure the six vibration modes simultaneously with optimal signal-to-noise ratios, the laser beam must be focused in one of the regions indicated by Figure 2d. Figure 2d represents the product of the square of the six partial slopes represented in Figure 2c. In Figure 2a, we show the four regions (red points) where our laser was arbitrarily focused to measure the six resonance frequencies. The frequency sensitivity of the membrane resonator may be estimated by the Allan deviation. The minimum detectable mass, calculated from the Allan deviation of the (1,1) vibration mode, is of about 0.7 fg (1 fg=10^{-15} g) for a particle located at the center of the membrane [10].

The rectangularity of the membrane, with an aspect ratio of about 1.14, makes the resonator different enough from a perfectly square membrane with aspect ratio of one to avoid frequency degeneration. It plays an important role in the frequency density of the vibration modes, since the pairs *(i,j)* and *(j,i)* split into two close, but uncoupled, modes. The frequency ratio between the modes (3,2) and (1,1), seventh and first modes in frequency respectively, is only of about 2.5, which facilitates the tracking of a large number of vibration modes simultaneously.

Measurement of the mass heterogeneity of cells
The adsorption of a particle in a highly tensioned membrane causes changes in the kinetic energy due to the added mass of the particle m_p, while the potential energy is negligibly affected by the particle's stiffness [10]. Thus, the relative frequency shift of a vibration mode due to a particle on the membrane at position *(x₀, y₀)* is given by equation 2 [10].

$$\frac{\Delta v_{ij}}{v_{ij}} = -\frac{m_p}{2M}\psi_{ij}(x_0,y_0)^2 \quad (2)$$

where M is the mass of the membrane resonator and Δv_{ij} is the frequency shift induced by the cell.

A real time record of the fractional changes of the frequency of the selected vibration modes of the membrane during ESI of *S. epidermidis* cells is shown in Figure 3a. The landing events of individual bacterial cells cause abrupt 'jumps' in the tracked eigenfrequencies. The inset of Figure 3a shows a close-up view to 60 seconds of the measurement, where more than 30 jumps can be seen,

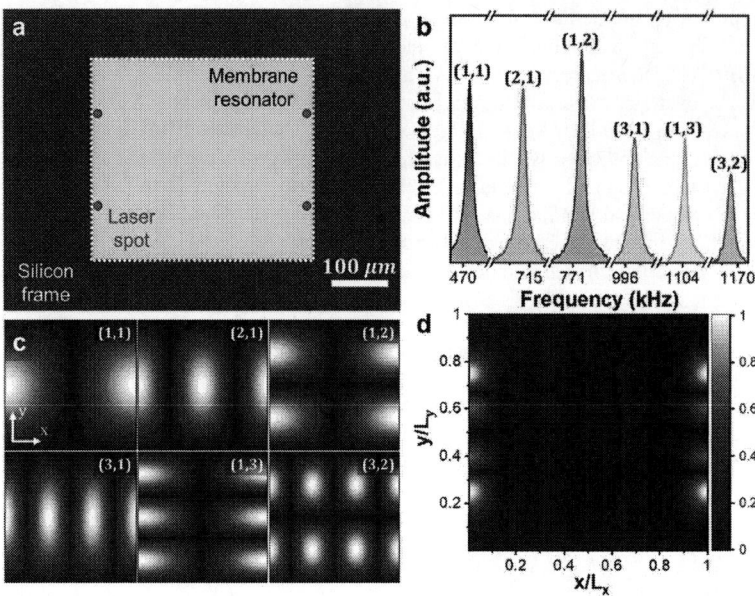

Figure 2: Silicon nitride membrane resonators. a) Optical microscope image of the resonator (400×350×0.05 μm) where the red points indicate the optimal measurement positions. b) Frequency spectra of the vibration modes (1,1), (2,1), (1,2), (3,1), (1,3) and (3,2). c) Squares of the theoretical partial slopes along x direction of the six vibration modes, respectively. d) Product of the square of the six partial slopes, which indicate the optimal positions for the laser beam.

which translates into around a jump each 2 seconds.

From the tentative events of *S. epidermidis* adsorptions of Figure 3a, represented by frequency shifts, about 70% are considered as valid events. We consider a valid event when: i) the obtained position of the particle by our inverse problem method is distanced more than 6% of the membrane length from the physical clampings and ii) the Pearson correlation coefficient between the experimental frequency 'jumps' and the square of the eigenmodes at the highest probability particle position is greater than 0.9. Further details about the inverse problem method used in this work can be found elsewhere [10].

Figure 3b shows the evolution of the normalized probability distribution function (PDF) of the dry mass as a function of the number of *S. epidermidis* cells detected. The final PDF of the dry mass has a mean and a standard deviation of 153±69 fg. The mass resolution of our

measurements has a median of 3 fg, which can determine variations of 2% in the mass of single *S. epidermidis* cells.

We have also examined the dry mass heterogeneity of the bacterial cells. Here, we consider the model of cell division processes proposed by Koch and Schaechter [12]. In this model, the deterministic distribution for the mass of the steady-state populations of exponentially growing cells is given by $\theta(m)=m_d/m^2$, for $m_d/2 \leq m \leq m_d$ and zero otherwise, with m_d the mass at the instant of division. Equation 3 defines the effect that the stochasticity of the cell division has on the mass distribution.

$$\theta(m) = \frac{A}{m^2} \int_m^{2m} g(x)dx \qquad (3)$$

where A is a normalization constant and $g(x)$ is the probability density of the mass of the cells at division.

In Figure 4, we apply the Koch and Schaechter model to the mass PDF of *S. epidermidis* cells (in yellow) and of

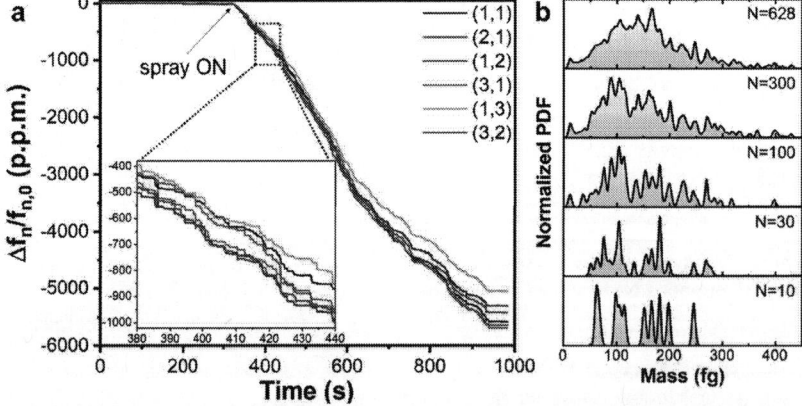

Figure 3: Nanomechanical spectrometry of S. epidermidis cells. a) Real time record of the fractional changes of the frequency of the selected vibration modes (inset: close-up view to 60 seconds of the measurement). b) Evolution of the normalized probability distribution function of the dry mass for the first 10, 30, 100, 300 and 628 cells.

E. coli K-12 cells (in green). The mass PDF of *S. epidermidis* is constructed with the 628 events of Figure 3b and the mass PDF of *E. coli* K-12 is constructed with 685 events. We have assumed normal distribution for the division mass of Equation 3 to obtain a coefficient of determination R-squared of about 0.96 for both bacteria species [10]. Koch and Schaechter model allows inferring the mean dry mass for both bacterial cells: 153 fg for *S. epidermidis* and 469 fg for *E. coli* K-12. The coefficient of variation can be also measured and is 0.46 for *S. epidermidis* and 0.37 for *E. coli.*

Koch and Schaechter model provides the mean mass at the instant of division of 248 fg for *S. epidermidis* cells and 763 fg for *E. coli* K-12 cells, and a CV of the division mass of 0.31 and 0.27 for the bacteria species, respectively.

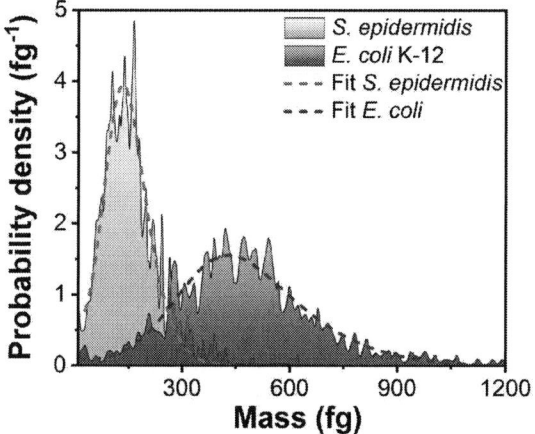

Figure 4: Stochasticity of the dry mass of the cells. Probability distribution functions of the mass of S. epidermidis cells (filled curve in yellow), N=628 events, and E. coli K-12 cells (filled curve in green), N=685 events, measured by our silicon nitride membrane resonators. The dashed lines represent the fitting to the Koch and Schaechter model, which leads to infer growth stochasticity parameters.

CONCLUSIONS

Cell growth and cell division are stochastic processes, and the mechanisms of cell size regulation are still not clear. Here we present a technique that can measure the dry mass of bacterial cells with high efficiency. Our membrane resonators can measure the dry mass of cell populations in a time span of few minutes with an unparalleled mass resolution of about 1%. We have applied the Koch and Schaechter model to obtain important parameters of the stochasticity of the cell growth. The extraordinary performance of our resonators leads to a better understanding on the mechanisms behind cell growth.

ACKNOWLEDGEMENTS

This work was supported by the European Union's Horizon 2020 Research and Innovation Program under Grant Agreement No. 731868-VIRUSCAN and by the ERC CoG Grant 681275 "LIQUIDMASS". We acknowledge the service from the Micro and Nanofabrication Laboratory an X-SEM laboratory at IMN-CNM funded by the Comunidad de Madrid (Project S2018/NMT-4291 TEC2SPACE) and by MINECO (Project CSIC12-4E-1794 with support from FEDER, FSE). E. G. S. acknowledges financial support by the Spanish Science and Innovation Ministry through Ramón y Cajal grant RYC-2019-026626-I.

REFERENCES

[1] A. Amir, "Cell Size Regulation in Bacteria", *Phys. Rev. Lett.,* 112, 208102, 2014.

[2] R. Milo, "What is the total number of protein molecules per cell volume? A call to rethink some published values", *BioEssays*, 35, 1050-1055, 2013.

[3] Y. Park, C. Depeursinge, G. Popescu, "Quantitative phase imaging in biomedicine", *Nat. Photon.*, 12, 578-589, 2018.

[4] N. Cermak *et al.,* "High-throughput measurement of single-cell growth rates using serial microfluidic mass sensor arrays", *Nat. Biotechnol.*, 34, 1052-1059, 2016.

[5] D. Z. Keifer, E. E. Pierson, M. F. Jarrold, "Charge detection mass spectrometry: weighing heavier things", *Analyst*, 142, 1654-1671, 2017.

[6] M. S. Hanay *et al.,* "Single-protein nanomechanical mass spectrometry in real time", *Nat. Nanotechnol.,* 7, 602-608, 2012.

[7] S. Domínguez-Medina *et al.,* "Neutral mass spectrometry of virus capsids above 100 megadaltons with nanomechanical resonators", *Science*, 362, 918-922, 2018.

[8] R. T. Erdogan *et al.,* "Atmospheric Pressure Mass Spectrometry of Single Viruses and Nanoparticles by Nanoelectromechanical Systems", *ACS Nano*, 16, 3821-3833, 2022.

[9] O. Malvar *et al.,* "Mass and stiffness spectrometry of nanoparticles and whole intact bacteria by multimode nanomechanical resonators", *Nat. Commun.*, 7, 13452, 2016.

[10] A. Sanz-Jiménez *et al.,* "High-throughput determination of dry mass of single bacterial cells by ultrathin membrane resonators", *Commun. Biol.*, 5, 1227, 2022.

[11] J. Tamayo *et al.,* "Imaging the surface stress and vibration modes of a microcantilever by laser beam deflection microscopy", *Nanotechnology*, 23, 315501, 2012.

[12] A. L. Koch, M. Schaechter, "A Model for Statistics of the Cell Division Process", *J. Gen. Microbiol.*, 29, 435-454, 1962.

CONTACT

*J. Tamayo, tel: +(34)918060700; javier.tamayo@csic.es

MICROFABRICATED ISOTHERMAL EG-FET SENSOR FOR LAMP MEDIATED CRISPR/CAS12A DETECTION OF HEPATITIS C VIRUS

Hsin-Ying Ho, Wei-Sin Kao, Piyush Deval, Ling-Shan Yu and Che-Hsin Lin
National Sun Yat-sen University, TAIWAN

ABSTRACT

A highly sensitive and rapid platform for the detection of hepatitis C virus (HCV) has been developed by combining loop-mediated isothermal amplification (LAMP) with the use of clustered regularly interspaced short palindromic repeats of associated protein 12 (CRISPR/cas12a). An isothermal sensor which serves as an indium tin oxide (ITO) based extended gate field-effect transistor (EG-FET) and is equipped with a polymethyl methacrylate (PMMA) microwell, is used to detect the potential change on the sensing surface, induced by the LAMP/CRISPR reaction. The immobilized signal reporter, which is single-strand deoxyribonucleic acid (ssDNA) on the ITO sensing surface, is cleaved by the activated CRISPR/cas12a. This trans-cleavage occurs due to the highly specific base-pairing of a nucleic acid target to a complementary guide ribonucleic acid (gRNA). The result is a significant potential change within four minutes. The diagnostic platform that is established in this study can be applied to the detection of other virus/bacteria through the use of different pathogen-specific LAMP/CRISPR assays. Therefore, this technique provides a rapid yet high performance platform for the detection of label-free pathogens.

KEYWORDS

EG-FET, HCV, ITO, LAMP, CRISPR/cas12a

INTRODUCTION

Infection by the HCV is a major cause of death around the world. HCV can be transmitted through the transfusion of infected blood, the use of contaminated syringes for drug intake, unprotected sex with infected people, and a lack of sterile medical devices [1, 2]. It is a major concern because there is no vaccine [3]. HCV infection is typically asymptomatic but may non-specifically increase the risk of complications of chronic liver disease in about 10-20% of the infected patients [4]. Therefore, the existence of a rapid and sensitive diagnostic method would enable early detection of HCV, which could lead to control its spread and prevention of its proliferation to a critical stage.

Currently, a polymerase chain reaction (PCR) method is considered the gold standard for HCV detection. However, PCR is a complex procedure to use for diagnostic assays [5]. Although some studies have reported successful HCV diagnosis based on an electrochemical method with the use of cyclic voltammetry [6, 7]. However, the device fabrication is complex and their use requires sophisticated and expensive instruments, presenting barriers to their utilization in low resource countries.

To bridge the gap, a LAMP method combined with the use of CRISPR/Cas12a has been developed to provide a sensitive, rapid, low-cost and robust detection of HCV. The method utilizes a sensor that is an ITO-based EG-FET modified with the addition of a ssDNA signal reporter.

Compared to current diagnostic systems, the developed platform requires no fluorescence label. It works faster than extant gold standard PCR methods and achieves HCV RNA detection levels as low as one genomic copy/reaction. The developed method will give substantial impact on the rapid and label-free genomic detections.

DESIGN AND FABRICATION

Principle of CRISPR mechanism

CRISPR/cas12a is an RNA-guided DNA-cutting nuclease [8, 9]. Upon target DNA binding, cas12a cleaves both targeted double-strand target DNA (dsDNA) in the *cis-configuration* and non-target ssDNAs in the *trans-configuration*. Figure 1 elucidates the molecular basis of both DNase cleavage modes [10, 11].

First, dsDNA targets the recognition initiates with recognition of the protospacer adjacent motifs (PAMs) by the wedge (WED) and PAM-Interacting (PI) domains. This process promotes dsDNA target unwinding. Second, positive-sense DNA binds with the CRISPR (cr) RNA and the ssDNA to form the R-loop. Once the crRNA-DNA hybridizes, conformational changes are induced in the cas12a recognition (REC) lobe, and these changes result in allosteric unblocking of the catalytic site in the recombination UV C (RuvC) domain. Third, the negative-sense DNA is guided towards the RuvC catalytic site, and this results in cis-cleavage of the negative-sense DNA. Subsequently, further unwinding of the PAM-distal DNA duplex allows the positive-sense DNA to enter the RuvC catalytic site, which results in cis-cleavage of the positive-sense DNA. The PAM-distal dsDNA is released. At the end of the process, the PAM-proximal dsDNA remains bound to the cas12a-crRNA complex, which holds the cas12a in a catalytically activated conformation. This conformation enables trans-cleavage of non-target ssDNAs [12].

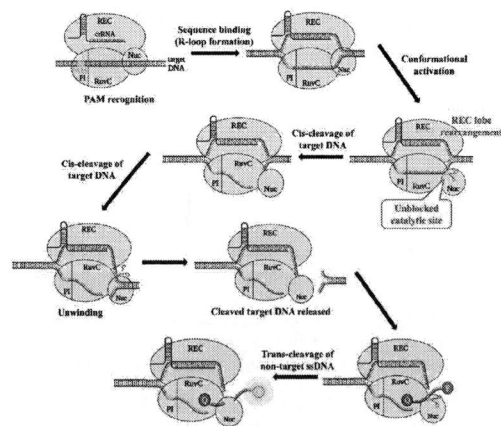

Figure 1: Principle of CRISPR/cas12a reaction. In the figure, recognition lobe is abbreviated as REC, Recombination UV C is abbreviated as RuvC, and Nuc is the abbreviation of nuclease.

ITO modification and EG-FET device fabrication

Figure 2 shows the steps in the fabrication of an ITO-based EG-FET sensor for detecting the potential change of the CRISPR/Cas12a reaction. The ITO surface was modified according to a previous protocol with slight changes [13]. ITO glass slide was cleaned ultrasonically with a series of chemical reagents, ethanol, acetone and deionized (DI) water, each of which was used for seven minutes. The cleaned ITO surface was then coated with 3-aminopropyltriethoxysilane (APTES) by immersion of the ITO slides in 5% APTES in absolute ethanol for eight hours. The surface was washed thoroughly with absolute ethanol and dried in air overnight in absence of humidity. The APTES-modified ITO slides were then immersed in 2% glutaraldehyde (GA) solution in phosphate-buffered saline (PBS) for two hours. The GA-functionalized surfaces were washed thoroughly with PBS followed by DI water to remove any free GA on the surfaces, followed by curing in a ventilated oven at 40°C for one hour.

EG-FET devices were fabricated on a printed circuit board (PCB) as described in our previous work [14, 15]. After modification of the ITO with APTES and GA, the ITO was mounted on the PCB board using epoxy, followed by assembling the PMMA microwell. The ssDNA was then immobilized on the ITO surface by pouring 10 µM of ssDNA in microwell through Schiff base chemistry. After 20 hours of reaction at room temperature (RT), surface was washed gently with nuclease-free water and it was kept wet until use.

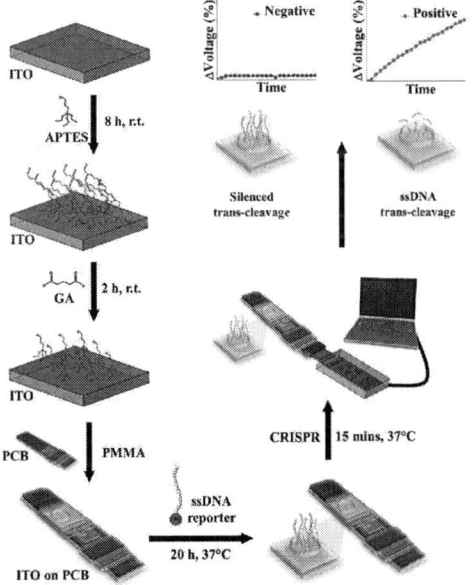

Figure 2: Schematic diagram of surface modification of ITO and fabrication of EG-FET device.

LAMP based CRISPR/cas12a design

HCV RNA was pre-amplified by performing a LAMP experiment [16], which involved the application of the following: 1x isothermal buffer (pH8.8), 4 mM MgSO$_4$, 1 mM deoxynucleotide diphosphates, 1 ng/mL bovine serum albumin, 0.8 M betaine, 1x primer mix (2.5 µmol/L each of F3 and B3 primers, 20 µmol/L each of forward inner primer

and backward inner primer, 10 µmol/L each of loop forward and loop backward primers), 0.636 U/µL *Bacillus stearothermophilus* 2.0 DNA polymerase, 0.2 U/µL avian myeloblastosis virus reverse transcriptase, 0.2 U/µL RNasin® plus RNase inhibitor, synthesized RNA solutions that ranging from 10^6 to one genomic copy/reaction, and enough nuclease-free water to bring the volume to 10 µL. The LAMP reaction is then initiated by addition of the DNA polymerase and carried out at 63°C for 30 minutes [17].

The CRISPR solution was prepared at the same time. The solution comprised 3x NEBuffer r2.1, 0.1 µM EnGen® Lba cas12a (Cpf1), 1.5 µM ssDNA reporter (5'-FAM/TTTTTT/3'-BHQ), 90 nM crRNA, and enough nuclease-free water to bring the volume to 10 µL. Next, the pre-amplified LAMP product was mixed in a ratio of 20:10 with the CRISPR solution. The intensity of the fluorescence was monitored in QuantStudio™ 5 Real-Time PCR system at 37°C for 30 minutes. As the cas12a-crRNA complex recognized the amplified HCV DNA products that were complementary to the crRNA, trans-cleavage by the cas12a enzyme was induced to cleave the ssDNA reporter in the solution for the fluorescence-based detection, or to cleave the ssDNA probe, which was immobilized on the ITO surface. Note that the fluorescence detection was used to validate the CRISPR/Cas12a reaction and compare the results with the produced chip device.

EXPERIMENTAL RESULTS
Characterization of modified ITO

To confirm the surface modification and ssDNA immobilization, atomic force microscopy (AFM) was used to observe the surface topography and roughness of the sensor surface at different stages. A minor roughness incensement from 0.434 nm, 0.625 nm and 1.064 nm was for bare ITO surface, after APTES and GA functionalization and after ssDNA immobilization, respectively, as shown in Fig. 3. The result confirmed the successful immobilization of the ssDNA.

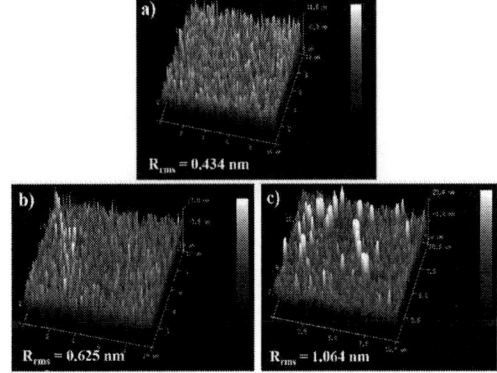

Figure 3: AFM images of modified substrates of ITO (a) bare ITO, (b) ITO modified by APTES+GA, and (c) ITO modified by APTES+GA+ssDNA.

LAMP-based CRISPR/cas12a design for HCV detection

The amplification efficiency of LAMP was first verified to ensure that all concentrations of dilution reached plateaus within 30 minutes. Figure 4a shows the mean time-to-positive (TTP) for different HCV RNA concentrations, which ranged from 10^5 to one genomic copies/reaction. The TTPs were 4.98 ± 0.007 minutes for the 10^5 genomic copies per reaction concentration, 5.34 ± 0.553 for the 10^4, 6.51 ± 0.567 for the 10^3, 7.22 ± 0.687 for the 10^2, 8.94 ± 1.640 for the 10 genomic copies per reaction, and 9.85 ± 1.890 for the one genomic copy per reaction concentration.

The sensitivity of the LAMP-based CRISPR/cas12a method was evaluated through the study of serial 10-fold dilutions, 10^6 to one genomic copies/reaction, of the synthetic HCV RNA template. As is shown in the Figure 4b, the positive signal come out in around four minutes, and the limit of detection (LOD) was one genomic copy per reaction.

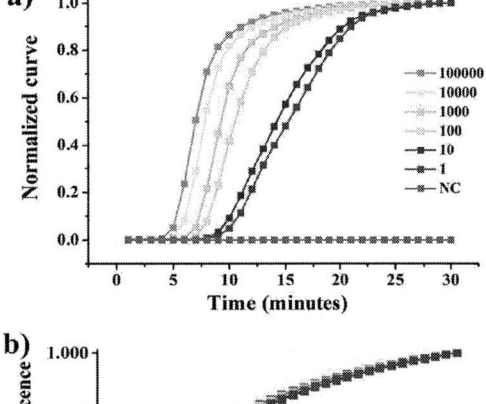

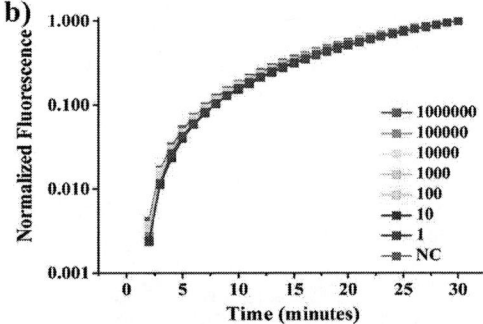

Figure 4: (a) Real-time LAMP amplification for the detection of different concentrations of synthetic HCV RNA templates. (b) Normalized fluorescence of real-time LAMP-CRISPR/cas12a detection of HCV RNA templates, achieved through the use of a PCR machine. NC: nuclease-free water as input instead of HCV RNA dilution.

Detection of HCV through use of the EG-FET device

This LAMP-based CRISPR/cas12a-based method for the detection of HCV was successfully tested through the use of a PCR machine, as shown in Figure 4. As the goal of this research was to detect HCV RNA through the use of an electrochemical platform that involved the application of a LAMP-based CRISPR/cas12a method, the design was transferred to an EG-FET device. In the case, the ssDNA was immobilized on the EG-FET device, whereas for PCR method, it was suspended in solution. A volume of nuclease-free water equivalent to that of the ssDNA that had been in the PCR based CRISPR/cas12a reaction mixture was added to keep the design volume. The

temperature control was achieved in a home-built copper heating plate for establishing the isothermal reaction environment. The experimental setup for electrochemical sensing the potential change of the LAMP-based CRISPR/cas12a trans-cleavage of ssDNA on the ITO based EG-FET is shown in Figure 5.

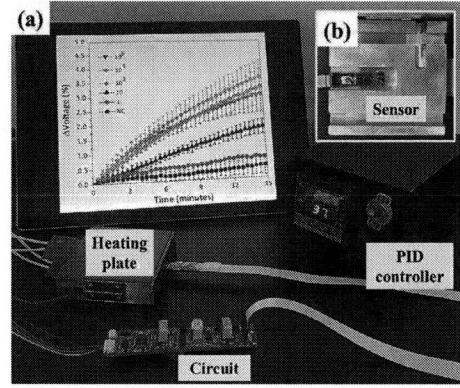

Figure 5: (a) Experimental setup of the EG-FET based biosensor showing the heating plate with sensor underneath, circuit, and PID controller. (b) Sensor under the heating plate during the analysis.

For chip-based detection, the LAMP reaction mixture was amplified to saturation at 63°C for 30 minutes prior to the detection, after which it was mixed with the CRISPR/cas12a reaction mixture. This LAMP-amplified CRISPR/cas12a reaction mixture (30 μL) was then poured into the PMMA microwell on the EG-FET device and sealed with PCR tape to retain a closed environment and avoid contamination. The sensor chip was placed on the designed heater and incubated at 37°C. The PCB was connected to the computer system flowing by the circuit and the real-time voltage change was recorded over 15 minutes.

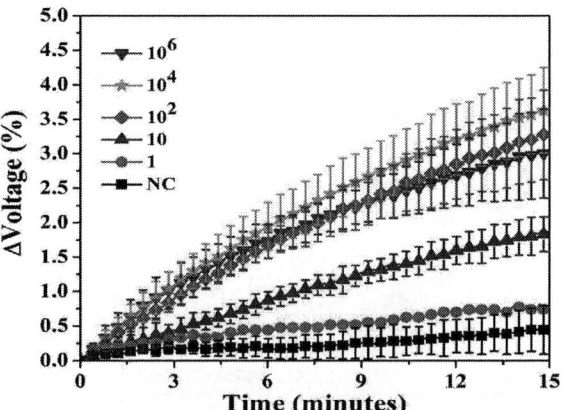

Figure 6: Real time voltage change on the EG-FET device due to the trans-cleavage of immobilized ssDNA on the ITO surface. (For NC and concentrations of genomic copies per reaction of 10^6–10^1, n=5; for one genomic copy per reaction, n=2).

Figure 6 presents the measured voltage responses on the EG-FET platform due to the trans-cleavage of immobilized ssDNA on the ITO surface. CRISPR trans-cleavage cut the ssDNA into several fragments and suspended in the solution which increased the surface negative charge density and in turn resulted in the voltage change. After 15 minutes of reaction there was a clear cut-off for positive and negative controls.

CONCLUSION

We have developed an ITO-based EG-FET platform that is easily fabricated and offers sensitive and quick virus detection. The proposed use of a LAMP-mediated CRISPR/cas12a reaction was validated through its monitoring in a fluorescence-based PCR machine prior to its use for detection on a chip. The HCV LAMP assay was found to amplify the initial RNA template successfully within 15 minutes and to achieve an LOD of one genomic copy per reaction. The sensitivity of the CRISPR/cas12a assay was validated through the addition of LAMP amplicons to the CRISPR/cas12a reaction mixture. The results revealed that the rapid and highly sensitive collateral cleavage of the ssDNA reporter was completed within four minutes. The combination of the use of this technique and an ITO-based EG-FET sensor enabled us to demonstrate successful real-time detection of both synthetic and clinical HCV samples. When the HCV-positive sample was used, a significant increase in voltage was observed within four minutes due to the collateral cleavage of immobilized ssDNA. In contrast, with the use of the negative control, no collateral cleavage occurred and only a negligible voltage change was reported. These results suggest that the developed diagnostic platform could offer a sensitive and robust way to accelerate the screening of HCV patients and could be applied in resource-limited settings.

ACKNOWLEDGEMENTS

The financial supports from the Ministry of Science and Technology of Taiwan and the Southern Taiwan Science Park Bureau (108CB03) are greatly acknowledged. (MOST109-2221-E-110-019-MY3, MOST110-2222-E-110 -005 -MY3).

REFERENCES

[1] M. Jefferies, B. Rauff, H. Rashid, T. Lam, and S. Rafiq, "Update on global epidemiology of viral hepatitis and preventive strategies," *World journal of clinical cases,* vol. 6, no. 13, p. 589, 2018.

[2] N. Brunner and P. Bruggmann, "Trends of the Global Hepatitis C Disease Burden: Strategies to Achieve Elimination," Journal of Preventive Medicine and Public Health, vol. 54, no. 4, p. 251, 2021.

[3] W. A. El-Said and J.-w. Choi, "High selective spectroelectrochemical biosensor for HCV-RNA detection based on a specific peptide nucleic acid," Spectrochimica Acta Part A: Molecular and Biomolecular Spectroscopy, vol. 217, pp. 288-293, 2019.

[4] A. Alberti, F. Noventa, L. Benvegnu, S. Boccato, and A. Gatta, "Prevalence of liver disease in a population of asymptomatic persons with hepatitis C virus infection," Annals of internal medicine, vol. 137, no. 12, pp. 961-964, 2002.

[5] W. Liu, J. Das, A. H. Mepham, C. R. Nemr, E. H. Sargent, and S. O. Kelley, "A fully-integrated and automated testing device for PCR-free viral nucleic acid detection in whole blood," Lab on a Chip, vol. 18, no. 13, pp. 1928-1935, 2018.

[6] A. Roohizadeh, A. Ghaffarinejad, R. Salahandish, and E. Omidinia, "Label-free RNA-based electrochemical nanobiosensor for detection of Hepatitis C," Current Research in Biotechnology, vol. 2, pp. 187-192, 2020.

[7] A. Venkatesh, H. Brickner, D. Looney, D. Hall, and E. Aronoff-Spencer, "Clinical detection of Hepatitis C viral infection by yeast-secreted HCV-core: Gold-binding-peptide," Biosensors and Bioelectronics, vol. 119, pp. 230-236, 2018.

[8] J. P. Broughton et al., "CRISPR–Cas12-based detection of SARS-CoV-2," Nature biotechnology, vol. 38, no. 7, pp. 870-874, 2020.

[9] B. Paul and G. Montoya, "CRISPR-Cas12a: Functional overview and applications," biomedical journal, vol. 43, no. 1, pp. 8-17, 2020.

[10] D. C. Swarts and M. Jinek, "Mechanistic Insights into the cis-and trans-Acting DNase Activities of Cas12a," Molecular cell, vol. 73, no. 3, pp. 589-600. e4, 2019.

[11] S.-Y. Li et al., "CRISPR-Cas12a-assisted nucleic acid detection," Cell discovery, vol. 4, no. 1, pp. 1-4, 2018.

[12] S.-Y. Li, Q.-X. Cheng, J.-K. Liu, X.-Q. Nie, G.-P. Zhao, and J. Wang, "CRISPR-Cas12a has both cis-and trans-cleavage activities on single-stranded DNA," Cell research, vol. 28, no. 4, pp. 491-493, 2018.

[13] C.-H. Lin, H.-P. Cheng, C.-B. Yang, and C.-N. Yang, "Solving satisfiability problems using a novel microarray-based DNA computer," Biosystems, vol. 90, no. 1, pp. 242-252, 2007.

[14] W.-S. Kao, Y.-W. Hung, W.-H. Liao, Y.-C. Chou, and C.-H. Lin, "Experimental Validation on the Industrial Panel-Level Process for Producing Solid-State EGFET Sensor Chips," IEEE Transactions on Electron Devices, vol. 69, no. 11, pp. 6304-6309, 2022.

[15] W.-S. Kao, Y.-W. Hung, and C.-H. Lin, "Solid-State Sensor Chip Produced with Single Laser Engraving for Urine Acidity and Total Dissolved Ion Detections," ECS Journal of Solid State Science and Technology, vol. 9, no. 11, p. 115016, 2020.

[16] S. Hongjaisee et al., "Rapid visual detection of hepatitis C virus using a reverse transcription loop-mediated isothermal amplification assay," International Journal of Infectious Diseases, vol. 102, pp. 440-445, 2021.

[17] T. Notomi et al., "Loop-mediated isothermal amplification of DNA," Nucleic acids research, vol. 28, no. 12, pp. e63-e63, 2000.

CONTACT

* Che-Hsin Lin, tel: +886-7-5252000 #4275
E-mail: chehsin@mail.nsysu.edu.tw
Ling-Shan Yu, tel: +886-7-5252000 #7204
E-mail: lingshanyu@mail.nsysu.edu.tw

SMART ELECTRODE ARRAY FOR COCHLEAR IMPLANTS

Ahmad Itawi[1], Sofiane Ghenna[1], Guillaume Tourrel[2], Sébastien Grondel[1], Cédric Plesse[3],
Tran Minh Giao Nguyen[3], Frédéric Vidal[3], Yinoussa Adagolodjo[4], Lingxiao Xun[4], Gang Zheng[4],
Alexandre Kruszewski[4], Christian Duriez[4], and Eric Cattan[1]

[1]Univ. Polytechnique Hauts-de-France, CNRS, Univ. Lille, UMR 8520 - IEMN, Valenciennes,
France.
[2]Oticon Medical, Research & Technology, Vallauris, France.
[3] CY Cergy Paris Université, LPPI, EA2528, Cergy-Pontoise, France and
[4]DEFROST, Univ. Lille, Inria, CNRS, Centrale Lille, UMR 9189 CRIStAL, Lille, France.

ABSTRACT

Cochlear implants made of standard silicone electrode array (EA) are currently used to stimulate the auditory nerve of patients' cochlea. The implants have a proximal diameter of 0.5mm and 2~3cm long composed of 20 bulk platinum electrodes and connection wires (Ø 25μm). Due to their stiffness and passive nature, the most difficult task during implant surgery is inserting the EA properly into the tympanic ramp of the patient's cochlea, often leading to trauma or incomplete insertion. In this work, we developed an original smart EA for efficient insertion. This prototype has a lower stiffness and functionalized with an electronic conducting polymer based micro-actuators able to bend under low electrical voltage stimulation. This prototype is expected to reduce the friction forces during insertion, allow better control of the insertion process, facilitate the work of the surgeon and decrease the probability of trauma.

KEYWORDS

Cochlear electrode array, electronic conducting polymer, actuator

INTRODUCTION

At present, commercially available cochlear implant electrode arrays are all manually assembled, consisting of a wire-bundle design with 12–26 electrodes, a length of 16–30mm, and an electrode surface area of 0.12–1.5mm². Platinum or platinum-iridium alloy electrode arrays are hand-fabricated and assembled from a bundle of fine wires. The most important factor when designing an electrode array is to provide an atraumatic insertion solution in order to prevent causing damage to any of the intra-cochlear structures. For example, the damage caused by the insertion of the cochlear implant may be a translocation from the scala tympani (ST) to the scala vestibuli. The implementation of EAs is particularly challenging because of the friction between the conventional electrode and the lateral wall of the ST. This is partly due to the mechanical mismatch between the stiffness of the materials, e.g., silicone (Young's modulus E ≈ 1 - 10MPa), platinum-iridium (Young's modulus E ≈ 170GPa), constituting such probes and the softness of the walls of the cochlea (E ≈ 10kPa). This mechanical mismatch, which can reach seven orders of magnitude, leads to irreversible tissue damage. It is also necessary that the structure of the EA facilitates a complete insertion, therefore having the array homogeneously placed along the lateral wall in close proximity with the basilar membrane is essential (Figure

1). A motorized insertion tool, external to the ear, is currently used to push a blunt pin into an insertion tube and eject the EA into the cochlea.

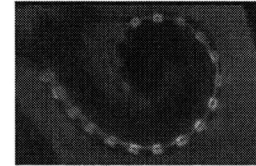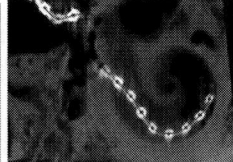

Figure 1: Computed tomography scan post-insertion of a cochlear implant; Left: successful insertion allowing activation of all electrodes; Right: incomplete insertion.

Research work is trying to get out of traditional manufacturing methods by moving towards the use of skills in the field of flexible electrodes made from microfabrication technologies. This evolution is driven in large part by the progress made in the field of stimulation electrodes for the brain [1], [2]. The use of flexible probes achieves better compliance with the surrounding tissues of the cochlea whereas simultaneously minimal friction improves the ST-EA interface. Thin-film EAs have already been developed from boron diffusion silicon substrate designs [3], [4] and flexible polymer substrate designs [5]. The state-of-the-art cochlear thin-film EAs developed by Johnson and Wise contains 32 electrodes on an 8-mm-long array in a curve-controlled shape with a system-level ASIC [6]. Despite these significant advances, none of the MEMS electrodes is currently used in commercial cochlear implants applications for humans. Notice however that friction along the lateral wall inducing jamming upon entry into the cochlea is not discussed nor considered for very flexible EA. Moreover, to our knowledge, micro-actuated EA for microfabricated thin cochlear implants do not exist. Consequently, electronic conducting polymer (ECP) based micro-actuators have been chosen because of their high stress per unit mass at low voltage (2V) [7, 8]. For completeness, it should also be recalled that current catheters fabricated with ECP [9, 10] are not suitable for integrating an EA thin film.

RESULTS AND DISCUSSION

The bending stiffness of a traditional implant EA results primarily from the platinum electrode and the electrical connection wires. The measured stiffness at 3.5mm of the clamping point is close to 60μN/μm (Figure 2a). This stiffness is 300 times lower and reduced to 0.2μN/μm, at the same distance, by removing bulk platinum electrodes and wires and by keeping only the silicone carrier. In this last situation, the blocking forces produced by poly(3,4-ethyledioxythiophene):poly(styrene sulfonate) (PEDOT:PSS) based micro-actuator are optimized by changing the thicknesses of each layer of the micro-actuator (Figure 2b).

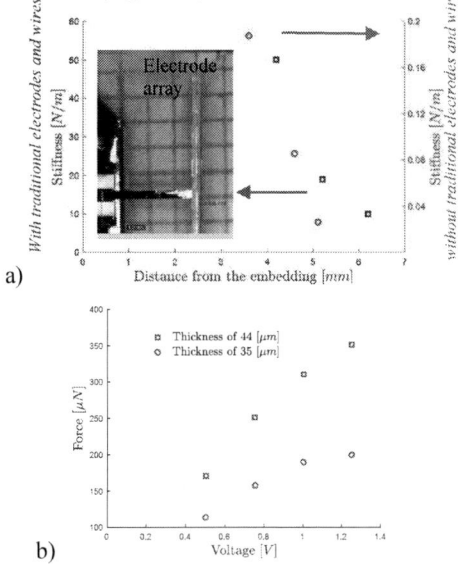

a)

b)

Figure 2: a) Comparison of the stiffness for a round silicon bar with (blue points)/without (red points) traditional platinum electrodes and connection wires, b) Optimization of the micro-actuator thickness to obtain enough force to bend the EA.

The micro-actuators composed of a nitrile butadiene rubber/poly(ethylene oxide) (NBR-PEO) ion storage membrane sandwiched between two electroactive PEDOT:PSS-PEO layers were fabricated using the layer stacking method [11], [12]. The PEDOT:PSS-PEO casting solutions were obtained by mixing PEO (poly(ethylene oxide)) with an aqueous dispersion of PEDOT:PSS (40wt% with respect to the final electrode). The resulting solution was cast ($0.1ml.cm^{-2}$) onto a glass slide and placed on a heating plate at 50°C to evaporate the water. The ion storage membrane is based on a semi-interpenetrating polymer network layer composed of a PEO network (50wt%) and linear NBR (50wt%). First, the NBR solution was prepared by dissolving NBR in cyclohexanone to obtain a concentration of 20wt%. The PEO is added to the NBR solution. The radical initiator DCPD (3wt% with respect to the PEO network) was then added to the solution. The final solution was stirred until complete homogenization and degassed. During the next step, the reactive mixture was spin coated (1500rpm – 1000rpm/s – 30s) onto the first PEDOT:PSS-PEO electrode layer and pre-polymerized in a closed annealing chamber under continuous nitrogen flow for 45min at 50°C to initiate the formation of the PEO network. The second PEDOT:PSS-PEO electrode was fabricated on top of the PEDOT:PSS-PEO/NBR-PEO bilayer in the same way as the first electrode: new solutions were prepared, cast, and solidified at 50°C by evaporating the water. The resulting trilayer micro-actuators were then placed in a closed annealing chamber and the final heat treatment was carried out at

50°C for 3 h followed by post-curing at 80°C for 1 h under a continuous nitrogen flow. Finally, the micro-actuator is swollen in 1-ethyl-3-methylimidazolium bis(trifluoro-methanesulfonyl) -imide electrolyte.

The micro-actuator force can reach 350µN at a distance of 2mm from the embedding point under 1.25V (Figure 2b). Depending on the applied voltage, the induced displacement is enough to move the silicone carrier to a few hundred micrometers in order to separate the implant from the lateral wall and reduce friction and capillary effect.

The overall design of the traditional stimulation of EA has been fully amended to reduce the EA stiffness. The proposed microsystem is an addition of a very thin and flexible EA, one or several distributed ECP based micro-actuators and a Polydimethylsiloxane (PDMS) carrier (Figure 3a). For new EA, a SU-8 thin-film was microfabricated with a density of 20 electrode contacts along a 25mm in length. (Figure 3c). The neutral line is used to run the 10µm wide gold lines.

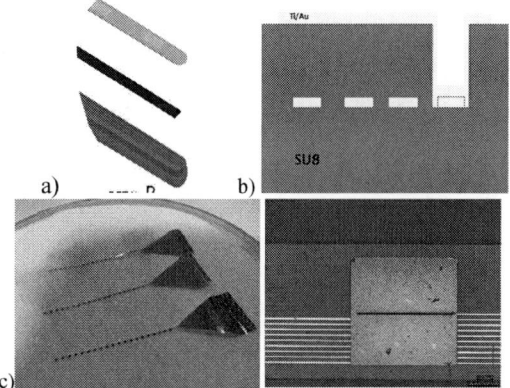

c)

Figure 3: a) Exploded view of the new prototype, b) cross-sectional drawing showing microfabrication steps c) SU-8 electrode arrays with detailed view of an electrode contact and connecting lines.

The microfabrication process is done as follows: first, Omnicoat is spun coated to create a sacrificial layer to remove flexible EA from the silicon substrate at the end of the process. Then 7 µm thickness of SU-8 photoresist is deposited and UV insolated to define the shape of the EA. The first metallic layer (Ti/Au - 20nm/0.5µm) was deposited by evaporation. Gold and titanium layers circuit were defined by chemical etching: $KI:I_2:H_2O$ and BOE 7.1 respectively. The second SU-8 layer with the same thickness is spin-coated to place the metal circuit in the neutral plane. The opening of SU-8 is made and the second metal layer is processed like the previous one to make the interconnect. Each site comprises a 0.45mm×0.40mm square gold electrode to maximize the surface in front of the basilar membrane (Figure 3b). With a total thickness of 14µm the new EA stiffness is 0.6µN/µm at 2mm, which corresponds to a decrease of a factor of 100 compared to traditional electrode arrays, and can be reduced more by decreasing the thicknesses of the two SU-8 layers: for example, the stiffness is reduced to 0.2µN/µm when the thickness is decreased at 10µm. The measured resistances at the first and last electrodes are around 45±5Ω and 300±20Ω respectively which is consistent with the theoretically calculated value.

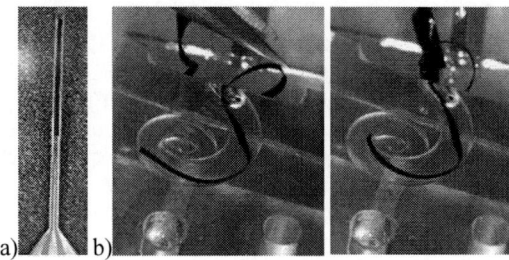

Figure 4: a) ECP-based micro-actuators clamped in a SU-8 micro-structure with integrated gold remote contacts. The actuator is a beam of 6 mm length and 300 μm wide. b) Actuation principle (without EA) inside a transparent 3D printed cochlea model. Actuator "off" on the left and "on" on the "right".

However, its insertion into a transparent 3D printed cochlea model does not exceed 180°due to the friction of the SU-8 beam edges on the lateral wall of the 3D printed cochlea model. Indeed, this friction causes an "accordion" effect of the EA at the entrance of the cochlea preventing adequate thrust. To actuate the EA in three different points, the smallest ECP-based micro-actuators (6x0.3mm^2) integrated ever used were clamped to a SU-8 micro-structure with integrated gold remote contacts by using a microfabrication method described in [13] (Figure 4a). The tests with this integrated micro-actuator inside the cochlea are currently in progress. The micro-actuator can also be envisaged as a laser cutting beam of 2.5cm with a width of 0.5mm used for an actuation along all the EA. With the appropriate electrical excitation, we have already demonstrated that the laser cutting micro-actuator can move its tip from a position near the lateral wall of the cochlea model to the inner wall, when it is full of glycerin to simulate the perilymph (Figure 4b).

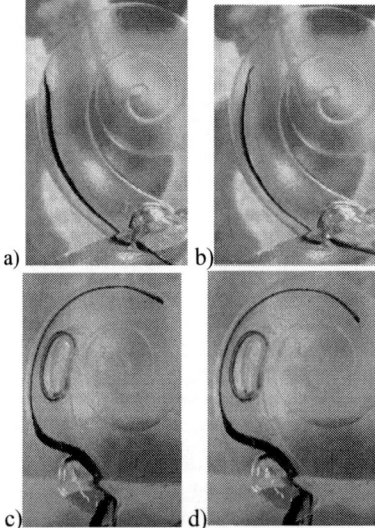

Figure 5: Micro-actuator added along an EA a) and b) and PDMS U-shape carrier without reservoir c) and d) to detach them from the lateral wall a) and d) off, b) and c) on.

It is then easier to push and move forward the actuator in

the cochlea but its flexibility also produces the accordion effect. To limit this effect, the thickness of the micro-actuator has been increased to 70μm. Tests performed with this micro-actuator added along an EA and PDMS carrier (Figure 5a, b, c and d) have demonstrated that it could be successfully detached from the lateral wall.

At the same time, a new PDMS U-shape carrier has been prepared with small reservoirs inside the PDMS. The latter have been filled with SU-8 in order to add rigidity along the thrust axis while keeping bending flexibility thanks to the PDMS reservoir separations (Figure 6a). Note also that SU-8 added prevents the PDMS carrier from folding like an accordion. With this solution, a successful insertion in the 3D printed cochlea model with an angle of 490° has been obtained (Figure 6b).

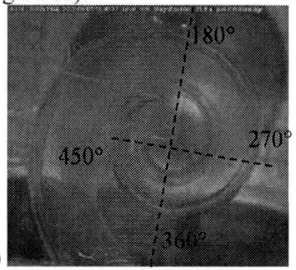

Figure 6: a) PDMS U-shape with reservoirs, b) PDMS U-shape carrier with SU-8 added. b) Insertion into 3D printed cochlea model with a progression to an angle of 490°.

In parallel, using the SOFA framework, we performed a simulation of the process of EA insertion into the cochlea based on the finite element (FE) method and Cosserat's theory [15]. To manage the different interactions that may appear between the EA and the cochlea, we have chosen a formulation based on Lagrangian constraints using Signorini's law for contact and Coulomb's law for friction. Figure 7 shows the intensity of the frictional force as the EA progresses through the cochlea. The calculation of this intensity depends on the friction parameter between the EA and the cochlea. A first empirical study was therefore carried out by comparing the forces obtained at the base of the EA during the simulation and that recorded by a sensor placed on the real electrode. The friction parameters that offer a force profile close to that recorded by the sensor, throughout the insertion procedure, were considered for the simulation of this new EA. We saw a 30% drop in max strength using these settings on the new EA.

The objective of this simulation is therefore to determine the regions of high friction between the EA and the cochlea. Then provide an optimal distribution of micro-actuators in the EA to further reduce the frictional force and avoid blockages during insertion.

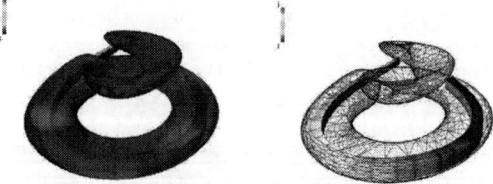

Figure 7: Simulation of cochlear implant insertion to identify contact areas in the cochlea at different

stages of the process.

CONCLUSIONS

The main aim of this study was to propose a solution to reduce the friction forces against the lateral cochlea wall during implant insertion by the surgeons. To solve this problem, the first idea has been to reduce the stiffness of the cochlear implant. However, it should be noted that by reducing drastically the stiffness, the implant will set in the accordion and lock at the entrance to the cochlea. To counterbalance this drawback, we have thereafter proposed to integrate ECP micro-actuators on the cochlear implant in order to allow its bending during the insertion. Using such original approach, we have demonstrated that the EA stiffness of the cochlear implant could be reduced by 100 and that the forces necessary for its bending became compatible with forces developed by ECP micro-actuators. These forces are currently on the order of $400\mu N$ at 2mm from the clamping point. Next, on the one hand, the EAs have been designed to put gold electrode as close as possible to the basilar wall of the cochlea, with large electrode surfaces of $0.2\mu m^2$ and a low resistance whereas on the other hand micro-actuators have been added to the EA and to the PDMS carrier. Then, it has successfully been shown that the use of micro-actuators allowed first the implant separation from the lateral wall of the cochlea under an electrical voltage of 1.5V and second improved its progression through the cochlea. To improve the efficiency of the proposed solution, an original PDMS U-shape holder with reservoirs has been finally designed. In fact, the addition of SU-8 in the reservoir as well the UV cure increase the EA stiffness and then prevent the implant from folding like an accordion. A successful insertion in the 3D printed cochlea model with an angle of 490° has been obtained which demonstrates the interest of the proposed solution. Future work will be focused on two aspects to improve the insertion efficiency. The first deals with the optimization of the microactuators distribution thanks to the simulation whereas the second concerns the possibility of using a closed loop control [14] during insertion.

ACKNOWLEDGEMENTS

This work was partly financially supported by the French National Research Agency with RENATECH, and ROBOCOP (ANR PRCE 19 CE19) projects.

REFERENCES

[1] E. Patrick, M. Ordonez, N. Alba, J. C. Sanchez, and T. Nishida, "Design and fabrication of a flexible substrate mieroelectrode array for brain machine interfaces," in *Annual International Conference of the IEEE Engineering in Medicine and Biology - Proceedings*, 2006, pp. 2966–2969.

[2] C. Cointe *et al.*, "Scalable batch fabrication of ultrathin flexible neural probes using a bioresorbable silk layer," *Microsystems Nanoeng.*, vol. 8, no. 1, Dec. 2022.

[3] J. Wang and K. D. Wise, "A hybrid electrode array with built-in position sensors for an implantable MEMS-based cochlear prosthesis," *J. Microelectromechanical Syst.*, vol. 17, no. 5, pp. 1187–1194, 2008.

[4] N. S. Lawand, P. J. French, J. J. Briaire, and J. H. M. Frijns, "Design and fabrication of stiff silicon probes: A step towards sophisticated cochlear implant electrodes," *Procedia Eng.*, vol. 25, pp. 1012–1015, 2011.

[5] Y. Xu, C. Luo, F. G. Zeng, J. C. Middlebrooks, H. W. Lin, and Z. You, "Design, fabrication, and evaluation of a parylene thin-film electrode array for cochlear implants," *IEEE Trans. Biomed. Eng.*, vol. 66, no. 2, pp. 573–583, Feb. 2019.

[6] A. C. Johnson and K. D. Wise, "An active thin-film cochlear electrode array with monolithic backing and curl," *J. Microelectromechanical Syst.*, vol. 23, no. 2, pp. 428–437, 2014.

[7] D. Zhou *et al.*, "Solid state actuators based on polypyrrole and polymer-in-ionic liquid electrolytes," *Electrochim. Acta*, vol. 48, no. 14-16 SPEC., pp. 2355–2359, 2003.

[8] J. D. Madden, D. Rinderknecht, P. A. Anquetil, and I. W. Hunter, "Creep and cycle life in polypyrrole actuators," *Sensors Actuators, A Phys.*, vol. 133, no. 1, pp. 210–217, 2007.

[9] A. Mazzoldi and D. De Rossi, "Conductive-polymer-based structures for a steerable catheter," *Smart Struct. Mater. 2000 Electroact. Polym. Actuators Devices*, vol. 3987, p. 273, 2000.

[10] T. Shoa, J. D. Madden, N. Fekri, N. R. Munce, and V. X. D. Yang, "Conducting Polymer Based Active Catheter for Minimally Invasive Interventions inside Arteries," *30th Annu. Int. IEEE EMBS Conf.*, pp. 2063–2066, 2008.

[11] K. Rohtlaid, G. T. M. Nguyen, C. Soyer, E. Cattan, F. Vidal, and C. Plesse, "Poly(3,4-ethylenedioxythiophene):Poly(styrene sulfonate)/Polyethylene Oxide Electrodes with Improved Electrical and Electrochemical Properties for Soft Microactuators and Microsensors," *Adv. Electron. Mater.*, vol. 5, no. 4, 2019.

[12] T. N. Nguyen *et al.*, "Ultrathin electrochemically driven conducting polymer actuators: fabrication and electrochemomechanical characterization," *Electrochim. Acta*, vol. 265, pp. 670–680, 2018.

[13] L. Seurre *et al.*, "Demonstrating Full Integration Process for Electroactive Polymer Microtransducers to Realize Soft Microchips," in *Proceedings of the IEEE International Conference on Micro Electro Mechanical Systems (MEMS)*, 2020.

[14] L. Xun *et al.*, "Modeling and Control of Conducting Polymer Actuator," *IEEE/ASME Trans. Mechatronics*, accepted for inclusion in a future issue, 2022.

[15] Y. Adagolodjo et al "Coupling numerical deformable models in global and reduced coordinates for the simulation of the direct and the inverse kinematics of Soft Robots" IEEE Robotics and Automation Letters, 2021.

CONTACT

*S. Ghenna, tel: +33(0)327 51 14 45;
Sofiane.ghenna@upf.fr

A THREE-DIMENSIONAL ARTIFICIAL INTESTINAL TISSUE WITH A CRYPT-LIKE INNER SURFACE

Shuma Tanaka[1], Shun Itai[2], and Hiroaki Onoe[1]
[1] Faculty of Science and Technology, Keio University, JAPAN and
[2] Graduate School of Biomedical Engineering, Tohoku University, JAPAN

ABSTRACT

Disruption of the microbiome in humans caused several diseases. To infer causality, the intestinal models are essential, but the commonly used model has still room for improvements such as biomimicry, productivity, and ability to co-culture with bacteria. This paper describes the artificial intestinal tissue by culturing intestinal cells (Caco-2) in the collagen gel tube with crypt-like cavities created by electrolysis-triggered microbubbles. This tube device has high biomimicry and productivity and also consists of glass/silicone tubes connector at both ends, which allows for connecting external pumps easily and running two types of flows: the inside of the tube (internal flow) and permeation from the outside of the tube to the inside (external flow). In order to show this high extensibility, we demonstrated the co-culture of Caco-2 cells and bacteria (*Bifidobacteria*). We believe that our device will be a platform for investigating complex intestinal diseases due to its high biomimetic properties.

KEYWORDS

Intestine, Intestinal tissue, Organoid, Collagen gel tube, Cell culture, Electrolysis, Co-culture, Bacteria

INTRODUCTION

The microbiome in the human intestine has important roles in the digestion of food and host immunity. Therefore, the disruption of the microbiome triggers digestive system diseases or chronic diseases[1][2][3]. Since the physiological interaction between intestinal cells and the microbiome affects these diseases, artificial intestinal models are required for understanding that relationship. Particularly, crypt-villus structures are known to play an important role to maintain homeostasis in the intestinal tissue (Figure 1 (a))[4]. Several reports have previously fabricated crypt-like structures *in vitro*, such as laser ablation[5] and elastomeric stamps[6]. However, these methods have poor biomimetic properties because they are conducted by a planar configuration. In addition, they required a long machining time with large and expensive equipment.

Here, we propose three-dimensional artificial intestinal tissue formed on the collagen gel tube with an uneven inner wall by microbubbles (Figure 1 (b)). During the fabrication process of a scaffold in the pre-gel condition, microbubbles are generated on a tungsten wire, the role of electrode and mold, to form cavities on the inner surface. The tube device has crypt-shaped cavities arranged three-dimensionally as observed *in vivo*. Besides, we can fabricate them simply and instantly by using a commonly used electric waveform generator. Furthermore, a perfusion culture can be easily performed by connecting an external pump system.

In this paper, we certified that our device maintains its cavities even though the gel is deformed by the traction force of the cells by culturing Caco-2 cells in the intestinal tube device. Moreover, by connecting that tube tissue to the perfusion device, we demonstrate the co-culture of intestinal and microbial cells (Figure 2 (a)). This demonstration indicates our culture system can reproduce the intestinal environment *in vivo*, the conditions of the intestinal human–microbe interface, by using two types of flow which are

(a) Intesinal epithelial tissue mechanism

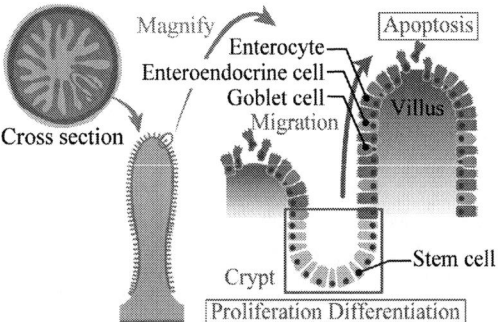

(b) Concept of this work
3D crypt structure formed by microbubbles

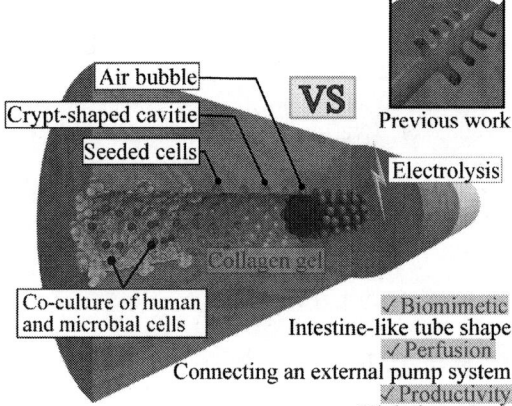

Figure 1: (a) schematic illustration of intestinal epithelial tissue. Intestinal stem cells proliferate and differentiate at the bottom of the crypt and undergo apoptosis at the tip of the villus. By constructing the crypt-like structure on the scaffold, we can reproduce the expression of a variety of cell types and turnover for homeostasis. (b) Conceptual illustration of the artificial intestinal tissue. The collagen gel tube with a crypt-like inner surface was fabricated by using electrolysis-triggered microbubbles. By culturing Caco-2 cells in the gel tube, The intestinal tissue was constructed.

external flow delivering cell culture medium and internal flow diverting microbial culture medium (Figure 2).

EXPERIMENTAL METHODS
Cell culture

A human colon adenocarcinoma cell line (Caco-2) was obtained from RIKEN BRC and cultured in Dulbecco's modified eagle medium (Sigma-Aldrich) containing 10% fetal bovine serum (FBS, Serana Europe) and 1% penicillin-streptomycin solution (Sigma-Aldrich) under a humidified atmosphere of 5% CO_2 at 37°C. The cells were passaged using 0.25% trypsin-EDTA solution (Thermo Fisher).

Fabrication of the collagen gel tube with a crypt-like inner surface

The overview of fabricating process is shown in Figure 3 (a). First, the aluminum mold, connectors made of a glass capillary (0.6×1.0 mm, NARISHIGE) and a silicone tube (1×2 mm, AS ONE), and tungsten wire (500 μm, Nilaco) were assembled and set to the voltage-applied base that was combined with an acrylic plate, carbon rod terminal (NaRika), and electric lead wires. Next, collagen pre-gel solution consists of 8 volumes of type-I collagen acidic solution (5 mg/mL, KOKEN), 1 volume of 10× Hanks' buffer (Sigma-Aldrich), and 1 volume of reconstitution buffer (containing 50 mM NaOH, 200 mM HEPES and 260 mM $NaHCO_3$) was poured into the mold (Figure 3 (b)). Then, collagen pre-gel solution was applied voltage (10 V, 150 ms) to fabricate microbubbles on the tungsten wire (Figure 3 (c)). Then, the pre-gel solution was heated for 1.5 h using an incubator (5 % CO_2 and 37°C) to solidify the gel. This allows cavities were maintained even when the

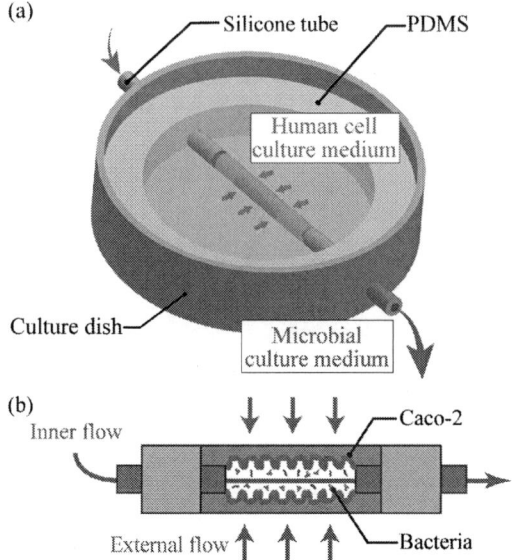

Figure 3: Co-culture of intestinal and microbial cells. (a) The image of the co-culture system. An artificial intestinal tube was connected to a co-culture device via a connector at both ends. (b) Cross-section view showing two types of flow. Inner flow supplies microbial culture medium and external flow supplies cell culture medium.

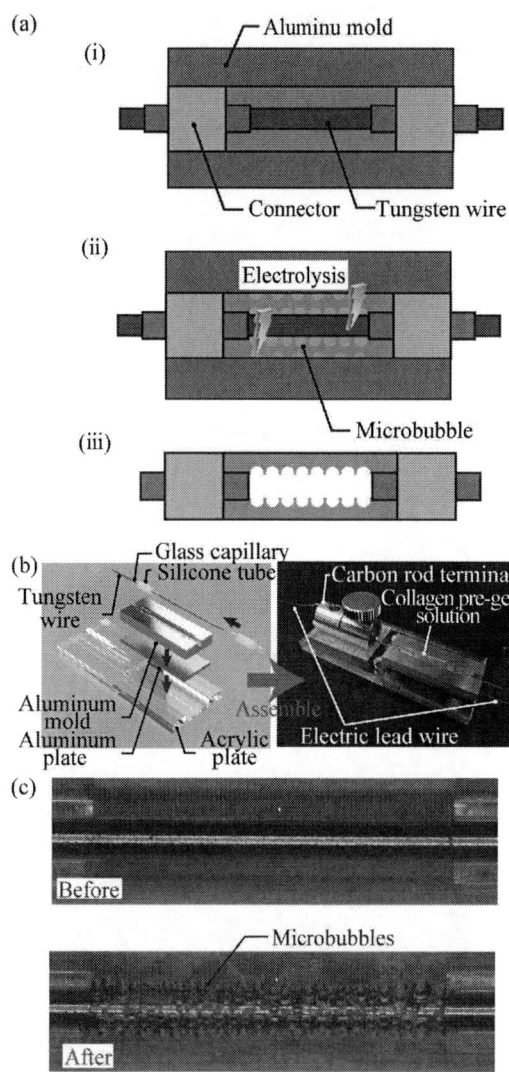

Figure 2: (a) The fabrication process of collagen gel tube with the crypt-like uneven inner wall from a top view. (b) Composition of a bubble-generation device. (c) Comparison images before and after electrolysis. After electrolysis, microbubbles were generated on the tungsten wire.

microbubbles were disappeared. Finally, the collagen gel was removed from the aluminum mold, and then the tungsten wire was withdrawn from the solidified collagen. As a result, the collagen gel tube with a crypt-like inner surface was obtained.

Intestinal tissue development and culture

The procedure of cell seeding and culture using the collagen gel tube prepared above is as follows. First, the Caco-2 cells were harvested and resuspended in a medium at a density of 4.0×10^6 cells per mL. Next, the suspension was loaded into the collagen gel tube by using a syringe (Figure 3 (a)) and then incubated (5 % CO_2 and 37°C) for 30 min to adhere cells to collagen. Then, after the collagen gel tube with the whole aluminum mold was upside-down, the suspension was loaded again and incubated (Figure 3

978-1-6654-9309-3/23 $31.00 © 2023 IEEE   457

(b)). This protocol allows seeding cells on the opposite side. Then, after the remaining suspended cells were flushed by loading the medium, the collagen gel tube was removed from the aluminum mold. Finally, the collagen tube was cultured under a humidified atmosphere of 5% CO2 at 37 °C for 5 days(Figure 3 (c)). The phase-contrast images were captured after flushing the medium to remove the remaining dead cells.

Demonstration of co-culture, intestinal and microbial cells, by using the perfusion system

We demonstrated the co-culture of intestinal and microbial (*Bifidobacteria,* MORINAGA MILK) cells. The perfusion device consists of poly-(dimethylsiloxane)

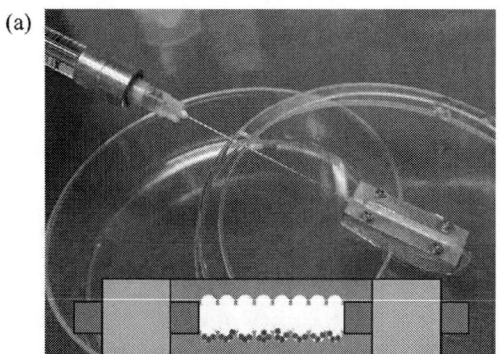

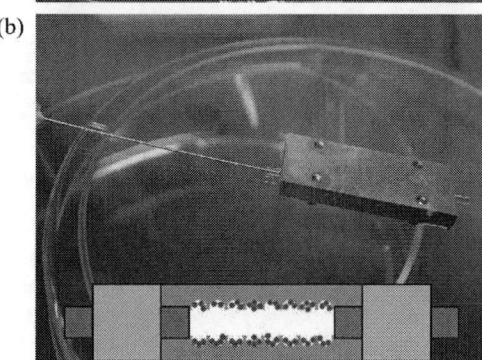

Figure 4: Intestinal cell culture in the tube device with an uneven inner wall. (a) First cell seeding. The operation allowed the cells to adhere the one side of collagen. (b) Second cell seeding. The intestinal tube with aluminum mold was flipped upside-down in order to seed on the opposite side. (c) Cell culture for 5 days, after tubes were removed from aluminum molds.

(PDMS), a 35 mm culture dish, and silicone tubes (1×2 mm). The intestinal tube fabricated in the previous section was connected to the perfusion device and then the cell culture medium was poured into a 35 mm dish and the bacterial culture medium supplemented with bacteria stained red (5×10^6 cells per mL) without antibiotics (PS) was perfused in a silicon/intestinal tube. After culturing under a humidified atmosphere of 5% CO2 at 37 °C for a day, live Caco-2 cells were stained (Calcein-AM, Life Technologies).

RESULT AND DISCUSSION

Fabricating the collagen gel tube with a crypt-like inner surface

The fluorescent images, captured by fluorescent particles adhering to the inner wall to observe, show that crypt-like cavities were conducted. The diameters of cavities were approximately 100 μm that matched those *in vivo* (Figure 5).

Intestinal tissue development and culture

By using the two steps of the cell seeding process, cells adhered to the whole inner surface. The Caco-2 cells, cultured for 5 days, proliferate and filled the luminal surface (Figure 5 (a)). Accordingly, the ductal diameter decreased by 14%, and the width and height of cavities were increased by 34% and decreased by 61% respectively (Figure 5 (b)(c)). The shape change of cavities was conducted in a day and hardly observed from day 2 to day 5. These results suggest that intestine-like tissue that mimics a tube shape and an uneven inner wall was fabricated.

Demonstration of co-culture, intestinal and microbial cells, by using the perfusion system

We demonstrated the co-culture of *Bifidobacteria* in the artificial intestinal tissue. The fluorescence image shows that bacteria were loaded into the artificial intestinal tissue without leaking and, nevertheless, many Caco-2 cells were alive (Figure 7 (a)(b)). These results suggest that our perfusion system can reproduce the *in vivo* intestinal environment divided into conditions for cells and for bacteria by using two types, external and inner, flows.

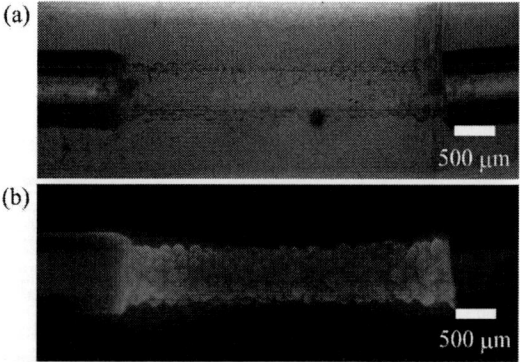

Figure 5: Collagen gel tube with a crypt-like uneven inner surface. (a) Phase-contrast image. (b) Fluorescent image observed by perfusing fluorescent microspheres.

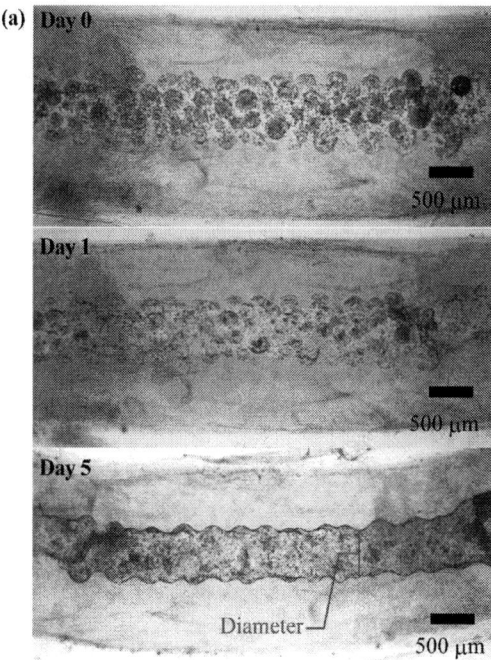

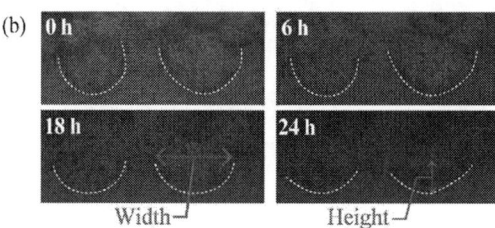

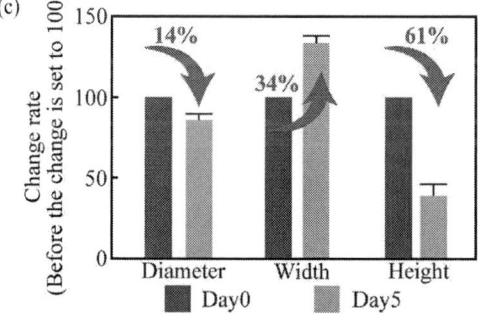

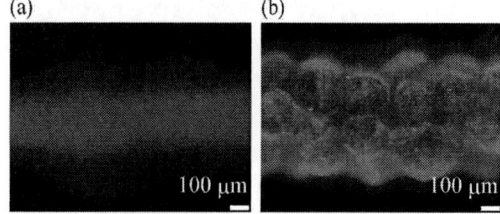

Figure 7: Co-culture of Caco-2 cells and bifidobacteria. (a) Fluorescent image of bifidobacteria (red). (b) Fluorescent image of Stained live Caco-2 cells (green).

culturing bacteria as an inner flow. We believe that our high biomimetic intestinal tube and perfusion system hold promise as a model for human development and diseases.

ACKNOWLEDGMENTS

This work was partly supported by Mitochondrial preemptive medicine, MOONSHOT research & development program from Japan Agency for Medical Research and Development (AMED).

REFERENCES

[1] A. Bergström *et al.*, "Establishment of intestinal microbiota during early life: A longitudinal, explorative study of a large cohort of Danish infants," *Appl. Environ. Microbiol.*, vol. 80, no. 9, pp. 2889–2900, 2014, doi: 10.1128/AEM.00342-14.

[2] KANAMORI and Y, "Abnormal intestinal microbiota in pediatric surgical patients and the effects of a newly designed synbiotic therapy," *Int J Prob Preb*, vol. 1, pp. 149–160, 2006, Accessed: Jan. 05, 2022. [Online]. Available: http://ci.nii.ac.jp/naid/10026623776/ja/.

[3] P. Panigrahi *et al.*, "A randomized synbiotic trial to prevent sepsis among infants in rural India," *Nature*, vol. 548, no. 7668, pp. 407–412, 2017, doi: 10.1038/nature23480.

[4] H. J. Snippert *et al.*, "Intestinal crypt homeostasis results from neutral competition between symmetrically dividing Lgr5 stem cells," *Cell*, vol. 143, no. 1, pp. 134–144, 2010, doi: 10.1016/j.cell.2010.09.016.

[5] M. Nikolaev *et al.*, "Homeostatic mini-intestines through scaffold-guided organoid morphogenesis," *Nature*, vol. 585, no. 7826, pp. 574–578, 2020, doi: 10.1038/s41586-020-2724-8.

[6] Y. Wang *et al.*, "A microengineered collagen scaffold for generating a polarized crypt-villus architecture of human small intestinal epithelium," *Biomaterials*, vol. 128, pp. 44–55, 2017, doi: 10.1016/j.biomaterials.2017.03.005.

CONTACT

*S. Tanaka, tel: +81-45-566-1507; ses@keio.jp

Figure 6: Intestinal tissue development. (a) Phase contrast images of the cultured Caco-2 cells in the tube. Caco-2 cells proliferate and filled the luminal surface on day 5. Time-lapse image of the crypt-like cavities. The images show the width is spread and the height is lower over time. Diameter, width, or height variation over time (n=10, error bars: S.D.).

CONCLUSIONS

By using a collagen gel tube with an uneven inner wall, we succeed in an artificial intestinal tissue mimicking tube shape and crypt-like structure. Furthermore, by connecting that tube tissue and a perfusion device, we were able to reproduce the intestinal environment divided into conditions for cells and for bacteria.

Future, we will replace Caco-2 cells with iPS-derived intestinal stem cells and perfuse an anaerobic medium for

TISSUE-ENGINEERED PENNATE MUSCLES ON A CHIP

Motoki Ito, Yuya Morimoto, and Shoji Takeuchi

Graduate School of Information Science and Technology, The University of Tokyo, JAPAN

ABSTRACT

In this paper, we report the device to culture skeletal muscle tissues modeling pennate muscles. The muscle tissues fabricated on the device have a different directed contractile force from their muscle fibers, by using two kinds of anchors to regulate both directions. Furthermore, we achieved to culture three pairs of the skeletal muscle tissues, pennate-like structured muscle tissue, on the device. Considering that it is possible to culture more pairs of tissues using the device, our proposed method will be useful to form large muscle tissues modeling pennate muscles.

KEYWORDS

Tissue Engineering, Pennate Muscle, Muscle Fibers Orientation, Contractile Force

INTRODUCTION

Recently, soft actuators, which produce force by their contractions, have been attracting attention since unlike conventional actuators such as mechanical motors, they are lightweight, portable, and can achieve flexible movement which mimics that of a living body [1]. Among soft actuators, those composed of materials from living organisms and synthetic materials are called as biohybrid actuators, which have attracted more and more attention in recent years. Mainly cardiac and skeletal muscle tissues are used in biohybrid actuators because they can easily produce contractile force by electrical or optical stimulations [2,3]. In particular, skeletal muscle tissues have high energy efficiency, the ability of self-repair, and higher controllability than cardiac ones, so many studies on biohybrid actuators composed of skeletal muscle tissues have made progress recently.

In previous researches, a number of biomimetic movements have been achieved by applying contractile force of skeletal muscle tissues to move various contrived structures, such as swimming on the air-liquid interface [4], crawling [5] and rotating its joint [6]. These movements are realized by contractile force of the skeletal muscle tissue, and thus it is thought that greater contractile force is required to move bigger structures. However, in these studies, muscle tissues modeling pennate muscle, which generates larger force in living organisms, have not been used.

Living organisms have two kinds of skeletal muscles - parallel muscles and pennate muscles (Fig. 1(a)). Parallel muscles are muscles in which the direction of muscle fibers and the direction of the contractile force are same. Pennate muscles have the direction of contractile force tilted with respect to the direction of the muscle fibers, so have more muscle fibers per unit volume than parallel ones. For this reason, the pennate muscles produce greater force than the parallel ones [7,8]. In spite of this advantage, previous research has not established how to make and utilize skeletal muscles modeling pennate muscles.

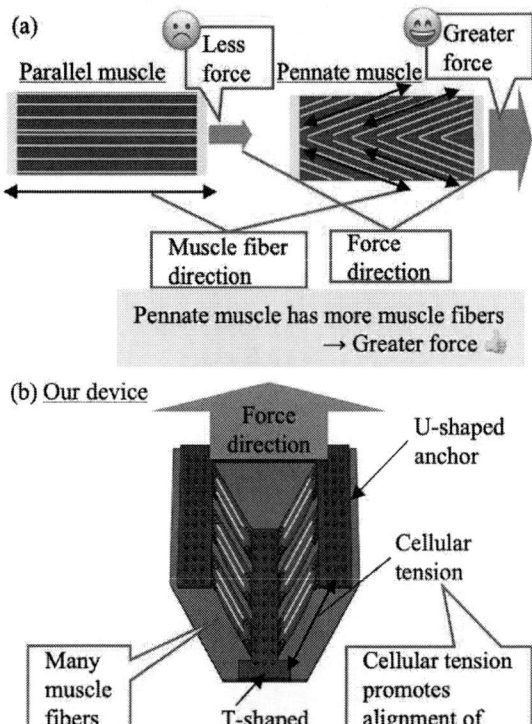

Figure 1: (a) Schematic illustrations of parallel muscle (left) and pennate muscle (right). (b) Conceptual illustration of three pairs of skeletal muscle tissues modeling pennate muscle cultured on our device.

In this paper, we propose a device to culture skeletal muscle tissues modeling pennate muscles. Cultured skeletal muscle tissues on the device have a different directed contractile force from their muscle fibers like pennate muscles, by regulating both directions according to design and dimensions of two kinds of anchors. Furthermore, we achieved three pairs of pennate muscles which are expected to produce larger force, using bigger anchors and devices (Fig. 1(b)).

EXPERIMENTAL METHODS

Device Preparation

The device was composed of four components: a 3D-printed T-shaped anchor, a 3D-printed U-shaped anchor, a polydimethylsiloxane (PDMS) mold, and a PDMS base. The anchors fixed the ends of skeletal muscle tissues and applied cellular tension to them. Both of the PDMS mold and the base had concavities which anchors were set on. The PDMS mold had grooves, and they were used to shape myoblast-laden hydrogel. The PDMS base was used to culture the myoblast-laden hydrogel to differentiate into

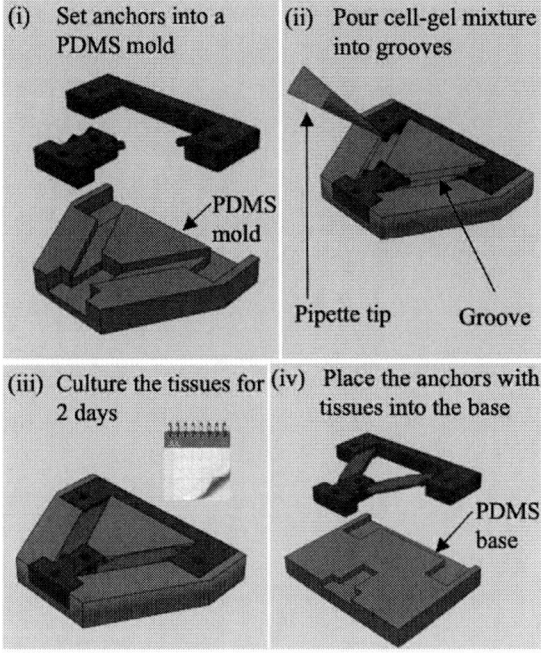

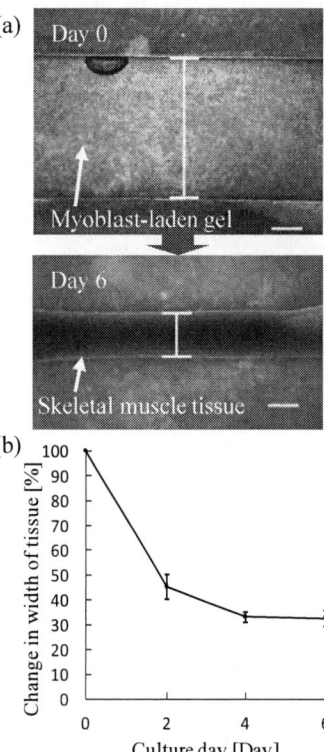

Figure 2: Procedure for culturing skeletal muscle tissue modeling pennate muscle.

muscle tissue. When fabricating skeletal muscle tissues, anchors were firstly placed on the PDMS mold (Fig. 2(i)).

Tissue Fabrication

To fabricate skeletal muscle tissues, a mixture of C2C12 mouse myoblast, Cultrex Basement Membrane Extract (BME), fibrinogen and thrombin was poured into grooves of the mold with two kinds of the anchors (Fig. 2(ii)). After 30 min of incubation at 37 °C for hydrogel gelation, the cell-gel mixture was put in the medium and cultured in the incubator for 2 days (Fig. 2(iii)). Subsequently, the muscle tissue with the anchors was moved to the PDMS base (Fig. 2(iv)) and cultured for additional days.

EXPERIMENTAL RESULTS
Measuring width of tissues

From the observation of skeletal muscle tissue during culture, we measured change in their width using ImageJ. As shown in Figure 3, the width of tissues became about 30 % from their initial width on day 6. This is a typical phenomenon in formation of skeletal muscle tissues with myoblast-laden hydrogel [9]. This result indicates that cellular tension was applied to the tissues in the direction perpendicular to that of tissue shrinking which is same with the direction of facing the anchors.

Calculation of contractile force

To calculate the contractile force of tissue, we put a pair of muscle tissues with the anchors cultured for 10 days on a device with a cantilever made of PDMS (Fig. 4). Then, we applied electrical stimulations to it and measured the displacement of the cantilever's tip. Using the theory of

Figure 3: (a) Shrinkage of muscle tissue in 6 days culture (scale bar; 500 μm). (b) Graph showing the change in width of skeletal muscle tissues (n=6).

mechanics of materials, we calculated its contractile force in a direction of the arrow in Fig. 4(a) from the displacement of the cantilever pushed by the muscle tissue. This result shows that the tissues had contractile force, which increased with the application of a larger electric field in a different direction from that of the cellular tension in the muscle tissues.

Evaluation of muscle fiber orientation

To evaluate muscle fiber orientation in the fabricated tissue, staining of longitudinal section of the tissues was conducted using Phalloidin, which stains actin filament. The stained image was analyzed using the FFT (Fig. 5), the result of which represents the existence probability of muscle fibers at each angle in a figure of the stained tissue. This result indicates that the muscle tissues had the orientation in a direction of cellular tension, which was different from that of the contractile force. These results suggest that our proposed method will be useful to make muscle tissues modeling pennate muscles.

Fabrication of the larger tissue

As a demonstration for fabrication of multiple pairs of skeletal muscle tissues to increase contractile force, we formed three pairs of the tissues using same method (Fig. 6). As a result of immunostaining (Fig. 6(b)), we confirmed that the tissues were enough matured because α-actinin, a protein related to muscle contraction, was expressed in them. This result indicates that it will be possible to make

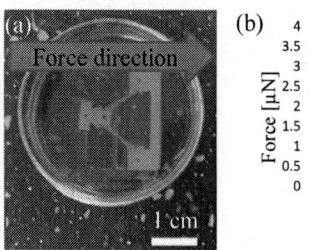 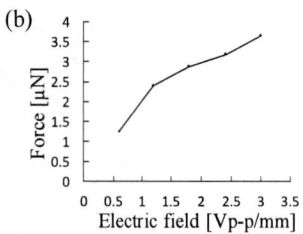

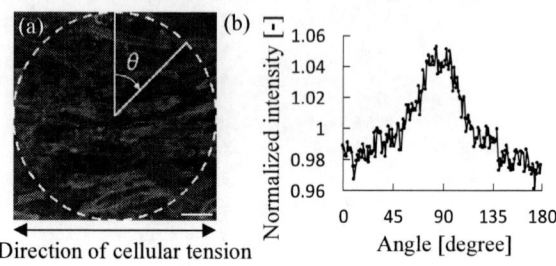

Direction of cellular tension

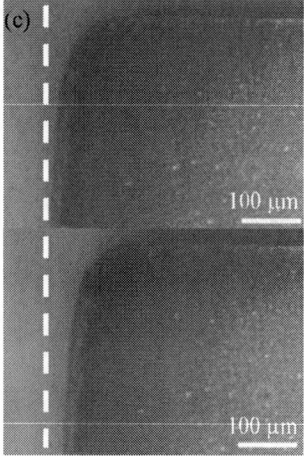

Figure 5: (a) Confocal microscopic image of F-actin in the muscle tissue after 10 days of differentiation (scale bar; 50 µm). (b) Result of the FFT analysis of (a). Values in the horizontal axis represent the angle θ. Values in the horizontal axis represent existence probability of muscle fibers at the angle.

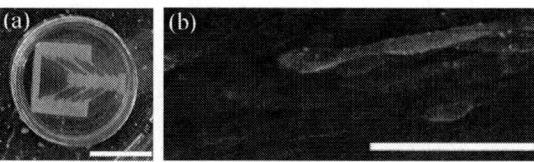

Figure 6: (a) Fabricated three pairs of skeletal muscle tissue (scale bar; 2 cm). (b) Confocal microscopic image of the immunostained skeletal muscle tissue (Green; α-actinin. Blue; nuclei. Scale bar; 100 µm).

Figure 4: (a) Anchors with tissues on the device consisting of a cantilever. (b) Graph showing the relationship between applied electric field and contractile force of the tissue in a direction of the arrow in (a) (n=1). (c) Displacement of the cantilever's tip pushed by the muscle tissue.

large and morphologically matured muscle tissues modeling pennate muscles in our proposed method.

CONCLUSION

In this study, we developed a device to fabricate skeletal muscle tissues modeling pennate muscles. The tissues made on it had the different directed contractile force from their muscle fibers, which is the feature of pennate muscles. We also achieved to fabricate three pairs of pennate muscle tissues which were enough matured by using this device. We believe that the muscle made by our proposed method could be used for biohybrid actuators with larger contractile force moving big structures.

ACKNOWLEDGEMENTS

This work was partially supported by JST-Mirai Program JPMJMI18CE and the JSPS Grants-in-Aid for Scientific Research (KAKENHI) (Grant Number 21H00321), Japan.

REFERENCES

[1] J. Kim, *et al.* "Review of Soft Actuator Materials," *int. J. Precis. Eng. Manuf.* vol. 20, pp. 2221-2241, 2019.

[2] Y. Morimoto, *et al.* "Biohybrid device with antagonistic skeletal muscle tissue for measurement of contractile force," *Advanced Robotics,* vol. 33, pp. 208-218, 2019.

[3] R. Raman, *et al.* "Optogenetic skeletal muscle-powered adaptive biological machines," *Proc. Natl. Acad. Sci. U.S.A.* vol. 113, pp. 3497-3502, 2016.

[4] M. Guix, *et al.* "Biohybrid soft robots with self-stimulating skeletons," *Science Robotics,* vol. 6, abe7577, 2021.

[5] C. Cvetokovic, *et al.* "Three-dimensionally printed biological machines powered by skeletal muscle," *Proc. Natl. Acad. Sci. U.S.A.,* vol. 111, pp. 10125-10130, 2014.

[6] Y. Morimoto, *et al.* "Biohybrid robot powered by an antagonistic pair of skeletal muscle tissues," *Science Robotics,* vol. 3, eaat4440, 2018.

[7] Y. Kawakami, "Morphological and functional characteristics of the muscle-tendon unit," *The Journal of Physical Fitness and Sports Medicine,* vol. 1, pp. 287-296, 2012.

[8] H. Degens, R. M. Erskine, and C. I. Morse, "Disproprtionate changes in skeletal muscle strength and size with resistance training and aging," *Journal of Musculoskeletal Neuronal Interactions,* vol. 9, pp.123-129, 2009.

[9] S. Hinds, *et al.* "The role of extracellular matrix composition in structure and function of bioengineered skeletal muscle," *Biomaterials,* vol. 32, no. 14, pp. 3575-3583, 2011.

CONTACT

*Motoki Ito, tel: +81-3-5841-6488; ito@hybrid.t.u-tokyo.ac.jp

WEIGHT TRAINING DEVICE TO PROMOTE MATURATION IN SKELETAL MUSCLE TISSUES

Kentaro Motoi, Byeongwook Jo, Yuya Morimoto, and Shoji Takeuchi

Graduate School of Information Science and Technology, The University of Tokyo, JAPAN

ABSTRACT

This paper proposes a weight training device to promote maturation in skeletal muscle tissues. The proposed device has a 3D-printed floating structure and a weight that enable the weight training movement inspired by muscle training. The floating structure allows free contractile movement in skeletal muscle tissues under electrical stimulation. Also, using the weight, mechanical stimulation can be applied to the tissues. Through the fabricated weight training device, thicker and more highly-aligned myotubes were formed compared to ones without training. This weight training device will be applicable to promote the maturation of skeletal muscle tissues for applications such as bioactuators, regenerative medicine, and cultured meat.

KEYWORDS

Tissue engineering, Electrical stimulation, Mechanical stimulation, Biofabrication, Myotube diameter, Myotube orientation

INTRODUCTION

Recently, *in vitro* skeletal muscle tissues have attracted attention due to their potential applications in the bioactuators and the food industry. Skeletal muscle tissues have been used as actuators for biohybrid robots and have succeeded in performing a variety of motions [1-3]. In addition, skeletal muscle tissues have been used in the construction of cultured meat, one of the promising candidates as environmentally friendly next-generation meat [4-6]. For the development of *in vitro* skeletal muscle tissues with these diverse application possibilities, training is essential since they are markedly inferior to living muscle tissues in terms of maturation, especially myotube thickness [7]. In previous studies, electrical stimulation (ES) has been applied to promote maturation. However, they still showed a lower maturation compared to living muscle tissues [8]. One of the challenges of previous studies is that the tissue is fixed at both ends and cannot change its length, limiting contractile movement. When training *in vivo* muscle, not only exercise that does not change its length (isometric exercise) but also exercise that changes its length (isotonic exercise) are widely used, each showing a different functional development on muscles [9]. To overcome the limitation of *in vitro* skeletal muscle tissues, it is necessary to achieve training with free contractile movement similar to that of *in vivo* muscle.

Here, we propose a weight training device that enables free contractile movement with adjustable loadings (Fig. 1). In the weight training device, one end of the skeletal muscle tissue is mounted to a floating structure, which serves to straighten the tissue horizontally without deflection. Also, the floating structure is not fixed, so it can move freely without friction at the bottom in a horizontal direction when the skeletal muscle tissue contracts under

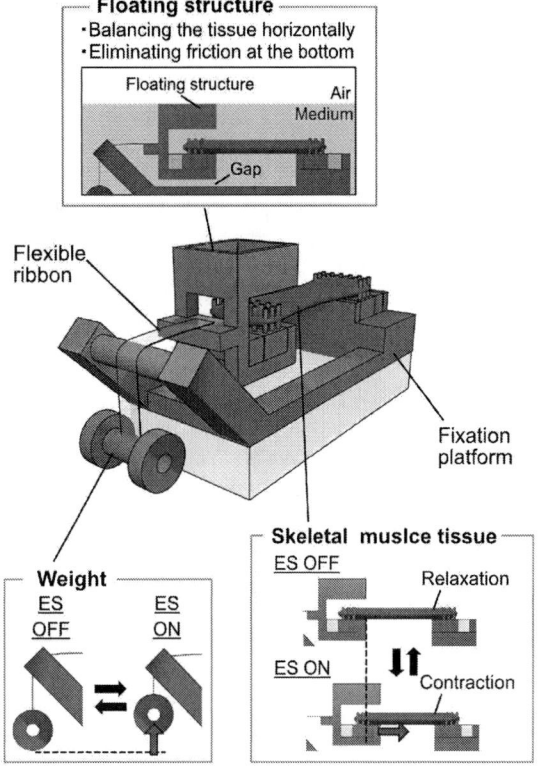

Figure 1: Concept of a weight training device. The weight training device with a floating structure and a weight allows free contractile movement in skeletal muscle tissue under electrical stimulation.

ES. The weight, connected to the floating structure by the flexible ribbon, moves vertically in accordance with the muscle contraction.

EXPERIMENTAL METHODS

Isolation of myoblasts

Skeletal muscle samples were obtained from hind limb of 1-year-old Wistar neonatal rat (Sankyo Labo Service Corporation, Inc.). Muscle tissue was mechanically minced using scissors and enzymatically digested with type-II collagenase (Invitrogen). Isolated cells were centrifuged and used to fabricate *in vitro* skeletal muscle tissues. All rats were maintained in accordance with the policies of the University of Tokyo Institutional Animal Care and Use Committee.

Fabrication of the skeletal muscle tissue

To fabricate the skeletal muscle tissue, first, we prepared a polydimethylsiloxane (PDMS) substrate with 3D-printed anchors. The PDMS substrate with grooves was fabricated by solidifying silicone elastomer base and curing

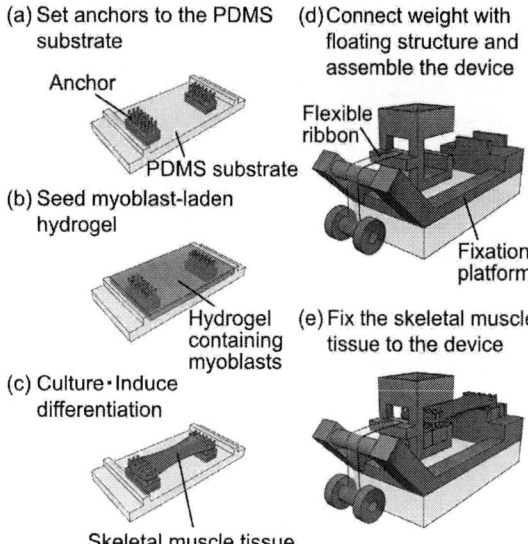

(a) Set anchors to the PDMS substrate

(b) Seed myoblast-laden hydrogel

(c) Culture·Induce differentiation

(d) Connect weight with floating structure and assemble the device

(e) Fix the skeletal muscle tissue to the device

Figure 2: Fabrication process of fabricating the weight training device. (a) Set anchors to the grooves on both sides of the PDMS substrate. (b) Seed myoblast-laden hydrogel on the PDMS substrate with anchors. (c) Culture the myoblasts and induce differentiation. (d) Connect the floating structure and the weight with the flexible ribbon, and arrange it to the device. (e) Fix the skeletal muscle tissue detached from the PDMS substrate to the device.

agent (SILPOT 184 Silicone Elastomer, Dow Toray Co., Ltd.) mixed with 10:1 ratio within a 3D-printed mold. Anchors were attached to the grooves on both sides of the PDMS substrate (Fig. 2(a)). The anchors serve to fix the ends of the skeletal muscle tissue and maintain tension during culture. It is essential to keep the tissue under tension to prevent shrinkage of the tissue leading to the loss of its contractility [10]. Then, hydrogel containing myoblasts was put on the PDMS substrate (Fig. 2(b)). After culturing the myoblasts in growth and differentiation medium, the skeletal muscle tissue was obtained as a result of differentiation of the myoblasts into myotubes (Fig. 2(c)).

Construction of the weight training device

To prepare the weight training device, the floating structure and the weight were connected with the flexible ribbon, and then arranged to the fixation platform (Fig. 2(d)). The flexible ribbon was fabricated by cutting and peeling off a 5μm thick parylene sheet. Finally, the fabricated skeletal muscle tissue detached from the PDMS substrate was mounted to the training device (Fig. 2(e)).

RESULTS AND DISCUSSION
Characterization of the skeletal muscle tissue

Fabricated skeletal muscle tissues are shown in Fig. 3(a, b). As shown in the microscopic images, the skeletal muscle tissue shrank by about 40% on day 7 in width since cells are densely packed together due to multinucleation during culture (Fig. 3(c)). When electrical pulses were applied using a contraction measurement device with a

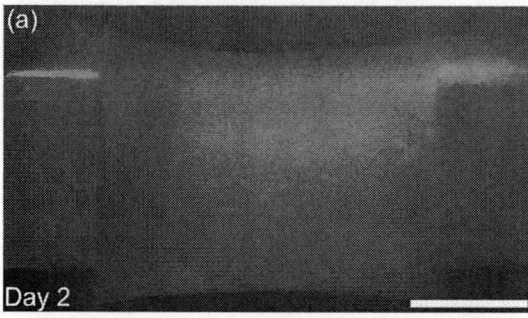

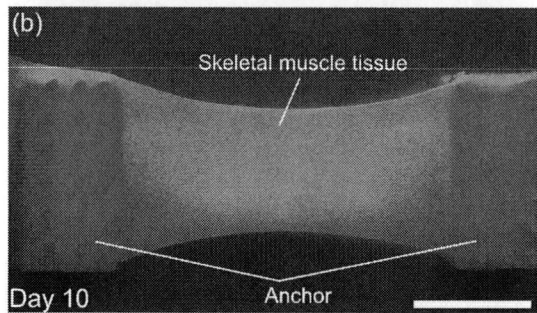

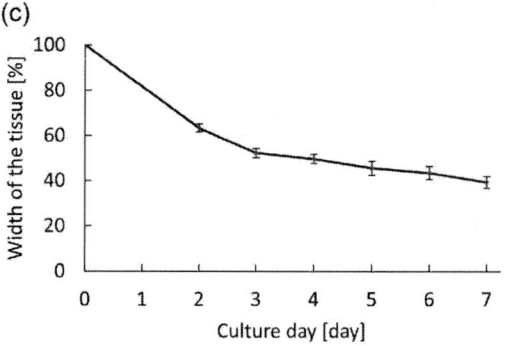

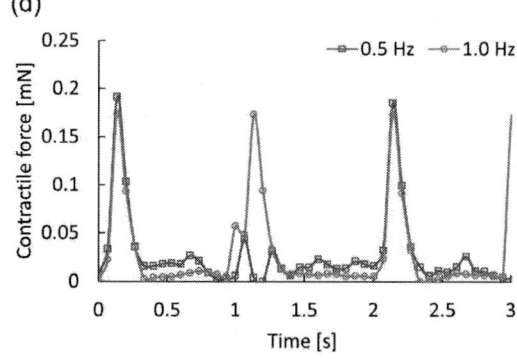

Figure 3: Characterization of the skeletal muscle tissue. (a), (b) Microscopic images of the skeletal muscle tissue on day 2 and day 10. (c) Change on the width of the skeletal muscle tissue during culture (n = 6). (d) Plots showing the contraction of the skeletal muscle tissue under electrical stimulation. Scale bars: (a), (b) 3 mm.

cantilever, contractile movement of the tissue in response to the applied electrical pulses was confirmed (Fig. 3(d)). This result suggests that contractile movement of the skeletal muscle tissue can be controlled by electrical stimulation.

978-1-6654-9309-3/23 $31.00 © 2023 IEEE

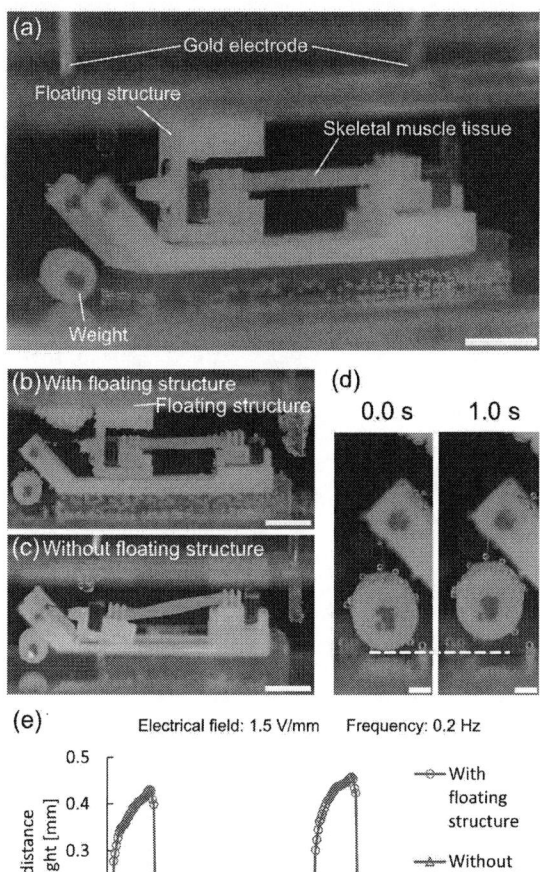

Figure 4: Evaluation of the weight training movement. (a) Photograph of the weight training device. (b), (c) Side views of the weight training device with and without floating structure. (d) Photographs showing the weight training movement with floating structure under ES. (e) Moving distance of the weight under ES. Scale bars: (a)-(c) 5 mm, (d) 1 mm.

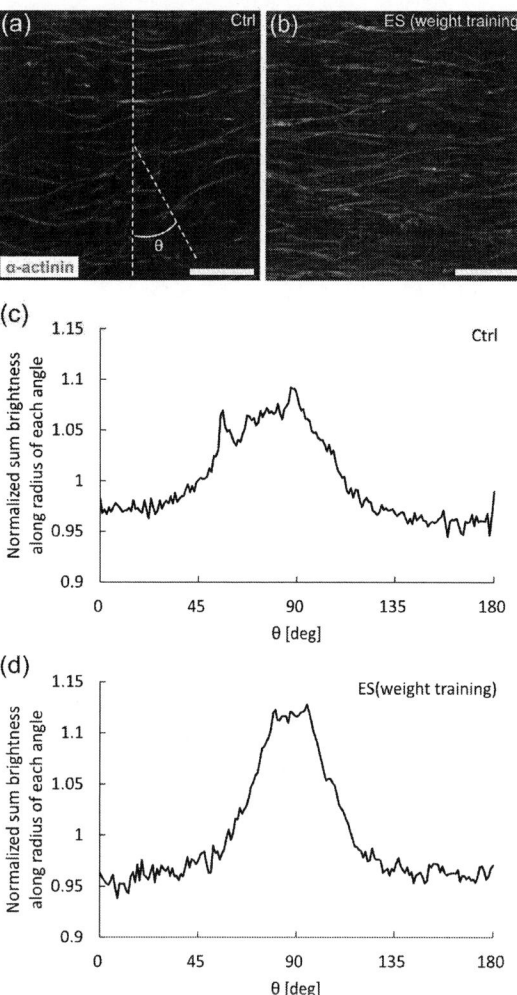

Figure 5: Orientation of myotubes. (a), (b) Confocal images of skeletal muscle tissues without ES as control and with ES (weight training). Green; α-actinin. (c), (d) Directional distribution of brightness of the tissue calculated by Fast Fourier Transform (FFT). Orientation of myotubes is quantified by previously proposed method [11]. Scale bars: (a), (b) 200 μm.

Verification of the weight training movement

Constructed weight training device is shown in Fig. 4(a). To verify whether the floating structure is necessary to achieve the weight training movement, two types of devices were compared: one with the floating structure and one without the floating structure. As shown in Fig. 4(b, c), the fabricated tissue succeeded to keep the balance along the horizontal direction owing to the floating structure. When electrical stimulation (electrical field, 1.5 V/mm; frequency, 50Hz; ON for 1 s and OFF for 4 s, repeated) was applied, the weight lifting movement was observed on the tissue with the floating structure, whereas significant movement was not observed on the tissue without the floating structure (Fig. 4(d, e)). Considering the loss of energy to straighten the deflected tissue, the proposed floating structure is effective to perform the weight training movement.

Evaluation of the effect of the weight training

Weight training was performed for 3.6 min each day from day 10 to day 12. To evaluate the effects of weight training on skeletal muscle tissues, immunostaining with anti-α-actinin antibody was conducted to the skeletal muscle tissue (Fig. 5(a, b)). The comparison of orientation between without ES as control and weight training under ES showed orientation values of 0.91 and 0.89 (the smaller numbers indicate that myotubes are more oriented), respectively (Fig. 5(c, d)). It is essential to align myotubes in the same direction for improving contractile properties. These results show that the skeletal muscle tissue with weight training under ES possesses higher orientation

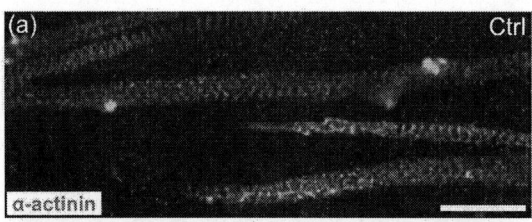

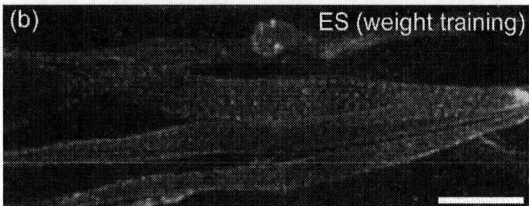

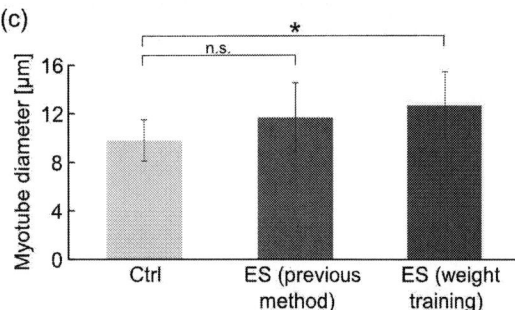

Figure 6: Thickness of myotubes. (a), (b) Confocal images of tissues without ES as control and with ES (weight training). (c) Quantification of myotube diameter (*p < 0.01, n = 50-51 myotubes from N = 5 tissues). Scale bars: (a), (b) 20 μm.

which indicates higher maturation of the skeletal muscle tissue.

Also, myotube thickness between without ES and weight training under ES was 9.8 ± 1.7 μm and 12.7 ± 2.8 μm, respectively (Fig. 6). Since the cross-sectional area of muscle is one of the determinants of its contractile force [12], it is necessary to increase myotube thickness. This significant increase (p < 0.01) means that the weight training induces hypertrophy in myotubes indicating maturation of the skeletal muscle tissue.

CONCLUSION

In this paper, we developed the weight training device to achieve free contractile movement inspired by *in vivo* muscle exercise. In addition, we showed that the weight training promoted the maturation in skeletal muscle tissues in terms of myotube orientation and diameter. We believe that this device can be a useful system to mature *in vitro* skeletal muscle tissues which possess high potential in applications for the fabrication of bioactuators and cultured meat.

ACKNOWLEDGEMENTS

This work was partially supported by JST-Mirai Program JPMJMI20C1, and the JSPS Grants-in-Aid for Scientific Research (KAKENHI) (Grant Number 21H01779).

REFERENCES

[1] R. Raman *et al.*, "Optogenetic skeletal muscle-powered adaptive biological machines", *Proc. Natl. Acad. Sci. U.S.A.*, vol. 113, no. 13, pp. 3497–3502, 2016.

[2] K. I. Kabumoto, T. Hoshino, Y. Akiyama, and K. Morishima, "Voluntary movement controlled by the surface EMG signal for tissue-engineered skeletal muscle on a gripping tool", *Tissue Eng. Part A*, vol. 19, no. 15–16, pp. 1695–1703, 2013.

[3] Y. Morimoto, H. Onoe, and S. Takeuchi, "Biohybrid robot powered by an antagonistic pair of skeletal muscle tissues", *Sci. Robot.*, vol. 3, eaat4440, 2018.

[4] R. Tanaka, K. Sakaguchi, A. Yoshida, H. Takahashi, Y. Haraguchi, and T. Shimizu, "Production of scaffold-free cell-based meat using cell sheet technology", *NPJ Sci. Food*, vol. 6, 41, 2022.

[5] B. Jo, M. Nie, and S. Takeuchi, "Manufacturing of animal products by the assembly of microfabricated tissues", *Essays Biochem.*, vol. 65, no. 3, pp. 611–623, 2021.

[6] M. Furuhashi, Y. Morimoto, A. Shima, F. Nakamura, H. Ishikawa, and S. Takeuchi, "Formation of contractile 3D bovine muscle tissue for construction of millimetre-thick cultured steak", *NPJ Sci. Food*, vol. 5, 6, 2021.

[7] L. Madden, M. Juhas, W. E. Kraus, G. A. Truskey, and N. Bursac, "Bioengineered human myobundles mimic clinical responses of skeletal muscle to drugs", *Elife*, vol. 4, e04885, 2015.

[8] A. Khodabukus *et al.*, "Electrical stimulation increases hypertrophy and metabolic flux in tissue-engineered human skeletal muscle", *Biomaterials*, vol. 198, pp. 259–269, 2019.

[9] A. F. Widodo, C. W. Tien, C. W. Chen, and S. C. Lai, "Isotonic and Isometric Exercise Interventions Improve the Hamstring Muscles' Strength and Flexibility: A Narrative Review", *Healthcare*, vol. 10, no. 5, 2022.

[10] Y. Yamamoto, A. Ito, H. Fujita, E. Nagamori, Y. Kawabe, and M. Kamihira, "Functional evaluation of artificial skeletal muscle tissue constructs fabricated by a magnetic force-based tissue engineering technique", *Tissue Eng. Part A*, vol. 17, no. 1–2, pp. 107–114, 2011.

[11] D. Kiriya *et al.*, "Meter-Long and Robust Supramolecular Strands Encapsulated in Hydrogel Jackets", *Angew. Chem. Int. Ed.*, vol. 51, no. 7, pp. 1553–1557, 2012.

[12] T. Fukunaga, M. Miyatani, M. Tachi, M. Kouzaki, Y. Kawakami, and H. Kanehisa, "Muscle volume is a major determinant of joint torque in humans", *Acta Physiol. Scand.*, vol. 172, no. 4, pp. 249–255, 2001

CONTACT

*K. Motoi, tel: +81-3-5841-6488; motoi@hybrid.t.u-tokyo.t.u-tokyo.ac.jp

MICROSYSTEM VIBRATING MESH ATOMIZER WITH INTEGRATED MICROHEATER FOR HIGH VISCOSITY LIQUID AEROSOL GENERATION

Pallavi Sharma, Irma Rocio Vazquez, and Nathan Jackson
University of New Mexico, Mechanical Engineering Department and Center for High Technology Materials, New Mexico, USA

ABSTRACT

Atomizers are used to generate an aerosol, which can be applied to numerous applications including inhaled drug delivery, ink jet printing, additive manufacturing, and spray cooling. However, current atomizing mechanisms either have poor droplet size uniformity or an inability to atomize liquids with viscosities >2 cP. This paper investigates a new MEMS based vibrating mesh technology with an integrated microheater to reduce the liquid viscosity and allow atomization to occur. The device aims to atomize high viscosity liquids while maintaining low span or narrow uniform droplet size distribution. The paper demonstrates the ability to atomize various high viscous liquids such as Su-8, PEDOT, and propylene glycol (PG).

KEYWORDS

Vibrating Mesh Atomizer, Aerosol, Microheater, Microelectromechanical Systems, Viscosity.

INTRODUCTION

Atomization is the process of generating micro-sized liquid droplets in the form of a spray. Atomizers are widely used in many industrial applications such as inhaled drug delivery (nebulizers), pharmaceutical manufacturing, coatings, spray cooling, fuel injection, pesticides and many more applications. Droplet size distribution is one of the most critical parameters that most of these applications desire.

There are various mechanisms used to atomize liquids with the three most common methods being air pressure through a nozzle, ultrasound based on cavitation, and vaporization based on Joule heating of the liquid (commonly used in vaping tools). However, each of these methods results in poor droplet size distribution [1, 2], and thus are typically undesired for most applications, but yet they are the most widely used methods because their mechanisms of operation are well understood and easily to manufacture. Recently two new methods have been developed i) vibrating mesh atomizer (VMA) and ii) surface acoustic wave (SAW) [3, 4]. However, VMA have demonstrated significant advantages in reducing span or generating a more uniform droplet size distribution compared to SAW and they avoid heating effects.

VMA's consist of a membrane with a mesh or array of nozzles and a piezoelectric film. The piezoelectric film vibrates the membrane causing it to oscillate and when operated in the (0,2) vibration mode enough force is generated to squeeze the liquid through the nozzles at a high frequency (~100 kHz) to create an aerosol [5]. All current VMA devices consist of a metal membrane and are manufactured using laser drilling or electroplating

techniques. Recently a new type of VMA made using MEMS fabrication techniques and a silicon membrane were developed [6]. The Si-VMA device has since been used to demonstrate significant advantages over metallic VMA [5, 7-9]. However, all current VMA devices are limited to low viscosity liquids of <2-3 cP. This has limited the number of applications where these devices can be used.

This paper aims to investigate methods of atomizing higher viscosity liquids using a MEMS VMA device. One of the advantages to manufacturing the VMA using microfabrication techniques is the ability to monolithically integrate various components into the system. This paper investigated the ability to integrate a micro-heater into the atomizer. The heater can then be used to reduce the viscosity of the liquid to meet threshold values of 2-3 cP thus allowing the liquid to be aerosolized. The paper describes the design and fabrication of the heater and integration into a MEMS VMA and validation of the device using various high viscosity liquids.

MATERIALS AND METHODS

VMA generate an aerosol based on mechanical vibrations, where a membrane with an array of nozzles (mesh) oscillates in the (0,2) vibration mode when excited by a piezoelectric actuator. The dynamics of a silicon MEMS based VMA system has been previously described [5]. The stiffness of the silicon and unique pyramidal nozzle shape contribute to the Si-VMA advantages. The concept of VMA aerosol mechanism of action has also been previously described [10].

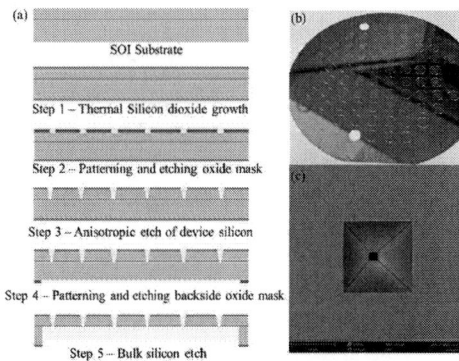

Figure 1: Fabrication of Silicon MEMS VMA (a) process flow, (b) fabricated wafer, and (c) individual nozzle.

Fabrication of MEMS Atomizer

The MEMS atomizer was fabricated using an SOI wafer with 25 µm thick (100) device silicon layer. The process flow is illustrated in Fig. 1 and has been previously described in more detail [6, 7]. Briefly, a thermal oxide layer was grown on the SOI wafer to act as a mask layer. The oxide layers were patterned using photolithography and reactive ion etching. Then the wafer was placed in KOH etchant at 90°C while the backside was protected. The anisotropic wet etching resulted in pyramidal shaped nozzles with a 54.74° angle as demonstrated in Fig. 1c. The membrane was created by etching the handle silicon using DRIE. The buried oxide layer was then removed using HF Vapor.

Design and Fabrication of Microheater

The microheaters function was to heat the liquid to reduce the viscosity and enable atomization. To accomplish this the heater needs to be low power (for portable applications), fast heat transfer rate, and uniform distribution of temperature. COMSOL Multiphysics finite element modelling (FEM) software was used to design the heating structure to meet specifications.

The microheater developed was based on Joule heating mechanism and consisted of polyimide substrates and insulating layers and a Ti/Pt conducting layer. The microheaters fabrication process is illustrated in Fig. 2. The process started with a clean silicon substrate. Then a thin (7μm) thick polyimide layer was deposited via spin coating (PI-2611, HD Microsystems). The film was then cured at 350°C with a 5°C/min ramp rate in a N_2 rich environment. The film was then micro-roughened to enhance metal adhesion using O_2 plasma. The metal layers Ti/Pt were then sputtered on the polyimide with a thickness of 25/250 nm and patterned using a lift off technique. Then a top photodefinable polyimide layer (HD4100, HD Microsystems) was spin coated and patterned using a puddle process. The bottom polyimide layer was etched using reactive ion etch with O_2 plasma and 15% CF_4 [11] to create the fluid passageway from the top surface to the surface of the atomizer. The microheaters were removed from the surface by removing the oxide layer on the silicon substrate.

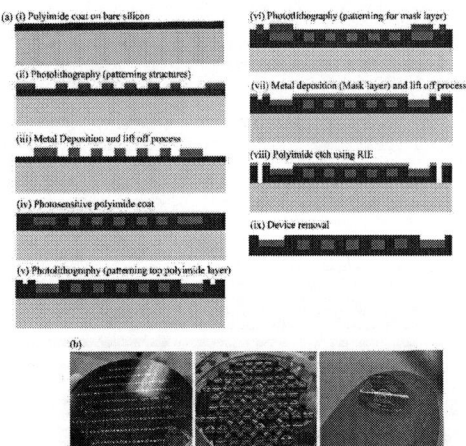

Figure 2: Fabrication process for the microheater with images of the fabricated devices demonstrated in (b).

Characterization

The microheaters were characterized experimentally and numerically. After fabrication the microheaters

impedance was measured (Fluke 6.5-digit multimeter) and the performance was determined by applying a voltage/current to the heaters. An infrared thermal imaging camera (PI 640i) was used to determine the spatial and temperature profile of the heaters. The heat transfer rate was determined by observing the time and temperature.

The atomizers were assembled using a piezoelectric film, a stainless-steel holder, MEMS VMA, and microheater as shown in Fig. 3. The atomizers were experimentally validated using water to demonstrate that an aerosol was generated. The frequency of the membrane in the (0,2) mode was measured using a laser interferometer (Picoscale, SmartAct). A signal generator and high frequency piezoelectric amplifier were used to excite the membrane at approximately 100 kHz.

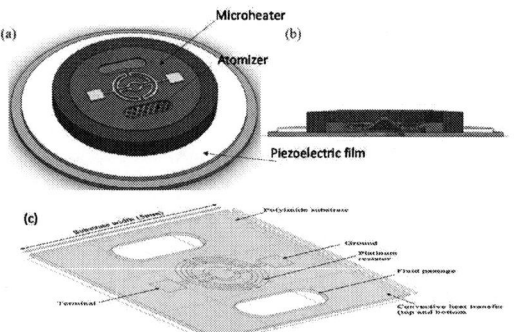

Figure 3: Overall assembly of the VMA device with microheater and (c) microheater design.

Droplet size of various liquids were measured using a laser particle analyzer (Sympatec Helos KF). The aerosols were placed close to the laser (1 cm). A high-speed camera (Nikon Nikkor) was used to image the aerosol and to investigate cone angle. In addition, viscosity of the liquids was measured using a cannon-Fenske viscosity tube with a variety of ranges.

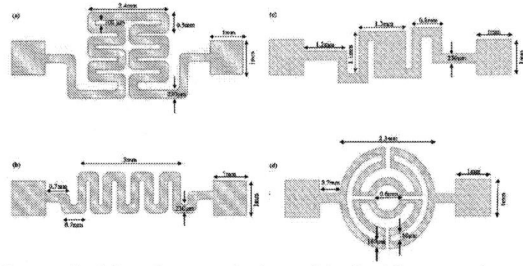

Figure 4: Microheater designs (a) double meander, (b) single meander, (c) S-shape, and (d) driving wheel.

RESULTS AND DISCUSSION

Four microheater designs were investigated as shown in Fig. 4. The heating rate and spatial distribution of heat were investigated by putting a drop of glycerol on the surface to determine how the heating was distributed in the liquid. The current as a function of voltage of the various designs of the fabricated heaters is shown in Fig. 5.

The driving wheel design was selected due to the low power consumption and the more uniform heat distribution shown in Fig. 6. Controlling the heating so that it is focused in the center of the heater was important as high temperatures on the outside could cause adhesion layers to fail, and atomization occurs at the center of the membrane. The fluid passageways allowed the heated fluid to interface with the atomizer. The gap spacing had to be large enough to not interfere with the dynamics of the membrane. The top surface of the heater was hotter than the bottom but due to the thin film properties of the polyimide the heat was able to transfer from the top side to the bottom thus allowing the fluid interfaced with the atomizer to be elevated. The heat transfer rate was quick as the features were on the micro-scale, where 95% of the final temperature reading was achieved within 2 seconds.

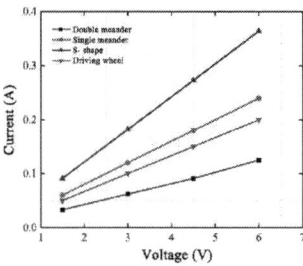

Figure 5: I-V characteristics for the microheater designs.

Once the heater was validated the complete system was tested using diluted glycerol up to 70% which had a viscosity of 29.8 cP, density of 1190 kg m^{-3} and a surface tension of 68 mN/m. The results shown in Fig. 7 demonstrate that the device was able to atomize the glycerol liquid. The atomizer without the heater was unable to create an aerosol from the 70% glycerol liquid. The outlet nozzle dimensions used in the device were 10 μm, but this could be changed to optimize droplet size by changing the mask layout. The volumetric median diameter (VMD) for water was 10.16 μm whereas for 60% glycerol the VMD was 12.88 μm and 70% glycerol had a VMD of 13.46μm. Although the VMD increased with viscosity the narrow size distribution is maintained compared to other atomizing methods such as ultrasound or air pressure [2, 9]. However, previous metallic VMA were not able to atomize liquids with > 2 cP [12, 13]. The ability to atomize higher viscosity liquids can lead to new applications or liquids that can be used to enhance performance.

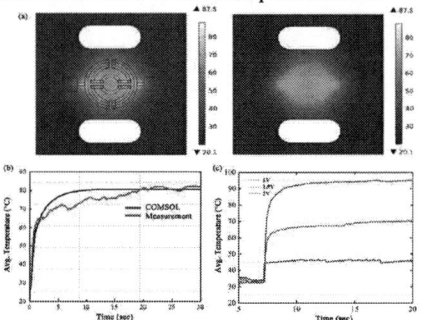

Figure 6: Driving wheel design (a) FEM model of top and bottom surface and (b) comparison of FEM and experimental time results with 1V applied, (c) measured temperature vs time as a function of voltage.

To validate the device and demonstrate the use of aerosols in a variety of applications we investigated the atomization of three liquids (PG, Su-8 2005, and PEDOT). PG is one of the base materials used in e-liquids and has a viscosity of 45.8 cP, whereas PEDOT has a viscosity of 30 cP and Su-8 2005 has a viscosity of 45 cP at 20°C. The device demonstrated the ability to atomize each liquid as shown in the droplet size distribution graphs shown in Fig. 8. SU-8 2005 had an increased average droplet size compared to PG or PEDOT likely due to the larger density and surface tension of the liquid. The ability to atomize Su-8 2005 could lead to new spin spray deposition to reduce waste and increase uniformity [7, 10].

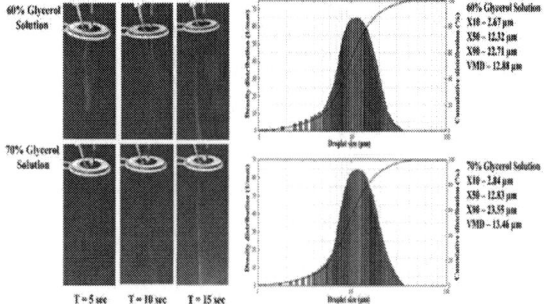

Figure 7: Aerosol high speed images of 60 and 70% glycerol and (right) droplet size distribution of the liquids.

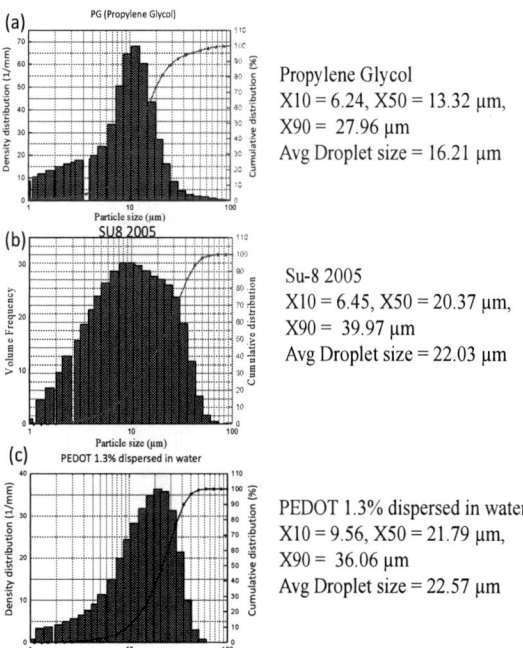

Figure 8: Droplet size distribution curves for three different high viscosity liquids (a) propylene glycol, (b) SU-8 2005, and (c) PEDOT.

CONCLUSIONS

The results demonstrate that the atomizing system was able to create an aerosol from relatively high viscosity liquids. Previous metallic VMA devices had a limitation of <2 cP, but the new silicon atomizer was able to generate an aerosol from liquids with a viscosity of 45 cP, by using low temperature heating of <100°C. We demonstrated that we could atomize up to 70% glycerol as well as some specialty liquids such as PEDOT, which could be used for additive manufacturing. Other specialty liquids investigated were Su-8 2005 and PG. The ability to atomize a liquid is not only dependent on the viscosity but is also dependent on the surface tension and density of the material. Therefore, future studies need to investigate all three parameters or develop new methods of optimizing these factors. However, the ability to atomize higher viscosity liquids can lead to new applications for these atomizers in the area of inhaled drug therapeutics as well as additive manufacturing.

ACKNOWLEDGEMENTS

The authors would like to thank the members of the Manufacturing Training and Technology Center at the University of New Mexico for their assistance. A special thanks to Prof. Pavan Muttil for his support and collaboration in using the particle size analyzer. This research was funded by the Research Allocation Committee Grant from University of New Mexico.

REFERENCES

[1] P. Sharma, M. Quazi, I. R. Vazquez, and N. Jackson, "Investigation of Droplet Size Distribution for Vibrating Mesh Atomizers," *Journal of Aerosol Science,* 2022.

[2] S. Kooij, A. Astefanei, G. L. Corthals, and D. Bonn, "Size distributions of droplets produced by ultrasonic nebulizers," *Scientific reports,* vol. 9, no. 1, p. 6128doi: 10.1038/s41598-019-42599-8.

[3] A. Qi, J. R. Friend, L. Y. Yeo, D. A. Morton, M. P. McIntosh, and L. Spiccia, "Miniature inhalation therapy platform using surface acoustic wave microfluidic atomization," *Lab on a Chip,* vol. 9, no. 15, pp. 2184-2193, 2009.

[4] M. Kurosawa, T. Watanabe, A. Futami, and T. Higuchi, "Surface acoustic wave atomizer," *Sensors and Actuators-A-Physical Sensors,* vol. 50, no. 1, pp. 69-74, 1995.

[5] P. Sharma and N. Jackson, "Vibration analysis of MEMS vibrating mesh atomizer," *Journal of Micromechanics and Microengineering,* vol. 32, no. 6, p. 065007, 2022.

[6] O. Z. Olszewski, R. MacLoughlin, A. Blake, M. O'Neill, A. Mathewson, and N. Jackson, "A Silicon-based MEMS Vibrating Mesh Nebulizer for Inhaled Drug Delivery," *Procedia Engineering,* vol. 168, pp. 1521-1524, 2016/01/01/ 2016, doi: https://doi.org/10.1016/j.proeng.2016.11.451.

[7] P. Sharma and N. Jackson, "Vibrating Mesh Atomizer for Spin-Spray Deposition," *Journal of Microelectromechanical Systems,* vol. 30, no. 4, pp. 582-588, 2021.

[8] P. Sharma, J. Ortega, I. R. Vazquez, and N. Jackson, "Spray Cooling Using Silicon Vibrating Mesh Atomizer," *Available at SSRN 4122013.*

[9] P. Sharma, M. Quazi, I. R. Vazquez, and N. Jackson, "Investigation of droplet size distribution for vibrating mesh atomizers," *Journal of Aerosol Science,* vol. 166, p. 106072, 2022.

[10] P. Sharma and N. Jackson, "Spin-spray deposition of spin on glass using MEMS atomizer," in *2021 IEEE 34th International Conference on Micro Electro Mechanical Systems (MEMS)*, 2021: IEEE, pp. 681-684.

[11] N. Jackson and J. Muthuswamy, "Flexible chip-scale package and interconnect for implantable MEMS movable microelectrodes for the brain," *Journal of microelectromechanical systems,* vol. 18, no. 2, pp. 396-404, 2009.

[12] S. C. Shen and Y.-J. Wang, "A Novel Handhold High Power MEMS Atomizer Using Micro Cymbal Shape Nozzle Plate for Inhalation Therapy," in *ASME 2009 International Design Engineering Technical Conferences and Computers and Information in Engineering Conference*, vol. 49033, pp. 375-382. 2009.

[13] T. Ghazanfari, A. M. A. Elhissi, Z. Ding, and K. M. G. Taylor, "The influence of fluid physicochemical properties on vibrating-mesh nebulization," *International Journal of Pharmaceutics,* vol. 339, no. 1, pp. 103-111, 2007/07/18/ 2007, doi: https://doi.org/10.1016/j.ijpharm.2007.02.035.

CONTACT

*N. Jackson, tel: +1-505-2727095; njack@unm.edu

SCALABLE MODULAR MEASUREMENT SYSTEM FOR CONTINUOUS BLOOD MONITORING WITH PIEZOELECTRIC MEMS RESONATORS

Michael Schneider[,1], Bernhard Kößl[1], Suresh Alasatri[1], Ingrid A.M. Magnet[2] and Ulrich Schmid[1]*
[1]Institute of Sensor and Actuator Systems, TU Wien, Vienna, Austria and
[2]Department of Emergency Medicine, Medical University of Vienna, Vienna, Austria

ABSTRACT

This paper reports on a novel scalable modular measurement system for continuous monitoring of the conductance spectrum of a piezoelectric MEMS resonator in blood either in standing or continuous flow conditions. Resonance frequency and quality factor of the resonator are extracted by automatic data fitting of the conductance spectrum. Both quantities can be monitored over extended periods of multiple hours and the sensor can be reused after cleaning. After careful sensor calibration, the dynamic viscosity can be extracted from the quality factor. This system allows for simultaneous dynamic viscosity measurements of multiple blood samples due to several modules arranged in parallel.

KEYWORDS

Piezoelectric MEMS, resonator, blood, continuous monitoring

INTRODUCTION

Aluminum nitride (AlN) based piezoelectric silicon MEMS sensors are a versatile technology platform for a multitude of current and future integrated sensor applications, including the measurement of physical parameters of liquids such as viscosity [1-4]. MEMS plate type resonators, especially when excited by tailored electrode designs in the so-called roof tile shaped (RTS) modes, offer high quality factors in liquids and are thus an ideal candidate for monitoring the dynamic viscosity μ of liquids due to the impact of viscous damping on the time evolution of the quality factor Q [5].

In medical treatments, there is often the need for artificial transportation of blood using mechanical pumps, *e.g.* in heart-lung machines or during extracorporeal membrane oxygenation (ECMO) [6]. The pumping process induces increasing shear forces with increasing pumping speeds in the red blood cells (RBC), which can result in the destruction of RBC if the shear forces exceed a critical threshold, causing hemolysis [7]. As this threshold is different for each patient, currently, the pumping speed is determined in a case-by-case trial-and-error approach. RBC contribute significantly to the visco-elastic behavior of blood [8] and thus it is reasonable to assume, that the onset of hemolysis would result in a detectable change in the viscous characteristics of blood.

The developed measurement system is a first step towards this goal by integrating a resonant MEMS viscosity sensor into a temperature-controlled liquid cell prepared for both standing and flowing liquid conditions. In addition, the system is modular and can be easily scaled up to multiple cells. Typical lab scale measurement setups only allow measurements of one liquid sample at a time, which is a problem when targeting biological liquids such

as blood, whose properties change with time (both *in vivo* and *ex vivo*). The developed measurement system is therefore also an invaluable tool for studying the statistical variations occurring in biological liquids as well as their change over time.

EXPERIMENTAL DETAILS

The key component of the MEMS resonator is a single-side clamped silicon plate with integrated piezoelectric transducers comprising a piezoelectric AlN thin film sandwiched between bottom and top electrodes. The electrode design is tailored to efficiently excite and measure the 17 RTS mode (Leissa's nomenclature for mode labeling is used [9]) [3]. The mode shape is indicated in the upper inset in Fig. 1.

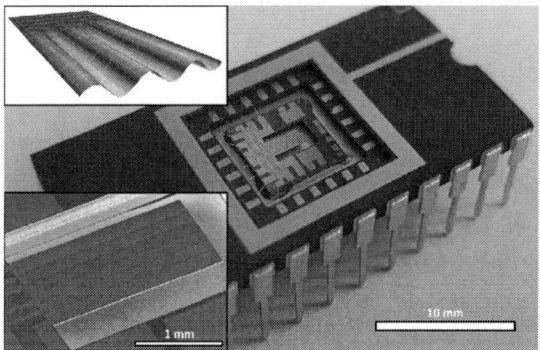

Figure 1: Piezoelectric MEMS sensor for blood analysis in dual inline package. The upper inset shows the mode shape of the 17 RTS mode. The lower inset shows an SEM image of the MEMS sensor.

The manufacturing process starts with a 100 mm silicon on insulator (SOI) wafer with a device layer thickness of 20 μm and a buried oxide thickness (BOX) of 1 μm. The SOI wafer is coated by the supplier with a stress compensated stack of 250 nm thermally grown silicon oxide and 80 nm LPCVD (low pressure chemical vapor deposition) silicon nitride for electrical insulation of the patterned bottom electrodes. The bottom electrode is deposited using evaporation of 50 nm chromium as adhesion promoter and 150 nm gold and patterned with lift-off. AlN is deposited by reactive DC sputter deposition in an industry-type Von Ardenne LS730S sputter tool and patterned with an in-house developed sputter lift-off process. A clamping holder ensures low substrate temperatures to avoid resist degradation [10]. The top electrode is deposited and patterned in the same way as the bottom electrode. The cantilever structure is defined by topside deep reactive ion etching (DRIE) and released by a subsequent DRIE backside etching step. The BOX is removed by hydrofluoric acid. The dies are separated with a dicing saw, glued into dual inline packages (DIP) and connected by wire bonding. Finally, the entire DIP is coated with amorphous PECVD (plasma enhanced chemical vapor deposition) silicon oxide to ensure a hydrophilic sensor surface to avoid air bubble inclusion below the plate structure. Fig. 1 shows a finished device. Details on the manufacturing process can be found in [11].

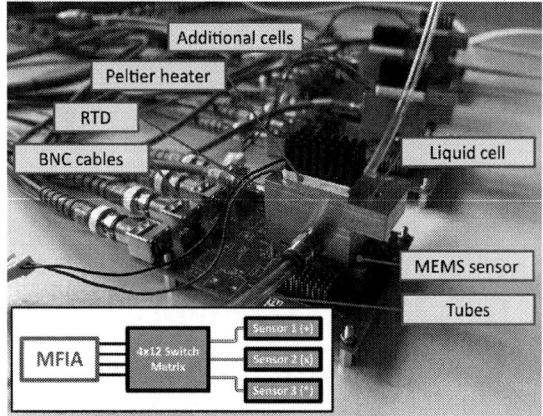

Figure 2: Liquid measurement cell with heater and MEMS sensor on interconnect PCB. The tubes can be used for both continuous and static flow conditions. The inset shows the measurement setup with impedance analyzer MFIA and switch matrix. Red double line connections indicate 4 BNC cables.

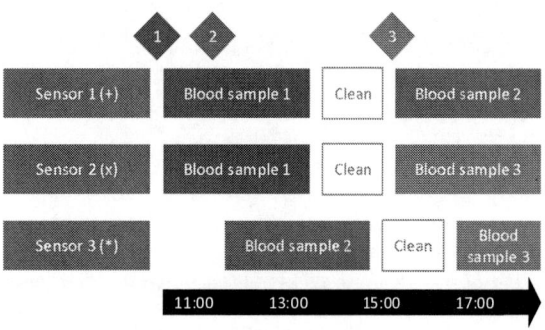

Figure 3: Measurement scheme showing which sensor is exposed to which blood sample during the measurement. Diamond symbols indicate 3 different times of blood draw. Cleaning step includes sensor disassembly and cleaning in cold water as well as subsequently in isopropanol.

Fig. 2 shows a single measurement module comprising a liquid cell with a resistance temperature detector (RTD), an individually controllable Peltier heater to maintain a constant temperature T of 37°C and a MEMS sensor in dual inline package (Fig. 1) mounted below the cell. Liquids are fed into the cell via tubes, allowing for both static and continuous flow conditions. The RTDs are measured using a Keithley 3706A switching multimeter and the Peltier elements are powered by a Hameg HM7044 programmable power supply. In this work, 3 modules are used and connected via 4 BNC cables through a Keithley 7173-50 4×12 switch matrix to an impedance analyzer (MFIA by Zurich Instruments) to continuously measure the conductance spectrum. This setup is shown schematically in the inset of Fig. 2.

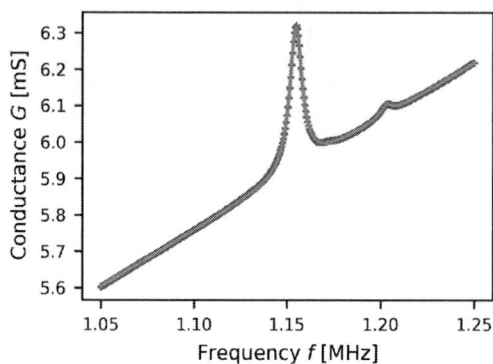

Figure 4: Exemplary conductance spectrum of a piezoelectric MEMS sensor immersed in blood. The fit is done based on the conductance spectrum of two parallel LCR equivalent circuits (corresponding to the two peaks).

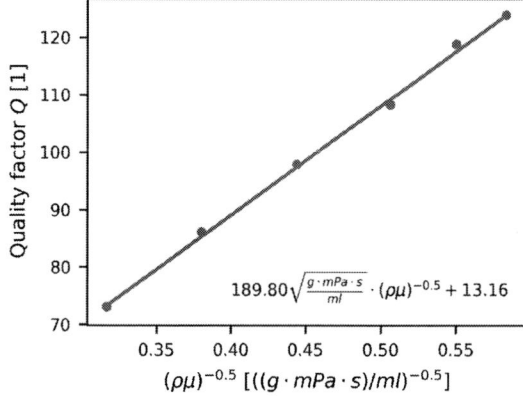

Figure 5: Q over $(\rho\mu)^{-0.5}$, measured in N10. Red line is the resulting linear regression needed for sensor calibration.

In this work, ~20 ml of blood is inserted into each liquid cell. Blood samples are drawn voluntarily from the lead author by a medical doctor and heparin is added to inhibit coagulation. Fig. 3 shows the measurement scheme indicating the times of blood draw (diamonds) and which sensor was exposed to which blood sample.

After each measurement, the sensor is cleaned with cold deionized water and isopropanol, dried and reassembled for the measurement of the next sample. Resonance frequency measurements in air are used to check, if the cleaning procedure was successful.

An exemplary conductance spectrum is shown in Fig. 4. Both the resonance of the 17 RTS mode at 1.16 MHz ($Q = 142.6$) and a smaller resonance at 1.21 MHz are fitted each by the conductance spectrum of a series LCR circuit

$$G(f) = G_0 \left(1 + \frac{Q^2 f_0^2}{f^2}\left(\frac{f^2}{f_0^2} - 1\right)^2\right)^{-1} \qquad (1)$$

with the conductance peak height G_0 at the resonance frequency f_0, frequency f and quality factor Q. The baseline is modelled by a linear function. Fig. 4 shows excellent agreement between Eq. 1 and the experimental data.

the day.

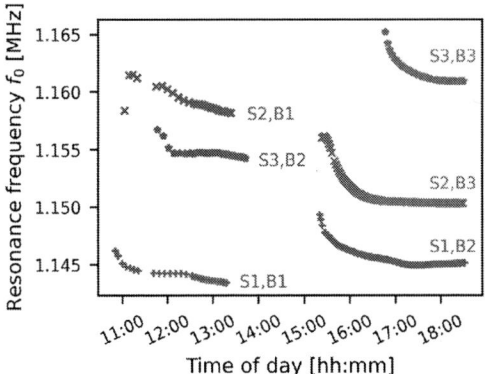

Figure 6: Resonance frequency of 17 RTS mode as function of time of day. Different colors correspond to different blood samples (B1, B2, B3). Symbols correspond to sensor 1 (S1, +), sensor 2 (S2, ×) and sensor 3 (S3, ∗).

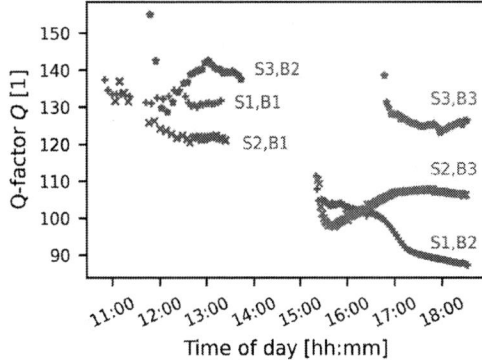

Figure 7: Quality factor of 17 RTS mode as function of time of day. Different colors correspond to different blood samples (B1, B2, B3). Symbols correspond to sensor 1 (S1, +), sensor 2 (S2, ×) and sensor 3 (S3, ∗).

For sensor calibration, the resonator is immersed in viscosity standard N10 by Paragon Scientific with known temperature dependency of density $\rho(T)$ and dynamic viscosity $\mu(T)$ and $\rho\mu$ is thus varied by changing the temperature from 30°C up to 73°C, providing Q as a function of $\rho\mu$. When plotting Q vs. $(\rho\mu)^{-0.5}$, a linear relation can be fitted, providing a calibration function to calculate the dynamic viscosity from Q for any liquid with known density. This is shown in Fig. 5.

RESULTS AND DISCUSSION

Fig. 6 shows f_0 as a function of the time of day for different sensors and blood samples. In general, all samples show the same behavior of decreasing f_0 with increasing time. This is most likely caused by a continuously increasing sensor mass due to adsorption of red blood cells on the sensor surface due to sedimentation. Different starting points can be attributed to slight variations in blood density for different blood samples and also to manufacturing tolerances (*e.g.* layer thickness inhomogeneities) of the different sensor dies.

Fig. 7 shows Q as a function of time for different sensors and blood samples. Sensors 1 and 2 are exposed to the same blood sample and show a similar behavior but with slight deviations. Sensor 3 is exposed to the blood sample drawn about 90 min after sample 1 and shows a different behavior with an initial decrease in Q, followed by an increase. Overall, for all samples and sensors, a similar value for Q is measured, which is reasonable given the chronological proximity of the blood draws. Sensor 1 is later exposed to blood sample 2 (re-homogenized by shaking after separation of the blood) after cleaning, demonstrating the strong change in Q due to changes in blood viscosity when the drawn blood is stored at room temperature for multiple hours compared to the fresh blood sample 2. Finally, sensors 2 and 3 are subsequently exposed to blood sample 3, again demonstrating a strong change in blood properties when comparing fresh to old samples. In addition, Q of the fresh sample is much lower, which can be attributed to the blood draw at a later time of

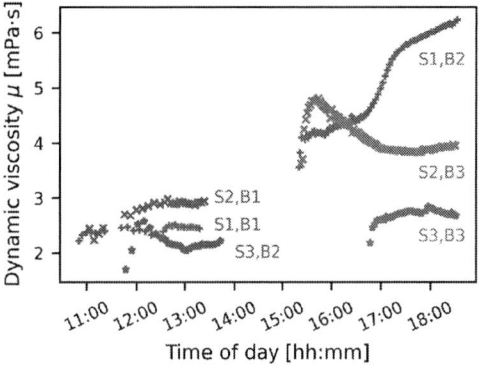

Figure 8: Dynamic viscosity μ as a function of time of day. Different colors correspond to different blood samples (B1, B2, B3). Symbols correspond to sensor 1 (S1, +), sensor 2 (S2, ×) and sensor 3 (S3, ∗).

Fig. 8 shows the dynamic viscosity μ of the different blood samples when applying the calibration function from Fig. 5 and a density value for blood of $\rho = 1.05$ g/ml [REF]. As is typical in biology, the viscosity of a donor's blood an indivual characteristic, which is strongly dependent on factors such as the donor's physique, if the donor exercised before blood draw (*e.g.* riding a bike), time of day and amount of food and water consumed prior to blood draw. Reference values for blood can therefore only be given to be in the range of 3.5 to 20 mPa·s [12]. Our measured values for the samples taken in the morning are slighly below this range, whereas the samples taken in the afternoon are in good agreement with it. Future studies will include reference measurements of the same blood samples performed in parallel to our sensor to allow for a direct comparison.

CONCLUSION

In this work, we presented a scalable and modular system to provide future inline blood viscosity measurements in *ex-vivo* applications such as ECMO using an integrated piezoelectric MEMS resonator. With this approach, we are able to monitor the viscosity of multiple blood samples over a period of multiple hours and could also demonstrate the reusability of the sensor after cleaning. In this concept study, the blood samples were kept static, but the next step will involve integration of the measurement system into a fluidic system with pump induced flow to test, whether the onset of hemolysis can be detected in the viscosity signal or not. Beyond this specific application, the system could also be valuable to monitor blood viscosity when blood-viscosity reducing drugs such as pentoxifylline (Trental) and oxypentifylline are administered to a patient.

ETHICS STATEMENT

Blood samples in this work were provided voluntarily by the lead author and drawn by a medical doctor. Samples were disposed of after the study concluded.

REFERENCES

[1] X. Le, Q. Shi, P. Vachon, E.J. Ng and C. Lee, "Piezoelectric MEMS—evolution from sensing technology to diversified applications in the 5G/Internet of Things (IoT) era", *Journal of Micromechanics and Microengineering*, vol. 32, pp. 014005, 2021.

[2] L. Huang, W. Li, G. Luo, D. Lu, L. Zhao, P. Yang, X. Wang, J. Wang, Q. Lin and Z. Jiang, "Piezoelectric-AlN resonators at two-dimensional flexural modes for the density and viscosity decoupled determination of liquids", *Microsystems & Nanoengineering*, vol. 8, pp. 38, 2022.

[3] G. Pfusterschmied, F. Patocka, C. Weinmann, M. Schneider, D. Platz and U. Schmid, "Responsivity and sensitivity of piezoelectric MEMS resonators at higher order modes in liquids", *Sensors and Actuators, A: Physical*, vol. 295, pp. 84-92, 2019.

[4] D. Platz and U. Schmid, "Vibrational modes in MEMS resonators", *Journal of Micromechanics and Microengineering*, vol. 29, pp. 123001, 2019.

[5] M. Kucera, E. Wistrela, G. Pfusterschmied, V. Ruiz-Diez, T. Manzaneque, J. Luis Sanchez-Rojas, J. Schalko, A. Bittner and U. Schmid, "Characterization of a roof tile-shaped out-of-plane vibrational mode in aluminum-nitride-actuated self-sensing micro-resonators for liquid monitoring purposes", *Applied Physics Letters*, vol. 104, pp. 233501-233501-5, 2014.

[6] A. Ahmed, X. Wang and M. Yang, "Biocompatible materials of pulsatile and rotary blood pumps: A brief review", vol. 59, pp. 322-339, 2020.

[7] I. Köhne, "Haemolysis induced by mechanical circulatory support devices: unsolved problems", *Perfusion*, vol. 35, pp. 474-483, 2020.

[8] G.B. Thurston, "Viscoelasticity of human blood", *Biophys J*, vol. 12, pp. 1205-17, 1972.

[9] A.W. Leissa, "The free vibration of rectangular plates", *J. Sound Vib.*, vol. 31, pp. 257-293, 1973.

[10] M. Fischeneder, E. Wistrela, A. Bittner, M. Schneider and U. Schmid, "Tailored wafer holder for a reliable deposition of sputtered aluminium nitride thin films at low temperatures", *Materials Science in Semiconductor Processing*, vol. 71, pp. 283-289, 2017.

[11] M. Kucera, E. Wistrela, G. Pfusterschmied, V. Ruiz-Díez, T. Manzaneque, J. Hernando-García, J.L. Sánchez-Rojas, A. Jachimowicz, J. Schalko, A. Bittner and U. Schmid, "Design-dependent performance of self-actuated and self-sensing piezoelectric-AlN cantilevers in liquid media oscillating in the fundamental in-plane bending mode", *Sensors and Actuators B: Chemical*, vol. 200, pp. 235-244, 2014.

[12] E. Nader, S. Skinner, M. Romana, R. Fort, N. Lemonne, N. Guillot, A. Gauthier, S. Antoine-Jonville, C. Renoux, M.-D. Hardy-Dessources, E. Stauffer, P. Joly, Y. Bertrand and P. Connes, "Blood Rheology: Key Parameters, Impact on Blood Flow, Role in Sickle Cell Disease and Effects of Exercise", *Frontiers in Physiology*, vol. 10, pp. 2019.

CONTACT

*Michael Schneider, tel: +43-1-58801-76636; michael.schneider@tuwien.ac.at

SILICON COMPATIBLE PROCESS TO INTEGRATE IMPEDANCE CYTOMETRY WITH MECHANICAL CHARACTERIZATION

Quentin Rezard[1,2], Faruk Azam Shaik [2,3], Jean Claude Gerbedoen[2,3], Fabrizio Cleri[1], Dominique Collard[2,3], Chann Lagadec[2,4] and Mehmet C. Tarhan[1,2,3]

[1]Univ. Lille, CNRS, Centrale Lille, Polytechnique Hauts-de-France, Junia, UMR 8520-IEMN, Villeneuve d'Ascq, FRANCE
[2]CNRS, IIS, COL, Univ. Lille, SMMiL-E Project, Lille, FRANCE
[3]LIMMS/CNRS-IIS, UMI 2820, The University of Tokyo, Lille, FRANCE
[4] Univ. Lille, CNRS, Inserm, CHU Lille, Centre Oscar Lambret, UMR9020 – UMR-S 1277 - Canther – Cancer Heterogeneity, Plasticity and Resistance to Therapies, F-59000 Lille, FRANCE

ABSTRACT

We introduced 3D silicon electrodes to perform impedance cytometry on single cells without compromising practical integration with sensors measuring complementary properties, *e.g.,* mechanical properties. Microfabricated from a highly-doped SOI wafer, some design modifications were made to improve their sensing performance. According to simulations, a trajectory-free measurement can be obtained by replacing the silicon backside under the sensing area with an insulating material. In addition, enlarging the distance between the electrodes and the surrounding silicon structure and filling them with some insulating material results experimentally in better signal quality and reduced parasitics. Combining these two device modifications improves the system frequency response and the signal quality at higher frequencies. The proposed process aims at creating silicon-based electrodes for impedance spectroscopy applications while providing opportunities to integrate them with MEMS sensors and actuators.

KEYWORDS

3D electrodes, impedance cytometry, single-cell analysis, microfabrication.

INTRODUCTION

Cellular systems exhibit a significant degree of heterogeneity. Among all the methods for quantifying cell heterogeneity, microfluidic impedance cytometry (MIC) provide biophysical analysis on single-cells in a label-free and high-throughput way [1]. MIC devices detect changes in the impedance signal due to a cell passing through an electric field zone between electrodes. Information regarding dielectric properties, cell size, membrane capacitance, and cytoplasm resistivity can be extracted [2]. Due to several reasons, *e.g.,* supreme electrical properties and bandwidth, 2D planar gold electrodes are widely used on a glass substrate. However, their inhomogeneous field line inside the channel results in a need for trajectory-specific measurements or more complex fabrication/assembly steps to achieve reliable measurements at the single-cell level. Then, 3D electrodes were created with an electric field distribution more homogeneous and suitable for highly sensitive impedance detection [3]. Nevertheless, these methods show difficulties in combining the sensitive electrical MIC analysis with other biophysical properties, *e.g.,* mechanical properties, to benefit more from MEMS

sensors and actuators in a high-throughput format. For example, highly-doped silicon has tremendous fabrication possibilities to integrate with various MEMS sensors/actuators providing multi-parameter analysis, *e.g.,* electrical and mechanical [4]. However, a silicon wafer requires each electrode to be adequately isolated. Also, the effect of nearby structures and parasitics to be minimized to ensure the most reliable measurement and broader bandwidth.

Here, we target improving silicon-based electrodes for impedance cytometry applications by providing trajectory-free measurements, better signal quality, and reduced parasitics without compromising opportunities to integrate them with MEMS sensors and actuators.

METHOD

Device description

The proposed device is composed of two layers: (i) a PDMS layer assembled on (ii) a MEMS layer. The PDMS layer has an inlet to inject a cell suspension and an outlet to connect to a vacuum pump for controlling the flow along the microfluidic channel embedded in the MEMS layer.

We use the MEMS layer, *i.e.,* the functional layer, by designing in three parts:

(i) A *microfluidic channel* is formed between silicon walls on the sides, the backside silicon layer or some insulating material as the bottom, and the PDMS slab as the top. The microfluidic channel handles the cells with an inlet on one side and an outlet on the other side of the PDMS layer.

(ii) *Electrical characterization area* is fabricated on the front side of an SOI wafer. It includes 3D electrodes facing each other at two sides of a microfluidic channel.

(iii) *Backside silicon layer* is critical for the designs we tested in this study, although it is mainly used for structural integrity and handling. We replaced silicon in some areas, *e.g.,* under the electrical characterization area, with a dielectric material.

Microfabrication

We used three different types of devices. The first type is an all-silicon device with two electrode pairs on the front side of an SOI wafer. The second type uses an insulating material on the sides of the electrode pairs to extend the surrounding silicon structures on the front side. The third type uses the same front side as the second type but has a modified backside by replacing different parts of the backside silicon with an insulating material.

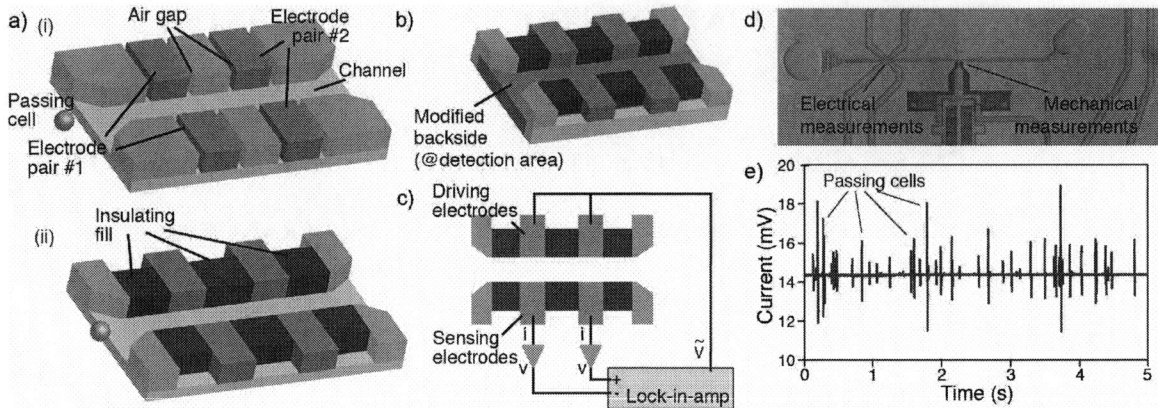

Figure 1: a) Two frontside designs are fabricated and tested: (i) narrow air gaps, (ii) insulating areas. b) Silicon and insulating backside devices are also compared. c) Measurements are performed using a lock-in-amplifier in differential mode connected to silicon 3D electrodes. d) 3D electrodes could be integrated with mechanical MEMS sensors. e) An example of detected cells flowing in the channel.

All types of devices are fabricated on SOI wafers (30/2/300 μm). Electrodes and the microfluidic channel are fabricated on the front side by a photolithography process followed by a 30-μm DRIE etching (Figure 1a-i). Devices with insulating front side regions have an additional step of insulating material filling and patterning (SU8 or CYTOP; Figure 1a-ii). One key point of filling those cavities is to keep the insulating material at the same level as the front side wall to have proper PDMS adherence on all the devices, especially in the detection area, and avoid any liquid leakage during the experiment. Similarly, devices with modified backside geometry received 350-μm DRIE etching of backside and insulating material filling steps (Figure 1b).

Working principle

The proposed device targets single-cell impedance spectroscopy in a continuous flow to achieve high throughput. We use either air-liquid interfaces or insulated walls at the electrical characterization area to keep the electrodes at minimal immersion in liquid for better performance. Cells are inserted via the inlet and flow through the 100-μm-width channel to reach the electrical characterization area where the channel width drops to 25 μm. Electrical characterization is performed using two pairs of 3D electrodes working in the differential mode. Being less conductive than the media, cells decrease the current passing between the electrodes while passing through. This change in the current can be detected in real-time.

All electrode designs are 25 μm in width. The 30-μm top layer of the SOI wafer defines the height of electrodes. The device without any insulating material at the front side has a 4-μm gap between the electrodes, forming a stable air-liquid interface due to the surface tension of the liquid preventing any leakage. The other designs have 25 μm space between the electrodes and surrounding silicon structures. This space is filled with an insulating material. A sinusoidal signal (1 Vp-p, 1 Mhz) is applied on driving electrodes with a lock-in-amplifier (Zurich Instruments HF2LI). Sensing electrodes are first connected to a trans-impedance amplifier (Zurich Instruments HF2TA), with a

gain of 10k, before being fed into the lock-in-amplifier in the differential mode (Figure 1c). This impedance cytometry setup allows us to simultaneously perform measurements at three different frequencies for providing information on cell size, membrane capacitance, and cytoplasm resistivity. The signal demonstrated here is at 1 MHz, related only to cell size measurements.

Setup

Experiments are performed on an upright microscope stage. The outlet of the PDMS slab is connected to a vacuum pump (Fluigent EZ), and the electrodes are connected to the lock-in-amplifier at different channels. Signals obtained from sensing electrodes are first amplified using trans-impedance amplifiers before feeding into the lock-in-amplifier.

Biological materials

SUM159-PT, a triple-negative breast cancer cell line, is tested with the proposed method. Cells are grown in monolayer conditions up to subconfluence. After being trypsinized to obtain a homogenous cell suspension solution, they are put in a PBS 1x solution and passed through a 40-μm filter before being used in the experiment.

Experimental procedure

We first assemble the PDMS slab on the MEMS device. The assembled device is attached to a PCB, and all electrical connections are established. Placing the device on the microscope stage, we complete the fluidic connections. A PBS solution is injected to fill the channel and the tubing until no bubbles remain. Then, 5 μl cell suspension is injected into the channel via the inlet. A constant flow speed of 2 μl/min is sustained inside the channel throughout the experiments by controlling the vacuum pump. Cells start moving with the flow and go through the electrical characterization area while continuously measuring the passing current. The frequency response measurements are performed by sweeping the frequency from 100 kHz to 50 MHz (2 Vp-p driving signal) with a gain of 1k.

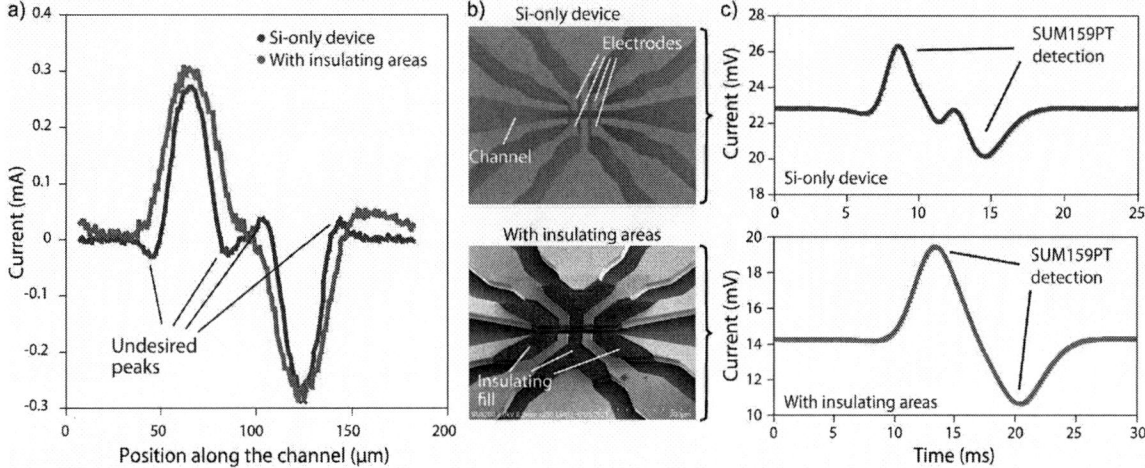

Figure 2: a) Simulations showed undesired peaks when cells are passing nearby silicon structures (blue) while a device with insulating areas showed an improved response (red). b) SEM images of fabricated devices used in the experiments. c) Similar to the simulations, cell experiments showed undesired peaks on Si-only devices but not on devices with insulating areas.

Simulations

A finite element model is used to investigate the sensitivity of a MEMS sensor over the detection and quantification of cells. In particular, we evaluated the sensitivity of the sensor according to different electrode geometries and the cell positions while flowing through a channel. Several sensor topologies are considered to define proper electrode design guidelines.

The model was built on COMSOL platform, referring to the Electric Current (EC) model, and given as,

$$-\nabla \cdot ((\sigma + j\omega\varepsilon_0\varepsilon_r)\nabla V - J_e = 0 \qquad (1)$$

where σ is the conductivity [S/m], ω is the angular frequency, ε_0 is the vacuum permittivity, and ε_r is the relative permittivity of the material, V is the applied potential, J_e is an externally generated density, defining the electric field by the constitutive law,

$$D = \varepsilon_0\varepsilon_r E \qquad (2)$$

where D is the electric displacement field, and E is the electric field.

The cell membrane was represented as contact resistances:

$$n.J_1 = \frac{1}{d_s}(\sigma + j\omega\varepsilon_0\varepsilon_r)(V_1 - V_2) \qquad (3)$$

where, d_s is the surface thickness. Analyses were solved in the frequency domain, setting the parametric solver to evaluate the response of the system at several working cell displacements. The input signal was given using a Voltage Terminal, and the current amplitude was evaluated as a function of cell displacement.

Results and Discussion

Two pairs of 3D-silicon electrodes (Figure 1d) were used in the differential mode to detect passing cells in a high-throughput way (Figure 1e). Air gaps on both sides of an electrode provided electrical insulation while the surface tension of the liquid protected the channel from leaking out. Both simulations and experimental results showed that the silicon structures surrounding electrodes significantly affect the detected signal as multiple additional peaks appear before, after, and in between the peaks corresponding to the electrodes. Those additional peaks altered measurement performance on the intrinsic cell

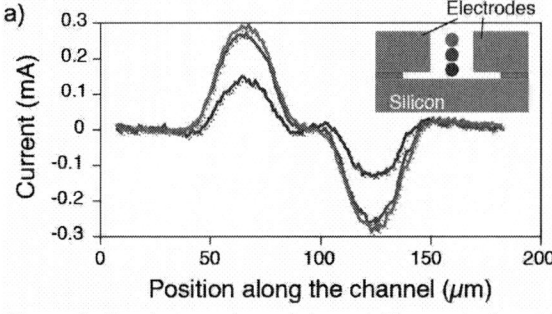
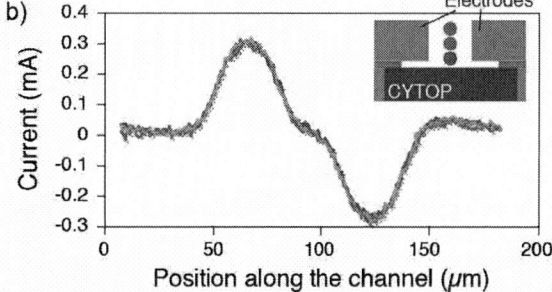

Figure 3: Simulations showed that: a) The vertical position of passing cells caused significant changes in the obtained current amplitude when the backside of the detection area was silicon. b) Removing the silicon and filling the detection area backside with a dielectric material such as CYTOP resulted in a trajectory-free measurement. Insets correspond to cross-sectional views.

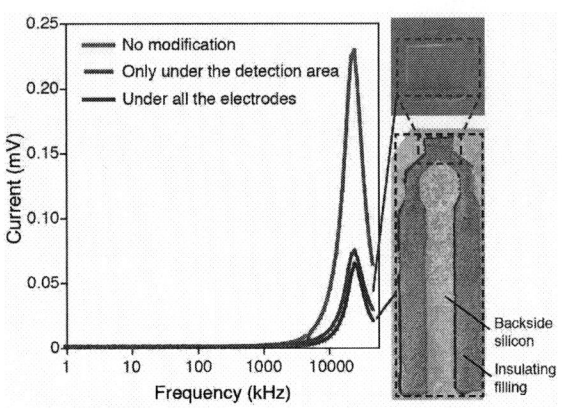

Figure 4: Three different backside designs (Red: Si-only, green: detection area-limited modification, and blue: extended to the electrode area modification) were fabricated and frequency responses of the devices were measured. The y-axis corresponds to the current value after the trans-impedance amplifier. Modified versions performed better at high frequencies by decreasing parasitics. Inset: The SEM image of the modified backside with PDMS filling. The green rectangle corresponds to the detection-area limited modification version.

properties. Those parasitics could be eliminated by extending the distance to the electrode and filling this hole with some insulating material, *e.g.*, SU8 and CYTOP, as demonstrated with simulations (Figure 2a). This measurement improvement was also shown experimentally between the two geometries, the Si-only device and the one with SU8 fillings between electrodes on the front side (Figure 2b,c).

The simulation of the effect of backside material on the measured current concerning the vertical position of cells showed significant differences (Figure 3). To obtain a trajectory-free measurement for single-cell analysis, we had to minimize the effect of cells' vertical position on the measured current. According to the simulations, a cell passing closer to the bottom surface results in a lower current passing between the electrodes (Figure 3a), which compromises reliable measurements. On the other hand, an alternative design with the backside silicon near the measurement area being replaced with an insulating material solves the problem. Simulations show identical measurement results regardless of the position of the passing cell, achieving a trajectory-free measurement (Figure 3b).

Frequency response analyses showed that the backside silicon also causes parasitics at high frequencies. Three different devices with different backside structures were tested for their frequency responses. All of these three designs had the same front side structures, *i.e.*, SU8-filled spaces between electrodes. The first type had an all-silicon backside. The second and third devices had PDMS-filled backside holes under only the electrical characterization area (green rectangle in figure 4 inset) and an area extended

along the electrodes (blue rectangle in figure 4 inset), respectively. The frequency response was swept from 100 kHz to 50 MHz (2 Vp-p) with a gain of 1k. As these measurements were taken in air, we expected complete insulation between the electrodes. However, all devices showed elevated current values at higher frequencies suggesting the parasitic currents. Among the tested devices, the all-silicon device performed poorly compared to the others. As a result, the modified backside devices show improved performance for trajectory-free cell measurements and frequency measurements with extended bandwidth.

CONCLUSION

Demonstrated fabrication steps significantly improved the use of 3D silicon electrodes for continuous flow single-cell impedance spectroscopy applications. Extending the space between electrodes and the surrounding silicon structures and filling those spaces with insulating materials significantly improved the cell signal quality by reducing parasitics and simplifying the detected signal geometry. The device showed improved bandwidth and potential trajectory-free measurements by replacing the backside silicon with an insulating material. Those optimizations suggest a reliable use of 3D silicon electrodes for impedance cytometry without compromising the integration possibilities with other MEMS sensors and actuators.

ACKNOWLEDGEMENTS

This work is in the framework of SMMiL-E activities (a joint project of CNRS, Institute of Industrial Science, Centre Oscar Lambret, and the University of Lille). The authors acknowledge the French State-Region plan (CPER IRICL, Lille Interdisciplinary research Institute against cancer) for financial support and IRCL for hosting SMMiL-E. M. C. Tarhan and Q. Rezard acknowledge I-SITE ULNE.

REFERENCES

[1] C. Honrado, P. Bisegna, N. S. Swami and F. Caselli, Single-cell microfluidic impedance cytometry: from raw signals to cell phenotypes using data analyticsLab Chip, 21, pp.22-54, (2021).

[2] Petchakup, C.; Li, K.H.H.; Hou, H.W., Advances in Single Cell Impedance Cytometry for Biomedical Applications, Micromachines 2017, 8, 87.

[3] Gawad S, Sun T, Green NG, Morgan H, Impedance spectroscopy using maximum length sequences: Application to single cell analysis (2007) Rev Sci Instrum 78:054301

[4] Q. Rezard, G. Perret, J.C. Gerbedoen, et al., IEEE Int. Conf. on MEMS (MEMS'21), pp. 494, (2021).

CONTACT

*M.C. Tarhan, cagatay.tarhan@junia.com

SORTING OF EXTRACELLULAR VESICLES BY USING OPTICALLY-INDUCED DIELECTROPHORESIS ON AN INTEGRATED MICROFLUIDIC CHIP

Wei-Jen Soong, Chih-Hung Wang, Yi-Sin Chen, Chihchen Chen and Gwo-Bin Lee*

Department of Power Mechanical Engineering, National Tsing Hua University, Taiwan

ABSTRACT

This study presents a new method to automatically sort nano-scaled extracellular vesicles (EVs) by using optically-induced dielectrophoresis (ODEP) techniques. An optimal intensity of green light at an optimal moving velocity (10 μm/sec) of the moving light patterns on an integrated microfluidic chip was explored. It successfully sorted EVs with three different sizes (small: 150 to 175 nm, middle: 175 to 200 nm, and large: 200 to 250 nm) within 30 mins. It may be useful for the diagnosis and risk assessment of cancer in future clinical applications.

KEYWORDS

extracellular vesicles (EVs), microfluidics, size-based sorting, optically-induced dielectrophoresis

INTRODUCTION

Extracellular vesicles (EVs) are composed of lipid bilayer membranes containing various components including proteins, nucleic acids (DNA and RNA), and metabolites, which are generally identified as the mediators in cell-to-cell communications [1]. EVs are heterogeneous and exhibit individual biophysical characteristics based on biogenesis [2]. Moreover, exosomes and exomeres, the subtypes in the EVs family, are derived from multi-vesicular bodies (MVBs) with distinct sizes ranging from 40 to 160 nm in diameter [3]. Exosomes can be divided into two subtypes according to the size of the particles including large exosomes (90 to 120 nm) and small exosomes (60 to 80 nm). Moreover, the size of exomeres is less than 50 nm [1]. Briefly, the size of exosome nanoparticles might alter their biological responses which are associated with immune responses, viral pathogenicity, pregnancy, metabolic and cardiovascular diseases, central nervous system-related diseases, and cancer progression [4]. Therefore, there is a great demand to sort EVs with different dimensions.

Conventional EVs isolation methods including ultracentrifugation centrifugation (UC), size-exclusion chromatography (SEC), and commercial immune-affinity kits are widely used to separate EVs. All the methods could be operated for various clinical samples. However, they may suffer from relatively low purity, lengthy and tedious processes, and risk of contamination by unidentified chemicals [5]. Recently, with the advancement of microfluidic platforms, the issues of conventional sorting methods can be properly addressed within a shorter process time in an automatic format while only consuming low-volume samples and reagents by a microfluidic assay [6]. For instance, viscoelastic flow could isolate different sizes of exosomes by Newtonian media and a co-flow of viscoelastic on a microfluidic chip [7]. It only required a simple chip fabrication. However, the viscoelastic flow

system could be limited by low throughput and specificity [7]. Moreover, immuno-acoustic sorting of EVs particles, which isolated different particles depending on the acoustic force by using an ultrasound waves system can achieve fast and high-resolution separating of intact EVs [8]. Nonetheless, the acoustic sorting system required relatively complicated microfabrication processes and may suffer from the risk of protein contamination. Furthermore, there are still some disadvantages such as low purity and time consumption.

In this work, an improved method to isolate nano-scaled EVs via ODEP techniques was reported. Compared to our previous work which used white light patterns [9], green light patterns were adopted due to their higher photoconduction efficiency on the photosensitive layer [9]. Moreover, optimal operating conditions including gradually decreasing light intensity of the green light and moving velocity of light patterns were explored for improving the sorting efficiency. Experimental results showed that the ODEP sorting method with the green light patterns successfully separated nano-scaled EVs, providing a fast, label-free, and high-sorting-efficiency approach.

MATERIALS AND METHODS

Experimental procedure of on-chip EVs sorting

A new EVs sorting approach was reported in this work, which used pure green light with gradually intensity-decreasing patterns and different moving velocities of the moving light patterns on a microfluidic chip. EVs and 40% iodixanol were first parallelly driven into the ODEP operation area (Figure 1(a)). Vertically dynamic light patterns generated a positive ODEP force (i.e. attractive) such that larger EVs in the upper flow, which experienced a stronger force than small EVs from the gradually decreasing light patterns, could be moved to the downflow (Figure 1(b)). After ODEP force was applied, EVs were sorted out by particle sizes and collected at outlets (Figure 1(c)).

The formula of ODEP force could be defined by equation (1), where r denotes the radius of a particle, ε_m is the permittivity of the solution, E represents the local electric field, and f_{CM} is the Clausius–Mossotti factor [10].

$$\underline{F}_{ODEP} = 2\pi r^3 \varepsilon_m Re(f_{CM})\nabla E^2 \qquad (1)$$

The larger particles then experience stronger forces because of the diameter of the particles (Figure 2(a)). The electric field could be manipulated by the moving velocity and the intensity of optical patterns such that small particles were remained in place under faster velocities and weaker intensity conditions (Figure 2(b)). With this new approach, EVs with different sizes could be sorted continuously and then collected in outlet channels.

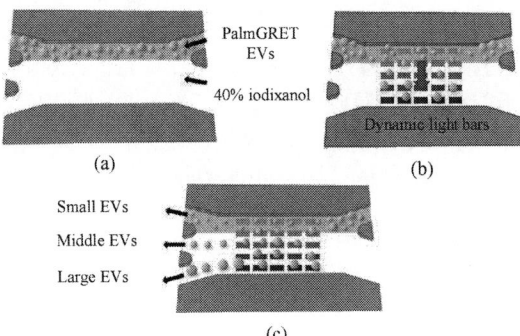

Figure 1: Schematic procedure for EVs sorting according to particle sizes. (a) EVs and 40% iodixanol were parallelly driven into the microchannel. (b) Dynamic light patterns, whose intensity was gradually decreased from the top to the bottom, generated ODEP force such that larger EVs in the upper flow could be attracted to the downflow. (c) Sorted EVs were collected in three-outlet chambers.

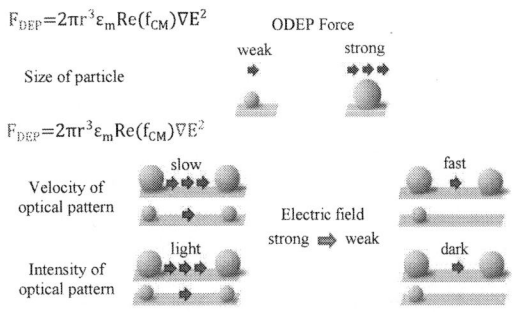

Figure 2: Schematic for manipulation of ODEP force. (a) ODEP force is proportional to the third power of radius of the EVs particles so that large particles experience a strong force. (b) ODEP force could be fine-tuned by light moving velocity and intensity. For a light pattern moving at a slow speed with a high intensity, EVs of all sizes can be dragged along. However, when the movement of the optical patterns increases and light intensity decreases while the electric field becomes weak, the particles in small size could not be dragged.

Chip design and microfabrication process

Figure 3(a) shows a schematic optically-induced EV sorting (OIES) chip. The dimensions of the OIES microfluidic chip were 70 mm × 30 mm (length × width) that was consisted of a polydimethylsiloxane (PDMS)-based (Sylgard 18A/18B, Dow Corning, USA) liquid channel layer for microfluidic control, an upper indium-tin-oxide (ITO) glass (200 mm x 200 mm x 1.1 mm, 6Ω, Onset Electro-Optics, Taiwan) with a 30-μm-thick double-side-tape (MSDS, Deer Brand, Taiwan) layer to form microchannels, and a bottom ITO glass deposited with photoconductive amorphous silicon (a-Si) layer for ODEP operation (Figure 3(b)).

The microfluidic platform was microfabricated by using an inverse micro-structure mold made by a computer-numerical-control (CNC) machining process (EGX-600, Roland DGA, USA). A standard PDMS

replication process by using a weight ratio of 10: 0.9 (184A:184B) for preparing the PDMS mixture was used. The baked PDMS replica was treated under oxygen plasma (CUTE-MPR, Femto Science, Korea) for 90 seconds and 90 W before bonding the upper ITO glass and PDMS layer. The double-sided tape layer was made by a laser cutting machine (VL-200, Universal Laser Systems, USA). Finally, the double-sided tape layer was used to bond the upper ITO glass and liquid channel layer with the ITO glass coated with amorphous silicon to form a complete OIES platform (Figure 3(c)).

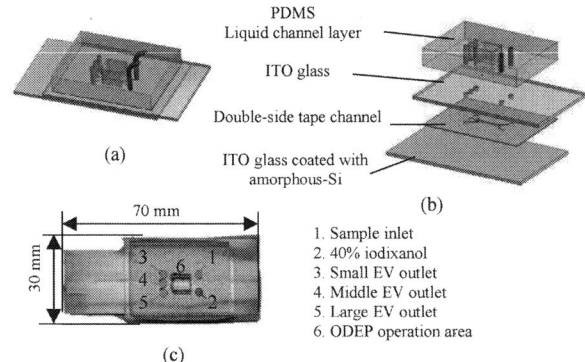

Figure 3: (a) Schematic of optically-induced EVs sorting (OIES) microfluidic platform. (b) Exploded view of the OIES platform. (c) A photograph of a prototyped chip (70 mm x 30 mm) presented the corresponding microfluidic components and functions.

RESULTS AND DISCUSSION
Optimal sorting conditions

Two EVs with different sizes were first prepared to explore the optimal sorting light velocity under the static situation. Sizes of small PalmGRET EVs (with green light) and large MDA-MB-23A EVs (with red light) were found to be 120.0 ±47.8 and 145.7 ± 52.6 nm, respectively, by a membrane-based chip, respectively [11]. These two kinds of EVs, small PalmGRET EVs and large MDA-MB-231 EVs were first loaded into a single-channel microfluidic chip [12] and concentrated to upwards to be a baseline (Figures 4(b) & 5(b)). In Figure 4(c), the merged results after applying light bars (230 to 70 G-value) at a moving velocity of 10 μm/s, these two EVs were manipulated to almost the same distance (5 light bars) from 230 to 150 G-value. After decreasing the value of each light bar, larger EVs were moved further to 7 light bars (140 G-value). However, small EVs were only carried to 4 light bars at 155 G-value (Figures 4(d)). It is then concluded that the optimal range of sorting intensity was from 170 to 130 (G-value).

After the optimal light intensities were found, in Figure 5(c), the continuous moving light bars with velocities of 5, 10, and 15 μm/s were used to sort these two EVs into different distances. In a moving velocity of 5 μm/s, two EVs were manipulated downwards and moved almost the same distance to 7 bars (140 G-value). After increasing the moving velocity to 10 μm/s, the large red EVs and the small green EVs were moved to 7 light bars (140 G-value) and 3 light bars (160 G-value), respectively, which demonstrated that this moving velocity (10 µm/s) could be the sorting threshold for these two EVs. It is then concluded that the optimal velocity condition was found to be 10 µm/s under a 130 to 170 (G-value) intensity (Figure 5(d)). After exploring the static sorting condition, the EVs were sorted under a horizontal-flowing condition to improve the sorting throughput while using vertically-moving light patterns.

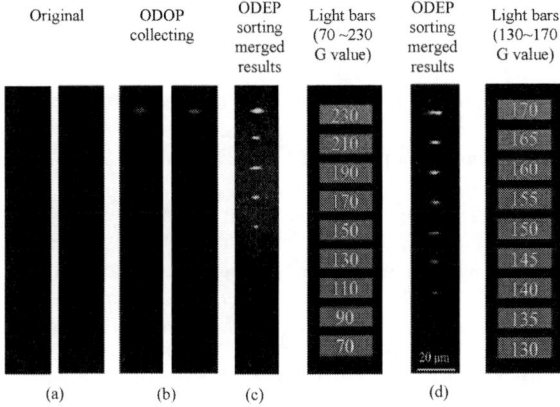

(a)   (b)   (c)   (d)

Figure 4: Statically optimal intensity conditions of green light bars. (a) Original graph of the microchannel with green PalmGRET EVs (mean size: 120.0 nm) and red MDA-MB-231 EVs (mean size: 145.7 nm), taken by two different microscopic filters, named U-FGW and U-FBWA, respectively. (b) Light bars with 255 green value (G-value) and a velocity of 10 µm/s were used to collect two EVs to the upper part of the microchannel as a baseline. (c) Merged results of two EVs after applying light bars with 70 to 230 G-value, revealing that two EVs were manipulated to almost the same position at 150 G-value. (d) After decreasing the value of light bars, the larger MDA EVs were moved to 140 G-value. However, smaller Palm EVs were only carried to the 155 G-value position.

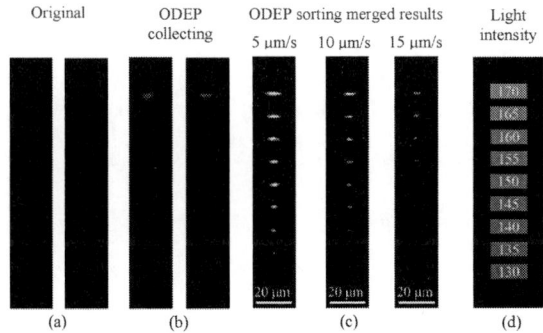

(a)   (b)   (c)   (d)

Figure 5: Statically optimal velocity conditions of green light bars. (a) and (b) were treated with the same conditions as explored in Fig. 4. (c) Under the same light intensity condition with 130 to 170 G-value; (d) The merged results revealed that the optimal sorting efficiency was 10 µm/s, which clearly separated two EVs better than the ones at 5 µm/s and 15 µm/s.

Dynamic sorting result

    Figure 6 demonstrated that EVs could be successfully sorted and collected based on qNano analysis (IZON, New

Zealand), in which samples were diluted three times with an electrolyte solution. The particle size of the inlet (original EVs) was distributed from 110 to 270 nm, and the average concentration was around 2×10^{11} particle/ml. After the ODEP sorting, the particle size of small EVs was measured to be around 150 to 175 nm with a peak located at 160 nm at a concentration of about 4.5×10^{11} particle/ml. Moreover, the particle size of middle EVs was quantified to be 110 to 200 nm with a peak at 185 nm with a concentration of 4.0×10^{11} particle/ml. Finally, the particle size of large EVs was determined to be 160 to 240 nm and had a peak at 215 nm at 1.0×10^{12} (particle/ml) concentration. The dynamic EVs sorting throughput was around 1.8 to 2.2 µL/min per group.

    Moreover, transmission electron microscope (TEM) images (Figure 7) demonstrated the shape and geometric size of sorted EVs, in which the average diameters of small, middle and large EVS are around 43 nm, 85 nm and 186 nm, respectively. Note that the sizes measured from the TEM images are not directly comparable with the qNano measurements because the EVs sample has to be fixed and dehydrated before being analyzed [13].

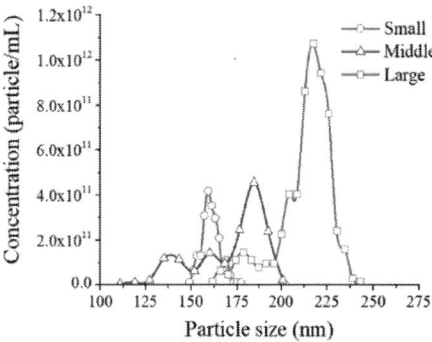

Figure 6: The qNano (TRPS) measurement results of the sorted EVs from three different outlets, which disclosed the small, middle and large sizes of EVs were around 150 to 175 nm, 175 to 200 nm, and 200 to 250 nm, respectively.

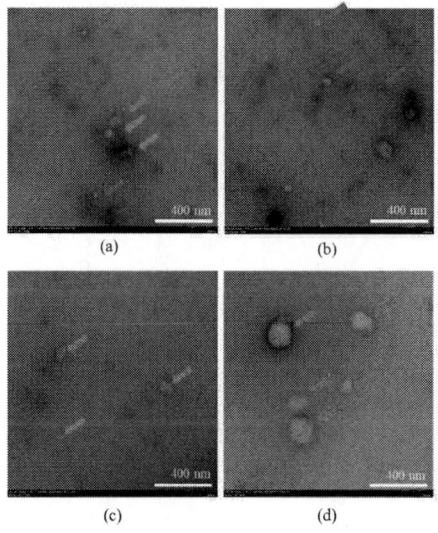

(a)   (b)

(c)   (d)

Figure 7: TEM images of sorted EVs. (a) The mixture of EVs shows the three size groups of EVs. Moreover, the TEM images demonstrated the shape and geometric size of EVs, in which the average diameters of small (b), middle (c) and large (d) sizes were around 43, 85, and 186 nm, respectively.

CONCLUSION

In this study, an automatic sorting of nano-scaled EVs on an integrated microfluidic chip via ODEP techniques has been demonstrated. The sorted EVs were measured by TEM and qNano, which demonstrated that this microfluidic platform successfully sorted EVs into three different sizes (small: 150 to 175 nm, middle: 175 to 200 nm, and large: 200 to 250 nm) and achieved high sorting efficiency. This technique can provide an efficient separation method, which may be useful for understanding the bioactive roles of the EV cargo, and immunomodulatory pathways for therapeutic treatments.

ACKNOWLEDGEMENTS

The authors appreciate the financial supports from 1) Ministry of Science and Technology (MOST) of Taiwan (MOST 109-2221-E-007-006-MY3 & MOST 110-2221-E-007-010-MY3), and 2) National Health Research Institutes of Taiwan (NHRI-EX110-11020EI). Authors also thank Dr. Charles Lai for providing PalmGRET EVs.

REFERENCES

[1] F. J. Verweij, L. Balaj, C. M. Boulanger, D. R. Carter, E. B. Compeer, G. D'angelo, S. El Andaloussi, J. G. Goetz, J. C. Gross, and V. Hyenne, "The power of imaging to understand extracellular vesicle biology in vivo", *Nature Methods*, vol. 18, pp. 1013-1026, 2021.

[2] H. Zhang, D. Freitas, H. S. Kim, K. F. A. Hoshino, I. Matei, C. M. Kenific, M. T. Mark, H. Molina, A. B. Martin, L. Bojmar, J. F. S. Rampersaud6, M. Nakajima, A. P. Mutvei, P. Sansone, L. Cohen-Gould, N. Paknejad, A. Magalhães, J. R. Cubillos-Ruiz, J. Blenis, H. Matsui, R. Schwartz, C. A. Reis, J. A. Ferreira, W. Buehring, M. B. H. Osório, D. Lyden, H. Wang, K. Manova-Todorova, A. M. Silva, J. P. Jimenez, G. G. P. Giannakakou, A. M. Cuervo, M. S. Brady, A. Massey, and H. P. J. Bromberg, "Identification of distinct nanoparticles and subsets of extracellular vesicles by asymmetric flow field-flow fractionation", *Nature Cell Biology*, vol. 20, pp. 332-343, 2018.

[3] D. E. Murphy, O. G. de Jong, M. Brouwer, M. J. Wood, G. Lavieu, R. M. Schiffelers, and P. Vader, "Extracellular vesicle-based therapeutics: natural versus engineered targeting and trafficking", *Experimental & Molecular Medicine*, vol. 51, pp. 1-12, 2019.

[4] H. Zhang, D. Freitas, H. S. Kim, K. Fabijanic, Z. Li, H. Chen, M. T. Mark, H. Molina, A. B. Martin, and L. Bojmar, "Identification of distinct nanoparticles and subsets of extracellular vesicles by asymmetric flow field-flow fractionation", *Nature Cell Biology*, vol. 20, pp. 332-343, 2018.

[5] C. Théry, S. Amigorena, G. Raposo, and A. Clayton, "Isolation and characterization of exosomes from cell culture supernatants and biological fluids", *Current protocols in cell biology*, vol. 30, pp. 3-22, 2006.

[6] S. C. Guo, S. C. Tao, and H. Dawn, "Microfluidics-based on-a-chip systems for isolating and analysing extracellular vesicles", *Journal of extracellular vesicles*, vol. 7, pp. 1508271, 2018.

[7] C. Liu, J. Guo, F. Tian, N. Yang, F. Yan, Y. Ding, J. Wei, G. Hu, G. Nie, and J. Sun, "Field-free isolation of exosomes from extracellular vesicles by microfluidic viscoelastic flows", *ACS nano*, vol. 11, pp. 6968-6976, 2017.

[8] J. Liu, Y. Qu, and H. Wang, "Immuno-acoustic sorting of disease-specific extracellular vesicles by acoustophoretic force", *Micromachines*, vol. 12, pp. 1534, 2021.

[9] W. J. Soong, Y. S. Chen, C. H. Wang, and G. B. Lee, "Size-based sorting of extracellular vesicles via optically-induced dielectrophoresis on a microfluidic chip", *IEEE 17th International Conference on Nano/Micro Engineered and Molecular System (IEEE NEMS)*, pp. 131-132, 2022.

[10] P. Y. Chiou, A. T. Ohta, and M. C. Wu, "Massively parallel manipulation of single cells and microparticles using optical images", *Nature*, vol. 436, pp. 370-372, 2005.

[11] Y. S. Chen, Y. D. Ma, C. c. Chen, S. C. Shiesh, and G. B. Lee, "An integrated microfluidic system for on-chip enrichment and quantification of circulating extracellular vesicles from whole blood", *Lab on a Chip*, vol. 19, pp. 3305-3315, 2019.

[12] P. F. Yang, C. H. Wang, and G. B. Lee, "Optically-induced cell fusion on cell pairing microstructures", *Scientific reports*, vol. 6, pp. 1-10, 2016.

[13] R. Szatanek, M. Baj-Krzyworzeka, J. Zimoch, M. Lekka, M. Siedlar, and J. Baran, "The methods of choice for extracellular vesicles (EVs) characterization", International *journal of molecular sciences*, vol. 18, pp. 1153, 2017.

CONTACT

*Gwo-Bin Lee, Tel: +886-3-5715131 ext. 33765; Fax: +886-3-5742495; E-mail: gwobin@pme.nthu.edu.tw

A REPROGRAMMABLE MEM SWITCH
UTILIZING CONTROLLED CONTACT WELDING

Tsegereda K. Esatu, Hei Kam, Lars P. Tatum, Xiaoer Hu, Urmita Sikder, Sergio Almeida,
Junqiao Wu, and Tsu-Jae King Liu
The University of California, Berkeley, CA, USA

ABSTRACT

The feasibility of controllably welding and unwelding the contacting electrodes of a micro-electro-mechanical (MEM) switch for non-volatile information storage is investigated via experimental study. It is demonstrated that MEM switches can be programmed and erased with relatively small voltage ($< 3\,V$) and that they have excellent retention characteristics at elevated temperature (200°C). This shows that it is possible to implement MEM switches for ultra-low-power computing with non-volatile memory "on-chip" at zero incremental fabrication cost.

KEYWORDS

Micro-electro-mechanical switches, ultra-low-power, reprogrammable, non-volatile, embedded memory.

INTRODUCTION

MEM switches have been proposed as alternatives to transistors for ultra-low-power digital computing applications because they have the ideal characteristics of zero OFF–state leakage current and abrupt switching behavior, enabling sub-50 mV operating voltage across a wide range of operating temperatures [1-2]. MEM switches also can be designed to operate as non-volatile (NV) switches, e.g., for reconfigurable interconnects [3], [4].

In this work, MEM switches designed for digital logic applications are demonstrated to be multi-time programmable via controlled contact welding and unwelding. This newfound capability provides for greater versatility of device operation, enabling non-volatile information storage to be embedded with digital logic circuitry with no incremental fabrication cost. Considering that MEM logic circuits can operate with high energy efficiency and zero standby power, and that the fabrication process for MEM switches is relatively simple and can be compatible with a variety of substrates including plastic [5], these results affirm that MEM switches are attractive for applications such as the Internet of Things and wearable or disposable electronics that require ultra-low power consumption as well as relatively low manufacturing cost.

MEM SWITCH DESIGN

Fig. 1 is a plan-view scanning electron micrograph (SEM) image of a fabricated MEM switch; it is designed for digital logic applications and comprises two electrical switches and hence has two pairs of source/drain contact electrodes [6]. The 2-contact (2C) device comprises a movable gate electrode suspended by four folded-flexure beams over a fixed body electrode; the drain electrodes are attached to and routed underneath the gate electrode and electrically insulated from it by an Al_2O_3 gate-dielectric layer. In the OFF–state (Fig. 2a), an air gap separates the conductive source and drain electrodes, so that no current

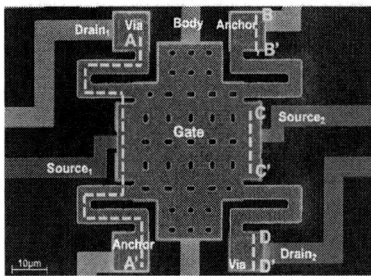

Figure 1: Plan-view scanning electron micrograph (SEM) image of the two-contact (2C) MEM switch.

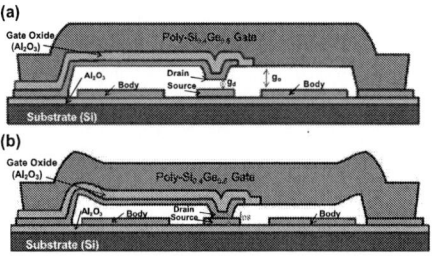

Figure 2: Schematic cross-sectional views along A-A' cutline in Fig. 1: (a) off–state (b) on–state. As-fabricated air-gaps $g_o = 220$ nm and $g_d = 60$ nm.

can flow between them. The movable gate is electrostatically actuated downward toward the body electrode when a voltage is applied between the gate and body (V_{GB}). When V_{GB} is larger than the turn-on voltage (V_{ON}), each of the drain electrodes comes into physical contact with its underlying source electrode, allowing current (I_{DS}) to flow under the influence of an applied voltage between the drain and source electrodes (V_{DS}). This state is referred to as the ON–state (Fig. 2b). To turn off the switch, V_{GB} is reduced toward 0V so that the spring restoring force (F_{spring}) of the folded flexure beams actuates the movable structure upward, causing the source and drain electrodes to break contact. The voltage at which I_{DS} drops back to zero is referred to as the release voltage (V_{RL}). Due to contact adhesive force, V_{RL} is always smaller than V_{ON}. The hysteresis voltage (V_H) is defined as $V_{ON} - V_{RL}$.

Fig. 3 illustrates key steps of the MEM switch fabrication process, which is described in detail in [6]. The structural material used for the gate electrode and suspension beams is 1.9 µm-thick boron doped polycrystalline silicon germanium (poly-$Si_{0.4}Ge_{0.6}$); the fixed body, source, and drain-anchor electrodes are formed from a sputtered layer of tungsten (W) deposited onto the Al_2O_3-coated silicon wafer substrate. The air gaps are formed by sequentially depositing and patterning two sacrificial layers of low-temperature oxide (LTO): the first layer is patterned to define the source contact regions;

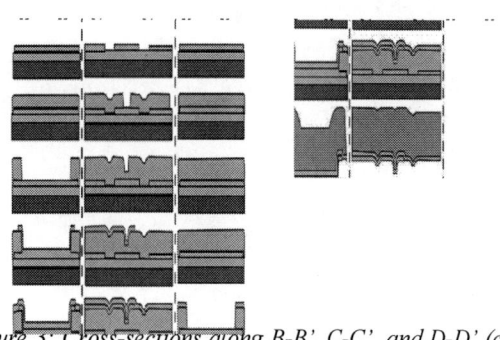

Figure 3: Cross-sections along B-B', C-C', and D-D' (cf. Fig. 1) illustrating key MEM switch fabrication steps.

the thickness of the second layer determines the as-fabricated air-gap thickness between the drain electrode and the source electrode in the contact regions (g_d) while the combined thickness of the two LTO layers determines the as-fabricated actuation air-gap thickness (g_o) between the movable structure and the body electrode. To minimize V_H, g_d is designed to be less than one third of g_o so that the MEM switch operates in non-pull-in mode [7]. Due to electrode surface roughness, physical contact between source and drain electrodes is made only at a small number of asperities, in the ON–state (Fig. 4a).

PROGRAM AND ERASE OPERATION

To program a MEM switch, one or more voltage pulses are applied between the drain and source electrodes while the switch is in the ON–state ($V_{GB} > V_{ON}$), as indicated in Table I. The resulting current flow causes Joule heating, which raises the local temperature at the contact point(s), T_C. If T_C is sufficiently high to soften the contacting material and allow atomic movement to occur, then the area of physical contact increases and new bonds can form between the contacting electrode surfaces; effectively welding the electrodes together (Fig. 4b). In the programmed state, the strength of the weld is greater than F_{spring} so that the switch remains ON when the gate voltage is removed, i.e., V_{GB} is reduced to 0V.

To erase the MEM switch, a voltage pulse is simply applied between the drain and source electrodes. If the Joule-heating induced temperature is sufficiently high to soften the contacting material, i.e., weaken the bonding strength between atoms, the spring restoring force can cause the contact to be broken (Fig. 4c) so that the switch turns OFF. The state of the MEM switch can be easily read by applying a small V_{DS} and measuring I_{DS}, as shown in Table I. If any current flows (i.e., $I_{DS} > 0$) then the switch is programmed.

Table I: Default parameters for program, erase, and read operations in this work.

Pulse Width	1ms	1ms	

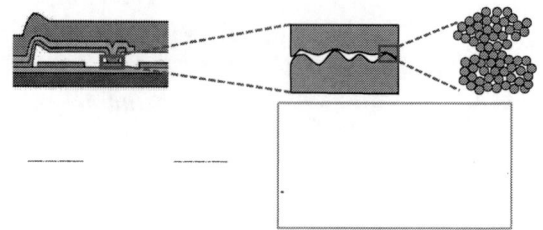

Figure 4: Contact welding and unwelding in a MEM switch: (a) Due to surface roughness, on-state conductance occurs through one or more contacting asperities. (b) During "Program" operation and after program showing effectively welded contacting electrodes. (c) During "Erase" operation and after erase showing opened contact.

RESULTS AND DISCUSSION

Fig. 5 shows measured I_D-V_{GB} characteristics for a MEM switch before and after the first program/erase (P/E) cycle. The observed reductions in V_{ON} by 2.5V and V_H by 0.3V can be qualitatively explained by asperity growth at the contact point(s), resulting in smaller effective contact air-gap thickness (g_d). This explanation is supported by atomic force microscopy (AFM) analyses of the source electrodes for a fresh switch and a cycled (programmed & erased) switch, in Figs. 6a and 6b, respectively. (The movable electrode layer stack – including the drain electrodes – was physically removed using scotch tape to expose these electrodes.) More pronounced asperities, approximately ~13 nm in height, are clearly seen for the cycled switch electrode vs. the fresh electrode. Also, the surface height distribution for the fresh electrode is Gaussian whereas it is skewed for the cycled electrode, indicative of atomic movement to grow asperities at the contact point(s). Based on the measured reduction in V_{ON}, the reduction in g_d is estimated to be 30 nm; therefore, similarly sized opposing asperities are expected on the surface of the cycled drain electrode. Since the average spacing between the contacting electrode surfaces is increased by ~30 nm due to asperity growth, Van der Waals force (which is the dominant component of the contact adhesive force [8]) in the ON–state is greatly reduced and hence V_H is greatly reduced.

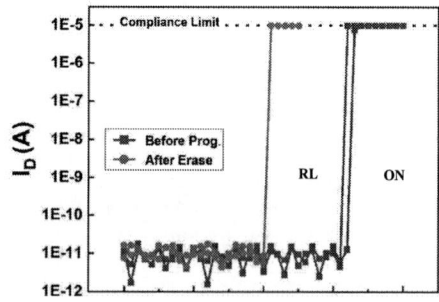

Figure 5: Measured I_{DS}-V_{GB} curves for a MEM switch before programming (blue) and after 1 P/E cycle (red).

978-1-6654-9309-3/23 $31.00 © 2023 IEEE   484

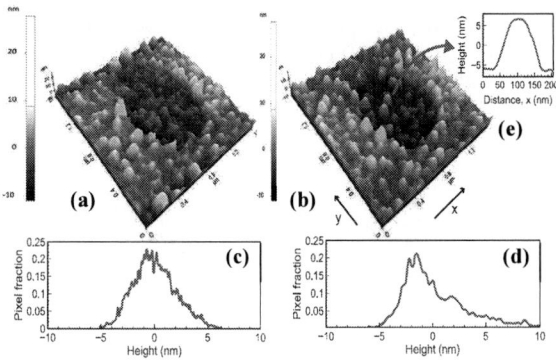

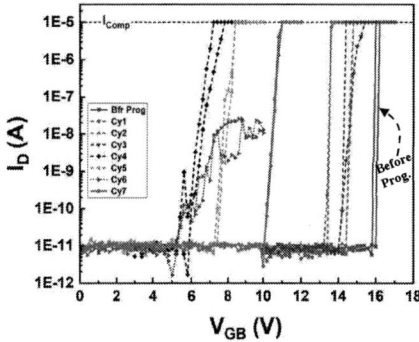

Figure 6: AFM analyses of MEM switch source electrode topography: Scans (1.6μm x 1.6μm) of electrode surfaces for (a) an unprogrammed contact, and (b) a P/E-cycled contact; height distributions within the contact dimple region for (c) the unprogrammed contact and (d) the P/E-cycled contact. (e) Height vs. distance for an asperity.

Reprogrammability is often desirable for embedded memory devices. Therefore, the program and erase operating conditions for reprogrammable memory operation were investigated; the experimental results are plotted in Figs. 7a and 7b, respectively. If the voltage and/or pulse width are too small, then the switch does not change state. If the voltage and/or pulse width are too large, then irreparable damage is caused to the contact so that the switch no longer functions properly, *i.e.*, they cease to conduct any current. The shaded regions in Fig. 7 indicate the voltage-time windows for reprogrammable operation.

Fig. 8 shows measured I_D-V_{GB} characteristics for a MEM switch before and after multiple program/erase cycles. It can be seen that V_{ON} varies substantially from cycle to cycle, indicating that the contact asperity height and shape change with each cycle. Nevertheless, low programmed state resistance (< 1kΩ) is maintained with each cycle, as can be seen from the endurance testing results in Fig. 9 (a). As shown in Fig. 9 (b), although the programmed state resistance is slightly larger at elevated temperature (200°C) due to degraded electron mobility, it does not change substantially over time; this indicates that the welded asperities are very stable, providing for excellent data retention.

Figure 8: Evolution of MEM switch I_{DS}-V_{GB} characteristic through multiple program/erase cycles. The change in V_{ON} from cycle to cycle is non-deterministic because it depends on the shape and height of the contacting asperities.

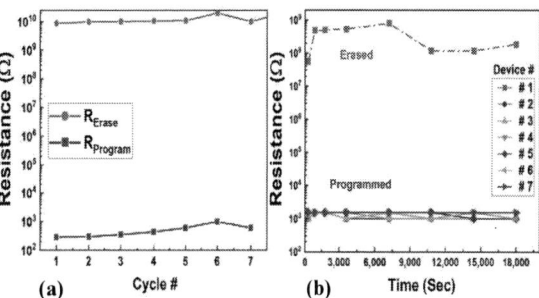

Figure 9: (a) Evolution of the measured resistance of a MEM switch in the programmed state ($R_{Program}$) and in the erased state (R_{Erase}), over multiple P/E cycles. R_{Erase} greater than ~10^9 Ω indicates an open circuit, due to the current measurement noise limit. (b) Data retention testing of multiple MEM switches (1 erased, 6 programmed) at high temperature (200°C) in vacuum (~1 μTorr). The current measurement noise floor is higher at elevated temperature, resulting in smaller apparent R_{Erase}.

THEORY AND SIMULATION

An analytical model for contact welding was derived in [9]; only the key results are presented herein. When a voltage pulse with amplitude V_{DS} and duration Δt is applied between the drain and source in the ON–state, heat is generated at a rate $\dot{Q} \approx \frac{V_{DS}^2}{R_{TOTAL}^2} \cdot R_C$ where $R_C = \rho_E/(2a_o)$ is the contact resistance, ρ_E is the electrical conductivity, a_o is the radius of the contacting asperity, and R_{TOTAL} is the total drain-to-source resistance. The peak temperature at the contact is

$$T_C = V_{DS} \cdot \frac{1}{4l} \frac{R_C}{R_{Total}} \sqrt{f(\tau)} \qquad (1)$$

where l = 2.44E-8 W·Ω/K² is the Lorenz number, $f(\tau) = 1 - e^{\tau}\left(1 - erf(\sqrt{\tau})\right)$, $\tau = \Delta t/t_{TH}$, $t_{TH} = \rho_E c a_o^2/\lambda$ is the thermal time constant, λ is the thermal conductivity, and c is the specific heat capacity. If atomic movement at the contacting points is diffusive, the rate of asperity growth should have an Arrhenius dependence on temperature, so that the increase in the radius of the contacting asperity is given by:

$$\Delta a_o \sim [D_o exp(-E_A/k_B T_C)]\Delta t \qquad (2)$$

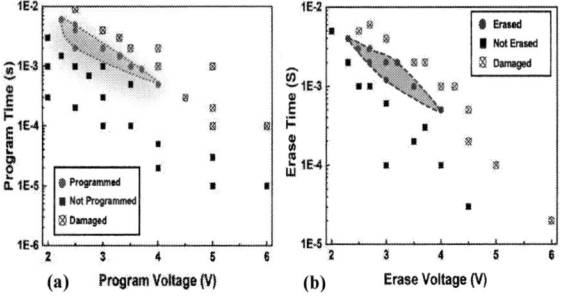

Figure 7: Operating windows for MEM switch (a) program operation and (b) erase operation. The shaded regions indicate the combinations of V_{DS} pulse duration and voltage for successful program/erase operation.

978-1-6654-9309-3/23 $31.00 © 2023 IEEE

where D_o is the diffusion coefficient, E_A is the activation energy, and k_B is the Boltzmann constant. Assuming that welding occurs when Δa_o reaches a critical value, and substituting Eqn. (1) into Eqn. (2), we obtain the following relationship between program time and program voltage: $\Delta t \sim A\exp(B/V_{DS})$ which is consistent with the results presented in Fig. 7.

COMSOL simulations were performed to confirm that the program and erase conditions used in this work result in contact temperature sufficient to soften the contact material. A welded-asperity structure with dimensions consistent with AFM measurements (cf. Fig. 6) and with electrical resistance consistent with measured values (cf. Fig. 9 (a)) was simulated. Since a thin native oxide forms on the surface of tungsten, the simulated structure comprises WO_3-coated asperities that are welded together.

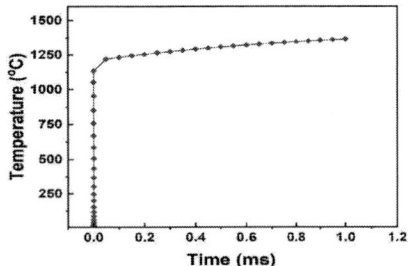

Figure 11: Evolution of peak contact temperature (cf. Fig. 12) during erase operation. The 2.7V, 1ms erase pulse is applied at time t = 0 with a ramp time of 0.1 μs.

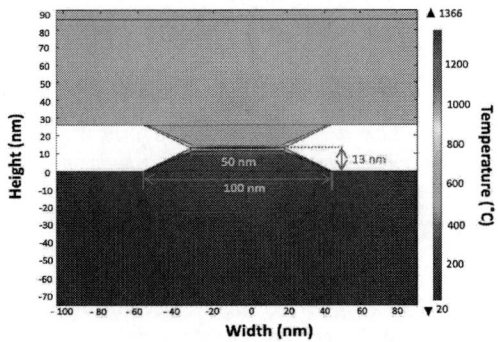

Figure 12: COMSOL simulation of Joule heating in a MEM switch during erase operation (cf. Table I). The W source/drain electrode contacting asperities are sized to match AFM measurements (cf. Fig. 6b) and are assumed to be oxidized, i.e., coated with 1.5 nm-thick WO_3.

Table II: Key material properties and values used for thermal simulation.

	Electrical Conductivity σ (S/m)	Thermal Conductivity κ (W/(m K))	Heat Capacity Cp (J/(Kg K))	Density ρ (kg/m³)
W	2×10^7	174	132	19,350
WO₃	1600	1.63	170	7,160
Al₂O₃	0	35	730	3,965
Poly-SiGe	167000	11	464	4,740
Si Substrate	4	131	700	2,329

The simulation results (Fig. 11) show that T_C exceeds 1366°C, which is approaching the melting temperature of WO_3 and is sufficient to cause atomic movement. Note from Fig. 12 that the heat is largely confined to the WO_3 contacting layers because the thermal conductivity of WO_3 is two orders of magnitude lower than that of W.

CONCLUSION

MEM switches designed for digital logic application can also be used as non-volatile memory devices, because contact welding and unwelding are controllable processes. Reprogrammability with consistently low programmed state resistance, and excellent (essentially infinite) retention time at elevated temperature, are experimentally demonstrated. Therefore, MEM switches are promising for low-cost implementation of ultra-low-power integrated systems.

ACKNOWLEDGEMENTS

The authors would like to thank Dr. M. Kim for useful discussions. Devices were fabricated in the UC Berkeley Marvell Nanofabrication Laboratory.

REFERENCES

[1] X. Hu *et al.,* "Ultra-Low-Voltage Operation of MEM Relays for Cryogenic Logic Applications", *IEEE International Electron Devices Meeting*, pp. 34.2, 2019.

[2] B. Osoba *et al.,* "Sub-50 mV NEM relay operation enabled by self-assembled molecular coating", *IEEE International Electron Devices Meeting*, pp. 26.8, 2016.

[3] K. Kato *et al.,* "Embedded Nano-Electro-Mechanical Memory for Energy-Efficient Reconfigurable Logic", *IEEE Electron Device Letters,* vol. 37, no. 12, pp. 1563-1565, Dec. 2016.

[4] U. Sikder *et al.,* "Vertical NV-NEM Switches in CMOS Back-End-of Line: First Experimental Demonstration and Array Programming Scheme", *IEEE International Electron Devices Meeting,* pp. 21.2, 2020.

[5] E. S. Park *et al.,* "Inkjet-printed micro-electro-mechanical switches", *IEEE International Electron Devices Meeting (IEDM)*, pp. 29.2.1-29.2.4, 2011.

[6] Z. A. Ye *et al.,* "Demonstration of 50-mV Digital Integrated Circuits with Microelectromechanical Relays", *IEEE International Electron Devices Meeting*, pp. 4.1.1-4.1.4, 2018.

[7] E. Falicov *et al.,* "Breakdown and Healing of Tungsten-Oxide Films on Microelectromechanical Relay Contacts", *in JMEMS*, pp. 265-274, 2022,

[8] R. Maboudian and R. T. Howe, "Critical review: Adhesion in surface micromechanical structures", *J. of Vacuum Science &Technology*, pp1–20, 1997.

[9] H. Kam *et al.,* "A predictive contact reliability model for MEM logic switches," *IEEE International Electron Devices Meeting* pp. 16.4, 2010.

CONTACT

*T.K. Esatu; t.esatu@berkeley.edu

MICROMECHANICAL RSSI BASED ON FORCE INTERACTION DERIVED TAPPING BANDWIDTH VARIATION IN VIBRO-IMPACT RESONATORS

Yi-Hsuan Huang, Hong-Sen Zheng, Chun-Pu Tsai, and Wei-Chang Li
Institute of Applied Mechanics, National Taiwan University, TAWIAN

ABSTRACT

This work demonstrates a micromechanical received signal strength indicator (RSSI) circuits based on CMOS-MEMS folded-beam comb-drive tapping-mode resonators. In particular, the tapping bandwidth of vibro-impact resonators that exhibits a linear growth as a function of input signal level when the dominant contact force transits from the attractive to repulsive force is used as RSSIs for ultra-low power indoor positioning reader applications. Based on CMOS-MEMS folded-beam comb-drive resonators, two designs are presented specifically for this purpose: (1) applying gap-narrowing technique to improve sensitivity, and (2) double-sided stoppers with different gap spacing to create digitized bandwidth vs. input signal strength for event-oriented positioning. These techniques would help to solve the power consumption and decoding problems of previous indoor localization method.

KEYWORDS

CMOS-MEMS, vibro-impact resonators, received signal strength indicator, indoor positioning

INTRODUCTION

Satellite-based positioning systems [1-3], such as the Global Positioning System (GPS), Galileo, and Beidou, are widely used for providing users geolocations and time information. These systems require an unobstructed line of sight between users' receivers and service satellite, making themselves impossible to be used for indoor environment.

In fact, indoor positioning technologies have been extensively developed for various purposes, such as navigation, location-based services, aged care, and life pattern study [4-10]. Among these applications, the indoor monitoring or tracking is probably the most important application for the aging society as the population of home rehabilitation is rapidly growing. For this particular application, of which the main purpose is to identify the personnel position and ensure the safety, the current popular solution of BLE beacons that in which continuously run and emit signal might not be the best solution in terms of energy efficiency.

To address this, this work introduces the use of tapping bandwidth of MEMS vibro-impact resonators as an indicator for receiving signal strength. The resoswitch-based receivers of [11] can be applied not only for wake-up receiver applications such as passive keyless entry (PKE) for ON/OFF signal-level indicator but also for RSSI using the dependency of tapping bandwidth on the input level. Fig.1 illustrates the tapping bandwidth of a tapping frequency response and its dependency on the input driving magnitude. The tapping bandwidth is by the attractive and repulsive interaction forces—the attractive force dominates with a small input driving voltage and as the input driving voltage increases, the repulsive force starts to dominate

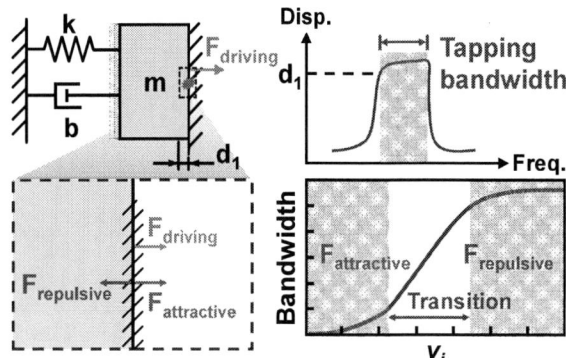

Fig. 1: Mathematical model for resonator, the tapping bandwidth is defined by the flat region of frequency response. The dependency is governed by attractive and repulsive forces and the transition region is between.

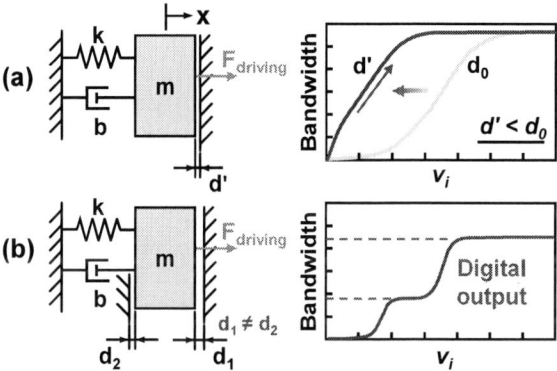

Fig. 2: Schematic of two designs and expected tapping - bandwidth graphs. (a) Combining the arc-beam gap-narrowing structure and (b) combining double-sided stoppers with different gap spacing.

[12].

RSSI CIRCUIT

To implement, this work introduces two designs for the RSSI purpose: (1) combining the arc-beam gap-narrowing structure [13] to improve the signal sensitivity (*cf.* Fig. 2 (a)) and (2) combining different stopper gap spacing in one device to realize multiple steps in the tapping bandwidth-input signal curve (*cf.* Fig. 2 (b)). Fig. 3 (a) illustrates the circuit that can be used to decode of the tapping bandwidth in a vibro-impact resonator based beacons—the pre-installed access point (AP) simply generates predefined sweeping frequency points and the beacon contains an impact resonator of the corresponding resonance frequency with a 1-bit ADC and a counter. The counter output represents the bandwidth magnitude as well as the RSSI level. Fig. 3 (b) and (c) depicts the utilization

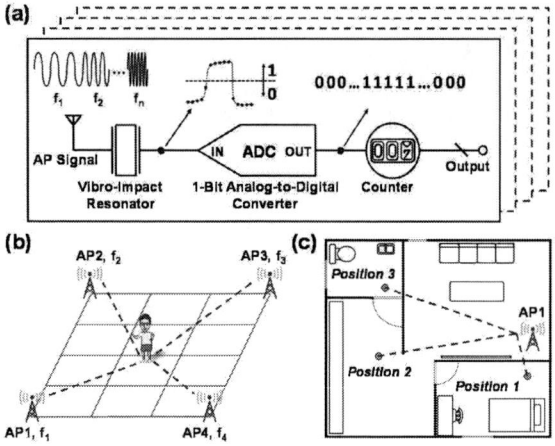

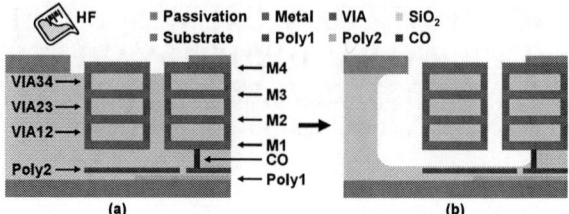

Fig. 5: Cross-section of the 0.35-μm CMOS-MEMS process platform used (a) before and (b) after the release step.

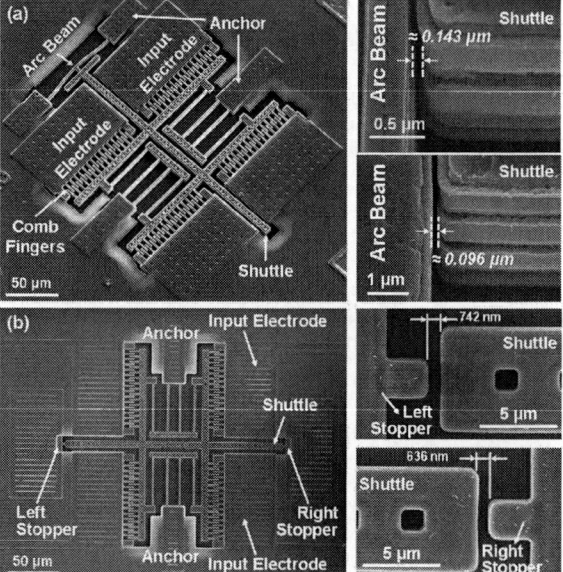

Fig. 3: Schematic of (a) wireless mechanical RSSI circuit that can be used for decode of the tapping bandwidth in a vibro-impact resonator, (b) and (c) shows the utilization scenarios for Fig. 2 (a) and (b).

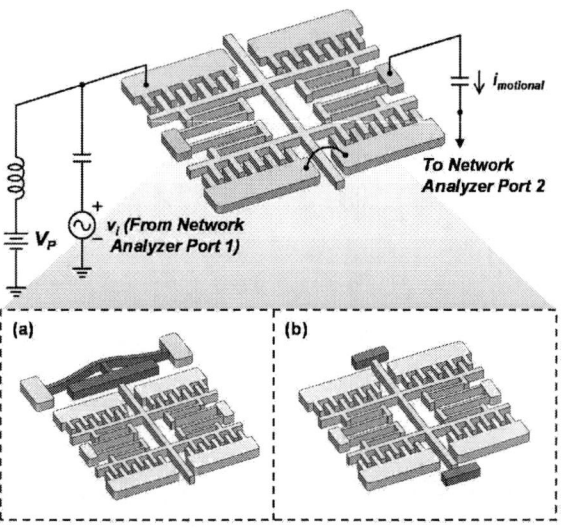

Fig. 4: The measurement set-up for the resonators and schematics of a regular CMOS-MEMS folded-beam comb-driven resonator with (a) a gap-narrowing arc-beam stopper and (b) double-sided stoppers of different gap spacings.

scenarios for Fig. 2 (a) and (b), respectively, where the positioning can be either coordination or a finite set of positions/events.

DEVICE STRUCTURE AND OPERATION

Fig. 4 illustrates the schematic of a regular CMOS-MEMS folded-beam comb-driven resonator with (a) a gap-narrowing arc-beam stopper and with (b) double-sided stoppers of different gap spacings. To operate, the dc bias V_P and the ac bias v_i, are applied on the input electrodes, yielding the driving force for operating the resonator. To measure the frequency transmission, i.e., S_{21}, *Port 1* of the network analyzer (KEYSIGHT E5080A) is applied to the input and *Port 2* to the output to pick up the motional current.

Fig. 6: SEM charts of (a) gap-narrowing arc-beam stopper with a final gap of 143 nm and 96 nm and (b) double-sided stoppers with 636- and 742-nm gaps.

EXPERIMENTAL RESULTS

This work uses a 2-poly-4-metal (2P4M) 0.35-μm CMOS-MEMS process platform to fabricate the micromechanical RSSI. Fig. 5 depicts the fabrication process before and after release step. The resonator structure and electrode consist of four metal layers with a total three oxide layers in between. Fig. 6 (a) shows the SEMs of devices with gap-narrowing arc-beam stoppers with a final gap of 143 nm and of 96 nm. Fig. 6 (b) shows the device with double-sided stoppers with 636 and 742 nm gap spacings, respectively. Fig. 7 shows the photo of measurement set-up. The device is placed in a customized vacuum chamber pumped down to below 7×10^{-5} Torr, and the curve is measured results of received signal strength vs. distance characterization based on LC coil antennas, which will be improved by parameter fine tuning and antennas with better performances. Fig. 8 plots the measurement results of (a) the tapping bandwidth vs. input voltage for Fig. 6 (a) with a final gap reduced to 96 nm and 143 nm, identifying the sensitivity has been improved by 39.93 dB; of (b) tapping bandwidth saturated at three input levels resulting from the resonator of Fig. 6 (b) with two different stopper gaps, all together indicating the vibro-impact resonators as a promising solution for ultra-low power RSSI receivers for AP-based indoor positioning.

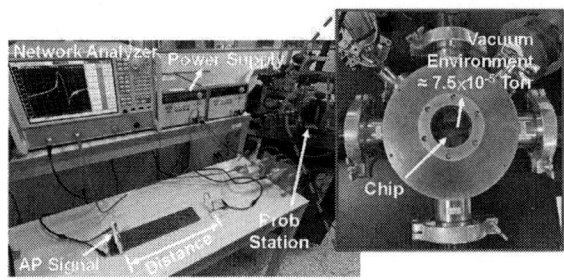

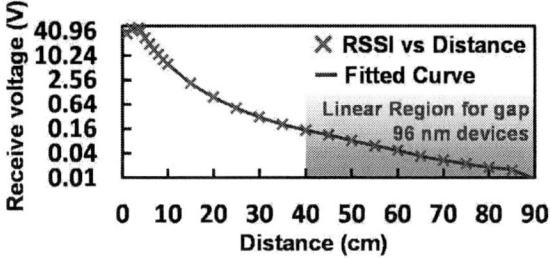

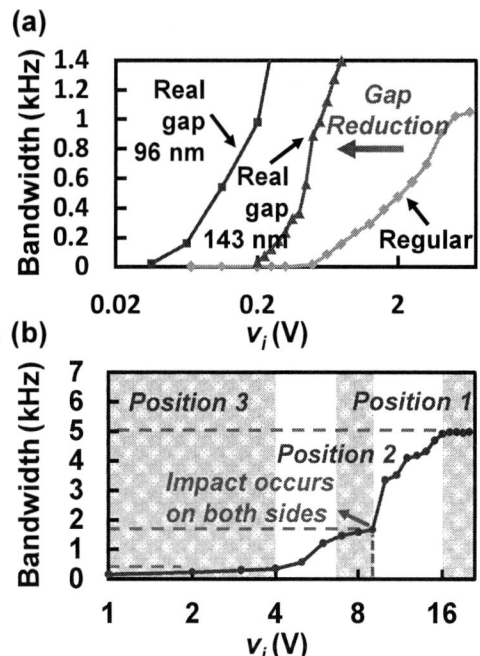

Fig. 7: Photo and received signal strength vs. distance characterization based on LC coil antennas. The distance will be further improved by parameter fine tuning and antenna replacement.

Fig. 8: Measured the results of (a) the tapping bandwidth vs. voltage for Fig. 4 (a), identifying the sensitivity has been improved 99.2× with (b) tapping bandwidth saturated at three input levels results from the vibro-impact RSSI resonator of Fig. 4 (b).

CONCLUSION

This work demonstrated two types micromechanical received signal strength indicator (RSSI) resonators for indoor positioning applications. In particular, the measurement results show that the vibro-impact resonators improved the sensitivity by 99.2× after replacing the stopper to gap-narrowing arc-beam structure. The version with double-sided stoppers of different gap spacings yields digitized bandwidth as a function of the input signal strength. However, in addition to reducing the sensitivity and increasing release yield, integration of the resonators with the backend circuit would be an important working item. Work towards these continues.

ACKNOWLEDGEMENTS

This research was funded by the National Science and Technology Council, Taiwan (MOST-109-2628-E002-004-MY3). The chip fabrication was supported by the Taiwan Semiconductor Research Institute (TSRI) and Taiwan Semiconductor Manufacturing Company (TSMC), Hsinchu, Taiwan. The authors would like to thank the staff in NEMS Research Center at NTU for providing technical support.

REFERENCE

[1] J. G. McNeff, "The Global Positioning System," *IEEE Transactions on Microwave Theory and Techniques,* vol. 50, no. 3, 645–652 Mar. 2002.

[2] A. Montesano, C. Montesano, R. Caballero, M. Naranjo, F. Monjas, L. E. Cuesta, P. Zorrilla and L. Martinez, "Galileo System Navigation Antenna for Global Positioning," *2nd European Conference on Antennas and Propagation (EuCAP),* 2007.

[3] C. Han, Y. Yang and Z. Cai, "BeiDou Navigation Satellite System and its Time Scales," *Metrologia,* vol. 48, no. 4, pp. S231-S218, Jul. 2011.

[4] A. Boukerche and E. Nakamura, "Localization Systems for Wireless Sensor Networks," *IEEE Wireless Communications,* vol. 14, no. 6, pp. 6-12, Dec. 2007.

[5] S. Sadowski and P. Spachos, "RSSI-Based Indoor Localization With the Internet of Things," *IEEE Access,* vol. 6, pp. 30149-30161, 2018.

[6] M. Sugano, T. Kawazoe, Y. Ohta and M. Murata, "Indoor Localization System using RSSI Measurement of Wireless Sensor Network based on ZigBee Standard," *Wireless and Optical Communications,* vol. 538, pp. 1-6, 2006.

[7] K. Langendoen and N. Reijers, "Distributed Localization in Wireless Sensor Networks: A Quantitative Comparison," *Computer Networks,* vol. 43, no. 4, pp. 499-518, Nov. 2003.

[8] A. A. Sohan, M. Ali, F. Fairooz, A. I. Rahman, A. Chakrabarty and M. R. Kabir, "Indoor Positioning Techniques using RSSI from Wireless Devices," *22nd International Conference on Computer and Information Technology (ICCIT),* Dec. 2019.

[9] I. Amundson and X. D. Koutsoukos, "A Survey on Localization for Mobile Wireless Sensor Networks," *Mobile Entity Localization and Tracking in GPS-less Environments,* pp. 235-254, 2009.

[10] A. Yassin, Y. Nasser, M. Awad, A. Al-Dubai, R. Liu, C. Yuen, R. Raulefs and E. Aboutanios, "Recent

Advances in Indoor Localization: A Survey on Theoretical Approaches and Applications," *IEEE Communications Surveys & Tutorials,* vol. 19, no. 2, pp. 1327-1346, Sec. 2017.

[11] C.-P. Tsai, Y.-Y. Liao and W.-C. Li, "A 125-KHZ CMOS-MEMS Resoswitch Embedded Zero Quiescent Power OOK/FSK Receiver," *the 33rd IEEE Int. Conf. on Micro Electro Mechanical Systems (MEMS'20),* pp. 18-22, Jan. 2020.

[12] K. M. Andersson and L. Bergström, "DLVO Interactions of Tungsten Oxide and Cobalt Oxide Surfaces Measured with the Colloidal Probe Technique," *Journal of Colloid and Interface Science,* vol. 246, no. 20, pp. 309-315, 2002.

[13] H.-S. Zheng, C.-P. Tsai, T.-Y. Chen and W.-C. Li, "CMOS-MEMS Resonators with Sub-100-Nm Transducer Gap Using Stress Engineering," in *the 35th IEEE Int. Conf. on Micro Electro Mechanical Systems (MEMS'22)*, Tokyo, Japan, Jan. 9-13, 2022.

CONTACT

W.C. Li, tel: +886-2-33665667; wcli@ntu.edu.com

WAKE-UP IOT WIRELESS SENSING NODE BASED ON A LOW-G THRESHOLD MEMS INERTIAL SWITCH WITH RELIABLE CONTACTS

Sagnik Ghosh[1], Duan Jian Goh[1], Yul Koh[1], Jaibir Sharma[1], Wei Da Toh[1], Weiguo Chen[1], Yao Zhang[1], Eldwin Ng[1], Amit Lal[2], and Joshua E.-Y. Lee[1]

[1]Institute of Microelectronics, A-STAR, SINGAPORE and
[2]Cornell University, USA

ABSTRACT

We demonstrate a wireless sensing node (WSN) that is woken-up by a Microelectromechanical systems (MEMS) inertial switch with low-g threshold of 9.4g for Internet-of-Things (IoT) applications based on zero-power event-driven sensing. By virtue of the device design and fabrication of the switch contacts, the inertial switch in the WSN maintains a contact time long enough to wake up a Bluetooth Low Energy (BLE) module. Upon wake-up, the BLE module initiates data transmission to a remote display, when the WSN senses an external acceleration above the threshold. We also demonstrate remarkable contact reliability in the MEMS inertial switch as a significant step towards long lifetime zero-power sensors. The inertial switch output shows little degradation even after 100 cycles of testing under ambient conditions with an unpackaged MEMS device; a notable first among MEMS inertial switches.

KEYWORDS

Low-g MEMS inertial switch, Event-driven zero-power IoT systems, switch contact time, Titanium Nitride (TiN) contact, zero-power sensor switch fabrication platform.

INTRODUCTION

Given the exponential growth in the number of sensors for Internet-of-Things (IoT), there is a mandate to drastically reduce power consumption of the wireless sensing nodes (WSNs). As such, there is a need to replace the always-on sensors with wake-up sensors [1-2] and mechanical sensor switches [3-6] to enhance battery lifetime of WSNs. Wake-up sensors and mechanical sensor switches get activated when a signal equals to or above the threshold is being detected and will remain dormant otherwise, thus reducing the power consumption during idle time. Wake-up sensors commonly employ piezoelectric sensing and require a charge amplifier which constantly consumes power [1,2]. Having no need for charge amplifiers, mechanical sensor switches [3-6] can be a better alternative to such wake-up sensors with an additional benefit of low leakage.

A MEMS inertial switch is triggered by an external acceleration beyond a certain predetermined value (known as the threshold value) and can be implemented in an IoT system as an alternative to conventional always-on MEMS acceleration sensors. Much progress has been made on the design of MEMS inertial switches, demonstrating switching at various threshold levels, most of which are on the high-g range of 100g [5,6]. There are few reports on low-g inertial switches (where the g value ranges from 1g to 40g) for use in environmental and building monitoring.

A class of low-g MEMS inertial switches was fabricated using multi-step metal electroplating processes [7,8]. Such complex fabrication platforms suffer from residual stress of individual metal layers, which is key to achieving high yield for low-g inertial switches.

Moreover, the reports on MEMS inertial switches have stopped well short of a full demonstration of a wake-up WSN, e.g., waking-up a WSN by an impulse input acceleration to validate the promise of event-driven zero-power IoT systems. One of the key bottlenecks is contact bouncing that prevents a sufficiently long contact time to wake-up the circuit and initiate data transmission. The switches also do not endure long enough for multi-cycle tests required in an actual demonstration of an event-driven wake-up WSN, degrading rapidly after a few cycles. In this report, we have successfully shown the full demonstration of waking-up a WSN (which consumes near-zero power during the idle time) by the virtue of sufficiently long contact time and contact reliability of our in-house fabricated low-g MEMS inertial switch.

DESCRIPTION OF THE SYSTEM

We deployed our in-house fabricated MEMS inertial switch to realize the wake-up of the WSN. The fabricated MEMS inertial switch was included in the BLE module beside other off-the-shelf components. In this section, we first describe our MEMS inertial switch and later provide a system level description of the WSN.

A. Description of MEMS inertial switch

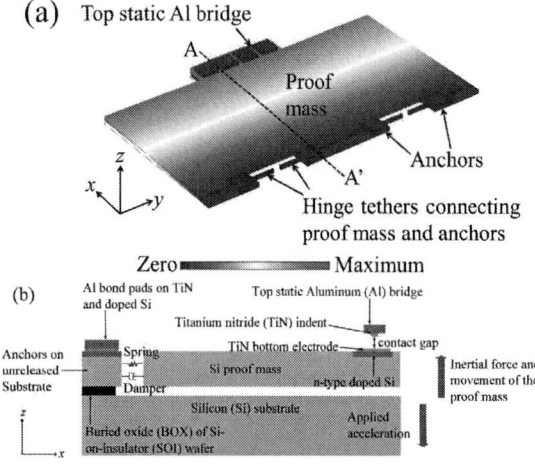

Figure 1: (a) Finite element simulated z-axis displacement contours under acceleration along z-axis; (b) Cross-sectional view along line-section AA' in Fig. 1a showing the torsional proof mass with bottom and top electrodes located near the tip of the proof mass (where displacement

is maximum).

Fig. 1a shows the out-of-plane (along z-axis) displacement contour of our MEMS inertial switch under 1g acceleration as simulated using finite element modeling (FEM) with COMSOL Multiphysics. As seen from Fig. 1a, the MEMS inertial switch comprises of a 30μm thick silicon (Si) proof mass (2200μm×1024μm) clamped at one end by four torsional hinges, forming a "simply clamped cantilever" to lower the spring constant. The cross-sectional view of the MEMS inertial switch is depicted in Fig. 1b. As shown in Fig. 1b, the top and bottom contact electrodes of the MEMS inertial switch are located along the free edge of the cantilever where the z-axis displacement is maximum.

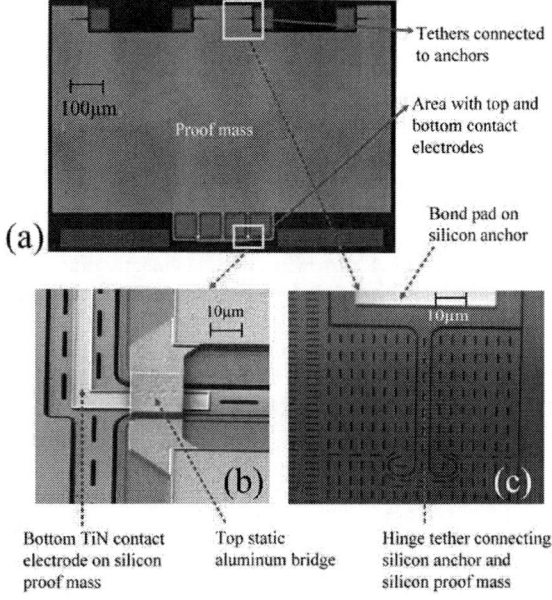

Figure 2: a) Top view optical micrograph showing the MEMS inertial switch with 30μm thick silicon proof mass; (b) Zoomed-in scanning electron micrograph (SEM) showing the bottom Titanium Nitride (TiN) contact with static Aluminum (Al) bridge on top; (c) Zoomed-in SEM showing a hinge tether between the silicon proof mass and the anchor. Each anchor has Al bond pads.

The fabrication process for our MEMS switch has been reported in [9]. An optical micrograph of the fabricated MEMS inertial switch is shown in Fig. 2a, while Figs 2b and 2c show the respective zoomed-in micrographs around the contact area and one of the torsional hinge-shaped tethers. The contact electrodes were fabricated with Titanium Nitride (TiN) to reduce wearing during the switching. As shown in Fig. 2b, the bottom TiN electrodes are directly placed on the Si proof mass, while an Aluminum (Al) bridge forms the static top electrode. It is worth noting that an array of TiN indents was fabricated on the underside of the top Al bridge to reduce stiction during switching. A single TiN indent on the underside of the Al bridge is shown in Fig. 1b. The compliance of the Al bridge eliminates bouncing, thus increasing the contact time (T_{ON}) upon impingement of the top and bottom electrodes.

Threshold-g is reduced by tether design and the nano-gap switch-contact as shown in Fig. 2. Given that the static contact electrode is placed on top of the dynamic bottom contact electrodes, the top and bottom electrodes will come into contact when an acceleration is applied downward along the z-axis. As such, the proof mass will be displaced upwards along the positive z-axis under an inertial force developed due to the external acceleration (given in Fig. 1b).

COMSOL Multiphysics was used to simulate the resonant frequency and static displacement of the device under applied acceleration. First, we used an eigenmode study to simulate the resonant frequencies for the fundamental mode (first resonance mode) and the 2nd vibration mode. The displacement profile corresponding to the fundamental mode is given in Fig. 1a. The hinge-shaped tethers result in a resonant frequency of 2.58kHz for the fundamental mode, while the 2nd vibration mode occurs at a resonant frequency of 21.89kHz (nearly 5 times higher than the fundamental mode). In addition, we have also simulated the cross-axis sensitivity by comparing the out-of-plane static displacements (along z-axis) around the contact regions while the MEMS inertial switch was subjected to accelerations along x, y and z axes respectively. We used a stationary study in the simulation, obtaining a sensitivity of 62.67nm/g when the proof mass is subjected to an acceleration along the z-axis, well over 10^5 times of sensitivities to acceleration applied along the x-axis (1.03×10^{-5} nm/g) and y-axis (3.60×10^{-6} nm/g). The near-zero cross-axis sensitivity can be attributed to the hinge-shaped tethers of the MEMS inertial switch.

B. Description of Bluetooth Low Energy (BLE) Module

Fig. 3 shows the key design blocks contained in a single compact package powered by a 3V coin battery. The fabricated MEMS inertial switch was wire bonded to a Leadless Chip Carrier (LCC) package and the LCC package was soldered to the Printed Circuit Board (PCB) of the WSN. The rest of the components within the WSN package were acquired off-the-shelf.

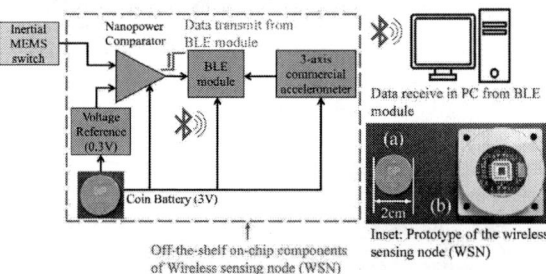

Figure 3: System level block diagram showing the active components of the wireless sensing node (WSN) that wirelessly communicates with a remote PC. The figure in the inset shows the compact form factor of the WSN prototype compared to a 2cm diameter coin cell battery.

As shown in Fig. 2a, the input voltage (V_{in}) from the coin battery was applied to one of the top Al bridges, while the output (V_{out}) from the MEMS inertial switch was fed to one of the inputs of the off-the-shelf comparator soldered to the WSN package. Upon detecting an acceleration impulse above threshold, the MEMS inertial switch is activated and V_{out} is compared against the reference voltage (which was set to 0.3V) of the comparator. The on-board components remain dormant until the voltage at the input of the comparator is at least 0.3V. After wake-up, the BLE module wirelessly transmits data from a built-in reference accelerometer to the BLE receiver at the remote PC, where the received acceleration data is displayed by a Guided User Interface (GUI) in real-time for validation.

MEASUREMENT RESULTS

Experimental validation was performed in two phases. The first phase focuses on the measurement of the MEMS inertial switch on its own. The MEMS inertial switch was later integrated to the BLE module to perform the second phase, where we woke up the WSN via V_{out} of the inertial switch.

A. Experimental validation of the MEMS inertial switch on its own

To measure the MEMS inertial switch, the fabricated die was mounted, and wire bonded to a LCC package. The LCC package was further clamped to the sockets soldered to a custom designed PCB. SMA connectors were included in the PCB for electrical connections to external devices. The PCB was then flipped upside down (as well as the MEMS inertial switch) and mounted on a shaker (Bruel and Kjaer, LDS V501) top. A reference accelerometer was mounted on the backside of the PCB to control the applied acceleration from the shaker and also tracks the acceleration experienced by the PCB as well as the MEMS inertial switch. A V_{in} of 3V was directly applied to one of top Al bridges, while V_{out} from one of the bond pads (that connects to the Si proof mass) was measured by an oscilloscope through a resistance load of $1M\Omega$ as set in the oscilloscope.

Experimental validation of threshold acceleration:

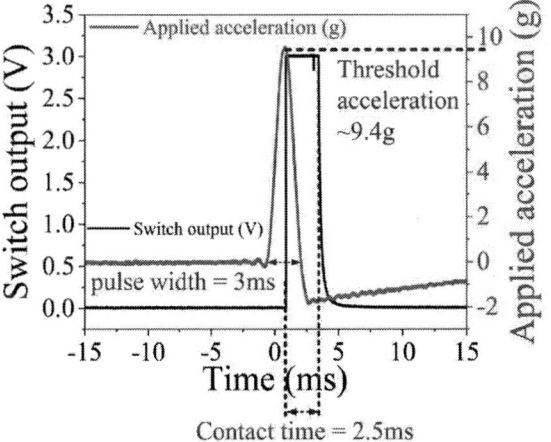

Figure 4: Time domain results showing threshold acceleration of 9.4g with contact time (T_{ON}) of 2.5ms when subjected to a maximum impulse of 10g acceleration with

3ms pulse width.

A measurable and consistent V_{out} was obtained from the MEMS inertial switch when an impulse with minimum peak acceleration of 10g with 3ms pulse width was applied from the shaker. A threshold acceleration of 9.4g with T_{ON} of 2.5ms was extracted from the time domain switching characteristics as shown in Fig. 4. Reduced bouncing from the compliant top Al bridge increases T_{ON}, which is more than 80% of the input acceleration impulse pulse width (3ms).

Measurement of contact reliability of the MEMS inertial switch:

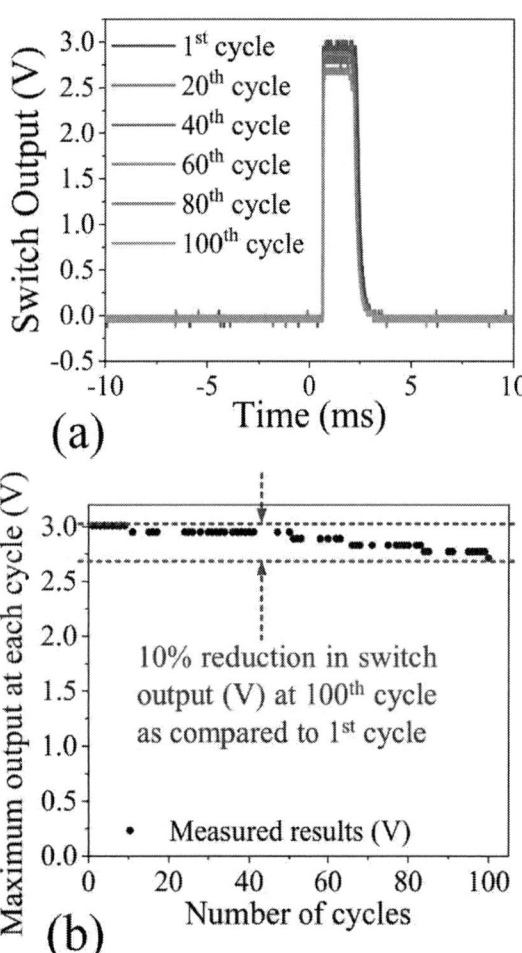

Figure 5: (a) Cyclic test results of the switch tested under ambient conditions showing time domain switch output response for the 1st, 20th, 40th, 60th, 80th and 100th cycles; (b) Output logged at each cycle showing only 10% reduction after 100 cycles under ambient conditions. Acceleration impulse applied: 40g at intervals of 5s.

The contact reliability of the MEMS inertial switch was electrically measured by applying 100 cycles of impulses with peak acceleration of 40g at intervals of 5s. It should be noted that the same measurement set-up was used for this study as disclosed before in this section. The cyclic test results under ambient conditions in air are shown in Fig. 5.

Fig. 5a shows the switching behaviors at different cycles, while the discrete points in Fig. 5b correspond to the maximum V_{out} recorded at each cycle. Comparing maximum V_{out} between 1st and 100th cycles from Fig. 5b, we can see that the drop in V_{out} is contained within 10%. This is a significant advancement in inertial switches given these tests were carried out under ambient conditions in air without packaging.

B. Wake-up of wireless sensing node

The prototype for the WSN (shown in the inset of Fig. 3) was flipped upside down and screw-mounted on the same shaker top. A reference accelerometer was also mounted on the backside of the prototype to control and record the acceleration experienced by the prototype. The prototype was exposed to 3 cycles of an impulse at 20g peak acceleration with 3ms pulse width at intervals of 5s. Fig. 6 shows the real time acquisition of the external acceleration applied from the shaker alongside the switch output and the reference acceleration (as recorded by the accelerometer mounted on the backside of the WSN prototype). The MEMS inertial switch woke up the BLE module in the first cycle. Given a finite time period required to establish the Bluetooth connection between the BLE module and Bluetooth of the PC, the input acceleration impulse of the first cycle has not been recorded in the GUI. But once the Bluetooth of the PC starts receiving data, the signal on the GUI accurately tracks the input acceleration applied to the WSN from the shaker as seen from the second and third cycles in Fig. 6 (bottom graph).

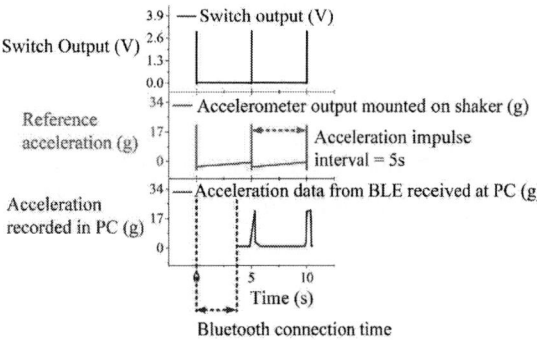

Figure 6: Real time results of MEMS switch output (top), output of accelerometer on shaker (middle) and output of accelerometer on BLE module as received remotely at PC (bottom) upon wake-up by the 1st cycle. Transmitted data of acceleration from WSN read out on the PC accurately tracks the reference accelerometer on the shaker.

CONCLUSIONS

We have successfully demonstrated the wake-up of a custom designed WSN by the switch output of a fabricated low-g MEMS inertial switch via Bluetooth transmit and receive between the BLE module of the WSN and the PC. Reduced bouncing during switching increases contact time which provides sufficiently long time to wake-up the WSN and initiate the transmission of applied acceleration data. Excellent contact reliability increases the probability to reproduce switch outputs through multiple cycles. Overall, the WSN prototype has the potential to be applied in environmental sensing and indoor condition monitoring.

ACKNOWLEDGEMENTS

This research work was supported by A*STAR under "Nanosystems at the Edge" programme (Grant number: A18A4b0055).

REFERENCES

[1] V. Pinrod *et al.*, "Zero-Power Sensors with Near-Zero-Power Wakeup Switches for Reliable Sensor Platforms", in *Digest Tech. Papers IEEE MEMS'17 Conference*, Las Vegas, USA, January 22-26, 2017, pp. 1236-1239.

[2] C. Stoeckel *et al.*, "Wake Up MEMS for Inertial Sensors", in *Digest Tech. Papers IEEE ISISS'15 Conference*, Hawaii, USA, March 23-26, 2015, pp. 1-4.

[3] L. J. Currano *et al.*, "3-Axis Acceleration Switch for Traumatic Brain Injury Early Warning", in *Digest Tech. Papers IEEE MEMS'12 Conference*, Paris, France, January 29 – February 2, 2012, pp. 484-487.

[4] Q. Xu *et al.*, "Multi-Threshold MEMS Shock Sensor for Quantitative Acceleration Measurements", in *Digest Tech. Papers Transducers'21 Conference*, Virtual, June 20-24, 2021, pp. 120-123.

[5] V. Singh *et al.*, "Design and Development of the MEMS-Based High-g Acceleration Threshold Switch", *J. Microelectromech. Syst.*, vol. 30, pp. 24-31, 2021.

[6] Z. Xi *et al.*, "High g MEMS Inertial Switch Capable of Direction Detection", *Sens. Actuators A Phys.*, vol. 296, pp. 7-16, 2019.

[7] Z. Yang *et al.*, "A Multidirectional-Sensitive Inertial Microswitch with Electrophoretic Polymer-Metal Composite Fixed Electrode for Flexible Contact", in *Digest Tech. Papers IEEE MEMS'12 Conference*, Paris, France, January 29 – February 2, 2012, pp. 504-507.

[8] Y. Gerson *et al.*, "Meso Scale MEMS Inertial Switch Fabricated Using an Electroplated Metal-on-Insulator Process", *J. Micromech. Microeng.*, vol. 24, pp. 025008, 2014.

[9] Y. Koh *et al.*, "Nano-Gap Contact MEMS Torsional Mode Acceleration Switch Wake-Up Sensor", in *Digest Tech. Papers IEEE SENSORS'22 Conference*, Texas, USA, October 30 – November 2, 2022 (to be published).

CONTACT

*S. Ghosh, Sagnik_Ghosh@ime.a-star.edu.sg

ARTIFICIAL INTELLIGENCE (AI)-ENHANCED E-SKIN WITH ARTIFICIAL SYNAPSE SENSORY OUTPUT FOR HUMANOID ROBOTIC FINGER OF MULTIMODAL PERCEPTION

Xinge Guo[1,2], and Chengkuo Lee[1]*

[1] Department of Electrical & Computer Engineering, National University of Singapore, Singapore

and

[2] Institute of Microelectronics (IME), Agency for Science, Technology and Research (A*STAR), Singapore

ABSTRACT

In this manuscript, we reported a self-generated e-skin with multimodal perception sensing enhanced by artificial intelligence. Inspired by the human skin, the proposed system contains the self-generated triboelectric sensor to mimic the fast adapting (FA) mechanoreceptors and the self-generated potentiometric sensor to mimic the slow adapting (SA) mechanoreceptors and thermoreceptors. By introducing deep learning with data fusion to extract features from multiple channels, the multimodal outputs can achieve more sensing functions and calibrate each other to overcome instability, such as applied force and speed. The surface temperature, roughness, texture, hardness, and material type measuring with high accuracy performed by a single element placed on the fingertip have been realized.

KEYWORDS

E-skin, Artificial Synapse, Multimodality, Artificial Intelligence.

INTRODUCTION

As one of the most essential sensory functions for humans to contact the environment, the human somatosensory system decodes a wide variety of external stimuli. Such complicated functions are achieved by two kinds of sensory neurons within the human skin, named mechanoreceptors and thermoreceptors [1]. As depicted in Figure 1, the mechanoreceptors, including SA and FA receptors, respond to innocuous mechanical stimuli such as pressure, stretch, and vibration, while the thermoreceptors respond to thermal stimuli in a range of 5 °C to 48 °C [2]. The valuable information sensed by such receptors endows us with a remarkable capacity for object recognition, texture discrimination, motion and temperature feedback, and social exchange. Since the 1980s, researchers have placed high expectations on bionic tactile sensors aiming to simulate the tactile functions of the human skin for their tremendous potential applications in fields including human-machine interfaces, humanoid robotics, wearable electronics, internet of things, etc [3]–[9].

Thanks to the burgeoning of functional materials and microfabrication technology, tactile sensors that can sense physical information, including pressure, temperature, and vibration, with high spatial resolution and sensitivity, have been developed based on piezoresistive, piezoelectric, capacitive, and pyroelectric principles [10]–[19]. However, previous works mostly either stay in signal mimicking without analysis in real applications and test the device with stable equipment or focus on tactile sensing without

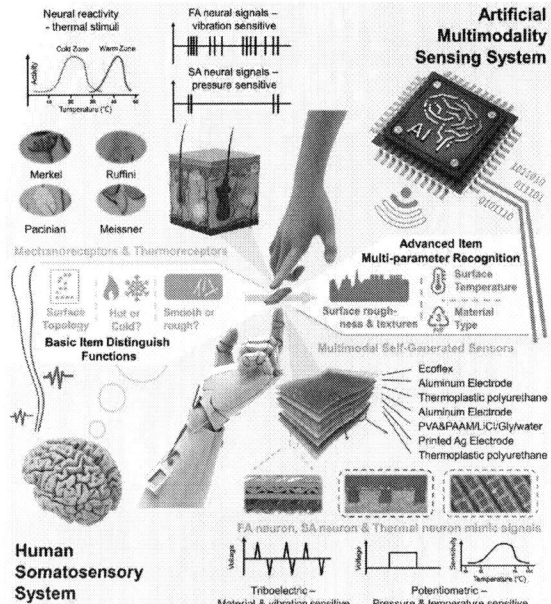

Figure 1: The structure and functions of the human somatosensory system, compared with the multilayer structures and advanced functions of the proposed multimodal sensor with FA, SA, and Thermal mimic sensing.

multimodality sensing demonstration. Besides, a complicated readout circuit and a continuous power supply are always required. While the instability of actual operations is a critical factor that can not be ignored, other sensing signals like temperature are vital for humans to avoid risks, and the circuit and battery also limit flexibility, cost, and lifetime [20]–[25]. Therefore, a self-generated multimodal sensing e-skin system with different types of signals as proper ways to self-calibrate, overcome instability, and provide more comprehensive information has been envisioned over a long period.

In this work, we have successfully developed a multimodal e-skin with self-generated FA neuron, SA neuron, and Thermal neuron mimic signals (Figure 1). The FA mimic function is achieved by a 250 μm thickness Ecoflex-0030 layer with patterned microstructures on top working based on the triboelectric mechanism. The SA and thermoreceptor mimic functions are achieved by the PVA&PAAM/LiCl/Gly/water material, which can also generate voltage responses with external physical stimuli based on the potentiometric mechanism [26]. Deep learning with data fusion is applied to extract features from multimodal signal channels. After the basic characterizations, the proposed e-skin is placed on the fingertip to perform multiple parameter measurements of various items, including the texture, surface roughness, hardness, temperature, and materials. The increased distinguish accuracy of multiple channels than the single channel proved the advancement of multimodal perception. And the nearly 100% accuracy for all the measured parameters shows the promising applications of the e-skin for robotics to achieve human-like item perceptions.

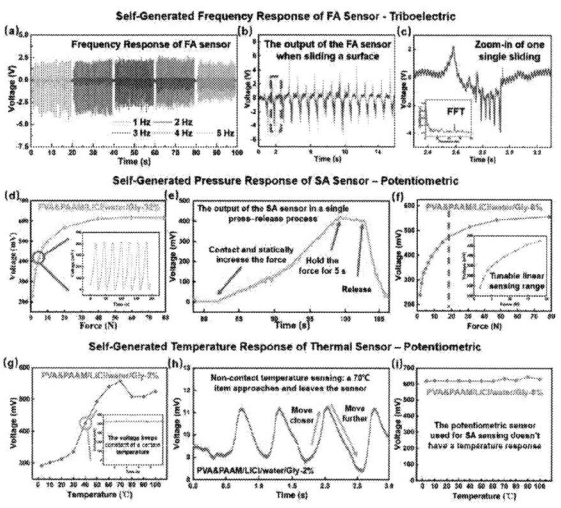

Figure 2: Characterization of applied multimodal sensors. (a-c) FA mimic sensor based on triboelectric; (d-f) SA mimic sensor based on potentiometric; (g-i) Thermal mimic sensor based on potentiometric

PERFORMANCE CHARACTERIZATION

Firstly, the performance of FA, SA, and temperature sensors are characterized (Figure 2). The triboelectric sensor with a rapid response (Figure 2(a)) is highly suited for mimicking the FA mechanoreceptors in the human skin for sensing the high-frequency vibration stimuli. When sliding through the item surface, the FA sensor with patterned microstructure on the surface leads to microscale continuous contact and separation and generates regular triboelectric signals (Figure 2(b, c)). The potentiometric sensor is composed of an electrolyte containing Lithium Chloride (LiCl) and two electrode layers with different reduction potentials, chosen as the Aluminum (Al) and Silver (Ag) for the proposed device. When the two electrode layers contact the electrolyte layer, a stable potential difference is developed between them, and the value is related to the difference in their reduction potentials. The conduction of the electrolyte, namely, its resistance, can be tuned by either changing the contact resistance with the electrode layer or modifying the electrolyte conductivity. And this characteristic makes it possible to apply it as a self-generated sensor for SA-mimic pressure sensing and thermoreceptor-mimic temperature sensing. To regulate the contact resistance variation under external pressure, the surface of the electrolyte layer is also patterned with microstructures. The increasing force applied to the device leads to an increase in the contact area and a decrease in the contact resistance. Therefore an

increased voltage can be sensed through the oscilloscope with a fixed resistance (Figure 2(d)), and a sensing range from 0 N to 10 N can be noticed. Besides, the fabricated device can also encode the entire process of applying force – holding at a particular value – releasing force (Figure 2(e)). Besides, the force sensing range is also tunable by tuning the mass fraction of the Glycerol to modulate the water content of the electrolyte, and the electrolyte with 8% Gly can reach 20 N force sensing (Figure 2(f)). Further decreasing the water content of the electrolyte to increase the resistance, it will reach the range that is dominated by the ionic conductivity, which can be varied by the temperature. For the electrolyte with 2% Gly, a higher temperature increases the ionic conductivity and leads to higher sensed voltage (Figure 2(g)). And thanks to the high sensitivity of the material, it can also achieve non-contact sensing for high-temperature items (Figure 2(h)). As a comparison, for the electrolyte with higher water content, since the resistance is not dominated by the temperature, it barely has a temperature response (Figure 2(i)).

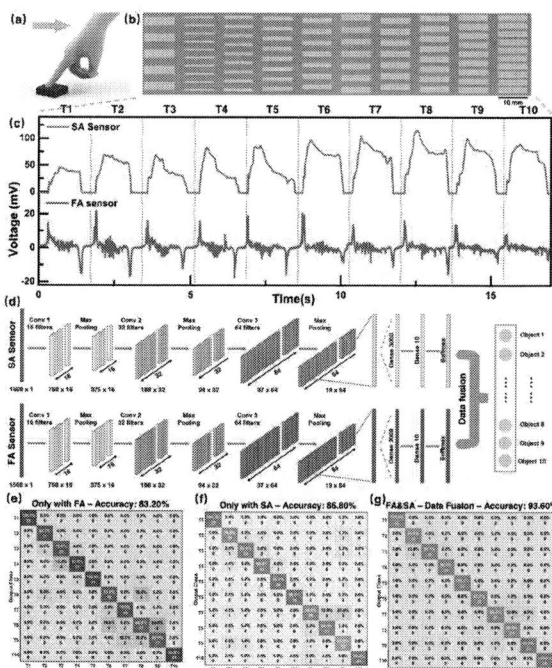

Figure 3: Surface texture recognition. (a-b) Schematic of testing method and tested ten different textures. (c) Output curves. (d) CNN with data fusion to extract multimodality features from each channel. (e-g) Machine learning results

AI-ENHANCED MULTIMODAL PERCEPTION

After that, the e-skin is attached to a fingertip to test ten different surface textures through sliding (Figure 3(a, b)). Both the SA and FA sensors can detect the texture to some extent, for one is sensitive to deformation and the other to vibration (Figure 3(c)). However, the accuracy is still limited due to the instability of the sliding process manually performed by the fingertip, such as the sliding speed and contact force. To overcome this deficiency, instead of applying a stable testing platform, which usually is infeasible for the actual applications, deep learning with data fusion is introduced (Figure 3(d)). By extracting the features from both SA and FA output and treating them as an entirety to achieve distinguishment, their features can complement and calibrate each other to overcome such instability. For example, the contact force and contact time encoded by the SA sensor are very important complementary information for the output amplitude and vibration frequency encoded by the FA sensor. It can be noticed that only with FA sensor or with SA sensor only achieve average accuracies of 83.20% and 86.80% for the identification of ten textures, while a higher accuracy of 93.60% is achieved through the data fusion.

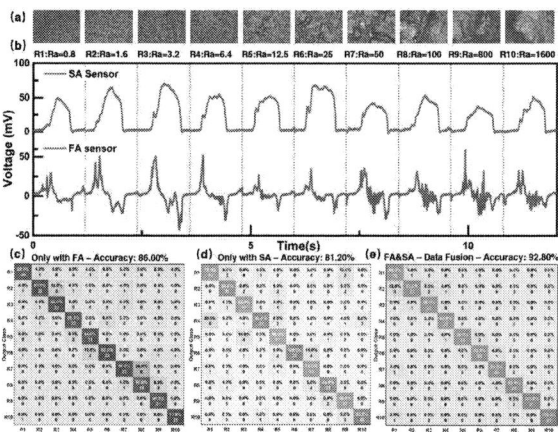

Figure 4: Surface roughness detection. (a) Optical images of ten items with different surface roughness. (b) Output curves. (c-e) Machine learning results with data fusion

Furthermore, the same device is utilized to measure objects with ten surface roughness ranging from 0.8 to 1600 Ra (Figure 4(a)), controlling the same material, hardness, and surface temperature for the testing samples. The Ra is used to describe the surface roughness by calculating the arithmetic average of the absolute values of the height deviations from the mean line, with μm as the unit. The FA sensor is the major component for roughness sensing as it generates distinct output corresponding to the surface (Figure 4(b)). At the same time, the bulges on the surface also lead to certain variations in the SA output due to their influence on the contact area. Similarly, average accuracies of 86.00% and 81.20% of solitarily working FA and SA sensors are obtained, which shows they have the capability for surface roughness sensing but still lack stability. Enhanced by data fusion, the multimodal sensing system can achieve an improved accuracy of 92.00%.

To comprehensively demonstrate the advancement of the proposed multimodal e-skin, measurements for all the other essential parameters for an item are performed, including the hardness, materials, and surface temperatures. An average accuracy of 99.58% is achieved for identifying eight different materials (Kapton, Aluminum, Copper, PVC, PTFE, Glass, Cloth, and PET). The testing materials are all attached to a 3D-printed platform to control the same surface roughness, hardness, and temperature. An average accuracy of 98.00% is achieved for identifying eight

different hardness (6HC, 27HC, 43HC, 70HC, 48HA, 83HA, 66HD, and 85HD). For each sample, the testing material is controlled as aluminum with the same surface roughness and temperature. At last, an average accuracy of 98.50% is achieved for identifying eight different temperatures (0°C, 15°C, 30°C, 45°C, 60°C, 75°C, 90°C, 105°C) with the multimodal e-skin composed by SA sensor and temperature sensor. Similarly, all the other parameters are controlled the same.

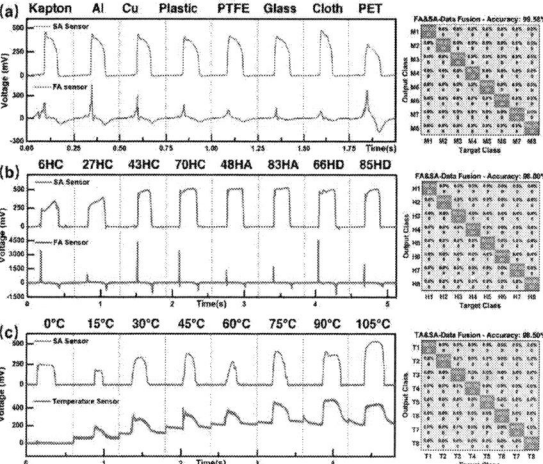

Figure 5: (a) Materials, (b) hardness, and (c) temperature sensing by the proposed multimodal e-skin with data fusion

CONCLUSION

In summary, a self-generated e-skin with multimodal sensors is proposed in this work. Triboelectric and potentiometric principles are applied to design sensors mimicking the FA and SA mechanoreceptors and thermoreceptors in human skins. Basically, they have the ability to endow the vibration, pressure, and temperature sensing functions to the robotics. With introducing the artificial intelligence, the information they encode can be further used to identify surface texture, roughness, hardness, materials, and temperature. Furthermore, enhanced by the multimodality and the data fusion, their features can be fully extracted and utilized to overcome the operation instability and achieve much higher accuracy for all the basic parameters for an item. The results show that the proposed multimodal e-skin is a promising sensing platform for humanoid robotics in applications like manufacturing, monitoring, assistance, etc.

ACKNOWLEDGEMENTS

This work was supported by the research grant of RIE Advanced Manufacturing and Engineering (AME) programmatic grant A18A4b0055 "Nanosystems at the Edge" at NUS, Singapore.

REFERENCES

[1] V. E. Abraira and D. D. Ginty, "The Sensory Neurons of Touch," Neuron, vol. 79, no. 4, pp. 618–639, Aug. 2013, doi: 10.1016/j.neuron.2013.07.051.

[2] M. Campero and H. Bostock, "Unmyelinated afferent s in human skin and their responsiveness to low temp

erature," Neurosci. Lett., vol. 470, no. 3, pp. 188–192, Feb. 2010, doi: 10.1016/j.neulet.2009.06.089.

[3] M. Zhu, Z. Sun, T. Chen, and C. Lee, "Low cost exoskeleton manipulator using bidirectional triboelectric sensors enhanced multiple degree of freedom sensory system," Nat. Commun., vol. 12, no. 1, p. 2692, Dec. 2021, doi: 10.1038/s41467-021-23020-3.

[4] Y. Yang, Q. Shi, Z. Zhang, X. Shan, B. Salam, and C. Lee, "Robust triboelectric information-mat enhanced by multi-modality deep learning for smart home," InfoMat, no. June, pp. 1–22, Aug. 2022, doi: 10.1002/inf2.12360.

[5] Z. Zhang et al., "Artificial intelligence of toilet (AI-Toilet) for an integrated health monitoring system (IHMS) using smart triboelectric pressure sensors and image sensor," Nano Energy, vol. 90, no. PA, p. 106517, Dec. 2021, doi: 10.1016/j.nanoen.2021.106517.

[6] M. Zhu et al., "Haptic-feedback smart glove as a creative human-machine interface (HMI) for virtual/augmented reality applications," Sci. Adv., vol. 6, no. 19, p. eaaz8693, May 2020, doi: 10.1126/sciadv.aaz8693.

[7] M. Zhu, Z. Sun, and C. Lee, "Soft Modular Glove with Multimodal Sensing and Augmented Haptic Feedback Enabled by Materials' Multifunctionalities," ACS Nano, vol. 16, no. 9, pp. 14097–14110, Sep. 2022, doi: 10.1021/acsnano.2c04043.

[8] Z. Zhang, F. Wen, Z. Sun, X. Guo, T. He, and C. Lee, "Artificial Intelligence-Enabled Sensing Technologies in the 5G/Internet of Things Era: From Virtual Reality/Augmented Reality to the Digital Twin," Adv. Intell. Syst., vol. 4, no. 7, p. 2100228, Jul. 2022, doi: 10.1002/aisy.202100228.

[9] S. Gao, T. He, Z. Zhang, H. Ao, H. Jiang, and C. Lee, "A Motion Capturing and Energy Harvesting Hybridized Lower-Limb System for Rehabilitation and Sports Applications," Adv. Sci., vol. 8, no. 20, p. 2101834, Oct. 2021, doi: 10.1002/advs.202101834.

[10] F. Wen et al., "Machine Learning Glove Using Self-Powered Conductive Superhydrophobic Triboelectric Textile for Gesture Recognition in VR/AR Applications," Adv. Sci., vol. 7, no. 14, p. 2000261, Jul. 2020, doi: 10.1002/advs.202000261.

[11] Z. Sun et al., "Artificial Intelligence of Things (AIoT) Enabled Virtual Shop Applications Using Self-Powered Sensor Enhanced Soft Robotic Manipulator," Adv. Sci., vol. 8, no. 14, p. 2100230, Jul. 2021, doi: 10.1002/advs.202100230.

[12] M. Zhu, T. He, and C. Lee, "Technologies toward next generation human machine interfaces: From machine learning enhanced tactile sensing to neuromorphic sensory systems," Appl. Phys. Rev., vol. 7, no. 3, p. 031305, Sep. 2020, doi: 10.1063/5.0016485.

[13] F. Wen, T. He, H. Liu, H.-Y. Chen, T. Zhang, and C. Lee, "Advances in chemical sensing technology for enabling the next-generation self-sustainable integrated wearable system in the IoT era," Nano Energy, vol. 78, no. July, p. 105155, Dec. 2020, doi: 10.1016/j.nanoen.2020.105155.

[14] Q. Shi, Z. Sun, Z. Zhang, and C. Lee, "Triboelectric Nanogenerators and Hybridized Systems for Enabling Next-Generation IoT Applications," Research, vol. 2021, pp. 1–30, Feb. 2021, doi: 10.34133/2021/6849171

.

[15] B. Dong et al., "Wearable Triboelectric–Human–Machine Interface (THMI) Using Robust Nanophotonic Readout," ACS Nano, vol. 14, no. 7, pp. 8915–8930, Jul. 2020, doi: 10.1021/acsnano.0c03728.

[16] F. Wen, Z. Zhang, T. He, and C. Lee, "AI enabled sign language recognition and VR space bidirectional communication using triboelectric smart glove," Nat. Commun., vol. 12, no. 1, p. 5378, Dec. 2021, doi: 10.1038/s41467-021-25637-w.

[17] M. Zhu, Z. Yi, B. Yang, and C. Lee, "Making use of nanoenergy from human – Nanogenerator and self-powered sensor enabled sustainable wireless IoT sensory systems," Nano Today, vol. 36, no. 800, p. 101016, Feb. 2021, doi: 10.1016/j.nantod.2020.101016.

[18] X. Le, Q. Shi, Z. Sun, J. Xie, and C. Lee, "Noncontact Human–Machine Interface Using Complementary Information Fusion Based on MEMS and Triboelectric Sensors," Adv. Sci., vol. 9, no. 21, p. 2201056, Jul. 2022, doi: 10.1002/advs.202201056.

[19] Z. Zhang et al., "Deep learning-enabled triboelectric smart socks for IoT-based gait analysis and VR applications," npj Flex. Electron., vol. 4, no. 1, p. 29, Dec. 2020, doi: 10.1038/s41528-020-00092-7.

[20] X. Guo et al., "Artificial Intelligence-Enabled Caregiving Walking Stick Powered by Ultra-Low-Frequency Human Motion," ACS Nano, vol. 15, no. 12, pp. 19054–19069, Dec. 2021, doi: 10.1021/acsnano.1c04464.

[21] L. Liu, X. Guo, and C. Lee, "Promoting smart cities into the 5G era with multi-field Internet of Things (IoT) applications powered with advanced mechanical energy harvesters," Nano Energy, vol. 88, no. July, p. 106304, Oct. 2021, doi: 10.1016/j.nanoen.2021.106304.

[22] L. Liu, X. Guo, W. Liu, and C. Lee, "Recent Progress in the Energy Harvesting Technology—From Self-Powered Sensors to Self-Sustained IoT, and New Applications," Nanomaterials, vol. 11, no. 11, p. 2975, Nov. 2021, doi: 10.3390/nano11112975.

[23] Z. Sun, M. Zhu, X. Shan, and C. Lee, "Augmented tactile-perception and haptic-feedback rings as human-machine interfaces aiming for immersive interactions," Nat. Commun., vol. 13, no. 1, p. 5224, Sep. 2022, doi: 10.1038/s41467-022-32745-8.

[24] T. He, X. Guo, and C. Lee, "Flourishing energy harvesters for future body sensor network: from single to multiple energy sources," iScience, vol. 24, no. 1, p. 101934, Jan. 2021, doi: 10.1016/j.isci.2020.101934.

[25] X. Guo et al., "Technology evolution from micro-scale energy harvesters to nanogenerators," J. Micromechanics Microengineering, vol. 31, no. 9, p. 093002, Sep. 2021, doi: 10.1088/1361-6439/ac168e.

[26] X. Wu et al., "A potentiometric mechanotransduction mechanism for novel electronic skins," Sci. Adv., vol. 6, no. 30, pp. 1–11, Jul. 2020, doi: 10.1126/sciadv.aba1062.

CONTACT
*C. Lee; elelc@nus.edu.sg

MULTI-MEMS DIFFERENTIAL PRESSURE SENSOR ELEMENTS-BASED AIRFLOW SENSOR WITH NEURAL NETWORK MODEL

Kotaro Haneda[1], Kenei Matsudaira[1], and Hidetoshi Takahashi[1]
[1]Keio University, JAPAN

ABSTRACT

This paper reports a compact spherical airflow sensor using multi-MEMS differential pressure (DP) sensor elements. Three built-in MEMS sensors simultaneously measure the DP around the spherical housing structure so that the measured DPs are converted into 2D wind direction and speed. The sensor outputs are converted into wind direction and speed by neural network. We attached the calibrated sensor to a toy drone as a demonstration. Then, it was confirmed that the output corresponding to wind direction and speed was measured when a crosswind was applied during flight.

KEYWORDS

Airflow sensor, Differential pressure sensor, Machine learning, Neural network

INTRODUCTION

Machine learning is useful in many fields, such as the medical, financial, and manufacturing industries. This technique also thrives in introducing smart sensors and is used to perform complex analyses in MEMS sensors. The research region is various, including analysis of the robot posture, simulation of human activity, gas composition test, and so on [1–3].

Previously, MEMS DP sensors have been widely used for airflow measurement. As one of the measurement methods, DP sensors are built in the housings of cylinder and sphere types[4–7]. Then, they measure the wind direction and speed from the DP of the inlets in the housing surface. Furthermore, DP sensors have high sensitivity and high time resolution, making it possible to detect low-speed wind quickly. However, wind speed is easy to measure, whereas wind direction measurement has a problem. When the device is placed in the flow field, the streamline separates, and the vortex is generated, resulting in a non-monotonic pressure gradient[8]. This effect causes the detection accuracy to be significantly poorer in specific wind direction areas. On the other hand, since machine learning is useful for measuring non-independent and nonlinear phenomena, it is applicable for DP sensors to improve the accuracy of wind direction detection.

In this paper, we propose a spherical airflow sensor with a neural network model for high wind direction accuracy (Fig. *1*). MEMS piezoresistive cantilever-type DP sensor chips are used as the built-in sensing element to achieve high sensitivity. Since the MEMS sensor chip is significantly small, the entire device becomes compact. Multiple inlets are formed three-dimensionally on the spherical housing surface. One sensor component comprises one DP sensor element and one pair of inlets. Using three sets of these components realizes 2D wind direction and speed detection corresponding non-monotonic pressure distribution around the sphere housing.

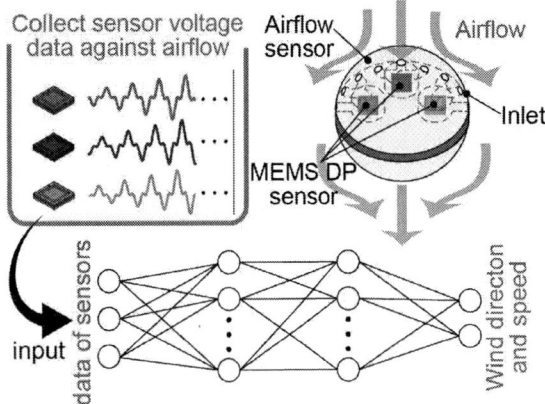

Fig. 1: Concept image of the proposed airflow sensor using a neural network model.

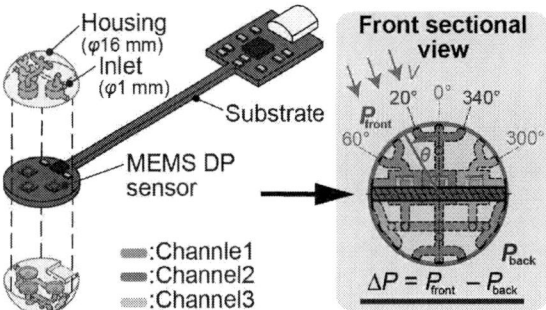

Fig. 2: Design diagram of the proposed airflow sensor.

We constructed a neural network model from the DP responses of the fabricated sensors to calibrate the wind direction and speed. Finally, the calibrated sensor was attached to a compact drone to demonstrate airflow measurement during flight.

DESIGN AND PRINCIPLE

Sensor Design

Fig. *2* shows the schematic diagram of the proposed airflow sensor. The sensor consists of two hemispherical housings and a substrate with three MEMS DP sensors. The spherical shape is developed by combining the housings with the circular substrate portion with the DP sensors. Air inlets are formed on the spherical surface, connecting to the corresponding inlets of the opposite side by each conduit. The MEMS DP sensors are located to block each conduit inside. Two pairs of three branched inlets, one channel, and one DP sensor comprise one channel. This channel is a total three. The airflow sensor redundantly measures the 2D wind direction and speed via the three DP sensor responses.

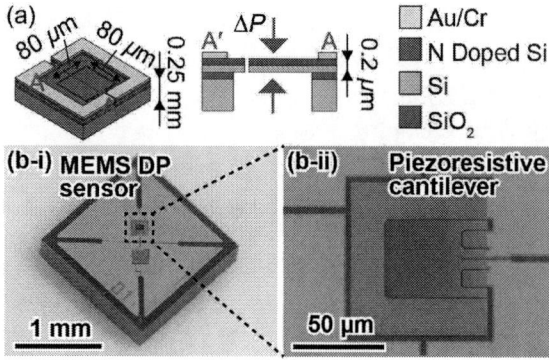

Fig. 3: (a) Schematic image of the piezoresistive cantilever. (b) Photograph of the (i) MEMS DP sensor and (ii) piezoresistive cantilever.

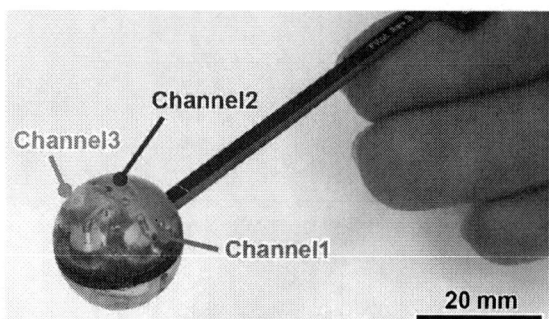

Fig. 4: Photograph of the fabricated airflow sensor.

Airflow Detection Principle

The detection principle of the proposed sensor is based on measuring DP. As shown in Fig. 2, it is assumed that the sphere symmetrizes by the middle inlet of channel2. Then, the wind direction θ is defined as 0°. When the airflow is applied from the θ direction, the DP ΔP is generated between each inlet pair and detected by the DP sensors. According to Bernoulli's equation, in the theoretical pressure distribution around a sphere with a constant wind direction, the DP is proportional to the wind speed square as

$$\Delta P \propto \frac{\rho v^2}{2} \qquad (1)$$

where ρ and v represent air density [kg/m^3] and wind speed [m/s], respectively. On the other hand, the ideal relationship between the DP and the wind direction is expressed by a sinusoidal equation as

$$\Delta P \propto C_p = 1 - \frac{1}{2}\sin^2\theta \qquad (2)$$

when the wind speed is constant. C_p represents the pressure drag coefficient, a one-to-one relationship with the angle. However, in Reynolds numbers from 10^3 to 10^4, which corresponds to the target sensor size and airflow range, a non-monotonic behavior occurs at $\theta = 60°$ to 120° and 240° to 300° [8]. Additionally, the wind direction and speed are not entirely independent because the pressure distribution relationship depends on the Reynolds number.

Based on these characteristics, we designed middle inlets of each channel at 60° intervals to measure the monotonic DP region in all directions. Moreover, three

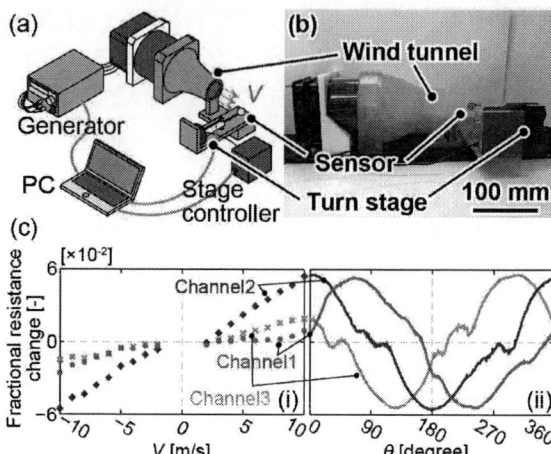

Fig. 5: (a) Concept diagram and (b) photograph of the wind tunnel experiment. (c)Relationship between the wind speed and direction, and the fractional resistance change when (i) θ is 0° and (ii) v is 10m/s.

inlets spaced at 20° intervals are connected to a single conduit to suppress the non-monotonic behavior [9]. Three sensor elements simultaneously measure the DP through the inlets when airflow is applied. The redundant measurement is realized since the wind direction and speed are converted from the three measured DPs.

FABRICATION

The sensor was developed by assembling three MEMS DP sensors, one substrate, and two hemispherical housings. The MEMS DP sensors were 1.5 mm × 1.5 mm × 0.25 mm in size. An 80 μm × 80 μm × 0.2 μm piezoresistive cantilever was formed on the center of the sensor chip as the sensing element [10]. Fig. 3 shows a schematic image and photograph of the cantilever.

Each DP sensor was fixed and wire-bonded to the substrate attachment port; the cantilever's position was aligned to the φ 0.45 mm penetrating hole of the attachment port. The housings were fabricated by a 3D printer (Form3, Formlabs). The inlet diameter on the spherical surface was approximately 1 mm. Fig. 4 shows a photograph of the developed sensor. The total size and weight became 85 mm and 4.3 g, respectively.

CALIBRATION USING NEURAL NETWORK

Data Collection

We conducted a wind tunnel experiment with the fabricated airflow sensor as shown in Fig. 5(a, b). A DC fan applied airflow with constant wind speed. The sensor was fixed on a turn stage. Then, the wind direction was controlled by rotating the stage. The wind speed was varied from 2 m/s to 10 m/s at 1 m/s interval. In addition, the rotation angle was changed from 0° to 359° at 1° interval. The wind direction was set as 0° when the middle inlet of channel2 was parallel to the airflow as defined in detection principal section. The sensor output, wind direction, and wind speed were measured simultaneously. First, the measurement was performed for 10 s with no airflow. Next,

978-1-6654-9309-3/23 $31.00 © 2023 IEEE

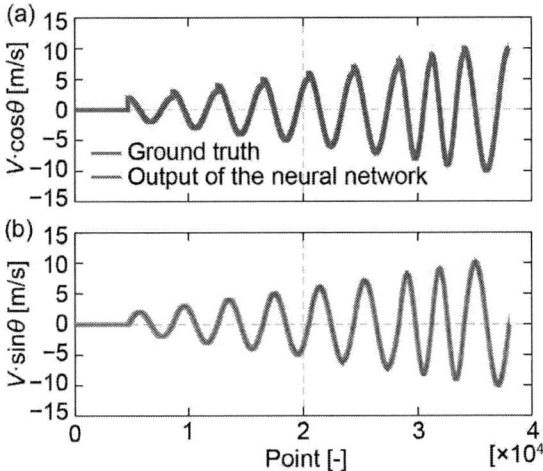

Fig. 6: Comparison learning results and ground truth of (a) $V \cdot \cos\theta$ and (b) $V \cdot \sin\theta$.

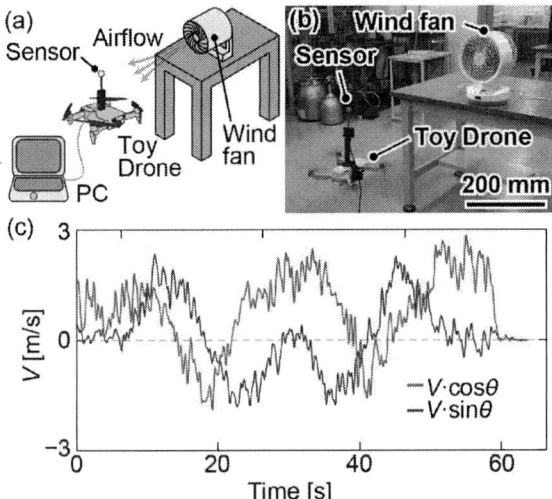

Fig. 7: (a) Concept diagram and (b) photograph of drone demonstration. (c) Measured $V \cdot \cos\theta$ and $V \cdot \sin\theta$ during drone flight.

the fan was turned on, and the measurements were conducted for 10 s at each wind direction and speed.

Fig. 5(c-i) shows the relationship between the wind speed and the fractional resistance change when channel2 was parallel to the wind direction. The relationship was linear for all three DP sensors. Similar results were obtained when channel1 and 3 were parallel to the airflow.

Fig. 5: (a) Concept diagram and (b) photograph of the wind tunnel experiment. (c)Relationship between the wind speed and direction, and the fractional resistance change when (i) θ is 0° and (ii) v is 10m/s.(c-ii) shows the relationship between the wind direction and the fractional resistance change in each channel when the wind speed was 10 m/s. It was observed that the response of each channel was approximately sinusoidal at each wind speed. Additionally, the non-monotonic pressure fluctuations were decreased at each channel, that region was 120° to

180° and 300° to 360° at channel 1, 60° to 120° and 240° to 300° at channel 2, and 0° to 60° and 180° to 240° at channel 3. This phenomenon appeared for other wind speeds.

Dataset And Training

The data count was a total of 252,000 samples. We also used the wind speed data of 0 m/s. The labels for each sensor output were $V \cdot \cos\theta$ and $V \cdot \sin\theta$, obtained from the wind speed V and wind direction θ at data collection. A 3-layer fully connected neural network was trained by these data.

Result For Test Data

The output results and ground truth of the neural network model for the test data are shown in Fig. 6. The noise of $V \cdot \cos\theta$ and $V \cdot \sin\theta$ for the output result of 0 m/s is 0.035 m/s and 0.0043 m/s, respectively, which is considered the noise level of this sensor when there is no wind. In total, the root mean square errors of the wind speed and angle were 0.24 m/s and 3.63°, respectively.

DRONE FROGHT DEMONSTRATION

We demonstrated the airflow measurement during drone flight (Fig. 7(a, b)). The fabricated airflow sensor was attached to a toy drone (DJI Mini 2, DJI). The attachment position was the center of the drone so that the sensor was not affected by wind turbulence from the propellers. The total weight of the sensor and jig was approximately 33 g, less than 20% of the drone. The experiment was conducted indoors, and a wind fan (YKAR-WEAD18, Yamazen) was placed on a desk 730 mm in height to ignore the ground effect. A hot wire anemometer (Climomaster, Kanomax) calibrated the spatial wind speed. We measured the sensor output via a cable. As a test procedure, we first moved the drone to a position where the drone was able to fly stably even when subjected to airflow, where the wind speed was 2 ~ 3 m/s. Then, the sensor response was measured when a drone rotated forward and counter on the spot for approximately 60 s.

Fig. 7(c) shows $V \cdot \cos\theta$ and $V \cdot \sin\theta$ calculated by the neural network from the sensor output. Both $V \cdot \cos\theta$ and $V \cdot \sin\theta$ had symmetrical sinusoidal responses with 33 s as the axis. In addition, the maximum wind speed became approximately 2.5 m/s. Thus, it was confirmed that the sensor could respond to the airflow current accurately.

CONCLUSION

We proposed the compact spherical airflow sensor using the neural network. The fabricated sensor was compact and lightweight, with a sphere diameter of 16 mm and a weight of 4.3 g. The wind tunnel experiment was conducted in all directions with a wind speed of 2-10 m/s. Then, a neural network model was built using the obtained data to derive the 2D wind direction and speed. Finally, we demonstrated that the toy drone with the fabricated sensor realized the crosswinds measurement during flight. It was confirmed that machine learning for DP sensors was useful for wind direction and speed measurement.

978-1-6654-9309-3/23 $31.00 © 2023 IEEE

ACKNOWLEDGEMENTS

This research was supported by the New Energy and Industrial Technology Development Organization (NEDO).

REFERENCES

[1] Nevlydov, I., Filipenko, O., Volkova, M., Ponomaryova, G., "MEMS-Based Inertial Sensor Signals and Machine Learning Methods for Classifying Robot Motion", *Proc 2018 IEEE 2nd Int Conf Data Stream Min Process DSMP 2018*, pp.13–16, 2018.

[2] Zhang, Y. S., Yang, T., "Modeling and compensation of MEMS gyroscope output data based on support vector machine", *Meas J Int Meas Confed*, vol.45-5, pp.922–926, 2012.

[3] Sitnik, S. P., "Нові Підходи Публічного Адміністрування Сфери Розширених Сімейних Відносин: Регіональний Аспект", *Theory Pract Public Adm*, vol.2-73, pp.86–92, 2021.

[4] Minh-Dung, N., Takahashi, H., Kuwana, K., Takahata, T., Matsumoto, K., Shimoyama, I., "3D airflow velocity vector sensor", *Proc IEEE Int Conf Micro Electro Mech Syst*, pp.513–516, 2011.

[5] Liu, C., Du, L., Zhao, Z., Fang, Z., Li, L., "A directional anemometer based on MEMS differential pressure sensors", *9th IEEE Int Conf Nano/Micro Eng Mol Syst IEEE-NEMS 2014*, vol.-April 2016, pp.517–520, 2014.

[6] Zhao, Z., Pan, Y., Zhao, R., Du, L., Fang, Z., Wu, H., Niu, X., "Design of a Wind Sensor Based on Cylinder Diametrical Pressure Differences for Boundary Layer Meteorological Observation", *Proceedings*, vol.2-13, p.1511, 2018.

[7] Haneda, K., Matsudaira, K., Noda, R., Nakata, T., Suzuki, S., Liu, H., Takahashi, H., "Compact Sphere-Shaped Airflow Vector Sensor Based on MEMS Differential Pressure Sensors", *Sensors*, vol.22-3, 2022.

[8] Hoerner, S.H., "Fluid - Dynamic Drag," 1965.

[9] Bruschi, P., Dei, M., Piotto, M., "A low-power 2-D wind sensor based on integrated flow meters", *IEEE Sens J*, vol.9-12, pp.1688–1696, 2009.

[10] Takahashi, H., Dung, N. M., Matsumoto, K., Shimoyama, I., "Differential pressure sensor using a piezoresistive cantilever", *J Micromechanics Microengineering*, vol.22-5, 2012.

CONTACT

*K. Haneda, tel: +81-070-1275-0056;
hane_taro@keio.jp

TRIAL-AND-ERROR LEARNING FOR MEMS STRUCTURAL DESIGN ENABLED BY DEEP REINFORCEMENT LEARNING

Fanping Sui[1,†], Wei Yue[1,†], Ziqi Zhang[2], Ruiqi Guo[1], and Liwei Lin[1,2]

[1]Department of Mechanical Engineering, University of California, Berkeley, USA
[2]Tsinghua-Berkeley Shenzhen Institute, Tsinghua University, Shenzhen, China
[†]Fanping Sui and Wei Yue contributed equally to this work.

ABSTRACT

We present a systematic MEMS structural design approach via a "trial-and-error" learning process by using the deep reinforcement learning framework. This scheme incorporates the feedback from each "trial" to obtain sophisticated strategies for MEMS design optimizations. Disk-shaped MEMS resonators are selected as case studies and three remarkable advancements have been realized: 1) accurate overall performance predictions (97.9%) via supervised learning models; 2) efficient MEMS structural optimizations to guarantee targeted structural properties with an excellent generation accuracy of 97.7%; and 3) superior design explorations to achieve one order of magnitude performance enhancement than the training dataset. As such, the proposed scheme could facilitate a wide spectrum of MEMS applications with this data-driven inverse design methodology.

KEYWORDS

Artificial Intelligence, MEMS Design, Design Space Exploration, Deep Reinforcement Learning.

INTRODUCTION

Artificial intelligence (AI) has shown prodigious success in solving complex real-world problems in interdisciplinary fields, such as Google's AlphaGo program [1], drug designing and development [2], material discoveries [3], protein engineering [4], and robotics [5]. In recent years, there has been an increasing interest in applying artificial intelligence to MEMS structural designs due to the complicated multi-physics coupling nature. The goal of such design methodologies is to improve or substitute the conventional time-consuming and compute-intensive design approaches such as finite element analysis (FEA) modeling. Several prior works have shown that artificial intelligence can be applied to predict the properties of MEMS structures accurately through deep neural networks [6, 7] and to customize MEMS devices based on targeted properties using conditional generative adversarial networks (CGAN) [8]. However, these methodologies are focusing on extracting underlying geometric features within the given training dataset and it is very challenging to explore and discover new designs with better performances out of the distributions of the training data.

On the other hand, it has been shown that humans can learn from the "trial-and-error" process [9]. Based on current knowledge, humans can develop strategies to make attempts that are most likely to result in success. By analyzing and summarizing the feedback obtained after each trial, humans can learn how to modify the strategies to improve the probabilities of success. Such a trial-and-error step can be repeated, and the corresponding knowledge is accumulated according to previous experiences. Finally, a sophisticated strategy can be established adaptively to handle the practical problems encountered by humans. Inspired by the "trial-and-error" scheme, a similar principle for MEMS structural design problems to explore the high-dimensional design space can be realized by using the deep reinforcement learning (DRL) algorithm.

In this work, case studies of disk-shaped MEMS resonators are used to demonstrate this trial-and-error-inspired methodology. Supervised learning (SL) is adopted to train several models, namely SL-based analyzers, which can predict the performance of arbitrary MEMS structures accurately (97.9%) and quickly (more than 10^4 times faster than FEA). Equipped with the SL-based analyzer, the DRL agent explores the design space efficiently and achieves a high generation accuracy of 97.7% based on prespecified targeted properties. The proposed DRL algorithm can also be used to find new MEMS designs with extreme physical properties that are out of the distribution of the training dataset for optimal performance, such as quality factors. Results show that MEMS resonators with remarkably high quality factors of one order of magnitude higher than those of both the training dataset and the CGAN approach [8] can be discovered through the proposed DRL scheme. Such methodology could be extended to other MEMS device design problems to open a new approach of using the DRL algorithm for data-driven inverse structural design.

SYSTEM ARCHITECTURE

The proposed DRL framework is illustrated in the flow chart as shown in Fig. 1. The step-by-step optimization strategy enabled by the DRL algorithm represents the core component of this framework. To overcome the low-sample-efficiency nature of reinforcement learning and accelerate the collection process of design-property pairs, supervised learning is utilized to train models that can capture the underlying essential physics for MEMS resonators and predict them accurately. The model is developed via deep residual neural networks as shown in the enclosed region in Fig. 1 and is adopted as the learning environment within which the reinforcement learning agent operates. Disk-shaped MEMS resonator devices [10] with three vibrational modes of interest as shown in Fig. 2 are chosen for case studies. MEMS resonator designs are translated into pixelated images with a resolution of 100 times 100 as free-form design representations while maintaining key geometric features. Over 100k cases of qualified pixelated images are initialized randomly from a topology generator with the depth-first-search (DFS) algorithm as the training data and topological constraints are always satisfied. Finite element analysis is used to numerically analyze these random designs for modal analysis to characterize their vibration responses and extract the associated parameters such as mode shapes, natural frequencies, and quality factors. The designs are subsequently labeled with the calculated vibrational characteristics. After sufficient training iterations in the form of the deep residual neural network, SL-based analyzers are obtained and they can be utilized to predict vibration responses with good accuracy and remarkably less amount of time.

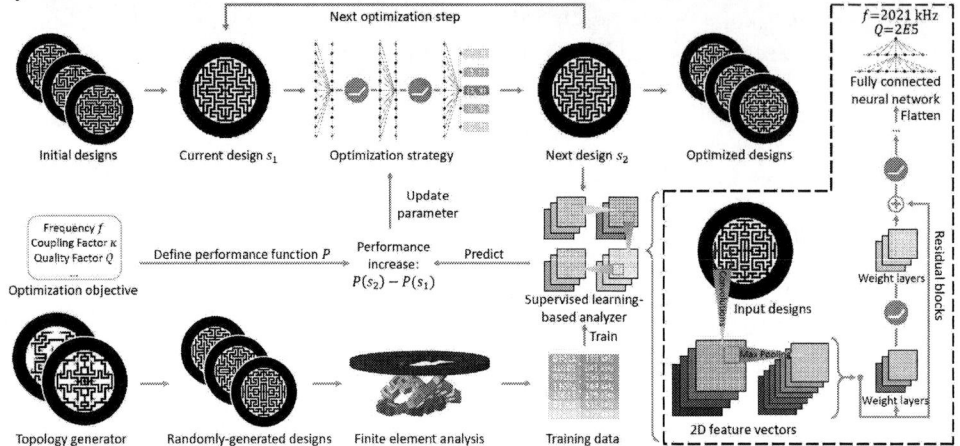

Figure 1: *The architecture of the reinforcement learning framework for MEMS structural designs. The region enclosed by dashed lines represents the architecture of the supervised learning-based analyzer to predict the properties of interest. After the design initialization, the proposed DRL agent starts to change the topologies of current designs by an optimization strategy to constitute new designs. The optimization strategy is updated by incorporating feedback (i.e., performance increases according to the optimization objectives, which are predicted from supervised learning-based analyzers) for implementing the design changes. The new optimization strategy is applied to new designs in the next cycle.*

With the fast prediction obtained from the SL-based analyzer, the DRL agent utilizes an optimization strategy (represented by the deep neural networks) to constitute new designs to achieve the optimization objectives through a step-by-step, trial-and-error manner. The optimization objective can be specified according to certain design tasks and a corresponding performance function that scores every design candidate from the whole design space can be subsequently determined by analyzing the relative distance between the current design and the desired designs with targeted performance. The reward criterion is defined as the performance value improvement after applying the design modification from the optimization strategy at the pixel level.

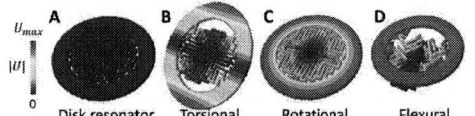

Figure 2: *A) Geometry and B-D) three vibrational modes of interest of a representative MEMS disk resonator.*

The operation pipeline for our reinforcement learning framework can be summarized as follows: In the first step, the MEMS resonator designs are initialized randomly to feed into the trial-and-error processes. Secondly, given the initial design as the current observation, the agent will change the properties (i.e. solid or void) of one set of specific pixels determined by the current optimization strategy to constitute a newly generated design as the next observation. Next, the newly created design is processed through a pre-trained SL-based analyzer to predict the overall performance of targeted properties. Afterward, the optimization strategy for the next iteration is updated by incorporating the reward signal (i.e., performance increment) from the environment as the feedback for implementing the previous modification decision. After several training iterations, a sophisticated optimization strategy would be produced, from which optimized designs with improved properties would be discovered after sufficient optimization steps.

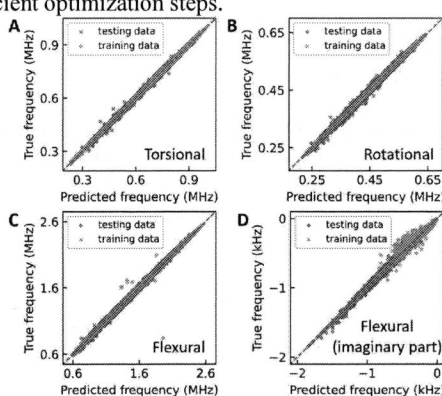

Figure 3: *Predicted frequencies derived by the SL-based analyzer with respect to true frequencies for A) torsional mode, B) rotational mode, C) flexural mode, and D) the imaginary part of the frequency for the flexural mode.*

RESULTS AND DISCUSSION

The SL-based analyzers are examined to predict frequencies (real part) for three modes of interest and the imaginary part of the frequency of the flexural mode for the disk-shaped MEMS resonators. The FEA simulation results are treated as ground truths and are compared with predictions from the SL-based analyzers. The accuracy is defined as how close the agreements between simulation results and the residual neural net outputs are. It is observed that all resulting points are located extremely close to the 45-degree line as shown in Fig. 3, indicating that high consistency has been achieved. Table 1 also shows that prediction results highly agree with FEA simulations with an averaged accuracy of 97.9%. For model validation purposes, SL-based analyzers are tested to show great agreements of ~99% with a previously published work [10]. Additionally, the SL-based analyzers are about 4×10^4 faster than that of the traditional FEA approaches, which is consistent with our previous results [6].

Table 1: The performance of the SL-based analyzers.

	Training error	Testing error	Accuracy
Torsional	0.2%	0.8%	99.2%
Rotational	0.1%	0.9%	99.1%
Flexural	0.5%	1.1%	98.9%
Flexural (Im)	2.3%	5.6%	94.4%

Two optimization objectives are further tested to demonstrate the feasibility of the DRL framework to discover MEMS structures with: 1) the multiple targeted modal frequencies (e.g., 0.51 MHz in rotational mode and 2.02 MHz in flexural mode); and 2) the highest quality factor in terms of anchor loss (e.g., in the flexural mode).

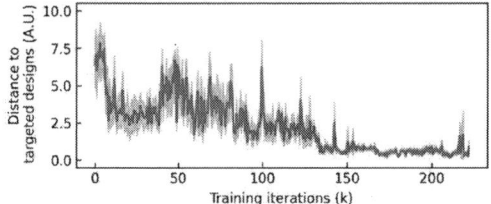

Figure 4: The relative distance to the targeted design with multiple desirable properties versus the training iterations. The red shaded areas represent the uncertainties of generated designs.

For the first optimization objective, it is important to show that the DRL agent can gradually learn how to find the desirable designs through the "trial-and-error" process by minimizing the relative distance to the targeted point. The relative distance, d, is defined as

$$d = \left(f_{flex} - f_{flex}^*\right)^2 + \left(f_{rot} - f_{rot}^*\right)^2 \quad (1)$$

where f_{flex} and f_{rot} represent the frequencies of flexural and rotational modes of the current design, respectively, and f_{flex}^* and f_{rot}^* represent the targeted frequencies of flexural and rotational modes of the desired design, respectively. The performance function is defined to be the negative of the relative distance to the targeted value. Afterward, the DRL agent explores non-trivial decisions for performance improvements in the global design space. After adequate parameter updating steps, a desirable optimization strategy is established which can put forward a series of effective modification steps toward the targeted natural frequencies (0.51 MHz, 2.02 MHz in this example). Figure 4 shows the learning curve of the optimization strategy for modifying designs toward the combination of targeted natural frequencies. The trained optimization strategy is evaluated by measuring the averaged deviation between the natural frequencies of generated designs to that of the targeted designs. Initially, the optimization strategy is far from effective and precise as it always has large fluctuations during the learning process. Gradually, the DRL agent discovers the underlying pattern behind the optimization problem and finally converges to a stable model that can minimize the relative distance to the desired design. Figure 5A shows the distribution evolutions of optimized designs after each optimization step in terms of two targeted variables. Initially, the pre-optimized designs are randomly produced such that they spread over a wide range of natural frequencies with a mean of (381.27 kHz, 1090.98 kHz) and a standard deviation of (60.80 kHz, 229.75 kHz). After applying the first optimization step, the resultant data points advance collectively towards the targeted point with a mean of (399.73 kHz, 1198.27 kHz) and a reduced standard deviation of (58.49 kHz, 261.66 kHz). By the same token, the corresponding mean values will further approach the targeted values and the data points will become less dispersed for every step. After being modified by three successive steps, the optimized designs start to be concentrated near the targeted values. With a seven-step optimization process, the majority of data points reside within the neighboring zone of the targeted location while some outliers that may contain some impeditive features remained to be further improved. After a 10-step optimization process, nearly all data points congregate within a tiny region centered at (513.88 kHz, 2030.90 kHz) with a standard deviation of (12.73 kHz, 42.20 kHz) which represents a high generation accuracy of 97.7%. This step-by-step optimization process verifies the effectiveness of every decision that our powerful DRL-based optimization strategy put forward at each step and further validates the feasibility of deep RL approaches for supervising the design process. A representative design satisfying the objectives of the targeted combination of natural frequencies is exhibited in Fig. 5A. Figure 5B shows the FEA simulation results of the optimized design. The calculated natural frequencies are very close to the targeted combination at 510.08 kHz and 2020.71 kHz, indicating our DRL approach is very effective at precisely achieving the targeted optimal designs.

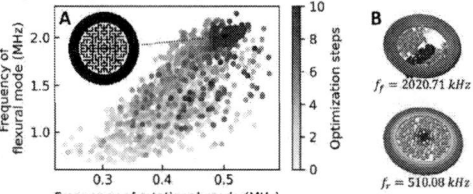

Figure 5: A) The optimization steps (color map) for multiple targeted frequencies (the flexural and rotational modes) and an optimized design example. By applying the optimization strategy, the generated designs collectively move toward the targeted points after each optimization step to result in high generation accuracy of 97.7% after 10 steps. B) Flexural (f_f) and rotational (f_r) mode frequencies and the shapes of the optimized design.

It is also important to show that the proposed DRL framework can be used to discover designs with high quality factors. High quality factor is highly desired for MEMS resonator devices since it represents the low dissipation rate of energy for high efficiency. The quality factor, Q, can be calculated as $Q = -f_r/2f_i$, where f_r is the real part of frequency (i.e. natural frequency), f_i is the corresponding imaginary part. In this study, design optimization of the flexural-mode quality factor is utilized to illustrate the essential ideas of the proposed methodology. The performance function is defined as the predicted quality factor based on SL-based analyzers. Figure 6 shows the learning process curve for the quality factor versus the training iterations, in which the DRL agents gradually produce high-performance designs with the resultant average quality factor at the order of 10^5.

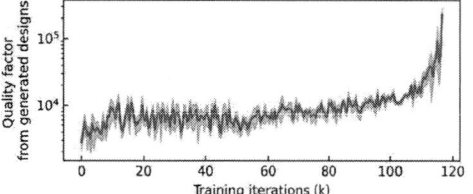

Figure 6: The obtained quality factor versus training iteration plot showing the optimization process. The DRL agent can learn the essential strategy to optimize quality factors which are progressively enhanced with respect to training iterations. The red shaded areas represent the uncertainties of generated designs.

Figure 7 shows the statistic distribution comparison between randomly generated designs and the high-performance designs produced from the RL algorithm. Most of the original designs reside in the lower performance range with the mean value of quality factor around 1.77×10^4, whereas most of the optimized structure designs obtained by our DRL method reside in the higher performance range, whose mean quality factor can be as high as 2.54×10^5, representing more than one magnitude performance improvement against the initial random designs and the CGAN scheme [8]. This shows that the DRL algorithm has indeed learned the underlying patterns of the top-ranked designs as well as how to improve them in an effective manner. As an application of our high-performance DRL algorithm, the trained neural networks can be served as a top-performed design generator, with much lower computational costs than those of traditional exhaustive approaches.

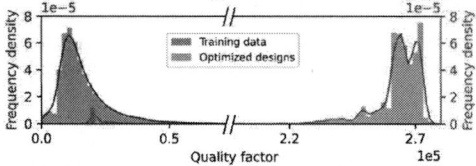

Figure 7: Performance distributions of training data and optimized designs with the high-quality factor objective. The performance of optimized designs can be one order of magnitude higher than that of the training data on average.

CONCLUSION

The framework for using deep reinforcement learning to optimize MEMS structural design in a free-form manner has been proposed and demonstrated. With a sufficient number of training iterations, the proposed DRL-based design methodology can successfully generate MEMS circular disk resonator designs with natural frequencies close to targeted frequencies for multiple modes and with

small natural frequency standard deviations. Furthermore, the quality factors can be successfully optimized, being more than one order of magnitude larger than the original randomized resonator designs. The results show great promise in demonstrating that DRL algorithms can be considerably helpful in designing MEMS devices that satisfy all design constraints and parameter requirements while being extremely time and energy efficient. We believe that with reasonable modifications, a similar approach can be developed for the automated design and optimization of other types of MEMS devices in the future.

ACKNOWLEDGEMENTS

The authors appreciate the helpful discussions with Prof. Grace X. Gu, Dr. Renxiao Xu, and Mr. Mingxin Jia.

REFERENCES

[1] J. X. Chen, "The Evolution of Computing: AlphaGo", *Computing in Science & Engineering*, vol. 18, pp. 4-7, 2016.

[2] R. Gupta, D. Srivastava, M. Sahu, S. Tiwari, R. K. Ambasta, and P. Kumar, "Artificial intelligence to deep learning: machine intelligence approach for drug discovery", *Molecular Diversity*, vol. 25, pp. 1315-1360, 2021.

[3] F. Sui, R. Guo, Z. Zhang, G. X. Gu, and L. Lin, "Deep Reinforcement Learning for Digital Materials Design", *ACS Materials Letters*, vol. 3, pp. 1433-1439, 2021.

[4] J. Jumper *et al.*, "Highly accurate protein structure prediction with AlphaFold", *Nature*, vol. 596, pp. 583-589, 2021.

[5] L. Kunze, N. Hawes, T. Duckett, M. Hanheide, and T. Krajnik, "Artificial Intelligence for Long-Term Robot Autonomy: A Survey", *IEEE Robotics and Automation Letters*, vol. 3, pp. 4023-4030, 2018.

[6] R. Guo *et al.*, "Deep learning for non-parameterized MEMS structural design", *Microsystems & Nanoengineering*, vol. 8, pp. 1-10, 2022.

[7] Q. Li *et al.*, "A Novel High-Speed and High-Accuracy Mathematical Modeling Method of Complex MEMS Resonator Structures Based on the Multilayer Perceptron Neural Network", *Micromachines*, vol. 12, p. 1313, 2021.

[8] F. Sui, R. Guo, W. Yue, K. Behrouzi, and L. Lin, "Customizing Mems Designs via Conditional Generative Adversarial Networks", in *2022 IEEE 35th International Conference on Micro Electro Mechanical Systems (MEMS)*, Virtual, January 9-13, 2022, pp. 450-453.

[9] H. P. Young, "Learning by trial and error", *Games and Economic Behavior*, vol. 65, pp. 626–643, 2009.

[10] F. Sui, W. Yue, R. Guo, K. Behrouzi, and L. Lin, "Designing Weakly Coupled Mems Resonators with Machine Learning-Based Method", in *2022 IEEE 35th International Conference on Micro Electro Mechanical Systems (MEMS)*, Virtual, January 9-13, 2022, pp. 454-457.

CONTACT

*F. Sui; +1-341-333-9262; fpsui@berkeley.edu

*W. Yue; +1-510-984-8328; wei_yue@berkeley.edu

FULLY MICROELECTROMECHANICAL NON-VOLATILE MEMORY CELL

Elliott Worsey, Mukesh K. Kulsreshath, Qi Tang, and Dinesh Pamunuwa
University of Bristol, U.K.

ABSTRACT

This paper reports the first non-volatile memory cell composed entirely of MEM relays that does not need CMOS circuitry for reading or writing. The cell uses a 7-terminal non-volatile relay for information storage and two 3-terminal relays for read and write access. This three-relay cell has been fabricated and characterised to demonstrate read and write operations. The cell architecture has been designed such that multiple cells can be tiled in combination with a relay-based multiplexer to make a digital all MEM reprogrammable non-volatile memory. Such a memory will have near zero standby power and ability to operate in harsh environments beyond the capabilities of conventional memory technologies.

KEYWORDS

MEMS, Microswitches, Nonvolatile memory, Radiation hardening (electronics), High-temperature

INTRODUCTION

Data storage in harsh environmental conditions characterised by high temperatures and high radiation levels is a requirement in many applications ranging from automotive and manufacturing to space and nuclear decommissioning. To cope with these conditions, conventional CMOS and Flash technologies require additional cooling, special packaging and custom design techniques [1]–[3]. While emerging MRAM and ReRAM memory technologies show potential for integrated non-volatile memory, they are still susceptible to radiation upsets and have similar or worse high temperature resilience as compared to CMOS [2], [4]. Further, they have unresolved development and integration challenges and depend on CMOS driving circuitry. By contrast, MEM-based data storage offers the promise of robust operation at high-temperatures as well as high-radiation levels with near zero standby power across all environmental conditions. However, to date, non-volatile memories either incorporate CMOS and charge storage [5] or focus on singular devices for proof of concept [6]–[9]. By contrast, this work uses a nonvolatile 7-terminal (7-T) rotational relay as the storage device [8], [9], and combines it with two 3-terminal (3-T) relays [10] to produce the first all MEM storage cell including read and write circuitry. As this cell does not depend on stored charge or CMOS addressing circuitry, it can be tiled with the aid of a relay-based multiplexer [11] to produce a fully MEM, addressable memory that consumes near zero standby power and is resilient to high temperatures and high levels of radiation.

DESIGN

Our memory cell uses a 7-T non-volatile rotational relay [8] as the storage element. The straight section of the relay beam (source) is anchored via a short soft hinge that allows the entire suspended beam to rotate anticlockwise to drain D1 by actuating gate pair PG1/AG1, or clockwise to drain D2 by actuating gate pair PG2/AG2 (see Fig. 1). Once rotated, the beam stays switched through surface adhesion forces and can be reprogrammed by actuating the opposing gate pair. The relative balance between the adhesion force, elastic force in the hinge and applied electric force dictates the voltages for rotating from neutral and switched states [9]. A logic '1' is stored by rotating the beam to drain D1, and a logic '0' by rotating it to D2. We use two conventional 3-terminal (3-T) relays [10] where the beam pulls out when the actuation voltage is removed, i.e. volatile behaviour, to enable digital reading and writing as explained below. Having this access circuitry contained within the cell itself allows for simplified control through a read/write line and a data-in signal.

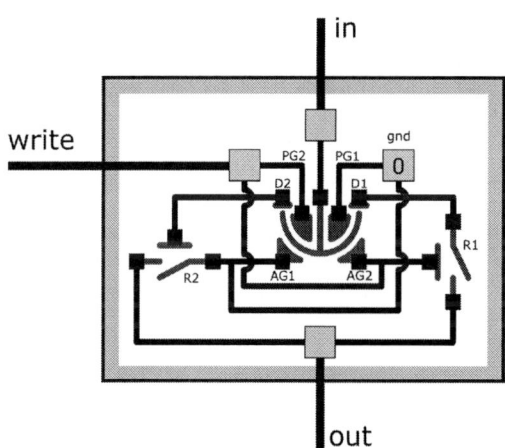

Figure 1: Schematic view of the proposed memory cell architecture consisting of one 7-T device [8] and two 3-T devices [10]. Two logical inputs ("in", "write") and one output ("out") are used for writing and reading the cell. The cell is programmed by switching the central 7-T relay to drain D2 to store a '0', or to drain D1 to store a '1'.

To store a value in the cell, "write" is asserted high, and the value to be stored is presented at "in". The potential difference between the gate pairs and the movable beam causes the beam to either rotate clockwise to contact the drain D2 (if a '0' is present on "in") or anticlockwise to drain D1 (if a '1' is present on "in"). The signal values for writing a '0' and '1' are shown in Fig. 2, top, where the grey cells labelled LUT (short for look up table) cell represent the circuit of Fig. 1. After contact, surface adhesion forces overcome the restorative spring force in the hinge to ensure non-volatility [9] until reprogramming occurs.

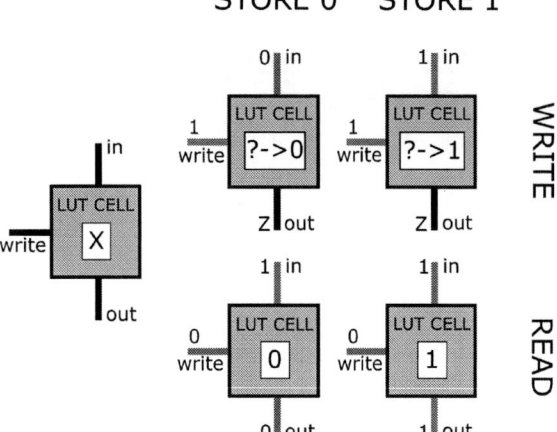

Figure 2: Operating principle of the memory cell showing signals required to read or write a value. Writing a value occurs by asserting "write" and applying the value to "in". Reading is achieved by applying a '1' to "in" and deasserting "write", after which the stored value will present at "out".

To read a stored value, "in" is driven high and "write" driven low, ensuring that all gates of the 7-T relay are grounded so that the clockwise and anticlockwise moments cancel, and the state remains unchanged (i.e. a read upset does not occur). The logic high on "in" propagates through the beam of the 7-T relay to a 3-T relay determined by the stored value. If a '1' was stored the beam of relay R2 in Fig. 1 is driven high, while its gate is driven low via "write". Thus, it pulls in and transmits the stored '1' to "out". Alternatively, if a '0' was stored, the logic high on "in" drives the gate of relay R1, which causes it to pull in as its beam is grounded, propagating a '0' to "out". The signal values for reading a stored '0' and '1' are shown in Fig. 2, bottom.

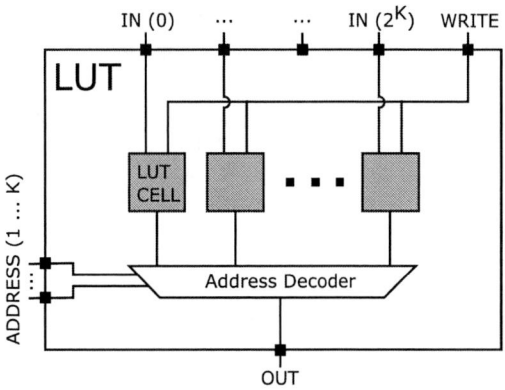

Figure 3: Combining a tiled array of memory cells with a multiplexer [11] to produce a look-up-table memory. All cells are simultaneously written to or read from through the shared "write" signal, with the multiplexer selecting a single value from the array.

This cell architecture has been designed such that multiple cells can be combined to provide a larger MEM non-volatile memory as shown in Fig. 3. Here, the grey cells represent the memory cell of Fig. 1, and all of the cells have a common "write" signal, so that all cells are written to or read from at the same time. With the use of a relay-based multiplexer (such as, for example, demonstrated in [11]) an exclusively MEM relay-based addressing scheme can easily be implemented, preserving the harsh-environment capability and zero standby power of the entire memory.

FABRICATION

The cell was fabricated with critical dimension of 2 μm on a silicon-on-insulator (SOI) chip with device and buried oxide (BOX) layer thicknesses of 5 μm and 4 μm respectively. E-beam lithography followed by deep reactive ion etching was used to pattern the silicon device layer. The moving parts were suspended by etching the BOX using vapour phase HF. Finally, a layer of gold was thermally evaporated to act as the relay contact material and provide a metal layer on top of the probe pads. Two additional pads were introduced into the design to compensate for the lack of a multi-layer interconnect stack, allowing each primary and auxiliary gate-pair connection to be made off-chip. An SEM of a cell after the silicon etch step can be seen in Fig. 4, showing the e-beam written pattern has been cleanly transferred to the silicon device layer.

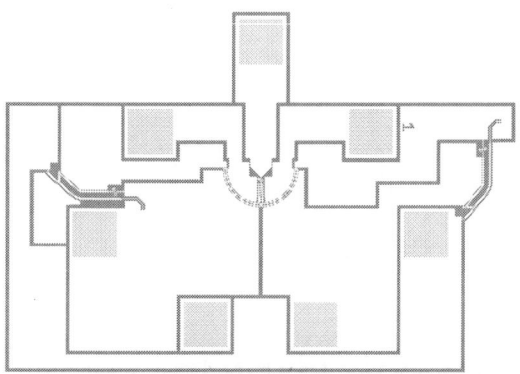

Figure 4: Original GDS design with pad locations marked (top) and SEM (bottom) of the fabricated memory cell before suspension. The circuit is patterned with ebeam lithography on an SOI wafer with 5 μm device layer and 4 μm BOX layer.

RESULTS

Electrical testing was carried out using a measurement setup equipped with six probes and four source measurement unit (SMU) channels. Each probe was placed on one of the designated pads "in", "out", "write", and "gnd" to either drive it to the requisite voltage or measure the voltage and current. To characterise the cell, first a '0' was written and subsequently a '1'. Due to the specific device layer and BOX layer thicknesses of the wafer used for prototyping (5 μm and 4 μm respectively) and the critical dimension used in the design (2 μm), we saw out-of-plane bending caused by the potential difference across the air gap between the substrate and the beam, when the beam was driven to the pull-in voltage. To minimise the risk of the beams collapsing onto the substrate, we modified the applied voltage patterns, taking advantage of the ambipolar nature of electrostatically actuated relays. Therefore, in order to write a '0', i.e. rotate the 7-T beam to drain D2, "in" was driven to '1' and gate pair 1 was driven to '0' as shown in Fig. 5a, top. At the same time, a bias was applied to the drain of relay R2, and its source grounded. After the 7-T relay completes switching and its beam contacts D2, the gate of relay R2 is driven high, and it pulls in. This event is shown in Fig. 5b, where the voltage on the x-axis is the value applied to the source of the 7-T. Next, the gate voltages were maintained and the voltage on the 7-T relay source ramped down, when the current continued to flow, demonstrating its non-volatile behaviour. The 3-T relay, though designed to be volatile, remained switched after its gate voltage was reduced initially. It was observed to eventually pull out after two days with the gate actuation removed. Following the initial storage of a '0' the cell was reprogrammed to '1', again by applying a modified pattern to avoid out-of-plane collapse, as shown in Fig. 6a, top. This operation was monitored via relay R1, in a similar fashion to the write '0' operation, shown in Fig. 6b. The pull-in was higher as only the principle gate (PG1) and not the auxiliary gate (AG1) of the 7-T relay was driven, due to the limitations imposed by a single layer of wiring and maximum of four SMU channels in our measurement setup. These reprogramming cycles were repeated several times before the gold contact failed and the current compliance of 10 nA could no longer be reached.

DISCUSSION

In this work we proposed a MEM only non-volatile memory cell combining two types of in-plane relays, a rotational 7-T relay as the storage device, and two 3-T relays for driving it. The cell was designed to be tiled to realise a MEM only memory using a standard word-line / bit-line arrangement. The main challenge we faced was caused by out-of-plane bending due to comparable in-plane and out-of-plane forces when a voltage was applied to the suspended structures. Nevertheless, we achieved proof of concept of the cell operation by utilising a modified access pattern, showing that MEM only memories without the need for CMOS drive circuitry or charge trapping schemes for information storage can be achieved. We expect to resolve the out-of-plane bending by ensuring the BOX layer thickness and critical dimension are such that there is an order of magnitude difference in the in-plane and out-of-plane forces for more robust operation.

Further challenges were encountered with the Au contact layer: high contact resistance; low number of switching cycles; and unpredictability of the surface adhesion force causing variation in switching event behaviour. We

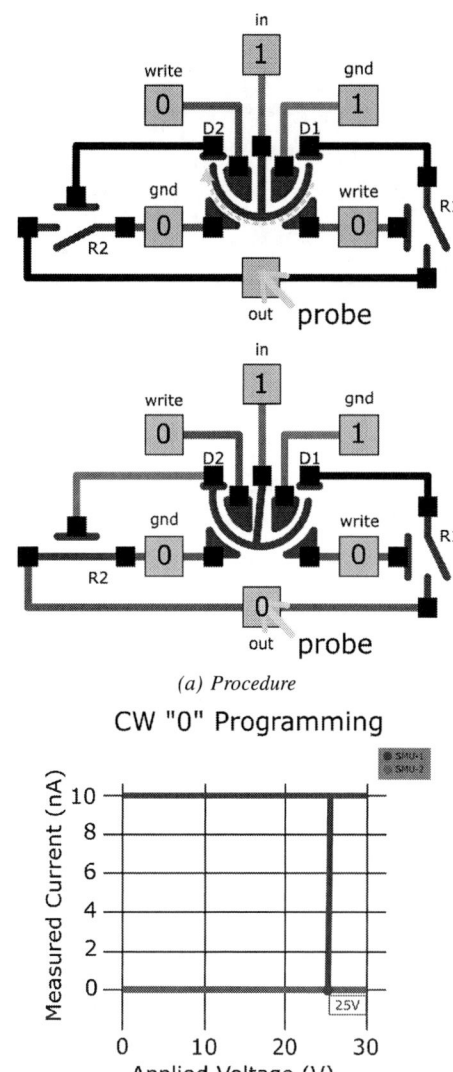

(a) Procedure

CW "0" Programming

(b) Characterisation

Figure 5: a) Probing procedure to validate functionality of the LUT memory cell being written to '0'. "In" is driven high, which along with "write" deasserted causes the 7-T to rotate clockwise to contact D2 (top). This state connects "in" to the gate of R2 to make it pull-in and allow a current to flow between "out" and "gnd" (bottom).
b) I-V characterisation ramps up the voltage at "in" to rotate the 7-T clockwise, close relay R2 and enable current to flow between the bottom left "gnd" pad and the "out" pad. Current starts to flow once the pull-in voltage for the 7-T is reached, and continues throughout ramp-down due to the nonvolatile nature of the devices.

expect the number of cycles to increase with a more suitable contact layer such as NCG [10], while a hermetic packaging solution will eliminate contaminants and reduce cycle-to-cycle variation in stiction. Finally, the lack of any additional routing layers can be addressed by a heterogeneous integration scheme [12]. With such an approach, MEM-based logic can be integrated with non-volatile memory on a single chip to provide processing alongside storage. Thus, this work has potential to yield

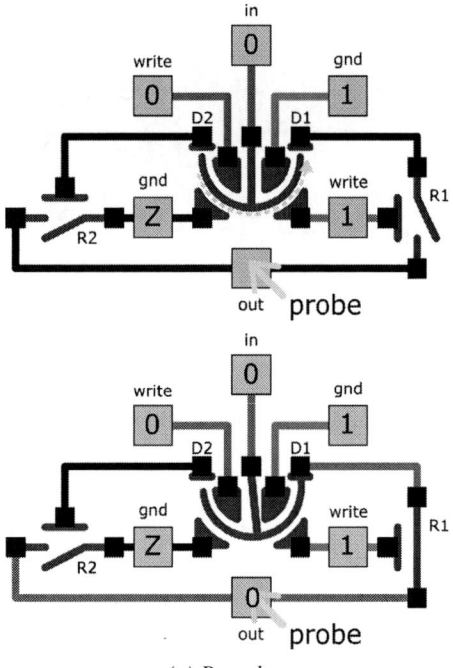

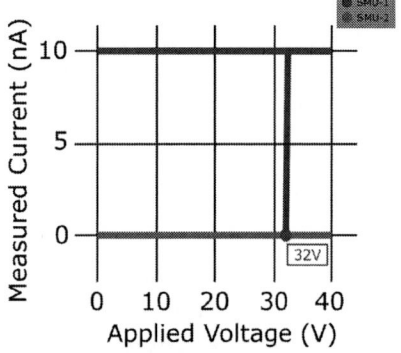

(a) Procedure

CCW "1" Programming

(b) Characterisation

Figure 6: a) Probing procedure to validate functionality of the LUT memory cell being written to '1'. "In" is driven low, which along with "gnd" forced high causes the 7-T to rotate counterclockwise to contact D1 (top). This state connects "in" to the source of R1 to make it pull-in and allow a current to flow between "in" and "out" (bottom). b) I-V Characterisation ramps up the voltage at "in" to rotate the 7-T counterclockwise, close relay R1 and enable current to flow between the "in" and "out" pads. An increased pull-in voltage comes from primary gate (PG1) actuation rather than using the gate pair (PG1 and AG1).

high-temperature, radiation-hard, zero standby power non-volatile memory to serve emerging edge-computing applications that have stringent environmental and power constraints.

ACKNOWLEDGEMENTS

This work received funding from the European Union's Horizon 2020 research and innovation programme under grant agreement No 871740 (ZeroAMP).

REFERENCES

[1] D. M. Mathew, H. Kattan, C. Weis, J. Henkel, N. Wehn, and H. Amrouch, "Thermoelectric cooling to survive commodity DRAMs in harsh environment automotive electronics", *IEEE Access*, vol. 9, p. 83950, 2021.

[2] J. Prinzie, F. Simanjuntak, P. Leroux, and T. Prodromakis, "Low power electronic technologies for harsh radiation environments", *Nature Electronics*, vol. 4, no. 4, pp. 243-253, 2021.

[3] S. Gesper, M. Weißbrich, T. Stuckenberg, P. Jääskeläinen, H. Blume, G. Payá-Vayá, "Evaluation of different processor architecture organisations for onsite electronics in harsh environments", *Int. J Parallel Programming*, vol. 49, no. 4, pp.541-569, 2021.

[4] M. J. Marinella, "Radiation effects in advanced and emerging nonvolatile memories", *IEEE Trans. Nuclear Science*, vol. 68, no. 5, pp. 546-572, 2021.

[5] P. Singh, D.S. Arya, and U. Jain, "MEM-FLASH non-volatile memory device for high-temperature multibit data storage", *App. Phy. Letters*, vol. 115, no. 4, p. 043501, 2019.

[6] K. Akarvardar, and H.-S.P. Wong, "Nanoelectromechanical logic and memory devices", *ECS Transactions*, vol. 19, no. 1, pp. 49, 2009.

[7] P.Singh, G.L. Chua, Y.S. Liang, K.G. Jayaraman, A.T. Do, and T.T.-H. Kim, "Anchor-free NEMS non-volatile memory cell for harsh environment data storage", *J Micromechanics and Microengineering*, vol. 24, no. 11, p. 115007, 2014.

[8] S. Rana, J. Mouro, S.J. Bleiker, J.D. Reynolds, H.M.H. Chong, F. Niklaus, and D. Pamunuwa, "Nanoelectromechanical relay without pull-in instability for high-temperature non-volatile memory", *Nature Communications*, vol. 11, no. 1, pp. 1-10, 2020.

[9] D. Pamunuwa, E. Worsey, J.D. Reynolds, D. Seward, H.M.H. Chong, and S. Rana, "Theory, design, and characterisation of nanoelectromechanical relays for stiction-based non-volatile memory", *IEEE J Microelectromechanical Systems*, vol. 31, no. 2, pp. 283-291, 2022

[10] S. Rana, J.D. Reynolds, T.Y. Ling, M.S. Shamsudin, S.H. Pu, H.M.H. Chong, and D. Pamunuwa, "Nanocrystalline graphite for reliability improvement in MEM relay contacts", *Carbon*, vol. 133, pp. 193-199, 2018.

[11] J.D. Reynolds, S. Rana, E. Worsey, Q. Tang, M.K. Kulsreshath, H.M.H. Chong, and D. Pamunuwa, "Single-contact, four-terminal microelectromechanical relay for efficient digial logic", *Advanced Electronic Materials*, p. 2200584, 2022.

[12] T. Qin, S. J. Bleiker, S. Rana, F. Niklaus, and D. Pamunuwa, "Performance analysis of nanoelectromechanical relay-based field-programmable gate arrays", *IEEE Access*, vol. 6, pp. 15997-16009, 2018.

CONTACT

E. Worsey, elliott.worsey@bristol.ac.uk

NONVOLATILE STATE CONFIGURATION OF NANO-WATT PARAMETRIC ISING SPINS THROUGH FERROELECTRIC HAFNIUM ZIRCONIUM OXIDE MEMS VARACTORS

Nicolas Casilli[1], Onurcan Kaya[1], Tahmid Kaisar[2], Benyamin Davaji[1], Philip X.-L. Feng[2], and Cristian Cassella[1]

[1]Northeastern University, Boston, MA 02115, USA and
[2]University of Florida, Gainesville, FL 32611, USA

ABSTRACT

This work reports on an Ising system formed by two artificial spins and relying, for the first time, on a ferroelectric MEMS device to permanently configure the spins' interactions to ferromagnetic or anti-ferromagnetic. These spins, consisting of two parametric oscillators (POs) built from off-the-shelf components, are driven by the same *2MHz* ("pump") signal and are coupled through a series LC-network with resonance frequency (f_C) formed by two inductors and a microfabricated hafnium zirconium oxide (HZO) MEMS varactor. We show that leveraging the nonvolatile ferroelectric memory characteristic of the HZO varactor allows a shift of f_C to a lower or to a higher value than *~1MHz* based on HZO polarization state, triggering nonvolatile ferromagnetic or anti-ferromagnetic interactions without consuming power for the spin-coupling during the system operation. These features, together with the exceptionally low power threshold of the POs, makes the reported Ising system consume the least power-per-spin (*~600nW*) reported to date.

KEYWORDS

MEMS, Ising Machines, Parametric Oscillators, Ferroelectric Devices, Ferromagnetic Spins

INTRODUCTION

Modern applications in science, economics, and engineering would benefit from solving a variety of nondeterministic polynomial-time (NP) hard problems [1]. For instance, there is a great interest in developing strategies to more efficiently allocate healthcare resources [2], maximize expected returns of investments [3], and allocate network resources in highly trafficked, rapidly changing channel configurations [4], [5]. These NP-hard problems are also used extensively in the algorithms underlying the capability of artificial intelligence (AI) and machine learning (ML) programs [6]. Unfortunately, many of these problems are left unsolved since the existing von-Neumann computing architectures are unable to solve them in a tractable amount of time. Quantum-inspired adiabatic hardware-solvers known as Ising machines (IMs) have emerged, generating new resources to find the solution for NP-hard problems by identifying the ground-state of a corresponding Ising Hamiltonian [1], [7]–[9]. Among the demonstrated IMs, those using resistively coupled sub-harmonic-injection-locked (SHIL) oscillators as spins [1], [7] are becoming popular because of robust manufacturability through complementary-metal-oxide-semiconductor (CMOS) fabrication processes and easier re-programmability through monolithically integrated CMOS-circuits. Nonetheless, the power-per-spin

consumed by SHIL oscillator IMs is set by the *mW*-range power levels needed to sustain the oscillation of their spins, trigger the injection-locking regime, and implement the spin-coupling, making SHIL-based IMs hardly usable to solve problems requiring *10³* spins or more. We addressed this limitation by demonstrating: i) CMOS-compatible Ising spins made of POs driven by the same *~2MHz* pump signal and engineered to consume only *nW*-power; and ii) a coupling network, unbiased during the spin operation, composed of two inductors and a ferroelectric HZO capacitor.

In this paper, a novel nonvolatile resonance coupling architecture capable of providing permanent state configuration to a system of two coupled Ising spins is presented. This architecture exploits the memory dynamics of ferroelectric HZO to create a permanent, re-configurable shift in the resonance frequency of the coupling network to set the interaction of the coupled POs as ferromagnetic or anti-ferromagnetic. In the first section, the principle of operation of the POs and the fabrication flow of the tunable HZO MEMS varactor are introduced. The operating principle, experimental methods, and experimental results of the system of two coupled POs, along with simulated solutions for several *4-node* Max-Cut problems, are reported in the following section. In the conclusions, future research directions are discussed.

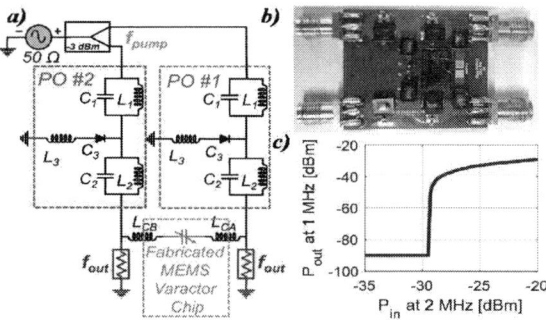

Fig. 1: a) Schematic view of the reported Ising system using two coupled POs and the HZO varactor in their coupling network; b) Top-view picture of the constructed PCB hosting the reported Ising system, showing the HZO device after wire-bonding; c) Measured trend of the POs' output power at 1MHz vs. their input power at 2MHz. The model numbers for the components are as follows: C_3: SMV1236-079LF (26.75pF, tuning range = 36%), L_1: 1812LS-334XLJC (330µH), L_2: 1812LS-474XLJC (470µH), L_3: 1812LS-334XLJC (330µH), C_1: GRM1555C1H750JA01 (75pF), C_2: GRM1555C1R70WA01 (0.7pF), L_{CA}: 1812LS-824XLJC (820µH), L_{CB}: 1812LS-684XLJC (680µH), HZO Varactor: 15.5pF, tuning range = 9%.

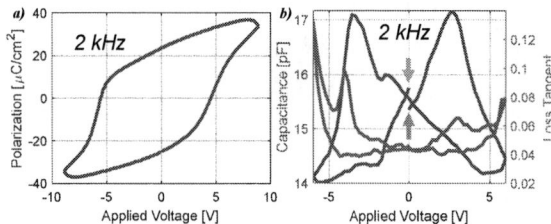

Fig. 2: Left) Measured polarization loop (PE-loop) by PUND method and C-V curve (right) of the HZO varactor. Both are measured at a frequency of 2kHz. Evidently, the device shows two different capacitance values at 0V in its two polarization states. These 0V capacitance values are marked with a blue and orange arrow to indicate whether the contribute to an f_C that results in a ferromagnetic or anti-ferromagnetic interaction, respectively.

METHODS

The coupled two-spin Ising system presented in this paper features PO designs relying on a varactor-based parametric frequency divider that exhibits a period-doubling mechanism under a particular operational region [10]. This region is activated through a supercritical bifurcation for an input power (P_{in}) exceeding a certain power threshold (P_{th}), resulting in an output frequency (f_{out}) that is one half of the frequency of the driving signal (f_{pump}) [11]. This subharmonic output signal provides the basis for the artificial spins, as the spins for this PO-based topology are represented by the subharmonic output phase of a single PO converging to 0 or π with respect to the output phase of a reference PO [1]. Recent theoretical studies have shown that the strategic selection of the components of a PO and its operational frequency greatly reduce the resultant P_{th} based on the satisfaction of four resonant conditions governing the equivalent impedance of the branches in the PO circuit topology [10].

In essence, these resonant conditions provide a structure for intelligently directing the emergent nonlinear subharmonic oscillations to the output of the oscillator. Particularly, i) the series of L_2, C_2, L_3, and C_3 must series-resonate at f_{out}; ii) the series of L_1, C_1, L_3, and C_3 must series-resonate at f_{pump}; iii) L_1 and C_1 must behave as a bandstop filter at f_{out}; and iv) L_2 and C_2 must behave as a bandstop filter at f_{pump} [10]. These conditions inform the design of the POs used in this paper to produce the lowest ever recorded P_{th} (-32dBm) for a PO constructed entirely with off-the-shelf lumped components and operating with f_{pump} = 2MHz. To create these POs, 3 inductors (L_1 = 330µH, L_2 = 470µH, and L_3 = 330µH), 2 capacitors (C_1 = 75pF and C_2 = 0.7pF), and 1 varactor (C_3 = 26.75pF, tuning range = 36%) are used. As observed, using POs with a P_{th} this low permits to create IMs with up to 10^6 spins while consuming less than 1W. This offers a marked improvement of scalability compared to the maximum IM-size that can be reasonably exploited with the available resistively coupled SHIL-based IMs.

To couple the POs, a network comprised of two off-the-shelf inductors and an in-house fabricated HZO MEMS varactor is employed [12]. This coupling network exhibits a resonance frequency (f_C) proximate to f_{out}. The coupled PO circuit topology, constructed system, threshold plot

characterizing the performance of the coupled POs, and a list of the lumped components are shown in Fig. 1. Within the coupling network, the HZO MEMS varactor was designed to operate at 15.5pF, and the lumped inductors are selected as 820µH and 680µH to set f_C to be as close as possible to the f_{out} of the POs. Using the positive-up, negative-down (PUND) method, the PE-loop and C-V curve of the HZO varactor have been measured and are reported in Fig. 2. The measured characteristics of the device demonstrate a clear difference in nominal capacitance and polarization, even at 0V, depending on the polarization state of the HZO film. This ferroelectric hysteresis behavior permits to exercise unique memory dynamics that allow our Ising system to operate without having to consume power due to leakage during the Ising system operation to maintain the desired programmable coupling weight between the spins.

By exploiting this phenomenon, we show that setting the polarization state of the HZO varactor before activating the spins generates a nonvolatile change of f_C, permanently varying the spin interaction between ferromagnetic and anti-ferromagnetic such that no power is needed for the spin coupling during the operation of the Ising system. Particularly, the HZO varactor exhibits different nominal capacitance values at 0V depending on its polarization (e.g., positively or negatively polarized), making f_C lower or higher than the POs' oscillation frequency of f_{out}. For $f_C < f_{out}$, the synchronization of the two POs exhibits out-of-phase output signals, realizing an anti-ferromagnetic interaction. The opposite case produces in-phase output signals, implementing a ferromagnetic interaction. As shown in the next section, the change in HZO polarization state produces a significant enough variation in the capacitance of the device at 0V that the resonance

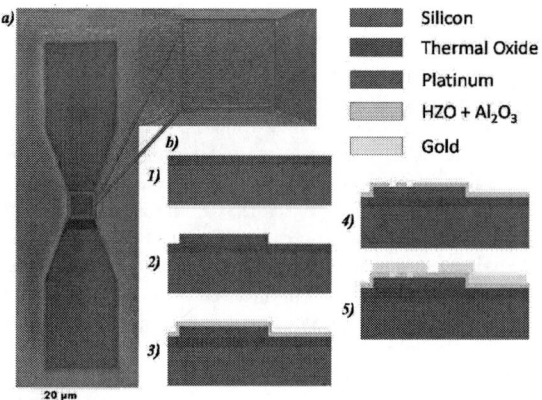

Fig. 3: a) Scanned-Electron-Microscope (SEM) picture of the fabricated HZO varactor device and magnification on the parallel plate capacitance structure. a) Fabrication flow we followed to build the HZO varactor used in this work. We started with 1) silicon wafer with a 150nm thermal oxide as the substrate; 2) sputtered and patterned 100nm platinum with lift-off; 3) deposited 20nm HZO and 3nm Al$_2$O$_3$ using ALD; 4) created vias using a dry plasma etch; and 5) deposited 150nm of gold using E-beam evaporation. After step 5), the deposition samples are annealed in a rapid thermal processor in an N$_2$ environment at 400°C for 40 seconds.

frequency of the coupling network can shift above or below f_{out} and permanently trigger a specific Ising coupling state without the need for an additional applied voltage.

The HZO MEMS varactor is designed as a typical parallel plate capacitor, with *100nm* of platinum and *150nm* of gold sandwiching a *20nm* thin film of HZO and *3nm* of Al_2O_3 deposited using atomic layer deposition (ALD). At such low thicknesses, the HZO exhibits the ferroelectric properties which enable the tuning of the capacitor and its consequent hysteresis relationship with applied voltage. In fact, at lower thicknesses (*10nm*), HZO becomes antiferroelectric, providing a rich set of functionalities in the thin-film regime, meaning that this material offers many exciting possibilities for future applications relevant to tunability [12]. The fabrication process used to create these varactors is shown and described in Fig. 3.

RESULTS AND SIMULATIONS

We have validated our theoretical predictions by fabricating an HZO varactor [12] and wire-bonding it onto a printed-circuit-board (PCB) hosting two identical coupled POs [10] driven by a *2MHz* input signal and exhibiting the lowest power threshold (*600nW*) ever demonstrated for electronic POs. Firstly, the HZO material is excited several times through the pads of the PCB using the PUND method to apply a *+5V 1.2kHz* pulse and ensure that the initial state of the HZO during experimentation is positively polarized. Then, the two POs are fed by a single signal generator (model-number: Tektronix TSG 4104A) providing pump power at *-20dBm*, above the system P_{th} of *-29dBm* which is *3dBm* higher than the threshold of a single PO since we are using the same pump signal to activate both POs. The POs' output waveforms are extracted in the time domain using an oscilloscope (model-number: InfiniiVision DSOX6004A). The driving signal is applied three separate times for each HZO polarization state to demonstrate permanence and consistency of the Ising behavior. As seen in Fig. 4, the POs exhibit in-phase output voltages, corresponding to a ferromagnetic interaction, when the HZO is positively polarized. The POs are then totally disconnected from the oscilloscope and signal generator and a *-5V 1.2kHz* signal is applied to the HZO varactor using PUND to negatively polarize the HZO device. After this step, a similar analysis is conducted and it is measured that the output phases of the POs are shifted by π with respect to each other, indicating an anti-ferromagnetic interaction. Again, the POs are fully disconnected from the measurement instruments and the ferroelectric varactor is polarized positively using a *+5V 1.2kHz* signal. The POs are driven again using the same signal at *2MHz* and their output phases are shown to be in-phase. Fig. 4 shows a system-level depiction of the transitions of the HZO polarization states performed in this experiment and the corresponding output waveforms of the POs for each of those states.

Furthermore, this experiment shows that the HZO polarization state, through the mechanism of changing f_C, offers a nonvolatile, permanent state configuration of the PO spins to ferromagnetic or anti-ferromagnetic without the need for a constantly applied bias voltage. Since the polarization state of the HZO is retained even at *0V* DC bias, the power cost of this system is only what is necessary

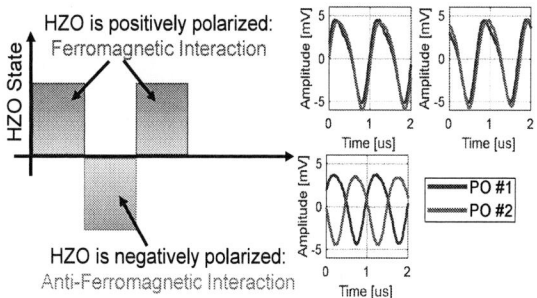

Fig. 4: Schematic view of the followed experimental procedure and measured output waveforms of the two POs for both polarization states of the HZO varactor. Two waveforms are reported for the case of the positively polarized HZO varactor in order to show the repeatability of the reported technique in controlling the interaction of the two Ising spins. With a positive polarization, the output waveforms of the coupled POs exhibit in-phase (ferromagnetic) behavior. When the HZO varactor is negatively polarized, the system exhibits anti-phase (anti-ferromagnetic) behavior.

to drive the POs above their input power threshold. This method of coupling offers exciting possibilities for the development of a larger, CMOS-compatible IM with an easily re-programmable binary-weighted coupling scheme consuming only RF power.

In order to benchmark the performance of our artificial spins compared to the state-of-the-art, it is useful to evaluate the cost-per-spin of the oscillators used in IM topologies. Furthermore, a table (Table 1) has been prepared delineating the operational frequency and power consumption per spin for several different types of oscillators used in existing IMs. It is worth emphasizing that this power per spin metric considers only the amount of power needed to excite the oscillators, and it does not consider any DC biasing voltage that needs to be applied to maintain any desired coupling weight. Upon a quick inspection, the RF power cost of using POs is one order of magnitude lower than the next best type of oscillator, demonstrating the immense potential of PO-based IMs to solve CO problems of larger complexity than those realistically solvable by IMs relying on other types of oscillators.

To show that PO-based systems, like the one presented in this work, can indeed be used as Ising solvers for NP-hard problems, we analytically show how a network of POs

Table 1. Comparison with other artificial spins.

Type of Oscillator	Operational Frequency	Power per Spin	Reference
LC	1 MHz	20.8mW	[1]
LC	50 kHz	/	[7]
Ring	1 GHz	23μW	[8]
Ring	118 MHz	41μW	[9]
PO	**2 MHz**	**600nW**	**This work**

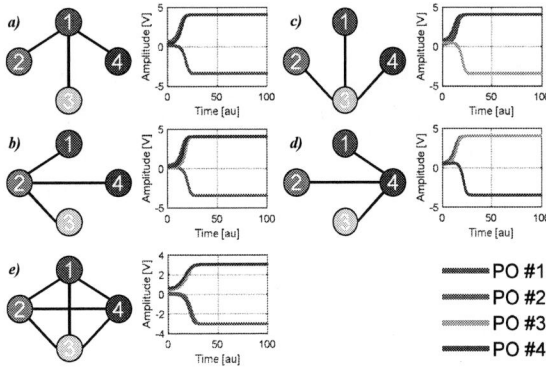

Fig. 5: Several 4-node graphs where a) all POs are connected to PO #1; b) all POs are connected to PO #2; c) all POs are connected to PO #3; d) all POs are connected to PO #4; and e) all POs are connected to all other POs. In each case, the simulated voltage outputs match with the correct solution for the Max-Cut problem, demonstrating the successful capability of POs-based Ising solvers to find the solution of NP-hard problems.

effectively solves several selected *4-node* Max-Cut problems [7] (Fig. 5). With our analytical model, we have verified the correctness of each spin phase solution through a comparison with solutions produced by other IMs and the solution derived using a Max-Cut approximate solution algorithm for the same set of problem graphs [7], [13]. Furthermore, it is demonstrated that by changing the value of f_C between two coupled POs, a specific Ising interaction can be permanently configured. A PO-based IM relying on this coupling architecture and using the type of POs reported in this paper is a promising path forward to addressing the problem-size, miniaturization, and re-programmability concerns currently plaguing other oscillator-based IMs and IMs in general.

CONCLUSIONS

In this paper, a novel approach to Ising spin state configuration is proposed based on the nonvolatile memory characteristics of the ferroelectric HZO MEMS varactor used for coupling the artificial spins. The proposed architecture is theoretically described and verified experimentally by measuring the output phases of the POs when driven with a 2MHz signal at different polarization states of the *20nm* HZO film at *0V* DC bias. It has been shown that the HZO film deterministically affects the phase convergence of the Ising system, triggering ferromagnetic or anti-ferromagnetic spin interactions depending on its current polarization state due to the resultant shift of the f_C of the coupling network relative to the output frequency of the POs. These experimental results, paired with a simulation framework for higher order Ising systems of coupled POs, will enable the development of exciting new capabilities for miniaturized, low power, CMOS-compatible, and re-programmable IMs.

ACKNOWLEDGEMENTS

We thank the support from the National Science Foundation CCF-FET program (Grants #2103351 & #2103091).

REFERENCES

[1] T. Wang and J. Roychowdhury, "Oscillator-based Ising Machine," 2017.

[2] Jeffrey. E. Arle and K. W. Carlson, "Medical diagnosis and treatment is *NP-complete*," *J. Exp. Theor. Artif. Intell.*, vol. 33, no. 2, pp. 297–312, Mar. 2021.

[3] R. Orús, S. Mugel, and E. Lizaso, "Quantum computing for finance: Overview and prospects," *Rev. Phys.*, vol. 4, p. 100028, Nov. 2019.

[4] A. Samuylov *et al.*, "Characterizing Resource Allocation Trade-Offs in 5G NR Serving Multicast and Unicast Traffic," *IEEE Trans. Wirel. Commun.*, vol. 19, no. 5, pp. 3421–3434, May 2020.

[5] C. She, R. Dong, W. Hardjawana, Y. Li, and B. Vucetic, "Optimizing Resource Allocation for 5G Services with Diverse Quality-of-Service Requirements," in *2019 IEEE Global Communications Conference (GLOBECOM)*, Waikoloa, HI, USA, Dec. 2019, pp. 1–6.

[6] C. J. Hillar and L.-H. Lim, "Most Tensor Problems Are NP-Hard," *J. ACM*, vol. 60, no. 6, pp. 1–39, Nov. 2013.

[7] J. Chou, S. Bramhavar, S. Ghosh, and W. Herzog, "Analog Coupled Oscillator Based Weighted Ising Machine," *Sci. Rep.*, vol. 9, no. 1, p. 14786, Dec. 2019.

[8] W. Moy, I. Ahmed, P. Chiu, J. Moy, S. S. Sapatnekar, and C. H. Kim, "A 1,968-node coupled ring oscillator circuit for combinatorial optimization problem solving," *Nat. Electron.*, vol. 5, no. 5, pp. 310–317, May 2022.

[9] I. Ahmed, P.-W. Chiu, W. Moy, and C. H. Kim, "A Probabilistic Compute Fabric Based on Coupled Ring Oscillators for Solving Combinatorial Optimization Problems," *IEEE J. Solid-State Circuits*, vol. 56, no. 9, pp. 2870–2880, Sep. 2021.

[10] H. M. E. Hussein, M. A. A. Ibrahim, G. Michetti, M. Rinaldi, M. Onabajo, and C. Cassella, "Systematic Synthesis and Design of Ultralow Threshold 2:1 Parametric Frequency Dividers," *IEEE Trans. Microw. Theory Tech.*, vol. 68, no. 8, pp. 3497–3509, Aug. 2020.

[11] H. M. E. Hussein, M. Rinaldi, M. Onabajo, and C. Cassella, "A chip-less and battery-less subharmonic tag for wireless sensing with parametrically enhanced sensitivity and dynamic range," *Sci. Rep.*, vol. 11, no. 1, p. 3782, Dec. 2021.

[12] V. Gund *et al.*, "Multi-level Analog Programmable Graphene Resistive Memory with Fractional Channel Ferroelectric Switching in Hafnium Zirconium Oxide," in *2022 EFTF/IFCS*, Paris, France, Apr. 2022, pp. 1–4.

[13] M. X. Goemans and D. P. Williamson, "Improved approximation algorithms for maximum cut and satisfiability problems using semidefinite programming," *J. ACM*, vol. 42, no. 6, pp. 1115–1145, Nov. 1995.

CONTACT

*N. Casilli, casilli.n@northeastern.edu

PHYSICAL RESERVOIR COMPUTING USING NONLINEAR MEMS RESONATOR HAVING HIGH MEMORY CAPACITY AT "EDGE OF CHAOS"

Hiroki Takemura, Takahiro Mizumoto, Amit Banerjee, Jun Hirotani, and Toshiyuki Tsuchiya
Department of Micro Engineering, Kyoto University, JAPAN

ABSTRACT

This paper reports physical reservoir computing (PRC) using a single nonlinear electrostatic resonator and demonstrates its high memory capacity at "edge of chaos. " The resonator is a simple doubly supported resonator fabricated from a silicon-on-insulator wafer. We proposed a PRC system without feedback loop, in which its memory capacity relies on the decay time of the high-Q resonator. The benchmark task results indicate that the system shows good linear and nonlinear memory capacities at the resonance and the maximum capacity was obtained at the vicinity of the instability edge of the frequency response.

KEYWORDS

Physical reservoir computing, electrostatic, nonlinear resonator, machine learning, silicon-on-insulator.

INTRODUCTION

Physical reservoir computing (PRC) is one of the machine learning algorisms where the inner layer of recurrent neural network (RNN) is implemented by nonlinear physical system. Figure 1 shows a human motion recognition system as an example. Vibration signal detected by accelerometer input to a recurrent neural network and outputs classifies person's activity, such as sitting, standing, walking, and running. Training of a conventional RNN is time- and power-consuming because of its complicated network structure and difficulties in optimizing all the node transfer functions and the connection weights, which makes difficult to implement it to portable systems. In the reservoir computing, the hidden (middle) layer of RNN is fixed (called as a reservoir) and only the weights of output layer are optimized by such as

linear and Ridge regression methods. PRC employs a nonlinear physical system for the reservoir, such as optical, spintronic, thermal, and electrical systems. A microelectro-mechanical nonlinear resonator is one of the candidates for the reservoir and experimental demonstration results have been reported [1-4]. One of the features of MEMS reservoir is that its integration to MEMS sensors [4] and mechanical signal processing is conducted before transduction to electrical signal, which greatly reduces the power consumption of remote sensing systems. We have utilized a frequency modulated capacitive accelerometer to a sensor-integrated physical reservoir. The doubly clamped beam resonator for acceleration detection was used as a physical reservoir with its enhanced nonlinearity and the delayed feedback loop. The vibration inputs were successfully processed by machine learning tasks and the results showed good performances [4]. However, the design strategy and parameter optimization procedure of MEMS resonator for PRC have not been investigated and clarified owing to its complicated output.

In this study, we use a simple nonlinear resonator driven by electrostatic force and detected by capacitance change and do not use the delayed feedback loop, which is often used in MEMS reservoir [1,4]. The memory capacity is obtained by controlling its Q-factor and time constant operated in a vacuum environment. Benchmark tasks were conducted by changing the oscillation frequency and sampling time to study its optimal memory capacities.

MEMS RESERVOIR

Figure 2 shows the nonlinear electrostatic resonator used as a reservoir in this study. The resonator has a pair of doubly clamped beams of 330 μm long and 3 μm wide, which shows the hard spring effect because of the axial force on the beams. There are two pairs of parallel plate

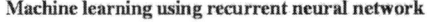
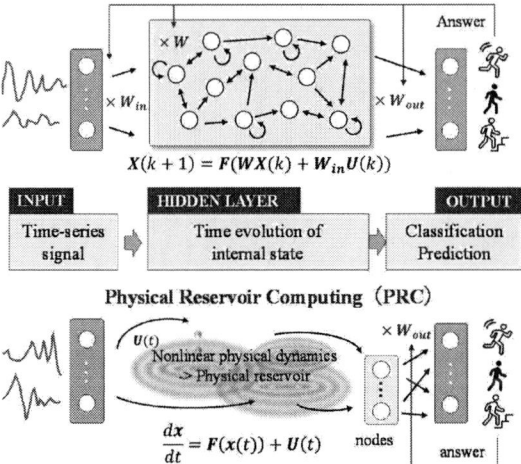

Figure 1: Concept of physical reservoir computing.

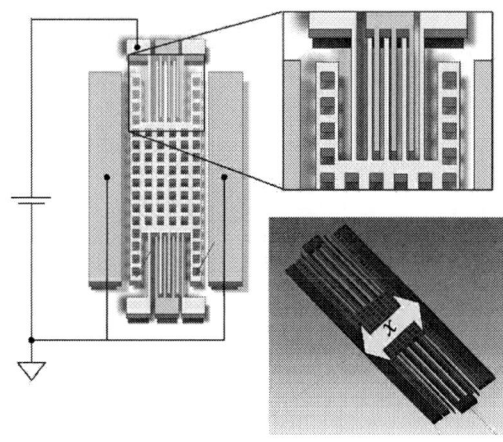

Figure 2: Electrostatic nonlinear SOI resonator as a physical reservoir.

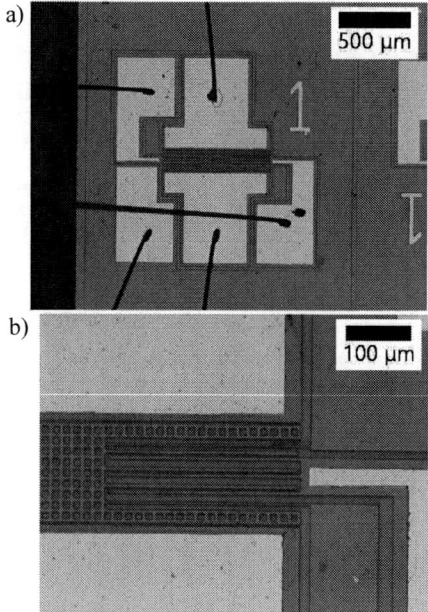

Figure 3: Fabricated SOI nonlinear resonator. a) Optical microscope image of resonator. b) Close-up view of doubly supported beams and parallel plate displacement sensor.

capacitances. The one pair on the left and right sides of the mass whose length is 700 μm is for actuation. The other pair on the top and bottom is for displacement detection. There are three capacitances on each side and the length is 275 μm. The gaps of these electrodes are 3 μm. The thickness is 15 μm. The resonant frequency at the first mode was designed and simulated as 11.96 kHz.

The resonators were fabricated on 4-inch silicon-on-insulator (SOI) wafer with 15-μm-thick device and 2-μm-thick buried oxide layers. A standard fabrication process using contact lithography and Bosch process (Samco, RIE-800iPB) for patterning the device layer and vapor HF etching (SPTS, MLT-SLE-Ox) for sacrificial etching. An Au/Cr layer is used for electrodes. A fabricated reservoir resonator is shown in Fig. 3.

RC ARCHITECHTURE

The reservoir computing architecture is shown in Fig. 4. The input is added as a bias voltage to the oscillation ac signal at a frequency near the resonant frequency. The oscillated vibration is detected at the displacement-detecting capacitance as a current by a biasing voltage applied on the mass. The reservoir output is just taken from the envelope of the vibration velocity signal as an output of transimpedance amplifiers (TIAs). The state variable vector X_k is acquired from the envelope waveform of the enveloped signal at a constant interval T. The vector dimensions were set to 100, which means that the hundred envelope values as virtual nodes were extracted at a constant interval at $T/100$, which are indicated as yellow points on the enlarged velocity plot in Fig. 4.

In the benchmark tasks, m random binary input data was input to the reservoir and the state matrix X was collected. The output data Y is calculated by multiplying the weight matrix W of a 100th order vector. In the teaching step, a certain number of inputs are processed in the RC. The weight values are determined by the Ridge regression;

$$W = \overline{Y}X^T(XX^T + \lambda I)^{-1} \qquad (1)$$

$\overline{Y}$ is the answer for the inputs and λ is a constant. In the testing step, The outputs $Y = XW$ were collected and compared with the answer $\overline{Y}$.

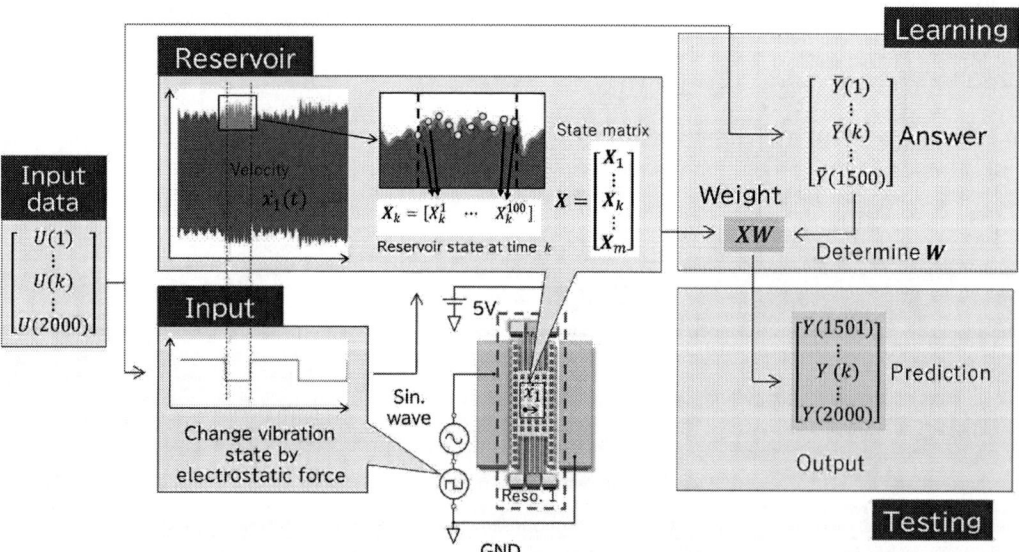

Figure 4: Schematic diagram of reservoir computing architecture for benchmark tasks.

978-1-6654-9309-3/23 $31.00 © 2023 IEEE

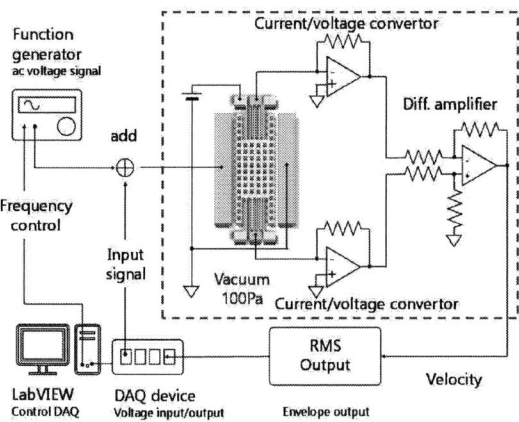

Figure 5: Experimental setups for RC.

EXPERIMENTAL

Data collection

Figure 5 shows the experimental setup for reservoir computing. The ac oscillation signal was generated using function generator controlled by LabVIEW and input signal was added by an op-amp and applied to the reservoir resonator for electrostatic oscillation. The displacement velocity was detected by the TIAs and differential amplifier. The rms-dc convertor (AD736, Analog Devices) was used to convert the velocity signal to the envelope with the time constant of 1 ms. The rms output of 2000 inputs was acquired and then the acquired data was processed offline.

Benchmark tasks

Benchmark tasks to evaluate memory capacity were conducted, as shown in the diagram in Fig. 4. We employ the parity check task (nonlinear memory) and the short memory task (linear memory). Main purpose of the benchmark tasks is to confirm machine learning capability. We defined the memory capacity MC;

$$ MC = \sum_{n=0}^{\infty} [\mathrm{Corr}(\overline{Y}_n, Y_n)]^2 \qquad (2) $$

where n is the order of the benchmark tasks, Y_n is the estimated answer of n-th-order task, and Corr( ) means the correlation coefficient. In this study, $m = 2000$ data is processed and the first 1500 data was used for teaching and the later 500 data was used for testing.

The tasks were conducted with the oscillation frequency and the input duration time T as parameters. The ac oscillation amplitude was 1 Vpp and the binary inputs were applied as ±1 V. The mass is biased at 5 V. The oscillation frequency was changed from 90% to 120% of the defined center frequency near the resonant frequency. The ambient pressure on the benchmark task was 100 Pa, so that the response time constant becomes 10 ms. The input holding time is from 2 to 30 ms with 2 ms interval.

RESULTS AND DISCUSSION

Nonlinear frequency response

Figure 6 shows the frequency responses at different oscillation amplitudes at 5 Pa. The resonant frequency was 9.74 kHz and the Q-factor was 3000 at small amplitude but the resonant frequency increased up to 12 kHz by the hard spring effect. The resonant frequency was lower than the design. The reason was fabrication error. The measured beam width was 2.6 μm.

Benchmark task

Figures 7 show the heat map of the correlation coefficient MC of each task as functions of normalized operating frequency Ω and the input holding time T. The square points in Fig. 7a show the rms output at the end of the benchmark task at the input duration of 4 ms. The frequency responses of upward and downward sweep are drawn with the continuous and broken curves. The plot indicates that the vibration kept at the upward sweep state during the task.

The memory capacity map of STM task shows a radial pattern centered at $\Omega =0.98$ and higher capacities are observed at $\Omega = 1.19$, just before the amplitude drops (Fig. 7b). The similar features are seen in that of PC task, but in the radial pattern, the point of high capacity in STM have low capacity in PC, and *vice versa*. These features indicate two rules of thumb about the performance of PRC. First, the memory capacities are complemental between the STM and PC tasks, at the parameters where the linear capacity is high, the nonlinear one is low. Second, the maximum capacities in both tasks were obtained near the unstable point in the frequency response at $\Omega = 1.19$, which is called as "edge of chaos." This is a well-known empirical feature of PRC that the performance become better near the boundary of chaotic responses.

To examine the reason of high capacity at the "edge of chaos" the transient responses at the holding time of 18 ms at the frequency was plotted in Fig. 8. Compared to the

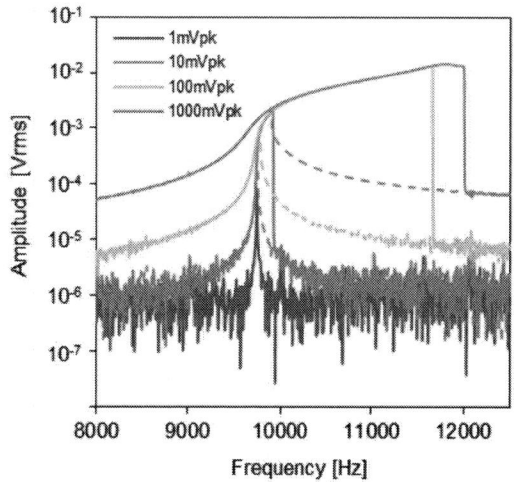

Figure 6: Measured frequency responses of reservoir resonator at different oscillation voltages. Measured at 5 Pa.

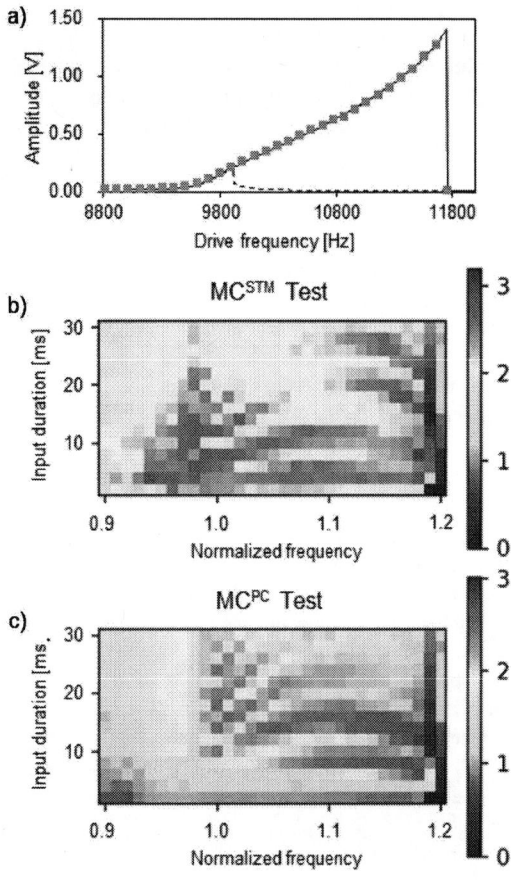

Figure 7: Benchmark results of linear and nonlinear task. a) Frequency response of reservoir resonator, b) heatmap of memory capacity in short term memory task and c) memory capacity in parity check task.

lower operating frequency, Ω=1.19 shows distorted and fluctuated waveforms, which means that the system has the proper complexity in the output signal.

CONCLUSION

We demonstrated that the simple nonlinear MEMS capacitive resonators can be applicable for PRC. The machine learning tasks were successfully conducted experimentally. The memory capacity of the device shows the "edge of chaos" empirical rules. To increase the performance, we are going to evaluate a nonlinear resonator array as a physical reservoir.

ACKNOWLEDGEMENTS

This work is supported by JSPS KAKENHI Grant Numbers JP22K18289, Tateisi Science and Technology Foundation, The Telecommunications Advancement Foundation, and Kyoto University Nanotechnology Hub in the "ARIM Project" sponsored by MEXT, Japan.

REFERENCES

[1] G. Dion, S. Mejaouri, and J. Sylvestrea, Reservoir

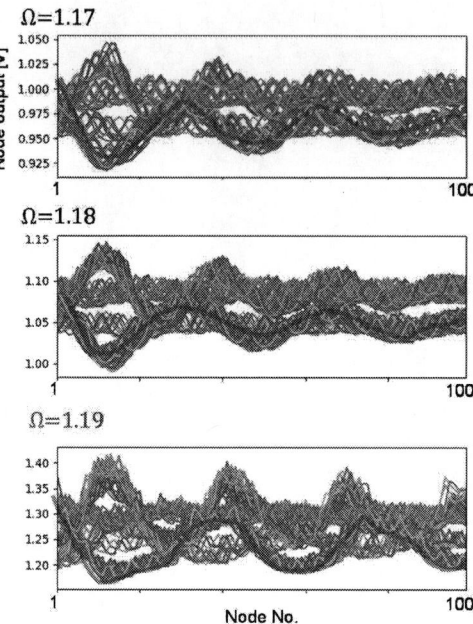

Figure 8: Envelopes of transient responses X_k at the oscillation frequencies Ω of 1.17, 1.18, and 1.19.

computing with a single delay-coupled non-linear mechanical oscillator, *Journal of Applied Physics*, 124 (2018), 152132.

[2] M. H. Hasan, A. Al-Ramini, E. Abdel-Rahman, R. Jafari, and F. Alsaleem, Colocalized Sensing and Intelligent Computing in Micro-Sensors, *Sensors*, 20 (2020), 6346

[3] J. Sun, W. Yang, T. Zheng, X. Xiong, Y. Liu, Z. Wang, Z. Li, and X. Zou, Novel nondelay-based reservoir computing with a single micromechanical nonlinear resonator for high-efficiency information processing, *Microsyst. Nanoeng.*, 7 (2021), 83.

[4] T. Mizumoto, Y. Hirai, A. Banerjee, and T. Tsuchiya, MEMS Reservoir Computing using Frequency Modulated Accelerometer, *IEEE MEMS2022*, Tokyo (2022) pp. 487-490.

CONTACT

*T. Tsuchiya, tel: +81-75-383-3690;

tutti@me.kyoto-u.ac.jp

PROGRAMMABLE FERROELECTRIC HZO NEMS MECHANICAL MULTIPLIER FOR IN-MEMORY COMPUTING

Shubham Jadhav, Ved Gund, and Amit Lal

School of Electrical and Computer Engineering, Cornell University, USA

ABSTRACT

This work presents the first-ever NEMS-based in-memory computing device that uses ultra-thin ferroelectric hafnium zirconium oxide (HZO) as the weight storage unit. The HZO unimorph's piezoelectric properties can be tuned by applying a poling voltage (V_p) from -8V to 8V across a 20-nm HZO film. The unimorph displacement, both at resonance and off-resonance, is found to be proportional to the product of V_p and V_{in}, where V_{in} is the input drive voltage when operating in compute mode. The NEMS device demonstrates analog multiplication, without electrical contacts, and consumes attoJoules of energy. The HZO-based unimorphs are scalable to nano-scale dimensions owing to the ultrathin HZO layer. This device provides the possibility of arraying of the unimorphs with capacitive readout for massively parallel MAC (Multiply and Accumulate) units.

KEYWORDS

HZO, ferroelectric material, in-memory computing, NEMS, programming

INTRODUCTION

With the potential slowing of Moore's law-led computational growth, many new accelerator architectures have been explored. Traditional von Neumann architecture suffers from large delays and significant energy consumption due to the transfer of data between storage bank and processing unit. To overcome these issues, in-memory computing technologies such as resistive random-access memory (RRAM), phase-change memory (PCM), and ferroelectric field-effect transistors (Fe-FETs) have been explored[1], [2]. Multi-level conduction modulation of HZO-based Fe-FET has been demonstrated in silicon and germanium channel devices[3], [4]. Though FET-based in-memory devices have a small footprint, further scaling of the gate dielectric <5nm is susceptible to high leakage currents [5]. Arrays of NEMS unimorphs offer an alternative architecture to perform in-memory computations with zero leakage, provided that the MAC operation can be demonstrated. Our work presented in a previous publication describes the detailed working of the FerroNEMS MAC unit [6]. It is shown that the displacement of the unimorph is the product of the programming and the input voltage ($\delta = \beta. V_p. V_{in}$), demonstrating analog multiplication. By operating many such unimorphs in parallel, an all-capacitive, non-current based DNN is possible, resulting in orders of magnitude higher computational capacity. This paper presents mechanical displacement measurements of a HZO-on-SiO$_2$ unimorph at two different frequency regions (on-resonance and off-resonance) to ensure working in both DC and AC modes.

DEVICE DESCRIPTION

The unimorph was fabricated by depositing HZO on 200nm sputtered platinum, with 1μm SiO$_2$ underneath which serves as the elastic layer (Fig. 1). Ferroelectric characterization with 1 kHz positive-up-negative-down (PUND) was performed with a programmable function generator, custom probe station, and digital oscilloscope (Fig. 2a,b) [7]. Poling at different values of V_p (-8V<V_p<8V with 0.5V increment) was done using the same setup using 3 kHz single-sided pulses for weight-storage. Following each poling step, unimorph displacements were measured using Polytec MSA-400 laser Doppler vibrometer with AC actuation ($V_{in} = 0.1V$), to prevent further polarization switching.

EXPERIMENTAL RESULTS

Input sinusoidal voltage (V_{in}) of 0.1V was applied in the frequency range of 450 to 460 kHz (corresponding to the fundamental mode f_{res}= 456kHz) to measure the maximum displacement δ_{max} at the center of the unimorph. The same measurement was performed in the frequency range of 45-50 kHz (corresponding to off-resonance) to plot RMS displacement δ_{RMS} shown in (Fig. 2c) which shows non-resonant programmability. To prove the multiplication functionality for MAC operations, V_{in} was varied from 0.02V to 0.2 V, and the corresponding δ_{max} and δ_{RMS} were plotted against V_p in the linear operating region (indicated by a blue ellipse in Fig. 2c) for

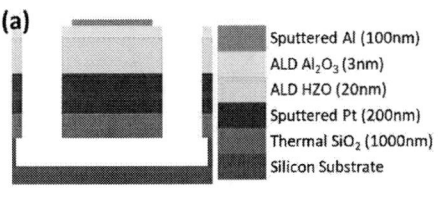

Figure 1: a) Schematic cross-section view of the ferroelectric unimorph b) SEM image of the unimorph showing released structure c) 3D interferometry profile of the released unimorph measured using Polytec topography measurement system (TMS) showing beam curvature due to residual stresses in the film-stack.

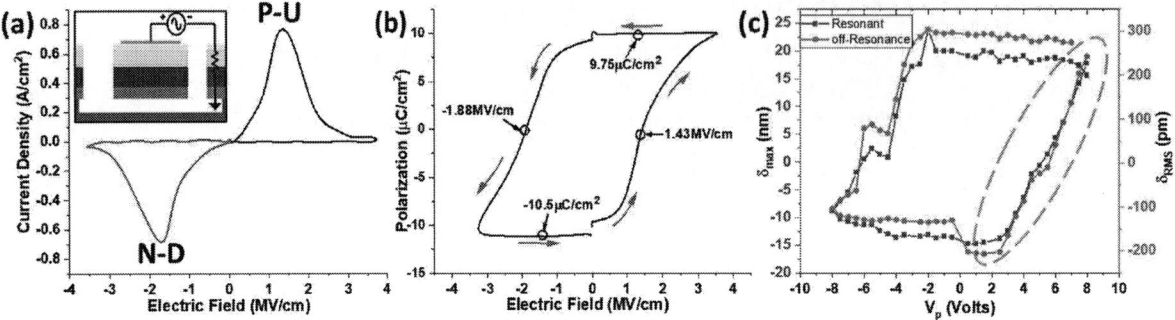

Figure 2: (a) Current density vs E-field plot showing polarization switching peaks. (b) Measured polarization vs. E-field characteristics of the released unimorph. c)Plot of unimorph displacement vs poling voltage V_p for resonant (δ_{max}) (black) and off-resonance (δ_{RMS}) (red) measurements at constant V_{in}=0.1V. The dashed blue ellipse indicates linear operating region for generating output displacements proportional to the product of V_p and V_{in}.

both on and off-resonances (Fig. 3a,b). These plots show that the output displacement is proportional to both the V_{in} and V_p, independent of actuation frequencies. Figure 3c inset shows the implementation of this unimorph with capacitive readout to detect beam motion. The beam displacement (δ) caused by the multiplication of V_p and V_{in} will change the gap between capacitive parallel plates. This change in gap results in capacitance change ΔC as plotted in Fig. 3c. By connecting these units in parallel and implementing a trans-impedance amplifier (TIA) at the output node, the motional current generated due to capacitance change can be read out as an output voltage (V_{out}). This V_{out} will be a scaled product of two inputs V_p and V_{in} serving a functional NEMS-based unit for MAC operation with a summation of capacitances in the form of individual displacement currents from each multiplier.

CONCLUSION

In conclusion, the proposed device concept was tested experimentally to show the displacement of the FerroNEMS beam is a scaled product of the multiplication of two inputs ($\delta = \beta.V_p.V_{in}$). The comparison of δ vs V_p at different V_{in} for on and off-resonance frequencies corroborated these results which ensure the working of a device in both DC and AC modes.

ACKNOWLEDGEMENTS

This work was performed in part at the Cornell NanoScale Facility, an NNCI member supported by NSF Grant No. NNCI- 2025233. Funding was provided by the DARPA TUFEN (No. HR0011-20-9-0048) program.

REFERENCES

[1] A. Chen, S. Datta, X. S. Hu, M. T. Niemier, T. Š. Rosing, and J. J. Yang, "A Survey on Architecture Advances Enabled by Emerging Beyond-CMOS Technologies," *IEEE Des. Test*, vol. 36, no. 3, pp. 46–68, 2019, doi: 10.1109/MDAT.2019.2902359.

[2] A. Sebastian, M. Le Gallo, R. Khaddam-Aljameh, and E. Eleftheriou, "Memory devices and applications for in-memory computing," *Nat. Nanotechnol.*, vol. 15, no. 7, pp. 529–544, 2020, doi: 10.1038/s41565-020-0655-z.

[3] M. Jerry *et al.*, "Ferroelectric FET analog synapse for acceleration of deep neural network training," in *2017 IEEE International Electron Devices Meeting (IEDM)*, Dec. 2017, pp. 6.2.1-6.2.4. doi: 10.1109/IEDM.2017.8268338.

[4] M. Si, X. Lyu, and P. D. Ye, "Ferroelectric Polarization Switching of Hafnium Zirconium Oxide in a Ferroelectric/Dielectric Stack," *ACS Appl. Electron. Mater.*, vol. 1, no. 5, pp. 745–751, May

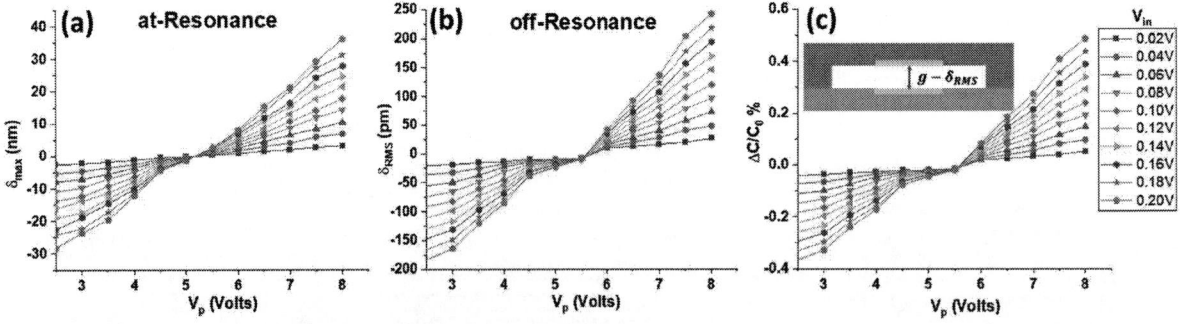

Figure 3: a) Unimorph displacement vs poling voltage V_p in the linear operating region for a) resonant (δ_{max}) and b) off-resonance (δ_{RMS}) measurements plotted at different input voltages V_{in} in the range of 0.02 to 0.2V. c) Extracted capacitance change calculations ($\frac{\Delta C}{C_0}\% = \frac{c-c_0}{c_0} \times 100$) for parallel plate readout assuming 50nm gap between plates. Where, C_0=original capacitance without beam displacement and C= capacitance due to change in parallel plate gap ($g - \delta_{RMS}$)

2019, doi: 10.1021/acsaelm.9b00092.

[5] S. S. Cheema *et al.*, "Enhanced ferroelectricity in ultrathin films grown directly on silicon," *Nature*, vol. 580, no. 7804, pp. 478–482, Apr. 2020, doi: 10.1038/s41586-020-2208-x.

[6] S. Jadhav *et al.*, "HZO-based FerroNEMS MAC for In-Memory Computing," *Appl. Phys. Lett.*, vol. 121, no. 193503, pp. 1–6, 2022, doi: 10.1063/5.0120629.

[7] V. Gund *et al.*, "Multi-level Analog Programmable Graphene Resistive Memory with Fractional Channel Ferroelectric Switching in Hafnium Zirconium Oxide," 2022.

CONTACT

*Shubham Jadhav, tel: +1-607-262-5059; saj96@cornell.com

STORING MEMS INTERFACES WITHOUT ELECTRICAL AUXILIARY ENERGY FOR LONG-TIME MONITORING

Martin Hoffmann[1], Philip Schmitt[1], Steffen Wittemeier[3], Falk Schaller[2], Alexey Shaporin[3], Chris Stöckel[2,3], Volker Geneiß[3], Roman Forke[3], Christian Hedayat[3], Ulrich Hilleringmann[4], Harald Kuhn[2,3], and Sven Zimmermann[2,3]

[1]Ruhr-Universität Bochum, GERMANY
[2]The Chemnitz University of Technology, GERMANY
[3]Fraunhofer ENAS, GERMANY
[4]Universität Paderborn, GERMANY

ABSTRACT

Two thoroughly integrated sensor node interfaces for the storage of mechanical and electrical measures are demonstrated. For long-time monitoring tasks, the continuous accessibility of electrical auxiliary energy is often critical, as the sensing device has to be on standby, even if the measures of interest - extremal values, integrals or total numbers of events - are retrospectively of interest, only. Novel mechanical or nanoionic interfaces allow storing these measures by using the inherently available energy from the measures. Any kind of acceleration, force or pressure provides sufficient mechanical energy to extract and store quantitative results in a mechanical MEMS storage whereas electrical energy, e.g. from piezo sensors or photoelectric elements can be stored in a nanoionic memory. Interfaces for both concepts exemplified for inertial sensors are described and demonstrated here.

KEYWORDS

Passive sensors, condition monitoring, predictive maintenance, nanoionic memory, ratcheting mechanism

INTRODUCTION

Although energy harvesting has extensively been investigated [1], it has found limited applications, as a constant accessibility to a non-electrical energy source is required and the efficiency is mostly low. On the other hand, many measures of interest come along with an appreciable energy content. The basic idea of this work was to investigate reliable interfaces that make directly use of this energy, without the loop of electrical auxiliary energy harvesting. Ratcheting mechanisms have already been used in MEMS [2, 3], but for precise data storage with high

memory depth, reliable MEMS interfaces with defined mechanical impedance and for electrical measures, memories with long-term stability (in contrast to e.g. common capacitors) are required. Here, we present an all-mechanical analog-to-digital MEMS memory interface and a long-term storage for the integration of electrical measures based on a nanoionic memory. Both interfaces can be reset by an external electrical signal.

METHOD

Fig. 1 illustrates the main steps of the proposed measurement acquisition chain. Two parallel paths are pursued: One for physical signals which can easily be transformed into force and displacement using membranes or proof masses (e.g. pressure or acceleration), and a second path for measures transformable into an electrical signal by means of a piezoelectric energy harvester or solar cells (e.g. vibration or light intensity). The energy provided by the measure itself causes a change of the non-volatile memory, available for read-out by RFID at any time.

The aim of the proposed measurement acquisition chain is to detect values of special interest, such as the maximum value of an amplitude or the integral of a measure over a long time span. The measurement itself is performed in the analog domain as indicated in Fig. 2.

Within a state-of-the-art digital data acquisition chain, measurements are performed at a defined sampling rate. For low sampling rates, short critical events can be missed, whereas a high sampling rate generates large amounts of unnecessary data that consume energy. The events of interest have to be extracted from this data.

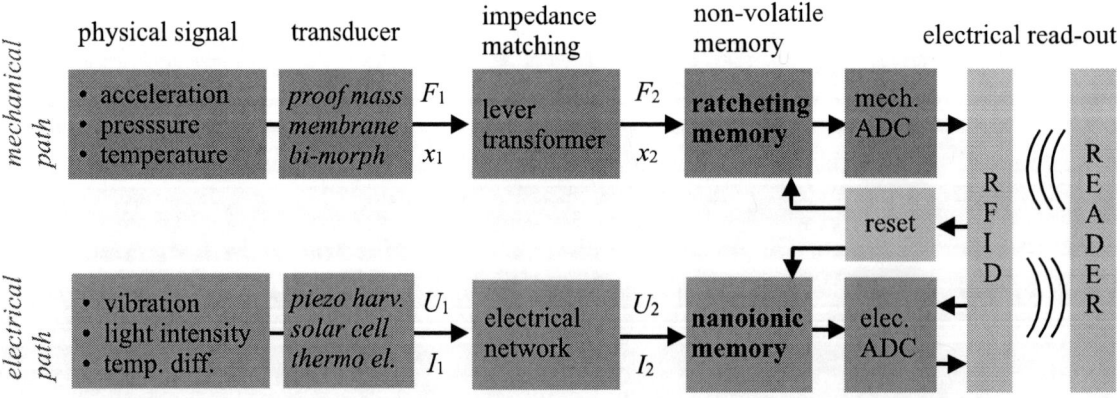

Figure 1: Concept of data acquisition for long-term monitoring.

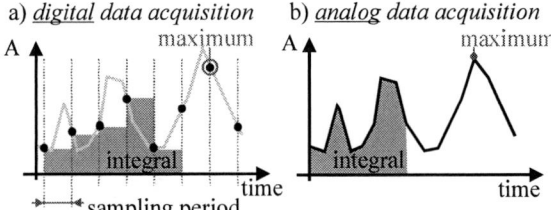

Figure 2: a) digital data acquisition with pre-defined sampling rate. b) analog data acquisition

Mechanical path

The use of MEMS components in sensor systems is often limited to serve as a transducer, which converts a physical measure into an electrical signal while the data acquisition and storage are performed in the electrical domain including amplification, filtering, digitization and storage and require a continuous amount of electrical auxiliary energy. Here, we perform the tasks amplification, selection of the maximum value, storage and digitization already in the mechanical domain. Physical measures are often also transformed into a mechanical displacement prior to generate a capacitive or resistive change based on the deflection.

Fig. 3 shows the working principle of a passive accelerometer that measures and stores the maximum value of an acceleration. The sensor consists of a primary transducer (a spring-guided proof mass), a lever-transformer, a mechanical memory and a mechanical-to-electrical AD-converter [4] resulting in an all-mechanical integrated sensor device with digital electrical output. As an external acceleration is applied at the proof mass, the induced displacement is first amplified by the lever transformer and subsequently conducted to the mechanical memory through a coupling.

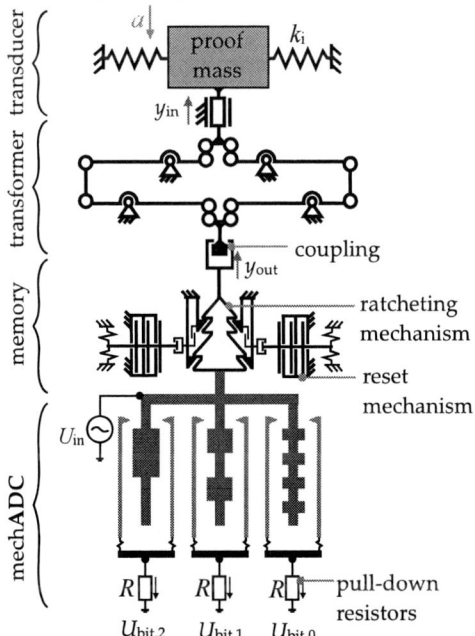

Figure 3: Mechanical setup of a passive accelerometer with mechanical memory

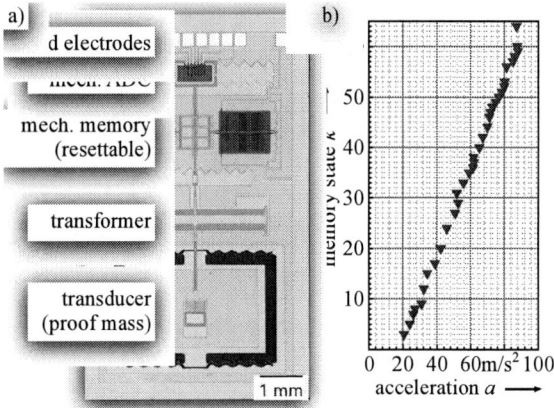

Figure 4: a) fabricated accelerometer based on SOI-technology and b) memory states as a function of imposed acceleration

The coupling compares the previously stored displacement at the memory with the displacement available at the output of the amplifier. If the current displacement exceeds the stored value, the new displacement is transmitted into the memory. This allows to overwrite the memory with the latest maximum value. The electrically readable memory is realized by a mechanical A/D converter (mechADC) that transforms the stored displacement at the memory into a binary electrical code as soon as a voltage is applied during read out. Fig. 4a shows the fabricated acceleration sensor based on SOI-technology. The mechanical memory features a ratcheting mechanism with 6 bit memory depth corresponding to 64 discrete states. Measurements in Fig. 4b show the memory states that correspond to the imposed acceleration. A reset of the mechanical memory is achieved by electrostatic actuators that allow to open the locked ratcheting position, as indicated in Fig. 5a and b. The required reset voltage strongly depends on the memory state, since the frictional force at the pawl-ratchet connection increases with the memory state. Fig. 5c shows the measured reset voltage as a function of the memory state.

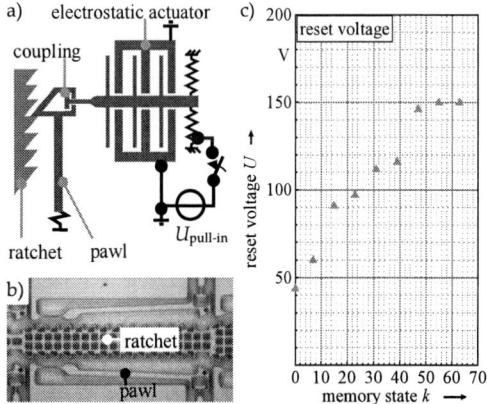

Figure 5: a) schematic of the reset mechanism at the memory. b) microscope image showing the reset at the memory. c) measured reset voltage as a function of the memory state.

Electrical path

The core of the electrical path concept is a combination of a piezoelectric MEMS for detecting inertial events such as shocks and vibrations and a nanoionic memory element whose internal resistance is reduced by the charge carriers provided by the transducer.

In order to obtain sufficient charge from vibration, we designed piezoelectric harvesters with a unique drone-like shape of the active elements. The MEMS are designed for a frequency of about 1.5 kHz. The first vibration mode, as well as design details are shown in the Fig. 6. This MEMS shape has sufficiently large gain to operate the nanoionic memory element. Moreover, it should be robust to withstand shocks and vibration.

Fabrication technology [5] and a resulting device are shown in the Fig. 7. The active layer thickness is about 20 μm, the AlN layer is about 600 nm. By coupling the piezo MEMS with the passive rectifier circuit, one obtains a passive vibration-to-voltage converter, which rectifies the low electrical AC signals from the MEMS to generate a sufficiently high DC output signal to write the nanoionic memories. Both functional units are matched to each other with a suitable coupling circuit.

The nanoionic memory element is based on a titanium dioxide Memristor. The memristive effect in rutile TiO_x layers leads to an increase in current or a reduction in internal resistance when an electrical voltage is applied, the writing process of the nanoionic memory element. If no electrical voltage is applied, the value of the internal resistance stabilizes in the direction of the starting value. With some nanoionic memory elements, a difference remains. This can be read out and thus qualitative information about the frequency and intensity of the writing processes can be derived. By applying a reversed electrical voltage, the resistance value is reduced enabling a reset of the memory. Various layer systems were designed and investigated for the nanoionic storage element. The layer structure creates a metal-semiconductor junction between the memristic TiO_x layer and the cover electrode. The Schottky barrier allows the current flow through the nanoionic storage element to be limited in a scalable manner and the energy requirement to be minimized. By defining the layer thickness of the TiO_x and the cover electrode area, the reverse current via the component can be selected in such a way that the nanoionic memory element can be connected to piezo-MEMS and can be read out via low-cost measurement electronics. Silicon with an additional 1 μm thermal SiO_2 layer serves as the substrate for the various TiO_x memristors. A 20 nm titanium layer is sputtered onto this as a bonding agent for the platinum layer as the back electrode of the memristor.

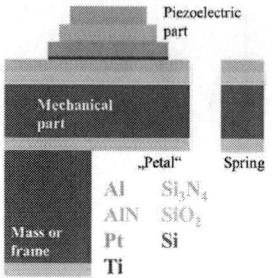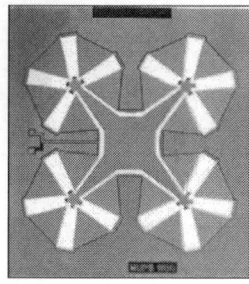

Figure 7: Cross-section of a fabrication technology for a piezoelectric MEMS and fabricated chip

The memristive TiO_2 layer was deposited and patterned using a PVD system CS400SR from Ardenne in a constant reactive gas flow. The electrically functional layer structure of the nanoionic memory element is a three-layer system with 100 nm Pt / 50 nm TiO_2 / 100 nm Pt. Between the deposition of the memristive layers and the deposition of the cover electrodes, the wafers are annealed at a temperature of 800 °C for 30 min. The Pt top electrode is necessary for the formation of a Schottky barrier and the 300 nm Al layer above it enables a downstream wire base.

To provide the functionality of the overall system, the challenge is to design a coupling circuit between the piezoelectric MEMS and the nanoionic memory element. The coupling circuit should be able to convert the voltage signal of the piezo MEMS into a DC voltage for the nanoionic memory cell. For this purpose, a Villard-Greinacher charge pump was designed and built.

Assembled systems, shown in the Fig. 8 generates electrical voltages of up to 3 V/g (peak-peak) at low accelerations of ≤ 1 g. Output voltages of >1.5 V/g measured on systems after passive rectification. The functionality of the system together with the nanoionic memory elements is demonstrated. Fig. 9 shows the read-out of the internal resistance value from the nanoionic memory element (blue area) with a voltage of 100 mV before a 30-minute write process (set process) with 1.5 V output voltage at the piezoelectric MEMS (green area).

After the set process, the second readout followed (blue area). The resistance value of the memristor reduced over the SET process from initially around 3.13 MΩ to around 2.79 MΩ. It was possible to prove that the resistance value of the nanoionic memory could be reduced or described by the output voltage of an excited wake-up MEMS. After a dwell time of about 18 hours, a third readout was started the following day and a resistance value of about 2.8 MΩ was determined.

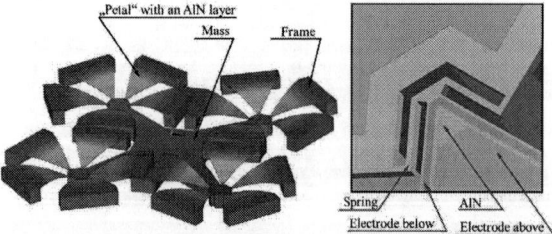

Figure 6: First mode of a piezoelectric MEMS as well as

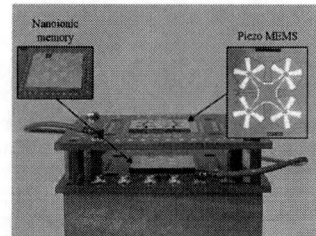

Figure 8: Assembled "electrical path" system

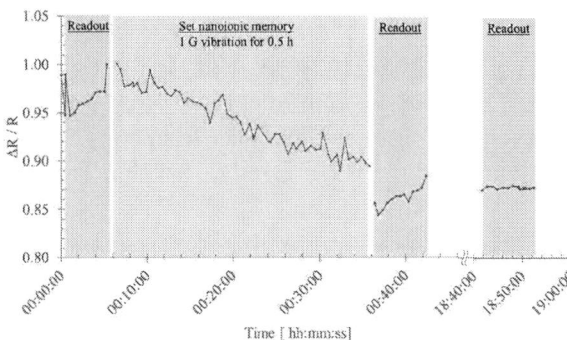

Figure 9: Obtained results

Electrical read-out by RFID

Both concepts have successfully been proven in combination with RFID interfaces for wireless read-out and reset while operating without any electrical auxiliary power during data recording.

Readout and reset of the MEMS:

The MEMS described previously have been bonded to a printed circuit board with readout electronics (Fig. 10). The assembly on the PCB can be divided into three parts: The communication interface with RFID/NFC (13.56 MHz, NTP53321G0JHKZ) interface with integrated harvester, a readout unit and a reset unit. The on-board communication is performed by an I2C interface. The state of the mechADC is read out via a digital interface (PCAL6408A). To reset the MEMS (Fig. 5a and b), a voltage >100 V (Fig. 5c) has to be applied to the electrostatic actuator of the MEMS. Here, a boost converter turned out to be the most suitable solution. The electrostatic actuator can be assumed to be an almost capacitive load. Thus, a theoretically infinite voltage can be generated at the output. The voltage is adjusted by the used diode and corresponds approximately to the maximum reverse voltage of this (~150 V). The RFID tag provides the control signals and can be read by a standard RFID reader or even a smartphone with a suitable interface.

Readout and reset of the nanoionic memory:

Similar to the MEMS, a readout and reset circuit for the nanoionic memory was developed (Fig. 11). The state of the nanoionic memory is stored as a resistance value. This is in the range of ~1MΩ. The resistance value is determined by measuring the resistance and applying a voltage to it. The current is measured using a shunt resistor with a relatively high impedance 100kΩ and allows accurate resistance measurement, even at only very low voltages (0.1V to 3V).

Figure 10: Readout and reset PCB for the "mechanical path". The NFC transponder can be addressed with a smartphone at a distance of up to 2 cm.

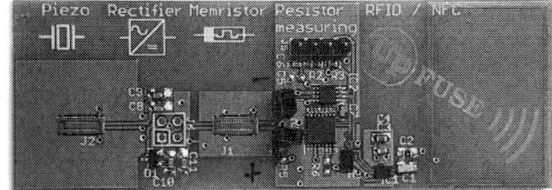

Figure 11: Readout and reset PCB of the nanoionic memory. Piezo and nanoionic memory are plugged onto the corresponding locations and are not bonded (not present in the picture).

To reset the nanoionic memory, a negative voltage is applied by applying the voltage in the reverse direction. For this reason, an ADC with two outputs is used. A characteristic feature of the circuit is its high impedance in the unpowered case.

CONCLUSION

In this contribution we presented two concepts to acquire and permanently store physical signals without auxiliary electrical energy using a mechanical ratcheting mechanism and nanoionic memories. Both types of memories could be read out by means of RFID. By changing the primary transducer, a multitude of other measures can be adapted to these concepts for control in logistic chains, preventive maintenance or general performance recording.

ACKNOWLEDGEMENTS

This research is funded by the German Federal Ministry of Education and Research within the ForMikro project upFUSE (contract number: 16ES1063). MEMS are partially realized at ZGH, Ruhr-Universität Bochum.

REFERENCES

[1] J. Singh, R. Kaur, and D. Singh, "Energy harvesting in wireless sensor networks: A taxonomic survey," International Journal of Energy Research, vol. 45, no. 1, pp. 118-140, 2021.

[2] R. R. Reddy, K. Komeda, Y. Okamoto et al., "A zero-power sensing MEMS shock sensor with a latch-reset mechanism for multi-threshold events monitoring," Sensors and Actuators A: Physical, vol. 295, pp. 1-10, 2019.

[3] L. J. Currano, M. Yu, and B. Balachandran, "Latching in a MEMS shock sensor: Modeling and experiments," Sensors and Actuators A: Physical, vol. 159, no. 1, pp. 41-50, 2010.

[4] P. Schmitt, and M. Hoffmann, "Direct Binary Encoding of Displacements on the Nano-Scale." pp. 677-680.

[5] Meinel, K., Melzer, M., Stoeckel, C., Shaporin, A., Forke, R., Zimmermann, S., Hiller, K., Otto, T., Kuhn, H.; 2d scanning micromirror with large scan angle and monolithically integrated angle sensors based on piezoelectric thin film aluminum nitride; (2020) Sensors (Switzerland), 20 (22), pp. 1-23.

CONTACT

*Martin Hoffmann, tel: +49 234 32-27700; Martin.Hoffmann-MST@RUB.de

A NEW FINDING ON NONLINEAR DAMPING AND STIFFNESS OF FLEXURAL MODE CAPACITIVE MEMS RESONATORS

Hung-Yu Chen[1], Ming-Huang Li[2], and Sheng-Shian Li[1,2]

[1]Institute of NanoEngineering and MicroSystems, National Tsing Hua University, Hsinchu, Taiwan
[2]Department of Power Mechanical Engineering, National Tsing Hua University, Hsinchu, Taiwan

ABSTRACT

This work reports a new resonant phenomenon of flexural mode capacitive MEMS resonators induced by nonlinear damping and nonlinear stiffness. The measured frequency response clearly demonstrates a variety of interesting resonant behaviors of a capacitive resonator when operated from its linear into nonlinear regime. As the input ac voltage increases (i.e., drive level dependency), the resonant frequency shifts with 6000 ppm while transmission magnitude also varies greater than 200%. Such strong nonlinear phenomenon mainly comes from the flexural mode capacitive resonator featuring excellent electromechanical coupling strength contributed by its narrow transduction gap spacing and large transduction area [1]. A behavior model for predicting nonlinear electrostatic force is built based on the measured linear response under different DC bias. By comparing the simulated result from the behavior model and the actual measurement data, both nonlinear stiffness term and nonlinear damping term can be successfully extracted.

KEYWORDS

CMOS, MEMS, capacitive transduction, resonator, electromechanical coupling, nonlinearity

INTRODUCTION

To enhance the electromechanical coupling capability of a capacitive transducer, minimizing the transduction gap spacing is the most efficient method. Narrower gap spacing efficiently helps the capacitive transduction to enjoy great electromechanical coupling strength under relatively lower DC bias voltage. Therefore, no matter it is flexural mode or bulk mode, many studies are pursuing the ultimate narrow gap [2][3] even towards nano scale for fabricating capacitive resonators. However, with the shrinkage of the gap spacing and the improvement of the electromechanical coupling, the deflection-dependent characteristics of the electrostatic force will become dominant and cannot be neglected especially for flexural mode resonator with lower structural stiffness, k_m. The system stiffness reduction caused by electrostatic force can be represented by adding a nonlinear stiffness term, k_e, into the system equation of motion:

$$F_{drive} = m\ddot{w} + b\dot{w} + (k_m - k_e(w))w \qquad (1)$$

where w is deflection of the beam in a simplified Single Degree of Freedom (SDOF) system. F_{drive} is the deflection-independent electrostatic force while m and b represent the equivalent mass and damper of the system, respectively.

Such nonlinear characteristic has been proved to reduce the equivalent stiffness of the resonant system which is called spring-softening Duffing nonlinearity [4].

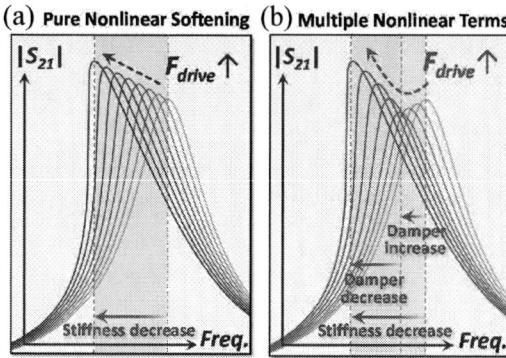

Fig. 1: Frequency response schematics of a nonlinear system with (a) pure nonlinear softening and (b) multiple nonlinear terms under different driving levels.

The resonant frequency shifts towards lower frequencies and S_{21} magnitude slightly increases as the input driving level increases [Fig. 1 (a)]. In addition, larger deflection of the structure would introduce a completely opposite stiffness-hardening phenomenon caused by the changes in material properties [5]. Most of the studies on the nonlinearity of capacitive resonators are quite limited to the discussion of the nonlinear stiffness, including softening and hardening. Besides, many studies can only be carried out using simulation because the generation of nonlinear phenomenon heavily relies on extremely high electromechanical coupling or unique material and structural properties, which are difficult to obtain.

In addition to the nonlinear stiffness terms contributed from electrostatic force, once the nonlinear damping term

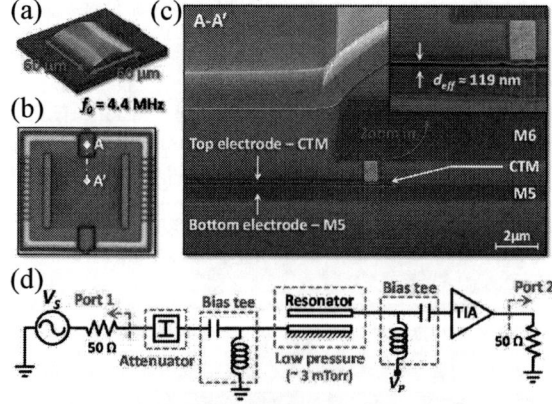

Fig. 2: (a) Finite element method (FEM) simulated mode shape and resonant frequency of the flexural-mode capacitive resonator. (b) Optical microscope picture of the resonator. (c) Focus ion beam (FIB) cut image of the effective 119nm capacitive transduction gap for the fabricated resonator. (d) Measurement setup for characterizing the nonlinear behaviors of the device.

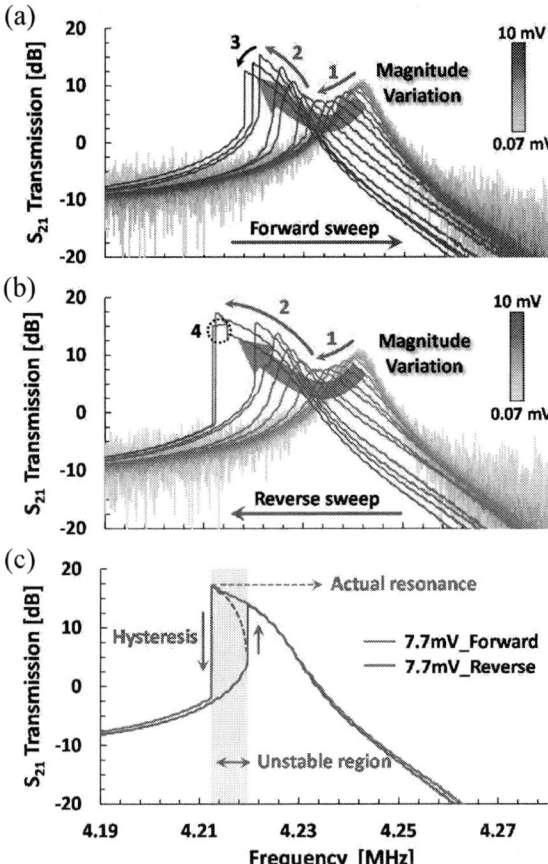

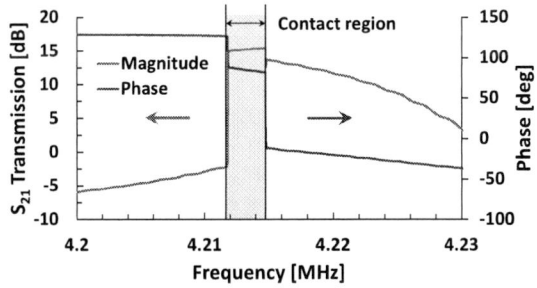

Fig. 3: Measured S_{21} transmission responses of the resonator in the setup depicted in Fig. 2(d) with various input AC voltage under $V_P = 8V$ using (a) forward and (b) reverse sweeps, respectively. (c) Hysteresis in frequency response with $V_P = 8V$ and input AC voltage V_{in} of 7.7mV.

Fig. 4: Measured S_{21} transmission response of the resonator with input AC voltage V_{in} of 10mV, shows a contact condition at resonance.

or other nonlinear stiffness term are considered, the frequency response and equation of motion of the system will become more complicated [Fig. 1(b)]:

$$\begin{aligned}F_{drive} \\ = m\ddot{w} + (b + b_n(w))\dot{w} + (k_m - k_e(w) + k_n(w))w\end{aligned} \qquad (2)$$

By building a behavior model for the nonlinear electrostatic force and comparing with the measurement results, we're able to extract the influence of each nonlinear term in equation (2) on the nonlinear system response.

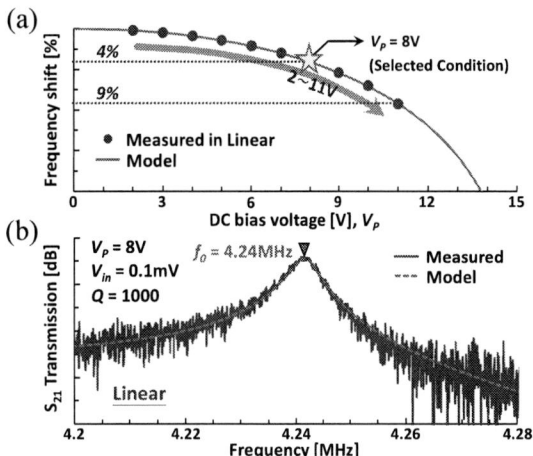

Fig. 5: (a) Measured and modeled resonant frequency shift in linear regime under different V_P. (b) Measured and modeled linear frequency response under $V_P=8V$.

MEASUREMENT RESULTS

To generate a sufficiently strong nonlinearity, a clamped-clamped beam capacitive resonator shown in Fig. 2 that features narrow transduction gap (119nm) and large electrode area is adopted as the experimantal target. The resonator is characterized using a network analyzer in low pressure chamber with forward and reverse frequency sweeps under different DC and AC voltage [Fig. 2(d)].

Fig. 3(a) and (b) show the measured S_{21} responses of the resonator characterized using forward and reverse sweeps, respectively. When the resonant frequency decreases with the increased input AC voltage as expected, the change in the magnitude of S_{21} shows a significantly nonlinear variation. The process from linear to nonlinear regime can be divided into 4 stages. In the first stage, once the input AC voltage reaches certain level, the stiffness of the system starts to decrease, thus leading to a resonant frequency reduction. At the same time, S_{21} magnitude drops of about 3dB. In the second stage, S_{21} magnitude sharply rises of about 10dB while the resonant frequency keeps decreasing. As the effect of nonlinear softening continuously increases, the system response exhibits hysteresis phenomena [Fig. 3 (c)] and can be observed from the magnitude decay in forward frequency response marked as third stage. Strong hysteresis implies that the actual resonant frequency with larger defection can only be found using reverse sweep. Thus, in forth stage, a structural collision to the bottom boundary happens during the reverse sweep with higher AC input voltage as shown in Fig. 4.

ANALYSIS OF NONLINEAR BEHAVIOR

Nonlinear Electrostatic Force

Fig. 5(a) presents the electrostatic spring softening effect under different DC bias voltage, V_P, caused by the deflection-dependent electrostatic force, F_e:

$$F_e = \frac{1}{2}\frac{dC}{dw}(V_P + V_{in})^2 \qquad (3)$$

where V_{in} is input AC voltage and $\frac{dC}{dw}$ is the source of

978-1-6654-9309-3/23 $31.00 © 2023 IEEE

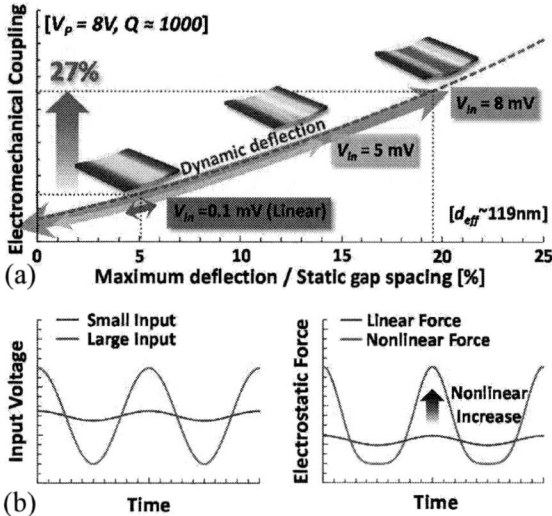

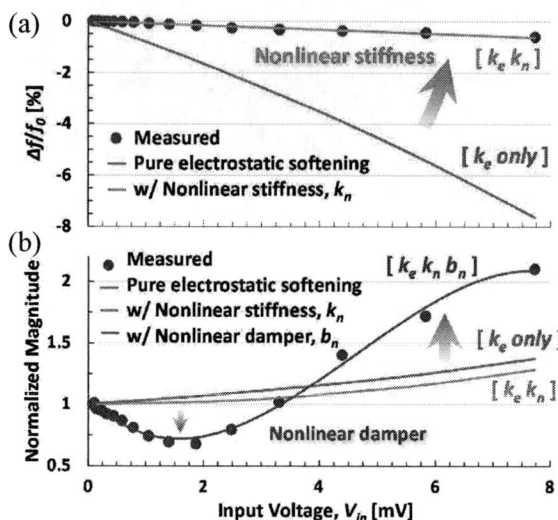

Fig. 6: (a) Modeled relationship between the electromechanical coupling capability and maximum structural deflection at the center of the resonator. Coupling strength is increased by 27% from linear state to 8mV drive voltage. (b) Time-domain schematic diagram of nonlinear electrostatic force enhancement generated by increasing small driving signal into large driving signal.

Fig. 7: (a) Measured and modeled resonant frequency shift with various input level based on considering mechanical nonlinear stiffness term, k_n, or not. (b) Measured and modeled transmission magnitude change with various input level based on considering different nonlinear terms.

nonlinearity that makes electrostatic force nonlinearly increases as the structural deflection increases. V_{in} is set low enough by using attenuator to ensure that the frequency response is in the linear regime and no other mechanical nonlinear terms, b_n and k_n, in equation (2) are introduced. To analyze the relationship between total capacitance and structural deflection, we first start with the simplified deflection profile, $w(x, t)$, of the clamped-clamped beam after applying voltage across the electrodes:

$$w(x,t) \approx w_p(t)\left[1-\left(\frac{x}{L}\right)^2\right]^2 \qquad (4)$$

where w_p is the displacement at the center of the beam, L is the half-side length of the beam, and x is the distance away from the center of the beam. The total capacitance, $C(t)$ of the deflected clamped-clamped beam can be written as

$$C(t) = \frac{W}{d_{eff}}\int_{-L}^{L}\frac{\varepsilon_0}{1-\frac{1}{d_{eff}}w(x,t)}dx$$
$$= C_0 g\left(\frac{x_p(t)}{d_{eff}}\right) = C_0 g(u) \qquad (5)$$

where W is the width of the beam, d_{eff} is the effective transduction gap spacing, ε_0 is the permittivity of free space, C_0 is the capacitance at zero deflection, and the function $g(u)$, which describes the shape of the capacitance curve, can be found by performing polynomial fitting. According to equation (5), the electrostatic force expressed in equation (3) can be rewritten as

$$F_e(t) = \frac{\varepsilon_0 WL}{d_{eff}^2}\cdot(V_P+V_{in}(t))^2\cdot g'(u) \qquad (6)$$

The only unknow variable, d_{eff}, can be extracted as

119nm by comparing the enhancement of electrostatic force in equation (6) to the measured frequency shift caused by different DC bias voltage in linear regime [Fig. 5(a)]. After obtaining the effective transduction gap spacing, we can establish the relationship between deflection and electrostatic force, and plot a curve as shown in Fig. 6(a). The linear frequency response shown in Fig. 5(b) is then used as a starting point for calculating the nonlinear response by iteration over the established curve. Note that the actual input AC voltage is calculated by the driving power of the network analyzer with considering the impedance mismatch between the resonator under test and ideal 50Ω port.

As depicted in Fig. 6(a), the electromechanical coupling strength with larger input AC voltage ($V_{in} = 8$mV) is increased by about 27% compared to the linear condition ($V_{in} < 0.1$mV). To better understand such nonlinear phenomena, a time-domain response of a nonlinear capacitive transducer can be simulated as shown in Fig. 6(b) by using harmonic balance method. An input AC voltage with larger amplitude will generate significant nonlinear variation on electrostatic force.

Extraction of nonlinear damping & stiffness

Such a nonlinearly rising electrostatic force will appear as a characteristic of resonant frequency drift during measurement with continuously increasing input AC voltage as shown in Fig. 7(a). After comparing the calculated resonant frequency shift due to nonlinear electrostatic forces with the measurement results, we can extract the mechanical nonlinear stiffness term, k_n, because another mechanical nonlinear damping term, b_n, is not related to the resonant frequency.

Finally, once all nonlinear terms (k_e, k_n, and b_n) are taken into consideration and the measured magnitude of S_{21} transmission frequency responses obtained with different

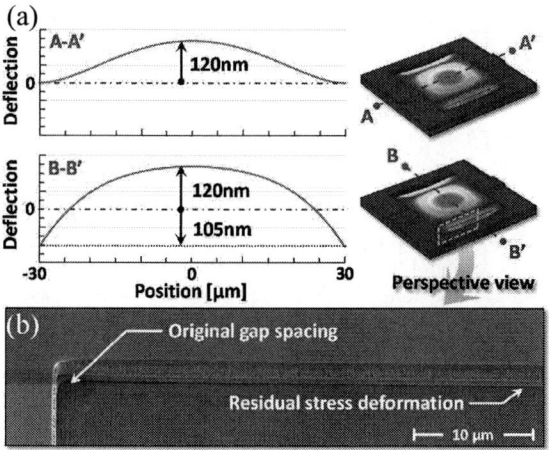

Fig. 8: (a) Optically measured static deformation in different axis with FEM simulated figures caused by residual stress and (b) SEM image of the transduction gap observed from the side of the resonator.

input AC voltages are used as the fitting target, the last nonlinear term, b_n, which represents nonlinear damper, can be extracted as shown in Fig. 7(b).

Nonlinear Frequency Response

After the extraction process of nonlinear terms, we can clearly identify the various steps where the capacitive resonator encounters in the process of nonlinearization during the measurement [Fig. 3]. According to the nonlinear stiffness extraction result shown in Fig. 7(a), the drift of the resonant frequency is mainly caused by the nonlinear electrostatic force, k_e. However, due to the addition of extra nonlinear mechanical stiffness term, k_n, the frequency drift is not as much as expected.

As for the measured variation of S_{21} magnitude, it can be seen from Fig. 7(b) that it is dominated by nonlinear damping term, b_n. Unlike k_e and k_n terms, which continue to grow with increasing deflection, b_n term exhibits higher-order variation behavior. This might reveal that at least two physical mechanisms involved during the mechanical motion.

Air Damping & Residual-stress Induced Deformation

It is speculated that the source of such peculiar nonlinear damping is air. The flow and squeezing of air between narrow transduction gaps contribute additional damping and stiffness to the mechanical system [6].

In addition to the small gap, the reason for significant air damping is that the deformation caused by residual stress restricts the flow of gas inside the gap, thereby causing such a special damping characteristic. Fig. 8(a) shows the deformation profile from optical measurement and FEM simulated 3D view. It can also be seen from the SEM image in Fig. 8(b). The upwardly bulging beam and the downwardly curved structural side edges make a large amount of air trapped in the transduction gap, and finally, with the appropriate ambient pressure, produce such a peculiar damping effect.

CONCLUSIONS

By characterizing a flexural-mode capacitive resonator featured with narrow transduction gap and large electrode area, this study reports a special nonlinear frequency response caused by multiple nonlinear sources, including nonlinear electrostatic force, nonlinear mechanical stiffness, and nonlinear mechanical damping, for the first time. Except for the 6000ppm resonant frequency drifting, the S_{21} transmission magnitude varies up and down for more than 200%.

To understand the effect of each nonlinear term on the overall system response, a behavior model for predicting nonlinear electrostatic force is built according to the measured electrostatic spring softening effect in linear condition. By comparing the calculated behavior with only nonlinear electrostatic force being considered to the measurement result in both resonant frequency drift and transmission magnitude, the nonlinear stiffness term (k_n) and the nonlinear damping term (b_n) can be extracted.

Finally, from the extraction results, we are able to clearly understand which nonlinear term dominates each stage of the device from linear to nonlinear regime.

ACKNOWLEDGEMENTS

The CMOS chip fabrication was supported by the Taiwan Semiconductor Research Institute (TSRI) and TSMC, Hsinchu, Taiwan. This work was supported by MOST 111-2221-E-007-135-MY3.

REFERENCES

[1] H.-Y. Chen, S.-S. Li, and M.-H. Li, "A low impedance CMOS-MEMS capacitive resonator based on metal-insulator-metal (MIM) capacitor structure," *IEEE Electron Device Letters*, vol. 42, no. 7, pp. 1045-1048, July 2021.

[2] A. Ozgurluk, K. Peleaux, and C. T.-C. Nguyen, "Single-digit-nanometer capacitive-gap transduced micromechanical disk resonators," *2020 IEEE 33rd International Conference on Micro Electro Mechanical Systems (MEMS)*, 2020, pp. 222-225.

[3] A. Ozgurluk, K. Peleaux, and C. T.-C. Nguyen, "Widely tunable 20-nm-gap ruthenium metal square-plate resonator," *2019 IEEE 32nd International Conference on Micro Electro Mechanical Systems (MEMS)*, 2019, pp. 153-156.

[4] Y.-W. Lin, S. Lee, S.-S. Li, Y. Xie, Z. Ren, and C. T.-C. Nguyen, "Series-resonant VHF micromechanical resonator reference oscillators," *IEEE Journal of Solid-State Circuits*, vol. 39, no. 12, pp. 2477-2491, Dec. 2004.

[5] A. M. Elshurafa, K. Khirallah, H. H. Tawfik, A. Emira, A. K. S. Abdel Aziz, and S. M. Sedky, "Nonlinear dynamics of spring softening and hardening in folded-MEMS comb drive resonators," *Journal of Microelectromechanical Systems*, vol. 20, no. 4, pp. 943-958, Aug. 2011.

[6] M. Bao and H. Yang, "Squeeze film air damping in MEMS", *Sens. Actuators A: Phys.*, vol. 136, no. 1, pp. 3-27, May 2007.

CONTACT

*H.-Y Chen, Email: neilswords@gmail.com.

EXPLOITING PARAMETRIC INSTABILITY IN BISTABLE MEMS ACTUATORS

Daniel Platz[1], Johannes Fabian[1], Elisabeth Samm[1], Mahdi Mortada[1], Michael Schneider[1], and Ulrich Schmid[1]

[1]TU Wien, Institute of Sensor and Actuator Systems, Gußhausstraße 27-29, 1040 Vienna, AUSTRIA

ABSTRACT

Parametric instability in elastic structures is mostly considered as danger to the structures' integrity. In this paper, we show how parametric instability can be exploited to implement buckled bistable MEMS actuators which reach exceptionally large actuation potentials. We introduce a model for efficiently simulating snap-through dynamics between different buckling states in bistable MEMS. Simulations of the snap-through initiation time in piezoelectrically excited devices show instability regions which are typical for parametric instability. Regions of parametric instability are also found in experiments with bistable MEMS proving that parametric excitation can be exploited for implementing novel bistable MEMS actuators.

KEYWORDS

Actuator, nonlinear dynamics, bistability, snap-through, parametric instability, parametric resonance, plates.

INTRODUCTION

During the last decades, microelectromechanical systems (MEMS) became an integral part of the semiconductors industry's product portfolio. At the same time, MEMS also evolved into a versatile technology platform for basic science in various fields. Despite the matureness of MEMS technology that was achieved during this development, MEMS still face fundamental limitations. One of these limitations are the small mechanical displacements in MEMS. Even resonant MEMS often do not reach motional amplitudes above a few micrometers. This limitation of mechanical displacements severely limits the application of MEMS as actuators. An example application that would greatly benefit from larger motional amplitudes in MEMS are loudspeakers. MEMS loudspeakers would enable the design of more compact mobile devices at lower cost. However, reaching a sufficient sound pressure level in MEMS loudspeakers remains a challenge due to the limited mechanical displacements.

A promising concept to overcome current limitations of mechanical displacements in MEMS is bistability [1]. In bistable MEMS, not only one but two (and possible more) static equilibrium mechanical configurations exist. A way to implement bistable MEMS is the application of a compressive pre-stressing to an elastic structure. Beyond a critical stress, the structure buckles and typically at least two stable buckling configuration exists for a given pre-stress [2]. It has been demonstrated that bistable MEMS reach buckling amplitudes up to tens of micrometers [3-5]. Switching between different stable buckling configurations results in motional amplitudes that are difficult to reach in conventional monostable MEMS. However, these large motional amplitudes come at the price of highly nonlinear switching dynamics. Moreover, only very little is known about the efficient and controlled initiation of the switching process, the so-called snap-through, since researchers focused mostly on methods for avoiding snap-through in elastic structures.

Here, we discuss how parametric instability can be exploited to efficiently initiate the snap-through process in bistable MEMS. We focus our analysis on bistable MEMS with integrated actuation by a layer of piezoelectric material. Previous experiments demonstrated successfully that piezoelectric excitation of bistable MEMS can initiate the snap-through between two buckled static configurations [6] but the mechanism of snap-through initiation and the optimal parameters for most efficient snap-through initiation remained unclear. We introduce a model for the efficient study of snap-through initiation in bistable piezoelectric MEMS. Using this model, we show that piezoelectric actuation allows for exploiting parametric instability for energy efficient snap-through initiation. The model reproduces well the essential features in experimental data that were obtained from bistable MEMS with similar geometry.

THEORETICAL MODEL

We consider a buckled elastic structure with a square geometry as shown in Fig. 1.

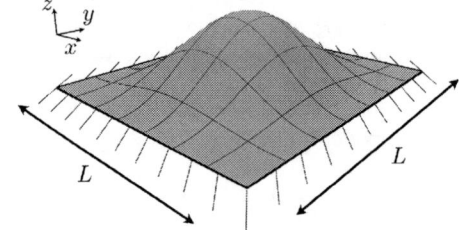

Figure 1: Plate geometry considered in the model. The same static in-plane pre-stress is applied on all four sides. Beyond a critical stress, the plate buckles either in positive or negative z-direction.

The four sides with length L are aligned with the x- and y-axes of the coordinate system. The structure is perfectly clamped at all sides whereas motion in z-direction is unconstrained in the structure's interior. A static pre-stress acts on the structure in the xy-plane such that above a critical stress the structure buckles in z-direction. To model the static buckling and the snap-through dynamics, we assume that the deflections in z-direction of the structure are moderately large and that no significant in-plane displacements occur in the xy-plane. An ad-hoc model which is compatible with these assumptions is the von Karman plate in which the equation of motion for the deflection w in z-direction is given by

$$\mu w'' + \mu \kappa w' + D \nabla^4 w$$
$$-\left(N_{xx}w_{,xx} + N_{yy}w_{,yy} + N_{xy}w_{,xy}\right) = 0, \qquad (1)$$

where μ is the aerial mass density, κ is the damping coefficient, D is the flexural rigidity and N_{xx}, N_{xx} and N_{xx} are the xx, yy and xy-in-plane stress resultant. The prime denotes differentiation with respect to time and the comma indicates spatial differentiation with respect to the coordinates following the comma. We assume that the in-plane stresses can be decomposed into three different contributions such that

$$N_\alpha(t,x,y) = N_\alpha^{(b)}(t,x,y) + N_\alpha^{(p)}(t) + N_\alpha^{(0)},$$
$$\alpha \in \{xx, yy, xy\}. \qquad (2)$$

Here, $N_\alpha^{(b)}$ represents the in-plane stress due to bending, $N_\alpha^{(p)}(t)$ is a homogeneous piezo-stress with a time-dependence that is controlled by the voltage applied to the piezo-material and $N_\alpha^{(0)}$ is a static homogeneous pre-stress that can be engineered during device fabrication. To efficiently solve equation (1), we use an Airy function ansatz for the in-plane stress resultants and a Galerkin method in which we approximate the deflection in z-direction as a superposition of the first four free vibrational modes,

$$w(t,x,y) \approx \sum_{n=1}^{4} q_n(t)\phi_n(x,y), \qquad (3)$$

where ϕ_n is the mode shape of the n-th mode and q_n is the time-dependent deflection of the n-th mode.

SIMULATION RESULTS

Simulation Parameters

For simulations, we choose parameters which are comparable to the parameters in recent experiments [6]. In detail, we consider a square membrane with a side length of 700 μm and a thickness of 3.5 μm. However, we do not include details of the piezoelectric material stack in the model and assume that the essential features of the dynamics are well reproduced using an isotropic linear material model. The Young's modulus in this model has a value of 165 GPa and Poisson's ratio is 0.22.

Static Analysis

To determine the critical stresses of the different buckling modes of the structure, we set all time derivatives and excitation terms in equation (1) to zero and solve the nonlinear algebraic systems of equations resulting from the Galerkin approach described above. The system has multiple solutions depending on the static pre-stress. In Fig. 2, the solutions in which only one of the modes exhibits a nonzero deflection are shown.

Below the critical stress of N_{cr} = -66.55 Nm^{-1}, only trivial solutions with zero deflection exist. Above this pre-stress value, the zero-deflection solution becomes unstable and the ϕ_1 mode begins to buckle. This buckling solution is stable and the buckling amplitude increases with increasing pre-stress. Other buckling modes involving the

ϕ_2 and ϕ_3 modes appear at a pre-stress of 1.79 N_{cr}. However, these buckling modes are unstable since the pre-stress is applied symmetrically on all four sides of the structure while the ϕ_2 and the ϕ_3 mode do not possess this symmetry. The ϕ_4 mode exhibits an appropriate symmetry for becoming a stable buckling mode and it starts to bifurcate from the zero-deflection solution at 2.48 N_{cr}. However, the solution remains unstable until a pre-stress of 3.03 N_{cr} is reached. Between this point and a pre-stress of 3.70 N_{cr} both the ϕ_1 and the ϕ_4 are stable buckling modes until the ϕ_1 mode becomes unstable.

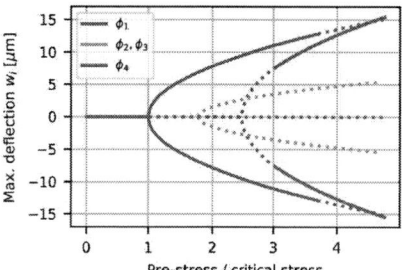

Figure 2: Static monomodal solutions of equation (1). The solid lines indicate a stable solution whereas dashed lines mark unstable solutions.

The results indicate that multiple buckling modes can exist simultaneously in a pre-stress regime that can be reached in experiments.

Snap-Through Dynamics

The use of bistable MEMS in actuator applications requires a controlled initiation of the snap-through between different buckling states. Those snap-through events can involve different states of the same buckling mode or the transition between different buckling modes. Here, we consider snap-through involving only the ϕ_1 mode. While in recent experiments [6], different waveform types have been used for initiating snap-through in bistable piezoelectric MEMS, we focus on homogeneous isotropic piezo-excitations which are purely sinusoidal and are characterized by a frequency f and a piezo-stress amplitude $N^{(p)}$.

A key figure of merit is the time T_s it takes from a static buckling equilibrium until the first snap-through event when an excitation is applied to the system. We study how this snap-through initiation time depends on the frequency and the amplitude of the piezo-stress. For doing so, we simulate the dynamics of the nonlinear system of coupled ordinary differential equations which stems from equation (1) by applying the Galerkin ansatz in equation (3). As initial condition, we choose the static positive-deflection buckling configuration of the ϕ_1 mode shown in Fig. 1 with zero velocity as the piezoelectric excitation sets in at $t_0=0$. We define a snap-through event as the sign change of the ϕ_1 mode deflection and measure the time between t_0 and the first snap-through event. The results are shown in Fig. 3 for a static pre-stress of 1.1 N_{cr} which is just above the bifurcation point of the ϕ_1 buckling mode.

In the white areas, no snap-through event occurred during the simulation period of 400 μs. The colored areas exhibit one or more snap-though events. The excitation parameter space in which snap-through events can be initiated is segmented into three tongue-like areas with tips at frequencies of 35.5 kHz, 297 kHz and 489 kHz. These frequencies correspond to the resonance frequency of the ϕ_1 mode and to twice the resonance frequencies of the ϕ_2/ϕ_3 mode and the ϕ_4 mode. This indicates that especially in the center and right tongue the snap-through is initiated by a parametric instability of the ϕ_2/ϕ_3 modes and the ϕ_4 mode, respectively.

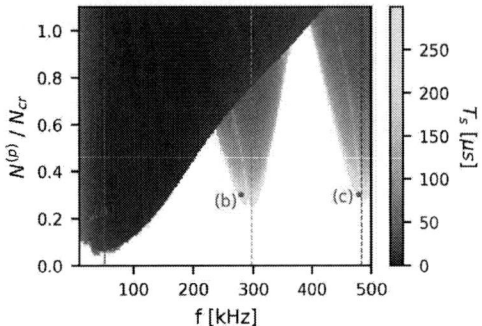

Figure 3: Simulated snap-through initiation time as a function of piezo-stress amplitude and frequency. The solid line indicates the resonance frequency of the ϕ_1 mode, the dashed lines are located at twice the resonance frequencies of the ϕ_2/ϕ_3 and ϕ_4 modes, respectively.

This interpretation is further strengthened when investigating the time-dependent deflection of the individual modes. In Fig. 4, time series of the simulated dynamics are shown for the excitation parameters which are marked by red points in Fig. 3.

The time series in Fig. 4a is located in the left tongue of Fig. 3 and is monomodal and only the ϕ_1 mode contributes to the dynamics of the system. In contrast, the dynamics shown in Fig. 4b involves multiple modes. The ϕ_1 mode exhibits small initial oscillations around the static buckling configuration at the beginning of the simulation. Additionally, an exponential increase of the oscillation amplitudes of the ϕ_2 and the ϕ_3 mode can be observed. During the later stage of this amplitude increase, the ϕ_1 mode leaves the stable buckling configuration. However, mode ϕ_1 does not reach the buckling configuration with inverted deflection sign. Instead, it oscillates around the zero-deflection line. The dynamics shown in Fig 4c are similar except that the ϕ_4 mode experiences an exponential amplitude increase and the post-snap-through oscillations of the ϕ_1 mode are significantly smaller.

The time required for snap-through initiation is not the only figure of merit for actuator applications. Another important quantity is the actuation potential which increases for larger buckling amplitudes. As shown in Fig. 2, the buckling amplitude increases with increasing pre-stress. Thus, it is generally advantageous to operate at larger pre-stresses. However, the clear segmentation of the excitation parameter space into regions in which only one mode dominates the snap-through dynamics is desirable as well since monomodal dynamics are significantly simpler to control. We investigate if the segmentation of the excitation parameter space also occurs at larger pre-stress values by introducing a modal participation factor

$$\lambda_i = \left. \langle q_i(t) \rangle_t \middle/ q_1(t_0) \right. \tag{4}$$

The participation factor λ_i measures the time-averaged participation of the i-th mode in the dynamics in relation to the initial static buckling deflection. We focus on the ϕ_4 mode as a representative example and show colormaps of λ_4 as a function of the piezo-stress amplitude and frequency in Fig. 5 for different pre-stress values.

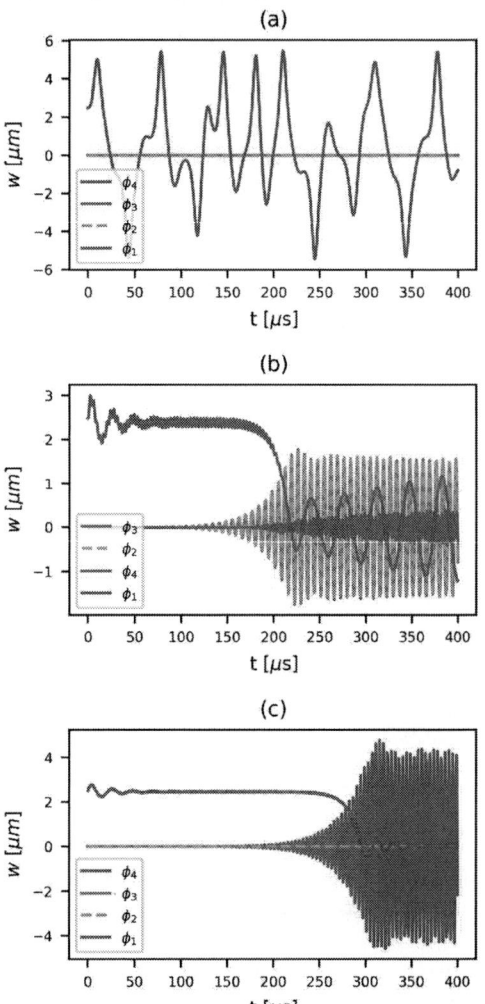

Figure 4: Simulated time-dependent max. displacements of the first four modes for the excitation parameters marked in Fig. 3.

Fig. 5a shows the modal participation factor λ_4 for the snap-through initiation time map in Fig. 3 where the pre-stress is 1.1 N_{cr}. In Fig. 5b, the pre-stress is increased to a value of 2.0 N_{cr}. It is apparent that the tongue-shaped areas from Fig. 3 have merged into a larger area. However, the tips of this area are still aligned with the resonance frequencies or twice the resonance frequencies of the system. The dynamics are predominantly still monomodal which is reflected in the participation factor λ_4 reaching significant values only in the right part of the snap-through area. This situation changes as the static pre-stress reaches 2.5 N_{cr} in Fig. 5c. The ϕ_4 mode participates in the snap-through dynamics for all excitation amplitudes and frequencies and the dynamics and no region with only mono-modal dynamics is left.

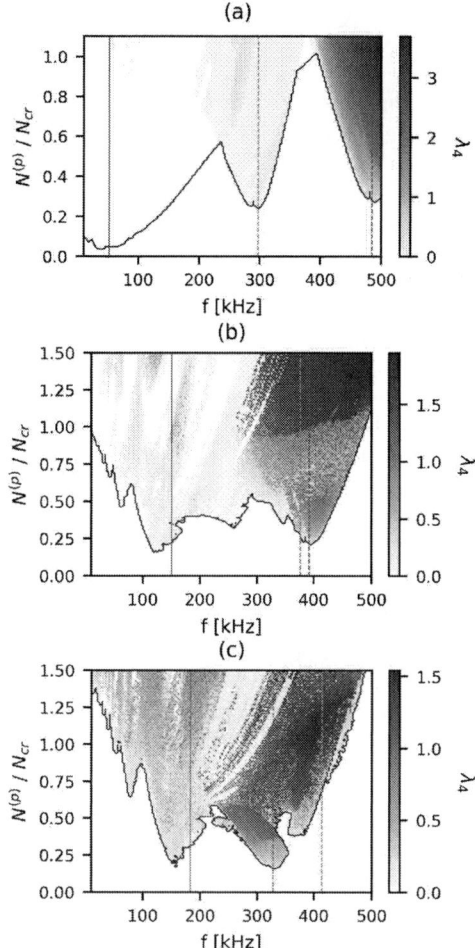

Figure 5: The modal participation factor λ_4 of mode ϕ_4 for pre-stress values of $N^{(0)}=1.1\ N_{cr}$ (a), $N^{(0)}=2.0\ N_{cr}$ (b) and $N^{(0)}=2.5\ N_{cr}$ (c).

EXPERIMENTAL RESULTS

One of the most prominent features of the simulated dynamics are the tongue-shaped snap-through areas in Fig. 3. These features are not only very prominent in the simulated data. Their presence is also of great practical importance since they hint at parametric instability as a mechanism for snap-through initiation at low piezo-stresses. However, the model underlying the simulations incorporates various simplifications like a homogeneous isotropic linear material model. To validate that parametric instability is also present in real devices, we performed experiments with a piezoelectric device that features similar geometry parameters as used in simulations. A photograph of the measured device is displayed in Fig. 6.

We measure the number of snap-through events as a function of the piezo-voltage amplitude and frequency using laser Doppler vibrometry. As found in the simulated data, there are clear tongue-like areas in the experimental data in Fig. 6 which indicates that parametric instability is the dominant mechanism for snap-through initiation. Moreover, the right area in Fig. 7 is located around 90 kHz which is close to the first resonance of the devices as predicted by simulations. Another smaller tongue appear

around 30 kHz. This tongue might be due to sub-resonance instabilities or quasi-static mode switching.

Figure 6: Photograph of the measured device.

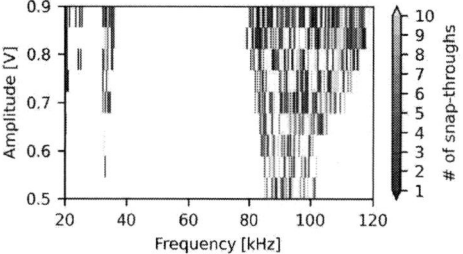

Figure 7: Measured number of snap-through events as a function of the piezo-voltage amplitude and frequency. The number of excitation pulses is the same for each frequency.

CONCLUSION

We presented a model for the efficient simulation of snap-through dynamics in bistable MEMS. The simulations results indicate that parametric instability can be exploited for efficiently initiating snap-through between buckling states at low piezo-stresses. At pre-stresses just above the critical buckling stress, the snap-through dynamics are monomodal. As the pre-stresses increases, the dynamics become increasingly multimodal. Parametric instability is found also in experimental data. These results demonstrated that parametric instability can be exploited as an efficient mechanism for snap-through initiation in buckled bistable MEMS with integrated piezo-actuation enabling novel MEMS actuators.

REFERENCES

[1] N. Hu and R. Burgueño, *Smart Mater. Struct.* **24**, 063001 (2015).

[2] M. T. A. Saif, *J. Microelectromech. Syst.* **9**, 157 (2000).

[3] M. Capanu, J. G. Boyd, and P. J. Hesketh, *J. Microelectromech. Syst.* **9**, 181 (2000).

[4] J. Qiu, J. H. Lang, A. H. Slocum, and A. C. Weber, *J. Microelectromech. Syst.* **14**, 1099 (2005).

[5] M. Dorfmeister, M. Schneider, and U. Schmid, *Sens. Actuator A Phys.* **282**, 259 (2018).

[6] M. Dorfmeister, M. Schneider, and U. Schmid, *Sens. Actuator A Phys.* **298**, 111576 (2019).

CONTACT

D. Platz, tel: +43 1 58801 76654;
daniel.platz@tuwien.ac.at

FIRST PROTOTYPE OF POLYMER MICROMACHINED FLAPPING WING NANO AIR VEHICLE

Rashmikant, Ryotaro Suetsugu, Minato Onishi, and Daisuke Ishihara
Kyushu Institute of Technology, JAPAN

ABSTRACT

The novelty of this study includes the development of an insect-inspired flapping wing nano air vehicle (FWNAV) using polymer micromachining or MEMS flyer and its computational flight performance using a fluid-structure interaction (FSI) analysis. The present FWNAV consists of a micro transmission with a support frame, a micro wing, and a piezoelectric bimorph actuator. This FWNAV can be easily fabricated using polymer micromachining and its flight performance can be accurately predicted using the FSI analysis. Hence, this study will lead toward the development of tethered and flyable FWNAVs with the size of the smallest flying natural insects.

KEYWORDS

MEMS flyer, flapping wing nano air vehicle, insect-inspired, flight performance, fluid-structure interaction, polymer micromachining.

INTRODUCTION

Insects have evolved sophisticated flight maneuverability and hovering using flapping wings [1], so many researchers have been developing flapping wing nano air vehicles (FWNAVs) for applications such as earthquake disaster management [2]. However, none of them have developed FWNAVs with a minimum insect size of about 1mm. Most recently, a few researchers have developed the FWNAVs with a size of about a few cm [3-5] and they produce enough lift-to-drag ratio for hovering [3,5], but their further miniaturization is difficult, because of the complicated fabrication process and the post-assembly. The development of FWNAVs at the insect scale has two difficulties; the primary issue is the design difficulty because of strongly coupled multi-physics like electro aeromechanical coupling coming from scale effects [6,7], and the secondary issue is the fabrication and assembly of components. The design difficulties can be overcome by a design window search methodology [8] where a design solution will satisfy all the design requirements while fabrication difficulty can be overcome by the microfabrication method [9].

We have developed the 2.5-D micro-transmissions for FWNAVs to produce the desired flapping motion transmitted from a small bending displacement of the piezoelectric bimorph actuator [9-11]. The 2.5-D structure induces ease in miniaturization and no post-assembly. In the development, we demonstrated the design of a 2.5-dimensional FWNAV using an iterative design window search method to solve the design difficulty [12,13].

We also demonstrated that the computational fluid-structure interaction (FSI) analyses can be used for the accurate estimation of the flight performance of the FWNAV because of the strong coupling between the flapping wings and the surrounding air [7, 14, 15].

Figure 1 shows the conceptual view of a polymer micromachined FWNAV which consists of a micro-transmission, a pair of the micro wing, and a piezoelectric bimorph actuator. This study presented a prototype of this FWNAV, and its computational flight performance was evaluated using nonlinear structural dynamic and FSI analyses. Since our FWNAV has a 2.5-dimensional structure that can be easily fabricated using polymer micromachining, this study will lead to the further miniaturization of FWNAVs equivalent to the smallest natural insects.

FLAPPING WING NANO AIR VEHICLE

Figure 2 shows the design of the first prototype of a polymer micromachined FWNAV in plan and sectional views, indicating that micro transmission is a central and essential component to get the desired flapping motion whereas a pair of micro wings exerts the flight simulation of the FWNAV. Our FWNAV was designed using the iterative design window search methodology where conflicting design requirements, nonlinear and unsteady were being satisfied [13]. The development of our FWNAV has been carried out in two steps: (1) The fabrication of micro-transmission and micro-wing, and (2) the assembly of these components along with the bimorph actuator. In our previous study [11], we proved that our polymer micromachining process is feasible for the fabrication of this kind of structure, since the dimensional precision of the fabricated one compared to the design was very high about 75% and the static performance error between the experiment and numerical was about 10% only.

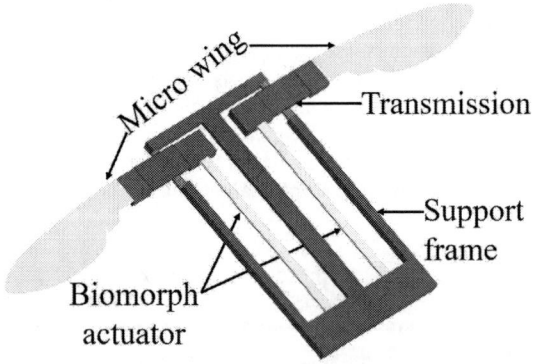

Figure 1: Concept of polymer micromachined flapping wing nano air vehicle or MEMS flyer.

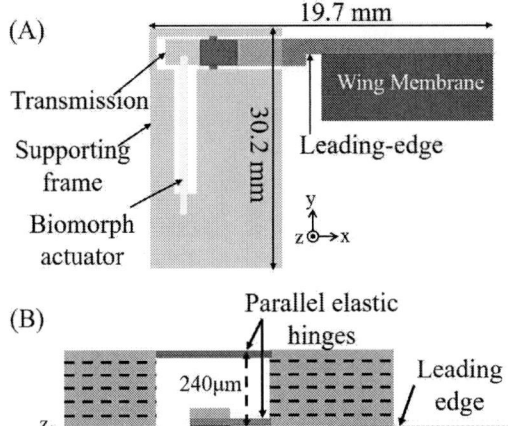

Figure 2: Actual design of FWNAV in plan (A) and section views (B). Note: Figures are the best views in color prints.

Transmission

In our FWNAV, the transmission is a key component and it has a 2.5-dimensional structure. It transduces the small translational displacement of the bimorph actuator into the large rotational displacement using the simple cantilever bending mechanism [9]. Initially, it was designed using nonlinear static analysis, and later it was designed using nonlinear dynamic analysis and it was observed that the stroke angle is also increasing from 40° to 60° due to the inertial effect. The transmission with the supporting frame is fabricated using standard microfabrication steps: the lamination, the exposure, the development, and the curing process [9–11]. Polyimide adhesive sheets with a thickness of 40μm are used as a material for microfabrication.

Micro Wing

The micro wing of our FWNAV is initially designed using the morphological and kinematic parameters of dipteran insects. Later its detailed design is done using the linear static stress analysis and mass of micro wing flapped by the transmission. It is also fabricated using the standard microfabrication steps and polyimide materials, of which material properties are given in Table 1.

Figure 3 shows the developed first prototype of the polymer micromachined FWNAV. The numerical mass (density × volume) of the transmission, micro wing, and FWNAV are respectively 76.30mg, 1.88mg, and 467.18mg because the actuator's mass is 289mg.

NUMERICAL FSI ANALYSIS

In the numerical FSI analysis, the coupled physics between the flapping wings and the surrounding air is incorporated by solving the following equations: an incompressible Navier-Stokes equation for the fluid (the ALE method is used), the equation of motion for the elastic body (the total Lagrangian method is used), and the coupled conditions on the fluid-structure interface are solved using the projection method [14]. These equations are discretized using the finite element method, and the monolithic equation system for the FSI can be obtained in the matrix-vector form as follows:

$$_L\mathbf{Ma} + \mathbf{Cv} + \mathbf{N} + \mathbf{q(u)} - \mathbf{Gp} = \mathbf{g}, \qquad (1)$$
$$_\tau\mathbf{Gv} = \mathbf{0}, \qquad (2)$$

where $\mathbf{M}$, $\mathbf{C}$, and $\mathbf{G}$ denote the mass, diffusive, and divergence operator matrices, respectively, and $\mathbf{N}$, $\mathbf{q}$, $\mathbf{g}$, $\mathbf{a}$, $\mathbf{v}$, $\mathbf{u}$, and $\mathbf{p}$ denote the convective term, elastic internal force, external force, acceleration, velocity, displacement, and pressure vectors, respectively, and the sub-scripts L and τ indicate the lumping of the matrix and the transpose of the matrix, respectively. This monolithic equation system is solved by the projection method using algebraic splitting in the parallel computation environment.

FLIGHT PERFORMANCE

The flight performance of the developed FWNAV has been evaluated using a pair of two distinct computational tools, that is, the nonlinear structural dynamic analysis of the FWNAV and the FSI analysis of the micro wing. Firstly, the former evaluates the flapping amplitude Φ_0 as an output of the nonlinear dynamic behavior of the FWNAV. Then, it is used for the sinusoidal flapping angular displacement $\Phi = \Phi_0 \times \sin 2\pi f t$ as an input for the FSI analysis of the flapping micro wing.

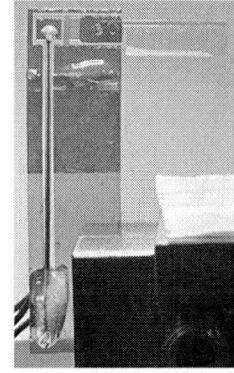

Figure 3: A prototype of the proposed FWNAV.

Table 1: Properties of polyimide materials.

Materials	Young's modulus	Mass density	Poisson ratio
Leading edge	2.5 GPs	1420 kg/m³	0.289
Wing membrane	3.5 GPs	1420 kg/m³	0.30

Nonlinear Structural Dynamic Analysis of FWNAV

The problem setup for the nonlinear structural dynamic analysis of the FWNAV along with the material distribution is shown in Fig. 4. The material properties of two polyimides are given in Table 1. In the problem setup, the piezoelectric actuator's output is considered as the forced displacement $u_x = U_x \times \sin 2\pi f t$, and it is applied to an area of the transmission, where the free end of the bimorph is attached. The evaluation of the amplitude of the forced displacement $U_x = 81\mu m$ has been given in our previous study [11]. In order to avoid sticking between elastic hinges during the flapping motion, the flapping frequency $f = 50$ Hz is selected [13]. In the problem setup, the lower part of the supporting frame is fixed, where the fixed end of the bimorph actuator is attached. The finite element mesh with 19,582 elements and 98,451 nodes is employed. During the dynamic analysis, the time increment $\Delta t = T/4000$ is used, where $T = 1/f$ is the flapping period.

Figure 5 shows the time history of the flapping angular displacement and the stroke angle from this time history is 43.80°, which will be used as an input for FSI analysis.

FSI Analysis of FWNAV

FSI analysis has been carried out using the in-house code, which was already validated [15]. The problem setup for the FSI analysis is shown in Fig. 6. In this setup, the sinusoidal flapping displacement $\Phi = \Phi_0 \times \sin 2\pi ft$ was applied to the wing base, where $\Phi_0 = 43.80°$ and $f = 50$ Hz were given from the nonlinear structural dynamic analysis of FWNAV. In this setup, the stroke axis coincides with the wing attachment tip line LL' as shown in Fig. 4. The time increment is $\Delta t = T/5000$.

The mesh for the fluid domain is shown in Fig. 7, and the section is shown in Fig. 8, where the mesh for the structural domain as well as the fluid domain is shown. The stabilized linear equal-order-interpolation velocity-pressure elements are used for the fluid mesh with the numbers of nodes and elements are 50,460 and 271,922, respectively, while the MITC shell elements are used for the micro wing with the numbers of nodes and elements are 143 and 120, respectively. In the analysis, the material properties of the fluid (air) are taken as follows; the mass density $\rho_a = 1.205$ kg/m³ and the viscosity $\mu_a = 1.837 \times 10^{-5}$ kg/(m sec), while the material properties of the micro wing are shown in the above sections.

The feathering motion is essential for the lift generation during the insect flapping flight. In the FSI model considered here, this motion occurs passively because of the FSI. Figure 9 shows the time history of the feathering motion. From this figure, the value of the mean feathering angle is evaluated as $\psi = 7.47°$. Figure 9 also shows the time history of the lift force, of which mean value is 0.0025mN, and it is less than the gravity force acting on the total weight of FWNAV $w = 4.58$mN. Hence, we will conduct the optimization of FWNAV such that it can generate the sufficient lift to hover.

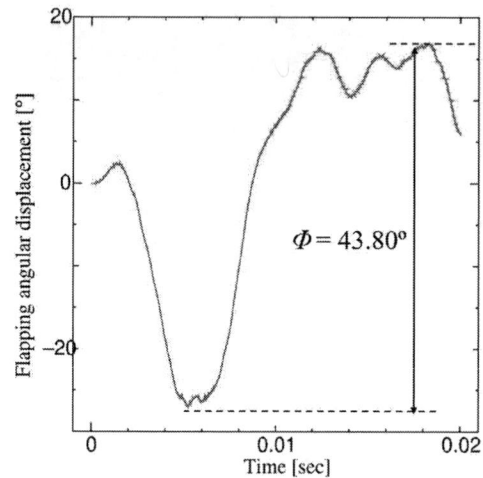

Figure 5: Time history of flapping angular displacement.

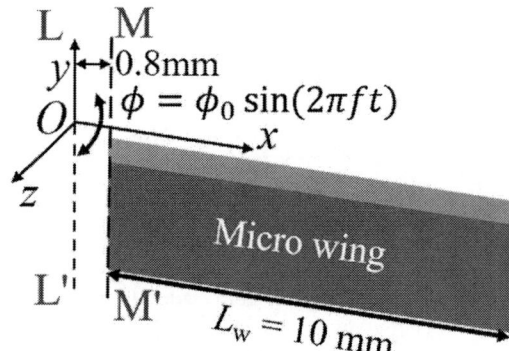

Figure 6: Problem setup for the fluid-structure interaction analysis of FWNAV.

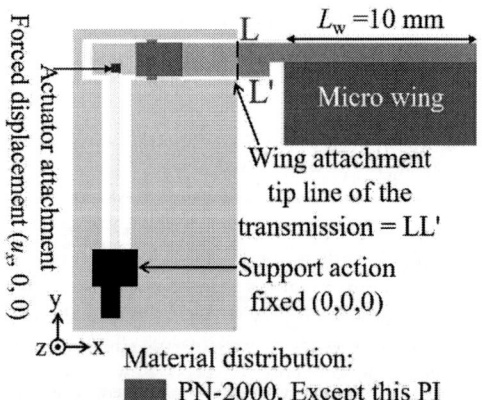

Figure 4: Problem setup for nonlinear structural dynamic analysis of FWNAV.

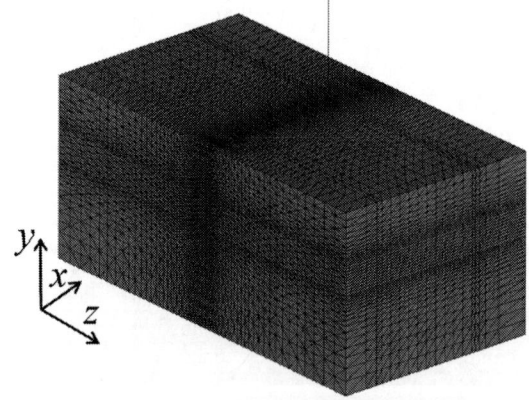

Figure 7: 3-D view of applied fluid mesh during FSI analysis.

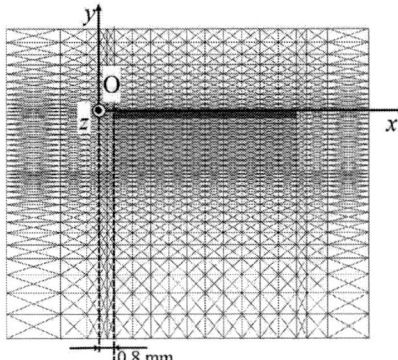

Figure 8: Applied fluid and wing mesh during FSI analysis.

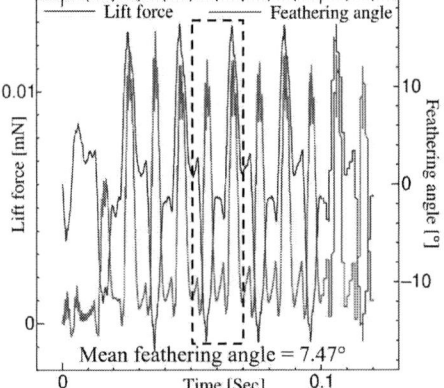

Figure 9: Flight simulation and performance of first prototype polymer micromachined FWNAV

CONCLUSIONS

In this study, we presented a prototype of the polymer micromachined developed FWNAV. Along with this, we have demonstrated the flight simulation and performance evaluation of the developed FWNAV using the nonlinear structural dynamic analysis of FWNAV and the FSI analysis of the micro wing. The flight performance of the prototype has shown the small mean feathering angle and the low mean lift force because of the small stroke angle and the low flapping frequency. Although the mean lift force is less than the weight, it can be increased by optimizing and miniaturizing FWNAV to develop a flyable one. So, this research will edge toward the development of tethered and flyable FWNAVs with the size of the smallest flying natural insects.

ACKNOWLEDGEMENTS

This work was supported by JSPS KAKENHI Grant Number 20H04199. We appreciate the support from Toray Industries, Inc.

REFERENCES

[1] A. R. Ennos, "The Kinematics and Aerodynamics of the Free Flight of some Diptera", *J. of Exp. Bio.*, vol. 142, pp. 49–85, 1989.

[2] C. Galinski, "Influence of MAV Characteristics on their Applications", *Aviation*, vol. 9, pp. 16–23, 2005.

[3] R.J. Wood, "The First Takeoff of a Biologically Inspired at-Scale Robotic Insect", *IEEE Trans. Robot.*, vol. 24, pp. 341–347, 2008.

[4] A. Bontemps, et. al., "Design and Performance of an Insect-Inspired Nano Air Vehicle", *Smart Mater. Struct.*, vol. 22, pp. 014008–20, 2013.

[5] T. Ozaki, K. Hamaguchi, "Bioinspired Flapping-Wing Robot with Direct-Driven Piezoelectric Actuation and its Takeoff Demonstration", *IEEE Robot. Autom. Lett.*, vol. 3, pp. 4217–4224, 2018.

[6] D. Ishihara, R. Takata, P. C. Ramegowda, N. Takayama, "Strongly Coupled Partitioned Iterative Method for the Structure–Piezoelectric–Circuit Interaction using Hierarchical Decomposition", *Comput. Struct.*, vol. 253, pp. 1–17, 2021.

[7] D. Ishihara, "Computational Approach for the Fluid-Structure Interaction Design of Insect-Inspired Micro Flapping Wings", *Fluids*, vol. 7, pp. 1–20, 2022.

[8] D. Ishihara, M. J. Jeong, S. Yoshimura, G. Yagawa, "Design Window Search using Continuous Evolutionary Algorithm and Clustering –its Application to Shape Optimization of Microelectrostatic Actuator", *Comput. Struct.*, vol. 80, pp. 2469–2481, 2002.

[9] D. Ishihara, S. Murakami, N. Ohira, J. Ueo, M. Takagi, "Polymer Micromachined Transmission for Insect-Inspired Flapping-Wing Nano Air Vehicle", *Proc. IEEE-NEMS '15 Conference*, September 27-30, 2020, pp. 176–179.

[10] Rashmikant, D. Ishihara, R. Suetsugu, S. Murakami, P. C. Ramegowda, "Improved Design of Polymer Micromachined Transmission for Flapping Wing Nano Air Vehicle", *Proc. IEEE-NEMS '16 Conference*, April 25-29, 2021, pp. 1320–1325.

[11] Rashmikant, D. Ishihara, R. Suetsugu, P. C. Ramegowda, "One-Wing Polymer Micromachined Transmission for Insect-Inspired Flapping-Wing Nano Air Vehicle", *Eng. Res. Exp.*, vol. 3, pp. 045006–25, 2021.

[12] Rashmikant, D. Ishihara, "A Design Window Search using Nonlinear Dynamic Simulation for Polymer Micromachined Transmission in Insect Inspired Flapping Wing Nano Air Vehicles", *Proc. IEEE-ICRAE '06 Conference*, November 19-22, 2021, pp. 162–167.

[13] Rashmikant, "Development of Polymer Micromachined Flapping Wing Nano Air Vehicle using Iterative Design Window Search Methodology", *Doctoral Dissertation Depository Kyushu Institute of Technology*, pp. 1–157, August 2022.

[14] D. Ishihara, T. Horie, "A Projection Method for the Monolithic Interaction System of an Incompressible Fluid and a Structure using a New Algebraic Splitting", *Comput. Model. Eng. Sci.*, vol. 101, pp. 421–440, 2014.

[15] D. Ishihara, T. Horie, T. Niho, "An Experimental and Three-Dimensional Computational Study on the Aerodynamic Contribution to the Passive Pitching Motion of Flapping Wings in Hovering Flies", *Bioinspir. Biomim.*, vol. 9, pp. 046009–22, 2014.

CONTACT

*D. Ishihara, ishihara.daisuke399@mail.kyutech.jp

ITERATIVE LEARNING CONTROL FOR QUASI-STATIC MEMS MIRROR WITH SWITCHING OPERATION

Matthias Macho[1], Han Woong Yoo[1], Richard Schroedter[2], and Georg Schitter[1]
[1]Automation and Control Institute, TU Wien, AUSTRIA and
[2]Fundamentals of Electrical Engineering, TU Dresden, GERMANY

ABSTRACT

This paper reports an iterative learning control (ILC) to compensate for the errors by the switching operation and the modeling inaccuracies for a quasi-static (QS) MEMS mirror. The modeling errors and uncertainties in dynamics with the switching operation between electrodes result in undesirable oscillations in beam positioning. A wideband frequency-domain ILC is proposed for a QS MEMS mirror with a flatness-based feedforward control. The improvement of the residual oscillations is demonstrated by reduced root mean square (RMS) errors for a 2 Hz and a 2-degree-amplitude sawtooth reference with a factor 69.9.

KEYWORDS

Iterative Learning Control, Quasi-Static MEMS Mirror, Electrostatic Actuation, Switching Operation

INTRODUCTION

Electrostatically actuated quasi static micro-electro-mechanical systems (QS MEMS) mirrors enable high quality arbitrary beam positioning with low power consumption via precisely aligned comb drive actuation, typically manufactured by CMOS processes [1]. QS MEMS mirrors require a special design of actuation such as staggered vertical comb (SVC) drives [2], providing beam positioning a wide control bandwidth or tracking a linear scan motion that open vast applications [1,3]. Due to the innate high Q factor, however, control of the QS MEMS mirror is essential to keep the desirable scanning trajectory without overshoot or oscillations. A switching operation for bi-directional scanning can also cause this oscillations.

Various types of controls are applied for QS MEMS mirrors to improve the precision scanning [4-6]. A flatness-based control is designed for QS MEMS mirror with staggered vertical comb (SVC) drives and successfully demonstrates significant reduction of root mean square (RMS) errors down to a few millidegrees [1]. Learning controls such as repetitive control (RC) and iterative learning control (ILC), which improve the control by learning from errors of repeating tasks in the previous trial, are investigated for target scan trajectories to reduce the control errors further with feedback controls [5, 6], while the reductions by the learning controls are mainly noticable in low frequency distortions. For a galvanometer scanner, an ILC strategy with an accurate model demonstrates wideband error compensation beyond two resonances [7]. For QS MEMS mirrors, the compensation solely using ILC for dynamic errors mainly due to switching operations and model inaccuracy has not been studied so far.

PROBLEM DESCIPTION
Model of Quasi-static MEMS Mirror

Figure 1a illustrates the actuation principle of the QS

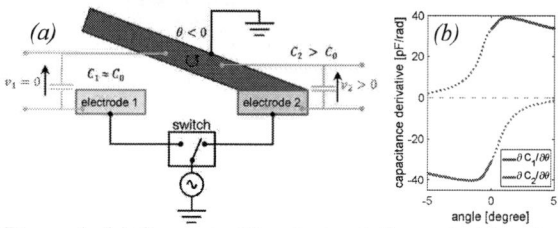

Figure 1: (a) Concept of the electrostatic comb drives for a QS MEM mirror. A voltage is applied between the stator electrode (red and blue rectangles) and the rotor electrode (dark gray), electrostatic torque rotates the mirror toward the turn-on side. The springs on the rotor apply a counter torque. (b) Angular derivative of capacitance for two comb drives along the deflection angle. The sign of $\partial C_1(\theta)/\partial \theta$ and $\partial C_2(\theta)/\partial \theta$ represents the torque direction. Dotted lines represent unused region by the switching operation.

MEMS mirror with two stator electrodes and a rotor electrode [1]. The used QS MEMS mirror is designed by the staggered vertical comb drive technique, allowing the vertical displacement between stator and rotor parts [2]. The deflection θ of the QS MEMS mirror is described by

$$ I\ddot{\theta} + b\dot{\theta} + k_m\theta = \frac{1}{2}\frac{\partial C_1(\theta)}{\partial \theta}v_1^2 + \frac{1}{2}\frac{\partial C_2(\theta)}{\partial \theta}v_2^2, \quad (1) $$

where I denotes the inertial of the mirror and rotor, b denotes the damping coefficient, and k_m denotes the mechanical stiffness of the mirror to the frame. Since the mirror and rotor part are suspended by a torsion bar to the frame under the atmospheric air pressure, the mechanical structure is mainly linear with a high Q factor about 90. C_1 and C_2 are the capacitance of the each comb drives and v_1 and v_2 are corresponding stator voltages while rotor is set to ground. When a voltage is applied at either stator electrodes, the mirror rotates toward to the stator by the electrostatic force. Since the torque is generated only by the pulling force, total control u_{tot} can be defined as a normalized torque by the inertia. The sign of the total control can be simply obtained by switching electrodes, i.e.

$$ u_{tot} = \begin{cases} \dfrac{1}{2I}\dfrac{\partial C_1(\theta)}{\partial \theta}v_1^2 & u_{tot} > 0, \\[3mm] \dfrac{1}{2I}\dfrac{\partial C_2(\theta)}{\partial \theta}v_2^2 & u_{tot} \le 0. \end{cases} \quad (2) $$

The direction of the torque is determined by the sign of angular derivative of capacitances as shown in Figure 1b. The total control is then scaled by the angular derivative at the current angle. Due to angular dependency, however, the dynamics become nonlinear in operation and the small model inaccuracy acts as impulse switching, exciting the main eigenmode of the mirror and resulting in oscillations.

Variation of Dynamics by Electrostatic Stiffness

The nonlinear angular dependency of the comb drive capacitance causes well-known electrostatic softening or hardening [8]. Assume that only the electrode 1 is activated, i.e. $v_1 > 0$ and $v_2 = 0$, which causes a constant deflection of the mirror at steady-state. Negative voltages of v_1 and v_2 provide the same torque due to the square function. The local dynamics at the operation angle θ_{op} and the stator voltage v_{op} can be approximated as

$$I\Delta\ddot{\theta} + b\Delta\dot{\theta} + k_m\Delta\theta = k_{el}\Delta\theta + K_{el}v_1, \quad (3)$$

$$k_{el} = \frac{1}{2}\frac{\partial^2 C_1(\theta)}{\partial \theta^2}v_1^2\Big|_{\theta=\theta_{op},\, v_1=v_{op}}, \quad (4)$$

$$K_{el} = \frac{1}{2}\frac{\partial C_1(\theta)}{\partial \theta}v_1\Big|_{\theta=\theta_{op},\, v_1=v_{op}}, \quad (5)$$

where Δ denotes small changes of the consecutive parameters. The nonlinear torque is represented in a linear manner by the local electrostatic stiffness k_{el} and the local torque constant K_{el}. This leads to the linearized transfer function at the operation point as [1]

$$G_{\Delta\theta/v_1} = \frac{K_{el}}{Is^2 + bs + k_m - k_{el}}. \quad (6)$$

As the electrostatic stiffness is changed by the electrode voltages with the operational angle, the eigenfrequency, of the local dynamics indeed varies by the operation angle. Figure 2a illustrates measured local frequency responses of the dynamics at -1° and -4°, showing discrepancy of local dynamics by a different operational angle [9]. Figure 2b describes the variation of the eigenfrequency and the electrostatic stiffening parameter along the deflection angle. The trend is similar according to the absolute deflection angle while it is not perfectly symmetric. The eigenfrequency varies according to this electrostatic stiffness by the operation angle, which varies most around 0° due to switching operations and the comb configuration.

Flatness-based Feedforward Control and Switching Function

To generate the inputs for the reference trajectory, flatness-based feedforward control is used [1]. The reference trajectory is generated as a smooth jerk limited trajectory, providing reference trajectory with their first and second order differentiation, i.e. $(\ddot{\theta}_{ref}, \dot{\theta}_{ref}, \theta_{ref})$ [10]. Then the flatness-based feedforward control input u_{ff} is obtained by applying the reference trajectory as

$$u_{ff}(\ddot{\theta}_{ref}, \dot{\theta}_{ref}, \theta_{ref}) = \ddot{\theta}_{ref} + \frac{b}{I}\dot{\theta}_{ref} + \frac{k_m}{I}\theta_{ref}. \quad (7)$$

Substitute (2) with (7) as the feedforward control is the total control, the electrode voltages are defined by the switching function as [1]

$$v_{1,ff} = \begin{cases} \sqrt{2Iu_{ff}\left(\frac{\partial C_1(\theta_{ref})}{\partial \theta_{ref}}\right)^{-1}} & u_{ff} > 0, \\ 0 & u_{ff} \leq 0, \end{cases} \quad (8)$$

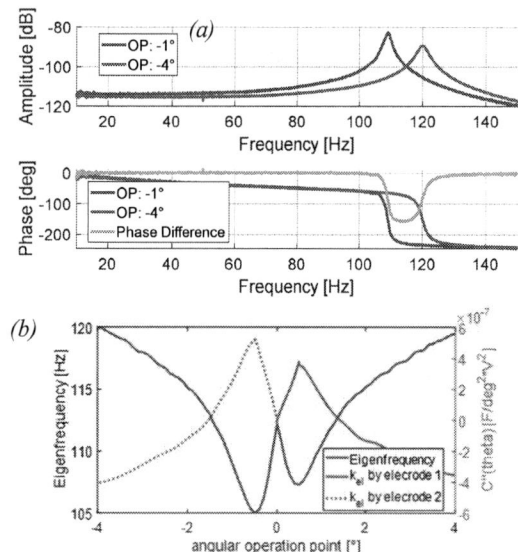

Figure 2: (a) Measured frequency responses of the QS MEMS mirror for different operational angles. (b) Measured eigenfrequency chances over the deflection angle and the estimated electrostatic stiffness k_{el} from the capacitance measurements.

$$v_{2,ff} = \begin{cases} 0 & u_{ff} > 0, \\ \sqrt{2Iu_{ff}\left(\frac{\partial C_2(\theta_{ref})}{\partial \theta_{ref}}\right)^{-1}} & u_{ff} \leq 0. \end{cases} \quad (9)$$

The trajectory tracking by the feedforward control is highly sensitive to modeling inaccuracies such as the errors in identification of (1), variation of the dynamics of (6), and parasitic motions of the QS MEMS mirror due to imbalanced actuation. With switching, this can lead to large residual errors in scanning trajectories.

ITERATIVE LEARNING CONTROL

Design of Iterative Learning Control

The update equation of frequency-domain ILCs [11], also called IIC [12], is defined by simple multiplications in the Fourier domain via discrete Fourier transform (DFT)

$$\boldsymbol{U}_{j+1}[n] = \boldsymbol{Q}[n]\big(\boldsymbol{U}_j[n] + \boldsymbol{L}[n]\boldsymbol{E}_j[n]\big), \quad (10)$$

$$\boldsymbol{E}_j[n] = \tilde{\boldsymbol{\Theta}}_j[n] - \boldsymbol{\Theta}_{ref}[n], \quad (11)$$

$$\boldsymbol{U}_j[n] = \mathcal{F}\{u_j[k]\} = \sum_{k=0}^{N-1} u_j[k]W_N^{kn}, \quad (12)$$

where $\mathcal{F}$ denotes the Fourier operator, defined with a weights of $W_N = e^{-i2\pi/N}$ and $i = \sqrt{-1}$, and the bold capital notations of $\boldsymbol{U}$, $\tilde{\boldsymbol{\Theta}}$, $\boldsymbol{\Theta}_{ref}$, $\boldsymbol{E}$, $\boldsymbol{L}$, and $\boldsymbol{Q}$ are Fourier coefficients of the ILC input u, measured output $\tilde{\theta}$, reference θ_{ref}, tracking error e, the learning filter, and the Q filter, respectively. The size N is chosen an integer multiple of the sample number in a scanning period to avoid spectral leakage, i.e. coherent sampling, and n defines the index of harmonic frequency component. For simplicity, $\boldsymbol{Q}$ is chosen as a unity, i.e. $\boldsymbol{Q}(n) = 1, \forall n$, and the learning filter is set to a frequency dependent learning gain ρ with an inversion of estimated QS MEMS mirror dynamics $\widehat{\boldsymbol{G}}$, i.e. $\boldsymbol{L}[n] = \rho[n]\widehat{\boldsymbol{G}}^{-1}[n]$. The learning gain is typically below 1 due to convergence, and a smaller learning gain allows more tolerance of modelling errors such as model uncertainty e.g. in (6) and nonlinear switching in (2) at a cost of a slower convergence speed. In this work, the learning gain is chosen as a sharp lowpass filter as $\rho[n] = \rho_0$, for $n \leq n_c$, where ρ_0 denotes a constant learning gain and n_c is the index of the cutoff frequency. For $n > n_c$, $\rho[n]$ is set to 0.

ILC with Switching Function

The ILC input u_{j+1} is added to the given feedforward input u_{ff} and together forms the total input, cf. Figure 3, i.e. $u_{tot,j+1} = u_{ff} + u_{j+1}$. As (8) and (9), the electrode voltages of the ILC are written by

$$v_{1,ILC} = \begin{cases} \sqrt{2 I u_{tot} \left(\frac{\partial c_1(\theta_{ref})}{\partial \theta_{ref}} \right)^{-1}} & u_{tot} > 0, \\ 0 & u_{tot} \leq 0, \end{cases} \quad (13)$$

$$v_{2,ILC} = \begin{cases} 0 & u_{tot} > 0, \\ \sqrt{2 I u_{tot} \left(\frac{\partial c_2(\theta_{ref})}{\partial \theta_{ref}} \right)^{-1}} & u_{tot} \leq 0. \end{cases} \quad (14)$$

EXPERIMENTAL RESULTS

Experimental Setup and ILC Implementation

Figure 3 describes the experimental setup and the control structure with the QS MEMS mirror. An optical readout by a one dimensional position sensitive device (1D PSD) is used as accurate angle measurements in degree via a calibration procedure by a linear motorized stage [13]. The measured PSD signals are recorded by a dSpace MicroLabBox. The ILC is calculated by Matlab in the PC and generates the electrode voltages by the dSpace MicroLabBox for each electrode via high voltage amplifiers, deflecting the laser beam that shines the PSD.

Figure 3 also illustrates a block diagram of the flatness-based feedforward control and the ILC in the dSpace MicroLabBox. The parameters in (1) for flatness-based feedforward control in (8) and (9) are identified as [1]. The model $\widehat{\boldsymbol{G}}$ is identified based on empirical transfer function estimate (ETFE) via a chirp input signal at the operation angle [14]. The constant learning gain ρ_0 in the learning filter $\boldsymbol{L}$ is set to 0.1, considering model uncertainty and switching nonlinearity of the QS MEMS mirror. The cutoff frequency of $\boldsymbol{L}$ is set to 200 Hz, which is much higher than the eigenfrequency around 114 Hz at 0° to compensate for the oscillations by switching operations. The sampling frequency of the dSpace MicroLabBox is set to 50 kHz.

Tracking Results

The flatness-based feedforward control with ILC of (13) and (14) are evaluated and are compared with the feedforward only case of (8) and (9) for sawtooth trajectories with 2° amplitudes for scan rates of 1, 2 and 4 Hz. Figure 4 illustrates the scanning trajectories, input trajectories, and error trajectories of a 2 Hz sawtooth reference trajectory in both cases. The error trajectory of the feedforward control clearly shows undesirable

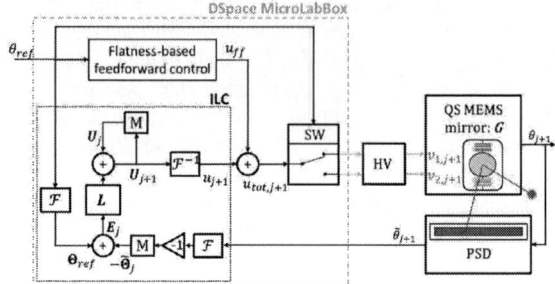

Figure 3: Block diagram of the experimental setup with a frequency-domain ILC. The feedforward input u_{ff} is generated by the reference trajectory θ_{ref}. With the ILC, the total input u_{tot} is applied to the QS MEMS mirror via a switch, steering the laser beam by θ. Then a PSD records the deflection angle of the QS MEMS mirror. With discrete Fourier transform (block $\mathcal{F}$ and $\mathcal{F}^{-1}$), the ILC updates the compensation $\boldsymbol{U}_{j+1}$ from the measured errors of the previous trial $\boldsymbol{E}_j$ from the memories (block M) via the learning filter $\boldsymbol{L}$ for improved next trial, $j+1$.

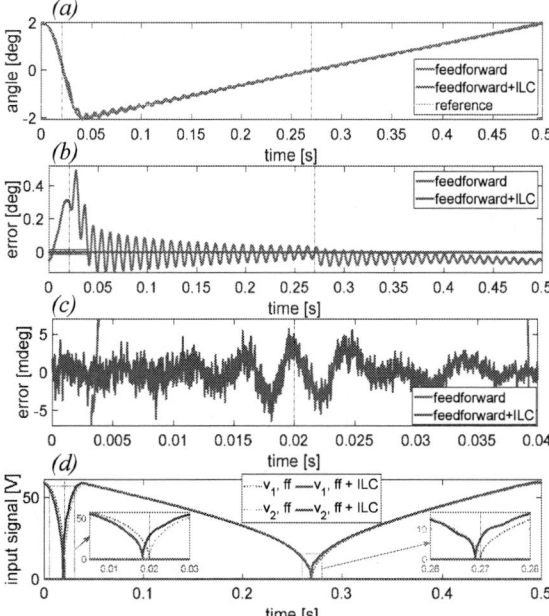

Figure 4: Experimental results of the feedforward control only (red solid lines) and feedforward control with the ILC (blue solid lines) for a 2 Hz sawtooth trajectory with 2° amplitude. (a) Scanning trajectories, (b) residual errors and (c) its zoomed plot near the switching operation (the violet box in (c)). (d) The input trajectories for each electrodes and zoomed plots near the switching operations. Original switching timing is drawn by vertical thin black dashed-dot lines.

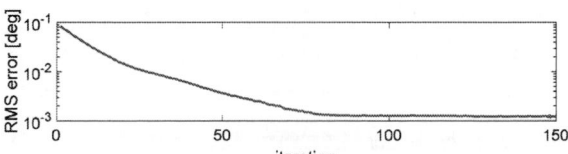

Figure 5: RMS errors along the iteration. The red dot represents the minimum of the RMS error.

Table 1: RMS errors of a sawtooth reference of 2° amplitude for scan rates of 1, 2, and 4 Hz

Scan rate	RMS Error (mdeg)		Error Ratio
	FF	FF+ILC	(FF/ FF+ILC)
1 Hz	43.9	1.3	35.0
2 Hz	84.8	1.2	69.9
4 Hz	151.2	3.8	39.7

oscillations triggered at the zero angle, where the switching operation between the electrodes happens. The impact of the switching operation is severe at the fast turnaround, generating large oscillations over the linear scanning region of the trajectory by the high Q factor. The switching operation at the slow linear scan region still adds oscillations while the impact is insignificant. The ILC reduces the errors in both phase and the oscillations at the eigenfrequency. The resulting RMS errors are 1.2 millidegrees with the ILC and 84.8 millidegrees in case of the flatness-based feedforward control only, showing benefits as a precise and accurate tracking control. The electrode voltages by the ILC also show the phase correction and oscillations for the compensation.

Figure 5 illustrates the learning transient of the ILC. The main error reduction takes about 80 iterations due to the small learning gain and the minimum is found at 148 iteration. Since ILC can recall the best correction afterwards once the best input is known, the performance can be kept the minimal by the use of the best input. Even if the ILC needs to run again, the number of iteration can be reduced by starting with the best input unless dynamics change significantly [11]. Table 1 describes improvements of the sawtooth references for other scan rates. In case of 1 and 4 Hz scan rates significant RMS error reduction less than 3.8 millidegrees can be achieved. These results demonstrate the feasibility of the proposed ILC for the error compensation of the QS MEMS mirror mainly due to innate nonlinear switching and model inaccuracy, showing potential as a simple and accurate control for QS MEMS mirrors.

CONCLUSION

This paper discusses a frequency-domain iterative learning control to compensate for undesirable oscillations caused by nonlinear switching and model inaccuracy of a QS MEMS mirror. QS MEMS mirrors allow arbitrary scanning based on CMOS process compatible electrostatic actuation while they require switching to change the torque direction by the zero angle, which can cause unwanted oscillations at the eigenfrequency. A frequency-domain ILC is designed with a low learning gain to cope with these nonlinearities and is implemented with a flatness-based feedforward control to compensate for undesirable oscillations. The tracking results for sawtooth reference trajectories of 2° amplitude demonstrate the ILC as a highly precise and accurate tracking control, which achieves the RMS error of 1.2 millidegrees from 84.8 millidegrees for the 2 Hz scan rate, leading to an improvement of a factor 69.9 than the flatness-based feedforward control.

ACKNOWLEDGEMENTS

This work has been supported in part by the Austrian Research Promotion Agency (FFG) under the scope of the AUTOScan project (FFG project number 884345). The authors would like to thank David Brunner of Infineon for fruitful discussions.

REFERENCES

[1] R. Schroedter et al., "Flatness-based open-loop and closed-loop control for electrostatic quasi-static microscanners using jerk-limited trajectory design," *Mechatronics*, vol. 56, pp. 318–331, Dec. 2018,

[2] T. Sandner et al., "Microscanner with vertical out of plane comb drive," in *16th Int. Conf. on Optical MEMS and Nanophotonics*, pp. 33–34, Aug. 2011.

[3] T. Sandner et al., "Quasistatic microscanner with linearized scanning for an adaptive three-dimensional laser camera," *J. Micro/Nanolith. MEMS MOEMS*, vol. 13, no. 1, pp. 011114, Feb. 2014,

[4] Y. Zhao, et al., "Fast and precise positioning of electrostatically actuated dual-axis micromirror by multi-loop digital control," *Sens. Actuator A Phys.*, vol. 132, no. 2, pp. 421–428, 2006.

[5] R. Schroedter et al., "Repetitive nonlinear control for linear scanning micro mirrors," in *MOEMS and Miniaturized Syst. XVII*, pp. 35, Feb. 2018.

[6] V. Milanović et al., "Iterative learning control (ILC) algorithm for greatly increased bandwidth and linearity of MEMS mirrors in LiDAR and related imaging applications," in *MOEMS and Miniaturized Syst. XVII*, vol. 10545, pp. 1054513, Feb. 2018.

[7] H. W. Yoo, S. Ito, and G. Schitter, "High speed laser scanning microscopy by iterative learning control of a galvanometer scanner," *Control Eng. Pract.*, vol. 50, pp. 12–21, May 2016,

[8] A. M. Elshurafa et al., "Nonlinear dynamics of spring softening and hardening in folded-mems comb drive resonators," *J. Microelectromech. Syst.*, vol. 20, No. 4, pp. 943–958, 2011,

[9] K. Janschek et al., "Adaptive Prefilter Design for Control of Quasistatic Microscanners." IFAC Proc. Volumes, vol. 46, no. 5, pp. 197–206, 2013,

[10] R. Schroedter et al., "Rasterscan with jerk-limited trajectories for quasi-static/resonant microscanners", *Tagungsband VDI Mechatronik*, pp. 155-160, 2019.

[11] Y. Li and J. Bechhoefer, "Model-free iterative control of repetitive dynamics for high-speed scanning in atomic force microscopy," *Rev. of Sci. Inst.*, vol. 80, no. 1, pp. 013702, Jan. 2009.

[12] K.-S. Kim and Q. Zou, "A Modeling-Free Inversion-Based Iterative Feedforward Control for Precision Output Tracking of Linear Time-Invariant Systems," *IEEE/ASME Trans. on Mechatronics*, vol. 18, no. 6, pp. 1767–1777, Dec. 2013.

[13] H. W. Yoo et al., "MEMS Test Bench and its Uncertainty Analysis for Evaluation of MEMS Mirrors," in *8th IFAC Symp. on Mechatronic Syst.*, vol. 52, no. 15, pp. 49–54, 2019.

[14] L. Ljung, *System Identification: Theory for the User*, 2nd Edition, Prentice-Hall, 1999

CONTACT

*H. W. Yoo, yoo@acin.tuwien.ac.at

M_z ATOMIC MAGNETOMETER USING A 3D MEMS GLASS ALKALI VAPOR CELL WITH VERTICAL SIDEWALLS

Jin Zhang, Jianfeng Zhang, Wenqi Li, Ziji Wang, and Jintang Shang

The Key Laboratory of MEMS of the Ministry of Education, Southeast University, Nanjing, China

ABSTRACT

This paper reports a novel M_z atomic magnetometer based on a low-cost 3D MEMS glass alkali vapor cell with vertical sidewalls for optical pumping/probing transmission. By designing vertical side walls, the 3D MEMS vapor cell provides high-quality centimeter-level optical paths for atom-photon interactions. The wafer-level high-quality multiple optical paths are also acquired for flexible M_z atomic magnetometer designs. Experiments show that the sensitivity of the proposed Mz magnetometer is 3.7 pT/Hz$^{1/2}$ at 1 Hz under 14000 nT.

KEYWORDS

3D MEMS glass alkali vapor cell, vertical sidewalls, atomic magnetometer, multiple optical paths.

INTRODUCTION

Miniaturized atomic magnetometers for bio-magnetometer and geomagnetic field measurement of high-precision based on MEMS alkali cells have attracted a lot of attention due to their low cost, small size, and acceptable performance [1]- [5]. The traditional MEMS alkali vapor cell is a glass-silicon-glass sandwich structure with an optical path perpendicular to the substrate plane [6]. However, the optical path length depends on the silicon wafer thickness, which limits the magnetic resonant signal.

Much research has focused on providing compact alkali vapor cells with a high-quality optical path to enhance the interaction between photons and atoms, thereby improving the signal-to-noise ratio of atomic magnetometers. A typical scheme is to extend the optical path in the direction perpendicular to the cell substrate, such as spherical MEMS alkali vapor cells and multilayer bonded alkali cells [7]- [9]. The optical path achieved under this design concept is around 5 mm. Another scheme is the reflection-type alkali vapor cell with the optical path parallel to the cell substrate [10], [11]. The optical length of the cell can reach a centimeter level. However, reflection-type cells require additional gratings or built-in micro-mirrors to adjust the orientation of the light path. Besides, the shaped rubidium vapor cell with a centimeter-level optical path for miniaturized atomic magnetometers is proposed in our previous work [12]. Nevertheless, there is an angle of 57.4, caused by silicon wet etching, between the cell sidewall and the optical path.

This paper proposes a novel Mz magnetometer with a centimeter-level optical path using a 3D MEMS glass alkali vapor cell with vertical sidewalls. This work builds on a fabrication process reported at IEEE NEMS 2022 [13]. The fabrication process is modified by changing the reflow time and the release process. The work also exhibits the performance of the M_z atomic magnetometer under different operating temperatures.

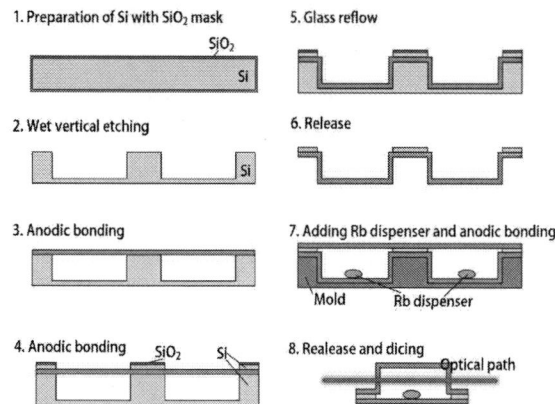

Figure 1: Fabrication process of the 3D alkali vapor cell with vertical sidewalls.

DESIGN AND FABRICATION

Cell design and fabrication

The 3D alkali vapor cell fabrication with vertical sidewalls is based on silicon micromachining and glass thermoforming. The key part of the fabrication is to form the vertical sidewalls by silicon anisotropic etching, refer to [7]. Due to the limitation of the anode bonding, the main materials of the 3D alkali vapor cell are glass and silicon. The whole preparation process consists of two parts: the first part is the preparation of vertical sidewalls in a (100) silicon wafer. The second part is cell shaping and sealing. Considering that the laser beam waist in subsequent tests was 1.5 mm, the sidewall height was set to 1.7 mm.

Fig. 1. Shows the fabrication process of the 3D glass alkali vapor cell with vertical sidewalls. The fabrication started with preparing and cleaning a 2-mm-thick (100) double-polished silicon wafer. In Step 1, silicon dioxide was thermally grown on the thick silicon wafer. In Step 2, a square opening was aligned in the primary flat, then the vertical wet etching was performed in a 90 °C water bath. The etching depth is around 1.7 mm. In Step 3, a 500-μm-thick glass wafer was bonded to the thick silicon wafer to seal the blind holes. The bonding voltage and temperature are 800V and 300 °C, respectively. In Step 4, the second anodic bonding was performed between a 500-μm-thick silicon wafer and the bonded wafer in Step 3 to protect the surface of the glass wafer during the following reflow process. Note that there is a silicon dioxide layer on the upper surface of the silicon wafer. In Step 5, the obtained multiple-bonded wafer is heated to the glass softening point for glass reflow. The glass reflow time is set to 10 ~ 20 minutes to prevent bubbles from forming in the glass. In Step 6, the 2-mm-thick silicon portion of the multi-bonded wafer is released by wet etching. In Step 7, the alkali dispenser was filled into glass cavities. After that, the last anodic bonding was completed under mold protection. In Step 8, the mold was released and the obtained wafer was diced.

Compared to the previous method, the fabrication is improved by changing the release process to solve the problem of overstressing of the reflowed glass, which will cause local cracking of the wafer.

M_z magnetometer design

We design an M_z atomic magnetometer based on the 3D glass alkali vapor cell with vertical sidewalls, as shown in Fig. 2. The magnetometer prototype is composed of a 3D alkali vapor cell, a coil set, optics, a PTC heater, and a photodiode. The housing of the magnetometer is 3D printed with grooves that allow it to hold the coils in place. The 3D alkali vapor cell is placed in the middle of the housing and aligned with the center of the coil set axis. The coil set consists of static magnetic field coils and RF magnetic field coils.

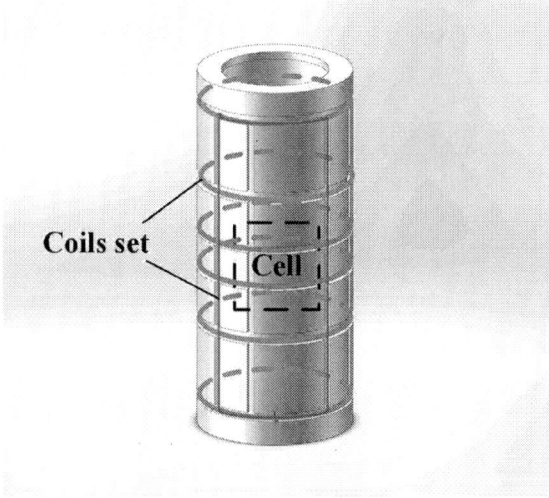

Figure 2: Schematic of the M_z atomic magnetometer prototype using the 3D alkali vapor cell.

EXPERIMENTAL RESULTS
Fabricated cell

The 3D MEMS glass vapor cells with a glass-silicon-glass three-layer structure and 4 vertical sidewalls are seen in Fig. 2(b). The size of the cell is $11 \times 11 \times 1.7$ mm³. The cell is filled with natural rubidium and 900 mbar N_2 as the buffer gas. The fabricated cell has a transparent and clear appearance compared to the previous 3D alkali vapor cell, which can be seen in Fig. 3(a). This proves that the problem of bubbles produced in the glass layer has been solved by improving the preparation process. However, unnecessary transparent areas appear around the glass shell of the fabricated cell. This is due to the destruction of the silicon dioxide layer of the thin silicon wafer in the process of wet etching for removing the thick silicon portion (Step 5 in cell fabrication). Therefore, it is necessary to thicken the silicon dioxide layer or replace the mask material to fabricate a more compact cell.

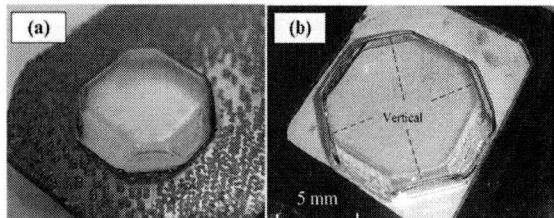

Figure 3: The fabricated 3D MEMS glass alkali vapor cell, (a) the previous cell with lots of bubbles in the glass, (b) the improved cell in this work.

M_z magnetometer test

The M_z magnetometer prototype is tested in a five-layer magnetic shield. Fig. 5 shows the schematic of the Mz magnetometer measurement system. A 795 nm laser beam emitted from a 5 mW DFB laser (NTG nanoplus, linewidth < 10 MHz) is introduced into the magnetometer prototype. The laser beam is circularly polarized by a polarizer and a quarter-wave plate, passing through the vertical sidewalls of the alkali vapor cell. The laser power is monitored through a photodetector (PD) as well as a trans-impedance amplifier circuit. A laser controller is used to adjust the laser power to an appropriate value. A 500 kHz alternating current (AC) is applied to the PTC heater to increase the rubidium vapor density inside the cell. Once the cell temperature is stable, it is necessary to lock the laser frequency to the rubidium D1 line due to the pressure broadening and the pressure shift of the optical absorption spectrum. After interacting with D1 circularly polarized laser, the rubidium atom will be pumped to a polarized state. Next, a 50 mV RF signal is applied on the RF coils to generate a small RF magnetic field at a frequency of ω along the X-axis. The frequency ω is modulated at a frequency of 89 Hz. After that, the static magnetic field coil is activated by a constant current source to produce a 14000 nT ambient static magnetic field along the Z-axis as the field to be measured. Following, the frequency ω is swept in the range of 2 kHz to 200 kHz. During the sweeping, the PD signal is demodulated by a lock-in amplifier (LiA, SRS 830) to obtain the magnetic resonant signal. Finally, an oscilloscope is utilized to monitor the output of the LiA.

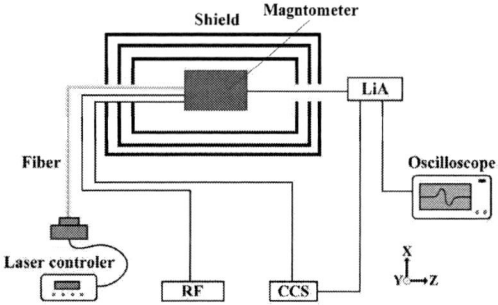

Figure 4: Experimental setup for the M_z atomic magnetometer. RF —— RF source; CCS —— constant current source; LiA —— lock-in amplifier.

The magnetic resonant signal has a dispersive lineshape. The difference in frequency between the maximum and minimum of the dispersive curve is defined as the linewidth. Moreover, the zero-crossing point of the dispersive curve corresponds to the Larmor precession frequency ω_0 of the rubidium atom. The ambient magnetic field can be calculated by the following formula:

$$B = \frac{\omega_0}{\gamma} \qquad (1)$$

Where ω_0 is the measured Larmor frequency, and γ is the gyromagnetic ratio of the rubidium atom. Besides, the Mz magnetometer is most sensitive at the zero-crossing point due to the largest slope S, i. e. the scale factor. The magnetic field sensitivity of the magnetometer can be calculated as [14]:

$$\delta B = \frac{(N/f)^{1/2}}{S} \qquad (2)$$

where $(N/f)^{1/2}$ is the magnetic noise density.

We have tested the performance of the magnetometer prototype at different operating temperatures. The operating temperature is set within 90~105 °C by a PID temperature control system. For each temperature, the 50-second time-domain signal at the zero-crossing point is recorded, as well as the slope at this point. The linewidth and magnetic noise floor of the M_z atomic magnetometer under different operating temperatures are recorded in Fig. 5. Under 90~105 °C, the magnetic noise floor is less than 6 pT/Hz$^{1/2}$ at 1 Hz, while the linewidth is less than 3.6 kHz. Results show that the optimal operating temperature is 105 °C with a sensitivity of 3.7 pT/Hz$^{1/2}$, as shown in Fig. 6.

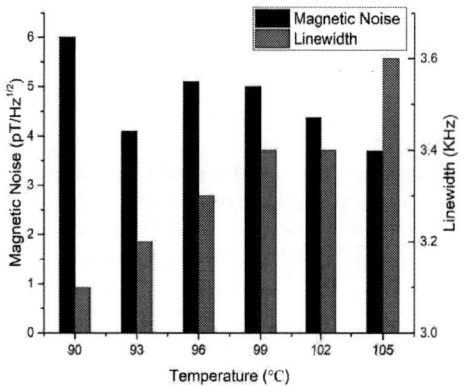

Figure 5: Performance of the M_z magnetometer under different operating temptempe

CONCLUSION

We have demonstrated a novel M_z atomic magnetometer based on a MEMS 3D alkali vapor cell with vertical sidewalls for the first time. Experiments show that the sensitivity of the proposed M_z magnetometer is 3.7 pT/Hz$^{1/2}$ at 1 Hz under 14000 nT, compared to 36.2 pT/Hz$^{1/2}$ in [15], 15 pT/Hz$^{1/2}$ in [16], and >3 pT/Hz$^{1/2}$ in [17] (not given, speculated from the noise spectrum) of state-of-art scalar atomic magnetometers. However, there is still room for improvement in the fabricated 3D alkali vapor cell. For example, the 3D alkali vapor cell can be made more compact by getting rid of unnecessary etched volume.

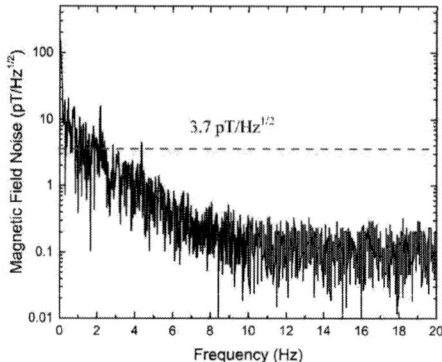

Figure 6: The magnetic noise spectrum of the magnetometer operated at 105 °C.

ACKNOWLEDGEMENTS

This work was supported by the National Science Foundation of China under Grant 51675102. The authors would like to thank G. Sun for their efforts in the probe design. We also thank B. Luo for the advice on the fabrication process.

REFERENCES

[1] V. Shah, S. Knappe, P. Schwindt, and J. Kitching, "Subpicotesla atomic magnetometry with a microfabricated vapour cell," *Nat. Photon.*, vol. 1, pp. 649–652, Nov. 2007.

[2] R. Mhaskar, S. Knappe, and J. Kitching, "A low power, high-sensitivity micromachined optical magnetometer", *Appl. Phys. Lett.*, vol. 101, pp. 241105-1–241105-4, 2012.

[3] J. Kitching, "Chip-Scale Atomic Devices," *Appl. Phys. Rev.*, vol. 5, no. 3, Aug. 2018, Art. no. 031302.

[4] S. Knappe *et al.*, "Cross-validation of microfabricated atomic magnetometers with superconducting quantum interference devices for biomagnetic applications," *Appl. Phys. Lett.*, Vol. 97, no. 13, Sep. 2010, Art. no. 113703.

[5] V. Shah, and R. T. Wakai, "A Compact, High Performance Atomic Magnetometer for Biomedical Applications," *Phys. Med. Bio.*, vol. 58, no. 22, pp. 8153-8161, Nov. 2013.

[6] L. Liew *et al.*, "Microfabricated alkali atom vapor cell," *Appl. Phys. Lett.*, vol. 84, no. 14, pp. 2694-2696, Apr. 2004.

[7] E. J. Eklund *et al.*, "Glass-blown spherical microcells for chip-scale atomic devices," *Sens. Act. A.*, vol. 143, no. 1, pp. 175-180, May 2008.

[8] Y. Ji *et al.*, "Improvement of Sensitivity by Using Microfabricated Spherical Alkali Vapor Cells for Chip-Scale Atomic Magnetometers," *IEEE Trans. Comp. Pack. Manu. Tech.*, vol. 8, no. 10, pp. 1715-1722, Oct. 2018.

[9] Y. Petremand *et al.*, "Microfabricated rubidium vapour cell with a thick glass core for small-scale atomic clock applications," *J. Micromech. Microeng.*, vol. 22, no. 2, Feb. 2012, Art. no. 025013.

[10] R. Chutani *et al.*, "Laser light routing in an elongated micromachined vapor cell with diffraction gratings for atomic clock applications," *Sci. Rep.*, vol. 5, Sep. 2015, Art. no. 14001.

[11] Y. Ji *et al.*, "Micro-fabricated shaped rubidium vapor cell for miniaturized atomic magnetometers," *IEEE Sensor Lett.*, vol. 4 no. 2, Feb. 2020.

[12] H. Nishino *et al.*, "Reflection-type vapor cell for micro atomic clocks using local anodic bonding of 45° mirrors," *Opt. Lett.*, vol. 46, no. 10, pp. 2272-2275, 2021.

[13] J. Zhang, J. Zhang, and J. Shang, "Preparation of 3D Alkali Vapor Cell with Vertical Sidewalls," in *Proc. 17th IEEE International*

Conference on Nano/Micro Engineered and Molecular Systems (NEMS), 2022.

[14] A. K. Vershovskii and A. S. Pazgale v, "Optimization of the Q factor of the magnetic Mx resonance under optical pump conditions," *Tech. Phys.*, vol. 53, pp. 646–654, May 2008.

[15] H. Ke *et al.*, "Parameters optimization of optical pumped Mz/Mx magnetometer based on rf-discharge lamp," *Optik*, vol. 223, 2020, Art. no. 165510.

[16] H. Korth et al., "Miniature atomic scalar magnetometer for space based on the rubidium isotope [87]Rb" *J. Geophys. Res-Space Phys.*, vol. 121, pp. 7870-7880, 2016.

[17] R. Zhang et al., "Subpicotesla scalar atomic magnetometer with a microfabricated cell" *J. Appl. Phys.*, vol. 126, 2019, Art. no. 124503.

CONTACT

*J. Shang, tel: +86-13913869603; jshang@seu.edu.cn

ON-CHIP HEATING NOISE SUPPRESSION OF 3D CHIP-SCALE ATOMIC MAGNETOMETER USING SINGLE-LAYER SHIFTED HEATER

Ziji Wang[1], Junming Wu[1], Jin Zhang[1] and Jintang Shang[1]*
[1] Key Laboratory of MEMS of Ministry of Education, Southeast University, Nanjing, China

ABSTRACT

This paper proposes a novel 3D MEMS atomic chip with single-layer on-chip heater for atomic magnetometry. A micro spherical vapor cell and an on-chip micro heater are monolithically integrated onto a single 3D atomic chip for the first time. Meanwhile, the impact of heating noise on magnetometer performance is effectively suppressed by lateral shifting the on-chip heater from probing region. Comparing to atomic chip heated by unshifted heater, experiments demonstrate that shifted heater reduces sensor noise by more than 90%, and the sensor sensitivity is increased by more than 12 times while reducing fabrication cost and improving device integration.

KEYWORDS

3D MEMS atomic chip, single-layer on-chip heater, heating noise suppression.

INTRODUCTION

As the core of atomic devices, alkali atomic vapor cell fabricated through microelectromechanical system (MEMS) technology holds the potential for low cost, small size, low power atomic clocks, atomic magnetometers and atomic gyroscopes [1-3]. Apart from the vapor cell, typical atomic devices also contain other key modules such like optical, modulation, thermal management and electronic components [4][5]. Developing these key components through CMOS/MEMS technique and heterogeneously integrating them with MEMS vapor cell can break dimension limitations and realize high integration atomic devices [6-8]. However, utilizing miniaturized components may affect device performance, therefore it is necessary to optimize the components design according to the characteristics of specific devices. A typical case is MEMS electrical heater used for atomic vapor cells' thermal management. Since stray magnetic noise is generated by heating current, non-magnetic design is specifically required to reduce the impact from heating noise on atomic magnetometer performance. To this end, discrete MEMS heater with double-layer or even multi-layer structure is investigated [9-11]. However, using separate MEMS heater and adding heating film layer directly increases fabrication complexity and cost. Compared with multi-layer MEMS heater, single-layer heater has the characteristics of simple structure, low fabrication cost and high reliability. Nevertheless, due to its insufficient magnetic field suppression ability, this kind of heaters have not yet been used in atomic magnetometers.

Differs from previous solution, a 3D atomic chip with a single layer shifted heater is proposed in this paper. Directly integrating MEMS heater onto vapor cell improves the heating efficiency and device integration. Meanwhile, the on-chip heater is shifted 7.5mm away from the probing region to reduce the heating noise impact on device performance. To verify the heating noise reduction

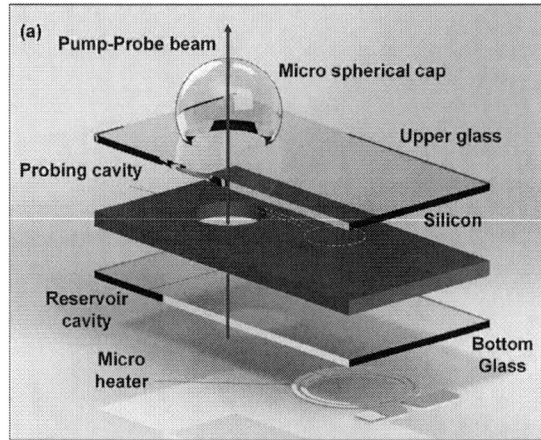

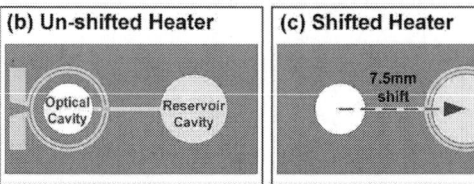

Figure 1: Schematic view of proposed 3D atomic chip and its (a) exploded view. (b) bottom view with un-shifted heater and (c) shifted heater

effect of shifted heater, a M_z atomic magnetometer uses the atomic chip as sensitive element is built and characterized. Experiments demonstrate that the sensitivity of magnetometer using shifted heater is effectively improved compared with using the un-shifted one.

DESIGN AND FABRICATION

Figure 1(a) shows the exploded view of the proposed 3D atomic chip, consisting of a glass-silicon-glass structure micro spherical vapor cell and an on-chip heater. A thermally foamed micro glass bubble is integrated onto the upper glass to improve the atom-light interaction path and atomic relaxation time [12] [13]. To realize the on-chip manipulation of atomic density, single-layer micro heater is integrated on the backside of atomic chip, as shown in figure 1(b). In order to suppress the stray heating noise without increasing heater layer, the on-chip heater is shifted 7.5mm away from the probing region, which is exhibited in figure 1(c). To verify design, magnetic flux density generated by heating current is calculated through the finite element method (FEM). The simulation results are shown in Figure 2, which plots the calculated 2D magnetic field distribution of the shifted heater in the MEMS vapor cell. The on-chip heater is located directly above the cell reservoir cavity, and the optical cavity is 7.5mm away from the reservoir cavity. When applying 100mA current onto the thin-film heater, magnetic field at the center of reservoir and probing region and is 1873nT and 85nT, respectively. Under different heating current,

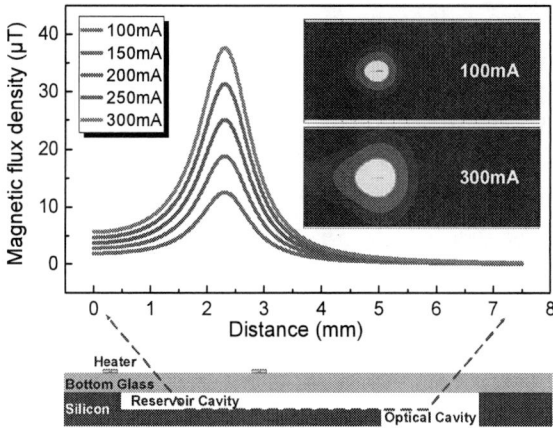

Figure 2: Calculated 2D heating magnetic field distribution in vapor cell. The right two figures demonstrate magnetic field distribution of thin-film heater section at 100mA and 300mA.

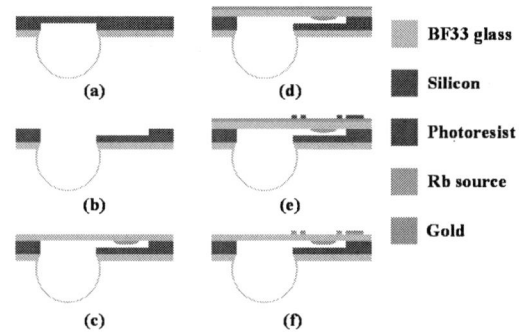

Figure 3: Wafer-level fabrication process of the proposed 3D atomic chip.

magnetic field at probing region decreases by more than 90% compared with that in reservoir region.

Based on the design and numerical analysis, the proposed 3D atomic chip is fabricated. The wafer-level fabrication process is compatible with standard CMOS/MEMS process and illustrated in Figure. 3. A 500µm thickness Si is etched to obtain the probing cavity and reservoir cavity. Followed by anodic bonding and chemical foaming process, micro spherical vapor cell is prepared. After this, bottom glass of the vapor cell is used as substrate, and Au layer with thickness of ~200nm is deposited through magnetron sputtering. Then one-step lithography and wet etching are performed to form the single-layer on-chip micro heater. Figure 4 shows the micrograph of the fabricated atomic chip and the details of its micro bubble and on-chip heater. The resistance of on-chip heater is measured to be 14.5Ω at room temperature.

EXPERIMENTAL RESULTS

The fabricated 3D atomic chip is mounted and wire bonded on a print circuit board (PCB). The sub-mount is then assembled with optical module (lens, polarizer, wave plate), RF coils and photodiode in a 3D-printed resinous sensor head. To characterize the fabricated atomic chip, a total-field atomic magnetometer is constructed based on it. The experimental setup is shown schematically in figure.

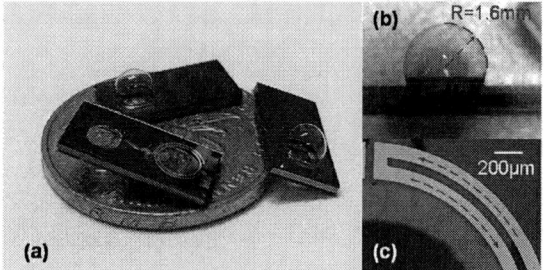

Figure 4: (a) Fabricated 3D atomic chips. (b) Side view of micro glass bubble. (c) Microscope image of on-chip heater.

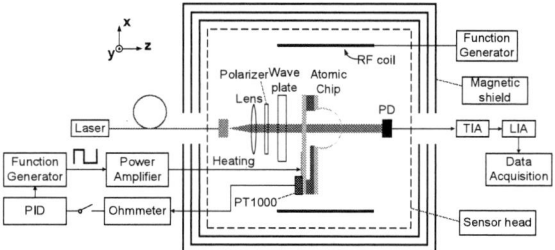

Figure 5: Experimental setup to test the proposed atomic chip, which is under the typical single-beam absorption – type M_Z magnetometer configuration.

5, which is under the typical single-beam absorption-type M_Z configuration. A 795nm DFB laser beam is circularly-polarized through optical module and then passes through the atomic chip. The transmitted light is received by a photodiode. A pair of RF coils generates modulation magnetic field on y-direction, while measured magnetic field B is generated by coils on z-direction. A square wave with 8Vpp amplitude is applied onto the heater to modulate the atomic density. To avoid low-frequency noise from heating current, frequency of square wave is tuned to 500kHz. After introducing the PID control, chip temperature is then stabilized at 110°C with a fluctuation less than 0.08°C. By sweeping the frequency of modulation signal applied on RF coils, dispersive signal is obtained. The relationship between amplitude of ambient magnetic field B and the modulation frequency corresponding to the zero-crossing point of the dispersive curve ω_0 can be determined by:

$$B = \omega_0 / \gamma_{Rb} \qquad (1)$$

where γ_{Rb} is the gyromagnetic ratio of rubidium atoms. The amplitude of the bias magnetic field can be changed by tuning the current applied on the Z-axis coil. Increase the excitation current applied on Z-axis coils from 200mA to 210mA, the zero-crossing frequency increases with the ambient magnetic field. Figure. 6 records dispersive signal response of the magnetometer under different bias magnetic field. Detail of zero-crossing points are shown on the right. Measured Larmor frequencies and their corresponding magnetic intensities at four resonance points are shown in Table I. Measured magnetic field is proportional to the stimulate current on Z-axis coils, showing an agreement with related theory.

978-1-6654-9309-3/23 $31.00 © 2023 IEEE   547

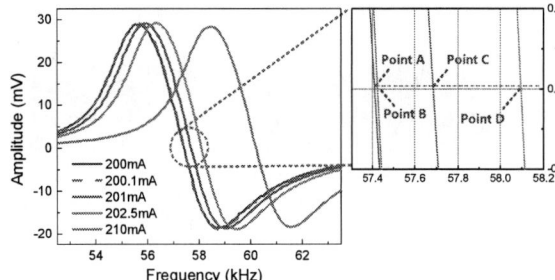

Figure 6: Dispersive signal of atomic magnetometer under measured magnetic field generated by different coil currents. Details of signal around zero-crossing point are shown on the right figure.

TABLE I
INFORMATION OF RESONANCE POINTS

Points	Frequency[kHz]	Magnetic intensity[nT]
A	57.411	12293.58
B	57.425	12296.57
C	57.686	12352.44
D	58.097	12440.43

To compare and characterize the effect of shifted heater, a 3D atomic chip which integrated an un-shifted and a shifted on-chip heater at the same time is designed and fabricated. Figure. 7 compares the output signal of magnetometer at the zero-crossing point when atomic chip is heated to 110°C by shifted and un-shifted heater, respectively. When the 3D atomic chip is heated by un-shifted heater, the output signal standard deviation (STD) is over 0.7mV. When the chip is heated by shifted heater, standard deviation of zero-point signal is reduced to 0.05mV. The zero-point output signal fluctuation of the magnetometer is suppressed by more than 93% after shifting heater 7.5mm from probing region.

Finally, sensitivity of magnetometer is obtained through magnetic noise spectrum density [14]. Figure 8 compares the magnetic field noise density of magnetometer with shifted and un-shifted heater. The magnetic noise density of magnetometer at 1Hz is 892.3pT/Hz$^{1/2}$ when chip is heated by un-shifted heater. After changing to shifted heater, the 1Hz noise density is reduced to 73.6pT/Hz$^{1/2}$, which shows a more than 12 times improvement. Combined with other optimization methods such as in-cell atmosphere and temperature control, the sensor performance can be further improved.

CONCLUSIONS

In summary, the design, fabrication, and characterization of a novel 3D atomic chip adopting single-layer on-chip MEMS heater is proposed for the first time. Compared with current multi-layer MEMS heaters, the proposed on-chip heater reduces fabrication complexity and packaging cost while improving the device integration. To avoid the heating noise generated by the on-chip heater, the heater is shifted from the probing region. The experimental results indicate that when shifting heater 7.5mm from probing cavity, heating noise is suppressed by

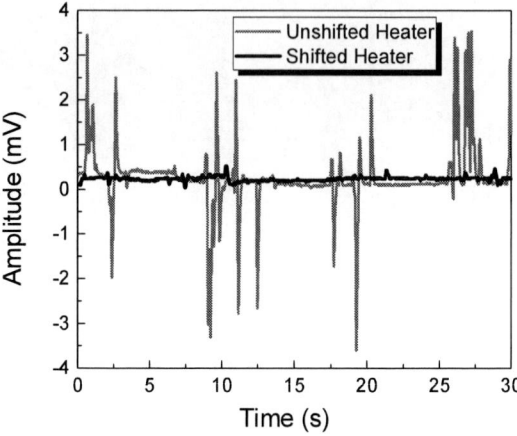

Figure 7: Measured time domain output signal of magnetometer at the zero-crossing point.

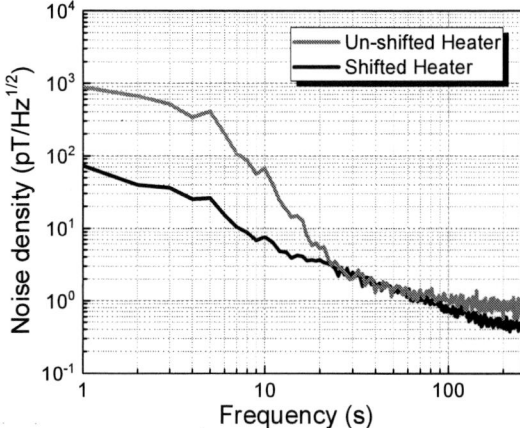

Figure 8: Magnetic field noise spectrum of the atomic magnetometer.

over 90% and the sensitivity of the chip-based magnetometer is improved by over 12 times.

ACKNOWLEDGEMENTS

This work is supported by National Science Foundation of China (No. 51675102). The authors would also like to thank J. Zhang, W. Li and G. Li for their cooperation during the experiments.

REFERENCES

[1] M. Hasegawa et al., "Microfabrication of cesium vapor cells with buffer gas for MEMS atomic clocks," *Sensors Actuat. A, Phys.*, vol. 167, no. 2, pp. 594–601, 2011.

[2] P. D. D. Schwindt et al., "Chip-scale atomic magnetometer," *Appl. Phys. Lett.*, vol. 85, no. 26, pp. 6409–6411, Aug. 2004.

[3] S. Karlen et al., "MEMS atomic vapor cells for gyroscope applications," *2017 Joint Conference of the European Frequency and Time Forum and IEEE International Frequency Control Symposium (EFTF/IFCS)*, 2017, pp. 315-316.

[4] D. Budker and M. Romalis, "Optical magnetometry", *Nature Phys.*, vol. 3, pp. 227-234, 2007.

[5] J. Kitching, "Chip-scale atomic devices", *Appl. Phys. Rev.*, vol. 5, no. 3, pp. 1-38, 2018.

[6] J. Kitching, S. Knappe and L. Hollberg, "Miniature vapor-cell atomic-frequency references", *Appl. Phys. Lett.*, vol. 81, no. 3, pp. 553-555, 2002.

[7] M. J. Mescher, R. Lutwak and M. Varghese, "An ultra-low-power physics package for a chip-scale atomic clock", *Proc. 13th Int. Conf. Solid-State Sens. Actuators Microsyst.*, pp. 311-316, 2005.

[8] J. Park, H. -G. Hong, T. Y. Kwon and J. -K. Lee, "Flexible Hybrid Approach for a 3D Integrated Physics Package of Chip-Scale Atomic Clocks," in *IEEE Sensors Journal*, vol. 21, no. 5, pp. 6839-6846, 1 March1, 2021.

[9] P. D. D. Schwindt, B. Lindsetha, S. Knappe, V. Shah, and J. Kitching, "Chip-scale atomic magnetometer with improved sensitivity by use of the Mx technique," Applied Physics Letters, vol. 90, no. 8, p. 081102, Feb. 2007, DOI: 10.1063/1.2709532.

[10] S. H. Yim, Z. Kim, S. Lee, T. H. Kim, and K. M. Shim, "Note: Double-layered polyimide film heater with low magnetic field generation," *Rev Sci Instrum*, vol. 89, no. 11, p. 116102, Nov. 2018.

[11] X. Liang et al., "A Quadra-Layered Multipole Moment Heating Film With Self-Cancellation of Magnetic Field," in *IEEE Transactions on Magnetics*, vol. 56, no. 12, pp. 1-11, 2020.

[12] J. Shang et al., "Preparation of wafer-level glass cavities by a low-cost chemical foaming process (CFP)," *Lab Chip*, vol. 11, no. 8, pp. 1532-40, 2011.

[13] Y. Ji, Q. Gan, L. Wu and J. Shang, "Geometry influence of the micro alkali vapor cell on the sensitivity of the chip-scale atomic magnetometers," *2017 IEEE 30th International Conference on* Micro *Electro Mechanical Systems (MEMS)*, 2017, pp. 342-345.

[14] A. K. Vershovskii and A. S. Pazgalev, "Optimization of the Q factor of the magnetic Mx resonance under optical pump conditions," *Tech. Phys.*, vol. 53, no. 5, pp. 646–654, 2008.

CONTACT

*Jintang Shang, tel: +86 13913869603; Email: Jshang@seu.edu.cn

LABOR-SAVING PLATFORM FOR CHARACTERIZATION OF MEMBRANE PROTEINS BY AUTOMATED MONITORING AND DATA REPORTING

Kazuto Ogishi[1], Toshihisa Osaki[2], Yuya Morimoto[1], and Shoji Takeuchi[1,2]
[1]Graduate School of Information Science and Technology, The University of Tokyo, JAPAN and
[2]Kanagawa Institute of Industrial Science and Technology, JAPAN

ABSTRACT

This paper describes an automated platform for lipid bilayer studies, which dramatically reduces manual labor. Lipid bilayers are widely used to investigate the functions of membrane proteins *in vitro*. However, the experimental processes, such as monitoring of bilayer functionality during the experiment and analysis of protein properties after the experiment, have relied on manual labor. In this study, we propose a platform that automates the monitoring and post-analysis, which are the most critical and labor-intensive processes. The platform integrates three developed functions: (1) an acquisition system that obtains transmembrane current in real-time, (2) digital filters for the analysis of the current, and (3) a bilayer reformation system that immediately activates when the measurement failure is detected. Combining all, the platform enables automated monitoring of bilayer status, allowing us to continue the measurement for a long period without human supervision. It also allows us to automatically create reports of the analyzed data, saving the trouble of manual post-analysis. We applied the system to gather a large amount of data of α-hemolysin (αHL) to create a histogram of their electrophysiological conductance values. Not only were the values well-estimated, but the time of manual labor throughout the experiment was also reduced to 1 min, which was far less than the 2 h or more required in a conventional procedure.

KEYWORDS

Artificial cell membrane, Lipid bilayer, Nanopore, Ion channel recording, Conductance, Droplet contact method, Real-time, Automation, Restoration

INTRODUCTION

Planar lipid bilayers, also known as artificial cell membranes, have been studied in various research fields, such as the investigation of drug candidates [1-2] and the development of bio-hybrid sensors [3-4], to characterize or utilize membrane proteins *in vitro*. Typical lipid bilayer systems comprise membrane proteins reconstituted into the lipid bilayer, electrophysiological equipment to acquire a transmembrane current, and digital analysis software to quantitatively characterize the properties of the proteins (Fig. 1a). However, these systems are not automated, and require constant monitoring throughout the experiment to ensure that the lipid bilayer is appropriately formed and measured, as well as manual analysis of the signals from a large number of log files after the experiment [5] (Fig. 1b). These labor-intensive and time-consuming processes have hindered the deployment of lipid bilayer technologies into more industrialized fields.

There have been several attempts to automate the lipid

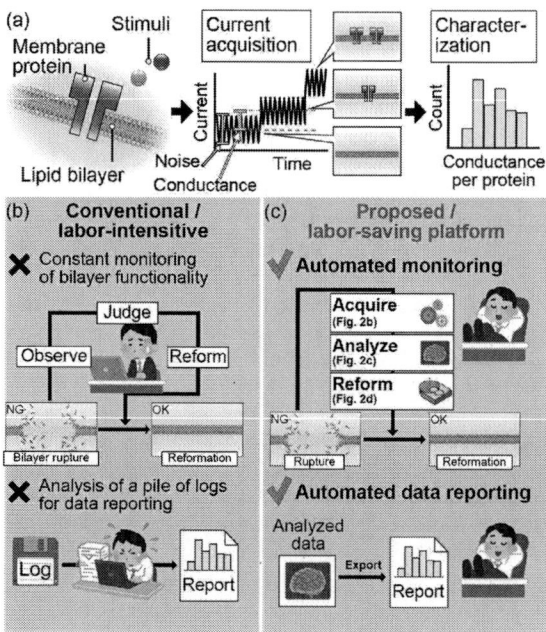

Figure 1: Concept of this work. (a) Experimental procedures for characterization of a membrane protein. Transmembrane current is acquired, and then protein feature is extracted for characterization. (b) Conventional manual labors throughout the lipid bilayer experiment. (c) Proposed labor-saving platform using the developed system, which automates most of the labors.

bilayer systems, but all have been only partially successful. For example, studies have been proposed to automate solution injection and bilayer preparation in combination with pipetting robots [6] or microfluidic channels [7], but none of them have been able to automate the most critical and time-consuming part of the process, *i.e.*, monitoring during the experiment and analysis after the experiment.

Here, we propose a platform that automates these critical processes, thereby reducing the amount of manual labor throughout the experiment (Fig. 1c). By integrating functions of current acquisition, analysis, and feedback into a single application, it is now possible to (1) automatically restore the lipid bilayer when the measurement fails, and (2) automatically create the analysis report of protein properties right after the measurement finishes. The function of the system is verified by applying the system to the long-term data collection of the conductance of α-hemolysin (αHL), nanopore-forming membrane proteins to investigate how much manual labor has been reduced.

EXPERIMENTAL METHODS

The developed automation platform integrates the processes of (1) ionic current acquisition across a lipid bilayer, (2) current analysis, and (3) bilayer restoration upon detecting the measurement failure, into a single PC application and peripherals.

First, a lipid bilayer chamber with rotational capability is fabricated (Fig. 2a). After surveying several previous studies, we consider that a rotational structure would be most convenient for quickly manipulating and restoring the state of lipid bilayers, and thus a rotational lipid bilayer chamber [8, 9] is employed in this study. The chamber is separated into an upper rotational table and a lower base, and the table is connected to a shaft of a stepper motor to allow accurate rotation relative to the base. The structure is fabricated by a 3D printer (microArch, BMF, U.S.A.) and coated with a fluorine coating agent (SF Coat, AGC, Japan) for reproducible lipid bilayer formation [10]. A lipid bilayer is then prepared by contacting a pair of aqueous droplets (28 μL each, 1 M KCl in ultrapure water) in phospholipid-dispersed oil (200 μL in total, 20 mg/mL DPhPC in *n*-decane). To verify the electrophysiological measurement ability, αHL nanopores are incorporated into the lipid bilayer at the bias voltage of +50 mV.

An important step in the automation of lipid bilayer experiments is the acquisition of ionic current in real-time. Unlike a conventional method, which requires the current to be read from a logger after the experiment, we exploit the API of the patch-clamp amplifier (Pico2, Tecella, U.S.A.) to obtain digitized current every second (Fig. 2b).

The current is then analyzed to characterize αHL nanopores reconstituted into the lipid bilayer (Fig. 2c). Since electromagnetic noise is overlapped, the current is first processed with average and Prewitt filters [11] for idealization, in which the number of αHL nanopores is quantified. The current is further processed to estimate a conductance increase for each nanopore incorporation. The analyzed data is immediately appended to a report.

The platform emits an actuation signal to peripherals if necessary (Fig. 2d). In this study, the signal is issued when the current deviates from the measurable range of the amplifier due to the rupture of a lipid bilayer or the incorporation of too many αHL nanopores. The signal actuates a stepper motor to rotate the rotational table of the lipid bilayer chamber. The separation and re-contact of droplets by the rotation allows the system to automatically reset the state of a lipid bilayer and restart the measurement from its beginning.

RESULTS AND DISCUSSION

Validation of the functionality for the automated monitoring and data reporting

Images of the developed platform are shown in Fig. 3a to Fig. 3c, and the workflow of the automated lipid bilayer monitoring and protein property reporting is depicted in Fig. 3d as time-lapse images. For each αHL incorporation, the electrical conductance of the nanopore was quantified immediately after the incorporation (i, ii). When the current overflow was detected (iii), the upper

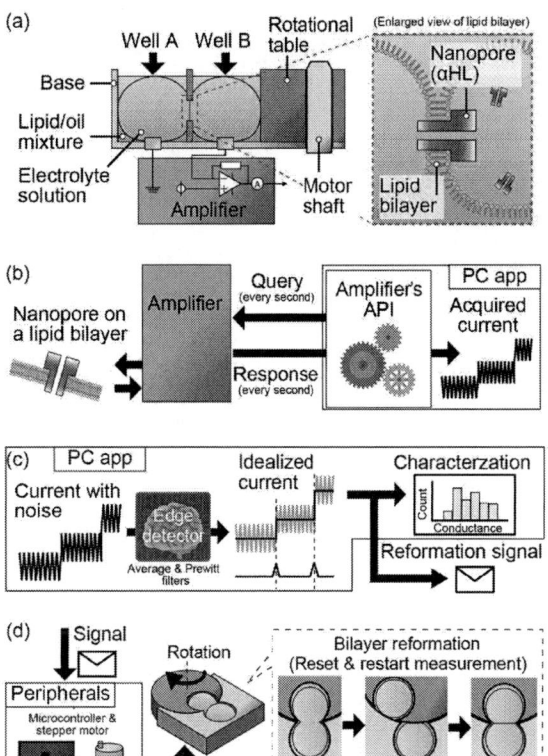

Figure 2: Description of the proposed system. (a) Preparation of a planar lipid bilayer. (b) Real-time current acquisition using the developed software. (c) Real-time analysis by designing optimal digital filters. Reformation signal is issued as needed. (d) Real-time lipid bilayer reformation by the developed system using a rotational chamber, a motor, and peripherals.

table rotated to separate and re-connect the droplets (iv), and we confirmed that this actuation successfully restored the current from the deviation from a measurable range (v).

Validation of the analysis accuracy

We then investigated the consistency between the results of the analysis obtained by the proposed system and those of conventional analysis. Even though the most important point of this study is to automate the monitoring and data reporting processes, the accuracy of the reported data is also of high importance. Therefore, we compared the values of conductance of several αHL nanopores estimated by the proposed platform with those obtained by manual analysis (Fig. 4). The results show that the conductance value of each αHL incorporation was well-estimated with an error of as small as 0.3% in terms of the root-mean-square-error (RMSE), confirming that the platform was able to extract protein features as accurately as the conventional method. In addition, the estimated conductance itself was roughly 1 nS per nanopore, which was in good agreement with the results of a previous study that investigated the conductance of αHL in a 1 M KCl environment [12]. Combined with the result of the previous section, we consider that we have successfully

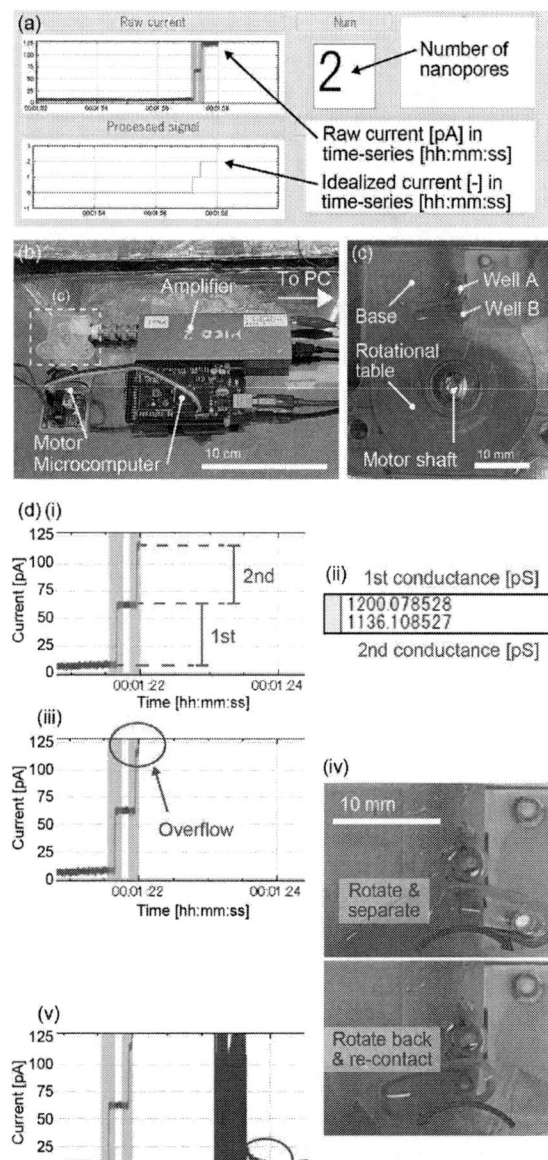

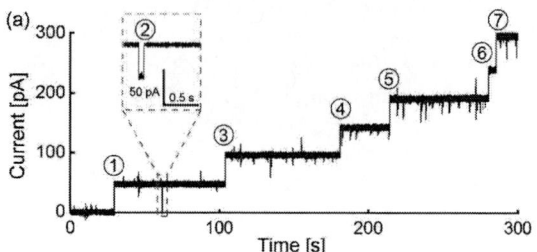

Figure 4: Quantitative comparison between analysis results of the conventional and proposed methods. (a) A representative signal of αHL nanopores. Applied voltage: 50 mV, [αHL]: 10 nM. (b) Comparison of the estimated conductance of each step in (a).

	Conventional method	Proposed method
①	941.6	941.6
②	905.1	903.4
③	968.2	967.5
④	903.6	903.8
⑤	958.7	958.1
⑥	966.7	961.5
⑦	1117.9	1113.0

(Unit: [pS])

Normalized RMSE = **0.3 [%]**

Figure 3: Images of the proposed system. (a) Screenshot of the PC application. (b) Photograph of the peripheral devices. (c) Enlarged view of the rotational lipid bilayer chamber. (d) Time-lapse images of real-time monitoring and in-situ reformation of a lipid bilayer. (i) Real-time conductance estimation when the lipid bilayer is successful. (ii) The quantified conductance for the two αHL molecules in (i). (iii) Beginning of the in-situ reformation. The overflow of current from the measurable range was detected. (iv) Rotation of the lipid bilayer chamber. The two aqueous droplets are separated and re-contacted. (v) End of the bilayer reformation process. The current came back to the measurable range.

constructed a system that can automatically characterize membrane proteins and report the results as accurately as a human, without human supervision.

Long-term, unmonitored lipid bilayer measurement enabled by the autonomous restoration ability

Lastly, as one application of our platform, we verified a long-term lipid bilayer measurement without human supervision and manual analysis to easily obtain a large amount of data of αHL. The platform was configured so that the lipid bilayer should be restored not only when it ruptured but also when a certain number of αHL nanopores were incorporated into the lipid bilayer, allowing for efficient collection of the αHL conductance data. Fig. 5a shows the raw current obtained in the measurement for around 1 h. In this case, the platform autonomously repeated the current acquisition, analysis, and bilayer restoration processes to finally obtain 370 data points of αHL nanopore conductance. The conductance histogram, which was also automatically generated by the platform, is shown as the red bars in Fig. 5b. The blue ones show the result of manual analysis, which was created for reference. Both graphs were in good agreement, indicating that the developed platform was capable of collecting data about membrane proteins with good accuracy as in humans even in continuous measurements over a long time.

Moreover, the amount of manual labor, which was saved by using this platform, was also investigated. Fig. 5c shows the time of manual labor required in the entire experiment to obtain the histograms in Fig. 5b. While the conventional method operated by the authors required 133 minutes of manual monitoring and data reporting to accomplish the experiment, the approach using our platform required only the one-minute of human intervention, except for a 1-hour break to wait for collecting the 1-hour data.

CONCLUSION

In this paper, we proposed an integrated platform that

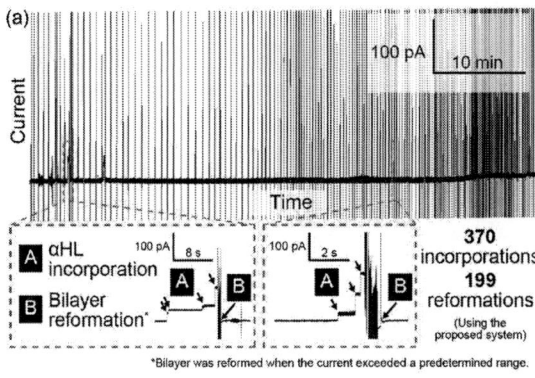

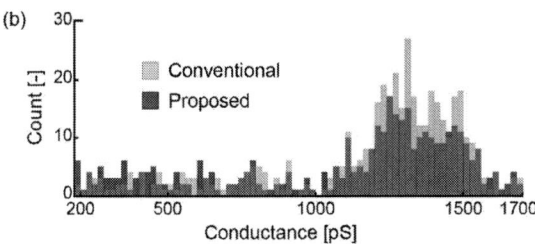

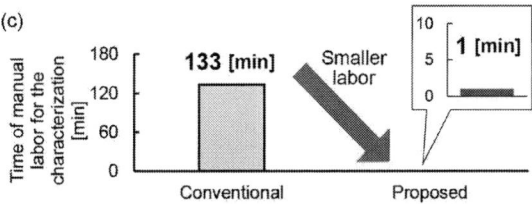

Figure 5: Demonstration of long-term measurement (around 1 h, 3356 s) without manual labor. (a) A representative signal (top) and two small excerpts (bottom). Applied voltage: 50 mV, [αHL]: 100 nM. (b) Comparison of the conductance histograms of αHL nanopores in (a). The red is created by the proposed system, and the blue is manually created for reference. (c) Comparison of the time of the manual labor throughout the experiment.

automates two of the most time-consuming and labor-intensive processes in lipid bilayer measurements: monitoring the functionality of a lipid bilayer during the experiment and analyzing protein features after the experiment. By integrating the acquisition of currents across a membrane protein and a lipid bilayer, analysis of the currents using optimized digital filters, and motor actuation based on the analysis results, the platform enabled the automated restoration of the lipid bilayer and the automatic generation of highly accurate reports about the protein features. Furthermore, the platform was applied to αHL data collection, and we verified that the total manual labor time was dramatically reduced to 1 minute for the continuous measurement around 1 h while maintaining the accurate quantification results as manual analysis. Therefore, we believe that this platform would provide great efficiency for practical applications of lipid bilayer systems, where large data sets must be collected efficiently, or where data analysis must be performed *in situ*.

ACKNOWLEDGEMENTS

This work was partly supported by JSPS KAKENHI Grant Number 21H05013.

REFERENCES

[1]. R. Santos, O. Ursu, A. Gaulton, A. Bento, R. Donadi, C. Bologa, A. Karlsson, B. Al-Lazikani, A. Hersey, T. Oprea, and J. Overington, "A comprehensive map of molecular drug targets", *Nature Reviews Drug Discovery,* vol. 16, pp. 19–34, 2017.

[2]. S. A. Portonovo, C. S. Salazar, and J. J. Schmidt, "hERG drug response measured in droplet bilayers", *Biomedical Microdevices*, vol. 15, pp. 255-259, 2013.

[3]. F. Mazur, M. Bally, B. Städler, and R. Chandrawati, "Liposomes and lipid bilayers in biosensors", *Advances in Colloid and Interface Science*, vol. 249, pp. 88-99, 2017.

[4]. C. Zhu, K. Huang, Y. Wang, K. Alanis, W. Shi, and L. A. Baker, "Imaging with Ion Channels", *Analytical Chemistry*, vol. 93, no. 13, pp. 5355-5359, 2021.

[5]. T. Yamada, H. Sugiura, H. Mimura, K. Kamiya, T. Osaki, and S. Takeuchi, "Highly sensitive VOC detectors using insect olfactory receptors reconstituted into lipid bilayers", *Science Advances*, vol. 7, no. eabd2013, 2021.

[6]. M. Rossi, F. Thei, and M. Tartagni, "A Parallel Sensing Technique for Automatic Bilayer Lipid Membrane Arrays Monitoring", *Sensors & Transducers Journal*, vol. 14-1, pp. 185-196, 2012.

[7]. M. A. Czekalska, T. S. Kaminski, S. Jakiela, K. T. Sapra, H. Bayley, and P. Garstecki, "A droplet microfluidic system for sequential generation of lipid bilayers and transmembrane electrical recordings", *Lab on a Chip*, vol. 15, pp. 541-548, 2015.

[8]. M. Gotanda, K. Kamiya, T. Osaki, N. Miki, and S. Takeuchi, "Automatic generation system of cell-sized liposomes", *Sensors and Actuators B: Chemical*, vol. 292, pp. 57-63, 2019.

[9]. Y. Tsuji, R. Kawano, T. Osaki, K. Kamiya, N. Miki, and S. Takeuchi, "Droplet Split-and-Contact Method for High-Throughput Transmembrane Electrical Recording", Analytical Chemistry, vol. 85, pp. 10913-10919, 2013.

[10]. K. Ogishi, T. Osaki, Y. Morimoto, and S. Takeuchi, "3D printed microfluidic devices for lipid bilayer recordings", *Lab on a Chip*, vol. 22, pp. 890-898, 2022.

[11]. S. Bhardwaj and A. Mittal, "A Survey on Various Edge Detector Techniques", *Procedia Technology*, vol. 4, pp. 220-226, 2012.

[12]. R. Kawano, A. Schibel, C. Cauley, and H. White, "Controlling the Translocation of Single-Stranded DNA through α-Hemolysin Ion Channels Using Viscosity", *Langmuir*, vol. 25, no. 2, pp. 1233–1237, 2009.

CONTACT

*K. Ogishi, tel: +81-3-5841-6488; ogishi@hybrid.t.u-tokyo.ac.jp

MODELLING IMPACT OF VISCOELASTIC PROPERTIES OF DIE-ATTACH MATERIAL ON THE BIAS RESPONSE OF RESONANT INERTIAL SENSORS

Theo Miani[1], Lokesh Gurung[1], Guillermo Sobreviela-Falces[1], Douglas Young[1], Colin Baker[1], and Ashwin A. Seshia[2]

[1]Silicon Microgravity Ltd., Cambridge Innovation Park, Waterbeach, Cambridge, UK
[2]Nanoscience Centre, Department of Engineering, University of Cambridge, UK

ABSTRACT

The method discussed in this paper is based on the viscoelastic-induced bias of a resonant accelerometer due to polymer-based die attach material. Based on a new physical model, the residual bias of two identical accelerometer designs employing two different polymer die-attach materials are compared in the temperature range of -40°C to +40°C. A description of the physical model is provided considering the packaging geometry and viscoelastic properties of the polymer die-attach material. The model results are compared to the bias response of two resonant accelerometers for an equivalent package. The comparison highlights the differences in the mechanical properties of the polymer materials over the temperature range and predicts a gain by using the adhesive with the most advantageous properties.

KEYWORDS

Vibrating beam accelerometer, viscoelastic-induced bias, polymer-based die attach package.

INTRODUCTION

In the recent years, rapid strides have been made in the field of microelectromechanical systems (MEMS) accelerometers including to high-performance applications such as inertial navigation [1], gravimetry [2] and seismology [3]. Specifically, the resonant beam architecture has demonstrated some of the best performance figures combining metrics such as dynamic range, bandwidth, noise floor, output stability and robustness to shock and vibration [1, 4].

However, high-performance resonant accelerometers can be highly sensitive to thermal and mechanical stresses that are mediated by the package. Considering typical operating temperature range requirements for emerging high-performance applications, the packaging must be carefully integrated with the device in order to address key specifications such as total bias error. Polymer-based materials are often used as an adhesive intermediary to mount a MEMS device onto a substrate. The properties of these polymer-based materials can significantly improve the package stress. Previous work has modelled the die-attach as an elastic material to improve the temperature sensitivity due to the package stress [5]. In this way, the residual bias response due to packaging can be compensated using calibration methods [6]. More recent work has highlighted the time dependence of the polymer-based material on temperature [7], which results in an abrupt change in the bias trend over small temperature intervals. The calibration is not sufficient to correct time dependent temperature sensitivity. Artificial neural network calibration [8] provides an approach to address the thermal calibration of the bias response without predicting the effects determined by the viscoelastic properties of the die-attach. In parallel, the viscoelasticity of adhesive materials has been studied to evaluate the long-term drift of micro-accelerometers [9] without proposing a method to improve the residual bias. In separate work, FEA modeling has been applied to model die warpage and induced die stresses for capacitive micro accelerometers to optimize temperature dependence of the offset by considering the material properties of the package stack [10].

In this paper, a new physical model is developed to predict the residual bias response of resonant accelerometers as determined by the properties of the die-attach material and further improvements required in the packaging process are outlined. The results of the model are compared to measurements of residual bias conducted on two identical device designs employing different polymeric die-attach materials for integration into a standard ceramic package.

PHYSICAL MODEL

The new method is based on a 2D geometric model of the packaged chip as schematically outlined in Figure 1(a). The package consists of a chip mounted onto a substrate using a polymer adhesive. In this first approach, the model

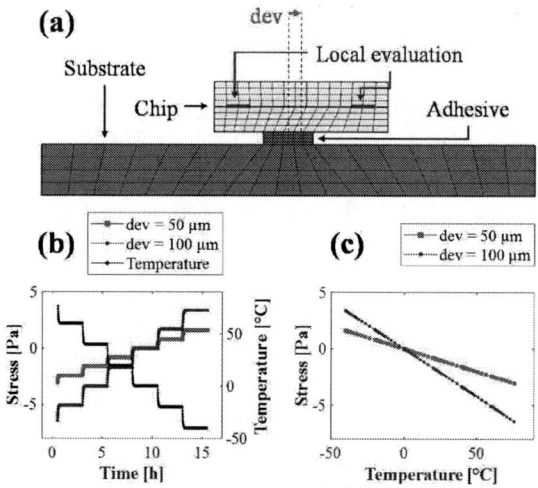

Figure 1: (a) Schematic illustrating the substrate, adhesive, and chip. The offset of the die relative to the adhesive is depicted along with the local evaluation regions. The output is proportional to the differential stress between the local evaluations shown. (b) The differential stress response for two offsets as a function of input temperature. (c) is the differential stress as function of temperature. Two different package process deviation, dev = 50 μm and 100 μm, are considered in (b) and (c).

evaluates the stress distribution in a homogeneous volume of silicon. This ideal case does not consider the detailed geometry of the accelerometer and therefore does not predict the absolute stress induced in the resonant beams. However, it does allow for the prediction of the relative improvement in bias response based on the mechanical properties of the adhesive. This is a particularly useful tool for improving the packaging process without the need for lengthy experiments. The package stress is evaluated differentially between two locations on the chip representing the positions of the resonators. Ideally, the symmetry of the model ensures a cancellation of the package stresses in a differential configuration. Therefore, the offset of the chip center from the center of the die-attach is considered as a key geometric parameter. The initial approach considers only the elastic properties of the adhesive layer. The temperature sensitivity is mainly due to the coefficient of thermal expansion (CTE) mismatch between chip and substrate. By calibrating the sensitivity in the operating temperature range, the effect can be cancelled. The efficiency of the correction depends on the accuracy of the temperature measurement. Adhesives with low Young's Modulus reduces the temperature sensitivity and thus, the requirement in temperature accuracy. This work studies two different polymer-based adhesives, henceforth labelled as glue 1 and glue 2, with Young's modulus of 2 MPa and 2.77 MPa respectively. However, the calibration is not sufficient to cancel the induced temperature sensitivity due to viscoelastic properties. The viscoelasticity introduces time dependence as a function of the temperature in the response. This dynamic behavior corresponds to the slow stress relaxation resulting from the CTE mismatch mediated by the viscoelastic nature of the polymer. To predict the residual bias error after calibration, viscoelastic properties are integrated in the model by considering the Standard Linear Solid (SLS) model of

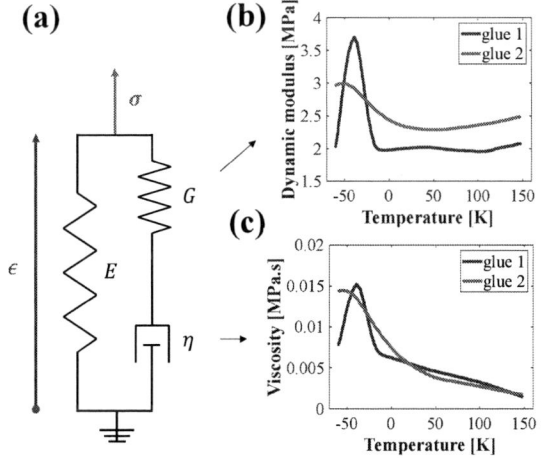

Figure 2: (a) Standard Linear Solid (SLS) model of the polymer-based material depicting the key model parameters. Where E represents the Young's Modulus, G represents the dynamic modulus and η represents the viscosity. (b) and (c) represent respectively experimentally measured dynamic modulus and viscosity for the two different types of adhesives.

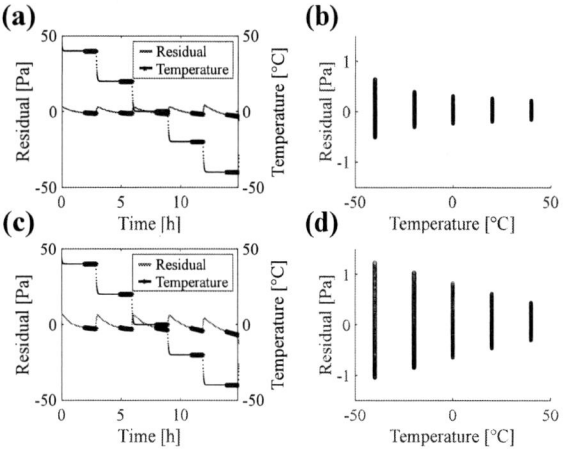

Figure 3: Overview of the physical model results under the - 40°C to + 40°C temperature range. (a) and (c) are the viscoelasticity-induced residual stresses corresponding to adhesives 1 and 2 respectively. The black curve represents the temperature input and the color curves (blue and red) represent the differential stress after temperature correction. Only the last hour of the result for each temperature step is considered. (b) and (d) show the residual stresses as a function of temperature.

Figure 2(a), i.e. representing the relaxation time through a spring-dashpot system. The coefficients of the system are deduced from Dynamic Mechanical Analysis (DMA). This experimental measurement method [11] consists of measuring the dynamic modulus and viscosity over the temperature range of interest. Figure 2 (b) and (c) shows the dynamic modulus and viscosity of the two used polymer respectively. Although their Young's moduli are similar, their viscoelastic behaviors are different in the specified temperature range. The physical model evaluates the impact of the two polymers. Figure 3 (a) and (c) shows the differential stress after temperature correction. The comparison is based on the 2D model of Figure 1 (a) considering the same geometry but using the two different adhesive properties.

In order to quantify the residual bias after temperature correction, a temperature-based criterion is defined. The criterion considers the last hour of measurement for each temperature step. It is highlighted as black-dotted line over the plots of Figure 3 (a) and (c), which represent respectively the residual stresses corresponding to adhesives 1 and 2. Figure 3 (b) and (d) plot respectively the criterion corresponding to the cases for adhesives 1 and 2 as a function of the temperature. For both adhesives considered, the residuals decrease for higher temperatures. The DMA results shown in Figure 2 (b) and (c) highlight the glass transition temperature is close to -40°C for both adhesives. In this configuration, the relaxation time is longer as indicated by a higher residual. The model quantifies (i) the residuals in the temperature range and (ii) compares the residuals corresponding to devices mounted using different adhesives.

EXPERIMENTAL RESULTS
Results from experiments performed using the same

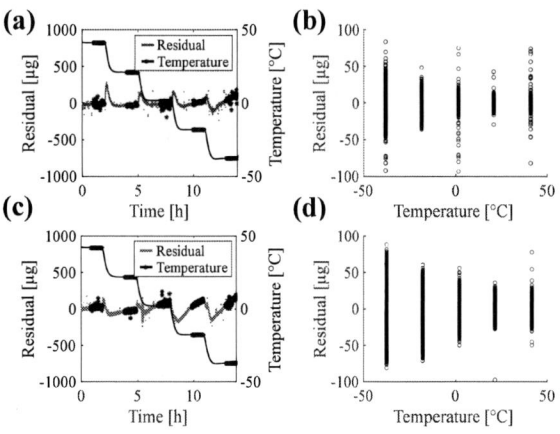

Figure 4: Overview of the experimental results over the -40°C to + 40°C temperature range. (a) and (c) are the residual bias of the accelerometers mounted using adhesives 1 and 2 respectively. The black curve represents the temperature input and the colored curves (blue and red) represent the equivalent acceleration after temperature correction. The criterion considers only the last hour of measurement for each temperature step. The criterion is highlighted in (a) and (b) in black. (b) and (d) plot the criterion as a function of temperature respectively for the cases of adhesives 1 and 2.

accelerometer architectures are compared to the model predictions. Two devices, one for each adhesive, were prepared using the same packaging process. A digital resonance tracking unit is implemented to track the resonant frequencies and their differential frequency represents an output reading proportional to the acceleration measurement. The differential frequency is measured over a temperature range from -40°C to + 40°C with 3 hours of data logged at each temperature step. The scale factor measurement allows for deducing the equivalent acceleration. A first calibration allows canceling the temperature sensitivity due to the package stress. The residual bias is shown in Figure 4 (a) and (c), respectively for die-attach adhesives 1 and 2. The same criterion used for the model is plotted as a function of temperature in Figure 4 (b) and (d), respectively for adhesive 1 and 2.

Table 1: Comparison between the physical model and the experiments. The two residuals for the two devices (packaged using different die-attach adhesives) are compared for the -40°C and + 40°C temperature steps.

Bias response of physical model		
Adhesive	**Adhesive 1**	**Adhesive 2**
+ 40°C	0.11 Pa	0.21 Pa
- 40°C	0.33 Pa	0.66 Pa
Bias response of resonant accelerometer		
Adhesive	**Adhesive 1**	**Adhesive 2**
+ 40°C	4.20 μg	7.93 μg
- 40°C	20.34 μg	36.10 μg

Table 1 shows the model and experimental comparison of the residual bias errors of both adhesives. As the model

does not specifically consider the accelerometer architecture, the simulation results are not directly comparable to experimental results. One is a function of the axial stress on the beam, whereas the other is the local stress evaluated in a homogeneous silicon chip. However, the physical model shows a relative comparison that follows the experimental results. By quantifying the viscoelastic-induced bias using the criterion, glue 1 yields a factor of 2 improvement in the bias response compared to glue 2.

CONCLUSION

The SLS model of die-attach polymers provides good insight in the prediction of thermal stress coupling to MEMS devices due to CTE mismatches between the die and package. The use of a 2D-homogeneous silicon volume demonstrates its ability to predict the bias of a resonant inertial sensor. The 2D FE model outputs provide a means of evaluating and optimizing the properties of die-attach adhesives and package geometries with low computational overhead. A more detailed 3D model can be considered a follow-on step to provide increased accuracy and predicting the absolute stress and frequency variations for specific device configurations informed by optimization performed using the parameter sweeps conducted under the 2D model.

REFERENCES

[1] G. Sobreviela-Falces *et al.*, 'A Navigation-Grade Mems Vibrating Beam Accelerometer', in *2022 IEEE 35th International Conference on Micro Electro Mechanical Systems Conference (MEMS)*, Tokyo, Japan, Jan. 2022.

[2] A. Mustafazade *et al.*, 'A vibrating beam MEMS accelerometer for gravity and seismic measurements', *Sci Rep*, vol. 10, no. 1, p. 10415, Dec. 2020.

[3] G. Sobreviela-Falces *et al.*, 'A Mems Vibrating Beam Accelerometer for High Resolution Seismometry and Gravimetry', in *2021 IEEE 34th International Conference on Micro Electro Mechanical Systems (MEMS)*, Gainesville, FL, USA, Jan. 2021, pp. 196–199.

[4] R. Dejaeger, O. Lefort, M. Jeanneteau, and J.-M. Muguet, 'A Low-Noise Mixed Signal ASIC for Navigation-Grade Resonant MEMS Accelerometer', in *2022 IEEE International Symposium on Inertial Sensors and Systems (INERTIAL)*, Avignon, France, May 2022.

[5] G. Li and A. A. Tseng, 'Low stress packaging of a micromachined accelerometer', *IEEE Trans. Electron. Packag. Manufact.*, vol. 24, no. 1, pp. 18–25, Jan. 2001.

[6] P. Aggarwal, Z. Syed, and N. El-Sheimy, 'Thermal Calibration of Low Cost MEMS Sensors for Land Vehicle Navigation System', in *VTC Spring 2008 - IEEE Vehicular Technology Conference*, Marina Bay, Singapore, May 2008, pp. 2859–2863.

[7] S.-H. Chae, J.-H. Zhao, D. R. Edwards, and P. S. Ho, 'Characterization of Viscoelasticity of Molding Compounds in Time Domain', in *ASME 2009 InterPACK Conference, Volume 1*, San Francisco, California, USA, Jan. 2009, pp. 435–441.

[8] R. Fontanella, D. Accardo, R. S. Lo Moriello, L. Angrisani, and D. De Simone, 'MEMS gyros temperature

calibration through artificial neural networks', *Sensors and Actuators A: Physical*, vol. 279, pp. 553–565, Aug. 2018.

[9] W. Zhou, P. Peng, H. Yu, B. Peng, and X. He, 'Material Viscoelasticity-Induced Drift of Micro-Accelerometers', *Materials*, vol. 10, no. 9, p. 1077, Sep. 2017.

[10] X. Zhang, S. Park and M. W. Judy, 'Accurate Assessment of Packaging Stress Effects on MEMS Sensors by Measurement and Sensor–Package Interaction Simulations,' in *Journal of Microelectromechanical Systems*, vol. 16, no. 3, pp. 639-649, June 2007.

[11] K. P. Menard, 'Dynamic Mechanical Analysis: A practical introduction', *CRC press* (2nd ed.), 2008.

CONTACT

Colin Baker, tel: +44 1 223 421 911 ; cbaker@silicong.com

CMOS-EMBEDDED 3D MICRO/NANOFLUIDICS EMPLOYING TOP-DOWN BEOL SINGLE-STEP WET-ETCHING TECHNIQUE

Wei-Yang Weng[1], Hung-Yu Hou[1], Yueh-Jung Chao[2], Shwu-Jen Liaw[2], Jun-Chau Chien[1]

[1] Graduate Institute of Electronics Engineering, National Taiwan University, Taipei, Taiwan and
[2] Department of Clinical Laboratory Sciences and Medical Biotechnology, National Taiwan
University, Taipei, Taiwan

ABSTRACT

This paper presents CMOS-embedded 3D micro/ nanofluidics for next generation point-of-care (POC) diagnostic devices. We take advantage of the fine line metallization in modern CMOS technology and create 3D micro/nanofluidics through the removal of the metals from the routings using a single-step post-CMOS wet-etching process. We demonstrate the method in both 180-nm and 65-nm CMOS chips with aluminum and copper back-end-of-line (BEOL), respectively. Various fluidic geometries are investigated including (1) a four-channel fluidic splitter; (2) a 3D fluidic channel; and (3) a width-varying channel for resistive pulse sensing (RPS). We also study the impact of etchants on the transistor characteristics and demonstrate functional circuit operation using a 32-bit shift register after prolong etching. Last, we demonstrate resistive pulse sensing in our CMOS-embedded microfluidics. This work presents the first demonstration of CMOS-embedded micro/nanofluidics for biomedical applications.

KEYWORDS

CMOS, microfluidics, nanofluidics, top-down BEOL etching, cellular detection, resistive pulse sensing.

INTRODUCTION

Over the past decades, many micro/nanofluidics devices have been developed for Point-of-Care (PoC) applications. Though the devices are well miniaturized, systems often still require a benchtop instrument for reading out the bio-signal responses, e.g., electrochemical currents, fluorescent lights, etc. Therefore, integrating millimeter-size CMOS chips with micro/nanofluidics has gained significant interest as it offers the potential for reader-less operation [1-3]. In addition, embedding CMOS electronics in close proximity to the samples-to-be-detected improves the accuracy as the unwanted routing parasitics can be largely eliminated.

Generally, the fluidics and CMOS electronics are integrated through a modular assembly [4-5]. Though flexible, such an approach is limited by the achievable alignment accuracy, thus putting constraints on the fluidics designs. On the other hand, [6] presents a wafer-scale bottom-up fabrication technique for implementing nanofluidics channels. The main idea is to etch away a patterned silicon layer encapsulated inside silicon oxides to create hollow channels. Channel dimensions at nanometers are achieved, and the paper demonstrates the detection of a stretched DNA optically using a lab microscope. However, the complexity of the fabrication flow makes its compatibility with the standard CMOS process challenging. [7] utilizes deep reactive ion etching (RIE) to pattern "open" fluidic channels on a BiCMOS chip. They also take advantage of the ion-etching selectivity and integrate the electrodes within the channel to trap a single cell using dielectrophoresis (DEP). [8] adopts both gas- and liquid-based etching techniques to remove the polysilicon and metal routings from a 2.0-μm CMOS chip for creating hollow channels with depths ranging from 0.5 to 100 μm. They demonstrate its integration with on-chip thermopiles for a liquid flow rate sensor. [9] introduces nanofluidics on top of CMOS photodetectors by sacrificing thin polysilicon layer in a CMOS chip. However, the developed process requires deep reactive ion-etching to create through-hole access from the passivation, which can lead to severe unwanted over-etching at lower controllability.

Inspired by the above pioneer works, we present a CMOS-embedded 3D micro/nanofluidics for biosensing in advanced CMOS technology. Compared to the conventional CMOS Lab-on-Chip systems where the fluidics and the electronics are modularly integrated together (Figure 1(a)), the proposed method, as shown in Figure 1(b), utilizes the metallization in the back-end-of-line (BEOL) of a CMOS process to create fluidic channels of desired geometry followed by a one-step wet etching process. As such, hollow channels are formed within the passivation and the inter-metal dielectrics of a CMOS die and the samples can be in closer interface with the detection electronics. Moreover, complex 3D fluidic structures can be implemented by utilizing multi-layer metallization offered by the foundry at fine resolution and accuracy. This work studies its feasibility and various fluidic geometries.

DEVICE DESGIN AND FABRICATION

The implementation flow starts with the layout of the fluidic structures using metal and pad layers. Figure 2 exhibits a few example structures. The designer has the freedom to choose channel widths and gaps as long as they meet the foundry design rules; on the other hand, the channel height is generally fixed by the chosen CMOS technology and the metal layers. After chip fabrication, we sacrifice these metals by immersing the chips in metal etchants at an elevated temperature (55°C). The types of etchants depend on the metal material. After several investigations, we select Type A aluminum etchant from Sigma Aldrich and Cu-129 copper etchant from Chemleader Corporation. Figure 2 shows the channel geometries and the optical images of 180-nm and 65-nm CMOS fluidic structures. Etch fronts are observed, and the estimated etch rates are 111 nm/min for the aluminum in 180-nm CMOS and 104 nm/min for the copper in 65-nm CMOS, respectively. Interestingly, we observe that the etching of the aluminum is more anisotropic in the 180-nm CMOS chip than the 65-nm counterpart, and the phenomenon is repeatable in multiple runs. The exact

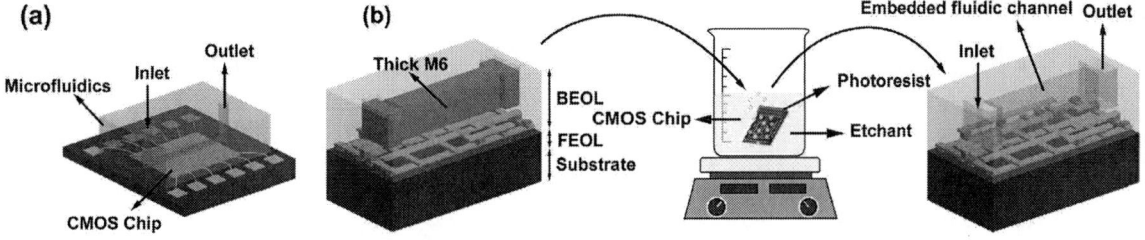

Figure 1: CMOS/microfluidics integration: (a) the conventional approach; (b) the proposed method.

reason is not yet known, and studies are in progress. The etch rate is highly dependent on the temperature and is ~4× slower than those from the datasheets. We believe this is due to the limited etchant diffusion rates.

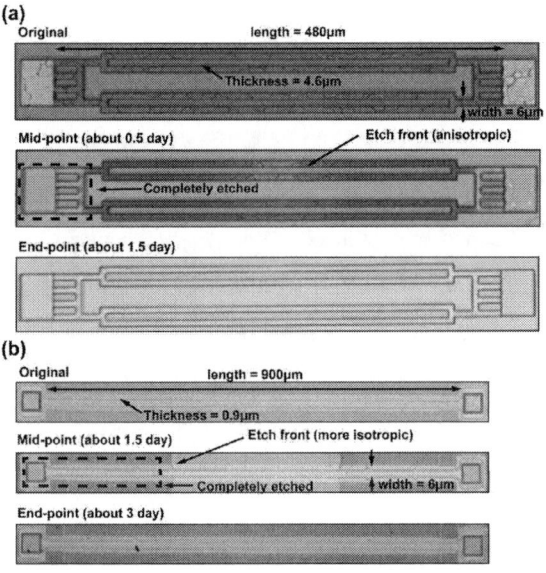

Figure 2: Etching status of (a) 180-nm and (b) 65-nm CMOS structures under an optical microscope.

We investigate the cross sections of the on-chip fluidic channel using the focus ion beam (FIB) and scanning electron microscope (SEM). Figure 3 shows the SEM images from the topmost metal layer in the 180-nm CMOS chip tilted at 60°. Complete removal of the aluminum is observed, and the dimensions match our layout structure.

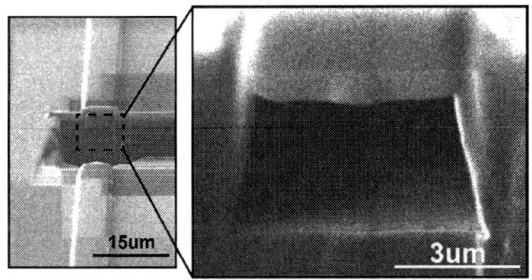

Figure 3: The SEM image of the on-chip fluidic channel from a 180-nm chip.

EXPERIMENTAL RESULTS
Channels Characterization

Various fluidic geometries are investigated, including (1) a four-channel fluidic splitter (Figure 2(a) and Figure 3(a)); (2) a 3D fluidic channel (Figure 3(b)); and (3) a width-varying single-layer channel (Figure 3(c)). The four-channel splitter is a widely used fluidic structure for enhancing assay throughput. We inject 1μm beads as a proxy for cells and observe successful flowing without clogging. We further explore forming 3D fluidic channels using multiple metal layers connected with vias. Remarkably, we find the same etchants can remove the CMOS vias at a size of 260×260 nm^2. Using the CMOS vias as pores can also open up new opportunities in pore-based sensing [10].

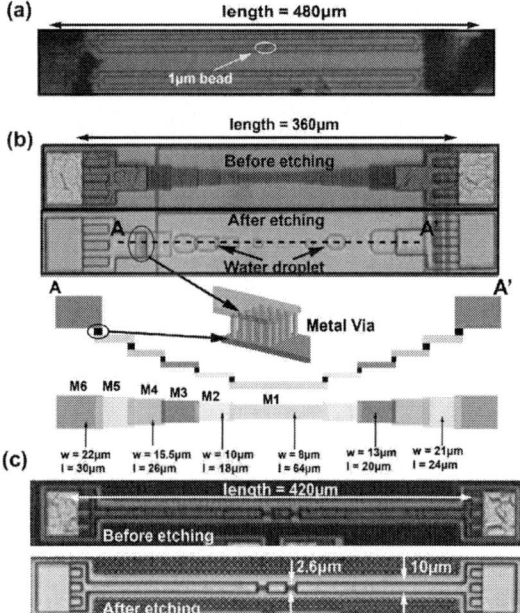

Figure 4: Investigated fluidics structures: (a) a 4-channel fluidic splitter; (b) 3D fluidics using multi-layer metallization; (c) a width-varying channel for resistive pulse sensing.

The last structure is a width-varying channel designed for resistive pulse sensing (RPS). RPS is a technique for detecting and measuring the size of the flowing particles from changing ionic resistances when a particle

978-1-6654-9309-3/23 $31.00 © 2023 IEEE   559

interrogates through the constricted channels [10]. To achieve this, two 2.6-μm wide, 10-μm long channel segments are inserted within a 10-μm wide fluidic channel. Figure 3(c) demonstrates complete removal of the metals.

Circuit Functions Verification

We study the impact of etchants on the transistor characteristics after prolonged immersion of the CMOS chip in the etchant medium. This is crucial to validate the feasibility of future fluidics/electronics integration. Figure 4(a) compares the measured I_D-V_{GS} of an NMOS transistor (W/L = 4 μm/ 0.18 μm) from two CMOS dies, with and without the metal removal process. The results show an average difference of 1.5 %, comparable to the degree of process variation. We also test the circuit operation post-etched using a 32-bit shift register. Figure 4(b) shows correct output bit sequences. Note that the small time delay between the input and output of the shift register arises from the interconnects and the wires in the measurement setup. Given these studies, we believe the immersion of CMOS dies in etchants for ~25 hrs has unnoticeable impacts on both the analog and digital circuits.

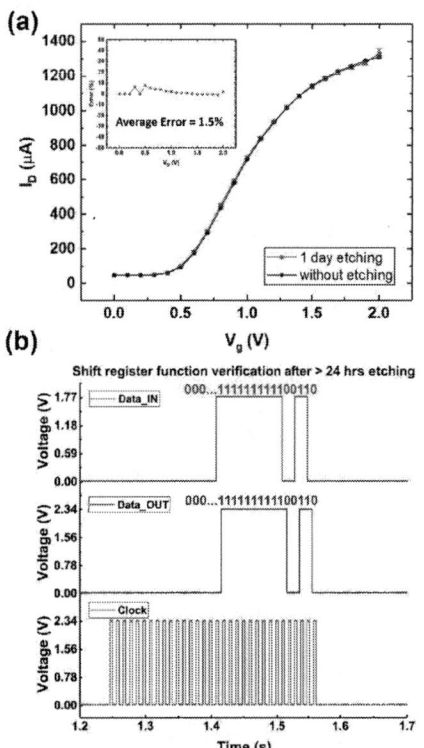

Figure 5: Circuit function before and after metal etching: (a) I_D-V_{GS} comparisons of a NMOS transistor (4 μm/0.18 μm) and (b) the functional demonstration of a 32b shift register.

Resistive Pulse Sensing (RPS) of bacteria cells

We perform RPS of bacteria cells using our CMOS-embedded microfluidics. Figure 6 shows the schematic and our testing setup. Two 250-μm diameter Ag/AgCl electrodes are inserted at both the inlet and the outlet, and the DC ionic resistance across the fluidic channel is measured through a transimpedance amplifier (IV204F3)

at a gain of 4MΩ. A flowing particle through the constriction zones creates an increase in the ionic resistance, and is registered as voltage pulses whose pulse-widths depend on the flowing velocity.

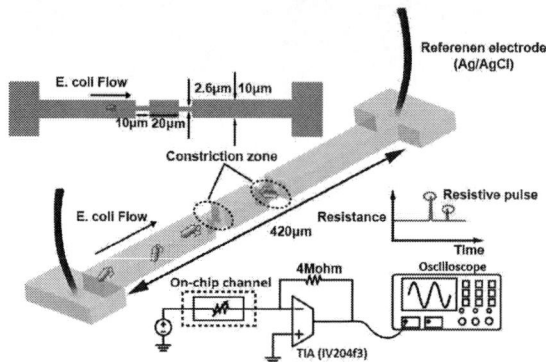

Figure 6: Resistive pulse sensing system schematic and its sensing mechanism.

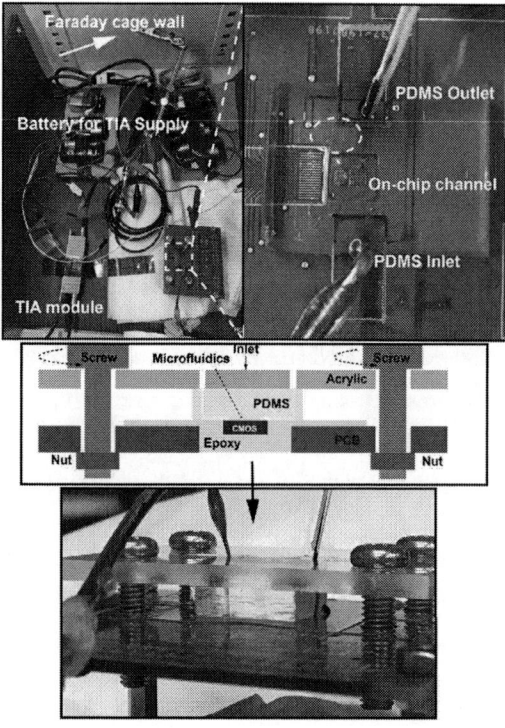

Figure 7: The measurement setup and device packaging for resistive pulse sensing.

Figure 7 shows the device packaging and our measurement setups. First, we use biocompatible epoxy (Epo-Tek 302-3M) to surround and planarize our CMOS chip into a printed circuit board (PCB). Next, we sandwich a sample-delivering PDMS between the PCB and an acrylic board. The PDMS is molded from a SU-8 patterned silicon wafer having wide channels that overlay with the pads of the fluidic structures. The electrodes for RPS are inserted into the tygon tubing through catheters. All measurements are performed inside a Faraday cage to minimize environmental noise, including those from the

60-Hz power lines.

Figure 8 shows the RPS results from flowing 1-μm diameter polystyrene beads (Figure 8(a)) and E. Coli cells (strain pir-116) (Figure 8(b)) in PBS medium. The particle entering the two constriction zones blocks the ionic flows, elevates the channel resistance, and causes a current reduction. Each pair of pulsing signals in the time-domain waveforms represents one particle flowing through the two constriction zones. The measured pulsewidth is ~80 μsec, corresponding to 125 mm/sec flow velocity.

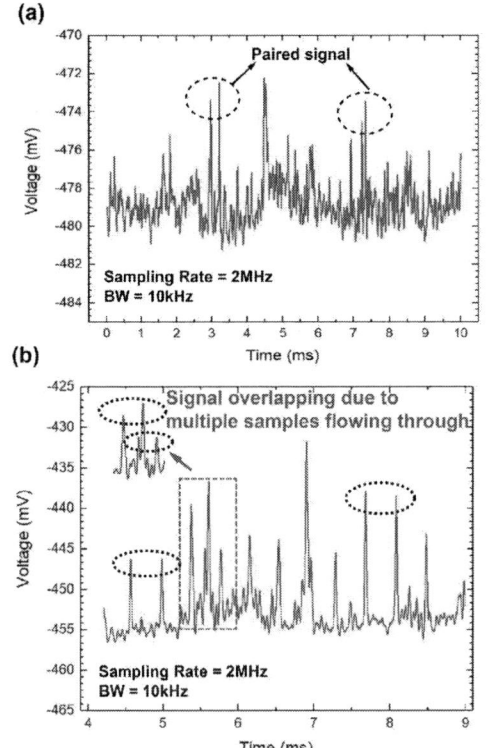

Figure 8: Resistive pulse sensing (RPS) of (a) 1-μm beads and (b) E. Coli bacteria cells.

CONCLUSIONS

This work presents a CMOS-embedded micro/ nanofluidics platform suitable for next-generation lab-on-a-chip devices. By sacrificing the BEOL metallization using a single-step wet-etching technique, sub 10-μm fluidic channels embedded in a CMOS chip are batch fabricated at low cost. Utilizing the proposed method, we demonstrate various fluidic structures and study the impact of the etchants on transistors and circuits. We also present the detection of a single E. Coli cell using resistive pulsing sensing. In the future, we aim to develop a fully-integrated system incorporating fluidics and electronics on the same CMOS chip.

ACKNOWLEDGEMENTS

The authors thank Yushan Young Scholar Program from the Taiwan Ministry of Education (NTU-110VV001 and NTU-111VV001), Taiwan National Science and Technology Council NSTC 110-2222-E-002-010-MY3, Taiwan Semiconductor Research Institute (TSRI) for

measurement supports, National Center for High-performance Computing (NCHC) for computational resources, NTU SoC Center, TSMC University Shuttle, Prof. Wei-Cheng Li and C. P. Tsai of NTU Institute of Applied Mechanics for insightful discussions, and Ms. C.-Y. Chien of Ministry of Science and NTU for FIB and SEM measurement supports.

REFERENCES

[1] Y. H. Ghallab and Y. Ismail, "CMOS Based Lab-on-a-Chip: Applications, Challenges and Future Trends," *IEEE Circuits and Systems Magazine,* vol. 14, no. 2, pp. 27-47, 2014

[2] S. M. Khan, A. Gumus, J. M. Nassar, and M. M. Hussain, "CMOS Enabled Microfluidic Systems for Healthcare Based Applications," *Adv Mater,* vol. 30, no. 16, p. e1705759, Apr 2018

[3] M. Lindsay *et al.*, "Heterogeneous Integration of CMOS Sensors and Fluidic Networks Using Wafer-Level Molding," *IEEE Trans Biomed Circuits Syst,* vol. 12, no. 5, pp. 1046-1055, Oct 2018

[4] Y. Huang and A. J. Mason, "Lab-on-CMOS integration of microfluidics and electrochemical sensors," *Lab Chip,* vol. 13, no. 19, pp. 3929-34, Oct 7 201

[5] J. C. Chien, A. Ameri, E. C. Yeh, A. N. Killilea, M. Anwar, and A. M. Niknejad, "A high-throughput flow cytometry-on-a-CMOS platform for single-cell dielectric spectroscopy at microwave frequencies," *Lab Chip,* vol. 18, no. 14, pp. 2065-2076, Jul 10 2018.

[6] C. Wang *et al.*, "Wafer-scale integration of sacrificial nanofluidic chips for detecting and manipulating single DNA molecules," *Nat Commun,* vol. 8, p. 14243, Jan 23 2017

[7] F. Hjeij *et al.*, "UHF dielectrophoretic handling of individual biological cells using BiCMOS microfluidic RF-sensors," in *2016 46th European Microwave Conference (EuMC)*, 4-6 pp. 265-26. Oct. 2016

[8] A. Rasmussen, M. Gaitan, L. E. Locascio, and M. E. Zaghloul, "Fabrication techniques to realize CMOS-compatible microfluidic microchannels," *Journal of Microelectromechanical Systems,* vol. 10, no. 2, pp. 286-297, 2001

[9] H. Meng, M. I. o. T. D. o. E. Engineering, and C. Science, *CMOS Nanofluidics*. Massachusetts Institute of Technology, Department of Electrical Engineering and Computer Science, 2018

[10] J. K. Rosenstein, M. Wanunu, C. A. Merchant, M. Drndic, and K. L. Shepard, "Integrated nanopore sensing platform with sub-microsecond temporal resolution," *Nat Methods,* vol. 9, no. 5, pp. 487-92, Mar 18 2012, doi: 10.1038/nmeth.1932.

CONTACT

* J.-C. Chien, jcchien@ntu.edu.tw
* W.-Y. Weng, r09943060@ntu.edu.tw

IMPLEMENTATION OF A MONOLITHIC SOC ENVIRONMENTAL SENSING HUB USING CMOS-MEMS TECHNIQUE

Ya-Chu Lee[1], Tung-Lin Chien[1], Chi-Te Fang[1], Yuanyuan Huang[1],
Wei-Lun Sung[2], Yen-Chang Chu[2], Rongshun Chen[1] and Weileun Fang[1,3]
[1]Dept. of Power Mechanical Engineering, National Tsing Hua University, Hsinchu, Taiwan
[2]PixArt Imaging Inc., Hsinchu Science Park, Hsinchu, Taiwan
[3]Institute of NanoEngineering and MicroSystems, National Tsing Hua University, Hsinchu, Taiwan

ABSTRACT

This study demonstrates a novel monolithic environmental sensing hub using CMOS MEMS processes. Merits of the proposed design are: (1) SoC environmental sensing hub monolithically integrated the gas sensor, anemometer, humidity sensor, visible light sensor and thermometer on a single chip, and (2) the compatibility of post-CMOS processes to implement a SoC environment sensing hub. The environmental sensing hub has been successfully implemented using TSMC 0.18μm 1P6M CMOS process and in-house post-process. Measurement results show that sensitivities of gas sensor, anemometer, humidity sensor and thermometer are respectively 0.057 %/% (O_2), 0.5 Ω/ms^{-1}, 1.6 pF/%RH and 0.6307 mV/°C. The sensitivities of visible-light sensor for each wavelength are 2.47 mV/μW (443 nm), 5.21 mV/μW (522 nm) and 8.94 mV/μW (634 nm). The fabricated environmental sensing hub reveals the highly integrated monolithic CMOS-MEMS system-on-chip (SoC) applications.

KEYWORDS

Environment sensing hub, CMOS-MEMS, Gas sensor, Thermometer, Anemometer, Humidity sensor, Visible light sensor.

INTRODUCTION

After the COVID-19 pandemic, environmental quality has become a significant global health and safety issue. Environmental sensors are key components to monitor environment conditions. Compared with traditional environmental sensors, the micro-electrical-mechanical-systems (MEMS) technology is considered as a promising solution for implementing environmental sensors. Hence, MEMS environmental sensors have been extensively investigated and applied in various fields [1-3]. Presently, the integration of many environmental sensors (as the sensing hub) have been commercialized [4-5]. For the demand of commercial products, "size" and "cost" are two critical considerations. As examples of the aforementioned considerations, BOSCH (BME680) sensing hub integrates a pressure sensor, thermometer and gas sensor through system-in-package (SiP) technology [4]. Multiple environmental informations can be obtained through the Sensirion (SEN5x) with the integration of particulate matter, volatile organic compounds (VOCs), oxidizing gases, humidity and temperature sensors [5]. However, in these examples, a SiP solution is required for the integration of multiple sensors. In contrast, the mature CMOS platform offer various inherent materials to develop environment sensors [3,6-8], which can be utilized for system-on-chip integration. Thus, this study will leverage the CMOS technology to implement and monolithically integrate environmental sensors.

DESIGN CONCEPT

Fig.1 shows the design (based on the 0.18μm TSMC 1P6M CMOS process) of the monolithic CMOS-MEMS SoC environmental sensing hub which contains gas sensor, anemometer, humidity sensor, visible light sensor, and thermometer. The vertical integration of MOS gas sensor with heater is enabled by the multi-layer stacking of CMOS process. In particular, the poly-Si heater is located between the interdigitated electrodes of the gas sensor to improve the heating efficiency of the gas sensing material. Thus, the gas sensor could effectively reach the target working temperature. The humidity sensor consists of interdigitated electrodes with sub-micron gaps (0.6μm) is designed to offer reasonable sensing signals without moister absorbing material. Note that the sensing performance may be reduced due to the absence of humidity sensing materials. However, the interference between the aforementioned gas sensing materials can be avoided. The visible light sensor and thermometer are formed by diodes for size reduction. The output voltage will be generated as the light sensor is irradiated by the incident light. For the thermometer, temperature fluctuations will change the threshold voltage of the diode, consequently affecting the detected voltage of the thermometer. Note the metal layer above thermometer (diode) can avoid the influence of external light. The resistance change on hot-wire (poly-Si layer) is employed to measure the wind speed (anemometer). Leveraging the advantages of CMOS platform consisting of sub-micron linewidth and multi-layer stacking, a compact CMOS-MEMS sensing chip with 5 different sensors could be implemented together with in-house post-CMOS processes.

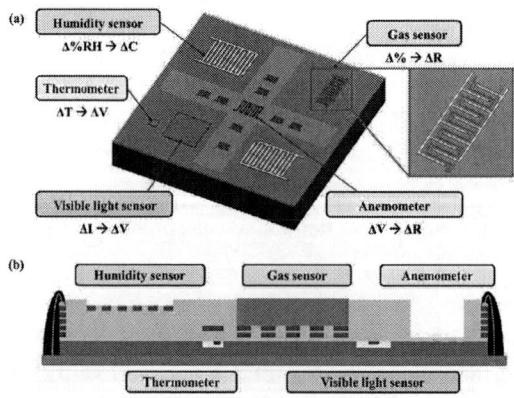

Figure 1: Design concept of CMOS-MEMS environmental sensing hub

FABRICATION PROCESS AND RESULT

Fig.2 shows the process flow. Fig.2a displays the CMOS chip fabricated by TSMC 0.18μm 1P6M CMOS process. Fig.2b-d shows post-CMOS processes. Fig.2b-c indicate the metal wet-etching process to define sensor structures, and the following reactive-ion etching (RIE) process were performed to expose the bond-pads and sensing electrodes. Finally, Fig.2d displays the wire-bonding process that the device was attached to the PCB board and was protected by the epoxy compounds. Then, the gas-sensing film was dispensed onto the gas-sensing area and then baked at 60°C for 24 hours. Note that the detailed preparation methods of the gas sensing material in this study can be obtained in [8].

Micrographs in Fig.3a show the typical fabricated environmental sensing hub including humidity sensor, thermometer, visible-light sensor, gas sensor, and anemometer. Details of each sensor is shown in their respective zoom-in micrographs. The heater of the gas sensor is located at the gap of the interdigital electrodes, and the heater of the anemometer is laid at the center of the

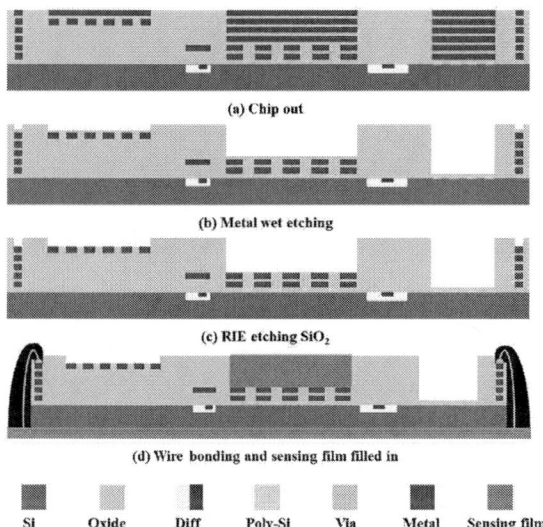

Figure 2: Fabrication process steps

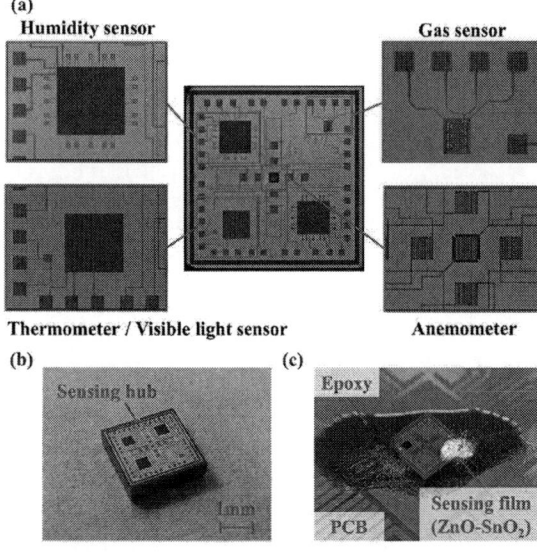

Figure 3: The OM micrograph of typical fabrication results.

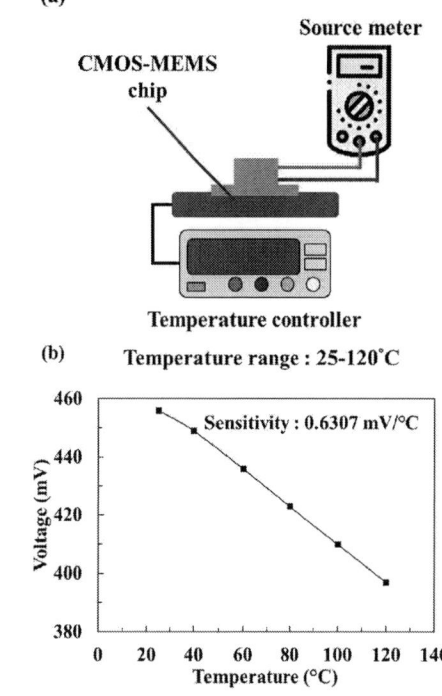

Figure 4: Experimental setup and response for the thermometer.

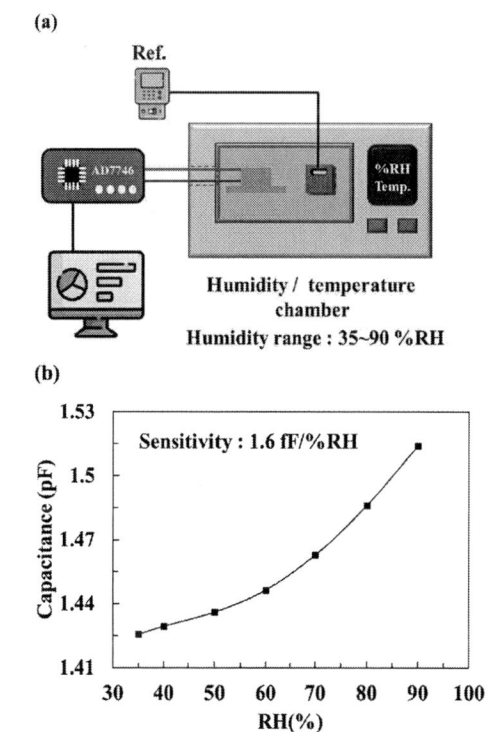

Figure 5: Experimental setup and response for the humidity sensor

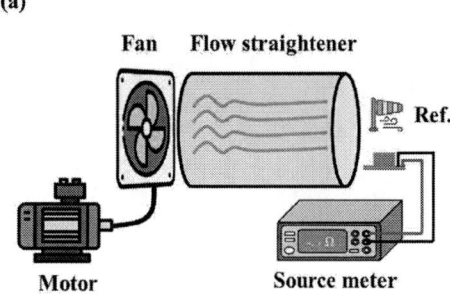

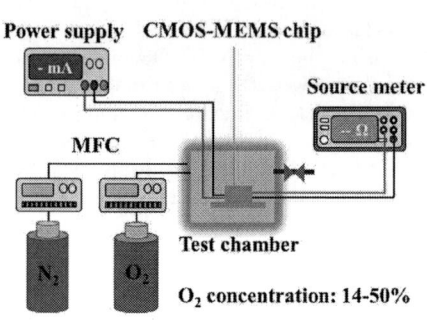

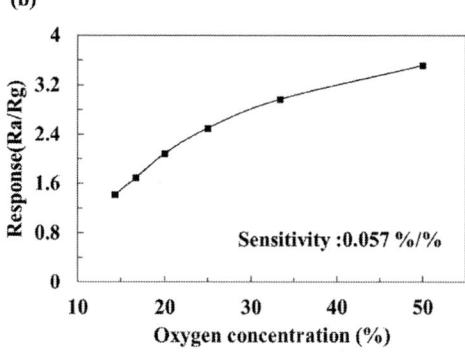

Figure 6: Experimental setup and response for the anemometer.

Figure 7: Experimental setup and response for the gas sensor.

sensing chip. As shown in the micrographs, the sub-micron gap electrodes of the humidity sensor, the visible light sensor, and the thermometer were successfully defined through the post-CMOS processes. Micrographs show the CMOS-MEMS chip after fabrication (Fig.3b), then wire-bonding on PCB and deposited with gas-sensing films to realize the device-under-test (DUT) (Fig.3c).

MEASUREMENT SETUP AND RESULT

Fig.4 displays the thermometer diode test to measure voltage changes when the chip was heated within 25~120°C. The voltage change of the thermometer diode was characterized by a source meter. Measurements indicate the sensitivity of thermometer is 0.6307mV/°C. Measurements in Fig.5 depict the capacitance change of humidity sensor when specifying chamber humidity from 35%RH to 90%RH at 25°C. A commercial readout circuit (Analog Devices, AD7746) was utilized to output the capacitance change under the variation of relative humidity. A commercial humidity sensor (Sensirion, SHT21) was used as a reference sensor to monitoring the humidity/temperature control chamber for calibration (with accuracy of ±3%RH). The sensitivity of humidity sensor is 1.6pF/%RH from 35%RH to 90%RH with intervals of 10%RH. Fig.6 displays the setup for anemometer. A motor was used to control fan speed, and the wind was rectified using the flow straightener. The wind speed is controlled to be between 0.5~5m/s, and the resistance change of the anemometer was characterized by the source meter. A commercial anemometer (YK-2005AH) was used as a reference sensor to monitor the wind speed (the wind speed

range of reference sensor is 0.2-20.0m/s±5%). Measurements show the sensitivity of the anemometer is 0.5Ω/(m/s). Fig.7 displays the gas sensor tests. A power supply was used to drive the heater for temperature control, while the source meter was obtained to measure the resistance change of the sensing film. The test chamber was purged with nitrogen for each measurement point. Then pure oxygen was injected to the chamber for testing. The oxygen concentration changes between 14-50% by mass flow controllers (MFC), and the heater on chip was driven by a power supply. By measuring resistance changes of the gas sensor, its sensitivity was characterized to be 0.057 %/%. Fig.8 displays the measurement setup for visible light sensor. The R/G/B lasers are used to emit lights of different wavelengths. The voltage change of the visible light sensor is recorded by a source meter. The sensitivities of visible light sensor for three different wavelengths are respectively: 2.47mV/μW (443nm), 5.21mV/μW (522nm), and 8.94mV/μW (634nm). Measurements summarized in Table1 shows the feasibility and performances of presented monolithic CMOS-MEMS environmental sensing hub.

Table 1: Specifications summary of environmental sensing hub.

Gas sensor (O$_2$)		Humidity sensor		Thermometer	
Sensitivity (%/%)	0.057	Sensitivity (pF/%RH)	1.6	Sensitivity (mV/°C)	0.6307
Anemometer		Visible light sensor			
Sensitivity (Ω/ms^{-1})	0.5	Sensitivity (mV/μW)	B:2.47 @ 443 nm G:5.21 @ 522 nm R:8.94 @ 634 nm		

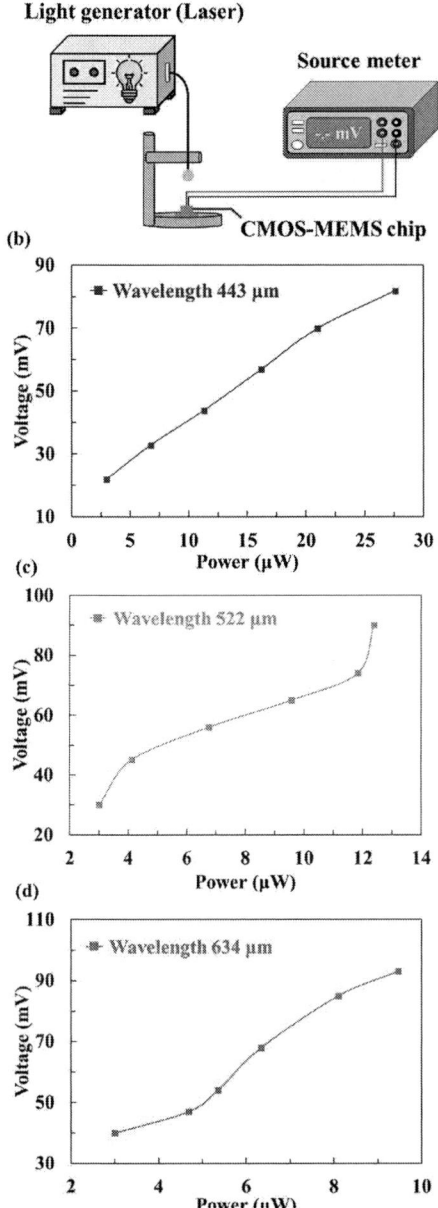

Figure 8: Experimental setup and response for the visible light sensor

CONCLUSION

This study demonstrates an environmental sensing hub implemented by using the TSMC 0.18μm 1P6M standard process platform. By leveraging the characteristics of the CMOS process platform, the integration of multiple sensors to form a compact SoC environmental sensing hub can be achieved. Table 1 summarizes the measurement results among different sensing units of the proposed environmental sensing hub. With proper post-CMOS processes arrangement, the integration with additional environmental sensors or inertial sensors in a single chip could offer more robust and comprehensive information for users in the near future.

ACKNOWLEDGEMENTS

This project was supported by the National Science and Technology Council under grant number NSTC 111-2923-E-007-003-, NSTC 111-2221-E-007-070-MY3, NSTC 111-2218-E-007-014-MBK, NSTC 111-2923-E-007-004-MY2, NSTC 110-2926-I-007-506-, and NSTC 110-2218-E-007-032-. The authors would like to appreciate the TSMC and the Taiwan Semiconductor Research Institute (TSRI), for the supporting of CMOS chip manufacturing. The authors also appreciate the Center for Nanotechnology, Materials Science and Microsystems (CNMM) of National Tsing Hua University for providing the process tools. The authors are also grateful to Prof. Chih-Yung Huang of National Tsing Hua University, and Prof. Cheng-Hao Ko of National Taiwan University of Science and Technology, Taiwan for providing the measurement facilities.

REFERENCES

[1] Y.-C. Lee, M.-L. Hsieh, P.-S. Lin, C.-H. Yang, S.-K. Yeh, T. T. Do and W. Fang, "CMOS-MEMS technologies for the applications of environment sensors and environment sensing hubs." *Journal of Micromechanics and Microengineering*, vol. 31, 074004, 2021.

[2] U. Kang and K. D. Wise, "A high-speed capacitive humidity sensor with on-chip thermal reset", *IEEE Transactions on Electron Devices*, vol. 47, pp.702-710, 2000.

[3] Y. -C. Lin, Y. -C. Lee, C. -H. Yang and W. Fang "Vertical integration of pressure/humidity/temperature sensors for CMOS-MEMS environmental sensing hub," *IEEE Sensors*, pp. 1-4, 2021.

[4] BOSCH (Available at: https://www.bosch-sensortec.com/products/environmental-sensors/gas-sensors/bme680/)

[5] Sensirion (Available at: https://developer.sensirion.com/sensirion-products/sen5x-environmental-sensor-node/)

[6] Y.-C. Lin, P.-H. Hong, S.-K. Yeh, C.-C. Chang and W. Fang, "Monolithic integration of pressure / humidity / temperature sensors for CMOS-MEMS environmental sensing hub with structure designs for performances enhancement," *IEEE MEMS*, Vancouver, Canada, pp. 54-57, 2020.

[7] C. -C. Chang, P. -H. Hong, S. -K. Yeh, Y. -C. Lin, M. -F. Lai and W. Fang, "Environmental sensing hub on single chip using double-side post-CMOS processes," *IEEE MEMS,* Vancouver, Canada, pp. 877-880, 2020.

[8] T.-L. Chien, Y.-C. Lee, T. Chou, Y.-Y. Lin, H.-Y. Chen and W. Fang, "Fabrication and integration of SOC environment sensing hub with gas/pressure/temperature sensors," *IEEE MEMS*, Tokyo, Japan, pp. 138-141, 2022.

CONTACT

* Weileun Fang; fang@pme.nthu.edu.tw

MONOLITHICALLY AND VERTICALLY INTEGRATED ENVIRONMENTAL SENSING HUB WITH NOVEL AIR-BASED HUMIDITY SENSOR DESIGN

Tung-Lin Chien[1], Yuanyuan Huang[1], Fuchi Shih[2], and Weileun Fang[1,2]

[1] Dept. of Power Mech. Eng., National Tsing Hua University, Hsinchu City, Taiwan
[2] Inst. of NanoEng. and MicroSyst., National Tsing Hua University, Hsinchu City, Taiwan

ABSTRACT

This study presents the vertical integration of humidity sensor, pressure sensor and thermometer on a single chip (Fig.1) based on the TSMC 0.18 μm 1P6M CMOS platform. Merits of the proposed sensing hub are: (1) Vertical SoC: vertically integrated environmental sensing hub consisting capacitive humidity sensor with microheater integrated, capacitive pressure sensor, and diode thermometer, (2) Novel humidity sensor: fast response and easily fabricated air-based capacitive humidity sensor. Measurements demonstrate the feasibility and performances of proposed environmental sensing hub (400 μm×400 μm): humidity sensor with sensitivity of 3.32 fF/%RH and response time of 2.48 sec, pressure sensor with sensitivity of 0.22 fF/kPa, and thermometer with response of 0.64 mV/°C.

KEYWORDS

Environment sensing hub, CMOS-MEMS, Pressure sensor, Humidity sensor, Thermometer

INTRODUCTION

The demand for environmental monitoring has gradually increased to improve the quality of life. The sensing hub consists of various environmental sensors could achieve better environmental monitoring. Through the approaches of SoC (system on chip) and/or SiP (system in packaging), the entire sensing system can be integrated to meet the requirements of various applications such as industries, home, personal and others. Recently, standard CMOS process has become a promising technology to realize the integration of environmental sensors [1-2], such as the integration of humidity, pressure, and temperature sensors [3-4], and the additional gas sensor integration [5].

In addition to sensing performances, process compatibility and size are also critical concerns during integration. In view of this, the multilayer stacking of commercial CMOS platform provides the design flexibility to fabricate several sensors on a single chip, especially for the environmental sensors. In general, capacitive humidity sensor with sensing material is commonly applied to increase the dielectric constant between electrodes, so as to improve the sensitivity of humidity sensing [6-7]. However, the sensing material needs to be further defined through post-processes such as dispensing, photosensitive pattern or etching definition after sensing chip fabricated [8]. Moreover, the additional process would further enlarge the performance variation of each sensor, reduce the yield rate, or encounter the incompatible process.

To solve aforementioned issues, this study presents a vertically integrated humidity/pressure/temperature sensors for size reduction, in which the novel air-based humidity sensor without moisture absorption material requirement is also presented.

DESIGN CONCEPT

Fig.1a shows the proposed sensing chip consisting of a capacitive humidity sensor with microheater, a capacitive pressure sensor and a diode thermometer. Leveraging the advantage of multi-layer stacking provided by standard TSMC 0.18 μm 1P6M (one polysilicon and six metal layers (M1-M6)) CMOS process platform and the in-house post CMOS processes, the humidity sensor is vertically integrated with pressure sensor to offer benefits of small footprint and less packaging effort. Moreover, a diode thermometer is monolithically integrated to monitor the surrounding temperature by its semiconductor characteristics [9]. The A-A' cross-section illustrates the vertical integration design of humidity/pressure sensors. In addition, the gap defined by M4 sacrificial layer between two sensing units not only serve as thermal isolation layer, but also provides the space for ambient pressure exposure. The top humidity sensor (Fig.1b) is formed by M5 interdigitated electrodes (IDE), where a set of sensing-electrode (in red, the same metal layer with blue one) is also exploited to act as the microheater. By leveraging the capability of sub-micron gap and linewidth offered by CMOS platform, the air-based humidity sensor is

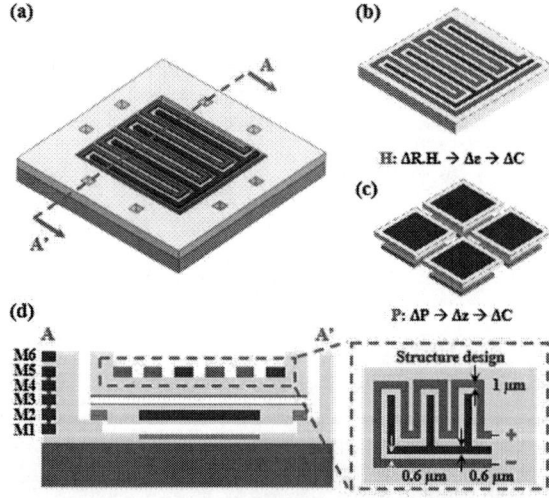

Figure 1: Design concept, (a) the proposed design consists of capacitive pressure sensor vertically integrated with capacitive humidity sensor with heater integrated and diode thermometer, (b) schematic and sensing principle of capacitive humidity sensor, (c) schematic and sensing principle of capacitive pressure sensor, (d) cross-sectional view and structure design of proposed humidity sensor.

proposed. Since no polymer-based sensing material is needed, the advantages of fast response time, simple fabrication with better process uniformity, easy integration, and no need of additional packaging effort are achieved. As the relative humidity increased in ambient environment, the dielectric constant (ε) of the air between interdigitated electrodes will be changed, and further caused the change in capacitance (ΔC). Typical humidity sensors often need long recover time when it comes to high humidity circumstance [10]. Therefore, the integration design of microheater in this study can provide Joule heating to prevent water vapor condensation. The proposed relative humidity sensor is designed at 400 μm × 400 μm, where the total resistance of the microheater is designed at ~5 kΩ.

An absolute type capacitive pressure sensor is designed below the humidity sensing unit for footprint reduction. Through the chemical vapor deposited (CVD) Parylene-C, the bottom pressure sensor (Fig.1c) consists of 2×2 sensing capacitor array are sealed in ~10Pa. The deformable diaphragm has an embedded sensing electrode (M2 layer), and the reference electrode (polysilicon layer) is anchored to the substrate. The sensing gap between two electrodes is defined by metal sacrificial layer (M1). The diaphragm will be deformed by the ambient pressure to change the gap as well as sensing capacitance of electrodes. Fig.1d illustrates the cross-sectional view of the vertical integration design, where the detail parameters of the proposed humidity sensor design with microheater are also demonstrated.

FABRICATION

The presented environmental sensing hub is implemented by the standard TSMC 0.18 μm 1P6M CMOS platform together with the in-house post-CMOS processes. Fig.2 shows fabrication steps. Fig.2a indicates the CMOS chip fabricated by TSMC. In Fig.2b, metal sacrificial layers connected with tungsten via were etched by H_2SO_4/H_2O_2 solution through the designed etching channels to release the suspended structures. After that, CVD Parylene-C was utilized to seal the etching channels to implement the pressure sensor (Fig.2c), the designed z-shaped etching channel was sealed under ~10 Pa vacuum condition to realize the absolute type capacitive pressure sensor. In Fig.2d, the Parylene-C was patterned by O_2-plasma. Then, the Al electrodes was exposed after RIE oxide etching for humidity detection, as indicated in Fig.2e. Finally, as shown in Fig.2f, the proposed chip was wire bonded onto PCB for electrical connections and additional measurements.

Micrographs display a typical fabricated sensing chip (Fig.3a) as the device-under-test (DUT) and the humidity/pressure sensing unit after RIE process (Fig.3b). The SEM micrograph (Fig.3c) depicts the geometry and submicron linewidth of proposed humidity sensor. FIB cross-section micrograph in Fig.3d further shows the vertical integration of sensors. The stacking of multi-layers for humidity and pressure sensors can be clearly observed. Furthermore, the micrograph also indicates that the initial deformation of the suspended structure is smaller than the initial sensing gap (0.53 μm) defined by metal sacrificial layer.

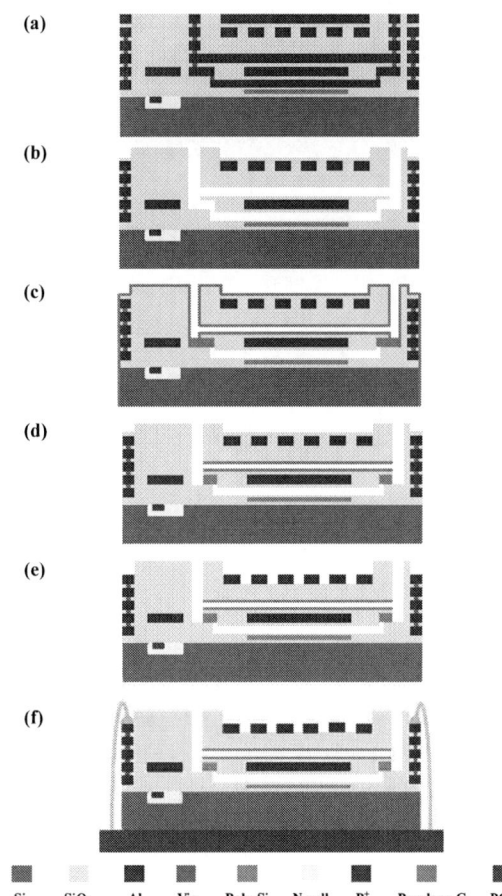

Figure 2: Fabrication process steps, (a) chipout prepared by TSMC, (b) metal wet etching, (c) polymer deposition, (d) O_2-Plasma patterning (e) RIE etching, (f) wire bonding.

MEASUREMENT RESULTS

Several experiments were conducted to characterize the performance of the environmental sensing hub. Fig.4a

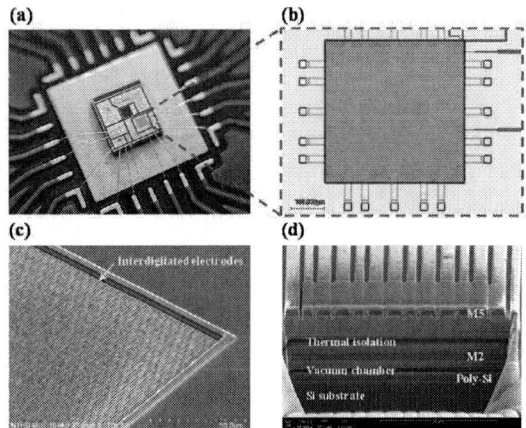

Figure 3: Micrographs of fabrication result, (a) overview of fabricated chip, (b) top view of the proposed design, (c) zoom-in micrograph of humidity sensor with interdigitated electrodes, (d) FIB micrograph of the proposed design.

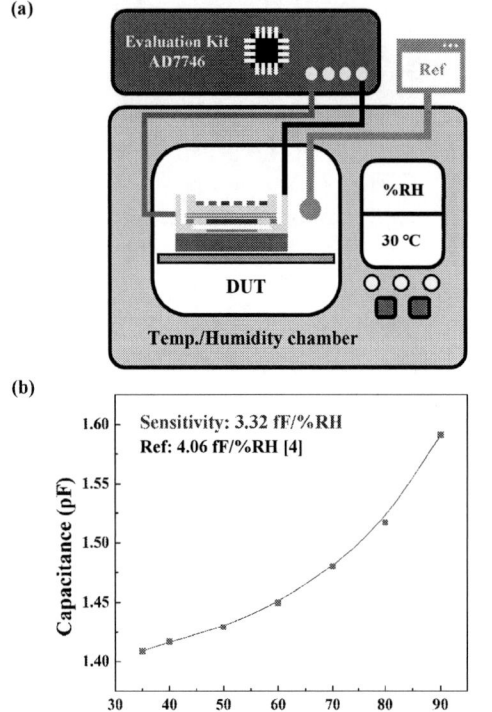

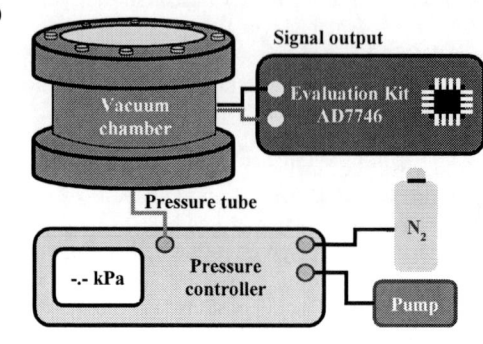

Figure 4: (a) Measurement setup for sensitivity measurement of capacitive humidity sensor, (b) measured capacitance versus relative humidity difference.

Figure 6: (a) Measurement setup for capacitive pressure sensor, (b) measured capacitance change versus chamber pressure difference.

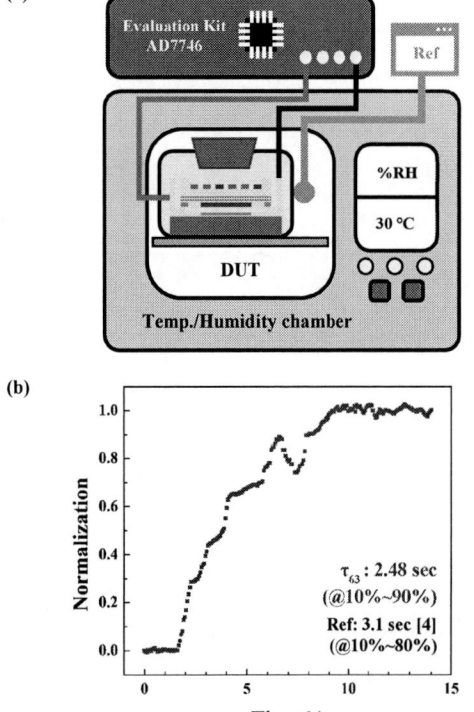

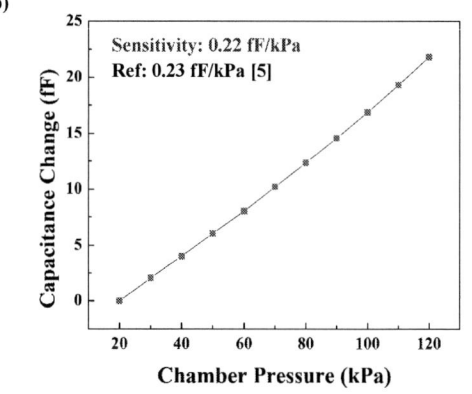

Figure 5: (a) Measurement setup for response time of capacitive humidity sensor, (b) measurement result of humidity response time.

depicts the DUT inside the programmable temperature/humidity chamber for humidity response measurement. The relative humidity inside the chamber is varied under a constant temperature, where the output signal of capacitive humidity sensor will be readout by a commercial evaluation kit (AD7746). Results in Fig.4b show the proposed humidity sensor without moisture-absorption materials has a sensitivity of 3.32 fF/%RH from 35%RH to 90%RH at 30 °C. The measurement setup for humidity response time is illustrated in Fig.5a. The DUT was placed inside a testing chamber where the anhydrous calcium sulfate was utilized to reach the required low humidity condition (~10%RH). Meanwhile, the outer chamber was set at high humidity condition (90%RH). The response time of humidity sensor can be measured through the sharply rise of the humidity from 10%RH to 90%RH. Measurements (Fig.5b) indicate the response time of proposed humidity sensor (τ_{63}) is only 2.48 sec at a constant temperature of 30 °C. Fig.6 shows the setup and measurements for pressure sensor. The DUT was placed inside the vacuum chamber where the chamber pressure will be controlled from 20 kPa to 120 kPa through a programmable pressure controller (PACE5000), as shown in Fig.6a. Moreover, the output capacitance change was recorded by the commercial evaluation kit (AD7746) which was connected to the DUT. Measurements in Fig.6b indicate the sensitivity of the pressure sensor is 0.22 fF/kPa between 20~120 kPa pressure range. Measurements in Fig.7 show the response of diode thermometer is 0.64 mV/°C within the 25~120 °C temperature range.

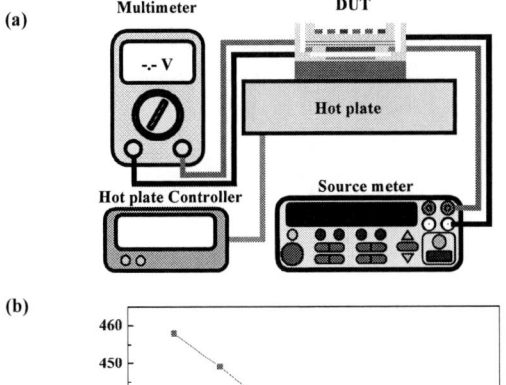

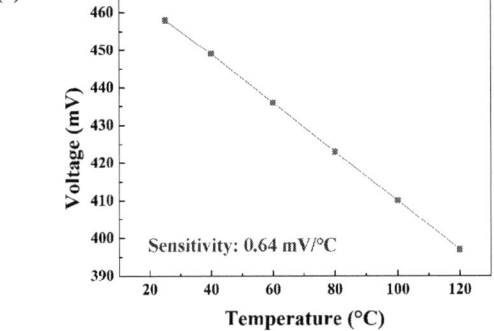

Figure 7: (a) Measurement setup for diode thermometer, (b) measured voltage versus setting temperature for diode thermometer.

CONCLUSION

In summary, this study exhibits a monolithically and vertically integrated environmental sensing hub (400 μm × 400 μm) consisted of the humidity (including heater), pressure, and temperature sensors. The proposed sensing hub is designed and implemented on the standard TSMC 0.18 μm 1P6M process together with the in-house post-CMOS processes. Furthermore, a fast response and easily fabricated air-based humidity sensor is also demonstrated. Benefits from the design flexibility provided by CMOS platform, a compact SoC environmental sensing hub with the integration of humidity (with heater), pressure, and temperature sensors has been successfully realized. Table 1 summarizes the characteristics of the proposed sensing units in this study. Measurements indicates the pressure sensor has a sensitivity of 0.22 fF/kPa, the sensitivity of humidity sensor is 3.32 fF/%RH with a response time of 2.48 sec, and results shows a response of 0.64 mV/°C for thermometer.

Table 1: Specifications summary of proposed design

CMOS-MEMS Environment Sensing Hub	
Humidity sensor	
Size	**400 × 400 μm²**
Sensing material	**Air**
Sensitivity	**3.32 fF/%RH**
Pressure sensor	
Size	**275 × 275 μm²**
Sensitivity	**0.22 fF/kPa**
Thermometer	
Sensitivity	**0.64 mV/°C**

ACKNOWLEDGEMENTS

This project was supported by the National Science and Technology Council under grant number NSTC 111-2923-E-007-003-, NSTC 111-2221-E-007-070-MY3, NSTC 111-2218-E-007-014-MBK, NSTC 111-2923-E-007-004-MY2, NSTC 110-2926-I-007-506-, and NSTC 110-2218-E-007-032-. The authors would like to appreciate the Taiwan Semiconductor Manufacturing Co., Ltd. (TSMC) and the Taiwan Semiconductor Research Institute (TSRI), for the supporting of CMOS chip manufacturing. The authors also appreciate the Center for Nanotechnology, Materials Science and Microsystems (CNMM) of National Tsing Hua University for providing the process tools.

REFERENCES

[1] Y.-C. Lee, M.-L. Hsieh, P.-S. Lin, C.-H. Yang, S.-K. Yeh, T.-T. Do, and W. Fang, "CMOS-MEMS technologies for the applications of environment sensors and environment sensing hubs," *J. Micromech. and Microeng.*, vol. 31, pp. 074004 (41), 2021.

[2] W. Fang, S.-S. Li, Y. Chiu, and M.-H. Li, *3D and Circuit Integration of MEMS.* 1st Ed., Weinheim, Germany: Wiley-VCH, 2021. (ISBN: 978-3-527-34647-9)

[3] Y.-C. Lin, P.-H. Hong, S.-K. Yeh, C.-C. Chang, and W. Fang, "Monolithic integration of pressure/humidity/temperature sensors for CMOS-MEMS environmental sensing hub with structure designs for performances enhancement," *IEEE MEMS*, Vancouver, Jan. 18-22, 2020, pp. 54-57.

[4] C.-C. Chang, P.-H. Hong, S.-K. Yeh, Y.-C. Lin, and W. Fang, "Environmental sensing hub on single chip using double-side post-CMOS processes," *IEEE MEMS*, Vancouver, Jan. 18-22, 2020, pp. 877-880.

[5] T.-L. Chien, Y.-C. Lee, T. Chou, Y.-Y. Lin, H.-Y. Chen, and W. Fang, "Fabrication and integration of SoC environment sensing hub with gas/pressure/temperature sensors," *IEEE MEMS*, Tokyo, Jan. 9-13, 2022, pp. 138-141.

[6] BOSCH BME280. [Online]. Accessed: Nov. 10, 2022. Available at: https://www.bosch-sensortec.com/products/environmental-sensors/humidity-sensors-bme280/

[7] Sensirion SHT41. [Online]. Accessed: Nov. 10, 2022. Available at: https://sensirion.com/products/catalog/SHT41/

[8] N. Lazarus and G. K. Fedder, "Designing a robust high-speed CMOS-MEMS capacitive humidity sensor," *J. Micromech. and Microeng.*, vol. 22, pp. 085021 (7), 2012.

[9] F. Udrea, S. Santra, and J. W. Gardner, "CMOS temperature sensors - concepts, state-of-the-art and prospects," *IEEE Proc. Int. Semiconductor Conf. CAS*, Sinaia, Oct. 13-15, 2008, pp. 31-40.

[10] U. Kang and K. D. Wise, "A high-speed capacitive humidity sensor with on-chip thermal reset," *IEEE Transactions on Electron Devices*, vol. 47, no. 4, pp. 702-710, 2000.

CONTACT

* W. Fang, tel: +886-3-5742923; fang@pme.nthu.edu.tw

A SELF-CORRECTED, SELF-CLEANED MEMS AND SUITABLE FOR ADVANCED FOUNDRY MULTI-PROJECT WAFER (MPW)

Sushil Kumar*, Dhairya Singh Arya, Manu Garg, and Pushpapraj Singh*

Centre for Applied Research in Electronics,
Indian Institute of Technology, NEW-DELHI

ABSTRACT

This work simplifies sacrificial-oxide (SOX) etching related implications (undercut and stress) in released MEMS that remains a leading process concern. Herein, SOX etch-time and stress-part get relaxed simultaneously simply by exposing MEMS chips to nanosecond (ns) laser before final SOX-etching (releasing). Therefore, > 80% stress-recovery in residue bending-correction, 6× extended etching window in undercut-correction, and for robust SOX etch-front and the direct cleaning feasibility of SOX's residues is the first of its kind. Applied method yields ~ 66% direct reduction in pull-in voltage (21-to-7 volt). Verified self-corrected/self-cleaned feature added advantage to released MEMS as no extra fabrication preventing measures (deposition/patterning) required, unlike past fabrication attempts. The presented method can be applied directly in MPW like foundry manufacturing to reduce undercut/stress and related yield loss.

KEYWORDS

MEMS, Residual Stress, Out-of-Plane Actuation, Refractory Transition Metal, Multi-Project Wafer

INTRODUCTION

Suspended structures are the essence of MEMS technology and require concluding SOX-releasing. Their actuating characteristics implies a large set of applications starts from signaling [1-3] to data storage [4], and in sensing [5]. All these everlasting applications of MEMS readily depends on the successful sacrificial release steps (surface and bulk micromachining) and require concluding SOX-releasing.

MEMS process integration offers various choices of sacrificial materials, including oxide, metal, and even the substrate Si. In the most MEMS-based micromachined process, sacrificial silicon oxide, SiO_x ($x \approx 2$) is widely used and can be selectively removed with respect to a large variety of semiconductor materials [1, 6]. Metal-MEMS in which the structural material is metal, provide advantageous properties compared to all those of the dominant MEMS structural material like polysilicon. These include higher electrical conductivity, higher optical reflectivity, higher mechanical ductility and better tribological resistance to wear [6]. Due to these properties, metal MEMS platforms are applied directly in implementing all metal ohmic switches [6-8] and non-volatile memories for high temperature electronics [4, 9]. Another significant benefits of metal MEMS are that processing is feasible after CMOS electronic wafer [1].

When building all-metal MEMS switching devices and memories, the etching step of SOX layer typically involves acids, with the common etchants including hydrofluoric acid (HF), mixed NH_4F/HF solutions

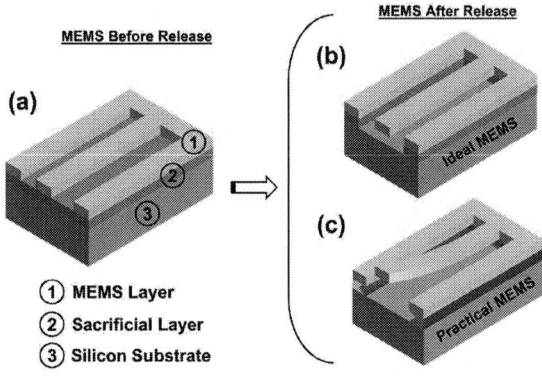

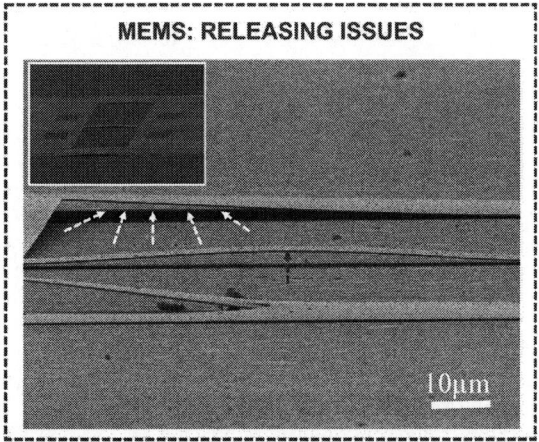

Figure 1: Post-release MEMS implications (a-c); residual stress and undercut issues. Thickness dependent SOX etch-time causes excessive undercut and designed air-gap gets modify due to residual bending which directly affects pull-in voltage (inset).

(buffered HF or BHF) and HNO_3/HF. However, removing these thin-to-thick SiO_2 SOX-layer in the final micromachining step is not straight forward. In one hand, thickness dependent etch-time in the release step causes insufficient or excessive sacrificial etch (Fig. 1) [6], on the other hand, actuation gap gets modify (Fig. 1 inset) due to non-zero residual stresses which finally affects actuation (pull-in) voltage [6]. As a consequence, the etch time cannot be arbitrarily, otherwise, unwanted release of some MEMS features would eventually take place. In fact, such effects result in non-functional devices and consequent yield loss. These issues become more prominent when variations (that appear either due to process or varying device geometries) could not be truly minimised. Such issues are always dealt separately (either using extra prevention/lithography step [10-11] or using high-temperature annealing [6]) and never be simultaneously control. Producing MPW in that varying pitch geometry are embedded, such effects become critical as variations (that appear either due to process or varying device geometries) could not be truly minimised. Reported method aids benefits here.

This work presents a viable solution and simplifies the release step that remains a leading process concern. Applied method introduces a trustworthy step for state-of-art sacrificial silicon oxide, SOX that correct the time-and solution condition-sensitive oxide release step with no extra caping layers deposition/etching unlike past methods. Results state both the etch-time and stress components get relaxed simply by exposing chips to nanosecond laser before final release step. Etch-time (6× extension) and stress recovery (> 80%) property correction and the direct cleaning feasibility of sacrificial residues is the first of its kind, and the reported fabrication is within CMOS compatibility limit.

EXPERIMENT DESIGN

We focus on all metal-based MEMS platform and its fabrication for low-voltage using refractory transition metals (RTMs) enabled directly by wafer-scale manufacturing. Efficacy of using molybdenum (Fig. 2), here, relaxed the restrictions on the application and could be translated into high temperature (T >150°C) MEMS devices that operate at high temperatures (and other harsh environments) [9]. Most part of the work presented in this paper has been realized on test devices with only the Mo electro-mechanical part featuring 1 μm suspended gap. In this way the effects of the single parameter, as well as the time required for release-etch could be better analyse limiting the attention only to metal release part. Since the process integration needs only one MEMS structural material therefore it offers a robust fabrication process against thermal stress.

Preferably, SOX-etch related characterization should be quick and contactless to allow many testing to be made efficiently. Therefore, a set of multi-pitched floating metal posts (FMP) structures included within the chip. Across the chip, FMP size varies from array to array, ranging from 2.5 to 50 μm, in fine increments. Herein, a worst-case situation (i.e. maximum etch time when FMPs is over-etched) can be easily detected using regular optical microscope. Fabrication starts by depositing Al_2O_3 insulating dielectric layer using electron beam deposition (ALD). A ~1 μm SiO_2 deposition is performed as a sacrificial layer for sacrificial etch experiment. Then, metallic part is formed of molybdenum (~300 nm). The entire active materials (oxide and metal) deposition is done using sputtering and the maximum substrate temperature is taken lower value to 200 °C [2]. After final lithography patterning of the test devices photoresist is stripped off and chips are ready for laser heating before release step. For heating a nanosecond excimer laser (Coherent: LS670K) is used. Laser allows heating treatment in selective chip area while keeping the underneath structures at much lower temperatures therefore suitable for CMOS wafer integration [12-13]. After selective heating, finally, chips are placed in a vapor-hydrofluoric acid (HF) chamber for release. Fig. 3 illustrate

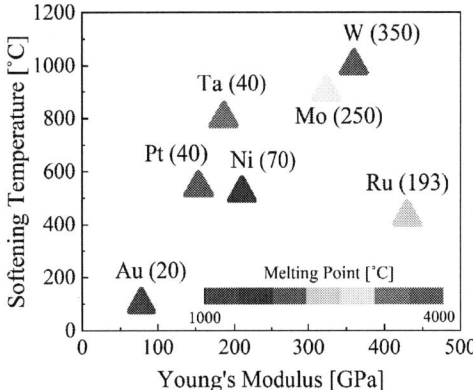

Figure 2: Beyond 'soft'; RTMs merits over conventional MEMS materials. Marker coordinates shows softening temperature and Young's modulus of each metals, and the marker colour represents its melting point and bracket values correspond to Brinell hardness. Stability of RTMs under rigid conditions is found to be larger than those of the noble metals [9].

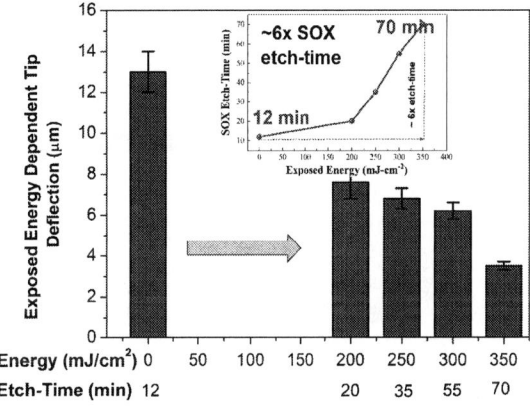

Figure 3: MEMS deflection control and ns-laser exposure. Results confirms ns-heating relaxed SOX etching window (12-to-70 min) and stress-part both in one go which is simply missing in past fabrication attempts.

the efficacy of laser heating. We compare merits of laser exposure (with and without) on timing related effects (insufficient/excessive undercut and stress). Herein, in one hand, the low-resolution FMPs characterizes the timed etch release process accurately (extended etch-time with no under-cut), on the other hand, fabricated metal part allow to analyze the stress effect altogether.

RESULTS AND DISCUSSION

A new etch-time landscape is realized upon localized laser heating which is exclusively determined by exposed energy presented in Fig. 3. Exposure before release changes the regular SiO_2-HF-reaction and correct the time and solution condition sensitive oxide wet release process, doing so 6× extension in release window is achieved as

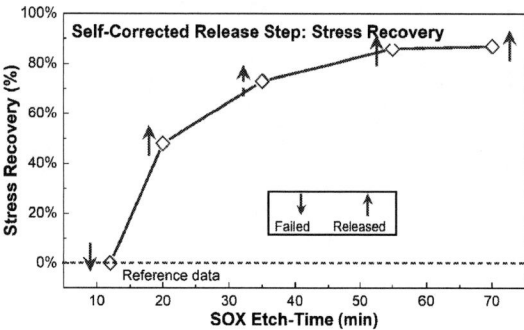

Figure 4: > 80% stress recovery (deflection reduces from 12.9 to 3.2 μm) compared to chips released without laser exposure.

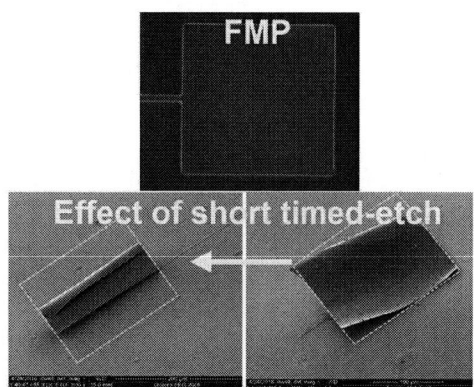

Figure 5: Effects of short timed-etch (both stress and undercut) in SOX-etching and characterized using integrated FMPs.

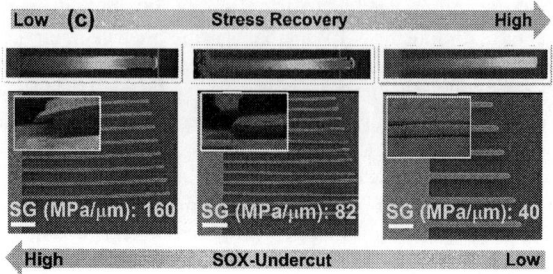

Figure 6: Self-corrected MEMS featuring tight controlled over post-releasing (stress/undercut) issues.

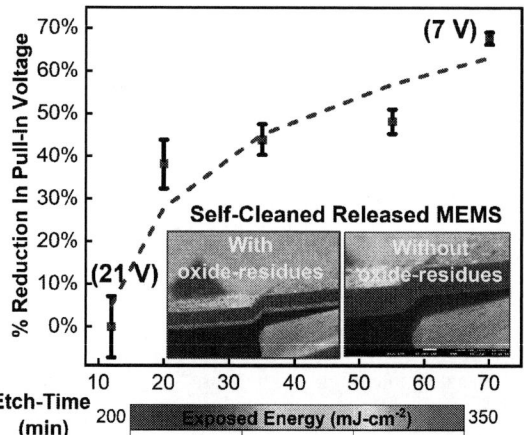

Figure 7: Direct pull-in reduction (∝ exposed energy/SOX etch-time). No post-cleaning required & described method would apply equally well to other MEMS systems/devices.

shown in Fig. 3 inset curve. This newly detected large etch-time behavior corresponds to oxide densification [12-13]. This can have short etch-time of 12 min or if needed it extended to 70 min. Much large release window (t = 12 min to 70 min) here not only covering all size FMPs (essential to compensate temporal effect that appear due to varying embedded geometry in MPW), but it also improved post-release deflection behavior. Recorded deflection results (Vecco model: 3300-NT) explained that the chips exposed to lesser energy deflects more (7.9 μm at 200 mJ/cm²) and at energy = 350 mJ/cm² relaxation for both etch-time and deflection maximizes presented in Fig. 3. Relaxing properties in the released MEMS have a common physical origin–suggesting self-corrected controlled release step. Enormous deflection (= 12.9 μm) with undercut in sacrificial oxide is detected for those chips which released without laser exposure (see Fig. 1 inset).

Fig. 4 plots the effectiveness of newly timed etch release-step. We record > 80% stress recovery compared to reference test chip results (deflection reduces from 12.9 μm to 3.2 μm) implies less stress sensitive wet release. We assume stress gradient value with no laser exposure (deflection = 12.9 μm) is 100%. Stress gradient data (Frontier Sem, model: 128-NT) measured at E (mJ/cm²) = 200, 300 and 350 to released MEMS. The best exposed energy is noted to be 350 mJ/cm² (t = 70 min) which

reduced the residual stress to 40 MPa (compared to 160 MPa for 200 mJ/cm²) per unit length offers lesser undercut. It is found that the stress gradient changes its sign to compressive for > 350 mJ/cm² and released structures stick to underline substrate. A shifted short etch-time (t < 10 min) under-etched FMP that released with no exposure in is presented in Fig. 5 implies large stress gradient values and summarizes in Fig. 6. Which further decreases by lasing action. Results show that the stress relaxing attribute in the employed timed etch release (∝ laser energy) ensures a stress controlled etch-release with no under-cut, which is essential for verity of MEMS applications. As discussed earlier, large etch-time yield a self-corrected release step– almost flat released MEMS with no undercut. Undercut distances directly measured by FE-SEM (JSM-7800F Prime) which found to correlate well with the size of the largest shifted FMPs. Measured interferometry fringing results (Interferometry, Vecco model: 3300-NT) and cross-sectional SEM results confirms both the stress gradient, and the undercut.

A non-zero stress gradient in the released MEMS significantly increase the pull-in voltage [7], since tensile stress changes the air gap to large values. Here our method aids benefits. Therefore, limiting temporal effect related penalties (undercut, stress, and actuation gap) merits in the self-corrected time etch process yields ~66 % direct reduction (21 V to 7 V) in pull in voltage identified in Fig. 7 results. Described results although targets clamed clamed/clamped free MEMS, this concept would apply equally well to other systems/devices. Notably, compensation to timing related effect, essential for multipitch embedded geometries in MPWs, and sacrificial residues cleaning with no post-cleaning step is unique and missing in regular release process.

CONCLUSION

In conclusion, we present a verified self-corrected solution for sacrificial etch in the released MEMS. We explained time-dependent etch-release implications (insufficient/excessive undercut and stress) that remains a leading process concern. Applied laser exposed method that correct the actual condition sensitive oxide-etch process implies insensitive release step with no temporal effects. >80% stress-recovery for bending-correction, 6× extended etching window for undercut-correction and direct cleaning feasibility of SOX's residues due to ns-laser exposure is the first of its kind and missing in regular SOX releasing (wet/dry) process. We believe our method can be applied directly in MPW and also in regular MEMS device production (those developed in lab rather than cleanroom environment) where controlling processes variations is not direct.

ACKNOWLEDGEMENTS

The authors would like to acknowledge the financial support from the IMPACT Research Innovation and Technology grant (IMPRINT), MHRD India (IMP/2018/600869). We also would like to thank Nano Research Facility (NRF) and Central Research Facility (CRF), IIT Delhi for fabrication and characterization facilities.

REFERENCES

[1] O. Y. Loh and H. D. Espinosa, "Nanoelectromechanical contact switches" *Nature Nanotechnol.*, vol. 7, no. 5, pp. 283–295, 2012.

[2] Sushil Kumar, D.S. Arya, and Pushpapraj Singh "Volatile or non-volatile switching? Establishing design parameters for metal-contact relays using ON/OFF hysteretic behavior (RT to 300° C)." Applied Physics Letters 118, no. 1, 013505, 2021.

[3] A. Levy, M. Oduoza, Akhilesh Balasingam, Roger T. Howe, and Priyanka Raina, "Efficient Routing in Coarse-Grained Reconfigurable Arrays Using Multi-Pole NEM Relays", Proc. Asia and South Pacific Design Automation Conf. (ASP-DAC), 978-1 6654-2135-5, 2022.

[4] Dhairya S. Arya, Sushil Kumar, Manu Garg and Pushpapraj Singh, "Piezoelectric Deactuation-Based Bi-Stable MEMS Switch: MEM-Z NVM", J. Microelectromech. Syst., vol. 30, no. 5, 2021.

[5] Manu Garg, D.S. Arya, S Sharma, Sushil Kumar, W. Uddin, Samaresh Das, Yi Chiu, and Pushpapraj Singh, "Highly Responsive Metal Oxide (V_2O_5)-Based NEMS Pirani Gauge for In-Situ Hermeticity Monitoring", J. Microelectromech. Syst., vol. 30, no. 3, 2021.

[6] Gabriel M. Rebeiz, RF MEMS: Theory, Design, and Technology, John Wiley & Sons, Inc., Hoboken, New Jersey, pp. 30-80, 2003.

[7] D. Bansal, A. Bajpai, A., P. Kumar, M. Kaur, A Kumar, "Effect of Stress on Pull-in Voltage of RF MEMS SPDT Switch", IEEE Trans. Electron Devices, 67(5), 2147–2152, 2020.

[8] Ankit Jain, Pradeep R. Nair, and Muhammad A. Alam, "Strategies for dynamic soft-landing in capacitive microelectromechanical switches", Appl. Phys. Lett., 98, 234104, 2011.

[9] Sushil Kumar, Dhairya S. Arya, Manu Garg, and Pushpapraj Singh, "Adhesion-Limit in Refractory Transition Metal (Mo) Contact Relay Operation at 300 °C—Avoiding Overestimation for Modern ICs", IEEE International Reliability Physics Symposium (IRPS), 978-1-6654-7950-9, 2022.

[10] Juan Valle, D. Fernández, and Jordi Madrenas, "Experimental Analysis of Vapor HF Etch Rate and Its Wafer Level Uniformity on a CMOS-MEMS Process", J. Microelectromech. Syst., vol. 25, no. 2, 2016.

[11] Ratul Majumdar, and Igor Paprotny, "Configurable Post-Release Stress Engineering of Surface Micro-Machined MEMS Structures", J. Microelectromech. Syst., vol. 26, no. 3, 2017.

[12] L. Grenouillet, T. Francois, J. Coignus, S. Kerdilès, N. Vaxelaire, C. Carabasse, F. Mehmood, S. Chevalliez, C. Pellissier, F. Triozon, F. Mazen1, G. Rodriguez, T. Magis, V. Havel, S. Slesazeck, F. Gaillard, U. Schroeder, T. Mikolajick, E. Nowak, "Nanosecond Laser Anneal (NLA) for Si-implanted HfO_2 Ferroelectric Memories Integrated in Back-End of Line (BEOL)", Symp. on VLSI Tech. Digest 2020.

[13] P. Singh and Sushil Kumar "Laser-assisted stress reduction in molybdenum microstructures for CMOS compatible MEMS integration." Sensors and Actuators A: Physical, 295, 523-531, 2019.

CONTACT

*Sushil Kumar; email: sushilsoi.2014@gmail.com.
Pushpapraj Singh; email: prsingh@care.iitd.ac.in.

MONOLITHIC INTEGRATION OF HUMIDITY/FLOW/TEMPERATURE SENSORS AS ENVIRONMENT SENSING HUB FOR APPARENT-TEMPERATURE DETECTION

Yu-Hsuan Li[1], Tung-Lin Chien[2], Fuchi Shih[1], Yuanyuan Huang[2], and Weileun Fang[1,2]
[1]Inst. of NanoEng. and MicroSyst., National Tsing Hua University, Hsinchu City, Taiwan
[2]Dept. of Power Mech. Eng., National Tsing Hua University, Hsinchu City, Taiwan

ABSTRACT

This study presents the monolithic integration of humidity/flow/temperature sensors to realize an environmental hub for apparent-temperature detection. The design is based on TSMC 0.18μm 1P6M CMOS platform. Features of the proposed design are: (1) For system: SoC environmental hub for apparent-temperature detection, (2) For component: capacitive humidity sensors without moisture sensitive filler (polyimide), and (3) For component: suspended hot-wire flow sensors for thermal-isolation. Measurements demonstrate performances of the proposed environmental hub: humidity sensor with the highest sensitivity of 2.57fF/%RH, flow sensor with the highest normalized sensitivity of 0.95V/W/(m/s), and temperature sensor with a sensitivity of 0.66mV/°C. The apparent-temperature measurement is also demonstrated.

KEYWORDS

CMOS-MEMS, capacitive humidity sensor, hot-wire flow sensor, apparent temperature.

INTRODUCTION

The ambient temperature usually refers to the atmospheric temperature that records from the thermometer screen, while apparent temperature is used to reflect the human body's feelings through calculation. A formula is introduced to determine the apparent temperature by humidity (RH, %RH), wind speed (V, m/s), and air temperature (T, °C) [1],

$$\text{Apparent temperature} = 1.04T + 0.2e - 0.65V - 2.7 \quad (1)$$

$$e = \frac{RH}{100} \times 6.105 \times \exp(\frac{17.27T}{237.7+T}) \quad (2)$$

the parameter e represents the vapor pressure (hPa), which can be extracted from temperature and relative humidity. In short, the apparent temperature varies with relative humidity, temperature, and wind speed. Thus, the increase in relative humidity will increase apparent temperature. Conversely, the increase in wind speed will decrease apparent temperature. According to Eqs.(1)-(2), at a wind speed of 2.5m/s (level 2 wind) with 33°C ambient temperature, if the relative humidity is 85%RH, the apparent temperature can reach 39°C (Table.1).

Apparent temperature is a useful information for the public to better understand environmental conditions. The typical measurement of apparent temperature is from a dry and wet bulb thermometer, a mercury thermometer, and a pitot tube anemometer. The proposed chip monolithically integrated with humidity, flow, and temperature sensors [2] can greatly reduce the size. Also, through the advantages of CMOS platform e.g. process compatibility of several different sensors and analog signal processing circuits, it could enable the apparent-temperature measurement on portable devices. Therefore, this study leverages CMOS technology to implement a sensing hub for apparent-temperature detection.

SENSOR DESIGN

The proposed environment sensing hub (Fig.1) is designed based on the standard TSMC 0.18μm 1P6M (1-polysilicon and 6-metal (M1~M6)) CMOS process [3-5]. As shown in Fig.1a, the proposed design consists of capacitive humidity sensors monolithically integrated with hot-wire flow sensors and a diode temperature sensor. This study extends the concept of pillar electrode array [6] to design a capacitive humidity sensor without polyimide (PI) filler (Fig.1b). The design without PI filler (air-type humidity sensor) can save polymer dispensing processes, and also shorten the response time. Moreover, since no PI filling is required, the air gap between pillar electrodes can be reduced to increase electrode number and sensing capacitance. The suspended hot-wire flow sensor is formed by the metal layer (M5) stacked with upper/lower SiO_2 films (Fig.1c). Thus, the flat surface on chip could ensure a smooth airflow [7]. In addition, the suspended structure is realized by removing sacrificial metal layers (M1-M3) to enhance thermal isolation and further increase sensitivity. The temperature sensor is formed by a Schottky diode [8].

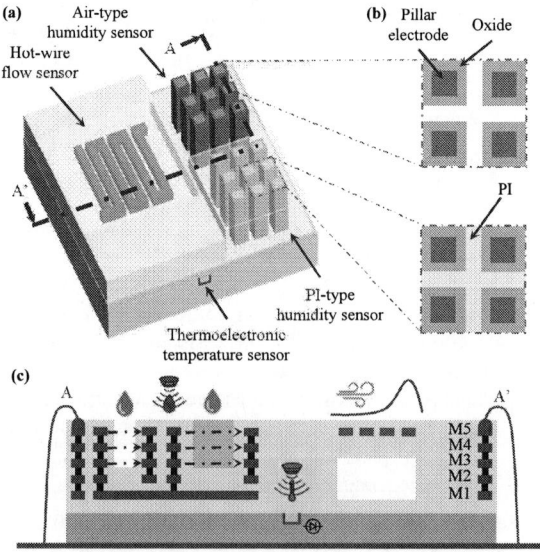

Figure. 1: Design concept:(a) the proposed design, (b) zoom-in top view, (c) sensing mechanism of the proposed design.

For capacitive humidity sensors, pillar electrodes are formed by 3 metal layers (M3, M4, and M5), each pillar electrode is designed with a minimum area of 1.65×1.65 μm^2, where the metal area is 0.45×0.45 μm^2 and the thickness of SiO_2 around electrodes is 0.6 μm and 0.8 μm. Electrical connection of each pillar is formed by M1 and M2 metal layers in this study. Without the need of PI dispensing, the proposed design (air-type humidity sensor) can realize 0.48 μm sensing gap distance between the pillar electrodes under 400×400 μm^2 sensing area. In comparison, the distance between electrodes in PI-type humidity sensor is designed to be 1 μm (by considering the dispensing of PI into the gap).

For hot-wire flow sensor, a meandering wire is defined by M5 metal layer, the wire width is 0.6 μm, and suspended by 3 sacrificial metal layers (M1, M2, and M3), and the intermetal dielectric layer (IMD4) is reserved to increase the stiffness of the suspended structure. The fabricated flow sensor has a measured resistance of 3.59 kΩ and a good linear TCR of $\alpha = 3.40 \times 10^{-3}$. The proposed flow sensor approximately occupied a 200×200 μm^2 area.

FABRICATION AND RESULTS

The proposed environment sensing hub is implemented by the TSMC 0.18 μm 1P6M CMOS platform and the in-house post-CMOS processes. Fabrication steps are shown in Fig.2. First, Fig.2a displays the chip fabricated by standard TSMC 0.18 μm 1P6M process. Fig.2b presents the metal wet etching process by H_2SO_4/H_2O_2 to release sensor structures of metal/via layers. The sensing electrodes were protected by SiO_2 films during metal wet etching. The pillar electrodes of humidity sensors and the suspended flow sensor structure were defined in this step. In addition, the temperature sensor was connected to the ambient simultaneously. In Fig.2c, PI dispensing gaps were defined after oxide wet etching by silox. Fig.2d illustrates the bonding pads were opened through laser-drilling for electrical connection as the device-under-test (DUT). Finally, in Fig.2e, sensing material (PI) was dispensed into sensing gaps for humidity sensors and then evacuated for 2 hours. Afterward, the device was baked for 3 hours to complete the whole fabrication process.

Micrographs in Fig.3 display the fabrication results. Fig.3a shows the typical fabricated sensing chip, including eight humidity sensor designs, two flow sensors with and without thermal isolation, and a temperature sensor. Zoom-

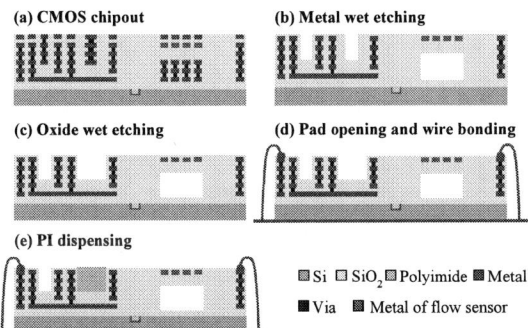

Figure. 2: Fabrication process steps.

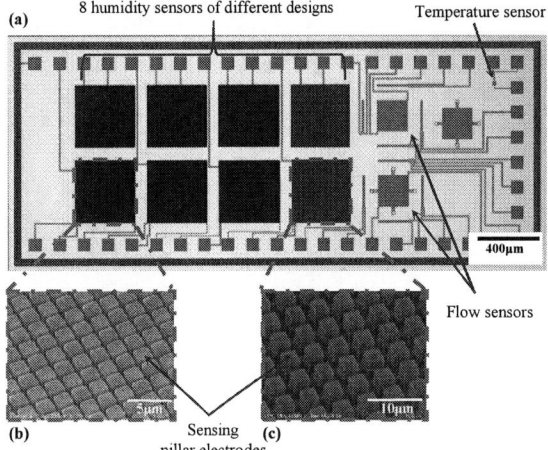

Figure. 3: Fabrication results, (a) overview, (b) zoom-in of air-type humidity sensor, (c) zoom-in of PI-type humidity sensor.

in micrograph (Fig.3b) shows a pillar-electrode array for air-type humidity sensor. The sensing gaps as discussed in previous section have been successfully defined, with a gap distance of 0.48 μm. Fig.3c exhibits the sensing pillar electrode design with a larger gap distance for PI-type humidity sensor, where the gap distance is 1 μm.

MEASUREMENT RESULTS

This study has demonstrated several tests to characterize the environmental hub. Fig.4a indicates the DUT inside temperature/humidity chamber for humidity sensor testing. A commercial humidity sensor (Sensirion, SHT21 with accuracy of ±3%RH) was placed beside the DUT as the reference signal for chamber condition monitoring. Evaluation kit (AD7746) was used as the signal readout for the humidity sensor. The humidity was set from 30%RH to 90%RH under a constant temperature of 30°C.

Measurements (Fig.4b) show the sensitivities of air-type humidity sensors with two different electrode gaps (g) are respectively 0.44 fF/%RH (g = 2.08 μm) and 0.86 fF/%RH (g = 1.68 μm). The results show that the reduction of air-gap between pillar electrodes can increase the number of electrode as well as the sensing capacitance. Note that the electrode gaps (g) here represents the distance between metal structures which were embedded in the SiO_2. In other words, the electrode gaps (g) of the air-type humidity sensor (1.68 μm) includes the sensing gap distance (0.48 μm) and the thickness of SiO_2 on both sides (1.2 μm). In comparison, the design with PI filler (g = 4.48 μm) has a sensitivity of 1.50 fF/%RH. Thus, even without PI filler, a reasonable sensitivity is achieved by leveraging the small line-width CMOS process.

Fig.5a indicates the setup for measuring humidity sensors' response time. The DUT was measured in a sealed chamber with anhydrous calcium sulfate to create a humidity of approximately 10%RH. The chamber was placed in the chamber which was set at 90%RH. After the humidity in the chamber became stable, quickly removed the plug of the chamber, which could instantly create a step

978-1-6654-9309-3/23 $31.00 © 2023 IEEE

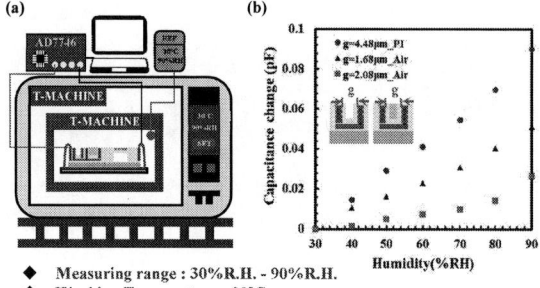

♦ Measuring range : 30%R.H. - 90%R.H.
♦ Working Temperature : 30°C

Figure. 4: (a) Measurement setup for humidity sensors, (b) measurement result of air-type humidity sensors with two different electrode gaps and PI-type humidity sensor.

change of about 80%RH. At the same time, the capacitance signal of humidity sensors was measured so as to determine the response time. The time for the sensor to reach 63% of the final steady-state capacitance value was defined as the response time (τ_{63}) [9]. Fig.5b-c depicts the response time of the air-type humidity sensor is only 0.19s, whereas the one with PI filler is 15.75s. Hence, the proposed design has a massive improvement in the response time of nearly 83-fold.

Fig.6a shows the setup in wind tunnel, the wind is generated by fan blades and a commercial anemometer was placed above the sensor for flow sensor calibration. In this measurement, fan blades were driven from 0 Hz to 60 Hz, which could provide a flow rate ranging from 0m/s to nearly 9m/s. Measurements in Fig.6b indicate responses of flow sensors increased with the operating current. The response increases from 0.03 V/(m/s) to 0.07 V/(m/s) when applied current increase from 4mA to 5mA (w/o cavity). In addition, the suspended thermal-isolation structure shows better performance. The response with a thermal-isolation

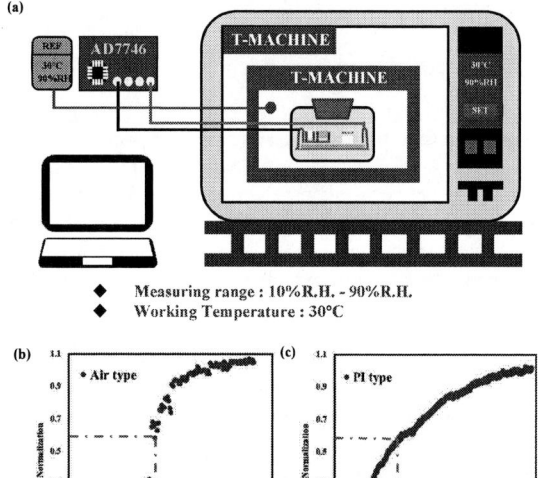

♦ Measuring range : 10%R.H. - 90%R.H.
♦ Working Temperature : 30°C

Figure. 5: (a) Measurement setup for response time, (b) measurement result of air-type humidity sensor, (c) measurement result of PI-type humidity sensor.

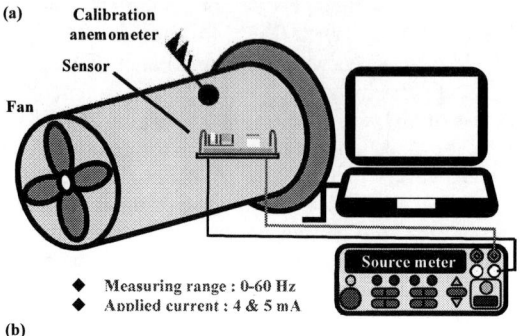

♦ Measuring range : 0-60 Hz
♦ Applied current : 4 & 5 mA

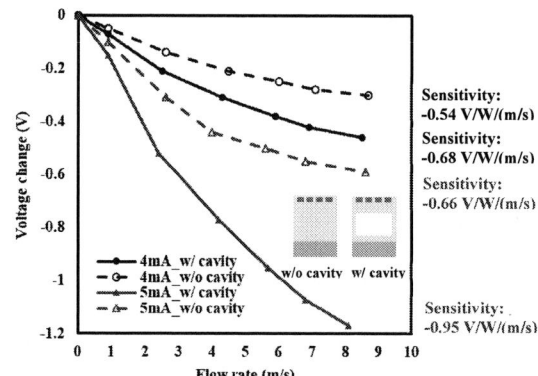

Figure. 6: (a) Measurement setup for flow sensor, (b) measured voltage change under different flow rate.

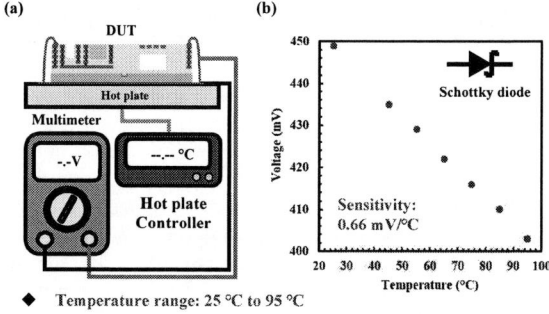

♦ Temperature range: 25 °C to 95 °C

Figure. 7: (a) Measurement setup for the temperature sensor, (b) measurement result.

structure (w/ cavity) increases from 0.07 V/(m/s) to 0.14 V/(m/s) under 5mA current applied. As a result, the response of 5mA_w/ cavity is 2.8 times larger than 5mA_w/o cavity. Furthermore, the proposed flow sensor shows an excellent normalized sensitivity of 0.95 V/(m/s)/W as compared to the one without thermal-isolation structure (0.66 V/(m/s)/W), which has a 144% improvement.

Fig.7a depicts the measurement setup for the temperature sensor. The temperature measured on the hot plate varied from 45°C to 95°C with intervals of 10°C and was also measured at room temperature ~25°C. The voltage response was measured by a multimeter. Fig.7b presents the sensitivity of the temperature sensor is 0.66 mV/°C.

978-1-6654-9309-3/23 $31.00 © 2023 IEEE

Table 1. Performance of each sensor in the proposed design.

CMOS-MEMS Apparent Temp. Sensing Hub		
Humidity sensor		
Size	400×400 μm^2	
Sensitivity	0.86 fF/%RH	
Response time	0.19 s	
Flow sensor		
Size	200×200 μm^2	
Sensitivity	0.95 V/(m/s)/W	
Temperature sensor		
Sensitivity	0.66 mV/°C	
Apparent Temp. demonstration		
85 %RH	2.5 m/s	33 °C
Calculated result: 39 °C		

In summary, Table.1 demonstrates the performance of each sensor and apparent-temperature measurement, the proposed humidity sensor has a sensitivity of 0.86 fF/%RH and a response time of 0.19s, the proposed flow sensor has a sensitivity of 0.95 V/(m/s)/W, and the response of the temperature sensor is 0.66 mV/°C. Through these three parameters (relative humidity, wind speed, and air temperature), the calculation of apparent temperature can be implemented.

CONCLUSION

In this study, a CMOS-MEMS monolithic integration of humidity/flow/temperature sensors as an environment sensing hub for apparent-temperature detection is presented, including a capacitive humidity sensor, a hot-wire flow sensor, and a diode-based temperature sensor. This chip was realized by TSMC 0.18 μm 1P6M CMOS platform and the in-house post-CMOS processes. First, by the design without moisture-sensitive filler, the response time of the humidity sensor can be enhanced by nearly 83-fold (the response time has been improved from 15.75s to 0.19s). Moreover, the flow sensor with a thermal-isolation structure can enhance 144% of response in a normalized sensitivity (the normalized sensitivity has been improved from 0.66 V/(m/s)/W to 0.95 V/(m/s)/W). Table.1 summarizes the results of this study.

ACKNOWLEDGEMENTS

This project was supported by the National Science and Technology Council under grant number NSTC 110-2926-I-007 -506 -, NTSC 110-2218-E-007-032-, NSTC 111-2923-E-007-003-, NSTC 111-2221-E-007-070-MY3, NTSC 111-2218-E-007-014-MBK, NTSC 111-2923-E-007-004-MY2. The authors would like to appreciate the TSMC and the Taiwan Semiconductor Research Institute (TSRI), for supporting of CMOS chip manufacturing. The authors also appreciate the Center for Nanotechnology, Materials Science and Microsystems (CNMM) of National Tsing Hua University for providing the process tools.

REFERENCES

[1] R. G. Steadman, "A Universal Scale of Apparent Temperature," *Journal of Climate and Applied Meteorology*, vol. 23. no. 12, pp. 1674-1687, 1984.

[2] Izhar, W. Xu, H. Tavakkoli, J. Cabot, X. Zhao, M. Duan, and Y.-K. Lee, "Single-Chip Integration of CMOS Compatible Mems Temperature/Humidity and Highly Sensitive Flow Sensors for Human Thermal Comfort Sensing Application," *Transducers'21 Virtual Conference*, June 20-25, 2021, pp. 1219-1222.

[3] Y.-C. Lee, M.-L. Hsieh, P.-S. Lin, C.-H. Yang, S.-K. Yeh, T. T. Do., and W. Fang, "CMOS-MEMS Technologies for The Applications of Environment Sensors and Environment Sensing Hubs," *J. Micromech. Microeng.*, vol. 31, no. 7, 074004(41), 2021.

[4] C.-H. Yang, C.-C. Chang, Y.-C. Lee, Y.-L. Chen, and W. Fang, "Capacitive CMOS Humidity Sensor with Novel Fringe Electrodes and Polyimide-Pillars for Performance Enhancement," *Transducers'21, Virtual Conference*, June 20-25, 2021, pp. 66-69.

[5] T.-L. Chien, Y.-C. Lee, T. Chou, Y.-Y. Lin, H.-Y. Chen, and W. Fang, "Fabrication and Integration of Soc Environment Sensing Hub with Gas/Pressure/Temperature Sensors," *IEEE MEMS'22*, Tokyo, January 9-13, 2022, pp. 138-141.

[6] Y.-C. Lin, P.-H. Hong, S -K. Yeh, C.-C. Chang, and W. Fang, "Monolithic Integration of Pressure/Humidity/Temperature Sensors for CMOS-Mems Environmental Sensing Hub with Structure Designs for Performances Enhancement," *IEEE MEMS'20*, Vancouver, January 18-22, 2020, pp. 54-57.

[7] E. Yoon, and K. D. Wise, "An Integrated Mass Flow Sensor with On-chip CMOS Interface Circuitry," *IEEE Transactions on electron devices*, vol. 39, no. 6, pp. 1376-1386, 1992.

[8] M. Mansoor, I. Haneef, S. Akhtar, A. D. Luca, and F. Udre, "Silicon Diode Temperature Sensors—A Review of Applications," *Sensors and Actuators A: Physical*, vol. 232, pp. 63-74, 2015.

[9] N. A. David, P. M Wild, and N. Djilali1, "Parametric study of a polymer-coated fibre-optic humidity sensor," *Meas. Sci. Technol.*, vol. 23, no. 3, 035103(9), 2012.

CONTACT

* W. Fang, tel: +886-3-5742923; fang@pme.nthu.edu.tw

978-1-6654-9309-3/23 $31.00 © 2023 IEEE

PIEZORESISTIVE PRESSURE SENSOR WITH MONOLITHICALLY INTEGRATED AMPLIFIER BASED ON METAL-OXIDE TRANSISTORS

Runxiao Shi[1],*, Dequan Lin[1],*, Kevin Chau[1,2], and Man Wong[1,3]

[1]Department of Electronic and Computer Engineering,
The Hong Kong University of Science and Technology, Hong Kong, CHINA
[2]CAS Engineering Laboratory for Deep Resources Equipment and Technology,
Institute of Geology and Geophysics, Chinese Academy of Sciences, Beijing, CHINA
[3]Shenzhen Research Institute, The Hong Kong University of Science and Technology,
Shenzhen, CHINA (*equal contribution)

ABSTRACT

Presently described is a low-temperature technology for the implementation of circuits based on indium-gallium-zinc oxide (IGZO) thin-film transistors (TFTs) and their post-MEMS, monolithic integration with a micro-fabricated sensor. The feasibility of the technology is demonstrated with the design, implementation, and characterization of a bulk-type, dual-cavity piezoresistive pressure sensor integrated with a differential amplifier consisting of the IGZO TFTs. The resulting sensor-amplifier system exhibits an increase in sensitivity by more than 20 times and stable performance up to a hydrostatic pressure loading of 100 MPa.

KEYWORDS

IGZO TFT; MEMS; piezoresistive pressure sensor; monolithic integration.

INTRODUCTION

MEMS sensors and silicon (Si)-based peripheral signal-processing circuits such as filters, amplifiers and analog-to-digital converters are rarely monolithically integrated. Though desired, such integration is often hindered by difficult process compatibility issues such as the high processing temperature of the transistors [1]. This is also true of the recently reported, piezoresistive MEMS sensor for high-pressure applications [2].

Though mostly deployed in flat-panel displays [3] and flexible electronic systems [4], indium-gallium-zinc oxide (IGZO) thin-film transistors (TFTs) have been recently used to implement an inverter amplifier [5]. Together with a separately fabricated piezoresistive pressure sensor, the amplifier has been packaged on a printed-circuit board and tested. Insensitive to pressure loading even above 100 MPa [1], the amplifier exhibits a gain of ~3.4 times. Its stable performance under pressure load and its low processing temperature of ≤ 400 °C make IGZO TFT a promising candidate for post-MEMS monolithic integration with the pressure sensor.

Presently described is the monolithic integration of a Si-based, bulk-type, dual-cavity piezoresistive pressure sensor and a differential amplifier consisting of IGZO TFTs fabricated using a process with a maximum temperature not exceeding 300 °C. With stable performance under hydrostatic pressure load up to 100 MPa, the integrated differential amplifier exhibits a gain of ~23 times.

MONOLITHIC INTEGRATION PROCESS

The evolution of the cross-sections of the pressure sensor and IGZO TFT through the monolithic integration process is shown in Figure 1. The integrated system is constructed of three wafers: a device wafer containing the sensor, the TFTs, and the bottom cavity; a cap wafer containing the top cavity; and a bottom wafer sealing the bottom cavity.

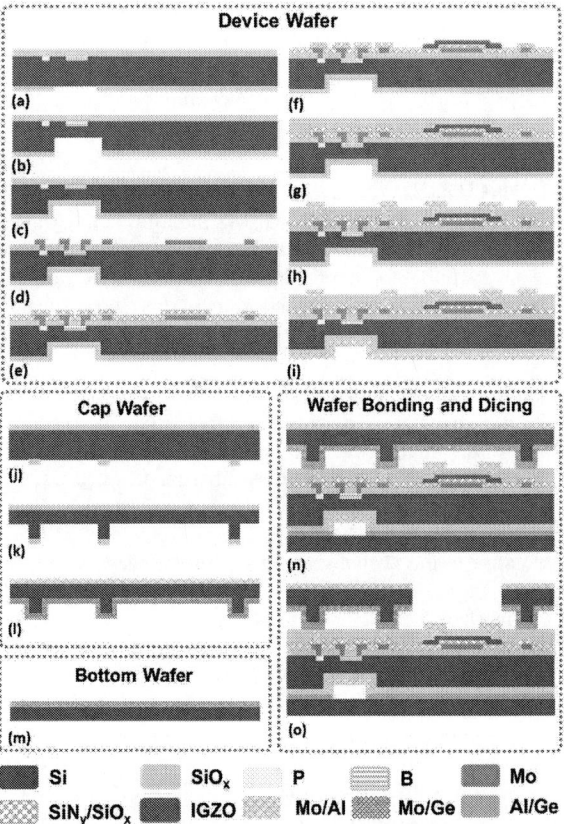

Figure 1. Evolution of the schematic cross-sections of the pressure sensor and IGZO TFT through the fabrication process. (a)-(i) Device wafer. (j)-(l) Cap wafer. (m) Bottom wafer. (n)-(o) Wafer bonding and dicing.

The fabrication of the device wafer starts with the implantation of boron (B) to form the p-type piezoresistors and phosphorus (P) to form the n+ contact for fixing the electrical potential of the substrate (Fig. 1a). Masked using a 3-μm low-temperature oxide (LTO), deep reactive ion etching (DRIE) is carried out to form the 250-μm-deep bottom cavity (Fig. 1b). After removal of the LTO mask, another 0.2-μm LTO is deposited. Densification of the LTO and activation of the implanted B and P are simultaneously accomplished at 900 °C for 30 min (Fig. 1c). After contact hole opening, 100-nm molybdenum (Mo) is sputter-deposited. The Mo layer is patterned to form the bottom gate of the IGZO TFTs and the first interconnection layer (Fig. 1d). The gate insulator stack, consisting of 75-nm silicon oxide (SiO_x) under 50-nm silicon nitride (SiN_y), is formed by plasma-enhanced chemical vapor deposition (PECVD) at 300 °C (Fig. 1e). A 20-nm IGZO acting as an active layer is sputtered at room temperature in a mixed atmosphere of 10% oxygen (O_2) and 90% argon at a total pressure of 3 mTorr. The active layer is patterned in a 1/2000 aqueous hydrofluoric acid solution (Fig. 1f). A passivation layer consisting of 250-nm PECVD SiO_x is deposited at 300 °C (Fig. 1g). Contact holes are opened to the Mo layer of the sensor and the TFTs before the deposition of a stack of 100-nm Mo under 1500-nm aluminum (Al). The stacked layers are patterned with a room-temperature mixture of phosphoric, nitric, and acetic acids (Fig. 1h), thereby forming the source/drain (S/D) electrodes of the TFTs, the second interconnection layer, and a "fence" for the subsequent Al-germanium (Ge) eutectic bonding with the cap wafer. A 100-nm Mo under 1500-nm Al is sputtered on the cavity side of the device wafer for the Al-Ge eutectic bonding with the bottom wafer (Fig. 1i).

The preparation of the cap wafer starts with a blanket P diffusion at 900 °C for 30 min. LTO is deposited as the hard mask (Fig. 1j) before DRIE is carried out to form a 160-μm-deep cap cavity (Fig. 1h). A 100-nm Mo under 1500-nm Al is blanket sputtered on the unetched side while the cavity is covered by 100-nm sputtered Mo under 500-nm evaporated Ge (Fig. 1l). The same stack of layers consisting of 100-nm Mo under 500-nm Ge also covers the bottom wafer (Fig. 1m).

After aligning and sandwiching the device wafer between the cap and the bottom wafers, all three are bonded by Al-Ge eutectic bonding under a clamping pressure of 0.4 MPa at 480 °C for 5 min (Fig. 1n). A shallow dicing process is carried out to expose the TFTs before the whole stack of wafers is sawn through to separate the sensor-amplifier dice (Fig. 1o). Finally, the whole system is annealed in O_2 at 300 °C for 4 hr to passivate the defects in the channel and to form the S/D regions of the TFTs [6].

SENSOR-AMPLIFIER SYSTEM

A schematic of the sensor-amplifier system is shown in Figure 2a. The pressure sensor is constructed of a half Wheatstone bridge connection of two sensing piezoresistors (R_1, R_4) and two relatively constant reference resistors (R_2, R_3). The two outputs (V_{PS_OUT+}, V_{PS_OUT-}) of the pressure sensor are respectively connected to the two inputs (V_{DA_IN+}, V_{DA_IN-}) of the differential amplifier. The TFT-level circuit diagram of the differential amplifier [7] consisting of 16 TFTs is shown in Figure 2b. The driver TFTs T1 and T3 and the load TFTs T2 and T4 form the basis pair to provide primary amplification. TFTs T5-8 are inserted to provide the gain-boosting feedback.

TFTs T9 and T10 as current sources are used to provide negative feedback for suppressing the common-mode signal. TFTs T11-14 are used to convert the differential outputs to a single-ended output. TFTs T15 and T16 provide the secondary amplification to generate the system output V_{DA_OUT}.

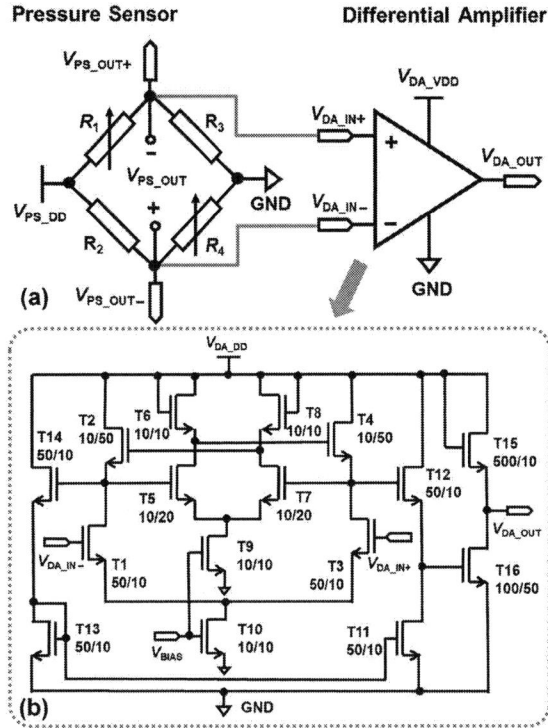

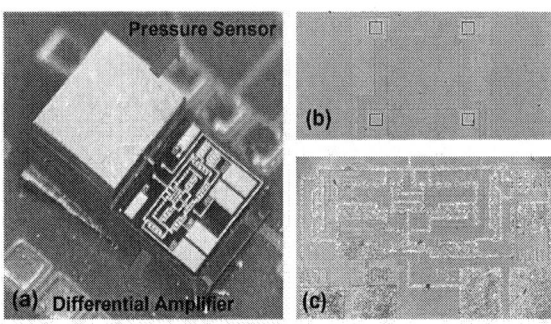

Figure 2. (a) Schematic of the sensor-amplifier system. (b) TFT-level circuit diagram of the differential amplifier.

The photograph of a fabricated sensor-amplifier system is shown in Figure 3a. The photographs of an uncapped pressure sensor and the differential amplifier are shown in Figures 3b and 3c, respectively.

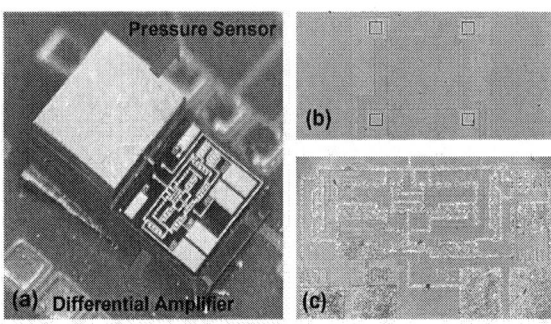

Figure 3. Photographs of the fabricated (a) sensor-amplifier system, (b) pressure sensor, and (c) TFT-based differential amplifier.

MEMS PRESSURE SENSOR

The pressure sensing structure consists of an n-type (110)-Si sensing plate sandwiched between the top and bottom cavities. Four identical piezoresistors are fabricated on the sensing plate. R_1, R_4 are aligned to the pressure-sensitive [1$\bar{1}$0] direction, while R_2, R_3 are along the

978-1-6654-9309-3/23 $31.00 © 2023 IEEE

pressure-insensitive [001] direction. The top cavity is deployed to eliminate the vertical pressure acting on the sensing plate, thus converting a hydrostatic pressure load (P) to a biaxial compressive stress field. The dual-cavity design further increases the lateral compression of the sensing plate, thereby increasing the sensitivity of the pressure sensor [2].

The anisotropic piezoresistive effects of Si induce a change in the differential output $V_{PS\_OUT} \equiv V_{PS\_OUT+} - V_{PS\_OUT-}$ of the Wheatstone bridge as P is changed. The measured dependence on P of V_{PS_OUT} at a bridge bias voltage of $V_{PS\_DD} = 5$ V and temperature (T) of 25 °C is plotted in Figure 4, exhibiting a sensitivity of ~0.5 mV/MPa.

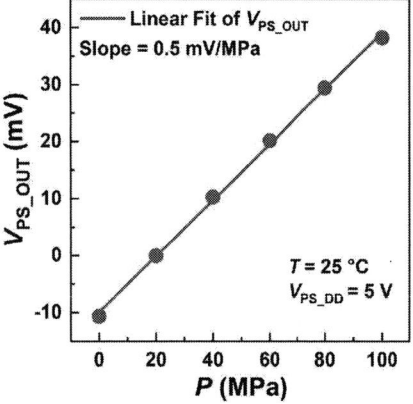

Figure 4. Dependence of V_{PS_OUT} on P at $V_{PS\_DD} = 5$ V and $T = 25$ °C.

CHARACTERISTICS OF IGZO TFT AND AMPLIFIER

The transfer characteristics, i.e., the dependence of the drain current (I_d) on the gate-to-source voltage (V_{gs}) of an IGZO TFT with channel width (W) and channel length (L) both equal to 20 μm are shown in Figure 5a. The gate-to-drain voltage (V_{ds}) is fixed at 0.5 or 5 V. A threshold voltage of −0.1 V, a subthreshold swing of 424 mV/dec and a mobility of 6.8 cm²/Vs are extracted. The output characteristics, i.e., the dependence of I_d on V_{ds} as V_{gs} is increased from 2 to 8 V are shown in Figure 5b.

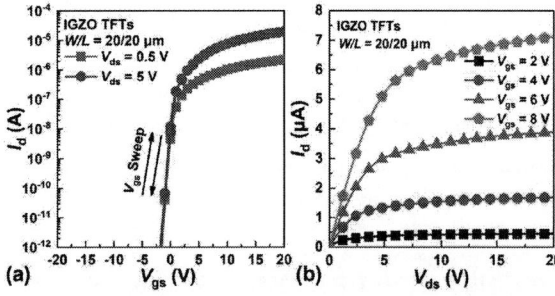

Figure 5. (a) The transfer characteristics and (b) the output characteristics of an IGZO TFT with $W/L = 20/20$ μm.

The voltage transfer characteristics (VTC) between V_{DA_OUT} and V_{DA_IN-} of the differential amplifier is shown

in Figure 6a. As V_{DA_IN+} is increased, the VTC shifts to the right. Defining the gain as the slope of the VTC, one obtains from Figure 6b a maximum gain of ~25. It can be seen in Figure 6c that the peak-to-peak value (V_{P-P}) of a 100-Hz sinusoidal signal is amplified from 0.1 V to 2.1 V, showing an amplification of ~22 times (or ~26.4 dB). The frequency response is shown in Figure 6d, which exhibits a bandwidth of ~1 kHz.

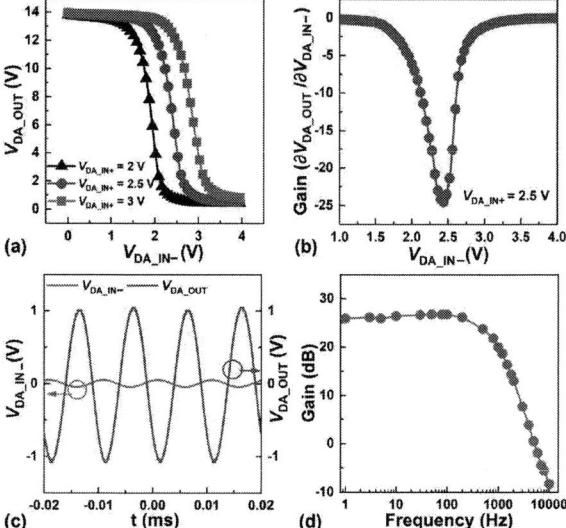

Figure 6. (a) The V_{DA_OUT} vs. V_{DA_IN-} VTC of a differential amplifier. (b) Extracted gain of the differential amplifier. (c) The V_{P-P} of a 100-Hz sinusoidal signal is amplified from 0.1 to 2.1 V by the amplifier. (d) Frequency response of the differential amplifier.

CHARACTERISTICS OF IGZO TFT AND AMPLIFIER UNDER PRESSURE

Shown in Figure 7 are the transfer characteristics of an IGZO TFT with $W/L = 20/20$ μm subjected to P between 0 and 100 MPa. The negligible change in the transfer characteristics indicates the insensitivity of the electrical properties of the TFT to hydrostatic pressure loading.

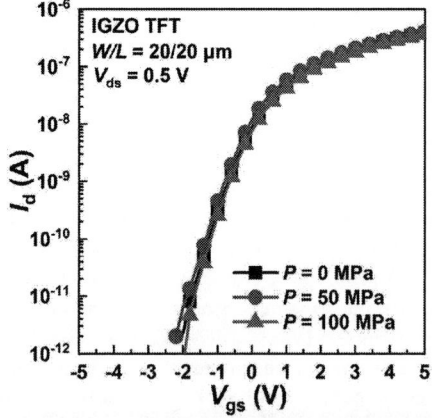

Figure 7. The transfer characteristics of IGZO TFT with $W/L = 20/20$ μm subjected to P between 0 and 100 MPa.

The performance of the differential amplifier under hydrostatic pressure loading is shown in Figure 8. As the pressure is increased to 100 MPa, the VTC (hence also the gain) of the amplifier exhibits negligible change. The stable performance of differential amplifier under hydrostatic pressure is reasonable as the constituent IGZO TFTs have been shown to be insensitive to pressure.

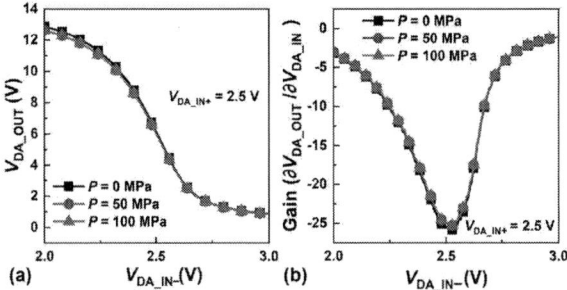

Figure 8. (a) VTC and (b) gain of the differential amplifier under hydrostatic pressure loading.

PERFORMANCE OF THE SENSOR-AMPLIFIER SYSTEM

The performance of the sensor-amplifier system is shown in Figure 9. The sensitivity of the raw sensor output V_{PS_OUT} at 25 °C is ~0.5 mV/MPa, whereas the slope of the amplifier output V_{DA_OUT} is ~11.3 mV/MPa, thus exhibiting an increase of around ~23 times. The good linearity of the sensor is maintained after the amplification.

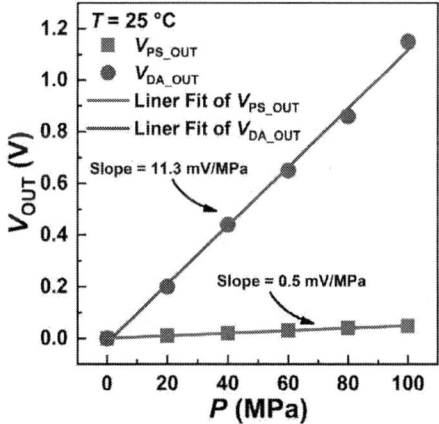

Figure 9. Comparison of the offset-compensated pressure sensor output $V_{OUT} = V_{PS\_OUT}$ and $V_{OUT} = V_{DA\_OUT}$ of the sensor-amplifier system.

SUMMARY

A process for the monolithic integration of a Si-based, bulk-type, dual-cavity piezoresistive pressure sensor and a differential amplifier made of IGZO TFTs is described. The resulting sensor-amplifier system exhibits good stability under pressure loading, good linearity and a sensitivity of ~11.3 mV/MPa. This is ~23 times larger than that of a standalone pressure sensor. These encouraging results demonstrate the potential of extending the proposed monolithic integration scheme to incorporate more complex TFT-based signal-processing circuits in a variety of micro-fabricated sensors.

ACKNOWLEDGMENT

This work was supported by the Fundamental and Applied Fundamental Research Fund of Guangdong Province 2021B1515130001 and in part by the Science and Technology Program of Shenzhen under Grant JCYJ20200109140601691.

The assistance of the Nanosystem Fabrication Facility (NFF) of The Hong Kong University of Science and Technology (HKUST) is gratefully acknowledged.

REFERENCES

[1] D. Lin, R. Shi, M. Wong, and K. Chau, "Metal-Oxide Thin-Film Transistor for Monolithic Integration with High-Pressure MEMS Pressure Sensor," *2022 IEEE 35th Int. Conf. Micro Electro Mech. Syst. Conf.*, vol. 2022-Janua, no. January, pp. 672–675, 2022.

[2] D. Lin, K. Chau, and M. Wong, "A Bulk-Type High-Pressure MEMS Pressure Sensor With Dual-Cavity Induced Mechanical Amplification," *J. Microelectromechanical Syst.*, vol. 31, no. 4, pp. 683–689, 2022.

[3] L. Lu, J. Li, H. S. Kwok, and M. Wong, "High-performance and reliable elevated-metal metal-oxide thin-film transistor for high-resolution displays," *Tech. Dig. - Int. Electron Devices Meet. IEDM*, no. December, pp. 32.2.1-32.2.4, 2017.

[4] R. Shi, S. Wang, Z. Xia, and M. Wong, "Low-temperature elevated-metal metal-oxide thin-film transistors and circuit building blocks on a flexible substrate," *J. Soc. Inf. Disp.*, vol. 30, no. 6, pp. 505–513, 2022.

[5] R. Shi, S. Wang, Z. Xia, and M. Wong, "Low-temperature elevated-metal metal-oxide thin-film transistors and circuit building blocks on a flexible substrate," *J. Soc. Inf. Disp.*, 2022.

[6] R. Shi, S. Wang, Z. Xia, L. Lu, and M. Wong, "Fluorinated Metal-Oxide Thin-Film Transistors for Circuit Implementation on a Flexible Substrate," *IEEE J. Flex. Electron.*, no. c, pp. 1–1, 2021.

[7] Y. C. Tarn, P. C. Ku, H. H. Hsieh, and L. H. Lu, "An amorphous-silicon operational amplifier and its application to a 4-bit digital-to-analog converter," *IEEE J. Solid-State Circuits*, vol. 45, no. 5, pp. 1028–1035, 2010.

CONTACTS

* Runxiao Shi, rshiab@connect.ust.hk
* Dequan Lin, dlinah@connect.ust.hk
* Kevin Chau, eekchau@ust.hk
* Man Wong, eemwong@ust.hk

A PERFORMANCE ENHANCEMENT METHOD FOR THERMOPILE SENSORS USING A CHIP PROBE TEST SYSTEM

Meng Shi[1,2], Mao Li[1,2], Yue Ni[3], Chenchen Zhang[1], Na Zhou[1,2], Haiyang Mao[1,2*], and Chengjun Huang[1,2]*

[1]Institute of Microelectronics of Chinese Academy of Sciences, Beijing 100029, CHINA
[2]University of Chinese Academy of Sciences (UCAS), Beijing 100049, CHINA and
[3]Jiangsu Hinovaic Technologies Co., Ltd, Wuxi 214135, CHINA

ABSTRACT

In this work, a conventional chip probe test system is not only adopted to evaluate functionality of thermopile sensors, but is also utilized to integrate ink dots on the devices. Compared with a pristine sensor, such devices integrated with ink dots are able to increase output by >200% in a wavelength range of 635-1550 nm, and by >40% in a temperature range of 30-450 °C. Moreover, the size of the ink dots can be well controlled in an 8-inch wafer, and the performance enhancement ability of this simple and low-cost approach has a satisfactory uniformity, thus it has great potential in mass preparation of thermopile sensors with high performance.

KEYWORDS

Thermopile sensors, chip probe (CP) test, ink dots, optical absorption enhancement

INTRODUCTION

Thermopile sensors are able to convert temperature differences between the hot and cold junctions to voltage signals. With advantages of simple structures, broadband response, easy fabrication and low cost, miniaturized thermopile sensors have been widely used in non-contact temperature sensing [1, 2], gas detection [3, 4], vacuum monitoring [5, 6], etc. Besides, recent development of internet in things (IoT), portable devices as well as the 5G technologies further requires optimization and upgrade of thermopile sensors with higher performance [7].

As is known, performance of a thermopile sensor is closely related with thermal conduction of thermocouple materials and the suspended membrane structures. Besides, performance of the sensor is also highly dependent on the radiation absorbing capability of its absorber [8]. Recently, different materials with low thermal conduction have been studied and utilized in thermopile sensors, and optimized structures of these devices have also been achieved, however, to meet certain demands, the performance of these devices still needs to be enhanced. In order to further improve performance, efforts have been made to increase absorption of thermopile sensors, which is also regarded as an effective way and has led to extensive research.

So far, various approaches for increasing absorption of thermopile sensors have been reported, and the integration of highly absorptive structures or materials on the devices is one of the resultful route. At present, different structures or materials that could be integrated have been investigated, such as the pyramidally-textured dielectric films [9], graphene oxides [10, 11], composite nanoforests [12, 13], black silicon [14], etc. However, these methods either require complicated micromachining steps or rely on expensive apparatuses which tend to cause damage to the pristine devices and increase manufacturing cost. In this sense, a simple method that is easy to implement is imminently required for integrating highly adsorptive materials or structures onto thermopile devices thus to achieve products with higher performance.

Normally, to launch thermopile sensors as a type of product, the related industry chain includes the following steps: device design, wafer-level fabrication, chip probe (CP) test, wafer dicing, device packing, final test and board assembly. Among them, the CP test is an indispensable wafer-level step to sort the fabricated chips according to their functionality. Specifically, this step is used to find out the unqualified chips and mark them with ink dots thus to avoid waste of cost in packaging those dies.

In this work, these ink dots generated in such a conventional CP test system are artfully adopted as the highly absorptive material to improve performance of qualified thermopile chips instead of marking those unqualified ones. Such ink dots are easy to prepare, with low cost and are controllable in size. Meanwhile, with this same CP test system, performance of the chips before and after integration of the ink dots can be detected, thus the enhancement ability of this approach can be evaluated directly. It is expected that such a simple and low-cost approach with size controlling capability is suitable for mass preparation of high-performance thermopiles.

FBRICATION

As illustrated in the schematic diagram (Figure 1), the CP test system mainly contains three installations. The first one is a module of functionality evaluation according to optical-electrical responses of the chips, the second one is with an ink-jet printhead for injecting ink dots. In addition, there is a 3D position controller to control the wafer movement in the X, Y and Z directions during the evaluation and printing process. The functionality evaluation module consists of a laser as the source of radiation, a probe supporter and an electrical testing equipment. For the probe supporter, it contacts with the wafer on one side and contacts with the electrical testing equipment on the other side. Thus, the output signals of thermopile chips can be obtained after exposure under the laser.

In our previous work, thermopile sensors with optimized structures and thermocouple materials have been fabricated on 8-inch Si wafers [13, 14]. Subsequently, the functionality evaluation module in the CP test system was utilized to estimated states of these wafer-level sensors. After functionality evaluation, a mapping file containing location information of the qualified chips on the whole wafer is generated, later on, with the help of the 3D position controller and the mapping file, precise printing of ink dots on the qualified chips can be directed. Herein, the size of the ink dots can be well controlled by diameters of the printheads, the injection speed as well as the period of time used for injection. In addition, after the ink dots are printed, the voltage signals of the new devices are tested by using the functionality evaluation module for the second time to check the new states and performance of the thermopiles with the ink dots.

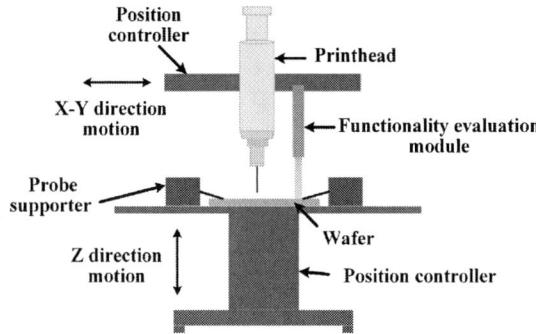

Figure 1: Diagram of the CP test system consisting of a printhead, a probe supporter, functionality evaluation module and a position controller.

Figure 2 (a) and (b) show the 3D structure and the cross-sectional view of the thermopile with an ink dot. As demonstrated, the ink dot is located on absorber of the thermopile sensor. With this design, the original absorber integrated with such an ink dot is regarded as the new absorber of the device.

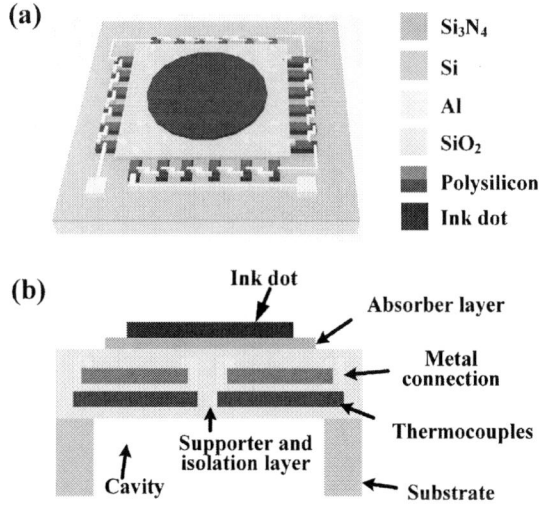

Figure 2: Diagrams of the thermopile with an ink dot: (a) 3D schematic diagram; (b) Cross-sectional view.

Figure 3 (a) and (b) illustrates microscope images of the thermopile sensors with ink dots. As illustrated, the black round ink dot covers the central region of the device, and the size and morphology of the ink dots on the adjacent devices are quite uniform. Figure 3 (c) shows a scanning

electron microscopy (SEM) image of the ink dot, from which it can be observed that the materials in the ink dot distribute evenly on the device, which is helpful for achieving stability of the ink dots as well as the devices. Besides, elemental components of the ink dots are investigated, as shown in Figure 3 (d). The results show that the main component of the ink dot is carbon whose atomic percent is 87.91%. As is known, black carbon-based materials that have wide-band and high-absorption properties have been widely used in photothermal conversion [15], thus it indicates that the ink dots have potential for photothermal conversion.

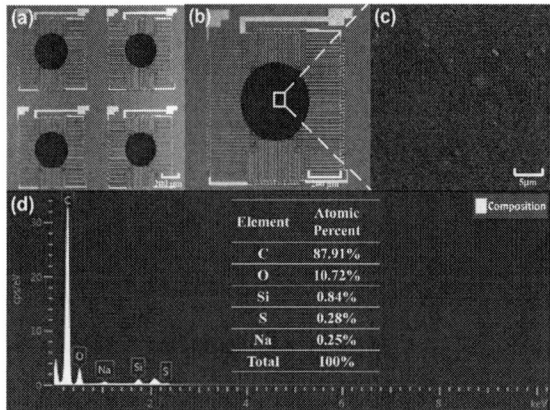

Figure 3: (a)-(b) Microscope images of thermopiles with ink dots;(c) A SEM image of the ink dot;(d) Elemental components of the ink.

RESULTS AND DISCUSSIONS

In order to verify the ability of the ink dots to enhance optical absorption, micro-region absorption spectra were detected in absorbers of a pristine thermopile device and the novel one integrated with an ink dot. As shown in Figure 4, in a wide wavelength range from 2.5 μm to 15 μm, the highest absorption of the pristine thermopile device is 68%, while for the new thermopile device, its lowest absorption is 70%, indicating that absorption of this novel thermopile device is entirely higher than that of the pristine one in a broad absorption region. Especially, in a narrower short-wavelength range from 2.5 μm to 7 μm that contains common gas absorption wavelengths, the absorption increases significantly, which implies the potential applications of the thermopile with an ink dot in the field of gas detection.

To further verify the performance improvement of the thermopile sensors integrated with ink dots in the short-wavelength range, the output voltages of the pristine thermopile and the thermopile with an ink dot were measured in a wavelength range from 635 nm to 1550 nm under a laser power of 100 mW. The test results are shown in Figure 5, as can be seen from this figure, the output voltages of the thermopile with such a new absorber are at least 200% higher than those of the pristine device, which demonstrates that the ink dot is able to significantly optimize performance of the devices in a short-wavelength band.

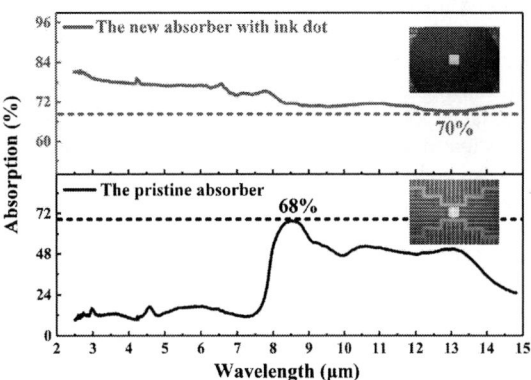

Figure 4: Micro-region absorption spectra of the new and the pristine absorbers in a wavelength range of 2.5 to 15 μm. Insets show the tested micro regions.

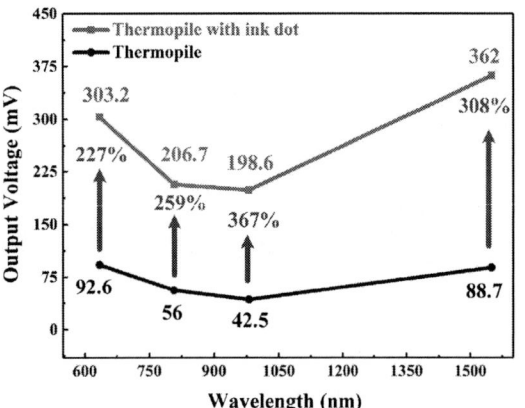

Figure 5: Voltage response curves of the two thermopile sensors with and without an ink dot in a wavelength range of 635-1550 nm.

Moreover, the voltage-temperature response curves of the sensors were recorded by gradually changing temperatures of a black-body from 30 °C to 450 °C, which was used as a source of infrared radiation. In this measurement, the dependence of device outputs on temperature could be observed. As is illustrated in Figure 6, when temperature of the blackbody increases, the output voltages of the thermopile with an ink dot increases much more rapidly than that of the pristine device, which is in accordance with the absorption results (Figure 4). Besides, this novel thermopile has a much larger slope and obtains at least 40% higher outputs than the pristine one, indicating that owing to the ink dot, sensitivity is the device is improved. This is because in the novel device, the ink dot functions as a part of the new absorber, when radiated by light, the ink dot helps to absorb more light than the original absorber of devices.

Furthermore, the controllability of the ink dots was also investigated. The size of the ink dots could be controlled by adjusting the injecting speed and the printing time. In our experiment, the injecting speed was set to a constant value, while the printing time could be altered. Accordingly, printing periods of 20 ms, 40 ms, 60 ms and 80 ms were chosen to print the ink dots on different

thermopile chips. Then, the diameters of these ink dots as well as the output voltages of the thermopile devices were measured under radiation of a 660 nm laser with power of 400 mW. As shown in Figure 7, with the printing time lasts, the diameter of the ink dots and the output voltages of the thermopile increase accordingly in an almost linear way, indicating that the printing periods of ink dots can well control the size as well as the output enhancement of devices. Moreover, the output of the pristine thermopile is much smaller than the one with an ink dot even when the printing time was only 20 ms, which further proves that the ink dots can greatly improve the outputs of thermopiles.

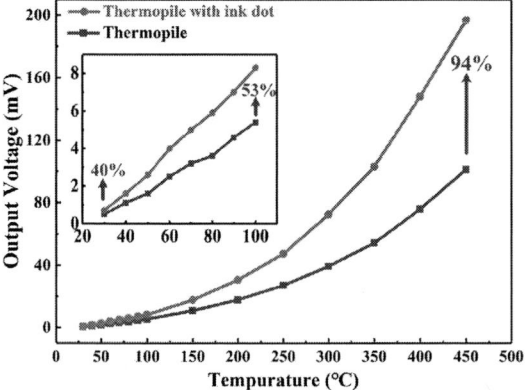

Figure 6: Voltage response curves of the two thermopile sensors with and without an ink dot in a temperature range from 30 to 450 °C. Inset shows the curves from 30 to 100 °C.

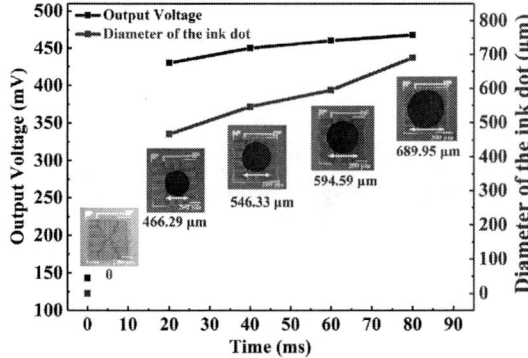

Figure 7: Diameters of ink dots under different printing periods, and the corresponding output of the thermopile chips radiated by a laser with a wavelength of 660 nm. The result of the pristine thermopile is also shown.

Finally, the uniformity of the output increment by injecting ink dots on the wafer is studied to verify the feasibility of mass preparation. After ink dot printing on the whole 8-inch wafer of thermopile sensors, at the concentric circle lines whose diameter are 20 mm, 80 mm and 170 mm, five thermopile chips at each circular lines were evenly chosen and tested by an 808 nm laser under the power of 100 mW. The output results of all these devices are depicted by boxplot as shown in Figure 8, which exhibits the maximum, minimum and average values of devices at positions of different concentric circles. As demonstrated, the average output increment at different positions is nearly

978-1-6654-9309-3/23 $31.00 © 2023 IEEE 584

the same. Moreover, for devices at the same circular line, the difference between a maximum and a minimum increment is all smaller than 15 mV. Thus, after calculation, the coefficient of variation (*CV*) is smaller than 2% on the whole wafer, which shows that the printed ink dots have good consistency and are suitable for mass preparation.

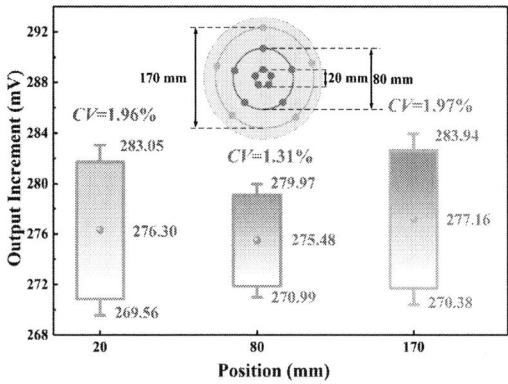

Figure 8: Output increment of thermopiles distributed at different positions on an 8-inch wafer under a 660 nm laser. Inset shows the tested thermopiles' positions, as well as the maximum, minimum, mean and CV values.

CONCLUSION

In this work, a simple, low-cost and size controllable method to enhance thermopile performance is proposed by using a conventional CP test system. Specifically, the CP test system not only can detect the device performance but also can print ink dots on the devices, so as to improve and evaluate the device performance in a convenient way. As a result, the performance of thermopile with an ink dot is able to be increased by >200% in a wavelength range of 635-1550 nm, and by >40% in a temperature range of 30-450 °C. Moreover, the ink dot printing process is controllable and has excellent uniformity, which is suitable for mass preparation. By this method, the thermopiles with ink dots are expected to satisfy more applications requiring high performance.

ACKNOWLEDGEMENTS

This work was supported by National Natural Science Foundation of China (Grant No. 62201567), Youth Innovation Promotion Association, Chinese Academy of Sciences (Grant No. 2022117), State Key Laboratory of Dynamic Test jointly built by Province and Ministry Open Fund (Grant No. 2022-SYSJJ-07), High Technology Research and Development Project of Guangdong (Grant No. 2019B010117001).

REFERENCES

[1] E. Sebastián, C. Armiens, and J. Gómez-Elvira, "Infrared Temperature Measurement Uncertainty for Unchopped Thermopile in Presence of Case Thermal Gradients", *Infrared Phys. Techn.*, vol. 54, pp. 75-83, 2011.

[2] A. Shajkofci, "Correction of Human Forehead Temperature Variations Measured by Non-Contact Infrared Thermometer", *IEEE Sens. J.*, vol. 22, no. 17, pp. 16750-16755, 2022.

[3] S. Zhang, W. Bin, B. Xu, *et al.*, "Mixed-Gas $CH_4/CO_2/CO$ Detection Based on Linear Variable Optical Filter and Thermopile Detector Array", *Nanoscale Res. Lett.*, vol. 14, 348, 2019.

[4] S. Cerimovic, F. Keplinger, and T. Sauter, "Calorimetric Flow Sensors Based on Thick-film Printed Thermopiles for Air Conditioning System Monitoring", in *Digest Tech. Papers IEEE IECON 2015*, Yokohama, November 09-12, 2016, pp. 005057-005061.

[5] S. Xu, M. Shi, N. Zhou, *et al.*, "A Performance Enhancement Method for MEMS Thermopile Pirani Sensors Through In-Situ Integration of Nanoforests," *IEEE Electron Device Lett.*, vol. 43, pp. 1752-1755, 2022.

[6] S. J. Chen and R. T. Ding, "Investigation on An Active Thermoelectric Vacuum Sensor with Low Frequency Modulation", in *Digest Tech. Papers 2019 IOP Conf. Ser.: Mater. Sci. Eng.*, Kenting, May 24-26, 2019, 012024.

[7] S.B. Mbarek, N. Alcheikh, and M.I. Younis, "Recent Advances on MEMS Based Infrared Thermopile Detectors", *Microsyst. Technol.*, vol. 28, pp. 1751-1764, 2022.

[8] C.N. Chen and W. C. Huang, "A CMOS-MEMS Thermopile with Low Thermal Conductance and A Near-Perfect Emissivity in The 8-14-μm Wavelength Range", *IEEE Electron Device Lett.*, vol. 32, pp. 96-98, 2011.

[9] Y. He, Y. Wang and T. Li, "Improved Thermopile on Pyramidally-Textured Dielectric Film", *IEEE Electron Device Lett.*, vol. 41, pp. 1094-1097, 2020.

[10] F. Liang, C. Cai, K. Zhang, *et al.*, "Infrared Gesture Recognition System Based on Near-Sensor Computing", *IEEE Electron Device Lett.*, vol. 42, pp. 1053-1056, 2021.

[11] S. Chen and B. Chen, "Research on A CMOS-MEMS Infrared Sensor with Reduced Graphene Oxide", *Sensors*, vol. 20, 4007, 2020.

[12] M. Li, M. Shi, B. Wang *et al.*, "Quasi-Ordered Nanoforests with Hybrid Plasmon Resonances for Broadband Absorption and Photodetection", *Adv. Funct. Mater.*, vol. 31, 2102840, 2021.

[13] M. Shi, X. Dai, Y. Liu, *et al.*, "Infrared Thermopile Sensors with in-Situ Integration of Composite Nanoforests for Enhanced Optical Absorption", in *Digest Tech. Papers IEEE MEMS 2021*, Gainesville, January 25-29, 2021, pp. 282-285.

[14] Q. Tan, L. Tang, H. Mao, *et al.*, "Nanoforest of Black Silicon Fabricated by AIC and RIE method", *Mater. Lett.*, vol. 164, pp. 613-617, 2015.

[15] A. Husnain, T. Asra, S. Amir, et al. "Stealth Technology: Methods and Composite Materials—A Review", *Polym. Compos.*, vol. 40, pp. 4457-4472, 2019.

CONTACT

*H.Y. Mao, tel: 86-010-82995934; maohaiyang@ ime.ac.cn

*N. Zhou, tel: 86-010-82995794; zhouna@ime.ac.cn

CHARACTERIZING INDUCTIVELY-COUPLED-PLASMA ETCHING OF SINGLE CRYSTALLINE LITHIUM TANTALATE FOR MICRO-ACOUSTIC APPLICATIONS

Yasaman Majd, Jorge Manrique Castro, Hakhamanesh Mansoorzare, and Reza Abdolvand

University Of Central Florida, Orlando, USA

ABSTRACT

In this paper, the etch characteristic of different crystalline planes (X-cut and 42 ° Y-cut) of lithium tantalate (LT) is studied by performing inductively coupled plasma reactive ion etching (ICP-RIE) with different CF_4/Ar gas mixtures and varying RIE powers using sputtered or e-beam evaporated Ni masks. Consequently, the influence of the etch parameters on the etch rate, selectivity, and roughness of the etched patterns was investigated. The best results in terms of sidewall roughness and selectivity are achieved at 28/30 SCCM of Ar/ CF_4 with RIE power of 80 W for sputtered Ni mask. These results will be used as a guideline for fabrication of micro-acoustic resonators on LT substrates.

KEYWORDS

Lithium Tantalate; Etch characteristic; Etch rate; ICP-RIE; Selectivity.

INTRODUCTION

Recently, thin-film lithium tantalate (LT) and lithium niobate (LN) crystals have attracted researcher's attention due to their remarkable physical properties such as large piezoelectric, acousto-optic, and electro-optic coefficients [1-5]. High piezoelectric and pyroelectric coefficients make LT a suitable candidate for acoustic devices and IR-sensors [3]. For instance, thin-film single crystalline LT-on-silicon platform has been suggested for realizing stable micro-resonators recently [6].

One of the important steps in fabrication of micromachined thin-film piezoelectric micro-resonators is to define resonator's in-plane boundaries through etching the piezoelectric films. To achieve high quality factor micro-acoustic resonators, the etch profile has to be as steep as possible and a large selectivity to the masking material is required for etching a relatively thick film.

Unlike LN, there have not been a substantial amount of work on characterization of the etch process for LT. Previous works report chemical wet etching of LT (in HF and HNO3 acids) [2], or purely physical ion beam etching (IBE) [3]. However, the processing accuracy in wet etching is poor due to the isotropic nature of the etch process while in purely physical etching, the selectivity and etch rate significantly deteriorates.

Previously, the surface evaluation of LN and LT crystal using flouring (CF_4) etchant with RIE has been reported [7]. However, in said work no mask has been used and sidewall roughness and slope has not been reported.

In this work, the etch rate, sidewall slope, and sidewall roughness of LT have been comprehensively studied and optimized for different ratios of Ar/CF_4 gas mixtures at different RIE bias voltage in an inductively coupled plasma etcher. Because of high cost of the thin film wafer, the study has been performed on die pieces of around 1cm^2

area of one-side polished X-cut 3000 um and 42 °Y-cut 5000 um thick LT.

Nickel (Ni) hard masks deposited through sputtering and e-beam evaporation are utilized here during the etch process and their selectivity are evaluated. The etched profile is evaluated by 3D optical profiler which could provide high resolution images without need for sample coating.

FABRICATION STEPS

The corresponding fabrication steps are outlined in Figure 1. First, the pieces of the saw-quality single-side-polished LT substrates are covered by a thick (~3 um) negative photoresist (NR9) to obtain a negative wall profile which is suitable for lift-off process and then patterned by UV-lithography (Figure 1.c).

660/20 nm of Ni/Ti is sputtered on LT with power of 180 W and low pressure of 2 mTor to yield better directionality (Figure 1.b). Ni is chosen here as a mask because of resistance to fluorine-based plasma etching. The patterns were subsequently formed by lift-off process (Figure 1.c). After the etching by Ar/CF_4 gas mixtures (Figure 1.d), the pieces were soaked into Ni and Ti etchants to remove the remained Ni mask (Figure 1.e).

It should be mentioned that all the pieces are soaked into the RCA-1 cleaning solution (5 parts H2O:1part NH4OH: 1part H2O2) to remove etched redeposited material on the side walls.

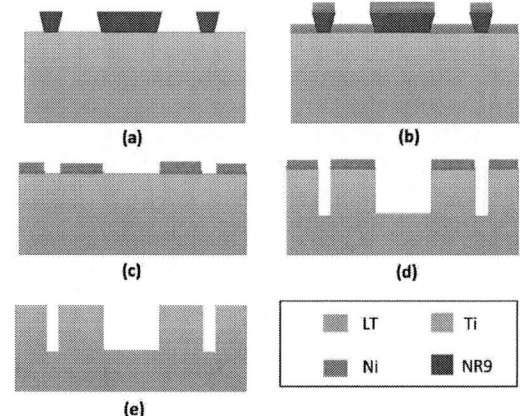

Figure 1: Fabrication process

Etching process

The etching steps were done in an APEX ICP-RIE machine which includes an ICP generator that produces the high density plasma, and a RF generator that accelerates the ions and promotes the directional ion bombardment. Here, a systematic parametric study has been performed in which the different parameters such as RIE power, gasses, and their flow rate are varied.

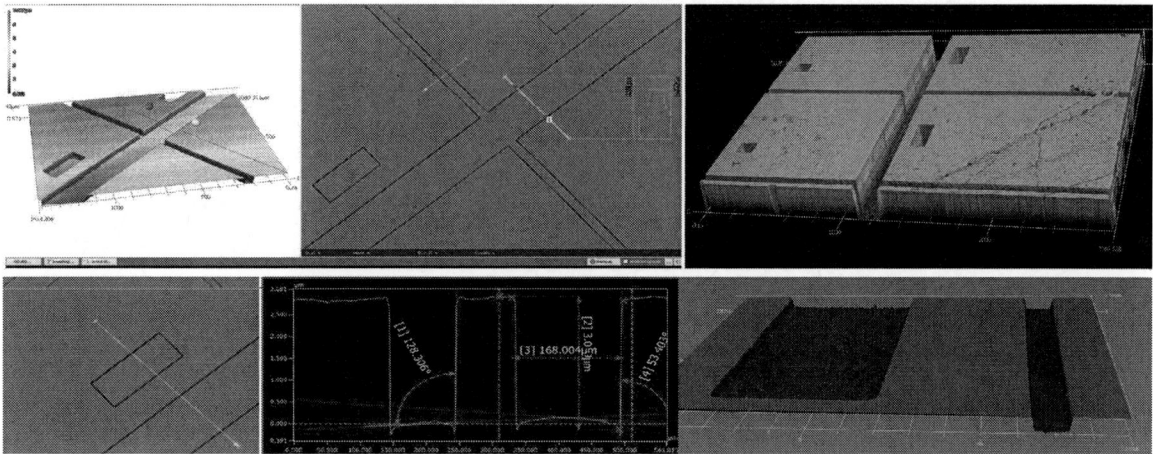

Figure 2: Etched Y surface profile of LT with Ni mask and Ar/CF4 gas obtained from 3D optical profiler

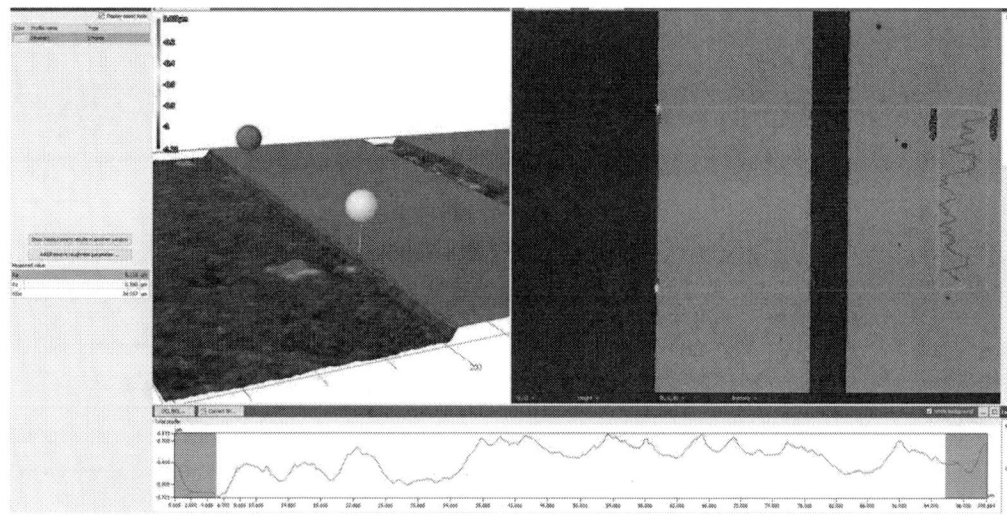

Figure 3: Measured sidewall roughness of Y-cut LT etched by Ar gas bombardment

First, the gas ratio of Ar/CF$_4$ is varied to investigate the effect of Ar gas on etch parameters. The etch was performed in Ar (physical etching with Ar$^+$ ions), in Ar/CF$_4$ gas mixtures (physically-assisted chemical etching) and finally by the pure CF$_4$ gas (mostly chemical etching). These experiments were done with RIE power of 100 W, ICP power of 500 W and low pressure of 1.3 Pa to improve the ion bombardment efficiency. Helium cooling (5 mTor) is used here to prevent the substrate from overheating, and to guarantee a sample process temperature of less than 25 °C.

The sidewalls and etch profile of samples are shown in Figure 2 measured by 3D profiler (VK-X3000, Keyence). In addition, the etch characteristic including etch rate, selectivity, and sidewall roughness (Figure 3), is reported in Table 1 at different gas ratio for different cuts of LT.

These experimental results show that by choosing the gas ratio effectively, both etch rate and selectivity could be enhanced. In fact, adding Ar gas could improve the etch rate and selectivity. Nevertheless, the ratio needs to be carefully optimized. As also suggested by [7], Ar gas ratio of 40% could result in best etch rate of single crystalline LT.

Table 1: Etch characteristic of LT at different gas mixtures.

AR/CF$_4$ (SCCM)	Cut of LT	Average etching rate (nm/min)	Average Selectivity (LT to Ni)	Average sidewall roughness (nm)
-/35	Y-cut	40	3	54
28/35	Y-cut	42	5.3	90
30/-	Y-cut	36	1	114.5
-/35	X-cut	39	3	43
28/35	X-cut	41	4.5	80
30/-	X-cut	33	1	100

Next, the Ar/CF$_4$ gas ratio is fixed at 28/35 SCCM, and the RIE power is varied from 80 W to 120 W to examine its effect on the etching process. Table 2 illustrates the relationship between RIE power and the etching results.

Table 2: Etching results of LT at different RIE power.

RIE power (W)	Cut of LT	Average etching rate (nm/min)	Average Selectivity (LT to Ni)	Average sidewall roughness (nm)
120	Y-cut	50	3.5	89
110	Y-cut	48	4.5	58
100	Y-cut	42	5.3	90
90	Y-cut	43	5	50
80	Y-cut	42.8	8.5	39
120	X-cut	48	4.5	115
110	X-cut	42	4.8	58
100	X-cut	41	4.5	90
90	X-cut	36	7	80
80	X-cut	39	9	40

For better comparison, the etch rate and selectivity graph of both X- and Y-cut samples are plotted in Figure 4 and Figure 5 respectively. As seen in Figure 4, Y-cut samples are etched slightly faster than X-cut LT samples. The best sidewall slope of 53 ° is achieved at RIE power of 100 W (shown in Figure 2).

From our observation, by increasing RIE power from 80 W to 120 W, the selectivity decreases by more than 50% while etching rate only increases by 20% for Y-cut.

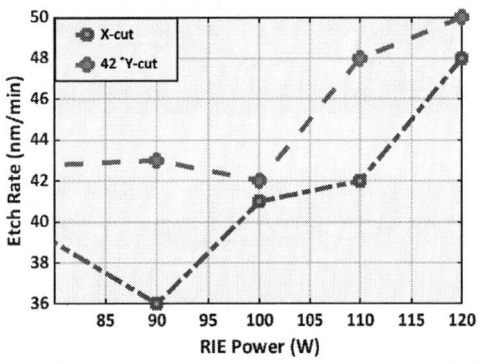

Figure 4: Etch rate of LT as a function of applied RIE power; the remaining etch parameters were kept constant.

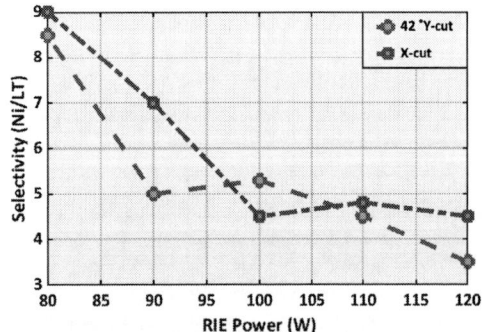

Figure 5: Selectivity value of LT as a function of applied RIE power; the remaining etch parameters were kept constant.

Based on the gathered data, increasing power also deteriorates the sidewall roughness which could be critical especially for optical applications.

Comparing Ni mask deposited by sputtering and E-beam Evaporation

280 nm Ni was sputtered on the top of 20 nm Cr with e-beam evaporation. The reason for depositing thin film of Ni stems from the high stress created during e-beam process on the film which cause metal peel off. Table 3 compares the etch rate, selectivity, and sidewall roughness of sputtered Ni and e-beam Ni on X-cut LT. Referring to the obtained data, the line roughness with e-beam deposition mask is better than sputtering deposition at the expense of worse selectivity. The line roughness enhancement might be attributed to the lift-off process. Usually, e-beam is more recommended for metal lift-off process because of its higher directionality.

Table 3: Comparison of etching results of LT with sputtered Ni and Ni deposited by e-beam.

AR/CF$_4$ (SCCM)	Method of deposition	Average etching rate (nm/min)	Average Selectivity (LT to Ni)	Average sidewall roughness (nm)
-/35	Sputtering	40	3	54
28/35	Sputtering	41	4.5	90
30/-	Sputtering	36	1	114.5
-/35	e-beam evaporation	39	2.8	55
28/35	e-beam evaporation	43	3	54
30/-	e-beam evaporation	37	1	50

Conclusion

To sum up, different plasma etching recipes are tested here employing ICP-RIE to etch X-plate and Y-plate of LT. First, the effect of gas ratio of Ar/CF$_4$ on etch parameters is analyzed and concluded that 40-44% percent of Ar gas could improve both etch rate and selectivity. Then RIE power were swept from 80 W to 120 W at the ICP power of 500 W. Based on the experimental data, it is found that selectivity could enhance significantly by lowering the RIE power.

The best results in terms of sidewall roughness and selectivity are achieved at 28/30 SCCM of Ar/ CF4 with RIE power of 80 W for sputtered Ni masks and the best sidewall slope (53˚) is achieved at RIE power of 100 W at 28/30 SCCM of Ar/ CF$_4$.

At the end the selectivity of Ni deposited through two methods have been compared. Data analysis reveals that e-beam Ni mask has weaker selectivity compared to sputtered Ni mask. This study could be applied to the realization of high-quality factor low-loss micro-acoustic resonators made from LT.

Finally, to compare the results of the previous research with this work, the experimental results of other works are brought in Table 4.

Table 4: Etch characteristic and recipe comparison between this study and previous works

Approach	[3]	[7]	This work
LT cut	Z-cut	Z-cut and X-cut	X-cut and Y-cut
Etch mask	Photoresist	N/A	Ni
Etch Technique	Reactive Ion beam etching	Plasma reactive Ion etching	ICP-RIE
Gas mixtures	Ar, C_2F_6	CF_4, Ar, H_2	CF_4, Ar
Maximum reported Etch rate	80 nm/min	13 nm/min for Z-cut	50 nm/min for Y-cut
Best Selectivity reported	1.8	N/A	9
Lowest Sidewall or surface roughness	N/A	Surface roughness: 3.1 nm	40 nm
Sidewall slope	N/A	N/A	45-50°

evaluation of LiNbO3 and LiTaO3 crystals etched using fluorine system gas plasma reactive ion etching." Electrical Engineering in Japan 149, no. 2 (2004): 18-24.

CONTACT
*Y.majd; majd92@knights.ucf.edu

REFERENCES

[1] Benchabane, Sarah, Laurent Robert, Jean-Yves Rauch, Abdelkrim Khelif, and Vincent Laude. "Highly selective electroplated nickel mask for lithium niobate dry etching." *Journal of Applied Physics* 105, no. 9 (2009): 094109.

[2] Gao, Z. D., Q. J. Wang, Y. Zhang, and S. N. Zhu. "Etching study of poled lithium tantalate crystal using wet etching technique with ultrasonic assistance." Optical Materials 30, no. 6 (2008): 847-850.

[3] Plehnert, Corinna, Volkmar Norkus, Silke Möhling, and Alan Hayes. "Reactive ion beam etching of lithium tantalate and its application for pyroelectric infrared detectors." *Surface and Coatings Technology* 74 (1995): 932-936.

[4] Ulliac, Gwenn, V. Calero, Abdoulaye Ndao, F. I. Baida, and M-P. Bernal. "Argon plasma inductively coupled plasma reactive ion etching study for smooth sidewall thin film lithium niobate waveguide application." Optical Materials 53 (2016): 1-5.

[5] Wang, Renyuan, Sunil A. Bhave, and Kushal Bhattacharjee. "Low TCF lithium tantalate contour mode resonators." In 2014 IEEE International Frequency Control Symposium (FCS), pp. 1-4. IEEE, 2014.

[6] Y. Majd, H. Kermani, and R.Abdolvand. "Temperature-Stable Thin-film Lithium Tantalite-on-silicon Resonators." In Hilton Head Workshop 2022: A Solid-State Sensors, Actuators and Microsystems Workshop. 2022. in press.

[7] Fujii, Takahiro, and Shinzo Yoshikado. "Surface

ROBUST POLYCRYSTALLINE 3C-SIC-ON-SI HETEROSTRUCTURES WITH LOW CTE MISMATCH UP TO 900 °C FOR MEMS

Philipp Moll[1], Georg Pfusterschmied[1], and Ulrich Schmid[1]
[1]TU Wien, Institute of Sensor and Actuator Systems, Vienna, AUSTRIA

ABSTRACT

In this paper we present for the first time polycrystalline cubic silicon carbide on monocrystalline silicon (3C-SiC-on-Si) heterostructures with very low coefficient of thermal expansion (CTE) mismatch at temperatures up to 900 °C. The use of different gas flow rates with alternating supply deposition (ASD) in a low-pressure chemical vapor deposition (LPCVD) system allows to tailor the CTE of the 3C-SiC thin films, resulting in thermal stress levels as low as 175 MPa at 900 °C (~300 MPa intrinsic stress at room temperature). This achievement unlocks robust 3C-SiC/Si interfaces for high temperature micro electromechanical systems (MEMS) applications by overcoming the well-known CTE mismatch of ~9 % between Si and 3C-SiC.

KEYWORDS

3C-SiC, LPCVD, CTE mismatch, thermal stress

INTRODUCTION

In the past 30 years silicon carbide (SiC) established as robust alternative in the field of silicon MEMS [1], as silicon lacks mechanical stability at temperatures above 500 °C [2, 3]. Especially as a structural material polycrystalline 3C-SiC is preferred compared to its silicon counterparts, as it features superior properties in the fields of chemical inertness, hardness and mechanical fracture strength, even at high temperatures [4, 5]. Even more, piezoresistive properties makes 3C-SiC a suitable material for the fabrication of *e.g.* strain-sensitive MEMS sensors [6]. Furthermore, polycrystalline 3C-SiC deposited by LPCVD enables process temperatures down to 800 °C and hence, facilitates the deposition on large-diameter Si substrates and therefore the cost-effective integration into standard Si-MEMS process flows [7]. In general, two different techniques established to deposit 3C-SiC successfully on Si substrates. Simultaneous supply deposition (SSD) provides a fast and simple approach, where all process gases are introduced into the reaction chamber at once. Alternating supply deposition (ASD) on the other hand, offers a more versatile deposition technique by introducing the process gases sequentially. The latter technique provides the deposition of high quality 3C-SiC thin films in terms of crystal quality, whereas even monocrystalline thin films are possible [8-11]. Independent of the selected deposition route, both feature the opportunity to tailor with the deposition parameters the resulting electrical, mechanical and thermal thin films properties, as demonstrated in this study for ASD layers. However, many approaches demonstrated the deposition of 3C-SiC-on-Si [7, 12, 13], the problem of the high coefficient of thermal expansion (CTE) mismatch between both materials still remains [14]. Especially in high temperature application, the CTE mismatch can cause many unwanted effects at SiC/Si interfaces [11, 15-17],

such as build-up of mechanical stress or even plastic deformation. This can result in material fatigue, accelerated aging effects, layer delamination or crack formation leading ultimately to thermally induced failure of the device. In this paper we present the impact of ambient temperature on the residual stress of polycrystalline 3C-SiC thin films-on-Si tailored for a low CTE mismatch up to 900 °C under vacuum.

EXPERIMENTAL DETAILS

For the deposition of our polycrystalline 3C-SiC thin films we performed ASD in a FirstNano© LPCVD system on n-doped 4 inch Si <100> wafers (thickness: 350 μm). As the first of three process steps hydrogen-cleaning was applied with 5 slm H_2 at 1000 °C for 10 min. Secondly, a carbonization step was performed at 1085 °C and 20 torr for 5.5 min with 1 slm H_2 and 100 sccm propane (C_3H_8). Hereby, a thin intermediate binding layer between the Si substrate and the SiC is formed, to overcome the problem of about 20 % lattice mismatch [18, 19]. This layer also functions as a protective layer to the substrate against the aggressive hydrogen during the high temperature deposition process and prevents the effusion of Si atoms from the surface [20, 21]. Subsequently, the actual SiC deposition process was done, whereas silane (SiH_4) and propane (C_3H_8) were introduced as precursor gases sequentially for 7 and 3.5 s, respectively, separated by a 5 s long pump out steps. Preliminary studies confirmed that the ratio of the precursor gas flows plays a significant role for the success of the 3C-SiC thin film deposition [22]. As a consequence, the ratio of the precursor flow rates was set to one part of SiH_4 to four parts of C_3H_8.

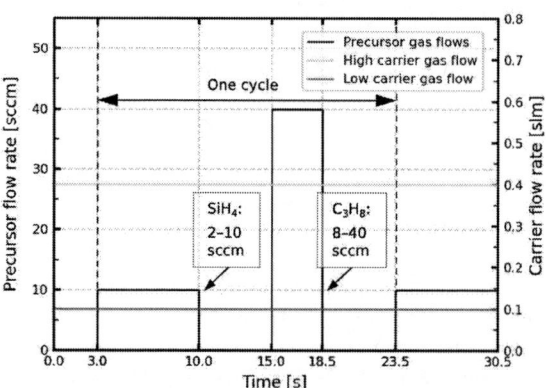

Figure 1: Deposition scheme of one ASD cycle. The precursor flow rates for SiH_4 and C_3H_8 were altered in five steps of 2, 4, 6, 8 and 10 sccm and 8, 16, 24, 32 and 40 sccm, respectively. Given flow rates were applied to a higher carrier gas flow of $H_2 = 0.4$ slm, as well as to a lower flow rate of $H_2 = 0.1$ slm.

To provide a sufficient flow regime during deposition [4, 23] the precursor gas flow is supported by H_2 as the carrier gas. In this study two different flow rates of $H_2 = 0.4$ slm and $H_2 = 0.1$ slm are investigated. A schematic representation of the deposition scheme is provided in Figure 1. The temperature and the chamber pressure were held constant at 1000 °C and 0.32 torr, respectively. To achieve a consistent layer thickness, the growth rate of each thin film composition was determined in preliminary deposition runs. Next, the thin films presented in this work were deposited with a layer thickness of 400 nm ± 5 %. In total, ten different 3C-SiC-on-Si samples were characterized under vacuum in terms of curvature with a thermal scan from K-Space Associates Inc.© kSA MOS Thermal Scan Model TS300-HT from room temperature (RT) to 900 °C with a hold time at 900 °C for each sample of 20 min before starting the cool-down phase. From the curvature measurements the corresponding thermal stress σ was determined using the Stoney-equation [24]

$$\sigma = \frac{E_s d_s^2}{6(1 - v_s)d_f} \frac{1}{R} \qquad (1)$$

as a function of temperature. Hereby E_S is the Young's modulus of the substrate, v_S the Poisson's ratio of the substrate, d_S and d_f the thicknesses of the substrate and the thin film, respectively and R the measured radius of the curvature. The coefficient of thermal expansion α is related to the thermal stress via [25]

$$\frac{\sigma}{(T_f - T_a)} = \frac{E_f}{1 - v_f}\left(\alpha_f - \alpha_S\right) \qquad (2)$$

whereas E_f is the Young's modulus of the thin film, v_f is the Poisson's ratio of the thin film and α_f and α_S are the CTE of the thin film and substrate, respectively. The term

$$\frac{\sigma}{(T_f - T_a)} \qquad (3)$$

represents the slope of the stress curve k. With

$$\frac{E_f}{1 - v_f} = E_{f,biax} \qquad (4)$$

as the biaxial Young's modulus, the temperature dependent CTE of the thin film $\alpha_f(T)$ can be expressed as

$$\alpha_f(T) = \frac{k}{E_{f,biax}} + \alpha_S(T) \qquad (5)$$

with $\alpha_S(T)$ as the temperature dependent CTE of the substrate. The temperature dependency of the Young's modulus was neglected, because of its comparable small impact compared to the thermal stress, whereas the Young's modulus of 3C-SiC at 900 °C is roughly $E_{900} = E_f*0.955$ [5]. In this analysis, the intrinsic layer stress of the thin films of about 300 MPa at RT was not considered, since it would only cause an offset shift and only the dependency of the thermal stress originating from the ambient temperature was calculated.

RESULTS

Using equation (1) on the measured temperature dependent radius of curvature R of the different 3C-SiC thin films the corresponding thermal stress values are determined. Hereby, a control of the thermal stress with the precursor and carrier flow rates is clearly visible and can be seen in Figure 2.

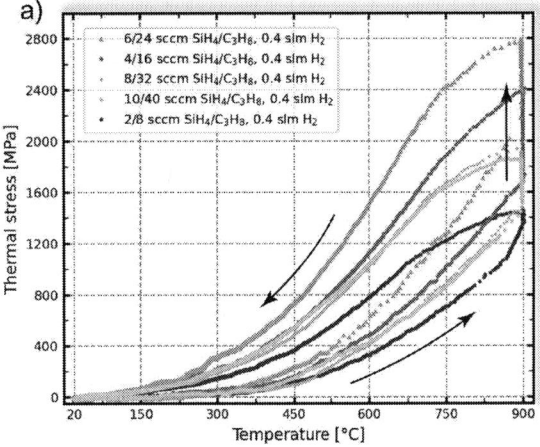

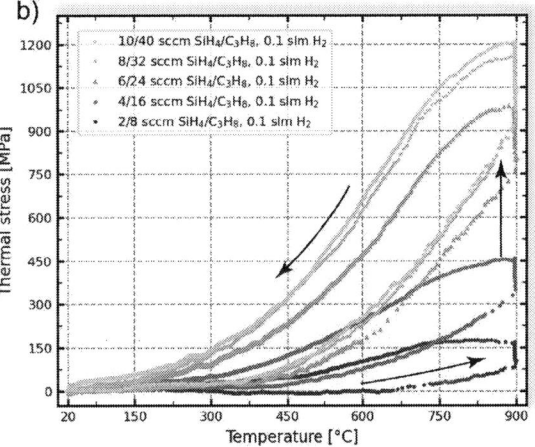

Figure 2: Temperature dependency of the thermal stress of poly 3C-SiC thin films on Si from 20 to 900 °C. In a) on average 4 times higher stress values for $H_2 = 0.4$ slm were measured compared to $H_2 = 0.1$ slm, given in b).

One outstanding result is that after cooling down to room temperature the initial stress value was reached again. This finding indicates that no stress relaxation, recrystallization or phase transitions took place during the heating process, confirming that only different CTEs of both materials caused the resulting stress. Different behaviors of the thermal stress with increasing precursor gas flow were found for the two different carrier flow rates. Figure 3 illustrates the maximum values of Figure 2 for each investigated thin film. Hereby, the thermal stress for the higher carrier gas flow rate increases at first and then decreases again with increasing precursor flow rate. However, the thermal stress for the lower carrier flow rate only increases with increasing precursor flow rate. Furthermore, the carrier flow rate of $H_2 = 0.1$ slm features much lower stress values compared to its higher counterpart. The measurements also revealed a 3C-SiC thin film (2/8 sccm SiH_4/C_3H_8, H_2 = 0.1 slm) featuring almost constant stress values of around 0 MPa up to 600 °C.

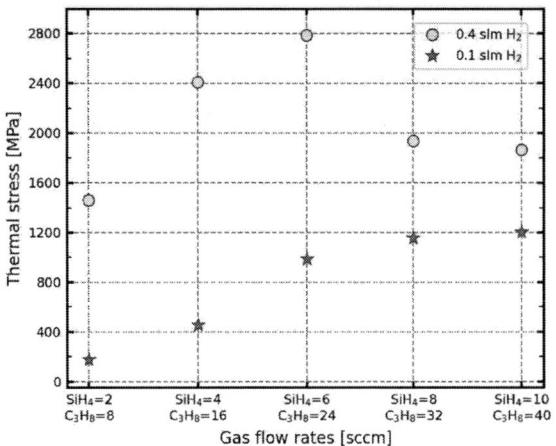

Figure 3: Maximum thermal stress values from Figure 2a and b. Hereby the maximum thermal stress of the higher carrier gas flow of H_2 = 0.4 slm increased at first with the precursor gas flow peaking at 6/24 sccm SiH_4/C_3H_8, followed by a decrease of the stress with further increasing of the precursor flow rates. Whereas for the thin films with lower carrier flow rate of H_2 = 0.1 slm the maximum values for the thermal stress solely increased with increasing precursor flow rates.

Above, a maximum of 175 MPa was determined at 900 °C. The thermally-induced stress results are based on varying CTE of our 3C-SiC thin films originated from different gas flow rates during deposition. For the calculation of the temperature dependent CTEs of each thin film we used (5) with $E_{f,biax}$ = 428.57 GPa [16, 26] and $\alpha_S(T)$ derived from [27]. The results of (5) for each thin film at RT and for 900 °C are listed in Table 1.

Table 1: Impact of the gas flow rates on the CTE of our poly 3C-SiC thin films with a thickness of 400 nm ± 5 % up to 900 °C.

	CTE [ppm/K] @ 20 °C	CTE [ppm/K] @ 900 °C
Si[a]	2.57	4.21
3C-SiC[a]	2.7 – 3.5	5.52

3C-SiC, this study				
Precursor flow rates	H_2 carrier flow rates [slm]			
[sccm]	0.4	0.1	0.4	0.1
2/8 SiH_4/C_3H_8	3.72	2.83	8.34	**4.46**
4/16 SiH_4/C_3H_8	3.82	2.94	13.12	**5.29**
6/24 SiH_4/C_3H_8	3.41	3.28	10.89	6.96
8/32 SiH_4/C_3H_8	3.99	3.47	9.95	6.95
10/40 SiH_4/C_3H_8	3.7	3.41	8.17	7.07

[a] Slack *et al* [27]

Two thin films (2/8 sccm SiH_4/C_3H_8, H_2 = 0.1 slm and 4/16 sccm SiH_4/C_3H_8, H_2 = 0.1) stand out with even lower values for the CTE than reported in literature for monocrystalline 3C-SiC at 900 °C [27]. On the other hand,

extraordinarily high CTE values for the higher carrier gas flow of H_2 = 0.4 slm were determined. The differences between the 3C-SiC thin films result from different growth mechanisms during deposition, which were caused by the ASD deposition schemes.

CONCLUSION

In the scope of this work we showed the impact of different precursor and carrier gas flow rates on the thermal stress of polycrystalline 3C-SiC thin films on Si substrates by using the alternating supply deposition technique. The wafer bow was measured up to a temperature of 900 °C and set in relation with different gas flow combinations. The corresponding thermal stress values were calculated with Stoney's equation and from that, the CTE of each thin film was determined. Hereby a minimum value of only 4.46 ppm/K was found, which is even lower than for monocrystalline 3C-SiC. However, the range of the different CTEs went up to 13.12 ppm/K giving the opportunity of tailoring the CTE depending on a specific application. This can be important when it comes to gas sensors, where heat induced vibrations causing accelerated aging are a common problem. A low CTE mismatch at high temperatures can minimize such unwanted side effects. On a wider scale this principle can be applied to 3C-SiC/Si interfaces, where a minimal vibrational behavior is preferred. Compared to that, a high CTE mismatch will cause high thermal stress in 3C-SiC/Si interfaces and for example high deflections in membranes or beams. Depending on the application this could be favored behavior, where high amplitudes of membrane deflections are required. The presented procedure of altered process gases using ASD allows to tailor the thermal stress behavior particularly to a specific application.

ACKNOWLEDGEMENTS

We want to share special thanking words for Barbara Schmid of the Department of Thin Film Material Science at TU Wien for providing know-how and support during wafer bow measurements with the K-Space.

CONTACT

*P. Moll, tel: +43-1-58801-76674; philipp.moll@tuwien.ac.at

REFERENCES

[1] W. J. Choyke, H. Matsunami, and G. Pensl, *Silicon Carbide: Recent Major Advances*, Switzerland: Springer Nature, 2004.

[2] P. French, G. Krijnen, and F. Roozeboom, "Precision in harsh environments," *Microsystems & Nanoengineering*, vol. 2, no. 1, pp. 16048, 2016/10/10, 2016.

[3] M. Mehregany, and C. A. Zorman, "SiC MEMS: opportunities and challenges for applications in harsh environments," *Thin Solid Films*, vol. 355-356, pp. 518-524, 1999/11/01/, 1999.

[4] M. B. J. Wijesundara, and R. Azevedo, *Silicon Carbide Microsystems for Harsh Environments*, Switzerland: Springer, 2011.

978-1-6654-9309-3/23 $31.00 © 2023 IEEE

[5] M. Pozzi, M. Hassan, A. J. Harris, J. S. Burdess, L. Jiang, K. K. Lee, R. Cheung, G. J. Phelps, N. G. Wright, C. A. Zorman, and M. Mehregany, "Mechanical properties of a 3C-SiC film between room temperature and 600 °C," *Journal of Physics D: Applied Physics,* vol. 40, no. 11, pp. 3335-3342, 2007/05/18, 2007.

[6] H.-P. Phan, D. Viet Dao, P. Tanner, L. Wang, N.-T. Nguyen, Y. Zhu, and S. Dimitrijev, "Fundamental piezoresistive coefficients of p-type single crystalline 3C-SiC," *Applied Physics Letters,* vol. 104, no. 11, pp. 111905, 2014.

[7] A. J. Steckl, C. Yuan, J. P. Li, and M. J. Loboda, "Growth of crystalline 3C-SiC on Si at reduced temperatures by chemical vapor deposition from silacyclobutane," *Applied Physics Letters,* vol. 63, no. 24, pp. 3347-3349, 1993.

[8] H.-P. Phan, D. V. Dao, P. Tanner, J. Han, N.-T. Nguyen, S. Dimitrijev, G. Walker, L. Wang, and Y. Zhu, "Thickness dependence of the piezoresistive effect in p-type single crystalline 3C-SiC nanothin films," *Journal of Materials Chemistry C,* vol. 2, no. 35, pp. 7176-7179, 2014.

[9] H. Nagasawa, and K. Yagi, "3C-SiC Single-Crystal Films Grown on 6-Inch Si Substrates," vol. 202, no. 1, pp. 335-358, 1997.

[10] L. Wang, S. Dimitrijev, J. Han, P. Tanner, A. Iacopi, and L. Hold, "Demonstration of p-type 3C–SiC grown on 150mm Si(100) substrates by atomic-layer epitaxy at 1000°C," *Journal of Crystal Growth,* vol. 329, no. 1, pp. 67-70, 2011/08/15/, 2011.

[11] T. Shimizu, Y. Ishikawa, and N. Shibata, "Epitaxial Growth of 3C-SiC on Thin Silicon-on-Insulator Substrate by Chemical Vapor Deposition Using Alternating Gas Supply," *Japanese Journal of Applied Physics,* vol. 39, no. Part 2, No. 6B, pp. L617-L619, 2000/06/15, 2000.

[12] M. Yazdanfar, H. Pedersen, P. Sukkaew, I. G. Ivanov, Ö. Danielsson, O. Kordina, and E. Janzén, "On the use of methane as a carbon precursor in Chemical Vapor Deposition of silicon carbide," *Journal of Crystal Growth,* vol. 390, pp. 24-29, 2014/03/15/, 2014.

[13] L. Wang, S. Dimitrijev, A. Fissel, G. Walker, J. Chai, L. Hold, A. Fernandes, N.-T. Nguyen, and A. Iacopi, "Growth mechanism for alternating supply epitaxy: the unique pathway to achieve uniform silicon carbide films on multiple large-diameter silicon substrates," *RSC Advances,* vol. 6, no. 20, pp. 16662-16667, 2016.

[14] D. N. Talwar, and J. C. Sherbondy, "Thermal expansion coefficient of 3C–SiC," *Applied Physics Letters,* vol. 67, no. 22, pp. 3301-3303, 1995/11/27, 1995.

[15] G.-S. Chung, and J.-M. Jeong, "Fabrication of 3C-SiC micro heaters and its characteristics," *Journal of the Korean Society of Sensors,* vol. 18, no. 4, pp. 311-315, 07/31, 2009.

[16] G.-S. Chung, and J.-M. Jeong, "Fabrication of micro heaters on polycrystalline 3C-SiC suspended membranes for gas sensors and their characteristics," *Microelectronic Engineering,* vol. 87, no. 11, pp. 2348-2352, 2010/11/01/, 2010.

[17] M. Hyung-Soo, C. Dongwon, and S. M. Spearing, "Development of Si-SiC hybrid structures for elevated temperature micro-turbomachinery," *Journal of Microelectromechanical Systems,* vol. 13, no. 4, pp. 676-687, 2004.

[18] S. E. Saddow, *Silicon Carbide Materials for Biomedical Applications,* Oxford: Elsevier, 2016.

[19] P. Becker, P. Scyfried, and H. Siegert, "The lattice parameter of highly pure silicon single crystals," *Zeitschrift für Physik B Condensed Matter,* vol. 48, no. 1, pp. 17-21, 1982/08/01, 1982.

[20] S. Roy, M. Portail, T. Chassagne, J. M. Chauveau, P. Vennéguès, and M. Zielinski, "Transmission electron microscopy investigation of microtwins and double positioning domains in (111) 3C-SiC in relation with the carbonization conditions," *Applied Physics Letters,* vol. 95, no. 8, pp. -, 2009.

[21] I. Kamata, H. Tsuchida, and K. Izumi, "The structure of 3C–SiC carbonized layer on Si substrate," *Microelectronic Engineering,* vol. 43-44, pp. 647-654, 1998/08/01/, 1998.

[22] L. Wang, S. Dimitrijev, J. Han, A. Iacopi, L. Hold, P. Tanner, and H. B. Harrison, "Growth of 3C–SiC on 150-mm Si(100) substrates by alternating supply epitaxy at 1000°C," *Thin Solid Films,* vol. 519, no. 19, pp. 6443-6446, 2011/07/29/, 2011.

[23] S. R. J. S. W. S. Choi D, and S. M. Spearing, "Residual stress in thick low pressure chemical vapor deposited polycrystalline SiC coatings on Si substrates," *J. Appl. Phys.,* vol. 97, no. 7, pp. 074904, 2005.

[24] G. G. Stoney, and C. A. Parsons, "The tension of metallic films deposited by electrolysis," *Proceedings of the Royal Society of London. Series A, Containing Papers of a Mathematical and Physical Character,* vol. 82, no. 553, pp. 172-175, 1909/05/06, 1909.

[25] J. A. Thornton, and D. W. Hoffman, "Stress-related effects in thin films," *Thin Solid Films,* vol. 171, no. 1, pp. 5-31, 1989/04/01/, 1989.

[26] T. Matsumoto, T. Nose, Y. Nagata, K. Kawashima, T. Yamada, H. Nakano, and S. Nagai, "Measurement of High-Temperature Elastic Properties of Ceramics Using a Laser Ultrasonic Method," *Journal of the American Ceramic Society,* vol. 84, no. 7, pp. 1521-1525, 2001/07/01, 2001.

[27] G. A. Slack, and S. F. Bartram, "Thermal expansion of some diamondlike crystals," *Journal of Applied Physics,* vol. 46, no. 1, pp. 89-98, 1975/01/01, 1975.

A 3D-PRINTED FUNCTIONAL MEMS ACCELEROMETER

Simone Pagliano[1], David E. Marschner[1], Damien Maillard[2], Nils Ehrmann[3], Göran Stemme[1], Stefan Braun[3], Luis Guillermo Villanueva[2], and Frank Niklaus[1]*

[1]KTH Royal Institute of Technology, SWEDEN
[2]École Polytechnique Fédérale de Lausanne, SWITZERLAND
[3]Hochschule Kaiserslautern, GERMANY

ABSTRACT

3D printing of MEMS devices could enable the cost-efficient production of custom-designed and complex 3D MEMS for prototyping and for low-volume applications. In this work, we present the first micro 3D-printed functional MEMS accelerometers using two-photon polymerization combined with the evaporation of metal strain gauge transducers. We measured the resonance frequency, the responsivity, and the signal stability over a period of 10 h of the 3D-printed accelerometer.

KEYWORDS

3D printing, two-photon polymerization, shadow masking, strain gauge transducer, Laser Doppler Vibrometer, custom MEMS, low-volume manufacturing.

INTRODUCTION

The success of MEMS devices in recent decades was fostered by their constant miniaturization and parallel fabrication using semiconductor micro-manufacturing technologies. This allows cost-effective production of large volumes of MEMS devices for large volume markets. However, the production of customized MEMS sensors and actuators at low- and medium-scale volumes does not benefit from the economy of scale in the same way. As a result, the development of novel commercial MEMS devices is often limited to devices that address very high-volume markets. 3D printing of MEMS devices could revert this trend and enable the realization of custom MEMS devices required in limited numbers.

Inertial sensors, such as accelerometers and gyroscopes, are among the most successful MEMS devices on the market today. In recent years, different 3D printing techniques have been used to fabricate functional inertial sensors [1]. Accelerometers have been fabricated using fused filament fabrication [2], laser powder bed fusion [3], and stereolithography [4]. However, since these 3D printing techniques have minimum resolutions in the order of several tens or hundreds of micrometers (μm), the resulting device footprints are of the order of several mm^2 or cm^2. Such devices are not suitable for applications where miniaturization is critical. Contrary to these techniques, 3D printing by two-photon polymerization is capable of achieving resolutions of below 1 μm in all spatial directions [1] and allows for freeform printing in the 3D space. Additionally, among 3D printing techniques with sub-μm resolution, 3D printing by two-photon polymerization has the highest volume printing rate [5]. Thus, two-photon polymerization is currently the best candidate for 3D printing truly miniaturized MEMS devices with well-resolved features at the μm-scale. However, the realization

of electrically functional components using two-photon polymerization remains challenging. 3D printing by two-photon polymerization has been used to manufacture electrostatic and thermomechanical microactuators [6], but its use for inertial sensors has been limited [7]. Here, we present a 3D-printed functional MEMS accelerometer using two-photon polymerization in combination with metal evaporation to form strain gauge transducers [8].

WORKING PRINCIPLE & DESIGN

The working principle of our 3D-printed MEMS accelerometer resembles that of a conventional piezoresistive MEMS accelerometer, where resistive strain sensors are fabricated on top of cantilevers connected to a proof mass. Any external acceleration applied to the proof mass results in the bending of the cantilevers and the consecutive straining of the resistive transducer element. The resistance change in the resistive transducer elements correlates linearly with the applied acceleration.

The mechanical accelerometer structure consists of a supporting pillar with two single-sided clamped horizontal cantilevers and a proof mass attached at the end of the two cantilevers (Fig.1). We patterned shadow-masking structures with T-shaped cross-sections on top of the cantilevers (Fig.1a) and the supporting pillar to define the areas of the strain gauge transducers and the probing electrodes, respectively.

The geometrical parameters of the accelerometer structure were determined based on the target measurement range (1-10 g), and a parametrized finite-element model in COMSOL®. We selected the width and the thickness of the cantilevers to be 20 μm, to ensure high-quality printing of the cantilevers and the shadow-masking features on top of them. The length of the cantilevers and the size of the proof mass was extracted from the parametric model that was run on COMSOL®. Based on the results of the simulation, we selected a cantilever length of 500 μm and proof mass dimensions of 350 μm × 300 μm × 210 μm (length × width × height).

FABRICATION METHODOLOGY

We 3D-printed the accelerometer structure by two-photon polymerization on an indium tin oxide-coated glass substrate using a Nanoscribe Photonic Professional GT2 3D printer (Nanoscribe GmbH, Germany) and commercial IP-S resin (Nanoscribe GmbH, Germany). The accelerometer structure was printed using the Dip-In Laser Lithography printing mode. The printing was performed using a 25x/NA 0.8 objective lens (Carl Zeiss AG, Germany) at a laser power of 50 mW with a scan speed of 100 mm/s. To ensure correct printing, the cantilevers and the proof mass were printed using the Solid printing mode

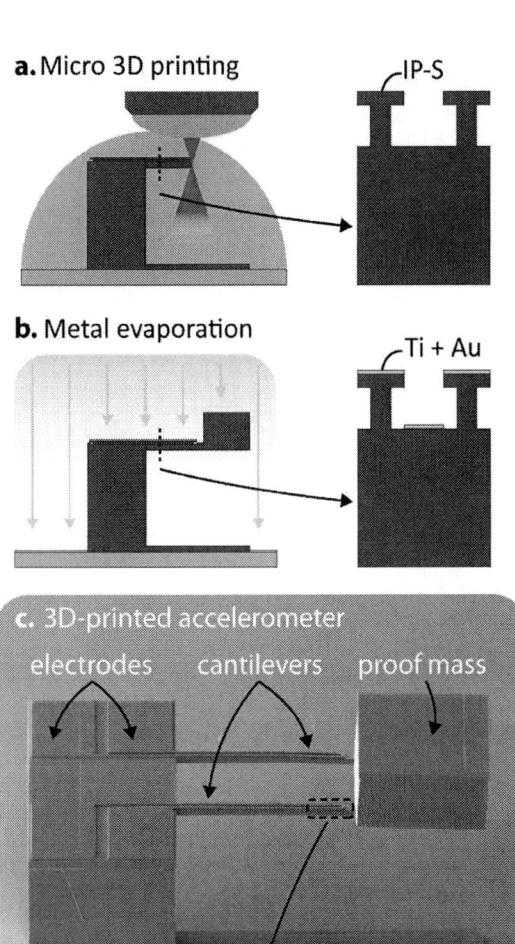

a. Micro 3D printing — IP-S

b. Metal evaporation — Ti + Au

c. 3D-printed accelerometer

electrodes   cantilevers   proof mass

supporting pillar

220 µm

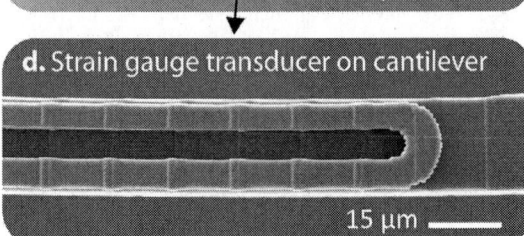

d. Strain gauge transducer on cantilever

15 µm

Figure 1: Fabrication of the 3D-printed accelerometer. *(a) Two-photon polymerization of the accelerometer structure , with details of the cross-section of the cantilever with shadow-masking features on top. (b) Directional evaporation of metal on the 3D printed structure with cantilever cross-section and isolation of the metal layer on top of the T-shaped structures for the strain gauge resistors. (c) SEM image with tilted view of the accelerometer. (d) Top view of the cantilever with the strain gauge resistor (transducer).*

with short stitching blocks of 13 µm, while the supporting pillar was printed in the Shell-&-Scaffold printing mode

with large stitching blocks. After printing, the structures were developed in propylene glycol methyl ether acetate (PGMEA) for 20 min and in isopropyl alcohol (IPA) for 5 min, then dried in air at room temperature. After drying, the structures were exposed to a UV flood for 5 min to cross-link all the internal volumes of the printed structure.

Next, the top surfaces of the 3D-printed structure were coated with Ti (10 nm) and Au (30 nm) using directional e-beam metal evaporation in direction perpendicular to the surface of the glass substrate (Fig.1b). The directional deposition on the T-shaped shadow-masking structures ensured the physical and electrical isolation of the metal on top of the shadow masking features from the rest of the deposited metal on the accelerometer, thereby forming the strain gauge resistors connected to the metal contact pads.

The 3D-printed accelerometer was inspected by SEM to ensure correct printing of the entire structure. The cantilever length was measured to be 480 µm, instead of 500 µm as designed, while all other geometrical parameters did not differ substantially from the intended size. Shrinkage of two-photon polymerized structures is well documented in literature and is related to the cross-linking level of the polymer [9]. The variation of the cantilever that we measured is within the range reported by the supplier of the resin [10].

EXPERIMENTAL RESULTS & VERIFICATION

Our measurement and characterization setup consisted of a piezoshaker (TA0505D024, Thorlabs, USA), a lock-in amplifier (H2FLI 50 MHz, Zurich Instruments, Switzerland) and a laser doppler vibrometer (OFV-551, Polytec, Germany) with the controller (OFV-5000, Polytec, Germany). We mounted the 3D-printed MEMS accelerometer on the piezoshaker together with the glass slide using double-sided tape. During each experiment, the piezoshaker driving frequency was swept between 1.4 kHz and 2 kHz, with voltage amplitudes ranging from 1 V_{rms} to 7 V_{rms}. At the same time, the signal applied to the strain gauge transduces was between 4 kHz and 4.6 kHz. The resistance change of the transducer was extracted by demodulating the signal at the downmixed frequency (2.6 kHz) obtained by multiplying the signal measured on the transducers and the one generated by the vibrometer. We calibrated the mechanical response of the piezoshaker by focusing the laser doppler vibrometer on the top surface of the supporting pillar and measured the oscillation velocity, from which the acceleration could be calculated by multiplying it with the angular frequency. The piezoshaker driving frequency was swept between 1.4 and 2 kHz, with voltage amplitudes from 1 V_{rms} to 7 V_{rms}. The piezoshaker could generate accelerations of up to 0.9 × g (8.8 m/s²).

First, we characterized the mechanical response of the accelerometer and compared it to the COMSOL® model. The resonance frequency of the accelerometer was measured at 1.78 kHz using a piezoshaker voltage of 7 V_{rms} (Fig. 2a). The Q-factor was calculated to be 35 by carrying out a Lorentzian fit of the measured data. By sweeping the piezoshaker voltage, we demonstrated the linear relation between the applied acceleration and the amplitude of

978-1-6654-9309-3/23 $31.00 © 2023 IEEE

oscillation of the proof mass (Fig. 2b). The resonance frequency of the structure extracted by the COMSOL® model matched the measured resonance frequency, assuming the polymer Young's modulus to be 6.5 GPa, instead of the 5.1 GPa specified by the material supplier [10]. To compare the simulated amplitude of the oscillation to the measured amplitude, we multiplied the applied acceleration in the model by the measured Q-factor. We found that this computed oscillation amplitude is consistent with the measured amplitude (Fig. 2b).

Next, we measured the electrical response of the strain gauge transducers. The relative resistance change ($\Delta R/R$) was measured at the resonance frequency (Fig. 3a). The off-resonance responsivity, defined as the relative resistance change as a function of the applied acceleration at standard testing frequencies (100-160 Hz), was calculated by dividing the values measured at resonance by the Q-factor. The accelerometer featured a linear response (Fig.3b). The responsivity was measured to be 420 ppm/g at resonance, with a Q-factor of 35. Thus, the responsivity at standard testing frequencies (100-160 Hz) was computed to be 11.2 ppm/g. From the extracted responsivity and the measured proof mass oscillation, we computed a gauge factor of 3.4 of the strain gauge transducer. The computed gauge factor is higher than the one extracted from our COMSOL® model, which only takes into account a resistance change due to geometrical deformation of the

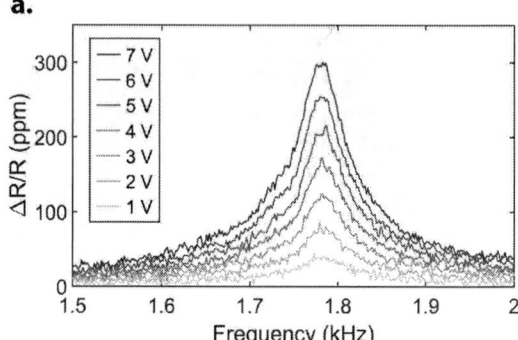

a.

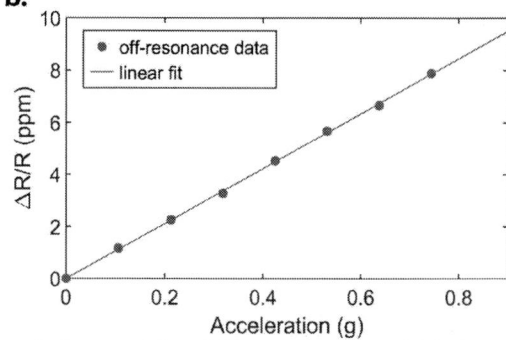

b.

Figure 3: Responsivity of the accelerometer. *(a) Relative resistance change measured with frequency sweeps at piezoshaker voltages from 1 V to 7 V. (b) Off-resonance relative resistance change as a function of acceleration, computed dividing the maximum values of relative resistance change measured at resonance by the Q-factor.*

resistor. Our measurement shows a contribution to the resistance change from a change in resistivity as well, which has been previously reported for sub-30 nm thin Au films [11].

A primary concern regarding polymeric micro-sensor structures, as the one presented here, is stability over time. To elucidate the stability of our sensor over time, we measured the stability of both the mechanical and electrical sensor response over a period of 10 h with an applied acceleration of 0.5 g. Therefore, frequency sweeps between 1.4 kHz and 2 kHz, with a voltage amplitude of 5 V_{rms} were continuously fed to the piezoshaker, and the maximum amplitude of the proof mass displacement and the relative resistance change of the transducer were recorded at each sweep and are plotted in Fig. 4. The measured resonance frequency remained within ±4 Hz of the average value during the entire duration of the experiment. In the same way, the measured relative resistance change remained within ±5 ppm of the average value. Temperature and relative humidity were monitored during this experiment to avoid any significant impact on the material characteristics.

Furthermore, we characterized the accelerometer behavior at resonance frequency to increase the signal-to-noise ratio at low applied accelerations (0.1 × g). The noise density was measured to be 90 nV/√Hz without a piezoshaker voltage applied, which corresponds to noise equivalent accelerations of 0.2 g/√Hz at standard testing frequencies (100–160 Hz). The noise density of the output signal can be decreased to 4 nV/√Hz with further

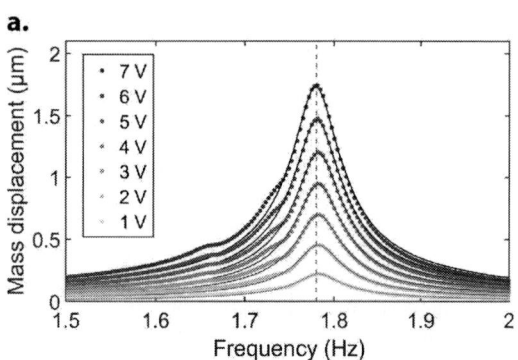

a.

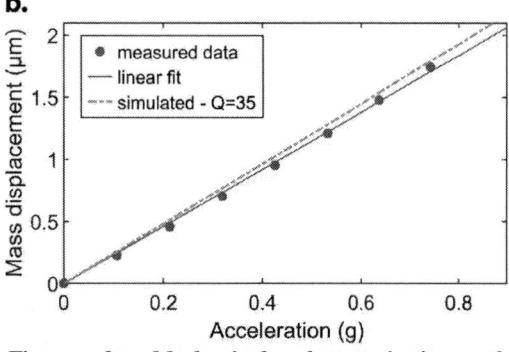

b.

Figure 2: Mechanical characterization of the accelerometer. *(a) Proof mass displacement measured with frequency sweeps at piezoshaker voltages from 1 V to 7 V. (b) Amplitude of oscillation of the proof mass at different accelerations. The measured data was extracted at the resonance frequency. The simulated data was computed using a Q-factor of 35 to match the behavior at resonance.*

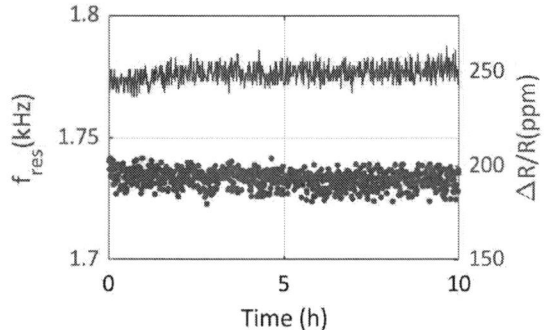

Figure 3: Stability of the accelerometer. *Measurements of relative resistance change and resonance frequency over a period of 10 h using a constant piezoshaker driving voltage of 5 V_{rms}. The resonance frequency remained stable within the interval 1.78 kHz ± 4 Hz. The relative resistance change at the resonance frequency remained stable at 193 ± 5 ppm.*

optimization of the read-out circuit, resulting in noise power densities of 10 mg/√Hz at standard testing frequencies, which is comparable to those of commercial MEMS accelerometers [12].

At the resonance frequency, we could achieve large proof mass displacements (1.9 μm) at low applied accelerations of 0.9 × g. Thus, we demonstrated a linear response of the accelerometer even at proof mass oscillations which otherwise would occur only at accelerations that are higher than 35 × g at standard testing frequencies.

The device and the measurements reported here show that functional 3D-printed MEMS accelerometers can be fabricated and have competitive performance. Additive manufacturing of MEMS has the potential to be applied to several types of MEMS sensors, including pressure sensors, microphones, flow sensors, and more. Moreover, 3D printing will enable the manufacturing of MEMS devices with innovative and complex device geometries that cannot be realized using conventional 2.5D silicon micromachining.

ACKNOWLEDGEMENTS

This work was supported by the Swedish Foundation for Strategic Research (SSF) (GMT14-0071) and the Wenner-Gren scholarship (UPD2020-0119).

REFERENCES

[1] T. Blachowicz, A. Ehrmann, "3D Printed MEMS Technology-Recent Developments and Applications", *Micromachines*, vol. 11, no. 4, 434, 2020.

[2] M. Arh, J. Slavič, M. Boltežar, "Design principles for a single-process 3d-printed accelerometer—theory and experiment", *Mech. Syst. Signal Process*, vol. 152, 107475, 2021.

[3] V. Zega, L. Martinelli, R. Casati, E. Zeppa, G. Langfelder, A. Cigada, A. Corigliano, "A 3D printed Ti6Al4V alloy uniaxial capacitive accelerometer", *IEEE Sens. J.*, vol. 21, no. 18, pp. 19640–19646, 2021.

[4] V. Zega, C. Credi, R. Bernasconi, G. Langfelder, L. Magagnin, M. Levi, A. Corigliano, "The First 3-D-Printed z-Axis Accelerometers With Differential Capacitive Sensing", *IEEE Sens. J.*, vol. 18, no. 1, 2017

[5] V. Hahn, P. Kiefer, T. Frenzel, J. Qu, E. Blasco, C. Barner-Kowollik, M. Wegener, "Rapid Assembly of Small Materials Building Blocks (Voxels) into Large Functional 3D Metamaterials". *Adv. Funct. Mater.*, vol. 30, pp. 1907795, 2020

[6] S. Kim, C. Velez, R. S. Pierre, G. L. Smith, S. Bergbreiter, "A Two-Step Fabrication Method for 3D Printed Microactuators: Characterization and Actuated Mechanisms", *J. of Microelectromechanical Syst.*, vol. 29, no. 4, pp. 544-552, 2020

[7] O. Tricinci, M. Carlotti, A. Desii, F. Meder & V. Mattoli, "Two-step MEMS microfabrication via direct laser lithography", *Advanced Fabrication Technologies for Micro/Nano Optics and Photonics XIV*, Proc. SPIE vol. 11696, 116960J, 2021

[8] S., Pagliano, D. E. Marschner, D. Maillard, N. Ehrmann, G. Stemme, S. Braun, L. G. Villanueva & F. Niklaus, "Micro 3D printing of a functional accelerometer", *Microsystems & Nanoengineering*, vol. 8, no. 105, 2022.

[9] J. Bauer, A. Guell Izard, Y. Zhang, T. Baldacchini, & L. Valdevit, "Programmable mechanical properties of two-photon polymerized materials: from nanowires to bulk", *Adv. Mater. Technol.*, vol. 4, pp. 1900146, 2019.

[10] [Online] Nanoscribe IP-S. https://support.nanoscribe.com/hc/en-gb/articles/ 360001750353-IP-S#.

[11] C. Li, P. J. Hesketh, and G. J. Maclay , "Thin gold film strain gauges", J. of Vac. Sci. & Tech. A, vol. 12, pp. 813-819 (1994)

[12] [Online] Devices Inc, A. Analog Dialogue 51-10, Devices Inc, A, 2017.

CONTACT

*Frank Niklaus, frank@kth.se

A FULL 3D PRINTING METHOD FOR MONOLITHIC INTEGRATION OF AN ACCELEROMETER AND A FORCE SENSOR

Guandong Liu[1,2], Changhai Wang[1], Kexin Wang[1], Zhili Jia[3], Ruiqi Luo[2], and Wei Ma[2,4]

[1]Institute of Sensors, Signals and Systems, School of Engineering & Physical Sciences, Heriot-Watt University, Edinburgh EH14 4AS, UK

[2]Intelligent Network Research Institute, Zhejiang Lab, Hangzhou 311100, China

[3]Center for Advanced Measurement Science, National Institute of Metrology, Beijing 100029, China.

[4]State Key Laboratory of Modern Optical Instrumentation, College of Information Science and Electronic Engineering, Zhejiang University, Hangzhou 310027, China.

ABSTRACT

This paper presents monolithic integration of an accelerometer and a force sensor based on a rapid 3D printing fabrication approach. With a triple-extruder 3D printer, for the first time, an insulating material, a conductive material, and a sacrificial material were printed alternately to fabricate a multifunctional MEMS device in a one-step printing process within an hour and without any additional metallization and assembly processes. Moreover, with the advantage of 3D printing in building geometrically complex structures, the accelerometer and the force sensor were easily integrated in vertical stacking using vertical feedthroughs in a TSV (Through-Silicon Via)-like design so that the multifunctional device could be directly mounted on a PCB substrate as a chip scale package. The performances of the multifunctional device indicate that the 3D printed sensors can be used in real-time motion and force monitoring for personalized customization and the monolithic integration method has potential applications in producing various integrated MEMS sensors.

KEYWORDS

Integrated 3D printing, monolithic integration, MEMS, accelerometer, force sensor

INTRODUCTION

Silicon-based MEMS (Micro Electromechanical System) sensors have received tremendous attention and have been widely used in many industrial and consumer applications over the past few decades [1-3]. Recently, with the development of the MEMS manufacturing techniques and the expansion of the consumer electronics market, monolithic integration of MEMS sensors on a single chip has become very important due to the advantages of smaller chip size and shorter electrical interconnections thereby reducing the manufacturing cost and improving the performance of the sensors. Wei *et al.* [4] reported an application of monolithic integration in TPMS (tire-pressure monitoring system). A pressure sensor and an accelerometer were monolithically integrated side-by-side on the silicon substrate. Gao *et al.* [5] reported monolithic integration of a pressure sensor and an accelerometer based on a cavity SOI (silicon on insulator) wafer. The accelerometer had a structure of beam-block-membrane design and the absolute pressure sensor was embedded in the mass block of the

accelerometer without increasing the chip size. Although the chip area was fully utilized, it was necessary to use a custom designed cavity SOI wafer. Wang *et al.* [6] reported vertical integration of a pressure sensor with a read-out chip. The pressure sensor chip with a TSV (through silicon via) structure and the read-out chip were fabricated separately, and then bonded together. In the field of the traditional sensor manufacturing, a large number of sensor dies have been manufactured on silicon wafers by using the conventional MEMS fabrication processes, such as thin film deposition, lithography, etching, metallization and bonding [1]. However, in the field of monolithic integration, both the dimensions of the sensors and the fabrication processes should be carefully designed to ensure process compatibility in order to protect the fabricated structures in the subsequent processes, which limit the wide application of the monolithic integration technique. Besides, it usually takes several months to fabricate the silicon-based sensors using the traditional MEMS fabrication processes, which cannot fulfill the increasing requirements of personalized customization and fast delivery time.

The three-dimensional printing (3D Printing) technique, also known as additive manufacturing, has the potential to meet the above personalized demands [7-9]. Different from the conventional subtractive methods, in 3D printing, the digital model of the 3D object is fabricated through an additive method in a bottom-up layer-by-layer process without drilling or molding. The primary advantage of 3D printing is its ability to create geometrically complex structures, which provides a possible solution for monolithic integration. In our previous work, several MEMS devices with complex suspended structures have been fully 3D printed [10-11]. However, the 3D printed MEMS devices were discrete, each device has only one function. Furthermore, the traditional wire bonding method and the flip-chip bonding method are difficult to be applied to connect the electrodes of the devices to the printed circuit board (PCB) since the pads of the electrodes were located inside or on the side wall of the devices.

In this paper, we report monolithic integration of an accelerometer and a force sensor based on a rapid, integrated 3D printing fabrication approach. With a triple-extruder 3D printer, a multifunctional device consisting of an accelerometer and a force sensor was fabricated by 3D printing in vertical stacking and without any additional metallization and assembling processes. With the TSV-like

978-1-6654-9309-3/23 $31.00 © 2023 IEEE

vertical feedthrough structure, the multifunctional device can be mounted on a PCB as a chip scale package.

SENSOR DESIGN

As shown in Fig. 1a and Fig. 1b, a capacitive force sensor and a capacitive accelerometer are vertically stacked in an integrated configuration. All the electrode plates of the force sensor and the accelerometer are extended horizontally firstly, then connected vertically to the corresponding conductive patterns in the bottom layer. The three-dimensional interconnection structure plays a similar function as the RDL (Redistribution Layer) and TSV in silicon-based devices.

The electrical model of the device is shown in Fig. 1c, the force sensor works as a single variable capacitor. While the accelerometer works as two variable parallel-plate capacitors for the purpose of differential capacitive sensing. The capacitance changes caused by a contact force or an external acceleration can be measured using an IC based readout circuitry. The sensitive structure of the force sensor and the accelerometer can be modeled as a suspended beam-plate or beam-mass. When an external force or an acceleration is applied to the device, the suspended plate or mass will deviate from its initial position thereby changing the capacitance values of the variable capacitors. Considering both the resolution of the current FDM (fused deposition modeling) 3D printer and the sensitivity of the device, the width of the beam spring was set to be 400 μm and the thickness of the beam spring was set to be 800 μm.

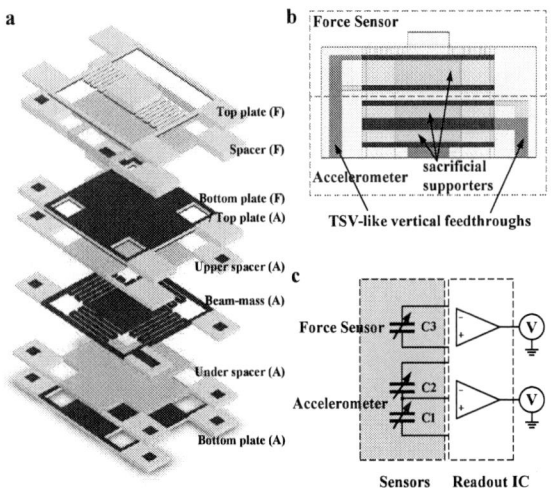

□ Normal PLA ■ Graphene PLA ▨ HIPS

Figure 1: Schematic view of the monolithic integration of a capacitive accelerometer (A) and a capacitive force sensor (F). (a) 3D structure of the device; (b) Cross section of the device; (c) Electronic model of the monolithic sensor system.

3D PRINTING METHOD

Although most of the available polymer materials for 3D printing are electrically insulating, the composite materials with conductive fillers such as carbon black, carbon nanotubes, or graphene exhibit electrical conductivity. In this work, the electrodes and the interconnections of the multifunctional device were fabricated by using a graphene-doped electrically conductive PLA (polylactic acid) polymer, which was obtained from Graphene 3D Lab Inc. The resistivity of the conductive PLA was 0.6 Ω·cm. The temporary supporting layers of the suspended sensitive structures were 3D printed by using a soluble polymer filament, HIPS (high impact polystyrene). The HIPS material and the associated solvent (D-Limonene), were both obtained from RS Components Ltd. The mechanical structures of the device were built by using a normally non-conductive PLA material. With a triple-extruder 3D printer, the electrically conductive material, the sacrificial material and the electrically insulating material can be alternately extruded in a single 3D printing process.

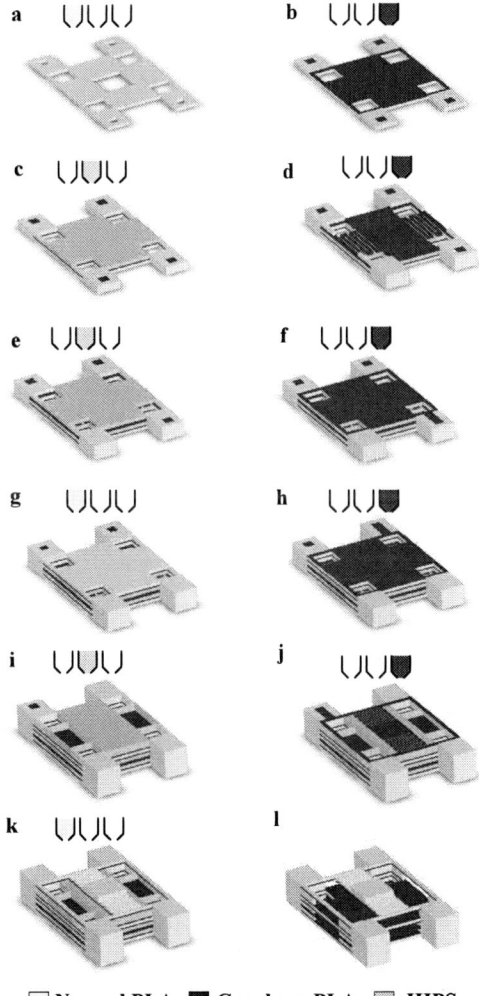

□ **Normal PLA** ■ **Graphene PLA** ▨ **HIPS**

Figure 2: The integrated 3D printing process. (a) Bottom plate printing; (b) Bottom electrode printing; (c) Under supporter printing; (d) Beam-mass printing; (e) Upper supporter printing; (f) Top electrode printing; (g) Top plate printing; (h) Bottom electrode printing; (i) Supporter printing; (j) Top electrode printing; (k) Top plate printing; (l) Sacrificial supporters removing.

978-1-6654-9309-3/23 $31.00 © 2023 IEEE

In the integrated 3D printing process, the three extruders of the 3D printer were operated alternately. The left extruder of the printer was loaded with the normal PLA filament. The graphene PLA filament was loaded into the right extruder and used to print the electrodes and the interconnections. The supporting structures of the monolithically integrated device were printed using the soluble HIPS filament, which was extruded from the middle extruder and used as sacrificial layers. The printing temperatures of the PLA, graphene PLA and HIPS were set to be 190°C, 240°C and 230°C, respectively. With a printing speed of 40 mm/s and a fill rate of 100%, the entire printing process was completed in approximately one hour. After 3D printing, the temporary HIPS layers were removed in an environmentally friendly solvent of D-Limonene at room temperature. Fig. 2 depicts the integrated 3D printing process for the monolithic integration of the accelerometer and the force sensor. For convenience, the printing steps of different structures are drawn in different illustrations. In the actual printing process, the three extruders of the 3D printer were operated alternately and the different structures at the same level were fabricated via a layer-by-layer process.

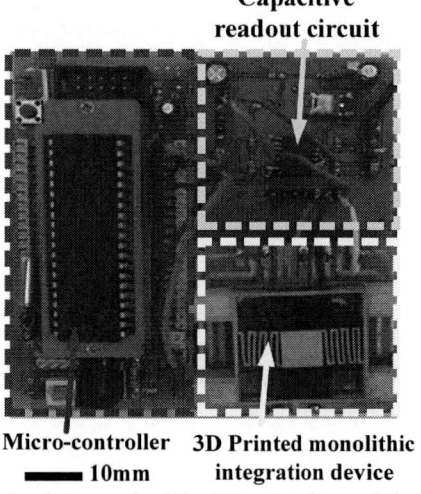

Figure 3: *photograph of the 3D printed monolithic device mounted on the capacitive readout circuit.*

Fig. 3 shows a photograph of the 3D printed monolithic device mounted on a PCB board and connected to the capacitive readout circuit MS3110, which has a capacitance resolution of 4.0 aF/rtHz and provides an output voltage proportional to the capacitance change [12].

The calibration results of the accelerometer in the range of ±1g gravity field are shown in Fig. 4, which were measured using a rotational mount. It can be seen that the accelerometer has a sensitivity of 146.8 mV/g and a nonlinearity error of 2.07%. The response of the force sensor as the applied force was measured by placing a plastic cap loaded with small screws of known weight on the moveable plate. The force was controlled by the number of screws. The sensitivity of the force sensor was approximately 1.4 V/N.

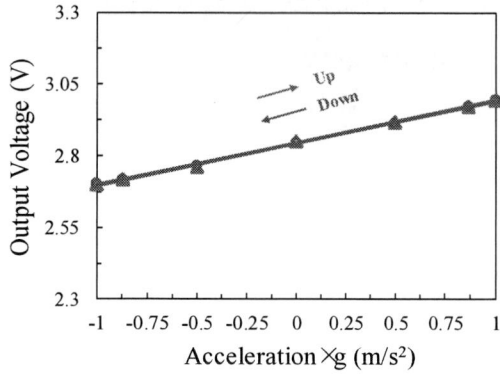

Figure 4: *Measurement results of the accelerometer of the fully 3D printed monolithic integration device.*

CONCLUSION

A 3D printing method has been developed for monolithic integration of an accelerometer and a force sensor. The MEMS devices were fabricated in a one-step printing process using a triple-extruder 3D printer, the insulating material, the conductive material and the sacrificial material were printed in alternate fashion to complete the device structure in one hour. The parallel electrode plates of the sensors were electrically connected to the pads at the bottom of the sensor stack using TSV-like vertical feedthroughs, which were made by a graphene PLA material. The 3D printed accelerometer has a sensitivity of 146.8 mV/g and a nonlinearity error of 2.07% and the 3D printed force sensor has a sensitivity of 1.4 V/N. The results show that MEMS sensors can be fully 3D printed by using the monolithic integration method.

ACKNOWLEDGEMENTS

This work was partly supported by the UK Engineering and Physical Sciences Research Council (Grant EP/R024502/1) and Research Project of Zhejiang Lab (No.2022QA0AL02). The authors would like to thank Mr Mark Leonard, and Dr Alissa Potekhina at Heriot-Watt University for useful discussions.

REFERENCES

[1] Q. Hu, C. Gao, Y. Hao, Y. Zhang, G. Yang, "Low cross-axis sensitivity micro-gravity microelectronmechanical system sandwich capacitance accelerometer." *Micro & Nano Letters*, pp. 510-514, 2011.

[2] F. Xiao, L. Che, B. Xiong, Y. Wang, X. Zhou, Y. Li, Y. Lin, "A novel capacitive accelerometer with an eight-beam-mass structure by self-stop anisotropic etching of (1 0 0) silicon." *Journal of Micromechanics and Microengineering*, 18 (7), P.075005, 2008.

[3] G. Liu, W. Cui, H. Hu, F. Zhang, Y. Zhang, C. Gao, Y. Hao, "Silicon on insulator pressure sensor based on a thermostable electrode for high temperature applications." *Micro & Nano Letters*, 10 (10), pp.496-499, 2015.

[4] C. Wei, W. Zhou, Q. Wang, X. Xia, X. Li, "TPMS (tire-pressure monitoring system) sensors: Monolithic integration of surface-micromachined piezoresistive pressure sensor and self-testable accelerometer." *Microelectronic Engineering*, 91, pp.167-173, 2012.

[5] Y. Zhang, C. Yang, F. Meng, G. Liu, C. Gao, Y. Hao, "A monolithic integration multifunctional MEMS sensor based on cavity SOI wafer." *IEEE SENSORS*, 2014, pp. 1952-1955.

[6] T. Wang, J. Cap, Q. Wang, H. Zhang, Z. Wang, "Design and realize of 3D integration of a pressure sensor system with through silicon via (TSV) approach." *12th International Conference on Electronic Packaging Technology and High Density Packaging*, 2011, pp. 1-4.

[7] L. E. Murr, "A Metallographic Review of 3D Printing/ Additive Manufacturing of Metal and Alloy Products and Components." *Metallography, Microstructure, and Analysis*, 7 (2), pp.103-132, 2018.

[8] J. Mai, L. Zhang, F. Tao, L. Ren, "Customized production based on distributed 3D printing services in cloud manufacturing." *The International Journal of Advanced Manufacturing Technology*, 84 (1), pp.71-83, 2016.

[9] M. Kreiger, J. M. Pearce, "Environmental Life Cycle Analysis of Distributed Three-Dimensional Printing and Conventional Manufacturing of Polymer Products." *ACS Sustainable Chemistry & Engineering*, 1 (12), pp.1511-1519, 2013.

[10] G. Liu, C. Wang, Z. Jia, K. Wang, "An Integrative 3D printing method for rapid additive manufacturing of a capacitive force sensor." *Journal of Micromechanics and Microengineering* 31.6, p.065005, 2021.

[11] G. Liu, C. Wang, Z. Jia, K. Wang, W. Ma, Z. Li. "A Rapid Design and Fabrication Method for a Capacitive Accelerometer Based on Machine Learning and 3D Printing Techniques." *IEEE Sensors Journal*, 21.16, pp.17695-17702, 2021.

[12] S. A. Bazaz, F. Khan, R. I. Shakoor, "Design, simulation and testing of electrostatic SOI MUMPs based microgripper integrated with capacitive contact sensor." *Sensors and Actuators A: Physical* 167 (1), pp.44-53, 2011.

CONTACT

*Guandong Liu, liuguand@outlook.com
*Changhai Wang, c.wang@hw.ac.uk
*Wei Ma, ma_wei@zju.edu.cn

AUTHOR INDEX

A

Abdolvand, Reza	586, 1183
Abdullah, Amjed	433
Abraham, Ebinesh R.	1049
Adagolodjo, Yinoussa	452
Agarwal, Ajay	665
Agarwal, Bhawana	1049
Agluschewitsch, Vladislav	1037
Aida, Yasuhiro	941, 957
Ainla, Alar	744
Al-Amidie, Muthana Last	433
Alasatri, Suresh	471
Alessandri, Anna	716
Alkorjia, Omar	433
Almasri, Mahmoud	433
Almeida, Sergio	483
Altmann, Robert K.	153
Amaya, Satoshi	57
Andreasen, Sune Zoëga	437
Andrianov, Nikolai	402
Ansari, Azadeh	1186
Anzinger, Sebastian	973
Aoyagi, Seiji	787
Arai, Fumihito	57, 99
Arakawa, Ryuichi	135
Archdeacon, Michael T.	235
Arnow, Hannah	728
Arya, Dhairya Singh	347, 570
Asai, Makoto	319
Asplund, Maria	421
Attanasio, Alaina G.	149
Azadi Keari, Shirin	767
Azadmehr, Mehdi	728

B

Bae, Kyubin	724
Bae, Yuna	651
Baglioni, Gabriele	131
Bahr, Andreas	370

Bahreyni, Behraad 779
Bainschab, Markus 1107, 1123, 1127, 1131
Baker, Colin 554
Banabathi, Naik T. 811
Bancora, Andrea 149
Banerjee, Amit 515
Bang, Dang Duong 437
Bao, Feihong 631
Barniol, Núria 933, 937, 1041
Basavanna, Abhiraj 973
Baumgart, Marcus 1107, 1123
Beigh, Faizan 343
Beigh, Nadeem Tariq 343
Berenschot, Erwin J.W. 639
Bhatta, Trilochan 217
Bhave, Sunil A. 149
Boehler, Christian 421
Borca-Tasciuc, Diana-Andra 728
Borrise, Xavier 1041
Bota, Sebastià 795
Bourouina, Tarik 1119
Bourrier, David 643
Bowen, Chris R. 697
Braun, Stefan 594

C

Cai, Bingyang 837
Cai, Junxiang 413
Calleja, Montserrat 153, 444, 1056
Cano, Álvaro 444
Cao, Entian 1037
Cao, Jiawei 384, 405
Cao, Yuan 1009
Cao, Zhen 103
Carminati, Marta 716
Carminati, Roberto 716
Carranza, Gabriel E. 259
Casilli, Nicolas 511
Cassella, Cristian 161, 511
Cattan, Eric 452
Ceyssens, Frederik 901
Chand, Rakesh 1198
Chang, Chin-Yu 953
Chang, Honglong 705, 740
Chang, Shu-Wei 123, 127

Chao, Yueh-Jung 558

Charlot, Samuel 643

Chau, Kevin 578, 748

Chau, Quy 961

Chen, Changnan 49, 881

Chen, Chihchen 479

Chen, Chun-Ming 135, 957

Chen, Guidong 229

Chen, Hao 157

Chen, Hung-Yu 526

Chen, Jianglong 189

Chen, Jianlin 865, 873

Chen, Lang 243

Chen, Liangqian 849

Chen, Minkan 115

Chen, Pai-Shan 1045, 1052

Chen, Pin-Chuan 1045, 1052

Chen, Qianhuang 606

Chen, Rongshun 562

Chen, Si-Han 720

Chen, Ting-Yi 185, 1151, 1155, 1206

Chen, Weiguo 491

Chen, Xi 606

Chen, Xiao-Wen 1021

Chen, Xing 1068

Chen, Yang 1175

Chen, Yen-Lin 791

Chen, Yi-Sin 479

Chen, Yih-Shurng 425

Chen, Ying 193, 255, 281, 969

Chen, Yu-Chen 685

Chen, Yung-Hsiang 1198

Chen, Yusa 814

Chen, Zhaohan 67, 247

Chen, Zhichao 157

Cheng, Hao-Chien 720

Cheng, Hsu-Hsiang 685

Cheng, Jiangong 157

Cheng, Ming-Ching 685

Cheng, Qian 33, 289

Cheng, Yu-Ting 425

Chidambara, Aaydha Chidambara 437

Chien, Jun-Chau 558

Chien, Tung-Lin 562, 566, 574

Cho, Incheol 799

Cho, Kyungsuk	811
Cho, Sung Kwon	1025
Cho, Wootaek	830
Choi, Kwang-Wook	993
Choi, Pan-Kyu	5
Chou, Tien	985, 1143
Choudhary, Chandrashekhar	235
Chu, Yen-Chang	562
Chung, Seokwhan	993
Cichon, Daniel	1099
Claar, Victor	409
Clausen, Niels	1099
Cleri, Fabrizio	475
Collard, Dominique	475
Colombo, Luca	161, 169
Comini, Elisabetta	251
Conédéra, Véronique	643
Costa, Antonio	795
Cui, Jian	833, 885
Cui, Tianhong	285

D

Da, Zhou	402
Dabas, Shaurya	1179
Dai, Ningxuan	53
Dalal, Mitul	853
Dao, Dzung Viet	366, 1064
Dao, Thang Duy	1139
Das, Samaresh	1033
Dau, Van Thanh	366, 1064
Davaji, Benyamin	161, 511
Decanini, Dominique	21
Degenfeld-Schonburg, Peter	877
Dehé, Alfons	973
Delgado, Rafael	153
Deshpande, Adwait	362
Deval, Piyush	448
Dinh, Thien Xuan	1064
Dinh, Toan	366, 1064
Dobson, Renwick C.J.	25
Dolamore, Fabian	25
Dostanic, Milica	374
Dou, Songtao	323
Dou, Wenkun	29
Du, Xu	99

Du, Zhiyuan 384
Duan, Mingzheng 1095
Duan, Xuexin 945
Duesberg, Georg S. 627
Dufour, Isabelle 643
Duriez, Christian 452

E

Ehrmann, Nils 594
Eovino, Ben 961
Erdale, Zeynep 433
Erkan, Derin 909
Esatu, Tsegereda K. 483
Escobar, Javier E. 1056
Esfahani, Kianoush S. 977

F

Fabian, Johannes 530
Fagnani, Andrea 989
Fahrbach, Michael 1037
Fang, Chi-Te 562
Fang, Mingdong 107
Fang, Weileun 123, 127, 143, 562, 566, 574, 685, 720, 791, 985, 1143
Fang, Xiaoli 177
Fang, Yanyan 841
Faraji, Mohammadmahdi 744
Fasel, Jens 370
Fastier-Woollel, Jarred 366
Favero, Ivan 153
Fee, Conan 25
Feng, Guo-Hua 239, 981
Feng, Philip X.-L. 511, 613, 897
Feng, Yongjian 763
Feng, Zhihong 1159
Fernandez-Cuesta, Irene 1041
Ferrarini, Paolo 716
Finkbeiner, Stefan 1
Fleury, Clement 1107, 1123, 1139
Floehr, Julia 1009
Forke, Roman 522
Frangi, Attilio A. 37, 989
Frigerio, Paolo 989
Fu, Sulei 1190
Fu, Yongqing 705, 740
Fujimoto, Kazuya 301, 309

Fujita, Takayuki 709
Furusawa, Gaku 1171

G

Galarza, Mathew 728
Gandotra, Rishabh 1001
Gang, Min-Ho 5
Gao, Chengchen 756
Gao, Le 837
Gao, Shupeng 606
Gao, Yanze 1147
Gao, Yunfei 115
García-López, Sergio 153, 444
Garg, Manu 347, 570
Garrill, Ashley 997
Gattere, Gabriele 37, 989
Gatti, Daniela A.L. 716
Ge, Yuqing 289
Gelaeschus, Anton 370
Geneiß, Volker 522
Gerbedoen, Jean Claude 475
Gesing, Andre L. 1072
Ghaderi, Erfan 779
Ghazinouri, Behrad 1163
Ghenna, Sofiane 452
Ghosh, Chayanjit 362
Ghosh, Sagnik 491
Gianola, Riccardo 716
Gidts, Michiel 901
Gil-Santos, Eduardo 153, 444, 1056
Giribaldi, Gabriel 169
Goh, Duan Jian 491
Gojdka, Björn 1099
Golabi, Mohsen 437
Golparvar, Ata 335
Gong, Tianjiao 602, 661
González, Pedro 744
Grondel, Sébastien 452
Grundmann, Annika 627
Gu, Chi 429
Gu, Jiebin 49
Gu, Yuandong Alex 413
Guerreiro, Sara 1107, 1123
Gund, Ved 519
Guo, Ruiqi 503

Guo, Wenlan	945
Guo, Xinge	495, 818, 1167
Guo, Yu	1190
Guo, Zhejun	635
Gurung, Lokesh	554

H

Hagiwara, Masaya	17
Haluska, Miroslav	925
Ham, Jimin	625, 651
Han, Jiangli	1068
Han, Xiaodong	763
Haneda, Kotaro	499
Hao, Yilong	756
Harouri, Abdelmounaim	21
Hartwig, Oliver	627
Hashimoto, Izumi	279
Hayakawa, Takeshi	17, 80
Hayashi, Tomohiro	1091
He, Kaixuan	845
He, Siyuan	1163
He, Wenzheng	285
He, Ying	433
He, Zongxing	205
Hedayat, Christian	522, 1099
Heeg, Jan	822
Hella, Mona M.	728
Heuken, Michael	627
Hierold, Christofer	925
Hiller, Karla	602
Hilleringmann, Ulrich	522
Hiraga, Hiroki	861
Hirayama, Takuto	760
Hirotani, Jun	515
Ho, Hsin-Ying	448
Ho, Yens	1198
Hoffmann, Martin	522
Hoffmann, Max	822
Hohe, Hans-Peter	1099
Holzmann, Dominik	1107, 1123
Hong, Sukjoon	625
Hong, Wen	405
Horsley, David A.	961
Hoshino, Ayuko	319
Hosokawa, Yoichiroh	1013

Hosseini-Pishrobat, Mehran 869
Hou, Hung-Yu 558
Hsiai, Tzung 53
Hsieh, I-Chieh 1155
Hsu, Tzu-Hsuan 1194, 1198
Hsu, Wei-Fan 901
Hu, Fangjing 841, 917
Hu, Senyong 814
Hu, Xiaoer 483
Hu, Yongbo 705
Hu, Yuqiang 157
Hu, Yushen 13, 259, 826
Hu, Zih-Song 127, 985
Hu, Zihsong 791
Hua, Chen 712
Huang, Chengjun 229, 582
Huang, Chung-Hao 981
Huang, Kai 1175
Huang, Lifeng 107
Huang, Linhai 1159
Huang, Nien-Tsu 1021
Huang, Siwei 107
Huang, Syuan-Rong 1021
Huang, Tony Jun 71
Huang, Wei-Chen 392
Huang, Wenjing 417
Huang, Xiaodong 736
Huang, Yi-Hsuan 487
Huang, Yu-Cheng 1143
Huang, Yuan 263
Huang, Yuanyuan 562, 566, 574, 791
Huang, Yuheng 771
Hulsey, Robert A. 433
Hunter, Gary L. 433
Huynh, Van Ngoc 437
Hwang, Gilgueng 21
Hwang, Jeonghyeon 327
Hyleme, George S. 433

I

Ichikawa, Keita 417
Ichiki, Masaaki 41
Ikeuchi, Shinsuke 941, 957
Ikezawa, Satoshi 1111
Irisa, Taiga 309

Ishida, Tadashi 1091
Ishihara, Daisuke 534
Islam, Sayemul 396
Isono, Yoshitada 701
Itai, Shun 305, 319, 456
Itawi, Ahmad 452
Ito, Motoki 460
Iwami, Kentaro 1111
Iwase, Eiji 221, 275, 339
Izhar 107, 1095
Izyumin, Oleg I. 961

J

Jackson, Nathan 467, 617
Jadhav, Shubham 519
Jafri, Ijaz 853
Jang, Gunyoung 625
Janssens, Yves L. 639
Jeon, Sungho 651
Jeong, Jinwon 693
Jeong, Seonghoon 217, 225, 732
Jeong, SeongHoon 197
Jia, Bin 736
Jia, Hao 119, 281, 1060
Jia, Yueyang 281
Jia, Yuyu 965
Jia, Zhili 598
Jiang, Chunpeng 355
Jiang, Jiaxin 235
Jiang, Wanqi 205
Jiao, Chenyin 189
Jiao, Rui 267, 752
Jin, Qiutong 173
Jin, Shengxiao 814, 1115
Jin, Yufeng 243
Jo, Byeongwook 378, 463
Jo, Eunhwan 45
Johnson, Isaac 315
Joshi, Khanjhan 347
Joshi, Sanjog V. 331

K

Kagawa, Gakuto 610
Kaisar, Tahmid 511
Kalisch, Holger 627

Kam, Hei	483
Kan, Tetsuo	1135, 1171
Kanamori, Yoshiaki	139
Kanda, Kensuke	709
Kanda, Natsuki	1171
Kaneda, Yuki	1135
Kaneko, Shingo	57, 99
Kang, Dongil	91
Kang, Dongwon	811
Kang, Mingu	799, 803
Kang, Minho	651
Kang, Sookyung	811
Kang, Tian	814, 1115
Kant, Rashmikant	534
Kao, Wei-Sin	448
Karkhanis, Mohit U.	362
Kataria, Satender	627
Kato, Aoi	313
Kaur, Navpreet	251
Kaya, Onurcan	161, 511
Khalil, Diaa	1119
Khan, Muhammad Jehanzeb	602, 661
Kim, Albert	396
Kim, Beomjoon	21
Kim, Dongjun	1029
Kim, Dongkyun	197
Kim, Eun S.	977
Kim, Hanseup	362, 657
Kim, Ilhwan	993
Kim, Jongbaeg	45, 724
Kim, Joonwon	327, 1080
Kim, Jungkwun JK	396
Kim, Jungwook	811
Kim, Sangmok	327
Kim, Su-Hyun	5
Kim, Tae-Soo	5, 76
Kim, Taejung	830
Kim, Taesung	1029
Kim, Yong-Jun	88
Kippenberg, Tobias J.	149
Kiya, Ryota	1013
Kline, Mitchell H.	961
Ko, Juhee	111
Kobayashi, Takeshi	359, 400
Koh, Yul	491

Koike, Yuha 17
Kong, David 53
Kooijman, Lucas J. 639
Kosaka, Priscila M. 153, 444, 1056
Kößl, Bernhard 471
Kounadis, Diamantis 153, 444
Kozasa, Mitsuki 775
Kraft, Michael 331, 865, 901, 917
Kraiem, Ines 627
Kruszewski, Alexandre 452
Kuhl, Matthias 370
Kuhn, Harald 522
Kulsreshath, Mukesh K. 507
Kumar, Manjeet 665
Kumar, Sushil 347, 570
Kumokita, Yuki 787
Kunwar, Deepak 617
Kuo, Feng-Chih 1001
Kuo, Justin 9
Kurashina, Yuta 305, 319, 1084
Kushwaha, Shashwat 901
Kwak, Jong-Hyun 830

L

Lagadec, Chann 475
Lagosh, Anton 1107, 1123, 1127, 1131
Lai, Mei-Feng 1143
Lal, Amit 9, 491, 519
Langfelder, Giacomo 873, 989
Lathia, Rutvik 1049, 1088
Lazarova, Borka 716
Lazzari, Carla M. 716
Ledesma, Eyglis 933, 937
Lee, Bong Jae 111
Lee, Cathy 961
Lee, Cheng-Hsun 95
Lee, Chengkuo 495, 818, 1167
Lee, Dong-Weon 61
Lee, Gwo-Bin 479, 1001
Lee, Hojoon 45
Lee, Jae-Ik 45
Lee, Jaehoon 977
Lee, Jaesung 613
Lee, Jeong Bong (JB) 693
Lee, Jongwan 1029

Lee, Joshua E.-Y. 491
Lee, Jungchul 111
Lee, Kichul 803
Lee, Mel S. 1001
Lee, Paul 440
Lee, Seung-Jun 76
Lee, Seyeon 811
Lee, So-Young 5, 76
Lee, Sueng Yoon 625, 651
Lee, Won Chul 625, 651
Lee, Wonhyung 1080
Lee, Wonjun 91
Lee, Ya-Chu 562
Lee, Yeonwoo 95
Lee, Yi-Kuen 107, 1095
Lee, Yong-Bok 5, 76
Lei, Tengteng 13
Lei, Wan-Lou 392
Leïchlé, Thierry 643
Lemme, Max C. 627
Lengert, Maren 822
Li, Bei 323
Li, Binghui 771
Li, Chengxin 917
Li, Defang 1147
Li, Dongxiao 1167
Li, Fangzheng 837
Li, Gang 763
Li, Hang 323
Li, Jiaqi 965
Li, Jinglun 728
Li, Jingzhen 763
Li, Junjian 849
Li, Lingyun 157
Li, Liye 814, 1115
Li, Mao 229, 582
Li, Meng 67, 205, 247, 297, 429
Li, Ming 119, 255
Li, Ming-Huang 143, 526, 1194, 1198
Li, Mingjie 893
Li, Na 1005, 1017
Li, Peng 881
Li, Qi 606
Li, Qingsong 849
Li, Sheng-Shian 143, 526, 905, 953

Li, Tao	235
Li, Teng	965
Li, Tingyu	381
Li, Wei	881, 969
Li, Wei-Chang	181, 185, 487, 1151, 1155, 1206
Li, Wenqi	542
Li, Xiaohui	209
Li, Xinxin	49, 119, 193, 255, 351, 881, 969, 1060
Li, Xinyu	193, 255
Li, Xuejiao	267
Li, Yu-Hsuan	574
Li, Yuankai	396
Li, Yung-Chen	1143
Li, Yunjia	740
Li, Zhuo	1147
Lian, Yujia	295
Liang, Jizhi	67, 247
Liang, Kai-Chih	720
Liaw, Shwu-Jen	558
Libaude, Guillaume	643
Lihachev, Grigory	149
Lim, Dohyun	625
Lim, Jaemook	625
Lim, Joowon	625, 651
Lin, Che-Hsin	448
Lin, Chen	921
Lin, Chia-Ying	235
Lin, Cong	1202
Lin, Dequan	578, 748
Lin, Guanzhou	814
Lin, Hung-Yu	720
Lin, Liwei	135, 503, 941, 957
Lin, Pen-Sheng	1143
Lin, Shihwei	791
Lin, Yang	285
Lin, You-An	1143
Lin, Zhong-Wei	905
Lin, Zude	929
Linh, Quyen Than	437
Liou, Ting-Jui	1206
Lisec, Thomas	1099
Liu, Chengze	673, 677
Liu, Chongbin	949
Liu, Chun-You	953
Liu, Guandong	598

Liu, Hai 965
Liu, Hanxiao 135, 941, 957
Liu, Huafeng 917
Liu, Huiliang 84
Liu, Jingquan 355, 384, 405, 635, 712, 783, 929
Liu, Kangfu 669, 969
Liu, Lin 440
Liu, Liying 285
Liu, MengWei 209
Liu, Min 881
Liu, Peisen 1190
Liu, Qihui 157
Liu, Ruichen 263
Liu, Shih-Chi 720
Liu, Song 965
Liu, Tsu-Jae King 483
Liu, Xiaojiang 893
Liu, Yi 1202
Liu, Yuncong 897
Liu, Yunfei 756
Liu, Yuxian 885
Liu, Zewen 84
Liu, Zhenhao 1068
Liu, Zhongyi 267, 752
Liu, Zong 259, 826
Liu, Zuheng 281
Llobet Sixto, Jordi 1041
Lo, Sung-Cheng 123, 685
Locquet, Jean-Pierre 901
Lotters, Joost C. 767
Lou, Liang 115, 413
Lu, Haojian 965
Lu, Yaoqing 669
Lukas, Sebastian 627
Luo, Bin 889
Luo, Ruiqi 598
Luo, Wenxin 893
Luo, Yuan 103
Lyu, Boming 705, 740

M

Ma, Wei 598
Ma, Xiao 323
Ma, Yibo 417
Macho, Matthias 538

Maenaka, Kazusuke	709
Magnet, Ingrid A.M.	471
Maharshi, Vikram	665
Mai, Luan Ngoc	366, 1064
Maillard, Damien	594
Majd, Yasaman	586
Mak, Daniel	25
Mallick, Dhiman	343, 1033
Malvar, Óscar	153, 444, 1056
Manrique Castro, Jorge	315, 586
Manrique Juarez, Dolores	643
Mansoorzare, Hakhamanesh	586, 1183
Mao, Haiyang	229, 582
Mao, Hongju	289, 1005, 1017
Maqsood, Waleed	402
Marschner, David E.	594
Martini, Irene	716
Maruyama, Hisataka	99
Mastrangeli, Massimo	374
Mastrangelo, Carlos H.	362
Masud, Mohammad Ayaz	2
Masuda, Akari	305
Masunishi, Kei	861
Mathieu, Fabrice	643
Matsudaira, Kenei	499
Matsunaga, Ryusuke	1171
Maynes, Jason	29
Mazenq, Laurent	643
Meffan, R. Claude	25
Meng, Ellis	388
Menges, Julian	25
Mercader, Anthony L.	1025
Miani, Theo	554
Miao, Jiahao	1202
Miao, Tongqiao	849
Miki, Norihisa	279, 293, 313
Mimura, Hisatoshi	65, 233, 279, 293, 313
Mingorance, Jesús	153, 444
Mirbakht, Sajjad	335
Mise, Nachi	775
Mita, Yoshio	21
Mitra, Bhaskar	665, 689
Miyazaki, Fumito	861
Mizumoto, Takahiro	515
Mizushima, Ayako	21

Mo, Dicheng 1179
Mo, Jiarui 72, 621
Modak, Chandantaru Dey 1049, 1088
Mohammadi, Ali 697
Molina, Juan 1056
Moll, Philipp 590
Møller, Jens Kjølseth 437
Moridi, Mohssen 402, 1107, 1123, 1139
Morimoto, Yuya 201, 381, 460, 463, 550
Morita, Tomohiro 378
Morita, Yuto 213
Mortada, Bassem 1119
Mortada, Mahdi 530
Motoi, Kentaro 463
Muhsin, Sura A. 433
Mukherjee, Dibyajyoti 343
Müller, Manuel 1041
Mummery, Christine L. 374
Murase, Hideaki 861
Murayama, Tomomi 1084
Mustafa, Muhammad 263

N

Nagpal, Satchit 1088
Nakajima, Hibiki 17
Nakamura, Nagi 221, 275
Nakane, Takuma 293
Nakano, Kyoka 80
Nakao, Kenji 233
Nakashima, Rihachiro 221
Napier, Cole 235
Nasri, Rukan 1041
Naval, Sourav 343
Ng, Eldwin 491
Nguyen, Clark T.-C. 173
Nguyen, Hong-Quan 366
Nguyen, Nam-Trung 1064
Nguyen, Thanh 366, 1064
Nguyen, Tran Minh Giao 452
Nguyen, Trieu 437
Nguyen, Tuan-Hung 366
Nguyen, Tuan-Khoa 366, 1064
Ni, Yue 582
Nicu, Liviu 643
Nie, Dezhi 705

Nie, Meng 771
Nie, Minghao 295, 378, 381
Nie, Ran 669
Niklaus, Frank 594
Ning, Bianca 440
Niroui, Farnaz 653
Nishimura, Akane 681
Nishiyori, Shusuke 701
Niu, Gaoqiang 826, 893, 913
Nock, Volker 25, 997
Noh, Seungbeom 657
Nomellini, Andrea 716
Nowbahari, Arian 728
Nunan, Kieran 853

O

Obispo, Meg 728
Oda, Haruka 295, 313
Odenthal, Marie C. 409
Ogawa, Etsuji 861
Ogawa, Jumpei 861
Oggioni, Laura 716
Ogishi, Kazuto 550
Okada, Hironao 41
Okamoto, Yuki 41
Onishi, Minato 534
Ono, Daiki 861
Ono, Takahito 1103
Onoe, Hiroaki 305, 319, 456, 1084
Osaki, Toshihisa 65, 233, 279, 293, 313, 550
Oshita, Masaaki 1135
Ou, Chuan-Hui 1103
Ou, Xin 177, 1175
Ouro-Koura, Habilou 728
Ouyang, Yifan 1115
Oyunbaatar, Nomin-Erdene 61

P

Pagliano, Simone 594
Pal, Sagnik 853
Pala, Sedat 135
Pamunuwa, Dinesh 507
Pan, Feng 1190
Pan, Pichao 49, 881
Pan, Xiaofang 107

Panagiotopoulos, Ilias 153, 444
Pang, Wei 673, 677, 945
Papanastasiou, Dimitris 153, 444
Park, Inkyu 799, 803
Park, Jae Yeong 197, 217, 225, 732
Park, Jiin 91
Park, Jongha 135
Park, Jungwon 651
Park, Jungyul 811, 1029
Park, Mingyo 1186
Park, Saeyoung 647
Park, Sung-Yong 95
Pathak, Pankaj 1033
Paul, Oliver 409, 421
Pedrini, Claudia 716
Peiner, Erwin 1037, 1147
Peng, Chih-Wei 392
Peng, Yande 135, 941, 957
Perelló-Roig, Rafel 795
Perez-Murano, Francesc 1041
Pezone, Roberto 131
Pfusterschmied, Georg 590
Phan, Hoang-Phuong 1064
Piazza, Gianluca 2
Pinto, Rui M.R. 744
Piot, Adrien 1107, 1123, 1139
Platz, Daniel 530, 1072
Plesse, Cedric 452
Pletka, Ryan J. 433
Pordeli, Yasser 639
Pradeep, Yelehanka Ramac R. 1198
Pradhan, Gagan Bahadur 217, 225, 732
Prechtl, Maximilian 627
Prelini, Carlo L. 716
Pribošek, Jaka 1123, 1127, 1131
Przybyla, Richard J. 961
Puff, Markus 1139

Q

Qamar, Afzaal 897
Qi, Jiali 267
Qi, Longheng 1095
Qi, Zhimei 1115
Qian, Hangyu 631
Qin, Mu 405

Qin, Nan 209, 807
Qiu, Anping 857
Qiu, Shihui 1005, 1017

R

Rahafrooz, Amir 853
Rais-Zadeh, Mina 897
Rajaraman, Swaminathan 315
Rana, S M Sohel 217, 732
Rassay, Sushant 1179
Ravi, Adarsh 9
Recaman Payo, María 901
Reddy, Bheema Sankar 1049, 1088
Redolfi, Sebastian 1139
Ren, Xinzhu 201
Ren, Zhihao 818, 1167
Reynaerts, Dominiek 901
Reza, Md Selim 197
Rezard, Quentin 475
Rhee, Joowon 1080
Riani, Manuel 37
Riemensberger, Johann 149
Rinaldi, Matteo 169
Rintaro, Yoshinaga 1013
Risquez, Sarah 1139
Rizzini, Francesco 37
Rodríguez-Tejedor, María 153, 444
Roman, Cosmin 925
Roshanghias, Ali 1139
Ruan, Tao 384
Ruther, Patrick 409, 421
Ruz, José J. 153, 444, 1056

S

Saadany, Bassam 1119
Sabry, Yasser M. 1119
Sadeghpour, Sina 331
Sadrafshari, Shamin 697
Sagi, H. Claude 235
Sahara, Yoshiki 301
Saito, Hiroki 1111
Saito, Shiro 1135
Sakamoto, Jinya 787
Sakuma, Shinya 271
Salvagnac, Ludovic 643

Samm, Elisabeth	530
San Paulo, Álvaro	153, 444, 1056
Sanders, Remco G.P.	767
Sanghvi, Rohan	9
Sano, Tomohiko G.	221
Sanz-Jiménez, Adrián	153, 444
Sarro, Pasqualina M.	131, 374
Sasaki, Takashi	1107, 1123, 1127, 1131
Sato, Ayane	1091
Sato, Takashi	339
Satterthwaite, Peter F.	653
Sbarra, Samantha	153
Schaller, Falk	522
Schitter, Georg	538
Schiwietz, Daniel	877
Schmid, Ulrich	471, 530, 590, 1072
Schmitt, Philip	522
Schmoltner, Kerstin	1139
Schnakenberg, Uwe	1009
Schneider, Michael	471, 530
Schröder, Dominik	1099
Schroedter, Richard	538
Schween, Oliver	370
Segovia-Fernandez, Jeronimo	165
Segura, Jaume	795
Sekiguchi, Takuma	139
Sen, Prosenjit	1049, 1088
Sentre-Arribas, Elena	153
Seo, Dongwoo	1029
Seo, Min-Ho	647
Sergovia, Karen Last	433
Serrano, Diego Emilio	853
Seshia, Ashwin	554
Shadymov, Vladimir	149
Shaik, Faruk Azam	475
Shang, Jintang	542, 546, 889
Shang, Kuang-Ming	53
Shankar, Shreyas	72
Shao, Shuai	969
Shaporin, Alexey	522
Sharifuzzaman, Md	197, 732
Sharma, Jaibir	491
Sharma, Kirti	421
Sharma, Pallavi	467
Sharma, Sudeep	225

Shaw, Steven W.	613
Shelton, Stefon E.	961
Shen, Gencai	355
Shen, Haixu	53
Shi, Meng	229, 582
Shi, Qin	857
Shi, Runxiao	578
Shi, Zhongyu	263
Shih, Fuchi	566, 574, 791, 1143
Shimokawa, Fusao	213, 775
Shin, Heungjoo	830
Shin, Yoo-Kyum	647
Shiratori, Toshihiro	787
Shokrani, Alborz	697
Siddharth, Anat	149
Sikder, Urmita	483
Sim, Sangjun	724
Simeoni, Pietro	169
Singer, Julian A.	370
Singh, Jujhar	440
Singh, Pushpapraj	347, 570
Snigirev, Viacheslav	149
Soberats, Bartomeu	795
Sobreviela-Falces, Guillermo	554
Song, Cheng	1190
Song, Hye Su	197, 732
Song, Ziliang	405
Soong, Wei-Jen	479
Spector, Sarah O.	653
Stahl-Offergeld, Markus	1099
Steeneken, Peter G.	131
Stemme, Göran	594
Stöckel, Chris	522
Stramm, Till	370
Streque, Jeremy	1139
Su, Rongxuan	1190
Su, Zhaoxi	889
Suetsugu, Ryotaro	534
Sugano, Koji	701
Sugiura, Hirotaka	57, 99
Suh, Bokyung	693
Suh, Seungbeum	91
Sui, Fanping	503, 941
Sun, Baoyun	621
Sun, Chen	945

Sun, Fengpei 1159
Sun, Hongshun 814, 1115
Sun, Jianwen 84
Sun, Ke 351, 881
Sun, Litao 771
Sun, Liuyang 33, 67, 205, 247, 297, 429
Sun, Mingchao 673, 677
Sun, Xiaopeng 845
Sun, Yi 351
Sun, Yiling 997
Sun, Yu 29
Sun, Zhenhuan 965
Sung, Gi-Bong 993
Sung, Wei-Lun 562
Suzuki, Masato 787
Suzuki, Yukio 602, 661

T

Tabrizian, Roozbeh 1179
Tabuchi, Ayumu 301
Tacchini, Riccardo 716
Tai, Yu-Chong 53
Takahashi, Haruna 271
Takahashi, Hidetoshi 221, 499, 610, 760
Takahashi, Tomokazu 787
Takamori, Sho 65, 233, 279, 293, 313
Takao, Hidekuni 213, 775
Takao, Yoshinori 681, 681
Takasato, Minoru 301
Takei, Yusuke 359, 400
Takemura, Hiroki 515
Takeshita, Toshihiro 359, 400
Takeuchi, Shoji 65, 201, 233, 279, 293, 295, 313, 378, 381, 460, 463, 550
Tamayo, Javier 153, 444, 1056
Tan, Jun Ying 396
Tanaka, Shuji 139, 602, 661, 865, 873
Tanaka, Shuma 456
Tanaka, Yo 1013
Tang, Gongbin 631
Tang, Qi 507
Tang, Qiankai 889
Tang, Yuan-Sin 425
Tang, Yue 752
Tao, Chen 205
Tao, Kai 705, 740

Tao, Tiger H.	**33, 67, 205, 209, 247, 297, 429, 807**
Tarhan, Mehmet C.	**475**
Tas, Niels R.	**639**
Tatar, Erdinc	**869, 909**
Tatum, Lars P.	**483**
Tavakkoli, Hadi	**1095**
Tayagui, Ayelen	**997**
Teng, Megan	**135**
Terao, Kyohei	**213, 775**
Teuber, Jeremy	**1041**
Thakkar, Pooja	**1127, 1131**
Thielen, Brianna	**388**
Tian, Hao	**149**
Tian, Ji'ao	**841**
Tian, Xuedi	**177**
Tian, Ye	**33**
Tian, Yuxin	**285**
Tichy, John A.	**728**
Tiggelaar, Roald M.	**639**
Toan, Nguyen V.	**1103**
Toh, Wei Da	**491**
Tohyama, Shugo	**305**
Tomizawa, Yasushi	**861**
Tong, Xing	**728**
Tope, Sayali	**657**
Torres Canals, Francesc	**1041**
Torres, Francesc	**933**
Tottori, Naotomo	**271**
Tourrel, Guillaume	**452**
Tran, Canh-Dung	**366, 1064**
Tran, Dang	**366**
Tran, Thomas	**1072**
Travnik, Aleš	**1107, 1123**
Tsai, Chia-Hsien	**1194**
Tsai, Chun-Pu	**181, 185, 487, 1155, 1206**
Tsai, Hsiao-En	**425**
Tsao, Peggy	**957**
Tsao, Pei-Chi	**135**
Tsuchiya, Toshiyuki	**515, 681**
Tsukamoto, Takashiro	**661, 865, 873**
Tu, Kejun	**405, 635**
Tu, Liangcheng	**837, 841**
Tung, Shao-Siang	**1198**
Turan, Bilal	**57**

U

Uchida, Kengo 861
Uesugi, Akio 701
Umar, Muhammad 335
Umezawa, Seiji 941, 957
Uranga, Arantxa 933, 937
Uzunoglu, Baha Erim 869

V

van der Heiden, Maurits 153
van Driel, Willem D. 621
van Meer, Berend J. 374
van Zeijl, Henk W. 621
Vazquez, Irma Rocio 467
Verd, Jaume 795
Verma, Satish K. 689
Vescan, Andrei 627
Veske, Tolga 909
Vidal, Frédéric 452
Villanueva, Luis Guillermo 594
Vimercati, Michele 716
Vinayakumar, K.B. 744
Vo, Tuan N.A. 1045
Vollebregt, Sten 72, 131
Vollmann, Morten 925
Voloshin, Andrey 149
Vu, Trung-Hieu 366, 1064

W

Waag, Andreas 1037
Wada, Hiroki 80
Wan, Xiu-Feng 433
Wang, Baosheng 413
Wang, Changhai 598
Wang, Chen 901, 917
Wang, Chih-Hung 479
Wang, Cong 1029
Wang, Daying 103
Wang, Fang 351
Wang, Fei 259, 826, 893, 913
Wang, Han 33
Wang, Huan 323
Wang, Jiachou 881
Wang, Jiachuang 209

Wang, Jie 756
Wang, Kexin 598
Wang, Linlin 917
Wang, Longchun 384, 405
Wang, Luming 281
Wang, Man 578
Wang, Ning 103
Wang, Peng 849
Wang, Qian 841
Wang, Rui N. 149
Wang, Ruoqin 267, 752
Wang, Wei 243, 323
Wang, Wenduo 783
Wang, Xiangyang 949
Wang, Xiaoyi 267, 752
Wang, Xin 1147
Wang, Xiner 67, 205, 247
Wang, Xiong 413
Wang, Xueying 33, 297
Wang, Yang 157, 433, 921
Wang, Yiwei 413
Wang, Yuan 917
Wang, Yudong 263
Wang, Yunong 613
Wang, Yuntong 323
Wang, Yuxi 669
Wang, Zenghui 189, 281
Wang, Zetian 243
Wang, Zhuangzhuang 355
Wang, Ziji 542, 546
Waquier, Louis 153
Wasisto, Hutomo Suryo 973
Wei, Jian 1076
Wei, Shuai 807
Wei, Ting-Chou 123, 127, 685
Wei, Xiaoling 33, 67, 205, 247, 297, 429
Weidner, Michael H. 822
Weig, Eva M. 877
Weng, Wei-Yang 558
Wiegerink, Remco J. 767
Wiendels, Maury 374
Wienecke, Marion 822
Windt, Laura M. 374
Wittemeier, Steffen 522
Wolff, Anders 437

Wong, Man 13, 259, 748
Worsey, Elliott 507
Wu, Cheng-Yen 905
Wu, Feng 263
Wu, Guan-Lin 1194
Wu, Guoqiang 949
Wu, Jinbo 177, 1175
Wu, Junming 546
Wu, Junqiao 483
Wu, Lang 913
Wu, Lixiang 402
Wu, Mingching 685, 720
Wu, Sheng 881, 969
Wu, Shuxian 631
Wu, Tao 177, 413, 669, 969
Wu, Wengang 814, 1115
Wu, Wenjie 837, 841
Wu, Xuezhong 845, 849
Wu, Yi-Xin 1052
Wu, Zhenhua 1005, 1017
Wu, Zhenyu 157
Wu, Zhipeng 115
Wu, Zonglin 631
Wulf, Matthias 1139

X

Xi, Jingqian 917
Xi, Xiang 849
Xi, Ye 384
Xia, Fan 135, 941
Xia, Guoming 857
Xiao, Dingbang 845, 849
Xiao, Yuhao 1159
Xie, Dongcheng 263
Xie, Fei 157
Xie, Huikai 267, 752
Xie, Maosong 281
Xie, Yong 949
Xing, Chong 263
Xing, Xiaoxing 1076
Xing, Yan 606
Xu, Bo 189
Xu, Cheng 818
Xu, Chengjian 33
Xu, Feihong 247

Xu, Feng 631
Xu, Han 243
Xu, Huiping 1190
Xu, Jinghui 1159
Xu, Jiushuai 1037, 1147
Xu, Lei 263, 917
Xu, Liangge 818
Xu, Linbing 673
Xu, Mengfei 384, 405, 635
Xu, Pengcheng 119, 193, 255
Xu, Qingda 384, 405
Xu, Qingmei 323
Xu, Wei 107
Xu, Xiaochao 841
Xue, Ying 323
Xun, Lingxiao 452

Y

Yabuuchi, Kensuke 301
Yadav, Vinit K. 1033
Yalikun, Yaxiaer 1013
Yamada, Genki 213
Yamagata, Chisaki 319
Yamaguchi, Masakazu 1111
Yamanishi, Yoko 271, 417
Yamaoka, Utana 1135
Yamashita, Yu 271
Yanez, Jesús 937
Yang, Bin 384, 405, 635, 712, 783, 929
Yang, Gai 267, 752
Yang, Hao 744
Yang, Heng 351
Yang, Huiran 205, 297, 429
Yang, Lujia 841
Yang, Rui 281
Yang, Shijia 103
Yang, Suhui 1147
Yang, Tung-Lin 425
Yang, Xiaopeng 673
Yang, Zhenchuan 756
Yao, Hulin 177, 1175
Yapici, Murat Kaya 335
Ye, Xiongying 285
Ye, Yifei 33
Yeh, Chun-Chen 1194

Yeh, Hung-Yu 239
Yen, Da-Jen 1143
Yen, Ernest T.-T. 165
Yin, Kuibo 771
Yokokawa, Ryuji 301, 309
Yokota, Takahito 709
Yokoyama, Yoshiyuki 17, 80
Yoo, Dongwoo 327
Yoo, Han Woong 538
Yoo, Seong-Jae 88
Yook, Se-Jin 993
Yoon, Jun-Bo 5, 76
Yoshida, Koki 1084
Yoshida, Shinya 139
You, Minmin 929
Young, Douglas 554
Younkin, Duane 853
Yousuf, Mujeeb 347
Yousuf, S M Enamul Hoque 613, 897
Yu, Haitao 193
Yu, Hongyu 267, 752
Yu, Lei 845
Yu, Ling-Shan 448
Yu, Xiaomei 1202
Yu, Zhenyi 1190
Yuan, Chao 1186
Yuan, Weizheng 740
Yue, Wei 135, 503, 941, 957
Yue, Xiawei 807

Z

Zahed, Md Abu 197, 732
Zaki, Muhammad Faizul 1052
Zamora, Iván 933, 937
Zappa, Dario 251
Zega, Valentina 37, 989
Zeng, Fei 1190
Zhai, Yanfen 402, 1107, 1123
Zhang, Bohan 205, 297, 429
Zhang, Chao 736
Zhang, Chenchen 582
Zhang, Dacheng 885
Zhang, Guoqi 72, 621
Zhang, Haitao 945
Zhang, Haochen 263

Zhang, Haoran	1060
Zhang, Haozhi	119
Zhang, Hemin	621, 865
Zhang, Jian	705, 740, 837
Zhang, Jianfeng	542
Zhang, Jin	157, 542, 546
Zhang, Jinwen	921
Zhang, Jinying	1147
Zhang, Kuikui	33
Zhang, Lejia	413
Zhang, Liping	177, 1175
Zhang, Meixuan	243
Zhang, Menglun	673, 677, 945
Zhang, Pengcheng	281
Zhang, Pingping	209, 807
Zhang, Qifu	285
Zhang, Shibin	177, 1175
Zhang, Shipeng	217
Zhang, Shuai	1190
Zhang, Xin	783
Zhang, Xinyuan	267
Zhang, Yan	263
Zhang, Yao	491
Zhang, Yi	323
Zhang, Yonggui	157
Zhang, Yucheng	402
Zhang, Yulong	84
Zhang, Yuyao	157
Zhang, Ziqi	503
Zhao, Chun	841, 917
Zhao, Jianlong	103, 289, 1005
Zhao, Lurui	965
Zhao, Nan	355
Zhao, Ning	783
Zhao, Qiancheng	833, 885
Zhao, Xiaomeng	177, 1175
Zhao, Xu	1095
Zhao, Yang	857
Zhen, Liyun	712
Zheng, Gang	452
Zheng, Hong-Sen	487, 1155
Zheng, Kevin H.	173
Zheng, Pengcheng	177, 1175
Zheng, Yue	1186
Zhou, Changdong	285

Zhou, Cunkai	33
Zhou, Hang	309
Zhou, Hong	818, 1167
Zhou, Lin	289
Zhou, Min	1175
Zhou, Na	229, 582
Zhou, Xiaoyong	107
Zhou, Xin	845
Zhou, Yangchao	945
Zhou, Yufan	255
Zhou, Zhanxuan	1202
Zhou, Zhitao	33, 67, 205, 247, 297, 429, 807
Zhu, Jiankai	189
Zhu, Linqian	889
Zhu, Xiantao	929
Zhu, Zhezheng	756
Zhu, Ziyi	205
Zhuang, Yi	826, 913
Zimmermann, Sven	522
Zou, Dujuan	205
Zou, Jie	631
Zymelka, Daniel	359, 400

IEEE
445 Hoes Lane
Piscataway, NJ 08854-4141

ISBN 978-1-6654-9309-3

2023 IEEE 36th International Conference on Micro Electro Mechanical Systems (MEMS 2023)

Munich, Germany
15-19 January 2023

Pages 602-1209

IEEE Catalog Number: CFP23MEM-POD
ISBN: 978-1-6654-9309-3

2023 IEEE 36th International Conference on Micro Electro Mechanical Systems (MEMS 2023)

Munich, Germany
15-19 January 2023

Pages 602-1209

IEEE Catalog Number: CFP23MEM-POD
ISBN: 978-1-6654-9309-3

**Copyright © 2023 by the Institute of Electrical and Electronics Engineers, Inc.
All Rights Reserved**

Copyright and Reprint Permissions: Abstracting is permitted with credit to the source. Libraries are permitted to photocopy beyond the limit of U.S. copyright law for private use of patrons those articles in this volume that carry a code at the bottom of the first page, provided the per-copy fee indicated in the code is paid through Copyright Clearance Center, 222 Rosewood Drive, Danvers, MA 01923.

For other copying, reprint or republication permission, write to IEEE Copyrights Manager, IEEE Service Center, 445 Hoes Lane, Piscataway, NJ 08854. All rights reserved.

*** This is a print representation of what appears in the IEEE Digital Library. Some format issues inherent in the e-media version may also appear in this print version.

IEEE Catalog Number: CFP23MEM-POD
ISBN (Print-On-Demand):   978-1-6654-9309-3
ISBN (Online):   978-1-6654-9308-6
ISSN: 1084-6999

Additional Copies of This Publication Are Available From:

Curran Associates, Inc
57 Morehouse Lane
Red Hook, NY 12571 USA
Phone: (845) 758-0400
Fax: (845) 758-2633
E-mail: curran@proceedings.com
Web: www.proceedings.com

TABLE OF CONTENTS

Monday, 16 January
All times are Central European Time (CET).

Welcome Address

08:00 **MEMS 2023 Conference Chairs**
Núria Barniol, *Universitat Autonoma de Barcelona, SPAIN*
Franz Lärmer, *Robert Bosch GmbH, GERMANY*

• IEEE Fellows Recognition in the Field of MEMS/NEMS
• IEEE Electron Devices Society Robert Bosch
Micro and Nano Electro Mechanical Systems Award

08:35 **IEEE Electron Devices Society Robert Bosch**
Micro and Nano Electro Mechanical Systems Award Recipient
John H. (Hal) Jerman will accept on behalf of the Gas Chromatograph on a Chip Project.

Plenary Presentation I

08:50 **FROM ETCH TO EDGE AI:**
OPENING NEW HORIZONS WITH SMART SENSOR TECHNOLOGIES 1
Stefan Finkbeiner
Bosch Sensortec GmbH, GERMANY

Session I - Novel MEMS/NEMS Devices for Computing/Imaging

09:35 **SUB-300 MILLIVOLT OPERATION IN NONVOLATILE 300 NM X 100 NM**
PHASE CHANGE NANOELECTROMECHANICAL SWITCH 2
Mohammad Ayaz Masud and Gianluca Piazza
Carnegie Mellon University, USA

09:50 **A FAST AND ENERGY-EFFICIENT NANOELECTROMECHANICAL**
NON-VOLATILE MEMORY FOR IN-MEMORY COMPUTING 5
Yong-Bok Lee[1], Min-Ho Gang[2], Pan-Kyu Choi[1], Su-Hyun Kim[1],
Tae-Soo Kim[1], So-Young Lee[1] and Jun-Bo Yoon[1]
[1]Korea Advanced Institute of Science and Technology (KAIST), KOREA and
[2]National NanoFab Center (NNFC), KOREA

10:05 **TOWARDS ULTRA-HIGH SPATIAL RESOLUTION SENSING OF GHZ**
ULTRASOUND USING STRAIN MODULATION OF FIELD EFFECT TRANSISTORS 9
Rohan Sanghvi[1], Justin Kuo[2], Adarsh Ravi[1], and Amit Lal[1]
[1]Cornell University, USA and [2]Geegah Inc., USA

10:20 **A TACTILE SENSOR ARRAY WITH A MONOLITHICALLY**
INTEGRATED NEURAL NETWORK FOR EDGE COMPUTATION 13
Tengteng Lei, Yushen Hu, and Man Wong
Hong Kong University of Science and Technology, HONG KONG

10:35   Break & Exhibit Inspection

Session II - BioMEMS I

11:05   EVALUATION OF LOCAL AND INTERNAL ELASTICITY OF HYDROGEL MATERIALS BY USING LIGHT-DRIVEN GEL ACTUATOR 17
Hibiki Nakajima[1], Yuha Koike[1], Yoshiyuki Yokoyama[2], Masaya Hagiwara[3], and Takeshi Hayakawa[1]
[1]Chuo University, JAPAN, [2]Toyama Industrial Technology Research and Development Center, JAPAN, and [3]RIKEN, JAPAN

11:20   3D PRINTED MINIATURIZED SOFT MICROSWIMMER FOR MULTIMODAL 3D AIR-LIQUID NAVIGATION AND MANIPULATION 21
Dominique Decanini[1], Abdelmounaim Harouri[1], Ayako Mizushima[2], Beomjoon Kim[2], Yoshio Mita[2], and Gilgueng Hwang[1,2]
[1]Paris-Saclay University, FRANCE and [2]University of Tokyo, JAPAN

11:35   SELF-DRIVEN CAPILLARIC VISCOMETER FOR DIRECT OR CASCADED BAR GRAPH READ-OUT OF RELATIVE SAMPLE VISCOSITY 25
Daniel Mak[1], R. Claude Meffan[1,2], Julian Menges[1], Fabian Dolamore[1], Conan Fee[1], Renwick C.J. Dobson[1], and Volker Nock[1]
[1]University of Canterbury, NEW ZEALAND and [2]Kyoto University, JAPAN

11:50   A FLEXIBLE BIOSENSING PLATFORM FOR HIGH-THROUGHPUT MEASUREMENT OF CARDIOMYOCYTE CONTRACTILITY 29
Wenkun Dou[1], Jason Maynes[2], and Yu Sun[1]
[1]University of Toronto, CANADA and [2]Hospital for Sick Children, CANADA

12:05   FLEXIBLE BI-DIRECTIONAL BRAIN COMPUTER INTERFACE FOR CONTROLLING TURNING BEHAVIOR OF MICE 33
Yifei Ye[1], Ye Tian[1,2], Han Wang[1], Qian Cheng[1], Kuikui Zhang[1], Xueying Wang[1,2], Cunkai Zhou[1], Chengjian Xu[1], Xiaoling Wei[1,2], Zhitao Zhou[1,2], Tiger H. Tao[1,2,3,4,5,6], and Liuyang Sun[1,2]
[1]Chinese Academy of Sciences, CHINA, [2]University of Chinese Academy of Sciences, CHINA, [3]ShanghaiTech University, CHINA, [4]Neuroxess Co., Ltd. (Jiangxi), CHINA, [5]Guangdong Institute of Intelligence Science and Technology, CHINA, and [6]Tianqiao and Chrissy Chen Institute for Translational Research, CHINA

12:20   Lunch & Exhibit Inspection

Session III - MEMS Inertial Sensors and Power MEMS

13:45   HIGH SENSITIVITY MEMS Z-AXIS ACCELEROMETER WITH IN-PLANE DIFFERENTIAL READOUT 37
Valentina Zega[1], Gabriele Gattere[2], Manuel Riani[2], Francesco Rizzini[2], and Attilio Frangi[1]
[1]Politecnico di Milano, ITALY and [2]STMicroelectronics, ITALY

14:00   TWO-AXIS ELECTROMAGNETIC SCANNER INTEGRATED WITH AN ELECTROSTATIC XY-STAGE POSITIONER 41
Yuki Okamoto, Hironao Okada, and Masaaki Ichiki
National Institute of Advanced Industrial Science and Technology (AIST), JAPAN

14:15   MEMS SHOCK ABSORBERS INTEGRATED WITH AL_2O_3-REINFORCED, MECHANICALLY RESILIENT NANOTUBE ARRAYS 45
Hojoon Lee[1], Eunhwan Jo[1], Jae-Ik Lee[2], and Jongbaeg Kim[1]
[1]Yonsei University, KOREA and [2]Harvard Medical School, USA

14:40   **HIGH-INDUCTANCE-DENSITY MEMS 3D-SOLENOID TRANSFORMERS WITH INSERTED THIN-FILM FERRITE MAGNETIC CORE FOR ON-CHIP INTEGRATED DC-DC POWER CONVERSIONS** 49
Changnan Chen[1,2], Pichao Pan[1,2], Jiebin Gu[1,2], and Xinxin Li[1,2]
[1]*Chinese Academy of Sciences, CHINA and* [2]*University of Chinese Academy of Sciences, CHINA*

Poster/Oral Session I

14:45   **Poster/Oral Session I**
Poster presentations are listed by topic category with their assigned number starting on Page 13.

16:15   **Break & Exhibit Inspection**

MEMS Community Announcement

16:45   Clark T.-C. Nguyen, *University of California, Berkeley, USA*

Session IV - BioMEMS II

16:50   **MICRON-SIZED PARYLENE-IN-OIL WATER PROTECTION LAYER** 53
Kuang-Ming Shang[1], Haixu Shen[1], Ningxuan Dai[1], David Kong[1,2], Tzung Hsiai[3], and Yu-Chong Tai[1]
[1]*California Institute of Technology, USA,* [2]*Harvard University, USA, and*
[3]*University of California, Los Angeles, USA*

17:05   **A PIPETTE TIP INTEGRATED WITH A CAPACITIVE MICROSENSOR FABRICATED BY COMBINED 3D PRINTING AND MEMS PROCESS FOR CELL DETECTION AND TRANSPORTATION** 57
Satoshi Amaya, Hirotaka Sugiura, Bilal Turan, Shingo Kaneko, and Fumihito Arai
University of Tokyo, JAPAN

17:20   **FOLDABLE POLYMER STENT INTEGRATED WITH WIRELESS PRESSURE SENSOR FOR BLOOD PRESSURE MONITORING** 61
Nomin-Erdene Oyunbaatar and Dong-Weon Lee
Chonnam National University, KOREA

17:35   **A DYNAMIC MICROARRAY DEVICE FOR SELECTIVE PAIRING AND ELECTROFUSION OF LIPOSOMES** 65
Sho Takamori[1], Hisatoshi Mimura[1], Toshihisa Osaki[1], and Shoji Takeuchi[1,2]
[1]*Kanagawa Institute of Industrial Science and Technology, JAPAN and* [2]*University of Tokyo, JAPAN*

17:50   **REAL-TIME FUNCTIONAL BRAIN MAPPING BASED ON HIGH-CHANNEL-COUNT, ULTRA-CONFORMAL NEURAL INTERFACE** 67
Xiner Wang[1,2], Zhaohan Chen[3], Jizhi Liang[1,2], Xiaoling Wei[1,2], Liuyang Sun[1,2],
Meng Li[1,2], Zhitao Zhou[1,2], and Tiger H. Tao[1,2,4,5,6]
[1]*Chinese Academy of Sciences, CHINA,* [2]*University of Chinese Academy of Science, CHINA,*
[3]*Shanghai Normal University, CHINA,* [4]*Neuroxess Co., Ltd. (Jiangxi), CHINA,*
[5]*Guangdong Institute of Intelligence Science and Technology, CHINA, and*
[6]*Tianqiao and Chrissy Chen Institute for Translational Research, CHINA*

18:05   **Adjourn for the day**

Tuesday, 17 January

All times are Central European Time (CET).

Plenary Presentation II

08:30 **ACOUSTOFLUIDICS: MERGING ACOUSTICS AND FLUID MECHANICS FOR BIOMEDICAL APPLICATIONS** 71
Tony Jun Huang
Duke University, USA

Session V - New Materials, Fabrication, and Packaging

09:15 **SILICON CARBIDE REINFORCED VERTICALLY ALIGNED CARBON NANOTUBE COMPOSITE FOR HARSH ENVIRONMENT MEMS** 72
Jiarui Mo, Shreyas Shankar, Guoqi Zhang, and Sten Vollebregt
Delft University of Techonology, NETHERLANDS

09:30 **A RELIABLE RELEASE METHOD FOR A BACK-END-OF-LINE NEMS SWITCH OF A MONOLITHIC THREE-DIMENSIONAL INTEGRATED CMOS-NEMS CIRCUIT** 76
Tae-Soo Kim, Yong-Bok Lee, So-Young Lee, Seung-Jun Lee, and Jun-Bo Yoon
Korea Advanced Institute of Science and Technology (KAIST), KOREA

09:45 **INCREASE OF EXPANSION RATE AND DIRECTION CONTROL OF MICROGEL ACTUATORS FOR SINGLE CELL MANIPULATIONS** 80
Kyoka Nakano[1], Hiroki Wada[1], Yoshiyuki Yokoyama[2], and Takeshi Hayakawa[1]
[1]Chuo University, JAPAN and [2]Toyama Industrial Technology Research and Development Center, JAPAN

10:00 **GENERALIZED-ACCUMULATED-TEMPERATURE PARAMETER FOR CHARACTERISTIC PREDICTION OF METAL-BASED MEMS CANTILEVER** 84
Yulong Zhang[1], Jianwen Sun[1], Huiliang Liu[2], and Zewen Liu[1]
[1]Tsinghua University, CHINA and [2]China Academy of Space Technology, CHINA

10:15 **Break and Exhibit Inspection**

Session VI - Micro- and Nanofluidics and Medical Applications

10:45 **MEMS-BASED WATER COLLECTION CONDENSATION PARTICLE COUNTER (WCCPC) OPTIMIZED FOR MULTI-POINT MONITORING OF AIRBORNE NANOPARTICLES** 88
Seong-Jae Yoo and Yong-Jun Kim
Yonsei University, KOREA

11:00 **RECONSTITUTING FUNDAMENTALS OF BACTERIA MEDIATED CANCER THERAPY ON A CHIP** 91
Wonjun Lee[1], Jiin Park[2], Dongil Kang[3], and Seungbeum Suh[4]
[1]Seoul National University, KOREA, [2]Ewha Womans University, KOREA, [3]Hanyang University, KOREA, and [4]Korea Institute of Science and Technology (KIST), KOREA

11:15 **3D SPATIAL FOCAL CONTROL BY ARRAYED OPTOFLUIDIC PRISMS** 95
Cheng-Hsun Lee, Yeonwoo Lee, and Sung-Yong Park
San Diego State University, USA

11:30 **HIGH-SPEED AND PINPOINT LIQUID EXCHANGE ON MICROFLUIDIC CHIP USING 3D PRINTED DOUBLE-BARRELED MICROPROBE WITH DUAL PUMPS** 99
Xu Du[1], Shingo Kaneko[2], Hisataka Maruyama[1], Hirotaka Sugiura[2], and Fumihito Arai[1,2]
[1]Nagoya University, JAPAN and [2]University of Tokyo, JAPAN

11:45 DESIGN OF A DNA SYNTHESIS CHIP FOR DATA STORAGE WITH ULTRA-HIGH
THROUGHPUT AND DENSITY FEATURING LARGE-SCALE INTEGRATED
CIRCUITS AND MICROFLUIDIC CONFINEMENT 103
Ning Wang[1,2,3], Shijia Yang[1,3], Dayin Wang[1,2,3], Zhen Cao[4], Yuan Luo[1,3], and Jianlong Zhao[1,3]
[1]Chinese Academy of Sciences, CHINA, [2]ShanghaiTech University, CHINA,
[3]University of Chinese Academy of Sciences, CHINA, and [4]Zhejiang University, CHINA

MEMS 2024 Announcement

12:00 **MEMS 2024 Conference Chairs**
Wen Li, *Michigan State University, USA*
Dana Weinstein, *Purdue University, USA*

12:15 **Lunch & Exhibit Inspection**

Session VII - MEMS Fluidic Sensors

13:15 A REAL-TIME WIRELESS CALORIMETRIC FLOW SENSOR SYSTEM
WITH A WIDE LINEAR RANGE FOR LOW-COST RESPIRATORY MONITORING 107
Lifeng Huang[1], Izhar[2,4], Xiaoyong Zhou[3], Mingdong Fang[3], Siwei Huang[1],
Yi-Kuen Lee[2], Xiaofang Pan[1], and Wei Xu[1]
[1]Shenzhen University, CHINA, [2]Hong Kong University of Science and Technology, CHINA,
[3]Mindray Medical International Limited, CHINA, and [4]University of Pennsylvania, USA

13:30 ADVANCED THERMOPHYSICAL PROPERTIES MEASUREMENTS
USING HEATER-INTEGRATED FLUIDIC RESONATORS 111
Juhee Ko, Bong Jae Lee, and Jungchul Lee
Korea Advanced Institute of Science and Technology (KAIST), KOREA

13:45 A MINIATURIZED TRANSIT-TIME ULTRASONIC FLOWMETER USING
PMUTS FOR LOW-FLOW MEASUREMENT IN SMALL-DIAMETER CHANNELS 115
Yunfei Gao[1,2], Zhipeng Wu[2], Minkan Chen[2], and Liang Lou[1,2]
[1]Shanghai University, CHINA and [2]Shanghai Industrial µ Technology Research Institute, CHINA

14:00 MEMS DIFFERENTIAL THERMOPILES FOR HIGHLY-SENSITIVE
HYDROGEN GAS DETECTION 119
Haozhi Zhang[1,2], Hao Jia[1,2], Ming Li[1,2], Pengcheng Xu[1,2], and Xinxin Li[1,2]
[1]Chinese Academy of Sciences, CHINA and [2]University of Chinese Academy of Sciences, CHINA

Poster/Oral Session II

14:15 Poster/Oral Session II
Poster presentations are listed by topic category with their assigned number starting on Page 13.

15:45 **Break & Exhibit Inspection**

Session VIII - Sonics & Ultrasonics MEMS

16:15 DOMAIN/BOUNDARY VARIATION IN CANTILEVER ARRAY FOR BANDWIDTH
ENHANCEMENT OF PZT MEMS MICROSPEAKER 123
Shu-Wei Chang[1], Ting-Chou Wei[1], Sung-Cheng Lo[2], and Weileun Fang[1]
[1]National Tsing Hua University, TAIWAN and [2]Transducer Star Technology Inc., TAIWAN

16:30   **ON THE DESIGN OF PIEZOELECTRIC MEMS MICROSPEAKER WITH HIGH FIDELITY AND WIDE BANDWIDTH** 127
Ting-Chou Wei, Zih-Song Hu, Shu-Wei Chang, and Weileun Fang
National Tsing Hua University, TAIWAN

16:45   **HIGH-PERFORMANCE WAFER-SCALE TRANSFER-FREE GRAPHENE MICROPHONES** 131
Roberto Pezone, Gabriele Baglioni, Pasqualina M. Sarro, Peter G. Steeneken, and Sten Vollebregt
Delft University of Technology, NETHERLANDS

17:00   **HIGH-SPL AND LOW-DRIVING-VOLTAGE PMUTS BY SPUTTERED POTASSIUM SODIUM NIOBATE** 135
Fan Xia[1,2], Yande Peng[1,2], Sedat Pala[1,2], Ryuichi Arakawa[1,3], Wei Yue[1,2], Pei-Chi Tsao[2], Chun-Ming Chen[2], Hanxiao Liu[1,2], Megan Teng[2], Jong Ha Park[1,2], and Liwei Lin[1,2]
[1]Berkeley Sensor and Actuator Center, USA, [2]University of California, Berkeley, USA, and [3]NGK Spark Plug Co., JAPAN

17:15   **EPITAXIAL $P_B(Z_R,T_I)O_3$-BASED PIEZOELECTRIC MICROMACHINED ULTRASONIC TRANSDUCER FABRICATED ON SILICON-ON-NOTHING (SON) STRUCTURE** 139
Takuma Sekiguchi[1], Shinya Yoshida[2], Yoshiaki Kanamori[1], and Shuji Tanaka[1]
[1]Tohoku University, JAPAN and [2]Shibaura Institute of Technology, JAPAN

17:30   **Adjourn for the day**

19:00   **Banquet at the Löwenbräu Keller**
- 22:00

Wednesday, 18 January

All times are Central European Time (CET).

Plenary Presentation III

08:30   **LEVERAGING SEMICONDUCTOR ECOSYSTEMS TO MEMS** 143
Weileun Fang, Sheng-Shian Li, and Ming-Huang Li
National Tsing Hua University, TAIWAN

Session IX - Optomechanics & Photonics Integration

09:15   **PROGRAMMABLE SILICON NITRIDE PHOTONIC INTEGRATED CIRCUITS** 149
Hao Tian[1], Alaina G. Attanasio[1], Anat Siddharth[2], Andrey Voloshin[2], Viacheslav Snigirev[2],
Grigory Lihachev[2], Andrea Bancora[2], Vladimir Shadymov[2], Rui N. Wang[2], Johann Riemensberger[2],
Tobias J. Kippenberg[2], and Sunil A. Bhave[1]
[1]Purdue University, USA and [2]Swiss Federal Institute of Technology Lausanne (EPFL), SWITZERLAND

09:30   **MULTIFREQUENCY NANOMECHANICAL MASS SPECTROMETER PROTOTYPE FOR
MEASURING VIRAL PARTICLES USING OPTOMECHANICAL DISK RESONATORS** 153
Oscar Malvar[1], Eduardo Gil-Santos[1], Jose J. Ruz[1], Elena Sentre-Arribas[1], Adrián Sanz-Jiménez[1],
Priscila M. Kosaka[1], Sergio García-López[1], Álvaro San Paulo[1], Samantha Sbarra[2], Louis Waquier[2],
Ivan Favero[2], Maurits van der Heiden[3], Robert K. Altmann[3], Dimitris Papanastasiou[4],
Diamantis Kounadis[4], Ilias Panagiotopoulos[4], Jesús Mingorance[5], María Rodríguez-Tejedor[5],
Rafael Delgado[6], Montserrat Calleja[1], and Javier Tamayo[1]
*[1]Instituto de Micro y Nanotechnologis, IMN-CSIC, CSIC (CEI UAM+CSIC), SPAIN, [2]Université Paris Cité,
FRANCE, [3]The Netherland Organization for Applied Scientific Research (TNO), NETHERLANDS,
[4]Fasmatech Science and Technology, GREECE, [5]Hospital Universitario La Paz, SPAIN, and
[6]Hospital Universitario 12 de Octubre, SPAIN*

09:45   **A MICROFABRICATED DIAMOND QUANTUM MAGNETOMETER
WITH PICOTESLA SCALE SENSITIVITY** 157
Fei Xie[1,2], Qihui Liu[1,2], Yuqiang Hu[3,4], Lingyun Li[1,2], Zhichao Chen[1,2], Jin Zhang[1], Yonggui Zhang[1,2],
Yuyao Zhang[3,4], Yang Wang[1,2], Jiangong Cheng[1,2], Hao Chen[1,2], and Zhenyu Wu[1,2,3,4]
*[1]Chinese Academy of Sciences, CHINA, [2]University of Chinese Academy of Sciences, CHINA,
[3]Shanghai University, CHINA, and [4]Shanghai Industrial µTechnology Research Institute, CHINA*

10:00   **Break & Exhibit Inspection**

Session X - RF MEMS Filters & Resonators (5G & 6G)

10:30   **A NON-VOLATILE THRESHOLD SENSING SYSTEM USING A FERROELECTRIC
$HF_{0.5}ZR_{0.5}O_2$ DEVICE AND A $LiNbO_3$ MICROACOUSTIC RESONATOR** 161
Onurcan Kaya, Luca Colombo, Benyamin Davaji, and Cristian Cassella
Northeastern University, USA

10:45   **RESONANT CONFINERS FOR ACOUSTIC LOSS MITIGATION
IN BULK ACOUSTIC WAVE RESONATORS** 165
Jeronimo Segovia-Fernandez and Ernest T.-T. Yen
Texas Instruments, Kilby Labs, USA

11:00   **HIGH-CRYSTALLINITY 30% SCALN ENABLING HIGH FIGURE OF MERIT
X-BAND MICROACOUSTIC RESONATORS FOR MID-BAND 6G** 169
Gabriel Giribaldi, Pietro Simeoni, Luca Colombo, and Matteo Rinaldi
Northeastern University, USA

11:15 FERRITE-ROD ANTENNA DRIVEN WIRELESS RESOSWITCH RECEIVER 173
Kevin H. Zheng, Qiutong Jin, and Clark T.-C. Nguyen
University of California, Berkeley, USA

11:30 ULTRA-WIDEBAND MEMS FILTERS USING LOCALIZED THINNED
128° Y-CUT THIN-FILM LITHIUM NIOBATE 177
Jinbo Wu[1,2,3], Shibin Zhang[1], Pengcheng Zheng[1,2], Liping Zhang[1,2], Hulin Yao[1,2],
Xiaoli Fang[1,2], Xuedi Tian[1,2], Xiaomeng Zhao[1], Tao Wu[3], and Xin Ou[1,2]
[1]Shanghai Institute of Microsystem and Information Technology, CHINA,
[2]University of Chinese Academy of Sciences, CHINA, and [3]ShanghaiTech University, CHINA

11:45 **Lunch & Exhibit Inspection**

Session XIa - MEMS/NEMS Resonators & Non-Linear Dynamics

13:00 ATTRACTOR EXCHANGER FOR OPEN-LOOP OPERATION OF MICROMECHANICAL
NONLINEAR RESONATORS USING GAP-SPACING CONTINUATION 181
Chun-Pu Tsai and Wei-Chang Li
National Taiwan University, TAIWAN

13:15 A CMOS-MEMS ULTRASENSITIVE THERMOMETER USING
INTERNAL RESONANCE INDUCED FREQUENCY COMBS 185
Ting-Yi Chen, Chun-Pu Tsai, and Wei-Chang Li
National Taiwan University, TAIWAN

13:30 ATOMICALLY THIN NEMS FREQUENCY COMB WITH BOTH FREQUENCY TUNABILITY
AND RECONFIGURABLE VIA SIMULTANEOUS 1:2 AND 1:3 MODE COUPLING 189
Bo Xu, Jiankai Zhu, Chenyin Jiao, Jianglong Chen, and Zenghui Wang
University of Electronic Science and Technology of China, CHINA

13:45 INSTRUMENTAL ANALYSIS OF ADVANCED CATALYSTS
BASED ON RESONANT MICROCANTILEVERS 193
Xinyu Li[1,2], Pengcheng Xu[1], Ying Chen[1], Haitao Yu[1], and Xinxin Li[1,2]
[1]Chinese Academy of Sciences, CHINA and [2]University of Chinese Academy of Sciences, CHINA

Session XIb - BioSensors I

13:00 A MULTIPLEXED BIOAFFINITY BIOSENSING PATCH FOR
POINT-OF-CARE CHRONIC ULCER MONITORING 197
Md Sharifuzzaman, Dongkyun Kim, Md Selim Reza, SeongHoon Jeong,
Hye Su Song, Md Abu Zahed, and Jae Yeong Park
Kwangwoon University, KOREA

13:15 3-DOF BIOHYBRID ACTUATOR WITH MULTIPLE SKELETAL MUSCLE TISSUES 201
Xinzhu Ren, Yuya Morimoto, and Shoji Takeuchi
University of Tokyo, JAPAN

13:30 A LOW NOISE MICROELECTRODE ARRAY FOR SPECIFIC
CELL ACTIVITY MODULATION FROM CELL TO TISSUE 205
Bohan Zhang[1,2], Huiran Yang[2], Xiner Wang[2,3], Ziyi Zhu[2,3], Zongxing He[1], Wanqi Jiang[2,3], Chen Tao[1,2],
Dujuan Zou[2,3], Meng Li[2,3], Zhitao Zhou[2,3], Liuyang Sun[2,3], Tiger H. Tao[1,2,3,4,5,6], and Xiaoling Wei[2,3]
*[1]ShanghaiTech University, CHINA, [2]Chinese Academy of Sciences, CHINA, [3]University of Chinese Academy
of Sciences, CHINA, [4]Neuroxess Co., Ltd. (Jiangxi), CHINA, [5]Guangdong Institute of Intelligence Science
and Technology, CHINA, and [6]Tianqiao and Chrissy Chen Institute for Translational Research, CHINA*

13:45   **BIONIC MECHANICAL HAND INTEGRATED WITH ARTIFICIAL OLFACTORY SENSOR ARRAY FOR ENHANCED OBJECT RECOGNITION** 209

Jiachuang Wang[1,2], Xiaohui Li[1,2], MengWei Liu[1,2], Pingping Zhang[3], Tiger H. Tao[1,2,4], and Nan Qin[1,2]
[1]Chinese Academy of Sciences, CHINA, [2]University of Chinese Academy of Sciences, CHINA,
[3]Suzhou Huiwen Nanotechnology Co., Ltd., CHINA, and [4]Neuroxess Co., Ltd. (Jiangxi), CHINA

Poster/Oral Session III

14:00   **Poster/Oral Session III**
Poster presentations are listed by topic category with their assigned number starting on Page 13.

15:30   **Break & Exhibit Inspection**

Session XIIa - Force & Displacement/ Tactile Sensors & Human-Machine

16:00   **HIGH RESOLUTION TACTILE SENSOR FOR MEASUREMENT OF A COMPLICATED TACTILE FEELING OF "*SHITTORI*" WITH MOISTNESS** 213
Genki Yamada, Yuto Morita, Kyohei Terao, Fusao Shimokawa, and Hidekuni Takao
Kagawa University, JAPAN

16:15   **PYRAMIDAL STRUCTURED MXENE/ECOFLEX COMPOSITE-BASED TOROIDAL TRIBOELECTRIC SELF-POWERED SENSOR FOR HUMAN-MACHINE INTERFACE** 217
Shipeng Zhang, Sm Sohel Rana, Gagan Bahad Pradhan,
Trilochan Bhatta, Seonghoon Jeong, and Jae Yeong Park
Kangwoon University, KOREA

16:30   **LIG-BASED TRIAXIAL TACTILE SENSOR UTILIZING ROTATIONAL ERECTION SYSTEM** 221
Rihachiro Nakashima[1], Nagi Nakamura[2], Tomohiko G. Sano[1], Eiji Iwase[2], and Hidetoshi Takahashi[1]
[1]Keio University, JAPAN and [2]Waseda University, JAPAN

16:45   **A STRETCHABLE STRAIN-INSENSITIVE SMART GLOVE FOR SIMULTANEOUS DETECTION OF PRESSURE AND TEMPERATURE** 225
Sudeep Sharma, Gagan Bahadur Pradhan, Seonghoon Jeong, and Jae Yeong Park
Kwangwoon University, KOREA

17:00   **A GESTURE RECOGNITION GLOVE ASSEMBLED WITH NANOFOREST-INTEGRATED INFRARED THERMOPILES** 229
Mao Li[1,2], Meng Shi[1,2], Guidong Chen[1,2], Na Zhou[1,2], Haiyang Mao[1,2], and Chengjun Huang[1,2]
[1]Chinese Academy of Sciences, CHINA and [2]University of Chinese Academy of Sciences, CHINA

Session XIIb - BioSensors II

16:00   **ONE PUSH MEMBRANE FORMATION FOR ITERATIVE MEASUREMENT OF ION CHANNEL ACTIVITY ON ARRAYED CHIP** 233
Hisatoshi Mimura[1], Toshihisa Osaki[1,2], Sho Takamori[1], Kenji Nakao[2], and Shoji Takeuchi[1,3]
[1]Kanagawa Institute of Industrial Science and Technology (KISTEC), JAPAN,
[2]Maqsys Inc., JAPAN, and [3]University of Tokyo, JAPAN

16:15   **AN IMPLANTABLE DIFFERENTIAL SENSOR WITH PASSIVE WIRELESS INTERROGATION FOR IN-SITU EARLY DETECTION OF PERIPROSTHETIC JOINT INFECTION** 235
Jiaxin Jiang, Cole Napier, Chandrashekhar Choudhary, H. Claude Sagi,
Chia-Ying Lin, Michael T. Archdeacon, and Tao Li
University of Cincinnati, USA

16:30   **MICROMACHINED PIEZOELECTRIC FILM-BASED FLEXIBLE ELECTRONICS WITH INTEGRATION OF FILM-SELF TEMPERATURE-DETECTING BREATH SENSOR AND ACETONE GAS SENSOR** 239
Hung-Yu Yeh and Guo-Hua Feng
National Tsing Hua University, TAIWAN

16:45   **FLEXIBLE TACTILE SENSING ARRAY WITH HIGH SPACIAL DENSITY BASED ON PARYLENE MEMS TECHNIQUE** 243
Meixuan Zhang[1], Zetian Wang[1], Han Xu[2], Lang Chen[1], Yufeng Jin[2,3], and Wei Wang[1,3,4]
[1]Peking University, CHINA, [2]Peking University Shenzhen Graduate School, CHINA,
[3]National Key Lab of Micro/Nano Fabrication Technology, CHINA, and
[4]Beijing Advanced Innovation Center for Integrated Circuits, CHINA

17:00   **SILK-ENABLED FOLDABLE AND CONFORMAL NEURAL INTERFACE WITH IN-PLANE SHIELDING FOR HIGH-QUALITY ELECTROPHYSIOLOGICAL RECORDINGS** 247
Jizhi Liang[1,2], Zhaohan Chen[1,3], Xiner Wang[1,2], Feihong Xu[1,2], Xiaoling Wei[1,2], Liuyang Sun[1,2], Meng Li[1,2,] Tiger H. Tao[1,2,4,5,6,7], and Zhitao Zhou[1,2]
[1]Chinese Academy of Sciences, CHINA, [2]University of Chinese Academy of Sciences, CHINA,
[3]Shanghai Normal University, CHINA, [4]ShanghaiTech University, CHINA, [5]Neuroxess Co., Ltd. (Jiangxi), CHINA, [6]Guangdong Institute of Intelligence Science and Technology, CHINA and
[7]Tianqiao and Chrissy Chen Institute for Translational Research, CHINA

17:15   **Adjourn for the day**

Thursday, 19 January
All times are Central European Time (CET).

Plenary Presentation IV

08:30 MATERIALS ENGINEERING FOR CHEMICAL SENSING ENHANCEMENT 251
Navpreet Kaur, Dario Zappa, and Elisabetta Comini
University of Brescia, ITALY

Session XIII - Gas & Flow Sensors

09:15 ON-DEMAND PREPARATION OF GAS-SENSING MATERIALS GUIDED BY
RESONANT CANTILEVER-BASED THERMOGRAVIMETRIC ANALYSIS 255
Yufan Zhou[1,2], Ming Li[1,2], Ying Chen[1,2], Xinyu Li[1,2], Pengcheng Xu[1,2], and Xinyu Li[1,2]
[1]Chinese Academy of Sciences, CHINA and [2]University of Chinese Academy of Sciences, CHINA

09:30 AN INTELLIGENT GAS ANALYSIS SYSTEM CONSISTING OF SENSORS AND
A NEURAL NETWORK IMPLEMENTED USING THIN-FILM TRANSISTORS 259
Zong Liu[1,2], Yushen Hu[1,2], Gabriel E. Carranza[1], Fei Wang[2], and Man Wong[1]
*[1]Hong Kong University of Science and Technology, HONG KONG and
[2]Southern University of Science and Technology, CHINA*

09:45 SINGLE-LAYER-ELECTRODE TEMPERATURE-MODULATED SNO_2 GAS SENSOR CELL
WITH LOW POWER CONSUMPTION FOR DISCRIMINATION OF FOOD ODORS 263
Chong Xing, Ruichen Liu, Yan Zhang, Dongcheng Xie, Yudong Wang, Yuan Huang,
Muhammad Mustafa, Haochen Zhang, Zhongyu Shi, Lei Xu, and Feng Wu
University of Science and Technology of China, CHINA

10:00 A PERFORMANCE ENHANCED THERMAL FLOW SENSOR WITH NOVEL DUAL-HEATER
STRUCTURE USING CMOS COMPATIBLE FABRICATION PROCESS 267
Zhongyi Liu[1], Ruoqin Wang[2], Gai Yang[1], Xinyuan Zhang[1], Rui Jiao[2],
Xuejiao Li[1], Jiali Qi[3], Hongyu Yu[2], Huikai Xie[1,4], and Xiaoyi Wang[1,4]
*[1]Beijing Institute of Technology, CHINA, [2]Hong Kong University of Science and Technology, HONG KONG,
[3]Hangzhou Dianzi University, CHINA, and
[4]BIT Chongqing Institute of Microelectronics and Microsystems, CHINA*

Session XIV - New Fabrication Techniques

10:45 LOCAL METAL DEPOSITION ON HYDROGELS USING MICRO-PLASMA-BUBBLES 271
Haruna Takahashi, Yu Yamashita, Naotomo Tottori, Shinya Sakuma, and Yoko Yamanishi
Kyushu University, JAPAN

11:00 FOLDING METHOD OF KIRIGAMI STRUCTURE WITH FOLDING LINES 275
Nagi Nakamura and Eiji Iwase
Waseda University, JAPAN

11:15 BUBBLE-ASSISTED RE-FORMATION OF INDIVIDUAL
LIPID BILAYERS IN ARRAYED DEVICE 279
Izumi Hashimoto[1,2], Toshihisa Osaki[2], Hisatoshi Mimura[2],
Sho Takamori[2], Norihisa Miki[1,2], and Shoji Takeuchi[2,3]
*[1]Keio University, JAPAN, [2]Kanagawa Institute of Industrial Science and Technology, JAPAN, and
[3]University of Tokyo, JAPAN*

11:30 **LARGE-SCALE ARRAYS OF TUNABLE MONOLAYER MoS$_2$ NANOELECTROMECHANICAL RESONATORS** 281

Zuheng Liu[1], Luming Wang[3], Pengcheng Zhang[1], Maosong Xie[1], Yueyang Jia[1], Ying Chen[4], Hao Jia[4], Zenghui Wang[3], and Rui Yang[1,2]

[1]*University of Michigan – Shanghai Jiao Tong University Joint Institute, Shanghai Jiao Tong University, CHINA, [2]Shanghai Jiao Tong University, CHINA, [3]University of Electronic Science and Technology of China, CHINA, and [4]Chinese Academy of Sciences, CHINA*

Awards Ceremony

11:45 **Awards Ceremony**

11:55 **Final Remarks**

12:00 **Conference Ajourns**

POSTER PRESENTATIONS
All times are Central European Time (CET).

M - Monday, 16 January - 13:45 - 15:45
T - Tuesday, 17 January - 13:30 - 15:30
W - Wednesday, 18 January - 13:30 - 15:30

Classification Chart
(last character of poster number)

a - Bio and Medical MEMS
b - Emerging Technologies and New Opportunities for MEMS/NEMS
c - Industry MEMS and Advancing MEMS for Products and Sustainability
d - Materials, Fabrication and Packaging for Generic MEMS and NEMS
e - MEMS Actuators and PowerMEMS
f - MEMS Physical and Chemical Sensors
g - Micro- and Nanofluidics
h - Optical, RF and Electromagnetics for MEMS/NEMS
i - Open Posters

a - Bio and Medical MEMS
Biosensors and Bioreactors

M01-a ANTIFOULING FOR ELECTROCHEMICALLY BIOSENSING IN BODY FLUIDS 285
Wenzheng He[1], Changdong Zhou[2], Yang Lin[2], Yuxin Tian[2], Liying Liu[2],
Qifu Zhang[2], Xiongying Ye[1], and Tianhong Cui[3]
[1]Tsinghua University, CHINA, [2]Jilin Cancer Hospital, CHINA, and [3]University of Minnesota, USA

T01-a ELECTRO-MAGNETIC SENSOR MEDIATED BY MAGNETIC BIOMOLECULES 289
Qian Cheng[1,2], Yuqing Ge[1], Hongju Mao[1,2], Lin Zhou[1], and Jianlong Zhao[1,2]
[1]Chinese Academy of Science, CHINA and [2]University of Chinese Academy of Sciences, CHINA

W01-a GAS-FLOW DEVICE FOR EFFECTIVE DISSOLUTION OF GAS-PHASE
ODORANTS UTILIZED FOR BIOHYBRID SENSORS 293
Takuma Nakane[1,2], Toshihisa Osaki[2], Hisatoshi Mimura[2], Sho Takamori[2],
Norihisa Miki[1,2], and Shoji Takeuchi[2,3]
*[1]Keio University, JAPAN, [2]Kanagawa Institute of Industrial Science and Technology, JAPAN, and
[3]University of Tokyo, JAPAN*

M02-a MULTIPLE WELLS ON A CMOS-MEA FOR CELL-BASED
BIOHYBRID ODORANT SENSORS 295
Yujia Lian, Haruka Oda, Minghao Nie, and Shoji Takeuchi
University of Tokyo, JAPAN

T02-a THE INTEGRATED RGO/PEDOT: PSS-MODIFIED ULTRAFLEXIBLE
MICROELECTRODES TOWARDS LONG-TERM NEUROPHYSIOLOGICAL
SIGNALING AND DOPAMINE SENSITIVE DETECTION 297
Xueying Wang[1,2], Huiran Yang[1], Bohan Zhang[1,3], Meng Li[1,2], Liuyang Sun[1,2],
Zhitao Zhou[1,2], Tiger H. Tao[1,2,3,4,5,6], and Xiaoling Wei[1,2]
[1]Chinese Academy of Sciences, CHINA, [2]University of Chinese Academy of Sciences, CHINA, [3]Shanghai Tech University, CHINA, [4]Neuroxess Co., Ltd. (Jiangxi), CHINA, [5]Guangdong Institute of Intelligence Science and Technology, CHINA, and [6]Tianqiao and Chrissy Chen Institute for Translational Research, CHINA

a - Bio and Medical MEMS
Devices & Systems for Cellular and Molecular Studies

W02-a COMPARISON OF SELECTIVE FILTRATION OF ON-CHIP GLOMERULUS
COMPRISED OF ORGANOID-DERIVED AND IMMORTALIZED PODOCYTES 301
Ayumu Tabuchi[1], Kensuke Yabuuchi[2,3], Yoshiki Sahara[2], Minoru Takasato[2,4],
Kazuya Fujimoto[1], and Ryuji Yokokawa[1]
[1]Kyoto university, JAPAN, [2]RIKEN, JAPAN, and [3]Osaka University, JAPAN

M03-a CONTROLLING FIRING POINT OF MICROFIBER-SHAPED HIPSC-DERIVED
CARDIAC TISSUE WITH LOCALIZED ELECTRICAL STIMULATION DEVICE 305
Akari Masuda[1], Shun Itai[1], Yuta Kurashina[2], Shugo Tohyama[1], and Hiroaki Onoe[1]
[1]Keio University, JAPAN and [2]Tokyo University of Agriculture and Technology, JAPAN

T03-a DEVELOPMENTAL PHASES OF ON-CHIP VASCULOGENESIS
CLASSIFIED USING A DEEP LEARNING VISUAL MODEL 309
Taiga Irisa, Hang Zhou, Kazuya Fujimoto, and Ryuji Yokokawa
Kyoto University, JAPAN

W03-a HAND-DRIVEN DEVICE FOR PREPARATION
OF LINEARLY ALIGNED HYDROGEL SHEETS 313
Aoi Kato[1,2], Haruka Oda[3], Sho Takamori[2], Hisatoshi Mimura[2],
Toshihisa Osaki[2], Norihisa Miki[1,2], and Shoji Takeuchi[2,3]
[1]Keio University, JAPAN, [2]Kanagawa Institute of Industrial Science and Technology, JAPAN, and [3]University of Tokyo, JAPAN

M04-a MICROFABRICATION AND CHARACTERIZATION OF MICRO-
STEREOLITHOGRAPHICALLY 3D PRINTED, AND DOUBLE METALLIZED
BIOPLATES WITH 3D MICROELECTRODE ARRAYS FOR
IN-VITRO ANALYSIS OF CARDIAC ORGANOIDS 315
Jorge Manrique Castro, Isaac Johnson, and Swaminathan Rajaraman
University of Central Florida, USA

T04-a OIL-SEALED RGD-MODIFIED HYDROGEL MICROWELL ARRAY WITH SIZE-
SELECTIVE PERMEATION FOR ANALYSIS ON EXOSOMES FROM SINGLE CELLS 319
Chisaki Yamagata[1], Shun Itai[1], Yuta Kurashina[2], Makoto Asai[1], Ayuko Hoshino[3], and Hiroaki Onoe[1]
[1]Keio University, JAPAN, [2]Tokyo University of Agriculture and Technology, JAPAN, and [3]Tokyo Institute of Technology, JAPAN

W04-a PICKING SINGEL CELLS FROM 10 ML SAMPLE BASED ON A
MICROFILTRATION- LIFT COMBINATION PLATFORM 323
Qingmei Xu[1], Yuntong Wang[2,3], Xiao Ma[4], Hang Li[5], Ying Xue[5],
Yi Zhang[1], Songtao Dou[1], Huan Wang[2], Bei Li[2,5], and Wei Wang[1,6,7]
[1]Peking University, CHINA, [2]Chinese Academy of Sciences, CHINA, [3]University of Chinese Academy of Sciences, CHINA, [4]Hangzhou Branemagic Medical Technology Co. Ltd., CHINA, [5]Hooke Laboratory, CHINA, [6]National Key Lab of Micro/Nano Fabrication Technology, CHINA, and [7]Beijing Advanced Innovation Center for Integrated Circuits, Beijing, CHINA

a - Bio and Medical MEMS
Flexible and Wearable Devices and Systems

M05-a   **A TRANSFER METHOD FOR EMBEDDING CONDUCTIVE FILLERS ON THE SURFACE OF MULTI-SCALE STRUCTURES FOR 3D FLEXIBLE CONDUCTORS** 327
Dongwoo Yoo, Sangmok Kim, Jeonghyeon Hwang, and Joonwon Kim
Pohang University of Science and Technology (POSTECH), KOREA

T05-a   **FABRICATION OF HIGH FREQUENCY 2D FLEXIBLE PMUT ARRAY** 331
Sanjog V. Joshi, Sina Sadeghpour, and Michael Kraft
KU Leuven, BELGIUM

W05-a   **FLEXIBLE SILK-BASED GRAPHENE BIOELECTRONICS FOR WEARABLE MULTIMODAL PHYSIOLOGICAL MONITORING** 335
Sajjad Mirbakht[1], Ata Golparvar[1,2], Muhammad Umar[1], and Murat Kaya Yapici[1,3]
[1]Sabanci University, TURKEY, [2]École Polytechnique Fédérale de Lausanne (EPFL), SWITZERLAND, and [3]University of Washington, USA

M06-a   **HIGHLY ACCURATE MEASUREMENT OF CONTACT RESISTANCE BETWEEN GALINSTAN AND COPPER USING TRANSFER LENGTH METHOD** 339
Takashi Sato and Eiji Iwase
Waseda University, JAPAN

T06-a   **MACHINE LEARNING ENABLED HIND FOOT DEFORMITY DETECTION USING INDIVIDUALLY ADDRESSABLE HYBRID PRESSURE SENSOR MATRIX** 343
Nadeem Tariq Beigh, Faizan Beigh, Sourav Naval, Dibyajyoti Mukherjee, and Dhiman Mallick
Indian Institute of Technology, Delhi, INDIA

W06-a   **MULTI-MODE E-SKIN INTEGRATING CAPACITIVE-PIEZOELECTRIC SENSORS FOR STATIC-DYNAMIC MECHANORESPONSE WITH WIDE SENSING RANGE** 347
Mujeeb Yousuf[1], Sushil Kumar[1], Dhairya Singh Arya[2], Manu Garg[1], Khanjhan Joshi[1], and Pushpapraj Singh[1]
[1]Indian Institute of Technology, Delhi, INDIA and [2]CSIR-Central Scientific Instruments Organisation (CSIO), INDIA

M07-a   **NON-INVASIVE INSTANT MEASUREMENT OF ARTERIAL STIFFNESS BASED ON HIGH-DENSITY FLEXIBLE SENSOR ARRAY** 351
Fang Wang[1,2], Heng Yang[1,2], Ke Sun[1], Yi Sun[1], and Xinxin Li[1,2]
[1]Chinese Academy of Sciences, CHINA and [2]University of Chinese Academy of Sciences, CHINA

T07-a   **SUPPRESSION OF BIOELECTRICAL NOISE SIGNALS IN MOTION STATE BY LOW-COST MICROPILLAR HYDROGEL ELECTRODE** 355
Gencai Shen, Nan Zhao, Chunpeng Jiang, Zhuangzhuang Wang, and Jingquan Liu
Shanghai Jiao Tong University, CHINA

W07-a   **ULTRA-THIN MEMS PACKAGING BASED ON AUXETIC STRETCHABLE STRUCTURES FOR APPLICATIONS IN WEARABLE ELECTRONICS** 359
Daniel Zymelka, Toshihiro Takeshita, Yusuke Takei, and Takeshi Kobayashi
National Institute of Advanced Industrial Science and Technology, JAPAN

M08-a   **ULTRALOW POWER FLEXIBLE OCULAR MICROSYSTEM FOR VERGENCE AND DISTANCE SENSING BASED ON PASSIVE DIFFERENTIAL MAGNETOMETRY** 362
Adwait Deshpande, Mohit U. Karkhanis, Chayanjit Ghosh, Hanseup Kim, and Carlos H. Mastrangelo
University of Utah, USA

a - Bio and Medical MEMS
Manufacturing for Bio- & Medical MEMS

T08-a   ELECTROHYDRODYNAMIC NEBULISER (eNEB) FOR DIRECT
PULMONARY DRUG DELIVERY APPLICATION 366
Trung-Hieu Vu[1], Luan Ngoc Mai[2,3], Tuan-Hung Nguyen[1], Dang Tran[1], Tuan-Khoa Nguyen[1],
Thanh Nguyen[4], Jarred Fastier-Woollel[1,5], Canh-Dung Tran[4], Toan Dinh[4], Hong-Quan Nguyen[1],
Dzung Viet Dao[1], and Van Thanh Dau[1]
*[1]Griffith University, AUSTRALIA, [2]Ho Chi Minh City University of Technology (HCMUT), VIETNAM
[3]Vietnam National University, VIETNAM, [4]University of Southern Queensland, AUSTRALIA, and
[5]University of Tokyo, JAPAN*

W08-a   FLEXIBLE POLYMER OPTICAL WAVEGUIDES FOR
INTEGRATED OPTOGENETIC BRAIN IMPLANTS 370
Julian A. Singer[1], Till Stramm[2], Jens Fasel[2], Oliver Schween[2], Anton Gelaeschus[1],
Andreas Bahr[1,3], and Matthias Kuhl[4]
*[1]Hamburg University of Technology, GERMANY, [2]TU Dortmund University, GERMANY,
[3]University of Kiel, GERMANY, and [4]University of Freiburg, GERMANY*

M09-a   HIGHLY REPRODUCIBLE TISSUE POSITIONING WITH TAPERED
PILLAR DESIGN IN ENGINEERED HEART TISSUE PLATFORMS 374
Milica Dostanic[1,2], Laura M. Windt[2], Maury Wiendels[2], Berend J. van Meer[2],
Christine L. Mummery[2,3], Pasqualina M. Sarro[1], and Massimo Mastrangeli[1]
*[1]Delft University of Technology, NETHERLANDS, [2]Leiden University Medical Center, NETHERLANDS, and
[3]University of Twente, NETHERLANDS*

T09-a   IN VITRO ASSEMBLY OF MUSCLE RINGS AND BIOPRINTED
HYDROGEL FOR BRANCHING TUBULAR TISSUE CONSTRUCTS 378
Tomohiro Morita, Byeongwook Jo, Minghao Nie, and Shoji Takeuchi
University of Tokyo, JAPAN

W09-a   MICROELECTRODES FABRICATED BY VACUUM FILLING WITH
LOW MELTING-POINT ALLOY FOR MUSCLE TISSUE STIMULATION 381
Tingyu Li, Minghao Nie, Yuya Morimoto, and Shoji Takeuchi
University of Tokyo, JAPAN

M10-a   OPTOELECTRONIC INTEGRATED ULTRAMICROELECTRODE FOR OPTICAL
STIMULATION AND ELECTRICAL RECORDING OF SINGLE-CELL 384
Qingda Xu, Ye Xi, Zhiyuan Du, Longchun Wang, Tao Ruan, Mengfei Xu,
Jiawei Cao, Bin Yang, and Jingquan Liu
Shanghai Jiao Tong University, CHINA

T10-a   THERMOFORMING OF PARYLENE C TO FORM HELICAL STRUCTURES 388
Brianna L. Thielen and Ellis Meng
University of Southern California, Los Angeles, USA

a - Bio and Medical MEMS
Materials for Bio- and Medical MEMS

W10-a   FABRICATION OF BIODEGRADABLE SOFT TISSUE-MIMICKED MICROELECTRODE
ARRAYS FOR IMPLANTED NEURAL INTERFACING 392
Wei-Chen Huang[1], Wan-Lou Lei[1], and Chih-Wei Peng[2]
[1]National Yang Ming Chiao Tung University, TAIWAN and [2]Taipei Medical University, TAIWAN

a - Bio and Medical MEMS
Medical Microsystems

M11-a   AN OPTIMIZATION OF PERFORATION DESIGN ON A PIEZOELECTRIC-BASED SMART STENT FOR BLOOD PRESSURE MONITORING AND LOW-FREQUENCY VIBRATIONAL ENERGY HARVESTING 396
Jun Ying Tan[1], Sayemul Islam[2], Yuankai Li[3], Albert Kim[2], and Jungkwun "JK" Kim[1]
[1]University of North Texas, USA, [2]University of South Florida, USA, and [3]Kansas State University, USA

W11-a   DEVELOPMENT OF AN ELECTRICAL-STIMULATION-INDUCED MECHANOMYOGRAM PROBE FOR MUSCLE CONTRACTION CHARACTERISTICS EVALUATION 400
Yusuke Takei, Toshihiro Takeshita, Daniel Zymelka, and Takeshi Kobayashi
National Institute of Advanced Industrial Science and Technology (AIST), JAPAN

M12-a   DUAL-FREQUENCY PIEZOELECTRIC MICROMACHINED ULTRASONIC TRANSDUCERS FOR FUNDAMENTAL AND HARMONIC IMAGING 402
Yanfen Zhai, Waleed Maqsood, Zhou Da, Nikolai Andrianov, Yucheng Zhang, Mohssen Moridi, and Lixiang Wu
Silicon Austria Labs GmbH, AUSTRIA

T12-a   FRACTAL MICROELECTRODES INTEGRATED WITH THE CATHETER FOR LOW-VOLTAGE PULSED FIELD ABLATION 405
Mengfei Xu[1], Mu Qin[2], Ziliang Song[3], Wen Hong[1], Qingda Xu[1], Jiawei Cao[1], Kejun Tu[1], Longchun Wang[1], Bin Yang[1], and Jingquan Liu[1]
[1]Shanghai Jiao Tong University, CHINA,
[2]Shanghai Chest Hospital Affiliated to Shanghai Jiao Tong University, CHINA, and
[3]Shanghai General Hospital Shanghai Jiao Tong University School of Medicine, CHINA

W12-a   HIERARCHICAL BONDING YIELD TEST STRUCTURE FOR FLEXIBLE HIGH CHANNEL-COUNT NEURAL PROBES INTERFACING ASIC CHIPS 409
Marie C. Odenthal, Victor Claar, Oliver Paul, and Patrick Ruther
University of Freiburg, GERMANY

M13-a   MICROWAVE-INDUCED THERMOACOUSTIC IMAGING USING ALUMINUM NITRIDE PMUT 413
Yiwei Wang[1], Lejia Zhang[1], Junxiang Cai[1,2,3], Baosheng Wang[1,2,3], Yuandong Alex Gu[5], Liang Lou[5], Xiong Wang[1,2,3,4], and Tao Wu[1,2,3,4]
[1]ShanghaiTech University, CHINA and [2]Chinese Academy of Sciences, CHINA, [3]University of Chinese Academy of Sciences, CHINA, [4]Shanghai Engineering Research Center of Energy Efficient and Custom AI IC, CHINA, and [5]Shanghai Industrial µTechnology Research Institute, CHINA

T13-a   NEEDLE-FREE DRUG INJECTION USING A SHOCK WAVE FOCUSING SYSTEM WITH THE FUNCTION OF REAL-TIME MICROBUBBLE-BASED DISTANCE SENSING 417
Yibo Ma, Wenjing Huang, Keita Ichikawa, and Yoko Yamanishi
Kyushu University, JAPAN

W13-a   NEW WAFER-LEVEL FABRICATION OF ULTRATHIN SILICON INSERTION SHUTTLES FOR FLEXIBLE NEURAL IMPLANTS 421
Kirti Sharma[1], Christian Boehler[1], Maria Asplund[1,2], Oliver Paul[1], and Patrick Ruther[1]
[1]University of Freiburg, GERMANY and [2]Chalmers University of Technology, SWEDEN

M14-a   REAL-TIME DYNAMIC LACTATE DETECTION IN A PIPELINE USING A MICROSENSING NEEDLE FOR ICU PATIENT MONITORING APPLICATION 425
Yuan-Sin Tang[1], Tung-Lin Yang[2], Yu-Ting Cheng[1], Hsiao-En Tsai[2,3], and Yih-Shurng Chen[3,4]
[1]National Yang Ming Chiao Tung University, TAIWAN, [2]National Taiwan Hospital HsinChu Branch, TAIWAN, [3]National Taiwan University College of Medicine Graduate Institute of Clinical Medicine, TAIWAN, and [4]National Taiwan University Hospital, TAIWAN

T14-a THREE-DIMENSIONAL FLEXIBLE NEURAL OPTO-ELECTRONIC
ARRAY WITH SILK-BASED SHUTTLE-FREE IMPLANTATION 429
Chi Gu[2,3], Huiran Yang[2], Bohan Zhang[2,4], Zhitao Zhou[2], Liuyang Sun[2,3],
Meng Li[2,3], Xiaoling Wei[2,3], and Tiger H. Tao[1,2,3,4,5,6]
[1]Guangdong Institute of Intelligence Science and Technology, CHINA, [2]Chinese Academy of Sciences,
CHINA, [3]University of Chinese Academy of Sciences, CHINA, [4]ShanghaiTech University, CHINA, [5]Neuroxess
Co., Ltd. (Jiangxi), CHINA, and [6]Tianqiao and Chrissy Chen Institute for Translational Research, CHINA

a - Bio and Medical MEMS
MEMS & BioMEMS for Fighting COVID-19 & Future Pandemic

W14-a A MICROFLUIDIC BIOSENSOR FOR RAPID DETECTION OF COVID-19 433
Sura A. Muhsin[1], Ying He[1], Muthana Al-Amidie[1], Karen Sergovia[1], Amjed Abdullah[1],
Yang Wang[1], Omar Alkorjia[1], Robert A. Hulsey[2], Gary L. Hunter[2], Zeynep Erdal[2],
Ryan J. Pletka[2], George S. Hyleme[2], Xiu-Feng Wan[1,2], and Mahmoud Almasri[1]
[1]University of Missouri, USA and [2]Black and Veatch, USA

M15-a A LOOP-MEDIATED ISOTHERMAL AMPLIFICATION (LAMP)-BASED POINT-OF-CARE
SYSTEM FOR RAPID ON-SITE CLINICAL DETECTION OF SARS-COV-2 VIRUSES 437
Trieu Nguyen[1], Aaydha Chidambara Vinayaka[1], Van Ngoc Huynh[1], Quyen Than Linh[1],
Sune Zoëga Andreasen[1], Mohsen Golabi[1], Dang Duong Bang[1], Jens Kjølseth Møller[2], and Anders Wolff[1]
[1]Technical University of Denmark, DENMARK and [2]University Hospital of Southern Denmark, DENMARK

a - Bio and Medical MEMS
MEMS & BioMEMS for Healthcare and Public Health

T15-a A SOLAR-DRIVEN WEARABLE MULTIPLEXED BIO-SENSING SYSTEM
FOR NONINVASIVE HEALTHCARE MONITORING IN SWEAT 440
Jujhar Singh, Bianca Ning, Paul Lee, and Lin Liu
Seattle Pacific University, USA

W15-a HIGH-THROUGHPUT MASS MEASUREMENT OF SINGLE BACTERIAL
CELLS BY SILICON NITRIDE MEMBRANE RESONATORS 444
Adrián Sanz-Jiménez[1], Oscar Malvar[1], Jose J. Ruz[1], Sergio García-López[1], Priscila M. Kosaka[1],
Eduardo Gil-Santos[1], Álvaro Cano[1], Dimitris Papanastasiou[2], Diamantis Kounadis[2], Elias Panagiotopoulos[2],
Jesús Mingorance[3], María Rodríguez-Tejedor[3], Álvaro San Paulo[1], Montserrat Calleja[1], and Javier Tamayo[1]
[1]Instituto de Micro y Nanotecnología, SPAIN, [2]Fasmatech Science & Technology, Lefkippos TESPA,
Demokritos NCSR, Patriarchou Gregoriou & Neapoleos, GREECE, and
[3]Hospital Universitario La Paz, Madrid, SPAIN

M16-a MICROFABRICATED ISOTHERMAL EG-FET SENSOR FOR LAMP MEDIATED
CRISPR/CAS12A DETECTION OF HEPATITIS C VIRUS 448
Hsin-Ying Ho, Wei-Sin Kao, Piyush Deval, Ling-Shan Yu, and Che-Hsin Lin
National Sun Yat-sen Universit, TAIWAN

T16-a SMART ELECTRODE ARRAY FOR COCHLEAR IMPLANTS 452
Ahmad Itawi[1], Sofiane Ghenna[1], Guillaume Tourrel[2], Sébastien Grondel[1], Cedric Plesse[3], Tran Minh Giao
Nguyen[3], Frédéric Vidal[3], Yinoussa Adagolodjo[4], Lingxiao Xun[4], Gang Zheng[4], Alexandre Kruszewski[4],
Christian Duriez[4], and Eric Cattan[1]
[1]University Polytechnique Hauts-de-France, FRANCE, [2]Oticon Medical, FRANCE,
[3]CY Cergy Paris Université, FRANCE, and [4]University of Lille, FRANCE

a - Bio and Medical MEMS
Tissue Engineering

W16-a A THREE-DIMENSIONAL ARTIFICIAL INTESTINAL
TISSUE WITH A CRYPT-LIKE INNER SURFACE 456
Shuma Tanaka[1], Shun Itai[2], and Hiroaki Onoe[1]
[1]Keio University, JAPAN and [2]Tohoku University, JAPAN

M17-a TISSUE-ENGINEERED PENNATE MUSCLES ON A CHIP 460
Motoki Ito, Yuya Morimoto, and Shoji Takeuchi
University of Tokyo, JAPAN

T17-a WEIGHT TRAINING DEVICE TO PROMOTE MATURATION
IN SKELETAL MUSCLE TISSUES 463
Kentaro Motoi, Byeongwook Jo, Yuya Morimoto, and Shoji Takeuchi
University of Tokyo, JAPAN

a - Bio and Medical MEMS
Other Bio and Medical MEMS

W17-a MICROSYSTEM VIBRATING MESH ATOMIZER WITH INTEGRATED
MICROHEATER FOR HIGH VISCOSITY LIQUID AEROSOL GENERATION 467
Pallavi Sharma, Irma Rocio Vazquez, and Nathan Jackson
University of New Mexico, USA

M18-a SCALABLE MODULAR MEASUREMENT SYSTEM FOR CONTINUOUS
BLOOD MONITORING WITH PIEZOELECTRIC MEMS RESONATORS 471
Michael Schneider[1], Bernhard Kößl[1], Suresh Alasatri[1], Ingrid A.M. Magnet[2], and Ulrich Schmid[1]
[1]TU Wien, AUSTRIA and [2]Medical University of Vienna, AUSTRIA

T18-a SILICON COMPATIBLE PROCESS TO INTEGRATE IMPEDANCE
CYTOMETRY WITH MECHANICAL CHARACTERIZATION 475
Quentin Rezard[1], Faruk Azam Shaik[1,2], Jean Claude Gerbedoen[1,2], Fabrizio Cleri[1], Dominique Collard[1,2],
Chann Lagadec[1], and Mehmet C. Tarhan[1,2]
[1]University of Lille, FRANCE and [2]University of Tokyo, Lille, FRANCE

W18-a SORTING OF EXTRACELLULAR VESICLES BY USING OPTICALLY-INDUCED
DIELECTROPHORESIS ON AN INTEGRATED MICROFLUIDIC CHIP 479
Wei-Jen Soong, Chih-Hung Wang, Yi-Sin Chen, Chihchen Chen, and Gwo-Bin Lee
National Tsing Hua University, TAIWAN

b - Emerging Technologies and New Opportunities for MEMS/NEMS
Internet of Things (IoT) with MEMS/NEMS

M19-b A REPROGRAMMABLE MEM SWITCH UTILIZING
CONTROLLED CONTACT WELDING 483
Tsegereda K. Esatu, Hei Kam, Lars P. Tatum, Xiaoer Hu, Urmita Sikder,
Sergio Almeida, Junqiao Wu, and Tsu-Jae King Liu
University of California, Berkeley, USA

T19-b MICROMECHANICAL RSSI BASED ON FORCE INTERACTION DERIVED TAPPING
BANDWIDTH VARIATION IN VIBRO-IMPACT RESONATORS 487
Yi-Hsuan Huang, Hong-Sen Zheng, Chun-Pu Tsai, and Wei-Chang Li
National Taiwan University, TAIWAN

W19-b WAKE-UP IOT WIRELESS SENSING NODE BASED ON A LOW-G THRESHOLD MEMS INERTIAL SWITCH WITH RELIABLE CONTACTS 491
Sagnik Ghosh[1], Duan Jian Goh[1], Yul Koh[1], Jaibir Sharma[1], Wei Da Toh[1], Weiguo Chen[1], Yao Zhang[1], Eldwin Ng[1], Amit Lal[2], and Joshua E.-Y. Lee[1]
[1]Agency for Science, Technology and Research (A*STAR), SINGAPORE and [2]Cornell University, USA

---

b - Emerging Technologies and New Opportunities for MEMS/NEMS
Machine Learning (ML) & Artificial Intelligence (AI) Enhanced MEMS/NEMS Design, Manufacturing, and Applications

M20-b ARTIFICIAL INTELLIGENCE (AI)-ENHANCED E-SKIN WITH ARTIFICIAL SYNAPSE SENSORY OUTPUT FOR HUMANOID ROBOTIC FINGER OF MULTIMODAL PERCEPTION 495
Xinge Guo[1,2] and Chengkuo Lee[1]
[1]National University of Singapore, SINGAPORE and
[2]Agency for Science, Technology and Research (A*STAR), SINGAPORE

T20-b MULTI-MEMS DIFFERENTIAL PRESSURE SENSOR ELEMENTS-BASED AIRFLOW SENSOR WITH NEURAL NETWORK MODEL 499
Kotaro Haneda, Kenei Matsudaira, and Hidetoshi Takahashi
Keio University, JAPAN

W20-b TRIAL-AND-ERROR LEARNING FOR MEMS STRUCTURAL DESIGN ENABLED BY DEEP REINFORCEMENT LEARNING 503
Fanping Sui[1], Wei Yue[1], Ziqi Zhang[2], Ruiqi Guo[1], and Liwei Lin[1,2]
[1]University of California, Berkeley, USA and [2]TSinghua University, CHINA

---

b - Emerging Technologies and New Opportunities for MEMS/NEMS
New Computing Devices and Systems with MEMS/NEMS

M21-b FULLY MICROELECTROMECHANICAL NON-VOLATILE MEMORY CELL 507
Elliott Worsey, Mukesh K. Kulsreshath, Qi Tang, and Dinesh Pamunuwa
University of Bristol, UK

T21-b NONVOLATILE STATE CONFIGURATION OF NANO-WATT PARAMETRIC ISING SPINS THROUGH FERROELECTRIC HAFNIUM ZIRCONIUM OXIDE MEMS VARACTORS 511
Nicolas Casilli[1], Onurcan Kaya[1], Tahmid Kaisar[2], Benyamin Davaji[1], Philip X.-L. Feng[2], and Cristian Cassella[1]
[1]Northeastern University, USA and [2]University of Florida, USA

W21-b PHYSICAL RESERVOIR COMPUTING USING NONLINEAR MEMS RESONATOR HAVING HIGH MEMORY CAPACITY AT "EDGE OF CHAOS" 515
Hiroki Takemura, Takahiro Mizumoto, Amit Banerjee, Jun Hirotani, and Toshiyuki Tsuchiya
Kyoto University, JAPAN

M22-b PROGRAMMABLE FERROELECTRIC HZO NEMS MECHANICAL MULTIPLIER FOR IN-MEMORY COMPUTING 519
Shubham Jadhav, Ved Gund, and Amit Lal
Cornell University, USA

T22-b STORING MEMS INTERFACES WITHOUT ELECTRICAL AUXILIARY ENERGY FOR LONG-TIME MONITORING 522
Martin Hoffmann[1], Philip Schmitt[1], Steffen Wittemeier[3], Falk Schaller[2], Alexey Shaporin[3], Chris Stöckel[2,3], Volker Geneiß[3], Roman Forke[3], Christian Hedayat[3], Ulrich Hilleringmann[4], Harald Kuhn[2,3], and Sven Zimmermann[2,3]
[1]Ruhr-Universität Bochum, GERMANY, [2]Chemnitz University of Technology, GERMANY, [3]Fraunhofer Institute for Electronic Nano Systems ENAS, GERMANY, and [4]University of Paderborn, GERMANY

b - Emerging Technologies and New Opportunities for MEMS/NEMS
Nonlinear Dynamics in MEMS/NEMS

W22-b A NEW FINDING ON NONLINEAR DAMPING AND STIFFNESS
OF FLEXURAL MODE CAPACITIVE MEMS RESONATORS 526
Hung-Yu Chen, Ming-Huang Li, and Sheng-Shian Li
National Tsing Hua University, TAIWAN

M23-b EXPLOITING PARAMETRIC INSTABILITY IN BISTABLE MEMS ACTUATORS 530
Daniel Platz, Johannes Fabian, Elisabeth Samm, Mahdi Mortada, Michael Schneider, and Ulrich Schmid
TU Wien, AUSTRIA

T23-b FIRST PROTOTYPE OF POLYMER MICROMACHINED
FLAPPING WING NANO AIR VEHICLE 534
Rashmikant, Ryotaro Suetsugu, Minato Onishi, and Daisuke Ishihara
Kyushu Institute of Technology, JAPAN

W23-b ITERATIVE LEARNING CONTROL FOR QUASI-STATIC
MEMS MIRROR WITH SWITCHING OPERATION 538
Matthias Macho[1], Han Woong Yoo[1], Richard Schroedter[2], and Georg Schitter[1]
[1]TU Wien, AUSTRIA and [2]TU Dresden, GERMANY

b - Emerging Technologies and New Opportunities for MEMS/NEMS
Quantum Devices and Systems with MEMS/NEMS

M24-b *Mz* ATOMIC MAGNETOMETER USING A 3D MEMS GLASS
ALKALI VAPOR CELL WITH VERTICAL SIDEWALLS 542
Jin Zhang, Jianfeng Zhang, Wenqi Li, Ziji Wang, and Jintang Shang
Southeast University, CHINA

T24-b ON-CHIP HEATING NOISE SUPPRESSION OF 3D CHIP-SCALE ATOMIC
MAGNETOMETER USING SINGLE-LAYER SHIFTED HEATER 546
Ziji Wang, Junming Wu, Jin Zhang, and Jintang Shang
Southeast University, CHINA

c - Industry MEMS and Advancing MEMS for Products and Sustainability
Barriers to Commercialization & Research Needs for Future Products

W24-c LABOR-SAVING PLATFORM FOR CHARACTERIZATION OF MEMBRANE
PROTEINS BY AUTOMATED MONITORING AND DATA REPORTING 550
Kazuto Ogishi[1], Toshihisa Osaki[2], Yuya Morimoto[1], and Shoji Takeuchi[1,2]
[1]University of Tokyo, JAPAN and [2]Kanagawa Institute of Industrial Science and Technology, JAPAN

c - Industry MEMS and Advancing MEMS for Products and Sustainability
MEMS Packaging Techniques

M25-c MODELLING IMPACT OF VISCOELASTIC PROPERTIES OF DIE-ATTACH
MATERIAL ON THE BIAS RESPONSE OF RESONANT INERTIAL SENSORS 554
Theo Miani[1], Lokesh Gurung[1], Guillermo Sobreviela-Falces[1], Douglas Young[1], Colin Baker[1], and Ashwin A. Seshia[2]
[1]Silicon Microgravity Ltd., UK and [2]University of Cambridge, UK

c - Industry MEMS and Advancing MEMS for Products and Sustainability
MEMS/NEMS - CMOS Integration

T25-c   CMOS-EMBEDDED 3D MICRO/NANOFLUIDICS EMPLOYING
TOP-DOWN BEOL SINGLE-STEP WET-ETCHING TECHNIQUE 558
Wei-Yang Weng, Hung-Yu Hou, Yueh-Jung Chao, Shwu-Jen Liaw, and Jun-Chau Chien
National Taiwan University, TAIWAN

W25-c   IMPLEMENTATION OF A MONOLITHIC SOC ENVIRONMENTAL
SENSING HUB USING CMOS-MEMS TECHNIQUE 562
Ya-Chu Lee[1], Tung-Lin Chien[1], Chi-Te Fang[1], Yuanyuan Huang[1], Wei-Lun Sung[2],
Yen-Chang Chu[2], Rongshun Chen[1], and Weileun Fang[1]
[1]National Tsing Hua University, TAIWAN and [2]PixArt Imaging Inc., TAIWAN

M26-c   MONOLITHICALLY AND VERTICALLY INTEGRATED ENVIRONMENTAL
SENSING HUB WITH NOVEL AIR-BASED HUMIDITY SENSOR DESIGN 566
Tung-Lin Chien, Yuanyuan Huang, Fuchi Shih, and Weileun Fang
National Tsing Hua University, TAIWAN

c - Industry MEMS and Advancing MEMS for Products and Sustainability
New MEMS System Design and Integration Approaches

T26-c   A SELF-CORRECTED, SELF-CLEANED MEMS AND SUITABLE FOR
ADVANCED FOUNDRY MULTI-PROJECT WAFER (MPW) 570
Sushil Kumar, Dhairya Singh Arya, Manu Garg, and Pushpapraj Singh
Indian Institute of Technology, New-Delhi, INDIA

W26-c   MONOLITHIC INTEGRATION OF HUMIDITY/FLOW/TEMPERATURE SENSORS AS
ENVIRONMENT SENSING HUB FOR APPARENT-TEMPERATURE DETECTION 574
Yu-Hsuan Li, Tung-Lin Chien, Fuchi Shih, Yuanyuan Huang, and Weileun Fang
National Tsing Hua University, TAIWAN

M27-c   PIEZORESISTIVE PRESSURE SENSOR WITH MONOLITHICALLY
INTEGRATED AMPLIFIER BASED ON METAL-OXIDE TRANSISTORS 578
Runxiao Shi[1], Dequan Lin[1], Kevin Chau[1,2], and Man Wang[1]
*[1]Hong Kong University of Science and Technology, HONG KONG and
[2]Chinese Academy of Sciences, CHINA*

d - Materials, Fabrication and Packaging for Generic MEMS and NEMS
Advancement in Conventional Materials for MEMS & NEMS

T27-d   A PERFORMANCE ENHANCEMENT METHOD FOR THERMOPILE
SENSORS USING A CHIP PROBE TEST SYSTEM 582
Meng Shi[1,2], Mao Li[1,2], Yue Ni[3], Chenchen Zhang[1], Na Zhou[1,2], Haiyang Mao[1,2], and Chengjun Huang[1,2]
*[1]Chinese Academy of Sciences, CHINA, [2]University of Chinese Academy of Sciences, CHINA, and
[3]Jiangsu Hinovaic Technologies Co., Ltd, CHINA*

W27-d   CHARACTERIZING INDUCTIVELY-COUPLED-PLASMA ETCHING OF SINGLE
CRYSTALLINE LITHIUM TANTALATE FOR MICRO-ACOUSTIC APPLICATIONS 586
Yasaman Majd, Jorge Manrique Castro, Hakhamanesh Mansoorzare, and Reza Abdolvand
University of Central Florida, USA

M28-d   ROBUST POLYCRYSTALLINE 3C-SIC-ON-SI HETEROSTRUCTURES
WITH LOW CTE MISMATCH UP TO 900 °C FOR MEMS 590
Philipp Moll, Georg Pfusterschmied, and Ulrich Schmid
TU Wien, AUSTRIA

d - Materials, Fabrication and Packaging for Generic MEMS and NEMS
Digital Micromanufacturing

T28-d A 3D PRINTED FUNCTIONAL MEMS ACCELEROMETER 594
Simone Pagliano[1], David E. Marschner[1], Damien Maillard[2], Nils Ehrmann[3],
Göran Stemme[1], Stefan Braun[3], Luis Guillermo Villanueva[2], and Frank Niklaus[1]
*[1]KTH Royal Institute of Technology, SWEDEN, [2]École Polytechnique Fédérale de Lausanne (EPFL),
SWITZERLAND, and [3]Hochschule Kaiserslautern, GERMANY*

W28-d A FULLY 3D PRINTED METHOD FOR MONOLITHIC INTEGRATION
OF AN ACCELEROMETER AND A FORCE SENSOR 598
Guandong Liu[1,2], Changhai Wang[1], Kexin Wang[1], Zhili Jia[3], Ruiqi Luo[2], and Wei Ma[2,4]
*[1]Heriot-Watt University, UK, [2]Zhejiang Lab, CHINA, [3]National Institute of Metrology, CHINA, and
[4]Zhejiang University, CHINA*

d - Materials, Fabrication and Packaging for Generic MEMS and NEMS
Generic MEMS & NEMS Manufacturing Techniques

M29-d CHARACTERIZATION OF VAPOR HF SACRIFICIAL ETCHING THROUGH
SUBMICRON RELASE HOLES FOR WAFER-LEVEL VACUUM PACKAGING
BASED ON SILICON MIGRATION SEAL 602
Tianjiao Gong[1], Yukio Suzuki[1], Muhammad J. Khan[1], Karla Hiller[2], and Shuji Tanaka[1]
[1]Tohoku University, JAPAN and [2]Fraunhofer Institute for Electronic Nano Systems, GERMANY

T29-d DAMAGE PROFILE MODELING AND EXPERIMENT OF SILICON CARBIDE SUBSTRATES
IN MICRO-NANO STRUCTURE FABRICATED BY HELIUM FOCUSED ION BEAM 606
Shupeng Gao, Xi Chen, Qianhuang Chen, Qi Li, and Yan Xing
Southeast University, CHINA

W29-d LIQUID-IMMERSION INCLINED-ROTATED UV LITHOGRAPHY FOR
MICRO SUCTION CUP ARRAY 610
Gakuto Kagawa and Hidetoshi Takahashi
Keio University, JAPAN

d - Materials, Fabrication and Packaging for Generic MEMS and NEMS
New & Emerging Materials for MEMS/NEMS

M30-d PARAMETRIC AMPLIFICATION AND PHONONIC FREQUENCY COMB
GENERATION IN MoS_2 NANOELECTROMECHANICAL RESONATORS 613
S M Enamul Hoque Yousuf[1], Yunong Wang[1], Jaesung Lee[1], Steven W. Shaw[2,3], and Philip X.-L. Feng[1]
[1]University of Florida, USA, [2]Florida Institute of Technology, USA, and [3]Michigan State University, USA

T30-d PARYLENE-N AS A HIGH TEMPERATURE THIN FILM PIEZOELECTRIC MATERIAL 617
Nathan Jackson and Deepak Kunwar
University of New Mexico, USA

W30-d SILICON CARBIDE-ON-INSULATOR THERMAL-PIEZORESISTIVE
RESONATOR FOR HARSH ENVIRONMENT APPLICATION 621
Baoyun Sun[1,2], Jiarui Mo[1], Hemin Zhang[3], Henk W. van Zeijl[1], Willem D. van Driel[1], and Guoqi Zhang[1]
*[1]Delft University of Technology, NETHERLANDS, [2]China University of Petroleum, CHINA, and
[3]KU Leuven, BELGIUM*

M31-d SPIN COATING OF HIGHLY ALIGNED AGCN MICROWIRES
EPITAXIALLY GROWN ON 2D MATERIALS 625
Jimin Ham, Jaemook Lim, Joowon Lim, Gunyoung Jang, Sueng Yoon Lee,
Dohyun Lim, Sukjoon Hong, and Won Chul Lee
Hanyang Universit, Ansan, KOREA

T31-d SUSPENDED TWO-DIMENSIONAL MATERIAL MEMBRANES FOR SENSOR
APPLICATIONS FABRICATED WITH A HIGH-YIELD TRANSFER PROCESS 627
Sebastian Lukas[1], Ines Kraiem[1,2], Maximilian Prechtl[3], Oliver Hartwig[3], Annika Grundmann[1], Holger
Kalisch[1], Satender Kataria[1], Michael Heuken[1,4], Andrei Vescan[1], Georg S. Duesberg[3], and Max C. Lemme[1,2]
[1]RWTH Aachen University, GERMANY, [2]AMO GmbH, GERMANY,
[3]University of the Bundeswehr Munich, GERMANY, and [4]AIXTRON SE, GERMANY

W31-d TCF-IMPROVED SH_0 MODE ACOUSTIC RESONATORS BASED
ON 30°YX-LINBO$_3$/SIO$_2$ MEMBRANE 631
Shuxian Wu[1], Zonglin Wu[1], Hangyu Qian[1], Feihong Bao[1],
Gongbin Tang[2], Feng Xu[1], and Jie Zou[1]
[1]Fudan University, CHINA and [2]Shandong University, CHINA

M32-d WAFER SCALE MULTILAYER GRAPHENE BASED BRAIN PROBES BY
SPIN-SPRAYING METHODS FOR MAGNETIC RESONANCE IMAGING 635
Kejun Tu, Zhejun Guo, Mengfei Xu, Bin Yang, and Jingquan Liu
Shanghai Jiao Tong University, CHINA

d - Materials, Fabrication and Packaging for Generic MEMS and NEMS
New Fabrication Processes for Making MEMS/NEMS

T32-d 3D SELF-ALIGNED FABRICATION OF SUSPENDED NANOWIRES
BY CRYSTALLOGRAPHIC NANOLITHOGRAPHY 639
Erwin J.W. Berenschot, Yasser Pordeli, Lucas J. Kooijman, Yves L. Janssens,
Roald M. Tiggelaar, and Niels R. Tas
University of Twente, NETHERLANDS

W32-d A SIMPLE PROCESS FOR THE FABRICATION OF PARALLEL-PLATE ELECTROSTATIC
MEMS RESONATORS BY GOLD THERMOCOMPRESSION BONDING 643
Dolores Manrique Juarez[1], Fabrice Mathieu[1], Guillaume Libaude[1], David Bourrier[1],
Samuel Charlot[1], Laurent Mazenq[1], Véronique Conédéra[1], Ludovic Salvagnac[1],
Isabelle Dufour[2], Liviu Nicu[1], and Thierry Leïchlé[1,3]
[1]LAAS-CNRS, FRANCE, [2]Université de Bordeaux, IMS UMR-CNRS, FRANCE, and [3]Georgia Tech, USA

M33-d ELECTROMECHANICALLY STABLE INTERCONNECTION BETWEEN
LIG AND THICK DAM-SHAPED METALLIC ELECTRODE VIA STORED
AG MICROPARTICLE SOLUTION 647
Saeyoung Park, Yoo-Kyum Shin, and Min-Ho Seo
Pusan National University, KOREA

T33-d FREE-STANDING MEMBRANES WITH SELF-ASSEMBLED
NANOPORE ARRAYS FOR TEM OBSERVATION OF LIQUID SAMPLES 651
Joowon Lim[1], Jimin Ham[1], Sungho Jeon[1], Yuna Bae[2,3], Minho Kang[2,3],
Sueng Yoon Lee[1], Jungwon Park[2,3], and Won Chul Lee[1]
*[1]Hanyang University, KOREA, [2]Seoul National University, KOREA, and
[3]Institute of Basic Science (IBS), KOREA*

W33-d NONPLANAR NANOFABRICATION VIA INTERFACE ENGINEERING 653
Sarah O. Spector, Peter F. Satterthwaite, and Farnaz Niroui
Massachusetts Institute of Technology, USA

M34-d WAFER-LEVEL FABRICATION OF CONFORMAL SUB 10-NM NANOGAPS 657
Sayali Tope, Seungbeom Noh, and Hanseup Kim
University of Utah, USA

d - Materials, Fabrication and Packaging for Generic MEMS and NEMS
Packaging & Assembly

T34-d MEMS RESONATOR VACUUM-SEALED BY SILICON
MIGRATION AND HYDROGEN OUTDIFFUSION 661
Muhammad Jehanzeb Khan, Yukio Suzuki, Tianjiao Gong, Takashiro Tsukamoto, and Shuji Tanaka
Tohoku University, JAPAN

W34-d MEMS THIN-FILM VACUUM PACKAGE UTILIZING GLOW DISCHARGE GETTER 665
Vikram Maharshi[1], Manjeet Kumar[1], Ajay Agarwal[2], and Bhaskar Mitra[1]
[1]Indian Institute of Technology, Delhi, INDIA and [2]Indian Institute of Technology, Jodhpur, INDIA

e – MEMS Actuators and PowerMEMS
Actuator Components & Systems

M35-e LNOI THIN-FILM DUAL-AXIS RESONANT MICRO-MIRROR
WITH E16 TORSIONAL ACTUATION 669
Yaoqing Lu[1,2,3], Kangfu Liu[1,2,3], Yuxi Wang[1,2,3], Ran Nie[1], and Tao Wu[1,2,3,4]
[1]ShanghaiTech University, CHINA, [2]Chinese Academy of Sciences, CHINA,
[3]University of Chinese Academy of Sciences, CHINA, and
[4]Shanghai Engineering Research Center of Energy Efficient and Custom AI IC, CHINA

T35-e A PIEZOELECTRIC MEMS SPEAKER WITH STRETCHABLE FILM SEALING 673
Linbing Xu, Mingchao Sun, Menglun Zhang, Chengze Liu, Xiaopeng Yang, and Wei Pang
Tianjin University, CHINA

W35-e BROADBAND MEMS SPEAKER BY SINGLE-WAY MULTI-RESONANCE
ARRAY WITH ACOUSTIC DAMPING TUNING: A PROOF OF CONCEPT 677
Mingchao Sun, Menglun Zhang, Chengze Liu, and Wei Pang
Tianjin University, CHINA

M36-e IONIC LIQUID ELECTROSPRAY THRUSTER WITH TWO-STAGE
ELECTRODES ON GLASS SUBSTRATE 681
Akane Nishimura[1], Yoshinori Takao[2], Toshiyuki Tsuchiya[1], and Yoshinori Takao[2]
[1]Kyoto University, JAPAN and [2]Yokohama National University, JAPAN

W36-e MONOLITHIC INTEGRATION OF PZT ACTUATION UNITS OF VARIOUS ACTIVATED
RESONANCES FOR FULL-RANGE MEMS SPEAKER ARRAY 685
Hsu-Hsiang Cheng[1], Sung-Cheng Lo[1], Yu-Chen Chen[1], Ming-Ching Cheng[1],
Ting-Chou Wei[1], Mingching Wu[2], and Weileun Fang[1]
[1]National Tsing Hua University, TAIWAN and [2]CoretronicMEMS Co., Ltd., TAIWAN

M37-e PULL-IN VOLTAGE REDUCTION IN ELECTROSTATIC AIRGAP ACTUATOR
USING 12 NM-ULTRATHIN INTERNAL DIELECTRIC TRANSDUCTION 689
Satish K. Verma and Bhaskar Mitra
Indian Institute of Technology, New Delhi, INDIA

e – MEMS Actuators and PowerMEMS
Energy Harvesting Materials, Structures, and Transducers

T37-e A REVERSE ELECTROWETTING-ON-DIELECTRIC (REWOD) ENERGY
HARVESTER USING NONWETTING GALLIUM COATED ELECTRODE
AND ULTRATHIN GALLIUM OXIDE SHELL AS DIELECTRIC LAYER 693
Jinwon Jeong, Bokyung Suh, and Jeong Bong (JB) Lee
University of Texas at Dallas, USA

W37-e   ASYMMETRIC QUAD LEG ORTHOPLANAR SPRING FOR
WIDEBAND PIEZOELECTRIC MICRO ENERGY HARVESTING 697
Ali Mohammadi, Shamin Sadrafshari, Alborz Shokrani, and Chris R. Bowen
University of Bath, UK

M38-e   EVALUATION OF THERMOELECTRIC PROPERTIES OF
MONOLITHICALLY-INTEGRATED CORE-SHELL Si NANOWIRE BRIDGES 701
Akio Uesugi, Shusuke Nishiyori, Koji Sugano, and Yoshitada Isono
Kobe University, JAPAN

T38-e   GLAZE TILE-INSPIRED LIQUID-SOLID POWER GENERATOR
FOR CONTINUOUS WATER FLOW ENERGY HARVESTING 705
Dezhi Nie[1], Boming Lyu[1], Yongbo Hu[1], Jian Zhang[1], Yongqing Fu[2], Honglong Chang[1], and Kai Tao[1]
[1]Northwestern Polytechnical University, CHINA and [2]Northumbria University, UK

W38-e   MEMS CANTILEVERED ENERGY HARVESTER WITH TAPERED
THICKNESS FOR STRESS CONTROL 709
Takahito Yokota, Kensuke Kanda, Takayuki Fujita, and Kazusuke Maenaka
University of Hyogo, JAPAN

M39-e   TAPERED HELMHOLTZ RESONATOR WIND ENERGY
HARVESTER DRIVEN BY AEROACOUSTICS 712
Chen Hua, Liyun Zhen, Jingquan Liu, and Bin Yang
Shanghai Jiao Tong University, CHINA

e – MEMS Actuators and PowerMEMS
Manufacturing for Actuators & Power MEMS

T39-e   ANDROMEDA: A FLEXIBLE MEMS TECHNOLOGY PLATFORM FOR A
VARIETY OF PIEZOELECTRICALLY ACTUACTED MICROMIRRORS 716
Irene Martini, Anna Alessandri, Marta Carminati, Roberto Carminati, Paolo Ferrarini, Daniela A.L. Gatti,
Riccardo Gianola, Borka Lazarova, Carla M. Lazzari, Andrea Nomellini, Laura Oggioni, Claudia Pedrini,
Carlo L. Prelini, Riccardo Tacchini, and Michele Vimercati
STMicroelectronis, ITALY

W39-e   DESIGN OF BUTTERFLY PLATE PIEZOELECTRIC ACTUATOR
WITH DUAL DRIVING ELECTRODES FOR MEMS MICRO-MIRROR 720
Si-Han Chen[1], Shih-Chi Liu[1], Hao-Chien Cheng[1], Hung-Yu Lin[1],
Kai-Chih Liang[2], Mingching Wu[2], and Weileun Fang[1]
[1]National Tsing Hua University, TAIWAN and [2]Coretronic MEMS Corporation, TAIWAN

M40-e   FULLY-FLEXIBLE MICRO-SCALE ACTUATOR ARRAY
WITH THE LIQUID-GAS PHASE CHANGE MATERIALS 724
Sangjun Sim, Kyubin Bae, and Jongbaeg Kim
School of Mechanical Engineering, Yonsei University, KOREA

e – MEMS Actuators and PowerMEMS
Power MEMS Components & Systems

T40-e   A NOVEL COMB DESIGN FOR ENHANCED POWER AND
BANDWIDTH IN ELECTROSTATIC MEMS ENERGY CONVERTORS 728
Jinglun Li[1], Habilou Ouro-Koura[1], Hannah Arnow[1], Arian Nowbahari[2], Mathew Galarza[1], Meg Obispo[1],
Xing Tong[1], Mehdi Azadmehr[2], Mona M. Hella[1], John A. Tichy[1], and Diana-Andra Borca-Tasciuc[1]
[1]Rensselaer Polytechnic Institute, USA and [2]University of South-Eastern Norway, NORWAY

e – MEMS Actuators and PowerMEMS
Self-Powered Devices and Microsystems

W40-e A HYBRID NANOGENERTOR-DRIVEN SELF-POWERED
WEARABLE PERSPIRATION MONITORING SYSTEM 732
Md Abu Zahed, S M Sohel Rana, Md Sharifuzzaman, Seonghoon Jeong,
Gagan Bahadur Pradhan, Hye Su Song, and Jae Yeong Park
Kwangwoon University, KOREA

M41-e A MONOLITHIC INTEGRATED AND TRANSPARENT MICROSYSTEM
CONSTRUCTED BY USING AMORPHOUS INGAZNO FILM 736
Bin Jia, Chao Zhang, and Xiaodong Huang
Southeast University, CHINA

T41-e FLOWING WATER ENABLES STEERABLE CHARGE
DISTRIBUTION ON ELECTRET SURFACE 740
Boming Lyu[1], Jian Zhang[1], Yunjia Li[2], Yongqing Fu[3], Honglong Chang[1], Weizheng Yuan[1], and Kai Tao[1]
*[1]Northwestern Polytechnical University, CHINA, [2]Xi'an Jiaotong University, CHINA, and
[3]University of Northumbria, UK*

W41-e SELF-POWERED FLEXIBLE PIEZOELECTRET
ARRAY FOR WEARABLE APPLICATIONS 744
Hao Yang[1,2], Rui M.R. Pinto[1], Pedro González[1], Alar Ainla[1],
Mohammadmahdi Faraji[1], and K.B. Vinayakumar[1]
[1]International Iberian Nanotechnology Laboratory, PORTUGAL and [2]Xi'an Jiaotong University, CHINA

f - MEMS Physical and Chemical Sensors
Fluidic Sensors

M42-f A BULK-TYPE PRESSURE SENSOR WITH FULL-BRIDGE
IMPLEMENTATION ENABLED BY STRESS-MODIFYING TRENCHES 748
Dequan Lin[1], Man Wong[1], and Kevin Chau[1,2]
[1]Hong Kong University of Science and Technology, CHINA, [2]Chinese Academy of Science, CHINA

T42-f A CMOS COMPATIBLE MICRO PIRANI GAUGE WITH STRUCTURE
OPTIMIZATION FOR PERFORMANCE ENHANCEMENT 752
Rui Jiao[1], Gai Yang[2], Ruoqin Wang[1], Yue Tang[2], Zhongyi Liu[2],
Huikai Xie[2,3], Hongyu Yu[1], and Xiaoyi Wang[2,3]
*[1]Hong Kong University of Science and Technology, HONG KONG, [2]Beijing Institute of Technology, CHINA,
and [3]BIT Chongqing Institute of Microelectronics and Microsystems, CHINA*

W42-f A THERMAL AIRFLOW SENSOR BASED ON MN-CO-NI-O THIN FILM 756
Jie Wang, Yunfei Liu, Zhezheng Zhu, Chengchen Gao, Zhenchuan Yang, and Yilong Hao
Peking University, CHINA

M43-f HIGHLY SENSITIVE WAVE HEIGHT SENSOR WITH MEMS
PIEZORESISTIVE CANTILEVER AND WATERPROOF MEMBRANE 760
Takuto Hirayama and Hidetoshi Takahashi
Keio University, JAPAN

T43-f MEMS CAPACITANCE DIAPHRAGM GAUGE WITH
TWO SEALED REFERENCE CAVITIES 763
Xiaodong Han[1,2], Jingzhen Li[3], Gang Li[4], and Yongjian Feng[1]
*[1]Xiamen University, CHINA, [2]University of Twente, NETHERLANDS,
[3]Beijing University of Technology, CHINA, and [4]Lanzhou Institute of Physics, CHINA*

W43-f TOWARDS A GAS INDEPENDENT THERMAL FLOW METER 767
Shirin Azadi Keari[1], Remco J. Wiegerink[1], Remco G.P. Sanders[1], and Joost C. Lotters[1,2]
[1]University of Twente, NETHERLANDS and [2]Bronkhorst High-Tech BV, NETHERLANDS

f - MEMS Physical and Chemical Sensors
Force & Displacement Sensors

M44-f AN INTEGRATED MEMS DEVICE FOR *IN-SITU* FOUR-PROBE
ELECTRO-MECHANICAL CHARACTERIZATION OF PT NANOBEAM 771
Yuheng Huang, Meng Nie, Binghui Li, Kuibo Yin, and Litao Sun
Southeast University, CHINA

T44-f FINGERLIKE TACTILE TEXTURE INTEGRATED SENSOR WITH
COLD AND WARM SENSATIONS OF SUB-MM SPATIAL RESOLUTION 775
Nachi Mise, Mitsuki Kozasa, Kyohei Terao, Fusao Shimokawa, and Hidekuni Takao
Kagawa University, JAPAN

W44-f MODIFIED BEAM STRUCTURES FOR IMPROVED RESONANT SENSING 779
Erfan Ghaderi and Behraad Bahreyni
Simon Fraser University, CANADA

M45-f OCCLUSAL PAPER-BASED FLEXIBLE PRESSURE SENSOR FOR
IN SITU MEASURING ORAL OCCLUSAL FORCE 783
Wenduo Wang, Xin Zhang, Ning Zhao, Jingquan Liu, and Bin Yang
Shanghai Jiao Tong University, CHINA

T45-f SUCTION CUP ARRAY WORKING ALSO AS TACTILE SENSOR
TO DETECT CUPS DEFORMATION USING KCF AND CNN 787
Toshihiro Shiratori, Jinya Sakamoto, Yuki Kumokita, Masato Suzuki,
Tomokazu Takahashi, and Seiji Aoyagi
Kansai University, JAPAN

W45-f VERTICAL INTEGRATION OF FORCE TRANSMISSION STRUCTURE ON CAPACITIVE
CMOS-MEMS TACTILE FORCE SENSOR FOR SENSITIVITY IMPROVEMENT 791
Yuanyuan Huang, Yen-Lin Chen, Shihwei Lin, Fuchi Shih, Zihsong Hu, and Weileun Fang
National Tsing Hua University, TAIWAN

f - MEMS Physical and Chemical Sensors
Gas & Chemical Sensors

M46-f 1-OCTADECANETHIOL SAM ON CMOS-MEMS GOLD PLATED
RESONATOR VIA DIP-CAST FOR VOCs SENSING 795
Rafel Perelló-Roig[1,2], Jaume Verd[1,2], Sebastià Bota[1,2],
Bartomeu Soberats[1], Antonio Costa[1], and Jaume Segura[1,2]
[1]*University of the Balearic Islands, SPAIN and* [2]*Health Research Institute of the Balearic Islands, SPAIN*

T46-f APPLICATION OF DEEP LEARNING NETWORK FOR HUMIDITY
COMPENSATION OF SEMICONDUCTOR METAL OXIDE GAS SENSORS 799
Mingu Kang, Incheol Cho, and Inkyu Park
Korea Advanced Institute of Science and Technology (KAIST), KOREA

W46-f DEVELOPMENT OF MONOLITHIC MICRO-LED GAS SENSOR BASED
E-NOSE SYSTEM FOR REAL-TIME, SELECTIVE GAS PREDICTION 803
Kichul Lee, Mingu Kang, and Inkyu Park
Korea Advanced Institute of Science and Technology (KAIST), KOREA

M47-f ELECTRONIC-NOSE: AN ARRAY OF 16 MOS-GAS SENSORS INTEGRATED
WITH TEMPERATURE AND MOISTURE SENSING CAPABILITIES 807
Xiawei Yue[1,2], Shuai Wei[1,2], Pingping Zhang[3], Zhitao Zhou[1], Tiger Tao[1,2,4,5,6], and Nan Qin[1]
[1]*Chinese Academy of Sciences (CAS), CHINA,* [2]*University of Chinese Academy of Sciences, CHINA,*
[3]*Suzhou Huiwen Nanotechnology Co. Ltd., CHINA,* [4]*ShanghaiTech University, CHINA,* [5]*Shanghai Research
Center for Brain Science and Brain-Inspired Intelligence, CHINA and* [6]*Neuroxess Co., Ltd. (Jiangxi), CHINA*

T47-f ENHANCEMENT OF SENSITIVITY IN PHOTONIC CRYSTAL BASED
CHEMICAL SENSOR USING CHEMO-MECHANICAL BILAYER EFFECT 811
Seyeon Lee[1], Naik T. Banabathi[1], Dongwon Kang[3], Sookyung Kang[2],
Kyungsuk Cho[2], Jungwook Kim[1], and Jungyul Park[1]
[1]Sogang University, KOREA, [2]Iwha University, KOREA, and [3]University of California, Los Angeles, USA

W47-f METAL ION RECOGNITION SENSOR BASED ON RESISTANCE SWITCHING EFFECT 814
Tian Kang, Yusa Chen, Guanzhou Lin, Shengxiao Jin,
Liye Li, Hongshun Sun, Senyong Hu, and Wengang Wu
Peking University, CHINA

M48-f MULTI-HOTSPOT MID-IR NANOANTENNAS WITH MATCHED LOSS
AND HIGH-INTENSITY NEAR-FIELD FOR SUB-PPM-LEVEL GAS DETECTION 818
Hong Zhou, Zhihao Ren, Cheng Xu, Liangge Xu, Xinge Guo, and Chengkuo Lee
National University of Singapore, SINGAPORE

T48-f PALLADIUM BASED MEMS HYDROGEN SENSORS 822
Max Hoffmann[1], Marion Wienecke[1], Maren Lengert[2], Michael H. Weidner[2], and Jan Heeg[2]
*[1]Hochschule Wismar, Institut für Oberflächen- und Dünnschichttechnik, GERMANY and
[2]Materion GmbH, GERMANY*

W48-f SELECTIVE DISCRIMINATION OF PPB-LEVEL VOCS USING MOS GAS
SENSOR IN PULSE-HEATING MODE WITH THE MODIFIED HILL'S MODEL 826
Gaoqiang Niu, Yi Zhuang, Yushen Hu, Zong Liu, and Fei Wang
Southern University of Science and Technology, CHINA

M49-f THERMAL CONDUCTIVITY DETECTOR (TCD)-TYPE GAS SENSOR BASED
ON THE SUSPENDED 1D NANOHEATER FOR IOT APPLICATIONS 830
Wootaek Cho, Jong-Hyun Kwak, Taejung Kim, and Heungjoo Shin
Ulsan National Institute of Science and Technology (UNIST), KOREA

f - MEMS Physical and Chemical Sensors
Inertial Sensors

T49-f 120 PPM QUALITY FACTOR THERMAL STABILITY FROM -40°C TO +60°C OF A
DUAL-AXIS MEMS GYROSCOPE BASED ON JOULE EFFECT DYNAMIC CONTROL 833
Jian Cui[1,2] and Qiancheng Zhao[1,2]
[1]Peking University, CHINA and [2]Beijing Advanced Innovation Center for Integrated Circuits, CHINA

W49-f A FORCE-BANLANCE CAPACITIVE MEMS GRAVIMETER
WITH SUPERIOR RESPONSE TIME, SELF-NOISE AND DRIFT 837
Le Gao[1], Fangzheng Li[1], Jian Zhang[1], Bingyang Cai[1], Wenjie Wu[1], and Liangcheng Tu[2]
[1]Huazhong University of Science and Technology, CHINA and [2]Sun Yat-sen University, CHINA

M50-f A MEMS-BASED GRAVIMETER FOR SIMULTANEOUS VERTICAL
AND HORIZONTAL EARTH TIDES MEASUREMENTS 841
Lujia Yang[1], Xiaochao Xu[1], Qian Wang[1], Ji'ao Tian[1], Yanyan Fang[1],
Chun Zhao[1], Wenjie Wu[1], Fangjing Hu[1], and Liangcheng Tu[1,2]
[1]Huazhong University of Science and Technology, CHINA and [2]Sun Yat-sen University, CHINA

T50-f A NOVEL MULTIPLE FOLDED BEAM DISK RESONATOR FOR MAXIMIZING
THE THERMOELASTIC QUALITY FACTOR 845
Xiaopeng Sun[1], Xin Zhou[1], Lei Yu[1], Kaixuan He[2], Xuezhong Wu[1], and Dingbang Xiao[1]
*[1]National University of Defense Technology, CHINA and
[2]East China Institute of Photo-Electronic IC, CHINA*

W50-f A TIME-SERIES CONFIGURATION METHOD OF MODE REVERSAL MEMS GYROSCOPES UNDER DIFFERENT TEMPERATURE-VARYING CONDITIONS 849
Liangqian Chen, Tongqiao Miao, Qingsong Li, Peng Wang, Junjian Li, Xuezhong Wu, Dingbang Xiao, and Xiang Xi
National University of Defense Technology, CHINA

M51-f ACOUSTICALLY ISOLATED MEMS BAW GYROSCOPES 853
Diego Emilio Serrano, Amir Rahafrooz, Duane Younkin, Kieran Nunan, Mitul Dalal, Sagnik Pal, and Ijaz Jafri
Panasonic Device Solutions Laboratory of Massachusetts, USA

T51-f ACTIVE QUALITY FACTOR STABILIZATION OF MEMS RESONATOR UTILIZING ELECTRICAL DISSIPATION REGULATION 857
Yang Zhao, Qin Shi, Guoming Xia, and Anping Qiu
Nanjing University of Science and Technology, CHINA

W51-f DEMONSTRATION OF GYRO-LESS NORTH FINDING USING A T-SHAPED MEMS DIFFERENTIAL RESONANT ACCELEROMETER 861
Kei Masunishi, Etsuji Ogawa, Daiki Ono, Fumito Miyazaki, Hiroki Hiraga, Kengo Uchida, Jumpei Ogawa, Hideaki Murase, and Yasushi Tomizawa
Toshiba Corporation, JAPAN

M52-f ENHANCED STIFFNESS SENSITIVITY IN A MODE LOCALIZED SENSOR USING INTERNAL RESONANCE ACTUATION 865
Jianlin Chen[1], Hemin Zhang[2], Takashiro Tsukamoto[1], Michael Kraft[2], and Shuji Tanaka[1]
[1]Tohoku University, JAPAN and [2]KU Leuven, BELGIUM

T52-f MODELING STRESS EFFECTS ON FREQUENCIES OF A MEMS RING GYROSCOPE 869
Mehran Hosseini-Pishrobat, Baha Erim Uzunoglu, and Erdinc Tatar
Bilkent University, TURKEY

W52-f RATE INTEGRATING GYROSCOPE TUNED BY FOCUS ION BEAM TRIMMING AND INDEPENDENT CW/CCW MODES CONTROL 873
Jianlin Chen[1], Takashiro Tsukamoto[1], Giacomo Langfelder[2], and Shuji Tanaka[1]
[1]Tohoku University, JAPAN and [2]Politecnico di Milano, ITALY

M53-f TEMPERATURE DEPENDENCE OF QUALITY FACTORS AT HIGH FREQUENCIES IN MEMS GYROSCOPES 877
Daniel Schiwietz[1,2], Eva M. Weig[2], and Peter Degenfeld-Schonburg[1]
[1]Robert Bosch GmbH, GERMANY and [2]Technical University of Munich, GERMANY

f - MEMS Physical and Chemical Sensors
Manufacturing Techniques for Physical Sensors

T53-f 0.5MM×0.5MM 150KPA-MEASURE-RANGE HIGH-TEMPERATURE PRESSURE SENSOR WITH HIGH-PERFORMANCE AND LOW FABRICATION-COST 881
Peng Li[1,2], Wei Li[1], Changnan Chen[1,3], Ke Sun[1], Min Liu[1], Sheng Wu[1], Pichao Pan[1,3], Jiachou Wang[1,3], and Xinxin Li[1,2,3]
[1]Chinese Academy of Sciences, CHINA, [2]Fudan University, CHINA, and [3]University of Chinese Academy of Sciences, CHINA

W53-f AUTOMATIC PICO LASER TRIMMING SYSTEM FOR SILICON MEMS RESONANT DEVICES BASED ON IMAGE RECOGNITION 885
Yuxian Liu[1], Qiancheng Zhao[1,2], Dacheng Zhang[1], and Jian Cui[1,2]
[1]Peking University, CHINA and [2]Beijing Advanced Innovation Center for Integrated Circuits, CHINA

M54-f MICROMACHINING FUSED SILICA MICRO SHELL RESONATOR WITH QUARTZ GLASS MOLD BY THERMAL REFLOW 889
Zhaoxi Su, Bin Luo, Qiankai Tang, Linqian Zhu, and Jintang Shang
Southeast University, CHINA

T54-f WAFER-LEVEL PATTERNING OF TIN OXIDE
NANOSHEETS FOR MEMS GAS SENSORS 893
Mingjie Li, Wenxin Luo, Xiaojiang Liu, Gaoqiang Niu, and Fei Wang
Southern University of Science and Technology, CHINA

f - MEMS Physical and Chemical Sensors
Materials for Physical Sensors

W54-f AIR DAMPING EFFECTS ON DIFFERENT MODES OF AlN-on-Si
MICROELECTROMECHANICAL RESONATORS 897
Yuncong Liu[1], S M Enamul Hoque Yousuf[1], Afzaal Qamar[2], Mina Rais-Zadeh[2,3], and Philip X.-L. Feng[1]
[1]University of Florida, USA, [2]University of Michigan, USA, and [3]California Institute of Technology, USA

M55-f A NOVEL PIEZORESISTIVE PRESSURE SENSOR
BASED ON CR-DOPED V_2O_3 THIN FILM 901
Michiel Gidts, Wei-Fan Hsu, María Recaman Payo, Shashwat Kushwaha, Chen Wang,
Frederik Ceyssens, Dominiek Reynaerts, Jean-Pierre Locquet, and Michael Kraft
KU Leuven, BELGIUM

f - MEMS Physical and Chemical Sensors
Metrology and Measurement Techniques for MEMS/NEMS Sensors

T55-f A NOVEL FEEDTHROUGH CANCELLATION TECHNIQUE FOR
PIEZOELECTRIC MEMS RESONANT SENSORS IN IONIC LIQUID MEDIUM 905
Cheng-Yen Wu, Zhong-Wei Lin, and Sheng-Shian Li
National Tsing Hua University, TAIWAN

W55-f CHARACTERIZATION OF PACKAGING STRESS WITH
A CAPACITIVE STRESS SENSOR ARRAY 909
Tolga Veske[1], Derin Erkan[1], and Erdinc Tatar[1,2]
[1]Bilkent University, TURKEY and [2]National Nanotechnology Research Center (UNAM), TURKEY

M56-f MILLISECOND-LEVEL PULSE-HEATING SENSING SYSTEM
FOR MEMS-BASED GAS SENSORS 913
Yi Zhuang, Gaoqiang Niu, Lang Wu, and Fei Wang
Southern University of Science and Technology, CHINA

T56-f MULTIPLE PARAMETER DECOUPLING USING A SINGLE
RESONANT MEMS SENSOR VIA BLUE SIDEBAND EXCITATION 917
Jingqian Xi[1], Lei Xu[1], Yuan Wang[2], Fangjing Hu[1], Chengxin Li[4], Linlin Wang[4],
Huafeng Liu[1], Chen Wang[4], Michael Kraft[4], and Chun Zhao[3]
*[1]Huazhong University of Science and Technology, CHINA, [2]University of Macau, CHINA,
[3]University of York, UK, and [4]University Leuven, BELGIUM*

f - MEMS Physical and Chemical Sensors
Nanoscale Physical Sensors

W56-f DIAMOND NANOWIRES ARRAY PREPARED BY ANNEALING NANO-CRYSTALLINE
DIAMOND IN AIR AND ITS APPLICATION IN FIELD EMISSION 921
Yang Wang, Chen Lin, and Jinwen Zhang
Peking University, CHINA

M57-f QUANTIFIED STRESS RELAXATION IN CARBON NANOTUBE RESONATORS 925
Morten Vollmann, Cosmin Roman, Miroslav Haluska, and Christofer Hierold
ETH Zürich, SWITZERLAND

T57-f SELF-REFERENCED TEMPERATURE SENSORS
BASED ON CASCADED SILICON RING RESONATOR 929
Xiantao Zhu, Minmin You, Zude Lin, Bin Yang, and Jingquan Liu
Shanghai Jiao Tong University, CHINA

f - MEMS Physical and Chemical Sensors
Sonic & Ultrasonic MEMS Transducers

W57-f A 0.35 mm[2] SYSTEM ON CHIP LEVEL DETECTOR BASED
ON AN ANNULAR PMUT-ON-CMOS ARRAY 933
Eyglis Ledesma, Iván Zamora, Francesc Torres, Arantxa Uranga, and Núria Barniol
Universitat Autònoma de Barcelona, SPAIN

M58-f AN ALSCN PMUT-ON-CMOS SENSOR FOR MONITORING FLUIDS' DENSITY,
VISCOSITY, SOUND VELOCITY, AND COMPRESSIBILITY 937
Eyglis Ledesma, Iván Zamora, Jesús Yanez, Arantxa Uranga, and Núria Barniol
Universitat Autònoma de Barcelona, SPAIN

T58-f AUTO-POSITIONING AND HAPTIC STIMULATIONS
VIA A 35 MM SQUARE PMUT ARRAY 941
Wei Yue[1], Yande Peng[1], Hanxiao Liu[1], Fan Xia[1], Fanping Sui[1], Seiji Umezawa[2],
Shinsuke Ikeuchi[2], Yasuhiro Aida[2], and Liwei Lin[1]
[1]University of California, Berkeley, USA and [2]Murata Manufacturing Co., Ltd., JAPAN

W58-f BODY FORCE BASED DROPLET EJECTION
BY GHZ ACOUSTIC MICRO-TRANSDUCER 945
Haitao Zhang, Yangchao Zhou, Menglun Zhang, Wenlan Guo, Chen Sun, Xuexin Duan, and Wei Pang
Tianjin University, CHINA

M59-f BONE CONDUCTION PICKUP BASED ON PIEZOELECTRIC
MICROMACHINED ULTRASONIC TRANSDUCERS 949
Chongbin Liu[1], Xiangyang Wang[1], Yong Xie[2], and Guoqiang Wu[1]
[1]Wuhan University, CHINA and [2]Xidian University, CHINA

T59-f BREAKING THE DEAD ZONE LIMITATION OF PMUTS BASED ON A PHASE SHIFT OF
DRIVING WAVEFORM WITH WINDOW FUNCTION 953
Chun-You Liu, Chin-Yu Chang, and Sheng-Shian Li
National Tsing Hua University, TAIWAN

W59-f DRONE-MOUNTED LOW-FREQUENCY PMUTS
FOR > 6-METER RANGEFINDER IN AIR 957
Hanxiao Liu[1], Yande Peng[1], Wei Yue[1], Seiji Umezawa[2], Shinsuke Ikeuchi[2], Yasuhiro Aida[2],
Chunming Chen[1], Peggy Tsao[1], and Liwei Lin[1]
[1]University of California, Berkeley, USA and [2]Murata Manufacturing Co., Ltd., JAPAN

M60-f MASS PRODUCED MICROMACHINED ULTRASONIC TIME-OF-FLIGHT
SENSORS OPERATING IN DIFFERENT FREQUENCY BANDS 961
Richard J. Przybyla[1], Stefon E. Shelton[1], Cathy Lee[1], Ben Eovino[1], Quy Chau[1],
Mitchell H. Kline[1], Oleg I. Izyumin[1], and David A. Horsley[1,2]
[1]TDK Invensense, USA and [2]University of California, Davis, USA

T60-f MEMS FIRST-ORDER BESSEL BEAM ACOUSTIC TRANSDUCER
FOR PARTICLE TRAPPING AND CONTROLLABLE ROTATING 965
Jiaqi Li[1], Zhenhuan Sun[1], Yuyu Jia[1], Teng Li[1], Haojian Lu[2], Lurui Zhao[3], Hai Liu[3], and Song Liu[1]
*[1]ShanghaiTech University, CHINA, [2]Zhejiang University, CHINA, and
[3]University of Southern California, Los Angeles, USA*

W60-f NON-INVASIVE CAROTID ARTERY MONITORING BY USING ALUMINUM NITRIDE PMUT CLOSE-PACKED ARRAYS 969
Sheng Wu[1,2,3], Kangfu Liu[2], Shuai Shao[2], Wei Li[1,3], Ying Chen[1,3], Tao Wu[2], and Xinxin Li[1,3]
[1]Chinese Academy of Sciences, CHINA, [2]ShanghaiTech University, CHINA, and
[3]University of Chinese Academy of Sciences, CHINA

M61-f NON-LINEAR BEHAVIORAL MODELING OF CAPACITIVE MEMS MICROPHONES 973
Sebastian Anzinger[1,2], Hutomo Suryo Wasisto[1], Abhiraj Basavanna[1], and Alfons Dehé[2,3]
[1]Infineon Technologies AG, GERMANY, [2]University of Freiburg, GERMANY, and
[2]Hahn-Schickard-Gesellschaft, GERMANY

T61-f VORTEX-BEAM ACOUSTIC TRANSDUCER FOR UNDERWATER PROPULSION 977
Jaehoon Lee, Kianoush Sadeghian Esfahani, and Eun S. Kim
University of Southern California, USA

W61-f WIDEBAND AND HIGHLY SENSITIVE MICROMACHINED PZT FILM-BASED ULTRASONIC MICROPHONE WITH PARYLENE FILM AND FLEXIBLE HELMHOLTZ RESONATOR ENHANCEMENT 981
Chung-Hao Huang and Guo-Hua Feng
National Tsing Hua University, TAIWAN

f - MEMS Physical and Chemical Sensors
Other Physical Sensors

M62-f HALBACH-ARRAY MAGNETIC COIL ARRANGEMENT ON CMOS CHIP FOR SENSITIVITY ENHANCEMENT OF INDUCTIVE TACTILE SENSOR 985
Tien Chou, Zih-Song Hu, and Weileun Fang
National Tsing Hua University, TAIWAN

T62-f *ON-MEMS-CHIP* COMPACT TEMPERATURE SENSOR FOR LARGE-VOLUME, LOW-COST SENSOR CALIBRATION 989
Paolo Frigerio[1], Andrea Fagnani[1], Valentina Zega[1], Gabriele Gattere[2], Attilio Frangi[1], and Giacomo Langfelder[1]
[1]Politecnico di Milano, ITALY and [2]STMicroelectronics, ITALY

W62-f PARTICULATE MATTER SENSOR BASED ON TWO STAGE CASCADE VIRTUAL IMPACTORS AND THERMOPHORETIC MICROHEATERS 993
Kwang-Wook Choi[1], Ilhwan Kim[1], Seokwhan Chung[1], Gi-Bong Sung[2], and Se-Jin Yook[2]
[1]Samsung Advanced Institute of Technology, KOREA and [2]Hanyang University, KOREA

g – Micro- and Nanofluidics
Biological and Medical Microfluidics and Nanofluidics

M63-g A MICROFLUIDIC OXYGEN GRADIENT GENERATOR FOR THE STUDY OF AEROTROPISM IN HYPHAE OF OOMYCETES 997
Ayelen Tayagui[1,2], Yiling Sun[1,2], Ashley Garrill[1], and Volker Nock[1,2]
[1]University of Canterbury, NEW ZEALAND and
[2]MacDiarmid Institute for Advanced Materials and Nanotechnology, NEW ZEALAND

T63-g A PAPER-BASED DUAL APTAMER ASSAY ON AN INTEGRATED MICROFLUIDIC SYSTEM FOR DETECTION OF HNP 1 AS A BIOMARKER FOR PERIPROSTHETIC JOINT INFECTIONS 1001
Rishabh Gandotra[1], Feng-Chih Kuo[2], Mel S. Lee[3], and Gwo-Bin Lee[1]
[1]National Tsing Hua University, TAIWAN, [2]Kaohsiung Chang Gung Memorial Hospital, TAIWAN, and
[3]Paochien Hospital, TAIWAN

W63-g AN INTEGRATED MICROFLUIDIC PLATFORM FOR TUMOR CELL SEPARATION AND FLUORESCENCE IN SITU HYBRIDIZATION AT SINGLE CELL LEVEL 1005
Shihui Qiu[1,2], Na Li[1,2], Zhenhua Wu[1,2], Jianlong Zhao[1,2], and Hongju Mao[1,2]
[1]Chinese Academy of Science, CHINA and [2]University of Chinese Academy of Sciences, CHINA

M64-g CHARACTERIZATION OF OOCYTE HARDENING USING A MICROFLUIDIC ASPIRATION-ASSISTED ELECTRICAL IMPEDANCE SPECTROSCOPY SYSTEM 1009
Yuan Cao, Julia Floehr, and Uwe Schnakenberg
RWTH Aachen University, GERMANY

T64-g DOUBLE PULSE IRRADIATION OF FS LASER FOR ENHANCING THE PERFORMANCE OF PRECISE LASER SORTING METHOD 1013
Ryota Kiya[1], Yoshinaga Rintaro[1], Yo Tanaka[2], Yaxiaer Yalikun[1,2], and Yoichiroh Hosokawa[1]
*[1]Nara Institute of Science and Technology, JAPAN and
[2]Institute of Physical and Chemical Research (RIKEN), JAPAN*

W64-g DROPLET BASED HIGH THROUGHPUT SINGLE-SPERM CRYOPRESERVATION PLATFORM 1017
Na Li[1,2], Shihui Qiu[1,3], Zhenhua Wu[1,3], and Hongju Mao[1,3]
*[1]Chinese Academy of Sciences, CHINA, [2]ShanghaiTech University, CHINA, and
[3]University of Chinese Academy of Sciences, CHINA*

M65-g DUAL ION-SELECTIVE MEMBRANE DEPOSITED ION-SENSITIVE FIELD-EFFECT TRANSISTOR (DISM-ISFET) INTEGRATING WHOLE BLOOD PROCESSING MICROCHAMBER FOR IN SITU BLOOD ION TESTING 1021
Xiao-Wen Chen, Syuan-Rong Huang, and Nien-Tsu Huang
National Taiwan University, TAIWAN

g – Micro- and Nanofluidics
Generic Microfluidics & Nanofluidics

W65-g STRONG MICROSTREAMING FROM A PINNED OSCILLATING MEMBRANE AND APPLICATION TO GAS EXCHANGE 1025
Anthony L. Mercader and Sung Kwon Cho
University of Pittsburgh, USA

M66-g TUNABLE NANOPORE-INTEGRATED MICRO-/NANOFLUIDIC PLATFORM FOR ION TRANSPORT CONTROL IN THE PRESENCE OF CONCENTRATION AND TEMPERATURE GRADIENTS 1029
Dongwoo Seo[1], Dongjun Kim[1], Jongwan Lee[1], Cong Wang[2], Jungyul Park[2], and Taesung Kim[1]
[1]Ulsan National Institute of Science and Technology (UNIST), KOREA and [2]Sogang University, KOREA

g – Micro- and Nanofluidics
Integrated/Embedded Microfluidics and Nanofluidic Systems & Platforms

W66-g QUANTITATIVE ASSESSMENT OF CAPTURED MAGNETIC NANOPARTICLES USING SELF-POWERED MAGNETOELECTRIC PLATFORM FOR BIOLOGICAL APPLICATIONS 1033
Pankaj Pathak, Vinit K. Yadav, Samaresh Das, and Dhiman Mallick
Indian Institute Of Technology Delhi, INDIA

M67-g REAL-TIME OPERATION OF MICROCANTILEVER-BASED IN-PLANE RESONATORS PARTIALLY IMMERSED IN A MICROFLUIDIC SAMPLER 1037
Jiushuai Xu, Entian Cao, Michael Fahrbach, Vladislav Agluschewitsch, Andreas Waag, and Erwin Peiner
Technische Universität Braunschweig, GERMANY

T67-g SUSPENDED NANOCHANNEL RESONATORS MADE BY
NANOIMPRINT AND GAS PHASE DEPOSITION 1041
Manuel Müller[1], Jeremy Teuber[1], Rukan Nasri[1], Francesc Torres Canals[2], Núria Barniol[2],
Jordi Llobet Sixto[3], Xavier Borrise[3], Francesc Perez-Murano[3], and Irene Fernandez-Cuesta[1]
*[1]University of Hamburg, GERMANY, [2]Universitat Autónoma de Barcelona, SPAIN, and
[3]IMB-CNM CSIC, SPAIN*

g – Micro- and Nanofluidics
Manufacturing for Micro- and Nanofluidics

W67-g DEVELOPING AN EXTREMELY HIGH FLOW RATE PNEUMATIC PERISTALTIC
MICROPUMP FOR BLOOD PLASMA SEPARATION WITH INERTIAL PARTICLE
FOCUSING TECHNIQUE FROM FINGERTIP BLOOD WITH LANCETS 1045
Tuan N.A. Vo[1,2,3], Pin-Chuan Chen[1], and Pai-Shan Chen[4]
*[1]National Taiwan University of Science and Technology, TAIWAN, [2]Ho Chi Minh City University of
Technology (HCMUT), VIETNAM, [3]Vietnam National University, VIETNAM, and [4]National Taiwan
University, TAIWAN*

M68-g DIRECT PATTERNING ON POROUS SURFACE USING DROP IMPACT PRINTING 1049
Bheema Sankar Reddy[1], Chandantaru Dey Modak[1,2], Rutvik Lathia[1],
Bhawana Agarwal[1,3], Ebinesh Abraham R[1], and Prosenjit Sen[1]
*[1]Indian Institute of Science, Bangalore, INDIA, [2]CNRS - ESPCI PSL, FRANCE, and
[3]Johns Hopkins University, USA*

T68-g MANUFACTURING 3D-PRINTED PAPER MICROFLUIDICS INTEGRATED
WITH IONIZATION MASS-SPECTROMETRY FOR ILLICIT DRUGS
ANALYSIS AND ON-CHIP CHROMATOGRAPHY 1052
Muhammad Faizul Zaki[1], Pin-Chuan Chen[1], Yi-Xin Wu[2], and Pai-Shan Chen[2]
*[1]National Taiwan University of Science and Technology, TAIWAN and
[2]National Taiwan University, TAIWAN*

g – Micro- and Nanofluidics
Materials for Micro & Microfluidics

W68-g DETECTION LIMITS IN NANOMECHANICAL MASS FLOW SENSING
FOR NANOFLUIDICS WITH NANOWIRE OPEN CHANNELS 1056
Javier E. Escobar, Juan Molina, Eduardo Gil-Santos, José J. Ruz, Óscar Malvar,
Priscila M. Kosaka, Javier Tamayo, Álvaro San Paulo, and Montserrat Calleja
Instituto de Micro y Nanotecnología, IMN-CNM (CSIC), SPAIN

g – Micro- and Nanofluidics
Modeling of Micro & Nanofluidics

M69-g CONTROLLING PARTICLE AGGREGATION AND SEPARATION
IN LIQUID ON MEMBRANE RESONATORS 1060
Haoran Zhang[1,2], Hao Jia[1,2], and Xinxin Li[1,2]
[1]Chinese Academy of Sciences, CHINA and [2]University of Chinese Academy of Sciences, CHINA

T69-g DEVELOPMENT OF BOAT MODEL POWERED BY
ELECTRO-HYDRODYNAMIC PROPULSION SYSTEM 1064
Luan Ngoc Mai[1,2], Tuan-Khoa Nguyen[3], Trung Hieu Vu[3], Thien Xuan Dinh[4], Canh-Dung Tran[5], Hoang-
Phuong Phan[6], Toan Dinh[5], Thanh Nguyen[5], Nam-Trung Nguyen[3], Dzung Viet Dao[3], and Van Thanh Dau[3]
*[1]Ho Chi Minh City University of Technology, VIETNAM, [2]Vietnam National University Ho Chi Minh City,
VIETNAM, [3]Griffith University, AUSTRALIA, [4]Explosion Research Institute Inc., JAPAN, [5]University of
Southern Queensland, AUSTRALIA, and [6]University of New South Wales, AUSTRALIA*

W69-g HEMODYNAMIC ANALYSIS OF CARDIOMEMS:
ADVERSE HEMODYNAMIC EFFECTS 1068
Zhenhao Liu[1], Jiangli Han[2], and Xing Chen[1]
[1]Beihang University, CHINA and [2]Peking University Third Hospital, CHINA

M70-g MODAL QUALITY FACTOR INVERSION OF NON-SLENDER
MEMS RESONATORS BETWEEN GASES AND LIQUIDS 1072
Andre L. Gesing, Thomas Tran, Daniel Platz, and Ulrich Schmid
TU Wien, AUSTRIA

g – Micro- and Nanofluidics
Other Micro- and Nanofluidics

T70-g CLASSIFYING CELL CYCLE BY ELECTRICAL
PROPERTIES USING MACHINE LEARNING 1076
Jian Wei and Xiaoxing Xing
Beijing University of Chemical Technology, CHINA

W70-g HIGH-THROUGHPUT SPHERICAL SUPRAPARTICLE SELF-ASSEMBLY
BY ENHANCED EVAPORATION OF COLLOIDAL WATER DROPLETS
THROUGH THIN FILM OF WATER-SOLUBLE OIL 1080
Wonhyung Lee, Joowon Rhee, and Joonwon Kim
Pohang University of Science and Technology (POSTECH), KOREA

M71-g IN-ICE POLYMERIZATION FOR FUNCTIONAL HYDROGEL MICROBEAD
WITH FLASH FREEZING CENTRIFUGAL MICROFLUIDIC DEVICE 1084
Tomomi Murayama[1], Koki Yoshida[1], Yuta Kurashina[2], and Hiroaki Onoe[1]
[1]Keio University, JAPAN and [2]Tokyo University of Agriculture and Technology, JAPAN

T71-g TEMPERATURE-RESPONSIVE MICROCAPSULES MANUFACTURED BY PROMOTING
CONTROLLED CLOAKING WITH THE HELP OF MICRO/NANOPARTICLES 1088
Rutvik Lathia[1], Bheema Sankar Reddy[1], Chandantaru Dey Modak[1,2],
Satchit Nagpal[1,3], and Prosenjit Sen[1]
[1]Indian Institute of Science, INDIA, [2]CNRS - ESPCI PSL, FRANCE, and [3]Texas A&M University, USA

W71-g WATER VITRIFICATION IN A MICROCHANNEL AT LOW COOLING RATE 1091
Ayane Sato, Tomohiro Hayashi, and Tadashi Ishida
Tokyo Institute of Technology, JAPAN

h - Optical, RF and Electromagnetics for MEMS/NEMS
Electrical Field and Magnetic Field Sensors and Transducers

M72-h A HIGHLY SENSITIVE 3-AXIS MICRO SEARCH-COIL MAGNETOMETER
ENABLED BY HIGH DENSITY THROUGH-SILICON-VIA PROCESS 1095
Hadi Tavakkoli, Mingzheng Duan, Longheng Qi, Izhar, Xu Zhao, and Yi-Kuen Lee
Hong Kong University of Science and Technology, HONG KONG

T72-h FULLY INTEGRATED BACK-BIASED 3D HALL SENSOR WITH
WAFER-LEVEL INTEGRATED PERMANENT MICROMAGNETS 1099
Björn Gojdka[1], Daniel Cichon[2], Markus Stahl-Offergeld[2], Dominik Schröder[3],
Niels Clausen[1], Christian Hedayat[3], Hans-Peter Hohe[2], and Thomas Lisec[1]
*[1]Fraunhofer Institute for Silicon Technology ISIT, GERMANY, [2]Fraunhofer Institute for Integrated Circuits
IIS, GERMANY, and [3]Fraunhofer Institute for Electronic Nano Systems ENAS, GERMANY*

h - Optical, RF and Electromagnetics for MEMS/NEMS
Free Space Optical Components & Systems

W72-h A LARGE-STROKE TIP-TILT-PISTON MICROMIRROR WITH
ELECTROMAGNETIC ACTUATORS BASED ON METALLIC GLASS 1103
Chuan-Hui Ou, Nguyen V. Toan, and Takahito Ono
Tohoku University, JAPAN

M73-h ARBITRARY SHAPED BACKSIDE REINFORCEMENT FOR
TWO DIMENSIONAL RESONANT MICROMIRRORS 1107
Takashi Sasaki, Adrien Piot, Anton Lagosh, Clement Fleury, Markus Bainschab, Yanfen Zhai,
Marcus Baumgart, Sara Guerreiro, Dominik Holzmann, Aleš Travnik, and Mohssen Moridi
Silicon Austria Labs, AUSTRIA

T73-h HIGH TRANSMITTANCE METASURFACE HOLOGRAMS USING SILICON NITRIDE 1111
Masakazu Yamaguchi, Hiroki Saito, Satoshi Ikezawa, and Kentaro Iwami
Tokyo University of Agriculture and Technology, JAPAN

W73-h MULTIFUNCTIONAL OPTICAL METASURFACE FOR ANOMALOUS REFLECTION,
STRUCTURAL COLOR, AND SURFACE LATTICE RESONANCE 1115
Liye Li[1], Hongshun Sun[1], Yifan Ouyang[1], Shengxiao Jin[1], Tian Kang[1], Zhimei Qi[2], and Wengang Wu[1]
[1]Peking University, CHINA and [2]Chinese Academy of Science, CHINA

M74-h NOVEL WAVEFRONT-SPLITTING INTERFEROMETER FOR ULTRA-COMPACT
BROADBAND FT-IR SPECTROSCOPY EXTENDING TO VISIBLE RANGE 1119
Bassem Mortada[1], Yasser M. Sabry[1,2], Bassam Saadany[1], Tarik Bourouina[3], and Diaa Khalil[2]
[1]Si-Ware Systems, EGYPT, [2]Ain Shams University, EGYPT, and [3]Université Gustave Eiffel, FRANCE

T74-h PIEZOELECTRICALLY ACTUATED MICROMIRROR WITH
DYNAMIC DEFORMATION COMPENSATION MECHANISM 1123
Takashi Sasaki, Adrien Piot, Jaka Pribošek, Anton Lagosh, Clement Fleury, Markus Bainschab, Yanfen Zhai,
Marcus Baumgart, Sara Guerreiro, Dominik Holzmann, Aleš Travnik, and Mohssen Moridi
Silicon Austria Labs, AUSTRIA

W74-h RESONANT d_{33} MODE PZT MEMS MIRROR EXCITED WITH
DIRECTIONAL INTERDIGITATED ELECTRODES 1127
Pooja Thakkar, Anton Lagosh, Takashi Sasaki, Markus Bainschab, and Jaka Pribošek
Silicon Austria Labs GmbH, AUSTRIA

M75-h RESONANT PIEZOELECTRIC VARIFOCAL MIRROR WITH ON-CHIP INTEGRATED
DIFFRACTIVE OPTICS FOR INCREASED FREQUENCY RESPONSE 1131
Jaka Pribošek, Anton Lagosh, Pooja Thakkar, Takashi Sasaki, and Markus Bainschab
Silicon Austria Labs, AUSTRIA

T75-h UNIQUE DISPERSION RELATION FOR PLASMONIC
PHOTODETECTORS WITH SUBMICRON GRATING 1135
Yuki Kaneda[1], Masaaki Oshita[1], Utana Yamaoka[1], Shiro Saito[2], and Tetsuo Kan[1]
[1]University of Electro-Communications, JAPAN and [2]IMRA JAPAN Co., LTD., JAPAN

h - Optical, RF and Electromagnetics for MEMS/NEMS
Infrared (IR) Sensors and Imaging Systems

W75-h INTEGRATION OF A HIGH TEMPERATURE TRANSITION METAL OXIDE
NTC THIN FILM IN A MICROBOLOMETER FOR LWIR DETECTION 1139
Sarah Risquez[1], Sebastian Redolfi[2], Clement Fleury[1], Matthias Wulf[2], Ali Roshanghias[1],
Adrien Piot[1], Jeremy Streque[1], Kerstin Schmoltner[2], Thang Duy Dao[1], Markus Puff[2], and Mohssen Moridi[1]
[1]Silicon Austria Labs GmbH, AUSTRIA and [2]TDK Electronics GmbH & Co OG, AUSTRIA

M76-h PERIODIC CAVITIES ON THE IR-ABSORBER FOR RESPONSIVITY
ENHANCEMENT OF CMOS-MEMS THERMOELECTRIC IR SENSOR 1143
Yung-Chen Li, Tien Chou, Pen-Sheng Lin, Yu-Cheng Huang, Fuchi Shih,
You-An Lin, Da-Jen Yen, Mei-Feng Lai, and Weileun Fang
National Tsing Hua University, TAIWAN

T76-h ULTRA-LARGE PIXEL ARRAY PHOTOTHERMAL TRANSDUCER
AND ITS THERMAL PERFORMANCE PREDICTION STRATEGY 1147
Defang Li[1,3], Jinying Zhang[1,2], Jiushuai Xu[3], Erwin Peiner[3], Zhuo Li[1,2],
Xin Wang[1], Suhui Yang[1], and Yanze Gao[1]
*[1]Beijing Institute of Technology, CHINA, [2]Yangtze Delta Region Academy of Beijing Institute of Technology,
CHINA, and [3]Technische Universität Braunschweig, GERMANY*

h - Optical, RF and Electromagnetics for MEMS/NEMS
MEMS for Timing & Frequency Control

W76-h A CMOS-MEMS BEAM RESONATOR WITH $Q > 10,000$ 1151
Ting-Yi Chen and Wei-Chang Li
National Taiwan University, TAIWAN

M77-h GENERIC TEMPERATURE COMPENSATION SCHEME FOR
CMOS-MEMS RESONATORS BASED ON ARC-BEAM
DERIVED ELECTRICAL STIFFNESS FREQUENCY PULLING 1155
I-Chieh Hsieh, Hong-Sen Zheng, Chun-Pu Tsai, Ting-Yi Chen, and Wei-Chang Li
National Taiwan University, TAIWAN

T77-h HIGH-Q AND LOW-MOTIONAL IMPEDANCE PIEZOELECTRIC
MEMS RESONATOR THROUGH MECHANICAL MODE COUPLING 1159
Linhai Huang[1], Zhihong Feng[1], Yuhao Xiao[2], Fengpei Sun[1], and Jinghui Xu[1]
[1]Huawei Technologies Company Ltd., CHINA and [2]Wuhan University, CHINA

h - Optical, RF and Electromagnetics for MEMS/NEMS
Photonic Components & Systems

W77-h CROSSTALK-FREE LARGE APERTURE 2D GIMBAL MICROMIRROR 1163
Behrad Ghazinouri and Siyuan He
Toronto Metropolitan University, CANADA

M78-h INVERSE INTERFERENCE EFFECT-ENHANCED ULTRASENSITIVE
SENSING VIA MID-IR NANOANTENNAS 1167
Hong Zhou, Dongxiao Li, Xinge Guo, Zhihao Ren, and Chengkuo Lee
National University of Singapore, SINGAPORE

T78-h TWISTED AND CONTACTED AU MICRO-RODS 3D CHIRAL METAMATERIALS
WITH CIRCULAR DICHROISM VIA AN ABSORPTIVE ROUTE IN
LONG-WAVELENGTH INFRARED 1171
Gaku Furusawa[1], Natsuki Kanda[2], Ryusuke Matsunaga[2], and Tetsuo Kan[1]
[1]University of Electro-Communications, JAPAN and [2]University of Tokyo, JAPAN

h - Optical, RF and Electromagnetics for MEMS/NEMS
RF MEMS Components & Systems

W78-h 3D HYBRID ACOUSTIC RESONATOR WITH COUPLED FREQUENCY RESPONSES
OF SURFACE ACOUSTIC WAVE AND BULK ACOUSTIC WAVE 1175
Liping Zhang[1,2], Shibin Zhang[1], Jinbo Wu[1,2], Pengcheng Zheng[1,2], Hulin Yao[1,2],
Yang Chen[1,2], Kai Huang[1,2], Xiaomeng Zhao[1], Min Zhou[1], and Xin Ou[1,2]
[1]Chinese Academy of Sciences, CHINA and [2]University of Chinese Academy of Sciences, CHINA

M79-h A C/K_U DUAL-BAND RECONFIGURABLE BAW FILTER
USING POLARIZATION TUNING IN LAYERED SCALN 1179
Dicheng Mo, Shaurya Dabas, Sushant Rassay, and Roozbeh Tabrizian
University of Florida, USA

T79-h ACOUSTOELECTRIC-DRIVEN FREQUENCY MIXING IN
MICROMACHINED LITHIUM NIOBATE ON SILICON WAVEGUIDES 1183
Hakhamanesh Mansoorzare and Reza Abdolvand
University of Central Florida, USA

W79-h EFFECT OF SCANDIUM COMPOSITION ON THE PHONON SCATTERING LIFETIME
OF ALUMINUM SCANDIUM NITRIDE ACOUSTIC WAVE RESONATORS 1186
Yue Zheng[1], Mingyo Park[1], Chao Yuan[2], and Azadeh Ansari[1]
[1]Georgia Institute of Technology, USA and [2]Wuhan University, CHINA

M80-h LITHIUM NIOBATE THIN FILM BASED A_1 MODE RESONATORS WITH FREQUENCY
UP TO 16 GHZ AND ELECTROMECHANICAL COUPLING FACTOR NEAR 35% 1190
Rongxuan Su[1], Zhenyi Yu[2], Sulei Fu[1], Huiping Xu[1], Shuai Zhang[1], Peisen Liu[1],
Yu Guo[2], Cheng Song[1], Fei Zeng[1], and Feng Pan[1]
[1]Tsinghua University, CHINA and [2]Jiangnan University, CHINA

T80-h SUB-3 DB INSERTION LOSS BROADBAND ACOUSTIC DELAY LINES AND
HIGH FOM RESONATORS IN LINBO$_3$/SIO$_2$/SI FUNCTIONAL SUBSTRATE 1194
Chun-Chen Yeh, Chia-Hsien Tsai, Guan-Lin Wu, Tzu-Hsuan Hsu, and Ming-Huang Li
National Tsing Hua University, TAIWAN

W80-h SUPPRESSION OF SPURIOUS MODES IN ALUMINUM NITRIDE S_1 LAMB WAVE
RESONATORS USING A MECHANICAL SOFT-CONTACT SCHEME 1198
Shao-Siang Tung[1], Tzu-Hsuan Hsu[1], Yens Ho[2], Yung-Hsiang Chen[2],
Yelehanka R. Pradeep[3], Rakesh Chand[3], and Ming-Huang Li[1]
*[1]National Tsing Hua University, TAIWAN, [2]Vanguard International Semiconductor Corporation, TAIWAN,
and [3]Vanguard International Semiconductor Corporation Singapore PTE. Ltd., SINGAPORE*

h - Optical, RF and Electromagnetics for MEMS/NEMS
THz MEMS Components & Systems

M81-h TERAHERTZ REFLECTIVE METALENS FOR ARBITRARY
OFF-AXIS FOCUSING WITH LARGE DEPTH OF FOCUS 1202
Jiahao Miao, Yi Liu, Cong Lin, Zhanxuan Zhou, and Xiaomei Yu
Peking University, CHINA

h - Optical, RF and Electromagnetics for MEMS/NEMS
Other Electromagnetic MEMS/NEMS

T81-h TOWARDS A BETTER CMOS-MEMS RESOSWITCH USING
ELECTROLESS PLATING FOR CONTACT ENGINEERING 1206
Ting-Jui Liou, Chun-Pu Tsai, Ting-Yi Chen, and Wei-Chang Li
National Taiwan University, TAIWAN

i - Open Posters

W81-i **A MEMS-CMOS INFRA-RED MICROSYSTEM WITH IN-SENSOR MACHINE LEARNING CAPABILITIES**
Marco Castellano, Ugo Garozzo, Luca Gandolfi, Davide Ruggiero, and Giuseppe Bruno
STMicroelectronics, ITALY

M82-i **A NOVEL BAROMETRIC PRESSURE SENSOR WITH A CAPACITVE TRANSDUCER AND IMPROVED PERFORMANCE**
Thomas Friedrich[1], Volkmar Senz[1], and Ferenc Lukacs[2]
[1]Robert Bosch GmbH, GERMANY and [2]Robert Bosch Kft., HUNGARY

T82-i **A NOVEL CLASS OF MOTION SENSORS FEATURED WITH AN ELECTRIC POTENTIAL SENSING CHANNEL**
Enrico R. Alessi, Fabio Passaniti, and Emanuele Lavelli
STMicroelectronics, ITALY

W82-i **A STABLE MIR PHOTODETECTOR BASED ON 2D PTSI/P-SI NANOHOLE ARRAYS**
Ashenafi A. Elyas, Masahiko Shiraishi, and Tetsuo Kan
University of Electro-communications, JAPAN

M83-i **AN EQUIVALENT CIRCUIT MODEL FOR THE PHASE GRADIENT METASURFACE ANALYSIS IN VISIBLE BAND**
Liye Li[1], Senyong Hu[1], Yifan Ouyang[1], Yusa Chen[1], Meizhang Wu[2], and Wengang Wu[1]
[1]Peking University, CHINA and [2]University of Science and Technology Beijing, CHINA

T83-i **DETECTION OF MASS AND MATERIAL NATURE OF MICROPARTICLES BY A PIEZOELCTRIC MEMS**
Francesco Foncellino and Luigi Barretta
STMicroelectronics, ITALY

W83-i **ELECTRO-OPTICAL TESTING SOLUTION FOR TMOS MEMS SENSOR SENSITIVITY ASSESSMENT AT WAFER LEVEL**
Roberta Carbone, Dario Premi, and Marco Rossi
STMicroelectronics, ITALY

T84-i **HIGH PERFORMANCE SPUTTERED PZT PMUTS OPERATING IN THE ULTRASOUND IMAGING RANGE REPRODUCIBLE AT WAFER-SCALE**
Jihang Liu[1], David Sze Wai Choong[1], Duan Jian Goh[1], Merugu Srinivas[1], Qing Xin Zhang[1], Steven Lee Hou Jang[1], Huamao Lin[1], Fabio Quaglia[3], Domenico Giusti[3], Laura Castoldi[3], Claudia Pedrini[3], Luca Barabani[3], Annachiara Esposito[3], Luigi Barretta[3], Rossana Scaldaferri[3], Alberto Leotti[2], Adriyan Hidayat Mohamed Hamsah[3], Peter Chang Hyun Kee[1], and Lee En-Yuan Joshua[1]
[1]Institute of Microelectronics, SINGAPORE, [2]ST Microelectronics, SINGAPORE, and [3]ST Microelectronics, ITALY

W84-i **PIEZOELECTRIC ACTUATOR INTRODUCTION FOR ACCURATE POSITIONING READ/WRITE ELEMENT IN HARD DISK DRIVE (HDD)**
Domenico Giusti and Marco Ferrera
STMicroelectronics, ITALY

M85-i **PIEZOELECTRIC MEMS FOR MICROPARTICLES DETECTION: ALTERNATIVE READOUT FOR MASS DETECTION**
Luigi Barretta and Francesco Foncellino
STMicroelectronics, ITALY

T85-i **SIDE WALL DETECTION TYPE SPR SENSOR WITH GOLD GRATING ON GLASS**
Masaaki Oshita, Shinichi Suzuki, Kazuto Masamoto, and Tetsuo Kan
University of Electro-Communications, JAPAN

W85-i **SPUTTERED PZT AIR-COUPLED PMUTS WITH WIDE BANDWIDTH AND LONG DETECTION RANGE FOR RANGING APPLICATIONS**
Mantalena Sarafianou[1], David Sze Wai Choong[1], Duan Jian Goh[1], Jihang Liu[1], Joshua En-Yuan Lee[1], Srinivas Merugu[1], Qing Xin Zhang[1], Peter Hyun Kee Chang[1], Fabio Quaglia[2], Domenico Giusti[2], Laura Castoldi[2], Filippo D'Ercoli[2], Riccardo Tacchini[2], Alberto Leotti[3], and Dao Hao Sim[3]
[1]Institute of Microelectronics, SINGAPORE, [2]ST Microelectronics, ITALY, and [3]ST Microelectronics, SINGAPORE

M86-i **THERMOELECTRIC MIROPHONE**
Akash Gupta[1], Dr. Achim Bittner[1], Prof. Dr.-Ing, and Alfons Dehe[1,2]
[1]Hahn-Schickard Institute for Applied Research e.V., GERMANY and [2]University of Freiburg, GERMANY

T86-i **ULTRA-PRECISE DEPOSITION: DIGITAL MICROMANUFACTURING FOR ADVANCED PACKAGING**
Lukasz Witczak, Jolanta Gadzalinska, Iwona Gradzka-Kurzaj, Mateusz Lysien, Ludovic Schneider, Aneta Wiatrowska, Karolina Fiaczyk, Piotr Kowalczewski, Lukasz Kosior, and Filip Granek
XTPL SA, POLAND

W86-i **WAFER-LEVEL DEFECT CHARACTERIZATION AND POLARITY-DEPENDENT RESISTANCE DEGRADATION OF SPUTTERED SODIUM POTASSIUM NIOBATE THIN FILMS**
Kuan-Ting Ho[1], Daniel Monteiro Diniz Reis[1], and Karla Hiller[2]
[1]Robert Bosch GmbH, GERMANY and [2]Technical University Chemnitz, GERMANY

CHARACTERIZATION OF VAPOR HF SACRIFICIAL ETCHING THROUGH SUBMICRON RELASE HOLES FOR WAFER-LEVEL VACUUM PACKAGING BASED ON SILICON MIGRATION SEAL

Tianjiao Gong[1], Yukio Suzuki[1], Muhammad J. Khan[1], Karla Hiller[2] and Shuji Tanaka[1]
[1]Tohoku University, JAPAN and
[2]Fraunhofer Institute for Electronic Nano Systems, GERMANY

ABSTRACT

Vapor hydrogen fluoride (vHF) sacrificial SiO_2 etching is a crucial process for wafer-level packaging based on silicon migration seal (SMS) technology. In this study, the characteristics of vHF etching through release holes with a diameter of 0.5 μm were investigated by using a series of test patterns and structures. The etch rate shows some dependence on the number of release hole array, distance from the release hole, and dimension of sealed cavity. This work also reveals the role of the water as both a by-product and catalyst during through-hole vHF etching. The results of this study provide important design guidelines for SMS-based MEMS packaging as well as other similar vacuum packaging technology and improve the understanding of vHF etching mechanisms.

KEYWORDS

Vapor hydrogen fluoride etching, Silicon migration sealing, Through-hole etching, Wafer-level vacuum packaging

INTRODUCTION

High-vacuum wafer-level packaging is highly required especially for resonant-type MEMS devices to improve the performance by reducing air damping effect and fabrication cost by wafer batch process. Silicon migration seal (SMS) [1][2] is a new wafer-level vacuum encapsulation technique based on the theory of silicon migration [3][4]. It is possible to obtain various kinds of complex structure by utilizing silicon migration. For example, a through-hole can be transformed into a spherical cavity, and thus the hole is closed and can no longer conduct [5]. The SMS technique takes advantage of this phenomenon to achieve a hermitical seal by through-hole occluding without large strain (Figure 1). In the SMS process, two SOI wafers are used: one is fabricated with MEMS structures in the cavities (namely device wafer), and another is fabricated with release holes (namely cap wafer). The two wafers are directly fusion bonded, and then MEMS structures in the cavities are released by sacrificial layer etching through the release holes.

After structure release, the release holes are closed by silicon migration in hydrogen atmosphere under high temperature annealing, and as a result the cavities are sealed. The release holes are required sub-micron size to be occluded in a shorter time SMS process considering the deformation of inside MEMS structure. The remaining hydrogen gas inside the sealed cavity will be thermally diffused out, consequently a high-vacuum environment can be achieved. For this technology, vapor hydrogen fluoride (vHF) release via submicron-sized, deep release holes is a key process.

The release holes are designed to sub-micron in size to facilitate subsequent sealing process, but this also leads to a reduced conductance. The supply of vapor HF and catalyst and the removal of etch products by diffusion is limited by the release holes. In particular, water will be generated as a by-product with any type of catalyst and could be easily condensed especially when in a small cavity. These factors may make the through-hole etching result unique from normal vHF etching. This study investigated vHF release in such a special condition for the first time.

In this paper, the properties of vHF sacrificial etching through submicron holes are investigated. Firstly, the background of SMS and importance of through-hole vHF etching to SMS is introduced. Then, various test patterns are designed and the test samples are fabricated to learn about the properties. Experimental contents including the fabrication process of test sample, the elements that constitutes a test unit, and the vHF etching equipment as well as the etching recipe used in the study are described in detail. A reference group without release holes is also set for comparison. After vHF etching by a commercially available equipment, the results of test samples are presented and discussed. Finally, based on the above results and discussions, the through-hole vHF etching properties are concluded.

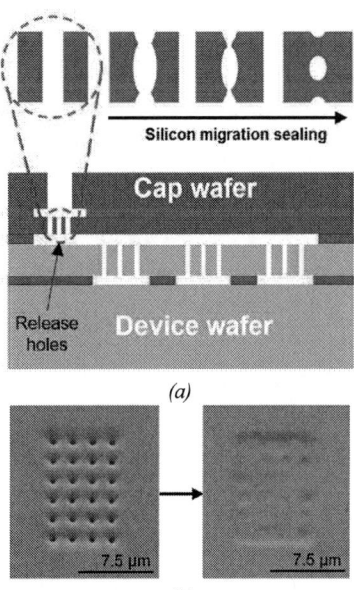

Figure 1: (a) Schematic figure of silicon migration seal (cross-section view). (b) The SEM image of release holes before and after silicon migration seal (top view).

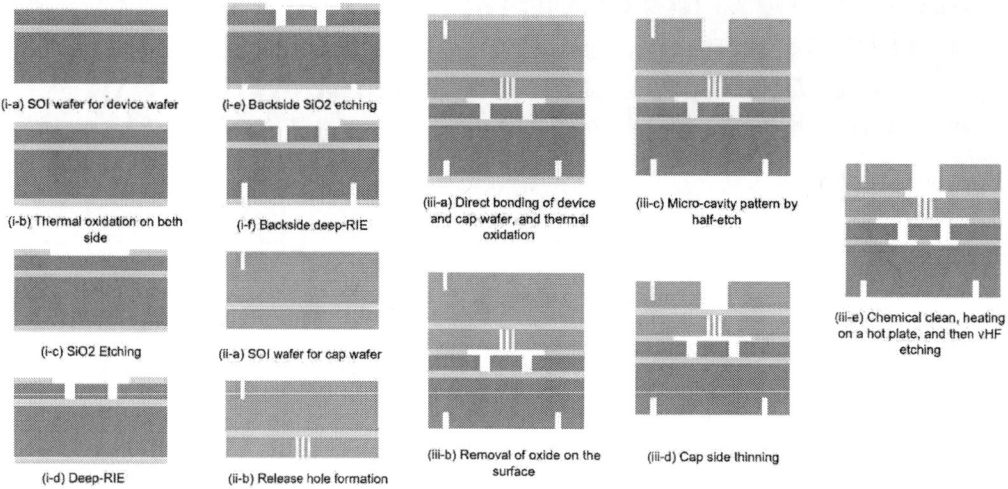

(i-a) SOI wafer for device wafer

(i-b) Thermal oxidation on both side

(i-c) SiO2 Etching

(i-d) Deep-RIE

(i-e) Backside SiO2 etching

(i-f) Backside deep-RIE

(ii-a) SOI wafer for cap wafer

(ii-b) Release hole formation

(iii-a) Direct bonding of device and cap wafer, and thermal oxidation

(iii-b) Removal of oxide on the surface

(iii-c) Micro-cavity pattern by half-etch

(iii-d) Cap side thinning

(iii-e) Chemical clean, heating on a hot plate, and then vHF etching

Figure 2: Fabrication process flow of the test samples

EXPERIMENTAL

Sample preparation

The fabrication process of test samples used for vHF etching study is shown in Figure 2. For the device wafer, firstly an SOI wafer (20/1/400 μm) is prepared (i-a) and thermal oxidized (i-b). Then oxide layer on the bonding interface side is partly opened by dry etching (i-c). In the oxide-opened area, test patterns as well as cavities are made by deep reactive ion etching (deep-RIE) (i-d). On the backside, dicing marks are made by SiO_2 dry etching (i-e) and deep-RIE (i-f). For the cap wafer, release holes pattern is firstly photolithographed on the active layer of SOI wafer by i-line stepper (Nikon, i14) and then the holes are formed by a 2-step deep-RIE method (ii-a) [6]. Some auxiliary marks are also made on another side of the cap wafer (ii-b). The two wafers are cleaned by piranha solution and RCA solutions. Bonding surface of the two wafers have an oxygen plasma activation. The cleaned wafers are then direct bonded by a manual wafer bonder. After bonding, the bonded wafer is put into a furnace for thermal oxidation at 1100°C (iii-a). The SiO_2 layer generated in last step on the wafer surface is removed by buffered hydrogen fluoride solutions (iii-b), and micro cavities are formed by half etching on the cap side of bonded wafer (iii-c). Then the cap side is thinned by deep-RIE until reaching to the BOX layer (iii-d), and subsequently the release hole arrays are exposed. Finally, the samples are cleaned and processed by vHF etching (iii-e).

The device wafer with test patterns but without bonded to another cap wafer is also prepared as the reference group, and the etching results will be used for comparison with through-hole etching results.

Test sample introduction

A test unit in the test samples includes release hole array, cavity, and test patterns (Figure 3(a)).

The release hole array is inside a micro-cavity which is fabricated on the opposite side of the cap wafer. A standard release hole array includes 4 × 6 release holes, 24 in total. Each release hole has a diameter of 0.5 μm as mask dimension and a depth of 5 μm, which is fixed in this study.

The diameter of a release hole needs to be submicron level to keep H_2 annealing time for sealing within a reasonable range, as the larger diameter requires much longer time for silicon migration.

The cavity is an independent unit that provides space for test patterns in the device wafer. The cavities are combined with different numbers of release holes array(s) after bonded to another cap wafer, so that the only passageway for inlet and outlet of the gases for a cavity is the release hole array(s). After the bonding, the cavities are supposed to be not affected by each other during vHF etching. The standard size of a cavity used in this study is 2510 μm × 2510 μm, except for some cavities with a size change to test the effect of cavity size.

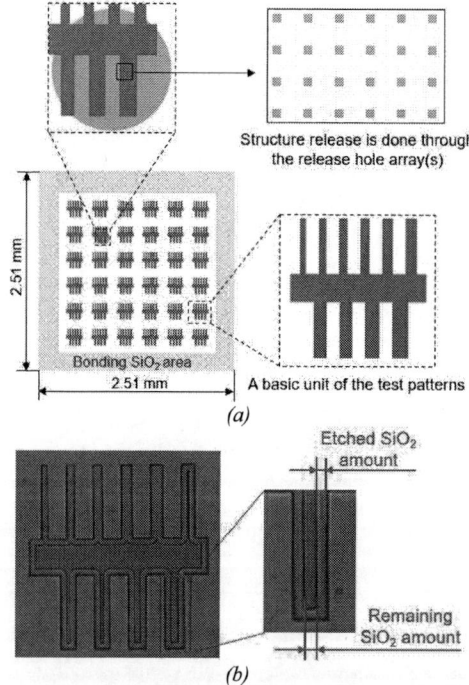

Figure 3: (a) Top view of one type of the test patterns on the device wafer and release hole array. (b) Infrared image of an etched small test pattern.

Table. 1 The vHF etching recipe used in this study

	Time/s	Pressure/Torr	HF/sccm	N2/sccm	Ethanol/sccm
Step 1: Stabilize	*120*	*50~70*	*0*	*1250*	*350*
Step 2: Etch	*600*	*110~120*	*310*	*1250*	*350*
Step 3: Pump	*30*	*10~30*	*0*	*0*	*0*

The evenly distributed small etch rate measurement patterns inside the cavity is used to show the etch rate distribution. When combined with different numbers of release hole array, such a cavity can be used to investigate the release hole array number dependency of etch rate. Normally, a cavity is divided into 6 × 6 blocks with small etch rate measurement pattern in each block. The etch rate of each block is measured and calculated based on the small patterns in that area, as shown in Figure 3(b), and subsequently the etch rate distribution in a cavity can be obtained.

VHF etching equipment introduction

The vHF etching equipment used in this study is Primaxx® uEtch (SPTS Technologies). The etching recipe used is shown in Table 1. For each etching cycle, the etching process includes 3 steps: stabilization, etch, and pump. In the etch step, vHF is carried by nitrogen to the etch chamber and SiO_2 is etched as the reactions above. As etching goes on, the etch rate becomes faster as more water is produced, so the etch time should be controlled. Additionally, the partial pressure of HF is also related to the etch rate within a certain range.

Test samples received an etching of 20 cycles by this recipe. After the vHF etching, the etch results are observed and the etch amount are measured by an infrared microscope (IRise, MORITEX Corporation).

RESULTS AND DISCUSSION

Position of release hole array

Figure 4 shows the dependence of etch rate on the distance from the release hole array. The result is based on test structure with only one release hole array. It is easy to find that the etch rate is higher in the area around the release hole array than in other areas. The etch rate is greatly increased as it becomes closer to the etch rate, especially within 1000 μm range. The etch rate shows some dependence up to 11%/mm regarding the distance from release holes, such a large value was never observed. However, if the distance is over a certain value, there will be almost no distance dependence, and the etch rate tends to be stable but low.

Number of release hole array

Figure 5 shows the average value and standard deviation of the etch rate with different number of release hole arrays. It is noted that the average value of etch rate increases as the number of release hole array increases, although there should be an upper limit. As for the standard deviation, it tends to decrease as the release hole array increases except in the cavities with only one release hole array. For the cavity with only one release hole array, the

standard deviation may vary in a relatively large range, which may indicate the etch rate distribution could be

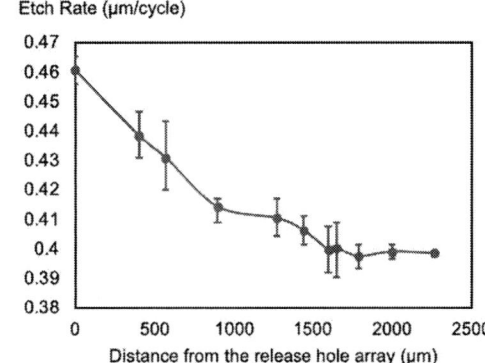

Figure 4. The dependence of etch rate on the distance from the release hole array.

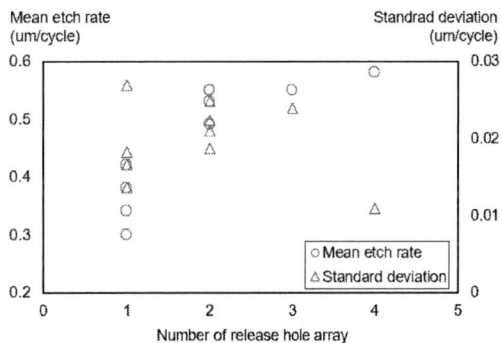

Figure 5: The mean etch rate and its standard deviation with different number of release hole array.

unstable and easily affected by other factors in such a case. The case with 4 arrays which have a distance with each other also proves that the evenly distributed array makes uniform etch rate distribution.

By contrast, the reference group which was not capped showed a general etch rate of 0.45~0.5 μm/cycle. This value is lower than most of the data collected from the capped samples, which revealed that the sealed cavity with some release holes has the potential to accelerate vHF etching. We believe that the difference in etch rate between capped and uncapped samples should come from the H_2O partial pressure. Limited by the conductivity of the submicron release holes, there is a higher partial pressure of H_2O (gas) generated by the reaction inside the capped cavity. Higher concentrations of H_2O (gas) near the release holes may result in higher etch rates near the release holes.

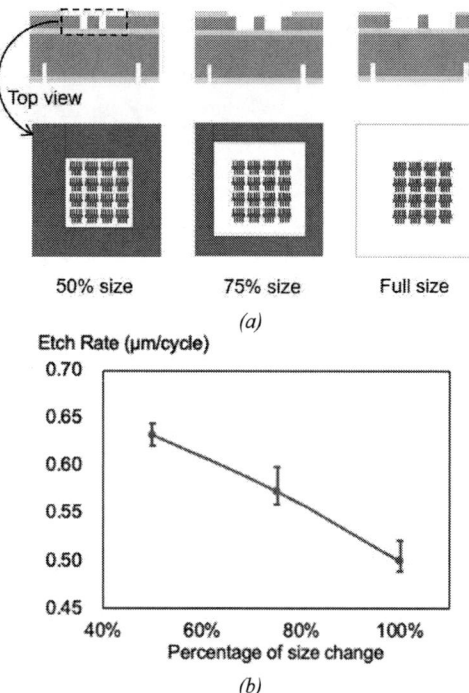

Figure 6. (a) Variations of cavity size with the same pattern (b) The dependence of etch rate on the percentage of size change

Cavity size effect

The varied cavity size with the same test patterns in some test samples is used to investigate the influence of cavity size. Unlike normal vHF etching, etching thorough submicron holes can easily cause the concentration of reactants as well as products in a small cavity. Therefore, the cavity size can affect the concentration, thereby affect the etching properties. As shown in Figure 6(a), the cavity size is set to 100%, 75%, and 50% of the default value by adjusting the remaining outer Si area, while the etch rate measurement patterns in the center remain the same. In the cap side, evenly distributed 3 release hole arrays are applied to eliminate the effect of the array number. Figure 6(b) shows the relation between etch rate and cavity size. The etch rate increases as the cavity size decreases. This result makes sense because water concentration occurs more in smaller cavity which accelerates the etching. We believe less conductance of the release hole will have the similar effect due to water concentration.

CONCLUSION

In this study, the characteristics of vHF etching through submicron holes are investigated. In general, through-hole vHF etching showed larger etch rate than normal vHF etching if the cavity size and release hole number is enough. The etching shows some dependence up to 11%/mm regarding the distance from release holes, although little dependence was observed if the distance exceeds a certain value. Generally, the average etch rate increases with the number of release hole array, and uniform distributed release hole arrays will lead to a uniform etch rate distribution. As for the cavity size,

etching in larger cavity size results in lower etch rates. As both a by-product and catalyst during through-hole vHF etching, water can be used to account for these characteristics for being easily condensed in a cavity. These characteristics will be meaningful for the design of MEMS devices applying SMS process for vacuum sealing.

ACKNOWLEDGEMENTS

This paper is based on results obtained from a project, JPNP19005, subsidized by the New Energy and Industrial Technology Development Organization (NEDO).

REFERENCES

[1] Y. Suzuki, V. Dupuit, T. Kojima, Y. Kanamori, and S. Tanaka. "Silicon migration seal wafer-level vacuum encapsulation." Electronics and Communications in Japan 104, no. 1 pp. 120-125. (2021)

[2] H. Suzuki, Y. Suzuki, Y. Kanamori, and S. Tanaka. "Improved Vacuum Level of Silicon-Migration-Sealed Cavity by Hydrogen Diffusion Annealing For Wafer-Level Packaging For Mems." In 2022 IEEE 35th International Conference on Micro Electro Mechanical Systems Conference (MEMS), pp. 565-568. (2022)

[3] T. Sato, K. Mitsutake, I. Mizushima, and Y. Tsunashima. "Micro-structure transformation of silicon: a newly developed transformation technology for patterning silicon surfaces using the surface migration of silicon atoms by hydrogen annealing." Japanese Journal of Applied Physics 39, no. 9R: 5033. (2000)

[4] T. Sato, I. Mizushima, S. Taniguchi, K. Takenaka, S. Shimonishi, H. Hayashi, M. Hatano, K.Sugihara, and Y. Tsunashima. "Fabrication of silicon-on-nothing structure by substrate engineering using the empty-space-in-silicon formation technique." Japanese Journal of Applied Physics 43, no. 1R: 12. (2004)

[5] J. Stehle, Vu A. Hong, A. Feyh, G. J. O'Brien, G. Yama, O. Ambacher, B. Kim, and Thomas W. Kenny. "Silicon migration of through-holes in single-and poly-crystalline silicon membranes." In Proc. Solid-State Sensors, Actuat., Microsystems Workshop, Hilton Head, pp. 32-35. (2014.)

[6] Y. Suzuki, K. Totsu, H. Watanabe, M. Moriyama, M. Esashi, S. Tanaka, "Low-Stress Epitaxial Polysilicon Process for Micromirror Devices", IEEJ Transactions on Sensors and Micromachine, Vol.133 No.6, 223-228 (2013)

[7] Y. Suzuki, M. Honda, H. Suzuki, H. Miyashita, S. Tanaka, "Optimization Development of Release Hole Geometry for Silicon Migration Seal (SMS) Lower Temperature", The 38th SENSOR SYMPOSIUM, 9P3-SS2-2 (2021)

CONTACT

*Y. Suzuki, tel: +81-22-795-6934;
suzuki@tohoku.ac.jp

DAMAGE PROFILE MODELING AND EXPERIMENT OF SILICON CARBIDE SUBSTRATES IN MICRO-NANO STRUCTURE FABRICATED BY HELIUM FOCUSED ION BEAM

Shupeng Gao, Xi Chen, Qianhuang Chen, Qi Li and Yan Xing

Jiangsu Key Laboratory for Design and Manufacture of Micro-Nano Biomedical Instruments,
Department of Mechanical Engineering, Southeast University, Nanjing, CHINA

ABSTRACT

The damage of silicon carbide substrates by the helium focused ion beam (helium FIB) process at the micro-nano scale is investigated. At the energy of $10\sim35\ KeV$ and the dose of $0.03\sim0.075\ nC/\mu m$, the helium FIB experiment was carried out on the silicon carbide substrate, and the evolution law of the damage of the silicon carbide substrate with the process parameters was explored. Then, an empirical equation for the damage profile of the amorphous region of silicon carbide substrates processed by the helium FIB is proposed and compared with that of silicon substrates.

KEYWORDS

helium focused ion beam; amorphous profiles; silicon carbide substrate

INTRODUCTION

In recent years, helium focused ion beam (helium FIB) has been widely used due to the high beam resolution, such as helium ion microscopy [1], nanolithography [2], ion beam induced deposition [3]. Current research on radiation damage of helium focused ion beams has only focused on a few materials, such as silicon [4], graphene [5] and diamond [6]. Silicon carbide is an important substrate material for high power devices due to its high voltage, high temperature and radiation resistance properties. Combined with previous work [7], this paper investigates the damage effect of helium FIB on silicon carbide substrates and compares them with silicon substrates.

The silicon substrate processed by the helium FIB exhibits typical amorphous damage region, transition damage region and helium bubble region [7]. However, experimental results of the helium FIB processing of silicon carbide at the same beam energy and ion dose show more complex damage region. Collisions between helium ions and substrate atoms create displacements and vacancies, leading to the transformation of the substrate from crystalline to amorphous. Stanford [8] presented the damage function in the simulation, which indicated that the damage of the crystal is cumulative, and the same lattice accumulated by 0.2 displacements will lead to the amorphization of this region. Based on the interaction mechanism between the incident ions and the substrate atoms, in the initial stage of the incident, the helium ions lose their energy mainly through electron stopping. Due to the smaller scattering angle, the helium ions continue to move deeper into the substrate along the incident direction. However, with increasing depth of incidence, the constant loss of its own energy reduces the role of electron stopping, and instead nuclear stopping dominates. At low

energies, nucleus collisions produce large scattering angles and the ions diffuse in a radial direction. Thus, both electron stopping and nuclear stopping together determine the shape of substrate damage, with electron stopping affecting the depth of high-energy damage regions and nuclear stopping affecting the radial extent of low-energy damage regions. At the same time, process parameters also affect the evolution of damage morphology. Compared with the damage of the silicon substrate, under the same beam energy and ion dose, the radial extent of the amorphous region of silicon carbide is larger because of the larger angle scattering of helium ions in the substrate, as shown in Figure 2. However, the area of the silicon carbide transition region is significantly reduced and the distribution of the helium bubble region is wider and closer to the substrate surface. In addition, the swelling of the surface of the silicon carbide substrate is more prominent.

It can be seen that the area of damage induced by helium FIB processing varies widely depending on the physical properties of the substrate, including elemental ratio and mass density. However, the form of damage determines the electrical, thermal and mechanical properties of the substrate material. Therefore, it is essential to establish a mathematical model of the relationship between process parameters and damage profiles for different substrates.

Based on the interaction mechanism between helium ions and silicon carbide substrates, this study proposed a mathematical model for the damage profile of silicon carbide substrate materials processed by helium focused ion beams, and established the relationship between the amorphous damage profile and the beam energy and ion dose through artificial intelligence algorithms. Finally, this work analyzed and compared the experimental results and simulation results in detail.

EXPERIMENTS

In this study, a 2" n-type silicon carbide (0001) wafer with a resistivity of $0.018\ \Omega\cdot m$ was chosen as the substrate. The experimental procedure includes the helium FIB processing and transmission sample preparation. First, the samples were ultrasonically cleaned in acetone solution for 15 minutes, then rinsed with negatively ionized water, and finally dried at 120 °C for 30 minutes. The ZEISS Orion NanoFab system implants helium ions into silicon carbide substrates in a line scan. The system can be equipped with three ion beam systems of Ga, Ne, and He at the same time, and has imaging and processing capabilities from micrometer to nanometer. This experiment is divided into 4 groups, the beam energies are 10, 15, 25 and 35 KeV, and the ion doses in each group are 0.03, 0.04, 0.05, 0.06 and 0.075 $nC/\mu m$. According to the line scan processing method, helium ions are implanted into the substrate with a dwell time of 0.1 μs and a constant current of 1.6 pA as shown in Figure 1. In order to improve the processing efficiency, the Ga ion beam with higher washing efficiency is used to process the T-shaped experimental mark.

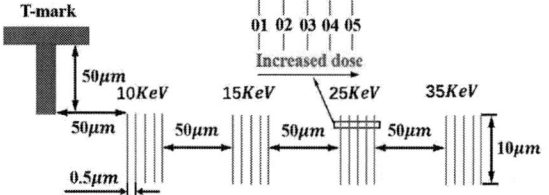

Figure 1: Schematic diagram of the experimental processing scheme.

After fabrication, the Zeiss Crossbeam 540 focused ion beam system was used to prepare transmission electron microscope (TEM) samples. A 1 nm layer of platinum was deposited to protect the sample surface, including 200 nm electron beam deposition and 800 nm gallium ion beam deposition. The sample was then washed by the gallium ion beam into a thin sheet and picked up by a robotic arm onto a copper mesh for observation. Finally, the damage cross-section was observed and photographed by the FEI Talos F200X transmission electron microscope system.

As shown in Figure 2(b), the cross-sectional image divides the damage into four regions: the amorphous region within the red line, the crystalline region outside the yellow line, the transition region between the two, and the helium bubble region within the blue line. The characteristic length of the damaged area of the silicon carbide amorphous region is marked, including the surface damage width, the maximum damage width and the maximum damage depth.

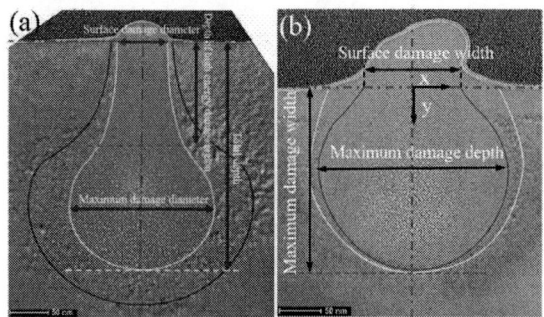

Figure 2: Damage profile and dimensional calibration of different substrates processed by Helium FIB;(a) Silicon substrate;(b) Silicon carbide substrate.

Figure 3 shows the TEM cross section at 35 KeV beam energy. In this study, the boundary of the damage region of the amorphous region was manually calibrated, clearly revealing the effect of beam energy and ion dose on amorphous damage. It is evident that with the increase of the ion dose, the depth of the high-energy damage region is decreasing, and the radial extent of the low-energy damage region is increasing. Figures 3(a)–3(d) show the change of the damage profile (characteristic

length) of the amorphous region at a fixed beam energy of 35 KeV, with the dose increasing from 0.03 $nC/\mu m$ to 0.06 $nC/\mu m$. It can be clearly seen that the maximum damage width of the amorphous region increases by approximately 13.5%, from 222 nm to 252 nm. The surface damage width increased by approximately 50%, from 37 nm to 54 nm. However, there was no significant increase in depth, and the larger surface expansion may have influenced this. In the previous study [7], the surface damage width, maximum damage width and maximum damage depth of amorphous regions in silicon substrates all increased with ion dose.

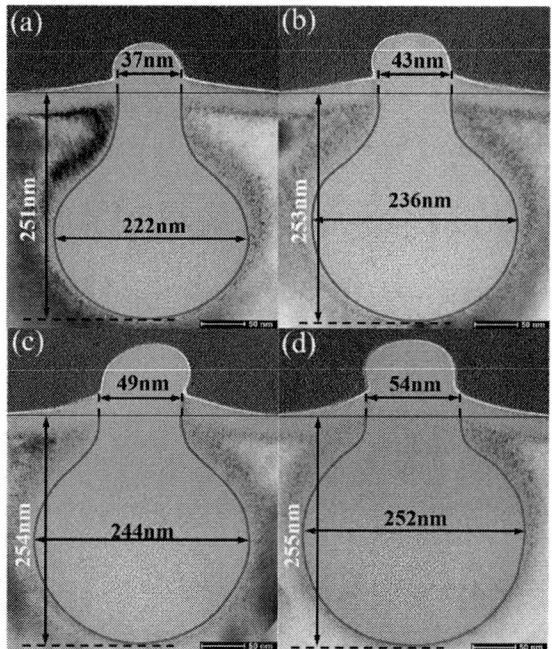

Figure 3: TEM images of transmission sample cross-section of silicon carbide damage processed with beam energy of 35KeV;(a) TEM image at ion dose of 0.03nC/μm;(b) TEM image at ion dose of 0.04nC/μm; (c) TEM image at ion dose of 0.05nC/μm; (d) TEM image at ion dose of 0.06nC/μm.

However, at the same ion dose, the maximum damage width of the amorphous region in the silicon carbide substrate increases with the increase of energy, which is not seen in the silicon substrate. Therefore, it is inferred that both the ion dose and the beam energy affect the maximum damage width of the amorphous region.

METHODS

Based on the variation of the damage profile of the amorphous region with the ion dose and beam energy in silicon carbide substrates, an empirical equation is proposed to simulate the amorphous damage profile. The defined coordinate system takes the center of the ion beam as the origin, the incident direction is the y-axis, the substrate surface is the x-axis, and the positive direction is shown in Figure 2(b). According to the symmetry of the scattering of incident ions in the substrate, only the profile to the right of the y-axis is simulated.

As a function of the damage radius x of the amorphous region relative to the incident depth y, the ion dose D, and the beam energy E, the empirical equation is as follows:

$$x(y,D,E) = \alpha_2 \left(1 - \frac{y+\delta}{\alpha_1}\right) \cdot \left(\frac{y+\delta}{\alpha_1}\right)^n + r. \quad (1)$$

Here, α_2 is the maximum damage width coefficient; δ is the damage depth indentation coefficient; α_1 is the damage depth coefficient; n is the depth influence coefficient of the maximum damage width; r is the surface damage width coefficient.

At the same energy and different doses, the damage depth of the silicon carbide substrate does not change significantly, so the damage depth coefficient α_1 is constant with a value of 243.8. Compared with the silicon substrate, under the same ion dose, with the increase of the beam energy, the radial extent of the low-energy damage region also increases in the silicon carbide substrate. Therefore, the coefficient m affects the depth at which the maximum damage width is located, and is related to the beam energy D and the ion dose E. At the same dose and different energies, it is found that the depth of the high-energy damage region decreases with decreasing energy. This is due to the fact that at low energies, nuclear stopping dominates. Therefore, the damage depth indentation coefficient δ is introduced to correct the empirical equation which is related to the beam energy D. Based on the experimental results, the surface damage width coefficient r is related to the ion dose D.

Table 1: Comparison of coefficient fits to empirical equations for silicon and silicon carbide substrates

	SiC	Si
α_2	$-4714.8 \times D + 680.13$	$16380 \times D + 82.68$
α_1	243.8	$1472 \times D + 273.5$
n	$-1.148 \ln D - 1.224 \ln E + 2.4973$	$-88.2 \times D + 9.504$
r	$629.2 \times D + 15.97$	$450.82 \times D + 24.008$
δ	$-5.5627 \times E + 193.96$	none

The equation coefficients are established by artificial intelligence algorithms. Compared with the empirical equations of the damage profile of the amorphous region of the silicon substrate, the relationship between the coefficients $\alpha_1, \alpha_2, n, r, \delta$ and the process parameters D

and E is shown in Table 1.

For silicon carbide substrates, beam energy D and damage depth indentation coefficient δ are introduced to correct the simulation accuracy. The fitting relationship between the coefficients and the process parameters is finally verified by experiments. Figure 4 shows the linear fitting results of the coefficients α_2, r and the ion dose D. So far, the empirical equation for predicting the profile of amorphous region damaged by focused helium ion beam processing silicon carbide is established by beam energy and ion dose.

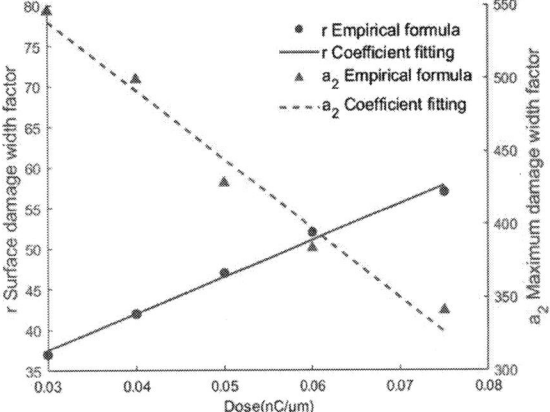

Figure 4: Variation of surface damage width coefficient d and maximum damage width coefficient α_2 with ion dose.

RESULTS

Figure 5 compares the experimental and simulated results for the characteristic length of the amorphous region profile at beam energies of 15, 25 and 35 KeV. It can be clearly seen that the simulation results at beam energies of 15 and 35 KeV can conform to the variation of the surface damage width radius of the amorphous region profile with the ion dose presented in the experiment, as shown in Fig 5(a). The error of beam energy of 35 KeV is all below 3.45 nm, much smaller than 9 nm of silicon substrate [7]. However, at beam energy of 25 KeV, the simulation results show large errors, all below 6.58 nm. Figure 5(b) shows the experimental and simulated results of the maximum damage width of the amorphous region profile. The simulation results show that the maximum damage width

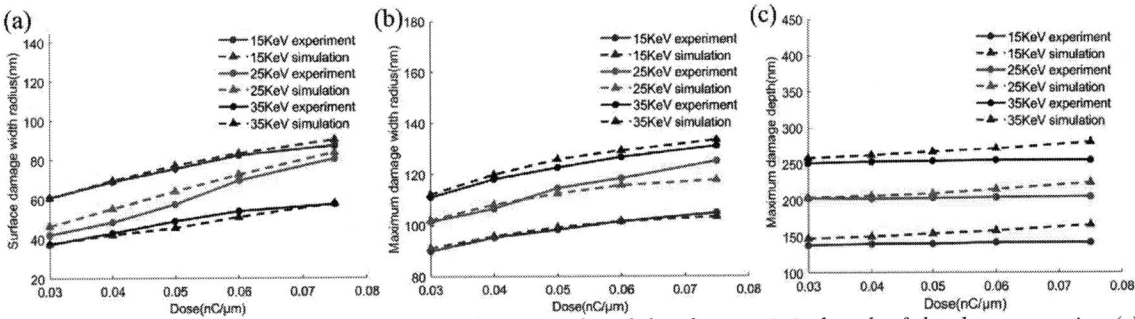

Figure 5: Comparison of experimental and simulation results of the characteristic length of the damage section;(a) Surface damage width radius;(b) Maximum damage width radius;(c) Maximum damage depth.

increases linearly with the ion dose, which is basically consistent with the experimental results. At the same time, beam energy is introduced to improve the accuracy of the simulation. Except for the results of beam energy of 25 KeV and ion dose of 0.075 $nC/\mu m$, the errors of other simulation results are all below 3.5 nm. The above errors are all affected by the accuracy of manual calibration and the degree of fitting between coefficients and process parameters.

Figure 5(c) shows the experimental and simulated results for the maximum damage depth of the amorphous region profile. The experimental results for the damage depth of the amorphous region profile are smaller than the simulation results and do not change much with the ion dose. The calculation error increases with increasing ion dose, with a maximum of about 25 nm. In addition to calibration error and fitting error, more process parameters can be used to improve the simulation accuracy of empirical equations.

CONCLUSION

In this study, the evolution law of the damage of amorphous region of carbonized substrate processed by helium FIB with process parameters was quantitatively analyzed and compared with that of silicon substrate. Based on the interaction mechanism and evolution law of helium ions and silicon carbide substrates, an empirical equation for calculating the damage of amorphous regions of silicon carbide substrates based on beam energy and ion dose is established in this work. Then, the helium FIB experiments were carried out on silicon carbide substrates at $10\sim35\ KeV$ energy and $0.03\sim0.075\ nC/\mu m$ dose. Comparing the experimental results, the calculated results of the empirical equations can accurately predict the damage distribution in the amorphous region of the silicon carbide substrate at different beam energies and ion doses. And the empirical equation can describe the change of amorphous region damage with ion dose and beam energy. The empirical equation has demonstrated the ability to apply to substrates of different materials, including silicon substrate and silicon carbide substrate. In addition, the method can be optimized to predict other forms of damage distribution on substrates processed by the helium FIB, such as helium bubble regions, transition regions.

ACKNOWLEDGEMENTS

We acknowledge financial support by the National Natural Science Foundation of China No.51875104.

REFERENCES

[1] J. Notte, B. Ward, N. Economou, R. Hill, R. Percival, L. Farkas, S. McVey, "An Introduction to the Helium Ion Microscope", *AIP conf. Proc.*, vol. 931, pp. 489-496, 2007.

[2] V. Sidorkin, E. van Veldhoven, E. van der Drift, P. Alkemade, H. Salemink, D. Mass. "Sun-10-nm Nanolithography with a Scanning Helium Beam", *J. Vac. Sci. Technol. B*, vol. 27, pp. L18-L20, 2009.

[3] HM. Wu, LA. Stern, JH. Chen, M. Huth, CH. Schwalb, M. Winhold, F. Porrati, CM. Gonzalez, R. Timilsina, PD. Rack, "Synthesis of Nanowires Via Helium and Neon Focused Ion Beam Induced Deposition with the Gas Field Ion Microscope", *Nanotechnology.*, vol. 24, 2013.

[4] D. Fox, YH. Chen, CC. Faulkner, HZ. Zhang, "Nano-Structuring, Surface and Bulk Modification with a Focused Helium Ion Beam", *Beilstein. J. Nanotechnology.*, vol. 3, pp. 579-585, 2012.

[5] S. Hang, Z. Moktadir, H. Mizuta, "Irradiation induced tunnel barrier in side-gated graphene nanoribbon," *2014 Silicon Nanoelectronics Workshop (SNW)*, pp. 1-2, 2014.

[6] CS. Kim, RG. Hobbs, A. Agarwal, Y. Yang, VR. Manfrinato, MP. Short, J. Li, KK. Berggren, "Focused-Helium-Ion-Beam Blow Forming of Nanostructures: Radiation Damage and Nanofabrication", *Nanotechnology.*, vol. 31, 2020.

[7] T. Shao, Q. Chen, Y. Xing, X. Lin, C. Fang, Q. Chai, "An Experiment Based Damage Profile Function for Focused Helium Ion Beam Process in Fabrication of Micro/Nano Structures", *2020 IEEE 33rd International Conference on Micro Electro Mechanical Systems (MEMS)*, pp. 905-908, 2020.

[8] Stanford. MG, Mahady. K, Lewis. BB, Fowlkes. JD, Tan. SD, Livengood. R, Magel. GA, Morre. TM, Rack. PD, "Laser-Assisted Focused He+ Ion Beam Induced Etching with and without XeF2 Gas Assist", *Acs Applied Materials & Interfaces*, vol. 8, pp. 29155-29162, 2016.

CONTACT

*Y. Xing, tel: +86-025-52098413; xingyan@seu.edu.cn

LIQUID-IMMERSION INCLINED-ROTATED UV LITHOGRAPHY FOR MICRO SUCTION CUP ARRAY

Gakuto Kagawa and Hidetoshi Takahashi
Keio University, Japan

ABSTRACT

This paper reports liquid-immersion inclined/rotated UV lithography for fabricating micro suction cups. Microstructures such as micro suction cups have been applied in various applications. However, the conventional fabrication method had limitations in suction force due to the insufficient inclination angle. We proposed liquid-immersion to inclined/rotated exposure (IRE), enabling greater microstructure inclination angles. IRE equipment was developed to fabricate PDMS micro suction cup arrays. The experimental inclination angle reached 51°, nearly doubling the conventional IRE method.

KEYWORDS

Suction cup, Inclined/rotated lithography, Liquid-immersion lithography

INTRODUCTION

IRE is widely used to fabricate MEMS 3D microstructures, such as micro suction cups and fluidic channels [1], [2] (Fig. 1(a)). Previously, the IRE method has been performed by irradiating inclined UV light to a rotating substrate [3]–[5]. However, the conventional method had difficulty increasing the inclination angle of the structure. This is due to the total reflection caused by the difference in refractive indices of the materials used for exposure. There have been multiple approaches to address this issue, though such methods require extensive setups, hindering compatibility with other processes[6], [7]. As shown in Fig. 1(b), micro suction cups are known to increase their suction force as the inclination angle increases [8]. Therefore, developing an improved IRE method that allows a larger inclination angle is desired.

This paper proposes UV exposure equipment that combines liquid-immersion and IRE. Using liquid-immersion reduces the total reflection of UV light and increases the inclination angle. The equipment is composed of a rotating assembly and wafer holder. The former assembly consists of cubic glass, a chamber on top, and two mirrors. With our proposed equipment, a mold for micro suction cups was obtained. A micro suction cup array was fabricated by pouring PDMS into the mold. Three differently angled micro suction cup arrays were fabricated and evaluated through experiments.

DESIGN AND PRINCIPLE

We present exposure equipment with liquid-immersion and IRE using inclined mirrors. Fig. 2 shows the proposed exposure equipment comprised of cubic glass and two adjustable mirrors. Additionally, a water chamber is placed on top of a cubic glass. A patterned wafer is placed inside the wafer holder and submerged horizontally in the water. The UV light ray is irradiated uniformly from above and reflected by mirrors to the cubic glass. In this

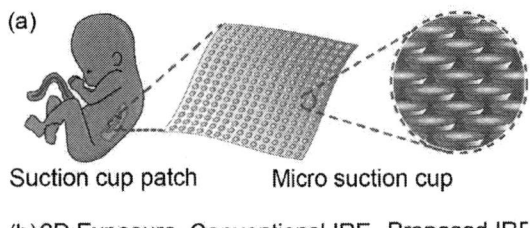

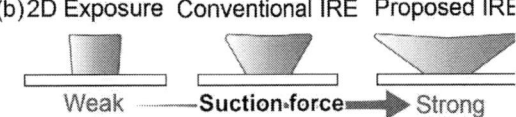

Fig. 1: (a) Concept and application examples of micro suction cups. (b) Comparison of suction force by the inclination angle of micro suction cups.

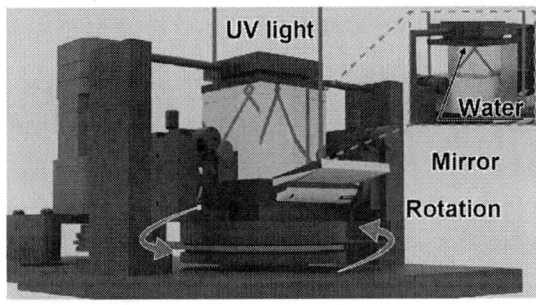

Fig. 2: Schematic image of the proposed inclined/rotated exposure (IRE) system.

moment, the patterned wafer is irradiated only from a reflected UV light ray. The rotating mirrors enable uniform exposure to the substrate. As presented in Fig. 3(a), this method can decrease the refractive index difference between pre-medium (water) and post-medium (photoresist), thus increasing the inclination angle of UV light to the substrate. Fig. 3(b) explains the relationship between the mirror angle and the inclination angle using air or water. Theoretical equations using Snell's law show that water provides a larger incidence angle for the same mirror angle compared to air. From theory, the conventional method's inclination angle covers 1° to 38°, whereas that in our proposed method covers between 45° to 72°. Therefore, we can obtain larger inclination angles by employing liquid-immersion for inclined/rotated UV lithography in microstructures.

DEVELOPMENT OF EQUIPMENT

During exposure, cubic glass (40 mm cubed) and both mirrors (20 mm x 30 mm) are rotated under UV exposure

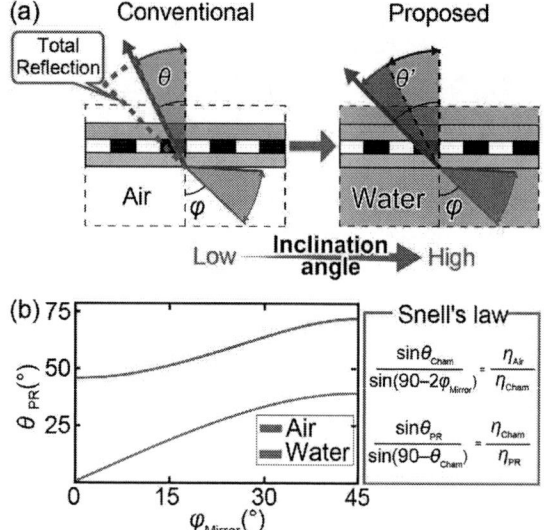

Fig. 3:(a) Difference in inclination angle between air and water as a median. (b)Relationship between the inclination angle and mirror angle using air or water as a median.

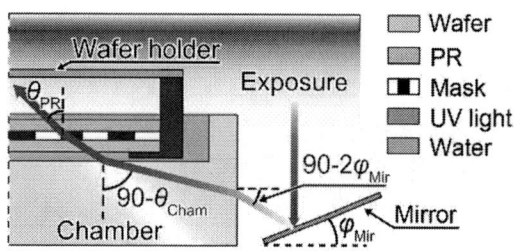

Fig. 4: Cross-sectional image of the proposed device's optical path.

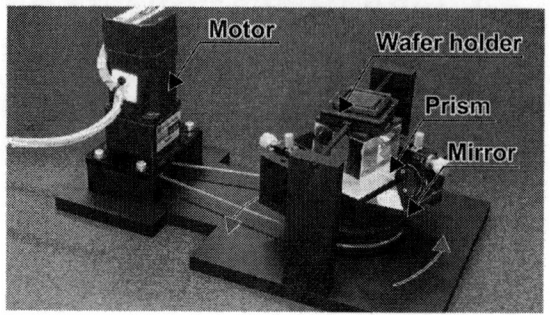

Fig. 5: Photograph of the developed IRE system. Rotational mechanism is actuated by external motor using rubber band.

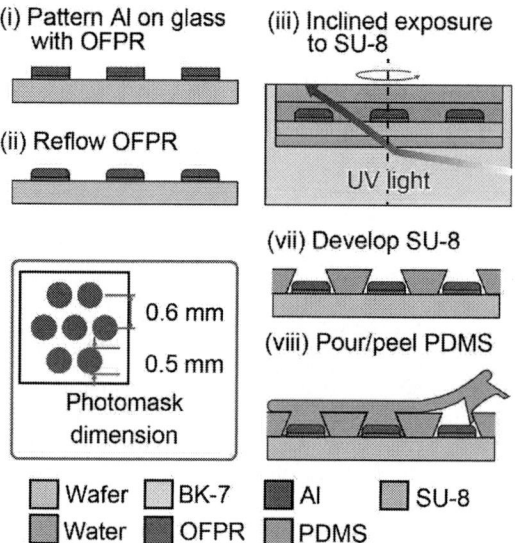

Fig. 6: Process flow for fabricating micro suction cups.

equipment (EMA-400, Union Optical Co.) with a valid exposure area of 100 mm diameter. The substrate is exposed to an inclined/rotated UV light ray through a patterned wafer (25 mm x 25 mm). A wafer holder is retained by two stands and designed to hold the wafer 0.5 mm above the cubic glass while storing water and the wafer. A rim is attached to prevent water from spilling from the cubic glass. Both cubic chamber and mirrors rotate around the wafer holder by bearing attached to the rotating base. As shown in Fig. 4, we can control the inclination angle of the micro suction cup mold by adjusting the mirror angle using the rotation stage (KSP-256, SIGMAKOKI Co). From the theoretical calculation, the angle of the mirrors can be adjusted from 11° to 20°. Any other angles would result in total reflection of the substrate.

We developed the equipment using aluminum as material (Fig. 5). The rotating stage is actuated by an external motor (0GN5K, ORIENTAL MOTOR Co.) using a rubber band. The motor speed can vary the rotation speed of the rotating equipment. All housings are attached to the aluminum base under a UV light source. All aluminum parts are anodized to matte black to reduce unnecessary UV reflection.

EXPERIMENT AND RESULT

We fabricated a PDMS micro suction cup array mold using the developed exposure system, which was placed under a UV light source. The fabrication process is explained in Fig. 6. SU-8 photoresist was used to develop the mold on the glass wafer. A mask pattern with a diameter and pitch of 500 μm and 600 μm, respectively, was used. The patterned wafers were made by aluminum deposition and lithography using OFPR-800 (TOKYO OHKA KOGYO CO., LTD). The wafers are reflowed to create a convex-shaped gap for suction. By developing the substrate, the patterned wafer is completely submerged in water inside the wafer holder. The experiment was performed with mirror angles set at 11°, 15°, and 20°. The rotation speed was set to 1.3 rps to account for water leakage due to centrifugal force. Considering the photomask shape, the inclination angle of UV light, and rotation speed, the exposure time was calculated to be 500 s. After exposure, the SU-8 mold was developed and baked using the normal process. Defoamed PDMS was poured into the SU-8 mold and baked at 75 degrees for an hour. After curing, the PDMS layer was separated from the SU-8 mold, revealing patterned micro suction cups.

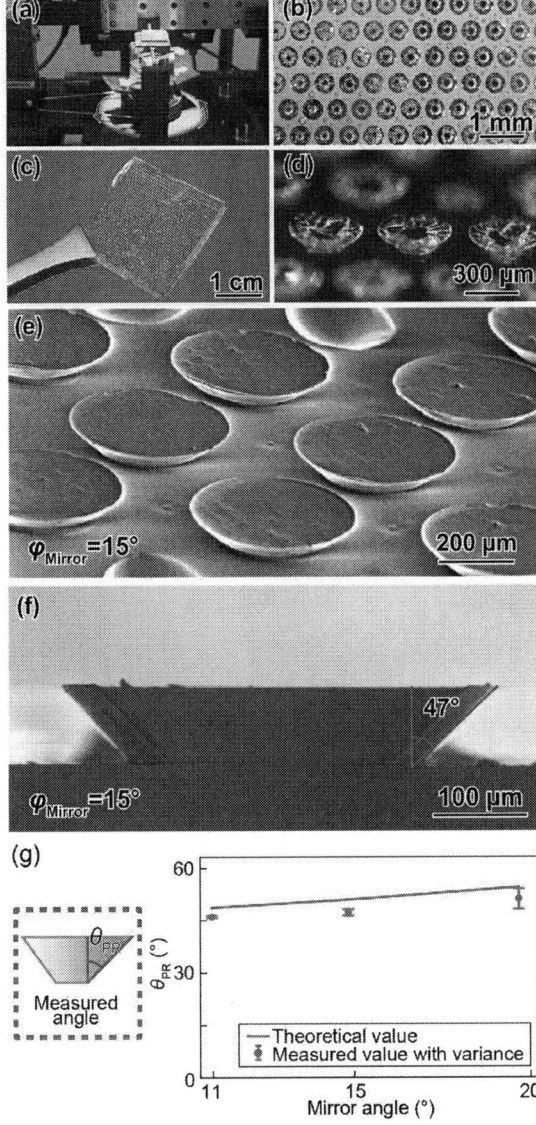

Fig. 7: Photographs of (a) equipment under UV light source, (b) SU-8 mold in array, (c) Fabricated micro suction cup, (d) close-up view of micro suction cups. SEM images of (e) micro suction cup, (f) cross- sectional view of the fabricated structure. (g) comparison of inclination angle in measured and theoretical values with variance.

Fig. 7(a) shows photographs of the experiment. Fig. 7(b) provides a photograph of the fabricated SU-8 mold. Fig. 7(c, d, e, f) shows close-up photographs and SEM images of the fabricated PDMS micro suction cup. As a result, the structures are uniformly formed across the wafer in all experiments. Fig. 7(g) compares the inclination angle of the fabricated PDMS micro suction cup and theoretical values. With the liquid-immersion, the inclination angle reached a maximum of 51°, nearly doubling the inclination angle conducted by a previous study using conventional inclined exposure [3].

Furthermore, as the mirror angle increased, the inclination angle of the micro suction cup increased. The difference between the theoretical and measured values was within 3%. This result suggested a high correlation with our proposed method. Thus, the proposed liquid-immersion IRE would further improve the performance of three-dimensional microstructural devices.

CONCLUSION

We proposed, designed, and developed exposure equipment for inclined/rotated lithography using liquid-immersion. The equipment increases the inclination angle of the structure while maintaining compatibility with other processes. As a result, we have managed to fabricate micro suction cups with a high inclination angle of up to 51°. For these reasons, it is expected to realize the performance of 3D microstructural MEMS devices.

ACKNOWLEDGEMENTS

This work was partly supported by Mitutoyo Association for Science and Technology (MAST).

REFERENCES

[1] R. Huang, et al., "Suction Cups-Inspired Adhesive Patch with Tailorable Patterns for Versatile Wound Healing," *Adv. Sci.*, vol. 8, no. 17, pp. 1–8, 2021.

[2] J. Zhou, et al., "Recent developments in PDMS surface modification for microfluidic devices," *Electrophoresis*, vol. 31, no. 1, pp. 2–16, 2010.

[3] T. Sugimoto et al., "Multidirectional UV lithography via inclined/rotated mirrors for liquid materials," *Appl. Phys. Express*, vol. 13, no. 7, 2020.

[4] M. Han, et al., "Fabrication of 3D microstructures with inclined/rotated UV lithography," *Proc. IEEE Micro Electro Mech. Syst.*, pp. 554–557, 2003.

[5] J. Kim, et al., "Fabrication of 3D nanostructures by multidirectional UV lithography and predictive structural modeling," *J. Micromechanics Microengineering*, vol. 25, no. 2, 2015.

[6] J. K. Kim, et al., "Adjustable refractive index method for complex microstructures by automated dynamic mode multidirectional UV lithography," *Proc. IEEE Int. Conf. Micro Electro Mech. Syst.*, pp. 733–736, 2009.

[7] T. Y. Chang, et al., "Complex Optical Microstructure Fabricated Using Inclined Immersion Lithography," *2009 Symp. Des. Test Integr. Packag. MEMS/MOEMS DTIP 2009*, pp. 4–7, 2009.

[8] N. Thanh-Vinh et al., "Micro suction cup array for wet/dry adhesion," *Proc. IEEE Int. Conf. Micro Electro Mech. Syst.*, pp. 284–287, 2011.

CONTACT

*Gakuto Kagawa, tel: +81-90-4670-9080; gakukagawa@keio.jp

PARAMETRIC AMPLIFICATION AND PHONONIC FREQUENCY COMB GENERATION IN MoS₂ NANOELECTROMECHANICAL RESONATORS

S M Enamul Hoque Yousuf[1*], Yunong Wang[1*], Jaesung Lee[1],
Steven W. Shaw[2,3], and Philip X.-L. Feng[1*]

[1]Department of Electrical & Computer Engineering, Herbert Wertheim College of Engineering,
University of Florida, Gainesville, FL 32611, USA
[2]Department of Mechanical & Civil Engineering,
Florida Institute of Technology, Melbourne, FL 32901, USA
[3]Department of Mechanical Engineering, and Department of Physics & Astronomy,
Michigan State University, East Lansing, MI 48423, USA

ABSTRACT

This paper reports on extraordinarily strong parametric amplification and spectral linewidth narrowing effects in atomically thin molybdenum disulfide (MoS₂) resonant nanoelectromechanical systems (NEMS), by applying the electrical pump signal at local gate. It also demonstrates phononic frequency comb (PnFC) formation based on 2:1 mode coupling during the degenerate parametric amplification. For a single-layer (1L) MoS₂ resonator, we measure the displacement spectral density without and with electrical parametric pumping and calculate the parametric gain and linewidth narrowing. We observe a parametric gain as high as ~10,000 (80dB) and spectral linewidth narrowing factor of ~5000 with 153mV pump voltage at $2f$. Since there is another mode present near $2f$, driving the device at $2f$ enables mode coupling between the fundamental mode and the mode near $2f$, and generates a PnFC with tunable comb spacing. The exceptional parametric amplification and spectral linewidth narrowing opens new possibilities towards building high performance atomically thin NEMS resonators for sensing applications.

KEYWORDS

Parametric amplification, parametric resonator, gain, phononic frequency comb (PnFC), two-dimensional (2D) materials, quality (Q) factor.

INTRODUCTION

Atomically thin NEMS resonators based on two-dimensional (2D) materials have demonstrated remarkable characteristics including broad dynamic range (DR) and ultrawide frequency tunability [1,2]. However, electrical readout of miniscule resonance motions of these devices is challenging because the intrinsic thermomechanical noise floor is overwhelmed by the noise from the electronics used at the front end of the measurement system. Quality (Q) factors of 2D resonators can be limited because of various extrinsic damping effects. To increase the signal amplitude and Q of 2D resonators, parametric amplification has been proposed to enhance the device motion at resonance frequency f with drive frequency of $f_p \approx nf$, where n is an integer ($n \geq 2$) [3-6]. Recent experiments on photothermal parametric pumping of undriven thermomechanical noise spectra have demonstrated giant parametric amplification with gains up to 71dB and spectral linewidth narrowing factor up to 1.8×10^5 [6]. Toward on-chip integration of parametric pumping, electrical parametric pumping is desired. Phononic frequency comb (PnFC), the phononic

analogue of optical frequency comb in the radio frequency (RF) domain has been reported in microelectromechanical systems (MEMS) [7] and NEMS [8]. In NEMS resonators, strong mode coupling is required to coherently transfer energy between different modes via internal resonances and thus to generate PnFCs [8]. Atomically thin circular drumhead resonators using 2D semiconductor membranes, with wide frequency tuning to satisfy internal resonance conditions, provide an ideal platform for realizing PnFCs.

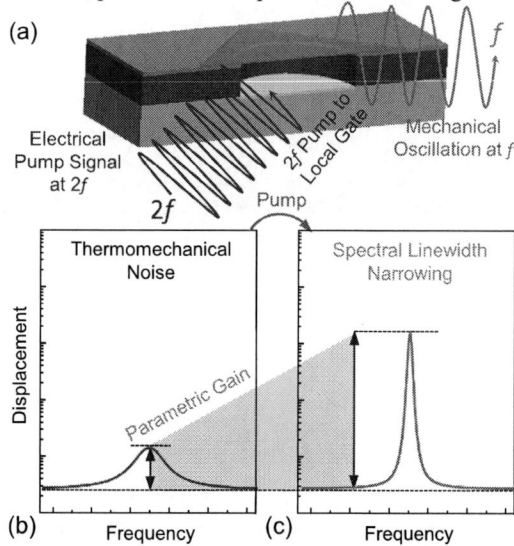

Figure 1: (a) Illustration of parametric amplification using time-domain signal in 2D NEMS resonator. Intrinsic thermomechanical noise spectral density, (b) before and (c) after electrical parametric pumping. The pump signal introduced at the local gate increases the signal amplitude with gain G along with spectral linewidth narrowing effect.

In this study, we realize highly efficient electrical parametric pumping via a local gate, and simultaneously measure the undriven thermomechanical noise spectra in a single-layer (1L) MoS₂ drumhead resonator. Undriven thermomechanical noise measurement enables exploring the full range of parametric amplification by directly amplifying the intrinsic noise floor. We experimentally demonstrate parametric gain up to ~10,000 before the device goes into parametric oscillation. The exceptional linewidth narrowing factor of ~5000 reveals the efficient electrical pumping in the sub-threshold regime (before saturation). We also observe a PnFC near the 1st mode by pumping near the 3rd mode to engender 2:1 mode coupling.

DEVICE DESIGN AND FABRICATION

The sapphire substrate for the electrostatically tunable MoS$_2$ 2D NEMS resonators is fabricated by using various photolithography and etching steps compatible with standard NEMS fabrication process (Fig. 2). An exfoliated 1L MoS$_2$ is transferred onto this substrate using an all-dry transfer method, to make a drumhead resonator with 3μm diameter. A local gate configuration with the sapphire substrate provides electrostatic control of individual drumheads located on the same chip with significantly reduced parasitic effects. The single-layer nature of the MoS$_2$ is confirmed by Raman spectroscopy showing two prominent peaks at E$^1_{2g}$=385.44cm^{-1} and A$_{1g}$=403.65cm^{-1}, with a separation of 18.21cm^{-1} between them.

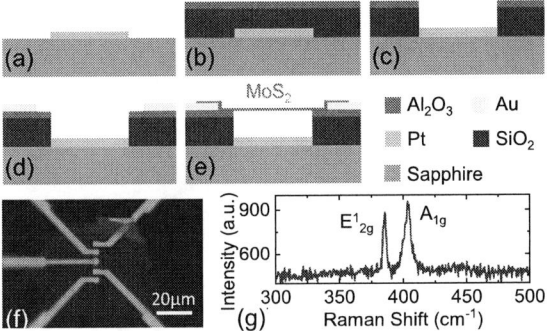

Figure 2: Fabrication process for the drumhead resonator. (a) Local gate patterning. (b) SiO$_2$ deposition and Al$_2$O$_3$ deposition. (c) RIE to form 290nm-deep trench. (d) Metal deposition for top electrodes. (e) All-dry transfer of the MoS$_2$ layer. (f) An optical image of the fabricated 1L MoS$_2$ drumhead resonator. (g) Raman signal showing E$^1_{2g}$ and A$_{1g}$ modes at 385.44cm^{-1} and 403.65cm^{-1}, respectively.

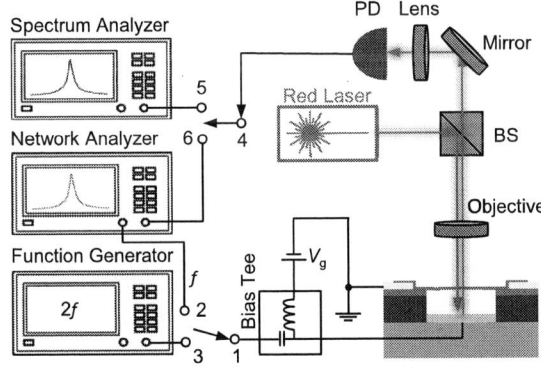

Figure 3: Illustration of the measurement scheme, showing how the driven resonance (connect nodes 1,2 and 4,6) and undriven thermomechanical noise spectral density (connect nodes 4,5) are measured. The pump signal is introduced to the local gate using a function generator (connect nodes 1,3). PD: photodetector, BS: beam splitter.

MEASUREMENT AND ANALYSIS

Resonance motion of the 2D NEMS resonator under electrical parametric pumping can be described by

$$m\ddot{x} + \frac{m\omega\dot{x}}{Q} + [k_1 + k_p\cos(2\omega t)]x + k_3 x^3 = F(t), \quad (1)$$

where x, t, m, ω, k_1, k_3, and $F(t)$ are displacement, time, effective mass, angular frequency at resonance ($\omega=2\pi f$), linear spring constant, third order spring constant, and

driving force, respectively. $F(t)$ can be either a harmonic drive with amplitude proportional to v_{drv} and frequency f_{drv} or thermal noise. The periodic modulation of linear stiffness at $2f$ is described by $k_p\cos(2\omega t)$, where k_p indicates the electrical pump strength. The atomically thin MoS$_2$ 2D

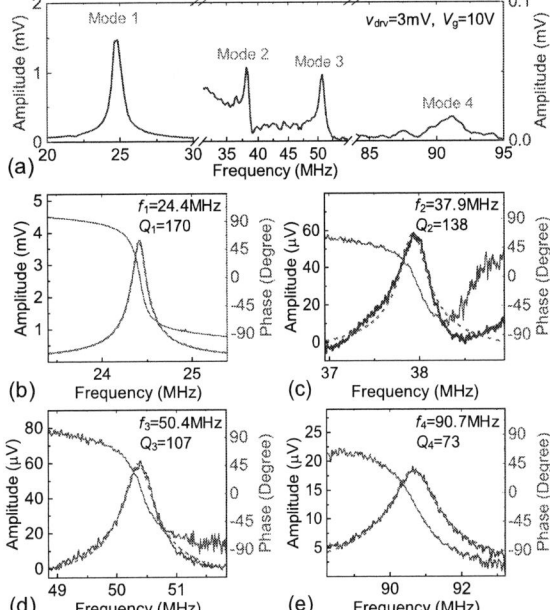

Figure 4: (a) Multimode resonances from a single-layer (1L) MoS$_2$ drumhead in a wide spectrum showing 4 distinct resonance modes measured. Driven resonance of the (b) 1st, (c) 2nd, (d) 3rd, and (e) 4th mode with Qs of 170, 138, 107, and 73, respectively. Blue and red curves show the measured data and fitting curves, respectively.

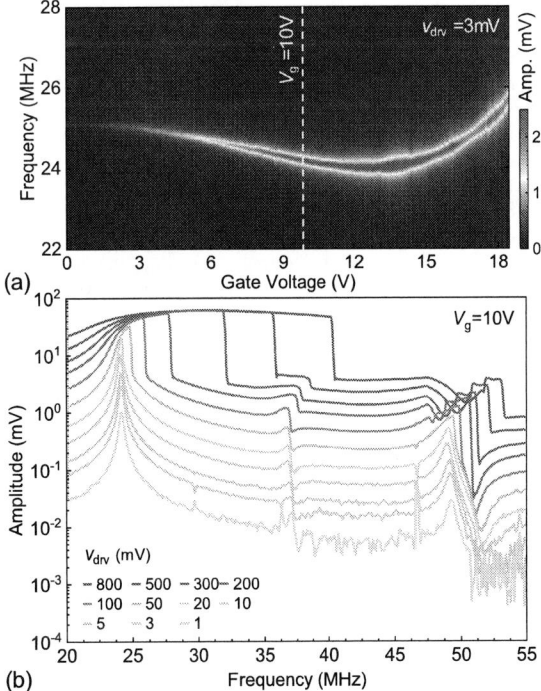

Figure 5: (a) Frequency tuning with V_g=0 to 18.5V, while at v_{drv}=3mV. (b) Duffing responses (up sweep) measured from the 1L MoS$_2$ device by varying v_{drv} (at V_g=10V).

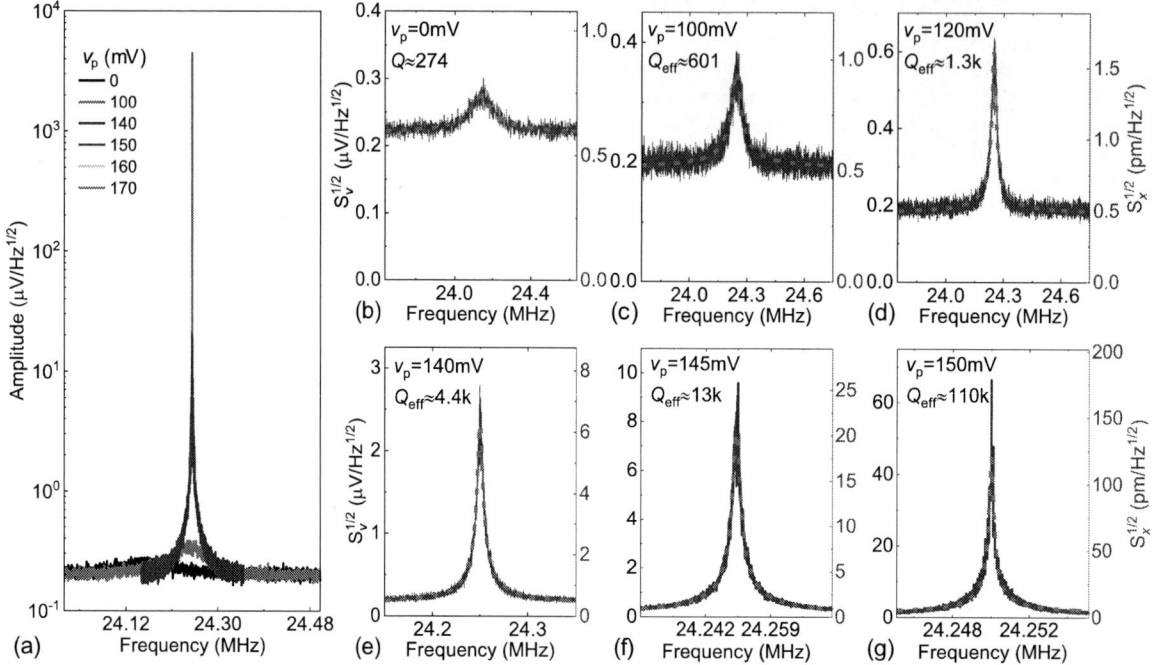

Figure 6: (a) Measured noise spectral density of the 1L MoS2 device at varying pump voltage. (b) Thermomechanical noise measurement without parametric pumping. Measured resonances with pump voltages at (c)100mV, (d) 120mV, (e) 140mV, (f) 145mV, and (g) 150mV, respectively. Red dashed curves are obtained by fitting the measured data (blue curves).

drumhead with low linear stiffness (k_1) operating in the tension dominated regime can be efficiently pumped with electrical pump voltage via the local gate.

Therefore, we electrically parametrically pump the 1L MoS2 resonator at twice the fundamental mode frequency using a function generator and record the corresponding spectrum using a spectrum analyzer (Fig. 3). A network analyzer is used to measure the driven resonance. Applied DC voltage (V_g) to local gate enables tuning the resonance frequency of the 2D NEMS resonator. The PnFC is generated by driving the device near 2f, establishing 2:1 mode coupling in the multimode MoS2 resonator.

RESULTS AND DISCUSSIONS

We first excite the device by applying a 10V DC gate voltage (V_g) and 3mV RF drive (v_{drv}) to the local gate and measure 4 driven resonance modes at 24.4MHz, 37.9MHz, 50.4MHz, and 90.7MHz, with Qs of 170, 138, 107, and 73, respectively (Fig. 4). Mode 1 shows clear frequency tuning for varying V_g from 0 to 18.5V. Capacitive softening dominates up to $V_g \approx 13$V showing a frequency downshift; tension-induced stiffening dominates when $V_g > 13$V, exhibiting frequency upshift. The device responses show clear Duffing nonlinearity with increasing v_{drv} (Fig. 5). The measured thermomechanical noise for mode 1 shows a resonance frequency of 24.14MHz with $Q \approx 274$. The responsivity of the system, defined as the ratio of the voltage domain thermomechanical noise spectral density $S_v^{1/2}$ to displacement domain thermomechanical noise spectral density $S_{th}^{1/2}$, is 0.371μV/pm at room temperature. After characterizing the undriven resonance of the drumhead, we pump the device at 2f using RF pump voltage (v_p) from a function generator, starting from 10mV to 170mV, and record the corresponding noise spectral

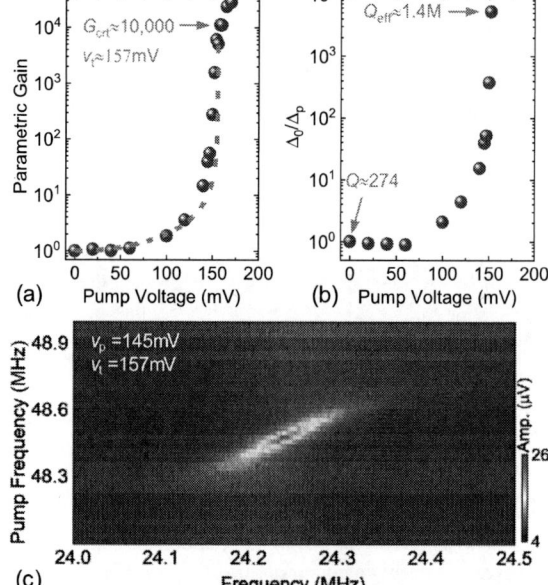

Figure 7: Measured (a) parametric gain and (b) Δ_0/Δ_p at varying pump voltage, showing giant gain and spectral narrowing due to electrical pump. (c) Measured pumped thermomechanical noise with varying f_p at v_p=145mV.

density near the resonance. Figure 6 shows a series of thermomechanical noise spectra measured at various pump voltages. At low v_p, the thermomechanical noise spectrum slowly starts to increase with increasing pump strength. When v_p approaches the threshold pump voltage (v_t), the thermomechanical resonance shows giant amplification with exceptional spectral linewidth narrowing. Parametric pumping below v_t can be described by the equation

$$G = \frac{S_{x,\text{pump}}^{1/2}(\omega_0)}{S_x^{1/2}(\omega_0)} = \sqrt{1 + \left(\frac{v_p^2}{v_t^2}\right)} \Big/ \left[1 - \left(\frac{v_p^2}{v_t^2}\right)\right], \quad (2)$$

where G is the parametric gain [6] and $S_{x,\text{pump}}^{1/2}(\omega_0)$ and $S_x^{1/2}(\omega_0)$ are the spectral peak with and without parametric pumping. Fitting the data to Eq. (2), we obtain a parametric threshold $v_t \approx 157\text{mV}$ and critical gain $G_{\text{crt}} \approx 10,000$ (80dB).

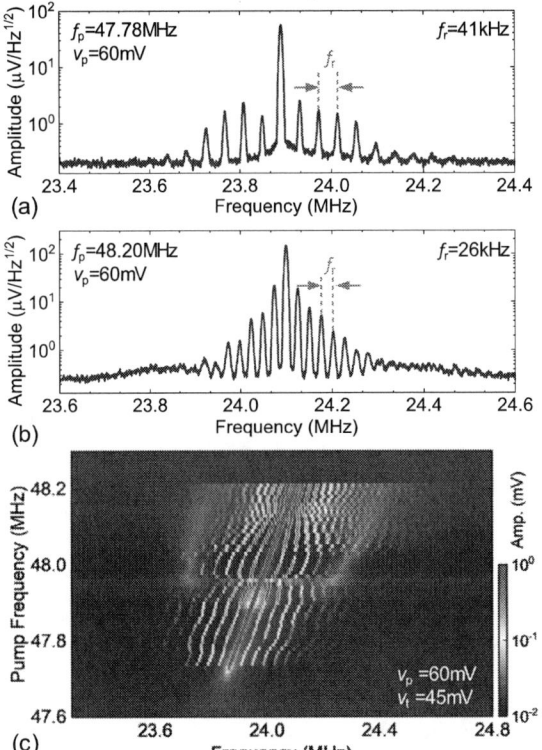

Figure 8: Phononic frequency comb generation by driving a new MoS₂ device at (a) 47.78MHz and (b) 48.2MHz, with comb spacing f_r=41kHz and 26kHz, respectively. (c) Evolution of phononic frequency comb by varying the pump frequency with v_p=60mV (for calibrated v_t=45mV).

The linewidth of noise spectral density also drops sharply near v_t due to spectral narrowing. With v_p=153mV, the parametric gain reaches up to ~10,000 (80dB) and the linewidth narrowing factor (Δ_0/Δ_p) reaches ~5000. We cannot reliably capture the linewidth narrowing beyond v_p=153mV due to the 1Hz resolution bandwidth limit imposed by the instrument. Sweeping the pump frequency (f_p) with fixed v_p=145mV reveals a range for f_p that can parametrically pump the resonator, as shown in Fig. 7c with spectrum analyzer measurements.

The multimodal MoS₂ drumhead resonator exhibits a 3rd mode (Fig. 4a) that lies within the parametric pump range near $2f$. As we vary the pump frequency, the 2:1 internal resonance condition is satisfied, leading to generation of phononic frequency comb in frequency domain with tunable comb spacing. As captured in careful spectrum analyzer measurements, pumping the device at f_p=47.78MHz with v_p=60mV generates a comb with spacing f_r=41kHz (Fig. 8a). When f_p=48.2MHz, the comb spacing changes to f_r=26kHz (Fig. 8b). Figure 8c shows the

evolution of the comb spacing and number of teeth with varying pump frequency. Generated PnFC spacing can be further tuned by controlling the pump strength.

CONCLUSION

In summary, we have experimentally demonstrated highly efficient electrical parametric amplification by directly pumping on undriven thermomechanical noise spectra. We have achieved giant parametric amplification up to ~80dB, and exceptional spectral linewidth narrowing factor ~5000. Driving the device parametrically near its 3rd mode (close to $2f$) to establish 2:1 internal resonance enables mode coupling, thus leading to PnFC generation. The observed frequency comb spacing can be tuned by varying pump frequency f_p and voltage v_p. The findings here shall contribute towards building high-performance NEMS for sensing in classical and quantum applications.

ACKNOWLEDGEMENTS

We thank the financial support from NSF (Grants: 2142552, 2103091) at University of Florida. S. W. Shaw is partly supported by BSF (Grant 2018041) at Florida Tech.

REFERENCES

[1] J. Lee, Z. Wang, K. He, R. Yang, J. Shan, P. X.-L. Feng, "Electrically Tunable Single- and Few-Layer MoS₂ Nanoelectromechanical Systems with Broad Dynamic Range", *Sci. Adv.* **4**, eaao6653, 2018.

[2] F. Ye, A. Islam, T. Zhang, P. X.-L. Feng, "Ultrawide Frequency Tuning of Atomic Layer van der Waals Heterostructure Electromechanical Resonators", *Nano Lett.* **21**, 5508-5515, 2021.

[3] P. Prasad, N. Arora, A. K. Naik, "Parametric Amplification in MoS₂ Drum Resonator", *Nanoscale* **9**, 18299-18304, 2017.

[4] R. J. Dolleman, S. Houri, A. Chandrashekar, F. Alijani, H. S. J. van der Zant, P. G. Steeneken, "Opto-Thermally Excited Multimode Parametric Resonance in Graphene Membranes", *Sci. Rep.* **8**, 9366, 2018.

[5] R. Singh, R. J. T. Nicholl, K. I. Bolotin, S. Ghosh, "Motion Transduction with Thermo-Mechanically Squeezed Graphene Resonator Modes", *Nano Lett.* **18**, 6719-6724, 2018.

[6] J. Lee, S. W. Shaw, P. X.-L. Feng, "Giant Parametric Amplification and Spectral Narrowing in Atomically Thin MoS₂ Nanomechanical Resonators", *Appl. Phys. Rev.* **9**, 011404, 2022.

[7] A. Ganesan, C. Do, A. Seshia, "Phononic Frequency Comb via Intrinsic Three-Wave Mixing", *Phys. Rev. Lett.* **118**, 033903, 2017.

[8] J. Lee, S. W. Shaw, P. X.-L. Feng, "Phononic Frequency Comb Generation via 1:1 Mode Coupling in MoS₂ 2D Nanoelectromechanical Resonators", in *Proc. of the 35th IEEE Int. Conf. on Micro Electro Mechanical Systems (MEMS 2022)*, Tokyo, Japan, January 9-13, 503-506, 2022.

CONTACT

*S M Enamul Hoque Yousuf: syousuf@ufl.edu
*Yunong Wang: yunongwang@ufl.edu
*Philip Feng: philip.feng@ufl.edu

PARYLENE-N AS A HIGH TEMPERATURE THIN FILM PIEZOELECTRIC MATERIAL

Nathan Jackson and Deepak Kunwar

University of New Mexico, Mechanical Engineering Department and Center for High Technology Materials, New Mexico, USA

ABSTRACT

Thin film piezoelectric materials are in high demand in microsystem devices, but common materials used are stiff ceramic materials. Flexible piezoelectrics are needed for low frequency applications or devices requiring large deformations, but they are often limited in microfabrication due to their low Curie temperatures. This paper investigates the development of a parylene-based high temperature piezoelectric material, and how crystallinity of the film affects the piezoelectric properties. Increasing the films crystallinity for Parylene-N resulted in an increase in d_{33} value of 12.6 pC/N when the film was annealed at 350°C and the sample maintained its piezoelectric properties when elevated to >250°C. The results demonstrate that modified Parylene-N thin films can be used a MEMS compatible piezoelectric material at elevated temperatures.

KEYWORDS

Parylene, Piezoelectric, Poling, Flexible Thin Film, MEMS.

INTRODUCTION

Piezoelectric films are commonly used as a transduction mechanism for a wide range of microelectromechanical systems (MEMS) devices [1]. However, most piezoMEMS devices use ceramic stiff piezoelectric materials such as AlN, ZnO, or PZT, but there is an increasing demand over the past decade to develop polymer flexible thin film piezoelectric materials for applications such as low frequency energy harvesting, wearable technologies, and biological sensors [2, 3].

Polyvinylidene Fluoride (PVDF) and its co-polymers is the most commonly used polymer piezoelectric material but its low Curie Temperature (~100 °C) limits is integration into standard MEMS processing [4, 5]. Recently there has been a significant increase in research focused on developing new polymer piezoelectric films. These include developing new composite films using piezoelectric nanoparticles [6-8], as well as creating new 2-2 hybrid materials consisting of a polymer substrate and a stiff thin film piezoelectric material [9-11]. Researchers have also investigated developing new polymer piezoelectric materials from bulk semi-crystalline polymers such as polyamide. However, currently the flexible piezoelectric thin films developed have poor temperature properties, as they have low melting or glass transition temperatures. This results in films losing their piezoelectric properties under elevated processing temperatures or operating temperatures.

Parylene is a unique polymer that is deposited using modified chemical vapor deposition techniques and it is widely used in various MEMS applications [12-14] due to its many advantageous properties including biocompatibility, high melting temperature, transparency, and deposition mechanism. Parylene also has a high crystallinity for a polymer, and recently this crystallinity along with electric poling has demonstrated that the material can be modified to generate piezoelectric properties [15-17]. There are various types of parylene which differ by its functional groups and each of these have unique properties. Parylene-C has a Cl functional group with an asymmetric molecular structure. Parylene-C is the only parylene film to date to demonstrate piezoelectric properties, but it has lower crystallinity and thermal properties than other parylene films.

This paper investigates methods of modifying both Parylene-C and Parylene-N thin films to alter their crystallinity in an attempt to increase their piezoelectric properties. Parylene-N is known to have higher crystallinity than Parylene-C and it has a higher melting temperature. In addition, it undergoes a phase transition from a monoclinic to a hexagonal structure which we hypothesize should result in increased piezoelectric properties. The goal of this study is to develop a flexible piezoelectric polymer that can be integrated into a standard MEMS process flow and be able to maintain its piezoelectric properties at elevated temperatures. The paper describes methods of depositing the films and annealing the films to alter their crystallinity, and to generate piezoelectric properties through high voltage poling.

MATERIALS AND METHODS

Parylene Deposition

Parylene-C and Parylene-N films were deposited on silicon (100) wafer substrate with a 1 μm thick thermally grown oxide layer, using the Gorham process (SCS Specialty Coatings, Labcoater). The parylene dimers were vaporized at 175°C and 160°C for Parylene-C and N respectively. The pyrolysis of the dimer occurred at 690°C and 650°C for Parylene-C and N type respectively. The films were deposited at a pressure of 3.33 and 7.33 Pa for C and N. Films were deposited to a desired thickness of approximately 20 μm.

After deposition, films were removed from the silicon substrate using a tweezer and cut into square pieces (2 cm x 2 cm). The samples were then annealed in a vacuum furnace (1.33 Pa) with N_2 gas flow to prevent oxidation. Annealing temperatures ranged from 25-350°C for Parylene-N and 25-250°C for Parylene-C. The ramp rate for heating and cooling was 5°C/min. The samples were annealed at their final temperature for 30 minutes and then allowed to cool to room temperature. Annealing was performed to alter the crystallinity of the film and to investigate if it had an impact on the piezoelectric properties.

Piezoelectric Poling

Standard parylene films do not have any piezoelectric properties, but the semi-crystalline structure of the films allows them to be poled to align the dipoles and create a film that has piezoelectric properties [15-17]. One side of the film was blanket coated with Ti/Au (10/100nm) to function as the bottom electrode. The films were then poled using a custom designed corona poling system as shown in Fig. 1, which consisted of an array of needles with spacing of 2 cm. The chamber was heated to 100°C and a voltage of -19 kV was applied between the needles for 10 minutes. The poling setup was validated using un-poled PVDF films, and was previously optimized for Parylene-C films [15].

Material Characterization

The parylene film thicknesses were validated using a Dektak 8 profilometer where a section of the wafer was masked off during deposition to obtain a topographical structure. The thickness of the parylene was held constant at approximately 20 μm.

The crystallinity of the films was measured using X-ray diffraction (XRD) 2θ scans using a (X-Pert Pro 45kV, 40 mA). The scanning angles were from 10-30°. The sharpness of the peaks was fitted using a Gaussian curve to determine the full-width-half-maximum (FWHM) value. A 2 mm mask was used to filter out the Cu K β radiation. SEM images of the cross-section view of the annealed parylene film was used to visually validate the crystallinity of the film.

The bulk piezoelectric film properties (d_{33}) were measured using a Berlincourt-based piezoelectric meter (PM300 Piezometer). A top Ti/Au electrode was deposited on the top surface after poling using a shadow mask with a diameter of 3 mm. Similar structures and procedures have been used to measure piezoelectric properties of polymer films [8, 18]. To determine the piezoelectric properties as a function of temperature the films were subject to varying temperatures during the testing.

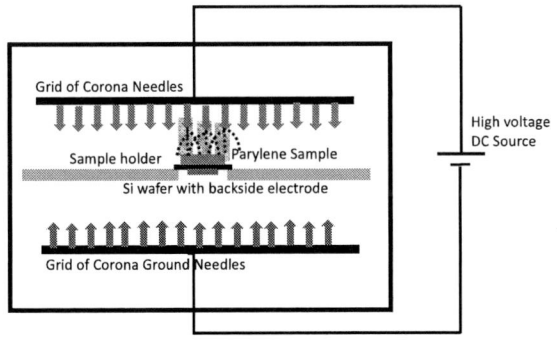

Figure 1: Schematic of the corona poling setup.

RESULTS AND DISCUSSION

The parylene films were able to be easily peeled off of the silicon substrate due to the SiO₂ surface. The annealed samples demonstrated an increase in crystallinity as shown in Fig. 2. Parylene-C films demonstrated a FWHM of 1.9°, 0.9°, and 0.55° for films with no annealing, 150°C, and 250°C respectively. The film annealed at 300°C which was

above the melting temperature (290°C) resulted in an amorphous material as the material recrystallized [19]. Parylene-N films demonstrated significantly higher crystallinity than Parylene-C films with FWHM values of 0.93°, 0.76°, 0.63°, 0.58°, 0.5°, and 0.3° for no annealing, 150°C, 200°C, 225°C, 250°C, and 300°C annealing temperatures respectively. The results are consistent with previous studies [20]. At temperatures of around 250°C the Parylene-N films undergo a phase transition from a monoclinic structure to a hexagonal structure (β-phase). The XRD measurements at 250°C illustrates a peak at 20.3° but a small peak at the original monoclinic angle (16.8°) can also be seen, demonstrating the transformation in crystal structure. Parylene-N demonstrated increased crystallinity at temperatures >300°C whereas Parylene-C illustrated amorphous structure at similar temperatures. Attempts to anneal Parylene-N above melting temperature (420°C) were not investigated. SEM scans of the cross-section image of the annealed Parylene-N film (300°C) illustrating the crystallin structure of the film is shown in Fig. 3.

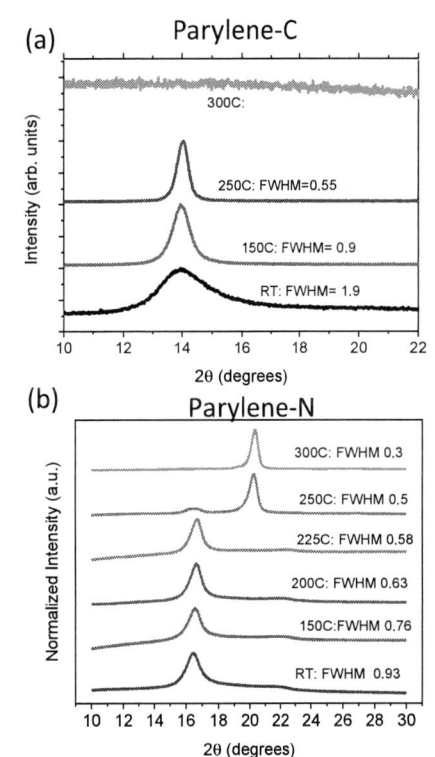

Figure 2: XRD scans for (a) Parylene-C films and (b) Parylene-N films annealed at various temperatures.

Both types of parylene films demonstrated piezoelectric properties after performing corona poling. To verify that the films were piezoelectric and not electrets measurements were made after poling and weeks after poling to verify no reduction in properties. Parylene-C films demonstrated a piezoelectric response of 2.94 pC/N with no annealing and a d_{33} value of 6.2 pC/N when annealed at 250°C. The piezoelectric values were significantly higher than previous reports of Parylene-C without annealing [17]. The amorphous film (300°C) had a piezoelectric value of 0, demonstrating a piezoelectric dependency on the crystallinity of the film. Therefore, the crystallinity was believed to be directly linked to the piezoelectric properties.

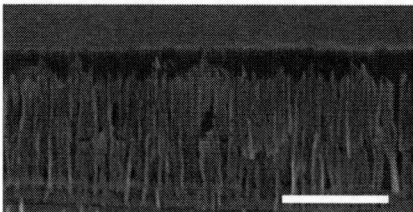

Figure 3: SEM image of cross section of the Parylene-N film with scale bar of 10 μm.

Parylene-N piezoelectric properties behaved similar to Parylene-C with annealing temperatures <250°C. However, they had a slight increase in values due to the lower FWHM, with d_{33} values of 7.57 pC/N for films annealed at 225°C, compared to 6.2 pC/N for Parylene-C annealed at 250°C. However, when the Parylene-N films underwent a phase transition to a hexagonal structure the piezoelectric properties increased significantly achieving a peak value of 12.6 pC/N when annealed at 350°C. The results illustrate that crystallinity and structure of the films have a significant effect on the piezoelectric properties of parylene films. The d_{33} values were averages of 10 films annealed and deposited in different batches. The mechanical properties of the films were impacted by the annealing as higher crystallinity will result in an increased elastic modulus as previously demonstrated for Parylene-C films [20]. The films were able to be patterned using reactive ion etching similar to un-annealed samples (data not shown).

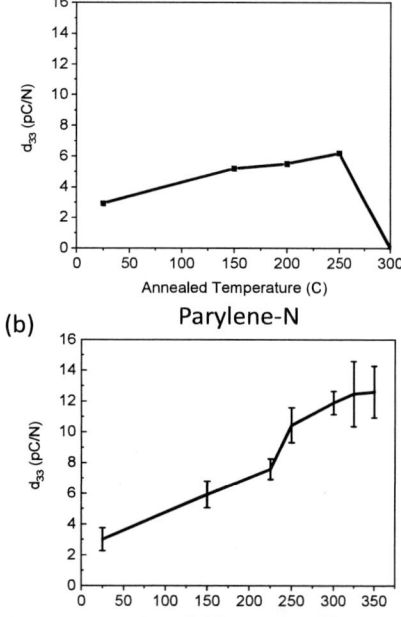

Figure 4: Piezoelectric d_{33} measurements for (a) Parylene-C and (b) Parylene-N with varying annealing temperatures prior to poling.

Next, we investigated how the Parylene-N d_{33} values would vary with an applied elevated temperature to determine if they maintained their piezoelectric properties. Parylene-N films annealed at 225°C (before the phase transition) were used in this experiment. The results shown in Fig. 5 demonstrate that the piezoelectric values remained relatively constant at temperatures <200°C. As the film went through the phase transition the piezoelectric properties decreased significant reaching a value of 1.15 pC/N at 350°C. However, films that were annealed above the phase transition >250°C and then annealed and poled demonstrated more stable piezoelectric values for higher elevated temperatures (data not shown). The results demonstrate that maximum operating temperature of the Parylene-N films was dependent on the annealing conditions, and if the applied temperature exceeded the annealing temperature, then there was a significant decrease in piezoelectric properties. Therefore, careful decision over the annealing temperature must be taken under consideration when using the film as a piezoelectric material.

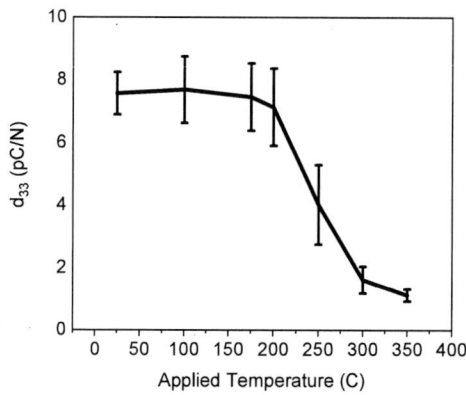

Figure 5: Piezoelectric values of Parylene-N films annealed at 225°C as a function of applied temperature.

CONCLUSIONS

The results in this paper have validated that Parylene-C can be converted to have piezoelectric properties via poling and enhancing crystallinity with similar properties as previously demonstrated [15-17]. The results also demonstrated that Parylene-N crystallinity can be increased through annealing to have higher crystallinity than Parylene-C, which can lead to increased piezoelectric properties. Parylene-N demonstrated significantly higher piezoelectric properties than Parylene-C, and the piezoelectric properties significantly increase after the film goes through a phase transition from monoclinic to hexagonal structure. The piezoelectric films were still able to be patterned using standard reactive ion etching methods. The Parylene-N films were able to maintain their piezoelectric properties at elevated temperatures, suggesting that these films could be a good material for MEMS applications that require elevated processing temperatures or elevated operating temperatures. Future work is needed to optimize the films and to determine long-term use.

ACKNOWLEDGEMENTS

The authors would like to thank the members of the SMART group at the University of New Mexico. The research was sponsored by the Army Research Office and was accomplished under Grant Number W911NF-20-1-0312. The views and conclusions contained in this document are those of the authors and should not be interpreted as representing the official policies, either expressed or implied, of the Army Research Office or the U.S. Government.

REFERENCES

[1] S. Trolier-McKinstry and P. Muralt, "Thin film piezoelectrics for MEMS," *Journal of Electroceramics,* vol. 12, no. 1, pp. 7-17, 2004.

[2] K. S. Ramadan, D. Sameoto, and S. Evoy, "A review of piezoelectric polymers as functional materials for electromechanical transducers," *Smart Materials and Structures,* vol. 23, no. 3, p. 033001, 2014.

[3] K. K. Sappati and S. Bhadra, "Piezoelectric polymer and paper substrates: a review," *Sensors,* vol. 18, no. 11, p. 3605, 2018.

[4] D. Kunwar, I. R. Vazquez, and N. Jackson, "Effects of solvents on synthesis of piezoelectric polyvinylidene fluoride trifluoroethylene thin films," *Thin Solid Films,* vol. 757, p. 139414, 2022.

[5] P.-H. Ducrot, I. Dufour, and C. Ayela, "Optimization of PVDF-TrFE processing conditions for the fabrication of organic MEMS resonators," *Scientific reports,* vol. 6, no. 1, pp. 1-7, 2016.

[6] K. K. Sappati and S. Bhadra, "Flexible piezoelectric 0–3 PZT-PDMS thin film for tactile sensing," *IEEE Sensors Journal,* vol. 20, no. 9, pp. 4610-4617, 2020.

[7] M. Kandpal, C. Sharan, P. Poddar, K. Prashanthi, P. R. Apte, and V. Ramgopal Rao, "Photopatternable nano-composite (SU-8/ZnO) thin films for piezo-electric applications," *Applied Physics Letters,* vol. 101, no. 10, p. 104102, 2012.

[8] I. Babu and G. de With, "Highly flexible piezoelectric 0–3 PZT–PDMS composites with high filler content," *Composites science and technology,* vol. 91, pp. 91-97, 2014.

[9] N. Jackson, L. Keeney, and A. Mathewson, "Flexible-CMOS and biocompatible piezoelectric AlN material for MEMS applications," *Smart Materials and Structures,* vol. 22, no. 11, p. 115033, 2013.

[10] N. Jackson and A. Mathewson, "Enhancing the piezoelectric properties of flexible hybrid AlN materials using semi-crystalline parylene," *Smart Materials and Structures,* vol. 26, no. 4, p. 045005, 2017.

[11] E. Smecca *et al.*, "AlN texturing and piezoelectricity on flexible substrates for sensor applications," *Applied Physics Letters,* vol. 106, no. 23, p. 232903, 2015.

[12] N. Jackson, O. Z. Olszewski, C. O'Murchu, and A. Mathewson, "Ultralow-frequency PiezoMEMS energy harvester using thin-film silicon and parylene substrates," *Journal of Micro/Nanolithography, MEMS, and MOEMS,* vol. 17, no. 1, p. 015005, 2018.

[13] B. J. Kim and E. Meng, "Micromachining of Parylene C for bioMEMS," *Polymers for Advanced Technologies,* vol. 27, no. 5, pp. 564-576, 2016.

[14] M.-N. Niu and E. S. Kim, "Piezoelectric bimorph microphone built on micromachined parylene diaphragm," *Journal of Microelectromechanical Systems,* vol. 12, no. 6, pp. 892-898, 2003.

[15] M. Duggina and N. Jackson, "Converting Parylene C into a Thin Film Piezoelectric Material," in *2021 IEEE 16th Nanotechnology Materials and Devices Conference (NMDC),* 2021: IEEE, pp. 1-5.

[16] J. Y.-H. Kim, "Parylene-c as a new piezoelectric material," California Institute of Technology, 2013.

[17] J. Y.-H. Kim, A. Cheng, and Y.-C. Tai, "Parylene-C as a piezoelectric material," in *2011 IEEE 24th International Conference on Micro Electro Mechanical Systems,* 2011: IEEE, pp. 473-476.

[18] N. Jackson, R. O'Keeffe, R. O'Leary, M. O'Neill, F. Waldron, and A. Mathewson, "A diaphragm based piezoelectric AlN film quality test structure," in *2012 IEEE International Conference on Microelectronic Test Structures,* 2012: IEEE, pp. 50-54.

[19] H.-w. Lo, W.-C. Kuo, Y.-J. Yang, and Y.-C. Tai, "Recrystallized parylene as a mask for silicon chemical etching," in *2008 3rd IEEE International Conference on Nano/Micro Engineered and Molecular Systems,* 2008: IEEE, pp. 881-884.

[20] N. Jackson, F. Stam, J. O'Brien, L. Kailas, A. Mathewson, and C. O'Murchu, "Crystallinity and mechanical effects from annealing Parylene thin films," *Thin Solid Films,* vol. 603, pp. 371-376, 2016.

CONTACT

*N. Jackson, tel: +1-505-2727095; njack@unm.edu

SILICON CARBIDE-ON-INSULATOR THERMAL-PIEZORESISTIVE RESONATOR FOR HARSH ENVIRONMENT APPLICATION

Baoyun Sun[1, 2], Jiarui Mo[1], Hemin Zhang[3*], Henk W. van Zeijl[1], Willem D. van Driel[1], and Guoqi Zhang[1]*

[1]Department of Microelectronics, Delft University of Technology, The NETHERLANDS
[2]Department of Engineering Mechanics, China University of Petroleum (East China), CHINA and
[3]Department of Electrical Engineering (ESAT-MNS), KU Leuven, BELGIUM

ABSTRACT

The thermal-piezoresistive effect in silicon (Si) has attracted great attention toward high-performance resonant devices but still faces major challenges for harsh environment applications. Instead of using Si, this paper, for the first time, reports a thermal-piezoresistive resonator based on a silicon carbide-on-insulator (SiCOI) platform. The resonance frequency simulation, CMOS-compatible fabrication, and thermoresistive properties characterization of the proposed SiCOI resonator are presented. The experimental results show linear current-voltage characteristics and a constant temperature coefficient of resistance (TCR) up to 200 °C.

KEYWORDS

Silicon carbide-on-insulator, thermal-piezoresistive, resonator, harsh environment.

INTRODUCTION

Since the first miniature Si piezoresistive heat engine was reported in 2011 [1], a variety of self-sustained thermal-piezoresistive resonators or oscillators have been developed for mass sensing [2, 3], Lorentz force measurement [4], gas detection [5], and signal amplification [6]. It was demonstrated that the thermal-piezoresistive effect is strong enough to initiate and maintain the mechanical oscillation of the resonator driven by direct current (DC), without the demand for any external alternating current (AC) source and amplifying circuitry. Moreover, the self-sustained oscillation enabled by this effect can significantly improve the quality factor of the resonator, resulting in an enhanced signal-to-noise ratio and sensitivity. However, the state-of-the-art thermal-piezoresistive resonators are mainly based on silicon-on-insulator (SOI) microelectromechanical systems (MEMS) technology. Owing to the intrinsic properties of Si, it is a major challenge to implement SOI resonant devices in harsh environments, like high temperatures and pressure, aggressive chemical corrosion, strong electric field, intense radiation, and high shock or vibration.

To overcome the limitations of Si devices, wide bandgap (WBG) semiconductor materials possessing a bandgap of 2.2 eV or higher [7], have recently attracted tremendous attention as possible MEMS platforms for extreme environment applications. Among all the WBG materials, silicon carbide (SiC) is one of the most appealing platforms for hostile environments compatible MEMS devices [8, 9], as its unique combination of excellent electrical, thermal, and mechanical properties. Furthermore, with the rapid progress in massive production and CMOS processing, SiC also enjoys the benefits of commercial availability of high-quality crystalline bulk wafers and compatibility with current CMOS nanofabrication processes.

SiC exists in more than 200 polytypes that are generally categorized into α-SiC and β-SiC [10, 11]. The most common α-SiC crystal structures are the hexagonal SiC including 4H- (bandgap 3.2 eV) and 6H-SiC (bandgap 3.0 eV) which are commercially available up to 200 mm in diameter. 3C-SiC (bandgap 2.3 eV) is the only cubic crystal of SiC, commonly known as β-SiC. As 3C-SiC is the lowest-temperature phase among all the SiC polytypes, it can be epitaxially grown on Si substrates by the chemical vapor deposition (CVD) method [12, 13]. Compared with Si, the large bandgap of SiC dramatically reduced the leakage current at elevated temperatures. The high ratio between Young's modulus and density enables high-frequency MEMS resonators and oscillators. In addition, the high thermal conductivity is especially crucial for quickly heating up and cooling down the beam in the thermal-piezoresistive resonator.

Here, we demonstrated a 3C-SiCOI thermal-piezoresistive resonator based on the internal thermodynamic feedback mechanism. The proposed structure and working principle of the resonator are described. The dependence of the resonance frequency on the piezoresistive beam dimension is determined by finite element analysis. A CMOS-compatible fabrication process for the SiCOI resonator is established. Experimental measurements are performed to characterize the thermoresistive properties of the resonator.

DESIGN AND SIMULATION

To implement the thermal-piezoresistive effect in SiC, we propose a resonator based on a SiCOI platform with a negative piezoresistive coefficient. The SiCOI thermal-piezoresistive resonator is comprised of two symmetrical suspended masses connected by a narrow piezoresistive beam and supported by four spring beams [14]. Figure 1(a) shows the finite element model of the SiC resonator oscillating in its in-plane mode. The two mass plates are free to move back and forth in the in-plane direction. The ends of the four spring beams are kept fixed to obtain eigenfrequency by simulation.

The operation mechanism of the thermal-piezoresistive resonator is attributed to the internal thermodynamic feedback illustrated in Figure 1(b). The device oscillates from its initial state (phase (1)). A constant DC (I_{dc}) passes through the piezoresistive beam, resulting in an increasing beam temperature T_b caused by the resistive heating power $P_h = I_{dc}^2 R_b$. The heat capacitance of the piezoresistive beam leads to a thermal delay between resistive heating power and temperature in the beam. Consequently, the beam expands at a higher temperature, and the strain in the beam is tensile. Meanwhile, the beam resistance decreases due to the negative piezoresistive coefficient. In phase (3), the beam resistance reaches its minimum value. The beam starts cooling and contracts to its initial state. In phase (5), the beam continues to contract because of the beam temperature is reduced, until it reaches phase (7) where the strain is compressive. Finally, the beam is heated and expanded again to phase (1). Thus, the beam expansion/contraction cycle forms an in-plane mechanical vibration. The piezoresistive beam is used as both a thermal actuator and a piezoresistive sensor in this structure. The oscillation of the beam can be detected by the alternating output voltage.

As a result of the geometry of the resonant structure, the heating power density and mechanical strain are concentrated in the narrow piezoresistive beam, but the resonance frequency of the structure is mainly determined by the masses and spring beams. The finite element analysis results shown in Figure 2 suggest that the 3C-SiC thermal-piezoresistive resonator can generate an in-plane mechanical resonance over 1 MHz. The eigenfrequency increases with increasing piezoresistive beam length and width. Scaling down the length-to-width ratio of the beam leads to a higher resonance frequency. Furthermore, scaling the beam dimension down reduces the power consumption of the resonator while having a higher resonance frequency. On the other hand, the effect of the beam thickness on the eigenfrequency can be neglected.

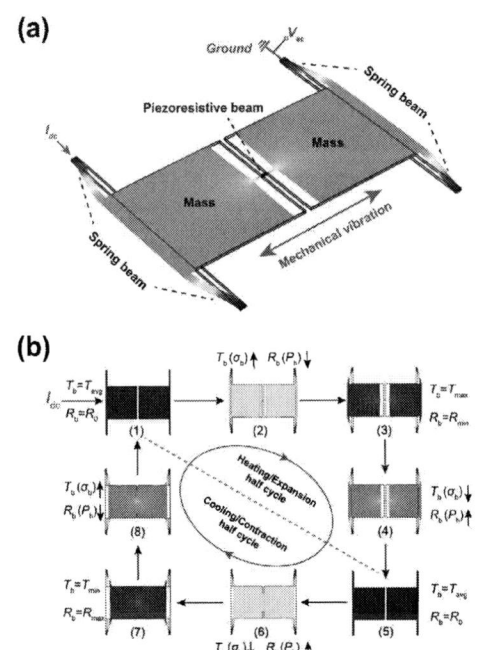

Figure 1: (a) Finite element model of the SiCOI piezoresistive resonator. (b) Schematic illustration of the thermodynamic cycle in the thermo-piezoresistive resonator. The heating power in the piezoresistive beam depends on its position as a result of the piezoresistive effect. Phases (1)-(4) indicate the heating and expansion half cycle, and phases (5)-(8) indicate the cooling and contraction half cycle.

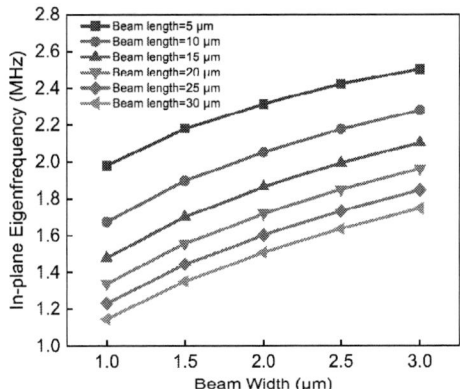

Figure 2: Finite element analysis results of the in-plane resonance frequency change with the piezoresistive beam length and width.

FABRICATION

The fabrication process of the proposed SiCOI thermal-piezoresistive resonator is schematically shown in Figure 3. The process begins with a double side polished 4-inch Si handle wafer with a thickness of 400 ± 8 μm. The Si wafer is thermally oxidized to provide a 2 μm thick buried SiO_2 layer on both sides. The 3C-SiC film is deposited on the Si wafer by low-pressure chemical vapor deposition (LPCVD) at 860 °C and 0.6 Torr. Dichlorosilane (SiH_2Cl_2) and acetylene (C_2H_2) with a flow rate of 80 sccm and 16 sccm are selected as precursor gases, respectively. Ammonia (NH_3) with a flow rate of 1.5 sccm diluted in hydrogen (H_2) is introduced as an N-type dopant during the deposition. As a result, a 1.7 μm thick N-type 3C-SiC film with an average resistivity of 0.02 Ω·cm and a negative piezoresistive coefficient [9] is obtained. The material stack of the created 3C-SiCOI platform is shown in Figure 3(a). The residual stress of the 3C-SiCOI platform is measured to be 6.8 MPa. Such small tensile stress enables flat multilayer thin films without buckling.

Starting with the 3C-SiCOI, the resonator structure is transferred into the top 3C-SiC device layer via lithography and chlorine (Cl_2)/ hydrogen bromide (HBr) plasma etching. A 500 nm thick titanium (Ti) metal layer is then sputtered onto the top 3C-SiC layer at 350 °C to form Ohmic contact with low contact resistance, followed by plasma etching. After finishing the frontside processing, the backside 3C-SiC and SiO_2 layers are opened using plasma etching in two steps. The 3C-SiC and SiO_2 films leave on the backside after etching act as hard masks for the subsequent Si deep reactive ion etching (DRIE). The "Bosch DRIE process" is employed to etch through the Si wafer. In the last step, the frontside SiO_2 layer underneath the resonator structure is removed using vapor hydrogen fluoride (HF) etching. Finally, the suspended 3C-SiCOI thermal-piezoresistive resonator is achieved.

The scanning electron micrograph (SEM) images of the resonator structure are shown in Figure 4. Two 400 μm × 400 μm suspended masses are connected by a 30 μm long, 1 μm wide piezoresistive beam. The masses are supported by four spring beams with the same length of 200 μm and width of 10 μm. Figure 4(b)-(c) show the zoom-in views of the piezoresistive beam and spring beam, respectively.

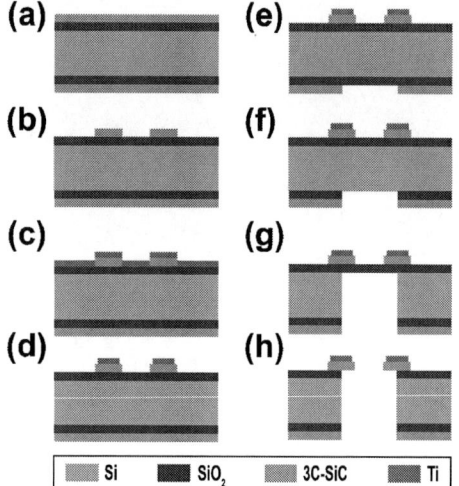

Figure 3: Schematic representation of the industry-standard, CMOS-compatible microfabrication process flow. (a) 3C-SiC thin film deposition and doping by LPCVD. (b) Frontside SiC resonator structure patterning via plasma etching. (c) Metal layer (Ti) deposition. (d) Metal layer patterning. (e) Backside 3C-SiC layer patterning. (f) Backside SiO_2 layer patterning. (g) Through-Si etching lands on the frontside SiO_2 sacrificial layer via the "Bosch process". (h) SiO_2 sacrificial layer removing in vapor HF.

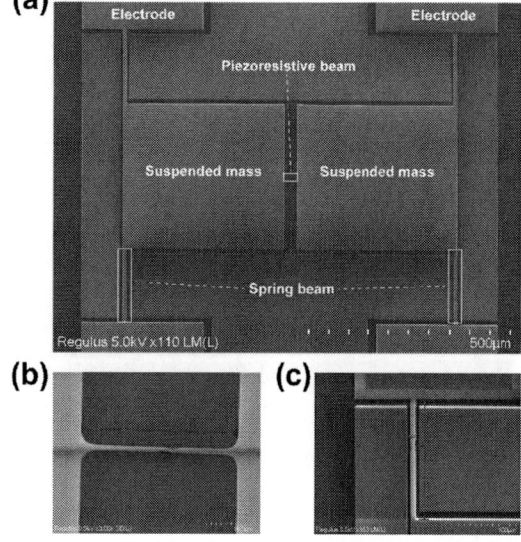

Figure 4: (a) Scanning electron micrograph (SEM) image of the SiCOI thermo-piezoresistive resonator. (b) Magnification of the piezoresistive beam. (c) Magnification of a spring beam.

THERMORESISTIVE CHARACTERISTIC

The thermoresistive effect of the 3C-SiC resonator was characterized by the CascadeMicrotech probe system. The applied current varied from -50 μA to 50 μA. Figure 5 shows the measurement results for the current-voltage (I-V) characteristics of the resonator at various temperatures ranging from 25 °C to 200 °C. The output voltage is linearly increased with the increasing current in the whole temperature range. The linear I-V curves indicate good Ohmic contact between the Ti electrodes and the 3C-SiC film is maintained at elevated temperatures. At a constant applied current, the measured voltage decreased with increasing temperature. The conduction of N-type 3C-SiC is thermally activated.

Figure 6 shows the relative resistance change of the resonator as a function of ambient temperature. The relative resistance change ($\Delta R/R$) linearly decreased with increasing temperature up to 200 °C. According to the definition of temperature coefficient of resistance (TCR),

$$TCR=(R_T-R_0)/[(T-T_0) R_0] \qquad (1)$$

where R_T is the resistance at temperature T, and R_0 is the reference resistance at room temperature T_0 (20 °C). The corresponding temperature coefficient of resistance (TCR) value is calculated to be -1670 ppm/°C. The constant TCR is of high interest for MEMS resonators working at elevated temperatures in terms of simplicity in the design and implementation of circuitry.

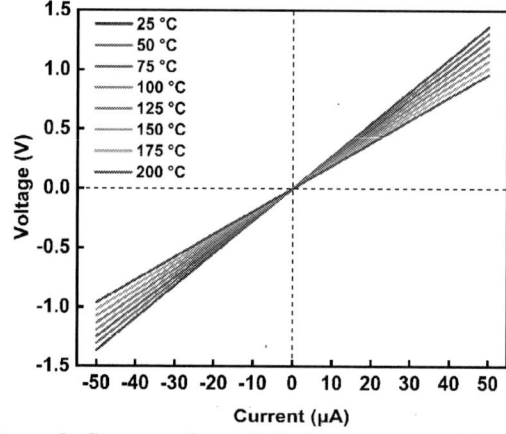

Figure 5: Current-voltage (I-V) characteristics of the 3C-SiC resonator at various temperatures ranging from 25 °C to 200 °C.

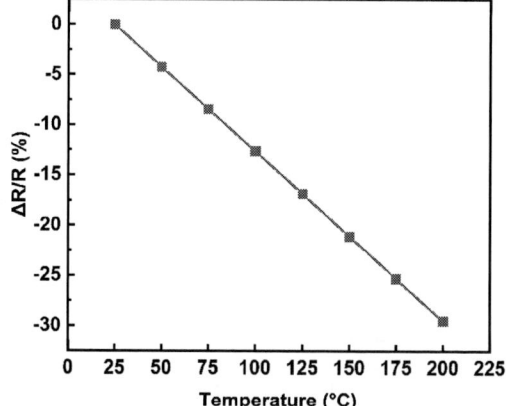

Figure 6: Relative resistance change as a function of the temperature.

CONCLUSION

In summary, we have demonstrated an N-type 3C-SiCOI thermal-piezoresistive resonator for high-temperature applications. The resonator structure design and operation mechanism were introduced in detail. The finite element analysis of the resonator indicates a high in-plane eigenfrequency over 1 MHz. We have developed a CMOS-compatible process for the SiCOI resonator fabrication. The as-fabricated device shows a linear I-V characteristic and constant negative TCR value up to 200 °C. Our work provides a viable route toward harsh environment-compatible, self-sustained SiCOI resonate devices. The thermal-piezoresistive effect at high temperatures and frequency responses of the SiCOI resonator in this work need to be further investigated and tested.

ACKNOWLEDGEMENTS

The authors thank the staff of Delft University of Technology Else Kooi Laboratory (EKL) for processing support. This work is supported by the European Union's Horizon 2020 research and innovation programme under grant agreement No 871741 and No 876659, ITEA-EUREKA project No 19037.

REFERENCES

[1] P. G. Steeneken, K. Le Phan, M. J. Goossens *et al.*, "Piezoresistive heat engine and refrigerator," *Nat. Phys.*, vol. 7, no. 4, pp. 354-359, 2011.

[2] A. Quan, H. Zhang, C. Wang *et al.*, "A Self-Sustained Mass Sensor With Physical Closed Loop Based on Thermal-Piezoresistive Coupled Resonators," *IEEE Trans. Electron Devices*, vol. 69, no. 10, pp. 5808-5813, 2022.

[3] A. Quan, C. Wang, H. Zhang *et al.*, "A Thermal-Piezoresistive Self-Sustained Resonant Mass Sensor with High-Q (>95k) in Air," in *21st International Conference on Solid-State Sensors, Actuators and Microsystems (Transducers)*, June 20-24, 2021, pp. 168-171.

[4] V. Kumar, A. Ramezany, M. Mahdavi *et al.*, "Amplitude modulated Lorentz force MEMS magnetometer with picotesla sensitivity," *J. Micromech. Microeng.*, vol. 26, no. 10, 2016.

[5] X. Guo, Y.-b. Yi, and S. Pourkamali, "Thermal-Piezoresistive Resonators and Self-Sustained Oscillators for Gas Recognition and Pressure Sensing," *IEEE Sens. J.*, vol. 13, no. 8, pp. 2863-2872, 2013.

[6] A. Ramezany, M. Mahdavi, and S. Pourkamali, "Nanoelectromechanical resonant narrow-band amplifiers," *Microsyst. Nanoeng.*, vol. 2, pp. 16004, 2016.

[7] M. N. Yoder, "Wide bandgap semiconductor materials and devices," *IEEE Trans. Electron Devices*, vol. 43, no. 10, pp. 1633-1636, 1996.

[8] J. Romijn, S. Vollebregt, L. M. Middelburg *et al.*, "Integrated 64 pixel UV image sensor and readout in a silicon carbide CMOS technology," *Microsyst. Nanoeng.*, vol. 8, pp. 114, 2022.

[9] L. M. Middelburg, H. W. van Zeijl, S. Vollebregt *et al.*, "Toward a Self-Sensing Piezoresistive Pressure Sensor for All-SiC Monolithic Integration," *IEEE Sens. J.*, vol. 20, no. 19, pp. 11265-11274, 2020.

[10] C. Eddy, and D. Gaskill, "Silicon carbide as a platform for power electronics," *Science,* vol. 324, no. 5933, pp. 1398-1400, 2009.

[11] A. L. Falk, B. B. Buckley, G. Calusine *et al.*, "Polytype control of spin qubits in silicon carbide," *Nat Commun*, vol. 4, pp. 1819, 2013.

[12] C. A. Zorman, A. J. Fleischman, A. S. Dewa *et al.*, "Epitaxial growth of 3C–SiC films on 4 in. diam (100) silicon wafers by atmospheric pressure chemical vapor deposition," *J. Appl. Phys.*, vol. 78, no. 8, pp. 5136-5138, 1995.

[13] L. Wang, S. Dimitrijev, J. Han *et al.*, "Growth of 3C–SiC on 150-mm Si (100) substrates by alternating supply epitaxy at 1000 °C," *Thin solid films,* vol. 519, no. 19, pp. 6443-6446, 2011.

[14] H. Zhang, A. Quan, C. Wang *et al.*, "A Fast-Startup Self-Sustained Thermal-Piezoresistive Oscillaror with >10^6 Effective Quality Factor In the Air," in *IEEE 35th International Conference on Micro Electro Mechanical Systems Conference (MEMS)*, January 9-13, Tokyo, 2022, pp. 142-145.

CONTACT

*Baoyun Sun, b.sun-3@tudelft.nl
*Hemin Zhang, 1989hemin@gmail.com

SPIN COATING OF HIGHLY ALIGNED AGCN MICROWIRES EPITAXIALLY GROWN ON 2D MATERIALS

Jimin Ham[1], Jaemook Lim[1], Joowon Lim[1], Gunyoung Jang[1], Sueng Yoon Lee[1], Dohyun Lim[1], Sukjoon Hong[1], and Won Chul Lee[1]

[1]Department of Mechanical Engineering, BK21FOUR ERICA-ACE Center, Hanyang University, Ansan, REPUBLIC OF KOREA

ABSTRACT

This study presents a solvent-controlled spin coating method that can uniformly synthesize a dense layer of epitaxial AgCN microwires onto various 2D materials. The combination of spin coating and an aqueous ethanol mixture promotes heterogeneous crystal nucleation on 2D material surfaces, which leads to highly aligned, uniform, dense growth of the microwires unlike the previous drop casting methods. Because the spin coated microwires are observable with conventional optical microscopy and removable with simple wet chemistry, we can apply our method to large-area crystallographic mapping of grains in polycrystalline graphene.

KEYWORDS

Crystallographic orientation, epitaxy, AgCN microwire, 2D materials, spin coating, inorganic material

INTRODUCTION

Identifying the crystallographic orientation of 2D materials is an important task for the reliable production of 2D materials. Atomic resolution imaging of 2D materials using transmission electron microscopy (TEM) and scanning tunneling microscopy is the most accurate method [1], but it is complex and time-consuming. Epitaxial growth of nanomaterials on 2D materials can offer a facile labeling technique to visualize the crystallographic structure and boundaries of 2D materials [2]. However, the label visualization requires scanning electron microscopy or atomic force microscopy, thus large-scale crystallographic identification is still challenging. Although a few cases based on optical microscopy are reported [3-5], poor uniformity and coverage of the epitaxial labels synthesized on 2D materials are still a limiting factor for large-area identification of crystallographic information.

Here, we present the solvent-controlled spin coating method for the uniform and dense growth of AgCN microwires (Fig.1), which are formed along the zigzag orientation of 2D materials. Spin coating of an aqueous ethanol mixture promotes the thin supersaturated solution layer on the 2D material surface and enables heterogeneous crystal nucleation [6]. As a result, spin coated microwires show improved uniformity and density compared with two types of previous drop casting methods [3,4] (Fig. 2). Drop casting of an aqueous AgCN solution (Fig. 1(a)) and an aqueous ethanol mixture (Fig. 1(b)) leads to sparse and nonuniform growth of microwires due to its low vapor pressure and inconsistent surface affinity. In addition, unavoidable limitations of drop casting method cause the formation of crystal residue from the bulk solution, which covers the microwires to prevent observation. Therefore, we propose that the presented spin coating method can be used as a practical tool for accurately measuring the crystallographic orientation of 2D materials.

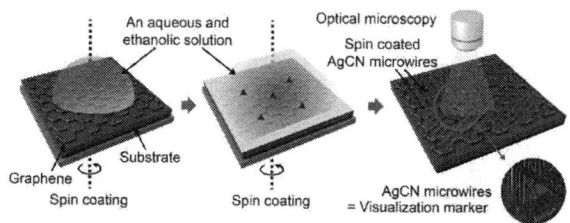

Figure 1: Spin coating process for the epitaxial growth of AgCN microwires on 2D materials.

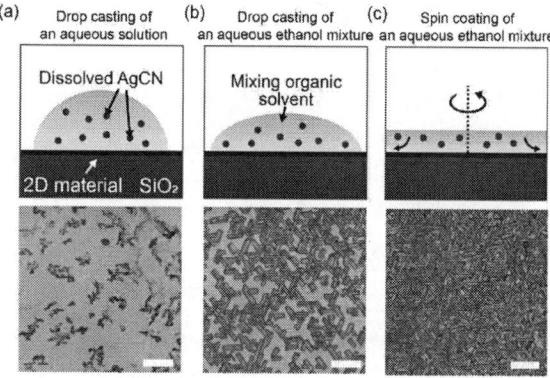

Figure 2: Growth methods and optical images of AgCN microwires grown on graphene by three synthesis methods. The scale bar is 10 μm.

The main claim of this study is that spin coating of an aqueous ethanol mixture can be applied to various types of 2D materials (Fig. 3). Although the 2D materials we used in this study have different surface properties, optical images of spin coated microwires synthesized on the five different substrates (Figs. 1(c) and 3) show similar growth shapes without noticeable differences. We assume that surface hydrophilicity can be adjusted by ethanol content and make surface affinity strong enough to form heterogeneous nucleation. Please note that this heterogeneous nucleation over the entire surface of 2D materials is generated on pristine 2D materials, not on graphene edges, amorphous carbon surfaces, and silicon oxides (Fig. 4).

978-1-6654-9309-3/23 $31.00 © 2023 IEEE

IEEE MEMS 2023, Munich, GERMANY
15 - 19 January 2023

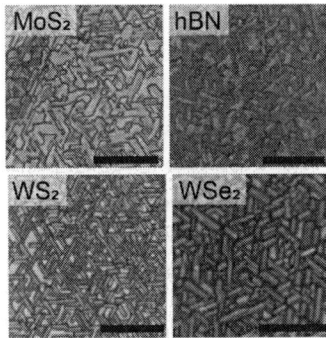

Figure 3: Optical images of spin coated AgCN microwires on four different substrates (MoS₂, hBN, WS₂, and WSe₂). The scale bar is 10 μm.

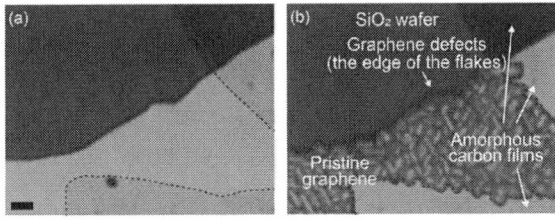

Figure 4: AgCN microwires spin coated on four different types of surfaces (pristine graphene, graphene edge, amorphous carbon, and silicon oxide). The scale bar is 2 μm.

The present spin coating method enables the large-scale optical identification of crystal structures in 2D materials because of the uniform and dense growth characteristics on a large area of 2D materials. We verify these abilities using single-layered polycrystalline graphene that is synthesized by CVD (chemical vapor deposition) and is transferred onto a SiO₂/Si substrate. After synthesizing AgCN microwires on the as-prepared graphene sample, we can map the crystal orientation in ~250 μm by ~250 μm areas (Fig. 5(a)) using an optical microscopic observation. Optical and SEM images clearly show that each microwire is aligned along the lattice orientation of the underlying graphene domain, and this microwire epitaxy enables the distinction of neighboring grains (Fig. 5(b)). The quantitative analysis from 346 grains and 930 grain boundaries in the map shows that the averaged grain size of 12.4 μm ± 5.3 μm is much smaller than the specific grain size expected by the manufacturer (Fig. 5(c)). The unsatisfied specification found in the commercially purchased graphene means that monitoring processes such as grain identification are mandatory for manufacturing commercial products from 2D materials. We also analyze grain shapes and lattice orientations and illustrate these relationships in a scatter plot and a histogram with a probability density function (Fig 5(d) and (e)).

In conclusion, we present that this presented spin coating method can form a dense layer of AgCN microwires and apply it to various types of 2D materials. The characteristics of high density and uniformity enable the large-area identification of polycrystalline graphene using optical microscopy. This practical application demonstrates that our method can be used as a useful tool to identify crystallographic orientations even in polycrystalline domains.

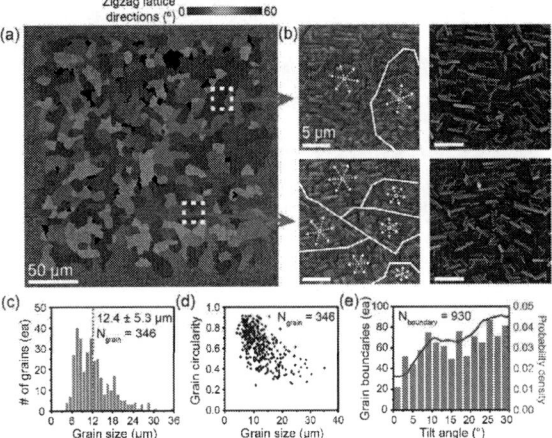

Figure 5: Large area identification of crystal structures in single-layered polycrystalline graphene based on conventional optical microscopy of AgCN microwires. (a) Crystal orientation map of 250 × 250 μm² graphene sample. (b) Optical and SEM images from the square areas in (a). (c-e) Analysis of grains and boundaries from orientation map in (a).

ACKNOWLEDGEMENTS

This work was supported by the National Research Foundation of Korea(NRF) grant funded by the Korea government(MSIT) (No. 2022R1A4A3031263 and No. 2021R1A2C1011797).

REFERENCES

[1] P. Y. Huang, "Grains and grain boundaries in single-layer graphene atomic patchwork quilts", *Nature*, 469, 389-392, 2011.

[2] W. C. Lee, "Graphene-templated directional growth of an inorganic nanowire", *Nature Nanotech.*, 10, 423-428, 2015.

[3] J. Ham, "Facile Identification of Graphene's Crystal Orientations by Optical Microscopy of Self-Aligned Microwires", in *IEEE 32ⁿᵈ MEMS*, Seoul, Jan 27-31, 2019, pp. 264-265.

[4] Y. Lee, "Universal Oriented van der Waals Epitaxy of 1D Cyanide Chains on Hexagonal 2D Crystals", *Adv. Sci.*, 7, 1900757, 2020.

[5] X. Cui, "Visualization of Crystallographic Orientation and Twist Angles in Two-Dimensional Crystals with an Optical Microscope", *Nano Lett.*, 20(8), 6059-6066, 2020.

[6] M. V. Kelso, "Spin coating epitaxial films", *Science*, 364, 166-169, 2019.

CONTACT

*W.C.L., tel: +82-10-6396-0309;
wonchullee@hanyang.ac.kr

SUSPENDED TWO-DIMENSIONAL MATERIAL MEMBRANES FOR SENSOR APPLICATIONS FABRICATED WITH A HIGH-YIELD TRANSFER PROCESS

Sebastian Lukas[1], Ines Kraiem[1,2], Maximilian Prechtl[3], Oliver Hartwig[3], Annika Grundmann[4], Holger Kalisch[4], Satender Kataria[1], Michael Heuken[4,5], Andrei Vescan[4], Georg S. Duesberg[3], and Max C. Lemme[1,2]

[1]Chair of Electronic Devices, RWTH Aachen University, Otto-Blumenthal-Str. 2, 52074 Aachen, GERMANY, [2]AMO GmbH, Advanced Microelectronic Center Aachen, Otto-Blumenthal-Str. 25, 52074 Aachen, GERMANY, [3]Institute of Physics, Faculty of Electrical Engineering and Information Technology (EIT 2) and Center for Integrated Sensor Systems, University of the Bundeswehr Munich, 85577 Neubiberg, GERMANY, [4]Compound Semiconductor Technology, RWTH Aachen University, Sommerfeldstr. 18, 52074 Aachen, GERMANY, [5]AIXTRON SE, Dornkaulstr. 2, 52134 Herzogenrath, GERMANY

ABSTRACT

This paper reports a novel dry-transfer method for suspending and patterning two-dimensional (2D) materials across closed (sealed) cavities on a target substrate. The process utilizes handling frames and heat to ensure sufficient adhesion and a multi-layer resist stack to pattern the suspended thin films. The yield of intact suspended monolayer graphene membranes is up to more than 75 %, higher than formerly reported [1]–[3]. The process also demonstrates high yields of intact membranes of double-layer graphene and of two different transition metal dichalcogenides. The process flow can be applied to the fabrication of various sensor applications, such as pressure sensors.

KEYWORDS

suspended 2D materials, graphene, $PtSe_2$, MoS_2, transition metal dichalcogenides, membrane-based sensors

INTRODUCTION

Very thin suspended membranes are essential for various sensor applications, e.g., pressure sensors or microphones [4]. The extreme thinness combined with a remarkable hermeticity [5], [6], high mechanical stability [7], and superior electronic and piezoresistive properties make 2D materials, particularly graphene, platinum diselenide ($PtSe_2$), molybdenum disulfide (MoS_2), and other transition metal dichalcogenides (TMDCs) promising candidates as membrane materials [8]–[12]. However, it remains challenging to transfer and suspend 2D materials across closed cavities etched into a target substrate (usually Si/SiO_2). In open-cavity or open-trench fabrication, methods like sacrificial layer under-etching are suitable [13], [14], in contrast to the fabrication of membranes across closed cavities. Additionally, commonly used wet-transfer methods have shown to be unfit due to liquid getting trapped inside the cavities or liquid pulling down and rupturing the 2D material membranes during the drying process [3]. Thermal-release-tape transfer, PDMS-stamp transfer, and similar techniques involving mechanical stress have also proven unsuitable for suspending 2D materials due to possible membrane rupture during the transfer process [3]. In several studies, the supporting poly(methyl methacrylate) (PMMA) layer was left on the 2D material membrane for improved mechanical stability [15], [16]. Although this

PMMA layer provides increased stability, it is also several orders of magnitude thicker than the 2D material it supports, making it a compromise that leads to a significantly reduced membrane flexibility and therefore sensitivity.

We demonstrate a dry-transfer method for various 2D materials, including the removal of the supporting PMMA layer. Handling frames are used, and high temperature ensures good adhesion to the target substrate. Compared to

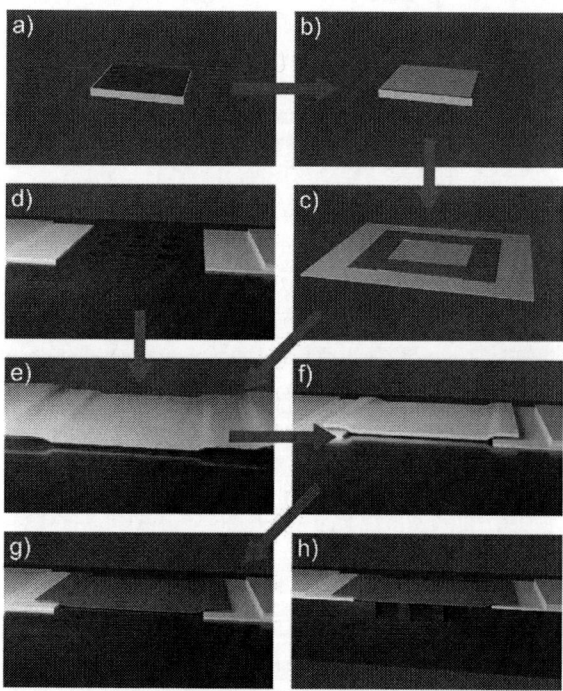

Figure 1: Sketch of the fabrication process. (a) 2D material on growth substrate. (b) PMMA spin-coating. (c) Prepared transfer frame with transparent foil and heat-resistant tape attached to PMMA/2D material on growth substrate. (d) Preparation of target substrate, cavity etching, metal contact deposition. (e) PMMA/2D material on target substrate after transfer on hotplate and frame removal. (f) Patterned PMMA/2D material stack. (g) 2D material on target substrate after PMMA removal. (h) Cross-section of the device.

the other described methods, our process results in high yields of intact 2D material membranes across closed cavities. We furthermore pattern the 2D materials after transfer and suspension using a three-layer photoresist stack. The process is applied to different large-area 2D materials: monolayer chemical-vapor-deposition-grown (CVD-grown) graphene, artificially stacked double-layer CVD-grown graphene, thermally-assisted-conversion-grown (TAC-grown) $PtSe_2$, and metal-organic chemical-vapor-deposition-grown (MOCVD-grown) MoS_2. Scanning electron microscopy (SEM) and Raman tomography [17] are utilized for evaluating the yield of intact suspended membranes.

METHODS

Commercial single-layer CVD graphene on copper (Cu) foil, 6-nm-thick TAC-grown layered nanocrystalline $PtSe_2$ on Si/SiO_2 (90 nm) substrates [18], [19], and few-layer MOCVD-grown MoS_2 on sapphire substrates [20], [21] were used. Artificially stacked double-layer graphene was created by etching the Cu from a piece of PMMA-covered graphene and wet-transferring it on top of another piece of graphene on Cu foil.

First, the 2D materials on their growth substrates were spin-coated with PMMA and baked (see Figure 1a-b) before attaching the transfer frames assembled from transparent plastic foil and heat-resistant polyimide tape (see Figure 1c and Figure 2a-b). Depending on the growth substrate, solutions of either $HCl/H_2O_2/H_2O$ (to etch Cu from graphene) or KOH/H_2O (to delaminate $PtSe_2$ and MoS_2 from SiO_2 and sapphire, respectively) were used to remove or detach the growth substrates. The transfer frames with the 2D material and PMMA were rinsed in deionized water and dried (see Figure 2c).

The target substrates were prepared from 6" Si wafers. Approx. 2-3 μm deep cavities were etched using a reactive ion etching (RIE) process. After thermal oxide growth (90 nm SiO_2), the wafers were diced, and metal contacts were deposited (see Figure 1d).

The target substrates were then placed on a hotplate at 115 °C and the dried transfer frames with the 2D materials was placed on top of them at the desired locations. While the temperature was ramped up to 180 °C, the 2D materials began adhering to the target substrates due to a softening of the PMMA support layer, indicated by a change of color (see Figure 2d). After full adhesion, the transfer frames were removed by cutting along their inner edge, leaving the stacks of 2D material and PMMA on the target substrates (see Figure 1e and Figure 2e). The PMMA was kept intact on the 2D materials for mechanical stability during further processing.

To pattern the transferred 2D material, a stack of LOR3A and AZ5214E photoresists was spin-coated on top of the PMMA, creating a three-layer resist stack. Here, LOR3A was used to prevent mixing of the PMMA and AZ5214E resist. After photolithography, the PMMA/2D material stack was etched using RIE with O_2 or O_2/CF_4 chemistry for graphene or $PtSe_2/MoS_2$, respectively. LOR3A and AZ5214E were then stripped using either dimethyl sulfoxide (DMSO) at room temperature or a TMAH-containing developer after a UV flood exposure (see Figure 1f). In both methods, the underlying PMMA layer is not affected. It was noted that DMSO heated to above 40 °C can remove PMMA, though. Additionally, it was observed that DMSO may lead to partial delamination of graphene, making the TMAH-based method more reliable for graphene.

In the final step, the PMMA was removed by acetone and subsequent careful isopropyl alcohol (IPA) rinsing. The samples were left to dry naturally, avoiding mechanical stress from blow-drying (see Figure 1g).

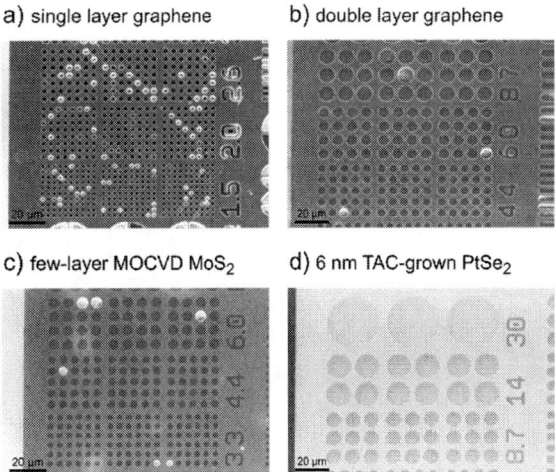

Figure 3: SEM images of suspended 2D material membranes. Intact membranes appear darker than entirely or partially ruptured membranes. Membrane diameters shown in units of micrometers. (a) Single-layer CVD graphene membranes. (b) Double-layer CVD graphene membranes. (c) Few-layer MOCVD MoS_2 membranes. (d) 6-nm-thick TAC-grown $PtSe_2$ membranes.

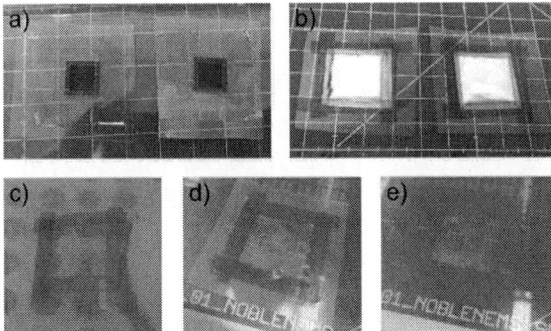

Figure 2: Photographs of the transfer process. (a) TAC-grown $PtSe_2$ on Si/SiO_2 and (b) CVD-grown graphene on Cu foil, all covered with PMMA and attached to transfer frames. 1 cm grid in background for scale. (c) Transfer frame with graphene/PMMA layer after Cu etching and rinsing, put up for drying. (d) Graphene transfer to target substrate on hotplate, during the process of adhering. (e) Graphene/PMMA on target substrate after complete adhesion and frame removal.

978-1-6654-9309-3/23 $31.00 © 2023 IEEE

RESULTS AND DISCUSSION

The suspended 2D material membranes were examined using SEM. The samples were slightly tilted to be able to image the cavity sidewall. Due to the high electrical conductivity of the 2D materials compared to the underlying SiO_2, the intact and broken membranes can be easily identified (see Figure 3). In case of broken membranes, the 2D material would either stick to the cavity sidewalls and bottom, be ripped off entirely, be folded on itself, or only feature small cracks or holes.

Additionally, using Raman tomography [17], i.e. linescans with a change of the focus point, cross-sectional images of suspended 2D material membranes across the Si/SiO_2 cavities were created from the peak intensities. This is shown for graphene and MoS_2 membranes in Figure 4. The measurement demonstrates that the 2D materials are truly suspended above the substrate across a circular cavity with a diameter of up to 30 μm.

The yield of intact 2D material membranes was analyzed from SEM images by evaluating several test areas with membranes of varying size (see Figure 3). The yield is plotted against the membrane diameter in Figure 5.

Our transfer and patterning method results in higher yields of intact single-layer graphene membranes across closed cavities than reported before [1]–[3]. Across ~ 2.5 μm wide circular cavities of > 2 μm depth, we can reach estimated yields of > 75 % of intact single-layer CVD-grown graphene membranes and of up to 90 % of intact double-layer CVD-grown graphene membranes (see Figure 5a-b). The generally higher yield of artificially stacked double-layer graphene membranes can be explained by a compensation of cracks and small holes in one graphene layer by the second one, thus eliminating predetermined breaking points. This effect becomes especially visible for membrane diameters > 5 μm, for which the yield using double-layer graphene is more than twice of that of single-layer graphene.

Additionally, high yields of intact suspended membranes of large-scale-grown transition metal dichalcogenides are observed (see Figure 5c-d). Intact MoS_2 and $PtSe_2$ membranes are found across cavities of up to 30 μm.

On the same substrates, several four-contact devices with a 2D material channel were fabricated, with graphene as an example, across an array of cavities and contacted by Ni electrodes (see Figure 6). These results demonstrate the suitability of the fabrication process for defining the 2D material structures in devices based on suspended membranes, such as pressure sensor devices.

Further experiments demonstrating the process on wafer-scale substrates with a much larger number of cavities and including electrical measurements of the pressure sensor devices are ongoing.

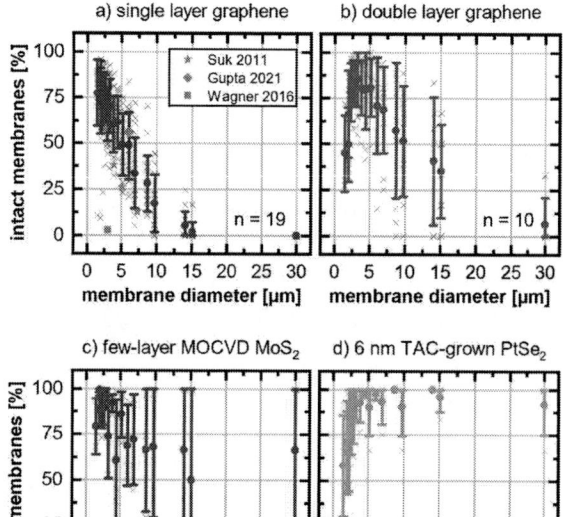

Figure 5: Yield of intact suspended 2D material membranes vs. the membrane diameter for single-layer CVD graphene, double-layer CVD graphene, MOCVD MoS₂, and 6-nm-thick TAC-grown PtSe₂. The number of counted test fields (see, e.g., Figure 3a) is indicated by n in each graph.

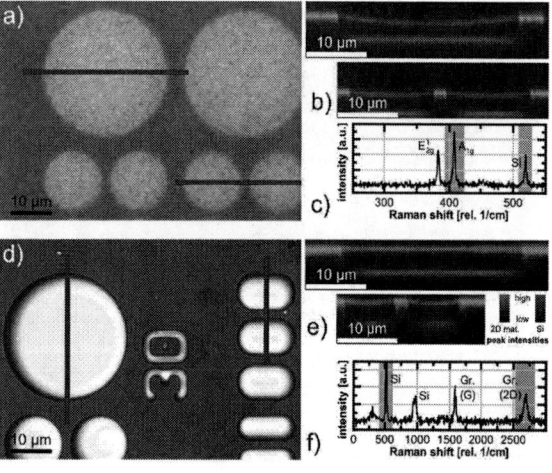

Figure 4: (a) Optical microscope (OM) image, (b) Raman tomography scans, and (c) single Raman spectrum of MOCVD MoS₂ membranes. (d) OM image, (e) Raman tomography scans, and (f) single Raman spectrum of double-layer CVD graphene membranes. The red lines in the OM images correspond to the x-axis in the tomography images. In the tomography images, blue color indicates a high Si peak signal and red color indicates a high MoS₂ A₁g or graphene 2D peak signal, respectively.

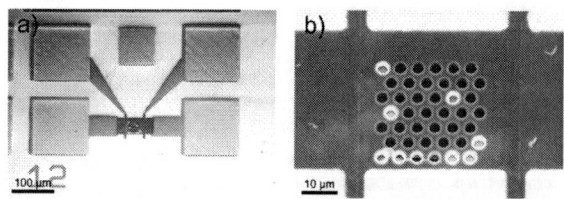

Figure 6: (a) Four-contact membrane device with single-layer graphene membranes, Ni bottom contacts, and Al pads for wire-bonding. (b) Zoom into (a).

CONCLUSION

The presented process for the transfer and patterning of suspended 2D material membranes across closed cavities has proven suitable for obtaining high yields of intact suspended membranes. For single-layer CVD-grown graphene, the observed yield is higher than in previous reports, and it is even higher for artificially stacked double-layer CVD-grown graphene. Suspended membranes of large-scale-grown MoS_2 and $PtSe_2$ with decent yields have additionally been demonstrated for the first time to the best of our knowledge.

The work shows that the developed versatile transfer and patterning process can be applied to a variety of 2D or other thin-film materials, thus opening new possibilities for the fabrication of a large range of membrane-based devices for electronics, sensing, and other fields.

ACKNOWLEDGEMENTS

This work has received funding from the German Ministry of Education and Research (BMBF) under grant agreements 16ES1121 (ForMikro-NobleNEMS), 16ME0399 (NEUROTEC 2), and 03ZU1106 (NeuroSys), and from the European Union's Horizon 2020 research and innovation programme under grant agreements 881603 (Graphene Flagship Core 3) and 825272 (ULISSES).

REFERENCES

[1] R. K. Gupta et al., "Suspended graphene arrays for gas sensing applications," *2D Mater.*, vol. 8, no. 2, p. 025006, Dec. 2020, doi: 10.1088/2053-1583/abcf11.

[2] J. W. Suk et al., "Transfer of CVD-Grown Monolayer Graphene onto Arbitrary Substrates," *ACS Nano*, vol. 5, no. 9, pp. 6916–6924, Sep. 2011, doi: 10.1021/nn201207c.

[3] S. Wagner, C. Weisenstein, A. D. Smith, M. Östling, S. Kataria, and M. C. Lemme, "Graphene transfer methods for the fabrication of membrane-based NEMS devices," *Microelectronic Engineering*, vol. 159, pp. 108–113, Jun. 2016, doi: 10.1016/j.mee.2016.02.065.

[4] P. Song et al., "Recent Progress of Miniature MEMS Pressure Sensors," *Micromachines*, vol. 11, no. 1, Art. no. 1, Jan. 2020, doi: 10.3390/mi11010056.

[5] M. Lee et al., "Sealing Graphene Nanodrums," *Nano Lett.*, vol. 19, no. 8, pp. 5313–5318, Aug. 2019, doi: 10.1021/acs.nanolett.9b01770.

[6] P. Z. Sun et al., "Limits on gas impermeability of graphene," *Nature*, vol. 579, no. 7798, Art. no. 7798, Mar. 2020, doi: 10.1038/s41586-020-2070-x.

[7] C. Lee, X. Wei, J. W. Kysar, and J. Hone, "Measurement of the Elastic Properties and Intrinsic Strength of Monolayer Graphene," *Science*, vol. 321, no. 5887, pp. 385–388, Jul. 2008, doi: 10.1126/science.1157996.

[8] S. Wagner et al., "Highly Sensitive Electromechanical Piezoresistive Pressure Sensors Based on Large-Area Layered PtSe2 Films," *Nano Lett.*, vol. 18, no. 6, pp. 3738–3745, Jun. 2018, doi: 10.1021/acs.nanolett.8b00928.

[9] M. C. Lemme et al., "Nanoelectromechanical Sensors Based on Suspended 2D Materials," *Research*, vol. 2020, Jul. 2020, doi: 10.34133/2020/8748602.

[10] A. D. Smith et al., "Electromechanical Piezoresistive Sensing in Suspended Graphene Membranes," *Nano Lett.*, vol. 13, no. 7, pp. 3237–3242, Jul. 2013, doi: 10.1021/nl401352k.

[11] V. Rana et al., "High-Performing Polycrystalline MoS2-Based Microelectromechanical Piezoresistive Pressure Sensor," *IEEE Sensors Journal*, vol. 22, no. 19, pp. 18542–18549, Oct. 2022, doi: 10.1109/JSEN.2022.3198761.

[12] X. Lin et al., "Polymer-Assisted Pressure Sensor with Piezoresistive Suspended Graphene and Its Temperature Characteristics," *NANO*, vol. 14, no. 10, p. 1950130, Oct. 2019, doi: 10.1142/S1793292019501303.

[13] J. Xu, Graham. S. Wood, E. Mastropaolo, Michael. J. Newton, and R. Cheung, "Realization of a Graphene/PMMA Acoustic Capacitive Sensor Released by Silicon Dioxide Sacrificial Layer," *ACS Appl. Mater. Interfaces*, Jul. 2021, doi: 10.1021/acsami.1c05424.

[14] R. Pezone, G. Baglioni, P. M. Sarro, P. G. Steeneken, and S. Vollebregt, "Sensitive Transfer-Free Wafer-Scale Graphene Microphones," *ACS Appl. Mater. Interfaces*, Apr. 2022, doi: 10.1021/acsami.2c03305.

[15] M. Šiškins et al., "Sensitive capacitive pressure sensors based on graphene membrane arrays," *Microsystems & Nanoengineering*, vol. 6, no. 1, Art. no. 1, Nov. 2020, doi: 10.1038/s41378-020-00212-3.

[16] C. Berger, R. Phillips, A. Centeno, A. Zurutuza, and A. Vijayaraghavan, "Capacitive pressure sensing with suspended graphene–polymer heterostructure membranes," *Nanoscale*, vol. 9, no. 44, pp. 17439–17449, Nov. 2017, doi: 10.1039/C7NR04621A.

[17] S. Wagner et al., "Noninvasive Scanning Raman Spectroscopy and Tomography for Graphene Membrane Characterization," *Nano Lett.*, vol. 17, no. 3, pp. 1504–1511, Mar. 2017, doi: 10.1021/acs.nanolett.6b04546.

[18] C. Yim et al., "High-Performance Hybrid Electronic Devices from Layered PtSe2 Films Grown at Low Temperature," *ACS Nano*, vol. 10, no. 10, pp. 9550–9558, Oct. 2016, doi: 10.1021/acsnano.6b04898.

[19] S. Lukas et al., "Correlating Nanocrystalline Structure with Electronic Properties in 2D Platinum Diselenide," *Advanced Functional Materials*, p. 2102929, 2021, doi: 10.1002/adfm.202102929.

[20] M. Marx et al., "Metalorganic Vapor-Phase Epitaxy Growth Parameters for Two-Dimensional MoS2," *J. Electron. Mater.*, vol. 47, no. 2, pp. 910–916, Feb. 2018, doi: 10.1007/s11664-017-5937-3.

[21] H. Cun et al., "Wafer-scale MOCVD growth of monolayer MoS2 on sapphire and SiO2," *Nano Res.*, vol. 12, no. 10, pp. 2646–2652, Oct. 2019, doi: 10.1007/s12274-019-2502-9.

CONTACT

*M. C. Lemme, max.lemme@eld.rwth-aachen.de

TCF-IMPROVED SH0 MODE ACOUSTIC RESONATORS BASED ON 30° YX-LINBO3/SIO2 MEMBRANE

Shuxian Wu[1], Zonglin Wu[1], Hangyu Qian[1], Feihong Bao[1,], Gongbin Tang[2], Feng Xu[1], and Jie Zou[1,*]*
[1]School of Information Science and Technology, Fudan University, CHINA and
[2]Institute of Novel Semiconductors, Shandong University, CHINA

ABSTRACT

In this paper, a 30° *YX*-Lithium Niobate (LN) 0-th shear horizontal (SH_0) plate acoustic wave (PAW) resonator is proposed. The SH_0 mode characteristics the superiority of interdigital transducer (IDT) in the frequency definition over most other plate modes. Using finite element analysis method, the rotation angle of LN and the thickness of each layer were optimized for large effective coupling coefficient (k^2_{eff}) and high acoustic velocity. The rotation angle and the thickness of LN membrane are optimized as 30° and 0.2λ, respectively. To improve the temperature stability of proposed PAW resonators, a SiO_2 film are added and the thickness is designed as 0.2λ. The measurement results derived a k^2_{eff} of 25.1%, a Bode-Q_{max} of 604, and a Figure of merit (*FoM*) of 151, which is higher than the reported similar-type PAW resonators. The measured first-order temperature coefficients of frequency at resonant frequency (TCF_{fs}) and anti-resonant frequency (TCF_{fp}) are -38ppm/°C and -26ppm/°C, suggesting the temperature stability improvement in comparison with only LN membrane-based resonators.

KEYWORDS

Large coupling, lithium niobate, plate acoustic wave, shear horizontal, temperature stability.

INTRODUCTION

In response to the rapidly increasing demands for 5G wireless connectivity, more and faster data transmission in wireless mobile, Internet-of-Things (IoT), autonomous vehicles [1, 2, 3], new systems currently are required wider and more frequency bands. With the increase of frequency bands dramatically, the complexity of radio frequency front-end (RFFE) system in mobile devices is also increased. To meet this challenge, a scheme of tunable acoustic filters with large bandwidth is desired [4, 5]. It is noteworthy that one of the vital technologies for the tunable filters is to design the acoustic resonators owing a large efficient coupling coefficient (k^2_{eff}) and a high quality factor (Q). So far, the widely adopted solutions of acoustic resonators are lithium niobate (LiNbO3/LN) or lithium tantalate (LiTaO3/LT)-based surface acoustic wave (SAW) devices and aluminium nitrogen (AlN)-based bulk acoustic wave (BAW) devices. However, the k^2_{eff} of commercial surface acoustic wave (SAW) or bulk acoustic wave (BAW) technology are 6%-13%, limiting the bandwidth to 6%. For the bandwidth wider than 6%, split-band method or inductor-aided technology would consume much more space and increase additional loss.

In the last decade, LN-based plate acoustic wave (PAW) resonators attract wide attention employing either the symmetric/asymmetric Lamb wave ($S_0, A_0, S_1, A_1, …$) or the plate shear horizontal wave ($SH_0, SH_1, …$), as they

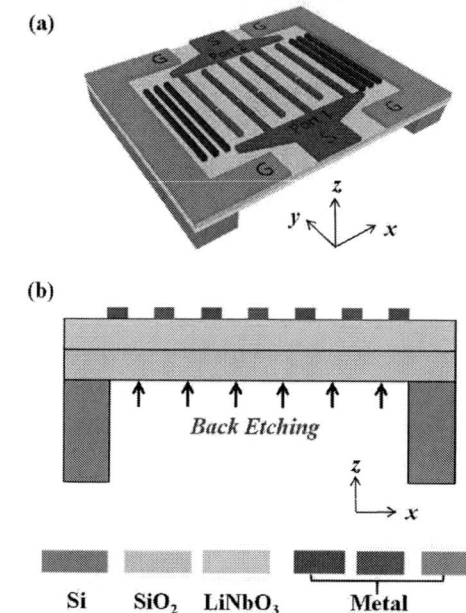

Figure 1: (a) 3-D geometry of plate acoustic wave (PAW) resonators using 30° YX- LiNbO3/SiO2 thin film. (b) 2-D cross-sectional view of PAW resonators.

provide an ultra-large electromechanical coupling coefficient (k^2) due to the large piezoelectric components of LN [6, 7, 8, 9]. Based on a 395nm *Z*-cut LN plate exploiting the A_1 mode, M. Kadota experimentally demonstrate a high resonant frequency, wide bandwidth, and high-impedance ratio resonator [6]. However, the measured Q was only 70, which is too low for practical applications. S. B. Gong designed a 5 GHz MEMS resonator exploiting A_1 Lamb wave and obtained a large k^2_{eff}, where the suspended plate structure was realized by etching LN grooves [7]. Recently, V. Plessky and co-workers proposed a laterally-excited bulk-wave resonators (XBARs) using *Z*-cut ion-sliced monocrystalline LN film (400nm), yielding k^2_{eff} of 26% and Q of 340 [8, 9]. Regrettably, the performance obtained by current studies is still inadequate for the requirements of commercial promotion.

In this work, a SH_0 plate acoustic wave (PAW) resonator using 800 nm LN thin film is proposed, and a SiO_2 layer is designed to improve the temperature stability. The released plate structure prevents the wave propagation loss leaking down into substrate, thereby it is characterized as low loss. According to our previous research, among all the vibration modes of LN-based PAW resonators, the shear wave (SH_0) PAW resonators have the largest k^2 up to 50% [10]. By finite element analysis (FEA) method, both the cut angle of LN layer and the thickness of each layer

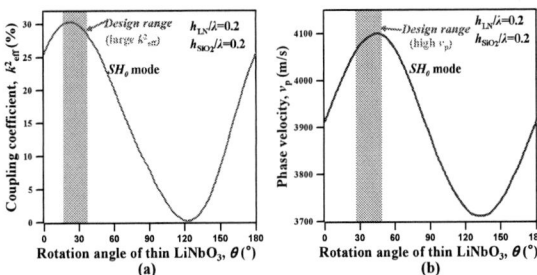

Figure 2: (a) k^2_{eff} and (b) v_p vary with different rotation angles of LiNbO$_3$ film in the case of h_{LN}=0.2λ, h_{SiO2}=0.2λ.

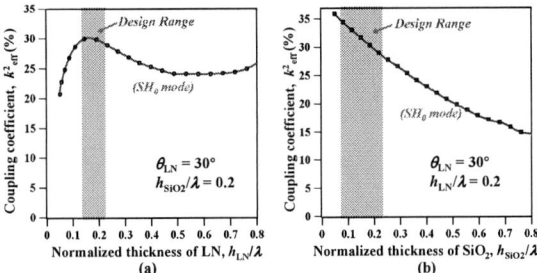

Figure 3: k^2_{eff} varies with different thicknesses of (a) LN layer and (b) SiO$_2$ layer, when the rotation angle of LN is 30°.

are optimized. The PAW resonators are fabricated by advanced smart-cut technology. Measured k^2_{eff} is extracted as 25.1%, and the Bode-Q_{max} is 604, showing that the overall performance is higher than other similar-type PAW resonators. The proposed PAW resonators can promise a solution to the emerging ultra-wide band devices and become a candidate for the next-generation tunable filters.

DESIGN AND SIMULATIONS

Fig.1 (a) illustrates the 3-D geometry of the 30° YX-LN/SiO$_2$ SH$_0$ PAW resonator, consisting of an IDT with metal material, a thin single crystal LN layer and a SiO$_2$ layer. There are two kinds of plate waves in LiNbO$_3$-based PAW resonators, Lamb and (S$_0$, A$_0$, S$_1$, A$_1$, …) or the plate shear wave (SH$_0$, SH$_1$, …), attract wide attention as they have ultra-large k^2 due to the large piezoelectric components of LiNbO$_3$ [11]. The release suspended plate structure prevents acoustic energy loss into the supporting substrate and SH$_0$ wave have a large k^2_{eff}. Some popular piezoelectric materials such as AlN and ZnO are hexagonal and isotropic in xy-plane, the SH plate modes cannot be excited by interdigital transducer (IDT) because of the lack of piezoelectric coupling component e_{21}. On the contrary, the e_{21} is very strong in Y-cut LiNbO$_3$; therefore, the SH plate modes can be excited as the main mode by the IDT. As shown in Fig. 1(a), N_t fingers of IDT is placed in the middle of the upper surface of resonator, and two grating reflectors (GRs) with N_r fingers are placed on both sides of the IDT to ensure the sufficient reflection of acoustic wave. The aperture (W) of IDT is set to 20λ, and the width (a) and pitch (p) of the electrode finger are set to $\lambda/4$ and $\lambda/2$, respectively, where the wavelength (λ) is set as 4 μm.

In FEA simulations, the Euler angle (ZXZ) of LN thin film is expressed as (0°, 90°−θ, 0°), where θ represents the

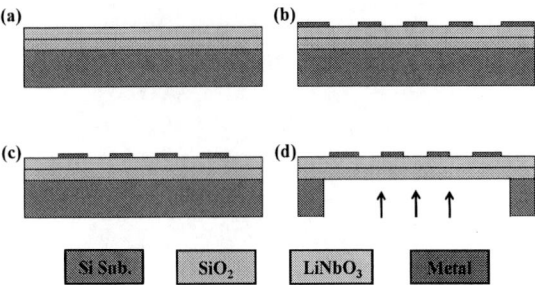

Figure 4: Fabrication process flow. (a) A piece of single crystal LN wafer implanted by He ions (He$^+$) is bonded to a piece of thermally oxidized Si substrate. (b) The bonded wafer is patterned by photolithography on the topside. (c) The electrodes are evaporated using lift-off process. (d) The silicon substrate is etched away from the backside.

rotation angle of LN. Fig. 1 (b) depicts the shows the 2-D cross-sectional view of PAW resonators. The thicknesses of Al, LN, and SiO$_2$ layers are denoted as h_{Al}, h_{LN}, and h_{SiO2}, respectively. To reduce computational complexity, periodic boundary conditions are imposed along the x-direction of propagation. According to IEEE standard [12], k^2_{eff} is calculated by the resonant frequency (f_s) and anti-resonant frequency (f_p) as follows,

$$k^2_{eff} = \frac{\pi}{2} \frac{f_s}{f_p} \frac{1}{\tan(\frac{\pi}{2} \cdot \frac{f_s}{f_p})} \cdot \tag{1}$$

The phase velocity (v_p) of SAW resonators is calculated using the following formula: v_p=$f_p\lambda$. By FEM simulations, both k^2_{eff} and v_p are as the functions of θ, h_{LN} and h_{SiO2} to optimize the stack of PAW resonators.

It is assumed that both the thicknesses of the LN layer and the SiO$_2$ layer are set to 0.2λ, and the rotation angle (θ) is swept from 0° to 180°. Fig. 2 shows the calculated k^2_{eff} with different rotation angles. It can be observed that when θ is around 30°, both the k^2_{eff} and v_p of SH$_0$ mode are extremely high. Based on the above results (θ =30°), the results of different LN layer thickness and SiO$_2$ layer thickness are illustrated in Fig. 3. As shown clearly in Fig. 3(a), the k^2_{eff} of SH$_0$ mode is large when h_{LN} is about 0.2λ. Considering the fabrication difficulty, 0.2λ is chosen as the thickness of SiO$_2$ layer to reduce the requirement of equipment accuracy.

MEASUREMENT RESULTS

Our wafers are fabricated using smart-cut technology, including ions implantation, wafer bonding and lift-off, and back etching, as shown in Fig. 4. First, a piece of single crystal LN wafer implanted by He ions (He$^+$) is bonded to a piece of thermally oxidized Si substrate. Second, the bonded wafer is patterned by photolithography from the top. Third, the metal electrodes are evaporated using lift-off process. In the end, the back etching is implemented by inductively coupled plasmas (ICP) dry etching using Bosch process, and the etching rate is about 4um/min. The S-parameters of resonators are measured by a GSG probe and an Agilent E5071C vector network analyzer (VNA). Fig. 5 shows the top view of a fabricated resonator by Scanning Electron Microscope (SEM), and the backside

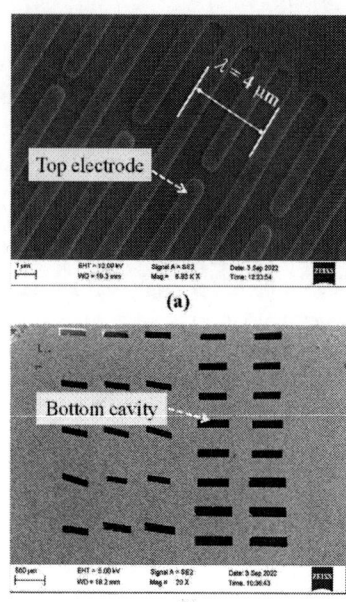

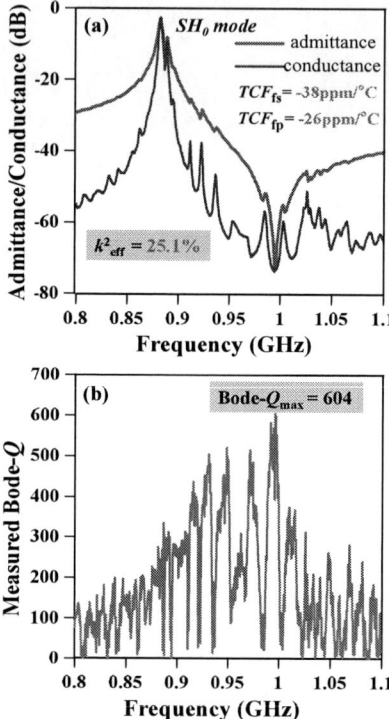

Figure 5: SEM images of fabricated PAW resonators. (a) Topography of the electrodes on the top. (b) Back cavities of multiple resonators on the bottom.

cavities are also presented. From Fig. 5(b), the profiles of the cavities are clearly visible and its inclination is controlled within 1°, indicating the high etching precision.

The measured admittance/conductance and the Bode-Q curves [13] derived from the probed S-parameters of resonators are plotted in Fig. 6. The derived k^2_{eff} of SH_0 mode is 25.1%, and the measured maximum of Bode-Q (Bode-Q_{max}) is calculated as 604, yielding the figure of merit ($FoM= k^2_{eff} \times$Bode-Q_{max}) of 151. It is observed that the FoM is higher than the ever-reported plate wave resonators. Moreover, the measured first-order temperature coefficients of frequency at resonant frequency (TCF_{fs}) and anti-resonant frequency (TCF_{fp}) are -38ppm/°C and -26ppm/°C, respectively.

CONCLUSION

In this paper, a SH_0 PAW resonator using 800nm *YX*-LiNbO$_3$/SiO$_2$ membrane has been proposed. According to FEM simulations, the SH_0 mode exhibits large k^2_{eff} and high acoustic velocity, and the cut angle of LN film and the thicknesses of layers were optimized as $\theta = 30°$, $h_{LiNbO3}=h_{SiO2}=0.2\lambda$, respectively. The fabricated resonators yielded a k^2_{eff} of 25.1%, a Bode-Q_{max} of 604, a FoM of 151, a TCF_{fs} of -38ppm/°C and a TCF_{fp} of -26ppm/°C, showing an attractive potential for wideband, low loss and low temperature sensitivity devices.

Figure 6: (a) Measured admittance/conductance curves of proposed PAW resonators. (b) Measured Bode-Q curve.

REFERENCES

[1] J. Lee et al., "Spectrum for 5G: Global status, challenges, and enabling technologies," *IEEE Commun. Mag.*, vol. 56, no. 3, pp. 12–18, Mar. 2018.

[2] S. Parkvall, E. Dahlman, A. Furuskar, and M. Frenne, "NR: The new 5G radio access technology," *IEEE Commun. Standards Mag.*, vol. 1, no. 4, pp. 24–30, Dec. 2017.

[3] W. Ejaz and M. Ibnkahla, "Multiband spectrum sensing and resource allocation for IoT in cognitive 5G networks," *IEEE Internet Things J.*, vol. 5, no. 1, pp. 150–163, Feb. 2018.

[4] M. Kadota, M. Esashi, S. Tanaka, Y. Ida, and T. Kimura, "Improvement of insertion loss of band pass tunable filter using SAW resonators and GaAs diode variable capacitors," in Proc. *IEEE Ultrason. Symp.*, 2013, pp. 1668–1671.

[5] M. Kadota and S. Tanaka, "Ultra-wideband ladder filter using SH$_0$ plate wave in thin LiNbO$_3$ plate and its application to tunable filter," *IEEE Trans. Ultrason., Ferroelect., Freq. Contr.*, vol. 62, no. 5, pp. 939–946, May. 2015.

[6] M. Kadota and T. Ogami, "5.4 GHz Lamb wave resonator on LiNbO3 thin crystal plate and its application," *Jpn. J. Appl. Phys.*, vol. 50, no. 7S, Jul. 2011, Art. no. 07HD11.

[7] Y. Yang, A. Gao, R. Lu, and S. Gong, "5 GHz lithium niobate MEMS resonators with high FoM of 153," in *Proc. IEEE 30th Int. Conf. Micro Electro Mech. Syst. (MEMS)*, Jan. 2017, pp. 942–945.

[8] V. Plessky, S. Yandrapalli, P. J. Turner, L. G. Villanueva, J. Koskela, and R. B. Hammond, "5 GHz

978-1-6654-9309-3/23 $31.00 © 2023 IEEE

laterally-excited bulk-wave resonators (XBARs) based on thin platelets of lithium niobate," *Electron. Lett.*, vol. 55, no. 2, pp. 98–100, Jan. 2019.

[9] S. Yandrapalli, S. E. K. Eroglu, V. Plessky, H. B. Atakan, and L. G. Villanueva, "Study of Thin Film LiNbO3 Laterally Excited Bulk Acoustic Resonators," *Journal of Microelectromechanical Systems*, vol. 31, no. 2, pp. 217–225, Jan. 2022.

[10] J. Zou, V. Yantchev, F. Iliev, V. Plessky, S. Samadian, R. B. Hammond, and P. J. Turner, "Ultra-Large-Coupling and Spurious-Free SH_0 Plate Acoustic Wave Resonators Based on Thin $LiNbO_3$," *IEEE Trans. Ultrason. Ferroelect. Freq. Control*, vol. 67, no. 2, pp. 374–386, Feb. 2020.

[11] M. Kadota, T. Ogami, and T. Kimura, "Ultra-wide band elastic resonators and their application: SH_0 mode plate wave resonators with ultra-wide bandwidth of 29%," in *Proc. ISAF-ECAPD-PFM*, Jul. 2012, Aveiro, Portugal, Jul. 2012, pp. 1–4.

[12] R. Aigner, "Bringing BAW technology into volume production the ten commandments and the seven deadly sins," *Intl. Symp. Acoustic Wave Devices for Future Mobile Communication Systems*. (2007) 85.

[13] D. A. Feld, R. Parker, R. Ruby, P. Bradley, and S. Dong, "After 60 years: A new formula for computing quality factor is warranted," in *2008 IEEE Ultrasonics Symposium*, Beijing, China, Nov. 2008, pp. 431–436.

CONTACT

*J. Zou, jiezou@fudan.edu.cn; and F.H. Bao, baofh@fudan.edu.cn.

WAFER SCALE MULTILAYER GRAPHENE BASED BRAIN PROBES BY SPIN-SPRAYING METHODS FOR MAGNETIC RESONANCE IMAGING

Kejun Tu[1,2], Zhejun Guo[1,2], Mengfei Xu[1,2], Bin Yang[1] and Jingquan Liu[1]
[1]National Key Laboratory of Science and Technology on Micro/Nano Fabrication,
Shanghai Jiao Tong University, CHINA
[2]DCI Joint Team, Collaborative Innovation Center of IFSA, Department of Micro/Nano Electronics,
Shanghai Jiao Tong University, CHINA

ABSTRACT

This paper proposes a convenient method for wafer scale multilayer graphene preparation by spin-spraying. Combining with the microfabrication process, graphene-based brain probes were acquired with graphene as both conducting traces and electrodes. Compared with metal ones, graphene electrodes show a nearly twelve-fold reduction in impedance and enhanced charge storage capacity, which is favorable for signal recording and brain stimulation. Specifically, they exhibit excellent biocompatibility, allowing safe implantation in the brain. Moreover, both simulations and in vitro experiments confirm their low artifacts during magnetic resonance imaging. Our studies provide a novel method for the fabrication of MR-safe probes.

KEYWORDS

Wafer Scale, Multilayer Graphene, Spin-spraying, Brain Probes, Magnetic Resonance Imaging

INTRODUCTION

Magnetic resonance imaging (MRI) compatible brain probes are of great significance in understanding complex brain connections and exploring the mechanism of clinical treatments [1-3]. They combine the advantages of high temporal resolution of electrophysiological measurements and high spatial resolution of MRI. Many applications can benefit from the use of MRI-compatible neural electrodes such as validating the position of implants [4], locating the epileptic foci [5], and mapping the brain network activity [6]. Commercial brain probes are mainly made of metal including platinum, iridium, gold, tungsten, nichrome and

so on [7], but they differ in magnetic susceptibility from brain tissue. This will lead to serious distortion of the magnetic field and produce artifacts or blind areas around the probes, damaging structural and functional imaging of the brain.

Recently, various carbon-based materials have proved to show fewer MRI artifacts compared with metal electrodes [8, 9], such as carbon fibers [10, 11], carbon nanotubes [2, 12], graphene [1, 3] and glassy carbon [13, 14]. However, the methods to prepare these carbon materials suffer from high temperature (800-1200 °C), small size (a few millimeters), complicated transfer process, and incompatible microelectromechanical systems (MEMS) fabrication [3, 8, 15, 16]. Sometimes, the as-prepared carbon materials show certain biological toxicity. Therefore, there is an urgent need for developing a low-cost, wafer scale, simple and MEMS-compatible method to prepare bio-safe carbon based probes.

In this work, we proposed an in-situ spin-spraying method combining subsequent polyimidization (350 °C) for wafer scale multilayer graphene fabrication. Based on this method together with further microfabrication process, graphene based brain probes are prepared. which not only show excellent electrochemical performance, but also are bio and MRI compatible. These studies will provide us with a novel method for MRI compatible brain probes with unique properties for both basic and applied neuroscience research.

DESIGN AND FABRICATION

Figure 1a shows a simple but novel method for the preparation of graphene film on the wafer. It combines

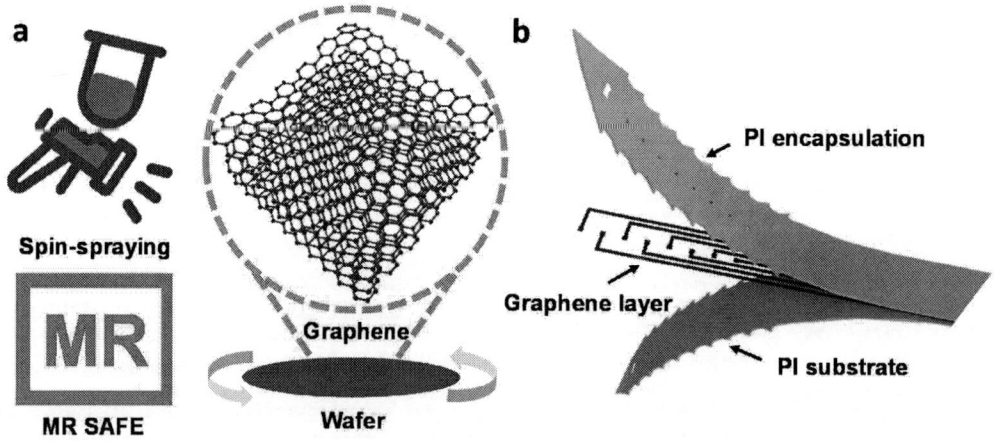

Figure 1: (a) Scheme of wafer scale multilayer graphene preparation; (b) structural diagram of the graphene-based brain probe.

spraying and spinning together. The graphene solution (10 wt %) is sprayed onto the wafer which is spun coated with polyimide (PI) before. To make the film uniform, the wafer is spun at a certain speed. Figure 1b shows the explosive view of graphene-based brain probes. It consists of three layers, The conductive carbon layer, which is made of graphene, is sandwiched between two flexible PI layers. To expose the electrodes, openings with an area of 20 μm ×20 μm are placed in the upper PI layer. The electrical connection to a flexible printed circuit is realized by an anisotropic conductive film for electrochemical measurements.

The graphene-based brain probes were fabricated by microfabrication processes (Figure 2). First of all, an aluminum sacrificial layer was sputtered on a clean wafer. Secondly, a flexible PI substrate was spun coated. Then, the graphene solution (10 wt% in water) was spun-sprayed onto the surface of the PI substrate which was prebaked at 100 °C for three minutes. Subsequently, they were cured at 350 °C in a nitrogen atmosphere for 1 hour to complete the imidization and adhesion process of the polyimide and graphene layers (Figure 2). Afterward, the graphene layers together with PI substrate were patterned by photolithography with hard mask technology and reactive ion etching. Finally, the probes were encapsulated by PI layer while the pads and electrodes were exposed. After releasing the sacrificial layer in diluted hydrochloric acid solution, the graphene-based brain probes were achieved.

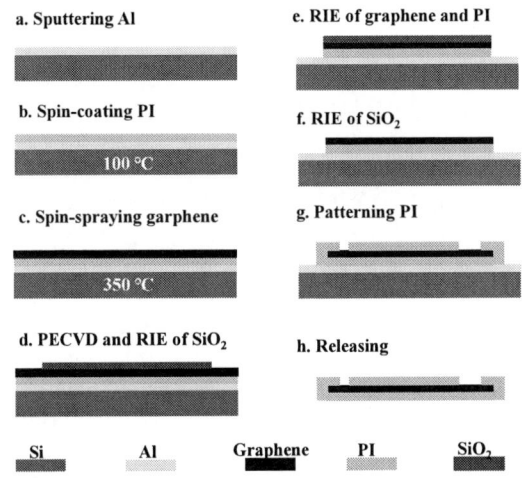

Figure 2: The microfabrication process of graphene-based brain probes.

RESULTS AND DISCUSSION

The in-situ spin-sprayed graphene layers and the released probes are shown in Figures 3a and b, exhibiting a dark black color of carbon materials. The detail of the electrodes' size and displacement is in the inset of Figure 3b, with the 20 μm line width of electrodes. The SEM images shown in Figures 3c and 3d illustrate the multilayer structure of graphene with a total thickness of nearly 1.5 μm and wrinkled surface of graphene.

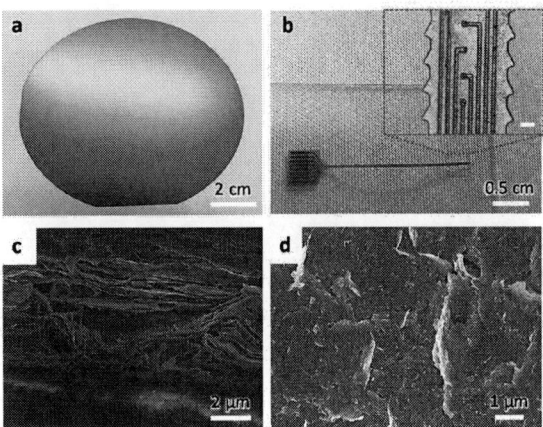

Figure 3: (a) Graphene layers sprayed on a 4-inch wafer; (b) optical photos of flexible graphene-based probe, inset is the zoomed-in graph of the dashed boxes, scale bar: 50 um; SEM of the (c) cross section and (d) surface of graphene layers.

The X-ray diffraction (XRD) patterns confirmed the presence of the carbon phase. The obvious peak at approximately 27 ° could be attributed to the (002) planes of graphene [17]. The graphitic degrees of the carbon layers were further confirmed through Raman spectroscopy (Figure 4b). There are mainly three characteristic peaks observed at about 1350, 1582, and 2692 cm^{-1}, which corresponds to D, G, and 2D band, respectively [18]. The intensity of the D peak is larger than the G peak, indicating more defective structures of graphene. While 2D band is much weaker and wider than the G band, reflecting the characteristic of multilayer graphene.

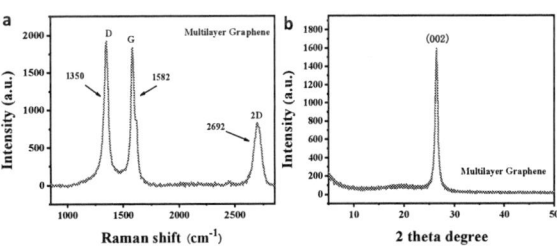

Figure 4: (a) Raman spectrum and (b) XRD pattern of the multilayer graphene.

For the safety of probes implanted in the brain, the biocompatibility of graphene-based probes is evaluated (Figure 5) by immersing the graphene wafer into the cell culture medium together with HT-22 neural cells. It can be seen that more neurons grow from dozens on the first day to hundreds a week in the same area, demonstrating the excellent biocompatibility of our probes. It can be attributed to the water solution of graphene and high cured temperature, which reduce the content of toxic organics in our probes.

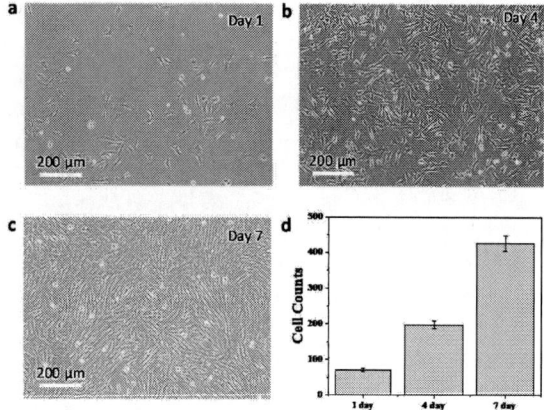

Figure 5: Biocompatibility test of graphene probes on (a) day 1, (b) day 4 and (c) day 7; (d) Corresponding cells counts in a-c.

Figure 6 is the comparison of electrochemical properties between graphene and bare gold electrodes. Cyclic voltammetry (CV) is commonly used to characterize the charge storage capacity (CSC) of electrodes. As shown in Figure 6a, graphene electrodes have a larger enclosed area than the bare gold ones, which means a better CSC for graphene. The value of CSC for graphene electrodes is 444.2 mC cm^{-2}, three times higher than that of bare gold ones, which is promising for brain stimulation. Figures 6b and 6c show the electrochemical impedance spectra (EIS). The impedance of graphene electrodes is 18.7 kΩ at 1 kHz, a nearly twelve-fold reduction compared to that of gold ones, and the phase angle decreases from 47.6 ° to 38.8 °. These results suggest the excellent recording ability of graphene electrodes. In a word, the graphene electrodes show outstanding performance both in stimulating and recording ability.

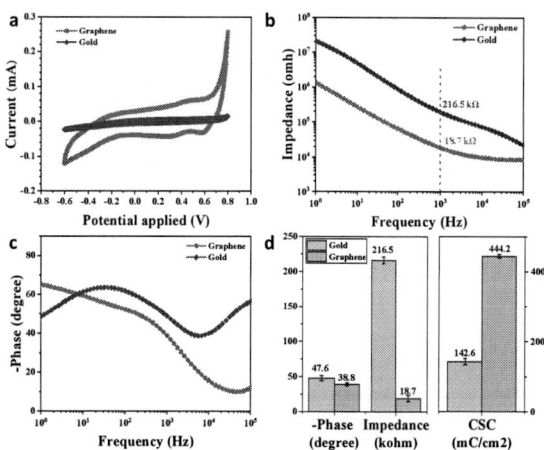

Figure 6: The electrochemical characterization of graphene and bare gold electrodes. (a) CV; (b,c) EIS and (d) Comparison of the phase, impedance and CSC.

To access the MRI compatibility of graphene-based probes, MRI image artifacts were compared with both simulations and experiments in a high-field 11.7T MRI scanner (Figure 7). Figures 7 a-c are the field variation maps calculated by a Fourier-based method [19, 20] for graphene and commercial platinum electrode respectively. From the simulation, it can be obtained that graphene-based probes exhibit nearly 40 times lower magnetic field perturbations than the commercial ones, making them better for MRI. In vitro MR artifact testing was then performed at an 11.7 T MRI scanner with a graphene probe and a platinum wire in agar phantoms. From the horizontal and coronal sections of the T2-weighted images (Figure 7d), the graphene probe shows an artifact of 0.38 mm while platinum wire displays a much larger artifact of nearly 1 mm. The results indicate that our graphene probes are highly MRI-compatible.

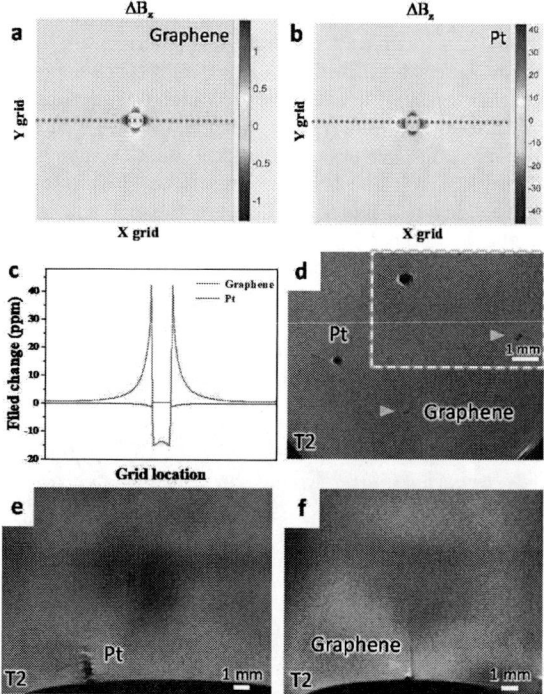

Figure 7: Comparison of the magnetic field perturbations of platinum and graphene models. (a-b) Simulation maps of the susceptibility-induced field variation (ppm); (c) variation of the z-component of the field perturbation along a line passing through the center in (a-b); (d) horizontal sections and (e-f) coronal sections of the T2-weighted images of an agar model implanted with (e) Pt wire and (f) graphene flexible electrodes.

CONCLUSION

In this paper, we propose a convenient method for wafer scale multilayer graphene preparation by spin-spraying. Combined with microfabrication technology, the as-prepared graphene-based brain probes show a nearly twelve-fold reduction in impedance at 1 kHz, and the CSC is also greatly enhanced compared to commercial metal ones without further electrochemical modification. These make them promising for both signal recording and deep brain stimulation. In particular, graphene-based probes have excellent biocompatibility, allowing safe implantation in the brain. Finally, the simulation of filed perturbations and in vitro tests confirm the low artifacts for our graphene probes, which is to some extent suitable for MRI of the brain.

ACKNOWLEDGEMENTS

This work was supported by the National Key R&D Program of China under grant (2022ZD0208601), the Strategic Priority Research Program of Chinese Academy of Sciences (Grant No. XDA25040100, XDA25040200 and XDA25040300), the National Natural Science Foundation of China (No.42127807-03), Project supported by Shanghai Municipal Science and Technology Major Project (2021SHZDZX), SJTU Trans-med Award (No.2019015, 21X010301627). The authors thank Zhangjiang Brain Imaging Center. The authors are also grateful to the Center for Advanced Electronic Materials and Devices (AEMD) of Shanghai Jiao Tong University.

REFERENCES

[1] S. Zhao, X. Liu, Z. Xu, H. Ren, B. Deng, M. Tang, *et al.*, "Graphene Encapsulated Copper Microwires as Highly MRI Compatible Neural Electrodes," *Nano Lett.*, vol. 16, pp. 7731-7738, 2016.

[2] L.-l. Lu, X. Fu, Y. Liew, Y. Zhang, S. Zhao, Z. Xu, *et al.*, "Soft and MRI Compatible Neural Electrodes from Carbon Nanotube Fibers," *Nano Lett.*, vol. 19 3, pp. 1577-1586, 2019.

[3] S. Zhao, G. Li, C. Tong, W. Chen, P. Wang, J. Dai, *et al.*, "Full activation pattern mapping by simultaneous deep brain stimulation and fMRI with graphene fiber electrodes," *Nat. Commun.*, vol. 11, p. 1788, 2020.

[4] I. Todt, G. Rademacher, P. Mittmann, J. Wagner, S. Mutze, and A. Ernst, "MRI artifacts and cochlear implant positioning at 3 T in vivo," *Otol. Neurotol.*, vol. 36, pp. 972-976, 2015.

[5] M. Stead, M. Bower, B. H. Brinkmann, K. Lee, W. R. Marsh, F. B. Meyer, *et al.*, "Microseizures and the spatiotemporal scales of human partial epilepsy," *Brain*, vol. 133, pp. 2789-2797, 2010.

[6] M. Figee, J. Luigjes, R. Smolders, C.-E. Valencia-Alfonso, G. van Wingen, B. de Kwaasteniet, *et al.*, "Deep brain stimulation restores frontostriatal network activity in obsessive-compulsive disorder," *Nat. Neurosci.*, vol. 16, pp. 386-387, 2013.

[7] P. Fattahi, G. Yang, G. Kim, and M. R. Abidian, "A Review of Organic and Inorganic Biomaterials for Neural Interfaces," *Adv. Mater.*, vol. 26, pp. 1846-1885, 2014.

[8] M. Devi, M. Vomero, E. Fuhrer, E. Castagnola, C. Gueli, S. Nimbalkar, *et al.*, "Carbon-based neural electrodes: promises and challenges," *J. Neural Eng.*, vol. 18, p. 041007, 2021.

[9] C. E. Cruttenden, J. M. Taylor, S. Hu, Y. Zhang, X.-H. Zhu, W. Chen, *et al.*, "Carbon nano-structured neural probes show promise for magnetic resonance imaging applications," *Biomed. Phys. Eng. Express*, vol. 4, p. 015001, 2017.

[10] X. Fu, G. Li, Y. Niu, J. Xu, P. Wang, Z. Zhou, *et al.*, "Carbon-Based Fiber Materials as Implantable Depth Neural Electrodes," *Front. Neurosci.*, vol. 15, 2021.

[11] Y. Huan, J. P. Gill, J. B. Fritzinger, P. R. Patel, J. M. Richie, E. Della Valle, *et al.*, "Carbon fiber electrodes for intracellular recording and stimulation," *J. Neural Eng.*, vol. 18, p. 066033, 2021.

[12] J. Feng, C. Chen, X. Sun, and H. Peng, "Implantable Fiber Biosensors Based on Carbon Nanotubes," *Acc. Mater. Res.*, vol. 2, pp. 138-146, 2021.

[13] D. Ashouri Vajari, M. Vomero, J. B. Erhardt, A. Sadr, J. S. Ordonez, V. A. Coenen, *et al.*, "Integrity Assessment of a Hybrid DBS Probe that Enables Neurotransmitter Detection Simultaneously to Electrical Stimulation and Recording," *Micromachines*, vol. 9, p. 510, 2018.

[14] S. Nimbalkar, E. Fuhrer, P. Silva, T. Nguyen, M. Sereno, S. Kassegne, *et al.*, "Glassy carbon microelectrodes minimize induced voltages, mechanical vibrations, and artifacts in magnetic resonance imaging," *Microsyst. Nanoeng.*, vol. 5, p. 61, 2019.

[15] M. J. Antonini, A. Sahasrabudhe, A. Tabet, M. Schwalm, D. Rosenfeld, I. Garwood, *et al.*, "Customizing MRI-Compatible Multifunctional Neural Interfaces through Fiber Drawing," *Adv Funct Mater*, vol. 31, p. 2104857, 2021.

[16] S. Ullah, X. Yang, H. Q. Ta, M. Hasan, A. Bachmatiuk, K. Tokarska, *et al.*, "Graphene transfer methods: A review," *Nano Res.*, 2021.

[17] K. Sun, Z. Wang, J. Xin, Z. Wang, P. Xie, G. Fan, *et al.*, "Hydrosoluble Graphene/Polyvinyl Alcohol Membranous Composites with Negative Permittivity Behavior," *Macromol. Mater. Eng.*, vol. 305, p. 1900709, 2020.

[18] Z. Wang, Y. Li, J. Liu, T. Gui, H. Ogata, W. Gong, *et al.*, "Facile synthesis of graphene sheets intercalated by carbon spheres for high-performance supercapacitor electrodes," *Carbon*, vol. 167, pp. 11-18, 2020.

[19] R. Salomir, B. D. de Senneville, and C. T. Moonen, "A fast calculation method for magnetic field inhomogeneity due to an arbitrary distribution of bulk susceptibility," *Concepts in Magnetic Resonance Part B: Magnetic Resonance Engineering*, vol. 19B, pp. 26-34, 2003.

[20] F. M. Martinez-Santiesteban, S. D. Swanson, D. C. Noll, and D. J. Anderson, "Magnetic field perturbation of neural recording and stimulating microelectrodes," *Phys. Med. Biol.*, vol. 52, pp. 2073-2088, 2007.

CONTACT

*J.Q. Liu, tel: +86-21-34207209; jqliu@sjtu.edu.cn

3D SELF-ALIGNED FABRICATION OF SUSPENDED NANOWIRES BY CRYSTALLOGRAPHIC NANOLITHOGRAPHY

Erwin J.W. Berenschot[1], Yasser Pordeli[1], Lucas J. Kooijman[1], Yves L. Janssens[1],
Roald M. Tiggelaar[2], and Niels R. Tas[1]

[1]Mesoscale Chemical Systems, MESA+ Institute, University of Twente, The Netherlands
[2]NanoLab, MESA+ Institute, University of Twente, The Netherlands

ABSTRACT

Known templating procedures mostly create out-of-plane nanowires where individual connections at both ends are complicated. Here we introduce a templating procedure for wafer scale fabrication of in-plane nanowires. The template fabrication process employs two simple interference lithography masking patterns and relies on self-aligned crystallographic processing. In-plane nanowires with diameters down to 10 nm can be fabricated wafer scale through this 3D templating procedure. As a first demonstration arrays of suspended silicon nitride wires have been created.

KEYWORDS

Corner lithography, nanowires, templating, 3D, silicon crystal

INTRODUCTION

Templated nanowire growth is a relative old technique, initially based on track-etched membranes [1] as well as anodic aluminum oxide (AAO) porous membranes [2]. Nanowires are typically oriented out-of-plane and connecting both ends is challenging. Typical diameters are at least tens of nm. Several procedures have been developed for *silicon* in-plane nanowires, typically using (advanced) lithography and silicon-on-insulator (SOI) substrates for defining the wire thickness and for release etching through oxide dissolution [3, 4]. Here we introduce an in-plane templating procedure with nm accuracy over the complete wafer for standard p-type silicon substrates, yielding nanowires in the 10 nm diameter range. More in detail, convex corner lithography [5, 6], an emerging self-aligned nanopatterning technique, is employed to create nano-cavities at the apex of silicon wedges [6]. Essential steps have been added to this basic scheme to create a template for suspended nanowires of ~100 nm length, and to create a strategy for controlled release etching. The complete procedure has been tested by creating arrays of suspended silicon nitride nanowires.

EXPERIMENTAL

Approach

Fig. 1 shows the basic steps for the template formation. Initial wedges are formed from a line pattern in silicon nitride, used as a hard mask for anisotropic etching of the (100)-silicon substrate, followed by LOCOS and another anisotropic silicon etching step [7]. Cavities are formed by "convex corner lithography" combined with anisotropic etching [6]. Nanowires are then formed in the cavities through "irreversible processing", i.e. conformal deposition of silicon nitride in confined space followed by isotropic etching (Fig. 2). This isotropic etching is performed through planar windows which are defined in a hard mask, as to define the finite length of the nanowires and their suspension. The hard mask stack consists of amorphous silicon on top of silicon nitride in which the pattern is created through interference lithography (DTL: displacement Talbot lithography). The same masking apertures are used to expose the silicon side walls. Exposed silicon is then etched in KOH to release the nanowires (fig. 2).

Details of template cavity fabrication

Boron doped (100) oriented silicon (Si) wafers (5-10 Ω cm, 100mm diameter, 525 μm thick, one-side polished, Okmetic, Finland) are used to fabricate wafer scale wedges combining DTL with reactive ion etching and anisotropic wet-chemical etching. Stoichiometric silicon nitride (Si_3N_4) is used as a hard mask for anisotropic etching of Si to form V-grooves followed by local-oxidation-of-silicon (LOCOS), stripping of Si_3N_4, and anisotropic etching of Si to form wedges, as shown in fig. (1a). Subsequently, low temperature (800°C steam) thermal oxidation is performed to grow SiO_2, as shown in fig. 1b. Timed isotropic thinning of SiO_2 is performed in 1% hydrofluoric acid (HF) to only expose the apices or the convex corners, shown in fig. 1c. Finally, the exposed apices are etched selectively in tetramethylammonium hydroxide (TMAH) to form cavities with free standing SiO_2 flaps, as shown in Fig. 1d [6].

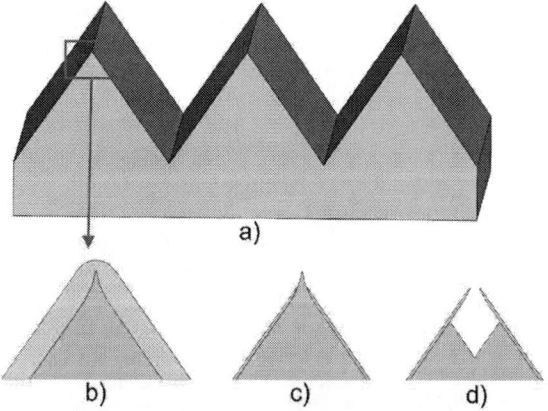

Figure 1: Process flow for template cavity fabrication. Starting from the silicon wedge array (a), a low temperature thermal oxidation follows (b), then HF thinning of SiO_2 (c) and finally TMAH etching of silicon (d). Adapted from [6] with permission.

Details of suspended nanowire formation

The substrate containing the wedge based cavities were standard pre-furnace cleaned by means of fuming 99% nitric acid (HNO_3) (2x 5 min) and boiling 69% HNO_3 (10 min). Low pressure chemical vapor deposition (LPCVD; Tempress horizontal diffusion system, type TS6604, 800 °C, 200 mTorr, 22 sccm SiH_2Cl_2, 66 sccm NH_3, 17 min) was carried out to conformally grow 13.1 ± 0.1 nm Si_3N_4 to embed the cavities. Next, amorphous silicon (a-Si) of 14.3 ± 0.8 nm was conformally deposited using LPCVD (Tempress horizontal diffusion system, 550 °C, 250 mTorr, 50 sccm $SiH4$, 6min 30s) to serve as a hard mask for patterning Si_3N_4. Next, a Si_3N_4 layer of 12.4 ± 0.2 nm was conformally deposited using LPCVD to be used as a hard mask for a-Si patterning. Subsequently, ~200 nm bottom anti-reflective coating (BARC, Barli-II) was spin coated at 3000 rpm for 45s followed by pre-exposure bake at 185 °C for 60s. Then ~160 nm positive-tone photo-resist (PFI:88, 1:1 PFI: PGMEA (propylene glycol monomethyl ether acetate – Sumitomo Chemical Co., Ltd.) was spin coated at 4000 rpm for 45s followed by pre-exposure bake at 90 °C for 60s. An advanced interference lithography technique DTL (PhableR 100C, Eulitha, Switzerland) was carried out. A phase-shift mask with gratings featuring a pitch of 500nm was aligned perpendicular to the substrate of nano- wedges. The photoresist was exposed at a wavelength of 375 nm with an intensity of 0.98 mW cm^{-2} for 75s at a Talbot distance of 3 µm and a gap spacing of ~65 µm. After DTL, the substrate was post-exposure baked at 110 °C for 60s followed by resist development in TMAH (OPD4252, Arch Chemicals) solution for 60s (substrate submerged two times for 30s in separate beakers). Next, the photoresist pattern was transferred into the BARC layer using a conductively coupled plasma RIE system (25W, 50 mTorr, 50 sccm N_2, 6 min 50s; TEtske home-built system). Prior to RIE, the chamber was pre-cleaned by wiping it with organic solvent, followed by oxygen plasma cleaning for 10 min (100W, 50mTorr, 50 sccm O_2). The etching was timed as such that the BARC layer was removed only from the apex of the nanowedges to expose Si_3N_4 layer whereas the concave corners were still protected. Next, the Si_3N_4 layer was etched for 60s in the same RIE chamber (25W, 10 mTorr, 25 sccm CHF_3, 5 sccm O_2; Tetske home-built system) to expose the a-Si layer, as shown in fig. 2a. The photoresist and BARC layer were stripped using an oxygen plasma (TePla 300) for 45 min, followed by 10 min Piranha cleaning (mixture (95 °C) of sulphuric acid (H_2SO_4) and hydrogen peroxide (H_2O_2) in a volumetric ratio of 3:1). Next, the substrate was etched in 1% HF at room temperature for 60s to remove native oxide from the surface of the exposed a-Si. The substrate was then placed in 20 wt.% potassium hydroxide (KOH) solution at 21 °C to selectively etch a-Si for 40s, as shown in fig. 2b. After KOH, the substrate was cleaned for 20 min to remove alkali-residue in RCA-2 (mixture (80 °C) of 36% hydrochloric acid (HCl), 31% hydrogen-peroxide (H_2O_2) and demineralized water (DI water) in a volumetric ratio of 1:1:5). Next, the substrate was etched in 1% HF at room temperature for 20s to remove native oxide on top of the exposed Si_3N_4 layer. The substrate was then placed in a solution of 85 wt.% H_3PO_4 at 140°C to isotropically thin

the Si_3N_4 layer at a pre-determined etch-rate of 2.0 nm/min. The substrate was timed for 7 min to leave the Si_3N_4 layer only in the cavities, shown in Figure 2c. Next, the exposed SiO_2 layer was etched in 1% HF at room temperature for 30s followed by etching of Si in 20% KOH at 21 °C for 8 min 30s to release the Si_3N_4 nanowires. The substrate was then carefully freeze dried to not damage the suspended Si_3N_4 nanowires. The freeze drying process includes rinsing in isopropanol and cyclohexane followed by freeze drying at -7 °C under a nitrogen flow.

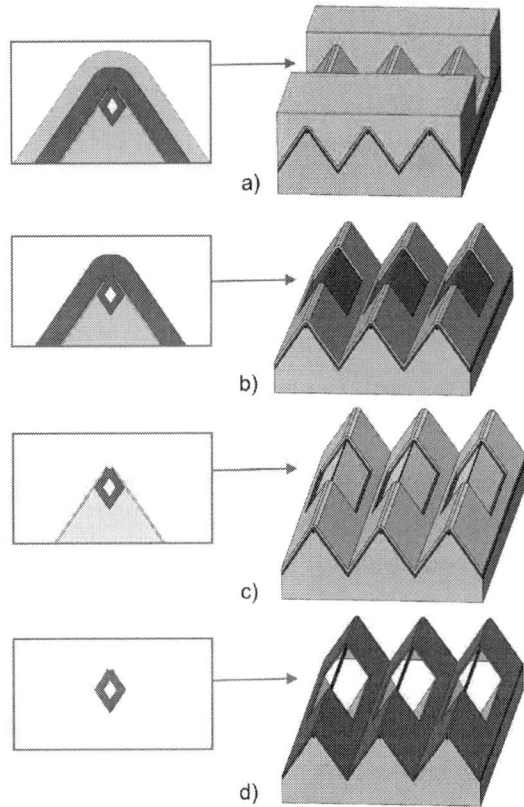

Figure 2: Process flow for finite length nanowire creation and release. a) After patterning the a-Si hard mask in 20% KOH, the silicon nitride is thinned in H_3PO_4 solution (b) -> c)). After HF removal of the SiO_2 stop layer, the exposed silicon is etched in 20% KOH at room temperature, followed by rinsing and freeze dying (d).

RESULTS AND DISCUSSIONS

The windows for removing silicon from underneath the nanowires by anisotropic etching, , are created by a hard mask layer stack composed of Si_3N_4/a-Si /Si_3N_4. Post to DTL and RIE (of BARC) the top layer of Si_3N_4, which is 12.4 nm thick, is opened with RIE. Via openings in this Si_3N_4 film, the underlying a-Si is selectively etched (patterned) using KOH. Post to cleaning steps, the bottom layer of Si_3N_4 (14.3 nm thick) is isotropically thinned, such that only in the cavities Si_3N_4 remained (i.e. non-suspended Si_3N_4-nanowires), followed by KOH-etching. By performing this sequence, the complete top layer of Si_3N_4 is removed during the HF thinning step, whereas the a-Si film and bulk-Si underneath the Si_3N_4-nanowires are removed during the KOH-step. Thus, the hard mask applied for realizing suspended Si_3N_4-nanowires in designated windows is a stack of three films, of which the top layer Si_3N_4 (deposited after a-Si) has to be thinner than the bottom layer Si_3N_4 (deposited at first).

Fig. 3 shows a TEM cross section of the templates after conformal filling with silicon nitride. The silicon oxide used for the convex corner lithography initially was grown at a thickness of about 8 nm on the {111}-planes (far away from the apex), of which about 2 nm remained after the HF etching employed to expose the silicon at the apex. As this is close to the minimum thickness for this procedure, the 6 nm slit at the apex is the lower limit for this material system and procedure. The cavity width is 12 nm and was determined to be rather uniform at five positions across the wafer [6]: 12.0 ± 0.5 nm (± 1 SD_N).

Fig. 4 shows SEM images and drawings illustrating the release at different times in the etching process. The top row illustrates when the hard masks are just opened and after 2.5 min of KOH etching (20% KOH, 21 °C). The middle shows the result of 5 min anisotropic etching of silicon is not enough because the template is still connected to the bottom of the nanowire. The bottom row shows the result after 8.5 min anisotropic etching of the template to create a cavity, yielding suspended Si_3N_4 nanowires.

Fig. 5 shows SEM images of fabricated suspended nanowires. The wires are about 10 nm in diameter and have a length slightly over 100 nm. Currently, the estimated yield is about 40%. The main loss of wires occurs in the final freeze drying release step, which has to be further optimized for the small scale of the nanowires.

CONCLUSIONS AND OUTLOOK

We have demonstrated the template formation of suspended nanowires with diameters close to 10 nm in a wafer scale process. The current procedure has now been demonstrated for silicon nitride wires and needs further optimization to increase the yield. It is expected that the presented method will enable wafer scale in-plane nanowire formation for a wide variety of materials including metals, semiconductors and piezoelectric ceramics.

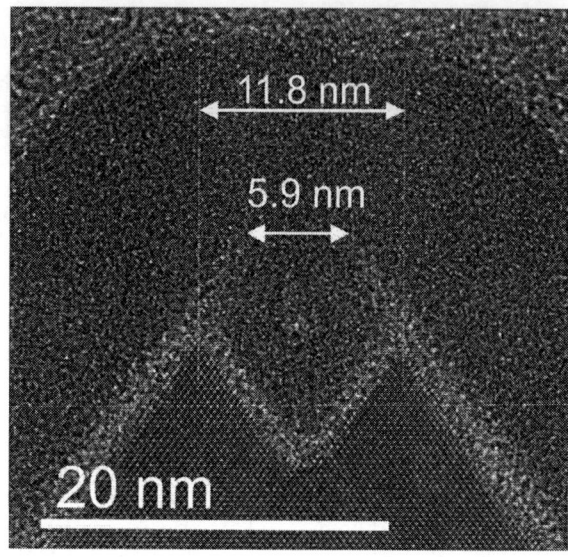

Figure 3: TEM image of a cross-slice of the nanowire template, showing the 12 nm wide cavity and the 6 nm top gap. The remaining thickness of free-standing silicon oxide flaps is about 2 nm.

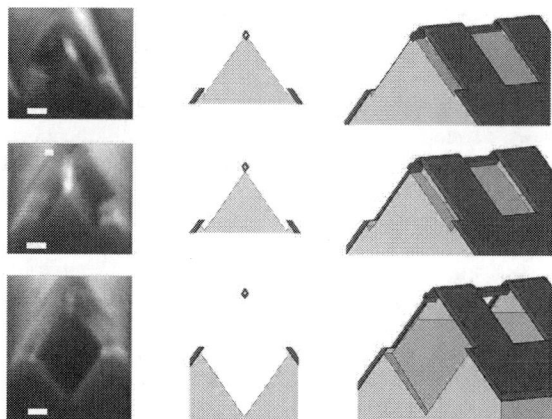

Figure 4: SEM photos and drawings illustrating the release etching at different stages. Top row: The hard masks have just been opened; Middle row: The silicon template still connects to the bottom side of the nanowires; Bottom row: A silicon cavity bound by slow etching crystal planes has formed beneath the nanowire, resulting in a suspended nanowire. Scale bars are 20 nm.

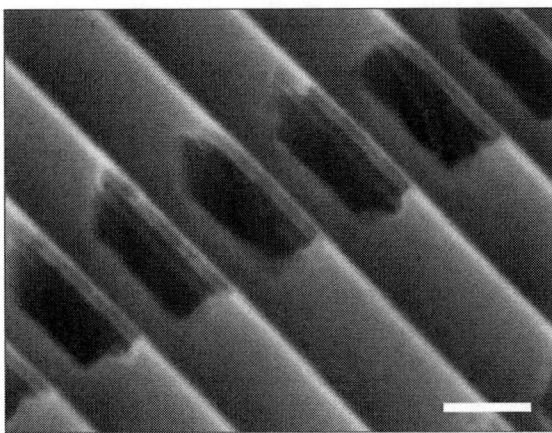

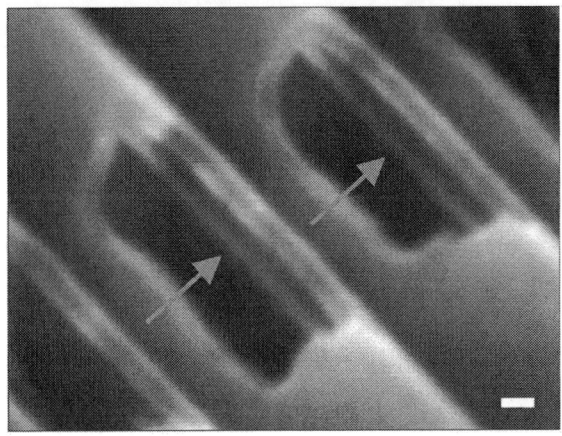

Figure 5: SEM photos showing finally produced suspended nanowires. Scale bars are 100 nm and 20 nm, respectively. It is noted that some suspended nanowires in the top image have collapsed during freeze drying.

ACKNOWLEDGEMENT

The authors would like to thank the MESA+ NanoLab staff for their continuous efforts in maintaining high standards in the facility.

REFERENCES

[1] C.R. Martin, "Membrane-based synthesis of nanomaterials", *Chem. Mater.*, vol. 8, pp. 1739-1746, 1996.

[2] D. Routkevitch, T. Bigioni, N. Moskovits, J.M. Xu, "Electrochemical fabrication of CdS nanowire arrays in porous anodic aluminum oxide templates", *J. Chem. Phys.*, vol. 100, pp. 14037-14047, 1996.

[3] S. Chen, J.G. Bomer, W.G. van der Wiel, E.T. Carlen, A. van den Berg, "Top-down fabrication of sub-30 nm monocrystalline silicon nanowires using conventional microfabrication", *ACS Nano*, vol. 3, pp. 3485-3492, 2009.

[4] K. Zhou, Z. Zhao, L. Pan, Z. Wang, "Silicon nanowire pH sensors fabricated with CMOS compatible sidewall mask technology", *Sens. Actuators B*, vol. 279, pp. 111-121, 2019.

[5] C. van Kampen, E.J. Berenschot, G.-J. Burger, R.M. Tiggelaar, R.G.P. Sanders, H.J.G.E. Gardeniers, N.R. Tas, "Massive parallel NEMS flow restriction fabricated using self-aligned 3D crystallographic nanolithography", in *Proc. 33rd IEEE MEMS*, Vancouver, Canada, January 18-22, 2020, pp. 1106-1109.

[6] E. Berenschot, R.M. Tiggelaar, B. Borgelink, C. van Kampen, C.S. Deenen, Y. Pordeli, H. Witteveen, H.J.G.E. Gardeniers, N.R. Tas, "Self-aligned crystallographic multiplication of nanoscale silicon wedges for high-density fabrication of 3D nanodevices", *ACS Appl. Nano Mater.*, vol. 5, pp. 15847-15854, 2022.

[7] G. Hashiguchi, H. Sakamoto, S. Kanazawa, H. Mimura, "Wedge-shaped silicon emitter fabricated by new method", *Jpn. J. Appl. Phys.*, vol. 32, pp. 6291-6292, 1993.

CONTACT

N.T., n.r.tas@utwente.nl
E.B., j.w.berenschot@utwente.nl

A SIMPLE PROCESS FOR THE FABRICATION OF PARALLEL-PLATE ELECTROSTATIC MEMS RESONATORS BY GOLD THERMOCOMPRESSION BONDING

Dolores Manrique Juarez[1], Fabrice Mathieu[1], Guillaume Libaude[1], David Bourrier[1], Samuel Charlot[1], Laurent Mazenq[1], Véronique Conédéra[1], Ludovic Salvagnac[1], Isabelle Dufour[2], Liviu Nicu[1], and Thierry Leïchlé[1,3], *

[1]LAAS-CNRS, Toulouse, FRANCE
[2]Université de Bordeaux, Laboratoire IMS UMR-CNRS 5218, Talence, FRANCE
[3]Georgia Tech-CNRS International Research Laboratory, School of Electrical and Computer Engineering, Atlanta, USA

ABSTRACT

The present work introduces a simple and quick process for the fabrication of flexural MEMS resonators with electrostatic actuation and capacitive detection. Gold thermocompression is used to anchor the free-standing structure, to provide electrical connections and to seal the chip to protect the structure to be released during wet etching, in a single step. The additional advantage of this process is the ability to simply adjust the gap of the parallel-plate capacitor through the thickness of electroplated gold. Squeeze damping is reduced by fabricating devices with increasing gaps.

KEYWORDS

Micromechanical device, Cantilever, Parallel-plate capacitor, Resonant frequency, Q-factor, Packaging, Thermocompression.

INTRODUCTION

Electrostatic actuation and capacitive read-out implemented in MEMS resonators are known to provide a combination of low-power consumption and high-level of integration. Even if electrostatic MEMS architecture remains relatively simple, the difficulty in their implementation is to precisely define the micrometric gap between the moving structure and the fixed actuation/sensing electrode. Parallel-plate electrostatic MEMS are commonly fabricated following two different approaches: silicon surface micromachining using a sacrificial layer [1], or transfer methods, such as wafer bonding techniques [2] and micro-masonry [3,4].

Thermocompression bonding is a technique widely used for MEMS packaging [5], that has so far only been exploited once for the fabrication of a free-standing structure, an electrostatic relay [6]. This is surprising since this approach offers the advantage of providing mechanical anchoring and electrical connection in a single step. Besides, shear tests have demonstrated that thermocompression bonds exhibit strength similar to those obtained by silicon/glass anodic bonding [7].

Here, thermocompression bonding was used to fabricate a parallel-plate mechanical resonator where the novelty of our work is the use of the gold seal formed on the edge of each chip to protect the resonator during its release by wet etching the handle wafer (which differs from the dry etch approach used in [6]), and to conveniently control the gap size through the thickness of the electroplated gold joint.

We evaluated the electro-mechanical performances of the fabricated MEMS that behave as theoretically expected: we measure a quadratic dependance of the gap and capacitance with applied voltage, a non-linear response regime at resonance is obtained with high AC voltage and the spring softening effect is observed with increasing DC voltage. The quality factor of the MEMS cantilever drops with increasing pressure and a strong squeeze film air damping is observed at atmospheric pressure. This effect is reduced by increasing the parallel-plate capacitor gap through the fabrication of a thicker thermocompression joint.

DESIGN AND FABRICATION

The thermocompression-based process developed to fabricate electrostatic MEMS resonators uses standard microfabrication techniques to create a cantilever with its top electrode and a bottom electrode on a SOI and a fused silica substrate, respectively. Gold is deposited on both substrates at the location of the cantilever anchor to provide mechanical and electrical connections to the cantilever with its substrate upon chip bonding by gold-gold thermocompression. Moreover, gold is deposited on the chip perimeter to serve as a sealing joint to protect the cantilever so the handle SOI layer can be removed and the cantilever released by a simple wet etch step. This unique process, on top of being straightforward, also enables to control the parallel-plate gap by simply adjusting the thickness of the gold layer.

We designed various shapes of cantilever structures inspired from our previous work on gas sensors [8]. However, the most studied devices consisted of 500µm or 250µm long, 500µm wide and 5µm thick silicon cantilevers. Cantilevers with a similar surface area, thus capacitance, but slightly different shape, thus resonant frequency, were added onto each chip to be used as an electrical reference for static capacitance compensation during dynamic characterizations. Each chip integrates 4 different devices and two reference cantilevers (figure 1).

The fabrication process is illustrated in figure 2 and consists of three main steps: i) processing the resonator and its electrode, ii) processing the bottom electrode on the receiving substrate, and iii) thermocompression bonding followed by a wet etch to release the resonator. A joint that seals the top and bottom substrates around the chip is integrated with the aim to protect the cantilever during its release by wet etching.

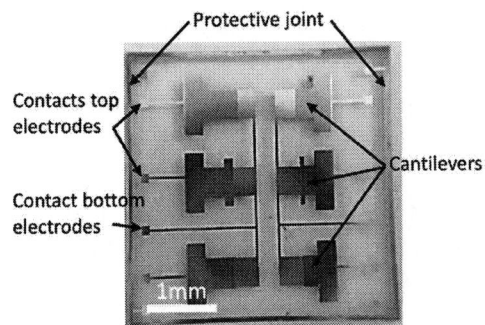

Figure 1: Optical image of a chip showing the various cantilever structures, the contacts for the top and bottom electrodes and the location of the gold joint used to protect the chip during wet etching.

The device process started with the 5µm thick dry etch of a 100mm SOI wafer (Si 5µm/SiO$_2$ 1µm/Si bulk 400µm) to create the shape of the cantilevers using AZ ECI 3012 photoresist (figure 2.a-b). Then, a 200nm thick silicon oxide layer was thermally grown to avoid gold diffusion into the silicon cantilever during the thermocompression process (figure 2.c). Electrodes were created by sputtering a 50nm thick Cr seed layer and a 100nm thick Au layer, which were patterned by photolithography (AZ 4999 photoresist) and wet etching (figure 2.d). The bonding pads were then created by a lift-off step using AZ nLOF 2035 photoresist and 500nm thick Au layer deposition (figure 2.e). The wafer was finally annealed at 250°C for 20min to decrease the residual stress within the metal layers.

The substrate hosting the fixed bottom electrode was processed from a 100mm fused silica wafer to ensure the isolation of the contact electrodes. The use of a silicon substrate for the bottom electrode was prohibited since the last fabrication step consisted in releasing the cantilever by wet etching the silicon handle wafer. A first method to set the gap between the top and bottom electrodes was the use of a cavity within the fused silica substrate (figure 2.f). This cavity was created by dry etching the glass substrate patterned by photolithography (AZ 15nXT photoresist). Integration of the bottom electrodes and the bonding pads were then carried out in a similar way as for the top electrode and the metals layers were annealed (figure 2.g-h). A second method to set the gap was to directly use the thickness of the bonding pad by implementing thicker gold joints, thus conveniently avoiding the substrate etching step and enabling large gaps. To this aim, thicker bonding pads were achieved through the electrochemical deposition of gold into a dedicated AZ 40XT photoresist mold (figure 2.h). For this modified process, we also switched to the use of Borofloat 33 substrates that exhibit a coefficient of thermal expansion closer to the one of silicon in order to avoid bonding failure due to layer delamination or substrate cracking that was sometimes observed with fused silica.

After processing the device and substrate wafers, both were diced into chips and cleaned using oxygen plasma. Thermocompression bonding was then conducted after aligning both chips using a FC150 flip-chip bonder from SET (figure 2.i-j). The bonding process was performed by applying 40KgF at 350°C during 10min. Finally, the

cantilevers were released by wet etching the device handle wafer and the buried silicon dioxide layer sequentially with TMAH and BHF (figure 2.k). To avoid stiction of the cantilever, sequential baths of ethanol-acetone-HFE7100-PF2 were used to rinse the device before drying it. Alternatively, supercritical CO$_2$ drying with a Tousimis 915B critical point dryer was also performed.

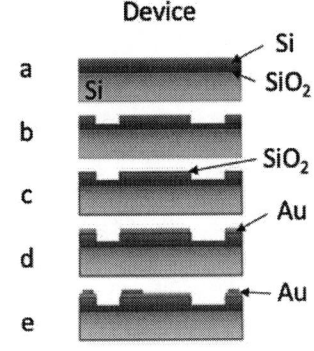

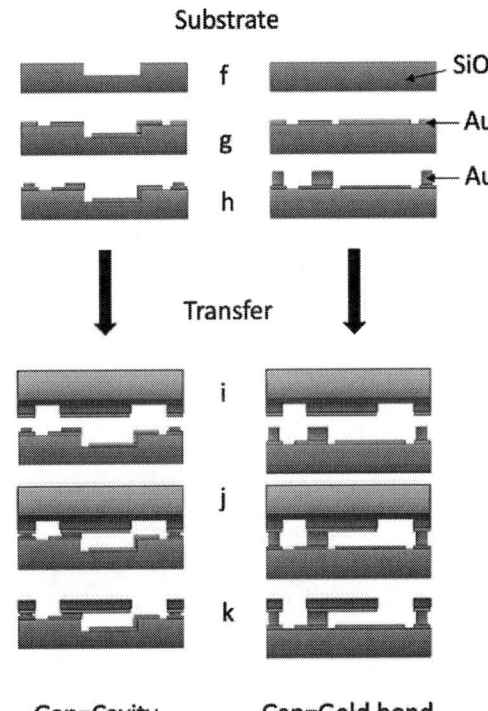

Figure 2: Thermocompression-based fabrication process of the MEMS resonator. The device is fabricated on a SOI wafer by silicon DRIE to form the cantilever and gold evaporation to create the top electrode (top). The bottom electrode and the electroplated gold joint are patterned onto a fused silica wafer that can be previously etched to create a cavity underneath the cantilever (middle, left) that defines the electrode gap. Alternatively, the gap can be set by the thickness of the gold bond, increased by electrochemical deposition of gold (middle, right). Fabrication of the device is finalized by gold-gold thermocompression to bond the 2 substrates and by wet etching the handle SOI layer to release the silicon cantilever (bottom).

978-1-6654-9309-3/23 $31.00 © 2023 IEEE

It is interesting to note that to save TMAH etching time, the backside of the device wafer could be mechanically thinned down (from 400µm to 200µm) with a dicing saw, thus demonstrating the thermocompression bond strength. Also, before the release step by wet etching, the seal was checked for leaks by immersing the chip in ethanol: test failure resulted in the etch/damage of the cantilever (figure 3).

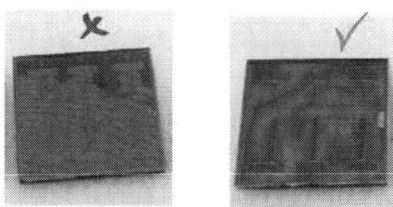

Figure 3: Optical picture of chips after immersion in an ethanol bath to test the joint sealing (Left: leaky device; Right: appropriate joint for subsequent wet etching).

RESULTS AND DISCUSSION

An example of a fabricated chip is shown in figure 1, where the protective joint at the periphery of the chip can clearly be seen. Figure 4 shows optical and SEM images of fabricated cantilevers where the gap of the parallel-plate capacitor, ranging from 3µm to 9µm, is set by a cavity etched in the silica substrate. The high-quality interface of the thermocompression bond can be seen on the SEM cross-section views. Figure 5 shows a SEM image of a device whose gap is defined by the thickness of the electroplated gold used for thermocompression. Structures with gaps up to 25µm were achieved with this process. Optical profilometry was used to image the surface of the fabricated cantilevers: as can be seen in figure 5, the cantilever surface is parallel to the substrate, indicating that the pressure is evenly distributed on the chip during the bonding process.

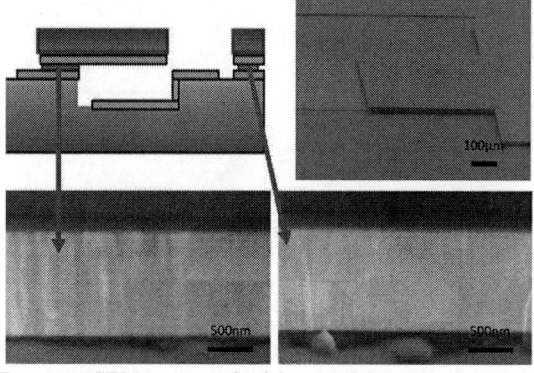

Figure 4: SEM images of a fabricated device with a cavity etched in the host glass substrate to create the gap; cross-section views of the gold bond.

The cantilevers were then characterized in static and dynamic modes. To this aim, we have used a home-made network analyzer to drive the excitation voltage and measure the gap capacitance using a charge amplifier. The deflection of the tip of the cantilever was additionally measured using a IFS2407-3 confocal microscope from micro-epsilon. Static characterizations are presented in

figure 6: the value of the gap size, at the tip of the cantilever, and the corresponding capacitance measured onto a 500µm long, 500µm width, 5µm thick cantilever with a ~22µm gap size exhibit a quadratic behavior with applied voltage, as theoretically expected.

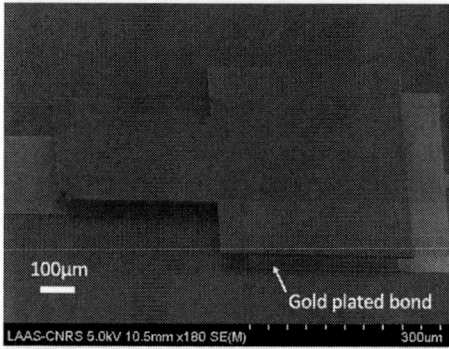

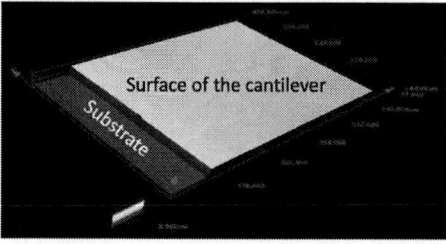

Figure 5: Top: SEM image of a cantilever with a large gap created by the electroplated gold. Bottom: Cantilever surface profile imaged with an optical profilometer.

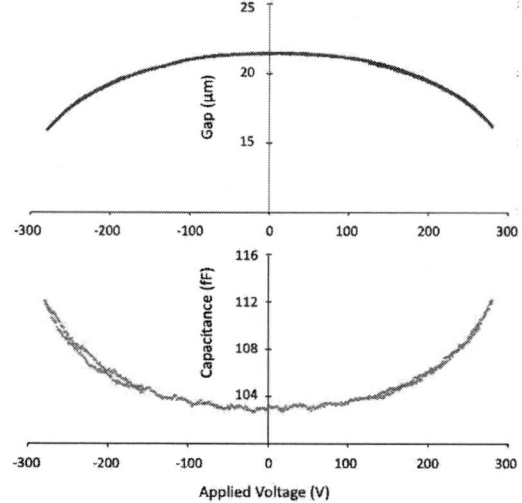

Figure 6: Static measurement of the cantilever gap and corresponding capacitance (after subtracting the static capacitance) as a function of applied DC voltage.

Last, we have studied the dynamic behavior of the cantilevers by driving them at resonance. The influence of the AC excitation and DC bias on resonant frequencies was studied and the expected non-linear response regime and the spring softening effect were observed with increasing AC and DC voltages, respectively (figure 7). The impact of environmental pressure on the quality (Q) factor of the mechanical resonators was studied using a Linkam vacuum chamber and a Cryoprober from SUSS MicroTec for medium and high vacuum measurements (figure 8). Q factors of up to 10^4 were obtained in high vacuum, indicating a limited dissipation due to the gold anchor. While devices with 3 µm gap experienced critically damped vibrations because of squeeze damping in air [9], a Q factor of ~45 was obtained by increasing the gap to 25 µm, thus demonstrating the interest of our fabrication technique for conveniently tunning the electrode gap.

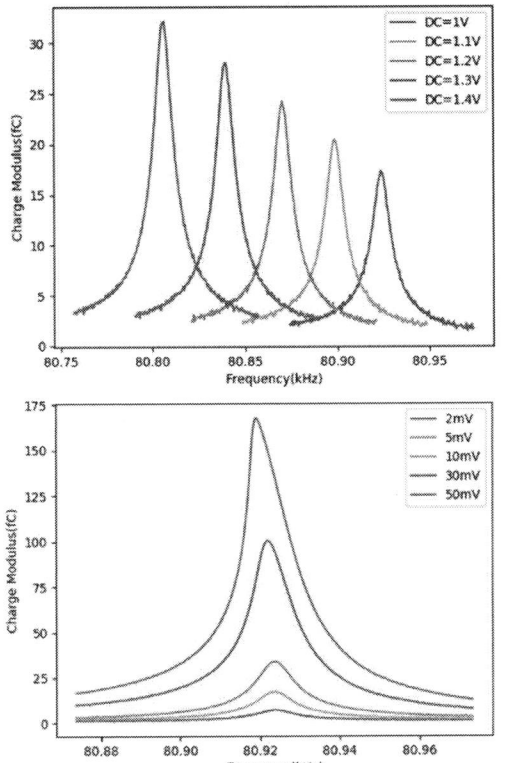

Figure 7: Characterization of a 3 µm gap cantilever in high vacuum. Resonance frequency as a function of DC bias for V_{AC}=50mV (top) and AC signal for V_{DC}=1V (bottom).

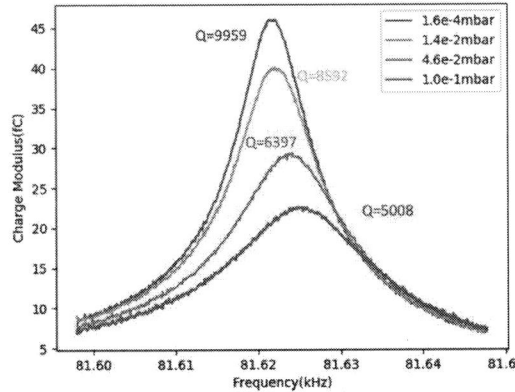

Figure 8: First mode resonance frequency spectrum of a 9 µm gap device as a function of pressure (V_{AC}=90mV and V_{DC}=2.5V).

CONCLUSION

In this paper, we present a straightforward process based on gold thermocompression bonding for fabricating suspended electrostatic flexural MEMS structures. Anchor of the free-standing structure and electrical connections are provided in a single step. Additionally, the parallel-plate gap size is controlled by simply adjusting the thickness of the electroplated gold layer to reduce the squeeze damping of cantilever resonators. This process is foreseen to be applicable to other geometries and material devices, e.g. SiC.

ACKNOWLEDGEMENTS

This research was performed thanks to the contest Programme d'Investissements d'Avenir of the French Government under the supervision of the French National Radioactive Waste Management Agency (ANDRA). This work was partly supported by the French RENATECH network.

REFERENCES

[1] M. I. Haller, and B. T. Khuri-Yakub, "A surface micromachined electrostatic ultrasonic air transducer", *IEEE Trans. Ultrason. Ferroelectrics. Freq. Contr.*, vol. 43, pp. 1-6, 1996.

[2] A. S. Erguri, Y. Huang, X. Zhuang, O. Oralkan, G. G. Yarahoglu, and B. T. Khuri-Yakub, "Capacitive micromachined ultrasonic transducers: fabrication technology", *IEEE Trans. Ultrason. Ferroelectrics. Freq. Contr.*, vol. 52, pp. 2242-2258, 2005.

[3] A. Bhaswara, H. Keum, S. Rhee, B. Legrand, F. Mathieu, S. Kim, L. Nicu, and T. Leïchlé, "Fabrication of nanoplate resonating structures via micro-masonry", *J. Micromech. Microeng.*, vol. 24, 115012, 2014.

[4] A. Bhaswara, H. Keum, F. Mathieu, B. Legrand, S. Kim, L. Nicu, and T. Leïchlé, "A simple fabrication process based on micro-masonry for the realization of nanoplate resonators with integrated actuation and detection schemes", *Front. Mech. Eng.*, vol. 2, pp. 1-7, 2016.

[5] C. H. Tsau, S. M. Spearing and M. A. Schmidt, "Fabrication of wafer-level thermocompression bonds", *J. Microelectromech. Syst.*, vol. 11, pp. 641-647, 2002.

[6] F. Copt, Y. Civet, C. Koechli and Y. Perriard, "Design and manufacturing of an electrostatic MEMS relay for high power applications", *Sens. Actuators A: Phys.*, vol. 321, 112569, 2021.

[7] S. Charlot, P. Pons, M. Dilhan, I. Vallet, and S. Brida, "Hermetic Cavities Using Gold Wafer Level Thermocompression Bonding", in *Proceedings of Eurosensors 2017*, Paris, France, September 3–6, 2017, 1(4), 607.

[8] M. T. Boudjiet, J. Bertrand, F. Mathieu, L. Nicu, L. Mazenq, T. Leichle, M. Heinrich, C. Pellet, and I. Dufour, "Geometry optimization of uncoated silicon microcantilever-based gas density sensors", *Sens. Actuators B: Chem.*, vol. 208, pp. 600-607, 2015.

[9] M. Bao and H. Yang, "Squeeze film air damping in MEMS", *Sens. Actuators A: Phys.*, vol. 136, pp. 3-27, 2007.

CONTACT

*T. Leïchlé, thierry.leichle@cnrs.fr

ELECTROMECHANICALLY STABLE INTERCONNECTION BETWEEN LIG AND THICK DAM-SHAPED METALLIC ELECTRODE VIA STORED AG MICROPARTICLE SOLUTION

Saeyoung Park[1†], Yoo-Kyum Shin[2†] and Min-Ho Seo[1]*

[1] School of Biomedical Convergence Engineering, Pusan National University, REPUBLIC OF KOREA

[2] Department of Information Convergence Engineering, Pusan National University, REPUBLIC OF KOREA

†These authors contributed equally to this work

ABSTRACT

This paper reports a method for forming an electromechanically stable interconnection between Laser-induced graphene (LIG) and metallic electrode exploiting silver (Ag) micro-particle (μ-P) solution (ink) and dam-shaped thick metallic electrodes. From visual inspection, we confirmed uniformly intertwined contact between the Ag μ-P and LIG. In practice, we confirmed that the proposed concept shows ohmic contact with high reproducibility (n=100). Finally, we adapted the proposed method to demonstrate a 3-electrode electrochemical biosensor, and it successfully detected various concentrations (0–5 mM) of ferrocene with high reproducibility (n=3)

KEYWORDS

Laser-induced graphene, Ag microparticle ink, dam-shaped electrode, Laser-induced graphene interconnection

INTRODUCTION

Laser-induced graphene (LIG) is known as three-dimensional (3D) porous graphene produced by scribing a CO_2 laser on a polyimide (PI) film in an atmospheric environment [1]. Compared to the conventional graphene manufacturing methods such as the chemical vapor deposition (CVD) technique and redox methods [2,3], LIG has huge advantages in terms of fabrication because it can be manufactured at atmospheric and room temperature without solvent, chemical treatment, and cleanroom process [4]. Moreover, the shape and dimension of LIG can be easily modified by scanning the laser path, implying high throughput, ease of fabrication, cost-effectiveness, and degree of design freedom [5].

Recently, LIG has gained considerable attention from various high-performance electrical devices because its 3D hierarchical porous structure is appropriate to demonstrate drastic performance-enhancement at the device-level. In practical, high-performance chemical sensors [6], biosensors [7], physical sensors [8,9], gas sensors [10], super-capacitors [11], and electro-actuators [12] have been developed using the LIG, and shown unprecedentedly high sensitivity and flexibility. However, LIG-based electrical devices still suffer from electrically and mechanically stable interconnection. Since LIG, the 3D nano-structure of carbon, is not appropriate for adapting conventional electrical interconnection, such as soldering, it is difficult to realize a stable interconnection with conventional metallic wire and electrode. Recently, researchers have demonstrated the electrical interconnection exploiting Ag paste and Ag ink hand-working [13,14]. However, these methods are not suitable to demonstrate the miniaturized and stable electromechanical interconnection, because the mechanically brittle LIG is easily broken during the process. Thus, researchers still seek for a method that can make the electromechanically stable and scalable interconnection method between the LIG and metal electrode. Here, we report a facile, miniaturized, and electromechanically stable interconnection method between LIG and the metallic electrode via a dam-shaped thick metallic electrode and a silver micro-particle solution (Ag μ-particle ink). Since the proposed method utilizes the micro-patterned (defined as "dam-shaped") 3D metallic electrode having the LIG inside the pattern, it can generate the localized and miniaturized electromechanical interconnection between the electrode and LIG through the dropped Ag ink. Using the proposed method, we have reproducibly demonstrated a bar-pattern device, composed of ~ 100 μm-thick-dam-shaped copper (Cu) electrodes and LIG with a 5 mm width and successfully confirmed a stable electromechanical interconnection between LIG and electrodes through visual inspection and electrical characteristics. We have adapted the proposed method to demonstrate a biosensor, and finally, confirmed that the various concentrations of ferrocene can be detected by the developed biosensor with reproducibility.

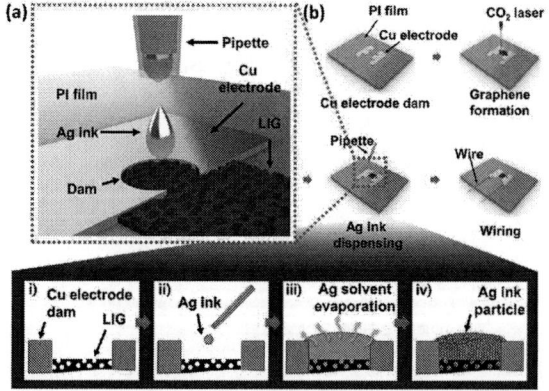

Figure 1: *Schematic illustration of the proposed concept. (a) Details of the proposed concept, "Ag μ-P (micro-particle) ink dispensing" step. Since the Cu electrode has a thick- thickness, the Ag ink solution does not overflow from the exquisite interconnection with LIG and Cu electrode. (b) The developed fabrication process.*

PROPOSED CONCEPT

This work proposes a method of realizing the electromechanically stable interconnection between the LIG and the dam-shaped metallic electrode using dropping Ag μ-particle ink (Figure 1a). The two important phenomena of this method are that the dam structure of the thick electrode, surrounding a part of LIG, traps the silver ink solution and the remaining Ag μ-P after the solvent evaporates physically connects the LIG and the metal electrode; First, when dispensing ink to the spot of LIG surrounded by the thick electrode, the ink is not spread out and is trapped in the electrode because of the shape of the electrode. In this state, when heat is further applied to the ink, the solvent of the ink evaporates, leaving only silver particles which became entangled with the LIG (Figure 1(a) i-iv)). The Ag microparticles not only generate the superior electrical interconnection due to the conformal contact between the LIG and electrode but also stabilize the mechanical durability of the contacting interface by means of the physical interlocking by agglomerates formation of the silver particles and resin remaining at the interface of the LIG and metal electrode.

To demonstrate the proposed concept, we developed a simple fabrication process as shown in Figure 1b. First, a ~100 μm-thick dam-shaped copper electrode is formed on the PI film using conventional UV laser machining. Then, LIG is formed by the CO_2 laser scribing process. In this step, the laser scans a wide range of the PI substrate including the electrode, but the Cu electrode does not react to the laser, and thus, the LIG has selectively formed only around and in the electrode on the PI substrate. After LIG formation, the Ag ink is quantitatively dispensed in a dam-shaped electrode with a micropipette. Since the Ag ink has enough viscosity only for filling the LIG in the dam structure, not for spread out, thus, when it is placed on a hot plate and heated, the solvent of the ink evaporates, locally leaving only Ag microparticles inside of the electrode.

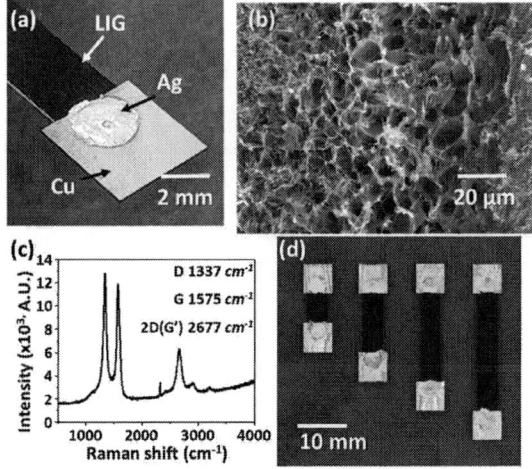

Figure 2: Result of Fabricated Ag μ-P interconnection. (a) Magnified optical image of the fabricated Ag μ-P interconnection with LIG and Cu. (b) SEM image of the fabricated LIG. (c) Raman spectra of the fabricated LIG. (d) Optical image of the samples with different lengths of LIG.

Figure 3: Visual analysis results of the contact interface. (a) Optical image of the top surface of the LIG-Ag μ-P contact interface after de-capping the Ag part. (b) SEM image of the top contact interface. (c, d) Cross-sectional SEM image of the LIG-Ag μ-P contact interface. The intertwined Ag μ-P and LIG are confirmed.

EXPERIMENTAL RESULT

The proposed method was confirmed experimentally. In the magnified optical image, the deposited ink is locally formed inside the dam-shaped metal electrode (Figure 2a). Scanning electron microscopy (SEM) confirmed that the LIG maintained its 3D porous structure even after the entire fabrication process was completed, which means that the solvent did not spread to the surrounding LIG by capillary force from the dam (Figure 2b). In addition, LIG of the fabricated device shows D (1337 cm^{-1}), G (1575 cm^{-1}), and 2D (2677 cm^{-1}) peaks in the Raman spectrum highly corresponding to the preceding research (Figure 2c) [15]. LIG devices with lengths of 5, 10, 15, and 20 mm were also fabricated, and a connection between the LIG and the metal electrode was formed by the proposed method. There was no noticeable morphological deformation or degradation in the fabricated device, so it can be concluded that the proposed concept can show high scalability and reliability (Figure 2d).

From further visual inspection, we verified the Ag microparticles successfully formed the mechanically stable contact between the LIG and the electrode. After removing the dried Ag μ-P in the fabricated device, the remaining dam part was confirmed from above through an optical microscope and SEM. As a result, it was found that the particles were locally embedded in the LIG inside the dam, and even when viewed in a magnified view, the LIG and silver particles were entangled (Figure 3a,b). In addition, the SEM was also confirmed in the cross-section, and as a result, it was confirmed that the LIG generated on the PI substrate was entangled with the Ag μ-P to form a composite layer between the PI substrate and the Ag particle cap (Figure 3c,d).

The electrical characterization of the proposed interconnection method was conducted using a conventional source-meter (Keysight B2902A, Keysight Technology) (Figure 4a). From the I-V characterization

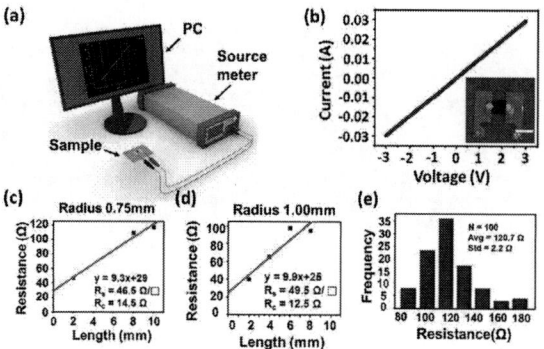

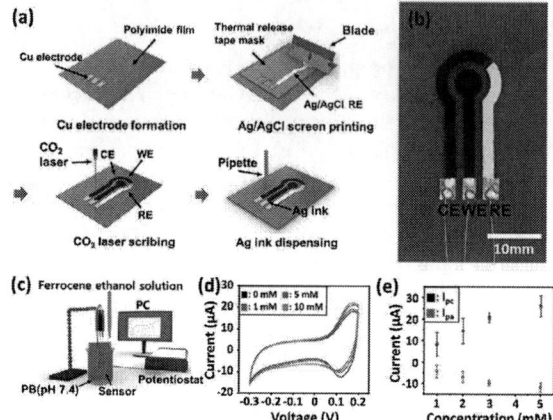

Figure 4: *Electrical characterization of the proposed interconnection. (a) Schematic illustration of the experimental set-up. (b) Measured I-V characteristics of the fabricated device. (Inset: optical image of the device, scale bar:5mm). (c, d) Measured transmission-line method (TLM) results of the different radius of round structure inside of the electrode (0.75mm(c), 1mm(d)) with different lengths of LIG. Rs and Rc indicates the sheet resistance of the LIG and contact resistance, respectively. (e) Reproducibility of the proposed method (total resistances of the 100-samples)*

Figure 5: *Biosensor Application. (a) Schematic illustration of the fabrication process of the biosensor with the developed method; (b) Optical image of the fabricated sensor. WE, CE, and RE means working, counter, and reference electrode, respectively. (c) Schematic illustration of the experimental set-up. (d) Measured cyclic voltammetry (CV) curve of the sensor with different concentration of ferrocene. (e) Results of the changes in the cathode and anode peak currents (I_{pc} and I_{pa}, respectively) with respect to the increasing ferrocene concentration.*

range from −3V to 3V, we confirmed that the device has an ohmic contact, which means that there is negligible parasitic capacitance at the interconnection (Figure 4b). In addition, the transmission line method (TLM) was performed to obtain the sheet resistance of LIG and contact resistance of the proposed method (The resistance value of metallic electrodes is omitted because they are relatively small compared to sheet resistance and contact resistance). When the radius of the dam shape is 0.75mm and 1.00mm, the contact resistance (R_c) is 14.5 **Ω** and 12.5 **Ω**, respectively, and the sheet resistance (R_s) is 46.5 **Ω/□** and 49.5 **Ω/□** respectively (Figure 4c,d). We confirmed that the contact resistance reasonably decreases as the contact area decreases, and the sheet resistance is similar to the results of previous research [16]. Also, as a result of measuring resistance by making 100 devices, it can be seen that the reproducibility is very high by confirming that it has a normal distribution with an average of 120.7 **Ω** and a standard deviation of 2.2 **Ω**.

We practically applied the proposed interconnection Since LIG provides high sensitivity as a sensor due to its large surface area, it can be used as a material for biosensors. The biosensor was manufactured in the following order (Figure 5a). First, an electrode designed in a dam shape is formed on the PI. Next, a reference electrode was fabricated through Ag/AgCl screen printing, and a counter electrode and a working electrode were fabricated by irradiating LIG. Immediately after that, Ag ink is dropped on the LIG inside the dam-shaped electrode (Figure 5b). The performance of the completed sensor was confirmed by dropping ferrocene, which is used as a sensing material for blood glucose measurement, into the phosphate buffer (pH 7.4) in which the sensor was dipped (Figure 5c). The results of the cyclic voltammetry (CV) curve were measured when ferrocene was injected at a concentration of 0, 1, 5, and 10 mM by applying a voltage

from −0.3V to 0.2V (scan rate: 20 mV/s). As a result, as the concentration increased, the peak value of the reduction current (I_{pc}) of CV gradually increased, which means that a more active electrochemical reaction between LIG and ferrocene of the sensor occurred (Figure 5d). In addition, the average standard deviation plot was drawn for the redox current peak values of the three sensors, and the corresponding increase in current was confirmed as the concentration increased to 1,2,3, and 5 mM (Figure 5e).

SUMMARY

We developed a simple method for electromechanical enhancement of the interconnection of LIG and metal electrodes by utilizing Ag ink and a dam-shaped copper electrode design. The core of this method is that the ink solution is confined in the dam structure, and the Ag particles and agglomerates remaining after the solvent evaporated from the solution become mediators that physically connect the LIG and the copper electrode, thereby improving electromechanically. To verify this, visual analysis and electrical property evaluation through optical and SEM images were conducted. As a result, it was confirmed that the ink particles were mechanically entangled with the LIG and the proposed concept shows it can be applied to miniaturization without deformation or deterioration. We also applied the method to an electrochemical biosensor where LIG is widely used and tested the three fabricated sensors. As a result, it was found that the sensing result of the sensor was well transmitted to the measuring device, which proves that the interface between the LIG and the metal electrode is stably interconnected and has high reproducibility.

ACKNOWLEDGEMENTS

This work was supported by the National Research

Foundation of Korea(NRF) grant funded by the Korea government(MSIT). (No. 2022R1A2C2091343)

REFERENCES

[1] J. Lin, Z. Peng, Y. Liu, F. Ruiz-Zepeda, R. Ye, E.L.G. Samuel, M.J. Yacaman, B.I. Yakobson, J.M. Tour, "Laser-induced graphene porous graphene films from commercial polymers", *Nat. commun.*, 5, 5714, 2014

[2] R. Muñoz, C. Gómez-Aleixandre, "Review of CVD Synthesis of Graphene", *Chem. Vap. Depos.*, vol. 19 (10-11-12), pp. 297–322, 2013.

[3] H. Jiang, "Chemical preparation of graphene-based nanomaterials and their applications in chemical and biological sensors," *Small*, vol. 7, no. 17, pp. 2413–2427, 2011.

[4] R. Ye, D.K. James, J.M. Tour, "Laser-Induced Graphene: From Discovery to Translation", *Adv. Mater.*, vol. 31, 1803621, 2018

[5] H. Wang, Z. Zhao, P. Liu, X. Guo, "Laser-Induced Graphene Based Flexible Electronic Devices", Biosensors, 12 (2), 55, 2022

[6] L. Huang, J. Su, Y. Song, R. Ye, "Laser-Induced Graphene: En Route to Smart Sensing", *Nanomicro Lett.*, 12, 157, 2020

[7] Z. Wan, N.-T. Nguyen, Y. Gao, Q. Li, "Laser induced graphene for biosensors", *Sustain. Mater. Technol.*, vol. 25, e00205, 2020

[8] W. Wang, L. Lu, z. Li, Z. Liang, X. Lu, Y. Xie, "Fingerprint-Inspired Strain Sensor with Balanced Sensitivity and Strain Range Using Laser-Induced Graphene", *ACS Appl. Mater. Interfaces*, 14, 1, pp. 1315–1325, 2022

[9] A. Kaidarova, N. Alsharif, B.N.M. Oliveira, M. Marengo, N.R. Geraldi, C.M. Duarte, J. Kosel, "Laser-Printed, Flexible Graphene Pressure sensors" *Global Challenges*, vol. 4 (4), 2000001, 2020

[10] M.G. Stanford, K. Yang, Y. Chyan, C. Kittrell, J.M. Tour, "Laser-Induced Graphene for Flexible and Embeddable Gas Sensors", *ACS Nano*, 13, 3, pp. 3474–3482, 2019

[11] F. Bu, W. Zhou, Y. Xu, Y. Du, C. Guan, W. Huang, "Recent developments of advanced micro-supercapacitors: design, fabrication and applications", *npj Flex. Electron.*, 4, 31, 2020

[12] L. Zhu, Y.-Y. Gao, B. Han, Y.-L. Zhang, H.-B. Sun, "Laser fabrication of graphene-based electrothermal actuators enabling predicable deformation", *Opt. Lett.*, vol. 44 (6), pp. 1363–1366, 2019

[13] A. Kaidarova, M. Marengo, G. Marinaro, N. Geraldi, C.M. Duarte, J. Kosel, "Flexible and Biofouling Independent Salinity Sensor", *Adv. Mater. Interfaces*. vol. 5 (23), 180110, 2018

[14] A.R. Cardoso, A.C. Marques, L. Santos, A.F. Carvalho, F.M. Costa, R. Martins, M.G.F. Sales, E. Fortunato, "Molecularly-imprinted chloramphenicol sensor with laser-induced graphene electrode" *Biosens. Bioelectron.*, vol. 124-125 (15), pp. 167–175, 2019

[15] W. Ma, J. Zhu, Z. Wang, W. Song, G. Cao, "Recent advances in preparation and application of laser-induced graphene in energy storage devices", *Mater. Today Energy*, vol. 18, 100569, 2020

[16] J.D. Kim, T. KIm, J. Park, "Fabrication and Transfer of Laser Induced Graphene (LIG) electrodes for Flexible Substrate-based Electrochemical Sensor Applications", *Trans. Korean Inst. Electr. Eng.*, vol. 67 (3), pp. 406–412, 2018

CONTACT

*M.-H. Seo, tel: +82-51-510-8558;
mhseo@pusan.ac.kr

FREE-STANDING MEMBRANES WITH SELF-ASSEMBLED NANOPORE ARRAYS FOR TEM OBSERVATION OF LIQUID SAMPLES

Joowon Lim[1], Jimin Ham[1], Sungho Jeon[1], Yuna Bae[2,3], Minho Kang[2,3], Sueng Yoon Lee[1], Jungwon Park[2,3], and Won Chul Lee[1]

[1]Department of Mechanical Engineering, BK21 FOUR ERICA-ACE Center, Hanyang University, Ansan, REPUBLIC OF KOREA,
[2]School of Chemical and Biological Engineering, Seoul National University, Seoul, REPUBLIC OF KOREA, and
[3]Center for Nanoparticle Research, Institute of Basic Science (IBS), Seoul, REPUBLIC OF KOREA

ABSTRACT

This work presents a method for creating free-standing, uniform nanopore arrays onto two-dimensional (2D) membranes such as silicon using AAO (anodic aluminum oxide) as an etching mask for jointly duplicating identical patterns, with an application of such silicon as holding chambers for *in-situ* liquid TEM (transmission electron microscopy) observations. By sandwiching the nanopatterned silicon between two sheets of graphene, liquid samples can be trapped and observed in real time. The given methods can provide guidelines in creating free-standing nanopatterned materials as well as applications for *in situ* liquid cell TEM observations.

KEYWORDS

AAO mask, reactive ion etching, liquid TEM, nanofabrication, *in-situ* electron microscopy.

INTRODUCTION

Free standing, 2D materials are used in a variety of applications for nanotechnology, such as graphene used as a trans-electrode membrane on a micro-fabricated chip [1]. Fabrication of such materials with identical patterns is a tedious and generally expensive process due to the patterning steps having to be repeated several times. Thus, methods to create identical patterns rapidly are of interest. Anodic aluminum oxide (AAO) is often used in nanofabrication due to easy availability and convenience of creating uniform pore arrays with tuned diameters, densities and thicknesses. Free-standing AAO has been used in nanofabrication for creating structures like nanopillars [2], or inside sensors [3]. Another use is for liquid-cells in TEM, as spacers in creating nanoscale chambers for trapping minute amounts of liquid [4], [5].

We present a self-assembled fabrication process for creating large-scale, free-standing, uniformly-patterned nanoscale membranes by utilizing AAO as an etching mask (Fig. 1), and demonstrate an application of one such patterned material, silicon, as a spacer material for liquid cells used in TEM observation (Fig. 2. a) with optically clear, mechanically strong graphene as a sealing and viewing window material. Liquid cells are required to seal liquid samples tightly to prevent evaporation and subsequent sample destruction from the TEM's interior vacuum, while having nanoscale (100nm~) thicknesses and optically clear windows for optimal observation resolution. Since silicon has a smooth surface, is readily available and is easily tuned to nanoscale thinness with reliable semiconductor fabrication methods, we imagined it to work well with graphene for use in observing *in-situ* liquid behaviors inside TEM (Fig. 2. b).

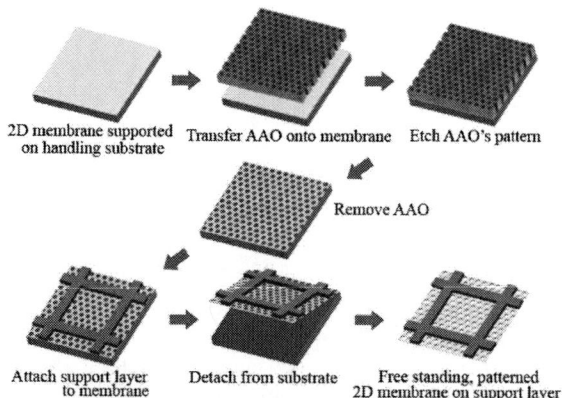

Figure 1. Schematic for creating free-standing, two dimensional patterned membranes.

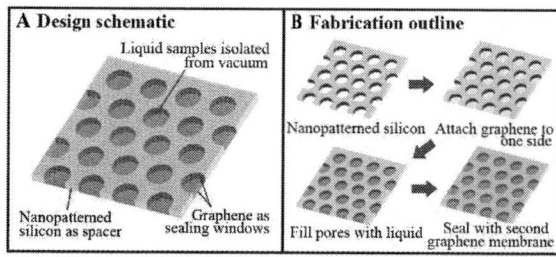

Figure 2. (a) Design schematic for using one nanopatterned material, silicon, for liquid-cell TEM observation. (b) Detailed fabrication process schematic for trapping liquid samples inside silicon pores.

FABRICATION

A SOI (silicon-on-insulator) wafer's top silicon layer of 88nm thickness is used as the membrane, and the mask AAO with pore diameters of 260nm and interpore (center to center) distances of 450nm is transferred onto the silicon and its pattern is emulated using reactive ion etching (RIE). The AAO is then removed, and the remaining nanopatterned silicon membrane is bonded to a TEM grid used for TEM observations, which is pre-coated with bilayer graphene. A gold ion-containing liquid sample, 1-10 mM $HAuCl_4$ in DI water, is dropped onto the silicon, and sealed inside by covering it with a second graphene-coated TEM grid. After the sample is dried overnight in a desiccator, it is placed in a TEM and observed.

RESULTS

Using AAO as an etching mask, we were able to duplicate its pattern with reactive ion etching (RIE) onto various 2D membranes such as graphene (Fig. 3. a), hexagonal boron nitride (hBN) (Fig. 3. b), thin carbon film (Fig. 3. c), and silicon (Fig. 3. d), and transferred the membranes onto support layers such as observation grids, where they exist in a free-standing state. Viewed through electron microscopes, numerous, uniform pore arrays conforming well to the mask's dimensions were shown to be emulated onto each material.

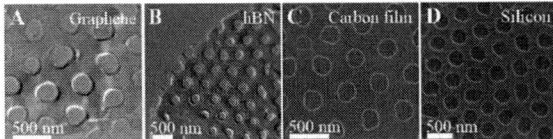

Figure 3. Electron microscope images of various nanomembranes patterned using AAO as an etching mask, existing free-standing on support grids. (a) Graphene, (b) hexagonal boron nitride (hBN), (c) thin carbon film (thickness 15nm), (d) silicon (thickness 90nm).

Before attempting to trap liquid samples inside the silicon nanopores, in order to check the initial viability of this method, we verified whether the silicon and graphene transfer well onto the TEM grid. The two materials were readily seen on the TEM grid (Fig. 4. a, b), and using fast Fourier transformation (FFT), we observed the lattice patterns of the viewed materials and confirmed that both were successfully transferred onto the TEM grid (Fig. 4. c).

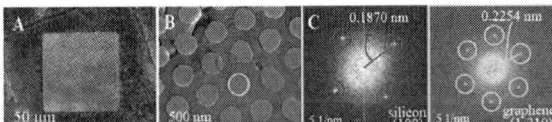

Figure 4. (a) Optical image of nanopatterned silicon and graphene on support grid. (b), (c) TEM images of graphene and silicon transferred onto observation support grid. (d) Fourier transformed images of lattice patterns respectively corresponding to silicon and graphene.

As we aim to observe *in-situ* liquid behaviors, prevention of evaporation and corresponding existence of liquid inside the silicon nanopores convey success of this method. Aqueous $HAuCl_4$ was chosen for the liquid sample as it has high reactivity to the TEM's electron beams and provides quick visual indicators of sample existence. We observed the results of sealing liquid samples with TEM, and discovered that nanoparticles had formed inside many of the silicon nanopores. Due to the high resolution of the images, owing to the thinness of the silicon (88nm~) and optical clarity of the graphene, we were able to track a single particle's movement over 21 seconds (Fig. 5. a). Using a tracking software, we followed the particle's trajectory using its center coordinates and calculated its displacement every 0.5 seconds (Fig. 5. b). The trajectory is reminiscent of a particle exhibiting Brownian motion inside a liquid medium, hence confirming successful liquid containment inside the silicon. Finally, similar particle movements and behaviors were observable a week after

initial trapping and observation, suggesting reliable liquid sample sealing and repeatability in the methods we used.

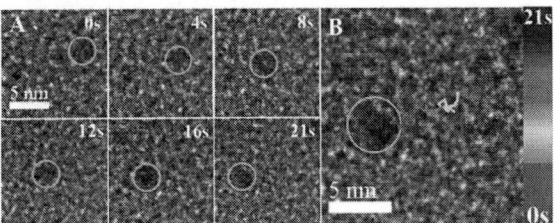

Figure 5. (a) Time series of in situ *particle movement in TEM. (b) Trajectory track of the observed particle over 21 seconds. Movement is reminiscent of Brownian motion exhibited in particles suspended in aqueous environments.*

CONCLUSIONS

In this research, we created various nanopore arrays from nano-thick 2D materials and showed an application of one material, silicon, as liquid cells for *in-situ* liquid TEM observations. The fabrication processes and results we procured are expected to provide guidelines for large-scale nano-level patterning techniques, as well as guidelines for simplified fabrication methods for creating liquid cells for *in situ* observations of nanoscale phenomena in TEM.

ACKNOWLEDGEMENTS

This work was supported by the National Research Foundation of Korea (NRF) grant funded by the Korea government (MSIT) (No. 2022R1A4A3031263 and No. 2021R1A2C1011797).

REFERENCES

[1] S. Garaj, W. Hubbard, A. Reina, J. Kong, D. Branton, J. A. Golovchenko, "Graphene as a Subnanometre Trans-Electrode Membrane", *Nature*, vol 467, pp. 190-193, 2010.

[2] T. Okamoto, T. Shimizu, K. Takase, T. Ito, S. Shingubara, "Formation of MoS_2 Nanostructure Arrays Using Anodic Aluminum Oxide", *Micro Nano Eng.*, vol 9, pp. 100071, 2020.

[3] K. Bae, J. Lee, G. Kang, D. S. Yoo, C.W. Lee, K. Kim, "Refractometric and Colometric Index Sensing by a Plasmon-Coupled Hybrid AAO Nanotemplate", *RCS Adv.*, vol 5, pp. 103052-103059, 2015.

[4] K. Lim, Y. Bae, K. Kim, S, Jeon, B. H. Kim, J. Park, W. C. Lee, "Self-Assembled Nanochamber Arrays for *in-situ* TEM Observation of Liquid-Phase Samples", in *MEMS 2019 Conference*, Seoul, January 27-31, 2019, pp 105-106.

[5] K. Lim, Y. Bae, S. Jeon, K. Kim, B. H. Kim, J. Kim, S. Kang, T. Heo, J. Park, W. C. Lee, "A Large-Scale Array of Ordered Graphene-Sandwiched Chambers for Quantitative Liquid-Phase Transmission Electron Microscopy", *Adv. Mater.*, vol 32, pp. 2002889, 2020.

CONTACT

*Won Chul Lee, tel: +82-31-400-5257; wonchullee@hanyang.ac.kr.

NONPLANAR NANOFABRICATION VIA INTERFACE ENGINEERING

Sarah O. Spector, Peter F. Satterthwaite, and Farnaz Niroui
Massachusetts Institute of Technology, Cambridge, Massachusetts, USA

ABSTRACT

We report an approach for scalable fabrication of suspended, ultrathin, nonplanar nanostructures without the use of sacrificial layers. We achieve this by engineering interfacial forces through a patterned self-assembled molecular monolayer to enable controlled delamination of a deposited oxide thin-film in predetermined locations. This allows formation of nonplanar structures with thicknesses < 10 nm and nanogaps reaching < 10 nm – features hard to achieve with conventional fabrication. This approach is versatile as it extends conventional, wafer-scale planar fabrication techniques to nonplanar designs. The resulting features are tunable in structure and compatible with various materials, enabling applications in miniaturized nanoelectromechanical devices including ultrathin mechanical resonators.

KEYWORDS

Nonplanar nanofabrication, Atomic layer deposition, Self-assembled molecular layer, Interface engineering, Nanoelectromechanical devices, Mechanical resonators.

INTRODUCTION

Nonplanar nanostructures, composed of suspended ultrathin films and nanogaps, are foundational for next-generation miniaturized nanoelectromechanical systems (NEMS), photonic elements, and metamaterials. However, instabilities due to nanoscale forces, including the van der Waals and capillary forces, challenge nonplanar fabrication through conventional top-down techniques, which often require the removal of a sacrificial support layer. These instabilities become more prominent as dimensions reduce to the few-nanometers regime, as is required for improved device efficiency, speed, and sensitivity, as well as new functionalities. Thus, it is imperative to overcome the inherent limits of top-down strategies to facilitate development of nonplanar designs.

Additive manufacturing using electron- and ion-beam-induced processes [1], two-photon lithography [2], and kirigami [3] provide alternative approaches for direct fabrication of three-dimensional nanostructures. However, these strategies typically remain limited to features > 100 nm that are often not mechanically-active, and are largely incompatible with conventional wafer-scale processes, limiting utility and integration opportunities.

To address these limitations, we propose a technique for stable and scalable fabrication of nonplanar nano-structures leveraging bottom-up engineering of surface interactions. In this approach, we deterministically modify the stress in a deposited thin-film and its local adhesion to a substrate, guiding delamination of planar films and their transformation into nonplanar architectures which can also be mechanically-active. Although thin-film delamination has been previously studied [4] and used for micron-sized features [5], this work leverages this principle to develop a deterministic fabrication strategy in the tens of nanometers.

FABRICATION METHODOLOGY

Our fabrication scheme is summarized in Figure 1a. We use lithographically-patterned self-assembled molecular monolayers (SAM) to selectively modify the surface properties of a substrate. The molecule's anchoring group allows for assembly onto the desired substrate while its functional chain is chosen to provide the desired surface interactions with a subsequently deposited thin-film.

To pattern the molecular layer, an Al_2O_3 hard-mask is used. In contrast to a polymer resist mask, this approach avoids contaminant residues that interfere with optimal SAM growth. To form the hard-mask, first, a 2 nm-thin film of Al_2O_3 is grown using atomic layer deposition (ALD). Then, a positive resist is lithographically patterned on top and developed with a tetramethylammonium

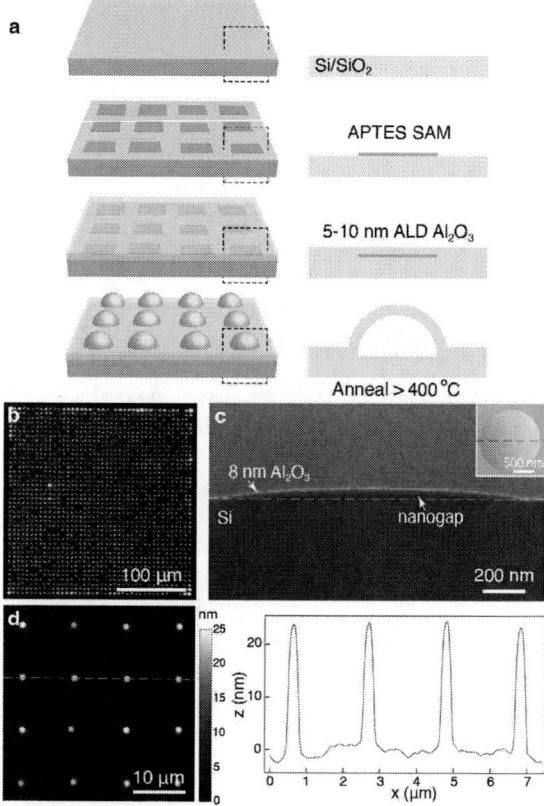

Figure 1: Fabrication methodology. (a) To fabricate the nonplanar structures, a patterned monolayer of APTES is self-assembled onto the substrate, followed by ALD of 5-10 nm of oxide. The sample is then annealed at > 400°C to induce selective delamination of the thin oxide film. (b) Dark-field microscope image of an example array of 8 nm-thin Al_2O_3 membranes fabricated in the form of nanoblisters. (c) Cross-sectional and top view (inset) scanning electron microscope images of a blister. (d) Atomic force micrograph (left) with line scan (right) of an array of blisters.

hydroxide (TMAH) developer. During development, the exposed Al₂O₃ is also etched by the TMAH, patterning the hard-mask. Resist is then stripped in acetone, followed by two minutes of oxygen plasma treatment. This results in an Al₂O₃ mask where the exposed regions of the substrate are free of polymer residues, having avoided direct contact with the resist. The SAM (3-aminopropyl)triethoxysilane (APTES) is then assembled onto the surface in vapor phase for 12 hours in a vacuum desiccator. Following assembly, the sample is sonicated in ethanol to remove the physisorbed molecules, and then the Al₂O₃ hard-mask is removed in TMAH developer, leaving the SAM selectively patterned on the surface.

Once the patterned SAM is formed, a thin layer of ALD oxide (5-10 nm) is deposited using alternating cycles of trimethylaluminum (TMA) and water precursors. The deposition is performed at 200°C, a temperature sufficiently low to avoid APTES degradation. Following oxide growth, the sample is annealed at 450°C for 5 minutes, inducing local oxide delamination into nonplanar nanostructures in the form of nanoblisters at sites defined by the SAM pattern. Figures 1b-d present examples of the resulting deterministic arrays of 8 nm-thin oxide blisters with controlled dimensions, characterized through optical microscopy (Figure 1b), scanning electron microscopy (SEM) (Figure 1c), and atomic force microscopy (AFM) (Figure 1d).

FABRICATION MECHANISM

The localized nonplanar transformation of the oxide film is driven by a combination of three effects: reduced oxide-substrate adhesion, compressive thermal stress buildup in the oxide film, and molecular layer degradation to guide film delamination.

Controlled Surface Adhesion

APTES was selected for its intermediary ALD-blocking properties. In addition to APTES, we repeated the experiments with a highly hydrophobic fluorosilane SAM (1H,1H,2H,2H-perfluorodecyltriethoxysilane, PFDTES) and a hydroxylated (OH) surface, shown schematically in Figure 2a. After 80 cycles of ALD on each molecular surface, the resulting Al₂O₃ film thickness is reported in Figure 2b. The hydroxylated surface readily grows the oxide with strong adhesion. The PFDTES blocks ALD, and as previously reported, precursor penetration through the molecular layer is needed to induce nucleation on the substrate to initiate the oxide growth [6], ultimately resulting in a strong adhesion. Being less hydrophobic, APTES displays intermediary ALD blocking performance. We hypothesize that its terminal amine group can facilitate precursor nucleation on the SAM surface. Under such condition, the film/substrate adhesion would be controlled by the molecular layer. This is consistent with our experimental observation that only the APTES-terminated surfaces allow for controlled film delamination.

Thermal Stress Buildup

The ALD film experiences thermal stress at any temperature other than the 200°C growth temperature due to the mismatch in the thermal expansion coefficients between the film and the substrate. Equation 1 [7] gives the

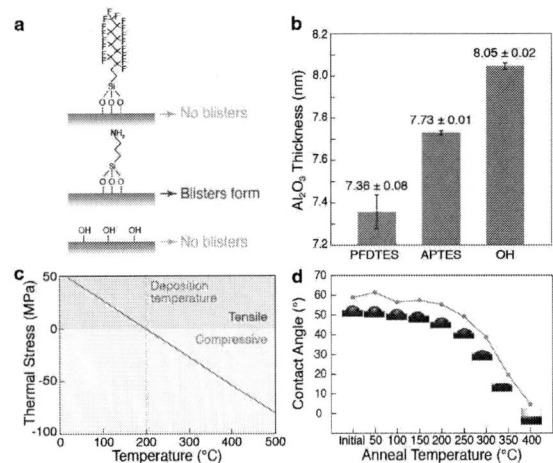

Figure 2: Fabrication mechanism. (a) A PFDTES or OH-terminated substrate does not cause controlled delamination in an 8 nm Al₂O₃ film, but an APTES-terminated substrate does. (b) After 80 ALD cycles of TMA/H₂O, APTES grows more Al₂O₃ than a PFDTES surface and less than an OH surface. (c) Modeling of thermal stress in an Al₂O₃ film grown on a silicon substrate at 200°C, then subjected to annealing. (d) APTES degrades at high temperatures, indicated by its reduced water contact angle with anneal temperature.

thin-film thermal stress,

$$\sigma_{Al_2O_3} = \frac{E_{Al_2O_3}}{1 - \nu_{Al_2O_3}} \left(\alpha_{Si} - \alpha_{Al_2O_3} \right) (T_1 - T_0) \qquad (1)$$

where $\sigma_{Al_2O_3}$ is the film stress, $E_{Al_2O_3} = 170$ GPa is the elastic modulus of thin ALD Al₂O₃, $\nu_{Al_2O_3} = 0.24$ is its Poisson ratio, $\alpha_{Si} = 3.0 \times 10^{-6}$ °C⁻¹ and $\alpha_{Al_2O_3} = 4.2 \times 10^{-6}$ °C⁻¹ are the respective thermal expansion coefficients for Si and Al₂O₃, [8] T_0 is the deposition temperature, and T_1 is the ambient temperature. This model is valid if the film and the substrate have similar elastic moduli, which is the case here since $E_{Si} = 180$ GPa [9].

As plotted in Figure 2c, when heated above the deposition temperature during the anneal step, the film is expected to experience a compressive thermal stress.

Molecule Degradation

Additionally, the APTES SAM, which controls the adhesion between the oxide and the substrate, degrades during the anneal step, assisting local release of the oxide from the substrate. The decomposition of APTES at high temperatures is evidenced by the change in the water contact angle of an APTES-functionalized surface after anneals at increasing temperatures, shown in Figure 2d.

Because the film/substrate adhesion is mediated by the molecular monolayer, we expect that this molecular decomposition provides preferential low-adhesion sites for the ALD film to selectively release and buckle, relaxing the compressive thermal stress incurred during the anneal. The result is delaminated, suspended oxide membranes with dimensions determined by the size and pitch of the patterned molecular sites.

FABRICATION VERSATILITY

The dimensions of the resulting features can be tuned using the size of the surface-functionalized sites and the spacing between them. By altering these parameters, the stress buildup per buckling site and thus the resulting blister diameter is controlled. As evidenced in Figure 3a, for an 8 nm-thin Al_2O_3 film, increasing either the size of or the spacing between the SAM-modified low-adhesion regions yields larger blisters. Here, an approximate linear relation is observed between each of the design parameters and the blister diameter.

Figure 3b plots the relationship between blister diameter and height, where height determines the size of the nanoscale gap formed between the oxide and the substrate. The observed relationship is approximately linear, which is expected for circular buckling of an elastic film under compressive stress [10]. The resulting gap size is ~ 1/10 the nanoblister diameter, as determined by AFM characterization of arrays formed in 8 nm-thin Al_2O_3, with molecular regions patterned by photolithography for microscale features and electron-beam lithography (EBL) for submicron features (Figure 3b, inset). These EBL-patterned samples result in stable, suspended membranes with gaps down to 7 nm demonstrated, dimensions difficult to achieve through conventional top-down fabrication alone.

Our platform is not limited to the Al_2O_3-on-silicon materials system discussed so far, and can be extended to other substrates and deposited thin-films (Figure 3c). We have demonstrated this with diverse substrates including W, Si, SiN, and HfO_2. While in this work we used substrates that are compatible with silane-based molecular self-assembly, different SAM anchoring group chemistries could accommodate alternative substrates. For the oxide film, besides Al_2O_3, we have also demonstrated our fabrication process with ALD HfO_2. We expect that the thin-film can be further extended to other materials, as long as the deposited material forms a continuous film on the SAM surface without causing its degradation, and the resulting film has a larger coefficient of thermal expansion than the substrate to help initiate delamination during annealing. For this, alternative molecular layers that meet the needs of the desired materials system might be necessary.

APPLICATIONS IN NEMS

As the examples provided demonstrate, our proposed technique allows fabrication of ultrathin suspended membranes and nanoscale gaps. These features are foundational building blocks for miniaturized NEM devices, yet are conventionally challenging to fabricate. For practical applications, however, designs beyond nanoblisters are necessary. Further processing after film delamination is one approach to modify the nonplanar nanostructures. To demonstrate an example, we have fabricated ultrathin nanobridge mechanical resonators, shown in Figure 4a. In this process, nanoblisters of the desired size are first fabricated in 8 nm-thin Al_2O_3 as per the process outlined in Figure 1a. Then, using electron-beam lithography, a polymethylmethacrylate (PMMA) resist is patterned on the oxide surface and developed,

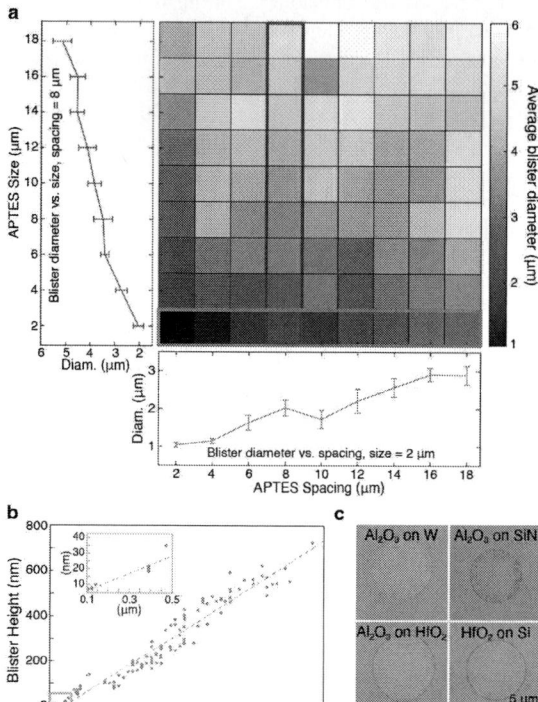

Figure 3: Fabrication versatility. (a) The average blister diameter achieved by a given lithographic pattern of APTES. Blister diameter depends on both the size of and spacing between patterned APTES sites due to increasing buildup of film stress with increasing area. (b) The relationship between blister height and diameter as measured by AFM, down to 7 nm gaps (inset). (c) The technique extends beyond Al_2O_3 films on a Si substrate. We have demonstrated Al_2O_3 features on W, SiN, and HfO_2 substrates, as well as HfO_2 features on a Si substrate, as shown in the optical micrographs.

exposing the sides of individual nanoblisters while masking off a central strip. A reactive-ion etch removes the exposed regions of oxide film. Lastly, the PMMA resist is stripped in acetone, using critical-point drying in isopropanol to avoid structure collapse due to capillary action during solvent drying. The resultant structures are stable, suspended 8 nm-thin double-clamped beams which can be fabricated in large arrays with wafer-scale techniques.

To evaluate the potential performance of such mechanical resonators, we used COMSOL Multiphysics to model the mechanical resonant modes of both the bridge and blister geometries. An ellipsoid shell was used to approximate the delaminated feature geometry, with the oxide modeled as a linear elastic material and fixed at the edges where the suspended feature intersects the substrate. Resonant vibrational modes were found and visualized using an eigenfrequency simulation.

The predicted fundamental frequency of our devices is highly size-dependent, but ranges from the low-MHz for larger features to low-GHz for smaller features. The bridge

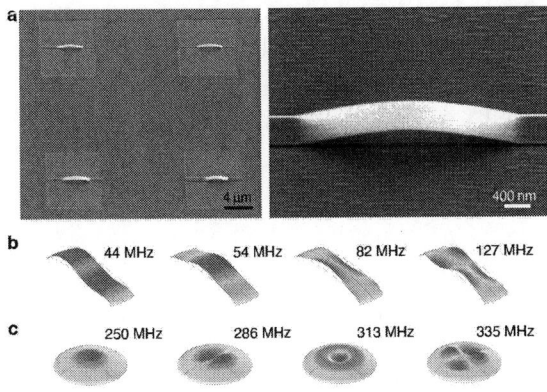

Figure 4: Application in mechanical resonators. (a) To achieve the desired electromechanical performance, the nonplanar structures are stable to undergo further lithography and etching. Tilted SEM images of example fabricated nanobridge mechanical resonators are shown. (b) An example 2 μm × 0.5 μm bridge resonator is simulated in COMSOL to find its first four vibrational resonant modes. (c) The same resonator as in (b) is simulated pre-etch to demonstrate the vibrational modes of the source nanoblister.

structures vibrate at a lower frequency than the equivalent blisters, as can be visualized in the comparison between Figure 4b and 4c. This is due to the decrease in overall spring constant when the structure edges are no longer fixed. We predict that built-in stress will be low in our structures due to the stress-relief-driven buckling mechanism, and therefore the vibrational properties should be largely determined by geometry.

CONCLUSION

We have demonstrated a fabrication technique to achieve nonplanar nanostructures composed of ultrathin suspended membranes (< 10 nm-thin) and nanogaps (< 10 nm), resulting in features that can also be mechanically-active. Core to our approach is bottom-up engineering of surface interactions to deterministically induce spontaneous transformation of planar features into nonplanar designs, providing a new dimension to the widely-used and wafer-scale fabrication steps. The resulting structures can be tuned in size, made of varying materials, and achieve alternative geometries through further processing. Using an example of ultrathin nanobridge mechanical resonators, we highlight the prospects offered by our platform for the development of miniaturized NEM devices. Applications can be further extended to diverse platforms requiring nanogaps or nonplanar nanostructures including in nanoelectronics, photonics and metamaterials.

ACKNOWLEDGEMENTS

S.O.S. and P.F.S. acknowledge support from the NSF Graduate Research Fellowship Program under Grant No. 1745302. S.O.S. further acknowledges support from the Samsung Semiconductor Fellowship and the MIT Alan L. McWhorter Fellowship. Fabrication and characterization were performed in MIT.nano and the Harvard University Center for Nanoscale Systems (CNS). CNS is a member of the National Nanotechnology Coordinated Infrastructure Network (NNCI), which is supported by the National Science Foundation under NSF ECCS Award No. 1541959.

REFERENCES

[1] P. Li *et al.*, "Recent advances in focused ion beam nanofabrication for nanostructures and devices: fundamentals and applications", *Nanoscale*, vol. 13, no. 3, pp. 1529-1565, 2021.

[2] S. Stassi, I. Cooperstein, M. Tortello, C. B. Pirri, S. Magdassi, C. Ricciardi, "Reaching silicon-based NEMS performances with 3D printed nanomechanical resonators", *Nat. Commun.*, vol. 12, no. 1, 6080, 2021.

[3] S. Chen, J. Chen, X. Zhang, Z.-Y. Li, J. Li, "Kirigami/origami: unfolding the new regime of advanced 3D microfabrication/ nanofabrication with 'folding'", *Light Sci. Appl.*, vol. 9, 75, 2020.

[4] B. Vermang *et al.*, "A study of blister formation in ALD Al$_2$O$_3$ grown on silicon", in *38th IEEE Photovoltaic Specialists Conference*, pp. 1135-1138, 2012.

[5] H. Liu, S. Guo, R. B. Yang, C. J. J. Lee, L. Zhang, "Giant blistering of nanometer-thick Al$_2$O$_3$/ZnO films grown by atomic layer deposition: mechanism and potential applications", *ACS Appl. Mater. Interfaces*, vol. 9, no. 31, pp. 26201-26209, 2017.

[6] X. Jiang, S. F. Bent, "Area-selective ALD with soft lithographic methods: Using self-assembled monolayers to direct film deposition", *J. Phys. Chem. C*, vol. 113, no. 41, pp. 17613-17625, 2009.

[7] S.-H. Jen, S. M. George, R. S. McLean, P. F. Carcia, "Alucone interlayers to minimize stress caused by thermal expansion mismatch between Al$_2$O$_3$ films and Teflon substrates", *ACS Appl. Mater. Interfaces*, vol. 5, no. 3, pp. 1165-1173, 2013.

[8] O. M. E. Ylivaara *et al.*, "Aluminum oxide from trimethylaluminum and water by atomic layer deposition: The temperature dependence of residual stress, elastic modulus, hardness and adhesion", *Thin Solid Films*, vol. 552, pp. 124-135, 2014.

[9] M. A. Hopcroft, W. D. Nix, T. W. Kenny, "What is the Young's Modulus of Silicon?", *J. Microelectromechanical Syst.*, vol. 19, no. 2, pp. 229-238, 2010.

[10] M. H. Bitarafan *et al.*, "Thermomechanical characterization of on-chip buckled dome Fabry-Perot microcavities," *JOSA B*, vol. 32, no. 6, pp. 1214-1220, 2015.

CONTACT

*F. Niroui, tel: +1-617-3247415; fniroui@mit.edu

WAFER-LEVEL FABRICATION OF CONFORMAL SUB 10-NM NANOGAPS

Sayali Tope, Seungbeom Noh, and Hanseup Kim

Electrical and Computer Engineering, University of Utah, Salt Lake City, Utah, USA

ABSTRACT

This paper reports successful wafer level fabrication of conformal sub-10-nm lateral nanogaps achieved mainly by finely optimizing the Atomic Layer Deposition (ALD) parameters in a so-called Exposure Mode to precisely deposit a sacrificial layer onto a high-aspect ratio sidewall. It is well-recognized that in a sub-10-nm thickness range, the ALD failed to maintain conformality especially onto a sidewall in a trench because of the relative inefficiency in gas transfer into tiny trenches. To resolve the issue, some ALD parameters were modified including (1) the use of an Exposure Mode where gas flow of pre-cursors was completely stopped in a chamber to let them take extended dosing/purging time to diffuse onto sidewalls (Figure 2-(a)) and (2) the number of cycles. SEM images showed that the optimized process improved the conformality in sub-10-nm ALD-deposited film thickness from 61.83 % to 90.54 % between at the top surface and at the sidewalls, enabling conformal coating of a sub-10-nm sacrificial layer onto a side wall in a trench. The max. deviation was only 0.7 nm in a 6.7 nm thick layer. Such a conformal sidewall ALD subsequently enabled the wafer-level fabrication of a nano gap arrays that would be utilized as a gas sensor. The precisely fabricated gaps successfully formed two isolated conductive electrodes with an initial 'off' resistance of 457 MΩ.

Index Terms—**ALD calibration, lateral nano-gap, sub-10-nm, Exposure Mode, Deep trench coating, conformality, uniformity**

INTRODUCTION

Sub-10-nm gap structures can be utilized as a critical component for various applications such as maximized inductor densities [1], capture of gas molecules [2] etc. However, they are challenging to microfabricate, especially on a wafer-level and for lateral (in-wafer) configurations. Several efforts have been reported, but with some limitations, as summarized in Table1. Electron Beam Lithography (EBL) and Focused Ion Beam Milling (FIB) are the most popular techniques, but they cannot pattern a feature below 20 nm unless with the help of extremely expensive equipment [4]. Methods like Mechanical cracking [6] and Self-Assembly [5] methods do not provide repeatable manufacturing [6], while nano-skiving method is not batch processible [7]. Recently, ALD sacrificial layer method [1] was utilized to show the potential of microfabricating a 5-nm gap between electroplated electrodes. However, it was not clear that the gap distance was maintained over a wafer level considering the known failure of ALD for sub-10-nm deposition. To achieve the wafer-level sub-10-nm uniformity, one possible way would be to provide enhanced period for molecules to cover a trench. In doing so, one can utilize the Exposure Mode, instead of the conventional Normal Mode, to deposit a sub-10 nm thin film utilizing an ALD. The Exposure Mode stops any gas flows in the chamber, thus giving the molecules a longer residence time, while the Normal Mode continuously flows gas molecules, as shown in Figure 1-(a). Typically, during the ALD operation, precursors of Trimethylaluminum (TMA) and Water are sequentially pulsed and quickly purged out of the chamber forming a monolayer in each cycle via thermal process. Thus, the extension of the dwell time between dosing and purging would enhance the adsorption in deep trenches as well. This paper reports the testing results of ALD uniformity by utilizing the Exposure Mode under various purging and dosing periods.

Table 1. Comparison of lateral nanogap fabrication techniques

Technique	ALD Layer Etching		Nano Skiving [7]	EBL/FIB Milling [3,4]	Mechanical Cracking [6]	Self-Assembly [5]
	Exposure Mode [This work]	Normal Mode [1]				
Wafer Level Fabrication	Yes	No	No	No	No	No
Sub 10-nm gap	Yes	Yes	Yes	No	Yes	No
Repeatability	Yes	Yes	Yes	Yes	No	No
Precision/Conformality	Yes	Yes	Yes	Yes	No	No
Area > 100 μm^2	Yes	Yes	Yes	Yes	Yes	Yes
Cost Effective	Yes	Yes	Yes	No	Yes	Yes

FABRICATION PROCESS FLOW

The Figure 1-(a) distinguishes the Normal mode from the Exposure mode. The key advantage of the adopted Exposure mode, in comparison to the Normal mode, is the fact that there is no gas flow in a deposition chamber, which enhances the chance of conformal deposition. Figure 1-(b) illustrates the fabrication process flow that utilized the Exposure mode, the fourth step. In the step 1, the photoresist (PR) was first spun on a glass substrate and developed to pattern the first electrode. On top of the PR, Chromium (Cr) and gold (Au) layers (10/20 nm) were deposited using Electron Beam evaporation. These Cr/Au layers were used as a seed layer for gold electroplating in the step 3. The PR was lifted off leaving the gold seed layer for electroplating (step 2). Gold electroplating was performed by utilizing a 10-mA current and resulting a deposition rate of 5 nm/min (step 3). The step 4 was the modified Exposure Mode ALD deposition for a 5.42-nm thick Al_2O_3 layer. This ALD layer was the sacrificial layer which was later etched to form a conformal air gap of precise thickness between the two electrodes. Then the second gold electrode was patterned through the liftoff process and electroplating (steps 5-7). The electroplating was performed to raise the second electrode to a desired height. All these processes were performed on a wafer level and the process success was verified after each step. Lastly, the Al_2O_3 was etched using AZ MIF 300 developer which has 2.38% Tetramethylammonium hydroxide (TMAH). The TMAH provides the selective etching of Al_2O_3 with an etch rate of 2 nm/min. As the post-fabrication verification, the electrical isolation,

between the two electrodes that were separated by only 7.93 nm, was measured by an electrical impedance measurement on equipment (4200A-SCS, Keithley). The impedance of the nanogap sensor device was found to be 457 MΩ.

ALD CALIBRATION METHODOLOGY

To calibrate the Exposure mode ALD parameters, first, the conformality was defined as the ratio of the deposited film thicknesses between at the side wall of a trench with a 5x5-mm² opening at the 600-nm depth and at the top surface as expressed in the following equation.

$$Conformality = \frac{Trench\ thickness}{Surface\ thickness} \times 100$$

Secondly, two parameters of the Exposure mode ALD operation were varied: dosing time and purging time. The dosing time is the wait time when the stop valve remained closed after precursors were pulsed into the chamber. The purging time is the wait time after the stop valve is opened and purge gas flows through the chamber before the next ALD cycle starts. The dosing time directly dictate the amount of precursors diffused in the deep trenches and thus the adsorption on the sidewalls. The insufficient dosing time can cause the discontinuous ALD deposition at the bottom of the trench. The purging time is responsible for the uniformity of the deposited layer because it removes excess deposition by CVD. This allows ALD monolayers of uniform and precise thickness to be formed in each cycle. Initial dosing and purging wait times were selected from the Cambridge NanoTech Fiji F200 ALD Systems manual [9]. As per the manufacturer standard recipe, the reaction temperature was set at 175°C for a thermal ALD process to deposit Al_2O_3. The Trimethylaluminum (TMA) pulse time was set at 0.4 sec, and water pulse time was set at 0.12 sec. The standard dosing time as per manual was 60 sec and standard purging time was 90 sec. The ALD recipe included these conditioning processes in sequence which were executed for the specified amount of time. The recipe modification was carried out on the interface software. The dosing and purging wait times were gradually increased from 60 to 180 sec in the ALD recipe while monitoring the resultant conformality in film thicknesses during Al_2O_3 deposition. The dosing and purging times were decided considering a general trend of deeper trenches needing longer times for the precursor to diffuse on all the surfaces. Thirdly, the conformality of the deposited ALD films was measured by taking an SEM image on the cross-section of a device achieved by the Dual Beam Focused Ion Beam Microscope (DBFIB).

Dosing Time Calibration

In the Exposure mode, the dosing time was varied from 60 to 90 sec by ALD recipe modification, and the film conformality was measured. Note that this dwell time allows pressure to build up in the deposition chamber and thus the precursor to diffuse into the deep trenches. Thus, it was believed that the increase in the dosing time

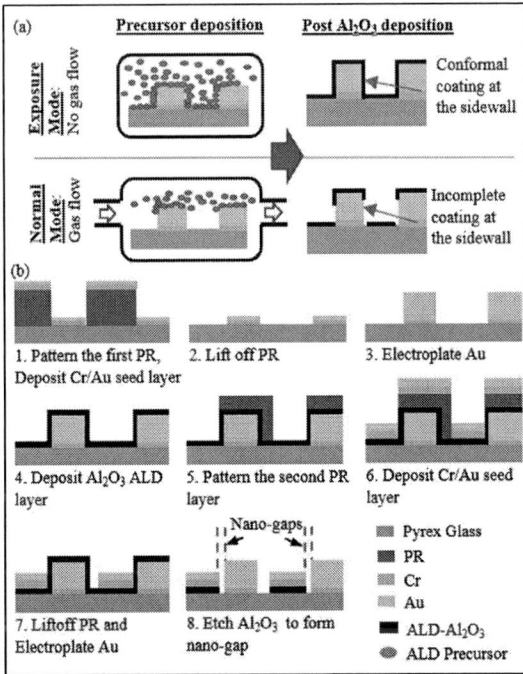

Figure 1. (a)ALD Exposure Mode mechanism for improved conformality (b)Cross section view of Lateral Nano-gap fabrication steps.

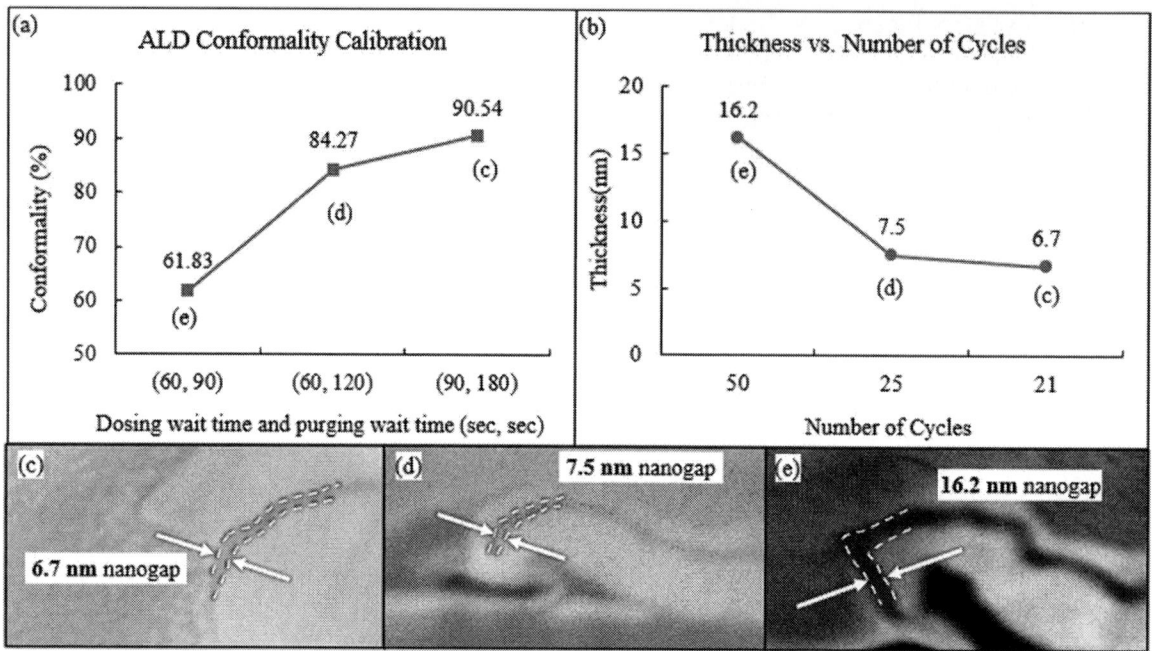

Figure 2. (a) Conformality trend as dosing and purging wait times are increased. Best value of 90.54% is observed at longest wait times. (b) Thickness variation with number of cycles. SEM of sample with measured conformality of (c) 90.54% (d) 84.27% and (e) 61.83%.

would result in higher conformality and better sidewall coverage in deep trenches [10, 11]. The initial dosing dwell time was selected to be 60 sec following the standard manufacturer's protocol.

Purging Time Calibration

The purging time was set to 90, 120 and 180 sec in the ALD recipe. Note that purge time needs to be sufficiently

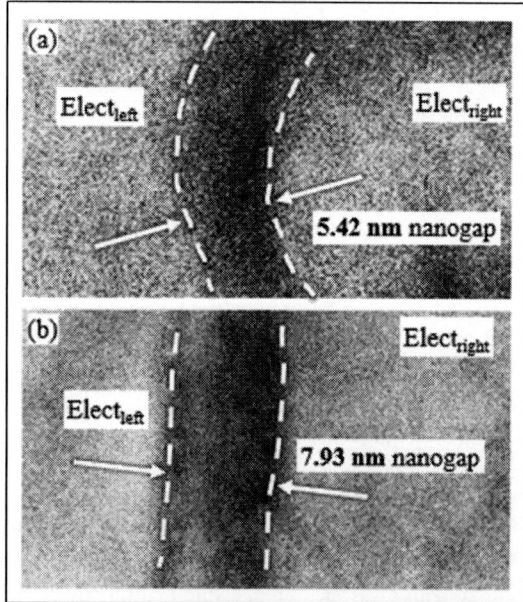

Figure 3. Cross section TEM of Lateral Nano-gap structure. (a) for 5.42nm gap size with 17 cycles and (b) for 7.93nm gap with 21 cycles.

longer to purge out all the byproducts of the reaction and excess precursor [8].

MEASUREMENT RESULTS
Dosing Time Calibration Results

The increase in dosing time from 60 to 90 sec enhanced the sub-10-nm film conformality from 84.27 % (sidewall deposition thickness of 7.5 nm and surface deposition thickness of 8.9 nm) to 90.54% (sidewall deposition thickness of 6.7 nm and surface deposition thickness of 7.4 nm). The thickness variation was mere 9.4% (0.7 nm variation for 7.4 nm thickness). For this work the dosing time of 60 sec was sufficient for the diffusion in 600-nm deep trenches and increasing it to 90 sec did not show any significant effect on the conformality. Figure 2-(c) shows the SEM image of a cross section of a 6.7-nm gap with 90 sec dosing time, and Figure 2-(d, e) shows the SEM images with 60 sec dosing time.

Purging Time Calibration Results

The increasing purging dwell time of 90, 120 and 180 sec resulted in the enhancing conformality of 61.83, 84.27 and 90.54 %, respectively, as shown in Figure 2-(c-e). Figure 2-(a) demonstrates the plot of conformality with respect to dosing and purging time variation. As purging time was increased from 90 to 180 sec, the conformality increased from 61.83 % to 90.54 %. Figure 2-(b) demonstrates that deposition thickness reduced as number of cycles were reduced. For all ALD runs, a constant deposition rate of 0.3 nm/sec was observed. It was believed that the increase in the purging dwell time removed all excess chemical vapors from the chamber and forming ALD layers with better uniformity. The increase in the purging time reduced the thickness variation from 10 nm to 0.7 nm resulting in a significant improvement of 93 % in the uniformity. Figure 3-(a) and -(b) illustrate the TEM images of a conformal 5.42-nm thickness obtained from 17 cycles and a 7.93-nm nanogap from 21 cycles.

CONCLUSION

This paper reported some testing results of the use of the Exposure mode ALD process for conformal deposition of sub-10-nm sacrificial layers. To ensure the conformality, during the dosing of precursor step, (1) the deposition chamber was isolated from the pump system (Exposure Mode), and (2) dosing and purging wait times were optimized. Compared to the conformality of nearly 0 % in the Normal mode, the Exposure mode successfully achieved that of 90.54 %.

ACKNOWLEDGEMENTS

This research work was generously supported by the cooperative agreement of DE-AR0001064 of the ARPAE OPEN 2018 program (Program Manager: Dr. David Babson). Microfabrication was performed at the state-of-the-art Utah Nanofabrication Facility in the University of Utah.

REFERENCES

[1]. Liu C, Schauff J, Lee S, Cho JH. Fabrication of Nanopillar-Based Split Ring Resonators for Displacement Current Mediated Resonances in Terahertz Metamaterials. J Vis Exp. 2017.

[2]. S. -u. H. Khan et al., "Characterization of a Wake-Up Nano-Gap Gas Sensor for Ultra Low Power Operation," in Journal of Microelectromechanical Systems, vol. 31, no. 5, pp. 791-801, Oct. 2022

[3]. Hammond, J.L.; Rosamond, M.C.; Sivaraya, S.; Marken, F.; Estrela, P. Fabrication of a Horizontal and a Vertical Large Surface Area Nanogap Electrochemical Sensor. *Sensors* 2016, *16*, 2128.

[4]. T Blom, K Welch, M Strømme, E Coronel and K Leifer "Fabrication and characterization of highly reproducible, high resistance nanogaps made by focused ion beam milling" 2007 Nanotechnology 18 285301

[5]. F. Niroui, M. Saravanapavanantham, T. M. Swager, J. H. Lang and V. Bulović, "Fabrication of nanoscale structures with nanometer resolution and surface uniformity," 2017 IEEE 30th International Conference on Micro Electromechanical Systems (MEMS), 2017, pp. 659-662.

[6]. Dubois V, Niklaus F, Stemme G. Design, and fabrication of crack-junctions. Microsyst Nanoeng. 2017 Oct 23; 3:17042.

[7]. Zhou, Z., Zhao, Z., Yu, Y., Ai, B., Möhwald, H., Chiechi, R. C., Yang, J. K. W., & Zhang, G. (2016). From 1D to 3D: Tunable Sub-10 nm Gaps in Large

Area Devices. Advanced materials, 28(15), 2956-2963.

[8]. M.F.J. Vos, A.J.M. Mackus, W.M.M. Kessels. Atomic Layer Deposition Process Development – 10 steps to successfully develop, optimize and characterize ALD recipes. 2019, 3. AtomicLimits

[9]. Cambridge NanoTech Fiji F200, 200mmThermal and Plasma ALD Systems, Installation and Use Manual CAW-02635 Rev. 0.6 13 March 2012

[10]. K. Arts, W.M.M. Kessels and H.C.M Knoops. Basic insights into ALD conformality – A closer look at ALD and thin film conformality. 2020, 1. AtomicLimits

[11]. Elam, J. W.; Routkevitch, D.; Mardilovich, P. P.; George, S. M. Conformal Coating on Ultrahigh-Aspect-Ratio Nanopores of Anodic Alumina by Atomic Layer Deposition. Chem. Mater. 2003, 15 (18), 3507–3517.

CONTACT

*H. Kim, +1-801-5879497; hanseup.kim@utah.edu

MEMS RESONATOR VACUUM-SEALED BY SILICON MIGRATION AND HYDROGEN OUTDIFFUSION

Muhammad Jehanzeb Khan[1], Yukio Suzuki[1, 2], Tianjiao Gong[1], TakashiroTsukamoto[1]
and Shuji Tanaka[1, 2]
[1]Department of Robotics, Tohoku University, JAPAN and
[2]Microsystem Integrated Center (μSIC), Tohoku University, JAPAN

ABSTRACT

MEMS resonators were vacuum-encapsulated by Silicon Migration Seal (SMS) technology. SMS is new wafer-level vacuum packaging technology, which utilizes silicon reflow phenomena to close release holes in hydrogen (H_2) environment at high temperature (>1000°C). In this study, we first demonstrated the encapsulation of a MEMS resonator made on an SOI wafer, which is one of the most standard structures for inertial sensors and timing devices. After the encapsulation, hydrogen trapped in the sealed cavity was diffused out by annealing at 430°C in nitrogen (N_2) environment for 27 hours. The resonator was capacitively driven and sensed, and the Q factor reached 6000. The sample after successful packaging was penetrated by focused ion beam (FIB) out of the resonating element area. Judging from the Q factor, the vacuum level of the sealed cavity is much better than that of the hydrogen annealing (10 kPa) and estimated ~60 Pa.

KEYWORDS

Resonator MEMS, Wafer-level Vacuum package, Silicon Migration Sealing (SMS), Q factor

INTRODUCTION

Wafer-level vacuum sealed packaging is critical process for micro-electro-mechanical systems (MEMS). The wafer-level process is advantageous for high yield, smaller device size and lower cost [1][2]. Resonant MEMS require high vacuum sealing to minimize air damping loss and achieve high Q factor. Epi-Seal technology is established for wafer-level packaging technology to realize stable vacuum sealing of single Pa or better [3]. In this technology, the release holes are closed by epi-poly-Si deposition. However, there are some problems such as complicated film stress control and inevitable regular cleaning of the deposition chamber [4]. Silicon Migration Sealing (SMS) technology [5] is expected to enable high vacuum packaging at wafer-level, and sealing is possible in a batch hydrogen annealing furnace, which does not need periodic cleaning. This idea was proposed by Howe *et al.* [6], but the demonstration for MEMS packaging was not reported. Our previous work confirmed the sealing vacuum level as low as 10 Pa, but there were no MEMS inside [7]. In this study, we successfully encapsulated MEMS resonators by SMS technology.

BACKGROUND
DEVICE DESIGN

A dual mass symmetric resonator with 4 anchors was designed. The tuning fork design can minimize anchor loss, which is necessary to measure the sealing pressure by squeezed film damping effect. The resonator is driven and

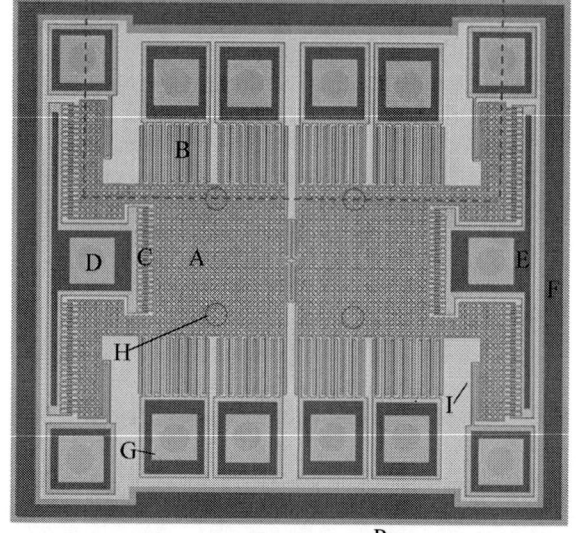

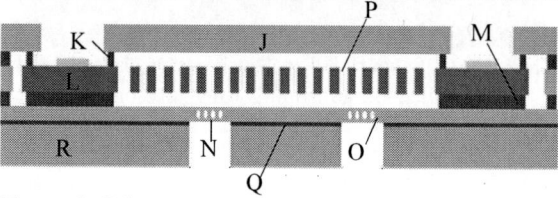

Figure 1: Schematic structure of Dual mass encapsulated resonator, (A) Dual mass resonator, (B) Sensing Comb fingers (w = 15 μm, l = 315 μm), (C) Drive comb fingers (w = 10 μm, l = 60 μm), (D) Metal electrode (Cr/Au: ~ 20 nm/250 nm, pattern size 130 μm diameter), (E) Bonding SiO_2 ring width (100 μm), (F) external sealing region width (150-200 μm), (G) Pad area for wire bonding (180×180 μm), (H) Cavity (100 μm diameter), (I) Flexural beam (w = 15 μm, l = 300 μm), (J) Device wafer handle layer (thickness (t) = 50 μm),(K) Device wafer BOX layer (t = 5 μm), (L) Device layer (t = 50 μm), (M) Bonding SiO_2 (t = 1.5 μm),(N) Vent holes after sealing (single vent hole area = 0.5×0.5 μm) and number of vent holes for each cavity =48, (O) CAP wafer device layer (t = 5 μm), (Q) CAP wafer BOX layer (t = 0.5 μm), (R) CAP wafer handle layer (t = 475 μm).

sensed using standard comb electrodes. The design details are presented in Figure 1. Gap between the driving and sensing combs is 5 μm, considering balance between sensitivity and fabrication yield. Release holes in the resonator are designed to be 20 × 20 μm. Full pad area was more than 350 × 350 μm to keep enough space for wire bonding after packaging. The total size for the packaged device is 3.31 mm square.

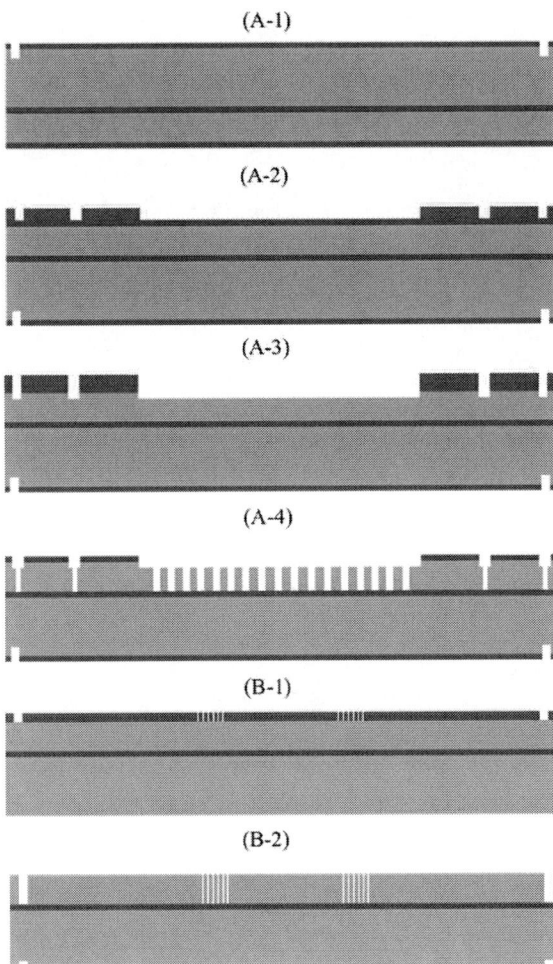

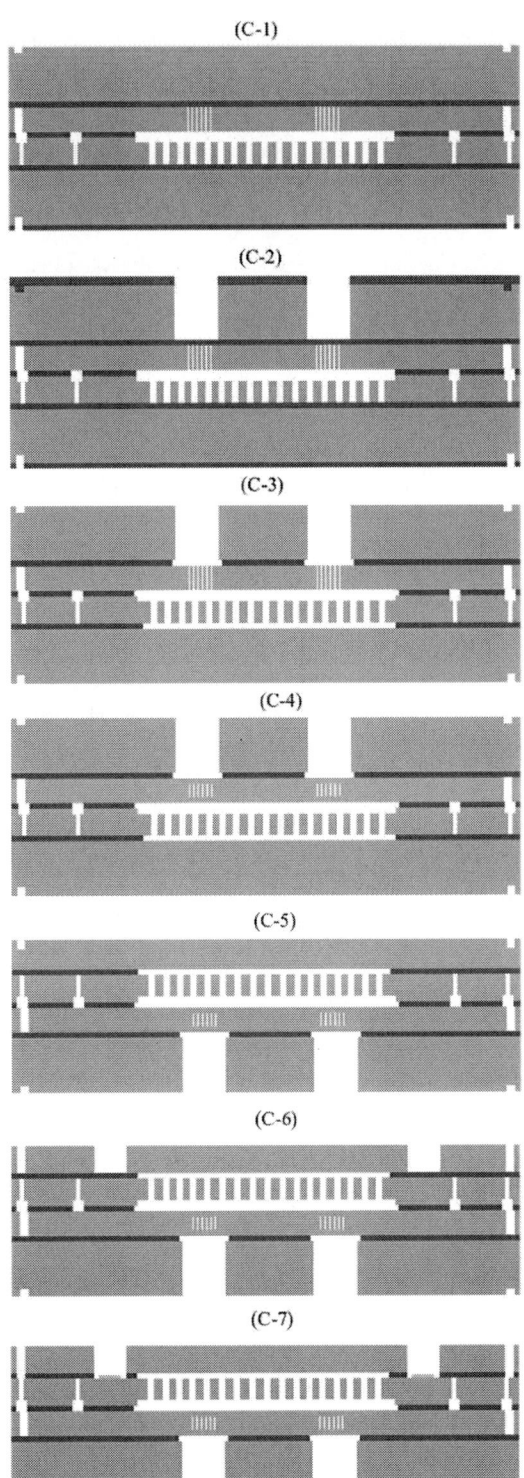

Figure 2: Fabrication process, (A-1) Device wafer after bonding SiO₂ (1.5 μm) growth by thermal oxidation and fabrication of reference backside alignment mark, (A-2) Photolithography for SiO₂ patterning and thinning (3.5 μm) silicon device layer by DRIE, (A-3) Photolithography for resonator pattern, DRIE and photoresist removal, (B-1) Cap wafer photolithography for submicron vent holes, (B-2) through vent holes DRIE and photoresist removal.

FABRICATION PROCESS

The starting materials are two SOI wafers. One is the device wafer, where the thicknesses of the device layer, BOX layer and handle layer are 50 μm, 5 μm and 490 μm, respectively. The other is the cap wafer, where the thicknesses of the device layer, BOX layer and handle layer is 5 μm, 0.5 μm and 475 μm, respectively. First, bonding SiO₂ of 1.5 μm thickness is grown by thermal oxidation, and alignment marks on the handle layer are patterned by photolithography and reactive ion etching (RIE) as shown in Fig. 2 A-1. In the next step, the thermal SiO₂ is pattered by photolithography (A-2) and RIE, and sequentially 3 μm thickness of silicon is etched down to make the gap between the cap and the resonator 5 μm (A-3). The resonator pattern is fabricated on the device layer by photolithography and deep reactive ion etching (DRIE) (A-4).

Figure 3: Fabrication process (continued), (C-1) Device and cap wafer after flip bonding at room temperature, (C-2) Photolithography for circular cavities to access vent holes, (C-3) Vapor HF (VHF) for the release of resonating element, (C-4) H₂ annealing at 1100°C to seal vent holes, (C-5) Device wafer handle layer thinning to 50 μm, (C-6) Photolithography and DRIE for pads, (C-7) VHF to remove BOX layer and metallization by stencil mask.

978-1-6654-9309-3/23 $31.00 © 2023 IEEE

In parallel, the submicron release holes are defined on the cap SOI wafer using an i-line stepper (B-1) and the device layer is penetrated by DRIE (B-2).

After fabricating the device and cap wafer separately, the next step was direct wafer bonding (C-1). Before bonding process, the wafer surfaces are cleaned to avoid bonding voids. RCA 1 and RCA 2 cleaning are deployed to remove organic and metallic contamination. The plasma surface activation and mega-sonic cleaning are applied for both wafers. The access paths for vent holes are made by photolithography and 475 μm deep DRIE (C-2). RCA1 and RCA 2 cleaning are done once again after photoresist removal. The sacrificial oxide layer of the resonator and the bonding interface oxide are etched by vapor HF (vHF) etching through the submicron release holes (C-3). The target SiO_2 undercut is 12 μm, such that resonating structure inside the cap is released. The high chemical bonding strength is crucial for controlled vapor HF etching. Then, the release holes are closed by SMS in pure hydrogen at 1100°C for 20 min (C-4). SMS completely cleans the silicon surface and removes any degas source by hydrogen reducing ambient.

The handle layer of the device wafer is thinned down by plasma dry etch from 490 μm to 50 μm in thickness (C-5). After that, DRIE followed by photolithography is carried out for pad opening (C-6). The BOX layer is removed by 10 cycles of vHF etching. After that heating process at 430°C for 27 h in N_2 atmosphere is done to diffuse out the residual hydrogen gas. Finally, Cr/Au is deposited on the bottom of the opened pad areas to create wire bonding pads (C-7).

EXPERIMENTAL RESULTS

Figure 4 shows the SEM images of the completed resonator chip with wire bondings. Figure 5 shows the magnified images of the submicron release holes before and after SMS. The encapsulated resonator was capacitively driven and sensed using a front-end circuit and a lock-in amplifier. The room temperature was kept 25°C. The input drive voltage was 0.3 V_{pp} over an offset voltage of 1.2 V. A modulation signal of 1 MHz and 4 V_{pp} was applied to the common electrode for capacitive sensing. Figure 6 shows the measured frequency characteristic.

The anti-phase mode was observed at 35.527 kHz and the measured Q factor was 6000. The other samples were also tested, and Q factors of 13000 and 19000 were measured.

After successfully measuring the Q factor without a vacuum chamber, a through-hole of several tenth micron diameter was created in the cap by focus ion beam (FIB). The sample with higher Q factors unfortunately showed charge up and failure during the FIB penetration process. Charge-up depended on the ion beam current. Higher ion current, faster the FIB penetration was possible, but the device showed stiction failure by charge-up. With low ion current, the FIB penetration was slower, but device showed no stiction failure.

The FIB penetrated sample was measured in the vacuum chamber by changing pressure. Figure 7 shows the relationship between the Q factor and the chamber pressure. From this relationship, the sample with a Q factor of 6000

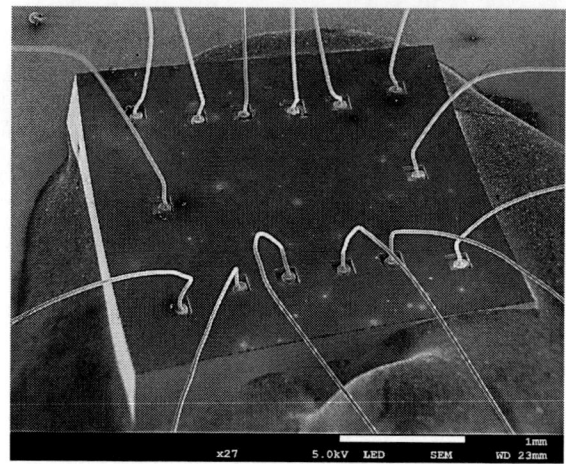

Figure 4: Completed MEMS resonator die after wire bonding.

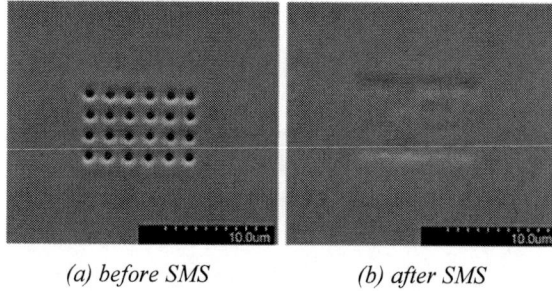

(a) before SMS   (b) after SMS

Figure 5: Submicron release holes on cap wafer.

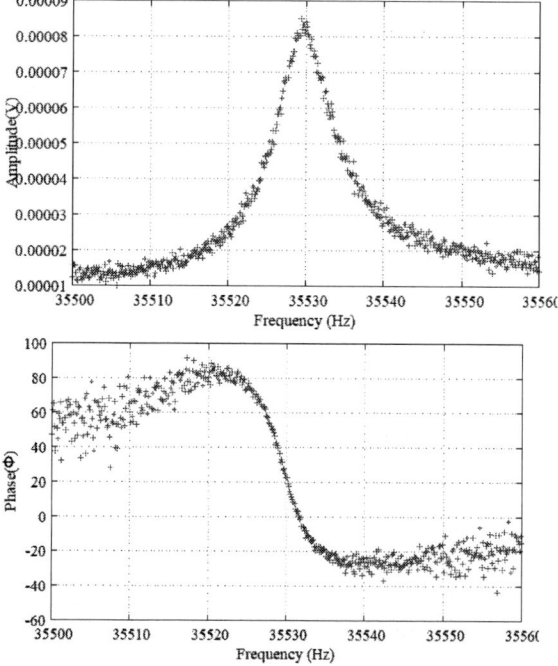

Figure 6: Frequency characteristic of packaged MEMS resonator, (a) Amplitude vs. frequency, (b) Phase vs. frequency.

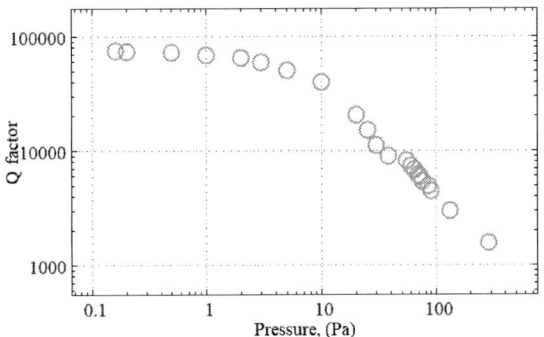

Figure 7: Relationship between Q factor and chamber pressure after FIB drilling.

was encapsulated at about 60 Pa, which was much better than the pressure of hydrogen annealing (10 kPa), i.e. sealing environment. This result successfully confirmed that the SMS technology worked for a typical MEMS resonator on the SOI wafer. Lower sealing pressure may be achievable by optimizing the hydrogen diffusion after SMS.

CONCLUSION

SMS was applied to the wafer level packaging of MEMS resonators fabricated on a SOI wafer. The dual mass resonators were successfully encapsulated and the Q factors were measured. The sealing pressure was estimated by measuring the Q factor of the same sample in a vacuum chamber after leaking the package by FIB drilling. Based on this method, the sealing pressure of a sample with a Q factor of 6000 was estimated at about 60 Pa. We obtained samples with higher Q factors, although the sealing pressure could not be confirmed due to the failure of FIB. A higher vacuum level of the sealed cavity may be obtained by optimizing the process conditions.

ACKNOWLEDGEMENTS

This presentation is based on results obtained from a project, JPNP19005, subsidized by the New Energy and Industrial Technology Development Organization (NEDO). A part of this work was supported by Advanced Research Infrastructure for Materials and Nanotechnology in Japan (ARIM) of the Ministry of Education, Culture, Sports, Science and Technology (MEXT), JPMXP12yyxx1234. The authors would like to acknowledge support by Institute of Materials Research, Tohoku University for FIB.

REFERENCES

[1] K. Najafi, "Micropackaging technologies for integrated microsystems: Applications to MEMS and MOEMS," Proc. SPIE, Vol. 4979 (2003) pp. 1–19.

[2] J. R. Vig, "Resonator aging," in Proc. Ultrason. Symp. 1977, pp. 848–850.

[3] R. Candler, M. Hopcroft, B. Kim, W. T. Park, R. Melamud, M. Agarawal, G. Yama, A. Partridge, M. Lutz, T. Kenny, "Long-Term and Accelerated Life Testing of a Novel Single-Wafer Vacuum Encapsulation for MEMS Resonators," J. Microelectromech. Syst., vol. 15, no. 6, pp. 1446–1456 (2006).

[4] Y. Suzuki, K. Totsu, H. Watanabe, M. Moriyama, M. Esashi, S. Tanaka, "Low-Stress Epitaxial Polysilicon Process for Micromirror Devices", IEEJ Trans. SM, Vol. 133, No.6 (2013) pp. 223–228.

[5] Y. Suzuki, V. Dupuit, T. Kojima, Y. Kanamori, S. Tanaka, "Silicon migration seal wafer-level vacuum encapsulation", Elect. Comm. Jpn., Vol. 104, No. 1 (2021) pp. 120–125.

[6] Kant Rishi, Roger Thomas Howe, "Deposition-free sealing for micro-and nano-fabrication," U.S. Patent 8,735,286, May 27, 2014.

[7] H. Suzuki, Y. Suzuki, Y. Kanamori, S. Tanaka, "Improved Vacuum Level of Silicon-Migration-Sealed Cavity by Hydrogen Diffusion Annealing for Wafer-Level Packaging for MEMS," 35th IEEE MEMS 2022, pp. 565–568.

CONTACT

*Yukio Suzuki
suzuki@tohoku.ac.jp, +81-22-795-6936

MEMS THIN-FILM VACUUM PACKAGE UTILIZING GLOW DISCHARGE GETTER

Vikram Maharshi[1], Manjeet Kumar[1], Ajay Agarwal[2] and Bhaskar Mitra[1]
[1]Indian Institute of Technology, Delhi, INDIA and
[2]Indian Institute of Technology, Jodhpur, INDIA.

ABSTRACT

This work demonstrates a thin-film encapsulated MEMS package integrated with a low-temperature plasma getter on the package lid. The thin-film encapsulated package was fabricated using anodized porous alumina as a capping layer and PECVD silicon dioxide as the sacrificial layer. A Silicon Pirani gauge was used to monitor the pressure changes inside the sealed cavity. The two titanium-gold film electrodes were separated by ~300 μm and placed above the capping layer. The package was sealed at 50 μTorr pressure using evaporated aluminum oxide. A glow discharge is formed in the cavity when a voltage >800V is applied at room temperature. The sputtered titanium from the discharge reacts with available reactive gas molecules and reduces the vacuum of the micro package. After the glow discharge getter activation, a decrease in the pressure of 50 μTorr from 2 μTorr was observed for the volume of $1.65 \times 10^{-6} cm^3$. The pressure in the micro package might be below the reported one since the limiting sensitivity of the MEMS Pirani gauge. The use of a glow discharge getter integrated with the package saves device space as well as avoids the use of high voltages near the device. EDX results indicate that the sputtered material is not deposited underneath the wafer.

KEYWORDS

MEMS, alumina, porous, pirani, getter, hermiticity.

INTRODUCTION

There is a need to maintain or generate the vacuum level inside the micro packages to avoid the degradation of the performance of the MEMS device. The vacuum of the micro package was ensured through the sealing process of the package. The sealing of the micro packages was performed using wafer capping by wafer-wafer bonding process and capping by thin-film. The outgassing and microleakages in the sealing process deteriorate the vacuum level of the micro package. The active vacuum generation inside the thin-film encapsulated package was required to prevent the device's performance deterioration. The thick or thin film NEG getter material is placed inside the microcavity to maintain the pressure level of the package. The activation of the thin film getter at higher temperatures leads to the degradation of several metals and polymer layers in the package. The glow discharge-based getter MEMS micropumps have been reported to generate a high vacuum level over a larger volume [1-8]. Getters have not been incorporated in thin-film packages; however, the large surface makes it compelling to place getters in the capping layer. Additionally, an active getter ion-sorption pump is likely to give additional control.

This work illustrates the integration of a low-temperature plasma getter on the package lid with a thin-film encapsulated MEMS package. The anodized porous alumina was used to provide the capping layer for the thin-film encapsulated package, while PECVD silicon dioxide was used as a sacrificial layer. Sputtered titanium from the discharge decreases the vacuum in the micro package by reacting with nearby reactive gas molecules. The variations in pressure inside the hermetically sealed package were tracked using characterizing a Silicon Pirani gauge.

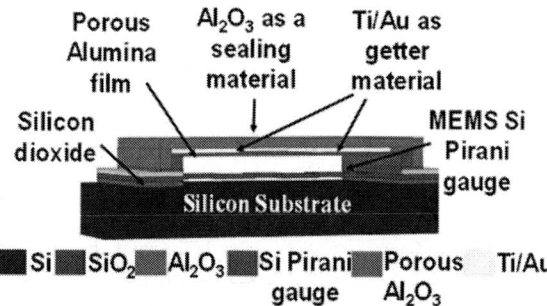

Figure 1: Cross-sectional schematic of Thin film Encapsulated MEMS vacuum Package utilizing glow discharge getter.

Figure 1 shows the package structure. The porous anodized alumina was used as a capping layer, and two Ti/Au metal electrodes were separated within 300 μm above the capping layer. The silicon MEMS Pirani was used to measure the pressure changes in a microcavity.

FABRICATION

The package was fabricated using three mask processes (Fig.2). First, the Pirani gauge was patterned and etched on an SOI wafer (10 μm device layer and 1 μm buried oxide layer) using a positive Photoresist (S1813) as masking and SF6 plasma for etching. Then the PECVD silicon dioxide (700 nm) was deposited conformally as a sacrificial layer of the package and patterned using buffered HF, keeping positive photoresist as the masking layer. The 500 nm thick aluminum was then deposited using sputtering. The aluminum's electrochemical anodization [9-10] was performed in oxalic acid with 20 V external applied voltage at 40°C applied temperature. The constant magnetic stirring was performed during the anodization process. The barrier layer of the anodized alumina layer was etched using phosphoric acid (5% wt). The sacrificial layer was etched through the nanopores using buffered HF for 20 minutes. The buried layer of SOI wafer (SiO_2) was also etched through the nanopores using a buffered hydro fluoride solution. The Ti/Au (20/200 nm) was deposited as metal electrodes using a shadow mask on the porous alumina layer. The micro package was vacuum sealed at 50 µtorr using evaporated aluminum oxide material (500 nm) to ensure the quality sealing of the package.

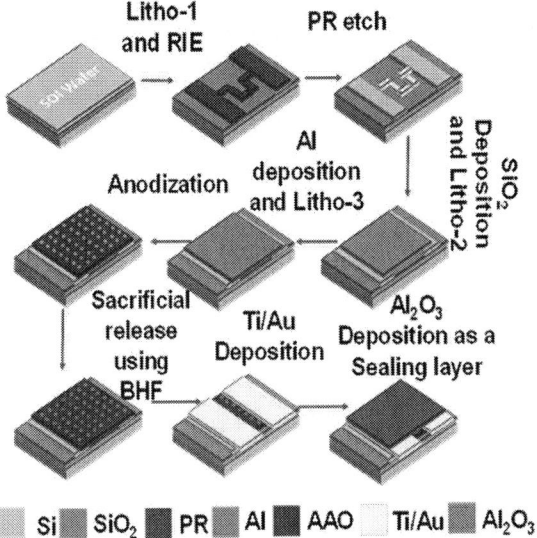

Figure 2: Cross-sectional schematic of Thin film Encapsulated MEMS vacuum Package utilizing glow discharge getter.

RESULTS AND DISCUSSION

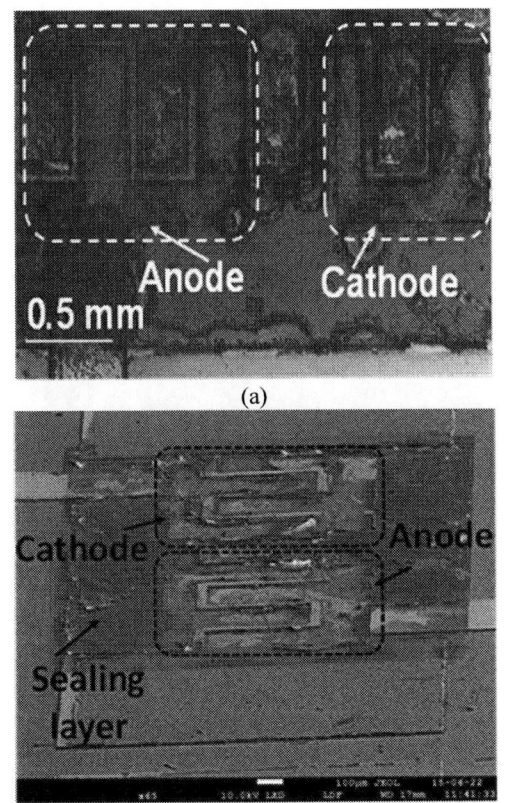

Figure 3: (a), (b) Optical and FESEM image of thin-film package, where sealing layer, pirani structure, and two metal electrodes can be seen.

Figure 4: (a) anodized porous alumina film, inset shows the magnified appearance of pores in alumina film, (b) cross-sectional view of anodized porous alumina film; inset shows the depth of the pore size. (c) FESEM image of a cross-sectional view of a ruptured micro package to ensure the etching of a sacrificial layer.

Figures 3a and b show the optical and FESEM images of the top of the fabricated device, which shows a thin film-encapsulated vacuum package consisting of two metal electrodes, the underlying Pirani, and the sealing layer be seen. The metal electrodes separated with the distance (300 µm) can be seen. Figure 4a, b shows a FESEM image of the anodized porous alumina after removing the barrier layer, showing a 50nm pore diameter with a significant amount of density and 500 nm pore depth; the inset shows the magnified appearance of the nanopores. The cross-sectional image confirms the complete anodization of the 500 nm thick aluminum layer. Figure 4c shows a cross-section of the microcavity, clearly showing a cavity being formed due to the removal of the sacrificial layer (silicon dioxide) through the nanopores of the anodized alumina membrane.

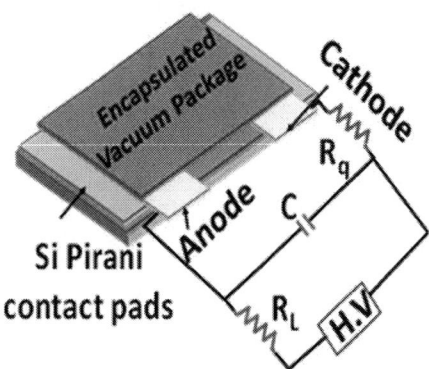

(a)

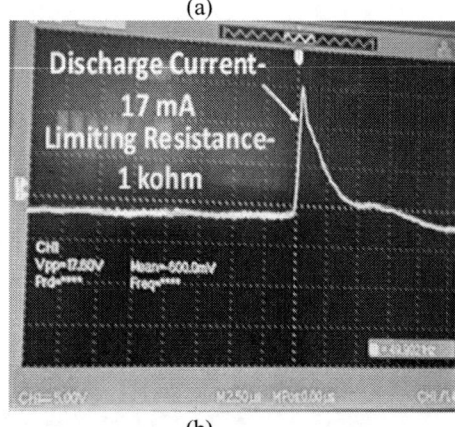

(b)

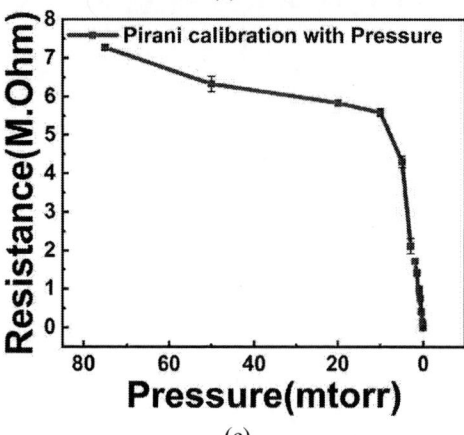

(c)

Figure 5: (a) Schematic of a thin-film package with the electrical circuit consisting 1kohm limiting resistor and the 11nF capacitor, (b) Discharge Current profile for discharge glow, (c) The calibration of Pirani gauge to different pressures using a cryogenic dc probe station (Lakeshore Cryotronics Probe Station), and the inset shows the specification of a MEMS package.

A high voltage (>800 V) between two metal electrodes was applied using the limiting resistor (1 Kohm) and discharge capacitor (11 nF). The schematic of a thin-film package with the electrical circuit is shown in Fig.5a. The ionized gas ion sputters the titanium, which adsorbs the gases that lead to reduced pressure inside the package.

Table. 1 MEMS Package Specifications.

Parameters	Design
L	2200 μm
W	1500 μm
Gap between electrodes	300 μm
Sealing Pressure	50 μTorr
Limiting resistance	1 kOhm
Discharge Capacitor	11 nF
Applied Voltage	>800 V

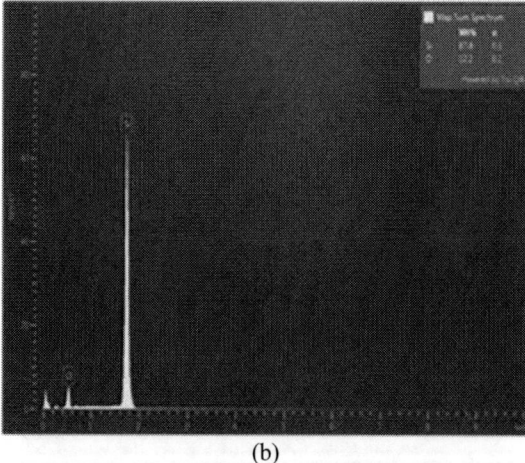

(a)

(b)

Figure 6: (a), (b) EDX mapping of the ruptured package below the cathode electrode after the glow discharge gettering process. The absence of a titanium peak in the mapping confirms that sputtered metal is not deposited underneath the device.

Fig.5b shows the current discharge profile at applied 800 voltage between two metal electrodes. The discharge comprises multiple such pulses. The pirani measurement done before the getter activation was 50.41 μtorr and 2.1

μtorr after. The pirani was calibrated separately (without package) using a vacuum dc probe station (Lakeshore Cryotronics Probe Station), as shown in Fig.5c. It is notable that the pirani gauge does not show high sensitivity at this pressure range, so the actual pressure is likely to be lower. The specification of the MEMS vacuum package and the parameters of electrical measurement is shown in Tab. 1. Fig.6a and b show the EDX mapping of the ruptured cavity below the cathode electrode, which does not show a peak for titanium. This indicates that the metal is not sputtered onto the device underneath but is likely redeposited on the porous alumina.

CONCLUSION

This work demonstrated a thin-film-based, hermetically sealed vacuum package for MEMS and microsensors. The novel aspect of the study is the integration of a glow discharge getter into a thin film-encapsulated vacuum package. The glow discharge getter integration enables vacuum reduction and stabilization inside the sealed micro package. The vacuum package was fabricated using anodization and surface micromachining techniques. The variations in pressure inside the hermetically sealed package were monitored using a Silicon Pirani gauge. The glow discharge forms when a voltage >800V is applied at two metal electrodes. The pressure of the sealed cavity was observed at 2 μTorr from 50 μTorr (sealed pressure) for a volume of 1.65×10^{-6} cm^3.

ACKNOWLEDGEMENTS

The authors acknowledge IIT Delhi's central research facility (CRF) and nano research facility (NRF) for their facilities. Vikram and Manjeet thanks the Ministry of Human Resources Development (MHRD) for providing financial support for his doctoral studies.

REFERENCES

[1] Grzebyk, T., Górecka-Drzazga, A., & Dziuban, J. A. (2014). Glow-discharge ion-sorption micropump for vacuum MEMS. Sensors and Actuators A: Physical, 208, 113-119.

[2] Grzebyk, T., & Górecka-Drzazga, A. (2018). Characterization of the ionization process inside a miniature glow-discharge micropump. Bulletin of the Polish Academy of Sciences. Technical Sciences, 66(2).

[3] Wright, S. A., & Gianchandani, Y. B. (2006, January). A micromachined titanium sputter ion pump for cavity pressure control. In 19th IEEE International Conference on Micro Electro Mechanical Systems (pp. 754-757). IEEE.

[4] Green, S. R., Malhotra, R., & Gianchandani, Y. B. (2012). Sub-Torr chip-scale sputter-ion pump based on a Penning cell array architecture. Journal of microelectromechanical systems, 22(2), 309-317.

[5] Grzebyk, T., & Górecka-Drzazga, A. (2018). Characterization of the ionization process inside a miniature glow-discharge micropump. Bulletin of the Polish Academy of Sciences. Technical Sciences, 66(2).

[6] Wright, S. A., & Gianchandani, Y. B. (2007). Controlling pressure in microsystem packages by on-chip microdischarges between thin-film titanium electrodes. Journal of Vacuum Science & Technology B: Microelectronics and Nanometer Structures Processing, Measurement, and Phenomena, 25(5), 1711-1720.

[7] Eun, C. K., & Gianchandani, Y. B. (2012). Microdischarge-based sensors and actuators for portable microsystems: Selected examples. IEEE Journal of Quantum Electronics, 48(6), 814-826.

[8] Grzebyk, T., Górecka-Drzazga, A., & Dziuban, J. A. (2017). Improved properties of the MEMS-type ion-sorption micropump. Journal of Vacuum Science & Technology B, Nanotechnology and Microelectronics: Materials, Processing, Measurement, and Phenomena, 35(6), 062001.

[9] Abd-Elnaiem, A. M., & Gaber, A. (2013). Parametric study on the anodization of pure aluminum thin film used in fabricating nano-pores template. Int. J. Electrochem. Sci, 8(7), 9741-9751.

[10] S. Dhahri, E. Fazio, F. Barreca, F. Neri, and H. Ezzaouia, "Porous aluminum room temperature anodizing process in a fluorinated-oxalic acid solution," Appl. Phys. A, Solids Surf., vol. 122, no. 8, pp. 1–7, Aug. 2016.

CONTACT

*Bhaskar Mitra. Public, tel: +91-11-2659-6074; bmitra@iitd.ac.in

LNOI THIN-FILM DUAL-AXIS RESONANT MICRO-MIRROR WITH E16 TORSIONAL ACTUATION

Yaoqing Lu[1,2,3], Kangfu Liu[1,2,3*], Yuxi Wang[1,2,3], Ran Nie[1], and Tao Wu[1,2,3,4]*

[1]School of Information Science and Technology, ShanghaiTech University, Shanghai, China and
[2]Shanghai Institute of Microsystem and Information Technology, Chinese Academy of Sciences, Shanghai, China and
[3]University of Chinese Academy of Sciences, Beijing, China and
[4]Shanghai Engineering Research Center of Energy Efficient and Custom AI IC, Shanghai, China

ABSTRACT

In this work, the dual-axis resonant micro-mirrors based on the 128°Y-cut lithium niobate on insulator (LNOI) have been experimentally demonstrated for the first time. We pioneered the utilization of the anisotropy of lithium niobate materials to simplify the control of micro-mirrors. Two pairs of e_{16} torsional actuators are used to drive torsion mode and rocking mode, respectively. High electro-mechanical coupling factor K^2_{16} in two orthogonal in-plane directions is required to achieve dual-axis actuation. The frequencies of the torsion mode and rocking mode are up to 60.034 kHz and 205.347 kHz, respectively. The corresponding optical angles of torsion mode and rocking mode are 13.04° and 9.16° in the air when the drive voltage is 10 V_{PP}. Our work provides a novel solution for dual-axis piezoelectric MEMS micro-mirrors, which is promising for application in miniaturized optical systems.

KEYWORDS

Dual-axis, Micro-mirror, LNOI, Torsional actuation

INTRODUCTION

The micro-electro-mechanical systems (MEMS) micro-mirrors are in great demand for numerous light scanning applications [1]. Recently, the development of LiDAR in micro-robotics and mobile devices has led to the requirement for ultra-small LiDAR systems [2]. The design of the micro-mirror focuses on three indicators: θ_{opt} (optical deflection angle, which is four times the mechanical deflection angle θ_{mech}), mirror size D (the scanning optical aperture size in the scanning direction), and f (the fast-axis scanning frequency) [3].

The piezoelectric micro-mirrors driven by the methods take advantage of compact size, low driven voltage, low power, and high energy density [4] among the main four actuation mechanisms: electrostatic [5], electromagnetic [6], thermoelectric [7], and piezoelectric [8]. Conventional piezoelectric micro-mirrors are mainly based on lead zirconate titanate (PZT) [9], zinc oxide (ZnO) [8], and aluminum nitride (AlN) [10]. However, PZT contains the toxic element lead (Pb) and is not compatible with the complementary metal-oxide-semiconductor (CMOS) process [11]; ZnO and AlN suffer from the limited piezoelectric coefficient.

The maturity of the lithium niobate on insulator (LNOI) process has led to a new choice of piezoelectric materials[12]. LNOI could be a novel platform that provides a high piezoelectric coefficient comparable to PZT and various orientations with more options for micro-mirror design.

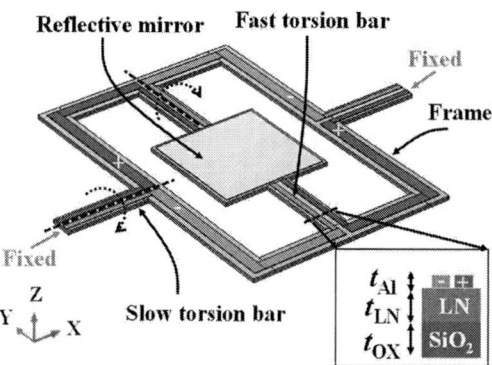

Figure 1: The schematic of the dual-axis mirror. (a) Mock-up view. (b) The cross-sectional view of the torsion bar.

Table 1: The Key Geometry Dimensions of the device

Parameter	μm
Width of reflective mirror	65
Width of fast torsion bar	8
Length of fast torsion bar	30
Width of slow torsion bar	8
Length of slow torsion bar	30

Single-axis micro-mirrors have been demonstrated using 36°Y-cut LNOI [13]. However, there is only one maximum piezoelectric stress constant e_{16} along the in-plane angle in this orientation. Compared with the single-axis micro-mirror, the dual-axis micro-mirror can effectively reduce the size and complexity of optical systems[14]. In order to drive the piezoelectric dual-axis micro-mirror effectively, the orientations with two electro-mechanical coupling factor K^2_{16} relatively large values in the plane should be explored.

In this work, we have first demonstrated the dual-axis mirror based on the LNOI platform. Such dual-axis mirror utilizes a specific orientation with dedicated electrode configuration and structural designs. The actuation method reported in this work differs from other piezoelectric dual-axis mirrors, which utilize a relatively large electro-mechanical coupling factor K^2_{16} in two vertical directions.

THEORY AND DESIGN

The schematic view of the dual-axis mirror is shown in Figure 1, which comprises a reflective mirror, a frame,

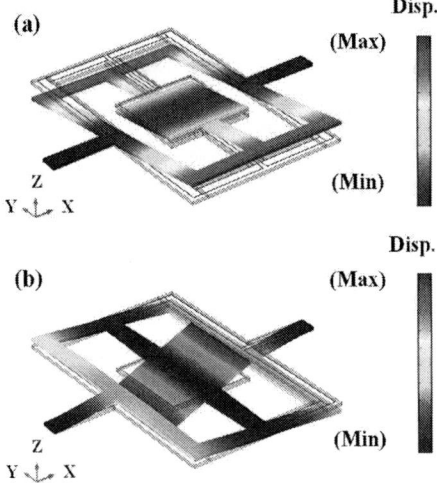

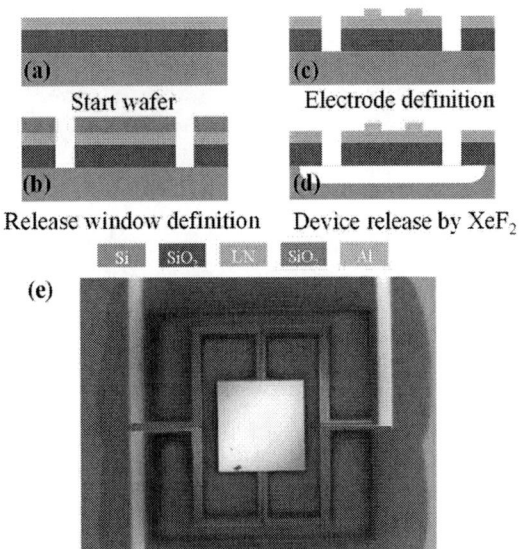

Figure 4: Process flow of the proposed device. (a) Start wafer with the LiNbO₃ thin film. (b) ICP-RIE of the device layer with SiO₂ hard mask. (c) Thermal evaporation and patterning of the aluminum electrode layer. (d) Releasing the device by etching the silicon layer using XeF₂. (e) Optical microscope image of the fabricated wafer.

Figure 2: The mode shapes for the dual-axis micro-mirror: (a) torsion mode (slow scanning mode), (b) rocking mode (fast scanning mode). Where cold and warm colors represent absolute minimum and maximum displacement, respectively.

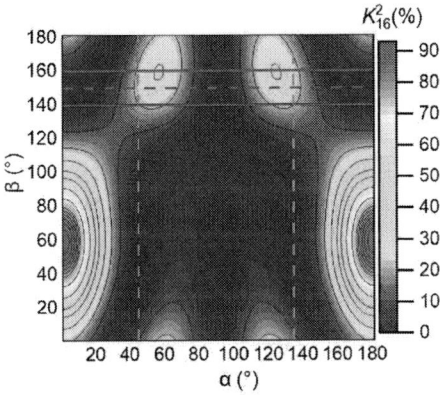

Figure 3: The contour plot of K^2_{16} as a function of Euler angles α and β. The red dotted line represents β = 150° (120°Y-cut).

and two pairs of torsional bars with driving electrodes on top. The electrode configuration in red and blue represents the positive and negative voltage, respectively. The inset shows the cross-sectional view of a torsion bar. The stack layer consists of a LiNbO₃ thin-film layer, a silicon dioxide (SiO₂) layer, and an Aluminum (Al) layer. The top Al electrode configuration, used for the driving electrodes of the torsion bar, simplifies the connection with control circuits. Two pairs of torsion bars that drive deflection in two directions are controlled by only one AC terminal. The key dimensions of the micro-mirror are listed in Table 1. The frequency of eigenmode and mechanical deflection angles are affected by the length and width of the torsion bars [15]. The total thickness and thickness ratio between LiNbO₃ and SiO₂ are optimized to achieve a high deflection angle. The dual-axis resonant micro-mirrors

require two different eigenmodes for the rotation in two directions.

Figures 2(a) and (b) show the mode shapes of slow and fast scanning modes simulated by COMSOL Finite Element Analysis, respectively. In the torsion mode (for the slow scan), the reflective mirror and the frame rotate as a whole body around the x-axis, requiring the torque generated by the slow torsion bar. In the rocking mode (for the fast scan), the reflective mirror rotates around the y-axis, and the outer frame performs a rocking motion out of phase with the reflective mirror. Thus, the torque generated by the fast torsion bar is needed. The resonant frequencies of the two eigenmodes are significantly different, so the torsion mode and rocking mode will not interfere with each other. The torques of the torsion and rocking modes are mainly generated by the slow/fast torsion bars, respectively.

The commonly used piezoelectric thin-film, such as PZT, AlN/AlScN[16], and ZnO, has a zero e_{16}, while LN has a strong e_{16} up to −4.5 C/m². With piezoelectric stress constant e_{16}, when the voltage is applied to the top electrode, the in-plane electric field ("1" direction) causes a pair of in-plane strain/stress ("6" direction), leading to the stress difference between the piezoelectric layer and elastic layer, resulting the rotation torque.

In order to effectively drive the dual-axis mirror, high electro-mechanical coupling factor K^2_{16} in two orthogonal in-plane directions are required. The K^2_{16} of LiNbO₃ with different orientations is calculated based on the Euler angle transformation, which follows the order of Z-X-Z[17], shown in Figure 3. The maximum value occurs at β =60° (30°Y-cut). However, the one maximum value of K^2_{16} at this cut can only drive a single-axis mirror. Note that there

978-1-6654-9309-3/23 $31.00 © 2023 IEEE

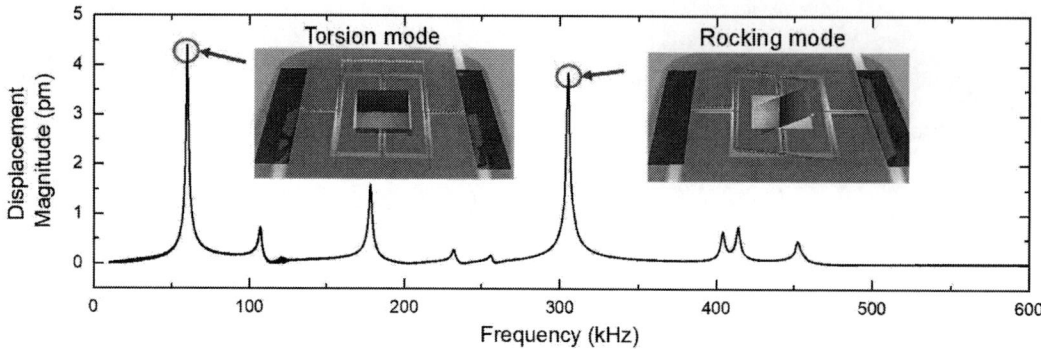

Figure 6: The LDV measurement of the dual-axis mirror in the air.

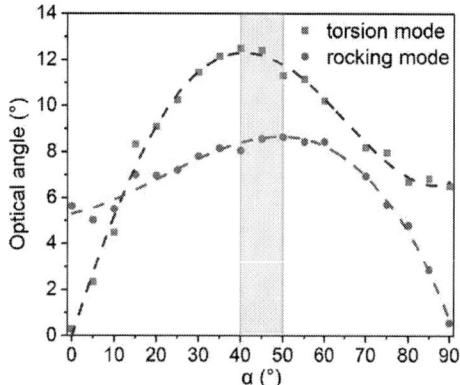

Figure 7: The optical deflection angle as a function of Euler angle α at 10 V.

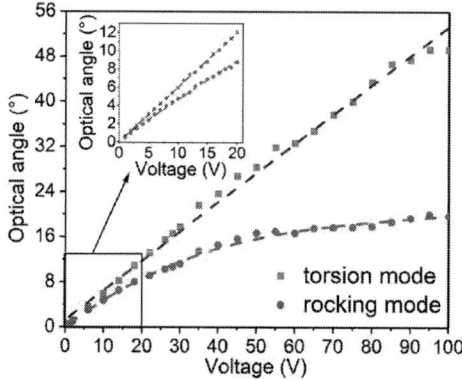

Figure 8: The optical deflection angle of two resonant modes under different driving voltages.

are two maximum values of K^2_{16} around the cut plane with β = 150° (120°Y-cut). The 128°Y-cut is close to the 120°Y-cut and is commercially available. The 128°Y-cut LiNbO₃ demonstrates a high K^2_{16} of 22% at both α = 45° and 135°. Thus, the 128°Y-cut LNOI thin film is selected for the design and analysis of dual-axis micro-mirrors.

FABRICATION AND MEASUREMENTS

The fabrication process is schematically shown in Figure 4. The fabrication process starts with a transfer-bonding 128°Y-cut single crystal, a 0.39 μm LiNbO₃ thin-film with a 0.685 μm SiO₂ layer. Next, a layer of SiO₂ is deposited by plasma-enhanced chemical vapor deposition (PECVD) as a hard mask. The photoresist SPR220-3 is used as an etching mask of SiO₂ for the release window. Then, inductively coupled plasma (ICP) with Cl₂-based reactive ion etching (RIE) is used to etch LiNbO₃ and SiO₂ and remove the SiO₂ hard mask. The electrodes are then defined by lifting off 100nm thermal evaporated aluminum. Finally, the micro-mirror is released by an isotropic XeF₂ dry etching.

The optical microscope image of the fabricated dual-axis micro-mirror is shown in Figure 5. The electrodes labeled in red and blue are connected to the positive and negative voltages, respectively. The electrodes on one pair of torsion bars need to be configured in opposite polarity to generate the torque in the same direction. The undercut at the bottom is due to the XeF₂ release.

Figure 6 shows the measured frequency spectrum of the laser Doppler vibrometer (LDV) for a micro-mirror in the air with sweeping signal excitation. The measured point is located at the corner of the mirror plate. The insets illustrate the mode shapes of the two highest peaks, which correspond to torsion mode and rocking mode, which are consistent with the mode shape shown in Figure 2. These results confirmed that the torsion mode and rocking mode work in two orthogonal directions, and they have a similar amplitude in excited deflection angle. The resonant frequencies of the torsion mode and rocking mode are 60.034 kHz and 205.347 kHz, respectively. The significant frequency difference between the two scan modes ensures that the scan modes are controlled by two different frequencies and that the driving signals for the two modes will not interfere with each other.

Figure 7 shows the relation between the measured optical angle and in-plane rotation angle (Euler angle α) when α in the range of 0°-90°. Define α = 0° as the positive direction of the x-axis. The curve at the interval of 90°-180° is antisymmetric with the curve at the interval of 0°-90°. Since the piezoelectric constants used in the fast and slow modes differ by 90° in the in-plane direction, their trends are also antisymmetric with each other. When α is in the range of 40-50°, the torsion mode and rocking mode achieves a relatively high optical deflection angle simultaneously, which is consistent with the trend of K^2_{16} illustrated in Figure 3.

978-1-6654-9309-3/23 $31.00 © 2023 IEEE

Figure 8 shows the dependency of optical angle with driving voltage. The torsion mode demonstrates an optical angle up to 48° at 100 V. From another perspective, it is also verified that the transferred single crystal film provides extremely high thin film quality, which is potential in terms of breakdown voltage, power handling, and electrostatic discharge (ESD). The curve for the torsion mode is almost linear. On the contrary, the rocking mode has a saturation voltage of around 30 V. The huge difference in saturation voltage between the two modes may require further study. The proposed dual-axis MEMS micro-mirror using 128°Y-cut LNOI provides an optical angle of torsion mode and rocking mode of 13.04° and 9.16° at 10 V_{pp} in the air when working at corresponding resonant frequencies.

CONCLUSIONS

We have successfully demonstrated dual-axis resonant micro-mirrors based on 128°Y-cut LNOI platform. Two pairs of e_{16} torsional actuators are used to drive torsion mode and rocking mode, respectively. The frequencies of the torsion and rocking mode are up to 60.034 kHz and 205.347 kHz, respectively. The corresponding optical angles are 13.04° and 9.16° in the air under a drive voltage of 10 V_{pp}. This work proposes and validates the feasibility of LNOI as a design platform for dual-axis micro-mirrors. Our design simplifies control and process complexity and may further inspire the application of LNOI in actuators and sensors.

ACKNOWLEDGEMENTS

This work was supported in part by the National Natural Science Foundation of China under Grant 61874073, and in part by the Lingang Laboratory under Grant LG-QS-202202-05. The device fabrication was performed at Soft Matter Nanofab (No. SMN180827) and ShanghaiTech Quantum Device Lab (SQDL).

REFERENCES

[1] E. Pengwang, K. Rabenorosoa, M. Rakotondrabe, and N. Andreff, "Scanning Micromirror Platform Based on MEMS Technology for Medical Application," *Micromachines*, vol. 7, no. 2, p. 24, Feb. 2016, doi: 10.3390/mi7020024.

[2] D. Wang, H. Xie, L. Thomas, and S. J. Koppal, "A Miniature LiDAR With a Detached MEMS Scanner for Micro-Robotics," *IEEE Sensors J.*, vol. 21, no. 19, pp. 21941–21946, Oct. 2021, doi: 10.1109/JSEN.2021.3079426.

[3] S. T. S. Holmstrom, U. Baran, and H. Urey, "MEMS Laser Scanners: A Review," *J. Microelectromech. Syst.*, vol. 23, no. 2, pp. 259–275, Apr. 2014, doi: 10.1109/JMEMS.2013.2295470.

[4] K. Uchino, "Piezoelectric actuators 2006," *J Electroceram*, vol. 20, no. 3, pp. 301–311, Aug. 2008, doi: 10.1007/s10832-007-9196-1.

[5] S. Ju, H. Jeong, J.-H. Park, J.-U. Bu, and C.-H. Ji, "Electromagnetic 2D Scanning Micromirror for High Definition Laser Projection Displays," *IEEE Photon. Technol. Lett.*, vol. 30, no. 23, pp. 2072–2075, Dec. 2018, doi: 10.1109/LPT.2018.2877303.

[6] Q. Wang, W. Wang, X. Zhuang, C. Zhou, and B. Fan, "Development of an Electrostatic Comb-Driven MEMS Scanning Mirror for Two-Dimensional Raster Scanning," *Micromachines*, vol. 12, no. 4, p. 378, Apr. 2021, doi: 10.3390/mi12040378.

[7] L. Zhou, X. Yu, P. X.-L. Feng, J. Li, and H. Xie, "A MEMS lens scanner based on serpentine electrothermal bimorph actuators for large axial tuning," *Opt. Express, OE*, vol. 28, no. 16, pp. 23439–23453, Aug. 2020, doi: 10.1364/OE.400363.

[8] M. Shkir *et al.*, "Investigation on structural, linear, nonlinear and optical limiting properties of sol-gel derived nanocrystalline Mg doped ZnO thin films for optoelectronic applications," *Journal of Molecular Structure*, vol. 1173, pp. 375–384, Dec. 2018, doi: 10.1016/j.molstruc.2018.06.105.

[9] U. Baran *et al.*, "Resonant PZT MEMS Scanner for High-Resolution Displays," *J. Microelectromech. Syst.*, vol. 21, no. 6, pp. 1303–1310, Dec. 2012, doi: 10.1109/JMEMS.2012.2209405.

[10] J. Shao, Q. Li, C. Feng, W. Li, and H. Yu, "AlN based piezoelectric micromirror," *Opt. Lett., OL*, vol. 43, no. 5, pp. 987–990, Mar. 2018, doi: 10.1364/OL.43.000987.

[11] J. R. Yuan *et al.*, "5I-1 Microfabrication of Piezoelectric Composite Ultrasound Transducers (PC-MUT)," in *2006 IEEE Ultrasonics Symposium*, Oct. 2006, pp. 922–925. doi: 10.1109/ULTSYM.2006.246.

[12] H. Hu, J. Yang, L. Gui, and W. Sohler, "Lithium niobate-on-insulator (LNOI): status and perspectives," presented at the SPIE Photonics Europe, Brussels, Belgium, Jun. 2012, p. 84311D. doi: 10.1117/12.922401.

[13] A. Emad, R. Lu, M.-H. Li, Y. Yang, T. Wu, and S. Gong, "Resonant Torsional Micro-Actuators Using Thin-Film Lithium Niobate," in *2019 IEEE 32nd International Conference on Micro Electro Mechanical Systems (MEMS)*, Seoul, Korea (South), Jan. 2019, pp. 282–285. doi: 10.1109/MEMSYS.2019.8870894.

[14] D. Wang, C. Watkins, and H. Xie, "MEMS Mirrors for LiDAR: A Review," *Micromachines*, vol. 11, no. 5, p. 456, Apr. 2020, doi: 10.3390/mi11050456.

[15] Y. Lu, K. Liu, and T. Wu, "Dual-Axis MEMS Resonant Scanner Using 128 ◦ Y Lithium Niobate Thin-Film," *Acoustics*, vol. 4, no. 2, Art. no. 2, Jun. 2022, doi: 10.3390/acoustics4020019.

[16] S. Shao, Z. Luo, Y. Lu, A. Mazzalai, C. Tosi, and T. Wu, "Low Loss Al0.7Sc0.3N Thin Film Acoustic Delay Lines," *IEEE Electron Device Letters*, vol. 43, no. 4, pp. 647–650, Apr. 2022, doi: 10.1109/LED.2022.3152908.

[17] L. Qin and Q.-M. Wang, "Mass sensitivity of thin film bulk acoustic resonator sensors based on polar *c* -axis tilted zinc oxide and aluminum nitride thin film," *Journal of Applied Physics*, vol. 108, no. 10, p. 104510, Nov. 2010, doi: 10.1063/1.3483245.

CONTACT

*Yaoqing Lu, luyq@shanghaitech.edu.cn
*Kangfu Liu, liukf@shanghaitech.edu.cn
First two authors contribute equally.
*Tao Wu, wutao@shanghaitech.edu.cn

A PIEZOELECTRIC MEMS SPEAKER WITH STRETCHABLE FILM SEALING

Linbing Xu, Mingchao Sun, Menglun Zhang*, Chengze Liu,
Xiaopeng Yang, and Wei Pang

State Key Laboratory of Precision Measuring Technology and Instruments, Tianjin University,
Tianjin, CHINA

ABSTRACT

This paper reports a solution to acoustic short circuit issue of piezoelectric MEMS speakers by sealing slits between actuators with a stretchable film. Validated by simulation and experiment results, sound pressure level (SPL) at low frequency range of a speaker could be improved significantly. With 15 μm thick stretchable film of polydimethylsiloxane (PDMS), the speaker SPL at 20 Hz increases by about 15 dB. Compared with alternative solutions of plastic film sealing, high sensitivity and resonant frequency of the speaker could be preserved since Young's modulus of the stretchable film is low enough for unconstrained bending of actuator diaphragms. The PDMS sealed MEMS speaker shows flat SPL curve at low frequency range with less than 5 dB variation from 20 Hz to 1 kHz, and excellent THD performance of lower than 1% over most of the whole audible frequency range. In addition, the PDMS sealed MEMS speaker is compatible with standard reflow process. The proposed stretchable film sealing solution is simple to implement and universally applicable to other MEMS speakers.

KEYWORDS

Piezoelectric; MEMS speaker; sound pressure level; acoustic short circuit; polydimethylsiloxane

INTRODUCTION

With growing demand of miniaturization, low cost and low power consumption for consumer electronics, microspeaker has been widely used in computers, smart phones, headphones, earbuds, high-fidelity hearing aids and Internet of things (IoT) [1]. In recent years, piezoelectric MEMS speakers have become a research hotspot. The advantages of piezoelectric MEMS speakers include small size, low power consumption, integrated manufacturing, fast response and large bandwidth [1-3].

For piezoelectric MEMS speakers, actuator diaphragms are usually decoupled or partially decoupled by slits [3], since a closed actuator diaphragm integrating piezoelectric films suffers from limited deflection and influences of residual stress. For these speakers, however, air leakage through the slits will cause sound pressure loss due to acoustic short circuit, especially at low frequency range. The wider the silts are, the more serious the issue is. Inevitable residual stress would cause decoupled actuator diaphragms to warp in fabrication and hence the slit widths are left larger than designed, which could not be well controlled. Therefore, current piezoelectric MEMS speakers are mostly associated with a reduced low frequency SPL, for example degrading the rhythmic feel of music.

To solve the issue, Tseng et al. [4] reported a design

adding a coupling mass to reduce the slit widths with out-of-phase driving. Wang et al. [5] reported a design adding structured Parylene film by rigid-flexible coupling mechanism. To reduce the restriction on the actuator movement by the added sealing, minimum etching gap needs to match the maximum gap distance. Hirano et al. [6] proposed a MEMS speaker adding a closed composite component by vertical stacking another wafer. All the solutions mentioned above aggravate complexity in either design or fabrication. The added part or structure would change sensitivity or resonant frequency of the original speaker, deviating from the target performance when trying to solve the acoustic short circuit issue.

In this work, an effective approach is proposed to solve the acoustic short circuit issue with negligible impact on speaker sensitivity and resonant frequency by adding a stretchable film sealing. The restriction of the sealing on actuator motion is released by its easy tensile deformation without any additional structuring. Therefore, the initial speaker design could be preserved fulfilling target specifications without further iterations. Design and fabrication simplicity leads to a lower cost and wider range of applications.

DESIGN CONCEPT

A thin film could be arranged to cover the whole actuator diaphragms or at least slits between diaphragms to block the acoustic short circuit for sealing. But the additional film would inevitably restrict bending motion of the actuators to some extent. When the actuators vibrate, the additional film is mainly subjected to tensile deformation. The key to minimize the restriction on bending motion by a sealing film is to enhance its deformation under tension.

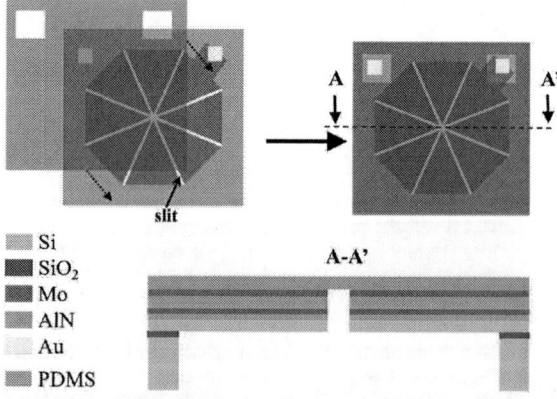

Figure 1: Schematic diagram of the proposed piezoelectric MEMS speaker with stretchable PDMS film sealing.

Young's modulus of polydimethylsiloxane (PDMS) is on the order of MPa, which is several orders of magnitude smaller than that of plastic films commonly used by electrodynamic microspeakers such as Polyimide and Parylene. PDMS as an additional film could endow actuators with minimally influenced deflection through its intrinsically low Young's modulus without further structuring, which simplifies the speaker design and fabrication.

As shown in Fig. 1, a PDMS film is implemented as the stretchable film on eight actuator diaphragms of a piezoelectric MEMS speaker, covering the slits between them. Fig. 2a shows simulation results where low-frequency SPL becomes flat after applying 15 μm PDMS or 1 μm Parylene. The speaker with PDMS film preserves the sensitivity of the original speaker, in contrast to the speaker with thinner Parylene. The strain fields of Fig. 2b-c show three times larger strain of PDMS compared to Parylene at slit locations, explaining the 10 dB higher SPL.

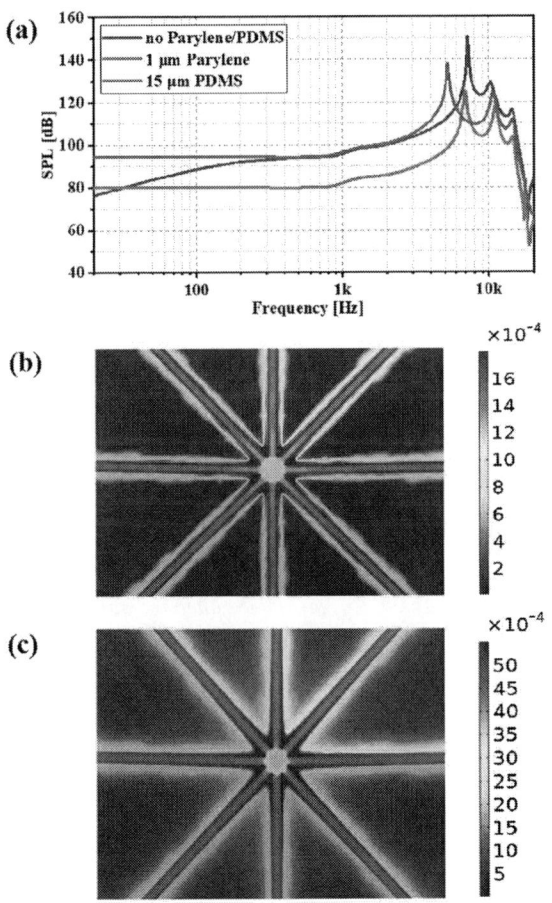

Figure 2: (a) Simulated SPL curves for different designs at 10 Vrms input, showing superiority of stretchable PDMS over flexible Parylene. Corresponding strain fields of 1 μm Parylene (b) and 15 μm PDMS (c).

FABRICATION

Fig. 3 illustrates the fabrication steps. Firstly, AlN seed layer was deposited on the silicon device layer of SOI wafers (Fig. 3a). Secondly, molybdenum with a thickness of 0.1 μm was deposited and patterned on the seed layer as the bottom electrode (Fig. 3b). Thirdly, AlN with a thickness of 0.25 μm as the piezoelectric layer was deposited on the bottom electrode (Fig. 3c). Next, molybdenum with a thickness of 0.1 μm was deposited and patterned on the piezoelectric layer as the top electrode (Fig. 3d). All molybdenum electrodes were dry-etched after photolithography. Then, the AlN and the silicon device layer with a thickness of 10 μm were etched to form slits between the actuators and expose the bonding area on the top and bottom electrodes (Fig. 3e). Then, TiW and Au were deposited and patterned for bonding pads on electrodes (Fig. 3f). Subsequently, the bulk silicon with a thickness of 400 μm of the SOI wafer was etched using DRIE to define the back cavity. Hydrofluoric acid solution was used to release the thermal oxide layer, which released the cantilever beams (Fig. 3g). Finally, PDMS with a thickness of 15 μm was applied on the cantilever beams (Fig. 3h).

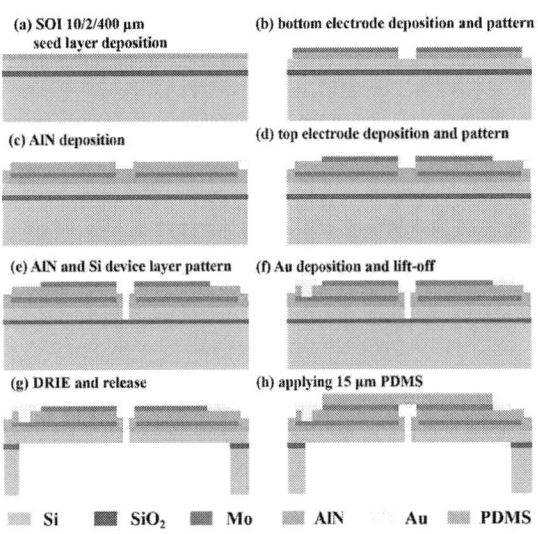

Figure 3: Fabrication process of the piezoelectric MEMS speaker.

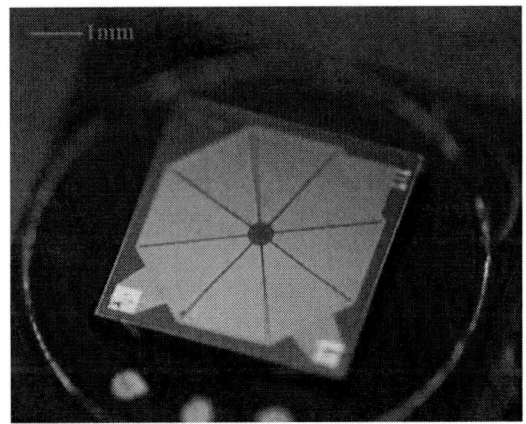

Figure 4: Photograph of the fabricated speaker with transparent 15 μm PDMS film on top.

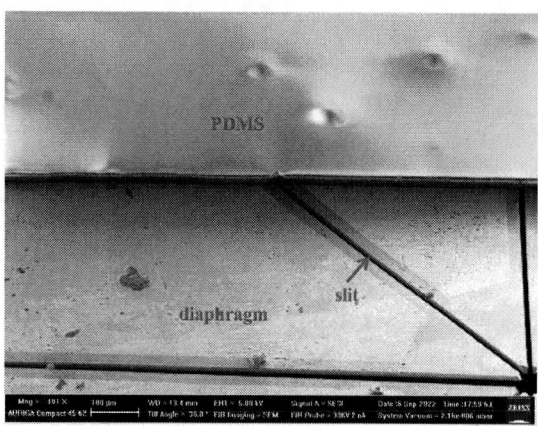

Figure 5: SEM photo with PDMS film partially removed to reveal actuator diaphragms and slits.

Fig.4 and Fig.5 show the photograph and the SEM photo of the fabricated speaker with PDMS film respectively.

ACOUSTIC CHARACTERIZATION

In order to characterize the acoustic performance of the piezoelectric MEMS speaker applied with PDMS film sealing, various electroacoustic measurements have been performed. During all measurements, the device under test (DUT) was installed on 711 ear simulator, which complies with IEC 60318-4 and simulates the acoustic load impedance of human ear canal in the frequency range of 100 Hz to 10 kHz. All measurements on SPL used a 10 Vrms sweep signal as input and THD was measured at 94 dB SPL at 1 kHz.

Experimental results shown in Fig. 6 correspond well to simulation in Fig. 2a. After sealing the slits using 15 μm PDMS, the speaker low frequency response is significantly improved with the SPL at 20 Hz increased by about 15 dB. And its SPL above 250 Hz is minimally influenced. Thus the difference in SPL ranging from 20 Hz to 1 kHz is kept within 5 dB, which is a quarter of that of the speaker without PDMS. The resonant frequency of the speaker is reduced by less than 1 kHz to a negligible extent, mainly due to mass of the PDMS film. This means that the high frequency response will not be degraded. In addition, the speaker with PDMS sealing withstands high temperature well, as its acoustic performance is well consistent before and after the 260 °C heat treatment, indicating its compatibility with standard reflow process.

As shown in Fig. 7, acoustic damping has been further added on the backside surface of the speaker to eliminate its resonance induced SPL peak at 5 kHz. The speaker applied with PDMS film sealing preserves its high SPL off resonance, in contrast to a degraded SPL curve when acoustic damping is added to a speaker without PDMS film sealing. The difference can be explained as follows. When acoustic damping is applied on one side of the speaker membranes, the sound pressure is pressed to the other side through the slits due to blocking effect of the acoustic damping. The sound pressures on both sides of the speaker membranes are out of phase and they cancel out, leading to

a low sound pressure output. Introduction of the PDMS film isolates the path of the cancellation.

Fig.7 illustrates that the resonant response of a speaker with stretchable film sealing can be flatten by acoustic damping without sacrificing sensitivity in the quasi-steady state. This is an important feature of speakers without acoustic short circuit issue. Fig. 8 shows excellent linearity performance of the speaker with measured THD lower than 1% over most of the whole audible frequency range.

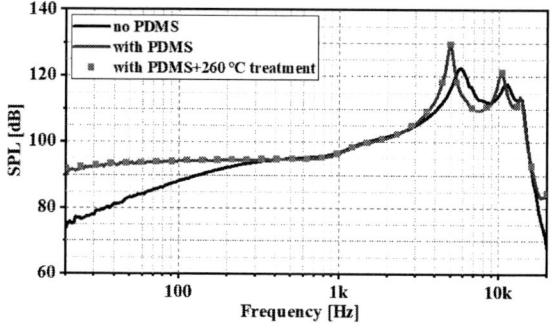

Figure 6: Measured SPL curves of the MEMS speaker before and after implementation of 15 μm PDMS, with following heating treatment at 260 °C for 1 min.

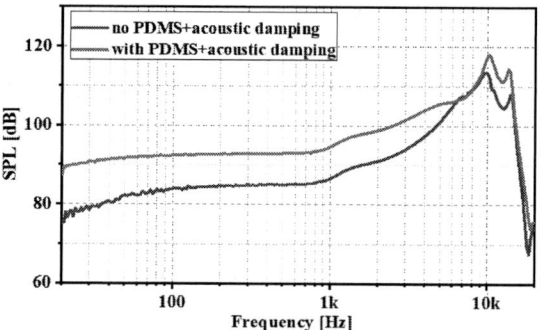

Figure 7: Measured SPL curves of the MEMS speakers with and without 15 μm PDMS, after the same acoustic damping is applied to both speakers to eliminate SPL peak at 5 kHz.

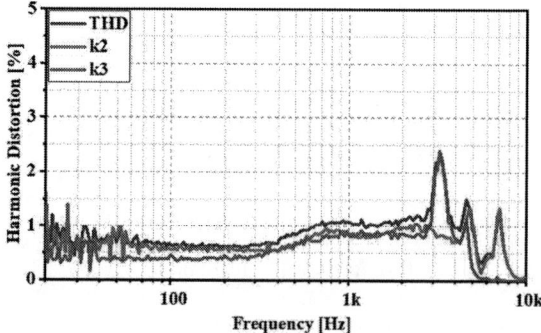

Figure 8: Harmonic distortion performance of the MEMS speaker with acoustic damping and 15 μm PDMS, measured at 94 dB SPL at 1 kHz.

CONCLUSION

This paper introduces a piezoelectric MEMS microspeaker sealed by stretchable PDMS film. Measurements of the piezoelectric MEMS speaker equipped with PDMS film show superior acoustic performance in terms of SPL and THD. After applying 15 μm PDMS, the speaker low frequency response is significantly improved with the SPL at 20 Hz increased by about 15 dB, and its SPL curve above 250 Hz is minimally influenced. In addition, the MEMS speaker with PDMS withstands high temperature well, indicating its compatibility with standard reflow process. Acoustic damping has been further added to eliminate its resonance induced SPL peak at 5 kHz while preserving its high SPL off resonance, in contrast to a degraded SPL curve when acoustic damping is added to a speaker without PDMS film. The proposed stretchable film sealing solution is simple to implement and universally applicable to other MEMS speakers.

ACKNOWLEDGEMENTS

This work was supported by funding provided by the Natural Science Foundation of China (NSFC Grant No.62001322) and National Key Research and Development Program (No. 2020YFB2008800).

REFERENCES

[1] C.Z. Liu, M.L. Zhang, M.C. Sun, L.B. Xu, Y.M. Lang, S.B. Gong, W.Pang, "Ultrahigh-Sensitivity Piezoelectric AlN MEMS Speakers Enabled by Analytical Expressions", *J. Microelectromech. Syst.*, vol. 31, pp. 664-672, 2022.

[2] Y.M. Lang, C.Z. Liu, A. Fawzy, C. Sun, S.B. Gong, M.L. Zhang, "Piezoelectric bimorph MEMS speakers", *Nano. Prec. Eng.*, vol. 5, pp. 033001(1-8), 2022.

[3] F. Stoppel, "NEW INTEGRATED FULL-RANGE MEMS SPEAKER FOR IN-EAR APPLICATIONS", *International Conference on Micro Electro Mechanical Systems' 31 Conference*, Belfast, JAN 21-25, 2018, pp. 1068-1071.

[4] S. H. Tseng, "PIEZOELECTRIC MICROSPEAKER USING NOVEL DRIVING APPROACH AND ELECTRODE DESIGN FOR FREQUENCY RANGE IMPROVEMENT", *International Conference on Micro Electro Mechanical Systems' 33 Conference*, Vancouver, JAN 18-22, 2020, pp. 546-549.

[5] Q. Wang, T. Ruan, Q.D. Xu, B. Yang, J.Q. Liu, "Obtaining High SPL Piezoelectric MEMS Speaker via a Rigid-Flexible Vibration Coupling Mechanism", *J. Microelectromech. Syst.*, vol. 30, pp. 725-732, 2021.

[6] Y. Hirano, "PZT MEMS SPEAKER INTEGRATED WITH SILICON-PARYLENE COMPOSITE CORRUGATED DIAPHRAGM", *International Conference on Micro Electro Mechanical Systems' 35 Conference*, Tokyo, JAN 09-13, 2022, pp. 255-258.

CONTACT

*M. Zhang, tel: +86-22-27401053; zml@tju.edu.cn

BROADBAND MEMS SPEAKER BY SINGLE-WAY MULTI-RESONANCE ARRAY WITH ACOUSTIC DAMPING TUNING: A PROOF OF CONCEPT

Mingchao Sun, Menglun Zhang*, Chengze Liu and Wei Pang
State Key Laboratory of Precision Measuring Technology and Instruments,
Tianjin University, Tianjin, China

ABSTRACT

This paper proposes a resonance synthesis (RS) method for broadband MEMS speaker. It involves multi-resonance actuator array with proper acoustic damping tuning. A broadband design for MEMS speaker is demonstrated based on the RS method and validated by simulation. Compared with state of the art methods, the proposed design achieves a much wider -3 dB SPL bandwidth from 2 kHz to 10 kHz, which is an improvement in bandwidth of ~250%. A proof of concept experiment is conducted by two piezoelectric MEMS speakers with different resonant frequencies of 3.6 kHz and 6 kHz. The experimental results agree with simulation. The RS method combines multiple mechanical resonances, forms highly sensitive and flat acoustic output, and yet only takes a single-way electrical signal as input.

KEYWORDS

Piezoelectric, MEMS speaker, broadband, SPL, resonance synthesis, acoustic damping

INTRODUCTION

Piezoelectric MEMS speaker has become a research hotspot with advantages of small size, low power consumption, low cost and batch fabrication ability [1-6]. However, its narrow bandwidth caused by non-flat frequency response around resonance is a remaining issue for piezoelectric MEMS speakers [1]. There have been a number of efforts to broaden the bandwidth. Stoppel et al. used electronic equalization to flatten the frequency response at the expense of abandoning the amplification effect of resonance, leading to a compromised sound pressure level (SPL) at resonant frequency [2]. They also proposed a two-way MEMS speaker with tweeter and woofer, driven by separate input signals below and above the crossover frequency [3]. Wang et al. proposed a multi-channel cantilever array by phase control with multi-way input [4]. Theoretically, for the methods proposed in [3,4], the more resonances or channels incorporated, the better frequency response will be. MEMS speaker is perfect for the multi-channel design thanks to easily obtained MEMS actuator array. However, its electrical input control would become formidably complex as the array number increases without CMOS integration.

In this work, a simple approach is proposed to fully utilize large actuator array with only one-way electrical input, breaking the above dilemma. The phase difference between different actuators is controlled by acoustic damping tuning to achieve a flat response curve without complex multi-way input.

CONCEPT

As shown in Fig. 1, the proposed resonance synthesis (RS) method combines multiple actuator resonances to form a flat response curve, leveraging resonance energy evenly distributed over a wide bandwidth. With the amplification effect of resonances, high SPL can be obtained along with wide bandwidth. In Fig. 1, $f_1 \sim f_n$ represent the resonance responses of the first to nth actuators, respectively. The target frequency response of actuator array is formed by combining multiple resonances.

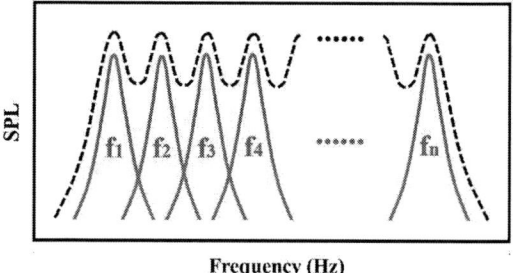

Figure 1: Target frequency response (dash line) by the resonance synthesis method.

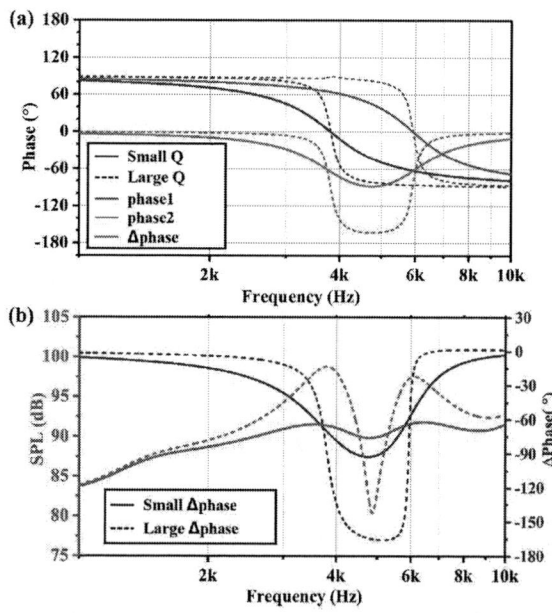

Figure 2: Simulation implemented with two actuators of different resonant frequencies. (a) Relationship of Q value of actuators and phase difference. (b) Relationship of phase difference and valley of SPL frequency response.

However, simply driving actuators of different resonant frequencies in parallel cannot achieve the ideal curve in Fig. 1. Piezoelectric actuators will undergo a sharp phase shift when they reach resonance and the different resonant frequencies of actuators mean different phase shift point. This leads to phase differences between the actuators. For instance, there will be a phase difference between two actuators if the one has exceeded its resonant frequency while the other is still far from reaching its resonant frequency. In fact, the phase shift is not abrupt but occurs within a frequency band; and the larger the Q value is, the narrower this frequency band is and the larger the phase difference is (dotted line in Fig. 2a). As piezoelectric MEMS speakers generally have a high Q value, large phase shift within a narrow band around resonance will cause nearly 180° phase difference between neighbor elements, leading to SPL valleys of a response curve. Deep valleys in frequency response curve lead to defects in broadband feature of multi-resonance array.

The simulation results of two actuators with different resonances driven by the same signal can be illustrative by Fig. 2. The 180°phase difference of two actuators means that the actuators vibrate in opposite directions to each other, leading to cancelled output pressure and thus a deep valley in SPL curve (dotted line in Fig. 2b). Therefore, reducing Q value and thus the phase difference is the key to solve the issue. When maximum phase difference is kept within 90°, the cancelled sound pressure will be greatly suppressed (solid line in Fig. 2b). The phase control is implemented by separating acoustic cavities of elements and then applying proper acoustic damping (Fig. 3b). As a result, phase shift will span a wider frequency band and therefore phase difference between elements will decrease. It is worth noting that the two conditions of acoustic damping and separated back cavities are indispensable, which will be discussed later. As this approach avoid deep SPL valleys by mechanical means, only one-way electrical input is required for the electrical signal control even with a large array, which is the key difference from other methods.

DEMONSTRATION DESIGN

Fig. 3 illustrates a demonstration design using the RS method, where the MEMS speaker is comprised of fifteen unimorph cantilevers with different lengths corresponding to different resonant frequencies. Each cantilever has its independent acoustic cavity covered by acoustic damping. Design parameters are shown in Table 1.

Table 1: Cantilever parameters of demonstration design and conventional/reference design

	Proposed	Reference
Length	965~1910 μm	1350 μm
Acoustic damping	70~200 Pa·s/m	0 or 750 Pa·s/m
Width	400 μm	
Piezoelectric layer	500 nm AlN	
Elastic layer	5 μm Si	
Electrode	100 nm Mo	
Cavity depth	400 μm	

To assess the acoustic performance of the proposed design in ear simulator, FEM simulation was implemented. Fig. 4 shows the simulation results of acoustic damping tuning influence on multi-resonance array. It can be seen that not only multi-resonance array without damping on separated cavities but also the one with damping on a shared cavity have deep valleys. Only when acoustic damping is applied on separated back cavities, the sound pressure cancellation can be greatly reduced.

This result can be explained as follows. In RS method, each actuator and its back cavity form a single unit and acoustic damping reduces the Q value of these units, so the phase difference between neighbor units becomes smaller without acoustic coupling to each other. While, actuators in a shared cavity are acoustically coupled to each other in the back cavity which causes a steep change in the phase. Fig. 4 demonstrates the necessity of independent acoustic cavities and proper acoustic damping for the RS method.

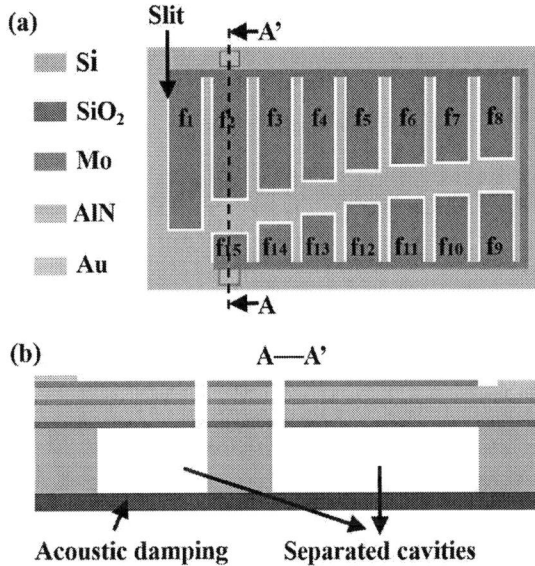

Figure 3: Structures schematic of demonstration design by RS method.

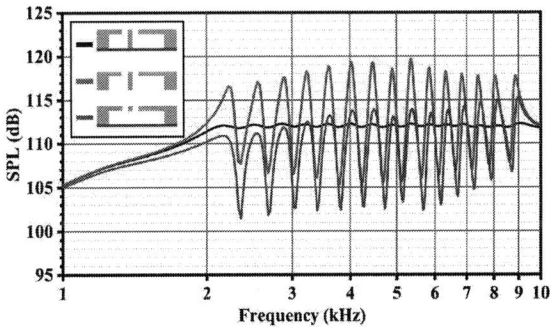

Figure 4: Simulation results of acoustic damping tuning influence on multi-resonance array. Black, red and blue lines correspond to proposed design, array without damping on separated cavities and array with damping on a shared cavity, respectively.

Fig. 5 shows the simulation results of multi-resonance array compared with a single resonance speaker. A flat SPL curve within -3 dB could be realized from 2 kHz to 10 kHz, which is improved by ~250% in bandwidth, compared with a single-resonance design whose maximum SPL is aligned with the multi-resonance array by acoustic damping. It is worth mentioning that damped single-resonance design need greater acoustic damping (as shown in Table 1) to achieve the same SPL. In other words, more energy is consumed in the damping. This can be interpreted as single-resonance design with resonant energy concentrated at one point, while the proposed design with resonant energy evenly distributed over a wide bandwidth, which proves that proposed design has better energy efficiency The single-resonance design without acoustic damping has a sharp frequency response at its resonant frequency, corresponding to a lower bandwidth.

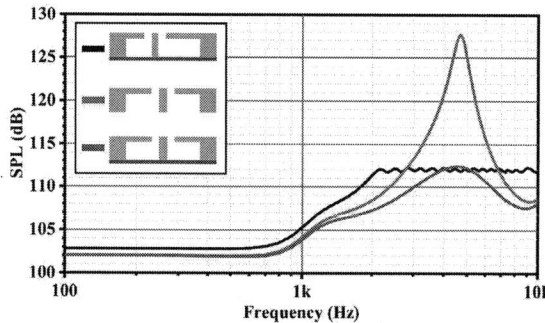

Figure 5: Simulation results of multi-resonance array influence. Black, red and blue lines correspond to proposed design, single-resonance without damping and single- resonance with damping.

PROOF OF CONCEPT

To prove the feasibility of the proposed RS method, experimental proof of concept was implemented with two different cantilever-type MEMS speakers. The schematic of experimental setup is shown in Fig. 6. Two speakers named *S1* and *S2* with respective resonant frequencies of 3.6 kHz and 6 kHz were arranged together to form a two-resonance speaker using RS method. Details are shown in Fig. 7. And a single-resonance speaker was set as a control group, which is comprised of two *S2* speakers.

The two-resonance speaker was wired bonded on PCB to form a DUT (device under test). The DUT was mounted on an ear simulator which conforms to IEC 60318-4 by an adapter. All measurement was carried out in anechoic chamber and the electrical input was a single-way 10Vrms frequency sweep signal.

Fig. 8 shows the SPL curve of RS method and control groups for comparison, including different acoustic cavity and damping conditions. The difference in response between the valley and the resonant peak is only 5 dB with RS method, while that without damping and with shared cavity are 26 dB and 10 dB respectively. The experimental result is the experimental counterpart of Fig. 4 and it confirms that two conditions of acoustic damping and separated back cavities are indispensable, otherwise deep valleys will inevitably appear.

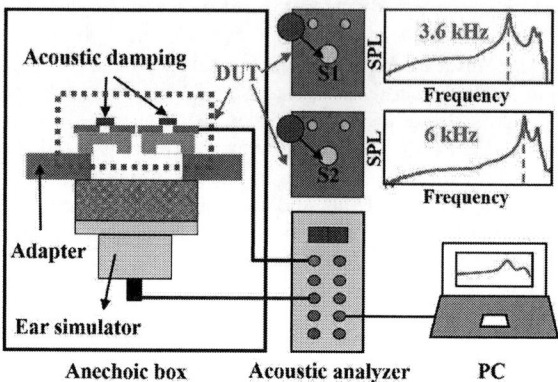

Figure 6: Schematic of experimental setup, with SPL curves of respective speaker S1 and S2.

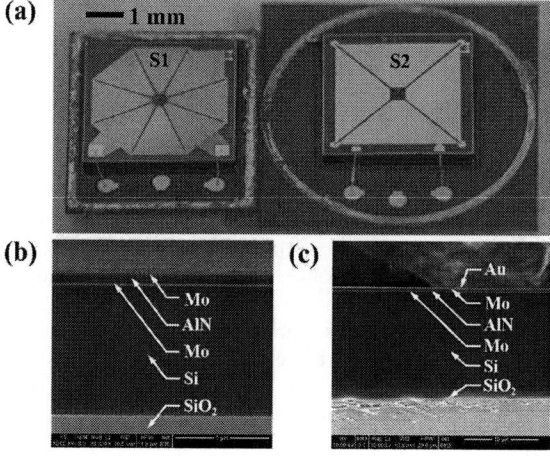

Figure 7: (a) Photo of S1 and S2 wired bonded on PCBs. (b)-(c) SEM photos of S1 and S2.

Similarly, the experimental result shown in Fig. 9 is the experimental counterpart of Fig. 5, demonstrating the superiority of multi-resonance array. The experiment results show that the RS method could achieve a wider bandwidth, as the proposed speaker has higher SPL from 1 kHz to 4 kHz than the control group of two *S2* speakers.

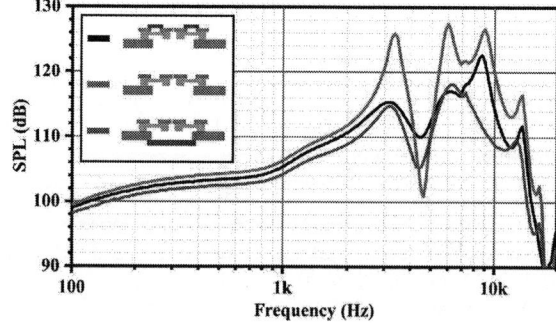

Figure 8: Experiment validation for acoustic damping tuning influence on multi-resonance array by speaker S1 and S2.

978-1-6654-9309-3/23 $31.00 © 2023 IEEE

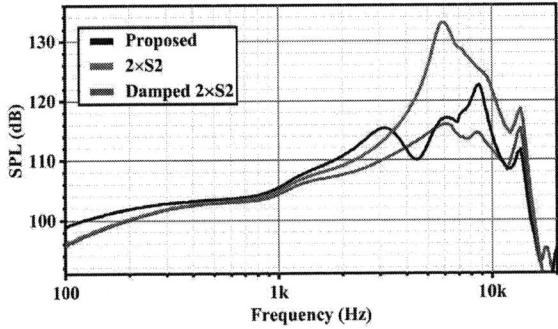

Figure 9: Experiment validation for multi-resonance array influence by speaker S1 and S2.

CONCLUSION

This paper proposes a broadband design for MEMS speakers using multi-resonance actuator array with proper acoustic damping tuning. The RS method shows advantages of high sensitivity, flat acoustic output and only one-way input. A flat SPL curve within -3 dB could be realized from 2 kHz to 10 kHz by simulation. Proof of concept experiments demonstrate feasibility of the RS method. Though the experimental results show limited improvement due to only two elements arranged, we believe that the result would improve steadily as the multi-resonance element number further increases.

ACKNOWLEDGEMENTS

This work was supported by funding provided by the Natural Science Foundation of China (NSFC Grant No.62001322) and National Key Research and Development Program (No. 2020YFB2008800).

REFERENCES

[1] H.R. Wang, Y.F. Ma, Q.C. Zheng, K. Cao, Y. Lu, H.K. Xie, "Review of Recent Development of MEMS Speakers", Micromachines, vol. 12, pp. 1-29, 2021.

[2] F. Stoppel, A. Mannchen, F. Niekiel, D. Beer, T. Giese, and B. Wagner "New integrated full-range MEMS speaker for in-ear applications", International Conference on Micro Electro Mechanical Systems' 31 Conference, Belfast, JAN 21-25, 2018, pp. 1068-1071.

[3] F. Stoppel, C. Eisermann, S. Gu-Stoppel, D. Kaden, T. Giese, B. Wagner, "Novel Membrane-Less Two-Way MEMS Loudspeaker Based on Piezoelectric Dual-Concentric Actuators", in Digest Tech. Papers Transducers'19 Conference, Kaohsiung, June 18-22, 2017, pp. 2047-2050.

[4] Y.J. Wang, S.C. Lo, M.L. Hsieh, S.D. Wang, Y.C. Chen, M. Wu, W. Fang, "Multi-way in-phase/out-of-phase driving cantilever array for performance enhancement of PZT MEMS microspeaker", International Conference on Micro Electro Mechanical Systems' 34 Conference, Online, JAN 25-29, 2021, pp. 83-84.

[5] C.Z. Liu, M.L. Zhang, M.C. Sun, L.B. Xu, Y.M. Lang, S.B. Gong, W. Pang, "Ultrahigh-Sensitivity Piezoelectric AlN MEMS Speakers Enabled by Analytical Expressions", J. Microelectromech. Syst., vol. 31, pp. 664-672, 2022.

[6] Y.M. Lang, C.Z. Liu, A. Fawzy, C. Sun, S.B. Gong, M.L. Zhang, "Piezoelectric bimorph MEMS speakers", Nano. Prec. Eng., vol. 5, pp. 033001(1-8), 2022.

CONTACT

*M. Zhang, tel: +86-22-27401053; zml@tju.edu.cn

IONIC LIQUID ELECTROSPRAY THRUSTER
WITH TWO-STAGE ELECTRODES ON GLASS SUBSTRATE

Akane Nishimura[1], Toshiyuki Tsuchiya[1], and Yoshinori Takao[2]
[1]Department of Micro Engineering, Kyoto University, JAPAN and
[2]Division of Systems Research, Yokohama National University, JAPAN

ABSTRACT

This paper reports ion emission of an ionic liquid electrospray thrustor with two-stage electrodes made on glass substrate having through hole for low-cost micro/nano satellites. By using the two-stage electrodes, one for ion extraction and the other for acceleration, high and stable ion emission and propulsion force is obtained. The emitter array was fabricated on a silicon wafer and the electrodes were fabricated on both sides of a glass substrate. The ion emission test was conducted, and the emission current was observed successfully. Almost no ions were collected on the accelerator electrode and reached to the collector electrode, which demonstrates the advantage of the two-stage configuration.

KEYWORDS

Electrospray thruster, ionic liquid, silicon emitter array, glass via hole.

INTRODUCTION

Development of low-cost miniature satellites, such as nano-satellites and CubeSat, is actively carried out. Even such a compact satellite, the attitude and orbital controls are demanded. Therefore, for these satellites compact electrical propulsion system is demanded. Among the electrical propulsion systems, ion emission thrusters using ionic liquid as a propellant is attracting more attentions because of its simple propellant supply system, no needs for neutralizer and large propulsion force [1]. There have been several reports on microelectromechanical system (MEMS) based ionic liquid electrospray thrustors (ILESTs) [2-5]. However, the emission stability is one of the major technical issues. The previous work measured current from silicon made emitters and metal extractor electrodes, but the emission is not stable because of the extraction voltage is too high [5]. The issues are poor alignment between emitters and extractor electrode and complicated fabrication process.

In this study, an ILEST was fabricated using microfabrication process, especially the electrodes are fabricated on the glass wafers with via holes using lift-off process with a dry thick photoresist film. In this paper, the fabrication process of both the emitter chips made from a silicon wafer and the electrode chips made from a glass substrate, and an assemble method with good precision to realize stable emission are reported.

OPERATION PRINCIPLE

Figure 1 shows the concept drawing of an ILEST with two-stage electrodes. We employed a glass substrate with via hole for ion emission and two electrodes on both sides for ion extraction and acceleration. By applying voltage between the bottom electrodes, the high concentrated electric field at the tip of the emitter generates the Tayler cone of ionic liquid propellant and the anions and cations are emitted by applying positive and negative voltages. The emitted ions are accelerated by the potential difference between the accelerator and emitter electrodes. By separating the functions of emission and acceleration, the stable and high thrust is expected because the emission requires a high electric field but the acceleration depends on the total voltage. The key technical challenge is the small gap between extractor electrodes and emitter tip. The low extraction voltage and large gap for accelerator electrodes avoids the electrical discharge. In addition, we can control the propulsion force with stable emission.

DESIGN AND FABRICATION
Device structure

Figure 2 shows the schematic design of the two-stage ILEST. The emitter chip is made from a 4-inch silicon wafer of 400 μm thick. The chip is 18 mm square. There are 57 emitters is arranged by 1 mm interval placed in the 12-mm-diameter circular reservoir cavity. The electrode

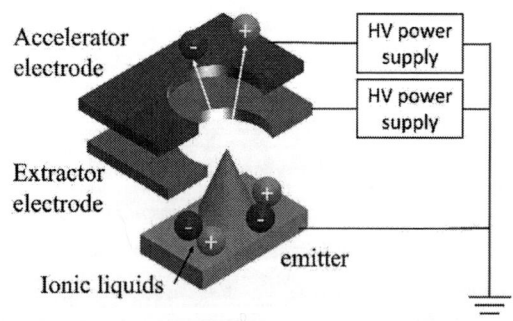

Figure 1: Concept of ionic liquid electrospray thruster with two-stage electrodes.

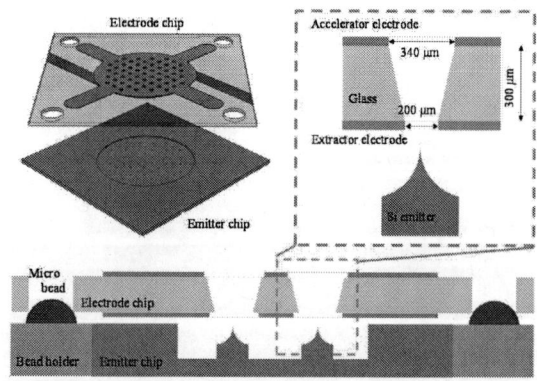

Figure 2: Device structure of ionic liquid electrospray thruster with double-side electrode.

978-1-6654-9309-3/23 $31.00 © 2023 IEEE

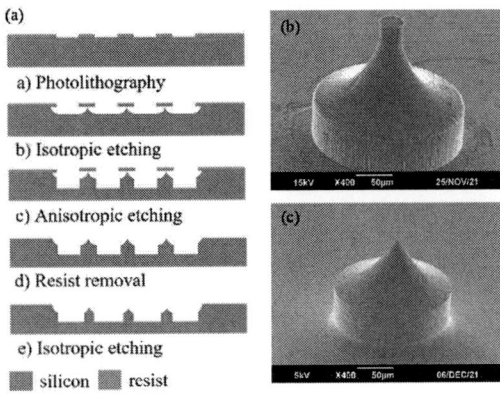

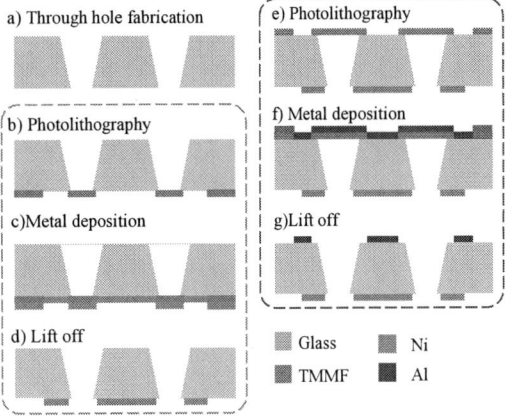

Figure 3: a) Fabrication process of silicon emitter electrodes, b) after deep RIE process and photoresist removal and c) after isotropic etching thinning.

Figure 5: Fabrication process of electrode chip.

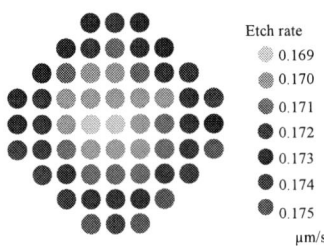

Figure 4: Etching rate distribution in an electrode chip at wafer center.

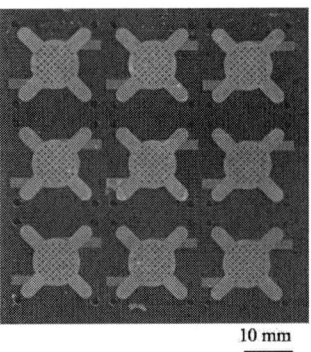

Figure 6: Fabricated electrode chip before dicing.

chip is made from a 4-inch glass wafer of 300 μm thick. 57 tapered via holes are arranged at the corresponding positions to the emitters. The diameters of the holes at bottom (emitter side) and top are 200 and 340 μm, respectively. The electrode chip is designed to be placed diagonal direction of the emitter chip to use the chip corners for the alignment described below.

To reduce the extracting voltage and to realize stable and uniform emission, the alignment accuracy between the two chips both horizontal and vertical directions is critically important. The alignment accuracy is ensured by the photolithography precision and the mechanical assembly using beads [2]. The emitter and electrode chips are fabricated with high precision because of semiconductor microfabrication technology. For the mechanical assembly, the four holes are made on the corners of the glass electrode chip and steel microbeads of 2 mm diameter is fit to the hole. The microbeads are placed in each bead holder.

Emitter chip

Figure 3a shows the fabrication process of the silicon emitter chip. Circular photoresist masks were patterned and SF_6 isotropic dry etching to form a conical emitter shape and then Bosch process to gain height was conducted using ICP-RIE successively. One of the fabricated emitters is shown in Fig. 3b. The diameter of the cylindrical part is 220 μm and the total height is about 150 μm. The diameter of the remained circle plane on the top is 30 μm and there

is a variation of about 7 μm in the chips. The difference is caused by the etching rate non-uniformity in each chip and the whole wafer. To sharpen the tip, the additional isotropic dry etching was conducted (Fig. 3c). The tips were sufficiently sharp, but the height of the emitter reduced by 50 μm.

The etching rate nonuniformity is evaluated. Figure 4 shows the etching rate map measured by observation of all the emitters on the chip that was taken from the wafer center. The rate at the center is 3% higher than the that at the edge. This is caused by the etching gas supply. In addition, the chip at the wafer edge showed higher etching rate of more than 5 % of that at the center. In this study, the insufficient etching shown in Fig. 3b is corrected by the additional isotropic etching. The additional etching makes the emitter height of smaller, which is not good to reduce the extractor voltage and we will consider the compensation by the mask diameter to mitigate the etching rate variation.

Electrode chip

Figure 5 shows the fabrication process of the glass chip. The via holes on the glass wafer were fabricated using laser-assisted chemical etching process. We also had examined dry etching of glass wafer after electrode fabrication. But the dry etching is too slow to make via hole on glass wafer of 300 μm thick. The technical problem for the current process is the deposition and patterning of the electrodes on both sides. To avoid electrical short circuit between electrodes, nickel and aluminum film electrodes were patterned by lift-off process. Dry film photoresist sheets were used to apply the photoresist on wafers with via holes. Fabricated electrode chips before blade dicing are shown in Fig. 6.

Assembly

The silicon emitter chip, glass electrodes chip, steel microbeads and bead holders were assembled using polytetrafluoroethylene (PTFE) parts, base plate, and top plate, as shown in Figure 7. The silicon chip was placed on the base plate and aligned by the four bead holders made of polyether ether ketone (PEEK). The alignment between the base and the glass chip is made by four steel microbeads of 2-mm diameter. The gap between the emitter tips and extractor electrode is controlled by the diameter of the holes on the corner of glass chip. In this experiment, the alignment hole is 1.48 mm in diameter and the gap between chips is designed as 400 μm. The electrode chip is fixed by placing the top plate using spring plungers. The spring plungers are used for electrical connection of accelerator

electrode.

Figures. 8a-8c show the assembling procedures. The bead holders were inserted to the holes of the base plate and the microbeads are put on the base plate (Fig. 8a). The emitter chip was put on the base plate aligned by the edge of the bead holders (Fig. 8b). The electrode chip was put on the microbeads (Fig 8c). The electrical connection of extractor electrode was made using a conductive tape as seen at the bottom left of the figure. The assembled device is shown in Fig. 8d. The top plate was fixed by PEEK screws.

EXPERIMENTS

1-Ethyl-3-Methylimidazolium Dicyanamide (EMI-DCA) was used as an ionic liquid propellant. The properties are listed in Table 1 [6, 7]. The propellant of about 1 μL was supplied to the reservoir of emitter chip before assembly. The emitter was processed by helium atmospheric pressure plasma to make the surface wettable to EMI-DCA. The assembled ILEST device was put in a custom-made vacuum chamber. The ion emission experiments were conducted in a vacuum at 3.0×10^{-3} Pa or lower.

The measurement setup was drawn in Fig. 9. To apply voltages to the emitter and extractor, high power source-measure units (SMU; 2657A, Keithley) were used. The currents of the collector and accelerator were measured by a SMU (2612B, Keithley). Bipolar square wave of 1Hz was applied and the emitter voltage was increased by 100V in the range from 0 to 1500 V and 10 V at 1500 V and higher. The extractor voltage was increased by 100 V to 500 V and fixed. Therefore, the voltage between accelerator and extractor was kept at 500 V, while the voltage between emitter and extractor was increased from 0 V to 2500 V.

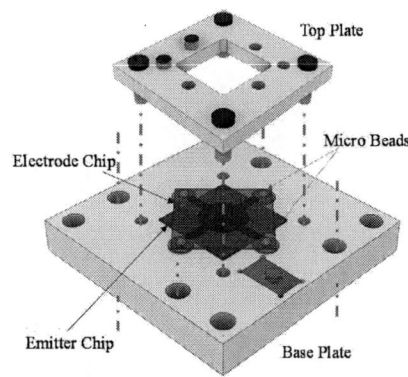

Figure 7: Schematic drawing of ILEST assembly.

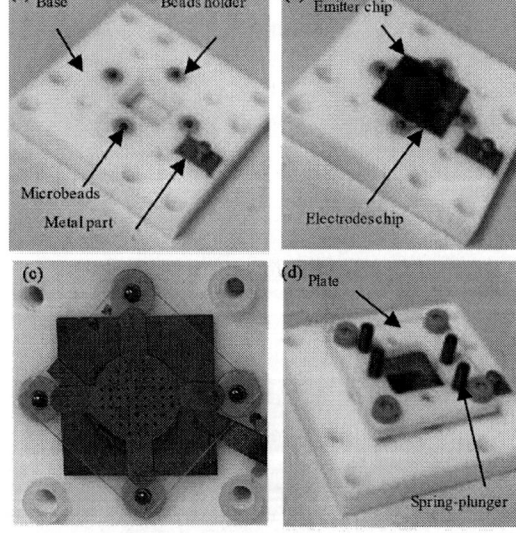

Figure 8: Fabricated jig parts and assembled ILEST.

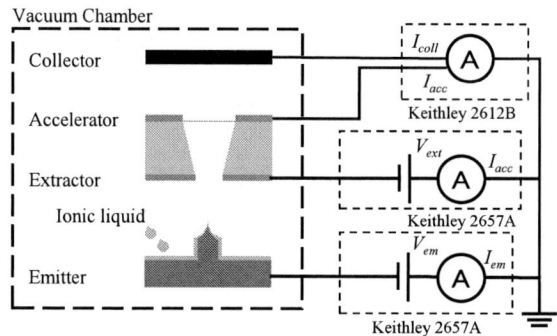

Figure 9: Ion emission test set-up.

Table 1: Propertiess of EMI-DCA (25 °C, 1 atm)

Electrical conductivity	2.8 S/m
Viscosity	0.021 Pa·s
Density	1080 kg/m³
Surface tension	0.04905 N/m
Cation mass	111.2 Da
Anion mass	66 Da

RESULTS AND DISCUSSION

Figures 10 show the measured current at each electrode when the distance between the emitters and extractor is about 400 μm. The horizontal axis indicates the voltage difference between emitter and extractor electrodes. The starting voltages of extracting cations and anions were 1810 V and −1860 V, respectively. The largest ion currents were 12 μA and −13 μA, respectively, in the stable emission range. Majority of the extracted ions collided to the extractor, and about 10% of extracted ions were detected on the collector. The accelerator current was almost zero, which shows the two-stage electrodes worked adequately. Unstable large current was detected over ±2 kV. This would be caused by droplets extraction. The droplets lead to lower specific impulse due to large mass flow rate.

We suspect that the main reason of low transmittance through via holes would be the misalignment of the electrode chip to the emitter chip. There was a significant rotational misalignment of the emitter chip since the chip position was guided by the edges of the bead holders and the chip often rotated during the assembly process. To increase the ion current on the collector, the hole diameter should be larger, and the electrode gap should be much smaller. But the former reduces the number of emitters per area and the total emissions will decrease. To reduce the gap between the emitter and glass electrode, ionic liquid should be supplied from the back side the chip to avoid unintended discharge during the operation.

CONCLUSION

The two-stage electrode ILEST using glass substrate was designed, fabricated, and tested. We successfully demonstrate ion emission but the emission to collector is small. The assembly methods should be improved to reduce ion corrosion to the extractor electrode and decrease the starting voltage. We also plan to employ a new propellant supply method from the backside of the emitter chip.

ACKNOWLEDGEMENTS

This study was funded by the Canon Foundation and Grant-in-Aid for Scientific Research of JSPS JP18H01623 and JP21H01530. A part of this work was supported by Kyoto University Nanotechnology Hub in "Advanced Research Infrastructure for Materials and Nanotechnology Project" sponsored by MEXT, Japan.

REFERENCES

[1] D. G. Courtney, H. Shea, K. Dannenmayer and A. Bulit, Charge Neutralization and Direct Thrust Measurements from Bipolar Pairs of Ionic Electrospray Thrusters, *J. Spacecraft and Rockets*, 55 (2018) pp. 54-65.

[2] M. A. C. Silva, D. C. Guerrieri, A. Gervone and E. Gill, A review of MEMS Micropropulsion Technologies for CubeSats and PocketQubes, *Acta. Astronautica.* 143 (2017) pp.234-243.

[3] S. Dandavino, C. Ataman, C. N. Ryan, S. Chakraborty, D. Courtney, J. P. W. Stark and H. Shea, Electrospray Emitter Arrays with Integrated Extractor and

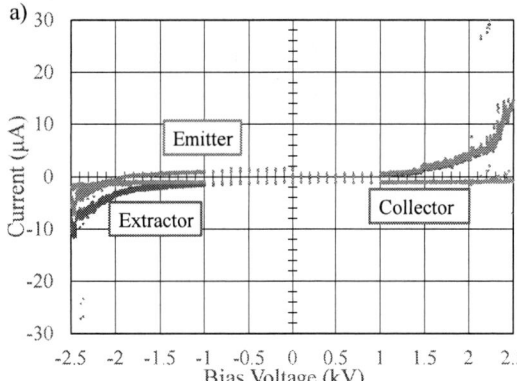

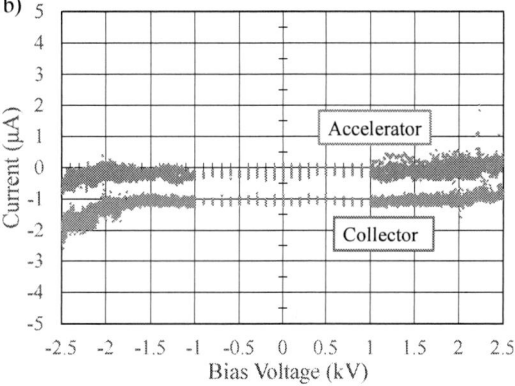

Figure 10: Measured current at each electrode as function of bias voltage between emitter and extractor electrodes. a) Currents at emitter, extractor, and collector electrodes. b) Magnified plots of collector and accelerator electrodes.

Accelerator Electrodes for the Propulsion of Small Spacecraft, *J. Micromech. Microeng.* 24 (2014) 075011.

[4] T. Henning, K. Huhn, L. W. Isberner and P. J. Klar, Miniaturized Electrospray Thrusters, *IEEE Trans. Plasma Science*, 46 (2018), pp. 214-218.

[5] K. Nakagawa, T. Tsuchiya, Y. Takao, Microfabricated Emitter Array for an Ionic Liquid Electrospray Thruster *Jpn. J. Appl. Phys.* 56 (2017), 06GN18.

[6] S. Zhang, N. Sun, X. He, X. Lu, and X. Zhang, Physical Properties of Ionic Liquids: Database and Evaluation, *J. Phys. Chem. Ref. Data*, 35 (2006) pp. 1475-1517.

[7] W. Martino, J. de la Mora, Y. Yoshida, G. Saito, and J. Wilkes, Surface Tension Measurements of Highly Conducting Ionic Liquids, *Green Chem.*, 8 (2006), pp. 390-397

CONTACT

*T. Tsuchiya, tel: +81-75-383-3690;

tutti@me.kyoto-u.ac.jp

MONOLITHIC INTEGRATION OF PZT ACTUATION UNITS OF VARIOUS ACTIVATED RESONANCES FOR FULL-RANGE MEMS SPEAKER ARRAY

Hsu-Hsiang Cheng[1], Sung-Cheng Lo[1], Yu-Chen Chen[1], Ming-Ching Cheng[1], Ting-Chou Wei[1], Mingching Wu[2], Weileun Fang[1]

[1]Power Mechanical Engineering, National Tsing Hua University, Hsinchu, TAIWAN
[2]CoretronicMEMS Co., Ltd., Hsinchu, TAIWAN

ABSTRACT

This study demonstrates the concept to monolithically integrate piezoelectric actuated structures of different natural frequencies on an SOI wafer to achieve the full-range MEMS speaker array. First of the two major points featured in this research is leveraging the characteristic of micromachining technology to design, fabricate, and integrate multiple PZT actuation units on a single chip to enhance the performance of microspeaker. The second is the design of novel driving electrodes on two bridge structures and one clamped diaphragm (5 mm by 1.5 mm) activating different resonant modes to increase SPL (Sound Pressure Level) at low, medium and high frequencies. As a result, the full-range MEMS speaker array can be realized. Measurements in the standard ear simulator illustrate that with 3.5 V_{rms} driving voltage, the proposed microspeaker has SPL higher than 81dB in the full audio range, and even larger than 90 dB from 60 Hz to 15 kHz.

KEYWORDS

MEMS speaker array, piezoelectric, PZT, full range.

INTRODUCTION

Along with the development over the past several years, the advantages and limitations of MEMS speakers have been well revealed in much research. In spite of smaller SPL at low frequencies compared to traditional dynamic speakers, MEMS speakers leveraging features of micromachining technologies have an upper hand in integration of various devices and cost reduction. In the condition of having smaller size and lower power consumption, MEMS speakers even provide a promising solution for consumer electronics, especially in-ear applications.

Among existing MEMS speakers, electrostatic type suffers lower electroacoustic efficiency although generally CMOS compatible. Not to mention the possible pull-in issue due to the initial gap between electrodes [1]. When it comes to remarkable capability at low frequencies, electrodynamic microspeakers may be preferred, yet the demand for magnet integration is a barrier to achieve smaller size and multi-unit integration on a single chip [2].

Piezoelectric MEMS speakers, taking benefits of high efficiency of electro-mechanical energy conversion and large driving force of piezoelectric films, possess advantages of low power consumption and high SPL. Some research of structure design to enhance SPL at low frequencies has been presented [3]. Recently, studies of bandwidth improvement to realize full-range devices also attract more interests [4-5].

This study presents the monolithic integration of piezoelectric MEMS speakers with three frequency units

for full-range applications. The PZT film is employed as the driving material in this work because of the highest transverse piezoelectric coefficient d_{31} among common piezoelectric materials [6].

DESIGN CONCEPT

To achieve a full-range MEMS speaker array, this study utilizes two bridges and one fully clamped diaphragm with the design of novel electrode pattern, as illustrated in Fig.1. Dimensions of all three diaphragms are 5 mm by 1.5 mm. As shown by A-A' cross-section view in Fig.1, three diaphragms have their own independent back chambers, so as to minimize the undesired decrease of out-of-plane displacement due to air coupling between each diaphragm. In addition, the top electrodes will be partitioned into two driving units: electrode A and electrode B to provide opposite bending deformation to enlarge out-of-plane displacement. Most importantly, the novel electrode pattern is designed to activate specific resonant modes of each diaphragm. The stimulated modes of three structures locate at low, medium, and high frequency range (around 1, 4, and 11 kHz respectively). As a results, it's expected that SPL in full audio range can be increased.

Table.1 depicts the simulation results of activated resonant frequencies of each frequency unit. Considering residual stress of -20 MPa in PZT film and 100 MPa in Cr/Au film, the resonances designed to stimulate will shift to higher frequencies, particularly the clamped diaphragm. Simulations in Fig.2 show the feasibility of the three-unit piezoelectric MEMS speaker. The active resonant modes to boost SPL at each frequency range are presented in

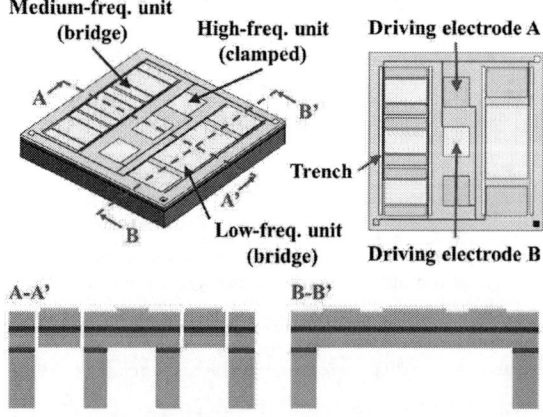

Figure 1: Design concept of the piezoelectric MEMS speaker with the structure of two bridge structures and one fully clamped diaphragm, and two separate driving electrode units providing opposite bending deformation.

Table 1: The activated resonant frequencies of each frequency unit with and without residual stress.

Frequency Unit	Low	Medium	High
Without residual stress	0.7 kHz	3.9 kHz	11.4 kHz
With residual stress	1.6 kHz	5.7 kHz	21.3 kHz

Fig.2a. These results verify the novel electrode pattern can drive higher modes of the structures. The frequency responses in Fig.2b further confirm the concept of SPL superposition of the three frequency units. Fig.2c exhibits the comparison between different driving methods in consideration of residual stress. The acoustic simulation indicates residual stress will not only cause higher resonant frequencies, but also induce a phase shift between low- and medium-frequency units. When electrode A and B are driven with out-of-phase signal, the phase shift will cause an out-of-phase movement between low- and medium-frequency units leading to destructive interference of sound wave, which will result in significant SPL drop. Hence, in-phase driving is expected to gain higher SPL at low and medium frequency range.

In order to realize a full-range MEMS speaker aimed to in-ear applications, the three-unit array will be fabricated on a single chip by utilizing micromachining technology. Besides, the PZT film with d_{31} coefficient of -39.8 pC/N is employed as the driving material.

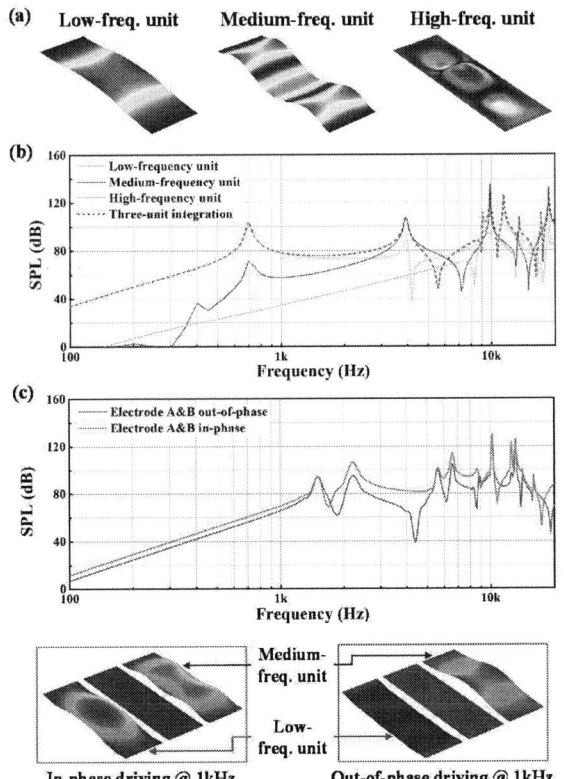

Figure 2: Simulations of (a) Activated resonant modes of three frequency units, (b) frequency responses of single unit and three-unit integration without residual stress, (c) frequency response and vibration modes of in-phase and out-of-phase driving with residual stress in PZT and Cr/Au thin films.

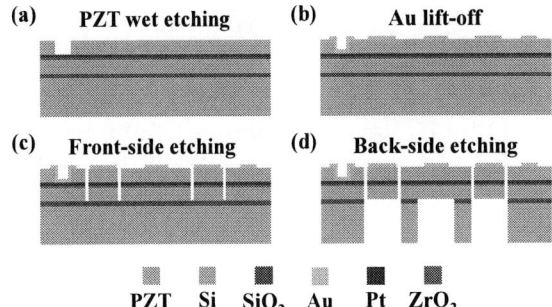

Figure 3: Fabrication process steps.

FABRICATION

Fig.3 illustrates the fabrication process steps of this work. First, 15 nm ZrO_2 insulation layer and 150 nm Pt bottom electrode were deposited on an SOI wafer with device layer thickness of 2 µm. Served as insulation layer to avoid leakage current, ZrO_2 was also applied as the adhesion layer for Pt bottom electrode. Then, 2 µm PZT film was sputtered and patterned by wet etching for bonding pads of the bottom electrode (Fig.3a). After that, Cr/Au (50 nm/300 nm) thin film was deposited by E-gun and patterned through lift-off to realize the routings and novel electrode design (Fig.3b). The front-side dry etching was then employed to define two bridge structures (Fig.3c) Subsequently, the silicon handle layer was etched by DRIE to define the back chambers of three driving units. Finally, the diaphragms of speaker array were released through removal of SiO_2 by RIE (Fig.3d).

Fig.4 displays a fabricated MEMS speaker chip wire-bonded on PCB served as the DUT (Device-Under-Test). An acoustic hole with diameter of 1 mm is in the center of PCB, as shown in the schematic view.

MEASUREMENT

To verify the residual stress of -20 MPa in PZT and 100 MPa in Cr/Au applied in simulation is reasonable, the warpage of low-frequency unit after structure releasing was measured by 3D optical profilometer. As demonstrated in Fig.5a, the maximum deformation (highest point minus lowest point) of bridge structure is around 14 µm, while the one calculated by simulation is about 12 µm. In addition, the surface profile of deformed structure is comparable between the measurement and simulation.

The acoustic performances of the MEMS speaker array were measured in the anechoic box. The setup of the standard ear simulator system (G.R.A.S. 43AG-6) is displayed by the photo and schematic view in Fig.6. The customized acrylic adapter was introduced to mount the DUT on the coupler (G.R.A.S. RA0401) flat and stably. Then, the signal input for the DUT was supplied by the

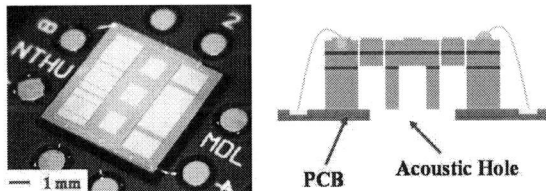

Figure 4: The micrograph and schematic view of wire-bonded MEMS speaker chip.

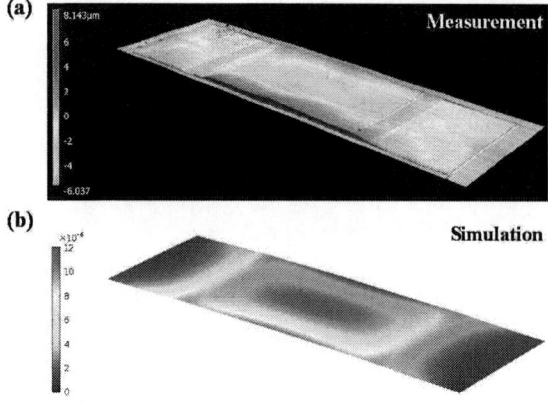

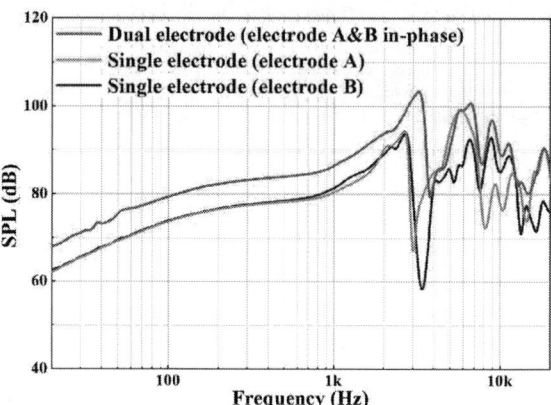

Figure 5: (a) Measurement, and (b) simulation results of low-frequency unit structure deformation under residual stress.

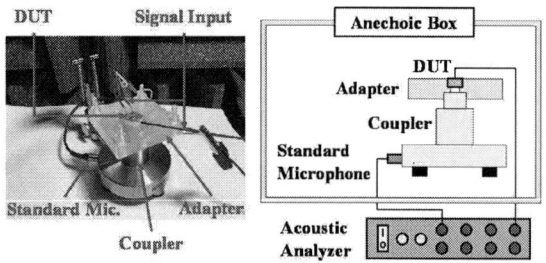

Figure 6: The setup of the standard ear simulator system.

B&K audio analyzer. At last, the standard microphone (G.R.A.S. 40AG) detected the output signal of DUT, which was also recorded by the analyzer.

Figs.7-10 depict the acoustic measurement results. Measurements in Fig.7 reveal the SPL variation with phase difference between AC voltage applied to the electrodes-A (in blue in Fig.1) and electrodes-B (in yellow in Fig.1). At 1 kHz, driven by 0.7 V_{rms}, the MEMS speaker array possesses the highest SPL under in-phase driving. On the other hand, out-of-phase driving shows the worst

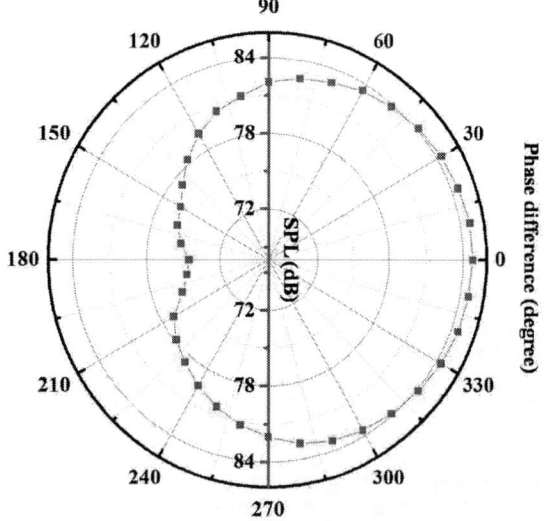

Figure 7: SPL at 1 kHz with various phase difference between AC voltage applied to electrodes A and B.

Figure 8: Comparison of frequency responses with different approaches from 20 Hz ~ 20 kHz.

performance in terms of SPL. These results are consistent with the expectation of simulation results in Fig.2c. Moreover, the phenomenon of residual stress induced phase shift can explain the contradiction results in the previous work [7].

Fig.8 shows the frequency responses of proposed MEMS speaker array with different driving approaches (0.7 V_{rms}). When the DUT is driven with single electrode A or B, the frequency responses demonstrate comparable SPL before the first mode of bridge structure. But the ones at medium to high frequency range are distinct from each other due to the different sensitivity to electrodes A and B of the higher resonant modes of bridge and clamped diaphragms. With single-electrode driving, the undesirable fluctuations of frequency responses larger than 30 dB SPL (after 100 Hz) are observed as well. In the case of dual-electrode actuation, there is around 6 dB enhancement of SPL before the first resonant peak. Furthermore, the characteristic of superposition of SPL not only illustrates better performance at low frequencies, but also benefits the high SPL ranges respectively contributed by electrodes A and B after the first resonance. The feature results in a flatter frequency response over the audio range. The full-range MEMS speaker is thus realized with SPL higher than 80 dB from 110 Hz to 20 kHz with only 0.7 V_{rms} input.

Fig.9 depicts the frequency responses of in-phase driving at different AC signal. With 3.5 V_{rms} driving voltage, SPL higher than 81 dB is achieved in the full audio range. Moreover, from 60 Hz to 15 kHz, the SPL of proposed design is even larger than 90 dB. The linearity at low frequency range is also favorable. Take the case of SPL at 1 kHz, the linearity reaches 0.996, as shown in Fig.10.

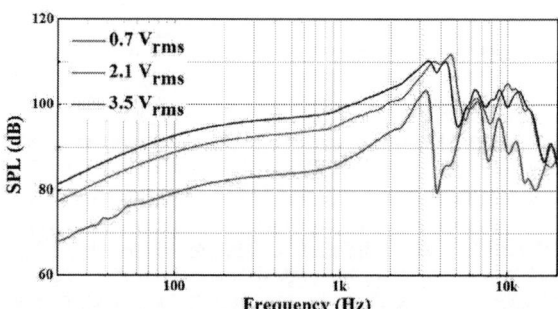

Figure 9: Frequency responses with different AC signal.

978-1-6654-9309-3/23 $31.00 © 2023 IEEE

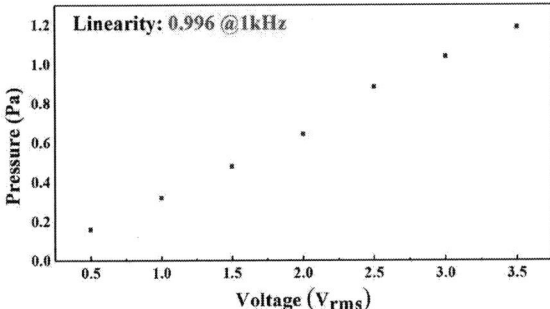

Figure 10: Linearity of proposed design at 1 kHz.

CONCLUSIONS

A full-range piezoelectric MEMS speaker was proposed and fabricated in this work. With the novel design of electrode pattern, the higher mode of bridge and clamped diaphragms were successfully activated to increase the SPL at low, medium, and high frequencies. Leverage the characteristic of micromachining technology, the designed chip monolithically integrated three frequency units to form a MEMS speaker array. Due to the phase shift caused by residual stress, in-phase actuation exhibits the best performance in terms of SPL. Driven at 0.7 V_{rms} without needing to apply DC voltage, the proposed MEMS speaker array demonstrates the SPL higher than 80 dB from 110 Hz to 20 kHz in the standard ear simulator. Moreover, with 3.5 V_{rms}, higher than 90 dB SPL from 60 Hz to 15 kHz is measured.

ACKNOWLEDGEMENTS

This project was supported by CoretronicMEMS Co. Ltd, and the National Science and Technology Council under grant number NSTC 111-2923-E-007 -003 -, NSTC 111-2221-E-007-069-MY3, NSTC 111-2221-E-007 -070 - MY3, NSTC 111-2218-E-007-014 -MBK, NSTC 111-2923-E-007-014 -MY2, NSTC 110-2926-I-007 -506 -, and NSTC 110-2218-E-007-032-. The authors would like to thank the Center for Nanotechnology, Materials Science, and Microsystems of National Tsing Hua University, and Taiwan Semiconductor Research Institute for providing the fabrication and measurement facilities.

REFERENCES

[1] B. Kaiser, S. Langa, L.Ehrig, M Stolz, H. Schenk, H.Conrad, H. Schenk, K. Schimmanz, and D. Schuffenhauer, "Concept and proof for an all-silicon MEMS micro speaker utilizing air chambers," Microsystems & Nanoengineering, 5 (2019), 43.

[2] I. Shahosseini, E. Lefeuvre, J. Moulin, E. Martincic, M. Woytasik, and G. Lemarquand, "Optimization and microfabrication of high performance silicon-based MEMS microspeaker," IEEE Sensors Journal, 13 (2013), pp. 273-284.

[3] H.-H. Cheng, S.-C. Lo, Z.-R. Huang, Y.-J. Wang, M. Wu, and W. Fang, "On the design of piezoelectric MEMS microspeaker for the sound pressure level enhancement," Sensors and Actuators A, 306 (2020), 111960.

[4] Y.-T. Lin, S. Lo, and W. Fang, "Two-way piezoelectric MEMS microspeaker with novel structure and electrode design for bandwidth enhancement," 2021 Transducers, Orlando, FL, USA, June 20-26, 2021, pp. 230-233.

[5] Y.-J. Wang, S. Lo, M.-L. Hsieh, S.-D. Wang, Y.-C. Chen, M. Wu, and W. Fang, "Multi-way in-phase/put-of-phase driving cantilever array for performance enhancement of PZT MEMS microspeaker," 2021 IEEE MEMS, Gainesville, FL, USA, January 25-29, 2021, pp. 83-86.

[6] S.-J. Chen, Y. Choe, L. Baumgartel, A. Lin, and E.S. Kim, "Edge-released, piezoelectric MEMS acoustic transducers in array configuration." Journal of Micromechanics and Microengineering, 22 (2012), 025005.

[7] H.-H. Cheng, S.-C. Lo, Y.-J. Wang, Y.-C. Chen, W.-C. Lei, M.-L. Hsieh, M. Wu, and W. Fang, "Piezoelectric microspeaker using novel driving approach and electrode design for frequency range improvement," 2020 IEEE MEMS, Vancouver, BC, Canada, January 18-22, 2020, pp. 513-516.

CONTACT

* W. Fang, Tel: +886-3-5742923; fang@pme.nthu.edu.tw

PULL-IN VOLTAGE REDUCTION IN ELECTROSTATIC AIRGAP ACTUATOR USING 12 NM-ULTRATHIN INTERNAL DIELECTRIC TRANSDUCTION

Satish K. Verma[1], and Bhaskar Mitra[1]
[1]Indian Institute of Technology Delhi, New Delhi, INDIA

ABSTRACT

This paper demonstrates a scheme for reduction of pull-in voltage issue in electrostatic actuator. It utilizes a hybrid actuation mode by combining airgap actuator with internal dielectric transduction in same device. A 12 nm ALD deposited ultrathin dielectric layer (UDL) is used as the internal dielectric actuator along with a 600 nm airgap actuator. The measured results show that the device has pull-in at 1.26 V for airgap actuation and reduces to 0.62 V for hybrid configuration (a 50% reduction) for the same air gap with off-to-on capacitance ratio of 27.

KEYWORDS

MEMS Actuation, Pull-In Voltage, Hybrid Actuator, Internal Dielectric.

INTRODUCTION

Airgap electrostatic actuation is commonly utilized in MEMS applications like RF switches, microwave varactors, microfluidics valves and optical micro-mirrors but it requires high voltages for actuation. The fundamental advantage of MEMS electrostatic actuator is negligible power consumption and ease of fabrication [1]. To reduce the pull-in voltage, gap reduction is usually employed, but reducing gaps beyond a point, leads to fabrication complexities and yield issues. Other clever approaches like curved electrode, nonlinear spring, and variable geometries have been reported to reduce the actuation voltage. But most of them require irregular shapes involving complex fabrication [2] [3]. Internal dielectric transducers (IDT) are studied and applied extensively for MEMS resonators but not for static actuation [4-6].

This work combines IDT with airgap actuator in a hybrid actuation scheme to achieve a lower pull-in voltage. An internal dielectric transduced folded cantilever beam shaped device along with airgap actuator is fabricated using CMOS compatible metals and 12 nm dielectric layer (Figure 1a). The IDT actuator provides additional force to airgap actuator in hybrid configuration (Figure 1b), which forces the device for pull-in at low voltage in comparison to only airgap actuation. Additionally, the fabrication process is discussed, and design principles are well summarized.

DESIGN AND FABRICATION

Design

The fabricated device combines an internal dielectric transducer along with airgap electrostatic device to form a hybrid actuator. The internal dielectric transducer is composed of 12 nm Al_2O_3 ultrathin dielectric layer (UDL) sandwiched between bottom folded cantilever shaped Al structure and top Au electrodes (Figure 1a). The electrostatic force between these electrodes compresses the

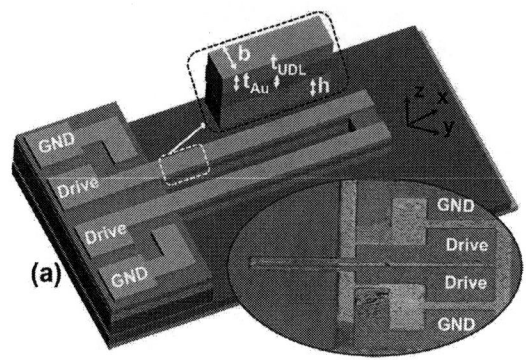

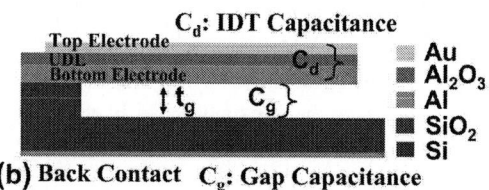

Figure 1: (a) Schematic of hybrid actuator with essential design parameters. Top view microscopic image of fabricated prototype (inset Bottom right). (b) Model of UDL filled hybrid device. The Image is not to scale.

UDL and elongates it due to the Poisson effect. The strain causes the beam to deflect in the transverse direction. The release gap between bottom electrode and substrate with back contact forms another electrostatic airgap actuator in same device (Figure 1b).

The stress and deflection profile of the device is analyzed using COMSOL Multiphysics. The FEM stress results show that the 400 nm Al beam anchors hold 992 kPa of normal stress at $V_{DC} = 1$ V with only airgap actuation. This stress significantly enhanced to 2.25 MPa for same beam thickness and with 10 nm UDL at lower voltage of $V_{DC} = 0.685$ V, when hybrid actuation is employed (Fig. 2a). The normal stress generated in z-direction (σ_z) within the UDL as a function of DC bias voltage is given by,

$$\sigma_z = \frac{\varepsilon_d}{t_{UDL}^2} v_{DC}^2 \qquad (1)$$

Where ε_d is the UDL permittivity. This normal stress translates into longitudinal strain along the x-direction due to the Poisson's effect which deflects the beam in transverse direction (Figure 2b).

The pull-in or actuation voltage for an airgap electrostatic actuator is given as,

$$V_{PI} = \sqrt{\frac{8k_z t_g^3}{27 A \varepsilon_0}} \qquad (2)$$

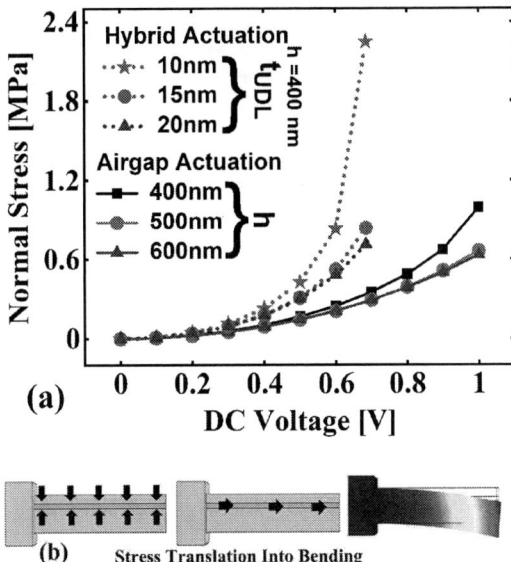

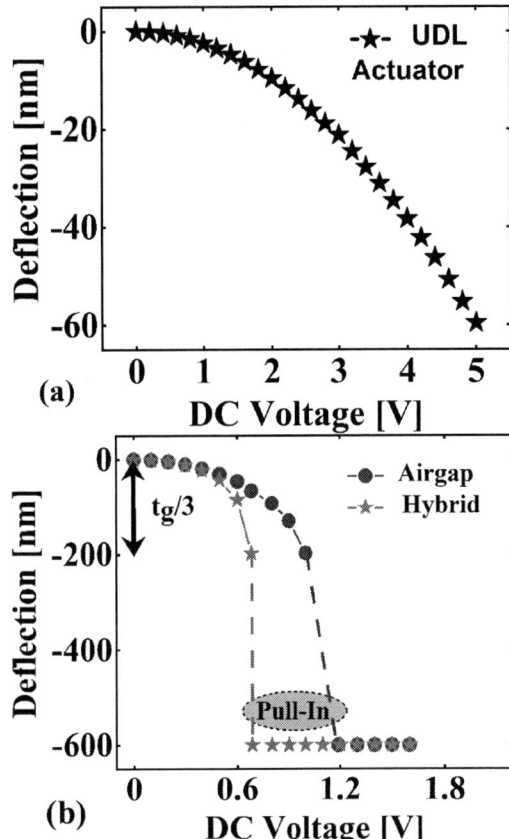

Figure 2: FEM Stress Analysis: (a) Normal stress experienced on beam anchor with DC voltage for airgap actuation (various beam thicknesses) and hybrid actuation (for h= 400 nm and various UDL thicknesses). (b) Illustration of electrostatic compression translation into transverse bending due to Poisson's effect in only UDL actuator. The Image is not to scale.

Where k_z is the spring constant, A is the device overlap area and ε_0 is the air permittivity. Equation (2) infers that a device with low spring constant has lower pull-in voltage.

The developed electrostatic force between top and bottom electrode due to applied voltage acts in the opposite direction of the mechanical restoring force [6]. It effectively softens the spring constant of the structure.

The deflection profile from FEM analysis using COMSOL Multiphysics shows that internal dielectric actuator provides a pull-in free deflection of up to 60 nm at $V_{DC} = 5$ V as shown in Figure 3a. The deflection behavior of the device for only airgap and hybrid actuation with dc voltage is shown in Figure 3b. This additional deflection due to UDL actuator along with the airgap actuator forces the device for pull-in even at lower voltage. Therefore, the simulated device has pull-in voltage of 1.19 V for only airgap actuator, and it reduces to 0.685 V in hybrid mode (Figure 3b). The device specifications are enlisted in Table I.

Table 1: Device Specifications.

Parameter	Value	Parameter	Value
Beam Length (l)	100 μm	UDL (t_{UDL})	12 nm
Beam Width (b)	10 μm	Electrode (t_{Au})	100 nm
Al Thickness (h)	400 nm	Airgap (t_g)	600 nm

Fabrication

The fabrication starts with a 2-inch p-type silicon wafer cleaning and 600 nm SiO₂ growth with thermal oxidation process. 400 nm Al bottom electrode is deposited using thermal evaporation process and it is patterned as folded cantilever shaped structure. 12 nm

Figure 3: (a) FEM predicted pull-in free, up to 60 nm of deflection at 5V dc, when only UDL actuator is active. (b) FEM results of deflection vs. dc voltage for airgap and hybrid actuator.

Al₂O₃ is deposited by atomic layer deposition (ALD) process as ultrathin dielectric layer and patterned for contact via opening for bottom electrode. The top electrode of the device is formed by depositing 100 nm of Au layer using thermal evaporation process followed by patterning. A 100 nm Al is deposited on backside of the wafer for back contact. Finally, device is stiction free released by etching sacrificial SiO₂ etching using vHF. The fabrication involves 2 dielectrics, 2 metals with 3 lithographic steps. The process is BEOL-CMOS compatible without any active materials (piezo/magnetic) or high-temperature processes (Figure 4a). The cross-sectional FESEM image of the fabricated device is shown Figure 4b.

EXPERIMENTAL RESULTS

Electrical stability of 12-nm UDL is of practical relevance since atomically thin dielectric films are prone to breakdown. To characterize dielectric quality, the device's pull-in behavior IV and CV measurement were performed using Keithley SCS-4200 parameter analyzer (Figure 6a).

The IV characteristics of MIM structure which forms the UDL actuator shows a breakdown voltage (V_{BD}) of ~7.7 V (Figure 5a). Additionally, we developed a FEM model to comprehend the device's safe working range for varied dielectric thicknesses. For 10-nm UDL ($v_{Drive} = 5\ V$), measured breakdown field 7.7 MV/cm > FEM predicted

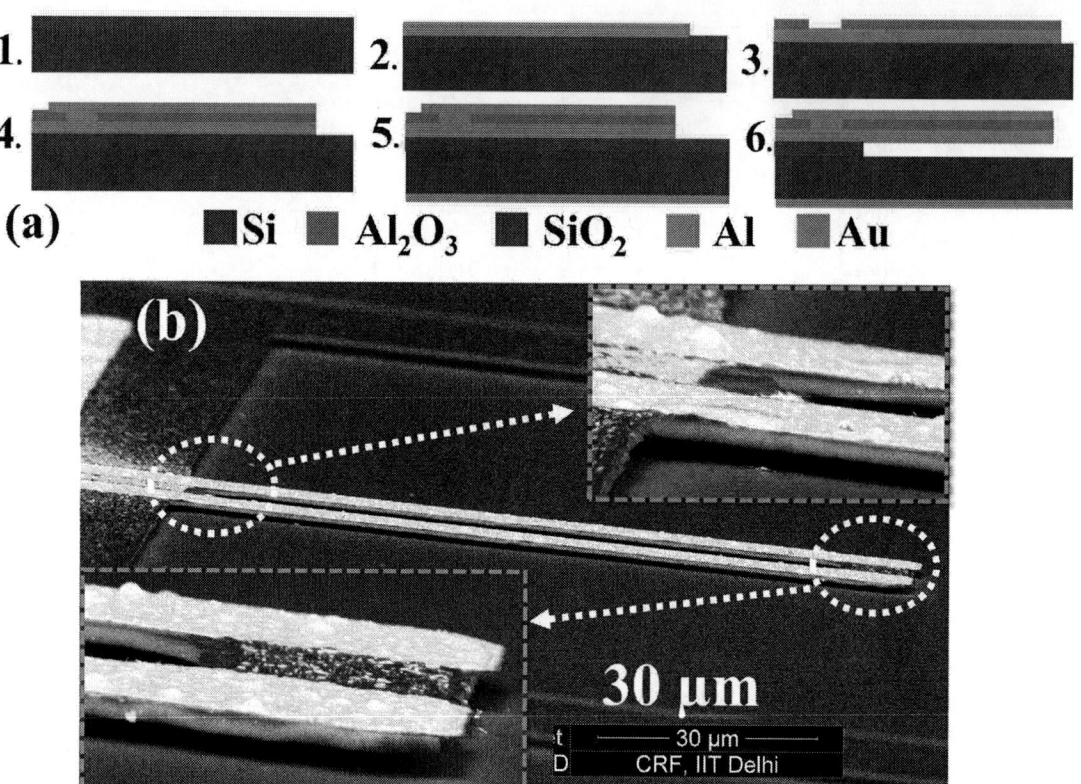

Figure 4: (a) The 6-step process fabrication process includes 1) Si oxidation, 2) Al deposition and bottom electrode patterning, 3) 12 nm UDL deposition using ALD, and via patterning for electrical connection, 4) Au deposition and top electrode patterning, 5) Backside Al deposition for back contacts, 6) Stiction free vHF device release. (b) FESEM image of fabricated folded cantilever with magnified view of anchor (top right) and free end (bottom left) of the device.

electric field is 2.5 MV/cm as shown in Figure 5b.

The CV measurement is carried using triaxial cables after performing standard instrument calibration to eliminate the parasitic effects. All the CV measurements (airgap and hybrid) are executed separately by varying the DC bias voltage and sensing airgap capacitance (C_g) with 30 mV high frequency ac signal. Figure 6a shows the measurement testbench with configuration setup for hybrid mode actuation measurement. The measured data shows that the device has pull-in voltage (V_{API}) of 1.26 V with only airgap actuation and it reduces to hybrid pull-in voltage (V_{HPI}) at 0.62 V in case of hybrid actuation with

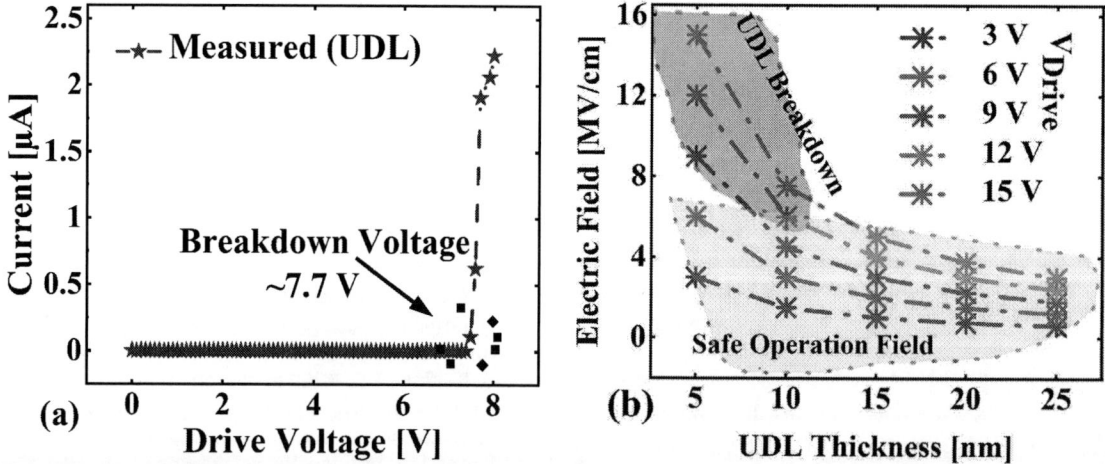

Figure 5: UDL thickness analysis (a) Measured I-V characteristics of MIM with 10-nm UDL (E_{BD} =7.7MV/cm). (b) Electrical stability of UDL is critical to prevent it from breakdown and to operate it into safe electric field: For 10-nm UDL (v_{Drive} = 5 V), measured breakdown field 7.7 MV/cm > FEM predicted electric field is 2.5 MV/cm.

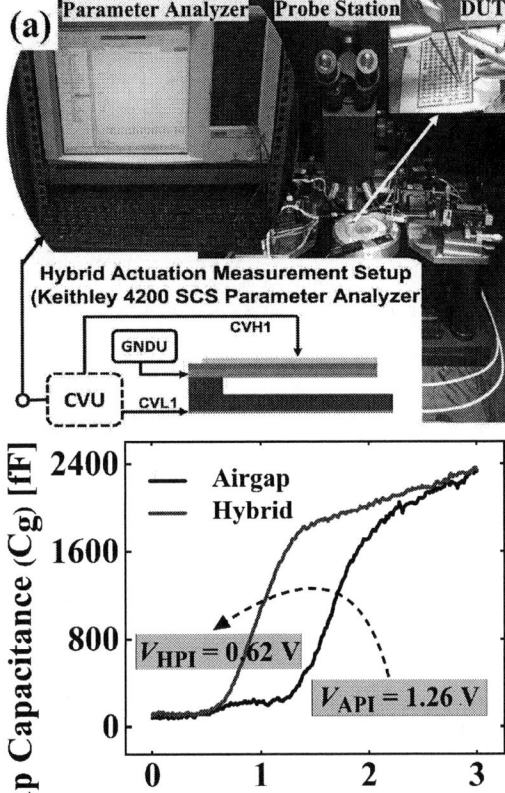

Figure 6: (a) Measurement testbench with Everbeing probe station, Keithley SCS-4200 parameter analyzer, DUT, and configuration setup for hybrid mode actuation measurements. (b) airgap capacitance vs. dc voltage for airgap and hybrid configurations, each measured in separate experiments.

off-to-on capacitance ratio of 27 (Figure 6b). The measured pull-in voltage drops by 50.8 % whereas FEM results predicts 42.2% of reduction. Therefore, the measured data is in good agreement with FEM results as listed in Table 2.

Table 2. FEM and Measured Results

	Pull-In Voltage (V)		
	Airgap	Hybrid	% Reduction
FEM	1.19	0.685	42.43
Measured	1.26	0.62	50.79

CONCLUSION

The paper experimentally demonstrated an effective technique to reduce pull-in voltage using electrical spring softening effect. An airgap electrostatic actuator and internal dielectric transducer is fabricated in same device. A 12 nm ALD deposited UDL is employed as the internal dielectric actuator along with a 600 nm airgap actuator as hybrid mode. We show that the device has pull-in at 1.26 V for airgap actuation and reduces to 0.62 V for hybrid configuration (a 50% reduction) for the same airgap with off-to-on capacitance ratio of 27.

ACKNOWLEDGEMENTS

The device fabrication is supported by DST CRG/20191005777. The fabrication and characterization were performed at NRF and CRF IIT Delhi.

REFERENCES

[1] Pacheco, Sergio P., Linda PB Katehi, and CT-C. Nguyen. "Design of low actuation voltage RF MEMS switch." In *2000 IEEE MTT-S International Microwave Symposium Digest (Cat. No. 00CH37017)*, vol. 1, pp. 165-168, 2000.

[2] M. Shimofuri, A. Banerjee, Y. Hirai and T. Tsuchiya, "Observation of Pull-In by Casimir Force in MEMS-Controlled Nanogap Fabricated by Silicon Cleavage," 2022 IEEE 35th International Conference on Micro Electro Mechanical Systems Conference (MEMS), 2022, pp. 511-514.

[3] Shmulevich, Shai, Ben Rivlin, Inbar Hotzen, and David Elata. "A gap-closing electrostatic actuator with a linear extended range." *Journal of Microelectromechanical Systems* 22, no. 5 (2013): 1109-1114.

[4] Ransley, James HT, A. Aziz, C. Durkan, and A. A. Seshia. "Silicon depletion layer actuators." *Applied Physics Letters* 92, no. 18 (2008): 184103.

[5] D. Weinstein and S. A. Bhave, "Internal Dielectric Transduction of a 4.5 GHz Silicon Bar Resonator," 2007 IEEE International Electron Devices Meeting, 2007, pp. 415-418.

[6] Bhave, Sunil A., and Roger T. Howe. "Internal electrostatic transduction for bulk-mode MEMS resonators." In *Solid State Sensor, Actuator and Microsystems Workshop (Hilton Head 2004)*, pp. 59-60, 2004.

[7] Stephen D. Senturia," Dynamics," in Microsystem Design, 1st ed., Springer New York, NY, 2005, pp. 149-178.

[8] M. Shimofuri, A. Banerjee, J. Hirotani, Y. Hirai and T. Tsuchiya, "Nanometer Order Separation Control of Large Working Area Nanogap Created by Cleavage of Single-Crystal Silicon Along {111} Planes Using a MEMS Device," in Journal of Microelectromechanical Systems, 2022.

CONTACT

*Satish K. Verma, eez188151@ee.iitd.ac.in.

A REVERSE ELECTROWETTING-ON-DIELECTRIC (REWOD) ENERGY HARVESTER USING NONWETTING GALLIUM COATED ELECTRODE AND ULTRATHIN GALLIUM OXIDE SHELL AS DIELECTRIC LAYER

Jinwon Jeong, Bokyung Suh, and Jeong Bong (JB) Lee

Department of Electrical and Computer Engineering, The University of Texas at Dallas, Dallas, USA

ABSTRACT

We report a method of REWOD energy harvesting using a bumpy formation of gallium coated (~20 nm) electrode as a nonwetting surface and utilizes naturally formed oxide shell of gallium (~3 nm) as a dielectric layer, while using naturally oxidized Galinstan as a conductive liquid. Compared with Cytop and Parylene-C coated flat electrodes, all liquid metal based energy harvester was able to generate continuous power without wetting on the surface. Maximum power generation of 268 μW was achieved when the gap between two electrodes was 4.5 mm, 7 V_{pp} and 65 Hz applied to a woofer.

KEYWORDS

REWOD, liquid metal, gallium oxide, energy harvest

INTRODUCTION

With the increasing interests in portable devices or wearable sensors, energy harvesting or scavenging which can generate electrical energy by converting various forms of energy sources such as mechanical vibrations, thermal gradients, and lights in surrounding environments has been extensively investigated [1].

Among several methods to harvest energy including triboelectric generators (TENG), and piezoelectric energy harvesters (PEG), reverse electrowetting-on-dielectric (REWOD) energy harvesting has gained a lot of attention with the advantages of harvesting energy without resonance of a solid structure, high power densities, and availability of harvesting using a wide range of mechanical forces and displacements [2]. Even if energy harvesting based on REWOD has a lot of advantages compared to other methods, limitations like strong charge trapping at the interface and evaporation issues in electrolyte resulting from the use of ionic liquids make REWOD energy harvesting difficult to be widely used in various applications [3].

Instead of ionic liquids in REWOD energy harvesting, liquid metal can be considered as a promising charge carrier in harvesting energy because of superior electric conductivity affecting energy production and surface tension which can increase energy per unit area. Krupenkin *et al.* proposed REWOD energy harvester using mercury as conductive liquid with Cytop-coated Ta_2O_5 electrode [4] and showed great promise as an energy harvester using various environmental mechanical energy sources. Nevertheless, it is highly undesirable for mercury to be used in energy harvester as the conductive liquid due to its severe toxicity.

Gallium-based liquid metal alloys such as Galinstan or EGaIn were identified as better material than mercury because they are non-toxic and has higher thermal/electrical conductivity and higher melting point with lower vapor pressure compared to mercury [4]. However, owing to naturally-formed oxide layer such as Ga_2O or Ga_2O_3 [5] which is easily attached to any surfaces because of sticky characteristic, it is tricky to employ gallium-based liquid metal alloys in energy harvesting using REWOD [4]. Wei *et al.* reported that triboelectric nanogenerator using gallium showed ~5x reduction in charge density compared with when using mercury due to gallium wetting on the surface of the friction layer [5].

In our previous work, we showed that thin film (75−200 nm in thickness) of gallium by evaporation formed nanoscale uneven and rough surface through Ostwald ripening with its surface covered with oxide shell. Such nanoscale uneven surface made almost any surface nonwetting to gallium-based liquid metal alloys [6]. In this work, we utilized evaporated gallium as electrodes which are nonwetting to Galinstan (conductive liquid). A naturally formed ~3 nm gallium oxide shell on the surface of Galinstan & gallium electrodes is used as ultrathin dielectric layer in the REWOD energy harvester.

FABRICATION

Figure 1-a1 shows a conceptual schematic of a REWOD energy harvester using liquid metal as electrodes. First, 20 nm chromium as an adhesive layer and 100 nm gold as the electrode were deposited on 3-inch glass wafer using e-beam evaporator (BJD-1800, Temescal, Inc., USA). On top of the Cr/Au layer, ~20 nm gallium was deposited using e-beam evaporator. When the sample was taken out of e-beam evaporator, ~3 nm shell of Ga_2O_3 was immediately formed on the surface of gallium by reacting with oxygen in ambient environment. The deposited gallium layer makes nanoscale uneven surface which makes the surface nonwetting.

To compare, we also prepared another set of samples with Cr/Au electrodes coated with Cytop and Parylene-C. Cytop (CTL-809A, AGC inc., USA) was spin-coated with the thickness of approximately 1 μm followed by thermally-cured at 200 °C for 30 minutes. On the other hand, ~ 1 μm Parylene-C was deposited on the Cr/Au electrodes using Parylene coater (Parylene deposition equipment, SCS, Inc., USA) (Fig. 1-a2).

A ~3 μL Galinstan droplet was placed between top and bottom electrodes with a separation distance of 3.4~6 mm (Fig. 1-b/c). An acoustic wave generated by a woofer was applied on the bottom electrode resulting in vertical oscillation of Galinstan droplet (Fig. 2). The shape of Galinstan liquid metal droplet was constantly deformed resulting from the acoustic vibration of the bottom electrode. The differences in contact area between the deformed liquid metal droplet and the dielectric layer on the top electrode changed the capacitance while

IEEE MEMS 2023, Munich, GERMANY
15 - 19 January 2023

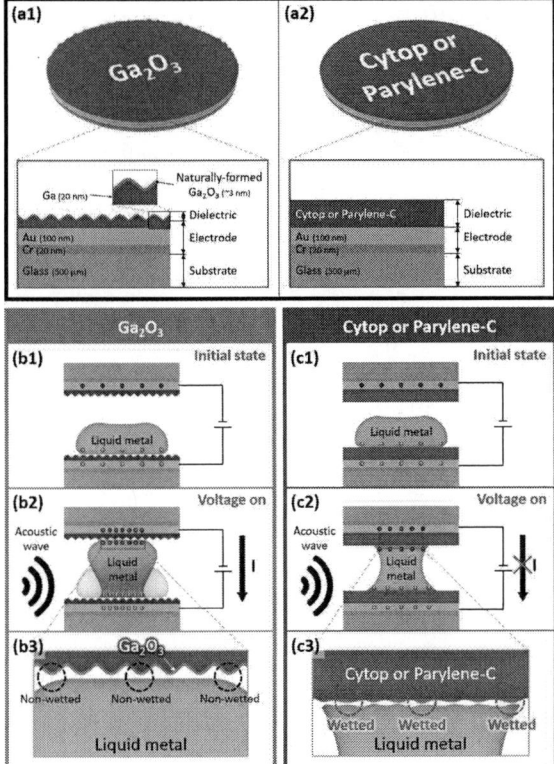

Figure 1: Conceptual schematic of a REWOD energy harvester: (a) gallium-coated nonwetting electrode vs. Parylene-C or Cytop coated electrodes; (b, c) schematics of oscillating Galinstan droplet resulting in nonwetting gallium-coated electrode vs. wetted Parylene-C or Cytop coated electrodes.

maintaining total charges resulting in the generation of an electrical current. The contact area of the Galinstan droplet further varies with variation of the separation distance and voltage applied on the woofer (Fig. 1-b2/c2, Fig. 2). When the woofer was turned off, Galinstan droplet returned to its initial state.

Since all liquid metal energy harvester is nonwetting to the Galinstan droplet, electrical current was continually

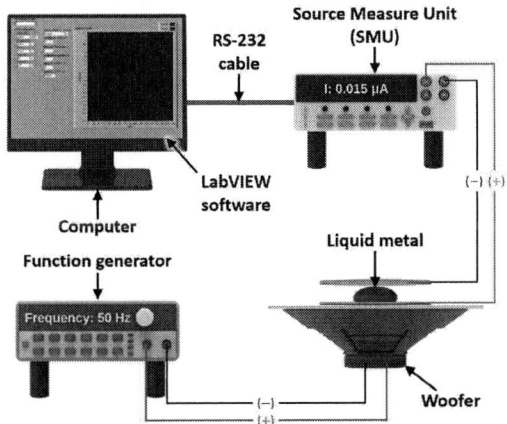

Figure 2: Measurement setup for acoustic-wave energy harvester.

generated (Fig. 1-b2/b3). On the contrary, the energy harvester with Cytop or Parylene-C dielectric layer, Galinstan liquid metal droplet was wetted on the top electrode which prevented it from working as energy harvester within a couple of minutes (Fig. 1-c2/c3).

EXPERIMENTAL RESULTS

Figure 2 shows a schematic diagram of the measurement setup used for energy harvester. A function generator (33120A, Agilent, Inc., USA) was used in order to apply AC voltage to the woofer (GW-210/8, Goldwood Sound, Inc., USA) which vibrates the energy harvester by acoustic wave. A Source Measure Unit (SMU) (Keithley 2400, Tektronix, USA) was used to measure the changes in the power of the energy harvester. The measurement result was recorded through the LabVIEW software.

A high speed camera (Mini CX, Photron, Inc., USA) was utilized to capture a series of optical images of the deformed liquid metal droplet by the vibration resulting from the acoustic wave from the woofer (Fig. 3a) which shows constant variation of contact area of the Galinstan droplet on the surface of the electrode. Figure 3b shows the captured optical images of the deformed liquid metal droplet with the maximum contact area to the top electrode at different heights (5.5 and 6 mm) while applying AC voltages (3 and 6 V_{pp}) to the woofer. From the optical images, we can verify that the large-sized contact area can be achieved with lower height between two electrodes coated with Ga_2O_3 and higher applied voltages to the woofer.

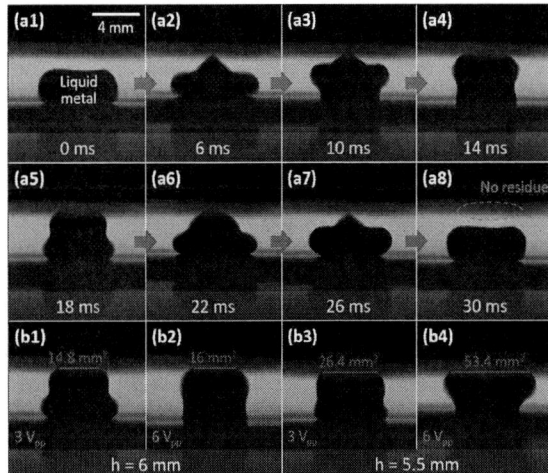

Figure 3: (a) High-speed camera images of the oscillating Galinstan droplet with applied acoustic wave through a woofer, (b) images of the Galinstan droplet showing varying areas of contact on the top electrode with different separation distance and voltages.

The measurement results of the output power by the control of applied AC voltage to the woofer with 5 seconds on/off period is shown in Figure 4. The applied voltage from the function generator was 7 V_{pp} with the frequency of 65 Hz. The gap between two electrodes was fixed to 4.5 mm. When the woofer was turned on, the output power started to be generated while showing similar values of power depending on the contact area between the deformed liquid metal droplet and the top electrode of the energy harvester.

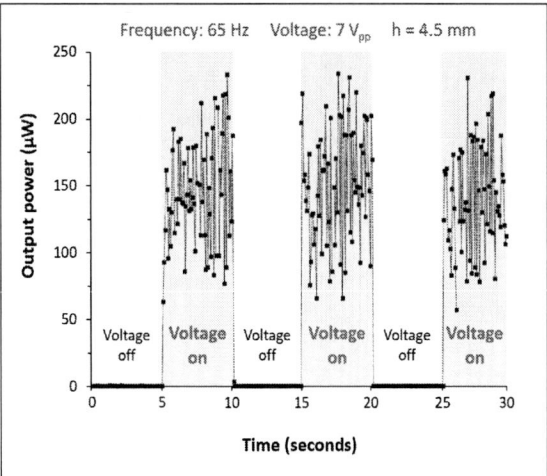

Figure 4: The output power generated from the REWOD energy harvester with sequential on/off acoustic wave generated by a woofer.

Figure 5 shows maximum output power from the fabricated energy harvester with different frequency and applied AC voltages to the woofer. For measurement of the frequency-dependent maximum output power, AC voltages of 7 V_{pp} was applied between two electrodes with the height of 4.5 mm. At the resonant frequency of 65 Hz in the energy harvester, maximum output power of around 268 μW resulting from the largest vibration of the energy harvester by acoustic wave was generated (Fig. 5a). In the case of measurements related to the maximum output power by applied voltages, with the height of 4.5 mm, 65 Hz AC voltages were utilized. From the graph, we can verify that the higher output power can be achieved with the increasing applied voltages by increasing amplitude of vibrations (Fig. 5b).

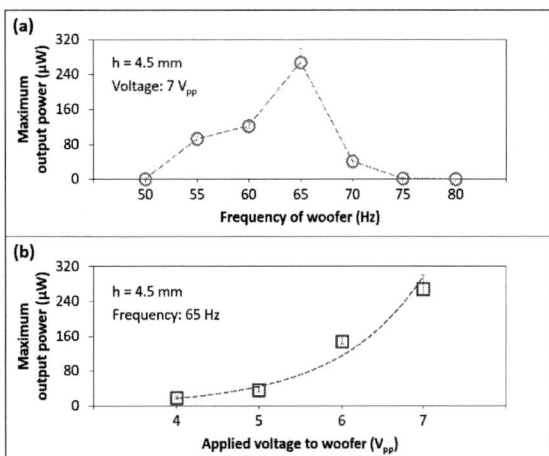

Figure 5: Output power of the REWOD energy harvester (a) as a function of the frequency of the woofer; (b) as a function of the applied voltage to the woofer.

To verify nonwetting characteristic of the nanoscale uneven gallium electrode compared with Cytop and Parylene-C coated flat electrodes, a series of optical images were taken. Figure 6 shows a series of images of surfaces of Cytop coated flat electrode (Fig. 6a), Parylene-C coated electrode (Fig. 6b) and nanoscale uneven gallium electrode with ultrathin Ga_2O_3 layer (Fig. 6c). For the energy harvester using Cytop as the dielectric layer, the liquid metal droplet started to be wetted on the surface of Cytop after 1 minute of operation. In the case of Parylene-C coated electrode, wetting occurred in the form of partial wetting after 2 minutes and the liquid metal droplet was fully wetted on the surface of the electrode at around 60 minutes of operation. In contrast, the energy harvester with Ga_2O_3 dielectric layer showed non-wetting characteristics of the liquid metal droplet even after 60 minutes of continuous operation.

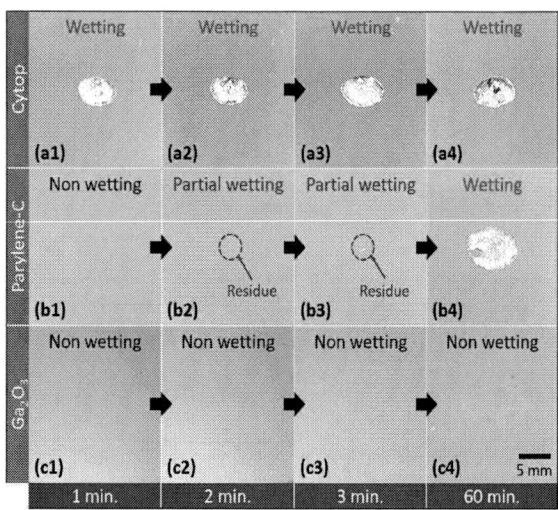

Figure 6: Optical images of the surfaces of (a) Cytop, (b) Parylene-C, and (c) Ga_2O_3 coated electrode showing rapid wetting on Cytop and Parylene-C coated electrodes.

Table 1 shows wetting states of the liquid metal droplet on the energy harvester using different dielectric layers: Ga_2O_3, Parylene-C, and Cytop depending on the height between two electrodes. The AC voltage of 6 V_{pp} was applied to the woofer for 5 seconds to test the wetting states with the range of height from 3.4 to 5 mm. Ga_2O_3 coated energy harvester can generate the electrical current continuously regardless of the changes in height from 3.4 to 5 mm. On the other hand, both energy harvesters using Parylene-C and Cytop as the dielectric layers showed wetting of the liquid metal droplet on the top electrode after heights of 4.2 and 4.6 mm respectively resulting in not operating as energy harvesters.

It should be noted that high dielectric constant is desirable for the dielectric layer in the REWOD, and the oxide shell of gallium (Ga_2O or Ga_2O_3) exhibits dielectric constant of ~11 [8] which is much higher than those of Cytop (~2) and Parylene-C (~3). We also measured the breakdown voltage of the gallium coted electrode-based energy harvester (280 V). This is well in line with high breakdown field of β-Ga_2O_3 (~8 MV/cm) [9].

Table 1: Near instant (~5 seconds) wetting on Cytop and Parylene-C surfaces in smaller separation distance between electrodes compared with nonwetting gallium coated electrode.

h (mm)	Insulator		
	Ga_2O_3	Parylene-C	Cytop
5	Non wetting	Non wetting	Non wetting
4.8	Non wetting	Non wetting	Non wetting
4.6	Non wetting	Non wetting	Wetting
4.4	Non wetting	Non wetting	Wetting
4.2	Non wetting	Wetting	Wetting
4	Non wetting	Wetting	Wetting
3.8	Non wetting	Wetting	Wetting
3.6	Non wetting	Wetting	Wetting
3.4	Non wetting	Wetting	Wetting

Voltage: 6 V_{pp} Applied time: 5 seconds

CONCLUSIONS

A REWOD energy harvester was demonstrated using nanoscale uneven formation of evaporated gallium as electrode and its naturally formed ultrathin gallium oxide as dielectric layer which is non-wetting Galinstan droplet. Compared to Cytop or Parylene-C as the dielectric layer, ~3 nm gallium oxide formed on gallium makes the surface nonwetting and suitable for energy harvester. This preliminary result of all liquid metal energy harvester may enable a highly efficient new energy harvester using REWOD principle.

ACKNOWLEDGEMENTS

This work was supported in part by the United States National Science Foundation grant NSF ECCS–1908779. The authors would like to acknowledge Dr. Xianming Dai, Arkadeep Mitra, Mohammad Salman Parvez, Muhammad Luqman Haider, and UT Dallas Clean Room staff for their support on this work.

REFERENCES

[1] S. P. Beeby, M. J. Tudor, N. M. White, "Energy Harvesting Vibration Sources for Microsystems Applications", *Meas. Sci. Technol.*, vol. 17, pp. R175, 2006.

[2] P. R. Adhikari, N. T. Tasneem, R. C. Reid, I. Mahbub, "Electrode and Electrolyte Configurations for Low Frequency Motion Energy Harvesting Based on Reverse Electrowetting", *Sci. Rep.*, vol. 11, pp. 1-13, 2021.

[3] J. Jeon, S.K. Chung, J.B. Lee, S.J. Doo, D. Kim, "Acoustic Wave-Driven Oxidized Liquid Metal-Based Energy Harvester", *Eur. Phys. J-Appl. Phys.* Vol. 81, pp. 20902, 2018.

[4] T. Krupenkin, J. A. Taylor, "Reverse Electrowetting as a New Approach to High-power Energy Harvesting", *Nat. Commun.*, vol. 2, pp. 1-8, 2011.

[5] D. Kim, P. Thissen, G. Viner, D.W. Lee, W. Choi, Y. Chabal, J.B. Lee, "Recovery of Nonwetting Characteristics by Surface Modification of Gallium-Based Liquid Metal Droplets Using Hydrochloric Acid Vapor" *ACS Appl. Mat. & Inter.* vol. 5, pp. 179-185, 2013.

[6] W. Tang, T. Jiang, F. R. Fan, A. F. Yu, C. Zhang, X. Cao, Z. L. Wang, "Liquid-metal Electrode for High-performance Triboelectric Nanogenerator at an Instantaneous Energy Conversion Efficiency of 70.6%", *Adv. Funct. Mater.*, vol. 25, pp. 3718-3725, 2015.

[7] Z. Chen, J. B. Lee, "Surface Modification with Gallium Coating as Nonwetting Surfaces for Gallium-based Liquid Metal Droplet Manipulation", *ACS Appl. Mater. Inter.*, vol. 11, pp. 35488-35495, 2019.

[8] A. Fiedler, R. Schewski, Z. Galazka, K. Irmscher, "Static Dielectric Constant of β-Ga2O3 Perpendicular to the Principal Planes (100), (010), and (001)", *ECS J. Solid State Sci. Tech.* vol. 8, pp. Q3083, 2019.

[9] M. Higashiwaki, K. Sasaki, A. Kuramata, T. Masui, S. Yamakoshi, "Gallium Oxide (Ga2O3) Metal-Semiconductor Field-Effect Transistors on Single-Crystal β-Ga2O3 (010) Substrates", *Appl. Phys. Lett.*, vol. 100, pp. 013504, 2012.

CONTACT

*Jeong Bong (JB) Lee, tel: +1-972-883-2893; jblee@utdallas.edu

ASYMMETRIC QUAD LEG ORTHOPLANAR SPRING FOR WIDEBAND PIEZOELECTRIC MICRO ENERGY HARVESTING

Ali Mohammadi[1], Shamin Sadrafshari[1], Alborz Shokrani[2], Chris R. Bowen[2],
[1] Department of Electronic and Electrical Engineering, and
[2] Department of Mechanical Engineering, University of Bath, UK

Abstract- **Piezoelectric energy harvesters (EH) generate their highest energy levels at the resonant frequencies of the transducer devices. To provide a wide-band EH solution, nonlinear mechanical resonators with multiple resonant modes can be used. We present a new asymmetric quad-leg orthoplanar spring (QOPS) EH microstructure to increase the harvesting bandwidth. The proposed design is implemented in CMOS compatible microfabrication processes. Finite element analysis show that the asymmetric designs increases the bandwidth by maximum 27% compared with symmetric designs. In order to measure the electrical output, one symmetric and three asymmetric piezoelectric devices implemented on the same microchip are exposed to mechanical vibrations over a wide frequency bandwidth. Experimental results approve the increase in the frequency bandwidth of resonators introduced by asymmetries added to the spring, as compared to the symmetrical configuration.**
Keywords: Piezoelectric Energy Harvester, Micro-electromechancial Systems (MEMS), Orthoplanar Spring.

INTRODUCTION

MICROELECTROMECHANICAL SYSTEMS (MEMS) provide a reliable and low-cost solution for integration of energy harvesting (EH) devices and sensor transducers. Vibration-based EH have demonstrated higher energy outputs compared with other mechanisms [1], such as radio frequency or pyroelectric approaches with similar transducer dimensions. Wireless power transfer [2], despite its straightforward implementation may have limitations in electromagnetically shielded environments. These EH systems have been implemented using piezoelectric, electromagnetic and capacitive transducers. Piezoelectric transducers have gained higher popularity among the other approaches since they offer the potential for high efficiency and autonomous solutions. In addition, compatibility of some piezoelectric MEMS processes with Complementary Metal Oxide Semiconductor (CMOS) processes offers a great advantage for integration of EH system within sensor nodes.

At the microscale, piezoelectric resonators generate high-quality factor (Q) vibrations, so that the power spectrum of the electrical output is concentrated at sharp resonance frequencies. However, the mechanical vibration sources to be harvested often have a wide frequency spectrum. Therefore, applying such wideband vibrational inputs to piezoelectric transducers results in low efficiency and output energy levels, due to the limited range of resonance frequencies.

Time-domain multiplexing of mechanical impacts, recently suggested in [3], increases the output energy of harvesters exposed to random vibrations by applying low frequency mechanical input to sequentially spaced multiple cantilever beams. However, most of the previous works are focused on frequency-domain approaches. Bandwidth widening of meso-scale piezoelectric transducers have been investigated by introducing a variety of nonlinearities, which leads to

additional resonant modes [4]. At the micrometre-dimension, a low frequency cantilever resonator can be used as a mechanical stopper against another high frequency cantilever beam [5]. This increases the bandwidth of the second resonator by a *scrape-through* effect. However, the reliability and repeatability of such an approach for fragile MEMS devices needs to be addressed. Permanent magnets in close vicinity of the free end of piezoelectric cantilever beams can also introduce nonlinearity by the magnetic coupling between the vibrating tip of the beam and stationary magnet in meso/macro systems [6]. However, the integration of permanent magnets in standard microfabrication processes is challenging [7]. In an alternative approach, a proof mass suspended from several linear cantilevers provide nonlinear behaviour [8-11]. For example in [11], a three-leg ortho-planar spring (OPS) system has been used with a suspended proof mass at the meso-scale. A comprehensive review of vibration energy harvesters with focus on planar spring systems is presented in [12]. The OPS system is best suited for smaller scales due to its compact size and planar structure, which is highly compatible with standard microfabrication processes.

We introduce here an alternative solution, where equal-sized asymmetries in a Quad-leg piezoelectric OPS system is introduced to further increase the bandwidth of an EH system. In the following sections we investigate the effect of asymmetric weight and stiffness added to specific sections of the OPS legs in a commercially available microfabrication process (Pz-MUMPS). The finite-element piezo-electromechanical analysis and measurement results are presented for different combinations of these asymmetrically located masses and a broadband response is demonstrated.

DEVICE MICROSTRUCTURE

The inherent wide bandwidth of OPS systems as a result of multiple resonant modes at close distances from each other, is

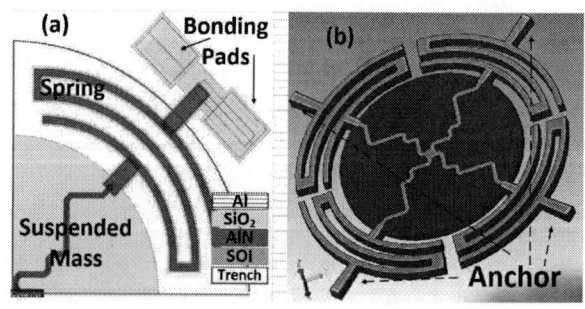

Fig. 1:. The proposed symmetric QOPS microsystem, (a) one leg layout that is rotated 90° four times to generate the full device layout, (b) 3D meshed model for symmetric structure in Coventor without the boning pads and the

a popular feature for EH applications. We show that shifting the resonance frequencies by adding asymmetric suspension beams in the microstructure can further increase the operational bandwidth. The proposed symmetric quad-leg ortho-planar spring (QOPS) system is illustrated in Fig 1. The quarter layout and the process cross-section are shown in Fig. 1(a). The active piezoelectric layer (AlN) is sandwiched between the metal (Al) and silicon (SOI) layers in this microfabrication process. The circular disc-shaped mass is etched out of the SOI layer and coupled to the substrate by S-shaped springs, which are anchored to the substrate by four cantilever beams. Etching the substrate (handle wafer) from the back-side leaves the mass suspended on the trench as shown in the three dimensional (3D) model in Fig 1(b). Coupling the mechanical vibration of the substrate to the suspended mass applies stress to the springs. The piezoelectric layer deposited on the springs converts this stress to a piezo-electric charge that can be collected from the bonding pads. The mechanical features of this structure are determined mostly by the silicon layer, which is an order of magnitude thicker (10μm) than the two other layers (<1μm).

The addition of weights to the symmetric mesoscale three-leg OPS systems is investigated for piezoelectric EH [13]. However, in standard microfabrication processes the only way to introduce asymmetries is to alter the 2D layout. In this work we have added three different asymmetries on the symmetric QOPS system above, where changes in the stiffness of the springs and the effective mass are used to affect the frequency

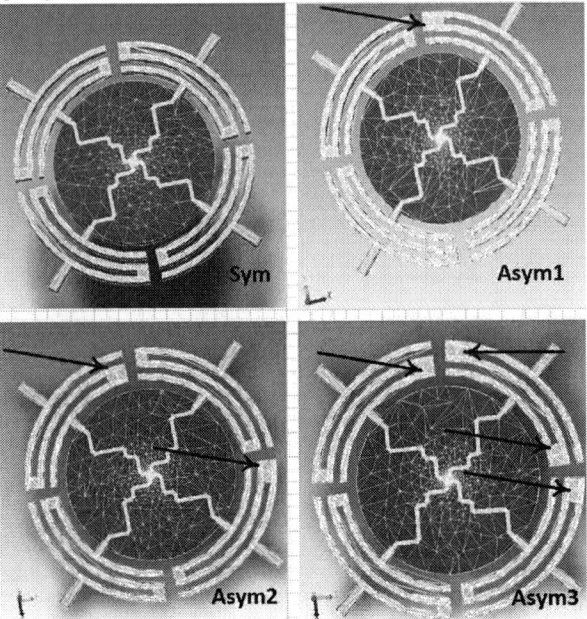

Fig. 2: Meshed model for the symmetric and asymmetric OPS in Pz-MUMP, (a) symmetric, (b-d) Asymmetric configurations with one (Asym1), two (asym2) and four (Asym3) asymmetries shown by arrows.

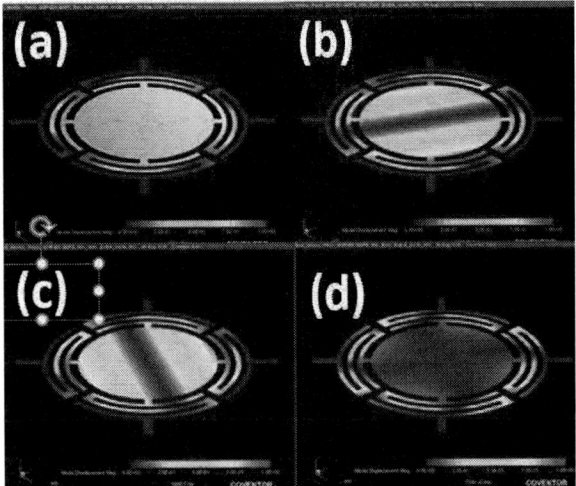

Fig. 3: Finite element analysis of the 3D meshed model for the symmetric COPS systems, (a) vertical mode, (b) and (c) torsion around symmetric access, (d) opposite torsion around the centre

of different resonant modes. The top view of 3D meshed model for these three different asymmetric designs are shown in Fig. 2. The S-shaped spring is widened as shown by arrows in one, two and four points for *Asym1*, *Asym2* and *Asym3*, respectively. Finite element analysis (FEA) has been applied to extract the mechanical resonance behavior of this QOPS system and Fig. 3 shows an animated view of first four resonant modes for the symmetric system. The resonance frequencies associated with each of these microstructures are reported in table 1 for the first four modes. Mode four can be ignored as its frequency is much higher than the others. The difference between first and third resonance frequencies is also reported as Δf_{31} in this table. The frequency difference between the first and third modes increases by adding the asymmetries,

Table 1: Resonant mode frequencies associated with the symmetric (Sym) and asymmetric (Asym) structures

Mode	Sym	Asym1	Asym2	Asym3
1	1209.8	1237.2	1266.1	1315.8
2	1672.4	1675.3	1716.9	1842.1
3	1686.7	1793.5	1869.2	1884.4
4	7230.4	7261.4	7137.4	7014.7
$\Delta f_{31} = f_3 - f_1$	476.9	556.3	603.1	568.6

which is expected to increase the bandwidth. The addition of asymmetries has a higher impact on the stiffness, rather than the effective mass of the structure, since the dominant suspended mass of this microstructure is determined by the circular disk. The addition of asymmetries has different impacts on the resonance behavior of *Asym1*, *Asym2* and

978-1-6654-9309-3/23 $31.00 © 2023 IEEE

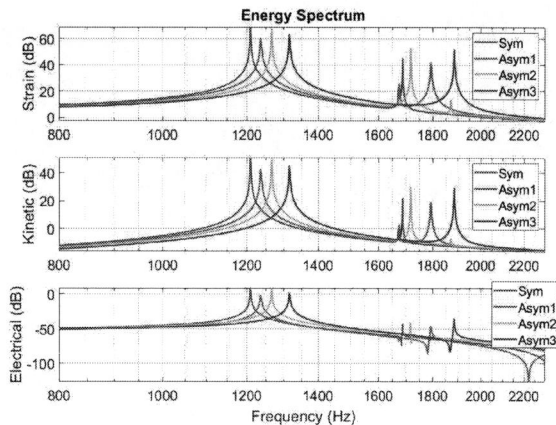

Fig. 4: Simulated harmonic energy spectrum for the proposed symmetric and asymmetric QOPS EH system, (a) Mechanical Strain Energy, (B) Electrical energy supplied to a 1MΩ resistor.

Asym3; see Table 1. For example, *Asym3* has experienced a higher increase in the spring stiffness of modes 2 and 3 compared with mode 1 as it can be seen in its resonance frequency shifts in mode 1 (Δf=13Hz), mode 2 (Δf=520Hz), and mode 3 (Δf=360Hz).

In addition, we have examined the frequency response of these structures by running direct harmonic analysis including piezoelectric and mechanical physics in Coventorware. The harmonic energy spectrum is simulated in 500 frequency points over 800Hz ~ 10 kHz bandwidth to highlight the critical frequencies only. The simulation results are shown in Fig 4. The strain energy and electrical output energy delivered to a

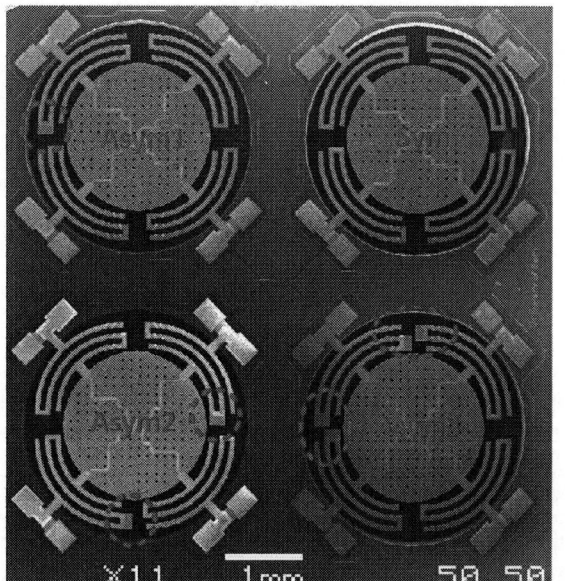

Fig. 5: SEM image of the chip manufactured in Pz-MUMP microfabrication process, with highlighted asymmetries in Asym1, Asym2 and Asym3.

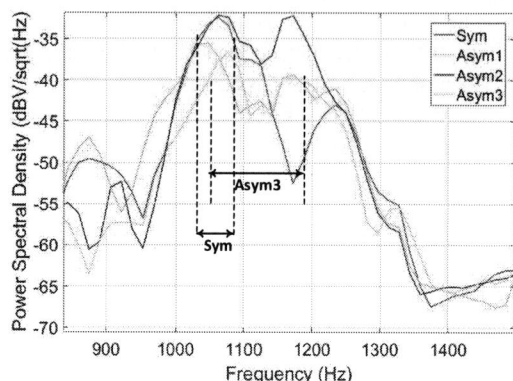

Fig. 6: Measured power spectrum of the symmetric and asymmetric configurations: adding asymmetric items increases the 3dB bandwidth.

1MΩ resistor are illustrated separately for symmetric and asymmetric modes. It can be seen that the addition of asymmetries shifts up the resonant modes due to increased stiffness. In addition, this results in higher magnitude of non-dominant resonant modes, but these are still much smaller than the dominant mode. The experimental verification of the fabricated microdevices are reported in the following section.

MEASUREMENT RESULTS

The chip is implemented in Pz-MUMP microfabrication process technologies, and wire-bonded using non-vacuum chip-holders in our in-house cleanrooms for characterization purposes. An SEM image of symmetric and three asymmetric versions of the microdevice on the same chip is illustrated in Fig. 5. The chip-holder is mounted on a shaker, which is actuated by a sinusoidal sweep signal generator. The signal generator sweeps 0.5 kHz – 2 kHz frequency range in 40 s and the voltage output of the four QOPS EH resonators are measured individually by dynamic signal analyzer (SR760). The power spectrum of the measured voltages are shown in Fig 6, where it can be seen that the bandwidth increases by adding the asymmetries in the symmetric QOPS structure. In particular, the symmetric structure shows a sharper tip at the dominant resonant frequency whereas the asymmetric structures spread the output power over a wider spectrum. Critically *Asym3* is shown to have highest bandwidth compared with the symmetric structure. This is due to the increased stiffness of asymmetric springs in *Asym3*, which shifts the resonance frequency of modes 2 and 3 further away from the original symmetric frequency, as seen in Table 1. The results for Asym2 also agrees with the finite element analysis wherein the stiffness of springs has increased symmetrically for all modes, which has resulted in increased bandwidth but to a smaller degree than *Asym3*. Differences between the resonance frequencies attained in simulation and measurement might be due to the incomplete release etch at the back side of the circular disk mass, which adds to the weight of the microfabricated sample compared with the simulated 3D model.

CONCLUSION

The conversion efficieny of piezoelectric vibration energy harvesters is limitted by sharp resonance behaviour especially in micro dimensions, which results in reduced ability to capture wideband meachnical vibrations. In this paper a arnge of nonlinear an dasymetric structures are investigated by both modelling and experimental charecterization to widen the bandwidth of these tansducers. The presented orthoplanar spring systems have potential for ease of integration in microfabrications processes. We investigated the effect of assymetric design in increasing the bandwidth. This approach provides a solution to realise broader band harvesting at the micro-scale using standard microfabrication processes.

ACKNOWLEDGMENT

This work has been partially supported by Engineering and Physical Science Research Council (EPSRC) grant (EP/V055011/1), University of Bath Alumni Fund, and Europractice MEMS Design Award.

REFERENCES

[1] R. Hamid, A. Mohammadi, and M. R. Yuce, "WE-Harvest: a wearable piezoelectric-electromagnetic energy harvester," in *Proceedings of the 10th EAI International Conference on Body Area Networks*, 2015, pp. 62-66: ICST (Institute for Computer Sciences, Social-Informatics and Telecommunications Engineering).

[2] A. Mohammadi, J.-M. Redoute, and M. R. Yuce, "Wireless power transmission for biomedical implants: The role of near-zero threshold CMOS rectifiers," in *2015 37th Annual International Conference of the IEEE Engineering in Medicine and Biology Society (EMBC)*, 2015, pp. 5453-5456: IEEE.

[3] A. Mohammadi, S. Sadrafshari, C. Bowen *et al.*, "Time Domain Multiplexing for Efficiency Enhanced Piezoelectric Energy Harvesting in MEMS," *IEEE Electron Device Letters*, vol. 41, no. 3, pp. 481-484, 2020.

[4] W. Wang, J. Cao, C. R. Bowen *et al.*, "Performance enhancement of nonlinear asymmetric bistable energy harvesting from harmonic, random and human motion excitations," *Applied Physics Letters*, vol. 112, no. 21, p. 213903, 2018.

[5] H. C. Liu, C. J. Tay, C. G. Quan *et al.*, "Piezoelectric MEMS Energy Harvester for Low-Frequency Vibrations With Wideband Operation Range and Steadily Increased Output Power," *Journal of Microelectromechanical Systems*, vol. 20, no. 5, pp. 1131-1142, Oct 2011. doi:10.1109/Jmems.2011.2162488

[6] Y. Zhu and J. W. Zu, "Enhanced buckled-beam piezoelectric energy harvesting using midpoint magnetic force," *Applied Physics Letters*, vol. 103, no. 4, p. 041905, Jul 22 2013. doi:10.1063/1.4816518

[7] A. Mohammadi, N. Karmakar, and M. Yuce, "A post-fabrication selective magnetic annealing technique in standard MEMS processes," *Appl. Phys. Lett.*, vol. 109, no. 22, p. 221906, 2016.

[8] M. Gratuze, A. H. Alameh, and F. Nabki, "Design of the Squared Daisy: A Multi-Mode Energy Harvester, with Reduced Variability and a Non-Linear Frequency Response," *Sensors (Basel)*, vol. 19, no. 15, p. 3247, Jul 24 2019. doi:10.3390/s19153247

[9] Q. He and T. Jiang, "Complementary multi-mode low-frequency vibration energy harvesting with chiral piezoelectric structure," *Applied Physics Letters*, vol. 110, no. 21, p. 213901, 2017.

[10] L. Zhang, R. Takei, J. Lu *et al.*, "Development of energy harvesting MEMS vibration device sensor with wideband response function in low-frequency domain," *Microsystem Technologies*, pp. 1-9.

[11] S. Dhote, J. Zu, and Y. Zhu, "A nonlinear multi-mode wideband piezoelectric vibration-based energy harvester using compliant orthoplanar spring," *Applied Physics Letters*, vol. 106, no. 16, p. 163903, 2015.

[12] A. Mohammadi, "Vibration-based energy harvesting for sensors," in *Reference Module in Biomedical Sciences*: Elsevier, 2021.

[13] S. Dhote, Z. B. Yang, and J. Zu, "Modeling and experimental parametric study of a tri-leg compliant orthoplanar spring based multi-mode piezoelectric energy harvester," *Mechanical Systems and Signal Processing*, vol. 98, pp. 268-280, Jan 1 2018. doi:10.1016/j.ymssp.2017.04.031

CONTACT

Dr Ali Mohammadi, a.mohammadi@bath.ac.uk,
+44 (0) 1225 383325

EVALUATION OF THERMOELECTRIC PROPERTIES OF MONOLITHICALLY-INTEGRATED CORE-SHELL Si NANOWIRE BRIDGES

Akio Uesugi, Shusuke Nishiyori, Koji Sugano, and Yoshitada Isono
Department of Mechanical Engineering, Kobe University, JAPAN

ABSTRACT

In order to clarify an effect of core-shell structuring on the thermoelectric properties of silicon nanowires (SiNWs), a new evaluation device with monolithic integration of bottom-up-grown SiNWs was developed, and Al_2O_3-shell-coated p-type SiNWs were evaluated. In fabrication, SiNWs bridge structures were successfully formed with a high yield of 61 % on desired local region, by employing VLS processes combined with surface nanoholes formation. Evaluated resistivity and Seebeck coefficient of the core-shell SiNW bridges were 7.6 $\Omega \cdot$cm and 0.61 mV/K, respectively. The Seebeck coefficient was compared with those of bulk and nano Si structures to discuss effect of core-shell structuring.

KEYWORDS

Thermoelectric conversion, Seebeck coefficient, Silicon, Nanowire, Crystal growth, Monolithic integration

INTRODUCTION

Thermoelectric generators directly convert heat into electrical energy, and are expected to be clean energy harvesters that effectively utilizes unused energy, such as waste heat from factories and high-density electronic devices. Pb-Te and Bi-Te materials have been studied as high thermoelectric conversion materials in past, but for wide applications and widespread usage, materials and elements with low cost, low environmental load, and high thermoelectric conversion properties are needed.

The conversion efficiencies are expressed in terms of figure of merit ZT and power factor PF by the following equations (1) and (2),

$$PF = S^2 \sigma \qquad (1)$$
$$ZT = S^2 \sigma T / \lambda \qquad (2)$$

where S is Seebeck coefficient, σ is electrical conductivity, and λ is thermal conductivity. The thermal conductivity λ is divided into a phonon-derived component λ_L and a carrier-derived component λ_e.

$$\lambda = \lambda_L + \lambda_e \qquad (3)$$

In order to improve thermoelectric conversion efficiencies, larger Seebeck coefficient S and electrical conductivity σ, and smaller thermal conductivity λ are necessary.

Silicon nanowires (SiNWs) and nanostructures have been investigated as thermoelectric conversion elements. Silicon is a material with a low environmental load, and many micro- and nanofabrication techniques have been studied. All of Seebeck coefficient S, electrical conductivity ρ, and thermal conductivity λ depend on carrier concentration, and a higher carrier concentration brought a lower Seebeck coefficient S, a higher electrical conductivity ρ, and a higher carrier-derived thermal conductivity λ_e. While thermal conductivity λ of the bulk

structure is 156 W/mK, that of nanostructures is decreased owing to phonon scattering. D. Li *et al.* (2003) reported thermal conductivity of less than 10 W/mK for nanostructures with a width of 22 nm [1]. Nanostructures with a high career concentration have been often studied because of their high electrical conductivity σ. A. I. Hochbaum *et al.* (2008) reported a high conversion efficiency of $ZT = 0.6$ [2], which still does not meet the practical level of $ZT = 1$. Therefore, further improvement of thermoelectrical properties of SiNWs are needed.

Core-shell structuring of SiNWs with a dielectric material film has a potential to improve thermoelectrical properties. Recently, we reported that carrier mobility of semiconductor nanowire of SiC with a low dopant concentration was significantly modified by core-shell structuring with dielectric films, SiO_2 and Al_2O_3 films [3]. That was because accumulation layer and depletion layer were formed at interfaces respectively, by electric field caused from fixed charges in the shell films. On measured p-type SiC nanowires, career mobility was enhanced with Al_2O_3-film shell which had a negative fixed charge. Career mobility change depending on surface films had been also reported on Si nanostructures [4]. Similarly, the core-shell structuring with dielectric films was expected to have a potential to modify relationship between electrical conductivity σ and Seebeck coefficient S in thermoelectric conversion. In addition, surface localization of carriers brought by the above accumulation formation has a potential to decrees thermal conductivity λ. A detailed description of prediction model is omitted here, but combination of thin surface layer with a high career concentration and core wire structure with a low career concentration was expected to decrease carrier-derived component λ_e. However, there has been no reports on effect of core-shell structuring on thermoelectric properties and experimental investigation and verification are needed.

In this study, we developed a thermoelectric evaluation device where a core-shell SiNWs with Al_2O_3 films can be integrated monolithically, and evaluations were conducted. It was one of efficient way for accurate measurement to integrate a countable number of SiNWs on evaluation devices. In this paper, processes for vapor-liquid-solid (VLS) growth of SiNWs and core-shell structuring were incorporated in fabrication process of the evaluation device for thermoelectric properties, and evaluation of monolithically-integrated core-shell p-type SiNWs bridges with Al_2O_3 shell was performed. Obtained Seebeck coefficient S was compared with those of bulk and nano Si structures to discuss effect of core-shell structuring.

EXPERIMENTS PROCEDURE

Evaluation devices for thermoelectric properties

Figure 1 shows schematics of a developed evaluation device for thermoelectric properties of core-shell SiNW

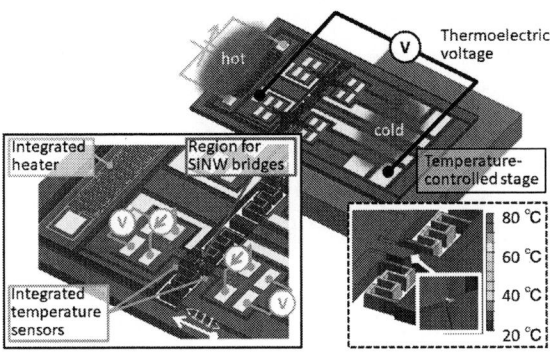

Figure 1: Schematics of an evaluation device of thermoelectric properties of core-shell SiNW bridges and measurement setup. Thin film heater and paired temperature sensors were integrated.

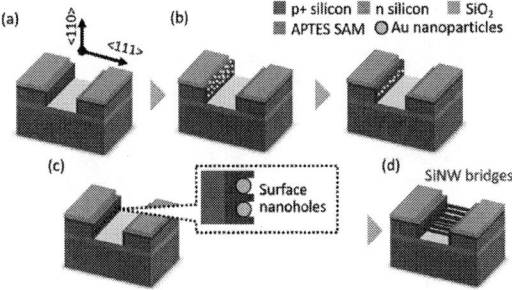

Figure 2: SiNW bridges formation process: (a)microtrench with smooth {111}-oriented sidewalls on silicon-on-insulator (SOI) substrate, (b)catalyst Au nanoparticles arrangement, and (c) metal-assisted chemical etching (MACE) process to form surface nanoholes, and (d) vapor-liquid-solid (VLS) process to form SiNW bridges.

bridges and measurement setup. The evaluation device was designed to be fabricated from 10 mm ×16 mm silicon-on-insulator (SOI) substrates with (110)-oriented surface device layer. The surface crystal orientation of the substrates was selected to grow SiNWs along in-plane <111> directions, as described in a next section. 10 μm-wide microtrenches were prepared on center area on the evaluation devices for integration regions of SiNW bridges and the sidewalls in the microtrenches were designed to be smooth {111}-oriented vertical surfaces formed using TMAH (Tetramethylammonium Hydroxide) wet etching process.

The evaluation device had an integrated metal-film heater and paired integrated metal-film temperature sensors near the microtrenches to monitor local temperatures near the both ends of the core-shell SiNW bridges. One side of the device was heated by an integrated heater, and the other side was fixed on a temperature-controlled stage including a Peltier device. As shown in FEM results in Fig 1, with design considering heat conduction of supporting beams and entire device, a large temperature difference could be given to the core-shell SiNW bridges with small heat supply. The temperature sensors had a four-terminal sensing configuration. A constant current was supplied using current regulating diodes and resistance change in response to temperature change was measured as a output voltage change.

In evaluations, *I-V* characteristics of the integrated core-shell SiNW bridges were measured using a semiconductor analyzer (Agilent, 4156C), and thermoelectric properties were measured using digital multimeters with the evaluation device placed in a vacuum chamber for a precise temperature control.

Formation of core-shell structured SiNWs bridge

Figure 2 shows processes for SiNWs bridge formation. In order to grow SiNWs perpendicular to {111}-oriented sidewall in microtrenches, metal-assisted chemical etching (MACE) process was employed before vapor-liquid-solid (VLS) crystal growth. MACE-formed surface nanoholes with adequate depths enabled to perpendicular growths in initial growth phase [5, 6].

Catalyst Au nanoparticles with a diameter of 60 nm from a colloidal solution were placed on one-side bank of

the mictotrenches, using UV-lithography patterning of 3-aminopropyltriethoxysilane (APTES) film. Next UV lithography filled in lower part of the microtrenches with photoresist and Au nanoparticles on upper side of the sidewalls were removed by Au wet etching.

The Au nanoparticles left on the middle part of the sidewalls were used as catalyst in following MACE process and VLS process. MACE process was conducted using a mixture solution of HF, H_2O_2 and H_2O with a volume ratio of 1:5:10. Surface of the sidewalls were slowly etched from small area the Au nanoparticles contacted, to form surface nanoholes with a diameter similar to the Au nanoparticles. Four-hours VLS process using diluted silane gas at 500 °C grow SiNWs to the opposite sidewalls, and formed SiNWs bridges with a length of 10 μm.

The formed SiNWs bridges went through a lightly-boron-doping process using phenyl boron acid (PBA). A slight oxide film was chemically formed on the SiNWs surface, followed by a 24-hour immersion in solution of PBA and tetrahydrofuran (THF) to modify the entire surface of the three-dimensional SiNW bridges. Thermal treatment at 950 °C in 5 sec caused a boron diffusion into SiNWs, and after cooling, surface oxide and residual of PBA films were removed with BHF solution. After the doping process, SiNWs were coated with 50-nm-thick Al_2O_3 film using thermal atomic layer deposition (ALD) process to form core-shell structures. In the core-shell configuration, accumulation layer of hole was formed at interface to shells due to a negative fixed charge in Al_2O_3 film.

During rest fabrication processes for the evaluation device, patterning metal film using a lift-off process, and deep RIE process on device and handle layers of the SOI substrates, the microtrenches with were fulfilled with a photoresist, to prevent damage on the SiNW bridges.

RESULTS AND DISCUSSIONS

Figure 2 shows SEM images of fabrication steps and results of core-shell SiNW bridges formation. As shown in Fig. 3(a), Au nanoparticles were successfully arranged on a local area on sidewalls, and embedded in MACE-formed surface nanoholes with estimated depths of around 100-150

978-1-6654-9309-3/23 $31.00 © 2023 IEEE

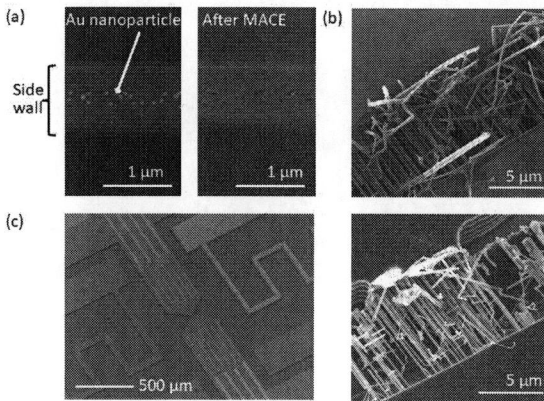

Figure 3: SEM images of SiNW bridges formation processes. (a) Au nanoparticle placed on sidewall and surface nanoholes containing the Au nanoparticles after MACE process. (b) SiNW bridges in microtrench, and (c) SiNW bridges after device-layer patterning.

nm. Many of VLS-grown SiNWs achieved to the opposite sidewall of the microtrenches. A yield of SiNW bridges which was ratio of a number of SiNW bridges to a number of arranged Au nanoparticles was 61%. This yield was sufficient high for the rest patterning and measurement of thermoelectric properties. Because a yield in case of SiNWs growth without surface nanoholes was only several percent in our previous experiments, surface nanoholes contributed to control growth directions of SiNWs and to form high-yield bridge SiNWs

In patterning of device layer of the SOI substrates, the microtrenches were also etched except for center region with an area of 10×8 μm^2. Core-shell SiNW bridges connected the microtrenches was counted after all fabrication process by SEM observation, and 18 core-shell SiNW bridges were subjected to following evaluations.

Temperature sensing and distribution on the fabricated evaluation device was confirmed as shown in Fig. 4. When the evaluation device was heated uniformly, the output from the temperature sensors showed a high and linear sensitivity, about 0.014 mV/°C as shown in Fig. 4 (a). Temperature destitution near the core-shell SiNW bridges was measured using the integrated temperature sensors with different input voltages to the integrated heater. As shown in Fig. 4(b), it was confirmed that temperature at the

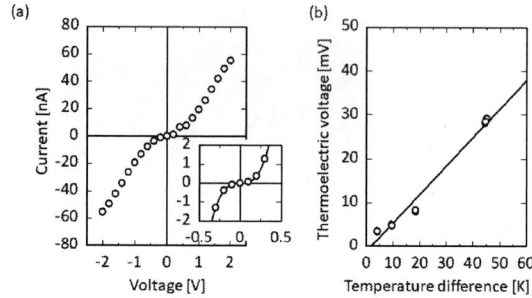

Figure 5: (a)I-V curves for single core-shell SiNW bridge. Solid line represents polynomial fitting near 0V. (b)Thermoelectric voltage as a function of temperature difference.

hot side could be controlled up to 70 °C with temperature at the cold side maintained by the temperature-controlled stage.

Figure 5(a) shows a I-V curve calculated as a single core-shell SiNW bridge. The resistance of single core-shell SiNW bridge was 27.3 MΩ, calculated from linear I-V change in range of V= 1.2 V to 2.0 V. A resistivity and carrier concentration calculated based on dimensions of the SiNW bridge were 7.6 Ω·cm and 2×10^{15} cm^{-3}, respectively. On the other hand, the I-V curve showed a slight nonlinearity near V = 0 V. This nonlinear I-V characteristic was attributed to the connection between the SiNWs and the evaluation device surface. Career concentration of the SiNW bridges were much smaller than that of the connected region in the evaluation device, which brought a Schottky-like contact.

Figure 5(b) shows the evaluation results of thermoelectric voltages. From the evaluation with the cold side temperature fixed at 25 °C, a thermoelectric voltage of 28.5 mV was obtained with a temperature difference of SiNW bridges of approximately 45 °C. From the measurement results, the Seebeck coefficient S was calculated to be about 0.61 mV, while some decrease was included due to the nonlinearity of the I-V properties curve near 0V. Seebeck coefficient S for the SiNWs without the contact regions was estimated to be higher than that for Si bulk structures with an identical career concentration, when the nonlinear I-V characteristics was used for the estimation.

Figure 6 shows relationship between experimentally-

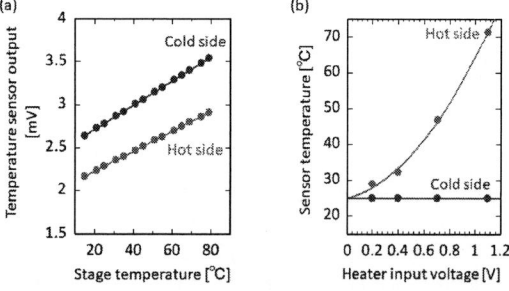

Figure 4: Temperature control of the fabricated evaluation device. (a) Relationship between temperature and sensor outputs, and (b)temperature distribution on the evaluation device.

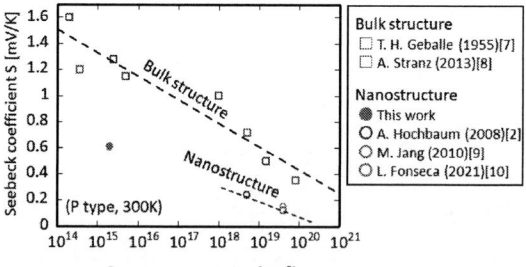

Figure 6: Comparison of Seebeck coefficient of p-type silicon as a function of career concentrations. When career concentration was not described, Career concentrations were estimated from electrical resistivities.

reported Seebeck coefficient S for p-type Si structures and career concentration. On p-type silicon structures, Seebeck coefficients S for nanostructures with a high career concentration [2, 9, 10] were often lower those of the bulk structures [7, 8]. Our obtained result of career concentrations and Seebeck coefficient S was consistent with trend for nanostructures, as shown in the graph. In addition, considering influence of the nonlinear I-V characteristic, the obtained data in this work could exceeded the trend for nanostructures, at least. On the other hand, some n-type Si nanostructures were reported to have similar or higher Seebeck coefficients S, compared with those for bulk structures [11, 12], which implied that not only structural factors but also the conductivity type of semiconductors affected an influence on Seebeck coefficients S. When increase in Seebeck coefficients S of the data obtained in this report is valid, it means that the surface potentials attracting majority carriers to surfaces affect generation of thermoelectric voltages. This is consistent with the above trend, because the native oxide film attracts electrons to the surface for n-type structures, and Al_2O_3 film deposited on the SiNW bridges attracted holes to surfaces for p-type structures. This effect has a potential to increase Seebeck coefficients S and figure of merit ZT. This effect of surface potential to Seebeck coefficients S will be investigated in detail in future evaluation on core-shell SiNW bridges.

CONCLUSION

In order to elucidate the effect of core-shell structuring on thermoelectric properties, a new thermoelectric evaluation device capable of a monolithic integration of core-shell SiNWs was developed, and thermoelectric voltages measurements for Al_2O_3-caoted core-shell SiNWs bridges were performed. SiNW bridges were successfully integrated in 10-μm-wide microtrench with a high yield of 61%, the evaluation device showed a sufficient temperature sensing and control properties. Obtained Seebeck coefficients S for the core-shell SiNW bridges of was found to be similar or higher, compared with experimentally-reported Seebeck coefficients S for p-type Si nanostructures. This result indicated a potential to increase Seebeck coefficients S by core-shell structuring.

Further investigation will be conducted on core-shell SiNWs with different conditions to clarify effects of core-shell structuring and surface potential on Seebeck coefficients S and to improve thermoelectric conversion performance of SiNWs.

ACKNOWLEDGEMENTS

This work was supported by JSPS KAKENHI Grant Number JP20K15150.

REFERENCES

[1] D. Li, W.Yiying, K. Philip, S. Li, Y. Peidong and A. Majumdar,"Thermal conductivity of individual silicon nanowires", *Applied Physics Letters*, vol.83, pp. 2934-2936, 2003.

[2] A. Hochbaum, R. Chen, R. Delgado, et al., "Enhanced thermoelectric performance of rough silicon nanowires", *Nature*, vol. 451, pp.163–167, 2008.

[3] A. Uesugi, S. Nakata, K. Inoyama, K. Sugano and Y. Isono, "Surface-potential-modulated piezoresistive effect of core–shell 3C-SiC nanowires", *Nanotechnology*, vol. 33, no. 50, 2022.

[4] K. Winkler, E. Bertagnolli, and A. Lugstein, "Origin of Anomalous Piezoresistive Effects in VLS Grown Si Nanowires", *Nano Letters*, vol.15 no.3, pp.1780-1785, 2015.

[5] A. Uesugi, T. Horita, K. Sugano, and Y. Isono, "Vapor-liquid-solid growth of silicon nanowires from surface nanoholes formed with metal-assisted chemical etching", *Jpn. J. Appl. Phys.*, vol. 60, no. 5, 055502, 2021.

[6] A. Uesugi, S. Nishiyori, T. Nakagami, K. Sugano, and Y. Isono, "Integration of silicon nanowire bridges in microtrenches with perpendicular bottom-up growth promoted by surface nanoholes", *Jpn. J. Appl. Phys.*, vol. 61, no. 7, 075502, 2022.

[7] T. H. Geballe and G. W. Hull,"Seebeck Effect in Silicon", Physical Review. vol.98, pp.940-947,1955.

[8] A. Stranz, J. Kähler, A. Waag, and E. Peiner, "Thermoelectric Properties of High-Doped Silicon from Room Temperature to 900 K", *Journal of Electronic Materials*, vol. 42, no. 7, 2013.

[9] M. Jang, Y. Park, M. Jun, et al. , "The Characteristics of Seebeck Coefficient in Silicon Nanowires Manufactured by CMOS Compatible Process", *Nanoscale Research Letters*, vol. 5, pp.1654–1657, 2010.

[10] L.Fonseca, I. Donmez-Noyan, M. Dolcet, et al., "Transitioning from Si to SiGe Nanowires as Thermoelectric Material in Silicon-Based Microgenerators", *Nanomaterials*, vol. 11, art. no.517, 2021.

[11] E. Krali and Z. A. K. Durrani, "Seebeck coefficient in silicon nanowire arrays", *Applied Physics Letters*, vol.102, artno.143102, 2013.

[12] N. S. Bennett, D. Byrne, and A. Cowley,"Enhanced Seebeck coefficient in silicon nanowires containing dislocations", *Applied Physics Letters*, vol.107, art.no.013903, 2015.

CONTACT

*A.Uesugi, tel: +81-78-803-6314;
uesugi@mech.kobe-u.ac.jp

GLAZE TILE-INSPIRED LIQUID-SOLID POWER GENERATOR FOR CONTINUOUS WATER FLOW ENERGY HARVESTING

Dezhi Nie[1,2], Boming Lyu[1,2], Yongbo Hu[1], Jian Zhang[1], Yongqing Fu[3],
Honglong Chang[1] and Kai Tao[1,2]

[1]School of Mechanical Engineering, Northwestern Polytechnical University, China
[2]Research & Development Institute in Shenzhen, Northwestern Polytechnical University, China
[3]Faculty of Engineering and Environment, Northumbria University, UK

ABSTRACT

This paper reports a cascaded liquid-solid power generator (CLPG) inspired by the Forbidden City's glazed tiles in ancient China. In order to solve the pain point that it is difficult to efficiently collect continuous water flow in the current micro water energy harvesting field, CLPG's breaking strategy is to convert water flow into droplets through a cascaded structure, and then collect energy discretely. In this way, the charge on the polymer surface can be fully accumulated, thus achieving an output efficiency 25 times higher than the existing planar structure. The CLPG provides an essential reference for micro water energy harvesting and will further promote the development of self-powered MEMS sensors.

KEYWORDS

Tile-inspired structure, micro water flow energy harvesting, liquid-solid contact electrification, electret energy harvesting

INTRODUCTION

With the development of Internet of Things (IOT), the demand for energy supply of wireless sensor networks is becoming increasingly prominent. Obtaining renewable energy from nature has been regarded as a sustainable solution to upgrade the current self-powered MEMS sensor networks [1-3].

The energy from water is currently in the spot-light because of the water's ubiquitous existence and abundance. The traditional water energy collection relies on the heavy and bulky electromagnetic generator (EMG), which is difficult to effectively collect small, low-frequency, random water energy, such as raindrop energy [4-9]. In recent years, contact electrification process was proposed as a potential way to harvest these random and low-frequency energies [10-14]. Z. K. Wang and coworkers reported a droplet-based electricity generator (DEG) [15], enhancing the instantaneous power density over water energy generator devices limited by the interfacial contact electrification. Conventionally, plane DEG devices only

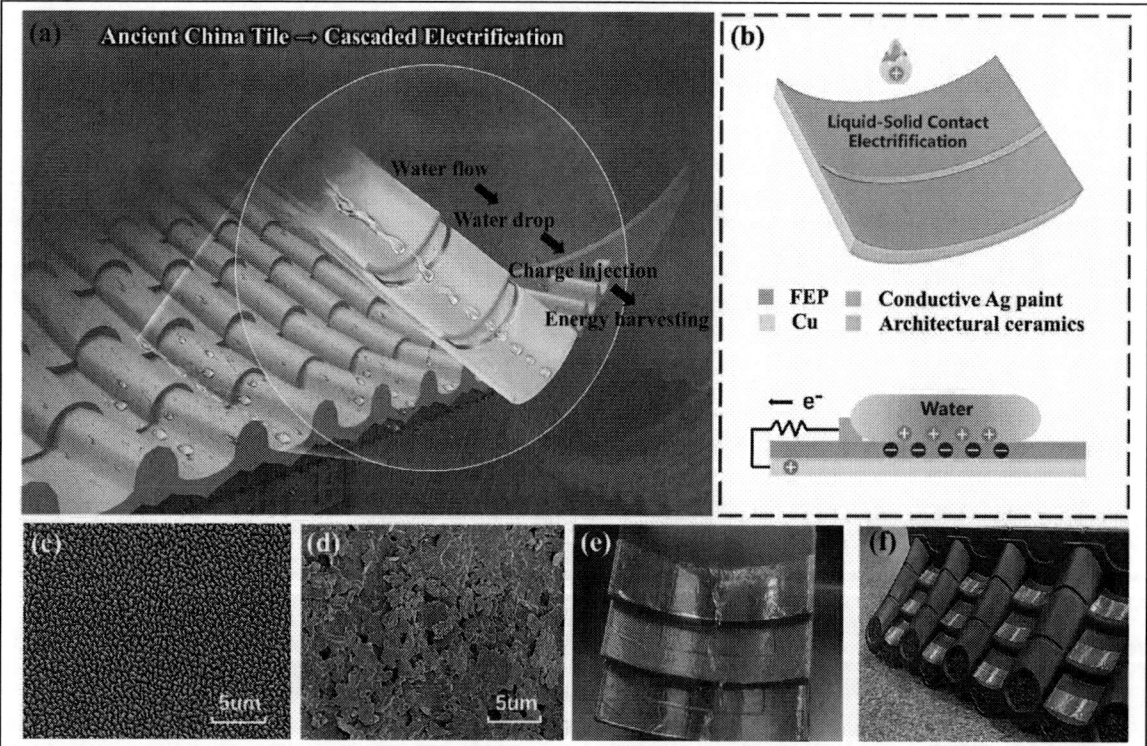

Figure 1: Glazed tile-inspired Cascaded liquid-solid power generator (CLPG). (a)Design concept; (b)Structure and mechanism of waterdrop energy harvesting; (c)The SEM of FEP; (d)The SEM of conductive Ag paint; (e)Structure unit; (f)Roof equipped with CLPG in large quantities.

show the ability to collect water droplets' energy, the same high electrical signal response was not obtained in the experiments of collecting water flows. As a result, the plane DEG structure has a limited Capacity Factor (CF) - the ratio of collected energy to the maximum energy theoretically obtainable [16,17]. In addition, previous studies are based on the ideal situation that water drops fall in order without interfering with each other. The reality is far more complex than this. The existing technology cannot effectively deal with the situation that a large number of water drops drop irregularly or even form water flow.

In this study, we draw inspiration from the overlapping shapes of glazed tiles and propose a cascade structure of a liquid-solid power generator. The core idea of this structure is to convert continuous water flow into water droplets and collect water energy discretely. After using surface potential scanning technology and high-speed camera technology, we demonstrated that the CLPG device increased the Capacity Factor (CF) of capturing solid-liquid electrification energy by utilizing the feature of the instability of water flow in gravitational traction and mechanical impact. Thus, the device can collect water energy in more prevalent forms. [18-20]

RESULTS AND DISCUSSION

The proposed tile-inspired liquid-solid power generator (CLPG) is shown in Fig.1. The basic idea is to convert continuous water flow into water droplets and collect energy through the droplet-based electricity generator (DEG) which includes Fluorinated Ethylene Propylene (FEP) film with strong hydrophobicity, conductive Ag paint, Cu electrode and tile substrate. The structure of the DEG module in CLPG is shown in Fig.1 (b). In this module, FEP has been proven to capture negative charges when contacting with water, as the electronegativity of FEP could perform an effective attraction to the negative ions or electrons in liquid or solid phases. The charge gradually accumulates as the droplet spreads on the FEP surface and is rapidly released within milliseconds when the droplet touches the top electrode,

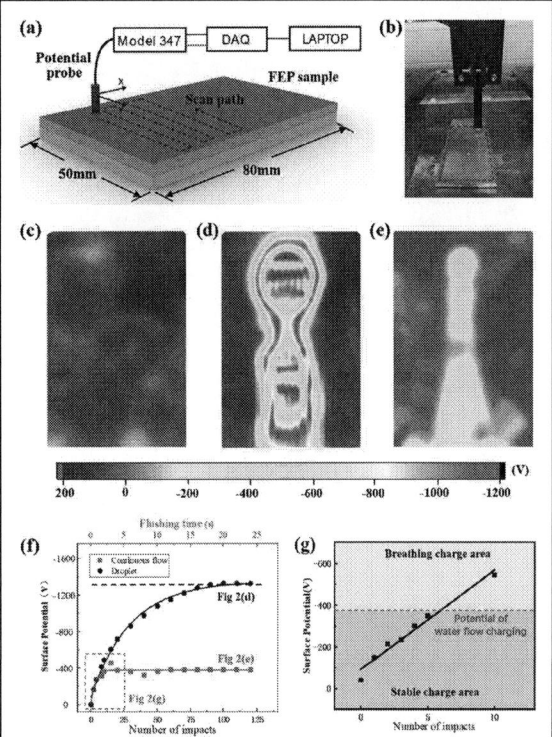

Figure 2: (a)Scanning of electret surface potential distribution; (b)The probe and the scanning area; Surface potential distribution for some cases: (c)before charging, (d)after droplet charging, (e)after water flow charging; (f) surface potential change in two cases; (g)The part where the surface potential is linearly related to the number of impacts

thus achieving ultra-high efficiency of droplet energy harvesting. The SEM images of the structure surface are shown in Fig.1(c-d), and the applications are shown in Fig.1 (e-f).

Figure 2 shows the surface potential scan method and

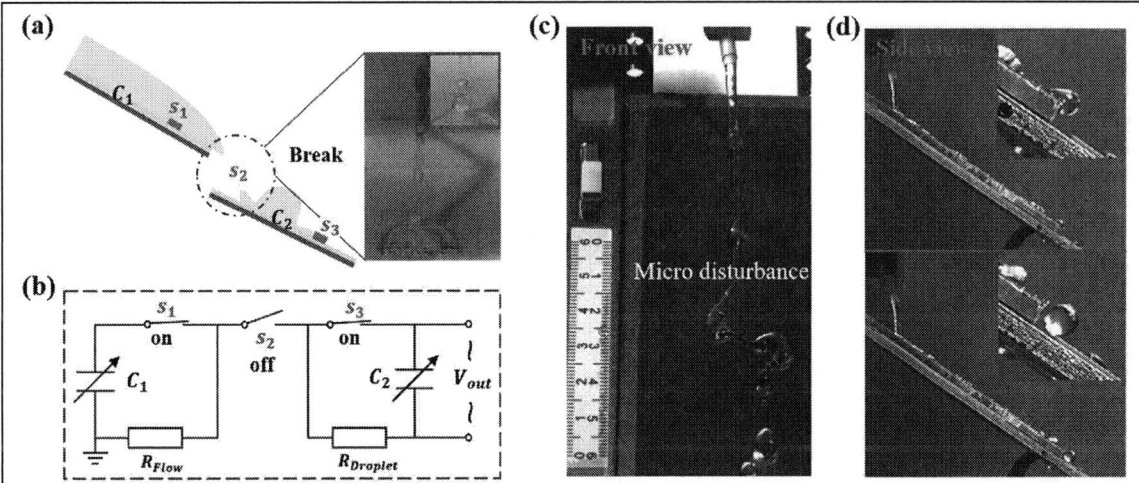

Figure 3: (a)The flow of water changes into impact; (b)Circuit abstraction of the influence of water flow splitting on power generation; (c)Front view and (d)Side view of the process of water splitting into droplets after flowing through the inclined wall, Plateau–Rayleigh instability promotes the transition from continuous flow to discrete droplets.

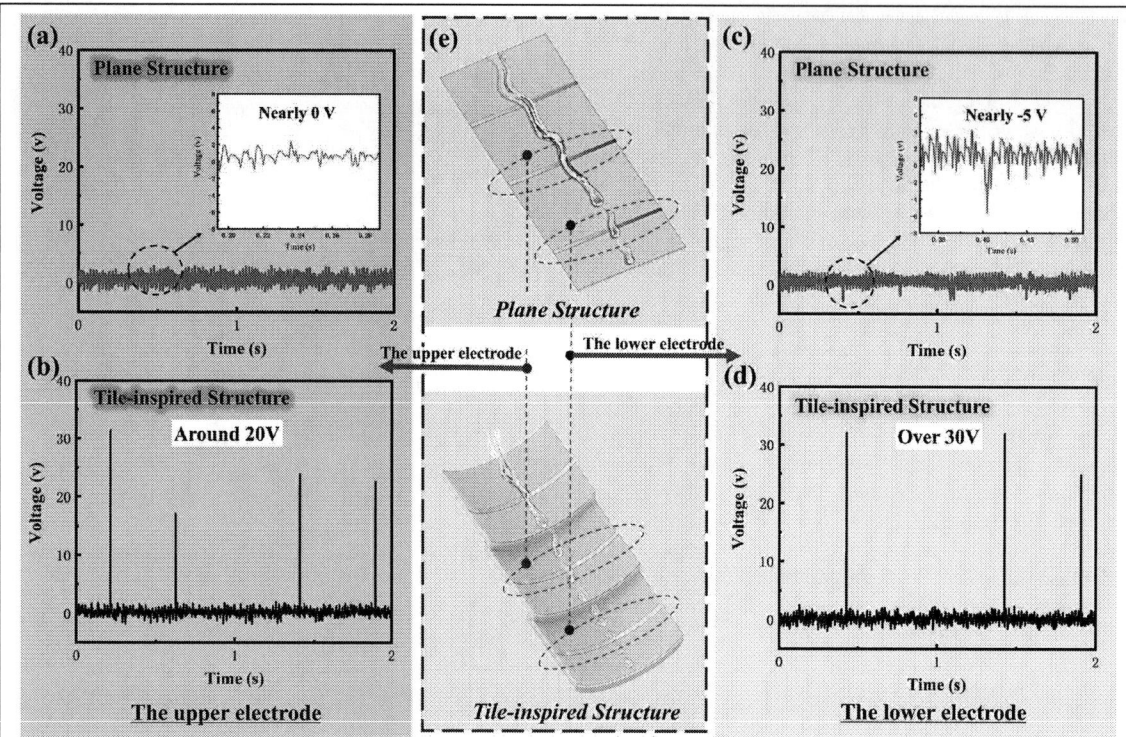

Figure 4: The comparison of device performance. (a-d) The output voltage signals of plane structure and tile-inspired structure, discussing electrodes at different positions; (e) Two experimental models of energy collection structure, whose impedance and external matching impedance are both 500 kΩ.

the experimental results, which proved that the water flow contact electrification process is difficult and inefficient compared to the water drop contact. As shown in Figure 2 (a-b), we used a surface potential scanner (Trek Model 347, USA) to scan the surface potential of FEP samples along a given path. The data were processed by a DAQ system (NI USB-6289 M series, USA), and then recorded by a computer. After data processing, we obtained the surface potential distribution of FEP after liquid-solid contact electrification. As shown in Fig.2 (c), the FEP surface after the impact of water droplets presents a ring-shaped high potential area, with the highest potential of about -1200V; in Fig.1 (e), after the water flow charging, the scan result of surface potential value is approximate -400V, and the pattern is consistent with the flow scouring area. The results of the potential distribution show that the charge effect of water droplets on the electret is more vital than that of continuous water flow. In addition, we focus on the highest surface potential in these two cases, as shown in Figure 2 (f). For water flow, after the electrification, the potential reached the saturation value indicated in Fig.2 (f) by the red dotted line and was not able to increase. We call the potential region below the red dotted line the "stable potential area". While intermittent water droplets can raise the FEP surface potential above the dashed line, but then fall with charge transfer (power generation), so we define it the "breathing charge area", as shown in Fig.2 (g). Because the amount of energy harvested by the water energy generator is positively correlated with the surface potential of the dielectric layer, devices with surface potential in the "breathing charge area" have higher power

output. Therefore, converting water flow into droplets will increase water energy harvesting efficiency.

The high-speed camera image shows how CLPG transformed water flow into water droplets. As shown in Fig.3 (a), when water flows down the slope, the water flow splits into droplets through the tile structure under the action of gravity and interfacial tension, then continues to hit the lower tile. This splitting process is called Plateau-Rayleigh instability. In Fig.3 (b), the break of water flow can be regarded as the disconnection of S2, which ensures the accumulation of charge on the surface of the FEP below and avoids the consumption of output electrical energy by the water flow itself. From the high-speed camera results in Fig. 3 (c-d), the conversion process of water flow to water droplets is observed from different viewpoints.

Finally, as shown in Figure 4, we compared the performance output of the plane DEG with CLPG. In the case of collecting water flow energy, the voltage output of the tile-inspired structure increases by a factor of nearly five. Considering the exact impedance matching, the tile structure increases the output power by a factor of 25, which proves the effectiveness of CLPG.

CONCLUSION

In this study, we proposed a tile-inspired cascaded liquid-solid power generator (CLPG) that can harvest continuous water flow energy. The scanning of electret surface potential illustrates the importance of converting water flow into water   droplets to collect energy. The measurement results of the high-speed camera image and the electrical output characteristics signal show that the

CLPG can convert water flow into droplets without external constraints by taking advantage of water instability and gravity, which increases the energy output efficiency of the water energy harvesting by 25 times.

ACKNOWLEDGEMENTS

This Research is supported by the National Natural Science Foundation of China Grant (No.51705429，No. 51735011), Science, Technology and Innovation Commission of Shenzhen Municipality (JCYJ20220530161809020), the Fundamental Research Funds for the Central Universities, 111 Project No. B13044.

REFERENCES

[1] Chu, S., Majumdar, A., " Opportunities and challenges for a sustainable energy future"，Nature 488, 294–303 (2012).

[2] Wang ZL，"Self-powered nanotech"，Sci Am. 2008 Jan;298(1):82-7.

[3] K. Tao, L. Tang et al., "Investigation of Multimodal Electret-Based MEMS Energy Harvester With Impact-Induced Nonlinearity," Journal of Microelectromechanical Systems, vol. 27, no. 2, pp. 276-288, Apr. 2018.

[4] Q. Liang, X. Yan, X. Liao, Y. Zhang, "Integrated multi-unit transparent triboelectric nanogenerator harvesting rain power for driving electronics"，Nano Energy 25 (2016) 18–25.

[5] C. Choi, D. W. Kim, D. Yoo, K. J. Cha, M. La & D. S. Kim, "Spontaneous occurrence of liquid-solid contact electrification in nature: Toward a robust triboelectric nanogenerator inspired by the natural lotus leaf", Nano Energy, 36, pp. 250-259, 2017.

[6] L. Zheng, Z. H. Lin, G. Cheng, W. Wu, X. Wen, S. Lee & Z. L. Wang, "Silicon-based hybrid cell for harvesting solar energy and raindrop electrostatic energy", Nano Energy, 9, pp. 291-300, 2014.

[7] Y. Liu, N. Sun, J. Liu, Z. Wen, X. Sun, S. T. Lee & B. Sun, "Integrating a Silicon Solar Cell with a Triboelectric Nanogenerator via a Mutual Electrode for Harvesting Energy from Sunlight and Raindrops", ACS Nano, 12(3), pp. 2893-2899, 2018.

[8] W. Tang, B. D. Chen, & Z. L. Wang, "Recent Progress in Power Generation from Water/Liquid Droplet Interaction with Solid Surfaces", Advanced Functional Materials, 29(41), p. 1901069, 2019.

[9] X. J. Zhao, S. Y. Kuang, Z. L. Wang, & G. Zhu, "Highly Adaptive Solid-Liquid Interfacing Triboelectric Nanogenerator for Harvesting Diverse Water Wave Energy", ACS Nano, 12(5), pp. 4280-4285, 2018.

[10] K. Tao, H. Yi et al., "Origami-Inspired Electret-Based Triboelectric Generator for Biomechanical and Ocean Wave Energy Harvesting," Nano Energy, vol. 67, p.

[11] Z. Zhao, H. Zhou et al., "Multiphase Bipolar Electret Rotary Generator for Energy Harvesting and Rotation Monitoring," Journal of Microelectromechanical Systems, 2022.

[12] K. Tao, B. Lyu et al., "Micro-Patterning of Electret Charge Distribution by Selective Liquid-Solid Contact Electrification," Journal of Microelectromechanical

Systems, vol. 31, no. 4, pp. 625-633, Aug. 202

[13] J. Zhang, B. Lyu et al., "High-Efficiency Raindrops Energy Harvester Using Interdigital Electrode," in Proc. IEEE 34th Int. Conf. MEMS, pp. 724-727, Jan. 2021.

[14] K. Tao, Z. Chen et al., "Ultra-Sensitive, Deformable, and Transparent Triboelectric Tactile Sensor Based on Micro-Pyramid Patterned Ionic Hydrogel for Interactive Human–Machine Interfaces," Advanced Science, vol.9, p. 2104168, Jan. 2022.

[15] Xu, W., Zheng, H., Liu, Y. et al. "A droplet-based electricity generator with high instantaneous power density"，Nature 578, 392–396 (2020).

[16] Wu, Hao et al., "Energy Harvesting from Drops Impacting onto Charged Surfaces"，Physical Review Letters 125, no. 7 2020 Aug: 078301.

[17] Riaud et al., "Hydrodynamic Constraints on the Energy Efficiency of Droplet Electricity Generators"，Microsystems & Nanoengineering 7, no. 1 2021 Dec: 49

[18] Zheng, Y., Liu, T., Wu, J., Xu, T., Wang, X., Han, X., Cui, H., Xu, X., Pan, C., Li, X., "Energy Conversion Analysis of Multilayered Triboelectric Nanogenerators for Synergistic Rain and Solar Energy Harvesting", Adv. Mater. 2022, 34, 2202238.

[19] Wu, H., Wang, Z., Zi, Y., "Multi-Mode Water-Tube-Based Triboelectric Nanogenerator Designed for Low-Frequency Energy Harvesting with Ultrahigh Volumetric Charge Density", Adv. Energy Mater. 2021, 11, 2100038.

[20] B. Kil Yun, H. Soo Kim, Y. Joon Ko, G. Murillo, & J. Hoon Jung, "Interdigital electrode based triboelectric nanogenerator for effective energy harvesting from water", Nano Energy, 36, 233-240, 2017.

CONTACT

*Kai Tao, tel: +29-88460434; taokai@nwpu.edu.cn

MEMS CANTILEVERED ENERGY HARVESTER WITH TAPERED THICKNESS FOR STRESS CONTROL

Takahito Yokota, Kensuke Kanda, Takayuki Fujita, and Kazusuke Maenaka
The Graduate School of Engineering, University of Hyogo, JAPAN

ABSTRACT

This paper reports on the improvement of the output power of piezoelectric energy harvesters (PEHs). A tapered thickness structure of a cantilever beam for uniform stress, tungsten proof mass, thick piezoelectric film, and series connection of the piezoelectric films are utilized for PEHs to increase the output power. The three-dimensional (3D) etching process of Si enables the formation of tapered thickness beam structure, which increases not only the output power, but the beam strength. The 3D shape allows to increase the weight of the proof mass, contributing to increase the output voltage and power. Furthermore, output voltage is enhanced by using series-connection of PZT cells to improve the circuit efficiency. The open-circuit output voltage and optimum power with resistance load of the fabricated harvesting device revealed 14.4 V_{0-P} and 92 μW at an acceleration of 9.8 m/s^2. The output power was improved by 8.9 times higher than that of the conventional model.

KEYWORDS

Uniform stress energy harvester, Series-connected piezoelectric film, Stress concentration relief, Fillet, Micro-loading effect, Thickness control

INTRODUCTION

Energy harvesting devices are essential for the autonomous sensors. Various devices have been developed in the last decade. Vibration is one of a promising candidate for the input energy of the harvesting device to power the sensors. A large number of researches for resonant-type vibration harvesters have been reported. Fundamental device design and fabrication technique are almost matured today. However, reliability and output performance are still important issues to miniaturize the harvesting devices. This research addresses the improvement of the endurance and output performance of silicon-based MEMS energy harvesters by using 3D fabrication process.

Piezoelectricity is a popular procedure to convert the mechanical energy into electrical one. The piezoelectric material generates electrical charge in proportional to the applied mechanical stress. Typical piezoelectric energy harvesters have bilayer cantilever structure consisting of piezoelectric and non-piezoelectric layers. A mass weight is usually attached to the tip of the cantilever. External vibration deforms the cantilever and the resulting internal stress is converted to electric charge. The internal stress of the bending beam structures is maximum at the fixed end. The stress concentration at the fixed end is critical to the fracture strength of the silicon based harvesters. Therefore the authors have applied a 3D fabrication technique of Si by using microloading effect, i.e. plasma etching rate dependent on the aperture-size (MEMSNAS process [2]) to a harvester structure for the improvement of the stress concentration at the fixed end. The fillet structures with a

tapered thickness have located at the fixed end and achieved to improve the fracture strength of the beam and output power. This means that a heavier proof mass can be used.

The output voltage is also an important feature of the MEMS energy harvesters. Since the output voltage from a harvester is bipolar voltage, rectification is required. Considering the voltage drop at the rectifier, a large output voltage is required. Although typical PZT thin films have a large piezoelectric constant, their large permittivity results in a small output voltage. For piezoelectric MEMS harvesters utilizing PZT, output voltage should be enhanced.

In this study, not only the tapered thickness structure, but the tungsten proof mass is attached for high power density. In addition, a thick PZT (lead zirconate titanate) film of 10 μm is used as a piezoelectric material, and the piezoelectric film is segmented and connected in series to achieve higher voltage and power.

DESIGN OF ENERGY HARVESTERS

Three devices are compared in this study as shown in Table 1, and the image of the third device is shown in Fig. 1.

Table 1: The three compared devices.

Device No.	Tapered thickness	PZT Series connection	Tungsten mass attachment
1	No	No	No
2	Yes	Yes	No
3	Yes	Yes	Yes

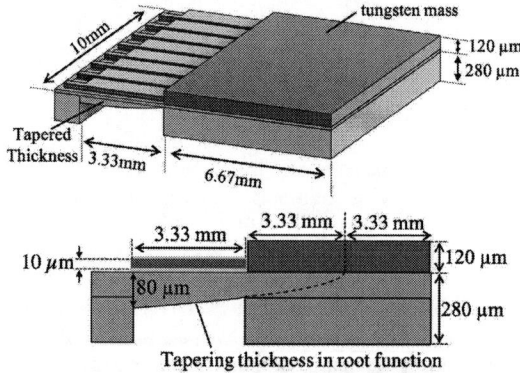

Figure 1: Tapered thickness, tungsten mass attached PEH.

The dimensions are based on a previous study [3], with the size of 10 mm square. The parameters are that the beam thickness is uniform or tapered, that PZT film is segmented and series-connected or not, and that a tungsten mass is attached on the mass or not. For all those devices, a cantilever beam supports a proof mass at the tip. The device with the tapered thickness can enhance the fracture strength and output power due to a fillet structure and a uniform stress structure. The segmentation and series connection of PZT thin films can increase output voltage proportional to the number of series connection. The tungsten mass induces large input energy into the device, resulting in large output power. Although the output voltage increases proportional to the number of cells, too large number of series connections are undesirable because of the effect of parasitic capacitance and too large output impedance [5]. Therefore, the number of series connections is determined to up to eight.

PROTOTYPE

The PZT thin film sandwiched by top and bottom metals is deposited on the beam. For the fabrication process, bottom metal, PZT, were sequentially sputtered on a Silicon-On-Insulator (SOI) wafer. Since a thicker piezoelectric thin film is necessary to obtain a larger output power, PZT thin film with the thickness of 10 μm was sputtered by using a multistep sputtering technique [4]. Next, the PZT and metals were dry-etched in chlorine-based gas-mixture plasma. The PZT thin films were segmented into multiple cells in this etching process. At last, device and handle layers of SOI are dry etched by using a Deep Reactive-Ion-Etching (DRIE) equipment to make the cantilever shapes.

Common MEMS structures derived from semiconductor processes are typically called 2.5D structure, i.e. freely designed planer shape and fixed layer thick. It is difficult to control the thickness of the beam freely. In this study, MEMSNAS process [2] is used to control the thickness. Already the authors have succeeded in fabricating the cantilever with controlled thickness of over 150 μm, and have also demonstrated the fabrication of a cantilever beam with tapered thickness and fillet structures. Furthermore, the thickness can be controlled with good accuracy within a maximum error of 10% [1].

The fabricated device is shown in Fig. 2. The device is assembled into a mount fabricated by an optical 3D printer, and a tungsten weight is attached on the tip mass with UV curable adhesive.

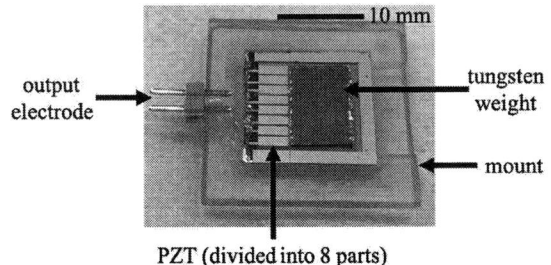

Figure 2: The image of the completed device with the tungsten mass attachment and PZT series connection.

EVALUATIONS

Comparisons of the generated voltage and power were conducted for the devices in Table 1. In the evaluations, the segmented 8 PZT cells are used as the device with the series connection number of 4 (4 pairs of cells are connected in series). Fig. 3 shows the measurement system.

Sinusoidal acceleration was applied by a shaker to the PEHs, and the output voltage and displacement of the mass center were measured.

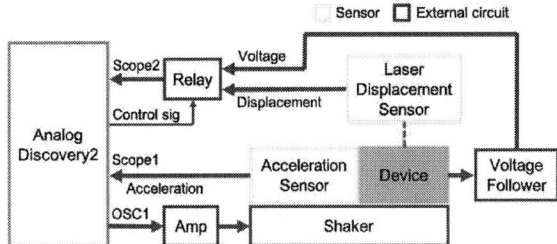

Figure 3: The measurement system.

Figure 4 shows the resonant characteristics of the three devices when the external acceleration of 1 G ($\approx$ 9.8 m/s^2) is applied. Due to the 4-series connection and tapered thickness beam for the uniform stress, the generated voltage was significantly increased, and the resonant frequency was decreased due to the tungsten mass attachments. From the viewpoint of practicality, these features are desirable because the higher output voltage can reduce the effect of voltage drop due to the rectifier circuit, and lower resonant frequency can be closer to environmental vibrations, which are relatively low in frequency [6][7].

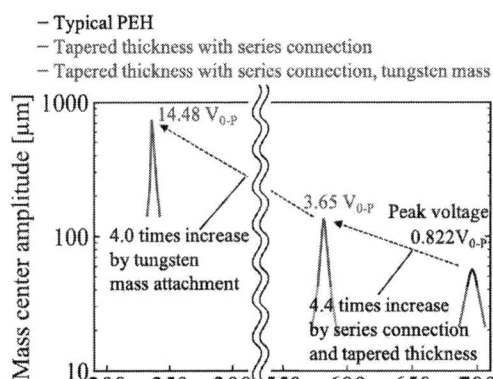

Figure 4: Measured displacement of mass center and maximum open circuit voltage.

Figure 5 shows the output power characteristics of each PEH at 1 G in resonance. The output power of the TTPEH with the tungsten mass was 91.7 μW at the optimum load resistance, which is 8.9 times higher than that of the typical PEH and 7.6 times higher than TTPEH without the tungsten mass.

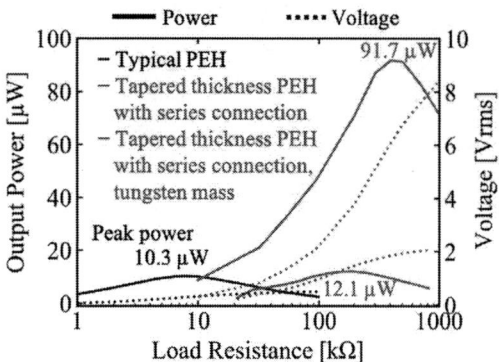

Figure 5: Measured mean output power and RMS voltage as a function of the load resistance.

CONCLUSIONS

In this study, it is shown that the uniform stress, thicker PZT film by sputtering, segmentation and series-connection of PZT, and tungsten mass attachment are significantly effective for increasing the generated voltage and output power compared to the conventional PEH. The fabrication process is very practical because it can be performed with general equipment and processes.

ACKNOWLEDGEMENTS

This research is supported by JST CREST, JPMJCR20Q2, and JSPS Grant-in-Aid for Scientific Research (B) 20H02120, JAPAN.

REFERENCES

[1] T. Yokota, K. Kanda, T. Fujita, and K. Maenaka, "Thickness Control of Cantilever Beam for Robust and High-power MEMS Energy Harvester", *Power MEMS2022*, Paper No. 0011, 2022.

[2] T. Bourouina, T. Masuzawa, and H. Fujita, "The MEMSNAS Process: Microloading Effect for Micromachining 3-D Structures of Nearly All Shapes", MEMS, Journal of 13, pp. 190-199, 2004.

[3] S. Hirai, K. Kanda, T. Fujita and K. Maenaka, "MEMS Energy Harvesting Based on Uniform-Stress Cantilever with Multilayer PZT Thin Films", J. Phys.: Conf. Ser. 1407 012081, 2019.

[4] K.Kanda, T. Koyama, T. Yoshimura, S. Murakami, and K.Maenaka, "Characteristics of Sputtered Lead Zirconate Titanate Thin Films with Different Layer Configurations and Large Thickness", *IEEE Trans. Ultrason. Ferroelectr.* Freq. Ctrl., Vol. 68 Issue 5, pp. 1988-1993, 2021.

[5] K. Kanda, T. Saito, Y. Iga, K. Higuchi, and K. Maenaka, "Influence of Parasitic Capacitance on Output Voltage for Series-Connected Thin-Film Piezoelectric Devices", *Sensors*, Vol. 12, pp. 16673-16684, 2012.

[6] S. Roundy, P. K. Wright and J. Rabaey, "A study of low level vibrations as a power source for wireless sensor nodes", *Computer Communications, Vol. 26*, Issue 11, pp.1131-1144, 2003.

[7] S. Roundy, Energy Scavenging for Wireless Sensor Nodes with a Focus on Vibration-to-Electricity Conversion, University of California, Berkeley, 2003.

CONTACT

*K. Kanda, Tel: +81-080-5445-8675; kanda@eng.u-hyogo.ac.jp

TAPERED HELMHOLTZ RESONATOR WIND ENERGY HARVESTER DRIVEN BY AEROACOUSTICS

Chen Hua[1,2], Liyun Zhen[1,2], Jingquan Liu[1], and Bin Yang[1]

[1]National Key Laboratory of Science and Technology on Micro/Nano Fabrication, Shanghai Jiao Tong University, CHINA

[2]Department of Micro/Nano Electronics, Shanghai Jiao Tong University, CHINA

ABSTRACT

This paper presents an aeroacoustics-driven piezoelectric energy generator (APEG) containing a tapered Helmholtz resonator (HR) and piezoelectric circular transducer (PCT) unit to scavenge wind energy. Through the tapered Helmholtz resonator will reduce the acoustic resistance and increase the highest sound absorption coefficient attainable to enhance the performance of Helmholtz resonator. With the help of the chemical mechanical thinning method and simulation, PCT unit is optimized to match the resonant cavity to achieve maximize power. At a flow rate of 170 L/min, the APEG delivers high output peak to peak voltage of 20 V, and large average output power of 21.3 mW at 1.5 kΩ.

KEYWORDS

Aeroacoustics, Wind Energy Harvester, Piezoelectric Circular Transducer, Tapered Helmholtz Resonator

INTRODUCTION

With the widespread use of new microelectronics and portable devices, the continuous power supply is paid more attention. Chemical energy battery is a common power supply solution for these low power consumption devices. However, Chemical energy has the disadvantages of short life, environmental pollution, inconvenience to carry and instability. In order to meet the fast and lasting energy demand of the devices, piezoelectric energy harvesters based on the piezoelectric effect have been widely developed. It can directly harvest energy from the surrounding environment, such as wind energy, to power electrical devices, which is an important exploration of renewable energy utilization and self-powered sensing [1].

As a renewable and widely distributed energy, wind energy has attracted more and more attention and plays an important role in power supply. At present, scientists report various wind energy harvesters based on electromagnetic, piezoelectric, electrostatic and triboelectric mechanisms [2]. The wind energy harvesters are usually composed of fixed parts and movable parts. In the process of wind energy utilization, movable parts are used as mechanical transducers to convert wind into mechanical rotation or vibration [3]. Therefore, based on the principle of aeroacoustics, wind can be used to induce sound field with stable frequency and large amplitude to power electronic devices. For example, Sun *et al.* designed a cross-junction configuration mean flow acoustic engine (MFAE), which converts wind energy in the pipeline into sound energy for driving power generation devices and thermoacoustic refrigerators [4]. After that, through the improvement of the pipe structure, Yu *et al.* built Pi-type MFAE and Jiang *et al.* used square cross-section pipe to improve the

efficiency of acoustic harvesting [5, 6]. In 2015, Zou *et al.* developed a piezoelectric energy harvester based on aeroacoustics principle and piezoelectric transducer mechanism, which can convert incoming wind energy into electric energy during flight [7]. On this basis, Li *et al.* studied the excitation conditions of the edge-tones generation by the similar jet resonator system and analyzed the influence of the structural parameters on the fluid dynamic sound sources, thus optimizing the design of the jet resonator system [8, 9]. Later, Huang *et al.* thinned the piezoelectric circular diaphragm to optimize the electrical performance. When the jet velocity was 79.6 m/s, the output power was 15.5 mW, which successfully provided real-time energy for different commercial electronic devices [10]. Although in the past studies, the performance of airflow-induced vibration piezoelectric generator has been studied thoroughly. However, the influence of resonator structure on the electrical performance of devices has not been studied. Some studies have shown that the tapered Helmholtz resonator structure can further amplify the sound field [11, 12]. Therefore, the tapered Helmholtz resonator structure is used to further amplify the bottom sound pressure, so as to achieve sustainable energy supply in the high-speed airflow environment.

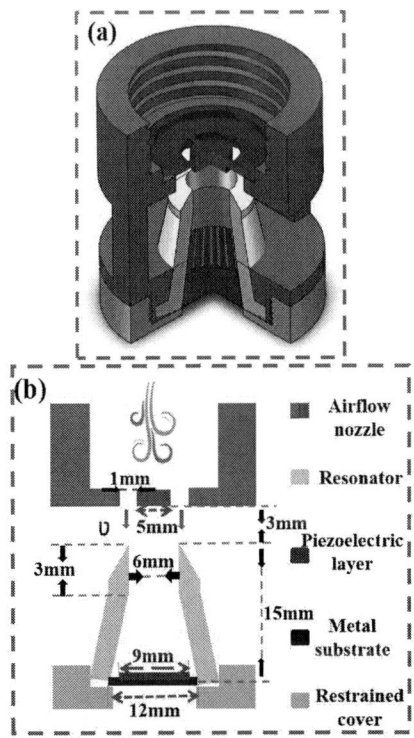

Figure 1: Schematic illustration (a) and the cross-section (b) of the APEG including tapered HR and PCT unit.

In this paper, we proposed a high-performance piezoelectric energy harvester with tapered Helmholtz resonator excited by airflow to meet the demand of sustainable energy supply in high-speed airflow environment. Based on the principle of aeroacoustics, using COMSOL simulation software to optimize the structure of the tapered Helmholtz resonator and to explore the influence of resonator structure on the electrical performance of devices. Through MEMS manufacturing process, the frequency of piezoelectric circular transducer unit is matched with one of the resonant cavity to improve the electrical output performance. Finally, the high-speed airflow simulation test platform is built and the electrical performance of devices is tested. It is enough to supply energy for commercial devices, and is expected to meet the demand for energy supply in high-speed environments such as unmanned aerial vehicles.

DESIGN AND FABRICATION

Figure 1a shows the designed structure including the tapered HR, PCT unit and the APEG. APEG is mainly composed of annular nozzle, open closed tapered Helmholtz resonator with edge and PCT unit. Jet-edge-resonator (JER) system aims to convert the airflow into edge-tones and generate stable standing wave for PCT. Figure 1b shows the corresponding parameters of the APEG. When the airflow through the pipe, the nozzle first converts the airflow into oscillating jet and collides with the edge, producing a dipole sound source called edge-tones. The edge-tones travel inside the tapered Helmholtz resonator. The air in the resonator cavity is like a spring. The pressure sound wave near the orifice forces the air mass to enter the resonator cavity, which will increase the pressure in the cavity. Then the increased pressure in the resonator pushes the air mass at the hole upward. When the APEG is coupled and resonant, it will reflect at the closed end to form a stable standing wave and be applied to PCT unit. Finally, PCT unit converts mechanical vibration into electrical energy based on piezoelectric effect by restraining the clamping of the cover and the resonator.

In Figure 2, the working mechanism of APEG in edge-tones generation is analyzed by using aeroacoustics principle, computational fluid dynamic simulation and the Navier-Stokes equation. The device states in the steady state and transient state are studied respectively. In figure 2a, velocity distributions for APEG under steady state is simulated. In Figure 2b-d, velocity distributions under transient state for APEG at three stages are simulated. At first, the jet generates reverse vortex to act on the edge at the opening end of the resonator, and then separates the jet, successively generating alternating vortices in opposite directions in Figure 2b. The vortex reacts on the flow field at the jet nozzle, reinforcing the vortex shedding to match with oscillation. After the oscillation system is stabilized in Figure 2c, a dipole sound source is generated at the edge, and the edge-tones are generated and traveled to the resonator, which is coupled with the air column of the resonator and gradually forms a standing wave. At this time, the resonator will occupy the dominant position, determine the oscillation frequency, and feed back to the edge and airflow nozzle in Figure 2d, finally forming a

stable self-oscillation system [10, 13].

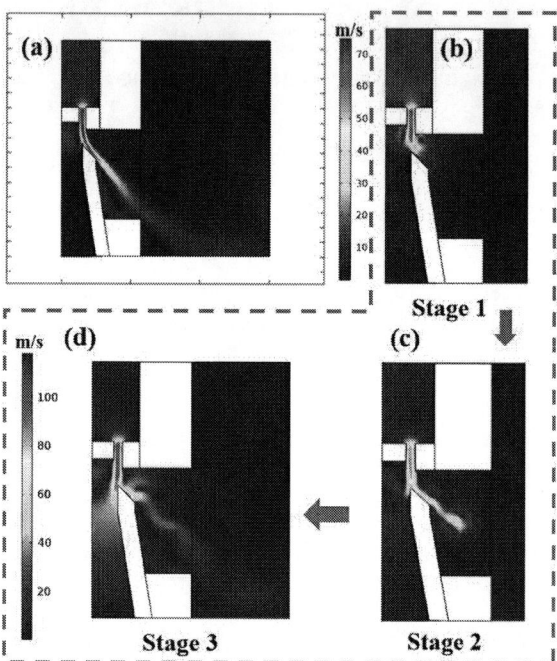

Figure 2: (a) Simulated velocity distributions under steady state. (b-d) Simulated velocity distributions under transient state at three stages .

By adding a probe to the bottom of the resonator through COMSOL simulation, the change of the device bottom force can be obtained, and the corresponding frequency change can be obtained through fast Fourier transform (FFT) spectrum analysis. As shown in Figure 3 a-b, the simulated pressure and frequency at the bottom of the resonant cavity are 5 kPa and 3322 Hz under the wind speed of 10 m/s.

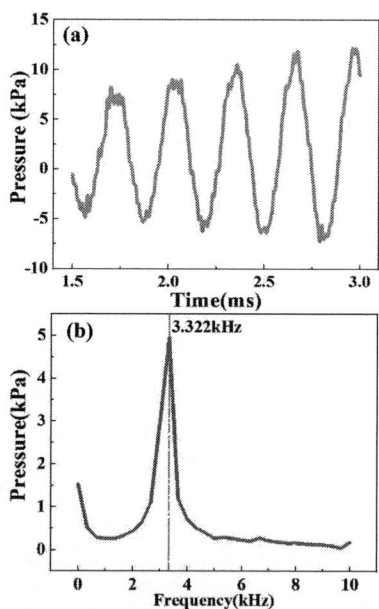

Figure 3: (a) Pressure simulation at bottom of resonator.(b) FFT spectrum analysis.

For transducer, PCT unit is mainly composed of stainless steel substrate and PZT piezoelectric layer. The detailed micro-fabrication of the PCT unit is depicted in Figure 4. First, the substrate and PZT are cut by laser to obtain the circle with the desired diameter. Then, Cr/Au electrode is sputtered on one side of PZT, and stainless steel is thermally bonded to PZT through conductive silver epoxy. The bulk PZT is thinned to a certain thickness by the chemical mechanical thinning method. Sputtering the Cr/Au electrode on the upper surface of PZT, and finally electrical performance are tested [14, 15].

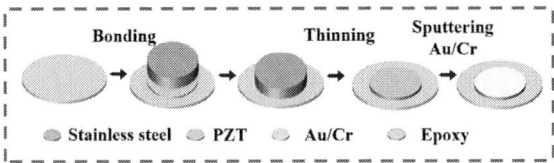

Figure 4:Fabrication process of the PCT unit.

Through COMSOL simulation, by controlling the diameter and thickness, the characteristic frequency of PCT unit with different diameters and thicknesses are obtained and matched with the previous simulation in Figure 5a. The bulk PZT with initial thickness of 300 μm was thinned down to the designed thickness through the chemical mechanical thinning method. In Figure 5b, PCT unit on 30 μm stainless steel substrate with 12 mm diameter and 30 μm thickness PZT with 9 mm diameter can be matched with the resonant cavity.

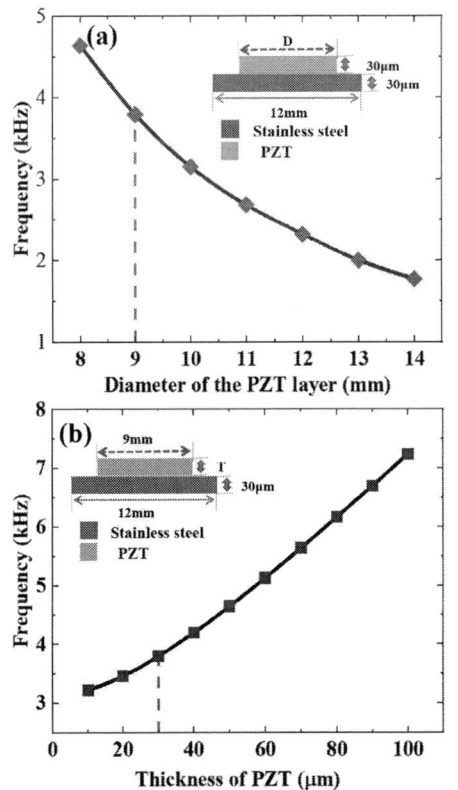

Figure 5: (a) Optimization of PZT diameter under both 30 μm substrate and PZT. (b) Optimization of PZT thickness under 30 μm substrate and 9mm diameter PZT.

Results and Discussion

The experimental platform for characterizing the APEG is shown in Figure 6a. In order to provide stable air flow to APEG, a gas tank is used to store and compress air. The incoming air is then purified through an air filter to ensure a stable output of the pressure regulating valve and the flow sensor. The flow sensor is used to monitor the change of air flow in real time. Finally, the output voltage is obtained by connecting the oscilloscope with the device lead. The APEG is mainly consisted of two parts, namely jet-edge-resonator system in Figure 6b by 3D-printing and PCT unit in Figure 6c.

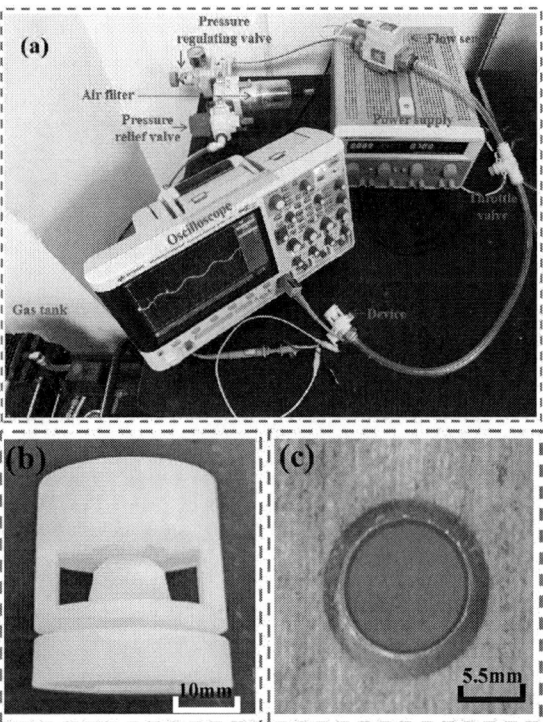

Figure 6: (a) Schematic diagram of the experimental setup for characterizing the APEG. (b) Schematic diagram of jet-edge-resonator system. (c) PCT unit

In Figure 7a, the peak to peak voltage of the APEG is tested at a flow rate of 80-220 L/min. It can be found that the correlation between voltage and flow is approximately a straight line. There is a critical jet velocity, which is divided output into two linear parts. Once the critical jet velocity is reached, APEG produces edge-tones and then the slope of voltage and jet velocity increases. With the increase of airflow, peak to peak voltage harvesting at 220 L/min is 34 V. Subsequently, in order to ensure the stability of the output, the device voltage and power are obtained at a flow rate of 170 L/min. The output peak to peak voltage waveform at the flow rate of 170 L/min is shown in Figure 7b and peak to peak voltage harvesting at 170 L/min is 20 V. The output peak to peak voltage change of the APEG obtained under different resistances of 100 Ω to 100 kΩ is shown in Figure 7c. In Figure 7d, the maximum root-mean-square (RMS) output power is 21.3mW at the optimal external resistance of 1.5 kΩ, which can meet the power requirement of some electronics devices or systems.

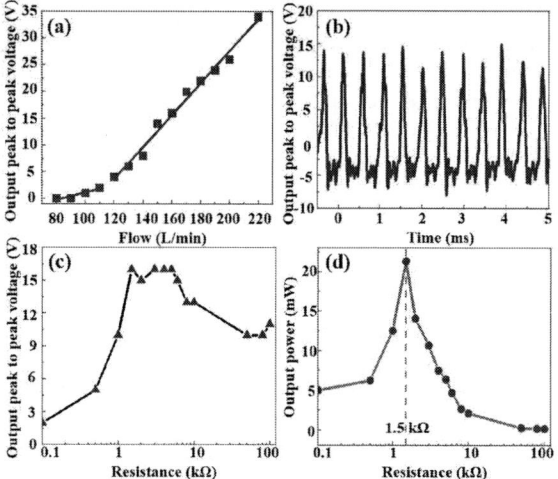

Figure 7: (a) The output peak to peak voltage of the APEG under various flow from 80 L/min to 220 L/min. (b) The output peak to peak voltage waveform under 170 L/min flow. (c) Peak to peak voltage under different load resistances at 170 L/min flow. (d) RMS Output power of APEG under different load resistances at 170 L/min flow.

Theoretically, milliwatt output power can effectively light up multiple LED lights and charge the capacitor in short time. The energy management circuit will be subsequently used to provide stable DC output to provide real-time energy for commercial electronic devices such as temperature and humidity sensors, wireless sensor networks, triple axis acceleration sensors, gyroscopes, etc. It is expected to meet the demand for energy supply in high-speed environments such as unmanned aerial vehicles and aircraft.

CONCLUSION

On the basis of the previous research, the tapered Helmholtz resonator is proposed to improve the performance of the device. Based on the principle of aeroacoustics, APEG composed of JER system and PCT unit has the advantages of small volume, simple structure and integrate design. In order to maximize the power, the frequency matching between the device and the PCT unit is carried out experimentally and theoretically. The APEG shows excellent output performance at 170 L/min flow, and provides 21.3 mW RMS Output power at 1.5 kΩ load resistance and 20 V peak to peak voltage. This research further provides a new structure improvement idea for the wind energy harvesting technology to supply power to commercial electronic devices in high-speed environments such as aircraft and unmanned aerial vehicles.

ACKNOWLEDGEMENTS

The authors grateful to the Center for Advanced Electronic Materials and Devices (AEMD) of Shanghai Jiao Tong University.

REFERENCES

[1] Z. Yang, S. Zhou, et al,. "High-performance piezoelectric energy harvesters and their applications", *Joule*, vol. 2, pp. 642-697, 2018.

[2] X. Fu, C. Zhang, et al., "Overview of micro/nano-wind energy harvesters and sensors", *Nanoscale*, vol. 12, pp. 23929-23944, 2021.

[3] X. Ma, S. Zhou, et al., "A review of flow-induced vibration energy harvesters", *Energy Conversion and Management,* vol. 254, pp. 115223, 2022.

[4] D. Sun, K. Wu, et al., "A mean flow acoustic engine capable of wind energy harvesting", *Energy Conversion and Management*, vol. 63, pp. 101-105, 2012.

[5] Y. Yu, D. Sun, et al., "Study on a Pi-type mean flow acoustic engine capable of wind energy harvesting using a CFD model", *Applied Energy*, vol. 189, pp. 602-612, 2017.

[6] Y. Jiang, H. Zhang, et al., "Numerical Simulation of Acoustic Resonance Enhancement for Mean Flow Wind Energy Harvester as Well as Suppression for pipeline", *Energies*, vol. 14, pp. 1725, 2021.

[7] H. Zou, H. Chen, X. Zhu, "Piezoelectric energy harvesting from vibrations induced by jet-resonator system", *Mechatronics,* vol. 26, pp. 29-35, 2015.

[8] Z. Li, J. Li, H. Chen, "Experimental research on excitation condition and performance of airflow-induced acoustic piezoelectric generator", *Micromachines*, vol. 11, pp.913, 2020.

[9] Z, Li, J. Li, H. Chen, "Sensitive Parameters of Dynamic Excitation on Fuze Airflow-Induced Acoustic Generator", *Micromachines*, vol.12, pp. 1033, 2021.

[10] Y. Huang, B. Yang, et al., "Aeroacoustics-driven jet-stream wind energy harvester induced by jet-edge-resonator", *Nano energy*, vol. 89, pp. 10644, 2021.

[11] Izhar Khan, Ullah Farid, "Three Degree of Freedom Acoustics Energy Harvester Using Improved Helmholtz Resonator", *Int. J. Precis. Eng. Manuf*, vol. 19, pp. 143-154, 2018.

[12] D. Li, A. Bao, et al., "Design of tunable low-frequency acoustic energy harvesting barrier for subway tunnel based on an optimized Helmholtz resonator and a PZT circular plate", *Energy Reports*, vol. 8, pp. 8108-8123, 2022.

[13] P.T. Nagy, A. Szabó, G. Paál, "A feedback model of the edge tone, using the adjoint Orr-Sommerfeld equation", *J. Fluid Mech*, vol. 915, pp. A13, 2021.

[14] Z. Yi, W. Zhang, B. Yang. "Piezoelectric approaches for wearable continuous blood pressure monitoring: a review", *J. Micromech. Microeng*, vol. 32, pp. 103003, 2022.

[15] Z. Yi, W. Zhang, B. Yang, et al., "Piezoelectric Dynamics of Arterial Pulse for Wearable Continuous Blood Pressure Monitoring", *Adv. Mater*, vol. 34, pp. 2110291, 2022.

CONTACT

*B. Yang, tel: +86-021-34206683; binyang@sjtu.edu.cn

ANDROMEDA: A FLEXIBLE MEMS TECHNOLOGY PLATFORM FOR A VARIETY OF PIEZOELECTRICALLY ACTUACTED MICROMIRRORS

Irene Martini*, Anna Alessandri, Marta Carminati, Roberto Carminati, Paolo Ferrarini, Daniela A. L. Gatti, Riccardo Gianola, Borka Lazarova, Carla M. Lazzari, Andrea Nomellini, Laura Oggioni, Claudia Pedrini, Carlo L. Prelini, Riccardo Tacchini, and Michele Vimercati

STMicroelectronics, AMS Group, Agrate Brianza (MB) and Cornaredo (MI), ITALY

ABSTRACT

The Andromeda platform is a novel and flexible thin film Piezo MEMS technology platform for the manufacturing of laser-beam scanning mirrors, which is capable to cover a wide range of requirements (quasi static and resonant operation, opening angle, resonant frequency, mirror dimension and wavelength range application) and to achieve the target performances of different market applications, from AR/VR projection to LiDAR. The results of the characterization of different Andromeda products and their performance improvement allowed by PZT technology evolution are presented in this paper.

KEYWORDS

MEMS scanner, MEMS mirror, PZT, piezoelectric, AR/MR, LiDAR.

INTRODUCTION

In literature dedicated process flows for different MEMS micromirror product designs are typically reported [1][2][3]. This approach has the drawback of a limited re-use of the same technology platform requiring continuous development efforts for each new design.

The peculiar and versatile process architecture presented here enables the integration directly on the MEMS chips of a piezoresistive sensing, for feedback control, together with thin film piezo actuation on MEMS mirror, with an area up to 8 mm^2, for high resolution projection. The very diverse electro-mechanical requirements for quasi static actuation and high frequency resonant operation are enabled by the definition of different technological perimeters which differ only for the active silicon thickness, respectively from tens up to hundreds of microns. Moreover, both the mirror and the cavity to accommodate its movements are integrated in the front-end processing. All these features allow a complete electro-optical characterization at wafer level testing [4], which in turns permits an early screening as well as an assembly cost reduction.

MANUFACTURING PROCESS

Figure 1 shows a cross section of the Andromeda architecture, where the main MEMS components are highlighted.

The manufacturing process is based on five different macro blocks as detailed hereafter.

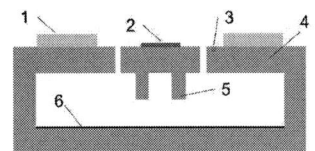

Figure 1: Schematic cross section of the Andromeda architecture: (1) PZT actuators, (2) mirror, (3) PZR sensor, (4) active silicon structure, (5) back-side reinforcement, (6) black cavity.

Piezoresistive sensor

Starting from an SOI (Silicon On Insulator) substrate, whose active layer thickness is selected based on the application (from 20 μm for slow scan application up to 200 of μm for the highest opening angle resonant mirror), the piezo resistive (PZR) sensor is defined through a sequence of lithography and ion implantation process steps. Dopant species are then activated and diffused by high temperature annealing, allowing the realization of a diffused PZR sensor in Wheatstone bridge configuration (sensitivity ranging from 0.7 mV/V/deg to 10 mV/V/deg, depending on the design).

Piezoelectric actuator

Piezoelectric actuation is attractive for micromirrors application because it allows larger actuation forces than electrostatic-based solution, offering better scanner characteristics with lower voltage. Starting from the consolidated ST proprietary technology PεTRA (Piezo-electric TRAnsducer) dedicated to PZT thin film MEMS actuators, many process improvements were done in the past years to move towards a higher actuation force (Table 1).

Table 1: Typical values of the absolute piezoelectric coefficient e_{31f}, as for the PεTRA piezo roadmap.

PZT-Gen1	PZT-Gen2	PZT-Gen3
14 C/m^2	16 C/m^2	17 C/m^2

The piezo stack (including barrier, bottom electrode, Lead zirconate titanate, top electrode) is defined on an insulating silicon dioxide layer. Depending on the piezo stack generation, the PZT layer deposition step may involve either a spin coating-based method or a PVD (Physical Vapor Deposition) process module. Once the metallic (and/or metallic oxide) layer accountable for the top electrode is deposited, the actuator is patterned through a sequence involving three different blocks of photolithographic mask, dry etching, resist stripping and cleanings. The Andromeda platform includes a low resistivity aluminum interconnection layer. This metallization is deposited on top of a pre-metal dielectric stack, acting as an insulator towards the actuators, and it is passivated by a moisture barrier layer. Pads can be finished afterwards with a gold-based stack to enable different backend packaging options.

Mirror and rotor

The core of the Andromeda platform is the wafer integrated mirror. Dielectrics are removed from the mirror area and the die region where torsional springs and movable structures (the rotor mask) are subsequentially defined via silicon etch. Reflective coating is chosen according to the type of application to match the highest reflection in the operating wavelength range: Aluminium for visible application (typical reflectivity R: 90% bare Al, typical R:82% with protection) and gold for infra-red operation (typical R:95%). In the micromirror application the "time zero distortion" must be limited as much as possible to increase the system resolution, that means achieving a low static mirror curvature. The mirror layers thickness is therefore optimized for each design to grant a good planarity (mirror radius of curvature from 1 m to 5 m depending on design). To protect the Al mirror from possible degradation phenomena (corrosion, hazing) a dielectric layer can be easily integrated at the expense of a reflectance spectrum reduction. Nonetheless, a multilayer dielectric stack, conveniently designed, can enhance the reflecting performance in a specific spectral range [5].

Backside wafer pattering

Beneficial to the proper handling of the sensor wafer during the backside manufacturing, a temporary bonding step is performed with a thermoplastic material and a carrier wafer. Wafer thinning (down to 160 µm) is achieved through a sequence of mechanical abrasion and polishing steps. The reinforcement ring, that aims to control the static mirror curvature and its dynamic deformation, particularly for fast scan application, is then defined in the residual silicon thickness through a sequence of photolithographic mask and dry etching.

Bonding and moving structure release

In the final process block, the sensor wafer is permanently bonded through a non-conductive die-attach adhesive glue with a cap wafer. The latest has been previously patterned by means of deep silicon etching to define the cavity as deep as needed to accommodate the mirror movement (cavity depth range 200-490 µm). A peculiar fabrication method [6] is used to blacken the above-mentioned cavity (R<1% in the whole wavelength range) and in turn avoid straylight during the MEMS operation inside the optical module. The moving structure is then released by detaching the temporary bonded carrier wafer and removing residues with a dedicated cleanings sequence. The final manufacturing procedure was developed to minimize mechanical stress on the suspended structure, that can be challenging for some layout (e.g. 20 µm thick, 10 mm span for design (e), Figure 3).

A SEM (Scanning Electron Microscopy) top view picture of an Andromeda MEMS scanning mirror (design (a), Figure 3) and its main features are shown in Figure 2.

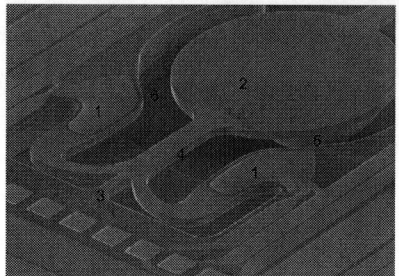

Figure 2: SEM image of a resonant mirror showing the actuators (1), the mirror (2), the PZR sensor (3), the active silicon structure (4), the backside reinforcement (5) and the black cavity (6).

EXPERIMENTAL RESULTS

MEMS mirrors validation has been done through electro-optical measurement and the preliminary results are reported in other papers [7][8][9][10]. The devices considered in this analysis, manufactured with the Andromeda platform are shown in Figure 3.

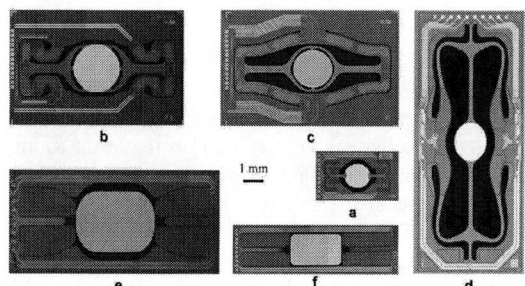

Figure 3: Optical image of MEMS micromirrors manufactured on the Andromeda platform. The different designs are labeled with a letter, which is used as a reference in this section.

The test setup used for the experimental measurements contains a laser source, which illuminates the mirror surface at normal incidence. The incident light is deflected by the reflective area of the device into a line on a perforated sheet. The image is collected by a camera and processed by a custom-made software with the goal of extracting the corresponding opening angle of the mirror. In addition, the output voltage from the PZR sensor is collected by means of an oscilloscope to characterize the behavior of the sensor, as it has been described in [8] [9].

The relation between the opening angle and the driving voltage is one of the most important design parameters for quasi-static mirrors since it describes the mirror actuation performance in operative conditions. In Figure 4 it is reported a comparison between two different designs of quasi-static mirrors in terms of peak mechanical scan angle collected at different values of the peak driving voltage.

The design (f) is a quasi-static mirror that has a reflective area of 2.45x1.44mm^2 in a die size of 7.94x2.34mm^2, which has been described in [9]. This

978-1-6654-9309-3/23 $31.00 © 2023 IEEE 717

mirror has a peak mechanical scan angle (θ) of more than 10deg at 40V, which means a total optical scan angle (TOSA) of more than 40deg. Similar values of the TOSA have been demonstrated also on MEMS mirrors with higher reflective area, such as the micro mirror design (e), which has a reflective area of 4x3mm² and overall die size of 11x5mm². In addition, higher TOSA can be reached by increasing the driving voltage of the mirror. As an example, considering the experimental data for design (e), by doubling the driving voltage TOSA of 60deg has been demonstrated.

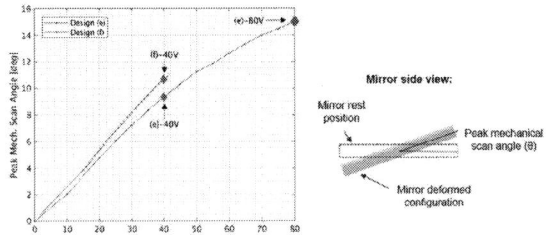

Figure 4: Peak mechanical scan angle vs peak driving voltage for two different designs of quasi-static mirrors. On the right it is shown a definition of peak mechanical scan angle θ, which corresponds to a TOSA of the projection equal to 4xθ.

In Figure 5 the frequency response function (FRF) curves of several designs of resonant micromirrors, collected at different operating angles are reported.

Design (a) has a reflective area of 1.1mm with a die size equal to 4x2.4mm² and it has been described in [9]. Its target operating TOSA is 56deg at 17V peak driving voltage. Even higher TOSA has been demonstrated in design (d), with a mirror diameter of 1.5mm and a full die size of 12.98x5.5mm². For this novel design, a TOSA of 96 deg has been demonstrated at 30V peak driving voltage.

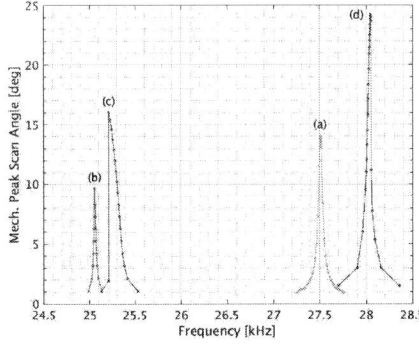

Figure 5: Experimental FRF curves collected at different operating angles for several designs of resonant micromirrors.

COMPARISON WITH THE STATE OF THE ART

The main challenge of new MEMS manufacturing technologies is the capability of being flexible enough to realize devices with very different requirements together with state-of-the-art performances. Andromeda platform achieves such a challenging result, as it can be demonstrated by comparing the performance of the devices

reported in the previous section with other results reported in the literature. Micromirrors are compared by looking at a Figure of Merit (FOM) defined as the product between mirror diameter, scanning angle and resonant frequency, following the approach described in [11].

For the sake of clarity, the comparison is divided into quasi-static mirrors and resonant mirrors, having the two families significantly different set of parameters.

The comparison of quasi-static micromirrors manufactured on Andromeda platform (represented by red diamonds) is reported in Figure 6, where it can be observed that Figures of Merit aligned to the best ones found in literature can be achieved.

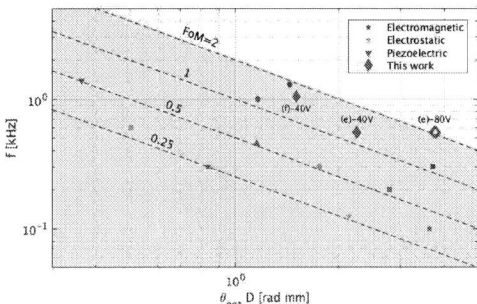

Figure 6: Comparison of the highest performing linear micromirrors found in the literature (see [12] [13] [14] [15] [16] [17] [18] [19]), following the approach described in [11]. The red shaded area represents the design space with same figure of merit, which can be covered by the presented technology, considering the highest performance micromirror manufactured on Andromeda platform.

Figure 7 reports instead the comparison of state-of-the-art high frequency resonant mirrors with the ones realized in Andromeda platform, for which the measurements have been reported in the previous section. Also in this case, it can be seen as the device (d) presented in this work well surpasses all the performances previously reported in literature for any kind of actuation technology.

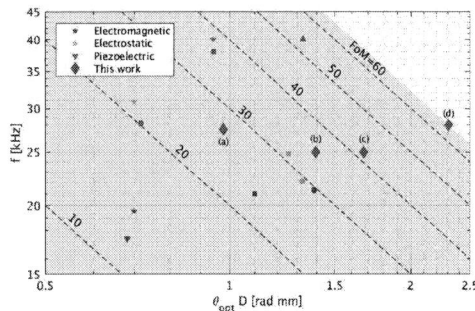

Figure 7: Comparison of the highest performing resonant micromirrors found in the literature (see [11] [20]), following the approach described in [11].

Different generations of PZT thin film layer have been applied to the same architecture and design, demonstrating also that piezoelectric coefficient increase reported in Table 1 translates into an effective increase of performance measured at device level (Figure 8).

978-1-6654-9309-3/23 $31.00 © 2023 IEEE

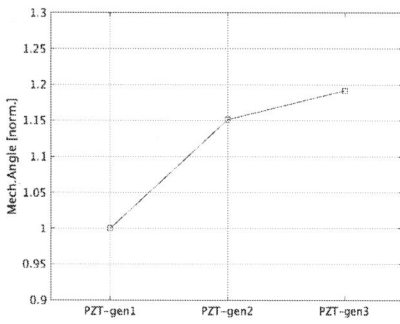

Figure 8: For the same design, the mechanical scan angle at the same drive voltage is improved by applying different piezo stack generation (experimental results).

CONCLUSION

Andromeda is a consolidated technology platform that enables the manufacturing of diverse micromirrors for a variety of applications. The newest design products surpassed the performance of the state of the art reported in literature, both for slow and fast scan. Based on the piezoelectric material roadmap, the MEMS actuation efficiency can be further improved.

REFERENCES

[1] U. Hofmann, J. Janes and H.J. Quenzer, *Micromachines* 2012, 3, 509-528.

[2] A. R. Cho, A. Han, S. Ju, H. Jeong, J.H. Park, I. Kim, J.U. Bu and C.H. Ji, *Optics Express* 16792, Vol. 23, No. 13.

[3] S.G. Stoppel, T. Lisec, S. Fichtner, N. Funck, C. Eisermann, F. Lofink, B. Wagner and A. Muller-Groeling, *Proc. of SPIE* Vol. 10931 1093102-4

[4] A. Della Bitta, and M. Rossi, 2022 *IEEE 35th International Conference on MEMS*, 2022, pp. 515-518.

[5] E. Cianci, A. Lamperti, G. Tallarida, M. Zanuccoli, C. Fiegna, L. Lamagna, S. Losa, S. Rossini, F. Vercesi, D. Gatti, and C. Wiemer, *Sensors and Actuators A: Physical*, p. 124–131, 2018.

[6] R. Somaschini, and P. Petruzza, U.S. Patent: 10364145B2, Jul. 30, 2019.

[7] G. Mendicino, M. Merli, R. Carminati, N. Boni, A. Opreni, and A. Frangi, *Proc. SPIE 11697, MOEMS and Miniaturized Systems XX, 1169715* (March 2021).

[8] N. Boni, R. Carminati, G. Mendicino and M. Merli, *Proc. SPIE 11697, MOEMS and Miniaturized Systems XX, 1169708* (March 2021).

[9] N. Boni, R. Carminati, G. Mendicino, M. Merli, D. Terzi, B. Lazarova and M.Fusi, *Proc. SPIE 12013, MOEMS and Miniaturized Systems XXI, 1201305* (March 2022).

[10] M. Merli, G. Mendicino, R. Carminati and N. Boni, *Proc. SPIE 12013, MOEMS and Miniaturized Systems XXI, 120130J* (March 2022).

[11] S. T. S. Holmström, U. Baran and H. Urey, *Journal of microelectromechanical systems*, vol. 23, no. 2, April 2014.

[12] https://www.maradin.co.il/wp-content/uploads/MAR1800-DataSheet20201207.pdf

[13] Marxer, C.; Herbst, P., U.S. Patent 8,482,833, 9 July 2013.

[14] Abhishek Kasturi, Veljko Milanovic, Frank Hu, Hong Joo Kim, Derek Ho, and Daniel B. Lovell, *Proc. SPIE 10931, MOEMS and Miniaturized Systems XVIII*, 109310L (4 March 2019).

[15] Dooyoung Hah, P. R. Patterson, H. D. Nguyen, H. Toshiyoshi and M. C. Wu, *IEEE Journal of Selected Topics in Quantum Electronics*, vol. 10, no. 3, pp. 505-513, May-June 2004.

[16] T. Sandner, T. Grasshoff, M. Schwarzenberg and H. Schenk, *2013 International Conference on Optical MEMS and Nanophotonics (OMN)*, 2013, pp.103-104.

[17] S. Gu-Stoppel, F. Senger, L. Wen, E. Yarar, G. Wille, and J. Albers, *Proc. SPIE 11697, MOEMS and Miniaturized Systems XX*, 116970F (5 March 2021).

[18] A. Piot, J. Pribošek and M. Moridi, 2021 *IEEE 34th International Conference on Micro Electro Mechanical Systems (MEMS)*, 2021, pp. 89-92.

[19] Takayuki Naono, Takamichi Fujii, Masayoshi Esashi, Shuji Tanaka, *Sensors and Actuators A: Physical*, vol.233, 2015, pp. 147-157, ISSN 0924-4247.

[20] Baran U., Homstrom S., Brown D., Davis W., Cakmak O., Urey H., *2014 International Conference on Optical MEMS and Nanophotonics (OMN)*, 2014, pp. 99-100.

CONTACT

*I. Martini, irene.martini@st.com

DESIGN OF BUTTERFLY PLATE PIEZOELECTRIC ACTUATOR WITH DUAL DRIVING ELECTRODES FOR MEMS MICRO-MIRROR

Si-Han Chen[1], Shih-Chi Liu[1], Hao-Chien Cheng[2], Hung-Yu Lin[1], Kai-Chih Liang[3], Mingching Wu[3] and Weileun Fang[1,2]

[1] Dept. of Power Mech. Eng., National Tsing Hua University, Hsinchu City, TAIWAN
[2] Inst. of NanoEng. and Microsyst., National Tsing Hua University, Hsinchu City, TAIWAN
[3]Coretronic MEMS Corporation, Hsinchu, TAIWAN

ABSTRACT

This study presents a novel butterfly-plate piezoelectric actuator to drive MEMS micro-mirror with decent scanning angle and frequency at low voltage. The micro-mirror is driven by four butterfly-plate actuators around the mirror plate. Merits of this study are, (1) butterfly-plate actuators with dual driving electrodes: butterfly-plate actuator has larger PZT area (than cantilever actuator) to contribute higher driving force, and the round corner on butterfly-plate is to remove the structure with larger bending caused by residual stress. The dual driving electrodes is exploited to further enhance the output displacement. (2) boundary of actuator: the proposed boundary condition of actuators could reduce the constraint of butterfly-plate to enhance its output displacement. Measurements demonstrate the proposed micro-mirror has 38-degree scanning angle with scanning frequency of 25.5 kHz when driving at 12 V_{pp}.

KEYWORDS

Piezoelectric MEMS Micro-mirror, PZT, Optical devices, AR-HUD

INTRODUCTION

MEMS micro-mirrors are extensively investigated and widely used in many applications. For instance, light detection and ranging (LiDAR) system [1], smart glasses [2], and medical imaging [3]. Recently, MEMS micro-mirror based augmented reality head-up display (AR-HUD) has gained attention in automotive industry. Compared to traditional HUD, which reflects the driving information on windshield. AR-HUD could offer more immersive experience by combing the information with surrounding environment, which could prevent driver from shifting visual focus while driving and provide a better driving experience.

AR-HUD could be realized by thin-film-transistor liquid-crystal display (TFT LCD), digital light processing (DLP), and laser beam scanning (LBS) [4-5]. The TFT LCD display takes the advantage of mature technology and high color saturation, but suffer insufficient brightness. The DLP technology features high color saturation and color contrast, however, rainbow effect [4] may occur according to vary users. Lastly, the LBS system gathers the merits of TFT-LCD/DLP and leverages the strong points of MEMS micro-mirror [5]. It could provide large field of view (FoV) in compact size. Nevertheless, the resolution and FoV are highly related to the scanning frequency and the scanning angle of micro-mirror. Hence, the performance enhancement of micro-mirror would directly lead to better specification of AR-HUD.

The electrostatic, electromagnetic and piezo-electric are three well-known driving mechanisms for MEMS micro-mirrors. Compared to electrostatic scanners, the piezoelectric one could drive in lower voltage and has no concern on pull-in effect [6]. Compared to electromagnetic scanners, complicated assembly process is not required for piezoelectric scanners [6]. Thus, to target on vehicle applications, this study employs the piezoelectric driving approach. In general, high scanning frequency and large scanning angle are design trade-offs for micro-mirror. This study proposes a novel actuator design to realize a micro-mirror with decent scanning frequency and angle at low driving voltage for future AR-HUD application.

DESIGN CONCEPT

Fig.1a-b presents the proposed and reference micro-mirrors consist of a 1.2 mm² circular mirror plate supported by two torsional bars, and driven by the four butterfly-plate actuators. Each butterfly-plate actuator consists two plates connected by a joint-bar, and plate-1 anchored to the substrate and plate-2 connected with the mirror by springs.

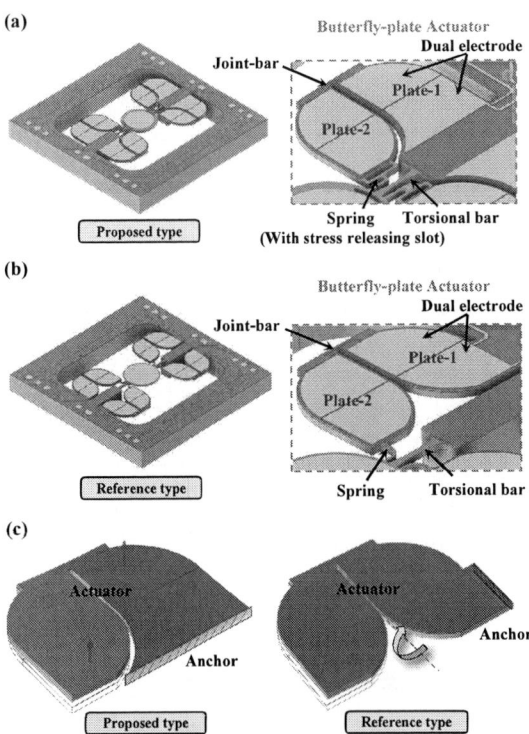

Figure 1: Design concept of micro-mirror, (a) proposed design, (b) reference design, (c) schema of proposed and reference boundary conditions of actuators.

Plate-1 and plate-2 are respectively driven by dual-electrodes with phase variation [3][6] to increase the output displacement of actuator. The reference micro-mirror in Fig.1b is employed to show the importance of actuator boundary design. In Fig.1c, as compare with the reference and existing micro-mirrors [3][7], the proposed boundary design could provide linear displacement and lower the constraint of the butterfly-plate to increase its output displacement.

Simulations in Fig.2a show that the torsional modes (optical scanning mode) of proposed and reference designs are their 1st resonant modes with frequencies at 27.1 kHz and 26.3 kHz respectively. Simulations in Fig.2b depict the proposed design has the mechanical angle of $\pm10°$ at 14 V_{pp} driving which is about 2-fold higher than reference one. The simulation results indicate that the proposed boundary design does play an important role to maximize the scanning capability.

(a)

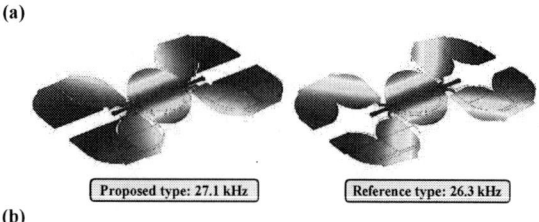

(b)

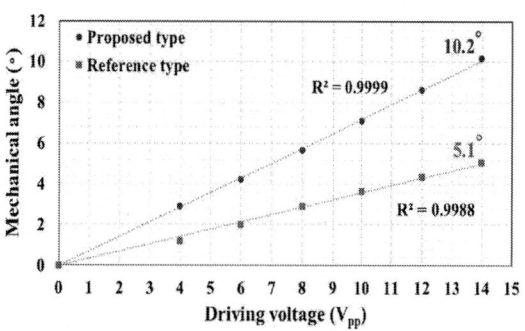

Figure 2: Typical simulation results, (a) torsional mode of proposed and reference designs, (b) performance comparison between proposed and reference designs.

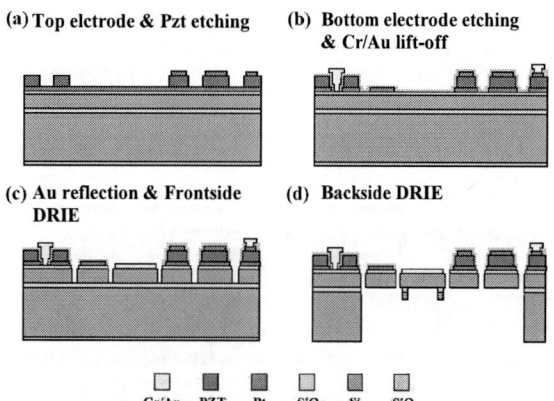

Figure 3: Fabrication process flow.

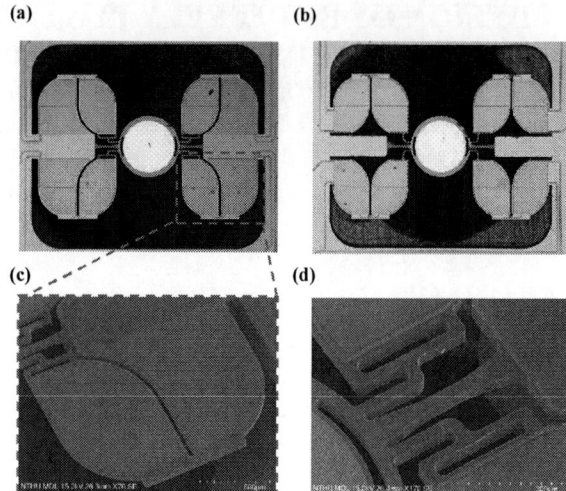

Figure 4: Fabrication results, (a) proposed design, (b) reference design, (c) butterfly-plate actuator, (d) spring with stress release slot.

FABRICATION

This section presents the process flow of this study. Both designs were realized on SOI wafer with 50 μm device layer. Pt bottom electrode layer and PZT film (2 μm) were deposited on SOI wafer. In Fig.3a, Pt top electrode layer was sputtered and then patterned by reactive ion etching (RIE). Following, PZT film was also patterned by RIE. Next, as shown in Fig.3b, bottom electrode was also patterned through RIE process. Subsequently, a thin SiO_2 passivation layer (~6000 Å) was deposited by plasma enhanced chemical vapor deposition (PECVD) so as to prevent the environment-sensitive PZT film from moisture, thereby retaining its piezoelectric performance [8]. After that, the SiO_2 covered on electrode contact area was removed for follow-up meatal pads lift-off process. In Fig.3c, Au reflection layer on mirror plate was deposited by shadow mask. After that, device silicon layer was patterned by deep Silicon reactive ion etching (DRIE) to define mirror structures. In Fig.3d, patterned backside silicon by two-step DRIE to define rib-structure and removed buried-oxide to suspend mirror.

Fabrication results is presented in Fig.4. Micrographs in Fig.4a-b respectively display typical fabricated proposed and reference micro-mirrors. Both designs exhibit negligible initial deformation induced by residual stress, which indicates that the rounded butterfly-plate actuators can effectively reduce the influence of residual stress. Zoom-in SEM micrographs in Fig.4c-d respectively depict proposed butterfly-plate actuator, and springs with stress releasing slot.

MEASUREMENT

The Fig.5a exhibits the micro-mirror wire-bonded on PCB as the device-under-test (DUT). Fig.5b-c shows the measurement setup and typical scanning pattern to measure scan angles of micro-mirror. The laser beam will incident on mirror plate and then reflect to the screen. When applying signals to induce torsional vibration of scanner, the mirror will rotate and steer the laser on screen (Fig.5b).

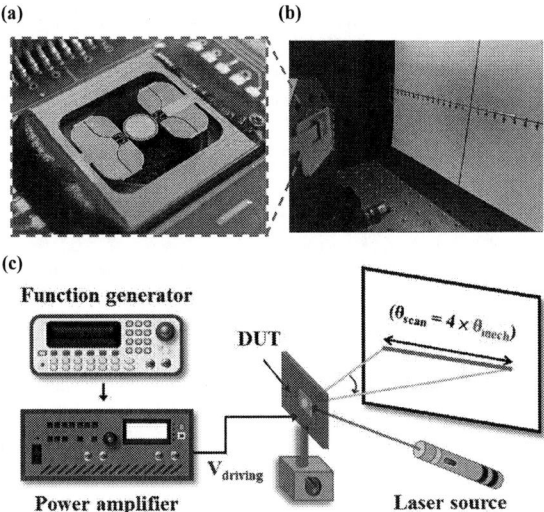

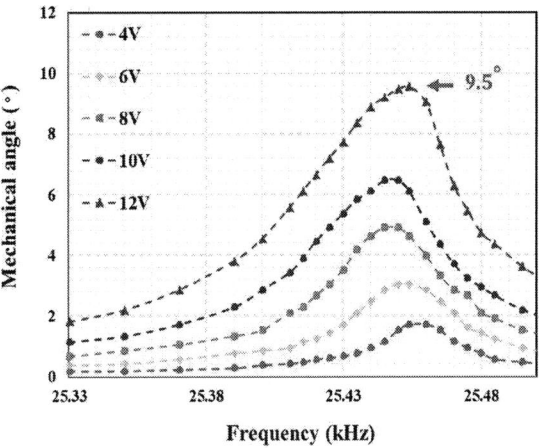

Figure 7: frequency response of proposed design under various driving voltages.

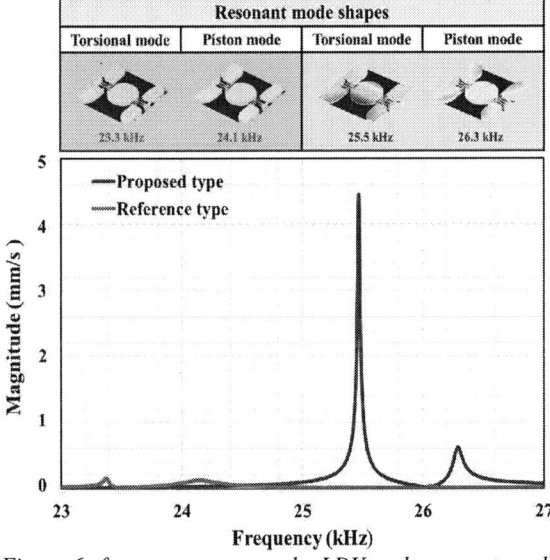

Figure 6: frequency response by LDV and resonant mode shape by DHM (Torsional mode: proposed: 25.5 kHz, reference: 23.3 kHz; piston mode: proposed: 26.3 kHz, reference: 24.1 kHz).

Figure 5: (a) The DUT for proposed design after wire bonding, (b) testing setup and typical scanning pattern, (c) schematic of optical measurement system.

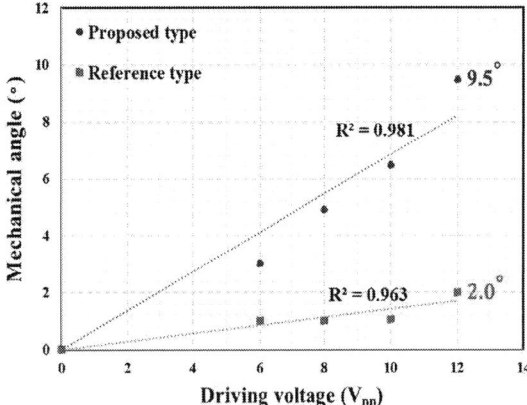

Figure 8: performance comparison between proposed and reference designs.

The measurement results will be classified as mirror dynamic analysis and mirror performance analysis. The former contains frequency response and resonant mode shape of mirror, while the latter includes the mechanical angle measurement.

Fig.6 shows the frequency responses of proposed (blue) and reference (red) designs measured using the commercial Laser Doppler Vibrometer (LDV). Resonant frequencies for scanning mode are 25.5 kHz (proposed) and 23.3 kHz (reference) respectively. Comparing with simulations, the frequency shift may result from the over etching during fabrication. The commercial digital hologram microscopy (DHM) was used to measure and confirm the resonant mode shapes, as displayed in figure.

Measurements in Fig.7 demonstrate the frequency responses of proposed design at different driving voltages. As depicted in figure, the mechanical scan angle of the mirror is increasing with the driving voltage, and the mechanical angle can reach ±9.5° at 12 V_{pp}. Besides, measurement results also depict decent linearity, which could be attributed to the proposed boundary design and the connecting springs with lower stiffness between mirror and actuators. Thus, with the contributions of boundary and spring designs, the performance of mirror could maintain linear under large deformation and high frequency.

Fig.8 exhibits the performance comparison between proposed and reference designs. Measurement results and simulations show good agreement. The proposed design has much higher mechanical angle than the reference design, which means that the proposed boundary condition has effectively reduce the constraint from anchor so that the energy could contribute to mirror's rotation efficiently.

Compared with designs proposed by other researchers summarized in Table1, the proposed micro-mirror has reasonable performances and successfully meets the specification of AR-HUD applications [2]. The proposed design will be further improved by structure designs to achieve a larger scanning angle and reduce the stress distribution at connecting springs as well.

978-1-6654-9309-3/23 $31.00 © 2023 IEEE

Table 1: Comparison of piezoelectric MEMS mirrors.

Type	Proposed	[2]	[9]	[10]
Scanning Frequency (kHz)	25.4	27.5	28.2	23.7
Optical Angle (°)	38°	56°	41°	40°
Driving voltage (V)	12	17	60	10

CONCLUSION

The piezoelectric actuation has shown its potential to realize MEMS micro-mirrors. This study proposes a novel butterfly-plate piezoelectric actuator to drive MEMS micro-mirror. Rounded actuator design has successfully mitigated bending induce by residual stress. While from measurement results, by modulating the boundary condition, the proposed design can attain the scanning angle of 38° at 12 V_{pp}, which is about five times larger than reference design. In summary, the proposed design meets the specification for AR-HUD applications by structure optimization and boundary condition modulation. Finally, in view of commercial use, long-term reliability is important and can be compensated by feedback control system. Therefore, integrated buried piezoresistors for mirror position sensing [11-12] will be take into consideration as well in future study.

ACKNOWLEDGEMENT

This project was supported by the National Science and Technology Council under grant number NSTC 111-2923-E-007 -003 -, NSTC 111-2221-E-007 -069 -MY3, NSTC 111-2221-E-007 -070 -MY3, NSTC 111-2218-E-007-014-MBK, NSTC 111-2923-E-007-004-MY2, NSTC 110-2926-I-007 -506 -, and NSTC 110-2218-E-007-032-. The authors are grateful to Coretronic MEMS Corporation, Taiwan, for providing the piezoelectric material, fabrication, and packaging.

REFERENCE

[1] K. Ruotsalainen, et al. " Resonating AlN-thin film MEMS mirror with digital control. " Journal of Optical Microsystems, *Journal of Optical Microsystems 2(1)*, 011006, 2022.

[2] N. Boni, R. Carminati, G. Mendicino, M. Merli, D. Terzi, B. Lazarova, M. Fusi, " Piezoelectric MEMS mirrors for the next generation of small form factor AR glasses," *Proc. SPIE, MOEMS and Miniaturized Systems XXI*, San Francisco, CA, Mar 1, 2022, 1201305.

[3] W. Liu, Y. Zhu, K. Jia, W. Liao, Y. Tang, B. Wang and H. Xie, "A tip–tilt–piston micromirror with a double S-shaped unimorph piezoelectric actuator," *Sensors and Actuators A: Physical, vol. 193*, pp.121-128, 2013.

[4] M. J. Baker, J. Xi, J. Chicharo, E. Li, "A contrast between DLP and LCD digital projection technology for triangulation-based phase measuring optical profilometers," *Proc. SPIE, Two- and Three-Dimensional Methods for Inspection and Metrology III*, Boston, MA, Nov 7, 2005, 60000G.

[5] J. Nakagawa, H. Yamaguchi, T. Yasuda, "Head up display with laser scanning unit," *Proc. SPIE, ODS 2019: Industrial Optical Devices and Systems*, San Diego, CA, Aug 30, 2019, 111250C.

[6] H.-H. Cheng, S. -C. Lo, Z. -R. Huang, Y. -J. Wang, M. Wu and W. Fang, "On the design of piezoelectric MEMS microspeaker for the sound pressure level enhancement," *Sensors and Actuators A: Physical, vol. 306*, 111960, 2020.

[7] M. Tni, M. Akamatsu, Y. Yasuda, H. Fujita and H. Toshiyoshi, "A Combination of Fast Resonant Mode and Slow Static Deflection of SOI-PZT Actuators for MEMS Image Projection Display," *IEEE/LEOS International Conference on Optical MEMS and Their Applications Conference*, Big Sky, MT, Aug 21-24, 2006, pp. 25-26.

[8] Dahl-Hansen, Runar Plünnecke, et al. "On the effect of water-induced degradation of thin-film piezoelectric microelectromechanical systems." *Journal of Microelectromechanical Systems vol. 30.1*, pp. 105-115, 2020.

[9] J.-H. Park, J. Akedo, H. Sato, " High-speed metal-based optical microscanners using stainless-steel substrate and piezoelectric thick films prepared by aerosol deposition method," *Sensors and Actuators A: Physical, vol. 135*, pp.86-91, 2007.

[10] S. Matsushita, I. Kanno, R. Yokokawa and H. Kotera, "Metal-based piezoelectric MEMS scanner mirrors composed of PZT thin films on titanium substrates," *Transducers*, Beijing, China, Aug 1, 2011, pp. 574-577.

[11] A. Vergara, T. Tsukamoto, W. Fang and S. Tanaka, "PZT MEMS Actuator with Integrated Buried Piezoresistors for Position Control," *IEEE MEMS*, Gainesville, FL, Mar 15, 2021, pp. 626-629.

[12] P. Frigerio, B. D. Diodoro, V. Rho, R. Carminati, N. Boni and G. Langfelder, "Long-Term Characterization of a New Wide-Angle Micromirror With PZT Actuation and PZR Sensing," *Journal of Microelectromechanical Systems vol. 30.2*, pp.281-289, 2021.

CONTACT

*W. Fang, tel: +886-3-5742923; fang@pme.nthu.edu.tw

FULLY-FLEXIBLE MICRO-SCALE ACTUATOR ARRAY WITH THE LIQUID-GAS PHASE CHANGE MATERIALS

Sangjun Sim[†], Kyubin Bae[†], and Jongbaeg Kim*

School of Mechanical Engineering, Yonsei University, Seoul, Republic of Korea

[†] These authors contributed equally to this work.

ABSTRACT

This paper reports a fully flexible micro-actuator array utilizing a liquid-to-gas phase change mechanism. A stretchable membrane of the actuator is deformed by volume expansion and contraction by inducing vaporization and liquefaction of the phase change materials. The phase change mechanism has a high energy density inducing large displacement even in a small volume at the microscale, which is advantageous for downscaling compared to other actuation mechanisms, such as electrostatic and pneumatic actuation. Additionally, this device could be a candidate for next-generation wearable devices due to the flexibility originating from polymer (chamber and membrane) and liquid (phase change material).

KEYWORDS

Phase change mechanism, micro-scale actuator, fully-flexible, MEMS fabrication process

INTRODUCTION

Human–machine interface (HMI) technology is the key element for exchanging information between users and robots. Therefore, this technology has been studied for use in fields such as robots and virtual/augmented reality (VR/AR) [1-4]. Previous studies have mainly developed focusing on technologies for delivering visual and auditory experiences to users [5, 6]. Recently, haptic interfaces consisting of tactile sensors [7-10] and displays [11-13] have been studied for more effective and intuitive manipulation by users.

In particular, flexible tactile displays can uniformly contact curved surfaces such as skin, enabling sophisticated tactile information delivery to users [14]. Various flexible tactile displays using actuator arrays with different mechanisms, including electromagnetic, electrostatic, and pneumatic actuation, have been developed [15-17]. Leroy et al. reported a flexible tactile display using hydraulic amplified electrostatic actuators [16]. The device achieved an ultrafast actuation time of about 5 ms and actuation forces of over 300 mN, which is sufficient to stimulate the palm. However, there has a limit to delivering complex tactile sensations to the fingertips because the size of a single actuator is as large as 6 mm. Carpenter et al. demonstrated the pneumatic actuated flexible haptic device [17]. This device exhibits a large actuation displacement of about 1 mm and stimulates the fingertip. However, additional devices, including a pump and valves to control each cell, are required, and thus it is practically difficult for the users to use without space constraints. Recently, a phase change actuator, which uses volume expansion during the phase transition, has been developed. It has considerable attention from many researchers owing to its simple operating mechanism and

inducing large force during phase change [18, 19]. Uramune et al. reported a wearable tactile display using a liquid-to-gas phase change actuator [20]. The authors demonstrated a device that can be attached to the fingertips with a large driving force of 1 N. However, the fabrication process of the device was complicated, and thus there was a limit to the scalable fabrication.

Here, we present a fully-flexible tactile display using micro-scale liquid-to-gas phase change actuators. Based on the batch fabrication process, the tactile display, which has a single cell diameter of 500 μm and device thickness of 500 μm, is demonstrated. The device shows a large actuation displacement of 70 μm even with a low power of 20 mW. Furthermore, the display exhibits a fast response time of 55 ms and relaxation time of 123 ms owing to the microscale-sized cells. To the best of our knowledge, this is the first microscale flexible tactile display fabricated in a batch process. This could be used as a tactile display that can deliver to users a complex and sophisticated tactile sense.

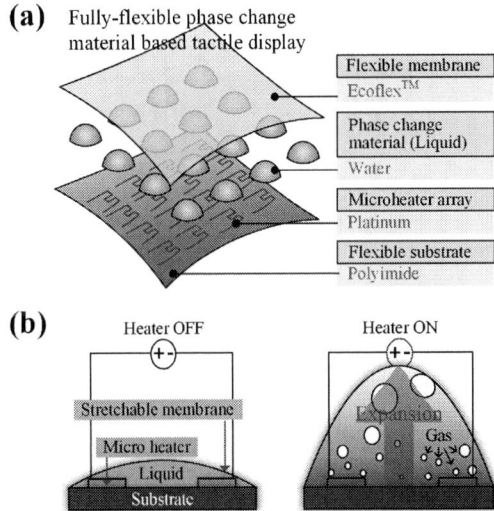

(a) Fully-flexible phase change material based tactile display

- Flexible membrane — Ecoflex™
- Phase change material (Liquid) — Water
- Microheater array — Platinum
- Flexible substrate — Polyimide

(b) Heater OFF / Heater ON

Figure 1: (a) Schematic diagram of the fully flexible actuator array using a liquid phase change material (PCM). All the materials consisting of the actuator are made of easily deformable materials to fabricate a flexible device. (b) Actuating mechanism of the actuator. When current flows through the heater, the volume expands as the liquid phase changes to gas due to Joule heating. This deforms the stretchable membrane, causing the actuator to change into a convex shape.

DESIGN AND FABRICATION

Figure 1 shows the schematics of a flexible micro-actuator array and the working mechanism. As shown in Figure 1a, the flexible polyimide (PI) film is used as a substrate, and micro-heater arrays are fabricated on the layer. After then, the phase change material (PCM) and the

(a)

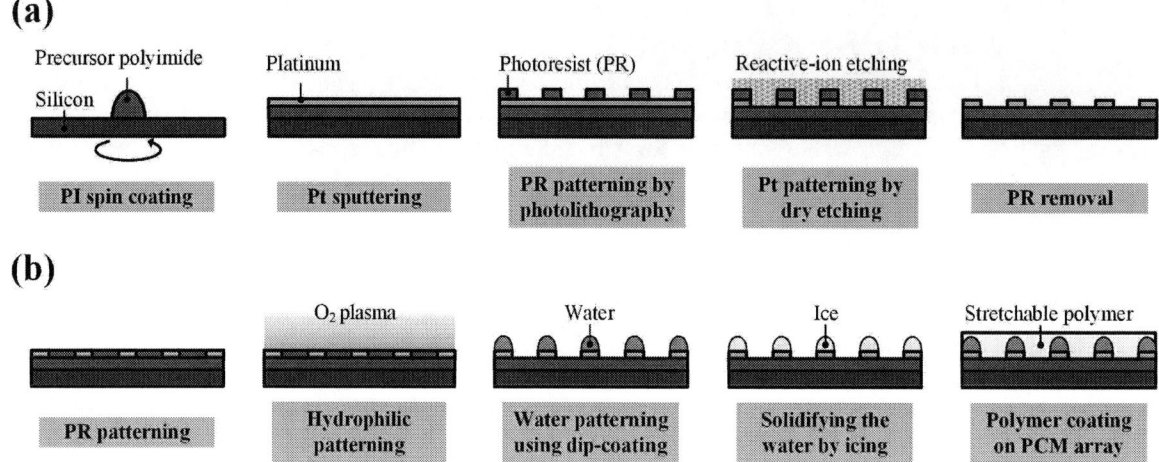

(b)

Figure 2: Schematic illustration of the fabrication process of the proposed flexible PCM actuator array. (a) micro heater array is patterned by a photo-lithography process. (b) Hydrophilic patterning on PI substrate using a photoresist (PR) mask and O_2 plasma surface treatment to pattern liquid PCM. The freezing method is a strategy for immobilizing the liquid PCM, which allows coating of stretchable polymer precursors on PCM array.

membrane layer are stacked to form the device structure. When the heater is off, the upper membrane does not deform, and while the heater is activated, the liquid PCM vaporizes and generates a deformation of the upper membrane (Figure 1b). When water vaporizes, the volume expands 1600 times [21], and the elastic membrane does not interfere with internal expansion, allowing the device to operate at high strain rates even in small volumes.

Figure 2 shows the fabrication process of the PCM-based micro-actuator array. As shown in Figure 2a, a PI substrate is fabricated by a spin coating process, and a micro-heater array is patterned by sputtering, photo-lithography, and reactive ion etching (RIE) processes. In this step, the heater is designed with a width of 10 μm and a total length of 15 mm to effectively generate heat up when applying a voltage. Continuously, a selective hydrophilic coating is applied only on the heater using an O_2 plasma treatment and photoresist (PR) mask. Then water is patterned on each micro heater array by a simple dip-coating process. In order to coat the precursor polymer on the phase change material, an uncured silicone elastomer (Ecoflex™) is spin-coated after liquid water turns into solid ice using a freezing process. Importantly, the membrane curing step is performed at a low temperature of 50 °C to prevent the phase change material from evaporating. Finally, after curing the membrane, the device is completed by peel-off on a dummy Si-wafer. The device (6×6 cells) is fabricated with a cell size of 500 μm, and the stretchable membrane is formed with a thickness of 100 μm.

Figure 3 depicts the experimental setup for driving the fully-flexible PCM actuator and evaluating its dynamic response. A microcontroller (Arduino Mega 2560) is used to drive target cells in 6 × 6 PCM actuator array while supplying power to the driver with the DC power supply (Keysight E3647a). When the target cell is operated, the displacement is measured with the laser displacement sensor (KEYENCE LJ-X8000) while checking that the cell is operating with a microscope.

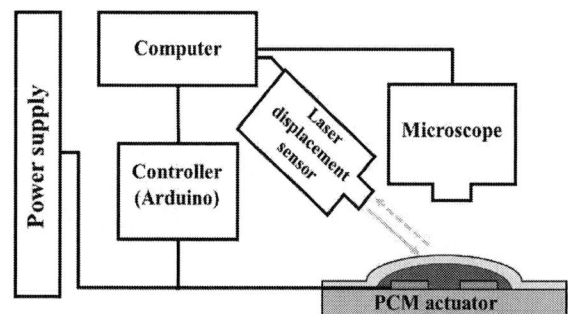

Figure 3: Experimental setup to measure vertical displacement and optically check when the device is activated simultaneously.

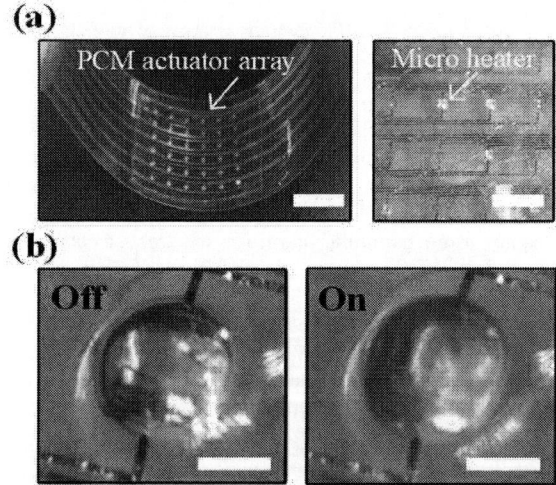

Figure 4: Optical images of the fabricated PCM actuator array. (a) fully flexible PCM actuator array and an enlarged image of the device. The size of each cell is 500 μm, and the distance between the cells is 5 mm (Scale bars: 1 cm (left), 5 mm (right), and (b) Single-cell when the heater is Off/On states (Scale bars: 500 μm).

Figure 4 presents optical images of the micro-actuator array and before and after actuation of the single cell. As shown in Figure 4a, the fabricated micro PCM actuator array shows flexibility under the bending state. In addition, the actuator is composed of the water and microheater array in a uniform size with a diameter of 500 μm. In Figure 4b, when the actuator is operated, the membrane is stretched as the water evaporates.

Figure 5a shows the driving displacement of the actuator measured by the laser displacement sensor. When a single heater is driven by 20 mW, the actuator shows a displacement of 70 μm. As shown in Figure 5b, this device exhibits an 'ON' response within about 55 ms and an 'OFF' response time of 123 ms. This is a significantly fast response time and sufficient displacement to stimulate the human tactile sense [22].

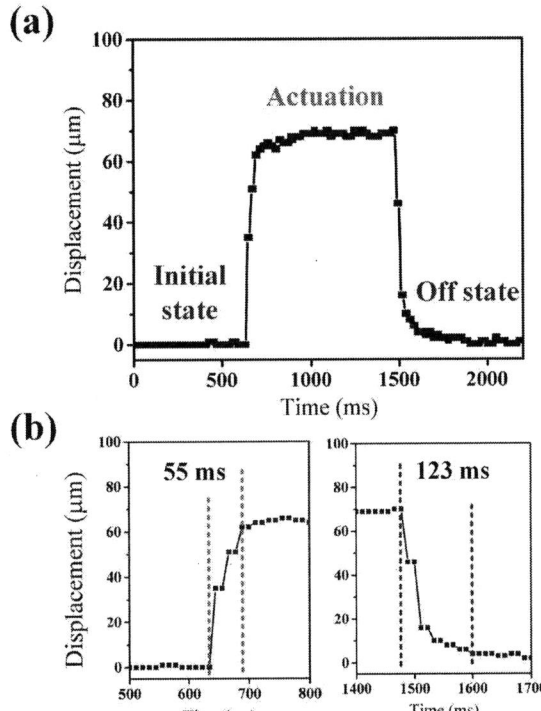

Figure 5: (a) Displacement measurement under PCM actuation. (b) Response time of the actuator at the loading/unloading processes. The response/relaxation time is 55 and 123 ms, respectively.

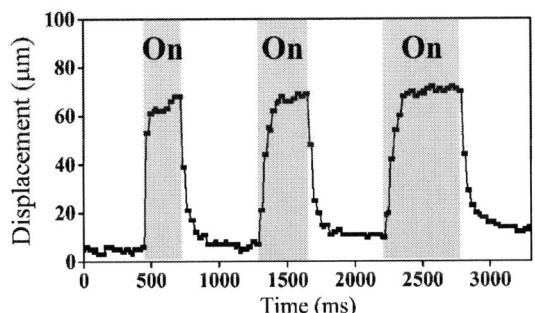

Figure 6: Repetitive test under 3 cycles. It exhibits stable activation even in repeated driving. It exhibits similar activating displacement even in repeated driving.

In Figure 6, it is experimentally demonstrated that the device can be operated stably after repetitive operations (3 times). In future works, the flexible actuator array can be used as a tactile display for realistic tactile transmission to a user through array driving.

CONCLUSION

We have developed a micro-scale, fully-flexible tactile display consisting of a liquid-to-gas phase change actuator array for the first time. All fabrication processes, including the selective water patterning process, are batch fabrication processes, and thus this device enables a highly uniform shape to form. The size of each phase change actuator is 500×500 μm². The actuator exhibits a large driving displacement of 70 μm even at a driving power of 20 mW, owing to the significant volume expansion during phase change from liquid to gas. Furthermore, the device shows a fast response time of 55 ms and a relaxation time of 123 ms. The proposed flexible tactile display could be utilized in various applications, including haptic feedback devices and virtual/augmented (VR/AR) communication systems.

ACKNOWLEDGEMENTS

S.S. and K.B. contributed equally to this work. This study was supported by a National Research Foundation of Korea (NRF) grant funded by the Korean government (MSIT) (No.2021R1A2B5B03002850).

REFERENCES

[1] Kim, Jiyong, et al. "Self-charging wearables for continuous health monitoring." Nano Energy 79 (2021): 105419.

[2] Khan, Salman, et al. "Review on the operation of wearable sensors through body heat harvesting based on thermoelectric devices." Applied Physics Letters 118.20 (2021): 200501.

[3] Lee, Taehoon, et al. "All Paper-Based, Multilayered, Inkjet-Printed Tactile Sensor in Wide Pressure Detection Range with High Sensitivity." Advanced Materials Technologies 7.2 (2022): 2100428.

[4] Pyo, Soonjae, et al. "Recent progress in flexible tactile sensors for human-interactive systems: from sensors to advanced applications." Advanced Materials 33.47 (2021): 2005902.

[5] Sato, Y., et al. "A 3D radiation image display on a simple virtual reality system created using a game development platform." Journal of Instrumentation 13.08 (2018): T08011.

[6] Cohen, Keren Shavit, and Elana Zion Golumbic. "The Dynamics of Attention Shifts Among Concurrent Speech in a Naturalistic Multi-Speaker Virtual Environment." bioRxiv (2019): 626564.

[7] Kim, Yunjeong, et al. "Self-powered Wearable Micropyramid Piezoelectric Film Sensor for Real Time Monitoring of Blood Pressure." Advanced Engineering Materials.

[8] Sim, Sangjun, et al. "Highly Sensitive Flexible Tactile Sensors in Wide Sensing Range Enabled by Hierarchical Topography of Biaxially Strained and Capillary-Densified Carbon Nanotube Bundles." Small 17.50 (2021): 2105334.

[9] Bae, Kyubin, et al. "Large-Area, Crosstalk-Free, Flexible Tactile Sensor Matrix Pixelated by Mesh Layers." ACS Applied Materials & Interfaces 13.10 (2021): 12259-12267.

[10] Bae, Kyubin, et al. "Dual-Scale Porous Composite for Tactile Sensor with High Sensitivity over an Ultrawide Sensing Range." Small 18.39 (2022): 2203193.

[11] Zhu, Minglu, et al. "Haptic-feedback smart glove as a creative human-machine interface (HMI) for virtual/augmented reality applications." Science Advances 6.19 (2020): eaaz8693.

[12] Jung, Yei Hwan, et al. "A wireless haptic interface for programmable patterns of touch across large areas of the skin." Nature Electronics (2022): 1-12.

[13] Frediani, Gabriele, et al. "A soft touch: wearable tactile display of softness made of electroactive elastomers." Advanced Materials Technologies 6.6 (2021): 2100016.

[14] Trase, Ian, et al. "Wearable Haptic Array of Flexible Electrostatic Transducers." International Conference on Human-Computer Interaction. Springer, Cham, 2021.

[15] Yu, Xinge, et al. "Skin-integrated wireless haptic interfaces for virtual and augmented reality." Nature 575.7783 (2019): 473-479.

[16] Leroy, Edouard, Ronan Hinchet, and Herbert Shea. "Multimode hydraulically amplified electrostatic actuators for wearable haptics." Advanced Materials 32.36 (2020): 2002564.

[17] Carpenter, Cody W., et al. "Electropneumotactile stimulation: multimodal haptic actuators enabled by a stretchable conductive polymer on inflatable pockets." Advanced materials technologies 5.6 (2020): 1901119.

[18] Yoon, Yeosang, et al. "Bioinspired untethered soft robot with pumpless phase change soft actuators by bidirectional thermoelectrics." Chemical Engineering Journal 451 (2023): 138794.

[19] Sanchez, Vanessa, et al. "Smart thermally actuating textiles." Advanced Materials Technologies 5.8 (2020): 2000383.

[20] Uramune, Ryusei, et al. "HaPouch: A Miniaturized, Soft, and Wearable Haptic Display Device Using a Liquid-to-Gas Phase Change Actuator." IEEE Access 10 (2022): 16830-16842.

[21] Weon, B. M., et al. "X-ray-induced water vaporization." Physical Review E 84.3 (2011): 032601.

[22] Arredondo, Luis T., and Claudio A. Perez. "Spatially coincident vibrotactile noise improves subthreshold stimulus detection." Plos one 12.11 (2017): e0186932.

CONTACT

*J. Kim, tel: +82-2-2123-2812; kimjb@yonsei.ac.kr

A NOVEL COMB DESIGN FOR ENHANCED POWER AND BANDWIDTH IN ELECTROSTATIC MEMS ENERGY CONVERTERS

Jinglun Li[1]*, Habilou Ouro-Koura[1]*, Hannah Arnow[1], Arian Nowbahari[3], Matthew Galarza[1], Meg Obispo[1], Xing Tong[2], Mehdi Azadmehr[3], Mona M. Hella[2], John A. Tichy[1], and Diana-Andra Borca-Tasciuc[1]

[1] Department of Mechanical, Aerospace, and Nuclear Engineering, Rensselaer Polytechnic Institute, USA

[2] Department of Electrical, Computer and System Engineering, Rensselaer Polytechnic Institute, USA

[3] Department of Microsystems, University of South-Eastern Norway, NORWAY

*Both authors contributed equally

ABSTRACT

Silicon-based kinetic energy converters employing variable capacitors hold promise as power sources for Internet of Things devices. However, they are plagued by low power output and a limited range for operation frequency. The objective of this work is to address these important challenges for a harvester based on interdigitated variable capacitors and gap-closing topography. The approach explored here employs non-uniform cross-section electrodes and a springless mass. The purpose is to enable frequency-up conversion following electrodes' impact and the impact of the springless mass with the shuttle mass. Frequency up-conversion is demonstrated experimentally and supported by the theoretical model.

KEYWORDS

ultralow frequency, energy harvester, comb design, up-conversion

INTRODUCTION

The growing need for wireless portable power sources for sensors is driving the development of MEMS energy harvesters [1]. The electrostatic energy harvester is gaining significant attention due to its compatibility with the majority of CMOS technology and fabrication techniques. It consists of a silicon-based interdigitated capacitor with electrodes on both sides of a large area shuttle mass. Despite its usefulness and wide range of applications, conventional electrostatic MEMS energy harvesters currently have limited output power, which is insufficient even for low duty cycle wireless sensors [2]. Thus, many approaches have been proposed to boost their power output [3-5]. Among those, it has been demonstrated that devices with a sloped sidewall electrodes have considerably increased output frequency and power. This comes from frequency up-conversion due to electrodes' impact under certain operating conditions [6-8]. However, fabricating electrodes with sloped sidewalls requires deep reactive ion etching processes that are not common in practice. To capitalize on the frequency up-conversion, a new electrode design that is compatible with the common fabrication technique is explored here. Specifically, the electrodes are designed to have vertical walls, but non-uniform width. This feature causes them to present a non-uniform gap, which was previously found to be critical for producing frequency up-conversion [6,8]. The devices are manufactured using a typical SOIMUMPS technique made available by MEMSCAP Inc. [9]. To further increase the output power of these devices a springless mass (a micro ball) was incorporated into a special cavity of the shuttle mass.

DEVICE DESIGN

A simplified schematic of the proposed in-plane electrostatic harvester is shown in Fig. 1(a). The device consists of a shuttle mass suspended by serpentine spring beams. The variable capacitor is formed by the two electrode pairs fastened to the anchor frame and to the shuttle mass. The shuttle mass contains a cavity that can host a micro ball. The direction of vibration is indicated by the arrow. The capacitance varies as the shuttle mass moves and the gap between the movable electrodes (in light blue) and the fixed electrodes (in gray) is changing. Two sets of soft stoppers on both ends of the fixed sections restrict the shuttle mass range of motion. The electrodes are separated by a trapezoidal-shaped gap due to their non-uniform cross-section. Pull-in is eliminated when the electrodes collide because of the irregular air gap. The electrodes are protected by 200 nm thick parylene C layer which acts as an electrical insulator and permits their impact without shorting.

Figure 1 (b) depicts the mounted device in cross-section and provides a detailed overview of the materials used in each layer. The device is mounted on a printed circuit board. A transparent poly (methyl methacrylate) (PMMA) cover is used to protect the harvester from dusting during testing or storage.

The MEMS energy harvester is fabricated on a silicon-on-insulator wafer via MEMSCAP SOIMUMPs process [9]. The device layer that includes the comb-like capacitor and springs is 25μm thick. The device is supported by a 2μm thick oxide layer and a 400μm thick handle silicon layer. Underneath the mobile components, the handle silicon and oxide layers are removed. The die size for this commercial process is restricted to 11.15mm x 11.15mm as is the device layer thickness (up to 25μm). Due to its size restriction, this proof-of-concept power harvester is not intended to demonstrate high power output. Instead, the objective is to demonstrate frequency-up conversion with this electrode geometry. However, the device performance could be compared qualitatively to a device with similar dimensions reported previously [4],[10].

Photos of the manufactured device are shown in Fig. 2. A view of the device mounted on A PCB with a dust cover is shown in Fig. 2(a). Figure 2(b) shows the flexure springs supporting the shuttle mass. Figures 2(c) and 2(d) illustrate the electrode sets at the minimum and maximum capacitance positions, respectively.

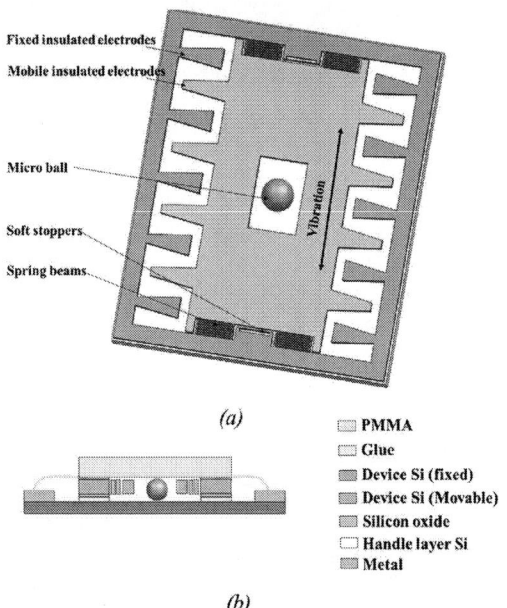

(a)

☐ PMMA
☐ Glue
☐ Device Si (fixed)
☐ Device Si (Movable)
☐ Silicon oxide
☐ Handle layer Si
☐ Metal

(b)

Figure 1: (a) A simplified schematic of the proposed device, (b) cross section view of the layer composition.

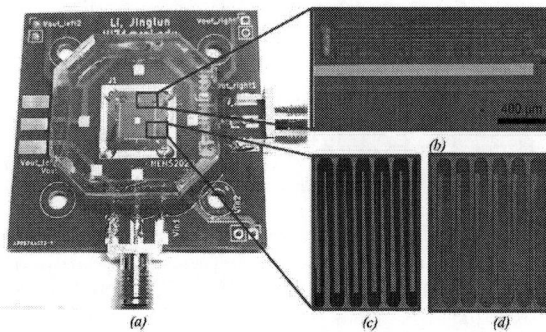

Figure 2: Photographs of the MEMS energy harvester: (a) The device mounted on a PCB, and close-up images of the structures; (b) the suspension spring and the trapezoidal electrodes at (c)minimum and (d)maximum capacitance positions.

The experimental setup used for the device characterization is shown in Fig. 3. The PCB hosting the device is mounted on a shaker table. A function generator creates a sinusoidal wave with a specified frequency. The signal is then amplified and used to power the shaker. An accelerometer is utilized to determine the excitation amplitude and frequency. The MEMS power harvester is connected in series with a load resistance and a DC voltage source. The output voltage across the load resistor along with acceleration frequency and output are measured using a NI Data Acquisition (DAQ) system. For visualization purposes, the voltage across the resistor is also monitored with an oscilloscope.

EXPERIMENTAL RESULTS

Figure 4 displays the measured power output as a function of load resistance. This power output was calculated using the RMS output voltage drop over the load resistance. The results were obtained using three distinct excitation frequencies as mentioned in the plot legend, a bias voltage of $6V$ and an excitation acceleration amplitude of $1.5g$. This result suggests that $0.5M\Omega$ provides the optimum condition for power harvesting and this resistance was used in all subsequent studies.

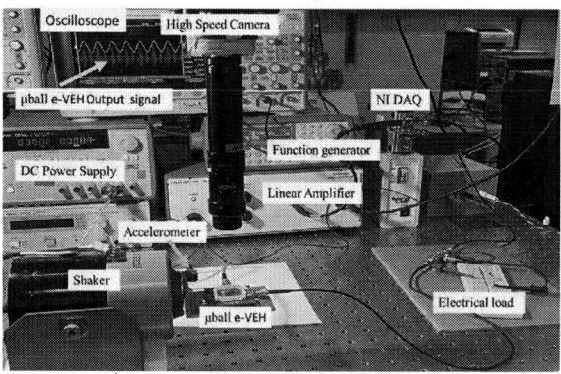

Figure 3: Testing setup of the MEMS characterization system.

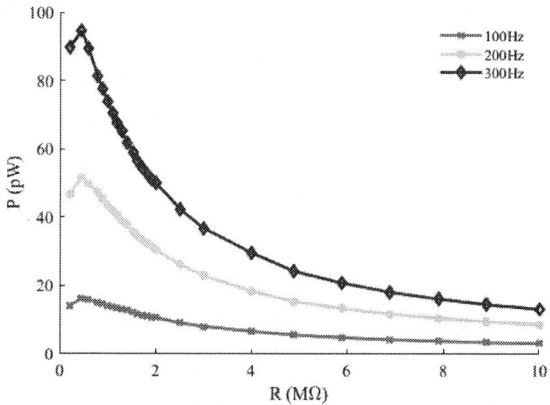

Figure 4: Measured output power of the MEMS device for various load resistances at 6 V bias voltage, 1.5g acceleration amplitude and multiple frequencies.

An example of frequency sweep is next illustrated in Fig. 5, which shows the $V_{out\,RMS}$ (measured across the series resistor) for upsweeps and downsweeps at three diferent excitation accelerations ($1g$, $1.5g$ and $2g$) and a bias voltage of 4 V. These results are obtained without a ball in the microcavity. The observed hysteresis at a given acceleration is typical to these devices [7] and is due to different initial conditions experienced by the device during each sweep direction. The hysteresis is seen for relatively larger values of acceleration amplitude and it vanishes at very low acceleration (50 mg and lower) due to small displacements and consequent linear response of the device [10].

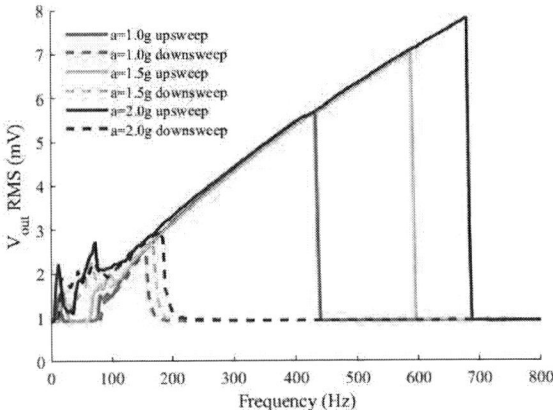

Figure 5: Measured RMS voltage from the device without a ball as a function of input vibration frequency under different amplitudes of input acceleration. The bias voltage is 4V.

Figure 6 (a) illustrate an example of device output as function of time at very low frequency of 20 Hz, an acceleration amplitude of 1*g* and a bias voltage of 4 V. The sinusoidal (orange) signal is represents the acceleration amplitude. The blue signal is the device output. Once again, these results are obtained in the absence of the ball. The multiple peaks grouped together are typical to devices that experience electrode impact and are caused by their secondary oscillations following the impact. The ftwo group of peaks have some dissimilarties, most likely to out-of-phase oscillation of the two fixed electrodes on each side of a mobile finger electrode. However, the sequence as shown in repeatable and stable in time.

Figure 6 (b) shows predictions of the device output as function of time. The predictions are based on a comprehensive lumped model [6] solving simultaneously motion equation and Kirchoff's law for electrical circuit. The model considers two masses (shuttle and electrode) and a set of three springs: the suspension beams, the soft stoppers and a third spring to account for electrode bending [6]. The measured (Fig.6(a)) and the simulation (Fig.6(b)) data for the device output voltage agree reasonable. Specifically, the voltage output is in the same order of magnitude and, as in experiments, the secondary oscillation is present.

However, there are several differences. Prediction shows identical groups due to identical oscillation following the impact of the mobile electrode with two fixed electrodes on its side. The predicted signal is also quickly damped. These differences can be explained taking a closer look at the electrode geometry. Due to their geometry, impact occurs at the base of the electrode attached to the shuttle mass (Fig. 2(d)). Since the base of this electrode is wide, this most likely causes the mobile electrodes to remain rigid and instead triggers oscillation of the fixed electrode which has a narrow base. As the anchored electrodes operate in pairs, only one is engaged by the mobile electrode during collision. The non-simultaneous electrode collision produces two distinct out of phase oscillation frequencies. Such disturbances in oscillation are exacerbated by the non-linearity of the device dynamics, tending towards chaotic electrode frequency-up conversion

as seen in Fig. 6(a). In contrast, the model considers oscillations of the mobile electrode. In addition, the out-of-plane bending of the shuttle mass, which has been previously documented in a similar device [10], and an assumption made in the damping calculation [11] that the plates are infinitely large, further contribute to the discrepancy between the simulation and measurement results.

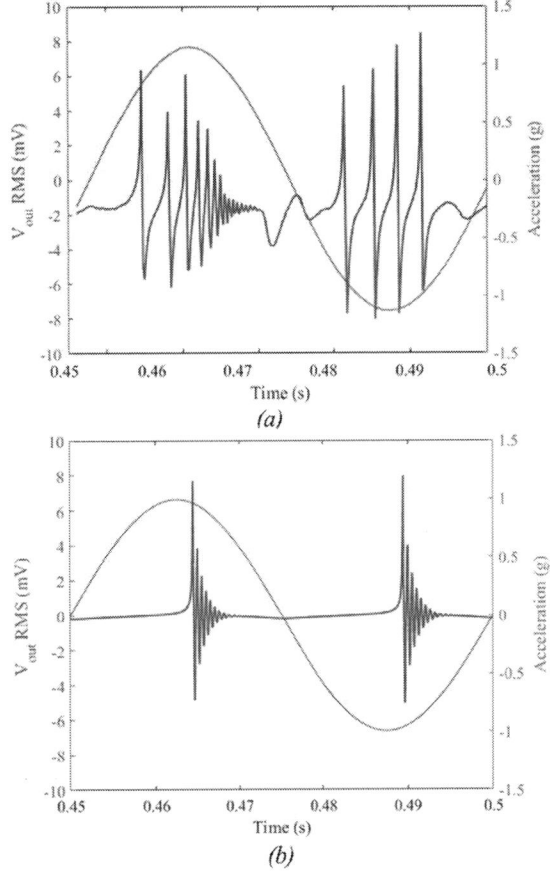

Figure 6: The RMS voltage output from the device without a ball under an input mechanical vibration of 20Hz and 1g acceleration, and a 4V bias voltage: (a) Measurement results; (b) Model prediction. (The acceleration amplitude is not shown to scale)

As previously noted, the device is designed to hold a micro ball in the cavity etched in the shuttle mass. The collision between the shuttle mass and the micro ball have been investigated on a device with identical design as the one discussed above. Examples of the $V_{out\ RMS}$ as function of time are shown in Fig. 7. for two distinct cases: 0.5 *mm* diameter and 0.8 *mm* diameter tungsten balls, respectively. In both cases the tests were performed at 4*V* bias voltage, 20*Hz*, and 1*g* acceleration. As compared to Fig. 6(a), both Fig. 7(a) and Fig.7(b) exhibit higher $V_{out\ RMS}$ amplitude. In addition to the individual ringing produced by the self-oscillation of the non-uniform electrodes, the impact between the shuttle mass and the springless ball is contributing to increase the output voltage. The comparison between Fig. 7(a) and Fig. 7(b) indicates that using a larger ball can help to increase both the amplitude and frequency of the output voltage.

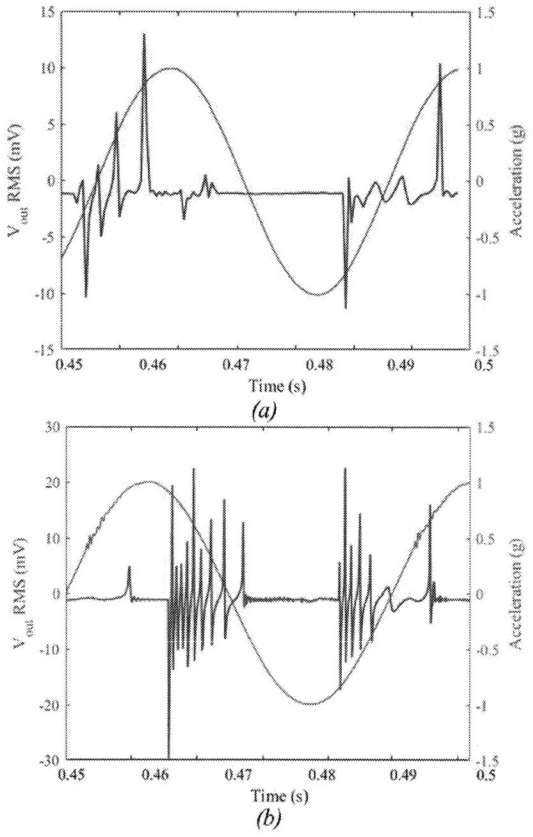

Figure 7: The measured RMS voltage output from the device with different balls under a 4V bias voltage and an input mechanical vibration of (a) 1g, 20Hz with a Tungsten micro ball of 0.5 mm diameter; (b) 1g, 20Hz with a Tungsten micro ball of 0.8 mm diameter.

CONCLUSIONS

The design, fabrication, and characterization of an electrostatic MEMS energy harvester with a non-uniform comb design are presented in this study. Previously it has been reported that electrodes with sloped walls and non-uniform gap may be used to increase the output frequency and power of electrostatic harvesters. However, creating sloping sidewalls requires specialized deep reactive ion etch processes, not common in practice. To capitalize on the effect of non-uniform electrodes' gap and avoid the practical limitation, a new electrode design was proposed here that also produces a non-uniform gap and is compatible with the standard MEMS fabrication methods. This work enhanced the state of the art in the following ways: 1) non-uniform cross-section electrodes are designed to prevent pull-in and to produce secondary oscillations at high frequencies following electrodes' impact; 2) frequency up-conversion is achieved, to the authors' best knowledge, for the first time using the commercial fabrication; 3) to further increase the bandwidth and power, the cutting-edge electrode design is integrated with springless micro balls. This proof-of-concept study opens new venues for enhancing the power output and frequency response of electrostatic harvesters bringing them closer to practical implementation.

ACKNOWLEDGEMENTS

The authors sincerely acknowledge the financial support from NSF ECCS 1609647. We thank Gavin Arthur Divincent for his help during lab testing.

REFERENCES

[1] D. J. Bishop, C. R. Giles, and G. P. Austin, "The lucent lambdarouter: MEMS technology of the future here today," *IEEE Commun. Mag.*, vol. 40, no. 3, pp. 75–79, Mar. 2002.

[2] E. Blokhina, D. Galayko, P. Harte, P. Basset, and O. Feely, "Limit on converted power in resonant electrostatic vibration energy harvesters," *Appl. Phys. Lett.*, vol.101, no. 17, Oct. 2012, Art. no. 173904

[3] J. Li, C. Xu, J. Tichy, and D.-A. Borca-Tasciuc, "A 1D model for design and predicting dynamic behavior of out-of-plane MEMS," *J. Micromech. Microeng.*, vol. 28, no. 8, May 2018, Art. no. 085021.

[4] J. Li, X. Tong, J. Oxaal, Z. Liu, M. Hella, and D.-A. Borca-Tasciuc, "Investigation of parallel-connected MEMS electrostatic energy harvesters for enhancing output power over a wide frequency range," *J. Micromech. Microeng.*, vol. 29, no. 9, Jul. 2019, Art.no. 09400.

[5] S. Meninger, J. O. Mur-Miranda, R. Amirtharajah, A. Chandrakasan, and J. H. Lang, "Vibration-to-electric energy conversion," *IEEE Trans Very Large Scale Integr VLSI Syst.*, vol. 9, no. 1, pp. 64–76, Feb. 2001.

[6] J. Li, "Electrostatic MEMS energy harvesting for sensor powering," *Ph.D. dissertation*, Rensselaer Polytechnic Institute, 2021.

[7] J. Oxaal, M. Hella, and D.-A. Borca-Tasciuc, "Electrostatic MEMS vibration energy harvester for HVAC applications with impact-based frequency up-conversion," *J. Micromech. Microeng.*, vol. 26, no. 12, Nov. 2016, Art. no. 124012

[8] J. Li, J. Tichy, and D.-A. Borca-Tasciuc, "A predictive model for electrostatic energy harvesters with impact-based frequency up-conversion," *J. Micromech. Microeng.*, vol. 30, no. 12, Nov. 2020, Art. no. 125012

[9] A. Cowen, G. Hames, D. Monk, S. Wilcenski, and B. Hardy, *SOIMUMPs design handbook*. Durham, NC: MEMSCAP Inc., 2011

[10] J. Oxaal, D. Foster, M. Hella, and D.-A. Borca-Tasciuc, "Investigation of gap-closing interdigitated capacitors for electrostatic vibration energy harvesting," *J. Micromech. Microeng.*, vol. 25, no. 10, Sep. 2015, Art. no. 105010.

[11] A. Moy, D.-A. Borca-Tasciuc, and J. Tichy, "Squeezing flow between rigid tilted surfaces: A general solution and case study for MEMS," *Lubr. Sci.*, vol. 29, no. 8, pp. 531–539, May 2017.

CONTACT

*D. Borca-Tasciuc, tel: +1-518-276 3385; borcad@rpi.edu

A HYBRID NANOGENERATOR-DRIVEN SELF-POWERED WEARABLE PERSPIRATION MONITORING SYSTEM

Md Abu Zahed, S M Sohel Rana, Md Sharifuzzaman, Seonghoon Jeong, Gagan Bahadur Pradhan, Hye Su Song, Jae Yeong Park*

Department of Electronic Engineering, Kwangwoon University, Seoul, 139-701, Republic of Korea

ABSTRACT

This paper reports a wireless and self-powered sweat biosensing system that consists of a novel, miniaturized hybrid nanogenerator with high output power (250.67 Wm^{-2}) and a microfluidic integrated stretchable biosensing patch. Polyvinylidene difluoride (PVDF)/Co$_3$O$_4$ nanofibers improve the triboelectric nanogenerator (TENG) performance whereas the Halbach magnet array in the electromagnetic nanogenerator (EMG) enhances the flux concentration in the coil. The nanoporous carbon (NPC) modified stretchable porous graphene patterned in Styrene-Ethylene-Butylene-Styrene (SEBS) (PCG-SEBS) demonstrated excellent linearity and sensitivities for Na$^+$, K$^+$, pH, and temperature sensing. Employing all-in-one system integration, we demonstrate a fully functional hybrid nanogenerator-driven system that can power multiplexed biosensors and transmit data wirelessly to user interfaces.

KEYWORDS

Self-Powered, Hybrid Nanogenerator, Wearable, Microfluidic, Biosensors.

INTRODUCTION

Self-powered wearable bioelectronics is greatly expanding their horizons due to their potential for personalized health monitoring [1]. The continuous operation of wearable electronics demands reliable sources of energy, currently met through Li-ion batteries. The recent nanotechnology advances have led to wearable energy harvesters that can convert biomechanical energy into electricity. Wei Gao et.al. reported a bulky flexible TENG-based energy harvester fabricated in complex microfabrication procedures that displayed a power output of ~416 mWm^{-2} [2]. Yansong Gai et. al. reported a unidirectional hybrid nanogenerator module for powering the real-time on-body sweat analyzing system [3]. Despite intensive research activities, most wearable energy harvesters suffer from complex fabrication procedures, low power density, and unidirectional movements making them unsuitable and limiting output power for wireless and continuous biosensing. On the other hand, perspiration-based biosensors are in demand for their potential applications in health management, wellness-tracking, and early warning for COVID-19 in a non-invasive manner [4]. However, currently reported sensors are largely focused on a limited number of analytes [5], not entirely self-powered, and fabricated in a sophisticated as well as expensive technique. As a result, the hybrid harvester, which harvests mechanical energy from a range of vibration sources generated by the human body, should be explored to improve its efficiency.

Sometimes overlooked aspect of wearable sweat sensing is sweat sampling. It is essential to collect sweat samples correctly to prevent measurement mistakes caused by evaporation and contamination. By employing microfluidics, it is possible to avoid these harmful impacts. By directing perspiration through a regulated channel, microfluidics allows continuous sampling and enhances sensing within a well-defined enclosed chamber by directing perspiration through a regulated channel. Few microfluidic chamber-based wearable sweat sensors for collecting and analyzing perspiration have been shown yet. In addition, existing fluidic integrated wearable biosensors require severe exercise to gather sufficient perspiration.

Utilizing transducing layers as immobilized matrices are one of the most effective ways to enhance the sensitivity of electrochemical sensors. Metal-organic frameworks (MOFs) are nanoporous materials with, a high specific surface area, an adaptable structure, ultrahigh porosity, and exceptional thermal and chemical resilience. At a high temperature of calcination, zeolite imidazolate framework-67 (ZIF-67) MOF-derived N-doped NPC demonstrate exceptional electrical conductivity, high nitrogen concentrations, and outstanding electrochemical activity[6]. In this work, we introduced PVDF/Co$_3$O$_4$ nanofibers for TENG and a miniaturized Halbach magnet array for EMG to convert biomechanical energy to electricity in an effective manner. Porous graphene electrodes were modified with NPC nanoparticles for sensitive sensing of Na$^+$, K$^+$, pH, and temperature. A newly designed sweat-collecting microfluidic channel was inserted into the patch. The developed self-powered and wireless sweat electrolyte monitoring system (SPEMS) is illustrated in Figure 1.

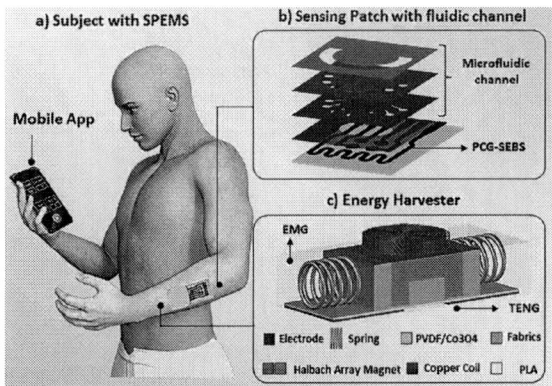

Figure 1: a) Schematic illustration of the SPEMS. b) Layer-by-layer illustration of the biosensing patch. c) Schematic drawing of the hybridized energy harvester.

MATERIALS AND METHODS
Reagents and Instruments

Ethylene glycol, Valinomycin, sodium tetrakis[3,5-bis(trifluoromethyl)phenyl] borate (Na-TFPB), potassium tetrakis (4-chlorophynyl) borate, poly (vinyl chloride), sodium ionophore X, bis(2-ethylehexyl) sebacate (DOS), high-molecular-weight polyvinyl chloride (PVC), tetrahydrofuran, potassium chloride (KCl), and sodium chloride (NaCl), polyvinyl butyral (PVB) were purchased from Sigma Aldrich. The Ag/AgCl paste was purchased from ALS Co., Ltd. (Japan). An electrochemical analyzer (CHI 660E, CH Instruments, Inc., USA) was employed for all electrochemical experiments.

Synthesis of Co_3O_4 and NPC nanoparticles

Following our previously optimized procedures [6], [7], Co_3O_4 and NPC nanoparticles were synthesized.

Fabrication of the Microfluidic Integrated Electrolyte Sensing Patch

Initially, polyimide sheets having a thickness of 125 μm were treated with acetone and methanol. After curing the polyimide sheet with nitrogen gas, a CO_2 laser was used to construct electrode patterns that included a temperature sensor. The geometric area of the working electrode is 7.07 mm². The graphene pattern was then transferred to a highly flexible and stretchable SEBS substrate. The SEBS solution was prepared by dissolving 3g of SEBS triblock copolymers in 20 mL toluene and magnetically stirring until a transparent homogeneous suspension was obtained. The graphene-patterned polyimide film was spin-coated with the SEBS suspension and dried at room temperature overnight. After that, the SEBS film with the graphene pattern was peel-off the polyimide film. The electrodes were subsequently modified with NPC particles. The NPC solution (5 mg mL^{-1}) in EG was prepared using probe sonication for 40 mins. 5 μL of the developed nanocomposite was drop cast on the top of the working electrodes of each sensor and dried overnight at 50 °C.

The Na$^+$ selective membrane cocktail consisted of Na ionophore X (1 mg), Na-TFPB (0.55 mg), PVC (33 mg), and DOS (66 μL). 100 mg of the membrane cocktail was dissolved in 660 μL of tetrahydrofuran.

The K$^+$ selective membrane cocktail was prepared by mixing (4 mg) valinomycin, (1 mg) potassium tetrakis (4-chlorophynyl) borate, (65 mg) poly (vinyl chloride), and (140 μL) DOS. The cocktail was dissolved in 700 μL of tetrahydrofuran through 15 minutes of magnetic stirring. 10 μL of the as-prepared cocktail was drop coated on top of the PCG-SEBS electrode.

To functionalize the working electrode of the pH sensor, aniline was polymerized in a 0.1 M aniline and 1 M HCl solution, and CV was conducted in the potential range of −0.2 to 1 V for 50 segments at a scan rate of 100 mVs^{-1}.

The serpentine-shaped PCG-SEBS electrode was used directly as a temperature sensor. The pattern was constructed in a serpentine approach to increase the initial resistance of the sensor, therefore, enabling it to exhibit improved sensitivity.

Thereafter, the microfluidic channel was fabricated on polyethylene terephthalate (PET) film and medical tape with the help of a CO_2 laser. First, a sweat accumulation layer was made of a half-circle pattern of water-proof, double-sided medical tape. Second, the inlet and outlet layers were subsequently created by cutting numerous 1-mm-diameter circles into a PET sheet. Third, several channels were designed into the channel layer of a medical tape. Fourth, PET film was laser-cut into an elliptical form to construct the sweat reservoir layer. Finally, the electrode substrate and fluidic channel were joined with a layer of medical tape cut in the same reservoir shape as the electrode substrate.

Fabrication of the Hybrid Nanogenerator

The hybrid nanogenerator's device frame, coil zig, and other components were fabricated using a 3D printer. EMG and TENG are two components of the hybrid nanogenerator. The EMG component consists of a stationary coil and a movable Halbach magnet array. Two non-magnetic coil springs were attached to both ends of the movable Halbach magnet array, and a fixed coil was attached to the center of the frame. Six NdFeB square magnets of the N52 class were used in each Halbach magnet array. The intensity of the magnetic field of the single magnet was 4.5×10^5 Am^{-1}. The coil and Halbach magnet are 1.5 mm apart based on COMSOL simulation. TENG was fabricated with nanofibers of PVDF/Co_3O_4 and nylon. The PVDF/Co_3O_4 nanofiber mat was attached to the bottom of the frame and the conductive fabric electrode was connected to the PVDF/Co_3O_4 nanofiber mat. The nylon films were attached to the bottom of the Halbach magnet, where they acted as triboelectric material with a positive charge. A case that had been produced via 3d printers was utilized to cover the final product.

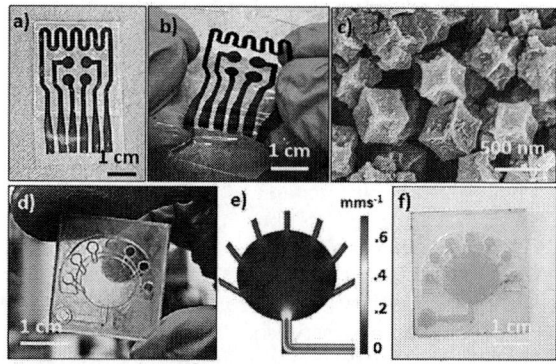

Figure 2: a) An optical image of the microfluidic integrated sensing patch. b) Stretched images of the developed patch, c) FESEM image of the NPC particles used as a transducing layer. d) Optical image of the adhesive microfluidic channel. e) Numerically simulated sweat distributions in the microfluidic reservoir. f) Green liquid flow through the microfluidic channel to check the efficiency.

RESULTS AND DISCUSSION
Physical Analysis of the transducing materials and Microfluidic Channel

An optical image and stretchability of the patch are shown in Figures 2a and 2b. The successful coating of NPC particles on top of the porous graphene is demonstrated in a FESEM image of PCG-SEBS in Figure 2c. An optical image and numerical simulation performed to optimize the geometric design of the microfluidic module are illustrated in Figure 2d, and Figure 2e, respectively. Green liquid flow through the module confirmed the feasibility of the microfluidic system (Figure 2f).

978-1-6654-9309-3/23 $31.00 © 2023 IEEE

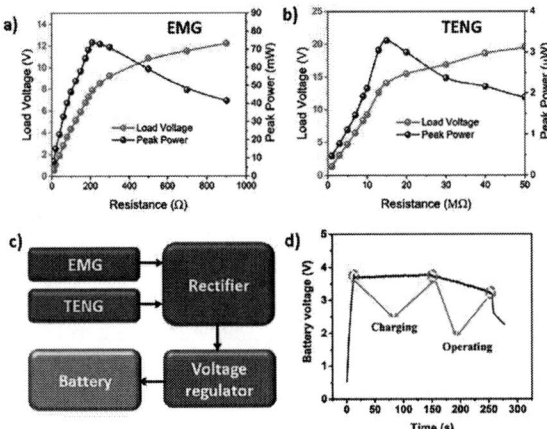

Figure 3: a, b) Electrical characterization of the TENG and EMG at 5 Hz frequency. c) Schematic diagram of the power management circuitry for battery charging. d) Battery charging curve and operating using hybrid nanogenerator.

Electrical Performance of the Hybrid Nanogenerator

The hybrid generator can deliver outstanding charging performance by applying a rational integration of electromagnetic and triboelectric generators. The relationship between the electrical output across various resistive loads was assessed and calculated for the TENG and EMG generator at 5 Hz and 1.5 g acceleration to evaluate the resistance matching and maximum output power offered by the proposed hybrid nanogenerator. As shown in Figures 3a and 3b, the output peak voltage increased as the external load resistance increased, when the internal load resistance was suited to the external load resistance, the maximum peak output power (250.67 Wm⁻²) was obtained. TENGs generate a high output voltage and a small output current at the load due to their high internal resistances, while EMGs generate a lower output voltage and a significant output current due to their lower internal resistance. We found that the TENG and EMG had distinguishable properties. To overcome this problem, we hybridized three low-power bridge rectifier integrated circuits (DF10S), LTC3331s, and PMC circuits. A complete power management circuit including rectifier units and a regulator was fabricated to charge the rechargeable battery as shown in Figure 3c. When storing harvested energy in a battery, the low output voltage of the EMG limits the maximum voltage that can be stored, which may be increased by combining the EMG with the TENG. The developed nanogenerator has a remarkable output power of 250 mWm⁻² at 5 Hz, allowing it to charge a 3.7 V, 100 mAh Li-ion rechargeable battery in 125s as shown in Figure 3d. The battery's charge level was enough to power the biosensing and data transmission unit.

Electrochemical Investigation of the Biosensing Patch

Electrolytes play an essential role in maintaining the balance of body fluids and regulating body functions. The concentration of Na^+ and K^+ in human sweat commonly varies from 10 to 100 mM and 4 to 8 mM, respectively. Sensitivity and linear response are important factors to ensure accurate measurements of all electrolytes. Under optimized conditions, the electrolyte sensors showed stable potentiometric responses with excellent linearities ($r^2 > 0.99$) and sensitivities of 80.7 mV/dec, and 50.3 mV/dec (Figure 4a-4d) in the physiologically relevant concentrations of 5–160 mM Na^+ and 1–128 mM K^+ for Na^+, K^+ sensors, respectively.

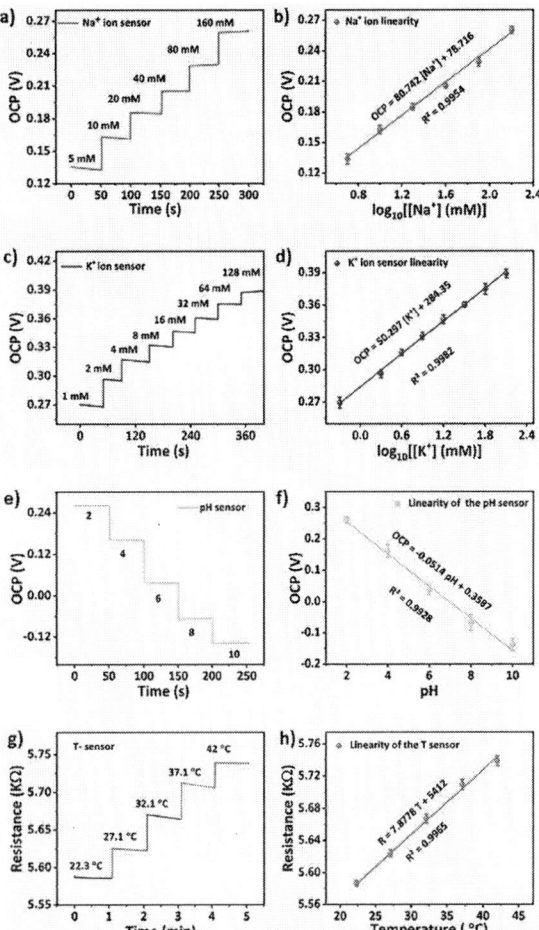

Figure 4: In-vitro characterization of the fabricated PCG-SEBS biosensing patch. a), c), and e) Potentiometric responses of the Na^+, K^+, and pH sensors. b), d), and f) Corresponding calibration curves of all ion sensors. g) Resistive responses of the graphene-SEBS-based temperature sensor. h) Calibration curve of the temperature sensor.

Similar to the Na^+ and K^+ sensors, the general performance of pH sensors was evaluated. Due to its ease of manufacture, reproducibility, and biocompatibility, PANI has been extensively used for measuring the pH of body fluids. Since the pH of human biological fluids often ranges between 3 and 8, sensors are categorized from pH 2 to 10 to accommodate this range. Figure 4e-4f illustrates the open circuit potentials of a PANI-based pH sensor with an outstanding sensitivity of 51.4 mV/dec when tested in commercial buffers.

Skin temperature is a reliable measure of a person's thermal condition and is instructive for a variety of skin-related illnesses, including ulceration. The outstanding thermoresistive effect of the PCG-SEBS electrode was used in the construction of the temperature sensor. Due to the thermal expansion of the polymeric SEBS electrode at an elevated temperature, which creates a crack between the hybrid scaffold and increases the phonon scattering in the sandwiched layers, its resistance increases as the temperature rises, thus demonstrating the positive temperature coefficient resistance behavior. The stepwise resistive responses and good linearity of the temperature sensors are depicted in Figures 4g and 4h, respectively. The developed resistive temperature sensors have a sensitivity of 0.141%/°C to the temperature variations over 22.3–42 °C with respect to its baseline resistance at room temperature.

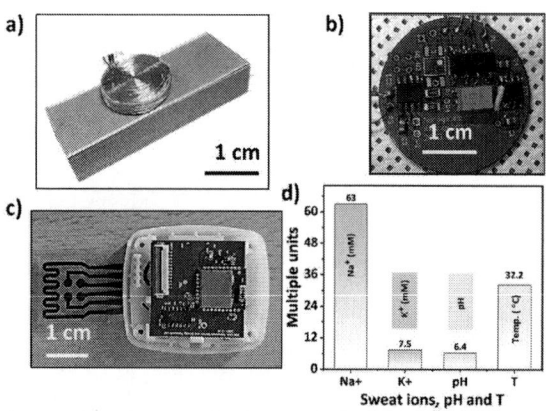

Figure 5: a) A real photo of the hybridized energy harvester. b) An optical image of the power management circuit required for battery charging. c) Optical image of the analog front-end, data transmission circuitry, and microfluidic integrated sweat biosensing patch. d) Sensors data found in the mobile application.

Epidermal Evaluation of the Self-Powered Perspiration Monitoring System

Eventually, after packaging a hybrid energy harvester (Figure 5a), power management circuit (Figure 5b), and data transmission system together with a biosensing patch (Figure 5c), data wireless transmission was successfully demonstrated on mobile interfaces. The operation of the whole SPEMS system was evaluated while the subject ran at a constant speed of 9 km/h. After 20 minutes of subject running, the microfluidic channel collected enough sweat, and the battery was charged sufficiently to transmit the first pair of data to the mobile application. Bluetooth enabled the wireless transmission of the physiological data collected by the sensors to a user interface for further analysis. The Na^+, K^+, pH, and temperature values recorded by the developed system were within the normal physiologically relevant range (Figure 5d), indicating that the SPEMS system functions well for perspiration analysis on the skin in real-time.

CONCLUSIONS

This research produced a self-powered, energy-harvester-driven sweat analysis device that is entirely powered by body movement. The developed hybrid nanogenerator has a remarkable output power of 250 mWm^{-2} at 5 Hz, allowing it to charge a 3.7 V, 100 mAh Li-ion rechargeable battery in 125s. The skin-conformable sensor patch based on microfluidic integrated NPC-SEBS automatically collected sweat and detected Na^+, K^+, pH, and body temperature. Finally, during on-body testing, the SPEMS system successfully transmitted sweat biomarkers and body temperature data to the mobile app to prove its potentiality in wireless personal health monitoring.

ACKNOWLEDGEMENTS

This research was supported by the National Research Foundation of Korea Grant funded by the Korean government (MSIT) (NRF-2020R1A2C2012820) and by the Technology Innovation Program (RS-2022-00154983, Development of Low-Power Sensors and Self-Charging Power Sources for Self-Sustainable Wireless Sensor Platforms) funded by the Ministry of Trade, Industry & Energy (MI, Korea). The authors are grateful to the group members of the Advanced Sensor and Energy Research (ASER) Laboratory of Kwangwoon University for their valuable suggestions and support.

REFERENCES

[1] W. Gao *et al.*, "Fully integrated wearable sensor arrays for multiplexed in situ perspiration analysis," *Nature*, vol. 529, no. 7587, pp. 509–514, 2016, doi: 10.1038/nature16521.

[2] Y. Song *et al.*, "Wireless battery-free wearable sweat sensor powered by human motion," *Sci. Adv.*, vol. 6, no. 40, pp. 1–11, 2020, doi: 10.1126/sciadv.aay9842.

[3] Y. Gai *et al.*, "A Self-Powered Wearable Sensor for Continuous Wireless Sweat Monitoring," vol. 2200653, pp. 1–11, 2022, doi: 10.1002/smtd.202200653.

[4] L. Manjakkal, L. Yin, A. Nathan, J. Wang, and R. Dahiya, "Energy Autonomous Sweat-Based Wearable Systems," *Adv. Mater.*, vol. 33, no. 35, 2021, doi: 10.1002/adma.202100899.

[5] T. Zhao *et al.*, "A self-powered biosensing electronic-skin for real-time sweat Ca2+ detection and wireless data transmission," *Smart Mater. Struct.*, vol. 28, no. 8, 2019, doi: 10.1088/1361-665X/ab2624.

[6] M. A. Zahed *et al.*, "A Nanoporous Carbon-MXene Heterostructured Nanocomposite-Based Epidermal Patch for Real-Time Biopotentials and Sweat Glucose Monitoring," *Adv. Funct. Mater.*, vol. 2208344, pp. 1–17, 2022, doi: 10.1002/adfm.202208344.

[7] M. T. Rahman *et al.*, "Silicone-incorporated nanoporous cobalt oxide and MXene nanocomposite-coated stretchable fabric for wearable triboelectric nanogenerator and self-powered sensing applications," *Nano Energy*, vol. 100, no. April, p. 107454, 2022, doi: 10.1016/j.nanoen.2022.107454.

CONTACT

*J. Y. Park, tel: +02-940-5113; E-mail: jaepark@kw.ac.kr

A MONOLITHIC INTEGRATED AND TRANSPARENT MICROSYSTEM CONSTRUCTED BY USING AMORPHOUS INGAZNO FILM

Bin Jia, Chao Zhang, and Xiaodong Huang*

Key Laboratory of MEMS of the Ministry of Education,
School of Electronic Science and Engineering, Southeast University, Nanjing, CHINA

ABSTRACT

Essential components for constructing a microsystem, including energy device (lithium-ion battery, LIB), electronic device (thin-film transistor, TFT) and sensing device (photodetector, PD), are successfully prepared on a single glass substrate by using amorphous InGaZnO film as the functional layers for the first time. The LIB presents a specific capacity of 9.8 μAh cm^{-2}, the TFT displays a carrier mobility of 3.3 cm^2 V^{-1} s^{-1} and the PD shows a responsivity of 0.35 A W^{-1}. Each device displays an acceptable performance and collaboratively works well to achieve systematical function. This integrated microsystem displays a relatively high transparency. Moreover, the fabrication processes of this microsystem can be significantly simplified owing to using the same film as the functional layers.

KEYWORDS

Integrated microsystem, transparent, InGaZnO (IGZO), on-chip energy device

INTRODUCTION

The rapid growing in many emerging fields such as the Internet of Things, artificial intelligence and smart healthcare, have driven the development of microsystems. Essential components in a microsystem include electronic device, energy device and sensing device, etc. [1]. Since various kinds of devices are involved, the microsystem has advantages of function diversity and high integration. As a kind of special microsystem, transparent microsystem has attracted increasing interest in next-generation consumer electronics and wearable equipments [2]. Unfortunately, no transparent integrated microsystems including all the above-mentioned essential components have been reported till now. Due to good uniformity, high carrier mobility and high optically transparency, amorphous InGaZnO (IGZO) has great potential in constructing a microsystem. So far, IGZO has already been widely used in transparent thin film transistor (TFT) [3] and sensor (e.g. photodetector, PD) [4]. Integration of PD with TFT is expected to construct an active-matrix sensor array with a quite high resolution, however, no work has been performed in this aspect [5]. In addition, there is still lack of IGZO-based transparent

energy device for constructing a complete integrated transparent microsystem.

In this work, a transparent LIB with IGZO as the anode is proposed as an on-chip power source for the first time. Then, TFT with IGZO as the channel layer and PD with IGZO as the photosensitive layer are also prepared. All the devices are fabricated on a single glass substrate for constructing a monolithic integrated and transparent microsystem. It is demonstrated that each device displays an acceptable performance and collaboratively works well to achieve systematical function.

EXPERIMENTAL

Figure 1 shows the schematic diagrams of this monolithic integrated microsystem and each essential competent. A thin-film LIB was prepared with a configuration of ITO electrode/V_2O_5 cathode/LiPON electrolyte/IGZO anode/ITO electrode as the on-chip power source. Although no work has been performed to explore the feasibility of IGZO used in LIB, the main compositions (ZnO [6], In_2O_3 [7] and Ga_2O_3 [8]) in IGZO have been well demonstrated to be good conversion-type LIB anode materials with a high specific capacity, suggesting the great potential of IGZO as the LIB anode. V_2O_5 is chose as the cathode mainly because of its good electrochemical performance in the amorphous state and good compatibility with the semiconductor processes [9]. TFT with a configuration of ITO gate/HfLaO dielectric/IGZO channel/ITO source and drain was also prepared. HfLaO is chose as the gate dielectric mainly because of its relatively high dielectric constant, good stability, and low processing temperature [10]. In addition, PD presents a two-terminal photoresistive structure and consists of ITO electrode/IGZO photosensitive layer/ITO electrode.

All the devices were fabricated on a single glass substrate (Kintec) for constructing a monolithic integrated microsystem by using physical vapor deposition (Hex Deposition System) and each layer is patterned by using shadow masks. The detailed deposition conditions as well as the deposition sequences of each layer are summarized in Table 1.

Table 1: Deposition conditions and sequences of each layer for constructing the microsystem.

Film	Sputtering Target	Sputtering Power (W cm^{-2})	Sputtering Mode	Sputtering Pressure (Pa)	Sputtering Ambient (sccm)	Nominal Thickness (nm)
ITO	In_2O_3/SnO_2, 90/10 wt%	4.2	RF	0.94	Ar=40	260
HfLaO	Hf/La, 50/50 at%	3.0	RF	0.69	Ar/O_2=24/4	80
V_2O_5	V_2O_5	4.4	RF	0.29	Ar/O_2=16/2	230
LiPON	Li_3PO_4	6.4	RF	0.59	N_2=13	660
IGZO	$InGaZnO_4$	4.0	RF	0.30	Ar/O_2=15/3	80
ITO	In_2O_3/SnO_2, 90/10 wt%	4.2	RF	0.94	Ar=40	260

The thickness of each layer was characterized by spectroscopic ellipsometry (Horriba UVISEL) and confirmed by scanning electron microscopy (SEM, Helios 5 CX). The transparency of each device and the integrated system was measured by ultraviolet and visible spectrophotometer (Yipu Instrument, U-T1810D). The performance of the LIB was characterized by battery test instrument (Lanhe, M340A). The performance of TFT and PD were measured by using Keithley 4200-SCS analyzer.

RESULTS AND DISCUSSION

The complete microsystem and the corresponding transmittance are displayed in the bottom right-hand corner of Figure 1. Both the optical image and the transmittance spectra suggest the high transparency of the microsystem as well as each component.

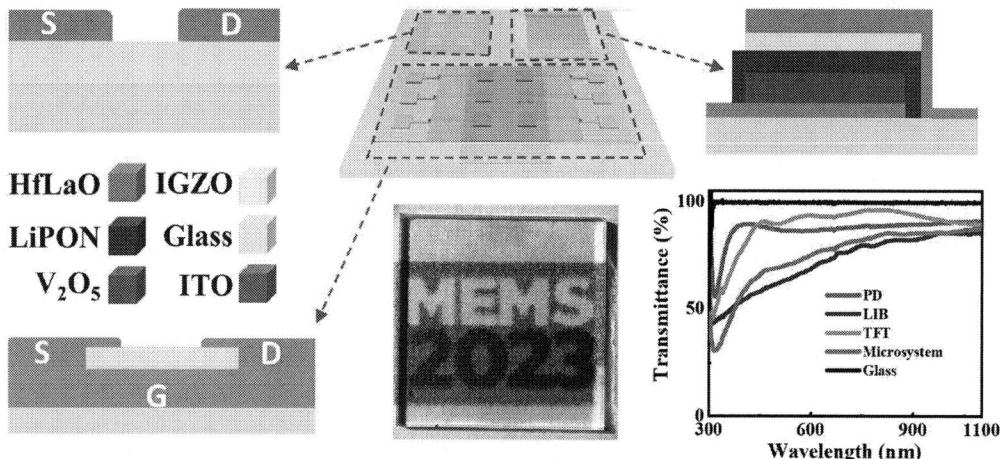

Figure 1: Schematic diagrams and transmittance of this monolithic integrated microsystem and each component. The two pictures in the bottom right-hand corner are the optical image of the microsystem placed on a MEMS 2023 logo and the transmittance of the microsystem and each component, respectively.

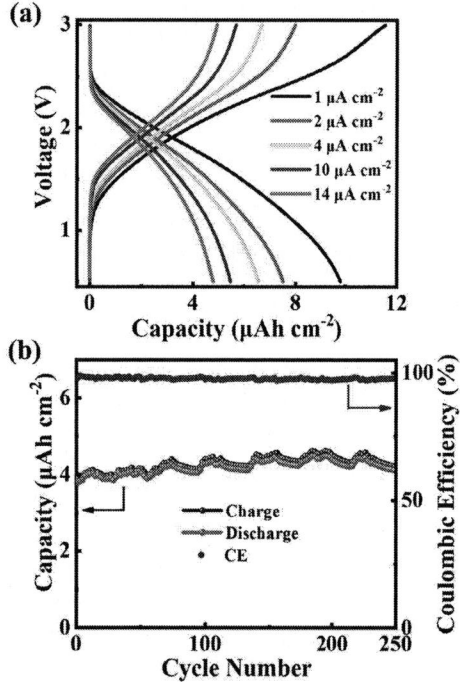

Figure 2: (a) Charge/discharge curves under different current densities and (b) cycling characteristics at a current density of 16 $\mu A\ cm^{-2}$ for the LIB.

Figure 2a shows the typical charge/discharge curves of the LIB under different current densities varying from 1μA cm^{-2} to 14 $\mu A\ cm^{-2}$. A relatively high specific capacity of

9.8 $\mu Ah\ cm^{-2}$ can be obtained at 1 $\mu A\ cm^{-2}$ for the LIB. It is also found that the capacity decreases gradually with increasing the current density, and a specific capacity of 4.8 $\mu Ah\ cm^{-2}$ can be still obtained even at a high current density of 14 $\mu A\ cm^{-2}$, suggesting good rate characteristics of this LIB. Figure 2b shows the cycling performance of this LIB under a current density of 16 $\mu A\ cm^{-2}$. The reversible capacity displays a negligible degradation even after 250 repeated charge/discharge cycles and also a relatively high coulombic efficiency (CE, $\geq$ 96%) is retained during the cycling, both of which demonstrate good cycling performance of this LIB. It is noted that the above-mentioned performance of this LIB is comparable to and even better than those of the V_2O_5-based thin-film LIBs reported in the literature [11], [12].

Figure 3a shows the typical transfer characteristic of the TFT in this work. The critical parameters, including the saturated carrier mobility (μ_{sat} ~ 3.3 $cm^2\ V^{-1}\ s^{-1}$), sub-threshold swing (SS ~ 486 $mV\ dec^{-1}$, turn-ON voltage (V_{ON} ~ 2.6 V) and on/off ratio (I_{on}/I_{off} ~ 10^3), can be extracted from the curve. In addition, by shorting the gate to drain terminal of the TFT, the TFT can act as a rectifier (called as TFTR). In this case, the gate to drain short terminal acts as the input terminal of the TFTR and the source terminal acts as its output terminal. Figure 3b shows the rectification characteristic of this TFTR and a rectification ratio of 10^3 can be obtained. In order to check the effectiveness of this rectifier, AC sinusoidal signals with different amplitudes are applied to the input of the rectifier by using a signal generator (Rigol) and the corresponding output signal of the rectifier across a load resistance (~ 50 kΩ) is recorded by using an oscilloscope

978-1-6654-9309-3/23 $31.00 © 2023 IEEE

(Rigol), as seen in Figure 3c inset. As shown in Figure 3c, the output displays half-wave DC signals in all the cases, demonstrating successful rectification.

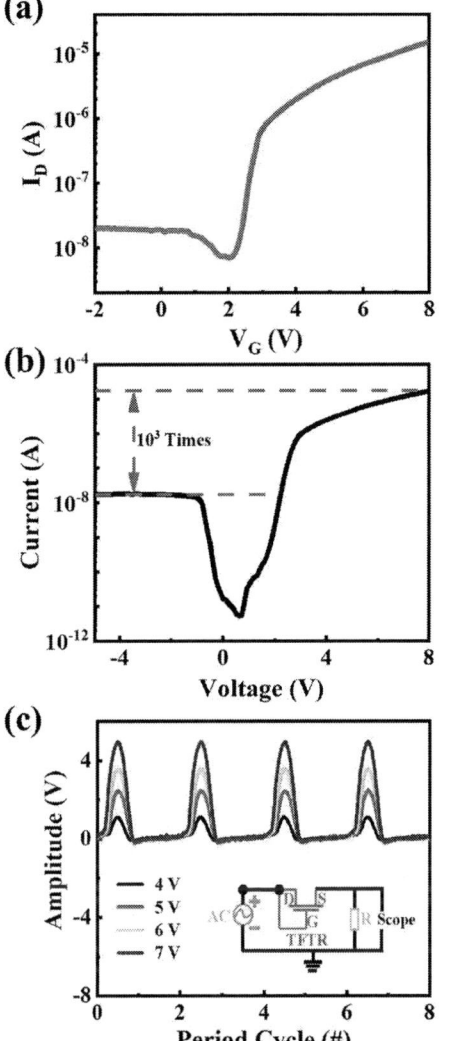

Figure 3: (a) TFT typical transfer characteristic. (b) TFTR DC rectification performance. (c) Corresponding half-wave DC output signals under AC sinusoidal input signals with different voltage amplitudes varying from 4 V to 7V and a fixed frequency of 300 Hz. The inset shows the schematic diagram of the rectification circuit based on the TFTR.

Figure 4a displays the PD current-voltage (*I-V*) characteristics measured under various power intensities of 405 nm light illumination and in the dark, respectively. The output current increases with increasing the light intensity. Moreover, there is a linear relationship of the *I-V* curves, indicating the ohmic contact between the electrode (ITO) and the photosensitive layer (IGZO), which is beneficial to the efficient collection of photogenerated carrier. The relationship between the photocurrent and light intensity is calculated and shown in Figure 2b. Its relationship can be well described by the following expression

$$I_P = AP^\theta \qquad (1)$$

where I_p is the photocurrent (difference between output current under light and in the dark), A is a constant, P is the light intensity, and θ is the empirical value. The θ value is estimated to be 0.79 by fitting, revealing the sublinear relationship between the photocurrent and light intensity, which is often observed in the metal-oxide-based photoconductors because of the complex processes of electron-hole generation, trapping, and recombination in the metal-oxide semiconductors. In addition, responsivity (*R*) is an important parameter for PD and can be defined as

$$R = \frac{I_P}{PS} \qquad (2)$$

where S is the light illumination area. The R value is found to be 0.35 A W^{-1} at a voltage of 5 V and a light intensity of 2.3 mW cm^{-2}.

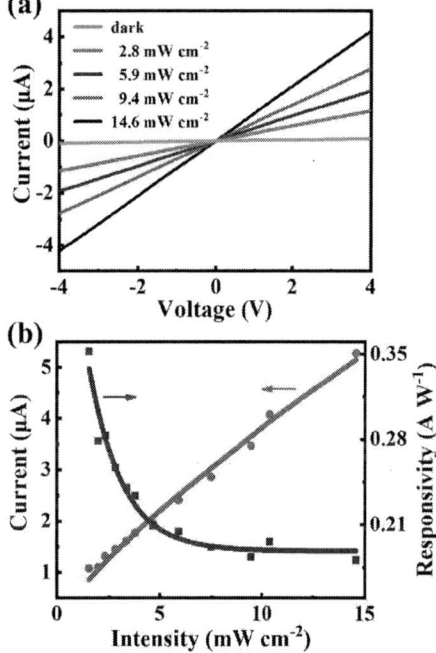

Figure 4: (a) I-V curves of the PD device measured under different power intensities of light illumination and in the dark. (b) Dependence of photocurrent and responsivity on the light intensity.

Figure 5a shows the LIB charge process by using the TFTR as the rectifier and the inset is the corresponding schematic diagram of the test setup. AC sinusoidal signals with different voltage amplitudes are applied to the input terminal of the TFTR, and the output of the TFTR is applied to LIB. The LIB voltage increases and then tends to saturate with charge time, suggesting that the energy is successful charged into the LIB. The input signal is intentionally shut down after charging for 200s. It can be found that the LIB voltage decreases at first and then tends to maintain constant, indicating the quite low self-leakage of this LIB.

978-1-6654-9309-3/23 $31.00 © 2023 IEEE

(a)

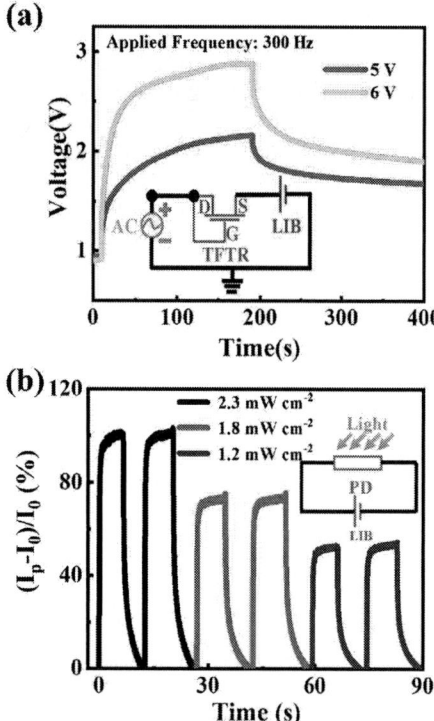

(b)

Figure 5: (a) LIB charge process by using TFTR as the rectifier with different voltage-amplitude sinusoidal AC signals as the input. The inset is the test setup by using the TFTR as the on-chip rectifier. (b) PD output under various power intensities of the light illumination. I_p and I_0 represent the PD current under light illumination and in the dark, respectively. The inset is the test setup by using the LIB as the on-chip power source.

Once the battery is charged, it can be used as an on-chip power source to drive the PD. The test setup is shown in the inset of Figure 5b and the test results are shown in Figure 5b. It can be found that the PD presents a stable output under a constant input light illumination, suggesting good reproducibility and stability of the PD. In addition, the output increases obviously with increasing the light illumination intensity, indicating high sensitivity of this PD. Therefore, the results in Figure 5 demonstrate that each component in this microsystem can collaboratively work well.

CONCLUSION

In summary, for the first time, a transparent thin-film lithium-ion battery with IGZO as the anode is prepared as the on-chip power source. Also, TFT with IGZO as the channel and PD with IGZO the photosensitive layer are also prepared. An integrated and transparent microsystem including energy device (LIB), information device (TFT) and sensing device (PD) are successfully developed. Because all the devices use IGZO as their functional layers, this transparent microsystem presents simple fabrication processes and compact structure.

ACKNOWLEDGEMENTS

This work was supported in part by the National Key R&D Program of China under Grants No. 2020YFB2007400 and in part by the National Natural Science Foundation of China under Grants No. 61974026.

REFERENCES

[1] J. Chen, and Z. L. Wang, "Reviving Vibration Energy Harvesting and Self-Powered Sensing by a Triboelectric Nanogenerator", *Joule*, 1(3), 480–521, 2017.

[2] Z. Y. Lin, Y. Huang, and X. F. Duan, "Van der Waals Thin-Film Electronics", *Nat. Electron.*, 2(9), 378–388, 2019.

[3] C. Zhang, D. Li, P. T. Lai, and X. D. Huang, "Effects of Back Interface on Performance of Dual-Gate InGaZnO Thin-Film Transistor with an Unisolated Top Gate Structure", *IEEE Electron Device Lett*, 42(8), 1176-1179, 2021.

[4] F. Z. Li, Y. Meng, R. T. Dong, S. P. Yip, C. Y. Lan, X. L. Kang, and J. C. Ho, "High-Performance Transparent Ultraviolet Photodetectors Based on InGaZnO Superlattice Nanowire Arrays", *ACS Nano*, 13(10), 12042-12051, 2019.

[5] Z. J. Pan, X. L. Zhao, W. B. Peng, X. M. Qi, and Y. N. He, "A ZnO-Based Programmable UV Detection Integrated Circuit Unit", IEEE Sens. J, 16(22), 7919-7923, 2016.

[6] V. K. H. Bui, T. N. Pham, J. Hur, Y. C. Lee, "Review of ZnO Binary and Ternary Composite Anodes for Lithium-Ion Batteries", *Nanomaterials*, 8, 2021.

[7] H. J. Xu, L. Wang, J. Zhong, T. Wang, J. H. Cao, Y. Y. Wang, X. Q. Li, H. L. Fei, J. Zhu, X. D. Duan, "Ultra-stable and High-rate Lithium Ion Batteries Based on Metal-organic Framework-derived In_2O_3 Nanocrystals/Hierarchically Porous Nitrogen-doped Carbon Anode", *Energy & Environmental Materials*, 2(9), 177-185, 2020.

[8] X. Tang, X. Huang, Y. M. Huang, Y. Gou, J. Pastore, Y. Yang, Y. Xiong, J. F. Qian, J. D. Brock, J. T. Lu, H. D. Abruna, L. Zhuang, "High-Performance Ga_2O_3 Anode for Lithium-Ion Batteries", *ACS Appl. Mater. Interfaces*, 10, 5519-5526, 2018.

[9] Y. Yue, H. Liang, "Micro- and Nano-Structured Vanadium Pentoxide (V_2O_5) for Electrodes of Lithium-Ion Batteries", *Adv. Energy Mater.*, 7(17), 1602545, 2017.

[10] H. Su, Y. X. Ma, P. T. Lai, W. M. Tang, "Influence of Gate Doping Concentration on the Characteristics of Amorphous InGaZnO Thin-Film Transistors With HfLaO Gate Dielectric", *IEEE Electron Device Lett*, 40(12), 1953-1956, 2019.

[11] C. Navone, R. Baddour-Hadjean, J. P. Pereira-Ramos, R. Salot, "Sputtered Crystalline V_2O_5 Thin Films for All-Solid-State Lithium Microbatteries", *J. Electrochem. Soc.*, 156(9), 763-767, 2009.

[12] S. H. Lee, P. Liu, C. E. Tracy, D. K. Benson, "All-Solid-State Rocking Chair Lithium Battery on a Flexible Al Substrate", *Electrochem. Solid-State Lett.*, 2(9), 425-427, 1999.

CONTACT

*Xiaodong Huang, xdhuang@seu.edu.cn

FLOWING WATER ENABLES STEERABLE CHARGE DISTRIBUTION ON ELECTRET SURFACE

Boming Lyu[1,2], Jian Zhang[1,2], Yunjia Li[3], Yongqing Fu[4], Honglong Chang[1], Weizheng Yuan[1], and Kai Tao[1,2]

[1]School of Mechanical Engineering, Northwestern Polytechnical University, China
[2]Research & Development Institute in Shenzhen, Northwestern Polytechnical University, China
[3]School of Electrical Engineering, Xi'an Jiaotong University, China and
[4]Faculty of Engineering and Environment, Northumbria University, UK

ABSTRACT

Contact electrification has been studied since as early as 2,600 years ago, but its mechanism is still unclear, especially for liquid-solid cases. In this paper, an innovative water charging method is proposed for the first time. The charge-transfer mechanism is suppressed by the steerable oxygen plasma treatment, and the 3D charge distribution map of NPU shape is obtained through the surface potential scanner. The method was further applied to the DEG and a self-powered thin-film sensor with droplet position monitoring function was prepared. This study demonstrates the feasibility of the proposed water charging method and proposes a new application in the field of water energy harvesting.

KEYWORDS

Contact electrification, oxygen plasma treatment, electret, water energy harvesting.

INTRODUCTION

Conventional energy collectors cannot obtain high-efficiency energy conversion from water [1-5], because its complex dynamic characteristics of irregular vibration and mixed frequencies and amplitudes [6-8]. In recent years, the generator based on liquid-solid contact (CE) mechanism has received a lot of attention by virtue of its unlimited generating area [5].

Compared with the solid-solid CE process [9-10], the CE process at the liquid-solid interface has not been studied extensively. Recently, many studies have explored the liquid-solid charging process. These studies show that there are two charge-transfer mechanisms in the liquid-solid CE process: electron transfer mechanism (ETM) and ion adsorption mechanism (IAM) [11-15]. In general, ETM dominates [12]. At the same time, liquid composition, pH value [15] and electronegativity of solid surface functional groups [16] have also been proved to significantly affect the charging performance. These studies provide the basis for understanding the CE process within the water & electret interfaces.

With regard to water/liquid energy collectors, existing technology relies on the random contact with water to generate uncontrolled output and signal waveforms. In this conditions, a high and stable surface potential on the electret is of great importance [17-19]. However, how to achieve steerable contact between liquid-solid interfaces remains a long-standing challenge, and it is difficult to control the contact between water and electret only in certain areas [20-21]. The controllable water charging method proposed in our paper is expected to solve this problem.

WATER CHARGING METHOD

In this paper, for the first time, we proposed a controllable water charging method, the charging process as following: first, etching the electret surface with the selective oxygen plasma treatment; second, washing the electret surface with flowing water. The oxygen plasma treatment replaces the untreated -C-F and -C-F$_3$ groups with a great deal of -C-O and -C=O groups on the FEP electret surface, which inhibited the CE mechanism on FEP surface. As shown in Fig. 1 (a), physical mask is not required during the liquid-solid CE process on account of the C and O surface groups can be considered as micron-masks grown on the electret surface directly.

In order to obtain a uniform charge distribution, flowing water is used to wash the electret surface directly, allowing a great deal of water to simultaneously contact the FEP surface. The CE mechanism occurs in the untreated "green" FEP area. On the contrary, the CE mechanism of the "purple" FEP area treated by plasma treatment is blocked. The different colors are just used to distinguish the plasma-treated FEP area from the untreated FEP area.

Fig. 1(b) illustrates the scanning electron microscope (SEM) image of the N-shaped plasma-treated FEP area. Connecting the FEP layer with an N-shaped hollow mask and processing them with 300W oxygen plasma treatment, 5 minutes. The size and line width of the N-shaped letter is around 2×2mm and 200μm, respectively. As is shown in Fig. 1(c), the boundary between the plasma-treated area and the untreated area is easily distinguished.

The SEM images and contact angles of the untreated FEP surface and the 300W oxygen plasma-treated FEP surface have been shown in Figs. 1(d) and 1(e). A series of "nano bun" microstructures with diameters ranging from 90 nm to 170 nm were prepared by 300W power oxygen plasma treatment. Meanwhile, after the oxygen plasma treatment, the contact angle of FEP surface decreased from $105 \pm 2°$ to $80 \pm 1.5°$, indicating that oxygen plasma treatment changed the hydrophobicity of FEP surface to hydrophilicity.

MEASUREMENT

To plot the charge distribution on electret surfaces after the water charging process, an experimental setup with a 3D surface potential scanner for measuring the charge distribution is set up. The stepper motor can be controlled to make snaking motion by programmable logic controller (PLC). During the motor process, the surface potential value is collected one time per second by a 3D surface potential scanner (Trek Model 347, USA), processed by a DAQ system (NI USB-6289 M, USA), and

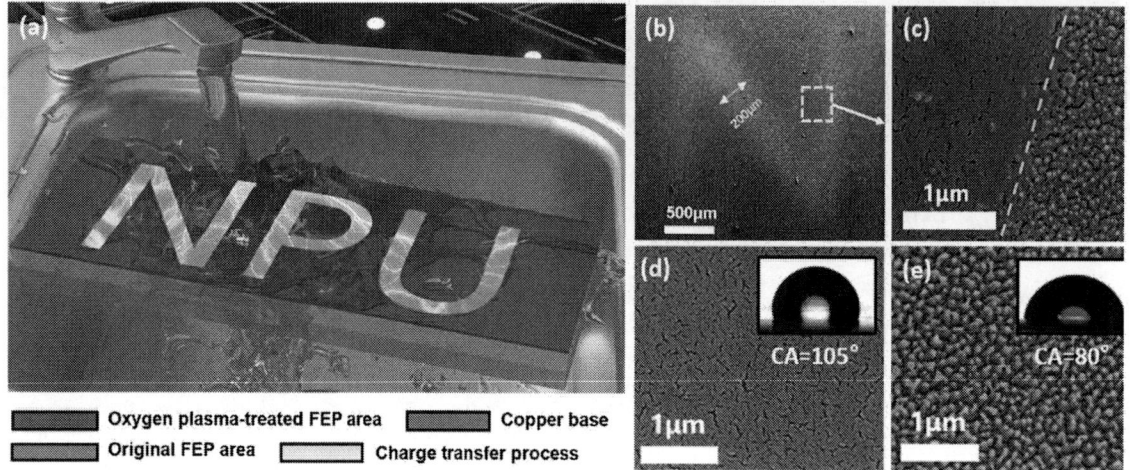

Figure 1: (a) Schematic diagram of forming an "NPU" charge distribution on the electret surface by flowing tap water; (b) The SEM image of N-shaped plasma-treated electret area; (c) The SEM image of the boundary between untreated area and plasma-treated area; (d)-(e) The SEM images and the contact angles of untreated FEP surface and 300W oxygen plasma-treated FEP surface, respectively.

then recorded by a computer. Finally, putting the combination of x position, y position and surface potential value into the Origin software to get potential distribution map. The charge density can be calculated by the following equation [8]:

$$\sigma_0 = \frac{Q}{A} = \frac{\varepsilon_r \varepsilon_0 V}{h} \qquad (1)$$

where Q is the stored charge, A is the surface area, ε_r is the relative dielectric constant of FEP, ε_0 is the dielectric constant of the vacuum, V is the surface potential, h is the thickness of the FEP layer, σ_0 is the surface charge density.

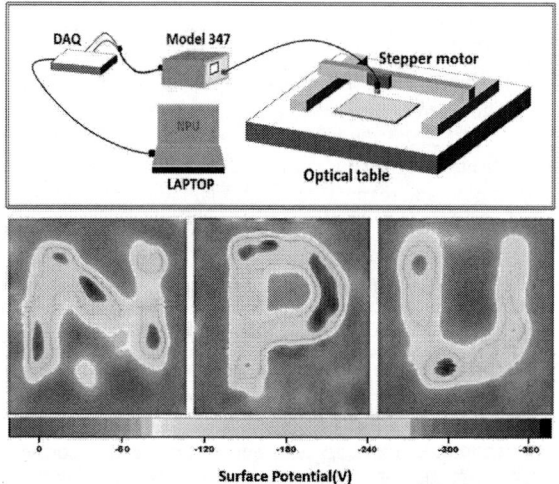

Figure 2: The NPU-shaped potential distribution map on the FEP surface was measured with a 3D surface potential scanner.

Fig. 2 shows the NPU surface potential diagram measured with a 3D scanner. After the NPU-shaped plasma-treated surface is charged by flowing water, the whole electret surface potential distribution map can be divided into "red" and "green" areas, representing neutral

and negative areas, respectively. It can be seen that the surface potential of the plasma-treated area is always about 0V, while the potential of the untreated area is about -200V to -380V.

WATER CHARGING PROCESS

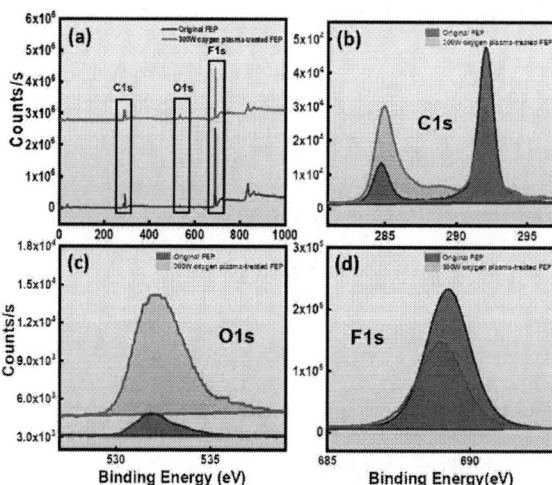

Figure 3: (a) The XPS full spectra of untreated FEP and the plasma-treated FEP; (b-d) The XPS C1s, O1s and F1s spectra contrast between untreated FEP and the plasma-treated FEP, respectively.

The X-ray photoelectron spectroscopy (XPS) full spectra of the untreated FEP surface and 300W oxygen plasma-treated FEP surface has been shown in Fig. 3(a). The detailed XPS spectra of C1s, O1s and F1s are shown in Figs. 3(b), 3(c) and 3(d), respectively. By comparing the spectra, we can infer the change of chemical groups on the surface of FEP. Based on the C1s spectrum, five peaks are displayed at 285.0eV, 286.5eV, 288.0eV, 291.6eV and 293.4eV, which correspond to five carbon environments on the surface of FEP electret, namely, carbon single-bonded carbon (C-C group), carbon single-bonded oxygen (C-O group), carbon double-bonded oxygen (C=O group), carbon single-bonded fluorine (C-F group), and carbon is bonded with three fluorine (C-F$_3$ group), single bonds respectively. It is found that the atomic counts of C element, O element, and F element have changed from 35.76%, 0.89%, 63.34% to 45.10%, 5.81%, 45.52% after the oxygen plasma treatment, respectively.

The XPS results show that there are many -C-F and -C-F$_3$ chemical groups on the untreated FEP surface, while a great deal of C and O chemical groups have superseded the untreated C and F chemical groups after the oxygen plasma treatment. The change of surface groups on the FEP surface inhibits the charging process within the water & electret interfaces.

(a) Charging process of original FEP

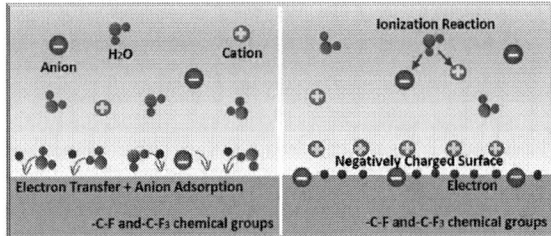

(b) Charging process of plasma-treated FEP

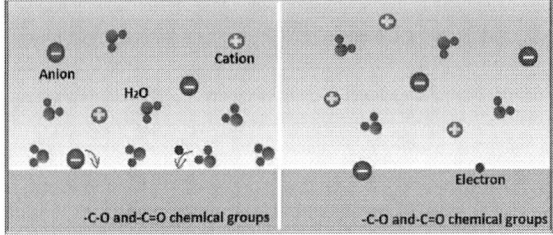

Figure 4: (a) The charging process within the water and untreated electret surface; (b)The charging process within the water and plasma-treated electret surface.

Fig. 4 (a) shows the water charge-transfer process on the untreated FEP surface. The untreated FEP surface is mainly took up by -C-F and -C-F$_3$ chemical groups. When water flushes on FEP surface, electron clouds overlap between water molecules and chemical groups. Owing to the utmost electronegativity of C and F chemical groups, the surface chemical groups of FEP can capture electrons in water molecules. At the same time, ionization reaction will occur to produce H$_3$O$^+$ ions and OH$^-$ ions, while negatively charged OH$^-$ ions will tend to be adsorbed on some areas. In the conditions, the electret surface has been full negative charged. Then, more positive ions in water tend to be attracted near the surface to form an electric double layer (EDL) [12].

The water charging process on plasma-treated FEP surface has been shown in Fig. 4(b). After the oxygen plasma treatment, a great deal of C and F functional groups on FEP surface are superseded by C and O groups. The common electron pairs in C and O chemical groups have no obvious deflection, so the electronegativity of C and O chemical groups will be markedly weakened. The ETM and IAM cannot occur normally when water rushes to the surface of plasma-treated FEP surface.

APPLICATION AND DISCUSSION

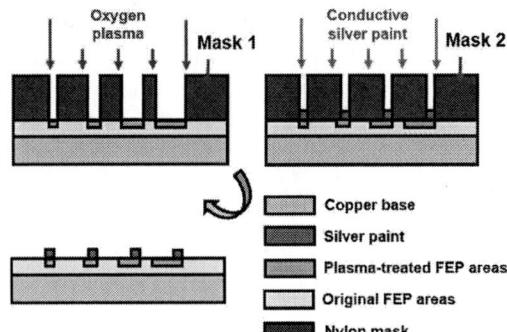

Figure 5: Process diagram of treating the gradient-distribution electret areas and preparing the surface electrode.

Fig. 5 describes the process diagram of the steerable oxygen plasma treatment and the preparation of integrated surface electrodes. The first step is to connect the copper substrate, FEP electret layer and mask 1 together, and place them in an oxygen plasma cleaner for 5 minutes at 300W power. Second, cool the sample under environmental conditions for 5 minutes. Third, replace mask 1 with mask 2 (mask 1 and mask 2 need to be aligned), and spray the simple with conductive silver paint in the shape of mask 2 hollow part.

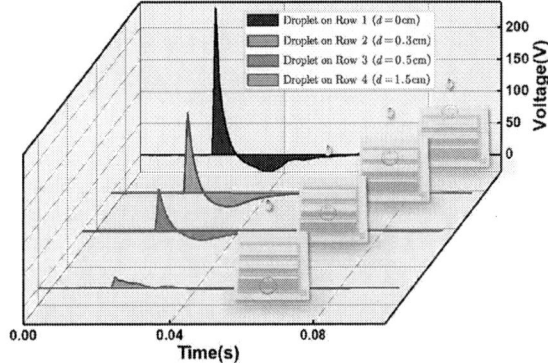

Figure 6: The signal waveform obtained when the droplets contact the different vertical positions of the self-powered sensor.

According to the processing flow in Fig. 5, a new self-powered sensor with the function of monitoring the droplets' vertical position is manufactured [22-24]. The line width and the parallel spacing of the interdigital electrodes are 3mm and 1.7cm, respectively. The widths of the gradient-distribution plasma-treated areas are 0 cm, 0.3 cm, 0.5 cm and 1.5 cm, respectively.

When the droplets released from the height of 8cm (~50μL) flushed on different rows of the sensor, different peak voltage amplitude and duration can be detected due to different sensing areas. When the water droplets contact with the first, second, third and fourth rows (with different plasma-treated widths: d=0cm; d=0.3cm; d=0.5cm; d=1.5cm) of the self-powered sensor, the peak voltages of about 220V, 130V, 65V and 10V can be obtained, respectively.

978-1-6654-9309-3/23 $31.00 © 2023 IEEE

CONCLUSIONS

In this paper, a steerable micro-patterning water charging method is innovatively proposed. By flushing the plasma treated FEP surface with flowing tap water, an NPU-shaped charge distribution has been mapped. Meanwhile, after the oxygen plasma treatment, there is a transition from C and F chemical groups to C and O chemical groups on the electret surface, which causes the attenuation of the charge-transfer mechanism. Based on this method, a self-powered sensor is successfully designed, manufactured and applied to monitor the location of droplets. The results of this work have promoted the application of micro-patterning charging method in the field of water energy collection.

ACKNOWLEDGEMENTS

This Research is supported by the National Natural Science Foundation of China Grant (No.51705429, No. 51735011), Science, Technology and Innovation Commission of Shenzhen Municipality (JCYJ20220530161809020), the Fundamental Research Funds for the Central Universities, 111 Project No. B13044.

REFERENCES

[1] A. Luo, Y. Zhang et al. "An inertial rotary energy harvester for vibrations at ultra-low frequency with high energy conversion efficiency," Applied Energy, vol. 279, p. 115762, Dec. 2020.

[2] N. Han, D. Zhao et al., "Performance evaluation of 3D printed miniature electromagnetic energy harvesters driven by air flow," Applied Energy, vol. 178, pp. 672-680, Sep. 2016.

[3] X. Zhang, P. Pondrom et al., "Ferroelectret nanogenerator with large transverse piezoelectric activity," Nano Energy, vol. 50, pp. 52-61, Aug. 2018.

[4] Y. Feng, K. Hagiwara et al., "Trench-filled cellular parylene electret for piezoelectric transducer", Applied Physics Letters, vol. 100, p. 262901, Jun. 2012.

[5] W. Xu, Z. Wang, "Fusion of Slippery Interfaces and Transistor-Inspired Architecture for Water Kinetic Energy Harvesting," Joule, vol. 4, no. 12, pp. 2527-2531, Dec. 2020.

[6] W. Xu, H. Zheng et al., "A droplet-based electricity generator with high instantaneous power density," Nature, vol. 578, pp. 392-396, Feb. 2020.

[7] H. Gu, N. Zhang et al., "A bulk effect liquid-solid generator with 3D electrodes for wave energy harvesting," Nano Energy, vol. 87, p. 106218, Sep. 2021.

[8] G. Zhu, Y. Su et al., "Harvesting water wave energy by asymmetric screening of electrostatic charges on a nanostructured hydrophobic thin-film surface," ACS Nano, vol. 8, no. 6, pp. 6031-6037, Jun. 2014.

[9] K. Tao, Z. Chen et al., "Ultra-Sensitive, Deformable, and Transparent Triboelectric Tactile Sensor Based on Micro-Pyramid Patterned Ionic Hydrogel for Interactive Human–Machine Interfaces," Advanced Science, vol.9, p. 2104168, Jan. 2022.

[10] K. Tao, H. Yi et al., "Origami-Inspired Electret-Based Triboelectric Generator for Biomechanical and Ocean Wave Energy Harvesting," Nano Energy, vol. 67, p. 104197, Oct. 2019.

[11] F. Zhan, A. C. Wang et al., "Electron Transfer as a Liquid Droplet Contacting a Polymer Surface," ACS Nano, vol. 14, no. 12, pp. 17565-17573, Dec. 2020.

[12] S. Lin, L. Xu et al., "Quantifying electron-transfer in liquid-solid contact electrification and the formation of electric double-layer," Nature Communication, vol. 11, no. 1, p. 399, Dec. 2020.

[13] C. Xu, Y. Zi et al., "On the Electron-Transfer Mechanism in the Contact-Electrification Effect," Adv. Mater., vol. 30, p. 1706790, Mar. 2018.

[14] J. Nauruzbayeva, Z. Sun et al., "Electrification at water-hydrophobe interfaces," Nature Communication, vol. 11, no.5285, Oct. 2020.

[15] J. Nie, Z. Ren et al., "Probing Contact-Electrification-Induced Electron and Ion Transfers at a Liquid-Solid Interface," Advanced Materials, vol. 32, Nov. 2019.

[16] S. Lin, M. Zheng et al., "Effects of Surface Functional Groups on Electron Transfer at Liquid-Solid Interfacial Contact Electrification," ACS Nano, vol. 14, no.8, pp. 10733-10741, Aug. 2020.

[17] K. Tao, L. Tang et al., "Investigation of Multimodal Electret-Based MEMS Energy Harvester With Impact-Induced Nonlinearity," Journal of Microelectromechanical Systems, vol. 27, no. 2, pp. 276-288, Apr. 2018.

[18] Z. Zhao, H. Zhou et al., "Multiphase Bipolar Electret Rotary Generator for Energy Harvesting and Rotation Monitoring," Journal of Microelectromechanical Systems, 2022.

[19] X. Cheng, Z. Song et al., "Wide Range Fabrication of Wrinkle Patterns for Maximizing Surface Charge Density of a Triboelectric Nanogenerator," Journal of Microelectromechanical Systems., vol. 27, no. 1, pp. 106-112, Feb. 2018.

[20] K. Tao, B. Lyu et al., "Micro-Patterning of Electret Charge Distribution by Selective Liquid-Solid Contact Electrification," Journal of Microelectromechanical Systems, vol. 31, no. 4, pp. 625-633, Aug. 2022.

[21] G. Zhu, Y. Su et al., "Harvesting water wave energy by asymmetric screening of electrostatic charges on a nanostructured hydrophobic thin-film surface," ACS Nano, vol. 8, no. 6, pp. 6031-6037, Jun. 2014.

[22] H. Wu, N. Mendel et al. "Energy Harvesting from Drops Impacting onto Charged Surfaces," Physical Review Letters, vol. 125, no. 7, p. 078301, Aug. 2020.

[23] M. H. Biroun, J. Li et al., "Acoustic Waves for Active Reduction of Contact Time in Droplet Impact," Physical Review Applied, vol. 14, Aug. 2020.

[24] J. Zhang, B. Lyu et al., "High-Efficiency Raindrops Energy Harvester Using Interdigital Electrode," in Proc. IEEE 34th Int. Conf. MEMS, pp. 724-727, Jan. 2021.

CONTACT

*Kai Tao, tel: +86-29-88460434; taokai@ nwpu.edu.cn

SELF-POWERED FLEXIBLE PIEZOELECTRET ARRAY FOR WEARABLE APPLICATIONS

Hao Yang[1,2#], Rui M. R. Pinto[1#], Pedro González[1], Alar Ainla[1], Mohammadmahdi Faraji[1], and K. B. Vinayakumar[1]*

[1]International Iberian Nanotechnology Laboratory, Av. Mestre Jose Veiga, 4715-330 Braga, Portugal
[2]Xi'an Jiaotong University, Xi'an 710049, China

ABSTRACT

In this manuscript, a flexible piezoelectret array is demonstrated for wearable applications. Piezoelectrets/ferroelectrets are voided polymers that exhibit piezoelectric effect after being subjected to a charging process that generates and orients charges in the voids. The developed piezoelectret array is composed of two fluorinated ethylene propylene (FEP) sheets bonded with a patterned adhesive layer. A piezoelectret array with optimized void geometry and a piezoelectric coefficient d_{33} = 74 ± 5 pC/N is tested as a flexible pressure sensor, allowing the measurement of the wrist flexors during contraction. The developed sensor is tested for the frequency and force response to understand the broad-band nature of the developed sensor. The sensor can be further miniaturized or fabricated in different shapes for particular applications.

KEYWORDS

Ferroelectret, pyroelectric, charging, flexible, wearable, sensor.

INTRODUCTION

Piezoelectrets are voided polymers that exhibit a piezoelectric effect due to charge trapping and alignment inside the voids [1, 2]. One advantage of piezoelectret devices is that contrary to resistive strain gauges, they don't consume any power and the output voltage or charge can be monitored over time in order to extract the physical parameters of interest (pressure or force applied to the device). Various techniques exist for the charging of ferroelectrets, including corona discharge, x-rays, and pyroelectric charging, but the underlying principle is the development of a high voltage across the ferroelectret, resulting in the ionization of gas molecules in the polymer voids and the orientation of the charged species, thus creating macroscopic dipoles that provide the material with a piezoelectric effect [1-4]. The piezoelectric coefficient (d_{33}) of the reported piezoelectrets is in the range of 10-4000 pC/N [1, 4]. Piezoelectrets can be prepared with a variety of materials such as polypropylene (PP), polyethylene (PE), polyethylene terephthalate (PET), polyethylene naphthalate (PEN)), cyclo-olefin polymer (COP) and copolymer (COC), polycarbonate (PC), and several fluoropolymers including polytetrafluoroethylene (PTFE), amorphous Teflon AF and FEP [5].

These novel low-cost sustainable sensors have been used to demonstrate applications such as force sensors, acoustic sensors, accelerometers, etc. [1-4]. Previously, we have demonstrated the pyroelectric charging approach to realize electrets and piezoelectrets for application in energy harvesting, storing electrostatic information and piezoelectret realization [6-11]. In our present study, we optimize the sensitivity and operating range by changing the void geometry, density, and number of layers.

EXPERIMENTAL METHODS

Piezoelectret Preparation

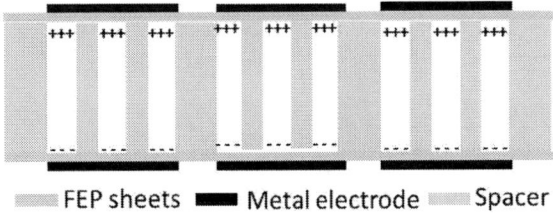

FEP sheets ▬ Metal electrode ▨ Spacer

Figure 1: Cross-section representation of the piezoelectret array.

The sensor is composed of two 25 μm-thick FEP sheets and a patterned adhesive layer (Tesa® 64621), as depicted in Figure 1.

To prepare the ferroelectret, first a 100 nm-thick film of Cr was sputter-coated on a FEP substrate, using a shadow mask with the shape of the electrodes. Then, a CO_2 laser cutter (Widlaser LS1390Plus) was used to pattern the 90 μm-thick adhesive layer. For each array element, the tape pattern was composed of four 400×400 μm squares with a separation of 400 μm. The assembled device was charged using the pyroelectric effect of a $LiNbO_3$ crystal, as described in [3]. Then, the 100 nm-thick Cr top electrodes were deposited by sputtering and the sample was characterized as a function of force and frequency. The final device is in Figure 2, both before and after the deposition of the top electrodes. Each piezoelectret in the array is 3×3 mm. The simple and low-cost fabrication process is scalable and it is possible to integrate it into roll-to-roll processes.

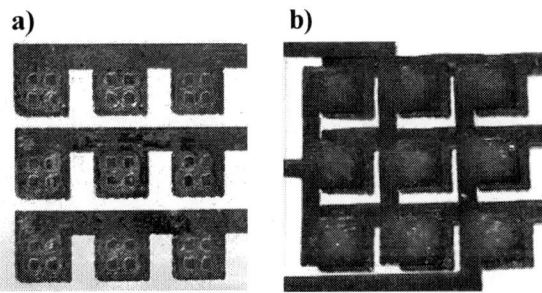

Figure 2: 9-Point piezoelectret array. a) Image of the patterned tape spacer and bottom electrodes. b) Image of the actual device with top and bottom electrodes.

RESULTS AND DISCUSSION

Ferroelectret Geometry Optimization

The number of FEP layers and the void geometry were optimized. Voids with diameters and pitches of 0.4, 0.6 and 1.5 mm were tested and the piezoelectric coefficient d_{33} measured. The geometry that yielded the highest sensitivity to low forces had a void size of 0.4 mm (Figure 3). The ferroelectret d_{33} was 74 ± 5 pC/N and the frequency dependency (20-200 Hz) shows a flat response, with variations within the error of the measurement (Figure 4). Then, the sensor was demonstrated in a few applications where the flexibility and spatial resolution are relevant.

Multi-Point Pressure Measurement using Charge Integration

A pressure chamber with a 2 cm-wide opening was fabricated and fitted with a 1 mm-thick PDMS membrane, to mimic the swelling caused by an edema (excessive fluid accumulation in the tissues). The array was mounted on top of the membrane (Figure 5) and used for mapping the pressure distribution, which was concentrated in a corner of the device (Figure 6). The charge (Q) was integrated over time using a precision switched integrator transimpedance amplifier (IVC102, by TI), when increasing the pressure by 30 mbar.

These results are promising for applications related to medical applications where swelling/edema need to be monitored over time.

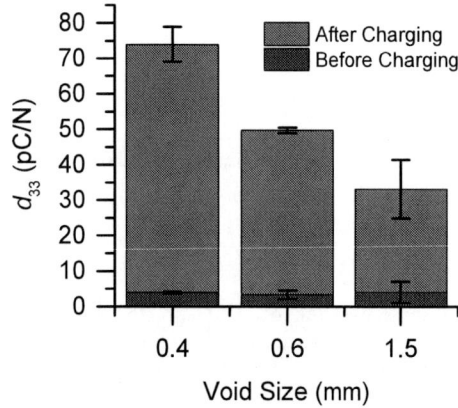

Figure 3: Piezoelectric coefficient as a function of the void size. The spacing between voids is equal to the void size (square-shaped).

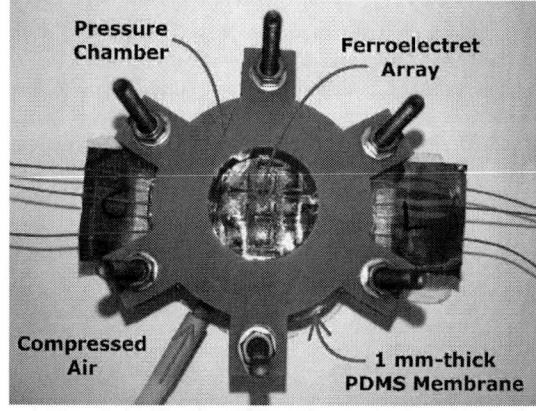

Figure 5: Pressure mapping using a ferroelectret array - pressure chamber used for the measurements.

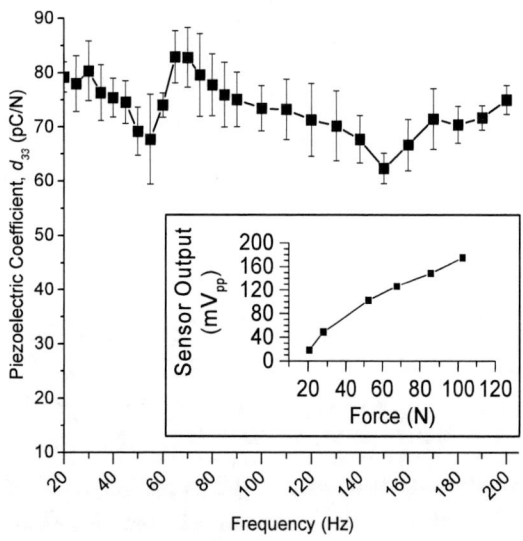

Figure 4: Frequency response of an element of the ferroelectret array, for a force of ~26 N. Inset: force response at 105 Hz.

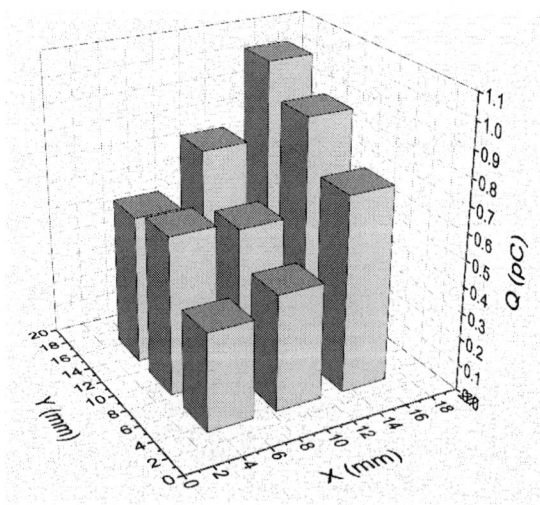

Figure 6: Charge generated by each of the array elements when applying a 30 mbar pressure (using Fig. 5 setup).

Ferroelectret Array as a Touch Screen/Keypad using Transient Voltage Measurements

To assess the possibility of using simple signal acquisition electronics (voltage reading without amplification), the signals were acquired from each of the rows and columns of the device, using a LabJack U3-HV, while pressing the different regions of the sensor (Figure 7). It is possible to identify the positive and negative current flows when pressing and releasing the array cell. By analyzing all the six voltage outputs the pressed cell can be identified. The example of pressing cell #3 for 3 times is plotted in Figure 8, where simultaneous voltage peaks with opposite polarities are visible on the corresponding row and column. The absolute polarity of the voltage peaks depends of the poling process, and can be oriented in either direction.

Figure 7: Ferroelectret array as a self-powered keypad.

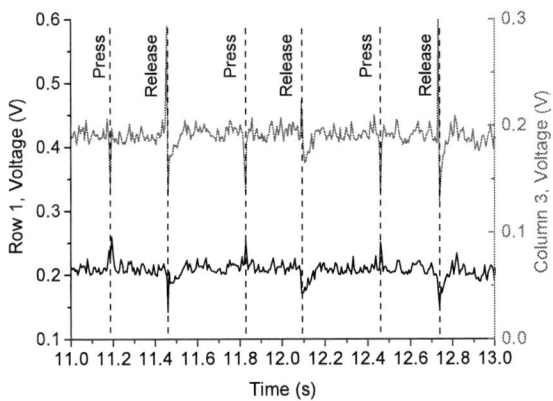

Figure 8: Voltage output when pressing number 3 for 3 times (column 3, row 1).

Ferroelectret Array as a Wearable Vibration, Bending or Pressure Sensor

To demonstrate the wearability of the flexible piezoelectret array, the device was fitted to the forearm and used to record the contraction of the wrist flexor muscles, as depicted in Figure 9. A time-domain direct voltage readout (no amplification) was also employed in this case, as in the previous sub-section. There is an immediate correlation between the movements and the signal readout, which shows the possibility of monitoring physical activity and eventually some physiological parameters such as heart or respiratory rate in wearable or medical applications.

a)

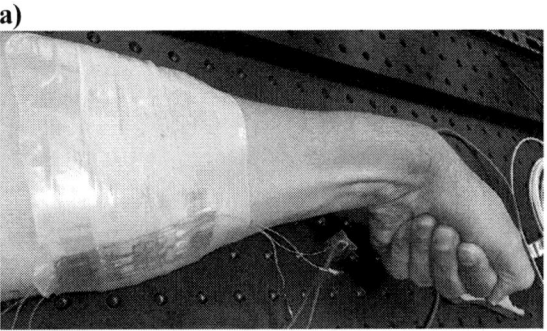

b)

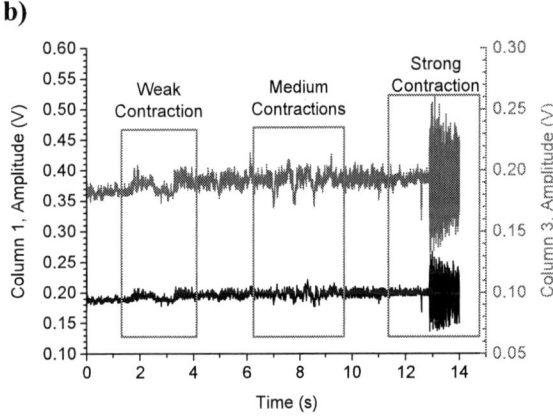

Figure 9: a) Sensor positioning in the forearm. b) Measurement of the contraction of the wrist flexors using the ferroelectret array (output voltage in the time domain).

CONCLUSIONS AND PERSPECTIVES

The manuscript discusses the design, development, and validation of a ferroelectret array suitable for wearable applications. The void geometry optimization allowed to maximize the piezoelectric coefficient of the sensor (d_{33} = 74 ± 5 pC/N). The frequency response of the developed piezoelectret shows a bandwidth in excess of 200 Hz, extending the applicability of the sensor for activity monitoring and vital sign measurements. The sensor was demonstrated as a pressure sensor, as a self-powered keypad and as wearable physical activity monitor, by measuring wrist flexor contraction. The demonstrated piezoelectret array concept will be used to realize acoustic and tactile devices. Further, we believe the demonstrated ferroelectret array concept will open new applications in robotics, sports, and in the health sector.

REFERENCES

[1] Zhong, Junwen, et al. "A flexible piezoelectret actuator/sensor patch for mechanical human–machine interfaces." ACS nano 13.6 (2019): 7107-7116.

[2] Dali, Omar Ben, et al. "Biodegradable 3D-printed ferroelectret ultrasonic transducer with large output pressure." 2021 IEEE International Ultrasonics Symposium (IUS). IEEE, 2021.

[3] González-Losada, P. et al. 2022. IEEE Transactions on Dielectrics and Electrical Insulation. 29, 3 (2022), 845–852.

[4] Zhukov, Sergey, et al. "Biodegradable cellular polylactic acid ferroelectrets with strong longitudinal and transverse piezoelectricity." Applied Physics Letters 117.11 (2020): 112901.

[5] Qiu, X., et al.: Ferroelectrets: Recent developments. IET Nanodielectr. 1–12 (2022). https://doi.org/10.1049/nde2.12036

[6] Vinayakumar, K. B., Gund, V., Lambert, N., Lodha, S., & Lal, A. (2016, October). Enhanced lithium niobate pyroelectric ionizer for chip-scale ion mobility-based gas sensing. In 2016 IEEE SENSORS (pp. 1-3). IEEE.

[7] Ni, D., Vinayakumar, K. B., Pinrod, V., & Lal, A. (2019, January). Multi Kilovolt Lithium Niobate Pyroelectric Cantilever Switched Power Supply. In 2019 IEEE 32nd International Conference on Micro Electro Mechanical Systems (MEMS) (pp. 970-973). IEEE.

[8] González-Losada, P., Alves, F., Martins, M., Mundy, S., Dias, R., & Vinayakumar, K. B. (2021, June). Pyroelectrically Rechargeable Electret for Continuous Vibration Energy Harvester. In 2021 21st International Conference on Solid-State Sensors, Actuators and Microsystems (Transducers) (pp. 928-931). IEEE.

[9] Pinto, R., González-Losada, P., Faraji, M., & Vinayakumar, K. B. (2022, January). Pyroelectrically-charged flexible piezoelectret sensors: route towards sustainable functional electronics. In 2022 IEEE 35th International Conference on Micro Electro Mechanical Systems Conference (MEMS) (pp. 547-550). IEEE.

[10] Sotgiu, E., González-Losada, P., Pinto, R. M., Yang, H., Faraji, M., & Vinayakumar, K. B. (2022). Pyroelectrically Charged Flexible Ferroelectret-Based Tactile Sensor for Surface Texture Detection. Electronics, 11(15), 2329.

[11] González-Losada, P., Yang, H., Pinto, R. M., Faraji, M., Dias, R., & Vinayakumar, K. B. (2022, July). Flexible ferroelectret for zero power wearable application. In 2022 IEEE International Conference on Flexible and Printable Sensors and Systems (FLEPS) (pp. 1-4). IEEE.

ACKNOWLEDGEMENTS

This work was supported in part by the PO Norte Agency (PT 2020: PROJETOS DE I&DT COPROMOÇÃO – Clube Fornecedores Bosch). Project under Contract POCI-01-0247-FEDER-04510. We would like to acknowledge the administrative support of Elisabete P. Fernandes.

CONTACT

* K. B. Vinayakumar, tel: +351-935 258 722; vinaya.basavarajappa@inl.int and vinayjgi@gmail.com
# Both authors contributed equally to the work

A BULK-TYPE PRESSURE SENSOR WITH FULL-BRIDGE IMPLEMENTATION ENABLED BY STRESS-MODIFYING TRENCHES

Dequan Lin[1], Man Wong[1], and Kevin Chau[1,2]

[1] The Hong Kong University of Science and Technology, Hong Kong, CHINA and
[2] CAS Engineering Laboratory for Deep Resources Equipment and Technology,
Institute of Geology and Geophysics, Chinese Academy of Sciences, Beijing, CHINA

ABSTRACT

Incorporating a half Wheatstone bridge consisting of two sensing piezoresistors and two relatively constant reference resistors to measure the biaxial compression induced by a hydrostatic pressure load, a bulk-type, dual-cavity pressure sensor exhibiting a sensitivity of 200 µV/V/MPa over a pressure range of 200 MPa has been reported. Presently described is a similar sensor further incorporating stress-modifying trenches in its cavity-sandwiched sensing plate. With all four piezoresistors being active and aligned to orientations with stress components exhibiting opposite signs, a full Wheatstone bridge can be realized. This leads to a significant 150% increase in the sensitivity of the pressure sensor to 513 µV/V/MPa over the same pressure range. Moreover, the resulting sensor can now be implemented on a more readily available (001) silicon substrate.

KEYWORDS

MEMS; pressure; sensor; piezoresistive; trench; stress-filtering.

INTRODUCTION

Realized on n-type (110) silicon (Si) and incorporating dual cavities sandwiching a sensing plate with its edges aligned to the orthogonal [001] and [$\bar{1}$10] crystalline orientations, a bulk-type sensor with a pressure range up to 200 MPa has been reported for applications such as heavy machineries, automobiles, and downhole exploration [1]. Placed around the center of the sensing plate, a pair of p-type sensing piezoresistors aligned to [$\bar{1}$10] and another pair of reference piezoresistors aligned to [001] are connected to form a half Wheatstone bridge. The reference pair is relatively insensitive to the biaxial stress field generated on the sensing plane by a hydrostatic pressure load.

Presently described is a pressure sensor realized on (001) Si and adopting a similar dual-cavity design but with trenches added on the sensing plate. Modified by the trenches, a biaxial stress field exhibiting compression and tension along orthogonal orientations is obtained. Consisting of appropriately aligned piezoresistors in such a stress field, a full Wheatstone bridge leading to higher sensitivity can be realized.

DESIGN AND SIMULATION

The relative change in the longitudinal resistivity $\Delta\rho/\rho$ of a piezoresistor subjected to a biaxial stress field is given by

$$\Delta\rho/\rho = \pi_l\sigma_l + \pi_t\sigma_t \qquad (1)$$

where $\pi_l(\sigma_l)$ and $\pi_t(\sigma_t)$ are the respective longitudinal and transverse piezoresistive coefficients (stress components). Shown in Fig. 1 is the orientation-dependence of π_l and π_t on the (001) plane of p-type Si, with respective values of 71.8×10^{-11} and -66.3×10^{-11} Pa^{-1} along the maximum-magnitude $\langle 110 \rangle$ directions.

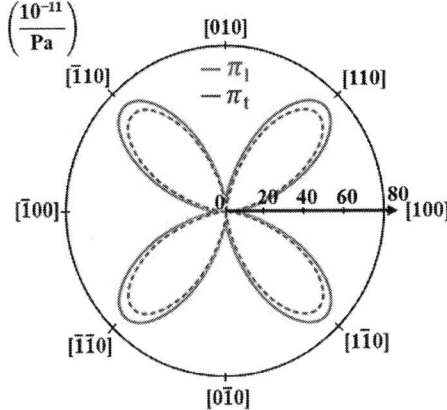

Figure 1: Polar plots of the piezoresistive coefficients π_l and π_t for p-type piezoresistors on Si (001) substrate, with the solid and dashed lines indicating respectively positive and negative values.

Since the signs of π_l and π_t are opposite but the magnitudes are roughly the same, (1) can be approximated as

$$\Delta\rho/\rho \approx \pi(\sigma_l - \sigma_t), \qquad (2)$$

where $\pi \equiv \pi_l \approx -\pi_t$. Consequently, $\Delta\rho/\rho \approx 0$ for a $\langle 110 \rangle$-oriented p-type piezoresistor on (001) Si subjected to a uniform biaxial stress field with $\sigma_l = \sigma_t$.

An important implication of (2) is that a full Wheatstone bridge leading to higher sensitivity can be constructed if the biaxial stress field can be manipulated such that piezoresistors can be deployed and aligned to experience σ_l and σ_t of opposite signs. Presently this is realized with the incorporation of stress-modifying trenches in the sensing plate.

The trench-modified stress field has been simulated using a 1/4-symmetry solid model consisting of a 400-µm-thick top cap and a 600-µm-thick bottom substrate. Top and bottom cavities with respective heights of 200 and 300 µm and similar cross-sectional areas of 400×400 µm² sandwiching a 100-µm-thick sensing plate are incorporated. Two parallel trenches stopping short of the bottom are dug from the top of the sensing plate. Between the two trenches is a 240×350 µm² sensing region for accommodating the piezoresistors.

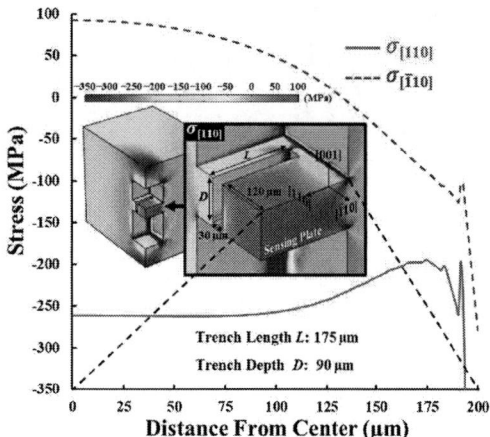

Figure 2: Plots of simulated distribution of $\sigma_{[110]}$ and $\sigma_{[\bar{1}10]}$ from the center to the edge of the sensing plane, with the sensor subjected to a 200-MPa hydrostatic pressure load. The 1/4-symmetry solid model for finite-element analysis is shown in the inset.

With the long axes of the trenches aligned along [110], the distribution of the stresses $\sigma_{[110]}$ and $\sigma_{[\bar{1}10]}$ along the respective [110] and [$\bar{1}$10] orthogonal directions on the sensing plane are shown in Fig. 2. With a magnitude 30% higher than the 200-MPa applied hydrostatic pressure due to the dual-cavity induced stress amplification [1], a compressive $\sigma_{[110]} \approx -260$ MPa is obtained at the center of the sensing plane, whereas a tensile $\sigma_{[\bar{1}10]} \approx 90$ MPa is obtained at the same location.

Shown in Fig. 3 is a squeezed sponge analogy for illustrating the origin of the tensile stress. When squeezed at the bottom, the trenches filter out the transverse compression on the top sensing plane. The corresponding relaxation results in a tensile stress along this direction on the sensing plane. In the orthogonal direction without trenches, the compression is retained. This results in a desired biaxial stress field exhibiting components of opposite signs along different orientations of [110] and [$\bar{1}$10].

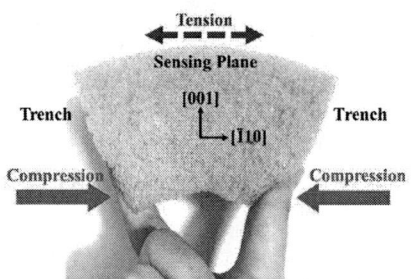

Figure 3: A squeezed sponge analogy explaining the origin of the tensile stress on the sensing plane.

Shown in Fig. 4 is the simulated dependence of $\sigma_{[110]}$ and $\sigma_{[\bar{1}10]}$ on the half-length L and the depth D of the trenches. While $\sigma_{[110]}$ changes by a small ~20 MPa when L is increased from 25 to 175 μm, $\sigma_{[\bar{1}10]}$ changes from a compressive −190 MPa to a tensile 90 MPa (Fig. 4a). Exhibited in Fig. 4b is the dependence of $\sigma_{[110]}$ and $\sigma_{[\bar{1}10]}$ on D. While $\sigma_{[110]}$ is relatively insensitive to D,

$\sigma_{[\bar{1}10]}$ changes significantly from −250 to 100 MPa when D is increased from 0 to 90 μm. The largest difference between $\sigma_{[110]}$ and $\sigma_{[\bar{1}10]}$ is obtained when $D = 90$ μm (90% of the thickness of the sensing plate) and $L = 175$ μm.

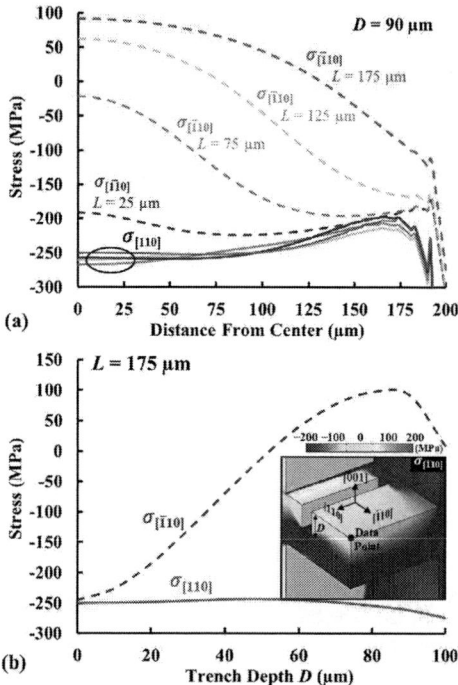

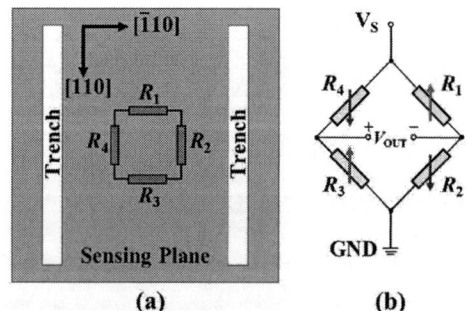

Figure 4: (a) Plots of the distribution of $\sigma_{[110]}$ and $\sigma_{[\bar{1}10]}$ from the center to the edge of the sensing plane with different trench half-length L. (b) $\sigma_{[110]}$ and $\sigma_{[\bar{1}10]}$ taken at the top center of the sensing plate plotted against the trench depth D. Solid and dashed lines indicate the respective $\sigma_{[110]}$ and $\sigma_{[\bar{1}10]}$.

End-to-end connected piezoresistors R_1 to R_4 forming a Wheatstone bridge are placed on the sensing plane. Their alignment to the trenches is shown in Fig. 5a. Summarized in Table 1 are the stress components acting on each of the piezoresistors and the calculated $\Delta\rho/\rho$ corresponding to $\sigma_{[110]} \approx -260$ MPa and $\sigma_{[\bar{1}10]} \approx 90$ MPa. A fully active Wheatstone bridge (Fig. 5b) is obtained because of the opposite signs of $\Delta\rho/\rho$ associated with the pair (R_1, R_3) and the pair (R_2, R_4).

Figure 5: (a) Layout of the piezoresistors R_1 to R_4 on a trench-sandwiched sensing plane. (b) Corresponding schematic of a full Wheatstone bridge.

Table 1: Calculated resistivity change of piezoresistors at 200-MPa applied pressure.

Piezoresistor	Longitudinal Stress	Transverse Stress	$\Delta\rho/\rho$ (%)
R_1, R_3	$\sigma_{[\bar{1}10]}$	$\sigma_{[110]}$	24.6
R_2, R_4	$\sigma_{[110]}$	$\sigma_{[\bar{1}10]}$	-23.7
$\Delta(\Delta\rho/\rho) = \Delta\rho_{1,3}/\rho - \Delta\rho_{2,4}/\rho = \textbf{48.3}$			

The layout of the device layer is shown in Fig. 6. Inside the bond ring is the 400×400-μm^2 vacuum-sealed cavity with two [110]-oriented trenches set 240-μm apart. Between the trenches is the Wheatstone bridge constructed of four 4-μm-wide and 110-μm-long end-to-end connected piezoresistors.

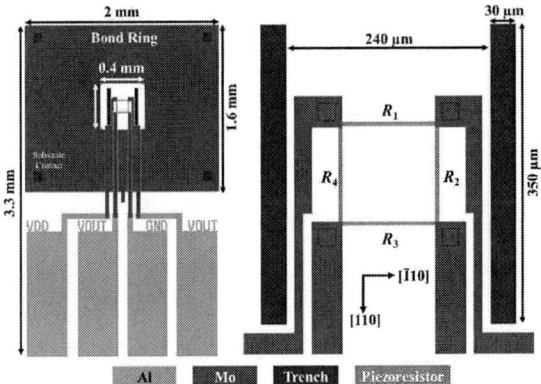

Figure 6: (Left) Layout of the sensor die. (Right) Piezoresisros R_1 to R_4 sandwiched by trenches.

DEVICE FABRICATION

The sensor was constructed using three wafers, with two 400-μm-thick n-type wafers for the cap cavity and the sensing device, and a 200-μm-thick wafer for sealing the bottom cavity. On the [001]-oriented device wafer, fabrication started with a 40-keV, 6×10^{14} cm^{-2} boron (B) implantation to form the p-type piezoresistors and a 50-keV, 3×10^{15} cm^{-2} phosphorus (P) implantation to form the n+ contacts for fixing the electrical potential of the substrate (Fig. 7a). This was followed by the deposition of a 3-μm-thick low-temperature oxide (LTO) layer as a hard mask for the 300-μm deep reactive ion etching (DRIE) to form the bottom cavity (Fig. 7b). The LTO layer was stripped before another 0.2-μm-thick LTO was deposited. The implanted impurities were activated during the densification of the LTO at 900 °C for 30 min. Contact holes were opened before a 0.1-μm molybdenum (Mo) was sputtered and patterned (Fig. 7c) to form the interconnects. An insulating 50/75-nm-thick silicon nitride/silicon oxide (SiN$_y$/SiO$_z$) stack was formed at 300 °C by plasma-enhanced chemical vapor deposition (Fig. 7d). Contact holes were opened before the sputtering and patterning of a 1.5-μm-thick aluminum (Al) layer to form the metal pads (Fig. 7e). Masked by photoresist, DRIE was used to form the 90-μm-deep stress-modifying trenches (Fig. 7f). Finally, a 1.5-μm-thick Al layer was blanket sputtered on the cavity side (Fig. 7g) for Al-germanium (Ge) eutectic bonding. Preparation of the cap wafer started with a 900-

°C, 30-min blanket P diffusion. Masked using a patterned 3-μm-thick LTO (Fig. 7h), DRIE was performed to form the 160-μm-deep cap cavity (Fig. 7i). After stripping the LTO, a 1.5-μm-thick Al layer was blanket sputtered on the unetched side and a 0.1-μm-thick Mo layer was blanket sputtered on the cavity side. This was followed by a 0.5-μm-thick Ge evaporation (Fig. 7j). The bottom wafer was prepared by the evaporation of a 0.5-μm-thick Ge on its top surface (Fig. 7k). The device wafer was attached to the bottom wafer to seal the bottom cavity and aligned to the cap wafer before they were eutectically bonded in vacuum under a clamping pressure of 1000 mbar at 480 °C for 5 min (Fig. 7l). A shallow dicing was first performed to form the Si cap and then the whole wafer was diced through to obtain an individual sensor die (Fig. 7m).

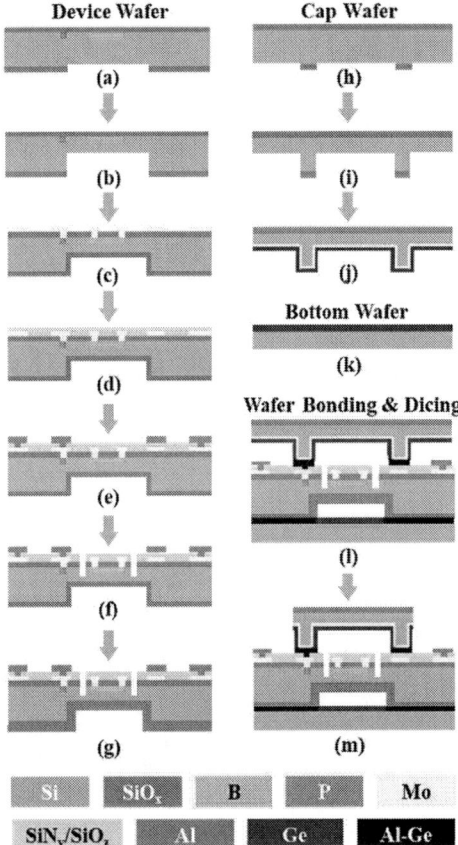

Figure 7: Schematic fabrication sequence of the pressure sensor. (a)-(g) Device wafer, (h)-(j) cap wafer, (k) bottom wafer, (l) wafer bonding, and (m) wafer dicing.

MEASUREMENT

A pressure sensor die (Fig. 8) is inspected using computed tomography (CT), revealing the internal cavities, stress-modifying trenches, and the Mo leads (Fig. 9). The sensor is characterized using a test system consisting of a thermal chamber, an HP4156 parameter analyzer and a Fluke 7615 hydraulic pressure controller.

The pressure dependence of the piezoresistance (Fig. 10) is characterized at room temperature by changing the hydrostatic pressure load P from 0 to 200 MPa. R_1 and

R_3 increase with increasing pressure and change by ~11% over the full pressure range, while R_2 and R_4 decrease by a similar magnitude of ~9% over the same pressure range. Both are less than half of the ~24% listed in Table 1. This is reasonable because the piezoresistive coefficients of a highly doped piezoresistor are roughly half of those of a lightly doped one [2]. The measured dependence on P and temperature T of the output voltage V_{OUT} of the Wheatstone bridge biased with a 5-V supply is plotted in Fig. 11. Ranging from -14 to 499 mV over 200 MPa and exhibiting a sensitivity of 513 μV/V/MPa at room temperature, this is 2.5 times the sensitivity previously reported for a dual-cavity design without trenches [1].

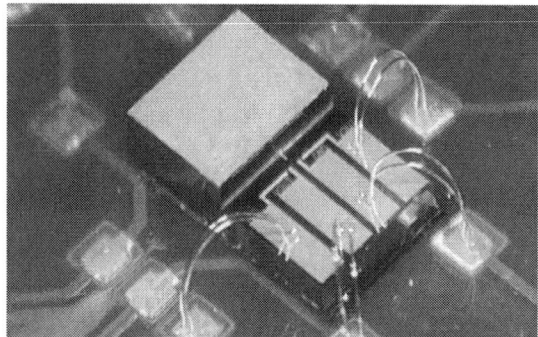

Figure 8: Photograph of a fabricated sensor die mounted on a printed circuit board and wire bonded to metal pads.

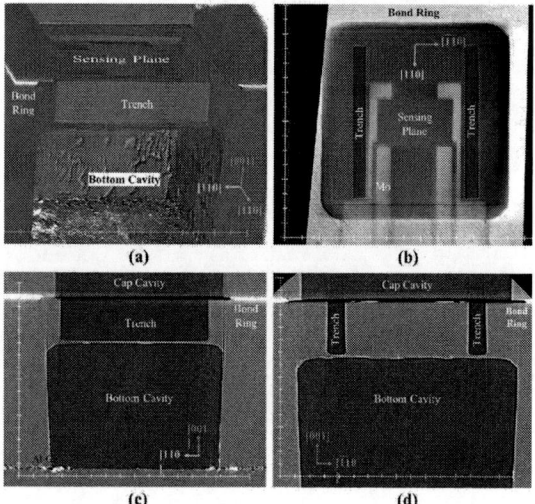

Figure 9: CT images of the sensor die. From left to right are the sectional views from the respective oblique, top, side, and front angle revealing the trenches and cavities.

DISCUSSION

A new bulk-type, dual-cavity pressure sensor with stress-modifying trenches has been realized and characterized. The dual-cavity construction, together with the trenches, modify the input hydrostatic pressure to a desired biaxial stress field exhibiting components of opposite signs along different orientations, enabling fully active Wheatstone bridge implementation and leading to a significant 150% increase in sensitivity.

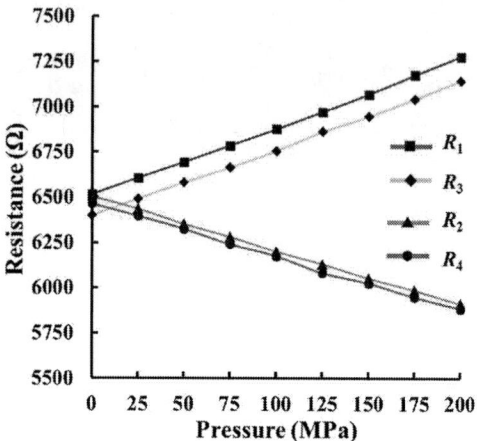

Figure 10: Measured dependence of the resistances of R_1, R_2, R_3, and R_4 on P at room temperature.

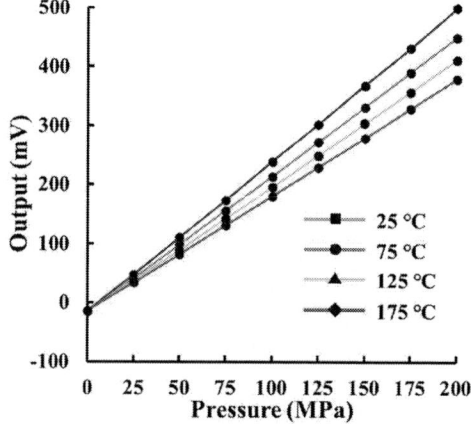

Figure 11: Measured dependence of the pressure sensor V_{OUT} on P and T at 5-V supply.

ACKNOWLEDGMENT

This work was supported by the Strategic Priority Research Program of the Chinese Academy of Sciences, Grant XDA14040202. Device fabrication was carried out at the Nanosystem Fabrication Facility of The Hong Kong University of Science and Technology. Prof. Chi Zhang at the Institute of Geology and Geophysics, Chinese Academy of Sciences, kindly provided the CT images.

REFERENCES

[1] D. Lin, K. Chau, and M. Wong, "A Bulk-Type High-Pressure MEMS Pressure Sensor with Dual-Cavity Induced Mechanical Amplification," IEEE JMEMS, vol. 31, no. 4, pp. 683-689, Aug. 2022.

[2] A. A. Barlian, W.-T. Park, J. R. Mallon, Jr., A. J. Rastegar, and B. L. Pruitt, "Review: Semiconductor Piezoresistance for Microsystems," Proc. IEEE, vol. 97, no. 3, pp. 513-552, Mar. 2009.

CONTACTS

*Dequan Lin, dlinah@connect.ust.hk
*Man Wong, eemwong@ust.hk
*Kevin Chau, zhouxl@mail.iggcas.ac.cn

A CMOS COMPATIBLE MICRO PIRANI GAUGE WITH STRUCTURE OPTIMIZATION FOR PERFORMANCE ENHANCEMENT

Rui Jiao[1], Gai Yang[2], Ruoqin Wang[1], Yue Tang[2], Zhongyi Liu[2], Huikai Xie[2,3], Hongyu Yu[1,*], and Xiaoyi Wang[2,3,*]

[1]Hong Kong University of Science and Technology, Hong Kong, China,
[2]Beijing Institute of Technology, Beijing, China, and
[3]BIT Chongqing Institute of Microelectronics and Microsystems, China.

ABSTRACT

In this paper, we reported a CMOS compatible micro-Pirani gauge using the CMOS compatible fabrication process. For the first time, we revealed that the tiny fluid conduction gap can not only enhance the sensitivity of the device but also can improve the dynamic measurement range, especially in the high-pressure region making it possible to exceed the 1 atm monitoring domain. Firstly, the CFD simulation was conducted to simulate the performance of devices with a tiny gap (T) and large gap (L) and validate the proposed discoveries. Then, a CMOS-compatible fabrication method was created with an overall dry etching method. Finally, the experimental results demonstrated the proposed idea that the T-type device could provide more than two folds higher sensitivity (4.01 μA/Torr) compared with the L-type one (1.63 μA/Torr), and a larger dynamic measurement range was achieved by the T-type gauge (0.01~760 Torr or more) compared with the L-type one (0.01~100 Torr). Besides, the device also showed high stability with a small variation value (3.8‰ @ 1.3e-4 Torr).

KEYWORDS

Pirani gauge, FEM, CMOS compatible, Performance Enhancement.

INTRODUCTION

Precise pressure monitoring is closely related to the development of automotive electronics, aerospace, and semiconductor industries [1-5]. For this matter, Pirani gauge is one of the most accurate methods to detect low and medium vacuum degrees. In addition, the Pirani gauge has a simple structure and no moving parts, so it has excellent mechanical strength and high reliability. Until recently, with the vigorous development of MEMS technology, the research of silicon-based Pirani gauge has also achieved a great breakthrough. By using mature silicon-based technology, the traditional macro pressure sensor can be transformed into micro-manufactured devices. Compared with traditional devices, miniaturized devices have advantages in size, power consumption, and cost. In addition, integrating the silicon-based Pirani gauge with the readout circuit can further reduce the size and improve the reliability of the entire system[6].

However, under the condition of high vacuum and near atmospheric pressure, the sensitivity will be greatly reduced due to the saturation effect of gas thermal conductivity [7]. To solve this problem, many miniature Pirani sensors have been reported, and different materials and sensing structures have been proposed to improve their performance. For example, Romihin [8] produced the world's first Pirani pressure sensor based on graphene, which greatly reduced the power consumption of traditional Pirani gauge; Moutaouekkil [9] designed a microbridge structure to extend the working range of the Pirani gauge from 1 kPa to atmospheric pressure or even more; Marconot [10] designed a Pirani pressure gauge array, which can measure the pressure in a wide range from atmospheric pressure to high vacuum (10e^{-3} Pa) without using a microcontroller. However, complex configuration, low sensitivity [11], and difficulty in integrating with circuits are the main disadvantages of these sensors.

In this study, we used CMOS compatible technology to manufacture two kinds of micro Pirani gauges with a tiny gap (T) and a large gap (L), as shown in Figure 1. The results of the FEM method are highly consistent with the experimental result of the actual device. The final results show that the Pirani gauge with a tiny gap structure can not only improve the sensitivity of the device to pressure detection, but also achieve high stability and a wider dynamic measurement range.

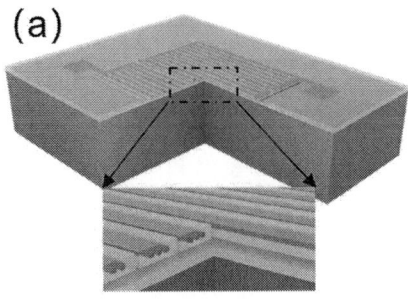

Tiny Gap Structure

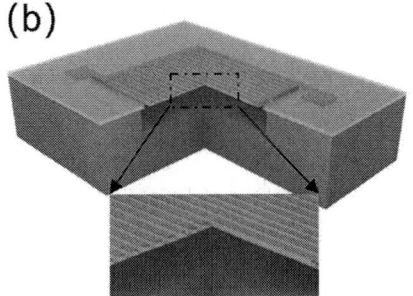

Large Gap Structure

Figure 1: Schematic of the Pirani gauge with (a)tiny gap structure and (b) large gap structure.

SIMULATION AND FABRICATION

Structure Design

As the micro Pirani gauge is based on the thermal sensing mechanism, the gap between the hot film and the substrate silicon is the significant factor that determines the performance of the devices. In this paper, two types of sensing gaps are designed for comparison, which is shown in Figure 1, one is with a tiny gap (T) and the other is with a large gap (L).

FEM Simulation

The heat producer of the miniature Pirani gauge is the suspended heater, which is affected by the ambient pressure because of the heat conduction of gas. When the power supply is injected into the suspended heater, heat is generated due to the Joule effect, which leads to the temperature rise of the device. When the system reaches thermal equilibrium, the temperature distribution will be formed on the heater. Generally, the total heat loss (Q_{total}) consists of four parts: thermal convection ($Q_{convection}$), thermal radiation ($Q_{radiation}$), solid heat conduction (Q_{solid}), and gas heat conduction (Q_{gas}). In the above heat transfer mode, Q_{gas} means that the gas under the heater conducts the heat generated by the heater, which is the only source of heat loss that depends on pressure. According to the above-mentioned mechanism, the finite element simulation was carried out, and the results showed that the T-type gauge could achieve higher sensitivity and larger dynamic range than the L-type gauge, as shown in Figure 2.

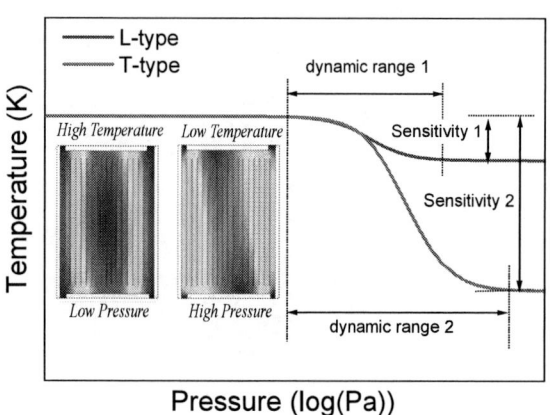

Figure 2: CFD simulation to compare the temperature change between two types of Pirani gauge versus pressure.

Fabrication Method

In this paper, polysilicon was used as the heater for both devices, and the main difference between these two structures was the cavity size. Moreover, the material and preparation process of the device was compatible with the CMOS process, making it possible to be integrated with an on-chip circuit for a monolithic design. The detailed fabrication process was introduced as follows.

Firstly, 0.5 μm silicon oxide was deposited on the wafer by LPCVD (figure3(i)). Secondly, 0.8μm polysilicon was deposited by LPCVD as a sacrificial layer, and patterned by the dry etching method(figure3(ii)). Thirdly,

1.5μm silicon oxide was deposited (figure3(iii)), followed by the deposition and patterning of 0.8μm polysilicon as sensing material (figure3(iv)). Then 1 μm silicon oxide was deposited on the top to protect the sensing structure, and contact holes were opened. Then, 2 μm of Al was sputtered and patterned for wire bonding (figure3(v)). After that, deep trenches were defined by DRIE (figure3(vi)). The last step is to release the sensing structure by XeF2 etch(figure3(vii)). The SEM image of the fabricated Pirani gauges is shown in Figure 4.

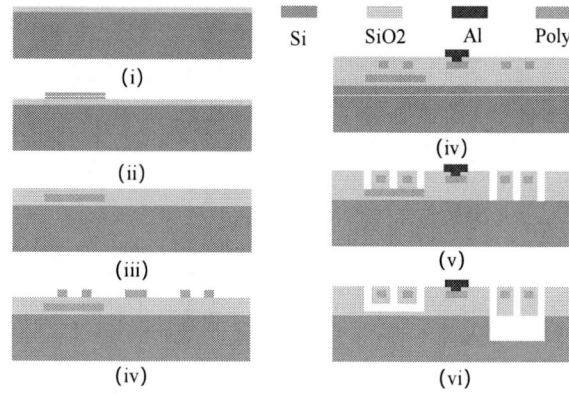

Figure 3: CMOS compatible fabrication process the micro-Pirani gauge.

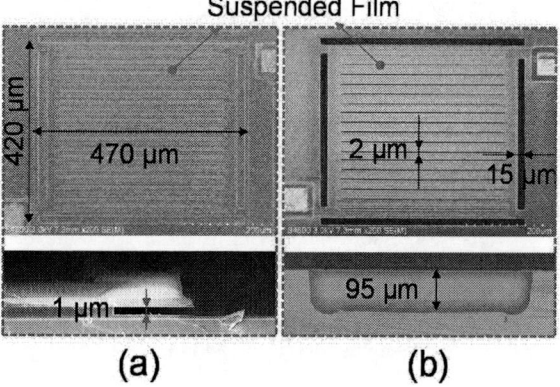

Figure 4: SEM image of the fabricated Pirani gauge with(a) tiny gap structure and (b) large gap structure using the CMOS compatible fabrication process.

EXPERIMENTAL METHOD

Test Setup Introduction

The experimental system is shown in Figure 5. The vacuum pump and pressure controller can accurately control the vacuum degree of the vacuum chamber, and the high-performance pressure gauge can accurately calibrate the vacuum degree of the chamber to ensure the accuracy of the test. Accurate output current information can be measured and recorded by high-performance current meter and computer. The vacuum test equipment and current meter are respectively Lake Shore TTPX probe station and NI9203 Current Meter.

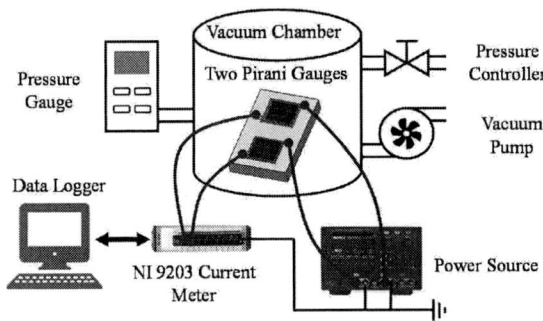

Figure 5: Test setup of the micro-Pirani gauge.

Initial Characterization

Before the vacuum degree test, the TCR of two types of Pirani gauges should be characterized to ensure that the TCR of the two periods was basically semblable, thereby eliminating interference with the subsequent performance analysis. There was a brief introduction to the TCR test process: stick the thermocouple on the back of the PCB with the sensor, put it into the oven, and record the temperature of the thermocouple and the resistance change of the sensor. The results were shown in figure 6(a). The results showed that the slopes of the two straight lines were identical, which indicates that the TCR of the Pirani gauge with two structures was similar. The slight difference in resistance value should be due to the difference in the inherent volume of the two devices, which did not influence the performance analysis of the devices.

RESULTS AND DISCUSSION

The proposed testing device characterized the sensing performance of the Pirani gauge. Figure6 (b) showed the test results of two types of Pirani gauges. The results showed that the pressure sensitivity of the T-type is higher (4.01 μA/Torr), which was almost twice compared with the L-type (1.63 μA/Torr). Moreover, the T-type would not only markedly improve the sensitivity but also enhance the dynamic range increasing from (0.01~100 Torr) to (0.01~760 Torr or more), which further expanded its application conditions of the Pirani gauge.

As shown in figure6(c), a greater bias voltage could achieve higher sensitivity needing more power consumption, and the dynamic detection range would shift towards high air pressure. Therefore, an appropriate bias voltage should be selected to balance power consumption and gauge performance in different application conditions.

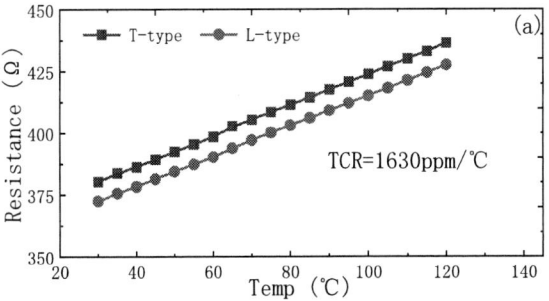

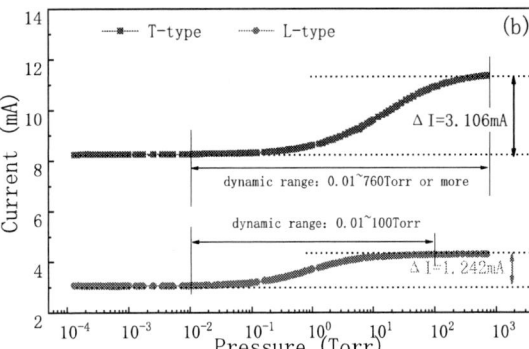

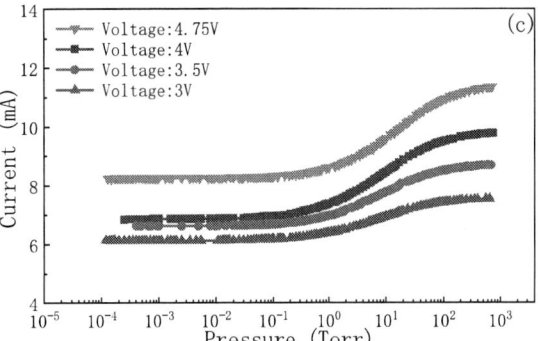

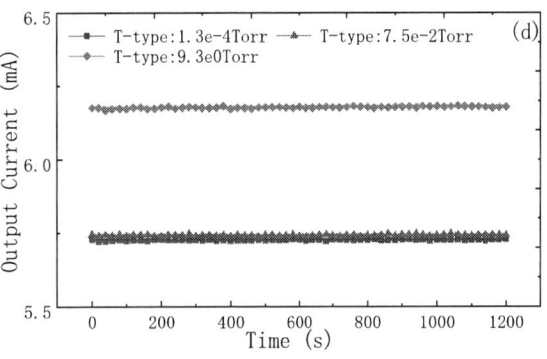

Figure 6: (a) TCR characterization of the sensing material; (b) performance comparison between these two types of structures; (c) performance characterization under different supply voltages; (d) stability test of the sensor with three pressure values.

In addition, stability is an important index to judge the performance of the gauge. As shown in figure6(d), the gauge kept working for 20 minutes, and its output showed good stability and the potential to detect air pressure continuously.

CONCLUSION

In this paper, two kinds of miniature Pirani gauges with different structures are manufactured by CMOS compatible manufacturing process. For the first time, we revealed that the fluid conduction gap can not only enhance the sensitivity of the device but also can improve the dynamic measurement range, especially in the high-pressure region making it possible to exceed the 1atm monitoring domain. Firstly, the performances of two types of Pirani gauges (L-type and T-type) are simulated by the FEM method. The results show that the T-type can not only improve the sensitivity of the device but also increase the dynamic measurement range of the gauge. Then, a CMOS-compatible fabrication method was created with an overall dry etching method. Finally, the experimental results proved the proposed idea. Compared with the L-type gauge (1.63 µA/Torr), the T-type gauge can provide more than twice the sensitivity (4.01 µA/Torr). Additionally, the T-type gauge can also achieve a larger dynamic measurement range (0.01~760 Torr) compared with the L-type gauge (0.01~100 Torr).

In addition, the device has excellent stability (3.8‰ @ 1.3e-4Torr). In this study, a Pirani gauge with high sensitivity, wide dynamic range, and good stability was manufactured, which provided a new idea for the improvement of the Pirani gauge, and proved the rationality of this idea, demonstrating its promising application in the field of accurate air pressure detection.

ACKNOWLEDGEMENTS

Rui Jiao and Gai Yang contributed equally to this paper. The authors would like to acknowledge the help of the staff from the Nanosystem Fabrication Facility (NFF) and Material Characterization and Preparation Facility (MCPF) of HKUST for the device fabrication. The work was supported in part by the Beijing Institute of Technology Research Fund Program for Young Scholars (XSQD-202206004), in part by the Natural Science Foundation of Chongqing (2022NSCQ-MSX5423), and in part by Shenzhen-Hong Kong-Macau S&T Program (Category C) Grant No. SGDX20210823103200004.

REFERENCES

[1] Mo, Jiarui, et al, "Surface-micromachined silicon carbide Pirani gauges for harsh environments." IEEE Sensors journal 21.2 (2020): 1350-1358.

[2] Garg, Manu, et al. "Stress engineered SU-8 dielectric-microbridge based polymer MEMS Pirani gauge for broad range hermetic characterization." Journal of Micromechanics and Microengineering (2022).

[3] Zhang, Guohe, et al. "Study of cavity effect in micro-Pirani gauge chamber with improved sensitivity for high vacuum regime." AIP Advances 8.5 (2018): 055131.

[4] Bosseboeuf, Alain, et al. "Calibration of micro Pirani vacuum gauges for internal pressure measurement in miniaturized vacuum chambers." 16th European Vacuum Confernce. 2021.

[5] Toto, Sofia, et al. "Correction: Design and Simulation of a Wireless SAW–Pirani Sensor with Extended Range and Sensitivity." Sensors 19.14 (2019): 3243.

[6] Xu, Wei, et al. "A Wafer-Level Packaged CMOS MEMS Pirani Vacuum Gauge." IEEE Transactions on Electron Devices 68.10 (2021): 5155-5161.

[7] Zhang, Ming, and Nicolas Llaser, "Exploiting a micro pirani gauge for beyond atmospheric pressure measurement." IEEE Transactions on Circuits and Systems II: Express Briefs 65.10 (2018): 1450-1454.

[8] Romijn, Joost, et al, "A miniaturized low power pirani pressure sensor based on suspended graphene." 2018 IEEE 13th Annual International Conference on Nano/Micro Engineered and Molecular Systems (NEMS). IEEE, 2018.

[9] Moutaouekkil, M., et al, "Elaboration of a novel design Pirani pressure sensor for high dynamic range operation and fast response time." Procedia engineering 120 (2015): 225-228.

[10] Marconot, Olivier, et al, "Dimensionless analysis of micro pirani gauges for broad pressure sensing range." IEEE Sensors Journal 20.17 (2020): 9937-9946.

[11] Piotto, Massimo, Simone Del Cesta, and Paolo Bruschi, "A CMOS compatible micro-Pirani vacuum sensor based on mutual heat transfer with 5-decade operating range and 0.3 Pa detection limit." Sensors and Actuators A: Physical 263 (2017): 718-726.

CONTACT

*Xiaoyi Wang, tel: +86-13611325054
E-mail: xiaoyiwang@bit.edu.cn

*Hongyu Yu, tel: +852-3469-2754
E-mail: hongyuyu@ust.hk

A THERMAL AIRFLOW SENSOR BASED ON MN-CO-NI-O THIN FILM

Jie Wang, Yunfei Liu, Zhezheng Zhu, Chengchen Gao, Zhenchuan Yang and Yilong Hao

School of Integrated Circuits, Peking University, Beijing, CHINA

ABSTRACT

A thermoresistive calorimetric airflow sensor with high sensitivity based on Mn-Co-Ni-O (MCN) thin film prepared by magnetron sputtering is reported. The microbridge structure of MCN thin film with a size of $900\mu m \times 20\mu m$ is fabricated by using bulk micromachining technology. The temperature coefficient of resistance (TCR) of the MCN thin film is approximately -34871 ppm/K at 293K and -21060 ppm/K at 373K. The fabricated sensor achieves an ultra-high sensitivity of 28.115 mV/(m/s)/mW with respect to input heating power and amplification factor, which is 10-fold better than the sensor with platinum (Pt) thin film. The proposed sensor demonstrates that the MCN thin film has promising applications in high-sensitivity thermal flow sensors.

KEYWORDS

Thermal airflow sensor, Mn-Co-Ni-O thin film, high temperature coefficient of resistance.

INTRODUCTION

MEMS thermoresistive flow sensors are widely used in medical equipment and air conditioning due to their low cost, small size, and stability [1]. However, fabricating such sensors with high sensitivity and accuracy is still challenging. The sensitivity and accuracy of the sensors mainly depend on the TCR of thermistors since the thermistors detect changes in the temperature field due to fluid flow. Traditionally, metal (e.g., aluminum [2], platinum [3], molybdenum [4]) and polysilicon [5] are employed as sensing elements due to their reliability and mature fabrication process. However, the relatively low TCR of these materials limits the sensitivity of such sensors. Therefore, a novel material with high TCR is required. A thermal flow sensor with 3C-SiC was proposed in [6]. The TCR of 3C-SiC is approximately -20716 ppm/K at 298K and -9367 ppm/K at 443K. Although the sensitivity of the sensor is improved, the fabrication process becomes complicated due to the requirement of chemical bonding between SiC and glass substrate. Moreover, the sensitivity and response time of the sensor is limited due to the difficulty of fabricating suspended SiC thermal structures on the glass substrate.

Recently, MCN thin film prepared by magnetron sputtering has shown excellent application prospects in miniature temperature sensors and uncooled bolometers due to their high TCR, excellent electrical stability, and wide operating temperature range [7, 8]. However, MEMS thermal flow sensors with MCN thin films are less reported due to the difficult micromachining process of such materials. In this paper, a thermal airflow sensor with the suspended microbridge structure based on MCN thin film is successfully fabricated by using bulk micromachining technology. The prominent normalized sensitivity of the sensor verifies that the MCN thin film has a broad application prospect in thermal flow sensors.

DESIGN

A schematic view of the proposed thermal airflow sensor based on MCN thin film is shown in Fig.1. The sensor consists of one suspended heater with Pt and two suspended thermistors with MCN thin film located symmetrically regarding the heater, where the heater has a size of $900\mu m \times 10\mu m$ (Length × Width). Each thermistor has a size of $900\mu m \times 20\mu m$. The Pt heater generates a temperature field by applying a constant voltage. The two thermistors with MCN thin film detect the upstream temperature T_u and the downstream temperature T_d, respectively. The temperature difference between the upstream and downstream thermistors increases with the flow velocity. In order to reduce the heat loss between the heater and substrate while improving the heat exchange between the heater and thermistors, the depth of the silicon cavity under heaters and thermistors is $120\mu m$. The output voltage of the sensor is acquired from the Wheatstone bridge formed by two thermistors with MCN thin film and two fixed resistors.

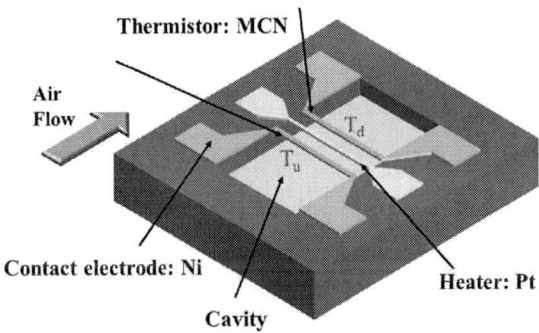

Figure 1: Schematic of the thermal airflow sensor based on MCN thin film.

FABRICATION

The fabrication of the thermal airflow sensor is based on bulk micromachining technology, and the key fabrication steps are shown in Fig. 2.

(a) Thermal oxidation of silicon oxide (SiO_2) is carried out on a silicon substrate as a buffer layer.

(b) A silicon nitride (Si_3N_4) layer is deposited as a supporting layer of the microbridge structure by low pressure chemical vapor deposition (LPCVD).

(c) The MCN layer is prepared by magnetron sputtering as thermal sensitive material.

(d) A chromium (Cr) layer as an adhesion layer and a Pt layer as heater material are sputtered and patterned by the lift-off process.

(e) A nickel (Ni) layer is sputtered and patterned to form the contact electrodes of the MCN thin film.

(f) MCN layer is etched by ion beam etching, and SiO_2/Si_3N_4 layers are etched by reactive ion etching (RIE) to define the sensor structure size.

(g) Silicon is etched by deep reactive-ion etching (DRIE) to fabricate a cavity that depth is approximately 90μm.

(h) The microbridge structure of heating and sensing elements is released by isotropic dry etching of silicon. The depth of the cavity is approximately 120μm.

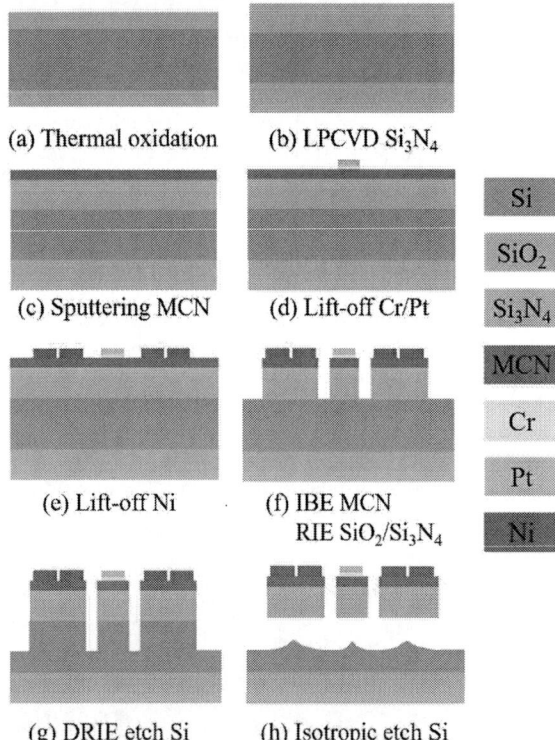

Figure 2: The key fabrication process of thermal airflow sensor based on MCN thin film.

The optical and the scanning electron microscope (SEM) images of the fabricated sensor are shown in Fig. 3. As shown in Fig. 3 (a), two rectangular electrodes with a certain distance are used as the contact electrode structure of the MCN thin film in order to reduce the readout resistance of the MCN thermistors. Fig. 3 (b) shows the overall view of the fabricated sensor. The etched cavity size is 1500μm × 900μm × 120μm (Length × Width × Depth). Fig. 3 (c) shows the close-up view of the fabricated microbridge structure of the Pt heater and MCN thermistors.

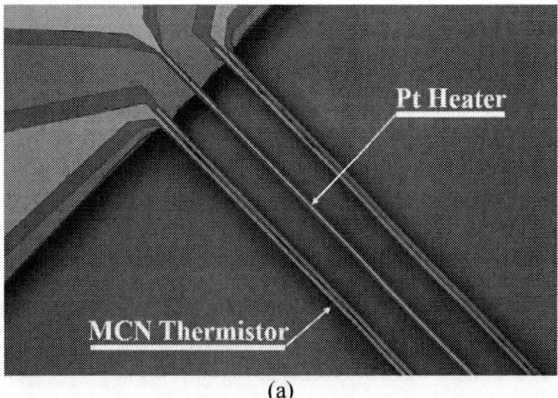

(a)

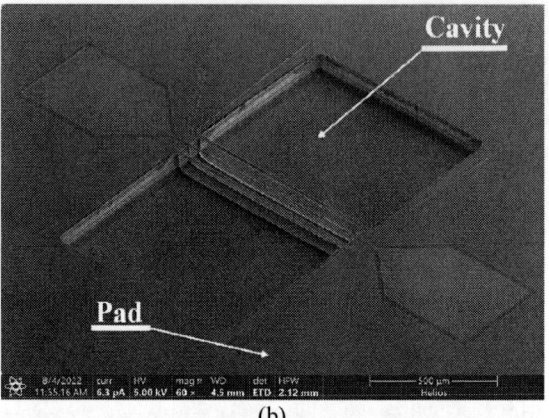

(b)

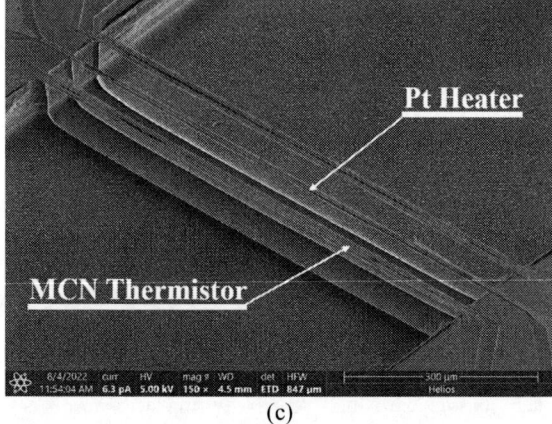

(c)

Figure 3: The images of the fabricated thermal airflow sensor based on MCN thin film. (a) optical microscope image of the heater and thermistors and (b)-(c) SEM images of the airflow sensor.

EXPERIMENT AND RESULT
Experimental Setup

The experiment setup and the packaged sensor are shown in Fig. 4. The fabricated sensor is attached on a PCB board and packaged in a PMMA flow channel with a size of 80mm × 10mm × 2mm (Length × Width × Height). Airflow is delivered to the sensor by a pneumatic pump and the pipe. The flow rate is evaluated by a reference volumetric flowmeter with a measuring range from 0.1~1.5L/min (8.3~125cm/s).

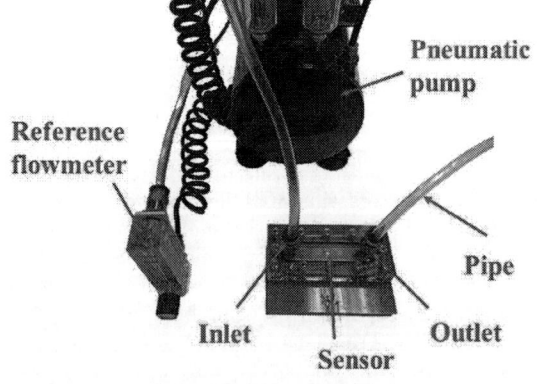

Figure 4: Experiment setup and the packaged sensor.

TCR Measurements

MCN has a negative temperature coefficient, which can be explained by the small polaron hopping theory [9]. The fabricated sensor was placed in an oven to evaluate the relationship between the resistance of the MCN thermistor and temperature. As shown in Fig. 5, the resistance of the MCN thermistor varies significantly over a temperature range from 293K to 393K. The TCR is calculated to range from -34871 ppm/K at 293K and -21060 ppm/K at 373K.

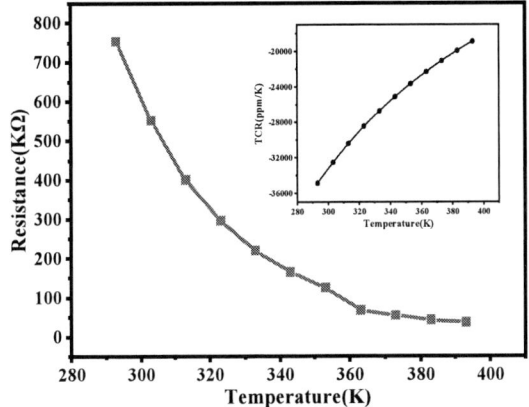

Figure 5: The relationship between the resistance of the MCN thermistor and temperature. The inset shows the calculated TCR value.

Sensitivity Measurements

The measured output voltage of the sensor versus the applied airflow is shown in Fig. 6. During the measurement, the sensor works in constant voltage mode, the supply voltage of the Pt heater is 2V, and the amplification factor of the circuit is 5. The output voltage of the sensor has a good linear relationship with flow velocity within the flow range of 8.3~125cm/s, and the sensor exhibits a high sensitivity of 1.76V/(m/s). Moreover, considering the input heating power and amplification factor, the sensor shows an ultra-high normalized sensitivity of 28.115mV/(m/s)/mW, which is 10-fold higher than the reported MEMS thermoresistive flow sensors, as shown in Table 1.

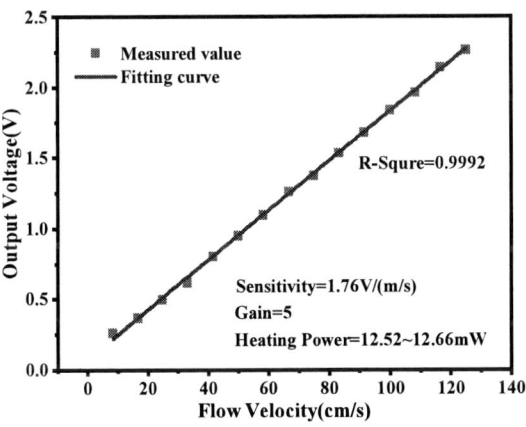

Figure 6: Measured output voltage as a function of airflow velocity of thermal airflow sensor based on MCN thin film (amplification factor of the circuit is 5).

Table 1: Comparison between the reported thermoresistive calorimetric flow sensor and our work.

Ref.	Thermal sensitive material	TCR of thermistor (ppm/K)	Sensitivity* mV/(m/s)/ mW
[2]	Al	2300	0.039
[3]	Pt	N/A	2.590
[4]	Mo	N/A	0.112
[5]	Polysilicon	903	0.228
Our Work	MCN	-34871@293K -21060@373K	28.115

Sensitivity normalized with respect to input heating power and amplification factor.

Noise Measurements

The inherent electronic noise of the thermal airflow sensor is mainly composed of the thermal noise of thermistors and heaters, flick noise, and natural convection caused by the overheated heater [10]. The ratio of noise voltage to sensitivity determines the resolution of the sensor.

In order to measure the inherent noise of the fabricated sensor accurately, we wrap the packaged sensor with a sponge and place the sensor on the vibration isolation pad to eliminate external flow perturbation and vibration interference, respectively. Moreover, the measured noise voltage of the sensor is amplified five times by the circuit. Fig. 7 shows a time diagram of the sensor output voltage measured over intervals of 6s at zero flow conditions. The calculated standard deviation (σ) of the noise voltage is 0.259mV. Considering the 95% of the population and the sensitivity of the sensor, the Minimum Detectable Flow Velocity (MDFV) can be obtained as [11]:

$$MDFV = 2\sigma/S \qquad (1)$$

The MDFV of the fabricated sensor at zero flow velocity is calculated as 0.294mm/s, which demonstrates that the sensor exhibits a high resolution.

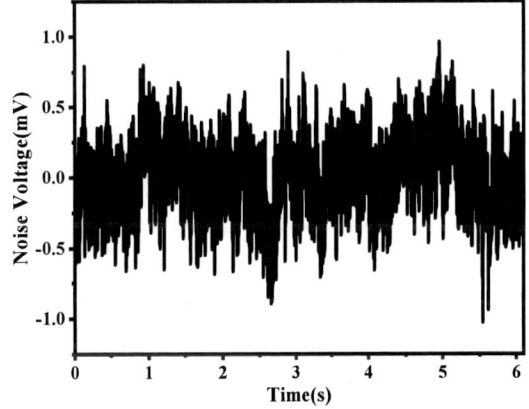

Figure 7: Output noise voltage of the sensor as time at zero flow velocity.

CONCLUSIONS

In this paper, a thermal airflow sensor based on MCN thin film has been designed and fabricated successfully. The suspended microbridge structures of MCN thermistors with a size of 900μm × 20μm have been successfully fabricated by using bulk micromachining technology. Benefiting from the high TCR of the MCN thin film, the fabricated sensor achieved an ultra-high normalized sensitivity of 28.115mV/(m/s)/mW within an airflow range of 8.3~125cm/s, which is 10-fold better than the sensor with Pt thin film. Moreover, the minimum detectable flow velocity of the sensor at zero flow velocity is 0.294mm/s. The proposed sensor verifies that the MCN thin film prepared by magnetron sputtering has promising applications in high-sensitivity thermal flow sensors.

REFERENCES

[1] F. Ejeian, S. Azadi, A. Razmjou, Y. Orooji, A. Kottapalli, M. E. Warkiani, M. Asadnia, "Design and applications of MEMS flow sensors: A review", *Sens. Actuator A-Phys.*, vol. 295, pp. 483-502, 2019.

[2] Z. Dong, J. Chen, Y. Qin, M. Qin, Q. A. Huang, "Fabrication of a micromachined two-dimensional wind sensor by Au–Au wafer bonding technology", *J. Microelectromech. Syst.*, vol. 21, pp. 467-475, 2012.

[3] D. Xue, W. Zhou, Z. Ni, et al., "A Front-Side Micro-Fabricated Tiny-Size Thermoresistive Gas Flow Sensor with Low Cost, High Sensitivity, and Quick Response", in *IEEE Transducers 2019*, Berlin, June 23-27, 2019, pp. 1945-1948.

[4] W. Xu, B. Gao, M. Ahmed, et al., "A wafer-level encapsulated CMOS MEMS thermoresistive calorimetric flow sensor with integrated packaging design", in *IEEE MEMS 2017*, Las Vegas, January 22-26, 2017, pp. 989-992.

[5] W. Xu, X. Wang, R. Wang, Izhar, J. Xu, Y. K. Lee, "CMOS MEMS thermal flow sensor with enhanced sensitivity for heating, ventilation, and air conditioning application", *IEEE Trans. Ind. Electron.*, vol. 68, pp. 4468-4476, 2020.

[6] V. Balakrishnan, T. Dinh, H. P. Phan, D. V. Dao, N. T. Nguyen, "Highly sensitive 3C-SiC on glass based thermal flow sensor realized using MEMS technology", *Sens. Actuator A-Phys.*, vol. 279, pp. 293-305, 2018.

[7] W. Zhou, X. F. Xu, C. Ouyang, J. Wu, Y. Q. Gao, Z. Huang, "Annealing effect on the structural, electrical and 1/f noise properties of Mn–Co–Ni–O thin films", *J. Mater. Sci. Mater. Electron.*, vol. 25, pp. 1959-1964, 2014.

[8] C. Wu, W. Zhou, Y. Yin, W. Ma, L. Jiang, Z. Huang, J. Chu, "Long wavelength infrared detection based on Mn-Co-Ni-O thin films with dielectric-metal-dielectric absorptive structures", *Infrared Phys. Technol.*, vol. 102, pp. 102987, 2019.

[9] R. Schmidt, A. Basu, A. W. Brinkman, "Small polaron hopping in spinel manganates", *Phys. Rev. B.*, vol. 72, pp. 115101, 2005.

[10] W. Xu, X. Wang, X. Zhao, et al, "An integrated CMOS MEMS gas flow sensor with detection limit towards micrometer per second", in *IEEE MEMS 2020*, Vancouver, January 18-22, 2020, pp. 200-203.

[11] S. Issa, W. Lang, "Minimum detectable air velocity by thermal flow sensors", *Sensors*, vol. 13, pp. 10944-10953, 2013.

CONTACT

*Chengchen Gao, tel: +86-10-62766598; gaocc@pku.edu.cn

HIGHLY SENSITIVE WAVE HEIGHT SENSOR WITH MEMS PIEZORESISTIVE CANTILEVER AND WATERPROOF MEMBRANE

Takuto Hirayama[1], Hidetoshi Takahashi
[1]Keio University, JAPAN

ABSTRACT

This paper reports a wave height sensor using a MEMS piezoresistive cantilever and a waterproof membrane. Conventional methods of wave height monitoring using GPS or accelerometers have problems of high-power consumption and integration errors. Therefore, measurement methods with simple principles and without integrals have been required to realize certain accuracy and lower power consumption. The proposed wave height sensor measures wave height from changes in atmospheric pressure because of its simple structure, which combines a chamber, a MEMS piezoresistive cantilever as a sensor element and a waterproof membrane. Here, we have theoretically and experimentally clarified the frequency response due to air leakage from the sensor element and the waterproof membrane. Actually, the height change could be measured with high sensitivity by the fabricated sensor. It is confirmed that a highly sensitive wave height sensor can be realized by adjusting the design parameters of the sensor.

KEYWORDS

Wave height sensor, piezoresistive cantilever, frequency response.

INTRODUCTION

It is essential to monitor wave heights at sea from the perspective of weather forecasting and the safe navigation of ships, so research has been conducted on robust, low-power and accurate wave height measurement. Currently, the main research into measuring local wave heights offshore involves using GPS (Global Positioning System) and accelerometers on buoys [1,2]. However, the GPS method has the disadvantages of high-power consumption and low temporal and spatial resolution because it measures the transmission time of reflected waves on the sea surface when microwaves are emitted from a satellite. The method using accelerometers is difficult to detect changes in wave height of several tens of centimeters at low frequencies, and the accelerometer data must be compensated for the rotational motion caused by a buoy by gyro-sensors. In addition, both methods have the disadvantage that, in principle, integration is required to measure changes in altitude, resulting in integration errors. Therefore, a wave height measurement based on a simple principle has been sought to realize low power consumption and certain accuracy.

Here, we propose a wave height sensor using a MEMS piezoresistive cantilever [3,4]and a waterproof membrane to measure pressure changes, as shown in Fig.1. The sensor has a cavity chamber and air leaks from the cantilever [5] and membrane. It has been suggested that the volume of the chamber and air leakage causes the sensor to have a unique frequency response. The wave height varies at a low frequency of response. The wave height varies at a low

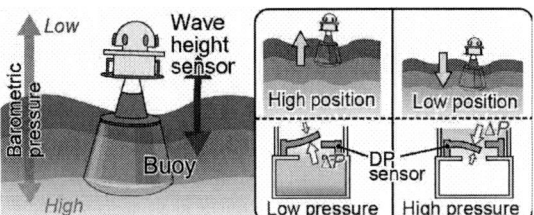

Figure 1 Concept sketch of the proposed wave height sensor.

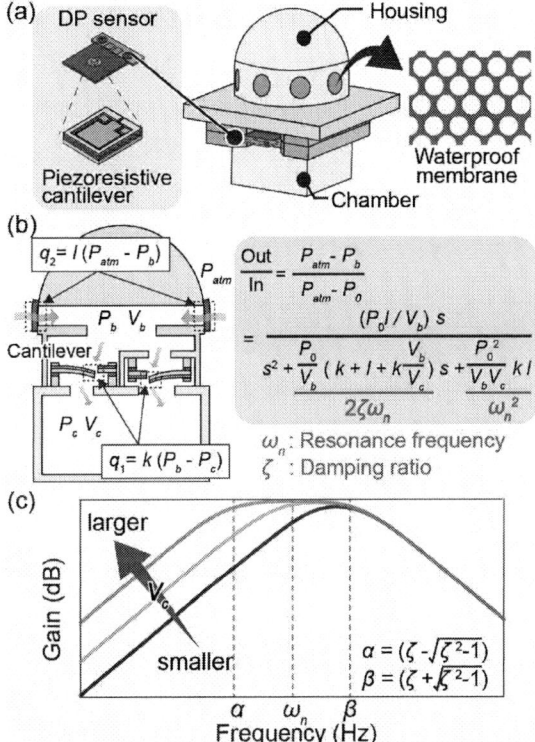

Figure 2 (a) Design, (b) principle, and (c) frequency response of the wave height sensor.

frequency of approximately 0.1~1 Hz, so it is necessary to design the sensor to meet the required specifications.

In this study, we clarified the sensor frequency response affected by various parameters theoretically and experimentally. We also demonstrated wave height change monitoring using the developed sensor.

DESIGN AND PRINCIPLE

Sensor design

The proposed sensor consists of a housing, a piezoresistive cantilever, and a cavity chamber as shown in Fig. 2 (a). The piezoresistive cantilever is sensitive enough to measure changes in air pressure. Two front and back reversed piezoresistive cantilevers are sandwiched between

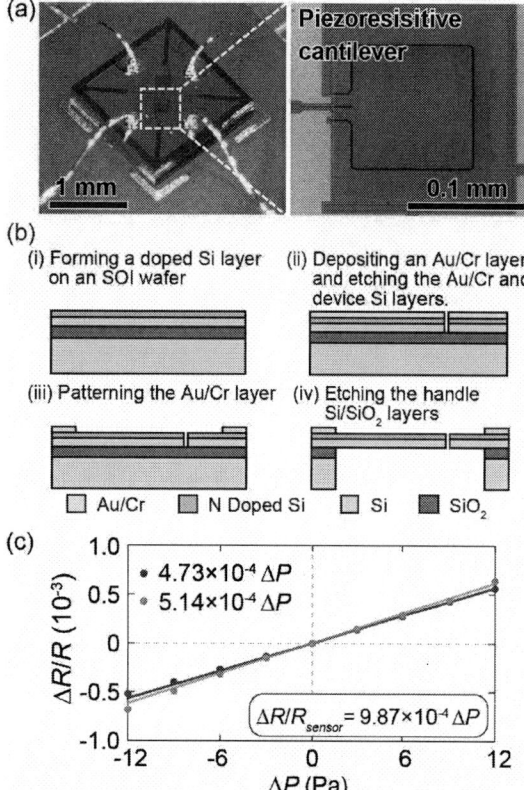

Figure 3 (a) Photograph and (b) fabrication process of the MEMS piezoresistive cantilever. (c) Response of the differential pressure sensor.

the housings. Micro-mesh filter membranes are attached to the air hole on the housing surface so that the sensor works well by splashing seawater. The housing is hemispherical to reduce vertical airflow dynamic pressure.

Sensing principle

When air pressure changes due to wave height, air passes through the micro-mesh filter so that the differential pressure on the piezoresistive cantilever changes. The piezoresistive cantilever has the characteristics of a high pass filter due to the small amount of air leakage [3]. This characteristic allows for the elimination of weather-induced pressure changes with a high differential pressure resolution. The waterproof membrane also limits air permeation slightly. Thus, air flows according to the respective pressure differences as shown in Fig. 2(b). The frequency response of the whole sensor system is derived as a transfer function using Laplace operator s. As shown in Fig. 2(c), the response uniquely has a low-pass filter component depending on the chamber volume from the air leakage of the cantilever and a high-pass filter component from the properties of the waterproof membrane. The cutoff frequencies are calculated from the co-evolution frequency ω and attenuation ratio ζ of the second-order delay system, which are determined from the chamber volume and the air transmission characteristics of the cantilever and waterproof membrane. Thus, it is necessary to design each parameter to adjust the frequency response suitably for the target wave motion.

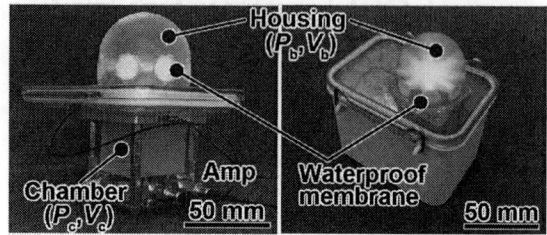

Figure 4 Photographs of the developed wave height sensor.

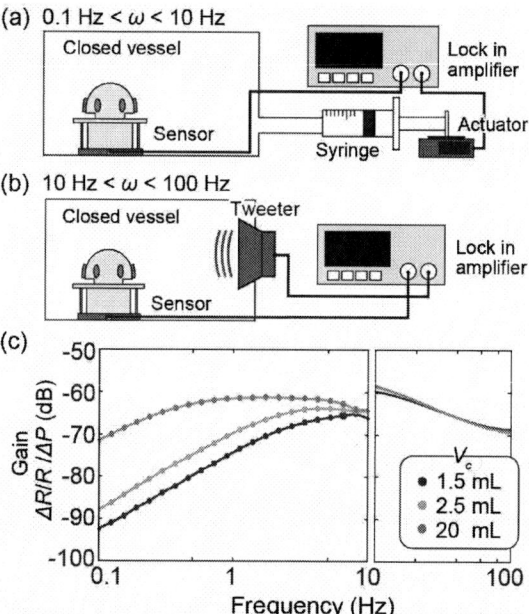

Figure 5 Experimental diagram for frequencies of (a) 0.1 Hz < ω < 10 Hz and (b) 10 Hz < ω < 100 Hz. (c) Frequency responses of the wave height sensor with a chamber of different volumes.

FABRICATION

Fig. 3(a,b) shows the photograph and fabrication process of the piezoresistive cantilever [4]. The piezoresistive cantilever is fabricated from a silicon-on-insulator (SOI) wafer. The detailed fabrication process is described in previous research. As shown in the figure, the cantilever is located on the center of the sensor element, with sizes of 1.5 mm × 1.5 mm × 0.25 mm. The cantilever size is 100 μm × 100 μm × 0.2 μm. The initial resistance of the cantilevers is approximately 2 kΩ.

The figure shows the relationship between the applied differential pressure and the fractional resistance change for the two cantilevers. The calibration results show that the cantilever responded linearly to the differential pressure and achieved a resolution of less than 0.1 Pa standalone. The fabricated wave height sensor is shown in Fig. 4. The housing and chamber were modeled using a 3D printer (Form3, Formlabs). A micromesh filter (S-NTF8031J, Nitto Denko Corporation) was attached to the air hole of the housing using adhesive as a waterproofing function. The sensor and amplifier were integrated. The sensor was designed to be combined into a waterproof box to prevent the sensor from getting wet as much as possible

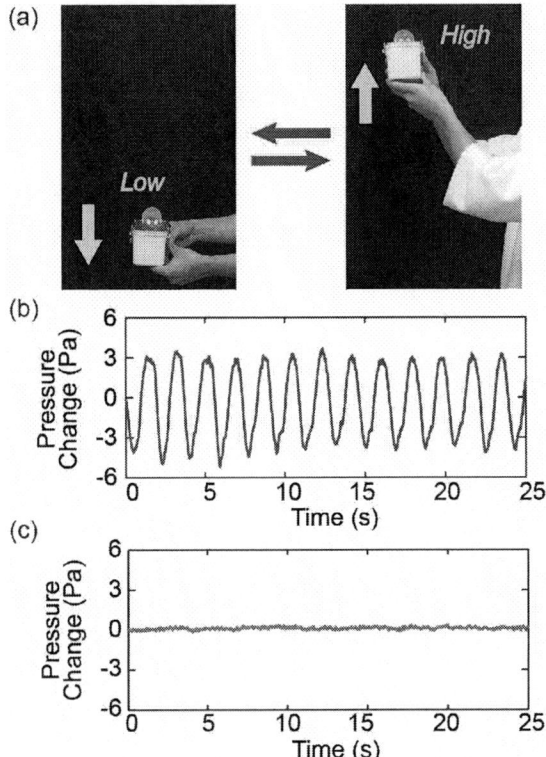

Figure 6 (a) Photographs of the demonstration. Pressure response (b) during the demonstration and (c) when not moving on the ground.

when installed on the buoy. Also, different volume chambers can be attached to the sensor.

EXPERIMENT AND RESULT
Frequency response experiment
Experiments were carried out with chamber volumes of 1.5, 2.5 and 20 mL to confirm the frequency response of the sensor as described in the principle. In order to measure a wide range of frequency responses, the syringe was moved by an actuator and measured with a lock-in amplifier at 0.1-10 Hz, as shown in the figure; at 10-100 Hz, the set-up used a tweeter instead of a syringe. As shown in Figure 6(c), the frequency response suggested that there were low-pass and high-pass filters similar to the theory. In addition, as the volume increased, the low-pass filter's cutoff frequency decreased.

Demonstration
Finally, we demonstrated wave height measurement using the developed sensor with V_b of 20 ml, of which frequency response was 0.1-10 Hz, moving up and down as shown in Figure 6(a). Figure 6(b, c) show the sensor response during the demonstration and when not moving on the ground, respectively. The sensor output were low-pass filtered at 10 Hz.

When not moving, there was almost no pressure change. On the other hand, during the demonstration, the height change was measured as the pressure change. The experimental result suggested that the proposed wave height sensor could easily detect wave height changes in the sea.

CONCLUSION
In this study, the high-sensitivity wave height sensor with MEMS piezoresistive cantilever was designed and developed. The sensor has a high-pass filter and a low-pass filter from its structure, and theory and experiment showed that the cut-off frequency of the low-pass filter depended on the chamber volume. Based on this experiment, the sensor was able to measure height change with high sensitivity. Therefore, by adjusting design parameters such as chamber volume, it is possible to realize a wave height sensor that can measure the target frequency band.

ACKNOWLEDGEMENTS
This study was partially supported by JSPS KAKENHI Grant Number 20H02102, the New Energy and Industrial Technology Development Organization (NEDO).

REFERENCES
[1] K. Komatsu and K. Tanaka, "Swell-dominant surface waves observed by a moored buoy with a GPS wave sensor in Otsuchi Bay, a ria in Sanriku, Japan," *J. Oceanogr.*, vol. 73, no. 1, pp. 87–101, 2017.

[2] T. H. C. Herbers, P. F. Jessen, T. T. Janssen, D. B. Colbert, and J. H. MacMahan, "Observing ocean surface waves with GPS-tracked buoys," *J. Atmos. Ocean. Technol.*, vol. 29, no. 7, pp. 944–959, 2012.

[3] H. Takahashi, N. M. Dung, K. Matsumoto, and I. Shimoyama, "Differential pressure sensor using a piezoresistive cantilever," *J. Micromechanics Microengineering*, vol. 22, no. 5, 2012.

[4] N. Minh-Dung, H. Takahashi, T. Uchiyama, K. Matsumoto, and I. Shimoyama, "A barometric pressure sensor based on the air-gap scale effect in a cantilever," *Appl. Phys. Lett.*, vol. 103, no. 14, pp. 1–5, 2013.

[5] R. Wada and H. Takahashi, "Time response characteristics of a highly sensitive barometric pressure change sensor based on MEMS piezoresistive cantilevers," *Jpn. J. Appl. Phys.*, vol. 59, no. 7, 2020.

CONTACT
*T. Hirayama, tel: +81-45-566-1847;
Hirataku99@keio.jp

MEMS CAPACITANCE DIAPHRAGM GAUGE WITH TWO SEALED REFERENCE CAVITIES

Xiaodong Han[1,2], Jingzhen Li[3], Gang Li[4], and Yongjian Feng[1]

[1]School of Aerospace Engineering, Xiamen University, CHINA

[2]MESA+ Institute, University of Twente, THE NETHERLANDS

[3]Key Laboratory of Optoelectronics Technology, Ministry of Education, Faculty of Information Technology, Beijing University of Technology, CHINA and

[4]Science and Technology on Vacuum Technology and Physics Laboratory, Lanzhou Institute of Physics, CHINA

ABSTRACT

A capacitance diaphragm gauge based on a glass-silicon-glass structure was designed to possessed two sealed reference cavities. Square pressure-sensing diaphragm with four corners chamfered was used as the sensitive element, and non-evaporation getter film was utilized to maintain the vacuum pressure in the reference cavities. The gauge with the size of $13\times8\times1.4$ mm^3 was manufactured by micromachining technology and its performance was studied systematically. Measurement rang, sensitivity and long-term stability of the gauge were tested and analyzed. The experimental results indicated that the gauge had an operating range of 1 to 8.5×10^5 Pa, and it can be used to measure accurate vacuum pressure ranging from 1 to 1000 Pa with a high sensitivity. Moreover, the long-term tests demonstrated that the gauge was stable enough for absolute vacuum pressure measurement.

KEYWORDS

capacitance, diaphragm gauge, vacuum, two cavities.

INTRODUCTION

Microscale sensors have become thriving with progress made in micro-electro-mechanical-system (MEMS) technology. These miniaturized devices have been widely participated in sectors of military industry, automobile industry, personal consumer electronics, medical treatment and implantable gauges inside human body [1-6]. Capacitance diaphragm gauge (CDG) is one of the major devices for the vacuum pressure measurement. CDGs fabricated based on MEMS have been a research hotspot due to small size, low power consumption and reasonable price [7,8]. Since 2017, we have set out the development of MEMS CDG [9-12]. Theoretical models considering size effect and edge field effect were established respectively to study mechanical and electric properties of microscale pressure-sensing diaphragms. The preparation of flat pressure-sensing diaphragm with large wide-thickness ratio and the development of nano getter film have also been tackled. The prototype of MEMS CDG has been developed and preliminary tests have been carried out.

The real superiority of CDGs manufactured by MEMS technology lies in absolute pressure type, which is highly required in the scenario of space tasks and implantable in situ vacuum measurements for minimal volume. Two major sticking points must be tackled to achieve the absolute type MEMS CDG: 1) the acquisition and maintenance of very low pressure in the reference cavity; 2) atmospheric pressure resistance and performance attenuation control during long term storage of pressure-sensing diaphragm in atmospheric environment. We reported a MEMS CDG with silicon block in 2022, which was capable of, for the first time, measuring absolute pressure ranging from 0.1 Pa to 84 kPa with high sensitivity [13]. However, the measurement results of the MEMS CDG would be affected by the silicon block, and its long-term stability is not good.

In this paper, we have focused on the design, fabrication and characterization of an optimized MEMS CDG with two sealed reference cavities, which has proved to be a good solution to above problems.

STRUCTURE AND PRINCIPLE

As shown in Fig. 1, the designed MEMS CDG has a glass-silicon-glass structure. The sensing capacitor is formed of the square diaphragm, the fixed electrode and the gap between them. The square diaphragm acting as a movable electrode of the sensing capacitor is fabricated in the silicon substrate, and its four corners are chamfered to eliminate stress concentration. The dimension of the diaphragm is $2250\times2250\times4$ µm^3.

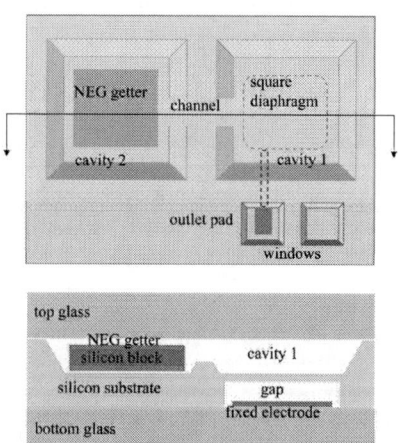

Figure 1: Structure of the MEMS CDG with two sealed reference cavities.

Two small windows are etched on the silicon substrate to extract the electrodes of the sensing capacitor, and one of the windows is also the inlet of the pressure to be measured. There are two cavities in the silicon substrate:

cavity1 is located above the diaphragm and cavity2 is beside it. The two cavities are connected by an etched channel. The cavity1 provides space for the diaphragm deformation, while the cavity2 is used to place the silicon block with non-evaporation getter (NEG) films on its surfaces. The two cavities are sealed in a vacuum environment where the pressure is lower than 10^{-4} Pa by anodic bonding of the silicon substrate and the top glass.

When the CDG is used to measure the vacuum pressure, the diaphragm will be deformed due to the applied pressure, causing an output capacitance variation of the sensing capacitor. There is a definite corresponding relationship between the output capacitance and the applied pressure, based on which the pressure can be calculated using the value of the output capacitance obtained by the detection circuit. The design of the MEMS CDG in this paper has several advantages: 1) the fixed electrode and the reference cavities are separated by the diaphragm, simplifying the electrode extraction process; 2) two cavities are manufactured to make the vacuum pressure of the reference cavities much easier to be obtained and maintained, and to avoid the effect of the silicon block with NEG films on the diaphragm' deformation; 3) the NEG in our design is deposited on the surfaces of silicon block in the form of thin film, and more than one silicon block with NEG films on two surfaces can be placed in the cavity2, which can significantly improve the absorb performance; 4) the small square diaphragm with corners chamfered is fabricated to serve as the pressure-sensing element, making the CDG have a strong overload resistance.

FABRICATION AND TEST

Manufacturing progress determines the performance of sensors to a great extent, especially for those based on MEMS technology. The main process steps of the designed MEMS CDG are shown in Fig. 2.

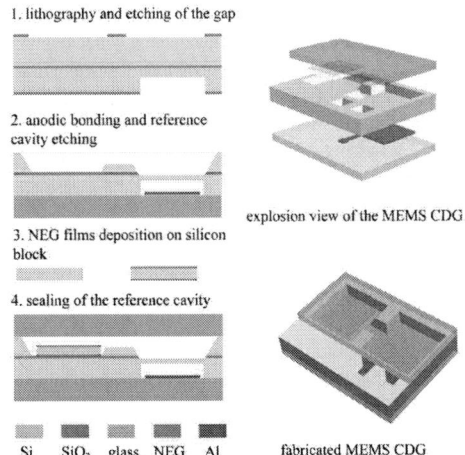

Figure 2: Fabrication progress of the MEMS CDG with two reference cavities.

The SOI (silicon-on-insulator) wafer which contains device, insulation and substrate layers was used due to its reliable fabrication steps and robust structures. Prior to

starting the processes, silicon oxide layers were formed on the both surfaces of the wafer as sacrificial layers. The first step is etching for the gap of the variable capacitor and lithography for the pattern of the reference cavities. Inductively coupled plasma (ICP) was used to etch the gap in the device layer of the SOI wafer, during which the shape and thickness of the diaphragm can be fixed. After the gap was fabricated, the pattern of the reference cavities was generated by lithography on the surface of the substrate layer, during which the alignment for the patterns of the gap and the reference cavities was required. The second step is anodic bonding between the glass with fixed electrode and the SOI wafer and etching for the reference cavities. The pattern of the fixed electrode was obtained by lithography on the BF33 glass, and then the magnetron sputtering was used to deposited an aluminum layer on the glass. The device layer of the SOI wafer was bonded to the glass by anodic bonding, forming the variable capacitor. The diaphragm was released by etching the substrate layer of the bonded wafer using the wet etching process, during which the reference cavities were also fabricated. The remaining steps concentrate on obtaining and maintaining a high vacuum in the reference cavities. In the third step, silicon block suitable for the cavity size was made to act as the substrate of the NEG films. In this paper, the Zr-Co-RE getter films was deposited on both upper and lower surfaces of the silicon block to exhibit an excellent gas absorption performance. The final step is the sealing of the reference cavities. Two silicon blocks with NEG films were put into the cavity2 and then the top glass was bonded to the substrate layer of the SOI wafer by anodic bonding. To keep a high vacuum in the reference cavities, the bonding system together with the wafer and top glass were heated at 300 °C for two hours under a high vacuum where the pressure was lower than 10^{-4} Pa. Then the temperature was increased to 350 °C and kept for 20 minutes to activate the NEG films, after which the temperature was increased to 360 °C for the anodic bonding. After applying a certain mechanical pressure and a DC voltage of 800 V on the wafer and the glass, the reference cavities were sealed and the fabrication of the MEMS CDG was preliminarily completed. Figure 2 also presents the explosion and the fabricated view of the MEMS CDG.

Figure 3: (a) Fabricated chip of the MEMS CDG with two sealed reference cavities; (b) diaphragm and fixed electrode of the MEMS CDG.

The fabricated MEMS CDG with two reference cavities is shown in Fig. 3 (a). Its dimension is $13\times8\times1.4$ mm^3. The diaphragm is deformed under atmospheric pressure. Figure 3 (b) illustrates the view of the diaphragm and fixed electrode seen through the bottom glass. All the four corners of the diaphragm are chamfered to reduce stress concentration.

A dedicated vacuum tests system was used to systematically verify the performance of the MEMS CDG with double reference cavities. With this system, measurement range, sensitivity and long-term stability of the CDG were evaluated comprehensively. The chip of the developed gauge shown in Fig. 3 was fabricated appropriately to be connected with the tests system. The vacuum chamber of the system can provide controllable and measurable pressure ranging from 10^{-4} to 8.5×10^5 Pa for the tests. The output capacitance of the MEMS CDG was extracted via two signal wires which are connected to the capacitance meter (Keysight-E4980AL). The capacitance-pressure characteristics of the MEMS CDG were obtained by recording the display values of the monitoring vacuum gauges and the capacitance meter when the pressure in the vacuum chamber was changed regularly within the rang.

RESULTS AND DISCUSSION
Measurement range

The capacitance-pressure curve is nonlinear due to the constrain of the diaphragm's working principle, but a linear voltage-pressure relation can be achievable through maneuvering in circuit. Generally speaking, a CDG can cover three orders of magnitude. For this MEMS CDG, there is no compulsion on the linear relation, so it is hopeful to exceed the typical measurement range. Thus, we first determined its lower and upper measurement limits using the tests system. The vacuum chamber was firstly pumped to 10^{-1} Pa, then the pumps were stopped and the pressure in the vacuum chamber was controlled to be increasing generally via the needle valve.

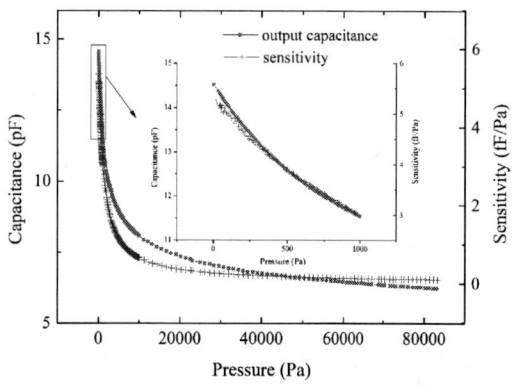

Figure 4: Capacitance-pressure characteristic of the MEMS CDG in the range of 1 to 8.5×10^5 Pa.

The capacitance-pressure characteristic of the MEMS CDG was obtained in the pressure range of 1 to 8.5×10^5 Pa, as depicted in Fig. 4. The capacitance-pressure curve shows nonlinearity as mentioned before. The maximum output capacitance, which can be regarded as an initial capacitance of the MEMS CDG, is 14.536 pF when the pressure is lower than 1Pa. In other words, the diaphragm

is in an undeformed state when the pressure is lower than 1Pa. The output capacitance decreases with the increased pressure, the total variation of the capacitance is 8.3 pF. The lower and upper measurement limits of the developed MEMS CDG are 1 Pa and 8.5×10^5 Pa, respectively.

The capacitance-pressure characteristic of the MEMS CDG in the range of 1 to 1000 Pa is shown in the inset in the Fig. 4. The variation of the output capacitance is approximately 3 pF, which accounts for 36% of the total variation. Considering the pressure range only accounts 1% of the whole pressure range, the ratio of capacitance to pressure is quite higher in the range of 1 to 1000 Pa compared with that in the whole range.

Sensitivity and resolution

The sensitivity of the MEMS CDG developed in this paper varies for different pressure points since the input-output curve is nonlinear. The sensitivity S can be characterized as the ratio of capacitance variation ΔC against pressure change Δp. Figure 4 also shows the sensitivity of the MEMS CDG. It is obvious that the sensitivity is reduced from 5.36 to 0.1 fF/Pa. This characteristic enables an enhanced ability for accurate lower pressure measurement, especially around the lower limit. It is an advantage of the developed CDG compared with the linear ones. As illustrated in the inset in Fig. 4, the sensitivity is diminished from 5.36 to 2.99 fF/Pa in the pressure range of 1 to 1000 Pa. It should be noted that the sensitivity of 2.99 fF/Pa is still sufficient for pressure measurement around the upper limit 1000 Pa. The change rate of the sensitivity is very small when the pressure exceeds 1×10^5 Pa, the sensitivity in this range is approximately 0.1 fF/Pa, which is suitable for occasions where the required measurement accuracy is not high. Therefore, the sensitivity of this gauge can meet the requirements of accurate pressure measurement in the lower pressure range and rough pressure measurement in the high-pressure range.

The resolution of the MEMS CDG is defined as the minimum pressure variation that can be distinguished by the capacitance meter. The minimum stable capacitance value that the capacitance meter can measure is 0.1fF. On the basis of the experimental results, the resolution of the MEMS CDG is obtained: 0.5 Pa in the pressure range of 1 to 1000 Pa and 5 Pa in the pressure range of 1000 to 8.5×10^5 Pa.

Long term stability

Long term stability is extremally vital for the absolute MEMS CDGs with vacuum sealed reference cavities. The measurement range of the MEMS CDG with two reference cavities depends on the vacuum pressure in the reference cavities to a great extent. The above discussion has demonstrated that the vacuum in the reference cavities is sufficient for the developed MEMS CDG because the lower measuring limit of 1 Pa was accomplished. In order to evaluate the outgassing of the materials such as silicon and glass, the sealing property of the reference cavities formed by anodic bonding and the gas adsorption capability of the NEG films, further tests performed between a long time are needed.

The MEMS CDG was fabricated in March 2021, when

the first capacitance-pressure curve was obtained. After being stored in the atmospheric pressure for six months, the MEMS CDG was tested under the same condition, and the second capacitance-pressure curve was acquired in August 2021. The results curves are plotted in logarithmic coordinates to vividly reveal the characteristics of the low-pressure range section, as illustrated in Fig. 5.

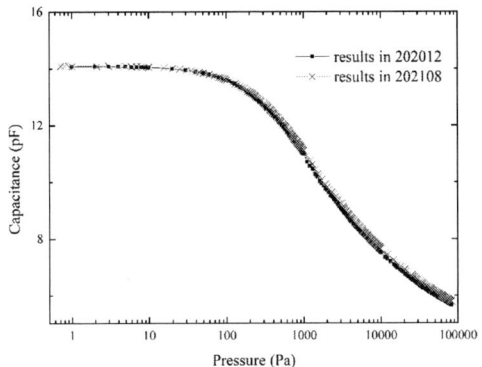

Figure 5: Long-term characteristic of the MEMS CDG with two sealed reference cavities.

There is no obvious difference between the two results. The relative discrepancies are: maximum 3.2%, minimum 1.3% and average value 2.2%. Within the pressure range from 1 to 10 Pa, the relative discrepancy of the output capacitance is less than 0.1%. The analysis data show that the performance of the MEMS CDG remains unchanged under the action of atmospheric pressure for 6 months, which means the developed MEMS CDG with two reference cavities has a better long-term stability compared to the MEMS CDG with one reference cavity [13]. On the one hand, the designed diaphragm is reliable enough to withstand atmospheric pressure; on the other hand, the reference cavities formed via anodic bonding have an excellent sealing performance and the NEG films is effective to eliminate the residual gas considering the inevitable outgassing and leakage of the materials.

CONCLUSION

An absolute MEMS CDG with two reference cavities was successfully fabricated and tested. The design of two cavities increases the amount of getter and eliminates the external influence on the diaphragm. The MEMS CDG has an operating range of 1 to 8.5×10^5 Pa. In the pressure range from 1 to 8.5×10^5 Pa, the sensitivity was reduced from 5.36 to 0.1 fF/Pa, while in the pressure range from 1 to 1000 Pa, the sensitivity was reduced from 5.36 to 2.99 fF/Pa. The resolution is 0.5 Pa in the pressure range from 1 to 1000 Pa, while 5 Pa in the range of 1000 to 8.5×10^5 Pa. The sensitivity and resolution indicate that the CDG has an enhanced ability for accurate measurement in the lower pressure range. The test results also show that the long-term stability of the MEMS CDG with two reference cavities is better than that of the MEMS CDG with one reference cavity. On the basis of the two cavities structure developed in this paper, MEMS CDGs with low measurement limit to 0.1 Pa and good long-term stability are expected to be realized.

ACKNOWLEDGEMENTS

The authors would like to acknowledge the support of the National Natural Science Foundation of China (Grant No. 61627805) and the China Scholarship Council.

REFERENCES

[1] Asri, Muhammad Izzudin Ahmad, et al. "MEMS gas sensors: a review." IEEE Sensors Journal (2021).

[2] Vivek K. Verma, and R. Yadava. "Stochastic resonance in MEMS capacitive sensors." Sensors and Actuators B: Chemical (2016).

[3] Grzebyk, T., and Anna Górecka-Drzazga. "MEMS type ionization vacuum sensor." Sensors and Actuators A: Physical 246 (2016): 148-155.

[4] Rafaie, Mostafa, Mohammad H. Hasan, and Fadi M. Alsaleem. "Neuromorphic MEMS sensor network." Applied Physics Letters 114.16 (2019): 163501.

[5] Varnava, Christiana. "MEMS sensor in flight control." Nature Electronics 2.8 (2019): 322-322.

[6] Li D, Wang Y, Zhang H, et al. "Applications of Vacuum Measurement Technology in China's Space Programs." Space: Science & Technology 2021 (2021).

[7] Miiller, A. P. "Measurement performance of high-accuracy low-pressure transducers." Metrologia 36.6 (1999): 617.

[8] Górecka-Drzazga, Anna. "Miniature and MEMS-type vacuum sensors and pumps." Vacuum 83.12 (2009): 1419-1426.

[9] Li G, Li D, Cheng Y, et al. "Design of pressure-sensing diaphragm for MEMS capacitance diaphragm gauge considering size effect." AIP Advances 8.3 (2018): 035120.

[10] Han X, Li D, Cheng Y, et al. "Analysis on edge effect of MEMS capacitance diaphragm gauge with square pressure-sensing diaphragm." Microsystem Technologies 25.7 (2019): 2907-2914.

[11] Xu M, Feng Y, Han X, et al. "Design and fabrication of an absolute pressure MEMS capacitance vacuum sensor based on silicon bonding technology." Vacuum 186 (2021): 110065.

[12] Han X, Xu M, Li G, et al. "Design and experiment of a touch mode MEMS capacitance vacuum gauge with square diaphragm." Sensors and Actuators A: Physical 313 (2020): 112154.

[13] Han X, Li G, Xu M, et al. "Miniature capacitance diaphragm gauge for absolute vacuum measurement." Measurement 194 (2022): 110851.

CONTACT

*Xiaodong Han, tel: +31645522909; hanxdcast@163.com

TOWARDS A GAS INDEPENDENT THERMAL FLOW METER

Shirin Azadi Kenari[1], Remco J. Wiegerink[1], Remco G.P. Sanders[1], and Joost C. Lotters[1,2]
[1]University of Twente, The Netherlands and
[2]Bronkhorst High-Tech BV, The Netherlands

ABSTRACT

We present a novel potentially gas-independent thermal flow sensor chip that contains three two-wire calorimetric flow sensors to measure the flow profile and flow direction inside a tube, and a single-wire flow-independent thermal conductivity sensor which detects the type of the gas through a simple DC voltage measurement. All wires have the same dimensions of 2000 μm in length, 5 μm in width and 1.2 μm in thickness. Four different gases Ar, N_2, Ne and He were used for the thermal conductivity measurement and the measured output voltage corresponds very well with a theoretical model.

KEYWORDS

Thermal flow sensor, thermal conductivity, calorimetric sensor, Wheatstone bridge.

INTRODUCTION

Thermal flow sensors are used to measure the flow rate of both gases and liquids. There are three types of thermal flow sensors: anemometric, calorimetric and time-of-flight. All three mentioned thermal flow sensor types are typically composed of a heater and temperature sensors; and follow a similar working principle, i.e., supplying power to the heater to elevate the temperature, and then measuring the change in temperature distribution over the sensor structure as a measure for the flow rate [1].

Thermal flow sensors have a simple working principle and low fabrication cost. However, they are dependent on the type of the flowing medium, more specifically the thermal properties of the gas or liquid. It means calibration of these sensors is required whenever the medium changes. There are different ways to make thermal flow measurements medium-independent. In [2], AC excitation was used to measure the voltage of a sensor close to the heater. They used the fact that by increasing the frequency, the thermal boundary layer will decrease and can be brought down close to the wall. Therefore, due to the non-slip condition on the wall, the thermal exchange between the heater and sensor is independent of the velocity, so the heat transfer is only affected by physical properties of the gas or fluid, and not by the flow. Two physical parameters, thermal conductivity κ and volumetric heat capacity ρc_p, are derived from the phase and amplitude of the third harmonic of the measured AC voltage. The flow rate itself is measured with DC excitation which is dependent on the κ and ρc_p. Therefore, by knowing these properties from the AC measurement, a gas independent flow rate measurement can be achieved. In [3], a thermal flow sensor was used consisting of a heater and a downstream temperature sensor to measure the thermal conductivity of the gas. The temperature of the downstream

temperature sensor in a specific flow region is only dependent on κ. Another technique is to measure the thermal conductivity by decreasing the dependency of the wire temperature on flow using different structures or implementing the sensor in a dead volume [4, 5, 6]. However, this technique is only suitable for flow sensors inside micromachined channels. In this paper, a novel design is proposed for a sensor probe that can be used to measure the flow rate in a larger tube. It can simultaneously measure the thermal conductivity by simply using a single heated wire with a DC voltage measurement. Moreover, the chip contains three pairs of flow sensing wires to measure the flow velocity at different locations in the tube.

DESIGN AND FABRICATION

Fig. 1(a) shows a schematic drawing of the sensor chip on a PCB, consisting of three pairs of wires realized at the front and back side of a silicon wafer to form calorimetric flow sensors, see Fig. 1(b), and two single wire structures to form thermal conductivity sensors, see Fig. 1(c). The distance between the wires in the flow sensor is 380 μm, which is defined by the wafer thickness. Fig. 2(a) shows the circuit schematic of the Wheatstone bridge. Resistors R_1 and R_4 are fixed resistors integrated on the silicon chip. Resistors R_2 and R_3 are the sensor wires. In no-flow condition, R_1 and R_4, and R_2 and R_3 have the same value, so the output signal of the Wheatstone bridge V_b is zero. When flow is applied, the heat will be transferred from the upstream wire to the downstream one. Therefore, there will be a positive or negative (depending on the flow direction) output voltage signal as a result of the temperature difference between the two wires R_2 and R_3.

An additional wire is suspended above a shallow V-groove cavity for thermal conductivity measurement. Fig. 2(b) shows a schematic cross-sectional drawing of V-groove with under etched beam. The V-groove cavity has a length of 2000 μm and a width of 40 μm. The temperature of this wire is dominated by the thermal conductivity of the gas inside the cavity and largely independent of the flow velocity. Hence, by monitoring the voltage drop over the wire at constant heating current, κ can be detected. Fig. 2(c) shows the circuit schematic of the thermal conductivity sensor.

Fig. 3 shows a summary of the fabrication process and photograph of the released chip. Because of the large aspect ratio of the beam length to the channel width, a good alignment to the crystal orientation {111} is required to minimize the under etch. Therefore, a Vangbo mask is used to obtain the required crystal orientation [7]. A layer of 150 nm LPCVD deposited SiRN is used as the etch mask to transfer the Vangbo pattern. By wet anisotropic etching in KOH (25 wt.% – 75 °C – etch time 10-15 minutes), the crystallographic orientation of a silicon wafer can be found easily and within an error of ± 0.05 degrees. Fig. 3(a) shows the Vangbo mask and the etch structure of the silicon. The second structure shows a perfect alignment while the right structure is not aligned parallel to the {111} crystal orientation, resulting in the unsymmetric under etch.

Fig. 3(b) shows the fabrication process. First, a layer of 1 μm SiRN is deposited by LPCVD (1). Then, a 20 nm Cr adhesion layer and 200 nm Pt layer are deposited and etched by sputtering and IBE etching, respectively, to pattern the wires and metal traces (2, 3). The IBE etching step is performed twice with two different masks. The first step is for transferring the metal pattern, the second one to narrow the beam width and define the pattern in the SiRN layer. In (4), SiRN is etched by plasma etching to open the window for etching the Si. All these steps are repeated for the backside of the wafer to have wires on both sides (5-7). Finally, Si is etched by KOH (KOH 1:3 DI-water) to realize a cavity inside the wafer between the two wires (8).

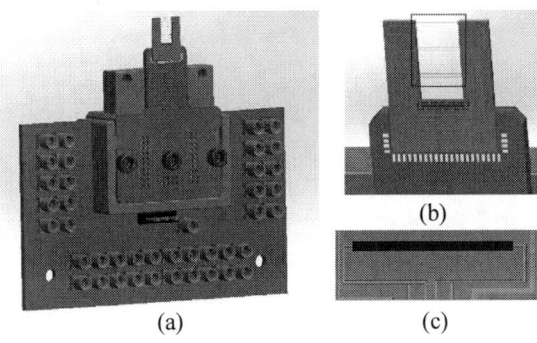

Figure 1: (a) Schematic drawing of the sensor on PCB, (b) chip (solid line and dashed line show the thermal flow sensors and thermal conductivity sensor, respectively), (c) a close-up image of thermal conductivity sensor with the cavity underneath. All wires are 2 mm long, 5 μm wide and 1.2 μm thick.

RESULTS AND DISCUSSION

Fig. 4 shows the experimental setup, consisting of a flow controller, pressure controller, digital voltmeter, and a 3D printed tube with the chip inside it. Each of the three wire pairs that are used as thermal flow sensor forms half of a Wheatstone bridge. The other half of the bridge is formed by fixed on-chip resistors. The sensor wires are heated by the power supply of the bridge, which is 2 V. A difference in temperature between the wires will result in an output voltage. At room temperature all wires have a resistance of approximately 300 Ω. Fig. 5(a) shows the measured output voltages of the three sensors as a function of volumetric flow rate of N₂. The sensitivity clearly depends on the location of the sensor wires, showing that the device can be used for measuring the flow profile. As can be seen, the upper wire is more sensitive to the flow in comparison to the other pairs. He and Ar are also used for thermal flow measurement, see Fig. 5(b). It shows that the voltage of the Wheatstone bridge is dependent on the type of the gas.

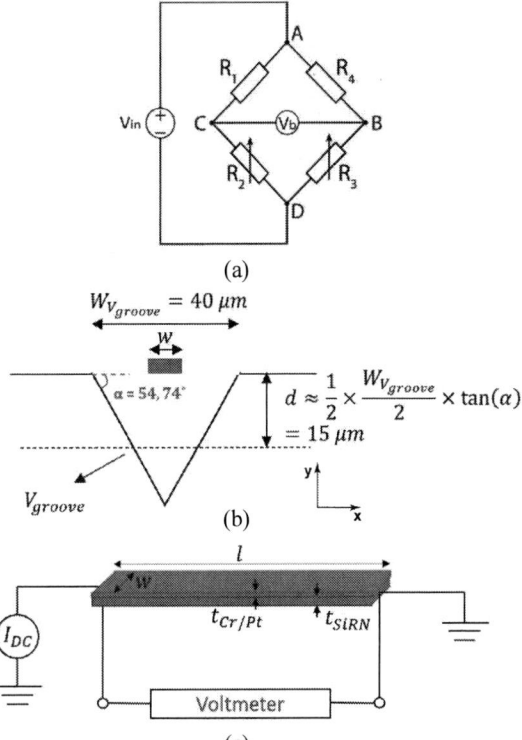

Figure 2: (a) Circuit schematic of the Wheatstone bridge (The bridge is fed by 2 V). (b) Schematic cross-sectional drawing of V-groove with under etched beam. (c) Circuit schematic of the thermal conductivity sensor. A 5 mA DC current is applied to the wire, and voltage of the wire is measured with a digital multimeter.

Next, the wire suspended above the V-groove cavity was used to measure the thermal conductivity of the gas simultaneously. The temperature profile $T(x)$ along the length of the beam can be expressed as [5]:

$$\frac{\Delta T(x_n)}{P'} = \frac{1}{G'_f}\left(1 - \sqrt{\frac{\cosh\left(x_n l \sqrt{R'_b G'_f}\right)}{\cosh\left(\frac{1}{2}l\sqrt{R'_b G'_f}\right)}}\right) \quad (1)$$

$$G'_f = k\frac{w}{d} \quad (2)$$

$$R'_b = \frac{1}{kA} \quad (3)$$

In the equations x_n is the dimensionless normalized position along the wire, l is the length of the wire, w and d are the width and depth of the cavity, and A is the cross-sectional area of the beam. P' is the electrical line power in [W/m] dissipated at position x_n, G'_f is the line conductance through the gas in [W/(Km)], and R'_b is the thermal line resistance of the beam in [K/Wm]. When the temperature is known, the voltage over the wire can be easily calculated ($V = R_a(1 + \alpha\Delta T)\times I$). Fig. 6(a) shows the calculated temperature distribution along the normalized wire length. Fig. 6(b) shows the measured voltage drop over the thermal conductivity sensor at 5mA heating current as a function of flow for He, Ne, N₂ and Ar. The measured voltage depends strongly on the thermal conductivity of the gas. At low flow levels we

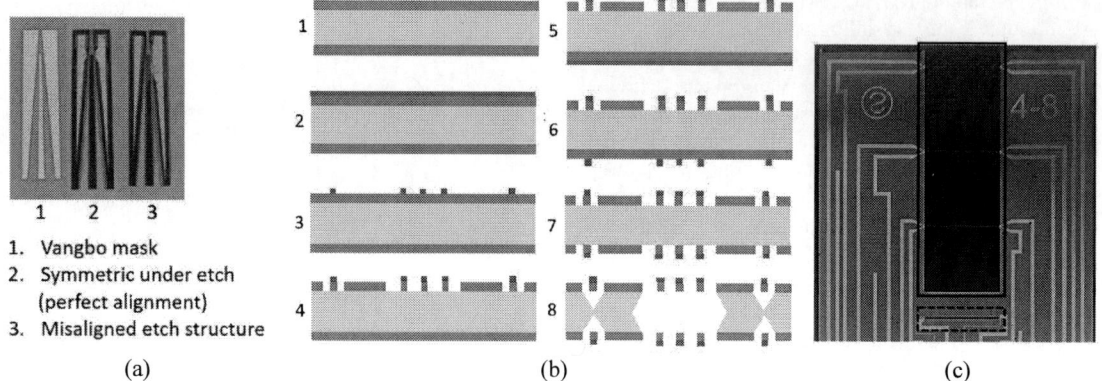

(a)   (b)   (c)

Figure 3: (a) Vangbo mask and the etch structure of the silicon. (b) Overview of the fabrication steps. Microscope photograph of the released (b) thermal flow and (c) thermal conductivity sensor. (1) LPCVD of 1 μm SiRN, (2) sputter a layer of Cr/Pt (20 nm/200 nm), (3) IBE etching of Cr/Pt, (4) SiRN layer is etched by plasma etching to open the window for etching the Si. (5-7) All these steps are repeated for the backside of the wafer, (8) Finally, Si is etched by KOH (KOH 1:3 DI-water) to realize a V-groove and a cavity inside the wafer between the wires. (c) Microscope photograph of the released sensor (solid-line: thermal flow sensor and dashed-line: thermal conductivity sensor).

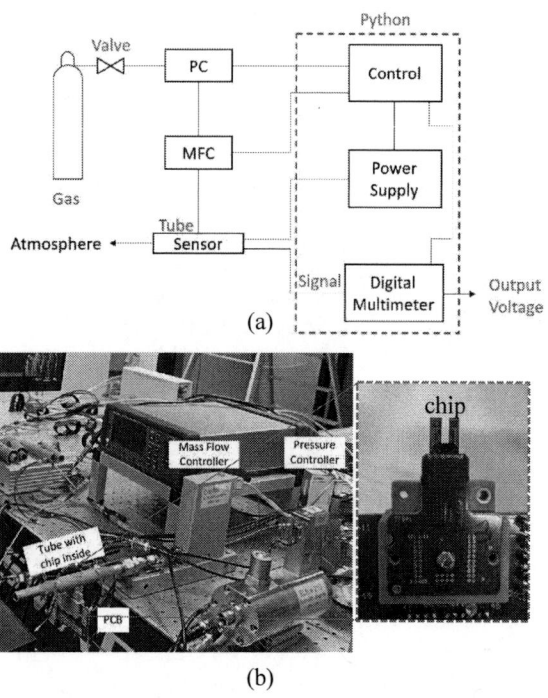

Figure 4: (a) Schematic drawing of the measurement setup. (b) Experimental setup. The setup consists of a pressure and flow controller for providing the gas flow to the 3D printed flow tube, excitation system for heating the wire, and data acquisition system for reading the voltage of the wire. Acquisition is implemented in a Python program. Red dashed line shows the PCB with the chip mounted on it.

see an influence of the flow because of outside air entering the tube. Fig. 7 shows the measured voltage as a function of thermal conductivity together with a theoretical curve that was calculated using the equation (1). The measured voltages correspond very well with the

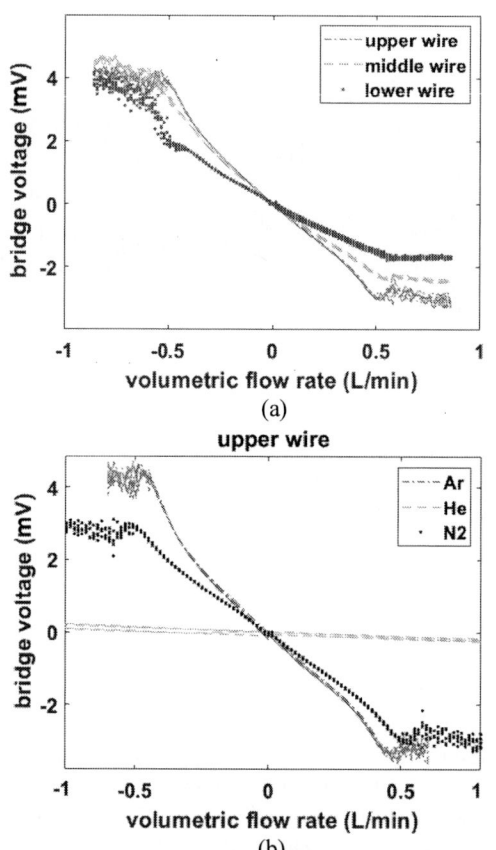

Figure 5: (a) Measured Wheatstone bridge voltage as a function of volumetric flow rate (L/min) for three pairs of wires with N_2. (b) Bridge voltage of upper sensor versus volumetric flow rate for three gases (N_2, He and Ar).

theoretical response. The output signal of the flow sensor can be adjusted with the measured thermal conductivity to be able to measure the flow rate independent of the gas type.

CONCLUSION

In this paper a novel potentially gas-independent thermal flow sensor chip is presented. It consists of three pairs of wires used as calorimetric flow sensors to measure the flow profile and flow direction inside a flow channel, and a flow-independent thermal conductivity sensor which detects the type of gas through a simple DC voltages measurement. Four gases, Ar, N_2, Ne and He were used for the thermal conductivity measurement and the measured output voltage corresponds very well with the theoretical model. The results show that the proposed thermal flow sensor is potentially capable of measuring the flow rate independent of the type of gas. Next steps will include the adjustment of the output signal of the flow sensor for the actually measured thermal conductivity to demonstrate the capability of the flow sensor to measure the flow independent of the gas type.

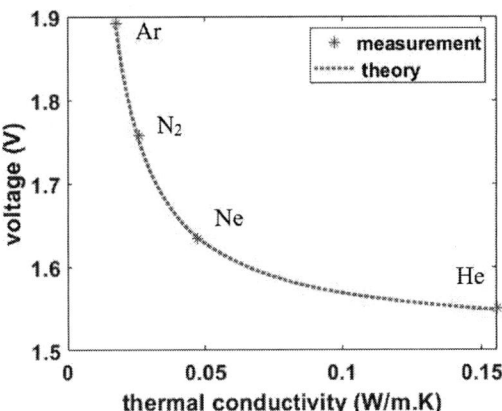

Figure 7: Theory and measurement results comparison. The dashed graph is derived from the stationary temperature distribution equation (1).

ACKNOWLEDGMENTS

The authors would like to thank Bronkhorst High-Tech B.V. and TKI for financially supporting this project.

REFERENCES

[1] Vivekananthan Balakrishnan et al., "Thermal Flow Sensors for Harsh Environments", Sensors 17, pp. 1-31, 2017.

[2] D. F. Reyes, "Measurement and simulation of the frequency response of a thermal flow sensor at different flow speeds", Sensors and Actuators A, 203, pp. 225–233, 2013.

[3] Y. Q. Zhu, "Modelling and simulation of a thermal flow sensor for determining the flow speed and thermal properties of binary gas mixtures", EUROSENSORS 2016, pp. 1028-1031.

[4] J. Wang, "Thermal Conductivity Gas Sensor with Enhanced Flow-Rate Independence", Sensors 2022.

[5] J. J. van Baar, "Micromachined structures for thermal measurements of fluid and flow parameters", J. Micromech. Microeng., 11, pp. 311–318, 2001.

[6] E. Wouden, "Multi Parameter Flow Meter for On-Line Measurement of Gas Mixture Composition", Micromachines, vol. 6, pp. 452-461, 2015.

[7] Mattias Vangbo, "Precise mask alignment to the crystallographic orientation of silicon wafers using wet anisotropic etching", J. Micromech. Microeng., vol. 6, 279–284, 1996.

CONTACT

E-mail: s.azadikenari@utente.nl

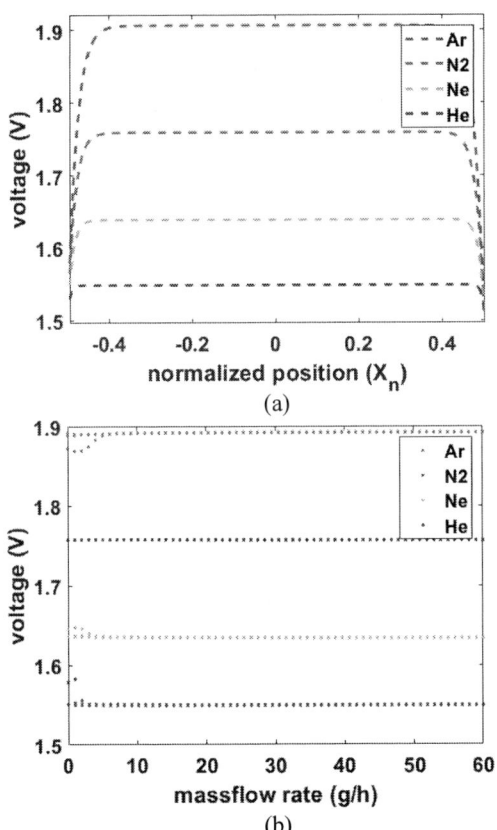

Figure 6: (a) Calculated temperature distribution along the beam of the thermal conductivity sensor for four different gases (He, N_2, Ne, and Ar). (b) Output voltage of the thermal conductivity sensor versus mass flow rate for He, N_2, Ne, and Ar. The wire is heated with a constant current value of 5mA, and the voltage of the wire is measured.

AN INTEGRATED MEMS DEVICE FOR *IN-SITU* FOUR-PROBE ELECTRO-MECHANICAL CHARACTERIZATION OF PT NANOBEAM

Yuheng Huang, Meng Nie, Binghui Li, Kuibo Yin and Litao Sun

SEU-FEI Nano-Pico Center, Key Laboratory of MEMS of Ministry of Education, School of Electronic Science & Engineering, Southeast University, CHINA

ABSTRACT

A miniaturized micro-electro-mechanical system (MEMS) device with a standard SOI process is presented, which integrates three parts, a thermal actuator, a four-probe electro-mechanical coupled test platform, and a capacitive sensor, to realize the four-probe electro-mechanical *in-situ* scanning electron microscopy or transmission electron microscopy (SEM/TEM) test for nanomaterials. Furthermore, it declares the electrical-mechanical properties of the Pt nanobeam fabricated by focused ion beam induced deposition (IBID) with the proposed device. The experimental results demonstrate that the resistance of IBID Pt increases with strain. Young's modulus and fracture strength are 84.24 GPa and 3.66 GPa, respectively. The results indicate that the proposed MEMS device will facilitate the *in-situ* test of 1D/2D nanomaterials.

KEYWORDS

Micro-electro-mechanical system; *In-situ* test; Electro-mechanical property; Pt nanobeam

INTRODUCTION

A variety of nanoscale materials and structures have emerged over the past decade, with wide-ranging applications in transistors, resonators, strain sensors, nanogenerators, and so on. The broad applicability of nanomaterials raises the demand for quantitative characterization of their mechanical and electro-mechanical. Their small-scale properties make characterization usually necessary with the assistance of SEM or TEM. Therefore, the *in-situ* structure-property relationship characterization technology of nanomaterials based on SEM/TEM has attracted extensive attention and developed rapidly.

Currently, two main techniques are used for *in-situ* electro-mechanical testing, TEM-STM (STM: Scanning-Tunneling-Microscopy) and MEMS [1-4]. However, TEM-STM techniques are difficult to perform quantitative mechanical measurements and often require atomic force microscopy (AFM) probes to be equipped. Moreover, such method is continuously affected by electron beam irradiation during testing, which severely affects its electro-mechanical properties. MEMS technology has the ability to integrate actuators and electro-mechanical test structures together, allowing *in-situ* quantitative electro-mechanical characterization in SEM/TEM to establish structure-property characterization of materials. However, most integrated MEMS devices are electrically tested with two probes, which cannot eliminate the influence of contact resistance. Bernal *et al.* reported a MEMS device which realized the four-probe electro-mechanical *in-situ* SEM/TEM test for nanomaterials [5]. However, the four electrodes of the reported device were fabricated by the focused ion beam (FIB) processing which was not suitable for large-scale industrial manufacture.

Herein, a novel integrated MEMS device with a standard SOI process for large-scale production is proposed for *in-situ* four-probe electro-mechanical coupling test. The integration of the drive structure and the test structure makes it easy to be used in SEM/TEM. The validity of the proposed test MEMS device is proved through the characterizations of the IBID Pt nanobeam.

EXPERIMENTAL METHODOLOGY

Device design

SEM image of the proposed MEMS device is shown in Fig. 1, with a close-up detail of the electro-mechanical coupled test platform. The device consists of three parts, a thermal actuator, a four-probe electro-mechanical coupled test platform, and a capacitive sensor. Thermal actuator, nanomaterial fixed on the platform, and capacitive sensor make up the lumped force-displacement model [6] during mechanical testing. When a voltage is applied to the thermal actuator, the V-beam moves forward due to the combined effect of Joule heating and thermal expansion [7]. The movement of the central shuttle of the thermal actuator generates tensile forces on the nanomaterial and capacitive sensor. During the tensile process, the strain of the nanomaterial is obtained by SEM with continuous photographs of the tensile section. Since the nanomaterial forms a series spring structure with the capacitive sensor, the stress of the nanomaterial can be obtained from the displacement change of the capacitive sensor, which can be obtained by SEM images or electronic sensing. Young's modulus and fracture strength are calculated through plotting the data points at different voltages into a stress-strain curve.

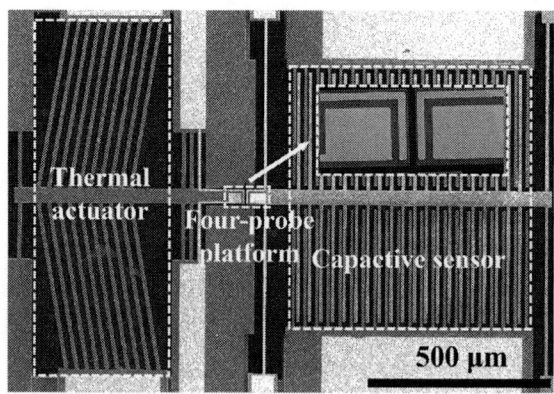

Figure 1: SEM image of MEMS device structure.

Fabrication process

Fig.2 shows the fabrication process of the MEMS

device. The device is fabricated on the 4-inch silicon on insulator (SOI) wafer, which has 30 μm thick device layer, 0.5 μm thick SiO_2 and 410 μm thick Si substrate. First, a 100-nm-thick silicon nitride (SiN) film was deposited by low-pressure chemical vapor deposition (LPCVD). Then, the actuator and sensor electrodes were exposed by reactive-ion etching (RIE) and electron beam evaporation was used to form mechanical and electrical test electrodes. The suspended structure is etched through a series of RIE and deep reactive-ion etching (DRIE). Finally, the SiO_2 layer below the suspended structure is removed by buffered oxide etch (BOE), completing the device fabrication. To realize the function of electro-mechanical coupling test, SiN insulating layer is designed to isolate the mechanical test signal and the electrical test signal, which eliminates the crosstalk from thermal actuators to electrical test electrodes.

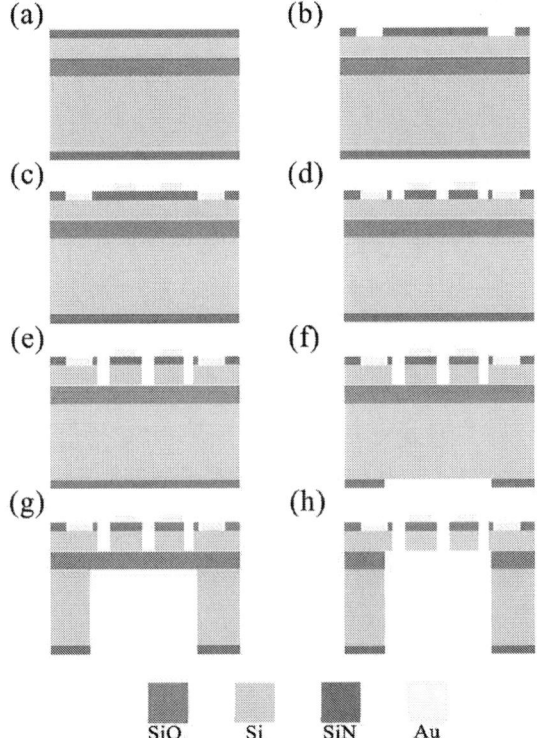

Figure 2: Fabrication process of the MEMS device.

RESULT AND DISCUSSION
Calibration of the device

The device electrodes were connected to the external test printed circuit board (PCB) by wire bonding. The PCB electrodes were electrically connected to the external circuit by the vacuum feedthrough in the SEM chamber. The thermal actuator and capacitive sensor should be calibrated before the mechanical test. A continuous voltage was applied to the thermal actuator and its displacement variation was obtained by SEM image measurement of the platform gap. Fig.3 shows the relationship between the applied voltage and the displacement. The deviation between the simulated results and experimental values may come from fabrication process errors and differences

between simulated and actual parameters [8].

A circuit based on the MS3110 (Irvine Sensors) capacitance reading chip was designed to test the capacitance sensor. A silicon block was transferred to the electro-mechanical test platform by FIB. Since the stiffness of the silicon block is much greater than that of the capacitive sensor, the displacement of the thermal actuator acts entirely on the capacitive sensor during static equilibrium. The calibration curves of the capacitive sensor are shown in Fig. 4. The difference between the theoretical and the experimental values may come from fabrication process errors and environmental noise [9].

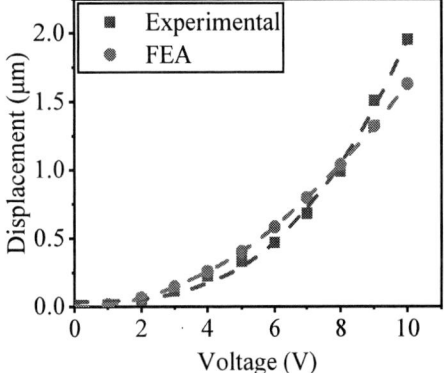

Figure 3: Calibration of the thermal actuator.

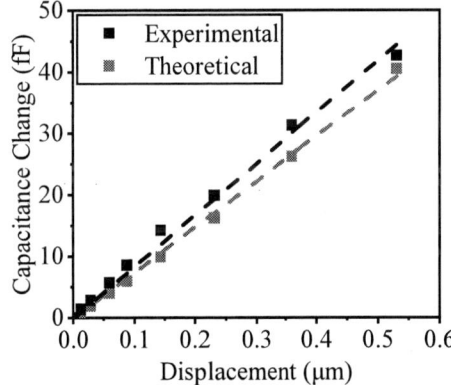

Figure 4: Calibration of the capacitive sensor.

Mechanical and electro-mechanical test of the Pt nanobeam

With the development of electron microscopic characterization, Pt fabricated by IBID is always used for *in-situ* fixation or electrode connection in nano-scale sample preparation or characterization [10]. Meanwhile, IBID Pt plays an important role in FIB-based 3D printing technology [11, 12]. Therefore, accurate characterization of its electro-mechanical properties is important for future electronic devices, especially NEMS.

The proposed device is utilized to characterize the electro-mechanical properties of IBID Pt in SEM. First, the mechanical stage inside the dual-beam microscope (Thermo Fisher Scientific, Helios 5CX) was rotated 52°. Then, the gas injection system (GIS) was inserted to spray the precursor into the target area. The adsorbed precursor gas was decomposed by ion beam, the volatile reaction products were desorbed and removed from the surface by a vacuum system, while the desired reaction product Pt was immobilized on the Si substrate. Both sides of the Pt block were etched with 5 μm deep concavity by FIB. Rotated the mechanical table back to 0° and used FIB to hollow out the bottom of the Pt block. The nanomanipulator was moved to the Pt block and fixed, then the junctions on both sides of the Pt were cut off. Lifted the Pt block and rotated the nanomanipulator to remove the Si attached to the bottom. Moved the nanomanipulator to the electro-mechanical test platform and used the same conditional of the Pt block for fixation and electrical connection. Finally, the suspended Pt is milled to a dog-bone-shape. Since the Pt block is made into a dog-bone-shape, the tensile deformation is concentrated in the middle nanobeam, and the change of resistance is also mainly caused by the deformation of the nanobeam. SEM images of the Pt nanobeam before and after stretching are demonstrated in Fig. 5 (a) and (b).

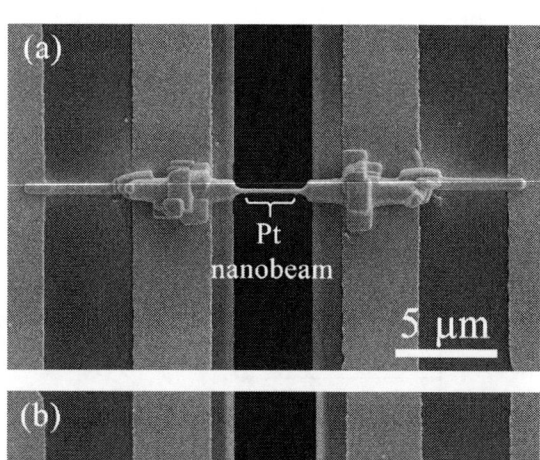
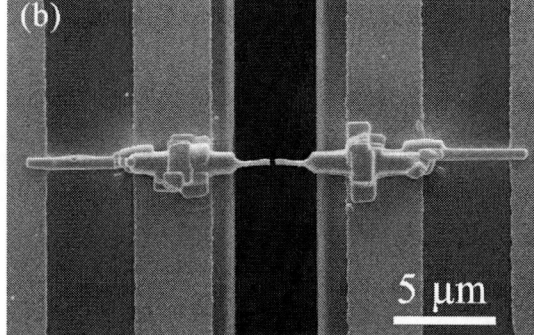

Figure 5: Pt nanobeam before and after tensile.

Using a multimeter to test the resistance of the actuator electrodes and the electrical test electrodes, the result shows an open circuit state, which proves that there is no crosstalk between the electrodes. After the Pt nanobeam was transferred to the electro-mechanical test platform, the same checking process was executed to confirm that the SiN film was not damaged during the transfer process. A 10 nA current was input between the outer electrodes, and the Pt nanobeam voltage changes of the inner electrodes were recorded. It was noted that electron beam irradiation would greatly affect the resistance of Pt nanobeam. Therefore, the electron beam was set to beam off or away from the test structure during the electro-mechanical test. Apply a continuous voltage to the thermal actuator and

calculate the strain of the Pt nanobeam from the platform gap. The results of the four-probe electro-mechanical test are shown in Fig. 6, indicating the resistance of the Pt nanobeam increased with strain.

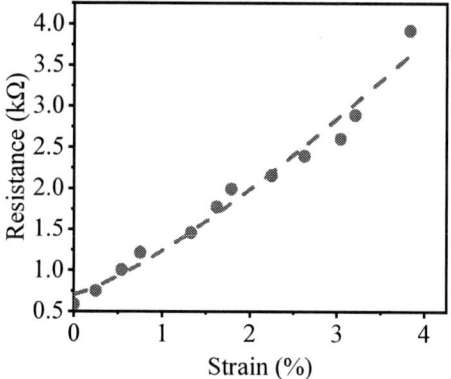

Figure 6: Four-probe electro-mechanical testing of Pt nanobeam.

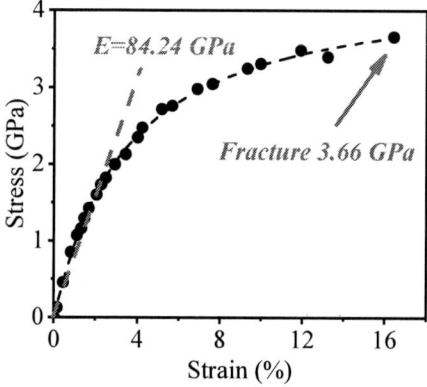

Figure 7: Stress-strain curve of Pt nanobeam.

It was found that the burning phenomenon of Pt nanobeam occurs even under extremely small current, which may lead to the degradation of its mechanical properties. To investigate the actual mechanical properties of IBID Pt, uniaxial mechanical tensile tests were subsequently performed on Pt nanobeams fabricated using the same fabrication method. Young's modulus and fracture strength of the Pt nanobeam are 84.24 GPa and 3.66 GPa, respectively, which demonstrated in the stress-strain curve of the Pt nanobeam (Fig. 7). The fracture strain was 16.39%, showing a superior ductility. The stretching strain range of the mechanical test was greater than that of 3.83% in the electro-mechanical test, which indicates the Pt nanobeam is extremely susceptible to current. The EDS energy spectrum revealed that IBID Pt was not pure, instead, it was a mixture consisting mainly of C, Pt, and Ga elements. The mass ratio of Pt was about 68.89%. The inherent impurities may undergo thermal expansion when an electric current was applied to the Pt nanobeam, thus affecting the overall structure and reducing its mechanical properties.

CONCLUSION

A novel MEMS device for *in-situ* four-probe electro-mechanical coupling testing was presented, which integrates a thermal actuator, an electro-mechanical coupling test platform, and a capacitive sensor. It can be easily incorporated into SEM/TEM to establish the *in-situ* structure-property relationship of nanomaterials. The Pt nanobeam fabricated by IBID was used to verify the functionality of the device. The experimental results illustrate electro-mechanical characteristics of the Pt nanobeam, and declare that IBID Pt is suitable to be used in nanostructure fabrication and strain engineering. Furthermore, the proposed MEMS device is fabricated by the standard SOI process, which can promote to the industrial field.

ACKNOWLEDGEMENTS

The authors acknowledge the funding provided by the National Natural Science Foundation of China (No. 62274031, 12174050, and 12234005), Jiangsu Provincial Natural Science Foundation of China (No. BK20201268), the Key Research and Development Program of Jiangsu Province (BE2021007-2), and the Fundamental Research Funds for the Central Universities.

REFERENCES

[1] Y. Zhang, X. Y. Liu, C. H. Ru, Y. L. Zhang, L. X. Dong, and Y. Sun, "Piezoresistivity characterization of synthetic silicon nanowires using a MEMS device," *Journal of Microelectromechanical Systems,* vol. 20, no. 4, pp. 959-967, Aug 2011.

[2] E. Ochoa, D. Alducin, J. E. Sanchez, C. Fernando, U. Santiago, and A. Ponce, "Semiconductor behavior of pentagonal silver nanowires measured under mechanical deformation," *Journal of Nanoparticle Research,* vol. 21, no. 7, 2019.

[3] Z. Fan, X. Tao, X. Fan, X. Li, and L. Dong, "Sliding Probe Methods for In Situ Nanorobotic Characterization of Individual Nanostructures," *IEEE Transactions on Robotics,* vol. 31, no. 1, pp. 12-18, 2015.

[4] Y. Yaakob *et al.*, "Study of structural and electrical behavior of silicon-carbon nanocomposites via in situ transmission electron microscopy," *Materials Today Communications,* vol. 32, p. 104081, 2022.

[5] R. A. Bernal *et al.*, "In situ electron microscopy four-point electromechanical characterization of freestanding metallic and semiconducting nanowires," *Small,* vol. 10, no. 4, pp. 725-33, Feb 2014.

[6] D. Zhang, J.-M. Breguet, R. Clavel, V. Sivakov, S. Christiansen, and J. Michler, "In situ electron microscopy mechanical testing of silicon nanowires using electrostatically actuated tensile stages," *Journal of Microelectromechanical Systems,* vol. 19, no. 3, pp. 663-674, 2010.

[7] H. D. Espinosa, Y. Zhu, and N. Moldovan, "Design and operation of a MEMS-based material testing system for nanomechanical characterization," *Journal of Microelectromechanical Systems,* vol. 16, no. 5, pp. 1219-1231, Oct 2007.

[8] M. F. Pantano, R. A. Bernal, L. Pagnotta, and H. D. Espinosa, "Multiphysics design and implementation of a microsystem for displacement-controlled tensile testing of nanomaterials," *Meccanica,* vol. 50, no. 2, pp. 549-560, Apr 2014.

[9] B. Pant, B. Allen, T. Zhu, K. Gall, and O. Pierron, "A versatile microelectromechanical system for nanomechanical testing," *Applied Physics Letters,* vol. 98, no. 5, p. 053506, 2011.

[10] D. Radic, M. Peterlechner, and H. Bracht, "Focused ion beam sample preparation for in situ thermal and electrical transmission electron microscopy," *Microsc Microanal,* vol. 27, no. 4, pp. 828-834, Aug 2021.

[11] C. Fang, Q. Chai, Y. Chen, Y. Xing, and Z. Zhou, "Pattern generation method and prediction model of nanohelices fabricated by focused ion beam induced deposition," *Precision Engineering,* vol. 77, pp. 241-250, 2022.

[12] I. Utke, J. Michler, R. Winkler, and H. Plank, "Mechanical Properties of 3D Nanostructures Obtained by Focused Electron/Ion Beam-Induced Deposition: A Review," *Micromachines,* vol. 11, no. 4, Apr 2020.

CONTACT

Meng Nie: m_nie@seu.edu.cn
Kuibo Yin: yinkuibo@seu.edu.cn

FINGERLIKE TACTILE TEXTURE INTEGRATED SENSOR WITH COLD AND WARM SENSATIONS OF SUB-MM SPATIAL RESOLUTION

Nachi Mise, Mitsuki Kozasa, Kyohei Terao, Fusao Shimokawa and Hidekuni Takao[1]
[1] Faculty of Engineering and Design, Kagawa University, Kagawa, JAPAN

ABSTRACT

In this study, a highly functional tactile texture sensor with cold and warm sensations similar to human finger tactile sensations has been realized. We report here the first successful results with quantitative measurement, and evaluation results of the delicate sensation of cold and warm with texture information of a target object at a high spatial resolution below 1 mm. This sensor device integrates a microheater on a fingerprint-like contactor to simulate the core temperature of a finger, and the temperature is kept at 35°C during the sensing operation. When an object comes into contact with the contactor tip, the heat flow rate from the microheater increases. The increased heat flow rate is quantitatively measured by the integrated temperature sensors around the contactor tip, which depends on the thermal conductivity of the material. Material warmth was measured by sweeping the device on the material at a speed of 100μm/s, and the sensor device successfully detected the spatial distributions of thermal conductivity, micro-surface shape, and friction force as waveforms at the same time at the same points.

KEYWORDS

tactile sensor, piezo resistor, surface shape, friction force, heat flex.

INTRODUCTION

In the evaluation of tactile sensations felt by the fingertips, cold and warm sensations are among the important tactile information that constitutes tactile perception, along with surface roughness and frictional force. The human fingertips acquire information on cold and warm sensation, surface roughness, and frictional force by touching an object, and by comparing the characteristics of these sensations, the fingertip remembers the specific tactile sensation of the object. In recent years, excellent tactile sensors [1] with diaphragm or cantilever structures have been developed, and most of them focus on "force sensation" including surface roughness and frictional force. No with cold and warm sensation with high spatial resolution have been realized silicon or non-silicon tactile sensors. We have so far developed a high-resolution tactile sensor that can acquire information on surface roughness and frictional force with a spatial resolution higher than that of fingertip skin [2-5]. However, it is essential to obtain the thermal properties of materials to reproduce the tactile discrimination ability of the fingertip since differences in thermal conductivity reflect differences in material warmness and coldness. In this study, we have successfully developed the first integrated tactile texture sensor with cold and warm sensation that acquires the thermal properties at a sub-mm spatial resolution.

SENSOR CONFIGURATION AND DETECTION PRINCIPLE

Figure 1 shows an overview of the cold and warm integrated tactile texture sensor in this study. The device has three contactors that simulate fingerprint ridges. The central contactor is equipped with a pair of pn-junction diode temperature sensor and a microheater for reproducing the thermal sensing capability of human fingers. Surface texture and frictional force are also detected by the same contactors at the same time.

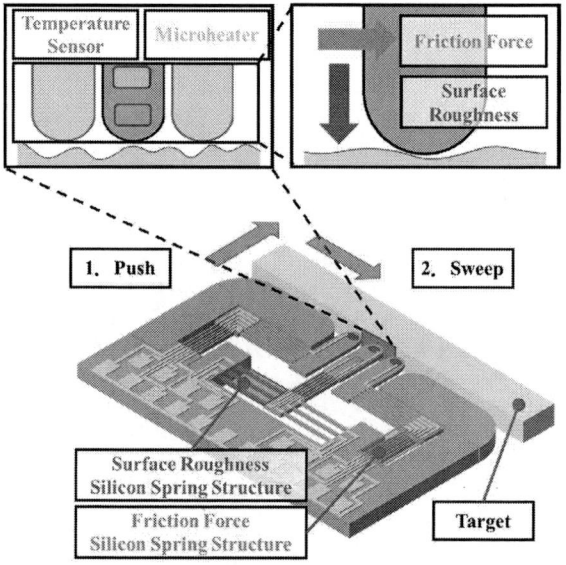

Figure 1: Overview of the cold and warm integrated tactile texture sensor in this study.

Figure 2 shows the measurement principle of the cold and warm sensation implemented for the first time in this device. Heat flux is a parameter that is highly correlated with the cold and warm sensation felt by humans [6] and our device measures its magnitude. The temperature of the microheater on contactor is kept at 35°C to simulate the core temperature of a finger during the sensing operation. By pressing the sensor contactors against an object in ambient temperature, a heat flux is generated from the contactor toward the object. The magnitude of the heat flux is measured by obtaining the difference temperature in between the microheater and the tip of the contactor using integrated temperature sensors. Tactile texture including the cold and warm sensation is measured by sweeping motion after pressing motion against the target object. During the sweeping motion, the contactor follows the surface of the object, and the silicon spring structure supporting the contactors is deformed, which results in a potential change in the output of a piezoresistive detection circuit to obtain the "surface roughness" and "friction force". By integrating multiple contactors with such a detection principle, we can simultaneously acquire information on the microscopic surface irregularities, friction force wave, and thermal characteristics of the target material in a single scanning measurement, as if a fingertip was acquiring them from the target material.

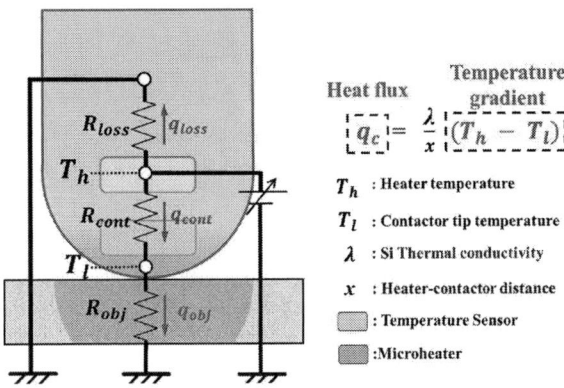

Figure 2: Principle of cold and warm sensory measurement

DEVICE FABRICATION

Figure 3 shows the fabrication flow of the device. The device uses a p-type SOI wafer with an active layer of 50 μm. First, a diffusion layer for the microheater section and piezoresistive circuit wiring is patterned on the active layer. The piezoresistive strain-sensing elements are formed by implanting phosphorus ions to make them n-type. In addition, a metal wiring (Al) film is deposited and patterned on the active layer, and circuit wirings for the microheater and temperature sensor are formed by the following etching process. After the wiring step, Cr films are patterned on the front and back sides of the active layer to form the hard mask for the following Deep-RIE process. The first Deep-RIE process forms the movable structure on the active layer side, and the second Deep-RIE process is performed to etch the handle layer under the movable structure. Finally, the buried oxide film is etched to release the movable structure.

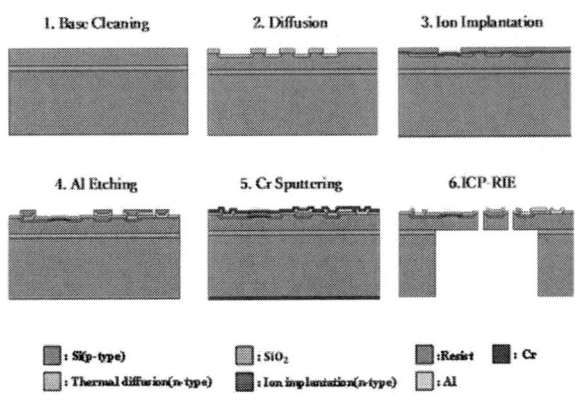

Figure 3: The fabrication flow of the device

Figure 4 shows a photograph of a sensor chip fabricated

using these processes and some details of the important parts.

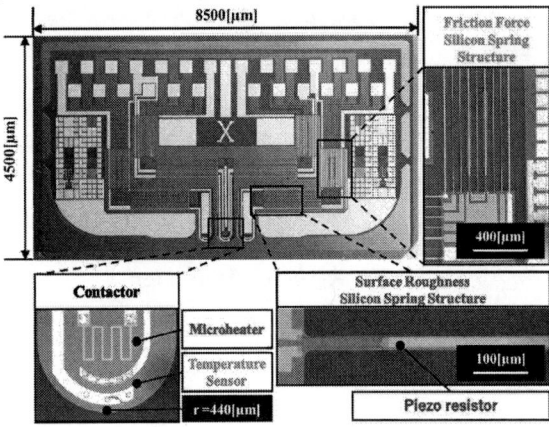

Figure 4: A photograph of a sensor chip fabricated and some details of the important part a photograph of a sensor chip. fabricated using these processes and some details of the important parts

DEVICE EVALUATION

The sensitivity and basic performance as a tactile sensor were evaluated with one-axis motorized stage, and obtained results of texture measurement are shown in Figure 5. The tactile sensitivities of the contactor are 56.7mV/μm for surface irregularities and 650mV/mN for the friction force. The sensitivity of the integrated temperature sensor was evaluated using an infrared thermal camera and a Peltier element at a constant temperature. Measured sensitivity and the transient response of temperature sensor are shown in Figure 6. The temperature sensitivity was 1.16mV/°C, and the response time constant the microheater in the structure is approximately 500ms. This finding implies that if we sweep the sensor device at a speed slower than 1mm/s, sub-mm spatial resolution can be realized since the width of contact area is less than 200μm.

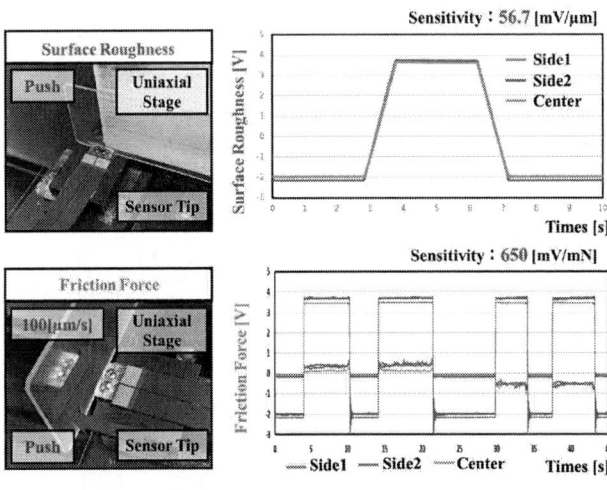

Figure 5: Evaluation experiment and the outputs of surface irregularities and friction force detection structure.

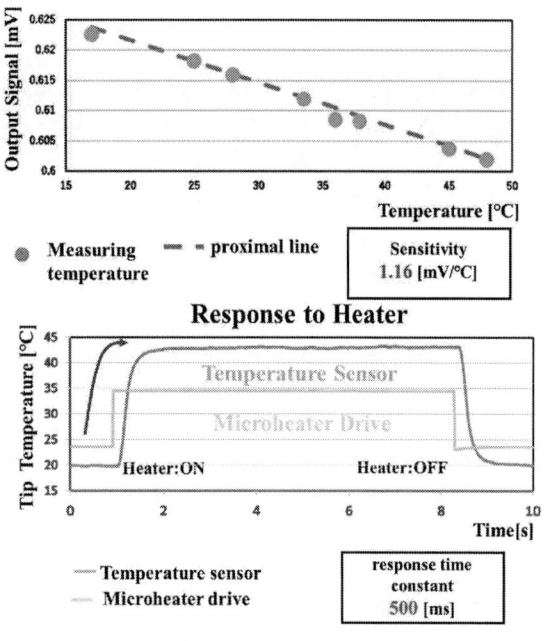

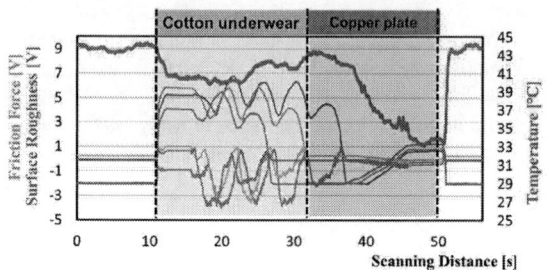

Figure 8: Output signals of surface roughness, friction force and the tip temperature under scanning from cotton underwear to copper plate.

Figure 6: Sensitivity calibration curve of temperature sensor and response of temperature sensor to the driven microheater.

DEVICE PERFORMANCES

Figure 7 shows an experimental setup of heat flux measurement using the device. After the device is pressed on a sample, it was held stationary for about 10 s, and then scanned for 15 seconds at a scanning speed of 100μm/s. In order to evaluate the spatial resolution of cold and warm sensation, a continuous sample of cotton underwear and a copper plate were evaluated as shown in Figure 8. Spatial distribution of surface irregularities, friction force, and temperature tip temperature of contactor were precisely obtained. When the contactor is running on the cotton, the textures of roughness, friction are clearly detected. As soon as the contactor reached the copper, the tip temperature drops suddenly, and settled at a temperature within 700μm. This result proves that the spatial resolution of cold and warm sensation is below 1mm in the device.

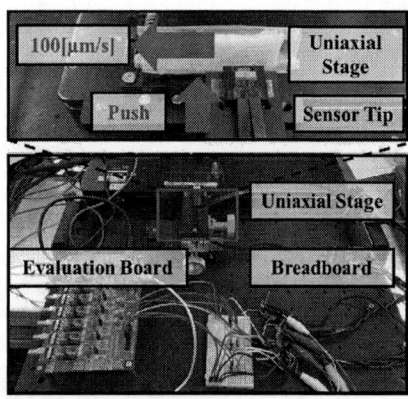

Figure 7: The experimental system of heat flux measurement using the fabricated device.

In the measurement data, the tip temperature just before pressing is extracted as the heater part temperature Th. Then it is used to calculate the difference with the contactor's tip temperature Tl, which corresponds to the temperature gradient proportional to the heat flow rate. The measurement results of various sample materials with the temperature gradient are shown in Figure 9. Temperature gradient is the largest on the copper plate, whereas the temperature gradient is much smaller for the cotton underwear material and cypress. This result clearly shows the device captures the difference in thermal heat flow depending on the thermal conductivity of each material with their surface roughness and friction characteristics.

To further demonstrate the performance of this sensor, we conducted an experiment to compare the cold and warm sensation of fabrics. The object that we used for measurement was a "cool-feeling mask" consisting of "polyester" for the outer layer and "rayon" for the inner "cool-feeling" layer. Rayon is a fiber material called "cool material" which has high hygroscopicity and thermal conductivity, and thus feels cool to the touch.Figure 10 shows the temperature change at the tip of the contactor when polyester and rayon were measured alternately. Surface unevenness and friction force vibration are detected with a temperature gradient at a very high spatial resolution at the same time. There is a clear difference in temperature change between rayon and polyester before and after pressing. The temperature gradient from the contactor to the object is larger for rayon, which gives a stronger cool feeling than polyester. This new sensor can numerically detect the small difference in "coolness" felt by the skin and was successfully used in the first integrated and quantitative measurement and evaluation of the delicate sensations of coolness, surface roughness, and frictional force perceived by the fingertips. Finally, the heat retention of each sample materials was measured and calculated its correlation with the temperature gradient, in order to verify the validity of using the temperature gradient. In the experiments, fingertips were pressed for 10 s against each sample to obtain heat retention. Their time constants of relaxation to return to ambient temperature were measured individually. The relaxation time constant was longer for the materials with high heat-retaining property, because they have a lower thermal conductivity. It is difficult to release the accumulated heat to the surrounding area in them. On the other hand, they are shorter for the material with a poor heat-retaining property

978-1-6654-9309-3/23 $31.00 © 2023 IEEE

owing to the higher thermal conductivities.

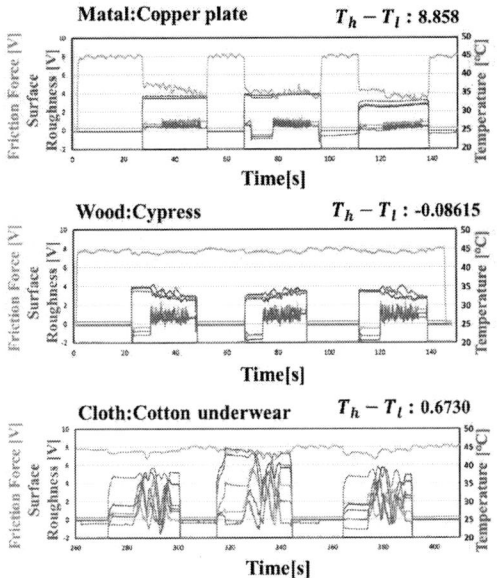

Figure 9: Temperature gradient and the output signals of surface roughness, friction force, and temperature for each material sample and temperature gradient. Samples are copper plates, cotton underwear material, and cypress.

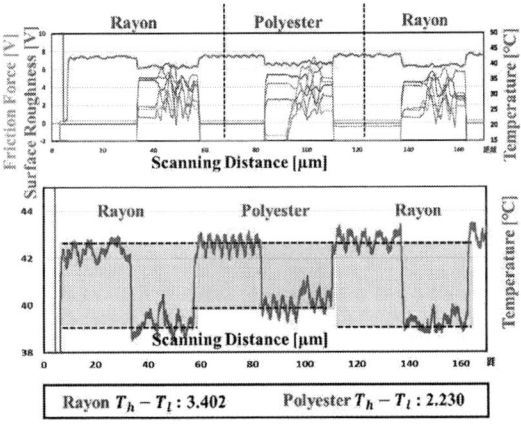

Figure 10: Comparison of temperature gradients and signals between the "cool-feeling side of rayon" and "polyester mask".

Figure 11 shows the relationship among thermal insulation, contactor, and temperature gradient. There is a clear inverse proportional relationship between heat retention and temperature gradient, and it is also theoretically rational. Cypress and copper plates, which are generally considered to have significantly different cold and warm sensations, are in opposite side positions, and the fabric samples are located in the middle. As a result, we succeeded in obtaining the temperature gradient that correlates exactly with the experimental on the results "heat retention" of the materials, i.e., the difference in the "warmth" of the materials detected using the new tactile-texture sensor.

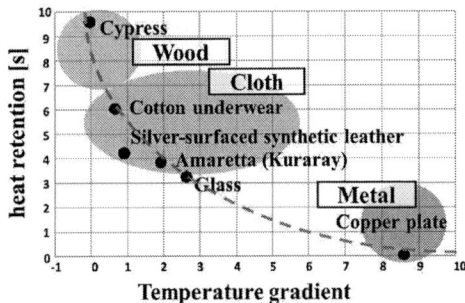

Figure 11: Correlation plot between measured temperature gradients and heat retention of the materials.

ACKNOWLEDGEMENTS

This research was partially supported by JST-CREST research funding program (Grant Number JPMJCR20C2), and JSPS Grant-in-Aid for Scientific Research (A) (Grant Number 22H00218).

REFERENCES

[1] For example, F Sato, T Shiwa, K Takahashi, T Abe, M Okuyama, H Noma, M Sohgawa "Texture measurement for fabrics including warm/cool and fluffiness sensation by multimodal MEMS sensor" TRANSDUCERS 2017 - 19th International Conference on Solid-State Sensors, Actuators and Microsystems pp.343 -346 ,2017

[2] K Watatani, K Terao, F Shimokawa, H Takao "A MEMS Tactile Sensor with Fingerprint-Like Array of Contactors for High Resolution Visualization of Surface Distribution of Tactile Information" Journal of Robotics and Mechatronics Vol.32 No.2, 2020

[3] K Watatani, K Terao, F Shimokawa, H Takao,"Planar-type MEMS tactile sensor integrating micro-macro detection function of fingertip to evaluate surface touch sensation", Japanese Journal of Applied Physics, Vol. 58, No.9, pp.097002-1-097002-9, 2019

[4] K Ando, T Yamamoto, Y Maeda, K Terao, F Shimokawa, M Fujiwara, H Takao,"Highly Sensitive Silicon Slip Sensing Imager for Forceps Grippers Used under Low Friction Condition", Technical Digest of IEEE International Electron Devices Meeting (IEDM2019), pp.IEDM19-422-IEDM19-425, 2019

[5] M Komatsubara, T Nakashima, K Terao, F Shimokawa, Y Matsui, and H Takao,"Development of a Minimally Invasive High-Resolution Tactile Sensor For Acquiring Delicate Haptic Changes in Hair", Proceedings of IEEE 35th International Conference on Micro Electro Mechanical Systems (IEEE MEMS2022), pp.699-702, 2022

[6] H.Liu and F.Chollet, "Layout controlled one-step dry etch and release of MEMS using deep RIE on SOI wafer ," Journal of Microelectromechanical Systems Vol 15, pp. 541-547, June 2006.

CONTACT

Prof. Hidekuni Takao, Tel:+81-87-864-2331;Email:takao.hidekuni@kagawa-u.ac.jp

MODIFIED BEAM STRUCTURES FOR IMPROVED RESONANT SENSING

Erfan Ghaderi, and Behraad Bahreyni
Simon Fraser University, Burnaby, Canada

ABSTRACT

This paper introduces simple structural designs to improve the sensitivity of micro-beam resonators to axial force. Herein we demonstrate that by modifying the boundary conditions of sensing beams, it is possible to significantly increase the sensitivity of beam resonators that are commonly used in resonant sensors. Devices with different boundary conditions were designed and simulated, and sample devices were fabricated and examined to validate the design principles. Experimental results validate significant improvements in the sensitivity of presented devices up to 100%. The work presents a versatile, zero-cost approach to notably improve the sensitivity of resonant force sensors and will be of interest to developers of various resonant microsensors. The proposed approach is applicable to virtually all manufacturing processes employed to fabricate resonant devices.

KEYWORDS

Force Sensors, Resonant Sensors, Microbeam, Boundary Conditions, Sensitivity.

INTRODUCTION

Resonant microsensors are attracting increasing interest in high-performance sensing due to their high dynamic range, good noise performance, and robustness against interference [1]. Moreover, batch microfabrication plus the possibility of on-chip electronics will reduce manufacturing costs and increase reproducibility. Resonant force sensors are at the heart of most resonant sensors, excluding gravimetric sensors. These force sensors consist of a resonating microbeam working as a sensing element in order to convert applied axial force to a change in the resonance frequency[2]. Microbeam resonators' relatively high axial stiffness has necessitated various force amplification mechanisms such as single and multi-stage micro-levers to improve the sensitivity of such sensors [3]–[5]. Earlier works were focused on increasing the sensitivity by optimizing the resonant beam dimensions[6]. Other studies have demonstrated improvements by operating in higher resonating modes [7].

Although the resonant sensing element can be designed in different shapes and operate in various vibrating modes [8], typically, the resonant elements were developed based on a clamped-clamped microbeam [9]. Herein, we propose a simple strategy to increase the sensitivity by changing the boundary conditions of the beam. By using Pinned-Pinned or Free-Free microbeams, we can significantly improve the performance of such sensors. In the next section, we introduce the fabricated MEMS devices, followed by an experimental setup and discussion of the results.

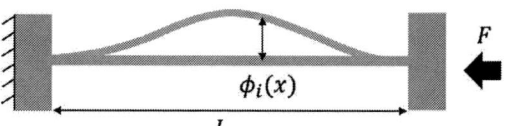

Figure 1 A vibrating beam under axial force

RESONANT SENSING ELEMENT

In a resonant force sensor, an input force leads to a change in the resonance frequency of a resonator [10]. Present designs of resonant force sensors are typically based on clamped-clamped beams excited in one of the flexural modes (Figure 1). The ith resonance frequency of a beam is given by $\omega_i = \sqrt{K_{eff}^i / M_{eff}^i}$, where the effective stiffness and mass of the beam, K_{eff}^i and M_{eff}^i, respectively, depending on the mode shape. The sensitivity of the beam response to an axial force mainly stems from the variations in the effective stiffness of the structure. From Euler-Bernoulli equations, we can write:

$$\rho A \frac{\partial^2 w}{\partial t^2} - F \frac{\partial^2 w}{\partial x^2} + EI \frac{4w}{\partial x^4} = P \qquad (1)$$

where I is the moment of inertia, E is Young's Module, ρ is the material density, A is the beam's cross-section, F is the axial force, P is lateral electrostatic force applied on the beam.

By assuming that the solution of the equation is a combination of orthogonal mode vibrations, ϕ_i, we know K_{eff}^i of a beam under the axial applied force F can be calculated from the equation shown below [11]:

$$K_{eff} = F \int_{x=0}^{L} \left(\frac{\partial \phi_i}{\partial x}\right)^2 dx + \int_{x=0}^{L} EI \left(\frac{\partial^2 \phi_i}{\partial x^2}\right)^2 dx \qquad (2)$$

where ϕ_i represents the ith beam mode shape, E is Young's module, and I is the moment of inertia. To improve the sensitivity of the sensor, we investigated methods to increase the first term in Equation 1 (i.e. $\int_{x=0}^{L} \left(\frac{\partial \phi_i}{\partial x}\right)^2 dx$) by changing the boundary conditions of the beam, and hence, influencing ϕ_i. The deflection equations of beams with Pinned-Pinned and Free-Free boundary conditions are such that they increase the sensitivity to the axial force.

Figure 2(a) shows a design of a Clamped-Clamped beam with a length $310\,\mu m$ and a width of $10\mu m$; this design acts as a reference design to compare the performance of resonators with different boundary conditions, and Figure 2(b) shows the optical image of the fabricated device.

Figure 2(c) shows the design of a Pinned-Pinned resonant beam; adding two beams at the two ends of the

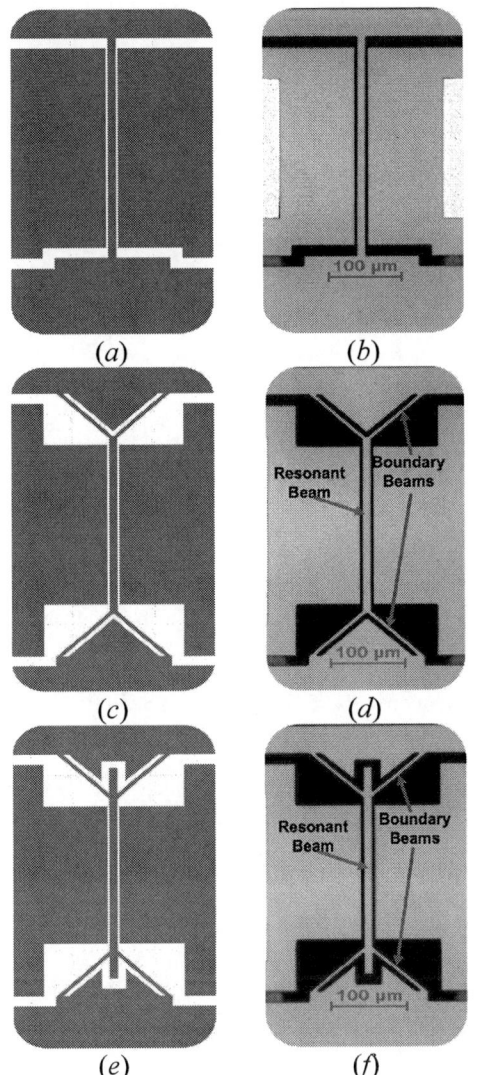

Figure 2(e) shows the design of a Free-Free beam; two boundary beams are attached to the resonant beam so that the beam's vibration mode is close to an ideal Free-Free beam. Similar to the Pinned-Pinned design, the angle of the boundary beams is 45 degrees. As the boundary beam width gets smaller, the vibration modes of the resonant beams get closer to an ideal Free-Free beam, as softer beams would have less impact on the resonant beam. The summary of the design performance is discussed in the result section. All the presented devices were fabricated using the standard PiezoMUMPs process with a 25 μm device layer. In the next section, we discuss the experimental setups.

EXPERIMENTS

All the fabricated devices introduced in the previous section have a large electrode at the top, as shown in Figure (3), for applying electrostatic force. Devices were put inside a vacuum chamber, and the vacuum level was kept around 20mTorr. We used a lock-in amplifier to measure the resonance frequency of the devices using the synchronous demodulation technique [12]. During measurement, all the resonant beams were kept at the same DC voltage to eliminate the effects of electrostatic voltages on the resonance shift. While keeping the electrostatic force zero, we measure the natural resonance frequency of each beam by applying the same AC voltage using the Lock-in amplifier.

In the next step, we increased the electrostatic force on the top electrode by applying voltage for all the resonant devices. We measured the resonance frequency one more time and reported the resonance shift. A low-noise transimpedance amplifier was used to amplify the output signal of the resonators.

The electrostatic force was applied to all test devices by using similar electrostatic force actuator electrodes and subjecting them to equal voltages, 50 Volts, which resulted in 34.5703 μN. Figure 3 shows the electrode topology and measurement setup.

Figure 2: Beams with different boundary conditions (a) Clamped-Clamped, (c)Pinned-Pinned, and (e)Free-Free (b,d,f) corresponding fabricated devices along with depiction of Boundary and Resonant. beams

Resonant beam with 45-degree angles (angle between boundary and resonant beam) allows the resonant beam to have a non-zero boundary condition at each end which results in change the vibration shape and as a result, based on equation (2), the sensitivity of the resonance beam will increase. Having 45-degree angle beams as the boundary beams (as shown in Figure 2), not only allows the resonant beam to experience Pinned boundary condition, but also 45-degree angle is not too wide to decrease the effective axial force on the beam resonators. Figure 2 (d) shows an optical microscope's fabricated Pinned-Pinned beam image. As the boundary beam becomes narrower, the resonant beam is closer to an ideal Pinned-Pinned beam. In this design, the length of the beam is ; this length is smaller than the Clamped-Clamped beam to keep their resonance frequency close to each other for comparison.

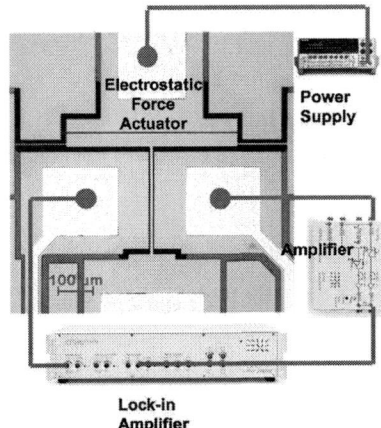

Figure 3: Electrode topology and measurement setup using a lock-in amplifier.

978-1-6654-9309-3/23 $31.00 © 2023 IEEE

Table 1. Summary of the performance and comparison to numerical simulations.

	Dimensions	Dimensions	Resonance Frequency [Hz]		Resonance Shift due to 34.57 μN force [Hz]	
	Resonant beam [μm]	Boundary beam [μm]	Simulation	Measurement	Simulation	Measurement
Clamped-Clamped	Length=310 Width=10	NA	829794	791485	97	91
Pinned-Pinned	Length=245 Width=10	Length=91 Width=5	842054	766992	175	180
Free-Free	Length=300 Width=10	Length=91 Width=5	1056970	977690	180	185

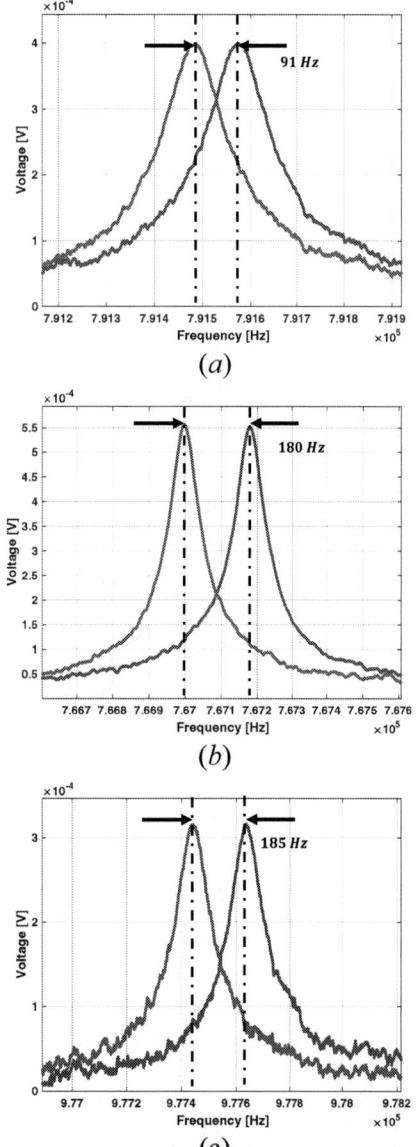

Figure 4: Experimental results and comparison of resonant shift between beams with different boundary conditions under the same axial force (34.57μN) (a)Clamped-Clamped (b)Pinned-Pinned (c)Free-free.

RESULTS

All the devices that were presented in the previous section have been simulated using the Finite Element simulator software CoventorWare. A summary of the simulations is shown in table 1. The simulation shows that by changing the boundary conditions, we can significantly increase the sensitivity of micro resonant force sensors to the axial force. Free-Free and Pinned-Pinned boundary conditions have better sensitivity to axial force. The sensitivity has increased from $97\,Hz$ for Clamped-Clamped beams to $175\,Hz$, and to $180\,Hz$. To validate simulation results, further tests have been done on the fabricated devices.

Figure 4 (a) shows the measured data for the Clamped-Clamped resonant beam, which depicts a 91 Hz resonance shift due 50Volts voltage difference on the electrostatic force electrodes; this $50\,V$ corresponds to $34.57\mu N$ according to the dimensions of the electrostatic electrode. Figure 4 (b) and (c) show the measured signals from the synchronous demodulator setup for resonance frequency measurement, showing a $180\,Hz$ shift in resonance frequency for the Pinned-Pinned resonant beam and a $185\,Hz$ shift in resonance frequency in the Free-Free resonant beam. These Experimental results validate simulation results that by changing the boundary condition of the resonant beams, the axial force sensitivity will significantly increase. Table 1 summarizes the simulation and measurement results.

CONCLUSION

This paper introduced a simple design approach to increase the sensitivity of micro-beam resonators to axial force. We demonstrated that by modifying the boundary conditions of sensing beams, it is possible to significantly increase the sensitivity of beam resonators that are commonly used in resonant sensors. Devices with different boundary conditions were introduced; simulation results have been presented showing the improvement in the performance, the resonance shift increased from 97Hz in the Clamped-Clamped beam to 175 Hz in Pinned-Pinned and to 180 using Free-Free boundary condition. Furthermore, experimental data were presented that validated the design principles, and the sensitivity increase by 100% due to changes in the boundary conditions. The resonance shift increased from 91 Hz in Clamped-Clamped to 180 Hz in Pinned-Pinned and 185 Hz in Free-Free resonant beams.

ACKNOWLEDGEMENTS

We would like to acknowledge CMC Microsystems for the provision of products and services that facilitated this research, including the fabrication of the devices presented in this paper.

REFERENCES

[1] G. Verma, K. Mondal, and A. Gupta, "Si-based MEMS resonant sensor: A review from microfabrication perspective," *Microelectronics Journal*, vol. 118, p. 105210, Dec. 2021, doi: 10.1016/j.mejo.2021.105210.

[2] C. Burrer and J. Esteve, "A novel resonant silicon accelerometer in bulk-micromachining technology," *Sensors and Actuators, A: Physical*, vol. 46, no. 1–3, pp. 185–189, 1995, doi: 10.1016/0924-4247(94)00887-N.

[3] X.-P. S. Su and H. S. Yang, "Single-stage microleverage mechanism optimization in a resonant accelerometer," *Structural and Multidisciplinary Optimization*, vol. 21, no. 3, pp. 246–252, 2001, doi: 10.1007/s001580050189.

[4] Y. Zhao *et al.*, "A sub-μg bias-instability MEMS oscillating accelerometer with an ultra-low-noise read-out circuit in CMOS," *IEEE Journal of Solid-State Circuits*, vol. 50, no. 9, pp. 2113–2126, Sep. 2015, doi: 10.1109/JSSC.2015.2431076.

[5] H. Ding, X. Le, Y. Ma, and J. Xie, "A biaxial resonant tilt sensor with two-stage microleverage mechanisms," in *2017 19th International Conference on Solid-State Sensors, Actuators and Microsystems (TRANSDUCERS)*, Jun. 2017, pp. 1013–1016. doi: 10.1109/TRANSDUCERS.2017.7994223.

[6] H. A. C. Tilmans, M. Elwenspoek, and J. H. J. Fluitman, "Micro resonant force gauges," *Sensors and Actuators A: Physical*, vol. 30, no. 1, pp. 35–53, Jan. 1992, doi: 10.1016/0924-4247(92)80194-8.

[7] M. Pandit *et al.*, "An Ultra-High Resolution Resonant MEMS Accelerometer," in *2019 IEEE 32nd International Conference on Micro Electro Mechanical Systems (MEMS)*, Jan. 2019, pp. 664–667. doi: 10.1109/MEMSYS.2019.8870734.

[8] G. Stemme, "Resonant silicon sensors," *Journal of Micromechanics and Microengineering*, vol. 1, no. 2, pp. 113–125, Jun. 1991, doi: 10.1088/0960-1317/1/2/004.

[9] H. Ding, W. Wang, B.-F. Ju, and J. Xie, "A {MEMS} resonant accelerometer with sensitivity enhancement and adjustment mechanisms," *Journal of Micromechanics and Microengineering*, vol. 27, no. 11, p. 115010, Oct. 2017, doi: 10.1088/1361-6439/aa8d99.

[10] R. Abdolvand, B. Bahreyni, J. Lee, and F. Nabki, "Micromachined Resonators: A Review," *Micromachines*, vol. 7, no. 9, p. 160, Sep. 2016, doi: 10.3390/mi7090160.

[11] B. Bahreyni, *Fabrication and design of resonant microdevices*. Norwich, NY: W. Andrew Inc, 2008.

[12] E. Ghaderi and B. Bahreyni, "Synchronous Demodulation for Low Noise Measurements," *IEEE Instrumentation & Measurement Magazine*, vol. 24, no. 2, pp. 72–78, Apr. 2021, doi: 10.1109/MIM.2021.9400956.

CONTACT

*Erfan Ghaderi, eghaderi@sfu.ca

OCCLUSAL PAPER-BASED FLEXIBLE PRESSURE SENSOR FOR IN SITU MEASURING ORAL OCCLUSAL FORCE

Wenduo Wang[1,2], Xin Zhang[1,2], Ning Zhao[3], Jingquan Liu[1], and Bin Yang[1]

[1]National Key Laboratory of Science and Technology on Micro/Nano Fabrication, Shanghai Jiao Tong University, CHINA

[2]Department of Micro/Nano Electronics, Shanghai Jiao Tong University, CHINA and

[3] Shanghai Ninth People's Hospital , CHINA

ABSTRACT

A pressure sensor based on occlusal paper is proposed for the first time to measure occlusal force at the given position in real time. A light-weight bacterial cellulose MXene (MX-BC) composite aerogel as piezoresistive sensing layer was prepared by liquid nitrogen assisted unidirectional freezing technology. The flexible pressure sensor was connected with the wireless test system based on Bluetooth module. The position and range of occlusion by coloring the occlusal paper acted on the teeth was observed, and the occlusal force at the corresponding location can be recorded through the fabricated pressure sensor. Meanwhile, and the waveform through the APP was displayed on the mobile equipment in real time.

KEYWORDS

Paper-based Flexible Sensor, Occlusal Force, In Situ Measurement, Interdigital Electrode.

INTRODUCTION

With the development of science and technology, people are paying more and more attention to physical health [1]. The development of implantable sensors has met people's demand for human internal health detection [2,3]. Among them, implantable sensors for oral health have been developed in recent years. In oral health studies, occlusal force has become an important therapeutic evaluation index for dental restoration, periodontal treatment and orthodontic treatment [4]. It is also an important parameter to evaluate the effectiveness of masticatory systems [5,6]. By measuring occlusal force and dynamic changes of occlusal contact at different stages of treatment, doctors can obtain the basis for occlusal adjustment and objectively evaluate the therapeutic effect. The function and integrity of the masticatory system has an important impact on a person's quality of life. Therefore, timely diagnosis and treatment of these diseases is essential to improve quality of life. At present, the principles of bite testing equipment are varied. Compared with other methods, pressure sensor has become the primary choice for the development of occlusal force sensor due to its simple principle and high measurement accuracy.

Nowadays, most of the clinical methods used to detect occlusal force are qualitative measurement of occlusal paper. Occlusal paper with different colors and thicknesses was used to mark the occlusal contact points in different contact states, and the position and range of occlusal contact were judged by the tooth surface coloring. The bite paper can be qualitatively measured, but not quantitatively. Most commercial pressure sensing devices used to detect bite force are mainly piezoelectric films, which have high sensitivity [7]. However, due to excessive hardness, it will interfere with the normal occlusal, after repeated using and its accuracy will be reduced. Therefore, it is urgent to develop new occlusal force testing equipment.

In this paper, we propose an occlusal paper-based flexible sensor (OPF) which can accurately and quantitatively measure the occlusal force through the sensing layer. Compared with commercial hard piezoelectric films, the flexible substrate does not affect the position and accuracy of the occlusal force. Compared with the clinical occlusal paper, OPF can not only mark the position, but also measure the occlusal force. Two-dimensional MXene [8,9] material with high electrical conductivity and bacterial cellulose with abundant surface functional groups are prepared composite aerogel as a

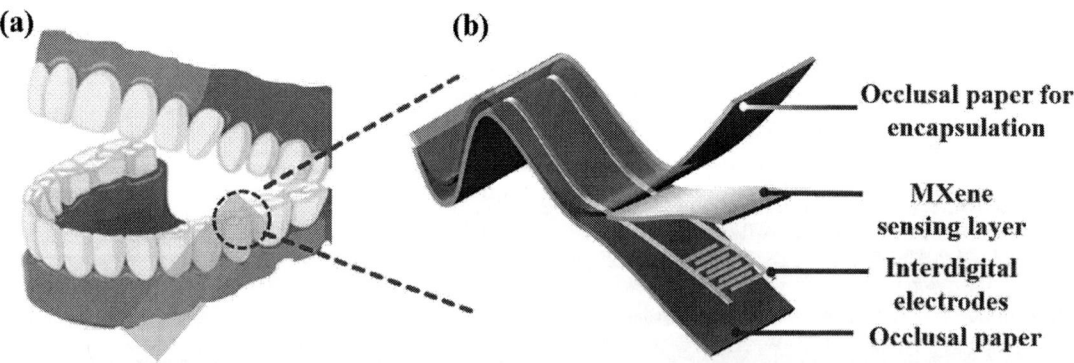

Figure 1: (a) Schematic diagram of flexible occlusal force measurement and positioning sensor (b) Structure of flexible occlusal force sensor.

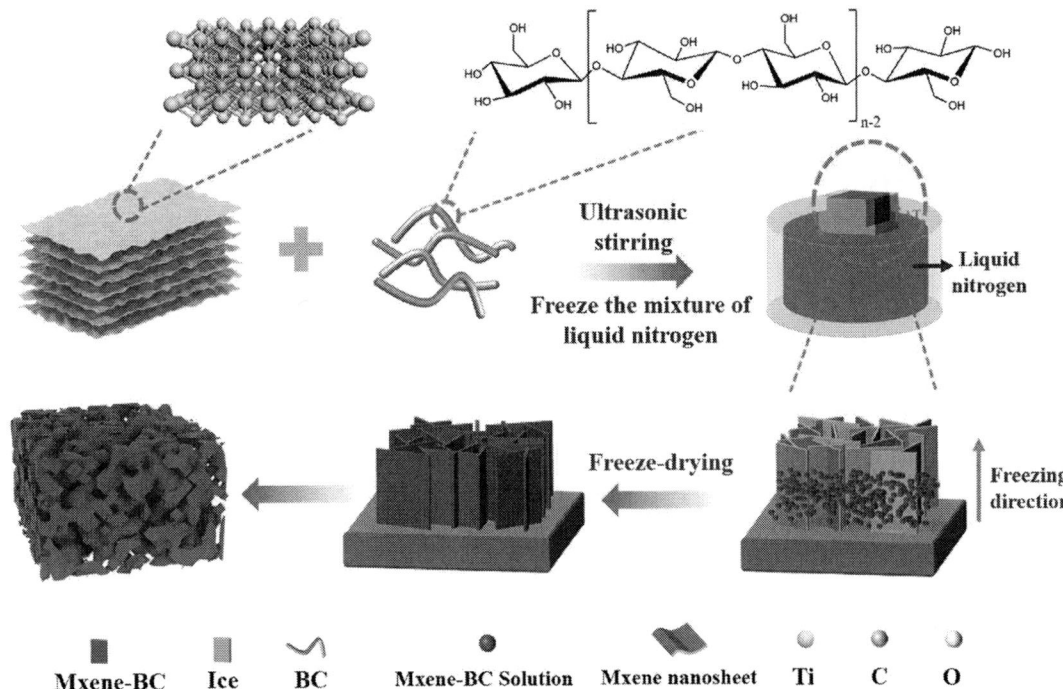

Figure 2: MXene-BC aerogel fabrication process by orientated freeze-vacuum drying.

piezoresistive sensing layer. Through Bluetooth real-time transmission, the force is transmitted to the APP, and can be shown in real-time. According to the current research, the occlusal paper-based flexible sensor is the first proposed.

DESIGN AND FABRICATION

Figure 1a shows the application scenario of the OPF sensor. Figure 1b shows the explosion view of the OPF sensor. The functional layers from top to bottom are the occlusal paper package layer, sensing layer, interdigital electrode and occlusal paper substrate, respectively. The interdigital electrode is directly deposited and patterned with hard mask on the flexible occlusal paper by electron beam evaporation. Then, the prepared composite aerogel sensing layer is fixed by double-sided tape and covered on the interdigital electrode. Finally, another layer of occlusal paper is used to wrap.

Figure 2 shows the preparation process of the sensing layer. The preparation method of MXene gives them abundant surface functional groups, large specific surface area and hydrophilicity. The microstructure of MXene can be directly adjusted and combined with other functional compounds to construct composite flexible materials for pressure sensors [10]. Aerogel is a three-dimensional porous solid material with high porosity and large surface area. The 3D conductive structure can withstand large deformation, thus greatly improving the pressure range of the sensor [11]. Aerogels were prepared by vacuum freeze-drying [12]. Bacterial cellulose with abundant surface functional groups was selected as the supporting material. Firstly, the bacterial cellulose and MXene solution were mixed uniformly by ultrasound for 15 minutes. Then the mixed solution was used for directional freezing with liquid nitrogen. Finally, put the frozen into the vacuum

freeze-dryer for 48 hours to get the composite aerogel.

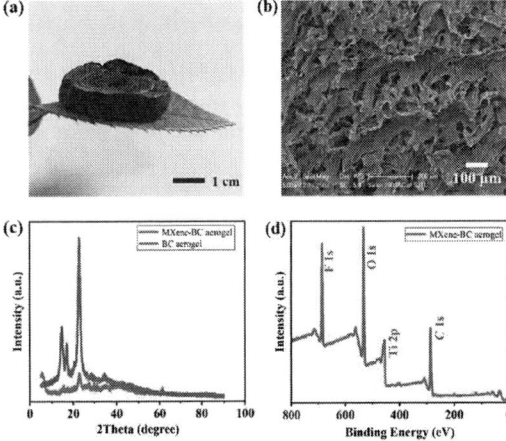

Figure 3: (a) Optical image of MXene-BC aerogel with lightweight. (b) SEM image of MXene-BC aerogel. (c) XRD patterns of MXene-BC aerogel and pure BC. (d) XPS patterns of MXene-BC aerogel.

The optical image of MXene-BC aerogel is shown in Figure 3a, the fabricated MXene composite aerogel is with the diameter of 3 cm , which is light enough to sit on leaves. Figure 3b shows its SEM image. It shows the porous structure of aerogel, and the pore size is uniform and about 100 μm, and the three-dimensional structure is clear. Figure 3c shows the XRD comparison between the composite aerogel and bacterial cellulose aerogel. It can be clearly seen that the structure changes after MXene material is involved. Figure 3d is an XPS image conforming to aerogel, in which carbon, oxygen, fluorine and titanium elements can be seen, confirming the successful incorporation of MXene material [13,14].

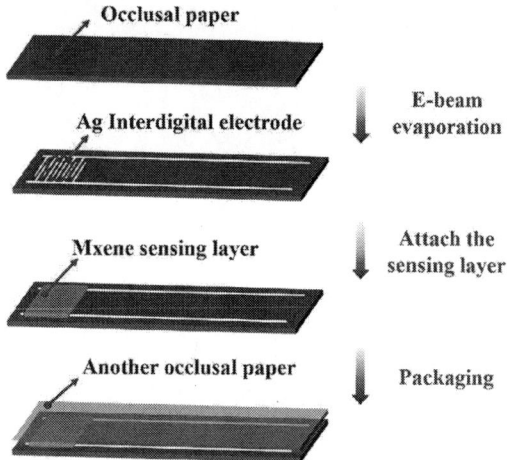

Figure 4: Fabrication process of flexible sensor.

Figure 4 shows the device preparation method. The occlusal paper is selected as the flexible substrate. The stainless steel interdigital structure is used as hard mask to pattern the interdigital metal electrodes by evaporation process. In order to improve the bonding force of metal, a layer of chromium metal is evaporated and then silver electrode is deposited. The aerogel sensing layer is fixed on the interdigital electrode, and finally the occluding paper is used for encapsulation, to ensure that the part deep into the mouth is waterproof, and at the same time to ensure that the leading electrode is connected to the voltage collector. The device can leave positioning information on the upper and lower teeth.

RESULTS AND DISCUSSION

Figure 5 shows the device test results. Considering the application scenario of the device, we pay more attention to the sensitivity, consistency and response time of the sensor. Figure 5a shows the test platform. Air compressor is used to apply force to the sensor and LCR meter is used to record the change of resistance value. Figure 5b is a schematic diagram of the device sensitivity. It can be seen from the diagram that the measuring range of this device is from 0 to 400 kPa and can be used to measure the occlusal force of teeth. The overall sensitivity is divided into four segments, each segment has good linearity. In the range of 0-10 kPa, the sensitivity of the device is 8.189 kPa^{-1}. The sensitivity is 1.678 kPa^{-1} from 10 to 50 kPa. With the increasing of applied force, up to 150kPa, the sensitivity reduces to 0.463 kPa^{-1}. At the last range of 150-400 kPa, the sensitivity is only 0.107 kPa^{-1}. Figure 5c shows the cyclic testing results under different pressures. It can be seen that the sensitivity is basically unchanged under the same pressures, and the consistency and repeatability of the device are good. Figure 5d shows the response and recovery time of the sensor, which are 163 ms and 232 ms, respectively.

The signal waveform of pressure can be visually displayed on the mobile phone by Bluetooth transmission module (Figure 6). Figure 6a and 6b are the picture of the sensor and the complete system with voltage acquisition and transmission circuit, respectively. Impressively, volunteers tested the occlusal force with the device and got satisfactory results. Figure 6c shows the staining marks left on the teeth of the volunteers after occlusion. The position

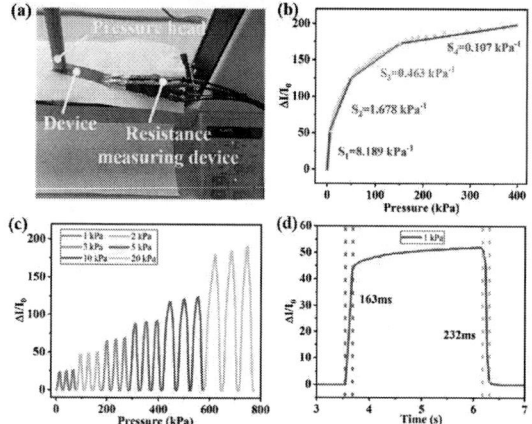

Figure 5: (a) Testing platform of the flexible sensor. (b) Sensitivity of MXene-BC aerogel flexible sensor. (c) Dynamic pressure response of MXene-BC aerogel flexible sensor under various pressures. (d) Response and recovery time of MXene-BC aerogel flexible sensor.

left by the occlusion of a single tooth can be seen. Figure 6d and 6e are the real-time display curve of occlusal force and enlarged images achieved through the acquisition circuit and Bluetooth transmission after the test respectively. In summary, it can be seen that the device has achieved both qualitative and quantitative test results.

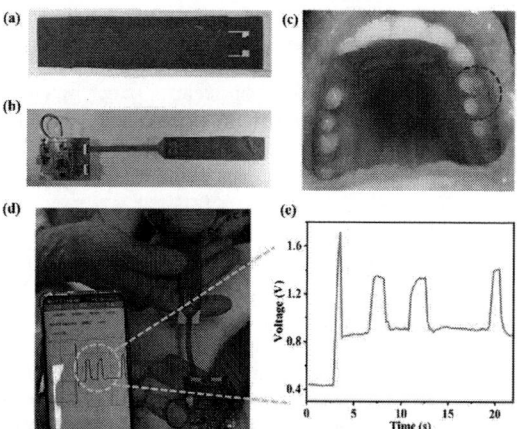

Figure 6: (a) Optical image of the whole sensor. (b) Test system integrated with Bluetooth wireless transmission. (c) The occlusal marks after the test. (d) Pressure signal shown in mobile phone. (e) Magnification of pressure signal.

CONCLUSION

This paper presents a paper based flexible

piezoresistive sensor for in-situ measurement of bite force, which can be transmitted to an APP via Bluetooth to display dynamic waveforms in real time. SEM and other images show that the three-dimensional structure of the composite aerogel is obvious, and it can withstand large deformation well, thus greatly improving the pressure range of the sensor. The test performance of the device shows that the device has excellent sensitivity, wide measuring range, good linearity, good repeatability. We also designed a voltage acquisition circuit and a Bluetooth transmission module to transmit the data measured by the occlusal force sensor to the APP for graphic display. In particular, we tested the device performance through volunteers, and the test results can well meet the requirements of in situ quantitative measurement.

ACKNOWLEDGEMENTS

The authors thank partly financial support from the National Natural Science Foundation of China (No. 12072189) The authors are also grateful to the Center for Advanced Electronic Materials and Devices (AEMD) of Shanghai Jiao Tong University.

REFERENCES

[1] Zhang, X., Lu, L., Wang, W., Zhao, N., He, P., Liu, J., & Yang, B., "Flexible Pressure Sensors with Combined Spraying and Self-Diffusion of Carbon Nanotubes", ACS applied materials & interfaces, vol. 14, pp. 38409–38420, 2022.

[2] Beg, S., Handa, M., Shukla, R., Rahman, M., Almalki, W. H., Afzal, O., & Altamimi, A., "Wearable smart devices in cancer diagnosis and remote clinical trial monitoring: Transforming the healthcare applications", *Drug discovery today*, vol. 27, pp. 103314, 2022.

[3] Kim, J. J., Stafford, G. R., Beauchamp, C., & Kim, S. A., "Development of a Dental Implantable Temperature Sensor for Real-Time Diagnosis of Infectious Disease", *Sensors (Basel, Switzerland)*, vol. 20, pp. 3953, 2020.

[4] Gu, Y., Bai, Y., & Xie, X., "Bite Force Transducers and Measurement Devices", *Frontiers in bioengineering and biotechnology*, vol. 9, pp. 665081, 2021.

[5] Bostancıoğlu, S. E., Toğay, A., & Tamam, E., "Comparison of two different digital occlusal analysis methods", *Clinical oral investigations*, vol. 26, pp. 2095–2109, 2020.

[6] Hashimoto, S.; Kosaka, T.; Nakai, M.; Kida, M.; Fushida, S.;Kokubo, Y.; Watanabe, M.; Higashiyama, A.; Ikebe, K.; Ono, T.;Miyamoto, Y., "A Lower Maximum Bite Force Is a Risk Factor for Developing Cardiovascular Disease: The Suita Study", *Sci. Rep.*, vol. 11, pp. 7671, 2021.

[7] Assery, M. K., Prosthodontics (Certified), Albusaily, H. S., Pani, S. C., & Aldossary, M. S., "Bite Force and Occlusal Patterns in the Mixed Dentition of Children with Down Syndrome", *Journal of prosthodontics : official journal of the American College of Prosthodontists*, vol. 29, pp. 472–478, 2020.

[8] Hang, G., Wang, X., Zhang, J., Wei, Y., He, S., Wang, H., & Liu, Z., "Review of MXene Nanosheet Composites for Flexible Pressure Sensors", ACS Applied Nano Materials, vol. 5, pp. 14191-14208, 2022.

[9] Zhao, L. J.; Wang, K.; Wei, W.; Wang, L. L.; Han, W., "High-Performance Flexible Sensing Devices Based on Polyaniline/MXene Nanocomposites", *InfoMat*, vol. 1, pp. 407−416, 2019.

[10] Wan, Y., Xiong, P., Liu, J., Feng, F., Xun, X., Gama, F. M., Zhang, Q., Yao, F., Yang, Z., Luo, H., & Xu, Y., "Ultrathin, Strong, and Highly Flexible $Ti_3C_2T_x$ MXene/Bacterial Cellulose Composite Films for High-Performance Electromagnetic Interference Shielding", *ACS Nano*, vol. 15, pp. 8439–8449, 2021.

[11] Wu, Z., Wei, L., Tang, S., Xiong, Y., Qin, X., Luo, J., Fang, J., & Wang, X., "Recent Progress in $Ti_3C_2T_x$ MXene-Based Flexible Pressure Sensors", *ACS Nano*, vol. 15, pp. 18880–18894, 2021.

[12] Zhu, M., Yan, X., Xu, H., Xu, Y., & Kong, L., "Ultralight, compressible, and anisotropic MXene@ Wood nanocomposite aerogel with excellent electromagnetic wave shielding and absorbing properties at different directions", *Carbon*, vol. 182, pp. 806-814, 2021.

[13] Jin, X., Li, L., Zhao, S., Li, X., Jiang, K., Wang, L., & Shen, G., "Assessment of Occlusal Force and Local Gas Release Using Degradable Bacterial Cellulose/$Ti_3C_2T_x$ MXene Bioaerogel for Oral Healthcare", *ACS Nano*, vol. 15, pp. 18385–18393, 2021.

[14] Chao, M. Y.; He, L. Z.; Gong, M.; Li, N.; Li, X. B.; Peng, L. F.;Shi, F.; Zhang, L. Q.; Wan, P. B., "Breathable $Ti_3C_2T_x$ MXene/ProteinNanocomposites for Ultrasensitive Medical Pressure Sensor with Degradability in Solvents", *ACS Nano*, vol. 15, pp. 9746−9758, 2021.

CONTACT

*B. Yang, tel: +86-21-34206683; binyang@sjtu.edu.cn

SUCTION CUP ARRAY WORKING ALSO AS TACTILE SENSOR TO DETECT CUPS DEFORMATION USING KCF AND CNN

Toshihiro Shiratori, Jinya Sakamoto, Yuki Kumokita, Masato Suzuki,
Tomokazu Takahashi, and Seiji Aoyagi
Mechanical Engineering, Engineering Science Major, Kansai University, JAPAN

ABSTRACT

We propose a micro-suction cup array device with tactile sensing capabilities. The suction cup array device is equipped with a camera for capturing images of the contact surface. The camera reads the deformation of the suction cups. Image processing estimates the displacement of each suction cup contact surface, its direction, and the contact force distribution. A preliminary experimental device has been fabricated. The master shape of the micro cup was fabricated by 3D laser lithography and UV nanoimprinting processes. A hole in the suction cup was opened by the femtosecond laser beam. Kernelized Correlation Filter was applied to the acquired images and the movement of the contact surface of each suction cup could be estimated. Convolutional Neural Networks (CNN) was used to estimate the pressing force. The correctness (accuracy) of the produced CNN model was 87%, which is a practical accuracy.

KEYWORDS

Suction cup array, CNN, tactile sensor, nanoimprint, femtosecond laser processing

INTRODUCTION

Robot grippers generally consist of two plates that sandwich the object between them, but fragile or non-flat objects are challenging to handle. Ordinary suction cup devices cannot handle small, grooved, or curved objects due to the size of the suction cups. Micro-suction cup arrays make it possible to handle such objects. Adding a suction cup array to the gripper ensures that such objects can be securely grasped and manipulated in the gripper, as small contact forces are achieved by moderate suction forces [1]. If contact conditions can be detected, nimbler handling becomes possible. GelSight [2] and Finger Vision [3], which detect tactile information by setting a transparent rubber sheet containing markers on a gripper and imaging the markers' movements, have been put to practical use and used in robot handling research. In this study, a new device integrating a suction cup array and tactile sensing function is proposed.

DEVICE CONCEPT

Our proposed device concept is illustrated in Fig. 1. Micro-suction cup arrays are placed on the device's surface. When an object is grasped, the contact surface of each micro-suction cup follows and deforms. As is the case with many vision-based contact sensors, this deformation is read from behind with a small camera. In this process, the micro-suction cups are constantly being suctioned. The shape of the suction cups that are captured depends on the vacuum suction, the pressing force, and the movement of the contact surface. By performing image analysis on the

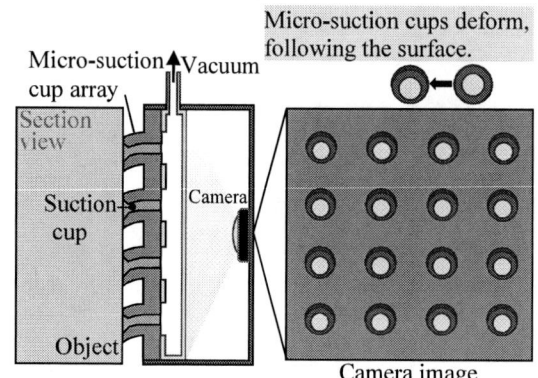

Figure 1: *The concept of a suction cup array device that also functions as a tactile sensor.*

images acquired from the camera, the real-time contact situation can be estimated and the object to be grasped can be vacuumed. This device is attached to the fingertip of the gripper.

MICRO-SUCTION CUP ARRAYS

Micro-suction cup arrays were fabricated using 3D laser lithography and nanoimprinting techniques. The fabrication method of the micro-suction cup array is shown in Fig. 2. Suction cups are made of polydimethylsiloxane (PDMS), and the ratio of the main ingredient to hardener is 5:1. A master (2×2) of micro-suction cups was fabricated using 3D laser lithography (Nanoscribe GmbH, resolution; 200 nm). As it is difficult to fabricate microstructures in large areas, nanoimprint technology was used to fabricate large areas. This master was replicated by UV nanoimprinting and manually assembled with epoxy glue. The processing conditions for nanoimprinting are: pressing pressure 0.35 MPa, pressure holding time before UV irradiation 200 s, and UV irradiation time 300 s. A 4×4 PDMS micro-suction cup array was fabricated by pouring and curing PDMS onto this assembled mold. FluroSurf NL-1(Fluoro Technology Ltd.) was used as a release agent in all the processes of releasing molds.

A through-hole was opened to the center of each suction-cup using a femtosecond laser (Cyber laser, IFRIT-D) for vacuum-sucking an object. The PDMS in the area outside the micro-suction cups is eliminated using a razor blade. As the PDMS pad is very thin, pressing it in this state causes excessive deflection and damage. A POM (polyoxymethylene) plate was used to prevent deflection. Opening holes in the POM plate was carried out by femtosecond laser. These fabrication processes are shown in Fig. 3. Details in the processing of PDMS using femtosecond lasers are shown in Fig. 4. It is difficult to hollow out PDMS in a cylindrical shape. The laser is

a) Master fabricated by 3D laser lithography

b) Pour PDMS and release

c) 2×2 micro-suction cups

d) UV nanoimprint

e) Cups made of UV curable resin

f) Assemble 4 molds manually

g) Pour PDMS and release

h) 4×4 micro-suction cups

Figure 2: Fabrication process of 4×4 PDMS micro-suction cup array using 3D laser lithography (Nanoscribe GmbH, resolution; 200 nm) and UV nanoimprint.

a) 4×4 micro-suction cups

b) Open hole by femtosecond laser beam

Hole

c) Eliminate extra PDMS by razor cutting

d) Bonding POM (polyoxymethylene) plate

Hole

4×4 micro-suction cups on POM plate

Figure 3: Opening holes by femtosecond laser and bonding POM plate.

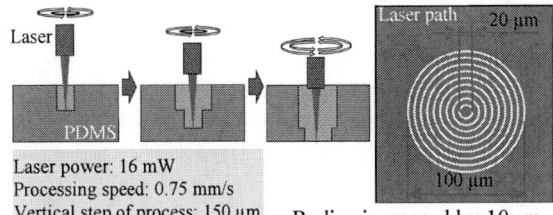

Laser power: 16 mW
Processing speed: 0.75 mm/s
Vertical step of process: 150 μm

Radius increased by 10μm

Figure 4: Path of the femtosecond laser beam for opening holes.

emitted in a circular pattern from the center while increasing the radius by 10 μm. This process is repeated twice, after which the laser depth of focus is lowered by 150 μm. This is one step. By repeating this step, the holes were successfully opened in the PDMS. The suction area of one suction cup is 0.26 mm² and the donut-shaped contact area is 0.23 mm².

PRELIMINARY EXPERIMENT

An experimental device was built to confirm the validity of the concept. An overview of the fabricated device is shown in Fig. 5. The device was fabricated using an optical 3D printer (formlabs, Form3). Four red LEDs illuminate the micro-suction cups at a 30° angle to ensure that the brightness of the external environment is not a disturbance factor. The inside of the device is evacuated by a suction pump. The deformation image of the suction cups was captured by a camera (WRAYCAM, 5,440×3,648 pixels) with a magnification lens (Kenko Tokina, KCM-Z6II). This device was used to lift a 0.14 g object with a single suction cup. The ultimate pressure of a pump used is 24.0 kPa. An industrial robot (FUNUC, LR Mate 200iD) and its end-effector, a commercial force sensor (BL AUTOTEC, MINI 2/10-A, resolution 19.6 mgf), were used to capture teacher images for the Convolutional Neural Networks (CNN) [4] and validation videos for the Kernelized Correlation Filter (KFC) method [5].

The steps in the capturing process are as follows. The device and camera are fixed to the desk and the camera is focused on the contact area. The force sensor of the robot

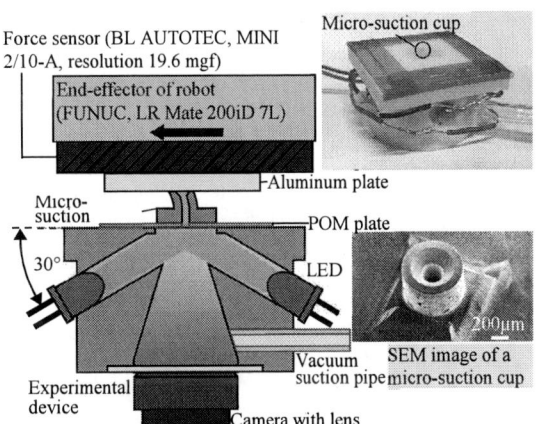

Figure 5: Device to observe deformation of micro-suction cups, while applied pressing and shearing forces by robot end-effector with commercial force sensor.

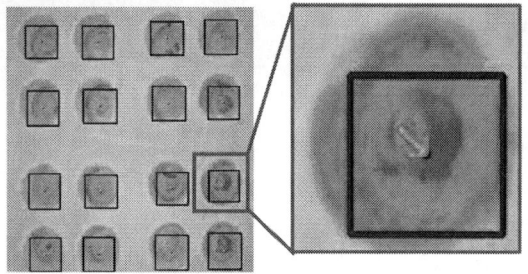

Pressing force: 2.5 N
Lateral movement: 500 μm

Figure 6: Movement detection of contacting surface of suction cups by KCF method.

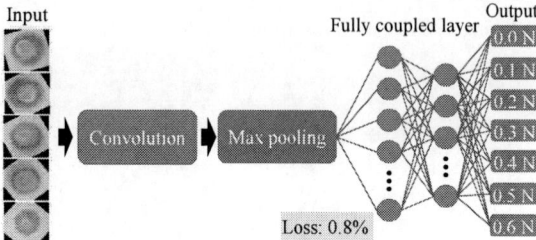

Figure 7: Input images and details of the structure of the CNN model.

arm's end-effector is pressed against a suction cup on the device. Depending on the pressing force, the suction cup follows and deforms, and the focus is shifted. With a constant pressing force, it also moves laterally. In this case, the doughnut-shaped contact area characteristically follows and moves. The camera's brightness and other factors, such as the zoom factor of the lens, are constant for all images captured.

IMAGE ANALYSIS

The A through-hole was opened to the center of each suction-cup using a femtosecond laser (Cyber laser, IFRIT-D) for vacuum-sucking an object. KCF method, a well-known Tracking API in OpenCV, was used to estimate the movement of the contact area of each suction cup. This is an algorithm that can track objects in a video. It can track objects in a bounding box. The results of adapting this algorithm to the fabricated 4×4 micro-suction cup array are shown in Fig. 6. The blue squares are the bounding boxes being tracked. The movement being estimated is indicated by the arrows. In Fig. 6, a vertical pressing force of 2.5 N is applied to the entire micro-suction cup array (0.16 N per suction cup). The movement in the lateral direction is approximately 500 μm.

A CNN model was created to enable the estimation of pressing forces from camera images. The teacher images and the structure of the CNN model are shown in Fig. 7. The teacher images were captured with a single micro-suction cup. This was done because the pressing force on the suction cup and its movement were stable. When capturing the teacher image, the pressing by the robot end-effector was from 0.0 N to 0.6 N in 0.1 N increments. At each pressing force, the robot end-effector moved from 0 μm to 250 μm in 50 μm increments. 540 images were captured. Lateral movements were made in one direction only. Image processing resulted in 2870 teacher images by randomly rotating the images from 360°. It is difficult to deal directly with large sized images. In CNN, convolution and Max pooling are used to reduce the image size while preserving the image features. The model was created by taking those images into A fully coupled layer. The image sources of the training and test images are different. As a result of the training, the loss of the created model was 0.8%.

The performance of the model created is shown in

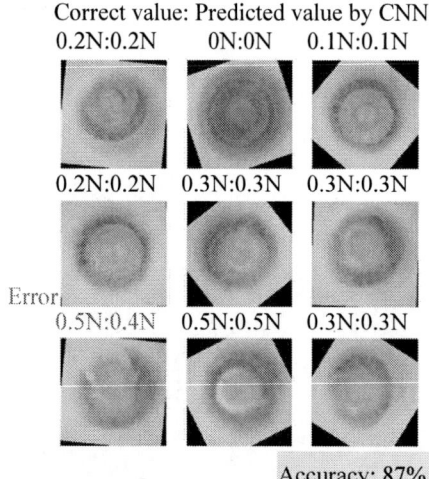

Correct value: Predicted value by CNN

0.2N:0.2N	0N:0N	0.1N:0.1N
0.2N:0.2N	0.3N:0.3N	0.3N:0.3N
Error 0.5N:0.4N	0.5N:0.5N	0.3N:0.3N

Accuracy: 87%

Figure 8: Performance evaluation of CNN model.

Figure 8. The correct (experimental) values are on the left and the values predicted by the CNN are on the right. An error is shown in red. The estimation was correct not only for simple pressing (top right) but also for pressing with lateral movement (bottom right). The accuracy (percentage of correct responses) was 87%.

DISCUSSION

In the KCF method, the estimation of direction was satisfactory, but the estimation of movement was not perfect. When large movements occur, the POM plate may obscure part of the contact area of the suction cup. If the feature points to be tracked are hidden, the movement is estimated to be small. A method that uses the KCF method to estimate the direction and a CNN to estimate the movement is under investigation.

It has been found that there are procedural problems when the molds replicated by UV nanoimprinting are assembled. The height and orientation of each suction cup are slightly different. This resulted in misaligned timing of follow-up deformation, even when in contact with a flat object, and in the diagonal movement of the suction cups, even when pressed in a vertical direction. This will be a major problem in the future as more and larger areas are developed. Assembly methods need to be improved.

In order to function as a practicable device, the camera and lens must be miniaturized. The bonding of the POM plates and suction cups to prevent deflection is done manually, so there are individual differences in the images obtained. In order to adapt this CNN model to the suction cup array video, it is necessary to take different teacher images of a larger number of suction cup samples.

SUMMARY

In this study, we developed a micro-suction cup array device that can detect contact with an object by judging its deformation captured by a camera by AI.

By 3D laser lithography and UV nanoimprinting processes, we were able to fabricate micro-suction cup arrays that can be scaled up to large areas. The suction cup array was formed by pouring PDMS into the mold and curing it. A femtosecond laser beam made a hole in each suction cup to suck an object. To confirm the usefulness of our proposed concept, experimental devices were fabricated. A camera captured the change in the contact area of the suction cup and the direction of deformation.

By applying a KCF to the acquired images, the movement of the contact surface of each suction cup was successfully estimated. CNN, one of machine learning, was used to estimate the pressing force. The force pressed against the micro-suction cup estimated by the created CNN model was compared with the force measured by the load cell. As a result, the accuracy of the pushing force estimated by the CNN model was 87%, which is a practical precision.

In future work, we will try to improve the estimation accuracy of pushing force using the CNN model.

ACKNOWLEDGEMENT

This work was financially supported in part by the Kansai University Fund for Supporting Research Group of ORDIST "Creation of nano and microdevices and their application to mechatronics, IoT, and medicine," 2020-2022. This research was conducted in part using the ORDIST clean room in Kansai University.

REFERENCES

[1] S. Nishita, H. Onoe, "Liquid-filled Flexible Micro-suction-Controller Array for Enhanced Robotic Object Manipulation," in *Journal of Microelectromechanical Systems*, vol. 26, no. 2, pp. 366-375, April 2017.

[2] R. Calandra, A. Owens, M. Upadhyaya, W. Yuan, J. Lin, E. H. Adelson, S. Levine, "The feeling of success: Does touch sensing help predict grasp outcomes?" in *CoRL*, 2017.

[3] A. Yamaguchi, C. G. Atkeson, "Implementing tactile behaviors using FingerVision," in *2017 IEEE-RAS 17th International Conference on Humanoid Robotics (Humanoids)*, 2017, pp. 241-248.

[4] M. Lambeta *et al.*, "DIGIT: A Novel Design for a Low-Cost Compact High-Resolution Tactile Sensor With Application to In-Hand Manipulation," in *IEEE Robotics and Automation Letters*, vol. 5, no. 3, pp. 3838-3845, July 2020.

[5] J. F. Henriques, R. Caseiro, P. Martins, J. Batista, "High-Speed Tracking with Kernelized Correlation Filters," in *IEEE Transactions on Pattern Analysis and Machine Intelligence*, vol. 37, no. 3, pp. 583-596, 1 March 2015.

CONTACT

*S. Toshihiro, k099383@kansai-u.ac.jp

VERTICAL INTEGRATION OF FORCE TRANSMISSION STRUCTURE ON CAPACITIVE CMOS-MEMS TACTILE FORCE SENSOR FOR SENSITIVITY IMPROVEMENT

Yuanyuan Huang[1], Yen-Lin Chen[1], Shihwei Lin[1], Fuchi Shih[2], Zihsong Hu[2] and Weileun Fang[1,2]
[1] Dept. of Power Mech. Eng., National Tsing Hua University, Hsinchu City, TAIWAN and
[2] Inst. of NanoEng. and Microsyst., National Tsing Hua University, Hsinchu City, TAIWAN

ABSTRACT

This study extends the capacitive tactile force sensor array to develop a novel force transmission structure for sensitivity enhancement (Fig.1). The sensor is designed and implemented using the commercial TSMC 1P6M CMOS process. Merits of the proposed design include: (1) Pillar-array force transmission structure: integrate with capacitive sensing array vertically to effectively transmit the load to each sensing elements to enhance the sensitivity; (2) Rigid force contact interface (i.e. glass bump): can be employed to further enhance the sensitivity. Measurements show that the sensitivity of proposed design with rigid glass bump is 5.04 fF/N (sensing range:5N). Compare with reference design having the same bump size, the sensitivity is increased for 2-fold.

KEYWORDS

Capacitive tactile force sensor, CMOS-MEMS, Vertical integration, Force transmission structure

INTRODUCTION

The tactile force sensors could serve as the human-machine interface (HMI) for the applications of industry 4.0, metaverse and consumer electronic product, etc. Tactile force sensor receives forces from surrounding environment and interprets the force data into various contact information. Currently, the sensing mechanisms of the tactile sensor include piezoresistive, capacitive, inductive and piezoelectric type, etc. [1-3]. Among them, piezoresistive and capacitive are more common sensing techniques. Furthermore, compared with the piezoresistive type, the capacitive one has the advantages of high sensitivity and low power consumption.

In recent years, commercial CMOS-MEMS platforms extends to sensor and actuator applications. By leveraging the inherent multi-layer thin films in CMOS process, various suspended structures and sensing mechanisms are realized through post-process. The capacitive CMOS-MEMS tactile force sensor has several advantages: sub-micron sensing gap, and the integration with sensing circuits and other sensors [4-6]. Moreover, for many existing tactile force sensors, the applied load will transfer to and then deform the sensing structure through the contact interface (e.g. tactile bump) and transmission structure (e.g. polymer). Thus, the contact interface and transmission structure are critical design issues for tactile force sensor.

Due to thin film residual stress and structure stiffness, large suspended membranes are typically not suggested on the CMOS-MEMS platform. For example, the array-type tactile force sensors with smaller sensing elements have been developed to improve the yield [7], [8]. However, for the array-type tactile force sensor, it is not straightforward to properly arrange the rigid contact interface. For example, the load applied on sensor would be interfered by rigid supports of each sensing elements (Fig.2). Most of the load was dispersed on anchor, leading to limited force transmitted to sensing structure. The flexible polymer is another option to transmit external load, yet sensitivity is reduced. Thus, this study proposes a novel force transmission structure to enchase sensitivity on CMOS-MEMS capacitive tactile force sensor.

DESIGN CONCEPT

Fig.1a illustrates the proposed design based on TSMC 1P6M 0.18μm CMOS process. The normal force sensor consists of 4×4 capacitive sensing array, the force transmission structure (including 4×4 pillar-array and a mounting plate) and tactile bump. In operation, the applied normal load will transmit to the capacitive sensing array through the tactile bump and force transmission structure. When the force is applied, the top electrode will approach the bottom electrode and leads to capacitance change. The 4×4 sensing units are connected (in parallel) to form an equivalent sensing capacitor. In addition, the etched channel between the sensing electrodes will be full of polymer, which act as a buffer. By the way, the polymer will be dispensed into the inlet hole beside the force sensor, and fill in the etched channel ultimately to form the buffer layer. Exploded views indicate each capacitive sensing unit (containing top deformable-electrode and bottom fixed-electrode) has a pillar attached to the top deformable-diaphragm. Moreover, another end of 4×4 pillar-array is fixed to the mounting plate which is the component to arrange tactile bump (e.g. rigid glass). In particular, the

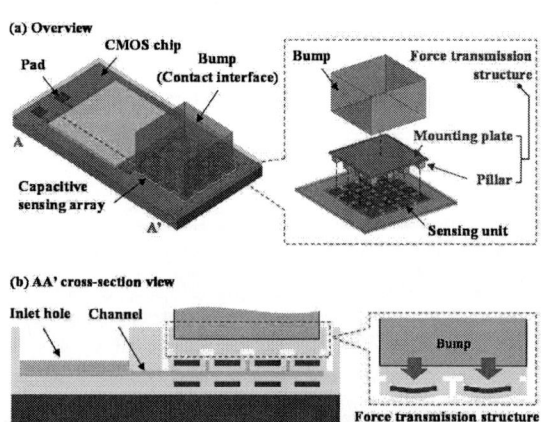

Figure 1: (a) 3D schema and (b) cross-section view of the proposed design concept.

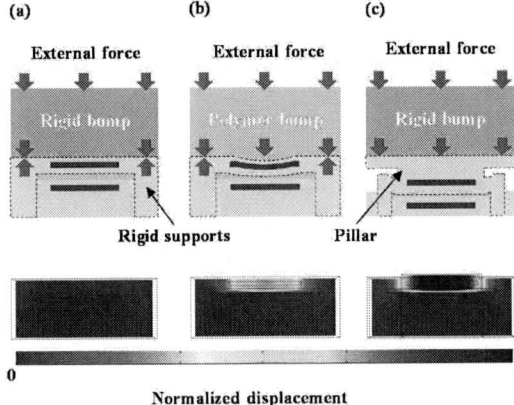

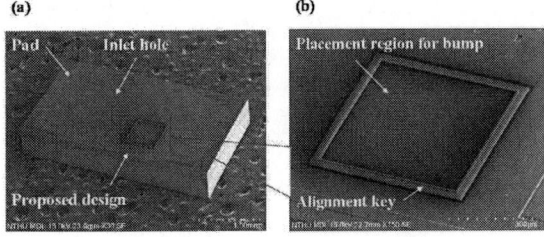

Figure.4: Fabrication results: (a) chip after wet etching process, (b) zoom-in SEM image.

Figure.2: Working principle :(a) reference type with rigid bump, (2) reference type with polymer bump, (c) proposed design

pillar size is smaller than the top membrane. Thus, forces applied on tactile bump could transmit to each sensing unit through the pillar-array without interfered by rigid supports. Fig.1b indicates the cross-section view of the proposed design. With the proposed force transmission structure, the external force can be efficiently transmitted to contribute larger deformation on the sensing electrode, as compared with the existing sensing structures [7], [8].

Fig.2 further depicts the difference between proposed and existing designs. For existing design (Fig.2a), the external load will be blocked by rigid supports when applied on rigid tactile bump, hence could not deform electrodes of capacitive sensing array. In this case, only flexible tactile bump can transmit the load (Fig.2b). With the proposed force transmission structure (Fig.2c), external loads could transmit to capacitive sensing array, even with rigid tactile bump. Simulations show proposed design causes larger deformation of sensing electrodes at a given load.

FABRICATION AND RESULT

The proposed CMOS process is illustrated in Fig.3. Fig.3a shows CMOS chip fabricated by TSMC 1P6M

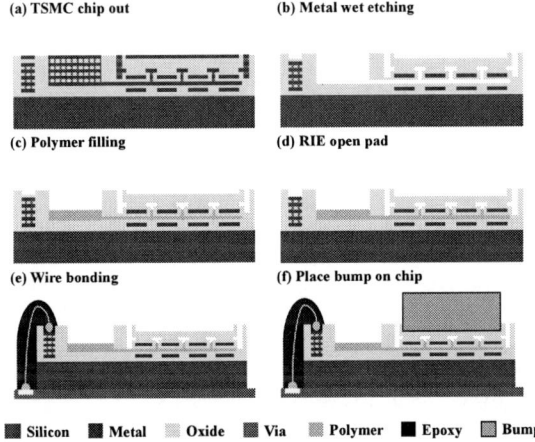

Figure.3: Fabrication process: (a) chip fabricated by TSMC, (b-d) post CMOS process steps.

CMOS-MEMS platform. As shown in Fig.3b, the sacrificial layers were removed through metal wet etching (H_2SO_4/H_2O_2) to define the sensing capacitor, force transmission structure, and inlet hole. As displayed in Fig.3c, the PDMS (SYLGARD 184, Dow Corning, USA, 10:1) was dispensed into inlet hole, and the polymer was completely filled into the gap between electrodes after placing in vacuum chamber. Next, PDMS was curing at 120 °C for 30 minutes. Fig.3d presents the bonding pads opened by the RIE. After that, the chip was attached and wire-bonded to PCB. Furthermore, the metal wires were protected with epoxy (Fig.3e). Finally, as shown in Fig.3f, tactile bumps (polymer or rigid glass) as discussed in Fig.2 were assembled on the mounting plate of the CMOS-MEMS chip.

Micrographs in Fig.4a-b show typical fabricated sensing chip and the zoom-in of mounting plate. After metal wet etching, the sensing structure is well defined. Among them, the outer frame of the mounting plate serves as the alignment mark, which is to improve the accuracy for tactile bump assembly. Fig.5a shows the CMOS chip wire-bonded on PCB and metal wires are protected by epoxy. Fig.5b displays rigid glass bumps and flexible PDMS bumps prepared are employed as force contact interfaced. Fig.5c-d demonstrates tactile force sensors assembled with bumps as the device-under-test (DUT). The top-view micrographs in Fig.5e-f depict that the PDMS and glass bumps are well attached to the mounting plate. Moreover, PDMS is served as an adhesive layer to fix the glass bump to prevent it from slipping.

MEASUREMENT RESULT

Fig.6 illustrates the setup for normal force measurement which consists of force gauge, tri-axis electrical control position stage and probe. Three different force sensors (reference design, proposed designs with PDMS/glass bumps) were prepared as DUTs and then fixed on position stage. During measurements, the DUT was moved by position stage to contact the probe on force gauge. Thus, by controlling the position stage, a normal load was applied on the DUT and then monitored by the force gauge. Sensing signals from DUT was recorded by commercial circuit (Evaluation Kit, AD7746).

Measurement results in Fig.7 show the capacitance change over normal force difference of three designs. Within the 0 N~5 N loading range, the sensitivities of three different sensors are: 1.69 fF/N for reference design, 3.67 fF/N for proposed design 1, and 5.04 fF/N for proposed

978-1-6654-9309-3/23 $31.00 © 2023 IEEE

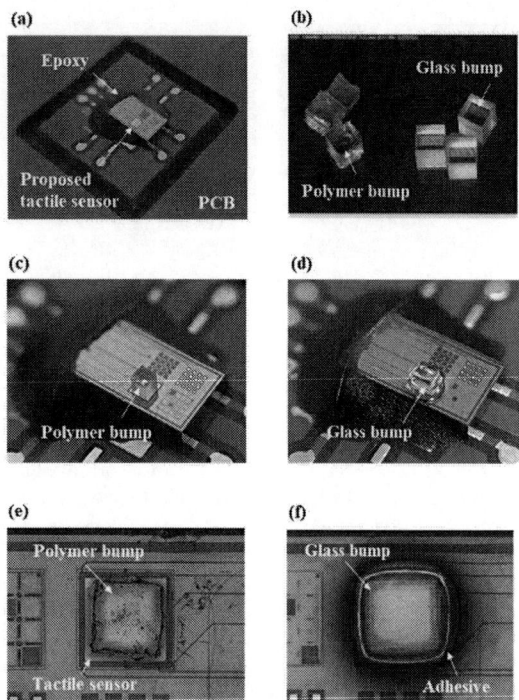

Figure.5: (a)Wire bonded for package and testing, (b) Force contact interface (Micro bump), (c-d) place bump on sensing chip, (e-f) top view of tactile sensor.

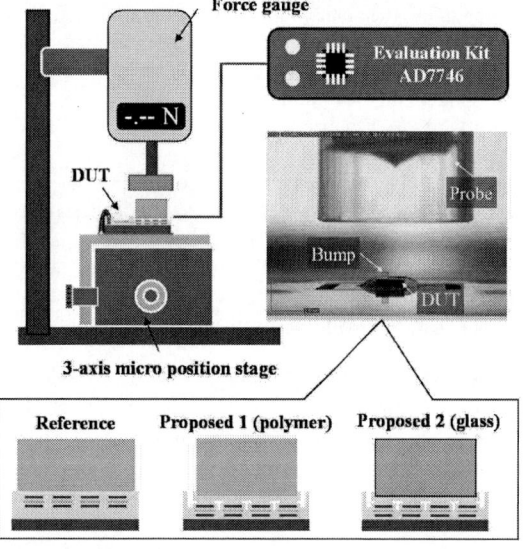

Figure.6: (a) Setup for force sensing measurements, and DUTs.

design 2. Compared with the reference design under same bump size condition, the sensitivity of proposed 2 is increased for 2-fold due to the contributions of both force transmission structure and rigid bump. Note: since the deformation of polymer may change the loading condition, the proposed design with flexible PDMS bump could only enhance the sensitivity for 40%.

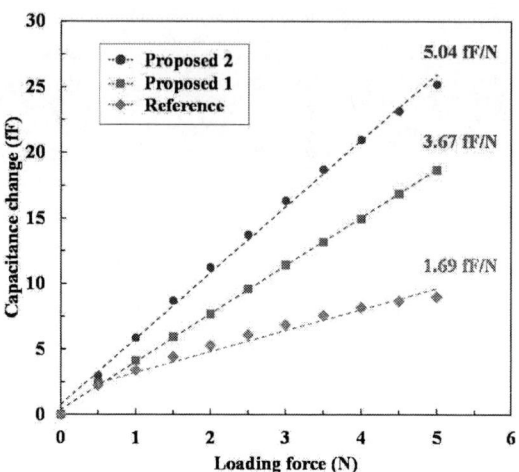

Figure.7: Measured capacitance changes versus normal force for three designs.

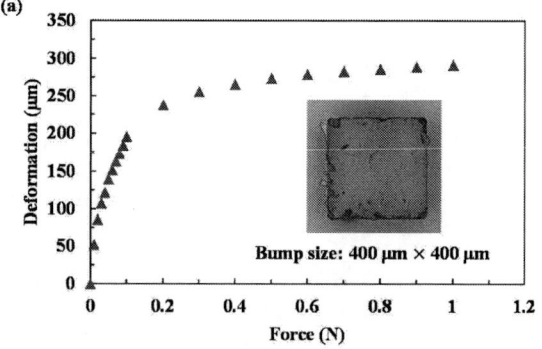

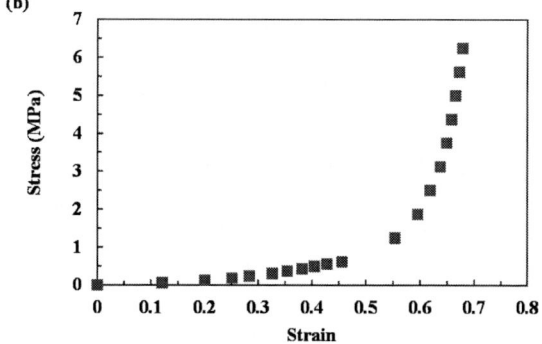

Figure.8: (a) Measured deformation changes versus normal force and (b) stress-strain curve of polymer bump

Nonlinearity behavior caused by PDMS hardening may occur when polymer was involved. Note that the turning point of proposed design 1 occurred before 0.01 N (limited by the current measurement resolution as shown in Fig. 8). Based on the aforementioned phenomenon, additional testing was performed to characterize the non-linearity load-deflection response of the bump. Fig.8 shows the load-deflection diagram and the stress-strain curve of the polymer bump, respectively. Results in Fig.8a verify that the polymer bump has spring hardening phenomenon when the applied force increased [9], [10]. Table 1 summarizes characteristics of three different tactile force sensors in this study.

978-1-6654-9309-3/23 $31.00 © 2023 IEEE

Table 1: The summary of tactile force sensor in this study

Type	Ref.	Proposed 1	Proposed 2
Bump size (μm^2)	500 × 500	400 × 400	
Material	Polymer		Glass
Array	4 × 4		
Sensitivity (fF/N)	1.69	3.67	5.04
Sensitivity (fF/MPa)	0.42	0.59	0.81

CONCLUSION

In this study, a tactile force sensor with vertical-integrated force transmission structure was successfully realized by the commercial CMOS-MEMS platform and in-house post-CMOS processes. With better force transmission efficiency, the proposed design with rigid glass bump (0.81 fF/MPa) can reach a 2-fold enhancement in sensitivity compared to the reference design (0.42 fF/MPa). In addition, the proposed design also solves the nonlinear issue caused by the polymer type.

ACKNOWLEDGEMENTS

This project was supported by the National Science and Technology Council under grant number NSTC 111-2923-E-007-003-, NSTC 111-2221-E-007-070-MY3, NSTC 111-2218-E-007-014-MBK, NSTC 111-2923-E-007-004-MY2, NSTC 110-2926-I-007-506-, and NSTC 110-2218-E-007-032-. The authors would like to appreciate the TSMC and the Taiwan Semiconductor Research Institute (TSRI), for the supporting of CMOS chip manufacturing. The authors also appreciate the Center for Nanotechnology, Materials Science and Microsystems (CNMM) of National Tsing Hua University for providing the process tools.

REFERENCES

[1] R. S. Dahiya, G. Metta, M. Valle, G. Sandini, "Tactile Sensing - From Humans to Humanoids", *IEEE Transactions on Robotics*, vol. 26, pp. 1-20, 2010.

[2] S.-K. Yeh, M.-L. Hsieh, W. Fang, "CMOS-based Tactile Force Sensor: A review", *IEEE Sensors Journal*, vol. 21, pp. 12563-12577, 2021.

[3] H. Yousef, M. Boukallel, K. Althoefer, "Tactile Sensing for Dexterous In-hand Manipulation in Robotics - A review", *Sensors and Actuators A: Physical*, vol. 167, pp. 171-187, 2011.

[4] W.-C. Lai, M.-L. Hsieh, W. Fang, "Electric Modulation on the Sensitivity and Sensing Range of CMOS-MEMS Tactile Sensor by Using the PDMS Elastomer Fill-In", *IEEE Sensors Journal*, vol. 21, pp. 5828-5835, 2021.

[5] S. Asano, M. Muroyama, T. Bartley, T. Kojima, T. Nakayama, U. Yamaguchi, H. Yamada, Y. Nonomura, Y. Hata, H. Funabashi, S. Tanaka, "Surface-mountable Capacitive Tactile Sensors with Flipped CMOS-diaphragm on a Flexible and Stretchable Bus Line", *Sensors and Actuators A: Physical*, vol. 240, pp. 167-176, 2016.

[6] T. Chou, Y.-C Lee, F. Shih, Y. Huang, S.-C. Lo, T.-L. Chien, C.-F. Hu, W. Fang, "Flip-Chip IR and Force Sensors for Both Touch and Touchless Elevator Buttons Applications", *IEEE MEMS'22 Conference*, Tokyo, January 9-13, 2022, pp. 523-526.

[7] M.-L. Hsieh, S.-K. Yeh, J.-H. Lee, M.-C. Cheng, W. Fang, "CMOS-MEMS Capacitive Tactile Sensor with Vertically Integrated Sensing Electrode Array for Sensitivity Enhancement", *Sensors and Actuators A: Physical*, vol. 317, pp. 112350(13), 2021.

[8] Y.-L. Chen, Y. Huang, F. Shih, T. Chou, T.-L. Chien, R. Chen, W. Fang, "A Dual Sensing Modes Capacitive Tactile Sensor for Proximity and Tri-Axial Forces Detection", *IEEE MEMS'22 Conference*, Tokyo, January 9-13, 2022, pp. 710-713.

[9] S.-K. Yeh, J.-H. Lee, W. Fang, "On the Detection Interfaces for Inductive Type Tactile Sensors", *Sensors and Actuators A: Physical*, vol. 297, pp. 111545(15), 2019.

[10] I. D. Johnston, D. K. McCluskey, C. K. L. Tan, M. C. Tracey, "Mechanical Characterization of Bulk Sylgard 184 for Microfluidics and Microengineering", *J. Micromech. Microeng.*, vol. 24, pp. 035017(7), 2014.

CONTACT

*Weileun Fang; fang@pme.nthu.edu.tw

1-OCTADECANETHIOL SAM ON CMOS-MEMS GOLD-PLATED RESONATOR VIA DIP-CAST FOR VOCs SENSING

Rafel Perelló-Roig[1,3], Jaume Verd[1,3]*, Sebastià Bota[1,3], Bartomeu Soberats[2], Antonio Costa[2] and Jaume Segura[1,3]

[1]Electronic Systems Group (GSE-UIB) at the University of the Balearic Islands, SPAIN,
[2]Department of Chemistry at the University of the Balearic Islands, SPAIN and
[3]Health Research Institute of the Balearic Islands, SPAIN

ABSTRACT

We present a novel chemical VOCs sensor based on a CMOS-MEMS gold plated micro-resonator functionalized with a self-assembled monolayer (SAM) of 1-octadecanethiol (ODT) via solution-casting being, in contrast to common techniques, fully compatible with batch production without additional lithographic steps requirement. The device operated at 2-MHz being fully integrated with an on-chip Pierce oscillator circuit for self-sustained operation. The MEMS resonator was fabricated using a commercial 0.35-μm CMOS technology and coated through electroless nickel immersion gold (ENIG) process providing thiol-based extended functionalization compatibility over standard CMOS BEOL aluminum layers. This CMOS-MEMS sensor platform opens several possibilities to multiple functional SAMs-based organic surfaces by synthetizing functionalized thiols and build a sensing matrix for breath pattern recognition, depicting tremendous potential low-cost non-invasive disease diagnosis applications.

KEYWORDS

MEMS resonators, VOCs, CMOS-MEMS, Lab-on-Chip, Self-Assembled Monolayer, Gas sensor, ENIG

INTRODUCTION

MEMS resonators, using sub-micrometer and nanometer scale structures, have been widely proposed for ultrasensitive mass sensing based on inertial mass detection with resolutions in the yoctogram (10^{-24} g) range [1]. Furthermore, surface functionalization has enabled gas sensing applications achieving, for example, reproducible measurements of NO_2 concentrations as low as 50 ppb [2] or 10 ppb H_2S concentrations [3]. While a large number of applications have been proposed and demonstrated including quality control [4] or pollutant monitoring [5], others like cancer biomarker detection applications [6] or volatile organic compound (VOCs) detection in exhaled human breath s for non-invasive disease diagnosis [7] have grown special interest in the biomedical domain [8].

Framed within the More-than-Moore paradigm, MEMS resonators offer system integration feasibility, miniaturization aligned with technological trends and low fabrication cost for mass production thanks to using standard micro-fabrication tools [9], [10]. Here, a CMOS-compatible technological approach provides a huge step forward for emerging applications [11] positioned to scientific and industry domains thanks to improved sensing capabilities. CMOS-MEMS brings a key added value for monolithically integrated single-chip systems [12]. MEMS resonators co-fabricated with a CMOS amplifier are able to operate as a self-sustained oscillator providing a quasi-digital output signal with frequency-encoded information, having demonstrated sub-ppm frequency stability [13]. However, to take full advantage of this System-on-Chip (SoC) approach, all post-CMOS fabrication steps must ensure batch compatibility.

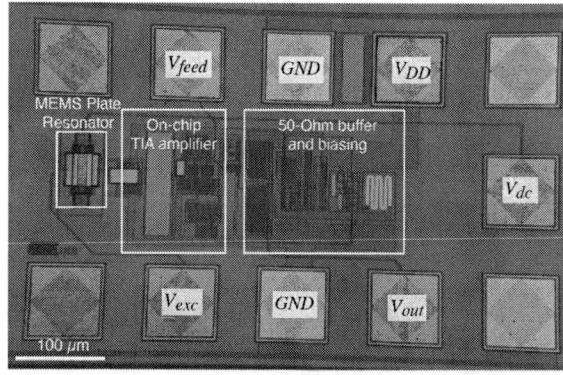

Figure 1: Optical image of the CMOS-MEMS chip after ENIG gold plating showing the process selectivity on exposed top-metal aluminum. The resonator is on-chip integrated with the CMOS sustaining amplifier.

Usually, in gas sensing applications, the resonator requires a relatively high capture area [14] or uses a specific fabrication process that limits the CMOS-compatible fabrication possibilities [2]. In addition, most reported works include sensing layers such as polymers [14], [15] or nanostructures [2], deposited by printing techniques [16] or other individual-assembly alternatives like shadow-mask-enabled spray-coating [14] or drop-based inkjet [7], [15] dramatically increasing fabrication process complexity and cost compared to the batch-compatible approach reported here. SAMs, besides attaining a maskless dip-cast assembly process, also offer much better sensor response time and reversibility over classical thick coatings thanks to ultra-fast surface interactions. Even CMOS-compatible solutions provide a huge step forward in emerging e-health applications, just few works have successfully integrated a fully CMOS-MEMS resonant gas sensor [15].

The monolithic CMOS-MEMS device presented in this work (Figure 1) combines a very high mass sensitivity CMOS-MEMS oscillator platform with a self-assembled monolayer of 1-octadecanethiol (SAM-ODT) functionalization step via solution-cast with all the post-CMOS fabrication processes being fully compatible with integrated circuit (IC) batch production. SAM-ODT compatibility has been accomplished thanks to a previous gold plating step on the CMOS standard aluminum layers based on an electroless nickel immersion gold (ENIG) process [17]. Alkanethiol chains offer a flexible and reliable functionalization approach that, while keeping the original process flow, includes various functional groups providing the MEMS resonator differentiated interaction with VOCs present in the exhaled breath. ODTs, as a starting point, show hydrophobicity [18] and larger interaction with non-polar analytes [19].

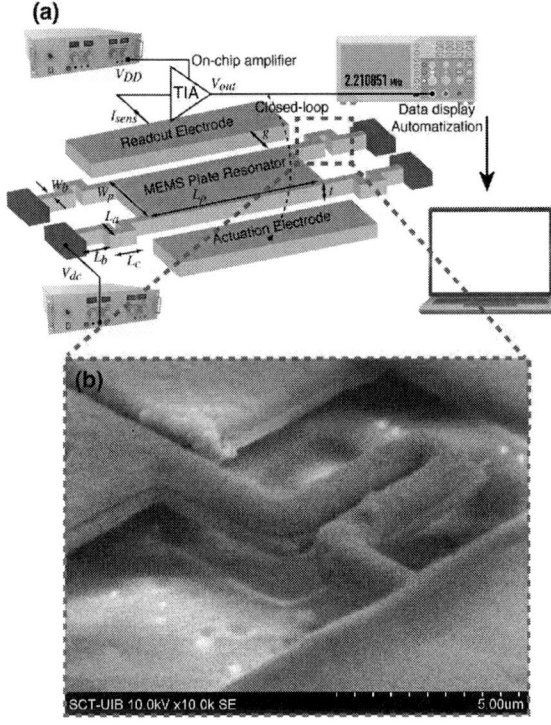

Figure 2: (a) Resonator schematic showing the design geometrical parameters and closed-loop configuration for self-sustained oscillator operation, and (b) SEM image detail of the CMOS-MEMS resonator anchor designed for VOCs sensing where the ENIG gold coating offers proper double-sided surface coverage and uniformity.

FABRICATED DEVICE AND POST-PROCESSING

The device reported in this work is a 4-anchored plate resonator with folded anchors optimized for VOC sensing (Figure 2a) [20], fabricated using the top metal layer of a commercial 0.35-μm CMOS technology followed by a post-CMOS wet-etching step performed without additional lithographic steps [7]. The samples are subsequently treated with a batch-fabrication compatible ENIG process [21] to deposit a high added-value gold layer (Figure 2b) for SAM alkanethiol anchoring [22] compared to standard CMOS aluminum-finish. Such gold plating step does not require any mask thanks to the high selectivity on the exposed metal layers. Only the top-level exposed aluminum (bonding PADs and MEMS structures) is coated as demonstrated in the optical image in Figure 1. The ENIG process flow is depicted and further detailed in

Figure 3: first, a double zincation sets up the aluminum surface for nickel adhesion, then, the nickel and gold plating steps follow.

ODT was assembled on the gold surface by immersion-based surface modification, a well-reported technique [19], [23], boosting non-polar molecules adsorption, opening a wide range of functional SAMs-based organic surfaces by modified alkanethiols. The method novelty relies on such functionalization stage consisting of a simple solution-cast followed by an ethanol cleaning without requiring any additional lithographic steps. Thus, all the post-CMOS processes required to fabricate a functional gas sensor device can be easily adopted at the wafer level in a batch fabrication process significantly reducing production costs.

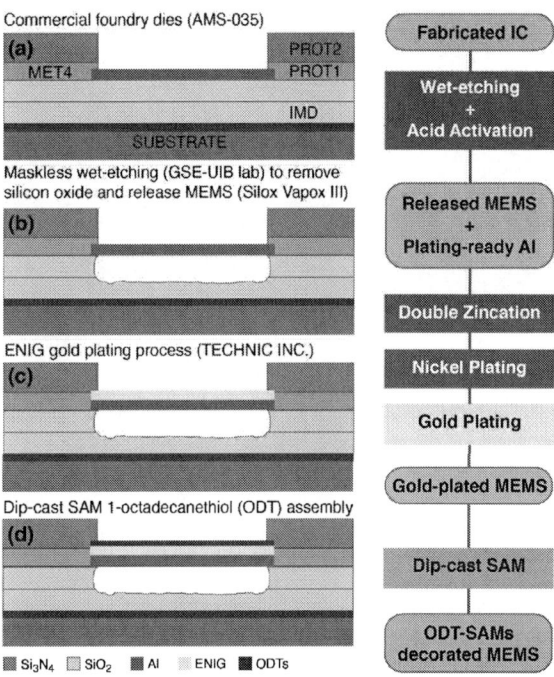

Figure 3: Fabrication process flow diagram run on commercial CMOS chips for having an ODT-decorated CMOS-MEMS resonator. The layers depicted are not to scale and the aluminum resonator underneath layers are a simplified version to improve the overall chip display.

EXPERIMENTAL CHARACTERIZATION

Extensive gas measurements with n-hexane as a non-polar analyte were performed using a two-valve gas flow controller setup while real-time tracking the sensor frequency (Figure 4a). A synthetic air stream was saturated with n-hexane using a bubbler at 22°C and mixed with the same synthetic air supply setting the analyte concentration (Figure 4b). Relative humidity (RH) characterization into a climate chamber at constant temperature demonstrated the ODT hydrophobic behavior (Figure 5).

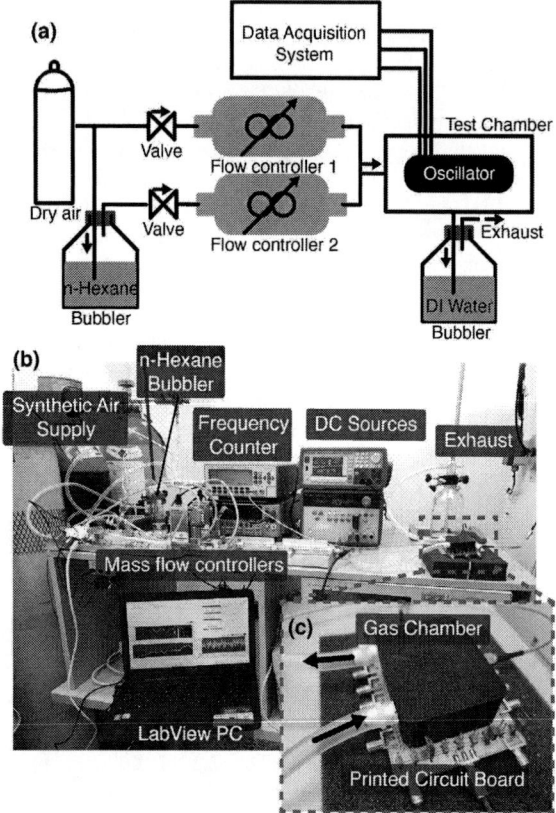

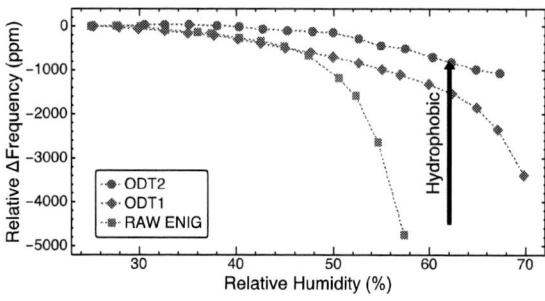

Figure 4: Automated gas sourcing and mixing experimental setup (a) schematic and (b) picture used to feed various n-hexane concentration to the device under test at a constant gas flow of 300 ml/min. (c) A custom-made gas chamber with inlet and outlet ports was designed to be attached to the interconnection PCB.

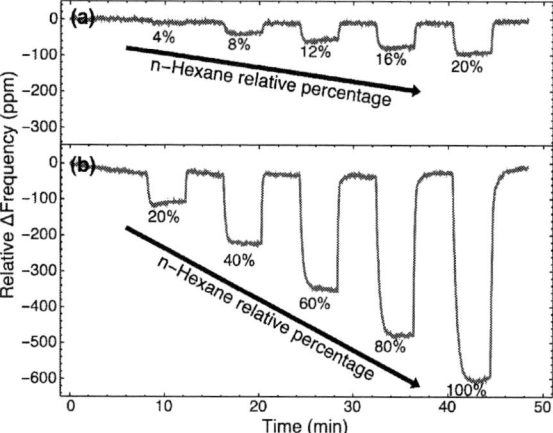

Figure 5: CMOS-MEMS resonator humidity response measured into a climate chamber after ENIG raw coating, 1st ODT and 2nd ODT, respectively, confirming the expected hydrophobic behavior. The resonance frequency was recorded as a function of the RH (±0.5%) at a constant temperature of 22°C (±0.1°C).

Measurements from 4% to 100% n-hexane saturated gas were performed (Figure 6) including a synthetic air cleaning phase after each analyte injection. In addition, the sensor response repeatability was confirmed by five consecutive exposures to n-hexane 100% and cleaning steps consecutively (Figure 7). Finally, Figure 8 plots the

steady state calibration curve for n-hexane showing a linear relationship for the whole range, exhibiting a sensitivity of 5.9 ppm/%. The sensitivity to the analyte together with the high frequency stability (ADEV = 0.35 ppm for 100 ms integrating time) provides a LOD of 0.16% (assuming a SNR = 3).

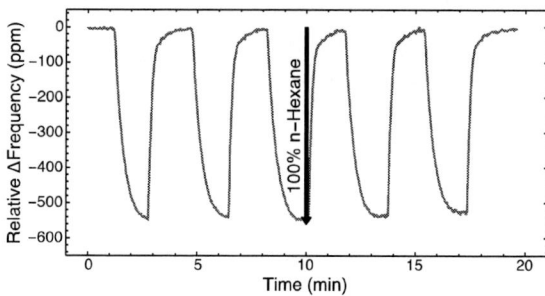

Figure 6: Measured sensor response to n-Hexane exposure at increasing concentration with 4-min period for (a) 0-20%, and (b) 0-100%. The temperature was 22°C.

Figure 7: Sensor resonance frequency decreases due to 100% n-Hexane-saturated synthetic air injections with 2-min period followed by a 2-min cleaning to test the sensor repeatability and response time. Temperature was 22°C.

CONCLUSIONS

This work reports an innovative solution-cast SAM-ODT functionalization process on monolithic CMOS-MEMS resonators, enabled by a batch-compatible ENIG gold plating step, for gravimetric VOCs sensing of human exhaled breath. The MEMS resonators are integrated with a sustaining amplifier for self-sustained oscillator operation and real-time frequency tracking. Experimental results demonstrate the increased hydrophobicity of the SAM-ODT layer by RH climate chamber characterization. Moreover, a proportional linear response to n-hexane injections by frequency downshift recording, with a sensitivity of 5.9 ppm/%, depicts tremendous potential low-cost non-invasive disease diagnosis applications in line with Lab-on-Chip and point-of-care trends in modern medicine.

978-1-6654-9309-3/23 $31.00 © 2023 IEEE

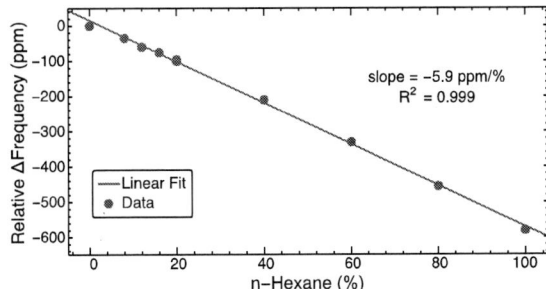

Figure 8: Measured sensor frequency change as a function of the n-Hexane mixing percentage revealing a linearly proportional relationship for the whole range.

ACKNOWLEDGEMENTS

This work is part of the projects TEC2017-88635-R funded by AEI/FEDER, UE and PID2021-122460OB-I00 funded by MCIN/AEI/10.13039/501100011033 and by "ERDF A way of making Europe". The authors thank Mr. Raúl Sánchez and Dr. Ferran Hierro, from SCT-UIB, for the test gas chamber and SEM images, respectively.

REFERENCES

[1] J. Chaste, A. Eichler, J. Moser, G. Ceballos, R. Rurali, and A. Bachtold, "A nanomechanical mass sensor with yoctogram resolution," *Nat. Nanotechnol.*, vol. 7, no. 5, pp. 301–304, 2012.

[2] J. Xu, A. Setiono, and E. Peiner, "Piezoresistive Microcantilever with SAM-Modified ZnO-Nanorods@Silicon-Nanopillars for Room-Temperature Parts-per-Billion NO2 Detection," *ACS Appl. Nano Mater.*, vol. 3, no. 7, pp. 6609–6620, 2020.

[3] L. Tang, P. Xu, M. Li, H. Yu, and X. Li, "Integrated Resonant Dual-Microcantilevers Combined Sensor with Accurate Identification and Highly-Sensitive Detection to H2S Gas," in *20th International Conference on Solid-State Sensors, Actuators and Microsystems & Eurosensors XXXIII*, 2019, pp. 338–341.

[4] I. Bargatin *et al.*, "Large-scale integration of nanoelectromechanical systems for gas sensing applications," *Nano Lett.*, vol. 12, no. 3, pp. 1269–1274, 2012.

[5] T. Y. Liu *et al.*, "Gated CMOS-MEMS thermal-piezoresistive oscillator-based PM2.5 sensor with enhanced particle collection efficiency," in *2018 IEEE International Conference on Micro Electro Mechanical Systems (MEMS)*, 2018, pp. 75–78.

[6] P. M. Kosaka *et al.*, "Detection of cancer biomarkers in serum using a hybrid mechanical and optoplasmonic nanosensor," *Nat. Nanotechnol.*, vol. 9, pp. 1047–1053, 2014.

[7] R. Perelló-Roig *et al.*, "CMOS–MEMS VOC sensors functionalized via inkjet polymer deposition for high-sensitivity acetone detection," *Lab Chip*, vol. 21, pp. 3307–3315, 2021.

[8] J. Dieffenderfer *et al.*, "Low-power wereable systems for continuous monitoring of environment and health for chronic respiratory disease," *IEEE J. Biomed. Heal. Informatics*, vol. 20, no. 5, pp. 1251–1264, 2016.

[9] G. K. Fedder, R. T. Howe, T. J. K. Liu, and E. P. Quévy, "Technologies for cofabricating MEMS and electronics," *Proc. IEEE*, vol. 96, no. 2, pp. 306–322, 2008.

[10] W. Fang *et al.*, "CMOS MEMS : A key technology towards the 'More than Moore' era," in *17th International Conference on Solid-State Sensors, Actuators and Microsystems*, 2013, pp. 2513–2518.

[11] S. S. Li, "A key more-than-moore technology: CMOS-MEMS resonant transducers," in *16th International Conference on Nanotechnology*, 2016, pp. 456–459.

[12] A. Uranga, J. Verd, and N. Barniol, "CMOS–MEMS resonators : From devices to applications," *Microelectron. Eng.*, vol. 132, pp. 58–73, 2015.

[13] J. Verd *et al.*, "Monolithic CMOS MEMS oscillator circuit for sensing in the attogram range," *IEEE Electron Device Lett.*, vol. 29, no. 2, pp. 146–148, 2008.

[14] L. A. Beardslee *et al.*, "In-plane vibration of hammerhead resonators for chemical sensing applications," *ACS Sensors*, vol. 5, pp. 73–82, 2020.

[15] K. L. Dorsey, S. S. Bedair, and G. K. Fedder, "Gas chemical sensitivity of a CMOS MEMS cantilever functionalized via evaporation driven assembly," *J. Micromechanics Microengineering*, vol. 24, no. 7, p. 075001, 2014.

[16] J. Dai *et al.*, "Printed gas sensors," *Chem. Soc. Rev.*, vol. 49, pp. 1756–1789, 2020.

[17] A. C. Sun, E. Alvarez-Fontecilla, A. G. Venkatesh, E. Aronoff-Spencer, and D. A. Hall, "High-Density Redox Amplified Coulostatic Discharge-Based Biosensor Array," *IEEE J. Solid-State Circuits*, vol. 53, no. 7, pp. 2054–2064, 2018.

[18] L. Xue *et al.*, "Hydrophobic 1-octadecanethiol functionalized copper catalyst promotes robust high-current CO2 gas-diffusion electrolysis," *Nano Res.*, vol. 15, no. 2, pp. 1393–1398, 2022.

[19] J. Wang *et al.*, "Ligand-assisted deposition of ultra-small Au nanodots on Fe2O3/reduced graphene oxide for flexible gas sensors," *Nanoscale Adv.*, vol. 4, no. 5, pp. 1345–1350, 2022.

[20] R. Perelló-Roig, J. Verd, S. Bota, and J. Segura, "Detailed analysis of flow-induced thermal mechanisms in submicron MEMS-based VOC biosensors: A design solution for the nanometer scale," in *IEEE 22nd International Conference on Nanotechnology (NANO)*, 2022, pp. 1–4.

[21] A. Bonanno *et al.*, "A multipurpose CMOS platform for nanosensing," *Sensors*, vol. 16, p. 2034, 2016.

[22] T. Ishida *et al.*, "High resolution X-ray photoelectron spectroscopy measurements of octadecanethiol self-assembled monolayers on Au(111)," *Langmuir*, vol. 14, no. 8, pp. 2092–2096, 1998.

[23] I. Hwang *et al.*, "Ultrasensitive Molecule Detection Based on Infrared Metamaterial Absorber with Vertical Nanogap," *Small Methods*, vol. 5, no. 8, 2021.

CONTACT

*J. Verd, tel: +34-971-172006; jaume.verd@uib.es

APPLICATION OF DEEP LEARNING NETWORK FOR HUMIDITY COMPENSATION OF SEMICONDUCTOR METAL OXIDE GAS SENSORS

Mingu Kang, Incheol Cho, and Inkyu Park
Korea Advanced Institute of Science and Technology (KAIST), REPUBLIC OF KOREA

ABSTRACT

In this research, we propose a humidity compensation method using a deep learning network for minimizing the effect of humidity on the semiconductor metal oxide (SMO) gas sensors. The SMO gas sensors were fabricated by depositing In_2O_3 on a suspended microheater platform, and the gas tests were conducted under various humidity conditions using NO_2 gas. The gas sensing data and humidity data were simultaneously used as input data of the deep learning network for real-time humidity compensation. Through the proposed method, we successfully improved the accuracy of predicting the concentration of NO_2 gas in different humidity conditions.

KEYWORDS

Semiconductor metal oxide gas sensor, Deep learning, Humidity compensation

INTRODUCTION

Nowadays, the need for gas monitoring systems has been increasing in terms of energy conservation and environmental protection. Especially, the detection of flammable and toxic gases is one of the most important issues in industrial fields because there have been many kinds of chemical gas leakage and related accidents. There are three main types of gas sensors electrochemical type [1], catalytic combustion type [2-3], and chemiresistive type [4-7]. Especially, as next-generation gas sensors, chemiresistive gas sensors are drawing attention since they have better properties in terms of sensitivity, response time, stability, and even cost [8]. In chemiresistive gas sensors, there are various types of sensing materials, and one of the most widely used sensing materials is semiconductor metal oxide (SMO).

The SMO gas sensors can detect target gases by measuring a change in the resistance of the metal oxide. When a target gas is adsorbed/desorbed on the surface of the metal oxide, the thickness of the electron depletion layer of the sensing material is changed, which changes the resistance of the metal oxide [9]. However, since the water vapor also can react with the sensing materials of the SMO gas sensors, the sensing performance of the SMO gas sensors is degraded depending on the humid condition, and thus it is difficult to use the SMO gas sensors in a humid environment [10]. To solve the problem of sensing performance degradation caused by humidity, various studies have been conducted and typical methods are using catalysts or filters to block the vapor gas from the sensing materials [11-12]. Nevertheless, the sensing performances of the gas sensors can be affected by the catalysts or filters, and also additional fabrication processes should be carried out.

In this study, to overcome these limitations, we propose a deep learning-based humidity compensation method that did not require additional fabrication processes

or moisture trap filters. Gas sensing data and humidity data were simultaneously used as the input data for the deep learning network, and the gas sensing data was collected from SMO gas sensors using nanocolumnar In_2O_3 thin films deposited by glancing angle deposition (GLAD) as a sensing material. The SMO gas sensors were fabricated using a micro-electro-mechanical-system (MEMS)-based suspended microheater platform to satisfy the high temperature conditions required for gas detection with low electrical power consumption. In the gas tests, NO_2 gas in the concentration of 1.0 ppm to 5.0 ppm was used as a target gas, and the relative humidity value in each gas test was set to 10%, 30%, and 50% using a humidity generator. The deep learning network for humidity compensation was composed of fully connected layers, and by applying preprocessed sensor response data with a moving time window for using the information in the transient region, it was possible to predict NO_2 gas concentration in real-time. Finally, it was possible to verify the possibility of humidity compensation through the deep learning network using humidity data by improving the accuracy of prediction of concentration of NO_2 gas even in an environment where the relative humidity changes.

FABRICATION

Deposition of sensing materials

An In_2O_3 sputtering target was used for the deposition of the sensing materials of the SMO gas sensors. In order to form the nanocolumnar thin films, the In_2O_3 was deposited using glancing angle deposition (GLAD) via RF sputtering at an angle of 85°. Ar gas flow rate, RF power, base pressure, working pressure, and rotation speed were 50 sccm, 250 W, 1×10^{-6} Torr, 4×10^{-3} Torr, and 4.0 rpm, respectively. To pattern the nanocolumnar In_2O_3 films, the photoresist (AZ5214, MicroChemicals, Germany) was used, and the lift-off process was carried out.

Suspended microheater platform-based gas sensors

On the Si wafer, a SiO_2 layer of 1 μm thickness was deposited by the plasma-enhanced chemical vapor deposition (PECVD) process. For the microheaters, the photoresist (AZ5214, MicroChemicals, Germany) was patterned by UV photolithography, and an electron beam (E-beam) evaporator was used to deposit a Ti/Pt (10 nm/200 nm) layer on the photoresist-patterned SiO_2 layer followed by lift-off. Then, a SiO_2 layer of 800 nm thickness was deposited by PECVD for an electrical insulation layer between and the microheaters the sensing electrodes. To make electrical contact of the heater pads, the SiO_2 layer on the heater pads was selectively etched by buffered oxide etchant (BOE). After the selective etching, interdigitated sensing electrodes of a Cr/Au (10 nm/200 nm) layer were deposited in the same way as the microheaters. To form the suspended platform by etching the Si substrate, the photoresist (AZ9260, MicroChemicals, Germany) was

978-1-6654-9309-3/23 $31.00 © 2023 IEEE

coated and patterned, and the SiO₂ layer was selectively etched using reactive ion etching (RIE) for the Si etching window. The nanocolumnar In_2O_3 films were deposited on the sensing electrodes as the sensing material through the GLAD via RF sputtering. For releasing the residual stress of the SiO₂ layer, the annealing was performed in a N₂ at 400°C for 2 hours. Finally, the Si substrate was etched by XeF₂ vapor etching, and the suspended microheater platform-based gas sensors were fabricated.

RESULTS

The sensing material of the SMO gas sensors was deposited through the GLAD which performs the physical vapor deposition on a wafer scale. When the substrate rotates at a certain angle with respect to the target materials, the target nanoparticles deposited on the substrate can generate a nanoscale shadow region. Due to this phenomenon, nanocolumnar thin films can be deposited on the substrate by the GLAD [6]. In this study, In_2O_3 was selected as the sensing material of the SMO gas sensors, and nanocolumnar In_2O_3 films were deposited through RF sputtering at a glancing angle of 85°. Fig. 1 shows the differences between the In_2O_3 films deposited by GLAD and the conventional sputtering, which provides that the surface-to-volume ratio of the films deposited by GLAD was larger than the films deposited through the conventional sputtering.

To promote the reaction with target gases generated on the surface of the metal oxide, SMO gas sensors generally require high temperature conditions (200°C to 400°C). Especially, MEMS-based microheater platforms that can generate high temperature conditions by the Joule heating have been widely used for SMO gas sensors to transfer the heat energy to the sensing materials with low electrical power consumption. Herein, a bridge type suspended microheater platform was fabricated based on our previous

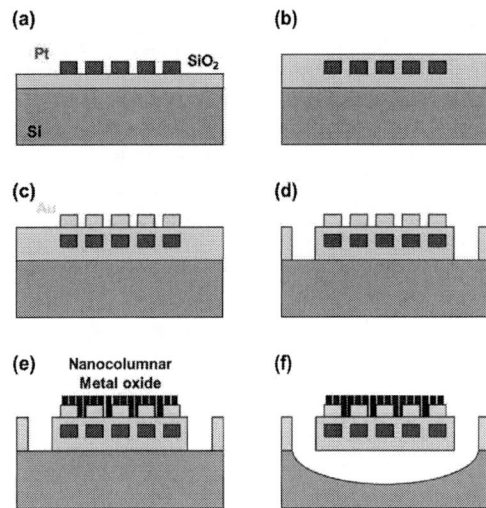

Figure 2: Fabrication process of the suspended microheater platform-based gas sensors. (a) Heater electrode (Ti/Pt) patterning by E-beam evaporation on the SiO₂ layer (PECVD). (b) SiO₂ layer deposition (PECVD) for electrical insulation layer. (c) Sensing electrode (Cr/Au) patterning by E-beam evaporation. (d) SiO₂ layer etching by RIE for Si etching window. (e) Nanocolumnar In_2O_3 films deposition by GLAD for sensing materials of the SMO gas sensors. (f) High temperature (400°C) annealing and Si etching by XeF₂ etcher for bridge type suspended microheater platform.

studies, and the In_2O_3 films were deposited on the microheater platform [4,6]. Fig. 2 shows the fabrication process of the suspended microheater platform-bases gas sensors, and in the Fabrication section, the detailed fabrication process of the suspended microheater platform-based In_2O_3 gas sensors is described. The fabricated microheater could generate a temperature condition of 250°C with low electrical power of 11 mW since the platform is a suspended bridge-shaped structure.

In the gas tests, NO₂ gas was used as the target gas under various humidity conditions, and the concentration range of the NO₂ gas and relative humidity were from 1.0 ppm to 5.0 ppm and from 10% to 50%, respectively. The concentration of the target gas and relative humidity could

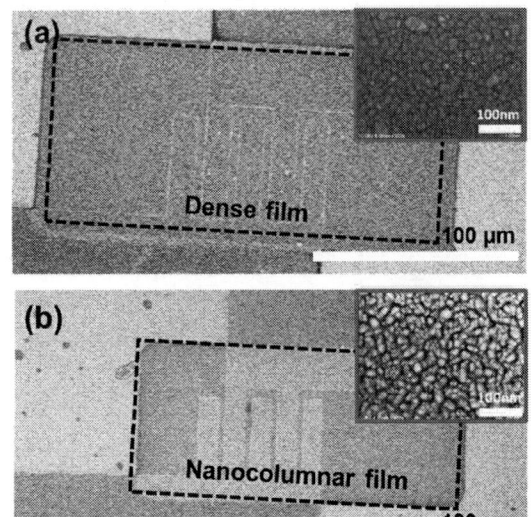

Figure 1: SEM images of In_2O_3 thin film deposited on the Au electrode by (a) conventional sputtering with a normal deposition angle and (b) glancing angle deposition via RF sputtering at an angle of 85°.

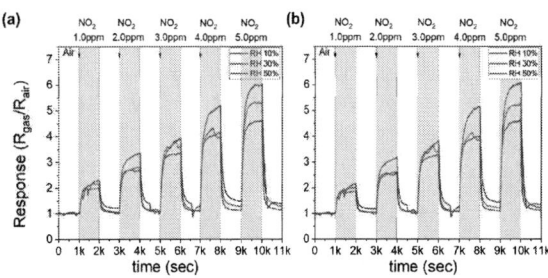

Figure 3: Dynamic responses of (a) #1 and (b) #2 suspended microheater platform-based In_2O_3 gas sensors to NO₂ gas in 10%, 30%, and 50% humidity conditions. For the high-temperature condition, an electrical power of 11 mW was applied to the microheater.

be controlled by adjusting the gas flow rate injected into the gas test chamber through the mass flow controllers (AFC500, Atovac, Korea). The gas tests were conducted with two types of microheater platform-based In_2O_3 gas sensors at the same time to make the configuration of the training data and test data different. While the gases are injected into the chamber, the current values through the nanocolumnar In_2O_3 films were measured by a bias of 2.5 V applied to the sensing electrode, and an electrical power of 11 mW was applied to the microheater. In this study, the ratio of resistance in the gas condition to resistance in the air condition (R_{gas}/R_{air}) quantified the sensor responses. Fig. 3 shows the dynamic responses of the In_2O_3 gas sensors to the NO_2 gas in 10%, 30%, and 50% relative humidity conditions. As shown in the gas test results, since the In_2O_3 is an n-type semiconductor and the NO_2 gas is an oxidizing gas, the resistance of the sensing materials of the gas sensors should increase when the gas sensor reacted with the NO_2 gas, and thus the response values of the gas sensors were larger than 1. However, it can be seen that the response of the gas sensors decreased under humid conditions because the sensing performance of the gas sensors degraded by the water vapor as the humidity increased. Therefore, in this study, the deep learning network was used to compensate for these humidity effects on the gas sensors.

For the humidity compensation in real-time, sensing data from the In_2O_3 gas sensors and humidity data were simultaneously used in the deep learning network, and Tensorflow (Google, USA) an open-source library was used to build the learning network. In this study, a deep neural network composed of 5 fully connected (Dense) layers was selected as the deep learning network for the humidity compensation, and the structure of the deep learning network is shown in Fig. 4. In the data preprocessing, a 10-second time window was applied to input the gas sensing data collected for 10 seconds into the learning network every second. In addition, to utilize the information about relative humidity, the humidity data was used with the gas sensing data simultaneously as the input data of the deep learning network. To minimize the vanishing gradient problem, the rectified linear unit (ReLU) was used as an activation function, and the batch normalization (BN) was carried out at every layer [13-14]. The adoptive moment estimation (Adam) was used as the optimizer of the learning network [15], and the network was trained for 200 epochs. In the last output layer, the concentration value of the NO_2 gas under various humidity conditions was predicted through the regression analysis. #1 sensor data was used to train the learning network, and #2 sensor data was used to test the trained network. Fig. 5a and b show the real-time prediction results of the concentration values of the NO_2 gas under various humidity conditions using the deep learning network with and without humidity data, respectively. An average error (7.11%) generated in the concentration prediction for the input data with humidity data was lower than half of the average error (16.58%) for the input data without humidity data. As a result, it could be demonstrated that humidity compensation is possible by adding humidity data to the input data of the deep learning networks without filters or any additional fabrication process.

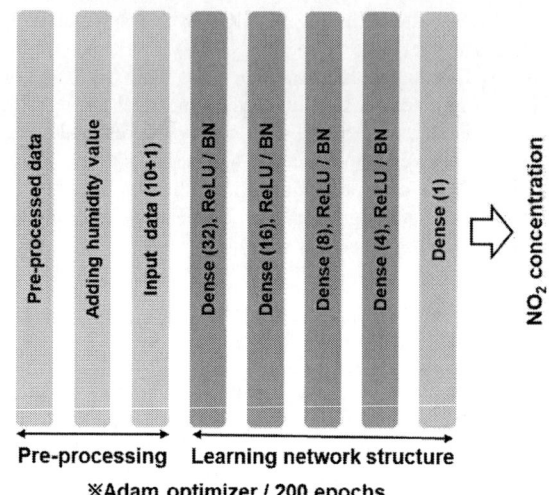

Figure 4: Structure of the deep learning network for quantifying the concentrations of NO_2 gas by compensating for humidity under various humidity conditions. The network consisted of five fully connected layers (Dense). An activation function of the network was the rectified linear unit (ReLU) function, and Batch normalization (BN) wass conducted in all layers.

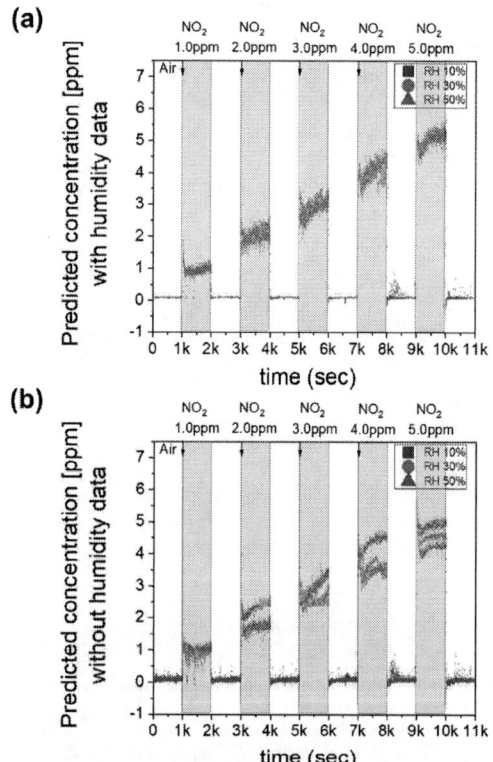

Figure 5: Real-time prediction results of NO_2 gas concentration under various humidity conditions (a) with humidity data and (b) without humidity data in input data of the deep learning network. An average error generated in the concentration prediction for the input data with and without humidity data was 7.11% and 16.58%, respectively.

CONCLUSION

In this study, the SMO gas sensors were manufactured by fabricating the MEMS-based suspended microheater platform and integrating nanocolumnar In_2O_3 films on the heater platform through GLAD. In the gas tests, NO_2 gas was selected as the target gas, and the gas sensing data of the gas sensors was collected under various humidity conditions. For humidity compensation, the deep neural network composed of 5 fully connected layers was used, and preprocessed gas sensing data and humidity data were used simultaneously as the input data of the deep learning network. Finally, we demonstrated the possibility of humidity compensation by improving the prediction accuracy of the concentration of the NO_2 gas in various humidity conditions using the deep learning network. In the future, by utilizing a gas sensor array composed of various types of SMO gas sensors and humidity data together, it is expected that a system that can identify many types of target gases even in an environment where the humidity changes can be implemented.

REFERENCES

[1] J. Zhang, G. Jiang, T. Cumberland, P. Xu, Y. Wu, S. Delaat, A. Yu, Z. Chen, "A Highly Sensitive Breathable Fuel Cell Gas Sensor with Nanocomposite Solid Electrolyte", *InfoMat,* 1 (2), 234–241, 2019.

[2] D. V. Del Orbe Henriquez, I. Cho, H. Yang, J. Choi, M. Kang, K. S. Chang, C. B. Jeong, S. W. Han, I. Park, "Pt Nanostructures Fabricated by Local Hydrothermal Synthesis for Low-Power Catalytic-Combustion Hydrogen Sensors", *ACS Appl. Nano Mater.,* 4, 7-12, 2021.

[3] D. V. Del Orbe Henriquez, H. Yang, I. Cho, J. Park, J. Choi, S. W. Han, I. Park, "Low-Power Thermocatalytic Hydrogen Sensor Based on Electrodeposited Cauliflower-like Nanostructured Pt Black", *Sensors Actuators, B Chem.* 329, 129129, 2021.

[4] K. Kang, D. Yang, J. Park, S. Kim, I. Cho, H. H. Yang, M. Cho, S. Mousavi, K. H. Choi, Park, I. "Micropatterning of Metal Oxide Nanofibers by Electrohydrodynamic (EHD) Printing towards Highly Integrated and Multiplexed Gas Sensor Applications", *Sensors Actuators, B Chem.* 250, 574–583, 2017.

[5] I. Cho, Y. C, Sim, M. Cho, Y. H. Cho, I. Park, "Monolithic Micro Light-Emitting Diode/Metal Oxide Nanowire Gas Sensor with Microwatt-Level Power Consumption", *ACS Sensors*, 5 (2), 563–570, 2020.

[6] M. Kang, I. Cho, J. Park, J. Jeong, K. Lee, B. Lee, D. V. Del Orbe Henriquez, K. Yoon, I. Park, "High Accuracy Real-Time Multi-Gas Identification by a Batch-Uniform Gas Sensor Array and Deep Learning Algorithm", *ACS Sensors*, 7 (2), 430–440, 2022.

[7] J. K. Han, M. Kang, J. Jeong, I. Cho, J. M. Yu, K. J. Yoon, I. Park, Y. K. Choi, "Artificial Olfactory Neuron for an In-Sensor Neuromorphic Nose", *Advanced Science*, 9, 2106017, 2022.

[8] G. Neri, "First Fifty Years of Chemoresistive Gas Sensors", *Chemosensors*, 3 (1), 1–20, 2015,.

[9] M. E. Franke, T. J. Koplin, U. Simon, "Metal and Metal Oxide Nanoparticles in Chemiresistors: Does the Nanoscale Matter?", *Small* 2 (1), 36–50, 2006.

[10] C. Wang, L. Yin, L. Zhang, D. Xiang, R.Gao, "Metal Oxide Gas Sensors: Sensitivity and Influencing Factors", *Sensors*, 10 (3), 2088–2106, 2010.

[11] J. W. Yoon, J. S. Kim, T. H. Kim, Y. J. Hong, Y. C. Kang, J. H. Lee, "A New Strategy for Humidity Independent Oxide Chemiresistors: Dynamic Self-Refreshing of In_2O_3 Sensing Surface Assisted by Layer-by-Layer Coated CeO_2 Nanoclusters", *Small*, 12 (31), 4229–4240, 2016.

[12] K. Hwang, J. Ahn, I. Cho, K. Kang, K. Kim, J. Choi, K. Polychronopoulou, I. Park, "Microporous Elastomer Filter Coated with Metal Organic Frameworks for Improved Selectivity and Stability of Metal Oxide Gas Sensors", *ACS Appl. Mater. Interfaces*, 12 (11), 13338–13347, 2020.

[13] V. Nair, G.E. Hinton, "Rectified Linear Units Improve Restricted Boltzmann Machines", *27th Int. Conf. Mach. Learn*, 807–814, 2010.

[14] S. Ioffe, C. Szegedy, "Batch Normalization: Accelerating Deep Network Training by Reducing Internal Covariate Shift", *32nd Int. Conf. Mach. Learn*, 1, 448–456, 2015.

[15] D. P. Kingma, J. L. Ba, "Adam: A Method for Stochastic Optimization", *3rd Int. Conf. Learn. Represent*, 1–15, 2015.

CONTACT

*Corresponding author:
Inkyu Park, tel: +82 42-350-7922; inkyu@kaist.ac.kr,

DEVELOPMENT OF MONOLITHIC MICRO-LED GAS SENSOR BASED E-NOSE SYSTEM FOR REAL-TIME, SELECTIVE GAS PREDICTION

Kichul Lee, Mingu Kang and Inkyu Park
Korea Advanced Institute of Science and Technology (KAIST)

ABSTRACT

As interest in environmental pollution, indoor air quality, and industrial safety increases, the demand for gas sensors is rapidly increasing. Until now, only high sensitivity and fast response speed were considered important, so there was a problem of high-power consumption and bulky size, and they were installed only in limited places. For higher utilization of gas sensors in the future, low power and mass production are required. In this study, a micro-sized, ultra-violet micro-LED gas sensor (μLG) that can operate in an ultra-low-power was fabricated, and using a convolutional neural network algorithm, an e-nose system was developed. As a result, it was possible to distinguish four different gas species and predict the concentration of each gas in real time.

KEYWORDS

Micro-LED, electronic nose, ultra-low-power, light-activated gas sensor, convolutional neural network

INTRODUCTION

Limitations of heater-based gas sensors

The global market for gas sensors is continuously growing due to global interest in environmental issues, climate change, and healthcare. Gas sensors were mainly used to detect harmful gases and air pollutants mainly generated in industrial facilities, but recently their use has expanded and is used in air quality monitoring, healthcare, and food monitoring. According to the principles of gas detection, gas sensors are classified into various types. Among them, semiconductor metal oxide (SMO)-type gas sensors have the advantages of small size, high sensitivity, and mass productivity. The SMO-type gas sensor uses a metal oxide as a gas sensing material and requires activation for gas sensing. Generally, Joule heating using an embedded heater has been used, but it has high power consumption and has the potential to ignite explosive gas such as hydrogen gas. Although microheaters and self-heating nanowires have been developed to minimize power consumption, their suspended structures have weak mechanical and thermal durability, and the fabrication process is expensive and complex. In addition, the SMO-type gas sensor has a non-selectivity problem with respect to target gases. To overcome this issue, several attempts have been made, and recently, an electronic nose (e-nose) system that mimics the human sense of smell is being studied. The e-nose system has heterogeneous gas sensors that have different sensor responses to target gases and a pattern recognition algorithm based on machine learning (ML). Recently, it is a trend to use a large number of sensors for higher accuracy of the e-nose system. As the number of sensors used in the e-nose system increases, reducing the size and power consumption of individual sensors becomes more important.

Monolithic micro-LED gas sensor (μLG)

Recently, gas sensing by light-activation method is of interest because it can be operated at room temperature (RT) and can safely measure any environmental circumstances. Early studies using light-activation mostly used external bulky light sources such as lamps or commercial LEDs. Therefore, the power consumption was several watts higher than that of microheater-based sensors. Recently, in 2020, a monolithic light-active gas sensor that can detect gas with ultra-low-power consumption of 0.1 mW was developed by using ultraviolet (UV) micro LEDs and hydrothermally grown ZnO nanowires as gas sensing materials [1]. This ultra-low-power gas detection was possible because a micro-sized light source was used and the light energy loss was reduced due to the close distance (less than 1 μm thickness) between the gas sensing material and a light source. However, in the previous study, since the sensor electrode and the micro-LED electrodes are on the same plane, defining the area where the ZnO nanowires are deposited was necessary. Otherwise, an electrical short circuit may occur.

In this study, the design of the micro-LED gas sensor (μLG) was modified and a double SiO_2 insulating layer was applied to avoid electrical short between the micro-LED electrodes and the sensor electrode. Therefore, various materials including SMO, Metal-Organic Frameworks, and MXenes can be used as gas-sensing materials. Additionally, the patterned sapphire substrate (PSS) enhanced the internal quantum efficiency and light extraction efficiency of micro-LEDs. PSS technology also lowers the dislocation density between the sapphire and the gallium nitride (GaN) layers. This improves light efficiency by reducing total internal reflection at the air-GaN interface and increasing the amount of upwardly directed light. The schematic diagram of the μLG developed in this study and the gas detection mechanism are shown in Fig. 1. First, an epitaxial GaN-grown 2-inch size wafer emitting UV light was used for micro-LED fabrication. After that, the top of the micro-LED was insulated by plasma-enhanced chemical vapor deposited (PECVD) SiO_2, and the sensor electrodes were patterned. After fabricating the micro-LED gas sensing platform, a porous In_2O_3 nanofilm was directly deposited on the micro-LED through a glancing angle deposition (GLAD). GLAD, a type of RF sputtering, is a method of physically depositing a target material while tilting the sputtering target. Porous columnar nanostructures are easily created by the nanoscale self-shadowing effect that occurs in the GLAD process [2]. The GLAD method has various advantages of a large surface-to-volume ratio, high throughput, and high uniformity between sensors. After the GLAD, the localized surface plasmon resonance (LSPR) was used by coating noble metal nanoparticles (NPs) on the metal oxide surface by electron beam (e-beam) evaporation to maximize the gas response of the sensor. The LSPR occurs when light is shined on plasmonic noble metal nanoparticles that are significantly smaller than the wavelength of light [3]. Due to oscillation brought on by the redistribution of the NPs' charge density, LSPR produces a strong electric field and a high concentration of energetic electrons close to the surface of NPs. Electrons with a high energy state are produced by the oscillations of electrons that undergo a non-radiation decay. The generated electrons then cross the energy barrier and are transferred from the NPs to the contacting SMO. As a result, the NPs coating on SMO surface generates additional carriers that can aid in gas sensing and contribute to the enhanced gas response and fast redox reaction between the SMO and the target gas.

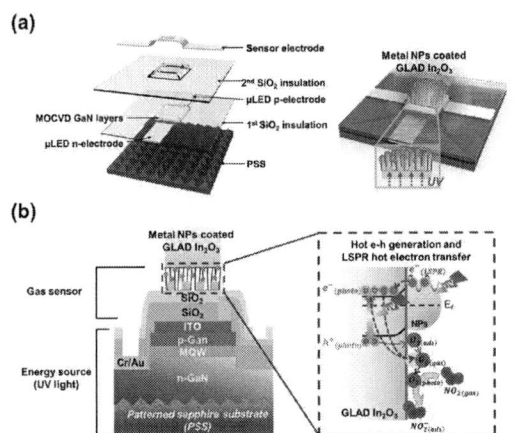

Figure 1: (a) Schematic images of design and concept of monolithic μLG that emits 395 nm wavelength ultra-violet light (b) Gas sensing mechanism of GLAD In_2O_3 coated with plasmonic noble metal NPs.

FABRICATION
Fabrication process

A comprehensive MEMS manufacturing process is shown in Fig. 2, which was used for the wafer-scale fabrication of μLG. A 6 μm-thickness n-GaN layer, a 50 nm-thickness multi-quantum well activation layer, and a 300 nm-thickness p-GaN layer were epitaxially grown by metal-organic chemical vapor deposition (MOCVD) method on a 600 μm-thickness PSS. First, a mesa structure was formed by vertically etching the GaN layer using inductively coupled plasma-reactive ion etching. Next, the deposition of the indium tin oxide (ITO) layer which is a current spreading layer was conducted on the GaN layer by e-beam. Then, a rapid thermal annealing process was followed to enhance the transparency of the ITO layer and reduce its electrical resistance. Third, gold patterning of the contact electrodes of micro-LED was performed by photolithography and lift-off process. Next, the first SiO_2 layer was deposited as an insulation layer using a PECVD method, and the deposited SiO_2 was then vertically etched using RIE to expose the gold contact electrodes. Then, the gold contact electrode was patterned using a lift-off process and connected to the previously patterned contact pad. Next, the second layer of SiO_2 was deposited using the PECVD and once again the second SiO_2 layer was etched

in a vertical direction using RIE to open the p-n contact pads. After that, photolithography and lift-off were used to pattern gold sensor electrodes on the micro-LED gas sensing platform. After patterning the sensor electrodes, an In_2O_3 gas sensing material was deposited by GLAD. Then, using an e-beam evaporation, silver, and gold which are two representative plasmonic metals, were coated onto the GLAD In_2O_3 surface. Finally, the fabricated μLG was diced into a sensor chip size of 5 mm by 5 mm.

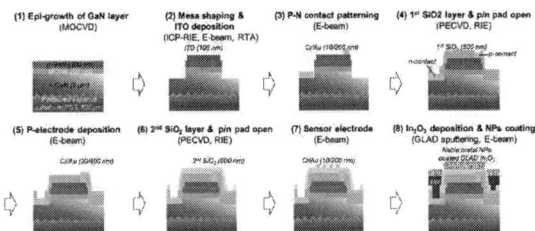

Figure 2: Schematics of a fabrication process of the monolithic light-activated μLG. An epitaxial GaN-grown PSS wafer emitting 395 nm wavelength UV light was used for μLG fabrication. The entire MEMS manufacturing process is carried out on a wafer scale.

Fabrication results

Fig. 3a shows the fabrication results of the μLG. The color (red, blue, and green) boxes show the pads of micro-LED-p, micro-LED-n, and sensor electrodes, respectively. The remaining areas are covered with a PECVD SiO_2 insulating layer. Scanning electron microscopy (SEM) image shows a 50 μm by 50 μm-sized μLG and the image on the right shows a UV light-emitting micro-LED at a forward bias of 3 V. Next, in Fig. 3b, the nanostructure of the noble metal NPs coated In_2O_3 film was analyzed by ultra-high-resolution SEM and transmission electron microscopy (TEM). The thickness of the In_2O_3 film was approximately 250 nm and the porosity was 35.0 %. The porous nanostructure increases the surface-to-volume ratio and improves gas response with the target gases. The size of silver and gold NPs coated on In_2O_3 gas sensing film were 30 nm and 10 nm, respectively.

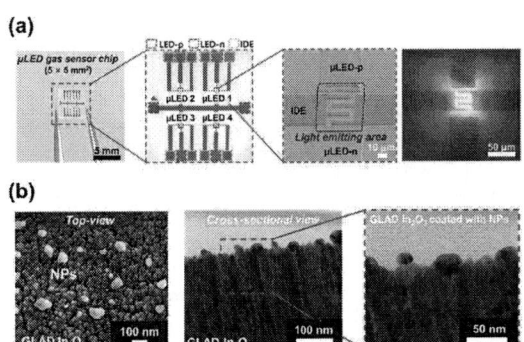

Figure 3: (a) Fabrication results of μLG. The size of the single sensor chip was 5 mm by 5 mm. There are multiple micro-LEDs in the form of a sensor array. (b) SEM and TEM images of GLAD In_2O_3 coated with plasmonic noble metal NPs.

RESULTS

Gas test results

After fabrication, the gas response to micro-LED power was examined to determine the μLG's ideal working power. Fig. 4a shows the gas response of bare In_2O_3, Ag NPs @ In_2O_3, and Au NPs @ In_2O_3 at various micro-LED electrical powers. The target gas used in the experiment was nitrogen dioxide at a concentration of 5 ppm, ethanol at a concentration of 100 ppm, and acetone at a concentration of 50 ppm. Since nitrogen dioxide gas is an oxidizing gas, it steals electrons from the In_2O_3 surface, which is an n-type semiconductor. Therefore, the electrical conductivity of In_2O_3 decreases, and the electrical resistance increases. Conversely, VOC gases such as ethanol and acetone react with adsorbed oxygen ions on the In_2O_3 surface and donate electrons and increase the electrical conductivity of the In_2O_3. The trend of gas sensitivity to micro-LED power was different for oxidizing and reducing target gases. For nitrogen dioxide gas, the sensitivity was maximum at a power consumption of nearly 1 μW, and for reducing VOC gases, as the power of the micro-LED increased, the gas sensitivity also increased monotonically. Through this experiment, the effect of metal NPs coating on the gas sensing material to improve the gas response was also confirmed. A forward bias of 3.0 V was the ideal operating condition for the micro-LED, which could detect both oxidizing and reducing target gases. In this state, the power consumption of sensors 1 and 2 (Ag NPs @ GLAD In_2O_3, Au NPs @ GLAD In_2O_3) was 0.18 mW and 0.20 mW, respectively, which means that the total power consumption of the e-nose system is only 0.38 mW which is ultra-low-power. In Fig. 4b, the gas test results of various concentrations of gases under the aforementioned conditions are shown.

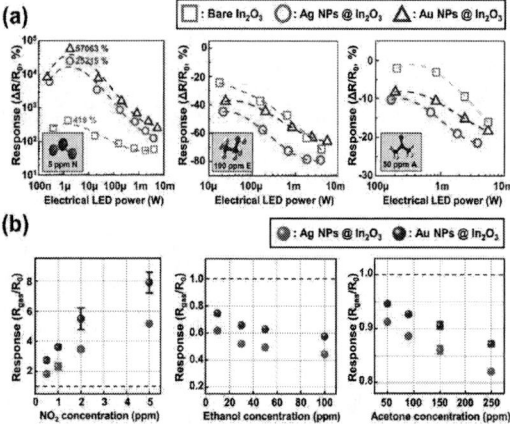

Figure 4: *Gas response of monolithic UV μLG. (a) Gas responses at various micro-LED powers for nitrogen dioxide at a concentration of 5 ppm, ethanol at a concentration of 100 ppm, and acetone at a concentration of 50 ppm. (b) Gas responses to various target gas concentrations. The total operation powers of sensors 1 (Ag NPs @ GLAD In_2O_3) and 2 (Au NPs @ GLAD In_2O_3) was 0.38 mW, which is ultra-low-power.*

Selective gas prediction of e-nose system

In Fig. 5a, the structure of the e-nose system based on

multi micro-LEDs and deep learning algorithm is described. In this study, the convolutional neural network (CNN) algorithm was used for pattern recognition of unique transient sensor responses caused by chemical reactions with each target gas. First, pre-processing was conducted on the gas test data measured by multi μLGs before training the CNN model. In order to increase the accuracy of the CNN model, a large number of data sets are required for training, and therefore, not only the original data but also artificially-made data were used for CNN training. The training data were normalized by log to prevent the CNN model from being trained by focusing on a target gas that has a high response. Next, data acquired by sensors 1 and 2 were concatenated in the form of a 2 × 30 (seconds) matrix and the sampling rate along the time direction was 1 Hz. A unique transient sensor signal generated by each target gas was captured by a sliding time window. The distinct transient sensor signal caused by each target gas was captured using CNN and a sliding time window size of 2 × 10 (seconds) matrix. When training the CNN model, the pre-processed gas response data and the corresponding true label (gas type and concentration) were input together. For gas type classification, air, ethanol, nitrogen dioxide, and acetone were labeled as 0, 1, 2, and 3, respectively. For regression of gas concentration, since the concentration values of each gas used in the gas experiment were different, normalization was conducted by dividing them by the maximum concentration value of each target gas (result value: 0 to 1). In the CNN model, six filters were used for the convolution layer and the stride interval was 5 seconds. The values calculated by convolution were then fed into a fully connected (FC) layer for calculation. During the training, batch-normalization and leaky-rectified linear unit (Leaky-ReLU) activation functions were applied to each hidden layer of CNN and FC. The final output layer has five nodes, four of which were used for gas classification and the other for regression of gas concentration. The classification was done by softmax function and four classification nodes output the gas label with the highest probability of being the corresponding target gas. The regression node outputs a normalized predicted gas concentration value. The categorical cross-entropy and mean-squared error were utilized in this study as the classification and regression loss functions, respectively. The total loss function is defined by simultaneously reflecting the two previous loss functions because the CNN model developed in this study conducts classification and regression simultaneously.

Fig. 5b-5d show the results of the e-nose system based on two different μLGs and a CNN. The confusion matrix presents the results of the gas type prediction for four different gases (air, ethanol, nitrogen dioxide, and acetone) (Fig. 5b). The average accuracy of classification was 99.5 %. The classification accuracy is calculated as the ratio of correct predictions to all the data. The results of the gas concentration regression are shown in Fig. 5c in normalized gas concentrations (zero to one). In regression, the error was 12.8 %. Fig. 5d describes the gas prediction results of the target gases in real-time. If data on more target gases are collected later and CNN models are trained, the performance and usefulness of the system will increase.

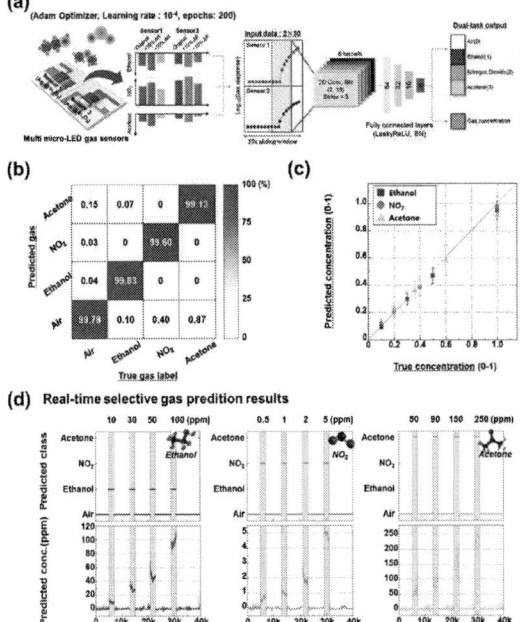

Figure 5: *(a) Structure of the e-nose system developed in this study. The e-nose system is based on two different µLGs and a CNN algorithm. It can selectively classify four target gases (air, ethanol, nitrogen dioxide, and acetone) and quantify their concentrations. Prediction results of (b) gas types summarized in a confusion matrix and (c) gas concentrations normalized in 0 to 1. (d) Gas prediction results of each target gas in real-time.*

CONCLUSIONS

In this study, an e-nose system that can be operated in ultra-low-power was developed based on the two different monolithic UV (395 nm) µLGs and a deep learning algorithm. The loss of light energy was reduced because the distance between the In2O3 and the light source (micro-LED) was micrometer scale (~ 1 µm). To increase the gas response, silver and gold which are plasmonic metal NPs were coated on porous In2O3 film by e-beam evaporation, utilizing the LSPR phenomenon. After several gas tests, the possibility of ultra-low power gas detection using micro-LED and the effect of NPs coating on gas sensing materials were confirmed. Next, the e-nose system that can selectively determine the gas type and gas concentration of four different gases (air, ethanol, nitrogen dioxide, and acetone) was developed based on µLGs with different sensing materials (Ag NPs @ In_2O_3, Au NPs @ In_2O_3) and CNN. As a result, real-time gas classification with a 99.5 % accuracy and regression with an error of 12.8 % was possible. Particularly, the overall power consumption of the µLG-based e-nose system was only 0.38 mW, which is less than one-hundredth of that of the heater-based e-nose system. Therefore, it can be used for a long time even when a battery is driven because it is ultra-low-power, has ultra-small size, and can be operated in room temperature conditions. So, the µLG-based e-nose system is expected to be highly useful as a mobile sensor in combination with IoT technology.

ACKNOWLEDGEMENTS

Facilities and materials for dicing sapphire wafers were supported by Disco Corporation (Japan).

REFERENCES

[1] I. Cho, Y. C. Sim, M. Cho, Y. H. Cho, I. Park, "Monolithic Micro Light-Emitting Diode/Metal Oxide Nanowire Gas Sensor with Microwatt-Level Power Consumption", *ACS Sens.,* vol. 5, pp. 563–570, 2020.

[2] C. Wongchoosuk, A. Wisitsoraat, D. Phokharatkul, M. Horprathum, A. Tuantranont, T. Kerdcharoen, "Carbon Doped Tungsten Oxide Nanorods NO_2 Sensor Prepared by Glancing Angle RF Sputtering", *Sens. Actuators, B,* vol. 181, pp. 388–394, 2013.

[3] X. C. Ma, Y. Dai, L. Yu, B. B. Huang, "Energy Transfer in Plasmonic Photocatalytic Composites", *Light: Sci. Appl.,* vol. 5, e16017, 2016.

CONTACT

1. Kichul Lee, tel: +82-10-7363-1542, kichul1106@kaist.ac.kr
2. Mingu Kang, tel: +82-10-4931-3780, kmg1994@kaist.ac.kr
3. Inkyu Park, tel: +82-10-2412-7337, inkyu@kaist.ac.kr

ELECTRONIC-NOSE: AN ARRAY OF 16 MOS-GAS SENSORS INTEGRATED WITH TEMPERATURE AND MOISTURE SENSING CAPABILITIES

Xiawei Yue[1,2], Shuai Wei[1,2], Pingping Zhang[3], Zhitao Zhou[1], Tiger H. Tao[1,2,4,5,6,7], and Nan Qin[1]*

[1]State Key Laboratory of Transducer Technology, Shanghai Institute of Microsystem and Information Technology, Chinese Academy of Sciences, Shanghai, CHINA and
[2]School of Graduate Study, University of Chinese Academy of Sciences, Beijing, 100049, CHINA and
[3]Suzhou Huiwen Nanotechnology Co. Ltd., Jiangsu, CHINA and
[4]Center of Materials Science and Optoelectronics Engineering, University of Chinese Academy of Sciences, Beijing, CHINA and
[5]School of Physical Science and Technology, ShanghaiTech University, Shanghai, CHINA and
[6]Shanghai Research Center for Brain Science and Brain-Inspired Intelligence, Shanghai, CHINA and
[7]Neuroxess Co., Ltd. (Jiangxi), 330029 Nanchang, Jiangxi, CHINA

ABSTRACT

This paper reports a new design of electronic-nose based on the Micro Electromechanical System (MEMS). An array of multiple metal-oxide semiconductor (MOS) gas sensors integrated with temperature and moisture sensing channels are fabricated on a single chip. Benefiting from the heat and humidity sensing capabilities, this electronic-nose exhibits an enhanced gas sensing property with improved accuracy. As low as 0.2 ppm of H_2S can be successfully identified, and the resistance of the gas sensor corresponding to the long-term stability drifts only about 3% after 100 days. It opens up new opportunities for fabricating bionic olfactory microsystems with multiple sensing channels and shows its potential for precisely detecting complex gases in practical applications.

KEYWORDS

Gas Sensor Array; Electronic-nose; Long-term Stability; Micro Electromechanical System;

INTRODUCTION

The biological sense of smell is based on multi-channel and multi-type olfactory receptor cells [1]. Human nervous system transports information about the odor to the brain, which is responsible for distinguishing different flavors. There are thousands of chemicals in the air, including CO, H_2S, CH_4, NO, NO_2, VOCs, etc. Many of them are harmful to human body and environment. Therefore, air quality monitoring is essential for safety and health conditions.

Recently, artificial olfaction has been developed for numerous industrial applications, such as indoor air quality monitoring [2], medical care [3], hazardous gas detection [4], environmental quality monitoring [5], food quality control [6-8], and military applications. Electronic-nose (e-nose) technology was first developed in 1964 by Wilkens, Hatman, and Buck, which is capable of detecting and distinguishing different species and concentrations of chemicals through a sensor array by mimicking the olfactory perception system of the organism [9]. Typically, the e-nose comprises a gas sensor array, electronic circuits, and algorithm corresponding to olfactory cells, the nervous system, and the brain of human beings. Each gas sensor exhibits a different response to exposed volatiles. With pattern recognition, the composition of odor is analyzed and reported [10].

However, the incorporation of sensor array, circuits, and algorithms increase the size and power consumption of the e-nose. The previous fabrication process of the electronic nose is quite complicated and accompanied by poor integration and high cost [11]. Moreover, humidity and temperature affect the MOS sensing sensitivity and accuracy of the e-nose. Herein, we propose a new design to integrate a low-power-consumption 16-channel gas sensor with temperature and humidity sensors to create an artificial electronic-nose principle similar to the biological version for characterizing the overall features of odors' information. It dramatically simplifies the process steps in subsequent applications and can be further 3D integrated with CMOS chips for hazardous gas detection, as well as improving the accuracy of complex gas identification and the long-term stability of the device [12].

EXPERIENT PROCEDURE

Figure 1 shows the design of the electronic-nose, which typically consists of an array of 16 MOS gas sensors, one temperature sensor and one moisture sensor. The total size is reduced to about 2.5×2.5 mm², as observed in the Scanning Electronic Microscope (SEM) image (Figure 1a). It is composed of four structural layers and one sensing layer from the top down, including an interdigital electrode, an insulated layer, a heating resistor, a substrate layer, the air cavity in silicon, and MOS sensing materials (Figure 1b and 1c). There are only two communal pads for heating circuits and gas-detecting circuits. Therefore, the lead frame is significantly simplified, and the number of pads is greatly lessened. A gas sensing circuit is located right behind the heating circuit to get a uniform thermal environment, and an air cavity is corroded beneath the circuits. The sensor array is decorated with different materials that operate at different temperatures. The working temperature of MOS sensing materials are ranging from 250 to 400 °C. 16 gas sensors are classified into four types according to the heater resistance, which determines the optimum working temperature. After loading with sensing nanomaterials and applying with suitable voltages, 16 gas sensors and 2 environment sensors show specific responses upon each type and concentration of gases.

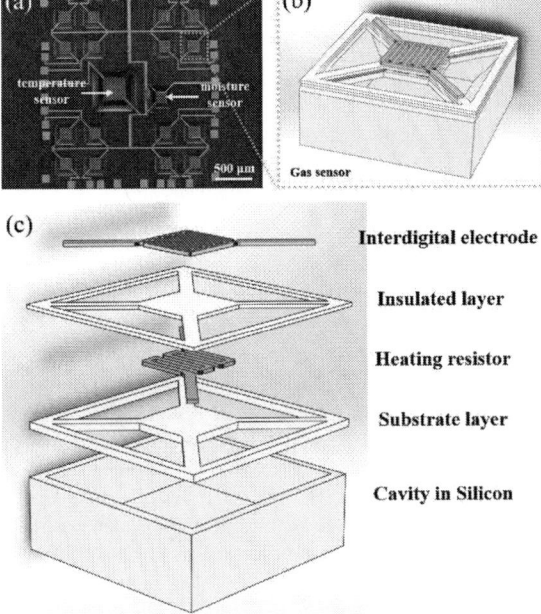

Figure 1: Structure of 16-gas-sensor array integrated with a temperature sensor and a humidity sensor. (a) SEM image of the sensor array. (b) Illustration of an individual gas sensor. (c) Explosive view of an individual gas sensor.

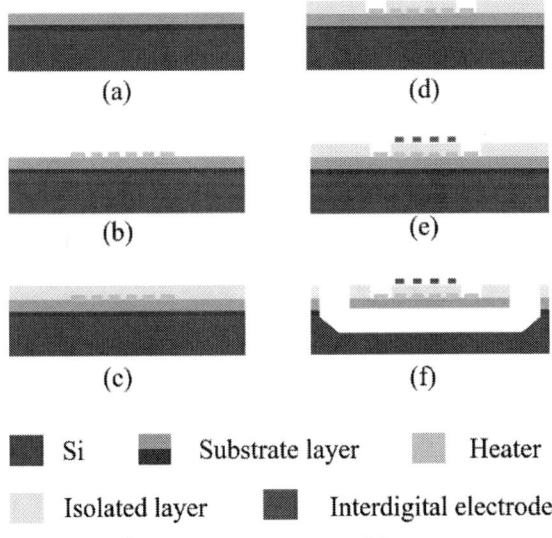

Figure 2: Fabrication process steps of the micro-sensor array in cross-sectional views.

Figure 2 shows the fabrication steps. Briefly, (a) a layer of 200 μm silicon oxide is deposited by thermal oxidation, and then a layer of 600 μm silicon nitride is deposited on (100) silicon wafer to balance the film stress. (b) 30 μm of adhesive layer Ta and 300 μm of Pt film are deposited and patterned as the heater. (c) insulative layer, which consists of 200 μm silicon oxide, 200 μm silicon nitride, and 200 μm silicon oxide, is deposited through Plasma Enhanced Chemical Vapor Deposition (PECVD). (d) pads of heating circuits are exposed by RIE (Reactive

Ion Etching) dry etching insulated layer. (e) 30 μm of adhesive layer Ta and 300 μm of Pt film are deposited and patterned as the interdigital electrode of temperature, moisture sensors, and 16 gas sensors. (f) etch the dielectric layer around each sensor, through where the silicon beneath is corroded after anisotropic wet etching in TMAH (Tetramethylammonium Hydroxide) solution for 4 hours at 81 °C to form suspension beam construction. Finally, MOS nanomaterials (e.g., ZnO nanoflowers consisting of nanorods with a diameter of ~100 nm) are loaded onto the interdigital electrode layer for gas sensing (Figure 3).

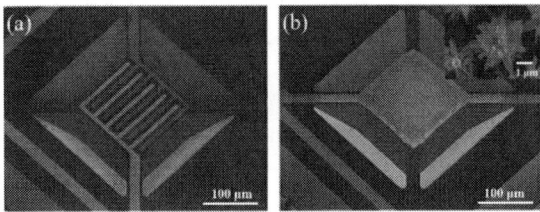

Figure 3: Structure of the electronic-nose. SEM image of an individual gas sensor (a) before and (b) after loading with gas sensing nanomaterials (the inset).

RESULTS

Figure 4 shows the detailed performance of the temperature sensor. Platinum is a temperature-sensitive metal and suitable for the temperature sensing. The serpentine resistance wire with a width of 8 μm is precisely fabricated on a 300×300 μm^2 suspended membrane (the inset). The corresponding temperature coefficient of resistance (TCR) reaches 0.00234/°C with an approximately proportional relation between 20 °C and 100 °C, indicating its reliable response under ambient temperature.

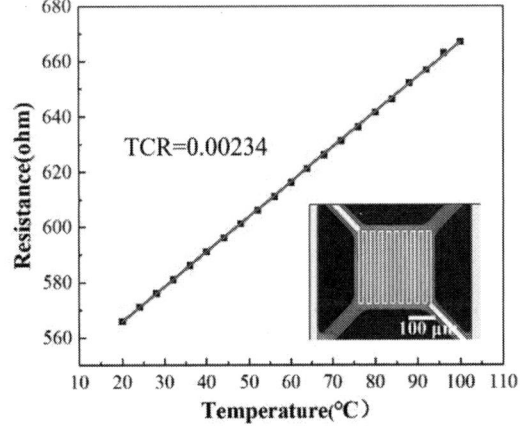

Figure 4: Temperature-Resistance relationship of temperature sensor.

Figure 5 shows the moisture sensing performance, which is evaluated by monitoring its relative resistance variation ($\Delta R/R_0\%$) upon exposure to test moisture. R_0 is defined as the initial resistance, and ΔR defined as resistance variation after vapor exposure. The quantitative response is defined as the normalized resistance difference ($\Delta R/R_0$) at the beginning and the end of the test moisture for each experimental cycle. The moisture sensor in this work is capable of detecting the relative humidity of 30%-70%.

For semiconductor materials, temperature exhibits a prominent sign on the movement of electrons and holes and water vapor in the air tends to bond on the surface of the sensing nanoparticles, thus affecting the electric conductivity. With the assistance of temperature and humidity sensors, the interference of environmental changes on gas response can be effectively eliminated.

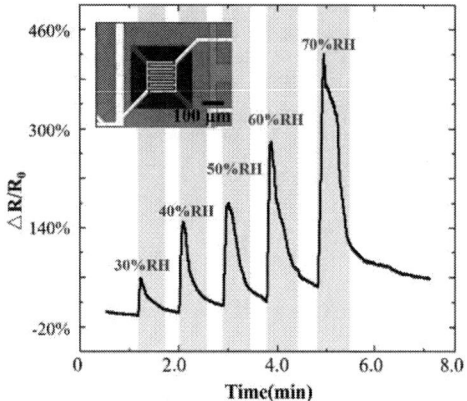

Figure 5: Response of humidity sensor between 30%RH and 70%RH.

Figure 6 shows the approximate linear temperature coefficient of heater resistance in the temperature-voltage (T-V) relationship. The designed heating resistance of sensor 1, sensor 2, sensor 3, and sensor 4 is 40 Ω, 50 Ω, 60 Ω and 70 Ω, respectively. Sensor 1 induces a higher working temperature while sensor 4 produces less heat. Under 300 °C, the power consumption of sensor 4 is ~46 mW. Suspension beam construction effectively reduces heat loss and promotes energy conservation. MOS nanomaterials exhibit varying responses at different operating temperatures. The ability of the heater to produce a steady amount of heat at the same voltage is necessary. As schemed in Figure 7, the resistance of the heater drifts only about 3% after 100 days, indicating the excellent long-term stability.

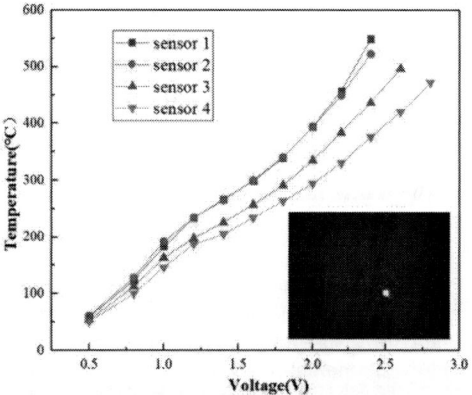

Figure 6: Temperature-Voltage relationship.

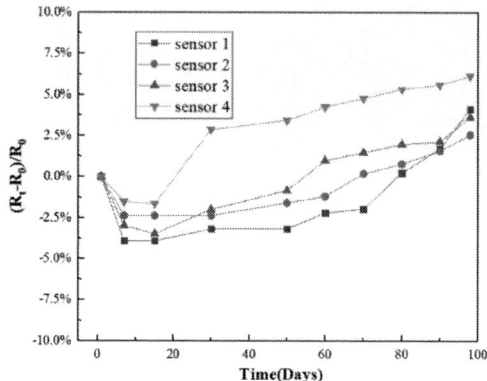

Figure 7: Resistance variation of sensors in 100 days.

Moreover, the sensor array is employed to test H_2S under 1.9 V (Figure 8). The detection limit of our electronic-nose toward H_2S reaches ~0.2 ppm and behaves well in several test cycles. All four types of sensors exhibit good response upon H_2S testing, further demonstrating their potential practical application in the hazardous gas sensing.

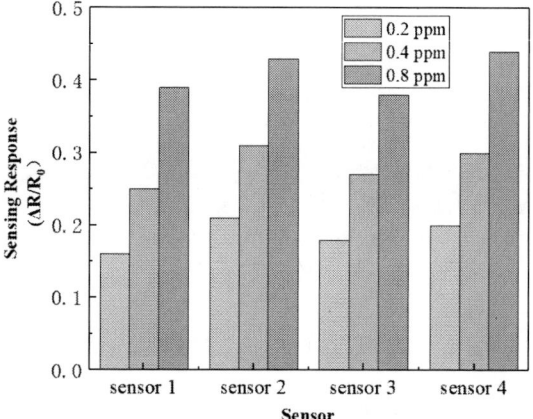

Figure 8: Response of H_2S with concentration of 0.2, 0.4 and 0.8 ppm.

CONCLUTION

In summary, this paper exploits a new design of low-power-consumption e-nose integrated with two environmental sensors and multiple MOS nanomaterial decorated gas sensors, creatively eliminating the effect of temperature and humidity and enhancing the accuracy of gas sensing. It demonstrates its ability to detect and distinguish multiple gases under complex environments and the potential of 3D integrated with CMOS chips, which brings enormous opportunities for environmental air quality monitoring, such as health monitoring, disease prediction, food quality control, and hazardous gas detecting.

ACKNOWLEDGEMENTS

Xiawei Yue and Shuai Wei contributed equally to this work. This work was supported by National Science and Technology Major Project from the Minister of Science and Technology of China (grant no. 2018AAA0103100),

National Natural Science Foundation of China (grant no. 62236005), National Science Fund for Excellent Young Scholars (grant no. 61822406), Key Research Program of Frontier Sciences, CAS (grant no. ZDBS-LY-JSC024) and Fund of Youth Innovation Promotion Association CAS (grant no. 2022234).

REFERENCES

[1] A. Rao, H. Long, A. Harley-Trochimczyk, et al. "In Situ Localized Growth of Ordered Metal Oxide Hollow Sphere Array on Microheater Platform for Sensitive, Ultra-Fast Gas Sensing", *J. ACS Applied Materials & Interfaces*, vol. 9, pp. 2634-2641, 2017.

[2] L Zhang, F Tian, H Nie, et al. "Classification of multiple indoor air contaminants by an electronic nose and a hybrid support vector machine", *J. Sensor Actuat B-Chem*, vol. 174, pp. 114-125, 2012.

[3] D Guo, D Zhang, N Li, L Zhang, J Yang. "A novel breath analysis system based on electronic olfaction", *IEEE Trans Biomed Eng*, vol. 57, pp. 2753-2763, 2010.

[4] Z Haddi, A Amari, H Alami, et al. "A portable electronic nose system for the identification of cannabis-based drugs", *J, Sensor Actuat B-Chem*, vol. 155, pp. 456-463, 2011.

[5] M Penza, D Suriano, G Cassano, et al. "A gas sensor array for environmental air monitoring: a study case of application of artificial neural networks", *AIP Conf Proc*, vol. 1362, pp. 205-206, 2011.

[6] C D Natale, A Macagnano, E Martinelli, et al. "The evaluation of quality of post-harvest oranges and apples by means of an electronic nose", *J, Sensor Actuat BChem*, vol. 78, pp. 26-31, 2001.

[7] I Concina, M Falas coni, V Sberveglieri. "Electronic noses as flexible tools to assess food quality and safety: should we trust them", *IEEE Sens J.* vol. 12 pp. 3232-3237, 2012.

[8] M Macías, A Manso, C Orellana, et al. "Acetic acid detection threshold in synthetic wine samples of a portable electronic nose", *J, Sensors*, vol. 13, pp. 208-220, 2013.

[9] S. Sankaran, L. R. Khot, S. Panigrahi. "Biology and applications of olfactory sensing system: A review" *J. Sensors and Actuators*, pp.1-17, 2012.

[10] S. Y. Park, Y. Kim, T. Kim, et al. "Chemoresistive materials for electronic nose: Progress, perspectives, and challenges", *J. InfoMat*, vol. 1, pp. 289-316, 2019.

[11] A D Wilson, M Baietto. "Applications and advances in electronicnose technologies", *J, Sensors*, vol. 9 pp. 5099-5148, 2009.

[12] Y. Chen, P. Xu, P Zhang, et al. "long-term stability improvement of micro-hotplate methane sensor product", *Proc. MEMS 2020,* pp. 1300-1303.

CONTACT

*N. Qin, tel: +86-21-62511070;
qinnan@mail.sim.ac.cn

ENHANCEMENT OF SENSITIVITY IN PHOTONIC CRYSTAL BASED CHEMICAL SENSOR USING CHEMO-MECHANICAL BILAYER EFFECT

Seyeon Lee[1], Naik T. Banavathi[1], Dongwon Kang[4], Sookyung Kang[3],
Kyungsuk Cho[3], Jungwook Kim[2] and Jungyul Park[1]
[1]Dept. of Mechanical Eng. Sogang Univ., [2]Dept. of Chemical Engineering. Sogang Univ.,
[3]Dept of Environmental Eng. Ihwa Univ. and [4]Dept of Mechanical & Aerospace Eng. UCLA

ABSTRACT

Photonic Crystals (PCs) enable the control of light through nanostructures and are used as colorimetric sensors. In this study, we propose a new mechanism for improving the sensitivity of the PCs based colorimetric sensor by inducing mechanical stress due to the difference from the relative swelling of the thin film on the nanostructure. Different types of polymer materials with a distinctive swelling ratio are patterned so that each polymer section shows a unique color. Using the color change combination that appears during chemical exposure makes possible a powerless color change sensor that can detect a specific target material.

KEYWORDS

Volatile Organic Compounds, Photonic Crystals, Chemo-mechanical Bilayer Effects, Diffusion

INTRODUCTION

Volatile Organic Compounds

Representative air pollutants, volatile organic compounds (VOCs), combine with oxides by photochemical reactions to be involved in smog and secondary organic aerosol formation [1]. VOCs are readily produced in gas form in indoor and outdoor industrial facilities, residential environments, and automobiles [2]. In particular, the human body is easily exposed through various cosmetics, cleaning tools, and furniture indoors, which cause serious health problems such as central nervous system diseases [3]. The need for technology to monitor the concentration of VOCs in real time is growing to inform workers of the risks [4]. Commercially available gas sensors such as gas chromatography and PID sensors have been employed, but they have drawbacks such as being difficult for non-experts to use or requiring expensive equipment, as well as, most crucially, requiring external power sources. Additionally, chemoresponsive dye-based colorimetric sensors were introduced, but because they can only be used once, it is tough to use them for continuous monitoring of VOCs.

Photonic Crystals

The epidermis of many animals and plants in nature exhibits structural colors based on nanostructures that can be visually responsive to outside stimuli. Inspired by this phenomenon, researchers have studied photonic crystals (PCs), which alter the inter-distance in response to chemical stimuli. Since PCs are made of dielectric materials with regular nanostructures, only a limited range of light wavelengths are reflected by diffraction and interference, which is based on the Bragg equation, and this is what is seen as the visible light region's color. The bandgap, which is the region where light cannot pass, shifts when the spacing between periodic nanostructures changes, and this color change may be seen with the human eye.

Colorimetric Sensor with the Integrated Thin Film

Polymers enable physical and chemical interactions by various external stimuli due to the nature of complex chain structures. By allowing chemicals to penetrate the molecular structure of the chain, the volume expands by the diffusion effect and then returns to the original state when the substance is removed. When PCs with regular nanostructures and polymer chain structures are made to interact chemically, the polymer expands when exposed to chemical stimulation and physically alters the spacing between the nanoparticles, causing overall color variations. When the chemical stimulus is removed, the enlarged polymer chain shrinks back to its original size, producing a reversible color change response. The range of color changes can be improved by causing more spacing changes between the PCs structures when a polymer material that has specific swelling capabilities is applied.

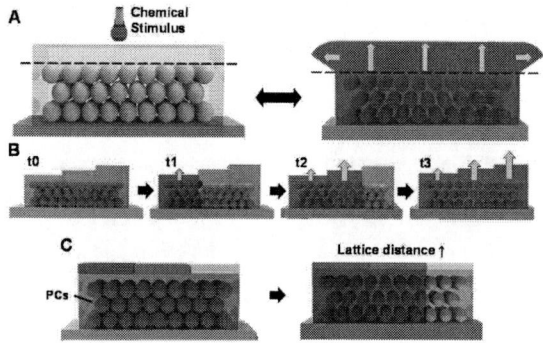

Figure 1: Spatial-temporal multiple color generation by chemo-mechanical effect induced from a single chemical stimulus (A) Thin layer swelling inducing the additional lattice displacement in the period nanostructures due to the chemo-mechanical effect (B) Timely regulated color formation controlled by thickness of thin layer (C) Multiple color formation with different swelling from different material properties

In this study, four polymers with high chemical affinity with VOCs are applied with PCs and a thin film composed of only polymers is additionally laminated thereon to induce mechanical stress. When an isotropic displacement of a polymer material occurs, the amount of polymer in a PCs structure is relatively much smaller than one in the thin film, and thus the chemo-mechanical bilayer effect is induced. This results in mechanical stress that lifts the nanostructure in the vertical direction, leading to the larger red shift of the bandgap and consequently, wider color changes are observed. Based on this principle, a polymer with a different degree of swelling depending on the type, mechanical stiffness, and concentration of VOCs can be controlled to show various color changes for one chemical stimulus, allowing selective and quantitative detection of VOCs through a combination of color changes when exposed to four chemical stimuli.

METHODS
Materials
Polystyrene 0.20μm (Bangs Laboratories Inc., USA), 3-(Trichlorosilyl) propyl methacrylate, Benzophenone, Acrylic Acid, Poly (ethylene glycol) diacrylate (PEGDA), Poly (ethylene glycol) methyl ether acrylate, Poly (propylene glycol) acrylate (PPG), Photo Initiator (Danocur 1173) (Sigma Aldrich, USA), Poly (propylene glycol) diacrylate, Poly (dimethylsiloxane) (PDMS) (Dow Corning. USA), Norland Optical Adhesive 144 (NOA144) (Edmund Optics, USA)

Preparation of Polystyrene Nanostructure
Polystyrene nanoparticles of uniform size are prepared to form opal structures on slide glasses prepared through several pre-processing steps. A method using a silane agent is applied as a process to increase the adhesion to the polymer, and then a mixed solution of acrylic acid, acetone, and benzophenone is sprayed to improve the coating performance of the nanoparticles. Then, monodisperse polystyrene nanoparticles with a diameter of 200 nm are self-assembled using Doctor blade method, which enables a regular and stable arrangement of particles as shown in Figure 2(A). 3D colloidal crystal nanostructures with Face Centered Cubic (FCC) structures are formed through the convective assembly.

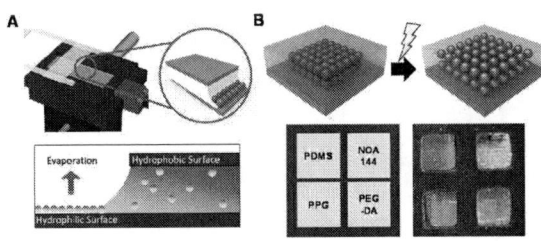

Figure 2: Polymer thin layer and photonic crystal-based color change sensor fabrication process (A) Self-assembly of polystyrene nanoparticles of monodisperse size (B) Process of making an array after applying PDMS, NOA144, PPG, and PEGDA through curing

Fabrication of Colorimetric Polymer Sensors
Figure 2(B) shows the fabrication process of PCs and swellable polymer composite. Applying a spacer with constant thickness beside the assembled nanostructure, PDMS, NOA144, PEGDA, PPG are fabricated through heat and photo-curing processes so that sensors can have specific thicknesses.

Initially, the sensor with four types of polymer thin

films after fabrication exhibits green in color. But a preferential difference in swelling occurs depending on the type of thin film, various color change combination can be obtained after exposure to VOCs.

RESULTS AND DISCUSSION
Chemo-mechanical Stress Generation Mechanism
In the absence of a thin film, AFM (NX10, Parksystems) analysis confirms that the stiffness of PDMS polymers embedded in photonic crystals differs about 10 times before and after swelling. This shows the possibility that the inter-distance change can be more easily caused by external mechanical stress when the photonic crystal-polymer are exposed to a chemical stimulus. The volume expansion of the thin film is relatively larger when the Young's modulus of the photonic crystal-polymer and the polymer thin film is significantly different, which leads to an improvement in the sensitivity of the entire sensor.

Bilayer Effects of PDMS Thin Layer
Figure 3(A) shows the verification of performance improvement when there is an additional thin film. It is confirmed that the bandgap shift by chemical stimulation is improved from about 20% to 100% by adjusting the Young's modulus by sequentially varying the ratio of PDMS and the curing agent from 3:1 to 20:1. In addition, when this sensor is exposed to 2000 ppm of acetone, it is experimentally verified that the performance is improved by about four times by the thin film.

In addition, when the sensor performance is measured by adjusting the thickness of the thin PDMS thin film, an additional wavelength change of 20 to 30 nanometers occurs when a thin film of about 20 to 40 micrometers is added (Fig. 3(B)). On the other hand, if the thickness of the thin film exceeds 60 micrometers, sufficient expansion of the thin film by chemicals does not occur, resulting in poor performance. It could be qualitatively verified that the relationship between the thickness of the thin film and the saturation time, which is a time when the reaction by chemical stimulation is completed, follows a trend of the diffusion equation.

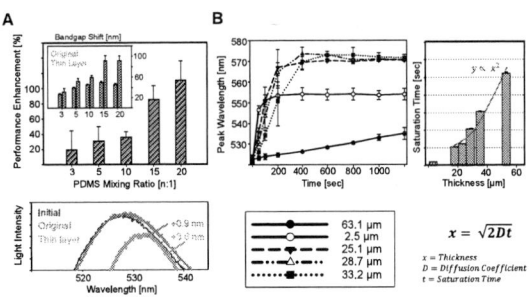

Figure 3: (A) Difference in color change according to mechanical stiffness of polymer in thin film (B) Relationship between reaction rate and saturation time according to the thickness of polymer thin layer

Exposure to Various Chemical Stimuli
When various types of VOC are exposed to the manufactured polymer-based PCs sensor, different color

change combinations can be obtained according to the preferred difference in swelling ratio depending on the type of thin film. Initially, sensors integrated with four types of polymer thin films have similar wavelengths of about 520 to 530 nm, but after the reaction, the degree of color change and bandgap shift varies depending on the type of chemical stimulus.

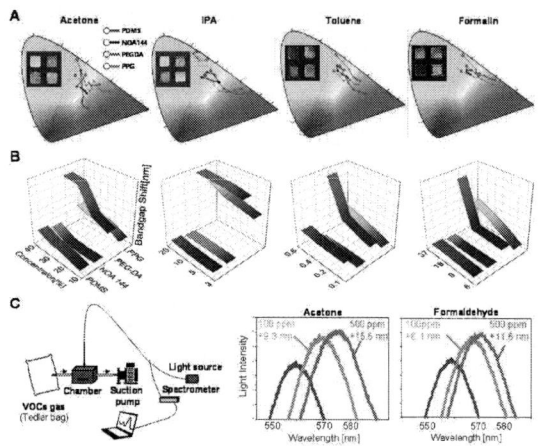

Figure 4: (A) Color change (CIE 1931) and (B) bandgap change of sensor according to polymer material when exposed to four aqueous solutions in liquid state (B) Variation of the wavelength of the PEGDA sensor when exposed to acetone, and formaldehyde 100 ppm and 500 ppm

Figures 4(A) and (B) illustrate the diversity of the degree of color change by varying the type and concentration of the VOCs aqueous solution. Figure 4(C) also shows the unique expansion characteristics of the polymer by showing different expansion rates when 2 types of gas concentration have concentrations of 100 ppm and 500 ppm. The measurement of wavelength shift was performed by a spectrometer (USB 2000, Ocean Optics).

CONCLUSION

By forming a polymer thin layer with a certain range of height in the PCs structure, the composite structure that can show a wider color change due to the relative swelling displacement difference between the thin film and the photonic crystal is presented. It was verified that, by applying a polymer that swells preferably for a specific chemical substance in a nanostructure, it is inducing different color combinations depending on the chemical substance, so it is possible to evaluate the type and concentration of VOCs quantitatively. To increase sensitivity to various types of chemicals, it is necessary to change the color more reliably. It can be solved by integrating a pre-concentrator capable of accumulating VOCs on the front end of the sensor. The developed sensor is expected to be used in various indoor environments because it is simple to manufacture and can selectively and quantitatively detect chemicals. Additionally, it can be used to create special cryptographic characters by adjusting specific patterns or color combinations at the desired time

when exposed to certain chemicals depending on the thickness and material type of thin film.

ACKNOWLEDGEMENTS

This work was supported by a National Research Foundation of Korea (NRF) grant funded by the Korean government (MSIP) (2020R1A2C2009093) and by the Korea Environment Industry & Technology Institute (KEITI) through its Ecological Imitation-based Environmental Pollution Management Technology Development Project funded by the Korea Ministry of Environment (MOE) (2019002790007).

REFERENCES

[1] R. Atkinson and J. Arey, "Atmospheric Degradation of Volatile Organic Compounds," *Chem. Rev.*, vol. 103, no. 12, pp. 4605–4638, Dec. 2003, doi: 10.1021/cr0206420.
[2] M. C. Janzen, J. B. Ponder, D. P. Bailey, C. K. Ingison, and K. S. Suslick, "Colorimetric Sensor Arrays for Volatile Organic Compounds," *Anal. Chem.*, vol. 78, no. 11, pp. 3591–3600, Jun. 2006, doi: 10.1021/ac052111s.
[3] K. Rumchev, H. Brown, and J. Spickett, "Volatile Organic Compounds: Do they present a risk to our health?," *Reviews on Environmental Health*, vol. 22, no. 1, Jan. 2007, doi: 10.1515/REVEH.2007.22.1.39.
[4] C. Peng, K. Qian, and C. Wang, "Design and Application of a VOC-Monitoring System Based on a ZigBee Wireless Sensor Network," *IEEE Sensors J.*, vol. 15, no. 4, pp. 2255–2268, Apr. 2015, doi: 10.1109/JSEN.2014.2374156.

CONTACT

*J. Park, tel: +82-2-705-8642; sortpart@sogang.ac.kr

METAL ION RECOGNITION SENSOR BASED ON RESISTIVE SWITCHING EFFECT

Tian Kang, Yusa Chen, Guanzhou Lin, Shengxiao Jin, Liye Li,
Hongshun Sun, Senyong Hu, and Wengang Wu*

National Key Laboratory of Science and Technology on Micro/Nano Fabrication, School of
Integrated Circuits, Peking University, Beijing 100871, P. R. CHINA

ABSTRACT

This paper reports a novel metal ion sensor composed of sandwiched Electrolyte-Oxide-Metal (EOM) structure. The EOM devices exhibit obvious resistive switching characteristics when excited by voltage. We test four electrolytes under different SiO_2 thicknesses and different electrolyte concentrations to research the resistive switching characteristics of the devices. The experiments show that the silicon oxide films fabricated by Plasma Enhanced Chemical Vapor Deposition can have relatively uniform turn-on voltage characteristics. In addition, we did more than 3000 tests for each electrolyte at 0.1M concentration and made a data set. A Deep Neural Network is constructed to train this data set, and the recognition accuracy reaches 99.6%. This demonstrates the potential of EOM devices in the field of identifying multiple metal ions.

KEYWORDS

Electrolyte-Oxide-Metal; Conductive filaments; Resistive switching; Ion sensing.

INTRODUCTION

RRAMs are widely developed by researchers for their good scalability [1], ultrafast operation speed [2], simple sandwiched-like structure [3], and CMOS compatibility [4]. RRAMs, which consist of a simple sandwiched meta-insulator-metal (MIM) structure, is based on the resistive switching (RS) effect [5], in which the resistance of an insulating layer sandwiched between metal electrodes can be changed by an externally applied electric bias.

The RS phenomenon has been observed in several materials, including chalcogenides, selenides [6], graphene oxide [7], organic polymer composites [8], and transition metal oxides. Among all these materials, transition metal oxide-based RS devices have been extensively researched in recent years. Several types of mechanisms such as (i) electrochemical metallization memory (ECM), (ii) thermochemical memory devices, and (iii) valence change memory have been proposed to explain the RS effect in these materials. In a low-resistance state (LRS) ohmic type conduction predominates, whereas in a high-resistance state (HRS) different conduction mechanisms such as (i) Poole Frenkel emission, (ii) Schottky emission, (iii) trap-assisted tunneling, and (iv) hopping conduction are observed.

Several methods have been used to improve the stability of RRAM devices, such as threading dislocations [9], voltage amplitudes [10], Ar plasma treatment [11], the coupling of oxygen and metal ions [12]. To obtain high speed and stable performance, most of the active

metal electrodes, inert metal electrodes, and insulating layers in the research are solid-state.

To explore the RS phenomenon of metal ions supplied by the electrolyte, a novel Electrolyte-Oxide-Metal (EOM) device was designed. Its RS mechanism was similar to that of metal conductive filament (CF) RRAM. There are some changes in EOM compared to the common RRAM structure. The metal active electrode was replaced by an electrolyte. The middle layer is the SiO_2 layer. In the RRAM structure, the metal on the active electrode needs to ionize at a positive bias to form a CF in the oxide layer. But in EOM the ions themselves are ionized. It is easier to form CF in an EOM device.

Previously, we also designed an ion sensing device composed of sandwiched Electrolyte-Oxide-Semiconductor (EOS) structure [13]. The EOS device uses a SiO_2 layer fabricated by the Low Pressure Chemical Vapor Deposition (LPCVD) process, and its bottom electrode uses a double-doped low resistance silicon wafer. Compared with EOS devices, EOM devices show more stable and uniform performance.

In this paper, EOM devices with three different SiO_2 thicknesses were used to measure four electrolytes separately. $CuSO_4$, $FeCl_3$, and $ZnSO_4$ show different RS characteristics respectively, while KCl electrolyte cannot produce RS phenomenon but can produce leakage current. The test results of different concentration and different thickness of SiO_2 shows that the devices can have a relatively stable Set voltage when using a different electrolyte, which can be regarded as a symbol of ion recognition. By comparing the switching success rates of three EOM devices with different thicknesses of SiO_2, the EOM device with a SiO_2 thickness of 50nm was selected for ion sensing data collection.

We did 4000 sets of experiments on four kinds of electrolytes with 0.1M concentration using a 50nm-SiO_2 EOM device. After data screening, the data sets of the four electrolytes are respectively 3319, 3732, 3791, and 3352. Three different types of neural network were used in the training test, and the prediction accuracy was higher than 99%. The results show that the EOM devices can realize the function of ion sensing and recognition.

STRUCTURE DESIGN

Figure 1 shows the fabrication process and structure diagram of the EOM devices. First, 10 nm Ti is sputtered onto wafer as an adhesive layer. Then 150 nm Pt was sputtered onto Ti as an inert metal electrode. Subsequently, SiO_2 of three different thicknesses was deposited onto Pt to make different EOM devices by by Plasma Enhanced Chemical Vapor Deposition (PECVD). Finally, a thick polydimethylsiloxane (PDMS) film with 6mm-diameter holes was fabricated. The PDMS film was then bonded on the surface of the SiO₂ layer as a solution tank.

To ensure accurate measurement in the liquid environment, we adopted a three-electrode measurement method to test the device. The counter electrode and the reference electrode were placed in the electrolyte, and the working electrode was connected to the Pt side. Due to the method of electrode connection, RS occurs when a negative voltage is applied to the device.

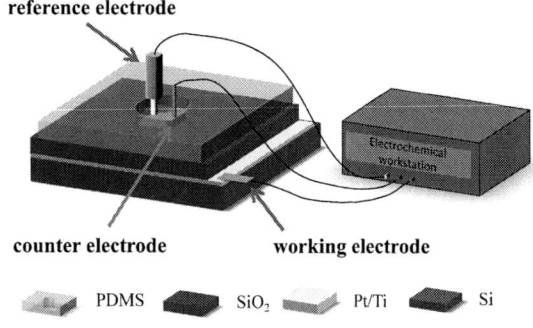

Figure 1: The schematic diagram of the EOM device structure and the three-electrode method for measuring the device.

RESULTS AND DISCUSSION

During the whole experiment, four electrolytes, $CuSO_4$, $FeCl_3$, $ZnSO_4$, and KCl, were used as the top electrode to test the performance of the EOM devices. The Cyclic Voltammetry (CV) method was used for testing, and the voltage was negatively scanned from 0V. In the IV curve shown in Fig. 2(a)-2(d), the thickness of the SiO₂ layer of the device is 50nm and the solution concentration is 0.1M. It can be observed that $CuSO_4$, $FeCl_3$, and $ZnSO_4$ electrolytes can produce RS phenomenon, indicating that these three electrolytes can provide the condition for the device to form stable metal CFs.

However, no RS phenomenon was observed in the test of the KCl solution. Compared with the other three electrolytes, the operating current of KCl is generated, but it is an order of magnitude smaller. This is due to the strong migration ability of K^+ in SiO₂, and a large number of K^+ distributed in the SiO₂ layer can cause the device leakage current. This indicates that the metal ion solution produced by extremely active metals cannot form RS phenomenon. This was also reflected in the test with NaCl solution, which showed a similar curve as KCl.

By observing the IV curves of $CuSO_4$, $FeCl_3$, and $ZnSO_4$ solutions, it is found that the RS voltages of the devices are different, and the curve morphologies are completely different. The Set voltage of $CuSO_4$ is about -0.3 V and the Reset voltage is about 2.2 V. The Set voltage of $FeCl_3$ is about -1.1V and the Reset voltage is about 0.8V. $ZnSO_4$ has a Set voltage of -1.5 V and a Reset voltage of 0.7 V. This indicates that the EOM devices can be used as ion sensors if they have relatively stable and uniform RS voltage characteristics when using different solutions.

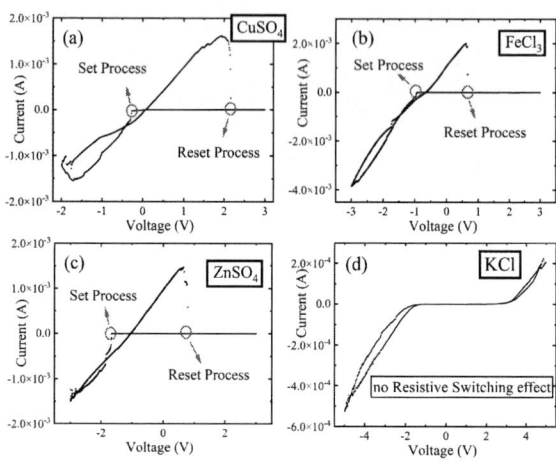

Figure 2: IV characteristic curves of four electrolytes. Initially, the voltage is scanned in a negative direction.

Firstly, three $CuSO_4$ solutions of different concentrations were used to test three EOM devices of different SiO₂ thicknesses, and each group of experiments was 100 cycles, as shown in Fig. 3. Figure 3 (a) shows the cumulative frequency diagram of voltage distribution of the 50nm-SiO₂ EOM device using three $CuSO_4$ solutions of different concentrations. When the solution concentration is 0.1M, the Mean Set voltage of the device is -0.174 V and the standard deviation (STD) is 0.083V. When the solution concentration is 0.01M, the Mean opening voltage of the device is -0.08 V and the STD is 0.047 V. When the solution concentration is 10^{-3}M, the Mean Set voltage of the device is -0.129 V and the STD is 0.046 V. Those EOM devices show relatively stable and uniform Setting performance at the three concentrations. The dispersion degree of the Reset voltage value is relatively large, but the Reset voltage of the device at 10^{-3} M concentration is concentrated and small, which is related to the fine formation of conductive filaments at this concentration.

Figure 3 (b) shows the box plot of the EOM devices with three different SiO₂ thicknesses using 0.1 M $CuSO_4$ solution. The experimental data volume is 75,178, and 137, respectively. The Mean and STD of the Set voltage of 10nm-SiO₂ EOM devices are -0.06V and 0.033V, respectively. The Mean and STD of the Set voltage of 30nm-SiO₂ EOM devices are -0.169V and 0.071V, respectively. The Mean and STD of the Set voltage of 50nm-SiO₂ EOM devices are -0.179V and 0.079V, respectively. It can be seen that the Set voltage tends to increase slightly with the increase of thickness, but the fluctuation range is very small, about 0V to -0.3V. The Reset voltage decreases gradually with the increase of the thickness of SiO₂, which is related to the breakage of different lengths of CFs. Figure 3 (c) shows the resistance distribution of HRS and LRS , indicating that the memory window of the device is about $6*10^2$. This also shows that the EOM devices have two distinct resistance states, which further proves that the RS occurs in the device.

978-1-6654-9309-3/23 $31.00 © 2023 IEEE   815

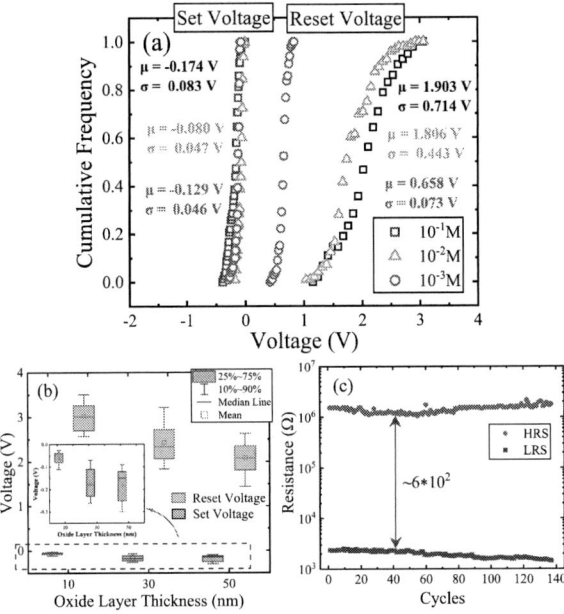

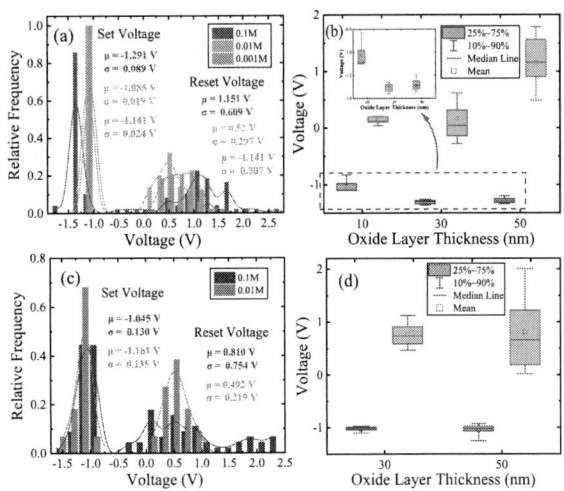

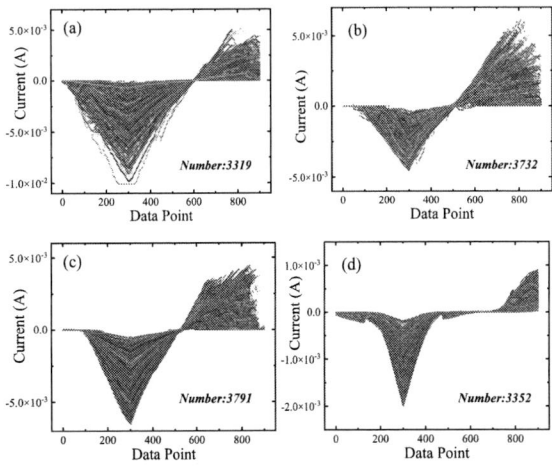

Figure 3: The experimental results of CuSO₄. (a) Cumulative probability of V_{set}/V_{reset} in the EOM of 50nm-SiO₂ layer with different concentration solution. (b) Statistical distribution of V_{set}/V_{reset} with different SiO₂ thickness at 0.1M concentration. (c) Resistance-state distribution over 100 cycles (50nm-SiO₂, 0.1 M concentration).

Figure 4 (a) shows the RS voltage distribution of a 50nm-SiO₂ EOM device using ZnSO₄ solutions of three different concentrations. When the solution concentration is 0.1M, the Mean and STD of Set voltage of the 50nm-SiO₂ EOM devices are -1.291V and 0.089V respectively. When the solution concentration is 0.01M, the Mean and STD of the Set voltage of 50nm-SiO₂ EOM devices are -1.085 V and 0.019 V respectively. When the solution concentration is 10⁻³M, the Mean and STD of the Set voltage of 50nm-SiO₂ EOM devices are -1.141 V and 0.024 V respectively. The Set voltage range of the three EOM devices with different SiO₂ thicknesses is concentrated between -1.05V and -1.3V. Using 0.1M ZnSO₄ solution, the box plot of RS voltage distribution of EOM devices with different SiO₂ thicknesses is shown in Fig. 4 (b).

Figure 4 (c) shows the RS voltage distribution of a 50nm-SiO₂ EOM device with three concentrations of FeCl₃ solution. When the solution concentration is 0.1M, the Mean and STD of the Set voltage of 50nm-SiO₂ EOM devices are -1.045 V and 0.13 V respectively. When the solution concentration is 0.01 M, the Mean and STD of the Set voltage of 50nm-SiO₂ EOM devices are -1.161 V and 0.135 V respectively. EOM devices no longer show RS properties when using a lower concentration of FeCl₃ solution. Using 0.1M ZnSO₄ solution, the switching voltage distribution boxes of EOM devices with different SiO₂ thicknesses are shown in Figure 4 (b). It can be seen from Fig. 3 and Fig.4 that the Set voltage of the three electrolytes is normally distributed.

Figure 4: The experimental results of ZnSO₄ (a-b) and FeCl₃ (c-d). (a-c) Statistical distribution of V_{set}/V_{reset} in the EOM of 50nm-SiO₂ layer with different concentration solutions. (b-d) Statistical distribution of V_{set}/V_{reset} with different SiO₂ thickness at 0.1 M concentration.

Based on the analysis of the switching performance and success rate of the device at three different SiO₂ layer thicknesses, the EOM device with 50nm-SiO₂ shows the most stable performance. Because the failure rate of RS is relatively high using 10nm-SiO₂ and 30nm-SiO₂ EOM devices. Therefore, we selected 50nm-SiO₂ EOM devices as metal ion sensors for data acquisition, to prepare for intelligent recognition of metal ions. We used the 50nm-SiO₂ EOM device to do 4000 experiments on each of the four electrolytes. After the unqualified data is screened out, the data volume of CuSO₄, FeCl₃, ZnSO₄, and KCl electrolytes is 3319, 3732, 3791, and 3352 respectively. We converted the voltage-current data into a one-dimensional array, as shown in Fig. 5 (a-d). Finally, the data volume of the four electrolytes reaches 14194, which can meet the data volume of DNN training.

Figure 5: One-dimensional data sets of four electrolytes. (a) CuSO₄, (b) ZnSO₄, (c) FeCl₃, (d) KCl.

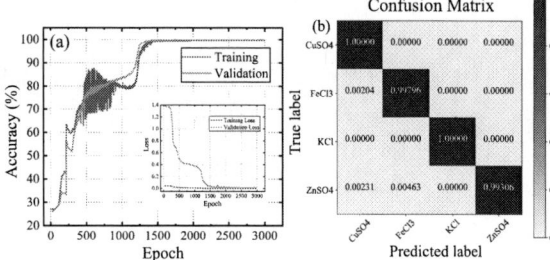

Figure 6:(a) Prediction accuracy of 99.6% was achieved by the trained DNN. (b) Confusion matrix of the predicted results.

Based on this data set, we adopted three kinds of neural networks for training, namely DNN, CNN, and LSTM. The predicted recognition rate of the three neural networks is all higher than 99%, among which the recognition rate of DNN is up to 99.6%, as shown in Fig. 6 (a). As shown in Figure 6 (b), it can be seen that $FeCl_3$ and $ZnSO_4$ have the possibility of misidentification. The prediction rate based on the neural network shows that the EOM devices can realize intelligent electrolyte recognition.

CONCLUSIONS

This paper introduces a RS device, EOM, which uses electrolyte as the top electrode. The device exhibit RS characteristics when tested with a variety of electrolytes. In this paper, four electrolytes are mainly selected for the test: $CuSO_4$, $FeCl_3$, $ZnSO_4$, and KCl. Three kinds of EOM devices with different SiO_2 thicknesses and three different electrolyte concentrations were used. The test results show that the Set voltage of the device is relatively concentrated for different electrolytes, which can be regarded as a marker for specific recognition. According to the success rate of RS and the dispersion degree of Set voltage, the EOM device with 50nm-SiO_2 was selected for data collection. The data sets of the four electrolytes were 14,194 in total. A deep neural network was used to train this data set. The final prediction accuracy reached 99.6%, indicating that the device has the function of ion sensing.

ACKNOWLEDGEMENTS

This work is supported by the National Key R&D Program of China (Grant No. 2021YFB3200100), and the National Natural Science Foundation of China (Grant No. 61974004 and 61931018).

REFERENCES

[1] K. Qian, V. C. Nguyen, T. P. Chen, and P. S. Lee, "Novel concepts in functional resistive switching memories", Journal of Materials Chemistry C, vol. 4, pp. 9637-9645, 2016.

[2] T. G. You, N. Du, S. Slesazeck, T. Mikolajick, G. D. Li, D. Burger, I. Skorupa, H. Stocker, B. Abendroth, A. Beyer, K. Volz, O. G. Schmidt, and H. Schmidt, "Bipolar Electric-Field Enhanced Trapping and Detrapping of Mobile Donors in BiFeO3 Memristors",

Acs Applied Materials & Interfaces, vol. 6, pp. 19758-19765, 2014.

[3] A. I. Ivanov, N. A. Nebogatikova, I. A. Kotin, S. A. Smagulova, and I. V. Antonova, "Resistive switching effects in fluorinated graphene films with graphene quantum dots enhanced by polyvinyl alcohol", Nanotechnology, vol. 30, 2019.

[4] Y. F. Chang, B. Fowler, Y. C. Chen, F. Zhou, C. H. Pan, T. C. Chang, and J. C. Lee, "Demonstration of Synaptic Behaviors and Resistive Switching Characterizations by Proton Exchange Reactions in Silicon Oxide", Scientific Reports, vol. 6, 2016.

[5] U. Das, S. Bhattacharjee, P. K. Sarkar, and A. Roy, "A multi-level bipolar memristive device based on visible light sensing MoS2 thin film", Materials Research Express, vol. 6, 2019.

[6] N. J. Lee, B. H. An, A. Y. Koo, H. M. Ji, J. W. Cho, Y. J. Choi, Y. K. Kim, and C. J. Kang, "Resistive switching behavior in a Ni-Ag2Se-Ni nanowire", Applied Physics a-Materials Science & Processing, vol. 102, pp. 897-900, 2011.

[7] K. C. Chang, T. C. Chang, T. M. Tsai, R. Zhang, Y. C. Hung, Y. E. Syu, Y. F. Chang, M. C. Chen, T. J. Chu,H. L. Chen, C. H. Pan, C. C. Shih, J. C. Zheng, and S. M. Sze, "Physical and chemical mechanisms in oxide-based resistance random access memory", Nanoscale Research Letters, vol. 10, 2015.

[8] B. Y. Mu, H. H. Hsu, C. C. Kuo, S. T. Han, and Y. Zhou, "Organic small molecule-based RRAM for data storage and neuromorphic computing", Journal of Materials Chemistry C, vol. 8, pp. 12714-12738, 2020.

[9] S. Choi, S. H. Tan, Z. F. Li, Y. Kim, C. Choi, P. Y. Chen, H. Yeon, S. M. Yu, and J. Kim, "SiGe epitaxial memory for neuromorphic computing with reproducible high performance based on engineered dislocations", Nature Materials, vol. 17, pp. 335-+, 2018.

[10] H. Garcia, O. G. Ossorio, S. Duenas, and H. Castan, "Controlling the intermediate conductance states in RRAM devices for synaptic applications", Microelectronic Engineering, vol. 215, 2019.

[11] M. Qi, Y. Tao, Z. Q. Wang, H. Y. Xu, X. N. Zhao, W. Z. Liu, J. G. Ma, and Y. C. Liu, "Highly uniform switching of HfO2-x based RRAM achieved through Ar plasma treatment for low power and multilevel storage", Applied Surface Science, vol. 458, pp. 216-221, 2018.

[12] G. Sassine, C. Nail, P. Blaise, B. Sklenard, M. Bernard, R. Gassilloud, A. Marty, M. Veillerot, C. Vallee, E. Nowak, and G. Molas, "Hybrid-RRAM toward Next Generation of Nonvolatile Memory: Coupling of Oxygen Vacancies and Metal Ions", Advanced Electronic Materials, vol. 5, 2019.

[13] T. Kang, X. Y. Chen, J. Zhu, Y. Huang, Z. J. Chen, G. Z. Lin, S. X. Jin, and W. G. Wu, "A resistive device with electrolyte as active electrode", International Journal of Modern Physics B, vol. 34, 2020.

CONTACT

*Wengang Wu, tel: +86-10-62767553;wuwg@pku.edu.cn

MULTI-HOTSPOT MID-IR NANOANTENNAS WITH MATCHED LOSS AND HIGH-INTENSITY NEAR-FIELD FOR SUB-PPM-LEVEL GAS DETECTION

Hong Zhou, Zhihao Ren, Cheng Xu, Liangge Xu, Xinge Guo, and Chengkuo Lee
Department of Electrical & Computer Engineering, National University of Singapore, SINGAPORE

ABSTRACT

Matched loss and high-intensity near-field are critical for plasmonic nanoantennas to achieve ultrasensitive molecular detection. However, the increment in losses during the loss-matching process weakens the intensity of the near-field due to the dissipation of the radiated field. Herein, we develop a multi-hotspot (up to 6) strategy by coupling dark-mode antennas to bright-mode antennas to, for the first time, achieve loss matching while keeping high-intensity near-field simultaneously, which provides ultrasensitive optical elements for molecular detection. Furthermore, amino groups are introduced into metal-organic frameworks (MOFs) through post-synthetic modification to expand the chemisorption function of the MOFs while maintaining their structural integrity for physisorption. Augmented by the physi-chemisorption of MOF-polymer films, our hybrid platform achieves sub-ppm level gas detection. This work provides a new methodology for nanoantenna design and ultrasensitive infrared spectroscopy.

KEYWORDS

multi-hotspot nanoantennas, metal-organic frameworks, gas detection

INTRODUCTION

Infrared (IR) fingerprint vibrations are directly related to the molecular components, chemical bonds, and their conformations [1-3]. The detection of fingerprint vibrations in infrared spectroscopy provides great opportunities for the non-invasive and non-destructive identification of various substances in nature [4-6]. However, it has encountered a technical bottleneck of low sensitivity due to the weak light-molecule interaction in the mid-IR (~10^{-20} cm^2 per molecule) [7], which limits its application in the detection of trace molecules [8, 9], such as trace gas sensing and monolayer molecular monitoring. Resonant nanoantenna [10-15] is an effective solution to enhance such light-molecule interaction by exciting collective oscillations of electrons at the surfaces [16, 17]. Although the concept of light-molecule interaction enhanced by nanoantennas is well-established [18, 19], the optimization of nanoantennas to maximally strengthen the interaction is critical and still presents a challenge. Currently, there are three methods of enhancing the plasmon-molecule interactions in the nanoantenna sensing technology. These methods include i) increasing the spatial overlapping of molecules to the near-field [20], ii) enhancing near-field intensity [21], and iii) loss optimization [18].

In terms of spatial overlapping, better spatial overlap means more analytes are located in the enhanced near-field, resulting in higher plasmon-molecule interactions. There are mainly two schemes that have been proven effective. Approaches from the antenna design side are to free up effective sensing space for molecules by undercutting the antenna structure or integrating microfluidic channels with the dielectric substrate [22]. The method from the molecule side is the introduction of enrichment materials in the near-field to concentrate molecules. Enrichment materials are critical for molecule detection at low concentrations [23, 24], especially gas molecules that are distributed loosely and flow freely in free space [25-27]. The widely used approach to increase the near-field intensity is the proximity of adjacent nanoantennas to form nanogaps, where strong near-field interaction occurs between the adjacent nanoantennas due to their capacitive coupling [28]. Additionally, the addition of an expansion structure at the end of the antenna (opposite to the gap) to increase the charge storage capacity can further improve the localized near-field intensity [29]. The plasmon-molecule interaction in the antenna is realized by the coupling of antenna resonance and molecular vibration. In the coupling system, the enhancement of the molecular vibrational signal is related to the system losses including external radiation loss, intrinsic absorptive loss, and molecular absorptive loss. The loss can be tuned to achieve loss matching [30]. To maximally strengthen the plasmon-molecule interactions, it is of great significance to integrate multiple optimization methods. However, the increment in losses during the loss-matching process weakens the intensity of the near-field due to the dissipation of the radiated field. Therefore, it is difficult to integrate all these optimization methods.

Here we develop a multi-hotspot (up to 6) strategy to achieve loss matching while keeping high-intensity near-field simultaneously by utilizing the increase of hotspot number to compensate for the intensity weakening of the near field.

METHOD AND FABRICATION

Design

The platform integrating MOFs hybrids and nanoantennas for gas detection consists of nanoantennas, calcium fluoride (CaF$_2$) substrate, and MOF hybrids (Figure 1). Up to 6 hotspots are excited by coupling dark-mode antennas to bright-mode antennas. The dark-mode antennas not only tune the loss and achieve a matching state, but they also can utilize the increase of hotspots to compensate for the intensity weakening of the near field in the loss-matching process. In gas detection demonstration, the gas trapping material MOFs mainly relies on the type of uptake gases by physisorption of MOFs through Lewis-acid open metal sites or molecular size confinement. This physisorption-induced variation in MOFs is small. Post-synthetic modification is used to improve it.

Figure 2 shows that the performance of nanoantennas

can be greatly improved by tuning the loss rate to the loss-matching state (Point F). As observed, as the loss rate increases, the changes in transmission and reflection are monotonic, while the enhanced signal intensity undergoes an evolution that first increases and then decreases. The trend of the intensity change of the enhanced signal is determined by loss rate γ_r/γ_a, which could weaken the near field of antennas. To achieve loss and near-field optimization at the same time, we propose a multi-hotspot strategy. More specifically, by increasing the hotspot number to compensate for the intensity weakening of the near-field caused by lossy coupling, we obtain a higher average field intensity than bright-bright mode coupling while implementing loss optimization (Figure 3). Then, we can achieve loss matching while keeping high-intensity near-field simultaneously.

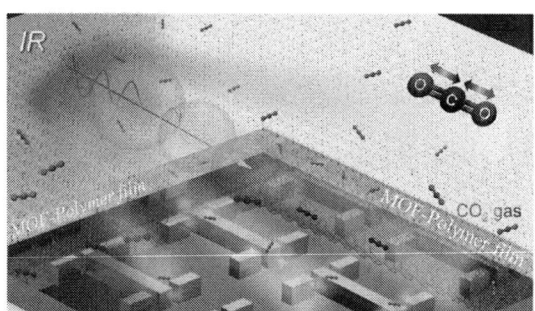

Figure 1: Design concept. The platform consisting of porous MOF-PEI hybrid film, CaF$_2$ substrate, and multi-hotspot nanoantennas. MOF-PEI hybrid film functions as gas-selective-trapping materials to adsorb CO$_2$ gases both physically and chemically. Nanoantennas provide a strongly enhanced multiple near fields for the detection of CO$_2$ captured in the hybrid film.

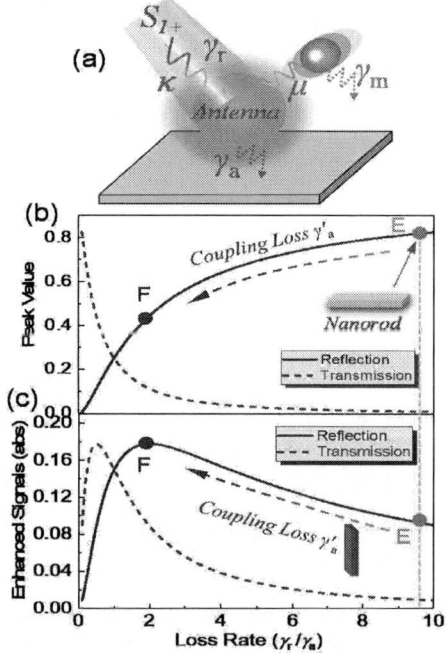

Figure 2: Loss-dependent signal enhancement. (a) Coupled-mode model of nanoantennas showing the

parameters that affect the performance of the antenna. (b, c) Optimization of enhanced vibrational signals through loss tuning. Nanorod-shaped antennas are at low levels of signal enhancement (Point E), and its signal enhancement can be improved (to Point F) by coupling appropriate loss.

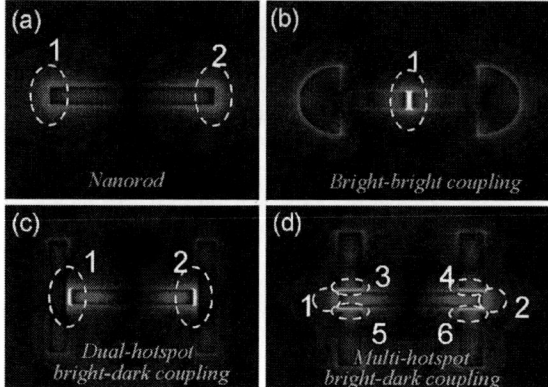

Figure 3: Multiple hotspot strategy. Hotspot distribution of (a) nanorod, (b) bright-bright mode coupling, and (c) dual-hotspot and (d) multi-hotspot bright-dark mode coupling. Multi-hotspot methods can excite new high-intensity hotspots and compensate for the near-field intensity weakening caused by loss tuning.

RESULTS AND DISCUSSIONS
Fabrication

We fabricated the EIA-EIT-integrated platform on 6-The nanofabrication began with the ultrasonic cleaning of CaF$_2$ wafer for 10 minutes in acetone. Then, it was rinsed through isopropanol (IPA) and dried with nitrogen, followed by the oxygen plasma treatment for 5 minutes. After cleaning, the CaF$_2$ wafer was spin-coated with a ~200 nm thick layer of PMMA e-beam lithography resist. After thermal bakes, a commercial electron-conducting polymer (Espacer 300Z from Showa Denko Singapore) was spin-coated at a speed of 2000 rpm to eliminate charge accumulation during e-beam exposure. Then, the nanoantenna pattern was exposed using e-beam lithography technique. After exposure, the sample was sequentially immersed in deionized water, MIBK/IPA (1:3) mixture, and IPA to complete the development of the resist. After development, Ti (5 nm) and Au (70 nm) were sequentially deposited on the surface of the developed sample using e-beam evaporation. Finally, according to the lift-off process, the sample was immersed in acetone for 24 hours to remove the unexposed resist and obtain nanoantenna patterns. In terms of the MOF modification, it began with the preparation of ZIF-8 solutions in methanol. Then, it was sonicated for 10 minutes at 50% energy in an ultrasonic cleaner to obtain fine crystals. After that, low-molecular-weight branched PEI was added to the MOF solution by using a pipette, followed by stirring with a magnetic stirrer for 4 hours to allow the amino group and MOF to interact sufficiently. Then, the modified MOF hybrids were spin-coated onto the device surface to complete the preparation of our MOF-PEI-functionalized multi-hotspot platform.

Gas detection demonstration

The SEM micrograph of the fabricated nanoantennas are shown in Figure 4a, indicating that the morphology and outline are clear and well-defined. After the loading of gas enrichment, the nanoantenna pattern is no longer observed in the SEM image (Figure 4b). However, rough surfaces with ZIF-8 particles inside are observed, which is conducive to the diffusion of gas into the film. Fig. 4d shows the map of enhanced molecular signal when film thickness and spectrum wavelength change. As observed, the IR peaks in ZIF-8-PEI are distinct in the map, and with increasing film thickness, the enhanced molecular signal between 5.5-7.5 μm first increases and then decreases slightly. According to the signal enhancement profile, 180 nm is selected as the optimal value of the film thickness.

Figure 5a shows the measured spectral response of the hybrid platform after reaching steady-state at various concentrations. The differential reflection spectra are extracted by taking the spectrum without CO_2 as a reference (Figure 5b). Figure 5c shows the nonlinear relationship between CO_2 concentration and spectral change. By plotting the total noise and the output signal of the platform at low concentrations together, we can obtain the limit of detection (LOD) of the platform, which reaches the sub-ppm level (Figure 5d).

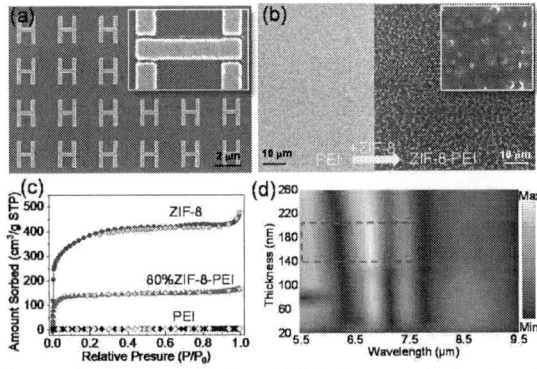

Figure 4: Characterization. SEM micrograph showing the details of multi-hotspot antennas (a) before and (b) after the functionalization with gas trapping materials. (c) N_2 adsorption-desorption isotherms. (d) Film thickness versus spectrum wavelength map revealing the optimal thickness of the gas-trapping film.

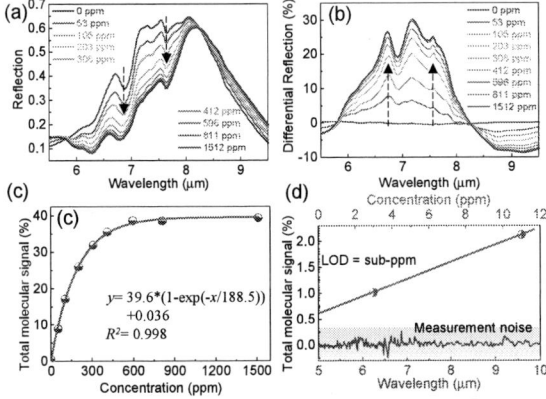

Figure 5: Demonstration of ultrasensitive CO_2 detection

using the multi-hotspot sensing platform. (a) Measured spectral response of functionalized multi-hotspot antennas when CO2 concentrations vary from 0 to 1512 ppm. (b) Corresponding differential signal with the measured spectrum of 0 ppm as reference. (c) Molecular signal versus CO_2 concentration profile showing the sensing behavior. (d) Limit of detection (LOD) of the hybrid platform

CONCLUSIONS

We have demonstrated a compatible all-in-one strategy integrating all major optimization methods to achieve a breakthrough in the limit of plasmon-molecule interaction in nanoantennas for ultrasensitive on-chip vibrational spectroscopy. Then trace carbon dioxide (CO2) gas detection is experimentally investigated to demonstrate the superior plasmon-molecule interaction capability of our all-in-one strategy. Additionally, porous MOFs are integrated with PEI as gas enrichments to improve the spatial overlapping of molecules to the near-field. The experimental results show that our all-in-one strategy possesses an overwhelming competitive advantage over any single method in ultrasensitive on-chip vibrational spectroscopy, achieving sub-ppm level gas detection.

ACKNOWLEDGEMENTS

This work is supported by the Advanced Research and Technology Innovation Centre (ARTIC) Project (WBS: A-0005947-20-00).

REFERENCES

[1] C. Xu, Z. Ren, J. Wei, C. Lee. "Reconfigurable terahertz metamaterials: From fundamental principles to advanced 6g applications", *iScience*, vol. **25**, pp. 103799, 2022.

[2] W. Liu, Y. Ma, X. Liu, J. Zhou, C. Xu, B. Dong, C. Lee. "Larger-than-unity external optical field confinement enabled by metamaterial-assisted comb waveguide for ultrasensitive long-wave infrared gas spectroscopy", *Nano Lett.*, vol. **22**, pp. 6112-6120, 2022.

[3] Q. Qiao, X. Liu, Z. Ren, B. Dong, J. Xia, H. Sun, C. Lee, G. Zhou. "Mems-enabled on-chip computational mid-infrared spectrometer using silicon photonics", *ACS Photonics*, vol. 9, pp. 2367–2377, 2022.

[4] D. Li, H. Zhou, X. Hui, X. He, X. Mu. "Plasmonic biosensor augmented by a genetic algorithm for ultra-rapid, label-free, and multi-functional detection of covid-19", *Anal. Chem.*, vol. **93**, pp. 9437-9444, 2021.

[5] D. Li, H. Zhou, X. Hui, X. He, H. Huang, J. Zhang, X. Mu, C. Lee, Y. Yang. "Multifunctional chemical sensing platform based on dual-resonant infrared plasmonic perfect absorber for on-chip detection of poly(ethyl cyanoacrylate)", *Adv. Sci.*, vol. **8**, pp. e2101879, 2021.

[6] X. Hui, C. Yang, D. Li, X. He, H. Huang, H. Zhou, M. Chen, C. Lee, X. Mu. "Infrared plasmonic biosensor with tetrahedral DNA nanostructure as carriers for label-free and ultrasensitive detection of mir-155", *Adv. Sci.*, vol. **8**, pp. e2100583, 2021.

[7] H. Zhou, D. Li, X. Hui, X. Mu. "Infrared metamaterial for surface-enhanced infrared absorption spectroscopy: Pushing the frontier of ultrasensitive on-chip sensing", *Int. J. Optomechatronics*, vol. **15**, pp. 97-119, 2021.

[8] X. Liu, W. Liu, Z. Ren, Y. Ma, B. Dong, G. Zhou, C. Lee. "Progress of optomechanical micro/nano sensors: A review", *Int. J. Optomechatronics*, vol. **15**, pp. 120-159, 2021.

[9] Y. Ma, Y. Chang, B. Dong, J. Wei, W. Liu, C. Lee. "Heterogeneously integrated graphene/silicon/halide waveguide photodetectors toward chip-scale zero-bias long-wave infrared spectroscopic sensing", *ACS Nano*, vol. **15**, pp. 10084-10094, 2021.

[10] Y.-S. Lin, Z. Xu. "Reconfigurable metamaterials for optoelectronic applications", *Int. J. Optomechatronics*, vol. **14**, pp. 78-93, 2020.

[11] H. Zhou, C. Yang, D. Hu, D. Li, X. Hui, F. Zhang, M. Chen, X. Mu. "Terahertz biosensing based on bi-layer metamaterial absorbers toward ultra-high sensitivity and simple fabrication", *Appl. Phys. Lett.*, vol. **115**, pp. 143507, 2019.

[12] Z. Ren, B. Dong, Q. Qiao, X. Liu, J. Liu, G. Zhou, C. Lee. "Subwavelength on-chip light focusing with bigradient all-dielectric metamaterials for dense photonic integration", *InfoMat*, vol. **4**, pp. e12264, 2021.

[13] H. Zhou, C. Yang, D. Hu, S. Dou, X. Hui, F. Zhang, C. Chen, M. Chen, Y. Yang, X. Mu. "Integrating a microwave resonator and a microchannel with an immunochromatographic strip for stable and quantitative biodetection", *ACS Appl. Mater. Interfaces*, vol. **11**, pp. 14630-14639, 2019.

[14] J. Xu, Y. Du, Y. Tian, C. Wang. "Progress in wafer bonding technology towards mems, high-power electronics, optoelectronics, and optofluidics", *Int. J. Optomechatronics*, vol. **14**, pp. 94-118, 2021.

[15] H. Zhou, D. Hu, C. Yang, C. Chen, J. Ji, M. Chen, Y. Chen, Y. Yang, X. Mu. "Multi-band sensing for dielectric property of chemicals using metamaterial integrated microfluidic sensor", *Sci. Rep.*, vol. **8**, pp. 14801, 2018.

[16] H. Zhou, L. Xu, Z. Ren, J. Zhu, C. Lee. "Machine learning-augmented surface-enhanced spectroscopy toward next-generation molecular diagnostics", *Nanoscale Adv.*, 2022.

[17] Z. Ren, Y. Chang, Y. Ma, K. Shih, B. Dong, C. Lee. "Leveraging of mems technologies for optical metamaterials applications", *Adv. Opt. Mater.*, vol. **8**, pp. 1900653, 2019.

[18] Z. Ren, Z. Zhang, J. Wei, B. Dong, C. Lee. "Wavelength-multiplexed hook nanoantennas for machine learning enabled mid-infrared spectroscopy", *Nat. Commun.*, vol. **13**, pp. 3859, 2022.

[19] Z. Ren, J. Xu, X. Le, C. Lee. "Heterogeneous wafer bonding technology and thin-film transfer technology-enabling platform for the next generation applications beyond 5g", *Micromachines*, vol. **12**, pp. 946, 2021.

[20] D. Hasan, C. Lee. "Hybrid metamaterial absorber platform for sensing of co2 gas at mid-ir", *Adv. Sci.*, vol. **5**, pp. 1700581, 2018.

[21] H. Aouani, H. Šípová, M. Rahmani, M. Navarrocia, K. Hegnerová. "Ultrasensitive broadband probing of molecular vibrational modes with multifrequency optical antennas", *ACS Nano*, vol. **7**, pp. 669–675, 2013.

[22] J. Xu, Z. Ren, B. Dong, X. Liu, C. Wang, Y. Tian, C. Lee. "Nanometer-scale heterogeneous interfacial sapphire wafer-bonding for enabling plasmonic-enhanced nanofluidic mid-infrared spectroscopy", *ACS Nano*, vol. **14**, pp. 12159−12172, 2020.

[23] H. Zhou, D. Li, X. Hui, X. He, H. Huang, X. Mu. "Mid-ir metamaterial absorber with polyvinylamine as a sensitive layer for on-chip sensing of carbon dioxide", in *2021 21st International Conference on Solid-State Sensors, Actuators and Microsystems (Transducers)*, Orlando, June 20-24, 2021, pp. 859-862.

[24] H. Zhou, D. X. Li, X. D. Hui, D. L. Hu, X. Chen, X. M. He, X. J. Mu. "Metamaterial gas sensing platform based on surface-enhanced infrared absorption", in *2020 33rd IEEE International Conference on Micro Electro Mechanical Systems*, Vancouver, January 18-22, 2020, pp. 717-720.

[25] H. Zhou, D. Li, Z. Ren, X. Mu, C. Lee. "Loss-induced phase transition in mid-infrared plasmonic metamaterials for ultrasensitive vibrational spectroscopy", *InfoMat*, pp. e12349, 2022.

[26] Y. Chang, D. Hasan, B. Dong, J. Wei, Y. Ma, G. Zhou, K. W. Ang, C. Lee. "All-dielectric surface-enhanced infrared absorption-based gas sensor using guided resonance", *ACS Appl. Mater. Interfaces*, vol. **10**, pp. 38272−38279, 2018.

[27] H. Zhou, X. Hui, D. Li, D. Hu, X. Chen, X. He, L. Gao, H. Huang, C. Lee, X. Mu. "Metal-organic framework-surface-enhanced infrared absorption platform enables simultaneous on-chip sensing of greenhouse gases", *Adv. Sci.*, vol. **7**, pp. 2001173, 2020.

[28] C. Huck, F. Neubrech, J. Vogt, A. Toma, D. Gerbert, J. Katzmann, T. Haertling, A. Pucci. "Surface-enhanced infrared spectroscopy using nanometer-sized gaps", *ACS Nano*, vol. **8**, pp. 4908-4914, 2014.

[29] L. V. Brown, X. Yang, K. Zhao, B. Y. Zheng, P. Nordlander, N. J. Halas. "Fan-shaped gold nanoantennas above reflective substrates for surface-enhanced infrared absorption (seira)", *Nano Lett.*, vol. **15**, pp. 1272-1280, 2015.

[30] J. Wei, Y. Li, Y. Chang, D. M. N. Hasan, B. Dong, Y. Ma, C. W. Qiu, C. Lee. "Ultrasensitive transmissive infrared spectroscopy via loss engineering of metallic nanoantennas for compact devices", *ACS Appl. Mater. Interfaces*, vol. **11**, pp. 47270–47278, 2019.

CONTACT
*C. Lee; elelc@nus.edu.sg

PALLADIUM BASED MEMS HYDROGEN SENSORS

Max Hoffmann[1], Marion Wienecke[1], Maren Lengert[2], Michael H. Weidner[2] and Jan Heeg[2]
[1]Hochschule Wismar, Institut für Oberflächen- und Dünnschichttechnik, GERMANY and
[2]Materion GmbH, GERMANY

ABSTRACT

The paper describes a novel approach of a new sensor based on a micro-electromechanical system (MEMS). These sensors utilise the volume change of Pd due to hydrogenation, thus they switch mechanically. Pd and Pd-alloy ultrathin films have been deposited on the membrane of Si-MEMS chips by magnetron sputtering. Our investigations revealed a reversible switching of MEMS-Pd-sensors without any drift. The sensor shows a fast response for hydrogen concentrations at and below the lower explosion limits up to 100%vol. hydrogen. Besides this, Pd based sensors show low cross sensitivities and particularly no cross sensitivity to methane (CH_4) since the switching mechanism is a physical one.

KEYWORDS

Hydrogen Sensor, Micro-Electro-Mechanical Sensor, Volume Change, Selectivity

INTRODUCTION

For more than 100 years, since hydrogen has been used technically, there has been a need to detect and reliably measure hydrogen concentrations. The risk of a hazardous event involving hydrogen can be mitigated through the use of reliable, robust and accurate hydrogen safety sensors which detect hydrogen before concentrations rise to hazardous levels [1]. The physico-chemical principles on which Pd-based MEMS sensors are based have been well researched for many decades. Because it features reversible absorption of H_2, palladium (Pd) is often used as active material in H_2 solid-state sensors. Recently researches focused on Pd-coated cantilever [2] and Pd-based MEMS resonant devices with increased sensitivity and low response time, together with high selectivity [3]. When it is mixed with air, hydrogen gas is highly flammable, thus made the importance of robust and fast hydrogen safety sensors for leak detection highly apparent and by such a high demand for highly sensitive and cost-efficient H_2 gas sensors.

A large number of H_2 sensors available on the market are based on a few measuring principles, including mainly electro-chemical sensors, catalytic pellistors, metal-oxide sensors, thermal conductivity sensors, and metal-oxide-semiconductor sensors. These sensors are mainly used for monitoring and explosion protection of stationary systems and meet the relevant standards. Gas sensors should fulfil requirements described in general in ISO/DIS 26142 [4].

Specific requirements arise for process control applications and in particular for new fields of application in the field of renewable energies [4], e.g. for monitoring fuel cells, in the automotive sector, when feeding regeneratively produced hydrogen into natural gas pipelines, as well as in medical technology for measuring H_2-proportions in the breathing gas to detect lactose intolerance [5]. These new requirements relate primarily to an extended measuring range for monitoring mobile fuel cells up to hydrogen concentrations of 100%vol, faster response times, in the automotive sector e.g. less than 1s and in the medical technology sector significantly lower detection limits of well below 100 ppm. A disadvantage of the above-mentioned commercially available H_2 sensors, especially for the new areas of application mentioned, is that all of them, with the exception of thermal conductive (TCD) sensors, have more or less strong cross-sensitivities to other combustible gases [6]. H_2 sensors that are based on the physical switching of Pd or Pd alloys and are therefore free of cross-sensitivities have recently been on the market. These include optical, resistive and metal-oxide (MOS) sensors with Pd-based thin films [7]. In this context, there are new requirements in terms of safety, energy consumption, miniaturization and price. Pd-based thin-film sensors also have potential in this regard, since they can be mass-produced using silicon technology. The purpose of this work is to investigate and describe the capabilities of a new sensor based on a micro-electromechanical system (MEMS). These sensors utilise the volume change of Pd due to hydrogenation, thus they switch mechanically. Pd and Pd-alloy ultrathin films have been deposited on the membrane of Si-MEMS chips by magnetron sputtering.

MATERIALS AND METHODS

PHYSICAL PROPERTIES OF THE PD-H SYSTEMS

The switchable physical properties are based on the effect of the atomic storage of hydrogen at interstitial sites in the metal lattice. Depending on its concentration, the hydrogen is first adsorbed when it hits the metal, dissociated and then dissolved in the metal lattice by diffusion. The α-phase, a solid solution, is formed. If a certain metal-specific hydrogen concentration is exceeded, the hydride phase (often referred to as β -phase) is formed. This process takes place at room temperature and is reversible. In Pd, the α-phase exists at room temperatures up to a hydrogen concentration of 1.68%. In the concentration range from 1.68 to 37.6% the material system is two-phase and consists of the α and the β phase. From 37.6% is only β-phase [8]. The solubility of hydrogen in palladium increases with increasing H_2 partial pressure. The α-phase is metallically conductive and opaque. The resistance increases with increasing concentration of H in the metal lattice due to electron scattering due to intercalation. The β-phase is semiconducting and transparent [9]. A change in volume occurs during the storage of hydrogen and in particular during the phase change. The lattice constant for pure Pd is 3.887 Å, up to a hydrogen content of approx. 2% it increases to 3.895 Å and for the β-phase with approx. 37.6% hydrogen the lattice constant is 4.02 Å [10].

This is associated with a volume expansion of 3.5% [8] according to the formula:

$$\Delta V/V = \Delta v/\omega \ C_H \qquad (1)$$
$$\Delta a/a = 1/3 \ \Delta v/\omega \ C_H \qquad (2)$$
$$\Delta a/a = \alpha_H \ C_H \qquad (3)$$

ΔV - volume change when n hydrogen atoms are dissolved
V - initial volume of the metal
Δv - characteristic change in volume per hydrogen atom
ω - atomic volume of a metal atom
C_H - ratio n/N, ratio of the number of hydrogen atoms to the number of metal atoms (at 50% is $C_H = 1$)
a - lattice constant
α_H - expansion coefficient - H intercalation is 0.063 for Pd

This volume change of Pd is the cause of poor adhesion of Pd thin films. Strategies to improve this consist of alloying the Pd or applying adhesion-promoting layers, are described by Fedtke et al. [11]. The hydrogen concentration in the α-phase is a function that is strictly dependent on the partial pressure. The storage of hydrogen in the metal is proportional to the square root of the hydrogen partial pressure (Sievert's law).

$$C_H = p(H_2)^{1/2}/K \qquad (4)$$

$p(H_2)$ - hydrogen partial pressure
K - Sieverts constant, for Pd = 0.12 at 300K.

MEMS BASED HYDROGEN SENSORS

Micro-electromechanical sensor structures for detecting hydrogen physically use the volume change of Pd or Pd alloys when hydrogen is stored. The measurement technique for detecting the change in volume can vary. Baselt et al. [12] report on a micro-cantilever which, like a bimetallic strip, is coated with Pd on one side and deforms under the influence of H_2. This deformation is measured capacitively and allows the detection of 1000 ppm to 100%vol. hydrogen. a response time of 90s is specified for 1%vol. by volume of hydrogen. Gurusamy et al. [13] also describe a microcantilever. 50 ppm H_2 can be reliably detected by optical reading of the deformation of the cantilever. Lee et.al [14] describe design proposals for so-called nanogap-based sensors. They are based on resistive brasses on Pd nanostructures, which percolate when they expand in volume and thus become more conductive. For a simple on-off arrangement, it was shown that 100 ppm H_2 can be reliably detected.

In this work, commercially available MEMS pressure sensors were coated with Pd/Au thin films (100 nm, magnetron sputtering with patchwork target). Figure 1 shows a schematic diagram. The MEMS pressure sensors are based on the piezoresistive properties of silicon. Resistance structures in the form of a Wheatstone bridge are detuned when the Si membrane deforms. MEMS pressure sensors are mass-produced using silicon technology. The output signal of the MEMS sensor is used in the following figures as a measure of the hydrogen concentration.

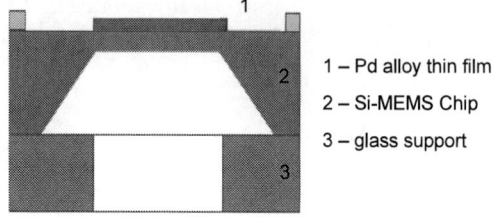

1 – Pd alloy thin film
2 – Si-MEMS Chip
3 – glass support

Figure 1: Sensor principle of a Pd modified MEMS sensor

METHODOLOGY FOR SENSOR PREPARATION

The device has been fabricated following a lithographic microfabrication process. Pd and Pd-alloy ultrathin films have been deposited on the membrane of Si-MEMS chips by magnetron sputtering. Before coating the sensors are mounted in the Mask frame to hold for coating and pressure equalization. The Argon gas flow was 50 sccm/min and the pressure was 4.7×10^{-3} mbar. The variation of the layer thickness was realized by different power and sputtering times. The deposition parameters were 40 W for the sputtering for 10 min coating process and for 40W for 5 min for a buffer layer. The thickness and material of this buffer layer can influence the internal stress in the functional thin film. The Pd coated MEMS chips have been bonded on widely used sensor sockets.

MEASUREMENTS SETUP

The measurement setup (see Figure 2) consists of a gas flow-controller that to generate a variation of the gas flow. It consists of a control unit and two valves with 100 sccm/min for H_2 and 100 sccm/min for Air, then the gas mixture is sent through a flow cell measuring chamber where the Pb MEMS are placed in such a way as the entire socket is also inside the chamber. Supply Voltage of 5V is applied. Resistance structures in the form of a Wheatstone bridge are detuned when the Si membrane deforms. All the data are processed with the coupled PC.

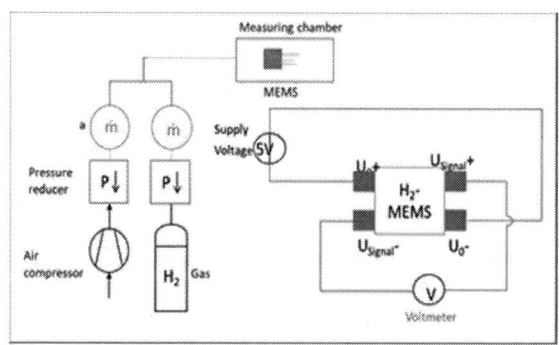

Figure 2: The measurement: - a gas flow-controller; - facility to generate a variation of the gas flow for H2 and Air; - an example of a thin film sensor; - Supply Voltage of 5V; - Wheatstone bridge; - Voltmeter which is connected with an IEEE connector to the PC with LabView

RESULTS

First results indicated a strong hysteresis. This effect was prevented with an additional buffer layer made of a ductile metallic thin film. The following figure shows the sensor performance up to the lower explosive limit (LEL; 4vol% H_2 in air) with and without different buffer layers.

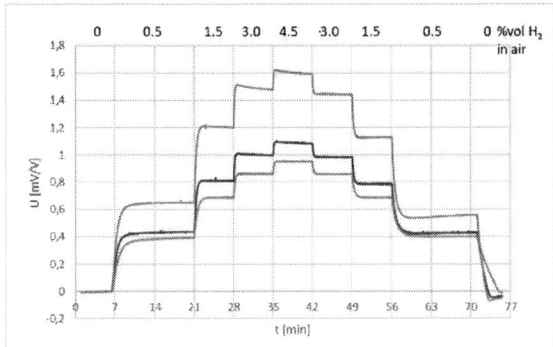

Figure 3: hysteresis reduction; green line represents no buffer layer, blue and orange line represent different buffer layers

Experimental results are presented in the Figure 4 as sensitivity versus time graph for Pd based MEMS sensors exposed to increase in H_2 concentration from 0 to 100% vol. in air.

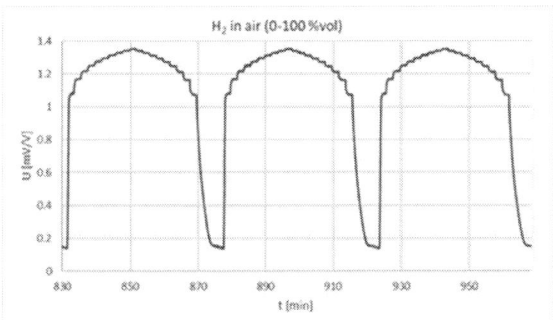

Figure 4: Supply voltage: 5V, flow rate: 100 sccm, increase in hydrogen concentration in air from 0 to 100% vol in 10% vol steps

The calibration curve for the measurements in Figure 4 is shown in Figure 5.

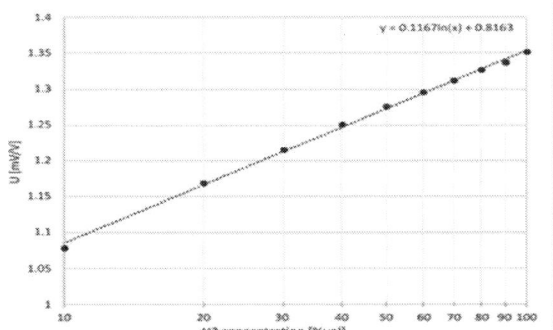

Figure 5: Logarithmic representation of the step test in Figure 4, interpolation points determined as the mean of 3 individual values; error bars are negligible small

In this graph the sensor signal is given over the logarithmic scale of the hydrogen concentration. It can be seen that the curve has linear behaviour.

Initial investigations into cross-sensitivity were carried out with the sensor structure described by exposing the sensor to CH_4-air mixtures and hydrogen air. Cross-sensitivity here refers to how sensitive a sensor is toward unwanted stimulus by another species and thus related to selectivity. A corresponding measurement with CH_4 concentrations in steps from 0 to 40%vol. is shown in Figure 6. It is clear that Pd based MEMS sensors do not exhibit any cross-sensitivity to methane (CH_4). In the explosive concentration region, the indicated hydrogen concentration was detected very fast. No hysteresis was observed.

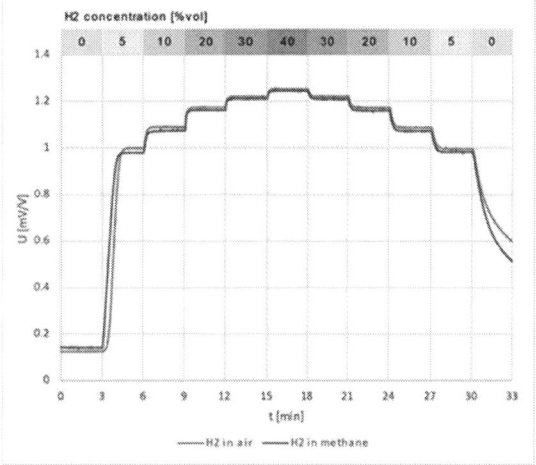

Figure 6: Increase in hydrogen concentration in air and methane from 0 to 40% vol% hydrogen, Supply voltage: 5V, flow: 80 sccm:

DISCUSSION

The developed sensor shows a fast and reliable response, a high selectivity to hydrogen, and particularly a high sensitivity up to the lower explosive limit as well as up to 100%. Experimental findings emphasise that, Pd based MEMS Hydrogen Sensors switch mechanically, because they utilise the volume change of Pd due to hydrogenation. Magnetron sputtering technique was used to deposit Pd and Pd-alloy ultrathin films on the membrane of Si-MEMS chips. The measurement results presented show that the sensory properties of Pd can be explained very well by physical properties of the Pd-H system. Under Hydrogen exposure, the deformation of the Si-Membrane occurs and cause a resistance change in a Wheatstone bridge structure in the piezoelectric Si-MEMS. The absorption of hydrogen by Pd is reversible, it follows the external partial pressure in the sample gas (Sievert's law). The hydrogen initially adsorbed on the surface diffuses into the metal lattice. The driving force is the concentration gradient; an equilibrium concentration of H atoms in the metal lattice is established, which corresponds to the phase equilibrium. Experimental investigations revealed a reversible switching of MEMS-Pd-sensors without any drift.

The sensor shows a fast response for hydrogen concentrations at and below the lower explosion limits up to 100% vol. hydrogen. Thus, the Pd-MEMS gas sensor fulfil the requirements of ISO/DIS 26142 [4] for control and explosion protection of stationary facilities. For hydrogen concentrations between 1%vol. and 100%vol, the t_{90} response time is about 30s or less. Further investigation showed that for small concentrations below 5000 ppm the response time is much larger than 30s, however, this is in agreement with data from literature, which reveal, that the response time increases for low hydrogen concentrations due to the diffusion driven process. Besides this, Pd based sensors show low cross sensitivities and particularly no cross sensitivity to methane (CH4) since the switching mechanism is a physical one.

Regarding the problem of the hysteresis a buffer layer proved to be a solution. The mechanical properties and effects providing this improvement have yet to be determined. The understanding of the theoretical foundation behind the elasto-mechanical system of MEMS in combination with hydrogen-sensitive thin films is crucial for the optimisation und reliable production of such sensors. Additional insight into this mechanical system will be established with numerical simulation.

CONCLUSION

The purpose of this work has been to demonstrate that the Pd-based MEMS sensors can reliably measure hydrogen concentrations from up to 100% vol. and have potential for new fields of application, since they are based on the physical switching of Pd at room temperature. Results show new potentials for hydrogen sensor applications like new application fields for in-situ control of hydrogen concentrations.

Pd-based MEMS systems have short response times of a few seconds in the concentration range 1-100 %vol, this fact make them suitable for applications in high hydrogen concentrations as well as application at and below the LEL. Applications for processes working in high hydrogen concentrations of up to 100 %vol, such as in metallurgy, for fuel cells, or chemical process engineering. Applications for process control at low hydrogen concentrations, widely below the explosive limit, the new sensor concept particularly show potentials because of their low cross sensitivity.

The most interesting application potential field for the MEMS hydrogen sensor is the injection of regenerative produced hydrogen into the natural gas system. It was demonstrated that the sensors have no cross sensitivity to methane and this type of Pd based MEMS Hydrogen Sensors has low power consumption, is miniaturised and can be produced in a mass production at low costs by silicon technology.

REFERENCES

[1] L. Boon-Brett, J. Bousek, G. Black, et. al. "Identifying performance gaps in hydrogen safety sensor technology for automotive and stationary applications". *International Journal of hydrogen energy.*, No. 35, pp. 373-384, 2010.

[2] D.R. Baselt, B. Fruhberger, E. Klaassen, et.al. "Design and performance of a microcantilever-based hydrogen sensor", *Sens. Actuators B.*, No. 88, pp. 120-131, 2003.

[3] T. Walewyns, D. Spirito, L.A. Francis, "A Tunable Palladium-based Capacitive MEMS Hydrogen Sensor Performing High Dynamics", *High Selectivity and Ultra-low Power Sensing. Procedia Engineering.*, No. 87, pp 268-271, 2014.

[4] ISO/DIS 26142. Hydrogen detection apparatus; 2009.

[5] S.C. Fleming, "Evaluation of a hand-held hydrogen monitor in the diagnosis of intestian lactase deficiency", *Ann Clin Biochem.*, No. 27, pp. 499-500; 1990.

[6] V. Palmisano, L. Boon-Brett, C. Bonato, et.al "Evaluation of selectivity of commercial hydrogen sensors", *International Journal of hydrogen energy.*, No. 39, pp. 20491-20496, 2014.

[7] P. Soundarrajan, F. Schweighardt, "Hydrogen Sensing and Detection", *Hydrogen fuel: Production, transport, and storage.*, CRC; Taylor & Francis, Boca Raton, Fla., London, 2009.

[8] A. Pundt, "Nanoskalige Metall-Wasserstoff-Systeme", Universitätsverlag Göttingen., 2005.

[9] A. Mandelis, J.A. Garcia, "Sensors and Actuators B", *Chemical.*, No. 49, pp.258-267, 1998.

[10] F.D. Manchaster, A. San-Martin, J.M. Pitre, "The H-Pd (Hydrogen-Palladium) System", *Journal of Phase Equilibria.*, No. 15, pp. 62-83, 1994.

[11] P. Fedtke, M. Wienecke, M.C. Bunescu et. al. "Hydrogen sensor based on optical and electrical switching", *Sensors and Actuators B: Chemical 100*, pp. 151-157, 2004.

[12] D.R. Baselt, B. Fruhberger, E. Klaassen et. al., "Design and performance of a microcantilever-based hydrogen sensor", *Sensors and Actuators B: Chemical.*, No. 88, pp.120-131, 2003.

[13] J.T. Gurusamy, G. Putrino, R.D. Jeffery et al., "MEMS based hydrogen sensing with parts-per-billion resolution", *Sensors and Actuators B: Chemical.* No. 281, pp. 335-342, 2019.

[14] J. Lee, W. Shim, J.S. Noh, W. Lee, "Design Rules for Nanogap-Based Hydrogen Gas Sensors", *ChemPhysChem.*, No. 13, pp. 1395-1403, 2012.

CONTACT

*M. Hoffmann; max.hoffmann@hs-wismar.de

SELECTIVE DISCRIMINATION OF PPB-LEVEL VOCS USING MOS GAS SENSOR IN PULSE-HEATING MODE WITH THE MODIFIED HILL'S MODEL

Gaoqiang Niu[1], Yi Zhuang [1], Yushen Hu[1], Zong Liu[1], and Fei Wang [1,*]

[1] School of Microelectronics, Southern University of Science and Technology, Shenzhen 518055, CHINA

ABSTRACT

This paper realizes the selective discrimination of ppb-level volatile organic compounds (VOCs) through a transient feature extraction method. The transient resistance of the micro-electromechanical systems (MEMS) based SnO_2 sensor in ethanol at various temperature were investigated in pulse-heating mode. Furthermore, we proposed the modified Hill's model to describe the transient resistance of the sensor when the pulse is on, and the experimental data was well fitted by the function of the modified Hill's model. Additionally, various VOCs gases with 500 ppb concentrations were distinguished with the principal component analysis (PCA).

KEYWORDS

Pulse-heating, MOSs gas sensor, the modified Hill's model, ppb-level, selectivity

INTRODUCTION

MEMS-based metal oxide semiconductors (MOSs) gas sensors exhibit low power consumption and low thermal inertia [1]. This capability of the MEMS-based MOSs gas sensors allows the working temperature changing instantaneously, which offers the possibility of pulse-heating operation [2, 3]. Meanwhile, the cross-sensitivity of the traditional MOSs gas sensors limits their further development in the field of gas recognition, especially at ppb-level concentration or below [4-6]. Based on the advanced technologies of MEMS for the fabrication of the micro-hotplate, MEMS gas sensors combined with the pulse-heating strategy is promising to address the issue of selective discrimination of various gases. Various kinetic models have been proposed in the past to describe the gas-sensing process at a fixed temperature, for instance, Freundlich isotherm, Langmuir model, and Eley−Rideal model, etc [7-9]. Nonetheless, these empirical models are insufficient to demonstrate the physicochemical process of the gas sensing in the pulse-heating mode. This is owing to several non-equilibrium processes involved in the gas-sensing in the pulse-heating mode, such as VOCs gases reactions, oxygen desorption/adsorption, and carrier transformation in the MOSs, etc [10]. To extract the kinetic features and fully understand the physico-chemical procedure in the pulse-heating mode, a quantitative kinetic model to describe the transient features of the gas-sensing process is of great interest, which is also essential for the further integration with the advanced algorithms.

In this work, we have proposed a modified Hill's model to interpret the transient resistance of the MEMS sensor in pulse-heating mode. The transient resistances of

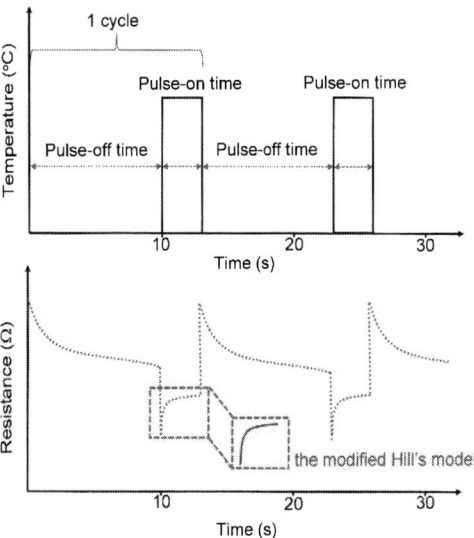

Figure 1: The set of pulse-heating strategy for the MEMS-based SnO_2 sensor. Corresponding transient resistance of the MEMS-based SnO_2 sensor, and the resistance was fitted with the modified Hill's model during the pulse-on period.

MEMS-based SnO_2 sensor were measured at various temperatures. Furthermore, experimental data was well fitted by the proposed model and the fitting parameters could be correlated with the gas-sensing kinetics in the pulse-heating mode. The effects of temperature on the parameters of the model were investigated and summarized. Finally, various VOCs gases (ethanol, toluene, acetone, amine and formaldehyde) at low concentration of 500 ppb could be distinguished successfully with the principal component analysis (PCA) based on the proposed model.

EXPERIMENT

Synthesis of SnO_2 nano-sheets

The SnO_2 nano-sheets were obtained via a hydrothermal method as described in our previous work.[11] 6 mmol $SnCl_2 \cdot 2H_2O$ was dissolved into 20 mL of deionized water. Then, the solution was adjusted to pH=13 with 0.4 M NaOH solution. The mixture was stirred for 30 min and transferred into a Teflon-lined stainless autoclave. The autoclave was sealed and kept in an oven at 180 °C for 12 h and cooled naturally to room temperature. The SnO_2 nanosheets were collected by centrifugation and washed with deionized water and absolute ethanol for

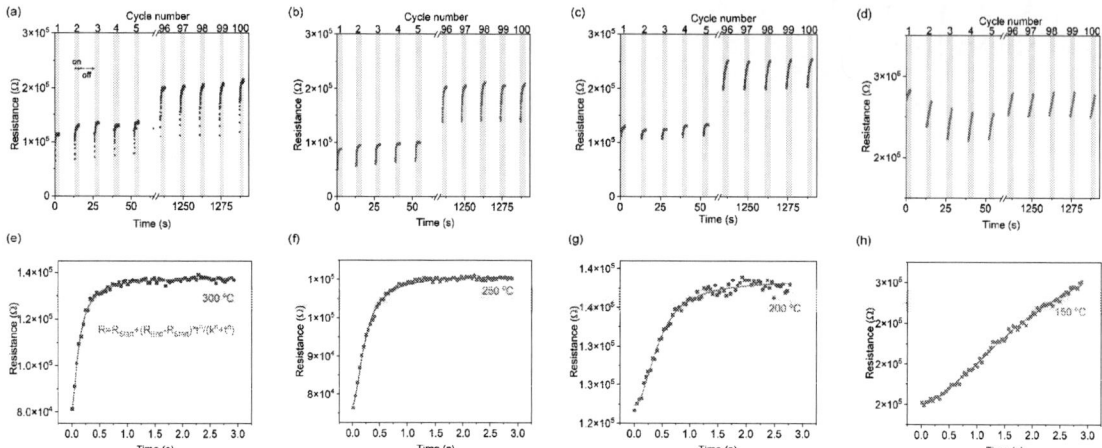

Figure 2: Transient resistance of the sensor in 100 cycles, which was operated in 100 ppm ethanol at (a) 300 °C, (b) 250 °C, (c) 200 °C and (d) 150 °C, respectively. The corresponding transient resistance in a cycle when the pulse on and their fitted lines with modified Hill's model at (e) 300 °C, (f) 250 °C, (g) 200 °C and (h) 150 °C, respectively.

several times, respectively, to remove any residual ions, and dried at 80 °C overnight. The final product was calcined at 500 °C for 2 h in air.

Pulse-heating measurements

The concentrations of VOCs during the pulse heating gas-sensing measurements were controlled by the commercial WS-30B system (Weisheng Instruments Co., Zhengzhou, China). The pulse-heating voltage was supplied by PWS2326 from Tektronix, Inc., and the pulse-heating voltage and pulse time were controlled by a Labview program based on the temperature coefficient of resistance (TCR) calibration of the sensors. Specifically, the temperatures of the micro-heater were set to 100, 150, 200, 250, and 300 °C by adjusting the applied voltages to 1.1, 1.3, 1.5, 1.7, and 1.9 V, respectively. The resistance of the sensor was monitored every 0.04 s, by measuring the voltage across a standard resistor under an applied voltage of DC 5 V. The voltage data was recorded by the Labview program through a data acquisition card (USB-6001 from National Instruments).

The pulse-heating strategy for the MEMS sensor is shown in Figure 1. The pulse-on time and pulse-off time are set as 3 s and 10 s, respectively. The transient resistances of the sensor during the pulse-on period are fitted with the modified Hill's model afterwards. Key parameters in the modified Hill's model are extracted according to the different sensing measurement temperatures. Finally, PCA is carried out based on the extracted parameters of the model and identification of target gases is realized at ppb-level.

RESULTS AND DISCUSSION

Figure 2 displays the transient resistance of the sensor in 100 cycles, which was operated in 100 ppm ethanol at (a) 300 °C, (b) 250 °C, (c) 200 °C and (d) 150 °C, respectively. The result only exhibits the transient resistance in 100 cycles when the pulse on for clarity. The resistance of the sensor increases at the pulse-on moment during all cycles. Compared with the initial five cycles, the resistances of the sensor in the last five cycles increase

slightly at each temperature. During the pulse-heating testing at each temperature, the resistance of the sensor at the pulse-on moment maintains basically stable and repeatable in each cycle at same temperature. At various temperature, the resistance curve of the sensor shows different characteristics, especially, when the sensor at 300 °C and 150 °C.

The corresponding fitted lines with modified Hill's model are shown in Figure 2(e-h), respectively. we have proposed a modified Hill's model, as described below:

$$R = R_{Start} + (R_{End} - R_{Start})\frac{t^n}{k^n + t^n} \qquad (1)$$

where R_{Start} is the resistance of the sensor when the pulse is ignited, R_{End} is the maximum resistance of the sensor during the pulse-on period, k reflects the rate of oxygen desorption (or adsorption) at which half the maximum resistance (or minimum resistance) is achieved and n is

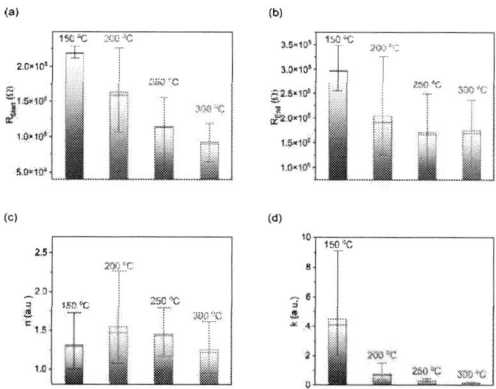

Figure 3: (a) R_{Start}, (b) R_{End}, (c) k and (d) n of transient resistance curves fitted by the modified Hill's model at different temperature. The sensor was exposed to 100 ppm ethanol. Values of the four parameters are summarized from 100 cycles when the pulse on.

978-1-6654-9309-3/23 $31.00 © 2023 IEEE

the Hill coefficient.[12] This nonlinear equation was initially used as the Hill's model to investigate the relationship between the oxygen tension and the saturation of hemoglobin, and then extended for the quantitative analysis of the drug–receptor in pharmacology as well as the gas adsorption onto homogeneous substrates.[13, 14] In this model, the gas adsorption is a cooperative process and the present gas species on the substrates would influence the further adsorption affinity of the rest sites. The Hill coefficient indicates the cooperative degree of gas adsorption, in which the larger n values reflects higher adsorption affinity of the sites. It should be noted that the resistance of the sensor is directly correlated with the surface oxygen coverage, which reflects the amount of adsorbed oxygen species on the surface of MOSs.[15]

The parameters extracted from the transient resistance curves with the modified Hill's model, are summarized in the Figure 3. The R_{Start} and R_{End} get smaller with the rise of temperature, which agrees well with the DC measurements. The n values with decrease with the increase of temperature, except for the sensor at 150 °C. And the k values decrease with the increase of temperature, which reflects that the adsorption rate increases with the rise of temperature.

The responses of the sensor to the various target gases in direct current (DC) mode at 300 °C are shown in Figure 4. The response of the sensor is defined as R_a/R_g (R_a: the resistance of the sensor in air, R_g: the resistance of the sensor in the target gases). Since only the responses of the sensor to the various target gases are provided in DC mode, it is difficult to distinguish the gases in ppb-level concentration, especially for toluene and amine with almost the same value. The poor selectivity of the MOSs based sensor is their inherent shortcoming in DC mode. To solve this problem, the extracted parameters of the sensor based on the Hill's model in the pulse-heating mode provide other choice. The extracted parameters from the resistance curves based on the modified Hill's equations reflect the gas-sensing kinetics of different target gases, which is useful for the gas selective discriminations with

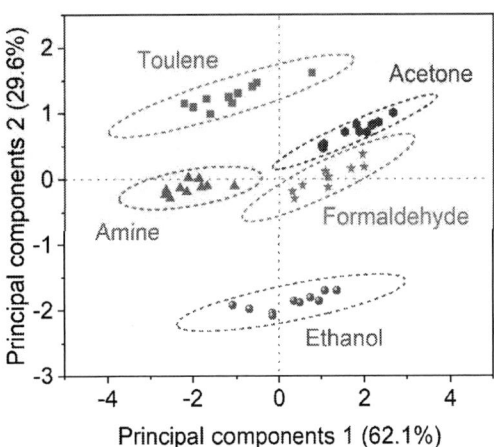

Figure 5: Scatter plot obtained through PCA performed over gas sensing kinetic parameter results for tested VOCs.

the advanced algorithms.

Recently, various pattern recognition tools, such as principal component analysis (PCA), BPNN, and ANN, have been employed to address the problems of gas classification and concentration prediction [16]. Herein, PCA is employed to realize a selective distinction among target gases with the extracted kinetic parameters (R_{Start}, R_{End}, k, n) as input features. Here, the four parameters of the target gases (ethanol, toluene, acetone, amine and formaldehyde) in 10 pulse-heating cycles are taken as a dataset (4 × 50 matrix). As shown in Figure 5, the test gases are presented in the same coordinate plane with the extracted parameters. The scatter plots of target points place in isolation, which facilitates the identification of tested gases selectively in 500 ppb concentration.

CONCLUSION

We have quantitatively analyzed the transient resistance of the sensor in pulse-heating mode using the modified Hill's model. The transient resistances of the sensors under different temperatures were well fitted by this model. Subsequently, the parameters of the modified Hill model were further investigated at various temperatures. Moreover, based on the parameters from the modified Hill's model via the PCA, various 500 ppb gases (ethanol, toluene, acetone, amine and formaldehyde) were discriminated. This pulse-heating approach combined with the modified Hill's model with the easy extraction of the MOSs gas sensors is promising for the other gas-sensing applications.

ACKNOWLEDGEMENTS

This work was financially supported in part by National Key R & D Program of China under Grant No. 2020YFB2008604, and in part by NSQKJJ under Grants K21799109 and K21799110. The authors would like to acknowledge the technical support from SUSTech CRF.

REFERENCES

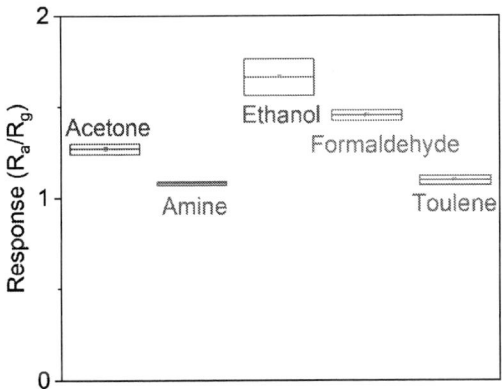

Figure 4: Responses (R_a/R_g) of the sensor to 500 ppb various gases (ethanol, amine, toluene, acetone and formaldehyde) at 300 °C in DC mode.

[1] G. Niu and F. Wang, "A Review of MEMS-based Metal Oxide Semiconductors Gas Sensor in Mainland China," *J. Micromech. Microeng.*, vol. 32, no. 5, p. 054003, 2022.

[2] Y. Chen, P. Xu, T. Xu, D. Zheng, and X. Li, "ZnO-nanowire Size Effect Induced Ultra-high Sensing Response to PPb-level H_2S," *Sens. Actuators B Chem.*, vol. 240, pp. 264-272, 2017.

[3] G. Niu, H. Gong, C. Zhao, and F. Wang, "H_2S Sensor Based on MEMS Hotplate and on-Chip Growth of $CuO-SnO_2$ Nanosheets for High Response, Fast Recovery and Low Power Consumption," in *2020 IEEE 33rd International Conference on Micro Electro Mechanical Systems (MEMS)*, 2020, pp. 799-802.

[4] D. Meier *et al.*, "The potential for and challenges of detecting chemical hazards with temperature-programmed microsensors," *Sens. Actuators B Chem.*, vol. 121, no. 1, pp. 282-294, 2007.

[5] S. Acharyya, S. Nag, S. Kimbahune, A. Ghose, A. Pal, and P. K. Guha, "Selective discrimination of VOCs applying gas sensing kinetic analysis over a metal oxide-based chemiresistive gas sensor," *ACS Sens.*, vol. 6, no. 6, pp. 2218-2224, Jun 25 2021.

[6] N. Barsan, M. Schweizer-Berberich, and W. Göpel†, "Fundamental and Practical Aspects in the Design of Nanoscaled SnO_2 Gas Sensors: A Status Report," *Fresenius' J. Anal. Chem.*, vol. 365, no. 4, pp. 287-304, 1999/10/01 1999.

[7] R. Afonso, L. Gales, and A. Mendes, "Kinetic Derivation of Common Isotherm Equations for Surface and Micropore Adsorption," *Adsorption*, vol. 22, no. 7, pp. 963-971, 2016.

[8] S. Acharyya, S. Nag, S. Kimbahune, A. Ghose, A. Pal, and P. K. Guha, "Selective Discrimination of VOCs Applying Gas Sensing Kinetic Analysis over a Metal Oxide-Based Chemiresistive Gas Sensor," *ACS Sens,* vol. 6, no. 6, pp. 2218-2224, Jun 25 2021.

[9] W. H. Weinberg, "Eley−Rideal Surface Chemistry: Direct Reactivity of Gas Phase Atomic Hydrogen with Adsorbed Species," *Acc. Chem. Res.*, vol. 29, no. 10, pp. 479-487, 1996/10/10 1996.

[10] C. Schultealbert, T. Baur, A. Schütze, S. Böttcher, and T. Sauerwald, "A novel approach towards calibrated measurement of trace gases using metal oxide semiconductor sensors," *Sens. Actuators B Chem.*, vol. 239, pp. 390-396, 2017.

[11] G. Niu, C. Zhao, H. Gong, Z. Yang, X. Leng, and F. Wang, "NiO Nanoparticle-decorated SnO_2 Nanosheets for Ethanol Sensing with Enhanced Moisture Resistance," *Microsyst. Nanoeng.*, vol. 5, no. 1, p. 21, 2019/05/20 2019.

[12] H. C. Leo Tsui *et al.*, "Graphene Oxide Integrated Silicon Photonics for Detection of Vapour Phase Volatile Organic Compounds," *Sci Rep,* vol. 10, no. 1, p. 9592, Jun 12 2020.

[13] D. I. Cattoni, O. Chara, S. B. Kaufman, and F. L. Gonzalez Flecha, "Cooperativity in Binding Processes: New Insights from Phenomenological Modeling," *PLoS One,* vol. 10, no. 12, pp. 1-14, 2015.

[14] M. I. Stefan and N. Le Novere, "Cooperative Binding," *PLoS Comput. Biol.*, vol. 9, no. 6, p. 1003106, 2013.

[15] N. M. Vuong, D. Kim, and H. Kim, "Surface gas sensing kinetics of a WO_3 nanowire sensor: Part 2—reducing gases," *Sens. Actuators B Chem.*, vol. 224, pp. 425-433, 2016.

[16] W. Ren, C. Zhao, G. Niu, Y. Zhuang, and F. Wang, "Gas sensor array with pattern recognition algorithms for highly sensitive and selective discrimination of trimethylamine," *Adv. Intell. Syst.*, 2200169, 2022.

CONTACT

* Fei Wang, Email: wangf@sustech.edu.cn

THERMAL CONDUCTIVITY DETECTOR (TCD)-TYPE GAS SENSOR BASED ON THE SUSPENDED 1D NANOHEATER FOR IOT APPLICATIONS

Wootaek Cho, Jong-Hyun Kwak, Taejung Kim, and Heungjoo Shin*
Ulsan National Institute of Science and Technology (UNIST), Korea

ABSTRACT

We developed a thermal conductivity detector-type gas sensor based on a suspended 1D nanoheater. The excellent thermal insulation and small size induced by the suspended nano-sized 1D architecture enabled ultrafast gas detection at ultralow power with pulsed power input. Due to these advantageous, thus, the nanoheater-based gas sensor is expected to be applicable to IoT applications. The nanoheater-based sensor chip was integrated with through-silicon vias adaptive to the heater fabrication for facile sensor packaging. Moreover, entire sensor chips were fabricated using wafer-level microfabrication processes, ensuring cost-effective manufacturing.

KEYWORDS

Thermal conductivity detector (TCD), Gas sensor, Pulse-width modulation, Suspended architecture, C-MEMS

INTRODUCTION

Various types of gas sensors have been widely studied for environmental monitoring in IoT applications. Among those, thermal conductivity detector (TCD)-type gas sensors have advantages for the detection of high-concentration gas because they detect the gas by measuring the heater resistance change through the thermal equilibrium process with the surrounding gas. However, conventional TCD-type sensors require large power consumption due to their relatively large heater size. To overcome this limitation, microscale bridge-type heater-based gas sensors have been developed using MEMS technologies, which could reduce the power consumption to 0.1–28 mW [1,2]. However, further power and size reduction require complex nanofabrication technologies, limiting mass production.

In this study, we developed a suspended 1D nanoheater-based TCD-type gas sensor. The suspended nanoheater was fabricated by selectively coating a thin metal layer on a suspended carbon nanowire backbone using a built-in shadow mask. The suspended carbon nanowires were fabricated at a wafer level using the carbon microelectromechanical systems (C-MEMS) technology [3,4]. C-MEMS enables batch fabrication of micro/nano or mixed-scale 3D carbon structures by pyrolyzing pre-patterned polymer structures at high temperatures [5, 6]. For practical applications, the developed sensors were packaged using through-silicon vias (TSVs) adaptive to the high-temperature pyrolysis process. Including the TSVs, the entire sensor was fabricated using only batch processes, ensuring cost-effective manufacturing. Furthermore, the small heater size enabled the sensor operation at low power (240 µW). Moreover, the fast gas response of the 1D nanoheater-based sensor resulting from the small size and

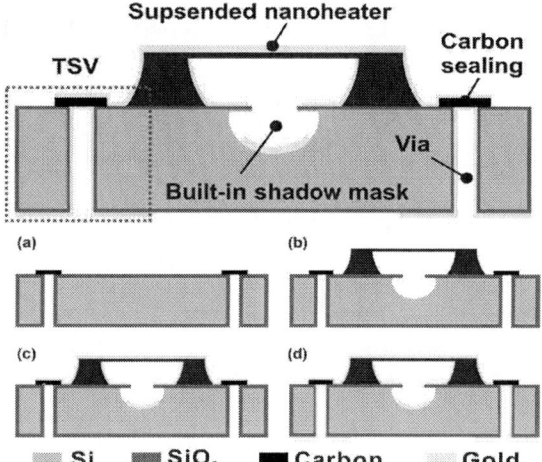

Figure 1: Schematic sectional view of a suspended 1D nanoheater with TSVs and corresponding fabrication steps: a) Patterning vias and carbon sealings. b) Fabrication of a built-in shadow mask and a suspended carbon nanowire. C) Deposition of a gold heater layer. D) Bottom electrode patterning.

suspended architecture allowed the sensor to be operated in a pulse-width modulation (PWM), leading to a power reduction of 1000 fold.

FABRICATION

Suspended 1D nanoheaters integrated with TSVs were fabricated by a four-step process, as shown in Figure 1. First, the via holes were drilled by the deep reactive ion etching (DRIE) and tetramethylammonium hydroxide (TMAH) silicon etching from the top and bottom sides of the wafer, respectively. Subsequently, the via holes were sealed with a thin carbon layer at the via hole top side using C-MEMS (Fig. 1a). Owing to the carbon sealing, the subsequent wet processes (e.g., photolithography) could proceed without leakage through the via holes. Next, the carbon sealing was thermally annealed by rapid thermal annealing at 1000 °C to increase the electrical conductivity [7]. Then, the built-in shadow mask was formed using silicon dioxide wet etching and isotropic DRIE silicon etching for selective metal coating. Subsequently, suspended carbon nanowire backbones were fabricated by pyrolyzing suspended polymer microwires at a low temperature (600 °C), ensuring low electrical conductivity (Fig. 1b) [7]. Then, a 50 nm thick gold heater layer was deposited using e-beam evaporation, resulting in the electrical connection only through the suspended carbon backbone due to the anisotropic e-beam evaporation and built-in shadow mask (Fig. 1c) [8]. The last step was a 300 nm thick copper bottom electrode deposition using e-beam evaporation (Fig. 1d).

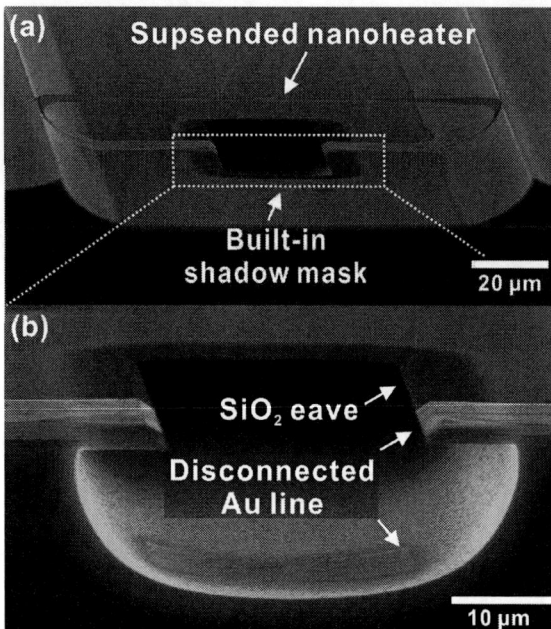

Figure 2: Scanning electron microscopy (SEM) images of the TCD-type gas sensor; a) Bird's eye view of the suspended nanoheater and built-in shadow mask. b) Cross-sectional view of the built-in shadow mask.

EXPERIMENTAL

The morphology and size of the suspended nanoheater were characterized using SEM (Quanta 200, FEI Co., USA). For the gas sensing test, the gas condition was precisely controlled using mass flow controllers (AFC 600, ATOVAC Co., Ltd., Korea). A current source meter (Keithley 6221, Keithley Instruments, Inc., USA) and a multimeter (Keithley 2401, Keithley Instruments, Inc., USA) were used to supply pulsed current signals to the nanoheater and measure the voltage output, respectively.

RESULTS AND DISCUSSION

The built-in shadow mask and suspended nanoheater were successfully fabricated, as shown in Fig. 2a. A suspended polymer wire (width: 1.3 µm, thickness: 2.5 µm, length: 50 µm) was converted to a suspended carbon nanowire (width: 200–300 nm, thickness: 300–400 nm, length: 80 µm) through a drastic volume reduction during the pyrolysis process. The gold heater layer was selectively coated only on top of the suspended carbon backbone. Although the gold heater layer was deposited wider than the width of the suspended carbon nanowire, the heater layer was electrically connected only through the suspended carbon nanowire due to the SiO_2 eave structure of the built-in shadow mask, as shown in Fig. 2b.

The developed sensor exhibited higher sensitivity than conventional micro-sized TCD-type gas sensors even at 240 µW due to its small size. For further power reduction, a current pulse (1.1 mA, 50 Hz, 0.1% duty) was applied to the nanoheater. Owing to the nanoscale heater, suspended architecture, and high aspect ratio, the nanoheater was instantly heated within 1 µs, ensuring a reduction in power consumption by 1/1000 through pulse power input. Fig. 3 shows the gas sensing result of the suspended nanoheater

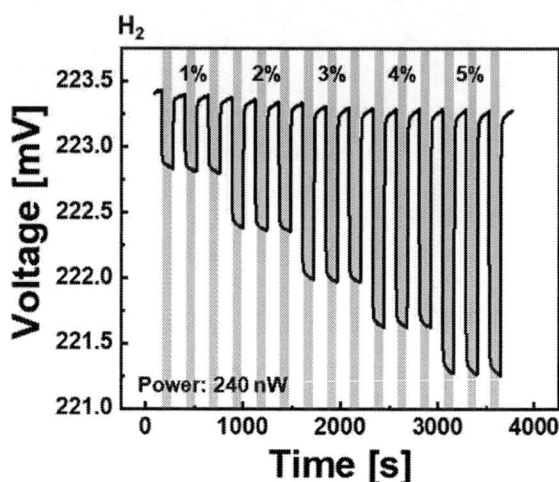

Figure 3: Output voltage of the suspended nanoheater upon repeated exposures to various concentrations of H_2 gas at 240 nW. The blue shaded areas in the graph represent the H_2 gas injection.

operated in the PWM mode (240 nW). H_2 has 7 times higher thermal conductivity than air. Therefore, more heat is dissipated to the surrounding gas, reducing the temperature of nanoheater. As a result, the voltage of the suspended nanoheater decreased with H_2 concentration.

CONCLUSION

In this study, we developed a suspended 1D nanoheater-based TCD-type gas sensor with TSV packaging for IoT applications. Due to the suspended architecture and built-in shadow mask structure, a metal heater layer was selectively deposited on the carbon nanowire backbone using conventional photolithography with micrometer-scale alignment accuracy. The nanoheater-based gas sensor showed the highest gas sensitivity per power compared with recently developed micrometer-sized TCD-based gas sensors. In addition, we developed the noble TSV packaging adaptive to the high-temperature process with carbon sealing. Thus, owing to this ultralow power consumption and the TSV suitable for C-MEMS, the developed sensor is expected to be applicable to various IoT applications.

ACKNOWLEDGEMENTS

This research was supported by Basic Science Research Program through the National Research Foundation of Korea (NRF) funded by the Ministry of Education (2020R1A6A1A03040570, 60%) and the Technology Innovation Program (00144157, Development of Heterogeneous Multi-Sensor Micro-System Platform, 40%) funded by the Ministry of Trade, Industry and Energy, Republic of Korea.

REFERENCES

[1] D. Berndt, J. Muggli, F. Wittwer, C. Langer, S. Heinrich, T. Knittel, R. Schreiner, "MEMS-based thermal conductivity sensor for hydrogen gas detection in automotive applications", *Sens. Actuator*

A Phys. vol. 305, 111670, 2020.

[2] A. Mahdavifar, M. Navaei, P.J. Hesketh, M. Findlay, J.R. Stetter, G.W. Hunter, "Transient thermal response of micro-thermal conductivity detector (μTCD) for the identification of gas mixtures: An ultra-fast and low power method", *Microsyst. Nanoeng.* vol. 1, 10525, 2015.

[3] J. Seo, Y. Lim, H. Shin, "High-performance hydrogen sensor based on an array of single suspended carbon nanowires selectively functionalized with palladium nanoparticles", *MEMS 2017 Conference*, Las Vegas, USA, January 22-27, 2017, pp. 1068-1070.

[4] J. Lee, D. Sharma, H. Shin, 1. "Biosensor platfrom based on sandwich carbon electrodes enabling enzymatic-electrochemical redox cycling", *MicroTAS 2017 Conference*, Savannah, USA, October 22-26, pp. 469-470, 2017.

[5] D. Sharma, J. Lee, H. Shin, "Electrochemical immunosensor based on 3D triple electrode system for sensitive detection of cardiac biomarker", *MicroTAS 2017 Conference*, Savannah, USA, October 22-26, pp. 491-492, 2017.

[6] J. Hong, B. Kim, H. Shin, "Diffusiophoresis-based particle entrapment in 3D microfunnels integrated in a mixed-scale PDMS fluidic device", *MicroTAS 2017 Conference*, Savannah, USA, October 22-26, pp. 1399-1400, 2017.

[7] Y. Lim, J. H. Chu, S. Y. Kwon, H. Shin, "Increase in graphitization and electrical conductivity of glassy carbon nanowires by rapid thermal annealing", *Journal of Alloys and Compounds*, vol 702, pp. 465-471, 2017.

[8] T. Kim, W. Cho, B. Kim, J. Yeom, Y. M. Kwon, J. M. Baik, J. J. Kim, H. Shin, "Batch nanofabrication of suspended single 1D nanoheater for ultralow-power metal oxide semiconductor-based gas sensors", *Small*, 2204078, 2022.

CONTACT

*Heungjoo Shin, tel: +82-52-217-2315;
hjshin@unist.ac.kr

120 PPM QUALITY FACTOR THERMAL STABILITY FROM -40 °C TO +60°C OF A DUAL-AXIS MEMS GYROSCOPE BASED ON JOULE EFFECT DYNAMIC CONTROL

Jian Cui[1,2] and Qiancheng Zhao[1,2]

[1]National Key Laboratory of Science and Technology on Micro/Nano Fabrication,
School of Integrated Circuits, Peking University, Beijing, China and
[2]Beijing Advanced Innovation Center for Integrated Circuits, Beijing, China

ABSTRACT

This paper presents a novel calibration method to improve the thermal stability of the quality factor of a MEMS gyroscope based on Joule effect dynamic control for the first time to the knowledge of the authors, which adjusts the dissipation energy of a resistive element connected to the mechanical structure with closed loop control by monitoring the driving voltage. The results show that the relative variation of the Q-factor dramatically reduced by more than 30000× down to ~120ppm from -40 °C to +60°C after applying the proposed means to a dual-axis gyroscope, which achieves only 30ppm of driving voltage stability over 100°C range with 3600× reduction. This technique is promising for enhancing the anti-shock performance and alleviating the bias thermal drift caused by the electrical coupling of the MEMS gyroscopes.

KEYWORDS

MEMS gyroscope, Quality factor tuning, Joule effect

INTRODUCTION

MEMS inertial sensors witness the tremendous emerging applications such as drones, automotive advanced driver-assistance systems, aeronautics in recent years owing to their low cost, compact size and increasing performances [1]. The in-phase electrical coupling results from drive mode to the sense mode via stray capacitances is considered as an important error source to the bias drift [2]. The parasitic capacitances are found to be inevitable in practice due to so many coupling paths exist in the complex sensor mechanical elements and also in circuit systems. Besides, the electrical coupling is in phase with the Coriolis force and thus cannot be separated from the hybrid signals [3]. Various methods are utilized to decrease this in-phase coupling effect. The amplitude of the driving voltage is used to compensate the bias drift [4]. A feed-forward coupling is utilized to cancel the electrical coupling by simply adding a scaled driving signal [5]. However, the thermal bias drift cannot be reduced since the Q-factor and the compensation gain change with the ambient temperature. Ascending frequency or half frequency driving method is used to separate the electrical signal from the mechanical vibration signal in the frequency domain [2], [6]. However, these methods complicated the circuit design.

Since this electrical coupling is relevant to the driving voltage amplitude that is inversely proportional to quality factor (Q-factor) value [7], an effective and direct approach to restrain the variation of the coupling signal is to stabilizing the Q-factor. Q-factor is a measure of the energy decay rate in each cycle of vibrations, which is limited by several energy mechanisms such as air damping, anchor loss, thermoelastic damping, surface loss, etc. Quite a few works have been reported to obtain a high Q-factor to reduce the driving voltage and mechanical thermal noise. However, the relative temperature variation of the Q-factor may be more essential because the electrical coupling from the driving voltage to the sense mode alters accordingly, thus giving rise to the bias drift.

The method reported here differs from previous works in two aspects: (1) directly stabilizing the Q-factor via Joule effect [8] to keep the driving voltage unchanged within the full temperature range; (2) having capability of precisely controlling the Q-factor to a reasonable value to enhance the anti-shock performance. The paper is organized as follows: section 2 describes the principle of the Q-factor control technique, section 3 shows the fabrication and packaging of the gyroscope, section 4 presents the experimental results and finally in section 5 conclusions are drawn.

METHOD DESCRIPTION

Figure 1 shows the schematic of the custom-designed dual-axis gyroscope with damping control via Joule effect. The gyroscope is a symmetrical wheeled horizontal dual-axis structure that mainly include a proof mass, eight driving beams, an outer frame and an inner frame [9]. The proof mass vibrates around Z-axis under an electrostatic force exerted on the driving combs. When an angular rate is generated about the X-axis, the proof mass swings by the Coriolis force around the Y-axis, introducing the variation of the capacitances between the proof mass and the substrate. The similar case also applies to the Y-axis.

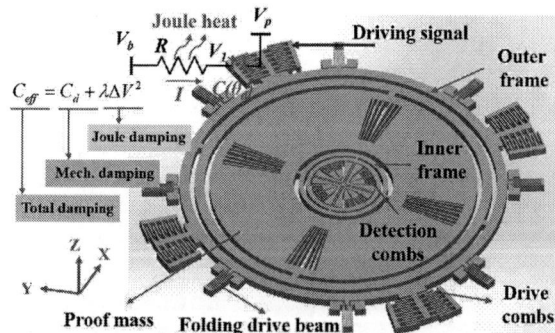

Figure 1: *The damping adjusting scheme via Joule effect.*

An adjusting voltage biased resistor is connected to the driving combs. A motion current is generated to flow through the resistor when the proof mass is actuated at drive mode as shown in Figure 2, which results a heat dissipation to lower the system energy. Therefore, the Q-factor of the gyroscope can be adjusted by adding this Joule damping depending on the resistor and the bias voltage. The overall damping of the system is written as.

$$C_{eff} = C_d + \lambda \Delta V^2 \qquad (1)$$

Where C_d is the mechanical damping of the drive mode, ΔV is the potential difference between the damping voltage V_b and V_p, λ is the converter coefficient described as follows.

$$\lambda = \frac{R\alpha^2}{1 + R^2 C_0^2 \omega^2} \qquad (2)$$

Here, R is the damping resistor, α is the gain from angular displacement to the capacitance variation, C_0 is the initial capacitance, ω_d is the resonant frequency of the drive mode. Equation (1) indicates that the system damping can be changed by adjusting the bias voltage.

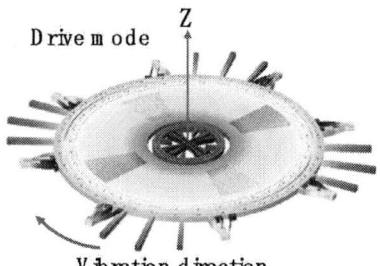

Figure 2: Drive mode analysis.

Since an extra damping caused by Joule effect is added to the dynamic gyro system, the Q-factor can only be decreased by this kind of method. Consequently, the mechanical damping should be designed as small as possible to prevent the increasing of the mechanical thermal noise. The tuning range is rather large as shown in Figure 3 which displays the simulated Q-factor varying with the damping control voltage and the resistor. The Q-factor is proportional to the square of the potential difference between the damping voltage and proof mass voltage. A 1000 kohm resistor is selected as the damping resistor.

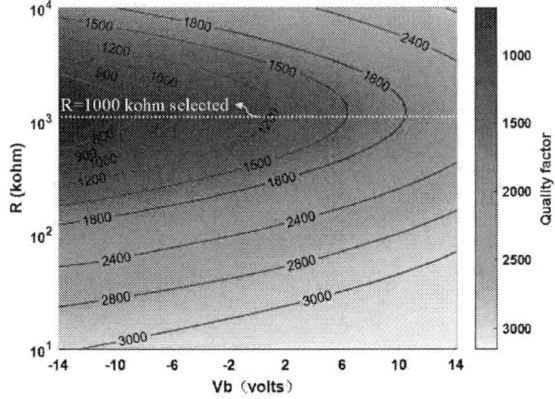

Figure 3: Q-factor varies with Joule resistor and adjusting voltage.

Based on this effect, a Q-factor stabilizing loop is proposed as shown in Figure 4. A self-oscillating loop is adopted to realize a stable vibrating with automatic gain control (AGC). Since the driving voltage is determined by

the Q-factor under closed-loop driving with which the target vibration amplitude is set to be V_r, the driving voltage magnitude can be kept to a stable value V_q by regulating the Joule damping with negative feedback. Thus, the electrical coupling caused by the variation of the driving signal amplitude can be suppressed.

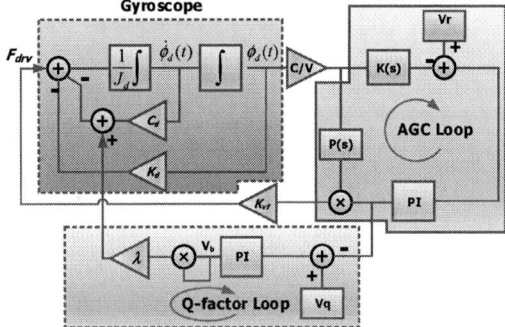

Figure 4: The schematic of the Q-factor dynamic control loop.

DEVICE FABRICATION

The gyroscope dies are fabricated using silicon-on-glass (SOG) processing. The anchor areas of the device are first defined by DRIE etching. Pyrex 7740 glass is selected as the substrate due to its similar thermal expansion coefficient with silicon material. The metal electrodes are patterned on the surface of the glass by lift-off technique. Then, the silicon wafer and the glass substrate are anodically bonded. The thickness of the device layer is adjusted using simple KOH etching. The mechanical structure is finally released through DRIE process. Figure 5(a) show the optical images of the fabricated wheeled horizontal dual-axis MEMS gyroscope.

The MEMS die is attached to a PCB soldered on a custom designed metal packaging and is moderately vacuum-sealed as shown in Figure 5(b). To test the characteristics of the Q-factor of the dual-axis gyroscope, the packaged device is mounted on a PCB with all the control circuits. The system circuit includes a closed driving loop and a Q-factor tuning loop for the drive mode with pure analog discrete electronics. Besides, the demodulations of the rate signal are realized by digital processing. The gyro system is installed in a test fixture as shown in Figure 6 and then put into a temperature chamber to evaluate the Q-factor thermal variation.

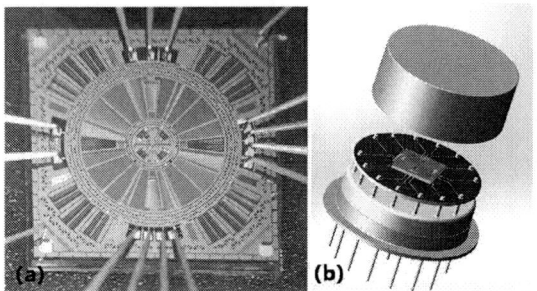

Figure 5: The photograph of the vacuum-sealed dual-axis gyroscope.

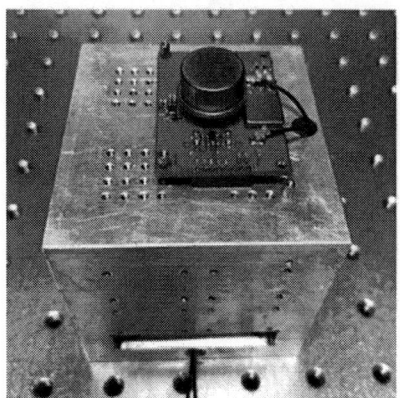

Figure 6: The photograph of the vacuum-sealed dual-axis gyroscope and the control circuit.

RESULT AND DISCUSSING

The Q-factor is firstly measured under different damping control voltages as illustrated in Figure 7. The Q-factor can be tuned from several hundred to ~2200 over about 30 volts range. The test results are in close agreement with the theoretical calculation results. In order to verify whether the driving voltage can be changed with the damping tuning voltage, the amplitudes of the driving voltage are adopted under different damping control voltages at room temperature. It indicates that the amplitude of the driving voltage changes from 0.3 to 1 volts corresponding to the Q-factor variation. Therefore, the driving amplitude can be viewed as a flag of Q-factor variation. It should be noted that the mechanical thermal noise may be increased because of the reduced Q-factor. This limitation can be easily solved by enlarging the initial Q-factor of the device, which can guarantee that the Q-factor value will not become too small after damping control.

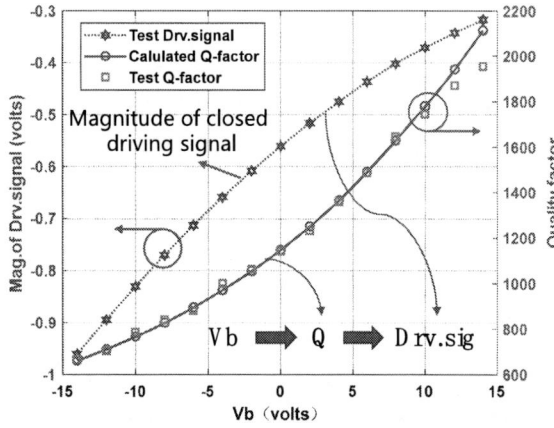

Figure 7: Magnitude of driving signal varies with adjusting voltage and tested Q-factor with calculated results.

The temperature dependences of the driving voltage amplitude with and without Q-factor control are illustrated in Fig.8. The variation range of the driving amplitude is 362 mV with 1.1×10^5 ppm stability without Q-factor control. When the dynamic damping control loop is employed, the peak range decreased sharply to ~0.17 mV with 30 ppm stability. The thermal variation of the electrical coupling of the gyroscope is expected to be reduced.

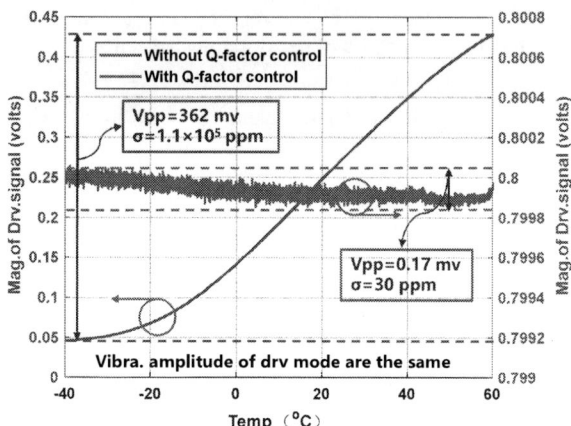

Figure 8: Magnitude of driving signal varies with temperature with and without Q-factor control.

To verify the Q-factor variation after damping control, the Q-factor of the drive mode are measured with and without closed-loop adjusting under no angular rate input when the proof mass vibrates. The Q-factor significantly reduced to ~120ppm from 3.88 over -40°C to +60°C as plotted in Fig.9.

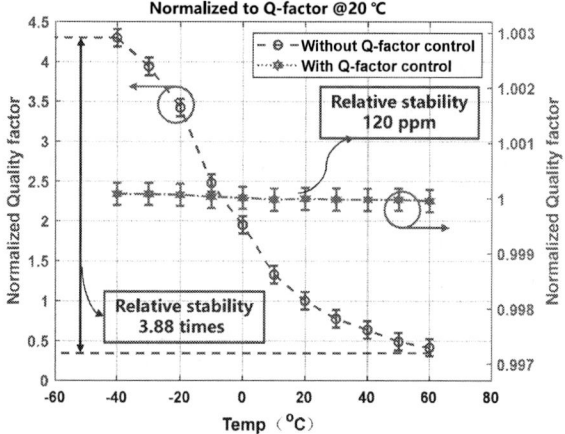

Figure 9: Q-factor varies with temperature with and without Q-factor control.

CONCLUSION

We have presented a novel method to improve the thermal stability of the quality factor of a dual-axis MEMS gyroscope based on Joule effect dynamic control. The amplitude of the driving voltage is served as a flag of the variation of the Q-factor. A negative feedback control loop is designed to adjust the dissipation energy of a resistive element connected to the mechanical structure. The relative variation of the Q-factor reduced to ~120ppm with 30000× improvement from -40°C to +60°C after applying the dynamic tuning for the system damping. Thanks to the Q-factor stabilization, a 30 ppm stability of the driving voltage is achieved over 100°C range with 3600×□ reduction. This technique can be utilized to improve the anti-shock performance and decrease the bias thermal drift caused by the electrical coupling of the MEMS gyroscopes.

ACKNOWLEDGEMENTS

The authors want to thank the technicians from the National Key Laboratory of Science and Technology on micro/nano fabrication for helping with the fabrication assistance. This work was supported by the Industrial Technology Development Program of China under Grant JCKY2021208B056.

REFERENCES

[1] Status of the MEMS Industry 2021, Yole Developpement, Lyon, France, 2021.

[2] M. Saukoski, L. Aaltonen and K. A. I. Halonen, "Zero-Rate Output and Quadrature Compensation in Vibratory MEMS Gyroscopes," in IEEE Sensors Journal, vol. 7, no. 12, pp. 1639-1652, Dec. 2007.

[3] M. S. Weinberg and A. Kourepenis, "Error sources in in-plane silicon tuning-fork MEMS gyroscopes," Journal of Microelectromechanical Systems, 15, pp. 479-491, (2006).

[4] K. Shcheglov, C. Evans, R. Gutierrez and T. K. Tang, "Temperature dependent characteristics of the JPL silicon MEMS gyroscope," 2000 IEEE Aerospace Conference. Proceedings (Cat. No.00TH8484), 2000, pp. 403-411 vol.1.

[5] M. J. Li et al., "Study on the influence induced by electrical coupling among interconnection lines in MEMS gyroscopes," 2017 19th International Conference on Solid-State Sensors, Actuators and Microsystems (TRANSDUCERS), 2017, pp. 1088-1091.

[6] J. Cui, Z. Y. Guo, Z. C. Yang, Y. L. Hao, and G. Z. Yan, "Electrical coupling suppression and transient response improvement for a microgyroscope using ascending frequency drive with a 2-DOF PID controller," Journal of Micromechanics and Microengineering, vol. 21, 2011.

[7] A. Walther, C. Le Blanc, N. Delorme, Y. Deimerly, R. Anciant and J. Willemin, "Bias Contributions in a MEMS Tuning Fork Gyroscope," in Journal of Microelectromechanical Systems, vol. 22, no. 2, pp. 303-308, April 2013.

[8] Jourdan, G., et al., Tuning the effective coupling of an AFM lever to a thermal bath. NANOTECHNOLOGY, 2007. 18(47).

[9] J. Cui and Q. Zhao, "A High-Performance Tactical-Grade Monolithic Horizontal Dual-Axis MEMS Gyroscope with Off-Plane Coupling Suppression Silicon Gratings," in IEEE Transactions on Industrial Electronics, vol. 69, no. 11, pp. 11765-11773, Nov. 2022.

CONTACT

* Q.C.Zhao, tel: +86-10-6274-5160; zqc@pku.edu.cn

A FORCE-BALANCE CAPACITIVE MEMS GRAVIMETER WITH SUPERIOR RESPONSE TIME, SELF-NOISE AND DRIFT

Le Gao[1], Fangzheng Li[1], Jian Zhang[1], Bingyang Cai[1], Wenjie Wu[1], and Liangcheng Tu[2]

[1]The MOE Key Laboratory of Fundamental, Physical Quantities Measurement, Hubei Key Laboratory of Gravitation and Quantum Physics, PGMF and School of Physics, Huazhong University of Science and Technology, Wuhan 430074, China and

[2] TianQin Research Center for Gravitational Physics and School of Physics and Astronomy, Sun Yat-sen University, CHINA

ABSTRACT

This paper reports a MEMS gravimeter with a capacitive displacement transducer and a force-balance system for superior response time, self-noise and drift compared to the state-of-the-art. It thus enables, for the first time, the measurements of low-frequency earth tide and high-frequency seismic motion simultaneously. More importantly, MEMS gravimeters with optimized dynamic response are crucial for a further wide field of applications, such as mobile gravity measurements.

KEYWORDS

electromagnetic force-balance, MEMS gravimeter, capacitive displacement transducer, earth tide, dynamic response

INTRODUCTION

Because variations in gravitational acceleration reflect changes in the density of the Earth's interior [1], precise local gravity measurements are critical for geophysics [2], resource exploitation [3], earthquake early warning [4], and gravitational passive navigation [5]. Traditional relative gravimeters as a tool for gravity measurement is large (~5 kilograms) and costly (more than 100,000 US dollars), which restricts its widespread application. Beneficial from its compact size and low cost, MEMS gravimeters have the potential to replace the traditional relative gravimeter.

A growing body of research have promoted the development of MEMS gravimeters and MEMS acceleration [6], [7] toward high accuracy and stability in recent years. Recently, open-loop MEMS gravimeters have been developed for static gravity measurements, featuring a quasi-zero stiffness spring-mass system for improving the sensitivity [8], [9], [10]. In order to prevent the proof-mass from being affected by the changes in air buoyancy, which results from the changes in atmospheric air density [11], [12], these MEMS acceleration sensing units are housed in a vacuum chamber. However, their high Q values due to working in open-loop mode diminished the responsiveness and limited their capacity for field gravity measurements. In addition, the resolution of these MEMS gravimeters was restricted by the low-precision optical shadow sensor with a resolution of 0.2 nm/Hz [9]. Generally speaking, the MEMS gravimeters have proved their stability and low self-noise in static testing. However, in order to make the MEMS gravimeter more practical, the issues of poor response time and low displacement sensitivity must be improved.

In this paper, we present a MEMS gravimeter utilizing a high-precision capacitive displacement transducer and an electromagnetic force-balance system for reduced response time, along with low self-noise and low drift. This constitutes a significant stride towards practical applications for MEMS gravimeters.

FORCE-BALANCE MEMS GRAVIMETER PRINCIPLE AND DESIGN

Principle

The force-balance capacitive MEMS gravimeter mainly consists of a spring-mass sensing element, arrayed area-changed capacitors for high-precision displacement sensing, and an electromagnetic force balance control system. The principle schematic of MEMS gravimeter is depicted in Figure 1. A positive-stiffness folded-beam spring and a negative-stiffness cosine beam connects the proof-mass on both sides, constituting a quasi-zero stiffness (QZS) system. Due to the low resonance frequency, this QZS system transfers the acceleration variation to the displacement of proof-mass with high sensitivity. The displacement is then detected by the arrayed area-changed capacitive displacement transducer [13]. Using the differential output, the displacement signal (Δx) is converted into a capacitance variation (ΔC) with an amplification factor of [14]

$$\frac{\Delta C}{\Delta x} = 2N\frac{\varepsilon b}{d}, \qquad (1)$$

where N is the number of periods, ε is permittivity, b is electrode length and d is the spacing of the capacitive transducer. The sensitivity is about two orders of magnitude larger than that of the optical shadow sensors in the previous works. The capacitance variation is detected by analog lock-in amplifying technique circuits, based on which an analog PID circuits calculate the proper feed-back voltage to be applied to the coils on the proof-mass. As the coils are in magnetic field produced by permanent magnets, it induced Ampere forces that are used to pull the proof-mass back to its balance position. Thus, the voltage-to-acceleration factor (H_a) can be written as

$$H_a = \frac{2nBL}{mR}, \qquad (2)$$

where n is the number of coil turns, B is the magnetic field strength of the permanent magnets, L is the length of the coils, m is the mass of the quasi-zero stiffness (QZS) system, and R is the resistance of the coils. This close-loop working mode is beneficial to improving the response time, linearity, and range of the MEMS gravimeter.

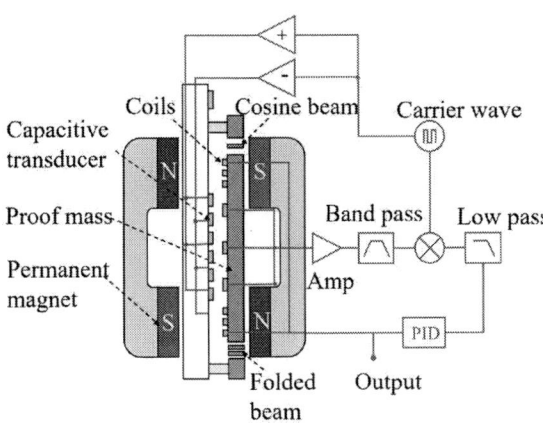

Figure 1: Design of the force-balance capacitive MEMS gravimeter

Fabrication and assembly

The chip of the MEMS gravimeter consists of a QZS silicon spring-mass system, and a glass cap which includes drive electrode arrays, spacing-control stoppers and bonding pads, as shown in Figure 2 (a). The QZS spring-mass system consists of five layers: a silicon spring-mass structure, the first insulator layer by a 400-nm-thick SiO_2, a 200-nm-thick gold layer for shielding and signal connections, a 2-μm-thick SiO_2 used as the second insulating layer in order to reducing the parasitic capacitance, a second gold layer with pick-up electrode arrays, coils for electromagnetic feedback, and bonding pads. The drive electrodes and pick-up electrodes form a precise area-changed capacitive displacement transducer after bonding the glass cap with the spring-mass system, as shown in Figure 2 (b). The dimensions of the fabricated MEMS gravity sensing chip are 31 mm×25 mm×3 mm.

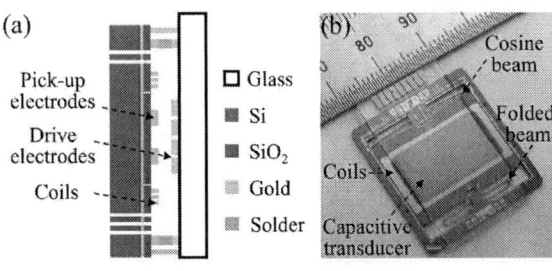

Figure 2: Schematic and photograph of the MEMS gravity sensor. (a) schematic of the MEMS gravimeter chip with coils for electromagnetic force balance. (b) MEMS gravity sensing chip bonded with the glass cap and the spring-mass system.

As shown in Figure 3 (a), the MEMS gravimeter chip was installed in a magnetic circuit system, which consists of permanent magnets and yokes assembled in a fixture for creating a steady and consistent magnetic field. The MEMS chip was electrically connected to the circuits via wire

bonding. In order to removing the disturbs from environmental pressure and temperature variations, the MEMS gravimeter is packaged inside a vacuum chamber with three-level temperature control (Figure 3 (b)).

Figure 3: The MEMS gravimeter chip (a) force-balance MEMS gravity sensing unit. (b) the packaged MEMS gravimeter.

TEST OF THE FORCE-BALANCE CAPACITIVE MEMS GRAVIMETER

Ring-down test

The packaged MEMS gravimeter calibrated with ring-down tests in both open-loop and closed-loop modes. The experimental results are shown in Figure 4. The response time of the gravimeter in the close-loop mode is 0.1 s, which is 2000 times faster than that in the open-loop mode. This response time is comparable to the commercial relative gravimeter (CG-6).

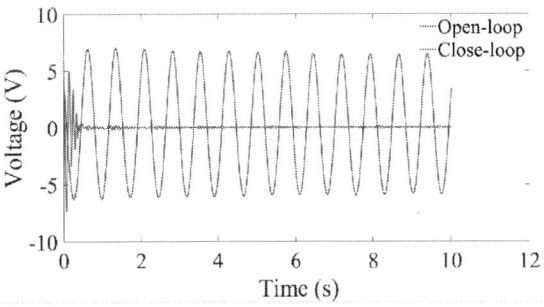

Figure 4: The response of the MEMS gravimeter working in open-loop (blue line) mode and in close-loop mode (orange line) with a knock at 0 s.

self-noise evaluation

The self-noise was calibrated in our cave lab experimentally with a commercial seismometer (Guralp-3ESPC) for reference. As shown in Figure 5, the earth micro-tremor can be detected by our device with high signal to noise ratio. Removing the environmental seismic signal, the self-noise of the proposed MEMS gravimeter is evaluated to be 1 μGal/$\sqrt{Hz}$ (1 μGal=10-9 g), which is 8 times lower than the previous works [9].

978-1-6654-9309-3/23 $31.00 © 2023 IEEE

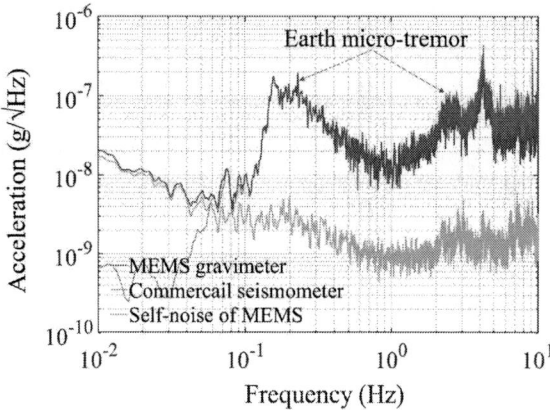

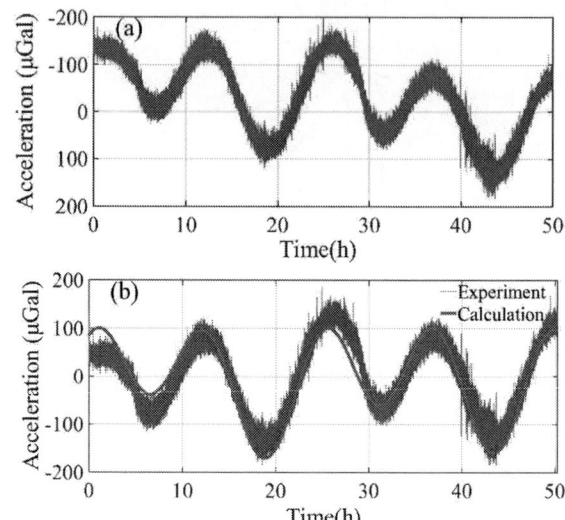

Figure 5: Power spectral density of the static output of the MEMS gravimeter (blue line), the referential commercial seismometer (orange line), and the self-noise of the MEMS gravimeter achieved by removing the environmental vibration signal from the blue line (yellow line).

Stability evaluation

The earth tide was measured in the cave laboratory for about two days, which was conducted to assess the stability of the MEMS gravimeter. Figure 6 (a) shows the data at a sampling rate of 10 Hz. Benefiting from the improved sensitivity of the displacement transducer, the drift of the proposed MEMS gravimeter was 60 μGal/day, which is about 4 times lower than that of previous works [9]. After linear correction of this drift（Figure 6 (b)), the output of the MEMS agrees well with the theoretical earth tide calculated by T-soft [15] with a coefficient of association of 0.92.

Figure 6: Long-term static output of the MEMS gravimeter (1 μGal=10^{-9} g). (a) the raw date with temperature correction, (b) the date of the MEMS gravimeter after removing a linear drift of 60 μGal/day (blue line). The calculated earth tide by T-soft was used as reference.

The key parameters comparison with previous published MEMS gravimeters and commercial bulk gravimeter (CG-6) are shown in Table 1. The proposed MEMS gravimeter was featured with lower noise level and static drift (see Table 1).

Table 1. Comparison with previous MEMS gravimeters and a commercial bulk gravimeter (CG-6).

Gravimeters	Technology	Working mode	Response Time (s)	Self-noise (μGal/√Hz)	Drift (μGal/day)
Ref [8]	MEMS/ optical shadow	Open-loop	—	40	268
Ref [9]	MEMS/ optical shadow	Open-loop	200	8	2400
CG-6 [16]	Fused Quartz/ capacitive	Close-loop	0.1	2	<200
This work	MEMS/ capacitive	Close-loop	0.1	1	60

CONCLUSIONS

It is worth noting that the proposed MEMS gravimeter in this paper is the first force-balance spring-based MEMS gravimeters reported to date. Benefiting from the high sensitivity of the capacitive displacement transducer, the proposed MEMS gravimeter achieved a self-noise floor of 1 μGal/√Hz and a static drift of 60 μGal/day, which are 8 times smaller and 4 times lower than that of the previous works, respectively. The performance is comparable to or even better than the commercial CG-6 relative gravimeter in terms of response time, self-noise and static drift. The MEMS gravimeter is being tested for practical applications in the field and is expected for mobile gravity measurement applications.

ACKNOWLEDGEMENTS

This research was funded in part by the National Natural Science Foundation of China under Grant 42274230, and in part by the National Key Research and Development Program of China under Grant 2021YFB3201601

REFERENCES

[1] M. Van Camp, O. de Viron, A. Watlet, et al.,

"Geophysics from terrestrial time-variable gravity measurements", *Reviews of Geophysics.*, vol. 55, pp. 938-992, 2017.

[2] S.C. Pearson-Grant, P. Franz, and J. Clearwater. "Gravity measurements as a calibration tool for geothermal reservoir modelling", *Geothermics.*, vol. 73, pp. 146-157, 2018.

[3] M. N. Nabighian, M. E. Ander, V. J. S. Grauch, et al., "Historical development of the gravity method in exploration", *Geophysics.*, vol. 70, pp. 63ND-89ND, 2005.

[4] J. P. Montagner, K. Juhel, M. Barsuglia, et al., "Prompt gravity signal induced by the 2011 Tohoku-Oki earthquake", *Nature communications.*, vol. 7, pp. 1-7, 2016.

[5] H. F. Liu, Z.C. Luo, Z. K. Hu, et al., "A review of high-performance MEMS sensors for resource exploration and geophysical applications", *Petroleum Science.*, 2022.

[6] G. Sobreviela-Falces, M. Pandit, A. Mustafazade, et al., "A Mems Vibrating Beam Accelerometer for High Resolution Seismometry and Gravimetry", in *2021 IEEE 34th Int. Conf. on Micro Electro Mech. Syst. (MEMS)*, Gainesville, FL, USA, January 25-29, 2021, pp. 196-199.

[7] M. Pandit, A. Mustafazade, C. Zhao, et al., "An Ultra-High Resolution Resonant MEMS Accelerometer", in *2019 IEEE 32nd Int. Conf. on Micro Electro Mech. Syst. (MEMS)*, Seoul, Korea (South), January 27-31, 2019, pp. 664-667.

[8] R. P. Middlemiss, A. Samarelli, D. J. Paul, et al., "Measurement of the Earth tides with a MEMS gravimeter", *Nature.*, vol. 531, pp. 614-617, 2016.

[9] S. H. Tang, H. F. Liu, S. T. Yan, et al., "A high-sensitivity MEMS gravimeter with a large dynamic range," *Microsyst Nanoeng.*, vol. 5, pp. 1-11, 2019.

[10] A. Prasad, R. P. Middlemiss, A. Noack, et al., "A 19 day earth tide measurement with a MEMS gravimeter", *Scientific reports.*, vol. 12, pp. 1-12, 2022.

[11] X. C. Xu , Q. Wang, L. J. Yang, et al., "On the Air Buoyancy Effect in MEMS-Based Gravity Sensors for High Resolution Gravity Measurements", *IEEE Sensors Journal.*, vol. 21, pp. 22480-22488, 2021.

[12] J. B. Merriam, "Atmospheric pressure and gravity", *Geophysical Journal International.*, vol. 109, pp. 488-500, 1992.

[13] W. J. Wu, D. D. Liu, H. F. Liu, et al., "Measurement of tidal tilt by a micromechanical inertial sensor employing quasi-zero-stiffness mechanism", *Journal of Microelectromechanical Systems.*, vol. 29, pp. 1322-1331, 2020.

[14] B. Bais, B. Y. Majlis, "Structure Design and Fabrication of an Area-changed Bulk Micromachined Capacitive Accelerometer", in *2006 IEEE Int. Conf. on Semicond. Electron.*, Kuala Lumpur, Malaysia, October 29 - December 01, 2006, pp. 29-34

[15] Van Camp, Michel, Simon DP Williams, and Olivier Francis. "Uncertainty of absolute gravity measurements", *Journal of Geophysical Research: Solid Earth.*, vol. 110, 2005.

[16] Scintrex, https://scintrexltd.com/wp-content/uploads/2018/04/CG-6-Operations-Manual-RevB.pdf

CONTACT

*Wenjie Wu, tel: +86 15871497300; wjwu@hust.edu.cn

*Liangcheng Tu, tel: +86 13971283278; tuliangch@mail.sysu.edu.cn

A MEMS-BASED GRAVIMETER FOR SIMULTANEOUS VERTICAL AND HORIZONTAL EARTH TIDES MEASUREMENTS

Lujia Yang[1], Xiaochao Xu[1], Qian Wang[1], Ji'ao Tian[1], Yanyan Fang[1], Chun Zhao[1], Wenjie Wu[1], Fangjing Hu[1] and Liangcheng Tu[1,2]

[1]MOE Key Laboratory of Fundamental Physical Quantities Measurement & Hubei Key Laboratory
of Gravitation and Quantum Physics, PGMF and School of Physics,
Huazhong University of Science and Technology, 430074 Wuhan, P. R. China and
[2]TianQin Research Center for Gravitational Physics and School of Physics and Astronomy,
Sun Yat-sen University (Zhuhai Campus), 519082 Zhuhai, P. R. China

ABSTRACT

This paper reports a MEMS gravimeter enabling simultaneous gravity measurements in both horizontal and vertical directions by combining two MEMS gravity sensors with an identical low-frequency linear spring structure. The proposed paired MEMS gravity sensors both experimentally demonstrate ultra-low self-noise floors of < 2 μGal/√Hz@1 Hz. The assembled gravimeter successfully measures long-term Earth tides signals with a high sensitivity and stability, showing the potential for three-axis gravity measurements.

KEYWORDS

Micro-electromechanical system (MEMS), gravimeter, Vertical and horizontal gravity measurements, Earth tides.

INTRODUCTION

Gravity measurement is one of the essential techniques for resource explorations [1]-[3]. Over the past years, due to the advancement of the micromachining technologies [4]-[6], Micro-Electro-Mechanical System (MEMS) based gravity sensors have been experimentally demonstrated. In 2016, Middlemiss *et.al* presented a MEMS device with a sensitivity of 40 μGal/√Hz using an anti-spring system. The MEMS successfully measured the Earth tides, indicating the capability for long term gravity measurements [7]. Tang *et al.* developed a more practical MEMS gravimeter with a sensitivity of 8 μGal/√Hz and a dynamic range of 8000 mGal by combining the folded beams with curved beams [8]. Fang *et al.* presented a micromachined resonant-type accelerometer with a noise floor of 75 ng/√Hz for moving-base gravimeter applications [9].

Compared with the single-axis gravity measurements in the vertical direction, tri-axial gravity measurements in the vertical and horizontal directions can reflect the longitudinal and lateral distribution characteristics of the underground minerals, respectively, promoting the spatial resolution of the gravity measurement [10]. However, the gravity sensor in the horizontal direction is sensitive to the change of the angle of the instrument, imposing an extremely high demand on the stability of the platform on which the gravity sensor is located. Therefore, the results on the long-term gravity measurement in the horizontal direction are rarely reported.

Recently, a MEMS-based three-axis sensor was proposed for borehole gravity survey by employing a MEMS vibrating beam accelerometer with a noise floor of 10 ng/√Hz@1 Hz [11], [12]. However, the result on the three-axis long-term gravity measurement has not been reported.

Although a MEMS gravity sensor using an anti-spring suspension has shown a noise floor of 18 μGal/√Hz@1 Hz and a good stability [13], the displacement of the proof-mass will be large owing to the ultra-low natural frequency of 7.35 Hz. Furthermore, due to the nonlinear characteristic of the spring, it is not applicable for such structure to measure gravity signals in the horizontal direction. In this paper, to improve the sensitivity of the MEMS gravity sensor both in horizontal and vertical directions, a linear spring with a low natural frequency is utilized. Through combining two identical MEMS gravity sensors, the proposed MEMS-based gravimeter is demonstrated for simultaneous Earth tides measurements in both horizontal and vertical directions.

MEMS DEVICE
MEMS Gravity Sensor Design

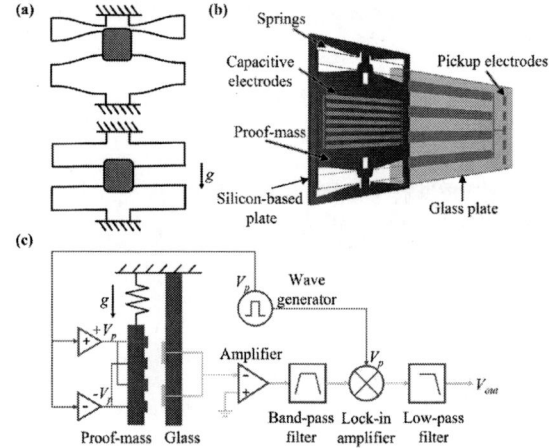

Figure 1: Design of the MEMS gravity sensor. (a) Schematic of the spring for the vertical MEMS gravity sensor with a designed initial offset. Under the 1 g gravitational field, the spring-mass will sag downward and operate in the "normal" linear spring state. (b) The spring-mass system with a capacitive transducer. (c) Readout circuitry.

In order to counteract the 1 g gravitational field on the vertical direction, the spring is expected to provide a sufficient restoring force. Therefore, the displacement of the proof-mass is carefully considered. Hereby, a linear spring with a natural frequency of ~14 Hz is designed. For the vertical MEMS gravity sensor, an initial offset is designed. Under the inertia force caused by 1 *g* gravitational field, the proof-mass will sag down to counteract the initial offset and centralize the structure, as shown in Fig. 1(a).

To obtain a high sensitivity, a capacitive transducer was employed for displacement detection, with 70 sets of pickup electrodes on the glass plate and 70 sets of driving electrodes on the silicon-based plate to achieve area-varying capacitive sensing, as shown in Fig. 1(b). The capacitance signal was further processed by a readout circuitry, as shown in Fig. 1(c). Two square carrier waves generated by a wave generator with a 180-degree phase difference were applied to the driving electrodes on the silicon-base plate, modulating the capacitance signal to a frequency of 50 kHz. Then, after a band-pass filter within a band from 38 kHz to 62 kHz, the signal was demodulated by a lock-in amplifier. Finally, the signal was fed to a low-pass filter with a cutoff frequency of 24 Hz to obtain the low-frequency gravity signal.

MEMS Device Fabrication

The fabrication process of the spring-mass system is shown in Fig. 2(a). The process of the silicon-based plate started from a 4-inch 500-μm-thick single-crystal silicon wafer with a crystal orientation of <100> and 300-nm-silicon dioxide layer on both sides. Firstly, followed by the cleaning of the wafer, the Metal 1 layer consisting of 40-nm-thick titanium and 400-nm-thick gold was patterned on the silicon wafer by a lift-off process [14]. Then, a 1.2-μm-thick SiO_2 layer was patterned as the insulation layer by plasma enhanced chemical vapor deposition (PECVD).

Afterwards, considering the compatibility during the fabrication, a layer with 40-nm-thick titanium and 200-nm-thick gold was evaporated as the conducting layer, followed by a 2-μm-thick electroplated gold layer to form the capacitive electrodes and bonding pads. In order to create an etch mask for the deep reactive ion etching (DRIE), a 13-μm-thick photoresist (AZ9260) was spun coated. In addition, a 200-nm-thick aluminum layer was deposited on the backside of the wafer after removing the SiO_2 layer to alleviate the notching effect of the through-wafer DRIE process. For protecting the fragile spring-mass structure, the silicon wafer was attached to a carrier wafer. Finally, the sample wafer was loaded into an Oxford Instruments PlasmaPro Estrelas100 to run a standard Bosch process for the release of the spring-mass structure [8].

The process of the glass cover started from a 500-μm-thick glass wafer. The fabrication process of the capacitive electrodes and the bonding pads on the glass wafer is similar to that of the silicon wafer. After the pattern of the metal layer, the solder ($Au_{80}Sn_{20}$) was reflowed on the bonding pads of the glass plate. Finally, the fabricated glass plate was bonded to the spring-mass structure by soldering reflow.

To ensure similar performances of the MEMS gravity sensors for gravity measurements in the vertical and horizontal direction, the devices were fabricated on the same silicon wafer, as shown in Figure 2(b). The photograph for the assembled MEMS gravity sensor is shown in Figure 2(c). The overall size of the MEMS gravity sensor is 20 mm × 20 mm × 0.5 mm.

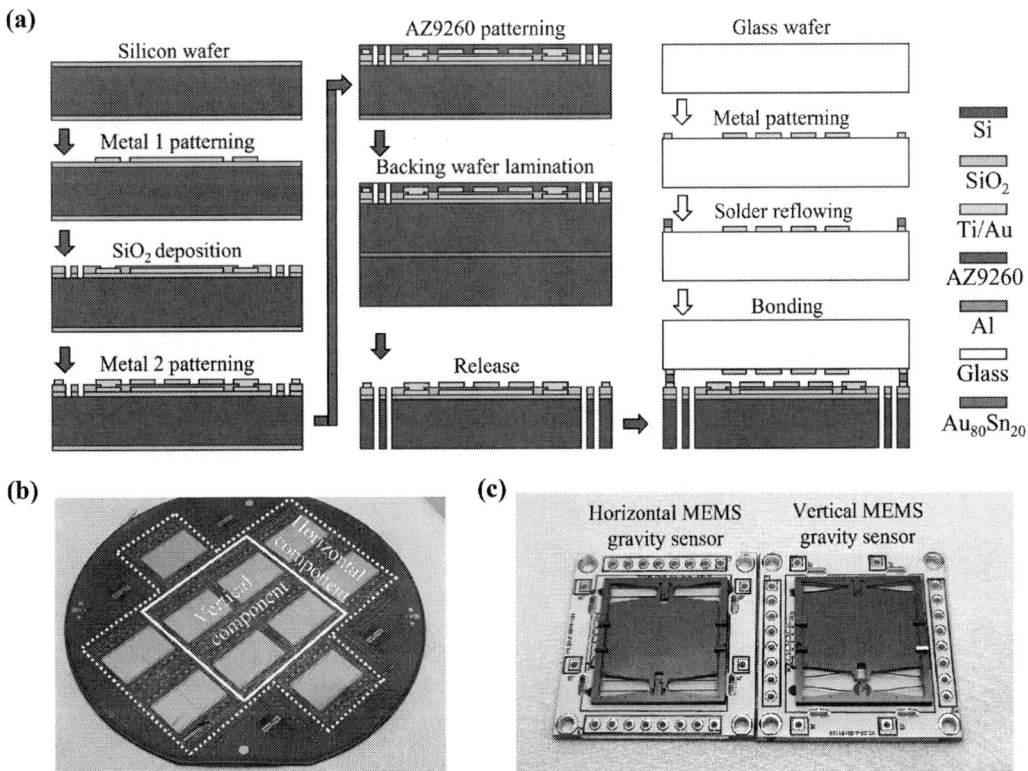

Figure 2: (a) Fabrication process of the spring-mass system. (b) MEMS gravity sensors on a silicon wafer showing four vertical components in the middle and six horizontal components in the remaining area. (c) Assembled MEMS gravity sensors for horizontal and vertical direction gravity measurements.

Self-noise Floor Measurement

In order to evaluate the self-noise floor of the MEMS gravity sensors in both horizontal and vertical directions, the MEMS gravity sensors were tested in a quiet laboratory with a commercial seismometer (CMG-3ESPC, Güralp) for reference. Since the predicted self-noise floor is lower than the ambient vibration according to the Earth's new high noise model (NHNM) [15], a coherence analysis [16] method was employed to remove the common-mode vibration signal. Fig. 3(a) shows the power spectral densities (PSDs) of the two vertical MEMS gravity sensors which are denoted as MEMS-1 and MEMS-2. The earth tremor peak at 0.2-0.4 Hz and 2-4 Hz can be clearly observed by all the devices, as seen from Fig. 3. After identifying the angular misalignment of the sensitive axis between the two vertical MEMS gravity sensors using the commercial seismometer, the calibrated self-noise floor of the vertical MEMS gravity sensor is <2 µGal/√Hz@1 Hz. Similarly, the self-noise floor of the horizontal MEMS gravity sensor is calibrated to be <2 µGal/√Hz@1 Hz, as shown in Fig. 3(b).

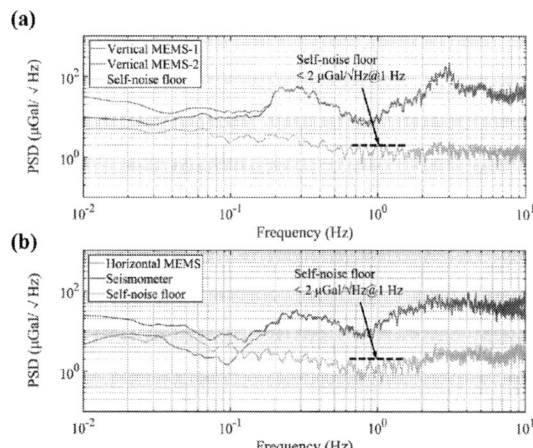

Figure 3: Self-noise floor of the MEMS gravity sensor in (a) vertical direction, and (b) horizontal direction.

SYSTEM ENVIRONMENTAL CONTROL

The assembled MEMS gravity sensors for vertical and horizontal gravity measurements are shown in Fig. 4(a). Generally, the MEMS gravity sensors are susceptible to the pressure changes, which will lead to a gravity-pressure coefficient of 501.5 µGal/hPa [17]. In order to isolate the effects of atmospheric pressure variation, the paired MEMS gravity sensors were sealed in a vacuum chamber. In addition, the stiffness of the springs is also temperature sensitive as Young's modulus of silicon is temperature dependent. To suppress ambient temperature variations, an active temperature control was therefore implemented. Polyimide film heaters were wrapped on the vacuum chamber, controlled by a commercial temperature control module (TCMM207, Oeshine). A two-stage passive thermal isolation scheme was introduced to further suppress the ambient temperature variations.

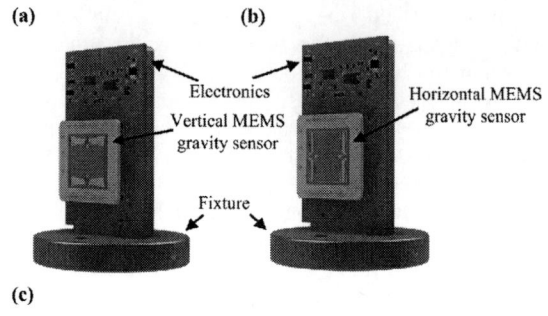

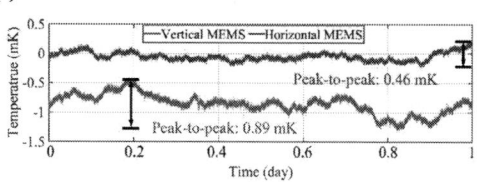

Figure 4: Schematics of the assembled MEMS gravity sensor for (a) vertical direction, and (b) horizontal direction. (c) Recorded temperature variations around the MEMS gravity sensors in the vertical and horizontal directions within a 1-day span.

The temperature around the MEMS gravity sensors is controlled at ~30°C. As shown in Fig. 4(c), the peak-to-peak of temperature variation around the MEMS gravity sensors in vertical and horizontal direction is about 0.89 mK and 0.46 mK within a 1-day span, respectively, showing a stable environment for long-term gravity observations.

EARTH TIDES MEASUREMENT

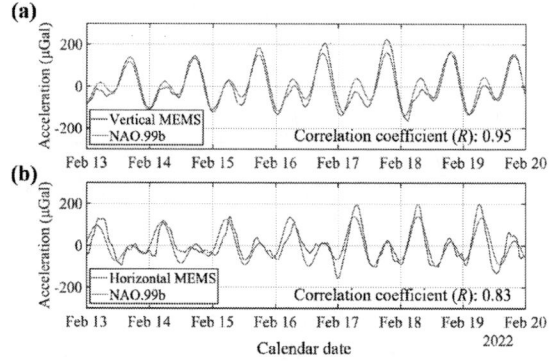

Figure 5: Earth tides measurements of the MEMS gravimeter within a time-span of 7 days from February 13, 2022 to February 20, 2022. (a) Data of the MEMS gravity sensor in the vertical direction and the theoretical result by NAO.99b. (b) Data of the MEMS gravity sensor in the horizontal direction and the theoretical result by NAO.99b.

The sensitivities and stabilities of the gravity sensors were verified via long-term Earth tides observations. Data from February 13 to February 20, 2022 were selected for further analysis. Firstly, the linear drift of raw data was removed by the 'detrend' function in MATLAB. As the frequency around 1×10^{-5} Hz is the frequency band of the Earth tides signal, a band-pass filter within a band from 1×10^{-5} Hz to 0.1 mHz was applied to the data to reject out-of-band noise. Fig. 5 shows the processed data for both the vertical and horizontal MEMS gravity sensors, giving a correlation coefficient of 0.95 and 0.83 with the theoretical results by NAO.99b [17], respectively. Our results indicate a high sensitivity and excellent stability of the proposed MEMS gravimeter.

CONCLUSION

In this paper, the MEMS gravity sensors with a linear spring structure were presented for simultaneous gravity measurement in both the vertical and horizontal directions. The proposed MEMS gravity sensors showed an ultra-low self-noise floor of <2 μGal/√Hz@1 Hz in both directions. The gravimeter has successfully observed the vertical and horizontal Earth tides signal over a 7-day span, giving a correlation coefficient of 0.95 and 0.83 with the theoretical results, respectively. It is arguably, to the best of our knowledge, the first MEMS-based gravimeter for simultaneous measurements of the Earth tides signals in both the vertical and horizontal directions. It is expected to be further applied to tri-axial gravity measurements for the exploration of detailed longitudinal and lateral distribution information of the underground structures.

ACKNOWLEDGEMENTS

This research was funded in part by the National Key Research and Development Program of China under Grant 2021YFB3201601, and in part by the National Natural Science Foundation of China under Grant 42274230. The authors thank Shasha Liu for the help on MEMS gravity sensor fabrication, and thank Fangzheng Li for the help on the circuits analysis.

REFERENCES

[1] M. N. Nabighian, M. E. Ander, V. J. S. Grauch, *et al.*, "Historical Development of the Gravity Method in Exploration", *Geophys.*, vol. 70, pp. 63ND-89ND, 2005.

[2] D. Sherlock, A. Toomey, M. Hoversten, *et al.*, "Gravity Monitoring of CO2 Storage in a Depleted Gas Field: A sensitivity study", *Explor. Geophys.*, vol. 37, pp. 37-43, 2006.

[3] L. Xing, X. Niu, L. Bai, *et al.*, "Monitoring Groundwater Storage Changes in a Karst Aquifer using Superconducting Gravimeter OSG-066 at the Lijiang Station in China", *Pure Appl. Geophys.*, pp. 1-18, 2022.

[4] Q. Lu, J. Bai, K. Wang, *et al.*, "Design, Optimization, and Realization of a High-performance MOEMS Accelerometer from a Double-device-layer SOI Wafer", *J. Microelectromech. Syst.*, vol. 26, pp. 859-869, 2017.

[5] M. S. Weinberg, J. J. Bernstein, J. T. Borenstein, *et al.*, "Micromachining Inertial Instruments", in *Proc. SPIE 2879*, Austin, TX, US., September 23, 1996, pp. 26-36.

[6] S. Finkbeiner, "MEMS for Automotive and Consumer Electronics", in *2013 Proc. of the ESSCIRC (ESSCIRC)*, Bucharest, Romania, September 16-20, 2013, pp. 9-14.

[7] R. P. Middlemiss, A. Samarelli, D. J. Paul, *et al.*, "Measurement of the Earth Tides with a MEMS Gravimeter", *Nature*, vol. 531, pp. 614-617, 2016.

[8] S. Tang, H. Liu, S. Yan, *et al.*, "A High-sensitivity MEMS Gravimeter with a Large Dynamic Range", *Microsyst. Nanoeng.*, vol. 5, pp. 1-11, 2019.

[9] Z. Fang, Y. Yin, C. Chen, *et al.*, "A Sensitive Micromachined Resonant Accelerometer for Moving-base Gravimetry", *Sens. Actuators A Phys.*, vol. 325, pp. 112694, 2021.

[10] H. Rim, Y. Li, "Advantages of borehole vector gravity in density imaging", *Geophys.*, vol. 80, pp. G1-G13, 2015.

[11] J. Lofts, A. Zett, P. Clifford, *et al.*, "Three-Axis Borehole Gravity Logging for Reservoir Surveillance", in *SPE Middle East Oil and Gas Show and Conf.*, Manama, Bahrain, March 18–21, 2019, pp. 194845.

[12] G. Sobreviela-Falces, M. Pandit, A. Mustafazade, *et al.*, "A Mems Vibrating Beam Accelerometer for High Resolution Seismometry and Gravimetry", in *2021 IEEE 34th Int. Conf. on Micro Electro Mech. Syst. (MEMS)*, Gainesville, FL, USA, January 25-29, 2021, pp. 196-199.

[13] A. Prasad, R. P. Middlemiss, A. Noack, *et al.*, "A 19 Day Earth Tide Measurement with a MEMS Gravimeter", *Sci. Rep.*,vol., 12, pp. 1-12, 2022.

[14] D. Liu, W. Wu, J. Liu, *et al.*, "A Force Balance Micromachined Accelerometer with a Self-noise of 1 ng/Hz1/2", in *2020 IEEE Int. Symp. on Inertial Sensors and Syst. (INERTIAL)*, Hiroshima, Japan, March 23-26, 2020, pp. 1-4.

[15] L. G. Holcomb, "A Direct Method for Calculating Instrument Noise Levels in Side-by-side Seismometer Evaluations", U.S. Geol. Surv., Albuquerque, NM, USA, Tech. Rep. 89-214, 1989.

[16] S. Yan, Y. Xie, M. Zhang, Z. Deng, and L. Tu, "A Subnano-g Electrostatic Force-rebalanced Flexure Accelerometer for Gravity Gradient Instruments", *Sensors*, vol. 17, pp. 2669, 2017.

[17] X. Xu, Q. Wang, L. Yang, *et al.*, "On the Air Buoyancy Effect in MEMS Based Gravity Sensors for High Resolution Gravity Measurements", *IEEE Sens. J.*, vol. 21, pp. 22480-22488, 2021.

CONTACT

*Fangjing Hu, tel: +86 15657859810; fangjing_hu@hust.edu.cn
*Wenjie Wu, tel: +86 15871497300; wjwu@hust.edu.cn

A NOVEL MULTIPLE-FOLDED BEAM DISK RESONATOR FOR MAXIMIZING THE THERMOELASTIC QUALITY FACTOR

Xiaopeng Sun[1], Xin Zhou[1,*], Lei Yu[2], Kaixuan He[2], Xuezhong Wu[1] and Dingbang Xiao[1]
[1] National University of Defense Technology, Changsha 410073, CHINA
[2] East China Institute of Photo-Electronic IC, Bengbu 233042, CHINA

ABSTRACT

This paper reports a micromachined multiple folded-beam disk resonator whose thermoelastic quality factor (Q_{TED}) can exceed 2 million owning to the topology improvement and parameters optimization for the (100) silicon resonant structure. The measured total quality factor (Q) reaches 710 k, surpassing the Q of most micromachined disk resonators. Moreover, the conducted temperature experiment indicates that the thermoelastic damping (TED) is no longer the only dominant damping mechanism.

KEYWORDS

Microelectromechanical systems (MEMS), quality factor, thermoelastic dissipation, anchor loss, disk resonator

INTRODUCTION

MEMS disk resonators are widely used for sensing applications due to the symmetrical structure, mature fabrication process and great performance potential [1-3]. Maximization the mechanical quality factor of the resonator is critical for improving the performance of the sensors. When air damping is minimized through high vacuum packaging, energy dissipation in vibratory MEMS resonators is mainly governed by the substrate dissipation and thermoelastic damping (TED) for the relatively low-frequency range [4, 5]. Although it is difficult to accurately model and analyze the anchor loss, many experimental results have indirectly shown that it is generally weaker than TED [4-6].

Employing the non-silicon material with smaller coefficient of thermal expansion (CET), such as fused silica [7], is a promising method to suppress resonator's TED, but facing manufacturing challenges. Thus, optimizing the structural design of the silicon MEMS resonators to improve the thermoelastic quality factor (Q_{TED}) is still attractive. The reported design strategies include optimizing structural parameters [1], placing thin slots [8], adding lumped masses and introducing other topological optimization methods [9,10]. Among them, adding lumped masses to realize stiffness-mass decoupling and the separation between resonance frequency f_0 and thermal relaxation rate f_z has demonstrated to be effective. However, the existing arrangement of masses in the disk resonators is limited to hanging or attaching to the flexible rings, which cannot achieve maximum benefit due to the negative impact on the mode shapes and the space for internal electrodes [10]. It is noteworthy that the radially pleated disk resonator reported in [11] shows great potential in enhancing Q_{TED} even without introducing lumped masses design. The pleated-beam design also shows advantages in vibration amplitudes over other topological forms, but there is still a lot of room for improvement.

Inheriting the advantages of the stiffness-mass decoupling strategy [9] and the pleated-beam design in ref. [11], this paper proposes a new kind of multiple folded-beam disk resonator. By carrying out the integrated design for the multiple folded beams and the lumped mass layers from the beginning, the resonator can gain great structure coherence and make full use of the stiffness-mass decoupling effect to realize greater Q_{TED}.

ANALYSIS AND DESIGN

Zener's standard model of TED reveals that [12]

$$Q_{TED} = \frac{C_v}{E\alpha^2 T_0}(\frac{2\pi f_0}{f_z} + \frac{f_z}{2\pi f_0}), \qquad (1)$$

where C_v is the heat capacity at constant volume, E is the Young's modulus, α is the linear coefficient of thermal expansion, T_0 is the equilibrium temperature, f_0 is the resonant frequency, and f_z is the thermal relaxation rate related to the thermal diffusivity χ and the strained beam width b with relation $f_z = \pi^2\chi/b^2$. Considering the limitations of the processing technology and the requirement of the resonator stiffness, the beam width b cannot be infinitely reduced, and there will be a lower limit. Then we can introduce lumped masses to further decrease f_0 without affecting f_z [9]. However, the maximization of this stiffness-mass decoupling effect is usually limited to the non-ideal compatibility between the frame and masses. Directly hanging masses on the outer rings also has a certain impact on the equivalent capacitance area of the resonator.

Here, we propose another design idea. We carry out the integrated design for the multiple folded beams and the lumped mass layers from the beginning to gain great structure coherence. Then further decrease the beam width and increase the mass width under the premise of ensuring rigidity. This trick can make full use of the stiffness-mass decoupling effect, thus providing lower f_0 and higher f_z to realize greater Q_{TED}. The designed multiple folded-beam disk resonator is shown in Figure 1, which consists of a central anchor, multiple folded beams extending radially and outer mass layers hanging between the beams. The inner folded beams with small width b_m can effectively separate resonance frequency and thermal relaxation frequency. With integral design, the outer mass layers help to further increase the effective mass and reduce the resonant frequency.

It should be noted that with the increase of Q_{TED}, the influence of the anchor loss would be more significant. Caused by the unbalanced out-of-plane vibration, the anchor loss in the (111)-single-crystal-silicon disk

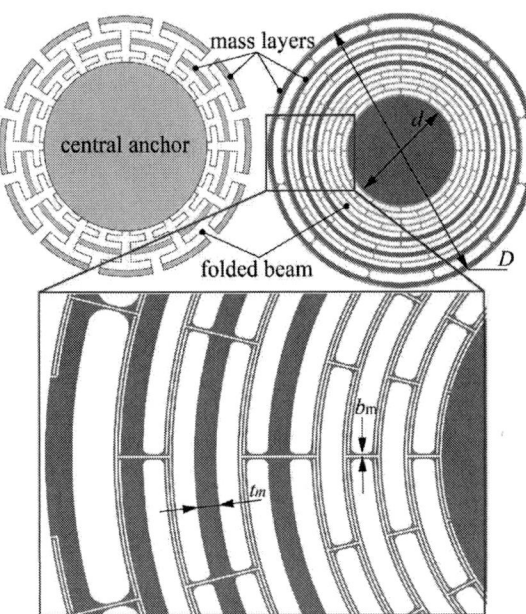

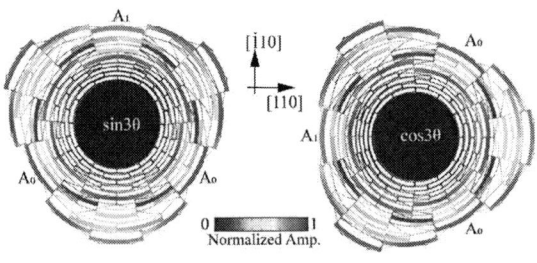

Figure 1: Design concept and structural design of the multiple folded-beam disk resonator.

Figure 2: The n = 3 mode shapes obtained by simulation.

resonator is usually difficult to control. Selecting (100) silicon can effectively suppress the out-of-plane displacement. However, it brings modal asymmetry concurrently resulting from the in-plane anisotropy of the material, which would destroy the balance of the centroid and aggravate the anchor loss as well [13]. Though the influences of the two above factors on Q are similar, the asymmetry in-plane vibration of (100) silicon resonator possesses the potential advantages for further regulation. And the 90° rotationally symmetric distribution of Young's modulus and Poisson's ratio in (100) silicon resonator ensures the degeneracy of the $n = 3$ modes [14]. Therefore, in this design, we use (100) silicon. The $n = 3$ mode shapes are illustrated in Figure 2, simulated with a density of 2330 kg/m³, thermal expansion coefficient of 2.6 ppm/K and the orthotropic stiffness matrix for (100) silicon with three axes at [100], [010], and [001] crystal orientations at 273 K. It can be seen that the amplitude of radial motion at the peaks along <110> orientation A_1 is smaller than that of the two symmetrical peaks A_0.

The outer diameter D of the resonator is designed to be about 6 mm. Accordingly, based on the optimal proportion ($D/d \approx 0.4$) obtained from preliminary design cases [1,10,11], the inner anchor diameter is designed to be 2.4

mm. The folded beam width b_m and mass layer width t_m, which show important impacts on f_0 and Q_{TED}, are analyzed in detail through finite element simulation.

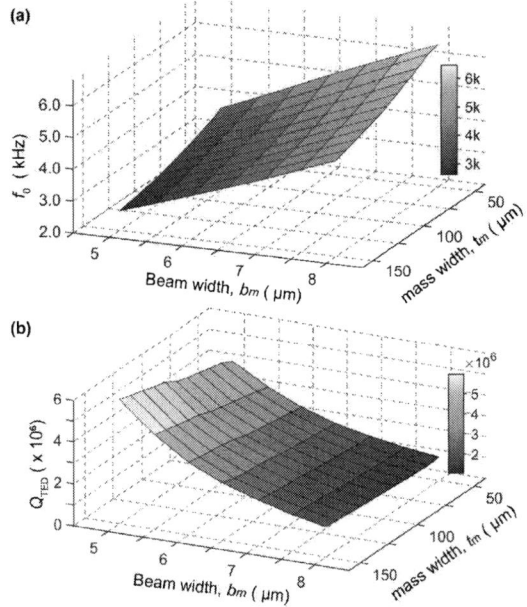

Figure 3: (a) Resonant frequency f_0 and (b) thermoelastic quality factor Q_{TED} as a function of beam width b_m and mass layer width t_m.

Figure 3 shows how the resonant frequency and Q_{TED} of the $n = 3$ modes vary as the beam width b_m and mass layer width t_m are varied with fixed outer diameter D and inner diameter d. Figure 3(a) indicates that low frequencies occur for thin beam and thick mass. High frequencies occur for thick beam and thin mass. As shown in Figure 3(b), for a given t_m, Q_{TED} increases as b_m is decreased. With a fixed b_m, Q_{TED} increases as t_m is increased. For small values of b_m, increasing t_m can significantly improve Q_{TED}. For example, when $b_m = 5$ μm, Q_{TED} changes from 4.1 million to 6.0 million as t_m varies between 50 μm and 150 μm. However, for large values of b_m, Q_{TED} is no longer sensitive to the change of t_m, which corresponds to the case that Q_{TED} just increases from 1.1 million to 1.4 million as t_m varies with a fixed $b_m = 8$ μm.

Table 1. Design parameters of the resonator.

Paremeter	Value
Outer diameter (D)	6080 μm
Inner diameter (d)	2400 μm
Structure height	100 μm
Beam width (b_m)	6.5 μm
Mass width (t_m)	100 μm
Electrode gap size	10 μm

Adopting smaller beam width can increase Q_{TED}, but it can also increase the risk of failure during manufacturing and result in a weaker susceptibility to the external interference. Under the trade-off between f_0 and Q_{TED}, we select $b_m = 6.5$ μm and $t_m = 100$ μm, which lead to the simulated f_0 = 4420 Hz and Q_{TED} = 2.2 million. The final structural design parameters of the multiple folded-beam disk resonator are summarized in Table 1.

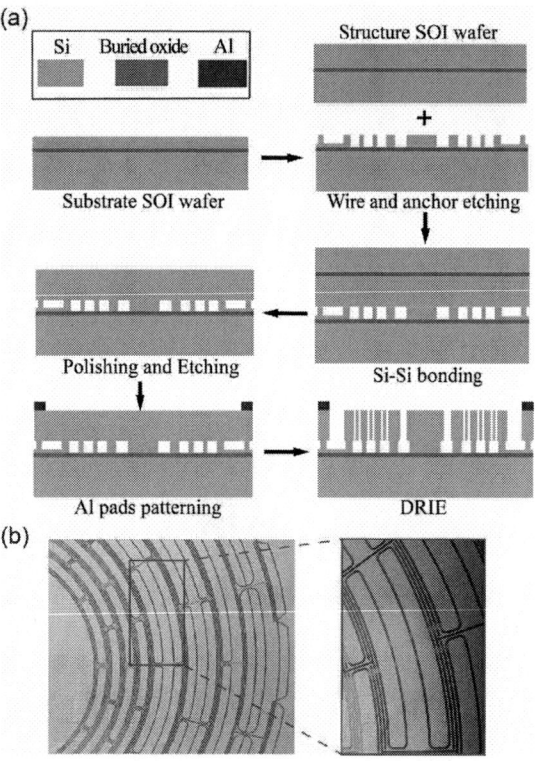

Figure 4: (a) Fabrication process. (b) Overview and zoomed-in view of the multiple folded-beam disk resonator.

FABRICATION PROCESS

The fabrication process of the resonator is shown in Figure 4(a). Firstly, we etch a substrate silicon-on-insulator (SOI) wafer for 10 μm to form the anchors and generate the thermal oxide layer on it with the thermal oxidation technology. Then, the substrate and another structure SOI are bonded together through the wafer fusion-bonding technology. The handle wafer is removed with the chemical and mechanical polishing subsequently. Next, the aluminum wire bonding pads are patterned, and the resonator and the electrodes are formed via deep reactive ion etching (DRIE) technology. Lastly, the device is diced by the laser stealth cutting, attached to a chip carrier, wire-bonded and then sealed in a metal shell with a high vacuum of 0.001 Pa [1]. The microscope images of the resonator are shown in Figure 4(b).

DEVICE CHARACTERIZATION

The characterization of the resonator is tested through a printed circuit board (PCB). The measured quality factors of the packaged resonators are between 550 k and 720 k. The frequency responses of the resonator with the largest quality factor are shown in Figure 5. The initial resonant frequencies of the n = 3 normal modes are extracted to be ω_1 = 2π × 4,116.55 Hz and ω_2 = 2π × 4,118.24 Hz, respectively, indicating an initial frequency split of only

1.69 Hz. The measured decaying time constants are τ_1 = 54.5 s and τ_2 = 55.4 s, corresponding to Q_1 = 704 k and Q_2 = 716 k, which are larger than the Q of most flexural MEMS disk resonators. However, there is a gap between the simulated Q_{TED} (2.2 million) and the experimental Q (0.7 million). It is inferred that there are other factors that significantly affect the energy dissipation of the resonator besides TED. With all loss mechanisms taken into account, the total quality factor of the resonator can be represented by

$$Q = 1/(1/Q_{TED} + 1/Q_{other}), \qquad (2)$$

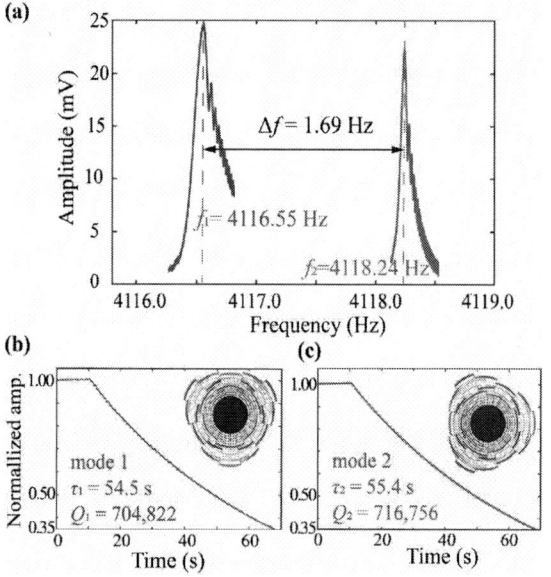

Figure 5: (a) Frequency responses of the packaged resonator. Ring-down signal of (b) $\sin 3\theta$ mode and (c) $\cos 3\theta$ mode.

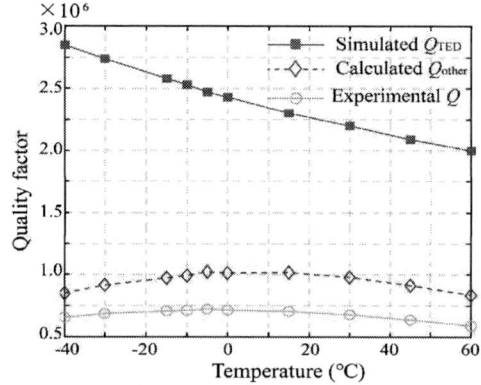

Figure 6: Relation between quality factor and temperature.

We conduct a temperature experiment for the resonator whose quality factor is Q = 716 k at room temperature. The relation between the quality factor and temperature is shown in Figure 6. The yellow dots are the measured quality factor, which increases before the temperature dropped to around −5 °C while gradually decreases with further cooling. The maximal Q is 732 k, close to that measured at room temperature. Based on Eq. (2) and the simulated Q_{TED} (blue dots), we can calculate Q_{other} shown by the red dots, which change between 835 k and 1.02 M. These results indicate that the dominant damping mechanism is not just TED. Some other factors, such as the anchor loss caused by the asymmetry in-plane vibration of the (100) silicon resonator, should be considered seriously.

CONCLUSION

In this paper, a multiple folded-beam disk resonator with high Q_{TED} (more than 2 million) was demonstrated. The $n = 3$ modes of the (100) silicon resonator show good degeneracy. The quality factor of the fabricated resonator reached 716 k, proving to be larger than Q of most flexural silicon MEMS disk resonators. Moreover, conducted temperature experiment indicates that the TED is no longer the only dominant damping mechanism. As several researches have mentioned, the mode asymmetry of the resonator caused by the in-plane anisotropy of the (100) silicon may lead to the anchor loss, which limit the overall quality factor of this resonator. Future study on verifying and compensating the influence of the in-plane mode asymmetry of this resonator is under way.

ACKNOWLEDGMENTS

This work was supported by the Microsystem Technology Laboratory, National University of Defense Technology and Hunan MEMS Research Center. We wish to thank Qingsong Li and Peng Wang of National University of Defense Technology for their assistance in device characterization. And we also thank National Natural Science Foundation of China (grant number 51905539 and U21A20505), Young Elite Scientist Sponsorship Program by CAST (grant number YESS20200127), and the Natural Science Foundation of Hunan Province for Excellent Young Scientists (grant number 2021JJ20049).

REFERENCES

[1] Li, Q.; Xiao, D.; Zhou, X.; He, Y.; Wu, X. "0.04 degree-per-hour MEMS disk resonator gyroscope with high-quality factor (510 k) and long decaying time constant (74.9 s)", *Microsyst. Nanoeng.* 2018, *4*, 1–11.

[2] Challoner, A.D.; Ge, H.H.; Liu, J.Y. "Boeing disc resonator gyroscope", In Proceedings of 2014 IEEE/ION Position, Location and Navigation Symposium—PLANS 2014, Monterey, CA, USA, 5–8 May 2014; pp. 504–514.

[3] Zotov, S.A.; Trusov, A.A.; Shkel, A.M. "High-range angular rate sensor based on mechanical frequency modulation", *J. Microelectromech. Syst.* 2012, *21*, 398–405.

[4] Trusov, A.A.; Prikhodko, I.P.; Zotov, S.A.; Shkel, A.M. "Low dissipation silicon MEMS tuning fork gyroscopes for rate and whole angle measurements", *IEEE Sens. J.* 2011, *11*, 2763–2770.

[5] Wong, S.J.; Fox, C.H.J.; McWilliam, S.; Fell, C.P.; Eley, R. A preliminary investigation of thermo-elastic damping in silicon rings. *J. Micromech. Microeng.*

2004, *14*, S108–S113.

[6] Zotov, S.A.; Simon, B.R.; Prikhodko, I.P.; Trusov, A.A.; Shkel, A.M. "Quality Factor Maximization Through Dynamic Balancing of Tuning Fork Resonator", *Sens. J. IEEE*, 2014, *14*, 2706–2714.

[7] Senkal, D.; Ahamed, M.J.; Ardakani, M.H.A.; Askari, S. "Demonstration of 1 million Q-factor on micro-glass blown wineglass resonators with out-of-plane electrostatic transduction", *J. Microelectromech. Syst.* 2015, *24*, 29–37.

[8] Candler, R.N.; Duwel, A.; Chandorkar, S.A.; Hopcroft, M.A.; Park, W.-T.; Kim, B.; et al. "Impact of geometry on thermoelastic dissipation in micromechanical resonant beams", *J. Microelectromech. Syst.* 2006, *15*, 927–934.

[9] Zhou, X.; Xiao, D.; Wu, Q.; Hou, Z.; He, K.; Wu, Y. "Mitigating thermoelastic dissipation of flexural micromechanical resonators by decoupling resonant frequency from thermal relaxation rate", *Phys. Rev. Appl.* 2017, *8*, 064033.

[10] Xu, Y.; Li, Q.; Wang, P.; Zhang, Y.; Zhou, X.; Yu, L.; Wu, X.; Xiao, D. "0.015 Degree-Per-Hour Honeycomb Disk Resonator Gyroscope", *IEEE Sens. J.* 2021, *21*, 7326–7338.

[11] Ren, X.; Zhou, X.; Tao, Y.; Li, Q.; Wu, X.; Xiao, D. "Radially Pleated Disk Resonator for Gyroscopic Application", *J. Microelectromech. Syst.* 2021, *30*, 825–835.

[12] Zener, C. Internal friction in solids. *Proc. Phys. Soc.* 1940, *52*, 152–166.

[13] Benvenisty, E.; Elata, D. "Frequency Matching of Orthogonal Wineglass Modes in Disk and Ring Resonators Made From (100) Silicon", *IEEE Sens. Lett.* 2019, *3*, 1–4.

[14] Chang, C.-O.; Chang, G.-E.; Chou, C.-S.; Chien, W.-T.C.; Chen, P.-C. "In-plane free vibration of a single-crystal silicon ring", *Int. J. Solids Struct.* 2008, *45*, 6114–61132.

CONTACT

*Xin Zhou, tel: +86-0731-84574958; zhouxin11@nudt.edu.cn

A TIME-SERIES CONFIGURATION METHOD OF MODE REVERSAL IN MEMS GYROSCOPES UNDER DIFFERENT TEMPERATURE-VARYING CONDITIONS

Liangqian Chen, Tongqiao Miao*, Qingsong Li, Peng Wang, Junjian Li, Xuezhong Wu,
Dingbang Xiao, and Xiang Xi
College of Intelligence Science, National University of Defense Technology, Changsha, CHINA

ABSTRACT

Mode reversal is a real-time online self-calibration method for gyroscopes. This paper reports a timing optimization method of mode reversal in MEMS gyroscopes. It provides guidelines for the timing configuration of mode reversal at different temperature change rates. Working time optimization can reduce the zero-rate output (ZRO) fluctuation of the gyroscope in the process of rapid temperature change, and transition time optimization can reduce the influence of the output when the gyroscope is unstable.

KEYWORDS

MEMS gyroscopes; Mode reversal; Timing configuration

INTRODUCTION

For the past few years, micro-electro-mechanical system (MEMS) gyroscope has become widely used in autonomous navigation, unmanned systems, precision guidance and other fields due to its low cost, low power consumption, and small size [1]. Because of the limitation of processing technology, the performance of the MEMS gyroscope still has room for improvement. Mode reversal is a real-time online self-compensation method for gyroscope ZRO, which can eliminate most of the ZRO drift of the gyroscope structure and significantly improve the long-term ZRO stability of the gyroscope [2]. Northrop Grumman [3] realized the application of mode reversal on hemispherical gyroscopes. Analysis of ZRO for disc resonator gyroscopes operating of mode reversal has been published elsewhere [4,5].

In fact, for different application scenarios, the timing configuration of mode reversal will have a great impact on its self-calibration effect. However, there have been no relevant reports on the timing configuration of mode reversal. This paper will present the theory and experimental results of timing optimization of mode reversal, and provide guidelines of mode reversal for timing configuration of application scenarios under different temperature-varying conditions.

THEORETICAL ANALYSIS AND MODEL
Design model of the HDRG

The study of this paper is conducted on honeycomb disk resonator gyroscope (HDRG), which is mainly composed of a resonant structure and electrodes, as shown in Figure 1(a). The resonant structure includes two parts, an anchor and a frame structure, wherein the anchor point is fixedly connected with the base and plays a supporting role, and the frame structure is suspended in the air and plays the role of sensitive angular velocity. The electrodes

are fixedly connected to the substrate through anchor points, including electrodes outside the resonance frame (outer electrodes) and electrodes inside the resonance frame (inner electrodes). An equivalent parallel plate capacitance is formed between the electrode and the adjacent resonant structure. The vibration of the gyroscope can be sensed by detecting changes in the capacitance. The SEM of the HDRG is shown in Figure 1(b), and the detailed process flow for the preparation of the HDRG is reported in [6,7].

The HDRG in this paper works in the $n = 2$ degenerate modes as shown in Figure 1(c). The working frequency and the quality factor (Q) are shown in Figure 1(d). Due to the resonator asymmetry caused by fabrication error, the initial frequency split is about 0.183 Hz.

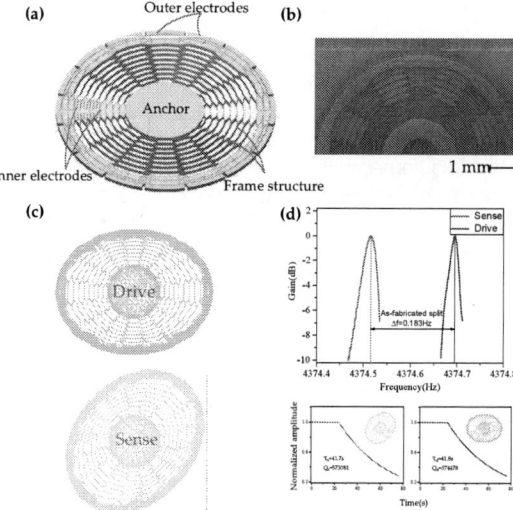

Figure 1: (a) The structure of HDRG. (b) SEM of HDRG. (c) Schematic diagram of the mode. (d) Resonant frequency and initial frequency difference of driving mode and sensing mode, and Ring-down curves of the driving and sensing modes of the HDRG.

Bias self-calibration based on the mode reversal

Mode reversal is a method for calibrating the gyroscope output by working on a periodic exchange between mode 1 and mode 2, as shown in Figure 2(a), α is the driving angle. When the resonator oscillates at the driving mode, its energy will be coupled through the Coriolis force into the sensing mode. And when the resonator oscillates at the sensing mode, the driving mode will sense the Coriolis force.

The two-degree-of-freedom equivalent system model of the HDRG with error is shown in Figure 2(b). The main error sources of HDRG are the structural imperfections which can cause the anisoelasticity and anisodamping. As shown in Figure 2(b), the anisoelasticity and anisodamping can lead to the unequal natural frequencies (ω_1 and ω_2) and principal damping time constants (τ_1 and τ_2). In the HDRG, when subjected to the angular velocity of the vertical plane, the direction of the Coriolis force and the vibration direction of the gyroscope have a determined orthogonal relationship. The action of the Coriolis force causes the sensing mode of the gyroscope to shift to achieve the detection of angular velocity. Therefore, the direction of the sensitive electrode of the sensing axis can be used as a reference. It is defined as the y-axis in Figure 2(b), and the x-axis orthogonal to it is the ideal driving axis direction. The stiffness and damping asymmetry of the resonant structure in the driving axis and the sensing axis make it have an orthogonal main stiffness axis ω_1-o-ω_2 and an orthogonal main damping axis τ_1-o-τ_2, which has a stiffness axis declination angle θ_τ and a damping axis declination angle θ_ω with the reference coordinate system x–o–y. Through the coordinate transformation between the main damping axis, the stiffness axis and the reference coordinate system, a centralized parameter dynamics model of the vibrating gyroscope can be established as [8]:

$$\begin{bmatrix} \ddot{x} \\ \ddot{y} \end{bmatrix} + \begin{bmatrix} \frac{2}{\tau} + \Delta\left(\frac{1}{\tau}\right)\cos 2\theta_\tau & \Delta\left(\frac{1}{\tau}\right)\sin 2\theta_\tau \\ \Delta\left(\frac{1}{\tau}\right)\sin 2\theta_\tau & \frac{2}{\tau} - \Delta\left(\frac{1}{\tau}\right)\cos 2\theta_\tau \end{bmatrix}\begin{bmatrix} \dot{x} \\ \dot{y} \end{bmatrix} +$$
$$\begin{bmatrix} \omega^2 + \omega\Delta\omega\cos 2\theta_\omega & \omega\Delta\omega\sin 2\theta_\omega \\ \omega\Delta\omega\sin 2\theta_\omega & \omega^2 - \omega\Delta\omega\cos 2\theta_\omega \end{bmatrix}\begin{bmatrix} x \\ y \end{bmatrix} =$$
$$\frac{1}{m_{eff}}\begin{bmatrix} \cos\alpha & -\sin\alpha \\ \sin\alpha & \cos\alpha \end{bmatrix}\begin{bmatrix} f_1 \\ f_2 \end{bmatrix} + 2nA_g\Omega\begin{bmatrix} \dot{y} \\ -\dot{x} \end{bmatrix} \quad (1)$$

The parameters of Eq. (1) satisfy the following relationship:

$$\omega^2 = \frac{{\omega_1}^2 + {\omega_2}^2}{2}, \quad \omega\Delta\omega = \frac{{\omega_1}^2 - {\omega_2}^2}{2},$$
$$\frac{2}{\tau} = \frac{1}{\tau_1} + \frac{1}{\tau_2}, \quad \Delta\left(\frac{1}{\tau}\right) = \frac{1}{\tau_1} - \frac{1}{\tau_2} \quad (2)$$

where m_{eff} is the equivalent mass of the HDRG, A_g is angle gain, Ω is the angular velocity exerted by the gyroscope in the z-axis direction, ω is the average natural frequency, ω_1 and ω_2 are the natural frequencies along the two stiffness axes, τ is the average attenuation time constant, τ_1 and τ_2 are the attenuation time constants along the two damping axes, Δ represents the difference.

The electronic and control scheme of mode reversal in the HDRG is shown in Figure 2(c). When HDRG is operating, the driving mode maintains a constant amplitude vibration along the driving axis through the automatic gain control (AGC) loop. The displacement along the driving axis can be expressed as :

$$x = |x|\cos\omega_d t \quad (3)$$

where $|x|$ and ω_d are the values of amplitude and frequency. The force-to-rebalance (FTR) loop [9] is designed for detecting angular velocity by solving the rebalanced force in the sensing mode. In the case where both the orthogonal force and the in-phase force are stabilized and closed-loop controlled by the feedback force, by substituting Eq. (3) into Eq. (1), the in-phase feedback force F_I can be obtained as:

$$F_I = f_2 = 2nA_g\Omega\omega_d|x|m_{eff} +$$
$$\left[-\Delta\left(\frac{1}{\tau}\right)\sin 2\theta_\tau\right]\omega_d|x|m_{eff} \quad (4)$$

Through Eq. (4), we can express the ZRO of HDRG as:

$$\Omega_{ZRO} = \frac{1}{2nA_g}\left[-\Delta\left(\frac{1}{\tau}\right)\sin 2\theta_\tau\right] \quad (5)$$

As shown in Figure 2(c), the virtual switches (VSs) in the FPGA are used to switch the AGC and FTR loops, thereby realizing the reversal of HDRG between mode 1 and mode 2. For the HDRG, if the driving axis is changed from x to y and the sensing axis is changed from y to x, the ZRO of the HDRG will be [10,11]:

$$\Omega_x = \frac{1}{2nA_g}\left[-\left(\Delta\frac{1}{\tau}\right)\sin 2\theta_\tau\right]$$
$$\Omega_y = \frac{1}{2nA_g}\left[\left(\Delta\frac{1}{\tau}\right)\sin 2\theta_\tau\right] \quad (6)$$

Adding the two ZRO gives:

$$\Omega_x + \Omega_y = 0 \quad (7)$$

Therefore, we believe that mode reversal suppresses most of the ZRO and its drift.

(a)
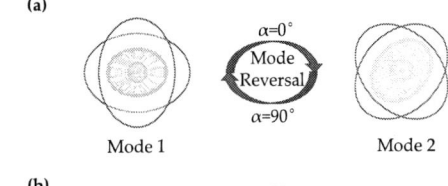

(b)
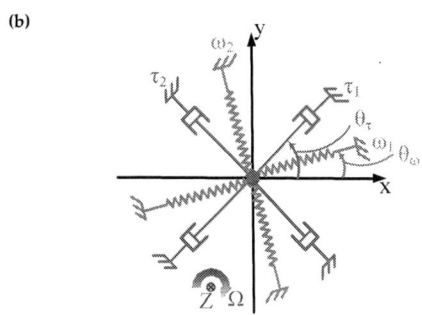

(c)
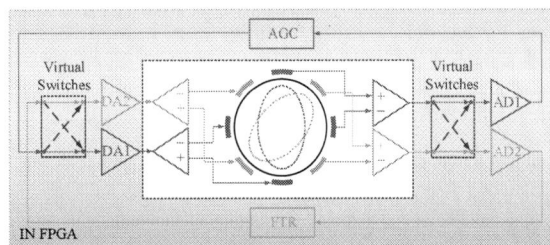

Figure 2: (a) Mode reversal Schematic. (b) Two-degree-of-freedom equivalent system model of the HDRG. (c) Electronic and control scheme of mode reversal in the HDRG.

For the actual ZRO and its drift of mode reversal, since the output of mode 1 and mode 2 cannot be obtained at the same time, there is a time-difference in the output between them, which we can express as [12]:

$$ZRO = \Omega_{x\_T1} + \Omega_{y\_T2}$$
$$ZRO_{drift} = \partial\Omega_{x\_T1} + \partial\Omega_{y\_T2} \quad (8)$$

where Ω_{x_T1} means the ZRO of the HDRG working in mode 1 when the time is $T1$, Ω_{y_T2} means the ZRO of the HDRG working in mode 2 when the time is $T2$.

This means that when the environment changes greatly in the time between T1 and T2, such as when the temperature changes rapidly, the self-calibration effect of mode reversal will deteriorate. Therefore, we need a method of selecting the appropriate timing configuration to ensure the effect of self-calibration.

Timing configuration of mode reversal

In practical applications, the timing configuration of mode reversal will greatly affect the effect of HDRG self-calibration. As shown in Figure 3(a), when the total test time length T_N is determined, the larger the value of the additional reversal times N, the more times the gyroscope can perform self-calibration, and the better the identification and compensation effect of mode reversal for rapid ZRO drift. However, the blind increase of the N value will introduce other measurement errors. The working time T_w represents the time when the gyroscope is working in mode 1 or mode 2 and can output normally. The transition time T_t represents the time when the output of the gyroscope is unavailable when switching from one mode to another. There are many ways to achieve the reversal of the two modes, but as shown in Figure 2(c), no matter which method, when switching VSs, the FTR and AGC loops need time to stabilize again. Therefore, the transition time is set up to ensure that the gyroscope can output normally after the mode is reversed.

Similarly, the ratio of the working time T_w and the transition time T_t will also affect the performance of the gyroscope. Therefore, we need to evaluate whether a timing configuration is appropriate through appropriate indicators. We can define the available data ratio (ADR) as:

$$ADR = \frac{T_w}{T_t} \qquad (9)$$

In practical applications, we generally require 1/ADR to be less than or equal to 1, to ensure that the gyroscope has sufficient output data. Similarly, we can express calibration time-difference (CTD) as:

$$CTD = T2 - T1 = T_w + T_t \qquad (10)$$

For a suitable timing configuration, we hope that the value of ADR is larger and CTD is smaller. Of course, we want the transition time T_t to be smaller, but due to the instability of the output of the gyroscope when the two modes are switched, we need to find a minimum T_t that does not affect the ZRO performance of the gyroscope. As shown in Figure 3(c), when T_t set too small, the gyroscope can't work stable before output. And when T_t set too large, the temperature field change will cause more ZRO drift. If we select an acceptable amount of drift, we can get a T_t selection interval, from T_{t_min} to T_{t_max}. Under normal circumstances, in order to obtain a larger ADR and a smaller CTD, we will choose T_{t_min} as the transition time.

When T_t is determined, then we need to select a suitable T_w. As shown in Figure 3(b), when the temperature changes rapidly, due to the CTD between the two modal outputs, the ZRO of the mode reversal will fluctuate. Of course, when the ADR is very small, the self-calibration effect of mode reversal will also deteriorate. As shown in Figure 3(d), if we select an acceptable amount of drift, we can get a T_w selection interval, from T_{w_min} to T_{w_max}. Next, in order to balance a larger ADR and a smaller CTD, we have to choose a suitable T_w as the working time according to the actual application needs. Usually in order to ensure the effect of self-calibration, the faster the

temperature change rates (TCR) is, the shorter CTD we want; and the slower the TCR is, the larger ADR we want.

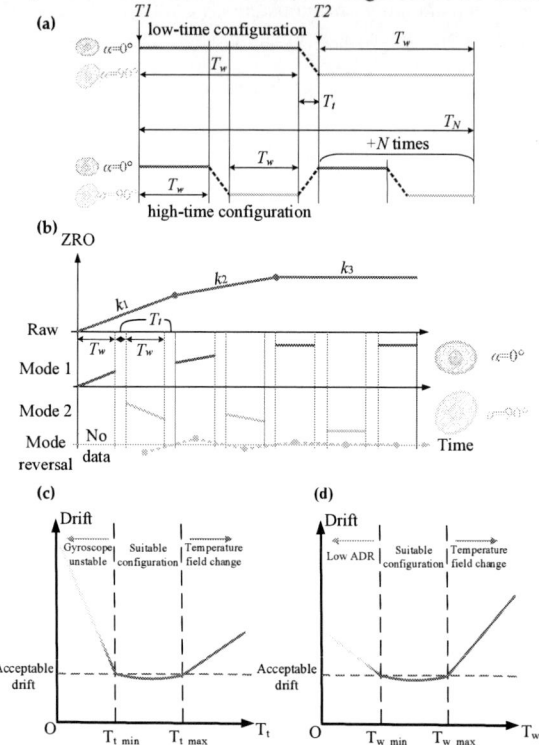

Figure 3: (a) Schematic diagram of mode reversal timing configuration principle. (b) Schematic diagram of rapid temperature drift effects on mode reversal. (c) Schematic diagram of the change of ZRO drift with the transition time and the configuration method. (d) Schematic diagram of the change of ZRO drift with the working time and the configuration method.

For different TCR, Schematic diagram of the configuration interval selection of T_t and T_w are shown in Figure 4(a) and Figure 4(b). It can be seen that the faster the TCR is, the shorter the time interval available for configuration, and the greater the possibility of incorrect output due to ADR and CTD.

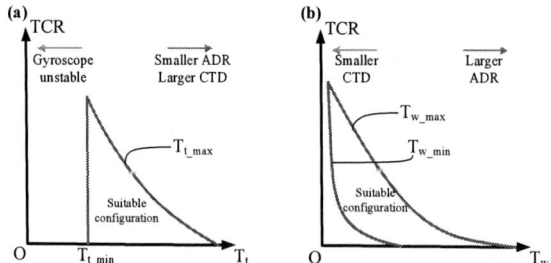

Figure 4: (a) Influence of different temperature change rates on T_t selection. (b) Influence of different temperature change rates on T_w selection.

EXPERIMENT

Experiment setting

According to the above timing configuration method, we optimize the timing configuration of the HDRG required by a given usage scenario. We will give the experimental results of the configuration of the transition time and working time respectively.

Transition time configuration

First configure the transition time, we set T_N and the N as fixed values, and we test the gyroscope with different transition times at room temperature.

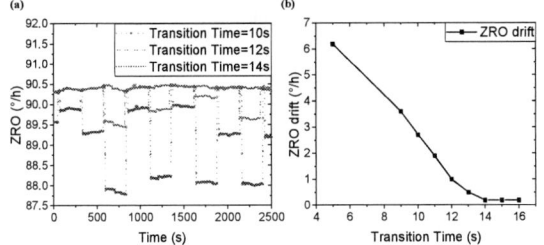

Figure 5: (a) The effect of different mode reversal transition times on ZRO. (b) The effect of different mode reversal transition times on ZRO drift.

Figure 5(a) shows ZRO with T_t of 10s, 12s, and 14s, respectively, and the data is averaged for 240s. Figure 5(b) shows the ZRO drift under different transition times, from which we can choose the optimal transition time T_{t_min} for this HDRG to be 14s.

Working time configuration

For working time configuration, we demonstrate the rapid cooling process from 30°C to 10°C using different timing configurations for the gyroscope with a temperature change rate of 5°C/min in temperature and humidity test chamber. As shown in Figure 6(a), we set T_t to 30s, and the figure shows the ZRO changes when T_w is 30s, 120s, and 210s. For these three different timing configurations, the relationship between 1/ADR, CTD and ZRO drift is shown in Figure 6(b). For such a fast rate of temperature change, the smaller the CTD, the smaller the ZRO fluctuation introduced. As we expected, the ZRO fluctuations were minimal at T_w of 30s.

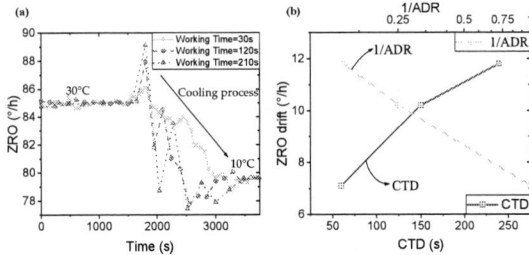

Figure 6: (a) The effect of different mode reversal working times on ZRO, the data is averaged for 150s. (b) When T_t as fixed values, the relationship between 1/ADR, CTD and ZRO drift.

CONCLUSION

In this paper, a timing optimization configuration method for mode reversal in MEMS gyroscopes is proposed. We not only establish the timing configuration model of the mode reversal at different temperature change rates, but giving evaluation indicators, optimized configuration methods and experimental verification. This method can improve the self-calibration effect of mode reversal in harsh environments, and providing guidance for the timing configuration of mode reversal in different practical application scenarios.

ACKNOWLEDGEMENTS

The authors would like to thank the National University of Defense Technology, China, for equipment access and technical support.

REFERENCES

[1] Wahlstrom, J., I. Fifteen Years of Progress at Zero Velocity: A Review. IEEE Sens. J., vol. 21, pp. 1139–1151, 2020.

[2] Miao, T.; Xu, Y.; Li, Q.; Sun, J.; Wu, X.; Xiao, D.; Hu, X. Temperature Drift Self-calibration for Honeycomb-like Disk Resonator Gyroscope. In Proceedings of the 2022 IEEE International Symposium on Inertial Sensors and Systems (INERTIAL), pp. 1–4, 2022.

[3] Meyer A. D. and Rozelle D. M., "Milli-HRG inertial navigation system," Proceedings of the 2012 IEEE/ION Position, Location and Navigation Symposium, pp. 24-29, 2012.

[4] Challoner A. D., Ge H. H. and Liu J. Y., "Boeing Disc Resonator Gyroscope," 2014 IEEE/ION Position, Location and Navigation Symposium - PLANS 2014, pp. 504-514, 2014.

[5] Miao T., et al., "Virtual Rotating MEMS Gyrocompassing With Honeycomb Disk Resonator Gyroscope," IEEE Electron Device Letters, pp.1331-1334, 2022.

[6] Xu, Y.; Li, Q.; Zhang, Y.; Zhou, X.; Wu, X.; Xiao, D. Honeycomb-Like Disk Resonator Gyroscope. IEEE Sens. J., pp.85–94, 2019.

[7] Xu, Y.; Li, Q.; Wang, P.; Zhang, Y.; Zhou, X.; Yu, L.; Wu, X.; Xiao, D. 0.015 Degree-Per-Hour Honeycomb Disk Resonator Gyroscope. IEEE Sens. J., pp.7326–7338, 2020.

[8] Lynch D., "Coriolis vibratory gyros," in Proc. Symposium Gyro Technology, pp. 1.0–1.14, 1998.

[9] Li, Q., et al. "Nonlinearity Reduction in Disk Resonator Gyroscopes Based on the Vibration Amplification Effect." IEEE Transactions on Industrial Electronics, vol. 67, pp. 6946-6954, 2020.

[10] Trusov, A.A., et al. Continuously self-calibrating CVG system using hemispherical resonator gyroscopes. In Proceedings of the 2015 IEEE International Symposium on Inertial Sensors and Systems (ISISS) Proceedings, pp. 1–4, 2015.

[11] Ge, H.H.; Liu, J.Y.; Buchanan, B. Bias self-calibration techniques using silicon disc resonator gyroscope. In Proceedings of the 2015 IEEE International Symposium on Inertial Sensors and Systems (ISISS) Proceedings, pp. 1–4, 2015.

[12] Chen L., Miao T., Li Q., Wang P., Wu X., Xi X., Xiao D., A Temperature Drift Suppression Method of Mode-Matched MEMS Gyroscope Based on a Combination of Mode Reversal and Multiple Regression. Micromachines, vol. 13(10), 2022.

CONTACT

*Tongqiao Miao, Email: miaotongqiao12@nudt.edu.cn

ACOUSTICALLY ISOLATED MEMS BAW GYROSCOPES

Diego Emilio Serrano[1]*, Amir Rahafrooz[1], Duane Younkin[1], Kieran Nunan[1], Mitul Dalal[1], Sagnik Pal[1], and Ijaz Jafri[1]

[1]Panasonic Device Solutions Laboratory of Massachusetts, Marlborough, MA, USA

ABSTRACT

This work reports on a new acoustically isolated MEMS bulk-acoustic wave (BAW) gyroscope, which provides an average improvement in zero-rate offset (ZRO) variation of ~7X (over a range of -55 °C to 110 °C) with respect to a baseline design with no isolation structure. The design consists of a wafer-level packaged resonant disk implemented in the device layer of a silicon-on-insulator (SOI) wafer, with a physically separated silicon region in the handle layer to prevent any coupling between the external package and the resonator. The isolation volume dimensions, defined by backside trenches, were optimized through a combination of finite element analysis (FEA) simulations and characterization results using a design of experiments (DOE). The fabricated baseline devices show great noise performance with angle white noise (AWN) of 0.021 "/√Hz, angle random walk (ARW) of 0.017 °/√h, and Allan Deviation (ADEV) values in the bias instability (BI) region as low as 0.25 °/h (0.4 °/h in average), which are projected to be the same for the isolated designs.

KEYWORDS

MEMS, Gyroscope, Bulk-Acoustic Wave.

INTRODUCTION

The success of fully autonomous driving systems will heavily depend on devices that can provide extremely accurate vehicle positioning information despite unfavorable environmental conditions such as weather, visibility, vibration/shock caused by terrain, and ambient or cabin temperature fluctuations. It is expected that global navigation satellite systems (GNSS) will be at the heart of these applications; however, limited signal coverage in densely populated and forested areas will require alternative systems such as ultrasound sensors, LiDAR, cameras, and inertial measurement units (IMUs) to serve as a reliable backup when satellite coverage is absent.

Given their excellent noise and vibration-rejection performance, MEMS BAW gyroscopes are well positioned to play a critical role in the implementation of IMUs for driverless cars. However, since these devices operate at high frequencies (in the MHz range), their modes of vibration can easily leak into the substrate and couple with the external package, which can adversely affect their temperature behavior [1]. Previously reported substrate-decoupled MEMS BAW gyros addressed this limitation by implementing a structure within the resonator to isolate the anchor with the substrate from the modes of vibration [1-3]. This has proven to be a very successful technique, but small imperfections during fabrication can cause residual amounts of energy to couple and propagate in the substrate, causing fluctuations in offset as a function of changes in the boundary conditions of the device.

For example, Figure 1 shows the substrate-decouple BAW gyro used in this work, which consists of a disk

implemented in a 40-μm thick (100) single-crystal silicon device layer, surrounded by polysilicon electrodes that provide capacitive electrostatic transduction via narrow capacitive gaps. This device was fabricated on silicon-on-insulator (SOI) wafers using the HARPSS process [4] with capacitive gaps between the resonator and the fixed electrodes of 270 nm. The trenches towards the center of the disk, which are self-aligned with its perimeter, correspond to the substrate decoupling structure that isolates the MEMS vibration from the underlying handle layer. The polysilicon anchor in the middle connects the resonator to the handle layer. Since this anchor is defined using a separate mask, there could be a slight misalignment with the disk, which can cause enough coupling with the substrate to affect the ZRO performance of the gyro, particularly over temperature. To mitigate this issue, in this paper we propose an additional level of isolation by separating the handle layer region right underneath the resonator using DRIE trenches in the backside of the device, and therefore preventing external disturbances from coupling with the acoustic wave of the resonance modes used for operation of the disk as a gyroscope sensor. The dimensions of the isolation region were optimized through a combination of FEA simulations and characterization results through a DOE that utilizes a simple square geometry that was an easy add-on to the fabrication process.

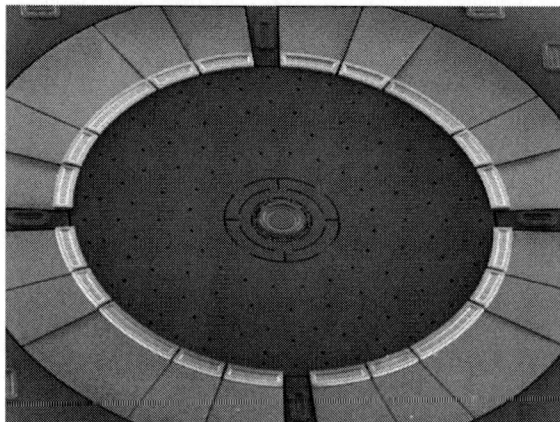

Figure 1. SEM image of bulk-acoustic wave (BAW) gyroscope. Trenches towards the center of the disk correspond to the decoupling structure that isolates the disk from its anchor.

The concept of modifying the handle layer to improve substrate dependencies for Q enhancement has been presented in SiC devices by use of a phononic crystal [5]. Our work builds on that idea by using a simpler and manufacturable approach in single-crystal silicon MEMS gyroscopes. Moreover, our solution focuses on improving how external stress affects product-specific parameters such as offset.

BACKSIDE TRENCH DESIGN

Figure 2 shows a 3D cross-section schematic of the SOI stack on which the MEMS BAW gyroscope is implemented, with and without the backside trench on the handle layer. In the case of the isolated design, the bottom attachment layer (represented in yellow) is only incorporated in the perimeter region of the die, limiting the surface area of handle layer that is physically attached to the underlying substrate.

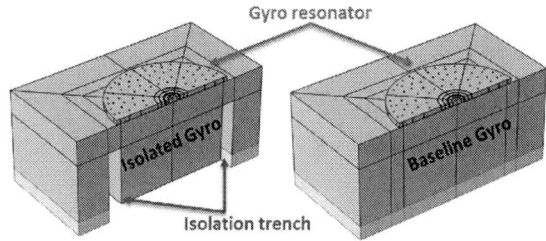

Figure 2. 3D cross-sectional schematic of MEMS gyro with and without a backside trench on the handle layer. Cap wafer that seals the device in a vacuum environment is not depicted in these diagrams.

For devices where the resonator decoupling structure does not fully isolate the design from the handle layer due to small process imperfections, the anchor acts as a source point of an acoustic wave that can travel throughout the substrate [6], reflect once it reaches the boundaries, and interfere with itself. This wave will be mostly generated by the displacement of the drive mode of the gyroscope, but since interference can happen in all directions throughout the bulk of the material, part of this signal will make it back to the source point and affect the sense mode of the resonator. If boundary conditions are changing unpredictably over temperature or stress, this behavior will be also observed in the ZRO at the output of the gyro sensor. By isolating the center region of the handle layer with a backside trench (as shown in left diagram of fig. 2), the interference effects experienced by the traveling wave will be contained within the single-crystal silicon region, making its behavior independent of the boundary conditions determined by both the non-linear material properties of the attachment, and conditions of the underlying layer to which the die is affixed to.

The efficiency of how well contained the acoustic wave is inside the center region is a function of the trench depth and the isolation area. FEA simulations were used to select the optimal trench parameters, and results were compared with the baseline model (no trench). In these simulations, the gyroscope drive mode is excited into oscillation and the ZRO from the sense mode is monitored as a function of temperature. Non-linear material properties were modeled between two extreme values; as seen in figure 3, an optimal isolated design shows significant reduction in hysteresis with respect to the baseline model.

IMPLEMENTATION

MEMS BAW gyroscopes with $n = 3$ degenerate modes at 4.9 MHz were implemented on 8" SOI wafers with a device layer thickness of 40 μm, buried oxide of 2 μm, and handle layer of 700 μm. The devices were fabricated using

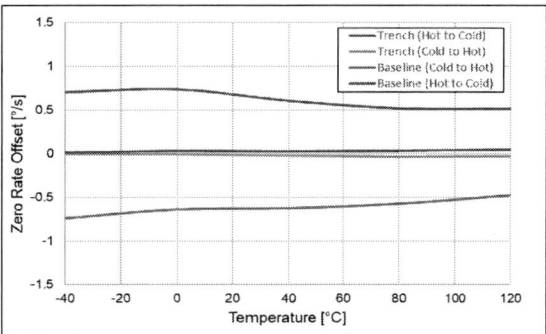

Figure 3. FEA simulation of ZRO vs. temperature (after linear compensation) for baseline model and a backside trench model with optimized dimensions. Hysteresis is significantly reduced when adding the isolation region.

the HARPSS process [4] with capacitive gaps of 270 nm. After processing the base wafers, these were wafer-level packaged with silicon caps using fusion bonding, which differs significantly from the gold-silicon eutectic bond method that has been used in previous generations of this type of devices [1-3]. The final cavity pressure level after bonding is below 0.5 Torr, making thermoelastic damping the limiting source of loss, which for this resonator design yields an average quality factor at room temperature of ~350,000 with standard deviation of only 1,600 across several wafers. Through-silicon vias (TSV) with very low parasitic resistance and capacitance carry the signals from the substrate electrodes to metalized pads on top of the packaged device, which are used to wirebond the device to the interface ASIC. The isolation region is then defined on the backside of the wafer using DRIE trenches with a depth of around 70% of the handle layer thickness. Figure 4 shows an optical picture of the front and back of the final packaged device. The front image (a) shows the metalized pads for electrical connection, and the back image (b) shows one of a series of backside trench designs used for the design of experiments (DOE), in which both the surface area and depth of the isolation region were varied. Figure 5 shows an infrared (IR) image of the same design where it can be seen where the isolation region sits with respect to the resonator design.

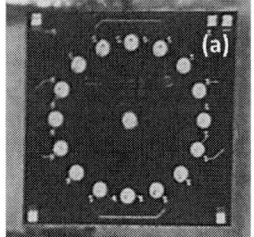 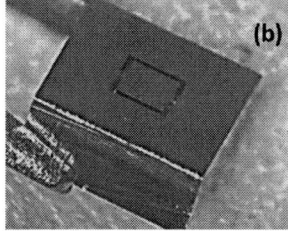

Figure 4. Optical image of (a) front and (b) back of one DOE design for the study of ZRO vs. temperature.

After fabrication, wafers were diced, and gyro sensors were packaged with the corresponding interface electronics. Figure 6 shows the full system, which includes an oscillator loop that regulates the amplitude of vibration of the drive mode utilizing an automatic gain control (AGC) circuit and a variable gain amplifier (VGA). The sense electronics include a frontend transimpedance amplifier (TIA), a programable gain amplifier (PGA) with

978-1-6654-9309-3/23 $31.00 ©2023 IEEE

Figure 5. IR image of fabricated BAW MEMS gyro with backside trench.

feedthrough (FT) cancellation capabilities, followed by a demodulator. The quadrature output of the demod is zeroed out by utilizing a quadrature cancellation loop (QCL). The in-phase output is regulated by means of force-to-rebalance (FTR), which also extends the open-loop bandwidth of the MEMS of about 7 Hz to 800 Hz. The inset in figure 6 shows the MEMS die stacked and wirebonded with its ASIC in the final package configuration (metal lid removed).

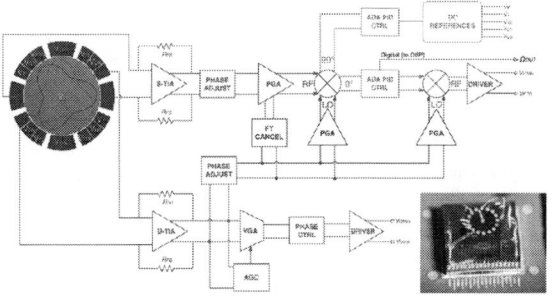

Figure 6. System block diagram of closed-loop BAW gyro system. Inset shows optical image of MEMS dice stacked on and wire-bonded to interface ASIC.

CHARACTERIZATION RESULTS

Figure 7 shows a comparison of the ZRO response of both baseline and isolated devices over the temperature range of -55 °C to 110 °C. It can be clearly seen how adding

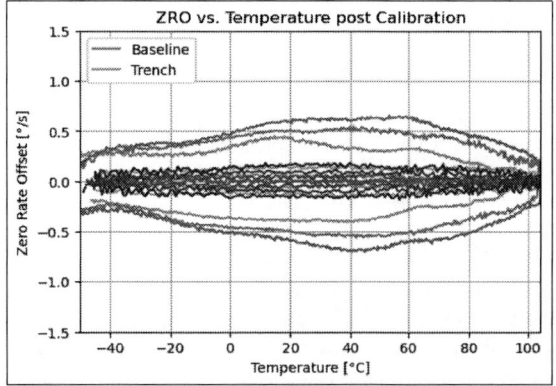

Figure 7. ZRO vs temperature comparison between baseline MEMS gyros (red-shaded traces) and MEMS gyros with isolation (blue-shaded traces).

the isolation region alleviates both the non-monotonic and hysteretic behavior that arises when the MEMS die is directly attached to the underlying layer. The calculated hysteresis (max - min ZRO value across temperature) for a total of 32 parts are summarized in figure 8. The results show that the baseline devices have average value and standard deviation of 1.5 °/s and 0.88 °/s, respectively. In contrast, gyros with isolation trench show an average of 0.22 °/s (~7X better than without isolation), and a deviation of 0.15 °/s (~6X better).

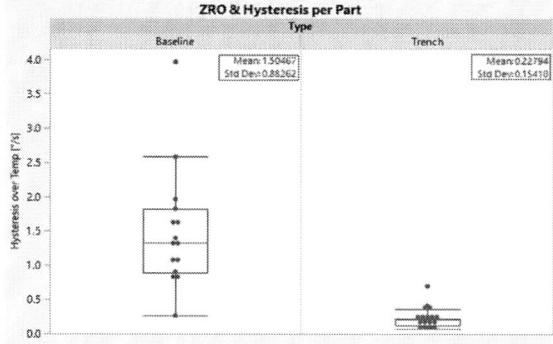

Figure 8. Summary of hysteresis over temperature results for 32 parts. Mean and standard deviation summarized inside green boxes. Devices with isolation show ~7X better average hysteresis, and ~6X better standard deviation.

Figure 9 shows a comparison of the Allan deviation (ADEV) for parts used in this temperature study. Unfortunately, these devices came from an earlier batch of wafers where the capping process was suboptimal. This resulted in higher pressure levels inside the MEMS cavity, which translates into lower Q factors. Therefore, the overall noise summarized in figure 9 is not on par with the latest generation of gyros. The critical conclusion to be drawn from this plot is that, on average, the noise for the isolated parts is identical to the baseline devices, which means that adding the isolation trench has no impact on noise performance. Furthermore, because the temperature

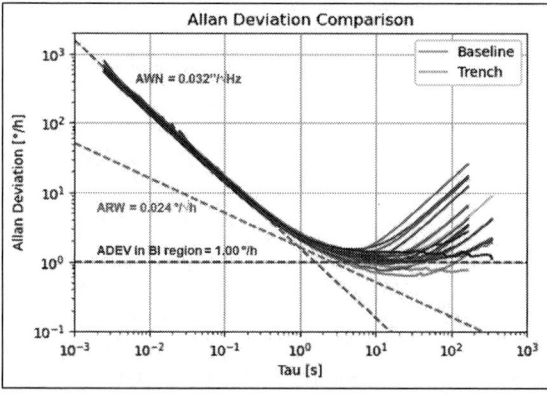

Figure 9. ADEV of baseline MEMS gyros (red-shaded traces) and isolated MEMS gyros (blue-shaded traces) for previous generation of capping process (lower Q values). Dashed lines represent the AWN (0.032 "/√Hz), ARW (0.024 °/√h) and ADEV in the BI region (1.0 °/h) for the averaged out over all parts.

stability of the isolated structures is significantly better than that of the baseline devices, figure 9 also shows that there is a large improvement in the drift ramp region for the new isolated designs.

To demonstrate the best possible noise performance these MEMS BAW gyros can achieve, the Allan deviation of 30 high Q baseline devices was collected at an unregulated room temperature and summarized in figure 10. The results show an angle white noise (AWN) of 0.021 $''/\sqrt{Hz}$, and an angle random walk (ARW) of 0.017 $°/\sqrt{h}$. The average Allan deviation in the bias instability (BI) region is around 0.4 $°/h$, but values as low as 0.25 $°/h$ have been recorded. Alas, wafers with both backside trenches and optimal bonding conditions had not been received at the time of writing, but as explained above, it is perfectly reasonable to expect that the isolated designs will have the same noise performance with improved drift.

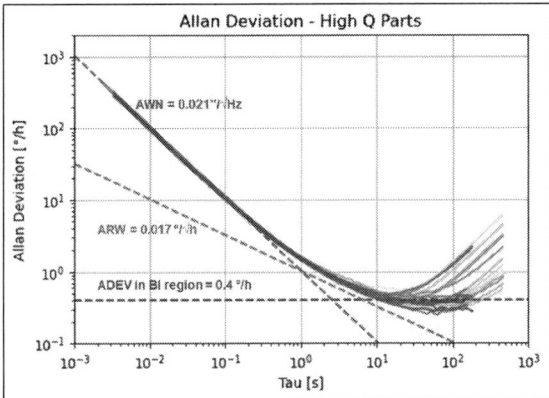

Figure 10. ADEV of high Q baseline MEMS gyros. Dashed lines represent the average AWN (0.021 ''/√Hz), ARW (0.017 °/√h) and ADEV in the BI region (0.4 °/h). Lowest recorded values around 0.25 °/h.

CONCLUSION

In this paper we demonstrated a new type of isolation method to improve the temperature performance of bulk-acoustic gyroscopes by adding a backside trench to the handle layer to contain any acoustic wave that can leak from the resonator and couple to the environment. We also showed that adding this trench has no impact on noise performance, providing an optimal solution for rugged, low noise gyroscope systems.

REFERENCES

[1] D.E. Serrano, M.F. Zaman, A. Rahafrooz, P. Hrudey, R. Lipka, D. Younkin, S. Nagpal, I. Jafri and F. Ayazi, "Substrate-decoupled, bulk acoustic wave gyroscopes: Design and evaluation of next-generation environmentally robust devices.", *Microsystems & Nanoengineering*, 2(1), 2016, pp.1-10.

[2] H. Wen, A. Daruwalla, C.S. Liu and F. Ayazi, "A Hermetically-Sealed 2.9 MHz N=3 Disk BAW Gyroscope with Sub-Degree-Per-Hour Bias Instability.", in *2020 IEEE 33rd International Conference on Micro Electro Mechanical Systems (MEMS)*, 2020, pp. 741-744.

[3] D.E. Serrano, A. Rahafrooz, R. Lipka, D. Younkin, K. Nunan, J. English, C. Chen, R. Hennessy, Y. Jeong, E. Ivanov, D. Sullivan and I. Jafri, "0.25 deg/h Closed-Loop Bulk Acoustic Wave Gyroscope", in *2022 IEEE International Symposium on Inertial Sensors and Systems (INERTIAL)*, 2022, pp. 1-4.

[4] F. Ayazi and K. Najafi, "High aspect-ratio combined poly and single-crystal silicon (HARPSS) MEMS technology," in *Journal of Microelectromechanical Systems*, vol. 9, no. 3, Sept. 2000, pp. 288-294.

[5] J. Yang, B. Hamelin and F. Ayazi, "Investigating Elastic Anisotropy of 4H-SiC Using Ultra-High Q Bulk Acoustic Wave Resonators," in *Journal of Microelectromechanical Systems*, vol. 29, no. 6, Dec 2020, pp. 1473-1482.

[6] Z. Hao, F. Ayazi. "Support loss in the radial bulk-mode vibrations of center-supported micromechanical disk resonators.", *Sensors and Actuators A: Physical* 134.2, 2007, pp 582-593.

CONTACT

*D.E Serrano, diego.serrano@us.panasonic.com
S. Hiraoka, tel: +81-80-9934-6216; hpg@ml.jp.panasonic.com

ACTIVE QUALITY FACTOR STABILIZATION OF MEMS RESONATOR UTILIZING ELECTRICAL DISSIPATION REGULATION

Yang Zhao, Qin Shi, Guoming Xia, and Anping Qiu

Nanjing University of Science and Technology, Nanjing, CHINA

ABSTRACT

This paper reports an active Q-factor stabilization approach utilizing electrical dissipation regulation for the first time. A bias current regulation loop is proposed to adjust the input impedance of the charge-sensitive-amplifier (CSA) automatically and stabilize the Q-factor of the MEMS resonator without altering its resonant frequency. After adopting the proposed stabilization loop, the Q-factor is stable at a relatively high point of 29.5k and its variation from -40°C to +60°C is reduced from 86.5% to 0.5%. The proposed approach requires no additional tuning structures or electrodes and applies to all MEMS resonators.

KEYWORDS

Quality factor, active stabilization, electrical dissipation, MEMS resonator, temperature variation.

INTRODUCTION

Q-factor is a fundamental parameter of the resonator, which suffers from strong and nonlinear dependence on temperature, whose typical TCO could be up to thousands of ppm/°C [1]. A stable and high Q-factor is essential for MEMS resonators and gyroscopes to maintain excellent frequency stability and low-drift under a wide temperature range.

On-chip ovenization is a common method in MEMS resonators to ensure a sub-ppm level resonant frequency stability [2], which also works for Q-factor stability. But it requires additional temperature sensing and adjustment elements as well as specific heat insulation mechanism, which consumes more power and cost.

Q-factor is traditionally limited by mechanical losses [3], such as air damping, anchor damping, and thermoelastic dissipation (TED). From this point of view, it's almost impossible to adjust the Q-factor after manufacturing unless focused ion beam trimming is employed [4]. On the other hand, electrostatic Q-factor tuning is emerging in MEMS gyroscopes to realize selective Q-factor tuning between its driving and sensing modes for drift suppression in the degenerate type or shock and vibration robustness in the nondegenerate type.

[5] utilizes mode coupling between anti-phase and in-phase modes to tune Q-factor. A DC voltage softens the inside proof mass and adds squeeze film damping (SFD) to realize Q-factor tuning. [6] reported an electrostatic anchor loss tuning method by introducing imbalanced stiffness between two proof masses via electrostatic softening effect. [7] introduce the electrical dissipation based Q-tuning method to MEMS rate integrating gyro. An external resistor in the motional current to voltage conversion path works as an electrical damper converting part of mechanical energy to heat. The polarization voltage is tuned to control the electrical dissipation for Q-factor

tuning. [8] present a similar method for disk resonator gyro.

The above electrostatic Q-factor tuning approaches provide a new idea for Q-factor stabilization. If the electrical dissipation can be properly and continuously regulated, then the temperature fluctuation of the Q-factor can be corrected automatically. [9] tunes the Q-factor via anchor loss modulation and an active temperature to tuning voltage loop is implemented reducing the Q-factor thermal variation from 25% to 0.3%. [10] presents a passive Q-stabilization utilizing electrical dissipation. Intrinsic Q-factor proportional to T^{-1} is carefully designed and the inherent sensing capacitor is employed as a Q-tuning capacitor. An external resistor with opposite TCO is integrated into the ASIC improving its thermal variation from 56.4% to 0.1%.

This paper reports an active Q-factor stabilization approach utilizing electrical dissipation regulation for the first time. The inherent input impedance of CSA is employed as the damping source instead of the additional damper in conventional approaches [7,8]. A bias current regulation loop is proposed to adjust the input impedance automatically and stabilize the Q-factor of the MEMS resonator without changing its resonant frequency. Compared to the passive stabilization we reported in [10], the specific TCO design of the Q-factor is not mandatory and the Q-factor reduction is avoided. Besides, no dedicated tuning structure and electrode is required in contrast to the active stabilization in [9], making the proposed method applicable to all kinds of MEMS resonators.

ENERGY DISSIPATION MECHANISMS IN MEMS RESONATOR

Q-factor of a resonator is defined as the ratio of the stored energy over the dissipated energy per vibration cycle. It suggests the Q-factor is inversely proportional to the energy loss in a resonant system. Various energy dissipation mechanisms exist in MEMS resonators, such as air damping, thermoelastic dissipation, anchor loss, and electrical dissipation, and the total Q-factor due to the above contributions becomes

$$\frac{1}{Q_T} = \frac{1}{Q_M} + \frac{1}{Q_E}, \qquad \frac{1}{Q_M} = \frac{1}{Q_{air}} + \frac{1}{Q_{TED}} + \frac{1}{Q_{anchor}} + \dots \quad (1)$$

where Q_{air} arises from the collision of gas molecules with the resonator, Q_{TED} is due to thermoelastic dissipation caused by strain-gradient-induced heat transfer, Q_{anchor} is caused by the energy dissipation via elastic wave propagation into the supporting bath and Q_E comes from the electrical dissipation when readout circuit and polarization voltage are considered.

Q_M in (1) is a well-known effect in MEMS resonators and is related to temperature-dependent parameters such as pressure, thermal conductivity, specific heat, thermal expansion coefficient, and so on. Fig.1 schematically presents the origin of Q_E, which is utilized to compensate the temperature variation of Q_M in (1).

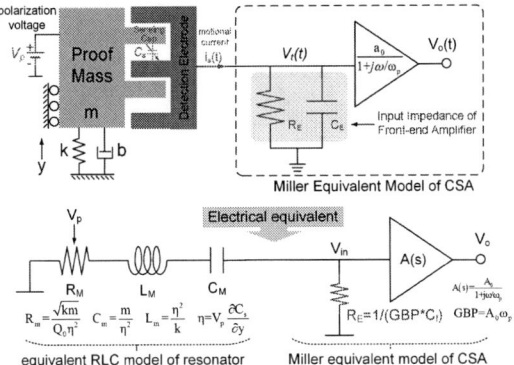

Figure 1: Schematic of electrical equivalent model of MEMS resonator with CSA.

The MEMS resonator can be modeled as a R_M-L_M-C_M network, with expressions in (2).

$$R_m = \frac{\sqrt{km}}{Q_M \eta^2} \quad C_m = \frac{m}{\eta^2} \quad L_m = \frac{\eta^2}{k} \quad \eta = V_p \frac{\partial C_s}{\partial y} \quad (2)$$

where Q_M is the intrinsic quality factor of the resonator without electrical dissipation and m, k, C_s, and V_p represent its mass, stiffness, capacitance, and polarization voltage, respectively.

R_E stands for the input impedance of CSA, which introduces additional resistive loading of the resonator and generates part of mechanical energy into heat, resulting in a loaded Q-factor as given in (3).

$$Q_T = Q_M \frac{R_m}{R_m + R_E} = \frac{\sqrt{km}}{\eta^2} \frac{1}{R_m + R_E} \propto \frac{1}{V_p^2} \quad (3)$$

Define A_0 and ω_p as the DC gain and dominant pole frequency of Opam in CSA. C_f is the feedback capacitance of CSA and R_b provides the DC feedback path for it, as shown in Fig.2. Using Miller's equivalent, the input impedance of closed-loop CSA can be written as

$$Z_{in}(s) = \frac{R_b}{1 + R_b C_f j\omega} \Big/ \frac{a_0}{1 + j\omega/\omega_p} \quad (4)$$

The real part in (4) stands for resistive impedance, R_E, which can be derived as

$$R_E = \frac{R_b \left(1 + R_b C_f \omega^2 / \omega_p\right)}{a_0 \left[1 + \left(R_b C_f \omega\right)^2\right]} \approx \frac{1}{a_0 \omega_p C_f} = \frac{1}{GBP \times C_f} \quad (5)$$

(5) and (3) indicate the input impedance of CSA and the Q_T can be regulated by the GBP of the Opam. Furthermore, the inherent temperature variation of Q_M can be compensated via a proper regulation of the GBP as long as $R_M + R_E$ maintains constant.

ACTIVE Q-FACTOR STABILIZATION APPROACH
GBP Regulated Opam in CSA

As shown in Fig.2, a typical two-stage Miller Opam is employed in CSA, in which the bias current of the input stage consists of two branches, i.e., I_{B1} and I_{B2}.

The GBP of two-stage Miller Opam is

$$GBP = \frac{g_{m1}}{C_c} = \frac{\sqrt{2\mu_n C_{ox}(W/L)_1 (I_{B1} + I_{B2})}}{C_c} \quad (6)$$

where I_{B1} provides the initial bias for the Opam and I_{B2} offers fine regulation of the GBP. Therefore, the Q-factor can be controlled by V_c to fulfill the proposed active stabilization.

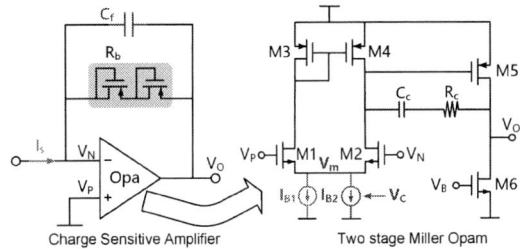

Figure 2: Schematic of CSA and two stage Miller Opam.

The R_E of CSA with different bias current, $I_B = I_{B1} + I_{B2}$ and GBP is simulated in Fig.3. The input impedance near DC is majorly decided by the feedback resistance R_B and the A_0 of the Opam. In the frequency range of 1~100kHz, the input impedance of the CSA is GBP and C_f dependent, which coincides with the theoretical analysis in (5). R_E at this frequency range varies from 300Ω to 2.4kΩ with a GBP from 3MHz to 24MHz. Since the resonant frequency of the studied MEMS resonator is 18kHz, such a GBP variation won't alter the CSA's I/V conversion gain.

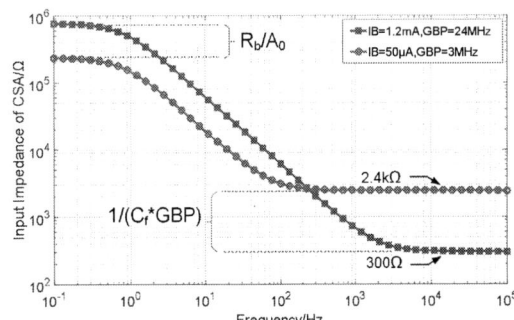

Figure 3: Simulated input impedance of CSA with 4pF feedback capacitance, C_f, versus bias current and GBP.

Active Stabilization Loop

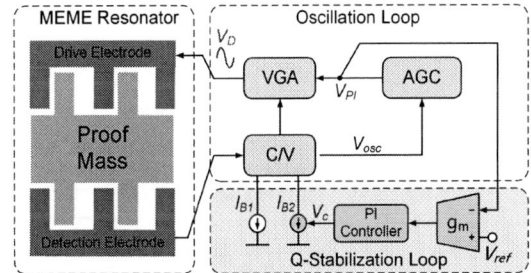

Figure 4: Block diagram of a MEMS oscillator with the proposed active Q-factor stabilization loop.

Fig.4 presents the proposed active Q-factor stabilization loop, which is embedded into a conventional amplitude-controlled MEMS oscillator. The output voltage of AGC, V_{PI}, is exploited as an instant measure of Q-factor, which is compared with a preset reference, V_{ref}. The error between V_{PI} and V_{ref} is converted into current and flows through a passive PI controller to generate a feedback control voltage, V_C, to adjust the bias current, I_{B2}, of Opam.

The implementation of the bias current in Opam is shown in Fig.5. A bandgap source provides the constant I_{B1} and I_{B2} is controlled by the error voltage, V_c.

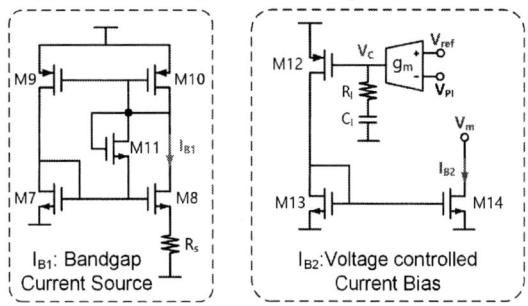

Figure 5: Schematic of two bias current branches in Opam.

Q-factor Tuning Range and Stable Point

The active Q-factor tuning range depends on the initial value of I_{B1} and the maximum current provided from I_{B2}, which decide the upper and lower limit of R_E for Q-factor tuning, i.e., R_{E1} and R_{E2}.

$$\Delta R_E = R_{E1} - R_{E2} = \frac{C_c/C_f}{\sqrt{2\mu_n C_{ox}(W/L)_1}}\left[\frac{1}{\sqrt{I_{B1}}}-\frac{1}{\sqrt{(I_{B1}+I_{B2})}}\right] \quad (7)$$

For a given bias current condition, the tuning range of R_E can be shifted by C_f. To compensate the inherent temperature variation of Q_M, ΔR_E shall be greater than the variation of R_m. Define α as the TCO of Q_M and ΔT as the temperature variation range, the following (8) shall be satisfied.

$$\Delta R_E > \left|\alpha R_m \Delta T\right| = \left|\alpha\frac{\sqrt{km}}{Q_M\eta^2}\Delta T\right| \quad (8)$$

To not introduce extra phase shift and gain loss, the input impedance of CSA typically needs to be designed below 10kΩ. For a given ΔR_E, (8) can be realized by selecting a proper V_p to decrease the electrical equivalent R_m to fall within the effective tuning range of ΔR_E.

Since the electrical dissipation generated by R_E always decreases the Q_M, the stable point of Q_T can only be close to its intrinsic value at the highest operating temperature point, T_{max}. As suggested by (3), the stable point of Q_T is

$$Q_{T,stable} = Q_{M,T_{max}}\frac{R_{m,T_{max}}}{R_{m,T_{max}}+R_{E2}} \quad (9)$$

Obviously, a smaller R_{E2} results in less loss of the intrinsic Q-factor, and the lower limit of R_{E2} depends on the maximum consumed current by $I_{B1}+I_{B2}$. In this work, the maximum bias current is designed to be 1.2mA and the initial value of I_{B1} is 50µA. The corresponding tuning range of R_E is thus 300Ω to 2.4kΩ, as simulated in Fig.3.

Fig.6 gives the simulated transient response of bias current when I_{B2} is enabled at 0.1s. The polarization voltage of the MEMS resonator is 10V and the corresponding R_m is 2kΩ with an intrinsic Q_M=50k. Since the designed tuning range of R_E is 2.1kΩ, a 105% temperature variation of R_m can be compensated. Before I_{B2} is enabled, the Q-factor is reduced to 23k with the loading effect of R_{E1}, and the total bias current converges to different points after 0.1s when setting different target Q-factor, i.e., V_{ref}. The simulation results prove the proposed active Q-stabilization loop is operational.

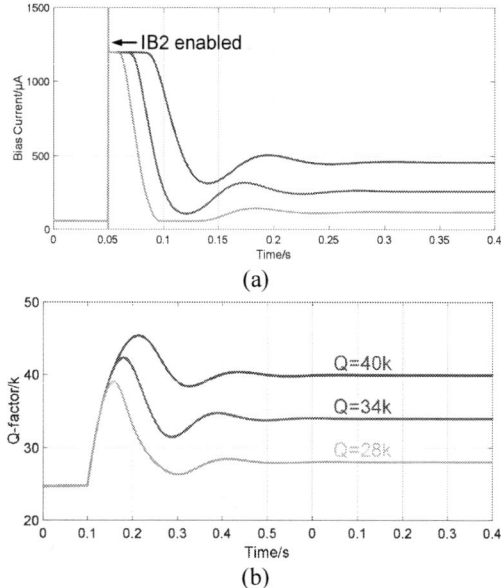

Figure 6: Simulated transient response of (a) bias current $I_{B1}+I_{B2}$ and (b) Q-factor when active stabilization is enabled at 0.1s with different target Q-factor.

EXPERIMENTAL RESULTS

The intrinsic temperature dependence of Q_M is evaluated first. Therefore, a low polarization voltage, V_p=5V, and R_E=0.3kΩ are adopted to reduce the electrical dissipation effect. The tested intrinsic Q_M is 46k with 86.5% thermal variation from -40°C to +60°C, as shown in Fig.7(a).

To implement the active stabilization, V_p=10V is then adopted to lower down its equivalent R_m to 2kΩ. As shown in Fig.7(b), the Q-factor variation is improved to 0.5% when the active stabilization loop is enabled. Moreover, the Q-factor stable point is 29.5k, which is close to the intrinsic value of Q_M at +60°C, which agrees with the design intent.

Fig.7(c) compares the resonant frequency difference versus temperature with and without the active stabilization loop with an identical V_p=10V. It proves the Q-factor tuning based on the regulation of bias current in Opam is frequency independent. The resonant frequency difference between these two conditions is less than 0.4Hz, which may attribute to the instability of environment temperature control. Because the TCO of resonant frequency in the studied MEMS resonator is 25ppm/°C, only a 1°C-temperature difference between the two experimental groups could result in a 0.45Hz frequency shift.

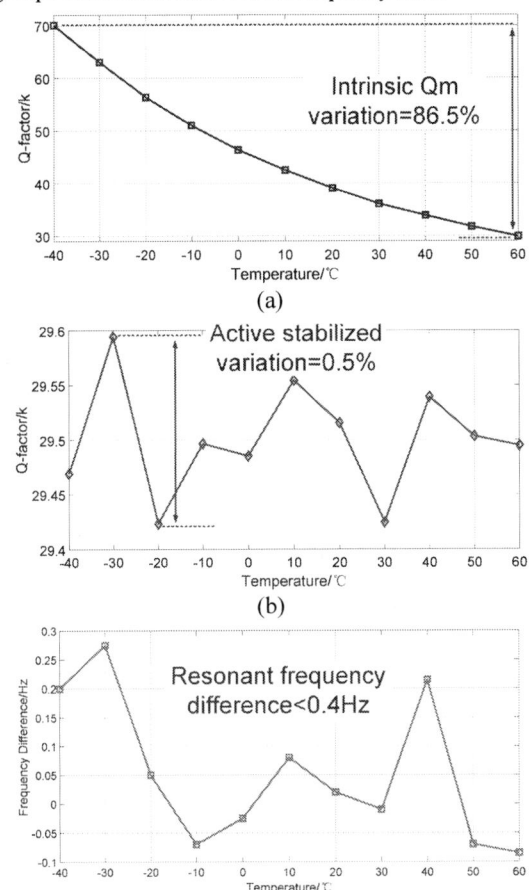

Figure 7: Tested (a) intrinsic Q_M with minor electrical dissipation (b) active stabilized Q-factor versus temperature from -40℃ to +60℃ and (c) resonant frequency versus temperature with and without active stabilization.

CONCLUSION

This paper reports an active Q-factor stabilization approach utilizing electrical dissipation regulation for the first time. A bias current regulation loop is proposed to adjust the input impedance of CSA to stabilize the Q-factor of the MEMS resonator. Compared to our previously reported passive stabilization approach, the specific TCO design of the MEMS resonator is not mandatory and the reduction of the Q-factor is also avoided. Besides, no dedicated tuning structure and electrode is required in contrast to available active stabilization.

ACKNOWLEDGEMENTS

This work was supported by the National Natural Science Foundation of China (NSFC) under Grant 62074078.

The authors would like to thank Zhi Yang, Chao Xu, Min Wang, Tianling Sun, Lixun Qian, and Songlin Guo from the CETC-13 institution for supporting the MEMS process and package development.

REFERENCES

[1] B. Kim et al., "Temperature Dependence of Quality Factor in MEMS Resonators," *Journal of Microelectromechanical Systems*, vol. 17, no. 3, pp. 755-766, June 2008.

[2] D. D. Shin, Y. Chen, I. B. Flader and T. W. Kenny, "Epitaxially encapsulated resonant accelerometer with an on-chip micro-oven," *2017 19th International Conference on Solid-State Sensors, Actuators and Microsystems (TRANSDUCERS)*, 2017, pp. 595-598.

[3] James M. Lehto Miller, Azadeh Ansari, David B. Heinz, Yunhan Chen, Ian B. Flader, Dongsuk D. Shin, L. Guillermo Villanueva, and Thomas W. Kenny, "Effective quality factor tuning mechanisms in micromechanical resonators", *Applied Physics Reviews* 5, 041307 (2018).

[4] H. Abdelli, T. Tsukamoto, M. Khan, and S. Tanaka, "A Novel Quality Factor Trimming Method for Multi-Ring MEMS Resonators Based on Thermoelastic Dissipation," *IEEE International Conference on Micro Electro Mechanical Systems Conference (MEMS)*, Jan. 2022, pp. 766-769.

[5] J. Chen, T. Tsukamoto, G. Langfelder and S. Tanaka, "Frequency and Quality Factor Matched 2-Axis Dual Mass Resonator," *2021 IEEE Sensors*, Oct. 2021.

[6] J. Chen, T. Tsukamoto, and S. Tanaka, "Triple Mass Resonator for Electrostatic Quality Factor Tuning," *in Journal of Microelectromechanical Systems*, vol. 31, no. 2, pp. 194-203, April. 2022.

[7] R. Gando, S. Maeda, K. Masunishi, et al, "A MEMS rate integrating gyroscope based on catch-and-release mechanism for low-noise continuous angle measurement," *IEEE Micro Electro Mechanical Systems (MEMS)*, Jan. 2018, pp. 944-947.

[8] M. Zhuo, Q. Li, Y. Zhang, Y. Xu, X. Wu, and D. Xiao, "Damping Tuning in the Disk Resonator Gyroscope Based on the Resistance Heat Dissipation," *2019 IEEE SENSORS*, Oct. 2019.

[9] J. Han et al., "Electrostatic stabilization of thermal variation in quality factor using anchor loss modulation," *IEEE SENSORS*, 2014, pp. 998-1001.

[10] Y. Zhao, Q. Shi, G. Xia, and A. Qiu, "A Compact Stabilization Scheme For Quality Factor in Nondegenerate Mems Gyroscope," *IEEE International Conference on Micro Electro Mechanical Systems Conference (MEMS)*, Jan. 2022, pp. 146-149.

CONTACT

*Yang Zhao, zhaoyang0216@njust.edu.cn

DEMONSTRATION OF GYRO-LESS NORTH FINDING USING A T-SHAPED MEMS DIFFERENTIAL RESONANT ACCELEROMETER

Kei Masunishi, Etsuji Ogawa, Daiki Ono, Fumito Miyazaki, Hiroki Hiraga, Kengo Uchida,
Jumpei Ogawa, Hideaki Murase and Yasushi Tomizawa
Corporate Research & Development Center, Toshiba Corporation, JAPAN

ABSTRACT

In this paper, we demonstrate the detection of true north by rotating a T-Shaped MEMS differential resonant accelerometer (DRA) at a constant angular velocity. In general, true north finding is performed by detecting the horizontal component of the Earth's rotational velocity. The projection of the Earth's rotation at 35.5°N for the experimental location is 12.2 deg/h. By rotating a small DRA module at a constant angular velocity (Ω_{table} = 1000–3000 dps), a sinusoidal Coriolis force with amplitudes corresponding to 37–124 μG (acceleration equivalent) was detected, with the upper and lower ends of the sinusoidal curves corresponding to the north and south directions, respectively. The unique feature of this achievement is that the north finder was realized without a gyroscope, using only an accelerometer. Compared with gyroscope-based north finders, the gyro-less device is expected to be easier to control and less expensive.

KEYWORDS

MEMS, differential resonant accelerometer, T-shaped MEMS DRA, north finding, Coriolis force, centrifugal force, table rotation, microcontroller module.

INTRODUCTION

Detecting true north is important for navigation and target orientation. For simple north detection, magnetic sensors are used to measure the direction of the Earth's magnetic field, which indicates magnetic north. Magnetic sensors are inexpensive and compact, but their accuracy can be affected by surrounding magnetic elements. In addition, there is a discrepancy between magnetic north and true north. Generally, gyroscopes are used for advanced true north detection. Gyroscopes measure angular velocity, and by rotating a single-axis gyroscope [1,2] or using a dual-axis gyroscope, for example, the horizontal component of the Earth's rotational velocity can be measured to detect true north. There are various types of gyroscopes based on different measurement principles, including fiber optic gyroscopes (FOG), ring laser gyroscopes (RLG), and MEMS gyroscopes [3,4], the latter of which use vibration. These gyroscopes are not affected by external magnetic fields and can therefore detect true north. However, they are expensive sensors with complex structures and controls.

This paper demonstrates the detection of true north by rotating a T-shaped MEMS differential resonant accelerometer (DRA) [5] at a constant angular velocity. This method differs from previous studies using conventional gyroscopes [1,2], because we achieved true north detection using only an accelerometer. It also differs from a previous theoretical study [6] as true north detection was confirmed by a demonstration involving actual rotation of the accelerometer.

MEMS AND MODULE

Figure 1a, b shows photographs of the small DRA module and the T-shaped MEMS DRA [5] used to detect true north. The analog front-end (AFE) circuitry and high-speed digital microcontroller unit (MCU) are the same as those used as in the miniature MEMS RIG (angular rate integral gyro) module [7]. The dimensions of the module are 5 cm × 5 cm × 1 cm. The DRA was mounted on a small module with only a few changes in the firmware. The principle of operation of the DRA [8] is as follows. When the inertia force due to acceleration acts on the proof mass, it rotates around a pivot (hinge) and the two beams are subjected to tensile and compressive stresses, respectively. As a result, the resonance frequencies of the two beams change in opposite directions. By taking the difference between these resonance frequencies, stable acceleration detection can be achieved. The resonant frequency of these beams is about 22 kHz, with a quality factor of 70 k-. A T-shaped electrode structure was added at the position of maximum displacement of the two resonant beams in order to improve the capacitance sensitivity, thereby achieving a high bias instability (BI) of 7.9 μG.

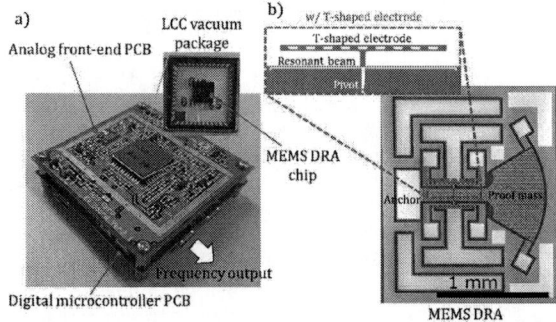

Figure 1: a) Photograph of a microcontroller-based module for the MEMS differential resonant accelerometer (DRA). b) SEM image of the T-shaped MEMS-DRA.

PRINCIPLE OF TRUE NORTH FINDING

Assume a case where an acceleration sensor is fixed on a rotary table with the axis of sensitivity in the direction of the axis of rotation (gravity direction) and the speed of rotation is constant (*Figure 2a, b*). In this case, the forces acting on the proof mass of the accelerometer include gravity, centrifugal force, and Coriolis force. If the accelerometer is placed at a distance r from the axis of rotation, the rotational speed of the table and the respective forces can be expressed by the following equation.

$$v_{table} = r\Omega_{table} \tag{1}$$

$$F_{centrifugal} = mr\Omega_{table}^2 \qquad (2)$$
$$F_g = mg \qquad (3)$$
$$F_{coriolis} = 2m\Omega_{e\_h}v_{table}$$
$$= 2mr\Omega_{e\_h}\Omega_{table} \qquad (4)$$

Here, v_{table} is the rotational speed of the table, $F_{centrifugal}$ is the centrifugal force, F_g is gravity, $F_{coriolis}$ the Coriolis force, Ω_{table} is the angular velocity of the table, m is the mass of the proof mass, g is gravitational acceleration (9.8 m/s^2), and Ω_{e_h} is the horizontal component of the rotational speed of the Earth.

If a horizontally installed rotary table rotates at a constant speed, gravity and centrifugal force will always be constant values. Gravity acts in the direction of the sensitivity axis and centrifugal force acts in the direction orthogonal to the sensitivity axis. Therefore, if the accelerometer is not fixed at an angle to the rotary table, centrifugal force will not be detected. Meanwhile, the Coriolis force generated by the horizontal component of the Earth's rotational velocity is detected in a sinusoidal shape. The upper and lower ends of the sinusoidal curve correspond to the north and south directions, respectively. Due to the constant rotational speed of the rotary table, the proof mass is instantaneously moving at a constant speed, so the Coriolis force acts on the proof mass, just like an oscillating MEMS gyroscope.

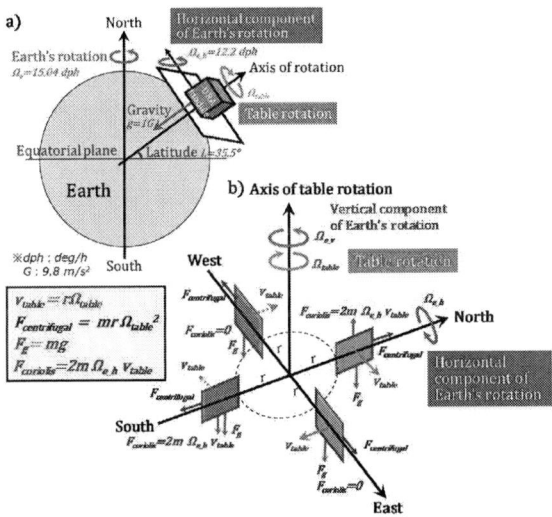

Figure 2: Principle of north finding with the DRA module.

EXPERIMENTAL SETUP

In this study, the DRA was mounted on a compact module to enable evaluation of actual rotation using a rotary table (1291BL, Ideal Aerosmith Inc.) (*Figure 3*). The rotary arm is fixed to the rotary table, and the DRA module can be fixed at several positions on the arm. The DRA module was fixed at a distance of r=15 cm from the rotary axis of the rotary table, standing upright so that the direction of the detection axis is toward the rotary axis (gravity direction). To reduce the load on the rotation axis, a counterweight of the same mass as the DRA module was installed symmetrically around the rotation axis. The rotary table was then rotated at a constant angular velocity (Ω_{table}

= 1000–3000 dps).

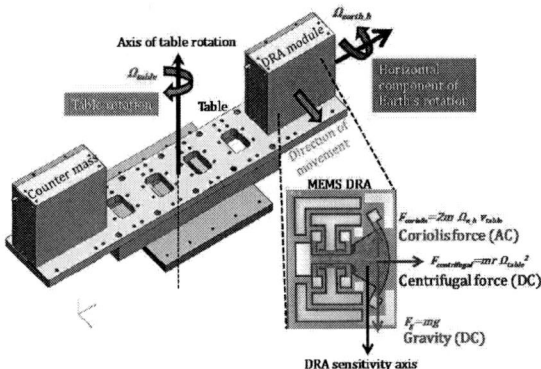

Figure 3: Experimental setup of north finding with the DRA module.

EXPERIMENTAL RESULTS

The projection of the horizontal component of the Earth's rotation at 35.5 degrees north latitude at the experimental site is Ω_{e_h}=12.2deg/h. By measuring this angular velocity, true north was detected. The DRA module was rotated at a constant angular velocity (Ω_{table}=2000dps) and the measurement data (600 s) was cyclically folded. The Coriolis force was detected by removing the influence of the DC component and analyzing only the AC component (*Figure 4*). Here, the Coriolis force is expressed as the acceleration equivalent divided by the mass of the proof mass m. The raw data (blue circles) show a very large scatter. Least squares Method fitting to a sine wave is shown by the red line. Fitting to a sine wave expressed by the following equation.

$$a_{coriolis} = \frac{F_{coriolis}}{m} = a \cdot cos(\phi - b) + c \qquad (5)$$

Here, $a_{coriolis}$ is the acceleration equivalent of the Coriolis force, and a, b, and c are the coefficients used in fitting the sine wave: a is the amplitude of the Coriolis force and b is the azimuth angle with true north and c is offset.

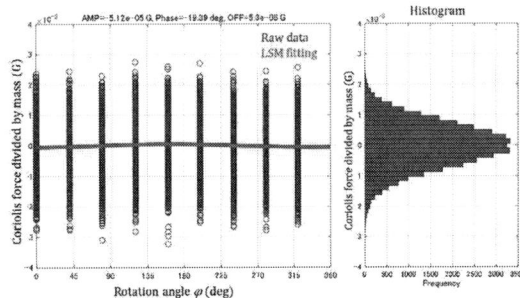

Figure 4: Experimental results of the Coriolis force (Ω_{table} = 2000 dps, r = 15 cm).

In addition, a comparison of the mean measured data at each rotation angle with sinusoidal LSM fitting is shown to verify the accuracy of the measurement (*Figure 5*). The upper and lower ends of the sine wave curves correspond to the north and south directions, respectively. The amplitude a of the sine wave is the amplitude of the

Coriolis force, corresponding to an acceleration of 51 μG. The accuracy was σ = 10 μG from the 95% confidence limit. The phase of the sine wave, *b*, is the azimuth angle, corresponding to –19 deg relative to true south. The precision was σ = 11 deg from the 95% confidence limit.

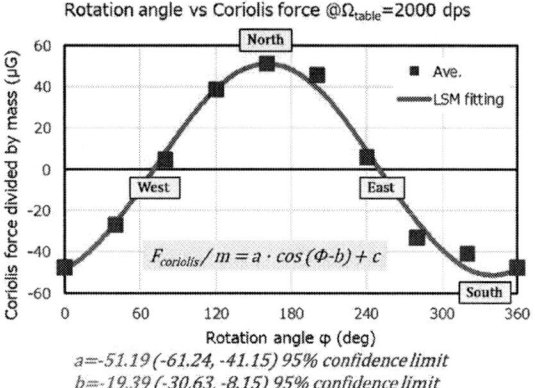

$a = -51.19\ (-61.24, -41.15)\ 95\%\ confidence\ limit$
$b = -19.39\ (-30.63, -8.15)\ 95\%\ confidence\ limit$

Figure 5: Experimental results of north finding (Ω_{table} = 2000 dps, r = 15 cm) based on sine wave fitting.

If the DRA module was not fixed at an inclination with respect to the rotary table, no centrifugal force would be detected. However, centrifugal force (in terms of acceleration) with a DC component corresponding to 38–294 mG was detected (*Figure 6a*), depending on the rotational speed (Ω_{table} = 1000–3000 dps). This suggests that the installation tilt of the DRA module was about 0.4 deg. The AC component was analyzed and a sinusoidal Coriolis force with an amplitude corresponding to 37–124 μG (acceleration equivalent) was detected, depending on the rotational speed (*Figure 6b*).

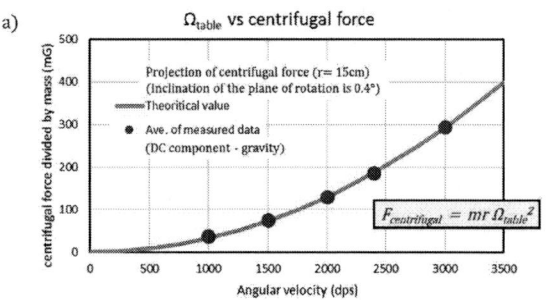

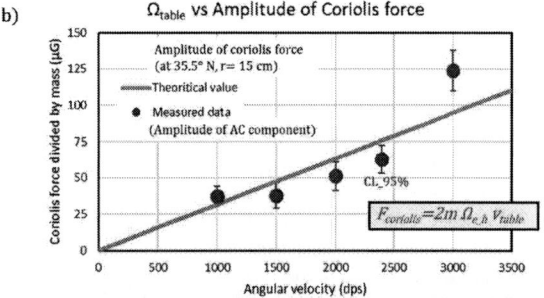

Figure 6: Experimental results of the dependence of angular velocity on a) centrifugal force and b) the Coriolis force.

From these results, we summarized the dependence of the measured S/N ratio of the Coriolis force and the precision of true north detection on the rotational speed (*Figure 7*). As the rotational speed increased, the S/N ratio of the Coriolis force improved, as did the precision of the true north detection. This is thought to be due to the fact that the Coriolis force is proportional to the rotational speed of the table. However, it may also be affected by rotational noise due to the increase in rotational speed. At an angular velocity of the rotary table Ω_{table} = 3000 dps, the S/N ratio of the Coriolis force was 9.7 (-) and the true north detection precision was 6.6 (deg).

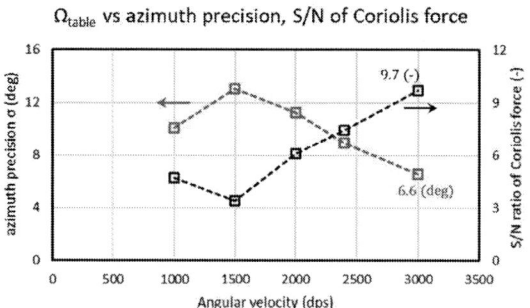

Figure 7: Experimental results of the dependence of angular velocity on azimuth precision and S/N ratio of Coriolis force.

SIMPLE S/N ESTIMATION

Finally, the results of a simple BI-based S/N estimation assuming true north detection are shown (*Figure 8*). Simple S/N estimations for the accelerometer and for the gyro sensor are represented by equations (6) and (7), respectively.

$$S/N_{acc} = (F_{coriolis}/m)/BI_{acc}$$
$$= 2r\Omega_{e\_h}\Omega_{table}/BI_{acc} \qquad (6)$$

$$S/N_{gyro} = \Omega_{e\_h}/BI_{gyro} \qquad (7)$$

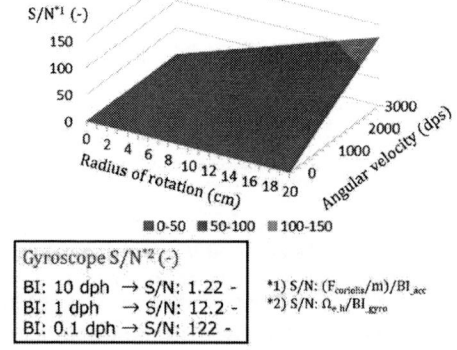

Figure 8: Results of a simple BI-based estimate of S/N for north finding. BI: bias instability.

Here, S/N_{acc} is the simple S/N estimate of Coriolis force detection by the accelerometer, S/N_{gyro} is the simple S/N estimate of Earth rotation detection by the gyroscope, BI_{acc} is the BI of the accelerometer, and BI_{gyro} is the BI of the gyroscope.

When using accelerometers, the Coriolis force increases in proportion to the rotation radius r and angular velocity Ω_{table}. Therefore, increasing the rotation radius r and angular velocity Ω_{table} or reducing BI_{acc} will improve the S/N ratio. In contrast, when using a gyroscope, reducing BI_{gyro} improves the S/N ratio. The simple S/N estimates are almost the same for the accelerometer condition ($BI_{acc} = 1$ μG, $r = 20$ cm, $\Omega_{table} = 3000$ dps) and the gyro sensor condition ($BI_{gyro} = 0.1$ dph).

CONCLUSIONS

A MEMS DRA with high precision achieved with use of T-shaped electrodes was mounted on a small module (5 cm × 5 cm × 1 cm) based on an MCU. By rotating the small DRA module at a constant angular velocity ($\Omega_{table} = 1000$–3000 dps), we detected a sinusoidal Coriolis force (acceleration equivalent) with amplitudes corresponding to 37–124 μG. The upper and lower ends of the sinusoidal curves correspond to the north and south directions, respectively. The unique feature of this achievement is that the true north finder was realized using only an accelerometer. Compared with conventional gyro-based north finders, our device is easier to control and less expensive and is thus expected to be a viable alternative. In addition, a simple BI-based S/N estimation was performed assuming true north detection. S/N estimation with accelerometers and gyroscopes were compared, and design guidelines for a true north finder were obtained.

ACKNOWLEDGEMENTS

This work was carried out with tremendous support and encouragement from H. Hirazawa and A. Kojima of Toshiba Corporation. The authors are grateful to Dr. R. Gando for his contributions during his tenure at Toshiba Corporation. This work was supported by the Innovative Science and Technology Initiative for Security Grant Number JPJ004596, ATLA, Japan.

REFERENCES

[1] I. P. Prikhodko, A. A. Trusov and A. M. Shkel, "North-finding with 0.004 radian precision using a silicon MEMS quadruple mass gyroscope with Q-factor of 1 million", Proc. MEMS2012, 2012, pp.164-167.

[2] Youngjian Zhang, Bin Zhou, Mingliang Song, Bo Hou, Haifeng Xing and Rong Zhang, "A novel MEMS gyro north finder design based on the rotation modulation technique", Sensors J., vol.17(5), 973, 2017.

[3] M. F. Zaman, A. Sharma and F. Ayazi, "High performance matched-mode tuning fork gyroscope", Proc. MEMS2006 Conference, 2006, pp.66-69.

[4] Sergei A. Zotov, Alexander A. Trusov and Andrei M. Shkel, "High-range angular rate sensor based on mechanical frequency modulation", Journal of Microelectromechanical Systems, vol.21, No.2, 2012, pp.398-405.

[5] Kei Masunishi, Etsuji Ogawa, Ryunosuke Gando, Daiki Ono, Shiori Kaji, Fumito Miyazaki, Hiroki Hiraga, Kengo Uchida and Yasushi Tomizawa, "A T-shaped MEMS differential resonant accelerometer with module-based demonstration of >134 dB dynamic range and <1mdeg absolute tilt angle precision", Proc. MEMS2022, 2022, pp.150-153.

[6] Guofu Sun and Qitai Gu, "Accelerometer based north finding system", Proc. of Position Location and Navigation Symposium (2000 IEEE), 2000, pp.399-403.

[7] Fumito Miyazaki, Ryunosuke Gando, Daiki Ono, Shiori Kaji, Hiroshi Ota, Hiroki Hiraga, Kei Masunishi, Etsuji Ogawa, Tetsuro Itakura and Yasushi Tomizawa, "A 0.1 deg/h module-level silicon MEMS rate integrating gyroscope using virtually rotated donut-mass structure and demonstration of the earth's rotation detection", Proc. Transducers 2021 Virtual Conference, 20-24 June, 2021, pp. 402-405.

[8] Dongsuk D. Shin, Chae Hyuck Ahn, Yunhan Chen, David L. Christensen, Ian B. Flader and Thomas W. Kenny, "Environmentally robust differential resonant accelerometer in a wafer-scale encapsulation process", Proc. MEMS 2017, 2017, pp.17-20.

CONTACT

*K. Masunishi, tel: +81-50-3191-0897; kei1.masunishi@toshiba.co.jp

ENHANCED STIFFNESS SENSITIVITY IN A MODE LOCALIZED SENSOR USING INTERNAL RESONANCE ACTUATION

Jianlin Chen[1], Hemin Zhang[2], Takashiro Tsukamoto[1], Michael Kraft[2] and Shuji Tanaka[1]
[1]Department of Robotics, Tohoku University, Sendai, JAPAN and
[2]Department of Electrical Engineering (ESAT-MNS), KU Leuven, Leuven, BELGIUM

ABSTRACT

In this paper, an enhanced stiffness sensitivity in a mode-localized sensor was demonstrated for 1:2 internal resonance (IR) for the first time. Two mode localized modes (anti-phase (AP) and in-phase (IP) modes, with frequencies around $2\omega_r$) could be simultaneously coupled with the rotational mode (ω_r) by a nonlinear interaction utilizing internal resonance (IR). When the nonlinear-actuated AP motion is coupled with the IP motion, an enhanced mode coupling can be observed at the IP mode. The concept was investigated by both theoretical model and experimental result. A balanced AP motion up to the jump frequency was generated under IR actuation. In contrast, a strong mode coupling was observed at the IP mode when the stiffness perturbation was introduced. By measuring the amplitude ratio of the IP mode for different stiffness perturbations, the stiffness sensitivity was found to be improved by an order of magnitude to $12400 /\Delta k/k$.

KEYWORDS

Mode localized sensors, Internal resonance, Enhanced stiffness sensitivity.

INTRODUCTION

Mode-localized MEMS sensors have been proved to have a two to three orders of magnitude higher parametric sensitivity compared with the frequency modulation method [1]. Besides, their amplitude ratio outputs are insensitive to variations in ambient temperature [2] and pressure [3], resulting in better sensing stability. Thanks to these merits, they have been applied to different kinds of sensors, like accelerometers [4], electrometers [5] and mass sensors [6].

High parametric sensitivity has been achieved in mode-localized sensors using a structure with more degree of freedom (DoF) and exploiting an electrically coupled design rather than a mechanically coupled design. A three DoF weakly coupled resonator [7] demonstrated to have a 49 times higher stiffness sensitivity compared with a two DoF structure. Besides, a novel triple mass resonator was proposed [8], whose mechanical system exploited a combination of the dual mass resonator and amplified dual mass resonator. It can achieve an extremely high and tunable stiffness sensitivity under a mechanical coupled structure since the frequency of the in-phase (IP) mode could be independently tuned to approach the anti-phase (AP) mode frequency, i.e. two frequencies are identical with each other.

In this paper, we firstly observed an enhanced stiffness sensitivity in a mode-localized sensor for 1:2 internal resonance (IR) actuation. The nonlinear interaction [9, 10] was applied in a mode-localized sensor and a new nonlinear interaction mechanism was found. When two

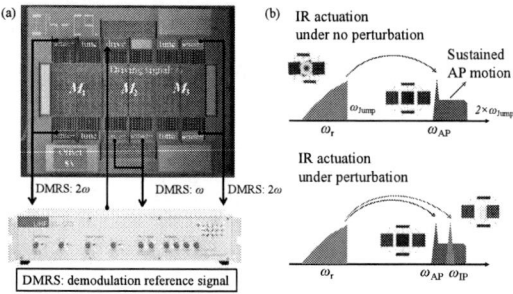

Figure 1: (a) Image of the fabricated devices and setup schematic for evaluating frequency response for IR actuation. (b) Concept of sensitivity enhancement. The structural perturbation causes the IR excitation of both AP and IP modes resulting in a strong mode coupling between AP and IP modes.

mode-localized modes are simultaneously driven by a low frequency mode by IR actuation, a stronger mode coupling can be acquired. Since the AP motion is extended in a wider frequency range by IR actuation and the frequency difference between AP and IP modes was decreased, an enhanced sensitivity in the IP mode can be achieved.

THEORETICAL ANALYSIS
Concept

In the proposed triple mass resonator, a rotational mode with a frequency ω_r and two mode-localized modes with frequencies around $2\omega_r$ can be excited (Fig. 1b). For the rotational mode, the center mass (M_2) and connection frames rotate around the center point and most of the vibration energy is concentrated on the rotatory components. Two mode-localized modes include the AP and IP modes. In the AP mode, the two outer masses (M_1 and M_3) oscillate in opposite direction and the inner mass remains static, assuming the structure is symmetric. In the IP mode, the two outer masses have an in-phase vibration and an anti-phase motion with the inner mass.

When the rotational mode is driven in the nonlinear regime, the response up to the jumping frequency ω_{jump} generates because of the amplitude-frequency (A-f) effect (Fig. 1b). Its nonlinear interaction with the AP and IP modes happens by the superharmonic under a frequency ratio as ~1:2, i.e. the IR actuation. The AP mode is firstly driven and a sustained motion up to $2\times\omega_{jump}$ can be acquired because of the continuous energy transfer with the rotational mode. However, no IP response is actuated by the nonlinear interaction under a symmetric resonator i.e. a balanced anti-phase motion without mode localization in two outer masses.

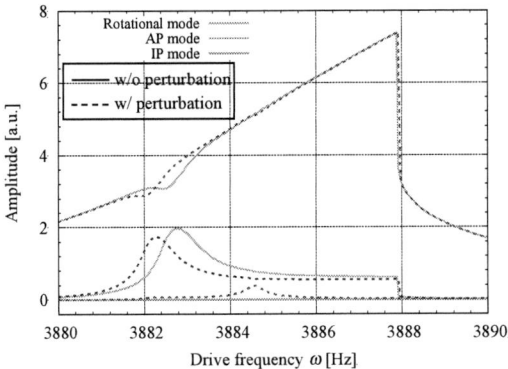

Figure 2: Nonlinear frequency response of the rotational mode (driving mode) and 2ω response of the AP and IP modes without (solid line) and with (dashed line) perturbation (numerical simulation).

When a perturbation is introduced, both of AP and IP modes can be observed and the nonlinear interaction of the rotational mode with both AP and IP modes occurs. Consequently, a strong mode coupling between the sustained AP motion and the IP motion takes place.

Nonlinear Dynamic

The nonlinear dynamic was analyzed by the equations of motion describing the rotational, AP and IP modes, respectively. The rotational mode (ω) is coupled with AP (2ω) and IP (2ω) modes by square and cubic coupling force. The AP and IP modes are coupled by a linear force relating to the stiffness perturbation. The equations of motion are:

$$\ddot{x}_{ro} + \mu_{ro}\dot{x}_{ro} + \omega_{ro}^2 x_{ro} + 2\beta x_{ro}x_{an} + 2\beta x_{ro}x_{in} + 2\gamma x_{ro}x_{an}^2 + 2\gamma x_{ro}x_{in}^2 + \gamma x_{ro}^3 = \frac{F}{M}\cos\Omega \quad (1)$$

$$\ddot{x}_{an} + \mu_{an}\dot{x}_{an} + \omega_{an}^2 x_{an} + \alpha x_{in} + \beta x_{ro}^2 + 2\gamma x_{ro}^2 x_{an} + \gamma x_{an}^3 = 0 \quad (2)$$

$$\ddot{x}_{in} + \mu_{in}\dot{x}_{in} + \omega_{in}^2 x_{in} + \alpha x_{an} + \alpha'\beta x_{ro}^2 + 2\alpha'\gamma x_{ro}^2 x_{in} + \alpha'\gamma x_{in}^3 = 0 \quad (3)$$

where x_i, μ_i and ω_i (i = ro, an, in) are the amplitudes, damping ratios and eigenfrequency of the rotational, AP and IP modes, respectively. x_{ro} is the displacement of Mass 2 (M_2) and connection frames, x_{AP} and x_{IP} are defined as the anti-phase motion $(x_1 - x_3)/2$ and the in-phase motion $(x_1 + x_3)/2$, where x_1, x_3 are the displacements of Mass 1 (M_1) and Mass 3 (M_3). $\alpha, \beta, \gamma, \alpha', F, M$ and Ω are the linear coupling parameters related to stiffness perturbation $\Delta k = k_1 - k_3$, square coupling parameter, cubic coupling parameter, the parameter related to Δk controlling the nonlinear coupling strength in IP mode, driving force, mass of each proof mass and the driving frequency, respectively.

The nonlinear dynamic equations were then solved by the perturbation method of multiple scales [11]. The amplitudes and phases of three coupled modes under steady state response could be obtained

$$-\frac{\partial\varphi_1}{\partial T_1}p_1\omega_{ro} - \frac{1}{2}\beta p_1 p_2 \cos\theta_2 - \frac{1}{2}\beta p_1 p_3 \cos\theta_3 + \frac{1}{4}\gamma p_1 p_2^2 + \frac{1}{4}\gamma p_1 p_3^2 + \frac{1}{8}\gamma p_1^3 - \frac{f}{2}\cos\theta_1 = 0 \quad (4)$$

$$\omega_{ro}\frac{\partial p_1}{\partial T_1} + \frac{1}{2}\mu_1\omega_{ro}p_1 + \frac{1}{2}\beta p_1 p_2 \sin\theta_2 + \frac{1}{2}\beta p_1 p_3 \sin\theta_3 - \frac{f}{2}\sin\theta_1 = 0 \quad (5)$$

$$-\frac{\partial\varphi_2}{\partial T_1}p_2\omega_{an} + \frac{1}{2}\alpha p_3 \cos(\theta_2 - \theta_3) + \frac{1}{4}\beta p_1^2 \cos\theta_2 + \frac{1}{4}\gamma p_2 p_1^2 + \frac{1}{8}\gamma p_2^3 = 0 \quad (6)$$

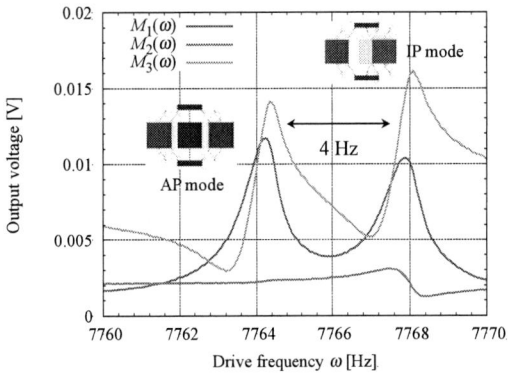

Figure 3: Frequency response of AP and IP modes under direct harmonic driving.

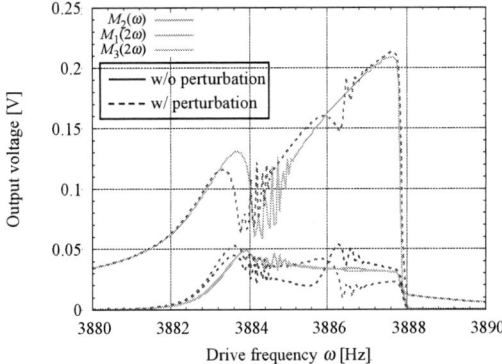

Figure 4: Nonlinear frequency response of M_2 with frequency ω and 2ω response of M_1 and M_3 for no perturbation (solid line) and stiffness perturbation (dashed line).

$$\omega_{an}\frac{\partial p_2}{\partial T_1} + \frac{1}{2}\mu_2\omega_{an}p_2 + \frac{1}{2}\alpha p_3 \sin(\theta_2 - \theta_3) + \frac{1}{4}\beta p_1^2 \sin\theta_2 = 0 \quad (7)$$

$$-\frac{\partial\varphi_3}{\partial T_1}p_3\omega_{in} + \frac{1}{2}\alpha p_2 \cos(\theta_3 - \theta_2) + \frac{1}{4}\alpha'\beta p_1^2 \cos\theta_3 + \frac{1}{4}\alpha'\gamma p_3 p_1^2 + \frac{1}{8}\alpha'\gamma p_3^3 = 0 \quad (8)$$

$$\omega_{in}\frac{\partial p_3}{\partial T_1} + \frac{1}{2}\mu_3\omega_{in}p_3 + \frac{1}{2}\alpha p_2 \sin(\theta_3 - \theta_2) + \frac{1}{4}\alpha'\beta p_1^2 \sin\theta_3 = 0 \quad (9)$$

where p_1, φ_1, p_2, φ_2, p_3, φ_3 are the displacement and phase of the rotational, AP and IP modes, respectively. Furthermore, $\theta_1 = \sigma_1 T_1 - \varphi_1$, $\theta_2 = 2\varphi_1 - \varphi_2 - \sigma_2 T_1 - \alpha T_1$ and $\theta_3 = 2\varphi_1 - \varphi_3 - \sigma_3 T_1 - \alpha T_1$. σ_1, σ_2, σ_3 are defined as the tuning parameters for the driving frequency, AP mode frequency and IP mode frequency.

The acquired polynomial equations Eq. (4)-(9) were solved by MATLAB and the displacements of the rotational, AP and IP modes for symmetric and asymmetric conditions are shown in Fig. 2. When the resonator structure is assumed to be symmetric, only an AP motion in M_1 and M_3 was excited and there was no linear and nonlinear interaction with the IP mode due to the parameter α' in the Eq. (3). On the other hand, not only AP motion but also IP motion were observed for an introduced perturbation due to the mode interaction between the IP mode with the rotational and AP modes, respectively. Therefore, mode localization phenomenon occurred, and a strong mode coupling could be observed at the IP mode since the sustained AP motion coupled with the actuated IP motion.

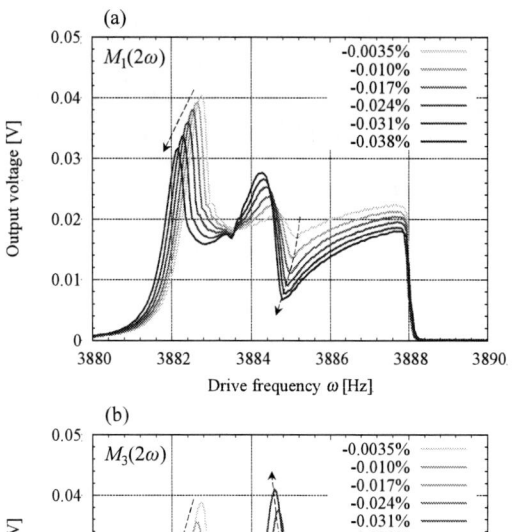

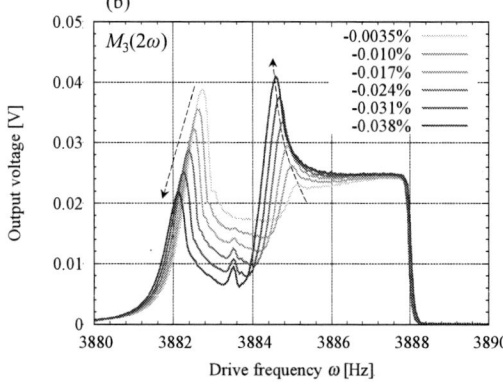

Figure 5: Frequency responses of the two outer masses for IR actuation with an increased negative stiffness perturbation. The vibrations of two outer mass in the AP mode decreased simultaneously, while the mass M_3 vibration increased and M_1 vibration decreased in the IP mode, resulting in a strong mode localization phenomenon.

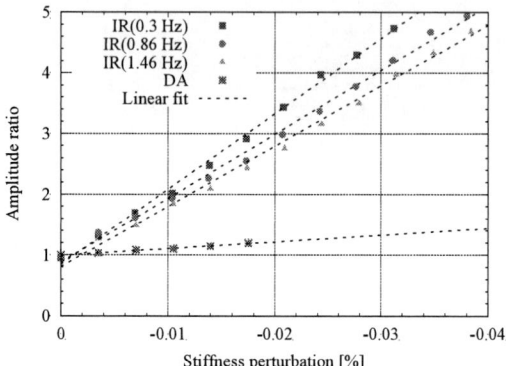

Figure 6: Stiffness sensitivity of amplitude ratio to stiffness perturbation under DA and IR driving, respectively. Under IR actuation, the amplitude sensitivity was enhanced under a smaller frequency difference ($\omega_{r-}\omega_{AP}/2$).

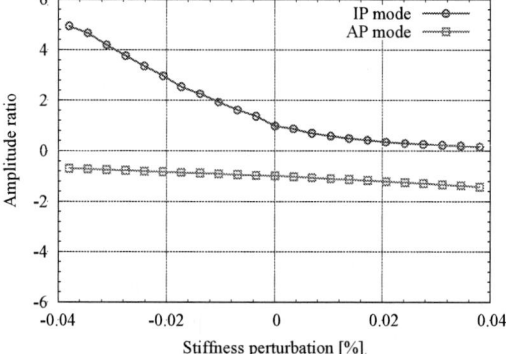

Figure 7: Amplitude ratio veering curve of IP and AP modes under IR actuation showing an enhanced sensitivity in the IP mode under negative perturbation.

EXPERIMENTAL RESULTS

Experimental setup

The device was fabricated by the standard SOI process and measured in a vacuum chamber at a pressure level of 25 Pa. For IR actuation, the driving signal was applied to the inner mass and the rotational mode was driven in the nonlinear regime. The motion signals were fed to the lock-in amplifier (Fig. 1), in which the inner mass signal (M_2) was demodulated by the reference signal with frequency ω and two outer masses signal (M_1 and M_3) were demodulated by the reference signal with frequency 2ω, i.e. the AP and IP modes were driven by nonlinear interaction under IR.

Frequency response

The AP and IP mode were first evaluated for direct harmonic actuation (DA), as shown in Fig. 3. The effect of the feedthrough signal was clearly observed in the response of M_3. The frequency of the AP (7764 Hz) and IP modes (7768 Hz) were tuned to be two times of the rotational mode frequency (3884 Hz) by applying a DC bias on M_1 and M_3. The response of the inner mass was almost zero in the AP mode, which means any structural perturbation was also compensated by a larger DC bias voltage on M_1.

Then, a driving signal was applied to M_2 and the rotational mode was driven in the nonlinear regime (Fig. 4). The resonant outputs of M_1 and M_3 at the AP mode could be clearly observed, while the IP mode was not excited by IR for no perturbation, which agreed well with the simulated results shown in Fig. 2. Instead, both AP and IP modes were clearly activated after introducing the stiffness perturbation by increasing the DC bias on M_1. A stronger mode localization phenomenon was observed at the IP mode, which meant the perturbation caused not only the mode coupling between the IP and AP modes but also the nonlinear interaction between the IP and rotational modes. The frequency difference $\omega_{IP}-\omega_{AP}$ was 5 Hz, which was larger compared with that for the linear actuation (Fig. 3) due to the nonlinear hardening effect.

Stiffness sensitivity

The amplitude ratio between M_1 and M_3 was analyzed for different stiffness perturbations and initial frequency differences between the rotational and AP modes, $\omega_{ro}-\omega_{AP}/2$. The introduced stiffness perturbation caused the frequency decrease of both AP and IP modes (Fig. 5), therefore the AP mode moved away from the rotational mode, while the IP mode got close to the rotational mode.

978-1-6654-9309-3/23 $31.00 © 2023 IEEE

Besides, the structural asymmetries caused an unbalanced motion between M_1 and M_3, i.e. mode localization phenomenon in both AP and IP modes. Especially, the amplitudes of M_1 decreased and whereas that of M_3 increased in the IP mode resulting in a larger unbalanced motion compared with that in the AP mode. Consequently, a larger stiffness perturbation caused a stronger nonlinear mode interaction between the IP and rotational modes, which enhanced the mode coupling between the IP motion and the sustained AP motion, i.e. a strong mode localization phenomenon.

Besides, the effect of the frequency difference $\omega_{ro} - \omega_{AP}/2$ was considered on the stiffness sensitivity. The DC bias applied on M_1 and M_3 exerted a 40 times larger effect on the frequency of the AP and IP modes compared with that of the rotational mode. Therefore, the frequency difference could be adjusted by adjusting the tuning bias. The frequency difference was tuned from 1.46 Hz to 0.3 Hz, i.e. the AP mode moved close to and the IP mode moved away from the rotational mode, respectively. Both AP and IP responses were improved thanks to coupling with a larger rotational mode output. In consequence, an improved sensitivity was acquired under the stronger nonlinear interaction.

The maximum amplitude ratio under IR actuation was calculated at the minimum value of M_1 (see in Fig. 5). The amplitude ratio under DA was measured after removing the feedthrough signal by data processing, which showed the stiffness sensitivity was $1100\ /\Delta k/k$ (Fig. 6). The maximum sensitivity of the IP mode for IR actuation was $12400\ /\Delta k/k$ (when $\omega_{ro} - \omega_{AP}/2 = 0.3$ Hz), which was enhanced by an order of magnitude compared with that under DA.

The amplitude ratio veering curve under IR driving was shown in the Fig. 7. The stiffness sensitivity of the AP mode for positive perturbation was approximated to that for the DA method as 1100, while the sensitivity of the IP mode for negative perturbation was considerably enhanced to 12400. This proved that the coupling between the sustained AP motion and excited IP motion could significantly improve the sensitivity for IR driving.

CONCLUSION

This paper presents a further enhancement in stiffness sensitivity of the proposed triple mass resonator enabled by IR actuation. Two mode-localized modes were driven by 1:2 IR and a strong mode localization phenomenon was observed in the IP mode thanks to the coupling with the sustained AP motion from the nonlinear oscillation. The concept was confirmed by theoretical and experimental studies. The results showed the stiffness sensitivity of the IP mode for IR was $12400\ /\Delta k/k$, revealing an improvement by one order of magnitude compared with the sensitivity of AP mode for IR and the sensitivity under DA.

ACKNOWLEDGEMENTS

This paper is based on results obtained from projects commissioned by the New Energy and Industrial Technology Development Organization (NEDO) and Japan Society for the Promotion of Science (JSPS) KAKENHI grant-in-aid for young scientist no. 21J11628.

REFERENCES

[1] P. Thiruvenkatanathan, J. Yan, J. Woodhouse, and A. A. Seshia, "Enhancing parametric sensitivity in electrically coupled MEMS resonators," *J. Microelectromech. Syst.*, vol. 18.5, pp. 1077-1086, 2009.

[2] J. Zhong, J. Yang, and H. Chang, "The temperature drift suppression of mode-localized resonant sensors", In 2018 *IEEE MEMS*, pp. 467-470, 2018.

[3] H. Zhang, J. Zhong, W. Yuan, J. Yang, and H. Chang, "Ambient pressure drift rejection of mode-localized resonant sensors". In 2017 *IEEE MEMS*, pp. 1095-1098, 2017.

[4] H. Zhang, B. Li, W. Yuan, M. Kraft, and H. Chang, "An acceleration sensing method based on the mode localization of weakly coupled resonators," *Journal of microelectromechanical systems*, vol. 25, no. 2, pp. 286–296, 2016.

[5] Y. Hao, J. Liang, H. Kang, W. Yuan, and H. Chang, "A micromechanical mode-localized voltmeter," *IEEE Sensors Journal*, 2020.

[6] J. R. Liu, C. P. Tsai, W. R. Du, T. Y. Chen, J. S. Chen and W. C. Li, "Vibration Mode Suppression in Micromechanical Resonators Using Embedded Anti-Resonating Structures," *J. Microelectromech. Syst.*, vol. 30.1, pp. 53-63, 2021.

[7] C. Zhao, G. S. Wood, J. Xie, H. Chang, S. H. Pu, and M. Kraft, "A three degree-of-freedom weakly coupled resonator sensor with enhanced stiffness sensitivity," *J. Microelectromech. Syst.*, vol. 25.1, pp. 38-51, 2015.

[8] J. Chen, T. Tsukamoto, and S. Tanaka, "A mechanical coupled three degree-of-freedom resonator with tunable stiffness sensitivity," *Sensors and Actuators A: Physical*, vol. 344, 113713, 2022.

[9] C. Xia, D. Wang, T. Ono, T. Itoh and R. Maeda, "A mass multi-warning scheme based on one-to-three internal resonance," *Mechanical Systems and Signal Processing*, 142, p. 106784, 2020.

[10] J. Chen, H. Zhang, T. Tsukamoto, M. Kraft and S. Tanaka, "A Mode Localized Force Transducer with Reduced Feedthrough via 1:2 Internal Resonance Actuation," in 2022 *IEEE MEMS*, pp. 743–746, 2022.

[11] G. Gobat, A. Frangi, C. Touzé, L. Guillot, and B. Cochelin, "Investigation of Quasi-Periodic Solutions in Nonlinear Oscillators Featuring Internal Resonance", *Advances in Nonlinear Dynamics*, pp. 797-806, 2022.

CONTACT

*J. Chen, Email: chenjl@tohoku.ac.jp

MODELING STRESS EFFECTS ON FREQUENCIES OF A MEMS RING GYROSCOPE

Mehran Hosseini-Pishrobat[1], Baha Erim Uzunoglu[1], and Erdinc Tatar[1,2]

[1]Department of Electrical and Electronics Engineering, Bilkent University, Ankara, Turkey and
[2]National Nanotechnology Research Center (UNAM), Bilkent University, Ankara, Turkey

ABSTRACT

We present, for the first time, an analytical model for the external stress effects on the frequencies of a vibrating ring gyroscope (VRG). The stress-induced anchor displacements cause gap changes in the electrodes and nonhomogeneous boundary conditions in the VRG's suspension structure. Stress stiffness arising from the geometric nonlinearity in the suspension is the principal mechanism affecting VRG's frequencies as it dominates variations of the electrostatic softening. We validate our model using external stress tests performed on a 57kHz VRG equipped with 16 symmetrically distributed, on-chip capacitive stress sensors, which provide anchor displacement measurements.

KEYWORDS

Analytical modeling; Extensible ring; Ring gyroscope; Stress sensing

INTRODUCTION

Long-term drift is one of the MEMS gyroscopes' main obstacles in achieving navigation-grade performance. Although temperature calibration is widely employed to suppress drift [1], it has been shown that on-chip stress calibration—in addition to temperature—could provide a superior drift performance [2], [3]. Accordingly, understanding how stress affects MEMS gyroscopes is crucial to their performance improvement. Zhang et al. [4] presented a finite element analysis of the thermomechanical stresses induced during packaging, and Schröder et al. [5] reported a new, adhesive-free wire bounding technique for low-stress packaging of MEMS gyroscopes. Even though the adverse effects of stress have been well-recognized, to the best of our knowledge, analytical modeling of the stress effects on MEMS gyroscopes is still missing in the literature.

As the main contribution, we present a new modeling approach that provides a rigorous, first-principle understanding of the stress effects in a VRG. Figure 1-(a) shows the SEM of our fabricated 57 kHZ VRG, which has two rings (Rings#1,2) and is equipped with eight inner and eight outer on-chip capacitive stress sensors. Any stress-related perturbation of the ideal arrangement of the anchors and electrodes poses two important ramifications: gap changes of the electrodes and nonhomogeneous (i.e., nonzero) boundary conditions for the displacement field of Ring#2, which acts as a suspension for Ring#1. To fully capture the stress stiffness caused by the latter effect, it is imperative to 1) take the extensibility of Ring#2's centerline into account and 2) use the higher-order, nonlinear Green-Lagrange strain. We note that this approach contrasts with the existing modeling techniques where inextensibility and linearity are fundamental

assumptions [6], [7]. We obtain the stress and electrostatic softening stiffness matrices from the strain and electrostatic potential energies of the VRG, respectively. We calculate the anchor and electrode displacements by interpolating the substrate's strain field using the stress sensors' outputs. The flowchart in Figure 2 summarizes our modeling approach.

ANALYTICAL MODELING

Our VRG is designed to operate at the $n = 2$ wineglass mode, and its main mechanical structure consists of Ring#1 (outer ring), Ring#2 (inner ring), eight connecting beams, and eight supporting beams (see Figure 1-(b)). We denote the mean radius and width of Ring#i ($i = 1,2$) by R_i and w_i, respectively. We denote the length and width of each connecting (supporting) beam by l_c (l_s) and w_c (w_s), respectively. Using the cylindrical coordinates $\{r, \theta, z\}$, we denote by $(u_{r,i}, u_{\theta,i})$ the displacement field (radial and tangential components) of Ring#i at its centerline and by ξ_i the distance from the centerline. The displacement field of Ring#i is then given by

$$\bar{u}_{r,i}(\xi_i, \theta, t) = u_{r,i}(\theta, t),$$
$$\bar{u}_{\theta,i}(\xi_i, \theta, t) = u_{\theta,i}(\theta, t) + \xi_i \gamma_i(\theta, t), \tag{1}$$

where

$$\gamma_i(\theta, t) = \frac{1}{R_i}\left(u_{\theta,i}(\theta, t) - \frac{\partial u_{r,i}(\theta, t)}{\partial \theta}\right) \tag{2}$$

is the rotation of the cross-section located at the angle θ and $t \geq 0$ is time. The extensional Green-Lagrange strain [8] in Ring#i is given by

$$\mathcal{E}_{\theta\theta,i} = \varepsilon_{\theta\theta,i} + \eta_{\theta\theta,i},$$
$$\varepsilon_{\theta\theta,i} = \varepsilon_{0,i} + \xi_i \kappa_i,$$
$$\eta_{\theta\theta,i} = \frac{1}{2}\left(\varepsilon_{\theta\theta,i}^2 + \left(1 + \frac{\xi_i}{R_i}\right)^2 \gamma_i^2\right), \tag{3}$$

where $\varepsilon_{\theta\theta,i}$ and $\eta_{\theta\theta,i}$ are the linear and nonlinear parts of the strain, respectively, and

$$\varepsilon_{0,i} = \frac{1}{R_i}\left(u_{r,i}(\theta, t) + \frac{\partial u_{\theta,i}(\theta, t)}{\partial \theta}\right),$$
$$\kappa_i = \frac{1}{R_i}\frac{\partial \gamma_i}{\partial \theta}, \tag{4}$$

are the extensional strain and curvature variation at the centerline, respectively [9]. Based on (3) and (4), the strain potential energy of Ring#i is given by

$$\mathcal{U}_i = \mathcal{U}_i^L + \mathcal{U}_i^{NL},$$

$$\mathcal{U}_i^L = \frac{1}{2}ER_i h \int_0^{2\pi} \int_{-\frac{w_i}{2}}^{\frac{w_i}{2}} \varepsilon_{\theta\theta,i}^2 \, \mathrm{d}\xi_i \mathrm{d}\theta,$$

$$\mathcal{U}_i^{NL} = \frac{1}{2}ER_i h \int_0^{2\pi} \int_{-\frac{w_i}{2}}^{\frac{w_i}{2}} \eta_{\theta\theta,i}^2 + 2\eta_{\theta\theta,i}\varepsilon_{\theta\theta,i} \, \mathrm{d}\xi_i \mathrm{d}\theta, \tag{5}$$

where $\mathcal{U}_i^L$ and $\mathcal{U}_i^{NL}$ are the strain energies associated with the linear and nonlinear strains, respectively, and h is the

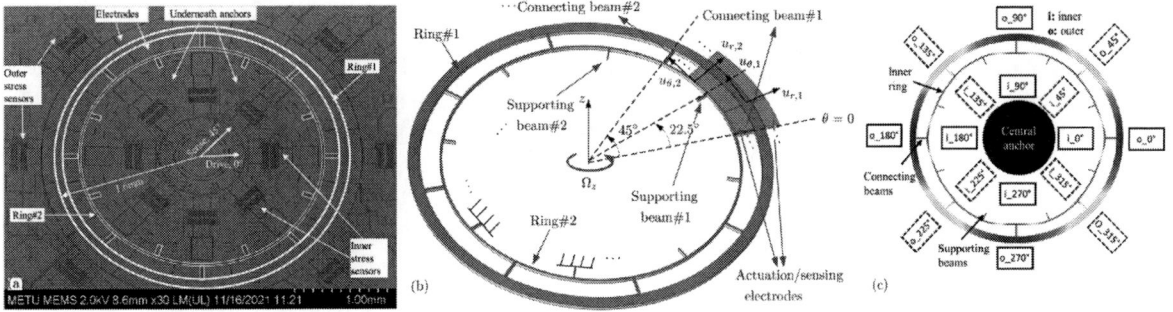

Figure 1: SEM of VRG (a), schematics of VRG (b), and arrangement of stress sensors (c).

structural thickness. We denote by $\vartheta_{c(s)}$ the deflection curve of the connecting (supporting) beam#i, $i = 1:8$, and use the coordinates x and y along and orthogonal to the beam's centerline, respectively. Following the Euler-Bernoulli beam theory and the *von Karman* strain analysis [8], the beam's strain energy is given by

$$\mathcal{U}_{c(s),i} = \mathcal{U}^L_{c(s),i} + \mathcal{U}^{NL}_{c(s),i},$$

$$\mathcal{U}^L_{c(s),i} = \frac{1}{2}Eh \int_0^{l_{c(s)}} \int_{-\frac{w_{c(s)}}{2}}^{\frac{w_{c(s)}}{2}} \left(y\frac{\partial^2\vartheta_{c(s)}(x,t)}{\partial x^2}\right)^2 dydx,$$

$$\mathcal{U}^{NL}_{c(s),i} = \frac{1}{8}Eh \int_0^{l_{c(s)}} \int_{-\frac{w_{c(s)}}{2}}^{\frac{w_{c(s)}}{2}} \left(y\frac{\partial^2\vartheta_{c(s)}(x,t)}{\partial x^2}\right)^4 dydx,$$

(6)

where $\mathcal{U}^L_{c(s),i}$ and $\mathcal{U}^{NL}_{c(s),i}$ are the strain energies associated with the linear and nonlinear extensional strains in the beam, respectively. The total strain potential energy of the VRG is given by the sum of strain energies of the rings and beams:

$$\mathcal{U} = \mathcal{U}^L + \mathcal{U}^{NL},$$

$$\mathcal{U}^L = \mathcal{U}^L_1 + \mathcal{U}^L_2 + \sum_{i=1}^8 \mathcal{U}^L_{c,i} + \mathcal{U}^L_{s,i},$$

$$\mathcal{U}^{NL} = \mathcal{U}^{NL}_1 + \mathcal{U}^{NL}_2 + \sum_{i=1}^8 \mathcal{U}^{NL}_{c,i} + \mathcal{U}^{NL}_{s,i}.$$

(7)

Electrostatic Potential Energy

There are 16 pairs of differential electrodes surrounding Ring#1 (see Figure 1). For the i-th pair, the capacitances between Ring#1 and the outer electrode, $C_{1,i}$ and the inner electrode, $C_{2,i}$ are given by

$$C_{1(2),i}$$
$$\approx \epsilon_0 h(R_1$$
$$\pm\frac{w_1}{2}) \int_{(2i-3)\pi/16}^{(2i-1)\pi/16} \left\{1 \pm \left(\frac{u_{r,1}(\theta,t)}{g_{out(inn)}(\theta,t)}\right)\right.$$
$$\left.+ \left(\frac{u_{r,1}(\theta,t)}{g_{out(inn)}(\theta,t)}\right)^2\right\}\frac{d\theta}{g_{out(inn)}(\theta,t)},$$

(8)

where $\epsilon_0 = 8.8542 \times 10^{-12}$ F/m is the permittivity of vacuum and g_{out} and g_{inn} are the outer and inner gaps, respectively. The (θ,t)-dependence of g_{out} and g_{inn} reflects the effects of stress-induced anchor displacements on the 3 μm nominal gap. The total electrostatic potential energy is then given by

$$\mathcal{U}_e = \frac{1}{2}\sum_{i=1}^{16} C_{1,i}V^2_{1,i} + C_{2,i}V^2_{2,i}.$$

(9)

where $V_{1,i}$ and $V_{2,i}$ are the voltage differences between Ring#1 and the outer and inner electrodes, respectively.

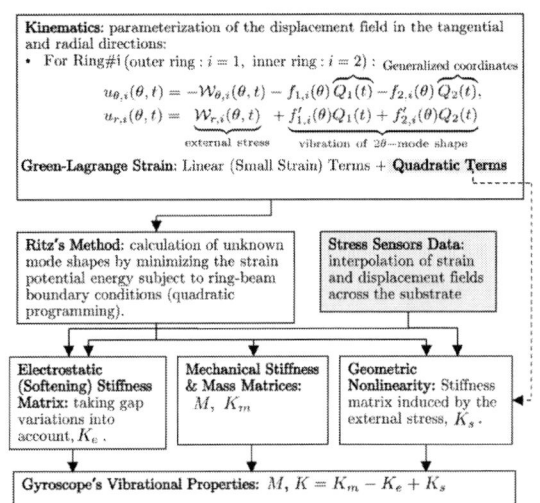

Figure 2: Modeling flowchart for stress effects.

Boundary Conditions

The axial rigidity of the connecting/supporting beams and the deflection/slope continuity at the ring-beam interconnections impose certain boundary conditions on the displacement fields [6]. To save space, we refer the reader to [6] for the details of these boundary conditions. However, a key difference of the current work with respect to [6] is the nonhomogeneous boundary conditions that the stress-induced anchor displacements impose on Ring#2 via supporting beams:

$$u_{r,2}(\theta_{s,i}, t) = \Delta_i(t), i = 1:8.$$

(10)

Here, $\theta_{s,i} = (2i-1)\pi/8$ is the angular position of the supporting beam#i and Δ_i is the displacement that the beam transmits to Ring#2.

Mode Shapes

The vibration of Ring#1 in the wineglass mode is governed by

$$u_{\theta,1}(\theta,t) = -\frac{\sin(2\theta)}{2}Q_1(t) + \frac{\cos(2\theta)}{2}Q_2(t),$$

$$u_{r,1}(\theta,t) = \cos(2\theta)Q_1(t) + \sin(2\theta)Q_2(t),$$

(11)

where Q_1 and Q_2 are the generalized coordinates. We have assumed in (11) that Ring#1's centerline is inextensible, i.e., $\varepsilon_{0,1} \equiv 0$. However, we take the centerline extensibility of Ring#2 into consideration as it is directly affected by the stress-induced anchor displacements. We therefore consider the following mode shape parameterization for Ring#2:

$$u_{\theta,2}(\theta,t) = -\mathcal{W}_\theta(\theta,t) - f_1(\theta)Q_1(t) - f_2(\theta)Q_2(t),$$

$$u_{r,2}(\theta,t) = \mathcal{W}_r(\theta,t) + f_1'(\theta)Q_1(t) + f_2'(\theta)Q_2(t),$$

(12)

where $f_1(\theta)$ and $f_2(\theta)$ are the orthogonal, wineglass-like mode shapes of Ring#2, and $\mathcal{W}_\theta(\theta,t)$ and $\mathcal{W}_r(\theta,t)$ account for the nonhomogeneous boundary conditions imposed by the stress-induced displacements. For the beams, we consider the mode shape parameterizations

$$\vartheta_{c(s),i}(x,t) = {}_1\varpi_{c(s),i}(x)Q_1(t) + {}_2\varpi_{c(s),i}(x)Q_2(t), \quad (13)$$

where ${}_1\varpi_{c(s),i}(x)$ and ${}_2\varpi_{c(s),i}(x)$ are the beam mode shapes corresponding to the two wineglass mode shapes of the rings. The procedure for calculating $f_1(\theta)$, $f_1(\theta)$, ${}_1\varpi_{c(s),i}(x)$, and ${}_2\varpi_{c(s),i}(x)$ is detailed in [6]. Hence, we focus on $\mathcal{W}_\theta$ and $\mathcal{W}_r$, for which we propose the following calculation algorithm: *Step I)* We consider the portion $\theta \in [0,\frac{\pi}{4}]$ of Ring#2, which involves connecting beams#1,2 and supporting beam#1, and assume that the supporting beam applies a unit displacement Δ_0 on Ring#2. *Step II)* We calculate the displacement field $(\overline{\mathcal{W}}_r(\theta), \overline{\mathcal{W}}_\theta(\theta))$, $\theta \in [0,\frac{\pi}{4}]$, that results from the static equilibrium in Step I. *Step III)* For any portion $\theta \in [\frac{\pi}{4}(i-1), \frac{\pi}{4}i]$, $i = 1:8$, of Ring#2 with the displacement $\Delta_i(t)$, we have

$$\mathcal{W}_\theta(\theta,t) = \frac{\Delta_i(t)}{\Delta_0}\overline{\mathcal{W}}_\theta\left(\theta - \frac{\pi}{4}(i-1)\right),$$
$$\mathcal{W}_r(\theta,t) = \frac{\Delta_i(t)}{\Delta_0}\overline{\mathcal{W}}_r\left(\theta - \frac{\pi}{4}(i-1)\right). \quad (14)$$

In Step II, based on the Ritz method, we calculate $\overline{\mathcal{W}}_r$ and $\overline{\mathcal{W}}_\theta$ by minimizing the strain potential energy subject to the ring-beam boundary conditions.

Stiffness Matrix

After calculating unknown mode shapes, Equations (11)–(13) give us a complete description of the VRG's displacement fields, and by substituting these equations into (7) and (9), we obtain the following components of the stiffness matrix:

Nominal mechanical stiffness matrix: $K_m = [k_{ij}^m] \in \mathbb{R}^{2\times2}$, where $k_{ij}^m = \frac{\partial^2 U^L}{\partial Q_i \partial Q_j}\Big|_{Q_1,Q_2=0}$; see [6] for the details.

Stress stiffness matrix: stiffness induced by the external stress, $K_s = [k_{ij}^s] \in \mathbb{R}^{2\times2}$:

$$k_{ij}^s = \frac{\partial^2 U^{NL}}{\partial Q_i \partial Q_j}\Big|_{Q_1,Q_2=0}$$
$$\approx \frac{Ehw_2}{R_2}\int_0^{2\pi}\left\{\varepsilon_{0,2}(\theta,t) + \frac{\varepsilon_{0,2}^2(\theta,t)}{2}\right.$$
$$\left. + \frac{3\gamma_{0,2}^2(\theta,t)}{2}\right\}\mathcal{F}_i(\theta)\mathcal{F}_j(\theta)\mathrm{d}\theta, i,j \in \{1,2\}, \quad (15)$$

where $\mathcal{F}_i(\theta) := f_i(\theta) + f_i''(\theta)$ and $\gamma_{0,2}(\theta,t) := \gamma_2(\theta,t)|_{Q_{1,2}} = 0$.

Electrostatic softening: $K_e = [k_{ij}^e] \in \mathbb{R}^{2\times2}$:

$$k_{ij}^e = \frac{\partial^2 U_e}{\partial Q_i \partial Q_j}\Big|_{Q_1,Q_2=0}$$
$$\approx \frac{\epsilon_0 h}{2}\sum_{i=1}^{16}\int_{(2i-3)\pi/16}^{(2i-1)\pi/16}\left\{V_{1,i}^2\frac{R_1 + \frac{w_1}{2}}{g_{out}^3(\theta,t)}\right.$$
$$+ V_{2,i}^2\frac{R_1 - \frac{w_1}{2}}{g_{inn}^3(\theta,t)}\right\}\left\{\delta_{ij}\right.$$
$$\left. + (-1)^{i-1}\cos\left(4\theta - \frac{\pi}{2}(1-\delta_{ij})\right)\right\}\mathrm{d}\theta, i,j \in \{1,2\}, \quad (16)$$

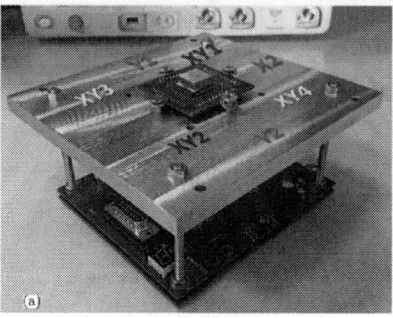

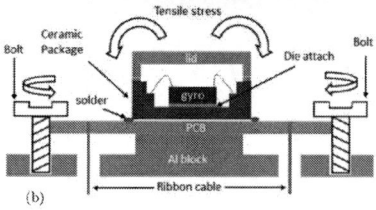

Figure 3: Stress test-bed (a) and the cartoon explaining the test concept (b).

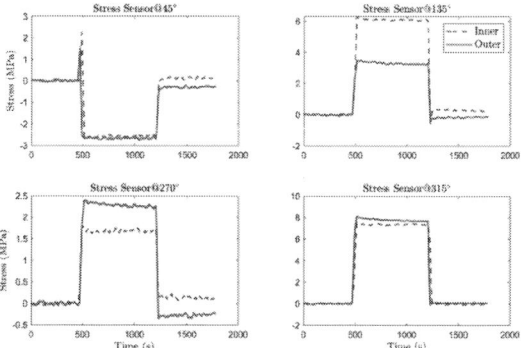

Figure 4: Sample stress sensor outputs for test#3.

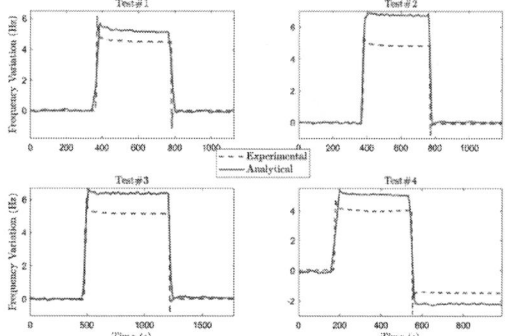

Figure 5: Analytical and experimental drive mode frequency variations for tests#1-4.

where δ_{ij} is the Kronecker delta. The total stiffness matrix is $K = K_m + K_s - K_e$. The external stress does not affect the mass matrix, which its calculation details can be found in [6].

STRESS SENSORS

Our VRG is equipped with 16 symmetrically distributed stress sensors as shown in Figure 1-(c). The stress sensor located at the polar coordinates (ϱ, φ) is a capacitive strain gauge that measures the local radial strain, $\tilde{\varepsilon}_{rr}(\varrho, \varphi, t)$ of the substrate via its two side anchors (see [3]). We use the outputs of the stress sensors to interpolate $\tilde{\varepsilon}_{rr}(.)$ as

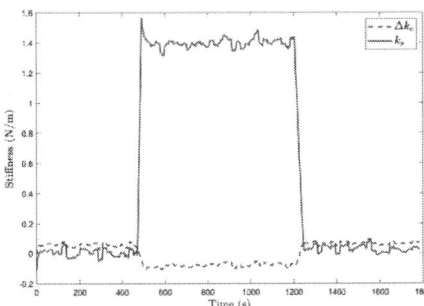

Figure 6: Stress stiffness (k_s) and variation of electrostatic softening (Δk_e) during test#3.

$$\tilde{\varepsilon}_{rr}(r,\theta,t) \approx \psi^{\mathsf{T}}(r,\theta)\alpha(t), \tag{17}$$

where $\psi(.) \coloneqq [\psi_1(.), \dots, \psi_N(.)]^{\mathsf{T}} \in \mathbb{R}^N$ is the vector of $N \geq 1$ suitable basis functions and $\alpha(.) \in \mathbb{R}^N$ is the vector of basis functions' coefficients. We obtain the optimal coefficients by solving the following least-squares problem:

$$\alpha^*(t) \coloneqq \arg\min_\alpha \sum_{i,j} \left(\bar{\varepsilon}_{ij}(t) - \psi^{\mathsf{T}}(\varrho_i, \varphi_j)\alpha(t) \right)^2, \tag{18}$$

where $\bar{\varepsilon}_{ij}$ is the output of the stress sensor located at (ϱ_i, φ_j). We then obtain the radial displacement field across the substrate as

$$\tilde{u}_r(r,\theta,t) \approx \int_0^r \psi^{\mathsf{T}}(\zeta,\theta)\alpha^*(t)\mathrm{d}\zeta, \tag{19}$$

which provides us with the stress-induced displacements, $\Delta_i(t)$ as well as electrostatic gap changes. A physically meaningful choice for the basis functions $\psi_i(r,\theta)$ is the functions that appear in *Michell's solution* to the biharmonic equation governing the static equilibrium of an elastic body in the polar coordinates [10]:

$$1, \ln(r), r\cos(p\theta), r\sin(p\theta), p = 1,2,\dots. \tag{20}$$

RESULTS

Figure 3 shows the aluminum stress test-bed where tensile stress was applied to the VRG by selectively tightening the bolt pairs. Ribbon cables connect the VRG's board to the main PCB. We performed four tests by tightening the bolt pairs: test#1 (bolts X1-X2), test#2 (bolts Y1-Y2), test#3 (bolts XY1-XY2), and test#4 (bolts XY3-XY4); see Figure 3-(a) for the bolt locations. The maximum tensile stresses in these tests were 4.41 MPa, 5.37 MPa, 8.10 MPa, and 5.82 MPa, respectively. Figure 4 shows the outputs of sample stress sensors during test#3, i.e., bending along 135°. We observed the maximum tensile stress at the stress sensors located at 135° and 315° (the tightening direction). A compressive stress was observed at 45° due to the Poisson effect. Figure 5 shows the comparative graph of the analytically estimated and experimentally observed variations of the drive mode's frequency for all the tests. The estimation errors are <1.21Hz. Apart from the modeling errors, four of the stress sensors were non-functional and contributed to the mismatch between the analytical and experimental results. During the experiments, we observed that decreasing the structural high voltage by 60% had a negligible effect on the frequency shift. As Figure 6 shows, the analytical model concurs with this observation as the stress stiffness is about 10X larger than the variation in electrostatic softening.

CONCLUSION

We presented and experimentally verified a new analytical model for the stress effects on ring gyroscopes. We showed that geometric nonlinearity and centerline extensibility of the inner suspension ring should be taken into account to identify stress-induced stiffness. We observed good agreement between the experimental results and the analytical predictions. Our method can be extended to the analysis of quadrature and in-phase errors that arise from external stress. Additionally, our model can accommodate fabrication-related structural imperfections and elastic anisotropy.

ACKNOWLEDGEMENTS

This work was supported by the Scientific and Technological Research Council of Turkey (TUBITAK) 2232 Program with the grant number 118C247. The ideas/claims presented here solely belong to the authors.

REFERENCES

[1] I. P. Prikhodko, A. A. Trusov, and A. M. Shkel, "Compensation of drifts in high-Q MEMS gyroscopes using temperature self-sensing," *Sens. Actuators Phys.*, vol. 201, pp. 517–524, Oct. 2013.

[2] E. Tatar, T. Mukherjee, and G. K. Fedder, "Stress Effects and Compensation of Bias Drift in a MEMS Vibratory-Rate Gyroscope," *J. Microelectromechanical Syst.*, vol. 26, no. 3, pp. 569–579, Jun. 2017.

[3] B. E. Uzunoglu, D. Erkan, and E. Tatar, "A Ring Gyroscope With On-Chip Capacitive Stress Compensation," *J. Microelectromechanical Syst.*, pp. 1–12, 2022.

[4] X. Zhang, S. Park, and M. W. Judy, "Accurate Assessment of Packaging Stress Effects on MEMS Sensors by Measurement and Sensor–Package Interaction Simulations," *J. Microelectromechanical Syst.*, vol. 16, no. 3, pp. 639–649, Jun. 2007.

[5] S. Schröder *et al.*, "Stress-Minimized Packaging of Inertial Sensors by Double-Sided Bond Wire Attachment," *J. Microelectromechanical Syst.*, vol. 24, no. 4, pp. 781–789, Aug. 2015.

[6] M. Hosseini-Pishrobat and E. Tatar, "Modeling and Analysis of a MEMS Vibrating Ring Gyroscope Subject to Imperfections," *J. Microelectromechanical Syst.*, vol. 31, no. 4, pp. 546–560, Aug. 2022.

[7] Z. Ma, X. Chen, X. Jin, Y. Jin, X. Zheng, and Z. Jin, "Effects of Structural Dimension Variation on the Vibration of MEMS Ring-Based Gyroscopes," *Micromachines*, vol. 12, no. 12, Art. no. 12.

[8] J. N. Reddy, *An introduction to nonlinear finite element analysis: with applications to heat transfer, fluid mechanics, and solid mechanics*, Second edition. Oxford: Oxford University Press, 2015.

[9] J. R. Barber, "Curved Beams," in *Intermediate Mechanics of Materials*, J. R. Barber, Ed. Dordrecht: Springer Netherlands, 2011, pp. 487–510.

[10] J. R. Barber, *Elasticity*, 2nd ed. Dordrecht: Kluwer Academic Publ, 2002.

CONTACT

*M. Hosseini-Pishrobat, tel: +90-312-2901219; mehran@ee.bilkent.edu.tr

RATE INTEGRATING GYROSCOPE TUNED BY FOCUS ION BEAM TRIMMING AND INDEPENDENT CW/CCW MODES CONTROL

Jianlin Chen[1], Takashiro Tsukamoto[1], Giacomo Langfelder[2] and Shuji Tanaka[1]
[1]Department of Robotics, Tohoku University, Sendai, JAPAN and
[2]Dipartimento di Elettronica, Informazione e Bioingegneria, Politecnico di Milano, Milano, ITALY

ABSTRACT

In this paper, structural asymmetries in a dual-axis symmetric resonator were reduced by FIB-trimming and electrostatic tuning and then fully compensated by the driving signal adjustment to achieve operation as a rate integrating gyroscope (RIG) for direct angle measurement. The as-fabricated frequency difference was minimized from 145 Hz to 24 Hz by applying symmetric focus ion beam (FIB) trimming on the suspension springs. The mode matching and quadrature null were further achieved by electrostatic softening the non-diagonal stiffness with a DC bias of 45 V. Finally, two degenerated modes were fully decoupled with a cross-coupling as small as 100 ppm enabled by adjusting the amplitude ratio and phase difference between the two-axis driving signals. The angle locking phenomenon was prevented and a 1.8 °/s self-precession rate was observed. The phase difference between clockwise (CW) and counterclockwise (CCW) modes, i.e. the RIG output, could be successfully modulated by a step-changed rotation. A rotation rate as small as 8.5 °/s could be detected.

KEYWORDS

Rate integrating gyroscope, Focus ion beam trimming, Mismatch compensation

INTRODUCTION

MEMS gyroscopes are essential components for many kinds of commercial electronics thanks to their low cost, small size and low power consumption [1]. To extend their application to navigation for automotive, the frequency modulated (FM) gyroscopes [2] and rate integrating gyroscopes (RIG) [3] were investigated towards further improvement of the scale factor stability over temperature variations, bandwidth and dynamic range. These gyroscopes require highly symmetric resonators with isoelasticity and isodamping matching to avoid a minimum rate threshold and angular dependent bias.

The anisoelasticity caused by fabrication errors and anisotropic material can be compensated by novel structure design [4], electrostatic tuning [5] and/or mechanical trimming [6]. The anisodamping can be independently compensated by air damping tuning [7], electrical damping tuning [8] and thermoelastic damping tuning [9]. These methods decrease the complexity of the frequency and Q-factor matching by independently tuning frequency and Q-factor.

In this paper, FIB trimming, electrostatic tuning and driving signals adjustment were combined to compensate the as-fabricated mismatches in a two-axis symmetric resonator. A coarse trimming was done by FIB and then the remaining mismatch was compensated by the electrostatic tuning. Finally, the driving signal tuning was exploited for

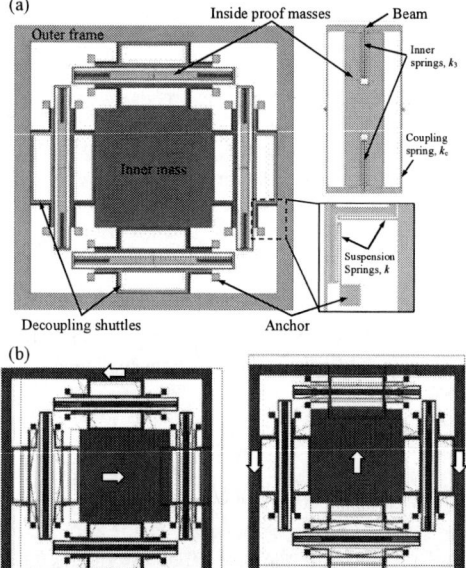

Fig. 1: (a) design of the dual-axis symmetric resonator with inside proof masses and (b) degenerated anti-phase modes along X and Y-axis.

a more precise matching.

DEVICE DESIGN

The proposed symmetric tuning fork resonator is shown in Fig. 1a. The resonator consists of two main proof masses (an *inner* nested mass and an *outer* frame), between which four *inside* proof masses are integrated symmetrically in X- and Y-axis. Such proof masses are connected to X- and Y-axis decoupling shuttles and anchored by suspension springs, which are defined as an outer tine and an inner nested tine. The inside proof masses are suspended from the connection beams by the inner springs, which can be treated as two additional one-mass systems integrated in a two-mass system. The resonator is symmetric in X- and Y-directions with an identical arrangement and the inner and outer tines (M_1 and M_2) are designed with an equivalent value, which improve the structural symmetries and mechanical Q-factors.

Figure 1b shows the simulated anti-phase modes in X and Y-axis (X-mode and Y-mode). Two orthogonal anti-phase modes are the desired modes, in which the outer frame and the inner mass vibrate in an opposite direction under structural balance and the inside proof masses keep static. Therefore, the frequency is dominant by the inner and the outer tine as well as the coupling springs. The X-axis and Y-axis motion can be decoupled by the shuttles mechanisms to decrease the cross coupling effectively.

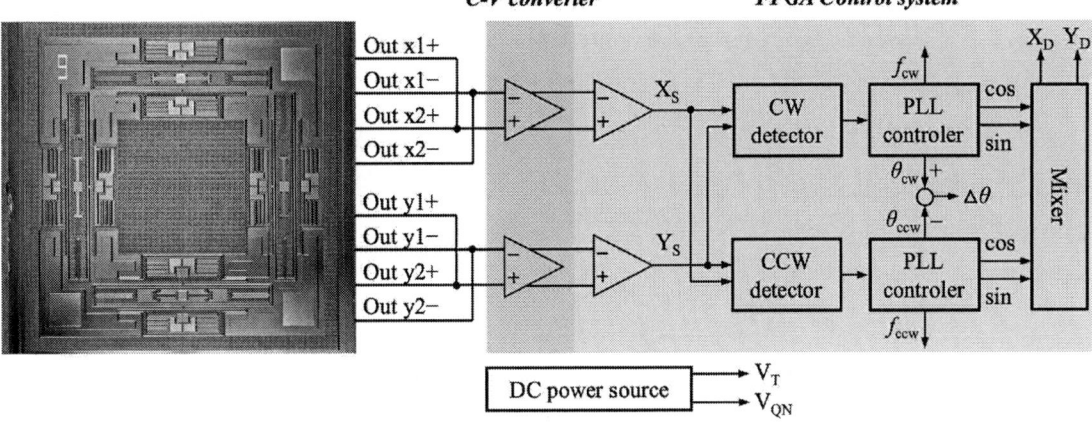

Figure 2: The fabricated device was connected to a C-V converter printed-circuit board (PCB). Then the output voltages were input in a FPGA board to achieve RIG operation.

EXPERIMENTAL SETUP

A control schematic is shown in Fig. 2. The device was mounted on a servo rotor and was connected to a C-V converter PCB in a vacuum chamber. Two pairs of parallel plate electrodes of the inner mass and outer frame were connected to the differential amplifier for reducing feedthrough signal and external acceleration effect. The motional signals were input to a FPGA control board for feedback control. The X- and Y-axis driving signal were generated by the control board and applied to the lateral comb electrodes along two orthogonal axes.

The motion signals were input in FPGA board and the mixed signals with CW and CCW modes were separated by CW/CCW mode detectors. Then CW and CCW modes can be locked to their resonant frequencies by PI controller and numerical controlled oscillators (NCOs). The *cos* and *sin* signals were synthesized by the mixer and applied to the actuating electrodes for feedback control. The frequencies and phase of CW and CCW modes can be acquired from the PLL controller based on synchronous demodulation. The phase difference was calculated by performing difference on the phases of CW and CCW modes, which was also the orientation angle of the vibration trajectory and was proportional to the rotation angle.

The DC bias was applied on the parallel-plate electrodes in the inner mass and outer frame, which electrostatic softened the suspension stiffness of two proof masses in primary axis of stiffness symmetrically. The quadrature cancellation was achieved by electrostatic softening the non-diagonal stiffness.

EXPERIMENTAL RESULTS
Fabrication and FIB trimming

The device was fabricated by the standard SOI process using a wafer with 60 μm device layer, 5 μm box layer and 500 μm handle layer, respectively. The as-fabricated frequency mismatch was 145 Hz, which was decreased to 24 Hz by FIB trimming on eight suspension springs (see in

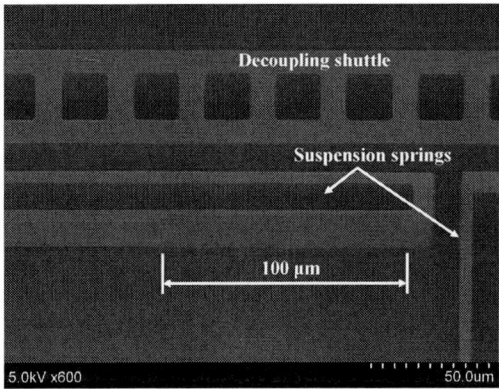

Figure 3: SEM image of a trimmed suspension spring in Y-axis.

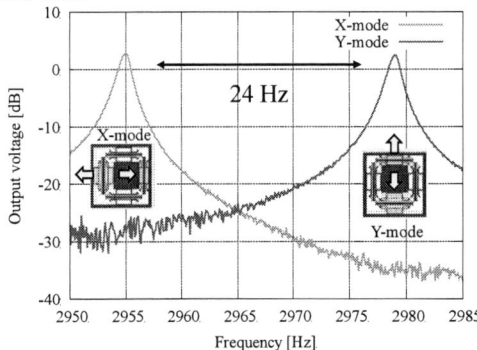

Figure 4: Frequency response of X- and Y-mode after FIB trimming. The frequency difference was decreased from 145 Hz to 24 Hz.

Fig. 3) (trimmed volume: $1.5 \times 100 \times 60 \ \mu m^3$) on Y-mode (Fig. 4). To further tune the frequency of Y-mode, a 35 V DC bias was applied on two pairs of parallel plate electrodes of inner mass and outer frame resulting a mode matching under 10 ppm. However, the ratio of quadrature error to the driving displacement was about 13%. To

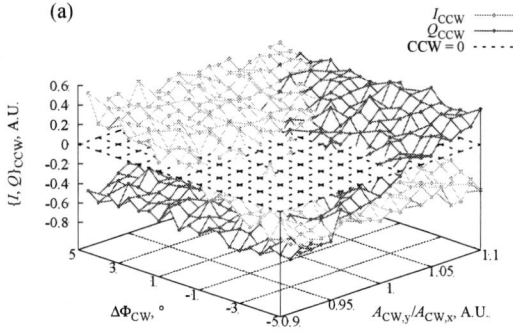

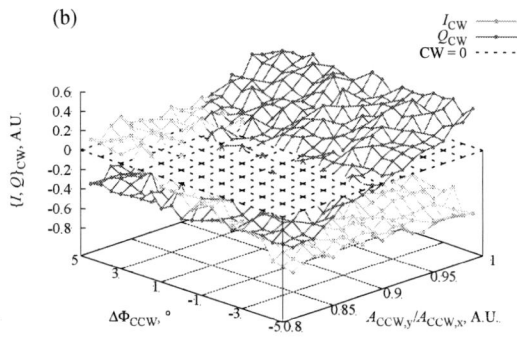

Figure 5: CCW (a) and CW (b) interfered components I, Q from CW and CCW oscillation, respectively. By sweeping the phase difference (ϕ_{CW}, ϕ_{CCW}) and amplitude ratio (A_Y/A_X) between two-axis driving signal, the mode cross-coupling could be calibrated.

simultaneously compensate frequency mismatch and quadrature error, a 45 V DC bias was applied on a pair of diagonal sensing electrodes in Y-axis [10], which achieved a frequency matching under 10 ppm and minimized the quadrature signal ratio under 3%.

Anisoelasticity and anisodamping compensation

Then the remained mismatches were precisely eliminated by adjusting the phase difference and amplitude ratio between the two-axis driving signals. The frequency and Q-factor mismatches could be compensated by adjusting the phase difference and the amplitude ratio of driving signals (A_Y/A_X), respectively. The calibration method was applied on CW/CCW mode, respectively. When the CW/CCW mode was actuated, the phase difference and amplitude ratio between the driving signals was adjusted to find the optimized parameters for the minimum cross-coupling components $I_{CW/CCW}$ and $Q_{CW/CCW}$. By adjusting the phase difference and voltage ratio of CCW mode to 0.9 and 3°, respectively, the cross coupling could be nearly compensated as shown in Fig. 5.

When both CW and CCW modes were simultaneously activated, the superposed oscillation of two degenerated modes changed to a linear trajectory. When the mismatches existed, the vibration trajectory could not freely precess under a rate threshold, i.e. locking to the direction with a higher Q-factor [11]. After mismatch compensation, a self-

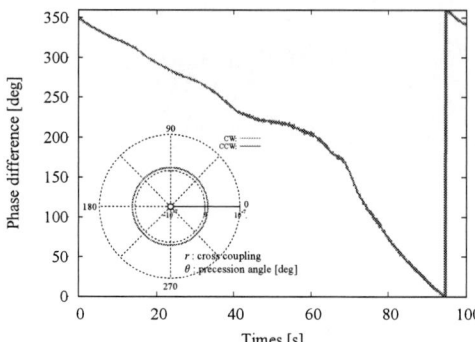

Figure 6: Self-precession of the oscillation trajectory after the frequency and Q-factor mismatch compensation. The cross-coupling terms were 100 ppm of the activated CW/CCW oscillation.

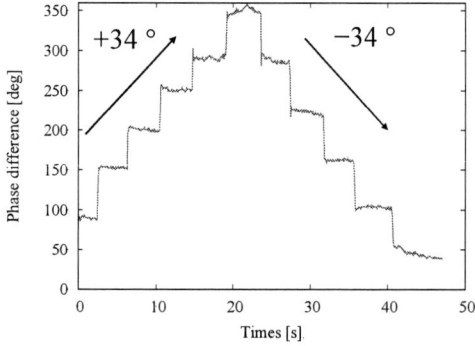

Figure 7: Phase difference was changed proportionally under a step-changed rotation with ±34° for each step, without angle locking phenomena.

precession of oscillation trajectory with a rate 3.7 °/s could be observed continuously as shown in Fig. 6. The interfered components of CW and CCW mode under different precession angles was collected showing a minor cross-coupling under 100 ppm (see in the polar graph of Fig. 6).

Rotation detection

Then a step-changed rotation with repeated ±34° (an increase/ a decrease of angle with a step of 34°, angle range: 0° to 170°) was applied on the device by the implemented rotor inside the chamber. The trajectory orientation was changed proportional to the rotation and stayed at the resulted angle without locking to the primary axis of damping (Fig. 7). The trajectory orientation shifted downward at the end of the return cycle (after steps of −34°), as a result of the self-precession observed in Fig. 6.

A rotation rate at 8.5 °/s was applied before and after the mismatches compensation, respectively. The mode matching and quadrature cancellation was firstly achieved by electrostatic tuning without compensating the Q-factor mismatch. The periodic nonlinear error was observed in the RIG output and the oscillation orientation was locked to 320° after 10 s as shown in Fig. 8a. On the other hand, when the remained mismatches were further compensated by

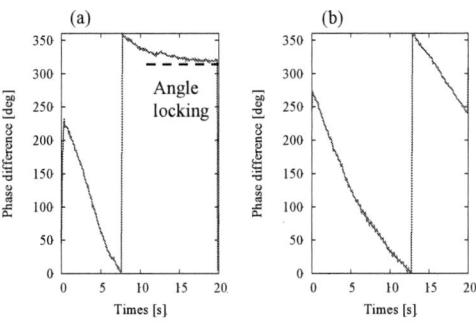

Figure 8: Phase difference under a rotation rate at 8.5 °/s (a) without and (b) with the mismatch compensation.

adjusting the driving signal parameters, the oscillation orientation was changed proportionally under the continuous rotation and a mixed self-precession rate could be observed (see in Fig. 8b).

CONCLUSION

This paper presents a two-axis symmetric resonator for RIG operation enabled by FIB trimming, electrostatic tuning and independent driving signal adjustment. The as-fabricated frequency difference was largely reduced by 6 times through the mechanical trimming. Then the remained anisoelasticity and quadrature error was compensated by the electrostatic softening method and was reduced to 10 ppm and 3%, respectively. Finally, a precise tuning was performed by adjusting the phase differences and amplitude ratios between two orthogonal driving signals, which successfully prevented the angle locking problems and enabled the vibration trajectory followed the expected precession under the applied rotation.

ACKNOWLEDGEMENTS

This paper is based on results obtained from projects commissioned by the New Energy and Industrial Technology Development Organization (NEDO) and Japan Society for the Promotion of Science (JSPS) KAKENHI grant-in-aid for young scientist no. 21J11628.

REFERENCES

[1] G. Langfelder, M. Bestetti, and M. Gadola, "Silicon MEMS inertial sensors evolution over a quarter century." *Journal of Micromechanics and Microengineering*, vol. 31, no. 8, pp. 084002, 2021.

[2] T. Tsukamoto and S. Tanaka, "Fully differential single resonator FM gyroscope using CW/CCW mode separator," *Journal of Microelectromechanical Systems*, vol. 27, no. 6, pp. 985–994, 2018.

[3] T. Tsukamoto and S. Tanaka, "Rate integrating gyroscope using independently controlled CW and CCW modes on single resonator," *Journal of Microelectromechanical Systems*, vol. 30, no. 1, pp. 15–23, 2020.

[4] S. Wang, J. Chen, T. Tsukamoto and S. Tanaka, "Mode-Matching Multi-Ring Disk Resonator Using (100) Single Crystal Silicon," in 2022 *IEEE MEMS*. IEEE, 2022, pp. 786-789.

[5] A. Efimovskaya, D. Wang, Y.-W. Lin, and A. M. Shkel, "Electrostatic compensation of structural imperfections in dynamically amplified dual- mass gyroscope," *Sensors and Actuators A: Physical*, vol. 275, pp. 99– 108, 2018.

[6] J. Chen, T. Tsukamoto, and S. Tanaka, "Quad mass resonator with frequency mismatch of 3 ppm trimmed by focused ion beam," *Journal of Microelectromechanical Systems*, vol. 30, no. 3, pp. 392–400, 2021.

[7] J. Chen, T. Tsukamoto, G. Langfelder, and S. Tanaka, "Frequency and quality factor matched 2-axis dual mass resonator," In *2021 IEEE SENSORS*. IEEE, 2021, pp. 1-4.

[8] R. Gando, D. Ono, S. Kaji, H. Ota, T. Itakura, and Y. Tomizawa, "A compact microcontroller-based MEMS rate integrating gyroscope module with automatic asymmetry calibration," in *2020 IEEE 33rd International Conference on Micro Electro Mechanical Systems (MEMS)*. IEEE, 2020, pp. 1296–1299.

[9] A. Hamza, T. Tsukamoto and S. Tanaka, "Quality factor trimming method using thermoelastic dissipation for multi-ring resonator," *Sensors and Actuators A: Physical*, vol. 332, pp. 113044, 2021.

[10] E. Tatar, S. E. Alper, and T. Akin, "Quadrature-Error Compensation and Corresponding Effects on the Performance of Fully Decoupled MEMS Gyroscopes," *Journal of Microelectromechanical Systems*, vol. 21, no. 3, pp. 656–667, 2012.

[11] P. Taheri-Tehrani, A. D. Challoner, and D. A. Horsley, "Micromechanical Rate Integrating Gyroscope With Angle-Dependent Bias Compensation Using a Self-Precession Method," *Journal of Microelectromechanical Systems*, vol. 18, no. 9, pp. 3533–3543, 2018.

CONTACT

*J. Chen, Email: chenjl@tohoku.ac.jp

TEMPERATURE DEPENDENCE OF QUALITY FACTORS AT HIGH FREQUENCIES IN MEMS GYROSCOPES

Daniel Schiwietz[1,2], Eva M. Weig[2], and Peter Degenfeld-Schonburg[1]
[1]Robert Bosch GmbH, GERMANY and
[2]Technical University of Munich, GERMANY

ABSTRACT

An efficient simulation approach for obtaining the temperature-dependent thermoelastic damping (TED) quality factors of the mechanical modes in complex microelectromechanical systems (MEMS) gyroscopes is reported. It is shown that the temperature dependence of TED in an application-relevant temperature range can be obtained by a Taylor expansion of the full solution around room temperature. Our approach is much faster than performing the full simulation at every temperature. We find good agreement of our simulation results with measured data and show that TED is highly relevant for the overall quality factors of higher order modes.

KEYWORDS

MEMS gyroscopes, numerical modelling, thermoelastic damping, quality factors

INTRODUCTION

Microelectromechanical systems (MEMS) gyroscopes are an integral part of modern automotive and consumer electronics [1,2]. In order to meet the ever-increasing performance requirements, the accuracy and completeness of the utilized simulation tools is crucial. At the same time, simulation methods must be fast enough to not slow down the development cycle.

In many applications, MEMS gyroscopes are faced with a wide range of external environmental conditions and disturbances. In automotive applications in particular, operability must be ensured over the temperature range from -40°C to 120°C [1]. Predicting the response of the gyroscope requires, among other quantities, the quality factors of the mechanical modes. The temperature dependence of the quality factors is governed by the underlying damping mechanisms.

The main damping mechanisms usually considered in MEMS gyroscopes are gas damping, thermoelastic damping (TED) and anchor losses [3]. The first two are temperature-dependent, while anchor losses are assumed to be independent of temperature. Gas damping arises from the interaction of the moving structure with the surrounding gas. At typical operational pressures of a few millibar, gas damping is known to be the dominant damping mechanism for the low order modes, such as the drive and sense modes [4]. However, for higher order modes, the gas damping contributions to the overall quality factors become less relevant compared to other damping mechanisms. Recently, it has been shown over a wide range of modes of an industrial MEMS gyroscope, that TED has a large impact on the overall quality factors of the higher order modes [5]. TED arises from the coupling of the stress and temperature fields and the resulting irreversible heat flows that occur during mechanical oscillations. TED is highly dependent on thermal material parameters, which show a much stronger temperature dependence than mechanical material parameters [6].

In the present work, we show that for our MEMS gyroscopes the temperature-dependent TED quality factors can be efficiently computed via a Taylor expansion of the finite element method (FEM) solution around room temperature. By that, the temperature dependence of the thermal material parameters can be accounted for, without having to rerun the simulation for every temperature. Our method is efficient and can be applied for predictive purposes in the development process of complex MEMS resonators. Comparisons to measured quality factors highlight the accuracy and validity of our method.

MODEL DESCRIPTION

The main damping mechanisms that are usually considered for MEMS resonators are gas damping, TED and anchor losses. The inverse of the total quality factor Q of a mode is the sum of the inverse quality factors of the individual damping mechanisms

$$\frac{1}{Q(T_0, p)} = \frac{1}{Q_{gas}(T_0, p)} + \frac{1}{Q_{TED}(T_0)} + \frac{1}{Q_{anchor}}, \quad (1)$$

with gas damping quality factor Q_{gas}, TED quality factor Q_{TED}, anchor loss quality factor Q_{anchor}, ambient temperature T_0 and ambient pressure p. This work is concerned with modelling TED. For that purpose, gas damping can be made negligible by reducing the pressure sufficiently, since $Q_{gas}^{-1} \propto p$ in the low-pressure regime [7]. Assuming that anchor losses are temperature-independent, the temperature dependence of the total quality factor is then governed by TED.

The origin of TED lies in the coupling of the mechanical equation of motion and the heat equation. Oscillating stress gradients give rise to oscillating temperature gradients, which induce irreversible heat flows, leading to dissipation of energy. Research on TED was sparked by Zener, who derived an analytic expression for the quality factor of a beam [8,9]. For arbitrary geometries FEM simulations are needed and TED can be described by the discretized equation of motion and heat equation [10]

$$M\ddot{u} + K^u u + \alpha K^{ut} \Delta T = f, \quad (2)$$

$$C_V C^t \dot{\Delta T} + \kappa K^t \Delta T = \alpha T_0 (K^{ut})^T \dot{u}, \quad (3)$$

with FEM system matrices M, K^u, K^{ut}, C^t and K^t, nodal displacement vector u, nodal temperature change vector ΔT and external force vector f. The temperature change vector is equal to the difference between the nodal temperatures and the ambient temperature. The thermal expansion coefficient α, specific heat C_V and thermal conductivity κ have been factored out of the matrices to make the dependence on thermal material properties more explicit. From Eqs. (2) and (3) the inverse TED quality factor of mode n can be derived as [5]

$$Q_{TED,n}^{-1}(T_0) = \mathrm{Re}\left\{\frac{\alpha^2 T_0}{\omega_n}\boldsymbol{\phi}_n^T \boldsymbol{K}^{ut} \boldsymbol{A}^{-1} (\boldsymbol{K}^{ut})^T \boldsymbol{\phi}_n\right\}, \quad (4)$$

with the matrix $\boldsymbol{A}$ defined as

$$\boldsymbol{A} = \kappa \boldsymbol{K}^t + i\omega_n C_V \boldsymbol{C}^t, \quad (5)$$

with the n-th mode's eigenfrequency ω_n and eigenvector $\boldsymbol{\phi}_n$ and the imaginary number i. The advantage of Eq. (4) is that the TED quality factor can be calculated by solving a linear equation system with as many unknowns as there are temperature degrees-of-freedom, in contrast to solving the full coupled equation system of Eqs. (2) and (3), e.g., by a complex modal analysis [11]. Nevertheless, solving Eq. (4) for potentially hundreds of modes at various temperatures still constitutes a significant computational effort. In order to reduce the computational complexity further, we linearize Eq. (4) around room temperature T_{RT}, leading to

$$Q_{TED,n}^{-1}(T_0) \approx$$
$$Q_{TED,n}^{-1}(T_{RT}) + (T_0 - T_{RT})\left.\frac{\partial Q_{TED,n}^{-1}}{\partial T_0}\right|_{T_0=T_{RT}}, \quad (6)$$

where the derivatives are calculated as

$$\frac{\partial Q_{TED,n}^{-1}}{\partial T_0} = \left(\frac{1}{T_0} + \frac{2}{\alpha}\frac{\partial \alpha}{\partial T_0}\right)Q_{TED,n}^{-1}$$
$$-\mathrm{Re}\left\{\frac{\alpha^2 T_0}{\omega_n}\boldsymbol{\phi}_n^T \boldsymbol{K}^{ut}\boldsymbol{A}^{-1}\frac{\partial \boldsymbol{A}}{\partial T_0}\boldsymbol{A}^{-1}(\boldsymbol{K}^{ut})^T\boldsymbol{\phi}_n\right\}, \quad (7)$$

$$\frac{\partial \boldsymbol{A}}{\partial T_0} = \frac{\partial \kappa}{\partial T_0}\boldsymbol{K}^t + i\omega_n \frac{\partial C_V}{\partial T_0}\boldsymbol{C}^t. \quad (8)$$

We neglect the temperature dependencies of Young's modulus and density as well as the temperature dependence of the eigenfrequency, as they are usually much weaker than the temperature dependencies of α, κ and C_V. The linearized equation allows us to take the temperature dependencies of all thermal material parameters into account, without having to solve a different linear equation system for each temperature, as Eq. (4) would require. The factorization of $\boldsymbol{A}$ must only be calculated once per mode. Evaluating $Q_{TED,n}^{-1}(T_{RT})$ and $\left.\frac{\partial Q_{TED,n}^{-1}}{\partial T_0}\right|_{T_0=T_{RT}}$ then requires a total of two forward and backward substitutions per mode. Therefore, the temperature-dependent TED can be computed efficiently.

DEVICES AND TESTING

We compare our simulations to measured data of two different industrial MEMS gyroscope designs. The quality factors of various modes were measured with scanning laser Doppler vibrometers (SLDV). The MEMS gyroscopes were made of isotropic polycrystalline silicon.

We have previously determined the temperature dependencies of α and κ with fit functions from measurements [5], which allow us to calculate their derivatives with respect to temperature analytically. The temperature dependence of C_V is calculated from the Debye model [5,12] and its derivative is estimated via finite differences.

Device 1 was actuated electrostatically by applying a broadband signal to various electrode pairs. The measurement was performed in a vacuum chamber on a thermal chuck, so that the temperature could be controlled. The measurement of device 1 was performed at a pressure of 10^{-3} mbar, where gas damping was found to be negligible [5]. In-plane and out-of-plane modes were measured for device 1. Device 2 was measured in a vacuum chamber at 1 mbar and 25 °C. The actuation of device 2 was realized with a piezo-shaker, that applied a broadband signal predominantly in out-of-plane direction. Only out-of-plane modes were measured for device 2.

RESULTS

In Fig. 1, the measured inverse quality factors for seven different modes of device 1 are compared to the simulation results based on Eq. (6) over an application-relevant temperature range. Furthermore, a constant offset was added to the result of Eq. (6) for each mode, to emulate the effect of temperature-independent anchor loss. The modes are labeled with letters A to G and their eigenfrequencies are also shown. The black dots show Eq. (6) with temperature-dependent α, κ and C_V. The blue circles show Eq. (6) with the room temperature value of κ and C_V assumed over the whole temperature range, i.e., where the second term in Eq. (7) was neglected. The red dots show the result of Eq. (6) where α, κ and C_V are all assumed as constant over temperature, so that only the temperature dependence due to the T_0 prefactor in Eq. (4) was considered. One can immediately see that the measured data is much steeper over temperature than the red dots, where the temperature dependence of the material parameters was neglected. This highlights the importance of incorporating the correct temperature dependencies of the material parameters into the simulation. The blue circles, which only include the temperature dependence of α, already show a much better agreement with the measured data. For our material and temperature range, the temperature dependence of α appears to be the most relevant out of all material parameters. Note that Eq. (4) scales quadratically with α. Considering also the temperature dependencies of κ and C_V leads to the black dots, which show a good agreement over the whole temperature range for all modes. For some modes the increase in accuracy between black dots and blue circles is negligible. However, for four of the measured modes one can see a noticeable impact of the temperature dependencies of κ and C_V on the inverse quality factor. Note that if one would consider a wider temperature range or a material with more drastic dependencies on temperature, the linearization in Eq. (6) could lead to

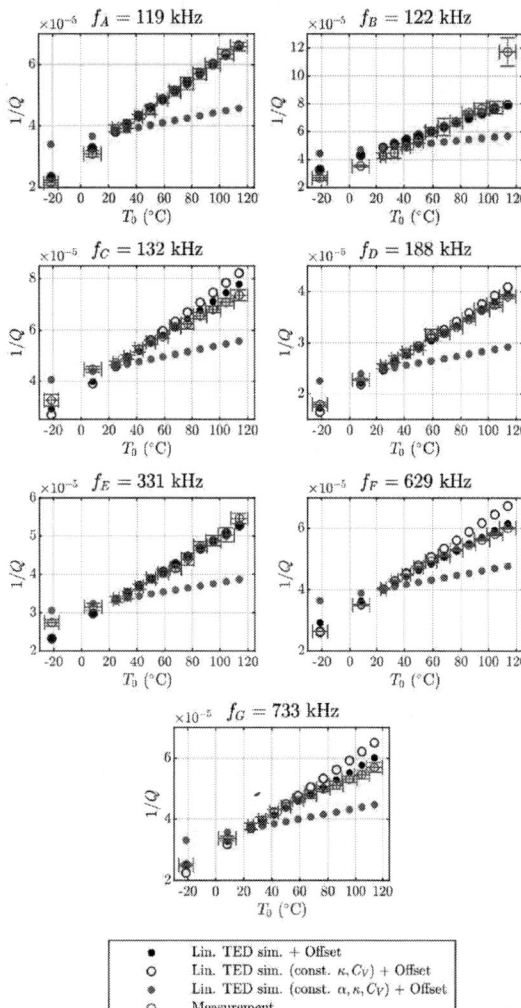

Figure 1: MEMS gyroscope design 1. Measured and simulated quality factors at a pressure of 10^{-3} mbar over temperature. The simulated quality factors are based on the linearized TED according to Eq. (6) and have a temperature-independent offset added to emulate anchor losses.

inaccurate results. A simple remedy would be to increase the order of the Taylor expansion, which would still only require one factorization of $\boldsymbol{A}$ per mode.

In the following, the temperature dependencies of all three thermal material parameters were incorporated into the simulations. Figure 2 shows measured quality factors over a wide range of out-of-plane modes for device 2. Furthermore, simulation results based on Eq. (6) are shown at -40 °C, +25 °C and + 120°C. Gas damping quality factors, based on molecular flow simulations, were also included in the total simulated Q. The gas damping quality factors were scaled as $Q_{gas}^{-1} \propto \sqrt{T_0}$ [6] from the room temperature value at 1 mbar. Additionally, a constant Q_{anchor} of 75000 was included for every mode to obtain a better fit to the measured data. One can see that the simulated quality factors show a good agreement with the measured trend. Looking at the results at -40 °C and

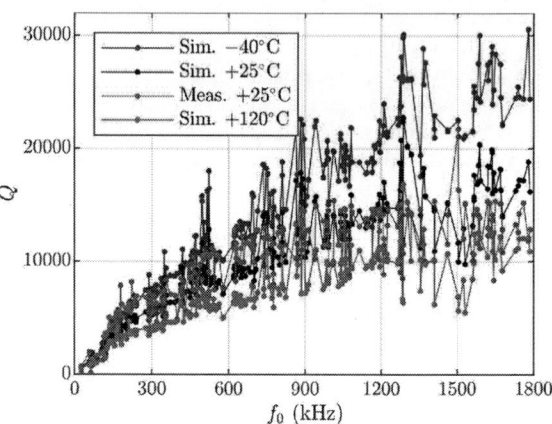

Figure 2: MEMS gyroscope design 2. Simulated quality factors based on gas damping, thermoelastic damping and a constant $Q_{anchor}=75000$ for different temperatures and a room temperature encapsulation pressure of 1 mbar over a wide range of modes. Also shown is the measured data at room temperature and 1 mbar.

+120 °C, one can see that the quality factors change by around 30%, compared to the values at +25 °C. The significant change within the application-relevant temperature range highlights the need for accurate and efficient simulation tools for the prediction of quality factors over temperature.

In Fig. 3 the ratio of the total simulated quality factors from Fig. 2 to the simulated pure gas damping quality factors is shown for the three different temperatures. One can clearly see that gas damping is dominant at the lower end of the spectrum. However, at the upper end, gas damping constitutes only 25% to 50% of the total damping. This highlights that the damping of the higher modes is dominated by TED. Furthermore, as $Q_{gas}^{-1} \propto \sqrt{T_0}$ and, for our devices, $Q_{TED}^{-1} \propto T_0$, one can see that the ratio Q/Q_{gas} decreases with increasing temperature. TED becomes more relevant at elevated temperatures, whereas gas damping gains relevance at lower temperatures.

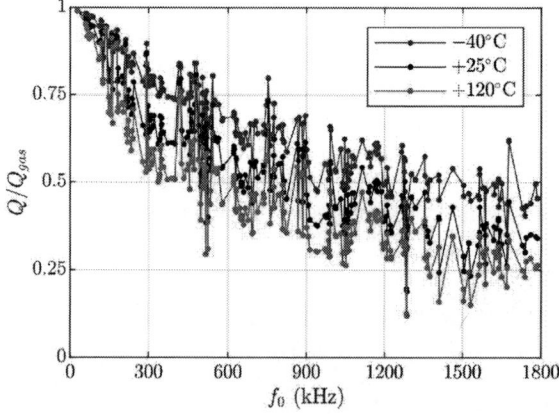

Figure 3: MEMS gyroscope design 2. Ratio of simulated total quality factors to pure gas damping quality factors for different temperatures over a wide range of modes.

CONCLUSION

We demonstrated an efficient method to simulate the temperature dependence of TED quality factors. By performing a linear Taylor expansion of the FEM closed-form expression around room temperature, we obtained good agreement of simulated and measured quality factors of an industrial MEMS gyroscope over the application-relevant temperature range. Our approach allows us to incorporate the full temperature dependence of the thermal material parameters, without having to simulate TED at every temperature.

Furthermore, we demonstrated the impact of TED over a wide range of modes. Our measurements and simulations show that TED is the dominant damping mechanism for high frequency modes. Based on our simulations, we found that the overall quality factor can significantly change over the application-relevant temperature range. Therefore, the simulation of the temperature-dependent quality factors is highly relevant. Our approach is efficient and suitable for complex industrial MEMS gyroscopes, where one often has to consider a wide range of modes.

ACKNOWLEDGEMENTS

The authors gratefully acknowledge technical support from Ulrich Kunz and Thomas Buck at Robert Bosch GmbH during the measurements of design 1. Furthermore, the authors would like to thank Christian Budak at Robert Bosch GmbH for providing the measurements of design 2.

REFERENCES

[1] R. Neul, et al. "Micromachined angular rate sensors for automotive applications", *IEEE Sensors Journal*, 7(2), pp. 302-309, 2007.

[2] D. K. Shaeffer "MEMS inertial sensors: A tutorial overview", *IEEE Communications Magazine*, 51(4), pp. 100-109, 2013.

[3] R. N. Candler, et al. "Investigation of energy loss mechanisms in micromechanical resonators", *Digest Tech. Papers Transducers'03 Conference*, Boston, June 8-12, 2003, pp. 332-335.

[4] A. Frangi, et al. "Near vacuum gas damping in MEMS: numerical modeling and experimental validation", *Journal of Microelectromechanical Systems*, 25(5), pp. 890-899, 2016.

[5] D. Schiwietz, E. M. Weig, P. Degenfeld-Schonburg "Thermoelastic Damping in MEMS Gyroscopes at High Frequencies", *arXiv preprint arXiv:2208.02591*, 2022.

[6] B. Kim, et al. "Temperature dependence of quality factor in MEMS resonators", *Journal of Microelectromechanical Systems*, 17(3), pp. 755-766, 2008.

[7] K. L. Ekinci, M. L. Roukes "Nanoelectromechanical systems", *Review of scientific instruments*, 76(6), 061101, 2005.

[8] C. Zener "Internal friction in solids I. Theory of internal friction in reeds", *Physical review*, 52(3), p. 230, 1937.

[9] C. Zener "Internal friction in solids II. General theory of thermoelastic internal friction", *Physical review*, 53(1), p. 90, 1938.

[10] R. Ardito, et al. "Solid damping in micro electro mechanical systems", *Meccanica*, 43(4), pp. 419-428, 2008.

[11] B. Antkowiak, et al. "Design of a high-Q, low-impedance, GHz-range piezoelectric MEMS resonator", *Digest Tech. Papers Transducers'03 Conference*, Boston, June 8-12, 2003, pp. 841-846.

[12] C. Kittel, *Introduction to solid state physics*, 8th ed., Wiley, Hoboken, NJ, 2005.

CONTACT

*D. Schiwietz, tel: +49-172-2971755; daniel.schiwietz@de.bosch.com

0.5MM×0.5MM 150KPA-MEASURE-RANGE HIGH-TEMPERATURE PRESSURE SENSOR WITH HIGH-PERFORMANCE AND LOW FABRICATION-COST

Peng Li[1,2], Wei Li[1], Changnan Chen[1,3], Ke Sun[1], Min Liu[1], Sheng Wu[1], Pichao Pan[1,3], Jiachou Wang[1,3], and Xinxin Li[1,2,3]

[1]State Key Laboratory of Transducer Technology, Shanghai Institute of Microsystem and Information Technology, Chinese Academy of Sciences, Shanghai 200050, China
[2]State Key Laboratory of ASIC and System, School of Microelectronics, Fudan University, Shanghai 200433, China
[3]University of Chinese Academy of Sciences, Beijing 100049, China

ABSTRACT

This paper reports a high temperature pressure sensor fabricated with TUB (thin-film under bulk) micromachining process from the front side of an ordinary (111)-SOI silicon wafer. Compared with the reported high temperature pressure sensor, the fabricated pressure sensor chip-size is as tiny as 0.5mm × 0.5mm with high sensitivity and low non-linearity. The testing results show that the 150-kPa-range pressure sensor has a full-scale output of 65.39mV/150kPa/3.3V, very high combined accuracy of 0.13%·FS within the range of -55°C to 300°C, and a low thermal hysteresis of 0.22%·FS at 300°C. In addition, the TCO (temperature coefficient of offset) is also tested to be as low as 0.01%/°C·FS for a full measure range of 150 kPa and the TCS (temperature coefficient of sensitivity) is about -0.07%/°C·FS.

KEYWORDS

pressure sensor, high temperature, tiny size, single side, TUB process

INTRODUCTION

With the rapid development of technology, the tiny-sized low-ranged high temperature pressure sensor has attracted more and more attentions and widely used in defense industry, aerospace application. Compared with most of the previously reported works [1-2], in this paper our previously developed TUB micromachining process [3] is employed on an ordinary SOI silicon wafer to successfully fabricate a tiny-sized low-ranged high temperature pressure sensor with high sensitivity, low non-linearity and high thermal-stability within the range from -55°C to 300°C.

SENSOR DESIGN

The 3-D schematic of our proposed high temperature pressure sensor is in Figure 1(a). The pressure sensor is prepared with single-side fabrication [4] from single ordinary (111)-SOI silicon wafer. The device layer of the SOI wafer is doped to the highest level to achieve a more stable, long-term electrical performance characteristics of the sensing elements [5]. The handle layer is used to fabricate 2μm-thick poly-Si diaphragm and SC-silicon (single-crystalline silicon) beam-island-reinforced structure in order to achieve both high sensitivity and low non-linearity [6]. The uniform and symmetrical Ti/Pt/Au-trace line is patterned to reduce the influence of the thermal

hysteresis phenomenon on the output signal of the sensor as shown in Figure 1(b).

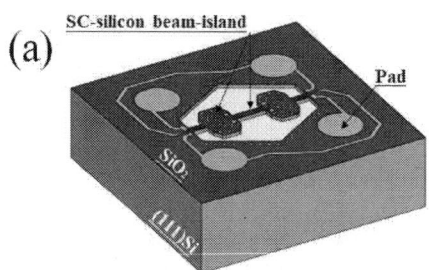

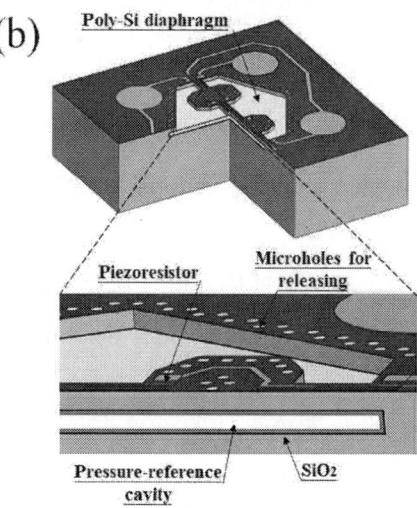

Figure 1: 3D sketch of the high temperature pressure sensor.

FABRICATION

Figure 2 shows the fabrication steps of the high temperature pressure sensor. It is worth pointing out that before LPCVD poly-Si sealing the microholes, a thin SiO_2 layer is thermally grown throughout the SOI silicon wafer in the step of (i). It acts as the etch-stop layer between the SC-silicon and the poly-Si for protecting the poly-Si diaphragm from over-etch when the beam-island-reinforced structure is formed by RIE process in the step of (l). When poly-Si is grown via microholes to form the poly-Si diaphragm for pressure sensing, the thickness of it is determined by the opening-size of the microholes in the step of (i).

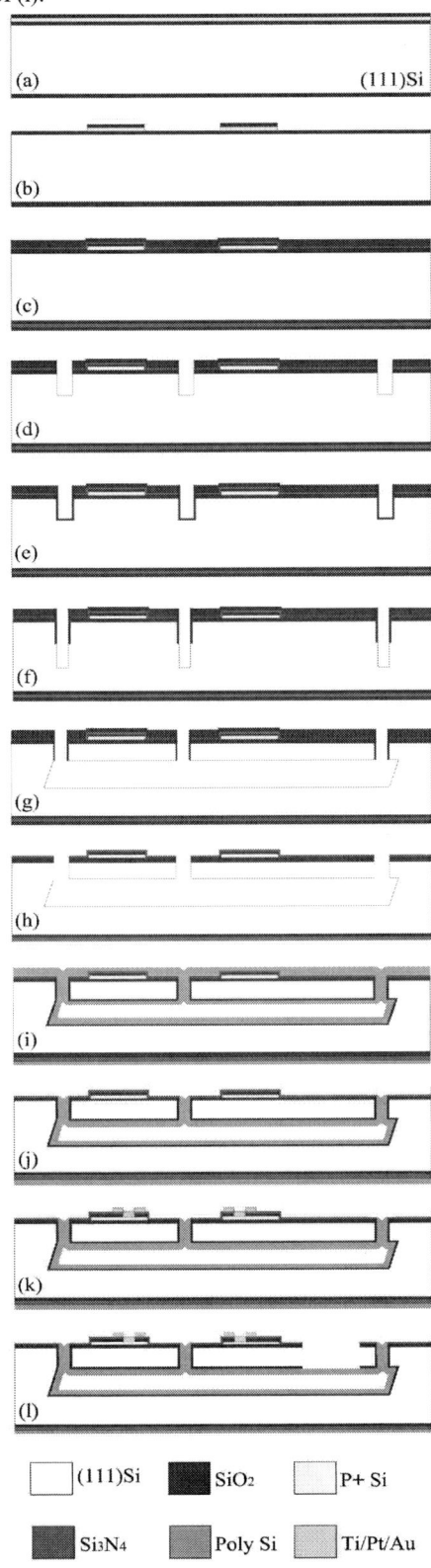

(a) (111)Si
(b)
(c)
(d)
(e)
(f)
(g)
(h)
(i)
(j)
(k)
(l)

☐ (111)Si ■ SiO₂ ▨ P+ Si

▨ Si₃N₄ ▨ Poly Si ▨ Ti/Pt/Au

Figure 2: Fabrication steps for the high temperature pressure sensors begin with N-type (111)-SOI silicon

wafers. *(a) Thermal oxidation and high dose boron ion implantation. (b) RIE for thermal SiO₂ and SC-silicon to pattern the piezoresistors. (c) LPCVD-deposition for Si₃N₄ and SiO₂. (d) RIE for SiO₂, Si₃N₄ and SC-silicon to pattern the micoholes. (e) LPCVD-deposition for SiO₂ to coat the sidewalls of the micoholes. (f) RIE for SiO₂ and SC-silicon to deepen the microholes. (g) Wet anisotropic etching of TMAH to form the pressure-reference cavity. (h) SiO₂ layer stripped in the BOE solution. (i) Thermal oxidation and LPCVD poly-Si are grown inside the cavity via microholes until they are sealed and then annealed. (j) Poly-Si etch by maskless RIE in the front side. (k) Contact holes are formed by RIE and Ti/Pt/Au-trace lines are patterned. (l) RIE for Si₃N₄, SiO₂ and SC-silicon is processed to form the strcture of beam-island reinforced diaphragm.*

TESTING RESULTS

Figure 3 shows the SEM and optical images of the fabricated high temperature pressure sensor. The pressure sensor is powered by a DC voltage of 3.3 V and measured by a digital pressure gauge and a hand-held pump.

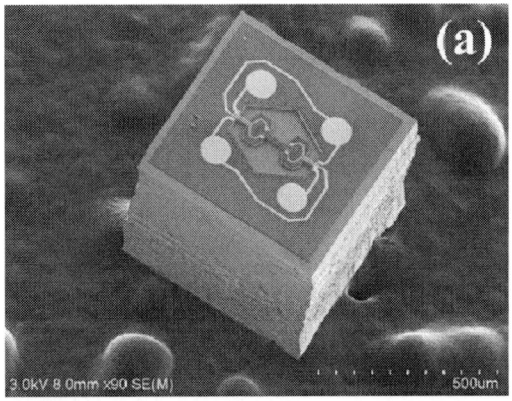

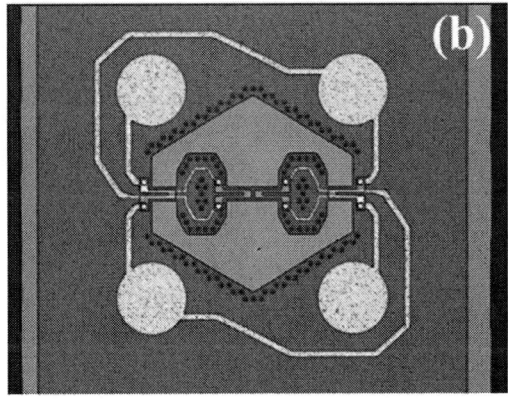

Figure 3: (a) SEM and (b) optical images of the fabricated sensor.

The tested output voltage is plotted against the applied pressure with a range from 20 to 150 kPa. At room temperature of 25 °C, the full-scale output voltage of the pressure sensor is about 65.39mV/150kPa/3.3V. Within the temperature range of -55°C to 300°C, the sensitivity is as high as 0.40 mV/kPa and the combined uncertainty is 0.13%·FS as shown in Figure 4.

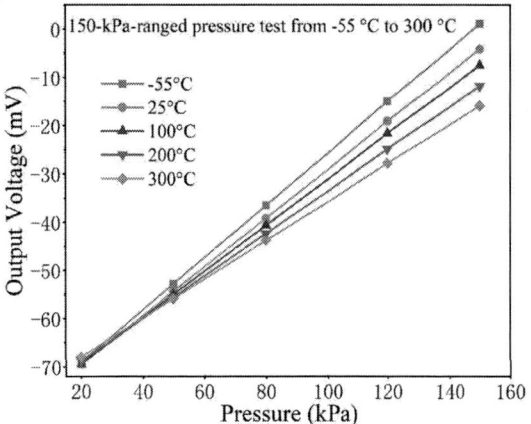

Figure 4: 150-kPa-ranged high temperature pressure sensor with a high combined uncertainty of 0.13 %·FS and a high sensitivity of 0.40 mV/kPa in the range of -55°C to 300°C.

The nonlinearity of the high temperature pressure sensor is better than 0.12% within the range of -55°C to 300°C as shown in Figure 5.

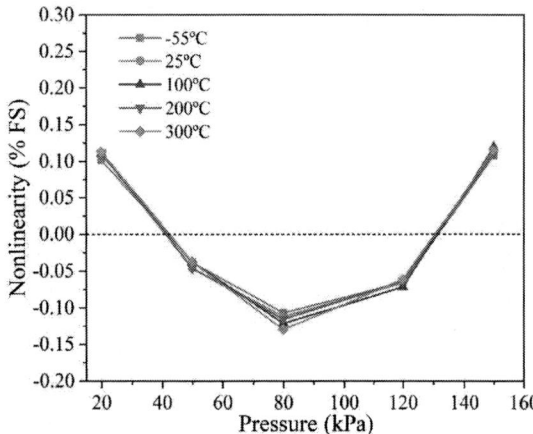

Figure 5: The high temperature pressure sensor with the nonlinearity of 0.12%.

The thermal hysteresis of the pressure sensor is as low as 0.22%·FS at 300 °C as shown in Figure 6. In addition, the TCO is tested to be as low as 0.01%/°C·FS for a full-scale pressure range of 150 kPa and the TCS is tested to be -0.07%/°C·FS as shown in Figure 7.

All the test results of the pressure sensor are obtained and listed in table 1.

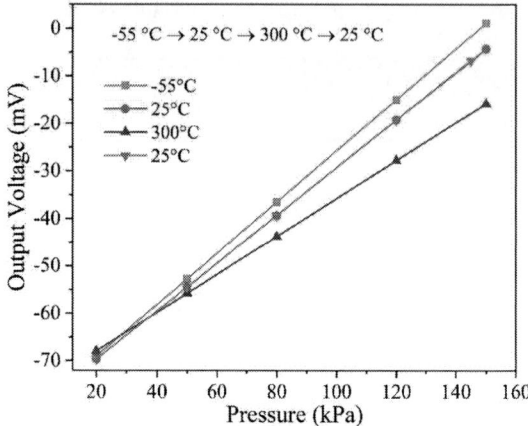

Figure 6: The thermal hysteresis is 0.22%·FS at 300 °C.

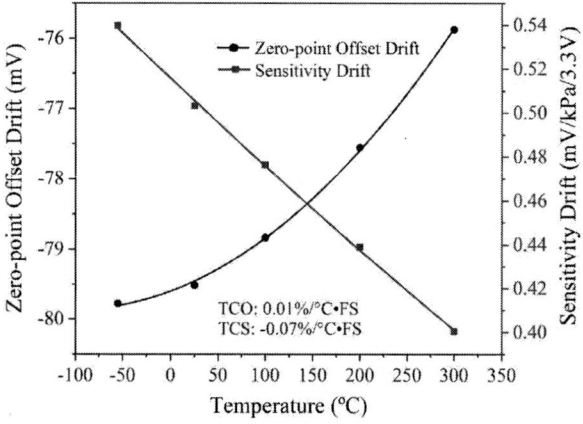

Figure 7: Tested TCO and TCS of the pressure sensor in terms of the temperature.

Table 1: Measurement results of the high temperature pressure sensor.

Parameters	Value
Sensitivity	0.40 mV/kPa/3.3V
Nonlinearity	0.12%·FS
combined accuracy	0.13%·FS
thermal hysteresis	0.22%·FS
TCO	0.01%/°C·FS
TCS	-0.07%/°C·FS

CONCLUSIONS

A high temperature pressure sensors with chip-size as 0.5mm × 0.5mm are reported in this paper. Combined SOI silicon wafer and TUB micromachining process, the tiny-sized pressure sensors perform high sensitivity and low non-linearity. Avoiding using wafer bonding process and taking use of the symmetrical and uniform layout of Ti/Pt/Au-trace line, the pressure sensors also exert a low thermal hysteresis and a high combined accuracy within the range of -55°C to 300°C. The fabricated sensor is promising for low-cost volume production and pressure measurements in harsh environment.

ACKNOWLEDGEMENTS

This work was supported in part by the National Science Foundation of China Projects (62074151, 61834007) and Innovation Team and Talents Cultivation Program of National Administration of Traditional Chinese Medicine (No: ZYYCXTD-D-202002, ZYYCXTD-D-202003).

REFERENCES

[1] Z. Yao, T. Liang, P. Jia, Y. Hong, L. Qi, C. Lei, B. Zhang, and J. Xiong, "A high-temperature piezoresistive pressure sensor with an integrated signal-conditioning circuit", *Sensors*, vol. 16, pp. 913, 2016.

[2] S. Guo, H. Eriksen, K. Childress, A. Fink, and M. Hoffman, "High temperature smart-cut SOI pressure sensor", *Sens. Actuators A Phys.*, vol. 154, no. 2, pp. 255-260, 2009.

[3] H. Zou, J. Wang, X. Li, "A novel TUB (thin-film under bulk) process for high-performance pressure sensors of sub-kPa measure-range", MEMS 2016 Conference, Shanghai, China, Jan 24-28, 2016, pp. 214-217.

[4] J. C. Wang, Xinxin Li, "Single-side fabricated pressure sensors for IC-Foundry compatible high-yield and low-cost volume production", IEEE Elec. Dev. Lett., vol. 32, 2011, pp. 979-981.

[5] A. D. Kurtz, A. A. Ned, and A. H. Epstein, "Ultra High Temperature, Miniature, SOI Sensors for Extreme Environments", *the IMAPS International HiTEC 2004 Conference*, Santa Fe, New Mexico, May 17-20, pp. 1-11, 2004.

[6] M. H. Bao, Micro Mechanical Transducers-Pressure Sensors, Accelerometers and Gyroscopes, Amsterdam: Elsevier, 2000.

CONTACT

*Jiachou Wang, tel: +86-62511070; jiatao-wang@mail.sim.ac.cn

*Xinxin Li, tel: +86-62131794; xxli@mail.sim.ac.cn

AUTOMATIC PICO LASER TRIMMING SYSTEM FOR SILICON MEMS RESONANT DEVICES BASED ON IMAGE RECOGNITION

Yuxian Liu[1], Qiancheng Zhao[1,2], Dacheng Zhang[1], and Jian Cui[1,2]

[1]National Key Laboratory of Science and Technology on Micro/Nano Fabrication,
School of Integrated Circuits, Peking University, Beijing, China and
[2]Beijing Advanced Innovation Center for Integrated Circuits, Beijing, China

ABSTRACT

This paper proposed an automatic laser trimming system for silicon MEMS devices based on picosecond laser source and image recognition technique. Through the design of laser focusing optical path and coaxial imaging optical path, the observation of laser machining process is realized. A set of picosecond laser trimming parameters are determined by experiments, which enables a 7um width trench on the silicon resonator. Compared with the femtosecond laser trimming systems, the cost of the pico-laser system is reduced by more than 60%. The shape matching algorithm based on OpenCV and the chord-to-point distance accumulation (CPDA) algorithm for MEMS resonators image measurement solve the problem that the inefficiency of manual trimming for complex resonators. The experimental results of dual-mass tuning fork resonator show that the trimming time of a single device using the automatic trimming method is reduced from more than 1 hour manual trimming to less than 1 minute and the accuracy consistency is also improved by 82.6%. Finally, the structural stiffness imbalance of tuning fork resonator caused by the manufacturing error is eliminated.

KEYWORDS

Silicon Resonator, Laser Trimming, Image Recognition.

INTRODUCTION

To alleviate the performance degradation of the MEMS devices caused by the fabrication errors [1], numerous trimming methods have been carried out including ion beam trimming [2], electrical tuning [3] and laser ablation [4]. As a mature method suitable for large-scale production, the core of laser trimming method is the laser source. The picosecond lasers can give consideration to the needs of precision and low cost. Compared with nanosecond laser, the thermal effect of picosecond laser on materials is greatly reduced. Compared with femtosecond laser, the cost of picosecond laser can be reduced by more than 60%-80%.

Fast and accurate positioning of the object area is a prerequisite to achieve efficient and precise trimming. Some articles are reported to adopt manual operation that usually takes a long time and relies heavily on the experience of operators [5], which undoubtedly leads to the cost rising. The picosecond laser trimming system for MEMS silicon resonators is rarely reported especially with automatic trimming design.

SYSTEM DESIGN

Laser source selection and the optical path design are two essential aspects of the trimming system. In order to realize the real-time observation of the laser trimming process, we designed a coaxial focusing optical path on the same side of the laser and illumination light. If the wavelength of the laser is not close to that of the illuminator, the image is blurred, which results that the laser processing cannot be monitored in real time. Therefore, the laser wavelength is set to be 532nm that lies in the visible light range. Besides, an achromatic objective (Mitutoyo, APO PLAN, 50X, NA 0.42, WD 17mm) is adopted to eliminate the imaging blur caused by chromatic aberration. In order to accurately locate the MEMS devices, a high-precision four-degree-of-freedom translation stage (Newport, 100 nm resolution) with the typical accuracy of ±0.37um is used for sub-micron positioning of the trimmed resonators. The automatic navigation of the local adjustment area depends on the construction of image recognition system.

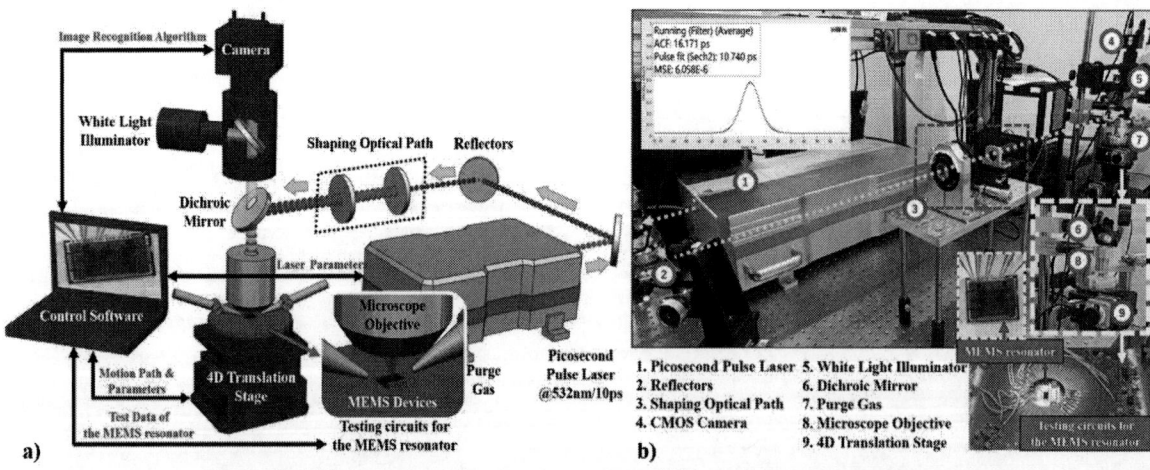

Figure 1: (a)Schematic of the trimming system. (b)View of the trimming system.

Thus, a COMS camera (DAHENG, MER-1220-32U3C) with 12 million pixels and 32 fps frame rate is used to identify and locate the area to be trimmed.

The process parameters of laser ablation are also the key to affect the precision of the trimming system. The pulse width of picosecond laser adopted in this paper is 10ps, which is close to the phonon relaxation time of the processed material. So the laser energy has almost no time to transfer to the surrounding material and there are few heat-affected areas.

IMAGE RECOGNITION TECHNOLOGY
Image Recognition of the Micro-structures

Image recognition can quickly locate the region to be trimmed from the complex structure of silicon MEMS resonant devices and greatly improves the laser trimming efficiency compared with manual operation. Traditional methods of image recognition include Hough transform, invariant moment, chain code feature etc. which have high recognition rate but slow recognition speed owing to a large amount of calculation [6]. Thus these algorithms are not suitable for rapid target area identification and localization. Corner detection algorithm has less computation burden and a high recognition rate and is normally applied to the regular geometric images. But it has poor anti-interference performance and strict image requirements.

To realize the rapid and precise image recognition of a microstructure, a shape matching algorithm based on OpenCV is adopted in the paper. There are three main procedures: (1) the images were preprocessed, (2) extracting the contours of the device structures from the images, (3) shape-matching algorithm is used to calculate the difference between the image moments of the image to be recognized and the template image to judge the similarity.

The Hu moment is an image feature with translation, rotation, and scale invariance. It introduces central moments and normalized central moments compared with the ordinary moments as indicated in formula (1) and (2),

$$\mu_{ij} = \sum_x \sum_y (x-\bar{x})^i (y-\bar{y})^j I\ (x,\ y) \tag{1}$$

$$\eta_{i,\ j} = \frac{\mu_{i,\ j}}{\mu_{00}^{(i+j)/2+1}} \tag{2}$$

$$
\begin{aligned}
H_0 &= \eta_{20} + \eta_{02} \\
H_1 &= (\eta_{20} - \eta_{02})^2 + 4\eta_{11}^2 \\
H_2 &= (\eta_{30} - 3\eta_{12})^2 + (3\eta_{21} - \eta_{03})^2 \\
H_3 &= (\eta_{30} + \eta_{12})^2 + (\eta_{21} + \eta_{03})^2 \\
H_4 &= (\eta_{30} - 3\eta_{12})(\eta_{30} + \eta_{12})\left[(\eta_{30} + 3\eta_{12})^2 - 3(\eta_{21} + \eta_{03})^2\right] + \\
&\quad (3\eta_{21} - \eta_{03})\left[3(\eta_{30} + \eta_{12})^2 - (\eta_{21} + \eta_{03})^2\right] \\
H_5 &= (\eta_{20} - \eta_{02})\left[(\eta_{30} + \eta_{12})^2 - (\eta_{21} + \eta_{03})^2 + 4\eta_{11}(\eta_{30} + \eta_{12})(\eta_{21} + \eta_{03})\right] \\
H_6 &= (3\eta_{21} - \eta_{03})(\eta_{30} + \eta_{12})\left[(\eta_{30} + \eta_{12})^2 - 3(\eta_{21} + \eta_{03})^2\right] + \\
&\quad (\eta_{30} - 3\eta_{12})(\eta_{21} + \eta_{03})\left[3(\eta_{30} + \eta_{12})^2 - (\eta_{21} + \eta_{03})^2\right]
\end{aligned}
\tag{3}
$$

Where μ_{ij} is the central moment, η_{ij} is the standardized moment, x- $\bar{x}$ and y- $\bar{y}$ respectively are the distances between a coordinate point and the coordinate center in the image, $i+j$ is moment of the order. Combining (1) and (2), seven values of the Hu moment can be obtained as follow

The smaller the difference in the seven values of Hu

moments, the higher the similarity of the images. The two images are identical when the difference is zero.

Image Dimension Measurement

Dimensional measurement in visual system is beneficial to improve the accuracy of the laser trimming. The Hough line detection algorithm is utilized to realize image measurement. The process flow is shown in Figure 2. First, the image to be measured is preprocessed. The contours of the microstructures are then extracted with chord-to-point distance accumulation (CPDA) algorithm that removes the wrong contour caused by the bulges or depressions of the silicon surface. The geometric dimension of microstructure, including the length, width and angles, are calculated using the Hough Line Detection algorithm which transforms the lines of the image space to the points of the parameter space.

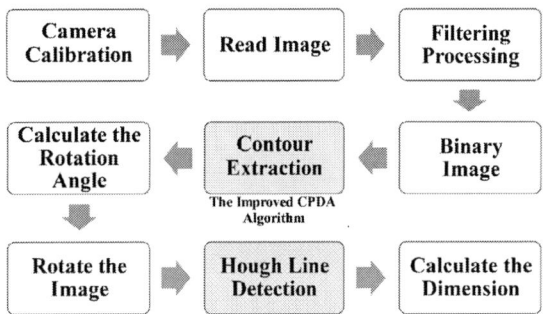

Figure 2: Flow chart of dimensional measurement.

Contour extraction and line detection are the key techniques of dimension measurement. Remove contour is significant for contour extraction, mainly to remove the wrong contour caused by impurities, such as bumps, depressions and so on. CPDA algorithm can produce the minimum positioning error when detecting the wrong contour, but its detection speed is slow. In this paper, we improve the CPDA algorithm from the speed, the result demonstrates that the improved algorithm takes 1-3s, which reduces by 200s compared to the CPDA algorithm.

The thought of Hough line detection algorithm lies in that the point set of a line in the image space and the line set in the parameter space can be converted to each other. By studying the line set in the parameter space, the information of line segment in the image space can be determined. The folding beams and comb fingers are widely used in MEMS design and are employed to verify the dimensional measurement as depicted in Figure 3. The results suggest that the measurement error is only ±0.5um compared with the microscopic observation.

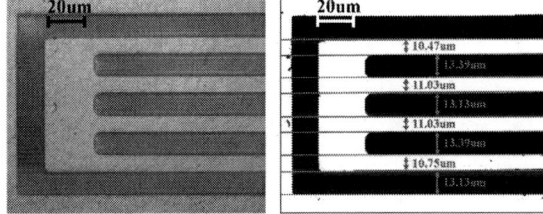

a) Original image of elastic beam b) Size measurement of elastic beam

Figure 3: Structural measurement result diagram.

EXPERIMENTS AND RESULTS

Experiment Setup

In order to test the trimming accuracy and efficiency of the system, the following two experiments are designed. Three silicon resonators are selected as the test samples in the experiment with the same structure. The frequency test circuit measures the resonator frequency with a measurement accuracy of ±0.05Hz. The process parameters are 1.18W laser power, 2MHz repetition frequency and 5um/s processing speed, which are obtained from the process experiment in Section 2.

System Performance Experiment

Each laser trimming process could be divided into three steps: the recognition of the trimming area, laser positioning, and the laser trimming. In order to test the fine-tuning efficiency of the system, two sample resonators (Sample1 and Sample2) were trimmed by manual trimming and automatic trimming method.

In the process of recognition, the method of manual operation need two operators who are in charge of the manipulation of the stage and recognize the location of the elastic beam with the aid of a CMOS camera respectively. This process time is generally more than 5 minutes and depends on the operator's proficiency. In order to coincide the starting position of the trimming with the position of the laser, skilled operators need to spend more than 5 minutes to move the translate stage. The automatic trimming system in this paper can achieve fast positioning at the magnitude of millisecond and ensure sub-micron positioning accuracy. In the laser trimming section, the automatic trimming system can automatically plan the laser machining path and control the cooperative operation of the translate stage and the laser. The time cost in this section is usually less than 50% of the time cost by manual trimming method. As shown in Table 1 and Figure 4, compared with the manual

operation which takes more than one hour, the time cost of the automatic trimming system is less than one minute, with a 75-fold increase in actual efficiency. The accuracy consistency of the automatic trimming system is also improved by 82.6%.

Table. 1 Accuracy consistency comparison between manual trimming system and automatic trimming system (*△f : Resonator frequency difference before and after trim).

Area	Sample1 (Manual Trim)		Sample2(Automatic Trim)	
	f/Hz	*△f /Hz	f/Hz	*△f /Hz
Org	7028.71	-	7162.7	-
L1	7027.81	-0.9	7161.98	-0.72
L2	7027.08	-0.73	7161.25	-0.73
L3	7025.88	-1.2	7160.47	-0.78
L3	7025.16	-0.72	7159.67	-0.8
	△f STD =0.224		△f STD = 0.039	

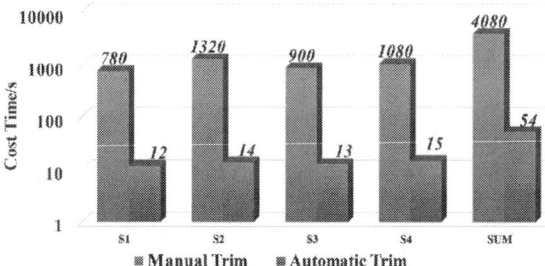

Figure 4: Time cost comparison between manual trimming system and automatic trimming system.

System Trimming Experiment

The existence of the 'double peak' deteriorates the ability of the resonator to suppress common-mode

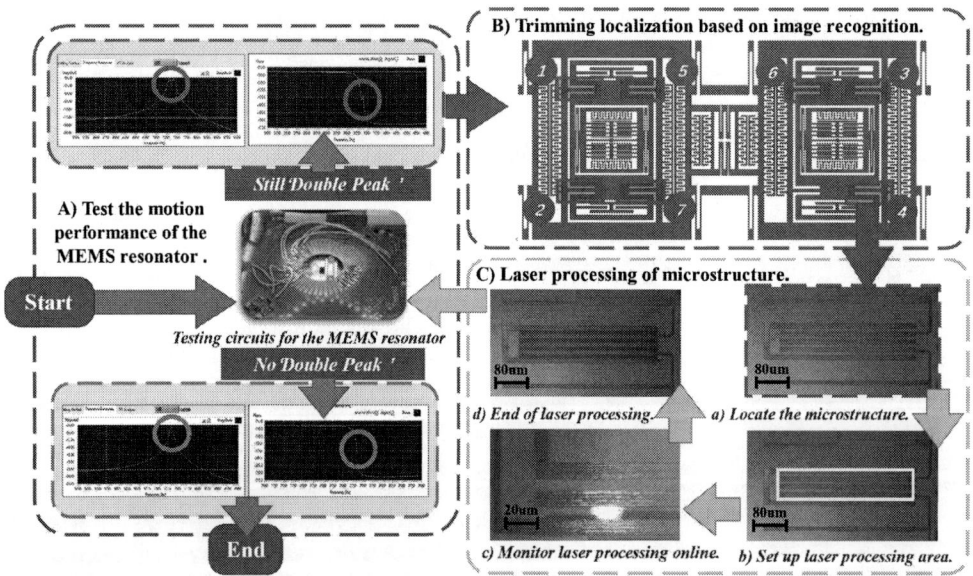

Figure 5: Flow chart of automatic laser trimming system. STEP A: Test whether the motion performance of the resonator meets the requirements. STEP B: The image recognition algorithm is used to locate the position of the elastic beam for machining. STEP C: Automatic laser machining of elastic beam. Repeat the process until the performance of the resonator meets the requirements.

vibration when the resonator works in the environment of vibration or impact. We try to use the laser trimming method to eliminate this double peak phenomenon with the automatic trimming system. Flow chart of automatic laser trimming system is shown in Figure 5.

Firstly, the Amplitude Frequency Characteristic (AFC) curves were detected by the testing circuit before each trimming, which is used to observe whether the double peak disappears. Due to the structural asymmetry error of the resonator, the AFC curve of the original resonator (Red Curve in Figure 6) has two peaks between 7300Hz and 7450Hz. Secondly, multiple areas of elastic beam structure are automatically identified through CMOS camera and the image recognition algorithm. After the trimming order of different beams is determined by the user, the identified elastic beam is successively moved within the view of the CMOS camera by the system via the 4D translation stage. After the shape and size of the area to be processed is defined by the user, the laser spot will automatically locate to the starting position of the processing area, and laser processing will be carried out according to the processing path and parameters automatically generated by the system. The process will be recorded by the CMOS camera in real time.

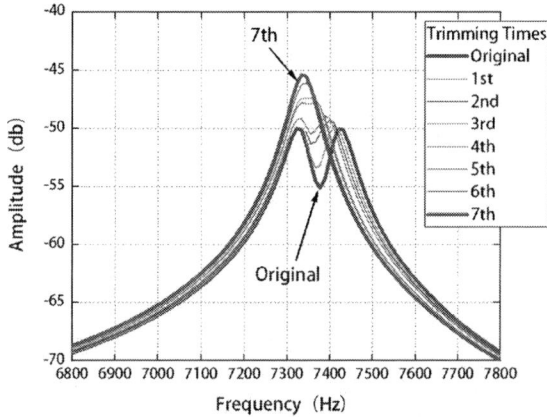

Figure 6: The amplitude-frequency characteristic curve of the resonator during 7 times laser modification.

The system repeated the above process until the end of the 7th laser process. The original 'double peak' disappeared and a new single peak appeared in the AFC instead, whose amplitude was 20% higher. This shows that the stiffness of the right mass gradually decreases after 7 times of continuous trim, making the two resonant frequencies coincide.

CONCLUSION

An automatic trimming system of silicon MEMS resonant devices based on picosecond pulse laser is realized. The shape matching algorithm based on OpenCV and the CPDA algorithm for MEMS resonators image measurement solve the problem that the inefficiency of complex resonators, which improves the size measurement accuracy by ±0.5um. The system is suitable for the tuning of various silicon resonators with complex structures. The efficiency of the system is 75 times higher than that of manual operation and the accuracy consistency is also improved by 82.6%. After 7 times of laser trims, the response amplitude of the tuning fork resonator is increased by 20%. This system is expected to solve the problem of resonator performance degradation caused by MEMS manufacturing process errors, and promotes the development of MEMS design and manufacturing technology.

REFERENCES

[1] M S Weinberg, A Kourepenis, et. al., "Error sources in in-plane silicon tuning-fork MEMS gyroscopes", *J. Microelectromech. Syst.*, vol.15(3), pp. 479-491, 2006.

[2] A. Efifimovskaya, D.Wang, et. al., "Mechanical trimming with focused ion beam for permanent tuning of MEMS dual-mass gyroscope", *Sensors and Actuators: A*, vol. 313, pp.112-189, 2020.

[3] B. J. Gallacher, J. Hedley, et al. "Electrostatic Correction of Structural Imperfections Present in a Microring Gyroscope", *J. Microelectromech. Syst.*, vol.14(1), pp. 221-234, 2005.

[4] S Hu, H Cui, et al., "A method of structural trimming to reduce mode coupling error for micro-gyroscopes". *NEMS Conference,* Suzhou, April 07-10, 2013, pp. 805-808.

[5] A. K. Samarao, F. Ayazi, et al., "Post-Fabrication Electrical Trimming of Silicon Bulk Acoustic Resonators using Joule Heating", *MEMS '09 Conference,* Sorrento, January 25-29, 2009, pp. 892-895.

[6] Z Zhang, et al. "A Flexible New Technique for Camera Calibration", *IEEE Trans. Pattern Anal. Mach. Intell*, vol. 11(22), pp. 1330-1334, 2000.

CONTACT

* J. Cui, tel: +86-10-6275-8911;eric.cuijian@pku.edu.cn
* Q.C.Zhao, tel: +86-10-6274-5160; zqc@pku.edu.cn

MICROMACHINING FUSED SILICA MICRO SHELL RESONATOR WITH QUARTZ GLASS MOLD BY THERMAL REFLOW

Zhaoxi Su[1], Bin Luo[1], Qiankai Tang[1], Linqian Zhu[1], and Jintang Shang[1]

[1]Key Laboratory of MEMS of Ministry of Education, Southeast University, Nanjing, CHINA

ABSTRACT

This paper demonstrates a low-cost process for fused silica micro shell resonators with quartz glass mold by thermal reflow under negative pressure. Wafer-level fused silica shell resonators with reduced size are prepared using this process. The results show that the fabricated resonator has a diameter of 2.7 mm with a Q-factor of 306k for n=2 wine-glass mode frequency at 63.698 kHz. The volume is reduced by at least 4 times compared to reported fused silica resonators. This work provides the potential for the preparation of sub-millimeter or even micron-fused silica shell resonators.

KEYWORDS

Micro Shell Resonator, Fused Silica, Thermal Reflow, Resonant Frequency, Quality Factor.

INTRODUCTION

The development of numerous new applications, including drones, submersible vehicles, self-driving automobiles, and augmented/virtual reality, has increased the demand for low-cost, high-performance MEMS gyroscopes. Hemispherical resonant gyroscopes （HRG） exhibit excellent performance with very low ARW and bias [1]. Researchers are striving to miniaturize 3D shell resonators in response to the success of HRG. The advantages of 3D shell resonators over their 2D counterparts are their robust construction, increased out-of-plane stiffness, and comparatively low operating frequency [2]. This not only improves the survivability of the sensor in extreme environments such as high-g shocks (>6000g) [3] or intense vibrations but also enables higher performance (BI ~ 0.0014 deg/hr) [4].

Fused silica exhibits excellent properties, including extremely low thermoelastic damping, high Q-factor, and thermal stability [5]. This makes it a fantastic choice for fabricating 3D shell resonators, promising navigation-grade performance at acceptable cost and power/size. At present, fused silica shell resonators with high Q-factor are prepared by glassblowing [6, 7] and blowtorch molding [8, 9]. Micro blown fused silica resonators provide the advantages of low cost, extremely smooth surface and potentially high structural symmetry [10]. However, the current situation is that although the Q-factor of some resonators has reached extremely high, their volume is still in the millimeter to centimeter region, and the reported resonators have diameters between 5 mm and 12 mm. Size reduction of fused silica shell resonators with low cost is attractive for future portable gyro applications. However, it is challenging to further shrink the current fused silica resonators.

Another way to fabricate micro 3D resonators is MEMS-based isotropic etching and deposition processes. The hemispherical mold is prepared by isotropic etching of silicon or glass. Then, isotropic materials such as polycrystalline diamond [11], polycrystalline silicon [12] and metallic glass [13] are deposited in the mold to obtain the micro-shell resonator. In most cases, the resultant resonators are on the micrometer size or smaller in diameter. However, the thermoelastic damping of materials places a limit on the superior Q-factor that may be achieved for these resonators. In addition, the poor relative processing tolerance results in non-negligible frequency mismatch and energy loss, limiting their further application in high-performance devices.

In conclusion, the thermal blowing-fabricated fused silica micro shell resonators exhibit good performance, but further size reduction is essential. Micro shell resonators fabricated by MEMS molding method are smaller in size but have poor performance. To achieve the goal of miniaturization while maintaining high performance, this paper proposes a novel thermal reflow process assisted by quartz glass mold to prepare wafer-level fused silica shell resonators with reduced size. And it is expected to maintain the performance of fused silica resonators. Further improvements are still in progress to achieve better performance and lower volume of resonators.

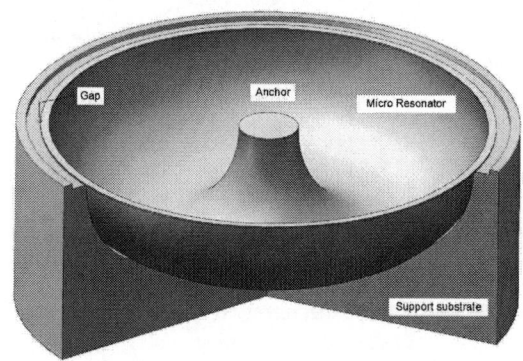

Figure 1: Schematic of the micro shell resonator with the support substrate.

ANALYSIS AND DESIGN

In this paper, the miniaturized fused silica shell resonator is designed by combining the thermal reflow process assisted by quartz glass mold and the MEMS isotropic etching process. The architecture of the micro resonator is shown in Fig. 1. It consists of a quartz glass mold substrate and a reflowed shell resonator with a self-aligning anchor. Instead of depositing a thin film in the cavity as the resonator structural layer, a thinned fused silica layer is used to form the resonator. The gap between the mold cavity and the resonator is defined by the sacrificial layer, allowing the resonator to move freely. The resonator is fixed to the self-aligning anchor in the center

978-1-6654-9309-3/23 $31.00 © 2023 IEEE

of the mold cavity, avoiding subsequent assembly errors.

Quartz glass wafer is used as the substrate, and hemi-toroidal cavities are etched as reflow molds of resonators. The etching of the cavities can be performed by isotropic wet etching or other methods. The aspect ratio (AR) obtained by isotropic wet etching is 1:2. Deeper AR can be achieved by other methods such as laser-induced etching. The softening point of the quartz glass substrate is higher than that of the fused silica device wafer, which can ensure that the substrate is not deformed during the thermal reflow process of resonators.

In our design, the size of the resonator is defined by the mold cavity, which theoretically can be as small as micron scale. The thickness of the resonator is determined by the fused silica device wafer's thickness, which can be thinned to the desired thickness. The resonant frequency of the resonator is determined by geometric parameters such as radius, thickness, and height. The desired operating frequency can be obtained by a suitable design.

FABRICATION

The wafer-level fabrication process flow is illustrated in Fig. 2.

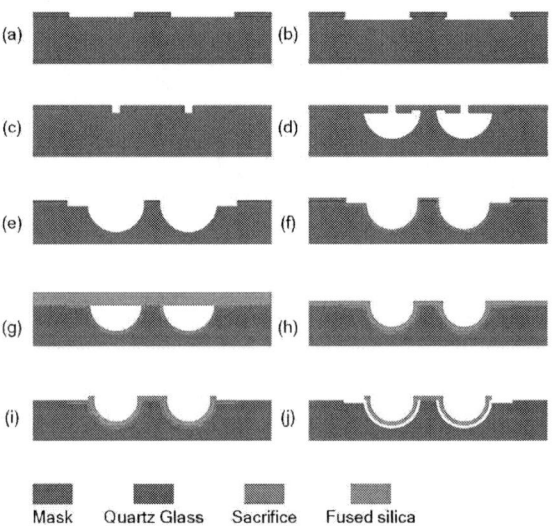

Figure 2: Fabrication process of fused silica micro shell resonator: (a) Deposition of Cr/Au and photoresist mask; (b) Time-Etching of quartz glass with HF 40% solution; (c) Deposition of Cr/Au and photoresist mask; (d) Time-Etching of quartz glass with HF 40% solution; (e) Photolithography to define the location of the sacrificial layer deposition; (f) Deposition of sacrificial layer; (g) Wafer bonding; (h) Thermal reflow; (i) Release of the resonator; (j) Removal of the sacrificial layer.

It starts with a 2mm thick 4-inch diameter quartz glass wafer coated with chrome/gold and photoresist layers mask on both sides. Define the annular pattern by photolithography and etching of the Cr/Au layer in preparation for isotropic wet etching, step (a). The quartz glass was then time-etched with an HF 40% solution to a depth of 10 µm, providing space for the subsequent deposition of the sacrificial layer, step (b). After etching, the photoresist layer and the Cr/Au layer are removed, and the wafer is cleaned. Then repeat Step (a) to define another ring pattern with a width of 200 µm by mask in the previous shallow etching cavity, step (c). Subsequently, the quartz glass is time-etched again with an HF 40% solution to obtain hemi-toroidal mold cavities, step (d). Figure 3 shows a quartz glass mold wafer etched with hemi-toroidal cavities without removing the Cr/Au and photoresist mask layers. Inset is a magnified image of one of the mold cavities. The cavity anchor diameter, cavity depth and cavity diameter are 400 µm, 520 µm, and 2.70 mm, respectively. After etching, remove the mask and clean the wafer. The photoresist is then sprayed on the wafer to ensure uniformity on the wafer with three-dimensional structures. Define the location of the sacrificial layer deposition by photolithography, step (e). This is followed by depositing a 2 µm thick layer of chromium as a sacrificial layer on the wafer, step (f). This sacrificial layer is used to define the spacing between the device and the mold cavity to ensure that they do not stick together during thermal reflow. After stripping the photoresist and cleaning the wafer, it is bonded to a 100 µm thick fused silica device wafer in the vacuum environment to maintain a vacuum inside the etched cavities, step (g). After bonding, the device wafer is thinned to the desired thickness, such as 70 µm in this paper, thereby defining the thickness of the resonator. The bonding wafer is then rapidly heated to the reflow temperature, which is 1500°C. The fused silica device wafer is softened and pressed into the quartz glass mold cavity driven by atmospheric pressure, the process takes no more than 1 minute, step (h). Subsequently, fused silica shell resonators are released by wet etching the rim, step (i). Finally, remove the chrome sacrifice layer and clean the devices thoroughly, step (j). 52 shell resonators are prepared on a 4-inch wafer.

Figure 3: A 4-inch quartz glass substrate wafer with wet-etched mold cavities. Inset: an enlarged image of a mold cavity.

Once the devices are prepared, they are removed from the 4-inch wafer by dicing to characterization. The inset of Fig. 6 is an optical photograph of a resonator under test.

DEVICE CHARACTERIZATION

The characterization of the resonator includes two parts: structural characterization and performance characterization.

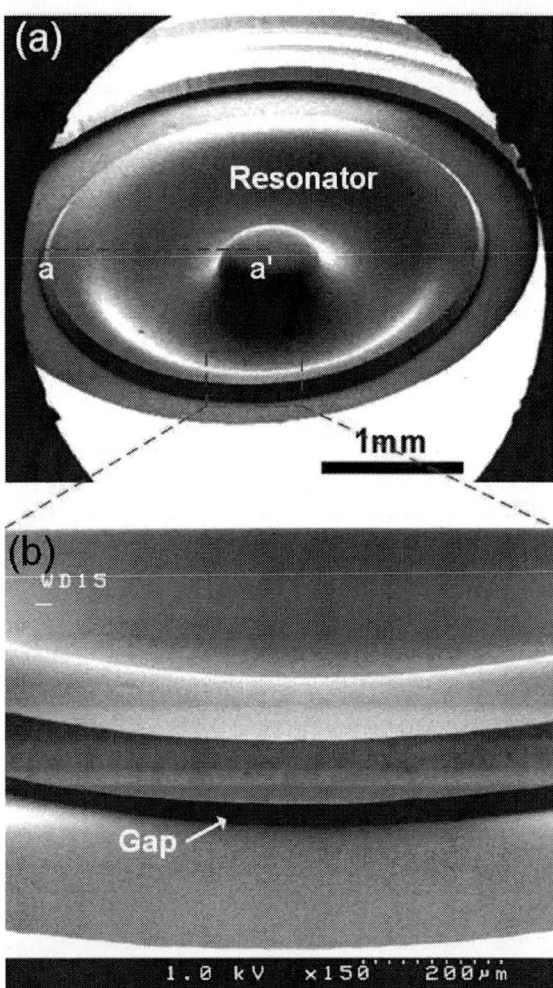

Figure 4: SEM images of the fused silica shell resonator. (a) Full image of a resonator. (b) Enlarged image of the gap between the resonator rim and the substrate.

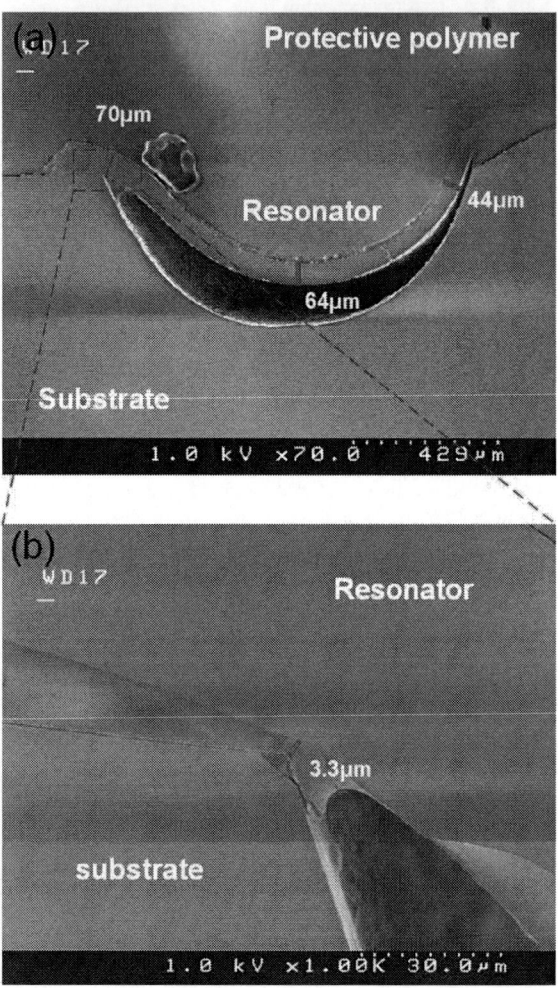

Figure 5: SEM images of the fused silica shell resonator. (a) Cross-section image of the aa' as shown in Fig. 4(a). (b) Enlarged image of the gap between the mold cavity edge and the resonator.

The structure of the resonator was characterized by SEM. Figure 4(a) shows the full picture of the resonator, and Figure 4(b) is a local close-up of the rim, showing the gap between the resonator rim and the substrate. Due to the thick device wafer, warping occurs at the edge of the cavity during reflow. Figure 5(a) is a cross-sectional view of the aa' of the resonator shown in Fig. 4(a). Polymer mounting is used for protection during sample preparation. Figure 5(b) is an enlarged image of the gap between the mold cavity edge and the resonator, which is about 3 μm. It indicates that the resonator is in a state where it can move freely. The geometry of the resonator was measured with a total diameter of 2.70 mm, shell height of 450 μm, anchor diameter of 620 μm and rim width of 170 μm. Further work can reduce the resonator volume by reducing the mold cavity size and the device layer thickness.

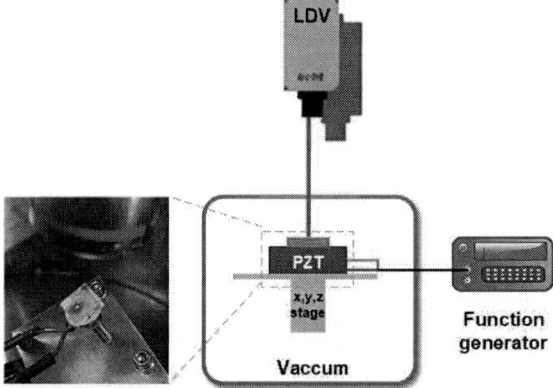

Figure 6: The schematic of characterization setup of the Q-factor measurement (Inset is a micro-shell resonator under test in the vacuum chamber).

The resonator performance, including the resonant frequency and Q-factor, was characterized by Laser

Doppler Vibrometer under a vacuum of 1×10^{-2} Pa, as shown in Figure 6. The test results are shown in Figure 7. The resonant frequency of the wineglass n=2 mode is 63.698 kHz and the ring-down time is 1.53s. The Q-factor reaches 306 k. Compared with existing fused silica shell resonators, the volume is reduced by at least 4 times with a considerate Q-factor.

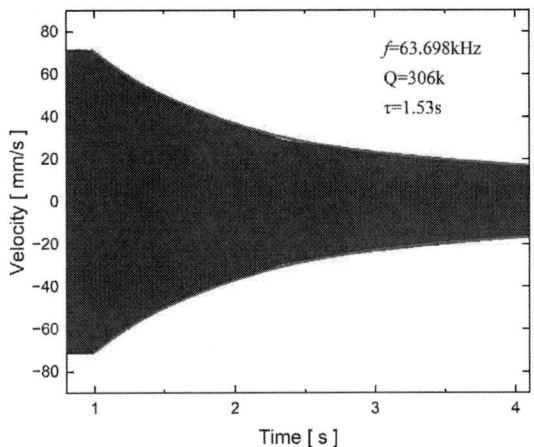

Figure 7: Ring-down time measurement of a fused silica shell resonator at wineglass n=2 mode under 1E-2 Pa.

CONCLUSION

This paper demonstrates a low-cost process for fused silica micro shell resonators with quartz glass mold by thermal reflow under negative pressure. Wafer-level fused silica shell resonators with reduced size were prepared using this process. The characterization results show that the resonant frequency of the n=2 wineglass mode of the prepared resonator is 63.698kHz, and the ring-down time is 1.53s. The Q-factor reaches 306k. The resonator has a diameter of 2.7 mm, and the volume is reduced by at least 4 times compared to the reported fused silica resonator. This work gives an idea of the preparation of sub-millimeter or even micron fused silica shell resonators while maintaining performance.

ACKNOWLEDGEMENTS

This work is supported by the National Natural Science Foundation of China (No. 51675102).

REFERENCES

[1] A. Trusov *et al.*, "mHRG: Miniature CVG with beyond navigation grade performance and real time self-calibration," in *IEEE International Symposium on Inertial Sensors and Systems (INERTIAL)*, Laguna Beach, USA, Feb. 22-25, 2016, pp. 29-32.

[2] D. M. Rozelle, "The hemispherical resonator gyro: From wineglass to the planets," in *Proc. 19th AAS/AIAA Space Flight Mechanics Meeting*, Savannah, USA, Feb. 8-12, 2009, vol. 134, pp. 1157-1178.

[3] S. Singh, A. Darvishian, J. Y. Cho, B. Shiari, and K. Najafi, "High-Q 3D micro-shell resonator with high shock immunity and low frequency mismatch for MEMS gyroscopes," in *IEEE 32nd International Conference on Micro Electro Mechanical Systems (MEMS)*, January 27-31, Seoul, Korea, 2019, pp. 668-671.

[4] J. Y. Cho, S. Singh, J.-K. Woo, G. He, and K. Najafi, "0.00016 deg/√ hr Angle Random Walk (ARW) and 0.0014 deg/hr Bias Instability (BI) from a 5.2 MQ and 1-cm Precision Shell Integrating (PSI) Gyroscope," in *IEEE International Symposium on Inertial Sensors and Systems (INERTIAL)*, Hiroshima, Japan, March 23-26, 2020, pp. 1-4.

[5] S. D. Penn *et al.*, "High quality factor measured in fused silica," *Review of Scientific Instruments,* vol. 72, no. 9, pp. 3670-3673, 2001.

[6] D. Wang, M. H. Asadian, D. Hii, and A. M. Shkel, "Fused silica dual-shell gyroscope with in-plane actuation by out-of-plane electrodes realized using glassblowing and thru-glass-vias fabrication," in *IEEE 35th International Conference on Micro Electro Mechanical Systems Conference (MEMS)*, Tokyo, Japan, January 09-13, 2022, pp. 154-157.

[7] S. Han, Z. Su, Y. Zhang, B. Luo, and J. Shang, "Three-Dimensional Digital Reconstruction Geometric Analysis for Fused Silica Microshell Resonators," *IEEE Transactions on Instrumentation Measurement,* vol. 71, pp. 1-7, 2022.

[8] B. Shiari, T. Nagourney, A. Darvishian, J. Y. Cho, and K. Najafi, "Simulation of blowtorch reflow of fused silica micro-shell resonators," *J. Microelectromech. Syst.*, vol. 26, no. 4, pp. 782-792, 2017.

[9] Y. Shi *et al.*, "Micro hemispherical resonators with quality factor of 1.18 million fabricated via laser ablation," in *IEEE 34th International Conference on Micro Electro Mechanical Systems (MEMS)*, Gainesville, USA, January 25-29, 2021, pp. 6-9.

[10] J. Y. Cho, J. Yan, J. A. Gregory, H. W. Eberhart, R. L. Peterson, and K. Najafi, "3-dimensional blow torch-molding of fused silica microstructures," *J. Microelectromech. Syst.*, vol. 22, no. 6, pp. 1276-1284, 2013.

[11] J. J. Bernstein *et al.*, "High Q diamond hemispherical resonators: fabrication and energy loss mechanisms," *J. Micromech. Microeng.*, vol. 25, no. 8, p. 085006, 2015.

[12] P. Shao, C. L. Mayberry, X. Gao, V. Tavassoli, and F. Ayazi, "A polysilicon microhemispherical resonating gyroscope," *J. Microelectromech. Syst.*, vol. 23, no. 4, pp. 762-764, 2014.

[13] M. Kanik *et al.*, "Metallic glass hemispherical shell resonators," *J. Microelectromech. Syst.*, vol. 24, no. 1, pp. 19-28, 2014.

CONTACT

*Jintang Shang, tel: +86 13913869603; jshang@seu.edu.cn

WAFER-LEVEL PATTERNING OF TIN OXIDE NANOSHEETS FOR MEMS GAS SENSORS

Mingjie Li[1], Wenxin Luo[1], Xiaojiang Liu[1], Gaoqiang Niu[1], and Fei Wang[1]
[1]Southern University of Science and Technology, Shenzhen 518055, CHINA

ABSTRACT

Gas sensing materials are typically loaded on designated area of micro-electrode via drop-casting for application in gas sensors. However, wafer-level patterning of gas sensitive materials with good reproducibility still faces challenge. In this work, we develop an effective 'top-down' approach to produce wafer-scale miniaturized gas sensing devices with high-throughput using standard photolithography of a mixture of photoresist and tin oxide (SnO_2) nanomaterials suspension, followed by calcination. The prepared gas sensing chips based on the proposed method exhibited satisfactory reproducibility and uniformity of sensing response toward ethanol detection, indicating promising application for large-scale production of MEMS gas sensors.

KEYWORDS

SnO_2 nanosheets, patterning, MEMS technology, ethanol sensors.

INTRODUCTION

With the increasing need for gas sensing [1-3], numerous types of gas sensors have been developed for environmental monitoring [4], toxic gas identification [5], medical diagnose applications [6], indoor air quality control [7], domestic appliances [8], etc. Metal-oxide-semiconductor (MOS) gas sensors, which were pioneered by Seiyama et al. [9] and Taguchi [10] in the 1960s, have many advantages including low-cost, high gas sensitivity, miniature size, and thus excellent compatibility with state-of-art electronic devices. The MOS gas sensors contain a sensing material layer to monitor the electrical resistance variations. In a typical fabrication procedure, MOS sensing materials are loaded on the specific area via drop-casting [11], spraying [12], or inkjet printing [13]. On the other hand, wafer-level patterning of MOS materials, which are highly desirable for the high-volume production, still encounters difficulties in terms of process incompatibility between the loading techniques and the microelectromechanical systems (MEMS) fabrication of the sensing chip [14].

Although SnO_2-based planar layers may be coated by sputtering [15] or atomic layer deposition [16], their sensing performance is handicapped due to the relatively small sensing surface area. Alternatively, porous metal oxide films can be directly coated on the sensing electrode via the 'bottom-up' strategy (e.g., solution-based synthesis). For instance, Gong et al. demonstrated on-chip growth process for α-Fe_2O_3/SnO_2 [17] and SnO_2/ZnO core-shell [18] nanoarrays for acetone and ethanol sensing, respectively. Guo et al. [19] developed a colloidal approach using an assembled polystyrene template to achieve bowl-like nanostructured SnO_2 films. In the contrary, the 'top down' strategy uses MEMS lithographic techniques for materials patterning. Chen et al. [20] patterned the sensor

with hydrophobic self-assembly monolayer, and drop-coated the sensing materials by a pipette. Whereas the nanomaterial dispersion spontaneously flows onto the hydrophilic sensing area.

More recently, the combination of 'bottom-up' and 'top down' techniques have been proposed for wafer-scale MOSs patterning. Niu et al. [21] patterned SnO_2 nanosheet array through standard lithography process, while unwanted sensing materials are removed by dry etching with SF_6 and BCl_3 gases. However, the reproducibility of micro-sensors diced from different locations on a wafer is not discussed. In addition, Liu et al. [22] reported a dry film photoresist template-guided de-wetting strategy to assemble a porous thermoplastic elastomer (TPE) mask for localized in-situ growth of NiO nanowalls. The prepared gas sensors show excellent reproducibility and uniformity; however, the concentration of TPE solution must be precisely controlled or peeling-off of the TPE mask could fail. Their method also requires multiple lithography steps, which may lead to low manufacturing yield. Despite the recent progress, engineers are still actively searching for a facile approach to resolve the bottleneck of the integration process of sensitive materials deposition and the wafer-scale fabrication of the miniaturized gas sensor.

Photoresists are common materials for mask in MEMS fabrication, and they can be easily removed from the substrate through calcination [23]. Herein, we implemented a 'top down' strategy for wafer-level fabrication of sensors with high-throughput and good reproducibility. The SnO_2 nanoparticles (NPs) and nanosheet (NS) particles are well-dispersed in the positive photoresist, followed by micro-patterning on the interdigital electrode (IDE) wafer through photolithography processes. High temperature calcination is carried out to remove the remaining photoresist after develop. The prepared gas sensors yield good reproducibility of sensing response for ethanol vapor detection, paving new routes for MEMS-compatible deposition of gas sensing materials and large-scale manufacture of miniature gas sensors

DEVICE FABRICATION AND SENSING MATERIAL PATTERNING

The schematic fabrication procedure is illustrated in Fig. 1. Firstly, Cr/Pt (15 nm/200 nm in thickness) IDEs were coated and patterned on a 4-inch (100) silicon wafer (with a 2 μm thermally-grown oxide) using a conventional lift-off process. Next, 0.6 mmol $SnCl_2 \cdot 2H_2O$ was dissolved into 20 mL of deionized water, while the solution pH was adjusted to ~13 by adding 6.7 mL 0.4 M NaOH solution. The mixture was magnetically stirred for 30 min and then transferred into a 50 mL Teflon-lined autoclave. The autoclave was sealed and subjected to annealing at 180 °C for 12 h, followed by furnace cool. The SnO_2 nanosheet products were then collected and washed by multiple

978-1-6654-9309-3/23 $31.00 © 2023 IEEE

centrifugations with deionized water and ethanol to eliminate any residual ions. The final particles were obtained after solvent evaporation at 80 °C for 12 h. Afterwards, 0.3 g of the synthesized nanomaterial was uniformly mixed with the positive photoresist (Ruihong RZJ 304.10, China) via stirring for 2 h. The acquired suspension was double spin-coated on the IDE wafer at 600 rpm for 60 s, followed by photolithography process. Lastly, as-prepared wafer was diced into individual devices, and calcinated at 500 °C for 5 h to remove the remaining photoresist. Besides nanosheets, commercial SnO_2 nanoparticles (99.9%, 50-70 nm diameter, Macklin, China) were also micro-patterned according to the above procedures to corroborate the coherence of the work.

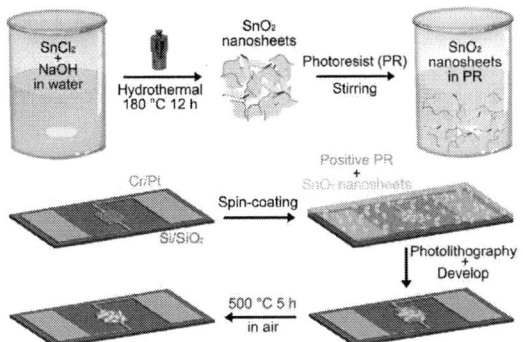

Figure 1: Schematic diagram of the fabrication procedure of the gas sensing chip with micro-patterned SnO_2 nanosheets sensing material.

The surface morphology of the prepared chips was characterized using a scanning electron microscope (SEM, Zeiss Merlin, Germany). Fig. 2a-c displays a photograph and SEM images of IDE wafer, which is covered by SnO_2 nanoparticles. A closer observation of the nanosheets (Fig. 2d-e) suggests the original flake-like nanostructures of the SnO_2 nanosheets is preserved after incorporation with the substrate. Further insights into the nanosheets are attained by a high-resolution transmission electron microscope (HRTEM, FEI Talos F200X G2, USA). The nanosheets have a size range of several hundred nanometers with a smooth surface morphology (Fig. 2f).

Figure 2: (a) The camera picture of wafer-scale MEMS chips patterned with sensing material. SEM images of (b) micro-patterned SnO_2 layer on platinum IDE, (c) SnO_2 NPs, and (d-e) SnO_2 NS. (e) TEM image of the SnO_2 NS.

GAS SENSING PERFORMANCE

Gas sensing performance was measured through a CGS-4TPs system (Beijing Elite Tech, China) at a relative humidity of 40%. The as-prepared sensing chip (3 mm × 5 mm in size) is placed on the test platform, where a hotplate was used to control the operating temperature, and two probes were connected to the bonding pads of a sensor chip [17, 18]. The atmospheric oxygen species may absorb on the SnO_2 grains and create an electron depletion layer [24, 25], while the chemisorbed oxygen species are removed when a reducing gas is in contact, causing reduction in both electron depletion region and the electrical resistance [24]. Monitoring the changes in the electrical properties could reflect the partial pressure of the target gas due to reduction in both electron depletion region. The sensor response is defined as the ratio of the sensor resistance in air (R_a) to that of in a mixture of target gas and air (R_g) [25].

Fig. 3a shows the dynamic response transients of two sensors when they are sequentially exposed to ethanol vapor at different concentrations spanning from 5 to 500 ppm. The response of SnO_2-based sensors increases quickly with lower ethanol concentration, and then become slower when ethanol concentration is above ~300 ppm (Fig. 3b), because ethanol adsorption gradually achieves saturation. The SnO_2 nanosheets based sensor exhibits higher response compared to that of the nanoparticles, possibly because of a larger number of effective pore channels and active sites of the nanosheets structure for

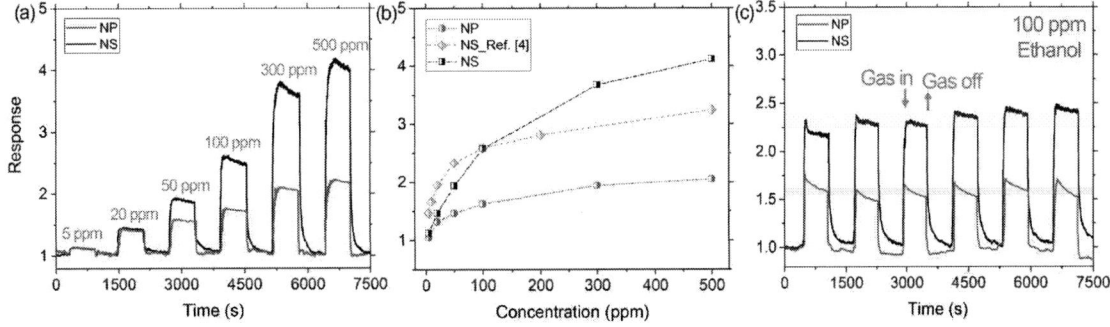

Figure 3: Gas sensing performance of sensors based on SnO_2 NPs and NS. (a) Dynamic response transients of the prepared chips to 5-500 ppm ethanol at an optimal work temperature of 300 °C. (b) Responses as a function ethanol concentration. (c) Reproducibility of the chips to 100 ppm ethanol at 300 °C for six cycles.

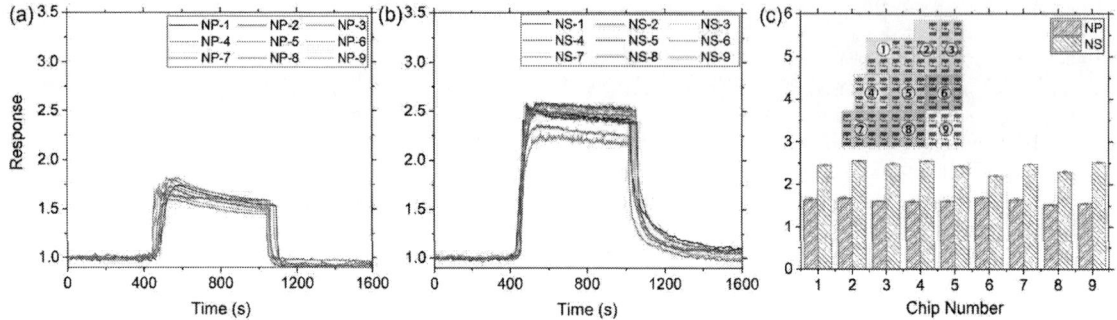

Figure 4: Gas sensing response of nine different gas sensors based on SnO$_2$ (a)NPs and (b)NS to 100 ppm ethanol. (c) The response deviations are ~3.3 % and ~4.5 % for NP and NS based sensors, respectively.

ethanol adsorption and desorption. The measured ethanol sensing properties are somewhat analogous to that of an on-chip grown SnO$_2$ NS based sensor [18], implying the proposed patterning method do not affect gas sensing functioning of assembled gas sensors. The response to 100 ppm ethanol was measured for 6 cycles (Fig. 3c), and both sensors yield repeatable sensing characteristics with relative standard deviations (RSDs) of ~2.3 % and ~3.5 %, respectively, suggesting good stability of fabricated devices.

Moreover, as presented in Fig. 4c insert, nine gas sensors were selected from different regions on two separate quarter wafers for gas sensing testing, respectively, i.e., one wafer was micro-patterned with SnO$_2$ NS, while another one was coated with NPs. The sensing responses of the selected sensors toward 100 ppm ethanol vapor are shown in Fig. 4a and Fig. 4b, respectively. The responses yield low RSDs of ~3.3 % and ~4.5 % for SnO$_2$ NP and NS based sensors (Fig. 4c), respectively, revealing high sensing performance uniformity throughout the quarter wafer.

CONCLUSION

In conclusion, we have developed a facile and low-cost fabrication procedure for wafer-scale micro-patterning of SnO$_2$ nanoparticles and nanosheets through standard photolithography followed by calcination. A uniform SnO$_2$ thin film patterned on a well-defined region are observed on the interdigital micro-electrodes on the silicon/oxide substrates. The reproducibility and uniformity of ethanol vapor response of prepared gas sensing devices were evaluated by picking nine gas sensors from various sites on a quarter 4-inch wafer. The acquired nine chips yielded response toward ethanol with a small RSD of ≤4.5%, inferring excellent potential to produce wafer-scaled MEMS gas sensors with excellent uniformity. In the future, further efforts will be made to enhance the gas sensing performance with both high response and good selectivity for highly efficient detection of ethanol and other gases, to meet the requirement for practical applications, such as breath analyzer.

ACKNOWLEDGEMENTS

This work is supported in part by the Shenzhen Research Foundation for Postdoc under grant K22797503, and in part by the National Key Research and Development Program of China under Grant 2020YFB2008604, and in part by NSQKJJ under Grant K21799109 and Grant K21799110.

Technical assistance provided by X. Ma and Y. Wang from the Core Research Facilities (CRF) at SUSTech is appreciated.

REFERENCES

[1] M. Li, W. Luo, X. Liu, G. Niu, and F. Wang, "Wafer-Level Patterning of SnO$_2$ Nanosheets for MEMS Gas Sensors," *IEEE Electron Device Lett.,* vol. 43, no. 11, pp. 1981-1984, 2022.

[2] W. Ren, C. Zhao, G. Niu, Y. Zhuang, and F. Wang, "Gas Sensor Array with Pattern Recognition Algorithms for Highly Sensitive and Selective Discrimination of Trimethylamine," *Adv. Intell. Syst.,* p. 2200169, 2022.

[3] G. Niu *et al.*, "Nanocomposites of pre-oxidized Ti$_3$C$_2$T$_x$ MXene and SnO$_2$ nanosheets for highly sensitive and stable formaldehyde gas sensor," *Ceram. Int.,* 2022.

[4] G. F. Fine, L. M. Cavanagh, A. Afonja, and R. Binions, "Metal oxide semi-conductor gas sensors in environmental monitoring," *Sens.,* vol. 10, no. 6, pp. 5469-5502, 2010.

[5] P. Tyagi, A. Sharma, M. Tomar, and V. Gupta, "Metal oxide catalyst assisted SnO$_2$ thin film based SO$_2$ gas sensor," *Sens. Actuators, B,* vol. 224, pp. 282-289, 2016.

[6] G. Gregis *et al.*, "Detection and quantification of lung cancer biomarkers by a micro-analytical device using a single metal oxide-based gas sensor," *Sens. Actuators, B,* vol. 255, pp. 391-400, 2018.

[7] E. J. Wolfrum, R. M. Meglen, D. Peterson, and J. Sluiter, "Metal oxide sensor arrays for the detection, differentiation, and quantification of volatile organic compounds at sub-parts-per-million concentration levels," *Sens. Actuators, B,* vol. 115, no. 1, pp. 322-329, 2006.

[8] N. Yamazoe, "Toward innovations of gas sensor technology," *Sens. Actuators, B,* vol. 108, no. 1, pp. 2-14, 2005.

[9] T. Seiyama, A. Kato, K. Fujiishi, and M. Nagatani, "A New Detector for Gaseous Components Using Semiconductive Thin Films," *Anal. Chem.,* vol. 34, no. 11, pp. 1502-1503, 1962.

[10] N. Taguchi, "Published patent application in Japan," *S37-47677,* 1962.

[11] Y. Chen, M. Li, W. Yan, X. Zhuang, K. W. Ng, and X. Cheng, "Sensitive and Low-Power Metal Oxide Gas Sensors with a Low-Cost Microelectromechanical Heater," *ACS Omega,* vol. 6, no. 2, pp. 1216-1222, 2021.

[12] M. Li *et al.*, "Sensitive NO_2 gas sensors employing spray-coated colloidal quantum dots," *Thin Solid Films,* vol. 618, pp. 271-276, 2016.

[13] B. Li *et al.*, "Inkjet printed chemical sensor array based on polythiophene conductive polymers," *Sens. Actuators, B,* vol. 123, no. 2, pp. 651-660, 2007.

[14] G. Niu and F. Wang, "A review of MEMS-based metal oxide semiconductors gas sensor in Mainland China," *J. Micromech. Microeng.,* vol. 32, no. 5, p. 054003, 2022.

[15] Y. Wang, W. G. Tong, and N. Han, "Co-sputtered Pd/SnO_2:NiO heterostructured sensing films for MEMS-based ethanol sensors," *Mater. Lett.,* vol. 273, p. 127924, 2020.

[16] A. J. Niskanen *et al.*, "Atomic layer deposition of tin dioxide sensing film in microhotplate gas sensors," *Sens. Actuators, B,* vol. 148, no. 1, pp. 227-232, 2010.

[17] H. Gong, C. Zhao, G. Niu, W. Zhang, and F. Wang, "Construction of 1D/2D α-Fe_2O_3/SnO_2 Hybrid Nanoarrays for Sub-ppm Acetone Detection," *Research,* vol. 2020, p. 2196063, 2020.

[18] H. Gong, C. Zhao, and F. Wang, "On-Chip Growth of SnO_2/ZnO Core–Shell Nanosheet Arrays for Ethanol Detection," *IEEE Electron Device Lett.,* vol. 39, no. 7, pp. 1065-1068, 2018.

[19] L. Guo *et al.*, "Response and stability improvement by fusing optimized micro-hotplatform and double layer bowl-like nano arrays," *Sens. Actuators, B,* vol. 231, pp. 450-457, 2016.

[20] Y. Chen, P. Xu, X. Li, Y. Ren, and Y. Deng, "High-performance H_2 sensors with selectively hydrophobic micro-plate for self-aligned upload of Pd nanodots modified mesoporous In_2O_3 sensing-material," *Sens. Actuators, B,* vol. 267, pp. 83-92, 2018.

[21] G. Niu, C. Zhao, and F. Wang, "Scalable Synthesis of SnO_2 Nanosheet Arrays on Chips for Ultralow Concentration NO_2 Detection," in *2021 IEEE 16th International Conference on Nano/Micro Engineered and Molecular Systems (NEMS),* 2021, pp. 820-823.

[22] L. Liu *et al.*, ""Top-down" and "bottom-up" strategies for wafer-scaled miniaturized gas sensors design and fabrication," *Microsyst. Nanoeng.,* vol. 6, no. 1, p. 31, 2020.

[23] D. Xia and S. R. J. Brueck, "Fabrication of enclosed nanochannels using silica nanoparticles," *J. Vac. Sci. Technol., B,* vol. 23, no. 6, pp. 2694-2699, 2005.

[24] N. Yamazoe and K. Shimanoe, "Theory of power laws for semiconductor gas sensors," *Sens. Actuators, B,* vol. 128, no. 2, pp. 566-573, 2008.

[25] N. Yamazoe, G. Sakai, and K. Shimanoe, "Oxide Semiconductor Gas Sensors," *Catal. Surv. Asia,* vol. 7, no. 1, pp. 63-75, 2003.

CONTACT

*Fei Wang, tel: +86-88018509; wangf@sustech.edu.cn

AIR DAMPING EFFECTS ON DIFFERENT MODES OF AlN-on-Si MICROELECTROMECHANICAL RESONATORS

Yuncong Liu[1†], S M Enamul Hoque Yousuf[1†*], Afzaal Qamar[2],*
Mina Rais-Zadeh[2,3], and Philip X.-L. Feng[1*]*

[1]Electrical & Computer Engineering, University of Florida, Gainesville, FL 32611, USA
[2]Electrical & Computer Engineering, University of Michigan, Ann Arbor, MI 48109, USA
[3]Jet Propulsion Laboratory, California Institute of Technology, Pasadena, CA 91109, USA

ABSTRACT

Piezoelectric thin films have enabled demonstrations of a wide range of piezoelectric-on-silicon (PoS) resonant transducers in microelectromechanical systems (MEMS) industry today. While the actual ambient pressure (p) levels in vacuum packages can greatly affect device performance, a comprehensive study and quantitative understanding of dynamical characteristics of multimode PoS resonators under varying pressure is still lacking and needs to be investigated. In this work, we report the first measurements and analyses of air damping effects on both bulk and flexural resonance modes of AlN-on-Si (AlN/Si) bulk acoustic resonators (BARs). Experimental results are presented for the resonant frequency (f) and quality (Q) factor with p varying from 760Torr down to 100μTorr. For AlN/Si resonator's bulk mode, f keeps a nearly constant value of 10.291MHz over the pressure range and the Q factor starts to decline when p is above 10Torr. For AlN/Si resonator's flexural mode, f is stable at 595.3kHz below 20Torr and drops to 595kHz at 760Torr. For the Si-only device's bulk mode, $f \sim 10.127$MHz is independent of p, and a high $Q \sim 257{,}000$ is maintained when $p < 3$Torr. The present work clearly reveals the different effects that air damping exerts upon the bulk and flexural modes of AlN/Si MEMS resonators, providing new quantitative knowledge and insight for device engineering.

KEYWORDS

Micromechanical resonator, air damping, aluminum nitride (AlN), piezoelectric resonator, quality (Q) factor.

INTRODUCTION

Thin film piezoelectric resonators have attracted a considerable amount of interest and are relevant to a broad variety of applications in communication, timing, sensing, and energy harvesting [1-3]. Among the important thin film materials with strong piezoelectricity, AlN is appealing for various radio-frequency (RF) applications also thanks to its high thermal conductivity, strong electric insulation properties, and ease of fabrication at scale [3]. Combining excellent electromechanical coupling of AlN with low acoustic loss of single crystal Si, AlN/Si BARs can exhibit high power handling and compatibility with mainstream CMOS wafer-scale manufacturing [4]. To maximize the benefits of the exceptional properties of laterally vibrating AlN/Si resonators, high-Q modes are preferable for more precise sensing and better frequency selection. Pursuing such high-Q devices requires carefully investigating and understanding the limits of loss mechanisms. Various energy loss mechanisms can contribute to the overall dissipation in such devices, which can be expressed as:

$$\left(\frac{1}{Q}\right)_{\text{tot}} = \left(\frac{1}{Q}\right)_{\text{med}} + \left(\frac{1}{Q}\right)_{\text{ted}} + \left(\frac{1}{Q}\right)_{\text{al}} + \left(\frac{1}{Q}\right)_{\text{others}}, \quad (1)$$

where it identifies three main dissipation mechanisms: medium (*i.e.*, air) damping Q^{-1}_{med}, thermoelastic damping Q^{-1}_{ted}, and anchor loss Q^{-1}_{al}. It has been shown that the thermoelastic damping from columnar boundaries in the AlN piezoelectric layer causes Q of AlN/Si device to be lower than that of Si-only device [5]. Further, inclusion of metal electrodes atop the AlN/Si heterostructure, as is typically needed for piezoelectric excitation and detection of resonant motions, can significantly lower the Q [6]. To address this issue and retain high Q, we have demonstrated a non-contact electrical excitation scheme recently [6], where an external electrode is devised to efficiently drive the length extensional (LE) and other vibrational modes of the electrode-less resonator, via inverse piezoelectric effect and gradient force from the electric field.

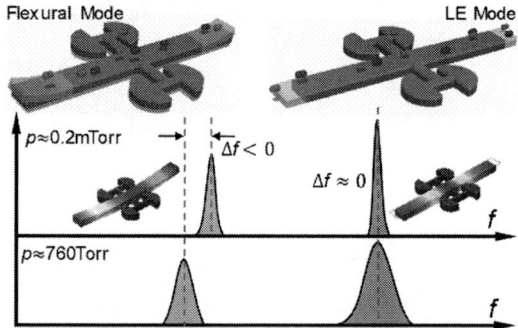

Figure 1: AlN/Si resonator vibrating at LE and flexural modes interacts with air molecules. The linewidths of both modes are broadened under atmospheric pressure. The frequency of flexural mode goes down while the frequency of bulk mode stays constant with increasing pressure level.

Another important factor in Eq. (1) is the medium loss. When the device is vibrating in moderate vacuum or near atmospheric pressure, the medium (air) damping can be an important dissipation mechanism which is heavily dependent on ambient pressure [7]. To date, many studies of pressure dependence and air damping have been conducted on traditional MEMS including cantilevers [7,8], doubly-clamped beams [9], and drumhead membrane resonators [10]. However, experimental investigation and analysis of pressure dependency of AlN/Si resonators is still lacking, especially for multiple modes. Moreover, a quantitative understanding of air damping effects in wider pressure range is of critical importance in designing and prototyping new MEMS technologies.

In this work, we present measurements and analyses of air damping effects on different modes of non-contact electrically driven MEMS resonators fabricated in different heterostructure stacks. We carefully measure Q and f values for both length extensional (LE) mode and flexural mode (Fig. 1), as functions of ambient pressure from relatively high vacuum (10^{-4} Torr) to atmospheric level (760 Torr) at room temperature. We find different modes of same device exhibit various tolerance to the air damping effects and provide quantitative evidences that could be helpful for future device engineering.

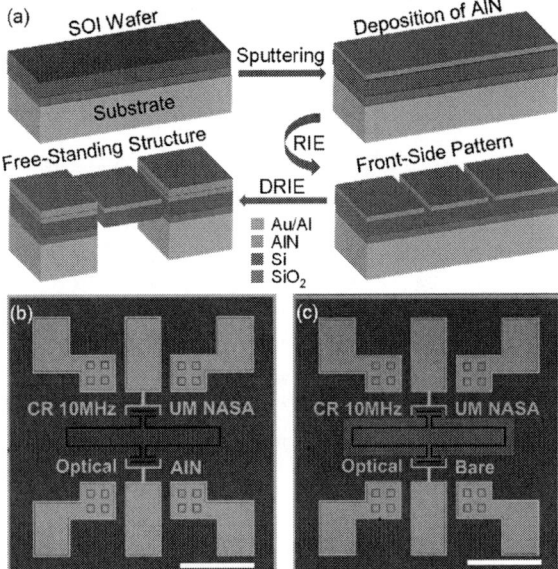

Figure 2: (a) Illustration of the AlN/Si device fabrication process using standard microfabrication techniques. RIE: reactive ion etching, DRIE: deep reactive ion etching. Optical images of (b) AlN/Si and (c) Si-only devices. Scale bar: 200 μm.

FABRICATION AND EXPERIMENT

As illustrated in Fig. 2a, we begin the fabrication with a Si-on-insulator (SOI) wafer comprising a 20 μm-thick n-type phosphorus (P) doped (at 4.6 ×10^{19} cm^{-3}) Si device layer and a 500 μm-thick Si substrate. Next, on top of the highly doped Si layer, a 1 μm-thick AlN layer is deposited to serve as the piezoelectric transducer. Then 100 nm of aluminum (Al) is chosen for metal routing and patterned by lift-off process. We use photolithography to pattern the front side of the wafer followed by reactive ion etching (RIE) through the buried oxide (BOX) layer. To create suspended devices, deep reactive ion etching (DRIE) is performed from the back side of the wafer. Si-only devices are made by removing the top deposited AlN piezoelectric layer and metal electrodes. Figure 2b & 2c present the optical images of AlN/Si and Si-only BARs, respectively. The main body has a length of 415 μm and a width of 43 μm, and its center is attached to two tethers each of which is 35 μm in length and 8 μm in width.

Figure 3 presents the experimental scheme for the characterization of the resonant properties of the devices under varying pressure level. To efficiently excite the resonance motion of the device, we employ a non-contact overhanging electrode [6] connected to the output of a network analyzer (NA). The multimode resonances are simultaneously read out by focusing a 633 nm red laser on the device employing ultrasensitive optical interferometry. The pressure is monitored via a pressure gauge connected to the chamber regulated by flow rate valves. We examine resonance behavior of the device from atmospheric pressure to high vacuum, then raise the pressure back to atmospheric conditions again. The measurements are carried out at room temperature.

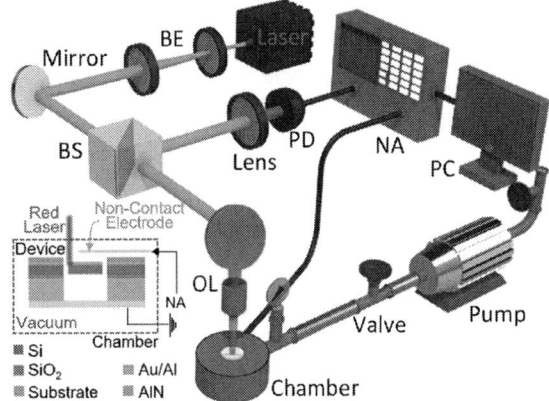

Figure 3: Schematic of measurement system using non-contact electric drive and optical detection. BE: beam expander, BS: beam splitter, NA: network analyzer, PD: photodetector, PC: personal computer, OL: objective lens.

RESULTS AND DISCUSSIONS

We first drive the LE bulk mode of the AlN/Si BAR and the resonance characteristics are clearly shown in Fig. 4a to 4f. To examine the pressure dependence of the damping, we measure and extract the f and Q at different pressure levels. Measured values of both parameters well overlap between sweeping-up and sweeping-down traces. As shown in Fig. 4a, the scheme resolves this mode at f=10.291 MHz which roughly remains constant during the experiment. This may be a consequence of inappreciable influence from mass loading due to a large spring constant of LE mode and indicates a high immunity of LE mode to external pressure in terms of f stability. Moreover, we observe a p-independent Q when p is below 10 Torr as the largest Q difference between Fig. 4d & 4f is less than 7%. This indicates that air damping impact is negligible in comparison to the intrinsic damping of the device itself. The independence of both f and Q on pressure within this pressure range suggests that dissipation is limited by other mechanisms (not air damping) for LE mode in p <10 Torr region. Beyond 10 Torr, Q declines with p and the measured tendency is consistent with the free molecule flow (FMF) damping which describes collisions of free air molecules with moving surface of the vibrating device. To examine if FMF is the primary dissipation mechanism, we have:

$$\left(\frac{1}{Q}\right)_{tot} = \left(\frac{1}{Q}\right)_{p\text{-indep}} + \alpha \left(\frac{1}{Q}\right)_{p\text{-dep}}, \quad (2)$$

where $Q^{-1}_{p\text{-indep}}$ ($Q^{-1}_{p\text{-dep}}$) is the pressure independent (dependent) dissipation and α is a fitting parameter associated with device structure. The pressure dependent Q [10,11] is given by:

$$Q_{p\text{-dep}} = \frac{\rho t \omega_0}{4} \left(\frac{\pi R T}{2m}\right)^{\frac{1}{2}} \frac{1}{p}, \quad (3)$$

where ρ is mass density of the device material, t is device thickness, R is the gas constant, T is temperature, ω_0 is the resonance frequency of the selected mode, and m is the mass of gas molecule. Equation (2) is valid when the FMF becomes predominant pressure dependent dissipation process. Using this relationship, we find good agreement between fitting (red dashed curve) and measurement as depicted in Fig. 4a, and we determine $\alpha=0.58$ through fitting to Eq. (2). This confirms that the air damping starts dominating the dissipation when p is above 10Torr for the LE bulk mode of AlN/Si device.

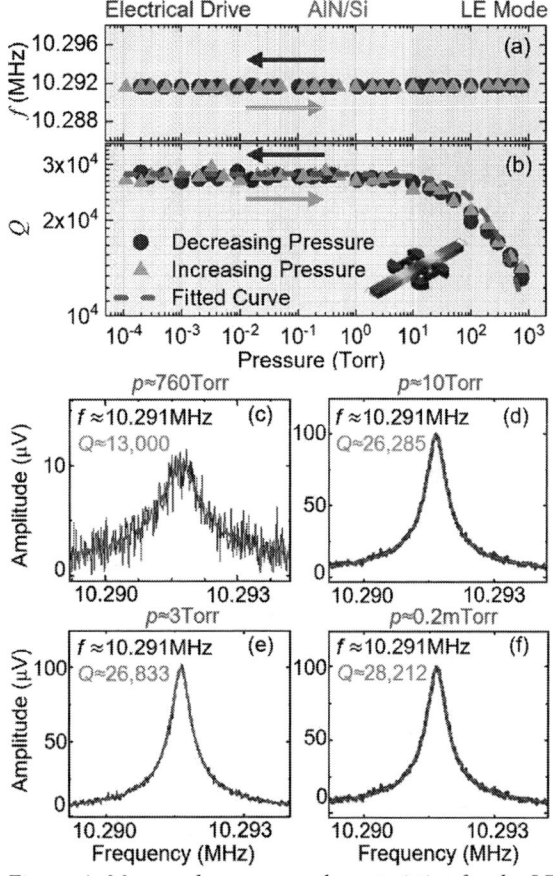

Figure 4: Measured resonance characteristics for the LE bulk mode of the AlN/Si resonator. (a)-(b): f and Q versus pressure. Red dashed curve shows the fitting to Eq. (2) with α=0.58. (c)-(f) Measured resonances with fitted Q values at p=760Torr, 10Torr, 3Torr, and 0.2mTorr.

We then excite the out-of-plane flexural mode of the AlN/Si device using non-contact electrical drive. As shown in Fig. 5, from 100μTorr to 1Torr, no clear influence from p on both f and Q is observed. The Q value extracted at 1Torr is only 0.8% lower than that extracted at 0.2mTorr. However, different from a stable f observed in the LE mode, the f of the flexural mode decreases as pressure goes to $p>$ 1Torr. This may arise from the fact that the extent of mass loading has more apparent impact on flexural mode f due to its comparatively much smaller spring constant ($k_{eff,flex}/k_{eff,LE}\approx0.002$). We apply Eqs. (2) and (3) to the measured trend of Q and the results agree well with fitting, which implies the air damping plays a role when p is above 1Torr for the flexural mode.

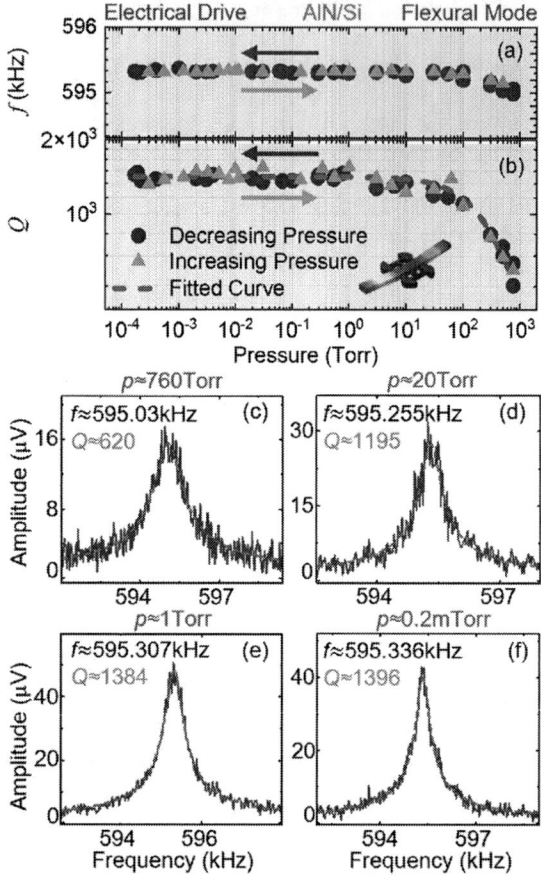

Figure 5: Measured resonance characteristics for the flexural mode of the AlN/Si device. (a)-(b): f and Q versus pressure. Red dashed curve shows the fitting to Eq. (2) with α=0.9. (c)-(f) Measured resonances at p=760Torr, 20Torr, 1Torr, and 0.2mTorr.

The Si-only device eliminates energy loss induced by loading and interfacial dissipation from the deposited metal electrode atop. Given its noticeably higher Q in the bulk mode [5,6] in comparison to that of the AlN/Si device, we also characterize the pressure dependence of its Q and f. As seen in Fig. 6, the LE bulk mode of Si-only device keeps a stable f of 10.127MHz with variations less than 0.001% throughout the measured pressure range. This LE mode maintains a high $Q\approx257,000$ with variation of 3% from the range $p=100\mu$Torr to 3Torr. As the Q starts declining at $p>$ 3Torr, the FMF model fits the data well with $\alpha=0.79$. Table 1 summarizes f and Q values for different modes in both devices at selected pressure levels.

CONCLUSION

In summary, air damping effects on both LE bulk and flexural modes of AlN/Si BAR, as well as LE mode of Si-only resonators, have been measured and analyzed. Using the non-contact electric drive for resonance excitation and a custom-built pressure-controlled apparatus, we have calibrated f and Q with p regulated from 760Torr to 100μTorr. We observe distinct features in the pressure-dependent resonance characteristics of dual modes: f of the LE bulk mode of AlN/Si resonator stays virtually constant

Table 1: Q factors and frequencies of the dual modes (LE & flexural) measured at various pressure levels.

Mode / Stacking		Length Extensional (LE) Mode				Out-of-Plane Flexural Mode			
		760Torr	10Torr	3Torr	0.2mTorr	760Torr	20Torr	1Torr	0.2mTorr
AlN/Si	f (MHz)	10.291	10.291	10.291	10.291	0.59503	0.595255	0.595307	0.595336
	Q	13,000	26,285	26,443	28,212	620	1195	1384	1396
Si-Only	f (MHz)	10.127	10.127	10.127	10.127	No Data Taken			0.59474
	Q	20,213	124,357	227,500	235,600				961

over the entire pressure range while Q begins to degrade above 10Torr; the flexural mode of AlN/Si device exhibits stable $f \approx 595.3$kHz below 1Torr but it falls to ≈ 595kHz at atmospheric pressure. The LE bulk mode of Si-only device shows stable $f \approx 10.127$MHz independence of p, while it offers high $Q \approx 257,000$ at $p<3$Torr. This work provides insight into distinct regimes in which air damping impacts the bulk and flexural modes of AlN/Si resonators, enhancing quantitative understanding that is beneficial for future device engineering and applications.

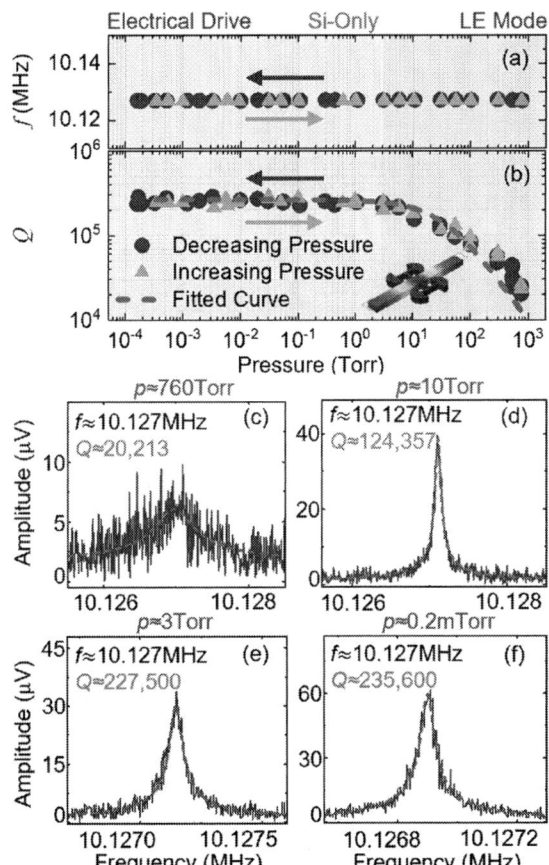

Figure 6: Measured resonance characteristics of the LE bulk mode of the Si-only resonator. (a)-(b): f and Q versus pressure. Red dashed curve shows the fitting to Eq. (2) with α=0.79. (c)-(f) Measured resonances at p=760Torr, 10Torr, 3Torr, and 0.2mTorr.

ACKNOWLEDGEMENTS

The effort at University of Florida (UF) is partly supported by National Science Foundation (NSF) CCF FET Program (Grant 2103091) and by the Margaret A. Ross Scholarship (S M Enamul Hoque Yousuf) in ECE at UF. The authors at UF thank X.-Q. Zheng and J. Lee for technical support and discussions.

REFERENCES

[1] R. Abdolvand, et al., "Thin-Film Piezoelectric-on-Silicon Resonators for High-Frequency Reference Oscillator Applications", IEEE Trans. Ultrason. Ferroelectr. Freq. Control, vol. 55, pp. 2596-2606, 2008.

[2] J. L. Fu, et al., "Dual-Mode AlN-on-Silicon Micromechanical Resonators for Temperature Sensing", IEEE Trans. Electron Devices, vol. 61, pp. 591-597, 2014.

[3] C. Fei, et al., "AlN Piezoelectric Thin Films for Energy Harvesting and Acoustic Devices", Nano Energy, vol. 51, pp. 146-161, 2018.

[4] G. K. Ho, et al., "Piezoelectric-on-Silicon Lateral Bulk Acoustic Wave Micromechanical Resonators", J. Microelectromech. Syst., vol. 17, pp. 512-520, 2008.

[5] A. Qamar, et al., "Study of Energy Loss Mechanisms in AlN-Based Piezoelectric Length Extensional-Mode Resonators", J. Microelectromech. Syst., vol. 28, pp. 619-627, 2019.

[6] S M E. H. Yousuf, et al., "Retaining High Q Factors in Electrode-Less AlN-on-Si Bulk Mode Resonators with Non-Contact Electrical Drive", Proc. of the 35th IEEE Int. Conf. Micro Electro Mech. Syst. (MEMS 2022), pp. 979-982, 2022.

[7] J. Yang, et al., "Energy Dissipation in Submicrometer Thick Single-Crystal Silicon Cantilevers", J. Microelectromech. Syst., vol. 11, pp. 775-783, 2002.

[8] J. E. Sader, et al., "Frequency Response of Cantilever Beams Immersed in Viscous Fluids with Applications to the Atomic Force Microscope", J. Appl. Phys., vol. 84, pp. 64-76, 1998.

[9] S. S. Verbridge, et al., "A Megahertz Nanomechanical Resonator with Room Temperature Quality Factor Over a Million", Appl. Phys. Lett., vol. 92, art. no. 013112, 2012.

[10] J. Lee, et al., "Air Damping of Atomically Thin MoS2 Nanomechanical Resonators", Appl. Phys. Lett., vol. 105, art. no. 023104, 2014.

[11] W. E. Newell, et al., "Miniaturization of Tuning Forks: Integrated Electronic Circuits Provide the Incentive and the Means for Orders-of-Magnitude Reduction in Size", Science, vol. 161, pp. 1320-1326, 1968.

CONTACT

†Equally Contributed Authors
*†Yuncong Liu: yuncong.liu@ufl.edu
*†S M Enamul Hoque Yousuf: syousuf@ufl.edu
* Mina Rais-Zadeh: mina.rais-zadeh@jpl.nasa.gov
* Philip Feng: philip.feng@ufl.edu

A NOVEL PIEZORESISTIVE PRESSURE SENSOR BASED ON CR-DOPED V$_2$O$_3$ THIN FILM

Michiel Gidts, Wei-Fan Hsu, Maria Recaman Payo, Shashwat Kushwaha, Chen Wang,
Frederik Ceyssens, Dominiek Reynaerts, Jean-Pierre Locquet and Michael Kraft
KU Leuven, B-3000 Leuven, Belgium

ABSTRACT

This paper reports the fabrication and characterization of a piezoresistive pressure sensor based on Cr-doped V$_2$O$_3$ thin film (Cr-V$_2$O$_3$TF). It is the first time that the piezoresistive effect of single crystalline Cr-V$_2$O$_3$TF is demonstrated experimentally and implemented as a pressure sensor. The Cr-V$_2$O$_3$TF piezoresistors on the membrane experience a stress change caused by a pressure input. This leads to a gradual phase transition of the material from an insulating phase to a metallic phase and results in a resistivity change. This new piezoresistive mechanism opens up the potential for developing highly sensitive piezoresistive sensors based on phase transition.

KEYWORDS

Piezoresistive pressure sensor, transition metal oxides, thin film, phase transition, sapphire membrane

INTRODUCTION

Piezoresistive MEMS sensors are among the earliest silicon micromachined devices. Successful integration of a piezoresistive material with micromachined flexure elements already enabled the implementation of the piezoresistive effect into various MEMS sensors and applications [1]. Though many materials like diamond [2], carbon nanotubes [3] and nanowires [4] show higher resistivity change with stress than silicon or germanium [1], they are harder to integrate in MEMS devices. However, thin film technology allows an easier integration of new piezoresistive materials in MEMS sensors [5,6]. Recently, vanadium oxide thin films have attracted much attention due to their high piezoresistive effect and integration feasibility to microfabrication [7,8]. Vanadium oxide materials are known for their interesting temperature-induced phase transition, termed as metal-insulator transition (MIT) [9,10]. This phase transition is associated with a three orders of magnitude resistivity change and has been successfully implemented in infrared microbolometer applications [11]. While most research about vanadium oxides focused on the resistivity change with temperature, vanadium oxide materials demonstrates great potential as piezoresistive material. The resistivity change with external stress has been previously investigated by Inomata et al. for sputtered VO$_x$ thin films (75% V$_2$O$_5$ and 25% Cr-VO$_2$) [7]. Though these VO$_x$ thin films demonstrated a high gauge factor, they did not show any obvious MIT property. Nevertheless, for single crystalline Cr-V$_2$O$_3$TF a gradual MIT transition at room temperature was observed when varying Cr doping level [12]. An in-plane lattice change caused by external stress was expected to cause a similar gradual MIT transition as a varying Cr doping level. Therefore, such a thin film material with a strain-caused phase transition could

potentially allow the development of highly sensitive piezoresistive sensors. To demonstrate the piezoresistive effect and integration of single crystalline Cr-V$_2$O$_3$TF in MEMS devices, a pressure sensor based on this material was fabricated and measured.

FABRICATION

Material fabrication

The piezoresistive material was epitaxially grown on a (0001)-oriented Al$_2$O$_3$ (sapphire) substrate by molecular beam epitaxy (MBE) [12,13]. Other techniques such as magnetron sputtering [14], pulsed laser deposition [15] and chemical vapor deposition [16] epitaxial thin films have also been used to successfully fabricate V$_2$O$_3$ thin films. However, MBE offers several advantages. The deposition is done at low pressures (8.5 10^{-6} Torr) reducing the number of impurities present in the thin film. The oxygen partial pressure in the MBE chamber and the substrate temperature can be precisely controlled allowing a good control of the stoichiometry and crystallinity of the growing layer. Finally, a low deposition rate by MBE allows a precise control of the thin film thickness.

Sensor fabrication

Figure 1 shows the process flow for sensor fabrication. At first the membrane was micromachined on a sapphire substrate with a high-precision milling machine. The total membrane size is 2.5mm with a measured membrane thickness of 96 μm. Then the sapphire wafer was cleaned with a RCA cleaning prior to the epitaxial growth of Cr-V$_2$O$_3$TF with MBE. The material was subsequently patterned and etched with reactive ion etching (RIE). Au and Cr (adhesion layer) were sputtered as metal tracks for electrical connection.

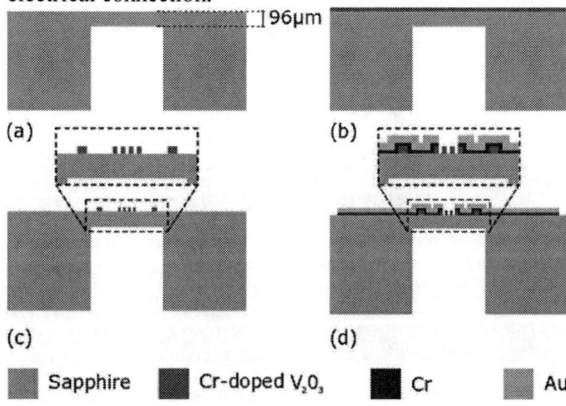

Figure 1: Process flow for the fabrication of the Cr-doped V2O3TF pressure sensor. (a) Micromachining of the sapphire membrane (b) Deposition of the Cr-V$_2$O$_3$TF (c) Patterning of the Cr-V$_2$O$_3$TF (d) Deposition of Au and Cr.

DESIGN AND CHARACTERIZATION
Sensor design

Figure 2 is a close-up top view of the sensor and sensor die bonded to a PCB. The total sensor die is 1x1 cm^2 with a frame thickness of 0.476 mm. The membrane is built on a sapphire substrate, which is a highly corrosion resistant material [17]. The elasticity modulus of sapphire is much bigger than silicon or stainless steel, resulting in a stiffer diaphragm with faster response time. Additionally, the piezoresistors are well isolated from the substrate, avoiding leakage currents that reduce the sensitivity of the pressure sensor. Four piezoresistors were fabricated on the membrane and connected in a Wheatstone bridge configuration through metal tracks and wire bonds. Two of the piezoresistors were placed at the center of the membrane where the stress change caused by a pressure input reaches to its maximum. The other two were placed more towards the edge of the membrane. The piezoresistors were designed with a meander shape to maximize the resistance in a limited membrane area. The piezoresistors at the center have a total length of 1.46 mm with a width of 20 µm. The piezoresistors on the outside have a length of 3.14 mm with a width of 43 µm. The bond pads are placed on the side of the sensor die Additional structures are placed on the die to characterize the material properties.

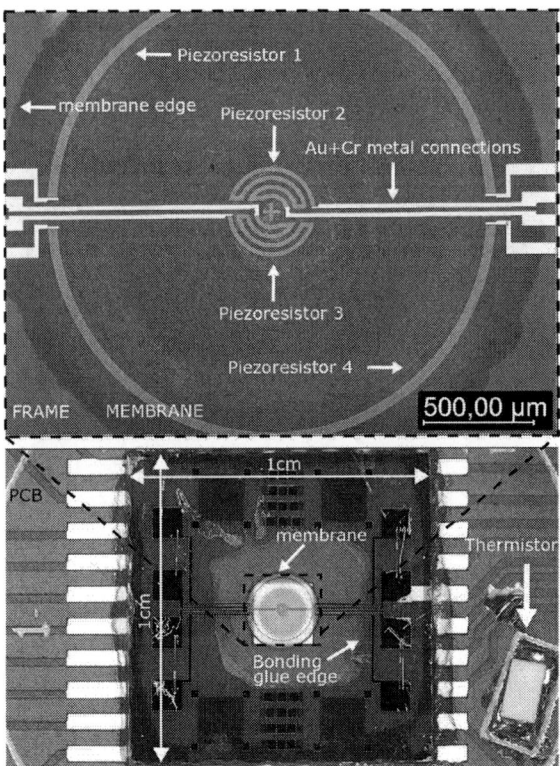

Figure 2: Picture of the sensor and sensor die bonded on the PCB for testing.

Material characterization

After the patterning step, the Cr-V$_2$O$_3$TF thickness measured with stylus profiling was 30 nm. The resistivity of the thin film has been determined by placing additional structures on the sensor die and using the Van der Pauw (VDP) method [18] and Transfer Length Method (TLM) [19]. Both structures are shown in Figure 3. The VDP structure consists of a Cr-V$_2$O$_3$TF square (1.5x1.5 mm^2) with a Au contact pad at each corner. The contact pads and square overlap 30x30 µm^2. The TLM structure consists of Cr-V$_2$O$_3$TF rectangles (0.3x1.4 mm^2) with three Au contact pads on top. The distance between the contact pads was measured with an optical microscope. These structures were placed on the sensor die close to (see Figure 2). A resistivity of respectively 0.00188 Ω cm and 0.00151 Ω cm was measured at room temperature (20°C) by each method.

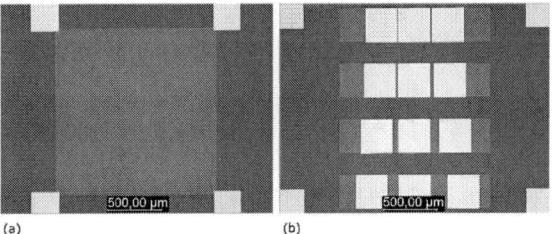

Figure 3: (a) VDP structure (b) TLM structure

Sensor characterization

The pressure sensor was finally glued on a PCB and the electrical connections between the sensor and the PCB were made with wire bonds. The glue was applied extensively between the sensor die and the PCB to create a good hermetic seal. Any glue under the sensor membrane was avoided. The PCB had a hole just underneath the membrane to allow applying pressure to the membrane. A thermistor was placed close to the sensor to measure the temperature (see Figure 2). This pressure sensor was characterized under different pressures and temperatures with a measurement setup as shown in Figure 4. This measurement setup includes an accurate reference pressure sensor, a PC, a source meter, a NI data acquisition card, a temperature sensor and a pressure vessel.

The PCB was connected via HDMI cables to an external circuit. The pressure vessel was made of aluminium, it covered the entire sensor die and sealed the sensor hermetically from the outside. The temperature was controlled by placing the pressure vessel under a heat gun in a temperature chamber. For each temperature, the differential pressure was increased from 0 to 1.8 bar with steps of 0.1 bar.

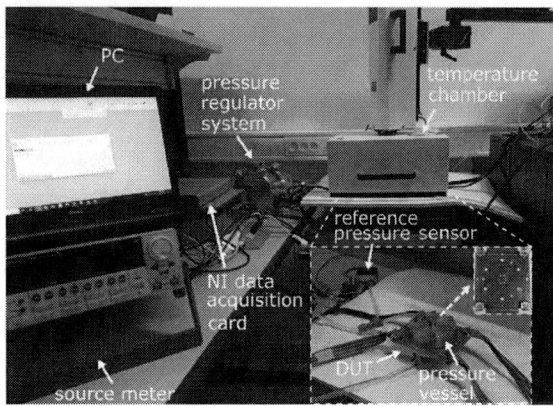

Figure 4: Measurement setup for the characterization of the pressure sensor system.

Figure 5 shows the output of the pressure sensor under differential pressures and temperature. At 20°C a sensitivity of 2.09 mV/V/bar was measured. The measured results showed that the pressure sensor demonstrated a linear response against pressure change. The larger the membrane size or smaller the membrane thickness are, the better the sensitivity of the pressure sensor is. The sensitivity and offset for different temperatures are listed in Table 1. With the increase of the temperature, the sensitivity was reduced while the offset was increased. The temperature coefficient of sensitivity (TCS) and offset (TCO) of the pressure sensor were -0.016 mV/V/bar/°C and 0.165 mV/V/°C respectively.

CONCLUSION

This work reports a novel piezoresistive pressure sensor based on Cr-V_2O_3TF. This is the first time that the Cr-V_2O_3TF piezoresistivity is observed experimentally and implemented as a piezoresistive sensor. The Cr-V_2O_3TF piezoresistors were deposited on a micromachined sapphire membrane and characterized under different pressures and temperatures. The piezoresistivity is caused by external stress inducing a gradual MIT phase transition. This new piezoresistive mechanism opens up the potential of developing highly sensitive piezoresistive sensors based on this phase transition.

Table 1: Measured sensitivity and offset of the pressure sensor.

Temperature	Sensitivity [mV/V/bar]	Offset [mV/V]
20°C	2.09	-36.71
30°C	1.77	-34.28
40°C	1.58	-32.25
50°C	1.53	-30.76
60°C	1.36	-29.42
70°C	1.29	-28.48

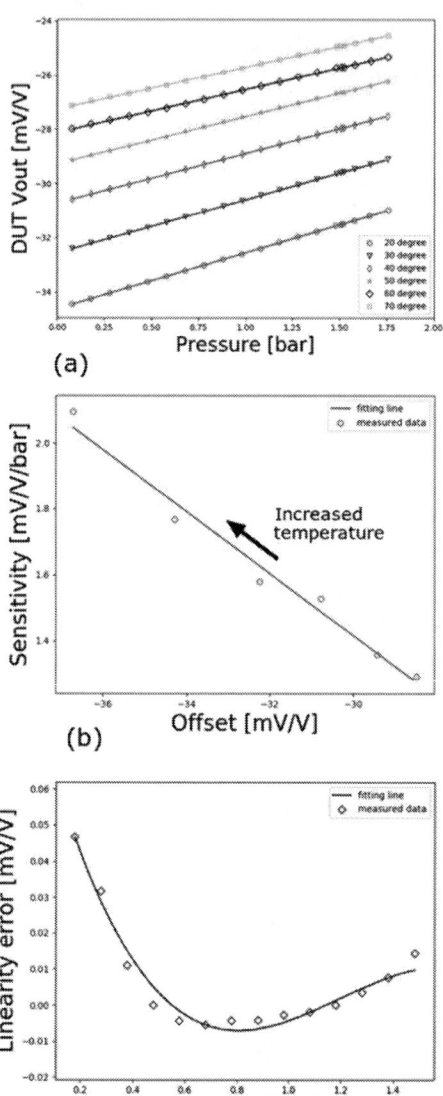

Figure 5: Measured output of the pressure sensor. (a) Output voltage under differential pressures and temperature. (b) Sensitivity and offset. (c) Linearity error for 60 °C.

REFERENCES

[1] A. Barlian, W. Park, J. Mallon, ... and B. Pruitt, "Review Semiconductor Piezoresistance for Microsystems", Proceedings of the IEEE, 97 (2009), pp. 513-552.

[2] S. Sahli, D.M. Aslam, "Ultra-high sensitivity intra-grain poly-diamond piezoresistors", Sensors and actuators. A. Physical, 71 (1998), pp. 193-197.

[3] P. Regoliosi, A. Reale, A. Di Carlo, ... and P. Lugli, "Piezoresistive behaviour of single wall carbon nanotubes", IEEE Conference on Nanotechnology, 2004, pp. 149-151.

[4] K. Reck, J. Richter, O. Hansen and E.V. Thomsen, "Piezoresistive effect in top-down fabricated silicon

nanowires", IEEE MEMS conference, 2008, pp. 717-720.

[5] M. Mehregany, C. Zorman, N. Rajan and C.H. Wu, "Silicon carbide MEMS for harsh environments", Proceedings of the IEEE, 86 (1998), pp. 1594-1609.

[6] H. Phan, D.V. Dao, K. Nakamura, S. Dimitrijev and N.T. Nguyen, "The Piezoresistive Effect of SiC for MEMS Sensors at High Temperatures: A Review", Journal of microelectromechanical systems, 24 (2015), pp. 1663-1677.

[7] N. Inomata, N. Van Toan, M. Toda and T.Ono. "Evaluation of piezoresistivity property of vanadium oxide thin film", IEEE sensors letters, 2 (2018), pp.1-4.

[8] P. Homm, M. Menghini, J.W. Seo, S. Peters and J-P. Locquet, "Room temperature Mott metal–insulator transition in V2O3 compounds induced via strain-engineering", APL Mater. 9, 021116 (2021).

[9] M. Imada, A. Fujimori and Y. Tokura, "Metal-insulator transitions", Reviews of modern physics, 70 (1998), pp.1039-1263.

[10] L. Dillemans, T. Smets, R.R. Lieten, ... and J-P Locquet, "Evidence of the metal-insulator transition in ultrathin unstrained V2O3 thin films", Applied physics letters, 104, 071902 (2014).

[11] R. Darling and S. Iwanaga, "Structure, properties, and MEMS and microelectronic applications of vanadium oxides", Sadhana, 34 (2009), pp. 531-542.

[12] P. Homm, L. Dillemans, M. Menghini. ... and J-P. Locquet, "Collapse of the low temperature insulating state in Cr-doped V2O3 thin films", Applied physics letters, 107 (2015).

[13] L. Dillemans, R.R. Lieten, M. Menghini, ... and J-P. Locquet, "Correlation between strain and the metal–insulator transition in epitaxial V2O3 thin films grown by Molecular Beam Epitaxy", Thin solid films, 520 (2012), pp. 4730-4733.

[14] G. Sun, X. Cao, S. Long, R. Li and P. Jin, "Optical and electrical performance of thermochromic V2O3 thin film fabricated by magnetron sputtering", Applied physics letters, 111 (2017).

[15] S. Majid, D Shukla, F. Rahman, … and D. Phase, "Characterization of pulsed laser deposition grown V2O3 converted VO2", Journal of Physics: Conference Series, 755 (2016)

[16] C. Piccirillo, R. Binions, I. Parkin, "Synthesis and Functional Properties of Vanadium Oxides: V2O3, VO2, and V2O5 Deposited on Glass by Aerosol-Assisted CVD", Chemical vapor deposition, 13 (2007), pp. 145-151.

[17] S. Fricke, A. Friedberger, H. Seidel and U. Schmid, "A robust pressure sensor for harsh environmental applications", Sensors and actuators. A. Physical, 184 (2012), pp. 16-21.

[18] L L. J. Van der Pauw, "A method of measuring specific resistivity and hall effect of discs of arbitrary shape". Philips Res. Rep, 13 (1958), pp. 1-9.

[19] D. Schroder, "Semiconductor material and device characterization", 2006

CONTACT

*Chen Wang: chen.wang@esat.kuleuven.be

A NOVEL FEEDTHROUGH CANCELLATION TECHNIQUE FOR PIEZOELECTRIC MEMS RESONANT SENSORS IN IONIC LIQUID MEDIUM

Cheng-Yen Wu[1], Zhong-Wei Lin[1], and Sheng-Shian Li[1,2]

[1] Institute of NanoEngineering and MicroSystems, National Tsing Hua University, Hsinchu, Taiwan
[2] Department of Power Mechanical Engineering, National Tsing Hua University, Hsinchu, Taiwan

ABSTRACT

This work presents a thin-film piezoelectric-on-silicon (TPoS) resonator and oscillator operating in an ionic (i.e., conductive) liquid environment without suffering from existing ionic background floor through *a novel feedthrough cancellation technique*. Such MEMS resonant transducers would serve as an important building block for mass sensing used in physical/chemical/bio applications demanding sensing in ionic liquid medium, such as saline solution. Resonators implementing both PZT and AlN piezoelectric materials are thoroughly explored in this work. With feedthrough cancellation, the PZT TPoS resonator features decent stopband rejection (SBR) of around 21 dB and also enables oscillation in an ionic solution, with an oscillating frequency of 5.75 MHz and Allan deviation (ADEV) of 2.05 ppm, which implies a mass resolution of only 3.7 pg. The PZT TPoS oscillator phase noise (PN) performance in the conductive liquid is -83.15 dBc/Hz and -127.04 dBc/Hz, respectively, at offset frequencies of 1 kHz and 100 kHz. To the best knowledge of the authors, this is the first real-time measurement of TPoS resonators and oscillators working in ionic circumstance without isolation between signal pads and liquids. Furthermore, the feedthrough cancellation makes the SBR of the resonator in ionic water (21 dB) even greater than that in DI (deionized) water (11 dB). This new technique has great potential in chemical and biosensing applications in future.

KEYWORDS

TPoS, ionic liquid, conductive medium, feedthrough cancellation, MEMS resonant transducers, PZT/AlN resonator, physical/chemical/bio applications, mass sensor.

INTRODUCTION

With the advantages of microelectromechanical systems (MEMS) devices, MEMS sensors are now widely used in a variety of applications. For instance, due to its compactness and high accuracy, MEMS sensors are very suitable for biological applications, known as bio-MEMS, and show great potential in various kinds of measurements. Once different biosensors can be integrated into single biomedical chip, it not only can replace large biochemical analysis instruments, but also reduce the consumption of biological reagents, the errors caused by human operation, the reaction time as well as the inaccuracy of detection. In addition, by integrating the entire inspection process including sample collection, transmission, reaction and detection into one chip, the ultimate scope of Lab-On-a-Chip (LOC) or even Micro Total Analysis System (μ-TAS) can be realized.

Many nanoscales biological or environmental sensing

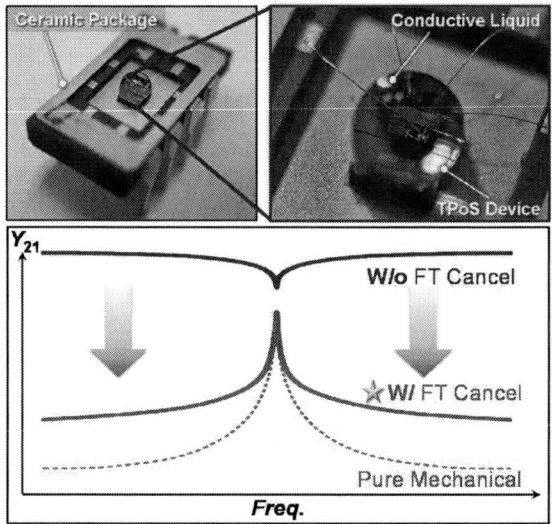

Figure 1: Photo of a PZT TPoS resonator under a conductive liquid droplet with the frequency schematic diagram with and without using the proposed feedthrough (FT) cancellation technique. Pure mechanical resonant behavior of the resonator is also provided for reference.

applications need to be operated in liquid or conductive medium environments, such as the interaction of protein antigens and antibodies, or the detection of drugs and metal ions. For MEMS sensors, there is a great quantity of research works related to liquid applications [1]. In literature, (i) resonant amplitude and Q reduction caused by liquid damping as well as (ii) high feedthrough floor due to dielectric constant of liquid, i.e., non-conductive medium, are major challenges in the field of MEMS resonators operating in liquids [2]. Therefore, mode shape exploration and design optimization are mostly discussed in deionized (DI) water to deal with the Q issue [3], while a fully differential transduction schematic usually used to handle the high feedthrough issue caused by DI water [4]. Unfortunately, in prior arts, real-time resonant signal outputs measured directly in conductive liquids are rarely investigated and reported. Since the electrical signal transduction is highly disturbed in conductive liquids, measurement after the liquid has evaporated [5] or the sensing device coated with an insulating layer [6] are the main methods often used to solve the issue.

As a result, a novel feedthrough cancellation technique that makes resonant sensors being satisfactorily measured without additional coating when working in ionic liquid medium has been successfully demonstrated in this work. As shown in Fig. 1, a TPoS resonator fully covered by an ionic droplet is adopted to illustrate the idea of how the resonant performance can be improved by using the proposed feedthrough cancellation technique. Once the feedthrough level is greatly reduced, the mechanical resonance of the resonator can be clearly seen for later oscillation and mass sensing. With this cancellation phenomenon, measuring resonant signal in ionic liquid becomes possible and such MEMS resonant transducers have a great potential to be used in physical/chemical/bio applications.

FABRICATION AND DESIGN CONCEPT
Resonator Fabrication

Both PZT and AlN based length-extensional (LE) mode resonators are designed, fabricated, and tested to verify the concept in this study. These resonators with different types of piezoelectric material are all fabricated by a similar four-masks process flow, as shown in Fig. 2. The fabrication process starts with a piezoelectric layer deposited on a 4" SOI wafer, followed by dry etching steps to define the contact via position. After that, top electrode is deposited and patterned through the lift-off technique. The resonator structure has been released through frontside dry etching and backside silicon DRIE. Fig. 3 shows the finite-element simulated mode shape of the resonator, the cross-section view, and the 3D confocal image of the

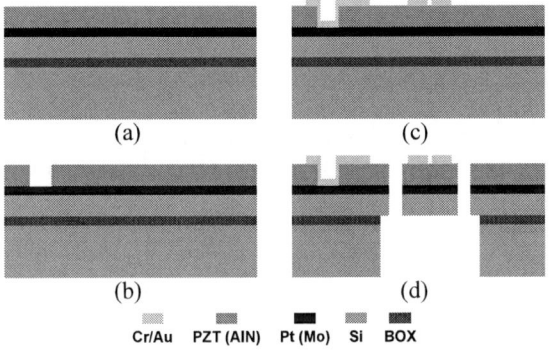

Figure 2: Process flow of the TPoS fabrication platform. (a) Piezoelectric material is firstly deposited on the 4" SOI wafer. (b) Contact via is defined by a dry etching process. (c) Top electrode is patterned by a lift-off process. (d) Structure is defined by front-side dry etching and released by backside silicon DRIE process.

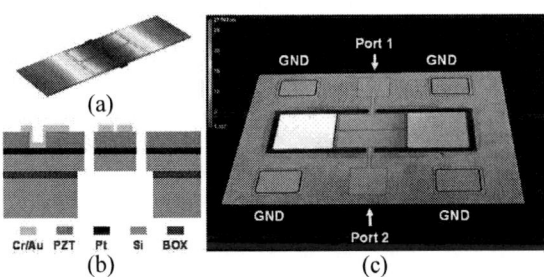

Figure 3: (a) Finite element simulated mode shape of the LE mode device. (b) Device cross-section view. (c) 3D confocal image of the LE mode resonator.

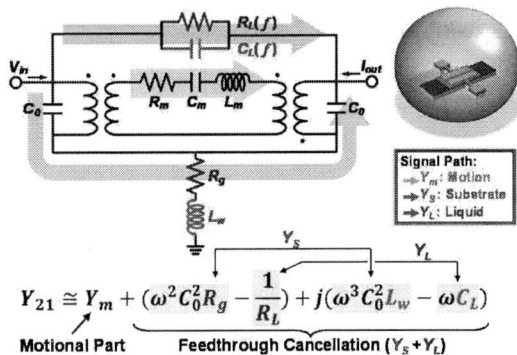

$$Y_{21} \cong Y_m + (\omega^2 C_0^2 R_g - \frac{1}{R_L}) + j(\omega^3 C_0^2 L_w - \omega C_L)$$

Motional Part | Feedthrough Cancellation ($Y_s + Y_L$)

Figure 4: Equivalent circuit model of the LE mode resonator in a conductive liquid environment. The admittance formula (Y_{21}) of this circuit model shows that signals from substrate (Y_S) and liquid (Y_L) would cancel out each other under certain conditions.

released device using KEYENCE confocal microscope.

Design Concept

With sufficiently high dielectric constant of the piezoelectric material and lack of a perfect signal ground, the feedthrough path through the bottom electrode (substrate) cannot be ignored. In this study, this feedthrough path from substrate is elegantly utilized to cancel the feedthrough caused by ionic liquid from the top electrodes, so that PZT is adopted as the desired piezoelectric material due to its high dielectric constant (compared to AlN). With this material property, the admittance at the substrate path becomes large enough, resulting in a strong feedthrough that can be utilized to cancel the feedthrough induced by the ionic medium.

The equivalent circuit model of this feedthrough cancellation technique is illustrated in Fig. 4. There are three admittance parts between the device I/O ports: motional path (Y_m), substrate path (Y_S), and liquid path (Y_L). The feedthrough Y_L from the ionic water is modeled as a parallel of $R_L(f)$ and $C_L(f)$ where f stands for operational frequency. After derivation and simplification of the Y_{21} formula, both real part and imaginary part of Y_S and Y_L would be able to cancel out each other under certain desirable conditions.

Typically, when a resonator (such as AlN) operates in a conductive liquid, the liquid will induce high feedthrough between the resonator I/O ports, which masks the desired motional signal Y_m. In this work, we adopt PZT resonator and carefully leverage Y_S and Y_L feedthroughs with opposite polarities and similar amplitudes to cancel out each other at a desired frequency range, making the oscillation in conductive liquid feasible.

In order to examine the idea of feedthrough cancellation, the constituent parameters in the equivalent circuit model using reasonable values based on FEM simulation and previous measurement data are implemented for quantitative verification. Fig. 5 shows the model simulation results of different combinations from the equivalent circuit model. It is implied that if the resonator in ionic liquid has no feedthrough cancellation, its frequency response behaves like a dip, as illustrated in

978-1-6654-9309-3/23 $31.00 © 2023 IEEE   906

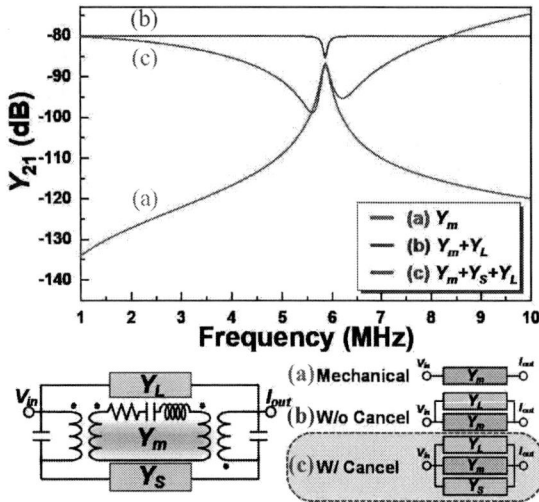

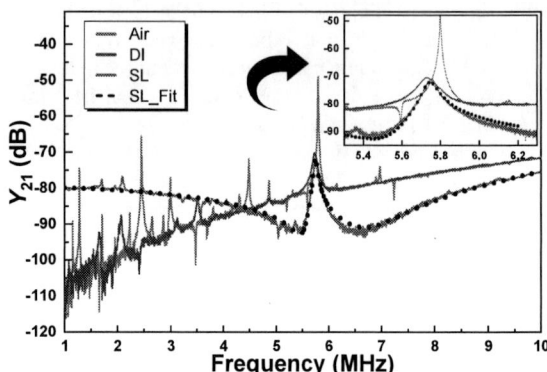

Figure 5: Simulation results of the equivalent circuit model. (a) Pure motional arm Y_m representing an ideal resonator behavior. (b) Ideal resonator working in conductive liquid, i.e., Y_m+Y_L, where motional signal is masked by liquid feedthrough. (c) Device response when feedthrough cancelation is implemented, i.e., $Y_m+Y_L+Y_S$.

Figure 7: Measurement results of PZT device in air, DI water, and saline (SL) solution. The SBR is greatly improved in SL even with better feedthrough cancellation as compared to DI. The fitting curve is also provided.

Fig. 5(b). When feedthrough cancellation is properly implemented, the resonant signal will be quite close to the pure motional arm within a certain frequency range as shown in Fig. 5(c).

EXPERIMENTAL RESULTS
Open-Loop Measurement

The frequency response of the PZT and AlN based LE mode resonators intentionally designed at similar resonant frequencies were measured by a network analyzer using the same setup and environmental condition. Fig. 6 shows comparison of the frequency characteristics between these two resonators working in a saline solution (ionic water).

Since the dielectric constant of AlN is not high enough, there is no significant feedthrough cancellation over the desired frequency span, and thus the frequency response behaves like a dip. On the contrary, the response of the PZT based resonator shows a noticeable peak thanks to the feedthrough cancellation effect.

Moreover, the PZT resonator was also measured in DI water and air for comparison and as a reference. The measurement results in Fig. 7 shows a decent SBR improvement by using the proposed technique in ionic water, which is even better than the 11-dB SBR performance in DI water. In addition, after the measurements are done for PZT resonator operated in air, DI water, and saline solution, all parameters can also be extracted by curve fitting, and Table 1 presents the comparison among these three conditions (air, DI, and SL). The fitting curve for the PZT resonator in this work is also provided in Fig. 7.

Closed-Loop Measurement

Apart from open-loop measurement, we use a lock-in amplifier (HF2LI+PLL) to form a PZT oscillator. Fig. 8 and Fig. 9 present the phase noise and Allan deviation performance of the PZT oscillator in air and in conductive liquid (saline solution). For mass sensing analysis, the mass resolution can be extracted from the measured Allan deviation results. In this work, the mass resolution of the PZT resonant sensor can be down to 3.7 pg. With this feedthrough cancellation technique, the resonance of the

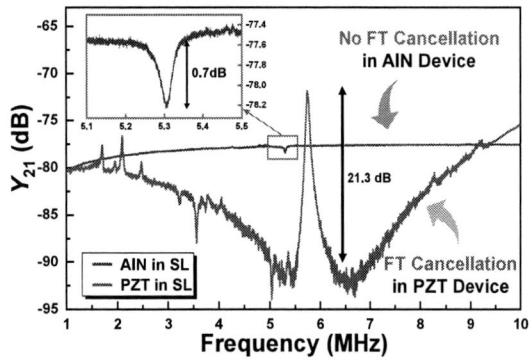

Figure 6: Measurement results of AlN and PZT LE mode resonators in saline (SL) solution. The measurement setup and environmental conditions are kept the same. Due to the lower dielectric constant of AlN, no obvious cancellation occurs in this frequency range. Feedthrough (FT) cancellation is attained when PZT is implemented thanks to its high dielectric constant, which leads to great SBR.

Table 1: Performance comparison of PZT device in air, DI water, and saline solution (FT cancellation).

Case	Air	DI	Saline Solution	Unit
f_r	5.79	5.73	5.75	MHz
Q	1071	62	76	-
R_m	0.27	4.68	3.82	kΩ
SBR	28.6	11.6	21.3	dB

*R_m: motional resistance; Q: quality factor.

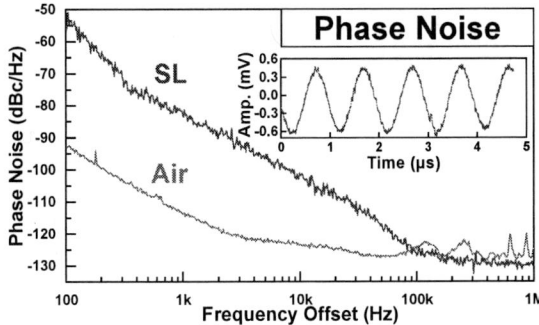

Figure 8: Phase noise measurement results of the PZT oscillator performance in air and saline (SL) solution. The inset shows time domain waveform of the oscillator working in SL.

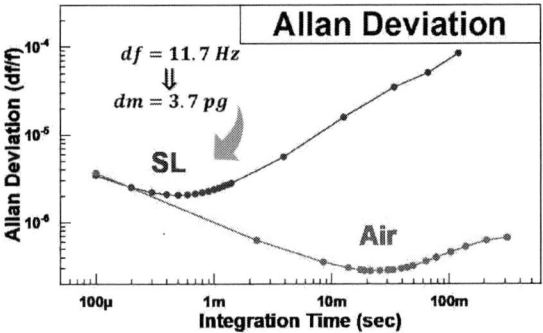

Figure 9: Allan deviation measurement results of the PZT oscillator working in air and saline (SL) solution. The mass resolution extracted from the measured Allan deviation in ionic liquid is around 3.7 pg.

MEMS device becomes readable and even has a better performance in conductive environment than in DI water. In future, we can use the PZT TPoS resonator and oscillator reported in this work to serve as a mass sensor for the applications in ionic liquid medium.

CONCLUSION

In this work, we successfully demonstrate the feedthrough cancellation technique for PZT MEMS resonant sensors operating in ionic liquid medium and validates the potential of this cancellation method for use in applications that need to accommodate conductive liquid environment. For comparison, both PZT and AlN based TPoS resonators are fabricated, characterized, and discussed in this research. Due to the sufficiently high dielectric constant of the PZT material, the feedthrough signal from substrate in PZT device becomes large enough, and the resonant signal is free from being masked by the high feedthrough introduced from the conductive liquid. As a result, the proposed method shows a significant enhancement for resonant sensor operated in conductive medium. Compared to the case without obvious cancellation effect (e.g., AlN device in saline solution or PZT device in DI water), the SBR performance has an enormous improvement, which is quite close to the value under air conditions.

With the help of the lock-in amplifier (HF2LI+PLL), the feedthrough-reduction based PZT resonator can stably oscillate in ionic liquid with an oscillating frequency of 5.75 MHz. The performance of the oscillator phase noise at offset frequencies of 1 kHz and 100 kHz is -83.15 dBc/Hz and -127.04 dBc/Hz, respectively. This PZT resonator working in ionic liquid medium in this work has an Allan deviation of 2.05 ppm, which implies a mass resolution of only 3.7 pg. Such MEMS resonant transducers would serve as a key building block in physical, chemical, and biological applications that require an ionic liquid medium environment.

ACKNOWLEDGEMENTS

The authors appreciate the financial support from the National Science and Technology Council (NSTC) of Taiwan under grant of NSTC 111-2223-E-007-007. We also appreciate Taiwan Semiconductor Research Institute (TSRI) and Center for Nanotechnology, Material science & Microsystem (CNMM) for providing fabrication facilities.

REFERENCES

[1] G. Pfusterschmied, C. Weinmann, M. Hospodka, B. Hofko, M. Schneider, and U. Schmid, "Sensing fluid properties of super high viscous liquids using non-conventional vibration modes in piezoelectrically excited MEMS resonators", in *Digest Tech. 32nd International Conference on Micro Electro Mechanical Systems* (MEMS), Seoul, Korea, Jan. 27-31, 2019, pp. 735-738.

[2] A. Ali and J. E.-Y. Lee, "Fully differential piezoelectric button-like mode disk resonator for liquid phase sensing", *IEEE Trans. Ultrason. Ferroelectr. Freq. Control*, vol. 66, no. 3, pp. 600-608, Mar. 2019.

[3] M. Mahdavi, A. Abbasalipour, and S. Pourkamali, "Thin film piezoelectric-on-silicon elliptical resonators with low liquid phase motional resistances," *IEEE Sensors Journal*, vol. 19, no. 1, pp. 113-120, Jan. 2019.

[4] A. Ali and J. E.-Y. Lee, "Piezoelectric-on-silicon square wine-glass mode resonator for enhanced electrical characterization in water," *IEEE Transactions on Electron Devices*, vol. 65, no. 5, pp. 1925-1931, May 2018.

[5] E. Mehdizadeh, J. Chapin, J. Gonzales, A. Rahafrooz, R. Abdolvand, B. Purse, and S. Pourkamali, "Direct detection of biomolecules in liquid media using piezoelectric rotational mode disk resonators," in *Digest Tech. IEEE Sensors Conference*, Taipei, Oct. 28-31, 2012, pp. 1-4.

[6] H. Mansoorzare, S. Shahraini, A. Todi, N. Azim, S. Rajaraman, and R. Abdolvand, "Liquid-loaded piezo-silicon micro-disc oscillators for pico-scale bio-mass sensing," *J. Microelectromech. Syst.*, vol. 29, no. 5, pp. 1083-1086, Oct. 2020.

CONTACT

*C.-Y. Wu, Email: tpmenthk24@gmail.com.

CHARACTERIZATION OF PACKAGING STRESS WITH A CAPACITIVE STRESS SENSOR ARRAY

Tolga Veske[1], Derin Erkan[1], and Erdinc Tatar[1, 2]

[1]Department of Electrical and Electronics Engineering, Bilkent University, Ankara, TURKEY and
[2]The National Nanotechnology Research Center (UNAM), Ankara, TURKEY

ABSTRACT

This paper presents a high-resolution capacitive stress sensor array to precisely characterize the distributed MEMS packaging stress, for the first time. The unit stress measurement cell utilizes a bridge-type mechanical amplifier that converts the substrate strain into capacitance variations. We have measured and compared the MEMS die stress over temperature for different die attaches with a custom designed PCB housing an on-chip heater. The proposed approach significantly simplifies the evaluation and selection of packaging materials. Comparing the temperature responses of a soft silicone-based and stiffer silver-filled epoxy reveals difficult to predict results.

KEYWORDS

Packaging stress, die-attach, capacitive stress sensor

INTRODUCTION

The choice of packaging materials is critical for MEMS sensors as the packaging determines the final stress, temperature and humidity response of the sensor. MEMS sensors are attached to a PCB through die-attach, package, and solder with different and not-ideal mechanical properties. As the temperature varies, the packaging materials expand and contract differently and non-repeatedly causing stress on the sensitive MEMS element. Among the packaging materials the die-attach requires special attention, since its properties might change by orders of magnitude over glass transition temperature, time, temperature and are not repeatable [1]. Therefore, a method precisely characterizing the distributed die-attach induced stress would benefit the MEMS development.

The importance of packaging stress is well known in the MEMS community even though distributed stress characterization efforts over temperature and time are limited. A recent study focuses on the transient offset behavior of plastic packaged MEMS devices in response to humidity [2]. It was shown that the package absorbs humidity resulting in stress, and the MEMS core could be optimized to minimize the moisture induced drift. Die attach stress was evaluated with an optical profilometer in [3], and thicker and compliant adhesive was found to result in the least stress on the MEMS die. The effects of moisture on flip chip packaging stress was measured with piezoresistive stress sensors in [4]. Compact piezoresistors could be built in IC processes with dedicated layers. However, piezoresistors with reasonable resistance consume large layout area in MEMS applications where the device layer is thick (>10μm) and highly-conductive, making fine spatial resolution stress measurements not-feasible [5]. In addition, piezoresistance coefficients of Silicon change over temperature requiring calibration. Capacitive stress sensors could also measure stress. Instead of utilizing a material property, anchor displacements due to strain are mechanically amplified and resulting capacitance changes are detected. Bent beams [6] and bridge-type [7] mechanical amplifiers were proposed for mechanical strain amplification. Compared with the piezoresistive counterparts, capacitive stress sensors don't consume DC power, are temperature insensitive, and can be designed compact for distributed on-chip stress measurements in high-conductive MEMS device layers. Towards that end, this paper reports distributed high-resolution (~kPa) stress measurements on two different die attaches over temperature with a modified capacitive stress sensor array, for the first time. We designed an on-PCB heater that enables heating the MEMS package independent from the electronics.

STRESS SENSOR AND THE ARRAY

Figure 1 presents the SEM image of the capacitive stress sensor (a), a finite element analysis (FEA) simulation result showing the mechanical strain amplification (b), and the equivalent circuit summarizing the electrical readout of the sensor (c). The overall sensor size is 480μmX386μm limited by the process. Fabrication was outsourced, and the devices were manufactured in a wafer level encapsulated process [8]. The device is attached to the substrate through the anchors located at the left and right sides. Anchors are labeled in Figure 1.b, and the rest of the device is suspended. When there is a strain in the substrate, the imbalanced bridge connections at the top and bottom of the device convert stress-induced anchor displacements in the x-direction to differential capacitive changes in the y-direction with amplification. The mechanical gain is $1/\tan(\alpha)$ and set by the imbalance angle (α) of the bridge. Smaller α results in higher gain at the cost of increased nonlinearity. We set $\alpha=8.5°$ in our design, corresponding to a mechanical gain of 6.7.

The mechanical design of the stress sensor is inspired by a cm sized bridge-type mechanical amplifier [7]. We have scaled down the macro device with significant modal enhancements. We have connected the suspended arms forming the sensing capacitors with additional beams (See "Added beam" in Fig.1.a). The connections move the out of plane rocking mode of the arms to high frequency making sure the resonant modes of the stress sensor are >300kHz. High modal frequencies combined with the differential motion of the arms with strain brings the acceleration sensitivity to negligible levels. Shown in Figure 1.c, the readout of the stress sensor is identical to an accelerometer readout. The stress variations are very slow in time so differential modulation clocks ($V_{mod+/-}$) are applied to move away from the 1/f noise corner of the electronics. A transimpedance amplifier converts the stress induced capacitive variations into voltage at the modulation frequency. Further amplification and final demodulation at the modulation frequency provides the stress measurement.

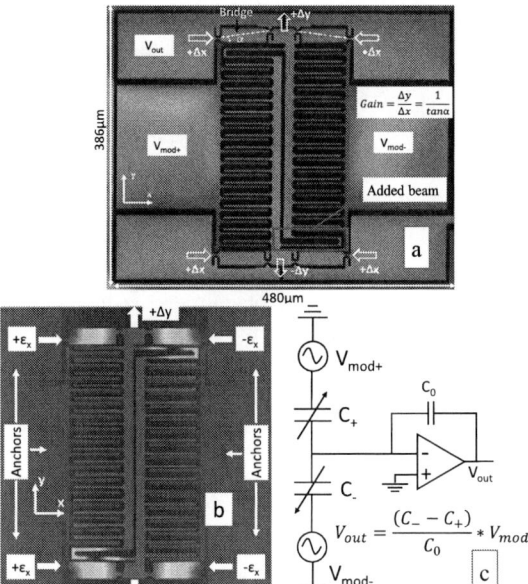

Figure 1: SEM image of the capacitive stress sensor (a), FEA image summarizing the sensor operation (b), and the equivalent circuit of the sensor with the front-end.

The individual stress sensors form a stress rosette that captures the in-plane normal x, y, and shear stress. Figure 2 presents the SEM photo for the two rows of the stress sensor array; the complete array consists of 3X3 stress rosettes. The stress rosettes are shown in dashed lines. 0° and 90° oriented sensors capture the x (σ_x) and y (σ_y) normal stress, respectively. The shear stress is measured by rotating a stress sensor with an angle of θ from the x-axis. The captured strain for a rotation angle of θ_a and θ_b are given by equations 1, and 2 [9]. ε_x, ε_y, and τ_{xy} are the in-plane x, y and shear strain, respectively. Setting θ_a=45° and θ_b=-45° and taking the difference of 1 and 2 cancels the normal stress and results in pure shear stress measurement. We implemented the subtraction operation by adjusting the clock polarities and shorting the outputs of the +/-45° oriented stress sensors. For area constraint applications, the shear stress measurement could be achieved with a single sensor by using equation 1 and the data from the normal stress sensors. The stress is found by multiplying the strain with the Young's Modulus.

$$\varepsilon_a = \frac{\varepsilon_x + \varepsilon_y}{2} + \frac{\varepsilon_x - \varepsilon_y}{2}\cos 2\theta_a + \frac{\tau_{xy}}{2}\sin 2\theta_a \quad (1)$$

$$\varepsilon_b = \frac{\varepsilon_x + \varepsilon_y}{2} + \frac{\varepsilon_x - \varepsilon_y}{2}\cos 2\theta_b + \frac{\tau_{xy}}{2}\sin 2\theta_b \quad (2)$$

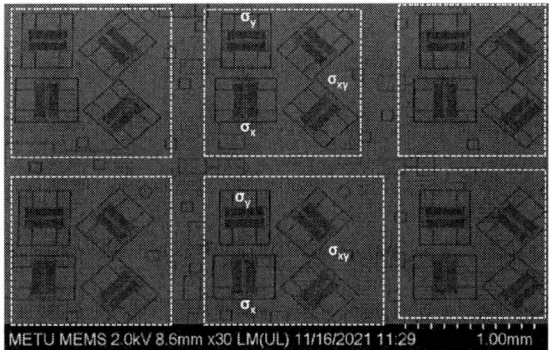

Figure 2: SEM photo of the stress sensor array, the stress rosettes are shown in dashed lines. The complete array consists of 3X3 array, 2 rows are shown.

TEST SETUP

Figure 3 presents our test setup that consists of the main PCB, the MEMS daughter board, and the voltage regulator board. The stress sensor array chip is wirebonded to a 44-pin ceramic LCC. The ceramic package is then soldered to the daughter board which acts as a socket and plugs into the main board. The daughter board houses an on-PCB heater covering the area underneath the ceramic package. Coupled with a temperature sensor the proposed setup allows heating of only the MEMS die independent of the readout electronics.

There are 9 stress rosettes in the array with a total of 27 individual stress sensors. By considering the slow variation of stress we designed a multiplexed readout circuit in Figure 3.b. All stress sensors share the differential clocks and the stress sensor outputs go through individual transimpedance amplifiers located on the main PCB. The amplified stress sensor outputs are then multiplexed, demodulated, and recorded one at a time. The digital control for the multiplexers, data recording, and demodulation were performed by a Zurich Instruments HF2LI digital lock-in amplifier. To utilize a single front-end amplifier, we tried moving the MUX right after the MEMS but the noise and drift got worse due to parasitics.

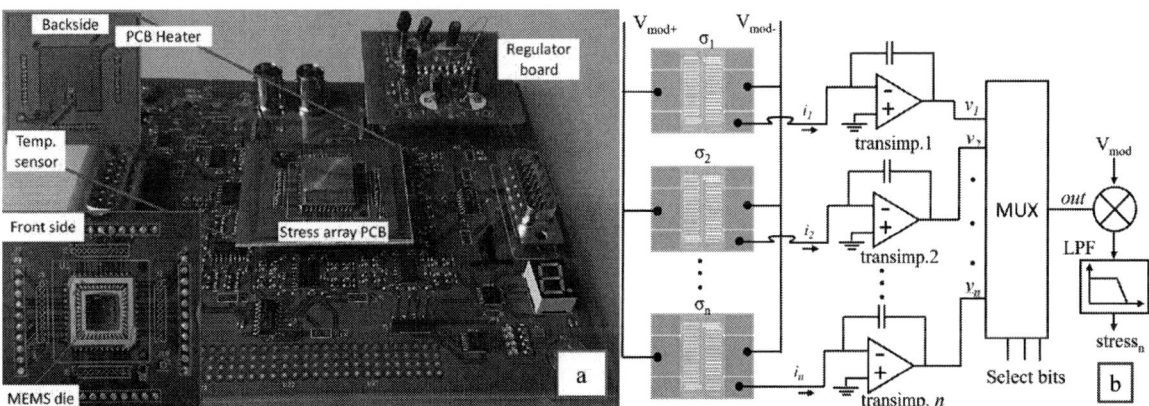

Figure 3: Photo of the test setup showing the main board, daughter board housing the MEMS die, the heater and temperature sensor (a), the block diagram showing the multiplexed measuring scheme (b).

978-1-6654-9309-3/23 $31.00 © 2023 IEEE

Two identical stress arrays are mounted to a 44-pin ceramic-LCC with a conductive soft silicone-based and non-conductive hard silver-filled epoxy. These epoxies represent typical MEMS die-attaches. Table 1 reports the epoxy properties. The Young's Modulus was calculated from the Shore Hardness. A temperature controller cycles the temperature between 25-100°C in 40mins (~2°/min) for two cycles, Figure 4.a. presents the temperature profile.

Table 1: Epoxy material properties

	Silicone-based	Silver-filled
Hardness	63 Shore A	75 Shore D
Young's Mod. (MPa)	6.92	198
CTE (ppm/°C)	140	31/158
Tg (°C)	-120	≥80
Resistivity (Ω-cm)	$\geq 10^{14}$	$\geq 4 \times 10^{-4}$

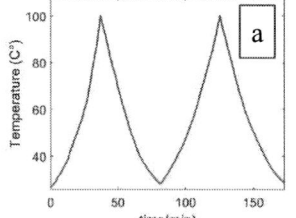 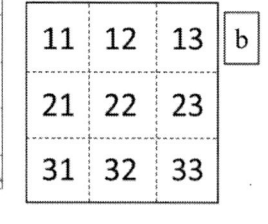

Figure 4: The applied temperature profile ~2°C/minute (a), and the labeling for the stress sensor rosettes (b). There is a total of 9 stress rosettes and 27 stress sensors.

MEASUREMENT RESULTS

Figure 5.a-b, 6.a-b, and 7.a-b present the distributed x, y, and shear stress measurements for two temperature cycles for the silicone-based and silver-filled epoxy, respectively. The labeling of the stress sensors is shown in Figure 4.b. While reporting the actual stress, we used calculated sensitivities for the stress sensors since applying a known stress to extract the sensitivity is not straightforward. We verified the functionality of the stress sensors first on a stress test-bed. The stress sensors responded negatively to tensile stress. x and y stress exhibit tensile stress with rising temperature as expected since Silicon is a low CTE material compared to the die-attach. However, a consistent tensile stress was not observed for the shear stress. Glass transition temperature for the silicone-based epoxy is listed as -120°C so its mechanical properties are expected not to change significantly. But a V-shaped stress response was observed in some stress components. The silver-filled epoxy is 30X stiffer than the silicone-based epoxy for T<80°C, and the as-born stress is higher in general. 80°C glass transition temperature can be observed with the hockey-stick curve shapes in silver-filled epoxy response. Overall the maximum stress is moderate with <10MPa including possible offsets.

The main discovery is the large hysteresis in the shear response of silver-filled epoxy in the 1st temperature cycle. The 1st cycle silver-filled epoxy response is repeatable when the sample waits at room temperature for 4-5 days. We believe the silver-filled epoxy absorbs moisture at room temperature and upon heating the moisture vaporizes returning to the original material properties. The silicone-based epoxy exhibits a smaller hysteresis which does not change drastically over temperature cycling thanks to the low water absorption of Silicone. The large hysteresis could cause serious sensor performance issues, and the die stress is not significantly different except the hysteresis for the two cases. The 30X stiffness difference between the epoxies does not result in order of magnitude stress difference. The results are difficult to predict especially for the shear stress and indicate that stress might have a large variation and different temperature response across the die. Measuring difficult to model on-chip stress could suppress the sensor drift for high performance gyroscopes [10].

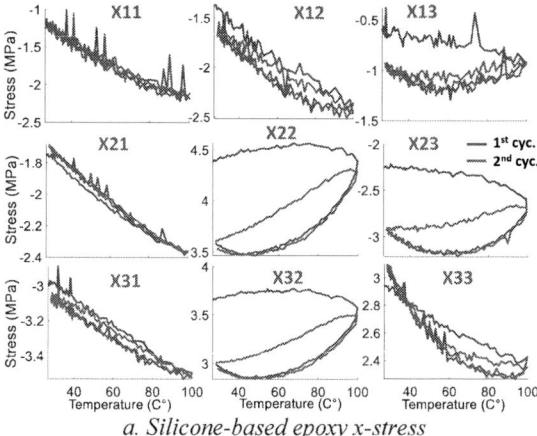

a. Silicone-based epoxy x-stress

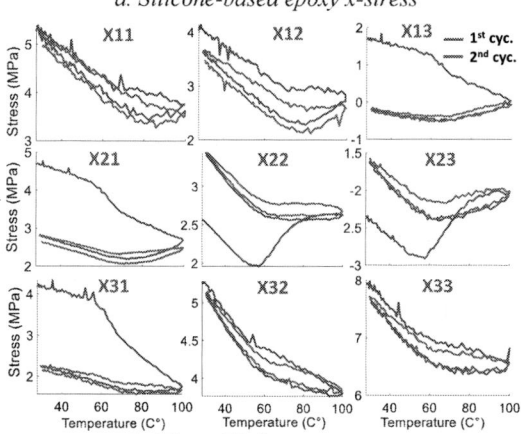

b. Silver-filled epoxy x-stress

Figure 5: Distributed x-stress for two temperature cycles.

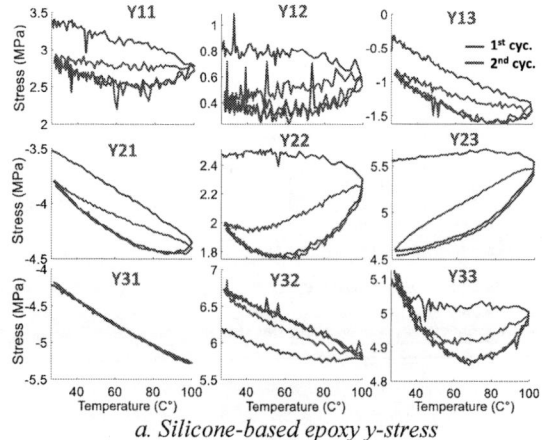

a. Silicone-based epoxy y-stress

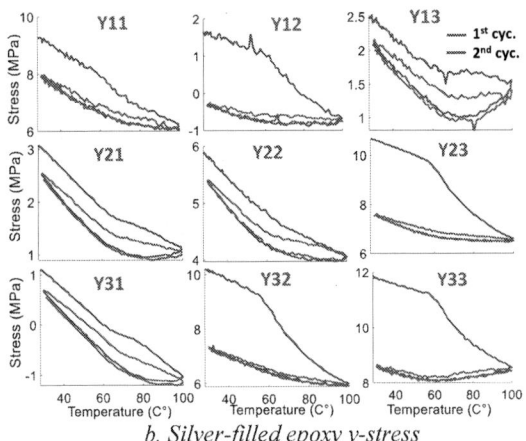

b. Silver-filled epoxy y-stress

Figure 6: Distributed y-stress for two temperature cycles.

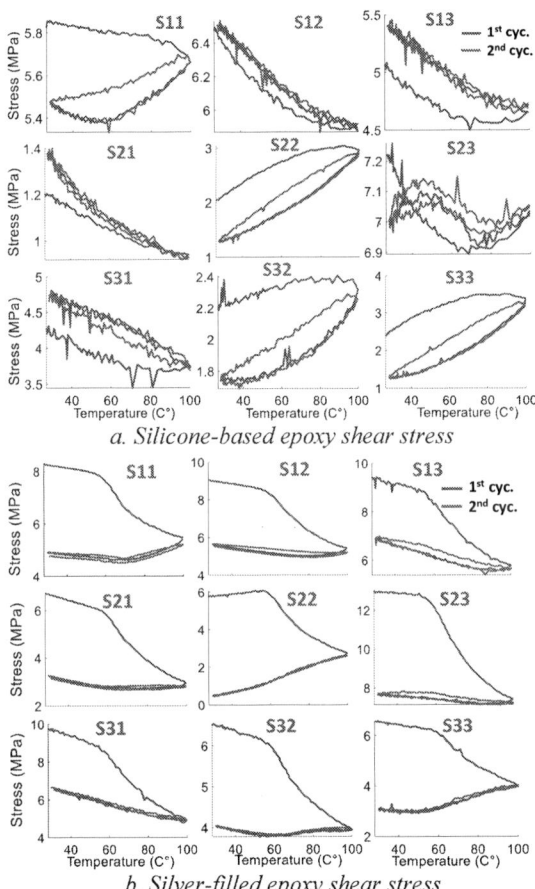

a. Silicone-based epoxy shear stress

b. Silver-filled epoxy shear stress

Figure 7: Distributed shear stress for two temp. cycles.

CONCLUSIONS

We have presented distributed MEMS die-attach stress measurements with a capacitive stress sensor array. The capacitive stress sensor amplifies the substrate strain with a bridge-type amplifier. Unlike piezoresistive stress sensors, capacitive stress sensors does not have temperature dependence. The distributed die stress has been characterized for a soft silicone-based epoxy and 30X stiffer silver-filled epoxy with a custom setup that only heats the MEMS die. The silver filled epoxy exhibited

significant hysteresis due to humidity absorption in the 1st temperature cycle after resting at room temperature. Apart from the hysteresis the stress responses of the two epoxies were in the same order. The distributed die stress has a large variation that might cause unexpected drift for the MEMS sensors. Extending the on-chip stress measurements for solder mounts could provide further understanding of packaging stress.

ACKNOWLEDGEMENTS

This work was supported by the Scientific and Technological Research Council of Turkey (TUBITAK) 2232 program under the Grant no. 118C247. The ideas/claims presented here solely belong to the authors.

REFERENCES

[1] A. Misrak *et al.*, "Impact of die attach sample preparation on its measured mechanical properties for MEMS sensor applicationa", *J. Microelectronics and Electronic Packaging*, 18(2021), pp. 21-28.

[2] J. Jurgschat, R. Scheben, I. Toth, T. Ohms, O. Kohn, A. Zimmermann, "Experimental investigation of moisture induced offset drifts on plastic packaged MEMS devices", *J. Microelectromech. Syst.*, vol. 31, no. 4, pp. 653-663, Aug. 2022.

[3] S. S. Walwadkar, J. Cho, "Evaluation of die stress in MEMS packaging: experimental and theoretical approaches", *IEEE Trans. Components and Packg. Tech.*, vol. 29, no. 4, pp. 735-742, Dec. 2006.

[4] Q. Nguyen, J.C. Roberts, J.C. Suhling, R.C. Jaeger, P. Lal, "A study on die stresses in flip chip package subjected to various hygrothermal exposures", *Proc. of IEEE 17th ITherm*, San Diego, May 29-June 01, 2018, pp. 1339-1350.

[5] E. Tatar, T. Mukherjee, G.K. Fedder, "Stress effects and compensation of bias drift in a MEMS vibratory-rate gyroscope", *J. Microelectromech. Syst.*, vol. 26, no. 3, pp. 569-579, June 2017.

[6] L.L. Chu, L. Que, Y.B. Gianchandani, "Measurements of material properties using differential capacitive strain sensors", *J. Microelectromech. Syst.*, vol. 11, no. 5, pp. 489-498, Oct. 2002.

[7] H.-W. Ma, S.-M. Yao, L.-Q. Wang, Z. Zhong, "Analysis of the displacement amplification ratio of bridge-type flexure hinge", *Sensors and Actuators A*, 132(2006), pp.730-736.

[8] M.M. Torunbalci, S.E. Alper, T. Akin, "Advanced MEMS process for wafer level hermetic encapsulation of MEMS devices using SOI cap wafers with vertical feedthroughs", *J. Microelectromech. Syst.*, vol. 24, no. 3, pp. 556-564, June 2015.

[9] Vishay Precision Group, "Plane-shear measurement with strain gauges," Tech. note TN-512-1, Nov. 2022.

[10] B.E. Uzunoglu, D. Erkan, E. Tatar, "A ring gyroscope with on-chip capacitive stress compensation", *J. Microelectromech. Syst.*, vol. 31, no. 5, pp. 741-752, Oct. 2022.

CONTACT

D. Erkan, tel: +90-312-290-1219; derin@ee.bilkent.edu.tr

MILLISECOND-LEVEL PULSE-HEATING SENSING SYSTEM FOR MEMS-BASED GAS SENSORS

Yi Zhuang[1], Gaoqiang Niu[1], Lang Wu [1] and Fei Wang [1]

[1]School of Microelectronics, Southern University of Science and Technology, Shenzhen 518055, CHINA

ABSTRACT

We have proposed a millisecond-level sensing system for MEMS-based SnO_2 gas sensors. Metal oxide semiconductors (MOSs) gas sensor normally works at high temperature, which is heated by a pair of electrodes on the micro-hotplate. In the conventional method, a constant/DC heating voltage is applied to establish a steady temperature, which demands high power consumption and long gas sensing times. MEMS-based MOSs gas sensors, with low power consumption and low thermal inertia, provide the possibility to employ a millisecond pulsed heating voltage. Herein, a 4×4 sensor array testing system has been prepared based on the pulse-heating strategy. By utilizing line scanning and parallel reading, the system realized the quality discrimination of the sensors with a home-made hardware. To address the instability of the sensing system in pulse-heating mode, Rg-50 ppm/Rg was used to describe the performance of the sensors and the reliability of the system was obviously improved.

KEYWORDS

Gas Sensor, Pulse-heating, Fast Detection, Sensing System

INTRODUCTION

Environmental problems have attracted increasing attention worldwide in recent decades. Gas sensors play an important role in our daily life for detecting various gases which have a negative effect on the environment. Metal oxide semiconductors (MOSs) gas sensors are widely used because of their low fabrication costs, high sensitivities, and long operational life [1-4]. With the development of MEMS technology, MOSs gas sensors are also evolving towards miniaturization and low power consumption. This gives it a very broad range of applications for the Internet-of-Things [5,6]. As the production scale of MEMS gas sensors increases, the need for a stable and efficient quality control system becomes more and more urgent for the sensor fabrication. However, in the conventional method, gas sensors need to be warmed up for a long time before testing and need to operate at very high temperatures, this results in inefficient testing and high energy consumption [7-9].

In this paper, we propose a millisecond pulse test method and build a demo test system. This system consists of a signal generation module, a sensor array module, and a data acquisition module. The signal generation module is used to generate pulse heating voltage with various amplitudes, pulse widths and intervals. Microcontroller unit (MCU) is used for logic control. Heating voltage was controlled by a direct current (DC) power supply.

Additionally, diodes were applied in order to prevent the crosstalk of measurement. Rapid performance

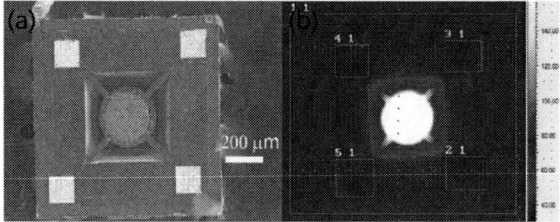

Figure 1: (a) SEM image of the MEMS gas sensor (b) Thermal infrared image under pulsed heating (voltage: 3.3V, pulse width: 300 ms).

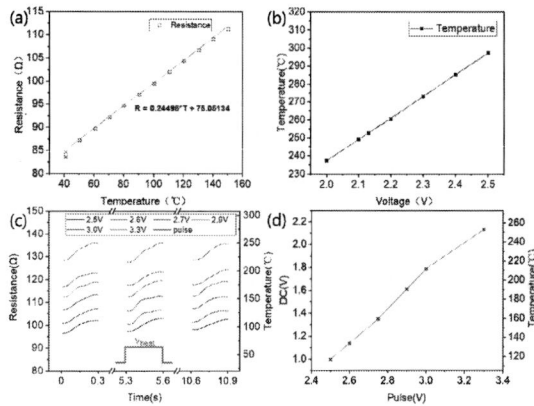

Figure 2: Relationship between (a) Heating electrode resistance vs. temperature; (b) DC heating voltage and sensor temperature; (c) Pulse voltage and sensor temperature (pulse width: 300 ms); (d) DC heating vs. Pulse heating.

judgment of the sensor and the positioning of corresponding sensor could be realized via a pulse-heating strategy. Meanwhile, the stability and reliability of the test were further improved by data processing. Therefore, the effectiveness of sensor evaluation could be significantly enhanced based on this strategy.

EXPERIMENT

The MEMS gas sensor contains sensing material and micro-hotplate. SnO_2 nano-sheets were obtained via a hydrothermal method as described in our previous work [10], and a commercialized micro-hotplate was purchased (SuZhou Huiwen Nano S&T). Fig.1(a) shows the SEM image of the gas sensor, which contains a pair of heating electrodes and a pair of test electrodes. The central circular area is the sensing layer modified by SnO_2. Fig.1(b) demonstrates a thermal infrared image of the sensor under pulsed heating at a voltage of 3.3V and a pulse width of 300 ms, with a maximum temperature up to 242°C for the sensing layer.

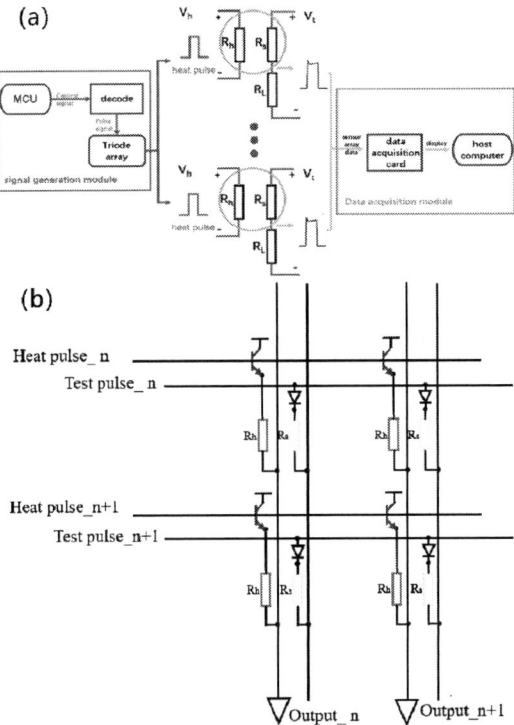

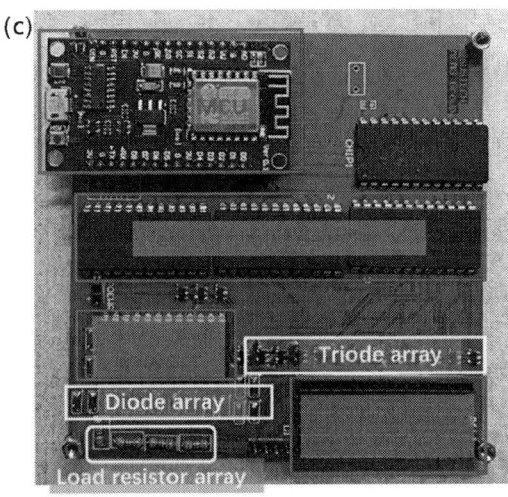

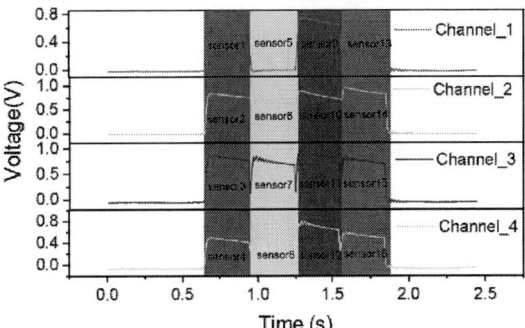

Figure 4: Voltage output of 16 sensors in different channels, which provides the feature of each sensor in the sensors arrays for the fast response.

DC voltage, the current-voltage (I-V) curve was measured on the heating electrode, and the resistance was introduced into equation (1). Fig.2(c) shows the transient response of resistance and temperature under different pulse heating voltage (V_{heat}). During the first 50ms of the pulse, temperature changes rapidly and then stabilizes at the end of the pulse width of 300 ms. Fig.2(d) shows the comparison between DC heating and pulse heating. The temperature of pulse heating is lower than that of DC heating under the same V_{heat}. Since the optimal working temperature corresponds to a DC voltage of 2.5 V and a pulse heating voltage of 3.3V could achieve the same temperature, all the subsequent experiments were conducted using a pulse voltage of 3.3V unless else mentioned.

Fig.3a shows the scheme of a test system designed to realize fast testing of sensor arrays. This system is composed of three modules: a signal generation module, a sensor array module, and a data acquisition module. The signal generation module generates pulses with varying amplitudes, widths and intervals for heating purposes. Timing sequences are generated by MCU. The decoder allows for multiplexed control and parallel control, and the triode is applied with the heating electrode to increase the driving capacity of the logic circuit. The sensor is tested by the classical series voltage division method [11-13]. The schematic depicting the sensor array is shown in Fig.3b, which we can use to design an array of N by N sensors.

In the array, a single pulse signal was applied to the n-th row. The sensor's measured signals on this line were read by channel n and channel n+1. In the next cycle, pulse signal is applied to row n+1 and all other row pulse voltage is set to zero. Fig.3c is a physical test circuit of 4 × 4 sensor array, and the output of this 4 × 4 sensor array is shown in Fig 4. Four sensing signals are received simultaneously every cycle of 300 ms. The number of channels and the sequence information allow us to determine where the sensor is located. This row scan-column read approach has well balanced the test speed and the hardware resources required for testing. Theoretically, this method can be applied to implement a rapid quality control system for large scale sensor arrays by reading data from any N by N sensor array.

Figure 3:(a) Schematic diagram of the test system structure (b) Schematic diagram of 2 ×2 sensors array (c) Physical circuits of test system.

Fig.2(a) demonstrates the relationship between the temperature and resistance of the heating electrode. Gas sensor was placed in the oven and the temperature was tuned. The heating electrode's resistance was measured using a source meter; therefore, a linear relationship between the temperature (T) and the resistance (R) was achieve as shown in equation (1).

$$R = 0.24497 \times T + 75.05134 \tag{1}$$

Fig.2(b) shows the relationship between the heating voltage and the working temperature under DC heating. To determine the relationship between temperature and

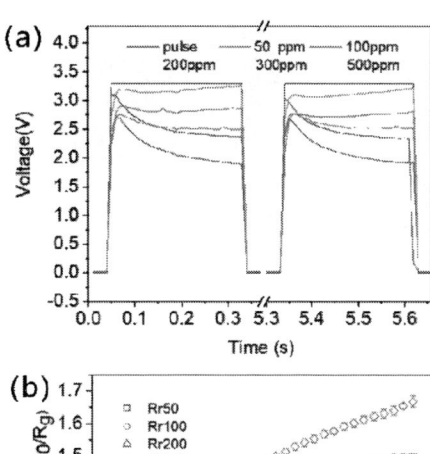

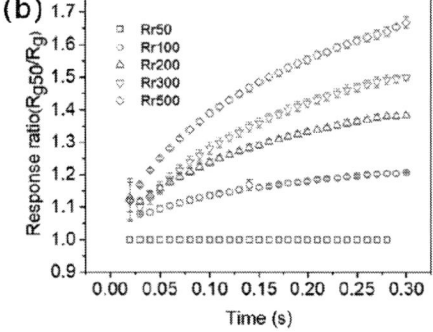

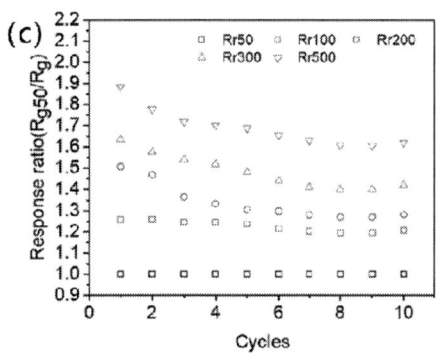

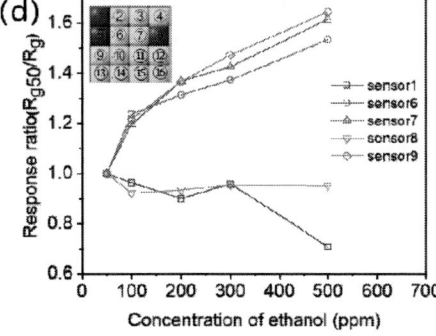

Figure 5: (a) Voltage output of sensor No. 7 at different concentrations under pulsed heating (b) the transient response during the 300 ms pulse-heating (c) repeatability of response at different cycles (d) the performance com- parisons of the sensors, and the location of the unqualified sensors in sensors array is exhibited in the upper left corner.

RESULTS AND ANALYSIS

The sensor array was placed in different atmospheres to obtain the transient response of the sensing performance under 300 ms pulse heating. The gas control system for sensing test was described as our previous work [14]. Using standard DC calibration, three failed devices were set as No.1, 5 and 8 sensors on purpose.

Fig.5(a) displays the results of the voltage output of one sensor No. 7 in the case of pulse heating at different concentrations (50 ppm, 100 ppm, 200 ppm, 300 ppm and 500 ppm) of ethanol. At low concentrations, the response tends to stabilize at a slow rate. Within the first 150 ms of the pulse, the output signal of the sensor is strongly influenced by the temperature. As the target gas concentration increases, the output signal tends to stabilize after 150 ms. Normally, the gas response is defined as Ra/Rg (Ra: sensor resistance in air, and Rg: sensor resistance in the target gas) [15-17]. Because of the large variation of the reference resistance in air, we have defined a parameter as the Response Ratio (Rr), which is the response resistance at 50 ppm ethanol divided by the response resistance at the certain concentration (Rr=R50/Rg). This parameter Rr is used as the quality control criteria for the sensors, which has shown more reliable and stable results compared with the traditional Ra/Rg. The response ratio Rr for sensor No. 7 during one pulse heating process was shown in Fig.5b. With all the different gas concentrations, Rr tends to be relatively stable in the second half of the temperature pulse.

Fig.5(c) presents the results of 10 cycles of device No. 7 under pulse heating conditions. Each point in Fig.5c represents the average value of Rr in Fig.5b, which gradually stabilize to a certain value after 10 cycles. From this plot, we can clearly see that the higher concentration of the target gas, the higher Rr value we can obtain for the good sensors. Fig. 5d shows the performance comparison of the sensors, and the failure sensors in the array have been successfully marked in the upper left corner in Fig. 5d. Based on this result, we believe that this N by N circuit provides a promising platform for the sensor array application, and a fast sorting strategy to mark the known-good-dies for the wafer-level fabrication of MEMS gas sensors.

CONCLUSION

We have proposed a millisecond-level pulsed gas test method and based on this method we have implemented a test system demo for fast sensor detection. A new parameter Rr was proposed to judge the performance of the sensor array and to improve the reliability and stability of the device quality control. This system can be extended to larger N by N arrays for wafer-level sensor sorting systems, which will result in significant improvements in the test efficiency and reduction of device cost.

ACKNOWLEDGMENT

This work was financially supported by the Shenzhen Science and Technology Innovation Committee under Grant JCYJ20170412154426330, in part by Foundation for Distinguished Young Talents in Higher Education of Guangdong, China under Grant 2018KQNX226, and Guangdong Natural Science Funds under Grants 2016A030306042 and 2018A050506001.

REFERENCES

[1] Z. Yuan, F. Yang, and F. Meng, "Research Progress on Coating of Sensitive Materials for Micro-Hotplate Gas Sensor," *Micromachines*, vol. 13, no. 3,pp. 491, 2022.

[2] G. Niu and F. Wang, "A review of MEMS-based metal oxide semiconductors gas sensor in Mainland China," *Journal of Micromechanics And Microengineering*, 32, 054003, May 2022.

[3] F. Palacio, J. Fonollosa, J. Burgués, J. M. Gomez, and S. Marco, "Pulsed-Temperature Metal Oxide Gas Sensors for Microwatt Power Consumption," *IEEE Access*, vol. 8, pp. 70938–70946, 2020.

[4] Y. Chen, M. Li, W. Yan, X. Zhuang, K. W. Ng, and X. Cheng, "Sensitive and Low-Power Metal Oxide Gas Sensors with a Low-Cost Microelectromechanical Heater," *ACS Omega*, vol. 6, no. 2, pp. 1216–1222, Jan. 2021

[5] K. H. Kim and H.-D. Kim, "Deep Sleep Mode Based Node MCU-Enabled Humidity Sensor Nodes Monitoring for Low-Power IoT," *Transactions on Electrical and Electronic Materials*,vol. 21,no. 6,pp. 617-620, 2020.

[6] W. Ren, C. Zhao, G. Niu, Y. Zhuang, and F. Wang, "Gas Sensor Array with Pattern Recognition Algorithms for Highly Sensitive and Selective Discrimination of Trimethylamine," *Advanced Intelligent Systems*, 2200169, 2022.

[7] H. Ji, W. Zeng, and Y. Li, "Gas sensing mechanisms of metal oxide semiconductors: a focus review," *Nanoscale*, vol. 11, no. 47, pp. 22664–22684, 2019.

[8] C. Zhao, H. Gong, G. Niu, and F. Wang, "Ultrasensitive SO_2 sensor for sub-ppm detection using Cu-doped SnO_2 nanosheet arrays directly grown on chip," *Sensors and Actuators B-Chemical*, vol. 324, pp. 128745, Dec. 2020.

[9] G. Niu, C. Zhao, and F. Wang, "Scalable Synthesis of SnO2 Nanosheet Arrays on Chips for Ultralow Concentration NO_2 Detection," *international conference on nano/micro engineered and molecular systems (NEMS)*, 2021, pp. 820–823.

[10] G. Niu, C. Zhao, H. Gong, Z. Yang, X. Leng, and F. Wang, "NiO nanoparticle-decorated SnO_2 nanosheets for ethanol sensing with enhanced moisture resistance," *Microsystems & Nanoengineering*, vol. 5, no. 1, p. 21, May 2019.

[11] H. Liu, G. Meng, Z. Deng, M. Li, J. Chang, T. Dai, "Progress in Research on VOC Molecule Recognition by Semiconductor Sensors," *Acta Physico-Chimica Sinica*, vol. 38, no. 5, pp. 2008018, 2022.

[12] N. M. Vuong et al., "Ni2O3-decorated SnO_2 particulate films for methane gas sensors," *Sensors and Actuators B-Chemical*, vol. 192, pp. 327–333, Mar. 2014..

[13] X. Kou et al., "High-performance acetone gas sensor based on Ru-doped SnO_2 nanofibers," *Sensors and Actuators B-Chemical*, vol. 320, Oct. 2020.

[14] Y. Hu, Y. Tian, Y. Zhuang, C. Zhao, and F. Wang, "Rapid Gas Sensing Based on Pulse Heating and Deep Learning," in *International conference on micro- electro mechanical systems (MEMS)*, 2021, pp. 438-441.

[15] H. Gong, C. Zhao, and F. Wang, "On-Chip Growth of SnO_2/ZnO Core–Shell Nanosheet Arrays for Ethanol Detection," *IEEE Electron Device Letters*, vol. 39, no. 7, pp. 1065–1068, 2018.

[16] M. Li, W. Luo, X. Liu, G. Niu, and F. Wang, "Wafer-Level Patterning of SnO_2 Nanosheets for MEMS Gas Sensors," *IEEE Electron Device Letters*, vol. 43, no. 11, pp. 1981–1984, Nov. 2022.

[17] C. Zhao, H. Gong, G. Niu, and F. Wang, "Electrospun Ca-doped In2O3 nanotubes for ethanol detection with enhanced sensitivity and selectivity," *Sensors and Actuators B-Chemical*, vol. 299, pp. 126946, Nov. 2019.

CONTACT

* Fei Wang, Email: wangf@sustech.edu.cn

MULTIPLE PARAMETER DECOUPLING USING A SINGLE RESONANT MEMS SENSOR VIA BLUE SIDEBAND EXCITATION

Jingqian Xi[1, §], Lei Xu[1, §], Yuan Wang[2,*], Fangjing Hu[1], Chengxin Li[4], Linlin Wang[4], Huafeng Liu[1,*], ChenWang[4,*], Michael Kraft[4], Chun Zhao[3,*]

[1] MOE Key Laboratory of Fundamental Physical Quantities Measurement, PGMF and School of Physics, Huazhong University of Science and Technology, Wuhan 430074, P. R. China
[2] Institute of Microelectronics, University of Macau, P. R. China
[3] Department of Electronics Engineering, University of York, YO10 5DD, U. K.
[4] ESAT Department, MNS, University of Leuven, Belgium

ABSTRACT

This work for the first time reports a simultaneous multiple parameter decoupling (MPD) technique using a single resonant MEMS sensor on the basis of a novel closed-loop blue-sideband excitation (BSE) scheme. Owing to the BSE, multiple vibration modes of the resonant MEMS sensor can be excited at the same time, enabling the implementation of the MPD algorithm. Experimental validation is conducted with a clamped-clamped beam resonator that concurrently undergoes temperature changes and either steady-state or periodic electrostatic perturbations. As a result, the output signals of the sensor that comprise both ambient temperature changes and the electrostatic perturbations can be separated in a single operation, with a cross-coupling suppression of up to 31 dB. In addition, the decoupled parameters are consistent with the stand-alone measurements, affirming the functionality of the MPD. The proposed technique is of great potential for practical applications such as suppressing temperature fluctuations in high-precision resonant sensors and multiple-parameter monitoring.

KEYWORDS

Multiple parameter decoupling, blue-sideband excitation, micro-resonator, parametric resonance

INTRODUCTION

MEMS resonators employed as sensors demonstrate prominent attributes in terms of sensitivity, resolution and stability. The measurand of resonant sensors ranges from physical forces, mass, physical fields, fluidic flow, acoustics and biological compounds. Despite the rapid development of micro-resonator-based sensors, the issue of cross-coupling sensitivity still remains, whereby resonant sensors are simultaneously sensitive to both the measurand (e.g., acceleration) and fluctuations in the ambient environment (e.g., temperature). To address this issue, environmental control [1] and frequency compensation [2] are the commonly adopted techniques. These methods typically require additional sensors and/or circuitry, which could limit the sensor resolution, and increase power consumption and form factors. Recently, a BSE scheme that originated from the principle of parametric resonance [3], has been demonstrated as suitable for sensor applications thanks to a dedicated closed-loop control system [4]. Distinct from the conventional driving scheme in which the excitation frequency is near or equal to a specific mode frequency of the resonator, the BSE exploits

a frequency equal to the sum of two modal frequencies, which is used in combination with a closed-loop system capable of tracking the frequency sum automatically. Such a mechanism also renders a feature that multiple vibration modes can be excited simultaneously, making it possible to simultaneously decouple multiple parameters from the frequency output of a single resonator.

In this work, a clamped-clamped beam resonator is employed to demonstrate the possibility of decoupling the cross-coupled parameters in the sensor readout. The first two flexural modes of the resonator, which are susceptible to both temperature changes and electrostatic perturbations, are used in this work. A closed-loop control system to track both modes' frequencies was implemented. A decoupling algorithm was then used to estimate the temperature and voltage changes based on the two modal frequencies. As a result, significant suppression of temperature effects in electrostatic voltage readout can be observed, and vice versa. This represents an important step toward (1) high-stability resonant sensors with significantly reduced ambient effects, and (2) in-situ multiple-parameter monitoring with a single resonator sensor.

THEORY

Unlike conventional excitation with a drive frequency equal to one particular modal frequency of the resonator, the BSE driving scheme utilizes an AC signal with a frequency equal to the sum of two vibration mode frequencies, i.e., blue-sideband. Based on the principle of parametric resonance [5], multiple modal frequencies of the resonator can be excited simultaneously, where the injected energy at the blue-sideband is distributed to the modes of interest, as depicted in Figure 1. Here, the first two modes of a resonator are selected as the BSE working modes, and the dynamics of the device subjected to BSE driving can be expressed as [4, 6]:

$$m_1\ddot{x}_1 + b_1\dot{x}_1 + k_1x_1 + k_{n1}x_1^3 + k_{n2}x_2x_1^2$$
$$+k_{n3}x_1x_2^2 + k_{n4}x_2^3 - k_{c1}x_2 \qquad (1)$$
$$= A_1\cos(2\pi f_{blue}t)x_1 + B_1\cos(2\pi f_{blue}t)x_2$$

$$m_2\ddot{x}_2 + b_2\dot{x}_2 + k_2x_2 + k_{n5}x_2^3 + k_{n6}x_1x_2^2$$
$$+k_{n7}x_2x_1^2 + k_{n8}x_1^3 - k_{c2}x_1 \qquad (2)$$
$$= A_2\cos(2\pi f_{blue}t)x_2 + B_2\cos(2\pi f_{blue}t)x_1$$

where m, b, k, k_c and x are the effective mass, damping, stiffness, inter-modal coupling stiffness, and the vibration amplitude of the resonator; k_n represents the nonlinear coefficient; A, B are the parametric excitation coefficient and inter-modal parametric coupling coefficient, respectively. The subscript numbers are the relevant vibration modes and higher nonlinear terms; f_{blue} is the BSE frequency.

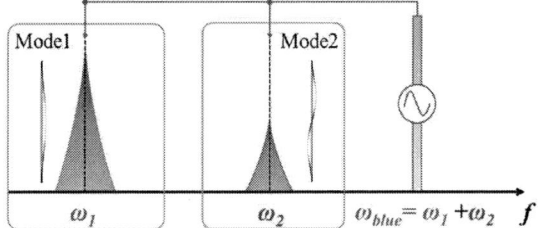

Figure 1: Fundamental of the blue-sideband excitation.

Given that the two vibration modes are simultaneously sensitive to temperature fluctuations and electrostatic perturbations, and assuming linear sensitivities, the modal frequencies can be written as:

$$f_1 = f_{01} + S_{v1}\Delta V + S_{t1}\Delta T \qquad (3)$$
$$f_2 = f_{02} + S_{v2}\Delta V + S_{t2}\Delta T \qquad (4)$$

where f_{01}, f_{02} and f_1, f_2 are the readout frequency before and after the input signal changes, of mode 1 and mode 2, respectively. S_{vi} and S_{ti}, $i=1,2$, are the sensitivity with respect to the electrostatic voltage perturbation and the temperature change. ΔV and ΔT are the input variations of the electrostatic voltage and the temperature, respectively. Equations (3) and (4) can be rewritten in matrix form, as shown in Figure 2. By multiplying the readout vector, which consists of both modal frequencies f_1 and f_2, with the inverse matrix consisting of the linear sensitivities, the voltage and temperature can be extracted and separated. This is the essence of the MPD algorithm.

$$\begin{pmatrix} S_{v1} & S_{t1} \\ S_{v2} & S_{t2} \end{pmatrix} \begin{bmatrix} v \\ t \end{bmatrix} = \begin{bmatrix} f_1 \\ f_2 \end{bmatrix} \quad \longmapsto \quad \begin{bmatrix} v \\ t \end{bmatrix} = \begin{pmatrix} S_{v1} & S_{t1} \\ S_{v2} & S_{t2} \end{pmatrix}^{-1} \begin{bmatrix} f_1 \\ f_2 \end{bmatrix}$$

Figure 2: A simplified illustration of the MPD algorithm to extract voltage and temperature based on the frequency of the first two modes of the resonator.

DEVICE DESCRIPTION

To verify the proposed MPD technique, a clamped-clamped beam resonator is chosen as the device-under-test (DUT). The dimensions of the DUT are given in Table 1. The device shown in Figure 3 is fabricated using a commercialized SOIMUMPS process from MEMSCAP. The DUT is mounted on an adapter carrier and then installed with the interface circuitry. A dedicated BSE closed-loop system comprising a digital lock-in amplifier, a phase-locked loop and a control module, as reported in [4], is employed here to execute the BSE driving scheme and maintain the oscillation of the DUT. Here, an extra phase-locked loop is also added, to extract the frequency of the second mode of the DUT. The block diagram of the closed-loop system is shown in Figure 4.

Table 1. The dimensions of the DUT.

Name	Parameter	Value
Clamped-clamped resonator	Beam length	800 µm
	Beam width	8 µm
Electrodes	Substrate length	394 µm
	Substrate width	329 µm

	Pad length	200 µm
Device thickness	Thickness	25 µm
Capacitive air gap	Width	2 µm

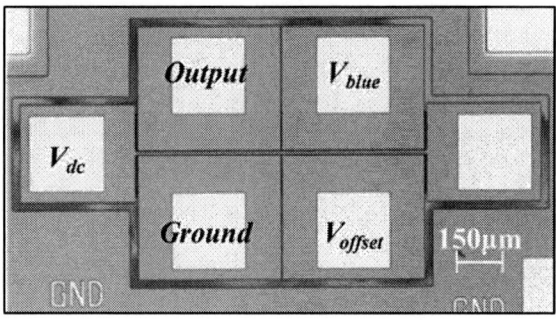

Figure 3: Image of the clamped-clamped beam resonator, the connections of the electrodes are included, in which the V_{offset} is the port of electrostatic voltage, and V_{dc} provides the polarization voltage. V_{blue} is the port for the BSE signal.

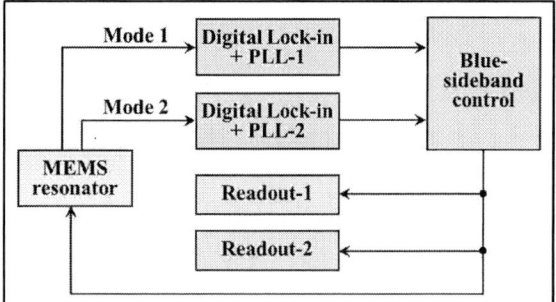

Figure 4: Schematic of the closed-loop system of BSE, containing a blue-sideband control that automatically locks on to and generates the sum frequency. The first two modes are both excited, hence two channels of digital lock-in and PLLs are implemented to extract the modal frequencies simultaneously. The digital system is implemented using an FPGA. More details of the blue-sideband control can be found in [4].

EXPERIMENT

The experiments are carried out in two stages: (i) the DUT is subjected to a step DC electrostatic perturbation, with continuous temperature changes; (ii) the DUT is subjected to a periodic AC electrostatic perturbation, with continuous temperature changes. The first experiment aims to investigate the possibility of decoupling multiple outputs at a quasi-steady-state, and the second experiment is intended to study dynamic decoupling. Regarding the experimental setup, a vacuum environment is provided with a vacuum level of 1m Pa. A temperature control (Stanford PTC 10) is used in the experiments to introduce a gradual temperature change from ~40.5℃ to ~43℃ in 10,000 s time. A commercial thermometer is used to monitor the temperature within the vacuum chamber. The BSE and associated lock-in amplifiers and PLL, as exhibited in Figure 4, are implemented based on a commercially available FPGA (Xilinx AX7102). The frequency readout is achieved using two synchronized frequency counters (Keysight 53230A). The experimental setup is shown in Figure 5.

978-1-6654-9309-3/23 $31.00 © 2023 IEEE

First, the stand-alone sensitivity of the resonator regarding electrostatic perturbations (to mimic stiffness perturbations) and temperature changes is characterized. The respective sensitivity is shown in Figure 6.

Figure 5: Photograph of the experimental setup.

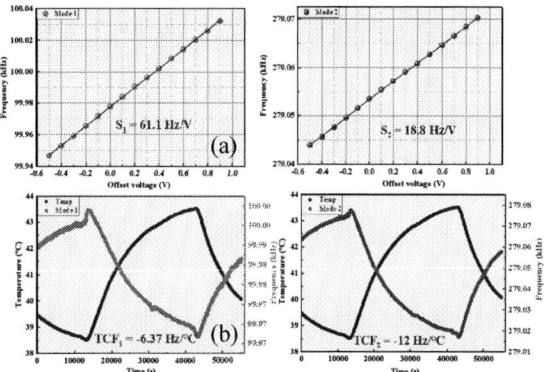

Figure 6: Stand-alone characterization of sensitivity of the first two modes of the DUT. (a) Sensitivity of the modal frequencies with respect to electrostatic perturbations. (b) Sensitivity of modal frequencies with respect to temperature changes.

Subsequently, a step variation experiment is carried out to observe the efficacy of the MPD approach in quasi-steady-state responses. The test is done with the BSE frequency controlled via the BSE control, near an initial value of $f_{blue} = f_1 + f_2 = 379.1$ kHz, and a BSE drive amplitude of 2.5 V. DC electrostatic perturbations are realized by tuning the value of V_{offset} (see Figure 3), ranging from 30 mV to 120 mV with a 30 mV step. This is done while the ambient temperature is rising (temperature changes are shown in black in Figure 7a and 7b). The directly measured responses of the two modes of the DUT are obtained and presented in Figure 7a and 7b. Unlike the tilted steps shown in Figure 7a and 7b due to temperature rising, the decoupled result obtained after applying the MPD algorithm (as shown in Figure 7c) demonstrates flat steps, implying the effect due to temperature rising is significantly suppressed with MPD. Further, the amplitudes of the decoupled signal are consistent with the source of the input, i.e., 30 mV to 120 mV.

Next, to demonstrate the feasibility of the simultaneous parameter decoupling technique with a dynamic input, an additional experiment is carried out with a slow changing (1 mHz, 0-0.1 V_{pp} sine wave) electrostatic perturbation and a gradual temperature rise introduced by

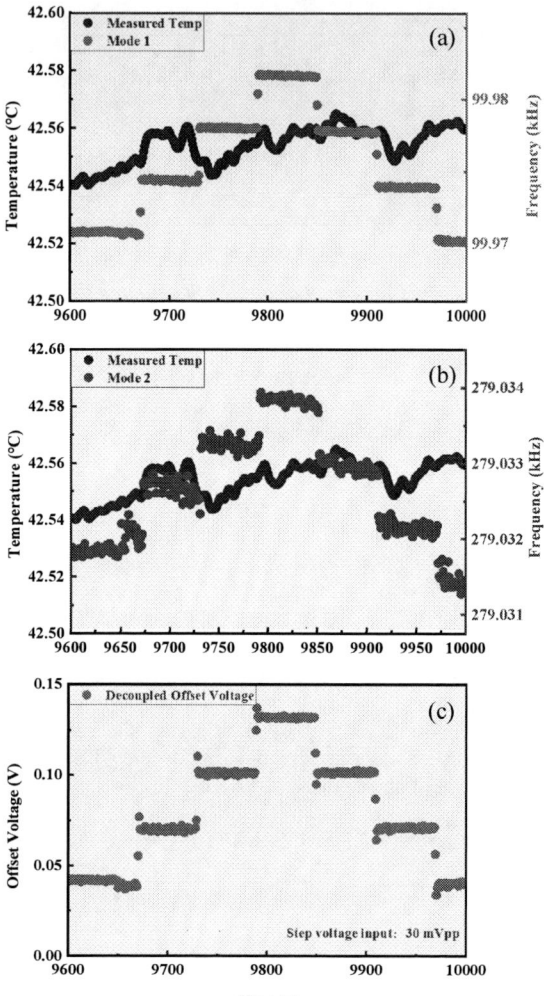

Figure 7: Step change experiment results. (a) and (b) Direct measurements of the frequencies (without MPD) of the first two flexural modes, containing fluctuations due to both the static offset voltage changes and monotonic trend of the temperature, along with the recorded gradual temperature changes. (c) Decoupled quasi-steady-state step offset voltage signal after applying MPD.

the temperature controller in place at the same time. The directly captured output signals (f_1 and f_2) and the decoupled data after applying the MPD algorithm, are depicted in Figure 8. It can be observed that the direct output changes contain both the voltage (periodic) and temperature (monotonic trend). After applying the MPD algorithm, the sine-wave fluctuations due to voltage changes are significantly suppressed in the temperature part. In addition, the decoupled temperature measurement follows closely to the temperature measurement using a thermometer (see Figure 8c). Furthermore, the frequency changes due to the monotonic temperature rise are significantly reduced in the voltage output (see Figure 8d). Some residual remains; this is likely due to small sensitivity errors obtained from Figure 6.

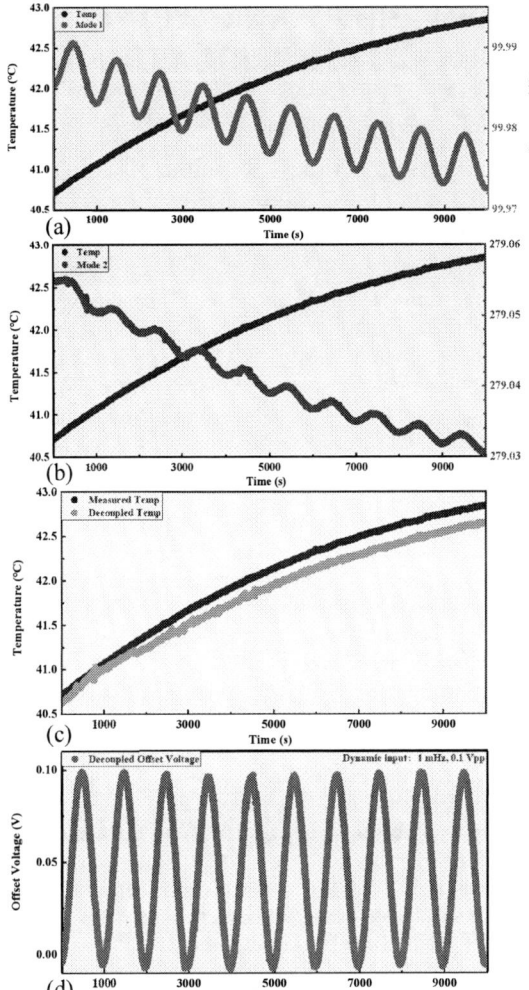

Figure 8: Dynamic experiment results. (a) and (b) Direct measurements of the frequencies of the first two flexural modes, containing fluctuations due to both the periodic voltage and monotonic trend of the temperature, along with the recorded gradual temperature changes. (c) The decoupled temperature in comparison with the recorded data. (b) Decoupled periodic voltage signal.

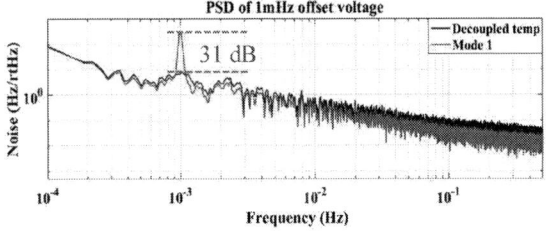

Figure 9: Delineation of cross-sensitivity suppression. It can be seen that within the decouple temperature data, the voltage changes are suppressed by 31 dB @ 1 mHz.

To preliminarily quantify the cross-sensitivity suppression, a PSD analysis is carried out, revealing that the peak at 1 mHz due to voltage changes is lowered by 31 dB within the decoupled temperature data, as shown in Figure 9. The results demonstrate the efficiency of the MPD technique for a single resonator.

CONCLUSION

In this study, a novel multiple-parameter decoupling technique featuring blue-sideband excitation is presented. Two simultaneous inputs of the resonant sensor, namely, the electrostatic perturbations and the temperature fluctuations, are successfully decoupled. A significant cross-coupling suppression of up to 31 dB is achieved, demonstrating considerable potential to differentiate and separate the coupled physical quantities. This subject has considerable potential to be applied in practical applications, in particular, the high-precision resonant sensors that demand the elimination of temperature fluctuations, as well as advanced resonant sensors with multiple-parameter monitoring capabilities without extra sensors.

ACKNOWLEDGEMENTS

This work was supported by the National Key Research and Development Program of China, Grant No. 2021YFB3201603.

REFERENCES

[1] A. Mustafazade, M. Pandit, C. Zhao, G. Sobreviela, Z. Du, P. Steinmann, X. Zou, R. T. Howe, and A. A. Seshia. "A vibrating beam MEMS accelerometer for gravity and seismic measurements." *Scientific reports*, vol. 10, no. 1, p. 1-8, 2020.

[2] S. Shin, H. K. Kwon, G. D. Vukasin, T. W. Kenny, and F. Ayazi, "A temperature compensated biaxial eFM accelerometer in Epi-seal process," *Sensors and Actuators A: Physical*, vol. 330, p.112860, 2021.

[3] T. Miao, X. Zhou, X. Wu, Q. Li, Z. Hou, X. Hu, Z. Wang, and D. Xiao, "Nonlinearity-mediated digitization and amplification in electromechanical phonon-cavity systems," *Nature Communications*, vol. 13, no. 1, pp. 1-8, 2022.

[4] L. Xu, J. Xi, L. Gao, F. Li, J. Pi, C. Li, K. Wang, X. Xiong, Y. Wang, and H. Liu, "A Closed-Loop System for Resonant MEMS Sensors Subject to Blue-Sideband Excitation," *Journal of Microelectromechanical Systems*, 2022.

[5] A. H. Nayfeh, and L. D. Zavodney. "The response of two-degree-of-freedom systems with quadratic non-linearities to a combination parametric resonance." *Journal of Sound and Vibration*, vol. 107, pp. 329-350, 1986.

[6] Y. Wang, X. Song, J. Xi, F. Li, L. Xu, H. Liu, C. Wang, S. Kuang, L.-C. Tu, and M. Kraft, "A Resonant Lorentz-Force Magnetometer Exploiting Blue Sideband Actuation to Enhance Sensitivity and Resolution," *Journal of Microelectromechanical Systems*, vol. 31, no. 3, pp. 402-407, 2022.

CONTACT

*wanglongyuan724@hotmail.com;
*huafengliu@hust.edu.cn;
*chen.wang@kuleuven.be;
*chun.zhao@york.ac.uk
§J. Xi and L. Xu contributed equally.

DIAMOND NANOWIRES ARRAY PREPARED BY ANNEALING NANO-CRYSTALLINE DIAMOND IN AIR AND ITS APPLICATION IN FIELD EMISSION

Yang Wang[1], Chen Lin[1], and Jinwen Zhang[1,2]*

[1]School of Integrated Circuits, Peking University, 100871, Beijing, People's Republic of China and
[2]National Key Laboratory of Micro/Nanometer Fabrication Technology, Peking University, 100871, Beijing, People's Republic of China

ABSTRACT

This paper reports the field emission (FE) properties of a novel array of diamond nanowires (DNWs). The DNWs structure was prepared by annealing the nano-crystalline diamond (NCD) films obtained by MPCVD in air, and the surface hydrogenation was realized by using hydrogen plasma. A flat anode was fabricated by micromachining, and assembled respectively with NCD film and nanowire cathodes to achieve a FE test structure with a gap of 1.03 μm. The FE properties of NCD films and DNWs were tested under a vacuum of 2.6×10^{-4} Pa. The results show that the FE current density of the DNWs array is as high as 174 mA/cm^2 and the turn-on field is as low as 2.3 V/μm, which is significantly better than that of the NCD film.

KEYWORDS

Diamond nanowires, nano-crystalline diamond, annealing, surface hydrogenation, field electron emission, field enhancement factor.

INTRODUCTION

FE can achieve stable emission of current at room temperature, with high efficiency, low power consumption, fast response rate and easy miniaturization, and has broad application prospects and great potential. The three allotropes of carbon, carbon nanotubes, graphene and diamond, have unique microstructures and excellent electron emission properties, which have attracted a lot of attention for application in display devices, microscopic imaging, switching devices, and electric thrusters [1]. Diamond surface shows significant negative electron affinity (NEA) after combining with hydrogen terminal, which greatly reduces the difficulty for electrons to leave the material surface. In addition, diamond has excellent stability, which is the key to long-term operation.

Micro-nanotip arrays with large aspect ratios is commonly used for FE enhancement. Diamond array processing methods applied to FE cathodes mainly include top-down reactive ion etching (RIE) [2], bottom-up diamond growth on patterned substrates [3], and CVD plasma growth regulated by RF power or magnetic field [4].

In addition, non-diamond phase carbon in NCD films can be selectively removed by annealing in oxygen-containing atmosphere, and various NCD porous structures can be prepared, including diamond microneedles [5], needle-like diamond skeleton structure [6], porous diamond film [7], diamond branch structure [8] and super hydrophobic nanoporous diamond [9]. This process has the same principle as maskless RIE etching, but due to the slower reaction rate, the diamond loss is smaller, and it is easy to prepare nano-tip structures with high aspect ratio and high distribution density [5]. On the one hand, a lower turn-on field and a larger FE current density can be achieved. On the other hand, NCD film with low requirements for growth conditions is adopted without complicated graphic mask and relatively expensive RIE processing technology, which greatly reduces the difficulty and cost of preparation and is conducive to the device and practical application of diamond FE. However, there are no reports on the FE cathode properties of nanoarray structures with large aspect ratio prepared by this method.

This paper systematically studies the FE properties of hydrogenated large aspect ratio DNWs. The DNWs structure with large aspect ratio, small radius of curvature and high distribution density was prepared by annealing in air for NCD films deposited by MPCVD. The surface hydrogenation was carried out by plasma. The structure of the high-field-strength flat-plate FE test anode was designed and fabricated, and a vacuum FE test system was built. The FE properties of NCD films and surface hydrogenated DNWs structures were tested and compared. Finally, the characteristic parameters of DNWs were extracted from the FE current-voltage curve and their enhancement effect on FE was analyzed.

EXPERIMENTAL

Preparation of DNWs

The single-cast n-type <100>Si was used as the substrate and placed in a 50 nm diameter diamond particle suspension for ultrasonic scratching for more than 1 h to form surface nanoparticle residues (as shown in Fig. 1(a)). NCD films were grown by MPCVD system with microwave power of 1800 W, air pressure of 6 kPa, temperature of 700 °C, atmosphere of CH_4(20 sccm)/H_2(80 sccm) and deposition time of 3 h.

The SEM image of the upper surface of the sample (as shown in Fig. 1(b)) indicates a large area of continuous film formation. The cross-section after splitting (as shown in Fig. 1(c)) shows that the thickness of the prepared diamond film is 0.98 μm. The 488 nm wavelength laser Raman spectrum (as shown in Fig. 1(d)) exhibits the 1332 cm^{-1} diamond characteristic peak, and a relatively sharp amorphous carbon characteristic peak appeared near 1500 cm^{-1}, which is the main filler in the NCD grain boundaries. In addition, the characteristic peak of polyacetylene at 1140 cm^{-1} is clearly visible, and polyacetylene is a typical structure at the junction of sp^3 diamond grains and sp^2 amorphous carbon in grain boundaries [10]. The two characteristic peaks at 1140 cm^{-1} and 1500 cm^{-1} appear at the same time, indicating that the sample is NCD film.

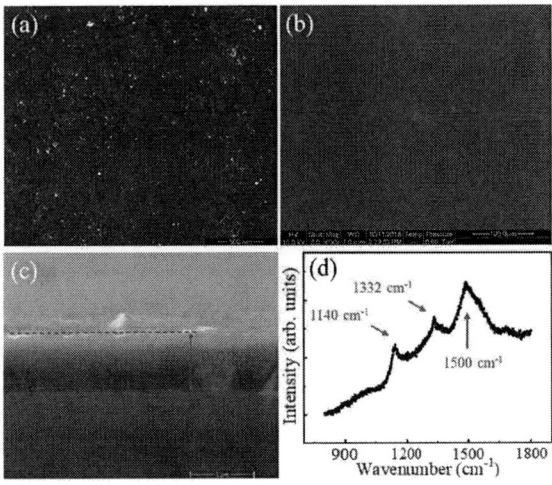

Figure 1: (a) Surface nanoparticle residue after ultrasonic scraping. Characterization of NCD films (b) upper surface (c) cross-sectional morphology (d) Raman spectrum under 488 nm laser light source.

The NCD samples were annealed in air at 500 °C (sample-A) and 600 °C (sample-B). The curves of the average top height $D_{o\text{-}avg}$, root height $D_{i\text{-}avg}$ and nanowire length d_{avg} of nanowires with time are shown in Fig. 2(a) and (b). The evolution of nanowires during the annealing process is divided into three stages, namely no nanowire stage, nanowire growth stage and nanowire length saturation stage. Finally, the $d_{avg\text{-}t}$ curve reaches the maximum value of 100.1 nm at 500 °C for 1020 min. At 600 °C, $d_{avg\text{-}t}$ curve reached its maximum value of 183.1 nm at 105 min.

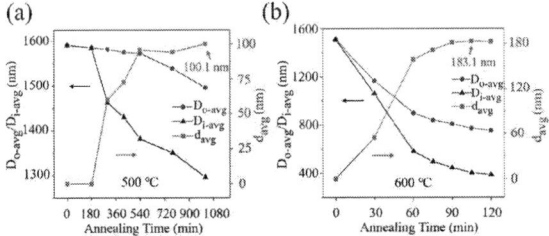

Figure 2: The average nanowire top height $D_{o\text{-}avg}$, root height $D_{i\text{-}avg}$ and nanowire length d_{avg} as a function of annealing time (a) sample-A at 500 °C (b) sample-B at 600 °C.

The SEM cross-sectional morphology observation photo of sample-B is shown in Figure 2(a). It can be seen that a porous structure layer is formed on the film, showing a large number of elongated DNWs arrays perpendicular to the substrate. Figure 2(b) shows a local high-magnification photo of the DNWs structure. The distribution of nanowires is very dense, with a diameter of about 10 nm, an aspect ratio greater than 18:1, and a distribution density of 4.6×10^{10} cm^{-2}.

Finally, the samples were put into an MPCVD chamber for hydrogenation under the following conditions: 100 sccm hydrogen, 1300 W microwave power, 3 kPa gas pressure, 770 °C substrate temperature and 20 min hydrogenation time.

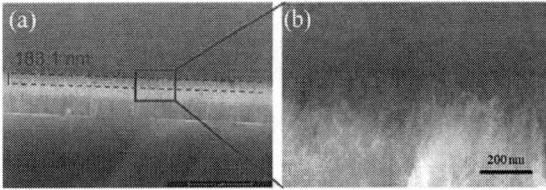

Figure 3: SEM image of sample B cross-section (a) diamond nanowires (b) partial enlargement.

Construction of FE Performance Test System

The FE anode was fabricated by micromachining. First, a thermally oxidized SiO$_2$ insulating layer was grown on a 4-inch single-cast n-type <100>Si, which not only precisely controlled the gap, but also served as the electrical isolation of the cathode and anode, with a thickness of 1.03 μm (as shown in Fig. 4(a)). Then a lithography plate as shown in Figure 4(b) was designed, photolithography and buffered hydrofluoric acid (BOE) wet etching the oxide layer to form a 1×1 mm^2 window and a large opening as the anode lead area, as shown in Figure 4(c). The measurement results of laser confocal microscopy show that the side length of the 1×1 mm^2 window is 1020.4 μm, as shown in Figure 4(d).

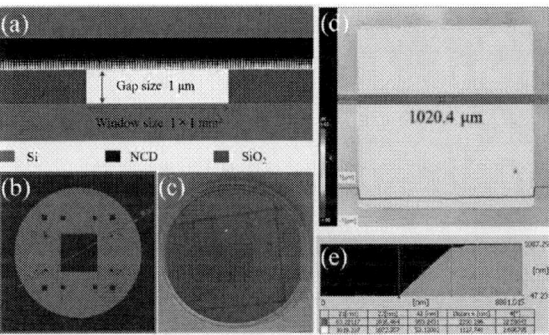

Figure 4: (a) Schematic diagram of the FE structure with 1 μm SiO$_2$ precisely controlling the cathode-anode spacing. Anode window (b) the lithography plate (c) processing results (d) actual area testing under laser confocal microscope (e) AFM measurement of oxide layer sidewalls.

The hydrogenated DNWs sample as the cathode covered upside down on the anode window, and was fixed firmly above the window with insulating tape. Finally, the anode and cathode terminals were drawn from the large window of the anode silicon wafer and the backside of the sample Si substrate with conductive glue. This formed flat FE test structure with a gap of 1.03 μm (as shown in Fig. 5(a)), which was mounted on the vacuum electrical test bench whose vacuum can reach 10^{-4} Pa. The FE test structure and electrical lead connection of the assembled sample are shown in Figure 5(b). The FE current was detected using the voltage scanning mode of the HP4156B semiconductor parameter analyzer, that is, the cathode FE was induced by applying a forward sweep voltage on the anode plate while the cathode of the sample was grounded.

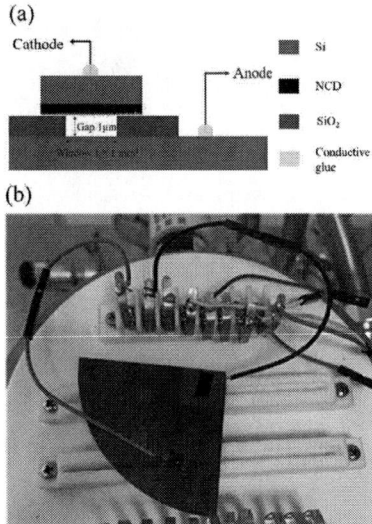

(a)

(b)

Figure 5: (a) Cross-section of schematic 1 μm plate FE testing structure (b) photo of the assembled FE testing structure mounted on the vacuum electrical test bench.

RESULTS AND DISCUSSION

The current characteristics of samples A and B were tested in a vacuum environment of 2.6×10^{-4} Pa and a scanning voltage of 0-60 V (as shown in Fig. 6). The curves showed an exponential upward trend. At 7.2 V, the FE current density of sample-A as a cathode reaches 10 μA/cm², indicating that its FE turn-on field is 7.2 V/μm, and the FE current density at 60 V is 61.8 mA/cm². The FE turn-on field of sample-B is 2.3 V/μm, and the FE current density at 60 V is 174 mA/cm².

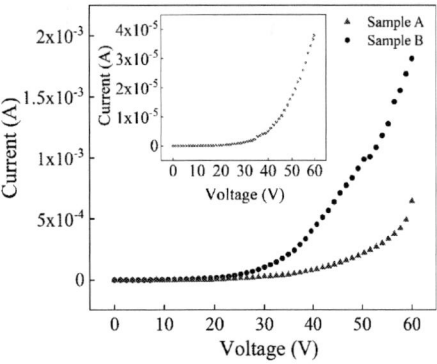

Figure 6: I-V curves at 0-60 V scanning voltage in vacuum of DNWs sample-A, sample-B, and NCD without nanowire structure (inset).

As a control group, the FE properties of unannealed NCD films (as shown in the inset of Fig. 6) were tested. The FE turn-on field is 19.4 V/μm, which is significantly higher than that of the nanowire samples. At 60 V, the FE current density is 3.50 mA/cm², which is 2 orders of magnitude smaller than that of the hydrogenated DNWs

sample. The nanowire arrays formed by annealing in air can significantly improve the FE performance of NCD films.

The classical FE phenomenon is described by Fowler-Nordheim theory [11]

$$J_0 = \frac{A E_t^2}{\Phi} \exp\left(-\frac{B\Phi^{1.5}}{E_t}\right) \qquad (1)$$

where J_0 is the FE current density, $A=1.54\times10^{-6}$ (A·eV/V²), $B=6.83\times10^7$ (V/((eV)$^{1.5}$)·m), Φ is the work function of the FE material, E_t is the local electric field at the cathode tip.

The relationship between the local electric field E_t at the cathode tip and the macroscopic electric field E is as follows

$$E_t = \beta E = \beta \frac{V}{d} \qquad (2)$$

where V is the applied voltage (V), d is the FE gap distance (m), and β is the field enhancement factor.

Considering the relationship between the current and the voltage obtained from the experimental data, the current is the product of the effective current density and the effective FE area S (m²), the following expression can be written:

$$I = \frac{SA(\beta V)^2}{\Phi d^2} \exp\left(-\frac{B\Phi^{1.5}d}{\beta V}\right) \qquad (3)$$

By removing V^2 on both sides and taking the logarithm, the following expression can be obtained:

$$\ln\left(\frac{I}{V^2}\right) = -\frac{B\Phi^{1.5}d}{\beta}\frac{1}{V} + \ln\frac{SA\beta^2}{\Phi d^2} \qquad (4)$$

The $\ln(I/V^2)$-$1/V$ curves of samples A, B and NCD films were obtained by processing the FE characteristic curves. The $\ln(I/V^2)$ is linear with $1/V$, which is divided into two stages, namely I and II. Staging is a normal phenomenon of FE characteristics, because the surface of the material cannot be absolutely smooth and flat, especially for the tip array. The slope is a quantity related to the work function Φ and the field enhancement factor β, while the intercept is a quantity related to β, Φ and the effective emission area S. Therefore, β can be obtained by calculating the slope of the $\ln(I/V^2)$-$1/V$ curve first, and bringing it into the intercept to obtain the effective emission area S. The parameter extraction is shown in Table 1.

It can be seen from Table 1 that the $\ln(I/V^2)$-$1/V$ curves of samples A, B and NCD films are divided into two stages. In stage I, the tips that dominate the emission has a smaller radius of curvature and a larger field enhancement factor, but a smaller effective emission area. While in stage II, the radius of curvature increases, the field enhancement factor decreases, but the effective emission area increases by several orders of magnitude. With the increase of the applied field strength, the field enhancement factors of some nanowires that dominate the FE gradually decrease, while the effective FE area increases by several orders of magnitude. This shows that during the FE process, some nanowires with small tip curvature radius and large field enhancement factor are the first to reach the open state and dominate the generation of FE current. With the further increase of field strength, more nanowires reach turn-on field and participate in the generation of dominant FE current.

Table 1: FE parameters extracted from relationship $ln(I/V^2)$ and $1/V$.

Sample	Stage	Slope	Intercept	β	S (m^2)
A	I	-43.48	-15.83	16.93	7.19×10^{-15}
	II	-111.92	-13.99	6.58	3.00×10^{-13}
B	I	-3.19	-16.95	230.81	1.26×10^{-17}
	II	-68.59	-13.49	10.73	1.86×10^{-13}
NCD	I	-138.75	-16.05	5.31	5.87×10^{-14}
	II	-535.40	-10.52	1.38	2.19×10^{-10}

For NCD films, the field enhancement factor in stage II is close to 1, indicating that there is no obvious sharp structure, and the electron emission mainly comes from the undulations of the film. For the nanowire samples, the β values are distributed between 6.58 and 230.81, with most of them in the order of 10^1, which is close to the nanowire size observed during the sample preparation stage. The field enhancement factor of sample-B in stage I is as high as 230.81, resulting in a very low turn-on field. Although the effective FE area of nanowire arrays is much smaller than that of NCDs, their FE currents are dozens of times larger, due to the advantages of large aspect ratio, small radius of curvature, and high distribution density of DNWs.

CONCLUSION

In this paper, NCD films obtained by MPCVD were annealed in air to prepare DNWs structures with large aspect ratio, small radius of curvature and high density. The surface hydrogenation was achieved by using hydrogen plasma. A flat anode was fabricated by micromachining, and assembled with NCD film and nanowire cathode to achieve a FE test structure with a gap of 1.03 µm. The FE properties of NCD films and DNWs were tested under a vacuum of 2.6×10^{-4} Pa. Based on the classical Fowler-Nordheim theory, relevant parameters were extracted and compared. The results show that the turn-on field of the NCD film is 19.4 V/µm, and the FE current density achieved at 60 V is 3.50 mA/cm^2. Under the same test conditions, the turn-on field of DNWs annealed at 500 ℃ and 600 ℃ are 7.2 V/µm and 2.3 V/µm, the FE current densities achieved at 60 V are 61.8 mA/cm^2 and 174 mA/cm^2, respectively, which are orders of magnitude higher than NCD films. The $ln(I/V^2)$-$1/V$ curve of both NCD films and DNWs has a staging phenomenon. With the increase of electric field, the field enhancement factor decreases, while the effective FE area increases. The extraction results of FE parameters show that the field enhancement factor of surface hydrogenated DNWs is in the order of 10^1 and the maximum value is as high as 230.81. On the other hand, the field enhancement factor of the NCD film in stage II is close to 1, and there is no obvious sharp structure. Compared with NCD films,

DNWs have larger field enhancement factor and smaller field emission area, which proves that DNWs have great application potential as high-performance FE cathodes.

ACKNOWLEDGEMENTS

This work is supported by the National Key R&D Program of China (Grant No. 2021YFB3200100).

REFERENCES

[1] M. L. Terranova, S. Orlanducci, M. Rossi, and E. Tamburri, "Nanodiamonds for field emission: state of the art," *Nanoscale*, vol. 7, no. 12, pp. 5094–5114, 2015.

[2] S. Kunuku *et al.*, "Investigations on Diamond Nanostructuring of Different Morphologies by the Reactive-Ion Etching Process and Their Potential Applications," *ACS Appl. Mater. Interfaces*, vol. 5, no. 15, pp. 7439–7449, Aug. 2013.

[3] D. Kim, H. L. Andrews, B. K. Choi, and E. I. Simakov, "Fabrication of Micron-Scale Diamond Field Emitter Arrays for Dielectric Laser Accelerators," in *2018 IEEE Advanced Accelerator Concepts Workshop (AAC)*, Breckenridge, CO, USA, Aug. 2018, pp. 1–3.

[4] J. Li *et al.*, "Plasma-enhanced synthesis of carbon nanocone arrays by magnetic and electric fields coupling HFCVD," *Surface and Coatings Technology*, vol. 324, pp. 413–418, Sep. 2017.

[5] A. Zolotukhin, P. G. Kopylov, R. R. Ismagilov, and A. N. Obraztsov, "Thermal oxidation of CVD diamond," *Diamond and Related Materials*, vol. 19, no. 7–9, pp. 1007–1011, Jul. 2010.

[6] S. Feng *et al.*, "Porous structure diamond films with super-hydrophilic performance," *Diamond and Related Materials*, vol. 56, pp. 36–41, Jun. 2015.

[7] J. Fecher, M. Wormser, and S. M. Rosiwal, "Long term oxidation behavior of micro- and nano-crystalline CVD diamond foils," *Diamond and Related Materials*, vol. 61, pp. 41–45, Jan. 2016.

[8] A. M. Alexeev, R. R. Ismagilov, and A. N. Obraztsov, "Structural and morphological peculiarities of needle-like diamond crystallites obtained by chemical vapor deposition," *Diamond and Related Materials*, vol. 87, pp. 261–266, Aug. 2018.

[9] Q. Wang *et al.*, "Morphology-controllable synthesis of highly ordered nanoporous diamond films," *Carbon*, vol. 129, pp. 367–373, Apr. 2018.

[10] H. Kuzmany, R. Pfeiffer, N. Salk, and B. Günther, "The mystery of the 1140 cm−1 Raman line in nanocrystalline diamond films," *Carbon*, vol. 42, no. 5-6, pp. 911-917, 2004.

[11] R.H. Fowler and L. Nordheim, "Electron emission in intense electric fields," *Proc. R. Soc. Lond.*, vol. 119, no. 781, pp. 173-181, May 1928.

CONTACT

*Jinwen Zhang, tel: +86-10-62766597;
zhangjinwen@pku.edu.cn

QUANTIFIED STRESS RELAXATION IN CARBON NANOTUBE RESONATORS

Morten Vollmann[1], Cosmin Roman[1], Miroslav Haluska[1], and Christofer Hierold[1]

[1] Micro and Nanosystems, Department of Mechanical and Process Engineering ETH Zurich, SWITZERLAND

ABSTRACT

We report an in-depth study of a suspended carbon nanotube resonator over the time span of more than 80 days. The resonator carries multiple carbon nanotubes, of which two show distinguishable frequency responses at different frequencies. The eigenfrequencies of one carbon nanotube ranges from 105 MHz and decays down to 30 MHz over time, indicating internal pre-stress relaxation from 90 MPa down to 10 MPa. With the application of new approaches in the modelling of carbon nanotube resonance currents, it is possible to quantify the time evolution of the internal pre-stress and the nonlinear spring constant.

KEYWORDS

Carbon nanotube resonator, NEMS, stress relaxation, harmonic balancing, contact slipping, nonlinear spring constant, tuning range

INTRODUCTION

Carbon nanotube (CNT) resonators can have versatile sensor applications for mass or force sensing [1][2]. Their extraordinary properties like the high yield strain and mechanical stiffness make them interesting for tunable resonators in particular. Resonance frequencies can range from several tens of MHz up to more than 2 GHz [3]. Advances in the controlled dry transfer of pristine CNTs [4] and the modelling of the CNT currents enable the extraction of the pre-stress or even gauge factors [5].

No studies of the long-term stability of such CNT resonators have previously been published. To establish the feasibility of industrial applications it is important to understand the degradation of both the contacts and the CNT. The electrical properties of CNT contacts with electrodes have previously been discussed in literature [6], however, the evaluation of the mechanical properties remains a challenge. Kumar et al. investigated the stress relaxation of adhesively clamped CNT resonators and linked the relaxation to actuation power high enough to trigger nonlinear responses [7]. Other research presented a tunable graphene resonator to investigate the shear stress present in the contacts between graphene and the electrodes [8].

In this paper we investigate the long-term behaviour of CNT resonators for 81 days and discuss the influence of stress relaxation on the frequency degradation and tuning range over time. Model parameters like the (internal) pre-stress and the nonlinear spring constant of the CNT mechanical response are extracted using harmonic balancing to solve the underlying differential equation of motion. The frequency response enables a distinction between different CNTs, which is impossible with pure electronic measurements. The device characterization is supported with Raman and SEM measurements.

METHODS AND MATERIALS

Device architecture

Our CNT resonator device layout consists of suspended CNTs that lay on top of Palladium electrodes in a field effect transistor configuration [9]. The device substrate is CNT dry transfer compatible [4] and its electrodes have been Argon sputtered and post-transfer-annealed to improve the electrical contacts [6].

During the transfer process six individual CNTs were transferred and identified by post-characterization using SEM imaging (Fig.1) after Raman spectroscopy (Fig. 2). The transfer characteristics are possibly a superposition of all electrically active CNTs. However, only two of them showed detectable and distinguishable mechanical resonance. The gate distance g_0, the suspended length l, and the radius of the CNT are presumably 2.0 µm, 2.1 µm, and 1.1 nm, respectively, based on previous measurements, as they could not have been evaluated precisely.

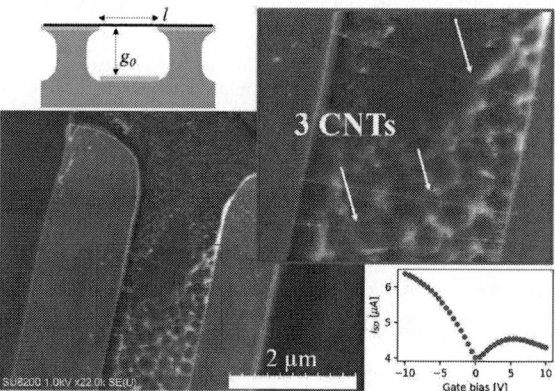

Figure 1: SEM picture in top view of the substrate architecture with Palladium electrodes. Red inset: 3 of 6 CNTs visible. Top left: side view of device architecture: Palladium/Chromium electrodes in yellow, Si and SiO2 in grey, CNT in black, geometry in table 1. Bottom right: Transfer characteristics with $V_{SD}=100$ mV on day 45.

Experimental setup

The resonance current has been detected using the common frequency down mixing setup with a gate actuation voltage $V_{GD}=V_{GD}^{DC}+V_{GD}^{AC} \cos(\omega t)$ [9]. The read-out signal through the CNT is $V_{SD}=V_{SD}^{DC}+V_{SD}^{AC} \cos(\omega t+\Delta\omega t)$ (Keithley 2400, Agilent N5181A). Due to the offset $\Delta\omega$, the mixed current has a frequency component at $\Delta\omega$, which is read out with a lock-in amplifier (Zurich instruments UHLFI, $\Delta\omega=10kHz$). A similar setup is presented in [5].

The measurement conditions are ambient temperature and vacuum ($<10^{-3}$ mbar). The measurements are repeated over a time-period of 81 days, at irregular intervals. One measurement set sweeps the gate bias and the actuation frequency and takes around 50 hours to complete. At the end of the time-period of 81 days, the sample was measured in a Raman setup.

Raman spectroscopy was performed with Renishaw Qontor inVia Raman confocal microscope utilizing the backscattering configuration with the microscope objective with 100x magnification and numerical aperture 0.85. For mapping, a 3 s accumulation time per spectrum was used in contrast to 120 s to obtain individual spectra of the suspended part of the CNT and (subtracted) background spectra. Two excitation wavelengths were used, 488 nm and 514 nm, respectively (Fig.2).

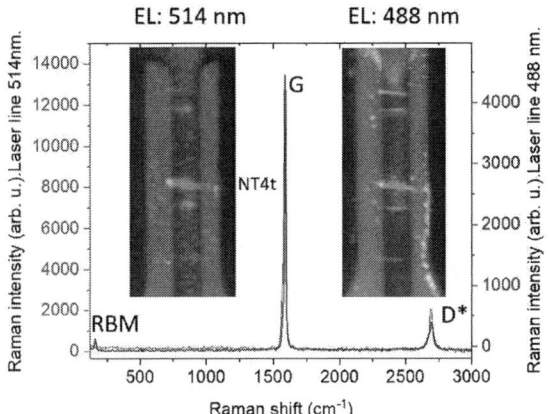

Figure 2: Raman spectra of the CNT (NT4t) obtained for laser excitation line 488 nm (blue curve) and 514 nm (green curve). Only one CNT shows a radial breathing mode (RBM) and no visible defect mode is present. Insets: Raman spectroscopy maps from the device measured by 488nm (left panel) and 514 nm (right panel) laser excitation lines. Light blue and green colors indicate CNT G modes (at around 1585 cm⁻¹) of suspended CNTs. Source and drain electrodes are visible in beige.

A typical electromechanical characterization set $(I_{SD}(f,V_{GD}))$ is shown in Fig. 3 a) and b). In the frequency response we can clearly distinguish between two different CNTs (as indicated by the white arrows in Fig. 3). Both show gate DC bias V_{GD} dependent eigenfrequencies with different current levels. The following analysis only addresses the CNT with the stronger signal ($f_0 > 35$ MHz in Fig. 3).

Model-based analysis

The model introduced in [5] was utilized in analyzing measurement sets (as in Fig. 3 a) and b)) as a function of time. The approach is based on the Duffing equation:

$$m\ddot{z}+b\dot{z}+\left(k_0+\sigma_0 A\int_0^l\left(\frac{\partial\phi}{\partial x}\right)^2 dx - k_{soft}\right)z+k_3z^3=F(t),\ (1)$$

where m is the effective resonator mass, b is the effective damping coefficient, k_0 is the linear stiffness, σ_0 is the pre-stress, l is the suspended CNT length, ϕ is the mode shape function, k_{soft} is the spring softening from the capacitive actuation force, and k_3 is a nonlinear spring constant describing the non-linearity in the restoring force.

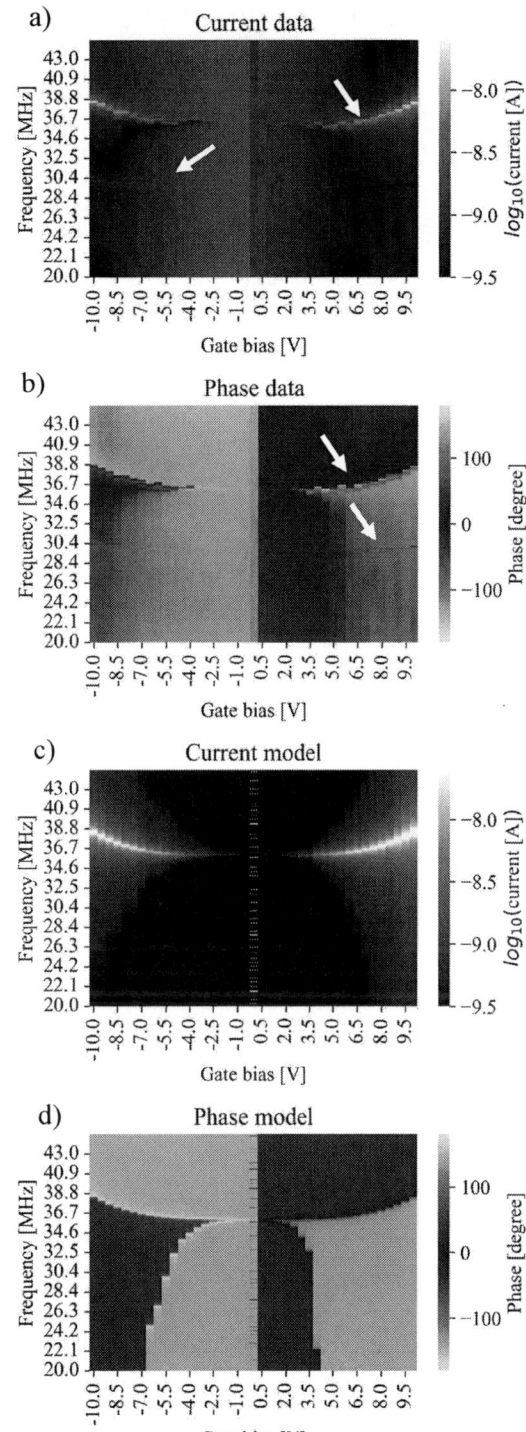

Figure 3: Typical current a) and phase b) of the resonator response at $\Delta\omega$ with two distinguishable and separate CNTs (white arrows) at different eigenfrequencies on day 45. c) and d) show the comparative model based results for current and phase with the fitted pre-stress and nonlinear spring constant. Only one CNT is modelled in c) and d).

Parameters such as the nonlinear spring constant k_3, the pre-stress σ_0, and the quality factor Q are fitted so that the eigenfrequencies fit the mechanical response of the device, throughout the range of V_{GD} biases. The solution for the fit is provided by a harmonic balancing solver of the first order [5]. Each set of measurements was fitted individually. Figure 3 c) and d) show the result of the model based fit. The model captures very well the current amplitude at $\Delta\omega$ and especially the non trivial phase information of the measurements.

The effective forces that act on CNTs can be described as [9]

$$F(t)=\frac{1}{2}C'0.62\cdot l\,V_{GD}^2(t) \qquad (2)$$

with

$$C'=\frac{2\pi e_0}{g_0\,\log\left(2\frac{g_0}{r}\right)^2} \qquad (3)$$

The effective forces acting on the here presented CNT are then F_{max}=26.2 pN.

Table 1: Estimation of the maximum acting force with assumed parameters for our CNT in comparison with different geometry from [7].

Device	l	r	g_0	V_{GD}	F_{max}
Fig.1	2.1 μm	1.1 nm	2 μm	10 V	26.2 pN
[7]	2.2 μm	1.2 nm	275 nm	1.9 V	13.3 pN

EXPERIMENTS AND RESULTS

The resonator was measured over a time-period of 81 days and showed eigenfrequencies ranging from 105 MHz at maximum V_{GD} = 10V on day one and 30 MHz on day 81 at V_{GD} = 10V. Both the maximum and the minimum eigenfrequencies ($f_{0,min}$, $f_{0,max}$) degraded with an exponential-like decay over time (Fig. 4a). The fitted exponential has a form of:

$$f(t)=a\cdot e^{\frac{t}{\tau}}+b \qquad (4)$$

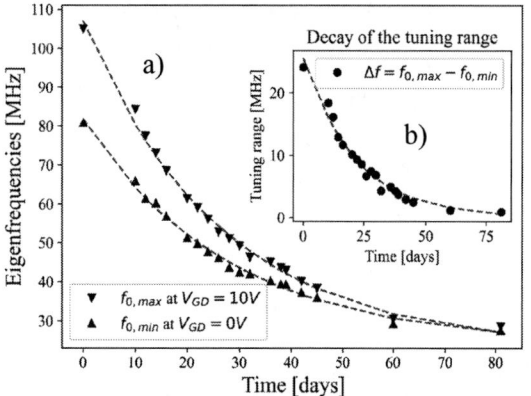

Figure 4: a) Reduction of the eigenfrequencies of the CNT ($f_0 > 35$ MHz) in Fig. 3 at different gate biases. Inset b): the corresponding tuning range Δf over time. The time constants as well as other fitting parameters are summarized in Tab 2.

The extracted τ values are 27.8 days and 26.1 days for $f_{0,min}$ and $f_{0,max}$ respectively with R^2=0.995 for both fits. Consequently, the tuning range by electrostatic actuation was extracted as the difference between the eigenfrequencies at 10 V and 0 V gate bias (Fig. 4b). The tuning range Δf=$f_{0,max}$-$f_{0,min}$ decays from 25 MHz down to 0.5 MHz with a τ value of 21.4 days and R^2=0.98. The τ values are similar in magnitude.

Table 2: Time constants and fitting parameters for the decay (eqn. 4) of different quantities.

Decay	a	b	τ [days]	R^2
$f_{0,min}$	57.7	24.0	27.8	0.995
$f_{0,max}$	83.7	23.3	26.1	0.995
Δf	25.6	0	21.4	0.980

The pre-stress σ_0, that is present in the CNT due to the fabrication, the nonlinear spring constant k_3 have been fitted to the experimental data (Fig. 5). Both the internal pre-stress and the nonlinear spring constant decay in time. They appear to have a different initial decay slope. After day 20, the actuation voltage was incrementally increased to trigger nonlinear mechanical response from the CNT resonator. There is no visible correlation between the actuation voltage and the eigenfrequency reduction.

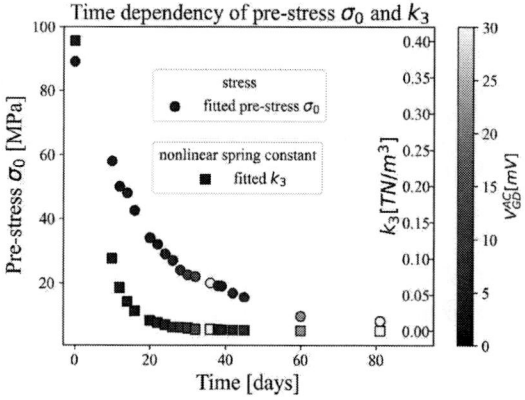

Figure 5: Relaxation of the fitted pre-stress and nonlinear spring constant from the model presented in [5]. The color of the data points represents the actuation amplitude.

Selected Duffing type resonances are presented in Fig. 6. Different read out voltages V_{SD} have been used, which explains the difference in the peak amplitudes. However, the peak shapes are not affected by the different read out voltages and they exhibit a nonlinear Duffing shape. No spring softening is visible, as expected, because CNT resonators form a small capacitance to the gate electrode. Fig. 6 shows once again the decay of the eigenfrequencies as well as the tuning range.

DISCUSSION AND CONCLUSION

Degradation/relaxation of the resonance frequency in carbon nanotube resonators has been observed previously by Kumar et al [7]. They attributed the relaxation to slippage of the CNT from the contacts due to large actuation voltages. Here, we observe a similar relaxation, but the actuation voltage appears not to be correlated to the relaxation (see Fig. 5). To estimate and compare the forces that act on the CNT with [7], we used equations (2) and (3) with their device parameters, yielding 13.3 pN, which is approximately half compared to the forces expected in our experiment (Tab. 1). We speculate that this relaxation is caused by slipping effects in the contacts.

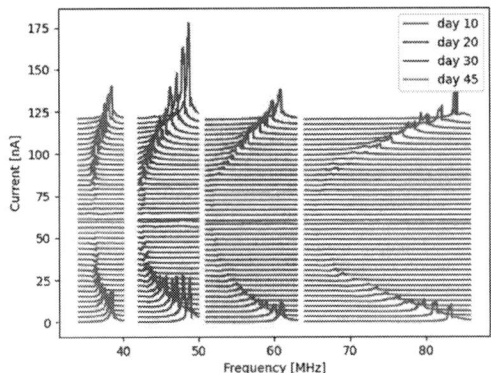

Figure 6: Selected frequency responses from the CNT resonator at different days. Each line corresponds to a gate bias and has an offset of 3 nA to improve visibility. Both the decay in eigenfrequency as well as tuning range is visible. The nonlinear Duffing shape of the resonance becomes more prominent at higher gate bias.

The nonlinear spring constant is relevant for the tuning range because of the large static deflections, but it is also relevant for the dynamics, as reflected in Duffing resonances in Fig. 6. In future investigations we will examine the apparent difference in the initial decay slopes of the internal pre-stress σ_0 and the nonlinear spring constant k_3.

One other interesting aspect is, that the frequency response of the multi-CNT resonator is able to distinguish between different mechanically/electrically active CNTs. The different current magnitudes can be explained by different contact resistances or different electrical properties of the tube itself.

The decay in resonance frequency from 105 MHz to 30 MHz is significant and seems to be caused by pre-stress relaxation and/or nonlinear spring constant relaxation. The relative impact of σ_0 and k_3 on f_0 remains to be investigated. Future research should also address the physical nano-tribological relaxation mechanism, and potentially technological means to control the nanotube electrode contact mechanics. Our setup and methodology could be used in the future to measure and quantify the quality of the mechanical contacts, where slower degradation rate indicates better contacts quality.

ACKNOWLEDGEMENTS

We thank the Cleanroom Operations Team of the Binnig and Rohrer Nanotechnology Center (BRNC) for their help and support, and Seoho Jung and Cristina Gentili for fabrication of the device.

REFERENCES

[1] G. Gruber, C. Urgell, A. Tavernarakis, A. Stavrinadis, S. Tepsic, C. Magén, S. Sangiao, J. M. de Teresa, P. Verlot, and A. Bachtold, "Mass Sensing for the Advanced Fabrication of Nanomechanical Resonators", *Nano Letters*, vol. 19, pp. 6987-6992, 2019.

[2] J. Moser, J. Güttinger, A. Eichler, M. J. Esplandiu, D. E. Liu, M. I. Dykman, and A. Bachtold, "Ultrasensitive force detection with a nanotube mechanical resonator", *Nature Nanotechnology*, vol. 8, pp. 493–496, 2013.

[3] X. Wang, D. Zhu, X. Yang, L. Yuan, H. Li, J. Wang, M. Chen, G. Deng, W. Liang, Q. Li, S. Fan, G. Guo, and K. Jiang, "Stressed carbon nanotube devices for high tunability, high quality factor, single mode GHz resonators", *Nano Research*, vol. 11, pp. 5812–5822, 2018.

[4] M. Muoth and C. Hierold, "Transfer of carbon nanotubes onto microactuators for hysteresis-free transistors at low thermal budget", *in Proceedings of the IEEE International Conference on Micro Electro Mechanical Systems (MEMS)*, pp. 1352–1355, 2012.

[5] M. Vollmann, C. Roman, and C. Hierold, "A Phase-Consistent Model for the Spectral Transfer Function of Carbon Nanotube Resonators", *in Proceedings of the IEEE International Conference on Micro Electro Mechanical Systems (MEMS)*, pp. 822–825, 2022.

[6] S. Jung, R. Hauert, M. Haluska, C. Roman, and C. Hierold, "Understanding and improving carbon nanotube-electrode contact in bottom-contacted nanotube gas sensors", *Sensors Actuators B Chem.*, vol. 331, p. 129406, 2021.

[7] L. Kumar, L. Jenni, M. Haluska, C. Roman, and C. Hierold, "Mechanical stress relaxation in adhesively clamped carbon nanotube resonators", *AIP Advances*, 8, 025118, 2018.

[8] Y. Ying, Z. Zhang, J. Moser, Z. Su, X. Song, and G. Guo, "Sliding nanomechanical resonators", *Nature Communications*, vol. 13, no. 6392, 2022.

[9] V. Sazonova, Y. Yalsh, I. Üstünel, D. Roundy, T. A. Arlas, and P. L. McEuen, "A tunable carbon nanotube electrochemical oscillator", *Nature*, vol. 431, no. 7006, pp. 284–287, 2004.

CONTACT

*M. Vollmann,+41446323826; mvollmann@ethz.ch

SELF-REFERENCED TEMPERATURE SENSORS BASED ON CASCADED SILICON RING RESONATOR

Xiantao Zhu[1,2], Minmin You[1], Zude Lin[1], Bin Yang[1] and Jingquan Liu[1]

[1]National Key Laboratory of Science and Technology on Micro/Nano Fabrication, Shanghai Jiao Tong University, CHINA and

[2] DCI Joint Team, Collaborative Innovation Center of IFSA, Department of Micro/Nano Electronics, Shanghai Jiao Tong University, CHINA

ABSTRACT

This paper proposed a novel self-referenced thermometer based on a silicon cascaded micro-ring resonator (CMRR). A tapered waveguide cascades the reference ring covered by the titanium dioxide (TiO_2) cladding with the sensing ring. Taking advantage of dual resonances derived from the two different waveguides, the simulation result shows the sensitivity of the CMRR is 63.4 pm/K utilizing the differential wavelength shifts ($\Delta\lambda=\lambda_{sensing}-\lambda_{reference}$) method. Moreover, the reference ring is temperature insensitive due to the thermo-optic effect of the silicon waveguide being offset out by the cladding with a negative thermo-optic coefficient (TOC). The work shows a new avenue for high-sensitivity and high-stability temperature sensing based on the silicon ring resonator.

KEYWORDS

Cascaded Micro-ring Resonator (CMRR), Titanium Dioxide (TiO_2), Reference Ring, Sensing ring, Temperature Insensitive

INTRODUCTION

Currently, temperature measurements play an important role in modern life, such as manufacturing, and artificial intelligence. In addition to the traditional resistance thermometers, optical temperature sensors based on different fiber sensor structures including the fiber Bragg gratings (FBG) [1], and Fabry-Perot interferometer (FPI) [2] have been widely studied. Employing electromagnetic immunity, complementary metal-oxide-semiconductor (CMOS) compatibility and high-quality factor, temperature sensors based on silicon micro-ring resonators (MRR) have also been demonstrated [3]. To improve the sensitivity of the MRR, different methods such as cascaded MRR for realizing the Vernier effect [4] and using the negative thermo-optic coefficient materials as a top cladding are proved [5].

Besides sensitivity, stability is another critical characteristic parameter for sensors. For other sensors like biosensors and gas sensors, the temperature is the main noise. To solve this problem, the cascaded MRR has been developed as one ring as the reference while the other for sensing. But for temperature monitoring, the noises arise from the power fluctuations and the output frequency drifts of the incident laser beside the intrinsic noises. Previous work has reported the self-referenced temperature sensors with a sensitivity of 3.0 GHz/K (24 pm/K) based on the thermo-optic birefringence effect of the lithium niobate [6]. Qiancheng Zhao, *etal* proposed a dual-mode resonance frequency difference temperature sensitivity of 188 MHz/K (1.5 pm/K) based on the same effect of silicon nitride [7]. The self-reference method based on the cascaded silicon-based MRR means the reference ring should be temperature insensitive. Researchers have realized the temperature-insensitive sensors based on a single MRR by covering the ring with negative TOC polymers such as SU-8 to compensate for the active TOC of the silicon. For silicon-based photonics, however, the polymers are typically not compatible with CMOS processes due to poor long-term operating stability and humidity sensitivity issues.

Titanium dioxide (TiO_2) due to its numerous desirable properties, such as CMOS-compatible and high transmission in the visible and near-infrared regions, has been widely used in various optoelectronic applications. Furthermore, TiO_2 which has a relatively strong negative TOC of approximately -1×10^{-4}/K ~ -2×10^{-4}/K around 1550 nm is an excellent candidate as an optical coating material to realize the temperature insensitivity of the silicon MRR and overcomes the polymer drawbacks of wetting and chemical instability. Researchers have realized the low sensitivity sensor distributed from -20 to 20 pm/K depending on the method of deposition such as sputtering, evaporation and atomic layer deposition (ALD) [8].

In this paper, we propose a self-referenced temperature sensor based on a silicon CMRR. The reference ring with the TiO_2 cladding achieved by the RF sputtering method is cascaded with the sensing ring by a tapered waveguide to materialize the self-referenced temperature monitoring. Exploiting the thermo-optic effect of the silicon waveguide being offset out by the cladding with a negative thermo-optic coefficient (TOC), the reference ring is relatively temperature insensitive. The simulation result shows the sensitivity of the CMRR is 63.4 pm/K utilizing the differential wavelength shifts method. Moreover, the stability can be further improved since some external interferences were excluded. The result indicates the great application potential for silicon-based CMRR in the field of extreme environments requiring high-sensitivity and high-stability temperature sensing.

EXPERIMENTAL METHOD

Sensor Design

Figure 1 shows the schematic diagram of the CMRR sensor. The waveguide width of the reference ring is different from the sensing ring to realize the temperature insensitive and is cascaded by a tapered waveguide in the design. The vertical grating couplers were designed at the waveguide ends to achieve an effective coupling between the fiber and the CMRR. The temperature can be determined from the resonant wavelength shift of the ring resonator. The shift of the resonance wavelength, $\Delta\lambda$,

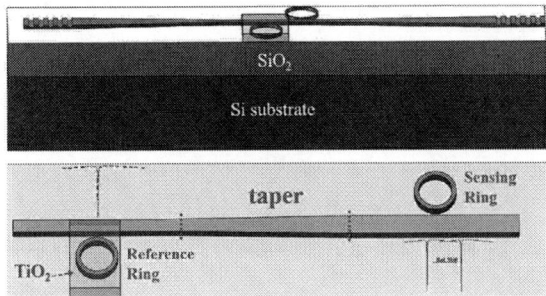

Figure 1: Schematic diagram of the CMRR.

induced by the thermal expansion and thermo-optic effects of silicon can be expressed as follows:

$$\Delta\lambda = \Delta\lambda_L + \Delta\lambda_T = \alpha_W \frac{n_{eff}}{n_g}\lambda\Delta T + \frac{\sigma_T}{n_g}\lambda\Delta T \quad (1)$$

where n_{eff} and n_g represent the effective index and group index of the waveguide, ΔT represents the temperature change, α_w and σ_T represent the thermo-expansion coefficient (TEC) and TOC, respectively. For the thermal-expansion effect, one can usually ignore it since the TEC of silicon (2.7×10^{-6} K) is smaller than the TOC (CTO,1.8×10^{-4} K) by two orders of magnitude. Therefore, the athermal condition of the waveguide means the TOC is zero, ie:

$$\sigma_T = \frac{dn_{eff}}{dT} = 0 \quad (2)$$

The reference ring segment is a sandwich structure consisting of TiO_2, silicon, and SiO_2, the equivalent TOC of which can be expressed as:

$$\frac{dn_{eff}}{dT} = \Gamma_{Si}\frac{dn_{Si}}{dT} + \Gamma_{SiO_2}\frac{dn_{SiO_2}}{dT} + \Gamma_{TiO_2}\frac{dn_{TiO_2}}{dT} \quad (3)$$

where Γ_{Si}, Γ_{SiO_2} and Γ_{TiO_2} represent the confinement factor of the silicon, SiO_2 and TiO_2, respectively. To ensure the positive TOC of the silicon is eliminated by the negative TOC of the cladding, the waveguide geometry of the

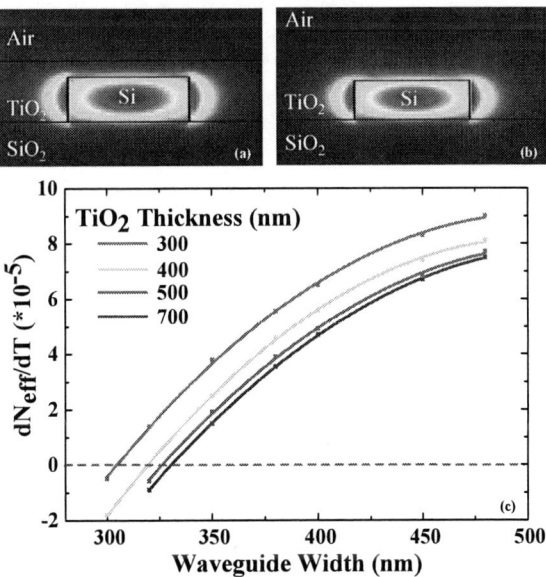

Figure 2: Optical field distribution of reference ring resonator with a cladding thickness of (a) 300 nm, (b)500 nm; (c) the TOC of the reference ring with different waveguide widths and different cladding thicknesses.

reference ring was designed carefully to satisfy the athermal condition. Utilizing the finite difference time domain (FDTD) simulation method, the reference rings with different waveguide widths and different cladding thicknesses are analyzed and optimized. The refractive index and TOC of the TiO_2 are set to 2.2 (data from ellipsometry) and $-1.8\cdot10^{-4}$ K^{-1}, respectively. Figure 2 (a)&(b) shows the distribution of the optical field within the reference ring. It can be seen that part of the optical field leaks in the air with 300-nm-thick TiO_2 cladding, while the 500 nm thickness is suitable. The simulated TOC of the reference ring at 1550 nm with different TiO_2 cladding thicknesses as a function of the waveguide core width is plotted in Figure 2 (c). With the waveguide width decreasing and the thickness of the TiO_2 cladding increasing, the TOC of the reference ring reduces continuously. However, the cladding thickness increases from 500 nm to 700 nm leading to a small modification of the TOC due to the limited propagation of the optical field outside the waveguides. Based on the calculated results, the thickness of the TiO_2 cladding is designed to be 500 nm, and the width of the waveguide is 330 nm for realizing the temperature insensitivity of the reference ring.

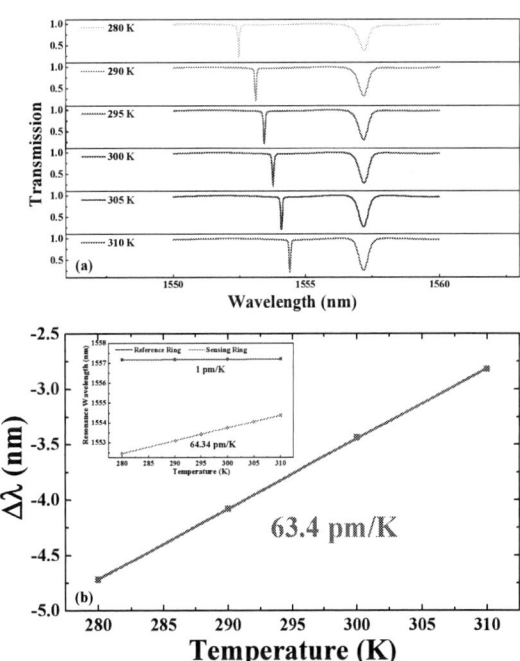

Figure 3: (a)Simulated transmission spectrum at different temperatures, (b)the sensitivity of the CMRR, and the inset figure shows the sensitivity of the reference ring and the sensing ring.

Using a three-dimensional FDTD simulation method, the transmission spectrum of the device is also calculated, as Figure 3 shows. The waveguide width of the reference ring is designed to be 330 nm, while the sensing ring is 480 nm. The 120-nm-wide gap between the bus waveguide and the circular resonator is designed to realize critical coupling between them. To avoid the transmission spectrum envelope of the traditional CMRR sensor, the diameters of the two rings are equally designed to be 20 μm. The normalized spectrum of the designed CMRR device at different temperatures from 280 K to 310 K is plotted in Figure 3 (a). As can be seen from the spectrum, the spaced resonance dips corresponding to different rings are generated. Clearly, the increase in temperature results in a redshift of the resonance wavelength. This shift is induced by the increased refractive index of the device as discussed above. However, due to the existence of the TiO_2 cladding, the redshift of the reference ring resonance is suppressed greatly and realizes the temperature insensitivity. The sensitivity of the two rings is calculated by using the least square method, as plotted in the inset figure in Figure 3 (b), showing a linear dependence on the temperature. The simulated results show the sensitivity of the sensing ring is 64.34 pm/K, while the reference ring is as low as 1 pm/K. So, the sensitivity of the device is about 63.4 pm/K, as Figure 3 (b) shows.

Fabrication Method

Figure 4 illustrates the device fabrication process using standard CMOS technology on silicon on insulator (SOI) wafer consisting of a top silicon layer of 220 nm and a silicon dioxide substrate of 2 μm. The device fabrication involves the 70-nm-depth grating couplers, the 220-nm-depth waveguide and the TiO_2 cladding film sputtering, the detailed process of preparation is as follows: Firstly, the SOI wafer was cleaned with acetone, isopropyl alcohol (IPA), and deionized water in an ultrasonic cleaner for 5 minutes, and dried with a stream of nitrogen gas. Secondly, oxygen plasma with a power of 500 W and a flow rate of 350 sccm was used to treat the wafer surface for 5 minutes to remove the residuals. Thirdly, a positive resist was coated on the wafer, and the device patterns were defined by e-beam lithography (EBL, Visten EBPG 5200+). Lastly, the patterns were transferred into the silicon layer by 70-nm-depth inductively coupled plasma (ICP) etching (SPTS DRIE-1, WF2SDSE01). The procedure was repeated as described above, except that the ICP etching depth was 220 nm for the fabrication of the waveguides. The TiO_2 film was fabricated using the lift-off process: the 1.3-μm-thickness PMMA photoresist was spun on the wafer, and the coverage area of the TiO_2 was defined by e-beam lithography. Then the TiO_2 cladding on the reference ring with a thickness of 500 nm is achieved using the RF sputtering technology. Lastly, the residual photoresist was cleaned with the acetone leaving the TiO_2 film on the wafer.

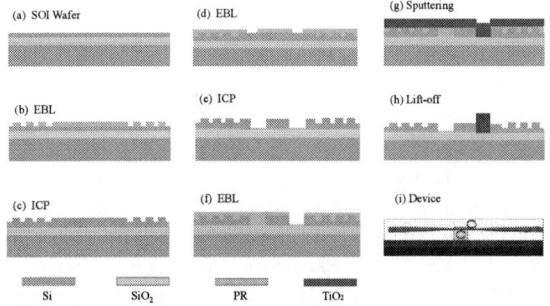

Figure 4: Fabrication process of the CMRR device. EBL: E-Beam Lithography; ICP: Inductively Coupled Plasma; PR: Photoresist

Experimental Setup

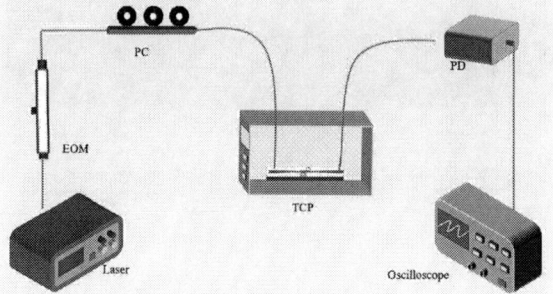

Figure 5: Schematic diagram of the experimental setup. PC: Polarization Controller; EOM: Electro-Optic Modulator, TCP: Temperature control Platform.

Figure 5 shows the experimental setup for temperature monitoring, the target temperature was provided by the temperature-controlled system and the temperature calibration was achieved by a commercial resistance thermometer. The system consists of a tunable laser (NEW PORT TLB-6728), an electro-optical modulator (THORLABS), a photodetector (NEW PORT 0901), and an oscilloscope (KEYSIGHT DSOX3034T). The tunable laser with an output from 1520 nm to 1570 nm was used to probe the sensor. An optical fiber-based polarization controller (PC) is used to adjust the polarization of the light from the laser. And the source was coupled into the waveguide in TE mode through the polarization controller. Finally, the output signal from the waveguide was detected by the photodetector and recorded by the oscilloscope to obtain the transmission spectrum.

RESULTS AND DISCUSSION

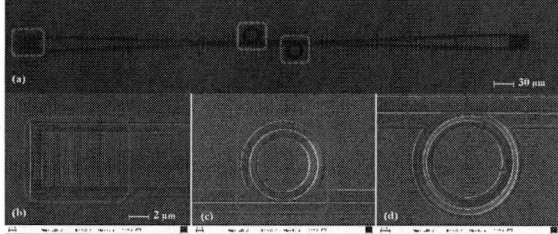

Figure 6: (a) Optical microscope image of the device; the SEM photos of (b) grating coupler(c) sensing ring and (d) reference ring.

Figure 6 (a) shows the optical microscope image of the fabricated device, which has a footprint of 716 μm × 45 μm. Figure 6 (b) shows the Scanning Electron Microscope (SEM) images of the grating couplers which have a period of 630 nm with a duty cycle of 50:50 to achieve the effective coupling of the optical signal between the fiber and the fabricated sensor. And the SEM images of the sensing ring and the reference ring covered by the TiO_2 are shown in Figure 6 (c) and Figure 6(d), respectively. Both of the rings have a similar diameter of 20 μm. A fine control on the TiO_2 cladding with a thickness of 500 nm was achieved through RF sputtering technique. The TiO_2 cladding is also a ring with a width of 2 μm to fully covered the 330 nm-width silicon waveguide.

The transmission spectrum of the fabricated cascaded

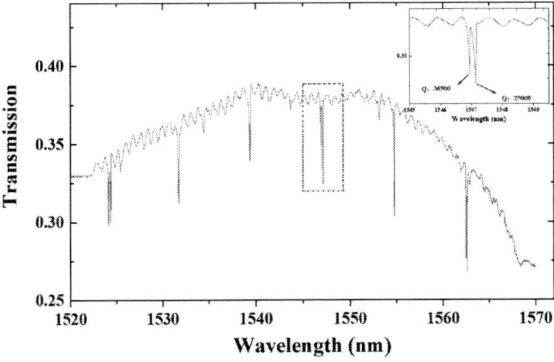

Figure 7: Transmission spectrum of the fabricated device, the inset figure shows the partial enlargement of the transmission spectrum.

MRRs was characterized. Figure 7 plots a wide-range transmission signal of the MRRs as a function of the wavelength measured at room temperature. A series of distinct absorption dips can be seen separated by the free spectral range (FSR) consistent with the simulation. Each dip consists of two resonant peaks corresponding to the cascaded MRRs. The inset in Figure 7 shows an enlarged view of one cascaded MRRs resonant wavelength with a separation of 0.21 pm corresponding to 26.3 MHz. Next, the quality factors (Q value) of the cascaded MRRs were calculated, and the results show that the Q values of the sensing ring and reference ring are 36500 and 25000, respectively. The smaller Q value of the sensing ring compared to the conventional silicon-based MRR is attributed to the optical field leakage caused by the narrow waveguide of the tapered waveguide and the reference ring. Moreover, the small Q value of the reference ring is mainly attributed to part of the optical field leaks in the TiO_2 cladding increasing the loss. However, both of these resonances with moderate Q factors also demonstrate the capability of the fabricated device to achieve high-resolution sensing with optical resonances.

CONCLUSION

This paper proposed a self-referenced temperature sensor based on a silicon CMRR. The reference ring was covered by a 500-nm-thickness TiO_2 cladding which has a negative TOC to achieve temperature insensitivity. The simulation result shows the sensitivities of the reference ring and the sensing ring are 1 pm/K and 64.34 pm/K, respectively. Utilizing the differential wavelength shifts method, the simulation result shows the sensitivity of the CMRR is 63.4 pm/K. The Q values are 36500 and 25000 for the fabricated reference ring and the sensing ring. It is a little smaller than the conventional silicon MRR due to the optical field leakage, but enough to achieve high sensitivity temperature monitoring. Moreover, through the sensing method, the stability of the sensor can be further improved since the external interference caused by the laser frequency drifts and some other noises is excluded. This paper paves a new avenue for silicon-based MRR for high-sensitivity and high-stability temperature monitoring in an extreme environment.

ACKNOWLEDGEMENTS

This work was partially supported by the National Key R&D Program of China under grant（2022ZD0208601）, the Strategic Priority Research Program of Chinese Academy of Sciences (Grant No. XDA25040100, XDA25040200 and XDA25040300), the National Natural Science Foundation of China (No. 42127807-03), Project supported by Shanghai Municipal Science and Technology Major Project（2021SHZDZX）, Shanghai Pilot Program for Basic Research - Shanghai Jiao Tong University (No. 21TQ1400203), the Oceanic Interdisciplinary Program of Shanghai Jiao Tong University (No.SL2020ZD205, SL2020MS017, SL2103), Scientific Research Fund of Second Institute of Oceanography, MNR (No.SL2020ZD205).

The authors are also grateful to the Center for Advanced Electronic Materials and Devices (AEMD) of Shanghai Jiao Tong University.

REFERENCES

[1] A. Arora, M. Esmaeelpour, M. Bernier, and M. J. F. Digonnet, "High-resolution slow-light fiber Bragg grating temperature sensor with phase-sensitive detection," Opt Lett, vol. 43, no. 14, pp. 3337-3340, 2018.

[2] C. E. Dominguez-Flores et al., "Real-Time Temperature Sensor Based on In-Fiber Fabry–Perot Interferometer Embedded in a Resin," J Lightwave Technol, vol. 37, no. 4, pp. 1084-1090, 2019.

[3] Minmin You, et al, "Chip-Scale Silicon Ring Resonators for Cryogenic Temperature Sensing", *Journal of Lightwave Technology*, vol. 38, no. 20, pp. 5768-5773,2020.

[4] T. Chen, H. Zhang, W. Lin, H. Liu, and B. Liu, "Highly Sensitive Refractive Index Sensor Based on Vernier Effect in Coupled Micro-Ring Resonators," J Lightwave Technol, vol. 40, no. 4, pp. 1216-1223, 2022.

[5] L. Y. Chiang, C. T. Wang, T. S. Lin, S. Pappert, and P. Yu, "Highly sensitive silicon photonic temperature sensor based on liquid crystal filled slot waveguide directional coupler," Opt Express, vol. 28, no. 20, pp. 29345-29356,2020.

[6] W. Weng, P. S. Light, and A. N. Luiten, "Ultra-sensitive lithium niobate thermometer based on a dual-resonant whispering-gallery-mode cavity," Opt Lett, vol. 43, no. 7, pp. 1415-1418, 2018.

[7] Q. Zhao et al., "Integrated reference cavity with dual-mode optical thermometry for frequency correction," Optica, vol. 8, no. 11, 2021.

[8] J.-M. Lee, "Influence of titania cladding on SOI grating coupler and 5 μm-radius ring resonator," Opt Commun, vol. 338, pp. 101-105, 2015.

CONTACT

*J.Q. Liu, tel: +86-021-34207209; jqliu@sjtu.edu.cn

A 0.35 mm² SYSTEM ON CHIP LEVEL DETECTOR BASED ON AN ANNULAR PMUT-ON-CMOS ARRAY

Eyglis Ledesma, Iván Zamora, Francesc Torres, Arantxa Uranga, and Núria Barniol
Electronics Engineering Department. Universitat Autonoma de Barcelona, 08193-Bellaterra, Barcelona, SPAIN

ABSTRACT

This paper presents a 0.35 mm² multi-element annular piezoelectric micromachined ultrasonic transducer array monolithically integrated on CMOS (PMUT-on-CMOS) capable of monitoring fluid level changes and evaporation rates in very small quantities and with high sensitivity in a non-invasive and non-intrusive way. Each individual PMUT consists of a 40 μm square AlScN PMUT with a resonance frequency of 8.6 MHz. Using a pulse-echo scheme, the high-frequency ultrasound system provides an interesting solution for the determination of fluid level changes and evaluation of evaporation rates in containers with low ultrasound reflection. Demonstrations of the evaporation rates of acetone and ethanol are provided as proof-of-concept. The thickness variations are 82 μm/min and 18 μm/min for acetone and ethanol, respectively; these values correspond to an evaporation rate of 3.9 kg/(m²h) (Acetone) and 0.9 kg/(m²h) (Ethanol).

KEYWORDS

MEMS-on-CMOS, PMUTs, ultrasound, fluid level sensor, evaporation rates, pulse-echo system, ring array.

INTRODUCTION

Nowadays, industrial control in the pharmaceutical, food, and chemical fields requires frequently liquid-based operations. The detection of fluid levels, leaks and evaporation rates are some of the tasks that must be constantly verified. In this context, ultrasound could be an interesting solution due to it is safe, non-invasive, and relatively inexpensive technology. Both invasive and non-invasive approaches could be implemented where the first one offers a fast solution, and the second one avoids the risk of fluid contamination.

Ultrasonic systems commercialized by [1], [2] show the capabilities of using ultrasound to estimate via non-invasive mechanism fluids levels in thin tubes and small containers for medical applications. In addition, capacitive micromachined ultrasonic transducers (CMUTs) have been recently reported as a high precision level and leak detection [3], [4]. These systems based on CMUTs require DC voltage bias, which is not suitable for low power demanding devices. In this context, Piezoelectric Micromachined Ultrasonic Transducers (PMUTs) could be an interesting alternative in this application due to their small area, low cost, and integration with CMOS circuitry providing a low power smarter solution.

In this work, we present a multi-element annular PMUT array monolithically integrated on CMOS (PMUT-on-CMOS) capable of monitoring fluid level changes and evaporation rates. Although only the results of a non-invasive approach are discussed, the presented tiny ultrasound sensor solution can work in contact mode (invasive) with different fluids such as was demonstrated in [5].

PMUT-on-CMOS SYSTEM DESCRIPTION

The PMUT-on-CMOS array consists of 10x10 PMUTs where the top electrodes are short-circuit forming five concentric irregular polygons similar to a conventional annular array (see the colored rings in Fig. 1b) with the advantage that the operation frequency operation is defined by the PMUTs [6]. The multi-element ring array has been monolithically fabricated on top of a 130 nm CMOS circuitry using the MEMS-on-CMOS technology developed by Silterra [5], [7]. In this case, the central ring (with 4 individual PMUTs) is only used to receive, and it is directly connected with a 1.5V CMOS low noise amplifier, whose topology is a voltage amplifier, and which allows the incoming ultrasound signal to be amplified with a gain of 25 dB. Each individual PMUT is an unimorph square device with a 40 μm side where the active layer consists of 0.6 μm $Sc_{9.5\%}Al_{90.5\%}N$ (AlN doped with a 9.5 % Sc concentration). The top and bottom electrodes are Al with thicknesses of 0.35 μm and 0.4 μm, respectively, and the passive layer, placed at the top of the PMUT, has 1 μm thick Si_3N_4 layer. Figure 1 shows the layout, optical image, and a schematic cross-section of the presented multi-element ring array showing the profile across two PMUTs elements.

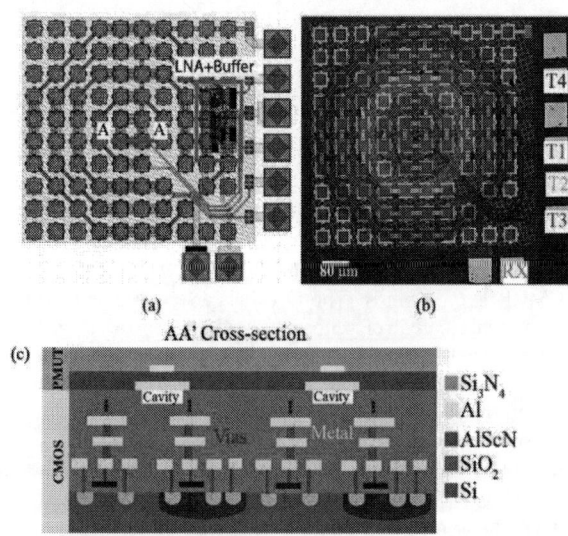

Figure 1: Multi-element PMUT-on-CMOS array (a) Layout with an LNA amplifier, (b) Optical image, and (c) Schematic cross-section.

The final system was coated with 700 μm of PDMS (10:1, Sylgard 184; ρ= 980 kg/m³ and c=1000 m/s) in order to improve the durability, and yield, and ensures waterproof protection through the total isolation of the bonding wires and an airtight seal of the cavity. In our case, the cavity is already sealed during device fabrication with the Si₃N₄ passive layer [5].

EXPERIMENTAL RESULTS

The acoustic characterization of the system was carried out through two scenarios. The first one allows characterizing the performance of the PMUT-on-CMOS array in terms of transmitting pressure and receiving sensitivity. Meanwhile, the second one was carried out in order to demonstrate the capabilities of the system to detect levels variations and obtain the evaporation rates of different fluids.

Ultrasound pulse-echo system:

The output pressure was measured in water (ρ = 1000 kg/m³ and c =1500 m/s) by placing a commercial hydrophone from ONDA (HNC-0200) at different heights along the axial direction with steps of 100 μm. The four transmitting rings were driven together with four cycles burst sine with 25 V_{pp} at 8.6 MHz using a function generator. Based on the pressure dependence in the far-field (1/z where z is the axial position), the normalized pressure with the distance can be obtained through the fitted curve, giving 28.2 kPa_{pp}*mm, being 5.5x times higher than one row of our previous linear array (2.55 kPa*2 mm) [7]. Considering the applied voltage, we obtain a sensitivity of 1.1 kPa/V at 1 mm, comparable with the PZT PMUT system reported in [8] but with an area around 2.8 times smaller.

On the other hand, the receiving sensitivity was measured by configuring the system in a pulse-echo scheme where a small Cu piece was used as a reflecting surface. The acoustic path was modified by lifting it each 50 μm which correspond to an increase of 100 μm (AP = 2*z). As in the previous experiment, the dependence of the received signal on the acoustic path was determined, giving a value of 19.7 mV_{pp}*mm. The receiving sensitivity of the PMUT-on-CMOS array can be evaluated based on both measured dependences, reaching approximately a value of 700 nV/Pa. Compared with the 1x128 PMUT array presented in [9], our system achieves a ~73 % improvement with a small area.

Fluid level detector:

The validation of the PMUT-on-CMOS array as a level detector was carried out based on the experimental set-up shown in Fig. 2 where a plastic cup is placed on the top with the testing liquid inside. As it is shown, the acoustic wave will propagate through three different media, causing reflected signals, and thus losses at each interface (PDMS/water, water/plastic, and fluid under test/air), degrading the receiving acoustic pressure and consequently the final Signal-to-Noise ratio.

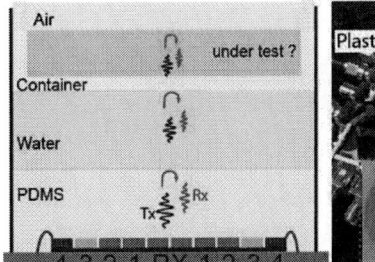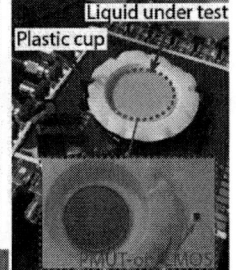

Figure 2: Experimental setup of pulse-echo measurements using different media (a) Schematic, and (b) Photograph.

Note that these losses can be minimized if the liquid under test is directly placed over the PDMS. In this experiment, the central row is used to receive and the three external rings (T4, T3, and T2) were excited with 25 V_{pp} four cycles at 8.6 MHz.

Figure 3 shows the reflecting signal coming from the different surfaces-involved in the measure. The initial time-of-flight (ToF) of 10 μs (green curve) represents an acoustic path (AP) of 14.3 mm (6.45 mm thick of water and 700 μm thick of PDMS) and it can be obtained from Eq. 1 where h_{PDMS} and c_{PDMS} are the thickness and sound velocity of PDMS, and h_{WATER} and c_{WATER} are the thickness and sound velocity of water, respectively. In this case, there is a total reflection at the water/ air interface due to the high acoustic impedance mismatch (Z_{Water}=1.5 MRayls and Z_{air}=345 Rayls).

$$AP = 2*ToF = 2*(h_{PDMS}/c_{PDMS}+h_{WATER}/c_{WATER}) \quad (1)$$

When the plastic container is placed into the water, it causes it spills out a bit until the water thickness is about 6 mm (blue curve). The temporal response shows two closer echoes due to the plastic container's thickness (~300 μm). The first maximum appears at the water/plastic interface, and, the second one is due to the plastic/air interface. The peak-to-peak amplitude decreases as a consequence of the new acoustic impedance and losses at these interfaces. Finally, once the container is filled, in this case with acetone (ρ = 784 kg/m³ and c = 1170 m/s), an echo coming from the fluid/air interface is detected (red curve), which based on the time-of-flight, an acetone thickness of 2 mm is obtained.

In order to obtain the sensitivity of the system to detect level variations with time, acetone and ethanol (ρ = 789 kg/m³ and c = 1160 m/s), were chosen and the received signal by the central ring was acquired in a 1-minute slot time. Figure 4 shows the acquired temporal response at three different times only for acetone: (a) initial measurement (reference signal, red curve), (b) after 5 min (magenta curve), and (c) after 13 minutes (blue curve), clearly showing how the time of flight decreases as the waiting time increases. This decrease in the time-of-flight corresponds to a 0.9 mm thickness reduction.

978-1-6654-9309-3/23 $31.00 © 2023 IEEE

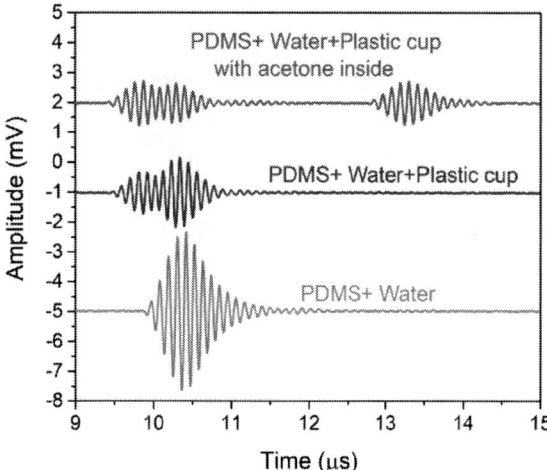

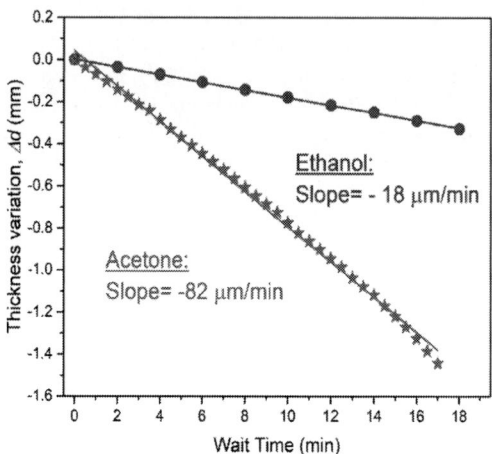

Figure 3: Temporal responses from the different experiments: green curve (PDMS+Water), echo is produced at the water-air interface; blue curve (PDMS+Water+Plastic cup), first echo is produced at the water-plastic cup interface and second echo at the plastic-air interface; red curve (PDMS+Water+Plastic cup+Acetone), first echo corresponds to water-plastic cup, second echo to plastic cup-acetone and third echo to acetone-air interface. Curves green, blue, and red are arbitrarily shifted for better visualization.

From these measurements, the cross-correlation between reference signal and the echo signal was used to obtain the difference in time between them, Δt, and from it, the estimation of the thickness variation, Δd, is determined ($\Delta d = c*\Delta t/2$, where c is the sound speed, 1170 m/s in acetone and 1160 m/s in ethanol). Figure 5 shows the experimental results for several waiting times for acetone and ethanol. Taking the slope of a linear fitting, the thickness variations are 82 μm/min and 18 μm/min for acetone and ethanol, respectively. These values ensure accurate control of the variation of the fluid level in micrometer order with slot times of a few minutes.

Figure 5: Experimental measurements of thickness variation versus waiting time, considering Acetone (red) and Ethanol (blue).

Based on these results and the density of the fluids, the evaporation rate can be obtained, which allows us to know how the mass of the liquid evaporates per unit of time in our small recipient. Equation 2 shows the evaporation rate, ER, as a function of the density (ρ), and thickness variation (slope, extracted from Fig. 5) of the fluid under test. From acetone and ethanol are obtained evaporation rates of 3.9 kg/(m²h) and 0.9 kg/(m²h), respectively. These values are in the same range as the reported ones for closed containers [10], which demonstrates the capability of the proposed system to characterize the evaporation rates of liquids.

$$ER\ (kg/(m^2h)) = slope\ (m/min) * 60 * \rho\ (kg/m^3) \quad (2)$$

Finally, Fig. 6 shows the incoming ultrasound signal with other liquids inside the plastic container (2-Propanol, methanol, and water). The temporal responses have been acquired at different time-of-flight. These experiments open the way to the characterization of different fluids inside a small plastic container without the need to directly immersed the PMUT in the liquid.

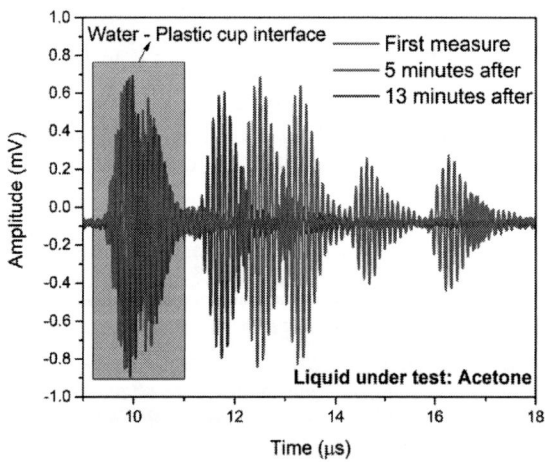

Figure 4: Temporal responses at different waiting times received by the central ring when the liquid under test is acetone.

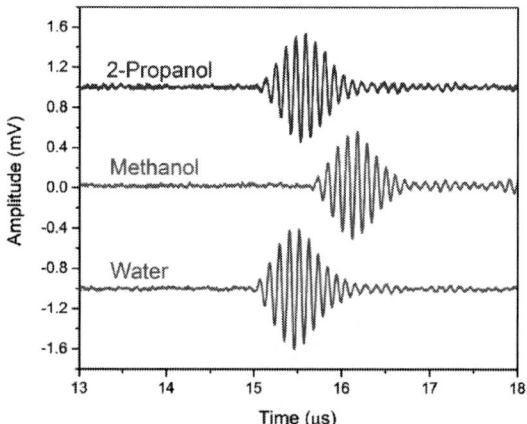

Figure 6: Temporal responses received by the central ring when the liquids under test are: (top-blue) 2-propanol, (middle-red) methanol, and (bottom-green) water at different ToF.

CONCLUSIONS

This work presents a tiny system-on-chip based on a multi-element ring PMUTs array as a level detector. The acoustic characterization as an actuator and as a sensor gives competitive sensitivities values, 1.1 kPa*mm/V as TX and 700 nV/Pa as RX, compared with the state-of-the-art. Pulse-echo experiments demonstrate the capability to use it as an accurate level detector, where the evaporation rates can be also determined through a non-invasive method. The monolithic integration allows edge processing directly at the chip level, providing smartness in a very compact system.

ACKNOWLEDGEMENTS

This research was partially funded by the Spanish Ministry of Science and Innovation and AEI with project PID2019-108270RB-I00

We would like to special acknowledge Eloi Marigó Ferrer and all the other members of the SilTerra's MEMS and SENSORS technology development team for supporting the fabrication of the PMUT-on-CMOS wafers.

REFERENCES

[1] "CeramTec." [Online]. Available: https://www.ceramtec-industrial.com.

[2] "SONOTEC GmbH." [Online]. Available: https://www.sonotec.eu.

[3] F. Merbeler, S. Anzinger, C. Bretthauer, and M. Kupnik, "Ultra-compact Clamp-on Liquid Level Sensor based on a Low-Voltage CMUT," in *2020 IEEE SENSORS*, 2020, pp. 1–4.

[4] N. Saeidi, K. Selvam, F. Tortato, M. Wiemer, and H. Kuhn, "High Precision Liquid Level and Leak Detection Based on Capacitive Micromachined Ultrasound Transducer," in *2022 Smart Systems Integration (SSI)*, 2022, pp. 1–5.

[5] E. Ledesma, I. Zamora, J. Yanez, A. Uranga, and N. Barniol, "Single-cell system using monolithic PMUTs-on-CMOS to monitor fluid hydrodynamic properties," *Microsystems Nanoeng.*, vol. 8, no. 1, p. 76, 2022.

[6] E. Ledesma, I. Zamora, A. Uranga, and N. Barniol, "Multielement ring array based on minute size pmuts for high acoustic pressure and tunable focus depth," *Sensors*, vol. 21, no. 14, p. 4786, 2021.

[7] I. Zamora, E. Ledesma, A. Uranga, and N. Barniol, "Phased array based on AlScN Piezoelectric Micromachined Ultrasound Transducers monolithically integrated on CMOS," *IEEE Electron Device Lett.*, vol. 43, no. 7, pp. 1113–1116, 2022.

[8] Z. Liu, S. Yoshida, D. A. Horsley, and S. Tanaka, "Fabrication and characterization of row-column addressed pMUT array with monocrystalline PZT thin film toward creating ultrasonic imager," *Sensors Actuators A Phys.*, vol. 342, p. 113666, 2022.

[9] S. Sadeghpour, S. V. Joshi, C. Wang, and M. Kraft, "Novel Phased Array Piezoelectric Micromachined Ultrasound Transducers (pMUTs) for Medical Imaging," *IEEE Open J. Ultrason. Ferroelectr. Freq. Control*, vol. 2, pp. 194–202, 2022.

[10] S. Akterian, "Evaluating the vapour evaporation from the surface of liquid pure organic solvents and their mixtures," *Food Sci. Appl. Biotechnol.*, vol. 3, no. 1, pp. 77–84, 2020.

CONTACT

* Núria Barniol; nuria.barniol@uab.cat

AN ALSCN PMUT-ON-CMOS SENSOR FOR MONITORING FLUIDS' DENSITY, VISCOSITY, SOUND VELOCITY, AND COMPRESSIBILITY

Eyglis Ledesma, Iván Zamora, Jesús Yanez, Arantxa Uranga, and Núria Barniol
Electronics Engineering Department. Universitat Autonoma de Barcelona, 08193-Bellaterra,
Barcelona, SPAIN

ABSTRACT

This paper presents a monolithic system-on-chip based on a single piezoelectric micromachined ultrasonic transducer (PMUT) and its CMOS analog front-end circuitry. This fabricated PMUT-on-CMOS system provides a novel solution for monitoring fluid hydrodynamic properties through very low amounts of liquid exploiting its characteristics as resonator and the acoustic wave propagation in the media. The PMUT-on-CMOS, based on an 80 μm two-port square Scandium-doped Aluminum Nitride ($Sc_{9.5\%}Al_{90.5\%}$) exhibits competitive sensitivity values (Transmitting: 9.4 kPa/V and Receiving: 3.5 V/MPa) in comparison with the PMUTs state of the art reported systems. The monolithic integration with CMOS enables on-chip further signal processing, providing smartness in a very compact system which could be beneficial in portable diagnosis devices.

KEYWORDS

PMUT-on-CMOS, AlScN, PMUT, fluid sensor, resonator, ultrasound system, pulse-echo system, acoustics.

INTRODUCTION

Hydrodynamic properties of a fluid such as density, longitudinal viscosity, compressibility, and sound velocity provide useful information about its quality and performance. Non-invasive ultrasonic monitoring and characterization of them can be a useful alternative in biomedical and industrial processes quality assessment [1], [2].

Nowadays, Micro-Electro-Mechanical systems (MEMS) have increased their use as fluids sensors taking advantage of their minute size and capabilities [3]–[5]. In this context, PMUT devices provide an attractive solution to characterize the hydrodynamic properties of fluids due to their small area, low power consumption and fabrication cost, as well as the capability to be monolithically integrated on CMOS substrate. Recently, PMUT-based microfluidic systems have been experimentally validated as fluid density sensors for low-viscosity liquids [6], [7].

In [8] we have presented a single-cell PMUT-on-CMOS system capable to provide main rheological liquid properties. In this work we provide complementary experimental characterizations of this single-cell PMUT-on-CMOS system reported in [8]. From this acoustic characterization as an actuator and as a pulse-echo system we demonstrate the excellent performance of this single PMUT-on-CMOS in comparison with the state-of-the-art, widening the usability of the system to other applications.

Figure 1: Schematic representation of the ultrasound system based on PMUT-on-CMOS. The inset image corresponds to an optical image of the PMUT device.

PMUT DESIGN AND FABRICATION

Figure 1 shows a schematic representation of the PMUT-on-CMOS system. It has been designed with the 130 nm CMOS HV technology developed by Silterra which has been reported in previous PMUT-on-CMOS ultrasound systems [8], [9], [10]. The single-cell is formed by an 80 μm square PMUT where the active layer consists of 0.6 μm AlScN (AlN with a 9.5% Sc concentration) sandwiched between three Al electrodes, two-top electrodes (0.35 μm thickness) and a common bottom electrode (0.4 μm thickness). As a final layer, all device is covered by 1 μm Si_3N_4 which acts as an elastic layer and allows the operation in liquid environment.

In the proposed ultrasound device, the output pressure is produced by the outer electrode, which is driven by a High Voltage (HV) CMOS Pulser that allows the application of 32 V monophasic pulses between it and bottom electrode. On the other hand, the incoming ultrasound wave is received by the inner electrode which is directly connected to a CMOS low noise amplifier (LNA). Additionally, two low-voltage switches are used to avoid amplifier damage during the transmission.

Flowchart fluid sensor:

The method to extract the fluid hydrodynamic properties (density (ρ), viscosity (η), sound velocity (c), compressibility (χ), and bulk modulus (K)) from this minute-size single cell acoustic system, exploits the PMUT as a flexural resonator and, also measures the acoustic wave propagation in the media (see schematic process, and main equations, in Fig. 2).

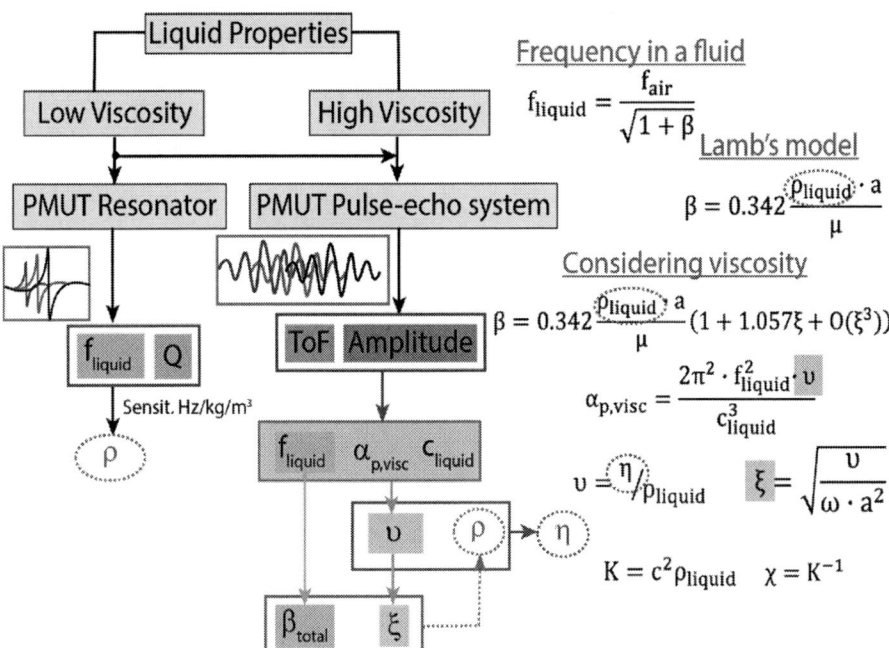

Figure 2: Schematic procedure to determine density, ρ, viscosity, η, sound velocity, c, compressibility, χ, bulk modulus, K.

The first step is to measure the frequency response (f_{liquid}), which could be measured as a resonator or in pulse-echo mode (sweeping the actuation frequency). From this measurement, the total added virtual mass (β) can be computed. However, density and viscosity affect the behavior of the PMUT as a resonator, being very difficult to extract the viscosity value from the resonator properties (frequency shift and quality factor), due to the dominant density dependence with respect to viscosity [8]. For this reason, in this case when the PMUT works as a resonator, only known low-viscosity liquids can be analyzed.

On the other hand, from pulse-echo measurements at different points information of the acoustic wave propagation can be extracted, allowing to determine experimentally the acoustic viscosity damping ($\alpha_{p,visc}$) and sound velocity (c_{liquid}). With these values and the resonance frequency in the liquid (f_{liquid}, $\alpha_{p,visc}$, and c_{liquid}), it is possible to estimate the kinetic viscosity (υ) and therefore the dimensionless parameter ξ, which allows us to see the influence of the viscosity on the added virtual mass. Density and the longitudinal viscosity are computed by using the previous estimated parameters (β, ξ, υ). Finally, compressibility and bulk modulus can be computed from sound velocity and density. All theoretical analyses are explained with details in [8].

EXPERIMENTAL RESULTS

Electrical Measurements:

Figure 3 depicts the electrical characterization in air performed between the outer and bottom electrodes, giving a resonance frequency for the first mode close to 4.2 MHz. Considering this and based on the equivalent circuit shown (top-left of Fig. 3), the static capacitance

(C_0, from the physical PMUT layout) gives 391.6 fF and the motional capacitance (C_M), resistance (R_M), and inductance (L_M) are 0.53 fF, 0.56 MΩ, and 2.66 H, respectively. These equivalent electrical components were used to compute the quality factor (Q) and the electromechanical coupling factor in air, giving 126.5 and 0.17 %, respectively.

Performance as an actuator and as a sensor:

The PMUT-on-CMOS transmitting and receiving sensitivities were measured in Fluorinert (FC-70: c = 685 m/s and ρ = 1940 kg/m^3) at 1.55 MHz. First, the output pressure was acquired by a commercial hydrophone from ONDA (HNC-0200) at 2 mm by driving the outer electrode with four 32 V monophasic pulses. Based on the hydrophone sensitivity, the output pressure gives 2760 Pa, and therefore, the normalized pressure and the transmitting sensitivity under these conditions give 136 Pa·mm/V and 9.4 kPa/V (considering the applied voltage 32V×1.27 and the Rayleigh distance R_0 = 14.5 µm).

On the other hand, to obtain the receiving sensitivity the experiment was performed in a pulse-echo scheme where the interface between liquid and air was used as a reflecting surface. An amplitude of 2.6 mV$_{pp}$ was measured when the acoustic path is close to 7.5 mm which represents a normalized value of 480V$_{pp}$·mm·V^{-1}. Taking into account this amplitude (2.6 mV$_{pp}$) and the generated output pressure at this distance (737 Pa$_{pp}$, the receiving sensitivity is around 3.5 V/MPa (2.6/0.737). Note that for this computation we have assumed no acoustics losses and perfect reflection from the interface liquid-air.

Table I summarizes a comparison between the performance of the presented single PMUT-on-CMOS and other PMUTs devices reported in the state-of-the-art that are operated in liquid. It can be stated that our PMUT-on-CMOS achieves high levels of transmitting

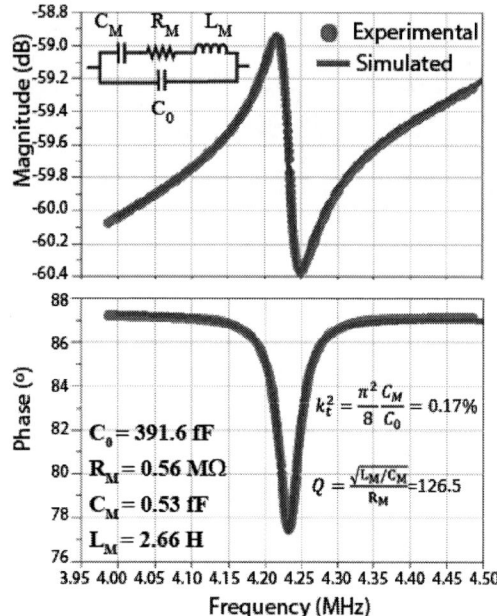

In the figure (Figure 3):

$$k_t^2 = \frac{\pi^2}{8}\frac{C_M}{C_0} = 0.17\%$$

$$Q = \frac{\sqrt{L_M/C_M}}{R_M} = 126.5$$

$C_0 = 391.6\ \text{fF}$
$R_M = 0.56\ \text{M}\Omega$
$C_M = 0.53\ \text{fF}$
$L_M = 2.66\ \text{H}$

Figure 3: Electrical PMUT characterization in air. Top: Magnitude and Bottom: Phase.

sensitivity (ST_0 in kPa/V) when compared with normalized single AlN PMUTs [11] and some PZT PMUTs [12]. In relation to the sensitivity as a sensor, it achieved a value around 3.5 V·MPa^{-1} a higher than other approaches either with AlN or PZT, and with almost the same value reached by PZT in array format [12]. Finally the PMUT-on-CMOS system can be compared for a pulse-echo system evaluating a figure of merit, defined as ST x SR. Picking the highest value (0.29 [13]) an improvement of 1.7x factor is achieved considering similar transducer area.

Fluid's characterization results:

Finally, single-cell PMUT-on-CMOS was tested as a fluid sensor using different water-glycerol mixtures as was explained in [8]. The sensitivity as a density sensor was determined by measuring the shift frequency when the fluid properties change (PMUT working as a resonator). Figure 4 shows the experimentally measured frequencies as a function of the density of the fluids. The linear fitted curve was performed to obtain the density sensitivity (excluding 100 % Glycerol) giving 482 ± 14 Hz/kg/m^3. This result demonstrates at least an enhancement of 64.7 % compared with the 250 μm diameter PMUT system presented in [6]. The bulk modulus, defined as K = c^2*ρ (where c is the sound velocity and ρ is the fluid density), is an important property of fluids that allows us to characterize their resistance to compression. Figure 5 shows the computed bulk modulus through the measured sound velocity reported in [8], demonstrating good correlation with the literature [3]–[5], and consequently validating the capabilities of the single-cell PMUT-on-CMOS system for fluid characterization.

CONCLUSIONS

In this paper, a single PMUT-on-CMOS has been characterized as an efficient ultrasound transducer capable of being used as a fluid sensor. The experimental verification shows that this tiny device, manufactured monolithically on a CMOS substrate, is an excellent candidate for a single measurement cell unit for use in microfluidic systems that require the characterization of the properties of small quantities of fluids. Additionally, the acoustic characterization as an actuator and as a sensor provides higher transmission and reception sensitivities in comparison with the state-of-the-art.

Table 1. PMUT comparison in liquid with the state-of-the-art.

Parameters	This work	2017 [11]	2021 [12]	2021 [14]	2022 [13]
Single/Array	single	6160 PMUTs arranged as 110x56 array	11776 PMUTs arranged as 1x64 array^2	6270 PMUTs arranged as 1x128 array^3	144 PMUTs arranged as 12x12 array
Size	80μm x 80μm	30μm x 43μm	300 μm pitch	160 μm diam.	85μm x 85μm
Piezoelectric Material	AlScN	AlN	PZT	PZT	PZT
Media	FC-70	FC-70	water	water	water
Frequency (MHz)	1.55	14	2.5	1.5	5
ST (kPa/V) @ distance	0.068 @ 2 mm	0.39 @ 220 μm^1	N/A	0.43 @ 30 mm	1 @ 4 mm^4
ST (kPa*mm/V) array//single PMUT	-// 0.14	0.086 // 1.5e-3	-	12.9^3 // 0.26	4 // 0.33
ST_0 array//single PMUT (kPa/V)	-//9.4	2.95//0.052	31^2//0.17	N/A	N/A
SR (V/MPa)	3.5	2	3.2	0.19	0.87
ST x SR //single PMUT	0.49	3e-3	-	0.05	0.29

^{1} Pressure emitted by one column of 56 PMUTs.
^{2} The pressure is generated by one element which is composed of 184 PMUTs connected in parallel.
^{3} The pressure is generated by one element of 49 PMUTs approximately (6270/128).
^{4} The pressure is generated by one element (1x12 PMUTs).

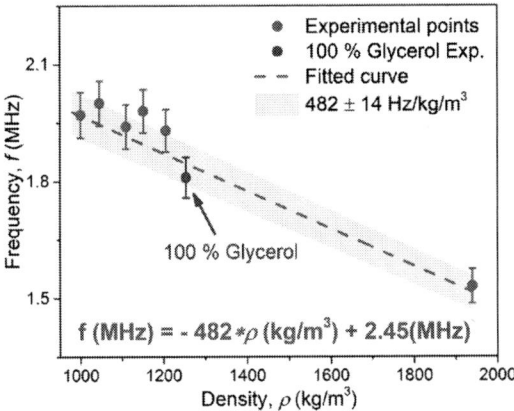

Figure 4: Frequency (MHz) vs. density (kg/m³) and the fitted curve excluding 100 % glycerol.

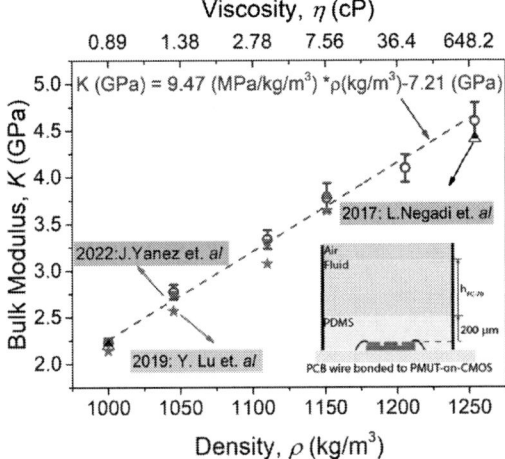

Figure 5: Bulk modulus experimental and published data in the literature. Inset: Experimental set-up.

ACKNOWLEDGEMENTS

This research was partially funded by the Spanish Ministry of Science and Innovation and AEI with project PID2019-108270RB-I00.

We would like to special acknowledge Eloi Marigó Ferrer and all the other members of the SilTerra's MEMS and SENSORS technology development team for supporting the fabrication of the PMUT-on-CMOS wafers.

REFERENCES

[1] R. Paxman, J. Stinson, A. Dejardin, R. A. Mckendry, and B. W. Hoogenboom, "Using Micromechanical Resonators to Measure Rheological Properties and alcohol content of model solutions and commercial beverages," *Sensors (Switzerland)*, vol. 12, no. 5, pp. 6497–6507, 2012.

[2] O. Cakmak, E. Ermek, H. Urey, G. G. Yaralioglu, and N. Kilinc, "MEMS based blood plasma viscosity sensor without electrical connections," in *Proceedings of IEEE Sensors*, 2013.

[3] J. Yanez, A. Uranga, and N. Barniol, "Fluid compressional properties sensing at microscale using a longitudinal bulk acoustic wave transducer operated in a pulse-echo scheme," *Sensors Actuators A Phys.*, vol. 334, p. 113334, 2022.

[4] L. Negadi *et al.*, "Effect of temperature on density, sound velocity, and their derived properties for the binary systems glycerol with water or alcohols," *J. Chem. Thermodyn.*, vol. 109, pp. 124–136, 2017.

[5] Y. Lu, M. Zhang, H. Zhang, Y. Jiang, H. Zhang, and W. Pang, "Microfluidic Bulk-Modulus Measurement by a Nanowavelength Longitudinal-Acoustic-Wave Microsensor in the Nonreflective Regime," *Phys. Rev. Appl.*, vol. 11, no. 4, p. 1, 2019.

[6] K. Roy *et al.*, "Fluid Density Sensing Using Piezoelectric Micromachined Ultrasound Transducers," *IEEE Sens. J.*, vol. 20, no. 13, pp. 6802–6809, 2020.

[7] K. Roy, K. Kalyan, A. Ashok, V. Shastri, and R. Pratap, "A PMUT Integrated Microfluidic System for Fluid Density Sensing," *J. Microelectromechanical Syst.*, vol. 30, no. 4, pp. 642–649, 2021.

[8] E. Ledesma, I. Zamora, J. Yanez, A. Uranga, and N. Barniol, "Single-cell system using monolithic PMUTs-on-CMOS to monitor fluid hydrodynamic properties," *Microsystems Nanoeng.*, vol. 8, no. 1, p. 76, 2022.

[9] I. Zamora, E. Ledesma, A. Uranga, and N. Barniol, "Phased array based on AlScN Piezoelectric Micromachined Ultrasound Transducers monolithically integrated on CMOS," *IEEE Electron Device Lett.*, vol. 43, no. 7, pp. 1113–1116, 2022.

[10] E. Ledesma, I. Zamora, A. Uranga, and N. Barniol, "Monolithic PMUT-on-CMOS Ultrasound System for Single Pixel Acoustic Imaging," in *Proceedings of the IEEE International Conference on Micro Electro Mechanical Systems (MEMS)*, 2021, pp. 394–397.

[11] X. Jiang *et al.*, "Monolithic ultrasound fingerprint sensor," *Microsystems Nanoeng.*, vol. 3, no. 1, p. 17059, Dec. 2017.

[12] A. S. Savoia *et al.*, "Design, Fabrication, Characterization, and System Integration of a 1-D PMUT Array for Medical Ultrasound Imaging," in *2021 IEEE International Ultrasonics Symposium (IUS)*, 2021, pp. 1–3.

[13] X. Jiang, V. Perrot, F. Varray, S. Bart, and P. G. Hartwell, "Piezoelectric Micromachined Ultrasonic Transducer for Arterial Wall Dynamics Monitoring," *IEEE Trans. Ultrason. Ferroelectr. Freq. Control*, vol. 69, no. 1, pp. 291–298, 2022.

[14] S. Sadeghpour, M. Ingram, C. Wang, J. D'Hooge, and M. Kraft, "A 128x1 Phased Array Piezoelectric Micromachined Ultrasound Transducer (pMUT) for Medical Imaging," in *2021 21st International Conference on Solid-State Sensors, Actuators and Microsystems (Transducers)*, 2021, pp. 34–37.

CONTACT

* Núria Barniol; nuria.barniol@uab.cat

AUTO-POSITIONING AND HAPTIC STIMULATIONS VIA A 35 MM SQUARE PMUT ARRAY

Wei Yue[1†], Yande Peng[1†], Hanxiao Liu[1†], Fan Xia[1], Fanping Sui[1], Seiji Umezawa[2], Shinsuke Ikeuchi[2], Yasuhiro Aida[2] and Liwei Lin[1]

[1]Department of Mechanical Engineering, University of California, Berkeley, USA, and
[2]Murata Manufacturing Co., Ltd., Japan
[†]Wei Yue, Yande Peng and Hanxiao Liu contributed equally to this work.

ABSTRACT

This work reports an engineered platform for the non-contact haptic stimulation on human skins by means of an array of piezoelectric micromachined ultrasonic transducer (pMUT) via the beamforming scheme. Compared to the state-of-art reports, three distinctive achievements have been demonstrated: (1) individual single pMUT unit based on lithium niobate (LN) with measured high SPL (sound pressure level) of 133 dB at 2 mm away; (2) a beamforming scheme simulated and experimentally proved to generate ~2.3x higher pressure near the focal point; and (3) the combination of auto-positioning and haptic stimulations on volunteers with the smallest reported physical device size to achieve haptic sensations. As such, this work could have practical applications in the broad areas to stimulate haptic sensations, such as AR (Augmented Reality), VR (Virtual Reality), and robotics.

KEYWORDS

Haptic Sensation, LN pMUT, Beamforming, Ultrasound

INTRODUCTION

Haptic feedback is an important tool for the human-machine interfaces in areas such as mobile phones, entertainment devices, … etc. [1] By applying force or displacement signals to human's mechanoreceptors, haptic sensation can effectively deliver information and bring different experiences in addition to the conventional visual and auditory ways currently used in the AR & VR systems. Recently, high intensity acoustic actuation has attracted attentions due to several key advantages. First, the acoustic actuation provides a non-contact way for haptic sensation, which does not require physical contact and can avoid hygiene issues. Second, the acoustic actuation can achieve high spatial resolutions when compared to other methods based on mechanical contacts [2] to enable more localized sensation patterns and provide various excitation features.

Several prior reports have demonstrated the feasibility of acoustic haptic sensations utilizing bulk piezoelectric transducers with a very large form-factor and high power-consumption [3,4]. Recent works have proposed the use of small-size and large array of piezoelectric micromachined ultrasonic transducers (pMUT) [5,6] with potential benefits in terms of size and ease of integration with other consumer electronics. However, the low piezoelectric coefficient of AlN used in the prior works produces low acoustic energy utilization efficiency to generate limited output pressure and vague sensation even with a large area of 256 mm².

Here, we report a pMUT array with an area of only 35 mm² by using lithium niobate (LN) as the piezoelectric

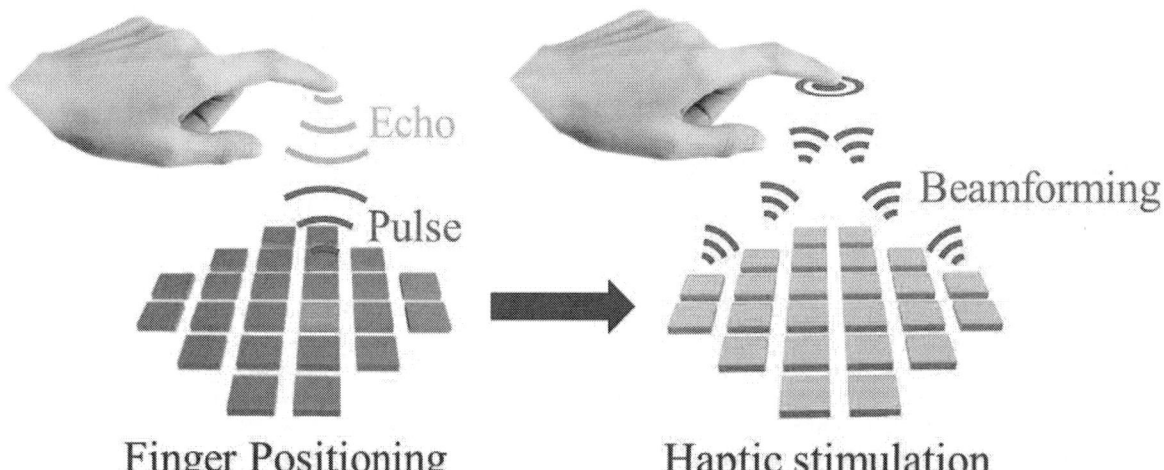

Figure 1: Illustration of the two-step haptic stimulation system. The array can locate the finger first, and generate haptic sensation with the beamforming scheme. In the finger position detection step, a transmitting pMUT unit shown as the green color emits a pulse signal, which is reflected by the surface of the finger. The echo signal can be received by the receiving units which are shown in the orange color. The position information can be analyzed with the obtained time-of-flight (ToF) information. Next, all pMUT devices are actuated with the beamforming scheme, to stimulate haptic sensations on the surface of the finger.

material. The single pMUT unit shows a high sound pressure level (SPL) of 133 dB at a distance of 2 mm benefiting from the better material property of LN due to the combination of piezoelectric coefficients (i.e., d_{31} and d_{11}) [7]. In addition, the pMUT array platform can detect the target and conduct beamforming based on the location to better utilize acoustic energy. As such, a high SPL of 156 dB has been achieved 2 mm away and haptic sensations have been experiences by volunteers at different heights. To the best of our knowledge, this is the smallest reported physical size to achieve haptic sensation.

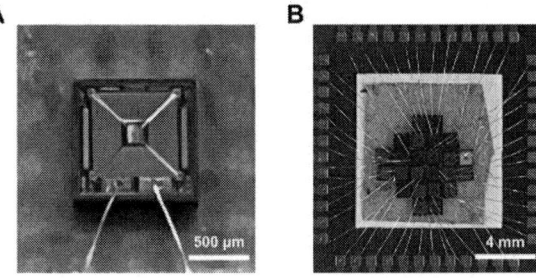

*Figure 2: **A)** The optical image of a single pMUT element. The size of the pMUT device has a rectangular shape with the size of 1.1 mm by 1.2 mm. **B)** The optical image of the 24-element array. The pMUT elements are placed on a PCB board, and the center-to-center distance between adjacent elements is 1.2 mm. The total area of the array is ~35 mm² and each element is connected to two different electrodes with aluminum wires.*

METHOD

The two-step haptic stimulation system is illustrated in **Fig. 1**, including the finger position sensing part and the haptic stimulation part. In the finger position sensing step, a transmitting pMUT emits a pulse wave, which is reflected by the finger to form an echo wave. The time-of-flight (ToF) information is extracted from the receiving pMUTs to analyze the location of the finger. This is accomplished by comparing the receiving signals without and with the finger. The haptic stimulation step is followed by exciting pMUT elements in the array with different phases for the resulting beamforming at the finger position. As such, the acoustic power is concentrated on the surface of the finger to stimulate haptic sensations. The excitation frequency of the acoustic wave is modulated to 200 Hz by pulse width modulation (PWM), for the optimal sensation results for mechanoreceptors in the skin [8]. The duty cycle in the PWM is set as 50%, which means the signal will be on for 2.5 ms and off for 2.5 ms in a cycle of 5 ms. Although the vibration frequency of the pMUT is still at 50.64 kHz, the modulated signal shows the behavior of a low frequency excitation of 200 Hz.

The optical photo of a single pMUT device used in this work is shown in **Fig. 2A**. The LN-based pMUT device has a rectangular shape with the size of 1.1 mm by 1.2 mm. AC signal can be applied to the two separate terminals to generate the desired ultrasound outputs. 24 identical pMUT devices are assembled as shown in **Fig. 2B** as the prototype. The pitch size between adjacent elements is 1.2 mm and the whole array has a total area of 35 mm². Each device has

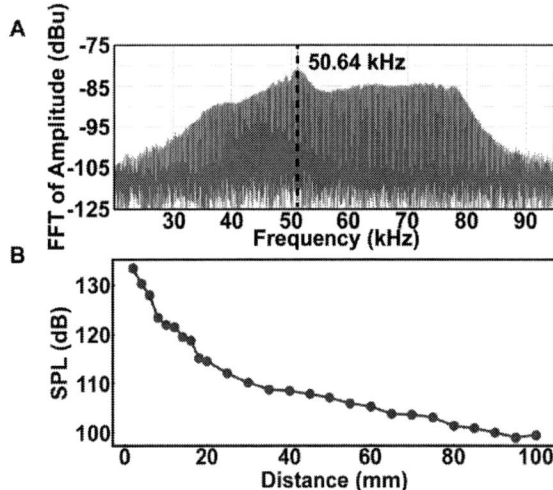

*Figure 3: **A)** Fast Fourier transform (FFT) of the measured output pressure from one device in air. The resonance frequency is observed at 50.64 kHz, which leads to the strongest pressure output. **B)** Measured sound pressure level (SPL) vs. distance curve at 50.64 kHz from one device. The SPL reduces from 133 dB at 2 mm to 100 dB at 10 cm.*

two aluminum bonding wires to electrodes on the PCB board and different signals can be applied on different pMUT devices to enable independent control.

RESULTS AND DISCUSSION
Characterization of a single pMUT

The characterization results of a single pMUT device are shown in **Fig. 3**. The resonance frequency is determined by applying a chirp signal with varying frequencies on the pMUT and measuring the pressure output at certain point by the microphone (Bruel & Kjaer 4136). The applied chirp signal plays the role of a frequency sweeping signal, containing equally strong components from different frequencies (20 kHz ~ 95 kHz). Therefore, the output sound wave generated by the pMUT can show frequency-varying patterns. The FFT of the amplitude obtained by the microphone is shown in **Fig. 3A**, from which the resonance

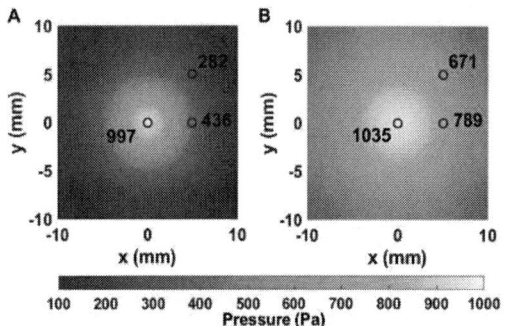

*Figure 4: Experimental and simulation results of pressure in a plane 5 mm above the array system: **A)** without and **B)** with beamforming. The circle symbols are experimental results at the three-dimensional Cartesian coordinates of (0 mm, 0 mm, 5 mm), (5 mm, 0 mm, 5 mm) and (5 mm, 5 mm, 5 mm), respectively.*

frequency is observed at 50.64 kHz. **Fig. 3B** shows results of the SPL of a single pMUT driven at resonance with respect to distance. The SPL declines monotonically with the distance, due to the spreading and attenuation [9]. At 2 mm above the surface, the SPL can reach ~133 dB, which is strong considering the area of a single unit is 1.32 mm^2.

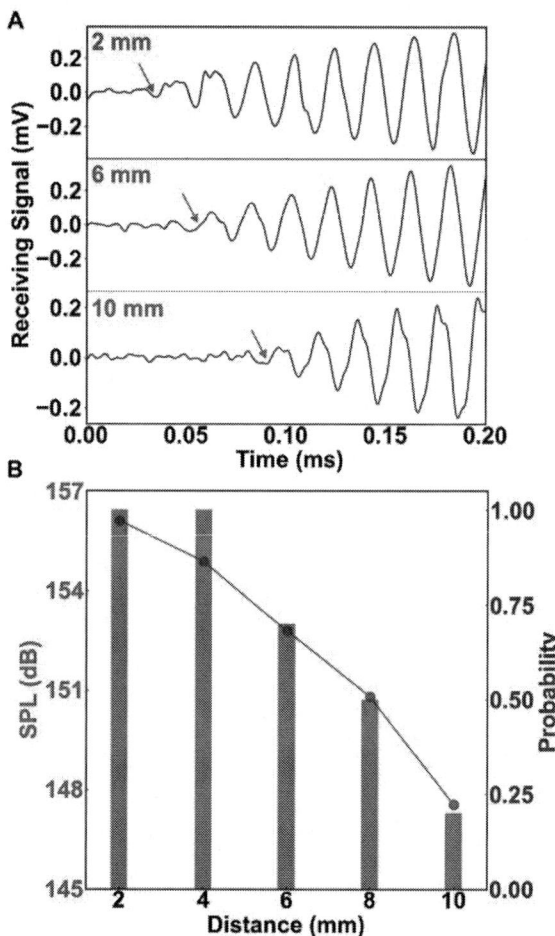

*Figure 5: **A**) The experimental setup for the finger position sensing test. The pulse signal emitted by the transmitting pMUT unit is reflected by the finger, and the echo signal is received by receiving pMUTs. The receiving signal is amplified by the charge amplifier first and collected by the oscilloscope for analysis. **B**) The optical photo showing the setup. The position of the finger can be located by the pulse-echo scheme. The beamforming scheme is utilized afterwards to stimulate haptic sensations by controlling the pMUT elements independently.*

Beamforming results

Low acoustic energy utilization is an important reason that a large array is needed for high pressure outputs. Here, we utilize the beamforming technique to concentrate the acoustic power by applying individual pMUT devices with input signals at the same frequency but different phase angles in the array to significantly increase the SPL at the focal point. The improvement on the 24-element array can be observed in **Fig 4**, showing the difference in the pressure output at 5 mm above the surface. The simulation results of the pressure output generated by the array without and with beamforming scheme together with several experimental measurements are shown in **Fig 4A** and **Fig 4B**, respectively. At each position in the plane, the pressure shown in **Fig 4B** represents the maximum pressure output obtained with the beamforming technique. Obviously, the beamforming technique strengthens the pressure output, especially in the region far from the center. Because at the location far from the center, the distance from this point to different pMUT elements can be different, which also leads to totally different phases of sound waves. When two sound waves have phase difference close to 180°, they can

counteract with each other and generate negligible pressure output. At the three-dimensional Cartesian coordinates of (0 mm, 0 mm, 5 mm), (5 mm, 0 mm, 5 mm) and (5 mm, 5 mm, 5 mm), the measured pressure output increases by 4%, 81% and 138%, from 997 Pa, 436 Pa and 282 Pa to 1035 Pa, 789 Pa and 671 Pa, respectively.

*Figure 6: **A**) Time-of-flight results at three distances of 2, 6 and 10 mm above the system. The beginning point of the echo signal indicates the distance of the finger. **B**) Measured SPL (red color dots) and human sensation probability results (blue color histogram) at 5 different distances of 2, 4, 6, 8, and 10 mm. The SPL reduces from 156 dB at 2 mm to 147 dB at 10 mm. All 10 volunteers report the haptic sensation 4 mm or closer to the system and two volunteers still have the sensation at 10 mm away.*

Experimental setup

The experimental setup diagram of the finger-position sensing experiment is illustrated in **Fig. 5A**. The transmitting and receiving pMUTs are assembled in the same board. When the target finger moves close to the system, the pulse signal sent by the transmitting pMUT unit will produce an echo signal from the surface of the finger to be received by each receiving pMUT unit due to different acoustic impedance. A charge amplifier is utilized to amplify the receiving signal before it is collected by the

978-1-6654-9309-3/23 $31.00 © 2023 IEEE

oscilloscope. The finger position is then analyzed and calculated from the receiving signal based on the Time-of-Flight (ToF). **Figure 5B** shows an optical image of the experimental setup. The 24-element array is fixed on a PCB board and each pMUT unit can be controlled independently with pre-defined phase angle by applying different input signals. Once the position of the finger is detected, pMUT units can be driven correspondingly to concentrate the acoustic power on the surface of the finger to generate strong haptic sensations.

Volunteer test results

The functionality of the two-step haptic stimulation system is demonstrated by volunteer tests with a total of 10 volunteers. They are asked to put their fingers at different locations, and the echo results at different distances of 2 mm, 6 mm and 10 mm are shown in **Fig. 6A**. The beginning point of the echo signal indicates the distance of the finger, which can be calculated based on the ToF. The echo signals can be clearly recognized so that the finger position can be precisely measured. This is accomplished by subtracting the background noises from the raw echo signal. When there is no finger over the array, the signal collected is defined as the background noises, including environmental noises such as wires and the terminals which may interfere the pulse-echo scheme. When the finger is present, echoes from these noise sources are still present and they will appear in the raw echo signals. By subtracting these noises, the signals representing the finger can be clearly presented. After identifying the finger position, the array is driven to stimulate haptic sensations by beamforming on the surface of the finger. The phase angles for different pMUT devices can be calculated with given finger positions and assigned separately to different elements. The haptic sensation at different distances from 2 mm to 10 mm are recorded and shown in **Fig. 6B**. An increased human sensation is observed as the finger is moving closer to the system as all volunteers report haptic stimulations 4 mm or closer to the system. Even at 10 mm away, two volunteers report clear sensations. The SPL at different distances by the beamforming scheme is also measured in **Fig. 6B**, reducing from 156 dB at 2 mm to 147 dB at 10 mm.

CONCLUSION

This paper demonstrates a two-step haptic stimulation system for both auto-positioning and haptic stimulation via a 24-element pMUT array. In the finger position sensing step, the pulse signal emitted by the transmitting pMUT can be reflected by the surface of the finger. The echo signal received by the receiving pMUTs is analyzed and the ToF information is used to determine the finger position. In the haptic stimulation step, the beamforming scheme is utilized to concentrate the acoustic power on the surface of the finger for haptic stimulations. With the beamforming scheme, the pressure output is significantly increased by up to ~138% at some points to make it possible to generate strong haptic sensations with a small array of only 35 mm². The SPL versus distance curve of the array is measured by a microphone, and the SPL reduces from 156 dB at 2 mm to 147 dB at 10 mm. All 10 volunteers report strong haptic sensations 4 mm or closer to the system, and two of them can still feel the sensations at 10 mm away. This proves the functionality of our proposed two-step haptic stimulation system, which can fine potential applications in mid-air haptic stimulations such as AR/VR.

ACKNOWLEDGEMENT

This work is supported in part by an NSF grant (ECCS- 2128311) and the membership support of Berkeley Sensor and Actuator Center. The authors would also like to thank Mingzheng Duan from UC Berkeley for his help in the PCB board design.

REFERENCES

[1] X. Yu, Z. Xie, Y. Yu, J. Lee, A. Vazquez-Guardado, H. Luan, J. Ruban, X. Ning, A. Akhtar, D. Li, B. Ji, Y. Liu, R. Sun, J. Cao, Q. Huo, Y. Zhong, C. M. Lee, S. Y. Kim, P. Gutruf, C. Zhang, Y. Xue, Q. Guo, A. Chempakasseril, P. Tian, W. Lu, J. Y. Jeong, Y. J. Yu, J. Cornman, C. S. Tan, B. H. Kim, K. H. Lee, X. Feng, Y. Huang, J. A. Rogers, "Skin-integrated wireless haptic interfaces for virtual and augmented reality", *Nature*, vol. 575, pp. 473–479, 2019.

[2] M. Ito, D. Wakuda, S. Inoue, Y. Makino, H. Shinoda, "High spatial resolution midair tactile display using 70 kHz ultrasound", in *Proc. EuroHaptics Part I LNCS 9774*, 2016, pp. 57–67.

[3] https://www.ultraleap.com/haptics/

[4] T. Carter, S. A. Seah, B. Long, B. Drinkwater, S. Subramanian, "UltraHaptics: multi-point mid-air haptic feedback for touch surfaces", in *Proc. UIST*, 2013, pp. 505-514.

[5] M. Billen, E. M. Ferrer, M. S. Pandian, X. Rottenberg, V. Rochus, "Mid-Air Haptic Feedback Enabled by Aluminum Nitride PMUTs", in *2022 IEEE 35th International Conference on Micro Electro Mechanical Systems Conference (MEMS),* January 2022, pp. 247-250.

[6] S. Pala, Z. Shao, Y. Peng, L. Lin, "Ultrasond-induced haptic sensations via PMUTS", in *2021 IEEE 34th International Conference on Micro Electro Mechanical Systems (MEMS)*, January 2021, pp. 911-914.

[7] F. Pop, B. Herrera, M. Rinaldi, "Lithium Niobate Piezoelectric Micromachined Ultrasonic Transducers for high data-rate intrabody communication", *Nature communications*, vol. 13, pp.1-12, 2022.

[8] P. J. J. Lamore, H. Muijser, C. J. Keemink, "Envelope detection of amplitude - modulated high - frequency sinusoidal signals by skin mechanoreceptors", *J. Acoust. Soc. Am. or JASA*, vol. 79, pp. 1082-1085, 1986.

[9] R.J. Przybyla, S.E. Shelton, A. Guedes, I.I. Izyumin, M.H. Kline, D.A. Horsley, B.E. Boser, "In-air rangefinding with an AlN piezoelectric micromachined ultrasound transducer", *IEEE Sens. J.*, vol. 11, pp. 2690-2697, 2011.

CONTACT

*W. Yue; +1-510-984-8328; wei_yue@berkeley.edu
*Y. Peng; +1-510-697-6951; yande_p@berkeley.edu
*H. Liu; +1-510-345-9811; liuhanxiao@berkeley.edu

BODY FORCE BASED DROPLET EJECTION BY GHZ ACOUSTIC MICRO-TRANSDUCER

Haitao Zhang, Yangchao Zhou, Menglun Zhang, Wenlan Guo, Chen Sun,
Xuexin Duan*, and Wei Pang*

State Key Laboratory of Precision Measuring Technology and Instruments, Tianjin University,
Tianjin, CHINA

ABSTRACT

This paper proposes a novel nozzle-free acoustic droplet ejection system based on GHz acoustic MEMS transducer. A strong body force could be generated to eject droplets when the GHz pulse signal is applied to the transducer. Compared with other acoustic ejection methods, the body force-based method generates focused radiation pressure without additional acoustic focus structures. Given its small size, nozzle-free, high acoustic radiation pressure and compatibility with complementary metal oxide semiconductor (CMOS) fabrication, the acoustic droplet ejection system has great potential in ink-jet printing, bioprinting and drug delivery applications.

KEYWORDS

Acoustic droplet ejection, GHz acoustic transducer, nozzle-free, body force, printing

INTRODUCTION

Droplet ejection has been widely used in bioprinting and biofabrication [1]. Existing ejection devices involve mechanic [2], electrical [3], optic [4] and acoustic [5] based approaches. Mechanic ejection may cause damage to cells because of the mechanical shear force. And they depend heavily on the nozzle structure which faces the challenge of clogging. The electrical and optic methods result different degrees of damages as well. Some studies have shown that acoustic flow fields affect little on cell activity [6]. The ejection system based on acoustics becomes gradually popular in bioprinting. kHz waves can propagate long distance in the liquid, which suits to accomplish the non-contact ejection. And the acoustic feedback system can measure the height of liquid in real time to adjust the power applied on the device, which avoids the generation of satellite droplets [7]. MHz surface acoustic waves travel along the surface of the medium, which result in the dispersive pressure in the liquid. In the past decades, researchers have designed devices to focus the surface acoustic waves, reducing the applied power and controlling the droplet diameter [8]. However, its compact integration is a serious issue. Similar to surface acoustic waves, Lamb wave devices requires liquid height control [9].

In this paper, we propose a novel droplet ejection system based on body force, different from conventional acoustic based methods requiring focus structures. The body force generating transducer is fabricated by MEMS technology and compatible with CMOS process. Our system has potential in bioprinting, inkjet printing and drug delivery applications.

DEVICE DESIGN AND FABRICATION

To more effectively utilize the energy, we designed the transducer working mainly in the longitudinal mode. As the core of transducer, the piezoelectric layer design including piezoelectric coupling coefficient, dielectric constant, and longitudinal wave velocity need to be optimized. Common piezoelectric materials include aluminum nitride, zinc oxide and lead zirconate-titanate (PZT), among which aluminum nitride is preferred in this work. The alternative layers of molybdenum and silicon dioxide were utilized as the Bragg reflectors, which the former is a high acoustic impedance material but the latter is low. The acoustic impedance ratio of the two is about 4.2:1. The thickness of each layer, d, was a quarter of the acoustic wavelength, λ. The thickness d is given by:

$$d = \frac{\lambda}{4} \tag{1}$$

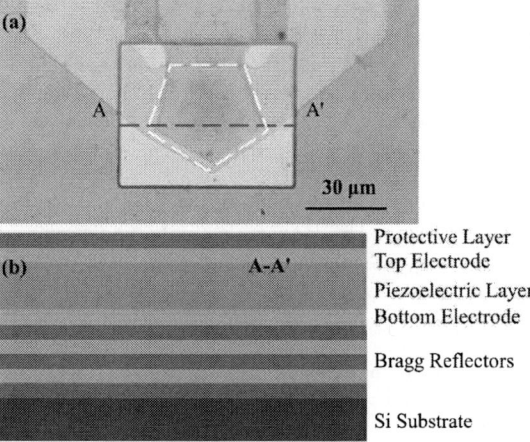

Figure 1: (a) Enlarged view of the acoustic transducer. The white line represents its vibration part. (b) Schematic sectional view of section A-A'.

The acoustic transducer was fabricated by MEMS process. Fig. 1a shows its top view. The pentagonal region is its vibration part. The transducer is composed of four parts: silicon substrate, Bragg reflective layer, piezoelectric structure, and protective layer (Fig. 1b). Molybdenum (0.64 μm) and silicon dioxide (0.65 μm) were alternately deposited as Bragg reflectors, which has different acoustic impedance coefficients to limit the loss of acoustic wave to the substrate. 0.17 μm and 0.15 μm thick molybdenum layers were used as the top electrode and bottom electrodes, respectively, between which there was the 1.1 μm thick aluminum nitride. The 0.2 μm thick silicon dioxide was deposited on top as the protective layer.

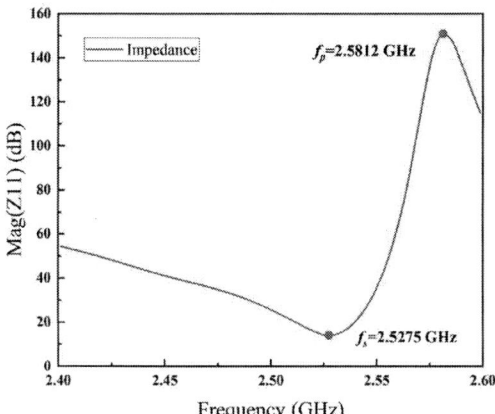

Figure 2: Impedance-frequency curve of transducer in air. f_s represents the series resonator frequency, f_p represents parallel resonator frequency.

The transducer was then connected to the evaluation board using gold wires. The impedance was measured by the network analyzer (E5061B, Agilent) in air, showing a high-quality factor (Fig. 2).

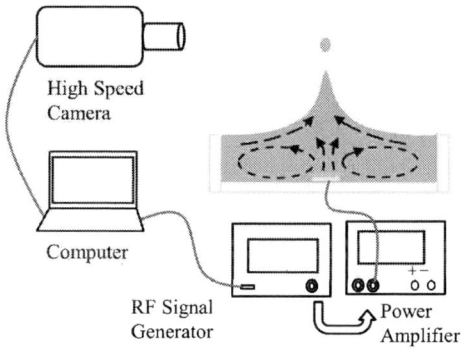

Figure 3: The diagram of acoustic droplet ejection system.

Fig.3 shows the diagram of acoustic droplet ejection system based on the GHz transducer. A sine signal was generated by the radio frequency (RF) signal generator (SSG-4000HP, Mini-Circuits) with input power from -50 dBm to 20 dBm. The signal was amplified by a power amplifier (ZHL-5W-422+, Mini-Circuits), which amplifies the power to 22 dB. Ultrapure water was dropped on tip of the transducer. To observe the change of liquid surface and the process of the ejection, a high-speed camera (FASTCAM SA-X2, Photron) was utilized and the video was captured at 12500 fps. The RF signal generator and the high-speed camera were controlled by the computer.

THEROY

By applying GHz electrical signal to the transducer, the high-frequency acoustic waves are generated due to the inverse piezoelectric effect and then coupled to the liquid. The acoustic waves will decay rapidly in the liquid and

induce fluid flow [10, 11]. The attenuation coefficient β is given by:

$$\beta = \frac{\omega^2}{2\rho c_L^3}\left(\frac{4}{3}\mu + \eta\right) \tag{2}$$

where ω is the angular frequency of the acoustic waves, ρ is the liquid density, c_L is the sound speed in the liquid, μ and η are the viscosity and bulk viscosity of the fluid, respectively. As the attenuation coefficient scales with square of waves frequency, the acoustic waves generated by GHz transducer attenuate in a rather short distance (around tens of micrometers).

The vibration displacement amplitude of the transducer in the z-axis is given by:

$$\xi(z) = u(x, y)e^{-\beta z} \tag{3}$$

where $u(x, y)$ is the vibration displacement magnitude at the direction of z, and z is the height from the transducer surface.

Due to the nonlinear attenuation of propagating acoustic waves, a finite time-averaged momentum flux is generated, exerting a body force on the liquid to drive a directed fluid motion along the propagation path of the acoustic waves [12]. The expression for the body force is:

$$F(x, y, z) = 2\rho\beta\omega^2 u^2(x, y)e^{-2\beta z} \tag{4}$$

The body force exponentially decays along z-axis. Higher-frequency acoustic waves provide stronger body force. The body force is also related to the vibration displacement amplitude of z-axis. Higher input power will induce larger vibration amplitude in the z axis, and result the stronger body force. The body force causes the liquid to flow, which overcomes the surface tension and forms the droplets. Fig. 3a and b show the vortex generated by GHz transducer and the schematic of ejection.

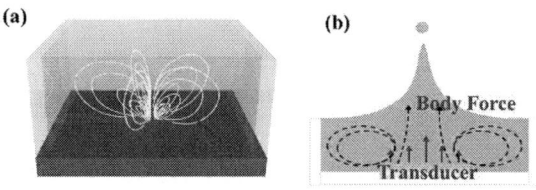

Figure 4: (a) Schematic of vortex generated by GHz transducer; (b) Schematic diagram of ejection.

RESULTS AND DISCUSSION

Before the experiment, hydrophilic treatment was applied to the transducer surface to obtain relatively low liquid level. We dropped a water droplet on the transducer surface and tested the series resonant frequency using network analyzer. The resonant frequency was 2.5063 GHz, which was slightly less than the result tested in air but with much lower factor (Fig. 5). A high-speed camera was utilized to observe the process of droplet ejection at 12500 fps. With a pulse RF signal (frequency=2.5063

GHz; duration=1 ms; period=10 ms; power=0.53 W), the obtained droplets were 170 μm in diameter under the interaction of surface tension and body force (Fig. 6a-c). Since the direction of the body force is perpendicular to the surface of transducer, the force used for ejection is more effective and concentrated.

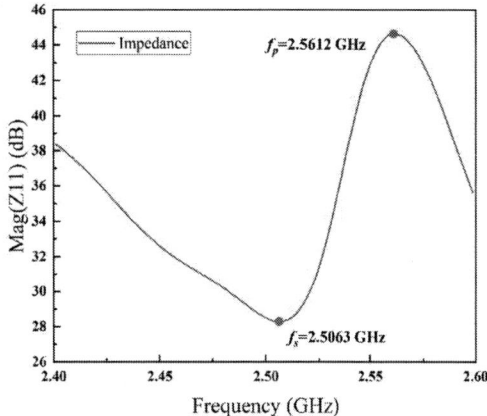

Figure 5: Impedance-frequency curve of transducer in water. fs represents the series resonator frequency, fp represents parallel resonator frequency.

We studied the effect of input power on the droplet ejection. The liquid needle was formed without droplet ejection when the power was low (0.3 W) (Fig. 6d). Until the power reached 0.5 W, the droplet ejection occurred (Fig. 6e). With the power gradually increasing, not only droplets but also satellite droplets could be observed which is undesirable during droplets ejection (Fig. 6f). In this experiment, body force-based acoustic droplet ejection system could eject single droplets at 0.5 W, lower than surface acoustic waves (2-5 W) and Lamb waves (5-8 W).

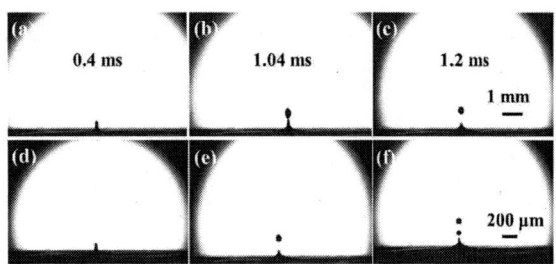

Figure 6: (a-c) Snapshot of droplet ejection process with pulse time of 1 ms (d-f) Different states with power increased. Unsuccessful droplet ejection with low power (d). Successful droplet ejection with appropriate input power (e). Satellite droplet ejection with high input power (f). Scale bar is 1 mm in (a-c) and 200 μm in (d-f).

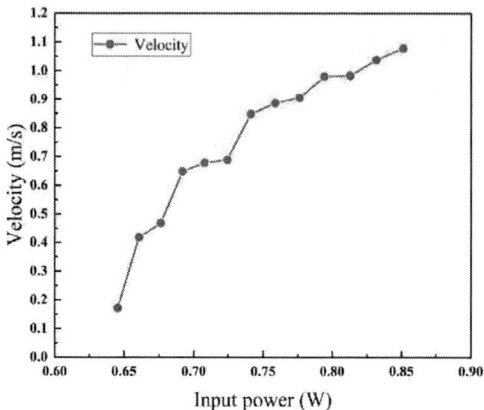

Figure 7: The relationship between input power and initial velocity of droplet. The duration and period of input power were 100 μs and 1 ms respectively.

In addition, we investigated the relationship between the input power and droplet initial velocity. According to the experiment above, the power was set to eject the droplets without satellite droplets. As shown in Fig. 7, the initial velocity increases with the power increases. Higher power results in larger amplitude, which causes stronger body force, inducing stronger fluid flow and thus higher ejection velocity. The velocity of droplet ejection reaches 1.1 m/s at 0.85 W input power.

CONCLUSIONS

In this paper, we proposed a novel nozzle-free droplet ejection system based on the GHz acoustic transducer. The GHz transducer is fabricated by standard MEMS process. Based on the inverse piezoelectric effect, the transducer produces high-frequency vibration with controlled input RF signal. The high-frequency acoustic waves caused by transducer vibration decay in liquid and generate strong body force, which causes the liquid to overcome the surface tension and gravity and then eject droplets. We studied the relationship between input power and initial velocity of the droplets. Given small size, nozzle-free and compatibility with CMOS technology, this work provides a new way for inkjet printing, bioprinting, and drug delivery applications.

ACKNOWLEDGEMENTS

The authors gratefully acknowledge financial support from National Key R&D Program of China (2018YFE0118700), the National Natural Science Foundation of China (NSFC No. 62174119), the 111 Project (B07014), and the Foundation for Talent Scientists of Nanchang Institute for Micro-technology of Tianjin University. Quanning Li, Xuejiao Chen, Bohua Liu, and Chongling Sun were thanked for the help on the device fabrication.

REFERENCES

[1] M. Śliwiak, R. Bui, M. A. Brook, and P. R. Selvaganapathy, "3D printing of highly reactive silicones using inkjet type droplet ejection and free space droplet merging and reaction," *Additive Manufacturing,* vol. 46, 2021.

[2] J. Luo, L. Qi, Y. Tao, Q. Ma, and C. W. Visser, "Impact-driven ejection of micro metal droplets on-demand," *International Journal of Machine Tools and Manufacture,* vol. 106, pp. 67-74, 2016.

[3] B. Zhang, J. He, X. Li, F. Xu, and D. Li, "Micro/nanoscale electrohydrodynamic printing: from 2D to 3D," *Nanoscale,* vol. 8, no. 34, pp. 15376-88, Aug 25 2016.

[4] J. Zhang *et al.,* "Single Cell Bioprinting with Ultrashort Laser Pulses," *Advanced Functional Materials,* vol. 31, no. 19, 2021.

[5] D. Foresti *et al.,* "Acoustophoretic printing," (in English), *Science Advances,* vol. 4, no. 8, Aug 2018.

[6] X. Guo *et al.,* "Controllable Cell Deformation Using Acoustic Streaming for Membrane Permeability Modulation," *Adv Sci (Weinh),* vol. 8, no. 3, p. 2002489, Feb 2021.

[7] Q. Guo, M. Shao, X. Su, X. Zhang, H. Yu, and D. Li, "Controllable Droplet Ejection of Multiple Reagents through Focused Acoustic Beams," *Langmuir,* Dec 13 2021.

[8] J. O. Castro, S. Ramesan, A. R. Rezk, and L. Y. Yeo, "Continuous tuneable droplet ejection via pulsed surface acoustic wave jetting," *Soft Matter,* vol. 14, no. 28, pp. 5721-5727, Jul 18 2018.

[9] Y. Ning *et al.,* "Mechanism and stability investigation of a nozzle-free droplet-on-demand acoustic ejector," *Analyst,* vol. 146, no. 18, pp. 5650-5657, Sep 13 2021.

[10] M. He *et al.,* "An on-demand femtoliter droplet dispensing system based on a gigahertz acoustic resonator," *Lab Chip,* vol. 18, no. 17, pp. 2540-2546, Aug 21 2018.

[11] H. Wu *et al.,* "Manipulations of micro/nanoparticles using gigahertz acoustic streaming tweezers," *Nanotechnology and Precision Engineering,* vol. 5, no. 2, 2022.

[12] D. J. Collins, Z. Ma, and Y. Ai, "Highly Localized Acoustic Streaming and Size-Selective Submicrometer Particle Concentration Using High Frequency Microscale Focused Acoustic Fields," *Anal Chem,* vol. 88, no. 10, pp. 5513-22, May 17 2016.

CONTACT

*X. Duan, xduan@tju.edu.cn

*W. Pang, weipang@tju.edu.cn

BONE CONDUCTION PICKUP BASED ON PIEZOELECTRIC MICROMACHINED ULTRASONIC TRANSDUCERS

Chongbin Liu[1], Xiangyang Wang[1], Yong Xie[2], and Guoqiang Wu[1]
[1] Institute of Technological Sciences, Wuhan University, Wuhan, 430072, CHINA
[2] School of Advanced Materials and Nanotechnology, Xidian University, Xi'an, 710071, CHINA

ABSTRACT

This paper reports a novel bone conduction pickup approach based on piezoelectric micromachined ultrasonic transducers (PMUTs) as the sound pickup element. The PMUT array is die-attached on an amplification circuit and encapsulated with acoustically transparent adhesive. The packaged PMUT array is attached on the throat of a human body to pick up the sound signals. Compared with the classical air conduction approach, the reported bone conduction pickup method demonstrates a signal-to-noise ratio (SNR) improvement of two times in noisy (with noise level of around 45 dB) and 5 times in very noisy (with noise level of around 65 dB) environments, respectively. This work explores a new application for PMUTs in the field of bone conduction pickups.

KEYWORDS

Bone Conduction, Microphone, Sound Pickup, Piezoelectric Micromachined Ultrasonic Transducers (PMUTs), Microelectromechanical System (MEMS).

INTRODUCTION

Compared with air conduction microphone, bone conduction microphone directly picks up sound signals through the vibration of vocal organs or related bones, which demonstrates a good isolation against external noises [1]. Therefore, bone conduction pickup has attracted lots of research attentions recently.

Firstly, the amplitude changes of bone conduction speech and air conduction speech have been compared and analyzed [2]. The studies showed that the bone conduction microphone has a good pickup effect in the noisy environment. But the results also showed that the individual performance in speech recognition of bone conduction speech was lower than that of air conduction speech [3,4]. Then, bone conduction microphone was used with the air conduction microphone for speech enhancement. It turned out that the verification performance using bone conduction to enhance air conducted speech can greatly reduce the equal error rate of speech [3,5]. Moreover, the algorithm and application of bone conduction microphone have also been studied. A novel person identification technique was developed based on bone conduction, which exploited bronchial breath sound and speech signal acquired by a stethoscope [6]. A real-time dual-microphone speech enhancement algorithm assisted by bone conduction sensor was proposed and the experimental results demonstrated that the speech sound quality and intelligibility in different noise environments is improved significantly when compared with the alternating current (AC) -only beamformer [7].

However, bone conduction microphones are generally based on accelerometers or capacitive microphones, which suffered from the complicated processing and packaging. The self-noise, falling resistance and service life of bone conduction microphone have always been problems in practical applications.

In this work, a novel bone conduction pickup approach based on piezoelectric micromachined ultrasonic transducers (PMUTs) as the sound pickup element is proposed and realized. Compared with typical bone conduction microphone, the current bone conduction PMUTs have distinguished features with direct collection of bone vibrations, simpler packaging, compact size, and more robust to mechanical shocks.

CHARACTERIZATION OF PMUTS

As essential pickup element in the reported bone conduction system, the PMUTs are designed as a bioinspired honeycomb structure [8,9], as shown in Fig. 1(a). The PMUTs are fabricated based on cavity silicon-on-insulator (CSOI) platform [10]. Fig. 1(b) shows the cross-sectional scanning electron scanning electron microscope (SEM) images (performed on a TESCAN MIRA3) of the sensing cell and diaphragm. The diaphragm of a sensing cell in the PMUT array consists of the bottom oxide, highly doped silicon (HDS), scandium-doped aluminum nitride ($Sc_xAl_{1-x}N$, $x = 9.5\%$) piezoelectric film, molybdenum (Mo) top electrode and top oxide, stacked from the bottom to the top.

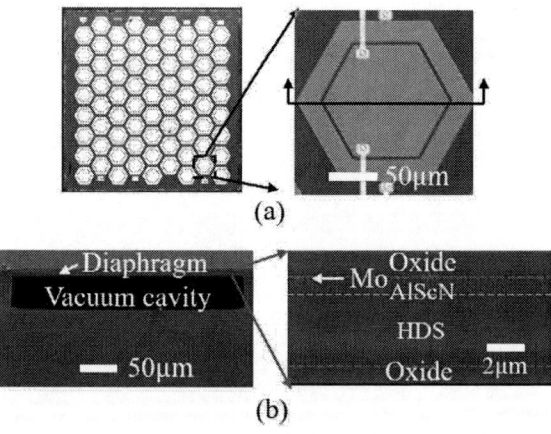

Figure 1: (a) optical micrographs of top view of a PMUT 8×9 array together with a magnified image of one sensing cell. (b) cross-sectional SEM images of the sensing cell and diaphragm.

The measured deflection versus frequency response and the mode shape of a single PMUT hexagonal sensing cell using laser doppler vibrometer (LDV) are shown in Fig. 2(a). Figure 2(b) illustrates the measured sensitivity of

978-1-6654-9309-3/23 $31.00 © 2023 IEEE

IEEE MEMS 2023, Munich, GERMANY
15 - 19 January 2023

the reported PMUTs in the frequency band required for human sound pickup. It can be seen that the PMUT array has a sensitivity of −154.5 dB±0.5 dB (re: 1 V/μPa) with a 60 dB circuit amplification. It shows a very flat response from 10 Hz to 2 kHz. Therefore, the sound signals of different frequencies collected by PMUTs have true proportions and will not be partially amplified.

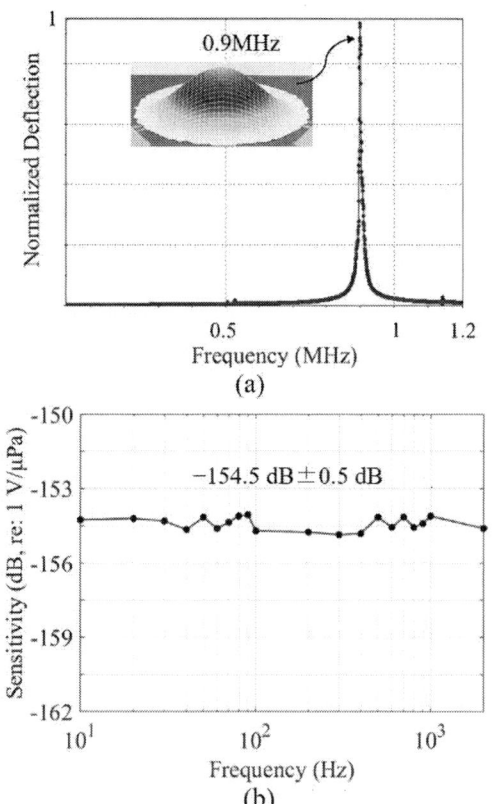

(a)

(b)

Figure 2: (a) Measured frequency response of a sensing cell in the PMUT array. (b) Measured acoustic pressure sensitivities of the reported PMUT array at different frequency.

DESIGN OF BONE CONDUCTION PICKUP SYSTEM

The traditional microelectromechanical system (MEMS) bone conduction microphone obtains the vibration signal by monitoring the pressure difference, $\Delta P(t)$, around the MEMS microphone, as shown in Fig. 3(a) [4]. On the contrary, the reported bone conduction PMUTs directly obtain the bone vibrations by measuring the induced charges on the surfaces of the piezoelectric layer as shown in Fig. 3(b) [11,12].

The schematic diagram of the reported PMUTs based bone conduction pickup system is illustrated in Fig.4(a) and the schematic diagram of the packaging method is shown in Fig. 4(b). The PMUT and the amplification circuit are integrated on a printed circuit board (PCB) and the PMUT is electrically connected using the bonded wires. Then, the assembled PCB with PMUT and circuit is packaged by an acoustically transparent polyurethane

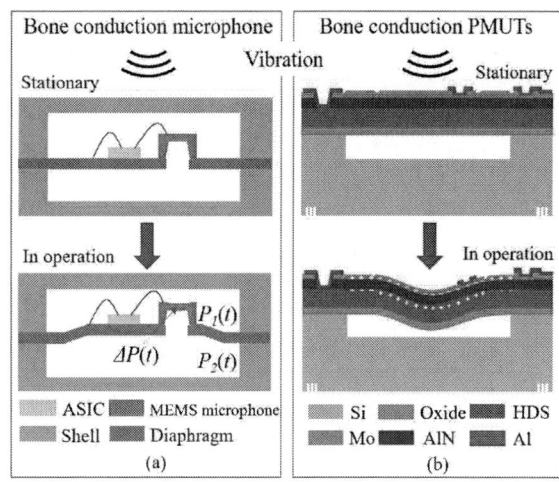

Figure 3: Comparison of working principles of bone conduction device based on (a) microphone and (b) PMUTs.

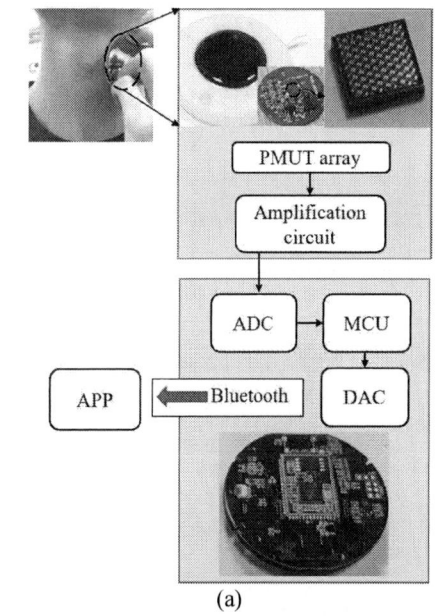

(a)

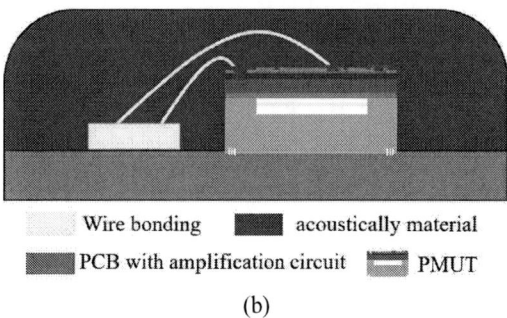

(b)

Figure 4: (a) Schematic diagram of the current bone conduction pickup system. The packaged PMUT based system is attached on the throat for sound collection. (b) The schematic diagram of packaging method.

material. The packaged system is attached on the throat for sound vibrations are collected and then amplified to be analog voltages. The obtained voltages are digitalized by an analog-to-digital converter (ADC) and then processed in a microcontroller unit (MCU). Finally, the processed signals are converted by a digital-to-analog converter (DAC) and transmitted to a mobile phone application via Bluetooth.

EXPERIMENTAL RESULTS

First, the measured voltages of the sound signals "Ni hao (hello in Chinese)" and "Wu han da xue (Wuhan University in Chinese)" are collected by the reported PMUTs based system and their frequency domain characteristics are analyzed, as shown in Fig.5. It can be seen that there are amplitude peaks at three frequencies of the two collected sound signals in the frequency domain. These three peak regions, P1, P2 and P3, can be defined as the features of the collected human voice to identify the speaker.

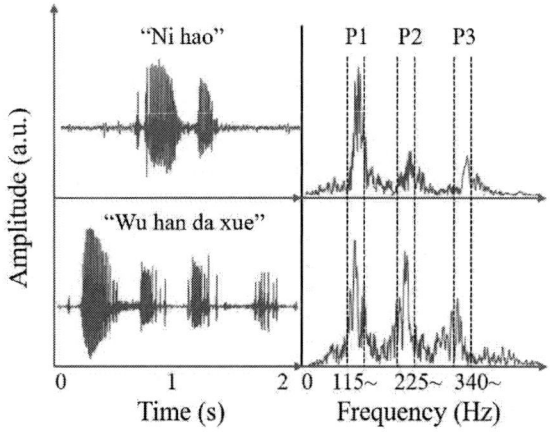

Figure 5: The measured voltages of the sound signals "Ni hao (hello in Chinese)" and "Wu han da xue (Wuhan University in Chinese)" collected by the reported PMUTs based system and their frequency domain characteristics.

Then, sound signals pickup are implemented using the reported PMUTs based system, as well as the commercial air conduction microphone in various environments with different noise levels. The measured voltages of the sound signals "Ni hao" collected by these two methods are compared in both time domain and frequency domain, as shown in Fig. 6(a) and Fig. 6(b), respectively. The signal-to-noise ratio (SNR) of the signals obtained by these two methods in different environmental noise level are listed in Table I.

The results show that the sound quality and SNR using the air conduction microphone are much better than the bone conduction PMUT in quiet environment (with noise level of around 15 dB). However, the SNR using the air conduction approach becomes worse in noisy environment (with noise level of around 45 dB). In contrast, the bone conduction PMUT system works well in all these environments, even with noise level around 65 dB.

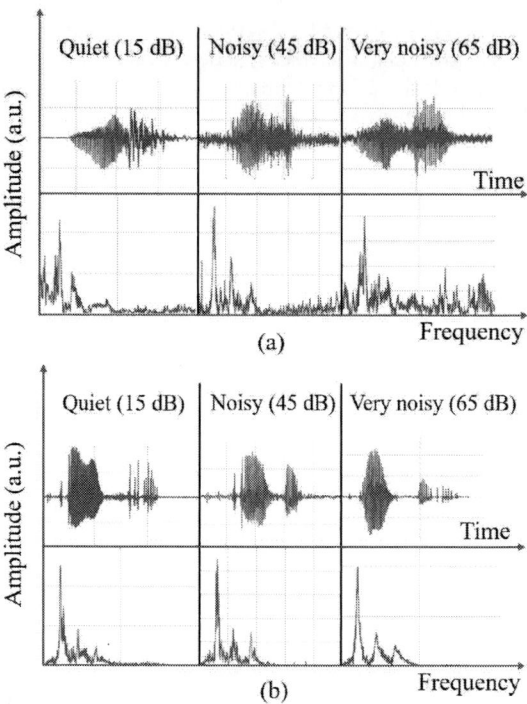

Figure 6: Collected sound signals obtained by (a) air conduction microphone and (b) bone conduction PMUTs.

Table I. Measured signal-to-noise ratios of the signals obtained by the commercial air conduction microphone and the reported bone conduction PMUTs in various environments with different noise level.

Noise level	Quiet (~15 dB)	Noisy (~45 dB)	Very noisy (~65 dB)
Air conduction microphone	150	20	10
Bone conduction PMUTs	70	55	50

DISCUSSION AND CONCLUSION

The basic flow of speech and voiceprint recognition is shown in Fig. 7 [13,14]. After enough sound signals are collected, they can be preprocessed, feature extracted and then used in training the template library. The library is the key to matching the sound signals to be recognized. In the future, the reported PMUTs based system can be used for speech and voiceprint recognition, which can prove the feasibility and accuracy of PMUTs in sound pickup.

In this work, a novel bone conduction pickup approach based on PMUTs as the sound pickup element is proposed. The PMUTs are designed as a bioinspired honeycomb structure and fabricated based on CSOI platform. The sensitivity of PMUTs in the frequency band required for human sound pickup and the mode shape of PMUTs are characterized. The packaged system is attached on the throat for sound collection including "Ni hao" and "Wu han da xue", which are analyzed in both time and frequency domain. Compared with the classical air conduction approach, the reported bone conduction pickup method collection. The generated charges caused by bone demonstrates a SNR improvement of two times in noisy (45 dB) and 5 times in very noisy (65 dB) environments, respectively. This work demonstrates the feasibility of PMUTs in the field of bone conduction pickups.

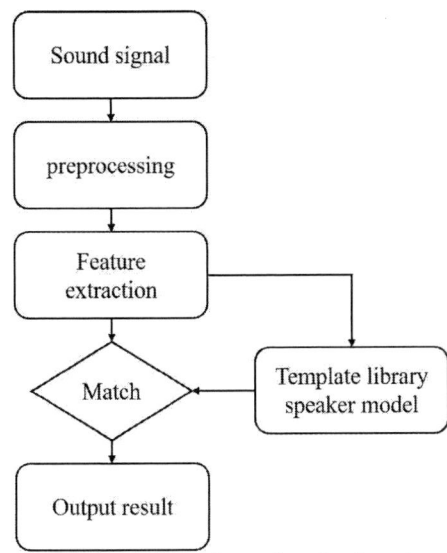

Figure 7: The basic flow of speech and voiceprint recognition

ACKNOWLEDGMENT

This work was supported in part by the Fundamental Research Funds for the Central Universities under Grant 2042021kf0191.

REFERENCES

[1] S. E. Ellsperman, E. M. Nairn, E. Z. Stucken, "Review of Bone Conduction Hearing Devices," *Audiology Research*, vol. 11, pp. 207-219, 2021.

[2] M. S. Rahman, T. Shimamura, "Amplitude Variation of Bone-Conducted Speech Compared with Air-Conducted Speech," *Acoust. Sci. Technol.*, vol. 40, no. 5, pp. 293-301, 2019.

[3] S. Tsuge, D. Koizumi, M. Fukumi, S. Kuroiwa, "Speaker Verification Method Using Bone-Conduction and Air-Conduction Speech," *Int. Symp. Intelligent Signal Process. Commun. Syst., Proc. (ISPACS)*, pp. 449-452, Dec. 2009.

[4] B. C. You, S. C. Lo, C. K. Chan, C. S. Li, H. L. Ho, S. C. Chiu, G. H. Hsieh, W. Fang, "Design and Implementation of Dual Pressure Variation Chambers for Bone Conduction Microphone," *J. Micromech. Microeng.*, vol. 30, no. 12, 2020.

[5] Y. Zhou, H. Wang, Y. Chu, H. Liu, "A Robust Dual-Microphone Generalized Sidelobe Canceller Using a Bone-Conduction Sensor for Speech Enhancement," *Sensors*, vol. 21, no. 5, p. 1878, Mar. 2021.

[6] V. Tran, W. Tsai, "Stethoscope-Sensed Speech and Breath-Sounds for Person Identification with Sparse Training Data," *IEEE Sensors J.*, vol. 20, no. 2, pp. 848-859, 2020.

[7] Y. Zhou, Y. Chen, Y. Ma, H. Liu, "A Real-Time Dual-Microphone Speech Enhancement Algorithm Assisted by Bone Conduction Sensor," *Sensors*, vol. 20, no. 18, p. 5050, Sept. 2020.

[8] L. Jia, L. Shi, Z. Lu, C. Sun, G. Wu "A High-Performance 9.5% Scandium Doped Aluminum Nitride Piezoelectric MEMS Hydrophone with Honeycomb Structure," *IEEE Electron Device Lett.*, vol. 42, no. 12, pp. 1845-1848, Dec. 2021.

[9] L. Shi, L. Jia, C. Liu, H. Yu, C. Sun, G. Wu, "Performance of Aluminum Nitride-Based Piezoelectric Micromachined Ultrasonic Transducers Under Different Readout Configurations," *J. Micromech. Microeng.*, vol. 32, no. 1, Nov. 2021.

[10] L. Jia, L. Shi, C. Liu, Y. Yao, C. Sun, G. Wu, "Design and Characterization of An Aluminum Nitride-Based MEMS Hydrophone with Biologically Honeycomb Architecture," *IEEE Trans. Electron Devices*, vol. 68, no. 9, pp. 4656-4663, Sept. 2021.

[11] L. Jia, L. Shi, C. Liu, J. Xu, Y. Xu, C. Sun, S. Liu, G. Wu, "Piezoelectric Micromachined Ultrasonic Transducer Array-Based Electronic Stethoscope for Internet of Medical Things," *IEEE Internet Things J.*, vol. 9, no. 12, pp. 9766-9774, Jun. 2022.

[12] L. Jia, L. Shi, C. Liu, C. Sun, G. Wu, "Enhancement of Transmitting Sensitivity of Piezoelectric Micromachined Ultrasonic Transducers by Electrode Design," *IEEE Trans. Ultrason. Ferroelectr. Freq. Control*, vol. 68, no. 11, pp. 3371-3377, Nov. 2021.

[13] H. Shen, B. Wang, J. Wang, "Research on Robustness of Voiceprint Recognition Technology," *ACM Int. Conf. Proc. Ser.*, Dec. 2018.

[14] Y. Xue, L. Wang, L. Li, Z. Liu, L. Liu, "Matlab-Based Intelligent Voiceprint Recognition System," *Proc. - Int. Conf. Instrum. Meas., Comput., Commun. Control (IMCCC)*, pp. 303-306, Dec. 2016.

CONTACT

*Guoqiang Wu, wuguoqiang@whu.edu.cn

BREAKING THE DEAD ZONE LIMITATION OF PMUTS BASED ON A PHASE SHIFT OF DRIVING WAVEFORM WITH WINDOW FUNCTION

Chun-You Liu[1], Chin-Yu Chang[2], and Sheng-Shian Li[1,2]

[1]Department of Power Mechanical Engineering, National Tsing Hua University, Hsinchu, Taiwan
[2]Institute of NanoEngineering and MicroSystems, National Tsing Hua University, Hsinchu, Taiwan

ABSTRACT

In this work, we demonstrate a simple and effective method for reducing the membrane ring-down time and suppressing the output interferences caused by nonlinearity to benefit axial resolution for piezoelectric micromachined ultrasonic transducers (PMUTs). This method breaks the dead zone limitation of PMUTs through *a phase shift of driving waveform* with a *window function*. The proposed phase shift waveform with window function can effectively minimize the ring-down time on the transmitting node (TX) and thus reduce the dead zone region in a pulse-echo measurement system. In addition, the unwanted ripples and interference signals of the receiving output (RX) caused by nonlinear behaviors of PMUTs, thereby resulting in worse axial resolution, can be significantly suppressed on the receiving node (RX). This technique shows a great performance enhancement for PMUTs with high accuracy and axial resolution during measurement.

KEYWORDS

Ring-down time reduction, axial resolution, PMUT, ultrasonic, dead zone, pulse-echo, time of flight, phase shift of driving waveform, window function.

INTRODUCTION

Ultrasonic technologies are widely used in industry, medical treatment, and consumer electronics for many decades. However, conventional bulk ultrasonic transducers with the disadvantages of large size and high fabrication cost have become a bottleneck in recent years for entering new and emerging markets necessitating miniaturization and integration. On the other hand, Microelectromechanical systems (MEMS) technology provides a solution to those problems. Through mature thin-film deposition technologies, the bulky piezoelectric ultrasonic transducers can be scaled down to small membrane transducers while reducing the power consumption as well as the manufacturing cost through wafer-level processing. With the advantage of low power consumption and small size, piezoelectric micromachined ultrasonic transducers (PMUTs) have been widely used for medical imaging techniques [1], fingerprint sensors [2], and rangefinders [3]. To realize these applications, the time-of-flight (ToF) method becomes a major characterization approach in ultrasonic technology. However, the ring-down time leading to an inevitable dead zone places a bottleneck for the minimum detectable range and limited resolution in axial direction.

To overcome the abovementioned issues, reducing the intrinsic quality factor of PMUT can directly minimize the ring-down time. However, the output pressure and sensitivity of PMUT will also be suppressed, resulting in a decrease in the overall performance of PMUT. As a result, providing a modification of driving waveform to actively eliminate the unwanted ring-down signals is a more appropriate solution for PMUT operating in pulse-echo mode. In literature, nonlinear pulse-shaping methods that exploit the nonlinear properties of PMUTs are used to rapidly increase the amplitude of the transmit output energy with complex chirp signals [4]. However, the excitation of mixing chirp signals is too difficult to achieve; especially the calibration of each device is required, and there is certainly a disadvantage of a long driving time. Another approach to modifying the driving waveform is to provide an inverting pulse [5][6]. By adjusting the inverting pulse, the amplitude of ring-down signals is significantly reduced. This approach is easier to implement and shows great performance on transmit output signals. Through this concept, a phase shift of driving waveform with window function is proposed in this work to break the dead zone limitation. The proposed modification to the driving waveform achieves significant ring-down time reduction and interference suppression for ultrasonic applications.

WORKING PRINCIPLE

To break the dead zone limitation and enhance the axial resolution of PMUTs, a phase shift of driving waveform with window function is proposed in this work. Fig. 1(a) illustrates the operation principle of a distance characterization system using two PMUTs for transmitting (TX) and receiving (RX), respectively. In general, the TX PMUT will be excited through a chirp signal with a TX output shown in Fig. 1(b). This TX output will propagate inside the medium (ex. air) and then be detected by an RX PMUT. However, the RX PMUT will be driven and triggered into its nonlinear state due to the strong TX output energy level in such a short distance and suffers unwanted ripples and interference at RX output signals, thus resulting in low accuracy and worse axial resolution. To overcome these issues, a phase shift of driving waveform with window function is proposed. Through modifying the excitation signal of the TX PMUT, not only the ring-down time as well as the dead zone region on TX output can be reduced, but also the interference signal on RX output can be suppressed, both of which significantly benefit high accuracy and axial resolution on RX output as shown in Fig 1(c).

Resonator Design

To maximize the transmitting efficiency, a PZT-based PMUT is used to serve as a TX PMUT thanks to its high piezoelectric coefficient, d_{31}. Note that the high dielectric constant of PZT will not influence the TX performance since TX serves as an actuator mainly depending on d_{31}.

978-1-6654-9309-3/23 $31.00 © 2023 IEEE

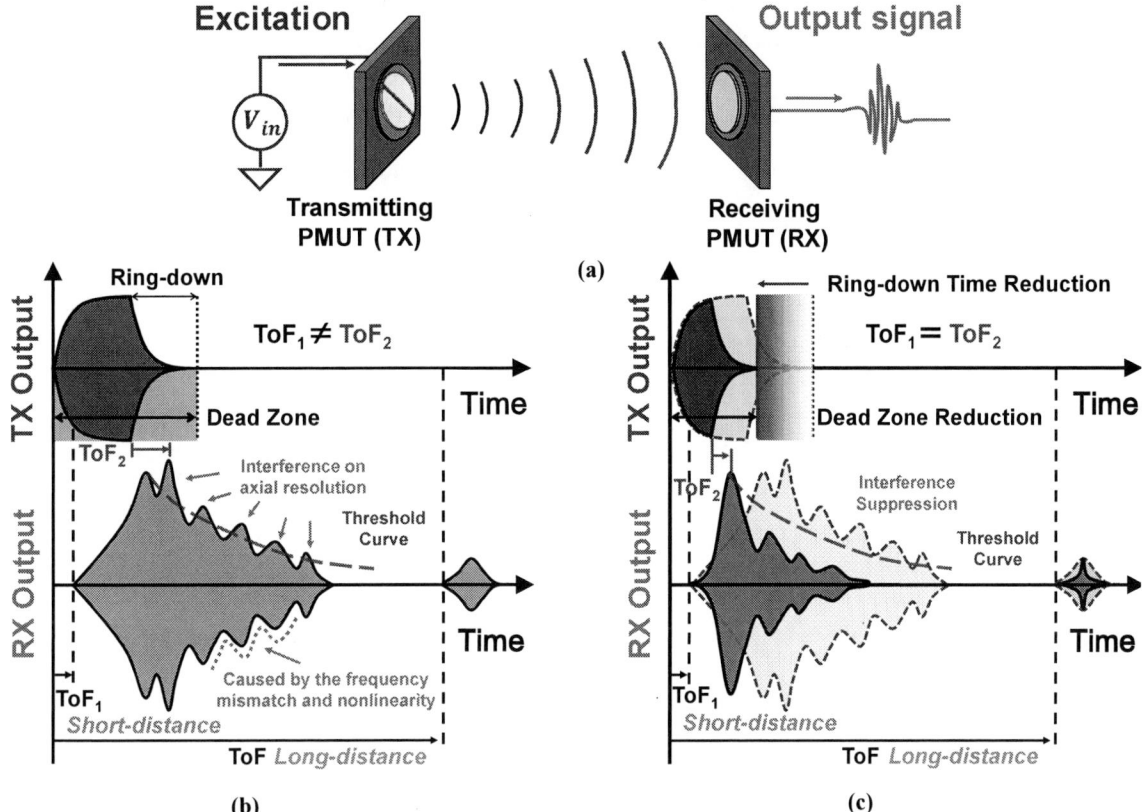

Fig. 1: (a) Operation principle of a distance characterization system based on time of flight (ToF). TX and RX output waveforms through (b) a chirp excitation and (c) a modified excitation based on the proposed methodology.

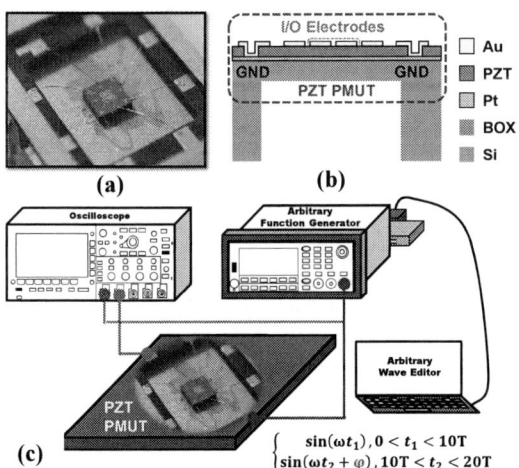

$$\begin{cases} \sin(\omega t_1), 0 < t_1 < 10T \\ \sin(\omega t_2 + \varphi), 10T < t_2 < 20T \end{cases}$$

Fig. 2: (a) 3D optical image of PZT TX PMUT. (b) PMUT cross-section view. (c) Measurement setup for the proposed phase shift of driving waveform.

The proposed TX PMUT is fabricated through a 4-mask fabrication process where a PZT layer is deposited on an SOI wafer followed by a wet etching process to pattern the contact via. Through the lift-off process, the top and bottom electrodes are patterned. The final MEMS structure is released by the DRIE process at the backside of the wafer. Fig. 2(a) presents the 3D optical image of PZT-based TX PMUT chip after fabrication and inside a ceramic package.

The details of each stacking layer and cross-section view of PMUT are also presented in Fig. 2(b).

EXPERIMENTAL RESULTS

The measurement setup of PZT TX PMUT is shown in Fig. 2(c) where the Arbitrary Function Generator (33250A, Agilent) provides a phase shift of driving waveform with window function while the time response of the output wave is measured in Oscilloscope (GDS-2104E, GW Instek). Note that 10 cycles with a designated phase shift are applied right after the original 10 cycles of the driving sinusoidal wave. The method of the phase shift is further elaborated in Fig. 2(c) and the excitation waveform is presented in Fig. 3(a).

Measurement results

To find out the most efficient way to suppress the ring-down effect by a phase shift function, the time responses of PZT PMUT under various phase shifts are carried out. Fig. 3(b) shows the measurement results under various phase shifts of 0°, 90°, 180°, and 270°, respectively. 90° is adopted as the excitation signal due to its significant reduction of the ring-down time. However, in short-distance measurement, the interference still needs to be suppressed for better axial resolution even at a low driving voltage of 0.1 V together with the phase shift method.

Fig. 4(a) shows the measurement setup of a short-distance (3 cm) measurement by an Aluminum Nitride (AlN) PMUT with a fixed gain (20 dB) voltage amplifier

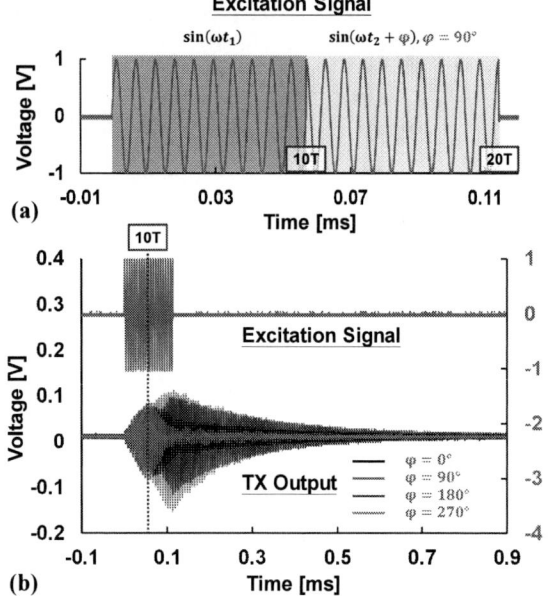

Fig. 3: (a) Phase shift ($\varphi = 90°$) of a driving waveform. (b) TX output measurements under various phase shifts.

as an RX PMUT for better signal-to-noise ratio (SNR). Note that, as compared to PZT, AlN features low dielectric constant as well as low dielectric loss, thus suitable to serve a receiving unit for PMUT. The TX and RX output measurements are shown in Fig. 4(b) and (c), respectively. According to the results, the ring-down effect of the output wave has been reduced through the phase shift (90°) of driving waveform and the ToF of 0.088 ms between the peaks of TX and RX outputs is measured, indicating the

extracted distance of 3 cm based on air sound speed of 340 m/s. However, the unwanted ripples and interferences, caused by the nonlinearity of RX PMUT and amplitude of ring-down signals, induce the worse axial resolution in short-distance measurement. Note that the threshold curve of the $1/x$ (x stands for distance) is used to identify the threshold of axial resolution.

To further reduce the ring-down time and enhance the axial resolution, window function is applied on excitation signal which contributes additional reduction on the ring-down time as well as shorter dead zone region as shown in Fig. 5. Through this window function adjustment (as presented in the inset of Fig. 5) in driving waveform, the measurement setup and results are shown in Fig. 6. The distance between TX/RX PMUTs is set to be 11 cm and the voltage of 5 V is used to excite PZT TX PMUT. Through this distance characterization system based on pulse-echo method, the receiving signals of 1st echo and 2nd reflecting echo are characterized in an oscilloscope while ToF_2 of 0.31 ms and ToF_3 of 0.94 ms are measured, respectively.

To verify the time difference (ToF_2) between the peaks of TX and RX outputs, the ToF_1 between starting points of excitation and RX output is also recorded. According to the result, ToF_1 and ToF_2 are almost the same, representing the measured distance of 11 cm (1st echo) and thus proving the efficacy of the proposed method. The ToF_3 of 0.94 ms representing the distance of 32 cm (2nd echo) is also verified. In summary, the method of phase shift with window function shows a great performance enhancement for PMUTs with high accuracy and axial resolution, which are beneficial to a variety of ultrasonic applications.

CONCLUSION

In this work, we demonstrate a simple and efficient

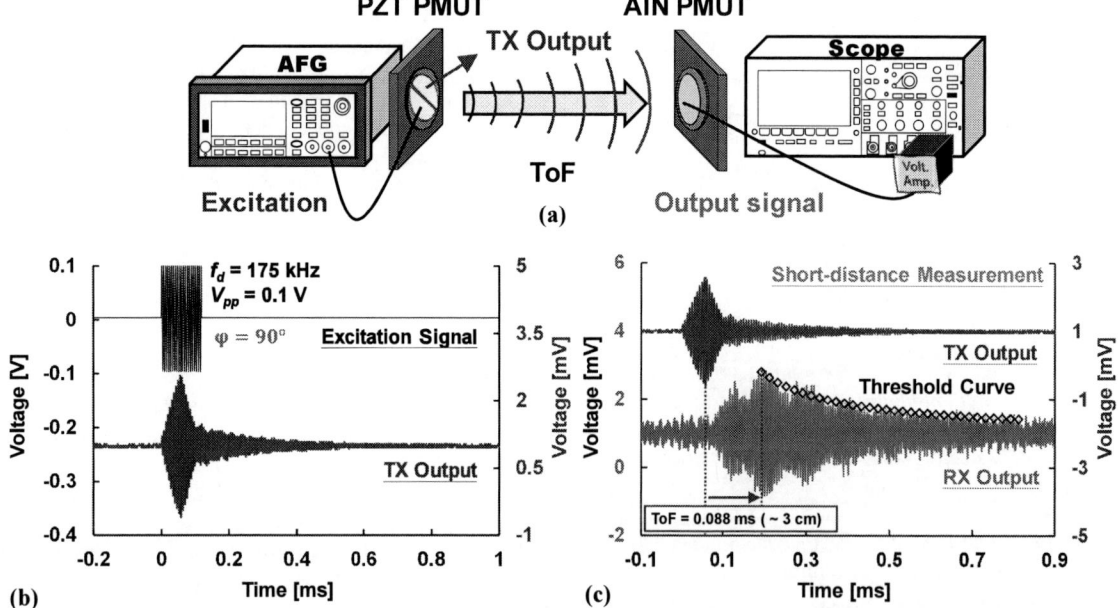

Fig. 4: ToF measurement (short distance) by the phase shift method. (a) Measurement setup of a distance characterization system. (b) TX output from the integrated electrodes of PZT TX PMUT. (c) RX output from the integrated electrodes of AlN RX PMUT. The unwanted ripples, resulting from the frequency mismatch between TX and RX, are mainly caused by nonlinearity of PMUTs.

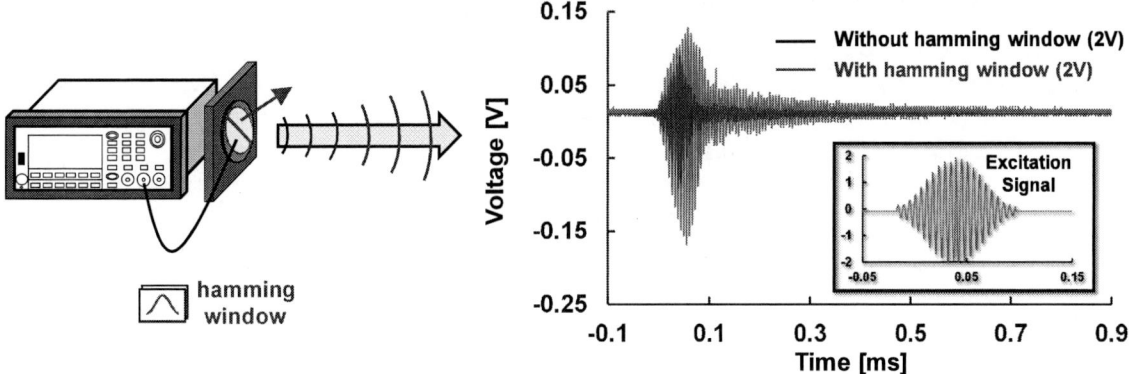

Fig. 5. Modification of the phase shift driving waveform by using a window function. The modification of excitation signal is presented in the inset while the comparison of TX output measurements w/wo modification is given here.

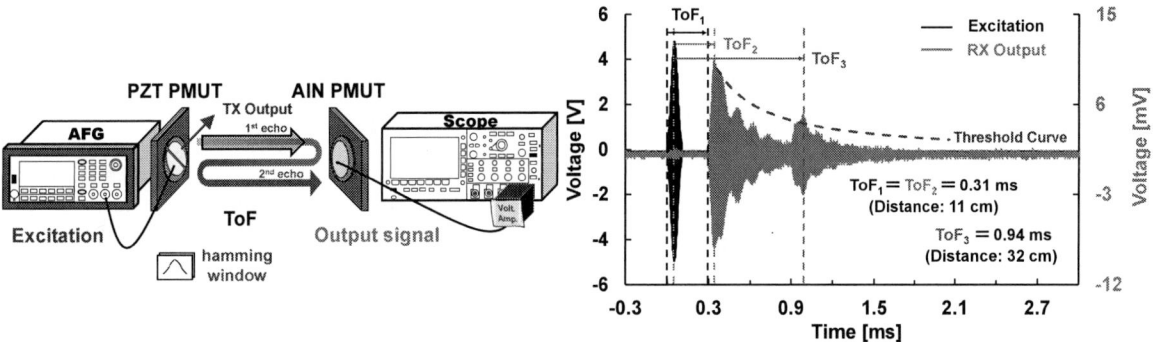

Fig. 6: ToF measurement through the modification of the driving waveform. The ToF_1 of 0.31 ms representing the 1st echo is same as ToF_2. The ToF_3 of 0.94 ms represents the 2nd echo. Note that ToF_1 is gauged on the starting points of the waveform response while ToF_2 is counted at the peak amplitudes of the waveform response.

way to eliminate the ring-down signals of PMUT in pulse-echo mode by implementing a phase shift of driving waveform with an additional window function. To maximize the transmitting efficiency and receiving sensitivity, PZT PMUT is used to serve as TX while AlN PMUT is used to serve as RX, respectively. A phase shift of 90° is adopted as the excitation signal due to the significant reduction of ring-down time. In addition, the ring-down time and amplitude will be further minimized by applying a window function on the driving waveform. Through the modification of driving waveform, the ring-down effect has been effectively reduced and the unwanted interference signal has been suppressed simultaneously. As a result, the proposed method shows a significant enhancement on axial resolution and leads to "breaking the dead zone limitation" for PMUTs.

ACKNOWLEDGEMENTS

The authors appreciate the financial support from the National Science and Technology Council (NSTC), Taiwan under Grant NSTC 111-2223-E-007-007. We also appreciate Coretronic MEMS (CMC) for the MEMS fabrication facility and the Center for Nanotechnology, Material science & Microsystem for fabrication of the dies.

REFERENCES

[1] K. M. Smyth, C. G. Sodini, and S.-G. Kim, "High electromechanical coupling piezoelectric micro-machined ultrasonic transducer (PMUT) elements for medical imaging," *Tech. Dig.*, 19th Int. Conf. on Solid-State Sensors & Actuators (Transducers'17), pp. 966-969, Kaohsiung, Taiwan, June 18-22, 2017.

[2] D. A. Horsley *et al.*, "Ultrasonic fingerprint sensor based on a PMUT array bonded to CMOS circuitry," *Proceedings*, IEEE International Ultrasonics Symposium (IUS), pp. 1-4, Tours, France, Sep. 18-21, 2016.

[3] G.-L. Luo, Y. Kusano, and D. A. Horsley, "Airborne piezoelectric micromachined ultrasonic transducers for long-range detection," *J. Microelectromech. Syst.*, vol. 30, no. 1, pp. 81-89, Feb. 2021.

[4] M. Gratuze, A.-H. Alameh, A. Robichaud, and F. Nabki, "A nonlinear pulse shaping method using resonant piezoelectric MEMS devices," *IEEE Transactions on Ultrasonics, Ferroelectrics, and Frequency Control*, vol. 69, no. 4, pp. 1515-1527, April 2022.

[5] S. Pala, Z. Shao, Y. Peng, and L. Lin, "Improved ring-down time and axial resolution of PMUTs via a phase-shift excitation scheme," *Proceedings*, 34th IEEE Int. Micro Electro Mechanical Systems Conf. (MEMS'21), pp. 390-393, online, Jan. 25-29, 2019.

[6] X. Liu, *et al.*, "Reducing ring-down time of PMUTs with phase shift of driving waveform," *Sensors and Actuators A: Physical*, vol. 281, pp. 100-107, 2018.

CONTACT

* S.-S. Li, Tel: +886-3-516-2401; ssli@mx.nthu.edu.tw

DRONE-MOUNTED LOW-FREQUENCY PMUTS FOR > 6-METER RANGEFINDER IN AIR

Hanxiao Liu[1†], Yande Peng[1†], Wei Yue[1†], Seiji Umezawa[2], Shinsuke Ikeuchi[2], Yasuhiro Aida[2], Chunming Chen[1], Peggy Tsao[1], and Liwei Lin[1]

[1]Department of Mechanical Engineering, University of California, Berkeley, USA
[2]Murata Manufacturing Co., Ltd., Japan
[†]Hanxiao Liu, Yande Peng and Wei Yue contributed equally to this work.

ABSTRACT

This paper reports a low-frequency piezoelectric micromachined ultrasonic transducer (pMUT) with a small attenuation coefficient to realize the long-distance range finding applications in air. Pulse-detection measurements of one pair of pMUT devices show a >6-meter traveling distance in air under a 37.32 kHz driving frequency. As an example, two pMUT chips are mounted on two drones as the transceiver and receiver, respectively. Measurements of the separation distance are conducted based on the time-of-flight (ToF) principle with up to 32 fps (frames per second) for real-time detections. This demo of drone-mounted pMUT system illustrates the advantages of pMUT in terms of compactness and low power consumption for applications in drones including obstacle avoidance, inter collision prevention, aerial coordination, and acoustic-based vision.

KEYWORDS

MEMS, pMUT, rangefinder, ultrasound, drone-mounted devices.

INTRODUCTION

Range finding is an important sensing technology in which sensors can detect the location of other devices or objects around the neighborhood areas [1]. Recent development in micromachining technologies have illustrated the new possibilities in range finding schemes using piezoelectric micromachined ultrasonic transducer (pMUT) for reduced size and power consumption when compared with those of bulk acoustic transducers, radars, and LIDAR for mobile drone and robotics applications [2]. Furthermore, pMUTs are also attractive in short-range navigation, obstacle avoidance, and robotic coordination applications in the uncertain or unknown environment [3,4], as compared with optical cameras and infrared transducer which requires clear optical paths [5].

Most current pMUT-based range finding systems have relatively short sensing distances. For example, a prior report shows a pMUT rangefinder reaches 4.8-meter in the traveling distance by using a 4×4 array [6] or >2-meter in traveling distance by using a single pMUT [7] due to the low coupling efficiency in air transmissions. Several approaches have been proposed to increase the sensing distance. In the area of the emission of ultrasound waves, the attenuation efficiency is related to the ultrasound frequency and the working frequency of pMUTs should be low to decrease the signal attenuation effect in air and there are several pMUT devices with relatively low operation frequencies in the 30-40 kHz range [8,9,10]. In the area of piezoelectric material property, the piezoelectric constant

(e_{31}) and electromechanical coupling coefficient (k_t^2) [11] should be high for high emission efficiency, while the dielectric constant (ε_{33}) should be low for high receiving sensitivity [12]. In terms of the structural characteristics, a large bandwidth is expected, which usually results in better coupling and efficient energy transfer.

Here, an air-coupled and wide bandwidth rangefinder is demonstrated based on the low- frequency lithium niobate (LN) pMUT (**Figure 1**). The rangefinder consists of two pMUTs as the transceiver and receiver, respectively, with a relatively low resonant frequency at 37 kHz. The system design enables a maximum of 6.4 m detection range in air. With the assistance of signal processing, the rangefinder can realize real-time distance detection up to 32 frames per second. Two mini drones (unfolded size with propellers of 245 mm × 289 mm × 55 mm and max speed in no wind condition of 4 m/s) are loaded with two pMUTs to demonstrate the range finding application. Experimental results show > 6-meter for the pMUT-based rangefinder with small form factor and low power consumption mounted on drones.

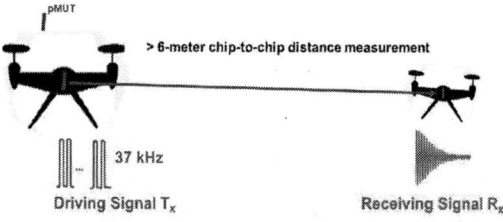

Figure 1: The low-frequency pMUT device for long-distance ultrasound transmissions in air for drone-mounted rangefinder applications. Inset table shows the influence of frequency for the detection range.

DESIGN

In this work, lithium niobate is used as the piezoelectric layer in the pMUT to enable a large electromechanical coupling coefficient, k_t^2, as compared to those of traditional piezoelectrical materials such as AlN, PZT and AlScN [13].

A packaged prototype device is shown in **Figure 2a** with a size of 7 mm ×10 mm × 1.0 mm. The two pins are used for input voltage signals. A well-designed acoustic hole (1 mm in diameter) is used to optimize the acoustic path (AP) for better wave propagations. **Figure 2b** shows the cross-section view of the packaged device. The pMUT is mound at the PCB board and aligned with the acoustic port. In the rangefinder application, two pMUTs are used as the transmitting and receiving unit in two drones and the time of flight (ToF) scheme is used as the sensing principle.

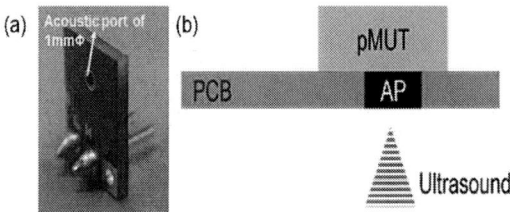

Figure 2: (a) The lithium niobate based pMUT device with package. Size: 7 mm ✕ 10 mm, thickness: 1.0 mm. (b) Cross-section schematic of the device.

DEVICE CHARACTERIATIONS

The transceiver function is characterized under the ambient environment by applying chirp signals with sweeping frequencies between 20-100 kHz and 12 V_{pp} in amplitude. The sound pressure is characterized by a microphone (Type 4138, Brüel & Kjær Type 4138-L-006). The received signal is analyzed with FFT as shown in **Fig. 3a** and results are shown in **Fig. 3b**. Two resonance peaks are identified at 37.32 kHz with a bandwidth of 2.53 kHz, and at 49.35 kHz with a bandwidth of 5.42 kHz. The broadened bandwidth ensures the efficient acoustic energy transfer between the pMUT and the medium. In this work, the working frequency is chosen at 37.32 kHz for high acoustic pressure outputs.

The measured sound pressure level (SPL) with respect to distance from 2 mm to over 1,100 mm is plotted in **Figure 3c**. At 2 mm, a high SPL of over 133 dB is obtained. The low operation frequency result in low attenuation lost such that the SPL remains over 70 dB at 1 m away. Analytically, the sound pressure reduces as the distance increases from the attenuation model [14]:

$$p_x \propto \frac{1}{D} 10^{-\alpha D}$$

where p_x is the sound pressure of a specific point in space; D is the distance, and α is the frequency-related attenuation coefficient. The relationship of sound pressure level (SPL) and distance is described in the following formula:

$$SPL \propto -\alpha D - m * log_{10}(D) + n$$

Where m and n are constants. The measurement results correspond well with the theoretical model as the blue curve in **Figure 3c**.

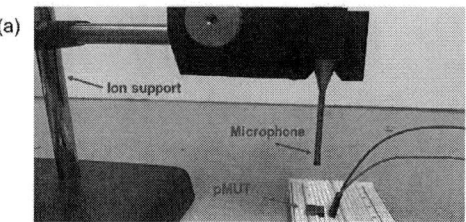

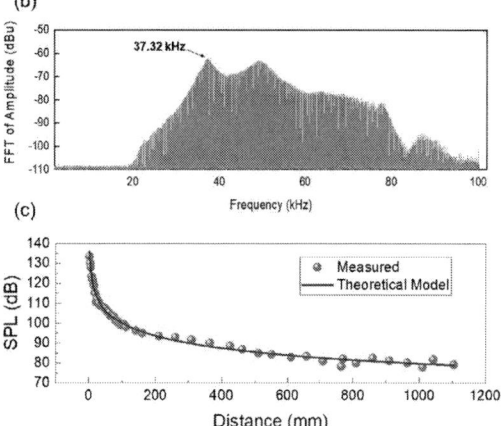

Figure 3: (a) The experimental setup for the PMUT Transceiver characterizations in air. (b) Fast Fourier transform (FFT) of the measured output pressure of a pMUT device in air. (c) Measured results and theoretical model of sound pressure level (SPL) vs. distance at 37.32 kHz under 12 V_{pp}.

MEASUREMENT SETUP AND RESULTS

The characteristics of a prototype pMUT rangefinder is evaluated by measuring the receiving signals amplitude with respect to the distance. As is shown in **Figure 4a**, two pMUTs are used as transmitter (T_x) and receiver (R_x), respectively. The transmitter is driven by a 10-cycle 12 V_{pp} 37.32 kHz square waves with 30 ms as the interval. The

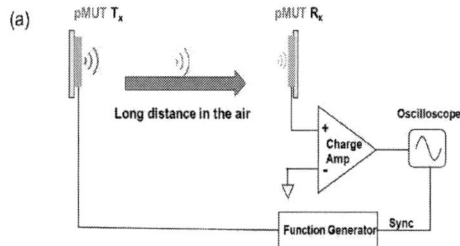

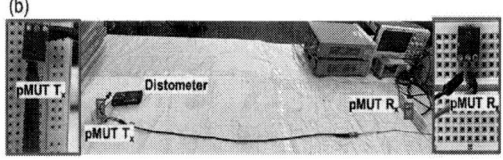

Figure 4: (a) The experimental setup for the distance measurement in air. (b) The optical photo showing the long-distance measurement setup. The insets are the T_x and R_x pMUTs.

978-1-6654-9309-3/23 $31.00 © 2023 IEEE   958

receiver is placed at different distances and the receiving signal is collected through the oscilloscope and amplified by a charge amplifier. **Figure 4b** shows the optical photo of the setup with the insets showing the pMUT mounted on two separated breadboards. A laser distometer is used to measure the distance from 0.4 m to over 6 m.

The receiving signals at the pMUT receiver are extracted and plotted in **Figure 5**. **Figure 5.a** shows the example of the measured receiving signal 5.46 m away from the T_x with good signal-to-noise ratio (SNR) of 22.79 dB. The ToF of 15.92 ms matches perfectly with the real distance by using the velocity of sound of 343 m/s.

The receiving signal amplitude with respect to the distance is plotted in **Figure 5b**. It is found that the signal can be detected up to 6.42 m away. Besides, there are no ring down effects due to the cross-talking phenomena of as the transceiver and receiver are separated (different from pulse-echo measuring principle). This means that there are no blind regions in the receiving signals.

The directional properties of the prototype rangefinder are characterized with a setup shown in **Figure 6a.** Under an excitation signal of 37.32 kHZ and 10 V_{pp} square waveform, the receiving voltage amplitude is measured 30 cm away at the same height. The transmitting pMUT is connected to a rotatable column, and the angle of rotation is measured by using the protractor at the bottom. **Figure 6b** shows the results of the normalized sound pressure level from -90 deg to 90 deg. Results show the range of the -6 dB sound pressure region relative to the maximum value is

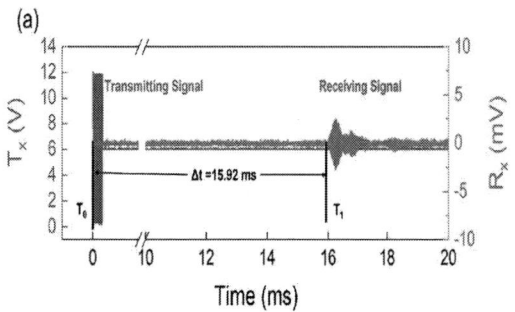

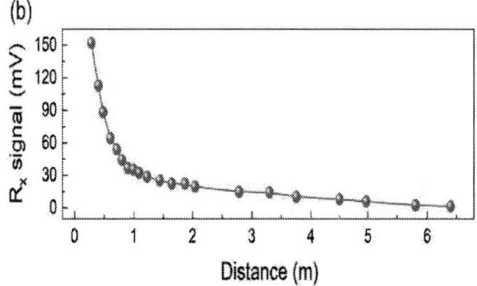

Figure 5: (a) Pulse detection results at a distance of 5.46 m. The transmitting pulse signal has 10 cycles of square waves with 20 ms intervals, and the receiving signal is 15.92 ms after the beginning of transmitting process. (b) Detected signal amplitude with respect to distance from 0.285 m to 6.42 m.

from -30 deg to 30 deg, which implies the rangefinder's field of view is around 60 deg.

One possible application of the rangefinder is for the inter drone collision prevention. In this test, T_x and R_x pMUTs are mounted to the two drones as shown in **Figure 7a.** The small form factor, light weight, and low power consumption of pMUTs enable this possible application for battery-powered drones. Experimentally, the two drones are hovering in air and the rangefinder is working

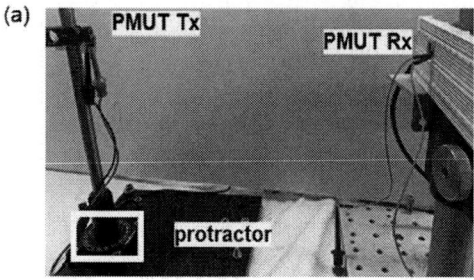

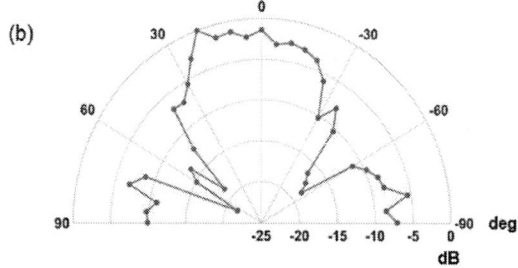

Figure 6: (a) An optical photo showing the experimental setup for the directional property characterization of the pMUT rangefinder. (b) Measured directional characterization results from -90 to 90 deg.

continuously. For the two examples, drones are separated with a distance of 4.38 and 6.05 m (**Fig. 7b**), respectively. The real-time receiving signals are clearly observed from the ToF information. The burst signals applied on the transceiver are continuous periodic signals with the same interval and the calculated distance corresponds well with the actual distance. As such, the distance perception information can be measured continuously which is important as drone-mounted sensing systems for collision prevention or machine vision applications.

CONCLUSION

In this work, a pair of well-packaged pMUTs are utilized for rangefinder applications based on the LN piezoelectric thin film and the application example to drones is demonstrated. The transceiver device is operated at 37.32 kHz while the sound pressure level at 2 mm is 133 dB. The low working frequency enables the low attenuation losses in air and long detecting range with measurements of over 6.4 m Two drone-mounted pMUTs are used as a demo to show the long-distance detection capability of pMUTs in the rangefinder application. The periodical burst signals with 0-12 V_{pp} are applied to the transmitter with a cycle of 30 ms, and the rangefinder system can operate at the 32-fps refresh rate. The pMUT rangefinder system has many advantages, including low power consumption with very small amount of computation power based on the ToF method, independent of lighting conditions, and small factor. The preliminary experiential results in this work also show the long-range detection potential of the pMUT rangefinder for flying drones.

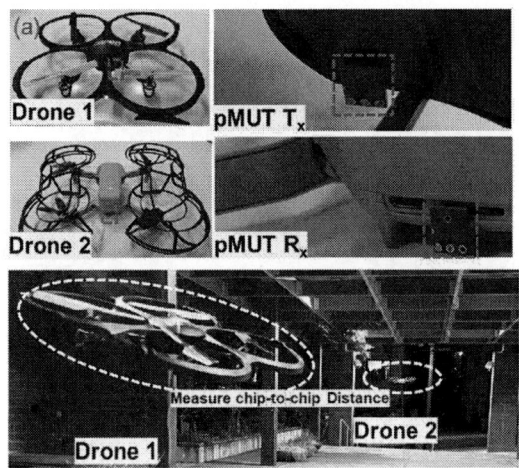

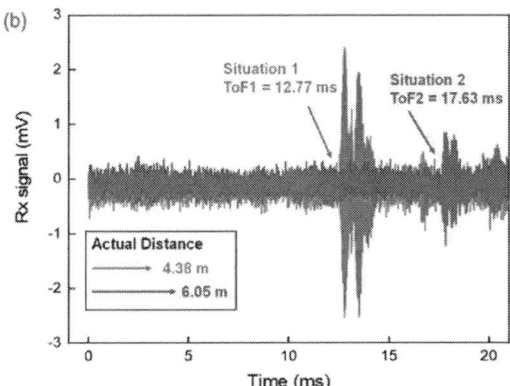

Figure 7: (a) Experimental setup of drone-mounted pMUTs for chip-to-chip distance measurement in real-time. (b) Time-of-flight results of two examples for the distances of two drones of 4.38 m and 6.05 m measured from the pMUT receiver.

ACKNOWLEDGEMENTS

This work is supported by the membership of BSAC (Berkeley Sensor and Actuator Center).

REFERENCES

[1] Przybyla, Richard J., et al. "3D ultrasonic rangefinder on a chip." IEEE Journal of Solid-State Circuits 50.1 (2014): 320-334.

[2] Joontaek J, et al. "Flexible soi-based piezoelectric micromachined ultrasound transducer (PMUT) arrays." Journal of Micromechanics and Microengineering 27.11 (2017): 113001.

[3] Zhou, Xin, et al. "Swarm of micro flying robots in the wild." Science Robotics 7.66 (2022): eabm5954.

[4] Richard J., et al. IEEE Sensors Journal, "In-Air Rangefinding with an AlN Piezoelectric Micromachined Ultrasound Transducer," Vol. 11 (2011): 2690-2697.

[5] Wu, Han, et al. "An ultrasound ASIC with universal energy recycling for> 7-m all-weather metamorphic robotic vision." IEEE Journal of Solid-State Circuits 57.10 (2022): 3036-3047.

[6] Shao, Zhichun, et al. "Bimorph Pinned Piezoelectric Micromachined Ultrasonic Transducers for Space Imaging Applications." Journal of Microelectromechanical Systems 30.4 (2021): 650-658.

[7] Shao, Zhichun, et al. "3D ultrasonic object detections with> 1 meter range." 2021 IEEE 34th International Conference on Micro Electro Mechanical Systems (MEMS). IEEE, 2021.

[8] Luo, Guo-Lun, et al. "Airborne piezoelectric micromachined ultrasonic transducers for long-range detection." Journal of Microelectromechanical Systems 30.1 (2020): 81-89.

[9] Simeoni, et al. "Long-range ultrasound wake-up receiver with a piezoelectric nanoscale ultrasound transducer (pNUT)." 2020 IEEE 33rd International Conference on Micro Electro Mechanical Systems (MEMS). IEEE, 2020.

[10] Simeoni, et al "A 100 nm thick, 32 kHz X-cut lithium niobate piezoelectric nanoscale ultrasound transducer for airborne ultrasound communication." Journal of Microelectromechanical Systems 30.3 (2021): 337-339.

[11] Pop, et al. "Laterally vibrating lithium niobate MEMS resonators with 30% electromechanical coupling coefficient."2017 IEEE 30th International Conference on Micro Electro Mechanical Systems (MEMS). IEEE, 2017.

[12] Smyth, et al. "Experiment and simulation validated analytical equivalent circuit model for piezoelectric micromachined ultrasonic transducers." IEEE Transactions on Ultrasonics, Ferroelectrics, and Frequency Control 62.4 (2015): 744-765.

[13] Pop, et al. "Lithium Niobate Piezoelectric Micromachined Ultrasonic Transducers for high data-rate intrabody communication." Nature communications 13.1 (2022): 1-12.

[14] Przybyla, Richard J., et al. "In-air range finding with an Aln piezoelectric micromachined ultrasound transducer." IEEE Sensors Journal 11.11 (2011): 2690-2697.

CONTACT

*H. Liu; +1-510-345-9811; liuhanxiao@berkeley.edu
*Y. Peng; +1-510-697-6951; yande_p@berkeley.edu
*W. Yue; +1-510-984-8328; wei_yue@berkeley.edu

MASS PRODUCED MICROMACHINED ULTRASONIC TIME-OF-FLIGHT SENSORS OPERATING IN DIFFERENT FREQUENCY BANDS

Richard J. Przybyla[1], Stefon E. Shelton[1], Cathy Lee[1], Benjamin E. Eovino[1],
Quy Chau[1], Mitchell H. Kline[1], Oleg I. Izyumin[1], and David A. Horsley[1,2]
[1]TDK InvenSense, Berkeley, USA and
[2]University of California, Davis, USA

ABSTRACT

This paper presents a new family of ultrasonic time-of-flight (ToF) sensors, which each incorporate AlN piezoelectric micromachined ultrasonic transducers operating at 175 kHz, 85 kHz, and 50 kHz. The work includes a theoretical model for the PMUT and the sensor's maximum range.

KEYWORDS

Ultrasound, piezoelectric, piezoelectric micromachined ultrasound transducer (PMUT), ultrasonic transceiver, rangefinder, time-of-flight

INTRODUCTION

Time-of-flight (ToF) sensors emit a stimulus into the environment and measure the propagation time of the signal as it travels to a target and back. While infrared (IR) optical time-of-flight sensors have been a big commercial success in the past 10 years, they suffer from fundamental problems including interference from sunlight and bright lighting and sensitivity to target color. By contrast, ultrasonic ToF sensors typically operate in a relatively interference-free acoustic band, and are insensitive to target color. In addition, the relatively slow speed of sound (343m/s compared to the 300×10^6 m/s speed of light) allows small changes in target range to be measured accurately without the use of high speed electronics.

Conventional ultrasonic ToF sensors use a piezoceramic transducer which typically has several nF of capacitance, leading to a relatively low impedance in the range of 100-1000Ω at the operating frequency. By contrast, MEMS piezoelectric micromachined ultrasonic transducers (PMUTs) using AlN [1],[2] have capacitance in the range of 10s of pF, with corresponding impedances in the range of 10k-100kΩ. While conventional ultrasonic sensors have high output pressure, they suffer from relatively low receive sensitivity; whereas AlN PMUTs have low transmit sensitivity and high receive sensitivity. As a result, at a given operating frequency, AlN PMUTs can be used to achieve similar maximum ranges compared with conventional ultrasonics, while consuming orders of magnitude lower power and volume [3], [4], [8],[9].

For these reasons, ultrasonic ToF sensors based on AlN PMUTs co-packaged with CMOS systems-on-chip (SoC) have been adopted in several applications including presence sensing, 6DoF tracking, and floor-type detection for robotic vacuum cleaners [4]. This work introduces a new line of ultrasonic ToF sensors operating at 175 kHz, 85 kHz, and 50 kHz.

This paper presents a model for the PMUT design and ultrasonic propagation. It develops a model for the maximum range and discusses SoC operation and shows

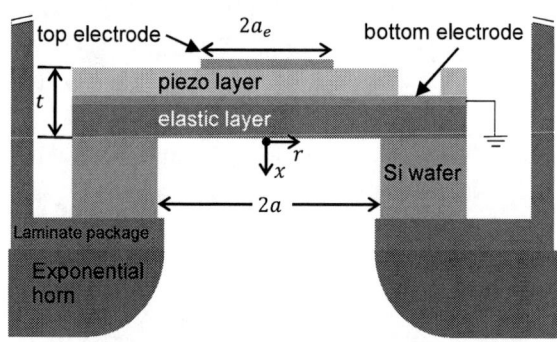

Figure 1. Cross section of a unimorph PMUT composed of a piezoelectric AlN layer sandwiched between top and bottom electrodes over a passive Si elastic layer and enclosed in a laminate package with an exponential horn.

experimental data from the three sensor products.

PMUT MODELING

A simplified cross-section of a basic PMUT design having a radius a is shown in Fig. 1. The sensor membrane comprises an active piezoelectric layer, sandwiched between two metal electrodes, deposited above a passive bending layer and suspended over a through-wafer etch in the Si substrate.

An applied voltage across the electrodes creates a vertical electric field which causes the piezoelectric material to expand in the horizontal plane according to the e_{31} piezoelectric constant of the material. The elastic layer creates a vertical stress gradient, causing the membrane to deflect vertically. Similarly, a pressure difference between the top and bottom of the membrane causes the membrane to deflect vertically, straining the piezoelectric material and causing a charge to develop in the piezoelectric layer.

For a circular membrane functioning as a clamped plate, the approximate modeshape, $d(r)$, is:

$$d(r) = d_0(1 - r^2)^2, \qquad (1)$$

where r is the position along the radius, d_0 is the displacement at the center of the membrane, and d is the vertical displacement. For simplicity, a piston-mode approximation is made, wherein the membrane is analyzed as if it had a uniform modeshape moving with a displacement of d_0, with a smaller area than πa^2 [5]. This allows the analytical solutions for a baffled piston to be used. Using (1) to calculate the effective area of the piston, it can be shown that $SA_{eff} = \frac{1}{3}SA_{PMUT} = \frac{1}{3}\pi a^2 = \pi a'^2$, where $a' = \sqrt{\frac{1}{3}}a$ is the effective (piston) radius. It follows that the modal mass $m_m = SA_{eff}\rho_m t$, where ρ_m is the

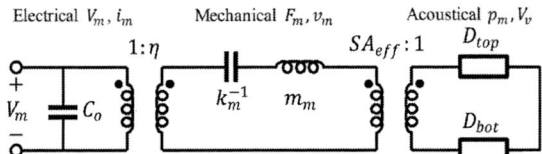

Figure 2. Equivalent circuit model of PMUT showing electrical, mechanical, and acoustical domains.

average density of and t is the thickness of the membrane. Neglecting film stress and acoustic loading effects, the resonant frequency of the membrane is

$$f_n \simeq \frac{10.2t}{2\pi a^2}\sqrt{\frac{E}{12(1-v^2)\rho_m}}, \qquad (2)$$

where E is the average Young's modulus of the membrane, and v is Poisson's ratio. It follows that the modal stiffness

$$k_m = \frac{0.65t^3}{\pi a^2}\frac{E}{(1-v^2)}.$$

The damping of the PMUT is dominated by the energy converted to sound through the acoustic impedance of the sensor. The top of the PMUT radiates into the package, while the bottom radiates into the environment through an exponential horn. The top side acoustic impedance (neglecting the package lid) is approximated using the equation for a baffled piston:

$$D_{top}(\lambda) = \frac{\rho c}{\pi a'^2}\left(1 - \frac{2\lambda J_1\left(\frac{4\pi a'}{\lambda}\right)}{4\pi a'} + j\frac{2\lambda K_1\left(\frac{4\pi a'}{\lambda}\right)}{4\pi a'}\right), \qquad (3)$$

where $\lambda = \frac{c}{f}$ is the wavelength of operation, ρ is the density of air, c is the speed of sound, J_1 is the first order Bessel function, and K_1 is the first order Struve function. A full model of the package acoustics requires finite element modeling and is beyond the scope of this paper.

At the bottom of the membrane we have an impedance D_{bot}. The straight sections of the acoustic path directly below the membrane can be modeled as transmission lines; for simplicity, we ignore those and approximate the horn's throat as starting at the effective radius of the membrane a', and so for an exponential horn with an exponentially increasing surface area $SA(x) = SA_{eff}e^{mx}$ [5],

$$D_{bot}(\lambda) = \frac{\rho c}{\pi a'^2}\left(\sqrt{1 - \frac{m^2\lambda^2}{16\pi^2}} + j\frac{m\lambda}{4\pi a'}\right). \qquad (4)$$

In practice, the real part of (5) dominates and approaches $D_{bot} = \frac{\rho c}{SA_{eff}}$ for a well-designed horn, while the real and complex parts of (4) will be significantly less than $\frac{\rho c}{SA_{eff}}$ for typical PMUT geometries where $a \ll \lambda$. Therefore, most of the generated sound will travel out the acoustic port.

The electrical characteristics of the PMUT depend on the electrode radius a_e and the piezo layer thickness t_p and the piezo dielectric constant ε_{33} and piezoelectric coefficient $e_{31,f}$. The parallel plate capacitance is

$$C_o = \frac{\varepsilon_{33}\pi a_e^2}{t_p}. \qquad (5)$$

The electromechanical coupling constant is

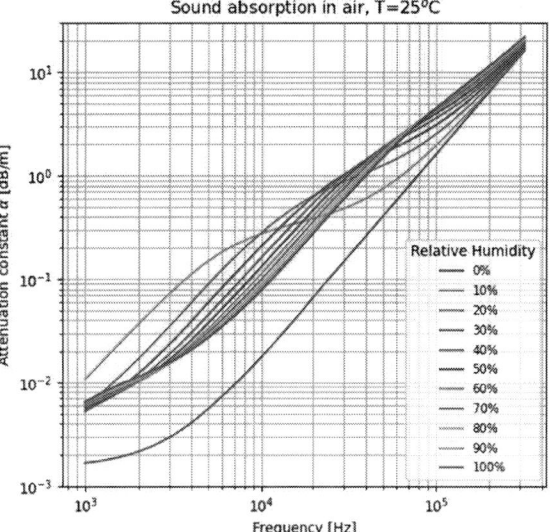

Figure 3. Sound absorption vs. frequency and humidity at standard temperature and pressure.

$$\eta = 12\pi\left(\frac{a_e}{a}\right)^2\left(\left(\frac{a_e}{a}\right)^2 - 1\right)e_{31,f}z_p, \qquad (6)$$

where z_p is the distance from the neutral axis of the membrane to the mid-plane of the piezo film.

Fig. 2 shows the 3 domain equivalent circuit model for the PMUT. The bandwidth of the PMUT at resonance is

$$BW = \frac{1}{2\pi}\frac{b_m}{m_m} = \frac{1}{2\pi}\frac{SA_{eff}^2(D_{top}+D_{bot})}{m_m} \simeq \frac{1}{2\pi}\frac{\rho c}{\rho_m t}. \qquad (7)$$

At resonance, the on-axis transmit sensitivity is

$$s_{tx} = \frac{p_{tx}}{V_m} \simeq \frac{\eta}{SA_{eff}}\frac{a_{th}}{a_{horn}}, \qquad (8)$$

where p_{tx} is the pressure at the mouth of the horn, and a_{th} and a_{horn} are the radii of the horn's throat and mouth, respectively. At resonance, the on-axis receive sensitivity is

$$s_{rx} = \frac{i_{rx}}{p_{rx}} \simeq \frac{a_{horn}}{a_{th}}\frac{SA_{eff}}{\eta D_{bot}\left(\frac{SA_{eff}}{\eta}\right)^2}, \qquad (9)$$

where i_m is the short-circuit receive current and we have assumed $D_{bot} \gg D_{top}$. Further simplification yields

$$s_{rx} \simeq \frac{a_{horn}}{a_{th}}\frac{\eta}{\rho c}. \qquad (10)$$

ACOUSTIC PROPAGATION MODELING

The factors that determine a ToF sensor's maximum range are the transducer's transmit and receive sensitivity, the shape of the target, the acoustic path loss, and the noise floor of the receiver. The transfer function through the transducer and the acoustic channel is:

$$\frac{i_{rx}}{V_{tx}} = s_{tx}s_{rx}G_{targ}G_{path}(R), \qquad (11)$$

where i_{rx} is the received short-circuit current through the PMUT, V_{tx} is the transmit voltage across the PMUT, s_{tx} [Pa/V] is the transmit sensitivity of the PMUT, and s_{rx} [A/Pa] is the receive sensitivity of the PMUT. The target reflectivity G_{targ} is dependent on the size, shape, and

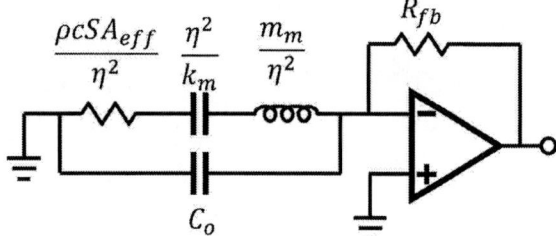

Figure 4. Electrical model for analog front-end amplifier for noise calculations.

texture of the target. For a large flat surface that is oriented towards the sensor, $G_{targ} \simeq 1$.

$G_{path}(R)$ is the acoustic path loss [5]:

$$G_{path}(R) = \frac{SA_{horn}/\lambda}{2R} 10^{-2R\alpha/20}, \qquad (12)$$

where R is the range to the target, SA_{horn}/λ is the Rayleigh range nearfield limit of the horn, and α is the sound absorption coefficient in dB/m, which is a strong function of frequency as shown in Fig. 3 [6].

MAXIMUM RANGE

Fig. 4 shows a simplified electrical model of the analog front end (AFE) for noise calculations. At resonance, the input referred current noise density of the transimpedance amplifier is [7]:

$$\frac{\overline{i_n^2}}{\Delta f} \cong \overline{v_a^2}\omega_o^2 C_o^2 + \frac{4kT}{R_{fb}} + \frac{4kT\eta^2}{\rho c S A_{eff}}, \qquad (13)$$

where $\overline{v_a^2}$ is the amplifier noise voltage, proportional to $\frac{4kT}{g_m}$ for a CMOS amplifier. The amplifier's spot noise is filtered by a subsequent bandpass filter, while the PMUT noise is filtered by the PMUT dynamics. Taking both filter noise bandwidths to be equal to $\frac{2\pi BW}{4}$ and using (7), we have an input referred current variance of

$$\overline{i_n^2} \cong kT \frac{\rho c}{\rho_m t}\left(\frac{\omega_o^2 C_o^2}{g_m} + \frac{1}{R_{fb}} + \frac{\eta^2}{\rho c S A_{eff}}\right). \qquad (14)$$

Combining (8), (10), (11), (12), and (14), we can write the SNR for a large target at a given range:

$$SNR(R) = \frac{\overline{i_{rx}^2}}{\overline{i_n^2}} \cong \frac{\left(\frac{V_{tx}\eta^2}{SA_{eff}\rho c}\frac{SA_{horn}/\lambda}{2R}10^{-\frac{2R\alpha}{20}}\right)^2}{kT\frac{\rho c}{\rho_m t}\left(\frac{\omega_o^2 C_o^2}{g_m} + \frac{1}{R_{fb}} + \frac{\eta^2}{\rho c S A_{eff}}\right)}. \qquad (15)$$

The maximum range occurs when the SNR falls to a value of about 12dB. Eq. (15) shows the importance of the frequency dependent parameter α on the SNR as R increases. In applications requiring fast sample rate and short range, high absorption is desirable because echoes from distant objects are faint and do not interfere with subsequent measurements. Long range applications are best suited to low-frequency transducers because the absorption loss is greatly reduced, as shown in Table 1.

Eq. (15) allows the AFE designer some tradeoffs due to the 3 terms in the denominator: the first term (amplifier noise) is dominant when power consumption and amplifier area are minimized and g_m is low, the 2nd term is dominant when R_{fb} area is minimized and R_{fb} is small, and SNR can

Table 1. Ultrasonic ToF sensor characteristics

Characteristic	Device		
	ICU-10201	ICU-20201	ICU-30201
Typical wavelength	2mm	4mm	7mm
Typical Operating Frequency	175kHz	85kHz	50kHz
Theoretical Absorption Coefficient α	$4.9-9.7$ dB/m	$1.2-3.7$ dB/m	$0.4-1.9$ dB/m
Large target typical max range	2m	5m	9m

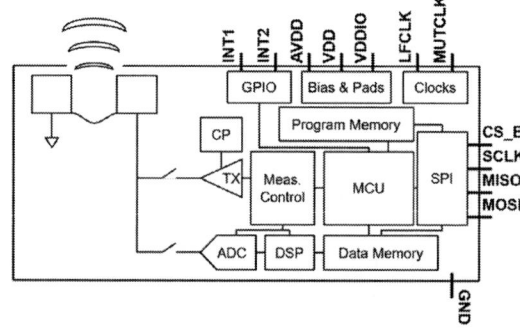

Figure 5. Block diagram of the ultrasonic ToF sensor [8]. A microcontroller (MCU) controls the measurement setup and provides on-board DSP. A measurement control state machine coordinates the ultrasonic measurement while the MCU is asleep. The transmitter includes an on-chip charge pump, while the receiver digitizes and writes the I/Q baseband data directly to a data SRAM.

be maximized by making the third term (air noise) dominant at the cost of AFE power consumption and area.

Eq. (15) is also instructive for PMUT design, as f_o, ρ_m, t, and SA_{eff} are typically design choices. While (15) can be maximized by letting $2\pi BW = \frac{\rho c}{\rho_m t}$ go to zero, this compromises the pulse-echo operation of the transceiver, so the MEMS design must make a tradeoff between bandwidth and SNR. Finally, when the amplifier noise is dominant, the SNR is proportional to η^4/C_o^2. This means that the piezoelectric material's ratio of $e_{31}^2/\varepsilon_{33}$ is critical for device performance.

SENSOR SYSTEM DESIGN

A block diagram of the ToF sensor is shown in Fig. 5 [3]. The programmable SoC contains mixed signal transceiver circuits, a measurement controller which handles the measurement timing in a cycle accurate way, and a 40MHz 16-bit MCU which is used for configuration and for processing the raw data generated by the transceiver. The SoC is manufactured in a 0.18µm 1.8V/3.3V/32V triple-gate oxide CMOS technology with deep N-well.

The transmitter includes an on-chip charge pump (CP)

978-1-6654-9309-3/23 $31.00 © 2023 IEEE

Figure 6: (Top) Isometric view of ICU-20201 ultrasonic ToF sensor, showing the circular acoustic port. (Bottom) Cross-section schematic of ultrasonic ToF sensor, showing MEMS transducer (top) and CMOS ASIC (bottom), enclosed in an open cavity package consisting of 3 laminate layers [10].

that generates 20V from the 1.8V supply without using any external components, which can drive PMUT electrodes with impedance as low as 30kΩ. The transmitter is isolated from the low-voltage AFE by high voltage switches; the transceiver cannot transmit and receive concurrently.

The AFE is a current-current amplifier with programmable gain and attenuator to accommodate the wide dynamic range of the current input from the PMUT. The current from the front-end is converted to a digital signal by a bandpass ΔΣ ADC similar to the one presented in [9], then converted to in-phase and quadrature (IQ) baseband signals by a digital mixer and CIC filters, and written directly to the SRAM for further processing by the CPU or readout via the serial peripheral interface (SPI).

The CMOS and MEMS dice are co-packaged in a custom laminate package shown in Fig. 6 [5]. Unlike capacitive microphones, where the MEMS die is wirebonded directly to the CMOS die to minimize stray capacitance, the lower impedance afforded by piezoelectric transduction reduces sensitivity to parasitics, enabling more flexible sensor packaging.

EXPERIMENTAL RESULTS

Three ultrasonic ToF sensor modules with operating frequencies of 178kHz, 88kHz, and 55kHz and horns with full-width half-max field of views of 20°, 45°, and 45°, respectively, were connected to a SAMG55 Cortex-M4 processor on a TDK SmartSonic™ 2 evaluation board, which was controlled by a PC. The ToF sensors were configured to take 6 measurements per second, with a maximum range of approximately 2.2m, 5.1m, and 9m, and an average current consumption of 90uA, 110uA, and 170uA, respectively, through 1.8V.

The ToF sensors were pointed at a large flat glass surface. An on-chip rangefinding algorithm returned the range and amplitude of the primary echo from the glass surface. The ΔΣ ADC's full-scale range limits the maximum amplitude to approximately 16kLSB; in order to determine the true amplitude at short ranges, the AFE gain was reduced; this reduction was measured, then compensated for. A fit was carried out using the propagation model presented above; the results show a

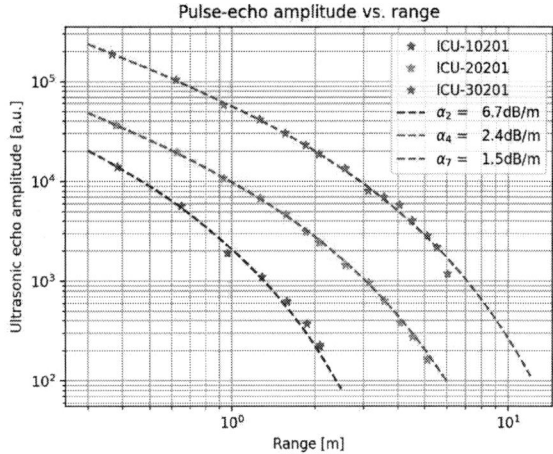

Figure 7: Measured pulse-echo amplitude vs. range for three ultrasonic transceivers (points) and fitted propagation equation (dashed lines).

good match to the theoretical values for the absorption coefficient α for the three transceivers tested.

CONCLUSION

PMUT-based ultrasonic time-of-flight sensors are in mass production. Three sensor products operating at three different frequencies have pulse-echo maximum ranges of 2m, 6m, and 10m.

REFERENCES

[1] S. Shelton et al., "CMOS-compatible AlN piezoelectric micro- machined ultrasonic transducers," in *Proc. IEEE Ultrasonics Symp.*, Oct. 2009, pp. 402–405.

[2] S. Shelton et al., "Improved acoustic coupling of air-coupled micromachined ultrasonic transducers," in *27th IEEE MEMS*, San Francisco, CA, 2014, pp. 753-756.

[3] R. J. Przybyla et al., "In-Air Rangefinding with an AlN Piezoelectric Micromachined Ultrasound Transducer," *IEEE Sensors Journal*, vol. 11, no. 11, pp. 2690-2697, Nov. 2011.

[4] D. A. Horsley et al., "Piezoelectric micromachined ultrasonic transducers in consumer electronics: The next little thing?," in *29th IEEE MEMS*, 2016, pp. 145-148.

[5] D. Blackstock, *Fundamentals of Physical Acoustics.* John Wiley & Sons, 2000.

[6] H.E. Bass et al., *"Atmospheric absorption of sound: Further developments"*, *The Journal of the Acoustical Society of America*, 1995, pp. 680-683.

[7] Richard J. Przybyla, *"Ultrasonic 3D Rangefinder on a Chip,"* *Ph.D. Dissertation, University of California,* Berkeley, CA, December 2013.

[8] TDK InvenSense, "Long Range Ultrasonic Time-of-Flight Range Sensor", *ICU-20201 datasheet*, Sept. 2022.

[9] Richard J. Przybyla et al., "3D Ultrasonic Rangefinder on a Chip," *IEEE Journal of Solid-State Circuits*, vol. 50, no. 1, pp. 320-334, Jan. 2015.

[10] TDK InvenSense, "CH101 Design Guide", *AN-000259*, Sept. 2022.

CONTACT

Richard J. Przybyla, rjprzy@gmail.com

MEMS FIRST-ORDER BESSEL BEAM ACOUSTIC TRANSDUCER FOR PARTICLE TRAPPING AND CONTROLLABLE ROTATING

Jiaqi Li[1†], Zhenhuan Sun[1†], Yuyu Jia[1], Teng Li[1], Haojian Lu[2], Lurui Zhao[3], Hai Liu[3], and Song Liu[1*]*

[1] School of Information Science and Technology
ShanghaiTech University, Shanghai 201210, China
[2] State Key Laboratory of Industrial Control and Technology
Zhejiang University, Hangzhou 310027, China
[3] Department of Electrical and Computer Engineering
University of Southern California, Los Angeles, CA 90089, USA
[†] Both authors contributed equally to the paper.

ABSTRACT

This paper reports a novel and easy-to-fabricate MEMS first-order Bessel beam acoustic transducer (BBAT) for particle trapping and electrically controllable rotating, in terms of the design, fabrication, characterization, and experiments. The BBAT is primarily composed of a bulk lead zirconate titanate (PZT) ceramic and polydimethylsiloxane (PDMS) membrane, bonded by SU-8 photoresist. The transducer encloses customized air-cavities in order to generate Archimedes' spiral acoustic source. We prototyped a 2.32 MHz BBAT with the design, which was capable of firmly trapping polystyrene (PS) particles as large as 800 μm diameter under 18 V_{pp}, and precisely steering the rotating velocity by controlling the driving voltage. A linear relationship between the rotating velocity and driving volage square was observed, which can reach as high as 138 rpm under 150 V_{pp}.

KEYWORDS

MEMS acoustic transducer, acoustic Bessel beam, soft lithography, PDMS, acoustic manipulation, BBAT

INTRODUCTION

Noncontact particle manipulation (NPM) is of significant value in biomedical field such as drug delivery, material characterization, and single cell analysis [1], etc. Typical NPM is by resorting to optical tweezers, magnetic tweezers and acoustic tweezers. Optical tweezers have negative effect to biostructures or living systems for its high light intensity [2], and magnetic tweezers are used in biological single molecule manipulation since it can only affect in nanoscale [3]. In the contrary, acoustic tweezers exhibit attractive advantages, like versatility to materials, good biocompatibility, propagation through opaque medium, adaptiveness to particle size [4-6], all of which makes acoustic tweezers promising potential in biological and medical fields [7].

Ideal Bessel beams are non-diffractive and self-healing, and thus can propagate to infinity. A first order Bessel Beam is characterized by its periodic phase dislocation along circular sound pressure profile with a null center. *N. Jiménez*, et al. [8] theoretically proved that Archimedes' spiral diffraction gratings can generate high order acoustic Bessel-like beams, while *T. Wang* and *R. Zhang* thereafter practically proved this theory by encoding Archimedes' spiral acoustic source through brass or silicon plate [9-10] fabricated by machining and wet etching.

This paper proposes a simple, compact and easy-to-fabricate MEMS first-order BBAT based on soft lithography with improved fabrication accuracy and time-efficiency, and better-quality control. The BBAT is capable of firmly trapping and rotating PS particles with electrical controllability. We anticipate that the BBAT can find potential applications in cell manipulation, particle analysis, and other biomedical scenarios, etc.

DEVICE DESIGN

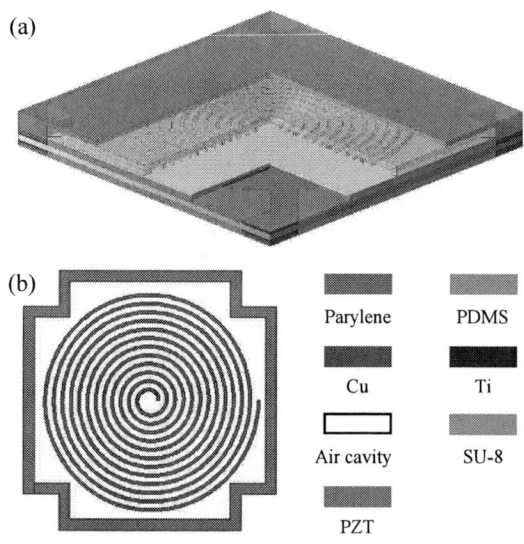

Figure 1: (a) Schematic overview of the first-order BBAT. The transducer consists of a bulk PZT plate and a PDMS film with Archimedes' spiral grating, bonded by SU-8 photoresist, and sealed by Parylene for waterproofing. (b) The contact profile between the PZT and the PDMS membrane, encoding the Archimedes' spiral acoustic source.

Air cavities within continuous solid medium can effectively block the propagation of acoustic waves due to the high acoustic impedance contrast (0.42 kRayl against >1 MRayl). Therefore, dedicatedly designed air cavities can stop part of plane waves in order to form customized acoustic source. In this work, the first-order BBAT is designed by bonding a soft PDMS membrane on top of a bulk PZT plate (Fig. 1a), within which air cavities are encapsulated, in order to generate acoustic source in Archimedes' spiral.

978-1-6654-9309-3/23 $31.00 © 2023 IEEE

IEEE MEMS 2023, Munich, GERMANY
15 - 19 January 2023

The Archimedes' spiral pattern (Fig. 1b) encoded by PDMS is characterized by three parameters in design, i.e., the maximum spiral radius $R = 15$ mm, spiral pitch $a = 1.2$ mm, and line width $w = 0.3$ mm. The line can't be too narrow to transmit enough energy or too wide to deform the acoustic field. The sound source is a 1-mm-thick PZT-5A plate with circular top Cu electrodes. The PZT vibrates in its fundamental thickness mode when 2.32 MHz sinusoidal voltage is applied on the electrodes.

DEVICE FABRICATION

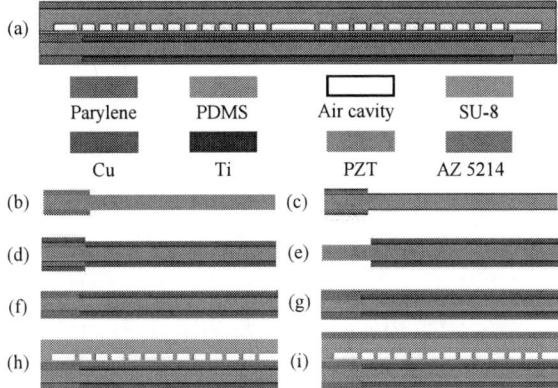

Figure 2: The first-order BBAT fabrication steps. (a) Cross-sectional view of the transducer structure, where the air-cavities block ultrasound to form binary amplitude source. (b) Naked PZT plate and AZ 5214 with electrode pattern, (c)-(d) Deposit 20 nm titanium and 100 nm copper by sputtering. (e) Remove photoresist by acetone and get PZT with circular electrode pattern. (f) Deposit Parylene. (g) Spin coat SU-8 as adhesive layer (600 rpm for 10 s and 2k rpm for 30 s). (h) Cover PDMS film onto PZT followed by over-exposure with 395 nm UV-light. (i) Deposit 9 μm thick Parylene for waterproofing.

The fabrication of the first-order BBAT primarily consists of three steps, i.e., preparation of the PZT plate, fabrication of the PDMS membrane, and further processing including bonding, soldering and waterproofing. Fig. 2a gives the overall layered structure of the BBAT.

The 1 mm thick naked PZT plate is diced into 36×36 mm² piece (Fig. 2b). Then sputter and pattern circular electrode (20 nm thick titanium and 100 nm copper) area on PZT with photolithography (Fig. 2c and d). After acetone rinsing, redundant metal and residual photoresist can be removed (Fig. 2e). The PZT plate preparation finally ends with 5.4 μm thick Parylene deposition (Fig. 2f) for tight bonding with PDMS membrane.

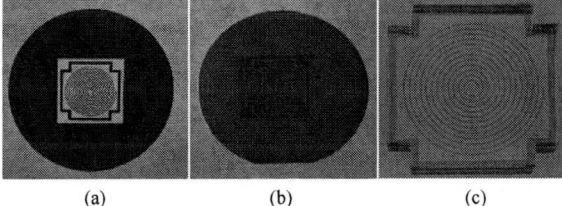

Figure 3: (a) Mask for mold fabrication with pattern in Fig. 2. (b). Mold with 170 μm thick SU-8. (c) the fabricated and tailored PDMS membrane.

The PDMS fabrication follows canonical soft lithography process. Mold of PDMS membrane is made by silicon wafer and SU-8 3025. First, spin coat 170 μm thick SU-8 on silicon wafer and prebake (5 mins at 65 °C on hot plate, 45 mins at 95 °C on hot plate and 2 hours at 95 °C in oven). Then transfer the mask pattern (Fig. 3a) to SU-8 followed by exposure for 60s under 365 nm UV light with intensity of 25 mW/cm². After develop and post bake (5 mins at 65 °C and 45 mins at 95 °C), the mold is fabricated (Fig. 3b). Finally, the mold is silanized by HMDS to reduce the adhesion between the mold and PDMS for easy peeling off. The mixture ratio of curing agent to pre-polymer is 1: 5. After 15 minutes stirring and vacuum degassing, the PDMS mixture is poured onto mold. Then cure the PDMS mixture at 85 °C for one hour. Finally, peel off the PDMS membrane and customize its shape to fit PZT size (Fig. 3c).

Then spin coat another SU-8 layer to PZT as adhesive layer (Fig. 2g) to bond the PDMS and PZT together with UV exposure (Fig. 2h). The last step is to solder Teflon wires on four corners of the transducer and deposit another Parylene layer with PECVD for waterproofing (Fig. 2i). The fabricated transducer is shown in Fig. 4.

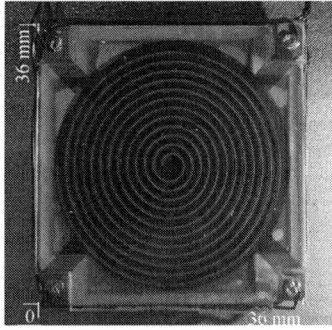

Figure 4: The fabricated First-order BBAT.

DEVICE CHARACTERIZATION

The fabricated first-order BBAT was immersed in DI water enclosed in $10 \times 10 \times 10$ cm³ water tank (Fig. 5). Then the BBAT was connected to power amplifier (Aigtek ATA-3415), which can provide as high as 150 V_{pp} and 105W power. The signal generator used was DG1022Z.

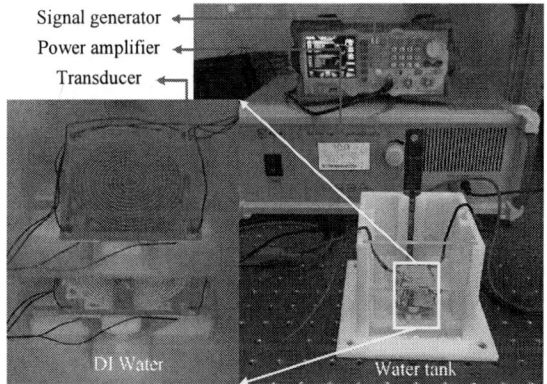

Figure 5: The experimental setup to characterize the fabricated first-order BBAT, including a signal generator DG1022Z and a power amplifier Aigtek ATA-3415.

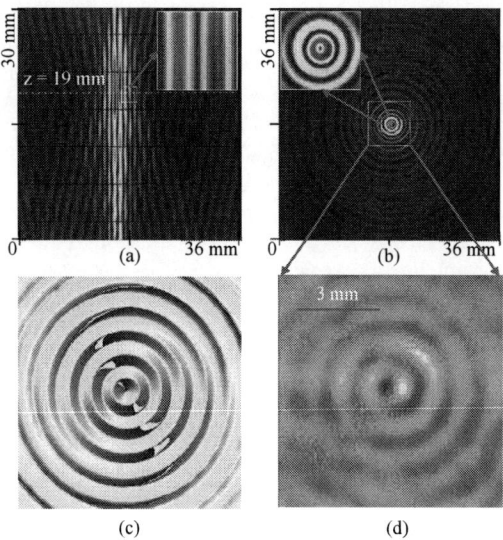

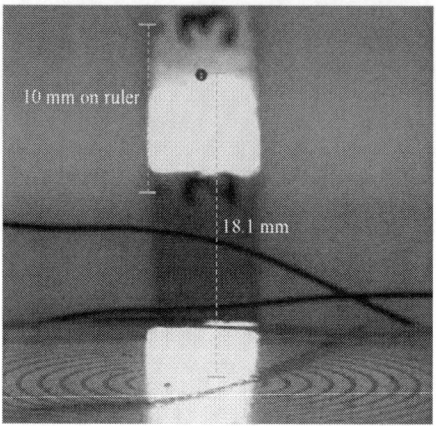

Figure 7: A PS particle was trapped amid water about 18.1 mm above the PZT source. Particle diameter is 800 μm.

Figure 6: Simulation and experiment characterization of both the vertical and transversal acoustic pressure profiles. (a) Simulated vertical acoustic pressure profile in XZ-plane. Maximum acoustic pressure appears at plane of z = 19 mm. (b) Simulated transversal acoustic pressure profile, (c) simulated periodical phase profile at the central region, and (d) scanned transversal acoustic pressure profile with 0.2 mm needle hydrophone (Precision Acoustics Ltd., UK), in XZ-plane at z = 19 mm.

The first-order BBAT was characterized by both simulation with COMSOL Multiphysics and experiment with the setup in Fig. 5. As can be seen from Fig. 6, the simulated vertical pressure profile (Fig. 6a) shows good propagation property along depth direction and reaches the maximum at z = 19 mm; while the simulated transversal pressure (Fig. 6b) and phase (Fig. 6c) profiles conform to the first-order Bessel function and would yield a vortex in the circular center. The fabricated first-order BBAT was actuated by 2.32 MHz sinusoidal voltage with 30 V_{pp}. The generated acoustic filed was scanned with a 0.2 mm needle hydrophone (Precision Acoustics Ltd., UK), and the transversal pressure profile (Fig. 6d) well agrees with the simulation, which validate the effectiveness of the fabrication recipe.

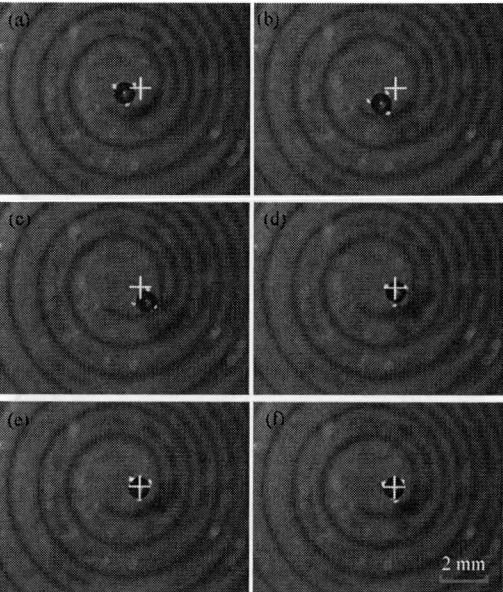

Figure 8: The PS particle trapping process recorded by digital microscope (Dino-Lite AM7115MZTL). The time interval from (a) to (f) is 0.3 s. Particle size is 800 μm.

EXPERIMENTS

Particle trapping and rotating experiments were conducted with PS particle, which has positive acoustic impedance contrast against DI water and thus can be trapped by the vortex beam generated from the first-order BBAT. A PS particle can both be trapped amid water (Fig. 7) and on water surface due to surface tension (Fig. 8f). With the experiment setup in Fig. 5, we first set the actuation voltage to be 60 V_{pp}, and then released an 800 μm diameter PS particle by a drop pipette to the BBAT focal zone, then the particle was firmly trapped amid water (Fig. 7). To better illustrate the trapping process, the PS particle was placed on water surface which initially floated around. Then moved the PS particle to central zone, and turn on power amplifier at 6 V_{pp}. Through Dino-Lite Edge digital microscope (AM7115MZTL), the PS particle trapping process was recorded (Fig. 8).

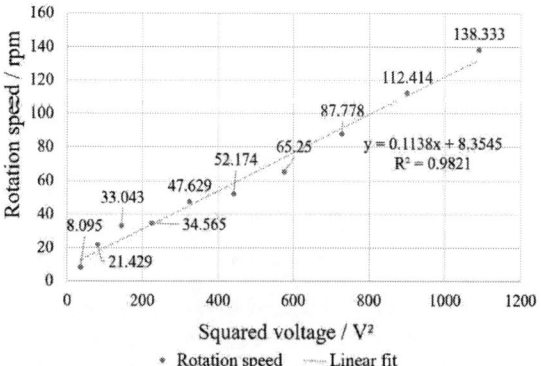

Figure 9: The measured PS particle rotating velocity against voltage square. The observations well fit to a linear relationship, conforming well to theoretical analysis.

After the PS particle was trapped on water surface (Fig. 8f), the actuation voltage was continuously increased from 6 V_{pp} to 33 V_{pp} in order to modulate the particle rotating velocity. We used Dino-Lite Edge digital microscope (AM7115MZTL) to record particle rotating in order to analyze rotating velocity. Fig. 9 shows that rotating speed of PS particle increased linearly against input voltage square, which confirms well to theoretical analysis [11].

CONCLUSION

This paper reported the design, simulation, fabrication, characterization and experiment results regarding a novel first-order BBAT. Both the scanned acoustic field profile and trapping experiments show that the BBAT successfully generated first-order acoustic Bessel beam, and also proved its capacity for particle trapping and electrically controllable particle rotating. The first-order BBAT has great application potentials in biomedical engineering, material science, etc.

ACKNOWLEDGEMENTS

This paper is based on the work supported by Shanghai Pujiang Talents Program under Grant 2021X0203-101-01.

REFERENCES

[1] Y. Ren, Q. Chen, M. He, X. Zhang, H. Qi, and Y. Yan, "Plasmonic optical tweezers for particle manipulation: Principles, methods, and applications", *ACS nano*, vol. 15, no. 4, pp. 6105-6128, 2021.

[2] A. B. Castro, "Optical tweezers: Phototoxicity and thermal stress in cells and biomolecules", *Micromachines*, vol. 10, no. 8, pp. 507, 2019.

[3] M. J. Shon, S. H. Rah, and T. Y. Yoon, "Submicrometer elasticity of double-stranded DNA revealed by precision force-extension measurements with magnetic tweezers", *Science advances*, vol. 5, no. 6, pp. eaav1697, 2019.

[4] M. Wiklund, "Acoustofluidics 12: Biocompatibility and cell viability in microfluidic acoustic resonators", *Lab on a Chip*, vol. 12, no. 11, pp. 2018-2028, 2012.

[5] K. H. Lam, Y. Li, Y. Li, H. G. Lim, Q. Zhou, and K. K. Shung, "Multifunctional single beam acoustic tweezer for non-invasive cell/organism manipulation and tissue imaging", *Scientific reports*, vol. 6, no. 1, pp. 1-7, 2016.

[6] M. Sundvik, H. J. Nieminen, A. Salmi, P. Panula, and E. Hæggström, "Effects of acoustic levitation on the development of zebrafish, Danio rerio, embryos", *Scientific reports*, vol. 5, no. 1, pp. 1-11, 2015.

[7] A. Ozcelik, J. Rufo, F. Guo, Y. Gu, P. Li, J. Lata, and T. J. Huang, "Acoustic tweezers for the life sciences", *Nature methods*, vol. 15, no. 12, pp. 1021-1028, 2018.

[8] N. Jiménez, R. Picó, V. Sánchez-Morcillo, V. Romero-García, L. M. García-Raffi, and K. Staliunas, "Formation of high-order acoustic Bessel beams by spiral diffraction gratings", *Physical Review E*, vol. 94, no. 5, pp. 053004, 2016.

[9] T. Wang, M. Ke, W. Li, Q. Yang, C. Qiu, and Z. Liu, "Particle manipulation with acoustic vortex beam induced by a brass plate with spiral shape structure", *Applied Physics Letters*, vol. 109, no. 12, pp. 123506, 2016.

[10] R. Zhang, H. Guo, W. Deng, X. Huang, F. Li, J. Lu, and Z. Liu, "Acoustic tweezers and motor for living cells", *Applied Physics Letters*, vol. 116, no. 12, pp. 123503, 2020.

[11] W. Li, M. Ke, S. Peng, F. Liu, C. Qiu, and Z. Liu, "Rotational manipulation by acoustic radiation torque of high-order vortex beams generated by an artificial structured plate", *Applied Physics Letters*, vol. 113, no. 15, pp. 051902, 2018.

CONTACT

*Song Liu, E-mail: liusong@shanghaitech.edu.cn
or Hai Liu, E-mail: hailiu@usc.edu

NON-INVASIVE CAROTID ARTERY MONITORING BY USING ALUMINUM NITRIDE PMUT CLOSE-PACKED ARRAYS

Sheng Wu[1,2,3], Kangfu Liu[2], Shuai Shao[2], Wei Li[1,3], Ying Chen[1,3], Tao Wu[2], Xinxin Li[1,3]

[1]Shanghai Institute of Microsystem and Information Technology, Chinese Academy of Sciences and
[2] ShanghaiTech University, CHINA and
[3]University of Chinese Academy of Sciences, CHINA

ABSTRACT

In this work, we present a prototype for the first time of using miniaturized aluminum nitride (AlN) micromachined piezoelectric ultrasonic transducer (PMUT) arrays for non-invasive carotid artery monitoring. The designed hexagonal close-packed PMUT arrays device is in size of $3\times3mm^2$, which contains four independent arrays. And any two of arrays can be used to accomplish pulse-echo method on skin tissue above carotid artery. The artery pulsation is successfully acquired with high signal-to-noise ratio and transformed to depth positioning image along time domain. Besides heart rate, the largest relative displacement of carotid artery wall to neck skin is also estimated via long time recording.

KEYWORDS

Artery monitoring, PMUT arrays, Pulse-echo, Pulsation wave, Portable medical ultrasonic system.

INTRODUCTION

As the most traditional and powerful auxiliary method for medical application, ultrasound imaging system has been used for decades and proved with advantages such as real-time response, non-invasion treat and safety. However, it is still difficult for people to get medical ultrasound diagnosis without the help of hospital equipment.

Handheld personal ultrasound devices have evolved with MEMS technology in recent years, especially for the capacitive micro-machined ultrasound transducer (CMUT) and PMUT devices. The actual production such as Butterfly-IQ with CMUT chips was released in consumer electronics market in 2018 for the first time. And it has been proved that with the help of these new smaller equipment, self-surveillance with ultrasound method is possible for people even at home.

Up to date, it is still less than common of non-invasive medical detecting using PMUT devices. In the previous related works, Alessandro Stuart Savoia group developed PZT PMUT arrays with relative large area to produce B-mode images on artery position [1, 2]. Xiaoyue (Joy) Jiang group proposed a 5-MHz PZT PMUT array with chip area of $1.5\times1.5\ mm^2$ for radial artery wall motion detecting [3]. Yande Peng group reports a subcutaneous blood pressure (BP) monitoring system in sheep [4]. Hong Ding group proposed a blood flowmeter concept validation experiment by the pulsed wave Doppler method [5]. Siyu Liu group reports miniaturized sensing system for non-invasive measurement of blood temperature [6].

In our previous work, an AlN PMUT with pre-concaved membrane was proposed and tested with excellent displacement performance and high quality factor

in high frequency from around five to nine megahertz in air environment. And the resonance frequencies of our devices after packaging are still proper for the medical ultrasonic application. Compared to PZT, AlN material is nontoxic and environment-friendly. The fabrication of AlN sputtering can be accomplished in a relative lower temperature, which is compatible for IC fabrication and with potential of COMS-MEMS chips. Therefor in this work, we further investigated the usage of these AlN PMUT array devices to the portable health surveillance field of superficial artery monitoring, which demonstrate the great potential of AlN PMUT devices.

DEVICE DESIGN AND TESTING SYSTEM

Device and Array Design

The designed PMUT structure is shown in Fig. 1(a), in which AlN is chosen as piezoelectric material for IC-compatible fabrication. Molybdenum as bottom electrodes helps better crystal orientation during AlN sputtering. Low pressure chemical vapor deposition (LPCVD) tetraethyl orthosilicate (TEOS) is used as sacrificial layer for vacuum cavity. Polysilicon served as elastic layer and thermal oxidation provides excellent isolation performance. The details of this fabrication has been discussed in [7]. The dynamic performance in high frequency in air median is shown in Fig. 1 (b), which also shows the reflection echo amplitude in water at distance of 7 mm. It is clear that the resonance frequency has an obvious drop and the bandwidth goes up to around 1MHz.

The designed PMUTs arrays are in the hexagonal geometry and all 91 units are connected in parallel in one array to enhance the acoustic output. The total capacitor of one PMUT array is about 23pF without considering fringe field effect.

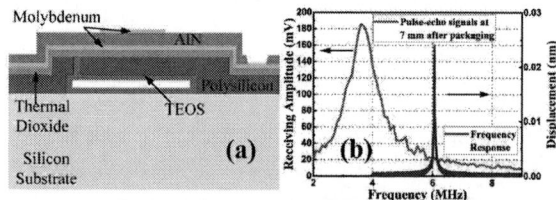

Figure 1: (a) Schematic of our PMUT device structure. (b) The displacement frequency response in air (the blue one) and pulse-echo reflection amplitude (the red one) in water after package.

Pulse-Echo Signal Capture System

The selected PMUT arrays device is wire-bonded and packaged to the customized printed circuit board (PCB) with epoxy adhesive. And for both membrane protection and sound wave transmitting functions, the human silica gel is spread and then cured on the top surface of device. And thickness of this silica gel is approximately 0.5 mm. Shown in Fig. 2 (a), there are four independent arrays on one chip and only two of them are utilized in this pulse-echo application. And any two of them can be used to accomplish pulse-echo function. For convenience, the top two arrays are put to use in this work.

The schematic of signals capture system is shown in Fig.2 (b). The arbitrary wave generator is used as stimulating signal source. And according to the experiment results shown in Fig.1 (b), 5 pulse of signal with the frequency of 3.75 MHz is chosen for larger receiving signal amplitude. Under excitation of the sinusoidal pulses, the corresponding ultrasound is generated towards carotid artery via superficial tissue. Due to the distinct sound impedance mismatch around artery vessel wall, the reflected signal happens and carries with the motion information. The reflection signal is captured via charge amplifier and high-pass filter circuit, and then can be displayed on the oscilloscope. Via the LABVIEW software, the exact signals of both excitation and reflection can be transferred to the personal computer for further processing.

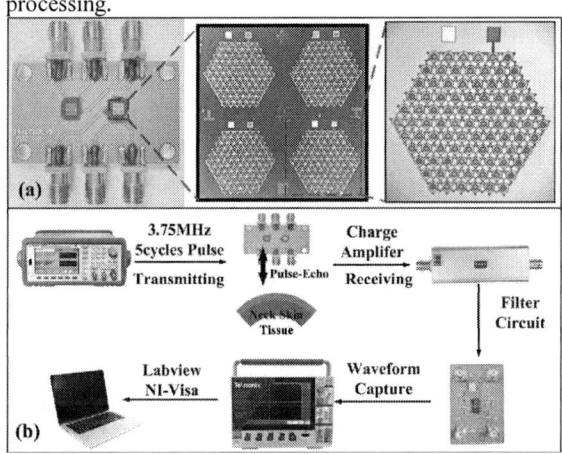

Figure 2: (a) PMUT arrays Device packaged on the customized PCB and its enlarged view. (b) The signals capture system in carotid artery monitoring application.

TESTING RESULTS

The excitation and reflection signal of once time acquisition during long time recording is shown in Fig. 3 (a). Since the transmitting and receiving PMUT arrays stay on the same chip, the coupling effect between them makes it possible to get both exciting and reflecting information via receiving array. The fast Fourier transform (FFT) results are shown in the inset (b), which shows with dissatisfactory frequency difference resolution between excitation and reflection part due to limited pulse width and data size. And shorter pulses can provide better axial space resolution along depth direction but cause simultaneous smaller reflection amplitude. On the other hand, the pulse-echo method gives the flight distance information. Here, we use 5 cycles of pulses with the frequency of 3.75MHz. The amplitude of excitation signals is 20V peak-to-peak value and leads to the more than 60mV peak-to-peak value

of receiving signals shown in the inset (c). For the subsequent processing in the software MATLAB, both upper and lower envelops of original and digital-filtered signals are displayed. The digit band-pass filter helps to improve the signal-to-noise ratio and to reveal clear peak boundary.

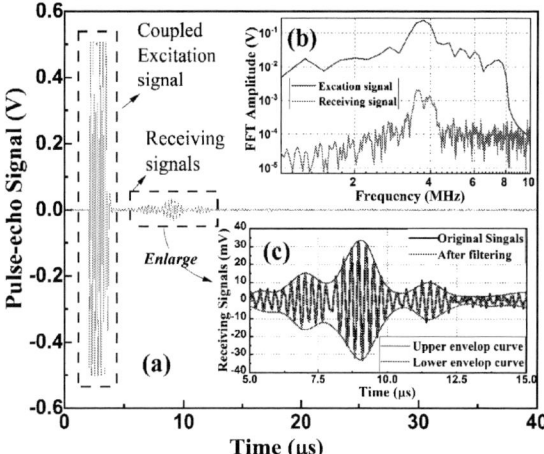

Figure 3: (a) Stimulating signals and reflection signals in once acquisition. (b) The FFT results are also displayed. (c) The enlarged view of reflection signals before and after filtering.

When the carotid artery moves with the pulse beat, the distance between vessel wall and skin above artery is also changed periodically. And this distance can be detected by using time of flight (ToF) of the reflection pulse signals via formula (1). Hereinto, d is the depth information of carotid vessel wall and c is the sound speed in human tissue (around 1584m/s [8]). Consequently, the average depth of upper artery wall below the skin is calculated about 4.7 mm.

$$d = c \cdot ToF /2 \qquad (1)$$

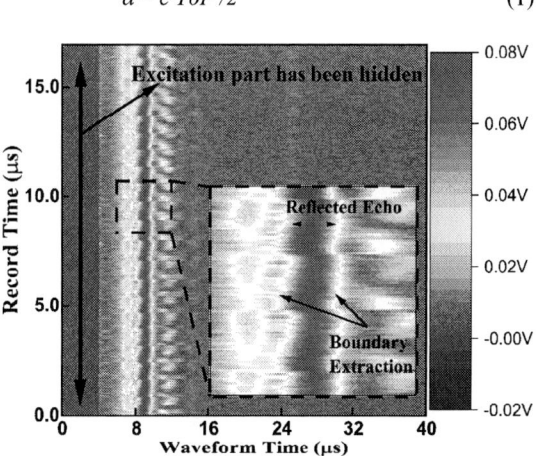

Figure 4: Reflection signals amplitude contour figure of long time recording.

Using the difference of two envelops shown in Fig. 3 (c), the reflection wave amplitude can be extracted. And to enlarge the pulsation feature of carotid artery, the contour figure via MATLAB is used and shown in Fig. 4. It can be demonstrated that during long time recording of more than 15 seconds, the reflection signals of artery wall exhibits regular pulsation motion.

The boundary value of the reflection signals can be extracted with the enlarged contour inset shown in Fig.4. Between two boundaries of reflection wave, the left one gives the more clear pulsation shape and can be used for pulse extraction.

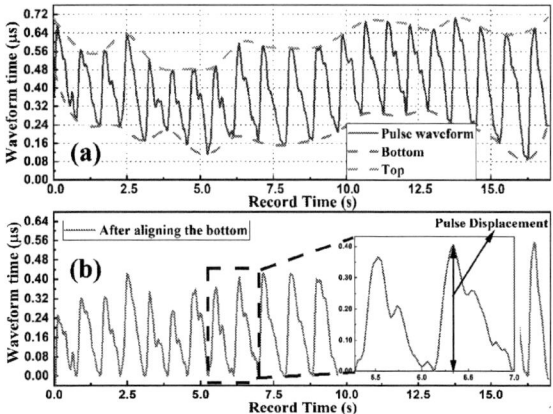

Figure 5: Extracted pulse wave of carotid artery for long time recording.

The pulsation of the TOF variation in several cardio cycles is shown is Fig.5 (a). Both top and the bottom envelops are displayed, which demonstrated that hand shaking during the measurement can also leads to reflection signals changing. After eliminating the bottom variation, the final pulsation signals of artery wall is shown in the inset (b). The heart rate is estimated as 76 times per minute and the largest relative displacement between skin and upper vessel wall is about 0.3mm.

DISCUSSION

The spatial resolution on the axial direction of this array can be calculated as 211 μm, which is not high enough for the distinction of each layer of vessel wall.

Compared to previous related work of ultrasound imaging in the artery wall extraction, this work used signals boundaries for identifying the vessel motion. And it is more direct and robust on the TOF working principle. For the pulsation information, the exact boundary value for each measurement is not necessary since the relative displacement is nearly equal in the reflection signals. And the grain pattern on the both side of the largest peak in reflection signals could be generated by other tissue such like muscles.

It is also possible to extract carotid artery wall velocity by using regular Doppler frequency spectrum method. But the resonance frequency should be larger to reach better resolution. Simultaneously, higher frequency introduces higher power dissipation in the human tissue, which may require larger device area. The single array used in this work is in the size of 1×1mm², which indicates that it is

possible to design ultra-small ultrasound transducer in practical situation.

This work demonstrates the feasibility of utilizing AlN for non-invasive medical ultrasonic application in miniaturized device size. It shows the great possibility of portable integrated CMOS-MEMS medical ultrasonic systems in the future.

The future work will be focused on the higher frequency PMUT array design with stronger output sensitivity for portable, non-invasive and high-resolution medical ultrasound application.

CONCLUSION

This work utilized miniaturized AlN PMUT arrays device as ultrasound transducer to capture carotid artery vessel pulsation signals. Apart from the hardware circuit processing, the signals are also analyzed within MATLAB software. The carotid artery pulse wave is successfully obtained with high signal-to-noise ratio. The depth and displacement of the upper vessel wall can be also estimated and are consistent with normal values. The results demonstrates the great potential of portable medical ultrasound system using PMUT devices. The future work would be focused on the performance improvement of this application.

ACKNOWLEDGEMENTS

This work was supported in part by the National Science Foundation of China Projects (62074151, 61834007), Innovation Team and Talents Cultivation Program of National Administration of Traditional Chinese Medicine (No: ZYYCXTD-D-202002, ZYYCXTD-D-202003) and the Lingang Laboratory under Grant No. LG-QS-202202-05. The authors also gratefully acknowledge ShanghaiTech Quantum Device Lab and ShanghaiTech Public Research Facilities Platform.

REFERENCES

[1] A. S. Savoia, R. Matera, F. Quaglia, and S. Ricci, "A feasibility study of a PMUT-based wearable sensor for the automatic monitoring of carotid artery parameters", in *2021 IEEE International Ultrasonics Symposium (IUS)*, 11-16 Sept. 2021, pp. 1-4.

[2] A. S. Savoia et al., "Design, Fabrication, Characterization, and System Integration of a 1-D PMUT Array for Medical Ultrasound Imaging", in *2021 IEEE International Ultrasonics Symposium (IUS)*, 11-16 Sept. 2021, pp. 1-3.

[3] X. J. Jiang et al., "Piezoelectric Micromachined Ultrasonic Transducers for Blood Vessel Motion Tracking", in *2021 IEEE 34th International Conference on Micro Electro Mechanical Systems (MEMS)*, 25-29 Jan. 2021, pp. 423-425.

[4] Y. Peng, S. Pala, Z. Shao, H. Ding, J. Xie, and L. Lin, "Subcutaneous and Continuous Blood Pressure Monitoring by PMUTs in an Ambulatory Sheep", in *2022 IEEE 35th International Conference on Micro Electro Mechanical Systems Conference (MEMS)*, 9-13 Jan. 2022, pp. 416-419.

[5] H. Ding et al., "A Pulsed Wave Doppler Ultrasound Blood Flowmeter by PMUTs", *Journal of*

Microelectromechanical Systems, vol. 30, pp. 680-682, 2021.

[6] S. Y. Liu, X. H. Feng, R. C. Zhang, and Y. J. Zheng, "Portable photoacoustic system for noninvasive blood temperature measurement", in *2018 Ieee International Symposium on Circuits and Systems (Iscas)*, MAY 27-30, 2018, pp. 1-5.

[7] S. Wu, W. Li, D. Jiao, H. Yang, T. Wu, and X. X. Li, "An Aluminum-Nitride Pmut with Pre-Concaved Membrane for Large Deformation and High Quality-Factor Performance", in *2021 21st International Conference on Solid-State Sensors, Actuators and Microsystems (Transducers)*, JUN 20-25, 2021, pp. 46-49.

[8] M. P. Brewin, P. D. Srodon, S. E. Greenwald, and M. J. Birch, "Carotid atherosclerotic plaque characterisation by measurement of ultrasound sound speed in vitro at high frequency, 20MHz", *Ultrasonics*, vol. 54, pp. 428-441, 2014.

CONTACT

*Tao Wu, tel: +86 (021) 20685357;
wutao@shanghaitech.edu.cn
*Xinxin Li, tel: +86-21-6213-1794;
xxli@mail.sim.ac.cn

NON-LINEAR BEHAVIORAL MODELING OF CAPACITIVE MEMS MICROPHONES

Sebastian Anzinger[1,2], Hutomo Suryo Wasisto[1], Abhiraj Basavanna [1], Marc Fueldner[1] and Alfons Dehé[2,3]

[1]Infineon Technologies AG, 85579 Neubiberg, Germany
[2]Department of Microsystems Engineering, University of Freiburg, 79108 Freiburg, Germany
[3]Hahn-Schickard-Gesellschaft, 78052 Villingen-Schwenningen, Germany

ABSTRACT

In this work, a non-linear behavioral lumped model applicable to commercial state-of-the-art single-backplate microelectromechanical system (MEMS) microphones is described. The model allows to simulate all relevant microphone key performance indicators (KPIs), including the bias voltage dependent sensitivity, signal-to-noise ratio (SNR), and input pressure dependent total harmonic distortion (THD) of the microphone output voltage. Non-linear behavioral descriptions of the capacitive transduction and mechanical properties of a membrane are considered, which can be implemented in common circuit simulators using industry standard hardware description languages (i.e., Verilog-A). While state-of-the-art lumped models focus on purely linear microphone investigations of the sensitivity and SNR, the proposed model inherits an improved accuracy and allows to consider the system non-linearity and THD in theoretical design optimizations. Measurement results of a commercial MEMS microphone were employed to validate the precision of the developed model.

KEYWORDS

MEMS, Microphone, Verilog-A, Modeling, SNR

INTRODUCTION

Capacitive microelectromechanical system (MEMS)-based microphones have reached a mature state of fabrication and nowadays form a multi-billion-dollar market. Their key performance indicators (KPIs) are primarily bound to their acoustical characteristics, which include acoustical sensitivity, signal-to-noise ratio (SNR) and output linearity quantified as the total harmonic distortion (THD). The sensitivity of a microphone is typically specified to be -38 dBV/Pa and reached by trimming the bias voltage applied to the MEMS chip. A high SNR particularly ensures low noise recordings of low sound-pressure-level audio signal and defines a bottom limit of the microphone dynamic range. At increasing sound pressures, the recording quality is limited by the microphone non-linearity, causing harmonically distorted output signals. Consequently, both high SNR and low THD are demanded nowadays in high-performance microphones to ensure a good audio recording quality and a high dynamic range [1,2].
The MEMS design is conventionally supported by lumped-element based simulations. Corresponding lumped models are typically linearized small-signal approximations and focus on simulating and optimizing the sensitivity and SNR, while the output linearity is neglected [3,4]. In this work, a fully non-linear behavioral simulation model of a MEMS microphone is presented, which allows for precise

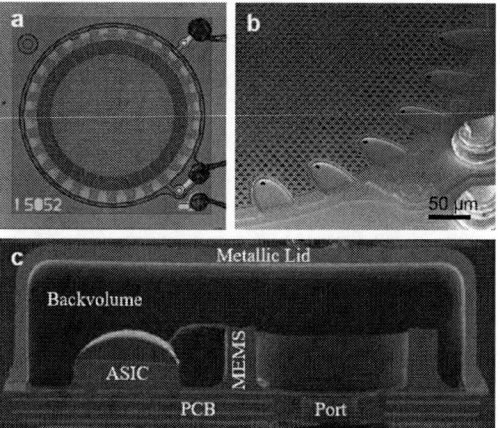

Figure 1: (a) Optical microscopy and (b) scanning electron microscopy (SEM) image of the microphone chip. (c) SEM image of an ASIC-integrated MEMS microphone.

simulations of all relevant microphone KPIs (e.g., sensitivity, SNR, and THD).

CAPACITIVE MEMS MICROPHONE

The subsequent modeling approach is shown and validated for a commercial capacitive MEMS microphone from Infineon Technologies AG (see Figs. 1(a)-(c)). The 0.97×0.97 mm² sized MEMS consists of a 330 nm thick circular polysilicon membrane, which is clamped at its perimeter. The tensile prestress in the material is lowered via six corrugation rings. A highly perforated counter-electrode (backplate) is spaced by a 2.0 µm air gap from the membrane, forming a capacitive system. An insensitivity of the microphone to low frequency pressures is ensured via 24 ventilation holes having a diameter of 5.5 µm (see Fig. 1(b)). The MEMS chip is placed in a $3.2 \times 1.9 \times 0.9$ mm³ sized package and wire bonded to an application specific integrated circuit (ASIC) enabling a constant charge-based readout of capacitance changes. The package comprises a printed circuit board (PCB) and a metal lid. While a circular hole in the PCB defines a Helmholtz-resonator shaped sound port, the metal lid encloses an air volume called backvolume.

NON-LINEAR BEHAVIORAL MODELING

A behavioral description of the MEMS can be deduced from the membrane's equation of motion

$$m\ddot{z} + d\dot{z} + k_1 z + k_3 z^3 = F_{ac} + F_{el} . \quad (1)$$

This equation describes membrane displacement z via a harmonic oscillator, defined by a mass m, a damping coefficient d, and a linear spring constant k_1. To account for mechanical non-linearities at large membrane displacements, a cubic spring constant k_3 is further added [5]. External acoustic F_{ac} and electrostatic F_{el} forces allow to drive the oscillator and ensure a coupling to the corresponding energy domains.

Components of the mechanical energy domain

For the mechanical energy domain, the membrane velocity in its center is chosen as the across quantity (voltage equivalent), while the force is assigned as the through quantity (current equivalent). Considering Eq. (1), the mass m, the inverse damping coefficient $1/d$, and the inverse linear spring constant $1/k_1$ correspond to a mechanical capacitance, a resistance, and an inductance, respectively.

The microphone membrane transduces incoming sound pressure into displacements and subsequently into a capacitance change. Its radial displacement profile $z(r)$ due to incoming sound pressure can be analytically approximated from a clamped and thin circular plate under a uniform pressure load [6]:

$$z(r) = z_{pk}\left(1 - \frac{r^2}{a^2}\right)^2 , \qquad (2)$$

where a and z_{pk} are the radius and center displacement of the membrane, respectively. Deflection can be described by a single amplitude parameter z_{pk}, enabling a single degree-of-freedom (DoF) mechanical modeling.

The lumped mass of the single DoF system can be calculated from energy conservation of its kinetic energy in lumped and continuous descriptions:

$$\frac{1}{2} m \left(\omega z_{pk}\right)^2 = \frac{1}{2}\int_0^a 2\pi t_m \rho \left[\omega z_{pk}\left(1 - \frac{r^2}{a^2}\right)^2\right]^2 r\,dr , \quad (3)$$

$$m = \frac{1}{5}\rho a^2 \pi t_m , \qquad (4)$$

where ρ, t_m, and ω are the membrane material density, thickness, and angular frequency, respectively.

The displacement amplitude of a thin tensile prestressed membrane due to a sound pressure P is given by [7]:

$$P = \frac{F}{A} = \frac{4\sigma t_m}{a^2}z_{pk} + (1 - 0.241\nu)\frac{8}{3}\frac{E}{(1-\nu)}\frac{t_m}{a^4}z_{pk}^3 , \quad (5)$$

where σ is the tensile membrane prestress, ν is the Poisson ratio, and E is the Young's modulus. Converting the sum terms of Eq. (5) into mechanical spring constants (k_1, k_3) requires to consider flux conservation in the single DoF description. Ensuring, that the lumped volume flow $Q = v_{Pk}A_{eff}$ created by deflections of a membrane's piston equivalent equals its continuous counterpart, requires to define an effective piston area A_{eff}:

$$\omega z_{pk} A_{eff} = \int_0^a 2\pi \omega z_{pk}\left(1 - \frac{r^2}{a^2}\right)^2 r\,dr , \qquad (6)$$

$$A_{eff} = \frac{1}{3}\pi a^2, \qquad (7)$$

The spring constants k_1 and k_3 finally result from multiplying Eq. (5) with the effective area A_{eff}.

The microphone damping coefficient is typically dominated by viscous damping, originating from squeeze-film damping and viscous flow through the backplate perforation holes. For the analytical description of damping

in perforated structures, the backplate can be split into unit cells containing a single perforation hole with a radius a_{perf}. The backplate perforation density Ψ can be defined from the ratio of the perforation hole and unit cell area A_{cell}

$$\Psi = \frac{a_{perf}^2 \pi}{A_{cell}} . \qquad (8)$$

Furthermore, the viscous dissipation in a perforated unit cell can be split into four regions [8], allowing to define damping coefficients for squeeze-film damping d_S, channel formed by the perforation hole d_C, acoustic end-effects of the channel d_E and a region connecting the channel and squeeze-film region d_I:

$$d_S = \frac{3\eta}{2\pi}\frac{A_{cell}^2}{t_0^3}\left[4\Psi - \Psi^2 - 3 - 4\ln\left(\sqrt{\Psi}\right)\right] , \qquad (9)$$

$$d_c = \frac{8\eta}{\pi}\frac{t_{bp}}{a_{perf}^4}(1 - \Psi)A_{cell}^2 , \qquad (10)$$

$$d_E = \frac{3\eta}{2a_{perf}^3}(1 - \Psi)A_{cell}^2 , \qquad (11)$$

$$d_I = \frac{3\eta}{2a_{perf}}\frac{0.8}{t_0^2}\left[(1 - \Psi)A_{cell}\right]^2 , \qquad (12)$$

where t_{bp} and η correspond to the backplate thickness and the dynamic viscosity of air, respectively. The lumped damping coefficient of the total membrane backplate systems can finally be determined as

$$d = \frac{A_{eff}}{A_{cell}}(d_S + d_C + d_E + d_I) . \qquad (13)$$

Coupling of mechanical and electrical energy domains

In a microphone operation, membrane displacements due to incident sound pressure lead to capacitance changes, which can be further converted into an output voltage. This capacitive transduction can be generally described by a two-port component defined by the following non-linear set of equations:

$$I = C\frac{dV}{dt} + V\frac{dC}{dt} , \qquad (14)$$

$$F_{el} = \frac{1}{2}V^2\frac{dC}{dz} . \qquad (15)$$

Beyond their mechanical and electrical DoFs (I, V, F, v), Eqs. (13) and (14) require a description of the microphone's variable capacitance C and its spatial derivative $\frac{dC}{dz}$. Detailed analytical descriptions allowing for behavioral implementations in circuit simulators are derived and given in [9]. Considering a membrane deflection profile given in Eq. (2), the variable capacitance of the microphone can be formulated to

$$C = \frac{\varepsilon_0 \pi a^2}{\sqrt{z_{pk}}}\left(\frac{(1 - \Psi)\,\text{atanh}\left(\sqrt{\frac{z_{pk}}{t_0}}\right)}{\sqrt{t_0}} + \frac{\Psi\,\text{atanh}\left(\sqrt{\frac{z_{pk}}{t_0 + n}}\right)}{\sqrt{t_0 + n}}\right) . \quad (16)$$

The first sum term accounts for the parallel-plate-like capacitance formed where the membrane opposes the backplate. The second sum term defines capacitances from electrostatic fringing fields forming inside the perforation holes. The parameter $n = 1\,\mu\text{m}$ is extracted from finite element modeling (FEM) and ensures an applicability to backplate geometries of $0.5\,\mu\text{m} \leq a_{perf} \leq 6\,\mu\text{m}$ and $0.5\,\mu\text{m} \leq t_0 \leq 10\,\mu\text{m}$. The capacitance's spatial derivative

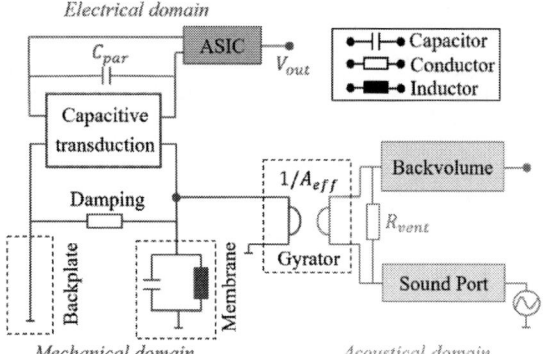

Figure 2. Behavioral system-level model of a capacitive MEMS microphone implementable using Verilog-A.

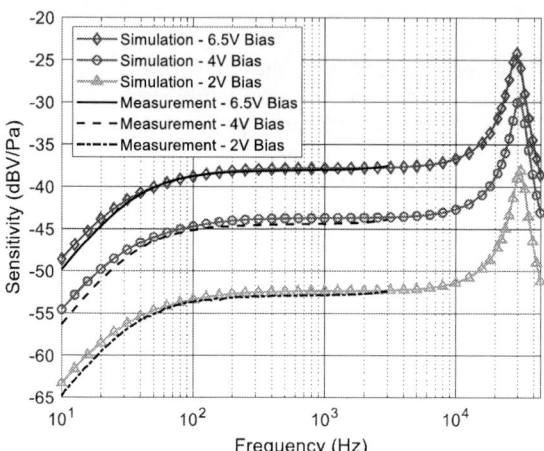

Figure 3. Measured and simulated microphone acoustic receive sensitivities for different applied bias voltages.

is analytically given in [9] and completes the description of the capacitive transductions

Components of the acoustical energy domain

The acoustical energy domain primarily describes the interaction of the MEMS with its package. A coupling of the acoustical domain to the mechanical domain is given by a two-port component

$$P = \frac{F}{A_{eff}} \,, \tag{17}$$

$$v_{pk} = \frac{Q}{A_{eff}} \,. \tag{18}$$

Eqs. (17) and (18) describe a linear gyrator component with gyration constant $1/A_{eff}$.

The package further consists of the sound port and the backvolume. The Helmholtz-resonator shaped sound port is formed by the cylindrical MEMS front-cavity and the opening in the PCB. Modeling them via coupled acoustic transmission lines [10], allows to account for wave propagation and thermo-viscous dissipation and provides precise estimation of the Helmholtz-resonance frequency. Acoustic transmission lines are moreover applied to model the backvolume. Approximating the backvolume by a cylinder of equivalent volume accounts for the compliance of the enclosed air volume, its frequency dependent transition from an isothermal to adiabatic state, and thermal dissipation [10]. The sound port and backvolume are acoustically connected through the membrane ventilation holes. The acoustical resistance of a single circular perforation hole can be derived from the acoustical resistance of a circular channel and an aperture defining end-effects. The acoustic resistance R_{vent} of N ventilation holes of radius a_{vent} can be formulated to

$$R_{\text{vent}} = \frac{1}{N}\left(\frac{8\eta}{\pi}\frac{t_m}{a_{vent}^4} + \frac{3\eta}{a_{vent}^3}\right). \tag{19}$$

Behavioral implementation in circuit simulators

The previous models can be implemented into common circuit simulators and linked to construct a system-level description of the microphone. While most previously described models can be implemented as conventional lumped components, the capacitive transduction and cubic spring constant particularly require non-linear behavioral implementations using hardware

description languages (HDL). In this work, these models are implemented using Verilog-A and placed in the commercial circuit simulator SIMetrix (see Fig. 2).

The backplate is here approximated to be perfectly stiff. In the electrical energy domain, parasitic capacitances of the MEMS C_{par} and a circuit description of the constant-charge-based readout circuit are implemented.

NON-LINEAR SIMULATION OF A CAPACITIVE MEMS MICROPHONE

The modeling approach can finally be applied to simulate the acoustic KPIs of the previously described commercial MEMS microphone. Fig. 3 shows the measured and simulated acoustic receive sensitivities of the microphone at different applied bias voltages (i.e., 2, 4, and 6.5 V). While the sensitivity is calculated from a working point linearized small-signal analysis, non-linearities are still considered in the computation of the DC working point. The Verilog-A implementation of the capacitive transduction therefore allows to precisely calculate the sensitivity at different DC bias voltages and moreover accounts for bias voltage dependent resonance frequency shifts caused by electrostatic spring softening. The measured sensitivity of -37.7 dBV/Pa at 6.5 V bias very well corresponds to its simulated counterpart of -37.8 dBV/Pa.

To calculate the SNR, noise analyses need to be performed. Resistive circuit components are thereby replaced by thermal white-noise voltage sources of amplitude $V_n = \sqrt{4k_bTR}$. This allows to extract the system output noise and separate the noise contributions of subsystems. Fig. 4 shows the measured and simulated spectral noise densities of the microphone at 6.5 V bias. The simulated noise is moreover split into thermal noise contributions of the ASIC, MEMS, backvolume, ventilation hole, and sound port. The measured and simulated total noise values again show a precise matching in the whole audio frequency band. The measured A-weighted noise in the audio band is -104.7 dBV and exactly meets the simulated value of -104.7 dBV. The sensitivity and noise finally conclude the microphone's SNR. The measured and simulated SNR values are 67.0 dB(A) and 66.9 dB(A),

978-1-6654-9309-3/23 $31.00 © 2023 IEEE

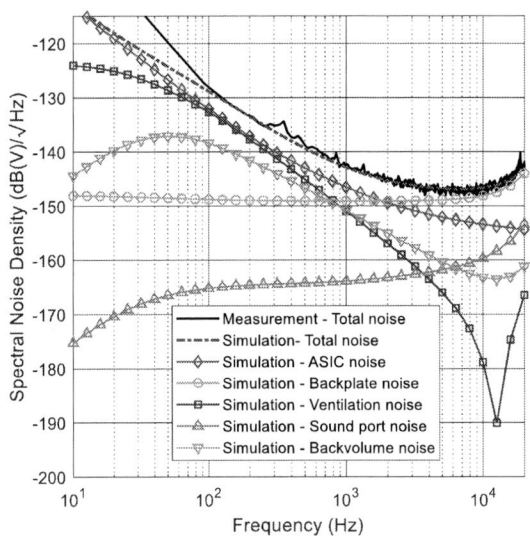

Figure 4. Measured and simulated spectral noise densities of the capacitive MEMS microphone.

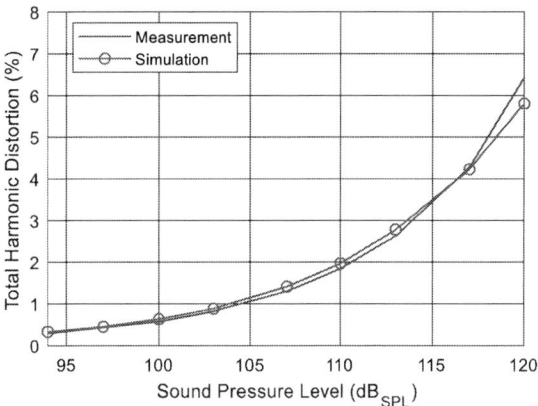

Figure 5. Measured and simulated total harmonic distortion (THD) of the capacitive MEMS microphone.

respectively, confirming an excellent small-signal precision of the model. The non-linear mechanics and capacitive transduction become major importance when large-signal analyses of the microphone's output linearity are considered. Even for a perfectly sinusoidal input pressure signal of frequency $f_0 = 1$ kHz, the system non-linearities cause a harmonic distortion of the output signal, which then contains higher order harmonic frequencies. The THD is then defined from the ratio of the output voltage amplitude V_0 at the input frequency $f_0 = 1$ kHz and the cumulated amplitudes V_i at the higher order harmonic frequencies $f_i = i \cdot f_0$, where i is integer valued . The THD considering k harmonic frequencies results to

$$THD = \frac{\sqrt{\sum_{i=1}^{k} V_i^2}}{V_0}. \quad (18)$$

In Fig. 5 the measured and simulated THD values of the microphone output voltage are compared at different input sound pressure levels ranging from 94 to 120 dB$_{SPL}$. Measurements and simulations finally show an excellent agreement. The THD of the microphone is here dominated by the non-linearity of the capacitive transduction and highly depends on the parasitic capacitances (e.g. C_{par}) of the MEMS and ASIC.

CONCLUSION

In this work, a non-linear system-level model of a capacitive MEMS microphone has been developed, which allows for a Verilog-A based behavioral implementation in commercial circuit simulators. The system-level model was validated by measurements of a commercial MEMS microphone from Infineon Technologies AG. The model is able to simulate all microphone key performance indicators, including the sensitivity, SNR, and THD, in which an excellent agreement between simulation and measurement results has been demonstrated. As all sub-models were analytically formulated, the model finally scales with the MEMS microphone's geometrical design parameters. Hence, this model can be further employed for multi-parameter optimizations of various next-generation capacitive microphones.

ACKNOWLEDGEMENTS

This work is fully funded by Infineon Technologies AG. Authors acknowledge technical support from Christian Bretthauer.

REFERENCES

[1] M. Fueldner and A. Dehé, "Dual back plate silicon MEMS microphone: balancing high performance." *Proc. DAGA*, 2015.

[2] L. Sant, et al. "A 130dB SPL 72dB SNR MEMS Microphone Using a Sealed-Dual Membrane Transducer and a Power-Scaling Read-Out ASIC." *IEEE J Sensors* 22(8): 7825-7833, 2022.

[3] S. Shubham, et al. "A Novel MEMS Capacitive Microphone with Semiconstrained Diaphragm Supported with Center and Peripheral Backplate Protrusions." *Micromachines* 13(1): 22, 2021.

[4] S. Anzinger, et al. "A comb-based capacitive MEMS microphone with high signal-to-noise ratio: modeling and noise-level analysis." *Proc.MDPI* 1.4 (2017): 346.

[5] R. Williams, et al. "Multidegree-of-Freedom State-Space Modeling of Nonlinear Pull-in Dynamics of an Electrostatic MEMS Microphone." *JMEMS* 31(4): 589-598, 2022.

[6] S. Timoshenko, et.al. *Theory of plates and shells*. Vol. 2. New York: McGraw-hill, 1959.

[7] R.P. Vinci, et al. "Mechanical behavior of thin films." *Ann. Review of Mat. Sc.* 26(1): 431-462, 1996.

[8] R. Sattler, et al. "Analytical compact models for squeeze-film damping." *DTIP 2004*. 2004.

[9] S. Anzinger, et al. "A non-linear lumped model for the electro-mechanical coupling in capacitive MEMS microphones." *JMEMS* 30(3): 360-368, 2021.

[10] S. Anzinger, et al. "Acoustic transmission line based modelling of microscaled channels and enclosures." *JASA*, 145(2), 968-976, 2019.

CONTACT

*S. Anzinger; sebastian.anzinger@tum.de

H.S. Wasisto; hutomosuryo.wasisto@infineon.com

VORTEX-BEAM ACOUSTIC TRANSDUCER FOR UNDERWATER PROPULSION

Jaehoon Lee, Kianoush Sadeghian Esfahani, and Eun S. Kim

Department of Electrical and Computer Engineering, University of Southern California, USA

ABSTRACT

This paper presents vortex-beam acoustic transducers (with air-cavity lens) for underwater acoustic propulsion. Acoustic propulsions generated by vortex-beam acoustic waves with Spiral-arm Vortex-beam Acoustic Transducer (SVAT), annular-ring-based Self-Focusing Acoustic Transducer (SFAT), 18-sectored SFAT, and combinations of SVAT and SFAT have been compared and analyzed. Experimental results show that a 1-arm SVAT can generate 70.7µN/W propulsion force per applied electrical pulse, 1.3 times higher than that of SFAT (53.7µN/W), while a transducer which combines SVAT and SFAT generates 130µN/W, 2.4 times higher than that of the SFAT. When the number of arms used in SVAT is increased to 4, the 4-arms SVAT generates acoustic propulsion force 173µN/W and 2.28µN/W/mm² (per transducer area). The acoustic propulsion forces per transducer area of the combined transducer, the one-arm SVAT, and SFAT are measured to be 2.33, 1.53, and 1.19 µN/W/mm², respectively.

KEYWORDS

Acoustic transducer, Vortex beam, Acoustic propulsion, Self-focusing acoustic transducer

INTRODUCTION

Acoustic propulsion generated by a MEMS-fabricated self-focusing acoustic transducer (SFAT) has been reported and analyzed in the previous work [1]. The advantage of the SFAT-based underwater propulsion is that no mechanical moving part is needed for propulsion, and it can propel robot or submarine without causing turbulence or low-frequency sound that propagates a long distance. However, the SFAT-based propeller produced a relatively low acoustic propulsion force per electrical power. In a wireless stand-alone propulsion system powered by a battery or a lithium-ion capacitor, its volume and weight is mainly determined by the power source. Consequently, by increasing the acoustic propulsion force per applied electrical power, the system can be made smaller, lighter, and more efficient.

This paper shows that the propulsion force can be increased through using vortex-beam wave. Different designs of Spiral-arm Vortex-beam Acoustic Transducers (SVAT) have been investigated and fabricated in a lead zirconate titanate (PZT) substrate to compare its acoustic propulsion with those of SFAT-based propellers. The experimental results showed that SVAT can generate at least 2.4 times higher acoustic propulsion per electrical power. Furthermore, a more number of arms in SVAT is shown to lead to a stronger acoustic propulsion force.

Also described is a new design of combining single-focusing and vortex-beam acoustic waves which enhances the acoustic pressure at the central axis. The combined transducer is shown to be capable of generating the highest acoustic propulsion force per applied electrical power per transducer area.

DEVICE DESIGN AND FABRICATION

Transducer Designs

When a sinusoidal electrical signal with its frequency matched to PZT's thickness-mode resonance is applied to between the top and bottom electrodes sandwiching a PZT thin substrate, the PZT vibrates and generates acoustic waves most effectively. With SFAT, the acoustic waves are designed to constructively interfere at a focal point through patterning air-cavity lens (composed of Parylene-sealed air-cavity rings and non-air-cavity rings) into Fresnel half-wavelength bands (Fig. 1), according to the following equation for the ring radii [2]:

$$R_n = \sqrt{n\lambda \times \left(f + \frac{n\lambda}{4}\right)} \qquad (1)$$

where n, λ, and f are the nth ring, acoustic wavelength in liquid, and focal length, respectively. The waves are blocked by the air cavity due to the large acoustic-impedance mismatch between air (0.4 kPa·s/m) and solid (over 1 MPa·s/m), while propagating through the non-air-cavity area.

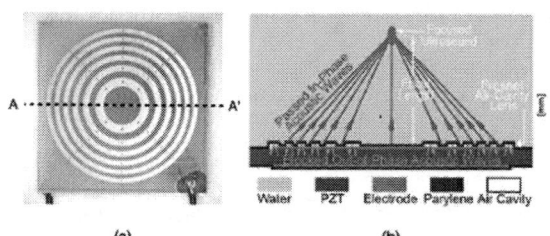

Figure 1. (a) Top-view photo of SFAT showing air-cavity Fresnel rings on a 1-mm thick PZT substrate. (b) Cross-sectional schematic across the dashed line (A-A') in (a) of SFAT, showing the focusing principle of SFAT

Vortex beam acoustic wave can be generated by logarithmic spiral helix electrode design, according to the following equation for radius r in polar coordinate [3]:

$$r = ae^{b\theta} \qquad (2)$$

where a, b, and θ are the initial radius, azimuth coefficient, and angle, respectively. The azimuth coefficient b determines the growth rate of the radial distance from the center with the angle increment. A larger azimuth coefficient results in larger gaps between the spiral arms and larger width at the outer electrode. A spiral arm electrode (e.g., Fig. 2a or 2b) can be defined with two logarithmic helixes with initial radii (a_1 and a_2) and origin and termination angles (θ_1 and θ_2) with one azimuth coefficient b as follows:

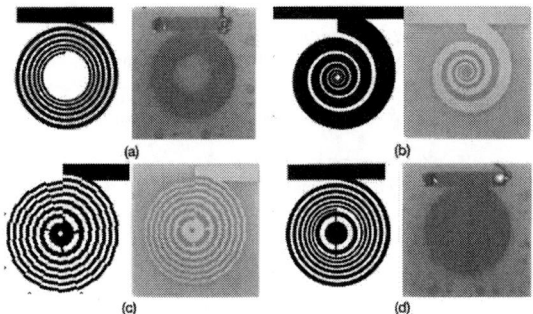

Figure 2. The designs (left) and photos of the fabricated transducers (right) of (a)1-arm Spiral-arm Vortex-beam Acoustic Transducer (SVAT), (b) a variation of 1-arm SVAT (A_2), (c) 18-sectored Self-Focusing Acoustic Transducer (SFAT) based on annular Fresnel rings and (d) 1- arm SVAT and SFAT with two rings in the center.

$$r = a_1 e^{b\theta} \text{ and } r = a_2 e^{b\theta}, \; \theta \in [\theta_1, \theta_2] \quad (3)$$

The a_1 and a_2 determine the smallest width of the arm, while the widest width can be calculated as $(a_2 - a_1)e^{b(\theta_2-\theta_1)}$. As the available frequency bandwidth of SVAT depends on the electrode width, the spiral arm electrode design can produce the vortex beams over the wavelength between $2d_{min}$ and $2d_{max}$, where d is the width of the electrode arm [3]. In other words, the operating frequency range for the vortex-beam acoustic wave is limited to

$$f_{min} = \frac{c}{2(a_2 - a_1)e^{b(\theta_2-\theta_1)}} , f_{max} = \frac{c}{2(a_2 - a_1)} \quad (4)$$

where c is the speed of sound in liquid. For example, A_0 design in Table 1 is designed to have $a_1 = 2.3 \, mm$, $a_2 = 2.5 \, mm, b = 0.090, \theta_1 = 0°, and \, \theta_2 = 12\pi$, which gives $d_{min} = 0.2$ mm and $d_{max} = 5.95$ mm. In water, where the speed of sound is 1480 m/s, the operating frequency range is from 124 kHz to 3.7 MHz, which includes the thickness-mode resonant frequency of 1-mm-thick PZT, 2.32 MHz. Table 1 shows other designs with the number of spiral arms varying from 1 to 4; while the termination angle × number of arms is kept to be 12π, the initial position of each arm is rotated by 180°, 120°, and 90° for the 2, 3, and 4-arms design, respectively.

Another method of generating vortex beam is to use sectored SFATs where each sector is pie-shaped with angle of 20° for a total of 18 sectors (Fig. 2c). The focal length

(F) of each sector increased with the step of 100 μm from 2 mm with the focal length of the ith sector being $F_i = 2 + 0.1 \times i \, [mm]$. The origin of each sector is displaced by 100 μm from that of its previous sector, forming a vacancy in the center of transducer $(D_{i+1} = D_i + 100\mu m)$. Furthermore, we have designed a transducer (Fig. 2d) by combining SVAT having a large empty space in the center (such as A0 in Table 1) with SFAT having two rings (for a focal length of 2 mm) in the center to enhance the acoustic pressure along with the z-axis.

Finite Element Method (FEM) Simulation

Vortex-beam acoustic distribution in liquid has been FEM-simulated with COMSOL in a three-dimensional cone space (10-mm base diameter with 15 mm height). The acoustic pressure p at a point in liquid can be calculated by integrating the pressure contributions from acoustically active transducer surface with Rayleigh-Sommerfeld diffraction integral [3]:

$$p(\rho, \phi, z) = \rho_0 c_0 \int_{\theta_1}^{\theta_2} \int_{a_1 e^{b\theta}}^{a_2 e^{b\theta}} \frac{i\omega\rho_0}{2\pi h} e^{i(\omega t - kh)} r dr d\theta \quad (5)$$

where ρ_0 and c_0 are the density and the acoustic velocity in liquid, respectively, while $\omega, k, and \, h$ are the angular frequency, wave number ($k = 2\pi/\lambda$) and distance between observation point in liquid and the surface element ($dS = r dr d\theta$) in the electrode. Figure 3 shows the simulation results of SFAT (Fig. 3a), various SVATs (Fig. 3b-d), 18-sectored SFAT (Fig. 3e), and combined transducer (Fig. 3f) under the condition of 1-μm vibration on the surface. The results show that A1 design among the SVATs generates the longest helix length in the medium. Interestingly, the combined transducer generates 1.5 times higher peak absolute acoustic pressure than that of SVAT (A0). In addition, the simulated acoustic pressure distributions over the xy-plane at z = 7mm produced by multi-arms A2-design SVAT (Fig. 4) clearly show that multiple acoustic vortices are produced at the observation plane.

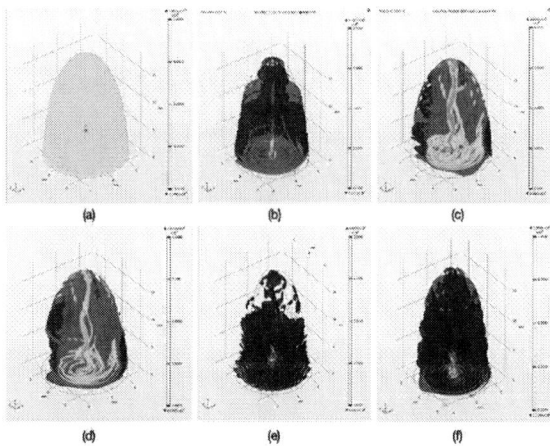

Figure 3. Three-dimensional COMSOL simulations with 1μm displacement on the transducer: (a) SFAT with focal length of 4.5 mm, (b, c, d) 1-arm SVAT (A0, A1, A2), (e) 18-sectored SFAT, and (f) 1- arm SVAT and SFAT with two rings in the center.

Table 1 Various Designs of Spiral-arms Vortex-beam Acoustic Transducers (SVATs)

Design (a_1, a_2, b)	1-Arm SVAT	2-Arms SVAT	3-Arms SVAT	4-Arms SVAT
A0 Design (2.3mm, 2.5mm, 0.090)				
A1 Design (0.5mm, 0.7mm, 0.090)				
A2 Design (0.2mm, 0.3mm, 0.083)				

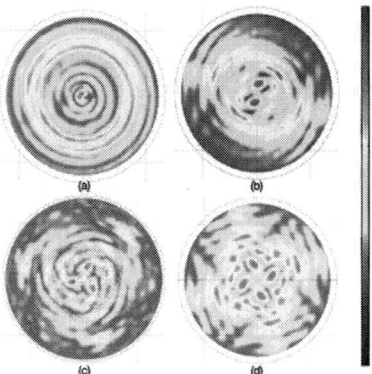

Figure 4. Normalized absolute acoustic pressure distributions over the xy plane at an observation point (z= 7mm) produced by the A2 Design (a) 1-arm SVAT, (b) 2-arms SVAT, (c) 3-arms SVAT and (d) 4-arms SVAT.

Fabrication Steps

In the fabrication steps of the air-cavity-lens transducers (Fig. 5), the top and bottom nickel electrodes over 1-mm-thick PZT substrate are first patterned into a large circle. The bottom circle is intentionally designed to be 200 μm larger in diameter than the top electrode to compensate any potential top and bottom misalignment. Fresnel half-wavelength band annular rings are patterned with 3μm thick AZ5214 photoresist on the top electrode as a sacrificial layer (Fig. 5e) and 4μm thick Parylene D layer is coated. The sacrificial layer is removed by acetone through the release holes patterned with photoresist and etched by RIE (Fig. 5g). Finally, additional 10μm thick Parylene D layer is coated to seal the release holes. The SVATs also are fabricated similarly without air-cavity lens.

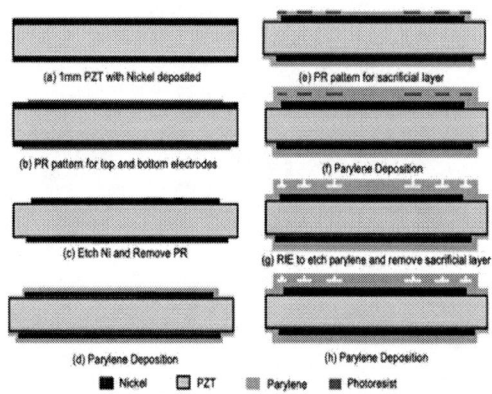

Figure 5. Brief fabrication process of the air-cavity lens transducers.

For underwater acoustic propulsion, a laser-machined air-cavity reflector made with 2-mm-thick acrylic sheet is attached on the other side of the Fresnel air-cavity lens to constrain the acoustic wave in one direction, as illustrated in Fig. 6. In the figure, the propulsion is upward, with the air-reflector at the top, as the focused acoustic beam (or vortex acoustic beam) works only on the bottom side of the transducer.

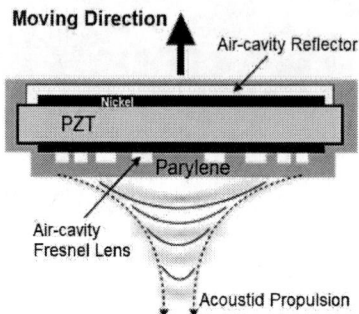

Figure 6. Conceptual diagram of the underwater acoustic propeller.

EXPERIMENTS AND RESULTS

Acoustic propulsion force generated by the transducer is measured by quantifying the acceleration of a pendulum swing, as the transducer is suspended with flexible wires. With $100V_{pp}$ sinusoidal wave applied (continuously) to the transducer, the propelling motion is observed and captured with a video camera of 30 frames per second (fps), and its traveling distance, velocity, and acceleration are obtained every frame (33ms) by a fine precision ruler immersed in the water, as shown in Figs. 7 and 8. By multiplying the acceleration with the transducer mass (2.86 grams in average including the mass of the air-cavity reflector on one side), the acoustic propulsion forces are obtained and summarized in Table 2.

The electrical power P_E consumed by the transducers is obtained with

$$P_E = \left(\frac{V_{rms}}{|Z|}\right)^2 \times R \tag{6}$$

where V_{rms}, Z, and R are the root-mean-square (rms) voltage, SFAT impedance, and real part of the SFAT impedance, respectively. The SFAT impedance is calculated by reflection coefficient, S_{11}, measured with a network analyzer:

$$Z = R + jX = 50\ \Omega \times \frac{1 + S_{11}}{1 - S_{11}} \tag{7}$$

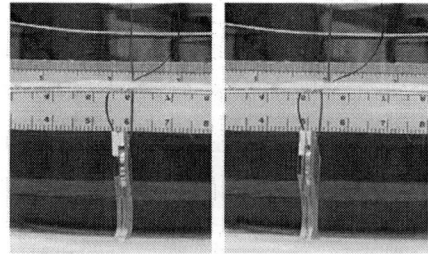

Figure 7. Photos of the propeller (the six-ring SFAT with 4.5 mm focal length) moving at (a) t=0 and (b) t = 660 ms

The SFAT (focal length = 4.5 mm) having 6 rings (active area = 75 mm²) is measured to produce 214 μN propulsion force under $100V_{pp}$ continuous sinusoidal voltage (Fig. 7). The acoustic propulsion force per electrical power by the SFAT is 53.7 μN/W, which is close

978-1-6654-9309-3/23 $31.00 © 2023 IEEE

Table 2 Summary of Fabricated and Tested Underwater Acoustic Propellers

	SFAT	1-Arm SVAT (A0 Design)	1-Arm SVAT (A2 Design)	4-Arms SVAT (A2 Design)	18-Sectored SFAT (Fig. 2c)	SFAT+SVAT (Fig. 2d)
Total Weight [g]	2.88	2.86	2.84	2.88	2.93	3.05
Acceleration [mm/s²]	74.28	98.50	218.72	238.78	112.69	170.30
Activated Area [mm²]	45.00	45.95	77.02	75.60	56.75	55.90
Propulsion per Power [μN/W]	53.68	70.69	155.87	172.56	82.56	130.34
Propulsion per Power per Active Area [μN/W/mm²]	1.19	1.53	2.02	2.28	1.46	2.33

to the previously reported value (52.07μN/W) in a wireless stand-alone system [4].

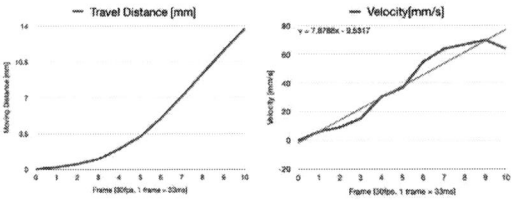

Figure 8. (a) Measured traveling distance (in mm) and (b) velocity (in mm/s) of the 4-arms SVAT (A2 design) acoustic propeller. The traveling distance is measured in every frame (33ms) and the averaged acceleration is calculated to be 238.8 mm/s²

Experimental results among the 1-arm SVATs show that the one with a larger transducer area (A2 design in Table 1) generates higher propulsion force than other designs. Also, as the number of arms increases, higher acoustic propulsion force is obtained. For example, the acoustic propulsion force from the 4-arms A2-design SVAT (Fig. 8) is measured to be 711.6 μN, while the 1-arm SVAT (same design) generates 621.2 μN, even though the active electrode area of the 4-arms SVAT (75.60 mm²) is smaller than that of the 1-arm SVAT (77.02 mm²). Since the acoustic pressure at the focal point depends on the total active electrode area of the transducer, the acoustic propulsion forces per electrical power per transducer area also are compared in Table 2. Although 4-arms A2-design SVAT generates the highest acoustic propulsion per electrical power, 172.6 μN/W which is 3.2 times larger than that of the SFAT (53.68 μN/W), the highest acoustic propulsion force per power per transducer area is obtained from the combined design of the SVAT (A0 design) with SFAT (2.33 μN/W/mm²), and is almost double of that of the SFAT-based propeller (1.19 μN/W/mm²). As can be seen in Fig. 9, the SVAT/SFAT combination produces much stronger propulsion force than the SFAT alone (Fig.

7). However, the vortex beam from the SVAT produces forces that are not perpendicular to the transducer surface, resulting in a twisting motion as the transducer propels itself in the direction perpendicular to the transducer surface. This phenomenon is observable in other SVAT-based propellers.

SUMMARY

This paper presents new designs of spiral-arms vortex-beam acoustic transducers (SVATs) which generate helical acoustic pressure distribution in liquid. Various designs with varying number of arms (1 to 4) have been investigated and tested. Furthermore, alternative method of generating vortex beam using an 18 pie-sectored SFAT is introduced, and a new design of combining SVAT and SFAT is implemented and compared with other designs. The acoustic propulsion force generated by the SVAT is measured to be up to 3.2 times larger than that by the SFAT-based propeller, while the highest acoustic propulsion force per electrical power per transducer area is obtained with the combined design of SVAT and SFAT. From the wired experiments, the 4-arms A2-design SVAT is measured to produce 711.6 μN acoustic propulsion force, 3.35 times larger than the SFAT-based propeller.

ACKNOWLEDGEMENTS

This paper is based on the work supported by National Science Foundation under grant ECCS2017926.

REFERENCES

[1] L. Zhao and E.S. Kim, "Subminiature Underwater Propeller with Electrical Controllability of Steering," 2021 IEEE International Ultrasonics Symposium, 2021 pp. 1-4, doi: 10.1109/IUS52206.2021.9593504.

[2] J. Lee and E.S. Kim, "Phase Array Ultrasonic Transducer Based on a Flip Chip Bonding with Indium Solder Bump," IEEE International Ultrasonics Symposium, Virtual Symposium, September 11 - 16, 2021

[3] Han Zhang and Yang Gao 2019 Chinese Phys. Lett. 36 114302

[4] J. Lee and E.S. Kim, "Wireless and Stand-Alone Submarine Propeller Based on Acoustic Propulsion", Solid-State Sensor and Actuator Workshop, Hilton Head Island, SC, June 5-9, 2022

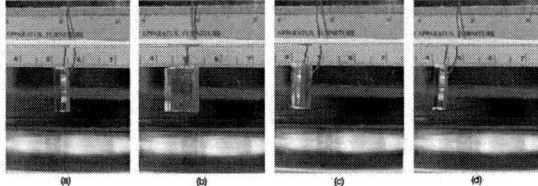

Figure 9. Photos of the propeller (1- arm SVAT(A₀) + two-ring SFAT) moving at (a) t = 0, (b) t = 330 ms, showing a twisting motion, (c) t = 660 ms, and (d) t = 1 s.

CONTACT

*Jaehoon Lee tel: +1-206-9028988; lee172@usc.edu

WIDEBAND AND HIGHLY SENSITIVE MICROMACHINED PZT FILM-BASED ULTRASONIC MICROPHONE WITH PARYLENE FILM AND FLEXIBLE HELMHOLTZ RESONATOR ENHANCEMENT

Chung-Hao Huang[1], and Guo-Hua Feng[1,2]

[1]Department of Power Mechanical Engineering, National Tsing Hua University, Hsinchu, TAIWAN
[2]Institute of Nano Engineering & MicroSystems, National Tsing Hua University, Hsinchu, TAIWAN

ABSTRACT

The innovative piezoelectric micromachined ultrasonic microphone fabricated by a bottom-up scheme with wideband operation frequency and sensitivity higher than a commercial MEMS microphone is presented for the first time. The device is constructed of a freestanding triangular cantilever beam array fabricated by titanium foil and quality hydrothermal PZT film of residual stress less than 60 MPa. A unique method is applied to deposit parylene film, fully covering the surface of triangular cantilever beam array and sealing the gaps between beams for enhancing the performance of the ultrasonic microphone. The deposited 0.5 μm-thick parylene film increases 3 dB sensitivity compared to that without parylene film sealing. The sensitivity further increases after the fabricated microphone integrated with a parylene flexible Helmholtz resonator cavity.

KEYWORDS

Piezoelectric, micromachining, Helmholtz resonator, paylene, ultrasonic microphone.

INTRODUCTION

There are many usages for microphones, including communication purpose, hearing aids, underwater ultrasonic and acoustic recognition and surveillance [1]. The miniature MEMS microphone becomes promising technology in our daily life, such as medical, automotive and recreational industrial applications. Compared to conventional electret condenser microphones, micromachined microphones possess many advantages of compatibility with microelectronic circuitries, temperature stability, and better immunity from surrounding vibration interference. Miniature micromachined microphones can also reduce the weight and size, enabling the creation of small products when they employed in the form of arrays of multiple microphones [2].

The demand for MEMS microphones is mostly driven by mobile devices, particularly the usage of smartphones microphone surging from a decade ago. MEMS microphones generate a sizable amount of revenue from applications besides the smartphones, including computers, earbuds, hearing aids, and smart speakers. The global microphone market reaches $1.6 billion in 2020 and is expected to reach 2.3 billion by 2025 [3]. Especially, the increasing use of wireless microphones in concerts, broadcasting studios and sporting events, is driving revenue growth in the wireless microphone market.

Three transduction mechanisms used for developing micromachined microphones have been broadly investigated: piezoresistive, capacitive and piezoelectric types. The critical characteristics of microphones include sensitivity, bandwidth, power consumption, fabrication cost and stability.

In terms of sensitivity, capacitive microphones exhibit the best sensitivity among the three types, which can reach 400-1000 μV/Pa and the sensitivity of the piezoelectric microphone is reported within 10-500 μV/Pa. The piezoresistive microphones display a sensitivity which is approximately an order of magnitude less than the piezoelectric microphones [3]. For the dynamic range of operation, the capacitive microphone exhibits a narrow bandwidth and the piezoelectric microphone has the widest working frequency range. Both piezoresistive and capacitive microphones required input power for operation. The piezoresistive microphone typically uses silicon piezoresistors constructed as a Wheatstone bridge configuration for sensing resistance change due to acoustic pressure. The capacitive microphone typically requires a bias voltage (several to tens of volts) between the rather rigid backplate and movable diaphragm for sensing the capacitance change. This increases the overall power consumption of the sound detection. In addition, capacitive microphones possess high output impedance, which complicates the design of readout circuits. On the other hand, piezoelectric microphones are relatively simple in structure and have low output impedance without the necessity of additional biasing voltage. In addition, compared to capacitive microphones, piezoelectric microphones offer higher reliability because of its relatively uncomplicated working configuration.

Airborne ultrasonic microphones exhibit broad applications in 3C electronic products such as proximity sensing and gesture recognition [4]. Recently the PMUT device with Helmholtz resonantor and active backing plate has shown the increase of output acoustic pressure. Although PZT possesses attractive piezoelectric effect, the challenging of quality PZT film deposition could lower the researchers' interest in developing PZT film based PMUT.

Here, we propose a bottom-up fabrication scheme to achieve a lower-stressed PZT film on a patterned free-standing thin film. The complete basic microphone unit possesses superior performance with a sensitivity of 20 dB greater than the commercial microphone and can be further enhanced by coating parylene film and integration of flexible Helmholtz resonant cavity.

DEVICE DESIGN AND FABRICATION

Design of ultrasonic microphone

The major goal for the ultrasonic microphone design in this study is to set the operation frequency at ultrasonic frequency band and utilized the proposed novel structure and fabrication schemes to investigate the performance of the complete device. Since we have developed the PMUT as ultrasonic emitter operated in a frequency range of approximately 25 kHz [5], we will design the fundamental frequency of the ultrasonic microphone to match this frequency range so that we can construct an ultrasonic emitter-receiver pair for better performance in applications. The configuration of the proposed ultrasonic microphone module was composed of two main parts: basic functional ultrasonic microphone unit and parylene flexible Helmholtz resonator cavity.

The former part was formed by micromachined titanium foil with double-side coated PZT film and aluminum electrode to construct a sandwich structured active layers and secured on a SU-8 plate, which allowed us to acquire the piezoelectric signal from the vibration of active layers due to ultrasonic pressure. The geometry design of the titanium foil and thicknesses of active layers dominated the resonant frequency of the basic functional ultrasonic microphone unit, which affected the dynamic response of the microphone. Thus, the finite element analysis software COMSOL was employed for simulation. The thicknesses of top aluminum electrode, PZT films, and titanium foil were reasonably designed as 0.175 μm, 7 μm, and 4 μm. We chose the geometry of radial triangular cantilever array around a circle for the shaped titanium foil substrate. With different cantilever numbers and circle size as the design parameters for simulation, we finally selected 10 cantilevers and circle diameter of 1.6 mm as the active region of the single ultrasonic microphone. The resulting fundamental resonance of the triangular beam was about 28 kHz.

The purpose of the parylene flexible Helmholtz resonator cavity aimed at increasing the sound pressure levels inside cavity during microphone detection. The air volume inside the cavity could increase its pressure due to the air around the neck portion compressing this air volume inside the cavity. The geometry design selection of the flexible resonator cavity mainly focused on its function demonstration in this study.

Device fabrication

The bottom-up fabrication started with using a 4 μm-thick titanium foil as a substrate and bottom electrode of the microphone. Micromachining was applied to obtain the designed triangular-beam array. After annealing, titanium dioxide layer was formed as the seed layer of PZT film deposition. Modified hydrothermal process was followed to conformally grow PZT film on the array [5, 6]. Aluminum electrode was selectively sputtered on the PZT film through a micromachined shadow mask. Individual microphone-film chips were diced from the processed titanium foil. Each chip was bonded to a 250 μm-thick patterned SU-8 plate as a basic functional ultrasonic microphone unit (Fig. 1).

Two enhanced fabrication processes significantly strengthened the sensitivity of the microphone. One is anchoring the microphone unit on the bonding wax to deposit parylene film [7]. After releasing wax, the parylene film fully covered the gaps and the surface of triangular beam array of the microphone. Besides that, a flexible parylene bellow cavity made by 3D stereolithography and bonding wax technology was integrated with the microphone unit to further increase the sensing performance.

Figure 2 shows that fabrication results. The fabricated triangular sensing beam exhibited a low residue stress. We employed the confocal laser scanning microscope (VK-X3000, Keyence Co., Japan) to measure the profile height of the triangular cantilever beam from the fixed and to the front tip end. The tip displacement is 25 μm for the measured beam length of 700 μm.

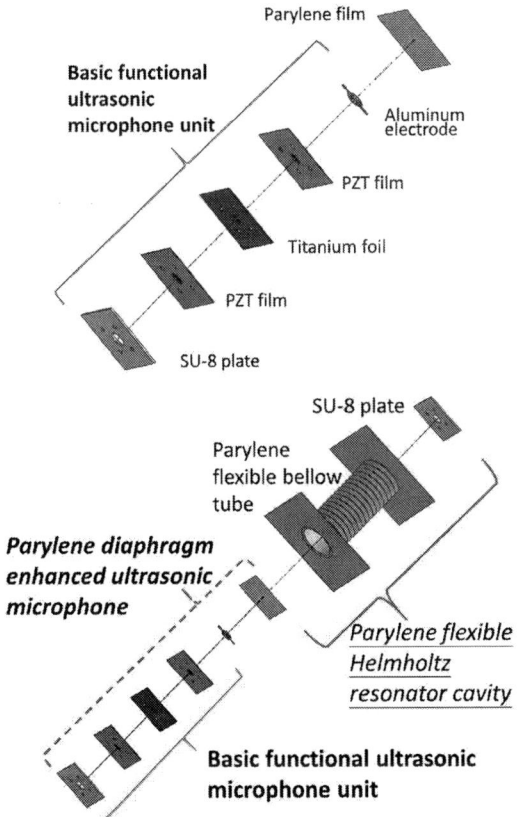

Figure 1: Configuration of bottom-up scheme fabricated PZT-film ultrasonic microphone with parylene film (top) and Helmholtz resonant cavity enhancement (bottom).

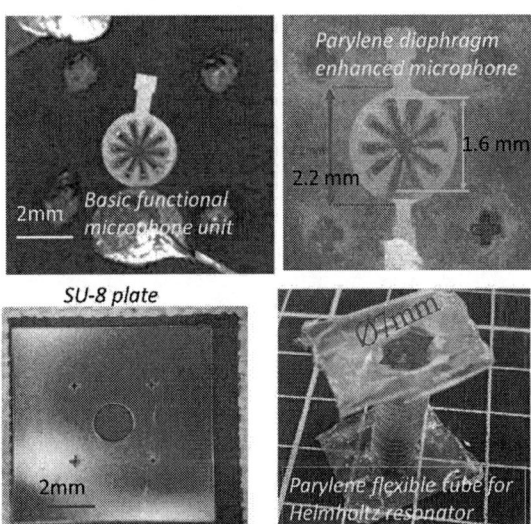

Figure 2: Fabricated results of the complete ultrasonic microphone and its major components.

Using the Stoney's equation to estimate the residue stress of the fabricated triangular cantilever beam with the layers of aluminum/PZT/titanium/PZT (Fig. 1), the resulting residual stress was calculated about 58 MPa.

EXPERIMENTAL SETUP AND RESULTS
Experimental setup

The same experimental setups were applied to characterize the commercial MEMS microphone (SPU0410LR5H-QB, Knowles, Taiwan) and the fabricated microphones (Fig. 5). To perform the experiment, we utilized the commercial speaker (Pettersson L60, Sweden) with operation frequency band 2-60 kHz as the ultrasonic source. The sinusoidal wave with a 12 Vpp amplitude generated from the function generator was amplified by Piezo Linear Amplifier EPA-104 (Piezo.com, MA, USA) with a magnification of 13x and fed into the ultrasound speaker to emit ultrasonic waves.

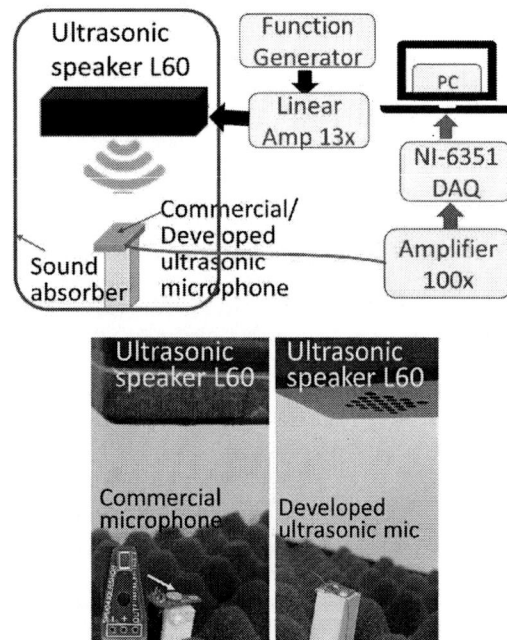

Figure 5: Experimental setup for testing and comparing the commercial and developed microphones.

To setup the microphones for receiving ultrasonic wave, two 3D printed hollow columns was created working as the stands so the development board of the commercial MEMS microphone would be fixed to the hollow column through the center hole at the development board by screw. The fabricated basic functional ultrasonic microphone unit was secured on to the top surface of the hollow column by glue. Both commercial and fabricated microphones were set the distances of 13 mm to the ultrasonic speaker and kept the sound receiving surfaces were normal to the emitting plane of the ultrasonic speaker.

Both acquired electrical signals from the commercial and fabricated microphones were amplified using CN0350 board (Analog Device, USA) with a magnification of 100x. respectively. Then, the processed signals were went through the data acquisition board (NI USB-6351) and

recorded by computer for further time-domain or frequency-domain analysis.

In order to minimize the sound reflection to increase the fidelity of the received signal, we covered the sound-absorbing cotton around the microphones. In addition, a home-made sound attenuator box was housed the entire experimental setup to reduce the likely sound wave inference as much as possible.

Results and discussion

Figure 6 shows the experimental results of the commercial MEMS microphone and the fabricated basic functional ultrasonic microphone unit to be operated at the identical three fixed frequencies and wideband frequency range (5 kHz -55 kHz).

The obtained signals for three fixed frequencies (21 kHz, 22 kHz and 23 kHz) displayed almost the same amplitudes with a slightly reducing trend as frequency increased for both commercial and fabricated microphones.

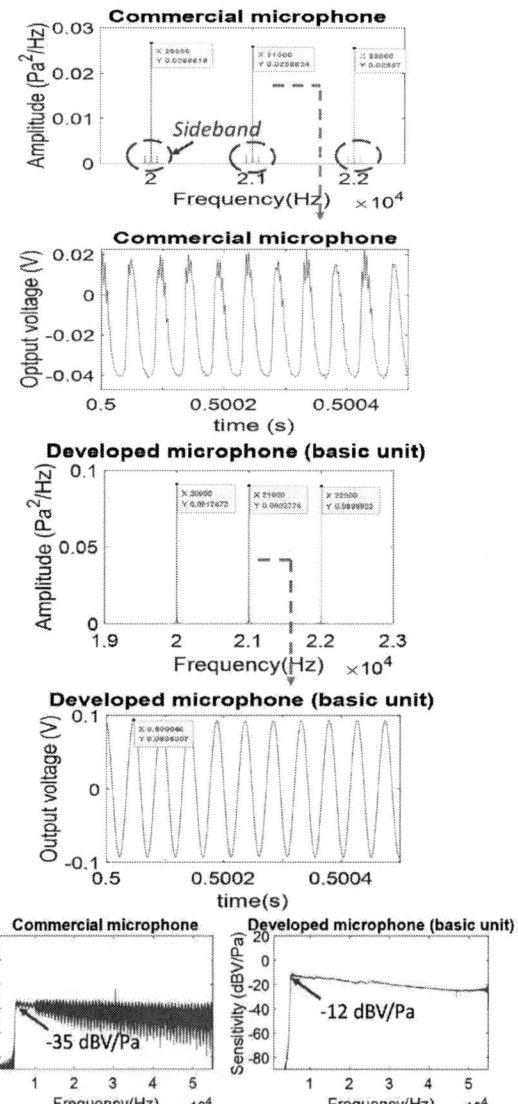

Figure 6: Experimental setup for testing and comparing the commercial and developed microphones.

978-1-6654-9309-3/23 $31.00 © 2023 IEEE

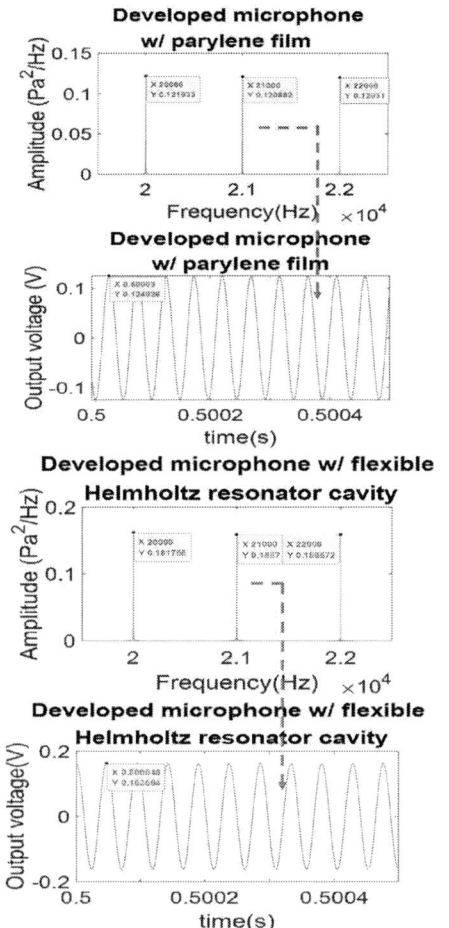

Figure 7: Experimental results of the developed microphone with parylene film and further enhanced by flexible Helmholtz resonant cavity.

An interesting phenomenon was observed for the commercial microphone, which upper and lower sideband signals existed for each single frequency. The time-domain signal also found the main wave was modulated by another wave.

The fixed frequency operation of our fabricated basic microphone unit exhibited a pure sinusoidal wave output without no sideband. It also shows approximately 3 times larger voltage output compared to the commercial microphone at a fixed frequency. As for the sensitivity performance, the sensitivity of the fabricated microphone unit possessed 20 dBV/Pa higher that of the commercial one while operated in a wide range of frequencies (5 kHz-55 kHz). This demonstrates the superiority of the proposed bottom-up fabrication scheme.

In addition, the basic microphone unit covered with parylene film displayed a significant improvement of the sensing performance. The fixed frequency operation at 22 kHz indicated a time-domain amplitude of 0.25 Vpp for the fabricated microphone with parylene film fully covered the gaps between the triangular cantilever beams and the surface of the beam array (Fig. 7). When this parylene film enhanced microphone further integrated with the flexible resonator cavity, the time-domain voltage output increased to approximately 0.33 Vpp at the same frequency. Hence, The sensitivity of the flexible resonator cavity integrated microphone was considerably boosted compared to that without resonator cavity.

SUMMARY

We successfully developed the ultrasonic microphone of high sensitivity by proposed bottom-up fabrication technique. The designed and fabricated PZT thin-film based triangular cantilever beam array exhibits low residual stress and wide bandwidth response. With parylene film coating and flexible cavity enhancement, the device performance can be even strengthened.

ACKNOWLEDGEMENTS

This work was supported by Ministry of Science and Technology in Taiwan under grant No. 109-2221-E-007-112-MY3. We also thank Prof. W. Fang at NTHU for providing confocal laser scanning microscope to facilitate the device measurement.

REFERENCES

[1] S. Anzinger, C. Bretthauer, J. Manz, U. Krumbein, A. Dehé, "Broadband acoustical MEMS transceivers for simultaneous range finding and microphone applications", in *Digest Tech. Papers Transducers'19 Conference*, Berlin, June 23-27, 2019, pp. 865-868.

[2] W.Mickiewicz, M. Raczyński, A. Parus, "Performance Analysis of Cost-Effective Miniature Microphone Sound Intensity 2D Probe", *Sensors,* vol. 20, no. 1, 2020, pp. 271.

[3] A. Kumar, A. Varghese, A. Sharma, M. Prasad, V. Janyani, R. P. Yadav, K. Elgaid, "Recent development and futuristic applications of MEMS based piezoelectric microphones", *Sensors and Actuators A: Physical*, vol. 347, no. 1, pp. 113887, 2022.

[4] G. H. Feng, G. R. Lai, "Hand Gesture Detection and Recognition Using Spectrogram and Image Processing Technique with a Single Pair of Ultrasonic Transducers", *Applied Sciences,* vol.11, no. 12, 2021, pp. 5407.

[5] G. H. Feng, W. S. Chen, "Piezoelectric Micromachined Ultrasonic Transducer-Integrated Helmholtz Resonator with Microliter-Sized Volume-Tunable Cavity", *Sensors*, vol. 22, no. 19, pp. 7471, 2022.

[6] G. H. Feng, K. Y. Lee, "Hydrothermally synthesized PZT film grown in highly concentrated KOH solution with large electromechanical coupling coefficient for resonator", *Royal Society open science*, vol. 4, no. 12, 2017, pp. 171363.

[7] G. H. Feng and E. S. Kim, "Universal concept for fabricating micron to millimeter sized 3-D parylene structures on rigid and flexible substrates," *in Digest Tech. Papers MEMS'03 Conference*, Kyoto, Jan. 19-23, 2003, pp. 594-597.

CONTACT

*G.-H. Feng, tel: +886-3-571-5131; ghfeng@pme.nthu.edu.tw

HALBACH-ARRAY MAGNETIC COIL ARRANGEMENT ON CMOS CHIP FOR SENSITIVITY ENHANCEMENT OF INDUCTIVE TACTILE SENSOR

Tien Chou [1], Zih-Song Hu [2], and Weileun Fang [1,2]
[1] PME, [2] iNEMS, National Tsing Hua University, Hsinchu, Taiwan

ABSTRACT

This study exploits the concept of Halbach array magnets and also leverage the multiple metal layers in the standard CMOS process to design and implement magnetic-coil distributions on the CMOS chip to concentrate and enhance magnetic fields. In this study, four vertical magnetic coils are designed to surround the four edges of traditional horizontal magnetic coil to reduce the magnetic flux loss. Thus, the sensitivity of inductive tactile force sensor (CMOS chip with magnetic bump and polymer filler is improved. Measurements indicate the proposed Halbach array coil design could enhance the sensitivity of tactile force sensor for more than 2-fold, as compare with the one with planar spiral coil.

KEYWORDS

Inductive sensing, tactile force sensor, CMOS chip, Halbach array.

INTRODUCTION

Tactile force sensors have wide applications in consumer electronics, robotics, etc. In general, physical contact with the users or sensing objects is required for the tactile force sensors. To use medical application as an example, the tip of the scalpel is equipped with a high-sensitivity tactile force sensor, which enables medical personnel to control surgical instruments at the same time [1]. In terms of consumer electronics applications, as the finger of users touches the sensing interface of tactile force, the contact information can be exploited to achieve richer command control, such as volume, strokes, color depth, etc. for the headphone, electronic brush, and so on [2-3].

Many approaches have been reported to achieve the tactile force sensors. Depending on the purpose of different applications, tactile sensors can be fabricated with silicon substrates, polymer materials, soft electronic materials, 3D printing technology, and more. In addition, tactile sensors with different sensing mechanisms, such as capacitive type, piezoresistive type, optical type, inductive type, etc., can be achieved by using different fabrications and designs [4-5]. The inductive type tactile force sensor typically consists of magnetic bump, polymer buffer, and sensing coil [6-8], could offer various advantages such as the relatively simple fabrication process and no fragile suspended structures are required during sensing. Thus the inductive sensing is a promising approach to realize the chip-scale tactile force sensor.

As reported in [9], the commercially available CMOS processes could provide various inherent materials and thin film layers to offer the design flexibility for the design of tactile force sensing chip. Moreover, various sensors have also been demonstrated using the processes, so that the monolithic sensors integration can be achieved. In [10], the inductive type tactile force sensor with no suspended thin film structures is realized using the CMOS chip. The small linewidth metal film in CMOS process enables the dense metal windings to increase the inductance effect. In addition, the multi-layer metal films offer the possibility to realize the coils in both planar and vertical directions. Thus, this study exploits this characteristic to design and integrate different sensing coils on the CMOS chip to control the direction of magnetic field lines, so as to enhance the sensitivity of the inductive type tactile force sensors.

DESIGN CONCEPT

Fig.1a presents the arrangement of magnetic-pole directions for the Halbach array magnets to concentrate the magnetic flux [11]. The magnetic flux from the N-pole will be attracted by the surrounding S-pole, forming the effect of magnetic field convergence. In this case, the arrangement of magnet array could enhance the planar magnetic field. Based on the concept of Halbach array arrangement, this study exploits the six metal-layers in the

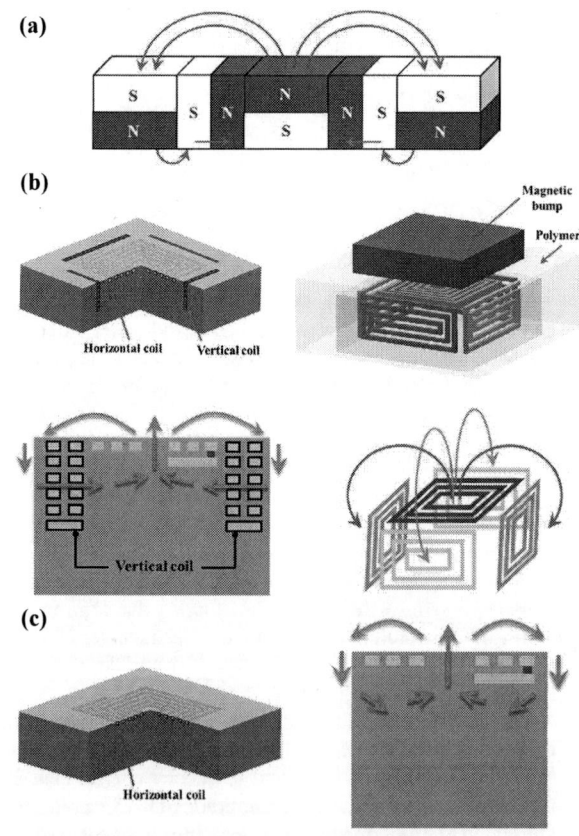

Figure 1: Scheme of the direction of the magnetic field, (a) Halbach array magnets, (b) proposed Halbach array coil design, (c) referenced horizontal spiral coil design.

standard TSMC CMOS process to design horizontal and vertical magnetic coils in Fig.1b to concentrate and enhance the magnetic field. In addition, the characteristics of CMOS multi-layer stacking offer the possibility to easily design coils. In short, the top metal layer is used to realize the horizontal planar spiral coil, and the six metal layers together with the five tungsten-vias are used to implement four vertical spiral coils surrounded the horizontal spiral coil. As indicated in figure, the magnetic flux generated by the horizontal spiral coil will be collected and re-directed by the vertical spiral coils. Thus, the escape and dissipate of magnetic flux from the sensing coils will be reduced. Moreover, the CMOS chip with Halbach array coil will be encapsulated by the polymer (Ecoflex 00-30) as the force transmitting layer and then covered with a soft magnet to form the tactile force sensor. In comparison, the conventional planar horizontal spiral coil in Fig.1c is also fabricated on the sensing chip. This reference design is a flat single-layer square coil with a planar dimension of 680×680 μm².

Fig.2 demonstrates the differences of the magnetic flux density between referenced and proposed type through 2D finite element simulation. Simulations in Fig.2a depict the distribution of magnetic flux for the planar spiral coil. In comparison, as indicated in Fig. 2b, the combination of the horizontal coil and vertical coil design will effectively enhance the magnetic flux. The magnetic flux at the edge has been collected and re-directed by the vertical magnetic coils.

FABRICATION AND RESULTS

Fig.3 shows process steps to fabricate the tactile force sensor. Fig.3a shows the CMOS chip from the TSMC with a square plane coil surrounded by four vertical coils. After that, the CMOS chip was encapsulated by the polymer through the molding process, and then a magnetic bump was bonded to the polymer, as shown in Fig. 3b. Micrographs in Fig.4a-b respectively show the reference and proposed sensing coil designs on the CMOS chip.

The optical micrographs in Fig.4a-b display different coil structures. Fig.4a is the single layer spiral coil, and Fig.4b is the single layer spiral coil surrounded by four

vertical coils. Fig.4c further shows different colors marked on the micrograph of Fig.4b to depict the positions of the horizontal coil (in light gray) and vertical coils (in orange). The vertical coils located on the periphery of the horizontal planar coil only occupy 7% of the footprint of chip. Therefore, the design of the vertical coil does not significantly enlarge the overall footprint of the sensor. Fig.5a displays typical fabricated CMOS chips, the overall chip size is about 2×2 mm², which contains the Halbach array coil designed in this study (The area for Halbach array coil is only about 0.73×0.73 mm²). Fig.5b is the chip after molding and assembly as the tactile force sensor (device-under-test, DUT). In this study, the 430 stainless steel with dimensions 1×1×0.35 mm³ was used as the magnetic bump. In addition, the material of polymer filler was Ecoflex 00-30 with 1:1 mixing ratio.

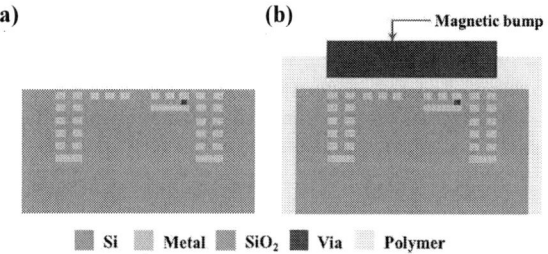

Figure 3: Fabrication steps, (a) the CMOS chip with coil designs fabricated by the TSMC, (b) the in-house post-CMOS processes for polymer molding and magnetic bump assembly.

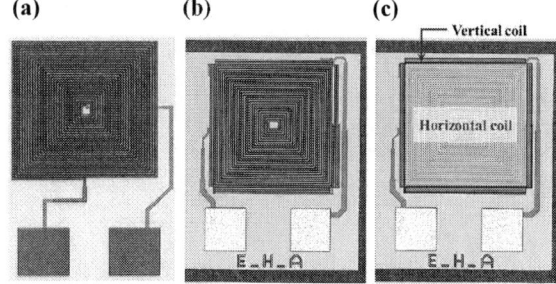

Figure 4: Micrographs of typical fabricated chips, (a) chip with the reference horizontal spiral coil design, (b) chip with the proposed Halbach array coil design, (c) vertical and horizontal coils of the proposed Halbach array design.

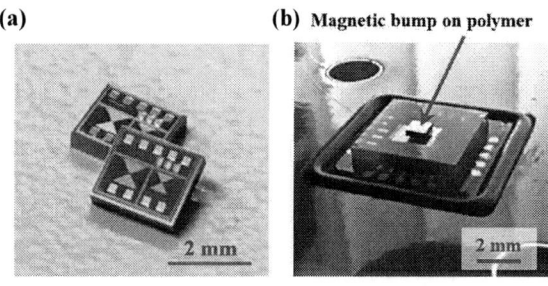

Figure 5: Micrographs of proposed inductive tactile force sensor, (a) the CMOS chip, (b) the chip after polymer molding and assembly with magnetic bump as the DUT.

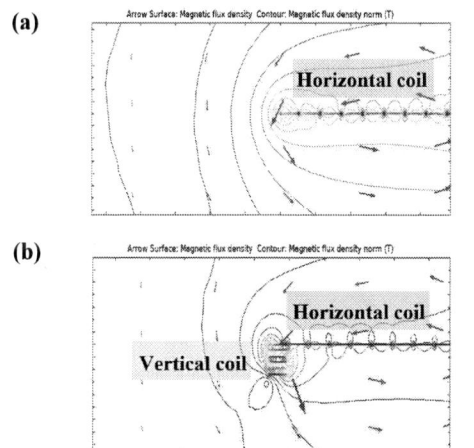

Figure 2: 2D simulation of two coil designs, (a) horizontal spiral coil design, (b) proposed Halbach array coil design.

MEASUREMENT RESULTS

Fig.6a shows measurement setup to characterize the inductance change of two coil designs with the magnet distance d. The soft magnet was adhered to the probe of the force gauge, and then the position stage was used to control the distance d between the magnet and the DUT with coils. Meanwhile, the inductance change generated by the coil will be determined by the LCR meter. Results in Fig.6b indicate the inductance change of the reference and proposed coil designs varying with the distance d. The inductance change of Halbach array design is 3-fold higher than the reference design with only horizontal spiral coil. Fig.6c depicts the layout of proposed and reference coils.

Moreover, Fig.7a shows measurement setup to measure the inductance change of tactile force sensors of different coil designs with the external load F. The force gauge applies and records loads on DUT. In more detail, the computer controlled position stage was employed to move the sensing chip to contact with the force gauge. The tactile load on DUT can be specified

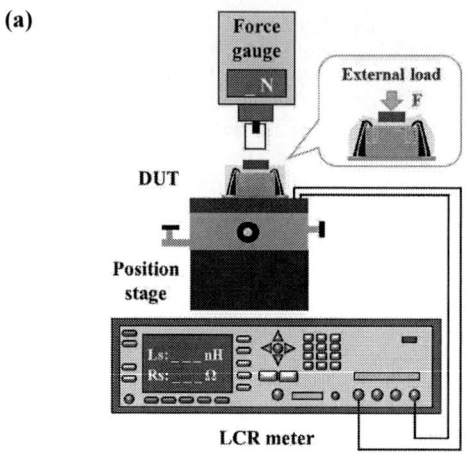

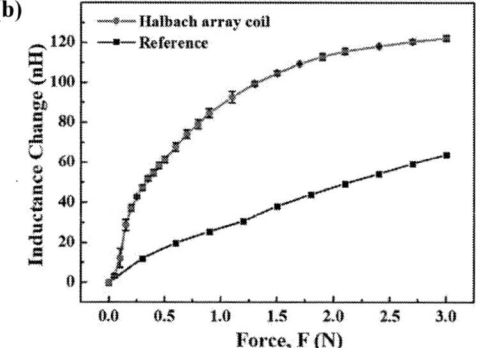

Figure 7: Inductance change vs the external load tests for the DUT in Fig.5b, (a) measurement setup, (b) measurement results.

by the position stage. Moreover, the tactile loads applied on DUT are monitored by the force gauge and the corresponding inductance change can be measured by the LCR meter as well. Measurements in Fig.7b demonstrate the design with Halbach array coils exhibiting a double inductance change compared with the reference design. However, the footprint is only increased for 7%. As the force applied to the DUT increases, the inductance changes of the two coils will gradually saturate. This is the well-known behavior due to the material properties of polymer layer (the force transmission layer) [12].

In summary, the multiple metal layers available in CMOS process can be exploited to realize the Halbach array magnetic coils to enhance the magnetic field, and further improve the performance of inductive tactile sensor.

CONCLUSIONS

In this study, the characteristics of the multi-layer stack of the CMOS standard process platform and the magnetic field concept of the Halbach array are used to design and implement the integration of a middle planar coil and four surrounding vertical coils. Thus, by changing the direction of the magnetic field lines, the effect of magnetic field convergence can be achieved,

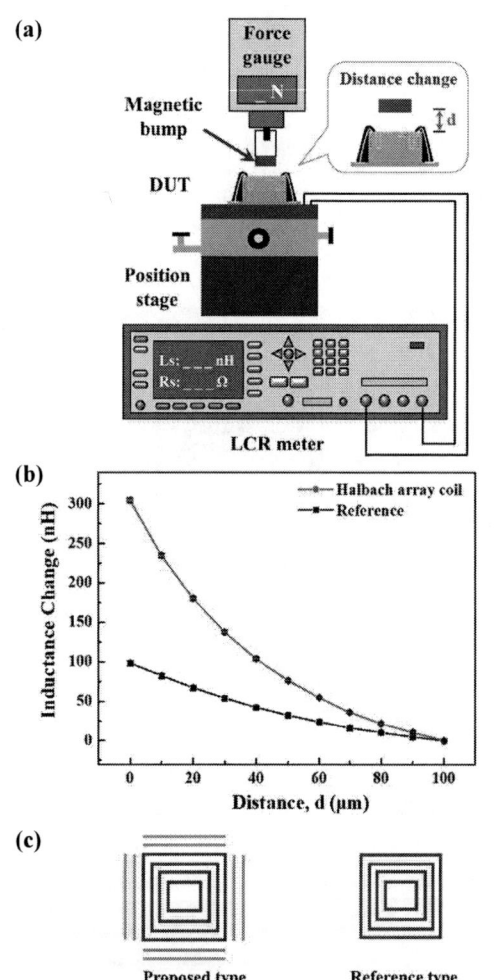

Figure 6: Inductance change vs the distance of magnetic bump tests for the CMOS chip in Fig.5a, (a) measurement setup, (b) measurement results, (c) scheme of coils.

and the occurrence of coil flux leakage can be reduced. As a result, the magnetic flux can be enhanced and the sensitivity of the inductive type tactile force sensor can be improved. Pure magnetic tests indicate the inductance change of Halbach array design is 3-fold higher than the reference design with only horizontal spiral coil. Moreover, the force sensing tests show that the presented tactile force sensor with Halbach coil array design could enhance the sensitivity for near 2-fold, as compared with the reference one.

ACKNOWLEDGEMENTS

This project was supported by the National Science and Technology Council under grant number NSTC 111-2923-E-007-004-MY2, NSTC 111-2923-E-007-003-, NSTC 111-2221-E-007-070-MY3, NSTC 111-2218-E-007-014-MBK, NSTC 110-2926-I-007-506-, and NSTC 110-2218-E-007-032-. The authors would like to appreciate Taiwan Semiconductor Manufacturing Co., Ltd. (TSMC) and Taiwan Semiconductor Research Institute (TSRI), for supporting the IC manufacturing and fabrication process. The authors are also grateful to Center for Nanotechnology, Materials Science, and Microsystems of National Tsing Hua University (CNMM) for providing measurement facilities.

REFERENCES

[1] K. Miller and M. Curet, "Intuitive surgical: an overview," in Robotic-Assisted Minimally Invasive Surgery: A Comprehensive Textbook, S. Tsuda and O. Y. Kudsi eds. Cham: Springer International Publishing, 2019, pp. 3-11. (ISBN: 978-3-319-96865-0)

[2] V. G. Chouvardas, A. N. Miliou, and M. K. Hatalis, "Tactile displays: a short overview and recent developments," *Proceedings of the ICTA*, Thessaloniki, Oct., 2005, pp. 246-251.

[3] O. Oliver, and R. Dahiya, "Smart tactile gloves for haptic interaction, communication, and rehabilitation," *Advanced Intelligent Systems*, vol. 4, pp. 2100091(22), 2022.

[4] Y. Liu, R. Bao, J. Tao, J. Li, M. Dong and C. Pan, "Recent progress in tactile sensors and their applications in intelligent systems," *Science Bulletin*, vol. 65, pp. 70-88, 2020.

[5] A. Tandon, P. Shukla, and H. K. Patel, "Review of transduction techniques for tactile sensors and a comparative analysis of commercial sensors," *2nd International Conference on Multidisciplinary Research & Practice*, Gujarat, Dec., 2015, pp. 133-138.

[6] T. Kawasetsu, T. Horii, H. Ishihara, and M. Asada, "Flexible tri-axis tactile sensor using spiral inductor and magnetorheological elastomer," *IEEE Sensors Journal*, vol. 18, pp. 5834-5841, 2018.

[7] L. Wang, D. Jones, G. J. Chapman, H. J. Siddle, D. A. Russell, A. Alazmani and P. Culmer, "A review of wearable sensor systems to monitor plantar loading in the assessment of diabetic foot ulcers," *IEEE Transactions on Biomedical Engineering*, vol. 67, pp. 1989 - 2004, 2020.

[8] H. Wanga, J. Kow, N. Raske, G. De Boer, M. Ghajari, R. Hewson, A. Alazmani, and P. Culmer, "Robust and high-performance soft inductive tactile sensors based on the Eddy-current effect," *Sensors and Actuators A: Physical*, vol. 271, pp. 44-52, 2018.

[9] S.-K. Yeh, M.-L. Hsieh and W. Fang, "CMOS-based tactile force sensor: A review," *IEEE Sensors Journal*, vol. 21, pp. 12563 - 12577, 2021.

[10] S.-K. Yeh, J.-H. Lee and W. Fang, "On the detection interfaces for inductive type tactile sensors," *Sensors and Actuators A: Physical*, vol. 297, pp. 111545(15), 2019.

[11] J. Masi, "Overview of Halbach magnets and their applications," *Electrical Manufacturing and Coil Winding Conference*, Dallas, Oct., 2010, pp. 134-139.

[12] I. D. Johnston, D. K. McCluskey, C. K. L. Tan and M. C. Tracey, "Mechanical characterization of bulk Sylgard 184 for microfluidics and microengineering," *Journal of Micromechanics and Microengineering*, vol. 24, pp. 035017(7), 2014.

CONTACT

* Weileun Fang; fang@pme.nthu.edu.tw

ON-MEMS-CHIP COMPACT TEMPERATURE SENSOR FOR LARGE-VOLUME, LOW-COST SENSOR CALIBRATION

Paolo Frigerio[1], Andrea Fagnani[1], Valentina Zega[1], Gabriele Gattere[2], Attilio Frangi[1], and Giacomo Langfelder[1]

[1]Politecnico di Milano, Milano, ITALY and
[2]STMicroelectronics, Cornaredo, ITALY

ABSTRACT

The work introduces a MEMS temperature sensor for use in the same module of other sensors, to provide optimal measurements for their thermal calibration/compensation purposes. The sensor, made of 25-µm-thick, lowly-doped, epitaxial polysilicon, relies on a multi-mode resonator, where the frequency ratio is used to extract relative temperature changes without the need of an external absolute time reference. Measurements across the [5-85]°C range validate the predictions of the frequency behavior, enabling to estimate ±50 mK$_{rms}$ resolution at 4-Hz data rate.

KEYWORDS

MEMS, multi-mode resonators, temperature sensors, temperature coefficient of frequency.

INTRODUCTION

The performance of several MEMS sensors is determined by stability over temperature T [1]. Though efforts are put to understand underlying phenomena [2], the behavior vs temperature remains perturbed by nth-order effects (e.g. parasitics drift and package stress, among the others) which are hard to model and predict. A need for calibration vs T is thus often mandatory. Several times the T sensor is aboard the integrated circuit or the printed circuit board (PCB): this means that, under spatial or temporal temperature gradients, the T estimate is inaccurate due to steady-state temperature offsets between MEMS and circuit, and/or due to their different thermal constant, yielding sub-optimal compensation. A solution could be a local T measurement, close to the MEMS sensor.

In this context, it is known that the natural frequency of MEMS resonators changes with T in a well repeatable manner, especially at low doping values [3]. However, not always a sensor has a self-sustained mode in operation (e.g. accelerometers); sometimes the sensor has one resonant mode (e.g. the gyroscope drive); even if more modes are forced (e.g. in FM gyroscope), they have nominally identical temperature coefficients of frequency (TCf). Can one infer relative temperature changes $\Delta T = T - T_0$ from a single mode with frequency $f(T)$, properly calibrated at a reference temperature T_0? Given the equation:

$$f(T) = f(T_0)(1 + \alpha \Delta T) \Rightarrow \Delta T = \frac{f(T)-f(T_0)}{\alpha f(T_0)} \quad (1)$$

where α is the linear TCf, the answer apparently seems: *yes*. However, to measure a frequency, one needs a frequency/time reference, which is itself an oscillator, which may itself drift. As this typically occurs, the issue is not solved. The key-point would be to avoid the need for an absolute, accurate frequency/time reference. This work

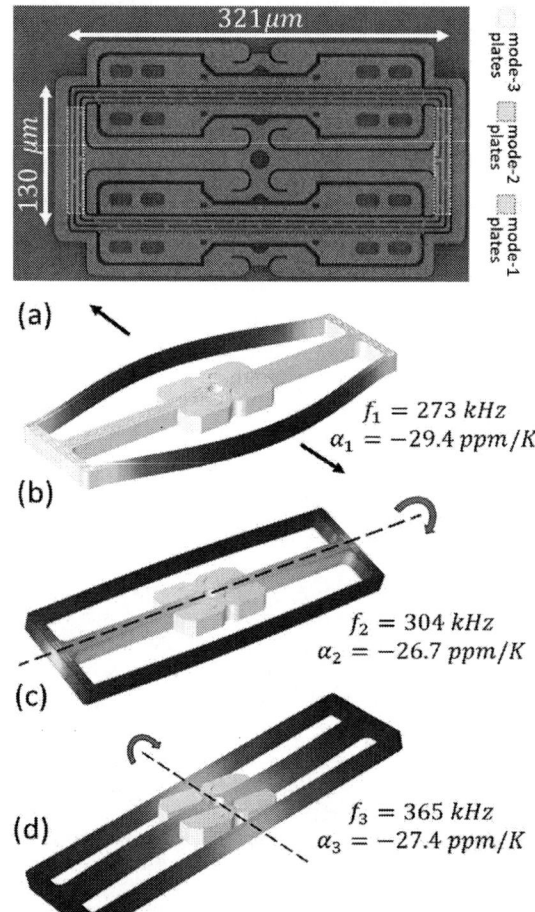

Figure 1: scanning electron microscope image of a 3-mode resonator (a) with the electrodes to drive and sense the three modes highlighted in colors. The modal shapes, frequencies and predicted TCfs are shown in plots (b-d).

proposes a T sensor based on a multi-mode resonator fabricated in a large-volume MEMS technology [4], compatible with low-cost applications and free of the need for an absolute time/frequency reference.

THEORY OF OPERATION

The sensor is based on the simultaneous oscillation of both flexural and torsional modes in a MEMS structure. The T dependence of the density and of the two elastic constants describing the constitutive laws of low-doped polysilicon, assumed as linear isotropic elastic material, can be derived from [5]: a different TCf can be then expected for different modes, as long as they are characterized by different distributions of flexural, shear

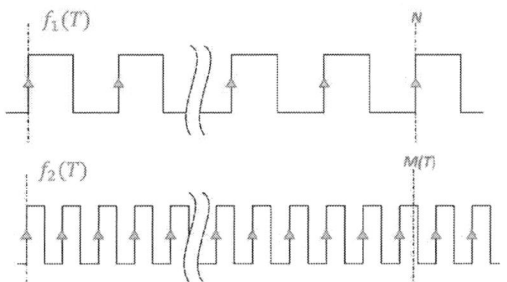

Figure 2: scheme of the temperature measurement method of this work: with two modes having different TCf, the number of periods M of the fast mode $f_2(T)$ that fits in a prescribed number N of periods of the slow mode $f_1(T)$ is a function of T and is used as a sensing method independent of an absolute time/frequency reference.

and torsional strains. If a difference can be observed in the temperature dependence of different modes of a single MEMS resonator, then the simultaneous actuation of at least two of its modes would enable the implementation of the self-referenced temperature sensor as described below.

The reference structure for the experiments is the one reported in Fig. 1a. The device, fabricated by STMicroelectronics using the ThELMA process [6], is anchored only in the center, and its main resonant mode is the flexural one whose mode shape is represented in Fig. 1b, based on the anti-phase bending of two silicon beams. The nominal resonance frequency is about 273 kHz. The other two resonant modes that can be exploited for the sensor are the torsional ones represented in Fig. 1c-d. These modes show a resonance frequency of about 304 kHz and 365 kHz, respectively. The device is actuated via parallel plates: Fig. 1a reports the electrodes arrangement for each mode. While the flexural mode can be actuated and sensed differentially, via the four electrodes highlighted in blue, the other two are inherently single-ended due to the arrangement of the electrodes.

Fig. 1 also reports the expected temperature dependence of each resonance frequency as calculated by FEM analysis: the average value of the TCf α is in the order of -30 ppm/K, as expected for silicon, however the coefficients of the three modes show a difference of a few ppm in their absolute value, corresponding to ~ 6%. Such small difference is enough to enable the sensor realization.

Its basic working principle is represented in Fig. 2 using the waveforms associated to just two modes, labeled $f_1(T)$ and $f_2(T)$. The idea is to adopt a counter to measure the number of cycles of both modes that fall within a given observation interval. However, the duration of such interval is not fixed as a value in seconds by an absolute reference; rather, it is given by a number N of cycles of the slowest of the two modes, i.e., $f_1(T)$ in the example. Thus, as temperature changes, the number of cycles N remains constant, but both the length of the measurement window and the number of cycles M of the second mode that fall within such frame change according to temperature. The two frequencies are expressed as:

$$f_1(T) = f_1(T_0)(1 + \alpha_1 \Delta T)$$
$$f_2(T) = f_2(T_0)(1 + \alpha_2 \Delta T) \qquad (2)$$

where $T_0 = 25\ °C$ is a reference temperature chosen for the initial frequency calibration, α_1 and α_2 are the temperature coefficients, and $\Delta T = T - T_0$ is the deviation of T from the reference temperature. The temperature variation can be measured by taking the ratio of the frequencies:

$$R(T) = \frac{f_2(T)}{f_1(T)} \qquad (3)$$

and evaluating its variation with respect to the reference temperature, i.e., $R(T) - R(T_0)$, resulting in the theoretical formula:

$$\Delta T = \left(\frac{R(T_0)}{R(T)} (\alpha_2 - \alpha_1) - \alpha_1 \right)^{-1}. \qquad (4)$$

An estimate of such exact quantity can be inferred from direct measurement of the number of cycles, minus a quantization error. Expressing the frequency ratio as a function of the number of counted cycles, the temperature estimate is written as:

$$\widehat{\Delta T} = \left(\frac{N}{M(T) - M(T_0)} R(T_0)(\alpha_2 - \alpha_1) - \alpha_1 \right)^{-1}. \qquad (5)$$

For small temperature-induced frequency variations, i.e., if $\alpha_1 \Delta T \ll 1$, the expression can be simplified, resulting in:

$$\widehat{\Delta T} \approx \frac{\Delta M(T)}{N} \frac{f_1(T_0)}{f_2(T_0)} \frac{1}{\alpha_2 - \alpha_1} \qquad (6)$$

where $\Delta M(T) = M(T) - M(T_0)$.

In (6) the value of N is chosen accounting for the bandwidth of the measured temperature variations. Given the desired output data rate (ODR) of the sensor, the value of N is chosen according to:

$$N < \frac{f_1(T_{max})}{\text{ODR}} \qquad (7)$$

where $T_{max} = 85\ °C$ is the maximum temperature that corresponds to the minimum value of f_1.

The number of cycles $M(T)$ is measured in real-time. Thus, there are four parameters that require an initial calibration, namely the nominal resonance frequencies at T_0, and the temperature coefficients of the two modes. The value of $M(T_0)$ derives from N and the nominal frequencies. As such, a printed circuit board (PCB) based system was developed to both characterize these parameters, as well as to sustain the simultaneous oscillation of two modes and thus to implement the sensor.

EXPERIMENTAL SETUP

Fig. 3a shows the circuit architecture adopted to keep each of the two modes in self-sustained oscillation. The MEMS displacement is sensed via a charge amplifier (CA) followed by gain stages that raise the output voltage to the level of ~ 1 V. A phase-shifter (90D) adjusts the phase-lag of the loop transfer function to 0° to meet the oscillation condition. A variable-gain amplifier (VGA) adjusts the gain of the loop transfer function according to the output of an automatic gain control (AGC) circuit to keep the

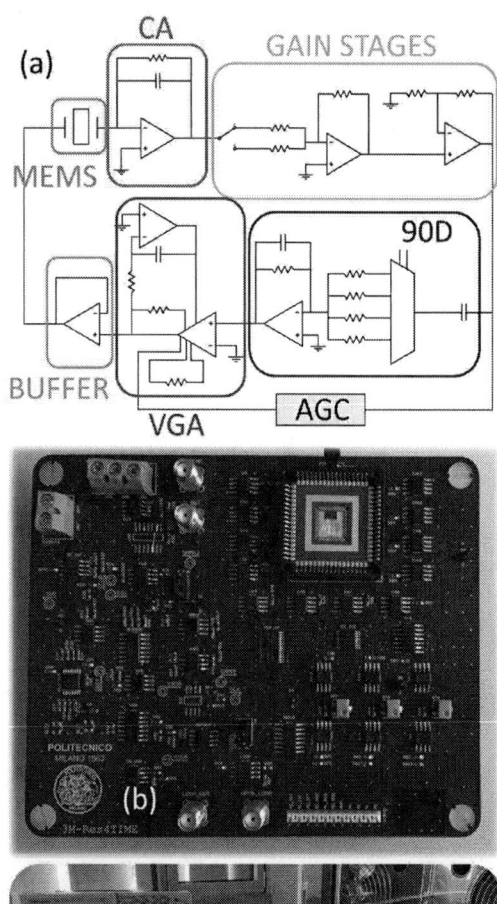

Figure 3: scheme of the oscillators (a) to sustain the modes. In (b-c) the board and climatic chamber setup are shown.

maximum displacement limited to about 200 nm, set using an external voltage reference. Finally, a buffer stage drives the MEMS actuators.

A switch is placed before the actuators to allow both open- and closed-loop operation. The former is used for preliminary characterization of the frequencies and temperature coefficients, while the latter is used for the sensor operation. The prototype PCB is shown in Fig. 3b.

Fig. 3c shows the whole setup, comprising the DC voltage generators for the ±5 V power supply, the rotor bias of about 20 V, and the AGC reference voltage, a frequency counter (Keysight 53230A), an oscilloscope and the climatic chamber to perform temperature tests.

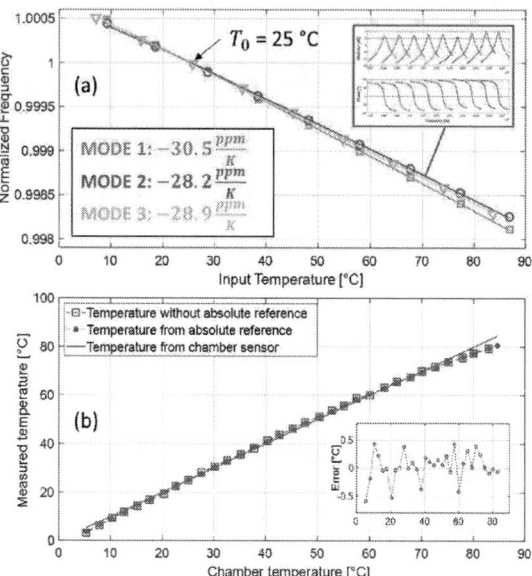

Figure 4: validation of TCf predictions through peak measurements vs T (a): results on the 3 modes show TCf trends in line with custom code predictions. Closed-loop operation (b) matches the use of an absolute reference (Keysight 53230A) within quantization (see the inset).

OPEN-LOOP CALIBRATION

Open-loop characterization was performed by sweeping the temperature from 85 °C to 5 °C with 10 °C steps and acquiring for each set-point the transfer function of the MEMS device on a frequency range centered on the resonance frequency. From such data, both the resonance frequency $f(T_0)$ at the reference temperature and the TCf can be extracted.

Fig. 4a shows the obtained TCf for the three modes. The inset shows a sample set of transfer function measurements, both magnitude and phase, used to extract the frequency data. Results well validate predictions, apart from an offset observed on all the modes: TCfs are all slightly larger than prediction, by 1.1 ppm for the flexural mode, and by 1.5 ppm for both the torsional ones. These results were repeatably obtained on eight different sets of resonators, belonging to two different MEMS dies.

CLOSED-LOOP OPERATION

Based on the results discussed in the previous section, Fig. 4b compares T measurements with the proposed method against the use of an absolute reference, i.e., the frequency counter which is kept outside the climatic chamber. The loop was closed for two modes, f_1 and f_2, corresponding to Fig. 1b and 1c, adopting two copies of the circuit of Fig. 3a. The nominal resonance frequencies for the device under test (DUT) are 276 kHz and 296 kHz, respectively.

The counter measures the oscillation frequency of both modes, and temperature is computed according to (4) – thus exploiting an absolute reference. The counter is also used to count the number of cycles of both modes, measuring over a gating time which varies as a function of temperature to keep the count N for the reference mode f_1 constant –

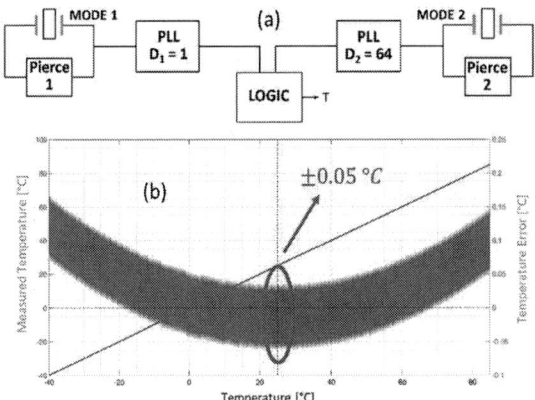

Figure 5: system-level scheme to reduce quantization error, while holding similar frequencies between modes. Pierce oscillators followed by PLLs with different DIV factor (a) yield ±50 mK quantization (b). A logic block implements the digital calculation of the ΔT equation in the text.

thus avoiding the absolute reference. Temperature is swept from 85 °C to 5 °C in 2.5 °C steps. The temperature calculation is then performed offline by a custom MATLAB code.

Measurement accuracy is dominated by quantization error, whose experimental value is shown in the inset of Fig. 4b, within about ±0.5 °C. The value chosen for N is $68 \cdot 10^4$, which enables a 0.4-Hz ODR. From a theoretical standpoint, the quantization error corresponds to the rounding error in counting M: the maximum value of such error thus corresponds to missing an entire period of f_2. Thus, it can be shown that the error improves by either increasing the observation interval, so the number of cycles N (which trades-off with the ODR); or increasing the ratio of frequencies R. The latter is the only viable solution to attain an acceptable ODR, i.e., in the order of a few Hz.

SYSTEM ARCHITECTURE

Attaining largely different frequencies, however, would require significantly different resonator topologies, inducing circuital differences which may be a source of mismatches and drifts. Therefore, it is preferred to keep a moderate frequency split between modes, use identical (well matched) oscillators, and further use a phase-locked-loop (PLL) after each oscillator with division moduli such to attain an additional frequency multiplication for mode f_2 (Fig. 5a). With this strategy, adopting two PLLs with division factors $D_1 = 1$ (buffer) and $D_2 = 64$ on f_1 and f_2 respectively, raising $f_{2,PLL}$ to 18.94 MHz, the quantization error drops below ±0.05 °C at a 4-Hz ODR, i.e., $N = 68 \cdot 10^3$, as shown by system-level simulation. Fig. 5b illustrates the overall temperature estimation error (orange curve). The quadratic dependence is related to the simplification introduced in equation (6) and can be eliminated by using the exact formula in equation (5), reducing the theoretical error down to ±0.05 °C due to quantization only. This is an acceptable value as it lies below the noise introduced by low-power Pierce oscillators.

CONCLUSION

This paper demonstrated the implementation of a temperature sensor based on the temperature-induced resonance frequency variations of multi-mode MEMS resonators. The sensor exploits the different temperature coefficients of two structural modes, enabling a self-referenced temperature measurement that does not rely on using a calibrated timing reference to readout the temperature-induced frequency variations. Results based on direct counting of the oscillators outputs shows that a 0.4-Hz ODR can be obtained with an accuracy of about ±0.5 °C. Increasing by a factor 64 the frequency of the "slave" mode using an integer-N PLL and performing counting on such upscaled reference should theoretically enable a 4-Hz ODR with a measurement accuracy within ±0.05 °C. Future work will focus on implementing the remaining parts of the system, namely the PLLs and a custom logic to perform temperature calculation, in order to implement a sensor with fully-digital output.

REFERENCES

[1] D. D. Shin, Y. Chen, I. B. Flader and T. W. Kenny, "Temperature compensation of resonant accelerometer via nonlinear operation," *2018 IEEE Micro Electro Mechanical Systems (MEMS)*, 2018, pp. 1012-1015, doi: 10.1109/MEMSYS.2018.8346730.

[2] B. Kim *et al.*, "Temperature Dependence of Quality Factor in MEMS Resonators," in *Journal of Microelectromechanical Systems*, vol. 17, no. 3, pp. 755-766, June 2008, doi: 10.1109/JMEMS.2008.924253.

[3] G. Mussi, P. Frigerio, G. Gattere and G. Langfelder, "A MEMS Real-Time Clock With Single-Temperature Calibration and Deterministic Jitter Cancellation," in *IEEE Transactions on Ultrasonics, Ferroelectrics, and Frequency Control*, vol. 68, no. 3, pp. 880-889, March 2021, doi: 10.1109/TUFFC.2020.3013976.

[4] G. Mussi, M. Bestetti, V. Zega, A. Frangi, G. Gattere and G. Langfelder, "Resonators for real-time clocks based on epitaxial polysilicon process: A feasibility study on system-level compensation of temperature drifts," *2018 IEEE Micro Electro Mechanical Systems (MEMS)*, 2018, pp. 711-714, doi: 10.1109/MEMSYS.2018.8346654.

[5] V. Zega, A. Frangi, A. Guercilena, & G. Gattere, "Analysis of frequency stability and thermoelastic effects for slotted tuning fork MEMS resonators," *Sensors*, vol. 18, no. 7, July 2018, doi: 10.3390/s18072157.

[6] C. Comi, A. Corigliano, G. Langfelder, V. Zega and S. Zerbini, "Sensitivity and temperature behavior of a novel z-axis differential resonant micro accelerometer", *Journal of Micromechanics and Microengineering*, vol. 26, n. 3, 2016, doi: 10.1088/0960-1317/26/3/035006.

CONTACT

*P. Frigerio, tel: +39-02-2399-3744; paolo.frigerio@polimi.it

PARTICULATE MATTER SENSOR BASED ON TWO-STAGE CASCADE VIRTUAL IMPACTORS AND THERMOPHORETIC MICROHEATERS

Kwang-Wook Choi[1], Ilhwan Kim[1], Seokwhan Chung[1], Gi-Bong Sung[2] and Se-Jin Yook[2]
[1]Samsung Advanced Institute of Technology, Suwon, KOREA
[2]Hanyang University, Seoul, KOREA

ABSTRACT

A particulate matter (PM) sensor that simultaneously measures the concentrations of $PM_{2.5}$ and $PM_{1.0}$ is reported in this study. Two serially connected virtual impactors (VIs) separate aerosol particles according to their sizes, which are subsequently collected on the surface of the gravimetric sensor through thermophoresis. The fabricated device achieved a mass sensitivity of 11 Hz/min per $\mu g \cdot m^{-3}$ and 10 Hz/min per $\mu g \cdot m^{-3}$ for $PM_{2.5}$ and $PM_{1.0}$, respectively, which is the highest among the gravimetric PM sensors reported to date. This result is attributed to the precise fluid-dynamic design of the VIs and the development of efficient microheaters that can provide localized heat and corresponding forces to the aerosol particles.

KEYWORDS

Particulate matter sensor, Gravimetric sensor, Virtual impactor, Thermophoresis, Airborne particle

INTRODUCTION

Airborne particles have different effects on the human body depending on their size, and continuous exposure to particulate matter (PM) can cause respiratory distress and cardiovascular diseases [1,2]. Particularly, short-term exposure to low concentrations of small-sized particles, such as $PM_{2.5}$ and $PM_{1.0}$, can be critical because they penetrate deep into the lungs. Consequently, the real-time monitoring of ultrafine particles has attracted much attention, and the accurate measurement of their concentration according to their size is crucial. Among various technologies, gravimetric methods are well suited for providing real-time data for size-discrete PM concentrations because they measure the direct mass of ultrafine particles regardless of their aerodynamic size. Generally, the aerosol particles entering the device are preprocessed to separate them according to their sizes and are subsequently collected on a mass sensor. As a precise mass detection sensor, a surface acoustic wave (SAW) sensor has proven its sensing capabilities, which provide a proportional resonance frequency shift to mass loading [3,4].

Herein, we report a gravimetric PM sensor with improved particle collection efficiency. The two-stage cascade virtual impactor (VI) is designed based on the computational fluid dynamics simulation, and a thermophoretic microheater is developed to optimize the particle collection on the surface of the SAW sensor.

DESIGN

Particle separation of VIs is based on the inertia of the particles [5]. When the aerosol flows into the entrance of the VI, the high-speed aerosol is divided into two parts, as shown in Fig. 1; the major flow takes 90% of the total flow, whereas the minor flow takes 10%. Small particles have low inertia and properly follow the curved streamlines of the major flow. However, large particles with large inertia deviate from curved streamlines and flow with the minor flow. Suppose 50% of the particles with a particular diameter flow into the minor flow channel and another 50% flows into the major flow channel. In that case, the corresponding diameter is called the cutoff diameter. Thus, particles larger than the cutoff diameter are concentrated and contained in the minor flow, whereas those smaller are separated and contained in the major flow.

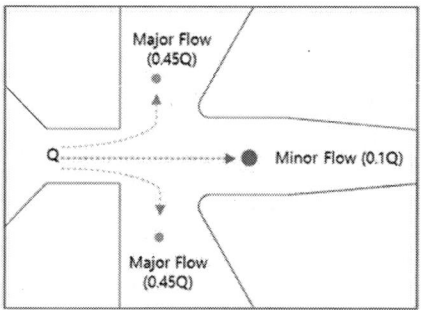

Figure 1: Inertial particle separation in a nozzle of the virtual impactor.

Schematics of the PM sensor based on the two-stage cascade VIs are shown in Fig. 2(a) and (b). The device is composed of VIs, microheaters, and SAW sensor dies integrated into a microfluidic channel. The SAW sensor die was embedded in the lower part of the major flow channel and aligned with the microheater in the upper part. The microheater applied a thermophoretic force to the separated particles. Thermophoresis is the drift of particles in a fluid medium in the direction of decreasing temperature. Because the temperature of the SAW sensor surface is maintained lower than that of the microheater, the particles passing through the space between the microheater and SAW sensor are supposed to drift toward the SAW sensor. Once the particles are deposited on the surface of the SAW sensor, the mass loading on the sensor changes the SAW velocity. The synchronous frequency shift induced by the SAW velocity change is utilized for $PM_{2.5}$ or $PM_{1.0}$ mass detection.

The two-stage VIs were designed to have a cutoff diameter of 2.5 μm in the first stage and 1.0 μm in the second stage, based on computational fluid dynamics simulations. The simulation results of $PM_{2.5}$ and $PM_{1.0}$, in the microfluidic channel, are shown in Fig. 2(c). As designed, a significant portion of $PM_{2.5}$ enters the side channel (major flow) of the first stage VI and the middle channel (minor flow) of the second stage VI. In contrast, a significant portion of $PM_{1.0}$ enters the side channels (major flow) of the first- and second-stage VI. That is, the major flow of the first-stage VI contains $PM_{2.5}$ (which includes $PM_{1.0}$), whereas the major flow of the second-stage VI contains only $PM_{1.0}$. Therefore, by placing the SAW sensors and microheaters in the major flow channels, $PM_{2.5}$ particles were collected on the SAW sensor of the first-stage VI, whereas $PM_{1.0}$ particles were captured on the SAW sensor of the second-stage VI, as shown in Fig. 2(a).

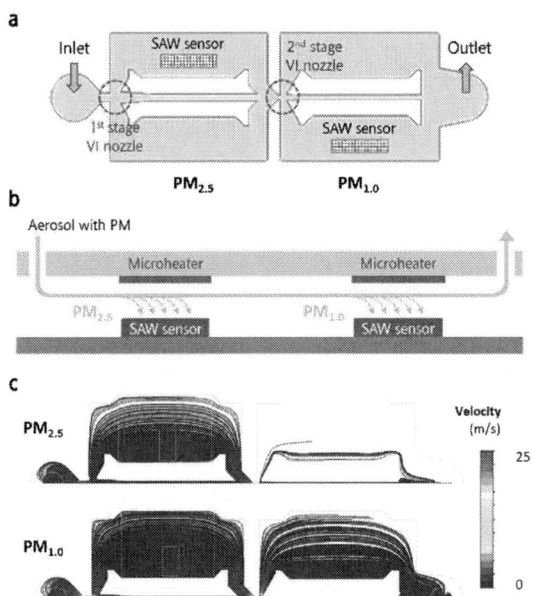

Figure 2: Schematics of a two-stage VI-based PM sensor. (a) Top and (b) Cross-sectional view. (c) Simulated particle trajectories in microfluidic channels.

FABRICATION

A three-dimensional schematic model of the proposed PM sensor is shown in Fig. 3. Each component was fabricated on 8-inch wafers using a micro-electromechanical system process and was assembled into a single module after being diced into individual dies. The detailed fabrication process of each component is described in the following subsections.

Microheater

The required characteristics of a microheater for use in a PM sensor include low power consumption and localized heating areas. In this study, a silicon-on-glass (SOG) substrate was employed to fabricate membrane-type microheaters. An SOG wafer with a 100-μm thick glass and a 500-μm thick Si layer was prepared. Furthermore, a lift-off patterning process was performed to define the heating material (Au, 100 nm) on the glass side of the wafer. A thin layer of Ti (10 nm) was used as the adhesion layer before Au deposition. Subsequently, a trench was formed on the backside Si under the heating area by a deep reactive-ion etching process (90% SF_6 10% O_2 etch, C_4F_8 passivation).

VI channel

The VI channel is the main component of the PM sensor, which separates aerosol particles according to their size. A silicon-on-insulator substrate with a 200-μm thick device layer, 1-μm thick buried oxide layer, and 400-μm thick handling layer was used to fabricate the VI channel. The Si device layer was patterned using the deep reactive-ion etching process and served as a microfluidic channel where size-separated aerosol particles flow. The Si-handling layer was laser-drilled to open windows to insert the SAW sensor. A silicon-on-insulator substrate is advantageous because the height of the microfluidic channel can be defined accurately based on the thickness of the device layer.

SAW sensor die

Size-separated aerosol particles accumulated on a SAW sensor to measure their mass. The fabrication of the SAW sensor die commenced by depositing a 0.5-μm thick SiO_2 layer on bare Si wafers using the plasma-enhanced chemical vapor deposition technique. Subsequently, the signal line was formed by a lift-off patterning process (Ti/Au 30 nm/400 nm) on the insulated substrate. After being diced into individual dies, the SAW devices were mounted on a predefined position using epoxy adhesive and wire bonded to the signal line. Because the thickness of the SAW device was 350 μm and that of the epoxy adhesive was 50 μm, a thermophoresis distance of 200 μm was achieved after assembly with the VI channel. Finally, all components were assembled into a single module using a manual clamping jig, which ensured complete sealing. An assembly image before the microheater was attached is shown in Fig. 4(b).

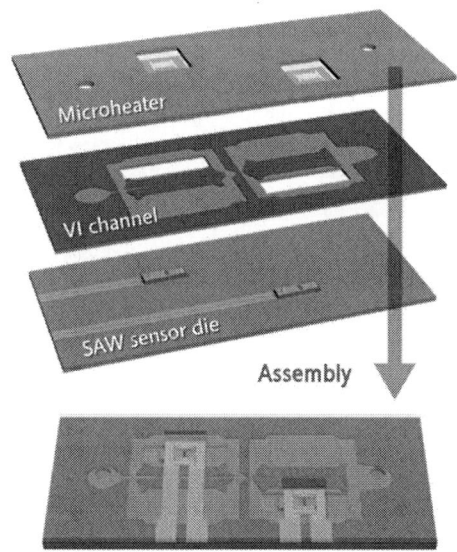

Figure 3: A 3D schematic model of the proposed two-stage PM sensor. (The substrate of the microheater is made transparent in the assembly image for clarity)

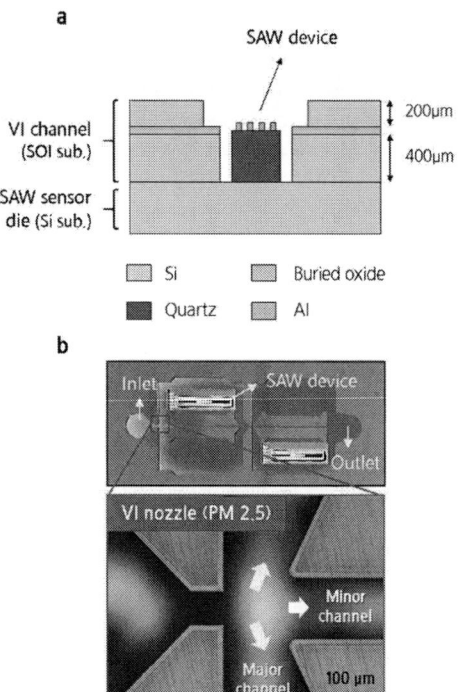

Figure 4: Assembly of the VI channel and SAW sensor die. (a) Schematic of the cross-section and (b) Optical images of the fabricated device.

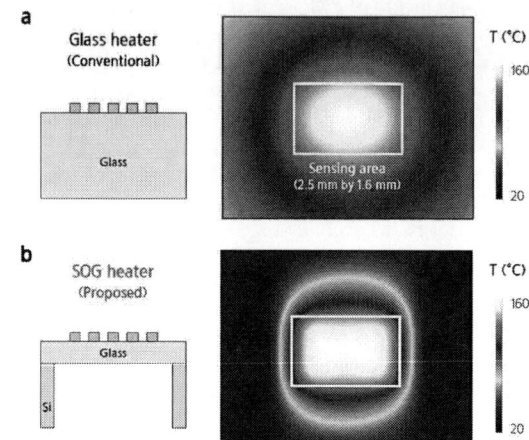

Figure 5: Schematic of the cross-section and obtained infrared thermal images of (a) conventional glass heater and (b) proposed SOG heater.

RESULTS

The SOG heater, which applied thermophoretic forces to aerosol particles in the VI channel, was first evaluated (Fig. 5). The proposed structure significantly reduced the heat conduction loss through the substrate, thereby lowering the power consumption and localizing the heat in the specified region. The localized heat created over the mass-sensing SAW sensor induced a large temperature gradient and the corresponding thermophoretic force to the aerosol particles passing through the VI channel. The developed SOG heater provided a concentrated force to the aerosol particles and improved the collection efficiency of the PM sensor compared with the conventional glass heater.

PM sensor measurements were performed using a custom-made PM chamber. The chamber can maintain a uniform particle concentration, and the size distribution of the particles ranged from submicron to 10 μm. Two small fans were placed in the chamber to ensure continuous air circulation within the chamber while maintaining a uniform particle concentration. Mass loading was performed on the surface of the $PM_{2.5}$ and $PM_{1.0}$ SAW sensors by repeated sampling at various PM mass concentrations to evaluate the PM sensor. A resonator-type SAW sensor with a synchronous frequency of 433 MHz was used for mass sensing, and each sampling time was 10 min. The temperature of the microheater used to generate the thermophoretic force was 160 °C. The resonance frequency change for each sampling time was measured by comparing the synchronous frequency of the SAW sensor before and after mass loading using a network analyzer.

The optical microscopy images of the surface of the $PM_{2.5}$ and $PM_{1.0}$ SAW sensors after a few sampling cycles are shown in Fig. 6. The measured size of the particles on each sensor surface was less than 2.5 μm and 1.0 μm, respectively, proving that designed VI is sufficiently accurate to separate the targeted PM size.

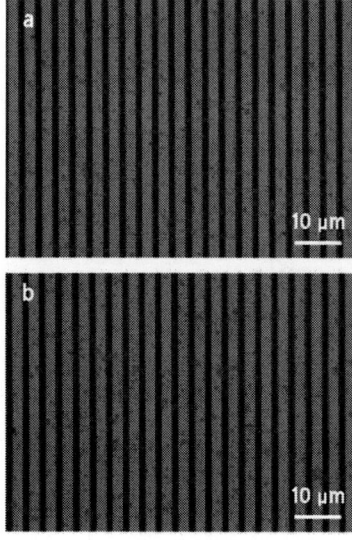

Figure 6: Scanning electron microscopy images of the surface of the SAW sensor after the PM collection. (a) $PM_{2.5}$ (b) $PM_{1.0}$

The resonance frequency shift of the SAW sensor measured during repeated PM loading cycles is shown in Fig. 7. Data were recorded after 10 min of sampling, and the same process was repeated five times. The average frequency shifts to PM concentrations ranging from 25 μg·m⁻³ to 200 μg·m⁻³ were linear, indicating that the developed PM sensor could reliably distinguish various concentrations of ultrafine particles. The mass sensitivity was calculated as 11 Hz/min per μg·m⁻³ and 10 Hz/min per μg·m⁻³ for $PM_{2.5}$ and $PM_{1.0}$ (the insets of Figs. 7(a) and (b)), respectively. To the best of our knowledge, this is the highest value achieved for gravimetric PM sensors [6]. The ability to sensitively detect the mass of ultrafine particles of small sizes up to $PM_{1.0}$, can be attributed to the precise fluid-dynamic design of the VIs and efficient microheaters developed to provide concentrated thermophoretic forces to the aerosol particles.

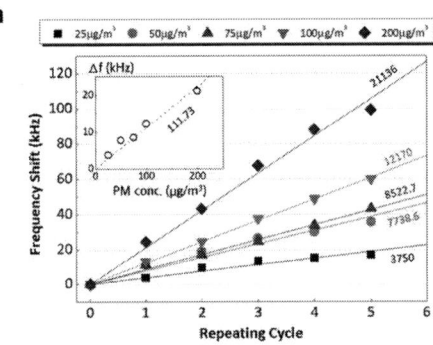

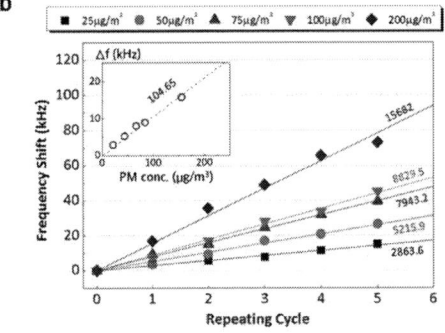

Figure 7: Frequency shift of the SAW sensor at various PM mass concentrations. (a) $PM_{2.5}$ (b) $PM_{1.0}$. (The sampling time for each repeating cycle is 10 min.)

CONCLUSION

This study designed a two-stage cascade VI and developed an SOG microheater to efficiently collect size-separated ultrafine particles on a mass-sensing SAW sensor. The aerodynamically designed VIs successfully separated $PM_{2.5}$ and $PM_{1.0}$, and the sensor provided a reliable and sensitive frequency response to various PM mass concentrations. The calculated mass sensitivity was the highest among the gravimetric PM sensors reported to date. The results demonstrated the possibility of real-time monitoring of small-size ultrafine aerosol particles and low-cost gravimetric PM sensors. In the future, we will focus on further improving the collection, which will lead to the detection of smaller particles, such as $PM_{0.3}$.

ACKNOWLEDGEMENTS

This work was supported by BK21 program from National Research Foundation.

REFERENCES

[1] H. Lin et al., "Particle size and chemical constituents of ambient particulate pollution associated with cardiovascular mortality in Guangzhou, China," *Environ. Pollut.* **208**, 758–766 (2016)

[2] R. Chen et al., "Beyond PM2.5: The role of ultrafine particles on adverse health effects of air pollution," *Biochim. Biophys. Acta (BBA) Gen. Subj.* **1860**, 2844–2855 (2016)

[3] W. Hao, J. Liu, M. Liu, S. He, "Development of a new surface acoustic wave based PM 2.5 monitor," *Proceedings of the 2014 Symposium on Piezoelectricity, Acoustic Waves, and Device Applications (SPAWDA), Beijing, China,* 52–55 (2014)

[4] S. Thomas, M. Cole, "High frequency surface acoustic wave resonator-based sensor for particulate matter detection," *Sens. Actuators A Phys.* **244**, 138–145 (2016)

[5] J. Zhao, M. Liu, W. Wang, J. Xie "Airborne Particulate Matter Classification and Concentration Detection Based on 3D Printed Virtual Impactor and Quartz Crystal Microbalance Sensor," *Sens. Actuators A Phys.* **238**, 379–388 (2016)

[6] D. Fahimi, O. Mahdavipour, J. Sabino, R. M. White, I. Paprotny, "Vertically-stacked MEMS $PM_{2.5}$ sensor for wearable applications," *Sens. Actuators A Phys.* **299**, 111569 (2019)

CONTACT

*K.-W. Choi, tel: +82-10-4022-8395, k.-w.choi@samsung.com

A MICROFLUIDIC OXYGEN GRADIENT GENERATOR FOR THE STUDY OF AEROTROPISM IN HYPHAE OF OOMYCETES

*Ayelen Tayagui[1,2,3], Yiling Sun[1,3], Ashley Garrill[1,2], and Volker Nock[*1,3]*

[1]Biomolecular Interaction Centre, Electrical & Computer Engineering, University of Canterbury, NZ
[2] School of Biological Sciences, University of Canterbury, NZ and
[3]The MacDiarmid Institute for Advanced Materials and Nanotechnology, Wellington, NZ

ABSTRACT

This paper reports a microfluidic platform that enables the study of aerotropism in hyphal microorganisms, such as fungi and oomycetes. While oxygen concentration is likely a key factor influencing the direction of growth of these organisms, how they sense and respond to O_2 is poorly understood. Our device generates O_2 gradients in a central growth channel in which hyphae have been inoculated. The gradient is generated via diffusion of gas through gas-permeable polydimethylsiloxane from two adjacent channels running in parallel to the growth channel. The resulting O_2 gradient perpendicular to the axis of growth is monitored via an integrated fluorescent sensor film. Our results show that the device enables hyphal growth, that the directionality of growth can be characterized and that hyphae of the oomycete *Achlya bisexualis* tend to grow away from low O_2.

KEYWORDS

Oomycetes, Aerotropism, Hyphae, Oxygen Gradient, Oxygen Sensor, Lab-on-a-Chip.

INTRODUCTION

Oomycetes and fungi play critical roles in ecosystems, recycling dead and decaying organic matter. Many species also break down living material, causing disease in their plant and/or animal host [1]. Their predominant vegetative structure is a mycelium, a network of branching, filamentous cells called hyphae. To be able to locate organic matter and hosts hyphae extend by a process called tip growth. The directionality of tip growth can be dictated by environmental cues, such as nutrients or gases [2]. One such gas is oxygen and hyphae can show aerotropic growth towards, or away from, oxygen [2,3]. Little is known about the cellular mechanisms that underlie the sensing and response to these O_2 gradients, although pathogenic fungi adapt to low oxygen conditions by altering their gene expression [4].

Microfluidic devices are able to generate precisely defined O_2 gradients, in contrast to conventional experimental systems [2]. They may therefore provide a means to study the mechanisms that underlie aerotropism in oomycetes and fungi. We have previously demonstrated the integration of polystyrene-encapsulated optical oxygen sensors into microfluidic devices [5] to enable the control and visualization of O_2 gradients in microfluidic flow [6], as well as their use to simulate the oxygen microenvironment in cancer cell cultures [7]. In this paper we show how the integration of a polydimethylsiloxane (PDMS)-based oxygen sensor into a microfluidic device can be used for the study of aerotropism in an oomycete. In particular, we show that the microfluidic platform

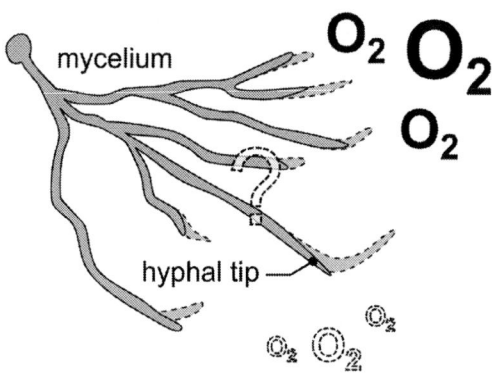

Figure 1: Aerotropism is hypothesized to involve the sensing of oxygen (O_2) concentration by fungi and oomycetes, and subsequent re-orientation of hyphal tips during polarized growth.

facilitates the observation of changes in hyphal tip orientation and growth in response to O_2 gradients (Fig. 1). As such, the platform provides a useful tool to improve our understanding of the underlying cellular mechanisms of aerotropism.

EXPERIMENTAL METHODS

Platform Design

O_2 gradients were generated across a media-filled microchannel by gas diffusion through PDMS channel walls [8] from two gas channels located on each side of the central channel (Fig. 2(a)). A fluorescence-based O_2 sensor was integrated into the channels in the form of a thin layer coated onto a glass substrate (Fig. 2(b)). Once filled with media, hyphae of the oomycete *Achlya bisexualis* were seeded into the central channel inlet (Fig. 2(c)). Oxygen gradients were then generated by flowing O_2, N_2 or air through the parallel gas channels, and hyphal growth and tip orientation was recorded using bright-field microscopy.

Chip Fabrication

PDMS microfluidic chips and the oxygen sensor substrate were fabricated using a combination of photo- and soft-lithography [9,10]. The design for the fluidic and two adjacent gas channels was created using L-Edit (Mentor Graphics). A laser mask writer (µPG101, Heidelberg Instruments) was used to produce a 4" chrome-on-glass photomask (Nanofilm) for the PDMS mold. Negative-tone dry film photoresist (SUEX 100, DJMicrolaminates) was laminated onto a 4" silicon wafer, which had been dehydrated in a 185 °C oven for 2 h and then cleaned using O_2 plasma (Tergeo, PIE Scientific) at 100 W for 10 min. Patterns were transferred from the

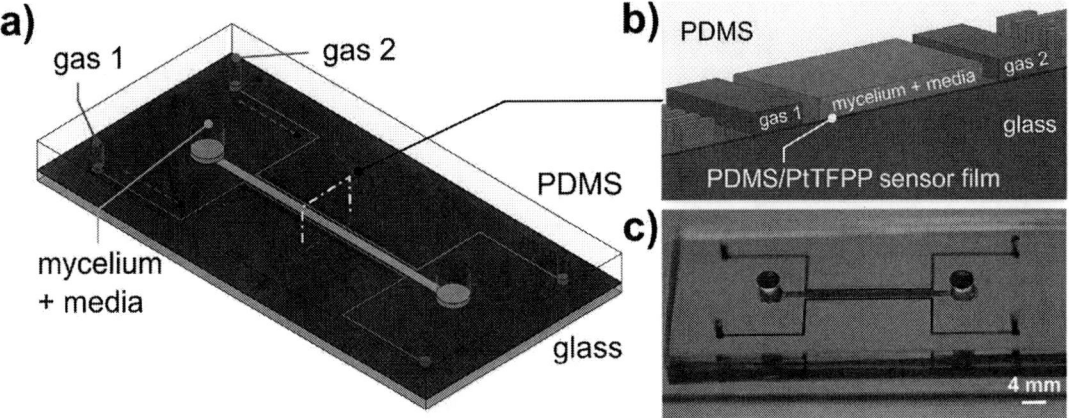

Figure 2: Oxygen gradient generation chip used to study hyphal aerotropism. (a) Schematic of the chip illustrating the central channel (mycelium + media) for mycelial culture and two adjacent parallel gas channels (gas 1 & gas 2) made from gas-permeable PDMS. (b) Schematic of the device cross-section and (c) photograph of a completed chip with the central observation channel colored in blue. The PDMS channels were bonded to a thin PDMS/PtTFPP O_2 sensor film spin-coated onto a glass microscope slide and visible as purple hue through the PDMS.

template mask onto the photoresist using UV lithography (MA-6, Suss) in low-vacuum-contact mode and a post-exposure bake. The resist was then developed in propylene glycol methyletheracetate (PGMEA), rinsed with isopropanol and dried with N_2, before a post-exposure bake was performed.

PDMS (10:1 w/w, Sylgard 184, Dow Corning) was replica-cast off the mold after treatment with trichloro (1H, 1H, 2H, 2H-perfluorooctyl) silane (Sigma-Aldrich). Premixed and degassed PDMS was poured onto the mold, and degassed again. After baking for 2 hours at 80°C on a hotplate, the PDMS was carefully peeled off and baked for a further 2 hours at 80°C. Once cured, the PDMS was cut to the appropriate size for the individual devices and inlet and outlet holes were punched with 1- and 3-mm hole punches (ProSciTech) [10]. For visualization, chips were filled with dye-colored epoxy as described previously [11].

Oxygen sensor films

Oxygen sensor films were prepared by mixing 11.7 mg of Platinum(II)-5,10,15,20-tetrakis-(2,3,4,5,6-pentafluorophenyl)-porphyrin (PtTFPP, Sigma-Aldrich) with 9.7 g of PDMS (10:1 w/w). The mixture was degassed for 30 minutes. The PtTFPP-PDMS was then spin-coated onto the glass slides at 800 rpm for 30 seconds using a Laurell WS-650 spin coater and left in the fume hood for 4 hours in order for the toluene, in which the dye was suspended, to evaporate. Finally, the oxygen sensor films were cured by baking the glass slides at 80° C on a hotplate for 2 hours [12].

For the experiments, PDMS microfluidic chips were bonded to the PtTFPP-PDMS-coated glass slide, using an O_2 plasma asher (Emitech k1050X) at 100 W for 50 seconds. These were then baked for 2 hours at 80°C on a hotplate to strengthen the bonding. Finally, they were degassed for 2 hours and vacuum sealed prior to use [13].

Biological Sample Preparation

Stock cultures of the oomycete *A. bisexualis* were grown on peptone yeast glucose agar media (PYG -

containing [in % w/v] peptone [0.125], yeast extract [0.125], glucose [0.3], and agar [2]) (all ThermoFisher) on Petri dishes and incubated for 48 hours at 26°C.

Experimental Setup

To inoculate the chips, a 2.5 mm agar plug was cut from the edge of the growing culture and placed upside down onto the chips containing PYG broth (containing [in % w/v] peptone [0.125], yeast extract [0.125] and glucose [0.3]). These were incubated for 12 hours at 26°C. A rotameter (Dwyer Instruments, PTY, LTD) was used to maintain a constant flow of O_2 and N_2 at 10 cc/min throughout the experiments, through the parallel gas channels. The growth and the tip orientation of the hyphae were recorded using an upright light microscope (Nikon Eclipse 80i) and a digital camera (ORCA-Flash4.0 V2, Hamamatsu). Image acquisition was controlled by a PC running HCImageLive (Hamamatsu). Image sequences were imported into ImageJ (V1.8_172, FIJI) [14] and the growth angle of the tip-most 100 μm of each hypha, relative to the horizontal, was determined using the angle measurement tool.

EXPERIMENTAL RESULTS

Oxygen gradients across the central channel were simulated as a function of supplied O_2 using COMSOL (Fig. 4(a)). For this, a *Transport of Diluted Species* model was used with O_2 diffusion coefficients of 3.25×10^{-9} m²/s in PDMS [15] and 2.14×10^{-9} m²/s in media/water [16]. Simulation results were compared to gradients established on chip via the fluorescent response of the integrated sensor film (Fig. 4(b)). For each measurement point, a set of fluorescence and brightfield images was recorded, the former to visualize the O_2 gradient, and the latter to track hyphal growth and tip orientation. To determine whether *A. bisexualis* hyphae reoriented their growth direction in response to the O_2 gradients, the central channel was split horizontally into two halves and the angle of growth of the tip-most 100 μm of each hypha was measured relative to

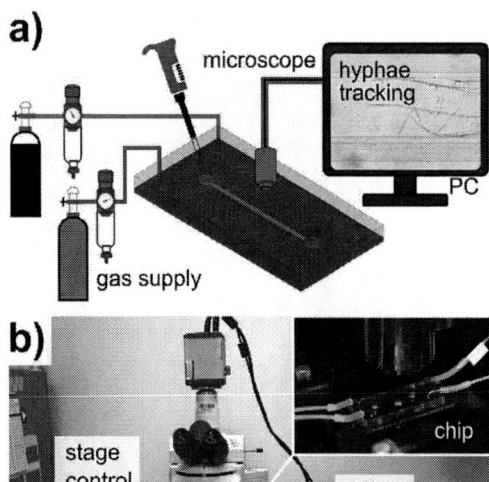

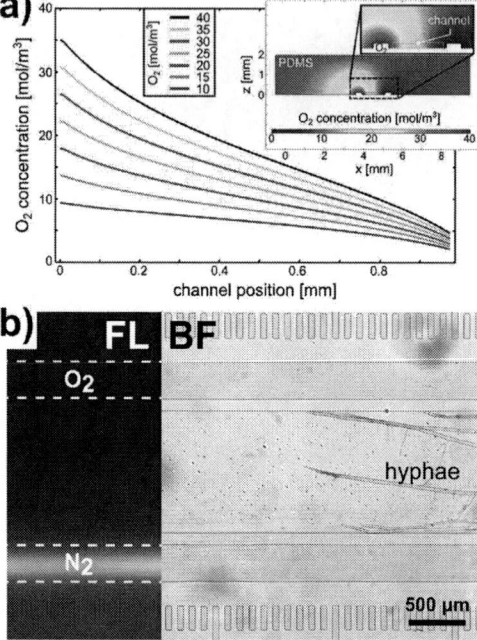

Figure 4: Experimental setup used for aerotropism experiments. (a) Schematic of the supply of gradient generating gases (air, N_2 and O_2) and seeding of the microorganisms into the central media-filled observation channel. (b) Changes in hyphal orientation in response to O_2 gradients generated in the central channel were recorded using an upright epi-fluorescence microscope with automated stage control. Inset shows the chip interfaced with tubing for gas supply on the microscope stage.

Figure 3: Oxygen gradients inside the observation channel. (a) COMSOL simulation results for gradients generated across the width of the central channel as a function of O_2 concentration in the gas channels. Inset shows a plot of the cross-section for 40 mol/m^3 O_2 in the left, and 0 mol/m^3 in the right gas channel, respectively. (b) Example composite of fluorescence (FL) and bright field (BF) microscopy images of the chip with A. bisexualis hyphae growing inside the central channel from right to left. The FL image visualizes the O_2 gradient, with a bright signal indicating low O_2 concentration. In this example, gaseous O_2 was applied to the top channel, N_2 to the bottom channel.

the horizontal using ImageJ, as illustrated in Fig. 5(a). Growth towards the channel edge was labelled positive and growth away from it negative. Hyphal branches, which tended to grow perpendicular to the axis of the channel, were excluded from analysis.

A combination of gradients was investigated, with air flowing through both gas channels acting as control. The influence of O_2 concentration on hyphal tip orientation and growth was studied for static and dynamic conditions by applying O_2 gas (gradient source) to one and N_2 gas (gradient sink) to the other gas channel. In the dynamic case, gases flowing through the two channels were swapped during an experiment to test for tip re-orientation in response to changing gradients. Examples of the different conditions, including repetitions and gradient reversal, can be found online in the following video: https://youtu.be/RZiS8A7xsGA.

In the absence of a gradient, no significant difference was found between angle of growth in the two halves (Fig. 5(b)). When a gradient was applied, a significant difference was observed in the angle of growth, furthermore the average angle was positive in the O_2 half and negative in the N_2 half, consistent with growth of the hyphae along the O_2 gradient toward the higher O_2 concentration. In experiments where the gradient was

dynamically reversed, hyphae reoriented their growth towards O_2, demonstrating that *A. bisexualis* hyphae sensed and used O_2 concentration as a directional cue.

CONCLUSIONS

In this paper we report the first demonstration of a microfluidic platform that enables the study of aerotropism in hyphal microorganisms. Our initial experiments show that is possible to quantify the gradient through the use of dyes and that oomycete hyphae are able to sense and grow towards areas of higher O_2 concentration. The platform provides a tool for future studies on the cellular mechanisms that underlie aerotropic responses in oomycetes and fungi.

ACKNOWLEDGEMENTS

The authors would like to thank Linda Chen and Gary Turner of the University of Canterbury Nanofabrication Laboratory for technical support. Financial support was provided by Rutherford Discovery Fellowship RDF-19-UOC-019, the Biomolecular Interaction Centre and the

978-1-6654-9309-3/23 $31.00 © 2023 IEEE

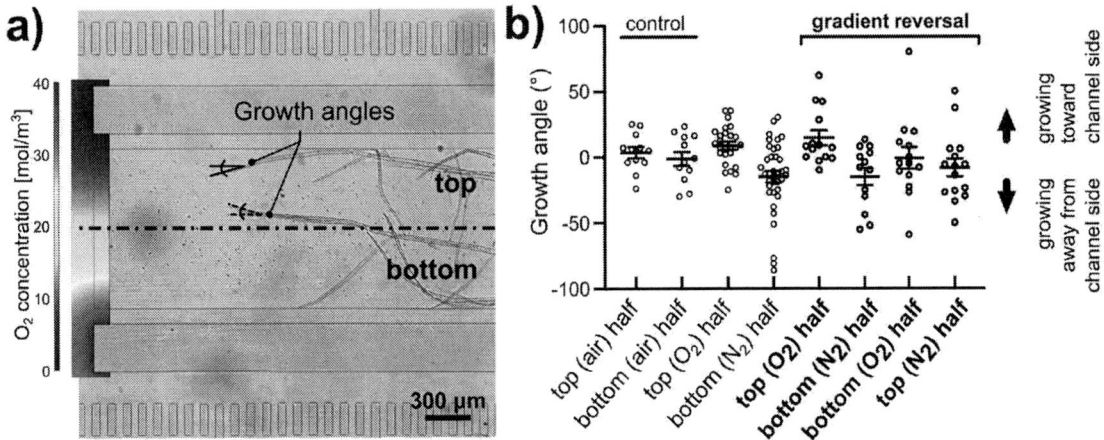

Figure 5: Investigation of aerotropism in hyphae of the oomycete Achlya bisexualis. *(a) For analysis, the channel was split horizontally into two halves (top and bottom) and the growth angle of the tip-most 100 μm of each hypha was measured relative to the horizontal, and as function of the respective O_2 gradient. Color overlay shows an example gradient as simulated using COMSOL. (c) Plot of the measured growth angle in response to different O_2 gradients, with the gas used in the respective gas channel given in brackets, demonstrating O_2 sensing and resulting tip direction changes in hyphae of* A. bisexualis.

MacDiarmid Institute for Advanced Materials and Nanotechnology.

REFERENCES

[1] Y. Sun, A. Tayagui, S. Sale, D. Sarkar, V. Nock, and A. Garrill, "Platforms for High-Throughput Screening and Force Measurements on Fungi and Oomycetes," *Micromachines*, vol. 12, no. 6, p. 639, 2021.

[2] A. Brand and N. A. R. Gow, "Mechanisms of hypha orientation of fungi," *Curr. Opin. Microbiol.*, vol. 12, no. 4, pp. 350-357, 2009.

[3] H. Chung and Y.-H. Lee, "Hypoxia: A Double-Edged Sword During Fungal Pathogenesis?," *Front. Microbiol.*, vol. 11, pp. 1920, 2020.

[4] F. Hillmann, E. Shekhova, and O. Kniemeyer, "Insights into the cellular responses to hypoxia in filamentous fungi," *Curr. Genet.*, vol. 61, no. 3, pp. 441-455, 2015.

[5] V. Nock, R. Blaikie, and T. David, "Patterning, integration and characterisation of polymer optical oxygen sensors for microfluidic devices," *Lab Chip*, vol. 8, no. 8, 1473-0189, pp. 1300-1307, 2008.

[6] V. Nock and R. Blaikie, "Spatially Resolved Measurement of Dissolved Oxygen in Multistream Microfluidic Devices," *IEEE Sens J*, vol. 10, no. 12, 1530-437X, pp. 1813-1819, 2010.

[7] L. Orcheston-Findlay, A. Hashemi, A. Garrill, and V. Nock, "A microfluidic gradient generator to simulate the oxygen microenvironment in cancer cell culture," *Microelectronic Eng.*, vol. 195, pp. 107-113, 2018.

[8] Y.-H. Chen, C.-C. Peng, Y.-J. Cheng, J.-G. Wu, and Y.-C. Tung, "Generation of nitric oxide gradients in microfluidic devices for cell culture using spatially controlled chemical reactions," *Biomicrofluidics*, vol. 7, no. 6, p. 064104, 2013.

[9] A. Tayagui, Y. Sun, D. A. Collings, A. Garrill, and V. Nock, "An elastomeric micropillar platform for the study of protrusive forces in hyphal invasion," *Lab Chip*, vol. 17, no. 21, pp. 3643-3653, 2017.

[10] Y. Sun, A. Tayagui, A. Garrill, and V. Nock, "Microfluidic platform for integrated compartmentalization of single zoospores, germination and measurement of protrusive force generated by germ tubes," *Lab Chip*, vol. 20, pp. 4141 - 4151, 2020.

[11] R. Soffe, A. J. Mach, S. Onal, V. Nock, L. P. Lee, and J. T. Nevill, "Art-on-a-Chip: Preserving Microfluidic Chips for Visualization and Permanent Display," *Small*, vol. 16, p. 2002035, 2020.

[12] Y. Gao, G. Stybayeva, and A. Revzin, "Fabrication of composite microfluidic devices for local control of oxygen tension in cell cultures," *Lab Chip*, vol. 19, no. 2, pp. 306-315, 2019.

[13] Y. Sun, A. Tayagui, A. Garrill, and V. Nock, "Fabrication of In-Channel High-Aspect Ratio Sensing Pillars for Protrusive Force Measurements on Fungi and Oomycetes," *J. Microelectromech. Syst.*, vol. 27, no. 5, pp. 827-835, 2018.

[14] J. Schindelin et al., "Fiji: an open-source platform for biological-image analysis," *Nat. Meth.*, vol. 9, no. 7, pp. 676-682, 2012.

[15] D. A. Markov, E. M. Lillie, S. P. Garbett, and L. J. McCawley, "Variation in diffusion of gases through PDMS due to plasma surface treatment and storage conditions," *Biomed. Microdev.*, vol. 16, no. 1, pp. 91-96, 2014.

[16] P. Han and D. M. Bartels, "Temperature Dependence of Oxygen Diffusion in H_2O and D_2O," *J. Phys. Chem.*, vol. 100, no. 13, pp. 5597-5602, 1996.

CONTACT

*V. Nock, tel: + 64 3 3694303;
volker.nock@canterbury.ac.nz

A PAPER-BASED DUAL APTAMER ASSAY ON AN INTEGRATED MICROFLUIDIC SYSTEM FOR DETECTION OF HNP 1 AS A BIOMARKER FOR PERIPROSTHETIC JOINT INFECTIONS

Rishabh Gandotra[1], Feng-Chih Kuo[2], Mel S. Lee[3] and Gwo-Bin Lee[1]

[1]National Tsing Hua University, Hsinchu, TAIWAN
[2]Kaohsiung Chang Gung Memorial Hospital, Kaohsiung, TAIWAN
[3]Paochien Hospital, Pingtung, TAIWAN

ABSTRACT

This work demonstrated a novel dual-aptamer assay performed on a paper-based membrane composed of nitro-cellulose (NC) on an integrated microfluidic platform for detection of a biomarker, human neutrophil peptide 1 (HNP 1) for periprosthetic joint infection. A fully automated device involved a single loading process (< 5 min) for the newly developed sandwich assay for HNP 1 quantification. The primary aptamer was immobilized on NC membrane where HNP 1 was captured and detected using fluorescent-labelled secondary aptamer. The developed assay is faster (~ 30 min) and required less volume (only 50 µL) than traditional assays.

KEYWORDS

Dual-aptamer assay, paper-based assay, microfluidics, HNP 1, PJI

INTRODUCTION

The recent emergence of total joint arthroplasty (TJA) increases dramatically in the last decade, thus causing many challenges in diagnosis of periprosthetic joint infection (PJI) that is a major post-operative complication. It is worth noting that PJI is a painful and damaging surgical condition that requires multiple surgeries to eliminate the complete infection [1]. Based on data for globally increasing ageing population [2], it can be rational to say that the cases of TJA and PJI may increase accordingly. Undiagnosed PJI cases can be fatal and the treatment cost with prolonged stay in hospitals can cause significant economic distress to patients and their families. However, the diagnostic methods for PJI with higher accuracy are still in great need and therefore constructively newer, faster and cheaper approaches are of much importance at this pivotal point.

The International Consensus Meeting by Muscoskeletal and Infection Society on orthopedics have discussed several aspects of PJI subdivided into pre-operative, intra-operative and post-operative categories [3]. The highest score based on a log diagnostic ratio was set for pre-operative criteria for alpha-defensins, i.e. human neutrophil peptide 1-3 (HNP 1-3). Previous work reported that HNP 1 accounted for 50 % of HNP 1-3 [4] and has been tested with high specificity and sensitivity (>95%) for PJI detection [5]. There exist several conventional immunoassays or enzyme linked immunosorbent assay (ELISA) kits for HNP 1 detection; however, they possess challenges like longer time (24~48 hr), higher reagent consumption (> 100 µL) and manual errors with high dependence on antibodies, which may suffer from being sensitive to temperature and humidity, batch-to-batch variations and high cost.

Alternatively, aptamers have demonstrated various advantages over antibodies like temperature/humidity stability and 100% accuracy of chemical synthesis; therefore, conventional antibody-based immunoassays may be replaced with aptamer-based assays. Aptamers screened through a process called systematic evolution of ligands by exponential enrichment (SELEX) [6] possess high specificity and affinity. For instance, our previous work has screened aptamers against HNP 1 on a microfluidic platform [7]. In this work, a faster, cheaper, dual-aptamer assay on a nitro-cellulose (NC) membrane as a paper-based substrate has been incorporated into a microfluidic chip. The aptamers possess high affinity to capture the target HNP 1 protein in synovial fluid (SF) and could be used to detect positive and negative cases of PJI. The approach for aptamer-based sandwich assay removed antibody reliability completely. The primary aptamer captured HNP 1 and secondary aptamer was fluorescent labeled to for fluorescent measurement. An image-based quantification of HNP 1was employed. This is the first time that a paper-based, dual-aptamer assay for HNP 1 with image-based detection has been performed for PJI detection.

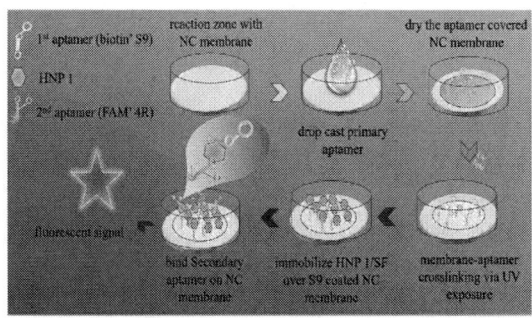

Figure 1: Schematic of the paper-based, dual-aptamer assay. The process involved the crosslinking of a primary aptamer onto a NC membrane via UV crosslinking, followed by HNP 1 capture, secondary aptamer binding and finally fluorescent signals detection.

MATERIALS AND METHODS
Paper-based dual-aptamer assay

A stepwise illustration of paper-based, dual-aptamer assay is shown in Fig 1; starting from crosslinking of primary aptamer on NC membrane to capture HNP 1 protein and finally biding of secondary aptamer for fluorescent signals and its quantification, it could be used for quantification of HNP 1 in SF.

Figure 2a describes the sample loading procedure on chip. First, a previously screened primary aptamer S9 (10 µL, 5 µM) [7] synthesized by Genomics (Taiwan) was immobilized onto the NC membrane (11306-41BL, 0.45 µm Nitrocellulose Blotting Membrane, Sartorius Stedim Biotech, Germany) was exposed to an ultraviolet (UV) lamp (250 nm) for crosslinking for 5 min and was blocked using bovine serum albumin (BSA; 0.1% w/v); it was left to dry to minimize non-specific binding and was later washed twice using 100 µL 0.01 M phosphate buffer saline (PBS). Next, the target protein HNP 1 (AS-60743, HNP 1, 0.1 mg, Anaspec, USA) in 0.01 M PBS at concentrations ranging from 1 to 25 mg/L were prepared. For the clinical tests or calibration curve development, 50 µL of HNP 1 (or SF sample) was used and transported via a transportation unit (i.e., micropump) from a storage chamber towards the reaction zones and was gently washed by using the micropump for 5 min for HNP 1 capture with S9 and was then washed thrice with 100 µL of 0.01 M PBS. Finally, 50 µL, 500 nM of the previously screened secondary aptamer i.e., FAM labeled 4R (synthesized by Genomics, Taiwan) was transported to the reaction zones and was incubated again gently for 5 min to bind with HNP 1. Once the process was completed, it was then washed again as mentioned above and finally the chip was placed under a fluorescent microscope (FITC filter, BX43, Olympus LS, Japan). The recorded images (Fig. 6) were then analyzed by using an image analysis software (ImageJ).

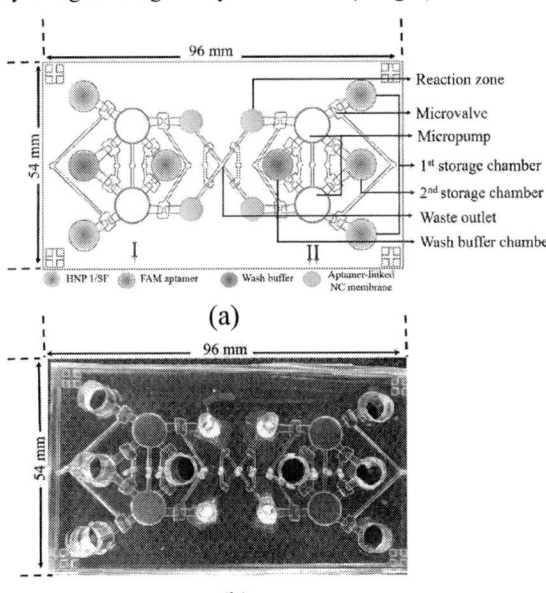

Figure 2: (a) Detailed design and functions of the microfluidic chip with the micro-components of the microfabricated chip and the two identical patterns (1 and 11). (b) Photograph of the chip. The chip was composed of 12 EMV ports, 1 micropump, 8 reagent chambers, 4 NC membrane chambers and 1 waste outlet.

Chip design and microfabrication

The microfluidic chip was designed with specific dimensions of 9.6 cm (L) x 5.4 cm (W) (Figs. 2a and 2b) in order to carry out the automated the paper-based, dual-aptamer assay for faster diagnosis. In order to analyze for points simultaneously or two positive and two negative samples together, 2 identical patterns of all micro-components were designed and pneumatically controlled by using just 12 inlets for electromagnetic valve (EMV; SMC 2070B-sbg-05; Wei-Chia Electro Materials, Taiwan) controller for applying negative (vacuum; DF-506K vacuum pump, Doctor's Friend Medical Instrument, Taiwan) and positive (compressed air, S-101 oil-free air compressor, Sharp Gun, Taiwan) gauge pressures. The microfluidic device was actuated by a flow control module [8]. Each EMV controlled two microvalves and two micropumps for each pattern One waste outlet for four reaction zones were compactly designed to eliminate waste and reduce chip size.

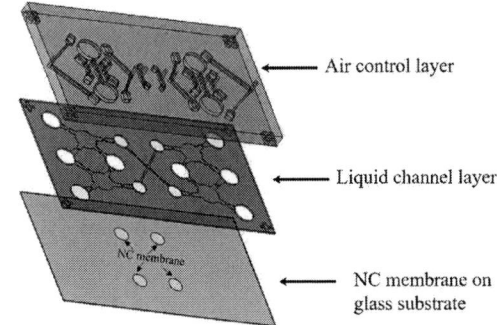

Figure 3: Exploding view of the microfluidic chip with 3-layer structures. The first air control layer was used to supply EMVs controls for actuation via vacuum and compressed air; the second liquid channel layer was for flow of liquid inside the microchannels; the third layer was a glass substrate composed of 4 paper substrates (NC membranes) bound with the glass substrate via double-sided tape. All 3 layers were bonded via plasma treatment.

The chip was designed as a 3 layered structure (Fig. 3) with polydimethylsiloxane (PDMS)-based (Sylgard 184 A: B (10:1), Dow Corning, USA) air control top layer and liquid channel middle layer. A tertiary layer of paper-based NC membrane bonded with a glass substrate by using double-sided tape was used. The PDMS mixture was poured over the inverse mold for chip structure designed using computer-numerical-control (CNC) machine (EGX-600, Roland DGA, USA) and were baked for 2 hr at 80 °C. In order to assemble the complete microfluidic device, all three layers were stepwise bonded together by exposing under oxygen plasma (CUTE-MPR, Femto Science, Korea) at 90 W for 45 sec.

RESULTS AND DISCUSSION
Optimization of assay conditions

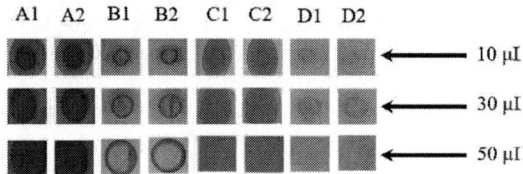

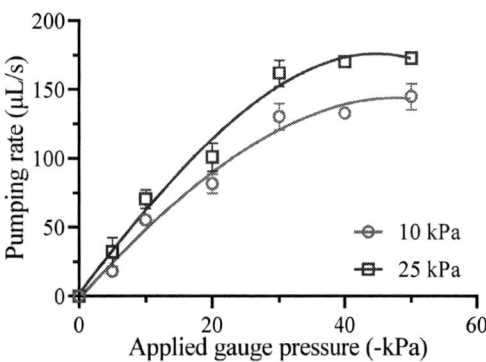

Figure 4: Characterization of absorption capacity of the NC membranes with 4 different dye colors (N=2) with loading volumes of 10, 30 and 50 µL. The absorption capacity of 50 µL showed the highest coverage using water and oil-based dye as simulation for clinical samples and assay buffer conditions and hence was utilized for further experiments.

Based on the selection of NC membrane as a paper-substrate, the characterization of absorption capacity was first carried out within the specifications of the microfluidic device. The test was based on previous works for similar conditions presented by our group [9]. The substrate was loaded with 4 different colored dyes (water and oil based) with varied volumes (10 µL, 30 µL and 50 µL) to simulate the clinical sample conditions over the membrane during the assay (N=2), as shown in Fig. 4. It appeared that 50 µL showed the highest coverage area in all significant dyes, which is important in our case due to fluid coverage over the NC crosslinked primary aptamer and hence it was used for all subsequent experiments.

Figure 5: Pumping rate of the transport unit under a driving frequency of 1 Hz to characterize the pumping volume of the micropump at variable negative gauge pressures for two different positive gauge pressure conditions of +10 kPa and +25 kPa (N=3).

Once the optimal required volume was concluded, the pumping volume of the micropump for transporting liquid from storage chambers to reaction zones as well as for required volume for washing buffer conditions was explored. The relationship between pumping rate and applied negative gauge pressure was explored at variable pressures driving at fixed driving frequency of 1 Hz. As shown in Fig 5, the pumping rate increased with the applied gauge pressure, which is similar to the one reported

previously [10] and a saturation was observed at +25 kPa/-40kPa with better stability and higher volume transport as compared with +10 kPa positive gauge pressure. A single pumping iteration was then found to be sufficient for +25 kPa/-20 kPa to transport a resulting volume of 50 µL for clinical samples and reagents from the storage chambers as well as for 100 µL washing buffer.

Calibration curve and clinical sample assessment

The paper-based, dual-aptamer assay was performed based on the above-mentioned conditions and a calibration curve was established (N=3) between the digitally analyzed images (Fig. 6) for fluorescence intensity and HNP 1 protein (Fig. 7). A dynamic range of 0.5 to 25 mg/L was spiked for the calibration. To cover a threshold region of HNP 1 of 2.6 mg/L [11], several points were added near the threshold region to identify PJI positive cases

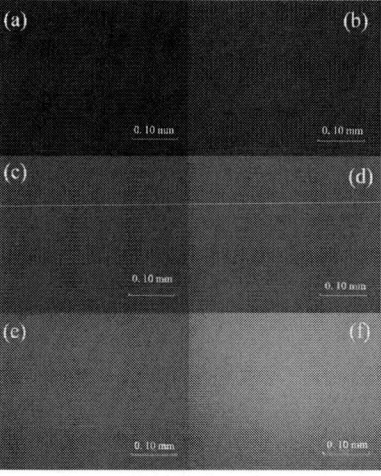

Figure 6: Images observed under a fluorescence microscope for 5 different conditions of HNP 1. (a) Negative control (0 mg/L), (b) HNP 1 (1 mg/L), (c) HNP 1 (2 mg/L) (d) HNP 1 (5 mg/L), (e) HNP 1 (10 mg/L) and (f) HNP 1 (25 mg/L)

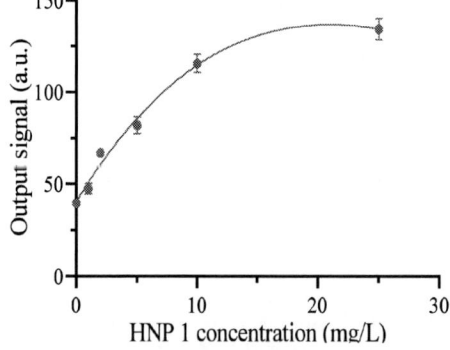

Figure 7: Output signal quantification using image analysis software. The output signals were quantified for 5 different concentrations of HNP 1 from 1 mg/L – 25 mg/L (N=3). Values below 2.6 mg/L can be set as a threshold to identify PJI positive case.

accurately.

For the clinical sample assessment, the storage chamber for HNP 1 was replaced with clinical samples (SF from patients provided by Chang Gung Memorial Hospital, Kaohsiung, Taiwan and Institutional review board (IRB) guidelines were followed) and the same assay conditions as mentioned above were performed (IRB No. 201802127B0C502). The collected fluorescent images under a fluorescent microscope were then quantified using ImageJ and the HNP 1 concentration was estimated using the calibration curve. Out of 5 clinical samples, 2 positive cases of PJI and 3 negative samples were correctly identified (Table 1).

Table 1. Clinical sample results

Sample No.	HNP 1 concentration (mg/L)	Result	Image Under Microscope
Sample 1	1.5	Negative	
Sample 2	4.7	Positive	
Sample 3	7.5	Positive	
Sample 4	0	Negative	
Sample 5	0.8	Negative	

CONCLUSION

A new paper-based, dual-aptamer assay on a microfluidic platform was successfully developed. The aptamer-based sandwich assay increased affinity and specificity for targeting PJI biomarker HNP 1 without antibody reliability. A fluorescent image-based quantification was used to develop the calibration curve that could be used to analyze the clinical samples. The developed assay is cost effective and can be performed within 30 min where two positive and three negative samples were analyzed simultaneously on a single chip. The developed microfluidic device can be used for further prognosis and diagnosis of PJI.

ACKNOWLEDGEMENTS

The authors would like to thank the 1) Ministry of Science and Technology (MOST) of Taiwan (MOST 109-2221-E-007-006-MY3 & MOST 110-2221-E-007-010-MY3), 2) National Health Research Institutes of Taiwan (NHRI-EX110-11020EI), and 3) Chang Gung Memorial Hospital, Kaohsiung (CMRPG8K0501) for funding this work.

REFERENCES

[1] J. A. Meyer, M. Zhu, A. Cavadino, B. Coleman, J. T. Munro, and S. W. Young, "Infection and periprosthetic fracture are the leading causes of failure after aseptic revision total knee arthroplasty," *Arch Orthop Trauma Surg*, vol. 141, pp. 1373-1383, 2021.

[2] C. H. Chang, S. H. Lee, Y. C. Lin, Y. C. Wang, C. J. Chang, and P. H. Hsieh, "Increased periprosthetic hip and knee infection projected from 2014 to 2035 in Taiwan," *J Infect Public Health*, vol. 13, pp. 1768-1773, 2020/11/01/ 2020

[3] J. Parvizi and T. Gehrke, "Definition of Periprosthetic Joint Infection," *J Arthroplasty*, vol. 29, pp. 1331-1331, 2014.

[4] X. Gao, J. Ding, C. Liao, J. Xu, X. Liu, and W. Lu, "Defensins: The natural peptide antibiotic," *Adv Drug Del Rev*, vol. 179, pp. 114008-114022, 2021.

[5] J. Bingham, H. Clarke, and M. Spangehl, "The alpha-defensin-1 biomarker assay can be used to evaluate the potentially infected total joint arthroplasty," *Clin Orthop Relat Res*, vol. 472, pp. 4006-4009, 2014.

[6] A. D. Ellington and J. W. Szostak, "In vitro selection of RNA molecules that bind specific ligands," *Nature*, vol. 346, pp. 818-822, 1990,

[7] R. Gandotra, H. B. Wu., P. Gopinathan, Y. C. Tsai, F C. Kuo, M. S. Lee and G. B. Lee "Aptamer selection against alpha-defensin human neutrophil peptide 1 on an integrated microfluidic system for diagnosis of periprosthetic joint infections," *Lab Chip*, vol. 22, pp. 250-261, 2022.

[8] C. Y. Lee, G. B. Lee, J. L. Lin, F. C. Huang, and C. S. Liao, "Integrated microfluidic systems for cell lysis, mixing/pumping and DNA amplification," *J Micromech Microeng*, vol. 15, pp. 1215-1223, 2005

[9] C. H. Wang, J. J. Wu, and G. B. Lee, "Screening of highly-specific aptamers and their applications in paper-based microfluidic chips for rapid diagnosis of multiple bacteria," *Sens Actuators B Chem*, vol. 284, pp. 395-402, 2019.

[10] C. W. Huang, S. B. Huang, and G. B. Lee, "Pneumatic micropumps with serially connected actuation chambers," *J Micromech Microeng*, vol. 16, pp. 2265-2272, 2006.

[11] P. Melicherčík, E. Klapková, K. Kotaška, D. Jahoda, I. Landor, and V. Čeřovský, "High-performance liquid chromatography as a novel method for the determination of α-Defensins in synovial fluid for diagnosis of orthopedic infections," *Diagnostics*, vol. 10, pp. 33-43, 2020.

CONTACT

*Gwo-Bin Lee, Tel: +886 35715131 ext. 33765; Fax: +886 35742495; E-mail: gwobin@pme.nthu.edu.tw

AN INTEGRATED MICROFLUIDIC PLATFORM FOR TUMOR CELL SEPARATION AND FLUORESCENCE IN SITU HYBRIDIZATION AT SINGLE CELL LEVEL

Shihui Qiu[1,2], Na Li[1,2], Zhenhua Wu[1,2]*, Jianlong Zhao[1,2]*, Hongju Mao[1,2]*

[1] State Key Laboratory of Transducer Technology, Shanghai Institute of Microsystem and Information Technology, Chinese Academy of Sciences, Shanghai 200050, China
[2] Center of Materials Science and Optoelectronics Engineering, University of Chinese Academy of Sciences, Beijing 100049, China

ABSTRACT

This platform realizes tumor cells' isolation from WBCs and conduct fluorescence in situ hybridization (FISH) in only one chip. This platform enjoys the following advantages: (1) It could realize precise cell isolation based on differences in cell size and protein biomarkers; (2) Contamination can be avoided because cells are analyzed in situ; (3) Single-cell level FISH analysis for CTCs (circulating tumor cells) is also realized on this chip; (4) This platform could save reagents and time; (5) Low costs and automation prospects. Therefore, the platform has the potential for FISH assay at single-cell level.

KEYWORDS

Fluorescence In Situ Hybridization (FISH), Single-cell Analysis, Circulating Tumor Cells (CTCs)

INTRODUCTION

CTCs are tumor cells that shed from the primary tumor of the patient, enter the human peripheral blood circulation, and finally take root in the distal end and develop into tumor tissue [1-3]. CTCs contain complete information on tumor lesions, it can provide important information for medical teams and researchers, so it is one of the important markers for clinical research and liquid biopsy. Its biggest disadvantage is its low abundance [4]. There is only a few CTCs among 10^6 peripheral blood mononuclear cells (PMBC) in real samples. Therefore, the successful isolation of CTCs from peripheral blood is the most basic requirement for all CTCs-based studies.

Fluorescence in situ hybridization (FISH) is one of the gold standards for clinical gene analysis [5]. For example, EGFR gene-amplification is an important instruction for the prognosis of patients [6]. However, traditional platforms even microfluidic platform could not perform FISH on CTCs [7]. This limits its application in liquid biopsies. Because of CTCs' low abundance, the enrichment and manipulation are still challenging. Fortunately, CTCs could be isolated according to its size differences (CTCs have larger size than other blood cells) [8-11]. Here we propose an integrated platform for CTCs' enrichment and FISH analysis. This platform integrates CTCs enrichment with FISH detection. The potential cell loss during export of CTCs is reduced and the consumption of reagents is reduced. Also, it provides a new idea for in situ detection of CTCs and other types of rare cell samples.

EXPERIMENT
Integrated Microfluidic Platform Establishment

The microfluidic chip is composed of PDMS functional layer, PDMS micro-valve layer and glass substrate. The PDMS functional layer contains ten single cell capture units for the capture of tumor cells in the sample, and the downstream capture array for FISH detection. PDMS release-valves realize the releasing of cells. Inlet and outlet valves control the channels state of opening or closing. The FluidClab pressure controller (Prinzen Biomedical, China) is used for sample and buffer injection. The inlet and outlet valves are controlled by syringe while the release valves are controlled by a solenoid valve.

Chip Design

The design of CTCs' capture unit is based on the Bypass principle. Each unit is designed with bypass channel and capture channel. By precise design of the flow resistance relationship between the two channels, we can control the path that cells with different sizes choose to pass through, to achieve the enrichment of tumor cells.

The channels are designed according to the flowing resistance formula:

$$R = \frac{12\mu l}{wh^3\left(1-0.63\frac{h}{w}\right)} \tag{1}$$

where w, h and l denote the width, height and length of the channel respectively; R and μ are the flow resistance and coefficient of viscosity.

The width of the capture channel and bypass channel were designed to allow only one single cell to pass through. The width of the capture channel is slightly smaller than the minimum size of tumor cells, which is about the diameter of WBC ~7.5 μm, and a capture port with a width of 15 μm is designed at the intersection of the two channels. When the cells are not captured, the flow resistance of the bypass channel is bigger than that of the capture channel. The cells will go through the capture channel and the white blood cells will pass through the channel without blocking. Once tumor cells pass through, they will be captured at the capture port. At the same time, the cells will go through the bypass channel to the next CTC capture unit. When the 10 capture units were filled with tumor cells, the enrichment process was finished.

In real PMBC samples, a large number of WBC clusters or larger WBCs, which were close to the size of CTCs, would be inevitably captured. In order to improve the enrichment purity, it is necessary for controllable release of non-CTC cells. In this platform, we have introduced a micro valve based on PDMS film to achieve cell trapping, channel washing, cell release and exportation.

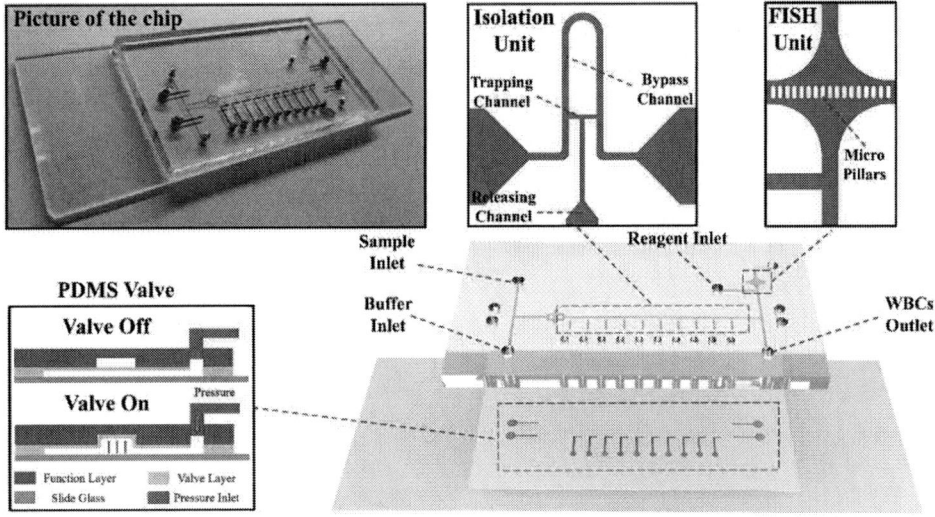

Figure 1: The schematic diagram of the whole system. The function layer is composed of 10 isolation units, one FISH analysis unit, sample inlet, buffer inlet, reagent inlet and WBCs outlet. The valve layer consists of 2 inlet/outlet valves and 10 cell release valves.

Figure1 is the schematic of the micro-valve. For the inlet and outlet valves, when the pressure is applied, the

membrane will be jacked up and blocked the crisscross channel and realize channel's opening or closing. For the release valve, the buffer in crisscross channel would be pumped out and squeeze out the captured cell. The controllable cell release finished.

Chip Fabrication

All chip manufacturing processes are based on microfabrication technology. Photolithography and Deep RIE were applied to manufacture the silicon mold of functional layer and micro-valve layer. All the manufacturing process is shown in figure2. After the silicon molds are prepared, the surface of them were modified by trichloro-(1H,1H,2H,2H-perfluorooctyl) silane vapor under vacuum to reduce the surface energy and facilitate the peeling of the PDMS layer. PDMS were prepared by mixing pre-polymer and cross-linking agent. For the functional layer, the mass ratio of the two components is 10:1. The PDMS mixture was poured on the silicon mold and the thickness of this layer was 2–3 mm suitable for the fixation of inserted tubes. It was then placed in a horizontal oven and left for at least 1 hour at 65°C for crosslinking. For the micro-valve layer, the mass ratio of the two components is 20:1. The PDMS mixture was poured on the mold and spinning coated at a speed of 1300rpm for 40 seconds. After the spin coating, the mold was also placed in a horizontal oven and left for at least 30min at 65°C for crosslinking. After that, the PDMS were peeled off and cut into proper size. After holes were punched at the corresponding positions, the PDMS layers were treated with alignment bonding by plasma to complete the chip manufacturing. Prior to experiments, to prevent the non-specific adsorption of cells and reagents in PDMS chip channels, PBS with 0.2% Tween was introduced into the channels for half an hour of surface treatment.

Control of the platform

The pressure source of the release-valves and sample injection in inlets is a nitrogen gasholder. The controller of the sample injection is FluidiClab pressure injection pump, the pressure accuracy of which is 0.1mBar. The controller of release valve is solenoid valve.

The inlet and outlet valves were controlled by syringes. We rotate the screw of the syringe to control the jacking-up of the valve layer thus control the opening or closing of the channels.

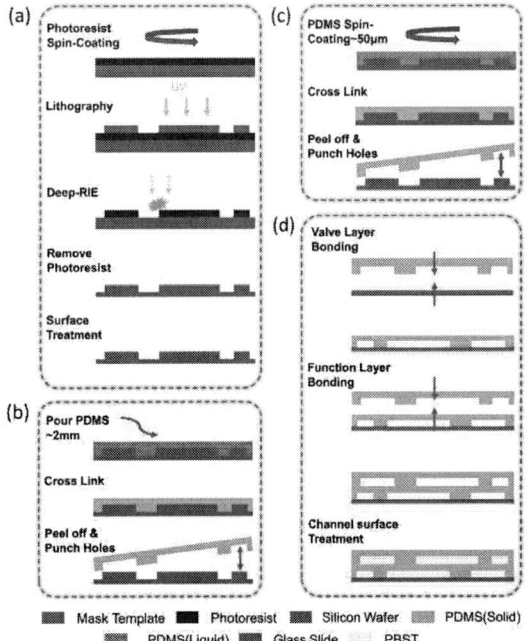

Figure 2: a. The fabrication of function and valve layer's silicon mold. b. The fabrication of PDMS function layer. c.

978-1-6654-9309-3/23 $31.00 © 2023 IEEE

The fabrication of PDMS valve layer. d. The bonding of different layer.

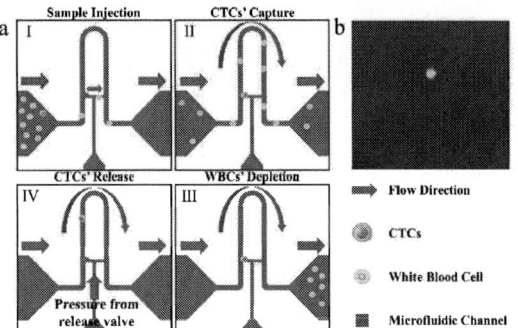

Figure 3: a. The basic mechanism of Bypass structure that applied to isolate cancer cells (bigger cells) and the way to release these captured cells. b. experiment results that A549 cell line were successfully captured.

Tumor Cells' Enrichment

The buffer's injection channel, reagent's injection channel and FISH unit's outlet channel are sealed by the valves. Cell samples are added from the sample inlet and the condition of the capture cell is observed in real time. If the cells captured in the capture ports are cancer cells, no operation is required. If cells captured are WBCs, the corresponding solenoid valve at the unit is activated, and the captured cells are released after pressurization. Until all the ten capture units are filled with tumor cells, the enrichment is finished.

As showed in figure4, after all the samples have been injected, seal the sample injection channel and the leukocyte sampling channel. Open the buffer injection channel and the FISH unit injection channel. The solenoid valves are triggered one by one to release the captured cancer cells while the buffer is being injected. After all cells have been flushed to the FISH unit and immobilized, the buffer injection would stop. After that, the buffer injection channel is sealed, and the reagent injection channel is unsealed. The next step is to perform FISH detection on the captured cells.

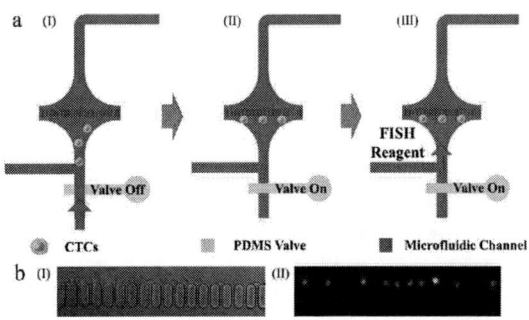

Figure 4: a. Experiment flow of FISH performed on chip. I. Cells were fixed on the pillars in FISH analysis module. II. The outlet valve is on to disconnect FISH analysis module and isolation chip. III. Reagents' injection and start FISH assay. b. Cells' fixation onto the micro pillars.

(I)Captured cells under bright view(II) Captured cells with DAPI stained.

Process of Fish in the chip

1. Inject pepsin solution, digest the cells at 37°C for about 3 min;
2. Rinse cells with PBS for about 5 min;
3. Wash cells with PBS for 10min;
4. Gradient dehydrate the cells with 70%, 85%, and 100% ethanol for 3 min respectively;
5. Vacuum the chip for at least 10 min;
6. The FISH probe solution are sucked into the chip by the negative pressure generated by vacuum until the chamber is filled with reagents;
7. Put the chip into a wet box and hybridize at 73°C for 3 min and keep it at 40°C for overnight;
8. After the hybridization is completed, a pre-warmed 73°C washing buffer is injected into the chip to wash off the excess probes;
9. The chamber was rinsed with deionized water, and then the chip was placed in a vacuum desiccator for 10 min. Then counterstained with DAPI for 10 min;
10. Observe the results of on-chip FISH under a fluorescence microscope with a 60x objective lens.

In order to optimize the flow velocity, the flow rate should be tested. When ethanol is injected, the injection pressure of the selected pump is controlled at 20 mbar. In the case of PBS, washing buffer and other reaction reagents, the pressure is selected as 100 mbar. The FISH detection kit was purchased from Henan Sinot Biological Co., Ltd. The testing NSCLC cell lines A549 and HCC-827 were purchased from China Typical Cell Culture Bank.

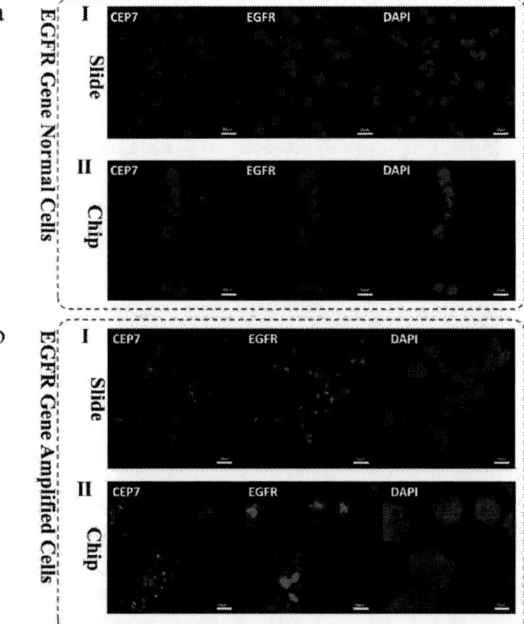

Figure 5: a. FISH results of cell line-A549. Both detection on slide and chip shows the average signal's amount of CEP7 and EGFR gene is 2 which means EGFR gene amplification doesn't take place in this cell line and the other group is the normal cell line-A549. b. The results of cell line-HCC-827. The average signal's amount of CEP7

and EGFR gene is over 2 which means EGFR gene amplified. The results one both slide and chip correspond with each other. Scalebar= 10 μm.

RESULTS

Results of Tumor Cells' Enrichment

The microfluidic chip was able to capture all cancer cells in simulated sample of 10^5 leukocytes doped with 10 A549 or HCC-827 cells in 1 mL. At the same time, the integrated PDMS microvalve can realize the control of the flow channel and finally collect the cells. The release valve can well achieve the release of cancer cells. As shown in Figure 3, the first row of arrays can intercept and immobilize all captured cancer cells for downstream FISH detection.

Results of FISH Analysis

The FISH kit contains two probes: EGFR (red) gene and CEP7 (green) gene; FISH operation supporting reagents. The results evaluation criteria are as follows:

1. Normal cells: 2 red and 2 green signals in a single interphase nucleus;

2. Abnormal EGFR gene amplification cells: The red signal in a single interphase cell nucleus is greater than 2, and the green signal is not less than 2.

As Figure 5.a(I)and 5.b(I)showed, for A549 and HCC-827 cell line, the FISH results on slides indicate that A549 cell line did not have EGFR gene amplification, while the HCC-827 cell line had a very strong amplification phenomenon which called cluster amplification.

The on-chip FISH results showed in Figure 5.a(II) and 5.b(II) indicated that all A549 cells were normal, while all HCC-827 cells had cluster amplification of EGFR gene. The results of FISH performed in-chip were consistent with those of traditional methods.

CONCLUSION

In this work, we introduced a single-cell level CTCs enriching and manipulating platform. We realized high recovery and purity for CTCs enrichment from blood cells. With integrated pillars, we realized in situ FISH analysis of captured tumor cells. The results are the same as that of traditional FISH assay.

In our future work, we need to further improve the flux for CTCs' enrichment and improved the FISH analysis unit for better efficiency of hybridization and observation.

ACKNOWLEDGEMENTS

This study was supported by National Natural Science Foundation of China (Nos. 62231025, 61801465, 61971410, 61801464 and 62001458), Shanghai Sailing Program (No. 20YF1457100), Shanghai Engineer & Technology Research Center of Internet of Things for Respiratory Medicine (No. 20DZ2254400), and the Science and Technology Commission of Shanghai Municipality (No. 19511104200).

REFERENCES

[1] K. Pantel, C. Alix-Panabieres, "Circulating tumour cells in cancer patients: challenges and perspectives", *J.* Trends. Mol. Med., vol. 16, pp.398–406, 2010.

[2] M. Cristofanilli, G.T. Budd, M.J. Ellis, A. Stopeck, J. Matera, M.C. Miller,J.M. Reuben, G.V. Doyle, W.J. Allard, L.W.M.M. Terstappen, D.F. Hayes, "Circulating tumor cells, disease progression, and survival in metastatic breast cancer", *J.* New. Engl. J. Med. Vol. 351, pp. 781–791, 2004.

[3] C. Wittekind, M. Neid, "Cancer invasion and metastasis", *J.* Oncology., vol. 69, pp. 14–16, 2005.

[4] Y.L. Song, T. Tian, Y.Z. Shi, W.L. Liu, Y. Zou, T. Khajvand, S.L. Wang, Z. Zhu,C.Y. Yang, "Enrichment and single-cell analysis of circulating tumor cells", *J.* Chem. Sci., vol.8, pp. 1736–1751, 2017.

[5] Song, Y.L., et al., "Enrichment and single-cell analysis of circulating tumor cells". *J.* Chem. Sci., vol. 8(3): pp. 1736-1751, 2017.

[6] Swennenhuis, J.F., et al., "Efficiency of whole genome amplification of single circulating tumor cells enriched by CellSearch and sorted by FACS". *J.* Genome Med., vol.5, 2013.

[7] Kao, K.J., et al., "A fluorescence in situ hybridization (FISH) microfluidic platform for detection of HER2 amplification in cancer cells". *J.* Biosensors & Bioelectronics., vol. 69, pp. 272-279, 2015.

[8] Wang, K., et al., "A microfluidic platform for high-purity separating circulating tumor cells at the single-cell level". *J.* Talanta., vol. 200, pp. 169-176, 2019.

[9] J. Bu, Y.T. Kang, Y.S. Lee, J. Kim, Y.H. Cho, B.I. Moon, "Lab on a fabric: Mass producible and low-cost fabric filters for the high throughput viable isolation of circulating tumor cells", *J.* Biosens. Bioelectron. vol. 91, pp. 747–755, 2017.

[10] X. Fan, C. Jia, J. Yang, G. Li, H. Mao, Q. Jin, J. Zhao, "A microfluidic chip integrated with a high-density PDMS-based microfiltration membrane for rapid isolation and detection of circulating tumor cells", *J.* Biosens. Bioelectron. vol. 71, pp. 380–386, 2015.

[11] T. Huang, C.P. Jia, Y. Jun, W.J. Sun, W.T. Wang, H.L. Zhang, H. Cong, F.X. Jing, H.J. Mao, Q.H. Jin, Z. Zhang, Y.J. Chen, G. Li, G.X. Mao, J.L. Zhao, "Highly sensitive enumeration of circulating tumor cells in lung cancer patients using a size-based filtration microfluidic chip", *J.* Biosens. Bioelectron. vol. 51, pp. 213–218, 2014.

CONTACT

*Z. Wu, tel: +86-21-62511070-8705;
  wuzhx@mail.sim.ac.cn.
*J. Zhao, tel: +86-21-62511070-8701;
  jlzhao@mail.sim.ac.cn.
*H. Mao, Tel: +86-21-62511070-8701;
  hjmao@mail.sim.ac.cn.

CHARACTERIZATION OF OOCYTE HARDENING USING A MICROFLUIDIC ASPIRATION-ASSISTED ELECTRICAL IMPEDANCE SPECTROSCOPY SYSTEM

Yuan Cao[1], Julia Floehr[2], and Uwe Schnakenberg[1]

[1]Institute of Materials in Electrical Engineering 1, RWTH Aachen University, GERMANY and
[2]Helmholtz-Institute for Biomedical Engineering, RWTH Aachen University, GERMANY

ABSTRACT

This paper reports a microfluidic aspiration-assisted electrical impedance spectroscopy system (MAEIS) with the aim of characterizing oocyte hardening for in-vitro fertilization (IVF) purposes. A microfluidic-based cell trap device is utilized, which offers simple oocyte handling through hydrodynamic trapping. The aspiration of the zona pellucida (ZP), which is a gelatinous outer layer of extracellular matrix, at an orifice of the microfluidic channel was measured observer-independently by applying electrical impedance spectroscopy (EIS). Impedance spectra obtained from three different oocyte genotypes were fitted to an appropriate equivalent electrical circuit (EEC) model and correlated with the aspiration length changes via a novel analytical model. Therefore, Young's moduli of the zona pellucidae were calculated and the different oocyte stages were successfully determined.

KEYWORDS

Microfluidics, lab-on-a-chip, oocyte, zona pellucida, electrical impedance spectroscopy, Young's modulus, equivalent electrical circuits.

INTRODUCTION

Artificial fertilization of oocytes is a well-established method in human reproductive medicine for preserving and generating breeding and laboratory animals. The success depends critically on the quality of oocytes, in particular on the hardening state of the zona pellucida (ZP), which is a gelatinous outer layer of extracellular matrix. Several techniques were already proposed to characterize the elasticity of ZP [1-3]. However, these techniques are suffering from challenging cell handling and/or observer-dependent readouts. Therefore, Cao et al. introduced a microfluidic lab-on-a-chip system with hydrodynamic single microparticle trapping capability [4]. The chip with its microfluidic channel and orifice in the coverlid is schematically shown in Figure 1. The suction is controlled by the corresponding microfluidic channel. Two ring-shaped electrodes are arranged around the intermediate orifice. Trapped oocytes can therefore be characterized by electrical impedance spectroscopy (EIS). The system was further improved to a microfluidic aspiration-assisted electrical impedance spectroscopy (MAEIS) system for the characterization of the hardening status of mouse oocytes ZP [5]. The chip was connected to a custom-made precise pressure management unit. By varying the suction pressure to a trapped oocyte, the aspiration-related changes in the electrical impedance were measured.

In this work, the MAEIS system was applied to three genetically different mouse oocyte models owing a super soft, a normal, and a stiff ZP, respectively. The impedance spectra were fitted to an EEC model. The suction pressure depending circuit element was correlated with the aspirated ZP depth via an analytical model. Young's moduli of all three ZPs of the investigated oocyte genotypes were determined.

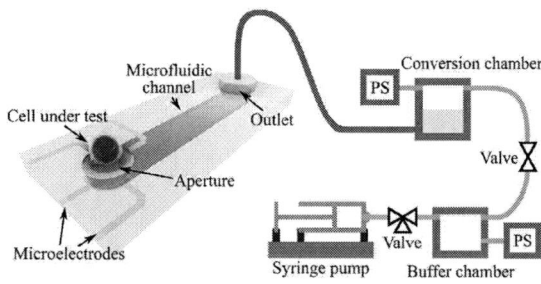

Figure 1: Schematic drawing of the microfluidic aspiration-assisted electric impedance spectroscopy (MAEIS) system comprising a microfluidic chip and a pressure regulation module. PS: pressure sensors. Drawing not to scale.

EXPERIMENTAL

The fabrication process of the microfluidic chip was described in detail in [4,5,6]. Briefly summarized, three SUEX® dry film resist (DFR) layers (micro resist technology) were laminated and patterned by standard UV-lithography. A two-step dry etching process in an inductively coupled plasma reactive ion etching tool structured the orifice. The intermediate electrodes were manufactured by deposition of a thin-film plating base (titanium and gold), standard UV-lithography, and subsequent gold electroplating. A replica-molded poly(dimethylsiloxane) (PDMS) interposer was bonded on top of the chip to connect the channel via tubing to the pressure regulation unit and to provide a chamber above the orifice. The chamber was conditioned by filling with EmbryoMax Advanced KSOM Embryo Medium (Sigma-Aldrich) before the experiments. An Agilent 4294A precision impedance analyzer interconnected the electrodes. The chip was mounted under an inverted microscope with a hotplate. For the experiments, the KSOM medium was heated to 36 °C. The fabricated chip and a PCB board with the inserted chip are shown in Figure 2.

Three different mouse models owing oocytes in MII stage with super soft (DKO oocytes), normal (WT oocytes), and stiff ZP structures (KO oocytes) [7,8] served as models to study the range of ZP elasticities by the MAEIS system. Oocytes in metaphase II (MII) state were isolated after hormonal treatment on the mice. All animal experiments were in accordance with the German Animal Welfare law and approved by the State Governments in North Rhine Westphalia, Germany. Maintenance, handling, and treatment of the mice were performed according to the Federation for Laboratory Animal Science Associations (FELASA) recommendations.

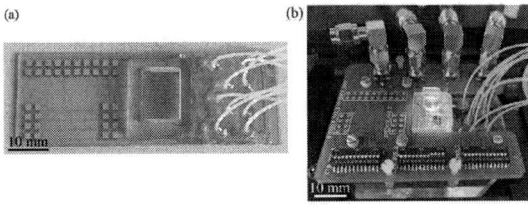

Figure 2: Microfluidic chip: (a) Photo of the microfluidic chip composed of three sections: contact pad area on the left, a PDMS frame with the reaction chamber well (middle section), and tubes to connect the microfluidic channels at the right. (b) PCB board with inserted microfluidic chip. The chamber is closed with a PDMS cover to prevent evaporation of the medium.

An oocyte was manually positioned to the orifice area with a micropipette and was trapped hydrodynamically at the orifice by application of a minimum suction pressure of 1 mbar ensuring that the oocyte sealed the orifice and stayed stable at the orifice. This minimum fixation pressure was defined as 0 mbar in the following as reference with regard to the applied suction pressure. EIS measurements were carried out between 1 kHz and 10 MHz with a 25 mV oscillation voltage every 5 s period, while the suction pressure was subsequently increased from 1 mbar to 8 mbar in 1 mbar steps adjusted by the pressure regulation unit. Impedance values were evaluated after the signal stabilization just before the next pressure step.

RESULTS AND DISCUSSION

Figure 3 presents the normalized impedance magnitude $\Delta Z/Z_0 = (Z - Z_0)/Z_0$ with respect to the applied suction pressure for the three different mouse oocyte genotypes, in which Z indicates the impedance magnitude with a trapped oocyte at the orifice and Z_0 the impedance magnitude measured at an open aperture (without oocyte), respectively. The impedance ratio rises with pressure, which indicates an increasing aspiration of the ZP into the orifice. The observed behavior is linear. The impedance values obtained from the oocyte with soft ZP show a significantly higher slope than those obtained from the oocytes with the hard ZP. The slope of the impedance measured at the normal ZP lies between the two other ones.

Such a linear dependency was also obtained when not the impedance change but the aspiration length L of the aspired tissue as measured with regard to the applied pressure. This behavior can be described with the well-established theory of Alexopoulos *et al.*, from which Young's moduli can be calculated [9]. Recently, Cao *et al.* developed an analytical model to correlate the slope of the aspiration length L with the impedance change [10]. In a

first step, the impedance spectra were fitted with an appropriate EEC model, as depicted in Figure 4. *CPE* refers to the constant phase elements, which is related to the roughness of the electroplated electrodes, C_E to the cross-talk capacitance

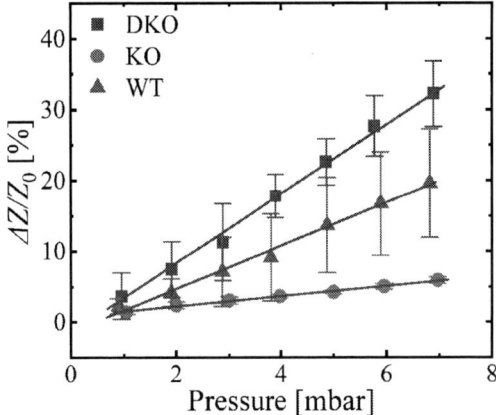

Figure 3: Normalized impedance magnitude at the frequency of highest impedance sensitivity (30 kHz) with respect to applied suction pressure at the orifice for KO (n = 4), WT (n =5) and DKO (n = 4) oocytes, respectively. Error bars represent standard deviations.

between the electrodes, which is dominant in the high-frequency range, and R_S to the solution resistance. R and C presents the resistance and the capacitance at the orifice, respectively.

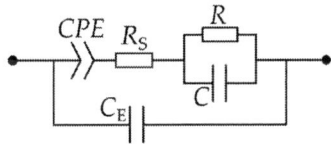

Figure 4: EEC model for fitting the electrical impedance spectra. CPE: constant phase element, C_E: cross-talk capacitance between the electrodes, R_S: solution resistance, R: resistance and C: capacitance of the orifice.

Cao *et al.* showed that the resistance R represents the circuit element, which depends on the suction pressure [10]. During the experiment, the ZP was pressed to the orifice. The ZP was thinned at the rim of the orifice. The resistance R increased by increasing pressure. Young's moduli E_{ZP} of the zona pellucida were calculated according to equation (1) with $C(h_{ZP}^*)$ a factor which was defined in ref. [9], γ_{ZP} the Poisson ratio of ZP (0.04 [3]), r_i the radius of the orifice, R_2 the resistance of a sealed orifice with applied pressure, $R_{2,0}$ the resistance of the sealed orifice with the minimum fixation pressure, R_0 the resistance of an empty orifice, r_c the radius of the oocyte, Δp the suction pressure difference, and R the resistance obtained from the EEC model in Figure 4, respectively.

$$E_{\text{ZP}} = r_{\text{i}} C(h_{\text{ZP}}^{*})(1 - v_{\text{ZP}}^{2}) \frac{R_{2,0}}{R_0} \left(\frac{1}{r_{\text{C}}} + \sqrt{\left(\frac{1}{r_{\text{C}}}\right)^2 + \frac{2}{r_{\text{C}}\left(r_{\text{C}} - \sqrt{r_{\text{C}}^2 - r_{\text{i}}^2}\right)}\left(\frac{R_2}{R_{2,0}} - 1\right)} \right) \frac{\Delta p}{\Delta R / R_0} \qquad (1)$$

Young's modulus E_{ZP} of zona pellucida of the WT oocyte genotype was calculated to be 3.6 kPa. In future applications, the oocyte maturation needs to be investigated with regard to the detection of the best window of fertilization. During oocyte maturation, the ZP passes through different steps of hardening. At the time of full maturity, the oocyte undergoes a first hardening step, which is important for the robustness of the oocyte during ovulation. At this time, the ZP remains permeable to sperm. Only after fertilization, a definitive ZP hardening occurs that no longer allows sperm binding and penetration. In this context, relative values of ZP hardening are of interest and not absolute values. Therefore, it is obvious to define a stiffness factor. In our case, the stiffness factor is defined as the ratio of KO or DKO Young's modulus with regard to E_{ZP} of WT oocytes. The values listed in Table 1 are well comparable to values obtained with a nanoindentation approach [11]. The stiffness factor can therefore be used to quantify different ZP hardening stages.

Table 1: Stiffness factor for three oocyte genotypes.

Oocyte genotype	Stiffness factor with respect to WT
MII-WT	1
MII-KO	6.51
MII-DKO	0.33

MII: metaphase II, WT: wild type, KO: Fetuin-B/single deficient, DKO: Fetuin-B/ovastacin double deficient. Errors are standard deviations.

CONCLUSION AND OUTLOOK

The proposed MAEIS system provides a non-destructive and label-free assessment to indicate oocyte quality. The setup is user-friendly because the oocytes can be easily handled due to the hydrodynamic trapping capability at an orifice. The system offers a user-independent electrical readout and is therefore not limited to feature resolution of an optical system.

In the future, the setup will be improved to a portable MAEIS system with the option to characterize multiple oocytes in parallel, which are desired for an in-vitro fertilization (IVF) procedure. Besides the application in the field of human and animal reproductive medicine as well as for the development of contraceptives, the technique provides, in general, an innovative approach to characterize the rheological properties of any kind of individual cells and tissues.

ACKNOWLEDGEMENTS

The project was funded by the Deutsche Forschungsgemeinschaft (DFG, German Research Foundation) - 422444193. YC and US express their sincere thanks to Dorothee Breuer, Jochen Heiß, Sascha Neske, Ewa-Janina Sekula, Erkan Yilmaz for their valuable assistance in processing the microfluidic chips, and Linda Wetzel (all with Institute of Materials in Electrical Engineering 1, RWTH Aachen University) for carefully proof-reading.

REFERENCES

[1] Y. Murayama, J. Mizuno, H. Kamakura, Y. Fueta, H. Nakamura, K. Akaishi, K. Anzai, A. Watanabe, H. Inui, S. Omata, "Mouse zona pellucida dynamically changes its elasticity during oocyte maturation, fertilization and early embryo development", *Human Cell*, vol. 19, pp. 119-125, 2006.

[2] Y. Sun, K.-T. Wan, K. P. Roberts, J. C. Bischof, B. J. Nelson, "Mechanical property characterization of mouse zona pellucida", *IEEE Trans. Nanobiosci.*, vol. 2, pp. 279-286, 2003.

[3] M. Khalilian, M. Navidbakhsh, R.M. Valojerdi, M. Chizari Mahmoud, P.E. Eftekhari, "Estimating Young's modulus of zona pellucida by micropipette aspiration in combination with theoretical models of ovum", *J. R. Soc. Interface*, vol. 7, pp. 687-694, 2010.

[4] Y. Cao, J. Floehr, S. Ingebrandt, U. Schnakenberg, „Dry film resist laminated microfluidic system for electrical impedance measurements", *Micromachines*, vol. 12, 632, 2021.

[5] Y. Cao, J. Floehr, D. Azarkh, U. Schnakenberg, „Mouse Oocyte Characterization by Electrical Impedance Spectroscopy", in *Digest Tech. Papers IEEE Sensors 2022*, Dallas, TX, USA, October 30 - November 2, 2022.

[6] Y. Cao, J. Floehr, E. Yilmaz, T. Kremers, U. Schnakenberg, „Microfluidic-based electrical impedance spectroscopy system using multilevel lamination of dry Film Photoresist", in *Digest Tech. Papers 14th Int. Workshop on Impedance Spectroscopy IWIS 2021*, Chemnitz, Germany, September 29 - October 01, 2021, pp. 11-14.

[7] J. Floehr, E. Dietzel, C. Schmitz, A. Chappell, W. Jahnen-Dechent, "Down-regulation of the liver-derived plasma protein fetuin-B mediates reversible female infertility", *Mol Hum Reprod.*, vol. 23, pp. 34-44, 2017.

[8] E. Dietzel, J. Wessling, J. Floehr, C. Schafer, S. Ensslen, B. Denecke, et al., "Fetuin-B, a liver-derived plasma protein is essential for fertilization", *Dev. Cell*, vol. 25, pp. 106-112, 2013.

[9] L.G. Alexopoulos, M.A. Haider, T.P. Vail, F. Guilak, "Alterations in the mechanical properties of the human chondrocyte pericellular matrix with osteoarthritis", *J. Biomech. Eng.*, 125, pp. 323-333, 2003.

[10] Y. Cao, J. Floehr, D. Azarkh, U. Schnakenberg, "Microfluidic Aspiration-Assisted Electrical Impedance Spectroscopy System is a Reliable Tool for the Characterization of Oocyte Hardening". *Sensors and Actuators B*, in revision, October 2022.

[11] C. Schmitz, S.Z. Sadr, H. Korschgen, M. Kuske, J. Schoen, W. Stocker, et al., "The E-modulus of the oocyte is a non-destructive measure of zona pellucida hardening", *Reproduction*, 162, pp. 259-66, 2021.

CONTACT

* U. Schnakenberg, schnakenberg@iwe1.rwth-aachen.de

DOUBLE PULSE IRRADIATION OF FS LASER FOR ENHANCING THE PERFORMANCE OF PRECISE LASER SORTING METHOD

Ryota Kiya[1], Yoshinaga Rintaro[1], Yo Tanaka[2], Yaxiaer Yalikun[1,2], and Yoichiroh Hsokawa[1].
[1]Division of Materials Science, Nara Institute of science and Technology, Japan and
[2]Center for Biosystems Dynamics Research, RIKEN, Japan.

ABSTRACT

Cell sorting systems are necessary for large amount cell analysis in biological, medical, and clinical applications. As one of the highest throughput cell sorting technologies (100 kilohertz (kHz)), femtosecond (fs) laser-based cell manipulation systems have the potential for single-step multi-sorting. To further improve the performance, instead of single pulse irradiation mode of laser, double pulses irradiation mode was developed. In this research, after experimental investigations and numerical analysis, we found that a pulse interval of 10 µs could achieve larger displacement of the sample than single ones.

KEYWORDS

Multiple pulse irradiation, femtosecond laser, high-throughput, cell sorter

INTRODUCTION

Cells contain countless important information in life sciences as the most basic structural and functional single position of the human body. Due to its heterogeneity, the same type of cells may express different characteristics and various biomolecules even in the same environment, which is related to the mechanism of cell development, distribution, and cancelation. The ability to analyze cells is critical for drug screening [1], stem cell research [2], tissue and organ regeneration [3], cancer diagnosis and treatment [4-9], In the process of cell analysis, sorting of target cells efficiently and quickly is an important issue.

FACS (Fluorescence-actived cell sorter) is most popular cell sorting system. The FACS has two functions, a detection system and a manipulation system.

A detection system can detect many types of cell quickly. Stained cells exhibit fluorescence with a specific wavelength. Cell fluorescence varies depending on cell physical characteristics [10] viability [11], and cell cycle characteristics [12-13]. Apart from fluorescence, even the same cell line may differ in several aspects, such as dielectric properties [14-15], deformability [16-18]. Therefore, by using indices such as fluorescence and dielectric properties as well as the presence or absence of fluorescence, it is possible to distinguish several types of cells even if they are the same cell line.

Manipulation system is developed many types, Those cell manipulation methods are divided into two types: passive-method and active-method [19].The passive-method are mainly using the geometry of channel and gravity to effect on flow path. Then a secondary flow will be generated, and the flow force works on target cell, separate them by their size and mass. On the other hand, the active-method actively applied the external forces such as electric fields, magnetic fields, surface acoustic waves, and flow drag force to separate the target cell from a large population, by its components, size, and mass.

Among above methods, these is a trade-off between throughput and the ability of selection. Which means, only binary selection could be applied on the original large population at a high-throughput. For example, our group have realized a binary selection sorting system with high real-time throughput (100,000 pieces/s) by using impact force generated when a short pulse laser is irradiated in water [20](Fig.1(a)). Because it is fluoresce triggered system, only the target cell having a fluorescent value over a fixed threshold will be selected. However, cell with other fluorescent value will be lost with non-fluorescent cells. Since the cell's heterogeneity, binary selection limits its potential to be widely used in the field of biology. Other method capable of multi-selection have a low throughput and time consuming, which limits its clinical application.

To increase the selecting ability to current binary laser sorter, (for example triplet sorter, Fig.1(b)) a new methodology/mechanize is required. This report say that displacement of the micro bead by double pulses irradiation is larger than displacement of the micro bead by single pulse irradiation. This mechanism was investigated with numerical analysis, and we found that a pulse interval of 10 µs could achieve larger displacement of the bead than single ones.

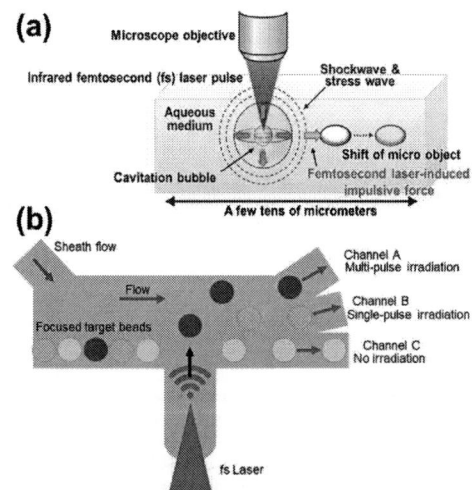

Fig.1 (a) Schematic diagram of manipulation of a micro object by femtosecond laser, and (b)triplet cell sorter.

MATERIAL AND METHOD

1. Experimental Samples

First, we diluted blocking reagent (Blocking Regent-N102, NOF Co.) in the ultrapure water (Milli-Q, EMDMillipore Co.) and prepared a 20% blocking reagent solution for preventing the adhesion between particles. Then we added beads (φ10 μm fluorescent polystyrene microspheres, Fluoresbrite Yellow Green Mailboxylate Microspheres) into the prepared solution.

2. System Configuration

Fig. 2 shows the experimental system in this report. fs-laser (generated by Yb femtosecond laser, Spirit One, Spectra-Physics, 1040 nm, < 400 fs, 2 μJ/pulse, laser power measured immediately after output from the amplifier) is introduced to the upright microscope (BX-53, Olympus). Concentrated irradiation was performed underwater with a water-immersion 20x object lens (N.A.:0.8). Pulse extraction was performed by inputting a trigger generated by a function generator to the pulse picker built into the laser. Single signal generated by function generator input to pulse picker, it extraction a single pulse from the laser pulse row, also two signal input to the pulse picker, it extraction double pulses from the laser pulse row. Laser pulses irradiate into the water, then it displace a bead. This displacement is observed by high speed camera with a microscope.

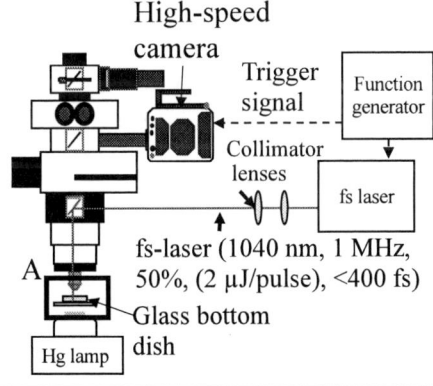

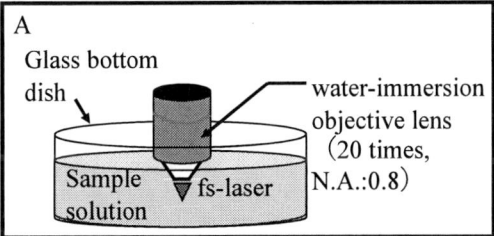

Fig.2 Experimental system.

RESULT AND DISCUTTION
Result

We adjusted that focal point and a micro bead distance is 15 μm. When the single fs-laser pulse is irradiated, a cavitation bubble was generated from the focusing point (Fig.1(A)). The cavitation bubble expanded, displacing the micro bead away from the focusing point. A cavitation bubble then contracted, displacing the micro bead closer from the focusing point. A cavitation bubble then burst, again displacing the micro bead away from the focusing point. When the double fs-laser pulses are irradiated, 2 cavitation bubbles were generated from the focusing point (Fig.1(B)). The 1st cavitation bubble expanded, displacing the micro bead away from the focusing point. The 1st cavitation bubble then contracted, displacing the micro bead closer from the focusing point. The 1st cavitation bubble then burst, again displacing the micro bead away from the focusing point. Then, the 2nd cavitation bubble expanded, displacing the micro bead away from the focusing point. The 2nd cavitation bubble then contracted, displacing the micro bead closer from the focusing point. The 2nd cavitation bubble then burst, again displacing the microbeads away from the focusing point.

In Fig. 3 (C), plots show the bead displacement when irradiated with single and double fs-laser pulses. The vertical axis indicates displacement of the micro bead based on the initial bead position before fs-laser pulse irradiation. The horizontal axis is a time change based on the 1st fs-laser pulse irradiation timing. The maximum displacement of bead when single fs-laser pulse irradiated was 26 μm. The maximum displacement of bead when double fs-laser pulses irradiated was 40 μm.

In Fig.3 (C), two lines shows numerical analysis value (1) [21 e.q. 6]. Equation 1 say that the displacement of a micro bead due to fluid movement associated with the cavitation bubble. We used this equation to analyze the displacement of bead when each fs-laser pulse irradiated.

Table 1 shows the value of water flow from equation 1. t_1 is the time of water flow from the focal point toward the bead, t_2 is the time of the water flow from the bead toward the focusing point. v_1 is the expanded volume of the water flow speed from the focal point toward the bead, and v_2 is the volume of the water flow speed from the bead toward the focal point. The upper right indent 1 and 2 shows 1st cavitation bubble and 2nd cavitation bubble.

When single fs laser pulse irradiation, it was found that micro bead were displaced by the expansion and contraction of 1 cavitation bubble. By 1st cavitation bubble, each volumes speed were (v_1^1): 1.5×10^{-7} m³/s and (v_2^1): 2.6×10^{-7} m³/s. By 2nd cavitation bubble, each volumes speed were (v_1^2): 5.0×10^{-7} m³/s and (v_2^2): 5.0×10^{-7} m³/s.

When irradiated with a double fs-laser pulses, it was found that 2 cavitation bubbles displaced the micro bead. It was found that the 1st cavitation bubble is equivalent to the cavitation bubble that occurs during single fs-laser pulse irradiation. The volume velocity and time of expansion of the 2nd cavitation bubble was 5.0×10^{-7} m³/s, and 2.2×10^{-6} s. The volume velocity and time of contraction of the 2nd cavitation bubble was 5.0×10^{-7} m³/s, and 4.0×10^{-6} s (Table 1).

Discussion

A micro bead of distance is 15 μm away from the focusing point. Irradiated with single and double fs-laser pulses. Cavitation bubble generated displaced a micro bead. Displacement of micro bead was observed with a high-speed camera. When a single fs laser pulse was irradiated, a cavitation bubble was generated, and the micro bead was displaced through a process of expansion and contraction (Fig.3 (A)). When double fs laser pulses were irradiated, two cavitation bubbles were generated (Fig.3 (B)). With the expansion and contraction of the 1st cavitation bubble, the micro bead was displaced. The displacement was equivalent to that of the cavitation bubble generated when single fs-laser pulse irradiation (Fig.3 (C)). Due to the expansion and contraction of the 2nd cavitation bubble, the micro bead was displaced more than single fs-laser pulse ones (Fig.3 (C)).

The results obtained from the experiment were numerically analyzed using the numerical analysis equation (e.q.1) (Fig.3 (C)). The cavitation bubble generated by single fs laser pulse irradiation. When double fs laser pulses irradiation, two cavitation bubbles generated. The 2nd cavitation bubble generated when double fs laser pulses irradiation had a larger volume expansion and contraction rate than the 1st cavitation bubble (=single cavitation bubble). The fluid flow was outward from the focusing point when the 2nd pulse irradiation of double fs laser pulses irradiation (10μs). Therefore, it is considered that the pressure at the focusing point is lower than when a single pulse is irradiated when a cavitation bubble is generated.

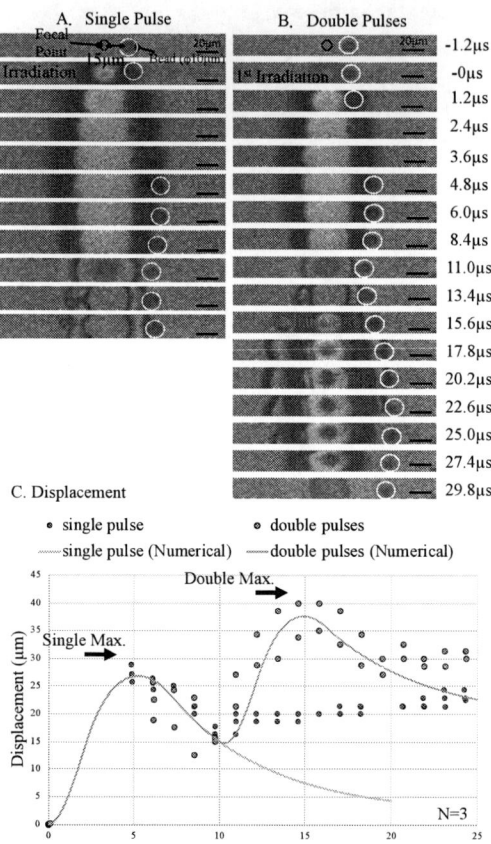

Fig.3 Bead displacement and expansion/contraction of the cavitation bubble when (A) irradiated with a single fs laser pulse and (B) irradiated with a double fs laser pulses. (C) Bead displacement when irradiated with single and double fs laser pulse.

$$m\ddot{x} + 6\pi\eta r(\dot{x} - v_w(x,t)) = 0 \qquad (1)$$

CONCLUTION

In this research, to conceptually propose a multi selectable cell sorter, we developed a controllable multi-pulse fs-laser irradiation system to achieve different shift distances for different type of target (micro bead). The shift process happens in a 10 μs, which means these is a possibility of achieving a high through-put over 100,000 event/s for cell sorting. By further optimizing the design of the microfluidic channel, and utilizing biological samples, a multi-selectable cell sorter is realizable.

REFERENCES

[1] P.M. Haverty, E. Lin, J. Tan, Y. Yu, B. Lam, S. Lianoglou, R.M. Neve, S. Martin, J. Settleman, R.L. Yauch, R. Bourgon, Reproducible pharmacogenomic

Table 1. Behavior of water flow associated with the generation of 1st cavitation bubble and 2nd cavitation bubble obtained from numerical analysis.

	1st CB	2nd CB
t_1 (s)	2.2×10^{-6}	2.2×10^{-6}
t_2 (s)	5.0×10^{-6}	4.0×10^{-6}
v_1 (m³/s)	1.5×10^{-7}	5.0×10^{-7}
v_2 (m³/s)	2.6×10^{-7}	5.0×10^{-7}

profiling of cancer cell line panels, Nature 533 (2016) 333–337.

[2] Y. Chen, T.-H. Wu, A. Chung, Y.-C. Kung, M.A. Teitell, D. Di Carlo, P.-Y. Chiou, Pulsed laser activated cell sorter (PLACS) for high-throughput fluorescent mam- malian cell sorting, Proc. SPIE - Int. Soc. Opt. Eng. 9164 (2014).

[3] A. Scialdone, Y. Tanaka, W. Jawaid, V. Moignard, N.K. Wilson, I.C. Macaulay, J.C. Marioni, B. Göttgens, Resolving early mesoderm diversification through single- cell expression profiling, Nature. 535 (2016) 289–293.

[4] S. Takaishi, T. Okumura, S. Tu, S.S.W. Wang, W. Shibata, R. Vigneshwaran, S.A.K. Gordon, Y. Shimada, T.C. Wang, Identification of gastric cancer stem cells using the cell surface marker CD44, Stem Cells 27 (2009) 1006– 1020.

[5] D.T. Miyamoto, Y. Zheng, B.S. Wittner, R.J. Lee, H. Zhu, K.T. Broderick, R. Desai, D.B. Fox, B.W. Brannigan, J. Trautwein, K.S. Arora, N. Desai, D.M. Dahl, L.V. Sequist, M.R. Smith, R. Kapur, C.L. Wu, T. Shioda, S. Ramaswamy, D.T. Ting, M. Toner, S. Maheswaran, D.A. Haber, RNA-Seq of single prostate CTCs implicates noncanonical Wnt signaling in antiandrogen resistance, Science 349 (2015) 1351–1356.

[6] S.B. Cheng, M. Xie, J.Q. Xu, J. Wang, S.W. Lv, S. Guo, Y. Shu, M. Wang, W.G. Dong, W.H. Huang, High- efficiency capture of individual and cluster of circulating tumor cells by a microchip embedded with three- dimensional poly (dimethylsiloxane) scaffold, Anal. Chem. 88 (2016) 6773–6780.

[7] D. Ren, B. Wang, C. Hu, Z. You, Quantum dot probes for cellular analysis, Anal. Methods 9 (2017) 2621–2632.

[8] D. Ren, J. Wang, B. Wang, Z. You, Probes for biomolecules detection based on RET-enhanced fluorescence polarization, Biosens. Bioelectron. 79 (2016) 802–809.

[9] D. Ren, Y. Xia, B. Wang, Z. You, Multiplexed analysis for anti-epidermal growth factor receptor tumor cell growth inhibition based on quantum dot probes, Anal. Chem. 88 (2016) 4318–4327.

[10] S. Hamann, J. F. Kiilgaard, T. Litman, F. J. Alvarez-Leefmans, B. R. Winther, T. Zeuthen, Measurement of Cell Volume Changes by Fluorescence Self-Quenching. *Journal of Fluorescence*. **12**, 139–145 (2002).

[11] A. Andrzejewska, A. Jablonska, M. Seta, S. Dabrowska, P. Walczak, M. Janowski, B. Lukomska, Labeling of human mesenchymal stem cells with different classes of vital stains: robustness and toxicity. *Stem Cell Research & Therapy*. **10**, 187 (2019).

[12] V. Roukos, G. Pegoraro, T. C. Voss, T. Misteli, Cell cycle staging of individual cells by fluorescence microscopy. *Nature Protocols*. **10**, 334–348 (2015).

[13] A. Ferro, T. Mestre, P. Carneiro, I. Sahumbaiev, R. Seruca, J. M. Sanches, Blue intensity matters for cell cycle profiling in fluorescence DAPI-stained images. *Laboratory Investigation*. **97**, 615–625 (2017).

[14] T. Tang, X. Liu, R. Kiya, Y. Shen, Y. Yuan, T. Zhang, K. Suzuki, Y. Tanaka, M. Li, Y. Hosokawa, Y. Yalikun, Microscopic impedance cytometry for quantifying single cell shape. *Biosensors and Bioelectronics*, 113521 (2021).

[15] C. Honrado, P. Bisegna, N. S. Swami, F. Caselli, Single-cell microfluidic impedance cytometry: From raw signals to cell phenotypes using data analytics. *Lab on a Chip* (2021), , doi:10.1039/d0lc00840k.

[16] R. Huisjes, A. Bogdanova, W. W. van Solinge, R. M. Schiffelers, L. Kaestner, R. van Wijk, Squeezing for Life – Properties of Red Blood Cell Deformability. *Frontiers in Physiology*. 0, 656 (2018).

[17] J. Kim, H. Lee, S. Shin, Advances in the measurement of red blood cell deformability: A brief review. *Journal of Cellular Biotechnology*. **1**, 63–79 (2015).

[18] Y. Hao, S. Cheng, Y. Tanaka, Y. Hosokawa, Y. Yalikun, M. Li, Mechanical properties of single cells: Measurement methods and applications. *Biotechnology Advances*. **45** (2020), doi:10.1016/J.BIOTECHADV.2020.107648.

[19] Yigang Shen, Yaxiaer Yalikun, Yo Tanaka, Recent advances in microfluidic cell sorting systems, Sensors & Actuators: B. Chemical 282 (2019) 268–281.

[20] T. iino et al. High-speed microparticle isolation unlimited by Poisson statistics, Lab chip, 19, 2669-2677. (2019)

[21] Yoichiroh Hosokawa, " Applications of femtosecond laser-induced impulse to cell research," JJAP invited review, vol. Volume 58, Number 11, pp110102, 31 (2019)

CONTACT

*kiya.ryota.kj8@ms.naist.jp

DROPLET BASED HIGH THROUGHPUT SINGLE-SPERM CRYOPRESERVATION PLATFORM

Na Li[1,2], Shihui Qiu[1,3], Zhenhua Wu[1,3], Hongju Mao[1,3], *

[1] State Key Laboratory of Transducer Technology, Shanghai Institute of Microsystem and Information Technology, Chinese Academy of Sciences, Shanghai 200050, China

[2] ShanghaiTech University, Shanghai 201210, China

[3] Center of Materials Science and Optoelectronics Engineering, University of Chinese Academy of Sciences, Beijing 100049, China

ABSTRACT

With the popularization of assisted reproduction technology (ART), as a key technology in ART, cell cryopreservation still has great space for improvement. At present, there are few cryopreservation technologies suitable for rare cells. And these technologies have high requirements for operators. Therefore, there is a lack of easy-to-use single sperm cryopreservation techniques for oligospermia patients.

Here we present a cryopreservation system that composed of thousands of pico-liter-volume droplets [1]. This platform enjoys the following advantages (1) Single cell level manipulation: The platform could realize single sperm cryopreservation. (2) High throughput: This chip could preserve 104 sperms separately in thousands of droplets. (3) Handleability: Droplets in microfluidic chips are easy to generate and extract. (4) Great performance: faster cryoprotectant agents (CPAs) loading and unloading, smaller cryopreservation volume could improve sperms survival rate [2,3]. (5) Customizability: Different individuals, even sperms, could have different responses to CPAs. This platform could generate CPAs with gradient concentration to determine the optimum reagent formula.

KEYWORDS

Azoospermatism, Cryopreservation, Vitrification, Single Cell, Droplet Microfluidics

INTRODUCTION

Nowadays, as affected by habits and environmental changes, the probability of males wordwide suffering from oligospermatism and asthenospermiakeeps rising [4]. This is the key factor that influencing the fertility.

With the development of Intracytoplasmic Sperm Injection (ICSI)technology, infertile patients with oligospermia and asthenospermia may be able to have children.

However, severe oligospermia is always the challenges against sperms' cryopreservation. Patients who are non-obstructive azoospermia (NOA) have active sperms in semen need sperm cryopreservation for storage. Those still want a child after sperm duct removal surgery; Others have cancer or need radioactive therapy need preserve their sperms before they lost their reproductive ability [5], sperm cryopreservation is of vital importance to them. ICSI is a technology to help those patients and they need cryopreservation to store the germ cells for a better opportunity for ICSI.

As to cryopreservation, two methods are developed: programmable slow freezing method and vitrification [6]. Slow freezing typically applied cryoprotective agents (CPAs) of low concentration to cells and froze them at a rate of 1°C/min. With such an optimized frozen rate, the extra- and intracellular ice formation (EIF and IIF) are minimized which would decrease the survival of cells. The strategy of vitrification is to avoid ice formation, which needs high concentration CPA or ultrafast cooling rate to promoting the formation of a non-crystalline glassy state.

The cryopreservation study based on microfluidic mainly focused on loading and unloading of CPAs because reducing the contact opportunity of living cells with CPAs could reduce the damage brought by its toxicity.

Vitrification has a better performance while it needs ultrafast cooling rate and high CPAs concentration. Theoretically, concentration of CPAs is in reverse proportion to the cooling rate. Based on the toxicity of CPAs, ultrafast cooling is preferred. To realize the ultrafast cryopreservation, the larger specific surface area is the key. Zou et al. [5] proposed a microfluidic device for ultra-fast and low-temperature preservation of a small amount of sperm without the use of cryoprotectants. When the chip height and width are both 10μm, there is no obvious difference compared with the traditional preservation method. It has been proved that the advantages of microfluidic small freezing volume and fast operation can achieve rapid freezing without cryoprotectant.

Droplets microfluidic provides thousands of independent reaction volume. As for cryopreservation, droplets were unstable and could fused thus could not be suitable for clinical usage in the cooling process. In this work, we designed a structure to achieve rapid and ordered cryopreservation for single sperm by wrapping single sperm into a droplet. The device has a better survival rate for frozen sperms because it could realize fast loading and unloading of CPAs, a pico-liter level cryopreservation volume to eliminate the generation of ice-crystal and CPAs toxicity. We also tried non-CPAs fast cryopreservation which proved to be feasible. The platform was friendly for clinical because technicians could directly snap the droplets of sperms for ICSI and could be a standard cryopreservation platform enjoys lower costs and storage space.

EXPERIMENT
Chip Design

When being heated, frozen and thawed, droplets would easily fused together. In order to cope with the phenomenon that droplets are prone to fuse in media such as PDMS, we designed microchannels to immobilize droplets and then the droplets were isolated. After isolation, droplets are difficult to get fused. When droplets

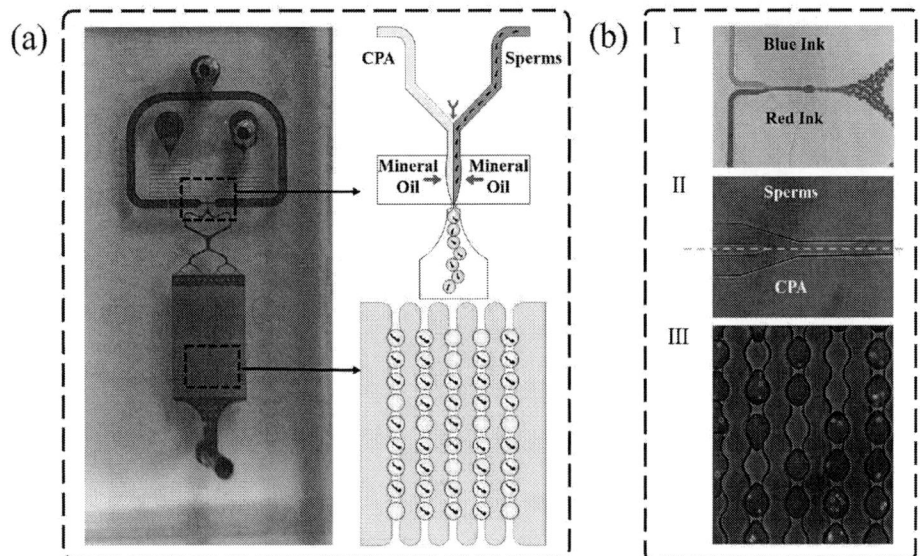

Figure 1: a. Schematic diagram of the platform; b. Demo of droplet generation and single-sperm encapsulation.

get deformed, they must restore to the state with lowest potential energy. When they became spherical again, they would be trapped in the channel and realize the immobilization and isolation. These droplets will not fuse under freezing and freezing-thawing conditions, furthermore, realize the freezing of single sperm. Due to the diameter of the droplet in this work (40μm) which can achieve high freezing and thawing speed. This helps to realize ultra-rapid freezing and provides a new idea for vitrification cryopreservation.

Droplets' Generation Module

The droplet storage channel imitates the shape of the gourd. The wider part is a circle with a diameter of 40μm and the narrower one is 20μm wide. The repetition period is 60μm which means the gap between the 40μm circle is 20μm.

When the channel parameters are decided, the droplets generation port should be designed. In this work, three inlets were designed. two dispersion phase inlets-the sample inlet and the cryoprotectant inlet. The third inlet is the continuous phase inlet. Flow Focusing structure was determined for single sperm encapsulation.

$$P = R \times Q \quad (1)$$

Here P, R and Q denotes pressure, flow resistance and flux respectively. For the convenience of droplets generation, here $P_1 = P_2$

This means: $R_1 \times Q_2 = R_2 \times Q_2$. Here the index 1 and 2 denotes different phases. We calculated the volume of the dispersion phase and the continuous phase in the channel which we called V_1 and V_2. The specific value of the volume should be proper, and the droplets could be stuck in the wider part of the channel.

$$V_1 : V_2 = Q_1 : Q_2 = R_2 : R_1 \quad (2)$$

According to the following equation for flow resistance:

$$R = \frac{12\mu l}{wh^3 \left(1 - 0.63\frac{h}{w}\right)} \quad (3)$$

where w, h, l denote width, height and length of the channel respectively; μ is the coefficient of viscosity.

In this chip, the height of the channel was set as 40μm. After, we just set the proper parameter of the channel length and width.

Chip Fabrication

All chip's manufacture is based on microfabrication. Photolithography and Deep RIE were applied to manufacture the silicon mold of the droplet chip. The specific process is shown in figure2. After the molds were manufactured, they were silanized by trichloro-
(1H,1H,2H,2H-perfluorooctyl) silane vapor under vacuum to reduce the surface energy and ease the peeling of the PDMS layer. PDMS were prepared by mixing pre-polymer and cross-linking agent. For the functional layer, the mass ratio of the two components is 10:1. The PDMS mixture was poured on the silicon mold and the thickness of this layer was 2–3 mm suitable for the fixation of inserted tubes. It was then placed in a horizontal oven and left for at least 1 hour at 65°C for crosslinking.

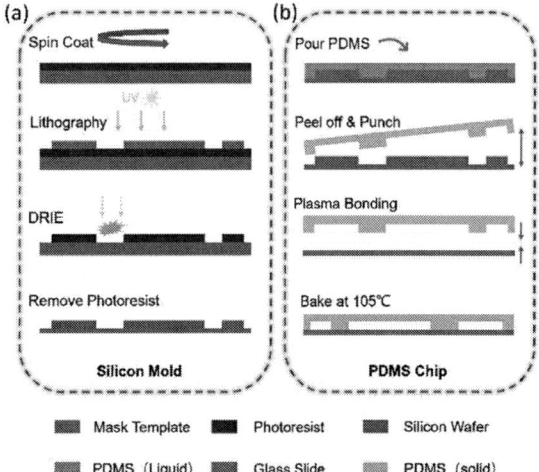

Figure 2: The whole manufacture. a. The fabrication of the silicon mold. B. The fabrication of the PDMS chip.

After punched holes at the corresponding positions, the PDMS layers were treated with alignment bonding by plasma to complete the chip manufacturing. Before use, the chips had to be baked in a hot oven at 105°C overnight for hydrophobization.

Single Sperm Encapsulation and Cryopreservation

We provide two strategies for single-sperm level cryopreservation. One was the cryopreservation system containing CPA, in which we used double dispersed phase for the encapsulation of single sperm. After the confluence of the two dispersed phases, they are immediately cut into droplets by the continuous phase. The other one is the CPA-free cryopreservation system, which only requires one dispersed phase injection port. After pressurization, the dispersed phase is cut into droplets by the continuous phase.

After the droplets were generated, they are squeezed into the downstream channels. The pressure was stopped after the droplets filled the entire channels. The chip needed no seal and was fumigated directly on liquid nitrogen for about five minutes. Finally, it was immersed into liquid nitrogen for preservation.

Thawing and observation

Take out the chip from the liquid nitrogen tank and immediately place them in the oven at 37°C for 5 minutes. When fully thawed, the chip was gently placed under an inverted microscope to observe the droplet status.

Droplets exportation and sperm characterization

One of the injection ports of dispersed phase and the one of continuous phase were sealed with adhesive tape. After that, PBS buffer was added into the unsealed dispersed phase injection port, all droplets were exported and collected with a pipette gun.

Eosin solution was added to the collected sperm sample, incubated for about 2 minutes, and then it was dripped onto a glass slide. After covering the glass slide, the stained sperm sample was observed under an inverted microscope. Dead sperm are stained distinctly red due to disruption of membrane integrity. Living sperm cells were either clear or light red. Count the total sperm number as well as live and dead sperm number and calculate the living sperm ratio.

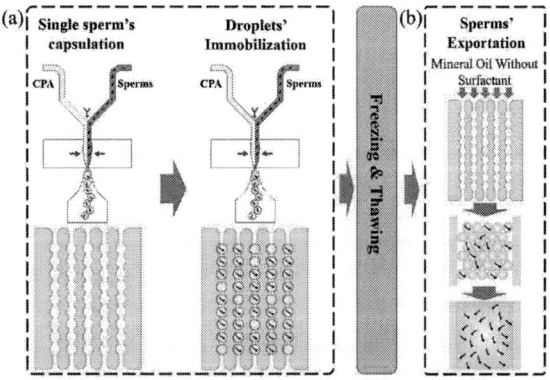

Figure 3: a. The basic experiment flow on the system. 1. Generate droplets that contains sperms (each droplet

contains no more than 5 sperms); 2. Trapping these droplets into the arrays. After droplets' trapping, freezing the chip into liquid nitrogen; b. After thawing in 37°C water bath, inject mineral oil without surfactant to push the droplets out of the arrays. Collecting the sample after the droplets' demulsification.

RESULTS

Improvement To Avoid droplet fusion

As shown in Figure 3, all droplets stored in the microfluidic chamber will fuse after freezing and thawing. However, the droplets in the gourd-shape microchannel have been completely preserved. The filled-up ratio of the droplets in the microchannel is over 80%, which can achieve high throughput single sperm freezing.

Sperms' Survival Statistics

We compared sperm cryopreservation with a conventional cryopreservation tube and cryopreservation on a single-sperm chip.

As showed in Figure6, the proportion of living sperms was about 50% with conventional method, while the proportion of living sperms with single sperm chip was about 90% with 10% CPA formula, and about 68.25% with no CPA.

Compared with traditional methods, single sperm chip has certain advantages in sperm viability.

Less Damage Brought by Simplifying Operation

For the chip with a droplets' storing-scale of 10^5,100 seconds was enough for droplets' filling-up with a generation ratio about 100 per second. The sperm encapsulation was automatic and no manual monitoring and operation. Simplified operation steps and high throughput reduce the requirements for operators and shorten the loading time of CPA at room temperature. This helps to reduce the damage to sperms [7,8].

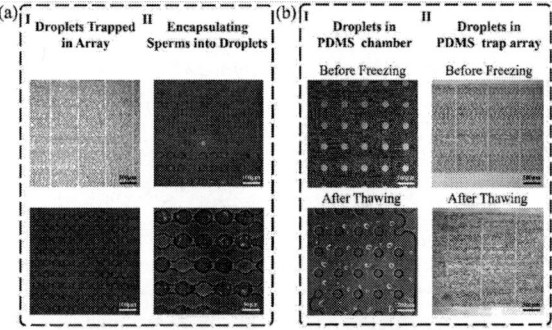

Figure 4: a. I. Real image of droplets' trapping in the array; II. Real image of sperms' encapsulation in droplets; b. Comparison between droplets before freezing and those after thawing. I.Droplets in PDMS chamber; II. Droplets in PDMS trap Array.

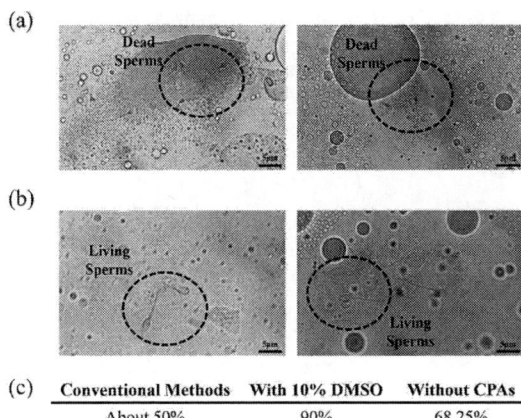

(a)

(b)

(c)

Conventional Methods	With 10% DMSO	Without CPAs
About 50%	90%	68.25%

Figure 5: a. Deep pink staining illustrates dead sperms; b. Light pink staining indicates living sperms; c. Statistics of sperm survival rate. The original samples are centrifugated, and almost 100% were survived. The gradient CPAs system has a 90% survival rate. The non-CPA system has a 68.25% survival rate.

CONCLUSION

In this work, we introduced a single-sperm level cryopreservation chip based on droplet microfluidics. The chip enjoys high throughput and simple fabrication. It can provide a pico-liter level size droplet as cryopreservation media and has great advantages in rapid freezing and thawing over traditional media. The freezing system with 10% CPA and without CPA through this platform, the survival rate of sperm before and after freezing has certain advantages compared with that of traditional media.

In our future work, we need to further improve the flux and reducing the flow resistance of the chip to improve the droplets' generating speed and the 'traffic Jam' phenomena in droplets' filling-up in the channels.

ACKNOWLEDGEMENTS

This study was supported by National Natural Science Foundation of China (Nos. 62231025, 61971410, 61801465, 61801464 and 62001458), Shanghai Sailing Program (No. 20YF1457100), Shanghai Engineer & Technology Research Center of Internet of Things for Respiratory Medicine (No. 20DZ2254400), and the Science and Technology Commission of Shanghai Municipality (No. 19511104200).

REFERENCES

[1] Christian H J, Schmitz, Amy C , "Dropspots: a picoliter array in a microfluidic device", *J. Lab. Chip.*, vol.9, pp.44-49, 2009.

[2] Deutsch M, Afrimzon E , Namer Y , "The individual-cell-based cryo-chip for the cryopreservation, manipulation and observation of spatially identifiable cells. I: Methodology", *J. BMC. Cell. Biol.*, vol. 11(1):pp. 54-54, 2010.

[3] Michael T , Steffen H , Martin G , "The individual-cell-based cryo-chip for the cryopreservation, manipulation and observation of spatially identifiable cells. II: Functional activity of cryopreserved cells",*J. Bmc. Cell. Biol.*, vol. 11(1), pp.1-13, 2010.

[4] Merzenich, H, H. Zeeb, M. Blettner, "Decreasing sperm quality: a global problem?", *J. Bmc. Public. Health.*, vol. 10, 2010.

[5] Zou, Y, "On-Chip Cryopreservation: A Novel Method for Ultra-Rapid Cryoprotectant-Free Cryopreservation of Small Amounts of Human Spermatozoa", *J. Plos. One.*, vol. 8(4), 2013.

[6] He, X.M, "Vitrification by ultra-fast cooling at a low concentration of cryoprotectants in a quartz micro-capillary: A study using murine embryonic stem cells", *J. Cryobiology.*, vol. 56(3), pp. 223-232, 2008.

[7] Wheeler, M.B, E.M. Walters, and D.J. Beebe, "Toward culture of single gametes: The development of microfluidic platforms for assisted reproduction", *J. Theriogenology.*, vol. 68, pp. S178-S189, 2007.

[8] Li, S, W. Liu, L.W. Lin, "On-Chip Cryopreservation of Living Cells", *J. Jala.*, vol. 15(2), pp. 99-106, 2010.

CONTACT

*H. Mao, tel: +86-21-62511070-8701;
hjmao@mail.sim.ac.cn.

DUAL ION-SELECTIVE MEMBRANE DEPOSITED ION-SENSITIVE FIELD-EFFECT TRANSISTOR (DISM-ISFET) INTEGRATING WHOLE BLOOD PROCESSING MICROCHAMBER FOR IN SITU BLOOD ION TESTING

Xiao-Wen Chen[1], Syuan-Rong Huang[1] and Nien-Tsu Huang[1,2]*

[1]Graduate Institute of Biomedical Electronics and Bioinformatics, National Taiwan University and
[2]Department of Electrical Engineering, National Taiwan University

ABSTRACT

Blood ion testing is one of the methods that is commonly used as of monitoring immune status and providing lots of physiological information for disease diagnosis. We proposed a microfluidic device integrating dual ion-selective membranes and dual-gate ion-sensitive field-effect transistor (DISM-ISFET) for whole blood processing and blood ion concentrations monitoring. In this system, the Anti-D-coated trench array and filter membrane embedded microchamber greatly shorten the serum extraction time to 4 min. Besides, the DISM-ISFET array sensor enables simultaneous Na+ and K+ ion sensing in 10 min. The above features make this fully integrated and portable system a rapid and sensitive tool for in situ blood ion concentration diagnosis.

KEYWORDS

blood ion concentration test, ion-selective membrane (ISM), ion-sensitive field-effect transistor (ISFET)

INTRODUCTION

Blood is one of the most important body fluids, containing blood cells, nucleic acids, metabolites, enzymes, antibodies, and electrolytes. Among them, electrolytes are substances with positive or negative charges dissolved in the blood, which can help the body to regulate chemical reactions and maintain the fluid balance between cell membranes. Since electrolytes play a critical role in maintaining body function, the electrolyte imbalance may be a sign of abnormal health conditions, including nausea, malaise [1], and cardiovascular and kidney diseases [2, 3]. For example, acute kidney injury (AKI) is defined as an abrupt decrease in kidney function, leading to electrolyte and acid-base disturbances, and may cause a high mortality rate (50% – 80%) [4]. To treat AKI, patients usually rely on intermittent hemodialysis (IHD) to maintain the electrolyte balance and continuously monitor blood ion (electrolytes) levels to ensure successful therapy [5]. However, the current blood ion measurement methods usually require prolonged and labor-intensive serum extraction process from whole blood using bulky instruments, which makes real-time and continuous blood ions measurement difficult [6].

To address the above issues, we propose a dual-gate ion-sensitive field-effect transistor (ISFET), a rapid and sensitive array-based technology, for in situ and continuous blood ion monitoring with less blood sample requirement. Our previous research has shown ISFET can be applied for real-time H$^+$ monitoring for bacterial antimicrobial susceptibility tests. [7] The working principle is measuring the source-drain current difference (ΔIds) caused by various ion concentrations in the solution. The current limitation of using ISFET for blood ion testing is the requirement of prolonged sample preparation and non-specific ion detection. Therefore, we integrated a microchamber device embedding trench array and filter membrane to perform quick serum extraction and deposit Na/K dual ion-selective membrane (DISM) onto ISFET array sensors to achieve real-time and continuous blood ion sensing.

MATERIALS AND METHODS

Whole blood sample preparation

In each whole blood sample test, two individual 9-mL clinical blood samples were collected with the blood collection tube without anticoagulants coating. One blood tube is processed by standard centrifugation, served as the control sample, and another blood tube is processed and tested using the DISM-ISFET system. All human whole blood samples were stored at 4 °C and used within 24 h after blood collection. The Ethics Committee of National Taiwan University Hospital (NTUH) approved this research, and informed consent was obtained before collecting blood samples.

DISM preparation and deposition process

The composition of ISM contains (1) a neutral carrier, to select specific ions; (2) a polymer matrix with plasticizer, to solidify the ISM structure, and (3) tetrahydrofuran (THF), to serve as a solvent. To prepare the Na ISM, we used 2.12 g of sodium ionophore X (Sigma-Aldrich®) as the neutral carrier; 0.1 g of poly (vinyl chloride) (PVC) and 0.2 g of bis (2-ethylhexyl) sebacate (DOS) as the polymer matrix and plasticizer. Finally, the above components were dissolved in 2 mL THF.

For K ISM, we used 0.01 g of valinomycin (Sigma-Aldrich®) as the neutral carrier; 0.16 g of poly (vinyl chloride) (PVC), and 0.33 g of bis (2-ethylhexyl) sebacate (DOS) as the polymer matrix and plasticizer. Finally, the above components were dissolved in 10 mL THF. The above ISM solution were stored in a 4 °C refrigerator to maintain the chemical stability. Before ISM deposition, the ISFET surface was cleaned by DI water, dried by N2 gas and placed in 65 °C oven for 60 min to prevent any hydroxyl group on the surface. After each experimental round, we used the tweezer to carefully replace ISM from the ISFET surface and deposit the new ISM for next experiments.

DISM-ISFET device

The schematic of the DISM-ISFET system is shown in Figure 1A. Briefly, it can be divided into three sections:

(1) Na/K ISM deposited ISFET sensor; (2) the PMMA-made microchamber embedded with the trench array and a filter membrane and (3) a portable reader for real-time and continuous electrical signal recording. The photograph of the portable reader and the DISM-ISFET device is shown in Figures 1B and 1C. The device dimension is 35 mm × 60 mm. The microchamber is fabricated by a computer numerical control (CNC) machine (EGX - 400, Roland, CA) on a 5-mm PMMA board. The microchamber was attached to the printed circuit board (PCB) and sealed by liquid phase PDMS to prevent any solution leakage. Figure 1D showed the microscopic image of the DISM-ISFET sensor deposited Na and K ISM. The DISM-ISFET sensor is fabricated by Taiwan Semiconductor Manufacturing Company (TSMC) using the 0.18 μm silicon-on-insulator (SOI) standard technology.

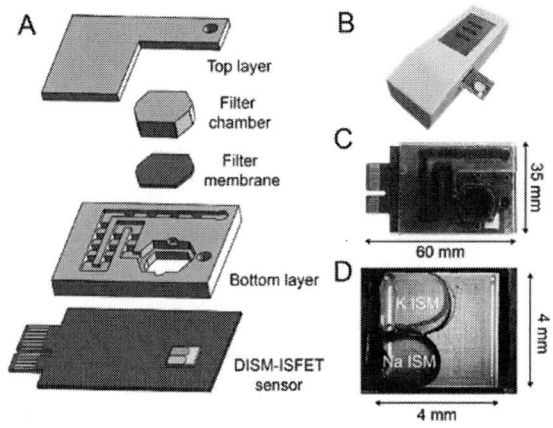

Figure1: (A)The schematic of the DISM-ISFET system (B) Photograph of the portable reader to record and process Ids signal. (C) The photograph of the DISM-ISFET device. (D) Microscopic images of the DISM-ISFET sensor deposited Na and K ISM.

The operational procedure of the DISM-ISFET system

The PMMA microchamber was assembled from three regions: (1) RBCs sedimentation region, (2) filtration region, and (3) detection region (Figure 2A). To accelerate RBCs sedimentation time, we coated anti-D on the surface of each trench (13 trenches in total). The porous filter membrane (0.22 μm pore size, hydrophilic PVDF, Durapore®) was placed at the end of the channel to further enhance serum purity. The operational procedure of the DISM-ISFET system includes six steps, shown in Figure 2B. First, 500 μL of whole blood was injected from the inlet into the RBCs sedimentation region and waited 3 min for RBCs sedimentation (step 1). Next, 200 μL buffer ($10^{-1.5}$ M NaCl + 10^{-3}M KCl) was injected into the detection region from the outlet and measured 3-min Ids as the signal baseline (step 2). Then, the buffer was withdrawn from the outlet (step 3). Next, the filter chamber was attached to the bottom layer to seal the filtration region. The air was then introduced from the inlet and pushed the purified serum into the filtration region (step 4). Then, the serum was withdrawn and passed through the filter to the detection region (step 5). A significant Ids signal fluctuation was shown during the liquid/gas transition step (steps 3 and 5). Once the serum was fulfilled in the detection region, Ids from Na/K ISM deposited ISFFET sensor were simultaneously recorded for another 3 min (step 6). The complete processes including on-chip serum purification and in situ Na+/K+ ion concentration measurement can be done in a single device in 10 min.

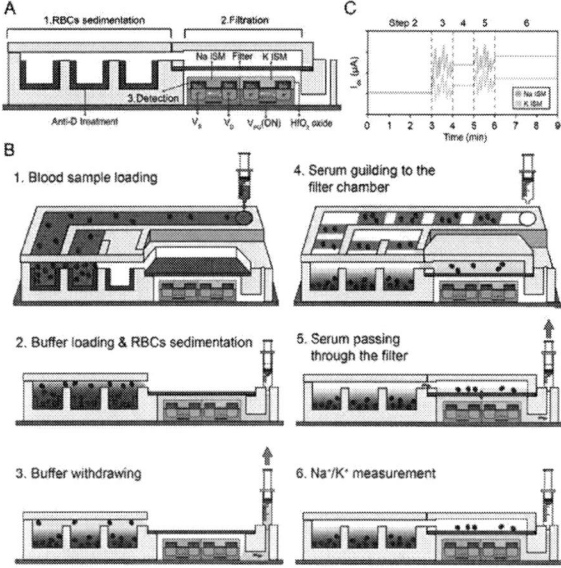

Figure 2: (A) The cross-sectional view and (B) operational procedure of the DISM-ISFET device. (C) The continuous Ids profile in steps 2 to 6 represents in situ Na+/K+ ions sensing features.

RESULTS AND DISCUSSION
The sensing performance of the ISFET sensor

Before performing specific ion measurements, we first evaluate the pH sensitivity of the bare ISFET sensor without ISM deposition. Here, we prepared three pH solutions (pH = 8, 7, and 6) and sequentially dropped 200 μL of each solution on the ISFET sensor for 3-min I_{ds} sensing, followed by the rinse step of the next solution twice. The continuous I_{ds} profile was recorded by the portable reader, shown in Figure 3A. To quantify the ion sensitivity, $I_{ds, initial}$ at the initial point served as the signal baseline and subtracted from the averaged $I_{ds, signal}$ at last 1 min to obtain ΔI_{ds}. In each sample, a complete concentration round was repeated three times to get the average ΔI_{ds} value with the standard deviation. As shown in Figure 3B, the pH (H^+) sensitivity of the bare ISFET sensor is 3.13 μA/pH with high linearity and stability. Next, to evaluate the bare ISFET sensitivity to specific ions, we prepared five sodium chloride (NaCl) concentrations ($10^{-1.5}$, $10^{-1.25}$, 10^{-1}, $10^{-0.75}$, and $10^{-0.5}$ M, equivalent to 31.6, 56.2, 10, 177.8, 316.2 mmol/L), and five potassium chloride (KCl) concentrations (10^{-3}, $10^{-2.75}$, $10^{-2.5}$, $10^{-2.25}$ and 10^{-2} M, equivalent to 1, 1.78, 3.16, 5.62, and 10 mmol/L). The concentration range is determined based on the normal Na^+ (135 - 145 mmol/L) and K^+ (3.5 – 5.5 mmol/L) levels in the blood. If the sodium and potassium level are below or above the normal range, it can be considered as hypokalemia (low sodium), hypernatremia (high sodium), hypokalemia (<3 mmol/L, low potassium), or hyperkalemia (>6.0 mmol/L, high potassium). To evaluate Na^+ and K^+ sensitivity and specificity, 200 μL of five NaCl and KCl solutions were sequentially dropped on the ISFET sensor and continuously recorded the I_{ds} profile, accordingly (Figure 3C). As shown in Figure 3D, ΔI_{ds} did not change along with the NaCl and KCl concentration, indicating a low Na^+/K^+ ion sensitivity and specificity. The above results indicate that the ISM deposition on ISFET is required to detect specific ion concentrations.

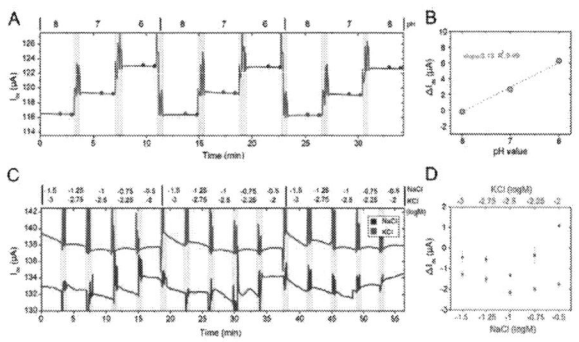

Figure 3: (A) The continuous I_{ds} profile with sequentially loading of three pH solutions (pH 8, 7, 6). (B) Corresponded ΔI_{ds} of pH solutions. (C) The continuous I_{ds} profile with sequentially loading five NaCl solutions ($10^{-1.5}$, $10^{-1.25}$, 10^{-1}, $10^{-0.75}$, $10^{-0.5}$ M) and five KCl solutions (10^{-3}, $10^{-2.75}$, $10^{-2.5}$, $10^{-2.25}$, 10^{-2} M). (D) Corresponded ΔI_{ds} of NaCl and KCl solutions.

ISM deposited ISFET performance for a mixed solution

To perform the signal compensation of Na and K ISM, we prepared three samples: (1) pure NaCl (2) pure KCl, and (3) mixed ion solution with NaCl and KCl at different concentrations. Then, we deposited 2 μL Na and K ISM solution at a 0.6 mm gap on the top and bottom half of the ISFET sensor, respectively. Based on ΔI_{ds} readout on the Na/K ISM, we can then generate the calibration curve of specific ions. As shown in Figure 4, we recorded the continuous I_{ds} profile and ΔI_{ds} of pure NaCl (red curve), KCl (blue curve), and mixed solution (purple curve) at different NaCl/KCl concentrations in Na ISM (Figure 4A and B) and K ISM (Figure 4C and D), respectively. For the Na ISM case, ΔI_{ds} of NaCl and the mixed solution were almost identical, indicating a minimum effect of K^+ ion in the mixed solution (Figure 4B). Instead, ΔI_{ds} of the mixed solution were higher than NaCl and KCl solution, indicating the signal was contributed by both Na^+ and K^+ ions (Figure 4D). Based on the calibration curve of three solutions in Na and K ISM, we can get the correlation of ΔI_{ds} with Na^+ and K^+ ions as follow:

$$\Delta I_{ds, \text{Na ISM}} = 8.09\,([Na^+] + 1.5) + 0.01\,([K^+] + 3) \quad (1)$$
$$\Delta I_{ds, \text{K ISM}} = 1.4\,([Na^+] + 1.5) + 4.58\,([K^+] + 3) \quad (2)$$

The above two equations allowed us to quantify Na^+ and K^+ levels in the clinical samples based on $\Delta I_{ds, \text{Na ISM}}$ and $\Delta I_{ds, \text{K ISM}}$.

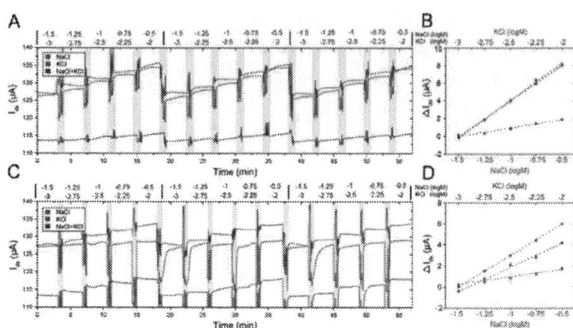

Figure 4: (A) Continuous Ids profiles and (B) ΔIds of standard NaCl, KCl, and mixed solution in Na ISM. (C) Continuous Ids profiles and (D) ΔIds of standard NaCl, KCl, and mixed solution in K ISM.

Beads trapping efficiency

We aim to evaluate the RBC trapping efficiency of the microchamber. To simulate red blood cells in the trapping process, we applied 6μm size beads modeling the RBCs and quantified the capture efficiency of the trench array and filter membrane by calculating the number of beads at the area of 0.12 mm². First, we injected 500 μL of beads solution at 10^8 mL^{-1} into the channel for 3 min sedimentation. After sedimentation and filtration, 4.5 μL of suspended microbeads were collected from different regions (inlet, filter, and outlet) and loaded into the other channel for shooting (Figure 5). From the counting results, we know that the trapping efficiency ($P_{\text{trapping efficiency}}$) of the trench array and filter membrane is about 54.1 % and 45.9 %. Although there were still some suspended microbeads flowing to the filter membrane, after the second interception, the trapping efficiency could be nearly 100%. The trapping efficiency was calculated by equation (3). Therefore, we can verify that our device could avoid RBCs escaping from the interspace of the trench array and the filter membrane to interfere with the sensing result.

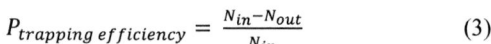

$$P_{\text{trapping efficiency}} = \frac{N_{in} - N_{out}}{N_{in}} \quad (3)$$

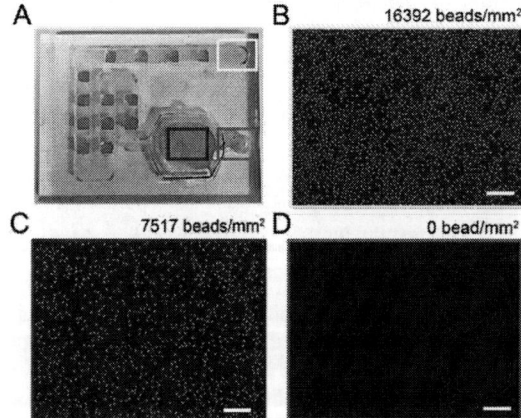

Figure 5: (A) Photograph of the PMMA microchamber filled with 6 μm beads at 10^8 beads/mL; the fluorescent images of beads seeding at (B) the inlet, (C) the top of the

filter membrane, and (D) the outlet. The scale bar is 50 µm.

On-chip blood ions measurement using the DISM-ISFET system

Once the off-chip serum extraction and filter membrane trapping efficiency was confirmed, whole blood processing and ion concentration detection with multiple ISMs and microchamber integrate trench and filter were then conducted. To match the calibration curve converting the current to ion concentration, the calibration solution we injected at the beginning of each whole blood test was the same solution used in the previous experiment that optimized the thickness of the membrane as a baseline value. So far, we have performed 10 clinical samples using our DISM-ISFET platform. (Figures 6A and B) and compared the result with the commercial Na/K meter (Figure 6C), and the overall error is under 10% for Na ISM-ISFET and 15% for K ISM-ISFET.

Therefore, we proved that this portable detection system could perform in-situ blood Na^+/K^+ ion concentration measurements within 10 min.

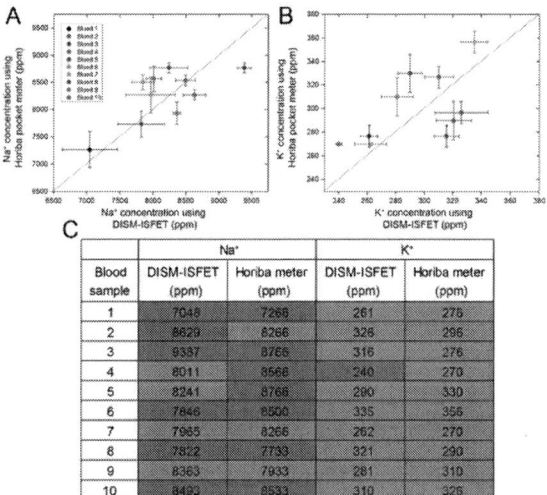

Figure 6: On-chip blood ions measurement results using DISM-ISFET system and Horiba pocket Na/K meters. The corresponded (A) Na+ ion and (B) K+ ion comparison of two devices. Each point represents an individual clinical blood sample. (C) detailed Na+ and K+ ion values of ten clinical blood samples. Red box and green box indicate abnormal and normal value, respectively.

CONCLUSION

In this paper, we demonstrated a DISM-ISFET system consisting of a microchamber for on-chip serum extraction from whole blood sample and a dual ISM deposited ISFET sensor for in situ Na+ and K+ measurements. Overall, three main features were demonstrated. First, the ISM thickness can be precisely adjusted to achieve dual ISM deposition and enable the best ion sensitivity and selectivity. Second, a microchamber embedded with anti-D-coated trench array and filter membrane enables a rapid (3 min) whole blood processing with an excellent RBC removal rate and low hemolysis. Finally, the above features were validated by 10

clinical whole blood samples and compared with the commercial Na/K pocket meter.

Compared to the commercial ion meter, the DISM-ISFET system can simultaneously measure two blood ions with a similar ion sensing performance. Furthermore, the system can directly process clinical whole blood samples without any sample pretreatment processes, which can simplify the whole blood ions sensing procedures and lessen the user's operational difficulty. Based on the above features, we envision that this platform could apply in an intensive care unit (ICU) or point-of-care (POC) environment in the future.

ACKNOWLEDGEMENTS

This work was supported by the Ministry of Science and Technology, Taiwan, under the grant "MOST 110-2221-E-002-009-MY3" and Taiwan Semiconductor Manufacturing Company (TSMC) under the University JDP project.

REFERENCES

[1] R. M. Reynolds, P. L. Padfield, and J. R. Seckl, "Disorders of sodium balance," BMJ, vol. 332, no. 7543, p. 702, 2006, doi: 10.1136/bmj.332.7543.702.

[2] W. E. Much and C. S. Wilcox, "Disorders of body fluids, sodium and potassium in chronic renal failure," The American Journal of Medicine, vol. 72, no. 3, pp. 536550,1982/03/01/1982,doi:https://doi.org/10.1016/0 002-9343(82)90523-X.

[3] E. P. Philippe Bu¨hlmann, and Eric Bakker, "Carrier-Based Ion-Selective Electrodes and Bulk Optodes. 2. Ionophores for Potentiometric and Optical Sensors," Chem. Rev., 1998.

[4] D.Patschan and G. A. Muller, "Acute kidney injury,"J Inj Violence Res, vol. 7, no. 1, pp. 19-26, Jan 2015,doi: 10.5249/jivr.v7i1.604.

[5] F. Locatelli, V. La Milia, L. Violo, L. Del Vecchio, and S. Di Filippo, "Optimizing haemodialysate composition," Clin Kidney J, vol. 8, no. 5, pp. 580-9, Oct 2015, doi: 10.1093/ckj/sfv057.

[6] D. W. C. Melissa K. Tuck, David Chia,Andrew K.Godwin,William E. Grizzle,, W. R. Karl E.Krueger, Martin Sanda,O Lynn Sorbara, Sanford Stass,Wendy Wang,, and a. D. E. Brenner, "Standard Operating Procedures for Serum and Plasma Collection: Early Detection Research Network Consensus Statement Standard Operating Procedure Integration Working Group," ACS Journal of Proteome Research, 2009.

[7] C.-Y. Hsieh and N.-T. Huang, "A proton-selective membrane (PSM)-deposited dual-gate ion-sensitive field-effect transistor (DG-ISFET) integrating a microchamber-embedded filter membrane for bacterial enrichment and antimicrobial susceptibility test," Sensors and Actuators B: Chemical, vol. 359, 2022, doi:10.1016/j.snb.2022.131580.

CONTACT

Corresponding Authors:
*nthuang@ntu.edu.tw; +886-2-33661775

STRONG MICROSTREAMING FROM A PINNED OSCILLATING MEMBRANE AND APPLICATION TO GAS EXCHANGE

Anthony L. Mercader and Sung Kwon Cho*

University of Pittsburgh, Pittsburgh, Pennsylvania, USA

ABSTRACT

We present a novel configuration to generate strong acoustic streaming vortices by a pinned oscillating membrane in a microchannel, its characterizations via advanced measurement techniques, and initial studies in application by augmentation of gas exchange across a permeable membrane towards microfluidic artificial lung technology. The configuration is stable over time and does not create any obstruction in flow passages. For an audible-frequency 20 V_{pp} input to a piezo buzzer, streaming velocity was measured up to 47 mm/s. Mixing from helical flow patterns in the microchannel augments gas transfer rate across the membrane up to 3.4 times compared to no actuation, allowing larger channel dimensions (better facilitation of scale-up manufacturing) and reduced shear (more hemocompatible) in microchannel-based artificial lung systems.

KEYWORDS

Acoustofluidics, Microstreaming, Artificial Lung

INTRODUCTION

Acoustic microstreaming in microfluidic devices is often used as a means to disrupt the typically laminar flows seen at the micro-scale for various purposes such as mixing and propulsion [1-3]. In this phenomenon, an oscillating acoustic input can interact with certain features in a flow field to generate time-averaged flows, typically in the form of vortices. It overcomes the limits of low Reynolds number flows due to generation of high local velocity which allows inertial forces to become significant in relation to viscous forces [4, 5].

Many works studying this phenomenon use ultrasound as the input [6, 7] since the effect can scale with frequency due to higher oscillatory velocity [8]. The work here, however, focuses on audible frequency effects where there exists a more limited number of transduction methods to generate the flows because the acoustic wavelength is much longer than any characteristic dimension in the system, meaning one cannot rely on any mechanism which occurs at the scale of that wavelength. Despite the scaling of the strength of the effect, acoustic streaming at audible frequency can carry some benefits including being cheaper to generate with commercial actuators without requiring precise design of an interdigitated electrode, for example. Among the methods which still work at audible frequency is bubble microstreaming, where a gas bubble resonates and the streaming flow is driven by oscillation of the liquid gas interface [9]. Though the effect is strong, it suffers critical disadvantages including stability due to growth of dissolution of the bubbles and limited applicability to medical devices where contact of bubbles with blood is highly undesirable. More recently discovered is sharp edge streaming where acoustic velocity oscillations in the presence of a sharp tip generates a pair of counter rotating vortices as fluid particles driven in the direction of that tip are directed away by centrifugal effects due to the sharp radius of curvature [10]. While this phenomenon does not suffer from the stability issue, the strength of the effect can be lower and obstruction in a microchannel is a new disadvantage.

This paper presents a novel configuration to generate acoustic microstreaming flows at audible frequency by a pinned, oscillating membrane. A microchannel is formed with polymethylsiloxane (PDMS) as the bottom wall, which is bonded to a glass substrate so that the channel 'hangs' off exposing the membrane to open air. When an acoustic wave is sent through the rigid glass substrate, it is focused on the glass edge creating strong flexural oscillations and thus strong vortex flow patterns above the membrane (Fig 1 (a)).

The focused application here is the microfluidic artificial lung, a device which seeks to oxygenate and/or decarbonate blood by blood and gas microchannels separated by a gas permeable PDMS membrane [11]. Applying active mixing by microstreaming enables both better gas transfer performance and larger channel sizes, greatly reducing shear and potential clotting response in blood. Using the newly described microstreaming concept, a gas transfer device is designed which shows up to 3.4 times improvement in gas exchange efficiency compared to no acoustic actuation.

EXPERIMENTAL
Fabrication

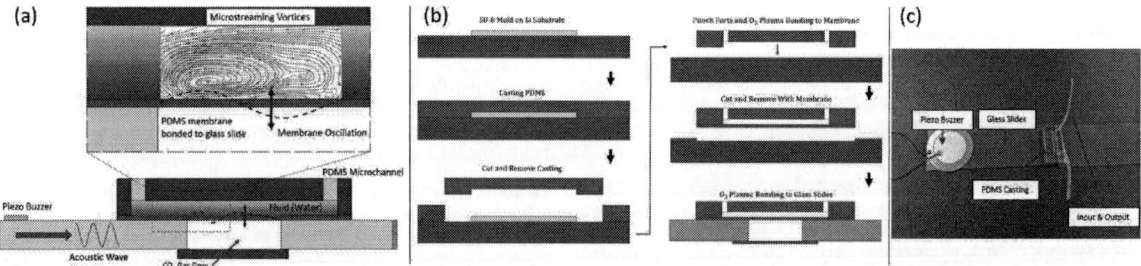

Figure 1: (a) Schematic diagram of the new configuration for microstreaming by membrane; (b) Fabrication process flow; (c) Sample fabrication.

Fabrication details for the base configuration as shown in Fig 1 (b) are as follows. The standard soft lithography process is used to create a mold in the shape of the microchannel. First, a spin coater is used to spread SU-8 2075 negative photoresist in an even layer on top of a silicon wafer up to a maximum of 300 μm depending on spin speed. The resin is cured on a hot plate. The process is repeated in multiple layers for channel heights above 300 μm. The photoresist is exposed to UV light using a mask aligner in an area defined by a photolithography mask in the shape of the top view of the channel. The design used for the initial characterization of the flow to be described used a channel cross section of 600 x 600 μm and length around 1 cm. The exposed photoresist is submerged in SU-8 developer chemical to remove the area of SU-8 which was not exposed to UV light, leaving a negative mold of the channel. PDMS elastomer base and curing agent are mixed in the standard 10:1 ratio, poured over the mold, and cured. The casting is then cut and removed from the mold by hand and the inlet and outlet holes are created by a 1 mm biopsy punch.

The membrane is fabricated by spin coating PDMS in a 20 μm layer curing as mentioned above. The bottom face of the channel cutout and the membrane are then chemically bonded using a handheld corona air plasma generator. The plasma head is held above the surfaces to be bonded together for 1 minute each, then the channel cutout is pressed onto the membrane surface and left overnight. The membrane with bonded channel cutout is then cut and peeled away from the Si substrate with a razor blade. The bonding process is then repeated as described to bond the underside of the membrane between two glass slides with a desired gap.

23 gauge PTFE tubing is inserted into the inlet and outlet holes and the channel is filled with liquid by a syringe. A piezo buzzer element is attached to one of the glass slides by epoxy to transmit the acoustic waves given a sinusoidal voltage input from a function generator and amplifier. An image of a completed device of this type is shown in Fig 1 (c) The system is viewed with a microscope and high speed camera to confirm the streaming flows. The frequency input is swept to find the frequency which leads to the strongest streaming flow for each sample fabricated, typically between 5 and 7.5 kHz due to sample-to-sample variation.

Laser Doppler Vibrometry (LDV)

In order to characterize its oscillation pattern and investigate the driving mechanism behind the streaming flows, LDV measurement was performed on the section of the membrane exposed to open air. LDV is a measurement technique which shines a laser onto the sample and measures the reflection in such a way that it can characterize the magnitude and phase over a specified frequency spectrum. Because the measurement requires a reflective surface, a slightly different PDMS membrane is fabricated with two layers of spin coating for the same total thickness and a 50 nm layer of Ti in sandwiched in the middle deposited by e-beam evaporation, thin enough to have minimal effect on the oscillation characteristics. The underside of the membrane is aligned with the view of the microscope objective and a grid of measurement points is

specified over the area of interest. The oscillation is measured on the range of 2.5 to 20 V_{pp} at the resonance frequency of 5.5 kHz. Displacement magnitude and phase at each measurement point are exported for later use.

Particle Image Velocimetry (PIV)

Micro PIV (μPIV) is used to measure the velocity field of the streaming effect. Laser light is guided through a filter cube and microscope objective to excite fluorescent particles seeded in the fluid field. Two images are captured with a specified time delay and correlated by computer software to track the particle motions and calculate a velocity field. In this particular case, the light is guided by a right-angle prism to capture a side view of the streaming, and the time delay was specified at integer multiples of the acoustic period to cancel any phase effects. Measurements were performed on the same input range as with LDV.

Computational Fluid Dynamics (CFD) Simulation

CFD simulation using the CFX solver in ANSYS is performed to confirm the membrane oscillation as the driving mechanism of the streaming flows. Geometry was created to match the dimensions of the fabricated channels. The inlet and outlet were left open, and the top and side walls were rigid, no-slip walls. The membrane oscillation magnitude and phase from LDV measurement were applied as a boundary motion to the bottom wall. The transient simulation is performed with 64 time steps per period of oscillation for a number of periods until the solution between periods does not vary. The streaming flow is visualized by calculating the time average of the velocity field over all the time steps in the final period.

Gas Exchange

For gas exchange, a separate channel geometry was designed so that the glass edge is aligned with the channel length so that the axis of rotation of the vortices is aligned in the same direction and a streamwise flow leads to helical patterns. The new channel width is 1.6 mm and the height varied from 250 μm to 1 mm. A PDMS slab is bonded to

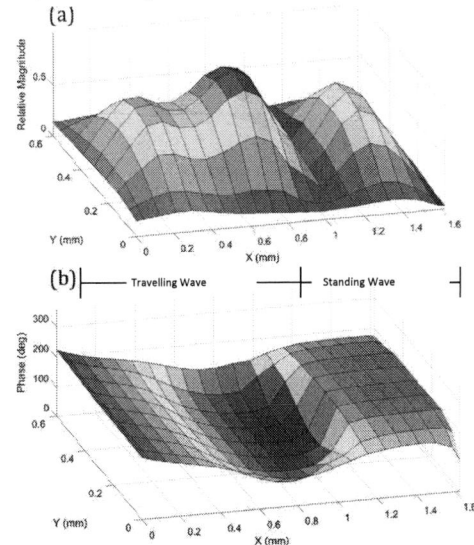

Figure 2: Experimental measurement of membrane oscillation by Laser Doppler Vibrometry (LDV): (a) magnitude and (b) phase.

the underside of the substrate to seal a gas channel. Deionized (DI) water is input by a syringe pump at flow rates of 0.2 and 0.4 mL/min and CO_2 is input by a separate syringe pump to the gas channel at a constant 0.4 mL/min. The pH of the output is measured by a pH meter with a micro pH electrode to characterize the CO_2 transfer into the liquid channel.

RESULTS AND DISCUSSION

Membrane Oscillation

As seen from the magnitude and phase plots in Fig 2, the observed pattern is a superposition of travelling and standing waves starting at the glass edge from which the acoustic waves originate. The travelling wave component dominates near that same edge and gives way to a standing wave dominant portion close to the opposite glass edge, The largest velocity gradient and magnitude appear nearest the primary glass edge, which corresponds to the main vortex and fastest velocity observed in the streaming flow field. On the 2.5 to 20 V_{pp} range of input voltage, the maximum displacement amplitude in the membrane ranged from 2.3 µm to 18.5 µm with approximately linear scaling with voltage.

Velocity Field Measurement

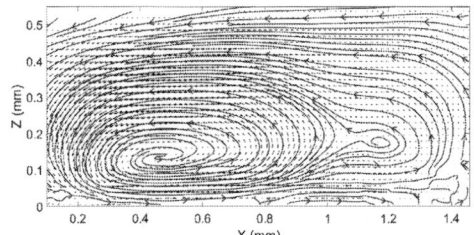

Figure 3: Experimental measurement of velocity fieled and streamline by Particle Image Velocimetry (PIV) (1.5 mm membrane length).

The PIV data give a clear picture of the streaming patterns. For the representative configuration presented in Fig 3, primary vortex with maximum velocity at the bottom is seen corresponding to the leading glass edge and first membrane amplitude peak. A second, corotating vortex is seen, and a third, counterrotating vortex is observed visually outside of the PIV field of view. Maximum velocity ranged from 0.41 mm/s to 47 mm/s on the same input voltage range.

CFD Simulation and Comparison

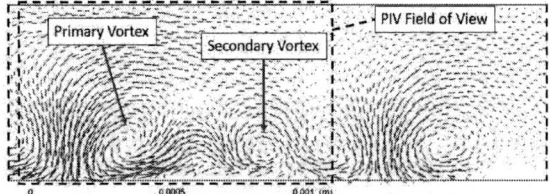

Figure 4: CFD results given membrane oscillation as input.

With the displacement magnitude and phase from the LDV data applied to CFD, the simulation results show good agreement in pattern and magnitude to the PIV results, providing evidence that the mechanical oscillation

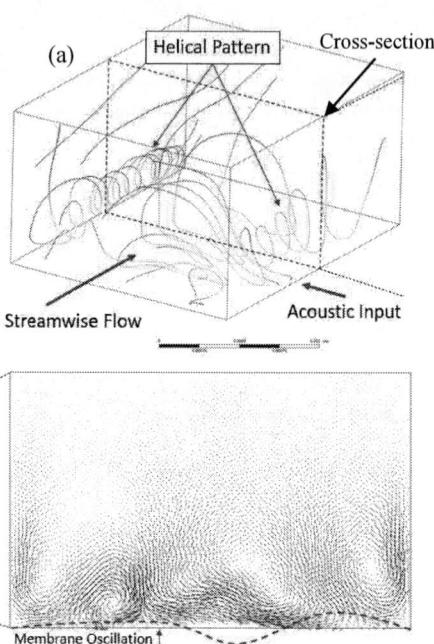

Figure 5: CFD simulation with infused streamwise flow shows (a) representative helical streamlines and (b) cross-sectional velocity field. Note that the membrane oscillating wave is in the cross-channel direction.

of the membrane is the driving mechanism of the acoustic streaming vortices. Especially for the representative 1.5 mm membrane length case, the whole three-vortex system is correctly predicted by simulation, including the fine detail of the very slow corotating vortex within the primary vortex, as seen in Fig 4.

The input amplitude of displacement for the simulation is related to the input voltage in the PIV experiment through the LDV results. In both cases tested, the velocity was predicted well, though slightly higher than experiment by close to 15% at higher input amplitudes. The scaling was approximately quadratic just as in experiment.

A similar simulation was performed on the previously mentioned perpendicular orientation using the centerline LDV magnitude and phase applied as a constant profile along the streamwise direction. The cross-sectional velocity profile appeared similar to the velocity field before, as expected considering the same membrane oscillation profile was used, just turned 90°. Adding channel flow as a constant velocity at the inlet and outlet shows the helical streamlines as desired as shown in Fig 5.

Augmentation of Gas Exchange

The data for each liquid flow rate are presented in Fig 6. Comparing the channel heights, at no actuation, CO_2 transfer through the permeable membrane was greater or equal for the shorter channel at any flow rate. This is in line with expectation due to the fact that the taller channel will tend to be more diffusion limited, although there is the competing effect of lower velocity and therefore larger residence time of the liquid over the membrane at a given flow rate with the taller channel. At any level of actuation applied, the taller channel outperforms the shorter channel, up to 3.4x compared to no actuation. This is in line with the original goal for the experiment, demonstrating the dual benefit of this method: the gas transfer performance is increased outright and favors a larger channel height which may lead to increased hemocompatibility due to highly reduced shear. Comparing the flow rates, output concentration decreased with increasing flow rate, though the total volume output of liquid at that concentration is increased leading to potentially higher magnitude of gas transfer. Due to these competing effects, the optimal flow rate will likely depend on the needs of the target application.

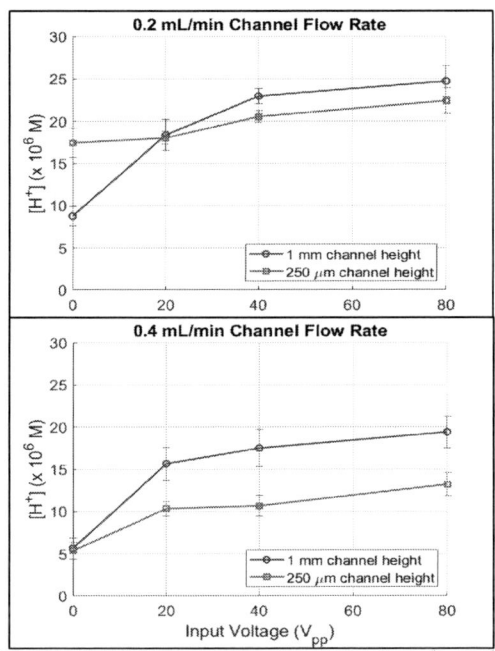

Figure 6: CO_2 concentration in microchannel by measuring pH for probing gas transfer rate at two water flow rates (CO_2 gas flow rate: 0.4 ml/min).

CONCLUSION

A novel configuration to generate audible frequency acoustic streaming flows in a microchannel is presented. A PDMS microchannel with a membrane as the bottom wall is bonded to a glass substrate extending over the edge so that acoustic input sent through the substrate is focused into a flexural oscillation in the membrane, which drives streaming flows in the form of vortices in the channel. The membrane oscillation and streaming flow field were measured by LDV and PIV, respectively, and CFD simulation using the LDV results as a boundary condition shows good agreement, indicating that the membrane oscillation is the driving mechanism and the shorter wavelength of that mechanical vibration compared to the acoustic wavelength allows the phenomenon to operate at audible frequency. Due to the vertical orientation of the mixing and requirement of membrane interface with open air, gas exchange towards microfluidic artificial lung technology is seen as the target application. A gas exchange device was fabricated with a gas channel formed on the underside of the membrane. With DI water input to the

liquid side and CO_2 input to the gas side, pH measurement characterizes the magnitude of CO_2 transfer into the liquid side. Results show a taller channel outperforming a shorter channel at any level of actuation tested demonstrating a dual benefit of increased gas transfer efficiency and allowing for larger dimensions for reduced shear. Such a concept may eventually benefit blood flow on the liquid side.

ACKNOWLEDGMENTS

This work was supported in part by the NSF grant (ECCS-1951051).

REFERENCES

[1] J. Feng, J. Yuan, and S. K. Cho, "Micropropulsion by an acoustic bubble for navigating microfluidic spaces," *Lab on a Chip,* vol. 15, no. 6, pp. 1554-1562, 2015.

[2] J. Feng, J. Yuan, and S. K. Cho, "2-D steering and propelling of acoustic bubble-powered microswimmers," *Lab on a Chip,* vol. 16, no. 12, pp. 2317-2325, 2016.

[3] F.-W. Liu and S. K. Cho, "3-D swimming microdrone powered by acoustic bubbles," *Lab on a Chip,* vol. 21, no. 2, pp. 355-364, 2021.

[4] S. Sadhal, "Acoustofluidics 13: Analysis of acoustic streaming by perturbation methods," *Lab on a Chip,* vol. 12, no. 13, pp. 2292-2300, 2012.

[5] M. Wiklund, R. Green, and M. Ohlin, "Acoustofluidics 14: Applications of acoustic streaming in microfluidic devices," *Lab on a Chip,* vol. 12, no. 14, pp. 2438-2451, 2012.

[6] M. K. Tan, L. Yeo, and J. Friend, "Rapid fluid flow and mixing induced in microchannels using surface acoustic waves," *Europhysics Letters,* vol. 87, no. 4, p. 47003, 2009.

[7] G. Destgeer, S. Im, B. H. Ha, J. H. Jung, M. A. Ansari, and H. J. Sung, "Adjustable, rapidly switching microfluidic gradient generation using focused travelling surface acoustic waves," *Applied Physics Letters,* vol. 104, no. 2, p. 023506, 2014.

[8] J. Friend and L. Y. Yeo, "Microscale acoustofluidics: Microfluidics driven via acoustics and ultrasonics," *Reviews of Modern Physics,* vol. 83, no. 2, p. 647, 2011.

[9] P. Tho, R. Manasseh, and A. Ooi, "Cavitation microstreaming patterns in single and multiple bubble systems," *Journal of Fluid Mechanics,* vol. 576, pp. 191-233, 2007.

[10] P.-H. Huang *et al.*, "An acoustofluidic micromixer based on oscillating sidewall sharp-edges," *Lab on a Chip,* vol. 13, no. 19, pp. 3847-3852, 2013.

[11] J. A. Potkay, "The promise of microfluidic artificial lungs," *Lab on a Chip,* vol. 14, no. 21, pp. 4122-4138, 2014.

CONTACT

* S.K. Cho, tel: +1-412-624-9798; skcho@pitt.edu

TUNABLE NANOPORE-INTEGRATED MICRO-/NANOFLUIDIC PLATFORM FOR ION TRANSPORT CONTROL IN THE PRESENCE OF CONCENTRATION AND TEMPERATURE GRADIENTS

Dongwoo Seo[1], Dongjun Kim[1], Jongwan Lee[1], Cong Wang[2], Jungyul Park[2], and Taesung Kim[1]
[1]Ulsan National Institute of Science and Technology, KOREA and
[2]Sogang University, KOREA

ABSTRACT

Ion transport at nanoscales plays an important role in the human body, including neural signal transmission. We describe a self-assembled particle membrane (SAPM)-based tunable nanopore-integrated micro-/nanofluidic platform that can manipulate ion transport through nanofabricated nanopores. With facile fabrication and easy tuning of channel characteristics, the platform employs various nanoparticles with different material properties and diameters, making it possible to conduct unprecedented nanofluidic transport experiments in the presence of multi-physical fields conditions. From the experimental results under various conditions, we suggest theoretical models considering the ion transports via diffusioosmosis (DO) and diffusion, which would be useful to understand and control the nanoscopic ion transport phenomena governing neural signaling at low temperature conditions. Lastly, we anticipate that the platform would contribute to understanding the underlying mechanism for the neural signal transmission and its cut-off at low temperature conditions, further optimizing various medical protocols such as cryo-anesthesia.

KEYWORDS

Nanofluidics, Neural Signaling, Diffusioosmosis, Ion transport, Nanofabrication, Nanopore/Nanostructure.

INTRODUCTION

Nanoscopic ion transport plays an important role in various natural systems, including neural signal transmission in the human body [1]. Especially, controlling neural signal transmission and retardation through temperature change can be widely applicable to low-temperature medical procedures such as cryo-anesthesia. However, neural signal retardation and/or cut-off mechanisms have not yet been fully revealed because of the lack of proper nanopore fabrication and experimental quantification techniques. To investigate these phenomena, previous studies rely on the conventional experimental platforms and nanofabrication techniques while applying single physical fields [2] which seem to be inappropriate for the precise measurement of ion transport. To resolve these problems, recent studies have focused on the fabrication and manipulation of the structures with novel unconventional methods and materials, enabling the control of the dimension of nanostructures as well as their properties [3-5]. Although these works have successfully demonstrated the manipulation of both structural and material characteristics, developing simple and facile fabrication methods is still necessary to the perspective of applicability. To tackle these problems, we employ a novel micro-/nanofluidic platform and facile fabrication method

enabling to not only integrate material property-tunable nanopores with different pore-size but also apply multi-physical fields at the same time. With the platform and following results obtained by applying various concentration and temperature conditions to the system, we explain the characteristics of concentration and temperature-dependent ion transport in the nanochannel. Furthermore, we anticipate that the platform would contribute to understanding the underlying mechanism for the neural signal transmission and its cut-off at low temperature conditions, as well as optimizing various low temperature-based medical protocols and treatments such as cryo-anesthesia.

EXPERIMENTAL METHODS

To investigate the ion transport under various conditions, we developed an artificial nanopore-integrated micro-/nanofluidic platform that can partially mimic the neural signaling as illustrated in Figure 1.

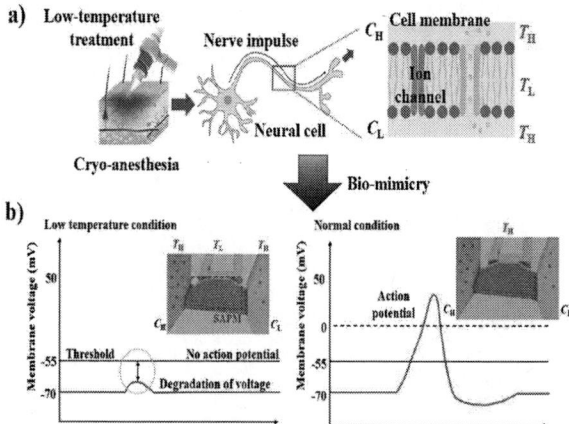

Figure 1: a) Neural signal transmission cut-off at low temperature and its bio-mimicry using the SAPM-integrated micro-/nanofluidic platform. b) Neural signal cut-off due to the ion transport retardation by lowering temperature.

We have developed a SAPM-integrated microfluidic device as shown in Figure 2 in our previous study [6]. Simply, an SAPM is constructed by arranging nanoparticles through evaporation. As described in Figure 2, injected nanoparticles with various properties are used for forming SAPMs, which in turn act as a bundle of nanopores with different electrical and material properties, and pore-sizes. For this work, nanoparticles of 50, 143, and 444 nm diameter are used to modify the pore-size of SAPM. The concentration and temperature gradient are acquired by loading NaCl solution with different concentrations

978-1-6654-9309-3/23 $31.00 © 2023 IEEE

(i.e., higher concentration of C_H and lower one of C_L) to the main channels and using the Peltier-based temperature controller. Electric signals are measured by using Ag/AgCl electrodes and a source meter (Keithley 2635A, Tektronix, Inc., USA) to characterize ion transport along the SAPMs in the presence of various temperature and concentration gradient conditions at the same time.

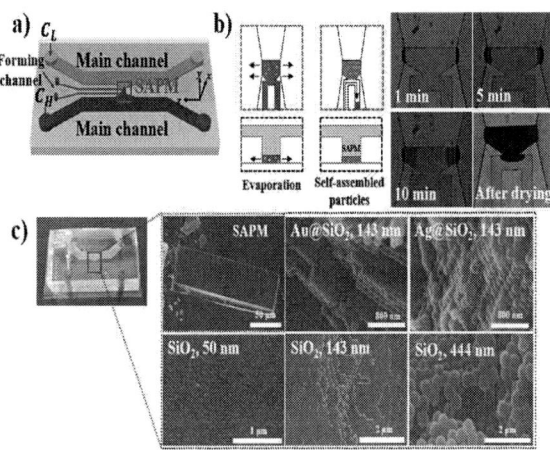

Figure 2: a) Schematic of the SAPM-integrated micro-/nanofluidic platform. b) Fabrication process of the SAPM into microfluidic channel network. c) Various SAPMs with different electrical and material properties, and diameter, determining the resulting zeta potential and pore-size of the nanopores/nanochannels.

THEORETICAL MODELING

Figure 3 shows theoretical models describing the relationship between the thickness of electric double layer (EDL) λ_D (Debye layer) and the pore-size (d_p). As the ion transport in the nanochannel can be divided by the ratios of the channel size (h) to the λ_D, three representative theoretical models can be established. First, for a low concentration, the l_D is longer than the pore-size ($2\lambda_D > d_p$), resulting in the overlapped EDL. In this case, with example of Figure 3a, cations mainly occupy the nanochannel and eventually go through the channel. However, anions cannot pass through the nanochannel due to the electric repulsive force from the wall. Therefore, DO (or Chemiosmosis, CO)-dominant selective ion transport occurs. In addition, ion selectivity can be seen for the low concentration because of the cation-selective ion transport by the negative zeta potential of the SAPMs as mentioned. For a high concentration, the effect of selective ion transport is diminished as the l_D decreases because diffusion gets more dominant. This is because the size of EDL decreases as the concentration increases as illustrated in Figure 3b ($2\lambda_D < d_p$). Lastly, when the nanochannel size and the EDL thickness are at the same order of magnitude ($2\lambda_D \sim d_p$), ion transport is governed by the combined effects of DO and diffusion (Figure 3c).

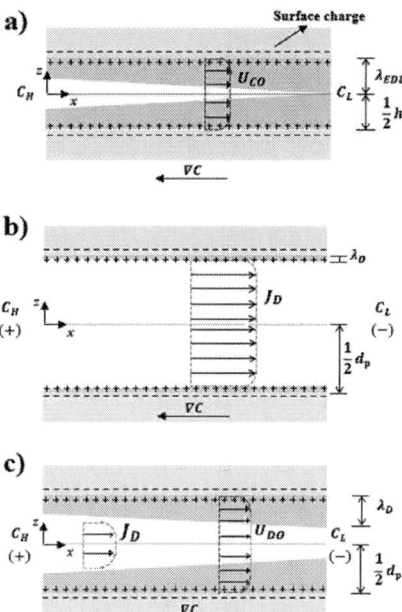

Figure 3: a) DO-dominant ion transport model ($2\lambda_D > d_p$) for a small pore-size and a low concentration. b) Diffusion-dominant ion transport ($2\lambda_D < d_p$) for a large pore-size and a high concentration. c) Ion transport based on the combined effect of DO and diffusion ($2\lambda_D \sim d_p$) for an intermediate pore-size and concentration. U_{DO} and J_D are velocity of DO and diffusion flux, respectively.

RESULTS AND DISCUSSION

The amount of net ion transport and cation selectivity can be explained by not only the nanochannel-based models but also the experimental results quantified by short circuit current (SCC) and open circuit voltage (OCV) as shown in Figure 4. Figure 4a shows the amount of net ion transport depending on the concentration and the pore-size of SAPM. For a low concentration, DO-driven selective ion transport is mainly affected by EDL, so the SCC values are inversely proportional to the pore size of the nanochannel. For a high concentration, the diffusion-dominant ion transport occurs, showing an opposite trend to that of the low concentration as λ_D decreases. Figure 4b shows the cation selectivity through the concentration and the pore-size. For a low concentration, ion selectivity can be seen because of the cation-selective DO by the negative zeta potential of the SAPMs. An opposite trend of ion selectivity can be seen at the high concentration as the effect of EDL decreases and the diffusion-driven ion transport gets dominant. Figure 4c and 4d shows the amount of net ion transport and the ion selectivity depending on the temperature and concentration, respectively. At both concentrations, SCC values increase with the temperature regardless of the concentration. This is because both the DO mobility and the diffusion coefficient are proportional to the temperature. Ion selectivity at the low concentration increases with the temperature because DO is more dominant at the condition, while the change of ion selectivity with the temperature is relatively small for high concentration.

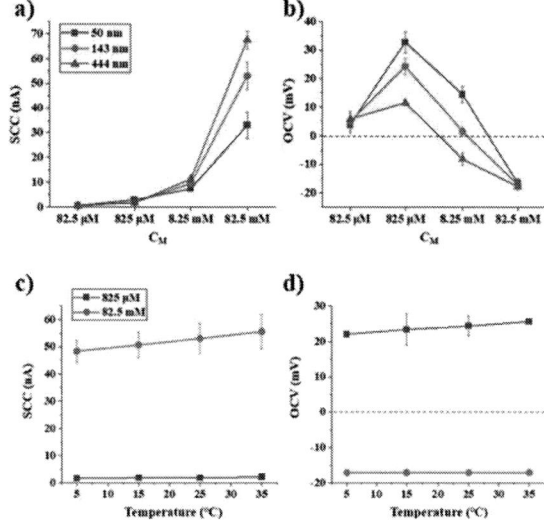

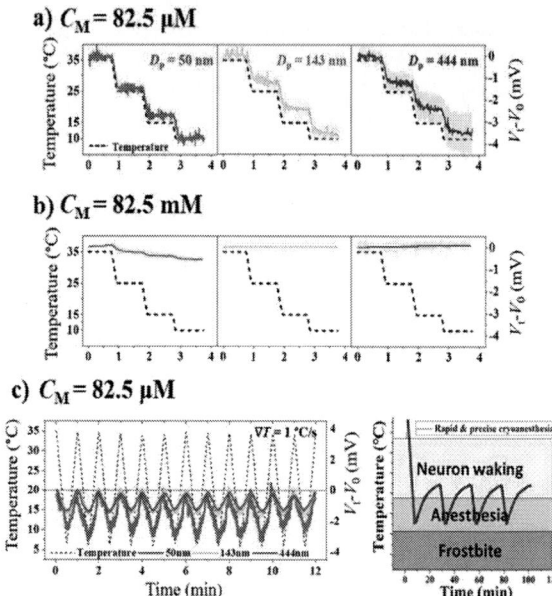

Figure 4: a) SCC and b) OCV results depending on the concentration and the pore-size that is about 15% of the nanoparticle diameters (50, 143, and 444 nm). c) SCC and d) OCV results depending on the concentration and temperature. C_M is the mean concentration, which expressed as $C_M = (C_H + C_L)/2$.

Figure 5 shows the differential OCV values between the reference (V_0 at 37°C) and the cooling temperatures ranging from 35°C to 10°C (V_t). The OCV values are calculated by subtracting the OCV value at reference temperature (i.e., 35°C) from the value at cooling temperature. For the low concentration, the differential OCV values generated at cooling conditions decrease more than those for the high concentration. As DO-driven selective ion transport is dominant for a low concentration, decrease of DO mobility due to low temperature leads to weaken the effect of selective ion transport in the channel, resulting in differential OCV values shown in Figure 5a. In addition, differential OCV values decrease as the pore-size increases, resulting in weakened effect of DO. Figure 5b shows the results for a high concentration. Unlike the values from a low concentration, the differential OCV values are relatively indistinctive regardless of the pore-size at this condition. The results of differential OCV generated by the cooling temperature under dynamic temperature change is shown in Figure 5c. The overall trend of temperature effect on the ion transport and OCV generation is similar to that of static temperature change shown in Figure 5a and 5b. Since the relation between the ion concentration across the neural cell and the size of the ion channel is similar with the low concentration condition in this work [7], the results not only prove that the cryo-anesthesia can be achieved by applying those temperature controls, but also provide an adequate and efficient mechanism for the temperature control which ensures the effect of cryo-anesthesia while avoiding the possible damages such as frostbite. Furthermore, these overall results may imply that the ion transport through the nanochannel decreases with cooling, emulating the decrease in ion transport along the ion channel across cell membranes, which is one of the main reasons contributing to the retardation of neural signal transmission.

Figure 5: Effect of temperature change on ion transport characterized by differential OCV values. a)-b) Differential OCV values from different pore sizes and concentrations under static temperature change. c) Differential OCV values from different pore sizes under dynamic temperature change and low concentration comparing with cryo-anesthesia operation mechanism.

CONCLUSION

We developed a tunable SAPM-integrated micro-/nanofluidic platform enabling the experimental characterization of ion transport through various nanochannels in the presence of concentration and temperature gradients by analyzing generated electric signals. The platform demonstrated that the nanoscopic ion transport is retarded at low temperature conditions, reminiscing the retardation and cut-off of neural signal transmission. We also established theoretical models to support the experimental results by describing the fundamental ion transport mechanisms. Hence, we anticipate that the introduced micro-/nanofluidic platform in conjunction with the theoretical models and overall experimental results would be very useful for not only developing and optimizing low-temperature medical applications including cryo-anesthesia, but also understanding the temperature effect on neural signal transmission through cell membrane.

ACKNOWLEDGEMENTS

This work was supported by National Research Foundation of Korea grants funded by the Korean government (NRF-2020R1A2C3003344 and NRF-2020R1A4A2002728).

REFERENCES

[1] C. C. Moser, J. M. Keske, K. Warncke, R. S. Farid, P. L. Dutton, "Nature of biological electron transfer", *Nature*, vol. 355, pp. 796-802, 1992.

[2] J. Yang, X. Hu, X. Kong, P. Jia, D. Ji, D. Quan, L. Wang, Q. Wen, D. Lu, J. Wu, L. Jiang, W. Guo, "Photo-induced ultrafast active ion transport through graphene oxide membranes", *Nat. Commun.*, vol. 10, pp. 1171, 2019.

[3] G. Liu, W. Jin, N. Xu, "Two-Dimensional-Material Membranes: A New Family of High-Performance Separation Membranes", *Angew. Chem., Int. Ed.*, vol. 55, pp. 13384-13397, 2016.

[4] G. Yang, D. Liu, C. Chen, Y. Qian, Y. Su, S. Qin, L. Zhang, X. Wang, L. Sun, W. Lei, "Stable Ti3C2Tx MXene-Boron Nitride Membranes with Low Internal Resistance for Enhanced Salinity Gradient Energy Harvesting", *ACS Nano*, vol. 15, pp. 6594-6603, 2021.

[5] K. Xiao, P. Giusto, L. Wen, L. Jiang, M. Antonietti, "Nanofluidic Ion Transport and Energy Conversion through Ultrathin Free-Standing Polymeric Carbon Nitride Membranes", *Angew. Chem., Int. Ed.*, vol. 57, pp. 10123-10126, 2018.

[6] J. Lee, K. Lee, C. Wang, D. Ha, G.-H. Kim, J. Park, T. Kim, "Combined Effects of Zeta-potential and Temperature of Nanopores on Diffusioosmotic Ion Transport", *Anal. Chem.*, vol. 93, pp. 14169-14177, 2021.

[7] M. E. Tagluk, R. Tekin, "The influence of ion concentrations on the dynamic behavior of the Hodgkin-Huxley model-based cortical network", *Cogn. Neurodyn.*, vol. 8, pp. 287-298, 2014.

CONTACT

* T.Kim; Phone: +82-42-217-2313; tskim@unist.ac.kr

QUANTITATIVE ASSESSMENT OF CAPTURED MAGNETIC NANOPARTICLES USING SELF-POWERED MAGNETOELECTRIC PLATFORM FOR BIOLOGICAL APPLICATIONS

Pankaj Pathak[1], Vinit K. Yadav[1], Samaresh Das[2] and Dhiman Mallick[1]*

[1]Department of Electrical Engineering, Indian Institute of Technology Delhi, INDIA and
[2]Centre for Applied Research in Electronics, Indian Institute of Technology Delhi, INDIA

ABSTRACT

This work demonstrates a self-powered, Ni/PMN-PT based magnetoelectric platform for rapid assessment of captured magnetic nanoparticles (MNPs) concentration. Injected MNPs are captured by the generated field gradient from Ni thin film, and the optothermal-pyroelectric property of PMN-PT is used to quantitative assess the captured MNPs concentration. Under the incident infrared pulse at zero bias, the device exhibits different photoresponse with varied injected MNPs concentrations. The transient photocurrent shows nearly linear relationship to the injected MNP concentration. The fabricated device exhibits high sensitivity of $0.29 nA.mg^{-1}.ml$, a short response time (< 2 sec), excellent stability and selectivity towards MNPs. Such a self-powered ME platform may open up avenues for new applications in areas such as controlled drug delivery, hyperthermia, cell labelling, etc., where a single device can capture as well as quantitatively assess the captured MNPs efficiently.

KEYWORDS

Self-powered, Magnetoelectric, Magnetic Nanoparticles, Photocurrent, Optothermal, Pyroelectric.

INTRODUCTION

Several attempts have been strived to exploit magnetically labelled cells and magnetic nanoparticles (MNPs) in the lab-on-a-chip devices for various biotechnology applications. These applications include cell delivery, biomedical diagnostics, cell targeting, sensing etc. [1]. In preceding studies, conventional methods incorporating high-power electromagnets, magnetic microstructures and current-based methods are investigated widely for such applications [2]. Unfortunately, these methods are energy-inefficient and spatially inaccurate at the micro and nanoscale. Recently, it has been shown that magnetoelectric (ME) devices can provide a prevalent energy-efficient route in scalable devices without using external, high-powered electromagnets [3]. Although preceding research attempts using ME devices have mainly focused on memory [4] and logic applications [5] only, recently researchers have demonstrated magnetic cell separation, controlled cell capture/release [6] and manipulation of captured MNPs [7] using ME devices with microscale precision.

Despite the great potential, ME devices have limited capability for critical biotechnology applications such as cancer prognosis, drug regulation etc. This is because such applications require a sensitive and precise quantitative assessment of the captured or released MNPs for drug regulation [8], which has not been explored till date using ME devices. This is crucial to extend their scope of applications so that a single device can capture, release, manipulate as well as quantitatively assess the magnetically labelled cells or drug coated MNPs for excellent therapeutic efficacy.

To circumvent this issue, we propose a self-powered ME platform based on a Nickel/PMN-PT (ferromagnet/piezoelectric) heterostructure. MNPs (average diameter= $500nm$) of varying concentrations ($C_{MNP} = 0.0 - 0.8 mg/ml$) are injected on the ME platform at a low flow rate ($0.001mm/sec$). Due to the generated field gradient from Nickel (Ni) thin film, injected MNPs are captured independently without using an external magnetic field. Finally, the optothermal-pyroelectric property of PMN-PT [9] is used to quantitatively assess the captured MNPs concentrations.

DEVICE FABRICATION AND CHARACTERIZATION

The proposed ME platform is fabricated by depositing Cr/Au planar electrodes on top and bottom of the unpoled single-crystal [011]-cut PMN-PT $[Pb(Mg, Nb)O_3 - PbTiO_3]$ substrate ($5mm \times 5mm \times 0.5mm$), followed by Ni thin film ($3mm \times 5mm \times 60nm$) deposition on the top-electrode by magnetron sputtering. X-ray Diffraction (XRD) is used to characterize the crystal structure and phase of the device. Magnetic force microscopy (MFM) is used to characterize the magnetic domain configuration and magnetic state of the Ni thin film in its demagnetization state, i.e. without using an external magnet. To illustrate the cross-sectional morphology of the fabricated device, Field emission scanning electron microscopy (FESEM) is utilized. The transient photocurrent response is recorded using the parameter Analyzer unit (KEITHLEY 4200A-SCS) for the ME platform under IR irradiation ($100W/m^2$) at zero bias voltage. Top and bottom electrodes are used to obtain the photoresponse of the device under an exposed IR pulse. To quantitatively assess the captured or released MNPs using the ME platform, iron oxide (Fe_3O_4) MNPs with varying concentrations ($C_{MNP} = 0.0 - 0.8 mg/ml$) are utilized. Fe_3O_4 MNPs are considered as they have high drug loading capacity and are biocompatible, chemically stable, non-toxic, and inexpensive [10]. The underlying physics of the MNP capture and magnetic field distribution using Ni thin film are envisaged using an analytical model. In an analytical model, spherical Fe_3O_4 MNP is considered, which has a density $5000kg/m^3$, volume susceptibility $800kAm^{-1}T^{-1}$ and saturation magnetization $4.78 \times 10^5 A/m$, same as the MNPs used in experiments. Water is used as the fluid in the experiment, which is nonmagnetic with density $1000kg/m^3$, permeability $4\pi \times 10^{-7} H/m$ and viscosity $0.001kg/ms$.

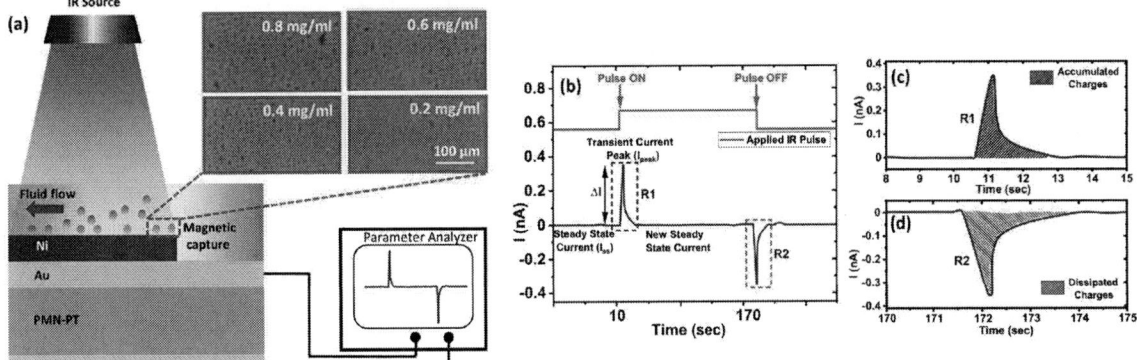

Figure 1: (a) Conceptual illustration of the self-powered magnetoelectric platform for capturing and sensing magnetic nanoparticles (Not to scale). Inset shows the microscopic image of captured magnetic nanoparticles of various concentrations by Ni magnet at one fixed edge. (b) Transient photocurrent recorded for magnetoelectric platform in absence of magnetic nanoparticles injection. When Infrared (IR) pulse is switched on, a sharp positive transient current peak (I_{peak}) is observed (region R1) before achieving a steady state photocurrent. Once IR pulse is switched off, a negative transient current peak is observed (region R2) before reaching to the new steady state. (c) and (d) illustrating accumulated and dissipated charges due to positive and negative transient photocurrent at region R1 and R2, respectively.

WORKING PRINCIPLE

Capturing of Magnetic Nanoparticles

Several forces govern the capturing of injected MNPs, which include magnetic force due to deposited magnetic film, viscous drag force that depends on fluid viscosity and injected velocity of the MNPs, surfactant force, gravitational, inertia force on MNPs etc. However, the magnetic force on MNP due to deposited magnetic film (F_m) and viscous drag force (F_d) are the dominant contributions of the total force, and other forces are negligible [11]. F_m is given as

$$F_m = \mu_o V[M_p . \nabla]H_m \qquad (1)$$

Where μ_o, V and M_p are free space permeability, MNP volume and magnetization respectively. H_m is magnetic field generated by the deposited magnetic thin film. F_d is given by

$$F_d = -6\pi\eta r_p(v_p - v_f) \qquad (2)$$

Where η and v_f are the fluid viscosity and velocity respectively. r_p and v_p are radius and injected velocity of the MNPs, respectively. For MNPs injected in non-flow regime, fluid velocity (v_f) is 0mm/s. Thus, for fixed fluid viscosity ($\eta=0.001 kg/m-s$) and average MNP radius ($r_p = 250nm$), $|F_d|$ depends on injected velocity of MNP (v_p) only, as described by Equation (2). To capture the MNPs, magnetic force on MNP due to deposited magnetic film ($|F_m|$) should surpass the drag force ($|F_d|$), i.e. $|F_m| > |F_d|$. This condition predicts that as long as F_m surpasses F_d, injected MNPs bind to the magnetic thin film, as shown in Fig 1(a). Inset of Fig. 1(a) shows the microscopic image of captured MNPs of various concentrations by Ni nanomagnet at one fixed edge. On the

other hand, if $|F_d|$ overcomes $|F_m|$, the MNPs do not bind to the magnetic thin film, causing release of the MNPs.

Quantitative Assessment of Captured Magnetic Nanoparticles

The optothermal-pyroelectric property of the PMN-PT substrate is utilized to assess the captured MNPs quantitatively. Fig. 1(b) shows the photoresponse of the device under an exposed IR pulse at zero bias. At room temperature, dipole arrangement in PMN-PT causes bound charges at the surface of PMN-PT. When the IR pulse is switched on, heat is transferred to the PMN-PT substrate. Due to this, a transient photocurrent (I) with a positive peak (I_{peak}) is obtained. This peak is observed as heating causes a reduction in bound charges and excess charges flow until a static charge balance is achieved. Finally, at the thermal equilibrium, I drops to the new steady state value, as shown in region R1 of Fig. 1(b). By integrating positive transient photocurrent with respect to time, accumulated charge at the substrate surface is obtained, as shown in Fig. 1(c). On the other hand, a transient photocurrent with a negative peak is obtained when the IR pulse is switched off, as shown in region R2 of Fig. 1(b). A negative peak is observed due to the cooling effect. By integrating negative transient photocurrent with respect to time, dissipated charge at the substrate surface is obtained, as shown in Fig. 1(c). In both cases, transient photocurrent is given by

$$I = Ap \frac{dT}{dt} \qquad (3)$$

Where A, p and $\frac{dT}{dt}$ are the surface area of the ME platform, pyroelectric coefficient of PMN-PT and first order time derivative of temperature, respectively. When injected MNPs are captured, and the device is exposed to the IR pulse, the effective heat transferred to the PMN-PT substrate is reduced. The reduced heat transfer to the

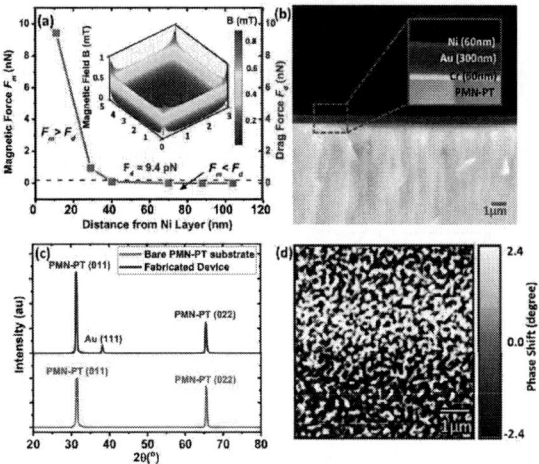

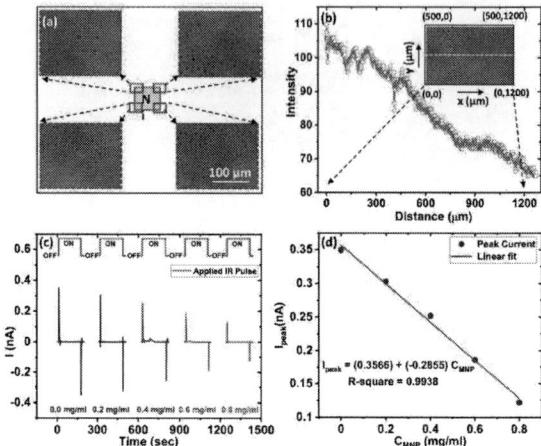

Figure 2: (a) Left ordinate shows exerted magnetic force on Fe_3O_4 nanoparticles due to Ni thin Film with respect to change in the distance from the upper surface of the Ni thin film. Right ordinate shows the drag force for injected nanoparticles for flow velocity $0.001mm/s$. Inset figure shows the numerical simulation of magnetic field intensity of Ni Thin film at a height of $50\mu m$. (b) cross-sectional FESEM image of the Ni thin film on the PMN-PT substrate. (c) X-ray diffraction (XRD) pattern of Bare PMN-PT substrate and fabricated device to obtain the lattice structure and (d) Magnetic force microscopy (MFM) phase shift in degree of nickel thin film in its demagnetized state.

Figure 3: (a) The magnetic field intensity is maximum at the edges of the Ni Thin Film leads to a high magnetic field gradient. Consequently, most MNPs trap corresponding to these locations. (b)Trapped MNPs distribution with respect to the location. Trapping minimizes from the edge to the center of the Ni thin Film. Corresponding change in particle trapping is measured using light intensity of microscope for a fix portion, which is shown in inset. (c) Photoresponse recorded against time under different injected nanoparticle concentration. (d) Calibration plot between positive transient current peak (I_{peak}) and injected nanoparticle concentration ($0.0 - 0.8mg/ml$).

substrate is attributed to the fact that MNPs in an aqueous solution scatter as well as absorb the incident IR light [12]. As a result, I_{peak} decreases with increasing injected MNP concentration.

RESULTS AND DISCUSSION

A schematic of the ME device fabricated is illustrated in Fig. 1(a), with Fig. 2(b) showing the cross-sectional FESEM image of the Ni thin film on the PMN-PT substrate. The lattice structure of the fabricated device is confirmed using XRD, as shown in Fig. 2(c). It only indicates firm (011) and (022) peaks originating from the [011]-cut PMN-PT substrate. A small peak of (111) orientation of the Au is also observed. No Ni peak is observed, which may be due to the fact that Ni thickness being much less than that of Au and PMNPT substrate. Hence, to confirm the deposited Ni film and its magnetic nature, MFM is performed. The magnetic domain configuration of Ni thin film in its demagnetized state is shown in Fig. 2(d). Without applying an external magnetic field, a dipolar contrast is observed, suggesting a magnetic nature of the Ni thin film.

Next, MNPs of varying concentrations (C_{MNP}) are injected on the upper surface of the ME platform. Maximum magnetic field intensity is observed at the edges of the Ni thin film, which decreases as we move beyond the edges, as shown in the inset of Fig. 2(a). This results in a large spatial variation of the magnetic field component at the film edge and corresponds to a higher magnetic field gradient. Such localized magnetic field gradient is

responsible for generating the required magnetic force for trapping the injected MNPs, which is calculated using equation (1). Consequently, most MNPs trap at the film edges, as shown in Fig. 3(a). This is also experimentally verified with the microscope light intensity plot. Fig. 3(b) shows trapped particle distribution with respect to the location. MNP trapping minimizes from the edge to the center of the Ni film, which is measured using the light intensity of the microscope for a fixed portion, as shown in the inset of Fig. 3(b). The magnetic force (F_m) acting on the MNPs due to Ni thin Film with respect to change in the distance from the upper surface of Ni film is also calculated, as shown in the left ordinate of Fig. 2(a). This clearly illustrates the maxima of magnetic force near the film surface. As already described, the trapping of MNPs also depends on viscous drag force; F_d is calculated at different MNP velocities. With the change in the injected velocity of the MNPs (v_p), the drag force varies according to equation 2. In contrast, the magnetic force remains the same, as no modifications are implemented to the deposited thin film. To ensure 100% capture probability, MNPs are injected at a low flow rate ($v_p = 0.001mm/s$), for which $|F_m| > |F_d|$, as shown in Fig. 2(a).

Once the injected MNPs are captured, analytical parameters (sensitivity, response time and linearity) of the device at different MNPs concentrations are obtained. Fig. 3(c) shows the photoresponse of the device. As the device is exposed to the IR pulse, the effective heat transferred to the PMN-PT substrate is reduced as MNPs in an aqueous solution scatter as well as absorb the incident IR radiation [12]. Consequently, the transient peak current reduces upon increasing the MNPs concentration, as shown in Fig. 3(c). In each case, when the IR pulse is switched on, the transient photocurrent exhibits a sharp rise before an exponential decay to reach a new steady state value. Response time, i.e. time to reach I_{peak} from I_{ss}, upon IR irradiation once injected MNPs are captured, is observed to be less than 2 sec. The transient photocurrent with a positive peak reduced linearly with increasing MNP concentration with a correlation coefficient (R^2) of 0.9938, as shown in the calibration plot in Fig 3(d). Additionally, the ME platform shows the sensitivity of $0.29 nA.mg^{-1}.ml$, calculated by dividing the slope of the calibration plot with the IR irradiation ME area.

CONCLUSION

In conclusion, we have demonstrated a self-powered ME platform for quantitatively assessing captured MNPs concentration. The combination of generated field gradient from a deposited magnetic thin film and the optothermal-pyroelectric property of the piezoelectric substrate allows the capture and enumeration of the captured MNPs. The ME platform exhibits a sensitivity of $0.29 nA.mg^{-1}.ml$ under the incident IR pulse without using external bias voltage. The platform also shows a fast response time of less than 2 sec with excellent stability and selectivity towards MNPs. The present study provides flexibility to the current ME platform, where a single device can capture, release, manipulate as well as quantitatively assess the MNPs. The self-powered ME platform described above can be implemented for future energy-efficient compact lab-on-a-chip applications such as controlled drug delivery, hyperthermia, cell labelling, etc.

ACKNOWLEDGEMENTS

This work is financially supported by the International Bilateral Corporation Division of Department of Science and Technology (DST), Government of India, under Indo-Norway joint project (Project No: INT/NOR/RCN/NS/P-06/2019).

REFERENCES

[1] N. Murali, S.K. Rainu, N. Singh and S. Betal, "Advanced Materials and Processes for Magnetically Driven Micro- and Nano-Machines for Biomedical Application", *Biosens. Bioelectron. X*, vol. 11, pp. 100206, 2022.

[2] M. Donolato, A. Torti, N. Kostesha, M. Deryabina, E. Sogne, P. Vavassori, M. F. Hansen, and R. Bertacco, "Magnetic domain wall conduits for single cell applications", *Lab Chip*, vol. 11, pp. 2976-2983, 2011.

[3] Y.-H. Chu, L. W. Martin, M. B. Holcomb, M. Gajek, S.-J. Han, Q. He, N. Balke, C.-H. Yang, D. Lee, W. Hu, Q. Zhan, P.-L. Yang, A. Fraile-Rodríguez, A. Scholl, S. X. Wang, and R. Ramesh, "Electric-field control of local ferromagnetism using a magnetoelectric multiferroic", *Nature Mater.*, vol. 7, pp. 478–482, 2008.

[4] P. Pathak and D. Mallick, "Size-Dependent Magnetization Switching in Magnetoelectric Heterostructures for Self-Biased MRAM Applications", *IEEE Trans. Electron Devices*, vol. 68, pp. 4418-4424, 2021.

[5] P. Pathak and D. Mallick, "Straintronic Nanomagnetic Logic Using Self-Biased Dipole Coupled Elliptical Nanomagnets", *IEEE Trans. Magn.*, vol. 58, pp. 1-6, 2022.

[6] R. Khojah, Z. Xiao, M. K. Panduranga, M. Bogumil, Y. Wang, M. Goiriena-Goikoetxea, R. V. Chopdekar, J. Bokor, G. P. Carman, R. N. Candler, and D. Di Carlo, "Single-Domain Multiferroic Array-Addressable Terfenol-D (SMArT) Micromagnets for Programmable Single-Cell Capture and Release", *Adv. Mater.*, vol. 33, pp. 2006651, 2021.

[7] H. Sohn, M. E. Nowakowski, C.-yen Liang, J. L. Hockel, K. Wetzlar, S. Keller, B. M. McLellan, M. A. Marcus, A. Doran, A. Young, M. Kläui, G. P. Carman, J. Bokor, and R. N. Candler, "Electrically Driven Magnetic Domain Wall Rotation in Multiferroic Heterostructures to Manipulate suspended on-chip magnetic particles", *ACS Nano*, vol. 9, pp. 4814-4826, 2015.

[8] D. Yoo, J.H. Lee, T.H. Shin, J. Cheon, "Theranostic Magnetic Nanoparticles", *Acc. Chem. Res.*, vol. 44, pp. 863–874, 2011.

[9] H. Fang, C. Xu, J. Ding, Q. Li, J.-L. Sun, J.-Y. Dai, T.-L. Ren, and Q. Yan, "Self-powered Ultrabroadband Photodetector Monolithically Integrated on a PMN–PT Ferroelectric Single Crystal", *ACS Appl. Mater. Interfaces*, vol. 8, pp. 32934–32939, 2016.

[10] E. P. Furlani and K. C. Ng, "Analytical model of magnetic nanoparticle transport and capture in the microvasculature", *Phys. Rev. E*, vol. 73, pp. 061919, 2006.

[11] Y. Sahoo, A. Goodarzi, M. T. Swihart, T. Y. Ohulchanskyy, N. Kaur, E. P. Furlani, and P. N. Prasad, "Aqueous Ferrofluid of Magnetite Nanoparticles: Fluorescence Labeling and Magnetophoretic Control", *J. Phys. Chem. B*, vol. 109, pp. 3879-3885, 2005.

[12] N. J. Hogan, A. S. Urban, C. Ayala-Orozco, A. Pimpinelli, P. Nordlander, and N. J. Halas, "Nanoparticles Heat through Light Localization", *Nano Lett.*, vol. 14, pp. 4640-4645, 2014.

CONTACT

* Dhiman Mallick, tel: +91-11-2659-1102; dhiman.mallick@ee.iitd.ac.in

REAL-TIME OPERATION OF MICROCANTILEVER-BASED IN-PLANE RESONATORS PARTIALLY IMMERSED IN A MICROFLUIDIC SAMPLER

Jiushuai Xu, Entian Cao, Michael Fahrbach, Vladislav Agluschewitsch,
Andreas Waag and Erwin Peiner

Technische Universität Braunschweig, Institute of Semiconductor Technology, Braunschweig,
GERMANY and Laboratory for Emerging Nanometrology (LENA), Braunschweig, GERMANY

ABSTRACT

This paper reports a real-time liquid-phase quantitative analysis system, where in-plane flexural piezoresistive Si cantilever resonators are utilized as microsensors. The new-developed system comprises a 3D-printed sampler, which enables a controllable setting of the immersion depth of the thermally self-actuated cantilevers of rectangular and triangular shapes into a microfluidic volume. By electromechanical amplitude modulation (EAM) of the supply voltage of the piezoresistive strain gauge, parasitic actuator-detector coupling effects were removed from the read-out signal enabling real-time tracking of resonant-frequency (rf) shifts using a phase-locked loop (PLL) circuit. High quality factors (Q: ~100-680) and low rf reduction (max. ~2.5%) with respect to operation in air were observed with cantilevers triangular-tip immersed into DI water. A real-time limit of detection (LOD) of ~1.0 ng in liquid and quasi-real-time LOD of ~150 pg have been obtained.

KEYWORDS

Piezoresistive microcantilever, electrothermal actuation, liquid-phase detection, electromechanical amplitude modulation

INTRODUCTION

Real-time quantitative measurements in liquid are preferably required for, e.g., point-of-care (PoC) applications, in situ environmental monitoring, and cancer cell detection [1]. Conventionally applied analytical methods for small molecule detection including gas chromatography-mass spectrometers (GC-MS), high-pressure liquid chromatography (HPLC) and UV spectrophotometers are specific and sensitive, but require expensive equipment and technical expertise, therefore are difficult to be adapted to the aforementioned applications. CMOS-compatible MEMS resonators are an attractive solution being extremely sensitive to the addition of mass. Among them, microcantilevers are one of the simplest and most common MEMS for mass sensing and biosensing [2]. Detection of single virus particles (~femtogram masses) in air [3], NO_2 gas molecules of part-per-billion by volume in air [4] and prostate-specific antigen in liquids at ng/mL concentrations [5] using MEMS cantilevers have been reported.

Miniaturized Lab-on-a-Chip (LoC) systems comprised of microfluidic systems and microsensors are naturally suited for PoC applications and biochemical detection. Here, reduced sample and reagent volumes are used by microfluidics and label-free detection is realized by microsensors. The limit of detection (LOD) with MEMS biosensors is identical or even better than the other label-free techniques, such as surface plasma resonance (SPR), quartz crystal microbalance (QCM) and electrochemical detection, indicating the potential of using MEMS sensors in pharmaceutical and biochemical diagnosis [6], [7]. However, liquid-phase sensing poses significant challenges for dynamic-mode microsensors due to severe viscous damping. Not only the resonance frequency (rf) is reduced due to mass loading/liquid dragging, but also the quality factor (Q), i. e., mass sensitivity and LOD of the resonator system are degraded by orders of magnitude compared to these of operating in air or vacuum, which may inhibit stable close-loop operation [8].

Compared to cantilevers vibrating in their first out-of-plane flexural mode, in-plane modes provide much lower damping and reduced mass loading by a surrounding fluid. Electrothermally self-actuated in-plane cantilevers with piezoresistive read-out are miniaturized resonators CMOS compatible strategy which can be embedded into microfluidic channel devices. However, continuous tracking of a cantilevers' resonant frequency during liquid-phase operation is still a great challenge. Low Q factor in liquid, as well as non-Lorentzian amplitude shape and reversing-phase characteristic around resonance hindering unambiguous resonance-phase tracking using a phase-locked loop (PLL)-based circuit have to be solved [9].

EXPERIMENTAL

Sensor fabrication

Microcantilevers (R1, T1 and T2) of different shapes, geometries and thicknesses (Figure 1) have been fabricated

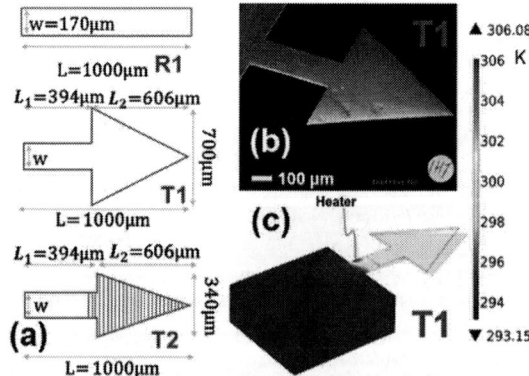

Figure 1. (a) different cantilever designs, (b) SEM graph of cantilever T1, (c) Temperature distribution across the heating resistor of T1 by finite-element modelling.

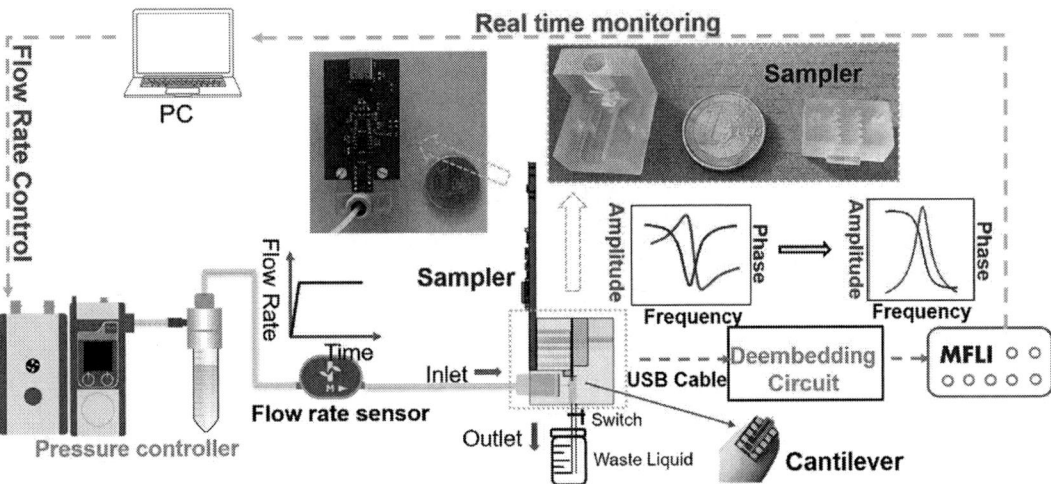

Figure 2. Schematic of real-time liquid-phase quantitative analysis system, including: Pressure/flow-rate controller, 3D-printed sampler, de-embedding (EAM) circuit board, PLL-based circuit for real-time resonance-frequency monitoring. Inset photos are the EAM circuit board (left) and 3D-printed sampler (right, above), cantilever on finger-tip (right, below).

from *n*-type bulk silicon substrates ((100), 1-10 Ω×cm). The heating and Wheatstone bridge resistors were embedded into the cantilevers' respective clamped-end by boron thermal diffusion. The detailed fabrication steps have been described in our previous work [2], [10].

Set-up of system

The whole system is comprised of three parts based on their corresponding functionalities, which are a pressure controller (LU-FEZ-2000, LINEUP FLOW EZ™) for generation of microfluidics, a sampler with a reaction-chamber allowing for controllable contact and immerse/ascend of cantilever beam to the liquid, and a circuit for de-embedding the cantilevers' true mechanical deflection signal from the measurement thus enabling the real-time PLL-based frequency tracking. The sampler (Fig. 2, inset photo upper right, size: 20×20×25 mm³) made of transparent resin has been fabricated using 3D-printing using a desktop 3D-Printer-Form from Formlabs. The reaction chamber of 2 mm in diameter and 6.6 mm in height (diameter of inlet: 500 μm, diameter of outlet: 2 mm) enables replicable insertion of a cantilever chip actuated laterally by providing a sinusoidal signal through a heating resistor located at its clamped-end (Fig. 1(c)), causing periodical heating, deformation and thus deflection of the cantilever. The inserted cantilever chip is connected to the de-embedding-circuit (Fig. 2 inset photo upper left) and a PLL-circuit (MFLI lock-in amplifier, Zurich Instruments) via needle-spring-contacts.

RESULTS

In order to investigate the effect of beam dimensions on the *rf* and *Q* of cantilevers operated in their fundamental in-plane flexural mode in liquid-phase, cantilevers of the same beam length and different geometrical shapes and thicknesses (Fig. 1 and Table 1) have been tested. Previous to the in-liquid measurements, the correspondence of immersion depth of cantilever beam to the given volume of microfluidic have been verified by immersing the cantilever along its axis downwards into a staining solution

consisting of blue triphenylmethane dye, sucrose and DI water, in which dye particles adsorb on the cantilever surface. The adsorption of dye particles can be observed under an optical microscope as shown in Fig. 3 (b), the immersion depth *h* was defined as the distance of the observed interface (Fig. 3 (b)) to the beam free-end. In addition, leaving the reaction chamber topside open to air

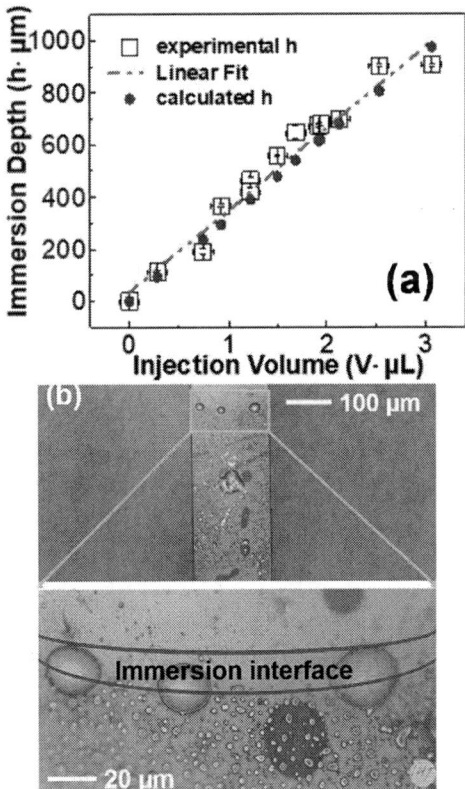

Figure 3. (a) Immersion depth h of a rectangular cantilever in staining liquid, vs. injection volume and (b) observed h observed under microscope.

Table 1: Geometries of the cantilevers measured in this work.

Microcantilevers	R1	T1-1	T1-2	T2
Thickness (μm)	60 ± 2	22 ± 2	61 ± 2	15 ± 2
Width of dipped-end (μm)	170 ± 2	606 ± 2	606 ± 2	340 ± 2

allows to ensure that there are no air bubbles trapped next to the cantilevers, which could lead to inaccurate results. The consistent fitting (Fig. 3(a)) of the calculated immersion depth according to the given injection volume and measured immersion depth by verifying adsorption of dye particles provides further confirmation that there is no air trapped beneath the resonator, and the microfluidic-sampler will be reliable for further experiments.

Cantilever mechanical properties were checked in DI water with various immersion depths, for T1 and T2, the complete immersion of triangular tips (~606 μm) was considered. Values of rf and Q are calculated from resonance-frequency-sweep curves using the 3-dB method in MATLAB (Mathworks, Natick, MA) [11]. Changes of rf and Q of the tested cantilevers in dependence of immersion depth (compared to their respective values in air) are depicted as Figs. 4 (a) and (b), respectively. Different cantilever geometries have been considered and utilized in this work, according to the previous reports and recommendations [8], [12], i. e., shorter and wider cantilevers for better Q and sensing characteristics, narrower beams for larger deflections. Therefore, cantilevers (T1 and T2) with wider and shorter sensing heads (equilateral triangle (T1) and isosceles triangle (T2)) as well as narrow beam fixed-end have been designed, fabricated and tested. Compared to the rectangular shape microcantilever, whose rf and Q decreased 8% and 94%, respectively, the triangular cantilevers showed less relative rf decrease (Fig. 4). Further, the shorter and wider T1-2

showed lowest rf reduction (by ~2.5%, from ~180 kHz) and highest Q of ~102 (decrease by 85%). The high value of Q is a prerequisite for real-time measurements in liquid.

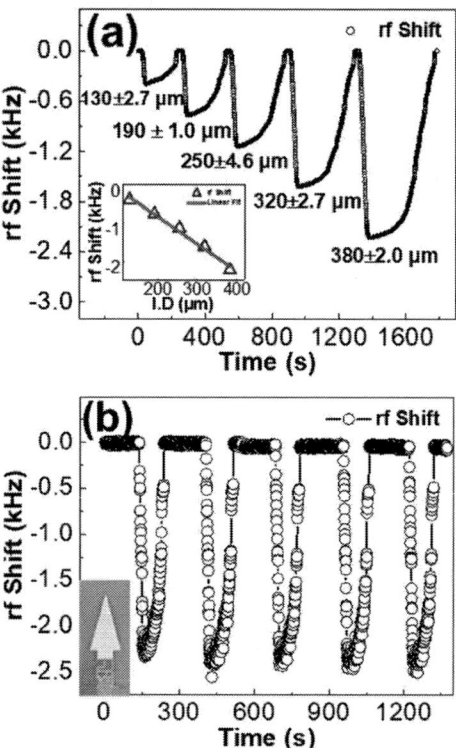

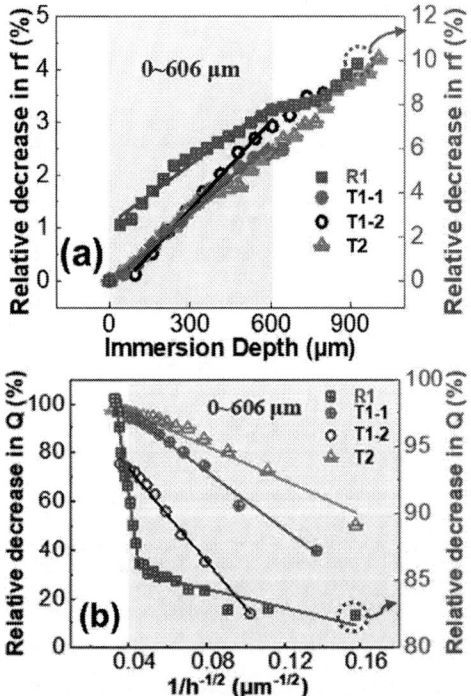

Figure 4. rf (a) and Q (b) related to respective values in air of different cantilevers.

Figure 5. Real-time responses of the cantilevers (a) T1-2 at various immersion depths h and (b) T2-1 repeatedly at h = 160 μm.

Real-time or quasi-real-time monitoring is crucial for bio- and chemical sensing, however, due to the near-field thermal radiation from the heating resistor, the Wheatstone bridge resistors have parasitical resistivity aberrations, leading to asymmetric resonance and non-monotonic phase spectral line shapes, which disable resonance-phase-change-based real-time measurements. Therefore, in our work, a home-built de-embedding circuit has been employed. By electromechanical amplitude modulation (EAM) of the supply voltage of the piezoresistive strain gauge, herein, the WB supply is AC-modulated (sinusoidal voltage of 3 V_{pp}) to separate and remove the parasitic actuator-detector (the heating resistor was driven by 3 V AC excitation superimposed on a 6 V bias voltage) feedthrough from the mechanical deflection signal in the frequency domain, enabling tracking and recording cantilever response in real-time. For cantilevers fabricated using the same process, the applied voltage to obtain an optimized output signal differs slightly. Typical resistance values of both the heating resistors and the piezoresistors are ~1 kΩ.

Real-time measurements of rf shifts of cantilever T1-2 at varying immersion depth are shown as Fig. 5(a), where we can find a linear decrease of *rf* with the immersion depth. Real-time monitoring of an immersing/ascending cantilever (T2) into/from DI water of $h = 160$ μm, respectively, in 5 repetitions, are depicted as Fig. 5(b), confirming their good repeatability (stability in 60 s: 0.13 ± 0.03 %) and feasibility in liquid detection. Furthermore, the real-time mass resolution of the cantilevers has been characterized in the stain solution. The *rf* shifts observed in real-time tests (Fig. 6) of T1-1 before and after been immersed into the stain solution for ~60 s reveals a mass of adsorbed-dye-particles of ~5.6 ng, corresponding to a mass sensitivity of 143 pg/Hz. The limit of detection was calculated to be ~1.0 ng (triangular-tip totally immersed in liquid):

$$LOD = \frac{3\Delta f_{min}}{\Delta f} \Delta m$$

where Δf_{min} represents the minimum detectable resonance frequency change, which was measured to be ~ 2.27 ± 0.07 Hz in DI water at $h \sim 606$ μm, Δm is the change of adsorbed mass, and Δf is the corresponding resonance frequency change. In addition, since the system allows real-time measurements for the cantilever in liquids for detection and in air for calculation (Fig. 6, ~20s for dumping liquids is required), wherein a higher Q (~680) can be obtained, realizing a quasi-real-time LOD of mass improved to ~150 pg.

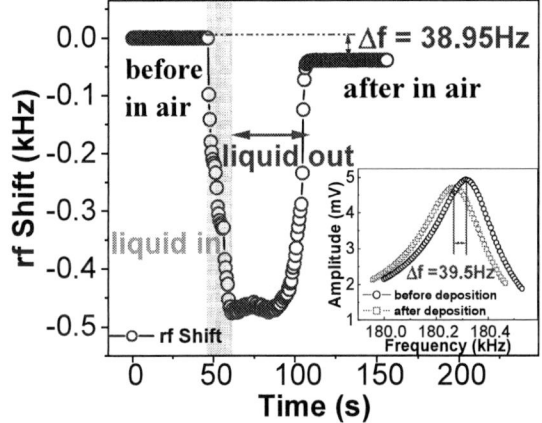

Figure 6. Real-time and spectroscopic (inset) adsorption monitoring of liquid-borne dye-particles from staining microfluidics to T1-1.

CONCLUSIONS

Microcantilevers of different geometries have been electrothermally actuated to their fundamental in-plane resonance mode and tested in liquids at various immersion depths in a microfluidic sampler for liquid-phase biochemical sensing applications. The shorter and wider equilateral-triangular cantilever beams exhibit best sensing characteristics, showing large Q (~102 to 680) as necessary for real-time particle detection, i. e., a limit of detections of mass in liquids from ~1.0 ng (real-time) to ~150 pg (quasi-real-time). As well known, cantilever deposited with gold layer on top surface allows the immobilization of thiol-modified DNA probes, therefore, the Lab-on-a-Chip (LoC) systems comprised of microfluidic systems and

microcantilever in this work has potentiality for real-time biochemical detection.

REFERENCES

[1] H. Jia, Y. Chen, X. Wang, T. Xu, and X. Li, "IN-PLANE MODE ENCASED CANTILEVERS FOR CANCER CELL DETECTION IN LIQUID," in Transducers 2021 Virtual Conference, 2021, no. June, pp. 735–738.

[2] J. Xu, M. Bertke, H. S. Wasisto, and E. Peiner, "Piezoresistive microcantilevers for humidity sensing," J. Micromechanics Microengineering, vol. 29, p. 053003, 2019.

[3] A. Gupta, D. Akin, and R. Bashir, "Single virus particle mass detection using microresonators with nanoscale thickness," Appl. Phys. Lett., vol. 84, no. 11, pp. 1976–1978, 2004.

[4] J. Xu, A. Setiono, and E. Peiner, "Piezoresistive Microcantilever with SAM-Modified ZnO-Nanorods@Silicon-Nanopillars for Room-Temperature Parts-per-Billion NO2 Detection," ACS Appl. Nano Mater., vol. 3, no. 7, pp. 6609–6620, 2020.

[5] C. Vančura, Y. Li, J. Lichtenberg, K. Kirstein, A. Hierlemann, and F. Josse, "Liquid-Phase Chemical and Biochemical Detection Using Fully Integrated Magnetically Actuated Complementary Metal Oxide Semiconductor Resonant Cantilever Sensor Systems," Anal. Chem., vol. 79, no. 4, pp. 1646–1654, 2007.

[6] M. Alvarez and L. M. Lechuga, "Microcantilever-based platforms as biosensing tools," Analyst, vol. 135, no. 5, pp. 827–836, 2010.

[7] R. M. R. Pinto, V. Chu, and J. P. Conde, "Label-Free Biosensing of DNA in Microfluidics Using Amorphous Silicon Capacitive Micro-Cantilevers," IEEE Sens. J., vol. 20, no. 16, pp. 9018–9028, 2020.

[8] J. A. Schultz et al., "Lateral-Mode Vibration of Microcantilever-Based Sensors in Viscous Fluids Using Timoshenko Beam Theory," J. Microelectromechanical Syst., vol. 24, no. 4, pp. 848–860, 2015.

[9] A. Setiono et al., "Improvement of frequency responses of an in-plane electro-thermal cantilever sensor for real-time measurement", J. Micromech. Microeng. 29 (2019) 124006 (12pp).

[10] J. Xu et al., "Fabrication of ZnO nanorods and Chitosan@ZnO nanorods on MEMS piezoresistive self-actuating silicon microcantilever for humidity sensing," Sensors Actuators, B Chem., vol. 273, pp. 276–287, 2018.

[11] M. Bertke et al. "Analysis of asymmetric resonance response of thermally excited silicon micro-cantilevers for mass-sensitive nanoparticle detection", J. Micromech. Microeng. 27 (2017) 064001 (10pp).

[12] L. A. Beardslee, F. Josse, S. M. Heinrich, I. Dufour, and O. Brand, "Geometrical considerations for the design of liquid-phase biochemical sensors using a cantilever's fundamental in-plane mode," Sensors Actuators, B Chem., vol. 164, no. 1, pp. 7–14, 2012.

CONTACT

*J. Xu, tel: +49 5313913820; jiushuai.xu@tu-bs.de

SUSPENDED NANOCHANNEL RESONATORS MADE BY NANOIMPRINT AND GAS PHASE DEPOSITION

Manuel Müller[1], Jeremy Teuber[1], Rukan Nasri[1], Francesc Torres Canals[2], Núria Barniol[2], Jordi Llobet Sixto[3], Xavier Borrisé[3], Francesc Perez-Murano[3] and Irene Fernandez-Cuesta[1,4]
[1]University of Hamburg, Hamburg, 22765, GERMANY
[2]Universitat Autónoma de Barcelona, Bellatera, 08193, Catalonia, SPAIN
[3]IMB-CNM CSIC, Bellatera, 08193, Catalonia, SPAIN and
[4]Hamburg Centre for Ultrafast Imaging, CUI

ABSTRACT

In this work, we report on the fabrication of fluidic devices with hollow, suspended nanochannels, made by a combination of nanoimprint lithography and gas-phase deposition. We fabricated and characterized complete fluidic devices with arrays of nanochannels in the range of 500 nm x 500 nm to 700 nm x 700 nm, 30 µm long, with wall thickness just few tens of nm thick. We also developed a COMSOL model to predict the motion and resonance frequency range of the hollow beams with different materials, geometries, and investigate the relevant damping conditions.

KEYWORDS

Nanofluidics, suspended nanochannels, resonators, ALD, nanoimprint, COMSOL

INTRODUCTION

Measuring the mass of single molecules, one by one, is one of the biggest challenges in biomedical technology. It would allow to identify and quantify a variety of biomarkers, for example, for early disease detection. We work towards weighting single molecules in their natural liquid environment. For this, we will use hollow nano-resonators as nanochannels, which are suspended and free to resonate in vacuum, while the liquid and molecules flow inside. Previous works on suspended microchannels have shown the versatility of the idea to measure cells and particles [1, 2]. But a crucial factor to measure single biomolecules is that the nanochannel has to be extremely light, to ensure the necessary sensitivity to measure the small masses. However, the challenges for making such devices with very light, suspended nanochannels are huge. In addition, the usually complicated process leads to low throughput and low yield in the device fabrication. Here, we use nanoimprint lithography [3], which is a parallel nanofabrication method, compatible with mass production, and our newly developed methodology for a gas-phase material deposition in a "*flow-through*" mode [4], to conformally coat the inner walls of the channels to define the hollow resonator. This combination allows to fabricate hollow, suspended nanochannels, with control on the wall thickness and with flexibility on the design.

FABRICATION

Figure 1 (a) shows the step-by-step fabrication process. We combine UV nanoimprint lithography (UV-NIL) as a high throughput nanostructuring technique to define the nanochannels and microchannels in just one step in a polymer [3]. And our newly developed methodology

for a gas-phase material deposition in a "*flow-through*" mode [4], to conformally coat their inner walls to define the hollow resonators, embedded in a complete, functional fluidic device. For this, first, we make a silicon stamp with micro and nanochannels using electron beam lithography and photolithography, followed by reactive ion etching (RIE). Then, after coating the stamp with a monolayer of fluorosilanes to avoid adhesion, we perform a double replication process [5] to define the fluidic system in a UV-curable polymer (step 1). We have used several different functional materials to define the micro- and nanochannels in the final imprinting step, including purely organic polymers, or hybrid (inorganic-organic) ones, like Ormostamp® [6]. After defining the micro- and nanochannels by UV-NIL, and to seal the fluidic device, a 3 µm thick polymer coverslip with a stabilizing and sacrificial transfer layer is covalently bonded to the surface (step 2). The device is then placed in the dedicated gas-phase deposition reactor to conformally coat the inner walls of the micro and nanochannels with a thin layer of Al_2O_3 (step 3). The gas phase deposition process can be tuned from atomic layer deposition characteristic growth rates (in the order of Å/cycle) to homogenous and conformal multilayer condensation growth rates with up to 70 nm deposited in a one-cycle process. This allows to overcome the limits of slow ALD processes in high-aspect-ratio-structures at the cost of thickness gradients of around ten percent.

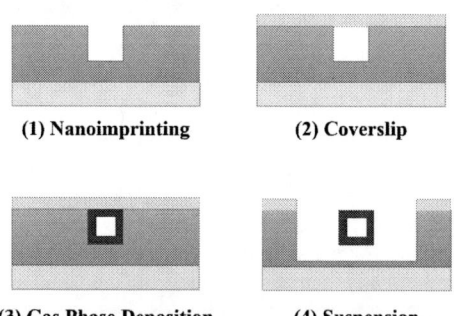

(1) Nanoimprinting   **(2) Coverslip**

(3) Gas Phase Deposition   **(4) Suspension**

Figure 1: Fabrication process: The micro- and nanostructures are defined by a UV NIL into a UV-curable polymer, drop casted on a substrate (step 1). The nanochannel is enclosed by bonding a coverslip with sacrificial layer (step 2). The inner walls are conformally coated with Al_2O_3 in a self-developed, gas-phase deposition process (step 3). The hollow nanochannel is suspended by selectively etching trough the coverslip and locally into the polymer around the Al_2O_3 nanochannel by RIE (step 4).

In a final step, the nanochannels are released by selectively etching through the coverslip and into the imprinted polymer around and below the nanochannel using RIE (4). A different etching process is used depending on the specific polymer used for the imprinting (O_2-based for the purely organic, and $SF_6 + O_2$ based for the hybrid material). The width of the trench used to release the beams can be easily varied, and allows for adjustable resonator lengths for the same imprinting and deposition process without modification of the basic device design. The underetching of the nanochannel can be further enhanced by tilting the sample 15 degrees inside the RIE chamber.

DEVICE

Figure 2 shows a photo of a polymer device after the gas phase deposition process, where the inlet holes and microchannels (which connect them to the nanochannels) can be seen. The access ports for the precursor and vacuum connectivity can be differentiated by the selective deposition of the "*flow-through*" process. In the fluidic experiments these ports will be used for the liquid input and pressure control for flow manipulation.

Figure 2: The photo shows a polymer fluidic chip after the gas phase deposition, where the inlet holes and microchannels can be seen. The deposition around the bottom inlet holes shows the result of the "flow-through" deposition as these inlets are selectively exposed to the precursor pulse while maintaining vacuum at the other two holes at the top.

Figure 3 shows a SEM picture of an Al_2O_3 suspended, hollow resonator, after the imprinting, Al_2O_3 deposition and selective coverslip removal. The channel is 700 nm wide, 700 nm deep, 50 μm long, and the suspended part is 35 μm. The walls have been coated with few tens of nanometers of Al_2O_3. The inset shows a channel which broke during characterization (in a different sample), where it can be seen that they are indeed hollow. Systematic characterization by FIB confirmed this, and shows that the wall thickness along the channel is very homogeneous and stays within ± 10% along the suspended part. Different device configurations can be easily adapted and fabricated with this process, to obtain, for example, single resonators for individually studying them, or nanochannel arrays, for higher throughput flow devices.

Figure 3: Scanning Electron Microscope (SEM) image, false colored, showing a suspended, hollow nanochannel, 700 nm outter dieameter. The inset shows a detail of a nanochannel, broken with a micromanipulator, where the hollow interior can be seen. The nanochannels have also been characterized by cross-sectioning using a FIB, to measure the homogeneity of the wall thickness (not shown here).

RESONANCE

The mechanical motion of the suspended nanochannels has been measured with an optical interferometer. For this, the sample is placed in a vacuum chamber, and excited using a piezoelectric actuator, wire-bonded to a printed circuit board to facilitate the connection to external equipment.

The nanochannels show typical resonances in the Megahertz regime, with quality factors in the order of 10^2. Figure 4 shows an exemplary resonance peak for a 500 nm x 500 nm nanochannel, 30 μm long. It has the center at 3MHz, and a quality factor, Q, of 274. Other samples with nanochannels with slightly different geometries and wall thicknesses show resonances around 1.77 MHz, and quality factors of 160.

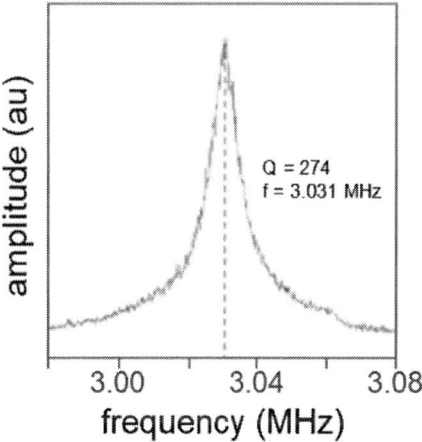

Figure 4. Resonance of a hollow Al_2O_3 nanochannel, 500 nm x 500 nm in cross section, 30 μm long. The resonance was measured optically by interferometry/reflectance, actuating the resonator with a piezoelectric crystal.

COMSOL MODELING

We modelled the hollow nanoresonators using COMSOL, to simulate their mechanical motion and to predict the trends for the resonant peak of the different modes for nanochannels with different geometries, coated with reflective layers, and made of and embedded in different materials.

Figure 5 (a) shows the first eigenfrequency of hollow nanoresonators as a function of their wall thickness for nanochannels with external diameters of 500 nm (blue dots) and 700 nm (grey dots), for a fix beam length of 30 μm. Figure 5 (b) shows the eigenfrequency for those two same nanochannel diameters as a function of the beam length, for a fix wall thickness of 70 nm.

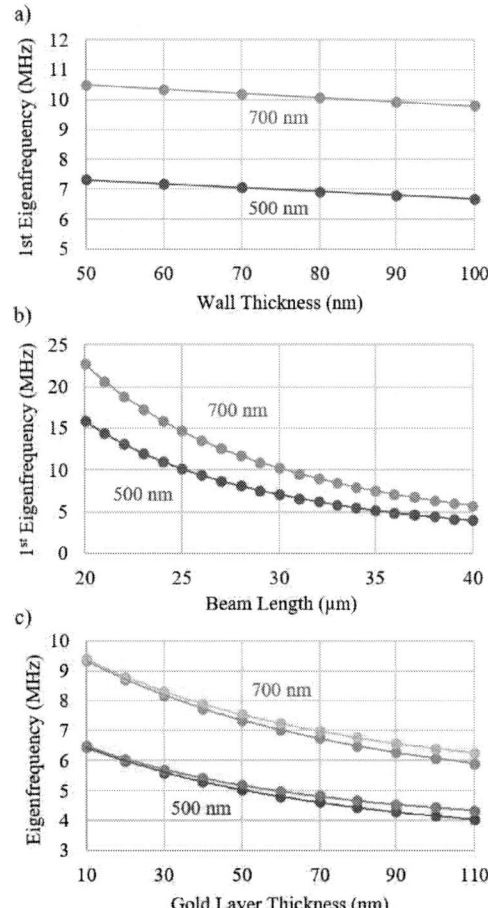

Figure 5. Position of the resonances (1st Eigenfrequencies) modelled by COMSOL for different types of hollow resonators. (a) shows the dependence of the resonance on the beam length for two different nanochannel diameters (700 nm, in grey, 500 nm in blue). (b) shows the dependence on the wall thickness for the same two nanochannel systems. (c) shows the effect of evaporating a thin layer of gold on the resonators: in the graph the position of the eigenfrequencies is shown as a function of the thickness of the gold layer, for a 700 nm wide nanochannel (yellow and green dots) and 500 nm wide (blue and orange). As a result of the simetry breaking, we observe a splitting of the mode.

The resulting eigenfrequencies are in agreement with numerically calculated values for these systems. And the observed trends show the flexibility of the sample design of our devices: with one nanochannel design the resonance frequency of the first mode can be varied (i) by tunning the suspended length (width of the trench etched into the polymer, used to release the nanochannel), and/or (ii) by varying the wall thickness by controlling the cycle number or cycle length during the gas-phase deposition. Thanks to this, an eigenfrequency range from few to tens of MHz can be achieved without complicated modifications in the device design. This allows to tune the frequency for optimization of the readout electronics and the sensor sensitivity.

For the experimental optical readout, a reflective gold layer is typically evaporated or sputtered on top of the channels. We also used the COMSOL model to evaluate the effect of this layer. The thickness of the gold layer is usually just a few nanometers, but since the density of gold is high, it can result in a significant added mass. The simulation results for the position of the eigenfrequency for an increasing layer thicknesses are shown in Figure 5 (c), again for a 500 nm wide nanochannel (orange and blue dots) and a 700 nm wide nanochannel (yellow and green dots). A significant shift of the eigenfrequency position towards lower frequencies can be observed in both cases. This was expected, since, in some cases, the mass of the gold layer can be similar to the mass of the cantilever itself. And, also interestingly, the added gold layer results in the splitting of the first and second eigenfrequency, as shown in the graph. This is a consequence of the symmetry breaking, since now the mass distribution is not equal along all the axes of the beam.

ACKNOWLEDGEMENTS

This project has received funding from the European Research Council (ERC) under European Union's Horizon 2020 research and innovation program (grant agreement No 714073). This work is also supported by the Cluster of Excellence 'CUI: Advanced Imaging of Matter' of the Deutsche Forschungsgemeinschaft (DFG) - EXC 2056 - project ID 390715994. In addition, the authors would like to thank R. Zierold for fruitful discussions and feedback about the ALD and gas deposition process.

REFERENCES

[1] T. Burg, M. Godin, S. M. Knudsen, W. Shen, G. Carlson, J. S. Foster, K. Babcock and S. R. Manalis, "Weighing of biomolecules, single cells and single nanoparticles in fluid", *Nature* 446, 1066-1069 (2007)

[2] A. De Pastina and L. G. Villanueva, "Suspended micro/nano channel resonators: a review" *J. Micromech. and Microeng* 30 043001 (2020)

[3] F. M. Esmek, P. Bayat, F. Pérez-Willard, T. Volkenandt, R. H. Blick and I. Fernandez-Cuesta, "Sculpturing wafer-scale nanofluidic devices for DNA single molecule analysis", Nanoscale 11 (28), 13620-13631 (2019)

[4] M. Müller, R. Zierold and I. Fernandez-Cuesta, Patent application DE 102020102076A1 (2020)

[5] I. Fernandez-Cuesta, A. L. Palmarelli, X. Liang, J. Zhang, S. Dhuey, D. Olynick and S. Cabrini, "Fabrication of fluidic devices with 30 nm nanochannels by direct imprinting" Journal of Vac Sci and Tech B, 29, 6, 06F801 (2011)

[6] A. Schleunitz, M. Vogler, I. Fernandez-Cuesta, H. Schift and G. Gruetzner, "Innovative and tailor-made resist and working stamp materials for advancing NIL-based production technology", J. Photopol. Sci and Tech., 26, 1, 119 (2013)

CONTACT

*ifernand@physnet.uni-hamburg.de
*manuel.mueller@physnet.uni-hamburg.de

DEVELOPING AN EXTREMELY HIGH FLOW RATE PNEUMATIC PERISTALTIC MICROPUMP FOR BLOOD PLASMA SEPARATION WITH INERTIAL PARTICLE FOCUSING TECHNIQUE FROM FINGERTIP BLOOD WITH LANCETS

Tuan N.A. Vo[1,2,3], Pin-Chuan Chen[1], and Pai-Shan Chen[4]

[1]Department of Mechanical Engineering, National Taiwan University of Science and Technology, Taipei, TAIWAN
[2]Faculty of Mechanical Engineering, Ho Chi Minh City University of Technology (HCMUT), 268 Ly Thuong Kiet Street, District 10, Ho Chi Minh City, VIETNAM
[3]Vietnam National University Ho Chi Minh City (VNU-HCM), Linh Trung Ward, Thu Duc City, Ho Chi Minh City, VIETNAM
[4]Institute of Toxicology, College of Medicine, National Taiwan University, Taipei, TAIWAN

ABSTRACT

This paper proposes an on-chip for blood plasma separation based on a pneumatic peristaltic micropump (PPM) linked to a spiral channel with a trapezoidal cross-section, which is used as an inertial particle focusing technique. The concept was based on the excellent bonding strength of PMMA/PDMS/PMMA and developed the PPM capable of delivering an extremely high flow rate of 3,500 µL/min. The device rapidly extracts blood plasma from diluted blood within 3 minutes for four rounds, with an efficiency of 97%. The total volume needed for each extraction process was only 4µL whole blood drawn from a fingertip with a lancet.

KEYWORDS

Micropump, Inertial Microfluidics, Blood Plasma Separation

INTRODUCTION

Human blood is one of the essential biological fluids utilized in medical laboratory diagnostics, and 45% of the whole blood includes red blood cells (RBCs), white blood cells (WBCs), and platelets. The remaining 55% of whole blood is blood plasma, which contains proteins, antibodies, DNA, and RNA fragments [1]. Blood cells are typically eliminated prior to diagnostic testing because the biological components of blood can easily interfere with the operation of biosensors, hence influencing the final assay results. The conventional blood processing approach is centrifugation at high rotation speed. However, the centrifugation method is challenging to be well-trained, time-consuming, requires bulky equipment, and difficult to be integrated on microfluidic platforms.

In recent years, inertial particle focusing has attracted great attention in a subfield of microfluidics, and the inertial platform would be suitable for blood plasma separation. Several groups have proved that a spiral microfluidic device with a trapezoidal cross-section could efficiently isolate plasma from diluted blood [2-5]. However, to achieve the required flow velocities (from 1,100 µL/min to 1,700 µL/min), the separating system typically uses an external syringe or a peristaltic pump to achieve sufficient blood flow rates inside the microchannel. Although these external systems are effective, they are not ideal for miniaturized total analysis systems (µTASs) or lab-on-a-chip (LoC) systems that utilize relatively small volumes of blood. In addition, with an external syringe pump and associated connection tubings, the blood volume used for a single test would be greater than the volume required for the microfluidic device. An on-chip micropump is ideal for reducing the volume of blood lost in the tubing and minimizing the blood volume required for a single test.

This research aims to develop a microfluidic device capable of transporting a precise and high throughput of a tiny of blood and separating blood plasma using the inertial focusing approach without external pumps or valves. Pneumatic peristaltic micropump (PPM) combined with spiral microchannel-based inertial microfluidics is capable of providing these inevitable requirements. However, due to the limitation of the fabrication process, the working pressure of reported PPMs are less than 200 kPa, affecting the flow rate of less than 900 µL/min [6, 7].

Recently, we have developed a novel boding method to maximize the interfacial bonding strength between polydimethylsiloxane (PDMS) and polymethylmethacrylate (PMMA) by using simple methods at room temperature and achieving the highest bonding strength [8]. Based on this bonding technique, we propose a PPM fabrication method that combines the elasticity of PDMS and the rigid of PMMA that could be operated under high pneumatic pressure and high frequency. The PPM can deliver an extreme flow rate of 3,500 µL/min, and a spiral microchannel with a trapezoidal cross-section area directly linked with an on-chip micropump enables rapidly isolating blood cells from only 4 µL human blood drawn by punching fingertips with commercial lancets.

EXPERIMENT METHOD

Fabrication process

The design of the PPM includes a top PMMA layer for pneumatic channels, an elastomer PDMS membrane in the middle, and a bottom PMMA substrate for liquid microchannels. The principle working of the PPM is based on the concept of pressurizing and depressurizing the pneumatic chambers, and fluids are typically driven by subsequent elastic membranes actuated by their corresponding pneumatic chambers. The fabrication process of the hybrid PMMA/PDMS/PMMA PPM is shown in Fig. 1. The pneumatic channel and liquid channel were fabricated by micromachining (see Fig. 1(a-i)) on the PMMA substrates. In Fig. 1(b), the liquid channel of the PPM has dimensions such as: depth of channel dc=300 μm, depth of the spherical chamber dp= 400 μm, while the width of the channel and diameter of the chamber was dw=500 μm, and the curvature of the liquid chamber was db=2,000 μm. While the dimension of the pneumatic channel was designed such as: depth of 500 μm, width of 1,000 μm, and a curvature of the pneumatic chamber of 2,100 μm. The design consisted of triple-channel pumping, with each channel having eight microchambers. The distance of parallel channels was 3,500 μm. The flexible PDMS membrane was employed in the architecture of the PPM. The PDMS membrane was prepared by combining elastomer and curing agent at a volume ratio of 10:1. The thickness of the PDMS membrane was 200 μm which was prepared by spin coating. The membranes then underwent thermal curing at 80 °C for 2 hours.

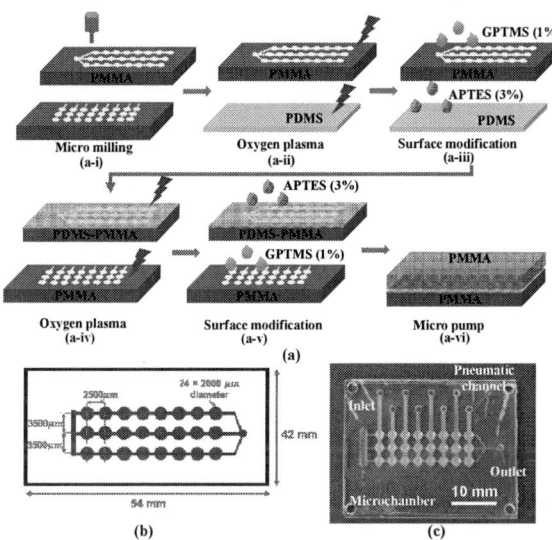

Figure 1: (a) Schematic illustrations of the fabrication process: (a-i) Fabrication of microchannel on PMMA substrates by using CNC machine; (a-ii) Oxygen plasma treatment on PMMA substrate and PDMS membrane; (a-iii) PMMA surface modification with GPTMS; PDMS surface modification with APTES; (a-iv) Oxygen plasma treatment on PMMA substrate and PDMS/PMMA chip; (a-v) PMMA surface modification with GPTMS; and PDMS/PMMA surface modification with APTES; (a-vi) Micropump; (b) Dimension of the liquid channel; (c) Bonded PMMA-PDMS-PMMA PPM.

For the bonding process, the bottom PMMA substrate and the PDMS membrane were directly bonded by using our previously reported method to achieve high bonding strength [8], in which the PMMA and PDMS surfaces were treated with oxygen plasma to form hydroxyl groups, shown in Fig. 1(a-ii). Immediately after plasma treatment, the modified PDMS substrate was soaked in an aqueous solution of 3% 3-Aminopropyltriethoxysilane (APTES), whereas the modified PMMA was soaked in an aqueous solution of 1% 3-glycidoxypropyltrimethoxysilane

(GPTMS), as shown in Fig. 1(a-iii). After thorough drying, the two substrates were brought together with a 1 kg weight at room temperature to form a tight amine–epoxy bond in 1 hour. At this point, fluidic access holes were punched on the PDMS membrane by using a Uni-Core™ Puncher. After successfully bonding PDMS and bottom PMMA substrate, then the top PMMA substrate was brought in contact and bonded by following the same process shown in the previous study and shown in Fig. 1 (a-iv)~(a-vi). Fig.1(c) shows the visible of the bonded PPM.

Experimental setup

In the previous study, we proved that by applying higher pressure, number of chambers, and depth of chamber, flow rate of the PMM would be increased [9]. In particular, when infusing high pressure of 300 kPa into the pneumatic chambers, the PDMS membrane always contacts the spherical microchamber, increasing pump flow, even at high frequencies. Combining the previous analysis procedures, we developed the parallel PPM with triple-channel to achieve an extreme flow rate.

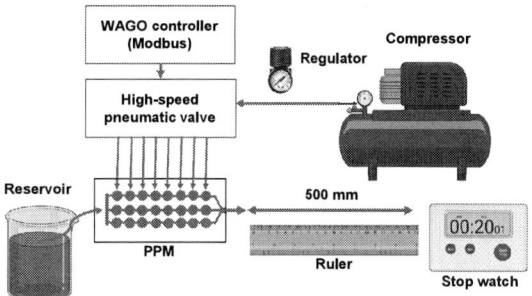

Figure 2: Schematic of the experimental setup for the measurement flow rate of the PPM.

Fig. 2 shows a schematic experiment setup to characterize the performance of the PPM. The compressed air system consists of an air compressor that supplies compressed air to the pneumatic actuation chamber and the high-speed valves (MHA2-MS1H, FESTO, USA) in which its operation could be modulated by a programmable WAGO controller (750-362, Germany). A software interface was developed in LABVIEW (National Instruments, Texas, USA) for the control frequency of actuators. Compressed air was routed to the solenoid valves, which were then connected via tubing to the fabricated chips containing functional elements. A precise gas regulator was used to adjust the actuation of compressed air pressure. To determine the generated flow rates of the PPM under different conditions, the volume of liquid passing through the PPM was estimated by recording the time when the liquid reached 500 mm in a tubing with an inner diameter of 0.38mm.

RESULTS

Fig. 3 shows the flow rates of different applied frequencies from 1.25 Hz to 10 Hz. Conclusions can be drawn, including: (1) if the same operating conditions were applied to the chamber with an applied pressure of 300 kPa, total chamber number of 24, and depth of chamber dp=400 μm, it is clear that increasing applied frequency increased the flow rates significantly. For example, the flow rate of 528.02±20.64 μL/min was achieved with the applied frequency of 1.25 Hz, while the flow rate of 3510.13±50.73 μL/min was achieved with the applied frequency of 10 Hz; (2) the performance of PPM developed herein has high reliability and repeatability since the error bar is small; (3) high performance can be potentially realized in a small footprint by integrating more PPM on a single chip because the distance, center lines between microchambers, is only 2,500 μm apart. Here, flow rate of PPM has provided sufficient magnitude to apply inertial microfluidics technique.

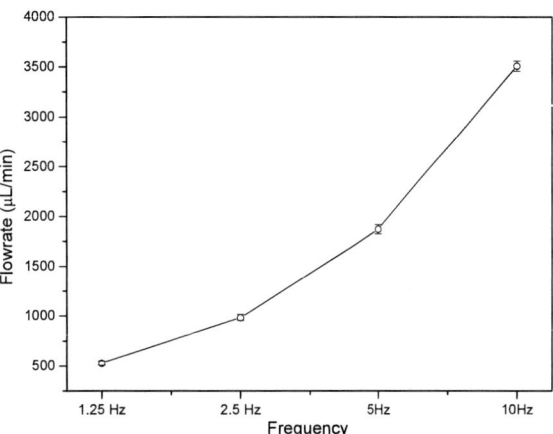

Figure 3: Pumping flow rate of the PPM with different frequencies under air pressure of 300 kPa.

BLOOD PLASMA SEPARATION
An integrated inertial microfluidic device

We characterized the performance of the integrated inertial microfluidic device for blood plasma separation only in a single device without any external pumps or valves. The integrated micropump consists of the PPM and a spiral microchannel with a trapezoidal cross-section area. The PMMA mold was manufactured using the micro-milling technique, which precisely milled a positive print of the spiral microchannels. The slanted five-circular-loops-spiral channels have a trapezoidal cross-section with a width of 500 μm, and inner and outer heights of 70 μm and 40 μm, respectively [10]. The spiral PDMS channel was bonded with the PPM following the process of bonding mentioned above. Fig. 4(a) shows the integrated microfluidic chip and the experiment setup of blood plasma separation, and Fig. 4(b)~(c) show the SEM of a trapezoidal cross-sectional view of spiral microchannels (section A-A') and interfacial of the microchamber (section B-B'), respectively.

The blood sample preparation process is shown in Fig.5 (a~c). To understand the performance of using integrated microfluidics to extract blood plasma from the diluted blood, 4 μL whole blood was obtained from the fingertip by using a commercial lancet, diluted with 45× phosphate buffered saline (PBS), and a total of 180 μL of diluted blood was loaded in the reservoir of blood shown in Fig. 5(d). A flow rate of 1,500 μL/min was generated by controlling the frequency and pneumatic pressure to push

the diluted blood through the microfluidics, the extracted blood plasma and cells were collected individually from different outlets (i.e., reservoir (1) and (2) in Fig. 4(a)), and a hemocytometer was used to quantify the blood cell concentrations in the collected plasma and understand the performance of this integrated microfluidics. The separation efficiency is calculated by dividing the measured number of RBCs in the collected plasma sample by the measured number of RBCs in the original sample.

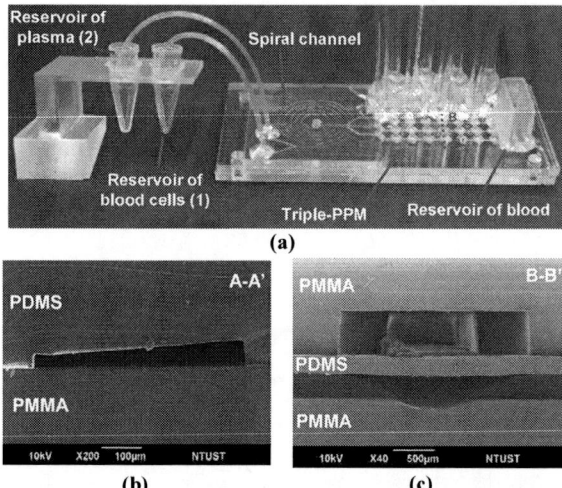

Figure 4: (a) The preparation for blood plasma separation with the hybrid device; (b) SEM of trapezoidal microchannels (section A-A'), (c) SEM of PMMA/PDMS/PMMA interfacial of the microchamber (section B-B').

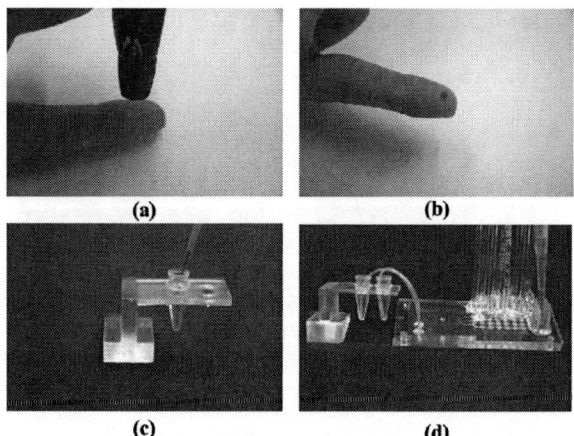

Figure 5: (a) Punching finger with the lancet, (b) A 4μL droplet of whole blood, (c) Diluting whole blood 45× with PBS into 180μL diluted blood, (d) Loading diluted blood into the reservoir.

Efficient blood plasma extraction

Fig. 6 shows the experiment results, in which the inner centrifuge tube (remark No. 1) has the blood cells while the outer centrifuge tube (remark No. 2) has the separated blood plasma. The experiment results clearly show that this blood plasma separation can be realized on-chip with inertial microfluidics, and the on-chip micro pump can provide a sufficient flow rate. As depicted in Fig. 6, blood plasma collected from the outside branch is clear; the blood cells collected from the inner branch are dark red, indicating a high concentration of blood cells that are predominantly blood cells. Moreover, no phenomenon of RBCs logging or leakage of microchannels was observed. Most importantly, the separation efficiencies were close to 97% after four rounds. Note that the separation process takes less than 3 min with a tiny volume of diluted blood.

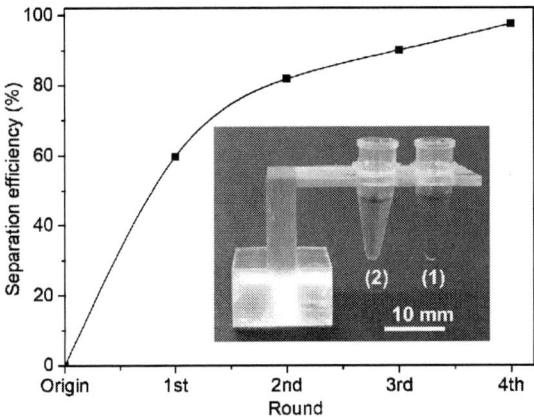

Figure 6: Separation efficiencies of the diluted blood: picture of samples collected from the inner outlet (No. 1) for blood cells and the outer outlet (No. 2) for blood plasma after the first round; and separation efficiencies of the diluted blood after 4 rounds of separations.

CONCLUSIONS

In this study, we successfully developed a novel microfluidic device to separate blood plasma from human blood without external valves or pumps. The concept is based on using a PPM integrated with a trapezoidal cross-section for ultra-fast separation. First, we proposed a fabrication process of PPM with a high pumping rate by combining rigidly of PMMA and elasticity of PDMS. Following the procedure to achieve excellent bonding strength of PDMS/PMMA, the PPMs could be operated under high pressure. We proved that the PPM achieved the highest pumping rate of 3,500 µL/min. Then the micropump was integrated with a spiral microchannel with a trapezoidal cross-section area and used to rapidly extract plasma from human blood within 3 minutes and with a small blood volume of 200 µL with separation efficiency up to 97%.

ACKNOWLEDGEMENTS

This research was funded by National Science and Technology Council (Taiwan) with grant numbers of NSTC 110-2628-E-011-006, and Mechanical Engineering Department of National Taiwan University of Science and Technology (NTUST).

REFERENCES

[1] W. S. Mielczarek, E. A. Obaje, T. T. Bachmann, and M. Kersaudy-Kerhoas, "Microfluidic blood plasma separation for medical diagnostics: is it worth it?," *Lab on a Chip,* vol. 16, no. 18, pp. 3441-3448, 2016.

[2] N. Nivedita, and I. Papautsky, "Continuous separation of blood cells in spiral microfluidic devices," *Biomicrofluidics,* vol. 7, no. 5, Sep, 2013.

[3] M. E. Warkiani, G. F. Guan, K. B. Luan, W. C. Lee, A. A. S. Bhagat, P. K. Chaudhuri, D. S. W. Tan, W. T. Lim, S. C. Lee, P. C. Y. Chen, C. T. Lim, and J. Han, "Slanted spiral microfluidics for the ultra-fast, label-free isolation of circulating tumor cells," *Lab on a Chip,* vol. 14, no. 1, pp. 128-137, 2014.

[4] L. D. Wu, G. F. Guan, H. W. Hou, A. A. S. Bhagat, and J. Han, "Separation of Leukocytes from Blood Using Spiral Channel with Trapezoid Cross-Section," *Analytical Chemistry,* vol. 84, no. 21, pp. 9324-9331, Nov 6, 2012.

[5] G. F. Guan, L. D. Wu, A. A. S. Bhagat, Z. R. Li, P. C. Y. Chen, S. Z. Chao, C. J. Ong, and J. Y. Han, "Spiral microchannel with rectangular and trapezoidal cross-sections for size based particle separation," *Scientific Reports,* vol. 3, Mar 18, 2013.

[6] Y. N. Yang, S. K. Hsiung, and G. B. Lee, "A pneumatic micropump incorporated with a normally closed valve capable of generating a high pumping rate and a high back pressure," *Microfluidics and Nanofluidics,* vol. 6, no. 6, pp. 823-833, Jun, 2009.

[7] H. Y. Tan, W. K. Loke, and N. T. Nguyen, "A reliable method for bonding polydimethylsiloxane (PDMS) to polymethylmethacrylate (PMMA) and its application in micropumps," *Sensors and Actuators B-Chemical,* vol. 151, no. 1, pp. 133-139, Nov 26, 2010.

[8] T. N. A. Vo, and P. C. Chen, "Maximizing interfacial bonding strength between PDMS and PMMA substrates for manufacturing hybrid microfluidic devices withstanding extremely high flow rate and high operation pressure," *Sensors and Actuators a-Physical,* vol. 334, Feb 1, 2022.

[9] T. N. A. Vo, P. C. Chen, and Y. H. Chen, "Applying Hybrid Bonding Technique to Manufacture A Peristaltic Micropump With Extremely High Flow Rate." 2022 IEEE 17th International Conference on Nano/Micro Engineered and Molecular Systems (NEMS), pp. 1-5.

[10] M. Rafeie, J. Zhang, M. Asadnia, W. H. Li, and M. E. Warkiani, "Multiplexing slanted spiral microchannels for ultra-fast blood plasma separation," *Lab on a Chip,* vol. 16, no. 15, pp. 2791-2802, 2016.

CONTACT

* Prof. Pin-Chuan Chen, Tel: +886-2-2737-6456; Email: pcchen@mail.ntust.edu.tw

DIRECT PATTERNING ON POROUS SURFACE USING DROP IMPACT PRINTING

Bheema Sankar Reddy[1], Chandantaru Dey Modak[1,2], Rutvik Lathia[1], Bhawana Agarwal[1,3],
Ebinesh Abraham R[1], and Prosenjit Sen[1,*]
[1]Indian Institute of Science, Bangalore, INDIA,
[2]CNRS - ESPCI PSL, Paris, FRANCE, and
[3]The Johns Hopkins University, Baltimore, Maryland, USA.

ABSTRACT

This work reports the direct printing of highly concentrated nanoparticle inks on porous surfaces using a simple, low-cost sieve-based technique. The formulated ink has a seven-fold higher concentration as compared to conventional printer inks. On porous surfaces, highly concentrated ink reduces drop spreading by ~ 33% and penetration by ~ 55%. Printing such high-concentration inks exhibits extremely high conductivity of 3.7×10^3 S/m for a single pass print. This work demonstrates novel high-conductive printing possibilities on porous surfaces for developing flexible/wearable sensors and electronic applications. The technique further removes the constraints in ink palettes and substrate dependency.

KEYWORDS

Superhydrophobic Sieve; Droplet Impact; Printing; Mass loading

INTRODUCTION

Conductive printing has been widely explored over decades for printed-electronic applications due to several benefits such as efficient usage of materials, reduced energy consumption both in the manufacturing and utilization phases, reduced usage of hazardous substances, and better overall recyclability [1], [2]. The choice of substrates for printing is much higher for printed electronics which provides favorable physical characteristics to the products, such as low weight, stretchability, resistance to bending, etc. [3], [4]. Inkjet printing has proven to be superior to conventional methods of developing flexible and wearable devices and systems.

Printing on porous flexible substrates has been of interest due to its application in flexible electronics. However, it comes with many challenges. The porosity of the substrate makes the printing difficult as the ink spreads and penetrates into the surface. This significantly degrades the line width and conductivity [5]. In general, printing is carried out by coating the porous substrate to reduce its porosity, which requires an additional step [6]. Further, the low concentration of inks that most printers are able to handle requires multiple passes to achieve reasonable conductivity. This reduces the throughput of the printing process.

We used a sieve-based technique to print ink droplets in the present paper. This technique is unlike conventional printers such as inkjet, EHD and acoustic, that use nozzles to generate droplets on demand. The technique known as Drop impact printing [7] uses a sieve that is fabricated using an etching technique followed by Teflon coating. The superhydrophobic sieve acts as a nozzle that generates tiny droplets when a mother drop is impacted from an optimized height (Figure 1A). This mechanism allows us to print droplets of high particle concentrations (~ 70% mass loading (w/w)). In contrast to other techniques, clogging is not a problem as the droplet remains in contact with the sieve for only ten milliseconds. High concentration reduces the spreading and penetration of the ink. Thus, allowing us to print on porous substrates without an additional coating layer.

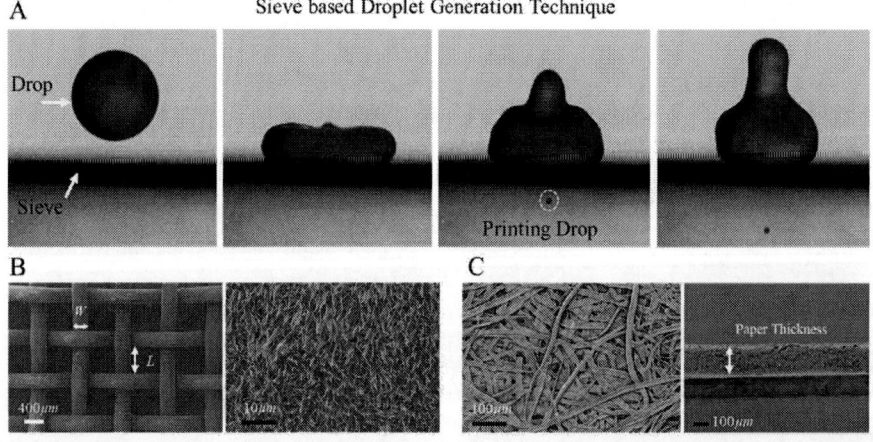

Figure 1: Time-lapse image of a drop (diameter~2.55mm) impacting on a superhydrophobic sieve generating a single droplet. (B) SEM image of sieve #0.012 showing sieve structure and grown nanowires using the etching technique, (C) SEM image of paper (top and cross-sectional view)

METHODOLOGY

All the chemicals used in the sieve preparation as well as the ink formulation, are purchased from Sigma-Aldrich unless mentioned otherwise.

Sieve Preparation

The superhydrophobic sieve used for printing the droplets is made by wet etching [8] the copper sieve #0.012 with a pore opening size of ~ 533 μm. The copper sieve (area = 6 cm^2) is initially cleaned by sonicating with acetone, isopropyl alcohol, and deionized (DI) water for 5 minutes each. The sieve was blow-dried using a nitrogen gun and placed in 33% (v/v) sulphuric acid for about 30 seconds to remove any oxides present on the surface. The sieve was cleaned again with DI water and dried with a nitrogen gun to remove residues.

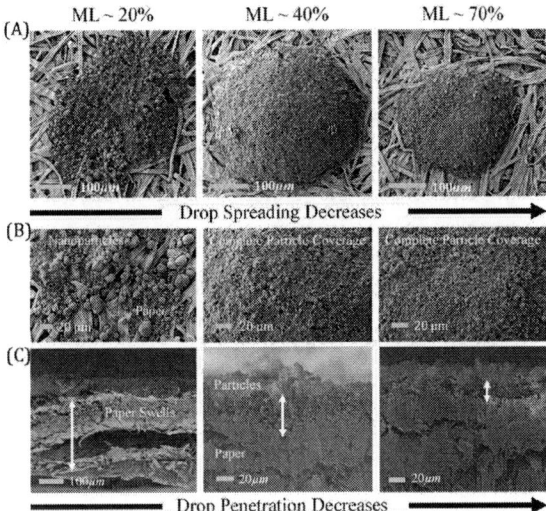

Figure 2: SEM images of deposited droplets. (A) Different mass-loaded drops printed on paper substrate using sieve #0.0045 with pore opening 139μm. (B) Zoomed in top view image of the deposited droplets. (C) Cross-section of the deposited droplets.

The etching is carried out by placing the sieve in a mixture of two solutions. The first solution is prepared by adding 5 gram sodium hydroxide in 50 mL of DI water. Simultaneously, another solution is prepared by adding 1.34 gram of ammonium persulfate in 50 mL of DI water. Both solutions were stirred thoroughly using magnetic stirrers for about 10 minutes. They are then mixed properly, and the sieve was dipped vertically in the etch solution. After 15 minutes, the sieve turned light bluish color due to the growth of copper hydroxide (Cu(OH)$_2$) nanowires (tip diameter ~ 220 nm) and was removed from the solution. Finally, it was cleaned with DI water, blow-dried with a nitrogen gun, and coated with Teflon by soaking it in Teflon solution for about 5 minutes followed by heating at 120 °C for 10 minutes to make it superhydrophobic.

Ink Preparation

Initially, a solution of 25% (v/v) ethanol-water is prepared. To make the conducting ink for printing

purposes, a solution containing 5% Ethylene Glycol, 0.5% Ethanol Amine and 94.5% of Ethanol-Water mixture prepared earlier is mixed thoroughly using a magnetic stirrer for several minutes. The silver nanopowder (purchased from BioChemazone) is then added to this solution in the required proportions to obtain the final conducting ink of varying mass loadings (ML) or concentrations (i.e., 4%, 20%, 40% and 70% ML (w/w)).

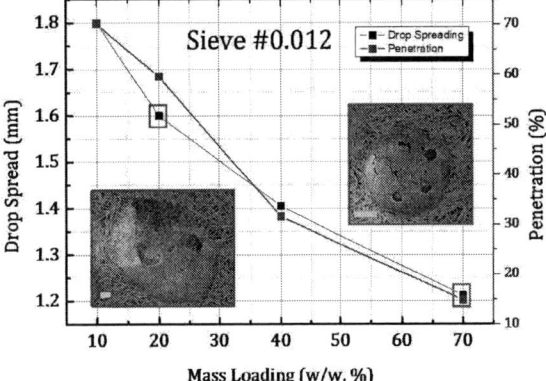

Figure 3: Plot showing spreading diameter and penetration versus mass loading for sieve #0.012 with pore opening ~533 μm. Scale = 100 μm.

RESULTS

The study was carried out using a sieve of different pore openings (#0.0045 with 139 μm pore size, and #0.012 with 533 μm pore size, Figure 1B). Different particle ink concentrations (4%, 20%, 40%, and 70% Mass loading, ML) were printed on A4 sheets of thickness ~ 150μm (Figure 1C). Figure 2 shows the SEM image of printed drops for different mass-loading inks. The spreading factor and penetration dynamics of ink is shown with varying mass loading. The increase in mass loading results in less spreading and less penetration (with respect to 4% mass loading ink) without any substrate swelling.

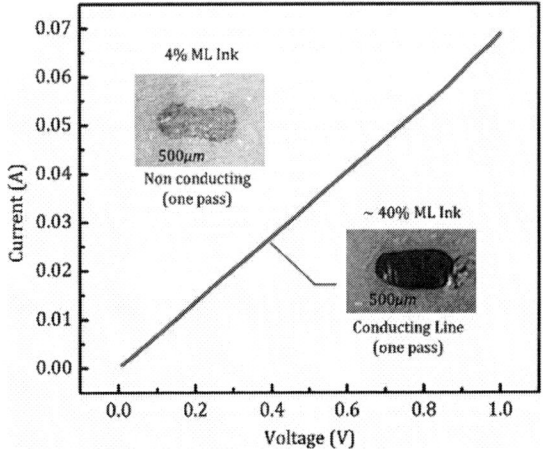

Figure 4: IV Characteristics of printed line for two different mass loading ink.

For low mass loading (20%), printed drop cause

978-1-6654-9309-3/23 $31.00 © 2023 IEEE

swelling of the paper (Figure 3). These results indicated that printing with high mass loading proves to be advantageous as it will leave more particles at the surface and prevent swelling. Further, we printed a line using 4%- and 40%-ML ink to check for conductivity. The high mass loaded ink (40%) exhibits extremely high conductivity (3.7 x 10^3 S/m) on paper with a single pass printing (Figure 4). Further, the conductivity can be increased by using higher-loading inks. To achieve similar conductivity, inkjet printers need 6-8 passes [9]. The process is time consuming and multiple passes compromise the resolution. The sieve-based printing can solve the present challenges in paper-based circuit printing and open an avenue for advanced flexible electronic printing applications.

CONCLUSION

We report a novel way of printing nanoparticle inks with high concentrations on porous surfaces using drop impact printing technique which is not only simple but also a low-cost sieve-based technique. Due to its ability to print inks with high nanoparticle concentrations, we could use an ink that is seven times the concentration of inks used in a conventional inkjet printer hence, reducing the number of passes required to print a conducting line with the ink to achieve a particular conductivity significantly making it an efficient and economical method.

ACKNOWLEDGEMENTS

The authors would like to thank the Department of Science and Technology and Ministry of Electronics and Information Technology, Government of India for financial support. The authors also acknowledge the Prime Minister's Research Fellowship for the support.

REFERENCES

[1] Kunnari, E.; Valkama, J.; Keskinen, M.; Mansikkamäki, P. Environmental evaluation of new technology: Printed electronics case study. J. Clean. Prod. 2009, 9, 791–799.

[2] Dimitriou E, Michailidis N. Printable conductive inks used for the fabrication of electronics: an overview. Nanotechnology. 2021 Oct 13;32(50). doi: 10.1088/1361-6528/abefff. PMID: 33735843.

[3] Cheng, I.C.; Wagner, S. Overview of flexible electronics technology. In Flexible Electronics; Springer: Berlin/Heidelberg, Germany, 2009; pp. 1–28.

[4] Nathan, A.; Ahnood, A.; Cole, M.T.; Lee, S.; Suzuki, Y.; Hiralal, P.; Bonaccorso, F.; Hasan, T.; Garcia-Gancedo, L.; Dyadyusha, A.; et al. Flexible electronics: The next ubiquitous platform. Proc. IEEE 2012, 100, 1486–1517.

[5] Asai A. et. al., "Impact of an Ink Drop on Paper", *J. Img. Sc. Tech.* **1993**, 37, 205-207.

[6] Lessing J. et. al., "Inkjet Printing of Conductive Inks with High Lateral Resolution on Omniphobic "R F Paper" for Paper-Based Electronics and MEMS", *ACS Appl. Mater.* **2014**, 26, 4677-4682.

[7] Modak, C. et. al., "Drop Impact Printing", *Nat. Comm.* 2020, 1-11.

[8] Kumar, Arvind, et al., "Designing assembly of meshes having diverse wettability for reducing liquid ejection at terminal velocity droplet impact", *Journal of Microelectromechanical Systems* 27.5 (2018): 866-873.

[9] Patil, P. et.al. Inkjet printing of silver nanowires on flexible surfaces and methodologies to improve the conductivity and stability of the printed patterns. *Nanoscale Adv.*, **2021**, 3, 240-248

CONTACT

*Prosenjit Sen, +917406890351, prosenjits@iisc.ac.in
Bheema Sankar Reddy, bheemareddy@iisc.ac.in
Chandantaru Dey Modak, chandantarum@iisc.ac.in
Rutvik Lathia, rutviklathia@iisc.ac.in
Bhawana Agarwal, bhawanaa@iisc.ac.in
Ebinesh Abraham R, ebineshar@iisc.ac.in

MANUFACTURING 3D-PRINTED PAPER MICROFLUIDICS INTEGRATED WITH IONIZATION MASS-SPECTROMETRY FOR ILLICIT DRUGS ANALYSIS AND ON-CHIP CHROMATOGRAPHY

Muhammad Faizul Zaki[1], Pin-Chuan Chen[1], Yi-Xin Wu[2], and Pai-Shan Chen[2]

[1]Department of Mechanical Engineering National Taiwan University of Science and Technology
[2]Graduate Institute of Toxicology, College of Medicine, National Taiwan University

ABSTRACT

This paper presents paper-based microfluidic devices fabricated by stereolithography 3D, enabling spray ionizations and chromatography mass-spectrometry analysis of illicit drugs. The fabrication techniques provide a rapid printing process (<1s) and precise control of channel structure that can be performed in a highly reproduced manner. The significant improvements in analytical performance on mass-spectrometry using this device are described such as the extended signal lifetime until 50 minutes and high stability of MS[1] signal can be achieved with CV<15%. A reliable signal of Methamphetamine via tandem mass-spectrometry can be produced with the Limit of Detection (LOD) of methamphetamine can be as low as 0.5 ppb using plasma as a matrix. Finally, the chromatography method was established by using two solvents during the spray which can separate two different drugs Methamphetamine and Morphine-G during the acquisitions. This 3D printed paper spray further simplifies the ionization methods and significantly enhanced the performance of the paper spray which allowed continuous spray with high intensity and reliable signal including chromatography of drugs mixture.

KEYWORDS

Paper-based microfluidic devices, stereolithography 3D printing, paper-spray ionization, chromatography, drug analysis

INTRODUCTION

Clinical drug testing becomes an important testing method since many fields are required to perform such as abuse drug testing, sports doping testing, drug treatment, and legal evidence. Accordingly, the demand for clinical drug testing increased over the year. Practically, clinical drug testing is consisting two steps a presumptive test and a confirmatory test. The presumptive test mainly used an immunoassay method. Meanwhile, the confirmatory test has to be conducted in the lab by employing ambient ionization methods such as gas chromatography or liquid chromatography mass-spectrometry analysis. However, current clinical drug testing was very expensive, labor-intensive, and time-consuming [1].

Currently, paper spray ionization (PSI) has been developed to simplify the ionization method. PSI became a promising technique for ambient ionization mass-spectrometry since it provides simple, low-cost, and the properties of paper itself naturally perform the extraction in the sample liquid. Moreover, the capillary force of paper becomes a major advantage that can be operated without external devices. Since the first development by Graham Cooks et.al [2], PSI becomes an emerging device in many

applications such as detection (e.g. blood, plasma, urine), Forensic toxicology, and environmental analysis [3]. Thus, the application promoted PSI that suitable method to detect complex samples and performed all compounds for a specific profile as well as chemical structure.

Conventionally, PSI is only plain paper to detect the sample as they are coupled with mass spectrometry. The geometry of the paper becomes an important factor to deliver the sample directly to the Mass-spectrometry. Therefore, the shape of the paper itself has been modified to enhance the spray ability as well as the Taylor cone spraying shape [4]. Several paper modification techniques have already been attempted. The key is to modify the porosity of the paper which aimed to decrease the matrix effect, improve the extraction and prevent the sample diffusion [5].

However, the analytical performance of paper spray ionization is still limited especially for chromatography. Chromatography is an important aspect of mass-spectrometry analysis which aimed to separate delicate and complex natural mixtures of raw samples. Commonly column chromatography is used to separate the drugs in the LC-MS process by allowing the sample through the compact column and applying continuous solvent as well as high pressure as a mobile phase separation. Unlike ESI, PSI could not apply a continuous solvent to separate. Instead, PSI used a dumping method in which, a few microliters of solvent were added to perform sample elution and ionizations. The reason is simply that the paper properties could not retain a large amount of solvent which makes the paper easily broken. Hence, the PSI was hardly performing chromatography since it required continuous solvent and high pressure

In this study, we reported a new approach for creating paper-based microfluidic devices fabricated by Stereolithography 3D printing for paper spray ionization and on-chip chromatography. The rapid printing process can proceed with <1s. The hydrophobic structure of the printed chip assists the flow of solvent directly toward the tip. In addition, precise control of the channel depths provided herein significantly reduced sample diffusion and improve the signal lifetime. A continuous flow of solvent was applied and the chip can maintain the spray for >60 minutes. Single drug analysis was performed and the chip can generate, high stability, intensity, and reliability mass-spectrometry signal. Finally, the chromatography of paper spray was established by utilizing two different solvents continuously that can separate two drugs simultaneously during the acquisitions.

EXPERIMENTAL METHODS
Material and Instruments

Photocurable resin (Durable⁺) purchased from FreEntity, Taipei City, Taiwan was used to print the structure. Alcohol 95% (ethanol) was purchased from Echo Chemical Co., Ltd. Cellulose filter paper (Advantec 5c) purchased from Toyo Roshi Kaisha Ltd (Japan) was used with 1 μm pore size. The digital mask format was in PNG with a resolution of 1920 x 1080 px. 3D print digital light processing (DLP) Stereolithography (SLA) purchased from FreEntity (Taipei, Taiwan) was used to print the channel. The specification of the 3D printing consists of UV light with 405nm wavelength, a resolution of 25 μm for each pixel and 48 mm x 27 mm printing areas. The system of the 3D print SLA DLP is shown in Figure 1(a). It is noteworthy that, the Digital Micromirror Device (DMD) has an important role to determine UV light patterns following the pattern of the digital mask (Figure 1 (b)). The black area of the digital mask was an uncured area and the white area is the area that was cured and exposed to UV light.

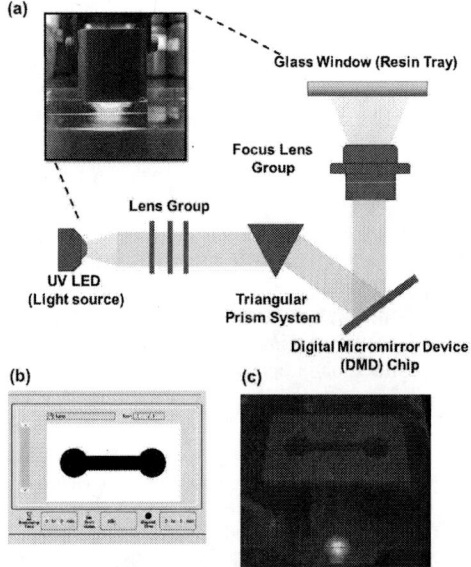

Figure 1: (a) Stereolithography Digital Light Processing 3D Printing system (b) digital mask on 3D print software(c) UV light pattern from the projection of digital mask on the 3D print platform

Fabrication Method

The fabrication process consists of three main parts: (1) preparation, (2) printing (3) cleaning. After the paper was soaked in the resin, the first printing process was performed to create the main channel structure by exposing the UV light at 0.2s exposure time (Figure 2(b)). After that, a second printing process was performed to generate a specific channel depth using a less exposure time (0.06s) shown in Figure 2(c). Then, the post-printing was performed including washing using alcohol, drying it in the oven at 80°C for 3 min, and the UV post-curing (Form 3 purchased from Formlabs Inc. Somerville, Massachusetts, USA) to stabilize the resin which is shown in Figure 2(d). The total time to fabrication of the μPADs was approximately < 10 minutes.

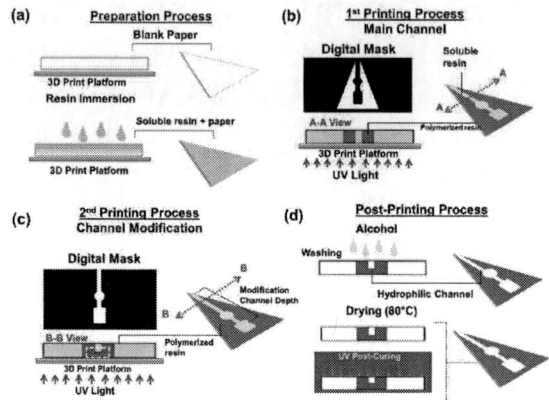

Figure 2: fabrication process of 3D printed paper spray (a) preparation process of resin and paper on the 3D print platform (b) 1st printing to form to print the main channel (c) 2nd printing to modify the channel depths to specific thickness (d) post-printing process including washing, drying, and post-curing.

3D Printed Paper-Spray Setup

All the experiments were carried out using TSQ Altis triple quadrupole MS (Thermo Fisher Scientific, MA, USA). 0.1% formic acid in acetonitrile was used as a spray solvent. Figure 3(a) presents the experimental setup of paper-chip mass-spectrometry analysis. The paper was placed on the crocodile clip which the clip was connected to the high-voltage source from the MS machine to generate electrical discharge. The preparation is described as follows: (1) The sample was dripped onto the channel and dried at room temperature. (2) After it dried, the paper chip was placed in the crocodile clip at the base channel. (3) Once the chip was ready the solvent was applied continuously to the channel followed by high-voltage electricity. The solvent flowed via capillary force through the channel while also eluting the samples. The solvent formed a spray at the paper tip and the compounds were ionized. A preliminary test was performed to evaluate the ideal MS parameters such as the flow rate and voltage which resulting the ideal flow rate being 20 μl/min (Figure 3(b)) and the ideal voltage being 3 kV (Figure 3(c)).

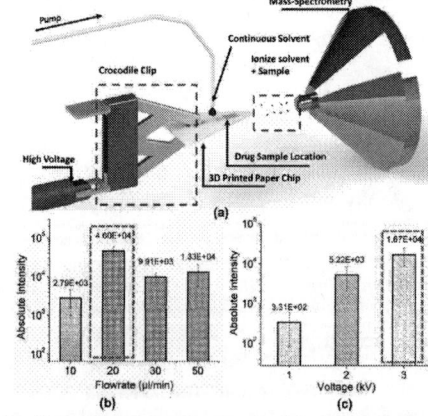

Figure 3: mass-spectrometry setup and parameter identification (a) 3D printed paper-spray mass-spectrometry setup (b) ideal conditions of solvent flow rate (c) ideal voltage conditions parameters

The data were recorded in selected ion monitoring (SIM) or selected reaction monitoring (MRM) mode and were analyzed with the Thermo Scientific Freestyle software. Three sample drugs such as Methamphetamine, Methamphetamine-d5, and Morphine-G were purchased from Cerilliant Corporation (Round Rock, TX, USA). LC-MS grade acetonitrile was obtained from Duksan Pure Chemicals. Formic acid was purchased from Sigma-Aldrich (St. Louis, MO, USA).

On-Chip Chromatography

Herein, we proposed a new method to conduct the chromatography analysis on a paper chip which is shown in Figure 4. Initially, a mixture of drugs consisting of methamphetamine and morphine-G was added to the channel. After that, the chip was placed on the high voltage clip and applied the continuous solvent flow as a mobile phase separation. Two types of a solvent such as 100% and 50% Acetonitrile were applied sequentially. In the first three minutes, 100% acetonitrile was applied and then switched to 50% acetonitrile which aimed to elute a specific drug.

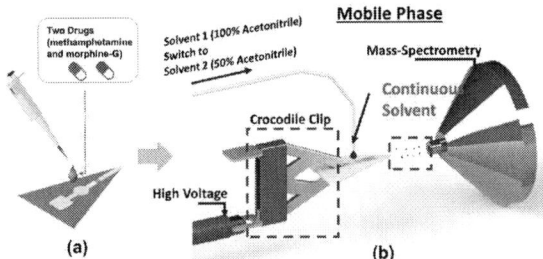

Figure 4: On-chip chromatography method (a) sample preparations on paper chip (b) the setup and MS parameters of on-chip chromatography

EXPERIMENTAL RESULTS
3D Print parameters and Device Characterization

The paper was printed using photocurable resin and exposed to UV light to lead polymerization of resin and form the hydrophobic structure. The exposure time becomes the main parameter to determine the thickness of the resin structure. Figure 5(a) shows that as exposure time increases, resin thickness also increases. Extremely fast printing can be proceeded to print the channel with an exposure time of only <1s. The design of 3D printed paper spray is shown in Figure 5(b) and consists of the electrode location area, solvent chamber, main channel, and tip. Channel depths were modified by employing shorter exposure time (0.06 s) to perform a half resin structure (110 µm) shown in Figure 5(c). The purpose of modifying the depth is to reduce the solvent diffusion, pointed the flow toward the tip and MS inlet, and concentrated the samples. Meanwhile, the tip was not modified, and maintain the porous substrate to get the optimized spray shape. The surface property namely the water contact angle was measured and it shows a significant difference between the resin surface and paper substrate with the angle was 79° and 0° respectively (Figure 5(d)). This can be guaranteed that the resin can prevent the diffusion of the samples and solvent.

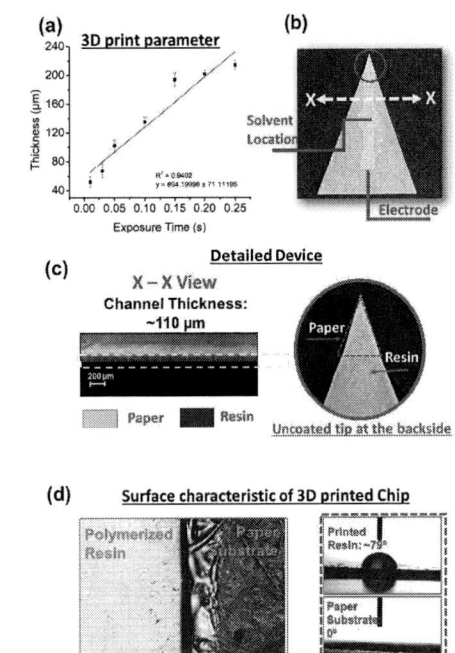

Figure 5: Printing parameter and device characterization (a) calibration curve of resin thickness and exposure time (b) the components of 3D printed paper-chip (c) Cross-sectional channel structure and the tip conditions of 3D printed paper spray (d) surface characterization of the resin surface and paper substrate

Spray Performance and Single Drug Analysis on 3D Printed Paper-Spray Mass-spectrometry

The spray lifetime was characterized by recording the background signal of the 3D-printed paper chip and blank paper. Figure 6(a) shows that the 3D print paper chip can maintain the spray for up to >50 minutes via continuous solvent. The paper chip was still in perfect condition after 50 minutes of spray with no change or damage in the channel structure. On the other hand, the blank paper was only able to spray for 5 minutes and the chip can not spray after 10 minutes due to the excessive amount of solvent that damaged the paper. The spray duration was also evaluated by using a limited solvent volume (10 µl) or dumping method. As a result, Figure 6(b) shows that the paper chip has a longer spray duration in different solvent conditions (Pure water, 50% acetonitrile, and 100% acetonitrile) compared to the blank paper. The reason is that the 3D-printed paper chip can significantly reduce the diffusion of the solvent hence the spray can last longer. The stability of the spray on the 3D printed paper chip was recorded and the result shows that the 3D print paper chip has good stability with the CV<15% (Figure 6(c)). Furthermore, the analytical performance of a 3D-printed paper chip was evaluated via tandem mass-spectrometry by adding methamphetamine as a targeted drug. Figure 6(d) shows the MRM chromatograms of methamphetamine. Three transitions of methamphetamine were obtained (Black: m/z $150.1 \rightarrow$ m/z 91.1; Red: m/z $150.1 \rightarrow$ m/z 119.1, Green: m/z $150.1 \rightarrow$ m/z 64.9) and the accurate ion ratio was also

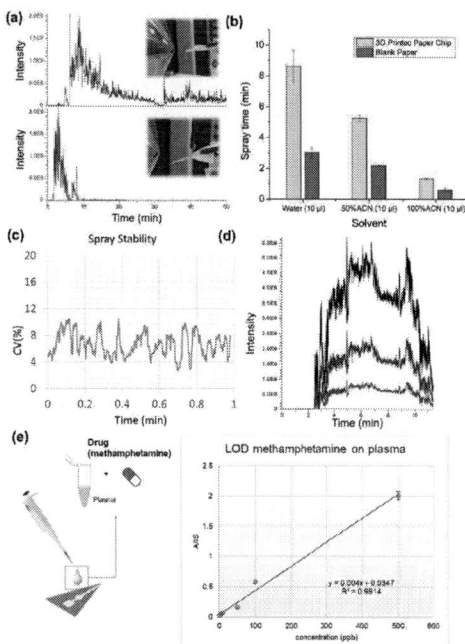

Figure 6: (a) Continuous spray evaluation of 3D printed paper chip and blank paper (b) The spray lifetime using 10 µl solvent volume (c) stability test of m/z 150.1 on 3D printed chip (Mean CV: 13.05 ± 5.17 (n=4)) (d) The MRM chromatograms (MS²) of the methamphetamine (Black: m/z 150.1 → m/z 91.1; Red: m/z 150.1 → m/z 119.1, Green: m/z 150.1 → m/z 64.9) (e) Limit of Detection (LOD) of methamphetamine on plasma (LOD: 0.5 ppb R²=0.9941).

obtained (ESI infusion as the reference) indicated that the signal was surely coming from methamphetamine. LOD was conducted for single drug analysis in the plasma matrix. The quantitation was based on the methamphetamine to methamphetamine-d5 ratio. As a result, 3D printed paper chip can achieve the LOD until 0.5 ppb in 0.1 µl volume or equal to 0.05 pg, with the linearity of the calibration curve, also achieved with $R^2 = 0.9914$ as shown in Figure 6(e).

Mixture Drug Analysis and On-Chip Chromatography on 3D Printed Paper-Spray Mass-spectrometry

The sample consists of 2 drugs such as methamphetamine and Morphine-G were mixed into the

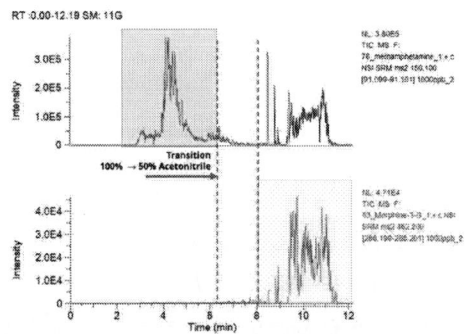

Figure 7: separated drug chromatograms of the methamphetamine (3 minutes – 6 minutes) and morphine-3-G (9 minutes – 11 minutes).

solvent. Then, 0.1 µl of the mixture of drugs was added to the channel following the protocols described in Figure 4. As a result, we successfully separated two different drugs during the acquisitions. Figure 7 shows the MS signal of each drug that showed up as a peak. The methamphetamine showed up earlier due to its hydrophobicity so it was eluted first by 100% acetonitrile. Then followed Morphine-G which showed up afterward when the more polar solvent (50% acetonitrile) was applied since Morphine-G is more hydrophilic

CONCLUSION

In summary, we developed paper-based microfluidic devices fabricated by stereolithography 3D printing for spray ionizations and on-chip chromatography analysis. Several advantages were provided using our chip in mass-spectrometry analysis: (1) The spray lifetime can significantly extend until 50 minutes via continuous flow solvents and also allowed a longer spray lifetime using limited solvent volumes compared to blank paper. (2) High stability (CV<15%) and high-reliability signal of the methamphetamine can be generated via tandem mass-spectrometry (3) The LOD for single methamphetamine can achieve low to 0.5 ppb on plasma as a matrix (4) Finally, chromatography can be performed by separating two drugs namely methamphetamine and Morphine-G during the acquisitions with clear distinguished peak signal.

ACKNOWLEDGEMENTS

This research was funded by National Science and Technology Council (NSTC) Taiwan with grant numbers of 111-2221-E-011-122-MY3

REFERENCES

[1] K.E. Moeller, K.C. Lee, and J.C. Kissack, "Urine drug screening: practical guide for clinicians", *Mayo Clin. Proc.*, Vol. 83, pp. 66-76, 2008.

[2] H. Wang, L. Jiangjiang, R. G. Cooks, and Z. Ouyang. "Paper spray for direct analysis of complex mixtures using mass spectrometry", *Angew. Chem. Int.*, Vol. 122, pp. 889-892, 2010.

[3] E. M. McBride, P. M. Mach, E. S. Dhummakupt, S. Dowling, D. O. Carmany, P. S. Demond, G. Rizzo, N. E. Manicke, and T. Glaros. "Paper spray ionization: Applications and perspectives" *Trac-Trends Anal. Chem.*, Vol. 118, pp. 722-730, 2019.

[4] Q. Yang, H. Wang, J.D. Maas, W. J. Chappell, N.E. Manicke, R. G. Cooks, and Z. Ouyang, "Paper spray ionization devices for direct, biomedical analysis using mass spectrometry", *Int. J. Mass Spectrom.*, Vol. 312, pp. 201-207, 2012.

[5] S. Chiang, W. Zhang, and Z. Ouyang, "Paper spray ionization mass spectrometry: recent advances and clinical applications", *Expert Rev. Proteomics.*, Vol. 15, pp. 781-789, 2018.

CONTACT

*P.C. Chen phone: +886-2-2737-6456; email: Pcchen@mail.ntust.edu.tw

DETECTION LIMITS IN NANOMECHANICAL MASS FLOW SENSING FOR NANOFLUIDICS WITH NANOWIRE OPEN CHANNELS

Javier E. Escobar[†], Juan Molina[†], Eduardo Gil-Santos, José J. Ruz, Óscar Malvar,
Priscila M. Kosaka, Javier Tamayo, Álvaro San Paulo*, and Montserrat Calleja
Instituto de Micro y Nanotecnología, IMN-CNM, CSIC (CEI UAM+CSIC), SPAIN
[†]These authors contributed equally.

ABSTRACT

Nanomechanical sensing is applied to measure the transport of liquid along the external surface of nanowire open fluidic channels. We integrate nanowire open channels with microcantilever resonators as sensors, enabling larger flow measurements; or use the nanowires simultaneously as open channels and resonators, for quantifying extremely small amounts of liquid. We study the detection limits of both systems and their implications for measuring the transport of ionic liquid from a microdroplet reservoir to Si nanowires when these are partially inserted into the microdroplet in the absence of any electrical signal. We obtain detection limits in the order of tens of femtogram and single attograms for each system, respectively, and a combined sensing range from 1 ag to beyond 10 pg, enabling the detection of transport of liquid volumes down to the zeptoliter scale.

KEYWORDS

Semiconductor nanowires, NEMS, nanomechanical resonators, nanofluidics, open fluidics, ionic liquids.

INTRODUCTION

Closed nanofluidic systems, such as channels, nozzles, or tubes, where fluids are enclosed by solid structures, are commonly used for the control of liquids [1]. Open systems, in which the liquid moves along the outer surface of a solid channel, are consolidating as an alternative bioinspired technological approach that provides otherwise impossible functionalities [2], especially for extremely miniaturized liquid handling [3]. Nanomechanical resonators [4], firmly established as ultra-high resolution mass sensors [5], can be integrated with open fluidics to obtain a quantitative measurement of flow rate in open systems with an unprecedented precision.

In this work we analyze the detection limits of two different realizations of this approach: first, a microcantilever-nanowire (MC-NW) device designed to allow larger flow measurements; and second, a nanowire-alone (NW-A) device that offers a much higher mass sensing resolution. Our study compares the fundamental mass detection limit of both devices with the actual resolution derived from their experimental frequency stability, and analyzes their sensing performance for

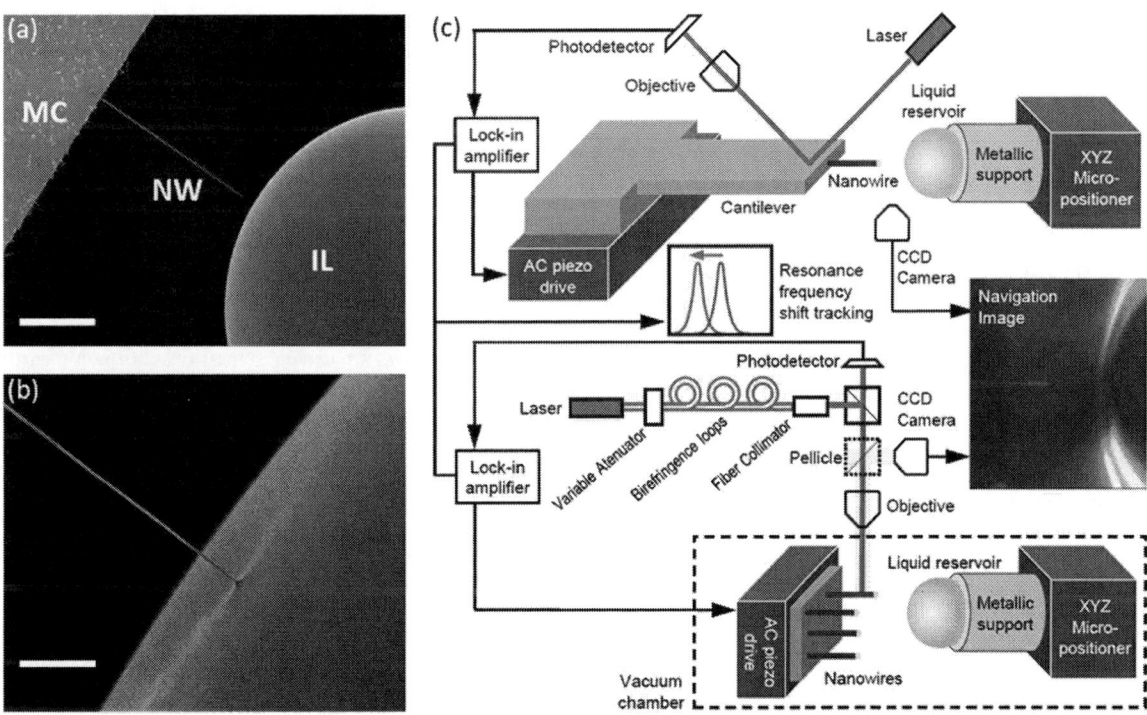

Figure 1: Nanomechanical mass flow sensing measurement setup. (a) SEM image of a MC-NW device placed in proximity of an ionic liquid microdroplet reservoir; scale bar: 20 μm. (b) SEM image showing a partial insertion of a nanowire into the ionic liquid microdroplet reservoir; scale bar: 2 μm. (c) Schematic depiction of experimental setup and resonance frequency tracking based on laser beam deflection transduction for the MC-NW device (upper section) and on laser beam scattering modulation for the NW-A device (lower section).

measuring the transport of ionic liquids (ILs) to the NWs when they are partially inserted into a microscopic IL reservoir in the absence of any electrical signal. We also determine the mass detection range of the devices by analyzing how the coupling between MC and NW in the MC-NW device limits its upper boundary.

RESULTS

Device design, fabrication and operation

The Si NWs used as open nanofluidic channels were grown by the VLS mechanism in an atmospheric pressure chemical vapor deposition reactor with $SiCl_4$ as precursor gas and 150-250 nm diameter colloidal Au nanoparticles (NPs) as growth catalyst [6]. Typical NW lengths are in the 30-50 μm range, base diameters of 150-470 nm and tip diameters of 30-270 nm. The MC-NW devices consist of a Si MC with its longitudinal axis oriented along a (111) crystallographic direction, where Si NWs are grown after random deposition of the Au catalyst NPs. These are conveniently diluted to obtain no more than single NW at the edge of the MC (Figure 1a). The MCs have a length of 500 μm, a width of 90 μm and a thickness of 1-2 μm. The NW-A devices consist of VLS Si NWs vertically grown on a raw Si (111) substrate, with a previous step in which Au catalyst NPs are placed at desired positions near one edge of the substrate by nano-pipetting.

The IL reservoir is a microdroplet of DMPI-TFSI or DEME-TFSI placed on a metallic support that is mounted on a micro-positioning stage, which enables controlled insertions of the NW tip of around 1 μm (Figure 1b). This insertion length is small compared to our typical device length, allowing to relate a shift in its resonance frequency, Δf_R, to a liquid mass deposited near its tip, Δm_L, by the expression

$$\Delta m_L = -2 m_R \, \Delta f_R / f_R, \qquad (1)$$

where m_R and f_R are, respectively, the effective mass and resonance frequency of the resonator device before the insertion [7]. The effective mass of the NW-A device can be calculated as described in [6]. A similar approach to that followed in [8] can be applied to obtain m_R for the non-uniform cross section MC-NW device: interestingly, for the typical dimensions specified above, the only significant contribution of the NW to the total MC effective mass is given by the position of the added mass, placed away from the free end of the MC to the tip of the NW. This results in a 20% decrease in m_R.

In order to determine Δf_R, the resonance frequency of the fundamental flexural mode is measured before and after each insertion. Laser beam deflection readout is used for the MC-NW implementation, whereas scattering modulation [9] is used for the NW-A resonator (Figure 1c). Piezo-acoustic driving is used in both cases. Resonance frequency tracking is performed via phase-locked loop. MC-NW devices are operated in air, whereas NW-A devices are placed in a high vacuum chamber at 10^{-5} mbar; both at room temperature conditions. Typical values of fundamental resonance frequency f_R and quality factor Q are around 5-15 kHz and 10-50 for the MCs, and around 200-300 kHz and 7,000-40,000 for the NWs.

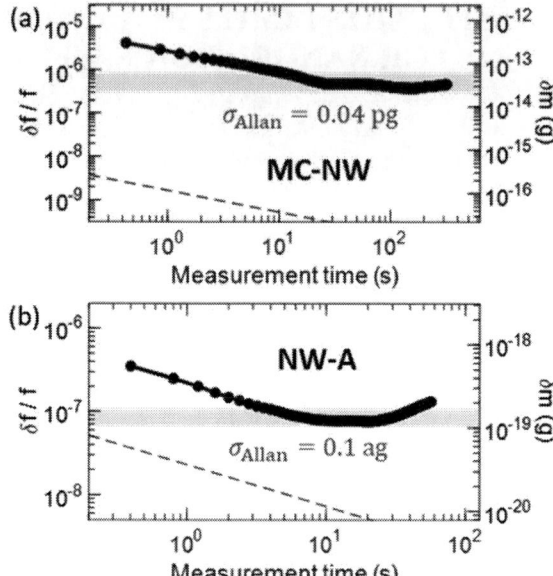

Figure 2: Experimental Allan deviation of resonance frequency as a function of measurement time (solid lines) for the MC-NW device (a) and for the NW-A device (b). The right axes use the effective mass of each device to convert frequency stability into minimum detectable mass. The dashed lines represent the fundamental limit.

Device performance and detection limits

Figure 2 shows measurements of Allan deviation performed to study the frequency stability $\delta f / f$ of our resonators [10]. At the measurement times of interest regarding our experiments (1-100 s), the Allan deviation takes values below 1 and 0.1 ppm for the MC-NW and NW-A devices, respectively; which can be converted into minimum detectable mass δm through equation 1. This results in experimental limits of 0.04 pg and 0.1 ag (see blue areas and right axes of Figure 2). Thus, the NW-A device provides a much higher mass resolution, mostly as a consequence of its much lower effective mass. The fundamental mass detection limit in δm was calculated as described in [6], and it is also plotted in the graphs for comparison (dashed lines). This theoretical value considers that the resonators are driven to their onset of nonlinearity and that thermomechanical noise is the dominant noise source. The calculation for the MC-NW considers that the added mass is located at the tip of the NW instead of the free end of the MC, which results into a 12% decrease in δm, mainly due to the 20% reduction in m_R discussed above. At short measurement times (< 10 s), the Allan deviation of the MC-NW device is three orders of magnitude higher than the fundamental limit. This difference is attributed to the fact that the full intrinsic mechanical dynamic range (DR) of the resonator is not accessed. At the lower DR boundary, the low Q resulting from operation in air makes instrumental noise dominate over thermomechanical noise. The upper DR boundary is also limited because the resonator is not driven up to its onset of nonlinearity, but around one order of magnitude lower due to piezo-drive power limitations. Vacuum operation to increase Q and instrumental improvements would reduce the difference between experimental and

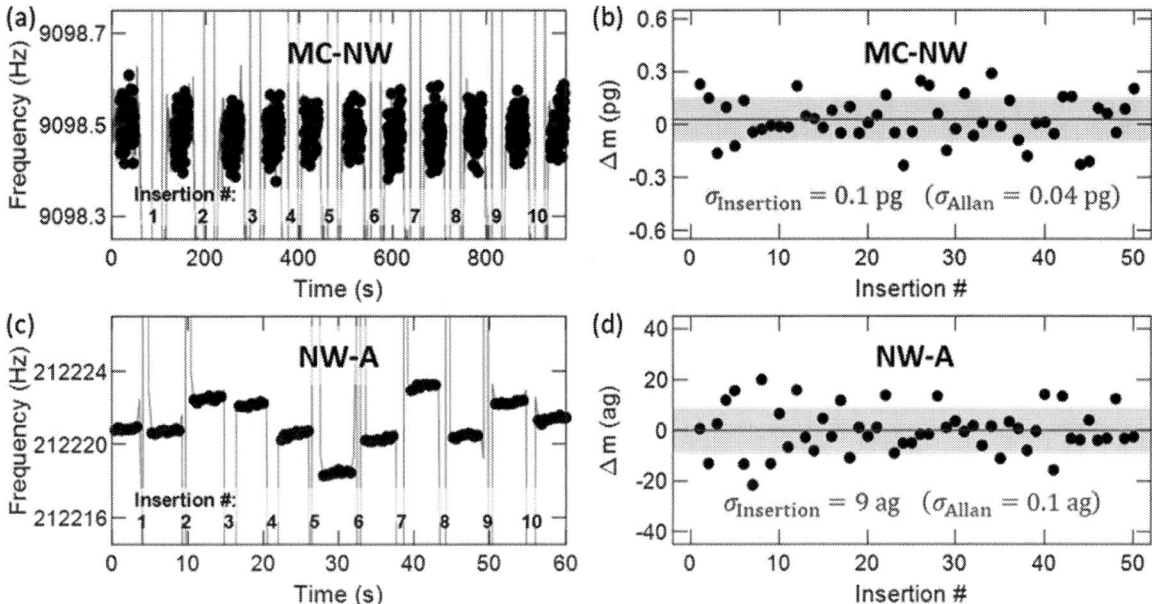

Figure 3: Effect of repeated insertions of NW tip into IL reservoir. (a) Example of resonance frequency tracking during a series of 10 iterated insertions of 20 s for the MC-NW device from Figure 2a, showing a negligible effect on the measured signal after each insertion event. The thin line represents all experimental values, including those for which the acquired signal is lost during insertion time, whereas dots only indicate the points from which the average frequency is calculated before and after each insertion. (b) Effective mass shift calculated from the tracked frequency of the MC-NW device after 50 iterated insertions. The red line indicates an average value of 0.03 pg and the red area represents a standard deviation of 0.1 pg. (c) Example of resonance frequency tracking during a series of 10 iterated insertions of 1 s for the NW-A device from Figure 2b, showing a clear effect of insertion events on the acquired signal after applying the same analysis as in (a). (d) Effective mass shift calculated from the tracked frequency of the NW-A device after 50 iterated insertions. The red line indicates an average value of -0.1 ag and the red area represents a standard deviation of 9 ag.

theoretical limits. Regarding the NW-A device, thermomechanical noise can be properly resolved in vacuum, pushing its Allan deviation only one order of magnitude above the fundamental limit. Increasing piezo drive up to the onset of nonlinearity would improve the experimental limit of both devices at short measurement times, but it could also compromise frequency stability at longer times.

We now analyze the implications of these detection limits for measuring the transport of IL from a microdroplet reservoir to the tip of NWs when these are repeatedly inserted into the IL reservoir in the absence of any applied electrical signal. Previous work has determined that in such conditions, a precursor film of a few nm thickness is expected to spread spontaneously on the NW surface, but no additional flow happens without applying a bias voltage between the IL reservoir and the NW [3]. Figure 3a shows that for the MC-NW device, the resonance frequency before each insertion does not significantly change when it is measured after the insertion (the signal is not tracked during insertion). If we convert these small frequency shifts into deposited mass Δm_L through equation 1, we obtain the series shown in Figure 3b (extended to 50 insertion events). The standard deviation $\sigma_{Insertion}$ with respect to the 0.03 pg average deposited mass obtained from this series is 0.1 pg, which is comparable to the detection limit established by σ_{Allan}. Since the average value lies below both σ_{Allan} and $\sigma_{Insertion}$, we conclude that any IL mass transferred during the insertions is not sufficiently above the detection

limit of the MC-NW device to provide meaningful measurements. We perform the same analysis on the NW-A system (Figure 3c-d), finding a remarkable difference. For this device, the signal is clearly shifted after each insertion, which translates into an average transported mass of -0.1 ag and a $\sigma_{Insertion}$ of 9 ag, almost two orders of magnitude above σ_{Allan}. On the one hand, the negligible average value is consistent with the expectation of no liquid flow at zero bias voltage [3]. But on the other hand, the increased dispersion well above σ_{Allan} is attributed to a real mass effect due to the adsorption and desorption of small amounts of liquid after each insertion. The mass resolution of the NW-A device allows to observe this effect, which is screened by frequency noise in the MC-NW device. We thus conclude that the NW-A device provides enough resolution to detect IL transfer in these conditions.

Figure 4a shows a characterization of $\sigma_{Insertion}$ as a function of tip diameter in 41 different NW-A devices, revealing a clear and steep positive correlation between both magnitudes. The observed correlation provides support for our interpretation of $\sigma_{Insertion}$ as a mass effect, as it is expected that wider NWs move larger IL masses. The thinnest NW tip diameter that we have tested is around 30 nm, providing a dispersion of around 1 ag. This value defines a lower experimental limit in the ability of our technology to measure liquid transport, which, converted into volume, lies in the single zeptoliter scale.

Finally, we estimate the upper limit of our sensing approach as the maximum added mass that the MC-NW

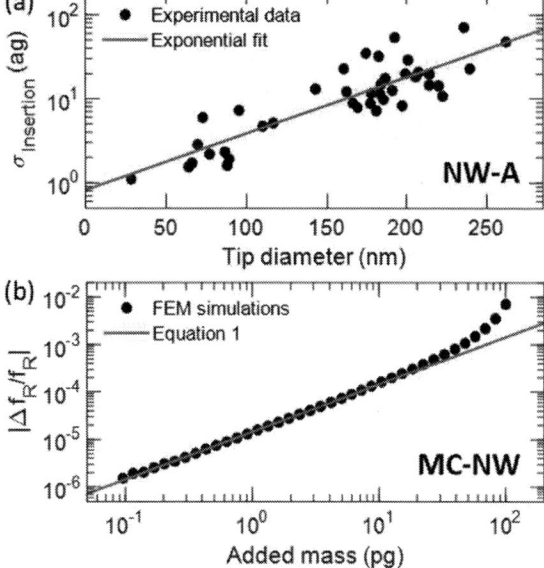

Figure 4: Liquid transport detection limits. (a) Insertion dispersion of 41 different NW-A devices as a function of tip diameter (black dots), showing a clear correlation between both magnitudes (an exponential fit is included in red as a guide to the eye). (b) Simulated frequency shift as a function of the amount of added mass at the tip of the NW in the MC-NW device (black dots), and comparison with the theoretical expression used in this work to convert frequency shift into effective mass (red line).

device can support within the framework of our analysis. Equation 1 does not have into account the coupling effect arising when, as a consequence of the added mass, the frequency of the fundamental flexural mode of the NW approximates that of the MC [11]. This phenomenon can be calculated by finite element simulations by varying the magnitude of a point mass located at the tip of the NW. Black dots in Figure 4b show the results of this simulation compared to the theoretical dependence expected if only equation 1 is considered (red line). The region where both curves diverge significantly, above a few tenths of pg, defines an upper limit for mass flow control where our method would not correctly determine the amount of liquid mass added. This upper limit could be further increased by designing MC and NW such that their fundamental flexural modes have more distant resonance frequencies.

CONCLUSION

The integration of nanomechanical resonators with open fluidics provides unprecedented tools for applications such as printing, patterning, constrained chemical reactions or bioassays. Our study of the mass detection limits of two particular implementations of these devices demonstrates their suitability for high resolution and high dynamic range mass flow sensing. After exploring the implications of such detection limits for measuring the transport of IL from a microdroplet reservoir to the tip of NWs as these are repeatedly inserted into the IL reservoir in the absence of any applied electrical signal, we find that the random adsorption and desorption of small amounts of liquid after each insertion can be resolved when the NWs are used both as channels and as sensors.

ACKNOWLEDGEMENTS

This work was supported by the ERC-CoG 681275 "LIQUIDMASS" and by the Spanish Science and Innovation Ministry through Projects "EXOFLUX" (PGC2018-101762-B-I00) and "MOMPs" (TEC2017-89765- R). We acknowledge the service from the MiNa Laboratory at IMN, and funding from CM (project S2018/NMT-4291 TEC2SPACE), MINECO (project CSIC13-4E-1794) and EU (FEDER, FSE). E.G.S. acknowledges financial support by the Spanish Science and Innovation Ministry through Ramón y Cajal grant RYC-2019-026626-I.

REFERENCES

[1] W. Sparreboom, A. van den Berg, and J. C. T. Eijkel, "Principles and applications of nanofluidic transport," *Nat. Nanotechnol.*, vol. 4, no. 11, pp. 713–720, 2009, doi: 10.1038/nnano.2009.332.

[2] C. T. Ertsgaard, D. Yoo, P. R. Christenson, D. J. Klemme, and S.-H. Oh, "Open-channel microfluidics via resonant wireless power transfer," *Nat. Commun. 2022 131*, vol. 13, no. 1, pp. 1–9, Apr. 2022, doi: 10.1038/s41467-022-29405-2.

[3] J. Y. Huang *et al.*, "Nanowire liquid pumps," *Nat. Nanotechnol. 2013 84*, vol. 8, no. 4, pp. 277–281, Mar. 2013, doi: 10.1038/nnano.2013.41.

[4] S. Schmid, L. G. Villanueva, and M. L. Roukes, *Fundamentals of nanomechanical resonators*, vol. 49. Springer, 2016.

[5] J. Chaste, A. Eichler, J. Moser, G. Ceballos, R. Rurali, and A. Bachtold, "A nanomechanical mass sensor with yoctogram resolution," *Nat. Nanotechnol.*, vol. 7, no. 5, pp. 301–304, 2012, doi: 10.1038/nnano.2012.42.

[6] J. Molina *et al.*, "High Dynamic Range Nanowire Resonators," *Nano Lett.*, vol. 21, no. 15, pp. 6617–6624, Aug. 2021, doi: 10.1021/acs.nanolett.1c02056.

[7] K. L. Ekinci, Y. T. Yang, and M. L. Roukes, "Ultimate limits to inertial mass sensing based upon nanoelectromechanical systems," *J. Appl. Phys.*, vol. 95, no. 5, pp. 2682–2689, Feb. 2004, doi: 10.1063/1.1642738.

[8] O. Malvar *et al.*, "Tapered silicon nanowires for enhanced nanomechanical sensing," *Appl. Phys. Lett.*, vol. 103, no. 3, Jul. 2013, doi: 10.1063/1.4813819.

[9] J. Molina *et al.*, "Optical Transduction for Vertical Nanowire Resonators," *Nano Lett.*, p. acs.nanolett.9b04909, Mar. 2020, doi: 10.1021/acs.nanolett.9b04909.

[10] M. Sansa *et al.*, "Frequency fluctuations in silicon nanoresonators," *Nat. Nanotechnol.*, vol. 11, no. 6, pp. 552–558, Jun. 2016, doi: 10.1038/nnano.2016.19.

[11] E. Gil-Santos *et al.*, "Optomechanical detection of vibration modes of a single bacterium," *Nat. Nanotechnol.*, vol. 15, no. 6, pp. 469–474, 2020, doi: 10.1038/s41565-020-0672-y.

CONTACT

*A. San Paulo, tel: +34 918060700;
alvaro.sanpaulo@csic.es

CONTROLLING PARTICLE AGGREGATION AND SEPARATION IN LIQUID ON MEMBRANE RESONATORS

Haoran Zhang [1,2], Hao Jia [1,2], and Xinxin Li[1,2]
[1]State Key Lab of Transducer Technology, Shanghai Institute of Microsystem and Information Technology, Chinese Academy of Sciences, Shanghai 200050, CHINA
[2]University of Chinese Academy of Sciences, Beijing 100049, CHINA

ABSTRACT

We design a dual-mode membrane resonator for non-invasively and fast manipulating microparticle aggregation and separation in liquid. By performing finite element modeling, we verify that the interaction between a pair of microspheres (d=20 μm, mimicking cancer cells) can be precisely controlled on our proposed rectangular membrane resonator. The model consists of Multiphysics coupling between mechanical vibration, fluid flow, and particle tracing. Stokes drag, acoustic radiation, and particle-particle interaction forces are analyzed to predict the microparticle trajectory (including aggregation and separation). When exciting the rectangular membrane (100×50 μm) at its 1st mode ($f_{1,1}$), a pair of microspheres are observed to aggregate at the device center (antinode of $f_{1,1}$). While switching to its 2nd mode ($f_{2,1}$), the 2 microspheres tend to separate towards the opposite directions to 2 antinodes of $f_{2,1}$. Our proposed membrane platform holds promise for high-precision, acoustically manipulating and analyzing cellular interactions in liquid.

KEYWORDS

membrane resonator, particle interaction, streaming flow, Chladni figure.

INTRODUCTION

The ability to control microscale objects and analyze their interactions is important for engineering functional interfaces, and sensor devices. Particularly, in biological research, techniques to control relative positions and interactions of biological species (*e.g.*, cells) play an important role in fundamental studies to better understand intercellular behaviors and cancer metastasis [1]. To date, micropipette aspiration [2], single-cell force microscopy [3], and optical tweezers [4] have been exploited for such purposes. They have shown limitations of invasiveness to cells sophisticated operations when quantifying intercellular interactions mechanically or optically [1]. More importantly, they measure cells pair by pair.

More recently, acoustic tweezers [5] (*e.g.*, SAWs) offer a versatile approach to manipulating cells to pressure nodes of ultrasonic waves (tens to hundreds of MHz) via acoustic radiation force.

In this work, we propose a dual-mode membrane resonator to control particle aggregation and separation, by exploiting the Chladni patterning effect induced by flexural resonances (hundreds of kHz) [6]. MEMS resonators have been widely studied for ultrasensitive chemical and biological sensing [7-9], especially for single microparticles and cells [10-11], while their capabilities for acoustic manipulation of microscale objects in liquid have remained largely unexplored. In our case, Stokes's drag force dominates the particle trajectory towards antinodes with minimized acoustic radiation and heating. As shown in Figure 1, when exciting the rectangular membrane at its 1st mode ($f_{1,1}$), two microparticles tend to aggregate at the device center (antinode of $f_{1,1}$). While switching to its 2nd mode ($f_{2,1}$), the 2 microspheres tend to separate towards the opposite directions to 2 antinodes of $f_{2,1}$.

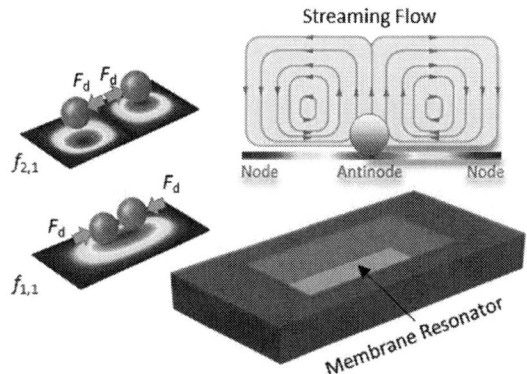

Figure 1: Concept of controlling aggregation and separation of 2 microspheres via a dual-mode membrane resonator ($f_{1,1}$ and $f_{2,1}$). The Chladni figure effect tends to drag the microspheres to the nearest antinode(s) of the mode shape by the vibration-induced streaming flows.

DEVICE DESIGN AND MODELING
Modeling of Multimode Membrane in Liquid

We first theoretically expect the first 3 modes of a multimode membrane in liquid based on the following equation,

$$f_{mn,l} = \frac{f_{mn,v}}{\sqrt{1+\gamma}}, \qquad (1)$$

Where $f_{mn,l}$ is the resonant frequency in liquid, $f_{mn,v}$ is the resonant frequency in vacuum, and γ is a nondimensionalized added virtual mass incremental (NAVMI) factor [12]. It explains the downward shift of the resonance frequency caused by hydrodynamic loads in the liquid environment compared to that in vacuum. $f_{mn,v}$ can be modeled by the following wave equation [13],

$$f_{mn,v} = \frac{1}{2}\sqrt{\frac{\sigma_0}{\rho_{3D}}}\sqrt{\left(\frac{m}{a}\right)^2 + \left(\frac{n}{b}\right)^2}, \qquad (2)$$

Where σ_0 and ρ_{3D} represent the initial stress and membrane mass density, m and n are the numbers of the half wave periods along the length and width, and a, b are the length and width of the rectangular membrane.

Figure 2 illustrates the multimode resonances of a rectangular-shaped membrane in liquid. Given the lateral dimensions $L \times W \sim (20\text{-}200) \times (10\text{-}100\ \mu m)$, we expect the first 3 modes $(f_{1,1}, f_{2,1}, f_{3,1})$ in the range of hundreds of kHz to a few MHz.

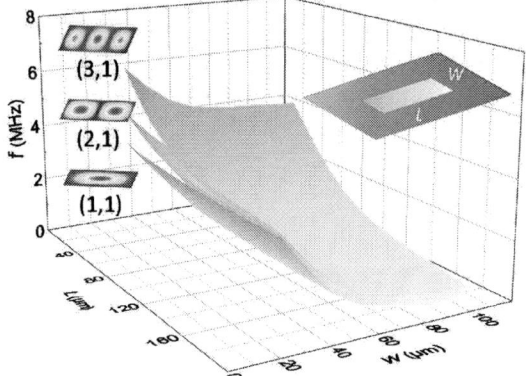

Figure 2: Design of multimode membrane resonator in liquid. Given the lateral dimensions $L \times W \sim (20\text{-}200) \times (10\text{-}100\mu m)$, we expect the first 3 modes $(f_{1,1}, f_{2,1}, f_{3,1})$ in the range of hundreds of kHz to a few MHz.

Modeling of Streaming Flow Patterns

According to the frequency range of the multimode resonance in Figure 2, we perform finite element modeling (FEM) to verify that our proposed membrane platform could manipulate the microparticle aggregation and separation, hence controlling the particle interaction. The model consists of Multiphysics coupling between mechanical vibration, fluid flow, particle-particle interaction, and particle trajectory tracing.

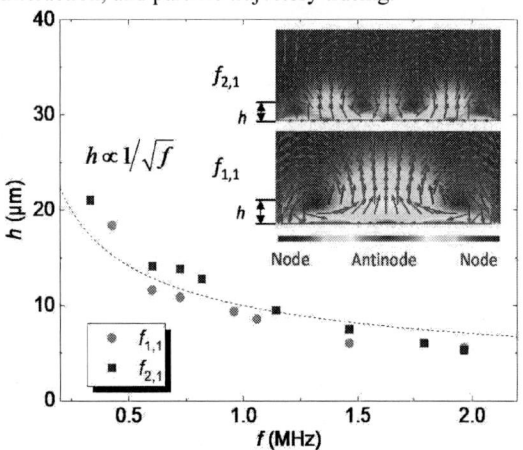

Figure 3: Modeling of streaming flow patterns induced by $f_{1,1}, f_{2,1}$ of membrane resonators with different dimensions in Fig. 2. The streaming flows dominate the microparticle movement on the device. The height from the device surface to the outer vortex eyes h is proportional to $\sqrt{f}$.

The streaming flow field, which is influenced by the fluid-resonator interaction, is derived from the Navier-Stokes (N-S) equation [14]:

$$\rho_0 \nabla \cdot \overline{u_2} = -\nabla \cdot \overline{\rho_1 u_1}$$
$$-\nabla \overline{p_2} + \mu \nabla^2 \overline{u_2} + \left(\mu_B + \frac{1}{3}\mu\right)\nabla\left(\nabla \cdot \overline{u_2}\right) = \overline{\rho_1 \frac{\partial u_1}{\partial t}} + \rho_0 \overline{(u_1 \cdot \nabla) u_1} \tag{3}$$

where ρ_i, p_i and u_i (i=0,1,2) are the initial, 1st- and 2nd-order fluid density, pressure, and velocity. Navier-Stokes

equation with periodic moving boundary (the mode shape, $A\cos(m\pi x/L)\cos(n\pi y/W)$) is solved to visualize the acoustic streaming flow field. As shown in Figure 3, outer vortices (h) are frequency dependent, $\propto 1/\sqrt{f}$, which are mostly considered given the microsphere diameter (d=20μm, mimicking cancer cells) and resonance frequencies ~hundreds of kHz to several MHz from rectangular membranes with lateral dimensions on the order of ~100μm.

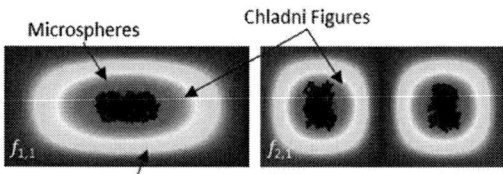

Figure 4: Micro Chladni figures in liquid on dual-mode membrane resonator. Given device dimensions of $L \times W = 100 \times 50 \mu m$, and $f_{1,1}$=1.17MHz and $f_{2,1}$=1.65MHz, tracing particles (d~1μm) are simulated to be dragged by the vortex flows towards the antinode(s) of the mode shapes, forming inverse Chladni figures.

RESULTS AND DISCUSSIONS

Micro Chladni Figures in Liquid

To verify the particle manipulation capability of the membrane resonator in liquid, we first simulate the Chladni figure formation using microspheres (PS beads, d=1μm). As shown in Figure 4, given the rectangular membrane platform with $L \times W = 100 \times 50 \mu m$, and the resonance frequencies of $f_{1,1}$=1.17MHz and $f_{2,1}$=1.65MHz, tracing particles are simulated to be dragged by the vortex flows towards the antinode(s) of the mode shapes, forming inverse Chladni figures.

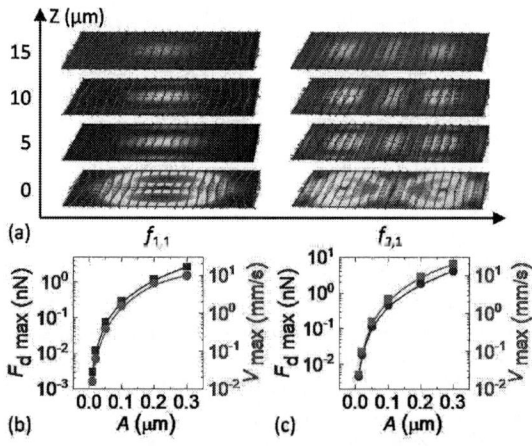

Figure 5: Fluid flow field and drag force on the membrane resonator ($100 \times 50 \mu m$). (a) Slides of velocity field when the membrane vibrates at $f_{1,1}, f_{2,1}$ with displacement $A = 100$nm. (b-c) corresponding maximum velocity and drag force ($d/2$=10μm) vs. vibrational amplitude A.

Fluid Flow Field and Drag Force

By varying vibrational amplitude (A), we expect to quantitatively control the streaming flow velocity, hence the drag force on the membrane resonators from pN to nN. The slides of flow patterns above the membrane surface are shown in Figure 5a. As the resonance amplitude increases from 0.02μm to 0.3μm, the maximum drag force and velocity on the membrane resonator raise about three orders of magnitude, as shown in Figure 5b. It can be inferred that we can further control the particle interaction via resonance amplitude and achieve fast manipulation of microparticle aggregation and separation in liquid.

Microparticle Aggregation and Separation

We further verify the capability of the dual-mode membrane resonator to manipulate the interaction between a pair of microparticles (d=20 μm, mimicking cancer cells). As shown in Figure 6a, when exciting the rectangular membrane (L×W ~100×50μm) at its 1st mode ($f_{1,1}$), the microspheres can be observed to aggregate at the antinode of $f_{1,1}$. While switching to its 2nd mode ($f_{2,1}$), the 2 microspheres tend to separate towards the opposite directions to 2 antinodes of $f_{2,1}$. The aggregation and separation forces (pN-nN) can be controlled via resonance amplitude, and the aggregation and separation are simulated to happen within 1 millisecond at A=100 nm (Fig. 6b). These simulation results also validate that the membrane resonator could be potentially used to quantify the particle-particle interaction, as the particle separation happens when the drag force is larger than the particle-interaction force till reaching a cut-off distance.

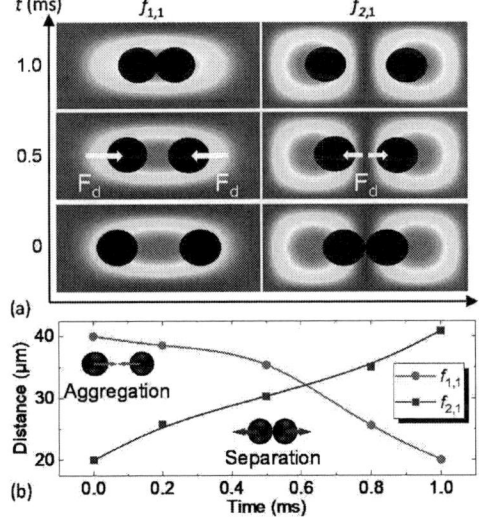

(a)

(b)

Figure 6: Microparticle aggregation and separation by controlling dual modes of the membrane resonator (100×50μm). The aggregation and separation are simulated to happen within 1 millisecond at A=100nm.

CONCLUSION

We report on the non-invasive, fast manipulation of microparticle aggregation and separation in liquid via a dual-mode membrane resonator. Using finite element simulation, we demonstrate for the first time that the interaction between a pair of microspheres (d=20μm, mimicking cancer cells) can be precisely controlled (at pN level) via exploiting flexural vibration-induced Chladni figures effect, *i.e.*, exciting a rectangular membrane (100×50μm) at its fundamental mode ($f_{1,1}$) brings microspheres in contact at the device center (antinode of $f_{1,1}$), while switching to its 2nd mode ($f_{2,1}$) tends to separate the 2 microspheres towards opposite directions (2 antinodes of $f_{2,1}$). The aggregation and separation forces can be further controlled via resonance amplitude. Our results can guide the design and realization of a new type of acoustofluidic device which can spatially reconfigure the particle assembly, and stimulate a versatile chip-scale platform for investigating intercellular interactions toward biophysical applications.

ACKNOWLEDGEMENTS

This work is supported by the National Natural Science Foundation of China (62227815, 61974155, 61874130, 61804156, 62104241), the National Key R&D Program of China (2021YFB3200800, 2020YFB2008603), Key Research Program of Frontier Sciences of the Chinese Academy of Sciences (QYZDJ-SSW-JSC001), Shanghai "Road and Belt" International Young Scientist Exchange Program (19510744600), Scientific Instrument Project of the Chinese Academy of Sciences (YJKYYQ20210024), Shanghai Pujiang Program (20PJ1415600). The authors also acknowledge the support of the Innovation Team and Talents Cultivation Program of the National Administration of Traditional Chinese Medicine (ZYYCXTD-D-202002, ZYYCXTD-D-202003).

REFERENCES

[1] J. Kashef, C. M. Franz, "Quantitative Methods For Analyzing Cell-Cell Adhesion In Development", *Developmental Biology*, vol. 401, pp. 165–174, 2015.

[2] Y.-S. Chu, W. A. Thomas, O. Eder, F. Pincet, E. Perez, J. P. Thiery, S. Dufour, "Force Measurements in E-Cadherin–Mediated Cell Doublets Reveal Rapid Adhesion Strengthened by Actin Cytoskeleton Remodeling Through Rac and Cdc42", *Journal of Cell Biology*, vol. 167, pp.1183–1194, 2004.

[3] M. Benoit, D. Gabriel, G. Gerisch, H. E. Gaub, "Discrete Interactions in Cell Adhesion Measured by Single-Molecule Force Spectroscopy", *Nature Cell Biology*, vol. 2, pp. 313–317, 2000.

[4] H. Zhang, K. K. Liu, "Optical tweezers for single cells", *J. R. Soc. Interface*, vol. 5, pp. 671–690, 2008.

[5] F. Guo, L. Peng, J. B. French, "Controlling Cell-Cell Interactions Using Surface Acoustic Waves", *Proceedings of the National Academy of Sciences of the United States of America*, vol. 112, pp. 43–48, 2014.

[6] H. Jia, H. Tang, P. X.-L. Feng, "Standard and Inverse Microscale Chladni Figures in Liquid for Dynamic Patterning of Microparticles on Chip", *Journal of Applied Physics*, vol. 124, art. no. 164901, 2018.

[7] H. Jia, P. Xu, X. X. Li, "Integrated Resonant Micro/Nano Gravimetric Sensors for Bio/Chemical Detection in Air and Liquid", *Micromachines*, vol. 12,

pp. 645, 2021.

[8] F. Yao, P. Xu, H. Jia, X. Li, H. Yu, X. X. Li, "Thermogravimetric Analysis on a Resonant Microcantilever", *Analytical Chemistry*, vol. 94, pp. 9380–9388, 2022.

[9] Z. Wang, B. Xu, S. Pei, J. Zhu, T. Wen, C. Jiao, J. Li, M. Zhang, J. Xia, "Recent Progress in 2D Van Der Waals Heterostructures: Fabrication, Properties, and Applications", *Science China Information Sciences*, vol. 65, art. no. 211401, 2022.

[10] H. Jia, P. X.-L. Feng, "Very High-Frequency Silicon Carbide Microdisk Resonators with Multimode Responses in Water for Particle Sensing", *Journal of Microelectromechanical Systems*, vol. 28, pp. 941–953, 2019.

[11] H. Jia, Y. Chen, X. Wang, T. Xu, X. X. Li, "In-Plane Mode Encased Cantilevers for Cancer Cell Detection in Liquid", *in Proceedings of 21st International Conference on Solid-State Sensors, Actuators and Microsystems (Transducers)*, Orlando, FL, Jun. 20-24, 2021.

[12] M. K. Kwak, "Hydroelastic Vibration of Rectangular Plates", *J. Appl. Mech.,* vol. 63, pp. 110-115, 1996.

[13] S. Timoshenko, *Vibration Problems in Engineering* (2nd Ed.), New York: D. van Nostrand Company, Inc., 1937.

[14] J. Friend and L. Y. Yeo, "Microscale Acoustofluidics: Microfluidics Driven via Acoustics and Ultrasonics", *Reviews of Modern Physics,* vol. 83, pp. 647-704, 2011.

CONTACT

*H. Jia, hao.jia@mail.sim.ac.cn
*X. X. Li, xxli@mail.sim.ac.cn

DEVELOPMENT OF BOAT MODEL POWERED BY ELECTRO-HYDRODYNAMIC PROPULSION SYSTEM

Luan Ngoc Mai[1,2,*], Tuan-Khoa Nguyen[3], Trung Hieu Vu[4,**], Thien Xuan Dinh[5], Canh-Dung Tran[6], Hoang-Phuong Phan[7], Toan Dinh[6], Thanh Nguyen[6], Nam-Trung Nguyen[3], Dzung Viet Dao[4], and Van Thanh Dau[4,8,***]

[1]Ho Chi Minh City University of Technology (HCMUT), Ho Chi Minh City, VIETNAM.
[2]Vietnam National University Ho Chi Minh City, Ho Chi Minh City, VIETNAM.
[3]Queensland Micro and Nano Technology Centre, Griffith University, AUSTRALIA.
[4]School of Engineering and Built Environment, Griffith University, AUSTRALIA.
[5]Explosion Research Institute Inc., R&D Division, Tokyo, JAPAN.
[6]School of Mechanical and Electrical Engineering, University of Southern Queensland, AUSTRALIA.
[7]School of Mechanical and Manufacturing Engineering, UNSW Sydney, AUSTRALIA.
[8]Centre for Catalysis and Clean Energy, Griffith University, AUSTRALIA.

ABSTRACT

This paper demonstrates the development of an ion-wind powered boat (iBoat) that offers a simple propeller-less design, trivial noise, and relatively low energy consumption. The apparatus of the iBoat includes a conducting wire working as the ion emitter and the water surface as the ground electrode, eliminating the need to carry a counter electrode essential in conventional ion wind generation. A miniaturized version of the iBoat and an electro-hydrodynamic propulsion system that can generate a thrust-to-power ratio up to 1.45 mN/W using a novel wire-dielectric-water configuration. Numerical simulations will also be formulated to shed light on the physical principles underlying our novel concept.

KEYWORDS

Boat model, electro-hydrodynamic propulsion, ion wind, numerical simulations, thrust-to-power.

INTRODUCTION

In the field of electro-hydrodynamic, the flow of air induced by corona discharge at elevated electric field has been researched intensively. The phenomenon states that intense electric field near an electrode can emits airflow from an electrode with a momentum conversion efficiency at 1% [1, 2]. Multiple ion wind based EHD techniques have been reported for various applications, such as inertial sensing application [3, 4], generation of particles [5, 6] propulsion systems [7]. The advantages of ion wind/flow compared to other methods of generating propulsion include less power consumption and simple structure with no requirement of moving components [8]. For some applications using high potential between electrodes, ion wind is an assisting factor that contributes to the final results, such as reducing total net charge [9, 10], improving sensor quality [11 - 13], or assisting jet flow in fluidic application [14]. Furthermore, the employment and optimization of large-scale ionic wind for propulsion systems in the last three decades have proven great advantages and potentials of this method. Christenson and Moller [15] presented the thrust to input electric power ratio, (θ) and the thrust to airflow area ratio (Φ) as the important parameters to evaluate the efficiency of the system. It is then shown that ion wind would produce adequate thrust for aerospace applications if θ and Φ reach up to 20 N/kW and 20 N/m^2, respectively. Various approaches have been proposed to achieve the above specifications, for example, multi-staged propulsion or alternating positive/negative discharge [8, 16]. Optimization of propulsion in such system is studied via numerous electrode arrangements, such as wire-to-inclined wing [17], rod-to-plate [18], and oppositely charged dual-pin [19 - 21]. In the aspect of aerospace propulsion, the first ionic device to be airborne and a self-enduring flight of EHD powered aeroplane were recently reported [7, 22]. These first-of-its-kind flights demonstrate the possibility of ionic wind in generating thrust density of about 7 N/m^2 and thrust-to-power ratio of 5 N/kW by aerodynamically optimizing the design of the miniaturized aircraft. In these in-atmosphere devices, both electrodes of EHD and conventional flaps are attached to their body to create and control the EHD flow. Another, perhaps, greater potential application of EHD propulsion is in maritime industry, i.e., ship propulsion, in which weight is not as critical of an issue as in aerospace. This technology offers replacement of propeller-based engines which normally generate noise and are not environmentally friendly and is ideal for military or wildlife surveillance missions.

This paper demonstrates the development of an ion-wind powered boat (iBoat) propelled by a novel ion wind-based wire-dielectric-water configuration. Our new structure conveniently exploits the water surface as the ground electrode whilst a dielectric barrier is employed to navigate the direction of the electric field, vectoring the thrust for the iBoat. This configuration eliminates the need of a ground electrode normally positioned on the device's part and utilizes a dielectric barrier as a flap for both electric and aerodynamic field to rectify the EHD flow. To prove the feasibility of our new concept, a miniaturized iBoat is constructed. This small-scale model successfully sails with several advantages: uncomplicated design, noiseless operation, and low energy use for a range of supplied voltages from 3 kV to 18 kV. Here, we achieve a sufficient thrust-to-power ratio of 1.45 mN/W, indicating the ability as well as the effectiveness of the proposed EHD system.

IEEE MEMS 2023, Munich, GERMANY
15 - 19 January 2023

*I*BOAT DESIGN

The schema of the iBoat is described in Fig. 1. At the boat's stern, a thin wire with a diameter of 20 μm is attached parallelly to the water surface. The distance between the wire center point and the water surface varies from 11 mm to 17 mm. The wire is connected to a high positive voltage, so electrons are repulsed from air molecules, leaving a domination of positive ions in the vicinity of the wire. Under the influence of electric field and the Coulombic force, positive ions travel in the direction of the electric field, colliding with neutral air particles and transfer momentum to them. This process drives the air and generates ionic wind downward the ground electrode which is, in this case, the water surface.

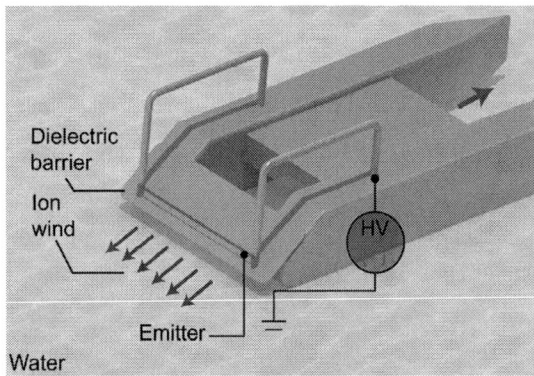

Figure 1: Fundamental schematic of ion-wind powered boat (iBoat).

As can be seen, our design does not include a second electrode downstream as used in most existing works, but instead utilizes the water surface as the collector. As ions are emitted out of the wire and induce airflow perpendicularly to the surface of water, an upward vertical reaction force (lift) is exerted on the wire. The installment of the dielectric barrier (hereinafter referred to as flap) converts lift into an effective horizontal force (thrust). Consequently, at an optimal position/angle of the flap, the lift can be sufficiently converted into thrust, propelling the iBoat forward. Experiments were conducted with various distances between the wire and water (d) and the dielectric inclination (by angle δ) to determine optimal thrust generation. Numerical simulations present the fundamental understanding over the generation of this propulsion.

SIMULATION MODEL AND RESULTS

Numerical EHD simulation depicts physical aspects of our novel design. Apparently, our concept is governed by fluid dynamic and electrostatic equations, commencing with the Navier-Stokes and continuity equations to describe fluid flow as follow

$$\nabla \cdot \left(\mathbf{UU} \right) - \nabla \cdot \left(\nu \nabla \mathbf{U} \right) = -\nabla p + \frac{q\mathbf{E}}{\rho}, \quad (1)$$

$$\nabla \cdot \left(\mathbf{U} \right) = 0, \quad (2)$$

where p is the air pressure; $\nu = 15.7 \times 10^{-5}$ m²/s the kinematic viscosity and $\rho = 1.204$ kg/m³ the air density.

For electrostatic system, Poisson and charge conservation equations calculate the electric field and govern the movement of charge in the domain,

$$\nabla \cdot \left(\mu \mathbf{E} q + \mathbf{U} q \right) = 0, \quad (3)$$

$$\nabla \cdot \left(\nabla V \right) = -\frac{q}{\varepsilon}, \quad (4)$$

where $\mu = 1.6 \times 10^{-4}$ m²/(Vs) is the mobility of ion; $\mathbf{U}$ is the velocity of air drifted by the charge movement; ε is the permittivity of air and $\mathbf{E}$ is the electric field determined as by $\mathbf{E} = -\nabla V$. The boundary condition on the electrodes' surface is set up by a density q_s calculated from the measured discharge current I

$$q_s = \frac{I}{\mu E_{on} A}, \quad (5)$$

where A is the area of the electrodes' surface, and $E_{on} = 3.23 \times 10^6$ V/m the onset electric field [18]. In this circumstance, Coulombic force is the only force exerting on electrical charge,

$$\mathbf{f}_e = q\mathbf{E}. \quad (6)$$

Thrust (T) generated by ion wind is estimated by [23]

$$T = \int_V qEdV = \int_0^d qEAdx, \quad (7)$$

where dV is the volume differential, d is gap length, A is a characteristic area, and E is the electric field intensity.

Simulation results show that, without the dielectric barrier, ion wind from the wire flows perpendicularly to the water surface, meaning only lift is generated (Fig. 2a). In contrast, when the dielectric is placed between the

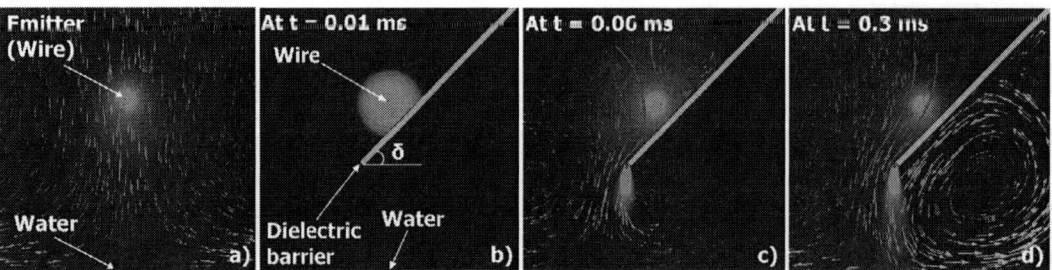

*Figure 2: a) 2-D numerical analysis of flapless configuration; b, c, d) 2-D transient numerical analysis of wire-dielectric-water configuration for interelectrode distance d = 15 mm and barrier angle δ = 45°, voltage applied V = 15 kV, current I = 50.41 μA, **vector**: ion wind velocity; **contour**: charge density (C/m³); **red arrows** generalize ion wind velocity.*

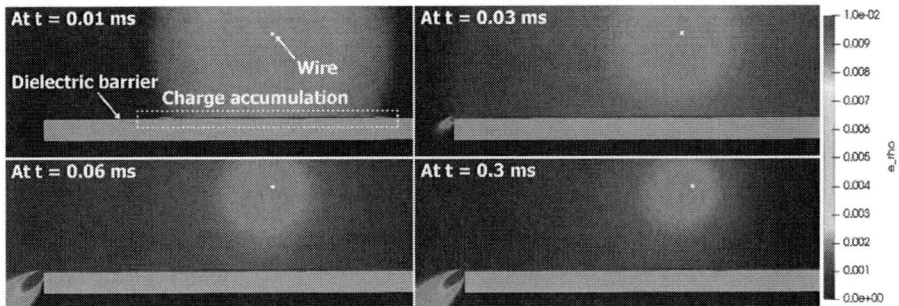

Figure 3: Simulation results of time-dependent charge accumulation on the upper surface of the flap, wire and flap positioned as annotated; **contour**: *charge density (C/m³).*

electrodes, the dielectric flap causes charge to accumulate on its upper surface, as can be observed in Fig 3. This charge build-up deflects the electrical field, diverting the force vector backwards and creating thrust. Our transient simulation demonstrates this progress in Fig. 2b, 2c, 2d. The charge at each time step is rendered together with ion wind velocity, showing the successful rectification of both electric field and ion wind direction.

EXPERIMENT RESULTS

The experiments are carried out in a large water tank where the lightweight miniaturized iBoat is manufactured and positioned in stationary state to eliminate the drag force. A force sensor is placed at the front-end of the iBoat and connected to a computer to measure thrust generated by the ion wind propulsion apparatus attached at the rear of the boat. This system includes an angle-adjustable polypropylene plate which would play as the electric field flap, an emitter (thin wire) powered by a high voltage supply unit which can generate up to 18 kV of potential difference. It is worth mentioning that the ground electrode of the power system is submerged in water.

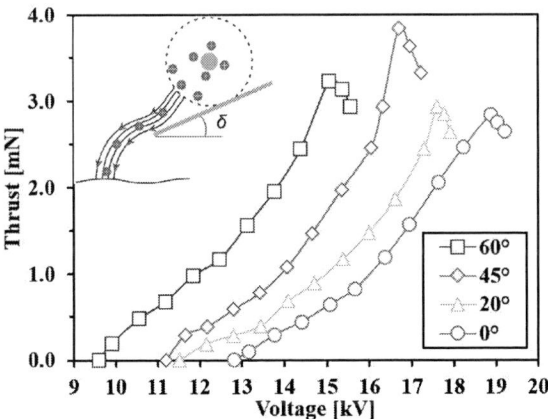

Figure 4: Experimental thrust plotted versus applied voltage for a range of the flap angles (δ) for the interelectrode (wire-water) distance d = 15 mm.

The relationship of thrust versus the electrode distance d and the dielectric inclination angle δ are investigated for a range of applied voltages from 3 to 18 kV. Our acquired experimental results show that at a certain emitter-water

distance d in the measured range, the horizontal thrust increases with the increasing of the applied voltage then decreases from a point when the spark occurs. The maximum thrust also increases from d = 11 mm then peaks at d = 15 mm followed by decreases when d larger than 15 mm. Therefore, for our existing configuration, the optimal electrode distance of 15 mm was registered to achieve the maximum thrust. Figure 4 shows the thrust (T) generated with d = 15 mm at different flap angles δ. Based on our results, the optimal flap angle was found to be at 45°.

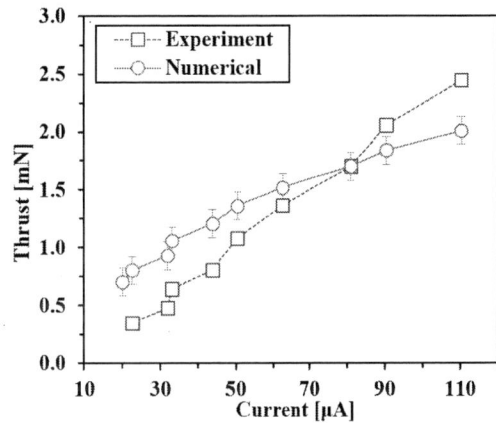

Figure 5: Thrust versus current comparison between experiment data and simulation result for case with flap angle δ = 45°, d = 15 mm.

Additionally, an average propulsion ratio of 1.45 mN/W is obtained which is comparable to reported data for other ion wind propulsions [2, 7, 15, 24] although our configuration utilizes the water surface as the reference electrode. We also include a quantitative comparison between empirical and simulated thrust results in Fig 5. Here, numerical thrust, calculated by Eq. 7, loosely agrees with measured data yet exhibits similar linear correlation. The relative inaccuracy of the numerical method can be caused by the interference of accumulated charge on the dielectric flap as Eq. 7 considers the whole volume of ions present in the gap [23]. Even so, the results of numerical method do show acceptable conformity with empirical data, thus the physics represented by simulations, to some extent, can explain the mechanism underlying our novel iBoat concept.

978-1-6654-9309-3/23 $31.00 © 2023 IEEE

CONCLUSION

We successfully presented the development of an ion-wind powered miniaturized boat (iBoat) using the novel wire-dielectric-water configuration. Empirical and simulation results of the deployed model confirms that the addition of a dielectric flap and the effect of subsequent surficial charge accumulation in adjusting electric field ionic wind direction allows propelling the iBoat within a range of applied voltages from 3 kV to 18 kV. An acquired thrust-to-power ratio of 1.45 mN/W is comparable to other two-electrode electro-hydrodynamic ion wind propulsion systems. This obtained proof of our concept would pave the way for further research and developments of highly efficient ion-wind powered boats for surveillance missions which is sensitive to noise.

ACKNOWLEDGEMENTS

Luan Ngoc Mai acknowledges the support of time and facilities from Ho Chi Minh City University of Technology (HCMUT), VNU-HCM for this study. This work has also been supported by Australian Research Council grant LP160101553 and was performed in part at the Queensland node of the Australian National Fabrication Facility (ANFF). V.T. Dau thanks Griffith University for the research grant: "Bipolar electrostatic atomizing system - fundamental study and application".

REFERENCES

[1] A. P. Chattock, "XLIV. On the velocity and mass of the ions in the electric wind in air," *Lond. Edinb. Dublin philos. mag. j. sci.*, vol. 48, no. 294, pp. 401-420, 1899.

[2] M. Robinson, "Movement of air in the electric wind of the corona discharge," *IEEJ Trans. Electr. Electron. Eng.*, vol. 80, no. 2, pp. 143-150, 1961.

[3] N. T. Van, T. T. Bui, T. X. Dinh, C. D. Tran, H. Phan-Thanh, T. C. Duc, and V. T. Dau, "A Circulatory Ionic Wind for Inertial Sensing Application," *IEEE Electron Device Lett.*, vol. 40, no. 7, pp. 1182-1185, 2019.

[4] N. T. Van, T. T. Bui, C. D. Tran, T. X. Dinh, H. P. Thanh, D. P. Van, T. C. Duc, and V. T. Dau, "Study on Point-to-Ring Corona Based Gyroscope," in *2019 IEEE 32nd International Conference on Micro Electro Mechanical Systems (MEMS)*, 2019, pp. 672-675.

[5] V. T. Dau, T. T. Bui, C. D. Tran, T. V. Nguyen, T. K. Nguyen, T. Dinh, H. P. Phan, D. Wibowo, B. H. A. Rehm, H. T. Ta, N. T. Nguyen, and D. V. Dao, "In-air particle generation by on-chip electrohydrodynamics," *Lab Chip*, vol. 21, no. 9, pp. 1779-1787, 2021.

[6] V. T. Dau, T.-H. Vu, C.-D. Tran, T. V. Nguyen, T.-K. Nguyen, T. Dinh, H.-P. Phan, K. Shimizu, N.-T. Nguyen, and D. V. Dao, "Electrospray propelled by ionic wind in a bipolar system for direct delivery of charge reduced nanoparticles," *Appl. Phys. Express*, vol. 14, no. 5, p. 055001, 2021.

[7] D. S. Drew, N. O. Lambert, C. B. Schindler and K. S. J. Pister, "Toward Controlled Flight of the Ionocraft: A Flying Microrobot Using Electrohydrodynamic Thrust With Onboard Sensing and No Moving Parts," *IEEE Robot. Autom. Lett.*, vol. 3, no. 4, pp. 2807-2813, 2018.

[8] V. T. Dau, T. X. Dinh, C.-D. Tran, T. Terebessy, and T. T. Bui, "Dual-pin electrohydrodynamic generator driven by alternating current," *Exp. Therm. Fluid Sci.*, vol. 97, pp. 290-295, 2018.

[9] V. T. Dau, T. X. Dinh, T. Terebessy, and T. T. Bui, "Bipolar corona discharge based air flow generation with low net

charge," *Sens. Actuator A Phys.*, vol. 244, pp. 146-155, 2016.

[10] T. H. Vu, H. D. Vu, N. L. Vu, H. T. Nguyen, D. V. Dao, and V. T. Dau, "Simultaneous Generation and Delivery of Neutral Polymeric Aerosol by Electro-Hydrodynamic Nebulizer," in *2022 IEEE 17th International Conference on Nano/Micro Engineered and Molecular Systems (NEMS)*, 2022, pp. 141-144.

[11] T.-H. Vu, H. T. Nguyen, J. W. Fastier-Wooller, C.-D. Tran, T.-H. Nguyen, H.-Q. Nguyen, T. Nguyen, T.-K. Nguyen, T. Dinh, T. T. Bui, Y. Zhong, H.-P. Phan, N.-T. Nguyen, D. V. Dao, and V. T. Dau, "Enhanced Electrohydrodynamics for Electrospinning a Highly Sensitive Flexible Fiber-Based Piezoelectric Sensor," *ACS Appl. Electron. Mater.*, vol. 4, no. 3, pp. 1301-1310, 2022.

[12] J. W. Fastier-Wooller, T.-H. Vu, H. Nguyen, H.-Q. Nguyen, M. Rybachuk, Y. Zhu, D. V. Dao, and V. T. Dau, "Multimodal Fibrous Static and Dynamic Tactile Sensor," *ACS Appl. Mater. Interfaces*, vol. 14, no. 23, pp. 27317-27327, 2022.

[13] V. T. Dau, T. T. Bui, T. X. Dinh, and T. Terebessy, "Pressure sensor based on bipolar discharge corona configuration," *Sens. Actuator A Phys.*, vol. 237, pp. 81-90, 2016.

[14] V. T. Dau, T. X. Dinh, T. T. Bui, and T. Terebessy, "Bipolar corona assisted jet flow for fluidic application," *Flow Meas. Instrum.*, vol. 50, pp. 252-260, 2016.

[15] E. A. Christenson and P. S. Moller, "Ion-neutral propulsion in atmospheric media," *AIAA Journal*, vol. 5, no. 10, 1967.

[16] C. Kim, K.-C. Noh, S.-Y. Kim and J. Hwang, "Electric propulsion using an alternating positive/negative corona discharge configuration composed of wire emitters and wire collector arrays in air," *Appl. Phys. Lett.*, vol. 99, no. 111503, 2011.

[17] A. Rashkovan, E. Sher and H. Kalman, "Experimental optimization of an electric blower by corona wind," *Appl. Therm. Eng.*, vol. 22, no. 14, pp. 1587-1599, 2002.

[18] H. Toyota, S. Zama, Y. Akamine, S. Matsuoka and K. Hidaka, "Gaseous electrical discharge characteristics in air and nitrogen at cryogenic temperature," IEEE *IEEE Trans Dielectr Electr Insul*, vol. 9, no. 6, pp. 891-898, 2002.

[19] V. T. Dau, T. X. Dinh, T. T. Bui and T. Terebessy, "Corona anemometry using dual pin probe," *Sens. Actuator A Phys.*, vol. 257, pp. 185-193, 2017.

[20] T. Xuan Dinh, D. B. Lam, C.-D. Tran, T. T. Bui, P. H. Pham and V. T. Dau, "Jet flow in a circulatory miniaturized system using ion wind," *Mechatronics*, vol. 47, pp. 126-133, 2017.

[21] V. T. Dau, T. X. Dinh, T. Terebessy and T. T. Bui, "Ion Wind Generator Utilizing Bipolar Discharge in Parallel Pin Geometry," *IEEE Trans. Plasma Sci.*, vol. 44, no. 12, pp. 2979-2987, 2016.

[22] H. Xu, Y. He, K. L. Strobel, C. K. Gilmore, S. P. Kelley, C. C. Hennick, T. Sebastian, M. R. Woolston, D. J. Perreault and S. R. H. Barrett, "Flight of an aeroplane with solid-state propulsion," *Nature*, vol. 563, p. 532-535, 2018.

[23] K. Masuyama and S. R. H. Barrett, "On the performance of electrohydrodynamic propulsion," *Proc. R. Soc. A.*, vol. 469, no. 20120623, 2013.

[24] V. T. Dau, T. X. Dinh, T. T. Bui, C.-D. Tran, H. T. Phan, and T. Terebessy, "Corona based air-flow using parallel discharge electrodes," *Exp. Therm. Fluid Sci.*, vol. 79, pp. 52-56, 2016.

CONTACT

*Luan Ngoc Mai; mnluan.sdh212@hcmut.edu.vn.

**Trung Hieu Vu; trunghieu.vu@griffithuni.edu.au.

***Van Thanh Dau; v.dau@griffith.edu.

HEMODYNAMIC ANALYSIS OF CARDIOMEMS: ADVERSE HEMODYNAMIC EFFECTS

Zhenhao Liu[1], Jiangli Han[2], and Xing Chen[1]*

[1]Beijing Advanced Innovation Center for Biomedical Engineering, School of Engineering Medicine, Beihang University, Beijing 100191, CHINA and
[2]Department of Cardiology, Peking University Third Hospital, Beijing 100191, CHINA

ABSTRACT

CardioMEMS is a renowned and representative MEMS sensor for medical uses, while its long-term safety after implantation is still unclear. This work has studied the hemodynamic changes caused by CardioMEMS sensors. Both simulation and experimental results have shown that the CardioMEMS sensor disturbs local hemodynamics of vasculature with strong correlation to the implantation angles, and have suggested that implantation of such a MEMS sensor could cause atherosclerosis. A streamlined profile is proposed to relieve the adverse effects. This study provides a design guideline for MEMS endovascular sensors with hemodynamic consideration, which is of clinical significance to medical devices.

KEYWORDS

CardioMEMS, hemodynamics, computational fluid dynamics, implantable sensor, endovascular sensor

INTRODUCTION

Heart failure is a chronic heart disease in which the pumping function or cardiac blood output of heart is impaired. Cardiac filling pressure is recognized as an accurate indicator to evaluate heart failure progress at its early stage. Abbott's CardioMEMS heart failure monitoring system is the first and currently the only FDA-approved implantable pulmonary artery pressure sensor to remotely measuring pulmonary artery pressures as a surrogate of cardiac filling pressures [1]. For heart failure patients, this wireless medical sensor has proved to reduce rates of hospital admissions by 48% in a RCT study. In other words, CardioMEMS can significantly reduce hospitalization rates and improve the life quality of heart failure patients [2].

However, CardioMEMS sensor's long-term safety beyond its follow-up period of clinical trial is unknown. One main concern is that the CardioMEMS sensor, like other endovascular implants (e.g., stent), can cause hemodynamics-related complications. It is well known that implanted stents have a restenosis rate of 12% because of altered hemodynamics. In comparison, CardioMEMS sensors are implanted at the lower lobe of pulmonary artery with blood flow velocity as high as 1.1 m/s [3]. Moreover, the tubular stent with thickness of 0.1 mm stays in a wall-fitted state, while the 2-mm-thick CardioMEMS sensor is suspended in the vascular lumen. Therefore, implanted CardioMEMs sensor certainly alters local hemodynamics and may result in long-term vascular complications.

In this study, we investigate the hemodynamic characteristics of CardioMEMS sensors using both computational fluid dynamics (CFD) and *in vitro* experiments. The hemodynamic parameters are interpreted to correlate with potential vascular complications.

MATERIALS AND METHODS
Computational Fluid Dynamics

In the CFD study, the CardioMEMS sensor and an optimized one proposed by this study have been studied and compared, as shown in Fig. 1.

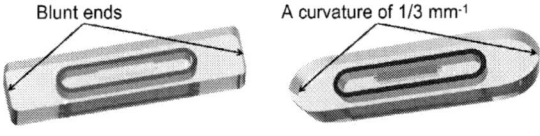

Figure 1: The 3D models of the original CardioMEMS sensor and the one with streamlined design.

We first 3D modeled a CardioMEMS sensor inside the blood vessel, in which all dimensions and anatomic structures were referenced to the datasheet of CardioMEMS sensor and CT tomography of the lower lobe branch of pulmonary artery. For the optimized sensor, two ends of it were designed with a curvature of 1/3 mm[-1] streamline considering both fabrication feasibility and hemodynamics.

During interventional operation, the CardioMEMS sensors are usually implanted with angles with respect to the arterial axis. According to the official data, the offset angle should be kept within 30° when working, otherwise the measurement is inaccurate. In order to cover all implantation scenarios, this work also studied hemodynamic characteristics of CardioMEMS sensors implanted at different angles, as shown in Fig. 2. For comparison, a control group was modeled in the same vessel but without sensor implanted.

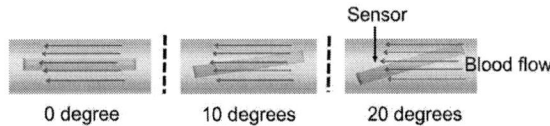

Figure 2: Schematics of implanted sensor with implantation angles of 0°, 10° and 20° with respect to the blood flow.

After that, we conducted finite element analysis on the above 3D model using ANSYS Fluent 14.5 software under pulsatile flow conditions. Blood was modeled as a Newtonian fluid, and was assumed to be homogeneous and incompressible [4, 5], and the vessel wall is set as a rigid wall without slip. The overall numerical simulation is based on the three-dimensional incompressible Navier-Stokes equations and the law of conservation of mass, with the expressions below.

$$\rho\left[\frac{\partial u}{\partial t}+(u\cdot\nabla)u\right]+\nabla p-\mu\nabla^2=0, \quad (1)$$

$$\nabla\cdot u=0, \quad (2)$$

where u and p denote the fluid velocity vector and pressure, respectively, and ρ and μ denote the density and viscosity of the blood (ρ =1050 kg/m^3, and $\mu = 3.5\times10^{-3}$ kg/m^{-s}) [6]. The SIMPLE algorithm was used to calculate the blood flow velocity, and a pressure-based solver was used to perform pressure correction and solve the momentum equation. The pulsatile inlet condition of blood flow velocity in Fig. 3 was extracted from clinical data of the pulmonary artery [7].

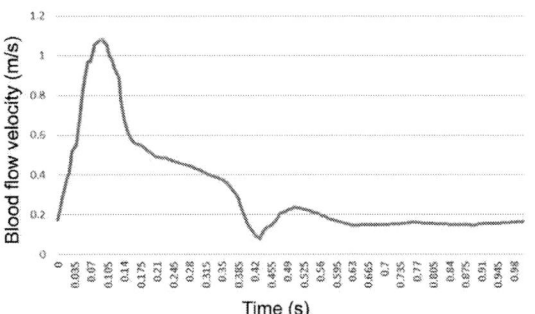

Figure 3: Inlet pulsating flow velocity. One cardiac cycle is 1 s and the peak flow rate is 1.08 m/s.

From the CFD results, the parameters of interest include wall shear stress (WSS), time-averaged ventricular wall shear stress (TAWSS), oscillatory shear stress index (OSI), relative residence time (RRT), and intravascular pressure gradient (IPG) [8].

WSS is defined as the product of fluid viscosity and the shear velocity of the adjacent vessel wall. A high WSS causes increased nitric oxide release from vascular endothelial cells and degenerative changes in the vessel wall. High WSS is a major factor in the development and progression of aneurysms. The vascular segment where maximum shear stress occurs is more susceptible to damage than elsewhere and often becomes a new site to cause aneurysms [9, 10]

TAWSS is the average of the WSS over the entire cardiac cycle and can be used to assess the pulsatile characteristics of the WSS. TAWSS is calculated as follows,

$$\text{TAWSS}=\frac{1}{T}\int_0^T|WSS(s,t)|\cdot dt, \quad (3)$$

where T is the duration of the cardiac cycle and s is the position on the vascular wall.

OSI is a dimensionless parameter to represent the magnitude of WSS fluctuations during the cardiac cycle, ranging from 0-0.5. A larger OSI indicates a more drastic oscillation of WSS and a greater impact on the vessel wall,

$$\text{OSI}=0.5\left[1-\left(\frac{\left|(1/T)\int_0^T WSS(s,t)\cdot dt\right|}{(1/T)\int_0^T|WSS(s,t)|\cdot dt}\right)\right]. \quad (4)$$

RRT is proportional to the OSI while is inversely proportional to TAWSS vector, which is used to characterize the near-wall material residence time, and OSI can modify the effect of TAWSS on RRT in specific endothelial regions,

$$\text{RRT}=\frac{1}{(1-2\cdot OSI)\cdot TAWSS}. \quad (5)$$

Studies show that RRT is a marker of disturbed blood flow, which is correlated to the pathology of atherosclerosis [11].

IPG is the pressure different across a blood vessel. According to the Poiseuille's Law, IPG is proportional to the flow resistance or the fourth power of radius at certain flow rate. From the hemodynamics perspective, a high IPG suggested the constriction of the vascular lumen.

***In Vitro* Experiments**

In the experimental setup, we built an *in vitro* blood circulation simulator, as shown in Fig. 4. The simulator included a peristaltic pump as the circulating source, silicone tubing to simulate vasculature, and multi-parameter sensing apparatus (flow sensor, pressure sensor, and data acquisition equipment) to real-time measure hemodynamics. A prototype of CardioMEMS and another streamlined one were 3D printed and tested in the blood circulation simulator, respectively. The pressure difference across the 3D-printed implants was measured to represent IPG and to quantitatively compare with the simulation results. A constant flow rate was set at 0.45 m/s, which is the average flow rate of the lower lobe branch of the pulmonary artery during one cardiac cycle.

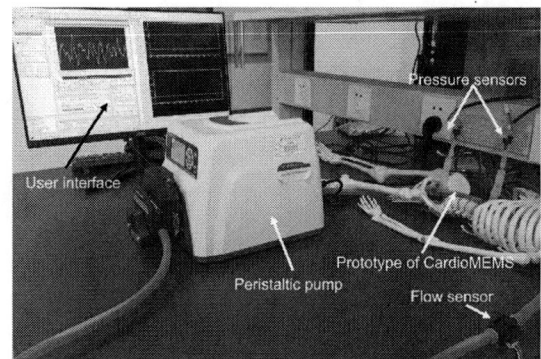

Figure 4. The setup of blood circulation simulator integrated with pumping and sensing functions.

We also studied the IPG of the CardioMEMS sensors implanted at different angles as simulated by CFD. Considering the accuracy to control the specific angle of the 3D-printed prototype inside the silicone tubing, only two experimental groups of angles (0° and 15°) were tested against the control group (without any implantation).

EXPERIMENTAL RESULTS
CFD Results of the CardioMEMS Sensor

The time with the peak flow rate in a cardiac cycle (t=0.095 s) was chosen to observe the CFD results, which can best illustrate what perturbations occur in the blow flow field around the sensor.

Fig. 5 shows that at the region around CardioMEMS sensor, the blood flow velocity was significantly higher than other domains of vascular lumen. The highest velocity occurred at the head, while the perturbation concentrated at the central body area. As the implantation angle ascended, the blood flow velocity along the sensor increased. At the ideal implantation angle of 0°, the blood flow velocity around sensor was higher than the one without sensor in the control group. At an implantation angle of 10°, the high-speed domain around the sensor increased. When the implantation angle reached 20°, the high-speed domain around the sensor became predominant, and the velocity streamlines got disturbed suggesting the occurrence of perturbation. The highest velocity reached 1.70 m/s, which was 70% faster than the blood flow velocity away from sensor. As the vascular WSS and TAWSS increased exponentially with the increased blood flow velocity, the implantation of CardioMEMS sensor led to higher WSS and TAWSS [12].

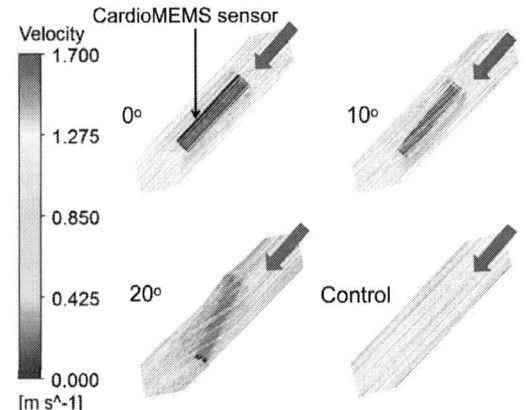

Figure 5. Simulated flow velocity streamline of a CardioMEMS sensor at different implantation angles. The red arrows indicate the direction of blood flow.

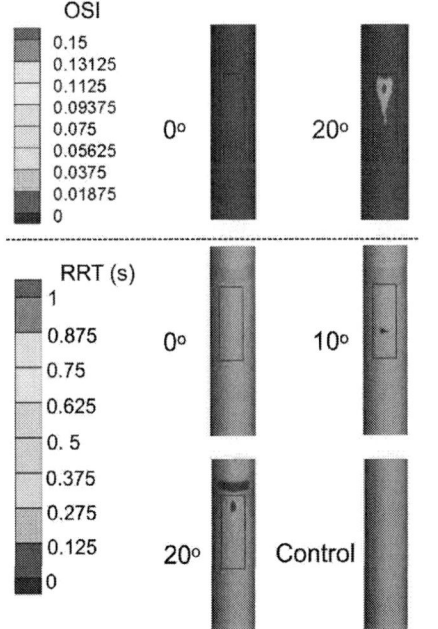

Figure 6. Simulated OSI (upper) and RRT (lower) of a CardioMEMS sensor at different implantation angles. The red boxes indicate the projection of the sensor onto the vascular wall.

Figure 6 shows that the implantation of CardioMEMS sensor and its implantation angles also affected both OSI and RRT, which in turn reflected the potential risk of degenerative vascular wall and atherosclerosis. Table 1

shows both the simulation and experimental results of IPG across the 3D-printed prototype of CardioMEMS sensor at different implantation angles. Both sets of results confirmed that the implantation of CardioMEMS sensor and increased implantation angles can raise IPG, which can be explicitly explained by the high flow resistance as a result of the reduced lumen radius. The increased vascular resistance is usually associated with circulatory pathology.

Table 1: Average values of IPG [mmHg] at different implantation angles.

Results	Control	0°	15°
Experiment	1.25	2.93	3.02
Simulation	0.70	1.21	1.43

Comparative Results between CardioMEMS Sensor and Streamlined One

As the current CardioMEMS sensor has been proved to alter local hemodynamics upon implantation, a streamlined design with a curvature of 1/3 mm⁻¹ was proposed in this study. The simulated OSI and RRT of the streamlined design was compared to the corresponding results of CardioMEMS, as shown in Fig. 7. With the same size but different profiles, the OSI up to 0.15 from the CardioMEMS sensor dropped to normal level at the implantation angle of 20°. For RRT, there was no significant difference between two profiles at the implantation angle of 0°. When the implantation rose to 20°, the RRT of CardioMEMS was reduced by 75% compared to the streamlined design, which helped to suppress disturbed blood flow.

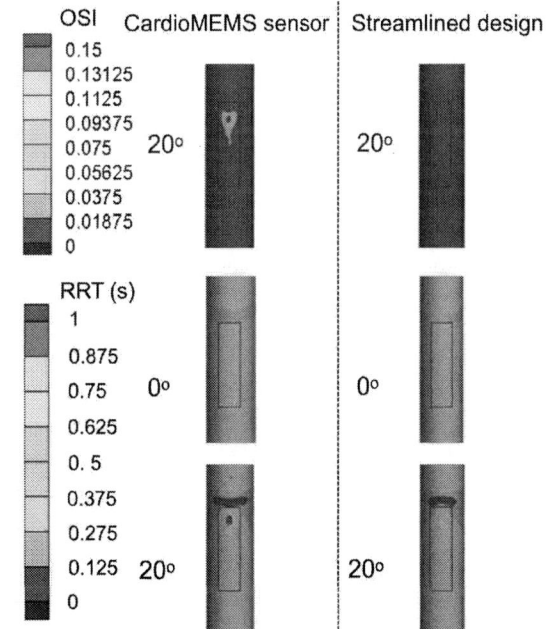

Figure 7. Simulated OSI (upper) and RRT (lower) of a CardioMEMS sensor (left column) and the streamlined one (right column) at different implantation angles. The red box indicates the projection of the sensor onto the vessel wall.

By comparing the blood flow pattern of the two profiles in Fig. 8, the streamlined sensor showed lower blood flow velocity than the CardioMEMS sensor at the implantation angle of 0°. The blood flow perturbation pattern became prominent for both of them at the implantation angle of 20°, but the streamlined design featured lower blood flow velocity at its head and less disturbance at the central body area.

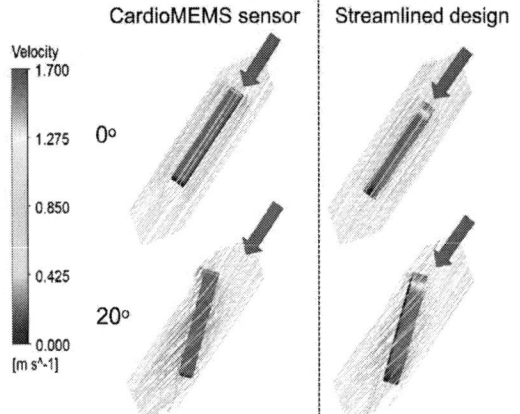

Figure 8. Simulated flow velocity streamline of a CardioMEMS sensor (left column) and the streamlined one (right column) at different implantation angles.

CONCLUSIONS

Long-term safety and efficacy are the two foundations for any medical devices. The purpose of this work aims to study the hemodynamic characteristics of CardioMEMS sensors by both computational fluid dynamics and *in vitro* experiments. A blood circulation simulator with implanted 3D-printed CardioMEMS sensor prototype was built to recapitulate the *in vivo* environment. Results have shown that the implanted CardioMEMS sensor has changed hemodynamic parameters including WSS, TAWSS, OSI, RRT and IPG, and thus caused perturbation of blood flow. The disturbed local hemodynamics poses potential risks of atherosclerosis and complications upon CardioMEMS sensor implantation. In the light of the current problem, the CardioMEMS sensor was optimized with a curvature of 1/3 mm^{-1} at two ends. This streamlined design has shown to improve hemodynamic characteristics and relieve adverse effects. This study provides a paradigm to design endovascular MEMS devices and evaluate their long-term safety.

ACKNOWLEDGEMENTS

This study was supported by the National Natural Science Foundation of China (No. 62074013), the National Key Research and Development Program (2020YFC0862900), and the Fundamental Research Funds of the Central Universities.

REFERENCES

[1] S. Ghio, A. Gavazzi, C. Campana, C. Inserra, C. Klersy, R. Sebastiani, E. Arbustini, F. Recusani, L. Tavazzi, "Independent and additive prognostic value of right ventricular systolic function and pulmonary artery pressure in patients with chronic heart failure", *J. Am. Coll. Cardiol.*, vol. 37, pp. 183-188, 2001.

[2] W. T. Abraham, P. B. Adamson, R. C. Bourge, M. F. Aaron, M. R. Costanzo, L. W. Stevenson, W. Strickland, S. Neelagaru, N. Raval, S. Krueger, S. Weiner, D. Shavelle, B. Jeffries, J. S. Yadav, "Wireless pulmonary artery haemodynamic monitoring in chronic heart failure: a randomised controlled trial", *Lancet*, vol. 377, pp. 658-666, 2011.

[3] X. Lei, Y. Guo, G. Xu, M. Chen, R. Li, J. Yang, "Study on blood flow of central pulmonary artery with MRI measurement" (in Chinese), *Chin. J. Med. Imaging. Technol.*, vol. 03, pp. 409-412, 2006.

[4] P. M. Walker, Basic Hemodynamics and its Role in Disease Processes, *Chest*, 1981.

[5] Z. Wang, A. Sun, Y. Fan, X. Deng, "Comparative study of Newtonian and non-Newtonian simulations of drug transport in a model drug-eluting stent", *Biorheology*, vol. 49, pp. 249-259, 2012.

[6] T. Chaichana, Z. Sun, J. Jewkes, "Computation of hemodynamics in the left coronary artery with variable angulations", *J. Biomech.*, vol. 44, pp. 1869-1878, 2011.

[7] K. Hilde, K. C. L. Carlsen, G. Haugen, "Doppler measures of blood flow in right and left branches of the fetal pulmonary artery", *J. Matern-fetal. Neo. M.*, vol. 35, pp. 2980-2983, 2020.

[8] G. Liu, J. Wu, D. N. Ghista, W. Huang, K. K. L. Wong, "Hemodynamic characterization of transient blood flow in right coronary arteries with varying curvature and side-branch bifurcation angles", *Comput. Biol. Med.*, vol. 64, pp. 117-126, 2015.

[9] T. Tanoue, S. Tateshima, J. P. Villablanca, F. Viñuela, K. Tanishita, "Wall shear stress distribution inside growing cerebral aneurysm", *Am. J. neuroradiol.*, vol. 32, pp. 1732-1737, 2011.

[10] A. Alaraj, S. F. Shakur, S. Amin-Hanjani, H. Mostafa, S. Khan, V. A. Aletich, F. T. Charbel, "Changes in wall shear stress of cerebral arteriovenous malformation feeder arteries after embolization and surgery", *Stroke*, vol. 46, pp. 1216-1220, 2015.

[11] A. Colombo, E. Bramucci, S. Saccà, R. Violini, C. Lettieri, R. Zanini, I. Sheiban, L. Paloscia, E. Grube, J. Schofer, L. Bolognese, M. Orlandi, G. Niccoli, A. Latib, F. Airoldi, "Randomized study of the crush technique versus provisional side-branch stenting in true coronary bifurcations: the CACTUS (Coronary Bifurcations: Application of the Crushing Technique Using Sirolimus-Eluting Stents) Study", *Circulation*, vol. 119, pp. 71-78, 2009.

[12] R. P. T. van Wijngaarden, J. A. Overbeek, E. M. Heintjes, A. Schubert, J. Diels, H. Straatman, E. W. Steyerberg, R. M. C. Herings, "Relation between different measures of glycemic exposure and microvascular and macrovascular complications in patients with type 2 diabetes mellitus: an observational cohort study", *Diabetes. Ther.*, vol. 8, pp. 1097-1109, 2017.

CONTACT

*X. Chen, tel: +86-10-82313101; xingc@buaa.edu.cn

MODAL QUALITY FACTOR INVERSION OF NON-SLENDER MEMS RESONATORS BETWEEN GASES AND LIQUIDS

Andre L. Gesing[1], Thomas Tran[1], Daniel Platz[1], and Ulrich Schmid[1]
[1]TU Wien, Institute of Sensor and Actuator Systems, Vienna, AUSTRIA

ABSTRACT

We present a semi-numerical method for determining the dynamics of non-slender structures in viscous fluids. An investigation of the quality factor (Q-factor) of the vibrational modes of non-slender structures in gases and liquids reveals that, in gases, the conventional Euler-Bernoulli (EB) modes exhibit the lowest Q-factors, while non-EB modes exhibit the highest Q-factors. In liquids, the opposite occurs. EB modes exhibit the highest Q-factors, and non-EB modes the lowest. This opposite Q-factor pattern in gases and liquids is the modal quality factor inversion. We expect the modal quality factor inversion of non-slender resonator geometries to enable novel MEMS fluid sensors.

KEYWORDS

MEMS; Fluid-structure interaction; Micro resonators; Plate theory; Stokes flow.

INTRODUCTION

The interaction between a MEMS resonator and a fluid environment is the most dominant dissipation mechanism experienced by a resonator outside vacuum. This fluid-structure interaction is well understood only for resonators with slender beam geometries. Slender beams vibrate in different vibrational modes as the out-of-plane transverse [1] and torsional modes [2], and the in-plane lateral and extensional modes [3] and are extensively used in MEMS. Recent experimental data show that resonators with non-slender geometries vibrating in out-of-plane transverse modes exhibit remarkably high quality factors in liquids [4] at frequencies in the hundreds of kilohertz frequency range as shown in Fig. 1.

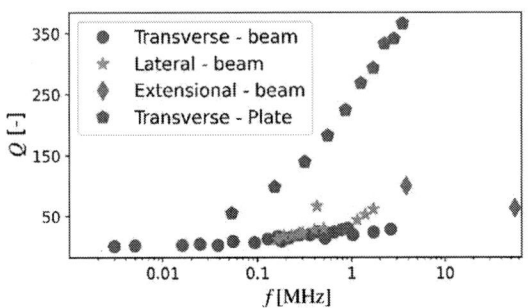

Figure 1: Experimental Q-factor of beams and plates in water. Beam data: transverse modes [5], lateral modes from [3] and extensional modes [6]. Plate data: transverse modes [4].

For slender beams, semi-analytic methods for determining the fluid-structure interaction exist [1, 2, 3]. For non-slender beams, only numerical studies exist, which due to the multi-scale characteristic of the problem, are

computationally costly. Here, we propose a semi-numerical method to efficiently investigate the dynamics of non-slender resonators in viscous fluids. The method is valid for gases and liquids, making it ideal for an investigation into the rich dynamics of micro-plates in these two fluid regimes. This analysis revealed an unexpected modal quality factor inversion which is here investigated.

SEMI-NUMERICAL METHOD
Microplate dynamics

Consider a microplate that moves only in the out-of-plane direction (z-direction) with a displacement ϕ, i.e., lateral displacements in x and y directions are neglected. The plate has a length l, width b and thickness h, as shown in Fig. 2.

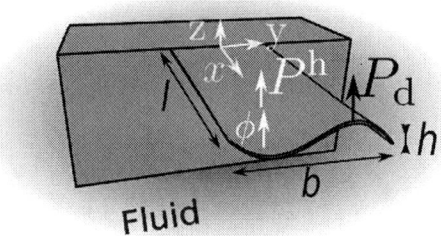

Figure 2: Microplate immersed in a viscous fluid. The plate has a length l, width b and thickness h and is clamped at one of its edges.

The equation of motion in the Fourier domain that governs the plate's dynamics in a fluid is

$$\frac{h^3}{12} C_{\alpha\beta\gamma\delta}\phi_{,\alpha\beta\gamma\delta} - \omega^2 \rho_p h\phi = P_d + P^h, \qquad (1)$$

where ω is the radial frequency, P_d is a drive pressure and P^h is the hydrodynamic pressure with which the viscous fluid acts back on the plate. Here, the material is anisotropic, described with a uniform density ρ_p and a fourth-order elastic tensor C. In Eq. 1, $C_{\alpha\beta\gamma\delta}$ are the elements of the elastic tensor C, and the indices $\alpha, \beta, \delta, \gamma$ represent x and y directions.

The discretized weak form of Eq. 1 is obtained by multiplying it with a test function w and integrating the equation over the plate domain Ω_p yielding

$$\int_{\Omega_p} \frac{h^3}{12} C_{\alpha\beta\gamma\delta}\phi_{,\alpha\beta\gamma\delta} w d\Omega - \omega^2 \rho_p h \int_{\Omega_p} \phi w d\Omega$$
$$= \int_{\Omega_p} P_d\, w d\Omega - \int_{\Omega_p} P^h\, w d\Omega. \qquad (2)$$

The test function w belongs to the function set v composed of the vacuum vibrational modes of the microplate. v is determined numerically using a finite element method for

solving the eigenvalue problem

$$\frac{h^3}{12} C_{\alpha\beta\gamma\delta} v_{i,\alpha\beta\gamma\delta} = \omega_i^2 \rho_p h v_i. \qquad (3)$$

Solution of Eq. 3 requires the implementation of a method for accounting the fourth order spatial derivatives of v_i and the cantilevered boundary conditions. We apply an interior-penalty method whose details can be found elsewhere [7]. The evaluation of the fourth integral (projection of the hydrodynamic pressure to v) in Eq. 2 requires an additional equation to determine the hydrodynamics pressure as a function of the plate's displacement.

Fluid flow dynamics

The fluid flow around the microplate is in the continuum regime, the fluid is incompressible, and the viscous forces dominate over inertial forces [7]. Hence, the fluid flow is determined by the Stokes equations. Furthermore, the volume of the fluid domain is much larger than the microplate's volume. Hence, a boundary integral formulation is an appealing method to determine the hydrodynamic pressure without the need to discretize the entire fluid domain. Assuming the no-penetration and no-slip boundary conditions on the plate's surfaces, an equation relating the fluid velocity in z-direction on the plate's surface u_z and the hydrodynamic pressure P^h is obtained as

$$u_z(x', y, 0) = j\omega\phi(x', y) =$$
$$\frac{1}{\mu_f} \int_{-b/2}^{b/2} P^h(x', y') \frac{\partial^2 \Psi(y, y', 0, 0)}{\partial y \partial y'} dy'. \qquad (4)$$

μ_f is the fluid's dynamic viscosity and Ψ is a fundamental solution to the Stokes equations given by

$$\Psi(y, y') = \frac{j}{2\pi} \frac{\nu_f}{\omega} \left[\log(|y - y'|) + K_0 \left(\sqrt{\frac{j\omega}{\nu_f}} |y - y'| \right) \right] \qquad (5)$$

where K_0 is the modified Bessel function of third kind and order zero. y' is the coordinate axis with coincident orientation to y. Considering the hydrodynamic pressure piece-wise constant in a quadrature scheme, Eq. 4 results in a matrix-vector product as

$$P = \mu_f j\omega A^{-1} \phi_{vec} \qquad (6)$$

where P and ϕ_{vec} are vectors, whose elements are the values of P^h and ϕ, respectively, at specific quadrature points. The quadrature points are defined in the y direction with a Chebyshev-Gauss quadrature scheme, and in the x direction the 1/3 Simpsons' rule. The value of each element of the A matrix is obtained from the integral of the second derivative of Ψ evaluated analytically around each quadrature point. For details of the implementation of this combination of quadrature schemes, we refer to [7].

RESULTS

To investigate how the non-slender structures interact with the fluid in different fluid regimes, we vary the density and viscosity of a fluid from a gas (air) to a liquid (water).

The object of the analysis is a silicon microplate with length equals 500 micrometers, the width is 250 micrometers, and the thickness is 5 micrometers. Fig. 3a shows the plate's spectral response ϕ_c at the plate's tip (x=l, y=b/2) in the 2 MHz frequency range in the fluid ranging from the air (top) to water (bottom) as a function of a fluid parameter η.

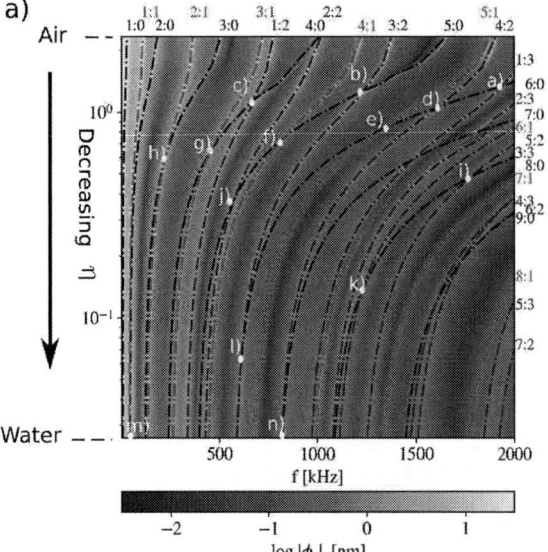

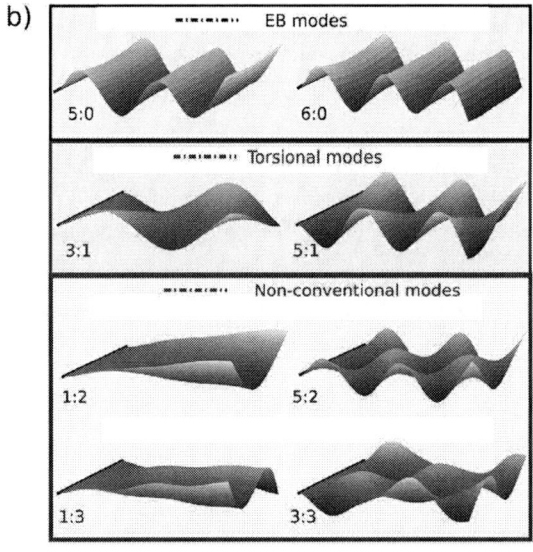

Figure 3: a) Absolute displacement spectrum of the plate in fluids between air (top) and water (bottom) as a function of η. The modes are identified according to Leissa's notation and color-coded in EB modes (black), torsional (pink) and non-conventional (blue). b) EB modes possess only nodal lines parallel to the clamped side. A torsional mode exhibits one nodal line orthogonal to the clamped edge and n_x parallel to the clamped edge. Modes with $n_y > 1$ are non-conventional modes.

The fluid parameter η is introduced as $\eta = log\,1/\sqrt{\rho_f \mu_f}$. For air, $\eta = 2.33$, while for water $\eta = 0.026$. In Fig. 3a, bright lines represent frequencies of maximum displacement associated with the different vibrational modes of plates. At the top of Fig. 3a the plate is considered in air, only fourteen vibrational modes can be found in the frequency range up to 2 MHz. The frequencies of these modes drop significantly as the fluid changes towards water. Additionally, other higher-order modes are present in the frequency range up to 2 MHz.

To classify the additional vibrational modes of the non-slender geometry, Leissa's notation is introduced, which categorizes the vibrational modes of plates according to their nodal lines in each direction as $n_x : n_y$. With the Leissa's notation, we define modes that also occur in slender beams as EB (from Euler-Bernoulli) modes $n_x : 0$, and torsional $n_x : 1$ modes as exemplified in Fig. 3b. Modes with $n_y > 1$ occur only in non-slender structures and are from hereon called non-conventional modes. The spectral response of the resonator changes continuously during the air-water transition. EB modes exhibit a more accentuated decrease in resonance frequency with increasing density and viscosity than other modes, which leads to two vibrational modes occurring with similar frequencies. For instance, at point a) in Fig. 3a the mode 6:0 and mode 1:3 occur at the frequency of 1925 kHz for a $\eta = 1.2$. Fourteen of these resonance frequencies on which two modes occur are highlighted in Fig. 3a from a) to n), following the order of decreasing η from air to water.

Quality factors are extracted for each mode in air and water by fitting a Lorentzian to the resonance peaks in the spectrum. The quality factors of all modes reduce significantly during the continuous sweep from air to water as seen in Fig. 4 for the Q-factor in air and in water in the 2 MHz frequency range. In air, non-conventional modes exhibit higher quality factors than flexural EB and torsional modes as shown in Fig. 4a. In water, the opposite occurs. The quality factors of EB modes are higher than those of torsional and of non-conventional modes as shown in Fig. 4b. We name this opposite Q-factor pattern in gases and liquids as the modal quality factor inversion.

To investigate the modal quality factor inversion, it is interesting to interpret the Q-factor of a vibrational mode in terms of its dissipated and stored energies, which yields

$$Q = \frac{2\pi f_n (m_p + M)}{c}. \qquad (7)$$

m_p is the plate mass, M is the fluid flow added mass and c is the modal damping and f_n is the mode's resonance frequency. For a proper meaningful comparison, we consider the pairs of modes that have similar resonance frequency in the air-water sweep. That is, the fourteen highlighted points in Fig. 3a. At these fourteen pairs of modes, we determine the ratio of quality factor Q^{EB}/Q^{other}, ratio of the damping coefficients c^{other}/c^{EB} and the ratio of fluid flow added masses M^{EB}/M^{other} between the Euler-Bernoulli modes and the other modes (torsional and non-conventional).

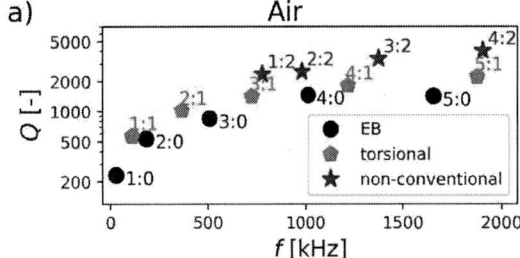

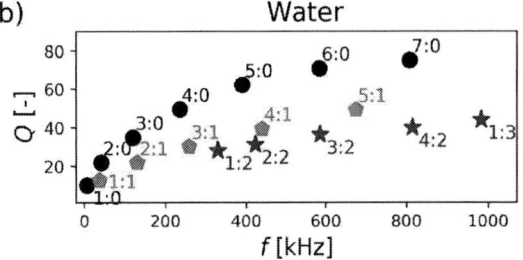

Figure 4: Q-factor of the vibrational modes of a microplate in a) air and b) water in the 2 MHz frequency range classified in EB modes, torsional modes and non-conventional modes.

Fig. 5a shows the quality factor modal inversion in terms of the ratios between Q-factors. In gases, this ratio is smaller than 1 because EB modes have lower Q-factor than other modes. In liquids, this ratio is larger than 1, since EB modes have in liquids higher Q-factors. Interestingly, the ratio of damping coefficients is approximately constant in the entire fluid range and smaller than 1, as seen in Fig. 5b. Thus, EB modes exhibit a larger damping coefficient independently of the fluid regime. Therefore, the explanation to the quality factor modal inversion lays on the fluid flow added mass of the different modes. As shown in Fig. 5c, the ratio of added mass is approximately 1 for gases, however, it continuously increases as the fluid is altered from a gas to a liquid, reaching a ratio higher than 3 in water. A higher added mass means the EB modes have more stored energy in the form of kinetic energy in the fluid than other modes. This in turn increases the EB modes' Q-factor in liquids, leading to the quality factor modal inversion.

CONCLUSIONS

The results show that vibrational modes of non-slender structures exhibit a modal quality factor inversion. EB modes exhibit the highest Q-factors in liquids, while non-conventional modes show the highest Q-factors in gases. These findings imply that new fluid-structure interaction effects can be found in MEMS with non-beam geometries and that non-slender resonator geometries can enable novel

MEMS fluid sensors.

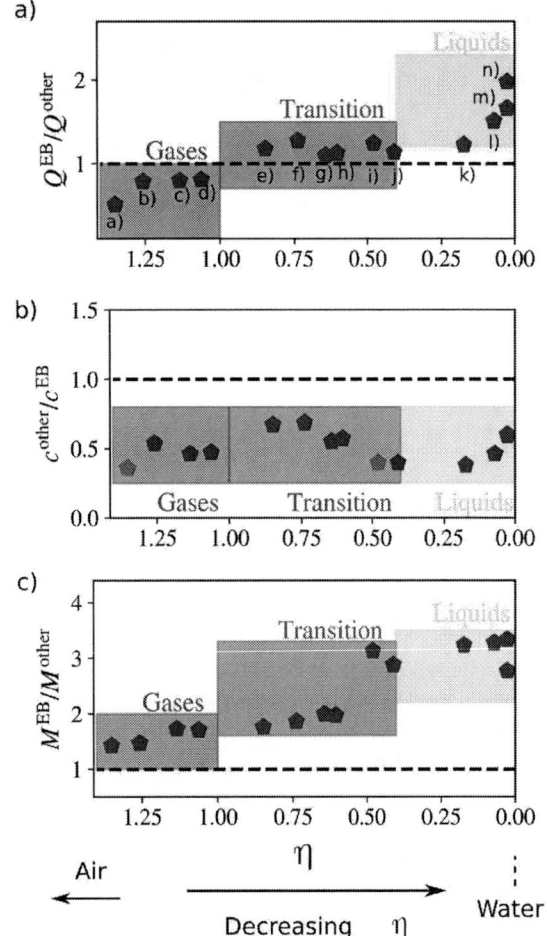

Figure 5: Ratios of the a) quality factor, b) damping coefficient and c) fluid flow added-mass of the EB-modes per other modes as the fluid properties range from air to water.

ACKNOWLEDGEMENTS

The authors gratefully acknowledge funding by the Innovative Projects program at TU Wien. The computational results presented have been achieved using the Vienna Scientific Cluster (VSC).

REFERENCES

[1] Sader, John Elie, "Frequency response of cantilever beams immersed in viscous fluids with applications to the atomic force microscope", *Journal of Applied Physics,* vol. 84, pp. 64-76, 1998.

[2] C. P. Green and J. E. Sader, "Torsional frequency response of cantilever beams immersed in viscous fluids with applications to the atomic force microscope," *Journal of Applied Physics*, vol. 92, pp. 6262–6274, 2002.

[3] Dufour, Isabelle, et al. "Unconventional uses of microcantilevers as chemical sensors in gas and liquid media", *Sensors and Actuators B: Chemical,* vol. 170, pp. 115-121, 2012.

[4] M. Kucera, E. Wistrela, G. Pfusterschmied, V. Ruiz-Dıez, J. L. Sanchez-Rojas, J. Schalko, A. Bittner, and U. Schmid, "Characterisation of multi roof tile-shaped out-of-plane vibrational modes in aluminium-nitride-actuated self-sensing micro-resonators in liquid media," *Applied Physics Letters*, vol. 107, no. 5, pp. 1–5, 2015.

[5] M. K. Ghatkesar, T. Braun, V. Barwich, J. P. Ramseyer, C.Gerber,M. Hegner, and H. P. Lang, "Resonating modes of vibrating microcantilevers in liquid," *Applied Physics Letters*, vol. 92, no.4, pp.10–13, 2008.

[6] T. Manzaneque, V. Ruiz, J. Hernando-Garcıa, A. Ababneh H. Seidel, and J. Sanchez-Rojas, "Characterization and simulation of the first extensional mode of rectangular micro-plates in liquid media," *Applied Physics Letters*, vol. 101, no. 15, p. 151904, 2012.

[7] Gesing, A., D. Platz, and U. Schmid, "A numerical method to determine the displacement spectrum of micro-plates in viscous fluids", *Computers & Structures*, vol. 260, pp. 106716, 2022.

CONTACT

*Andre Loch Gesing; andre.gesing@tuwien.ac.at

CLASSIFYING CELL CYCLE BY ELECTRICAL PROPERTIES USING MACHINE LEARNING

Jian Wei[1] and Xiaoxing Xing[1]

[1]College of Information Science and Technology, Beijing University of Chemical Technology,
P. R. of CHINA

ABSTRACT

Mitotic cell proliferation undergoes precisely controlled functional phases and results in two daughter cells divided from one parental cell. Some drugs interact with cell proliferation progression and lead to mitosis arrest at a certain phase, or cell death eventually. Combination of machine learning with impedance flow cytometry (IFC), an efficient tool for high-speed and label-free single cell analysis, allows linking of the cell physiology states post drug treatment with impedance data in a more accurate and convenient fashion. This work presents the use of different models of machine learning for classifying cell states, i.e. mitotic arrest at G1/S or G2/M phases and apoptosis, from the impedance data measured for drug-treated cells using microfluidic IFC chip.

KEYWORDS

cell cycle, machine learning, electrical characteristics, classification

INTRODUCTION

Flow cytometry plays an important role in high throughput single cell analysis [1]. However, fluorescence staining of cells required as prerequisite in running flow cytometry is a fairly complex task being time consuming and expensive. Moreover, the labeling of cell receptors may affect the biological activity of living cells [2]. As an alternative, microfluidic impedance flow cytometry (IFC) has emerged as a label-free technique for high throughput single cell analysis based on their electric properties. With single cells passing by precisely manufactured microelectrode pairs in a train, IFC chip allows measurement of the impedance amplitude and phase of each individual cell with high speed and accuracy [3]. In last two decades, microfluidic IFC chip has been exploited for detection as well as study of biophysical properties of varieties of cell types, focusing on which there has been very good review papers [4][5].

Recently, applying machining learning for data analysis of impedance flow cytometry has been demonstrated for being able to classify cells or their physiology states based on the electrical features at multiple frequencies measured by IFC [6][7]. With training on proper data amount, efficient prediction can be given with high accuracy. For example, Zhao et al. applied neural network (NN) model to classify H1299 cells and Hela cells using specific membrane capacitance and cytoplasm conductivity, and obtained high success rate of 90.9% [8]. Feng et al presented NN-assisted IFC in real time, which obtained classification of five cell types with accuracy up to 91.5% [6]. Researchers also demonstrated that combining the electrical features or even electrical feature with other biophysical properties, i.e. mechanical deformability and optical features, that gave rise to enhanced accuracy in classification [9]. Zhao et al. also demonstrated in their work that combining multiple electrical features could enhance the success rate for classification of A549 cells before and after TGF-β treatment [8]. Yang et al. demonstrated a significantly improved classification accuracy of 93.3% between normal MCF-7 cell and chemical-modified ones using NN, based on combination of electrical impedance, deformability and relaxation index [9]. D'Orazio et al used support vector machine (SVM) classifier for differentiation of eight types of pollen based on electrical and optical features, and obtained an accuracy of 88.3% [10].

Here we present our experimental results regarding applying machine learning models to IFC measured datasets for classification of cell status of cell cycle arrest and apoptosis. Tumor cells treated by anti-cancer drugs could exhibit multiple stage of states. For instance, the broadly used drugs, taxol and thymine, induce mitotic arrest of tumor cells at G2/M and G1/S phase [11][12]. Cells with abnormal mitosis could exit the division process and undergo apoptosis [13]. We used to demonstrate the examination of single-cell physiology states using IFC device on the MEMS 2022 last year [14]. By exploiting machine learning model here, it offers a more convenient way with high accuracy for cell state judgement, which potentially give insight into cell heterogeneity research as well as drug sensitivity assessment.

Figure 1a shows the flow chart of measuring cells using IFC chip and data processing by machine learning with example of using neural network model. Electrical signal trains at 500kHz and 10MHz frequencies are probed by the microelectrodes simultaneously, and subsequently acquired by the impedance spectroscope. The imaginary and real parts of the impedance data at dual frequencies are then extracted and applied as input features for the NN to classify cell states of mitosis arrest or apoptosis (e.g. G2/M phase vs. apoptosis, G1/S phase vs. apoptosis). Confusion matrix of classification results is obtained by properly splitting the training and validation datasets.

MATERIALS AND METHODS

Data acquisition

The cells tested here by IFC were human lung cancer cells (H1650) and cervical carcinoma cells (Hela). 2mM thymidine and 50nM taxol were respectively applied on H1650 cells for blocking the cells at G1/S or G2/M phases. Apoptosis state of H1650 cells was induced by taxol treatment with higher concentration of 800nM. Hela cells were treated by 800nM taxol for different time interval for inducing G2/M arrest and apoptosis. Cells post drug treatment were passed through the IFC chip for measuring impedance at 500kHz and 10MHz simultaneously.

IEEE MEMS 2023, Munich, GERMANY
15 - 19 January 2023

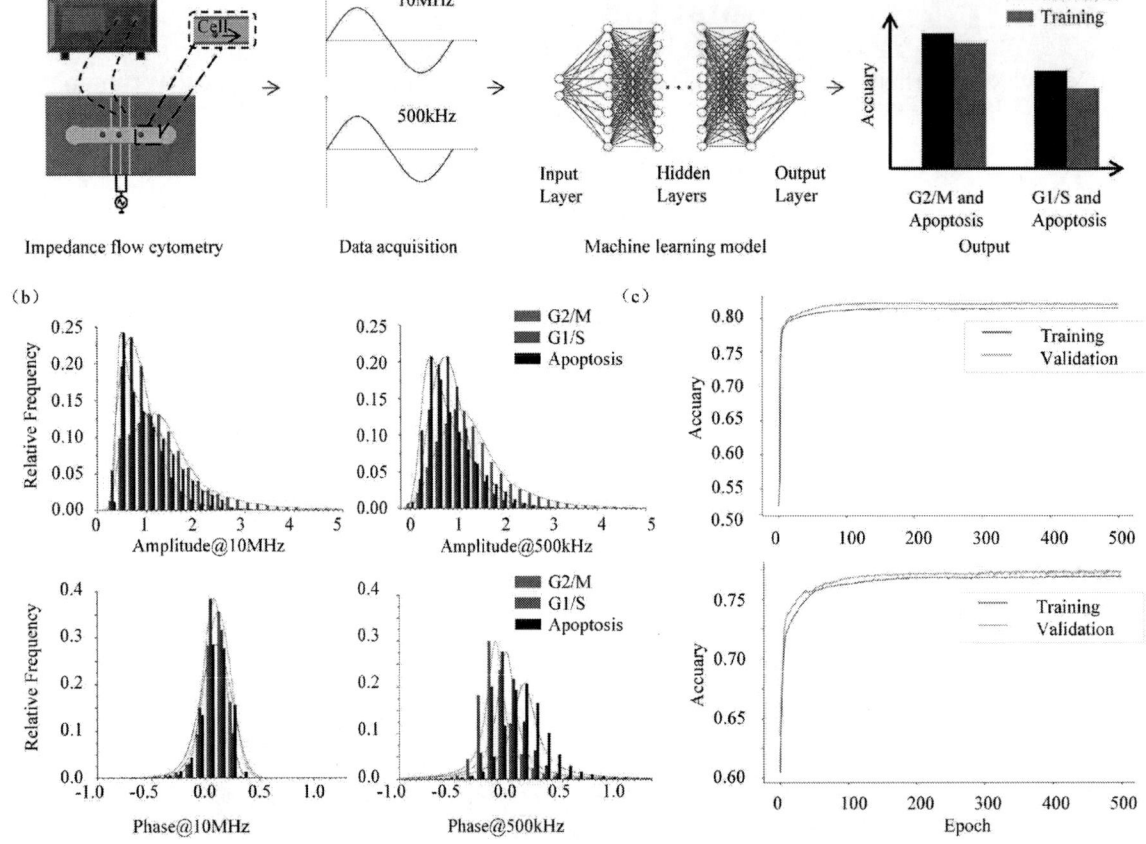

Figure 1: (a) Conceptual diagram of the process for classifying cells according to their electrical properties. The cell impedance information is obtained by flow cytometry, and then the amplitude and phase of the cell at 500kHz and 10MHz are obtained by data processing, and brought into the neural networks model to get an output that results in a confusion matrix. (b) Amplitude and phase distribution of human lung cancer cells in different cycles at 10MHz and 500kHz. (c) Learning curves of neural networks for different periodic cell (G2/M phase and Apoptotic phase, G1/S phase and Apoptotic phase) classification.

Impedance datasets were acquired using impedance spectroscope (HF2IS, Zurich Instrument) with both amplitude and phase being recorded.

Machine learning model

The machine learning models used here include deep neural network (DNN), random forest (RF) and support vector machine. We exploited DNN firstly for classification of H1650 cells and Hela cells at different states, i.e. G1/S, G2/M and apoptosis. DNN can be divided into input layer, hidden layer and output layer. The advantage of DNN is that each neuron has a weight value, and there is no connection between neurons in the same layer. We tried to improve the accuracy of the models for both types of cells by adjusting the number of hidden layers. Among them, the H1650 cells and the Hela cells had seven and three hidden layers in their DNN models, respectively.

In addition, we also tried to improve the classification accuracy of Hela cells with other types of models, such as RF and SVM. Random forest is a classifier that contains multiple decision trees, the number of which can be

adjusted to enhance the accuracy. The number of decision trees used for classification of Hela cells is 162. Support vector machine is built with Gaussian kernel function for Hela cells classification.

Four electrical features (i.e., amplitudes and phases at 500kHz and 10MHz) were applied for building these models. The electrical features were extracted by processing the raw data with a custom program. Cell states of blocked cycles as well as apoptosis were used as labels. The whole datasets were split by 7:3 for training and validation, respectively. The models returned binary classification of cells states (i.e., G1/S vs. apoptosis, G2/M vs. apoptosis) for the validation datasets. Confusion matrixes were then obtained. Both raw data processing and machine learning prediction were implemented with Python.

RESULTS AND DISCUSSION

The amplitude and phase distributions H1650 cells with different states are shown in Figure 1b, indicating that cell states can be classified according to the electrical feature of these cells. The data amount of sample points of

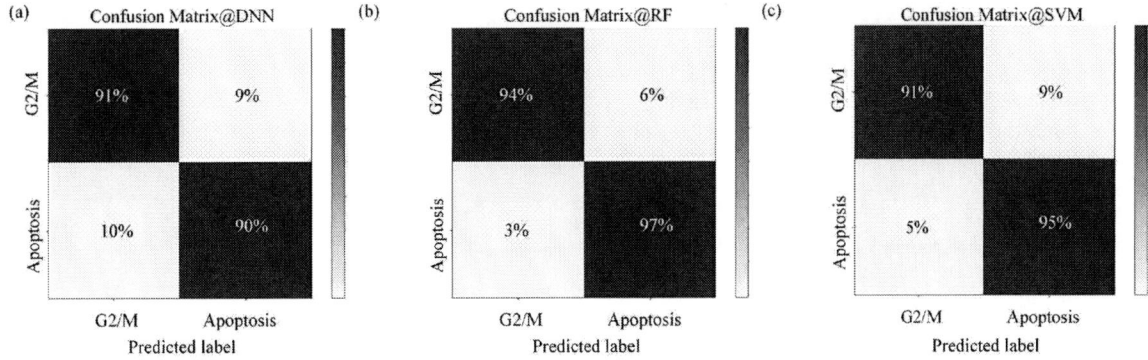

Figure 2: Confusion matrix from classifying Hela cells with different models: (a) DNN (93.29%), (b) RF (96.63%), (c) SVM (93.13%).

G1/S, G2/M and apoptosis are 30,000, 30,000 and 25,117, respectively. DNN model for classification of H1650 cells in G1/S and apoptosis states produces overall accuracy of 69.24% (76% for G1/S and 61% for apoptosis) for phase only, 69.21% (81% for G1/S and 57% for apoptosis) for amplitude only and 76.58% (80% for G1/S and 72% for apoptosis) for the combination. Using DNN for classifying G2/M and apoptotic states gives rise to success rate of 78.79% (82% for G2/M and 75% for apoptosis) for phase only, 71.84% (79% for G2/M and 63% for apoptosis) for amplitude only and 82.20% (81% for G2/M and 83% for apoptosis) for the combination. The learning curves characterized by both phase and amplitude datasets is shown in Figure 1c.

Hela cells at G2/M state and apoptosis state were also classified using trained DNN model. The confusion matrix in Figure 2a exhibit the binary states being correctly classified in diagonal units. In addition, we also trained random forest model and support vector machine model for classifying the cells states to improve the accuracy. The sample amount were 15,000 points for each state. The classification results are shown in Figure 2b-c, where the RF model exhibits higher compatibility with the datasets of Hela cells. It exhibits overall accuracy of 96.63% (94% and 97% successful rates for G2/M state and apoptosis state, respectively) as compared to 93.29% from DNN model and 93.13% from SVM model.

CONCLUSION

This paper presents a method for classifying cell states using machine learning models based on cell electrical features extracted by impedance flow cytometry. The DNN model gives rise to an overall accuracy of 76.58% and 82.20% for differentiation of G1/S and G2/M states of H1650 cells from their apoptosis state, respectively. The random forest model produces 96.63% accuracy for classification of G2/M state and apoptosis state of taxol-treated Hela cells. This label-free technique offers an effective solution for single cell analysis post anti-cancer drug treatment with high accuracy and easy operation.

ACKNOWLEDGEMENTS

Financial supports were supported by the Fundamental Research Funds for the Central Universities

(Grant No. BUCTRC201809 and Grant No. XK1802-4).

REFERENCES

[1] Müller S and Nebe-von-Caron G, "Functional single-cell analyses: flow cytometry and cell sorting of microbial populations and communities", *FEMS Microbiology reviews*, vol. 34, pp. 554-587, 2010.

[2] Adele De Ninno, Riccardo Reale, Alessandro Giovinazzo, Francesca R. Bertani, Luca Businaro, Paolo Bisegna, Claudia Matteucci and Federica Caselli, "High-throughput label-free characterization of viable, necrotic and apoptotic human lymphoma cells in a coplanar-electrode microfluidic impedance chip", *Biosensors and Bioelectronics*, vol. 150, 2020.

[3] S. Zhang, Z. Han, Z. Feng, M. Sun and X. Duan, "Deep Learning Assisted Microfluidic Impedance Flow Cytometry for Label-free Foodborne Bacteria Analysis and Classification", *2021 43rd Annual International Conference of the IEEE Engineering in Medicine & Biology Society (EMBC)*, pp. 7087-7090, 2021.

[4] Huang, L., Zhao, P., and Wang, W., "3D cell electrorotation and imaging for measuring multiple cellular biophysical properties", *Lab on a Chip*, vol. 18, pp. 2359-2368, 2018.

[5] Feng, Y., Huang, L., Zhao, P., Liang, F., and Wang, W., "A microfluidic device integrating impedance flow cytometry and electric impedance spectroscopy for high-efficiency single-cell electrical property measurement", *Analytical chemistry*, vol. 91, pp. 15204-15212, 2019.

[6] Y. Feng, Z. Cheng, H. Chai, W. He, L. Huang and W. Wang, "Neural network-enhanced real-time impedance flow cytometry for single-cell intrinsic characterization", *Lab on a Chip*, vol. 22, pp. 240-249, 2022.

[7] Caselli, F., Reale, R., De Ninno, A., Spencer, D., Morgan, H., and Bisegna, P., "Deciphering impedance cytometry signals with neural networks", *Lab on a Chip*, vol. 22, pp. 1714-1722, 2022.

[8] Yang Zhao, Ke Wang, Deyong Chen, Beiyuan Fan, Ying Xu, Yifei Ye, Junbo Wang, Jian Chen and Chengjun Huang, "Development of microfluidic impedance cytometry enabling the quantification of

specific membrane capacitance and cytoplasm conductivity from 100,000 single cells", *Biosensors & Bioelectronics*, vol. 111, pp. 138-143, 2018.

[9] D. Yang, Y. Zhou, Y. Zhou, J. Han and Y. Ai, "Biophysical phenotyping of single cells using a differential multiconstriction microfluidic device with self-aligned 3D electrodes", *Biosensors and Bioelectronics*, vol. 133, pp. 16-23, 2019.

[10] Travaglini and F. Caselli, "Electro-Optical Classification of Pollen Grains via Microfluidics and Machine Learning", *IEEE Transactions on Biomedical Engineering*, vol. 69, pp. 921-931, 2021.

[11] Schiff, P. B., and Horwitz, S. B., "Taxol stabilizes microtubules in mouse fibroblast cells", *Proceedings of the National Academy of Sciences*, vol. 77, pp. 1561-1565, 1980.

[12] Sherley, J. L., and Kelly, T. J., "Regulation of human thymidine kinase during the cell cycle", *Journal of Biological Chemistry*, vol. 263, pp. 8350-8358, 1988.

[13] K. Ahuja, G. M. Rather, Z. Lin, J. Sui, P. Xie, T. Le, J. R. Bertino and M. Javanmard, "Toward point-of-care assessment of patient response: a portable tool for rapidly assessing cancer drug efficacy using multifrequency impedance cytometry and supervised machine learning", *Microsystems & nanoengineering*, vol. 5, pp. 1-11, 2019.

[14] Yuan, X., Li, T., Yu, F., Han, C., and Xing, X., "Microfluidic Impedance Flow Cytometry For Single-Cell Physiology States Analysis of Tumor Cells Treated with Chemotherapy Drugs", *IEEE 35th International Conference on Micro Electro Mechanical Systems Conference (MEMS)*, pp. 880-883, 2022.

CONTACT

† X. Xing: xxing@mail.buct.edu.cn

HIGH-THROUGHPUT SPHERICAL SUPRAPARTICLE SELF-ASSEMBLY BY ENHANCED EVAPORATION OF COLLOIDAL WATER DROPLETS THROUGH THIN FILM OF WATER-SOLUBLE OIL

Wonhyung Lee, Joowon Rhee, and Joonwon Kim

Pohang University of Science and Technology (POSTECH), Republic of Korea

ABSTRACT

This paper proposes a microfluidic device that facilitates high-throughput and reliable spherical supraparticle self-assembly through enhanced evaporation of water droplets through thin film of water-soluble/low volatile olive oil. The olive oil, containing water droplets generated from a flow-focusing device, is spontaneously spread thinly throughout interpillar spaces of an open-microfluidic evaporator. The thin olive oil film on an evaporator offers enhanced water dissolution to oil and diffusion of dissolved water from oil to air, which can allow complete droplet evaporation within 10 min without any temperature and humidity control. Finally, high-throughput and reliable spherical supraparticle self-assembly are achievable, which can construct tens of micron-sized particles with nanoparticle building blocks.

KEYWORDS

Open-Microfluidics, Water-In-Oil Emulsification, Bottom-Up Fabrication.

INTRODUCTION

Supraparticles are three-dimensional structures that are composed of more minute building blocks such as micro/nanoparticles, being highlighted as versatile candidate for broad application spectrums by high surface-to-volume ratio and scalability [1]. Supraparticle formation has been acquired by drying suspension water droplets in immiscible phases (e.g., air and oil). During drying, capillary pressure around the interface brings micro/nanoparticles in contact, where van der Waals forces are dominant for interparticle cohesion [2]. Microfluidic emulsification has been successfully implemented, making it possible to produce intact spherical droplets at a high frequency with scalable droplet sizes, which are supported with continuous oil-phase [3]. However, three limitations have hindered the applicability of the emulsification methods for the supraparticle formation: (i) low water solubility that hinders the dissolution of water in oil, (ii) high volatility of oil that is diminished earlier than complete droplet evaporation and (iii) bulk oil layer formation that inhibits transfer of dissolved water in oil to air. Fluorinated oils, which is widely used in microfluidic water-in-oil (W/O) emulsification, inherently have low water solubility and high volatility. These properties hinder droplet evaporation and diminish the oil before supraparticle assembly. Moreover, bulk oil layer surrounding suspension droplets under sophisticated temperature/humidity control in bulk vials also hinders evaporation, which takes few hours to days until complete supraparticle assembly [3]. Therefore, the combination of water-soluble/low volatile oil and its thin film formation while having high droplet productivity would be beneficial for the supraparticle production in the microfluidic system [4].

In this paper, a high-throughput and reliable spherical supraparticle self-assembly manner is described. Water-soluble/low volatile olive oil with surfactant (continuous phase), and deionized (DI) water with various nanoparticles (dispersed phase) were utilized for W/O emulsification. The W/O emulsification is performed via a

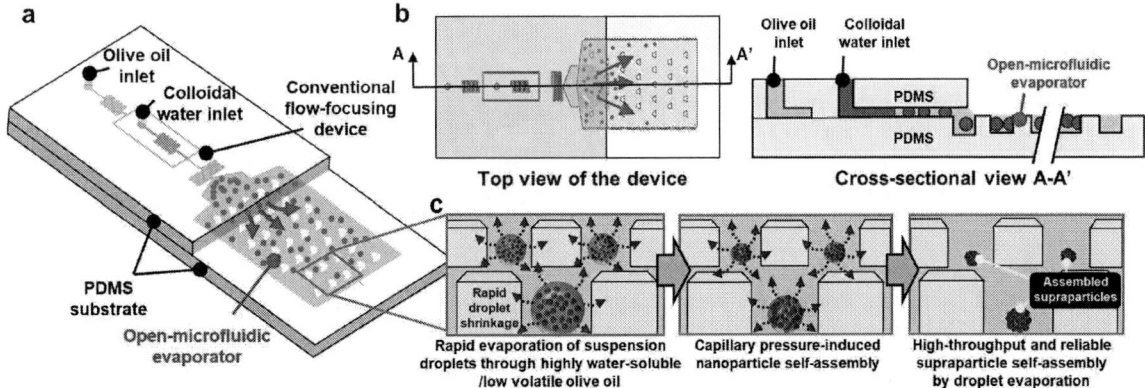

Figure 1: (a) Configuration of the hybrid microfluidic device for water-in-oil (W/O) emulsification and spontaneous and high-throughput supraparticle self-assembly via highly water-soluble/low volatile olive oil thin film. The generated W/O emulsion is spontaneously flowing through the micropillar array on the evaporator. (b) Top and cross-sectional view of the microfluidic device. (c) Through the even and thin olive oil layer, suspension droplets are uniformly and rapidly evaporated while assembling nanoparticles through interfacial capillary pressure. The dominant capillary pressure brings nanoparticles building blocks in contact and constructs supraparticles under complete droplet evaporation. All figures are not-to-scale.

conventional flow-focusing device (Figure 1a) [5]. The W/O emulsion can be continuously conveyed by the spontaneous capillary flow of the olive oil through the micropillar array on the open-microfluidic evaporator (Figure 1b). Then the open-surface evaporator can construct a uniform and thin oil layer which enhances the transfer of dissolved water molecules through the oil-air interface [4]. Therefore, the water droplets on the open-microfluidic evaporator can be rapidly evaporated and diminished by two synergetic characteristics of this system (i.e., thin film formation and properties of the olive oil). The effects of the two characteristics on water droplet evaporation were investigated with comparison tests on oil film thickness and types of oils. Finally, the TiO_2 or polystyrene (PS) nanoparticles in the water droplet were rapidly and reliably self-assembled as ~25 μm supraparticles within 10 min (Figure 1c).

MATERIALS AND METHODS

Reagents and Materials

Poly(dimethylsiloxane) (PDMS) Sylgard 184 monomer and curing agent were obtained from Dow Corning. Photoresist (PR) KMPR1025 and SU-8 developer were purchased from Kayaku Advanced Materials. NOVEC 7500 was purchased from 3M. Olive oil, TiO_2 nanoparticles (~21 nm in diameter), PS nanoparticles (~100 nm in diameter), and silicone oil AR 20 were purchased from Sigma-Aldrich. Polyglycerol polyricinoleate (PGPR) was kindly provided from Ilshin Wells. Abil EM 90 was obtained from Evonik. FluoSurf was purchased from Emulseo.

Fabrication of the Microfluidic Device

The microfluidic device consists of upper (flow focusing device; 70 μm in height) and lower (open-microfluidic evaporator; 100 μm in height) PDMS microchannels (Figure 1a). The PDMS microchannels were fabricated with standard soft lithography technique [6, 7]. The master molds were patterned using KMPR1025. The PR was spin-coated on a 4" silicon wafer, followed by soft-baking at 100°C. The designed microchannel patterns were lithographed by an ultraviolet exposure through a photomask, followed by post-exposure-baking at 100°C and development in the SU-8 developer. Then, the PDMS prepolymer solution mixed with 10:1 (w/w) ratio of monomer to curing agent, was poured and cured at 100°C for 20 min. The cured PDMS was peeled off and diced for proper sizes. The inlet holes at the flow-focusing device were punched with a biopsy punch. Both of the PDMS microchannels were treated with air plasma and irreversibly bonded. For strong bonding and hydrophobic recovery of the PDMS, the microfluidic device was incubated at 70°C for 24 hrs. Before use, the open-microfluidic evaporator was treated with air plasma to increase surface energy, which can enhance the capillary flow of the olive oil through the micropillar array. During the plasma treatment, other parts of the microfluidic device were masked with taping to block the unintended hydrophilic treatment on the closed channel (flow-focusing part), which can reversely generate oil-in-water droplets.

Experimental Setup

A custom-made vacuum chamber was used to degas the PDMS prepolymer solution. A lab-made pneumatic pressure pump was utilized to generate W/O droplets. The microscopic images were acquired through an inverted microscope (IX 73, Olympus) with a charge-coupled device camera (DP 80, Olympus). An ultrasonic homogenizer (VCX 130, Sonics & Materials) was used to disperse the nanoparticles. The plasma treatment was performed by a vacuum plasma system (CUTE-MP, Femto Science).

Droplet Generation

The W/O emulsification was performed with various combinations. DI water was used as the same as dispersed phase, while three oils and surfactants [i.e., olive oil with 2.5% (w/w) PGPR, NOVEC 7500 with 2% (w/w) FluoSurf and silicone oil AR 20 with 2% (w/w) ABIL EM 90] were used as the continuous phase. The input pressures of each phase were regulated to generate droplets of ~90 μm in diameter due to different viscosities of oils.

RESULTS AND DISCUSSIONS

Effect of Oil Thickness on Droplet Evaporation

The fast droplet evaporation through thin olive oil layer was demonstrated by comparing it with thick olive oil layer. Once the W/O emulsion was conveyed to the evaporator, the evaporation process was sequentially started. As shown in Figure 2a, the micropillar array can form uniform and thin oil film [8]. The thin film can offer same short distances (~10 μm) between droplets and air-oil interface, which enhances the transfer of dissolved water in

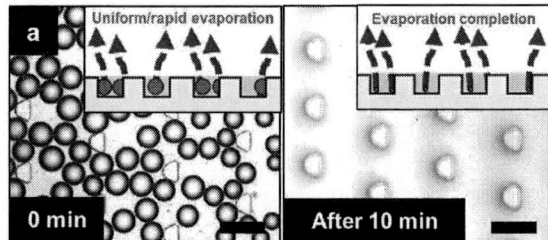

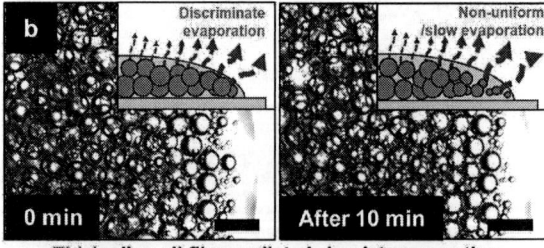

Figure 2: Comparison test on water droplets evaporation. (a) The uniform and thin olive oil film was formed via micropillar array. The droplets in this film were uniformly and rapidly evaporated. (b) The generated emulsion from the flow-focusing device was retrieved and loaded on a glass slide to form non-uniform and thick oilve oil film. The droplets in this film were discriminately and slowly evaporated. All scale bars are 200 μm.

Table 1: Water solubility and vapor pressure of oils for W/O emulsification.

Oil type	Water solubility @25°C (ppm)	Vapor pressure @25°C (mmHg)	Surfactant
Fluorinated oil NOVEC 7500	45 [9]	15.75 [9]	FluoSurf
Silicone oil AR 20	2902 [10]	<5 [11]	ABIL EM 90
Olive oil	>7000 [12]	<0.001 [11]	PGPR

oil toward the air. Therefore, droplets on the open-microfluidic evaporator were evenly and rapidly evaporated within 10 min. In the case of droplet evaporation in thick olive oil film, droplets were generated the same as the thin film test, while the emulsion was retrieved and dropped on a glass slide to form non-even and thick oil film (>1 mm). Because the thick oil film provided non-uniform and relatively long distance between droplets and air-oil interface, droplets were discriminately and slowly evaporated depending on the droplet's location (Figure 2b). The size of few droplets at the relatively thin oil film (i.e., the edge of oil film), had decreased, but the size of droplets at the thick oil part sustained. As a result, uniform and fast droplet evaporation in the thin oil film was confirmed.

Effect of Oil Types on Droplet Evaporation

Various oils have been incorporated to form W/O emulsions. Therefore, evaporation regimes on three

different oils (i.e., olive oil, fluorinated oil NOVEC 7500 and silicone oil AR 20) were investigated in the open-microfluidic evaporator to form the uniform and thin oil film. In Figure 2a, the droplets in the olive oil were successfully evaporated in the preserved oil film within 10-min. In the next case, droplets in the NOVEC 7500 were generated and incubated on the evaporator. As shown in Figure 3a, the droplets failed to evaporate in 10 min and merged together due to early evaporation of the supporting liquid. Because the NOVEC 7500 has low water-solubility and high vapor pressure compared to the olive oil, NOVEC 7500 was evaporated early by inhibiting the water dissolution in oil (Table 1). In the last case, the silicone oil AR 20 has relatively low water-solubility (Table 1). Therefore, the size of the droplets was slightly decreased after 10 min, while maintaining the spherical shapes in the preserved supporting liquid (Figure 3b). From these data, the formation of thin olive oil film could be the optimized combination for high-throughput and reliable supraparticle self-assembly.

Self-assembly of Nanoparticle Building Blocks for Tens of Micron Supraparticle Construction

Supraparticle self-assembly was demonstrated by using TiO_2 and PS nanoparticles. Colloidal droplets were generated with the same method as the previous experiment. The initial droplet size was set as ~90 μm in diameter. The nanoparticle concentrations were controlled to construct supraparticles ~25 μm in diameter, regardless of the nanoparticle types. Finally, both TiO_2- and PS-based supraparticles were successfully fabricated on the evaporator with the thin olive oil film (Figure 4). For further use, the supraparticles were washed in ethanol 3 times and resuspended in DI water. The washed supraparticles sustained themselves after the medium changes (Data are not provided).

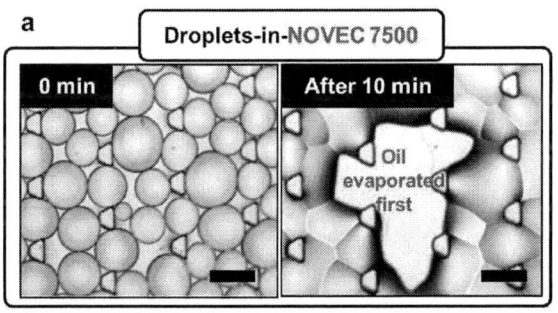

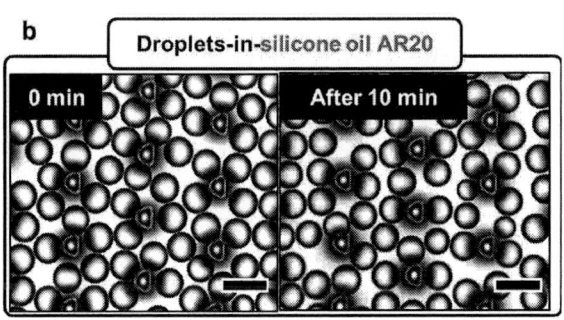

Figure 3: Droplet evaporation test on various oils for W/O emulsification. (a) NOVEC 7500 has highly volatile and poorly water-soluble properties. The supporting NOVEC 7500 evaporated faster than droplets inside, thereby the droplet merging occurred. (b) Silicone oil AR 20 is relatively low volatile and highly water-soluble. The size of droplets in silicone oil AR 20 slightly decreased after 10-min. All scale bars are 200 μm.

CONCLUSIONS

In this paper, a high-throughput and reliable spherical supraparticle self-assembly was proposed. To perform this method, open-microfluidics and advantageous properties of the olive oil were synergistically utilized. The olive oil, which has high water solubility and low volatility, was thinly formed through the open-surface micropillar array. Water droplets in the thin olive oil film were uniformly and rapidly evaporated in 10 min while sustaining the spherical shape until complete evaporation. Simply generating colloidal water droplets and drying, the TiO_2 or PS nanoparticle building blocks were successfully constructed into supraparticles due to dominant interfacial tension between water and oil, and van der Waals forces. As a

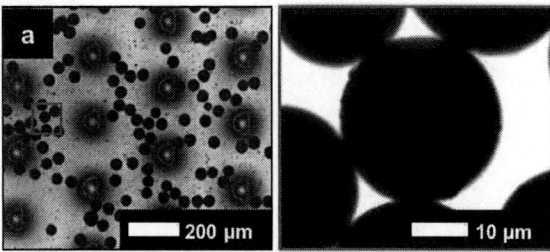

TiO₂ nanoparticle-based supraparticle self-assembly

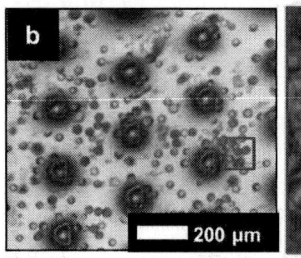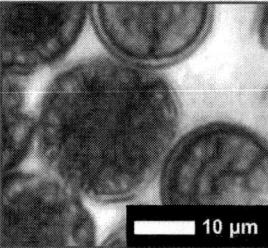

Polystyrene nanoparticle-based supraparticle self-assembly

Figure 4: Self-assembled supraparticles based on various nanoparticles. (a) TiO₂ nanoparticle-based and (b) PS nanoparticle-based supraparticles. The nanoparticle concentrations of suspensions were regulated to fabricate supraparticles with ~25 μm in diameter.

result, this method demonstrated its potential as an effective supraparticle production technique. Although further study is needed, the proposed system could potentially cover a broad range of applications for drug delivery systems (DDS), catalysis, and diagnostics with its versatility such as broad material selection and high-throughput manner [2, 13, 14].

ACKNOWLEDGEMENTS

There are no acknowledgements.

REFERENCES

[1] H. Tan, S. Wooh, H. J. Butt, X. Zhang, and D. Lohse, "Porous supraparticle assembly through self-lubricating evaporating colloidal ouzo drops," *Nat. Commun.*, vol. 10, no. 1, p. 478, 2019.

[2] S. Wooh, H. Huesmann, M. N. Tahir, M. Paven, K. Wichmann, D. Vollmer, W. Tremel, P. Papadopoulos, and H. J. Butt, "Synthesis of Mesoporous Supraparticles on Superamphiphobic Surfaces," *Adv. Mater.*, vol. 27, no. 45, pp. 7338-7343, 2015.

[3] J. Wang, C. F. Mbah, T. Przybilla, B. Apeleo Zubiri, E. Spiecker, M. Engel, and N. Vogel, "Magic number colloidal clusters as minimum free energy structures," *Nat. Commun.*, vol. 9, no. 1, p. 5259, 2018.

[4] H. Miyazaki and S. Inasawa, "Drying kinetics of water droplets stabilized by surfactant molecules or solid particles in a thin non-volatile oil layer," *Soft Matter,* vol. 13, no. 47, pp. 8990-8998, 2017.

[5] S. Lee, H. Kim, D.-J. Won, J. Lee, and J. Kim, "On-demand, parallel droplet merging method with non-contact droplet pairing in droplet-based microfluidics," *Microfluidics and Nanofluidics,* vol. 20, no. 1, 2016.

[6] W. Lee, L. Ha, D. P. Kim, and J. Kim, "Cytocompatible asymmetrical coating for Janus carrier synthesis through capillary wetting and ascending," *J. Colloid Interface Sci.*, vol. 623, pp. 54-62, 2022.

[7] W. Lee, H. Kim, P. K. Bae, S. Lee, S. Yang, and J. Kim, "A single snapshot multiplex immunoassay platform utilizing dense test lines based on engineered beads," *Biosens. Bioelectron.,* vol. 190, p. 113388, 2021.

[8] P. Baumli, H. Teisala, H. Bauer, D. Garcia-Gonzalez, V. Damle, F. Geyer, M. D'Acunzi, A. Kaltbeitzel, H. J. Butt, and D. Vollmer, "Flow-Induced Long-Term Stable Slippery Surfaces," *Adv. Sci.,* vol. 6, no. 11, p. 1900019, 2019.

[9] 3M™ NOVEC™ 7500 Engineered Fluid Product Information.

[10] T. Gerecsei, R. Ungai-Salánki, A. Saftics, I. Derényi, R. Horvath, and B. Szabó, "Characterization of the Dissolution of Water Microdroplets in Oil," *Colloids and Interfaces,* vol. 6, no. 1, 2022.

[11] J. Cahir, M. Y. Tsang, B. Lai, D. Hughes, M. A. Alam, J. Jacquemin, D. Rooney, and S. L. James, "Type 3 porous liquids based on non-ionic liquid phases - a broad and tailorable platform of selective, fluid gas sorbents," *Chem. Sci.,* vol. 11, no. 8, pp. 2077-2084, 2020.

[12] S. Saffar Taluri, S. M. Jafari, and A. Bahrami, "Evaluation of changes in the quality of extracted oil from olive fruits stored under different temperatures and time intervals," *Sci. Rep.,* vol. 9, no. 1, p. 19688, 2019.

[13] Y. Zhao, X. Zhao, C. Sun, J. Li, R. Zhu, and Z. Gu, "Encoded silica colloidal crystal beads as supports for potential multiplex immunoassay," *Anal. Chem.*, vol. 80, pp. 1598-1605, 2008.

[14] Y. Yu, X. Yang, M. Liu, M. Nishikawa, T. Tei, and E. Miyako, "Anticancer drug delivery to cancer cells using alkyl amine-functionalized nanodiamond supraparticles," *Nanoscale Adv.*, vol. 1, no. 9, pp. 3406-3412, 2019.

CONTACT

*J. Kim, tel: +82-54-279-2185; joonwon@postech.ac.kr

IN-ICE POLYMERIZATION
FOR FUNCTIONAL HYDROGEL MICROBEAD
WITH FLASH FREEZING CENTRIFUGAL MICROFLUIDIC DEVICE

Tomomi Murayama[1], Koki Yoshida[1], Yuta Kurashina[2], and Hiroaki Onoe[1]
[1] Department of Mechanical Engineering, Keio University, Kanagawa, JAPAN and
[2] Tokyo University of Agriculture and Technology, Tokyo, JAPAN

ABSTRACT

This paper describes a centrifuge-based fabrication method of microgel beads using flash freezing in liquid nitrogen and in-ice polymerization. Previously, materials for fabricating microgel beads with centrifugal microfluidic systems have been limited to only sodium alginate solution using rapid ionic cross-linking. By using our proposed method, the shape of beads can be defined by flash freezing of the ejected pre-gel microdroplets in liquid nitrogen and successive in-ice polymerization by UV irradiation. We succeeded in fabricating ethanol-responsive structural-color beads and confirmed their ethanol responsivity by exhibiting visible structural color change. Our fabrication method will expand the variety of materials for the centrifugal microfluidic microgel beads fabrication method and possible applications including biosensors and drug delivery systems.

KEYWORDS

Microgel beads, In-ice polymerization, Centrifugal force, Gelling by UV irradiation, Ethanol responsive hydrogel

INTRODUCTION

Microgel beads are microscale particles composed of a hydrogel. Hydrogels are three-dimensional polymer networks that are swollen by aqueous solvents. Because of their biocompatibility, hydrogels have been used in biochemical and medical applications such as biosensors [1]. In addition to this biocompatibility, microscale hydrogel beads have spherical, isotropic, and uniform shapes, showing highly responsive properties to changes in the external environment and to applied stimuli because of their rapid mass diffusion and rapid heat transfer due to the scaling effect. Based on these attractive properties, microgel beads have been applied in drug delivery systems, biosensors, and tissue engineering [2].

In previous studies, several methods have been proposed for the fabrication of microgel beads, including an emulsion method [3], a microfluidic method [4], and an electrojet method [5]. In addition, a centrifugal method [6][7][8] has recently been demonstrated for the fabrication of microgel beads: pre-gel solution filled in the micro-scale capillary was ejected by centrifugal force and then the ejected pre-gel solution formed droplets by surface tension. The shape of the fabricated droplets can be maintained due to instant gelation by the ionic cross-linking between sodium alginate and calcium ions. This centrifugal method has some advantages over other methods: uniformity of the microgel beads, oil-free processes, and easy-to-use simple setup. However, in this method, instantaneous gelation of the ejected pre-gel droplet is essential for the formation of hydrogel microbeads. In other words, the material for fabricating microgel beads is generally limited only to alginate-based hydrogels.

To solve this problem, here we propose a method for fabricating microgel by flash freezing in liquid nitrogen and in-ice polymerization by UV irradiation. This method is divided into two parts; a flash freezing part with centrifugal force and a gelation part by UV irradiation to the frozen pre-gel microdroplets (Fig. 1). The pre-gel solution is ejected from a pulled glass capillary by applying centrifugal force. The ejected pre-gel solution forms micro-scale droplets, followed by flash freezing in liquid nitrogen. Then frozen pre-gel droplets are gelled by UV irradiation

Freezing part

Pre-gel solution
Centrifugal force
Air
Flozen bead
Liquid N$_2$(-196°C)

Gelation part

Frozen bead | Polymerization | Melting ice

Monomer | Polymer | Gel

Figure 1: Concept of microgel beads fabrication method using flash freezing and in-ice polymerization.

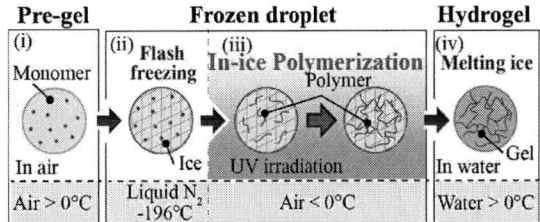

Figure 1: Fabrication principle of microgel beads gelatinized by UV. (i) Monomer droplets ejected by centrifugal force. (ii) Droplet instant frozen by liquid nitrogen. (iii) Polymerization under frozen conditions by UV. (iv) Thawing of frozen droplets.

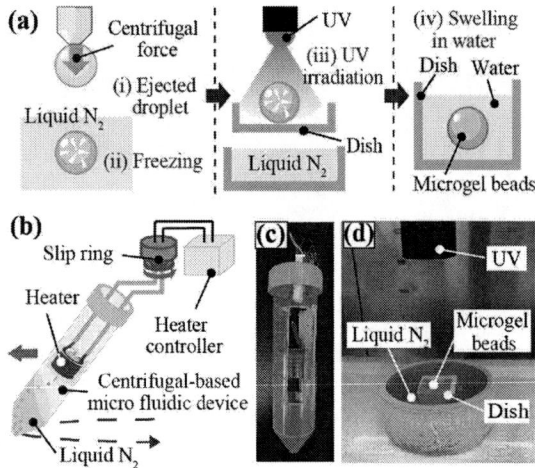

Figure 3: Fabrication method of microgel beads. (a) Fabrication procedure of microgel beads. (b) Schematic of the fabricated centrifugally driven microfluidic device. (c) Image of the device. (d) Image of device used for UV irradiation.

with maintaining the frozen state. To demonstrate this concept work, we chose *N*-(Hydroxymethyl)acrylamide (NMAM) as a material for microgel beads. NMAM is one of the common hydrogels that gel reversibly by UV irradiation [9]. NMAM has stimulus response to ethanol and its hardness could be controlled by UV. In this paper, we verify our fabrication method, confirm the gelation by temperature and UV irradiation time and demonstrate the NMAM microgel beads with the proposed in-ice gelation system.

PRINCIPEL

The principle and formation method of microgel beads using flash freezing and in-ice polymerization are as follows (Fig. 2); (i) pre-gel droplets are ejected by applying centrifugal force to the whole microfluidic device. (ii) Instant freezing of the droplet occurs by entering the droplets in liquid nitrogen at the bottom of the microtube (the water in the droplet becomes ice). (iii) The frozen pre-gel droplets are kept frozen and polymerized by UV irradiation at a temperature lower than 0°C. (iv) After UV irradiation, the frozen droplets are swollen in water to melt the ice (Fig. 3 (a)). For this process, we used a centrifugal microfluidic device and a setup for UV irradiation (Fig. 3 (b)(c)(d)). This system can fabricate microgel beads with various UV polymerizable materials while maintaining their spherical shape.

METHODOLOGY

Centrifugal microfluidic device with liquid nitrogen

For fabricating microgel beads, we fabricated flash freezing centrifugal microfluidic device (Fig. 3 (b)(c)). The device is composed of a pre-gel microdroplet ejection part with a heater system to keep the temperature of a pre-gel solution, and a freezing part filled with liquid nitrogen to achieve flash freezing of the pre-gel microdroplets. In fabricating microgel beads, we experimented to see if the

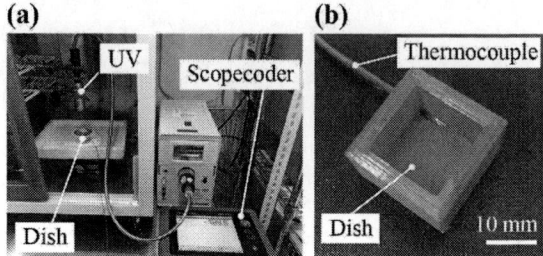

Figure 4: UV irradiation time device. (a) Set up the device. (b) Dish for gelation with temperature measurement.

temperature of the beads during gelation affected the UV irradiation time required for gelation.

Conditions for gelation by temperature of gel beads

We measured the temperature of the bottom of the dish in contact with the beads by connecting a thermocouple attached to the dish (Fig. 4 (a)(b)). First, we confirmed the temperature change at the bottom of the dish by taking measurements for 20 minutes with a 20 mm depth of liquid nitrogen added to the bottom dish. Next, we performed measurements for 10 minutes while adding liquid nitrogen to keep the dish at -10, -50, and -100°C. Finally, we confirmed the gelation of NMAM gels by irradiating the frozen pre-gel solution with UV light. The pre-gel solution was prepared by adding 10% NMAM, 0.33% BIS, 10 colloids, and 0.5% Irgacure1173 to a 1:1 mixture of water and glycerin. The pre-gel solution was dripped into liquid nitrogen with a pipette. Then, dripped frozen gel was irradiated for 10, 20, and 30 seconds at -10, -50, and -100°C.

Fabrication method of microgel beads

For fabricating NMAM microgel beads, a pre-gel solution was prepared by adding 10% NMAM, 0.33% BIS, 10 colloids, and 0.5% Ilgacure1173 to a 1:1 mixture of water and glycerin. The NMAM pre-gel solution was wrapped in aluminum foil to shield it from light up to the time of use. The inner diameter of the glass capillary was adjusted for 100 μm.

The fabrication process of NMAM microbeads are as follows (Fig. 3 (a)): the microfluidic device filled with liquid nitrogen was placed in a centrifuge. The glass capillary was filled with the NMAM pre-gel solution and inserted into the microfluidic device (Fig. 3 (b)). Then, the centrifugal force was applied to the microfluidic device by the rotation of the centrifuge. The NMAM pre-gel solution was ejected to liquid nitrogen by the centrifuge force (95 G) within in a minute. The ejected pre-gel droplets were frozen as soon as contact with the liquid nitrogen. The frozen NMAM pre-gel beads were moved to a liquid nitrogen-cooled dish and irradiated with UV light for 6 minutes while maintaining the frozen state. The intensity of UV irradiation was about 60.0 mW/cm². After UV irradiation, the microgel beads were swollen with pure water.

Ethanol responsivity of NMAM microgel beads

The ethanol responsivity was confirmed using the

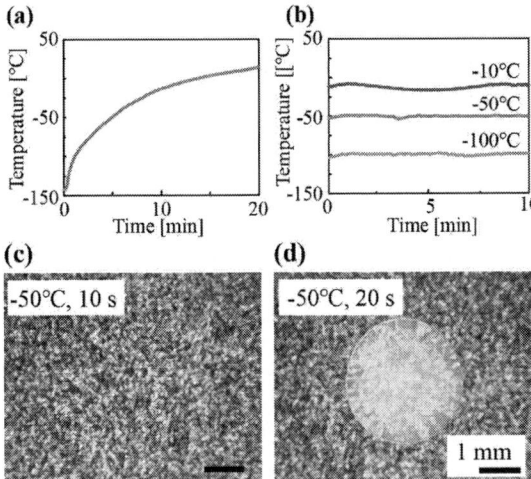

Figure 5: Relationship between gelation temperature and UV irradiation time during gel formation. (a) Temperature measurement at the bottom of the dish. (b) Temperature measurement while maintaining -10, -50, -100°C. (c) Image of no gelation. (d) Image of a NMAM gel by UV irradiation in 20 s at -50°C.

Table 1: Gels fabricated under various conditions.

		Irradiation time [s]		
		10	20	30
temp. [°C]	-10	○	○	-
	-50	×	○	-
	-100	×	×	○

○: Gelation, ×: No gelation, $n = 2$

fabricated NMAM microgel beads. After soaking the beads in each ethanol solution from 0 to 90% for 20 minutes, the beads were measured for changes in bead diameter and structural color using a microscope and a spectrometer.

EXPERIMENTAL RESULTS

Temperature measurement during UV irradiation

Figure 5 (a) shows a graph of the temperature at the bottom of the dish when liquid nitrogen was added. Immediately after starting the measurement, the temperature changes rapidly, but after some time has passed, the temperature changes slowly when the liquid nitrogen level lowers. This result shows that the temperature change was due to the difference in the temperature difference between the air above and below the bottom of the dish at the measurement point. When the liquid level of liquid nitrogen was high, the temperature difference between the temperature of the bottom of the dish and the air became large, and thus the rate of temperature change became large. On the other hand, when the liquid nitrogen level was low, the bottom surface of the dish was not sufficiently cooled. As a result, the temperature difference between the temperature of the bottom of the dish and the air became small, and the rate of temperature change became small.

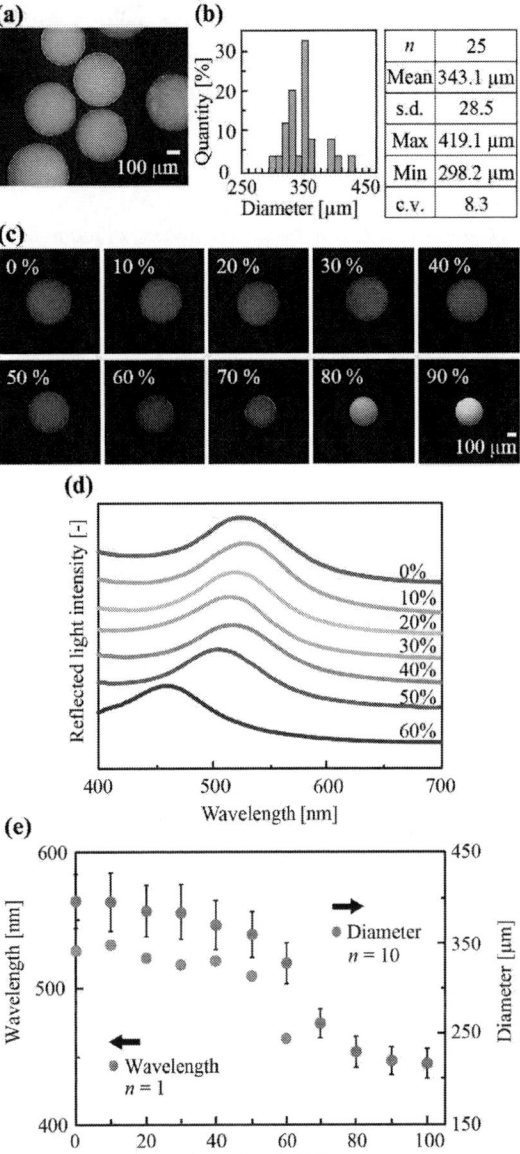

Figure 6: Confirmation of stimuli-responsivity of structural color microgel beads. (a) Microscopic images of beads. (b) Diameter distribution of beads. (c) Microscopic images of beads at various ethanol concentration. (d) Results of reflectance measurements at various ethanol concentration. (e) Relationship between maximum wavelength and beads diameter as a function of ethanol concentration. As the ethanol concentration rose, the maximum reflection wavelength was shortened and the diameters were shrunk.

Figure 5 (b) is a graph of the temperature measured while adding liquid nitrogen to the bottom of the dish at temperatures of -10, -50, and -100°C. There were no significant gaps at any of the temperatures of -10, -50, and -100°C.

Gelation by temperature and UV irradiation time

Table 1 shows the presence or absence of gelation when irradiated with UV for 10, 20, and 30 seconds respectively at -10, -50, and -100 °C. At -10°C, gelation occurred at an irradiation time of 10 seconds. On the other hand, gelation at -50°C took 20 seconds of UV irradiation time, and it took 30 seconds at -100°C. This result shows that the crosslinking reaction speed slowed down as the temperature became lower. After 10 seconds of UV irradiation at 50°C, the pre-gel solution dissolved in pure water without gelation (Fig. 5 (c)). On the other hand, after 20 seconds of UV irradiation, the pre-gel solution did not dissolve in pure water by gelation (Fig. 5 (d)).

Diameter distributions of NMAM microgel beads

NMAM microgel beads were successfully fabricated (Fig. 6 (a)(b)). The beads could be expressed in green color by structural color (Fig. 6 (a))

The diameters of 25 NMAM microgel beads made by the 100 μm glass capillary and centrifugal force of 95G were measured (Fig. 6 (b)). The average diameter was 343.1 μm with a maximum of 419.1 μm and a minimum of 298.2 μm. These results show that it is possible to produce a high amount of microgel beads at one time.

Measurement of diameter and peak wavelength by ethanol response

After 20 minutes of immersion in each ethanol solution from 0 to 90%, the beads decreased in diameter and changed in structural color from green to purple as the ethanol concentration increased (Fig. 6 (c)). Figure 6 (d) shows the result of the reflection spectra of microbeads at various ethanol concentrations. We confirmed that the peak wavelength of microbeads shifted to the shorter wavelength side as the ethanol concentration increased. As shown in Figure 6 (e), the diameter of microbeads decreased as the ethanol concentration increased, and a sharp diameter change was observed around 60%. In addition, the change in diameter of the microgel beads was linked to the change in peak wavelength. These results show that the shrinkage of the beads caused by ethanol caused the spacing between the colloids in the beads narrower, resulting in the reflection of visible light with shorter wavelength.

CONCLUSIONS

In this paper, we proposed a method of producing microgel beads by In-ice polymerization with centrifugal microfluidic device using flash freezing in liquid nitrogen. Our microfluidic device enabled to keep the shape of microgel beads, and UV irradiation device enabled to fabricate the gel during measuring the temperature of the pre-gel solution. We succeeded in fabricating NMAN microgel beads by instant freezing and in-ice polymerization. The fabricated microgel beads were spherical in shape and expressed structural color. NMAM microgel beads could change the diameter of the beads depending on the ethanol responsiveness. The beads were also able to change the wavelength of the structural color depending on the bead diameter. Our fabrication method will expand the variety of materials for the centrifugal microfluidic microbead fabrication method and possible applications including biosensors, tissue engineering and drug delivery systems.

ACKNOWLEDGEMENTS

This work was partly supported by Grant-in-Aid for Challenging Research (Pioneering) (21K18164), Japan Society for the Promotion of Science (JSPS).

REFERENCES

[1] Javad Tavakoli and Youhong Tang, "Hydrogel Based Sensors for Biomedical Applications: An Updated Review," *Polymers*, vol. 9, no. 364, pp. 1-25, 2017.

[2] X. Kang *et al.*, "Design and synthesis of multifunctional drug carriers based on luminescent rattle-type mesoporous silica microspheres with a thermosensitive hydrogel as a controlled switch," *Adv. Funct. Mater.*, vol. 22, no. 7, pp. 1470–1481, 2012.

[3] M. P. A. Lim, W. L. Lee, E. Widjaja and S. C. J. Loo, "One-step fabrication of core-shell structured alginate–PLGA/PLLA microparticles as a novel drug delivery system for water soluble drugs," *Biomater. Sci.*, vol. 1, pp. 486-493, 2013.

[4] M. Zhang, W. Wang, R. Xie, X. Ju, Z. Liu, L. Jiang, Q. Chen, L. Chu, "Controllable microfluidic strategies for fabricating microparticles using emulsions as templates," *Praticuology*, vol. 24, pp. 18-31, 2016.

[5] M. Ma, A. Chiu, G. Sahay, J. C. Doloff, N. Dholakia, R. Thakrar, J. Cohen, A. Vegas, D. Chen, K. M. Bratlie, T. Dang, R. L. York, J. Hollister-Lock, G. C. Weir, and D. G. Anderson, "Core–Shell Hydrogel Microcapsules for Improved Islets Encapsulation," *Adv. Healthc. Mater.*, vol. 2, pp. 667-672, 2013.

[6] K. Maeda, H. Onoe, M. Takinoue, and S. Takeuchi, "Controlled Synthesis of 3D Multi-Compartmental Particles with Centrifuge-Based Microdroplet Formation from a Multi-Barrelled Capillary," *Adv. Master.*, vol. 24, no. 10, pp. 1340-1346, 2012.

[7] M. Tsuchiya, Y. Kurashina, H. Onoe, "Eye-recognizable and repeatable biochemical flexible sensors using low angle-dependent photonic colloidal crystal hydrogel microbeads," *Sci. Rep.*, vol. 9, 17059, 2019.

[8] S. Yoshida, M. Takinoue, H. Onoe, "Compartmentalized spherical collagen microparticles for anisotropic cell culture microenvironments," *Adv. Healthc. Mater.*, vol. 6, 1601463, 2017.

[9] A. Toyotama, Toshimitsu Sawada, Jumpei Yamanaka, Kensaku Ito, and Kenji Kitamura, "Gelation of Colloidal Crystals without Degradation in Their Transmission Quality and Chemical Tuning", *Langmuir*, vol. 21, pp. 10268-10270, 2005.

CONTACT

*T.Murayama, tel: +81-070-2657-8845; t.murayama@keio.jp

TEMPERATURE-RESPONSIVE MICROCAPSULES MANUFACTURED BY PROMOTING CONTROLLED CLOAKING WITH THE HELP OF MICRO/NANOPARTICLES

Rutvik Lathia[1], Bheema Sankar Reddy[1], Chandantaru Dey Modak[1,2], Satchit Nagpal[1,3] and Prosenjit Sen[1,]*

[1]Indian Institute of Science, INDIA,
[2]CNRS – ESPCI PSL, FRANCE, and
[3]Texas A&M University, USA.

ABSTRACT

We demonstrate a simple and novel way of manufacturing temperature-responsive microcapsules by using wax-infused nanostructured surfaces. The infused wax in the nanostructured surface melts at higher temperatures, turning the surface slippery. When a liquid marble (LM), i.e., a droplet coated with hydrophobic particles/ is placed on the surface, the molten wax rises and cloaks the LM. Cooling down the LM below melting temperature solidifies wax, producing capsules. By varying particle size, the capsule shell thickness can be tuned over a wide range (5–100μm). Such capsules show stimuli-responsiveness, making them useful in chemical, drug, and pesticide delivery applications.

KEYWORDS

Cloaking; Liquid infused surface; Capsule; Liquid Marbles

INTRODUCTION

Encapsulations of chemicals are of utmost importance in many areas related to drugs, agriculture, food, cosmetics, and textiles [1]. Many techniques have been employed in the past for making such shells, such as blow molding, dip coating, and injection [2]. However, most of the encapsulation techniques are prone to the diffusion of inner liquid, especially for hydrophilic and lower molecular weight materials [3]. Non-porous and hard materials are often used to prevent undesirable diffusion. However, such materials are not useful in the on-demand and stimuli-responsive release of the inner liquid. Recently, a paraffin-wax encapsulation has been reported for hermetically sealed capsules with stimuli-responsive behavior [3]. However, the thickness and uniformity of the produced shells are limited to a millimeter scale (~ 0.5 mm) [2]–[4]. Microfluidic devices have also been used to create such encapsulations for the lower shell thickness coating. However, microfluidic devices are complex to fabricate, need time-consuming optimization, and are often limited by their shape and size [5]. Additionally, The microfluidics systems are limited to a smaller range of droplet size (10 – 200 μm) and shell thickness (7 – 50 μm) [6]–[8]. A wide range of thickness control is essential in many applications as it helps control the strength and dissolution of such capsules.

In the present paper, we discuss the simple and controlled fabrication of capsules with the help of micro/nanoparticles and phase change liquid-infused surfaces. The discussed technique produces stimuli-responsive, hermetically sealed capsules with a wide range of tunability in shell thickness (5 – 100 μm) and droplet size (> 0.5 mm). The shell thickness can also be further increased by taking larger particle sizes.

METHODOLOGY

Wax-infused surface fabrication

The nanostructured surface for the infusion of wax was prepared from a copper surface. The protocol of etching the copper surface for creating nanostructures has been previously reported [9]. Briefly, the clean copper surface of 3 cm × 2 cm is immersed into the aqueous solution of 2.5 mol/L sodium hydroxide and 0.1 mol/L ammonium persulfate for 20 min at room temperature. The solution produces nanowires of copper hydroxide on the copper surface. After cleaning the surface with DI water, it was dipped into a 5 % w/v solution of stearic acid in ethanol for 3 hours to make it superhydrophobic. The prepared surface shows a contact angle of 150°. The hydrophobicity of the surface is necessary to ensure non-wetting nature towards the encapsulated liquid.

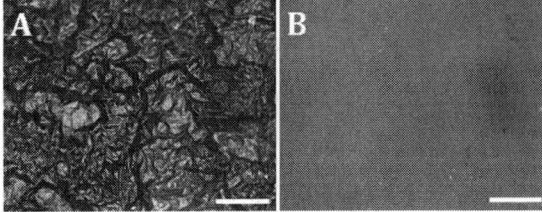

Figure 1: Microscopic images showing wax-infused surface at different temperatures (A) $T < T_m$ (melting temperature) where solid wax is seen and (B) $T > T_m$ (melting temperature) where wax is in liquid form. Scale bar = 200 μm.

The paraffin wax with a lower melting temperature ($T_m \sim 55$ °C) was used for the surface preparation. First, the wax was heated above the melting temperature, and the nanostructure surface was dipped inside the molten wax to infuse the nanostructure. Then, the surface is gently removed from the molten wax and placed at room temperature. This ensures the solidification of the wax infused in the nanostructures. The prepared surface shows a thermal phase change response. If the temperature of the surface is below the melting temperature, the infused wax is in the solid phase (Figure 1(A)). However, a phase change occurs above the melting temperature, and the wax is in liquid form. Thus, above the melting temperature of the wax, the surface behaves like a slippery surface (Figure 1(B)).

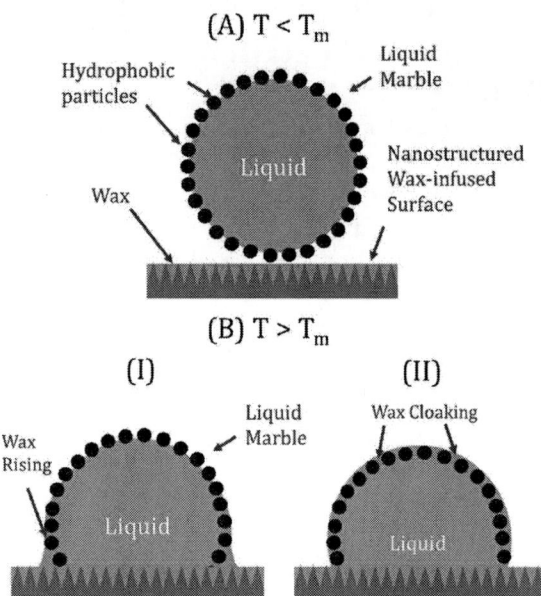

Figure 2: LM placed over the wax-infused surface for (A) T < T_m where LM retains its shape because of the solid wax phase (B) T > T_m where the wax melts and converts into liquid phase, which allows (I) wax to rise along LM interface because of capillary suction. (II) After sufficient time, the liquid wax covers the entire LM. Decreasing the temperature below T_m allows the wax to solidify, forming the capsule.

Capsule Fabrication

The droplet of desired encapsulation (~ 10 μL) was first rolled onto the hydrophobic particle (PTFE) bed. These particles settle on the outer surface of the droplet, making Liquid Marble (LM). On the other hand, the wax-infused surface was put onto the hot plate above the melting temperature (T_m). The wax-infused surface behaves as slippery above T_m and hydrophobic below T_m. LM was placed over a wax-infused surface. For temperature $T < T_m$, LM retains its shape as the wax is in a solid phase (Figure 2(A)). However, heating above T_m results in the rising of liquid wax along the LM surface because of the capillary forces of the porous particle coating (Figure 2(B)).

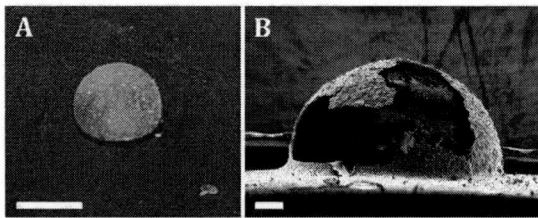

Figure 3: (A) Photograph showing a capsule with a solid wax shell and a liquid core. Scale = 2 mm. (B) SEM image of a wax capsule after cutting the capsule and complete drying of inner fluid. Scale = 400 μm.

The oleophilic nature and porous network of particles allow liquid wax to rise along the surface of LM through capillary suction. Particle-driven capillary suction ensures

coating thickness in the same order as particle size. After the liquid wax covers the entire LM, it is cooled below T_m. The wax solidifies and makes a solid encapsulation (Figure 3(A)). Figure 3(B) shows the thickness of the capsule wall as measured in SEM after cutting the capsule and drying the internal fluid.

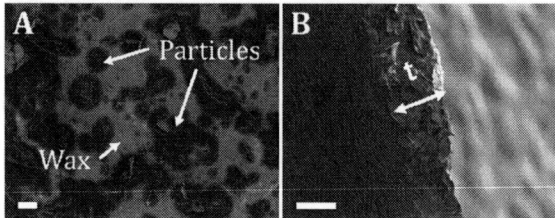

Figure 4: (A) SEM image of shell surface where particles can be seen embedded inside wax overlayer. Scale = 40 μm. (B) SEM image of the shell wall, where t is the shell thickness. Scale = 30 μm.

A similar kind of rise can also be seen if a bare droplet is placed onto the oil-infused surface. The cloaking of the oil over liquid droplets is feasible if the spreading factor of oil on the liquid marble (S_{ol}) is positive ($S_{ol} = \gamma_l - \gamma_o - \gamma_{ol} > 0$), where γ_l, γ_o, and γ_{ol} are the surface tension between LM-air, oil-air, and oil-LM interfaces, respectively. However, without particles, the thickness of the coating liquid is limited to a few hundred nanometers only [10]. Additionally, during the initial settling phase, the LM spreads on a wax-infused surface to achieve the minimum surface energy (Figure 2). Such spreading depends on the contact angle of the LM on the surface. For contact angles higher than the critical value (θ_c), the successful creation of a capsule is ensured. However, if the contact angle is lower than the critical value, there are higher chances of crack formation in particle coating because of larger spreading. Such cracks are undesirable as they create uneven coating and openings in the prepared capsules.

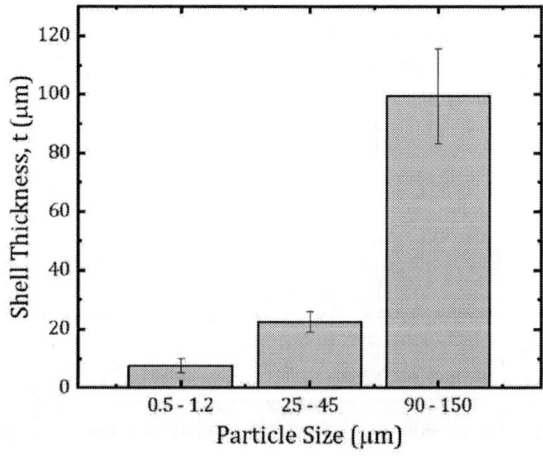

Figure 5: The shell thickness (t) variation with the particle size. The coating thickness remains nearly similar to the size of the particles.

RESULTS

The thickness of the produced capsule depends upon the size of the particles used in preparing LM. The mean particle size used in the experiments ranged from 0.8 μm to 100 μm. As shown in Figure 4(A), the particles are embedded in the solidified wax layer, which essentially dictates the thickness of the wax overlayer (Figure 4(B)). Thus, the thickness of the outer shell can be tuned by changing the size of the particles (Figure 5).

Because of the wax coating, the capsule can be made stimuli-responsive to the temperature. The shell of the capsule can be removed easily by temperature rise for the applications such as drug and pesticide release. The methylene blue dye-based capsule was prepared from PTFE particles (25-45 μm) to demonstrate the triggered release. Then it was dropped inside a water bath. If the temperature of the bath is below T_m, no diffusion of methylene blue can be seen for a prolonged duration (Figure 6(A)). However, if the bath temperature is raised above T_m, the outer wax shell melts, and the inner material is released. Figure 6(B) represents the mixing of methylene blue dye upon triggered release, which can be confirmed by a color change in the water.

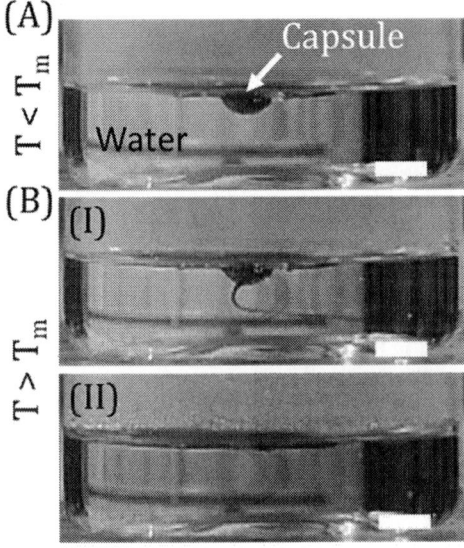

Figure 6: Encapsulated Methylene blue dye placed in water. (A) The temperature of the water is below T_m. Thus, the capsule stays intact over a longer duration. (B) The temperature rises above T_m, triggering the release of the inner material. (I) The wax shell started melting and releasing dye. (II) The dye is thoroughly mixed with water. Scale = 3 mm.

CONCLUSION

We report a novel and simple fabrication of a capsule with the help of a unique cloaking mechanism over droplets. The presented approach involves the use of wax-infused nanostructured surfaces and micro/nanoparticles. Due to particles at the droplet interface, molten wax rises from the infused surface and covers the entire droplet. Upon cooling, the wax solidifies, producing a capsule. The thickness of the capsule shell can also be varied by varying the particle size. Additionally, we demonstrated the triggered release of inner material upon applying temperature stimuli.

ACKNOWLEDGEMENTS

The authors acknowledge support from the Department of Science and Technology, Government of India. The authors acknowledge support from Prime Minister's Research Fellowship.

REFERENCES

[1] H. N. Yow and A. F. Routh, "Formation of liquid core–polymer shell microcapsules," *Soft Matter*, vol. 2, no. 11, pp. 940–949, Oct. 2006.

[2] A. Lee, P. T. Brun, J. Marthelot, G. Balestra, F. Gallaire, and P. M. Reis, "Fabrication of slender elastic shells by the coating of curved surfaces," *Nat. Commun.*, vol. 7, no. 1, pp. 1–7, Apr. 2016.

[3] J. P. Goertz, K. C. Demella, B. R. Thompson, I. M. White, and S. R. Raghavan, "Responsive capsules that enable hermetic encapsulation of contents and their thermally triggered burst-release," *Mater. Horizons*, vol. 6, no. 6, pp. 1238–1243, Jul. 2019.

[4] G. Beall, *Rotational Molding: Design, Materials, Tooling, and Processing*. Hanser/Gardner Publications, 1998.

[5] R. Lathia and P. Sen, "JMEMS Letters Fabrication of Self-Sealed Circular Microfluidic Channels in Glass by Thermal Blowing Method," *J. Microelectromechanical Syst.*, vol. 31, no. 2, pp. 177–179, Apr. 2022.

[6] E. M. Payne, D. A. Holland-Moritz, S. Sun, and R. T. Kennedy, "High-throughput screening by droplet microfluidics: perspective into key challenges and future prospects," *Lab Chip*, vol. 20, no. 13, pp. 2247–2262, Jun. 2020.

[7] F. Fontana, J. P. Martins, G. Torrieri, and H. A. Santos, "Nuts and Bolts: Microfluidics for the Production of Biomaterials," *Adv. Mater. Technol.*, vol. 4, no. 6, p. 1800611, Jun. 2019.

[8] P. W. Chen, R. M. Erb, and A. R. Studart, "Designer polymer-based microcapsules made using microfluidics," *Langmuir*, vol. 28, no. 1, pp. 144–152, Jan. 2012.

[9] A. Kumar, A. Tripathy, C. D. Modak, and P. Sen, "Designing assembly of meshes having diverse wettability for reducing liquid ejection at terminal velocity droplet impact," *J. Microelectromechanical Syst.*, vol. 27, no. 5, pp. 866–873, 2018.

[10] Q. Ge, A. Raza, H. Li, S. Sett, N. Miljkovic, and T. Zhang, "Condensation of Satellite Droplets on Lubricant-Cloaked Droplets," *ACS Appl. Mater. Interfaces*, vol. 12, no. 19, pp. 22246–22255, May 2020.

CONTACT

*Prosenjit Sen, +917406890351, prosenjits@iisc.ac.in
Rutvik Lathia, rutviklathia@iisc.ac.in
Bheema Sankar Reddy, bheemareddy@iisc.ac.in
Chandantary Dey Modak, chandantarum@iisc.ac.in
Satchit Nagpal, satchit@tamu.edu

WATER VITRIFICATION IN A MICROCHANNEL AT LOW COOLING RATE

Ayane Sato, Tomohiro Hayashi, and Tadashi Ishida
Tokyo Institute of Technology, JAPAN

ABSTRACT

Vitrification of water suppresses damages to cells in cryopreservation. It requires a high cooling rate (over 10^4 K/min) and the thickness is limited to 10 μm. This thickness is comparable to the size of microchannels, and high cell survival rate was achieved there. However, the cooling rate should not be sufficiently high due to low thermal conductivity of polydimethylsiloxane. To confirm the condition of water inside a low microchannel, we developed a 10-μm-high microchannel with a micro thermometer. According to fluorescent imaging and Raman spectroscopy, frozen water in the microchannel was not crystalline even though the cooling rate was low (~100 K/min). The Raman spectrum suggested that the frozen water in the microchannel was vitreous.

KEYWORDS

Cryopreservation, vitrification, microchannel, micro thermometer, Raman spectroscopy

INTRODUCTION

Cryopreservation is a method to preserve cells by storing them at low temperature. The low temperature stops their biological activities and maintains their structures for a long period [1]. At the freezing step of the cryopreservation, water is transformed into ice crystals and increases its volume. This damages cells and causes low cell survival rate. To avoid the damages, the formation of ice crystals should be suppressed. From the point of the formation of ice crystals, the cryopreservation is categorized into 3 types depending on its cooling rate; slow freezing (around 10 K/min), rapid freezing (10^2-10^4 K/min), and vitrification (over 10^5 K/min) [2].

In the case of slow freezing, the water outside of the cells is crystallized before that inside of the cells is. This leads the water inside the cells to move out from the cells by osmosis and the formation of ice crystals inside the cells are suppressed. However, the extracellular ice crystal somehow damages the cell and the concentrated electrolyte inside the cells is harmful to the cells. In the case of rapid freezing, the freezing duration is short. The intracellular water does not flow out from the cells and is transformed into ice crystals. This intracellular ice formation somehow damages the cells. In the case of the vitrification, the extracellular and intracellular water maintain random molecular alignment due to extremely short freezing duration, resulting in the suppression of the formation of ice crystals. The cells do not receive the damage from ice crystal formations because the volume increase of water and electrolyte concentration do not occur.

However, the vitrification is a promising method with low cellular damages, but is difficult for the cryopreservation. One of the conventional methods for the vitrification is slam freezing. The slam freezing method freezes specimens by the contact onto a metal block cooled at liquid nitrogen temperature. Although the specimen is cooled at a high cooling rate, the thickness of the vitrification is limited to only 10 μm from the interface between the specimen and metal block.

For the vitrification, microchannels, whose channel height can be controlled at the ten-micrometers scale, are promising technology to reduce the thickness of the specimen. For example, cryopreservation in microchannels of 10, 50, 100 μm in height dipped into liquid nitrogen showed different cell survival rates [3]. The best cell survival rate was about 50% in the case of the microchannel of 10 μm in height, which is comparable to that of using slow freezing with cryoprotectant agents. The authors explained the reason for such a high survival rate was the vitrification. However, the cooling rate should be low due to the low thermal conductivity of PDMS (polydimethylsiloxane) and nitrogen gas generated around the specimen.

In order to combine the technologies between the vitrification and microchannel, the mechanism of high cell survival rate should be studied. We developed microchannels with different heights with a micro thermometer inside the microchannel. In the microchannels, we observed the formation of ice crystals, measured temperature and Raman spectra of water at the freezing step.

MICROCHANNEL AND MICRO THEROMOMETER

Straight microchannels with micro thermometers were developed (Figure 1 (a)). The cross sections of the microchannels were rectangular and had different heights, 10, 30, 60, 90, 300 μm. We named the microchannel of 300 μm in height as Channel A, and that of 10 μm as Channel B. They were fabricated by the conventional soft lithography using PDMS.

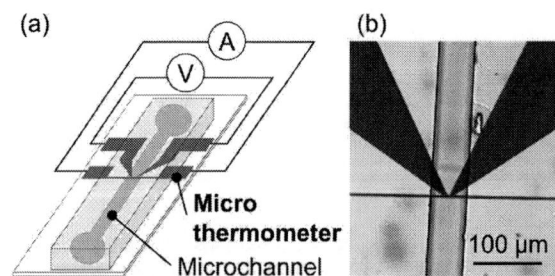

Figure 1: Microchannel with a micro thermometer. (a) Schematic diagram of micro thermometer for four-probe measurement in a microchannel. (b) Microscopic image of a fabricated microchannel with a thermometer.

The micro thermometer was designed for four-probe measurement precisely to measure the change of electrical resistivity against temperature. The temperature-measurement wire made of Pt/Ti on a slide glass was fabricated by lift off method. It was 2 μm in length, 4 μm in width, and 28 nm of Pt in thickness for high electrical resistance. To measure the temperature of water in the microchannel, the microchannels and micro thermometer were aligned (Figure 1 (b)). The micro thermometer was calibrated and the rate of the temperature against the electrical resistance of the fabricated micro thermometer was 0.001 K/Ω.

METHOD TO VISUALIZE ICE CRYSTALS

Water becomes polycrystalline when it is cooled under 273 K. To visualize ice crystals, fluorescent solution was used. At the freezing step, concentration of saturated solution decreases, resulting in the precipitation of the fluorescent dye out of ice crystals. Continuing the freezing step, the precipitated fluorescent dyes are concentrated at the grain boundaries of the ice crystals. By the end of the freezing step, the ice crystals are dark and the highly concentrated fluorescent dyes at grain boundaries are bright.

In our experiment, a fluorescein solution was introduced into the microchannels and cooled. When the solution was at room temperature, the microchannels seemed uniformly bright (Figure 2(a)). After freezing, the ice crystals seemed dark, surrounded by fluorescence from the concentrated dyes at grain boundaries (Figure 2(b)).

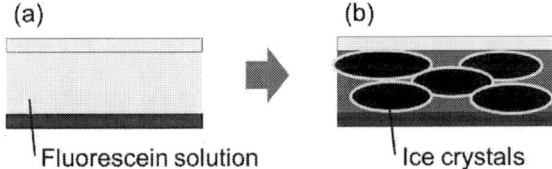

Figure 2: Visualization method of ice crystals by concentration of fluorescent dyes at grain boundaries. (a) Before and (b) after freezing.

METHOD TO ANALYZE CONDITION OF ICE

The conditions of water are analyzed by Raman spectrometry. Peaks of Raman spectra correspond to specific chemical bonds. According to Ref [4], ice crystals have a peak in 3100 cm^{-1} at 77 K, and vitrification shifts the peak towards 3110 cm^{-1} at 77 K. Furthermore, the position of the Raman peak of ice crystals depends on temperature [5]. Measuring the relation between the position of Raman peak and temperature is necessary. With the relation, we can analyze the condition of water.

EXPERIMENTAL SETUP

The experimental setup is shown in Figure 3. A microchannel filled with water was set on a stage of a confocal microscope. It was cooled by the contact of an aluminum block cooled in liquid nitrogen. Fluorescein solution of 3 mM in concentration was used to visualize ice crystals. In our microchannel, frozen water showed strip patterns of the fluorescein, and therefore the width of stripes (ice crystal width, ICW) was measured to compare the ice crystal size. In the case of Raman spectroscopy and temperature measurement, purified water was used to avoid the influence from other materials than water.

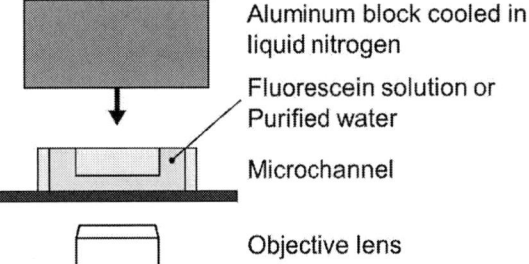

Figure 3: Experimental setup.

VISUALIZATION OF ICE CRYSTALS

Figure 4 shows the fluorescent images of water in a microchannel of 60 μm in height. The fluorescein solution showed the uniform fluorescent distribution in the microchannel before freezing (Figure 4 (a)). By cooling the solution in the microchannel, the fluorescent stripes appeared by the formation of ice crystals (Figure 4 (b)). The ICW in this case was 10.5 μm.

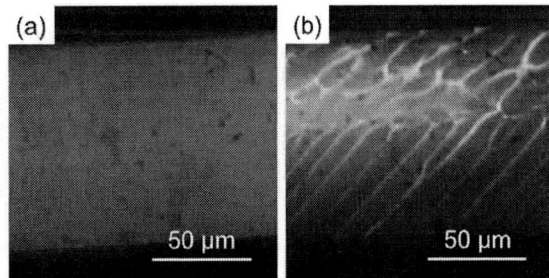

Figure 4: Fluorescent image of fluorescein solution inside a microchannel (a) before and (b) after freezing.

The relation between the ICW and channel height are shown in Figure 5. The ICW decreased as the channel height was lower. The ICW in Channel A was 15 μm (Figure 6 (a)) and it decreased to 9.5 μm in the microchannel of 30 μm in height. Furthermore, we could not observe the fluorescent stripes, but uniform fluorescent distribution after freezing in Channel B (Figure 6 (b)). This suggests that very small or no ice crystals were formed in Channel B.

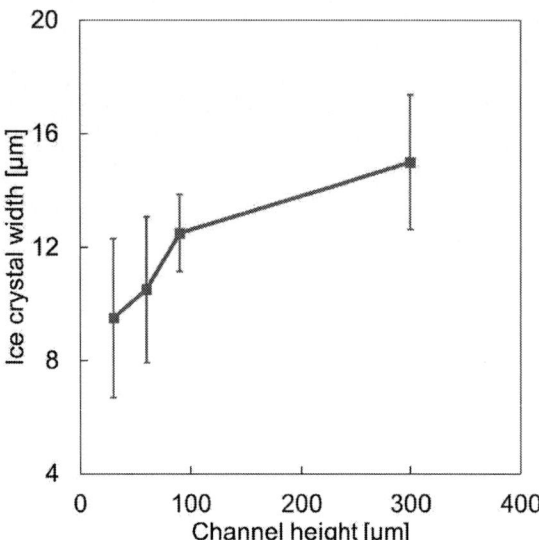

Figure 5: ICW as a function of channel height.

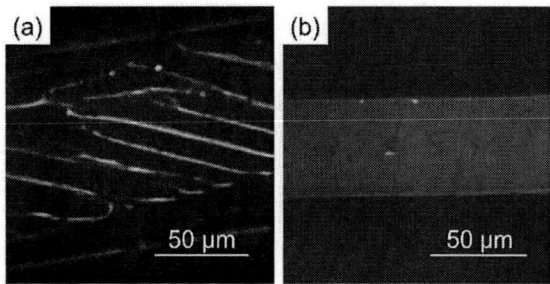

Figure 6: Fluorescent images of frozen fluorescein solution in microchannels, (a) Channel A and (b) Channel B.

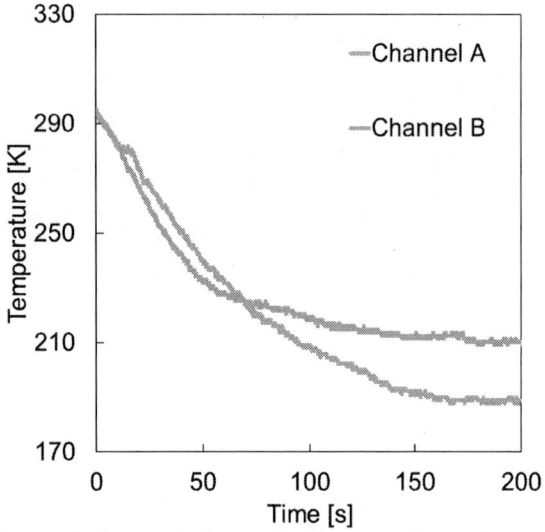

Figure 7: Temporal change of temperature during freezing water in Channels A and B.

TEMPERATURE MEASUREMENT OF FREEZING WATER IN MICROCHANNEL

Temperature changes of the water were measured in the microchannels of Channel A and Channel B at the freezing step (Figure 7). The cooling rates of water at the beginning (0-20 s from the contact) were 67.8 and 73.8 K/min in Channels A and B, respectively. The cooling rates were not largely different. From the point of the cooling rate, it was very low and vitreous ice should not be formed in both Channels A and B.

RAMAN SPECTRA OF FROZEN WATER

The Raman spectra of water changed by freezing in the microchannels. Figure 8 shows the Raman spectra of water before and after freezing. The Raman spectrum of the water before freezing had a broad peak from 3200-3500 cm^{-1} at room temperature. When the water in the microchannels were frozen, the Raman peak became sharp and shifted to 3100 cm^{-1} at 212 K. This sharp peak was applicable to distinguish the conditions of frozen water between ice crystals and vitreous ice.

The Raman spectra of the frozen water in Channels A and B at different temperatures are shown in Figure 9. The peak positions against temperature had linear trends. Using linear approximations, the peak positions of the frozen water in Channels A and B were estimated to be 3097 and 3110 cm^{-1} at 77 K, respectively. These peak positions matched well with the position of ice crystals, 3100 cm^{-1}, and vitreous ice, 3110 cm^{-1}, reported in Ref [4].

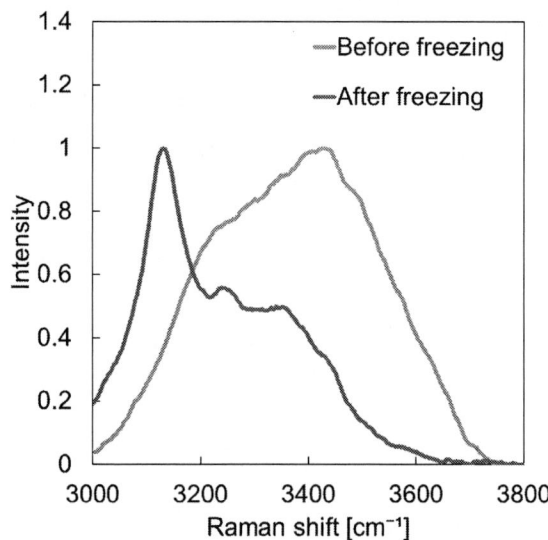

Figure 8: Raman spectra of water inside microchannels before and after freezing.

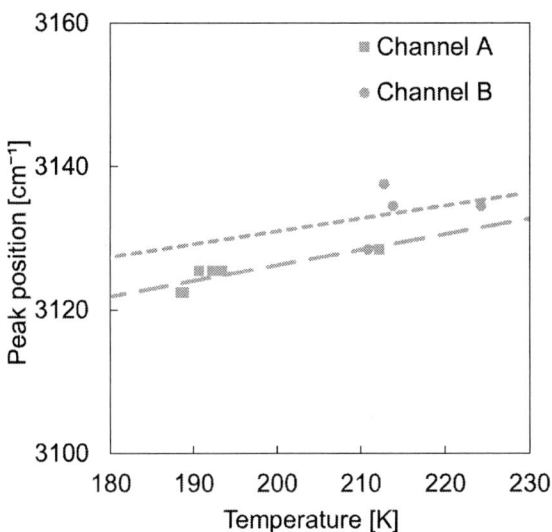

Figure 9: Relationship between the temperature and peak position in the spectrum of water in Channels A and B.

CONCLUSION

We analyzed the conditions of freezing water in microchannels. Ice crystals did not appear in low microchannel, whereas they appeared in high microchannel. The cooling rates were comparable and much lower than that required in vitrification. Furthermore, peaks of Raman spectrum at 77 K were estimated, and those of frozen water in high and low microchannels matched with that of ice crystals and vitreous ice, respectively. This result suggested that vitreous ice could be obtained in a low microchannel even at low cooling rate.

ACKNOWLEDGEMENTS

This project was supported in part by 2021 Female Student Support Publication Activity Support School of Engineering, Tokyo Institute of Technology supported by Micron Foundation.

REFERENCES

[1] T. H. Jang, S. C. Park, J. H. Yang, J. Y. Kim, J. H. Seok, U. S. Park, C. W. Choi, S. R. Lee, J. Han, "Cryopreservation and Its Clinical Applications," *Integr. Med. Res.*, vol. 6, no. 1, pp. 12–18, 2017.

[2] X. He, "Thermostability of Biological Systems: Fundamentals, Challenges, and Quantification," *Open Biomed. Eng. J.*, vol. 5, pp. 47–73, 2011..

[3] Y. Zou, T. Yin, S. Chen, J. Yang, W. Huang, "On-Chip Cryopreservation: A Novel Method for Ultra-Rapid Cryoprotectant-Free Cryopreservation of Small Amounts of Human Spermatozoa," *PLoS One*, vol. 8, no. 4, 2013.

[4] H. S. Cao, "Formation and Crystallization of Low-Density Amorphous Ice Topical Review Formation and Crystallization of Low-Density Amorphous Ice," *J. Phys. D Appl. Phys. J. Phys. D Appl. Phys*, vol. 54, p. 13, 2021.

[5] T. Zhan, Y. Xu, D. Wang, M. Cui, X. Li, X. Wang, "Interaction of Solute and Water Molecules in Cryoprotectant Mixture during Vitrification and Crystallization," *J. Mol. Liq.*, vol. 325, p. 114658, 2021.

CONTACT

*A. Sato, tel: +81-045-924-5450;
sato.a.ap@m.titech.ac.jp

A HIGHLY SENSITIVE 3-AXIS MICRO SEARCH-COIL MAGNETOMETER ENABLED BY HIGH-DENSITY THROUGH-SILICON-VIA PROCESS

Hadi Tavakkoli, Mingzheng Duan, Longheng Qi, Izhar, Xu Zhao, and Yi-Kuen Lee

Hong Kong University of Science and Technology, HONG KONG SAR

ABSTRACT

We report a highly sensitive 3-axis micro search-coil magnetometer (μSCM) fabricated in one single silicon die and enabled by an in-house developed through-silicon-via (TSV) process. The z-axis μSCM is realized by two planar spiral inductors on both sides of the silicon wafer. The x & y-axis μSCMs are realized inside a silicon wafer by 400 TSVs (an array of 20×20 TSVs with a diameter of 50 μm) for each axis. A conformal polysilicon deposition is utilized to achieve an electrical connection through the etch holes inside the silicon wafer. Although partial filling of TSVs with polysilicon has higher resistance (< 3Ω per TSV) compared to complete filling, the proposed fabrication method is simpler and free of grinding and polishing steps, making it a more cost-effective alternative. The sensitivity (S) of the 3-axis μSCM (S_x & S_y = 134 mV/mT, S_z = 3130 mV/mT at 100Hz) is better than existing μSCMs (13 mV/mT) by more than one order of magnitude.

KEYWORDS

3-axis search-coil magnetometer, MEMS, planar spiral inductor, through silicon via (TSV).

INTRODUCTION

Magnetic field sensors have a wide range of applications, such as brain function mapping, smart electrical power measurement, non-destructive evaluation, automotive applications, etc. There are different types of magnetometers; for instance, search-coil magnetometer (SCM), anisotropic magnetoresistance (AMR), giant magnetoresistance (GMR), and Tunneling Magneto-Resistance (TMR) [1]. Each type of magnetometer is suitable for a specific measurement range and selected applications. Magnetometers can measure a weak magnetic field from femto Tesla up to kilo Tesla range (SCM). Among these types of magnetometers, SCM has some unique advantages over other types, such as high power efficiency, the largest sensing range, and low-temperature dependency [2], [3].

SCMs are usually realized by wrapping a very long wire around a magnetic core. They are generally bulky and not CMOS compatible; therefore, they cannot take full advantage of the highly advanced silicon technology as these CMOS sensors [4–6]. The miniaturization of an SCM can be realized by a single-layer or multi-layer planar spiral inductor. Planar spiral inductors are widely used for GHz high-frequency applications. Few works have reported utilizing a planar spiral inductor as a magnetic field sensor. In [7], a low-cost four-layer printed circuit board (PCB) technology was employed to fabricate a μSCM. Stacking planar inductors enables on-chip calibration based on mutual inductance calculation [8]. Furthermore, the planar inductor can be employed for the on-chip calibration of a Hall effect sensor [4]. In our previous work [9], a novel nomogram was presented for the design optimization of a

μSCM, and the power consumption of a personal computer was measured by the μSCM. Eyre *et al.*[10] reported a 3-axis μSCM with post-CMOS processing. The permanent deformation of aluminum hinges was utilized to realize three perpendicular planar inductors, achieving a sensitivity of 1.3×10^{-4} V/T.Hz.

Nowadays, through substantial advances in microfabrication technology and the availability of through silicon via (TSV) process, inductors can be fabricated inside the bulk of the silicon wafer. In this work, for the first time, we utilized high-density TSVs to realize a 3-axis μSCM in a single die for a low-frequency magnetic field range.

μSCM DESIGN AND FABRICATION

The working principle of a μSCM is based on Faraday's law of induction. If the input magnetic field $B(t)$ is perpendicular to the surface of the μSCM and is equal to $B_m \sin(2\pi f t)$, the maximum output voltage (V_{om}) can be calculated by Eqn. (1):

$$V_{om} = 2\pi f A_e B_m \qquad (1)$$

where f, B_m, and A_e are the applied magnetic field frequency, the maximum amplitude of the magnetic field, and the effective area of the inductor, respectively. The effective area calculation is thoroughly described in [9] Therefore, the sensitivity (S) of a μSCM can be defined by Eqn. (2):

$$S = \frac{V_m}{B_m} = 2\pi f A_e \frac{V}{T} \qquad (2)$$

To increase the sensitivity, a higher effective area is preferred. Therefore, the resistance of the inductor can have a considerable value up to hundreds of kiloohms. Overall, the thermal noise generated by the inductor resistance is an important factor to determine the minimum detectable magnetic field.

The minimum detectable magnetic field of a μSCM, which is also called noise equivalent magnetic induction (*NEMI*) can be calculated by Eqn. (3):

$$NEMI = \frac{e_n}{S} \frac{T}{\sqrt{Hz}} \qquad (3)$$

where e_n is the total input referred noise of the μSCM and the amplifier circuit. e_n can be calculated by Eqn. (4):

$$e_n = \sqrt{2\left(i_{in}\frac{R}{2}\right)^2 + e_{nR}^2 + e_{in}^2 + \left(\frac{e_{out}}{G}\right)^2} \qquad (4)$$

where i_{in}, e_{in}, e_{out}, G, R, e_{nR} are input current noise, input voltage noise, output noise, the gain of the instrumentation amplifier, resistance of the μSCM, and thermal noise of the μSCM, respectively.

The entire structure of the presented 3-axis μSCM is shown in Fig. 1(a). The z-axis μSCM consists of two planar spiral inductors on both sides of the silicon wafer. These two inductors are connected in an in-phase configuration by TSVs. The utilization of TSV doubles the sensitivity of the z-axis μSCM within the same die size. The applied magnetic field in the x & y-axis is parallel to the surface of the planar inductors. A tiny area of the z-axis μSCM created by the two TSVs remains perpendicular to the x-axis magnetic field. The generated voltage in this small area due to the x-axis magnetic field is less than 0.2 % of the induced voltage due to the z-axis magnetic field.

The x & y-axis (in-plane) μSCMs are realized inside the bulk of the substrate by TSVs. The in-plane μSCMs can be realized inside a silicon wafer with two TSVs in each column, similar to a solenoid, as shown in Fig. 1(b). In this case, the in-plane μSCMs will be similar to Fig. 1(a). To increase the sensitivity of in-plane μSCMs, we utilized multiple TSVs in each column, as shown in Fig. 1(c). In other words, each column of the in-plane μSCMs is a planar spiral inductor inside the silicon wafer. By increasing the density of TSVs in each column, the number of turns of the spiral inductor in each column increases. Therefore, the effective area in each column increases and higher sensitivity can be achieved. The top and bottom metal tracks for in-plane μSCMs are designed in an anti-phase configuration to suppress the input magnetic field in the z-direction. Notably, the middle pins of all μSCMs are available by using the TSV process for differential output voltage.

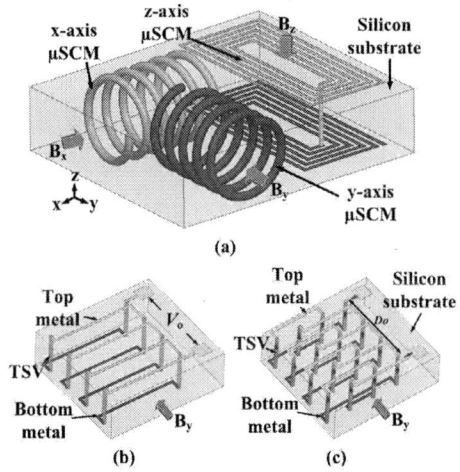

Figure 1: (a) The conceptual schematic of the 3-axis micro search-coil magnetometer (μSCM) in a silicon substrate; (b) Schematic of realization of y-axis μSCM inside a silicon wafer by low-density through-silicon-via (TSV) process; (c) Schematic of realization of y-axis μSCM inside a silicon wafer by high-density TSV process.

The fabrication processes of μSCMs are schematically shown in Figs. 2(a)-2(i). The fabrication started with 3μm thermal oxide growth on a 400 μm (100)-oriented n-type silicon wafer (Fig. 2(a)). This thick oxide layer acted as the mask layer during the deep reactive ion etching (DRIE) process. The first mask was designed for oxide patterning and defining the TSV locations on the wafer. HPR 504 photoresist was utilized as a mask layer for the dry oxide etching. After the lithography process, oxide etching was performed by dry method with an etch rate of 250 nm/min. Next, the remaining photoresist was removed by oxygen plasma (Fig. 2(b)). After photoresist stripping, the wafer

was bonded to a dummy wafer and placed in a DRIE machine to etch silicon and make etched-through holes with an etch rate of ~ 2.5μm/min. Then, the dummy wafer was detached from the silicon wafer, and the deposited polymer during the DRIE process was removed by a dry etching method (Fig. 2(c)).

The next step was TSV sidewalls electrical isolation. 500 nm thermal oxide was grown on the wafer (Fig. 2(d)). To achieve an electrical connection between two sides of the wafer, highly doped polysilicon deposition by low pressure chemical vapor deposition (LPCVD) method was employed.

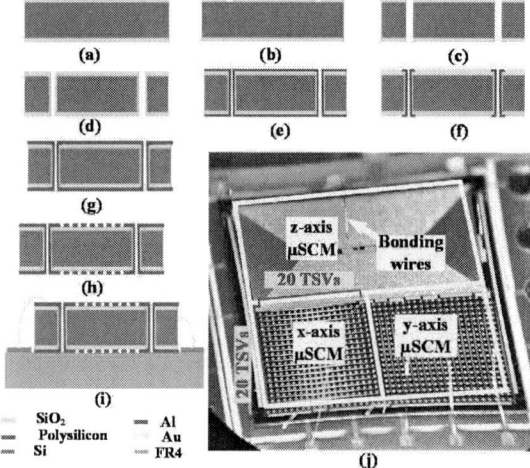

Figure 2: Fabrication processes: (a) Thermal oxidation; (b) Oxide dry etch; (c) Si etch through; (d) side-wall oxidation; (e)Polysilicon deposition; (f)Polysilicon dry etch; (g) Top and bottom Al deposition; (h)Al dry etch and forming gas annealing; (i) Wafer dice and wire bonding; (j) photo of wire bonded μSCMs.

The etched-through holes were partially filled by polysilicon deposition (Fig. 2(e)). It is noted that the sheet resistance of the deposited polysilicon was 3.2Ω. To remove the excess polysilicon, the second mask was applied to both sides of the wafer. Since there are many holes in the wafer at this step, the spray coating method was utilized for the second mask and all remaining lithography steps. During this process, excess polysilicon dry etching was performed on both sides of the wafer. Then, the photoresist was stripped from both sides of the oxygen plasma (Branson IPC 3000) (Fig. 2(f)).

After achieving an electrical connection between two sides of the wafer with highly doped polysilicon, top and bottom sides metallization was performed. Before metal deposition, the wafer was dipped inside a diluted HF acid solution for one minute to remove the native oxide, and then dried. Next, 1 μm aluminimum silicon (Al/Si (1%)) was immediately deposited on both sides of the wafer to reduce the contact resistance between the polysilicon and the metal layer (Fig. 2(g)). Then, top-side and bottom-side lithography were performed by third and fourth lithography masks. Afterward, the Oxford PlasmaPro 100 Cobra Aluminum dry etcher (Oxford Instruments, Oxfordshire, UK) was employed to etch the aluminum layer, followed by deionized water rinsing to eliminate chlorine corrosion.

After drying, oxygen plasma was utilized to strip the photoresist (Fig. 2(f)). The final step of the fabrication was forming gas annealing, which reduced the contact resistance between AlSi and polysilicon. Finally, the fabricated μSCM was mounted on the designed PCB and wire bonded to the designed pads (Fig. 2(i)). Although partial filling of holes with polysilicon leads to higher resistivity (< 3Ω per TSV) compared to the complete filling by copper electroplating, our method does not require grinding and chemical mechanical polishing (CMP) process to remove the excess polysilicon, which makes it a cost-effective alternative.

A picture of the wire-bonded die attached to a PCB is illustrated in Fig. 2(j). In the fabricated sample, 400 TSVs are connected in a series to realize the in-plane μSCMS, which proves the proposed fabrication process is a reliable method for realizing high-density TSVs. Furthermore, flip chip bonding can be utilized to eliminate the wire bonding, leading to a more compact design that will be reported in our future work.

EXPERIMENTAL RESULTS

The sensitivity of in-plane μSCMs can be calculated by Eqn. (2), and is a function of die size, substrate thickness, and TSV pitch. By decreasing the TSV pitch, the total effective area of the μSCMs increases, and higher sensitivity is achieved (S_x & $S_y \propto p^{-1.88}$), as shown in Fig. 3(a). Notably, the sensitivity of in-plane μSCMs is linearly proportional to the substrate thickness (S_x & $S_y \propto h^1$). Furthermore, the die size of the μSCM plays a pivotal role in the sensitivity calculation (S_x & $S_y \propto D_o^{2.86}$), as shown in Fig. 3(b). Therefore, the dependency of in-plane μSCMs sensitivity can be expressed by Eqn. (5):

$$S_x \, \& \, S_y \propto D_o^{2.86} \times p^{-1.88} \times h^1 \qquad (5)$$

The sensitivity of the z-axis μSCM is studied in reference [10] and our previous work [9]. The input-output response of μSCMs can be calculated by Eqn. (1). According to this equation, the output voltage is a function of both magnetic field frequency and amplitude. To measure the input-output response, the wire-bonded μSCMs were placed inside a uniform magnetic field generated by an electromagnet. To amplify the induced output voltage on the μSCMs, a micro-power instrumentation amplifier (INA126, Texas Instruments, TX, USA) was connected to the μSCMs as shown in Fig. 4. The calculated and measured output voltage as a function of input magnetic field amplitude at 10 Hz magnetic field for in-plane and out-plane μSCMs are illustrated in Fig. 4(a) and Fig. 4(b), respectively.

The sensitivity of the μSCMs is proportional to input magnetic field frequency as expressed in Eqn. (2). This unique feature of the μSCM enables the measurement of output voltage without any amplifier circuit [11]. Therefore, μSCM can be used as a passive and ultra-low-power sensor. To study the effect of input magnetic field frequency on the sensitivity of the fabricated μSCMs, a 1mT uniform magnetic field with a frequency range of 10~100 Hz was applied to the sensors. The calculated and measured sensitivity as a function of input magnetic field frequency for in-plane and out-plane μSCMs are shown in Fig. 5(a) and Fig. 5(b), respectively.

The noise measurement of the fabricated μSCMs was performed by a spectrum analyzer (HP DSA Dynamic Signal Analyzer 35665A) at a zero input magnetic. The *NEMI* of μSCMs was calculated by Eqn. (3).

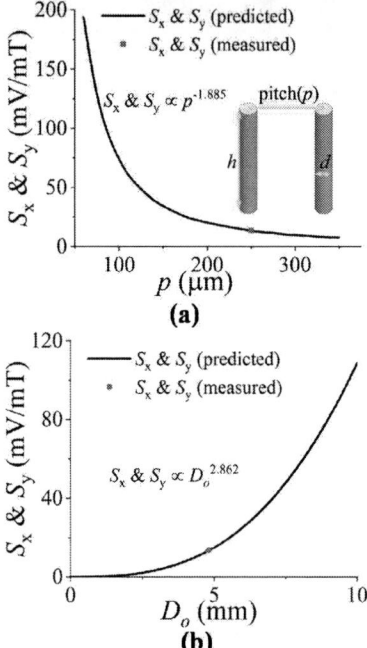

(a)

(b)

Figure 3: Sensitivity (S) at 10Hz of x & y axis μSCMs: (a) as a function of TSV pitch (d = 50 μm, die size = 4.8 × 4.8 mm², and h = 400 μm); (b) as a function of D_o (D_o is the length of the square side, p = 250 μm, and h = 400μm,).

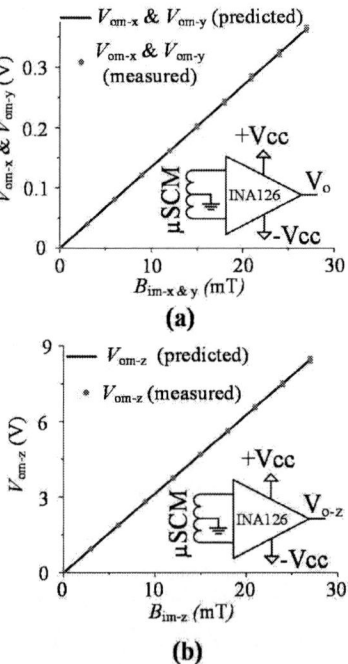

(a)

(b)

Figure 4: Input (maximum amplitude of input magnetic field (B_{im})) output (maximum output voltage (V_{om})) response at 10Hz magnetic field: (a) x&y-axis μSCMs; (b) z-axis μSCM.

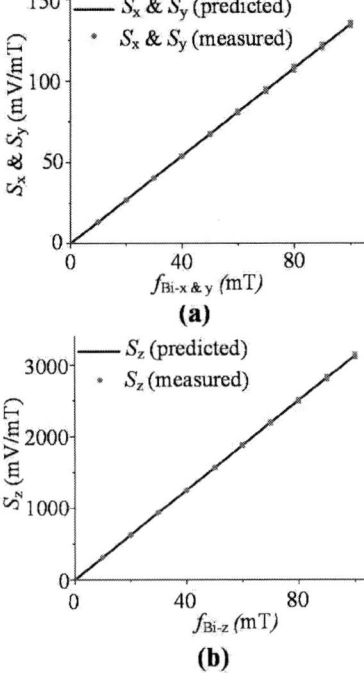

Figure 5: Measured and calculated sensitivity of the 3-axis µSCM as a function of input magnetic field's frequency: (a) x&y-axis µSCMs; (b) z-axis µSCM.

Finally, the performance of the fabricated 3-axis µSCM is compared with some available commercial Hall effect sensors and previously reported works, as listed in Table 1.

Table 1. Comparison between the presented 3-axis µSCM and other works.

Ref.	type	S@100Hz (mV/mT)	NEMI (nT/Hz$^{0.5}$)
[1]*TI	Hall sensor	200	130
[2]*AM	Hall sensor	40	230
[3]*ML	Hall sensor	280	125
[10]	3-axis µSCM	[4]*13	-
[4]	µSCM	[4]*10	-
This work	3axis-µSCM	$S_{x\,\&\,y}$=134 S_z=3130	$e_{n-x\,\&\,y}$ = 286 e_{n-z} = 12

[1]*TI-DRV5056A1/Z1; [2]*A1304ELHLX-T; [3]*MLX91205; [4]*the sensitivity was calculated with the amplifier gain = 1000.

CONCLUSION

To sum up, a 3-axis µSCM was realized in a single chip by a high-density TSV process. By use of the TSV process, the sensitivity of out-plane µSCM was doubled in the same die size. The in-plane µSCMs were fabricated inside the bulk of a silicon wafer. Each in-plane µSCM consisted of 400 TSVs in a series connection, which proves the reliability of the TSV fabrication process. The pitch of TSVs, and die size play a pivotal role in achieving higher sensitivity. The performance of fabricated µSCM is superior to reported works and certain selected commercial products. The TSV technology will be used for the other CMOS MEMS & 3D IC integration.

ACKNOWLEDGEMENTS

This work is partially financially supported by Hong Kong Center for Construction Robotics (Hong Kong ITC sponsored InnoHK center), Hong Kong RGC project (No. 16201220), Shenzhen Sci. Tech. Innovation Commission (Ref No. SZSTI21EG13) for partially sponsoring this research. The authors also acknowledge the technical support of all staff at HKUST NFF during the fabrication of the presented µSCM.

REFERENCES

[1] L. Francis, K. Poletkin, and K. Iniewski, Magnetic Sensors and Devices. CRC Press, 2017.

[2] A. Grosz, E. Paperno, S. Amrusi, and E. Liverts, "Integration of the electronics and batteries inside the hollow core of a search coil," J. Appl. Phys., vol. 107, no. 9, pp. 105–108, 2010.

[3] T. Hirota, T. Siraiwa, K. Hiramoto, and M. Ishihara, "Development of Micro-Coil Sensor for Measuring Magnetic Field Leakage," Jpn. J. Appl. Phys., vol. 32, no. 7 R, pp. L3328–L3329, 1993.

[4] H. Tavakkoli et al., "Dual-mode Arduino-based CMOS-MEMS Magnetic Sensor System with Self-calibration for Smart Buildings' Energy Monitoring," 17th IEEE Int. Conf. Nano/Micro Eng. Mol. Syst. NEMS 2022, pp. 91–94, 2022.

[5] M. Duan, X. Zhong, X. Zhao, O. M. El-Agnaf, Y. K. Lee, and A. Bermak, "An Optical and Temperature Assisted CMOS ISFET Sensor Array for Robust E. Coli Detection," IEEE Trans. Biomed. Circuits Syst., vol. 15, no. 3, pp. 497–508, 2021.

[6] W. Xu, M. Duan, M. Ahmed, S. Mohamad, A. Bermak, and Y. K. Lee, "A low-cost micro BTU sensor system fabricated by CMOS MEMS technology," TRANSDUCERS 2017 - 19th Int. Conf. Solid-State Sensors, Actuators Microsystems, vol. 1, pp. 406–409, 2017.

[7] Hadi Tavakkoli, Izhar, Xu Zhao, and Yi-Kuen Lee "Low-cost Micro Search-Coil Magnetic Sensor with Self Calibration for the Internet of Things," IEEE NEMS conference, 2021, pp. 813–817.

[8] H. Tavakkoli, E. Abbaspour-Sani, A. Khalilzadegan, G. Rezazadeh, and A. Khoei, "Analytical study of mutual inductance of hexagonal and octagonal spiral planer coils," Sensors Actuators, A Phys., vol. 247, pp. 53–64, 2016.

[9] H. Tavakkoli, K. Song, X. Zhao, M. Duan, and Y. Lee, "A Novel Design Nomogram for Optimization of Micro Search Coil Magnetometer for Energy Monitoring in Smart Buildings," Micromachines 13, no. 8 1342., vol. 13, no. 8, p. 1342, 2022.

[10] B. Eyre, K. S. J. Pister, and W. Gekelman, "Multiaxis microcoil sensors in standard CMOS," Micromachined Devices and Components, vol. 2642, p. 183, 1995.

[11] H. Tavakkoli et al. "A CMOS Hybrid Magnetic Field Sensor for Real-time Speed Monitoring," IEEE Sens. J., 2022.

CONTACT

*Yi-Kuen Lee, tel: +852 2358-8663; meyklee@ust.hk

FULLY INTEGRATED BACK-BIASED 3D HALL SENSOR WITH WAFER-LEVEL INTEGRATED PERMANENT MICROMAGNETS

Björn Gojdka[1], Daniel Cichon[2], Markus Stahl-Offergeld[2], Dominik Schröder[3], Niels Clausen[1], Christian Hedayat[3], Hans-Peter Hohe[2], and Thomas Lisec[1]

[1]Fraunhofer Institute for Silicon Technology ISIT, GERMANY
[2]Fraunhofer Institute for Integrated Circuits IIS, GERMANY and
[3]Fraunhofer Institute for Electronic Nano Systems ENAS, GERMANY

ABSTRACT

This paper reports on a 3D Hall sensor back-biased by rare-earth micromagnets which are integrated directly into the silicon substrate using a wafer-level fabrication process, called PowderMEMS. For the first time, the technique is used to realize a fully integrated back-biased magnetic field sensor. The feasibility of the approach is proven by measurements of the motion of a rotating gear wheel which represents a typical back bias application. Additionally, the ability of the approach to create optimal magnetic field geometries for magnetically biased sensors is demonstrated.

KEYWORDS

Integrated micromagnets, magnetic bias, back bias, wafer-level integration, Hall sensor, magnetic field shaping.

INTRODUCTION

Magnetic field sensors have been used for decades for the detection of rotational motion for example in the automotive industry. In a typical back-biased arrangement, a magnetic bias field is generated by a permanent magnet in close proximity to the sensor. The measuring signal results from the distortion of this field by a passing soft magnetic mechanical part. Although an established technology, both, sensors [1,2] and back bias magnets, are subject to continuous development. Commonly the magnet is mounted within the sensor package using pick-and-place techniques [3]. To overcome sequential and unprecise assembly of the back bias magnets, a polymer-based molding process was proposed to fabricate magnets on chip-level during the packaging process [3]. In addition, shaping of the magnetic field distribution is highly desirable to optimize the ratio of static bias field to signal modulation [3]. This work presents a fully integrated back bias sensor based on a novel approach on wafer-level, called PowderMEMS [4], which was already used to create micromagnets from NdFeB [5].

EXPERIMENTAL SETUP

Integrated back-biased Hall sensor

The demonstrator comprises a 3-axis Hall sensor (FH3D12) with two 3D Hall elements fabricated on an 8-inch silicon wafer. The latter is masked by photolithography on the backside to form a cavity in each chip with a targeted depth of 500 μm by DRIE underneath the 3D Hall elements. The cavities are dry-filled with NdFeB powder (Magnequench MQFP−B, D50=5 μm) using a dedicated filling process reported elsewhere [6]. Subsequently, the powder is solidified throughout the whole mold volume with a layer of approximately 75 nm Al_2O_3 using atomic layer deposition [5]. Finally, the solidified isotropic NdFeB powder is magnetized perpendicular to the wafer plane by application of an external field of 3.5 T across the whole wafer. The device wafer featuring the Hall sensors on the front and the magnets on the back is diced into chips as shown in Figure 1. Dicing was performed outside the dicing lanes across neighboring dies since an existing layout was used which was not adapted to accommodate the integrated magnets.

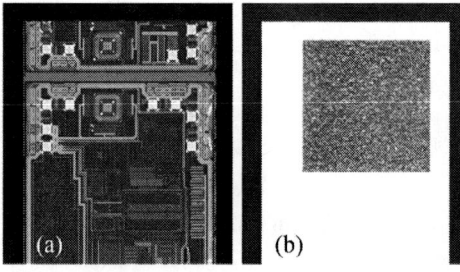

Figure 1: (a) Frontside of 3D Hall Sensor chip showing one of the Hall elements. (b) Backside of the same chip with integrated permanent NdFeB magnet with lateral dimensions of (1200 x 1200) μm² and a targeted thickness of 500 μm.

The back-biased Hall sensor is positioned opposite a gear wheel (13 mm diameter, 11 teeth, C45 steel) as depicted in Figure 2 to demonstrate a typical application scenario. In this basic proof-of-concept setup, the separation between gear wheel and sensor can only be manually adjusted. The estimated distance of 1400 μm is therefore subject to an error of ± 200 μm.

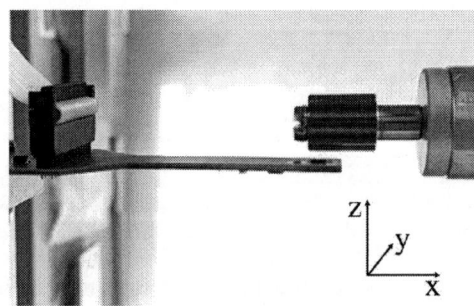

Figure 2: Demonstration setup with the back-biased sensor from Figure 1 and a gear wheel (13 mm diameter, 11 teeth, C45 steel) at a separation distance of (1400 ± 200) μm from sensor to tooth.

The gain of the sensor is configured to optimally adapt the measurement range to the magnitude of the magnetic flux density of the demonstrator. A microcontroller reads out the measurement values and transmits them to a LabView routine, which also controls the encoder (Physik Instrumente M-062).

Magnetic characterization setup

To locally characterize the magnetic field distribution of the integrated micromagnets a 3-axis Hall sensor (Melexis MLX90393) is utilized, which is positioned above a 3-axis portal robot. The samples are mounted on the surface of the robot with which they can be positioned in three dimensions with an accuracy of 1 μm. The magnetic field distribution can be recorded by varying the sample position. The minimum distance of the active Hall sensor element to the surface of the magnet is 160 μm, corresponding to contact between sample and sensor package. To characterize the three-dimensional magnetic field distribution, lateral scans in the x-y plane are performed at sensor-sample separations of z=160-260 μm in steps of Δz=20 μm.

As the size of the Hall element is comparable to lateral features of the integrated micromagnets, the measurement data need to be convoluted with the sensor element's lateral geometry. This is due to the fact, that the Hall sensor integrates the magnetic field over area. The presented measured magnetic field distributions are accordingly post-processed. The results are validated with finite element method (FEM) simulations of the respective magnetic fields.

FEM simulation and analytical calculations

The field distributions of differently shaped micromagnets are simulated using COMSOL Multiphysics® with the interfaces Magnetic Fields, No Currents. For demonstration purposes, the modulation of the fields in the presence of a single gear wheel tooth is investigated. Material parameters used are remanent flux density of the permanent magnet B_r=330 mT [5], and relative permeability of the soft magnetic tooth μ_r=2000.

Additional analysis was performed analytically to determine the actual distance between sensor and integrated magnet. For a rectangular prism magnet, the field on the axis of a square prism magnet is given by [7]

$$B(z) = \frac{B_r}{\pi}\left[\arctan\left(\frac{w^2}{2z\sqrt{4z^2+2W^2}}\right) - \arctan\left(\frac{w^2}{2(H+z)\sqrt{4(H+z)^2+2W^2}}\right)\right] \quad (1)$$

With B_r the remanence field, z the distance from the pole face on the symmetry axis, W the width of the block and H the height of the block. In the case of a frame-shaped magnet (hollow prism with a square base area), two fields for the full block and the inner one are calculated with Equation 1 and subsequently superposed.

RESULTS
Proof of concept

The sensor gain is set to a measurement range of 100 mT. A static bias field of 56.9 mT in z-direction is subtracted from the sensor signal. The rotation of the gear wheel is detected clearly by the integrated sensor setup as shown in Figure 3. B_z is modulated by (202 ± 31) μT peak-to-peak. As expected, B_x exhibits virtually no modulation. Additionally, a sinusoidal offset is observed which results from precession of the simple mechanical demonstration setup.

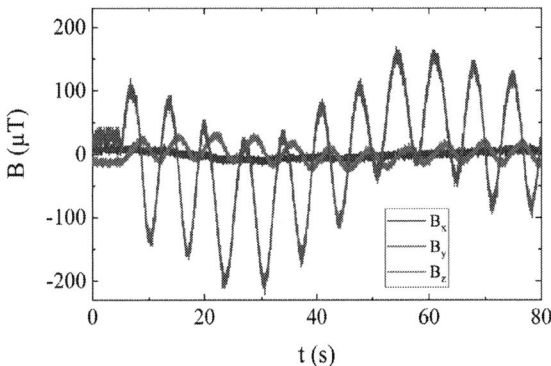

Figure 3: Variation of magnetic flux density during rotation of the gear wheel, yielding a peak-to-peak modulation of (202 ± 31) μT in B_z. The sinusoidal offset of the signal results from precession of the mechanical demonstration setup.

The unfavorable ratio of a large static bias field to a significantly smaller modulation of the field is a typical challenge in back bias setups. In the above case the ratio of the modulation versus the background is 0.2 mT / 56.9 mT = 3.5E-3. With the presented integration approach being able to create arbitrary lateral geometries across the substrate, an improved ratio can be achieved by tailoring the magnetic field distribution.

Magnetic field shaping

To demonstrate the field shaping ability of the solution, frame-shaped magnets are integrated into the backside of a device wafer as shown in Figure 4. All frames have outer edges of (1.2 x 1.2) mm² and varying frame thickness of 210-330 μm.

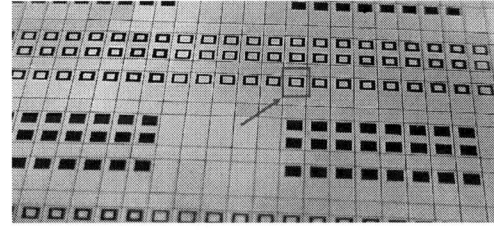

Figure 4: Backside of diced device wafer with wafer-level integrated NdFeB micromagnets of various geometries. Any lateral shape and custom variations can be realized to create custom magnetic field distributions. Highlighted: frame-shaped magnet with 260 μm frame thickness.

The field distribution of a frame-shaped magnet with 260 μm frame thickness (highlighted in Figure 4) is characterized by a Hall probe in various distances as depicted in Figure 5(a). At the position of the sensor a magnetic flux density of only B_z=6.5 mT is observed, compared to 56.9 mT in the previous case of the block-shaped magnet.

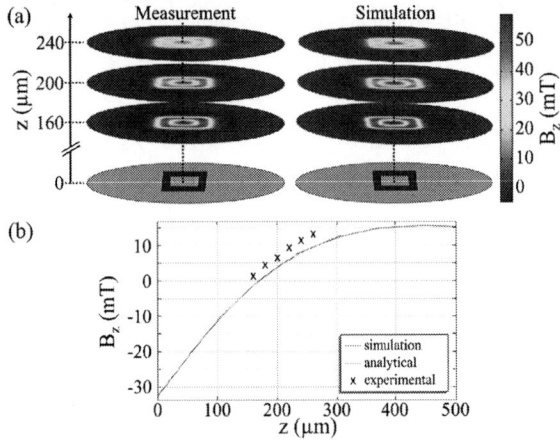

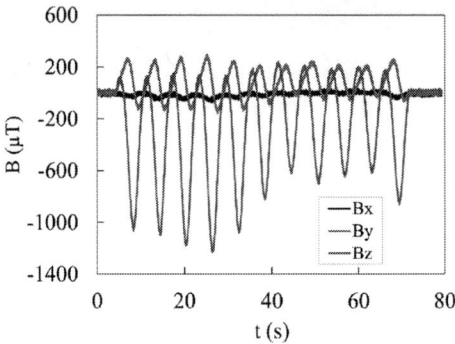

Figure 6: Modulation of magnetic flux density of a frame-shaped back bias magnet (1200 μm outer edge length, 260 μm frame thickness) during rotation of a gear wheel at a distance of 0.8 mm to the sensor. The variation in amplitude originates from a mechanical precession of the demonstration setup.

Figure 5: (a) Left: Measured x-y planes of magnetic flux density B_z for different distances z between Hall probe and magnet surface at z=0. Right: Corresponding FEM model in good agreement with the experimental data. (b) FEM Simulation, calculation (Equation 1) and measured data of $B_z(z)$ along the center axis dashed in (a).

A comparison of the FEM-simulation, an analytical calculation using Equation 1 and the Hall probe data along the z-axis is shown in Figure 5(b). As expected, the FEM simulation and analytical calculation are virtually identical. Measurement data lies 2.4-3.6 mT above theory. This apparently systematic error can result from the positioning error of ±1μm in z-direction of the Hall probe, the de-convolution of the sensor data with a generic sensor geometry and an uncertainty of ±10 mT in B_r.

A Hall sensor device with an integrated frame-shaped back bias magnet (frame thickness 260 μm) is assembled, yielding a static bias of 7.5 mT without the presence of the gear wheel. For this flux density and B_r=330 mT [5], Equation 1 yields a separation of 235 μm between sensor and magnet while the measurement data result in a separation of 207 μm. This is in good agreement with the DRIE target depth of 500 μm within the 725 μm thick substrate, leaving a targeted separation of 225 μm between substrate surface and integrated magnet. The deviation of 10-18 μm lies within the expected variation of the DRIE process across the 8-inch substrate without the use of a dedicated stop layer.

With this low static bias flux density, the modulation-to-bias ratio allows to increase the gain of the signal amplifier. Accordingly, a measurement range of 10 mT is chosen. Again, the rotation of the gear wheel was measured in the demonstration setup depicted in Figure 2. After subtraction of a static bias field of 8.8 mT in z-direction in presence of the gear wheel, B_z is modulated by (1050 ± 350) μT peak-to-peak during the rotation of the wheel in a distance of 0.8 mm. As noted before, a

sinusoidal offset is observed which results from precession of the simple mechanical demonstration setup. The magnetic field shaping does not change the general behavior of the measurement as shown in Figure 6.

Discussion

First measurements with wafer-level integrated permanent magnets demonstrate that back-biased sensor systems can be realized without discrete assembly of magnets. The measured gear wheel signals, in the range of 200 μT, can be measured amid a static back bias field of 56.9 mT. However, the ratio of 3.5E-3 between signal and bias forces the use of an unfavorable low sensor gain. This general challenge of back-biased Hall sensors can be improved using optimized magnetic field distributions as discussed in the following based on FEM simulations.

A single gear wheel tooth is placed opposite either a solid block magnet or a frame-shaped magnet, the latter being depicted in Figure 7(a). A sensor is supposed to detect the field 180 μm above the center of the magnet's surface (accounting for the substrate between magnet and sensor). The distance between the tooth and the sensor was chosen to be 300 μm at shortest distance. The tooth rotates on a circle with 13 mm diameter in front of the sensor, mimicking a gear wheel rotation.

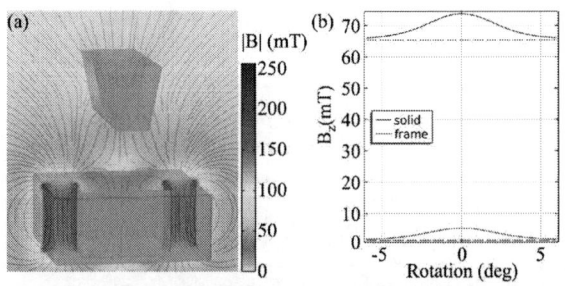

Figure 7: (a) FEM simulation of magnetic field distribution for one gear wheel tooth and a frame-shaped back bias magnet. (b) Resulting static bias field and field modulation upon rotation of the wheel tooth for a solid magnet and a frame-like magnet with B_r=300 mT and the same outer dimensions of (1.2 x 1.2) mm².

As shown in Figure 7(b), the solid block magnet delivers a static bias field of 65.4 mT at the location of the sensor. The rotation of the tooth within ±5° results in a modulation amplitude of 8.4 mT. In the case of a frame-shaped magnet, a bias magnetic flux density of 1.4 mT is observed with a modulation amplitude of 3.9 mT. Although the modulation is lower in absolute terms in the case of the frame-shaped magnet, the ratio of 2.7 between modulation and bias is advantageous compared to 0.1 for the solid magnet. Accordingly, the sensor can be configured for small magnetic input ranges which increases the sensor resolution and thus the accuracy of the sensor system.

COCNLUSIONS AND OUTLOOK

This work demonstrated the suitability of wafer-level fabricated rare earth micromagnets for fully integrated back bias sensor solutions. The precise field shaping ability of the PowderMEMS fabrication technique was shown by integration and characterization of different lateral magnet geometries within an 8-inch wafer. The measured and simulated field distributions were shown to deliver an improved signal-to-bias ratio allowing to use a favorable sensor gain.

In future work, an elaborate mechanical test setup will be used to precisely investigate the sensor performance with respect to air gap size and jitter. The more precise experimental procedures will be accompanied by refined FEM simulations of application specific setups. Adapted sensor layouts which cater for the specific design of optimal magnetic configurations will be realized. Future work will also aim at increasing the remanence of the PowderMEMS magnets to enable larger air gaps for Hall sensor configurations. Using the generic PowderMEMS approach, magnetic materials other than rare earth compounds will be integrated to cater for other types of sensors [2] and for industry demand to reduce rare earth consumption [8]. In addition, magnetic bias configurations other than back bias will be realized for other types of sensors.

REFERENCES

[1] S. Hainz, E. de la Torre, J. Güttinger, "Comparison of Magnetic Field Sensor Technologies for the use in Wheel Speed Sensors", in *2019 IEEE International Conference on Industrial Technology (ICIT)*, Melbourne, February 13-15, 2019, pp. 727-731.

[2] S. Hainz et al., Proc. AmE 12th GMM Symposium, pp. 24-27. S. Hainz, E. de la Torre, J. Güttinger, "New Magnetic Sensor Technology for Modern Speed Sensing Application of Rotating Shafts", in *AmE 2021-Automotive meets Electronics; 12th GMM-Symposium*, online, March 10-11, 2021, pp. 1-4.

[3] K. Elian, H. Theuss, "Integration of polymer bonded magnets into magnetic sensors", *Proceedings of the 5th Electronics System-integration Technology Conference (ESTC)*, September 16-18, 2014, pp. 1-5.

[4] T. Lisec, O. Behrmann, B. Gojdka, "PowderMEMS - A Generic Microfabrication Technology for Integrated Three-Dimensional Functional Microstructures", *Micromachines*, vol. 13, pp. 398-420, 2022.

[5] M. T. Bodduluri, B. Gojdka, N. Wolff, L. Kienle, T. Lisec, F. Lofink, "Investigation of Wafer-Level Fabricated Permanent Micromagnets for MEMS", *Micromachines*, vol. 13, pp. 742-759, 2022.

[6] C. Kostmann, T. Lisec, M. T. Bodduluri, O. Andersen, "Automated Filling of Dry Micron-Sized Particles into Micro Mold Pattern within Planar Substrates for the Fabrication of Powder-Based 3D Microstructures", *Micromachines*, vol. 12, pp. 1176-1190, 2021.

[7] J. M. Camacho, V. Sosa, "Alternative method to calculate the magnetic field of permanent magnets with azimuthal symmetry", *Rev. Mex. Fis. E*, vol. 59, pp. 8–17, 2013.

[8] D. Gielen, *Critical minerals for the energy transition*, International Renewable Energy Agency, Abu Dhabi, 2021

CONTACT

*B. Gojdka, tel: +49-4281-171412; Bjoern.gojdka@isit.fraunhofer.de

A LARGE-STROKE TIP-TILT-PISTON MICROMIRROR WITH ELECTROMAGNETIC ACTUATORS BASED ON METALLIC GLASS

Chuan-Hui Ou, Nguyen V. Toan, and Takahito Ono
Tohoku University, JAPAN

ABSTRACT

An electromagnetically driven 3-DOF (Degrees of Freedom) micromirror with a large stroke is presented. The stroke reaches the highest level of electromagnetic micromirrors with a tilting angle control mechanism, which allows spectrometers to achieve higher resolution. This research employs a novel actuation structure made of a unique material, metallic glass. It is the first attempt to integrate two functional elements, an electromagnetic actuation element and a mechanical supporting structure, into a single-layer spring made of metallic glass. The fabricated device achieves ±400 µm stroke, providing higher performance and robustness for micro-interferometer applications.

KEYWORDS

Micromirror, Electromagnetic, Metallic glass, Micro-interferometer.

INTRODUCTION

A translational micromirror is the core part of a spectrometer. The micromirror with a large stroke and a low tilting angle is required for high-resolution spectrums. Several micromirrors have been developed for such purposes. For instance, some micromirrors are actuated by depositing or assembling magnetic material under applying an external field [1, 2], while others are based on Lorentz force, in which electromagnetic coils are formed on the mirror, and an external field is applied [3]. Although many electromagnetic micromirrors have been demonstrated, the performances are limited by small strokes, the lack of a tilting angle control mechanism due to the high spring constants, and limited material and design for coil and magnet. A summarized comparison of the characteristics and the performance of the micromirror between previous works [1-3] and this research is shown in Table 1.

In this work, the key to developing a novel micromirror is the unique properties employing metallic glass material, a kind of amorphous alloy. Due to the amorphous phase, metallic glass possesses a high mechanical strength and high elasticities, which makes it an excellent spring material and has been successfully applied to various microdevices [4-7]. Since metallic glasses are conductive, they can be used as conductive wires for Lorentz force actuation. In this proposal, an actuator structure with four separate metallic glass springs was used to achieve large strokes and controllable angles. The actuation structure is based on driving principle using Lorentz force [8, 9]. The structure, material, and device fabrication are investigated to peruse the large strokes with controllable angles.

DESIGN OF MICROMIRROR

Device Structure

The micromirror supported by four metallic glass springs can be actuated by Lorentz force, as shown in figure 1 (a). Figure 1 (b) illuminates the cross-sectional view of the micromirror. As currents flow into the metallic glass springs under the application of a magnetic field in the in-plane direction, Lorentz forces are generated in the out-of-plane direction. The Lorentz force F in a current carrying straight wire can be expressed as

$$F = iL \times B, \qquad (1)$$

where i is current, L is the length of the wire, and B is external magnetic field, respectively. By controlling the currents of each metallic glass spring, the two-axes of tilting angles of the micromirror can be controlled.

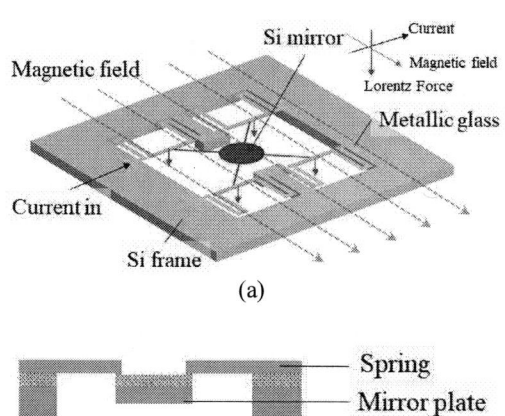

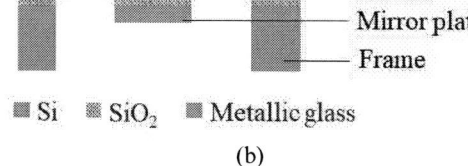

Figure 1. (a) Schematic of micromirror. (b)Cross-sectional view of the micromirror.

Table 1. Summary of translational electromagnetic micromirrors

Ref.	Actuation element	Stroke	Angle Control	Device footprint
[1]	Nickel film (Magnetization)	123 µm (static)	-	1 cm × 1 cm
[2]	Permanent magnet (Magnetization)	±62.5 µm (static)	-	Platform 1 mm × 1 mm Beam 2.43 mm × 25 µm
[3]	Cu coil (Lorentz force)	±162.8 µm (resonant)	-	7cm × 8cm
This work	Metallic glass spring (Lorentz force)	418 µm (static)	Possible	1 cm × 1 cm

Actuator Design

The key to meet the requirements of the novel spring-shaped Lorentz force-type actuator design lies in the use of metallic glass, a type of amorphous alloy. Metallic glasses with an amorphous phase have high mechanical strength and elastic limit, making them ideal candidates for spring structures. Since Pd-Si-Cu metallic glass has relatively high electrical conductivity compared to other metallic glass compositions, it can also be used as a coil material for electromagnetic actuators. In addition, Young's modulus of Pd-Si-Cu metallic glass is lower than the conventional structural materials, for instance, silicon. By using low Young's modulus materials on the springs, the performance of displacement can be increased significantly. Therefore, the single-layer metallic glass spring can be considered as a functional part that integrates the actuation element and supporting structure element.

The springs are designed with 200 μm in width, 1 μm in thickness, 1.8 mm in length of the folded part, and 1.9 mm in the straight part orthogonal to the magnetic field. The dimensions are summarized in table 2. The stroke of the micromirror was simulated by the finite element method, as shown in figure 2 (a). When 6.08 μN of force is applied, 1.04 mm of stroke can be obtained. The spring constant is estimated to be 0.023 N/m. Stress distribution was simulated to avoid the fracture of metallic glass during actuation, as shown in figure 2 (b). The maximum stress is 88.8 MPa, which is smaller than the tensile strength of a thin metallic glass film, 1.14 GPa [10]. Also, in the actual design, the corners are rounded to avoid the concentration of stress. Metallic glass, as an actuator material, can enhance mechanical strength and reduce the stiffness of the springs. Consequently, only several micronewtons force is enough to obtain a large stroke.

Table 2: Dimensions of micromirror

Parameters	Values
Device footprint	1×1 cm
Width of spring	200 μm
Length of the folded part of spring (parallel to magnetic field)	1.8 mm
Length of the straight part of spring (perpendicular to magnetic field)	1.9 mm
Thickness of metallic glass spring	1 μm
Diameter of mirror plate	1.5 mm

FABRICATION OF MICROMIRROR

The fabrication process is shown in figure 3. It starts with depositing SiO_2 on both sides of a Si wafer by CVD (Chemical Vapor Deposition). SiO_2 layer on the topside is used as an electrical insulation layer, and on the backside is used as the etching mask. After patterning the SiO_2 layer by wet etching, a photoresist layer is also patterned on the backside using photolithography. Next, deep reactive ion etching (DRIE) is conducted to create the thick mirror plate, and the metallic glass is formed by sputtering and patterned by the lift-off process. Finally, DRIE is conducted again to release the structure entirely. Figure 4 shows the fabricated micromirror.

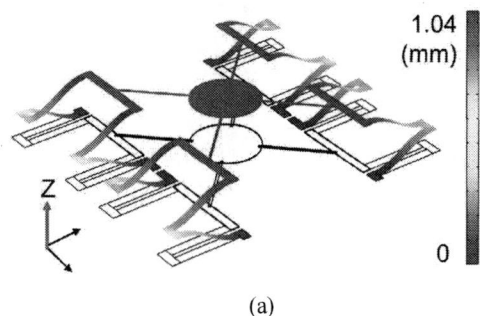

Displacement in Z direction

(a)

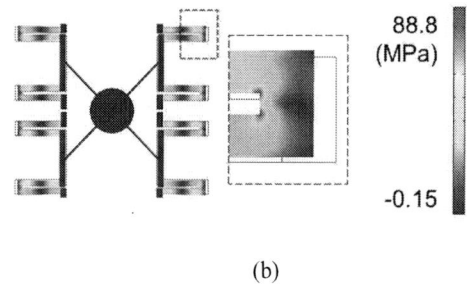

Stress distribution

(b)

Figure 2: Simulation results. (a) Displacement in Z direction. (b) Stress distribution.

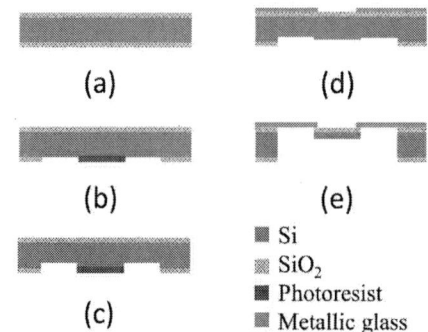

Figure 3: Fabrication process of the micromirror. (a) SiO_2 deposition. (b) Patterning of SiO_2 and photoresist. (c) 1st DRIE. (d) Patterning of metallic glass by lift off process. (e) 2nd DRIE

EVALUATION

Experiment Setup

The micromirror was placed in the magnetic field generator and was powered by a DC current source, as shown in figure 5. Between the DC current source and the device, there is a control circuit for distributing the current from the source to each metallic glass spring. The control circuit includes variable resistors in each output channel for the tilting angle control function. The actuated

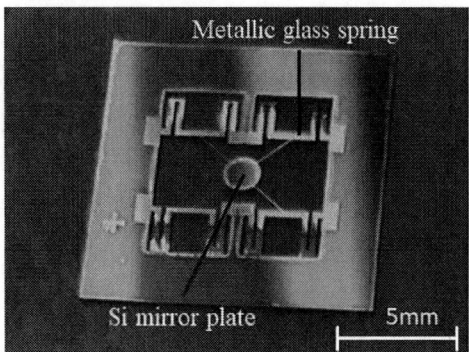

Figure 4: Fabricated micromirror

displacements are recorded by a CCD (Charged-Coupled Device) camera. This setup does not require a special power source or strong magnetic field. The resistance of each fabricated metallic glass spring is approximately 40 Ω. A 64 mW, 0.8 V, 80 mA power source is enough to actuate the device. As for the magnetic field, it is possible to generate a 160 mT magnetic field by a pair of $1.5 \times 1.5 \times 1$ cm^3 permanent magnets with an iron yoke. The simple setup requirements indicate that this device has high flexibility in being integrated into many systems.

Both translational displacement and angle tilting were observed. Measurement of out-of-plane displacements is conducted by applying 0~80 mA DC current to the four springs in a 160 mT magnetic field. Each of the springs will be powered by 0~20 mA DC current. Measurement of tilting angle is performed by applying 37.5 mA DC current on each two of the springs and -37.5 mA DC current on the other two springs for making the mirror have a rotational movement.

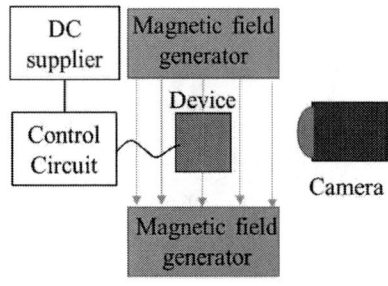

Figure 5: Experiment setup.

Results

The comparison of before and after translational actuation is shown in figure 6, which is a superimposed photo combining the photo of the before actuation (blue area) and the one after actuation (yellow area). The relationship between the applied current and the displacement is shown in figure 7, showing a large linear static stroke. At 80 mA, a maximum displacement of 418 µm was obtained with the Lorentz force of 24 µN. This stroke achieves the highest level in the translational electromagnetic micromirrors in comparison to other works [1-3].

The tilting angle was estimated with the displacement

of one edge point on the mirror plate observed by the CCD camera in figure 8. The maximum mechanical tilting angle is about 3.28° with 37.5 mA or -37.5 mA current in each spring under a 160 mT magnetic field.

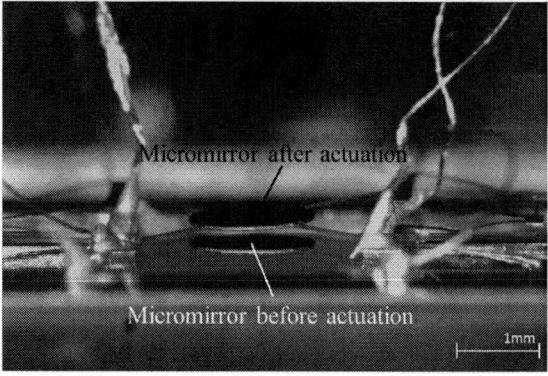

Figure 6: Comparison of before and after static actuation. Blue area: photo of initial position with no current; Yellow area: photo of under actuation at 80 mA.

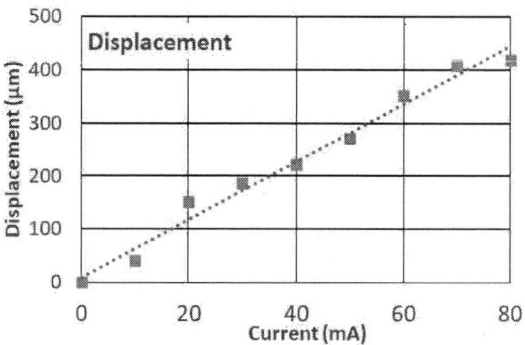

Figure 7: Measurement result of translational displacement during upward actuation. Dotted line: linear approximation.

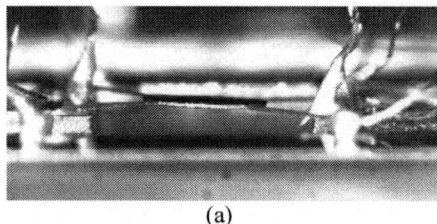

(a)

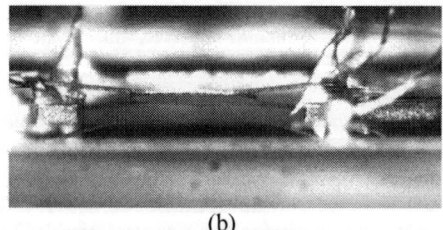

(b)

Figure 8: Tilting angle observation. (a) Position of micromirror before actuation. (b) position of micromirror after applying 37.5 mA DC current.

Discussion

The micromirror can be actuated in both upward and downward directions by only changing the direction of the actuation current. If applying a ±80 mA current, a stroke of ±400 μm will be achieved. The stroke of the micromirror is the key factor of the resolution in spectrometers. The theoretic resolution of an FTIR spectrum can be expressed as

$$\Delta v = \frac{1}{X}, \tag{2}$$

where X is the optical path difference (OPD). OPD is two times of the stroke. Theoretically, a ±400 μm stroke can ideally provide a resolution of 6.25 cm^{-1}.

In practical spectrum measurement, if the mirror is tilted, the resolution will be decreased significantly. Therefore, a control mechanism of tilting angle is necessary. The fabricated micromirror is capable of rotating in 3.28°. With changing the direction of the current, the rotation range could be ±3.28°. The tilting measurement result indicates that this micromirror is compactable to be integrated with the tilting angle control system, which is rarely achieved in electromagnetic micromirrors.

CONCLUSIONS

An electromagnetic micromirror employing metallic glass with a unique actuation structure was proposed and fabricated. The fabricated micromirror with a displacement of 418 μm had been achieved, and a maximum tilting angle of 3.28° had been obtained. This novel actuation structure brings more choices for micro-interferometers. Furthermore, the large displacement makes it possible to fulfill a high-resolution miniaturized interferometer.

ACKNOWLEDGEMENTS

Part of this work was performed in the Micro/Nanomachining Research Education Center (MNC) of Tohoku University, and Micro System Integration Center (μSiC), Tohoku University. This work was supported by JST SPRING, Grant Number JPMJSP2114.

REFERENCES

[1] Y. Xue and S. He, "A translation micromirror with large quasi-static displacement and high surface quality," Journal of Micromechanics and Microengineering, vol. 27, no. 1, p. 015009, 2016.

[2] Y. Hongbin, Z. Guangya, C. F. Siong, L. Feiwen, W. Shouhua, and Z. Mingsheng, "An electromagnetically driven lamellar grating based Fourier transform microspectrometer," Journal of Micromechanics and Microengineering, vol. 18, no. 5, p. 055016, Apr. 2008.

[3] U. Baran, K. Hedili, S. Olcer, and H. Urey, "FR4 electromagnetic scanner based Fourier Transform Spectrometer," Volume 3: 2011 ASME/IEEE International Conference on Mechatronic and Embedded Systems and Applications, Parts A and B, 2011.

[4] N. Van Toan, T. T. Tuoi, Y.-C. Tsai, Y.-C. Lin, and T. Ono, "Micro-fabricated pressure sensor using 50 nm-thick of pd-based metallic glass freestanding membrane," Scientific Reports, vol. 10, no. 1, 2020.

[5] Z.-Y. Wang, Y.-Y. Chen, Y.-C. Lin, T. Ono, M.-T. Lin, and Y.-C. Tsai, "Electrostatic metallic glass micro-mirror fabricated by the self-aligned structures," Japanese Journal of Applied Physics, vol. 59, no. SI, 2020.

[6] M. Toda, C. Li, N. Van Toan, Y.-C. Tsai, Y.-C. Lin, and T. Ono, "Torsional resonator of Pd–Si–Cu metallic glass with a low rotational spring constant," Microsystem Technologies, vol. 27, no. 3, pp. 929–935, 2020.

[7] Y.-F. Huang, C.-H. Tsou, C.-J. Hsu, Y.-C. Lin, T. Ono, and Y.-C. Tsai, "Metallic glass thin film integrated with flexible membrane for electromagnetic micropump application," Japanese Journal of Applied Physics, vol. 59, no. SI, 2020.

[8] E. Afsharipour, B. Park, and C. Shafai, "Large Tilt Angle Lorentz Force Actuated Micro-Mirror with 3 DOF for Optical Applications", 31st Eurosensors - European Conference on Solid-State Sensors, Paris, France, 3-6 September 2017.

[9] G. Kaltenboeck, M. D. Demetriou, S. Roberts, and W. L. Johnson, "Shaping metallic glasses by electromagnetic pulsing," Nature Communications, vol. 7, no. 1, 2016.

[10] Y. Liu, S. Hata, K. Wada, and A. Shimokohbe, "Thermal, mechanical and electrical properties of PD-based thin-film metallic glass," Japanese Journal of Applied Physics, vol. 40, no. Part 1, No. 9A, pp. 5382–5388, Sep. 2001.

CONTACT

*T. Ono, e-mail: takahito.ono.d4@tohoku.ac.jp

ARBITRARY SHAPED BACKSIDE REINFORCEMENT FOR TWO DIMENSIONAL RESONANT MICROMIRRORS

Takashi Sasaki, Adrien Piot, Anton Lagosh, Clement Fleury, Markus Bainschab, Yanfen Zhai,
Marcus Baumgart, Sara Guerreiro, Dominik Holzmann, Aleš Travnik, Mohssen Moridi
Silicon Austria Labs, Austria

ABSTRACT

This paper reports a backside reinforcement technique for two-dimensional resonant scanning mirrors. The backside reinforcement patterns are generated based on the product of the surface stress and the distance from the rotation axis of interest, and the patterns are combined for both axes. The piezoelectrically actuated two-dimensional micromirrors were fabricated inhouse. The novel fabrication approach makes it possible to obtain backside reinforcement of any desired thickness. It has been experimentally proven that the deformation is reduced by more than 50% for torsion mode and 40% for rocking mode without a significant increase in the moment of inertia: 3% and 9%, respectively.

KEYWORDS

Tow dimensional resonant micromirror, backside reinforcement, dynamic deformation, silicon on insulator wafer, piezoelectric actuator.

INTRODUCTION

In high-performance micromirrors, a low moment of inertia with low dynamic deformation is required [1-5]. The use of thick backside reinforcement from SOI wafer to meet the deformation requirement was reported by Ji et al. [6] and Ikegami et al. [7], etc. To our knowledge, no backside reinforcement has been specifically designed to mitigate dynamic deformation for both axes simultaneously. This paper will present the results of simulation and experiments on dynamic deformation of a two-dimensional piezoelectrically actuated resonant micromirror with backside reinforcement capable of two axes high-frequency operation.

PRINCIPLE AND DESIGN

The backside reinforcement (BSR) shape is generated based on the surface stress of mirror surface obtained at maximum rotation angle using finite element method. The reinforcement is divided into small silicon pixels that are enabled if the criterion value, calculated as the distance from the rotation axis multiplied by the surface stress, exceeds a certain value. The size of pixels is selected to be 14 um and the criterion value is set to be 8 MN/m in this study. The reinforcement is calculated for each rotation axis and superposed. Figure 1 shows the generated design of the backside reinforcement for this mirror. Figure 2 shows the calculated dynamic deformation for torsion and rocking modes with the backside reinforcement. The rocking mode and torsion mode can be found at 13.1 kHz and 21.1 kHz.

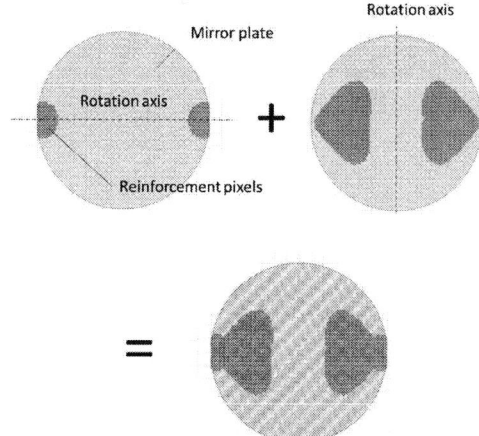

Figure 1: Designed arbitrary shaped backside reinforcement for two resonant axes

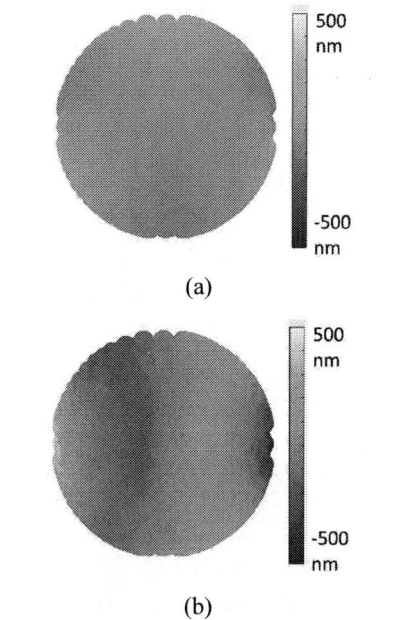

Figure 2: Calculated dynamic deformation with backside reinforcement at 28° optical angle (a) torsion mode (slow axis) (b)rocking mode (fast axis)

FABRICATION

The micromirror is fabricated from the SOI wafer with piezoelectric stack on top using developed and optimized multistep microfabrication technology as shown in Figure 2. The backside reinforcement is formed from the SOI handle layer. A novel approach with the combination of several dry etch techniques and independent lithography steps allows us to form the reinforcement of various shapes. An advantage of this microfabrication approach is grass- and residue free surfaces at the backside of the mirror. The thickness of the reinforcement shape is not limited to the thickness of the handle layer and can be up to any desired value. The BSR surface is flat, smooth and without eroded parts. This technology was implemented on 6-inch wafers.

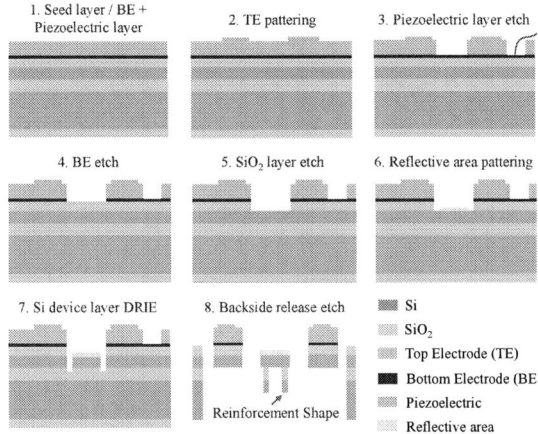

Figure 2: Process flow for fabricating a micromirror with backside reinforcement

RESULTS AND DISCUSSION

A fabricated 2D micromirror is shown in Figure 3. The almost all devices in 6-inch wafer were successfully fabricated using forementioned fabrication process. Figure 4 shows the optical image of the backside reinforcement of this device. The height of the backside reinforcement is about 50 um.

Figure 5 shows the measured micromirror frequency response for torsion mode and rocking mode. The rotation angle of mirror reaches more than 20° optical in both torsion and rocking mode at 4.8 Vpp. The deformation was measured using digital holographic microscope. The static deformation is shown in Figure 6, showing less than 20 nm bow. Although the silicon dioxide layer exists between top silicon layer and the reinforcement layer, a deformation high enough to degrade the optical quality was not observed.

Figure 7 and 8 shows the results of the dynamic deformation profile measurement for mirrors without and with the BSR. The static component was subtracted. With backside reinforcement, the deformation decreases by more than a factor of 2 as shown in Figure 7 for the torsion mode, and more than 30% for rocking mode shown in Figure 8. The resonant frequency change was about 900 Hz for rocking mode and 190 Hz for torsion mode. The obtained results show the effectiveness of the developed

and implemented backside reinforcement technique.

Figure 3: Fabricated two dimensional piezoelectrically actuated micromirror with backside reinforcement.

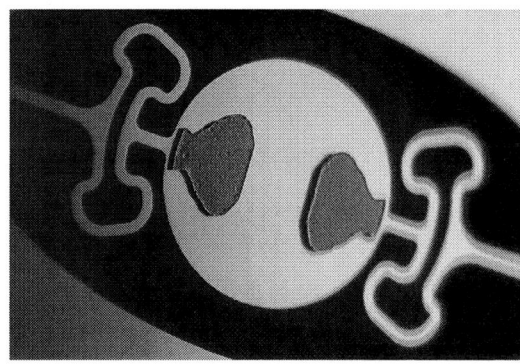

Figure 4: Arbitrary shaped backside reinforcement of two dimensional piezoelectrically actuated micromirror

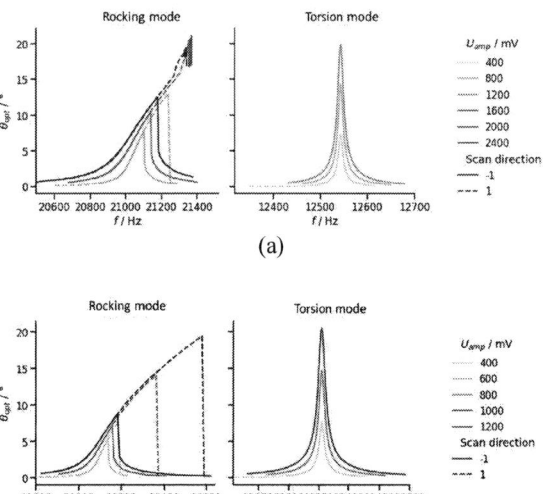

Figure 5: Frequency characteristic of mirror (a) without backside reinforcement and (b) with backside reinforcement

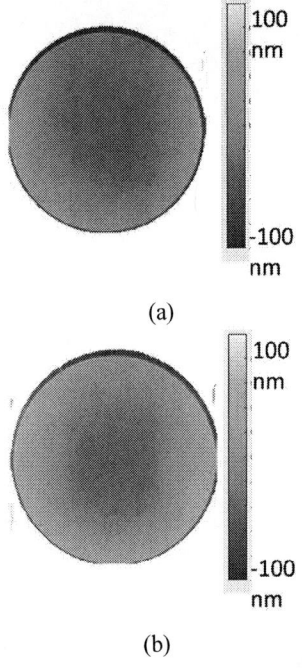

Figure 6: Static deformation of micromirror: (a) without backside reinforcement; (b) with backside reinforcement

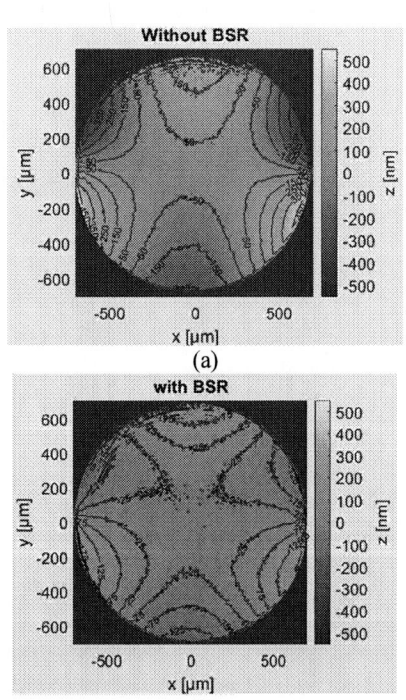

Figure 7: Dynamic deformation of torsion mode at 40° optical angle: (a) without backside reinforcement; (b) with backside reinforcement

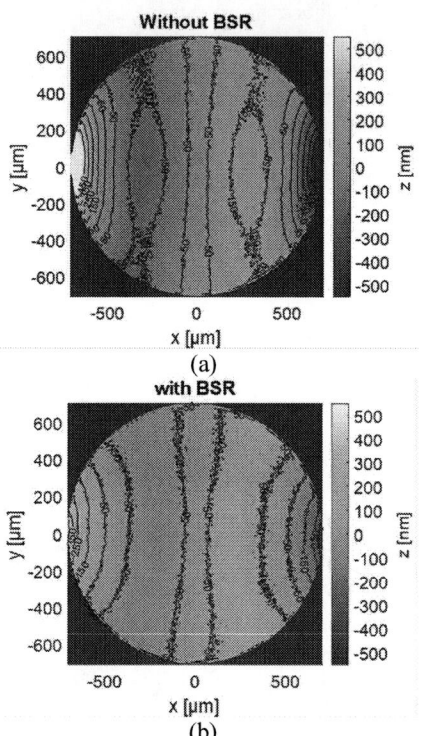

Figure 8: Dynamic deformation of rocking mode: (a) without backside reinforcement at 24° optical angle; (b) with backside reinforcement at 28° optical angle

CONCLUSION

The design procedure and fabrication have been shown for reducing the dynamic deformation with a small increase in moment of inertia. The design mirror with method reduces the deformation by 50% and 40 % for torsion mode and rocking mode with the moment of inertia increasing by 3 % and 9 %, respectively, for obtaining a high-performance two-dimensional micromirror.

ACKNOWLEDGEMENT

This work has been jointly supported by Evatec, EV Group, TDK Elec-tronics, ZKW and by Silicon Austria Labs (SAL), owned by the Republic of Austria, the Styrian Business Promotion Agency (SFG), the federal state of Carinthia, the Upper Austrian Research (UAR), and the Austrian Asso-ci-a-tion for the Elec-tric and Elec-tronics Industry (FEEI).

REFERENCES

[1] H. Urey, D. Wine, and J. Lewis, "Scanner design and resolution tradeoffs for miniature scanning displays," Proceedings of SPIE, vol. 3636, pp. 60–6860, Apr. 1999.

[2] A. Piot et al., "Resonant PZT MEMS mirror with segmented electrodes," in 2020 IEEE 33rd International Conference on Micro Electro

Mechanical Systems (MEMS), 2020, pp. 517–520.

[3] A. Piot, J. Pribošek, and M. Moridi, "Dual-axis resonant scanning mems mirror with pulsed-laser-deposited barium-doped PZT," in 2021 IEEE 34th International Conference on Micro Electro Mechanical Systems (MEMS), 2021, pp. 89–92.

[4] A. Piot et al., "Optimization of Resonant PZT MEMS Mirrors by Inverse Design and Electrode Segmentation," Journal of Microelectromechanical Systems, vol. 30, no. 2, pp. 216–223, Apr. 2021.

[5] P. Thakkar et al., "Measuring angle-resolved dynamic deformation of micromirrors with digital stroboscopic holography," in Optics and Photonics for Advanced Dimensional Metrology II, 2022, vol. 12137, pp. 112–119.

[6] C.-H. Ji et al., "An electrostatic scanning micromirror with diaphragm mirror plate and diamond-shaped reinforcement frame," Journal of Micromechanics and Microengineering, vol. 16, p. 1033, Apr. 2006.

[7] K. Ikegami, T. Koyama, T. Saito, Y. Yasuda, and H. Toshiyoshi, "A biaxial piezoelectric MEMS scanning

CONTACT

* T. Sasaki, tel: +43-664-88843732; takashi.sasaki@silicon-austria.com

HIGH TRANSMITTANCE METASURFACE HOLOGRAMS USING SILICON NITRIDE

Masakazu Yamaguchi, Hiroki Saito, Satoshi Ikezawa, and Kentaro Iwami
[1]Tokyo University of Agriculture and Technology, JAPAN

ABSTRACT

A metasurface hologram with high transmittance of 93% was achieved by using silicon nitride (SiN) as a meta-atom material, which far surpasses the conventional record of 80% reported using TiO_2 meta-atom [1]. This progress is because of the decrease in the Fresnel reflection due to the effective index matching between the substrate and metasurface hologram. The projection of an image and a cinematographic movie were demonstrated with improved brightness and color reproducibility.

KEYWORDS

Hologram, Metasurface, Silicon Nitride, Holography.

INTRODUCTION

Holography is a technology for recording and reproducing light waves, and holograms are the storage media. Holograms have attracted attention because they can display stereoscopic images without the use of external devices. Among them, metasurface holograms, which can obtain a wide viewing angle in a small nanoscale volume, are expected to be applied. In previous studies, colorization of the projected image [1][2][3], and movies based on a cinematographic approach [4][5] have been achieved. However, in these studies, the transmittance of the material used for the meta-atom was low and the projected images were dark. Table 1 summarizes the average transmittance of the meta-atom materials and the produced holograms in the previous studies. In terms of colorization, there are a few examples of color superposition. It is essential to obtain a projected image that accurately reproduces the brightness and color of the target image for viewing holograms that can be used for digital signage and other applications. In this study, we fabricated metasurface holograms with high transmittance by using silicon nitride (SiN), which has high transmittance in the visible light range, as the meta-atom material.

Table 1: Comparison of meta-atom material and transmittance reported in previous studies [1][2][3].

Material	Wavelength [nm]	Transmittance [%]
TiO_2	480, 532, 660	80
c-Si	445, 532, 633	61
a-Si	633	31

MATERIALS AND METHODS

A metasurface hologram is a medium that records the light itself as it propagates from an object. In this study, we adopted a phase hologram, which records the phase distribution of the object light. When the hologram is irradiated with a plane wave, the phase profile of the incident light is controlled on the hologram, and a projected image can be observed as if the object exists on the projection plane.

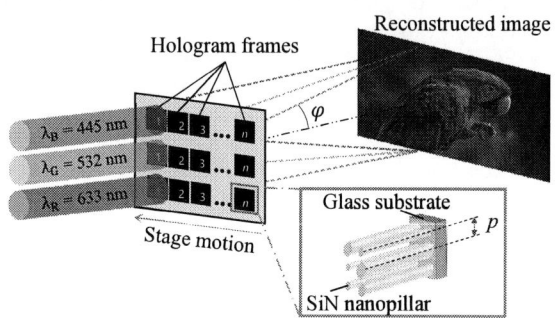

Figure 1: Schematic drawing of a multicolor metasurface hologram.

Figure 1 shows a schematic diagram of the metasurface hologram fabricated in this study. The hologram consists of three-color channels designed at the wavelengths of 445 (Blue), 532 (Green), and 633 nm (Red). For a cinematographic movie, each channel has 30 frames. Each hologram frame consists of a square lattice of octagonal nanopillars on a glass substrate. The light transmitted through each nanopillar has a different phase delay depending on the height and width of the pillar.

The viewing angle φ of the hologram shown in Figure 1 can be obtained as:

$$\varphi = 2\arcsin\left(\frac{p}{\lambda}\right), \qquad (1)$$

where p is the pixel period and λ is the wavelength of the incident light. In order to superimpose projected images of each wavelength channel on the identical size, we maintained the ratio p/λ constant. In addition, equation 1 indicates that a hologram with microscale and nanoscale size inter-pixel distance is necessary to obtain a sufficient viewing angle. Therefore, metasurface holograms are expected to achieve a wide viewing angle due to their subwavelength periods.

In this study, the height of the nanopillars was kept constant, and the phase delay of transmitted light is controlled only by changing the width of the nanopillars. As mentioned earlier, SiN, which has high transmittance in the visible range, will be used for the holograms produced in this study. Figure 2 plots the bandgap energy and refractive index of materials used for dielectric metasurface.

Figure 2 shows that the refractive index of SiN is lower than that of other materials. The phase delay Δ of light transmitted through a nanopillar is expressed as:

$$\Delta = \frac{2\pi}{\lambda}\left(n_{eff} - 1\right)h, \qquad (2)$$

where h and n_{eff} are the height and the effective refractive index of nanopillars, respectively. From equation 2, it can be seen that using SiN, which has a low refractive index, as a meta-atom to obtain a phase delay of 0~2π requires higher nanopillars.

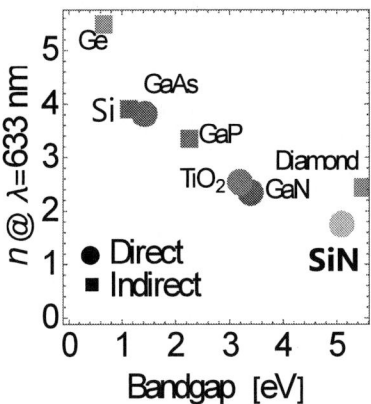

Figure 2: Bandgap energies and refractive indices of meta-atom materials.

Electromagnetic field analysis of SiN nanopillars

The transmittance and phase delay of the SiN nanopillars were calculated by a commercially available finite element method software COMSOL Multiphysics 5.6. Equation 3 derived from Maxwell's equations was used in the analysis. In Equation 3, the steady-state response $E(r, t) = E_0(r)\exp(j\omega t)$ due to monochromatic light with a single angular frequency ω is assumed as the solution.

$$-\nabla \times \frac{1}{\mu}\{\nabla \times E_0(r)\} - k_0^2\left(\epsilon_r - \frac{j\sigma}{\omega\epsilon_0}\right)E_0(r) = 0 \quad (3)$$

In this study, the height of the nanopillars was fixed at 1500 nm, and the width was varied from 50 nm to 300 nm in 10 nm increments. The wavelengths of the incident light corresponding to each color were 445 nm, 532 nm, and 633 nm, which are the wavelengths of lasers used for projection. The pitch was fixed at $p = 284$ nm, $p = 340$ nm, and $p = 404$ nm for each color, respectively. Figure 3 shows the calculation results. Based on the above simulation results, the nanopillar widths used in this study were 120~190 nm for the wavelength of 445 nm, 130~240 nm for 532 nm, and 70~280 nm for 633 nm.

Calculation of phase distribution

The phase distribution of the hologram surface was calculated using the iterative Fourier transform method (IFTM). The IFTM is a method in which the Fourier transform and inverse Fourier transform are repeated as shown in Figure 4, and the following procedures (a) through (d) were used.

(a) Combine the intensity and phase distribution of the target image. At the first step, a random phase is used.

(b) Inverse Fourier transform the combination obtained in step (a). The phase distribution obtained in this step becomes the phase distribution to be reproduced on the hologram surface.

(c) Combine the phase distribution obtained in step (b) and the incident light intensity of the laser used for projection and perform the Fourier transform. The calculated intensity becomes the projected image.

(d) Perform step (a) again using the phase distribution calculated in step (c).

The target image and the projected image obtained in step (c) are compared, and these operations are repeated until a sufficient matching is obtained.

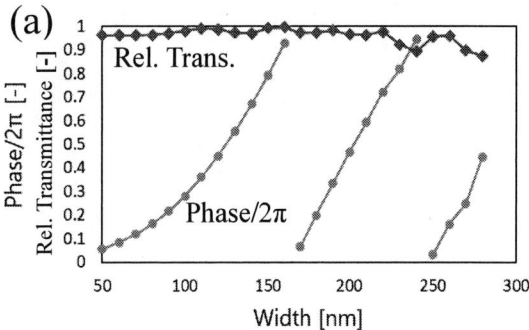

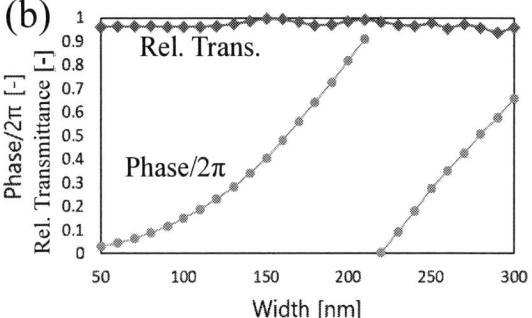

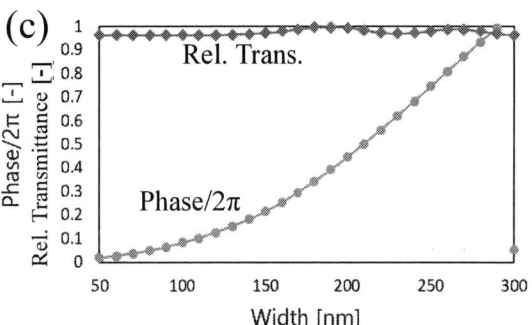

Figure 3: Calculated transmittance and phase delay. The incident light wavelengths are (a) 445 nm, (b) 532 nm, and (c) 633 nm, respectively.

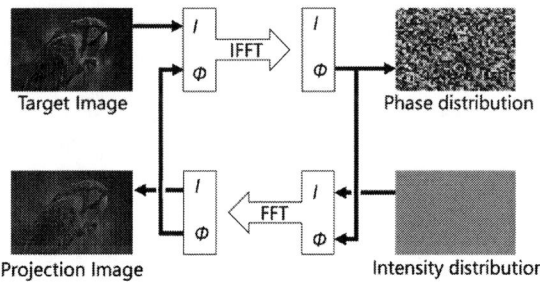

Figure 4: Schematic diagram of the IFTM.

The target image and some frames of the movie used in this study is shown in Figure 5(a) and (c)~(f). In order to avoid conjugate images and zero-order light in the projection, each image and frames are shifted to the left and padded with the black pixels as shown in Figure 5(b).

978-1-6654-9309-3/23 $31.00 © 2023 IEEE

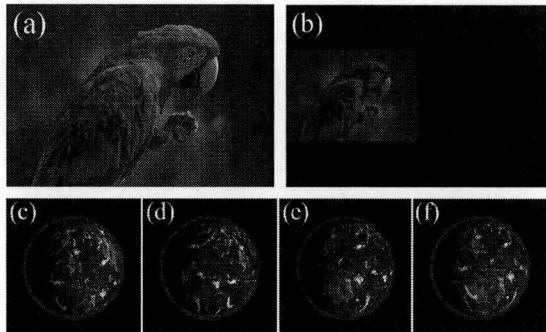

Figure 5: Original target image [6] (a), and modified target image of the green channel (b). Selected frames of the target movie (c)~(f).

Determination of production dimensions

Based on these calculated values, the design dimensions of the holograms for static images to be produced were determined as shown in Table 2. As mentioned earlier, different pitches are used for different wavelengths to keep the viewing angle constant. Therefore, if the resolution of the target image is kept constant, the hologram area and the size of the projected image will be different for each wavelength. Accordingly, target images of different resolutions were used for each wavelength to maintain the frame size constant.

Table 2: Dimensions of holograms for static images

λ [nm]	p [nm]	d [nm]	Resolution	Rel. Transmittance [%]
445	284	120–190	1920×1280	98
532	340	130–240	1606×1071	98
633	404	70–280	1350×900	98

Fabrication

The procedure for fabricating holograms is shown in Figure 6. A 2 cm square glass substrate deposited with SiN is coated with the resist, and the structure is transferred using an electron beam lithography apparatus. Chromium mask pattern is formed through the lift-off process. The SiN is then removed by reactive ion etching. Finally, the chromium is removed by wet etching.

RESULTS

Projection Evaluation

First, an optical system was constructed as shown in Figure 7 for projection evaluation. Lasers with wavelengths of 445 nm, 532 nm, and 633 nm were used for projection, and each structure on the same substrate was irradiated. The obtained projected images are shown in Figure 8. Figure 8 shows that the holograms fabricated in this study can obtain high brightness projection images. It was also shown that the images were in good agreement with the simulated images.

Transmittance and phase delay measurements using a microspectrometer

The transmittance and phase delay of the holograms produced in this study were measured using a

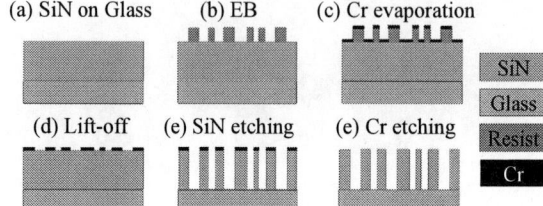

Figure 6: Schematic diagram of hologram fabrication procedure.

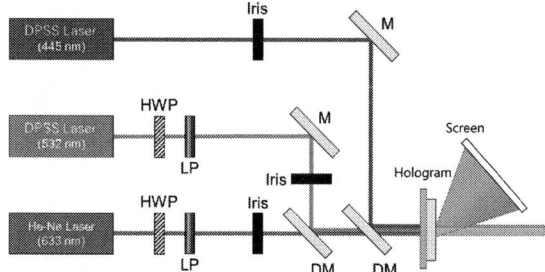

Figure 7: Schematic of the optics for projection.

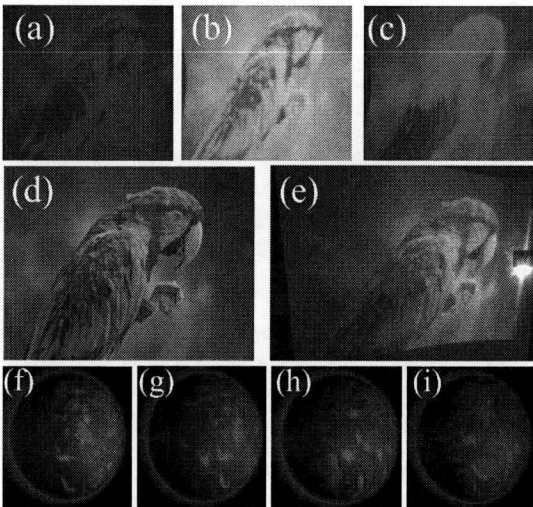

Figure 8: Projection images at 445 nm (a) 532 nm (b) 633 nm (c) and their superimposed projection (e). Also, simulated image (d). And some frames of movie projections (f)~(i).

microspectrometer. Test patterns were used for the measurements. The test pattern is an array of SiN nanopillars with diameters ranging from 50 nm to 300 nm, and was fabricated on the same substrate as the holograms.

In this measurement method, the intensity at each point is measured while moving the measurement point from the unmetasurfaced area (glass substrate) to the formed area as shown in Figure 9, and then the phase delay α is calculated using Equation 4.

$$I_2(x_d) = r^2 I_1 + (1-r)^2 I_3 + 2r(1-r)\sqrt{I_1 I_3} \cos\alpha \quad (4)$$

where I_1 through I_3 refer to the intensity of the unformed, boundary, and formed areas, respectively. Also, r is the percentage of glass substrate surface to measurement range. The transmittance was calculated by dividing I_3 by I_1.

The measurement results of holograms for static images are shown in Figure 10. Based on these results, the average transmittance of the holograms for static images fabricated in this study was 93%. Also, the average transmittance of the holograms for movie was 89%. These values are higher than the values in the previous study shown in Table 1, which indicates that this study successfully fabricated high transmittance metasurface color holograms using SiN.

CONCLUSION

In this study, metasurface holograms were fabricated using SiN, which has high transmittance in the visible light range, as a meta-atom material. The average transmittance of the fabricated holograms for static images was 93% and holograms for movie is 89 % as measured by a microspectrometer. This value was higher than the values in the previous studies [1][2][3]. In terms of color superposition, the projected images of the holograms produced in this study were in good agreement with the simulated images. These results indicate that this study succeeded in fabricating high transmittance metasurface color holograms.

ACKNOWLEDGEMENT

This work was supported in part by the Japan Society for the Promotion of Science (JSPS) KAKENHI under Grant 21H01781 and Grant 22K04894; in part by the Takeda Sentanchi Supercleanroom, University of Tokyo, through the "Nanotechnology Platform Program" of the Ministry of Education, Culture, Sports, Science, and Technology (MEXT), Japan, under Grant JPMXP1222UT1014. The authors would like to thank Prof. Y. Mita, Dr. A. Higo, Dr. E. Lebrasseur, and Mr. M. Fujiwara (The University of Tokyo) for their support during sample fabrication, also would like to thank Prof. Lucas Heitzmann Gabrielli (University of Campinas) for developing and maintaining gdstk, a Python library for creating and manipulating GDSII layout files, and also would like to thank a TSUBAME3.0 supercomputer at the Tokyo Institute of Technology for the numerical calculations.

REFERENCES

[1] R. C. Devlin: et al.: "Broadband high-efficiency dielectric metasurfaces for the visible spectrum", *Proc. Nat. Acad. Sci.*, vol. 113, pp. 10473-10478, 2016.

[2] A. Martins, et al.: "Highly efficient holograms based on c-Si metasurfaces in the visible range", *Opt. Exp.*, vol. 26, pp.9573-9583, 2018.

[3] Q.-T. Li., et al.: "Polarization-independent and high-efficiency dielectric metasurfaces for visible light", *Opt. Exp.*, vol. 24, pp.16309-16319, 2016.

[4] R. Izumi, et al.: "Metasurface holographic movie: a cinematographic approach", *Opt. Exp.*, vol. 28, pp. 23761-23770, 2020.

[5] N. Yamada, et al.: "Demonstration of a multicolor metasurface holographic movie based on a cinematographic approach", *Opt. Exp.*, vol. 30, pp. 17591-17603, 2022.

[6] "pixabay", https://pixabay.com/ja/photos/%e3%82%aa%e3%82 %a6%e3%83%a0-%e9%b3%a5-%e7%be%bd%e6% af%9b-%e7%be%bd-6342271/, accessed:2022-09-06.

CONTACT

*K. Iwami, k_iwami@cc.tuat.ac.jp

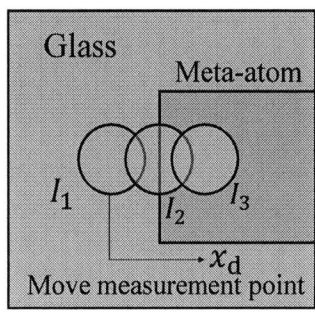

Figure 9: Measurement method for the transmittance and phase delay of meta-atom.

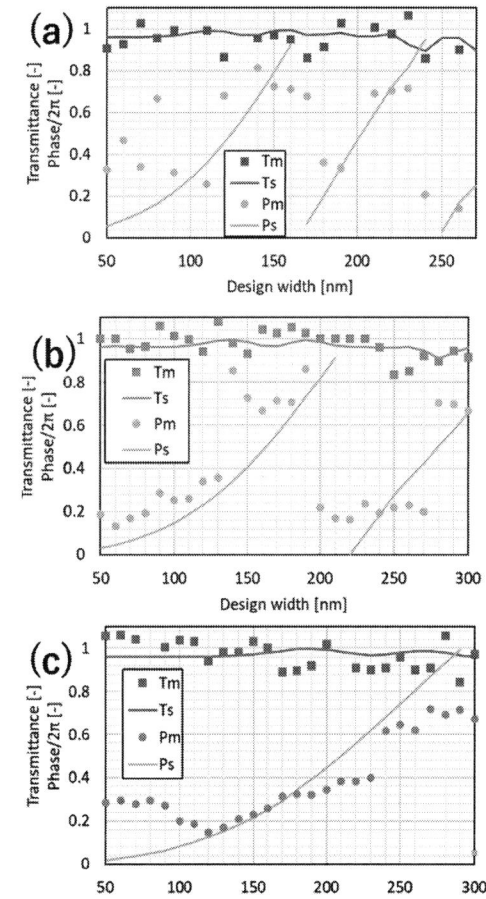

Figure 10: Measured transmittance and phase delay of silicon nanopillars at 445 nm (a), 532 nm (b), 633 nm (c). Solid lines represent simulated values and dots represent measured values.

MULTIFUNCTIONAL OPTICAL METASURFACE FOR ANOMALOUS REFLECTION, STRUCTURAL COLOR, AND SURFACE LATTICE RESONANCE

*Liye Li[1], Hongshun Sun[1], Yifan Ouyang[2], Shengxiao Jin[1], Tian Kang[1], Zhimei Qi[3], and Wengang Wu[1]**

[1]National Key Laboratory of Science and Technology on Micro/Nano Fabrication, School of Integrated Circuits, Peking University, Beijing 100871, P. R. China
[2]School of Electronics, Peking University, Beijing 100871, P. R. China
[3]State Key Laboratory of Transducer Technology, Aerospace Information Research Institute, Chinese Academy of Science, CHINA

ABSTRACT

Multifunction is one of the most important research directions for the optical metasurface.This paper creatively proposes a multifunctional optical metasurface based on polarization multiplexing, which is composed of periodic arrays of metal-insulator-metal structural meta-atom. In the y-polarization state, the metasurface realizes anomalous reflection and high-saturation structural color in the visible band. Besides, in the x-polarization state, the metasurface realizes refractive index sensing with a high sensitivity of 640.5 RIU/nm due to surface lattice resonance excited by the incident beam. The proposed metasurface has great potential applications in the field of spectrometers, displays, and sensors.

KEYWORDS

multifunctional metasurface, anomalous reflection, structural color, surface lattice resonance, polarization multiplexing

INTRODUCTION

Compared with traditional optical elements, the metasurfaces have attracted extensive attention for their unparalleled ability to modulate optical characteristics, including amplitude, polarization, phase, frequency, and so on [1-3]. It can control the spatial distribution of optical parameters with subwavelength resolution, and complish the optical modulation within a subwavelength distance. However, metasurface with only one single function can not meet the diversified needs of complex optical systems. Therefore, designing multifunctional metasurfaces is of great significance for practical applications of the metasurface in the future [4-7].

At present, parameter multiplexing is a general method for realizing multifunctional optical metasurfacaes: Fan et al. realized different structural colors by the incident direction multiplexing [8], and Chen et al. designed a metalens with multifocal focus based on polarization multiplexing [9]. Besides, the spatial multiplexing includes staggered, subarea and stacked arrangements, which will integrate multiple diversified functionalities into one single metasurface. For example, Genevet et al. spliced the meta-atom arrays with different functions to form a new metasurface for multiplex anomalous transmission beams [10]. Nevertheless, above methods increase the difficulty of the fabrication process, cause serious crosstalk, and reduce the efficiency of each function. Another common problem is that there is no obvious difference between multiple functions.

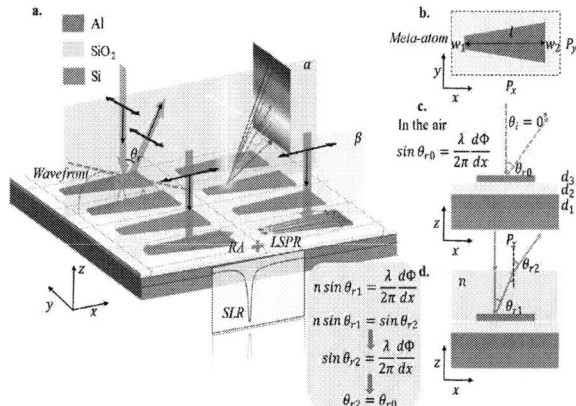

*Figure 1: **a.** The proposed multifunctional metasurface. **b.-c.** Top and cross-sectional views of a meta-atom. The geometric parameters are as follows: w_1=40 nm, w_2=160 nm, l=400 nm, d_1=150 nm, d_2=50 nm, d_3=35 nm, P_x=700 nm, P_y=200 nm. **d.** The sketch of refraction angle compensation at the air-medium interface and theoretical calculations.*

In this paper, we propose an optical metasurface for achieving anomalous reflection (AR), structural color (SC), and surface lattice resonance (SLR) functions based on polarization multiplexing. The structure and function of metasuface are shown in Figure 1a. The top metal is trapezoidal, the middle is a SiO_2 dielectric layer, and the bottom is thick metal to reflect all the light. On the one hand, the normal incident light with y-polarization will generate AR obeying the generalized Snell's Law. Considering that the reflection angle is a function of wavelength, AR can be seen as a kind of dispersion and can produce high saturation SCs in the visible band. On the other hand, in the x-polarization state, the metasurface produces SLR with a sharp resonance dip in the reflection spectrum, which is extremely sensitive to the environmental refractive index. It is worth noting that the refractive index matching layer has no effect on the AR direction due to the refraction angle compensation at the air-medium interface, which keeps the independence between the AR and SLR. In addition, the trapezoidal length l is a crucial parameter that influences the efficiency of three functions. The proposed metasurface is equivalent to the combination of blazed grating, prism, and refractive index sensors, which provides a strategy for optical component integration.

DESIGN AND SIMULATION

The proposed optical metasurface is composed of periodic arrays of metal-insulator-metal (MIM) meta-atom (Figure 1b-c). The y-polarized incident beam induces gap surface plasmons (GSP) in MIM structures which is a special surface electromagnetic wave from a couple of the surface plasmon polariton of two metal films. The GSP propagates along the polarization direction and is reflected at the structure terminations, forming two pairs of electric dipoles in opposite directions (Figure 2a). Further, they can be seen as an annular displacement current, which is approximately a magnetic dipole, causing the magnetic field to be localized in the SiO_2 medium (Figure 2b). Besides, the electromagnetic field energy distribution shifts to a narrower direction with decreasing wavelength (Figure 2c-d, taking λ=550 nm and λ=450 nm as examples). For the 400-650 nm visible band, the electric dipole, and magnetic dipole work together to result in the phase increasing linearly along the x-axis and covering 2π (Figure 2e). Therefore, the wavefront will be titled, and each reflected beam has a specific reflection angle θ_r instead of normal emergence (Figure 2f). The AR phenomenon can be described by the generalized Snell's Law, as follows:

$$nk_0 sin\theta_r - nk_0 sin\theta_i = \frac{d\varphi}{dx} \quad (1)$$

where n is the refractive index, k_0 is the wavevector in a vacuum, θ_r is the reflection angle, θ_i is the incident angle, and $d\varphi/dx = 2\pi/P_x$ is the phase gradient influenced by the P_x. According to Equation (1), the incident beams with different wavelengths have different θ_r, which proves that AR is a kind of dispersion and can produce high-saturation structural colors.

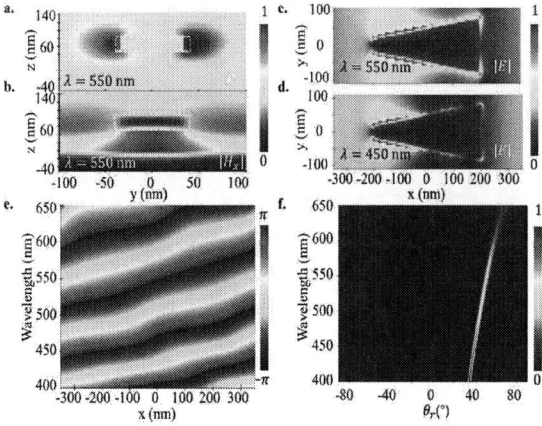

Figure 2: Numerical simulation results about AR. **a.-b.** The distribution of ρ and $|H_x|$ in the case of λ=550 nm (y-z plane, x=-50 nm). **c.-d.** The E-field intensity distribution on the trapezoidal surface at λ=550 nm and λ=450 nm (x-y plane, z=85 nm). **e.** Phase shift along the x-axis for the 400-650 nm band. **f.** Anomalous Reflection energy distribution in the far field, which is a function of incident wavelength λ and the anomalous reflection angle θ_r.

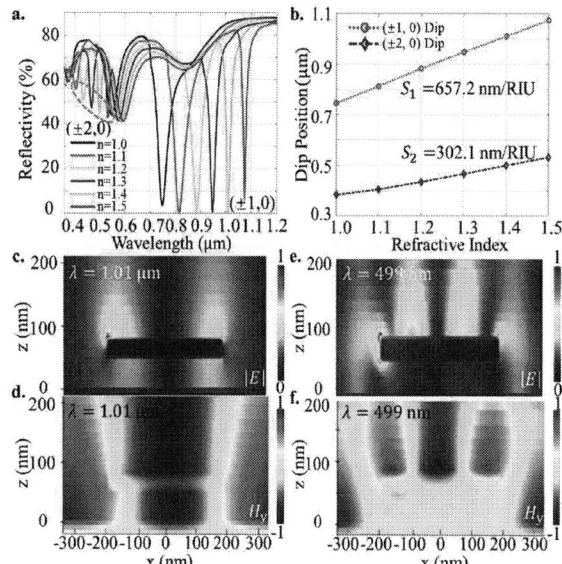

Figure 3: Numerical simulation results about SLR. **a.** The reflection spectra under different environmental refractive indices. **b.** (±1,0) and (±2,0) SLR dips shift with the increase of the refractive index. **c.-f.** The electromagnetic field distributions at SLR wavelengths (1.01 μm and 499 nm) in the case of n=1.4.

Moreover, the x-polarization beam excites the collective oscillation of free electrons in the top metal along the same direction, namely localized surface plasmon resonance (LSPR). What's more, the incidence, modulated by the periodic array, will produce the Rayleigh anomalies (RA) phenomenon, a set of diffraction plane waves parallel to the surface. The occurrence of RA at normal incidence conforms to the following formula:

$$\frac{2\pi}{n \cdot \lambda_{(p,q)}} = \sqrt{\left(p\frac{2\pi}{P_x}\right)^2 + \left(q\frac{2\pi}{P_y}\right)^2} \quad (2)$$

where the p and q are the diffraction orders in the x and y directions, and $\lambda_{(p,q)}$ is the corresponding wavelength. Obviously, only specific wavelengths can produce RA, so the spectrum of RA is a narrow band. On the contrary, due to Ohmic loss, the LSPR has a continuous broad spectrum. Therefore, the SLR, as a kind of Fano resonance, originates from the coupling of RA and LSPR, with a sharp resonance dip, which promotes the strong light-matter interaction because of the enhanced the localized electric field [11]. As shown in Figure 3a, there are two resonance dips in simulated reflection spectra. With the increase of the ambient index, the SLR wavelength increases linearly (Figure 3b), and the linewidth decreases gradually because of the reduced index contrast between the environment and SiO_2-substrate. Taking n=1.4 as an example, the electromagnetic field distributions at two resonance dips (λ=1.01 μm and λ=499 nm) have apparent plane wave characteristics, including directionality, periodicity, and non-locality (Figure 3c-f), which are different from GSP resonance completely (Figure 2a-b). The electric fields propagate periodically along the x-direction, so the P_x plays a major role in RA and P_y doesn't work, resulting that both resonance dips are originated from the (±1,0) order SLR and (±2,0) order SLR separately. We only concentrate on the (±1,0) order SLR due to the higher sensitivity and resonance strength, which is more practical and stable for refractive index sensing.

RESULTS AND DISCUSSION

The scanning electron microscope (SEM) image of the fabricated sample is shown in Figure 4a. Moreover, we take advantage of the atomic force microscope (AFM) to check the smoothness of the top trapezoidal metal (Figure 4b-c). A rough surface will cause extra optical loss, such as scattering and absorption, especially in the short band. The angular-resolved spectrometer was used to measure AR with an acceptance angle range from -90° to 90° at 1° interval. Perfectly, the measured results are consistent with the theoretical calculation and the simulated results (Figure 4d). The AR range is from 35°to 69°, and corresponding structural colors from dispersion can be observed clearly (Figure 4e). To analyze the structural colors quantitatively, we calculate the CIE 1931 chromaticity coordinates of each color at different reflection angles. As shown in Figure 4f, the structural color gamut can cover Adobe RGB completely [12], whose gamut area is 0.2682 about 173.8% of Adobe RGB. And the maximum excitation purity can reach 1.00, which is far more than that of structural colors generated by other metal metasurfaces [13].

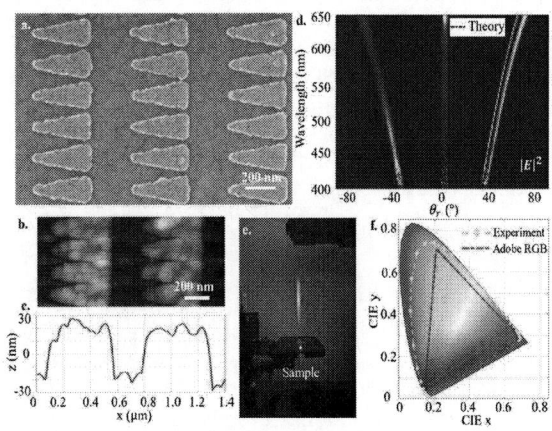

Figure 4: Sample characterization and measured results of AR. **a.** The SEM image of the proposed metasurface sample. **b.-c.** The AFM image and corresponding vertical thickness results. **d.** Measured AR results in the far field. **e.** The real scene photo of AR dispersion. **f.** Measured structural colors in CIE 1931 plot and Adobe RGB gamut.

Considering the narrow linewidth and strong local electric field properties of SLR, the sample is suitable for high-sensitivity refractive index sensing. Figure 5a shows the measured spectrum results with different environments, including air, deionized water, Ethanol, and PDMS. The spectra own obvious Fano lineshape, which proves that there is coupling between LSPR and RA. The experimental sensitivity is 640.5 nm/RIU with a relative error of 2.54% compared with the simulated sensitivity (Figure 5b). The quality factor (Q-factor) is also increasing, and the

maximal Q-factor is 43.3. Generally, the Q-factors of most LSPR are around 10-20. In contrast, the SLR is an efficient method to improve Q-factor [14].

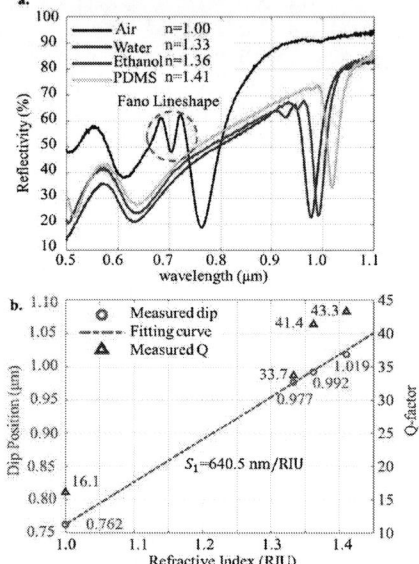

Figure 5: Measured Results for SLR. **a.** The reflective spectra of the metasurface sample immersed in liquids with different refractive indices. **b.** The measured SLR dip wavelength and Q-factor as an increasing function of the refractive index. And the refractive index sensitivity can reach 640.5 nm/RIU, which matches the simulation result.

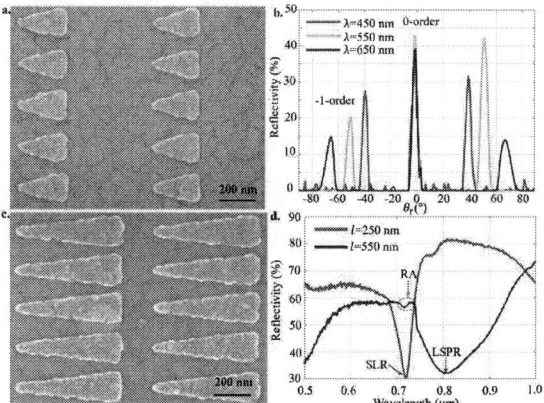

Figure 6: **a.-b.** The SEM of the sample with l=250 nm and measured AR reflectivity in the case of λ=450 nm, λ=550 nm, and λ=650 nm. **c.** The SEM of the sample with l=550 nm. **d.** Experimental reflection spectra of SLR for samples in **a.** and **c.**

It is worth mentioning that, when we make use of SLR to detect the refractive index, the medium layer covering the sample does effect the outgoing direction of AR. Although the medium layer with a larger refractive index than the air will reduce the AR angle θ_{r1}, additional refraction will occur at the interface between the medium and air to compensate the θ_{r1}, following the traditional Snell's Law:

$$n \cdot sin\,\theta_{r1} = 1 \cdot sin\,\theta_{r2} \qquad (3)$$

where θ_{r2} is the final outgoing angle, which is equal to the AR angle θ_{r0} without a medium layer. Figure 1c and Figure 1d present detailed theoretical derivations. This conclusion means both functions can work meanwhile.

The trapezoidal length l should be designed meticulously which decides the performance of three functions. If l is too small (Figure 6a, l =250 nm), the duty cycle of the top trapezoid metal will decrease, and part of the light will be reflected to 0-order and -1-order, resulting in a reduction in the energy of AR and SC (Figure 6b). However, the SLR dip will be sharper owing to the less Ohmic loss (Figure 6d, red line); If l is too large (Figure 6c, l =550 nm), the SLR can't be generated because of the mismatch between LSPR and RA. In that case, the spectrum mainly presents the LSPR features, and the RA strength is small (Figure 6d, blue line).

CONCLUSION

In summary, we propose a novel kind of multifunctional optical metasurface with MIM structure for achieving AR, SC, and SRL based on polarization multiplexing. The y-polarized beam excites GSP to increase the phase linearly and make the 400-650 nm visible band generate AR. The SCs from AR dispersion have high saturation and a large gamut of about 173.8% of Adobe RGB. The x-polarized beam excites SLR from the coupling between LSPR and RA, whose Q-factor can reach 43.3, higher than the Q-factors of resonance dip from most LSPR. the refraction angle compensation at the air-medium interface keeps the independence between RA and SLR. In application, we can use the above three functions separately to achieve light-splitting, display, and sensing, which promotes the practicality, miniaturization, and integration of optical elements.

ACKNOWLEDGEMENTS

This work is supported by the National Key R&D Program of China (Grant No. 2021YFB3200100), and the National Natural Science Foundation of China (Grant No. 61974004 and 61931018). The authors thank Jialei Zhu, Ling Li, and Zhenghong Li from Ideaoptics Inc. for their assistance in measurement.

REFERENCES

[1] F. Ding, Y. Q. Yang, R. A. Deshpande, and S. I. Bozhevolnyi, "A review of gap-surface plasmon metasurfaces: fundamentals and applications," *Nanophotonics*, vol. 7, pp. 1129-1156, 2018.

[2] N. F. Yu, and F. Capasso, "Flat optics with designer metasurfaces," *Nat. Mater.*, vol. 13, pp. 139-150, 2014.

[3] Y. Huang, J. Zhu, S. X. Jin, M. Z. Wu, X. Y. Chen, and W. G. Wu, "Polarization-controlled bifunctional metasurface for structural color printing and beam deflection," *Opt. Lett.*, vol. 45, pp. 1707-1710, 2020.

[4] L. G. Deng, J. Deng, Z. Q. Guan, J. Tao, Y. Chen, Y. Yang, D. X. Zhang, J. B. Tang, Z. Y. Li, Z. L. Li, S. H. Yu, G. X. Zheng, H. X. Xu, C. W. Qiu, and S. Zhang, "Malus-metasurface-assisted polarization multiplex-ing," *Light Sci. & Appl.*, vol. 9, 2020.

[5] M. Z. Liu, W. Q. Zhu, P. C. Huo, L. Feng, M. W. Song, C. Zhang, L. Chen, H. J. Lezec, Y. Q. Lu, A. Agrawal, and T. Xu, "Multifunctional metasurfaces enabled by simultaneous and independent control of phase and amplitude for orthogonal polarization states," *Light Sci. & Appl.*, vol. 10, 2021.

[6] C. Spagele, M. Tamagnone, D. Kazakov, M. Ossiander, M. Piccardo, and F. Capasso, "Multifunctional wide-angle optics and lasing based on supercell metasurfaces," *Nat. Commun.*, vol. 12, 2021.

[7] X. H. Yin, T. Steinle, L. L. Huang, T. Taubner, M. Wuttig, T. Zentgraf, and H. Giessen, "Beam switching and bifocal zoom lensing using active plasmonic metasurfaces," *Light Sci. & Appl.*, vol. 6, 2017.

[8] J. R. Fan, W. G. Wu, Z. J. Chen, J. Zhu, and J. Li, "Three-dimensional cavity nanoantennas with resonant-enhanced surface plasmons as dynamic color-tuning reflectors," *Nanoscale*, vol. 9, pp. 3416-3423, 2017.

[9] C. Chen, Y. Q. Wang, M. W. Jiang, J. Wang, J. Guan, B. S. Zhang, L. Wang, J. Lin, and P. Jin, "Parallel Polarization Illumination with a Multifocal Axicon Metalens for Improved Polarization Imaging," *Nano Lett.*, vol. 20, pp. 5428-5434, 2020.

[10] Q. Song, A. Baroni, P. C. Wu, S. Chenot, V. Brandli, S. Vézian, B. Damilano, P. D. Mierry, S. Khadir, P.Ferrand, andP. Genevet "Broadband decoupling of intensity and polarizationwith vectorial Fourier metasurfaces," *Nat. Commun.*, vol. 12, 3631, 2019.

[11] D. Khlopin, F. Laux, W. P. Wardley, J. Martin, G. A. Wurtz, J. Plain, N. Bonod, A. V. Zayats, W. Dickson, and D. Gerard, "Lattice modes and plasmonic linewidth engineering in gold and aluminum nanoparticle arrays," *J. Opt. Soc. Am.*, vol. 34, pp. 691-700, 2017.

[12] W. H. Yang, S. M. Xiao, Q. H. Song, Y. L. Liu, Y. K. Wu, S. Wang, J. Yu, J. C. Han, and D. P. Tsai, "All-dielectric metasurface for high-performance structural color," *Nat. Commun.*, vol. 11, 2020.

[13] M. Miyata, H. Hatada, and J. Takahara, "Full-Color Subwavelength Printing with Gap-Plasmonic Optical Antennas," *Nano Lett.*, vol. 16, pp. 3166-3172, 2016.

[14] V. G. Kravets, A. V. Kabashin, W. L. Barnes, and A. N. Grigorenko, "Plasmonic Surface Lattice Resonances: A Review of Properties and Applications," *Chem. Rev.*, vol. 118, pp. 5912-5951, 2018.

CONTACT

Wengang Wu,
tel: +80-010-62767553;
Email: wuwg@pku.edu.cn

NOVEL WAVEFRONT-SPLITTING INTERFEROMETER FOR ULTRA-COMPACT BROADBAND FT-IR SPECTROSCOPY EXTENDING TO VISIBLE RANGE

Bassem Mortada[1], Yasser M. Sabry[1, 2], Bassam Saadany[1], Tarik Bourouina[3] and Diaa Khalil[2]

[1]Si-Ware Systems, EGYPT

[2]Ain Shams University, EGYPT and

[3]Université Gustave Eiffel, ESIEE Paris, FRANCE

ABSTRACT

While MEMS interferometers reported so far rely on *amplitude*-splitting, this paper reports the first *wavefront*-splitting MEMS interferometer. The interferometer is ultra-compact and is part of a Fourier transform (FT) spectrometer fabricated using Deep Reactive Ion Etching (DRIE) of Silicon-on-Insulator (SOI). It enables the extension of the spectral range from Infrared to the visible range as validated from the experiments. The proposed interferometer architecture is based on spatial splitting and combination of optical beams using wavefront-splitting, where free-space propagation is only through air.

KEYWORDS

Interferometer, Wavefront splitting, Fourier transform and Spectrometer.

INTRODUCTION

Most of the reported micro-machined interferometers are based on *amplitude*-splitting of optical beams achieved by partial reflection and transmission at a dielectric interface [1]. In MEMS SOI-based interferometers, the beam splitter is usually an air-silicon interface, where light propagates through silicon [2, 3]. While silicon is translucent in most of the infrared range, it highly absorbs light at lower wavelengths (< 1100 nm wavelength), which prevents the operation in the visible and ultra-violate spectral ranges. Another challenge associated with propagation through silicon is chromatic dispersion; since silicon has a wavelength-dependent refractive index, it introduces wavelength-dependent phase errors [4]. A third challenge for conventional micro-machined interferometers is the need for selective metallization of micro-mirrors while keeping the beam splitter non-metalized [2]. A spatial-splitting interferometer based on Multi-Mode Interference MMI waveguides was reported to overcome these challenges [5], however the MMI waveguide is usually long, because its length is proportional to the square of the waveguide width, leading to a tradeoff between the interferometer compactness and optical etendue. The proposed interferometer structure herein is compact and overcomes these challenges opening the door for integration and fabrication of efficient wide spectral range MOEMS interferometers on a mass scale at a very low cost.

INTERFEROMETER ARCHITECTURE

The proposed interferometer (schematically shown in Fig. 1) is a Michelson-like interferometer based on a *wavefront* truncating beam splitter, which splits the input beam into a transmitted beam and a reflected beam. The two beams are directed towards two corner mirrors (retro-reflectors), one is moveable and one is fixed. An optical path difference between the two beams is introduced by a moving MEMS corner mirror. The split transmitted beam is reflected back from the moving retroreflector to the beam splitter, which directs it to a beam combiner. The split reflected beam is reflected back from the fixed retroreflector towards the beam combiner, which is implemented as a parabolic mirror that focuses the two output beams onto the output aperture.

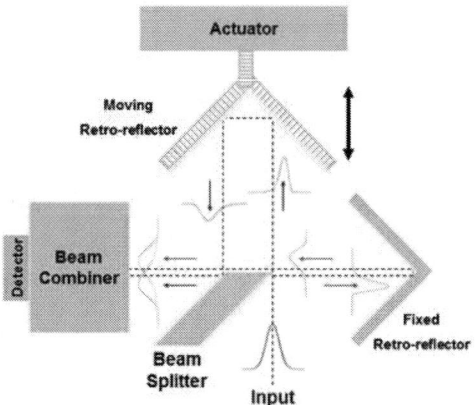

Figure 1: Wavefront splitting interferometer architecture showing splitting mechanism.

a) Wavefront spatial beam splitter

The spatial splitter divides the wavefront of the input incident beam into two skewed beams in the lateral direction, shifted from the optical axis with a spatial shift in-between, as shown in Fig. 2(a), where the split beam field distribution is simulated using plane wave expansion Fourier transform method for beam propagation [6]. For an incident beam field distribution $\psi_i(x, y) = \psi_{ix}(x) \psi_{iy}(y)$, for x-component of the filed, the angular spectrum of the transmitted beam $B(v_x)$ is given by

$$B(v_x) = \int_{-\infty}^{+\infty} \int_{-\infty}^{+\infty} T(x) \psi_{ix} e^{j2\pi v_x x} dx dy \qquad (1)$$

where $T(x) = 1$ for $x > 0$, $T(x) = 0$ for $x \leq 0$ and v_x is the lateral wavenumber. The field distribution after propagation of an optical path length d at wavenumber v, is calculated by applying the paraxial approximation and Fourier transform as,

$$\psi_x(x, y) = \int_{-\infty}^{+\infty} B(v_x) e^{-j\pi v_0 \left(1 - \left(\frac{v_x}{v}\right)^2\right) d} e^{-j2\pi v_x x} dv_x \qquad (2)$$

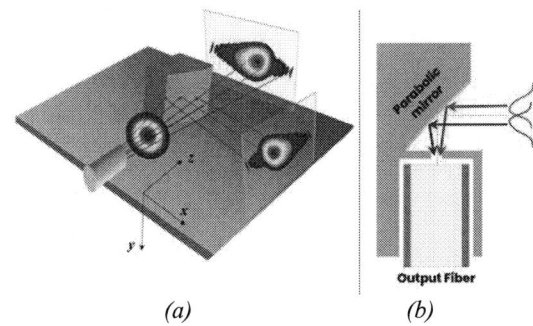

Figure 2: Beam splitting and combining: (a) spatial splitting beam profile. (b) Beam combining by a parabolic mirror, aperture and a multimode fiber.

b) Wavefront beam Combiner

The parabolic reflector focuses the two shifted beams onto the output aperture, where the interfering output beam is collected by an output multimode optical fiber such that the fiber facet is at the focus of the parabolic mirror, as shown in Fig. 2(b). The multimode fiber delivers the interfering beam to a photodetector.

c) Actuator

The light passing the interferometer is modulated by varying the optical path difference (OPD) between the two interferometer arms, where part of the light is directed by a beam splitter to a moveable retroreflector, while the other part is directed to a fixed retroreflector. The output interference pattern power P_o of the two arms beams versus OPD at wavenumber $v = 1/\lambda$ is given by [7]:

$$P_o(OPD, v) = \eta[1 + V\cos(2\pi v\ OPD)]P_i(v) \quad (3)$$

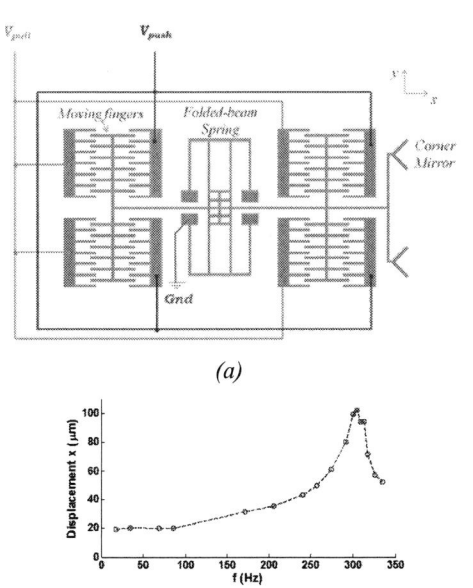

(a)

(b)

Figure 3: (a) Comb drive actuator driving a retroreflector (a corner mirror). (b) Measured frequency response of comb drive second order system.

where P_i is the power spectral density of the incident light from the input fiber and coupled to the output fiber, η is the coupling efficiency and V is the visibility of interference, which is a function of the lateral/angular shift between the two output beams and the beam width.

A double-sided electrostatic comb-drive actuator was designed to work stably in push-pull operation over a travel range of 130 μm (Fig. 3(a)). At the center of the actuator, a double-folded flexure is used to reduce the parasitic torque that is usually encountered due to fabrication tolerance [8]. The OPD between the interferometer arms is varied by attaching one of the retroreflectors to the actuator, such that the OPD is twice the displacement of the MEMS actuator. Measured displacement frequency response is shown in Fig. 3(b), where the resonance frequency of the actuator second order mass-spring system is around 300 Hz.

FABRICATION

A monolithically-integrated SOI optical bench of 80 μm device height, was designed and fabricated using deep reactive ion etching of silicon, with full metallization of the interferometer structure. An SEM photo of the fabricated interferometer structure with ray tracing is shown in Fig. 4, where the interferometer size is smaller than 0.5 mm × 0.5 mm. Indeed, high attention was given during the etching process to the quality of the optical surfaces, which highly impacts the interferometer efficiency. Multimode optical fibers are inserted into U-grooves, designed for passive alignment of the fibers with the interferometer mirrors, such that the light is in-plane coupled throughput the interferometer.

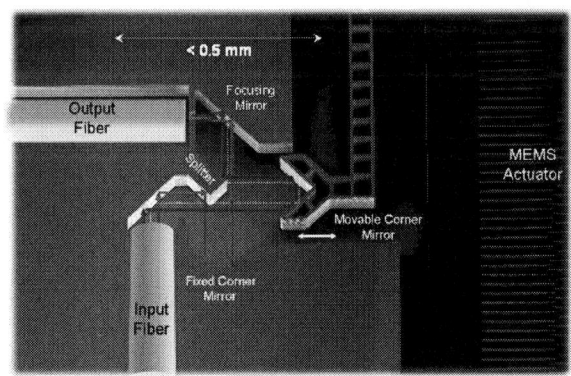

Figure 4: SEM photo of the interferometer showing the ray path coupled from the input fiber to the output fiber.

CHARACTERIZATION

Measurements carried out on the interferometer was focused on its operation as a core of a Fourier transform spectrometer operating both in Infrared and visible ranges. The testing setup used in the measurements, shown in Fig. 5, consists of two optical positioners for fiber injection inside the MEMS chip, an NIR tunable laser source working in C-band, a 635 nm red laser source, InGaAs photo-detector (for NIR range), silicon photo-detector (for visible range), 50 μm-core input multi-mode fiber, 105 μm-core output multimode fiber and an electronic control board with actuation, capacitive-sensing and photo-detection interface circuitry. An actuation voltage signal drives the actuator and consequently the movable mirror of the interferometer. Optical power of the output interfering beam coming out from the output fiber is measured by the photo-detector in the form of photo-current. An interferogram signal is measured by recording the optical path difference and the detector current simultaneously, as shown in Fig. 6(a). Capacitive sensing of MEMS actuator motion is employed to measure the OPD between the two interferometer arms. With such technique, extra laser source and reference interferometer are no longer needed [9]. Laser is only used to calibrate the capacitive sensing technique just one time and then OPD is always extracted from a stored pre-calibrated capacitance signal and the capacitance to OPD relation, shown in Fig. 6(b), which is measured using 1550 nm infrared laser. Fourier transform is then applied to the interferogram signal to calculate the spectrum of the detected light.

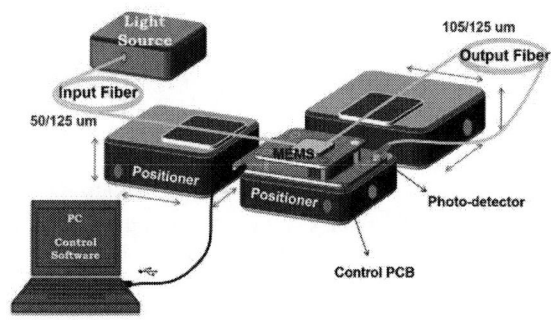

Figure 5: MEMS chip testing setup.

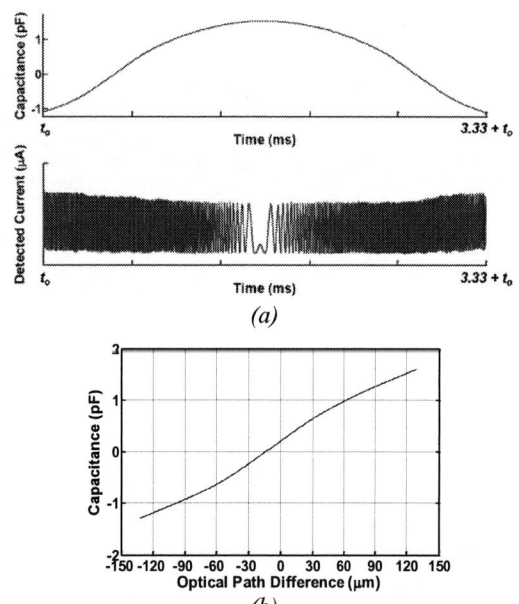

Figure 6: (a) Simultaneous measurement of capacitance and detector current signals (b) Measured capacitance to OPD relationship.

The novel wavefront-splitting interferometer configuration is validated and tested using the red laser at 635 nm and IR laser at 1480 nm in addition to the 1550 nm infrared laser used in capacitance to OPD calibration. The measured interferogram for red laser and the extracted spectra by Fast Fourier Transform FFT for both red laser and IR laser are shown in Fig. 7, indicating accurate peak positions with wavelength errors below 0.5 nm. A 2.15 nm Full Width at Half Maximum FWHM was measured at the 635 nm red laser line.

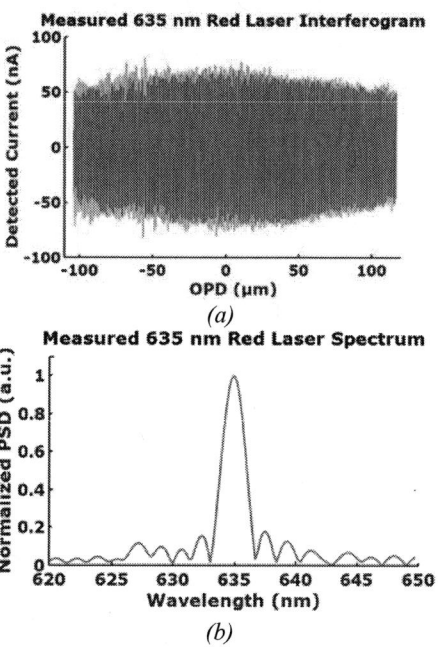

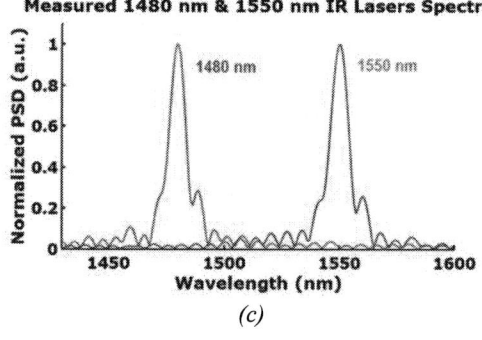

Figure 7: Interferometer measurements for (a) 635 nm visible laser interferogram and extracted spectrum, (b) IR laser wavelengths 1480 nm and 1550 nm.

CONCLUSION

Wideband optical MEMS interferometer based on spatial wavefront splitting onchip was presented. The light is coupled from the input fiber to the output fiber in-plane with respect to SOI substrate of the monolithic MEMS interferometer. Operating as a Fourier transform spectrometer, a resolution of 2.15 nm in the visible range was demonstrated. The presented interferometer opens the door for compact miniaturized and wider spectral range spectrometers extending from visible to mid-infrared range.

REFERENCES

[1] Y. M. Sabry, D. Khalil, and T. Bourouina, "Monolithic silicon-micromachined free-space optical interferometers onchip," Laser Photon Rev., vol. 9, no .1, pp. 1-24, 2015.

[2] B. Saadany, H. Omran, M. Medhat, F. Marty, D. Khalil, and T. Bourouina, "MEMS Tunable Michelson Interferometer with Robust Beam Splitting Architecture," In Optical MEMS and Nanophotonics, USA, 2009, pp. 49-50.

[3] H. Omran, M. Medhat, B. Mortada, B. Saadany and D. Khalil, "Fully Integrated Mach-Zhender MEMS Interferometer With Two Complementary Outputs," in IEEE Journal of Quantum Electronics, vol. 48, no. 2, pp. 244-251, Feb. 2012, doi: 10.1109/JQE.2011.2170825.

[4] T. A. Al-Saeed, and D. Khalil, "Dispersion compensation in moving optical wedge Fourier Transform spectrometer," Appl. Opts., vol. 48, no. 20, pp.3979-3987, July, 2009.

[5] B. Mortada, M. Erfan, M. Medhat, Y. M. Sabry, B. Saadany and D. Khalil, "Wideband Optical MEMS Interferometer Enabled by Multimode Interference Waveguides," in Journal of Lightwave Technology, vol. 34, no. 9, pp. 2145-2151, 1 May1, 2016, doi: 10.1109/JLT.2016.2531642.

[6] Bahaa E. A. Saleh, Malven Carl Teich, Fundamentals of Photonics, John Wiley & Sons, Inc., 2nd edition 2007.

[7] B. Mortada et al., "Ultra-Compact Fourier Transform Near-Infrared MEMS Spectral Sensor for Smart Industry and IoT," in IEEE Journal of Selected Topics in Quantum Electronics, vol. 27, no. 6, pp. 1-9, Nov.-Dec. 2021, Art no. 2700109, doi: 10.1109/JSTQE.2021.3091375.

[8] M. Medhat, Y. Nada, B. Mortada, and B. Saadany, "Long range travel MEMS actuator," US patent 8497619, issued 2013-07-30.

[9] B. Saadany, A. Hafez, M. Medhat, H. Haddara. "Technique to Determine Mirror Position in Optical Interferometers". US patent 20110222067. Sep., 15, 2011.

CONTACT

*B. Mortada, tel: +20 22 266 7344; bassem.mortada@si-ware.com

PIEZOELECTRICALLY ACTUATED MICROMIRROR
WITH DYNAMIC DEFORMATION COMPENSATION MECHANISM

Takashi Sasaki, Adrien Piot, Jaka Pribošek, Anton Lagosh, Clement Fleury, Markus Bainschab, Yanfen Zhai, Marcus Baumgart, Sara Guerreiro, Dominik Holzmann, Aleš Travnik, Mohssen Moridi

Silicon Austria Labs, Austria

ABSTRACT

This paper reports a piezoelectrically actuated micromirror with an active control of the dynamic deformation. The proposed approach allows a reduction of dynamic deformation without increasing the moment of inertia, for the first time in field. A 75 % dynamic deformation reduction at the center of mirror was calculated and the dynamic mirror shape correction was experimentally confirmed on devices fabricated.

KEYWORDS

Dynamic deformation, silicon on insulator wafer, piezoelectric actuator, compensation.

INTRODUCTION

Dynamic deformation is the inherent problem in high-performance micromirrors [1-8]. To keep the laser spot diameter reflected from the vibrating micromirror small, the mirror surface should be kept flat even in high-frequency operation. One way of reducing the dynamic deformation is by increasing the thickness of the mirror. To keep the device layer thickness low, the preferred approach is to introduce the backside reinforcement. Such structure is not only more difficult to fabricate but also increases the moment of inertia and will result in the high stress generation, that can exceed the yield stress of the material and lead to failure at small rotation angle [1]. Similar approaches have been actively studied for more than twenty years. For example, Nee et al. reported a drum-like mirror structure using tensile stress [3]. Ji et al. reported a mirror with a diamond-shaped backside reinforcement [4]. To push the performance further, a novel mechanism that mitigates this limitation is needed. In this article, for the first time in this field, a new approach based on active mirror compensation using a piezoelectric actuator will be shown. The approach allows to design a mirror with small dynamic deformation without increasing the moment of inertia.

PRINCIPLE AND DESIGN

Figure 1 shows the schematic of oscillating micromirror with a piezoelectric dynamic deformation compensator. The phase of rotation angle and voltage of compensator is operated in phase so that the mirror shape becomes flat. The compensators are placed on the left and right sides on the mirror and the 180° phase delayed voltage is applied to them, respectively. As thickness of the piezo layer is generally only a fraction of thickness of device layer, it can be considered negligible. As such, this mechanism fundamentally does not increase the moment of inertia. Figure 2 shows the designed micromirror with piezoelectric dynamic mirror shape correction. The device is fabricated from the Au/PZT/Pt/SiO$_2$ thin film on SOI wafer. The mirror with piezoelectric compensation mechanism is integrated in the two-dimensional piezoelectric micromirror platform. In this experiment, the rotation and compensation actuators are used. The width of the compensation electrode varies as a cubic function since the generated moment due to the dynamic deformation is changes as a cubic function [16-17]. The mirror size is 1.4 mm square, and thickness is 50 μm.

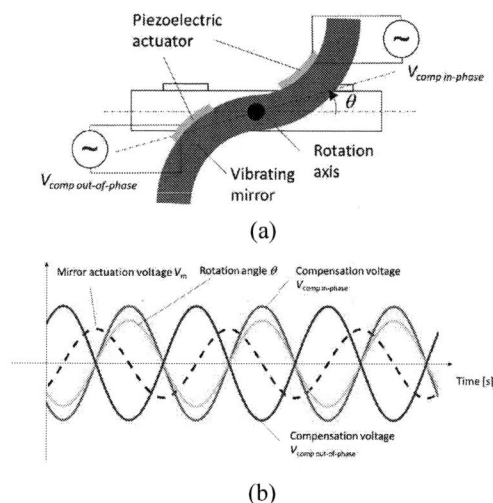

(a)

(b)

Figure 1: (a) Schematic of mirror with piezoelectric shape compensation mechanism (b) Rotation angle of mirror and applied voltage waveforms

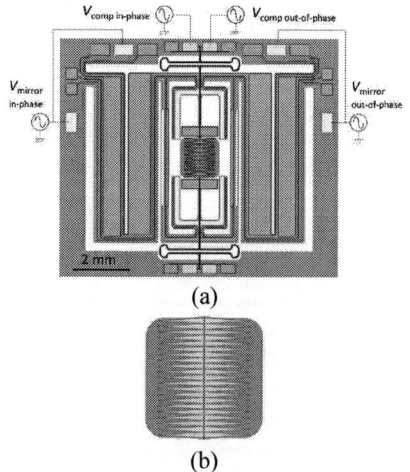

(a)

(b)

Figure 2: (a)Device design with piezoelectrically actuated micromirror with dynamic mirror shape correction. (b) Close up view of compensation electrode.

Figure 3 shows the simulated results of dynamic deformation. Without the compensation voltage applied, the mirror deforms by 22 nm, measured at 27 opt. deg. in the center of mirror at 12.6 kHz. The deformation is further reduced to 5 nm when compensation voltage is applied. The FWHM beam sizes calculated using the obtained mirror deformation in x-axis and y-axis are reduced by 6 % and 17 % due to compensation as shown in Figure 4.

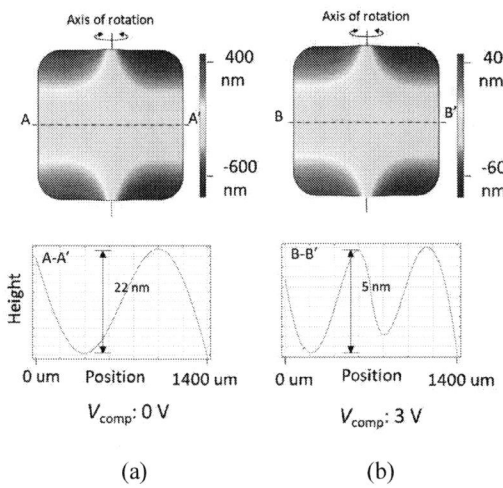

(a)   (b)

Figure 3: Simulated dynamic deformation of mirror at the maximum angle of mirror. (a) with, and (b) without compensation voltage.

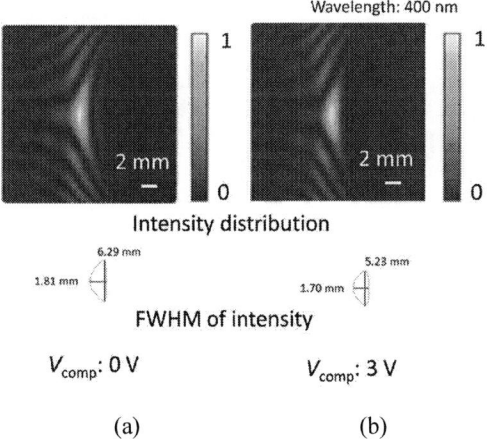

(a)   (b)

Figure 4: Simulated laser intensity distribution at 5 m far from mirror at maximum rotation angle based on diffraction theory(a) with, and (b) without compensation voltage.

FABRICATION

The device is fabricated from the Au/PZT/Pt/SiO$_2$ thin film on a 6-inch SOI wafer. The process flow consists of 5 lithography steps shown in Figure 5. Dry etch techniques based on ICP-RIE are used for patterning of the PZT stack, SiO$_2$ and Si device layer as well as for the backside release step. These techniques allowed us to resolve all salient features without any residues on the surfaces.

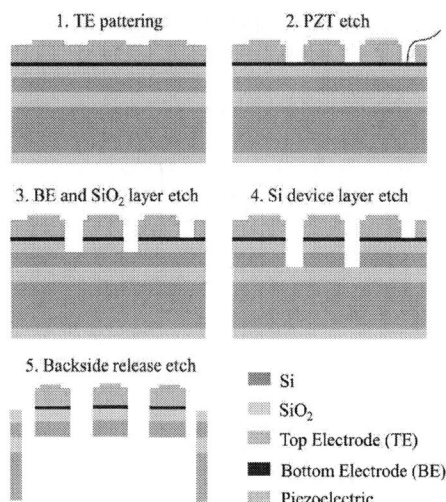

Figure 5: Step-by-step microfabrication technology for piezoelectrically actuated micromirrors.

RESULTS AND DISCUSSION

Figure 5 shows optical micrograph of fabricated micromirror. Figure 6 shows the static deformation measured from opposite side of compensation electrode. Digital holographic microscopy was used to measure the deformation [7]. There is a static spherical deformation about 300 nm in peak-to-peak due to film stress, which can be controlled by the voltage bias to the piezoelectric actuator or residual stress. The mirror rotation of about 24 opt. deg. was obtained at 9.4 Vpp with 4.7 V bias. Under this condition, mirror profiles at maximum angle were obtained. The measured results with the subtraction of the spherical components are shown in Figure 7.

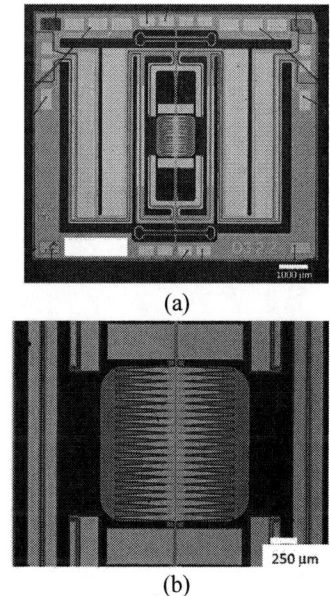

(a)

(b)

Figure 5: (a) Optical micrograph of fabricated micromirror. (b) Close up view of compensation electrode.

The symmetric shape with respect to the rotation axis was obtained. To see the amount of dynamic deformation from measured results, the coefficient of fitted 3rd order function is used [1], which is not affected by the initial deformation of the mirror, depending on the compensation voltage amplitude. The coefficient changed from negative to positive according to the applied compensation voltage. This shows that the fabricated device has compensation ability for dynamic deformation.

CONCLUSION

We report active control of dynamic deformation of MEMS mirrors, using structured piezoelectric actuators. Using the proposed approach, we report 75 % reduction of the dynamic deformation at the center of the mirror in simulation. The ability of the dynamic deformation compensation is also shown experimentally. The mechanism does not need to increase the moment of inertia fundamentally and has the potential to increase micromirror performance further.

ACKNOWLEDGEMENT

This work has been jointly supported by Evatec, EV Group, TDK Electronics, ZKW and by Silicon Austria Labs (SAL), owned by the Republic of Austria, the Styrian Business Promotion Agency (SFG), the federal state of Carinthia, the Upper Austrian Research (UAR), and the Austrian Association for the Electric and Electronics Industry (FEEI).

REFERENCES

[1] H. Urey, D. Wine, and J. Lewis, "Scanner design and resolution tradeoffs for miniature scanning displays," Proceedings of SPIE, vol. 3636, pp. 60–6860, Apr. 1999.

[2] T. Sasaki and K. Hane, "A confocal laser scanning endoscope using a varifocal scanning mirror," in 2013 Transducers & Eurosensors XXVII: The 17th International Conference on Solid-State Sensors, Actuators and Microsystems (TRANSDUCERS & EUROSENSORS XXVII), 2013, pp. 1412–1415.

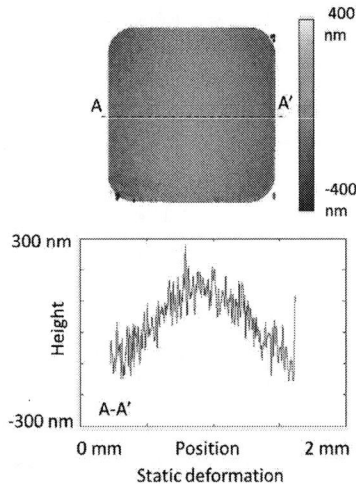

Figure 6: Surface profile in static condition

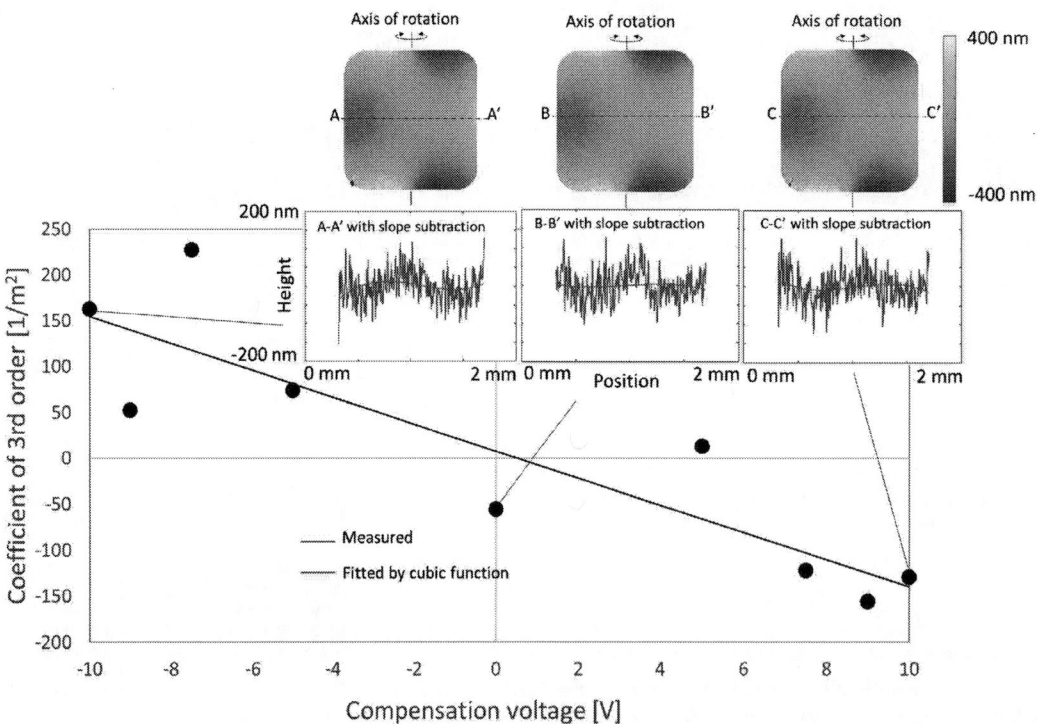

Figure 7: Compensation voltage as function of coefficient of fitted 3rd order function with inset of dynamic deformation profile and cross-sectional profile

[3] K. Nakazawa, T. Sasaki, H. Furuta, J. Kamiya, T. Kamiya, and K. Hane, "Varifocal scanner using wafer bonding," Journal of Microelectromechanical Systems, vol. 26, no. 2, pp. 440–447, 2017.

[4] A. Piot et al., "Resonant PZT MEMS mirror with segmented electrodes," in 2020 IEEE 33rd International Conference on Micro Electro Mechanical Systems (MEMS), 2020, pp. 517–520.

[5] A. Piot, J. Pribošek, and M. Moridi, "Dual-axis resonant scanning mems mirror with pulsed-laser-deposited barium-doped PZT," in 2021 IEEE 34th International Conference on Micro Electro Mechanical Systems (MEMS), 2021, pp. 89–92.

[6] A. Piot et al., "Optimization of Resonant PZT MEMS Mirrors by Inverse Design and Electrode Segmentation," Journal of Microelectromechanical Systems, vol. 30, no. 2, pp. 216–223, Apr. 2021.

[7] P. Thakkar et al., "Measuring angle-resolved dynamic deformation of micromirrors with digital stroboscopic holography," in Optics and Photonics for Advanced Dimensional Metrology II, 2022, vol. 12137, pp. 112–119.

[8] K. Mihara, K. Hanatani, T. Ishida, K. Komaki, and R. Takayama, "High Driving Frequency (>54 kHz) and Wide Scanning Angle (>100 Degrees) MEMS Mirror Applying Secondary Resonance For 2K Resolution AR/MR Glasses," in 2022 IEEE 35th International Conference on Micro Electro Mechanical Systems Conference (MEMS), Jan. 2022, pp. 477–482.

[9] J. T. Nee, R. A. Conant, M. R. Hart, R. S. Muller, and K. Y. Lau, "Stretched-film micromirrors for improved optical flatness," in Proceedings IEEE Thirteenth Annual International Conference on Micro Electro Mechanical Systems (Cat. No.00CH36308), Jan. 2000, pp. 704–709.

[10] M. Sasaki, T. Sasaki, K. Hane, and H. Miura, "An optically flat micromirror using a stretched membrane with crystallization-induced stress," Journal of Optics A: Pure and Applied Optics, vol. 10, no. 4, p. 044004, 2008.

[11] T. Sasaki and K. Hane, "Stress control of silicon membrane for optical scanner with deformable mirror," IEEJ Transactions on Sensors and Micromachines, vol. 129, no. 5, pp. 138–141, 2009.

[12] T. Sasaki and K. Hane, "Deformation of varifocal mirror with narrow frame by SOI wafer residual stress," Electronics and Communications in Japan, vol. 95, no. 8, pp. 26–33, 2012.

[13] T. Sasaki and K. Hane, "Varifocal micromirror integrated with comb-drive scanner on silicon-on-insulator wafer," Journal of microelectromechanical systems, vol. 21, no. 4, pp. 971–980, 2012.

[14] C.-H. Ji et al., "An electrostatic scanning micromirror with diaphragm mirror plate and diamond-shaped reinforcement frame," Journal of Micromechanics and Microengineering, vol. 16, p. 1033, Apr. 2006.

[15] K. Ikegami, T. Koyama, T. Saito, Y. Yasuda, and H. Toshiyoshi, "A biaxial piezoelectric MEMS scanning mirror and its application to pico-projectors," in 2014 International Conference on Optical MEMS and Nanophotonics, Aug. 2014, pp. 95–96.

[16] S. D. Senturia, Microsystem Design. Springer US, 2005.

[17] R. Farrugia, I. Grech, D. Camilleri, O. Casha, E. Gatt, and J. Micallef, "Theoretical and finite element analysis of dynamic deformation in resonating micromirrors," Microsyst Technol, vol. 24, no. 1, pp. 445–455, Jan. 2018, doi: 10.1007/s00542-017-3335-7.

CONTACT
*T. Sasaki, tel: +43-664-88843732; takashi.sasaki@silicon-austria.com

RESONANT d_{33} MODE PZT MEMS MIRROR EXCITED WITH DIRECTIONAL INTERDIGITATED ELECTRODES

Pooja Thakkar[*]*, Anton Lagosh, Takashi Sasaki, Markus Bainschab and Jaka Pribošek*
Silicon Austria Labs GmbH, AUSTRIA

ABSTRACT

This work reports a resonant piezoelectric mirror driven by rotated directional interdigitated electrodes exploiting the d_{33} mode. This work expands the prior work on directional IDEs where only small angle, quasistatic devices were reported. We, for the first time combine this actuation principle with mechanical amplification and demonstrate that it can be used to excite high-speed and high amplitude scanning mirrors. A torsional twist of 0.66°, measured at the end of the actuators was mechanically amplified by over 30x to achieve an optical scan angle (OSA) of 20° at 29.7 kHz, with a Q-factor of 2481. This work marks the highest OSA achieved with directional interdigitated electrodes so far.

KEYWORDS

Interdigitated electrodes, mechanical amplification, 1D PZT MEMS micromirror, d_{33} mode actuation

INTRODUCTION

Until now, piezoelectric d_{33} mode has been almost exclusively applied to surface acoustic waves (SAW) sensors, energy harvesting applications and active fiber composites, due to large produced displacements [1]. Recently, directional d_{33} piezoelectric actuators have emerged, that allow control of the in-plane stress distribution which can be leveraged to directly generate torsion. So far, these actuators have been applied to quasistatic applications [2]. Only one report can be found, where the d_{33} mode was used to induce torsion at resonance on micromirror, [3] reporting a torsional twist of 1.5° and a Q-factor of 27. Other work employs d_{33} mode for bending actuation or sensing. Table 1 provides the summary of the prior work using d_{33} for torsion and bending actuation. In the present paper, the directional electrodes are designed to induce in-plane stress distribution tuned to excite a torsional mode of the structure. For the first time, the directional IDE are applied to thin film PZT on silicon substrate combined with mechanical amplification to provide high-amplitude-high-frequency operation.

DESIGN AND FABRICATION

The micromirror was fabricated from an Au-PZT-Pt-SOI stack with 175 µm Si device layer thickness having 200 nm Au, 1.75 µm PZT, 100 nm Pt using the custom 6-litography step process reported in [4]. Fig. 1(a) shows the fabricated device. The optical part is suspended by two torsional springs, attached to the cantilevers. IDEs with fingers with 3.5 µm gap, rotated at 45° shown in Fig. 1(b) actuate the cantilevers (Fig. 1(b). Fig.1(c) shows classical 90° IDEs for comparison. When the directional fingers are excited, they induce directional in-plane stress and as a result a cantilever twists. The interdigitated actuators are excited with signals, offset by $+V_{DC}$ and $-V_{DC}$ respectively, as 4 cycle. The mirror design leverages mechanical amplification, whereby the system transfers the mechanical energy from large inertia actuators into the low inertia mirror through soft torsional springs. As a result, a small twist in the cantilevers induces a large deflection of the micromirror, shown in Fig 1. (e).

SIMULATIONS

The work from [5] has shown that strain ratio in the plane of strained membrane can be modelled independently of layer structure or material properties. By optimizing the width of electrodes, the gap between the electrodes and their placement on the on the membrane a highly directional strain profile can be engineered. In principle, transverse electric fields produce a spherical displacement profile and in-plane electric fields produce saddle shaped profiles. Considering a piecewise model, where the electric field components in *x, y, z* are E_x, E_y, E_z and assuming the approximation of $d_{31} = -\frac{1}{2}d_{33}$, the mean strain over a small region is given by [5],

$$s_{xx} \sim d_{33}\left(\langle\frac{E_x{}^2}{|E|}\rangle - \frac{1}{2}\left(\langle\frac{E_y{}^2}{|E|}\rangle + \langle\frac{E_z{}^2}{|E|}\rangle\right)\right) \qquad (1)$$

Table 1: Previous work employing d_{33} mode for torsion and bending actuation

Ref.	Applied Mode	Frequency (Hz)	Actuation - Bending(B)/Torsion(T)	PZT thickness (µm)	Bottom electrode	IDE	Tilt angle
[6]	d_{31}, d_{33}	2775	B	1.2	Yes	90°	-
[7]	d_{33}	528	B	1	No	90°	-
[5]	d_{31}, d_{33}	-	B+T	120	Yes	custom	-
[8]	d_{33}	13.9k, **21.9k**, 48.5k	B, **T**, B	0.48	No	90°	-
[2]	d_{31}, d_{33}	0	B, T	127	No	90°,45°	-
[3]	d_{33}	**6.1k**	**T**	127	Yes	45°	**1.5°**
This work	$(d_{31}), d_{33}$	**29.7k**	**T**	1.75	Yes	45°	**20°**

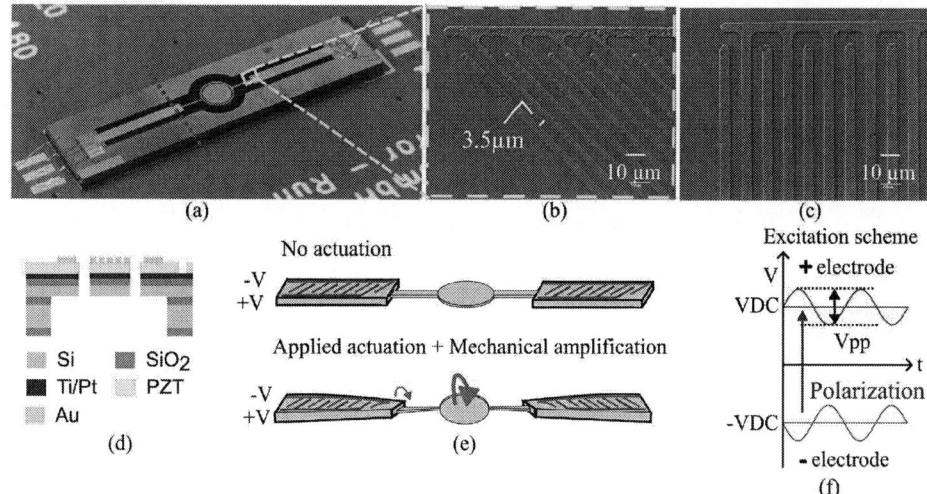

Figure 1: (a) Fabricated micromirrors having 45° directional IDEs on cantilevers for actuation promoting torsion mode, (b), (c) magnified optical SEM image cantilever with IDEs at 45° and 90° respectively. (d) fabricated cross section of micromirrors. (e) actuation mechanism of cantilevers with 45° IDEs showing twisting of edge and mechanical amplification achieved through springs (f) excitation scheme to maintain the polarization in PZT layer.

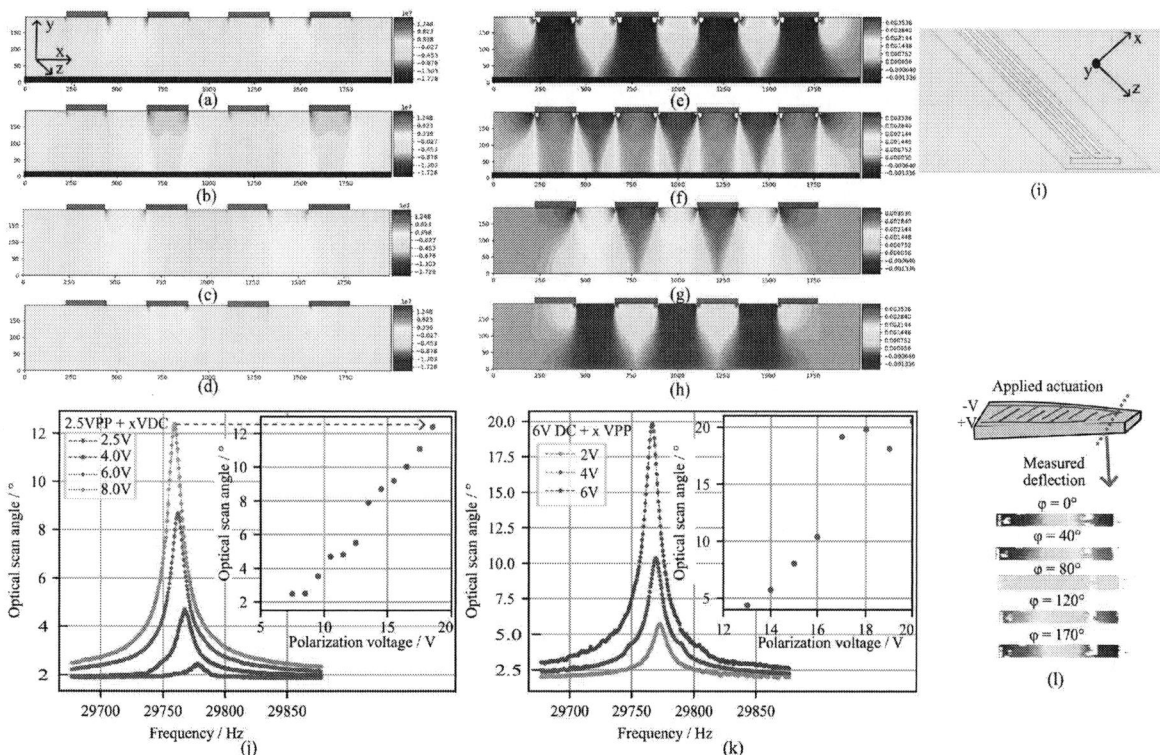

Figure 2: 16 V_{pp} applied on the electrodes, results in (a), (b) Electric field distribution in x-y plane observed at the cross section at 45° of the IDEs -(a) In-plane electric field distribution E_x (b) and transverse filed distribution E_y for PZT with passive bottom electrode, (c) E_x (d) E_y for PZT layer without bottom electrode. (e)-(h)-strain maps s_{xx}, s_{yy} with respect to the electric fields (a)-(d). (i) top view electric field distribution of the IDEs. (j) Frequency response of the micromirror at different DC bias applied to electrode pair $(+V_{DC}, -V_{DC})$, the optical scan angle represents peak-to-peak values. Inset-maximum optical scan angle achieved for different polarization conditions extracted from the frequency response. The jump in response at 5.5 V_{DC} might come from a nonlinear rise in strain due to increased polarization, followed by linear behavior. (k) Frequency response of micromirror at different VPP. Inset represents maximum optical scan angle achieved for different polarization condition extracted from the frequency response. (l) measured topography maps of deflection of cantilevers using DHM for complete harmonic cycle with phase steps of 10°, where maps are normalized (only 5 dissected phase maps shown)[9]. The mirror was operated at 6 V_{DC}, 4 V_{PP} at the resonant frequency of 29770 Hz.

978-1-6654-9309-3/23 $31.00 © 2023 IEEE

$$s_{yy} \sim d_{33} \left(\langle \frac{E_y^{\,2}}{|E|} \rangle - \frac{1}{2} \left(\langle \frac{E_x^{\,2}}{|E|} \rangle + \langle \frac{E_z^{\,2}}{|E|} \rangle \right) \right) \qquad (2)$$

where s_{xx}, s_{yy} are strain in x and y direction. In case, the polarizing field is not in the same direction as the actuation field, the projection of the field on the remanent polarization should be considered. Here we use d_{33} = 5.93e-10 m/V [10] for our computation.

To extract the electric field map, we performed a finite element method (FEM) analysis using the open-source software FEMM. The model consists of two pairs of electrodes at 45° IDE, excited at 16 V_{PP} Two types of electrode pair were used for the study: with and without bottom electrode as in Fig 2. (a). (c). In the first configuration, one gets a higher transverse electric field as compared to an in-plane electric field (E_x = 0.6M V/m and E_y = 2.5M V/m). This is due to the presence of the bottom electrode and rather low thickness of PZT as compared to the gap between the electrodes, as shown in Fig 2. (a), (b). Here, the electric field in the PZT layer is alternately dominated by E_x and E_y, resulting in change of ratio from -1 to 1 along the cantilever. Moreover, the average E_y is 4.1 times larger than E_x, indicating that a mixture of both d_{31} and d_{33} modes is responsible for the actuation of the device. Fig. 2 (e)-(h) shows the strain maps s_{xx}, s_{yy} with respect to two cases, where they are closely related to electric field distribution. In the second configuration without bottom electrode, higher in-plane electric field is observed, with E_x = 1.3M V/m and E_y = 0.3M V/m, which can potentially improve the performance of the device by using pure d_{31} mode. In the present report however, only the device with bottom electrodes were fabricated. The electric field distribution on the top surface of cantilever (Fig 2. (i)) shows strongly directional polarization with s_{zz}/s_{xx} ratio of 0, with respect to given co-ordinate system, This imposes an elliptic corrugation [5] oriented with the direction of the fingers, leading to torsional twist at the end of cantilevers.

EXPERIMENTAL RESULTS

To operate the mirror at optimized polarization to get maximum deflection, the DC sweep (5.5V_{DC} indicates +5.5V_{DC}, -5.5V_{DC} are applied to electrode pairs respectively), is performed at fixed modulation voltage of 2.5 V_{PP} (as in Fig. 2 (j)). The effect of a modulated polarization on the response is analyzed by measuring the frequency response of the mirror (as in Fig. 2(k)).

Maximum optical scan angles in excess of 20° were obtained with 6V_{DC}, 6V_{PP}. We report a quality factor of 2481 in air at ambient pressure for the device presented. We note that the applied voltage difference between the fingers is not yet reaching the breakdown limit. For 3.5 μm inter-distance, the breakdown is estimated to occur at of 175 V, which offers some room for further increase of the optical scan angle [11]. In addition to the directional fingers, presented in the study, another device with 90° interdigitated fingers was fabricated (not shown here) and its excitation was found to be highly selective to the bending mode only. In contrast, with directional torsional actuators oriented at 45° it was not possible to actuate the bending mode, which demonstrates the large modal selectivity of this principle.

In addition, a digital holographic microscope was used to observe the twist at the edge of the cantilever (Fig. 2(l)). The topography maps show a twist angle of 0.17° when operated at 6 V_{DC}, 4 V_{PP} at resonance (29770 Hz). At the same driving conditions, the measured OSA of the micromirror is 10.5° (Fig.2 (k) red curve), indicating a mechanical amplification factor of 30.1.

CONCLUSION

The presented actuation principles allow to design devices with slim form factor when compared to other devices exploiting d_{31} actuation, which is particularly well suited for application such as augmented reality. Further, the possibility to engineer the strain distribution in the preferred direction can be leveraged and tuned to specific requirements in both actuation and sensing, opening many other potential applications. The presented device achieves a resonant frequency of 29.7 kHz, a Q factor of 2481 and an optical scan angle of 20° for 1D torsional operation. When compared to state-of-the art devices, using direct torsion actuation [1,2], a 100-fold increase in quality factor, a 10-fold increase in optical scan angle and a 5-fold increase in frequency operation is observed. This work can be expanded to include 2D actuation of micromirrors for combined bending and torsion mode, where the bottom electrodes can also be patterned for improved performance.

ACKNOWLEDGEMENTS

Authors would like to thank PIEMACS Sàrl, Switzerland for the fabrication of piezoelectric MEMS micromirrors.

REFERENCES

[1] S. Priya *et al.*, "A Review on Piezoelectric Energy Harvesting: Materials, Methods, and Circuits," *Energy Harvesting and Systems*, vol. 4, no. 1, pp. 3–39, Jan. 2017, doi: 10.1515/ehs-2016-0028.

[2] I. (Hotzen) Grinberg, N. Maccabi, A. Kassie, S. Shmulevich, and D. Elata, "Direct Torsion of Bulk PZT Using Directional Interdigitated Electrodes," *Procedia Engineering*, vol. 168, pp. 1483–1487, 2016, doi: 10.1016/j.proeng.2016.11.429.

[3] I. Grinberg, N. Maccabi, and D. Elata, "A pure-twisting piezoelectric actuator for tilting micromirror applications," in *2017 19th International Conference on Solid-State Sensors, Actuators and Microsystems (TRANSDUCERS)*, Kaohsiung, Taiwan, Jun. 2017, pp. 2035–2038. doi: 10.1109/TRANSDUCERS.2017.7994472.

[4] J. Pribošek, A. Lagosh, P. Thakkar, T. Sasaki, and M. Bainschab, "Resonant piezoelectric varifocal mirror with on-chip integrated diffractive optics for increased frequency response," presented at the IEEE MEMS 2023, Munich, Feb. 2023, p. Accepted.

[5] M. C. Wapler, M. Stürmer, J. Brunne, and U. Wallrabe, "Piezo films with adjustable anisotropic strain for bending actuators with tunable bending profiles," *Smart Mater. Struct.*, vol. 23, no. 5, p.

055006, May 2014, doi: 10.1088/0964-1726/23/5/055006.

[6] D. Kim, N. N. Hewa-Kasakarage, and N. A. Hall, "A theoretical and experimental comparison of 3-3 and 3-1 mode piezoelectric microelectromechanical systems (MEMS)," *Sensors and Actuators A: Physical*, vol. 219, pp. 112–122, Nov. 2014, doi: 10.1016/j.sna.2014.08.006.

[7] J. C. Park, J. Y. Park, and Y.-P. Lee, "Modeling and Characterization of Piezoelectric d33 -Mode MEMS Energy Harvester," *J. Microelectromech. Syst.*, vol. 19, no. 5, pp. 1215–1222, Oct. 2010, doi: 10.1109/JMEMS.2010.2067431.

[8] Y. B. Jeon, R. Sood, J. -h. Jeong, and S.-G. Kim, "MEMS power generator with transverse mode thin film PZT," *Sensors and Actuators A: Physical*, vol. 122, no. 1, pp. 16–22, Jul. 2005, doi: 10.1016/j.sna.2004.12.032.

[9] P. Thakkar *et al.*, "Measuring angle-resolved dynamic deformation of micromirrors with digital stroboscopic holography," presented at the Optics and Photonics for Advanced Dimensional Metrology II, 2022, vol. 12137, pp. 112–119.

[10] Q. Guo, G. Z. Cao, and I. Y. Shen, "Measurements of Piezoelectric Coefficient d33 of Lead Zirconate Titanate Thin Films Using a Mini Force Hammer," *Journal of Vibration and Acoustics*, vol. 135, no. 1, p. 011003, Feb. 2013, doi: 10.1115/1.4006881.

[11] G. Meng and Y. Cheng, "Electrical Breakdown Behaviors in Microgaps," in *Electrostatic Discharge - From Electrical breakdown in Micro-gaps to Nano-generators*, S. H. Voldman, Ed. IntechOpen, 2019. doi: 10.5772/intechopen.86915.

CONTACT

*P. Thakkar, pooja.thakkar@silicon-austria.com, tel: +4366488843742

RESONANT PIEZOELECTRIC VARIFOCAL MIRROR WITH ON-CHIP INTEGRATED DIFFRACTIVE OPTICS FOR INCREASED FREQUENCY RESPONSE

Jaka Pribošek[1]*, Anton Lagosh[1], Pooja Thakkar[1], Takashi Sasaki[1], Markus Bainschab[1]

[1]Silicon Austria Labs, Villach, AUSTRIA

ABSTRACT

We report the design, fabrication, and characterization of a resonant varifocal micromirror excited with thin-film piezoelectric actuators. The mirror features a spiral suspension allowing both in-plane and out-of-plane motion which reduces the stress build up upon the membrane deformation and hence high shape fidelity. Driven at resonance, the mirror reaches an optical power of more than 27.5 diopters at 197 kHz and 7 V_{P-P} excitation. The unique feature of the mirror is a combination of large optical power with the integration of a binary Fresnel zone plate (BFZP) onto the flexible membrane. The BFZP provides a virtual and real focusing point, which, when combined with the main mechanical mode provides four focusing points per harmonic cycle, effectively doubling the resonant frequency to 394 kHz. Compared to state-of-the-art devices, this device features the highest figure of merit ever reported.

KEYWORDS

Varifocal mirror, diffractive optics, piezo actuators.

INTRODUCTION

Three-dimensional optical scanning is instrumental to many optical imaging, sensing, and laser-machining applications. While scanning in lateral direction can achieve a microsecond response time at a resolution of several thousand points, scanning in axial direction represents a bottleneck in systems' throughput, speed, and resolution [1]. Varifocal mirrors are known to provide a large axial resolution [2], but lack the speed and therefore still cannot compete with ultra-high speed focus tuning devices like acousto-optical and electro-optical lenses [1]. Over the past years, authors have tried to apply a number of different techniques to improve both axial resolution as well as speed. Currently, the highest speed of operation at 400 kHz was reported by Sasaki et al [3], but at only limited membrane deformation. In this paper, we report an integration of a diffractive optical elements directly onto the flexible membrane of a micromirror which, at sufficiently large membrane deflections increases the effective frequency response by a factor of two. This technique can be applied to any varifocal mirror, independent of the actuation type.

DESIGN AND FABRICATION

The micromirrors were fabricated from 6-inch (100) SOI wafer with 150 μm device layer thickness and 1.75 μm sol-gel deposited gradient-free PZT material with state-of-the art performance (e_{31} = -20 C/m²). Pt and Au layers are used as a bottom and top electrode, respectively. The full

stack and the corresponding process flow consists of the six lithography steps shown in Fig.1a. First, the fabrication starts by structuring the top electrode using a lift-off process. Second, we etch the PZT layer using an advanced ICP-RIE dry etch technique followed by dry-etch to structure the bottom electrode and the SiO₂ top layer. The device and handle layer are etched using a Deep Reactive Ion Etch (DRIE) process. Next, the wafer is protected with Parylene and the backside of the device is etched using DRIE to release the devices. Fig.1b shows the micrograph of the fabricated device. The mirror features a circular membrane with 3 mm diameter, suspended with six spiral springs. As opposed to our prior work, where the membrane was suspended by rather stiff spring structure thus introducing large nonlinearities [2], the springs of the current design allow both lateral and axial motion upon membrane deformation. This allows nearly free-edge boundary conditions and prevents stress build-up upon membrane deformation. As a result, large shape fidelity is obtained. The springs follow the Archimedean spiral formula, where centerline of each spring is given with the following analytical equation:

$$x_s = r \exp(\eta \, t) \cos(t) \qquad (1)$$
$$y_s = r \exp(\eta \, t) \cos(t)$$

Here $t \in (0, 2\pi)$, $\eta = 100$ and $r = 1600$. The spirals are then rotated and distributed around the center $(0,0)$ The coordinates of the k-th spiral are given as:

$$
\begin{aligned}
x_k &= x_s \cos\left(k\frac{\pi}{3}\right) - y_s \sin\left(k\frac{\pi}{3}\right) \\
y_k &= x_s \sin\left(k\frac{\pi}{3}\right) + y_s \cos\left(k\frac{\pi}{3}\right)
\end{aligned}
\qquad (2)
$$

with $k = \{0, 1, \dots 5\}$. Each centerline (x_k, y_k) is then offset on both sides to obtain the final spiral suspension with width of 120 μm. Suspension springs also serve as interconnection lines to the PZT, deposited at the optical membrane. The top electrode on the optical membrane is structured to form a binary Fresnel Zone Plate (BFZP), introduced by the difference in reflectivity between Au top electrode layer and PZT layer. The closeup in Fig.1b shows the binary Fresnel zone plate (BFZP). The n-th zone of the BFZP is defined as follows:

$$
r_n = \sqrt{n \, \lambda \, f + \frac{1}{4} \, n^2 \, \lambda^2 \, \lambda}
\qquad (3)
$$

Where f = 50 mm, λ = 633 nm and r_n < 1500 μm. The smallest feature size is limited by fabrication to 3 μm. Note,

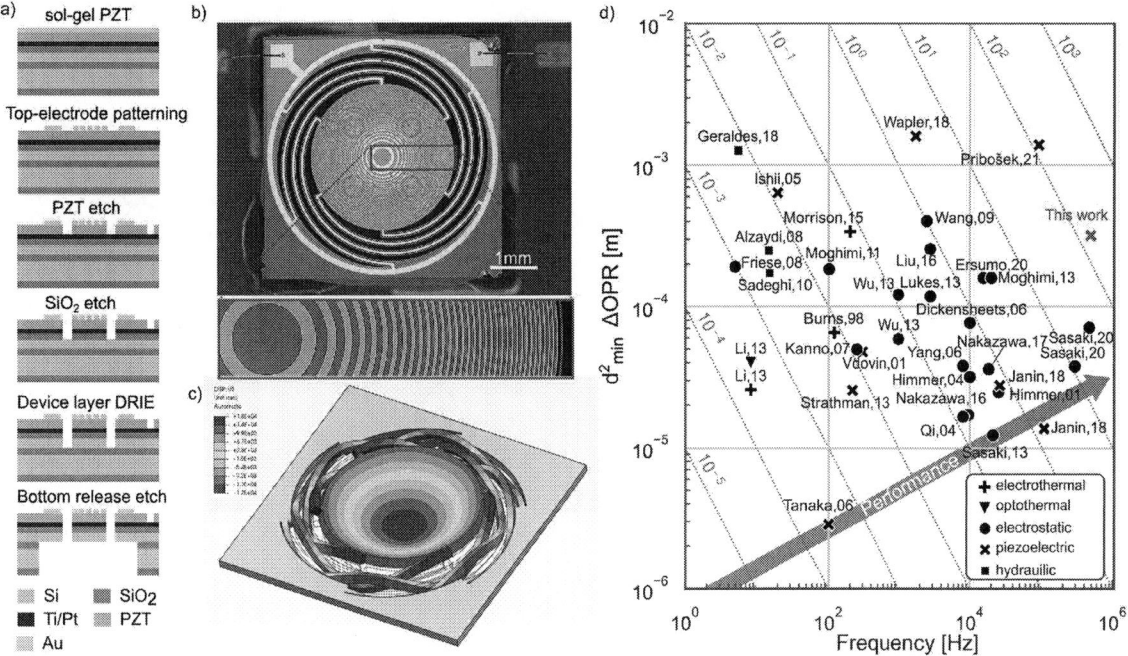

Figure 1: a) Microfabrication process flow b) Micrograph of the fabricated device bonded to the PCB substrate. Closeup shows the integrated binary Fresnel zone plate c) Parabolic mode, identified at 196 kHz using FEM software. d) Overview of state-of-the-art resonant varifocal mirrors, plotted in stroke vs frequency graph. Our device outperforms all devices from the literature.

that the zones of the BFZP are interconnected using 5 μm electrode to polarize PZT ceramics in the d_{31} mode. The device was modelled and optimized using CalculiX Finite interconnection lines to serve at the same time also as an lement Methods. We identified a parabolic mode around 196 kHz (Fig.1C). Parabolic membrane deflection is used to focus a collimated input beam into a focal spot, that is swept back and for that the resonant frequency. In typical fast focusing applications like laser micromachining and high-speed neural imaging, the membrane deflection provides two scans through the inspected volume per harmonic cycle. In this paper, we combine the main parabolic mode with the optical function of the BFZP to introduce two additional focusing points, one real and one virtual. Provided that the membrane deflection is stronger than the optical power of the BFZP, scanning of the membrane deflection in convex regime allows to also image the virtual focus of the BFZP in the real domain. Imaging virtual focus gives rise to two additional focusing points per harmonic cycle and hence twice the scanning speed. The functional schematic of the combined optical performance using both membrane deflection and BFZP is depicted in Fig.2B. Since these focal points are well separated in optical powers, and typically occur in different halves of the harmonic cycle, whenever real focus is used, almost no stray light arises from the virtual focusing point, and vice versa. This allows a simple single laser operation using a stroboscopic laser illumination and typical Pulse-On-Demand function. Hence individual focusing control at twice the speed can be obtained. With the current device, axial scanning at 394 kHz is reported.

EXPERIMENTAL RESULTS

We measured the frequency response function using a custom-built experimental setup (Fig.2a). A system consisting of a single mode fiber coupled laser diode (λ = 635 nm), collimated to form a TEM$_{00}$ beam with 3.5 mm diameter. The laser is pulsed synchronously to the excitation frequency with 100 ns pulse lengths. A CMOS camera (xiQ, Ximea) is used to capture the light intensity distribution, observed on the screen, located at a distance of 770 mm from the mirror (L in Fig.2a). By varying the pulse delay of the stroboscopic laser pulse, we captured the light intensity distribution at different membrane deflections. The lower part of the Fig.2b shows the measured intensity distribution plotted as a function of harmonic cycle's phase. The cycle comprises two focusing actions, first by focusing the real focus point of the BFZP, and second of focusing the virtual focusing point of the BFZP (RFP and VFP on Fig.2b). We note, that focusing of the virtual focusing point from the BFZP is only possible if the optical power provided by the membrane deflection is larger than the optical power of the BFZP. The total optical power of the membrane deflection was measured from the mirror backside and found to be 27.5 diopters at a peak frequency of 197 kHz and 7 V$_{P-P}$ and 5 V$_{DC}$ excitation. A frequency response of the device shows a Q-factor of 195 and slightly slanted response, indicating presence of modest spring stiffening (Fig.2c). Fig.2d shows the lateral point spread function, measured at the focus of the first real focal point is and compared to the theoretical point spread

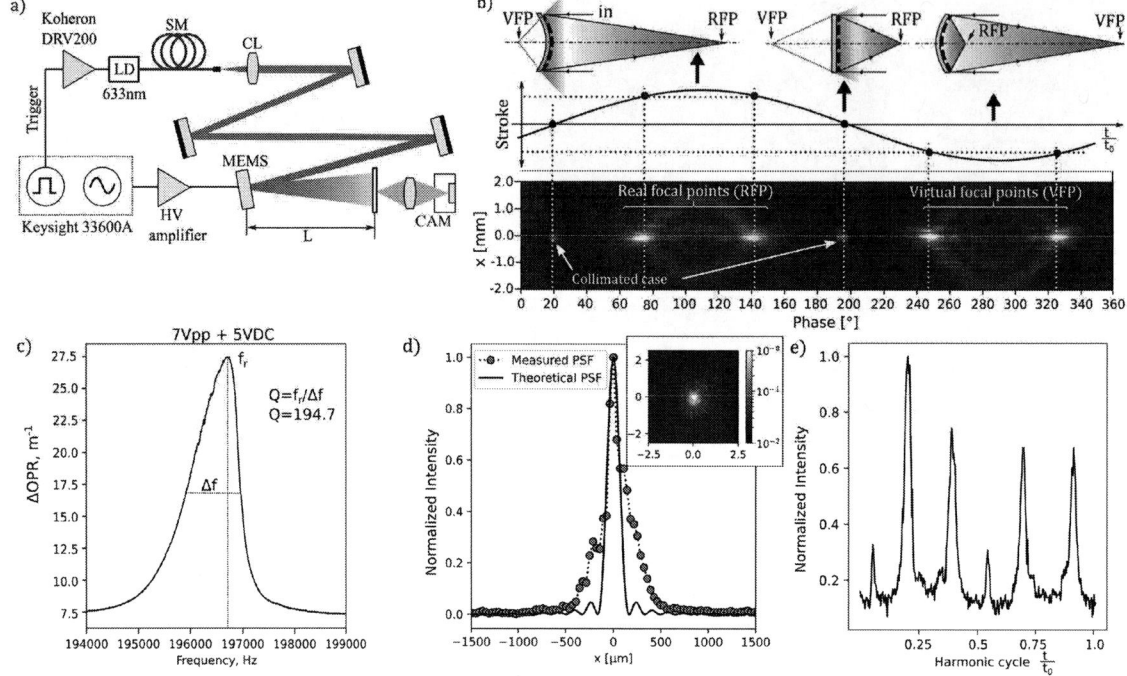

Figure 2: a) Experimental system for characterization, b) harmonic cycle with real and virtual focus points. c) Measured total optical power versus frequency. d) lateral point spread function of the first real focus point, compared to the theoretical point spread function obtained from diffraction theory. e) axial point spread function as a function of phase of harmonic cycle. Four focusing points per harmonic cycle are clearly visible.

function, calculated for the same aperture by the scalar wave beam propagation method. Slightly increased first order is indicating nearly diffraction limited performance and hence high shape fidelity of the parabolic mode. Fig.2e shows the axial point spread function over one harmonic cycle, as measured by changing the phase delay of the stroboscopic illumination. Clearly, four focusing points per harmonic cycle are visible, indicating two times increase in operation frequency of the main resonant mode. When compared against the state-of-the-art varifocal mirrors (Fig.1d) the proposed device delivers the highest figure-of-merit for varifocal mirrors [2], [4].

Finally, we investigated the shape fidelity of the proposed device. To avoid the effects of the topography on the top side, the membrane shape was measured from the backside through a hole in the PCB, where the device was mounted. Initial quasistatic deformation was found to be mostly parabolic with a sag of 75 nm and about 10 nm of astigmatism. Next, the membrane shape was recorded at different membrane deflections using stroboscopic digital holographic microscopy [5]. A deformation map obtained for maximum deflection is shown in the Fig.3b. Fig.3a shows the corresponding horizontal profile of the deformation, showing large parabolic mode with total p-v of 0.75μm. The fit of parabola reveals a total deviation of less than +/-50 nm. Next, a set of Zernike polynomials was fitted at each stroboscopic phase to identify the main structural modes at each position of the harmonic cycle. All Zernike terms were found to oscillate harmonically, with by far dominant deformation being the spherical

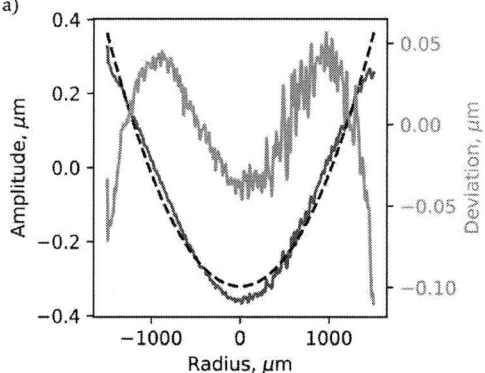

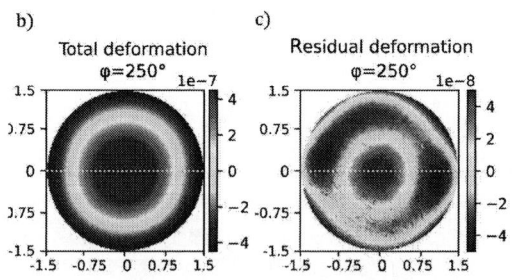

Figure 3: a) Horizontal profile of the deformation map obtained at $\varphi = 250°$ and its deviation from parabolic fit, b) measured total deformation map acquired at maximal deflection excited at 4 V_{p-p} and 6 V_{dc} c)the residual deformation map after the subtraction of parabolic mode

aberration, followed by astigmatism and tetrafoil (Fig.4). We note, that individual components do have different offsets. While Spherical aberration, Coma X, Coma Y and Tetrafoil 0 have almost zero offset, both astigmatisms were found to have vertical offset of roughly -10^{-8} m. As a result of this offset, the spherical aberration is somewhat compensated in the first half of the harmonic cycle, while the total aberration is more pronounced in the second half of the harmonic cycle. Presence of the astigmatism and tetrafoil are also responsible for slightly asymmetric map of residual deformations (Fig.3c). We believe one of the reasons for this can be attributed to the silicon crystal anisotropy.

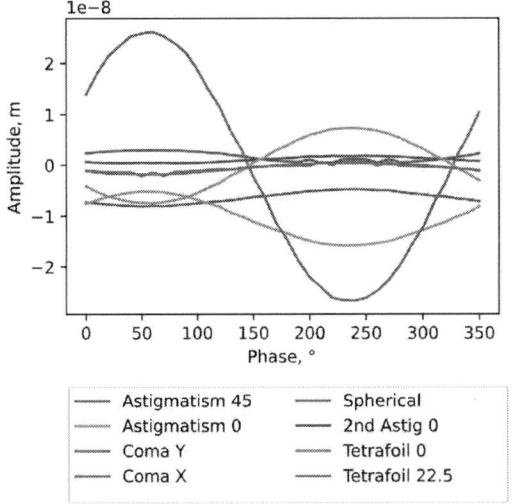

Figure 4: Most dominant Zernike polynomials, fitted to the residual deformation map, plotted as a function of harmonic phase

CONCLUSIONS

We report the design, fabrication, and experimental evaluation of a resonant varifocal micromirror with integrated diffractive optical element. The diffractive optical element produces two focal spots, virtual and real focus spot respectively. When combined with the membrane deformation from the structural mode, this results in four focusing points per harmonic cycle, and thus 2 times increase in resonant frequency. A successful operation at 397kHz and 27.5 diopters of total optical power is reported. This opens a novel way towards ultra-high-speed axial focusing. The presented device can be applied to a number of different applications, requiring ultra-high speed axial focusing, like laser machining [6], neural imaging [7], confocal microscopy or similar.

ACKNOWLEDGEMENTS

Authors would like to acknowledge Dominik Holzmann for the help with electronics.

REFERENCES

[1] S. Kang, M. Duocastella, and C. B. Arnold, "Variable optical elements for fast focus control," *Nat. Photonics*, vol. 14, no. 9, Art. no. 9, Sep. 2020, doi: 10.1038/s41566-020-0684-z.

[2] J. Pribošek, M. Bainschab, A. Piot, and M. Moridi, "Aspherical High-Speed Varifocal Piezoelectric MEMS Mirror," in *2021 21st International Conference on Solid-State Sensors, Actuators and Microsystems (Transducers)*, Orlando, FL, USA, Jun. 2021, pp. 1088–1091. doi: 10.1109/Transducers50396.2021.9495520.

[3] T. Sasaki, T. Kamada, K. Hane, and Tohoku University 6-6-01 Aoba, Aramaki, Aoba-ku, Sendai, Miyagi 980-8579, Japan, "High-Speed and Large-Amplitude Resonant Varifocal Mirror," *J. Robot. Mechatron.*, vol. 32, no. 2, Art. no. 2, Apr. 2020, doi: 10.20965/jrm.2020.p0344.

[4] J. Pribošek, M. Bainschab, and T. Sasaki, "Varifocal MEMS mirror for fast focus control: a review," 2022.

[5] P. Thakkar *et al.*, "Measuring angle-resolved dynamic deformation of micromirrors with digital stroboscopic holography," in *Optics and Photonics for Advanced Dimensional Metrology II*, Strasbourg, France, May 2022, p. 29. doi: 10.1117/12.2621325.

[6] M. Duocastella and C. B. Arnold, "Enhanced depth of field laser processing using an ultra-high-speed axial scanner," *Appl. Phys. Lett.*, vol. 102, no. 6, Art. no. 6, Feb. 2013, doi: 10.1063/1.4791593.

[7] M. Žurauskas, O. Barnstedt, M. Frade-Rodriguez, S. Waddell, and M. J. Booth, "Rapid adaptive remote focusing microscope for sensing of volumetric neural activity," *Biomed. Opt. Express*, vol. 8, no. 10, Art. no. 10, Oct. 2017, doi: 10.1364/BOE.8.004369.

CONTACT

*J.Pribošek, tel: +4366488200153
jaka.pribosek@silicon-austria.com

UNIQUE DISPERSION RELATION FOR PLASMONIC PHOTODETECTORS WITH SUBMICRON GRATING

Yuki Kaneda[1], Masaaki Oshita[1], Utana Yamaoka[1], Shiro Saito[2], and Tetsuo Kan[1]
[1]The University of Electro-Communications, JAPAN
[2]IMRA JAPAN CO., LTD., JAPAN

ABSTRACT

We proposed a plasmonic photodetector with a submicron grating pitch for accurate and robust near-infrared reconstructive spectroscopy. Miniaturized spectrometers have been intensively studied, and one promising candidate is the use of wavelength-specific plasmonic photodetector, which presents a correspondence between the detectable wavelength and the angle of incidence under measurements. However, the method fails to accurately reconstruct the spectrum because multiple wavelengths to be detected correspond to one angle. We, therefore, propose a plasmonic photodetector that has a one-to-one correspondence by shrinking the grating pitch to submicron, which was ~3 μm in previous reports. We compared reconstructed spectra with those with conventional photodetectors in the near-infrared range from 1150 to 1500 nm and confirmed the effectiveness of this method.

KEYWORDS

Spectroscopy, Plasmonics, Schottky photodetector, MEMS, Near-infrared

INTRODUCTION

Near-infrared (NIR) spectroscopy is a non-destructive or non-contact method for analyzing a chemical composition from its reflection and absorption spectrum and is widely applied in various fields [1]. Miniaturization of spectrometers realizes spectroscopic analysis on instruments with limited space and weight and is actively studied for applications such as remote sensing by exploration satellites [2]. Although various miniaturization methods have been proposed [3], these methods require multiple filters and photodetectors, limiting the miniaturization of the spectrometer footprint.

Recently, one promising method of spectroscopy using a plasmonic photodetector with a gold diffraction grating on an n-type silicon cantilever has been proposed [4-6]. The method reconstructs an incident spectrum from photoresponse spectra with photocurrent peaks at the corresponding incident angle to the grating using an inverse problem approach. Since the spectroscopy method acquires photoresponse spectra only by angle-scanning one photodetector, the method does not need any additional optical components. It is, therefore, expected to significantly miniaturize the footprint compared to conventional spectrometers. However, the measurable wavelength range was limited to 200 nm in the near-infrared, suffering from inaccuracy out of the range. This limitation is because the previous photodetector configuration exhibited that multiple wavelengths to be photo-detected correspond to an incident angle. This correspondence hinders the photodetector from distinguishing these wavelengths.

Here, we propose a configuration of a plasmonic photodetector that has a one-to-one correspondence between a wavelength and an incident angle in a near-infrared area from 1150 to 1500 nm. This unique, one-to-one correspondence was obtained by shrinking the grating pitch on the photodetector to submicron size, which was ~3 μm in previous reports. In this paper, we fabricated the photodetector with the submicron-sized grating and compared spectra reconstructed from the fabricated photodetector and the conventional structure.

METHOD

The plasmonic photodetector in this study consists of an n-type silicon (n-Si) grating evaporated a gold (Au) thin film and an aluminum (Al) electrode formed on the backside (Fig. 1(a)). The Au grating absorbs light of a specific wavelength at the corresponding angle of the incident light due to the surface plasmon resonance, and the photodetector converts the energy of the incident light into a photocurrent signal that has a peak at the corresponding angle (Fig. 1(c)). Fig. 2 shows relationships between the angle of incidence and the peak wavelength with two grating models. Fig. 2(a) presets the relationship of the previous plasmonic photodetector. In this case, the photodetector presents that multiple wavelengths λ_1 and λ_2 correspond to the incident angle θ_1 (Fig. 2(a)). If this spectra is used for the spectrum reconstruction, there are three possible possible spectra (Fig. 2(b)) from the peak of

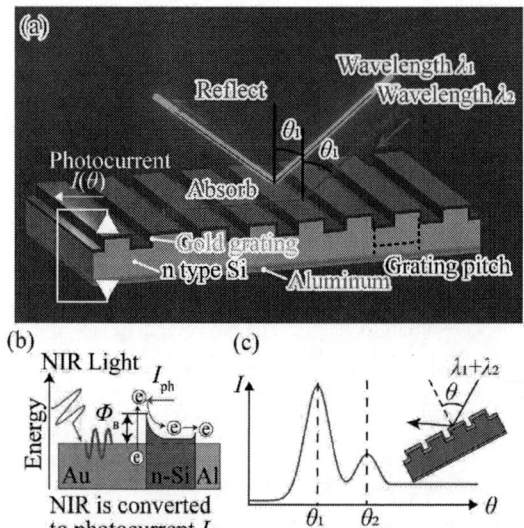

Figure 1: (a) Principle of plasmonic photodetector (b) Schottky barrier used in this device (c) Photocurrent measured when this device is rotated while irradiating light of wavelengths λ_1 and λ_2

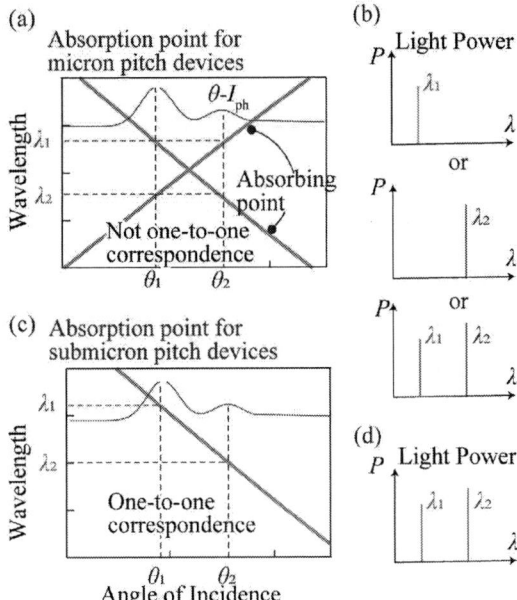

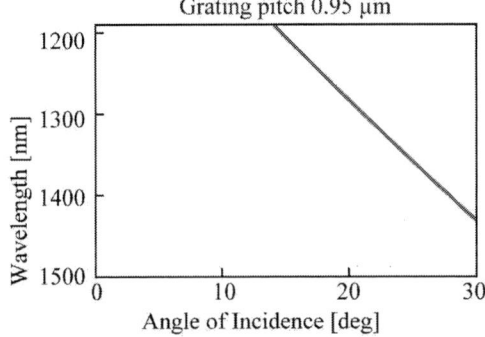

Figure 2: (a) A example of absorption relationship for a micron pitch device. (b) Possible incident spectra. (c) Example of absorption relationship for a submicron pitch device. (d) A possible incident spectrum.

Figure 3: Dispersion relation of the photodetector whose grating pitch is 0.95um.

$$\frac{\omega}{c}\sqrt{\varepsilon_{m}}\sin\theta + \frac{2m\pi}{a} = \frac{\omega}{c}\sqrt{\frac{\varepsilon_{m}\varepsilon_{Au}}{\varepsilon_{m} + \varepsilon_{Au}}} \quad (1)$$

where ω is the angular frequency of the incident light, c is the speed of light in vacuum, θ is the angle of incidence, m is the diffraction order, a is the pitch of the grating, ε_{m} is the dielectric constant of the sample, and ε_{Au} is the dielectric constant of Au. When SPR occurs, the energy of the incident light excites free electrons in Au. The excited electrons overcome the Schottky barrier formed at the interface between Au and n-Si. The Schottky barrier converts the electrons into a photocurrent signal (Fig. 1(b)). When we scan the incident angle to the grating on the photodetector, the peak of the photocurrent is observed at the angle where the SPR occurs (Fig. 1 (c)). Using the photocurrent signal, we calculated the incident spectrum using the method described in the reference [7].

Equation (1) indicates that a micron pitch grating has multiple absorption points within the near-infrared range for previous ~3-μm-pitch gratings, making it difficult to inverse calculate the incident spectrum uniquely. In contrast, the submicron pitch grating ($a = 0.95$ μm) exhibits a one-to-one correspondence in the calculated dispersion relation using Equation (1), as shown in Fig. 3. In this paper, we employed the pitch of the grating and compared the spectra obtained from the conventional micron pitch grating and the proposed submicron pitch grating.

FABRICATION

The process flow for the plasmonic photodetector with submicron pitch grating is described in Fig. 4 (a). First, the electron beam resist was patterned into the grating on an n-Si wafer by the electron beam (EB) lithography. Then, an n-Si wafer with a resistivity of 1-10 Ω-cm was spin-coated with EB resist ZEP520A (ZEON, Japan) at 6000 rpm for 60 seconds, and an EB lithography system (F5112, ADVANTEST, Japan) was used to draw the diffraction gratings with a dose of 104 μC/cm2 (Fig. 4. (a)(1)). The pattern was then developed by ZED-N50 for 4 minutes. After the diffraction grating was formed by etching the silicon with a deep reactive ion etching system (MUC21, SUMITOMO PRECISION PRODUCTS CO., LTD., Japan) (Fig. 4 (a)(2)), the EB resist was removed with a solution (Hakuri104, Tokyo Ohka Kogyo) (Fig. 4 (a)(3)). Next, a 50 nm gold film was evaporated, and aluminum was deposited on the backside of the device (Fig. 4 (a)(4)). Finally, the device was mounted on a printed circuit board to complete the device fabrication (Fig. 4(b)).

To investigate the effect of shrinkage on the grating pitch, we fabricated the photodetector with a pitch of $a =$ 0.95 μm and 3.4 μm, respectively. To check the pitch and depth of the fabricated device, we used an Atomic Force Microscope (AFM, AFM5500M, Hitachi, Japan) to scan the surface of the grating (Fig. 4(c)). As a result, the grating height and pitch were approximately 80 nm and 0.95 μm (Fig. 4(d)), indicating that the targeted structure dimensions were formed. For confirmation of whether the photodetector can detect light in the near-infrared range from 1150 to 1500 nm, the IV characteristics of the fabricated device were measured using a source meter (B1500A, Keysight, U.S.A.). In this study, the anode was

the photocurrent signal at the incident angle θ_1. Therefore, it is not possible to uniquely determine the incident spectrum from the photodetector signal in the previous configuration, causing noise on the reconstructed spectra. On the other hand, a one-to-one correspondence (Fig. 2(c)) between a wavelength and an incident angle realizes one possible spectrum (Fig. 2(d)) from the peak at the incident angle θ_1. If we obtain the configuration of the photodetector that has the one-to-one correspondence in the near-infrared area, we can expand the spectral range of the spectroscopy method, and also improve the accuracy and robustness of the measurement.

The absorption point of the plasmonic photodetector is caused by a coupling phenomenon between the collective vibration of free electrons (plasmons) on the metal surface and the incident light, called Surface Plasmon Resonance (SPR). SPR occurs when the wave number of the diffracted light along the surface of the metal matches the wave number of the plasmon. The resonance condition is shown in Equation (1).

978-1-6654-9309-3/23 $31.00 © 2023 IEEE

set on the Au electrode, and the cathode was set on the Al electrode. The measured IV characteristics are shown in Fig. 4(e). From these characteristics, we calculated the Schottky barrier height using an algorithm by Cheung *et al.* [8]. The height was calculated to be 0.71 eV, and the longest detectable wavelength (cutoff wavelength) was calculated to be 1750 nm, which is sufficient for the target wavelength range.

EXPERIMENT AND RESULT

Incident angle characteristic of the photodetector

To confirm whether the fabricated device exhibits an absorption point coherent with SPR theory, we constructed an experimental setup, as shown in Fig. 5(a). Since the generated photocurrent was as weak as 200 nA, we converted the photocurrent signal into the voltage signal using a trans-impedance amplifier. The amplified output was measured with a source meter (6242, ADCMT, Japan). The fabricated device and the amplifier were placed inside a shielded box to block external noise. The aperture is formed in front of the shield box to provide the incident light to the photodetector. For scanning an incident angle to the grating, the shielded box was placed on a rotating stage (OSMS-60YAW, Sigma Koki, Japan). The incident light was emitted from a supercontinuum (SC) light source (SC450-27, FIANIUM, U.K.), which outputs visible to near-infrared light. For filtering the spectrum of the SC light, we employed an acousto-optic tunable filter (AOTF) (Acousto-Optic Tunable Filter, FIANIUM, U.K.). The filtered light was passed through a polarizer (PBSW-10-10/20, Sigma Koki, Japan), which made the light TM-polarized. The TM-polarized light entered the pinhole (Φ = 2 mm) in front of the photodetector to prevent excitation

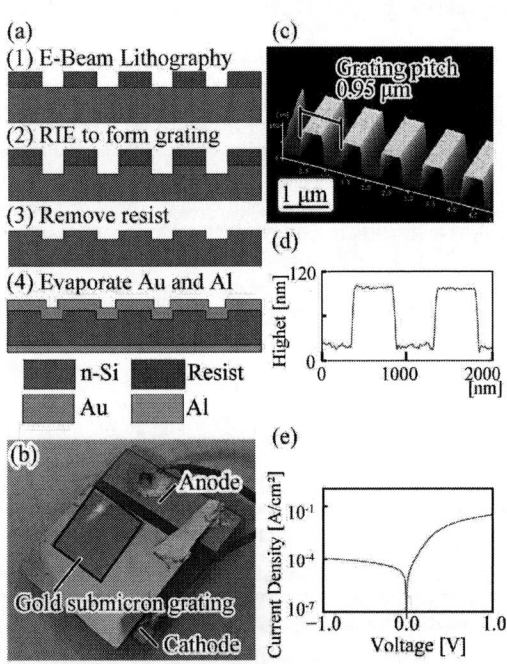

Figure 4: (a) Process flow (b) A photograph of the fabricated device (c) The surface of the garting measured by AFM (d) Cross-sectional view of the grating (e) IV characteristics

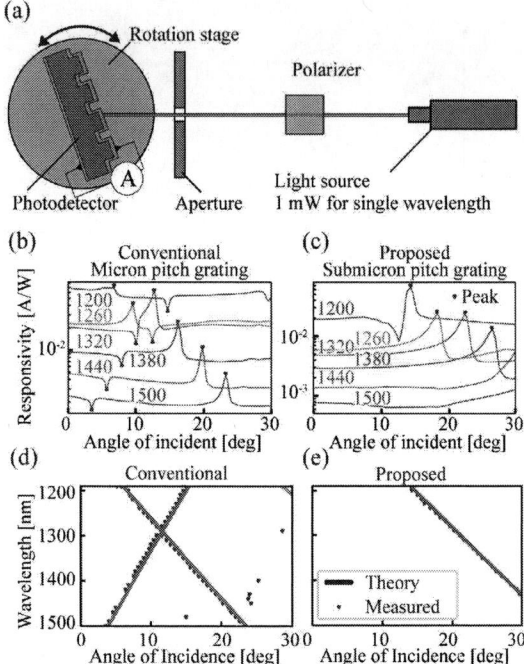

Figure 5: (a) Experimental setup for measuring responsivities at each incident angle. Measured responsivities from (b) conventional and (c) proposed device. Theoretical absorbing point and peaks of the measured signals obtained from (d) conventional and (e) proposed device.

light from reaching areas other than the diffraction grating. The photocurrent signals were measured in this configuration while changing the incident angle in 0.2° increments from 0° to 30°. As shown in Fig. 5 (b) and (c), the peak of responsivity shifted with changing the irradiation wavelength in each structure, and the peak angles were consistent with the theoretically derived angle in both structures (Fig. 5(d), (e)).

Spectral performance

Finally, we performed near-infrared spectroscopy using the fabricated photodetector to compare the spectra obtained from the conventional micron pitch grating and the proposed submicron pitch grating. The plasmonic photodetectors were irradiated with spectra that have peaks at multiple wavelengths (1230, 1280, 1330, and 1380 nm) filtered by AOTF, and the obtained photocurrent signals are shown in Fig. 6(a)(b). From these photocurrent signals, incident spectra were calculated using the method described in the reference [7]. As a reference, the incident spectrum was also measured with a commercial spectrometer (Sol2.2A, BWTEK, U.S.A.). The results are shown in Fig. 6 (c)(d). In these spectra, the blue lines show the spectra measured by the fabricated photodetector, and the orange lines show the spectra measured by a reference spectrometer. The reconstructed spectrum obtained from the conventional micro-sized grating structure showed peaks at wavelengths other than the irradiated wavelengths. Furthermore, the light intensity at 1170 nm was minimal, and the light intensity after 1400 nm was about 0.5, which is larger than the reference spectrum. The spectrum calculated from the conventional device did not reproduce the original spectrum. On the contrary, the reconstructed spectrum based on the proposed submicron grating structure showed peaks at the same wavelengths measured with the reference spectrometer. No peaks other than the incident spectrum were observed. Therefore, we concluded that the proposed photodetector could perform near-infrared spectroscopy comparable with a reference spectrometer, presenting highly accurate near-infrared spectrosocpy.

CONCLUSION

In this paper, we proposed a configuration of a plasmonic photodetector that has a one-to-one correspondence between a wavelength and an incident angle for expanding the spectral range of the reconstructive spectroscopy to a near-infrared range from 1150 to 1500 nm. We designed the pitch of the grating on the plasmonic photodetector to 0.95 nm, and the dispersion relation of the photodetector exhibits the one-to-one correspondence structure. The spectral shape was similar to the result of the reference spectrometers in the range of 1170 - 1500 nm. Since the spectral range has been limited to about 200 nm with the previous device configuration, the proposed submicron grating expanded the measurable wavelength range. In this study, we did not perform the miniaturization of the photodetector and angle-scanning mechanism, and the experimental setup was relatively large because we used the balk device (10 x 10 mm^2) and the commercial rotating stage. However, since a MEMS cantilever-type plasmonic photodetector has been reported [6], we expect this structure to be applied to MEMS devices.

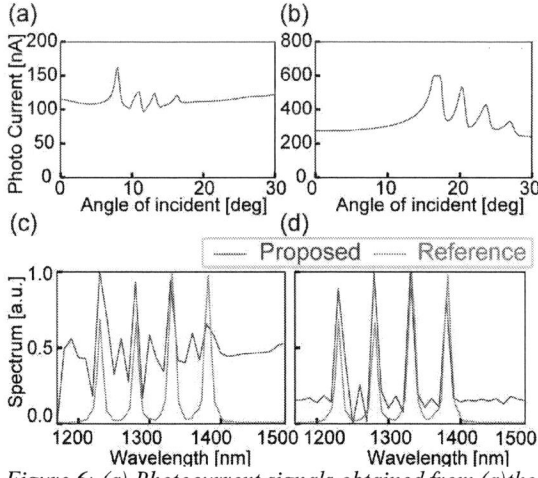

Figure 6: (a) Photocurrent signals obtained from (a)the conventional and (b) proposed structure. Calculated spectrum of the (c) conventional and (d) proposed structure. Spectra obtained from reference spectrometer is displaed in orange lines.

ACKNOWLEDGEMENTS

This research was financially supported by NEDO, Japan. This research was financially supported by IMRA JAPAN Co., Ltd. This work was also supported by "Advanced Research Infrastructure for Materials and Nanotechnology in Japan (ARIM)" of the Ministry of Education, Culture, Sports, Science and Technology (MEXT). Proposal Number JPMXP1222UT1021. Microfabrication was performed in a clean room of the Division of Advanced Research Facilities (DARF) of the University of Electro-Communications, Tokyo, Japan.

REFERENCES

[1] C. Bacon, *et al.,* "Miniature spectroscopic instrumentation: Applications to biology and chemistry," Review of Scientific Instruments, vol.75, no.1, pp.1–16, 2004.

[2] K. Kitazato, *et al.*, "The surface composition of asteroid 162173 Ryugu from Hayabusa2 near-infrared spectroscopy," Science, vol. 364, no. 6437, pp. 272275, 2019.

[3] Z. Yang, *et al.*, "Miniaturization of optical spectrometers," Science, vol. 371, no. 6528, p. eabe0722, 2021.

[4] W. Chen, *et al.*, "NIR spectrometer using a Schottky photodetector enhanced by grating-based SPR," Opt. Express, vol.24, no. 22, pp.25797–25803, 2016.

[5] M. Oshita, *et al.*, "Reconfigurable Surface Plasmon Resonance Photodetector with a MEMS Deformable Cantilever," ACS Photonics, vol. 7, no. 3, pp. 673–679, 2020.

[6] M. Oshita, *et al.*, "Reconstructive spectrometer based on plasmonic Schottky photodetector with MEMS angular modulator," Proceedings of Transducers 2021, pp. 1118-1121, 2021.

[7] Y. Suido, *et al.*, "Extension of the measurable wavelength range for a near-infrared spectrometer using a plasmonic au grating on a Si substrate," Micromachines, vol. 10, no. 6, 2019.

[8] S. K. Cheung, *et al.*, "Extraction of Schottky diode parameters from forward current-voltage characteristics," Applied Physics Letters, vol. 49, pp. 85–87, 1986.1989.

CONTACT

*Y. Kaneda, tel: +81-42-443-5423;
kaneda@ms.mi.uec.ac.jp

INTEGRATION OF A HIGH TEMPERATURE TRANSITION METAL OXIDE NTC THIN FILM IN A MICROBOLOMETER FOR LWIR DETECTION

Sarah Risquez[1], Sebastian Redolfi[2], Clement Fleury[1], Matthias Wulf[2], Ali Roshanghias[1], Adrien Piot[1], Jeremy Streque[1], Kerstin Schmoltner[2], Thang Duy Dao[1], Markus Puff[2], Mohssen Moridi[1]

[1]Silicon Austria Labs GmbH, AUSTRIA
[2]TDK Electronics GmbH & Co OG, AUSTRIA

ABSTRACT

In this paper, we report the integration of a non-CMOS transition metal oxide composite thin film with a high negative temperature coefficient resistance (NTCR) of 3.8 %/K on a silicon nitride membrane for uncooled infrared microbolometer working in the long wavelength infrared (LWIR) region. The NTC thin film is fabricated by a chemical solution deposition process requiring high crystallization temperature (>750°C). Different micromachined silicon-nitride membrane-supported bolometers (1, 2×2 and 4×4 pixels) with a 170×170 µm² pixel size are successfully fabricated and packaged in a vacuum-shielded TO housing. The fabricated devices exhibit a typical noise equivalent temperature difference (NETD) value down to 15 mK×Hz$^{-1/2}$ and a high specific detectivity characteristic of $1.4×10^8$ cm×Hz$^{1/2}$×W^{-1}.

KEYWORDS

Metal oxide thin film, negative temperature coefficient resistance thermistor, microbolometer, infrared detector micromachining

INTRODUCTION

The need for infrared detectors as single element or multi-elements array for thermal sensing and imaging keeps growing in many applications such as medical, agriculture, automotive, security and defense. Uncooled microbolometers take up the largest market for infrared cameras owing to CMOS compatibility and low fabrication costs. A microbolometer consists of a detector absorbing incident electromagnetic infrared radiation that causes temperature variation of a thermally isolated thermistor thin film. The thermal insulation of the device guarantees the maximum increase of temperature and is realized via a suspended arm, a suspended membrane, or a heat sink. The infrared absorption is realized through an absorber thin film stack targeting a maximal absorption in the 7 µm – 14 µm range, which corresponds to the maximum of emitted radiation of a black body at ambient temperature and a large window of the atmospheric transparency. The temperature change in the bolometer induced by the absorbed radiation, whose energy is converted to heat, can be measured by a change of electrical resistance of the thermistor material. Typically, a one-kelvin change in the object's temperature results in a temperature increase in the bolometer of a few millikelvins. The performance of the thermistor thin film is characterized by its temperature coefficient resistance (TCR) whose absolute value defines its sensitivity. Commercially, thermistor thin films can mainly be vanadium oxide (VO$_x$) or amorphous silicon

(a:Si) which present comparable performance in negative negative temperature coefficient resistance (NTCR) value ranging between 2.0 - 2.5 %/K [1]. Both materials are already widely available on the market. In the past decades, many efforts have been conducted to search for NTCR materials with increased sensitivity. Among them, transition metal oxide composites have attracted many interests from industry due to their high NTCR (> 3 %/K) and low fabrication cost [2-5]. It is believed that NTC detectors have the advantage over thermopiles with higher sensitivity [2]. However, integration of the transition metal oxide NTC thermistor with the long-wavelength infrared (LWIR) absorber in a micromachined thermal isolation structure still remains a veritable challenge as crystallization in spinel structure requires high deposition or annealing temperatures above 650°C [5,6]. To our knowledge, existing infrared detectors integrating a metal oxide NTC thermistor are presented in a single pixel form with a high pixel size (>0.1 mm²) [2-4] or with a heat sink [7]. In the present work, we demonstrate an uncooled infrared microbolometer working in the LWIR region (8 µm – 13 µm) using transition metal oxide composite thin film processed at an elevated temperature as the NTC thermistor material. The design and microfabrication as well as the characterization are presented.

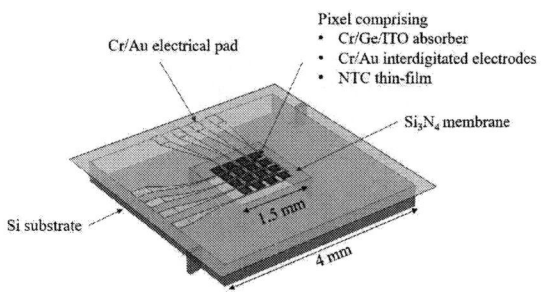

Figure 1: Sketch representing the proposed transition metal oxide thin film microbolometer with 4×4 pixels.

DESIGN AND FABRICATION

Figure 1 presents a sketch of the proposed NTC thin film microbolometer with 4×4 pixels. One pixel is composed of a transition metal oxide NTC thin film covered by interdigitated electrodes for the measurement of the pixel resistance and an absorber stack (Cr/Ge/ITO). The absorber is electrically isolated from the interdigitated electrodes by a SiO$_x$ thin film. All the pixels are resting on a thin Si$_3$N$_4$/SiO$_2$ stress compensated membrane for thermal insulation of the pixels. Membrane configuration was chosen due to the transition metal fabrication constrains requiring an annealing step at high temperature (>750°C). A minimal thickness of the thermal isolation layer with a 300-nm-thick Si_3N_4 and a 100-nm-thick SiO_2 films was chosen to keep the thermal capacitance as low as possible.

Prior to the fabrication, optimization of device parameters and expected device performances were estimated via an NETD analytical model [8] and thermal FEM simulations. The estimated total NETD of a pixel as a function of thermistor resistance for different 1/f constant of NTC material is presented in Figure 2. The details of the different NETD contribution (1/f-noise $NETD_{1/f}$, Johnson noise $NETD_{Johnson}$, thermal fluctuation noise including noise from radiation heat exchange $NETD_{thermal}$) show that the NETD is dominated by the 1/f noise. In addition to a high TCR, the thermistor must present a moderate resistance for reduced noise (Johnson, read-out circuit). Interdigitated electrodes configurations were designed to avoid metal/NTC interdiffusion that would take place in a sandwich or sandwich gap electrode configurations. Different interdigitated electrodes were designed to cover a thermistor resistance range between 25 kΩ and 100 kΩ.

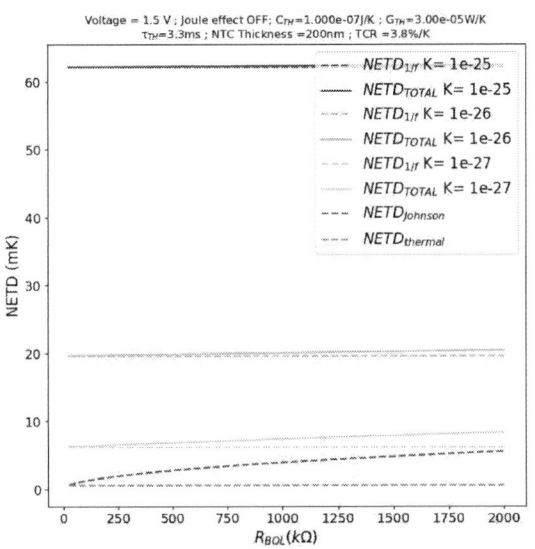

Figure 2: Simulated noise equivalent temperature difference (NETD) of a pixel as a function of thermistor resistance, considering several material 1/f constant K of NTC material.

Microfabrication description

Figure 3 shows the schematic of the process flow described as follows in detail. The fabrication of the microbolometer started from a 4-inch, 380-μm-thick Si(100) substrate including a 100-nm thermal SiO_2 and a 300-nm low stress LPCVD Si_3N_4 thin films. A total of six photolithography Cr masks were used for the realization of the devices. First, a 200-nm-thick transition metal oxide thin film was deposited via chemical solution deposition (CSD) with specific stoichiometry to obtain an optimized NTCR of 3.8 %/K and a resistivity of 4 Ω.m. The CDS process consists of the repetition of spin coating, pyrolyze and annealing steps. The annealing step is performed at 750°C. This temperature is required to obtain the spinel phase showing the NTCR behavior of the transition metal oxide. The film was subsequently patterned by ion beam etching (IBE) to define the 170×170 μm² pixel size. Then, the interdigitated electrodes made of evaporated Cr(50 nm)/Au(150 nm) were structured by a lift-off process.

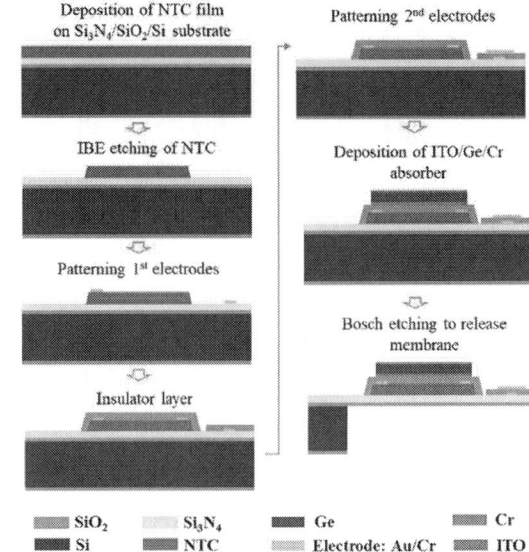

Figure 3: Fabrication process flow of the proposed transition metal oxide NTC thin film bolometer.

To isolate electrically the absorber from the electrodes and the column pads from the row pads, a thin SiO_x (375 nm) film was sputtered on the full wafer and structured by wet etching using BOE (10 HN_4F:1 HF). The electrical connection between columns was performed by a second lift-off of evaporated Cr(50 nm)/Au(250 nm). The IR absorber stack Cr(100 nm)/Ge(585 nm)/ITO(65 nm) was deposited by sputtering and structured by a lift-off process. Finally, the release of the membrane has been performed by back-side ICP-RIE dry etching using a photoresist mask. The backside etching also included chip separation design features to prevent any additional dicing step. An example of the microfabricated device (2×2 pixels) is depicted in Figure 4.

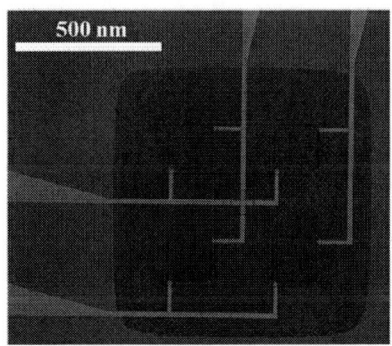

Figure 4: Optical microscope of a 2×2 pixels microbolometer

Packaged devices characterization

Bolometer pixels were packaged in vacuum shielded TO8 housing with an integrated germanium window (Figure 5).

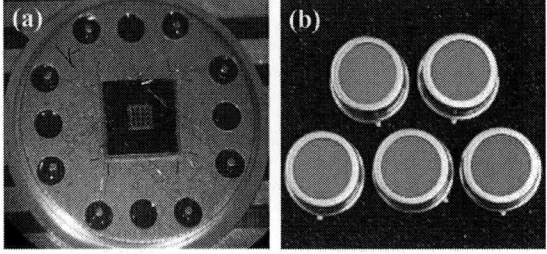

Figure 5: (a) Photo of a 4×4 pixel² NTC bolometer in a TO housing. (b) Vacuum-shielded TO-housing.

The fabricated devices were characterized using a thermal light source covering the LWIR region combined with a mechanical chopper as the excitation, equipped with a differential amplifier and an oscilloscope for the data acquisition and the dynamic spectral response. The noise was measured using the same setup, setting a low emissivity cover, and recording the signal in the same way. Checks with an equivalent resistor validated the setup.

RESULTS AND DISCUSSIONS
Microfabrication

Figure 6 shows a FIB-cut cross-sectional SEM image performed between two interdigitated electrodes after the deposition of the Cr/Ge/ITO absorber and before the membrane release.

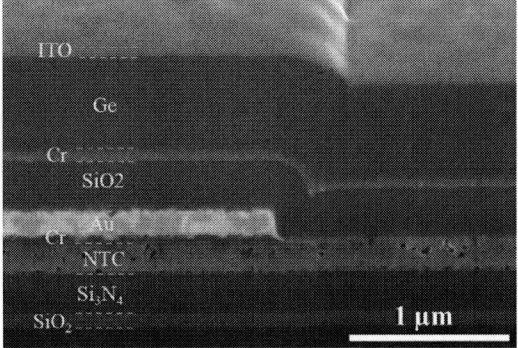

Figure 6: Tilted (52°) cross-sectional view after FIB cut of a NTC bolometer after Cr/Ge/ITO absorber deposition.

I-V characteristics

After patterning the NTC thin film and performing and the interdigitated electrodes, the extraction of the pixel resistance through I-V curve between [-1V;+1V] confirms the expected NTC thin film resistivity between 5.8 and 7.2 Ω.m. I-V curve present linearity up to 20V for high gap interdigitated electrodes (>10 μm). However, the resistance of the thermistor pixel was greatly impacted (> factor 6) by the overall microfabrication process: at the end of the microfabrication, the electrical contact nature becomes non-ohmic (rectifying). However, the sensitivity of the microbolometer to temperature was not changed (TCR=-3.8 %/K).

Absorber characteristics

The absorptivity spectrum of the fabricated microbolometer was measured by an FTIR microscope (LUMOS). The measured absorptivity spectrum shows a good agreement with the designated absorber covering a high absorptivity in the LWIR region.

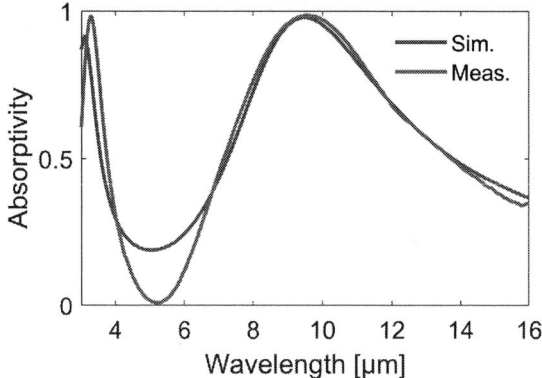

Figure 7: Simulated (blue) and measured (orange) absorptivity spectra of a fabricated 170×170 μm2 pixel bolometer working in LWIR. A gold mirror was used as the reference for the reflectance measurement. The absorptivity was calculated by 1 – reflectance

Microbolometer performances

The measured noise equivalent power (NEP) and noise equivalent temperature difference (NETD) of a fabricated 2×2 pixels bolometer under a bias of 1.5 V at 20 Hz was found to be as small as $1.3×10^{-10}$ W×Hz$^{-1/2}$ and 15 mK×Hz$^{-1/2}$, respectively, which results in a high detectivity D* of $1.4×10^{8}$ cm×Hz$^{1/2}$×W^{-1} (Figure 8).

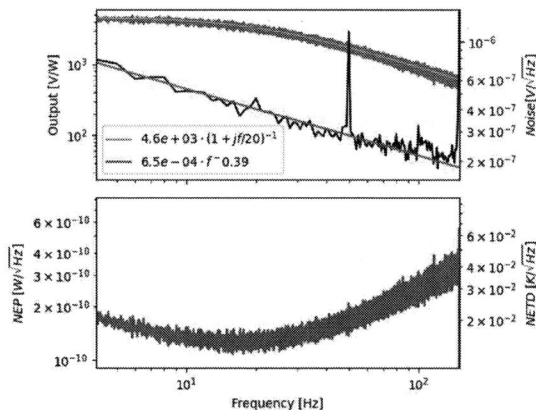

Figure 8: Measured responsivity and noise spectrum, and derived NEP and NETD of a fabricated 4×4 pixels bolometer. The NEP to NETD conversion assumes a lossless optics with a f/1.5 aperture.

The static responsivity (Figure 9) and the NETD (Figure 10) maps of a 4×4 pixels membrane were also derived. The noise has a characteristic 1/f power spectral density with a -0.78 exponent, which is expected for an NTC-based bolometer. The responsivity is slightly higher at the center of the array as a result of the higher insulation, at the cost of the cut-off frequency (not shown here).

978-1-6654-9309-3/23 $31.00 © 2023 IEEE

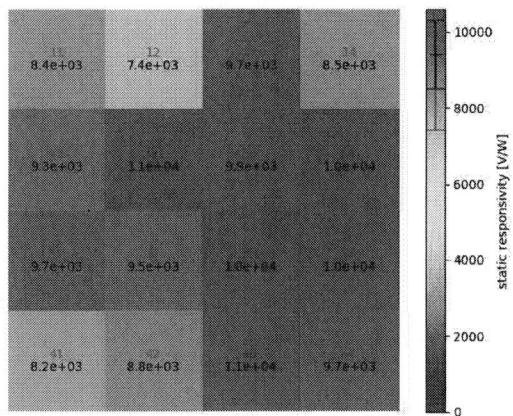

Figure 9: Measured responsivity map of a 4×4 pixels bolometer

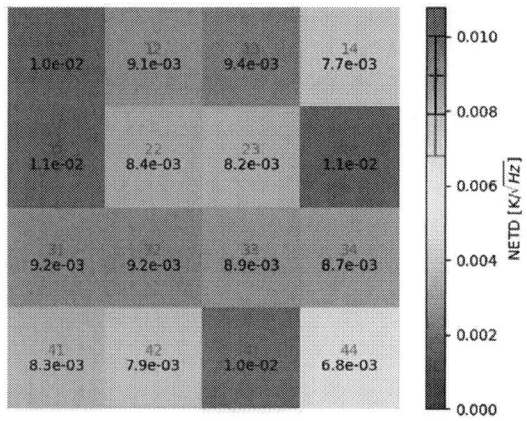

Figure 10: Measured NETD map of a 4×4 pixels bolometer

CONCLUSION

We have successfully demonstrated the integration of a high temperature processed metal oxide NTC into a membrane-supported thermal isolation for LWIR bolometers. Different number pixel arrays (1, 2×2 and 4×4 pixels) bolometers have been fabricated and characterized with a typical NETD of and 15 mK×Hz$^{-1/2}$ and detectivity of 1.4×10^8 cm×Hz$^{1/2}$×W^{-1}. The fabricated NTC thin film bolometers can be used for infrared sensing, thermal sensing, motion sensing and imaging applications. The fabrication process and micromachined thermal isolation structure can be applied to the integration of other thermistor materials.

ACKNOWLEDGEMENTS

This project was performed within the COMET Centre ASSIC Austrian Smart Systems Integration Research Center, run by FFG. The authors would like to thank the staff of the EPFL Center of MicroNanoTechnology (EPFL-CMI) for their support in microfabrication.

REFERENCES

[1] F. Niklaus, C. Vieider, H. Jakobsen, "MEMS-based uncooled infrared bolometer arrays: a review", *MEMS/MOEMS technologies and applications III*, 6836, pp. 125-139, 2008.

[2] L. Moonho, M. Yoo, "Detectivity of thin-film NTC thermal sensors", *Sensors and Actuators A: Physical*, vol 96, no. 2-3, pp. 97-104, 2002.

[3] S. Karanth, M. A. Sumesh, V. Shobha, H. G. Shanbhogue, C. L. Nagendra, "Infrared detectors based on thin film thermistor of ternary Mn–Ni–Co–O on micro-machined thermal isolation structure", *Sensors and Actuators A: Physical*, vol 96, no, pp. 69-75, 2009.

[4] C. OuYang, W. Zhou, J. Wu, Y. Hou, Y. Gao, Z. Huang, "Uncooled bolometer based on Mn1. 56Co0. 96Ni0. 48O4 thin films for infrared detection and thermal imaging", *Applied Physics Letters,* vol. 105, no. 2, p. 022105, 2014.

[5] C. Wu, W. Zhou, Y. Yin, W. Ma, L. Jiang, Z. Huang, J. Chu, "Long wavelength infrared detection based on Mn-Co-Ni-O thin films with dielectric-metal-dielectric absorptive structures", *Infrared Physics & Technology*, vol 102 , p. 102987, 2019.

[6] W. Zhou, Y. Yiming, C. Wu, W. Ma, N. Yao, J. Wu, Z. Huang, "Broadband infrared thermal detection using manganese cobalt nickel oxide thin film", *Optics Express,* vol 27, no. 11, pp. 15726-15734, 2019.

[7] T. Hu, W. Ma, Z. Zhang, J. Wu, L. Jiang, N. Yao, W. Zhou, Z. Huang, "Boosting performance of Mn1. 56Co0. 96Ni0. 48O4 film bolometer using randomly distributed nanostructures", *Japanese Journal of Applied Physics*, vol 60, no. 4, p. 040906, 2021.

[8] F. Niklaus, C. Jansson, A. Decharat, J. E. Källhammer, H. Pettersson, G. Stemme, "Uncooled infrared bolometer arrays operating in a low to medium vacuum atmosphere: performance model and tradeoffs", *Infrared Technology and Applications XXXIII* , vol. 6542, pp. 588-599, 2007.

CONTACT

*Sarah Risquez, sarah.risquez@silicon-austria.com

PERIODIC CAVITIES ON THE IR-ABSORBER FOR RESPONSIVITY ENHANCEMENT OF CMOS-MEMS THERMOELECTRIC IR SENSOR

Yung-Chen Li[1], Tien Chou[1], Pen-Sheng Lin[1], Yu-Cheng Huang[2], Fuchi Shih[2], You-An Lin[3], Da-Jen Yen[3], Mei-Feng Lai[1,2], and Weileun Fang[1,2]

[1]Dept. of Power Mechanical Eng., National Tsing Hua University, Hsinchu City, TAIWAN
[2]Inst. of NanoEngineering and MicroSyst., National Tsing Hua University, Hsinchu City, TAIWAN
[3]Dept. of Materials Science and Eng., National Tsing Hua University, Hsinchu City, TAIWAN

ABSTRACT

This study proposes a periodic cavities structure on the absorber of a CMOS-MEMS thermoelectric (TE) infrared (IR) sensor implemented through TSMC 0.18 μm 1P6M standard CMOS platform for absorption enhancement. Design proposed in this study features: (1) utilizing periodic cavities on the absorber to enhance IR absorption, (2) increasing cavity density to further improve IR absorption, (3) leveraging the advantages of CMOS platform to realize this sub-wavelength optical structure. The TE IR sensor in this study is aimed to be applied in human body temperature detections with a spectral sensing range from 8-14 μm wavelength. Absorption spectrum measurements show the enhancement within target band. Responsivity measurements also reveal cavity structures on the absorber can enhance the performance by 10.8%.

KEYWORDS

CMOS-MEMS, thermoelectric infrared sensor, IR-absorber, periodic cavities, responsivity.

INTRODUCTION

Infrared (IR) sensors have extensively appeared in various applications recently, such as thermometer, gas detection, thermal imager, human detection, etc. Especially due to the pandemic in the past few years, the IR sensors for body temperature detection are in great demand. IR sensor in general can be categorized into three types according to their sensing mechanisms [1], including bolometer [2], pyroelectric [3], and thermoelectric (TE) [4]. TE type exploiting the Seeback effect can directly convert temperature into voltage output [5]. The heat generated by incident IR-light causes redistribution of the charge carriers, which creates potential differences in the thermoelectric material, also known as thermopile. Their linear relationship makes TE type a decent mechanism in non-contact IR sensing. Furthermore, it also features few noise sources and no power consumption [6]. Those advantages have contributed to its growing prosperity in consumer market.

TE IR sensor is compatible with the CMOS processes [7]. The CMOS platform can leverage fabrication processes from the foundries, which is beneficial for batch production and precise fabrication results. It also offers stacking thin-film layers, including thermopile materials like Polysilicon, and dielectric materials like SiO_2 or Si_3N_4, which serve as outstanding absorbing materials for TE IR sensor. Its availability of sub-micron linewidth also provides a huge convenience for designing a MEMS structure. Multiple types of CMOS-MEMS TE IR sensors are studied to enhance the performances, especially for designs related to the interface of the absorber. Previous studies have successfully realized sensors integrated umbrella-like absorber [8], gold black coatings [9], plasmonic metamaterial absorber (PMA) [10], etc. In this study, a new form of the interface for IR-light receiving is proposed. By implementing periodic cavities structure on the IR-absorber, the performance enhancement in a CMOS-MEMS TE IR sensor is investigated.

DESIGN CONCEPT

Fig. 1 illustrates the proposed and the reference TE IR sensor designs based on the standard TSMC 0.18 μm 1P6M CMOS process. The CMOS platform offers multiple thin films to achieve the design flexibility. For example, the dielectric layer (SiO_2) is exploited as the IR-absorber. Noted the absorption band of dielectric material SiO_2 used in this study is located within target wavelength of 8-14 μm [11]. Moreover, by leveraging small linewidth process capability of the CMOS platform, this study presents the design to add periodic sub-wavelength cavities structure on the IR-absorber to create light-trapping effect [12], as shown in Fig. 1a. The design can strengthen the absorption of incident IR-light, so as to enhance the responsivity of the TE IR sensor.

Through the CMOS-MEMS processes, the sensing area is suspended and supported by four supporting beams where thermopile is embedded inside and simultaneously serves as heat-transfer path. In CMOS-MEMS component, the N-doped/P-doped Polysilicon pair are considered to be excellent materials for the implementation of TE effect because of their tremendous Seeback coefficient differences [13]. In addition, the two-dimensional sheet thermopile arrangement can decrease the electrical

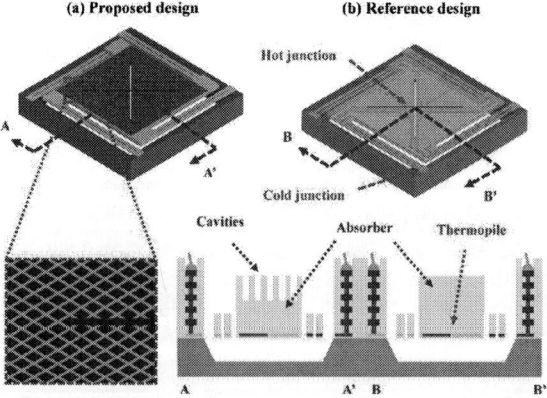

Figure 1: Design concept, (a-b) proposed and reference designs, and the zoom-in image of cavities on the absorber and cross-section view.

resistance and the noise, thus enhance the performances. The silicon substrate with good thermal conductivity will be removed by bulk Si etching to reduce the thermal loss.

This study also investigates the influence of the cavity density on the IR absorption. In Fig. 2, simulations demonstrate the absorption spectra can be improved by increasing the density of cavities at a given absorber area (220 μm×220 μm). Absorbers with four different cavity densities are respectively investigated. Compared with the reference type, significant absorption enhancement (particularly in the wavelength of 8-10 μm) is observed from the absorber with the highest cavity density (Type1).

FABRICATION

Fig. 3 displays the fabrication steps. First, the chip was fabricated by standard TSMC 0.18 μm 1P6M CMOS process (Fig. 3a). Next, the structure of the sensing area,

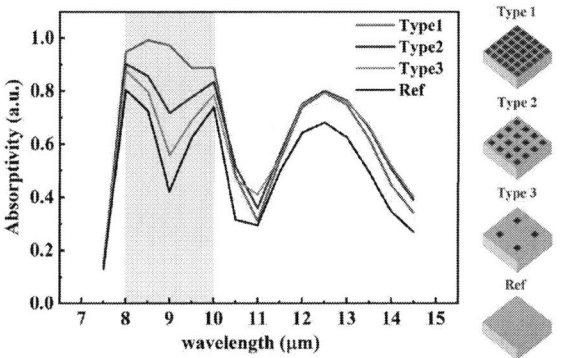

Figure 2: Simulations of IR-absorption in different wavelength for sensors of different absorber designs.

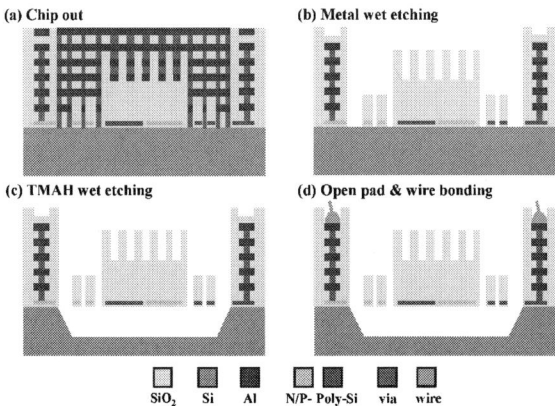

Figure 3: In-house post-fabrication process steps, (a) chip prepared by TSMC, (b) pattern definition by metal wet etching, (c) Si substrate removal, (d) wire bonding.

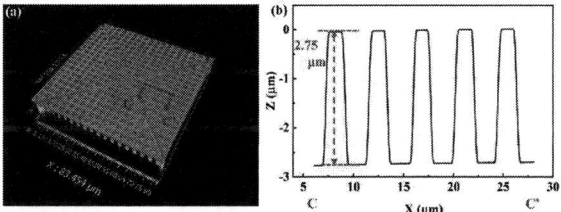

Figure 4: Surface topology measurements, (a) the IR-absorber, (b) CC' cross-section to show cavities.

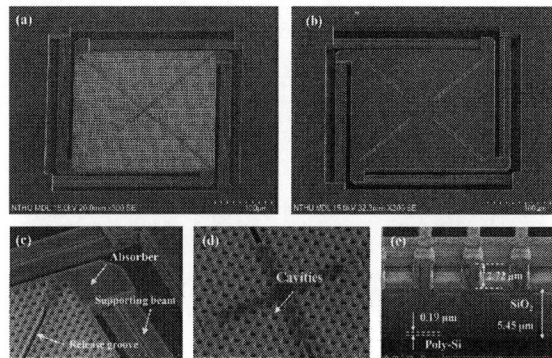

Figure 5: SEM micrographs of fabrication results, (a-b) proposed and reference designs, (c-d) zoom-in the supporting beams, cavities and etching release grooves on the absorber, (e) dimension measurements by FIB.

the supporting beams, and the cavities on the IR-absorber, were defined by metal wet etching process (Fig. 3b), of which the solution is a mixture composed of sulfuric acid (H_2SO_4) and hydrogen peroxide solution (H_2O_2). Third, the sensor was suspended by using TetraMethyl Ammonium Hydroxide (TMAH) anisotropic silicon etching processes (Fig. 3c). After that, the metal pads for electrical connection were exposed through Reactive Ion Etching (RIE), and the chip was interconnected to a printed-circuit-board (PCB) as the device-under-test (DUT) (Fig. 3d).

Fig. 4a presents the surface topology of the IR-absorber of the fabricated sensor measured by the optical interferometer. Fig. 4b further depicts the zoom-in graph of cross-section CC' to show the typical profile (depth/width) of periodic cavities on the IR-absorber. Measurements indicate the periodic cavities featuring 2.75 mm depth are well defined on the absorber by the CMOS process. SEM micrographs in Fig. 5a-b respectively display the entire proposed and reference sensor designs. The zoom-in micrograph in Fig. 5c demonstrates the supporting beams and etching release grooves on the IR-absorber. The supporting beam consists of two separate slender structures to enhance thermal isolation [14]. Fig. 5d shows periodic cavities and Fig. 5e presents the cross-section and the stacking layers by FIB (Focus-Ion-Beam) micrograph.

MEASUREMENT

The absorption spectrum with the feature of periodic cavities on the IR-absorber was measured by the Fourier-Transform InfraRed spectroscopy (FTIR) imager as shown in Fig. 6. The results in Fig. 7 indicate the IR-absorption at the target wavelength range (8-14 μm) for different absorber designs. As the cavity density increases, the gradual enhancement in absorptivity can be observed. Moreover, compared with the reference type, the proposed design with highest cavity density exhibits around 15% absorption improvement at 9 μm wavelength. Briefly speaking, the absorption behavior agrees well with simulations in Fig. 2.

Fig. 8 reveals the measurement setup for responsivity tests. The DUT was illuminated by using a blackbody radiator and the Ge/ZnSe filters were applied to attain an 8-12 μm band transmission. Noted the DUT was placed inside a vacuum chamber and the pressure was controlled

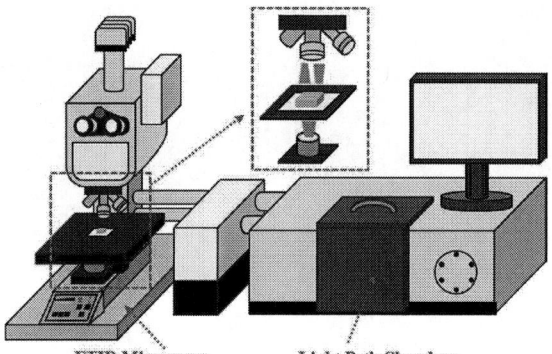

Figure 6: FTIR setup for absorption measurement.

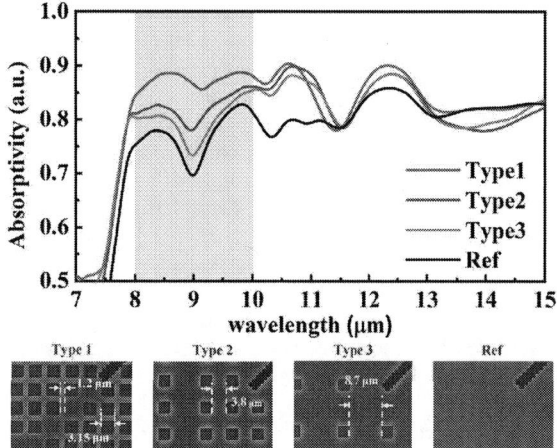

Figure 7: Measured absorption spectra of different absorber designs and their corresponding cavity specs.

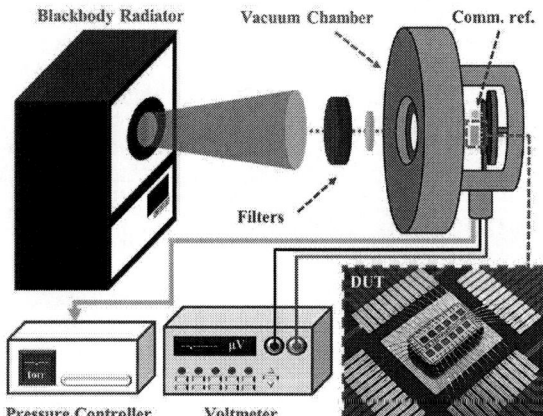

Figure 8: Schema of responsivity measurement setup and the chip on PCB as the DUT.

under 1 torr to reduce the heat loss from air. A commercial IR sensor was placed next to the DUT to monitor the input power. The output voltage was then measured by the voltmeter.

Measurements in Fig. 9 illustrate the responsivity for different designs. The responsivity of the proposed design with highest cavity density (Type1) is 157.33 V/W. As the cavity density gradually decreases for Type2 and Type 3 designs, their responsivities become 151.63 V/W and 142.86 V/W, respectively. In comparison, the reference

design with no cavity on the IR-absorber has a responsivity of 142 V/W. In short, the Type 1 design with highest cavity density (55.7% on the IR-absorber) has 10.8% responsivity enhancement compared with the reference design. Finally, the Type 1 design was measured under different ambient pressure. Measurements in Fig. 10 show the normalized output voltage varies from atmospheric pressure to 0.1 torr. The TE IR sensor performs better under higher vacuum environment due to the elimination of air conduction.

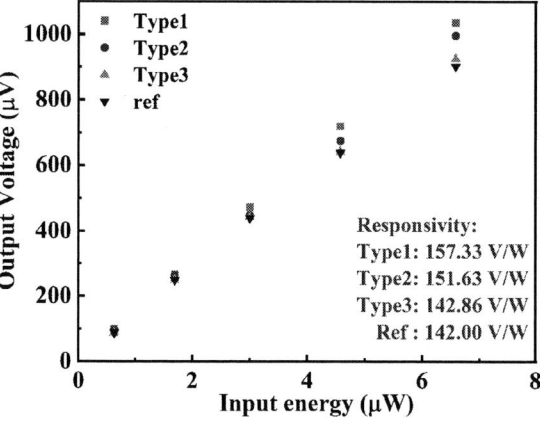

Figure 9: Responsivity measurements of different designs.

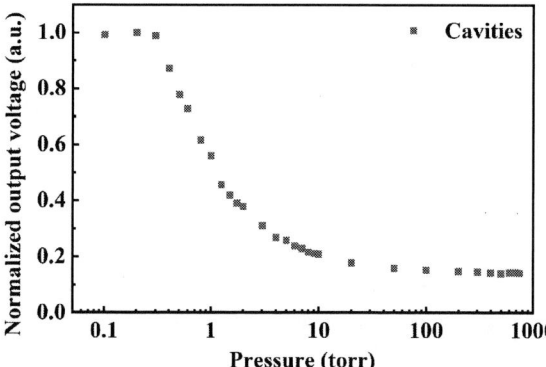

Figure 10: Measurement results of the proposed design under different pressure levels.

Table 1: Specs of IR sensors with different absorber designs.

	Type1	Type2	Type3	Ref
Area (μm²)	220×220			
Number of thermopile	4			
Resistance (kΩ)	190.6			
Cavity density on IR-absorber	55.7%	21.1%	4.2%	0%
Responsivity @1torr (VW⁻¹)	157.3	151.6	142.9	142.0
Detectivity (10⁸cmHz⁰·⁵W⁻¹)	0.62	0.60	0.56	0.56

978-1-6654-9309-3/23 $31.00 © 2023 IEEE

CONCLUSION

In this study, a CMOS-MEMS TE IR sensor with periodic cavities on the IR-absorber is successfully implemented. The CMOS platform simultaneously provides ideal thermoelectric materials and enables the realization of the sub-wavelength periodic structures. This study demonstrates the performance improvement of TE IR sensors by adding periodic cavity structures. The responsivity of the proposed design (with cavity density of 55.7% on the IR-absorber) is 157.33 V/W at 1 torr within target human body detection band 8-14 μm, which is 10.8% improvement in comparison with the reference type. Moreover, the simulations agree well with the absorption spectrum measurements. Table 1 concludes the specifications of different IR-absorber designs.

ACKNOWLEDGEMENT

This project was supported by the National Science and Technology Council under grant number NSTC 111-2923-E-007-004-MY2, NSTC 111-2923-E-007-003-, NSTC 111-2221-E-007-070-MY3, NSTC 111-2218-E-007-014-MBK, NSTC 110-2926-I-007-506-, and NSTC 110-2218-E-007-032-. The authors would like to appreciate Taiwan Semiconductor Manufacturing Co., Ltd. (TSMC) and Taiwan Semiconductor Research Institute (TSRI), for supporting the IC manufacturing and fabrication processes. The authors are also grateful to Center for Nanotechnology, Materials Science, and Microsystems of National Tsing Hua University (CNMM) for providing measurement facilities.

REFERENCES

[1] Y.-C. Lee, M.-L. Hsieh, P.-S. Lin, C.-H. Yan, S.-K. Yeh, T. Do, and W. Fang, "CMOS-MEMS Technologies for the Applications of Environment Sensors and Environment Sensing Hubs," in *Journal of Micromechanics and Microengineering*, vol. 31, no. 7, 2021, Art. no. 074004.

[2] A. Blaikie, D. Miller, and B. J. Alemán, "A fast and sensitive room-temperature graphene nanomechanical bolometer," *Nature Communications*, vol. 10, no. 1, 2019, Art. no. 4726.

[3] S. P. Gaur, K. Rangra, and D. Kumar, "MEMS AlN pyroelectric infrared sensor with medium to long wave IR absorber," *Sensors and Actuators A: Physical,* vol. 300, 2019, Art. no. 111660.

[4] M. J. Modarres-Zadeh and R. Abdolvand, "High-responsivity thermoelectric infrared detectors with stand-alone sub-micrometer polysilicon wires," in *Journal of Micromechanics and Microengineering,* vol. 24, no. 12, 2014, Art. no. 125013.

[5] A. Graf, M. Arndt, M. Sauer, and G. Gerlach, "Review of micromachined thermopiles for infrared detection," *Measurement Science and Technology*, vol. 18, no. 7, pp. R59-R75, 2007.

[6] C. Escriba, E. Campo, D. Estève, and J. Y. Fourniols, "Complete analytical modeling and analysis of micromachined thermoelectric uncooled IR sensors," *Sensors and Actuators A: Physical,* vol. 120, no. 1, pp. 267-276, 2005.

[7] R. Lenggenhager, H. Baltes, J. Peer and M. Forster, "Thermoelectric infrared sensors by CMOS technology," in *IEEE Electron Device Letters*, vol. 13, no. 9, pp. 454-456, 1992.

[8] T.-W. Shen, K.-C. Chang, C.-M. Sun, and W. Fang, "Performance enhance of CMOS-MEMS thermoelectric infrared sensor by using sensing material and structure design," in *Journal of Micromechanics and Microengineering*, vol. 29, no. 2, 2019, Art. no. 025007.

[9] H. Masaki and M. Shinichi, "Infrared sensor with precisely patterned Au black absorption layer," in *Proc. SPIE*, vol. 3436, pp. 623-634, 1998.

[10] P.-S. Lin, T.-W. Shen, K.-C. Chan and W. Fang, "CMOS MEMS Thermoelectric Infrared Sensor With Plasmonic Metamaterial Absorber for Selective Wavelength Absorption and Responsivity Enhancement," in *IEEE Sensors Journal*, vol. 20, no. 19, pp. 11105-11114, 2020.

[11] J. Kischkat, S. Peters, B. Gruska, M. Semtsiv, M. Chashnikova, M. Klinkmüller, O. Fedosenko, S. Machulik, A. Aleksandrova, G. Monastyrskyi, Y. Flores, and W. Ted Masselink, "Mid-infrared optical properties of thin films of aluminum oxide, titanium dioxide, silicon dioxide, aluminum nitride, and silicon nitride," *Appl. Opt.,* vol. 51, no. 28, pp. 6789-6798, 2012.

[12] J. Schieferdecker, R. Quad, E. Holzenkämpfer, and M. Schulze, "Infrared thermopile sensors with high sensitivity and very low temperature coefficient," *Sensors and Actuators A: Physical,* vol. 47, no. 1, pp. 422-427, 1995.

[13] Z. Yu, A. Raman, and S. Fan, "Fundamental limit of nanophotonic light trapping in solar cells," *Proceedings of the National Academy of Sciences*, vol. 107, no. 41, pp.17491-17496, 2010.

[14] Y.-C. Huang, P.-S. Lin, Y.-L. Chen, F. Shih, C.-F. Hu and W. Fang, "Novel High Thermal Resistance Structure Design for Responsivity and Detectivity Enhancements of Cmos Mems Thermoelectric Infrared Sensor," *35th IEEE MEMS*, Tokyo, Japan, Jan. 9-13, 2022, pp. 975-978.

CONTACT

*Weileun Fang; fang@pme.nthu.edu.tw

ULTRA-LARGE PIXEL ARRAY PHOTOTHERMAL TRANSDUCER AND ITS THERMAL PERFORMANCE PREDICTION STRATEGY

Defang Li[134], Jinying Zhang[12]*, Jiushuai Xu[34], Erwin Peiner[34], Zhuo Li[12], Xin Wang[1], Suhui Yang[1], and Yanze Gao[1]

[1] Beijing Key Laboratory for Precision Optoelectronic Measurement Instrument and Technology, School of Optics and Photonics, Beijing Institute of Technology, Beijing, 100081, P. R. China
[2] Yangtze Delta Region Academy of Beijing Institute of Technology, Jiaxing, 314001, P. R. China
[3] Institute of Semiconductor Technology (IHT), Technische Universität Braunschweig, Hans-Sommer-Straße 66, Braunschweig 38106, Germany and
[4] Laboratory for Emerging Nanometrology (LENA), Technische Universität Braunschweig, Langer Kamp 6, Braunschweig 38106, Germany

ABSTRACT

A photothermal (PT) transducer with beyond 4000 × 4000 pixels is proposed and fabricated. A thermal performance predication strategy for this ultra-large pixel array is proposed for the first time. Two different pixel patterns are fabricated and tested to explore their steady and transient thermal properties. Experimental results indicate this transducer shows great potential of high frame rate, and simulation results show that the microstructure design of pixel plays an important role in the thermal properties. The proposed PT transducer is promising in high resolution and its fabrication is compatible to low-cost and mass production.

KEYWORDS

Photothermal transducer, ultra-large pixel array, thermal performance predication strategy.

INTRODUCTION

Infrared emitter has been widely utilized in infrared scene generation and infrared detection fields [1,2]. With the development of infrared detection system, the performance index of infrared emitter is increasing [3,4]. The infrared generation system aims not only to provide convenient infrared source, but also to generate customized infrared pictures and videos. To generate high-precision infrared scene in the laboratory, the critical parameters of infrared emitters, such as pixel array and response time, are vital to the infrared scene generation. PT transducer is one of the key devices of the infrared emitters and it has gained great progress in recent years [5]. Benefit from the mature micro/ nano fabrication, the pixel array can be increased to million scale [6]. For the thin film PT transducer, the pixel array has achieved 1400 × 1400 pixels [7]. For the PT transducer based on silicon (Si), the pixel array is further increased to 2000 × 2000 pixels [8]. This type of transducer owns the merits of mature manufacture and low cost, but the suspended micro-bridge or micro-cavity structure limits its frame rate and larger pixel array is a hard nut to crack.

In this work, a PT transducer with ultra-large pixel array is proposed and fabricated. Based on the ingenious microstructure design, this transducer achieves the state of the art in pixel array up to 4000 × 4000 and the time response up to 2 ms. To value the thermal performance of PT transducer with ultra-large pixel array, a thermal performance predication strategy is proposed for the first time. Two different pixel patterns are designed and fabricated to explore the steady and transient thermal properties, and their corresponding FEM models are built to verify the accuracy of the proposed thermal performance predication strategy.

STRUCTURE OF THE TRANSDUCER

Figure 1 shows schematics of the previous PT transducer and the proposed PT transducer. Both of the PT transducers are based on the principles of photothermal conversion and thermal radiation. The PT transducers are driven by visible light. When the pixels of the transducer absorb the given light energy, their temperature increases to emit the corresponding infrared ray according to Plank's blackbody radiation law. Therefore, the distribution of infrared radiation field can be modulated by changing the distribution of given light energy. Through this method, various infrared pictures or infrared videos can be generated by the PT transducers.

As shown in Fig.1 (a), the previous PT transducer consists of four layers, including aluminum (Al) black, chromium (Cr), polyimide (PI) and silicon (Si). The pixel pattern is double S with period of 37 μm (including 34 μm side length and 3 μm spacing) and the supporting structure is a cavity [3]. For the proposed PT transducer, it contains Al black, silica (SiO_2) and Si. The pixel pattern is a square with period of 12 μm (including 9 μm side length and 3 μm spacing) and the supporting structure is a columnar support.

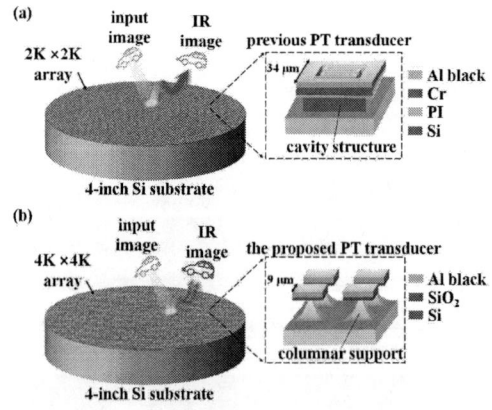

Figure 1: Schematics of (a) the previous PT transducer based on cavity structure and (b) the proposed PT transducer based on columnar support. IR image means infrared image.

To explore the influence of pixel pattern on the thermal performance, two kinds of pixel pattern are designed and fabricated. Fig.2(a) and (b) are scanning electron microscope (SEM) images of the two PT transducer before depositing Al black. One pattern is a circle with 7 μm diameter, and another pattern is a square with 8 μm side length. Both the spacing between adjacent pixel is 3 μm and the pixel patterns have the columnar support with the narrow top and wide bottom structure. As shown in the top views, the bright part is SiO_2 layer and the dark part is Si layer. It can be seen that the Si supporting structure is in the middle of pixel pattern.

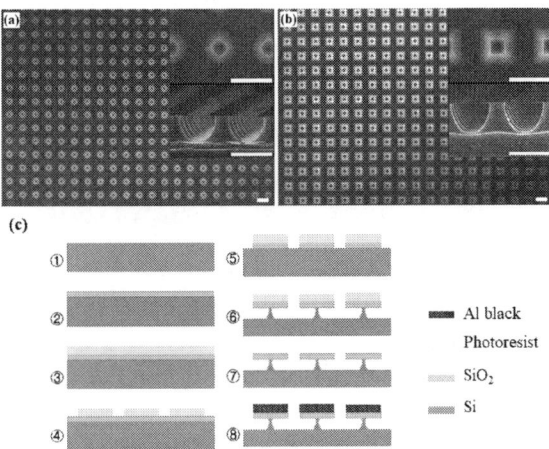

Figure 2: SEM images and processing flow of the two PT transducers. (a) pixels with circle pattern. (b) pixels with square pattern. Inserts are their corresponding magnified pictures with top view and sectional view. All scale bars are 10 μm. (c) processing flow of the proposed PT transducer.

The processing flow of the proposed PT transducer is shown in Fig. 2(c). First, a SiO_2 layer with 300 nm thickness is grown on a 4-inch clean Si wafer by thermal oxidation. Through sequential photolithography and dry etching, the pixel pattern is transferred from the photoresist to the SiO_2 layer. Then, Si is etched using the photoresist and SiO_2 layer as a mask until the photoresist is completely removed. Finally, a layer of Al black is deposited on the SiO_2 layer.

Fig.3 (a) and (b) are the SEM images of PT transducers in Fig.2 (a, b) after depositing Al black, and Fig. 3(d) is the photo of PT transducer with circle pattern. Fig.3 (c) is the SEM image of the previous PT transducer in the top view. The scale bars are 10 μm in Fig. 3(a-c). Since the proposed PT transducer possesses a single columnar support, its contact area (area between Si and SiO_2) is smaller than that of the previous PT transducer. The side length of single pixel is decreased by 76.5%. Therefore, the pixel array can be increased from 2000 × 2000 to 4000 × 4000 on the 4-inch silicon substrate.

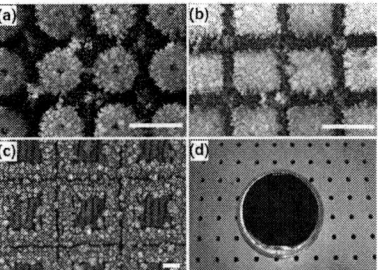

Figure 3: SEM images of the two PT transducers and previous PT transducer after depositing Al black. (a) pixels with circle pattern. (b) pixels with square pattern. (c) previous pixels with double S pattern. All scale bars are 10 μm. (d) photo of the fabricated PT transducer with circle pattern.

EXPERIMENTAL RESULTS

Fig.4 (a) and (b) are experimental setups of steady and transient thermal performance. For the steady thermal performance, a 532 nm laser is modulated by the signal generator. The laser beam is reflected by a reflecting visible light mirror and then passes through a reflecting visible light and transmitting infrared light mirror. Finally, the laser point is incident to the surface of PT transducer. The generated infrared radiation signal is collected by thermal imager (VarioCAM HO head) and the corresponding temperature data is recorded and saved in computer. For the transient thermal performance, an infrared detector (PVI-4TE-5, response wavelength: 3-5 μm) is used to receive the infrared signal and the signal is collected and saved by an oscilloscope. To explore the effect of pixel pattern on the steady thermal performance, the PT transducer is tested without Al black. Al black layer can improve absorption and emit strong infrared signal, but it is quite difficult to obtain a uniform and thickness-controllable Al black layer on different 4-inch samples by thermal deposition. However, in the test of transient thermal performance, Al black layer can be used to get a strong infrared signal, and the absorption difference can be eliminated by normalizing the radiation intensity.

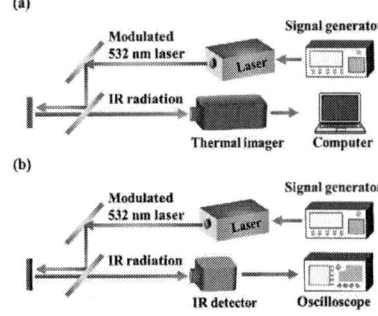

Figure 4: Schematics of the experimental setups of (a) thermal performance and (b) transient response.

The steady temperature result and transient response

time are shown in Fig.5 (a) and (b), respectively. The laser voltage is proportional to the laser power. When the voltage comes to its maximum value, the laser power is around 28 W. The two pixel patterns conform to a same trend that the temperature rises as the laser power increases. For the transient result presented in Fig. 5(b), the two pixel patterns both meet the response time within 2 ms. Their rising time (response time of 0.1 to 0.9 times of the maximum radiation intensity) and falling time (response time of 0.9 to 0.1 times of the maximum radiation intensity) are nearly 0.05 ms. This response time is 10 times faster than that of the previous PT transducer [5,6,9].

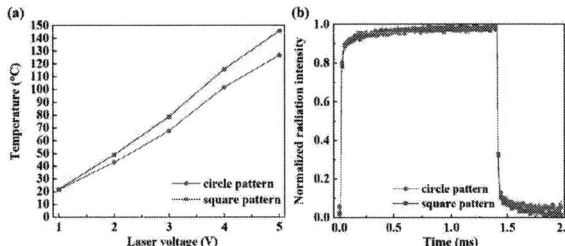

Figure 5: Measured results of (a) the steady temperature varied in different voltages and (b) the transient response for the two PT transducers.

SIMULATION OF THE TRANSDUCER

The main heat transfer methods of PT transducer include heating source, heat conduction, radiation and convention. The heating source applied to the pixel mainly depends on the laser power density and the absorptivity of the radiation layer. The heat conduction rate depends on the material parameters and contact thermal resistance between pixel and the supporting structure. Heat radiation is related to temperature and emissivity. Since the PT transducer is tested in the air environment, so all of the surfaces of the whole structure are set as air convection boundary condition. The formula (1) describes the heat transfer of the PT transducer in thermal steady state,

$$Q+kd\frac{\partial^2 T}{\partial r^2}-\sigma\varepsilon\left(T^4-T^4_{\text{amb}}\right)-h(T-T_{amb})=0 \qquad (1)$$

where Q is the heating energy applied on the surface of the radiation area, k is thermal conductivity, $\vec{r}$ is the inner normal direction, d and T are the thickness and temperature of the materials, respectively. For the radiation term, σ is the Stefan-Boltzmann constant of 5.67×10^{-8} W/(m²·K⁴) and ε is emissivity. T_{amb} is the ambient temperature. h is natural convection heat transfer coefficient. The terms in the formula (1) describe heating source, absorption, conduction, and radiation in turn. Since the order of magnitude of heat convection and heat radiation is much smaller than that of heat conduction, the thermal equation of steady state can be simplified as applied heating energy and heat conduction. It can be seen that when the applied heating energy and heat conduction change proportionally, this equation still holds. When a heating source (laser) with a certain power density is incident on the surface of PT transducer, a larger irradiated area indicates the device

absorbs more heat. Si substrate acts as a heat sink, so the larger volume indicates more heat is conducted. Since the thermal equation of steady state is constant, we can control the heating energy and heat conduction term equiproportionally by varying the irradiated area and the volume of Si, and further obtain the corresponding steady-state temperature.

Fig. 6 presents the principle of the proposed thermal performance predication strategy. Firstly, we calculate the dimension ratio of thickness of Si (T_{Si}) and diameter of Si substrate (D_{Si}) and laser point (D_{laser}), which conforms to $T_{\text{Si}} : D_{\text{Si}} : D_{\text{laser}} = 1 : 200 : 17.6$. The model is set as this dimension ratio to get the simulated result of steady temperature. The pixel array is built as 3×3 array, and then expand to 10×10, 20×20, ..., 50×50 array. Then linear fitting curve can be plotted by the simulated results of these arrays. Utilizing this linear fitting curve, interpolation and extrapolation can be calculated. Through positioning the point to the actual laser diameter (8.8 mm), the final deductive point can be obtained. Table 1 shows the material parameters used in FEM simulation.

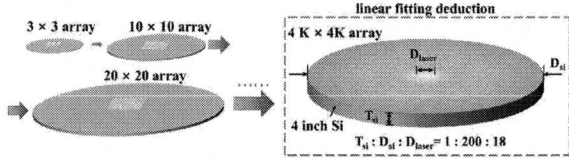

Figure 6: Schematic of the proposed prediction strategy based on the linear fitting deduction.

Table 1: Material parameters used in FEM simulation.

Parameter	SiO₂	Si
Density (kg/m³)	2200	2329
Specific heat capacity (J/ (kg · K))	730	700
Thermal conductivity (W/ (m · K))	1.2	130
Emissivity	0.6	/

Fig. 7(a) presents the simulated results of the two transducers with circle and square pixel patterns. The insets are the simulated results from 3×3 array to 50×50 array. As the side length of heating area expands to the diameter of laser (beyond 700×700 array), the simulated temperature of circle and square patterns are 121.81°C and 150.57°C, respectively. The measured temperature of circle and square patterns are 126.56°C and 146.01°C, respectively. The maximum percentage error of average temperature between simulated and measured results is within 5.22%.

Fig. 7(b) shows the simulated and measured results of the two pixel patterns varied in different laser voltages. It is obvious that the simulated points have nearly the same trend and value as the measured points. The average temperature of PT transducer increases with increasing laser power, regardless of the pixel pattern. Comparison of the measured and simulated results shows that the error of average temperature for the proposed prediction strategy is less than 6°C, which manifests the proposed thermal performance predication strategy has great predictability.

978-1-6654-9309-3/23 $31.00 © 2023 IEEE

The proposed thermal performance predication strategy not only provides design guidance for optimizing PT transducer, but is also useful for other ultra-large pixel array device that require thermal performance optimization. The wider universality of this thermal performance predication strategy will be further explored and proved in future studies.

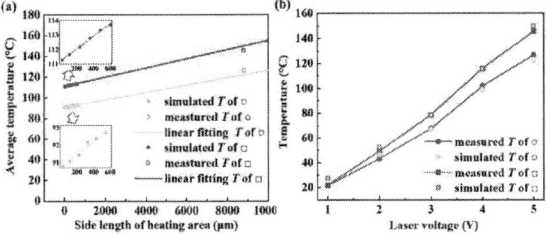

Figure 7: Comparison of measured and simulated results of the two PT transducer. (a) measured results of the two PT transducers and their corresponding simulated results based on the linear fitting deduction. Insets are the simulated results from 3 × 3 array to 50 × 50 array. (b) measured and simulated results of the two PT transducers varied in different voltages.

CONCLUSIONS

A PT transducer with more than 4000 × 4000 pixels is designed and fabricated. With the improved microstructure design, this transducer possesses excellent time response and state-of-the-art pixel scale. A thermal performance prediction strategy is proposed to study the influence of microstructure on the thermal performance of the ultra-large pixel array for the first time. Two different pixel patterns are fabricated and tested to explore their steady and transient thermal properties. Measured results indicate the two PT transducers show great potential of high frame rate with the time response of 2 ms. The average temperature of PT transducer increases with increasing laser power, regardless of the pixel pattern. Comparison of measured and simulated results shows that the proposed thermal performance prediction strategy has great predictability, and the maximum percentage error of average temperature is within 5.22%. The PT conversion efficiency of the proposed transducer will be further investigated in the future.

ACKNOWLEDGEMENTS

This work was partially supported by National Natural Science Foundation of China (No. 62174012, 61704166), National Key Research and Development Program of China (2018AAA0100301, 2018YFF01010304), and China Scholarship Council (student ID: 202106030163). Authors are thankful to Integrated Circuit Process and Test Laboratory, Prof. Jiang Yan and Prof. Jing Zhang in the North China University of Technology. Thanks are also given to Analysis & Testing Center in Beijing Institute of Technology for providing the fabrication and characterization facility for this study.

REFERENCES

[1] J. Singh, F. Kiamilev, T. Lassiter, A. Deputy and M. Joyce, "Fabrication of the Next Generation of Drive Electronics for Infrared Scene Projectors," in *2022 IEEE Research and Applications of Photonics in Defense Conference (RAPID)*, Miramar Beach, 12-14 September, 2022, pp. 1-2.

[2] N. Scherer-Negenborn and A. Schmied, "Semi-synthetic naval scene generation for infrared-guided missile threat analysis with separate setting of apparent temperatures for each target part," in *SPIE Security + Defence*, Strasbourg, 17 October, 2019, vol. 11158.

[3] J. Zhang, D. Li, Z. Li, X. Wang and S. Yang, "A Photothermal Transducer Based on 3D Thermal Management," in *2021 21st International Conference on Solid-State Sensors, Actuators and Microsystems (Transducers)*, Orlando, 20-24 June, 2021, pp. 601-604.

[4] Y. Gao, Z. Li, S. Zhang, T. Zhao, R. Shi, and Q. Shi, "Infrared scene projector (IRSP) for cryogenic environments based on a light-driven blackbody micro cavity array (BMCA)," *Opt. Express*, vol. 29, pp. 41428-41446, 2021.

[5] D. Li, J. Zhang, Q. Shi, X. Yuan, Z. Li, X. Wang, S. Yang and Y. Hao, "A Robust Infrared Transducer of an Ultra-Large-Scale Array", *Sensors*, vol. 20(23), pp. 6807, 2020.

[6] J. Zhang, D. Li, Z. Li, X. Wang and S. Yang, "Demonstration of thermal modulation using nanoscale and microscale structures for ultralarge pixel array photothermal transducers", *Microsystems & Nanoengineering*, vol. 7, pp. 102, 2021.

[7] L. Zhou, X. wang, S. Yang, J. Zhang, Y. Gao, C. Xu, D. Li, Q. Shi and Z. Li, "A self-suspended MEMS film convertor for dual-band infrared scene projection", *Infrared Physics & Technology*, vol. 105, pp. 103231, 2020.

[8] D. Li, J. Zhang, Q. Shi, Q. Li, D. Zhao and Z. Li, "A Robust Infrared Transducer Beyond 2K ×2K Pixels," in *2021 IEEE 34th International Conference on Micro Electro Mechanical Systems (MEMS)*, Gainesville, 25-29 January, 2021, pp. 923-926.

[9] L. Zhou, Z. Li, D. Li, C. Xu, Y. Gao, J. Zhang, X. Wang, S. Yang and K. Wang, "Large Scale Array Visible-Infrared Converter Based on Free-Standing Flexible Composite Microstructures," in *2019 20th International Conference on Solid-State Sensors, Actuators and Microsystems & Eurosensors XXXIII (TRANSDUCERS & EUROSENSORS XXXIII)*, Berlin, 23-27 June, 2019, pp. 2527-2530.

CONTACT

*Corresponding author: Jinying Zhang, jyzhang@bit.edu.cn

A CMOS-MEMS BEAM RESONATOR WITH $Q > 10,000$
Ting-Yi Chen and Wei-Chang Li
Institute of Applied Mechanics, National Taiwan University, Taipei, Taiwan

ABSTRACT

This work for the first time demonstrates a Q up to 11,450 in a flexural CC-beam resonator based on a CMOS-MEMS process platform. In particular, the CC-beam resonator employs the Q-enhancement stepped structure along the beam length to introduce stress dilution to address the low Q issue dominantly limited by the material loss in the CMOS-MEMS process. With a theoretically predicted dilution factor D_Q up to 3.8×, the measured Q of a stress-engineered stepped CC-beam resonator shows a similar improvement of 3.2× compared to that of regular counterparts. This technique not only avoids the typical post-fabrication annealing process but also overturns the low Q impression of CMOS-MEMS based devices.

KEYWORDS

Quality factor enhancement, stress dilution, CC-beam resonators, CMOS-MEMS resonators.

INTRODUCTION

While the popular demands for high data-rate communications keep pushing the development of micromechanical resonator devices towards higher frequencies [1], resonators operating at MHz or even kHz are still useful in abundant applications such as low-power sensors [2] and timing references [3]. In this context, beam resonators are one of the mostly used topologies [4]. Due to the use of materials and the operating frequency, Q's of these devices are typically limited by thermoelastic damping (TED) and anchor loss and therefore, Q raising techniques have been demonstrated to mitigate TED [5] and anchor loss [6]. For metal beam resonators [7], however, the material loss becomes rather the dominant factor, which requires other strategies to improve Q other than the abovementioned techniques. To address this, [7] introduces annealing induced tensile stress and successfully demonstrates an impressive Q improvement by 271.8×. [8] then avoids the use of post-fabrication annealing and introduces the Q-boosting stress dilution factor via structural design to improve Q by 2.4×. In a similar manner, this work employs the phononic crystal structure of [9] and implements in a CC-beam resonator to deal with the lossy materials in the CMOS-MEMS technology.

DEVICE STRUCTURE

Fig. 1 depicts the schematic of the beam resonator, which contains isometric stepped cells with dimensions labeled in the inset. To observe how the geometry affects the stress dilution technique, various beam lengths and numbers of cells are deployed in this work. Table 1 lists the geometric parameters of the device, where the length ratio l_3/l_2 of 1.2 is borrowed from [9]. In addition, the arc-shaped driving electrode is adopted to yield a gap spacing smaller than the nominal gap defined by the design rule of the standard 0.35-μm CMOS-MEMS process platform after

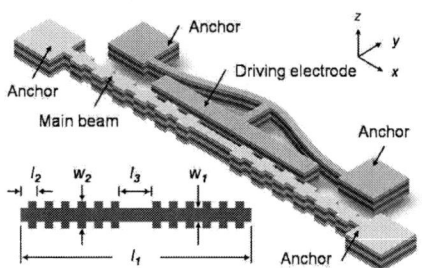

Fig. 1. Schematic of a stress-diluted stepped CC-beam resonator, showing the 1st mode shape and geometric parameters.

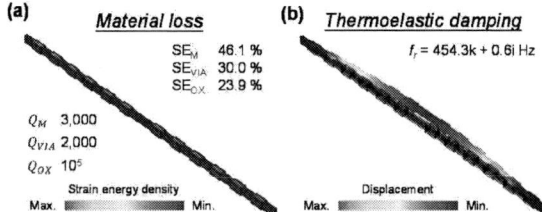

Fig. 2. (a) The material loss is calculated by considering the Q value and the strain energy (SE) stored in each material, where SE is obtained by integrating the FEA-derived strain energy density over the volume. (b) The FEA-derived TED, where the imaginary part indicates the lossy eigenfrequency.

Table 1. Geometric Parameters of the Device

Parameter	l_1	l_3	w_1	w_2	# Step
Value (μm)	160; 200	$1.2 \times l_2$	1.6	$2 \times w_1$	12; 16; 20; 24

release [10].

STRESS DILUTED CC-BEAMS

Borrowing the theory from [11], the intrinsic Q values in micromechanical beam resonators could be determined by considering TED, anchor loss, and material loss. To calculate, Fig. 2 (a) shows the TED derived from finite element analysis (FEA), which simply takes the form as

$$Q_{TED} = \frac{Re(f_r)}{Im(f_r)} \qquad (1)$$

where f_r and $Im(f_r)$ represent the resonance frequency and the lossy eigenfrequency, respectively. The anchor loss for the 1st flexural mode of a CC-beam resonator could be expressed as [12]

$$Q_{AN} = 0.638 \left(\frac{l}{w_{eff}} \right)^3 \qquad (2)$$

where l and w_{eff} are the length and effective width of the beam, respectively. The material loss could be extracted as

$$Q_{MAT}^{-1} = SE_{Al}Q_{Al}^{-1} + SE_W Q_W^{-1} + SE_{SiO_2}Q_{SiO_2}^{-1} \qquad (3)$$

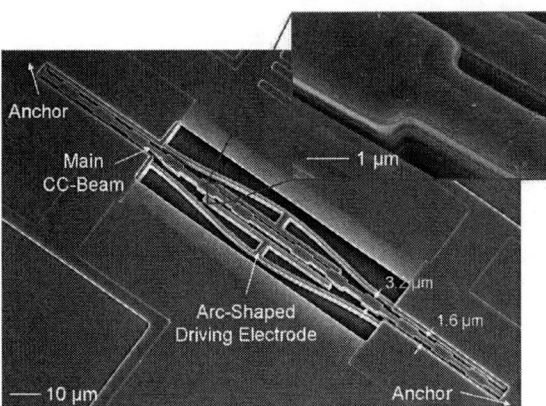

Fig. 3. The SEM photo of a releases 12-step CC-beam resonator, with a zoom-in to the step location.

Table 2. Stress Dilution Model Parameters

Parameter	Q_{TED}	Q_{AN}	Q_{MAT}	D_Q	Q_{ABS}
Value	$7.6×10^5$	$3.9×10^5$	3,570	*3.8	*13,380

*For the 16-step case

where SE_i and Q_i are the strain energy and the quality factor of ith material, respectively. Fig. 2 (b) shows the simulated strain energy density plot, along with the strain energy (SE) and Q of each material listed in the inset. Finally, combining all lossy mechanisms, the intrinsic Q value takes the form

$$Q_{IN}^{-1} = Q_{TED}^{-1} + Q_{AN}^{-1} + Q_{MAT}^{-1} \qquad (4)$$

Involving the stress dilution in [9],the overall Q takes the form

$$Q = Q_{IN} \times D_Q \qquad (5)$$

of which the dilution factor D_Q is expressed as [9]

$$D_Q = 1 + \frac{12\sigma_{eff}l^2}{Ew^2} \frac{\int_0^1 \big(u'(x)\big)^2 dx}{\int_0^1 v(x)\big(u''(x)\big)^2 dx} \qquad (6)$$

where σ_{eff}, E, w, $u(x)$, $v(x)$ are the effective deposition stress, Young's Modulus, width, mode shape function, and width ratio, respectively. In this work, σ_{eff} is determined by the comparing the difference between the simulated resonance frequency and the measured one due to the residual stress induced frequency shift [7], and $u(x)$ could be numerically calculated using the Adomian decomposition method (ADM) with material properties in [13]. Table 2 summarizes the stress dilution parameters derived in this work.

EXPERIMENTAL RESULTS
Fabrication Process and Measurement Set-Up

Both the regular and stepped CC-beam resonator are fabricated using the 0.35-μm CMOS-MEMS process platform in [14]. Fig. 3 depicts the SEM photo of a released stepped CC-beam resonator with 12 steps, along with a zoom-in to highlight the step location. Fig. 4 illustrates the typical measurement set-up to measure the frequency response, where an ac voltage v_{ac} from a network analyzer, combined with a DC bias V_P from a power supply, is applied across the driving electrodes and the main beam.

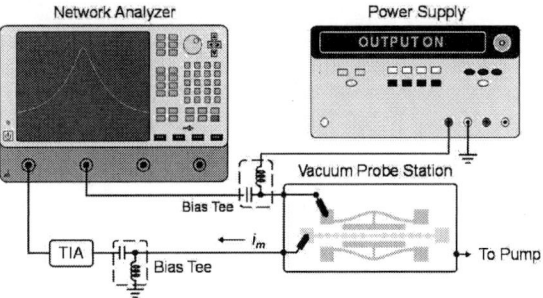

Fig. 4. Illustration of the measurement set-up. A network analyzer, a TIA, and a power supply are used for measuring the frequency responses.

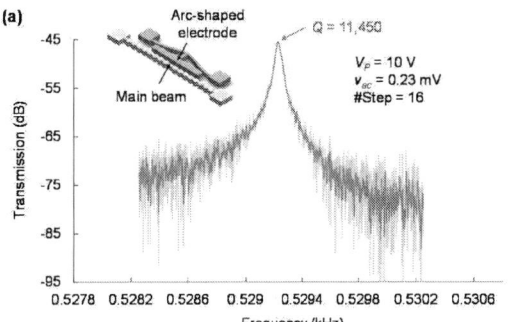

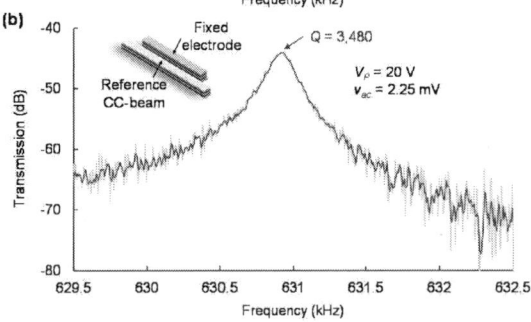

Fig. 5. Measured frequency responses of (a) the stress-engineered and (b) a regular CC-beam resonator, respectively. Compared to Q = 3,480 of a reference CC-beam resonator, this work achieves up to Q = 11,450 with a 16-step structure at V_P = 10 V and v_{ac} = 0.23 mV by simply introducing steps into the uniform profile and successfully shows Q enhancement of 3.2×.

The motional current, resulting from the time-varying electrostatic force of the device, is then picked up by the network analyzer to obtain the frequency response through a transimpedance amplifier (TIA).

Measurement Results

Fig. 5 (a) and (b) plot the measured frequency responses of a stress-engineered and a regular CC-beam resonator, respectively. Compared to Q of 3,480 of the reference one, the stepped CC-beam structure achieves a Q up to 11,450, which is a 3.2× enhancement by simply introducing steps into a uniform beam profile. Both the 3.2× enhancement and the reference Q of 3,480 reveal promising agreements to the theoretical derived results. Furthermore, to study Q variations as the beam geometry

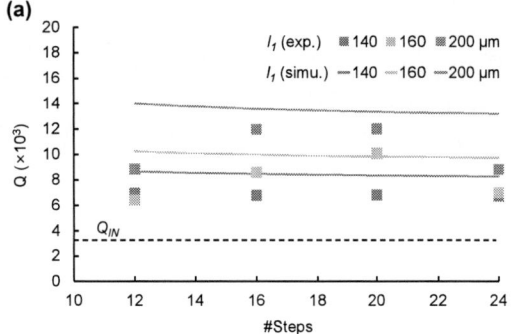

(a)

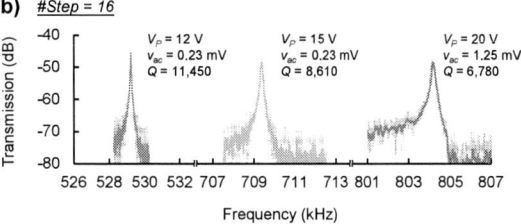

(b) #Step = 16

V_P = 12 V
v_{ac} = 0.23 mV
Q = 11,450

V_P = 15 V
v_{ac} = 0.23 mV
Q = 8,610

V_P = 20 V
v_{ac} = 1.25 mV
Q = 6,780

Fig. 6. (a) Theoretical and empirical geometric dependency of Q's of the stepped CC-beam resonator, showing that Q increases as the beam length increases, and all stress diluted Q's are higher than that without applying this technique. (b) The frequency responses of the 16-step structure.

changes, Fig. 6 (a) shows the measured and theoretically predicted Q's with varying beam lengths and the number of steps. All Q's overcome the intrinsic Q limit by employing the stress dilution technique. While some of the measured Q's fall short of the theory prediction, they still show the same tendency such that Q increases as the beam length increases. Zooming in to the highlighted 16-step case in Fig. 6 (a), Fig. 6 (b) shows the frequency responses of the 16-step structure with different beam lengths to examine this tendency. Finally, Table 3 finally compares the work with other Q-boosting micromechanical resonators.

CONCLUSION

This work demonstrates the stress dilution technique to enhance Q's of beam resonators with factors of 3.2× compared to regular CC-beam counterparts. Particularly, by introducing steps into a uniform profile to deal with the dominant material loss in the CMOS-MEMS process platform, the 16-step stress-engineered resonator for the first time achieves an even higher Q up to 11,450 in flexural modes. Indeed, a $Q > 10,000$, although still falling short of that of commercial timing resonators, already satisfies the requirement for low-cost applications.

ACKNOWLEDGEMENT

This research was funded by National Science and Technology Council, Taiwan (MOST-109-2628-E-002-004-MY3). The chip fabrication was supported by the Taiwan Semiconductor Research Institute (TSRI) and Taiwan Semiconductor Manufacturing Company (TSMC), Hsinchu, Taiwan. The authors would like to thank the staff in NEMS Research Center at NTU for providing technical

Table 3. Comparison of Q-boosting Techniques

Ref.	[2]	[7]	[12]	This work
Material & Process	Ru surface micro-machining	SOI epi-seal	SOI epi-seal	AlCu/W/SiO$_2$ CMOS-MEMS
Technique	Localized annealing	BPSO algorithm	Topology optimization	Stress engineering
Resonance frequency	12 MHz	400 kHz	~350 kHz	529 kHz
Measured Q	48,919	57,000*	~95,000*	11,450
Improvement	271.8×	1.4×*	1.75×*	3.2×

*Results are specifically for Q_{TED}.

support.

REFERENCES

[1] T. L. Naing, T. O. Rocheleau, Z. Ren, S.-S. Li and C. T.-C. Nguyen, "High-Q UHF Spoke-Supported Ring Resonators," *Journal of Microelectromechanical Systems*, vol. 25, no. 1, pp. 11-29, 2016.

[2] P.-C. Huang, T.-Y. Chen, C.-P. Tsai and W.-C. Li, "An Ultrasensitive Cmos-Mems Tuned-Mass-Damper (Tmd) Based Voltmeter," in *the 35th IEEE Int. Conf. on Micro Electro Mechanical Systems (MEMS'22)*, Tokyo, Japan, Jan. 9-13, 2022.

[3] S. Zaliasl, J. C. Salvia, G. C. Hill, L. Chen, K. Joo, R. Palwai, N. Arumugam, M. Phadke, S. Mukherjee, H.-C. Lee, C. Grosjean, P. M. Hagelin, S. Pamarti, T. S. Fiez, K. A. A. Makinwa, A. Partridge and V. Menon, "A 3 ppm 1.5 × 0.8 mm 2 1.0 µA 32.768 kHz MEMS-Based Oscillator," *IEEE Journal of Solid-State Circuits*, vol. 50, no. 1, pp. 291-302, 2015.

[4] J. Clark and G. Mostyn, "Microchip Oscillators and Clocks Using Microelectromechanical Systems (MEMS) Technology," 2017.

[5] J. Lake, E. Ng, C.-H. Ahn, V. Hong, Y. Yang, J. Wong and R. Candler, "Particle swarm optimization for design of MEMS resonators with low thermoelastic dissipation," in *the 17th International Conference on Solid-State Sensors, Actuators and Microsystems (TRANSDUCERS & EUROSENSORS XXVII)*, Barcelona, Spain, Jun. 16-20, 2013.

[6] A. H. Ghadimi, S. A. Fedorov, N. J. Engelsen, M. J. Bereyhi, R. Schilling, D. J. Wilson and T. J. Kippenberg, "Elastic strain engineering for ultralow mechanical dissipation," *Science*, vol. 360, no. 6390, pp. 764-768, 2018.

[7] A. Ozgurluk, R. Liu and C. T.-C. Nguyen, "Q-boosting of metal MEMS resonators via localized anneal-induced tensile stress," in *2017 Joint Conference of the European Frequency and Time Forum and IEEE International Frequency Control Symposium (EFTF/IFCS)*, Besançon, France, Jul. 10-13, 2017.

[8] M. J. Bereyhi, A. Beccari, S. A. Fedorov, A. H. Ghadimi, R. Schilling, D. J. Wilson, N. J. Engelsen

978-1-6654-9309-3/23 $31.00 © 2023 IEEE

and T. J. Kippenberg, "Clamp-Tapering Increases the Quality Factor of Stressed Nanobeams," *Nano Lett.,* vol. 19, no. 4, pp. 2329-2333, 2019.

[9] S. A. Fedorov, N. J. Engelsen, A. H. Ghadimi, M. J. Bereyhi, R. Schilling, D. J. Wilson and T. J. Kippenberg, "Generalized dissipation dilution in strained mechanical resonators," *Phys. Rev. B,* vol. 99, p. 054107, 2019.

[10] H.-S. Zheng, C.-P. Tsai, T.-Y. Chen and W.-C. Li, "CMOS-MEMS Resonators with sub-100-nm Transducer Gap Using Stress Engineering," in *the 35th IEEE Int. Conf. on Micro Electro Mechanical Systems (MEMS'22)*, Tokyo, Japan, Jan. 9-13, 2022.

[11] M.-H. Li, C.-S. Li and S.-S. Li, "Exploring the Q-factor limit of temperature compensated CMOS-MEMS resonators," in *the 28th IEEE International Conference on Micro Electro Mechanical Systems (MEMS)*, Estoril, Portugal, Jan. 18-22, 2015.

[12] Z. Hao, A. Erbil and F. Ayazi, "An analytical model for support loss in micromachined beam resonators with in-plane flexural vibrations," *Sensors and Actuators A: Physical,* vol. 109, no. 1-2, pp. 156-164, 2003.

[13] T.-Y. Chen, C.-P. Tsai and W.-C. Li, "1:6 Internal Resonance in CMOS-MEMS Multiple-Stepped CC-Beam Resonators," in *the 35th IEEE Int. Conf. on Micro Electro Mechanical Systems (MEMS'22)*, Tokyo, Japan, Jan. 9-13, 2022.

[14] J.-R. Liu, S.-C. Lu, C.-P. Tsai and W.-C. Li, "A CMOS-MEMS clamped–clamped beam displacement amplifier for resonant switch applications," *J. Micromech. Microeng. (JMM),* vol. 28, no. 6, p. 065001, 2018.

[15] D. D. Gerrard, Y. Chen, S. A. Chandorkar, G. Yu, J. Rodriguez, I. B. Flader, D. D. Shin, C. D. Meinhart, O. Sigmund and T. W. Kenny, "Topology optimization for reduction of thermo-elastic dissipation in MEMS resonators," in *the 19th International Conference on Solid-State Sensors, Actuators and Microsystems (TRANSDUCERS)*, Kaohsiung, Taiwan, Jun. 18-22, 2017.

CONTACT
*W.-C. Li, Tel: +886-2-3366-5636; wcli@iam.ntu.edu.tw

GENERIC TEMPERATURE COMPENSATION SCHEME FOR CMOS-MEMS RESONATORS BASED ON ARC-BEAM DERIVED ELECTRICAL STIFFNESS FREQUENCY PULLING

I-Chieh Hsieh, Hong-Sen Zheng, Chun-Pu Tsai, Ting-Yi Chen, and Wei-Chang Li
Institute of Applied Mechanics, National Taiwan University, Taiwan

ABSTRACT

This work demonstrates a generic temperature compensation scheme applicable for nearly *all* types of resonators in CMOS-MEMS technology without additional process steps. Particularly, based on the arc-beam structure used for gap narrowing, specific geometrical design yields autonomous temperature dependent gap spacing between the tuning electrode and the resonator body and in turn results in electrical stiffness necessary for counteracting resonance frequency variation at varying temperatures. Using a clamped-clamped beam (CC-beam) resonator in the 0.35-μm CMOS-MEMS process platform as a demonstration vehicle, the compensation scheme reduces the first-order temperature coefficient (TCF_1) from -394.16 ppm/°C to +14.18 ppm/°C with a TCF_2 of -1.5757 ppm/°C² from 0 °C to 90 °C. With the help of the gap variation prediction model derived and knowledge of stress information, the arc-beam structure offers an elegant solution of temperature compensation for CMOS-MEMS resonators with inherent yet thermally induced stress.

KEYWORDS

CMOS-MEMS resonator, temperature compensation, frequency pulling, electrical stiffness.

INTRODUCTION

Standard CMOS processes have been proved to be low-cost and fast prototyping platforms with circuit integration capability to realize resonant devices, such as oscillators [1], mass sensors [2], filters [3], and other resonator-based sensors [4]. For these resonant devices, the frequency stability against the ambient temperature changes remains one of the most important attributes that governs the precision of these applications and overall system performance.

While active temperature compensation techniques for micromechanical resonators, such as Joule-heating [5], a constant-R-algorithm [6] and phase-locked loop (PLL) based oscillators [7], have been shown to yield promising performances, they are not suitable for low-power applications due to the active power required to operate. In this context, passive temperature compensation methods that use material engineering [8], degenerate doping [9], manifest power-effective approaches of alleviating frequency drift to a certain level have been proposed to achieve so with no power consumed.

An electrical stiffness frequency pulling is the approach that can retain the original resonator material and structure. Fig. 1 depicts the previously demonstrated electrical stiffness compensation of using (a) heterogeneous material by additional process steps [10] and (b) *U*-shaped suspended structure [11] to generate a temperature-dependent gap-spacing. While the latter seems a better option without the need for added processes, it can serve probably best for the reported free-free beam (FF-beam) resonators only. Applying the technique to other resonator topologies such as CC-beams, for example, would require unavoidable modification to the *U*-shaped geometry, not to mention it would not be possible to use the compensation scheme for bulk resonators such as disk and Lamé-mode resonators.

DEVICE STRUCTURE AND OPERATION

To mitigate this, this work utilizes arc-beam structures as the frequency tuning electrodes that would move and then cause the gap-spacing to change in response to temperature variations. Fig. 1 (c) illustrates how the arc-beam structure can be used for various resonator topologies. Fig. 2 plots the schematic of the compensated

(a)

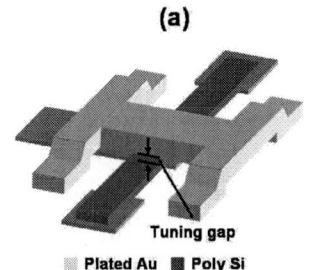

Plated Au Poly Si

× **Need additional processes**
× **For specific resonator types**

(b)

Tuning Gap

× **Large footprint**
× **For specific resonator types**

(c)

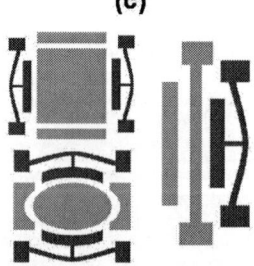

✓ **No additional processes**
✓ **Suitable for various resonator topologies**
✓ **Small footprint**

Fig. 1: Comparison of temperature compensation techniques of (a) heterogeneous of [10], (b)U-shaped of [11], and (c) this work which is suitable for various topologies.

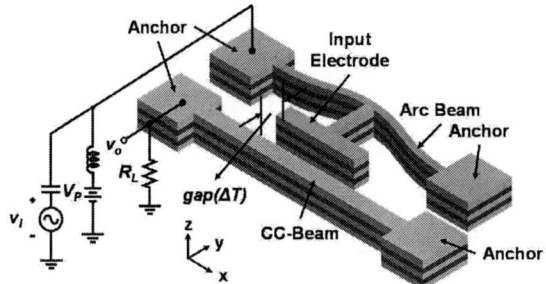

Fig. 2: Schematic of a clamped-clamped beam with the arc beam frequency tuning electrode and the illustration of the temperature dependent gap spacing.

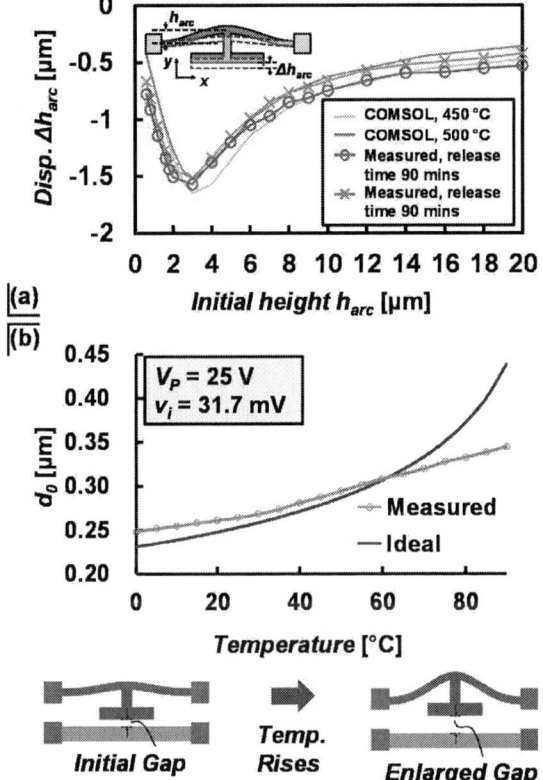

Fig. 3: (a) FEA simulation and measurement of test-key structures for the final displacement Δh_{arc} after release as a function of h_{arc}. (b) Ideal gap spacing as a function of temperature changed.

arc-beam structure together with a CC-beam resonator. In order to suppress the thermally dependent mechanical stiffness, the arc-beam is designed to deform and enlarge the gap spacing, thereby yielding temperature dependent electrical stiffness (k_e) to counteract the change of mechanical stiffness. The arc-beam is biased at V_P. The temperature-dependent electrical stiffness then takes the form [12]

$$k_e(T) = \int_0^{l_{arc}} V_P{}^2 \frac{\varepsilon_0 h}{(gap(T))^2} dx \qquad (1)$$

where l_{arc} and h are the length and thickness of the beam. As the ambient temperature rises, the arc-beam extends towards the $+y$ direction and results in a larger gap-

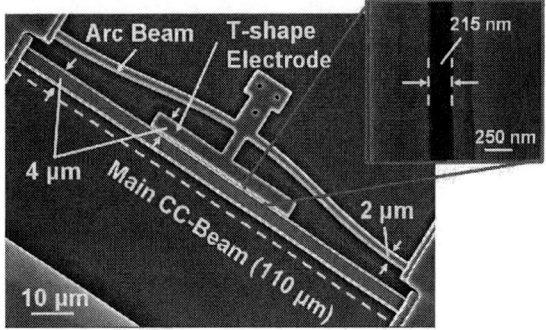

Fig. 4: SEM of a released temperature compensated CC-beam resonator with a zoom-in to the transducer gap.

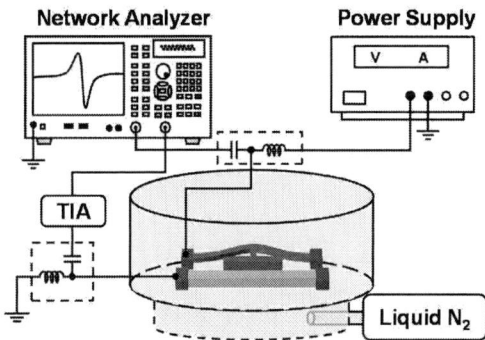

Fig. 6: Measurement set-up for frequency transmission, where power supply is used to bias the compensated arc-beam (V_P), and the network analyzer is used as the driving signal.

spacing. The temperature dependent gap spacing induced by the thermal stress of the arc-beam structure is key to generate corresponding electrical stiffness and yield a temperature insensitive resonance frequency.

Fig. 3 (a) plots the analytical model of [13] together with FEA simulation and measurement of test-key structures for the final displacement after release as a function of h_{arc}. Based on Fig. 3 (a), the gap-spacing as a function of temperature can be derived and tailored to match the required temperature temperature-dependent gap-spacing and resultant electrical-stiffness by varying the arc-beam geometry (cf. Fig. 3 (b)).

EXPERIMENTAL RESULTS

The fabrication of the temperature-dependent arc-beam structure and CC-beam resonator follows that of [14]. Specifically, the device is fabricated using a 0.35-μm 2-poly-4-metal CMOS-MEMS process platform followed by SiO_2 wet etching using Silox Vapox etchant III. Fig. 5 shows the SEM of a released CC-beam resonator, including a zoom-in to the transducer gap.

Measurement Set-Up

Fig. 6 depicts the experimental set-up for the device, in which the fabricated device is placed in a custom-built temperature-controlled vacuum probe station at a pressure below 10^{-4} Torr. To operate the device, a bias voltage, V_P, and an ac signal, v_i are applied on the arc-beam electrode

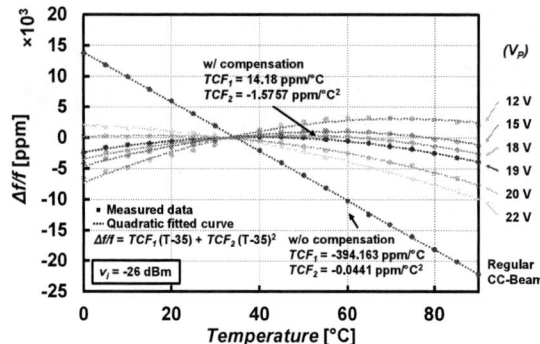

Fig. 7: Fraction of frequency change of the arc-beam equipped CC-beam resonator from 0 °C to 90 °C showing a $TCF_1 = 14.18$ ppm/°C is obtained with $V_P = 19$ V.

to generate an electrostatic force for driving the resonator. The motional current goes back to the network analyzer, yielding the transmission responses. Using liquid nitrogen and a PID temperature controller, the temperature of the chuck is varied and kept with a precision of ±0.1 °C at each temperature measurement point.

Compensated Resonance Frequency

Fig. 7 shows the fraction of frequency change over a temperature range from 0 °C to 95 °C with varying compensating voltages. At each temperature point, measurement is performed after a dwell time of 15 minutes. By adjusting the compensating voltage, the fractional frequency presents best compensation at a V_P of 19 V, corresponding to a TCF_1 of 14.18 ppm/°C. The proposed CMOS-MEMS resonator in this work shows a 27× improvements in terms of temperature-stability compared to regular CC-beam resonators. This technique not only presents a passive frequency compensation technique but also allows an area saving and more generic structure design compared to the previous *U*-shaped counterpart. *TABLE 1* compares the performances of this work with previously published temperature compensated resonators.

CONCLUSION

This work has verified the efficacy of the arc-beam structure for compensating frequency variations. This resonator yields a TCF_1 of +14.18 ppm/°C from 0 °C to 90 °C, which is 27× lower than the uncompensated one. The technique offers a flexible solution that can be applied literally to all resonator topologies. Of course, the performance of the arc-beam compensation scheme can be further improved to match with the performance of other techniques. Work towards this continues.

ACKNOWLEDGEMENTS

This research was funded by the National Science and Technology Council, Taiwan (MOST-109-2628-E002-004-MY3). The chip fabrication was supported by the Taiwan Semiconductor Research Institute (TSRI) and Taiwan Semiconductor Manufacturing Company (TSMC), Hsinchu, Taiwan. The authors would like to thank the staff in NEMS Research at National Taiwan University for providing technical support.

REFERENCES

TABLE 1: COMPARISON OF TEMPERATURE-COMPENSATED RESONATORS

	[8]	[10]	[11]	This work
Mechanism	Composite SiO_2 Structure	Temperature-Dependent Electrical Stiffness		
Frequency [MHz]	8.03	9.91	2.92	2.21
TCF_1 [ppm/°C]	N/A	-0.24	0.43	14.18
Max. $\Delta f/f_0$ from 0°C to 85°C [ppm]	3000	17.8	496	3751.6
CMOS Compatibility	✓	✗	✓	✓

[1] J. Van Beek and R. Puers, "A review of MEMS oscillators for frequency reference and timing applications," *Journal of Micromechanics and Microengineering,* vol. 22, no. 1, p. 013001, 2011.

[2] T.-Y. Liu, C.-C. Chu, M.-H. Li, C.-Y. Liu, C.-Y. Lo and S.-S. Li, "CMOS-MEMS thermal-piezoresistive oscillators with high transduction efficiency for mass sensing applications," in *2017 19th International Conference on Solid-State Sensors, Actuators and Microsystems (TRANSDUCERS),* 2017.

[3] F. Chen, J. Brotz, U. Arslan, C.-C. Lo, T. Mukherjee and G. K. Fedder, "CMOS-MEMS resonant RF mixer-filters," in *18th IEEE International Conference on Micro Electro Mechanical Systems, 2005. MEMS 2005.,* 2005.

[4] S.-C. Lu, C.-P. Tsai, Y.-C. Huang, W.-R. Du and W.-C. Li, "Surface condition influence on the nonlinear response of MEMS CC-beam resoswitches," *IEEE Electron Device Letters,* vol. 39, no. 10, pp. 1600-1603, 2018.

[5] G. Wu, J. Xu, E. J. Ng and W. Chen, "MEMS resonators for frequency reference and timing applications," *Journal of Microelectromechanical Systems,* vol. 29, no. 5, pp. 1137-1166, 2020.

[6] A. A. Zope, H. Ranjith, J.-H. Chang, C.-C. Chen, D.-J. Yao and S.-S. Li, "An effective temperature compensation algorithm for CMOS-MEMS thermal-piezoresistive oscillators with SUB PPM/° C thermal stability," in *2017 IEEE 30th International Conference on Micro Electro Mechanical Systems (MEMS),* 2017.

[7] J. C. Salvia, R. Melamud, S. A. Chandorkar, S. F. Lord and T. W. Kenny, "Real-time temperature compensation of MEMS oscillators using an integrated micro-oven and a phase-locked loop," *Journal of Microelectromechanical Systems,* vol. 19, no. 1, pp. 192-201, 2009.

[8] Y.-C. Liu, M.-H. Tsai, W.-C. Chen, M.-H. Li, S.-S. Li and W. Fang, "Temperature-compensated CMOS-MEMS oxide resonators," *Journal of microelectromechanical systems,* vol. 22, no. 5, pp. 1054-1065, 2013.

[9] A. K. Samarao and F. Ayazi, "Temperature compensation of silicon resonators via degenerate doping," *IEEE Transactions on Electron Devices*, vol. 59, no. 1, pp. 87-93, 2011.

[10] W.-T. Hsu and C.-C. Nguyen, "Stiffness-compensated temperature-insensitive micromechanical resonators," in *Technical Digest. MEMS 2002 IEEE International Conference. Fifteenth IEEE International Conference on Micro Electro Mechanical Systems*, 2022.

[11] J.-R. Liu and W.-C. Li, "A Temperature-Insensitive CMOS-MEMS Resonator Utilizing Electrical Stiffness Compensation," in *2019 IEEE 32nd International Conference on Micro Electro Mechanical Systems (MEMS)*, 2019.

[12] M. Akgul, L. Wu, Z. Ren and C. T.-C. Nguyen, "A negative-capacitance equivalent circuit model for parallel-plate capacitive-gap-transduced micromechanical resonators," *IEEE Transactions on Ultrasonics, Ferroelectrics, and Frequency Control*, vol. 61, no. 5, pp. 849-869, 2014.

[13] H.-S. Zheng, C.-P. Tsai, T.-Y. Chen and W.-C. Li, "Cmos-Mems Resonators with Sub-100-Nm Transducer Gap Using Stress Engineering," in *2022 IEEE 35th International Conference on Micro Electro Mechanical Systems Conference (MEMS)*, 2022.

[14] J.-R. Liu, S.-C. Lu, C.-P. Tsai and W.-C. Li, "A CMOS-MEMS clamped–clamped beam displacement amplifier for resonant switch applications," *Journal of Micromechanics and Microengineering*, vol. 28, no. 6, p. 065001, 2018.

CONTACT

*W.-C. Li, Tel: +886-2-3366-5636; wcli@iam.ntu.edu.tw

HIGH-Q AND LOW-MOTIONAL IMPEDANCE PIEZOELECTRIC MEMS RESONATOR THROUGH MECHANICAL MODE COUPLING

Linhai Huang[1#], Zhihong Feng[1#], Yuhao Xiao[2], Fengpei Sun[1] and Jinghui Xu[1]
[1] Huawei Technologies Company Ltd., Shenzhen, China
[2] The Institute of Technological Sciences, Wuhan University, Wuhan, China

ABSTRACT

A high-Q and low-motional impedance (R_m) resonator is achieved by mechanically coupling a low-impedance piezoelectric length extensional-mode resonator with two high-Q single-crystal silicon (SCS) Lame-mode resonators. The mass-weighted quality factor relationship between the mode-coupled resonator and the individual resonating elements has been demonstrated. By properly adjusting the mass ratio of the individual resonating elements, the mode-coupled resonator achieves a record high-quality factor of 172105 for piezoelectric resonators whereas the R_m is maintained as low as 350 Ω at 19.7 MHz. Our results offer opportunity to use MEMS resonators in high-performance temperature-compensated MEMS oscillators (TCMOs) and oven-controlled MEMS oscillators (OCMOs), making these devices a potential candidate for communication and intelligent vehicle applications.

KEYWORDS

Quality factor (Q), motional impedance, piezoelectric resonator, mode coupling.

INTRODUCTION

With the rapid development of 5G communication, wearable devices and intelligent vehicles, high-performance timing devices such as temperature-compensated oscillators and oven-controlled oscillators are highly required. For a long time, this market has been dominated by the conventional quartz crystal technology [1, 2]. But, with the ever-growing needs of timing devices for low cost, small size, shock resistance and wide temperature range compatibility, these requirements are coming to a limit that the conventional quartz industry cannot achieve.

Recently, Micro-electro-mechanical system (MEMS) timing devices have drawn extensive attention owing to their miniaturization capability and shock robustness, as well as high-temperature tolerance ability, making them attractive candidate to replace the quartz technology in the high performance timing market [3,4,5]. But, the performance of MEMS is limited by the phase noise level. The close-to-carrier phase noise is shaped by the quality factor of the resonator, while the phase noise floor at high-offsets is directly proportional to square of motional impedance R_m^2 [6]. Therefore, high quality factor and low motional impedance are important factors that determine the phase noise performance of resonators used as frequency references. To date, the piezoelectric resonators achieve R_m as low as 50 Ω [7] while the capacitive resonators achieve Q over 2e6 [8] at 10MHz. Unfortunately, these two merits are difficult to be achieved simultaneously in any single MEMS resonator design.

#These authors contribute equally to this manuscript.

In literature, approaches to improve the Q of piezoelectric MEMS resonators include using acoustic reflector [9] or PnC [10] as the anchor to suppress anchor loss, and using Q-boosting arrayed cells [11] and increasing the thickness of silicon device layer to reduce the AlN-to-Si ratio [12], but the highest Q achieved is limited to 74000 whereas the R_m is raised to 900 Ω. Methods to reduce the R_m of capacitive MEMS resonators include partial-gap filling to reduce the gap width [13] and using resonator array to increase the transduction area [14, 15]. However, the improvement is still unsatisfactory and it suffers from frequency fluctuations due to charging and electrical noise in DC BIAS [13].

In this study, we design a mode-coupled electromechanical resonator by mechanically connecting the piezoelectric resonator with SCS resonators. We demonstrate that the mode-coupled resonator inherits not only the merit of high Q from SCS resonators but also the merit of low R_m from piezoelectric resonators. We also explore the optimization methods of the performance of mode-coupling resonator by studying the relationship between the Q of coupled resonator and the Q and mass ratio of individual resonating elements.

DEVICE DESIGN AND FABRICATION

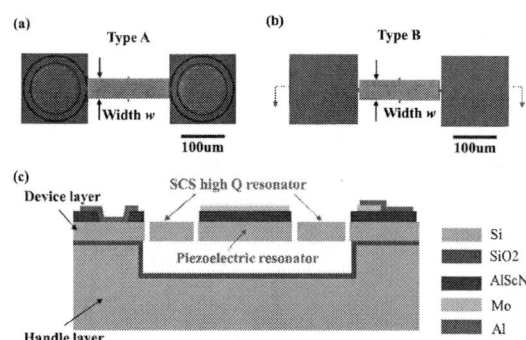

Figure 1: SEMs of the fabricated coupled resonators: (a) Type A; (b) Type B. (c) cross-sectional schematic view taken along the red dashed line in (b) together with the electrodes.

Fig. 1 shows the two types of resonators investigated in this work. Both resonators consist of two types of individual resonators mechanically connected together. The middle rectangular piezoelectric resonator serves as a high-electromechanical coupling transducer, while on its opposite sides two SCS resonators (Ring resonators for Type-A resonator in Fig. 1a and square plate resonators for Type-B resonator in Fig. 1b respectively) are connected to it. Owing to the absence of anchor loss and interfacial loss, these pure-SCS resonators serve as Q-enhancement element of the whole resonating system. To study the effect

IEEE MEMS 2023, Munich, GERMANY
15 - 19 January 2023

of mass ratio of individual resonating elements on Q of coupled resonator, the widths of piezoelectric resonators are varied in different designs.

The devices were fabricated in AlScN-on-CSOI MEMS process, as illustrated in Fig. 1c. First, 1um thickness of AlScN piezoelectric layer was sputtered on the highly doped SCS layer of CSOI wafer. Then, 0.15um thickness of Mo was deposited and patterned by dry etching on the AlScN piezoelectric layer, serving as the top electrodes. Next, the AlScN piezoelectric layer was etched both on the pure-SCS resonators regions and bottom electrodes regions to expose the SCS layer. The Al electrodes were then patterned on the bottom electrodes and top electrodes regions. Finally, the 40um thickness of SCS device layer was dry etched by deep reactive-ion etching (DRIE) process to form the resonating structures. The silicon device layer is highly doped and serves as the bottom electrode. The cross-sectional schematic view (Fig. 1c) shows that the piezoelectric layer and metal layer covers only the middle rectangular part forming the piezoelectric resonator while the other two are SCS high Q resonator.

ANALYSIS AND SIMULATIONS

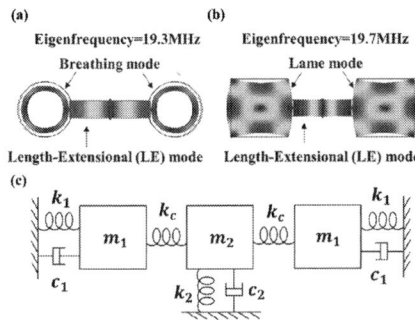

Figure 2: The mode shapes of the coupled resonators: (a) Type A; (b) Type B. (c) The mass-stiffness-damper model of coupled resonators. Resonator 1 and 2 represent the SCS resonator and piezoelectric resonator respectively.

Fig. 2a and 2b show the simulated mode shapes of the two types of coupled resonators. The rectangular piezoelectric resonators in both designs work in Length-Extensional (LE) mode. In Type-A coupled resonator, the SCS Ring resonators work in breathing mode while in Type-B coupled resonator, the SCS square plate resonators work in Lame mode. In both coupled resonators, mode coupling of the individual piezoelectric and SCS resonators is realized by exciting them in frequency-matched modes. To maximize the energy transfer efficiency, the individual resonators are designed to be connected at the maximum vibration amplitudes region. The strong coupling through the high-stiffness joints ensures that all the individual resonators in one coupled resonator system dissipate energy at same rate.

We analyze the effective quality factor of the coupled resonator by using the equivalent lumped-mass model shown in Fig. 2C. Parameters $k_{1,2}$ and $c_{1,2}$ represent the effective stiffness and damping factor of the individual resonators, and k_c the stiffness of the coupling joints. The effective mass $m_{1,2}$ is calculated from

$$m_{eff} = \int_V \rho \left(\frac{x}{x_{max}}\right)^2 dV, \qquad (1)$$

where x, x_{max} and ρ are the displacement, the maximum displacement and the density of the resonator. The quality factor Q of a resonator is defined by the ratio of energy stored and dissipated in one period by

$$Q = \frac{E_{stored}}{E_{dissipated}}. \qquad (2)$$

By taking into account the frequency matching condition $\sqrt{\frac{k_1}{m_1}} = \sqrt{\frac{k_2}{m_2}}$, the energy stored in the individual resonators follows the following relationship

$$\frac{E_1}{E_2} = \frac{\frac{1}{2} k_1 x_{1max}^2}{\frac{1}{2} k_2 x_{2max}^2} = \frac{k_1}{k_2} = \frac{m_1}{m_2}. \qquad (3)$$

For the whole coupled resonator system, the energy stored in the system and the energy the system dissipated are given by

$$E_{total} = 2E_1 + E_2,$$
$$E_{dissipated} = \frac{2E_1}{Q_1} + \frac{E_2}{Q_2} \qquad (4)$$

Therefore, the effective quality factor Q_c of the coupled resonator system is given by

$$\frac{1}{Q_c} = \frac{2m_1}{2m_1 + m_2}\frac{1}{Q_1} + \frac{m_2}{2m_1 + m_2}\frac{1}{Q_2}. \qquad (5)$$

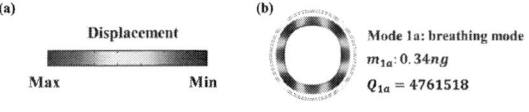
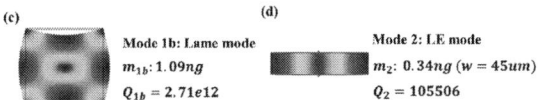

Figure 3: (a) Relative displacement scale of eigen mode simulations. The simulation results of quality factors and mode shapes of the individual resonators: (b) SCS ring resonator, (c) SCS square plate resonator and (d) piezoelectric rectangular resonator.

The mass-weighted Q relationship is investigated in simulation. The quality factor of individual resonators and coupled resonators are simulated by finite-element analysis in COMSOL Multiphysics™. Thermoelastic damping (TED) and anchor loss computed using a perfectly matched layer (PML) are taken into account in the simulation. As illustrated in Fig. 3, the simulated quality factor of the individual LE-mode piezoelectric resonator at w=45um is 1.05e5. The Q values and frequencies of the LE-mode resonators under different widths remain almost unchanged, so the frequency-matching condition of the coupled resonators can be maintained among different designs and the sole effect of mass ratio on Q_c can be extracted. In the coupled resonators, the SCS ring resonators and square plate resonators are hanged by the rectangular piezoelectric resonators with no anchor connected to the substrate, so anchor loss is not considered in the quality factor simulation of the individual SCS resonators in Fig. 3b and 3c. The effective mass values listed in Fig. 3 are calculated by volume integration in COMSOL using Eq. 1. The simulation results show that the SCS resonators have Q with orders of magnitude higher than that of the piezoelectric resonator and have effective mass comparable to that of the piezoelectric resonator, thus ensuring that the individual SCS resonators can contribute to considerable improvement in Q of the coupled resonators according to Eq. 5.

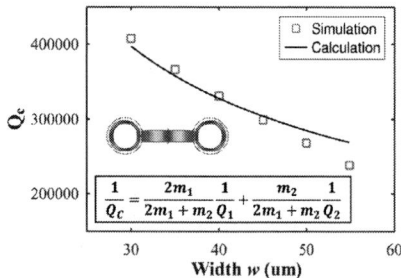

Figure 4: The simulation and calculation results of quality factors of Type-A devices under different widths.

Simulation results of quality factors of Type-A coupled resonators are illustrated in Fig. 4. As the width w of piezoelectric resonator decreases from 55um to 30um, the simulated Q_c increases by about 70% showing considerable enhancement ability of Q_c by adjusting the mass ratio. Calculation of Q_c by Eq. 5 is also performed based on the Qs and effective mass of individual resonators given in Fig. 3. It's seen that the simulation results show good consistency with theory which predicts that the Q of coupled resonator follows a mass-weighted relationship with the Qs of the individual resonators.

RESULTS AND DISCUSSIONS

Electrical characterization of the fabricated devices was conducted at room temperature in a vacuum probe station equipped with GSG probes. Measurements of S21 electrical transmission of the coupled resonators were performed by a network analyzer after a standard SLOT calibration. Fig. 5 shows the typical measurement results of the S21 transmission of both types of coupled resonator under different widths. The loaded Q_l is extracted from 3dB linewidth of series resonance peak. The unloaded Q_u and motional resistance R_m are obtained by the following equations to remove the loading effect of 50 Ω termination resistance R_0 of the network analyzer:

$$Q_u = \frac{Q_l}{1 - 10^{-\frac{IL}{20}}}, \qquad (6)$$

$$R_m = 2R_0 \left(10^{IL/20} - 1 \right).$$

It's seen from Fig. 5 that as the width w decreases, the unloaded Qs of both coupled resonators increase showing good agreement with both the analytical and also numerical predictions. It's worth noting that, for any of the widths, the R_m of Type-A coupled resonator is maintained below 300Ω. The Type-B resonator exhibits an apparently higher

Q compared with Type-A resonator at the same width w. This result is also consistent with Eq. 5 which predicts that higher Q of coupled resonator can be obtained by coupling to a resonator with both higher Q and higher mass ratio. For each resonator design involved in Fig. 5, measurements were performed on 5 samples and the distribution of results

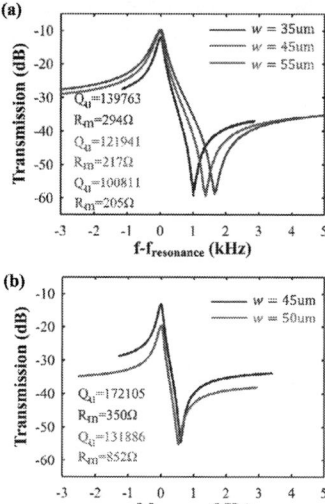

Figure 5: Typical measurements of the S21 transmission of the coupled resonators: (a) Type A; (b) Type B.

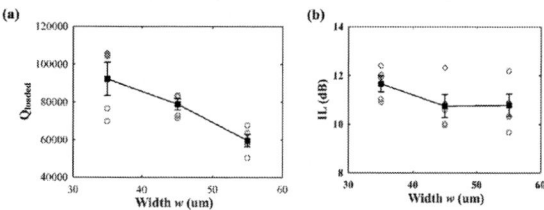

Figure 6: Measured (a) loaded Q and (b) insertion loss (IL) of the Type-A coupled resonators. Red circle: measured value of individual device sample. Black square and error bar: mean value and standard deviation of 5 measured samples.

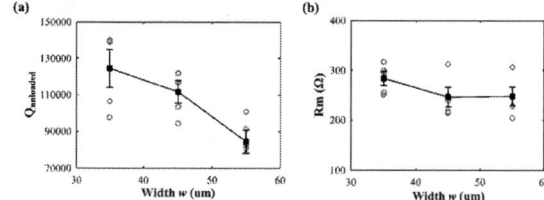

Figure 7: Extracted (a) unloaded Q and (b) motional resistance (R_m) of the Type-A coupled resonators. Red circle: extracted value of individual device sample. Black square and error bar: mean extracted value and standard deviation of 5 samples.

Table 1: The average and standard deviation of extracted unloaded Q and R_m of Type-A and Type-B coupled resonators over 5 samples.

	Width w (um)	Average $Q_{unloaded}$	Standard deviation of $Q_{unloaded}$	Average R_m (Ω)	Standard deviation of R_m (Ω)
Type A	45	111671	12221	246	40
Type B	45	157859	8527	391	31
	50	116084	15669	550	205

are shown in Fig. 6, Fig. 7 and Table I. The results demonstrate that the observed Q enhancement under smaller width w and the low R_m are repeatable. Fig. 8 compares the state-of-the-art performance of MEMS resonators in terms of Q and R_m. The mode-coupled piezoelectric resonator in this work achieves the highest figure of merit (FoM) Q/R_m among all the state-of-the-art piezoelectric and capacitive resonators [7, 10-14]. Notably, the FoM of our piezoelectric resonator is more than 5 times higher than other piezoelectric resonators in literature.

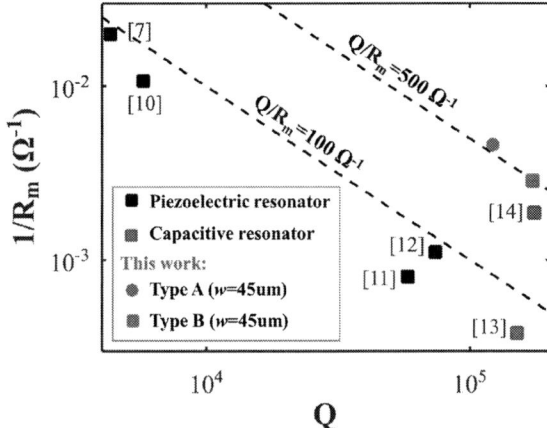

Figure 8: Performance comparison with state-of-the-art MEMS resonators. Red square and Red circle: this work. Black square and blue square: literature.

CONCLUSION

In this work, we have designed and fabricated mode-coupled piezoelectric resonators to improve their quality factors and motional resistances. We have demonstrated analytically and experimentally that the Q of the coupled resonator follows a mass-weighted relationship with the Qs of the individual resonators. The measurement results consistently show that considerable Q enhancement of the coupled resonator occurs when we improve the Q or mass ratio of the individual resonator with higher Q. The LE-Lame mode coupled (Type-B) resonator achieves a record high-quality factor of 172105 for piezoelectric resonators whereas the R_m is maintained as low as 350 Ω at 19.7 MHz. The findings of mass-weighted Q relationship may help guiding the development of high-Q and low-Rm resonators using mode-coupled method.

REFERENCES

[1] W. L. Hsieh, *et al.*, "The World's Smallest Quartz-Based OCXO for 5G Synchronization Applications", *2021 Joint Conference of the European Frequency and Time Forum and IEEE International Frequency Control Symposium (EFTF/IFCS)*, 2021, pp. 1-4

[2] R. L. Kubena, *et al.*, "A Fully Integrated Quartz MEMS VHF TCXO", *IEEE Trans. Ultrason., Ferroelect.,Freq. Contr.*, vol. 65, no. 6, pp. 904-910, Jun. 2018.

[3] H. K. Kwon, *et al.*, "An Oven-Controlled MEMS Oscillator (OCMO) With Sub 10mw, ± 1.5 PPB Stability Over Temperature," *2019 20th International Conference on Solid-State Sensors, Actuators and Microsystems & Eurosensors XXXIII (TRANSDUCERS & EUROSENSORS XXXIII)*, 2019, pp. 2072-2075

[4] C. Nguyen, *et al.*, "MEMS technology for timing and frequency control," *IEEE Trans. Ultrason, Ferroelectr., Freq. Control*, vol. 54, no. 2, pp. 251-270, Feb. 2007.

[5] Z. Wu, *et al.*, "A temperature-stable mems oscillator on an ovenized micro-platform using a PLL-based heater control system," *2015 28th IEEE International Conference on Micro Electro Mechanical Systems (MEMS)*, 2015, pp. 793-796

[6] S. Seth, *et al.*, "A −131-dBc/Hz, 20-MHz MEMS oscillator with a 6.9-mW, 69-kΩ, gain-tunable CMOS TIA," *2012 Proceedings of the ESSCIRC (ESSCIRC)*, 2012, pp. 249-252.

[7] G. Piazza, *et al.*, "Piezoelectric Aluminum Nitride Vibrating Contour-Mode MEMS Resonators," *IEEE/ASME J. Microelectromech. Syst.*, vol. 15, no. 6, pp. 1406-1418, Dec. 2006

[8] H. K. Kwon, *et al.*, "Crystal Orientation Dependent Dual Frequency Ovenized MEMS Resonator With Temperature Stability and Shock Robustness," *IEEE/ASME J. Microelectromech. Syst.*, vol. 29, no. 5, pp. 1130-1131, Oct. 2020.

[9] B. P. Harrington, *et al.*, "In-plane acoustic reflectors for reducing effective anchor loss in lateral–extensional MEMS resonators". *J. Micromech. Microeng*, 2011, 21(8): 085021.

[10] H. Zhu, *et al.*, "AlN piezoelectric on silicon MEMS resonator with boosted Q using planar patterned phononic crystals on anchors" *2015 28th IEEE International Conference on Micro Electro Mechanical Systems (MEMS)*. IEEE, 2015: 797-800.

[11] G. Pillai, *et al.*, "Quality factor boosting of bulk acoustic wave resonators based on a two dimensional array of high-Q resonant tanks", *Appl. Phys. Lett.* 116, 163502 (2020)

[12] W. Pan, *et al.*, "Thin-film piezoelectric-on-substrate resonators with Q enhancement and TCF reduction", *2010 IEEE 23rd international conference on micro electro mechanical systems (MEMS)*. IEEE, 2010: 727-730.

[13] L. Hung, *et al.*, "Capacitive transducer strengthening via ALD-enabled partial-gap filling", *Hilton Head. Citeseer*, 2008, 8: 208-211.

[14] H. Lee, *et al.*, "Low jitter and temperature stable MEMS oscillators," *IEEE International Frequency Control Symposium Proceedings*, 2012, pp. 1-5,

[15] Y. Lin, *et al.*, "Quality Factor Boosting via Mechanically-Coupled Arraying," TRANSDUCERS *International Solid-State Sensors, Actuators and Microsystems Conference*, 2007, pp. 2453-2456

CONTACT

*Jinghui Xu, xujinghui@huawei.com

CROSSTALK-FREE LARGE APERTURE 2D GIMBAL MICROMIRROR

Behrad Ghazinouri[1], and Siyuan He[1]
[1]Toronto Metropolitan University, CANADA

ABSTRACT

This paper presents a crosstalk-free, large aperture (19x19 mm^2) FPCB (flexible printed circuit board) 2D gimbal micromirror. The crosstalk-free is achieved through a novel arrangement of 6 brick shape permanent magnets and four layers of coils embedded in FPCB. The performance of the design was examined using FEM (finite element method) analysis and a prototype was built and tested. An optical FOV (field of view) of 36 degrees horizontal and 14 degrees vertical was obtained at 0.5V, 50Hz(resonance) and 2V, 1 Hz (quasi-static) respectively. The Large aperture size of the mirror makes it suitable for low vibration LiDAR (light detection and ranging) applications such as ground robotics.

KEYWORDS

Scanning micromirror, Crosstalk-free micromirror, Electromagnetic actuation, LiDAR (Light Detection and Ranging)

INTRODUCTION

Many of the micromirrors that are developed for applications such as projection display[1] and laser engraving[2], have an aperture size of ~1mm in diameter in order to steer a laser beam. But in LiDAR applications, the reflected light from the surface needs to be captured too. Since the received signal power is proportional to the square of receiver's aperture diameter, a small aperture size will limit the maximum detectable distance[3]. One solution is to use a small aperture size micromirror to steer the laser alongside solid-state receivers. In this case, in order to cover the whole FOV of the mirror, multiple receivers are required which will increase the cost and size of the LiDAR device. Another approach is to use a large aperture size micromirror to steer the laser and receive the reflected light back[4].

The 2D micromirror presented in this paper is diamond shaped with aperture size of 19x19 mm^2. The micromirror is electromagnetically actuated in 2 orthogonal directions with a novel magnetic orientation that eliminates crosstalk.

DESCRIPTION OF THE NEW DESIGN

One of the main problems with conventional electromagnetic 2D gimbal micromirrors is the crosstalk between the two orthogonal rotations. [5], [6] The most common magnet orientations used in conventional electromagnetically actuated 2D micromirrors are either radial[7] or 45-degrees[8]. When 45-degrees oriented magnetic field is used as shown in Figure 1(a), the Lorentz force is only applied to half of the coil due to the cross product between the magnetic flux density and the current. As a result, the Lorentz forces provide not only actuation torque around the y-axis but also an undesired torque component around the x-axis which causes the crosstalk. In the proposed design (Figure 1(b)), a novel arrangement of 6 brick shape permanent magnets was used to eliminate the

mechanical crosstalk. Moreover, the electrical crosstalk was eliminated using 4 layers of coils that allow separate electric paths for the frame and the mirror actuation.

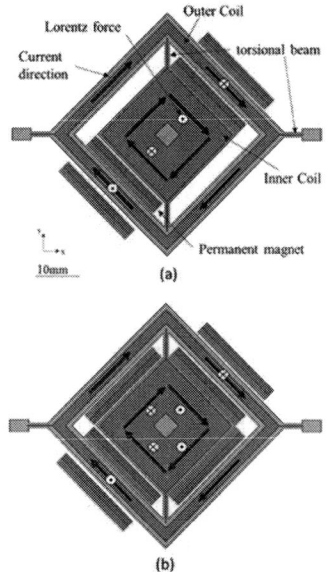

Figure 1: 2D micromirror with (a) Conventional 45-degrees magnetic field. (b) proposed magnet orientation.

As it's illustrated in Figure 1(b), the orientation of permanent magnets has been changed in a way that the inactive portion of the mirror coil has become active. The Lorentz force on the inactive parts contributes to the torque around the mirror torsional beam but cancels the undesired torque around the frame torsional beam. A diamond shape mirror is an ideal design for this magnetic orientation as it's easier to assemble rectangular magnets on the sides of the mirror with the specified orientation.

The 2D mirror was designed as a raster scanner which is preferred for LiDAR applications due to the even scanning pattern[9]. This means horizontal actuation a.k.a. fast axis will be at resonant and the vertical actuation a.k.a. slow axis will be quasi-static.

The high torque provided by the 4 layers of coil under the mirror a.k.a. inner coil and 2 layers of coil under the frame a.k.a. outer coil makes it possible to produce bigger micromirrors without sacrificing the scanning angle i.e. FoV. The large aperture size of this mirror makes it possible to use it for both emitting and receiving light.

Structural Parameters

FPCB manufacturing method[10] was used for the micromirror production. This method has two main steps: 1. FPCB and coil layer production which is outsourced to a commercialized FPCB manufacturing company. 2. Metal coated silicon layer which is added to the FPCB structure later.

The FPCB structure consists of 4 layers of copper coils which are sandwiched between 5 layers of polyimide. Structural parameters can be found in Table 1.

Table 1: Structural parameters of the micromirror

Parameter	Value (Unit)
Polyimide shear modulus	5.60E+08 Pa
Copper shear modulus	2.35E+10 Pa
Moment of inertia of the inner gimbal (mirror)	1.31E-08 kg.m^2
Silicon thickness	200 μm
Total polyimide thickness	267.5 μm
copper layer thickness (4 layers)	18 μm
Polyimide PI stiffener thickness	50 μm

Theoretical Analysis

The resonance frequency is one of the main factors to be considered in micromirror design. The proposed micromirror was designed to be used alongside a single point LiDAR with a refresh rate of 10kHz. As a result, the micromirror needed to have a resonance frequency of under 100Hz for the fast axis to achieve 100 points of measurement along the horizontal scan.

Since the aperture dimensions are a design input and thicknesses are all driven by the fabrication process, the moment of inertia of the mirror is known and according to equation (1), the only remaining variable for resonance frequency determination is the torsional stiffness.

$$f = \frac{1}{2\pi}\sqrt{\frac{k}{I}} \qquad (1)$$

Where f is the resonance frequency, k is the torsional stiffness and I is the moment of inertia.

The theoretical torsional beam width and length can be determined using rectangular section torsional beam equation (2)[11].

$$k = \frac{G}{L}ab^3\left(\frac{16}{3} - 3.36\frac{b}{a}\left(1 - \frac{b^4}{12a^4}\right)\right) \qquad (2)$$

Where G is the shear modulus, L is the length of the torsional beam, a is half width of the torsional beam, and b is half thickness of the torsional beam.

By inserting the parameter values of Table 1 in equation (2), the width and length of the polyimide torsional beam was calculated to be 1 mm and 5.5 mm respectively to achieve a resonance frequency of 45 Hz. The torsional stiffness of the copper wire inside the torsional beam can be neglected due to the small dimensions.

Modal Analysis

Modal analysis was done to analyze the resonance modes of the micromirror. Figure 2 shows the first 4 modes of the designed micromirror. Mode 2 is the desired mode of actuation for the fast axis.

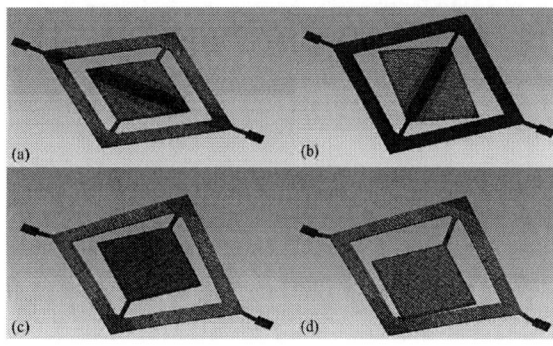

Figure 2: Ansys Workbench Modal Analysis of the design: (a): Mode 1:22.43Hz (b): Mode 2: 46.03Hz (c): Mode 3: 68.25Hz (d): Mode 4: 300.35Hz

Magnetostatic and Static Structural Analysis

In order to determine the performance of the micromirror before manufacturing, the Lorentz force on the inner coil and the outer coil had been calculated using Ansys Workbench Magnetostatic.

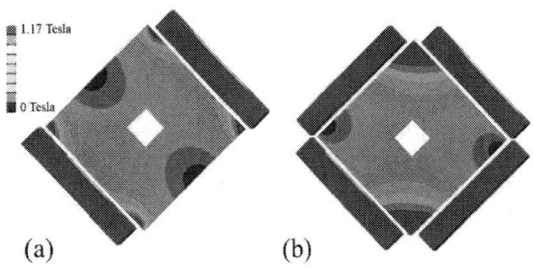

Figure 3: Magnetic flux density of the inner coil with (a):2 magnets (b): 4 magnets

The results in Figure 3 show that using 4 magnets on the inner coil causes an even distribution of the Lorentz force on the coil which would prevent the crosstalk.

The Lorentz forces were then applied to the design using Ansys Workbench Static Structural for 10 mA to 60 mA applied currents and the results were illustrated in
Figure 6.

FABRICATION

A prototype of the proposed design was manufactured using the FPCB manufacturing Method. The FPCB shown in

Figure 4(a) has 4 inner coils and 2 outer coils. The outer coils' surface and the screw holders were reinforced using PI stiffener. The inner coils require 2 layers of the outer coil dedicated to it for the current path. As a result, only the 2 remaining layers can be used for the outer coil. *The magnet housing was 3D printed using PLA (Polylactic Acid). And 6 bars of Neodymium N52 magnets (magnet dimensions: 19mm x 19 mm x 3.2mm) were used as shown in*

Figure 4(b).

978-1-6654-9309-3/23 $31.00 © 2023 IEEE

The reflective layer of the mirror is a gold coated silicon layer. A 101.6mm (4 inches) silicon wafer with 200 μm thickness was coated with 100 nm of Au using 10 nm of Cr as the adhesive layer. The wafer was then diced into 19x19 mm² pieces and glued to the inner gimbal as the reflective surface. (
Figure 4(c))

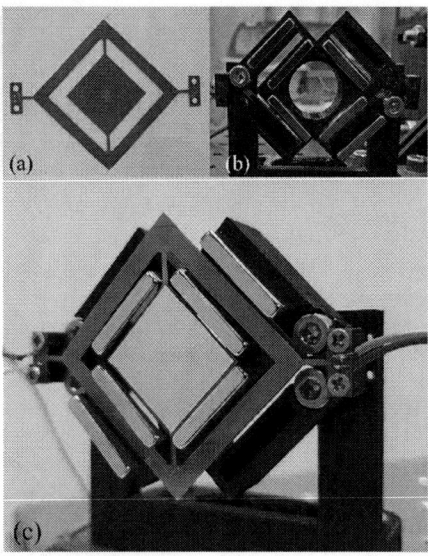

Figure 4: (a): FPCB gimbal (b): 3D printed magnets housing (c): assembled prototype

EXPERIMENTAL RESULTS

The prototype was then tested to verify the performance of the mirror.

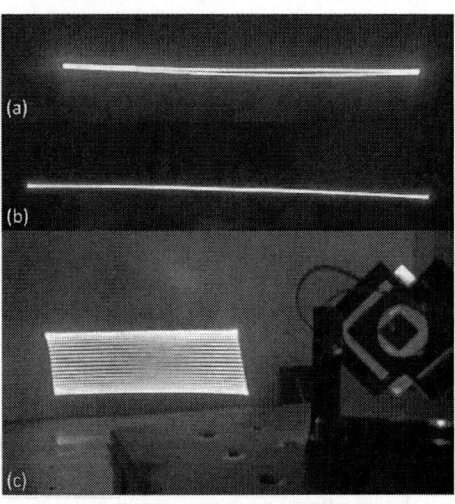

Figure 5: Scanning pattern of the 2D micromirror (a): horizontal actuation with Conventional 45-degrees magnetic field. (b): horizontal actuation with the hybrid-field. (c): 2D actuation with the hybrid field at 0.5V, 50Hz and 2V, 1 Hz for Horizontal and vertical actuation respectively

Figure 5(a) & (b) show a comparison between scanning lines generated by a traditional 2D micromirror with 45 degrees orientated magnetic field vs. the design proposed respectively. In both pictures, the mirror was actuated using a 0.5 V, 50Hz driving signal. According to Figure 5(b), It's clear that the crosstalk was eliminated using the 4 magnets on the inner coil and the max angle was also improved.

Figure 5(c) shows the full raster pattern of the mirror using driving signals of 0.5V, 50Hz, and 2V, 1 Hz for horizontal and vertical actuation respectively.

Figure 6 shows experimental and FEM results for static angle vs current diagram of the vertical and horizontal actuation. The resonance angle for the test shown in Figure 5(c) is 36 degrees optical using a driving signal of 0.5V, 6mA, 50Hz.

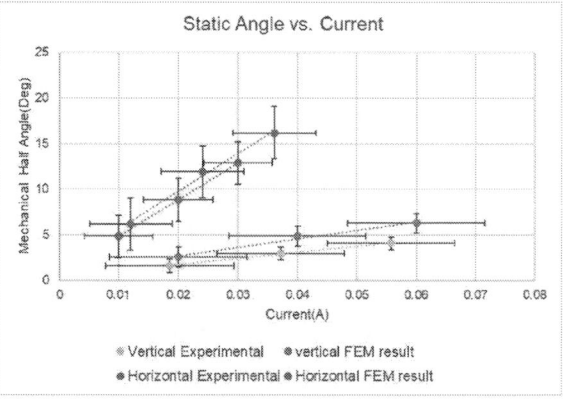

Figure 6: Static angle vs. current diagram for vertical and horizontal actuation

CONCLUSION

A large aperture, electromagnetically actuated 2D mirror was designed, analyzed, built, and tested. The 4 magnets' orientation on the mirror, eliminated the mechanical crosstalk and increased the FOV. Isolated layers of coil for inner and outer actuation eliminated the electrical crosstalk. The large aperture size of the mirror makes it suitable for LiDAR applications. Although the resonance frequency of the current prototype makes it suitable for low vibration applications e.g. ground robotics, the torsional beam design can be adjusted in order to achieve a higher resonance frequency application e.g. autonomous vehicle.

REFERENCES

[1] A. D. Yalcinkaya, H. Urey, D. Brown, T. Montague, and R. Sprague, "Two-axis electromagnetic microscanner for high resolution displays," *J. Microelectromechanical Syst.*, vol. 15, no. 4, pp. 786–794, 2006.

[2] K. G. Periyasamy, H. Zuo, and S. He, "Flexible printed circuit board magnetic micromirror for laser marking/engraving," *J. Micromechanics Microengineering*, vol. 29, no. 8, p. 085001, 2019.

[3] D. Wang, C. Watkins, and H. Xie, "MEMS mirrors for LiDAR: a review," *Micromachines*, vol. 11, no. 5, p. 456, 2020.

[4] T. Tai, S. He, and B. Ghazinouri, "2D FPCB micromirror for scanning LIDAR," *J. Micromechanics Microengineering*, 2022.

[5] Y. Park, S. Moon, J. Lee, K. Kim, S.-J. Lee, and J. H. Lee, "Via-less two-axis electromagnetic micro scanner based on dual radial magnetic fields," *IEEE Photonics Technol. Lett.*, vol. 30, no. 5, pp. 443–446, 2018.

[6] S.-Y. Kang, J.-H. Park, and C.-H. Ji, "Design optimization of a 6.4 mm-diameter electromagnetic 2D scanning micromirror," *Opt. Express*, vol. 28, no. 21, pp. 31272–31286, 2020.

[7] S.-H. Chung, S.-K. Lee, C.-H. Ji, and J.-H. Park, "Vacuum packaged electromagnetic 2D scanning micromirror," *Sens. Actuators Phys.*, vol. 290, pp. 147–155, 2019.

[8] A. R. Cho *et al.*, "Electromagnetic biaxial microscanner with mechanical amplification at resonance," *Opt. Express*, vol. 23, no. 13, pp. 16792–16802, 2015.

[9] B. Ghazinouri, S. He, and T. S. Tai, "A position sensing method for 2D scanning mirrors," *J. Micromechanics Microengineering*, vol. 32, no. 4, p. 045007, 2022.

[10] H. Zuo and S. He, "Extra Large Aperture FPCB Mirror Based Scanning LiDAR," *IEEEASME Trans. Mechatron.*, 2021.

[11] W. C. Young, R. G. Budynas, and A. M. Sadegh, *Roark's formulas for stress and strain*. McGraw-Hill Education, 2012.

CONTACT

* B. Ghazinouri; bghazinouri@ryerson.ca

INVERSE INTERFERENCE EFFECT-ENHANCED ULTRASENSITIVE SENSING VIA MID-IR NANOANTENNAS

Hong Zhou, Dongxiao Li, Xinge Guo, Zhihao Ren, and Chengkuo Lee

Department of Electrical & Computer Engineering, National University of Singapore, SINGAPORE

ABSTRACT

Electromagnetically induced transparency (EIT) and absorption (EIA) are quantum interference phenomena occurring in multilevel atoms and inducing a sharp peak or valley within a broad transmission dip. Herein, we realize both EIA and EIT effects in hybrid mid-IR nanoantennas, and, for the first time, utilize the two inverse effects to enhance the signal strength of nanoantennas, thereby making breakthroughs in ultrasensitive sensing without additional fabrication complexity and cost. As demonstrated through CO_2 gas detection, the performance of our EIA-EIT-integrated platform including signal strength, sensitivity, and limit of detection is comprehensively improved. Furthermore, the idea of using two inverse effects for sensing performance improvement is generally applicable to almost all EIA/EIT-like classical systems. Our work promises to have a profound impact on the spectroscopic sensing field.

KEYWORDS

electromagnetically induced transparency, electromagnetically induced absorption, gas detection

INTRODUCTION

Quantum coherence and interference, as one of the landmark fundamental features in quantum physics, have caused massive interesting and important phenomena that deviate from classical physics. Electromagnetically induced transparency (EIT) is such a destructive quantum interference phenomenon occurring in multilevel atoms and inducing a sharp peak within a broad transmission dip. This effect markedly transforms the anomalous dispersion (opaque) of the medium into normal dispersion (transparent), thereby inspiring substantial applications including optical switching, nonlinear optics, and slow light manipulation. Among them, slow light has proven its ability to enhance the light-matter interaction, which has a profound impact on the optical sensing field. For instance, the enhancement of light-matter interaction can be utilized to solve the low molecular absorption cross-sections in mid-infrared and then realize the detection and characterization of molecular substances with unprecedented sensitivity. However, in quantum systems, the implementation of these EIT phenomena must meet scathing experimental conditions including stable gas lasers and low-temperature environments, which hinders its wide applications. An effective way is to imitate the resonant quantum transition in the atomic system, and then reproduce the EIT concept in the classical system. Plasmonic metamaterials have attracted tremendous attention since they can easily achieve the EIT effect without the scathing experimental requirements in quantum optics via Fano-type linear destructive interference [1-5]. Because of its functional retention in enhancing light-matter interaction, the EIT effect using plasmonic metamaterials is widely used in the sensing field, such as dynamic reaction monitoring, hyperspectral infrared chemical imaging, and biochemical molecule detection [6-8]. With the rapid development of these applications, optimization of the light-matter interaction to obtain high signal enhancement becomes the developing main thread of EIT-based sensing technology.

Great efforts have been invested for EIT optimization ranging from structure design to device material [9-12]. For instance, when the plasmonic structure (nanoantenna) is designed to be close enough, strong mutual coupling between them occurs, resulting in a large near-field enhancement inside their gap region [13]. It is demonstrated that a nearly three orders of magnitude boost in enhancement factor (EF) is achieved when the gap is reduced from 20 to 3 nm [14, 15]. Another effective approach is to increase the spatial overlap between analytes and enhanced near-field by undercutting the antenna's structure or integrating microfluidic channels with the dielectric substrate [16, 17]. These design methods significantly improve the sensitivity, but it requires precise and complex photolithography and bonding processes. Furthermore, the development of fabrication technology is generally difficult and costly, so excessive reliance on the fabrication capability will hinder the optimization and application of EIT. From the perspective of device material, all-dielectric plasmonic metamaterials are a promising method of EIT optimization due to their characters in reducing the dissipative loss and strong heating in the metals [18]. However, when compared with metal antennas, all-dielectric devices suffers from a poor affinity for immobilizing some molecules, which is unfavorable for surface functionalization in biosensing. In response to the trade-off between performance and fabrication cost, we previously attempted to utilize loss engineering to optimize the enhanced molecule signals while maintaining the same fabrication process [19]. It is definitely effective, but its further improvement is limited. Based on this background, we notice a complementary effect of EIT, namely electromagnetically induced absorption (EIA), which is a constructive interference phenomenon and induces an enhanced absorption on top of the broad dipolar absorbance feature [20]. EIA is often ignored and rarely used in sensing applications because of its similar function to EIT [21-25]. However, their main difference, namely the inverse trend in spectral change, inspired us whether it could be used to address the above-mentioned issues and make a significant breakthrough in the field of sensing.

In this work, we demonstrate, for the first time to our knowledge, the integration of complementary EIA and EIT effects in classical systems to break through the performance limitations in the field of traditional spectroscopic sensing. More specifically, molecular vibration coupled in an EIT-state device exhibits a negative enhancement, and that coupled in an EIA-state device a positive enhancement. When the two inverse enhancements are innovatively and neatly combined, ultra-strong molecular vibration signals exceeding the limit of commonly used methods are expected. Considering that gas detection requires high sensor sensitivity, we demonstrate the improvement of sensing performance of our strategy by detecting trace CO_2 gas at mid-infrared. As a result, our strategy can comprehensively improve the sensing performance including signal strength, sensitivity, resolution, noise, and limit of detection (LOD). Furthermore, our strategy is generally applicable to the sensing performance improvement in almost all EIA/EIT-like classical systems without additional fabrication complexity and cost. This work opens a new avenue to achieve efficient sensing performance improvement from the perspective of fundamental physical effects, and it promises to have a profound impact on the spectroscopic sensing field.

METHOD AND FABRICATION
Design

We propose a metal-insulator-metal (MIM) based SEIRA platform with two independent enhancement areas, namely EIT Area and EIA Area, as shown in Figure 1a. The molecular vibration coupled in EIT Area is negative, and that coupled in EIA Area is positive (Figure 1b). EIA and EIT enhancements are determined and optimized by adjusting the dielectric insulator thickness. When the two reversal changes are extracted based on each other, ultra-strong molecular vibration signals are obtained (Figure 1c). Such a high signal strength is of great benefit to obtain excellent sensing performance, including sensitivity and LOD (Figure 1d). The realization of EIA and EIT depends on the external/intrinsic loss rate γ_r/γ_a of the whole system, which can be adjusted by adjusting the distance between the reflective layer and antennas (Figure 2). As observed, the increase in thickness T causes a large rise in γ_r and a slight decrease in γ_a.

Fabrication

We fabricated the EIA-EIT-integrated platform on 6-inch silicon using a CMOS-compatible process (Figure 3). The process began with the cleaning and drying of the silicon wafer. Then, Ti (10 nm) and Au (100 nm) were sequentially deposited on the silicon surface using an evaporation system. After that, 200nm and 500nm thicknesses of MgF2 were deposited into two separate areas by using an e-beam evaporator system. Then, the patterning of the metamaterials is performed. First, the PMMA photoresist was spincoated, baked, exposed, and developed sequentially through a mid-ultraviolet stepper lithography system. Then, Ti (10 nm) and Au (100 nm) were sequentially deposited on the MgF2 dielectric layer using the e-beam evaporator system. Finally, the device was etched by an IBE system. After patterning, the patterned two areas were bonded on a silicon wafer. PEI thick film was spin-coated onto the surface of the device to functionalize the platform. The ultimately achieved devices were stored in a drying and storage cabinet.

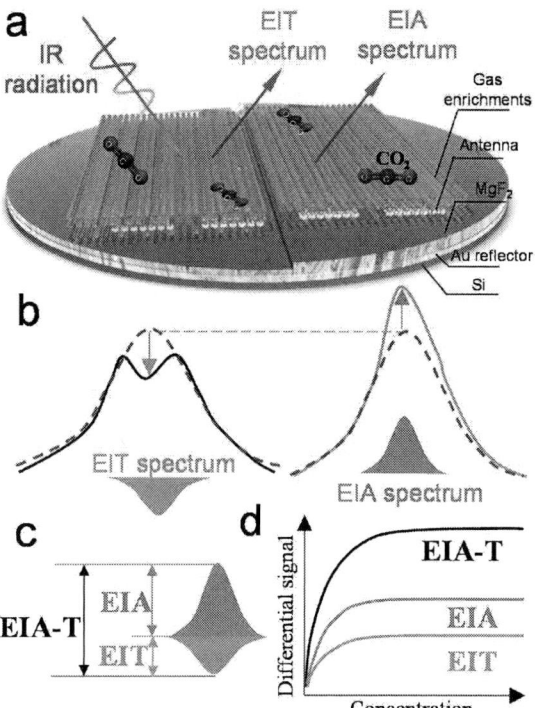

Figure 1: Design concept. (a) Nanoantennas with different configurations showing different spectral responses to the same vibrational signals under the irradiation of IR light. One is the EIT spectrum and the other is the EIA spectrum. The antenna-based platform consists of polyethylenimine (PEI) gas enrichments, nanoantennas, an MgF_2 spacer, an Au reflector, and Si substrate. (b) The spectral changes caused by the adsorption of CO_2 gases into gas enrichment materials on the antenna surface. For the same gas molecules, the caused change is negative in the EIT spectrum but positive in the EIA spectrum. (c) Ultra-strong molecular vibration signals are extracted when the two reversal changes are combined. (d) The EIA-T platform based on the two reversal effects exhibits higher sensing performance.

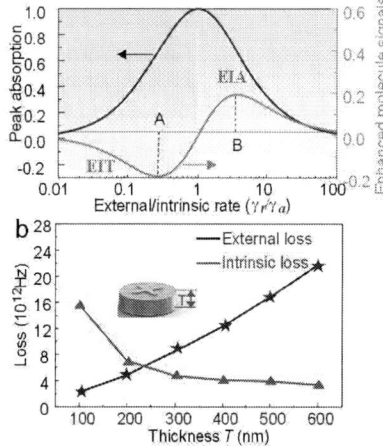

Figure 2: Thickness-dependent EIA&EIT transition. (a) Nanoantenna is in EIT or EIA state depending on the external/intrinsic loss rate γ_r/γ_a. (b) γ_r/γ_a further depends on the thickness of the spacer layer H. It means that we can

achieve both EIA and EIT by adjusting H.

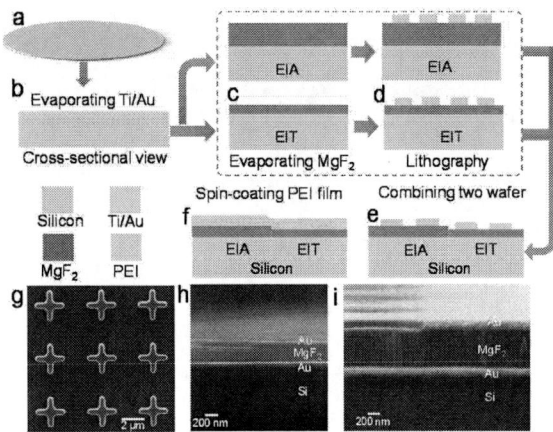

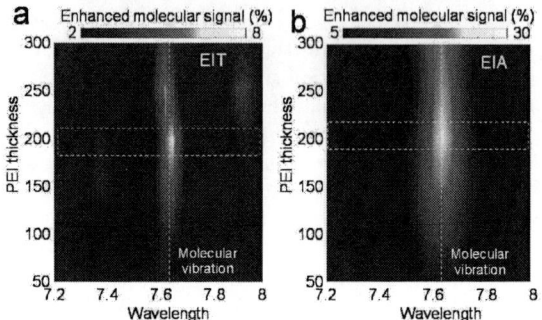

Figure 4: Thickness optimization of gas enrichment PEI. (a) PEI thickness versus spectrum wavelength map showing the optimal enhanced molecular signals at EIT and (b) EIA enhancement configuration.

Figure 3: Fabrication and Characterization. (a-f) Schematic diagram of processes to fabricate the EIA-EIT-integrated devices. The patterning of nanoantennas is performed using a mid-ultraviolet stepper lithography technique. (g) Top view SEM micrograph showing details of the nanoantenna. (h) Cross-sectional views of the EIT region and (i) the EIA region. All devices were fabricated on 6-inch silicon using a CMOS-compatible process.

RESULTS AND DISCUSSIONS

PEI is an excellent CO_2 gas-selective-trapping polymer that is often used for CO_2 adsorption and storage. During the adsorption process, the hard-acidic CO_2 molecules covalently react with the basic amine groups of PEI and are mainly converted into carboxyl groups -COOH and bicarbonate ions HCO_3^-. The reaction process is reversible, that is, the formed species can be recovered by heating, and the loss is negligible. The thickness of PEI is optimized. Thicker PEI certainly provides more binding sites for the gas, but a larger thickness also prevents the gas from entering the PEI interior, and vice versa. According to the PEI thickness versus spectrum wavelength (Figure 4), the optimal PEI thickness is about 200 nm.

In the gas detection demonstration, our device was placed in a gas cell with CO_2 concentrations changing from 0 to 1520 ppm. To void the interference of water vapor, the vibrational bands at 7.66 μm originating from the conformational change in HCO_3^- are chosen as the detection target for our platform. The measured spectral response of our EIA-EIT-integrated platform was depicted in Figure 5a. Clearly, the absorption in the EIA rose with the increase of CO2 concentration, while the changing trend of EIT is the opposite. Then, the difference absorption representing the enhanced molecular signals was extracted (Figure 5b). As observed, the signal intensity enhanced by our EIA-EIT-integrated platform is significantly larger than that of the EIA or EIT methods. The sensitivity and signal intensity of the EIA-EIT-integrated platform is about 4 times higher than the EIT method, which is critical in breaking the detection limit of gas sensors. When the maximum slope of the fitting curve is defined as the sensor sensitivity, the sensitivity of the EIA-T platform was calculated to be 0.141%/ppm, which was about 4 times higher than the EIT method.

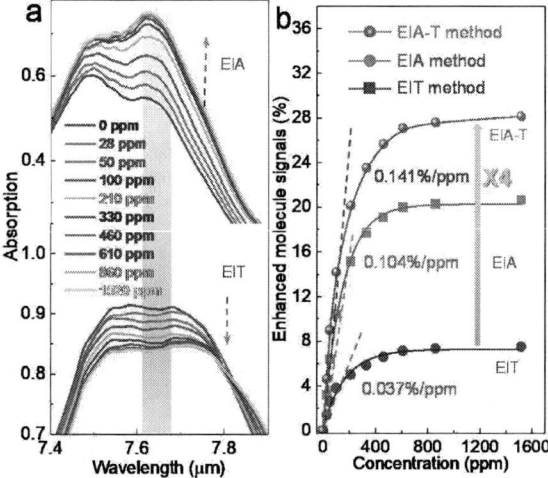

Figure 5: Demonstration of ultrasensitive detection of trace amounts of CO_2. (a) The measured reversal spectral response of the EIA-EIT-integrated platform when CO_2 concentrations change from 0 to 1520 ppm. (b) Enhanced molecular signal versus concentration profile showing a comparison of the sensitivity and signal intensity of the pure EIA device, pure EIT device, and our EIA-EIT-integrated platform.

CONCLUSIONS

We have demonstrated an ultrasensitive sensing strategy that innovatively combines the inverse enhancements of EIT and EIA in a coupled molecular and plasmonic SEIRA system. The combination of the two inverse enhancements into a hybrid EIA-T platform can further break the limit level achieved by the common improvement methods. Importantly, this breakthrough is the overall improvement of sensing performance including signal strength, sensitivity, resolution, noise, and LOD.

ACKNOWLEDGEMENTS

This work is supported by the Advanced Research and Technology Innovation Centre (ARTIC) Project (WBS: A-0005947-20-00).

REFERENCES

[1] Z. Ren, Z. Zhang, J. Wei, B. Dong, C. Lee. "Wavelength-multiplexed hook nanoantennas for machine learning enabled mid-infrared spectroscopy", *Nat. Commun.*, vol. **13**, pp. 3859, 2022.

[2] W. Liu, Y. Ma, X. Liu, J. Zhou, C. Xu, B. Dong, C. Lee. "Larger-than-unity external optical field confinement enabled by metamaterial-assisted comb waveguide for ultrasensitive long-wave infrared gas spectroscopy", *Nano Lett.*, vol. **22**, pp. 6112-6120, 2022.

[3] Y.-S. Lin, Z. Xu. "Reconfigurable metamaterials for optoelectronic applications", *Int. J. Optomechatronics*, vol. **14**, pp. 78-93, 2020.

[4] J. Xu, Y. Du, Y. Tian, C. Wang. "Progress in wafer bonding technology towards mems, high-power electronics, optoelectronics, and optofluidics", *Int. J. Optomechatronics*, vol. **14**, pp. 94-118, 2021.

[5] G. Zhou, Z. H. Lim, Y. Qi, F. S. Chau, G. Zhou. "Mems gratings and their applications", *Int. J. Optomechatronics*, vol. **15**, pp. 61-86, 2021.

[6] X. Hui, C. Yang, D. Li, X. He, H. Huang, H. Zhou, M. Chen, C. Lee, X. Mu. "Infrared plasmonic biosensor with tetrahedral DNA nanostructure as carriers for label-free and ultrasensitive detection of mir-155", *Adv. Sci.*, vol. **8**, pp. e2100583, 2021.

[7] D. Li, H. Zhou, X. Hui, X. He, H. Huang, J. Zhang, X. Mu, C. Lee, Y. Yang. "Multifunctional chemical sensing platform based on dual-resonant infrared plasmonic perfect absorber for on-chip detection of poly(ethyl cyanoacrylate)", *Adv. Sci.*, vol. **8**, pp. e2101879, 2021.

[8] D. Li, H. Zhou, X. Hui, X. He, X. Mu. "Plasmonic biosensor augmented by a genetic algorithm for ultra-rapid, label-free, and multi-functional detection of covid-19", *Anal. Chem.*, vol. **93**, pp. 9437-9444, 2021.

[9] H. Zhou, D. Li, X. Hui, X. Mu. "Infrared metamaterial for surface-enhanced infrared absorption spectroscopy: Pushing the frontier of ultrasensitive on-chip sensing", *Int. J. Optomechatronics*, vol. **15**, pp. 97-119, 2021.

[10] X. Liu, W. Liu, Z. Ren, Y. Ma, B. Dong, G. Zhou, C. Lee. "Progress of optomechanical micro/nano sensors: A review", *Int. J. Optomechatronics*, vol. **15**, pp. 120-159, 2021.

[11] Z. Ren, J. Xu, X. Le, C. Lee. "Heterogeneous wafer bonding technology and thin-film transfer technology-enabling platform for the next generation applications beyond 5g", *Micromachines*, vol. **12**, pp. 946, 2021.

[12] D. Hasan, C. Lee. "Hybrid metamaterial absorber platform for sensing of co2 gas at mid-ir", *Adv. Sci.*, vol. **5**, pp. 1700581, 2018.

[13] C. Huck, F. Neubrech, J. Vogt, A. Toma, D. Gerbert, J. Katzmann, T. Haertling, A. Pucci. "Surface-enhanced infrared spectroscopy using nanometer-sized gaps", *ACS Nano*, vol. **8**, pp. 4908-4914, 2014.

[14] L. Dong, X. Yang, C. Zhang, B. Cerjan, L. Zhou, M. L. Tseng, Y. Zhang, A. Alabastri, P. Nordlander, N. J. Halas. "Nanogapped au antennas for ultrasensitive surface-enhanced infrared absorption spectroscopy", *Nano Lett.*, vol. **17**, pp. 5768-5774, 2017.

[15] H. Zhou, L. Xu, Z. Ren, J. Zhu, C. Lee. "Machine learning-augmented surface-enhanced spectroscopy toward next-generation molecular diagnostics", *Nanoscale Adv.*, 2022.

[16] J. Xu, Z. Ren, B. Dong, X. Liu, C. Wang, Y. Tian, C. Lee. "Nanometer-scale heterogeneous interfacial sapphire wafer-bonding for enabling plasmonic-enhanced nanofluidic mid-infrared spectroscopy", *ACS Nano*, vol. **14**, pp. 12159−12172, 2020.

[17] T. H. H. Le, T. Tanaka. "Plasmonics-nanofluidics hybrid metamaterial: An ultrasensitive platform for infrared absorption spectroscopy and quantitative measurement of molecules", *ACS Nano*, vol. **11**, pp. 9780-9788, 2017.

[18] A. Leitis, A. Tittl, M. Liu, B. H. Lee, M. B. Gu, Y. S. Kivshar, H. Altug. "Angle-multiplexed all-dielectric metasurfaces for broadband molecular fingerprint retrieval", *Sci. Adv.*, vol. **5**, pp. eaaw2871, 2019.

[19] J. Wei, Y. Li, Y. Chang, D. M. N. Hasan, B. Dong, Y. Ma, C. W. Qiu, C. Lee. "Ultrasensitive transmissive infrared spectroscopy via loss engineering of metallic nanoantennas for compact devices", *ACS Appl. Mater. Interfaces*, vol. **11**, pp. 47270–47278, 2019.

[20] A. M. Akulshin, S. Barreiro, A. Lezama. "Electromagnetically induced absorption and transparency due to resonant two-field excitation of quasidegenerate levels in rb vapor", *Phys. Rev. A*, vol. **57**, pp. 2996-3002, 1998.

[21] H. Zhou, D. Hu, C. Yang, C. Chen, J. Ji, M. Chen, Y. Chen, Y. Yang, X. Mu. "Multi-band sensing for dielectric property of chemicals using metamaterial integrated microfluidic sensor", *Sci. Rep.*, vol. **8**, pp. 14801, 2018.

[22] H. Zhou, X. Hui, D. Li, D. Hu, X. Chen, X. He, L. Gao, H. Huang, C. Lee, X. Mu. "Metal-organic framework-surface-enhanced infrared absorption platform enables simultaneous on-chip sensing of greenhouse gases", *Adv. Sci.*, vol. **7**, pp. 2001173, 2020.

[23] H. Zhou, Z. Ren, C. Xu, L. Xu, C. Lee. "Mof/polymer-integrated multi-hotspot mid-infrared nanoantennas for sensitive detection of co2 gas", *Nano-Micro Lett.*, vol. **14**, pp. 207, 2022.

[24] H. Zhou, C. Yang, D. Hu, S. Dou, X. Hui, F. Zhang, C. Chen, M. Chen, Y. Yang, X. Mu. "Integrating a microwave resonator and a microchannel with an immunochromatographic strip for stable and quantitative biodetection", *ACS Appl. Mater. Interfaces*, vol. **11**, pp. 14630-14639, 2019.

[25] H. Zhou, C. Yang, D. Hu, D. Li, X. Hui, F. Zhang, M. Chen, X. Mu. "Terahertz biosensing based on bi-layer metamaterial absorbers toward ultra-high sensitivity and simple fabrication", *Appl. Phys. Lett.*, vol. **115**, pp. 143507, 2019.

CONTACT

*C. Lee; elelc@nus.edu.sg

TWISTED AND CONTACTED AU MICRO-RODS 3D CHIRAL METAMATERIALS WITH CIRCULAR DICHROISM VIA AN ABSORPTIVE ROUTE IN LONG-WAVELENGTH INFRARED

Gaku Furusawa[1], Natsuki Kanda[2], Ryusuke Matsunaga[2], and Tetsuo Kan[1]
[1]The University of Electro-Communications, JAPAN and
[2] The Institute for Solid State Physics, The University of Tokyo, JAPAN

ABSTRACT

In this study, we proposed a twisted and contacted Au micro-rods 3D chiral metamaterial that shows degree-order circular dichroism via an absorption route in the long-wavelength infrared region. The proposed Au chiral structure consisted of two layers of Au micro-rods whose second layer was oriented with 90-degree rotation relative to the first layer and arranged C_4 rotation symmetry. The Au chiral structure showed strong circular dichroism in absorption at a wavelength of 12.3 μm in simulation. The chiral structure was fabricated using a bi-layer lift-off process. The fabricated Au chiral structure showed 2.3 degrees circular dichroism at 12.2 μm wavelength.

KEYWORDS

Chiral metamaterial, optical region, microfabrication.

INTRODUCTION

Like the human right and left hands, symmetry structures with properties that do not overlap in translation or rotation operations are called chiral structures. Chiral structures show circular dichroism, an optical characteristic showing different permittivity for right and left circularly polarized light, respectively [1]. Chiral metamaterials, which are consisted of sub-wavelength scale metal chiral structures, have much stronger circular dichroism than that of chiral structures in natural materials. Therefore, chiral metamaterials are promising for applications to the circular polarizer and chiral sensing [2]. In principle, there are two routes to produce transmissive circular dichroism: one is a difference in reflections, and another is a difference in absorptions between two (right and left) circularly polarized lights. In previous studies, circular dichroism has been realized based on reflection, and there are a few examples of achieving circular dichroism via an absorption route [3], [4]. If circular dichroism via the absorption route is realized in the optical region, chiral metamaterials can be employed for many applications, such as a flexible anti-reflection film [5]. Previously, absorption-based circular dichroism chiral metamaterials have been reported in radio frequency [6]. However, a simple size reduction of this structure cannot produce circular dichroism in the optical region because of a considerable difference in the electromagnetic behavior of metals in optical and radio wave regions. Therefore, absorptive-type chiral metamaterials operating in the optical range are highly required.

This paper proposes a twisted and contacted Au micro-rods 3D chiral metamaterial that shows degree-order circular dichroism via an absorption route in the long-wavelength infrared region. We evaluated the proposed Au chiral structure's validity by simulation and measuring a

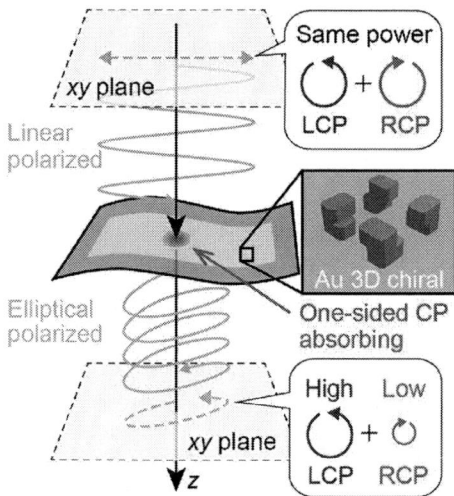

Figure 1: The proposed twisted and contacted Au micro-rods 3D chiral metamaterial that shows circular dichroism via an absorption route.

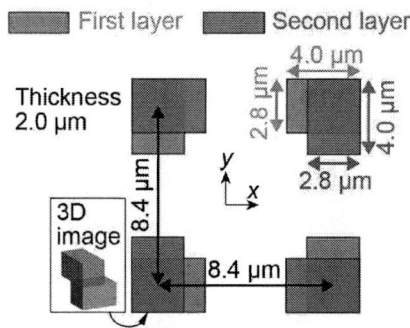

These units are periodically placed every 18 μm in the *x*- and *y*-directions.

Figure 2: The device design of the proposed Au chiral metamaterial. The proposed Au chiral structure was consisted of two layers of Au micro-rods whose second layer was oriented 90 degrees rotated with respect to the first layer and arranged C_4 rotation symmetry.

device's circular dichroism.

METHODOLOGY
Device design

The proposed chiral metamaterial is consisted of twisted and contacted Au micro-rods fabricated by forming thick rectangular two-layered Au (figure 1). In general, three-dimensional chiral structures exhibit strong circular dichroism. In order to fabricate three-dimensional chiral structures using planar processes such as photolithography, each layer of the Au pattern is designed as a thick film. Figure 2 shows the schematic diagram for the proposed chiral structure. A mask pattern of the chiral structure has rectangles of 4.0 μm × 2.8 μm and are placed 8.4 μm apart in C_4 rotation symmetry arrangement; four two-layered Au rods are placed on the device plane with 90-degree rotation with each other around the z-axis. The second layer of the chiral pattern is oriented 90 degrees rotation relative to the first layer. The pattern of each layer is composed of 2.0 μm thickness Au micro-rods patterns. The first and second layers of Au micro-rods are in contact.

In simulations using COMSOL Multiphysics 5.6, we found that contacting Au micro-rods of the first and second layers was essential to achieve circular dichroism via the absorptive route. The simulation conditions are shown in Figure 3. The simulation area was 18 μm × 18 μm × 30 μm. The dimensions of the Au chiral structure were the same as in Figure 2. The corners of the rods were rounded with a radius of 1.4 μm, considering the actual fabrication process by photolithography. The Au chiral structure was embedded in a thin SU-8 film with a thickness of 10 μm. The dielectric constant of Au was calculated with the Drude model [7]. Surrounding media of the chiral structure and the SU-8 film were air. Floquet periodicity was applied to the sides of the simulation domain. The transmittances of left and right circularly polarized light of the proposed chiral structure were 95% and 37%, and absorptions were 1.9% and 60%, respectively, at a peak wavelength of 12.3 μm.

Fabrication

The process flow of the proposed method is shown in Figure 4. Each layer of thick Au rods was patterned using a bi-layer lift-off process which uses two types of photoresist: positive photoresist and thickness-enhancing photoresist. An Al layer was evaporated on a Si substrate with a thickness of 100 nm. An anti-reflective film for i-line (light of 365 nm wavelength), XHRiC-307 (Brewer Science, USA), was spin-coated at 800 rpm for 3 s and 1500 rpm for 30 s on the substrate to reduce a reflection by the substrate at the exposure and to protect the Al layer. After spin-coating of XHRiC, the substrate was baked at 180°C for 1 min. A thickness-enhancing photoresist PMGI SF 13 (PMGI, Nippon Kayaku) was spin-coated at 800 rpm for 3 s and 3000 rpm for 30 s on the substrate baked at 180°C for 5 min. An i-line positive photoresist THMR-iP5720 HP 7 cP (TOKYO OHKA KOGYO, Japan) was spin-coated at 800 rpm for 3 s and 6000 rpm for 30 s on the substrate, and the substrate was baked at 110°C for 2 min. The THMR was exposed with an exposure energy of 50 mJ/cm² with a mask aligner (MA/BA6, SUSS, Germany). The substrate was baked at 110°C for 2 min and developed for 2 min with NMD-3 (TOKYO OHKA KOGYO, Japan). A 2.0 μm thickness Au was evaporated on the substrate. The THMR was removed with acetone, and the Au was lifted off. Here, the first layer of Au micro-rods patterning was completed. A g-line (light of 436 nm wavelength) positive photoresist OFPR-800 23cp (OFPR, TOKYO OHKA KOGYO) was spin-coated at 800 rpm for 3 s and 2000 rpm for 30 s, and the substrate was baked at 110°C for 2 min. Spin-coating and baking of the OFPR were done three times to flatten the first layer. A gap between the Au

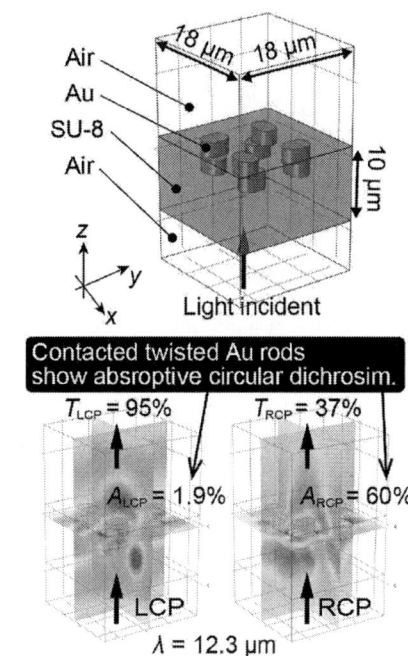

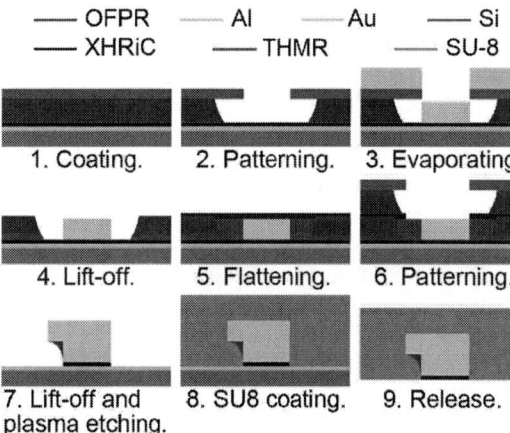

Figure 3: In simulation, the proposed Au chiral metamaterial showed strong circular dichroism at a wavelength of 12.3 μm.

Figure 4: The process flow of the proposed Au chiral structure. To make thick Au pattern, bi-layer lift-off process which uses two type resist, positive photoresist and enhancing thickness resist, was used.

and PMGI was filled with the OFPR. The OFPR was exposed with 180 mJ/cm² and developed for 3 min with NMD-3 to remove OFPR out of Au patterned area. After patterning OFPR, the substrate was baked at 180°C for 5 min. The OFPR was dry-etched by O₂ plasma at 10 W power and 10 sccm gas flow rate for 40 min to expose the top surface of Au. The XHRiC was spin-coated at 800 rpm for 3 s and 1500 rpm for 30 s on the substrate, and it was baked at 180°C for 1 min. The second layer of Au micro-rods pattering used the same procedure as the first. After the second layer lift-off, the OFPR was removed using O₂ plasma at 10 W power and 10 sccm gas flow rate for 90 min and completed to make the twisted and contacted Au micro-rods 3D chiral structures. SU-8 3010 (SU-8, Nippon

Kayaku) was spin-coated at 800 rpm for 3 s and 2000 rpm for 30 s, and the substrate was baked at 95°C for 10 min. The edges of the substrate were covered with aluminum foil and exposed to 270 mJ/cm². After exposure, the substrate was baked at 95°C for 5 min and developed for 5 min with propylene glycol monomethyl ether acetate (PGMEA, FUJIFILM Wako Pure Chemical Corporation). To transfer the Au chiral structures from Al evaporated Si substrate to SU-8 thin film, the Al layer of the substrate was removed using an Al etchant (H_2PO_4: CH_3COOH: HNO_3: H_2O = 4: 4: 1:1) for 1 hour on a 90°C heated hot plate. The Au structures were transferred into SU-8 thin film without destruction (Figure 5).

EXPERIMENTS AND RESULTS

The circular dichroism of the device was measured using terahertz time-domain spectroscopy (THz TDS) between the wavelength of 10.0 μm to 25.0 μm [8]. The device was irradiated with linearly polarized light, which is a superposition of left and right circularly polarized light of the same intensity. Since the device exhibits different transmittances for left and right circular polarized light, the linearly polarized light was converted into elliptically polarized light. The ellipticity η of the elliptically polarized light transmitted through the device was measured to evaluate circular dichroism. The measured ηs of the chiral and the reversed chiral structures were 2.3 degrees at a wavelength of 12.2 μm (Figure 6). The peak wavelength of circular dichroism exhibited by the fabricated device was close to the peak wavelength of 12.3 μm obtained in the simulation. Circular dichroism is known to be produced via an absorptive route if chiral structures placed on C_4 rotation symmetry arrangement [9]. It may be possible to obtain larger circular dichroism by improving the device's fabrication precision. In this work, we established a basic technology to provide a significant (deg scale) absorptive circular dichroism in the optical region.

CONCLUSION

In this paper, we proposed the twisted and contacted Au micro-rods 3D chiral metamaterial, and it showed deg order circular dichroism via the absorption route in the long-wavelength infrared region. The chiral structures, which consisted of single-micron-scaled thick Au patterns, were fabricated by the bi-layer lift-off process and formed 3D twisted shape. In the simulation, the proposed chiral structure showed that left and right circularly polarized light absorptions were 1.9% and 60%, respectively, at a peak wavelength of 12.3 μm. In the THz TDS measurement, the Au chiral structures showed ellipticity η = 2.3 degrees at a wavelength of 12.2 μm. Stronger circular dichroism could be achieved by improving the fabrication accuracy and optimizing the design of the chiral structures in this study.

ACKNOWLEDGEMENTS

This work was supported by JSPS KAKENHI Grant Number 20J23133. This work was partially performed using facilities of "Advanced Research Infrastructure for Materials and Nanotechnology in Japan (ARIM)" of the Ministry of Education, Culture, Sports, Science and

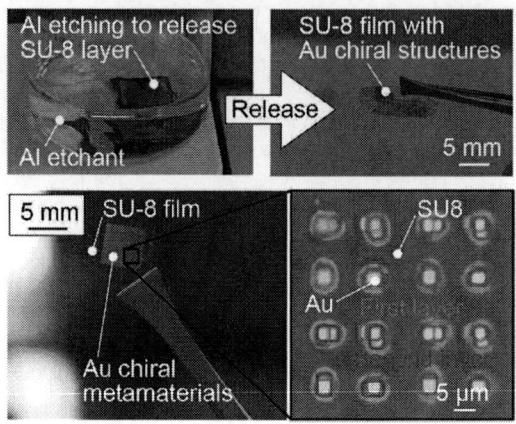

Figure 5: The Au chiral structures were transferred to SU-8 thin film by removing Al layer of the surface of the Si substrate which had the Au chiral structures on its surface.

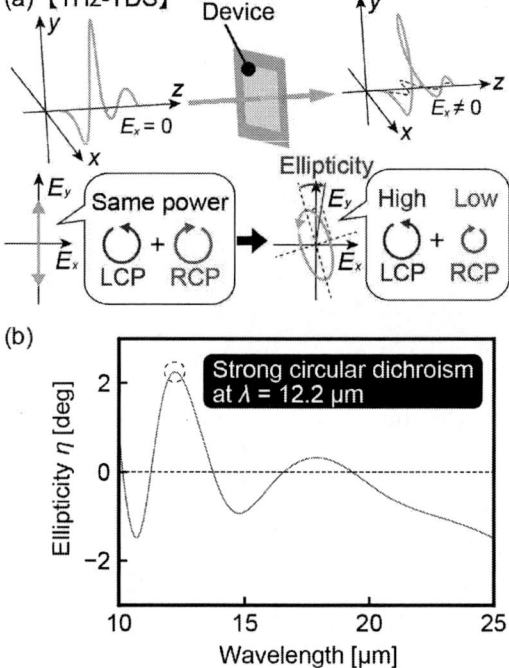

Figure 6: The Au chiral structures showed deg order circular dichroism at a wavelength of 12.2 μm. (a) A schematic of measurement using THz TDS. (b) The measured circular dichroism of the fabricated device.

Technology (MEXT) (Proposal Number JPMXP1222UT1030). Photomasks were fabricated in the Takeda Clean Room of d.lab, the University of Tokyo, Japan. The microfabrication was mainly performed in a clean room of the Advanced Research Section of the Research Facility Center of the University of Electro-Communications (UEC), Japan.

REFERENCES

[1] W. Wu and M. Pauly, "Chiral plasmonic nanostructures: recent advances in their synthesis and applications," *Mater Adv*, vol. 3, no. 1, pp. 186–215, Jan. 2022, doi: 10.1039/D1MA00915J.

[2] Y. Y. Lee, R. M. Kim, S. W. Im, M. Balamurugan, and K. T. Nam, "Plasmonic metamaterials for chiral sensing applications," *Nanoscale*, vol. 12, no. 1, pp. 58–66, Jan. 2020, doi: 10.1039/C9NR08433A.

[3] Y. Cheng, F. Chen, and H. Luo, "Plasmonic Chiral Metasurface Absorber Based on Bilayer Fourfold Twisted Semicircle Nanostructure at Optical Frequency," *Nanoscale Res Lett*, vol. 16, no. 1, p. 12, Dec. 2021, doi: 10.1186/s11671-021-03474-6.

[4] D. B. Stojanovic, G. Gligoric, P. P. Belicev, M. R. Belic, and L. Hadzievski, "Circular Polarization Selective Metamaterial Absorber in Terahertz Frequency Range," *IEEE Journal of Selected Topics in Quantum Electronics*, vol. 27, no. 1, pp. 1–6, Jan. 2021, doi: 10.1109/JSTQE.2020.3024570.

[5] R. Singh, K. N. Narayanan Unni, A. Solanki, and Deepak, "Improving the contrast ratio of OLED displays: An analysis of various techniques," *Opt Mater (Amst)*, vol. 34, no. 4, pp. 716–723, Feb. 2012, doi: 10.1016/j.optmat.2011.10.005.

[6] M. Li, L. Guo, J. Dong, and H. Yang, "An ultra-thin chiral metamaterial absorber with high selectivity for LCP and RCP waves," *J Phys D Appl Phys*, vol. 47, no. 18, p. 185102, May 2014, doi: 10.1088/0022-3727/47/18/185102.

[7] C. A. Ward *et al.*, "Optical properties of the metals Al, Co, Cu, Au, Fe, Pb, Ni, Pd, Pt, Ag, Ti, and W in the infrared and far infrared," *Applied Optics, Vol. 22, Issue 7, pp. 1099-1119*, vol. 22, no. 7, pp. 1099–1119, Apr. 1983, doi: 10.1364/AO.22.001099.

[8] N. Kanda, N. Ishii, J. Itatani, and R. Matsunaga, "Optical parametric amplification of phase-stable terahertz-to-mid-infrared pulses studied in the time domain," *Opt Express*, vol. 29, no. 3, p. 3479, Feb. 2021, doi: 10.1364/OE.413200.

[9] J. Kaschke, J. K. Gansel, and M. Wegener, "On metamaterial circular polarizers based on metal N-helices," *Opt Express*, vol. 20, no. 23, p. 26012, Nov. 2012, doi: 10.1364/OE.20.026012.

CONTACT

*G.Furusawa, tel: +81-42-443-5423;
gaku@ms.mi.uec.ac.jp

3D HYBRID ACOUSTIC RESONATOR WITH COUPLED FREQUENCY RESPONSES OF SURFACE ACOUSTIC WAVE AND BULK ACOUSTIC WAVE

Liping Zhang[1,2,#], Shibin Zhang[1,#,*], Jinbo Wu[1,2], Pengcheng Zheng[1,2], Hulin Yao[1,2], Yang Chen[1,2], Kai Huang[1,2], Xiaomeng Zhao[1], Min Zhou[1] and Xin Ou[1,2,*]

[1]State Key Laboratory of Functional Materials for Informatics, Shanghai Institute of Microsystem and Information Technology, Shanghai 200050, Chinese Academy of Sciences, China

[2]Center of Materials Science and Optoelectronics Engineering, University of Chinese Academy of Sciences, Beijing 100049, China

ABSTRACT

This work proposes a novel one-port 3D acoustic resonator based on the lithium niobate thin film on conductive silicon carbide substrate ($LiNbO_3$-on-SiC, LNCSiC). The fabricated resonator shows coupled frequency responses of the shear-horizontal surface acoustic wave (SH-SAW), the longitudinal leaky SAW (LL-SAW), and the high-overtone bulk acoustic waves (HBAWs). The HBAWs propagating in the thickness direction of LNCSiC show a wide frequency response span exceeding 4 GHz and an excellent maximum quality factor (Q_{max}) of 7980. The GHz SH-SAW propagating in the surface of LNCSiC show a large electromechanical coupling coefficient (k_t^2) of 25.95%, while the LL-SAW shows an extremely high velocity of ~6900 m/s. Such hybrid resonators could potentially open up new applications in radio frequency communications, 3D imaging, and sensing.

KEYWORDS

Hybrid Acoustic Resonator, Lithium Niobate Thin Film on Conductive Silicon Carbide, Surface Acoustic Waves, Bulk Acoustic Waves, Quality Factor, Electromechanical Coupling Coefficient, Phase Velocity.

INTRODUCTION

In the past decade, acoustic devices have received widespread attention for their profound applications in signal processing and sensing [1, 2]. With the vast expansion of wireless communication system, acoustic filters, sensors and imaging devices are recognized as indispensable components for mobile terminals and are now in aggressive and tough competition in radio frequency (RF) device markets [3-6].

At present, the most mainstream implementations of filters for mobile RF front-ends are surface acoustic wave (SAW) filters and bulk acoustic wave (BAW) filters [2-4, 7, 8]. Among them, SAW devices are increasingly demanded in the front-end module, thanks to their simple structures, lithography defined operating frequency and the low manufacturing cost. Recently, SAW filters based on the piezoelectric-on-insulators (POI) substrates have been investigated widely [1, 9-13]. For instance, the shear-horizontal surface acoustic wave (SH-SAW) filter based on the $LiNbO_3/SiO_2/poly$-Si/Si heterostructure shows a center frequency (f_c) of 3728 MHz and a 3-dB bandwidth (BW) of 1052 MHz [13], which meets the bandwidth requirements of the 5G n77 band. The longitudinal leaky

SAW (LL-SAW) filter based on the $LiNbO_3/SiC$ heterosubstrate shows a f_c of 6.12 GHz and a 3-dB BW of 483 MHz [14], showing the potential for enabling high-performance LLSAW devices for the 5G n79 band. In a word, these $LiNbO_3$-based POI-SAW filters with high frequency, large bandwidth and high quality (Q) factors are promising to make the 5G vision a reality.

The high-overtone bulk acoustic resonator (HBAR), another kind of acoustic device, consists of a piezoelectric thin film sandwiched by upper and lower electrodes acting as a transducer for acoustic wave generation, and the acoustic energy is mainly trapped in the supporting substrate [4, 15-18]. HBAR features multiple resonance peaks (high-overtone bulk acoustic waves, HBAW) in an extended frequency range and extremely high Q factors (up to 68000) at gigahertz [19]. As reported, the traditional AlN-based HBAR usually functions as wireless sensor and frequency sources (oscillators) for RF applications [7, 16], and also can be used as an alternative to SAW resonator for sensor development. Compared to the AlN-based HBAR, the novel $LiNbO_3$-based HBAR shows a wider frequency response span [18, 20, 21], which benefits from the larger electromechanical coupling coefficient (k_t^2). However, single-crystalline $LiNbO_3$ films cannot be directly prepared on metal electrodes or patterned bottom electrodes, leading to a more complicated fabrication process than that of SAW devices.

Then, if it is possible to build a new device with a simple structure but combines the merits of the SAW devices and the HBAR. In this work, the one-port 3D hybrid acoustic resonator with coupled frequency responses of the SH-SAW, LL-SAW and HBAWs is firstly demonstrated on the lithium niobate film on conductive silicon carbide ($LiNbO_3$-on-SiC, LNCSiC) substrate, showing a combination of high frequency, large bandwidth and high Q factors. The HBAWs show a broad frequency response span in the range of 1~5 GHz and a great $f \cdot Q$ product. The SH-SAW shows an extremely large k_t^2 of 25.95% and a resonance frequency (f_r) of 1.93 GHz, while the LL-SAW shows a f_r of 2.88 GHz and an ultra-high velocity of ~6900 m/s.

DEVICE DESIGN AND SIMULATION

As schematically shown in Figs. 1(a) and 1(b), thanks to the strong anisotropic piezoelectricity of $LiNbO_3$ and the high phase velocities of the bulk acoustic waves of SiC [22], the wideband SH-SAW and high-velocity LL-SAW can be excited by the lateral electric field induced by the

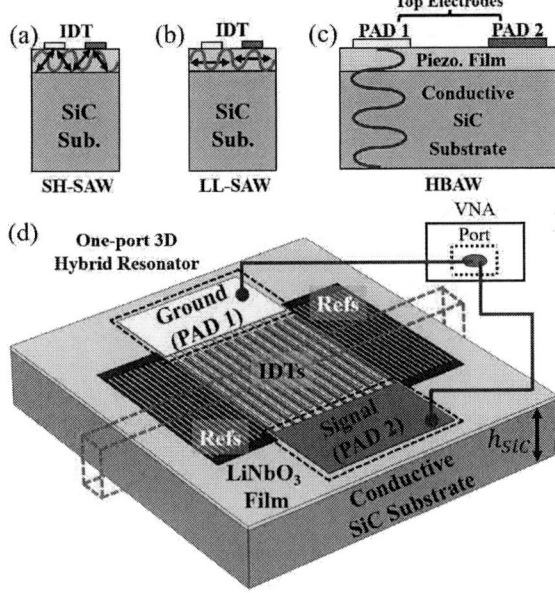

Figure 1: The schematic of (a) the SH-SAW resonator, (b) the LL-SAW resonator, and (c) the HBAW resonator. (d) The schematic of the proposed one-port 3D hybrid acoustic resonator based on the LNCSiC substrate.

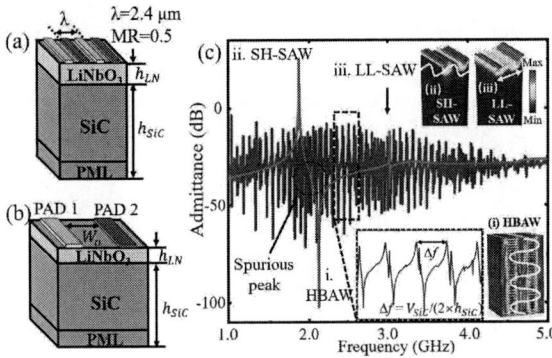

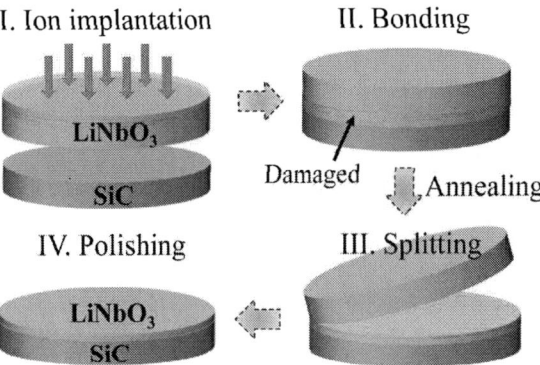

Figure 2: (a) The simulated model of SAW resonators. (b) The simulated model of HBAR. (c) The simulated frequency responses of the hybrid resonator, in which the HBAWs (i), SH-SAW (ii) and LL-SAW (ii) are marked, and their mode displacements are presented.

Figure 3: The process flow of transferring LiNbO₃ thin film to the conductive SiC substrate.

interdigital transducers (IDTs). The acoustic field of the above modes can be sufficiently confined in the LiNbO₃ thin film, which are suitable for high-frequency, large-bandwidth and high-quality SAW filters.

Meanwhile, unlike the traditional HBAR with a bottom electrode [4, 7, 18], the HBAR based on the LNCSiC heterosubstrate has the conductive SiC serving as the bottom electrode as well as the resonant cavity. The HBAWs can be excited by the bus-bar regions (PAD 1 and PAD 2) of the IDTs, and travel in the thickness direction of the SiC substrate, as shown in Fig. 1(c). Thanks to the large k_t^2 of the LiNbO₃ film and the low acoustic loss of the SiC substrate (resonant cavity), the wide frequency response span and the high Q are expected to be achieved.

Taken together, as schematically shown in Fig. 1(d), the unique one-port 3D acoustic resonator can be realized by the conventional SAW design consisting of the IDTs, reflectors (Refs) and bus-bar regions on the LNCSiC substrate. Then the SH-SAW, LL-SAW can propagate in the horizontal plane, while the HBAW can propagate in the thickness direction.

The frequency responses of the SAW and the HBAW parts are simulated individually based on the same LNCSiC substrate. For the SAW part, as shown in Fig. 2(a), the unit-cell model is adopted, in which the thickness of the LiNbO₃ film (h_{LN}) and SiC substrate (h_{SiC}) are 520 nm and 50 μm, respectively. The wavelength (λ) of the resonator is 2.4 μm, the metallization ratio (MR) is 0.5 and the aperture (W_a) is 50 μm. The in-plane orientation is set to be -12° to the +Y axis to optimize the k_t^2 of SH-SAW. For the HBAW part, as shown in Fig. 2(b), the distance between two bus-bars is 50 μm, the length and width of the bus-bar region are 55 μm and 30 μm. Besides, the electrical conductivity of SiC substrate is set to be 5000 S/m.

The simulated frequency responses and mode displacements of the SAW and BAW parts are presented in Fig. 2(c), respectively, in which (ii) and (iii) correspond to the in-plane SH-SAW and LL-SAW, respectively, while (i) corresponds to the HBAWs travelling in the conductive SiC. The simulated frequency spacing (Δf) of HBAWs is about 70 MHz, which is related to the velocity ($V_{SiC}\sim7100$ m/s) and thickness of the SiC substrate ($h_{SiC}\sim50$ μm), and $\Delta f = V_{SiC}/(2 \times h_{SiC})$. The SH-SAW shows a large k_t^2 of 35.9% and a f_r of 1.87 GHz. The LL-SAW shows an extremely high velocity of ~7200 m/s, a f_r of 3.02 GHz and a k_t^2 of 0.24%, However, the spurious peak marked by the black circle is Rayleigh SAW, which can be eliminated by adopting a suitable h_{LN}/λ or changing the in-plane orientation.

MEASUREMENT AND DISCUSSION

As shown in Fig. 3, the heterogeneous integration of a 4-inch X-cut LiNbO₃ thin film onto a 4H-SiC substrate (n-type, the electrical conductivity is about 5000 S/m) is prepared by ion-cutting and wafer bonding process. Based on the LNCSiC substrate, the one-port hybrid acoustic resonators are fabricated via electron-beam lithography, followed by Al evaporation and lift-off process. The

978-1-6654-9309-3/23 $31.00 © 2023 IEEE   1176

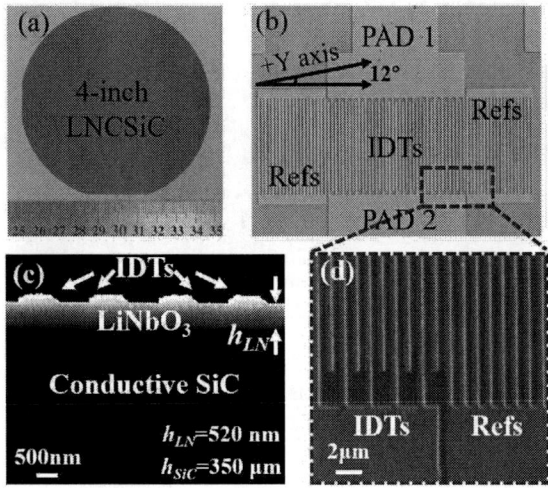

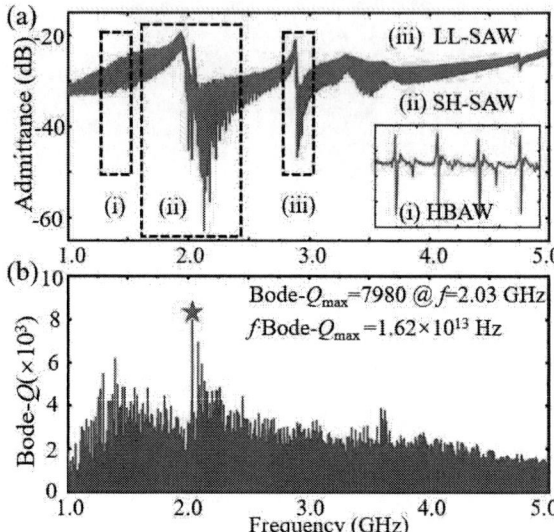

Figure 4: (a) The photo of the fabricated 4-inch LNCSiC substrate. (b) The optical microscope image and (d) the detailed scanning electron microscope (SEM) image of the IDTs and Refs of a fabricated one-port 3D hybrid resonator. (c) The cross-sectional SEM image of a one-port 3D hybrid resonator.

Figure 5: (a) The measured admittance curve of the fabricated hybrid resonator, with the black boxes marked by (i), (ii) and (iii) corresponding to HBAW, SH-SAW and LL-SAW, respectively. (b) The measured Bode-Q curve.

electrode thicknesses of the IDTs, Refs and bus-bar regions are all 130 nm.

Fig. 4(a) shows the photo of the fabricated LNCSiC substrate. Figs. 4(b)-(d) show the optical microscope image and the scanning electron microscope (SEM) images of a fabricated resonator and the detailed IDTs and Refs. The thickness of the LiNbO₃ thin film (h_{LN}) and SiC substrate (h_{SiC}) are 520 nm and 350 μm, respectively. The resonator is designed as followed: oriented at -12° to the +Y axis, the λ of 2.4 μm, the MR of 0.5. The profiles of IDTs are a little bit rough, and the fabrication process need to be optimized further.

The one-port hybrid resonator is characterized using a vector network analyzer (Keysight E5071C) with a terminal impedance of 50 ohm at room temperature in air, as schematically shown in Fig. 1(d). The measured admittance and Bode-Q curves are shown in Figs. 5(a) and 5(b), respectively. The hybrid resonator shows strong-coupled frequency responses of SH-SAW, LL-SAW and HBAW in the range of 1~5 GHz. The SAW peaks labelled as "ii" and "iii" corresponds to the SH-SAW and LL-SAW, respectively. The HBAWs indicate a broad frequency response span exceeding 4 GHz. As shown in Fig. 5(b), the maximum of the extracted Bode-Q (Bode-Q_{max}) of 7980 is achieved at 2.03 GHz, resulting in a great f·Bode-Q_{max} product of 1.62×10^{13} Hz.

The zoomed-in admittance curves of the SH-SAW and LL-SAW are plotted in Figs. 6(a) and 6(b), and the measured results are fitted by the modified Butterworth–Van Dyke (mBVD) model [23]. The SH-SAW shows a large k_t^2 of 25.95%, a f_r of 1.93 GHz and a relatively high velocity of ~4600 m/s. Compared with the simulated k_t^2 of SH-SAW in unit-cell model, the smaller k_t^2 factor may result from parasitic capacitance induced by the bus-bar regions of the IDTs. The LL-SAW shows an extremely high velocity of ~6900 m/s, a f_r of 2.88 GHz and a k_t^2 of

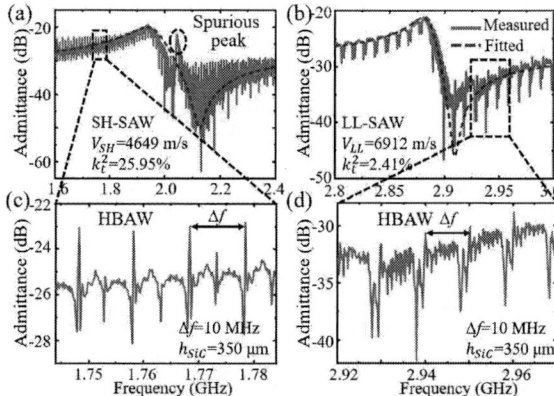

Figure 6: The zoomed-in measured and fitted admittance curve of (a) the SH-SAW and (b) the LL-SAW of the fabricated hybrid resonator. The zoomed-in coupled frequency response of (c) the SH-SAW and HBAWs, and (d) the LL-SAW and HBAWs.

2.41%, which also have some differences with the simulation results. The differences of f_r and phase velocity may be caused by the angular deviations between the fabricated resonator and the designed one. Other reasons are still under investigation. The further zoomed-in admittance curves in Figs. 6(c) and 6(d) show the coupled frequency response of SH-SAW and HBAWs, and the coupled frequency response of LL-SAW and HBAWs, respectively. Besides, the HBAWs present a Δf of 10 MHz, corresponding to the substrate thickness of 350 μm.

CONCLUSION

In this paper, we report a novel one-port 3D hybrid acoustic resonator based on the LNCSiC substrate. The LNCSiC substrate is obtained through ion-cutting and

978-1-6654-9309-3/23 $31.00 © 2023 IEEE

wafer bonding process. The fabricated resonator shows coupled frequency responses of the SH-SAW, the LL-SAW and the HBAWs in the range of 1~5 GHz. The multiple HBAWs show a wide frequency response span, and an excellent $f \cdot Q_{max}$ product. The SH-SAW present a large k_t^2 of 25.95%, while the LL-SAW shows an ultra-high velocity of ~6900 m/s. Further investigation of this hybrid technology may promote the progress of the 3D imaging, sensing and new applications in RF communications.

ACKNOWLEDGEMENTS

This work was supported by the National Key R&D Program of China (2022YFB3606701).

REFERENCES

[1] H. Zhou et al., "Surface Wave and Lamb Wave Acoustic Devices on Heterogenous Substrate for 5G Front-Ends," in *2020 IEEE International Electron Devices Meeting (IEDM)*, 2020, pp. 17.6.1-17.6.4.

[2] R. C. Ruby, P. Bradley, Y. Oshmyansky, A. Chien, and J. D. Larson, "Thin film bulk wave acoustic resonators (FBAR) for wireless applications," in *2001 IEEE Ultrasonics Symposium. Proceedings. An International Symposium (Cat. No.01CH37263)*, 2001, vol. 1, pp. 813-821 vol.1.

[3] R. Aigner, G. Fattinger, M. Schaefer, K. Karnati, R. Rothemund, and F. Dumont, "BAW Filters for 5G Bands," in *2018 IEEE International Electron Devices Meeting (IEDM)*, 2018, pp. 14.5.1-14.5.4.

[4] S. Ballandras et al., "High overtone Bulk Acoustic Resonators built on single crystal stacks for sensors applications," in *SENSORS, 2011 IEEE*, 2011, pp. 516-519.

[5] S. Zhang et al., "Surface Acoustic Wave Devices Using Lithium Niobate on Silicon Carbide," *IEEE Transactions on Microwave Theory and Techniques*, vol. 68, no. 9, pp. 3653-3666, 2020.

[6] Y. Yang, L. Gao, and S. Gong, "X-Band Miniature Filters Using Lithium Niobate Acoustic Resonators and Bandwidth Widening Technique," *IEEE Transactions on Microwave Theory and Techniques*, vol. 69, no. 3, pp. 1602-1610, 2021.

[7] S. Ballandras et al., "High overtone Bulk Acoustic Resonators: application to resonators, filters and sensors," 2012.

[8] R. Vetury et al., "High Rejection, 160MHz Bandwidth, High Q-factor 6 GHz RF Filters for Wi-Fi 6E manufactured in a Novel BAW Process," presented at the 2021 IEEE International Ultrasonics Symposium (IUS), 2021.

[9] S. Ballandras et al., "Development of Temperature-Stable RF Filters on Composite Substrates Based on a Single Crystal LiTaO3 Layer on Silicon," in *2019 Joint Conference of the IEEE International Frequency Control Symposium and European Frequency and Time Forum (EFTF/IFC)*, 2019, pp. 1-4.

[10] L. Zhang et al., "High Frequency, Low Loss and Low TCF Acoustic Devices on LiTaO3-on-SiC Substrate," presented at the 2021 IEEE International Ultrasonics Symposium (IUS), 2021.

[11] T. Takai et al., "I.H.P. SAW technology and its application to microacoustic components (Invited)," in *2017 IEEE International Ultrasonics Symposium (IUS)*, 2017, pp. 1-8.

[12] S. Inoue and M. Solal, "Layered SAW Resonators with Near-Zero TCF at Both Resonance and Anti-resonance," in *2019 IEEE International Ultrasonics Symposium (IUS)*, 2019, pp. 2079-2082.

[13] R. Su et al., "Over GHz bandwidth SAW filter based on 32° Y-X LN/SiO2/poly-Si/Si heterostructure with multilayer electrode modulation," *Applied Physics Letters*, vol. 120, no. 25, 2022.

[14] H. Zhou et al., "A 6.1 GHz Wideband Solidly-Mounted Acoustic Filter on Heterogeneous Substrate for 5G Front-Ends," in *2022 IEEE 35th International Conference on Micro Electro Mechanical Systems Conference (MEMS)*, 2022, pp. 1006-1009.

[15] S. S. Kongbrailatpam, J. P. Goud, and K. C. J. Raju, "The Effects of a Coated Material Layer on High-Overtone Bulk Acoustic Resonator and its Possible Applications," *IEEE Transactions on Ultrasonics, Ferroelectrics, and Frequency Control*, vol. 68, no. 4, pp. 1253-1260, 2021.

[16] T. Daugey, J. M. Friedt, G. Martin, and R. Boudot, "A high-overtone bulk acoustic wave resonator-oscillator-based 4.596 GHz frequency source: Application to a coherent population trapping Cs vapor cell atomic clock," *Rev Sci Instrum*, vol. 86, no. 11, p. 114703, Nov 2015.

[17] H. Tian et al., "Hybrid integrated photonics using bulk acoustic resonators," *Nat Commun*, vol. 11, no. 1, p. 3073, Jun 17 2020.

[18] M. Pijolat et al., "Mode conversion in high overtone bulk acoustic wave resonators," in *2009 IEEE International Frequency Control Symposium Joint with the 22nd European Frequency and Time forum*, 2009, pp. 290-294.

[19] K. M. Lakin, G. R. Kline, and K. T. McCarron, "High-Q microwave acoustic resonators and filters," *IEEE Transactions on Microwave Theory and Techniques*, vol. 41, no. 12, pp. 2139-2146, 1993.

[20] D. Gachon et al., "Fabrication of High Frequency Bulk Acoustic Wave Resonator Using Thinned Single-Crystal Lithium Niobate Layers," *Ferroelectrics*, vol. 362, no. 1, pp. 30-40, 2008/05/19 2008.

[21] J. Wu et al., "A New Class of High-Overtone Bulk Acoustic Resonators Using Lithium Niobate on Conductive Silicon Carbide," *IEEE Electron Device Letters*, vol. 42, no. 7, pp. 1061-1064, 2021.

[22] J. Yang, B. Hamelin, and F. Ayazi, "Investigating Elastic Anisotropy of 4H-SiC Using Ultra-High Q Bulk Acoustic Wave Resonators," *Journal of Microelectromechanical Systems*, vol. 29, no. 6, pp. 1473-1482, 2020.

[23] J. D. Larson, P. D. Bradley, S. Wartenberg, and R. C. Ruby, "Modified Butterworth-Van Dyke circuit for FBAR resonators and automated measurement system," in *2000 IEEE Ultrasonics Symposium. Proceedings. An International Symposium (Cat. No.00CH37121)*, 2000, vol. 1, pp. 863-868 vol.1.

A C/K_U DUAL-BAND RECONFIGURABLE BAW FILTER USING POLARIZATION TUNING IN LAYERED SCALN

Dicheng Mo, Shaurya Dabas, Sushant Rassay, and Roozbeh Tabrizian

Department of Electrical and Computer Engineering, University of Florida, Gainesville, USA

ABSTRACT

This paper reports, for the first time, on the scandium-aluminum nitride ($Sc_{0.28}Al_{0.72}N$) bulk acoustic wave (BAW) bandpass C/K_u dual-band filter with intrinsically reconfigurable center frequencies at 6.81 GHz and 13.29 GHz. The filter is created from electrical coupling of dual-mode BAW resonators with complementary-switchable operation in the first and second thickness-extensional modes ($TE_{1,2}$), defined by spatial poling in a layered $Sc_{0.28}Al_{0.72}N$ transducer. The transducer is created from alternate stacking of two ~150nm $Sc_{0.28}Al_{0.72}N$ films with three ~50nm molybdenum electrodes to enable independent control over polarization direction across the thickness, using ferroelectric behavior. Poling $Sc_{0.28}Al_{0.72}N$ layers in the same or opposite directions, through application of low-frequency pulses, enables complementary excitation of TE_1 and TE_2 BAW modes with similar electromechanical coupling but a frequency ratio nearing two. A filter prototype is presented with operating bands at 6.81 GHz and 13.29 GHz, bandwidths of 382MHz ($\approx$ 5.6%) and 809MHz ($\approx$ 6.1%), and insertion losses of 2.59dB and 2.62dB, respectively. The filter operation band is reconfigured through application of 79V triangular pulses at 20kHz to reverse polarization direction in one of the $Sc_{0.28}Al_{0.72}N$ layers. On-off isolation of ~7dB and 5dB are measured for C- and K_u-band filters, when switching between TE_1 and TE_2 operation modes of BAW resonators. The intrinsic reconfiguration of center frequency while sustaining similar fractional bandwidth, enabled by complementary ferroelectric poling in layered transducer, highlights the potential of C/K_u dual-band scandium-aluminum nitride BAW filters for creation of compact multi-frequency spectral processors in C/K_u bands.

KEYWORDS

Scandium Aluminum Nitride, ferroelectricity, intrinsic switchability, complementary switchable, periodically polled, BAW filter, super high frequency.

INTRODUCTION

Aluminum-nitride (AlN) BAW filters and duplexers have been pivotal constituents in radio-frequency front-end (RFFE) of current wireless systems, to enable integrated spectral processing within sub-6 GHz spectrum [1]. However, with the ever-increasing demand for wireless capacity and exponential increase in user number, there is an urgent need for RFFE hardware transformation, to accommodate communication in cm- and mm-wave regimes. Further, novel RFFE hardware should provide wideband configurability to enable efficient spectrum sharing and spread spectrum communication protocols in modern wireless generations.

Over the past decade, extensive research and development have been in progress for extreme frequency

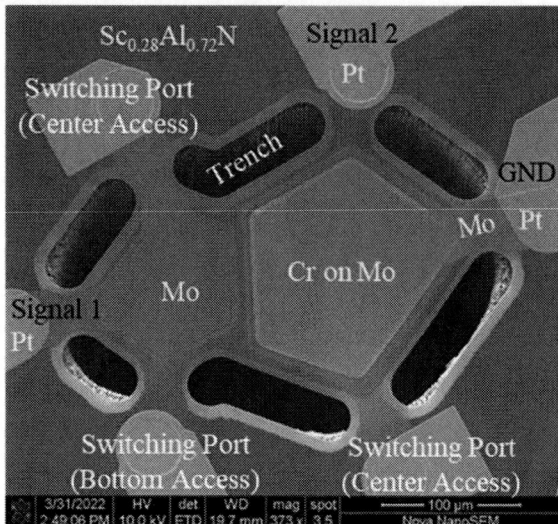

Figure 1: The scanning electron microscope (SEM) image of the intrinsically dual-band reconfigurable BAW single-stage L-ladder filter.

scaling of BAW filters beyond 6 GHz [2][3], enhancement of filter bandwidths beyond the limits set by the electromechanical coupling factor (k_t^2) in AlN [4], and development of multi-frequency technologies that enable single-chip arraying of reconfigurable filters over a wide spectrum [5].

These efforts include the use of scandium-aluminum nitride films ($Sc_xAl_{1-x}N$) that provide substantial enhancement of transducer k_t^2 and enable realization of filters with significantly larger bandwidths compared to their AlN counterpart. Further, the discovery of ferroelectricity in $Sc_xAl_{1-x}N$ has opened new opportunities through the use of polarization tailoring for intrinsic tuning or switching of resonators and filters. Specifically, the intrinsic switchability may enable multiband integration of filter arrays with a relieved need for external switches that impose excessive loss and footprint on RFFE. Furthermore, it is recently demonstrated that spatial polarization tailoring in Lamb-wave and BAW resonators enables the configuration of resonator operation in different acoustic modes, which may further reduce the number of individual filters in an array to cover a wide spectrum [5].

In this work, we present a novel intrinsically reconfigurable BAW filter technology with dual-band operation in C/K_u band. This filter in Figure 1 is created in a laminated transducer formed from alternate stacking of two $Sc_xAl_{1-x}N$ films with three electrodes to enable independent control over polarization direction across the thickness, using ferroelectric behavior. Poling $Sc_xAl_{1-x}N$ layers in the same or opposite directions through application of low-frequency pulses to an RF-floating middle electrode, enables complementary excitation of

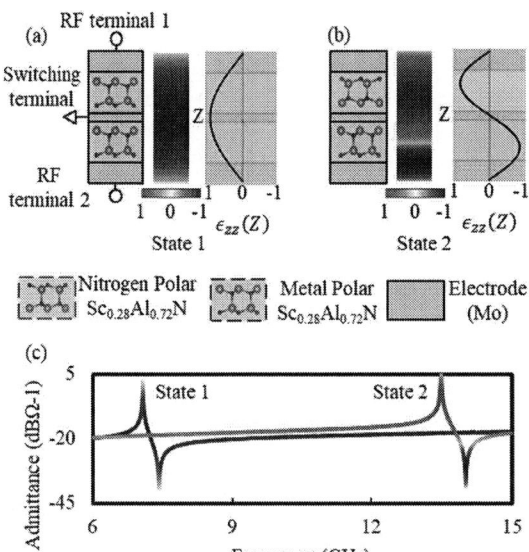

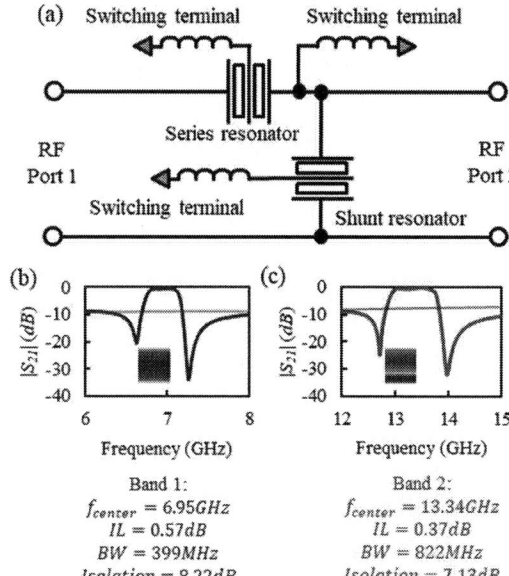

Figure 2: Transducer polarization states to excite (a) TE_1 mode for C-band, and (b) TE_2 mode for Ku-band operation. In each state, the mode-shape deformation and cross-sectional strain $\epsilon_{zz}(Z)$ are shown. The switching between two operation states is achieved by tuning polarization of constituent ScAlN layers in the same or opposite directions. (c) is the simulated admittance of the laminated resonator. In State 1 the $Sc_{0.28}Al_{0.72}N$ layers are uniformly poled, resulting in transduction of TE_1 and suppression of TE_2. In State 2 the $Sc_{0.28}Al_{0.72}N$ layers are poled in opposite direction, resulting in transduction of TE_2 and suppression of TE_1.

different thickness-extensional harmonics (*i.e.*, TE_1 and TE_2) with similar k_t^2 but a frequency-ratio nearing two. This enables realization of the first intrinsically reconfigurable C/K$_u$ dual-band BAW filter.

DUAL-BAND RECONFIGURABILITY IN SCALN FILTERS

To explore the intrinsic complementary switchability in laminated BAW resonator, one-port resonator models are constructed by stacking two $Sc_{0.28}Al_{0.72}N$ layers with top, middle, and bottom molybdenum (Mo) electrodes, as shown in Fig. 2(a). This architecture provides polarization direction in each $Sc_{0.28}Al_{0.72}N$ layer as a degree of freedom to define operating BAW harmonic. With RF electric field applied between the top and bottom electrode, and assuming a floating middle-electrode, the uniform polarization of $Sc_{0.28}Al_{0.72}N$ layers results in excitation and suppression of TE_1 and TE_2 modes, respectively. On the other hand, poling $Sc_{0.28}Al_{0.72}N$ layers in opposite directions, by applying switching voltage to the middle electrode, results in a suppression of TE_1 mode while enabling excitation of TE_2 mode in Figure 2(b). Figure 2(c) shows the COMSOL-simulated resonator admittance using the material property from Caro et al. [6], validating the resonator complementary switchability between the TE_1 and TE_2 modes through reconfiguration of polarization direction in $Sc_{0.28}Al_{0.72}N$ layers of the laminate transducer.

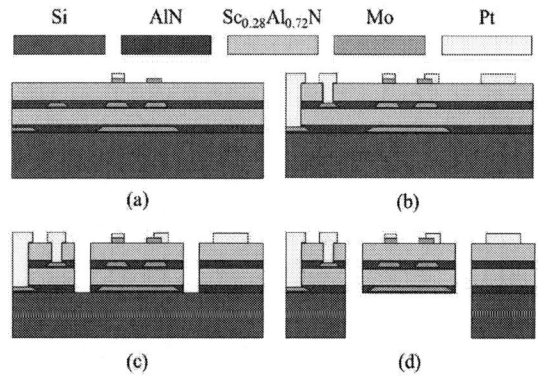

Figure 3: (a) The architecture of dual-band $Sc_{0.28}Al_{0.72}N$ TE BAW single-stage filters. The switching ports are also shown. The simulated $|S_{21}|$ of the dual-band filter when operating in (b) State 1 with uniform polarization in all $Sc_{0.28}Al_{0.72}N$ layers, and (c) State 2 corresponding to $Sc_{0.28}Al_{0.72}N$ layers polarized in opposite directions.

The simulated resonators are incorporated in a single-stage L-ladder filter model to enable the design of dual-band reconfigurable filter. Figure 3 (a) shows the filter architecture and the switching ports, connected through Bias-Tees. In the designed filter, the frequency offset between series and shunt resonators is achieved by adding a 19nm-thick chromium layer on the shunt resonator to load its frequency. Additionally, the transduction area of the shunt resonator (A_{shunt}) is chosen to be 2.5-times series resonators (*i.e.*, $A_{shunt} \approx 2.5A_{series}$) in the ladder filter. Further, the absolute value of transduction areas is chosen to hold the criteria for impedance matching given by:

$$A_{series}A_{shunt} = \frac{4H_{ScAlN}^2}{(2\pi\varepsilon_0\varepsilon_{ScAlN}R_{term})^2 f_{series,s}f_{shunt,p}} \quad (1).$$

Here, ε_{ScAlN} is the relative dielectric constant of $Sc_{0.28}Al_{0.72}N$. $f_{series,s}$ and $f_{shunt,p}$ are the series and parallel

Si | AlN | $Sc_{0.28}Al_{0.72}N$ | Mo | Pt

Figure 4: Fabrication process flow for the C/K$_u$ band reconfigurable $Sc_{0.28}Al_{0.72}N$ BAW L-ladder filter.

978-1-6654-9309-3/23 $31.00 © 2023 IEEE

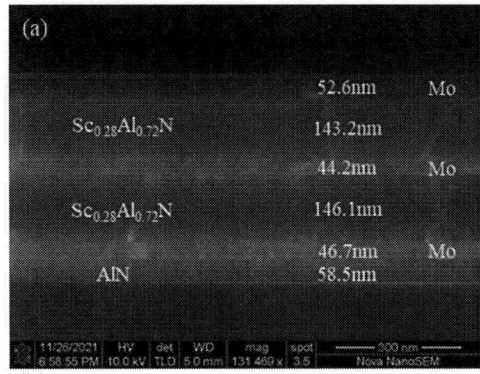

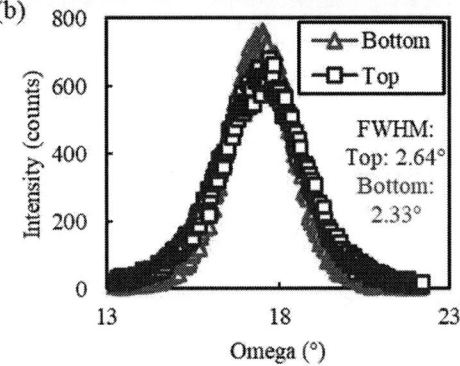

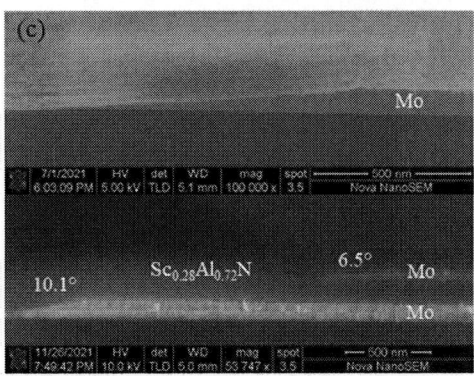

Figure 5: (a) Cross-sectional SEM of the $Sc_{0.28}Al_{0.72}N$ lamination with top and bottom FWHM of 2.64° and 2.33° shown in (b). The slanted bottom and middle Mo sidewalls in (c) contribute to the crack-less growth of $Sc0.28Al0.72N$.

resonance frequencies of series and shunt resonators in ladder filter, respectively. R_{term} is the termination impedance, which is taken as 50Ω. Fig. 3(b-c) shows simulated transmission using the resonator admittance extracted from COMSOL simulations.

FABRICATION PROCESS

Figure 4 illustrates the fabrication process flow for the complementary-switchable dual-mode BAW resonators and filters. Figure 1 shows the SEM image of the filter. DC and reactive magnetron sputtering are applied to deposit Mo and $Sc_{0.28}Al_{0.72}N$ layers [7]. A 58.5nm AlN seed layer is firstly deposited on the Si substrate to facilitate the (110)-texture growth of Mo film [8] and c-axis texture growth of $Sc_{0.28}Al_{0.72}N$ films. The device cross-sectional SEM in Fig. 5(a) highlights the crack-less growth of the $Sc_{0.28}Al_{0.72}N$

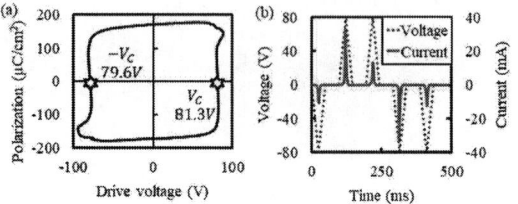

Figure 6: (a) Measured polarization hysteresis loop for the unreleased $Sc_{0.28}Al_{0.72}N$ layer, measured between the top and center electrode. (b) The polarization switching procedure for released resonators using triangular voltage pulses, and the corresponding polarization switching current, peaking at the transition of pulse sign.

lamination with FWHM of 2.64° and 2.33° for the top and bottom $Sc_{0.28}Al_{0.72}N$ layer shown in Fig. 5(b), respectively. This is contributed by the slanted bottom and middle Mo sidewalls in Fig. 5(c) from boron trichloride inductively coupled plasma reactive-ion-etching, as shown in [9].

DEVICE CHARACTERIZATION

The resonator's ferroelectric property is measured using a Radiant PiezoMEMS ferroelectric tester, to identify the switching/band reconfiguration voltage of the filter. Fig. 6(a) presents the polarization-voltage hysteresis loop of the top $Sc_{0.28}Al_{0.72}N$ layer, highlighting a coercive voltage of 80V. Fig. 6(b) shows the instantaneous current for the triangular pulse-train for polarization switching. The significant increase in instantaneous current when the pulse polarity flip indicates a large switching current, validating the desired ferroelectricity in $Sc_{0.28}Al_{0.72}N$ film that enables reversing polarization direction.

Fig. 7 here shows the wide-range measured admittance covering both the TE$_1$ and TE$_2$ modes of the laminated $Sc_{0.28}Al_{0.72}N$ BAW resonator, with the inter-state transition from State 1 to 2 detailed using a 20kHz 79V switching pulse. The admittance is extracted from the one-port resonator's reflection coefficient. In State 1, both $Sc_{0.28}Al_{0.72}N$ films are Metal-poled and TE$_1$ is excited at 6.96GHz with k_t^2 of 10.6% and quality factor (Q) of 109. With proper switching pulses, the top $Sc_{0.28}Al_{0.72}N$ film gradually poled to State 2, where the top $Sc_{0.28}Al_{0.72}N$ layer is Nitrogen-poled, resulting in the TE$_2$ at 13.4GHz with k_t^2 of 10.8% and Q of 130.

Fig. 8 (a) shows the measured transmission response ($|S_{21}|$) of the single-stage TE BAW ladder filter in two

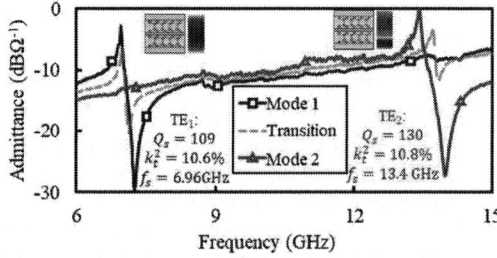

Figure 7: Measured wide-span admittance of the intrinsically switchable dual-mode $Sc_{0.28}Al_{0.72}N$ BAW resonator in the two complementary operation modes and the transition state.

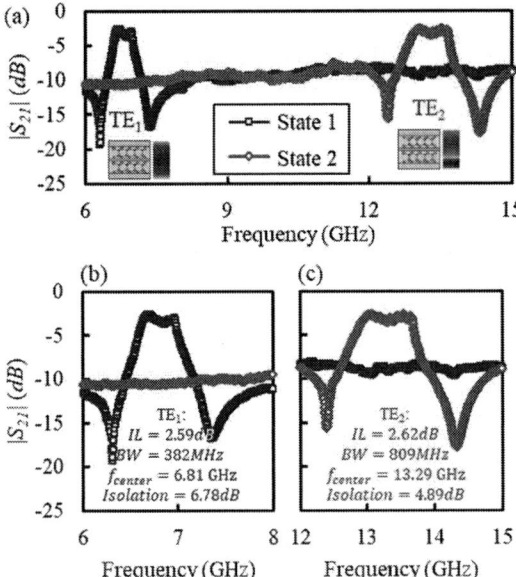

Figure 8: Measured (a) wide-range and (b, c) short-span $|S_{21}|$ of the single-stage intrinsically switchable dual-band $Sc_{0.28}Al_{0.72}N$ BAW filter in C- and K_u-band operation states.

different operation states. The shunt resonator frequency is shifted using a thin Chromium mass loading layer, and the transduction area ratio is 1:2.5. The interstate transition is achieved by applying triangular pulses between the top and center Mo electrodes of the series and shunt resonators in the filter. Fig. 8(b) details the filter $|S_{21}|$ around 6.81 GHz in two states. In State 1, a bandwidth of 382 MHz (fractional bandwidth of 5.61%) and an insertion loss of 2.59dB are measured. In State 2, this pass band is switched off with isolation of 6.78dB, set by the out-of-band rejection of the filter as defined by the relative area of series and shunt elements. Fig. 8(c) presents the focused filter $|S_{21}|$ around 13.29 GHz in two states, with a bandwidth of 809 MHz (fractional bandwidth of 6.09%) and an insertion loss of 2.62dB in State 2 and isolation of 4.89dB in State 1.

CONCLUSION

In conclusion, an intrinsically reconfigurable C/K_u dual-band TE BAW filter is demonstrated based on the use of ferroelectric polarization tuning in layered ferroelectric $Sc_{0.28}Al_{0.72}N$. The dual-band reconfiguration principle is formulated using qualitative discussion and finite-element modeling for different spatial polarization states. Resonator and filter prototypes with dual-frequency operation at 6.81 GHz and 13.29 GHz are implemented, and their intrinsic and complementary C/K_u dual-band reconfigurability is experimentally verified. The presented filter technology enables the creation of multiband spectral processors on a single chip and obviates the need for external switches in wireless systems, and avoids their excessive loss, latency, and footprint overhead.

ACKNOWLEDGEMENTS

This work was supported by the Defense Advanced Research Projects Agency (DARPA), Tunable Ferroelectric Nitrides (TUFEN) Program under Grant HR00112090049.

REFERENCES

[1] D. Kim et al., "Wideband 6 GHz RF Filters for Wi-Fi 6E Using a Unique BAW Process and Highly Sc-doped AlN Thin Film," *2021 IEEE MTT-S International Microwave Symposium (IMS)*, 2021, pp. 207-209.

[2] K. Karnati, M. Schaefer, W. Yusuf, R. Rothemund, M. Al-Joumayly and G. Fattinger, "5G C-V2X Filter Using BAW Technology," 2021 IEEE MTT-S International Microwave Filter Workshop (IMFW), 2021, pp. 109-111.

[3] M. Schaefer et al., "Process and Design Challenge for SMR-type Bulk Acoustic Wave (BAW) Filters at Frequencies Above 5 GHz," *2019 IEEE International Ultrasonics Symposium (IUS)*, 2019, pp. 1696-1699.

[4] Liu, Yan, et al. "Materials, design, and characteristics of bulk acoustic wave resonator: A review." *Micromachines* vol. 11, no. 7, pp. 630.

[5] S. Rassay et al., "Dual-Mode Scandium-Aluminum Nitride Lamb-Wave Resonators Using Reconfigurable Periodic Poling." *Micromachines,* vol 13, no.7, pp. 1003.

[6] S. Fichtner et al., "AlScN: A III-V semiconductor based ferroelectric," *J. Appl. Phys.*, vol. 125, no. 11, Mar. 2019, Art. no. 114103.

[6] M.A. Caro et al. "Piezoelectric coefficients and spontaneous polarization of ScAlN." *Journal of Physics: Condensed Matter.*, vol. 27, no. 24 May. 2015, pp. 245901.

[7] S. Rassay et al, "A Segmented-Target Sputtering Process for Growth of Sub-50 nm Ferroelectric Scandium– Aluminum–Nitride Films with Composition and Stress Tuning," *Physica Status Solidi (RRL)-Rapid Research Letters* vol. 15, no.5, pp. 2100087, May. 2021.

[8] V. Felmetsger et al, "Sputter Process Optimization for Al0.7Sc0.3N Piezoelectric Films," *2019 IEEE International Ultrasonics Symposium (IUS)*, 2019, pp. 2600-2603.

[9] S. Dabas et al, "Intrinsically Tunable Laminated Ferroelectric Scandium Aluminum Nitride Extensional Resonator Based on Local Polarization Switching," 2022 IEEE 35th International Conference on Micro Electro Mechanical Systems Conference (MEMS), 2022, pp. 1050-1053.

CONTACT

*Dicheng Mo, tel: +1-352-3288369; dicheng.mo@ufl.edu

ACOUSTOELECTRIC-DRIVEN FREQUENCY MIXING IN MICROMACHINED LITHIUM NIOBATE ON SILICON WAVEGUIDES

Hakhamanesh Mansoorzare and Reza Abdolvand
University of Central Florida, USA

ABSTRACT

This work reports on frequency mixing due to phonon-electron coupling (acoustoelectric effect) in lithium niobate (LN) on silicon (Si) micromachined Lamb mode waveguides. To study passive RF signal mixing in micro-acoustic domain, two-tone measurements are performed on said heterostructured waveguides. The results confirm the mixing is mainly due to the charge carriers in Si and is intensified as the input power level is increased. Furthermore, the mixed frequency components that lie outside of the design passband of the waveguide are also detected which is an indication of strong acoustic mixing. The preliminary results confirm the possibility of using the phonon-electron coupling in properly designed LN-on-Si waveguides for implementation of micro-acoustic mixers.

KEYWORDS

Acoustoelectric, frequency mixing, Lamb waves, lithium niobate, piezoelectric.

INTRODUCTION

The quest for realizing radio frequency (RF) frontend modules entirely in the micro-acoustic platform has intensified owing to the unprecedented efficiency and miniaturization that is inherent to this platform. Today, micro-acoustic filters have become a ubiquitous component in RF frontend modules. Said filters, which are passive devices, are currently integrated with active components to form the RF frontend module which comes at the cost of size, complexity, and performance, especially due to the multi-chip nature of these modules and the associated interconnects.

It has been shown that electrical or mechanical modulation of micro-acoustic devices could enable new functionalities that can be used for implementation of frequency tunable filters [1], non-reciprocal devices [2], and even amplifiers [3]. The amplification in the acoustic domain has been demonstrated using phonon-electron interactions known as the acoustoelectric effect (AE) [4]. This was discovered after initial observation of acoustic loss due to energy transfer to nearby charge carriers which later leveraged into acoustic wave amplification by using charge carriers to give up energy to the acoustic waves. First bulk waves in piezoelectric semiconductors and later surface acoustic waves (SAW) in layered piezoelectric-semiconductor structures were employed for said purpose. The latter allowed for tailoring the piezoelectric and semiconductor properties independently, thus improving the overall device performance. As a result of the promising results attained by layered SAW platform in conjunction with high performance piezoelectric materials such as lithium niobate (LN), AE SAW devices are even presently investigated by researchers [5, 6]. However,

SAW platform faces limitation in terms of efficiency, loss, bandwidth, and transition into higher frequencies. To overcome such problems, Lamb waves in micromachined waveguides made of piezoelectric on semiconductor have been explored as a potential solution [7]. In this platform, Lamb waves generated due to the piezoelectricity are coupled to charge carriers in the semiconductor, such as silicon (Si), leading to the generation of space-charge wave. Through application of a drift field to the space-charge wave so that their drift velocity exceeds the Lamb wave phase velocity, Lamb wave amplification could occur.

Another interesting aspect of this phenomenon is that the generated space-charge can interact with acoustic waves at other frequencies and produce strains at the sum or difference frequencies which could be significant once the nonlinear phonon-electron coupling is strong. This could enable implementation of frequency mixers within the micro-acoustic domain. Since LN-on-Si platform has shown promising results in terms of phonon-electron coupling for Lamb waves, it is expected to provide strong mixing of acoustic waves, therefore, it is investigated in this work. Previously frequency mixing in the micro-acoustic domain has been demonstrated using coupled mechanical resonators [8].

To investigate said electron-mediated mixing of acoustic phonons, ultra-high frequency LN-on-Si Lamb mode waveguides are fabricated and tested using a two-tone setup as described in following sections and the results are compared with that of identical LN-only waveguides to isolate the effect of phonon-electron coupling on mixing.

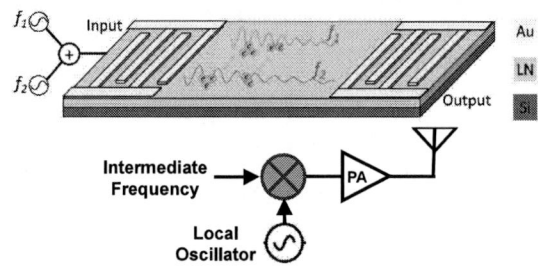

Figure 1: Conceptual schematic of wave mixing in phonon-electron coupled waveguide made of LN-on-Si and example application of a mixer in an RF transmitter.

DEVICE DESIGN & FABRICATION

The Lamb mode waveguide is fabricated on bonded LN on Si substrates having 1 μm LN on 1 μm Si with a resistivity of 5-10 Ω.cm which is chosen to enable strong phonon-electron interactions. X-cut LN is selected, and the waveguides are oriented along the ~30° off Y-axis of LN to harness the highest electromechanical coupling for fundamental symmetric (S0) mode [9]. This is crucial in maximizing the phonon-electron interactions and subsequently the frequency mixing. Interdigital transducers (IDT) with chirped design [10] are placed at the two ends of the waveguide to convert the RF signal from electromagnetic domain to acoustic domain at the input port and from acoustic domain back to the electromagnetic domain at the output port. Since the waveguides of this work are for proof-of-concept, they lack two separate ports for the input signal, instead the two tones are externally combined and fed into the input port. The waveguides have a footprint of ~300×800 μm^2 and a passband frequency of ~600 MHz as defined by the periodicity of the IDTs.

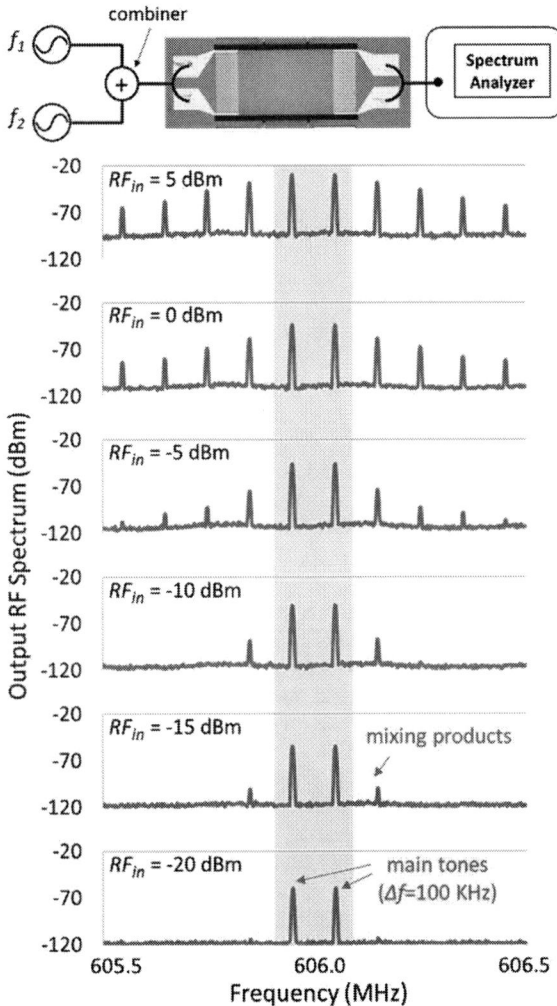

Figure 2: Micrograph of the waveguide under test and the test setup (top). Output frequency spectrum of the LN-on-Si waveguide under two-tone test as the input power is increased, resulting in stronger mixing (bottom).

EXPERIMENTAL RESULTS

The frequency mixing is demonstrated by performing two-tone tests on fabricated waveguides in room temperature and air. The test setup and the micrograph of the waveguide under test is shown in Figure 2. For a waveguide having a passband at ~600 MHz, the output of

the two RF generators (Rohde & Schwarz SMC100A) are set at 605.95 MHz and 606.05 MHz and the combined signal is fed to input port IDT. The amplitude of the input RF power is swept, and the output spectrum is measured by a spectrum analyzer (Rohde & Schwarz FSUP). The output spectrum for an input RF power sweep from -20 dBm to 5 dBm is shown in Figure 2. While at low levels of input RF power, the mixing products of the two tones are insignificant, as the input RF power is increased the mixing products start to intensify which can be attributed to increased electron bunching since increasing the power, up to a certain point, results in a stronger space-charge wave, therefore, intensified acoustic wave mixing. At this point the third order product can reach as high as -8.6 dBc.

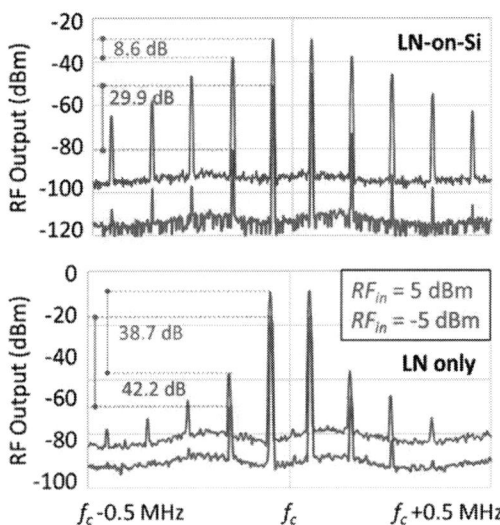

Figure 3: Output frequency spectrum of waveguides made of LN only and LN-on-Si at two input power levels (-5 and 5 dBm) showing much stronger mixing due to presence of Si.

The two-tone test is repeated after a backside etch step to completely remove the Si layer from underneath LN so that the phonon-electron mixing is isolated from other nonlinear processes which could cause frequency mixing. The output RF spectrum of the waveguide before and after the Si etching step is shown in Figure 3 top and bottom, respectively. The presence of Si results in -8.6 dBc and -29.9 dBc third order products for 5 dBm and -5 dBm input tones, respectively. On the other hand, the same for LN only waveguide is at -38.7 dBc and -42.2 dBc. This confirms that the origin of mixing is mainly due to the phonon-electron interactions within the heterostructure. The mixing products are not limited to only the design passband of the waveguide and can be observed at sum and difference frequencies of the two tones which are far from the passband of the IDT. This is shown in Figure 4 for two tones at 605 and 607MHz where the frequency mixing products at 2 MHz and 1212 MHz are present roughly at levels of -40 dBc and -50 dBc, respectively despite the large transducer loss at those frequencies.

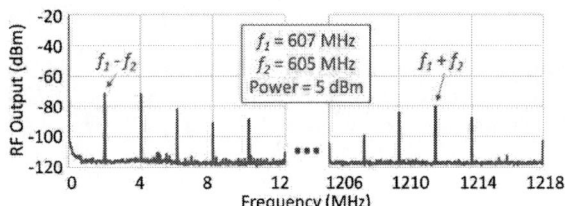

Figure 4: Output frequency spectrum of the LN-on-Si waveguide for two-tone test at 5 dBm input showing mixing products at the sum and difference frequencies.

The frequency mixing can be further enhanced by applying a bias voltage to Si layer for electron drift. This is shown in Figure 5 for a bias of 28 V (bias points marked); the transmission loss is improved through AE amplification and the mixing products are intensified, especially those that are further from the carrier. The reason for this is believed to be the saturation of AE amplification at higher acoustic powers.

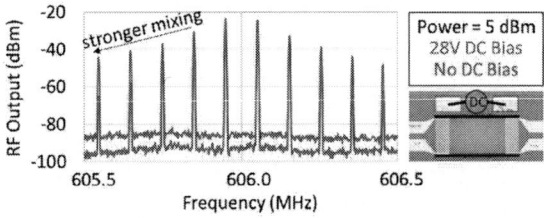

Figure 5: Output frequency spectrum of the LN-on-Si waveguide for two-tone test at 5 dBm without/with application of DC bias for electron drift; mixing is intensified at 28 V especially for the products that are further away from the carrier.

CONCLUSION

It was experimentally demonstrated that the phonon-electron coupling known as the acoustoelectric effect can lead to passive frequency mixing in lithium niobate on silicon Lamb mode waveguides. The frequency mixing is mediated by space-charge waves in the silicon layer and becomes stronger by increasing Lamb wave amplitude up to a certain point. The results indicate possibility of leveraging the phonon-electron coupling in such platform to realize RF mixers. By optimizing the transducers and propagation directions of waves, the performance of the mixer could be enhanced, thus enabling a micro-acoustic solution for this important RF communication component.

ACKNOWLEDGEMENTS

The authors would like to thank NGK Insulators for providing bonded lithium niobate on silicon substrates. This work was supported by the National Science Foundation under Grant 2122670.

REFERENCES

[1] A. Reinhardt, E. Defaÿ, F. Perruchot and C. Billard, "Tunable composite piezoelectric resonators: A possible "Holy Grail" of RF filters?," in *2012 IEEE/MTT-S IMS Digest*, 2012, pp. 1-3.

[2] Y. Yu et al., "Magnetic-free radio frequency circulator based on spatiotemporal commutation of MEMS resonators," in *2018 IEEE MEMS*, 2018, pp. 154-157.

[3] H. Mansoorzare and R. Abdolvand, "Micromachined Heterostructured Lamb Mode Waveguides for Acoustoelectric Signal Processing," in *IEEE Transactions on Microwave Theory and Techniques*, vol. 70, no. 11, pp. 5195-5204, Nov. 2022.

[4] R. H. Parmenter, "The acousto-electric effect." *Physical Review* 89.5, 1953.

[5] L. Hackett, et al. "Towards single-chip radiofrequency signal processing via acoustoelectric electron–phonon interactions." *Nature communications* 12.1, 2021.

[6] S. Ghosh, S. Cho and R. Lu, "Experimental Observation of Electron-Phonon Interaction in Semiconductor on Solidly Mounted Thin-Film Lithium Niobate," in *2022 IEEE MTT-S IC-MAM*, 2022, pp. 90-93.

[7] H. Mansoorzare and R. Abdolvand, "A Thin-Film Piezo-Silicon Acoustoelectric Isolator with More than 30 dB Non-Reciprocal Transmission," *2021 IEEE MEMS*, 2021, pp. 470-473.

[8] A. C. Wong and C. T. -C. Nguyen, "Micromechanical mixer-filters ("mixlers")," in *Journal of Microelectromechanical Systems*, vol. 13, no. 1, pp. 100-112, Feb. 2004.

[9] I. E. Kuznetsova et al., "Investigation of acoustic waves in thin plates of lithium niobate and lithium tantalate," in *IEEE Transactions on Ultrasonics, Ferroelectrics, and Frequency Control*, vol. 48, no. 1, pp. 322-328, Jan. 2001.

[10] H. Mansoorzare and R. Abdolvand, "Acoustoelectric Non-Reciprocity in Lithium Niobate-on-Silicon Delay Lines," in *IEEE Electron Device Letters*, vol. 41, no. 9, pp. 1444-1447, Sept. 2020.

CONTACT

*H. Mansoorzare, tel: +1-352-3464400; hakha@ucf.edu

EFFECT OF SCANDIUM COMPOSITION ON THE PHONON SCATTERING LIFETIME OF ALUMINUM SCANDIUM NITRIDE ACOUSTIC WAVE RESONATORS

Yue Zheng[1], Mingyo Park[1], Chao Yuan[2] and Azadeh Ansari[1]
[1]Georgia Institute of Technology, Atlanta GA, USA
[2]Wuhan University, Wuhan, China

ABSTRACT

In this work, we analyze the anharmonic phonon scattering time to project the acoustic loss limits in thin-film aluminum scandium nitride ($Al_{1-x}Sc_xN$) micromechanical resonators. Although electromechanical coupling coefficients have been improved significantly with Sc alloying, making $Al_{1-x}Sc_xN$ a promising material for wideband acoustic filters, lower quality factor (Q) have been observed when increasing Sc content in $Al_{1-x}Sc_xN$. This has often been attributed to poor crystallinity and surface roughness of the film. To investigate the intrinsic limiting factors of Q in $Al_{1-x}Sc_xN$ resonators, we summarize the figure of merits from the experimental works of $Al_{1-x}Sc_xN$ resonators from multiple research and industry groups in the past decades. We classify the major phonon scattering mechanisms present in $Al_{1-x}Sc_xN$ lattice and their theoretical expressions. The results suggest that while both AlN and $Al_{1-x}Sc_xN$ phonon lifetimes can be limited by grain boundary scattering, which depends on the crystal microstructure, as well as the film thickness, $Al_{0.7}Sc_{0.3}N$ suffers from a decrease of phonon lifetime by up to four orders of magnitude due to alloy scattering, conceivably responsible for the lower Q compared to AlN acoustic resonators.

KEYWORDS

Aluminum Scandium Nitride; Quality Factor; Phonon Scattering; Lattice Defect; Anelastic Damping; Energy Loss Mechanism; Phonon Relaxation Time.

INTRODUCTION

Aluminum scandium nitride ($Al_{1-x}Sc_xN$) has been a promising piezoelectric material for modern wide-band high-frequency acoustic filter applications [1]–[3]. Introducing Sc content to aluminum nitride (AlN) can boost the piezoelectric coefficient d_{33} up to five times than that of pure AlN. Enhanced effective electromechanical coupling coefficients (k_{eff}^2) are indeed reported for various $Al_{1-x}Sc_xN$-based acoustic resonators [4], [5]. However, there is a growing number of reports on quality factor (Q) degradation with increasing Sc concentration [6], [7]. This has often been attributed to degraded crystallinity and increased surface roughness of the film [8], while softening of the elastic constant c_{33} is also a popular explanation [6], [9]. Nevertheless, the underlying physical mechanisms behind the decrease of Q have not been thoroughly investigated to date. Studies of damping mechanisms generally begin by identifying the damping processes and categorizing them as intrinsic or extrinsic processes. Conventional intrinsic phonon dissipation analysis of micromechanical resonators considers Akhiezer [10], [11], Landau-Rumer [12] and thermoelastic damping [13], [14] as the ultimately dominating loss mechanisms [15] but

assumes that the losses are anelastic and the crystal lattice are in a perfect single crystal condition. Extrinsic losses, including interface reflection [16], electric loss [17], air damping and anchor loss [18], attenuate the acoustic wave by energy relaxation and can be minimized with proper resonator design. However, current theories do not provide a clear-cut solution and not all the works quantify the effect of material defects on the relaxation process [19]. In short, there has been a lack of literature discussing motional losses due to crystal imperfections (e.g., point defects, grain boundary, alloy scattering).

Crystal defects not only affect the thermodynamic properties [20] but also impact the kinetic process. Klemens derived the analytical relaxation time of the long-wavelength phonons perturbed by lattice defects due to Rayleigh scattering [21]. The theory has been served as the basis for thermal conductivity modeling of defect-constricted solid materials. Recently, analyzing the phonon scattering rate of $Al_{1-x}Sc_xN$ with Debye-Callaway model has revealed the phonon scattering mechanisms that contributes to thermal conductivity degradation and provided insights on quantifying loss limits in acoustic resonators [22]. Among various spectroscopy instrumentations for lattice defects characterization, assessing crystallinities using full width at half maximum (FWHM) read out from rocking curve (RC) in X-ray diffraction (XRD) scan has gained popularities in recent years [8]. However, it remains to verify the correlations between FWHM and the resonator performance, Q and k_{eff}^2, as well as the theory's applicability on wafers with different growth methods.

In our previous work [23], we computed the intrinsic limit of the frequency (f), quality factor product $f \cdot Q$ under Akhiezer, Landau-Rumer and thermoelastic damping of $Al_{1-x}Sc_xN$ using measured thermal conductivity values. In this work, we discuss the role of anharmonic phonon scattering time in thin-film $Al_{1-x}Sc_xN$ micromechanical resonators to show the acoustic loss limit under the presence of crystal defects. To investigate the intrinsic limiting factors of Q in $Al_{1-x}Sc_xN$ resonators, we, for the first time, confirm this Q degradation trend observed in lab by summarizing the figure of merits (FoMs) from the experimental works on $Al_{1-x}Sc_xN$ resonators. Furthermore, we identify the major phonon scattering mechanisms in $Al_{1-x}Sc_xN$ lattice and their mathematical formulations. Ultimately, the resonator Q is directly affected by the phonon lifetime.

SUMMARY OF PAST WORKS AlScN RESONATORS

To understand the impact of scandium concentration and material quality on the performance of $Al_{1-x}Sc_xN$ acoustic resonators, we gather the experimental works

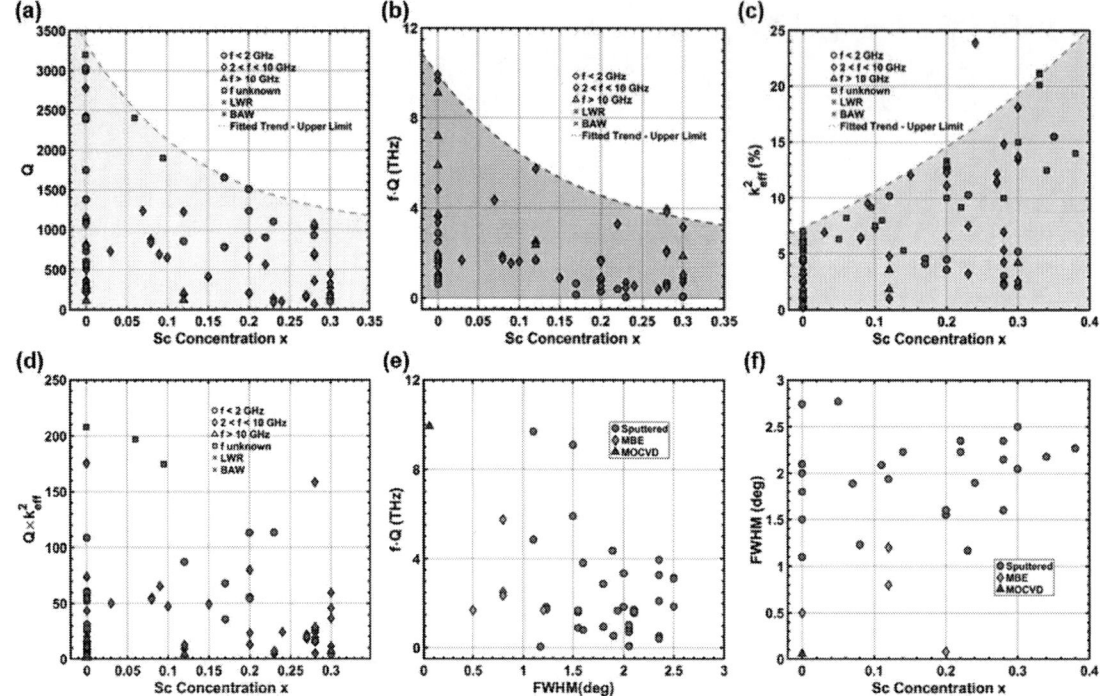

Figure 1: Comparisons of the measured device performance metrics and piezoelectric material quality of $Al_{1-x}Sc_xN$-based lamb-wave and bulk-acoustic-wave devices reported to date: (a) Q, (b) f·Q, (c) k_{eff}^2, and (d) $Q×k_{eff}^2$ vs Sc concentration x. (e) FWHM vs Sc concentration x and (f) f·Q vs FWHM of $Al_{1-x}Sc_xN$ film grown by Sputtering, MBE and MOCVD method.

from the past decade and record the reported Q, k_{eff}^2 and f, as well as the FWHM and growth method of the sample, and inspect their correlations. Fig. 1(a) and (b) summarizes the trend of Q and $f·Q$ of the previously reported $Al_{1-x}Sc_xN$-based Lamb-wave resonators (LWRs) and bulk acoustic wave (BAW) resonators with respect to the Sc content in the piezoelectric thin films. k_{eff}^2 and the FoM $Q×k_{eff}^2$ of these works are shown in Fig. 1(c) and (d). These trends suggest that there is a significant trade-off between Q and for AlScN acoustic resonators particularly when x is high. Considering the report of thermal conductivity κ degradation with increasing x of $Al_{1-x}Sc_xN$ [22], [24], the Q trend clearly indicates that the damping is not limited by the well-known thermally excited intrinsic damping mechanisms (Akhiezer damping and thermoelastic damping), under which Q is inversely proportional to κ of the material [11], [14], [15]. Fig. 1(e) shows the $f·Q$ correlations with FWHM from RC in XRD measurements. Fig. 1(f) presents the FWHM of the $Al_{1-x}Sc_xN$ thin films grown by sputtered, molecular beam epitaxy (MBE) and metalorganic chemical vapor deposition (MOCVD) methods reported to date. It should be noted that although lower FWHM is conducive to achieving better $f·Q$, AlN resonators and AlScN resonators made of thin films with similar FWHM could still yield drastically different $f·Q$, which requires further understanding from the crystal microstructure level.

DISSIPATIVE PHONON SCATTERING MECHANISMS IN ALSCN

We discuss the phonon dissipation mechanisms arising from the structure of the materials and present the anelastic phonon scattering relaxation time which is an indicator of loss in acoustic resonators. The fundamental thermal scattering relaxation time consists of the contributions from 3-phonon normal (τ_n) and umklapp (τ_u) scattering, phonon-point (τ_p) scattering and phonon-boundary scattering (τ_b). The expression for each scattering process derived in early theories [21], [22], [25], [26]. The theories are combined with virtual crystal (VC) assumption [27], wherein the material property a is treated as a weighted average by each alloy composition (Eq. 1), in order to be applied on a ternary alloy $A_{1-x}B_x$.

$$a_{A_{1-x}B_x} = (1-x)a_A + xa_B \tag{1}$$

Three-Phonon Normal and Umklapp Scattering

Vibrational energy can be transferred by phonon-phonon scattering. The scattering process can be elastic or anelastic: normal (N) scattering process preserve the phonon momentum while umklapp (U) process, which dominates in long-wavelength phonon scattering, destroys phonon momentum and restore equilibrium [28]. The corresponding longitudinal (^{L}) and transverse (^{T}) branches of relaxation time can be expressed as [25], [26]:

$$\tau_n^L = \frac{1}{B_N^L \cdot \omega^2 \cdot T^3}, B_N^L = \frac{k_B^3 \gamma_L^2 \delta^3}{M\hbar^2 v_L^5} \tag{2.1}$$

$$\tau_n^T = \frac{1}{B_N^T \cdot \omega \cdot T^4}, B_N^T = \frac{k_B^3 \gamma_T^2 \delta^3}{M\hbar^2 v_T^5} \tag{2.2}$$

$$\tau_u^L = \frac{1}{B_U^L \cdot \omega^2 \cdot T \cdot e^{-\theta_L/3T}}, B_U^L = \frac{\gamma_L^2 \hbar}{M\theta_L v_L^2} \tag{3.1}$$

$$\tau_u^L = \frac{1}{B_U^L \cdot \omega^2 \cdot T \cdot e^{-\theta_L/3T}}, B_U^L = \frac{\gamma_L{}^2 \hbar}{M\theta_L v_L^2} \qquad (3.2)$$

where ω is the angular phonon frequency, T is the temperature, k_B is the Boltzmann constant, $\hbar$ is the reduced Planck constant. In the following analysis, the Grüneisen parameter (γ), the average mass per atom (M), the average volume per atom (δ^3), the Debye temperature (θ) and the acoustic phonon velocity (v) are approximated using the VC model.

Phonon Scattering by Point Imperfections

A substitutional atom can perturb the vibration system by mass difference and strain difference of the neighbor linkages, resulting in unbalanced kinetic energy or potential energy of the lattice [21], [29]. Contributions of different types of point imperfections can be lumped into a scattering parameter Γ:

$$\Gamma = \Gamma_{impurity} + \Gamma_{isotope} + \Gamma_{alloy} \qquad (4)$$

where $\Gamma_{impurity}$ and $\Gamma_{isotope}$ are the represents the scattering due to impurities/interstitials/vacancies and isotopes. Γ_{alloy} describes alloy scattering which is induced by mass difference and strain difference and causes the constituent atoms to redistribute probabilistically [30]. It can be modeled as [27]:

$$\Gamma_{alloy} = \sum_i x_i \left\{ \left(\frac{M_i - M}{M}\right)^2 + \epsilon \left(\frac{\delta_i - \delta}{\delta}\right)^2 \right\} \qquad (5)$$

where subscript i denotes the index of each constituent, and ϵ is a fitting parameter that characterizes the alloy scattering strength. The phonon relaxation time due to point scattering is expressed as:

$$\tau_p^L = \frac{4\pi v_L^3}{\delta^3 \Gamma \omega^4}, \tau_p^T = \frac{4\pi v_T^3}{\delta^3 \Gamma \omega^4} \qquad (6)$$

It should be noted that point scattering exhibits a $\sim 1/\omega^4$ dependence, which should gradually dominate as the frequency grows.

Phonon-Boundary Scattering

We assume the grain dimension to be isotropic with an effective size of d and the scattering process to be independent of temperature and frequency. The phonon relaxation time due to boundary can be simply written as:

$$\tau_b^L = \frac{d}{v_L}, \tau_b^T = \frac{d}{v_T} \qquad (7)$$

With the major phonon scattering mechanisms identified, the overall phonon lifetime τ_c can be summed up using the Mathiessen's rule:

$$\tau_c^{-1} = \tau_n^{-1} + \tau_u^{-1} + \tau_p^{-1} + \tau_b^{-1} \qquad (8)$$

Fig. 2 compares the room-temperature frequency spectrum of the computed phonon relaxation time

contributed from each phonon scattering process of AlN with $Al_{0.7}Sc_{0.3}N$. It can be identified that alloy scattering and grain boundary scattering limit the overall phonon lifetime τ_c. The relaxation time due to Akhiezer damping is plotted using the method detailed in [23] for comparison. For $Al_{1-x}Sc_xN$, a unique scattering mechanism, alloy scattering, degrades the lifetime of high-frequency phonons by up to four orders of magnitude compared to AlN.

It can be observed that poor crystallinity limits of phonon lifetime through grain boundary scattering for both AlN and $Al_{1-x}Sc_xN$. The grain boundary sizes are usually determined from the high-resolution cross-sectional transmission electron microscopy (XTEM) image. [20], [22] and [31] present the XTEM images of sputtered- and MBE-grown AlN and $Al_{1-x}Sc_xN$ thin films, showing that small in-plane grain sizes can present in both AlN and $Al_{1-x}Sc_xN$ sputtered film. It should be noted that the in-plane and cross-plane grain boundary sizes can be different, and precise evaluation of τ_b and the phonon mean free path (MFP) depends on the acoustic mode shape and the actual grain geometry.

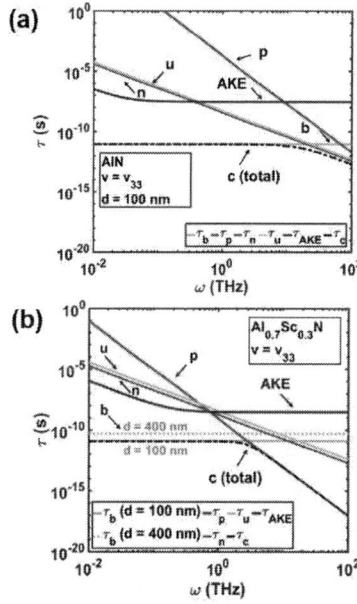

Figure 2: Phonon lifetime compositions at room temperature (292K) of (a) AlN compared with (b) $Al_{0.7}Sc_{0.3}N$. Even for the same grain boundary size (d = 100 nm), alloy scattering degrades the lifetime of high-frequency phonons for $Al_{0.7}Sc_{0.3}N$. Estimation of τ_b with d = 400 nm is plotted on (b) as a comparison.

From the referenced XTEM images, one can note that the grain boundaries of MBE films have a uniform columnar structure with a cross-plane size close to the film thickness. This suggests the possibility that as the film thickness thins down to meet the frequency requirement, the Q factor of BAW resonators could suffer as a result of increased boundary scattering. Another possibility is that the current Q trend of $Al_{1-x}Sc_xN$ could also be attributed by elastic scattering of acoustic wave with the thermal phonons and the energy dissipate out of the system via extrinsic process.

CONCLUSION

This work demonstrates the methodology of assessing phonon scatterings of alloy systems and theoretically model and compare the anharmonic phonon relaxation time due to phonon-phonon scattering and phonon-defect scattering of AlN and $Al_{1-x}Sc_xN$. We first confirm the degradation of Q factor with increasing x presenting in $Al_{1-x}Sc_xN$ resonators and analyze the correlation between the FWHM of the materials and FoMs of resonators. We identify the phonon scattering mechanisms arising from the material structures and present their mathematical formulations. With the computation the phonon lifetime of each scattering category of AlN and $Al_{0.7}Sc_{0.3}N$, it was found that the phonon lifetime of $Al_{0.7}Sc_{0.3}N$ is prone to alloy scattering. We also suggest future analysis of the Q factor limit should be in light of the grain boundary size of the thin film. With the estimation of phonon relaxation time, one can reasonably predict the intrinsic Q limit for $Al_{1-x}Sc_xN$-based acoustic resonators, with further assistance of ab initio study and first-principles calculations.

REFERENCES

[1] Y. Liu, Y. Cai, Y. Zhang et al., "Materials, Design, and Characteristics of Bulk Acoustic Wave Resonator: A Review," *Micromachines 2020*, vol. 11, no. 7, p. 630, Jun. 2020.

[2] A. Ansari, "Single Crystalline Scandium Aluminum Nitride: An Emerging Material for 5G Acoustic Filters," *Symp. IWS 2019 - Proc.*, May 2019.

[3] M. Akiyama, T. Kamohara, K. Kanon et al., "Enhancement of Piezoelectric Response in Scandium Aluminum Nitride Alloy Thin Films Prepared by Dual Reactive Cosputtering," *Adv. Mater.*, vol. 21, no. 5, pp. 593–596, Feb. 2009.

[4] J. Wang, M. Park, S. Mertin et al., "A Film Bulk Acoustic Resonator Based on Ferroelectric Aluminum Scandium Nitride Films," *J. Microelectromechanical Syst.*, vol. 29, no. 5, pp. 741–747, Oct. 2020.

[5] M. Park, Z. Hao, R. Dargis et al., "Epitaxial aluminum scandium nitride super high frequency acoustic resonators," *J. Microelectromechanical Syst.*, vol. 29, no. 4, pp. 490–498, 2020.

[6] M. Moreira, J. Bjurström, I. Katardjev, and V. Yantchev, "Aluminum scandium nitride thin-film bulk acoustic resonators for wide band applications," *Vacuum*, vol. 86, no. 1, pp. 23–26, 2011.

[7] A. Bogner, R. Bauder, H. Timme et al., "Enhanced Piezoelectric $Al_{1-x}Sc_xN$ RF-MEMS Resonators for Sub-6 GHz RF-Filter Applications: Design, Fabrication and Characterization," in *IEEE MEMS*, pp. 1258–1261, 2020.

[8] A. Bogner, H. Timme, R. Bauder et al., "Impact of High Sc Content on Crystal Morphology and RF Performance of Sputtered $Al_{1-x}Sc_xN$ SMR BAW," in *IEEE IUS*, pp. 706–709, Oct. 2019.

[9] M. Uehara, T. Mizuno, Y. Aida, et al., "Increase in the piezoelectric response of scandium-doped gallium nitride thin films sputtered using a metal interlayer for piezo MEMS," *Appl. Phys. Lett.*, vol. 114, no. 1, p. 012902, Jan. 2019.

[10] A. Akhiezer, "On the sound absorption in solids," *J. Phys.*, vol. 1, p. 277, 1939.

[11] T. O. Woodruff and H. Ehrenreich, "Absorption of sound in insulators," *Phys. Rev.*, vol. 123, no. 5, pp. 1553–1559, Sep. 1961.

[12] G. Landau, L; Rumer, "Absorption of sound in solids," *Phys. Z. Sowjentunion*, vol. 11, 1937.

[13] V. B. Braginskiĭ, V. P. Mitrofanov, and V. I. Panov, *Systems with small dissipation*, University of Chicago Press, 1985.

[14] R. Lifshitz and M. Roukes, "Thermoelastic damping in micro- and nanomechanical systems," *Phys. Rev. B*, vol. 61, no. 8, p. 5600, Feb. 2000.

[15] R. Tabrizian, M. Rais-Zadeh, and F. Ayazi, "Effect of phonon interactions on limiting the f.Q product of micromechanical resonators," in *Transducers 2009*, pp. 2131–2134, 2009.

[16] Z. Hao and B. Liao, "An analytical study on interfacial dissipation in piezoelectric rectangular block resonators with in-plane longitudinal-mode vibrations," *Sensors Actuators A Phys.*, vol. 163, no. 1, pp. 401–409, Sep. 2010.

[17] J. Rieger, A. Isacsson, M. J. Seitner et al., "Energy losses of nanomechanical resonators induced by atomic force microscopy-controlled mechanical impedance mismatching," *Nat. Commun.*, vol. 5, pp. 1–6, Mar. 2014.

[18] B. P. Harrington and R. Abdolvand, "In-plane acoustic reflectors for reducing effective anchor loss in lateral–extensional MEMS resonators," *J. Micromechanics Microengineering*, vol. 21, no. 8, p. 085021, Jul. 2011.

[19] R. Tabrizian and M. Rais-Zadeh, "The effect of charge redistribution on limiting the kt^2.Q product of piezoelectrically transduced resonators," in *2015 Transducers*, pp. 981–984, Aug. 2015.

[20] Y. Song, C. Zhang, J. Lundh et al., "Growth-microstructure-thermal property relations in AlN thin films," *J. Appl. Phys.*, vol. 132, no. 17, p. 175108, Nov. 2022.

[21] P. G. Klemens, "The Scattering of Low-Frequency Lattice Waves by Static Imperfections," *Proc. Phys. Soc. Sect. A*, vol. 68, no. 12, p. 1113, Dec. 1955.

[22] C. Yuan, M. Park, Y. Zheng et al., "Phonon heat conduction in $Al_{1-x}Sc_xN$ thin films," *Materials Today Physics*, vol. 21, pp. 100498, Feb. 2021.

[23] Y. Zheng, M. Park, A. Ansari et al., "Self-Heating and Quality Factor: Thermal Challenges in Aluminum Scandium Nitride Bulk Acoustic Wave Resonators," in *Transducers 2021*, pp. 321–324, Jun. 2021.

[24] Y. Song, C. Perez, G. Esteves et al., "Thermal Conductivity of Aluminum Scandium Nitride for 5G Mobile Applications and beyond," *ACS Appl. Mater. Interfaces*, vol. 13, no. 16, pp. 19031–19041, 2021.

[25] M. Asen-Palmer, K. Bartkowski, E. Gmelin et al., "Thermal conductivity of germanium crystals with different isotopic compositions," *Phys. Rev. B*, vol. 56, no. 15, p. 9431, Oct. 1997.

[26] R. Peierls, "Zur kinetischen Theorie der Wärmeleitung in Kristallen," *Ann. Phys.*, vol. 395, no. 8, pp. 1055–1101, Jan. 1929.

[27] W. Liu and A. A. Balandin, "Thermal conduction in $Al_{1-x}Ga_xN$ alloys and thin films," *J. Appl. Phys.*, vol. 97, no. 7, p. 073710, Mar. 2005.

[28] J. Callaway, "Model for Lattice Thermal Conductivity at Low Temperatures," *Phys. Rev.*, vol. 113, no. 4, p. 1046, Feb. 1959.

[29] R. Gurunathan, R. Hanus, M. Dylla et al., "Analytical Models of Phonon-Point-Defect Scattering," *Phys. Rev. Appl.*, vol. 13, no. 3, p. 034011, Feb. 2020.

[30] J. W. Harrison and J. R. Hauser, "Alloy scattering in ternary III-V compounds," *Phys. Rev. B*, vol. 13, no. 12, p. 5347, Jun. 1976.

[31] D. Wang, J. Zheng, Z. Tang et al., "Ferroelectric C-Axis Textured Aluminum Scandium Nitride Thin Films of 100 nm Thickness," in *IFCS-ISAF 2020*, Jul. 2020.

CONTACT

*Azadeh Ansari, Georgia Institute of Technology, Atlanta, GA, 30302; azadeh.ansari@ece.gatech.edu

LITHIUM NIOBATE THIN FILM BASED A_1 MODE RESONATORS WITH FREQUENCY UP TO 16 GHZ AND ELECTROMECHANICAL COUPLING FACTOR NEAR 35%

Rongxuan Su[1#], Zhenyi Yu[2#], Sulei Fu[1], Huiping Xu[1], Shuai Zhang[1], Peisen Liu[1], Yu Guo[2], Cheng Song[1], Fei Zeng[1], and Feng Pan[1]

[1] Key Laboratory of Advanced Materials (MOE), School of Materials Science and Engineering, Tsinghua University, Beijing, China and
[2] School of Internet of Things Engineering, Jiangnan University, Wuxi, Jiangsu, China
(# Authors contributed equally to the work)

ABSTRACT

The 5G or emerging sub-6G systems have turned to millimeter wave (mmw) frequency range with large bandwidth, so it is urgent to develop suitable filter to meet the spectrum requirements. In order to make acoustic platform still competitive in future communication, the first-order antisymmetric (A_1) lamb mode resonators were constructed on thin 128° Y-X $LiNbO_3$ (LN) film, while the high crystal quality and flat surface were ensured. By analyzing the propagation characteristics through calculation and experiment, it is found that the A_1 mode not only possesses high electromechanical coupling factor (K^2) around 35% but also can obtain frequency enhancement by thinning LN film. The measurement results show that as the LN film is trimmed to 110 nm, the center frequency of resonator can be extended 16.9 GHz, while the K^2 is as high as 34.2%. The achieved specifications demonstrate the potential of the acoustic platform to be applied in the mmw systems.

KEYWORDS

Lamb wave devices, high electromechanical coupling factor, high frequency, $LiNbO_3$

INTRODUCTION

Due to the demand for higher data rates in 5G system, researchers focus on scaling communication technologies to the millimeter wave (mmw) frequency range while large bandwidth is also required [1-3]. In current radio frequency front end, acoustic devices including the surface acoustic wave (SAW) and bulk acoustic wave (BAW) are the mainstream filtering technologies [4-7]. However, it is hard to directly scale current used SAW or BAW to mmw range due to several intrinsic restrictions.

In SAW field, the working frequency is inversely proportional to interdigital transducers (IDTs) width and the miniaturization of IDT is restricted by the resolution of the optical lithography system. The direct reducing width of IDT inevitably causes large ohmic loss, low yield and limited power durability [8-9]. Although the BAW shows certain advantages at high frequencies, the low electromechanical coupling factor (K^2) of AlN (near 6%) is difficult to support the desired bandwidth [10]. The Sc doped AlN films can raise K^2 to 12%, but it comes at the expense of the reduced quality factor (Q), leading to additional insertion loss [11].

As the combination of SAW and BAW, the first-order antisymmetric (A_1) lamb mode devices in suspended

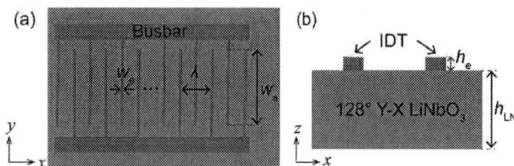

Figure 1: (a) and (b) Structure of the A_1 mode resonator

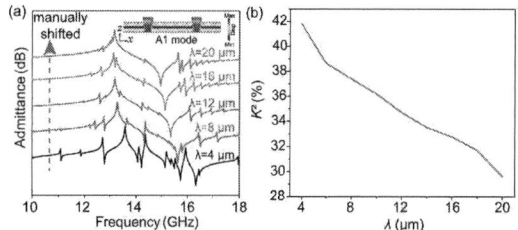

Figure 2: (a) Simulated admittance curves and (b) K^2 of A_1 mode with different λ.

$LiNbO_3$ (LN) film have shown great prospect of enabling high frequency filters with large bandwidth. Thanks to the ion-slicing technology, the LN cut with large piezoelectric component can be freely used without considering the growth conditions to maximize K^2, and high-quality single crystal films can also guarantee the Q value [12]. More importantly, the frequency of A_1 mode is mainly LN thickness defined, so the frequency enhancement can be easily achieved without reducing the IDT width. Kadota *et al.* [13] and Plessky *et al.* [14] demonstrated the A_1 resonator in Z-cut LN film with K^2 over 20% at frequency near 5 GHz. In order to increase the frequency to 10 GHz, Gong and his co-workers adopted high order modes such as A_3 or A_6, but the high order modes suffer a small K^2 less than 5%[15-16]. Recently, Link *et al.* successfully pushed up the frequency of A_1 mode to 11.7 GHz by LN thinning, but the K^2 and Q need further improvement [17].

In this work, through the theoretical and experimental exploration of A_1 mode resonator on 128° Y-X LN film, the quasi-mmw range resonators were designed and fabricated through LN thickness minimization. The measurement shows that when the LN film is 110 nm, the center frequency reaches to 16.9 GHz while a high K^2 of 34.2% is also ensured, which significantly exceeds the state-of-art of acoustic platform. The result of this work not only analyzes the properties of lamb mode resonators but also advances their potential applications in future mmw range.

Table I. Design Parameters of the Implemented A_1 mode Resonators

Device	h_{LN} (nm)	W_e (μm)	λ (μm)	h_e (nm)	W_a (μm)
A	155	1.5			
B	135	0.9	12	50	40
C	110	1.2			

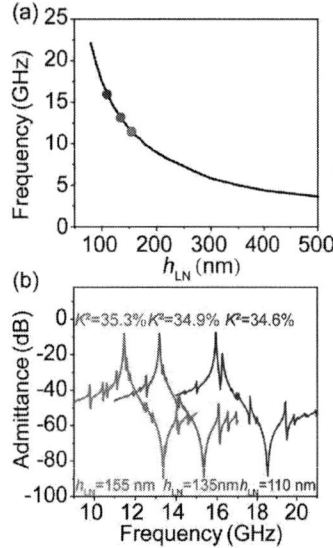

Figure 3: (a) Simulated f_r and (b) admittance curves of A_1 mode with LN thickness of 155, 135, 110 nm, respectively.

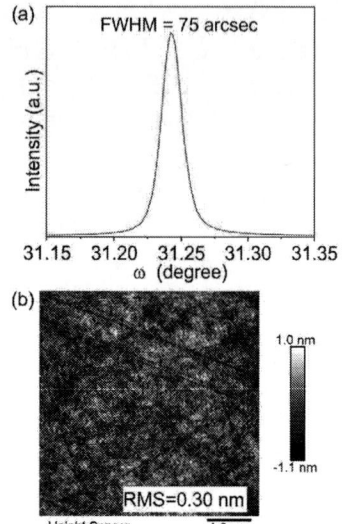

Figure 4: (a) XRD rocking curve of LN. (b)AFM image of LN film.

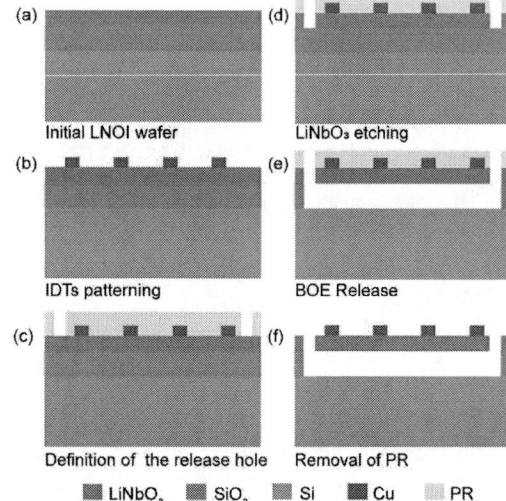

Figure 5: Fabrication process of LN film A_1 resonator

SIMULATION AND DESIGN

The basic structure of A_1 resonator is shown in Fig.1 (a) and (b). Several IDTs are placed on the top of suspended 128° Y-X LN film to generate transverse electric field, thereby exciting strong A_1 mode due to the high piezoelectric component e_{15}. Other piezoelectric components that may cause spurious mode are very small or even zero in 128 ° Y-X LN [18].

Finite element method (FEM) is used to capture the acoustic characteristics. In the following article, the resonance frequency and anti-resonance frequency of resonator are abbreviated as f_r and f_a, respectively. The K^2 is derived from $K^2 = \pi^2(f_a - f_r)/(4 \times f_a)$. We first analyze the influence of λ on propagation. The LN thickness is set as 135 nm. The admittance curves under λ from 4 to 20 μm are illustrated in Fig. 2(a), while the displacement is plotted in the inset. Even if the wavelength is reduced to one fifth, the frequency increase is very small, and f_r still remains at 13 GHz. As for K^2 in Fig. 2(b), although the smaller λ profit higher K^2, the very strong ripples at λ below 8 μm limit its actual application. Therefore, considering the K^2 and clean spectrum together, the λ is located at 12 μm in future.

Because reducing the λ in A_1 mode can no longer achieve the frequency improvement, we further studied the f_r under different LN thickness [shown in Fig. 3(a)]. It is found that the f_r decreases monotonously with the increase of LN thickness. This feature indicates that thickness reduction is one of the solutions to promote lamb wave platforms into the quasi-mmw field. Moreover, the simulated admittance curves with LN thicknesses of 110, 135 and 155 nm are presented in Fig. 3(b), while the corresponding f_r is also dotted in Fig. 3(a). These three resonators with different LN thickness are labelled as device A-C and follow the parameters in Table I. Apparently, all these resonator possess large span over 2 GHz between f_r and f_a leading to a high K^2 around 35%. When the LN thickness scales down to 110 nm, the f_r is as high as 16 GHz, while the large K^2 is also achieved of 34.6%.

FABRICATION OF A_1 MODE RESONATOR

For the implementation of A_1 resonator, three 4-inch wafers of 128°Y-X LN/SiO$_2$/Si (LNOI) were fabricated by NanoLN Inc. with initial LN thickness of 300 nm. To

978-1-6654-9309-3/23 $31.00 © 2023 IEEE   1191

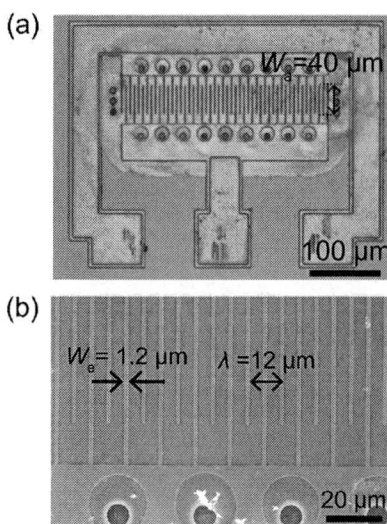

(a) W_a=40 μm

100 μm

(b) W_e = 1.2 μm λ =12 μm

20 μm

Figure 6: (a) optical image (b) SEM image of fabricated device.

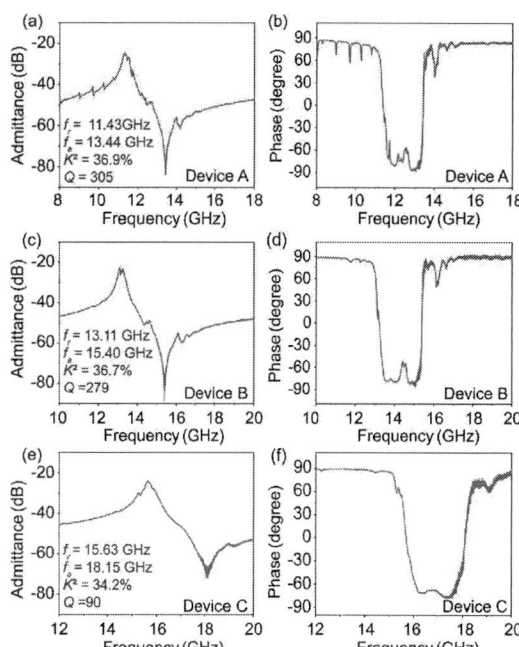

Figure 7: Measured admittance curves of A_1 mode with different LN thickness (a) and (b) 155 nm, (b) and (d) 135 nm, (e) and (f) 110 nm.

enhance the working frequency over 10 GHz, the LN film in three wafers were chemical mechanical polished (CMP) and trimmed to 155, 135 and 110 nm, respectively. The crystal quality of LN film with thickness 155 nm was characterized through the high-resolution X-ray diffraction (XRD) and shown in Fig. 4(a). The full width at half maximum (FWHM) of LN (300) plane is 75 arcsec, which is comparable to that of bulk LN [19]. The good LN crystal quality is conducive to high performance acoustic devices. Furthermore, through the measurement of atomic force microscope (AFM) in Fig. 4(b), flat surface with small root-mean-square roughness (RMS) of 0.3 nm is also realized. The high crystal orientation and smooth surface of LN film confirm the well-controlled transferring and trimming processes.

We then fabricated the resonators according to the parameters in Table I. In order to achieve the required suspended structure, the etching and releasing process were utilized and demonstrated in Fig. 5 (a)-(f). Firstly, Cu IDTs were placed on the top of LNOI through photolithography and lift-off processes. Next, the release holes were defined by overlay process and the thick photoresist served as mask for etching LN hole by SF_6 based inductive coupled plasma. Finally, to suspend the structure, the SiO_2 film under LN was removed through BOE to form the necessary cavity and the top photoresist is then plasma-removed.

The representative optical image of fabricated device C is shown in Fig. 6(a). It can be seen that the SiO_2 at the bottom of the resonator is successfully removed, thus yielding a complete suspended resonator. For detailed characterization, the resonator was observed through scanning electron microscope (SEM) and shown in Fig. 6 (b). The etched hole are circles with radius of 10 μm. The IDTs have clear pattern and are well-proportioned with λ of 12 μm and W_e of 1.2 μm, which is consistent with design.

DEVICE MEASUREMENT

The devices were measured by an Agilent E5071C network analyzer. To capture the acoustic performance, the

de-embedding process was utilized [18]. The measured admittance and phase curves of devices A-C are illustrated in Fig.7 (a)-(f). The obtained frequency, K^2 and 3-dB Q value are marked in the corresponding figure. Although some small spurious ripples exist in band, A_1 mode resonators with trimmed LN film can get large K^2 near~35% over 10 GHz. In device-C, the center frequency of resonator reaches 16.9 GHz, which is adaptable to electromagnetic waves with wavelength of 17.8 mm. This results indicate that acoustic devices are expected to remain competitive in future mmw wideband communications. The difference of frequency and K^2 between the measured and theoretical is mainly due to the thickness deviation from preset during trimming process. As for Q value, although not perfect and decreases with frequency enhancement, 3-dB Q around 300 at 11 GHz is slightly larger than pervious works by Link *et al.* [17] at similar frequency. In future, the major work is to suppress spurious mode and Q value improvement.

CONCLUSION

In this work, we theoretically and experimentally explore the lamb wave resonator to quasi-mmw range through LN thickness minimization. Through material characterization, the trimmed LN still maintains high crystal quality and flat surface. Under the FEM simulation guidance, A_1 mode devices were fabricated under different LN thickness. The resonator measurement shows that when the LN is trimmed down to 110 nm, A_1 mode possess high K^2 of 34.2% at center frequency of 16.9 GHz. Although there are still some rooms for improvement (small Q factor and in-band ripples), the demonstrations in this work still open a new avenue for acoustic devices for mmw wideband spectrum.

978-1-6654-9309-3/23 $31.00 © 2023 IEEE

ACKNOWLEDGEMENTS

This work was supported by the National Key Research Development Program of China (Grant No. 2022YFB3606700), the National Natural Science Foundation of China (Grant No. 52002205), the China Postdoctoral Science Foundation (Grant No. 2020M680557), the Key Research Development Program of Guangdong Province (Grant No. 2020B0101040002), and the Natural Science Foundation of Beijing Municipality (Grant No. JQ20010).

REFERENCES

[1] S. Mahon, "The 5G Effect on RF Filter Technologies," *IEEE Trans. Semicond. Manuf.*, vol. 30, no. 4, pp. 494–499, 2017.

[2] R. Su *et al.*, "Over GHz bandwidth SAW filter based on 32° Y-X LN/SiO₂/poly-Si/Si heterostructure with multilayer electrode modulation," *Appl. Phys. Lett.*, vol. 120, p. 253501, 2022.

[3] P. Popovski, K. F. Trillingsgaard, O. Simeone, and G. Durisi, "5G wireless network slicing for eMBB, URLLC, and mMTC: A communication theoretic view," *IEEE Access*, vol. 6, pp. 55765–55779, 2018.

[4] T. Kimura, M. Omura, Y. Kishimoto, and K. Hashimoto, "Comparative Study of Acoustic Wave Devices Using Thin Piezoelectric Plates in the 3-5-GHz Range," *IEEE Trans. Microw. Theory Tech.*, vol. 67, no. 3, pp. 915–921, 2019.

[5] R. Su *et al.*, "Wideband and Low-Loss Surface Acoustic Wave Filter Based on 15° YX-LiNbO₃/SiO₂/Si Structure," *IEEE Electron Device Lett.*, vol. 42, no. 3, pp. 438–441, 2021.

[6] H. Zhou *et al.*, "A 6.1 GHz Wideband Solidly-Mounted Acoustic Filter on Heterogeneous Substrate for 5G Front-Ends," *2022 IEEE 35th International Conference on Micro Electro Mechanical Systems Conference (MEMS)*, 2022, pp. 1006-1009.

[7] Y. Liu, Y. Cai, Y. Zhang, A. Tovstopyat, S. Liu, and C. Sun, "Materials, Design, and Characteristics of Bulk Acoustic Wave Resonator: A Review," *Micromachines*, vol. 11, no. 7, p. 630, 2020.

[8] S. Fu *et al.*, "High-frequency surface acoustic wave devices based on ZnO/SiC layered structure," *IEEE Electron Device Lett.*, vol. 40, no. 1, pp. 103–106, 2019.

[9] I. Ahmed, U. Rawat, J.-T. Chen and D. Weinstein, "GaN-on-SiC Surface Acoustic Wave Devices Up to 14.3 GHz," *2022 IEEE 35th International Conference on Micro Electro Mechanical Systems Conference (MEMS)*, 2022, pp. 192-195

[10] R. Aigner, G. Fattinger, M. Schaefer, K. Karnati, R. Rothemund and F. Dumont, "BAW Filters for 5G Bands," *IEEE International Electron Devices Meeting (IEDM)*, 2018, pp. 14.5.1-14.5.4

[11] M. Moreira, J. Bjurström, I. Katardjev, and V. Yantchev, "Aluminum scandium nitride thin-film bulk acoustic resonators for wide band applications," *Vacuum*, vol. 86, no. 1, pp. 23–26, 2011.

[12] E. Butaud *et al.*, "Innovative smart cut™ piezo on insulator (POI) substrates for 5G acoustic filters," *IEEE International Electron Devices Meeting (IEDM)*, pp. 34.6.1-34.6.4, 2020.

[13] M. Kadota and T. Ogami, "5.4 GHz Lamb wave resonator on LiNbO₃ thin crystal plate and its application," *Jpn. J. Appl. Phys.*, vol. 50, no. 7, pp. 6–10, 2011.

[14] V. Plessky, S. Yandrapalli, P. J. Turner, L. G. Villanueva, J. Koskela, and R. B. Hammond, "5 GHz laterally-excited bulk-wave resonators (XBARs) based on thin platelets of lithium niobate," *Electron. Lett.*, vol. 55, no. 2, pp. 98–100, 2018.

[15] Y. Yang, R. Lu, L. Gao, and S. Gong, "10-60-GHz Electromechanical Resonators Using Thin-Film Lithium Niobate," *IEEE Trans. Microw. Theory Tech.*, vol. 68, no. 12, pp. 5211–5220, 2020.

[16] R. Lu and S. Gong, "A 15.8 GHz A₆ Mode Resonator with Q of 720 in Complementarily Oriented Piezoelectric Lithium Niobate Thin Films," *2021 Joint Conference of the European Frequency and Time Forum and IEEE International Frequency Control Symposium (EFTF/IFCS)*, 2021, pp. 1-4.

[17] S. Link, R. Lu, Y. Yang, A. E. Hassanien, and S. Gong, "An A1 Mode Resonator at 12 GHz using 160nm Lithium Niobate Suspended Thin Film," *IEEE Int. Ultrason. Symp. IUS*, pp. 3–6, 2021.

[18] R. Lu, Y. Yang, S. Link, and S. Gong, "A1 Resonators in 128° Y-cut Lithium Niobate with Electromechanical Coupling of 46.4%," *J. Microelectromechanical Syst.*, vol. 29, no. 3, pp. 313–319, 2020.

[19] J. Shen *et al.*, "High-Performance Surface Acoustic Wave Devices Using LiNbO₃/SiO₂/SiC Multilayered Substrates," *IEEE Trans. Microw. Theory Tech.*, vol. 69, no. 8, pp. 3693–3705, 2021.

CONTACT

*Sulei Fu, suleifu@163.com

*Feng Pan, panf@tsinghua.edu.cn

SUB-3 DB INSERTION LOSS BROADBAND ACOUSTIC DELAY LINES AND HIGH FOM RESONATORS IN LINBO₃/SIO₂/SI FUNCTIONAL SUBSTRATE

Chun-Chen Yeh, Chia-Hsien Tsai, Guan-Lin Wu, Tzu-Hsuan Hsu, and Ming-Huang Li

Department of Power Mechanical Engineering, National Tsing Hua University, Hsinchu, TAIWAN

ABSTRACT

This work demonstrates high-performance shear-horizontal surface acoustic wave (SH-SAW) delay lines at 910 MHz with low insertion loss (IL) of 2.6 dB and wide fractional bandwidth (FBW) of 8.8% in thin film LiNbO₃/SiO₂/Si functional substrate. The SH-SAW acoustic delay lines (ADLs) based on X-cut LiNbO₃ thin film and gold (Au) electrodes show a group velocity (V_g) of 4,300 m/s and a propagation loss of 10.1 dB/mm, which yields an effective quality factor (Q_{PL}) of 675. To validate the Q_{PL} from the delay line measurement, a 969 MHz one-port SH-SAW resonator at a similar operation frequency was designed on the same chip and presents high effective electromechanical coupling factor (k_{eff}^2) up to 38.5% and maximum Bode-Q (Q_{max}) of 848, resulting in a high figure-of-merit (FoM) of 326. The low-loss behavior of SH-SAW devices illustrates the potential of LiNbO₃/SiO₂/Si functional substrates for RF signal processing and cross-domain applications.

KEYWORDS

MEMS, lithium niobite, acoustic delay lines, SAW resonators, and piezoelectric transducers.

INTRODUCTION

Elastic wave devices have long been an important part of radio frequency (RF) signal processing in mobile devices [1]. Over the past decade, research into new materials, innovative microstructures, and sophisticated fabrication processes has been a major driver of performance advancements in RF microelectromechanical systems (RF MEMS) to meet the ever-demanding specifications of wireless communication systems [2]. To provide efficient electromechanical energy conversion and low loss at UHF and SHF bands, thin-film piezoelectric MEMS resonators adapting critical materials such as aluminum nitride (AlN) [3], aluminum scandium nitride (AlScN) [4], lithium niobate (LiNbO₃) [5], and lithium tantalite (LiTaO₃) [6] have been intensively investigated and achieved commercial success. Although the suspended microstructure offers several advantages, such as low acoustic loss and good environmental isolation, it suffers from significant thermal nonlinearity and weak structure sturdiness [7]. To this end, acoustic devices designed in a solidly mounted configuration and equipped with acoustic reflectors are promising [8].

To address this issue, shear-horizontal surface acoustic wave (SH-SAW) devices using thin-film LiNbO₃ solidly mounted on substrates offer high electromechanical coupling factors (k_{eff}^2), wide fractional bandwidth (FBW), low insertion loss (IL), high linearity, ease of fabrication, and great mechanical robustness [9]-[11]. The acoustic impedance mismatch between the LiNbO₃ film and the substrate permits SH-SAW to propagate only in the

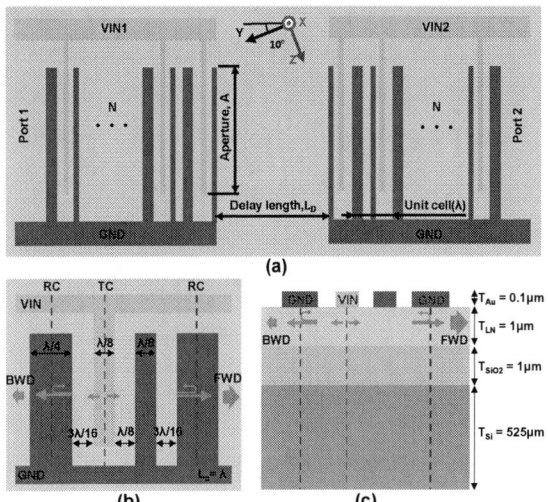

Figure 1: (a) Schematic of the SPUDT ADLs on LiNbO₃/SiO₂/Si platform. (b) Top view and (c) side view of an EWC unit cell.

piezoelectric waveguide, which is the key to maintaining the excellent electroacoustic behavior similar to the suspended thin-film MEMS structure. For the choice of the carrier substrate, high acoustic phase velocity materials such as sapphire [10] and silicon carbide (SiC) [11] have been reported to enhance acoustic energy confinement. However, those expensive materials may not be cost-effective for deploying devices on a large scale.

Based on the above considerations, this work again adopts the previously studied LiNbO₃/SiO₂/Si wafer configuration as a balanced solution between performance and cost [9][12]. Although several devices including acoustic delay lines (ADLs) [9] and one-port resonators [12] have been demonstrated before, the acoustic losses of the LiNbO₃/SiO₂/Si platform have not been systematically extracted. In addition, the metallization process at the time was not optimized, so the performance of the acoustic devices was underestimated. Therefore, further studies are required to reveal the true potential of the LiNbO₃/SiO₂/Si platform.

This work utilizes a group of low-loss and broadband unidirectional SH-SAW ADLs to extract the propagation loss in the LiNbO₃/SiO₂/Si platform. The fabricated ADL achieves a minimum IL of 2.6 dB and a 3-dB FBW of 8.8% with a group delay of 18.2 ns. The propagation loss extracted by SH-SAW is 10.1 dB/mm at 910 MHz, which is comparable to the LiNbO₃-Sapphire and LiNbO₃-SiC platforms. Finally, a one-port resonator with a high figure-of-merit (FoM) of 326 was fabricated and tested with ADLs, highlighting the great potential of LiNbO₃/SiO₂/Si platform for RF signal processing as well as cross-domain applications.

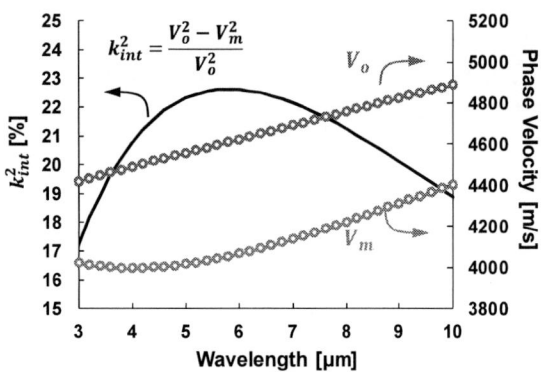

Figure 2: Simulated intrinsic coupling coefficient (k_{int}^2).

$$k_{int}^2 = \frac{V_o^2 - V_m^2}{V_o^2}$$

DESIGN AND SIMULATION

The ADL is a two-port RF passive component with two piezoelectric transducers located on opposite sides of the acoustic waveguide [9]-[11]. By adjusting the size of the transducer and the length of the delay, the desired transfer function from one port to another can be obtained. Fig. 1(a) shows the ADL schematic and the LiNbO₃ orientation. The SH-SAW is launched and received via a pair of single-phase unidirectional directional transducers (SPUDTs) to minimize insertion loss. The SH wave propagation direction is aligned with -10° to +Y axis to target the maximum coupling in X-cut LiNbO₃ [12]. The gap between the SPUDT is defined by L_D. Fig. 1(b) shows the zoomed view of the electrode-width-controlled (EWC) SPUDT unit cell. In either port, the SPUDT consists of 10 cascaded EWC cells with unidirectionality created by forward and backward asymmetry [10]. A customized LiNbO₃/SiO₂/Si wafer with stacking of 1 μm X-cut LiNbO₃ and 1 μm SiO₂ on top of a high resistivity Si substrate was chosen, as shown in Fig. 1(c). In this study, the piezoelectric transducers were made of 100 nm gold (Au) thin film to obtain high electromechanical coupling [12].

The intrinsic electromechanical coupling (k_{int}^2) along the SH-SAW propagation direction is extracted using finite element method (FEM) with a periodic unit cell model based on COMSOL Multiphysics, as shown in Fig. 2. The intrinsic coupling coefficient (k_{int}^2) is calculated based on the difference in the phase velocities of a free surface (V_o) and a metalized surface (V_m), showing a high $k_{int}^2 > 20\%$ for λ between 4 μm and 9 μm. The wavelength (λ) of 4 μm is selected in this work to operate the ADL around 900 MHz while considering the fabrication capability.

Fig. 3(a) and (b) show the simulated displacement profile and the transmission spectrum of the SH-SAW ADL, respectively. In this simulation, the SPUDT of the input and output ports are composed of 10 EWC unit cells [Fig. 1(b)] and the delay length L_D is set to 10λ (40 μm). Perfectly matched layers (PMLs) with a width of 1λ are placed at both ends outside the SPUDT and at the bottom of the model to absorb outgoing acoustic energy [9]. A low IL of 1.5 dB and wide bandwidth of around 8.9% is obtained for an 870 MHz ADL after conjugate impedance matching, as shown in Fig. 3(b). Note that the materials used in the simulations are assumed to be lossless, so the

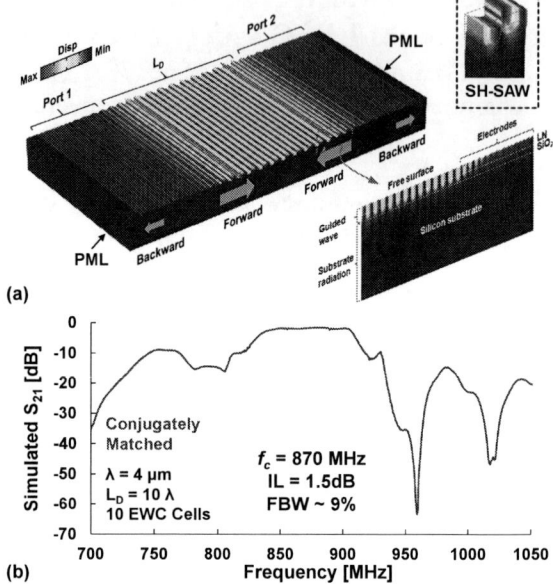

Figure 3: (a) Simulated displacement profile of the SH-SAW ADL. (b) S_{21} passband spectrum of the X-cut LiNbO₃/SiO₂/Si SH-SAW ADL with L_D = 10λ.

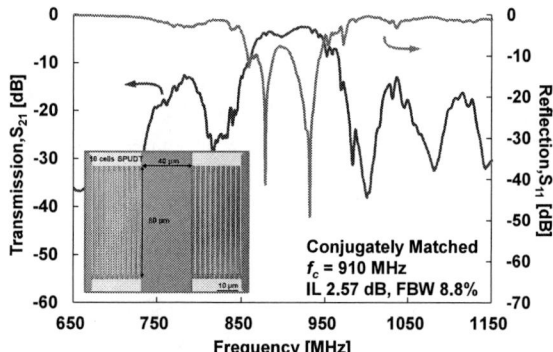

Figure 4: Measured S-parameters of the ADL with L_D=10λ with the optical image of the fabricated device.

acoustic loss only comes from the finite directionality of the SPUDT and the substrate radiation.

FABRICATION AND MEASUREMENT
Acoustic delay lines (ADLs)

The SH-SAW devices were fabricated by electron beam lithography (EBL), metal evaporation, and lift-off process with a minimum dimension of 0.5 μm. Fig. 4 shows the measured S-parameters for a prototyped ADL with λ of 4 μm, which features 10 pairs of EWC cells and a 10λ delay. After performing the conjugate matching at both ports (Z_1 = Z_2 = 105 + j138 Ω), a minimum IL of 2.6 dB, a wide FBW of 8.8%, a center frequency (f_c) of 910 MHz is obtained. Comparing the measurement results with the simulation results, the performance of IL and FBW is very similar. The center frequency increase shown by the measurement data is most likely caused by the variation in the thickness of the gold electrode. The electrodes of the fabricated ADLs might be thinner than 100 nm.

978-1-6654-9309-3/23 $31.00 © 2023 IEEE

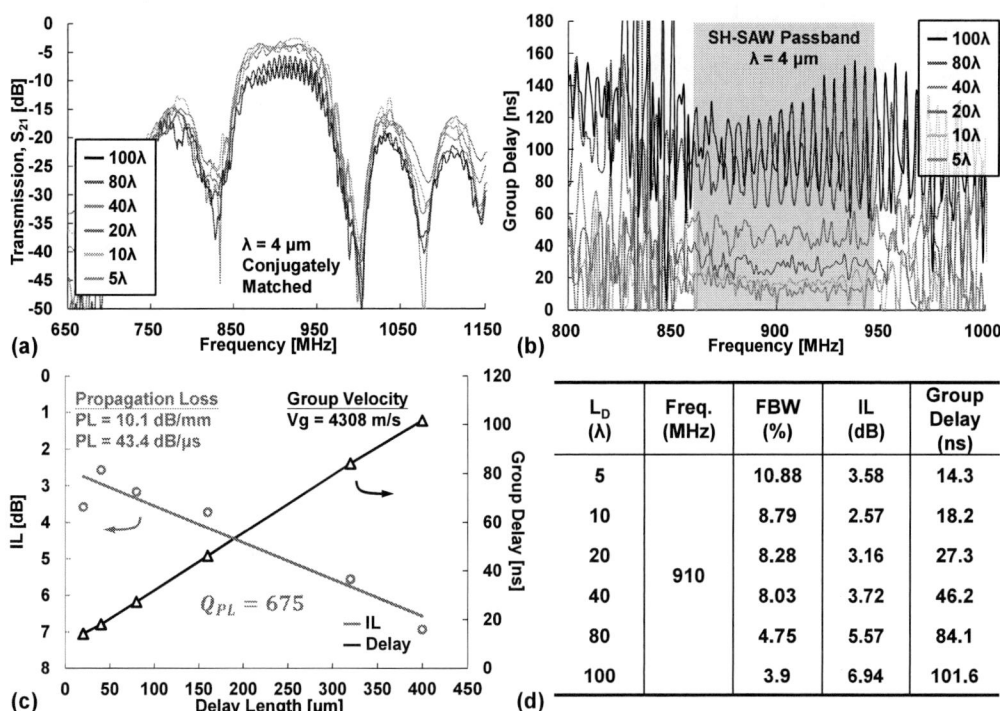

Figure 5: Measured ADL performance with λ = 4 μm, and L_D between 5λ and 100λ. (a) S_{21}, (b) group delay, (c) extracted propagation loss and group velocity, and (d) summary table.

L_D (λ)	Freq. (MHz)	FBW (%)	IL (dB)	Group Delay (ns)
5		10.88	3.58	14.3
10		8.79	2.57	18.2
20	910	8.28	3.16	27.3
40		8.03	3.72	46.2
80		4.75	5.57	84.1
100		3.9	6.94	101.6

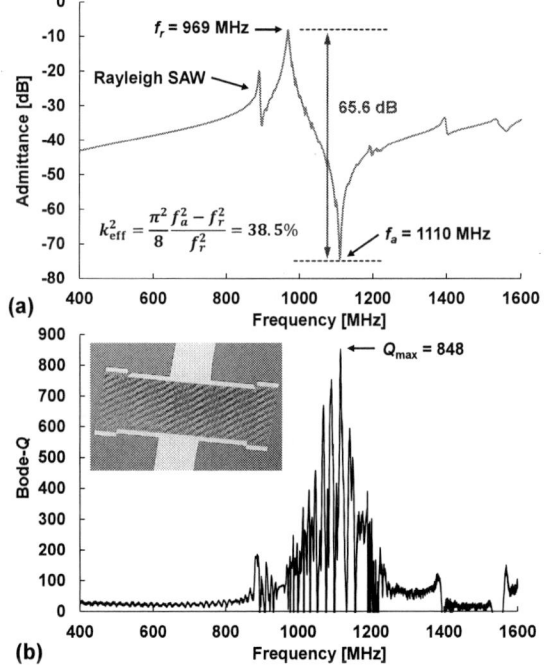

Figure 6: (a) Measured admittance response of the 1-port SH-SAW resonator and (b) extracted Bode-Q.

Furthermore, the propagation characteristics of the SH-SAW are extracted by a set of ADLs with a fixed λ of 4 μm at different delays from 5λ to 100λ (20 μm to 400 μm). Fig. 5(a) and (b) shows the measured transmission responses and the group delay of the ADLs. The finite directivity of the SPUDT transducer results in significant in-band ripples as L_D increases. The propagation loss (PL) and the group velocity (V_g) extracted from the measurement data are summarized in Fig. 5(c), showing a PL of 10.1 dB/mm (equivalence to 43.4 dB/μs) and V_g of 4,308 m/s, respectively. The maximum delay implemented in this work is 101.6 ns as $L_D = 100λ$.

Moreover, the effective quality factor extracted from the propagation loss (Q_{PL}) can be calculated by the equation of $Q_{PL} = \pi/PL$, where the unit of PL in the equation should be in Np/λ (1 Np = 8.686 dB) [10]. The extracted Q_{PL} based on Fig. 5(c) is 675, which is comparable to LiNbO₃-SiC and LiNbO₃-Sapphire platforms. The summary table is offered in Fig. 5(d).

One-port resonator

To validate the Q_{PL} extracted from the ADLs, we design a one-port SH-SAW resonator based on the same LiNbO₃/SiO₂/Si material stack with a series resonant frequency (f_s) of 969 MHz, as shown in Fig. 6(a). The wavelength of the resonator is designed to be 3.4 μm instead of 4 μm to achieve the targeted frequency. The number of interdigital transducer (IDT) pairs and reflective gratings is 57 and 20, respectively.

Based on Fig. 6(a) and (b), the SH-SAW resonator shows a high effective electromechanical coupling (k_{eff}^2) of 38.5% and a high Bode-Q (Q_{max}) of 848, resulting in a high figure-of-merit (FoM = $k_{eff}^2 \cdot Q_{max}$) of 326. Apparently, the measured $Q_{max} = 848$ is comparable to $Q_{PL} = 675$, which indicates a good correlation between different acoustic loss assessment methods. The admittance ratio of the resonator is 65.6 dB.

Comparison

Table 1: Comparison of SH-SAW ADLs around 1 GHz

REF	IL_{min} (dB)	FBW (%)	Freq. (MHz)	Platform	Category	Extract Q_{PL}
This work	2.6	8.8	910	LN/SiO$_2$/Si	SH-SAW SPUDT	PL = 10.1 dB/mm Q_{PL} = 675.4
[9]	11	9.5	640	LN/SiO$_2$/Si	SH-SAW IDT	N/A
[10]	2.8	6.1	1120	LN-on-Sapphire	SH-SAW SPUDT	PL = 6.73 dB/mm Q_{PL} = 878.3
[11]	8.1	4	1320	LN-on-SiC	SH-SAW SPUDT	PL = 3.16 dB/mm Q_{PL} = 1799
[13]	5	N/A	850	LN-on-Si	SH-SAW N/A	PL = 26.9 dB/mm Q_{PL} = 185

Table 2: Comparison of SH-SAW resonators

REF	Freq.	Platform	k_{eff}^2*	Q_{max}	FoM = $k_{eff}^2 \cdot Q_{max}$
This work	969 MHz	LN/SiO$_2$/Si	38.5%	848	326
[14]	1.3 GHz	LN-SiO$_2$-SiC	28%	330	92
[15]	2.2 GHz	LN-SiC	26.9%	1228	330
[15]	1.6 GHz	LN-SiC	22.5%	420	94
[16]	832 MHz	LN/SiO$_2$/Si	36.5%	251	91

*k_{eff}^2 in different references were re-calculated for fair comparison

Finally, the comparison of our work with the state-of-the-art ADLs [9]-[11][13] and the SH-SAW resonators [14]-[16] are shown in Table 1 and 2, respectively. The SH-SAW ADLs implemented the in LiNbO$_3$/SiO$_2$/Si platform shows very competitive performance to LiNbO$_3$-Sapphire [10] and LiNbO$_3$-SiC [11] in terms of PL and Q_{PL}. On the other hand, the one-port SH-SAW resonator built in this work not only exhibits a high FoM of 326, but also features a high electromechanical coupling coefficient of 38.5%. The remarkable performance validates the proposed LiNbO$_3$/SiO$_2$/Si functional substrate as a promising acoustic device platform.

CONCLUSION

In this work, we demonstrate low-loss broadband ADLs and high FoM resonators based on LiNbO$_3$/SiO$_2$/Si functional substrate. The ADL at 910 MHz achieves a low IL of 2.6 dB and a wide FBW of 8.8%. The corresponding propagation loss of 10.1 dB/mm is also reported for the first time, yielding an extracted Q_{PL} of 675. A one-port SH-SAW resonator at a similar operation frequency (969 MHz) reports a high admittance ratio of 65.6 dB, a high Bode-Q_{max} of 848, and a large k_{eff}^2 of 38.5%. This work suggests the strong potential of the LiNbO$_3$/SiO$_2$/Si platform for RF signal processing and cross-domain applications.

ACKNOWLEDGMENT

This research is supported by the NSTC of Taiwan under grants 110-2636-E-007-012 and 111-2221-E-007-074-MY2. The authors would like to appreciate the CNMM of NTHU for device fabrication.

REFERENCES

[1] K. Hashimoto, *Surface acoustic wave devices in telecommunications: Modelling and simulation.* Berlin, Germany: Springer, 2000.

[2] A. Hagelauer *et al.*, "Microwave acoustic wave devices: recent advances on architectures, modeling, materials, and packaging," *IEEE Trans. Microw. Theory Techn.*, vol. 66, no. 10, pp. 4548-4562, Oct. 2018.

[3] Y.-M. Huang *et al.*, "S-band high passive gain resonant transformers based on aluminum nitride FBAR resonators," *in Proc., IEEE MTT-S Int. Microw. Symp.*, June 2022, pp. 891-894.

[4] M. Park *et al.*, "Epitaxial aluminum scandium nitride super high frequency acoustic resonators," *J. Microelectromech. Syst.*, vol. 29, no. 4, pp. 490-498, Aug. 2020.

[5] R. Lu *et al.*, "A1 resonators in 128° Y-cut lithium niobate with electromechanical coupling of 46.4%," *J. Microelectromech. Syst.*, vol. 29, no. 3, pp. 313–319, Jun. 2020.

[6] H. Kando *et al.*, "Improvement in temperature characteristics of plate wave resonator using rotated Y-cut LiTaO$_3$/SiN structure," *in Proc., 24th IEEE Micro Electro Mechanical Systems*, 2011, pp. 768-771.

[7] R. Lu and S. Gong, "Study of thermal nonlinearity in lithium niobate-based MEMS resonators," *in Proc., 18th Int. Conf. on Solid-State Sensors & Actuators (TRANSDUCERS)*, 2015, pp. 1993-1996.

[8] T. Takai *et al.*, "High-performance SAW resonator with simplified LiTaO$_3$/SiO$_2$ double layer structure on Si substrate," *IEEE Trans. Ultrason., Ferroelectr., Freq. Contr.*, vol. 66, no. 5, pp. 1006-1013, May 2019.

[9] K.-J. Tseng and M.-H. Li, "Low loss acoustic delay lines based on solidly mounted lithium niobate thin film," *J. Microelectromech. Syst.*, vol. 29, no. 4, pp. 449-451, Aug. 2020.

[10] R. Lu *et al.*, "Gigahertz low-loss and high power handling acoustic delay lines using thin-film lithium niobate-on-sapphire," *IEEE Trans. Microw. Theory Tech.*, vol. 69, no. 7, pp.3246-3254, 2021.

[11] P. Zheng *et al.*, "Ultra-low loss and high phase velocity acoustic delay lines in lithium niobate on silicon carbide platform," *in Proc., IEEE Int. Conf. Micro Electro Mech. Syst.*, 2022, pp. 1030-1033.

[12] T.-H. Hsu, K.-J. Tseng, and M.-H. Li, "Thin-film lithium niobate-on-insulator (LNOI) shear horizontal surface acoustic wave resonators," *J. Micromech. Microeng.*, vol. 31, no. 5, pp. 054003, Apr. 2021.

[13] Y. Yang *et al.*, "Silicon-SAW resonators and delay lines based on sub-micron lithium niobate and amorphous silicon," *in Proc., 2022 IEEE MTT-S Int. Microw. Symp.*, June 2022, pp. 817-820.

[14] J. Shen *et al.*, "High-performance surface acoustic wave devices using LiNbO$_3$/SiO$_2$/SiC multilayered substrates" *IEEE Trans. Microw. Theory Tech.*, vol. 69, no. 8, pp. 3693-3705, Aug. 2021.

[15] S. Zhang *et al.*, "Surface acoustic wave devices using lithium niobate on silicon carbide," *IEEE Trans. Microw. Theory Tech.*, vol. 68, no. 9, pp. 3653-3666, Sept. 2020.

[16] T.-H. Hsu, K.-J. Tseng, and M.-H. Li, "Large coupling acoustic wave resonators based on LiNbO$_3$/SiO$_2$/Si functional substrate," *IEEE Electron Device Lett.*, vol. 41, no. 12, pp. 1825-1828, Dec. 2020.

CONTACT

*M.-H. Li, mhli@pme.nthu.edu.tw

SUPPRESSION OF SPURIOUS MODES IN ALUMINUM NITRIDE S₁ LAMB WAVE RESONATORS USING A MECHANICAL SOFT-CONTACT SCHEME

Shao-Siang Tung[1], Tzu-Hsuan Hsu[1], Yens Ho[2], Yung-Hsiang Chen[2], Yelehanka R. Pradeep[3], Rakesh Chand[3], and Ming-Huang Li[1]

[1]Department of Power Mechanical Engineering, National Tsing Hua University, Hsinchu, TAIWAN
[2]Vanguard International Semiconductor Corporation, Hsinchu, TAIWAN and
[3]Vanguard International Semiconductor Corporation Singapore PTE. Ltd, SINGAPORE

ABSTRACT

In this study, a novel mechanical soft-contact (MSC) scheme is proposed, for the first time, to enable selective mode suppression in aluminum nitride (AlN) S₁ Lamb wave resonators (LWR). For MSC-LWRs, the degree of energy dissipation introduced at the contact point depends heavily on the vibrational modes. We have demonstrated that in MSC-LWR, the ratio of the quality factor (Q) for the S₁ and S₀ Lamb modes can be as high as 37.7 (944 for Q_{S1} and 25 for Q_{S0}) to effectively eradicate the S₀ mode. The proposed resonators are fabricated via surface micromachining and exploit residual stress-induced structural deformation to create a mechanically soft contact between the resonator plate and the substrate. As a result, a spectrum-clean MSC-LWR achieves Q_{S1} of 1,381 and an effective electromechanical coupling coefficient (k_{eff}^2) of 2.72% at 2.19 GHz. Furthermore, we show that Q_{S1} can be further optimized by adjusting the tether width, which is presumably dominated by the normal force at the contact point induced by the supporting structure.

KEYWORDS

Aluminum nitride (AlN), microelectromechanical system (MEMS), resonator, Lamb wave, mechanical soft contact, spurious mode.

INTRODUCTION

Aluminum nitride (AlN) microelectromechanical systems (MEMS) resonators are promising solutions for radio frequency (RF) signal generation, selection, and control [1][2]. Among the resonator types that have been investigated so far, the Lamb wave resonators (LWRs) have attracted much attention due to their high quality factor (Q), lithographically defined resonance frequency, decent electromechanical coupling coefficient (k_{eff}^2), and fabrication compatibility with CMOS [3][4]. The lowest-order symmetric (S₀) Lamb wave has become the most preferred acoustic mode to realize LWR in AlN over the past decade due to its low dispersive phase velocity (v_p) characteristic [3]. However, its high frequency scaling is very difficult due to the limitation of $v_{p,S0}$ capped at 10,000 m/s and the fabrication challenges of realizing extremely short wavelength devices. Therefore, employing higher-order Lamb wave modes in AlN is one of the viable solutions adopted to overcome this issue by taking the advantage of their high phase velocity. In particular, the first symmetric (S₁) mode features even higher v_p and larger k_{eff}^2 than the S₀ mode, which is suitable for implementing filters, sensors, and oscillators at multi-GHz

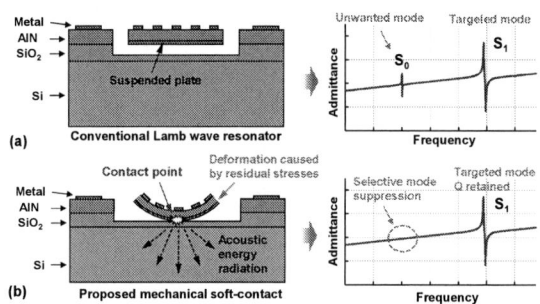

Figure 1: Conceptual schematic of (a) residual stress-free S₁ mode LWR with significant multi-mode resonance behavior and (b) spectrum-clean S₁ mode MSC-LWR.

[5]-[8]. Nevertheless, in addition to the target S₁ mode, other Lamb waves can also be excited and detected by the IDT and operate as spurious modes over a broad frequency band. This is also known as the multi-mode resonance behavior [9].

The strong presence of S₀ mode among typical S₁ LWR may hinder its application from practical use as it creates an unwanted passband for the filter and causes the oscillator to generate the wrong frequency. To address this issue, previous work utilized damped edge reflector [10] and reflector-less boundary [6] designs to suppress the S₀ mode by operating the S₁ mode under zero group velocity (ZGV) conditions. Since ZGV is associated with energy localization, the acoustic energy of the S1 mode is properly confined in the IDT region, while the other Lamb modes are suppressed by the damped edge reflectors or the reflector-less boundaries. However, this method requires a thick AlN plate (around 2.5 to 3.5 μm [6][10]) to enable ZGV operation of the S₁ mode, which needs more stringent process control in the fabrication process.

In this work, a novel mechanical soft contact (MSC) scheme to suppress the S₀ Lamb mode in S₁ LWRs without ZGV operation is proposed. We exploit the residual stress gradient in the thin films to deform the LWR structure and make it in contact with the substrate. Experiments have shown that the degree of energy dissipation introduced at the point of contact strongly depends on the vibration mode. As a result, a high performance S₁ MSC-LWR with $k_{eff,S1}^2$ of 2.72% and S₁ quality factor (Q_{S1}) of 1,381 is demonstrated with a suppressed S₀ mode ($Q_{S0} < 100$).

MSC-LAMB WAVE RESONATOR
MSC-LWR based on Stress-induced Deformation

As shown in Fig. 1, the mechanical soft contact (MSC) scheme is adopted in this work to perform selective Lamb

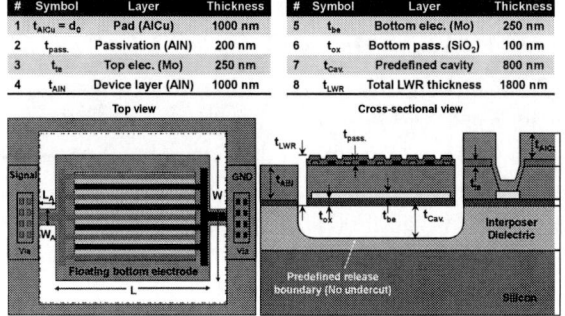

#	Symbol	Layer	Thickness
1	$t_{AlCu} = d_0$	Pad (AlCu)	1000 nm
2	$t_{pass.}$	Passivation (AlN)	200 nm
3	t_{te}	Top elec. (Mo)	250 nm
4	t_{AlN}	Device layer (AlN)	1000 nm

#	Symbol	Layer	Thickness
5	t_{be}	Bottom elec. (Mo)	250 nm
6	t_{ox}	Bottom pass. (SiO$_2$)	100 nm
7	$t_{Cav.}$	Predefined cavity	800 nm
8	t_{LWR}	Total LWR thickness	1800 nm

Figure 2: Illustrations of the AlN LWR platform with detailed thin film stacking configuration. The release cavity is defined by a shallow sacrificial layer with $t_{Cav.}$ of around 800nm.

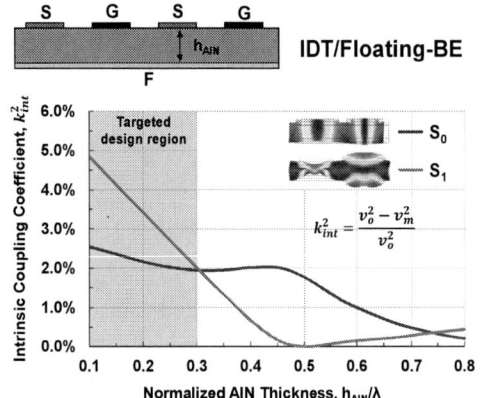

Figure 3: FEM-simulated dispersive intrinsic coupling coefficient (k_{int}^2) of the LWR.

Table 1: S$_1$ LWRs designed in this work

Device Index	A	B	C	D	E	Units
Device Type	No Cont.	MSC	MSC	MSC	MSC	–
Wavelength (λ)	12.4	12.4	10	10	10	μm
IDT Pairs	9	9	9	9	9	–
IDT Aperture	12	12	10	10	10	λ
IDT Duty	83	83	75	75	75	%
Device Width (W)	115.6	117.6	88.75	88.75	88.75	μm
Device Length (L)	200.4	229.2	137	133	129	μm
Tether Width (W$_A$)	34.1	9.3	10.5	8.5	6.5	μm
Tether Length (L$_A$)	6	6	6	6	6	μm

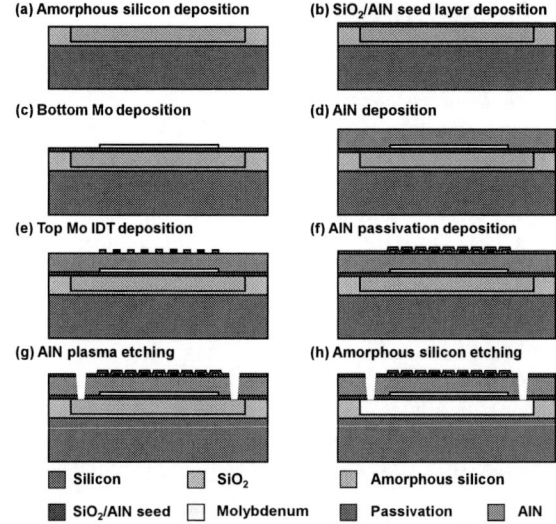

Figure 4: Fabrication process for AlN S$_1$ lamb wave resonator.

AlN plate and molybdenum (Mo) top/bottom electrodes of 250 nm. The height of the cavity is predefined by the thickness of the sacrificial layer ($t_{Cav.}$), which is around 800 nm in this work. In other words, the mechanically soft contact condition is achieved when the center of the LWR plate is deformed downward by about 800 nm. The equivalent contact force can be determined by changing the length (L) and width (W) of the resonator and the length and width of the support tether (L$_A$ and W$_A$), which will affect the Q_{S0} and Q_{S1}.

Resonator Design and Simulation

To determine the resonator design window, the dispersion characteristics of the S$_0$ and S$_1$ modes are simulated using finite element method (FEM), as shown in Fig. 3. A single-IDT electrode configuration with a floating bottom metal is chosen in this study. The intrinsic electromechanical coupling factor (k_{int}^2) of each mode is estimated by Alder's approach based on the difference in the phase velocities for a free surface (v_o) and metalized surface (v_m), given by $k_{int}^2 = (v_o^2 - v_m^2)/v_o^2$ [3]. The electrodes and passivation layers are omitted in k_{int}^2 simulation. As a result, the S$_1$ MSC-LWRs are designed with long wavelength (λ > 10 μm) to obtain a smaller normalized thickness (h_{AlN}/λ < 0.3), resulting in a higher k_{int}^2 than the S$_0$ mode.

To experimentally investigate the properties of the MSC-LWR, five different resonator designs are studied in this work, as shown in Table 1. For Device-A and B, the width (W) and length (L) were chosen to be about 120 μm

mode suppression. Fig. 1(a) shows a conceptual schematic of a residual stress-free LWR where the MEMS structure is suspended over a cavity through tiny tethers. In this ideal case, the resonance spectra of the S$_0$ and S$_1$ modes can be excited and detected simultaneously and they usually feature similar Q's [9]. However, in a practical case, due to the existence of high residual stresses in the thin film stacks, the MEMS structure deforms naturally after the thin sacrificial layer is etched, as shown in Fig. 1(b). The typical relative stress between the layers results in a concave resonator in which the sides of the device bend up and the center moves down. Therefore, if the cavity is shallow, the deformed LWR structure will come into contact with the bottom of the cavity [11]. As a result, this creates a new path for the acoustic energy radiation into the substrate through the contact point, causing additional acoustic energy loss. As shown in Fig. 1(b), we experimentally found that the S$_0$ mode is easily suppressed in MSC-LWR, but the S$_1$ mode is not, suggesting that the acoustic loss of mechanical contact is selective. Furthermore, the degree of plate deformation depends largely on the different tether widths of the Lamb wave resonator. Narrower tether widths lead to more severe deformation and thus increased loss of acoustic energy through the contact point. More details will be revealed in the Experimental Results section.

The information of the AlN LWR platform used in this work with detailed thin film stacking configuration is summarized in Fig. 2. The LWR is composed of a 1 μm

978-1-6654-9309-3/23 $31.00 © 2023 IEEE

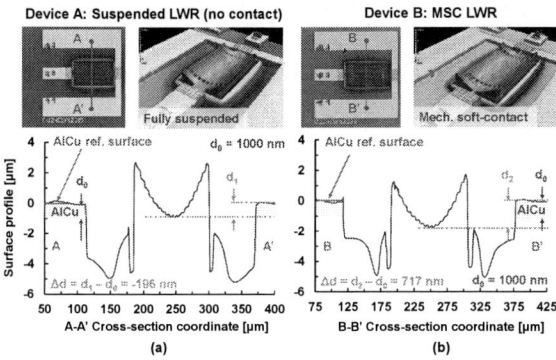

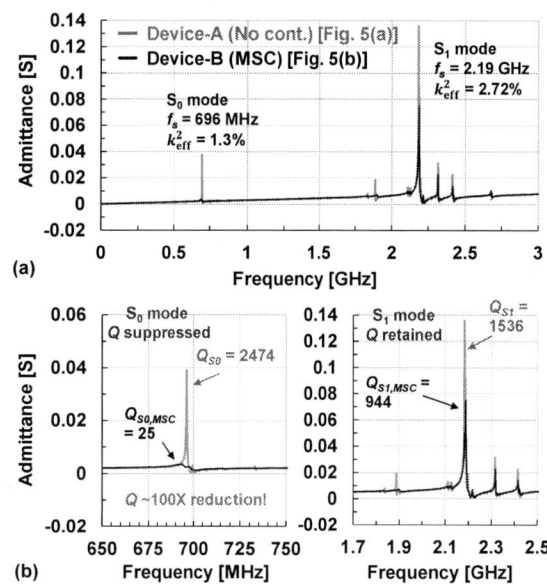

Figure 5: Optical profiles of (a) Device-A: wide-anchor conventional LWR (without physical contact), and (b) Device-B: narrow-anchor LWR (with physical contact).

and 200 μm, respectively, to intentionally create larger out-of-plane deformations, while the support tether width (W_A) was used to control whether mechanical contact occurred. On the other hand, Device-C to E are used to explore the relationship between Q_{S1} and W_A.

FABRICATION

The MEMS LWRs are fabricated on a custom AlN piezoelectric MEMS platform using 200 mm wafers [11]-[13], and the simplified process flow is shown in Fig. 4. In this process, the amorphous silicon (α-Si) of $t_{Cav.}$ = 800 nm is embedded in silicon oxide (SiO_2) as the sacrificial layer on a handling silicon wafer [Fig. 4(a)]. The resonator process starts with the deposition of bottom SiO_2 protective layer and AlN seed layer, followed by the deposition and patterning of bottom Mo electrode, AlN piezoelectric layer, top Mo IDT electrode, and top passivation layer [Fig. 4(b)-(f)]. The resonator geometry is then defined by a plasma-based dry etching process [Fig. 4(g)]. Finally, the gas-phase dry silicon etching removes the exposed α-Si and creates cavities underneath the MEMS LWRs. Other process steps used to make vias and thick aluminum copper (AlCu) probe pads are not detailed here.

EXPERIMENTAL RESULTS

Surface Profile Measurement

To study the deformation caused by the residual stress in our platform, a laser scanning confocal microscope (Keyence VK-X series) is introduced to characterize the surface profile of the Lamb wave resonators. Fig. 5 (a) and (b) depict the 3-dimensional images and the cross-sectional profiles of Device-A and B, respectively. To calculate the vertical deformation of the resonator center due to residual stress, the surface of the AlCu layer is set as the reference plane. The thickness of AlCu is d_0 = 1000 nm and the distance between the reference plane and the center of the lamb wave resonator is defined as d_1 and d_2 for Device-A and B, respectively. Therefore, the center deformation can be identified as $\Delta d_{1,2} = d_{1,2} - d_0$.

As shown in Fig. 5(a) and (b), the center of Device-A moves upward about 196 nm ($\Delta d_1 = d_1 - d_0 = -196$ nm) while center of Device-B moves downward about 717 nm ($\Delta d_2 = d_2 - d_0 = 717$ nm). It shows that the wider supporting tether is a key to suppress out-of-plane

deformation due to its higher stiffness. On the other hand, $\Delta d_2 = 717$ nm is very close to the thickness of the sacrificial layer ($t_{Cav.} \approx 800$nm), which indicates that a physical contact is occurred between the Lamb wave resonator and the bottom of the cavity.

Figure 6: Measured admittance plots of the conventional LWR and proposed MSC-LWR: (a) wide-span spectrum, and (b) zoom-in view. The Q_{S0} is suppressed 100-fold in MSC-LWR.

Admittance Measurement: Conventional vs. MSC

After confirming mechanical soft contact happened in Device-B, the devices were characterized in dry air using an RF probe station (MPI TS-150) and a network analyzer (Keysight E5071C). The measured admittance plots of Device-A (conventional LWR) and B (MSC-LWR) are shown in Fig. 6. The resonance frequency (f_o) and k_{eff}^2 of the S_0 mode and S_1 mode for both devices are 696 MHz and 2.19 GHz and 1.3% and 2.72%, respectively. This shows that f_o and k_{eff}^2 are hardly affected by the MSC boundary condition, but this is clearly not the case for Q.

As can be seen from Fig. 6(b), even though Device-A and B have similar geometry, the Q_{S0} ratio between the two reaches 100 times (2,474 and 25) while the Q_{S1} ratio is only 1.6 times (1,536 and 944). It is evident that mechanical soft contact affects different resonant modes to varying degrees. Furthermore, the minor spurious modes around the S_1 mode are also suppressed, which is helpful in designing oscillators and filters.

Admittance Measurement: Selective Q-tuning

As seen in the profile measurement (Fig. 5), the deformation of the LWR is highly dependent on the supporting tether width. In other words, the amount of deformation and contact force can be controlled by adjusting the width of the tether. We reasoned that a narrower W_A would result in more pronounced deformations and larger normal forces, resulting in increased acoustic losses (i.e., lower Q) when the structure

978-1-6654-9309-3/23 $31.00 © 2023 IEEE

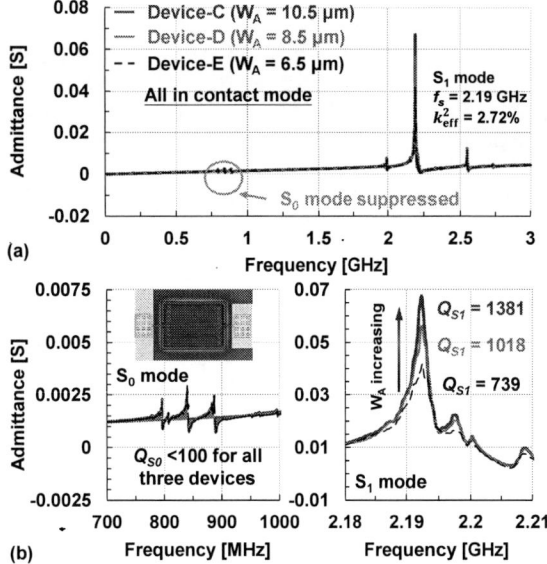

Figure 7: Measured admittance plots of MSC-LWRs with different W_A's (6.5 to 10.5 μm) for Q_{S1} tuning and Q_{S0} suppression: (a) wide-span spectrum, and (b) zoom-in view. Optical photo of the Device-E is given in (b).

contacts the cavity bottom. Therefore, we further investigated the Q-tuning effect in MSC-LWRs with different tether width devices based on Device-C to E in Table 1.

The measurement results of Device-C to E are provided in Fig. 7. The S_0 mode is suppressed for all devices regardless of W_A. The Q_{S1} is improved from 739 to 1,381 when W_A is increased from 6.5 to 10.5 μm while most of the resonator dimensions keep the same. This result is in good agreement with our predictions. We believed that the normal force controlled by W_A plays an important role in the Q-tuning effect, which paves a way for achieving Q_{S1} optimization in MSC-LWRs. As a result, we successfully demonstrate the first spectrum-clean S_1 Lamb wave resonator with superior performance (Q_{S1} > 1,300, Q_{S0} < 100, $k^2_{eff,S1}$ > 2.7%) at 2.19 GHz based on a novel mechanical soft-contact scheme.

CONCLUSION

In this study, a novel mechanical contact scheme is proposed to suppress the wide-band spurious modes. We exploit the deformation induced by residual stress to create a mechanically soft contact between the resonator and the substrate to control the damping of Lamb modes. Since the amount of acoustic energy dissipation at the contact point is strongly dependent on the vibration mode, most acoustic modes are suppressed except the S_1 mode. We experimentally demonstrate a spectrum-clean S_1 mode MSC-LWR at 2.19 GHz with a high Q_{S1} of 1,381 and high $k^2_{eff,S1}$ of 2.72% while the corresponding Q_{S0} is below 100. The proposed AlN S_1 mode resonator features high frequency operation, high electromechanical coupling, and a near-spurious-free frequency range up to 3 GHz, which enables a new frequency control solution for next-generation communication systems.

ACKNOWLEDGMENT

This work was supported in part by the Vanguard International Semiconductor Corporation Joint Development Project and in part by the NSTC of Taiwan under grant 111-2221-E-007-074-MY2.

REFERENCE

[1] G. Piazza *et al.*, "Piezoelectric aluminum nitride vibrating contour-mode MEMS resonators," *J. Microelectromech. Syst.*, vol. 15, no. 6, pp. 1406-1418, Dec. 2006.

[2] B. P. Otis and J. M. Rabaey, "A 300-μW 1.9-GHz CMOS oscillator utilizing micromachined resonators," *IEEE J. Solid-State Circuits*, vol. 38, no. 7, pp. 1271-1274, July 2003.

[3] C.-M. Lin *et al.*, "Micromachined one-port aluminum nitride Lamb wave resonators utilizing the lowest-order symmetric mode," *J. Microelectromech. Syst.*, vol. 23, no. 1, pp. 78-91, Feb. 2014.

[4] K. E. Wojciechowski *et al.*, "A fully integrated oven controlled microelectromechanical oscillator—Part I: design and fabrication," *J. Microelectromech. Syst.*, vol. 24, no. 6, pp. 1782-1794, Dec. 2015.

[5] J. Zou *et al.*, "High-frequency and low-resonance-impedance Lamb wave resonators utilizing the S1 mode," *in Proc. Int. Conf. Solid-State Sens. Actuators Microsyst. (Transducers)*, Anchorage, AK, Jun. 2015, pp. 2025-2028.

[6] V. Yantchev *et al.*, "Thin-film zero-group-velocity Lamb wave resonator," *Appl. Phys. Lett.*, vol. 99, 033505, Jul. 2011

[7] A. Gao, J. Zou, and S. Gong, "A 3.5 GHz AlN S1 lamb mode resonator," in *Proc. IEEE Ultrason. Symp.*, 2017, pp. 1-4.

[8] C. Zuo *et al.*, "Dual-mode resonator and switchless reconfigurable oscillator based on piezoelectric AlN MEMS technology," *IEEE Trans. Electron Devices*, vol. 58, no. 10, pp. 3599-3603, Oct. 2011.

[9] J. Zou *et al.*, "The multi-mode resonance in AlN Lamb wave resonators," *J. Microelectromech. Syst.*, vol. 27, no. 6, pp. 973-984, Dec. 2018.

[10] J. Zou *et al.*, "Spectrum-clean S1 AlN Lamb wave resonator with damped edge reflectors," *Appl. Phys. Lett.*, vol. 116, pp. 023505, Jan. 2020.

[11] T.-H. Hsu *et al.*, "On the geometry design of AlN Lamb wave resonators with predefined shallow release cavities," in *Proc., 2022 Solid-State Sensor, Actuator, and Microsyst. Workshop (Hilton Head)*, Hilton Head, South Carolina, June 2022.

[12] C.-Y. Chang *et al.* "On the spurious modes in FBAR resonators with quasi-free edges," in *Proc., 2022 Joint Conf. of Eur. Freq. Time Forum - IEEE Int. Freq. Contr. Symp.*, Paris, France, Apr. 2022, pp. 1-3.

[13] Y.-M. Huang *et al.*, "S-band high passive gain resonant transformers based on aluminum nitride FBAR resonators," *in Proc., IEEE MTT-S Int. Microw. Symp.*, June 2022, pp. 891-894.

CONTACT

*M.-H. Li, mhli@pme.nthu.edu.tw

978-1-6654-9309-3/23 $31.00 © 2023 IEEE

TERAHERTZ REFLECTIVE METALENS FOR ARBITRARY OFF-AXIS FOCUSING WITH LARGE DEPTH OF FOCUS

Jiahao Miao, Yi Liu, Cong Lin, Zhanxuan Zhou, and Xiaomei Yu[*]

School of Integrated Circuits, Peking University, National Key Laboratory of Science and Technology on Micro/Nano Fabrication, Beijing 100871, China

ABSTRACT

This paper reports a reflective metalens composed of fifteen rotationally symmetrical metal resonator-dielectric-metal structure with 24° step phase response by changing the geometry of the resonators and arranged by the derived phase distribution formula. An off-axis focusing with an angle of 1.20° is realized by the metalens at 2.52THz. The depth of focus (DOF) is at least 16mm. The off-axis metalens can reduce the interference between the incident and reflected wave, and arbitrarily control the spatial location of the focus, which remains stable while the position of the incident wave changes within a certain range.

KEYWORDS

Terahertz, reflective metalens, off-axis focusing, depth of focus, focusing efficiency.

INTRODUCTION

In recent years, terahertz (THz) waves have attracted much interest in fields of spectral imaging [1, 2], non-destructive detection [3], biomedicine [4, 5], broadband communications [6, 7], and other aspects because of the characteristics of low photon energy, strong penetration of non-polar materials, high absorption of polar molecules, and broad bandwidth [8], etc. Traditional THz components usually utilize materials like polymers or high-resistance silicon to process into specific shapes for manipulating the state of light with gradually accumulating changes in amplitude, phase, and polarization along the propagation path, which are bulky and thus limit the miniaturization of THz systems.

Metasurfaces are artificial electromagnetic materials composed of subwavelength microstructures that can modulate electromagnetic wave typically by the different response of resonators with varying shapes or sizes to incident wave. Therefore, metasurfaces are expected to replace traditional optical components in some fields due to their advantages of ultrathin and easy to be fabricated. Recently, many metasurface-based THz components have been proposed, such as beam deflector [9], metalens [10, 11], beam splitter [12], quarter-wave plate [13, 14], vortex beam generator [15], and so on [16].

As one of the important applications of the metasurface, metalens have reflective and transmissive propagation ways for incident electromagnetic waves. Many studies have been proposed to achieve better performance of metalenses such as achromatic [17], high efficiency [18], etc. However, most metalenses function well with on-axis focusing, which means in a small angle reflection mode, there will be interference between incident and reflected radiation. Meanwhile, some existing off-axis metalenses are polarization-sensitive [19, 20], which also limit the application of metalens to a certain extent.

In this paper, we present an off-axis metalens, which can precisely focus arbitrarily polarized THz wave to the designed location with a large DOF. Meanwhile, the off-axis focusing at same spatial position can be achieved when slightly shifting incident wave, which shows robustness and potential in compact, lightweight, and integrated THz systems.

DESIGN AND FABRICATION

The off-axis metalens was designed by imposing a discontinuous hyperbolic phase distribution on the metasurface. Figure 1(a) and (b) show the schematic of off-axis metalens with arbitrary focusing position. According to the principle of optical path reversibility, the process of reflecting a plane wave to a given focus position equals the focus emitting a spherical wave outward as a point source. Assuming that the phase shift of the projection position of the focus (F') on the metalens is zero, the phase delay at the point P on the metasurface is proportional to the distance PR, where R is the point of intersection of the line PF with the spherical surface with a radius of f_0. The off-axis angle (α) is defined as the angle between the line connecting the center of the metalens and the focus and the z-axis. Therefore, for a given position of the focus (x_0, y_0, z_0), the phase delay imparted on any point P (x, y) on the off-axis metalens must satisfy the following equation:

$$\varphi_{(x, y)} = \frac{2\pi}{\lambda} \left(\sqrt{(x-x_0)^2 + (y-y_0)^2 + f_0^2} - f_0 \right) \quad (1)$$

where λ is the wavelength in free space. Here, the radius of the metalenses was set to be 10mm, the off-axis angle was designed to be 1.15° while the off-axis value (dx, equals to x_0) and the distance between focus and metalens (f_0) were 2mm and 100mm, respectively. Note that $x_0=y_0=0$ represents the condition for on-axis metalens.

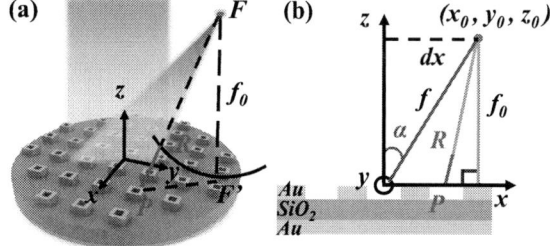

Figure 1: (a)Schematic diagram and (b) side view showing of the reflective metalens for off-axis focusing.

In order to provide a desired phase delay at each position of the metasurface, a metal resonator-dielectric-metal sandwich structure was utilized to respond to the THz wave at 2.52THz. As shown in Figure 1(b), the bottom layer of gold functions as a ground plane to eliminate the transmission of THz waves. The intermediate dielectric spacer is composed of silicon dioxide with permittivity ε of 3.9 and loss tan δ of 0.025. The top rotationally symmetrical square-shaped gold ensures insensitivity to the polarization of incident electromagnetic waves. The dimensions of the structure were optimized with a commercially available finite-difference time-domain (FDTD) software to obtain a relatively high reflection amplitude at 2.52THz. The thickness of the Au ground plane, the dielectric spacer, and the Au resonators were optimized to be 100nm, 5μm, and 100nm respectively. Figure 2(a) shows the simulated reflection magnitude and phase around 2π range with 24° step by modifying the geometrical parameters (l and w) of fifteen unit cells. Figure 2(b) and(c) depict the discretized phase distributions with fifteen levels of off-axis and on-axis metalenses, indicating that the closer to the focusing position, the larger the range between adjacent 2π.

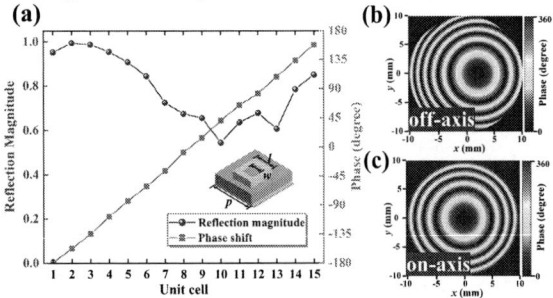

Figure 2: (a)Simulated reflection magnitude and phase around 2π range by optimizing geometrical parameters of fifteen unit cells. Discretized phase distributions of (b)off-axis and (c)on-axis metalenses.

The Fresnel-Kirchhoff diffraction theory was considered to calculate the light field intensity distribution of the THz beam reflected by the metalens in the focal plane. Each unit cell on the metasurface is regarded as a point source radiating spherical waves outward, and the intensity of any position on the focal plane can be obtained by superposing spherical waves radiated from all unit cells. Therefore, the field intensity distribution can be calculated by the following equation:

$$\tilde{E}(x_1,y_1)= \iint_{\Sigma} \frac{A(x_0,y_0)\exp(jk\varphi(x_0,y_0))}{i\lambda} K(\theta) \frac{\exp(jkr)}{r} dx_0 dy_0 \quad (2)$$

Where $A(x_0, y_0)$ and $\varphi(x_0, y_0)$ represent the amplitude and phase at the position (x_0, y_0) of metalens respectively. $K(\theta)=\frac{1+\cos(\vec{n}, \vec{r})}{2}$ is the inclination factor, and $r=\sqrt{(x_1-x_0)^2+(y_1-y_0)^2+f_0^2}$. Figure 3 shows the theoretically calculated field intensity distributions of off-axis and on-axis metalenses in the focal plane (a, b) and X-Z plane (d, e), and their normalized intensity distributions extracted along the center of the spots (c, f). It is worth noting that the theoretical full-wave half-maximum (FWHM) of foci for on-axis and off-axis metalenses are 649μm and 608μm, respectively. The off-axis value (dx) is 1.98mm due to the deviation between the simulated and ideal phases, which also proved the feasibility of the design method. The DOF of on-axis and off-axis metalenses were calculated to be 21.3mm and 17.2mm respectively, as shown in Figure 3(f).

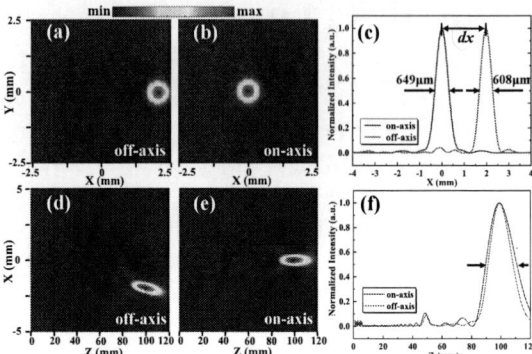

Figure 3: Calculated field intensity distributions of off-axis and on-axis metalenses in the X-Y focal plane (a, b) and X-Z plane (d, e) and their normalized intensity curves extracted along the center of the spots (c, f).

The fabrication process started with a 20/100 nm Cr/Au film sputtered on a silicon wafer, followed by a PECVD deposition of 5μm SiO$_2$. Then another 20/100 nm Cr/Au film was sputtered and finally, the metal was patterned by dry etching combined with photolithography. Figure 4 shows the optical images and SEM photograph of the fabricated off-axis metalens and on-axis metalens, in which the enlarged inset shows that the fabricated Au resonators have a relatively uniform morphology.

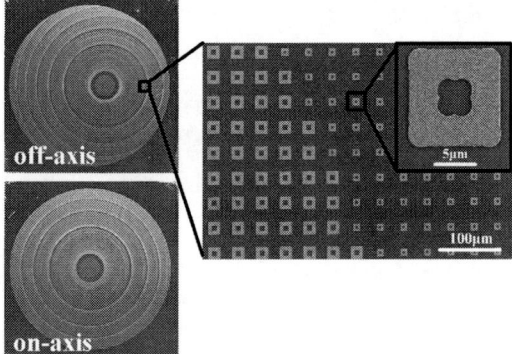

Figure 4: Optical images and SEM photograph of the fabricated off-axis metalens and on-axis metalens.

RESULTS AND DISCUSSION

The performance-verifying system was established to verify the focusing performance of the fabricated metalenses, as revealed in Figure 5. First, the THz wave with a frequency of 2.52THz was emitted by the THz laser, then was collimated and expanded by two parabolic mirrors (PM) with focal lengths of 50.8mm and 101.6mm successively. Eventually, the THz wave was incident on the metalens with a power of about 0.2mW, and the field intensity of the reflected beam was captured by a THz camera. Here, the aperture diaphragm (AD) was applied to restrict the size and position of the beam irradiated on the metalens.

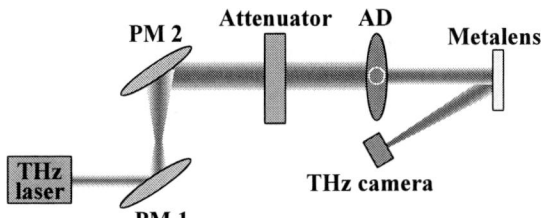

Figure 5: Experimental system established for verifying the focusing performance of metalenses.

To demonstrate the focusing performance of the fabricated off-axis metalens, the field intensity of off-axis metalens were captured with the performance-verifying system. During the experiment, we adjusted the angle of the THz camera's detection surface and moved the camera back and forth in different directions to ensure the recorded Z-axis was exactly at the actual reflection direction and the accuracy of distance along the Z-axis. Figure 6(a) shows the field intensity pictures of off-axis metalens near focus and their normalized intensity distributions extracted along the center at different Z-axis positions. Due to the influence of high THz laser power, there may be a saturated field intensity near the center of the spot, corresponding to a short flat curve at some certain position.

Figure 6(b) shows the experimental FWHMs of the spots and their Gaussian fitting curve, demonstrating that the focal length of off-axis metalens and FWHM of the focus are 97mm and 737μm respectively, which is slightly larger than the theoretical value due to the sensitivity of the camera to infrared radiation existing in the environment. The concentrated FWHMs indicate that the fabricated off-axis metalens had an excellent focusing within a range of 16mm along Z-direction, which is close to the theoretical value, showing the potential of large DOF. The focusing efficiency was measured to be 62.3% by metering the power of the incident THz beam and that reflected by off-axis metalens at the focus.

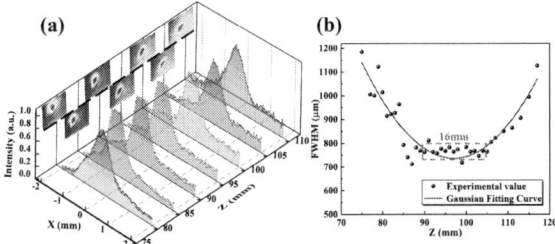

Figure 6: (a)Field intensity pictures of off-axis metalens near focus and normalized intensity distributions at different Z positions. (b)Experimental FWHMs of the spots and their Gaussian fitting curve, indicating a relative concentrated focusing within a range of 16mm along Z-direction.

Once the off-axis focus was captured by the camera, the off-axis metalens was replaced by the on-axis metalens under the circumstance that the positions of other optical components remain unchanged. In order to further investigate the function of off-axis focusing by comparing the distance between two spots, the reflection spot of on-axis metalens was pictured. Limited by the imaging area of the THz camera (4mm×4mm), we pre-adjusted the optical path so that the two foci can be completely imaged on two sides of the picture. The white dashed lines represent the position of X=0, as shown in Figure 7(a), which reveals the experimental off-axis value (Δx) is 2.04mm, and the off-axis angle was calculated to be 1.20°, which is close to the designed value of 1.15°, verifying the off-axis focusing ability.

The size and center position of the THz beam incident on off-axis metalens were determined by the diameter and the position of the aperture diaphragm in the direction perpendicular to the propagation path, which is also parallel to the X-direction of the metalens. Figure 7(b) and (c) show the intensity distributions of reflection focusing spots recorded when THz wave incident on the off-axis metalens at the position of X=0, ±2mm, and ±4mm, indicating that the off-axis focusing at the same spatial position can be achieved when slightly shifting incident wave. However, the intensity gets weaker when the size decreased incident beam moved along the X-direction, owing to the partially covered 2π phase range, especially near the off-axis position.

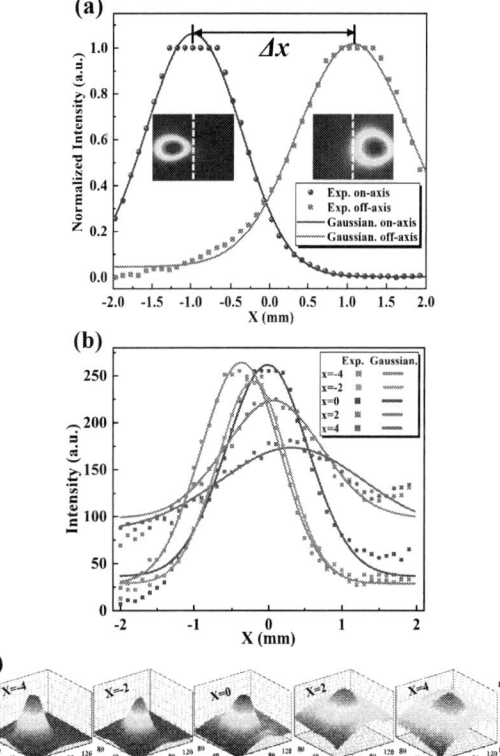

Figure 7: (a)Experimental normalized intensity of on-axis and off-axis metalenses obtained by successively placing two metalenses at the same position in the performance-verifying system. The white dashed line represents the position of X=0. (b)Intensity distributions extracted along the center of the spot and (c)3D intensity distributions of focusing spots recorded at the circumstances of THz wave incident on the off-axis metalens at the position of X=0, +2mm, and +4mm

CONCLUSIONS

In summary, we have designed and fabricated off-axis and on-axis metalenses with fifteen polarization-insensitive metal resonator-dielectric-metal structures which were arranged according to the derived formula. The phase response of the adopted unit cells to the 2.52THz wave can achieve 2π range coverage with a 24° step. The focusing performance of the fabricated metalens was verified with an established THz optical system. The experimental results demonstrated that an off-axis focusing with an angle of 1.20°, which is very close to the designed value, was obtained by the off-axis metalens. The depth of focus is at least 16mm and the focusing efficiency is 62.3%. This research provides a reference to design off-axis metalens and may find promising applications in the THz imaging system.

ACKNOWLEDGEMENTS

This research was funded by the National Key Research and Development Program of China (Grant No. 2020YFB200903) and the National Natural Science Foundation of China (Grant No. 61935001).

REFERENCES

[1] Y. Shen, Y. Yin, B. Li, C. Zhao, and G. Li, "Detection of impurities in wheat using terahertz spectral imaging and convolutional neural networks", *Computers and Electronics in Agriculture,* vol. 181, pp. 105931, 2021.

[2] A. Bandyopadhyay and A. Sengupta, "A review of the concept, applications and implementation issues of terahertz spectral imaging technique", *IETE Technical Review,* vol. 39, pp. 471-489, 2022.

[3] Y. H. Tao, A. J. Fitzgerald, and V. P. Wallace, "Non-contact, non-destructive testing in various industrial sectors with terahertz technology", *Sensors,* vol. 20, pp. 712, 2020.

[4] E. Pickwell-Macpherson, S. Huang, K. W. C. Kan, Y. Sun, and Y. T. Zhang, "Recent developments of terahertz technology in biomedicine", *Journal of Innovative Optical Health Sciences,* vol. 1, pp. 29-44, 2008.

[5] F. Sizov, "Infrared and terahertz in biomedicine", *Semiconductor physics, quantum electronics & optoelectronics,* pp. 273-283, 2017.

[6] R. Xu, S. Gao, B. S. Izquierdo, C. Gu, P. Reynaert, A. Standaert, G. J. Gibbons, W. Bösch, M. E. Gadringer, and D. Li, "A review of broadband low-cost and high-gain low-terahertz antennas for wireless communications applications", *Ieee Access,* vol. 8, pp. 57615-57629, 2020.

[7] C. Han and Y. Chen, "Propagation modeling for wireless communications in the terahertz band", *IEEE Communications Magazine,* vol. 56, pp. 96-101, 2018.

[8] D. Jia, Y. Tian, W. Ma, X. Gong, J. Yu, G. Zhao, and X. Yu, "Transmissive terahertz metalens with full phase control based on a dielectric metasurface", *Optics letters,* vol. 42, pp. 4494-4497, 2017.

[9] M. Liu, Q. Yang, A. A. Rifat, V. Raj, A. Komar, J. Han, M. Rahmani, H. T. Hattori, D. Neshev, and D. A. Powell, "Deeply subwavelength metasurface resonators for terahertz wavefront manipulation", *Advanced Optical Materials,* vol. 7, pp. 1900736, 2019.

[10] X. Jiang, H. Chen, Z. Li, H. Yuan, L. Cao, Z. Luo, K. Zhang, Z. Zhang, Z. Wen, and L.-g. Zhu, "All-dielectric metalens for terahertz wave imaging", *Optics Express,* vol. 26, pp. 14132-14142, 2018.

[11] Q. Cheng, M. Ma, D. Yu, Z. Shen, J. Xie, J. Wang, N. Xu, H. Guo, W. Hu, and S. Wang, "Broadband achromatic metalens in terahertz regime", *Science Bulletin,* vol. 64, pp. 1525-1531, 2019.

[12] R. Jia, Y. Gao, Q. Xu, X. Feng, Q. Wang, J. Gu, Z. Tian, C. Ouyang, J. Han, and W. Zhang, "Achromatic dielectric metasurface with linear phase gradient in the terahertz domain", *Advanced Optical Materials,* vol. 9, pp. 2001403, 2021.

[13] D. Liu, T. Lv, G. Dong, C. Liu, Q. Liu, Z. Zhu, Y. Li, C. Guan, and J. Shi, "Broadband and wide angle quarter-wave plate based on single-layered anisotropic terahertz metasurface", *Optics Communications,* vol. 483, pp. 126629, 2021.

[14] D. Wang, Y. Gu, Y. Gong, C.-W. Qiu, and M. Hong, "An ultrathin terahertz quarter-wave plate using planar babinet-inverted metasurface", *Optics express,* vol. 23, pp. 11114-11122, 2015.

[15] J.-S. Li and L.-N. Zhang, "Simple terahertz vortex beam generator based on reflective metasurfaces", *Optics Express,* vol. 28, pp. 36403-36412, 2020.

[16] X. Zang, B. Yao, L. Chen, J. Xie, X. Guo, A. V. Balakin, A. P. Shkurinov, and S. Zhuang, "Metasurfaces for manipulating terahertz waves", *Light: Advanced Manufacturing,* vol. 2, pp. 148-172, 2021.

[17] J. Ding, S. An, B. Zheng, and H. Zhang, "Multiwavelength metasurfaces based on single-layer dual-wavelength meta-Atoms: toward complete phase and amplitude modulations at two wavelengths", *Advanced Optical Materials,* vol. 5, pp. 1700079, 2017.

[18] T. Wang, R. Xie, S. Zhu, J. Gao, M. Xin, S. An, B. Zheng, H. Li, Y. Lin, and H. Zhang, "Dual-band high efficiency terahertz meta-devices based on reflective geometric metasurfaces", *IEEE Access,* vol. 7, pp. 58131-58138, 2019.

[19] M. Khorasaninejad, W.-T. Chen, J. Oh, and F. Capasso, "Super-dispersive off-axis meta-lenses for compact high resolution spectroscopy", *Nano letters,* vol. 16, pp. 3732-3737, 2016.

[20] M. Gong, Z. Zhang, P. Hu, H. Xiang, Q. Jiang, and D. Han, "Asymmetric off-axis focusing THz metasurface for circularly polarized light waves", *Results in Physics,* vol. 29, pp. 104815, 2021.

CONTACT

*Xiaomei Yu, tel: +86-10-62766592; yuxm@pku.edu.cn

TOWARDS A BETTER CMOS-MEMS RESOSWITCH USING ELECTROLESS PLATING FOR CONTACT ENGINEERING

Ting-Jui Liou, Chun-Pu Tsai, Ting-Yi Chen, and Wei-Chang Li
Institute of Applied Mechanics, National Taiwan University, Taipei, Taiwan

ABSTRACT

This work demonstrates a performance enhancement technique for both reliability and sensitivity using electroless plated contact for CMOS-MEMS based resoswitches with a factor of ~7× increase in cycle counts, from ~3×10^6 to 2.2×10^7 cycles, and shows integrity of the output voltage waveform—the initial and last cycles of output signal are almost the same, indicating an enhancement in the lifetime and signal stability of resoswitches. In addition, electroless nickel plating achieves the sub-micro gap spacing 0.1 μm and 3× decrease in the signal level, from 900 mV$_{pp}$ to 300 mV$_{pp}$ required for launching hot switching. In particular, applying electroless nickel plating on released CMOS-MEMS folded-beam comb-driven resoswitches to replace the two only contact materials of aluminum and tungsten available in standard CMOS yields not only a much reliable contact but also a reduced gap spacing in the same time.

KEYWORDS

Electroless nickel plating, contact materials, CMOS-MEMS resoswitches, reliability enhancement.

INTRODUCTION

Recent developments in MEMS resoswitches have shown a wide range of application [1-4]. Compared with conventional RF switches, resoswitches have multiple advantages such as ultra-lower stand-by power and much reduced parasitic capacitance. To push the resoswitch technology towards commercial applications, the reliability issue is one of the major issues needed to be solved. The reliability of resoswitch technology has been shown to yield more than 173 trillion cycles in a poly-silicon device [5]. However, unlike other process platforms [5-7], for which the process design can be tailored to incorporate materials for better contact, CMOS-MEMS resoswitches are hindered by limited available contacts materials. Take a 0.35-μm-CMOS for example, aluminum and tungsten are the two only available materials for contact. Aluminum has a low melting point and low oxidation resistance—undesirable as a contact material due to micro-welding and oxidation easily formed during contact [8]. Tungsten, on the other hand, although with a much higher melting point, requires a large force to deform and form better contact [8]. Plus, the tungsten layer has a fixed small size as defined by the design rule for VIA, with which the resulting tiny interfaces between the aluminum metal layers might become bottleneck handling the hot switching induced electrical current.

Inspired by [9], this work uses maskless, electroless nickel plating for released CMOS-MEMS resoswitches (*cf.* Fig. 1). There are many advantages to adopt electroless plating. For example, because of the high selectivity, electroless is a maskless process, plated nickel only goes on metal but not on the substrate—otherwise, the I/O signal

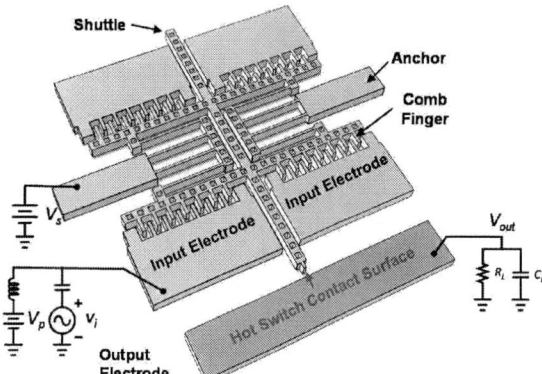

Fig. 1: Schematic of a comb-driven resoswitch and lifetime measurement set-up, where the device is biased and driven by V_P and v_i on the input electrode. The V_S on the structure is used to provide hot switching signal, and v_{out} is measured at output.

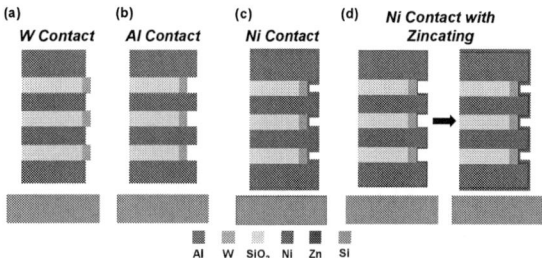

Fig. 2: Cross-sectional of four contact materials in this comparison. (a) W contact and (b) Al contact was based on CMOS-MEMS layout design. (c) Ni only (d) Ni plating with zincating.

and bias would be all shorted. In addition, compare to electroplating, electroless plating has a better coating uniformity [10]. Also, the thickness of plating film can be controlled by the plating time.

DEVICE STRUCTURE AND OPERATION

Fig. 1 presents the schematic of CMOS-MEMS-based comb-driven folded-beam resoswitches. The structure of the resoswitch and electrodes consist of the layer stacking from METAL1 to METAL4 of the CMOS process. The input electrodes are biased with a dc polarization V_p and driven with the input ac signal v_i, and the comb-driven structure is supplied with V_s for the hot switching operation.

FABRICATION PROCESS

Fig. 2 represents four contact materials compared in this work, namely the inherent (a) W and (b) Al, and electroless plated nickel (c) w/o and (d) w/ zincating. Fig. 3 presents fabrication steps of electroless nickel w/ zincating, the feature of this study. In order to enhance the adhesion between aluminum and nickel [11], the samples

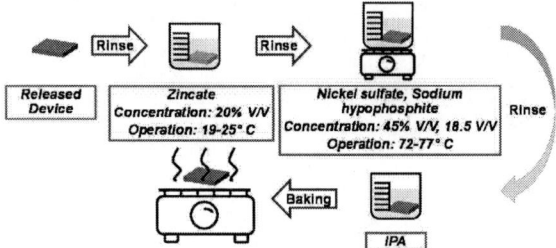

Fig. 3: *Fabrication process flow of nickel plating with zincating and bath condition.*

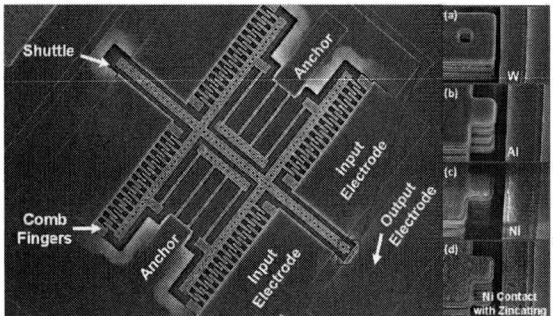

Fig. 4: *SEM of fabricated comb-driven resoswitches with different contact surfaces. (a) W contact geometry is different because of design rule of CMOS-MEMS platform (b) Al contact (c) Ni contact (d) Ni plating with zincating.*

were dipped in a 20 % zincating bath at 21 °C for 10-15 seconds, the extremely thin Zn film is deposited. In addition, zincating process activates the surface of plating seed layer and increases the reaction rate for the following plating process. It will be found later that zincating film enhances adhesion of plated Ni, which is key for reliability improvement of resoswitches. The plating solution contain 15 wt.% nickel and 6 wt.% phosphorus, the addition of phosphorus increases the hardness of nickel film [10], providing a stronger contact during continuous switching. The samples are dipped in the plating solution for 1 min at 75 °C, which is a critical parameter that prevents the resoswitches from short circuit.

EXPERIMENTAL RESULTS

The resoswitches are fabricated in a standard 0.35-μm CMOS-MEMS process platform. The release process utilizes SiO_2 etchant (Silox Vapox Ш, Transene Company, Inc.). Fig. 4 shows the SEM photos of the released resoswitches with different post processes and contact surfaces.

Measurement Set-Up

Fig. 5 illustrates the schematic of measurement set-up for the hot-switching lifetime, where V_p and the bias voltages are applied on input electrode, and the ac signal v_i is driven on the resonance frequency with OOK type of a 4-ms burst with a period of 8 ms. The testing samples are placed in a custom-built vacuum chamber that switching operating below 7.5×10^{-5} Torr. To fairly compare different contact cases of the switches, the bias V_p and driving voltage v_i are kept at the same. A data acquisition tool is connected to the output port and record the output waveform.

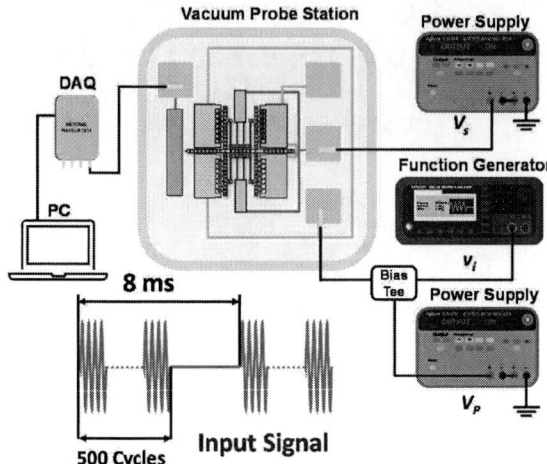

Fig. 5: *Illustration of the measurement set-up for hot switching lifetime measurement, the input signal v_i with resonance frequency OOK signal of a 4-ms burst with a period of 8 ms at 25 °C and 7.5×10^{-5} Torr, the output signal is recorded by a data acquisition tool.*

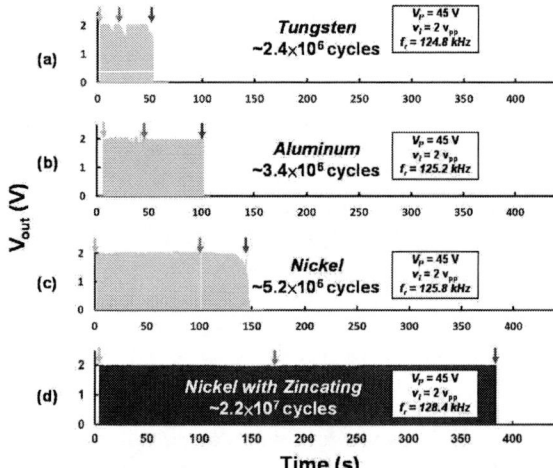

Fig. 6: *Lifetime measurement of the comb-driven resoswitches with different materials contact of (a) W (b) Al (c) Ni (d) Ni plating with zincating.*

Reliability Improvement

Fig. 6 plots the hot-switching voltage output for the four contact materials. To illustrate the details of output waveforms, Fig. 7 plots the output waveform comparison in 0.1 s time interval with three different periods—initial, middle, and last cycles towards to failure. As shown in Fig. 7 (a) and (b), the contact materials aluminum and tungsten based on the COMS-MEMS process exhibit the worst reliability—total cycles of 2.4×10^6 cycles and 3.4×10^6 cycles, respectively. In addition of total lifetime, the integrity of output waveform amplitude decreases during continuous contacts, and the contact resistance increases gradually. Compared to aluminum and tungsten, plated nickel shows a better reliability extended to 5.2×10^6 cycle counts. The hot switching pulse is missing after 1.7 million cycles for 3 seconds, but the signal does not degrade after recovery. Finally, nickel plating with zincating shows the

TABLE 1: LIFETIME COMPARISON FOR DIFFERENT MATERIAL CONTACT

	[12]	[13]	[14]	[15]	This Work
Type	RF Switches (Hot-Switching)	Resoswitches (Hot-Switching)	Resoswitches (Hot-Switching)	RF Switches (Hot-Switching)	Resoswitches (Hot-Switching)
Contact	Au-NP-CNT to Au	Poly-Si to Poly-Si	W to W	NCG (Nanocrystalline graphite)	Ni to Ni
Environment	Air	Vacuum	Vacuum	Air	Vacuum
Lifetime (cycle)	5×10^6	1.76×10^{14}	8×10^6	5×10^6	6.2×10^7

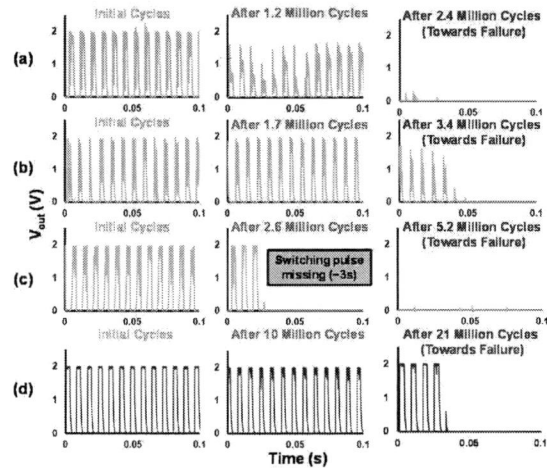

Fig. 7: Zoom-in to the initial, middle, and last periods for different.

best reliability. The total cycle counts extend to 2.2×10^7 cycles, revealing a ~9× improvement compared to the original Al and W and ~4× compared to Ni without zincating. The lifetime extended with zincating shows the adhesion improvement contributes greatly to reliability. In addition, plated Ni with zincating reveals the best output signal stability—the output signal does not decay after 21 million cycles.

Sensitivity Improvement

In addition to reliability, receiving sensitivity is highly dependent on the gap spacing. The electroless plating also serves as a gap narrowing technique. By controlling the plating time and temperature, the gap spacing is successfully reduced from 0.6 μm to 0.1 μm, which in turn decreases the minimum threshold voltage from 900 mV_{pp} to 300 mV_{pp} for hot switching.

CONCLUSION

A CMOS-MEMS folded-beam comb-driven resoswitches using electroless nickel plating with zincating to replace contact surface in a 0.35-μm CMOS platform has been successfully improved the reliability of resoswitches, extend the lifetime from ~3×10^6 to 2.2×10^7 cycle counts. In addition, plated Ni with zincating shows a better integrity in the output voltage waveform—the initial and last cycles are almost the same. Electroless Ni plating can also realize the gap narrowing from 0.6 μm to 0.1 μm, improve the sensitivity of resoswitches, 3× decrease in the signal level, from 900 mV_{pp} to 300 mV_{pp} required for lunching hot switching. TABLE 1 summaries the performances of this work and other MEMS switches. While performing promising results in terms of reliability and sensitivity, the electroless plating technique requires further process optimization and material characterization.

Work towards this continues.

ACKNOWLEDGEMENTS

This research was funded by the National Science and Technology Council, Taiwan (MOST-109-2628-E-002-004-MY3). The chip fabrication was supported by the Taiwan Semiconductor Research Institute (TSRI) and Taiwan Semiconductor Manufacturing Company (TSMC), Hsinchu, Taiwan. The solution of electroless process is supplied by LEADWAVE Industries Company, Taoyuan, Taiwan.

REFERENCES

[1] W.-C. Li, Y. Jiang, R. A. Schneider, H. G. Barrow, L. Lin and C. T.-C. Nguyen, "Polysilicon-filled carbon nanotube grass structural material for micromechanical resonators," in *2011 IEEE 24th International Conference on Micro Electro Mechanical Systems*, 2011, pp. 477-48.

[2] V. B. Sawant, S. S. Mohite and L. N. Cheulkar, "Comprehensive contact material selection approach for RF MEMS," in *Materials Today: Proceedings*, 2018 pp. 10704–10711.

[3] R. A. Moghadam, H. Saffari and J. Koohsorkhi, "Ni–P electroless on nonconductive substrates as metal deposition process for MEMS fabrication," in *Microsystem Technologies 27*, 2021, pp. 79-86.

[4] P. Ron, Properties and applications of electroless nickel, Toronto: Nickel Development Institute, 1997.

[5] J. Chen, G. Yu, B. Hu, Z. Liu, L. Ye and Z. Wang, "A zinc transition layer in electroless nickel plating," in *Surface and Coatings Technology*, vol. 201, no. 3–4, pp. 686–690, Oct. 2006.

[6] E. Jo, Y.-B. Lee, Y. Jung, S.-B. Kim, Y. Kang, M.-H. Seo, J.-B. Yoon and J. Kim, "Integration of Gold Nanoparticle–Carbon Nanotube Composite for Enhanced Contact Lifetime of Microelectromechanical Switches with Very Low Contact Resistance," *ACS Applied Materials & Interfaces*, 2021.

[7] Y. Lin, R. Liu, W.-C. Li, M. Akgul and C. T.-C. Nguyen, "A Micromechanical Resonant Charge Pump," in *the 17th Int. Conf. on Solid-State Sensors, Actuators, & Microsystems (TRANSDUCERS'13)*, Barcelona, Spain, Jun. 16-20, 2013, pp. 1727-1730.

[8] C.-P. Tsai and W. C. Li, "CMOS-MEMS Vibro-Impact Devices and Applications," in *Frontiers in Mechanical Engineering*, vol. 8, Jun. 2022.

[9] R. Sunil, J. D. Reynolds, T. Y. Ling, M. S. Shamsudin, S. H. Pu, H. M. Chong and D.

Pamunuwa, "Nano-crystalline graphite for reliability improvement in MEM relay contacts," in *Carbon*, vol. 133, pp. 193–199, Jul. 2018.

[10] R. Liu, J. N. Nilchi and C. T.-C. Nguyen, "RF-Powered Micromechanical Clock Generator," in *2016 IEEE Int. Frequency Control Symposium (IFCS'16)*, New Orleans, LA, USA, May 9-12, 2016, pp. 1-6.

[11] Y. Lin, R. Liu, W.-C. Li and C. T.-C. Nguyen, "Polycide Contact Interface to Suprress Squegging in Micromechanical Resowitches," in *IEEE 27th International Conference* , San Francisco, CA, Jan. 26-30, 2014, pp. 1273-1276.

[12] C.-P. Tsai, Y.-Y. Liao and W.-C. Li, "A 125-KHZ CMOS-MEMS Resoswitch Embedded Zero Quiescent Power OOK/FSK Receiver," in *2020 IEEE 33rd International Conference on Micro Electro Mechanical Systems (MEMS)*, Vancouver, BC, Canada, Jan. 2020, pp. 106–109.

[13] S.-C. Lu, C.-P. Tsai, Y.-C. Huang, W.-R. Du and W.-C. Li, "Surface Condition Influence on the Nonlinear Response of MEMS CC-Beam Resoswitches," in *IEEE Electron Device Letters*, vol. 39, no. 10, pp. 1600-1603, Oct. 2018.

[14] R. Liu, J. N. Nilchi and C. T.-C. Nguyen, "CW-powered squegging micromechanical clock generator," in *2017 IEEE 30th International Conference on Micro Electro Mechanical Systems (MEMS)*, 2017, pp. 905-908.

[15] C.-P. Tsai and W.-C. Li, "A Micromechanical Frequency Controlled Pulse Density Modulator," in *2022 IEEE 35th International Conference on Micro Electro Mechanical Systems Conference (MEMS)*, 2022, pp. 204-207.

[16] J. C. Salvia, R. Melamud, S. A. Chandokar, S. F. Lord and T. W. Kenny, "Real-Time Temperature Compensation of MEMS Oscillators Using an Integrated Micro-Oven and a Phase-Locked Loop," *Journal of Microelectromechanical Systems,* vol. 19, no. 1, pp. 192-201, Feb. 2010.

[17] Y.-C. Liu, M.-H. Tsai, W.-C. Chen, M.-H. Li, S.-S. Li and W. Fang, "Temperature-Compensated CMOS-MEMS Oxide Resonators," *Journal of Microelectromechanical Systems,* vol. 22, no. 5, pp. 1054-1065, Oct. 2013.

[18] W.-T. Hsu and C. T.-C. Nguyen , "Stiffness-Compensated Temperature-Insensitive Micromechanical Resonators," in *IEEE 15th International Conference on Micro Electro Mechanical Systems (MEMS'02)*, Las Vegas, NV, USA, 2002, pp. 731-734.

CONTACT
*W.-C. Li, Tel: +886-2-3366-5636; wcli@iam.ntu.edu.tw

AUTHOR INDEX

A

Abdolvand, Reza 586, 1183
Abdullah, Amjed 433
Abraham, Ebinesh R. 1049
Adagolodjo, Yinoussa 452
Agarwal, Ajay 665
Agarwal, Bhawana 1049
Agluschewitsch, Vladislav 1037
Aida, Yasuhiro 941, 957
Ainla, Alar 744
Al-Amidie, Muthana Last 433
Alasatri, Suresh 471
Alessandri, Anna 716
Alkorjia, Omar 433
Almasri, Mahmoud 433
Almeida, Sergio 483
Altmann, Robert K. 153
Amaya, Satoshi 57
Andreasen, Sune Zoëga 437
Andrianov, Nikolai 402
Ansari, Azadeh 1186
Anzinger, Sebastian 973
Aoyagi, Seiji 787
Arai, Fumihito 57, 99
Arakawa, Ryuichi 135
Archdeacon, Michael T. 235
Arnow, Hannah 728
Arya, Dhairya Singh 347, 570
Asai, Makoto 319
Asplund, Maria 421
Attanasio, Alaina G. 149
Azadi Keari, Shirin 767
Azadmehr, Mehdi 728

B

Bae, Kyubin 724
Bae, Yuna 651
Baglioni, Gabriele 131
Bahr, Andreas 370

Bahreyni, Behraad 779
Bainschab, Markus 1107, 1123, 1127, 1131
Baker, Colin 554
Banabathi, Naik T. 811
Bancora, Andrea 149
Banerjee, Amit 515
Bang, Dang Duong 437
Bao, Feihong 631
Barniol, Núria 933, 937, 1041
Basavanna, Abhiraj 973
Baumgart, Marcus 1107, 1123
Beigh, Faizan 343
Beigh, Nadeem Tariq 343
Berenschot, Erwin J.W. 639
Bhatta, Trilochan 217
Bhave, Sunil A. 149
Boehler, Christian 421
Borca-Tasciuc, Diana-Andra 728
Borrise, Xavier 1041
Bota, Sebastià 795
Bourouina, Tarik 1119
Bourrier, David 643
Bowen, Chris R. 697
Braun, Stefan 594

C

Cai, Bingyang 837
Cai, Junxiang 413
Calleja, Montserrat 153, 444, 1056
Cano, Álvaro 444
Cao, Entian 1037
Cao, Jiawei 384, 405
Cao, Yuan 1009
Cao, Zhen 103
Carminati, Marta 716
Carminati, Roberto 716
Carranza, Gabriel E. 259
Casilli, Nicolas 511
Cassella, Cristian 161, 511
Cattan, Eric 452
Ceyssens, Frederik 901
Chand, Rakesh 1198
Chang, Chin-Yu 953
Chang, Honglong 705, 740
Chang, Shu-Wei 123, 127

Chao, Yueh-Jung 558
Charlot, Samuel 643
Chau, Kevin 578, 748
Chau, Quy 961
Chen, Changnan 49, 881
Chen, Chihchen 479
Chen, Chun-Ming 135, 957
Chen, Guidong 229
Chen, Hao 157
Chen, Hung-Yu 526
Chen, Jianglong 189
Chen, Jianlin 865, 873
Chen, Lang 243
Chen, Liangqian 849
Chen, Minkan 115
Chen, Pai-Shan 1045, 1052
Chen, Pin-Chuan 1045, 1052
Chen, Qianhuang 606
Chen, Rongshun 562
Chen, Si-Han 720
Chen, Ting-Yi 185, 1151, 1155, 1206
Chen, Weiguo 491
Chen, Xi 606
Chen, Xiao-Wen 1021
Chen, Xing 1068
Chen, Yang 1175
Chen, Yen-Lin 791
Chen, Yi-Sin 479
Chen, Yih-Shurng 425
Chen, Ying 193, 255, 281, 969
Chen, Yu-Chen 685
Chen, Yung-Hsiang 1198
Chen, Yusa 814
Chen, Zhaohan 67, 247
Chen, Zhichao 157
Cheng, Hao-Chien 720
Cheng, Hsu-Hsiang 685
Cheng, Jiangong 157
Cheng, Ming-Ching 685
Cheng, Qian 33, 289
Cheng, Yu-Ting 425
Chidambara, Aaydha Chidambara 437
Chien, Jun-Chau 558
Chien, Tung-Lin 562, 566, 574
Cho, Incheol 799

Cho, Kyungsuk	811
Cho, Sung Kwon	1025
Cho, Wootaek	830
Choi, Kwang-Wook	993
Choi, Pan-Kyu	5
Chou, Tien	985, 1143
Choudhary, Chandrashekhar	235
Chu, Yen-Chang	562
Chung, Seokwhan	993
Cichon, Daniel	1099
Claar, Victor	409
Clausen, Niels	1099
Cleri, Fabrizio	475
Collard, Dominique	475
Colombo, Luca	161, 169
Comini, Elisabetta	251
Conédéra, Véronique	643
Costa, Antonio	795
Cui, Jian	833, 885
Cui, Tianhong	285

D

Da, Zhou	402
Dabas, Shaurya	1179
Dai, Ningxuan	53
Dalal, Mitul	853
Dao, Dzung Viet	366, 1064
Dao, Thang Duy	1139
Das, Samaresh	1033
Dau, Van Thanh	366, 1064
Davaji, Benyamin	161, 511
Decanini, Dominique	21
Degenfeld-Schonburg, Peter	877
Dehé, Alfons	973
Delgado, Rafael	153
Deshpande, Adwait	362
Deval, Piyush	448
Dinh, Thien Xuan	1064
Dinh, Toan	366, 1064
Dobson, Renwick C.J.	25
Dolamore, Fabian	25
Dostanic, Milica	374
Dou, Songtao	323
Dou, Wenkun	29
Du, Xu	99

Du, Zhiyuan 384
Duan, Mingzheng 1095
Duan, Xuexin 945
Duesberg, Georg S. 627
Dufour, Isabelle 643
Duriez, Christian 452

E

Ehrmann, Nils 594
Eovino, Ben 961
Erdale, Zeynep 433
Erkan, Derin 909
Esatu, Tsegereda K. 483
Escobar, Javier E. 1056
Esfahani, Kianoush S. 977

F

Fabian, Johannes 530
Fagnani, Andrea 989
Fahrbach, Michael 1037
Fang, Chi-Te 562
Fang, Mingdong 107
Fang, Weileun 123, 127, 143, 562, 566, 574, 685, 720, 791, 985, 1143
Fang, Xiaoli 177
Fang, Yanyan 841
Faraji, Mohammadmahdi 744
Fasel, Jens 370
Fastier-Woollel, Jarred 366
Favero, Ivan 153
Fee, Conan 25
Feng, Guo-Hua 239, 981
Feng, Philip X.-L. 511, 613, 897
Feng, Yongjian 763
Feng, Zhihong 1159
Fernandez-Cuesta, Irene 1041
Ferrarini, Paolo 716
Finkbeiner, Stefan 1
Fleury, Clement 1107, 1123, 1139
Floehr, Julia 1009
Forke, Roman 522
Frangi, Attilio A. 37, 989
Frigerio, Paolo 989
Fu, Sulei 1190
Fu, Yongqing 705, 740
Fujimoto, Kazuya 301, 309

Fujita, Takayuki 709
Furusawa, Gaku 1171

G

Galarza, Mathew 728
Gandotra, Rishabh 1001
Gang, Min-Ho 5
Gao, Chengchen 756
Gao, Le 837
Gao, Shupeng 606
Gao, Yanze 1147
Gao, Yunfei 115
García-López, Sergio 153, 444
Garg, Manu 347, 570
Garrill, Ashley 997
Gattere, Gabriele 37, 989
Gatti, Daniela A.L. 716
Ge, Yuqing 289
Gelaeschus, Anton 370
Geneiß, Volker 522
Gerbedoen, Jean Claude 475
Gesing, Andre L. 1072
Ghaderi, Erfan 779
Ghazinouri, Behrad 1163
Ghenna, Sofiane 452
Ghosh, Chayanjit 362
Ghosh, Sagnik 491
Gianola, Riccardo 716
Gidts, Michiel 901
Gil-Santos, Eduardo 153, 444, 1056
Giribaldi, Gabriel 169
Goh, Duan Jian 491
Gojdka, Björn 1099
Golabi, Mohsen 437
Golparvar, Ata 335
Gong, Tianjiao 602, 661
González, Pedro 744
Grondel, Sébastien 452
Grundmann, Annika 627
Gu, Chi 429
Gu, Jiebin 49
Gu, Yuandong Alex 413
Guerreiro, Sara 1107, 1123
Gund, Ved 519
Guo, Ruiqi 503

Guo, Wenlan	945
Guo, Xinge	495, 818, 1167
Guo, Yu	1190
Guo, Zhejun	635
Gurung, Lokesh	554

H

Hagiwara, Masaya	17
Haluska, Miroslav	925
Ham, Jimin	625, 651
Han, Jiangli	1068
Han, Xiaodong	763
Haneda, Kotaro	499
Hao, Yilong	756
Harouri, Abdelmounaim	21
Hartwig, Oliver	627
Hashimoto, Izumi	279
Hayakawa, Takeshi	17, 80
Hayashi, Tomohiro	1091
He, Kaixuan	845
He, Siyuan	1163
He, Wenzheng	285
He, Ying	433
He, Zongxing	205
Hedayat, Christian	522, 1099
Heeg, Jan	822
Hella, Mona M.	728
Heuken, Michael	627
Hierold, Christofer	925
Hiller, Karla	602
Hilleringmann, Ulrich	522
Hiraga, Hiroki	861
Hirayama, Takuto	760
Hirotani, Jun	515
Ho, Hsin-Ying	448
Ho, Yens	1198
Hoffmann, Martin	522
Hoffmann, Max	822
Hohe, Hans-Peter	1099
Holzmann, Dominik	1107, 1123
Hong, Sukjoon	625
Hong, Wen	405
Horsley, David A.	961
Hoshino, Ayuko	319
Hosokawa, Yoichiroh	1013

Hosseini-Pishrobat, Mehran 869
Hou, Hung-Yu 558
Hsiai, Tzung 53
Hsieh, I-Chieh 1155
Hsu, Tzu-Hsuan 1194, 1198
Hsu, Wei-Fan 901
Hu, Fangjing 841, 917
Hu, Senyong 814
Hu, Xiaoer 483
Hu, Yongbo 705
Hu, Yuqiang 157
Hu, Yushen 13, 259, 826
Hu, Zih-Song 127, 985
Hu, Zihsong 791
Hua, Chen 712
Huang, Chengjun 229, 582
Huang, Chung-Hao 981
Huang, Kai 1175
Huang, Lifeng 107
Huang, Linhai 1159
Huang, Nien-Tsu 1021
Huang, Siwei 107
Huang, Syuan-Rong 1021
Huang, Tony Jun 71
Huang, Wei-Chen 392
Huang, Wenjing 417
Huang, Xiaodong 736
Huang, Yi-Hsuan 487
Huang, Yu-Cheng 1143
Huang, Yuan 263
Huang, Yuanyuan 562, 566, 574, 791
Huang, Yuheng 771
Hulsey, Robert A. 433
Hunter, Gary L. 433
Huynh, Van Ngoc 437
Hwang, Gilgueng 21
Hwang, Jeonghyeon 327
Hyleme, George S. 433

I

Ichikawa, Keita 417
Ichiki, Masaaki 41
Ikeuchi, Shinsuke 941, 957
Ikezawa, Satoshi 1111
Irisa, Taiga 309

Ishida, Tadashi 1091
Ishihara, Daisuke 534
Islam, Sayemul 396
Isono, Yoshitada 701
Itai, Shun 305, 319, 456
Itawi, Ahmad 452
Ito, Motoki 460
Iwami, Kentaro 1111
Iwase, Eiji 221, 275, 339
Izhar 107, 1095
Izyumin, Oleg I. 961

J

Jackson, Nathan 467, 617
Jadhav, Shubham 519
Jafri, Ijaz 853
Jang, Gunyoung 625
Janssens, Yves L. 639
Jeon, Sungho 651
Jeong, Jinwon 693
Jeong, Seonghoon 217, 225, 732
Jeong, SeongHoon 197
Jia, Bin 736
Jia, Hao 119, 281, 1060
Jia, Yueyang 281
Jia, Yuyu 965
Jia, Zhili 598
Jiang, Chunpeng 355
Jiang, Jiaxin 235
Jiang, Wanqi 205
Jiao, Chenyin 189
Jiao, Rui 267, 752
Jin, Qiutong 173
Jin, Shengxiao 814, 1115
Jin, Yufeng 243
Jo, Byeongwook 378, 463
Jo, Eunhwan 45
Johnson, Isaac 315
Joshi, Khanjhan 347
Joshi, Sanjog V. 331

K

Kagawa, Gakuto 610
Kaisar, Tahmid 511
Kalisch, Holger 627

Kam, Hei **483**

Kan, Tetsuo **1135, 1171**

Kanamori, Yoshiaki **139**

Kanda, Kensuke **709**

Kanda, Natsuki **1171**

Kaneda, Yuki **1135**

Kaneko, Shingo **57, 99**

Kang, Dongil **91**

Kang, Dongwon **811**

Kang, Mingu **799, 803**

Kang, Minho **651**

Kang, Sookyung **811**

Kang, Tian **814, 1115**

Kant, Rashmikant **534**

Kao, Wei-Sin **448**

Karkhanis, Mohit U. **362**

Kataria, Satender **627**

Kato, Aoi **313**

Kaur, Navpreet **251**

Kaya, Onurcan **161, 511**

Khalil, Diaa **1119**

Khan, Muhammad Jehanzeb **602, 661**

Kim, Albert **396**

Kim, Beomjoon **21**

Kim, Dongjun **1029**

Kim, Dongkyun **197**

Kim, Eun S. **977**

Kim, Hanseup **362, 657**

Kim, Ilhwan **993**

Kim, Jongbaeg **45, 724**

Kim, Joonwon **327, 1080**

Kim, Jungkwun JK **396**

Kim, Jungwook **811**

Kim, Sangmok **327**

Kim, Su-Hyun **5**

Kim, Tae-Soo **5, 76**

Kim, Taejung **830**

Kim, Taesung **1029**

Kim, Yong-Jun **88**

Kippenberg, Tobias J. **149**

Kiya, Ryota **1013**

Kline, Mitchell H. **961**

Ko, Juhee **111**

Kobayashi, Takeshi **359, 400**

Koh, Yul **491**

Koike, Yuha 17
Kong, David 53
Kooijman, Lucas J. 639
Kosaka, Priscila M. 153, 444, 1056
Kößl, Bernhard 471
Kounadis, Diamantis 153, 444
Kozasa, Mitsuki 775
Kraft, Michael 331, 865, 901, 917
Kraiem, Ines 627
Kruszewski, Alexandre 452
Kuhl, Matthias 370
Kuhn, Harald 522
Kulsreshath, Mukesh K. 507
Kumar, Manjeet 665
Kumar, Sushil 347, 570
Kumokita, Yuki 787
Kunwar, Deepak 617
Kuo, Feng-Chih 1001
Kuo, Justin 9
Kurashina, Yuta 305, 319, 1084
Kushwaha, Shashwat 901
Kwak, Jong-Hyun 830

L

Lagadec, Chann 475
Lagosh, Anton 1107, 1123, 1127, 1131
Lai, Mei-Feng 1143
Lal, Amit 9, 491, 519
Langfelder, Giacomo 873, 989
Lathia, Rutvik 1049, 1088
Lazarova, Borka 716
Lazzari, Carla M. 716
Ledesma, Eyglis 933, 937
Lee, Bong Jae 111
Lee, Cathy 961
Lee, Cheng-Hsun 95
Lee, Chengkuo 495, 818, 1167
Lee, Dong-Weon 61
Lee, Gwo-Bin 479, 1001
Lee, Hojoon 45
Lee, Jae Ik 45
Lee, Jaehoon 977
Lee, Jaesung 613
Lee, Jeong Bong (JB) 693
Lee, Jongwan 1029

Lee, Joshua E.-Y.	491
Lee, Jungchul	111
Lee, Kichul	803
Lee, Mel S.	1001
Lee, Paul	440
Lee, Seung-Jun	76
Lee, Seyeon	811
Lee, So-Young	5, 76
Lee, Sueng Yoon	625, 651
Lee, Won Chul	625, 651
Lee, Wonhyung	1080
Lee, Wonjun	91
Lee, Ya-Chu	562
Lee, Yeonwoo	95
Lee, Yi-Kuen	107, 1095
Lee, Yong-Bok	5, 76
Lei, Tengteng	13
Lei, Wan-Lou	392
Leïchlé, Thierry	643
Lemme, Max C.	627
Lengert, Maren	822
Li, Bei	323
Li, Binghui	771
Li, Chengxin	917
Li, Defang	1147
Li, Dongxiao	1167
Li, Fangzheng	837
Li, Gang	763
Li, Hang	323
Li, Jiaqi	965
Li, Jinglun	728
Li, Jingzhen	763
Li, Junjian	849
Li, Lingyun	157
Li, Liye	814, 1115
Li, Mao	229, 582
Li, Meng	67, 205, 247, 297, 429
Li, Ming	119, 255
Li, Ming-Huang	143, 526, 1194, 1198
Li, Mingjie	893
Li, Na	1005, 1017
Li, Peng	881
Li, Qi	606
Li, Qingsong	849
Li, Sheng-Shian	143, 526, 905, 953

Li, Tao	235
Li, Teng	965
Li, Tingyu	381
Li, Wei	881, 969
Li, Wei-Chang	181, 185, 487, 1151, 1155, 1206
Li, Wenqi	542
Li, Xiaohui	209
Li, Xinxin	49, 119, 193, 255, 351, 881, 969, 1060
Li, Xinyu	193, 255
Li, Xuejiao	267
Li, Yu-Hsuan	574
Li, Yuankai	396
Li, Yung-Chen	1143
Li, Yunjia	740
Li, Zhuo	1147
Lian, Yujia	295
Liang, Jizhi	67, 247
Liang, Kai-Chih	720
Liaw, Shwu-Jen	558
Libaude, Guillaume	643
Lihachev, Grigory	149
Lim, Dohyun	625
Lim, Jaemook	625
Lim, Joowon	625, 651
Lin, Che-Hsin	448
Lin, Chen	921
Lin, Chia-Ying	235
Lin, Cong	1202
Lin, Dequan	578, 748
Lin, Guanzhou	814
Lin, Hung-Yu	720
Lin, Liwei	135, 503, 941, 957
Lin, Pen-Sheng	1143
Lin, Shihwei	791
Lin, Yang	285
Lin, You-An	1143
Lin, Zhong-Wei	905
Lin, Zude	929
Linh, Quyen Than	437
Liou, Ting-Jui	1206
Lisec, Thomas	1099
Liu, Chengze	673, 677
Liu, Chongbin	949
Liu, Chun-You	953
Liu, Guandong	598

Liu, Hai 965

Liu, Hanxiao 135, 941, 957

Liu, Huafeng 917

Liu, Huiliang 84

Liu, Jingquan 355, 384, 405, 635, 712, 783, 929

Liu, Kangfu 669, 969

Liu, Lin 440

Liu, Liying 285

Liu, MengWei 209

Liu, Min 881

Liu, Peisen 1190

Liu, Qihui 157

Liu, Ruichen 263

Liu, Shih-Chi 720

Liu, Song 965

Liu, Tsu-Jae King 483

Liu, Xiaojiang 893

Liu, Yi 1202

Liu, Yuncong 897

Liu, Yunfei 756

Liu, Yuxian 885

Liu, Zewen 84

Liu, Zhenhao 1068

Liu, Zhongyi 267, 752

Liu, Zong 259, 826

Liu, Zuheng 281

Llobet Sixto, Jordi 1041

Lo, Sung-Cheng 123, 685

Locquet, Jean-Pierre 901

Lotters, Joost C. 767

Lou, Liang 115, 413

Lu, Haojian 965

Lu, Yaoqing 669

Lukas, Sebastian 627

Luo, Bin 889

Luo, Ruiqi 598

Luo, Wenxin 893

Luo, Yuan 103

Lyu, Boming 705, 740

M

Ma, Wei 598

Ma, Xiao 323

Ma, Yibo 417

Macho, Matthias 538

Maenaka, Kazusuke 709

Magnet, Ingrid A.M. 471

Maharshi, Vikram 665

Mai, Luan Ngoc 366, 1064

Maillard, Damien 594

Majd, Yasaman 586

Mak, Daniel 25

Mallick, Dhiman 343, 1033

Malvar, Óscar 153, 444, 1056

Manrique Castro, Jorge 315, 586

Manrique Juarez, Dolores 643

Mansoorzare, Hakhamanesh 586, 1183

Mao, Haiyang 229, 582

Mao, Hongju 289, 1005, 1017

Maqsood, Waleed 402

Marschner, David E. 594

Martini, Irene 716

Maruyama, Hisataka 99

Mastrangeli, Massimo 374

Mastrangelo, Carlos H. 362

Masud, Mohammad Ayaz 2

Masuda, Akari 305

Masunishi, Kei 861

Mathieu, Fabrice 643

Matsudaira, Kenei 499

Matsunaga, Ryusuke 1171

Maynes, Jason 29

Mazenq, Laurent 643

Meffan, R. Claude 25

Meng, Ellis 388

Menges, Julian 25

Mercader, Anthony L. 1025

Miani, Theo 554

Miao, Jiahao 1202

Miao, Tongqiao 849

Miki, Norihisa 279, 293, 313

Mimura, Hisatoshi 65, 233, 279, 293, 313

Mingorance, Jesús 153, 444

Mirbakht, Sajjad 335

Mise, Nachi 775

Mita, Yoshio 21

Mitra, Bhaskar 665, 689

Miyazaki, Fumito 861

Mizumoto, Takahiro 515

Mizushima, Ayako 21

Mo, Dicheng 1179
Mo, Jiarui 72, 621
Modak, Chandantaru Dey 1049, 1088
Mohammadi, Ali 697
Molina, Juan 1056
Moll, Philipp 590
Møller, Jens Kjølseth 437
Moridi, Mohssen 402, 1107, 1123, 1139
Morimoto, Yuya 201, 381, 460, 463, 550
Morita, Tomohiro 378
Morita, Yuto 213
Mortada, Bassem 1119
Mortada, Mahdi 530
Motoi, Kentaro 463
Muhsin, Sura A. 433
Mukherjee, Dibyajyoti 343
Müller, Manuel 1041
Mummery, Christine L. 374
Murase, Hideaki 861
Murayama, Tomomi 1084
Mustafa, Muhammad 263

N

Nagpal, Satchit 1088
Nakajima, Hibiki 17
Nakamura, Nagi 221, 275
Nakane, Takuma 293
Nakano, Kyoka 80
Nakao, Kenji 233
Nakashima, Rihachiro 221
Napier, Cole 235
Nasri, Rukan 1041
Naval, Sourav 343
Ng, Eldwin 491
Nguyen, Clark T.-C. 173
Nguyen, Hong-Quan 366
Nguyen, Nam-Trung 1064
Nguyen, Thanh 366, 1064
Nguyen, Tran Minh Giao 452
Nguyen, Trieu 437
Nguyen, Tuan-Hung 366
Nguyen, Tuan-Khoa 366, 1064
Ni, Yue 582
Nicu, Liviu 643
Nie, Dezhi 705

Nie, Meng 771
Nie, Minghao 295, 378, 381
Nie, Ran 669
Niklaus, Frank 594
Ning, Bianca 440
Niroui, Farnaz 653
Nishimura, Akane 681
Nishiyori, Shusuke 701
Niu, Gaoqiang 826, 893, 913
Nock, Volker 25, 997
Noh, Seungbeom 657
Nomellini, Andrea 716
Nowbahari, Arian 728
Nunan, Kieran 853

O

Obispo, Meg 728
Oda, Haruka 295, 313
Odenthal, Marie C. 409
Ogawa, Etsuji 861
Ogawa, Jumpei 861
Oggioni, Laura 716
Ogishi, Kazuto 550
Okada, Hironao 41
Okamoto, Yuki 41
Onishi, Minato 534
Ono, Daiki 861
Ono, Takahito 1103
Onoe, Hiroaki 305, 319, 456, 1084
Osaki, Toshihisa 65, 233, 279, 293, 313, 550
Oshita, Masaaki 1135
Ou, Chuan-Hui 1103
Ou, Xin 177, 1175
Ouro-Koura, Habilou 728
Ouyang, Yifan 1115
Oyunbaatar, Nomin-Erdene 61

P

Pagliano, Simone 594
Pal, Sagnik 853
Pala, Sedat 135
Pamunuwa, Dinesh 507
Pan, Feng 1190
Pan, Pichao 49, 881
Pan, Xiaofang 107

Panagiotopoulos, Ilias 153, 444
Pang, Wei 673, 677, 945
Papanastasiou, Dimitris 153, 444
Park, Inkyu 799, 803
Park, Jae Yeong 197, 217, 225, 732
Park, Jiin 91
Park, Jongha 135
Park, Jungwon 651
Park, Jungyul 811, 1029
Park, Mingyo 1186
Park, Saeyoung 647
Park, Sung-Yong 95
Pathak, Pankaj 1033
Paul, Oliver 409, 421
Pedrini, Claudia 716
Peiner, Erwin 1037, 1147
Peng, Chih-Wei 392
Peng, Yande 135, 941, 957
Perelló-Roig, Rafel 795
Perez-Murano, Francesc 1041
Pezone, Roberto 131
Pfusterschmied, Georg 590
Phan, Hoang-Phuong 1064
Piazza, Gianluca 2
Pinto, Rui M.R. 744
Piot, Adrien 1107, 1123, 1139
Platz, Daniel 530, 1072
Plesse, Cedric 452
Pletka, Ryan J. 433
Pordeli, Yasser 639
Pradeep, Yelehanka Ramac R. 1198
Pradhan, Gagan Bahadur 217, 225, 732
Prechtl, Maximilian 627
Prelini, Carlo L. 716
Pribošek, Jaka 1123, 1127, 1131
Przybyla, Richard J. 961
Puff, Markus 1139

Q

Qamar, Afzaal 897
Qi, Jiali 267
Qi, Longheng 1095
Qi, Zhimei 1115
Qian, Hangyu 631
Qin, Mu 405

Qin, Nan **209, 807**
Qiu, Anping **857**
Qiu, Shihui **1005, 1017**

R

Rahafrooz, Amir **853**
Rais-Zadeh, Mina **897**
Rajaraman, Swaminathan **315**
Rana, S M Sohel **217, 732**
Rassay, Sushant **1179**
Ravi, Adarsh **9**
Recaman Payo, María **901**
Reddy, Bheema Sankar **1049, 1088**
Redolfi, Sebastian **1139**
Ren, Xinzhu **201**
Ren, Zhihao **818, 1167**
Reynaerts, Dominiek **901**
Reza, Md Selim **197**
Rezard, Quentin **475**
Rhee, Joowon **1080**
Riani, Manuel **37**
Riemensberger, Johann **149**
Rinaldi, Matteo **169**
Rintaro, Yoshinaga **1013**
Risquez, Sarah **1139**
Rizzini, Francesco **37**
Rodríguez-Tejedor, María **153, 444**
Roman, Cosmin **925**
Roshanghias, Ali **1139**
Ruan, Tao **384**
Ruther, Patrick **409, 421**
Ruz, José J. **153, 444, 1056**

S

Saadany, Bassam **1119**
Sabry, Yasser M. **1119**
Sadeghpour, Sina **331**
Sadrafshari, Shamin **697**
Sagi, H. Claude **235**
Sahara, Yoshiki **301**
Saito, Hiroki **1111**
Saito, Shiro **1135**
Sakamoto, Jinya **787**
Sakuma, Shinya **271**
Salvagnac, Ludovic **643**

Samm, Elisabeth	530
San Paulo, Álvaro	153, 444, 1056
Sanders, Remco G.P.	767
Sanghvi, Rohan	9
Sano, Tomohiko G.	221
Sanz-Jiménez, Adrián	153, 444
Sarro, Pasqualina M.	131, 374
Sasaki, Takashi	1107, 1123, 1127, 1131
Sato, Ayane	1091
Sato, Takashi	339
Satterthwaite, Peter F.	653
Sbarra, Samantha	153
Schaller, Falk	522
Schitter, Georg	538
Schiwietz, Daniel	877
Schmid, Ulrich	471, 530, 590, 1072
Schmitt, Philip	522
Schmoltner, Kerstin	1139
Schnakenberg, Uwe	1009
Schneider, Michael	471, 530
Schröder, Dominik	1099
Schroedter, Richard	538
Schween, Oliver	370
Segovia-Fernandez, Jeronimo	165
Segura, Jaume	795
Sekiguchi, Takuma	139
Sen, Prosenjit	1049, 1088
Sentre-Arribas, Elena	153
Seo, Dongwoo	1029
Seo, Min-Ho	647
Sergovia, Karen Last	433
Serrano, Diego Emilio	853
Seshia, Ashwin	554
Shadymov, Vladimir	149
Shaik, Faruk Azam	475
Shang, Jintang	542, 546, 889
Shang, Kuang-Ming	53
Shankar, Shreyas	72
Shao, Shuai	969
Shaporin, Alexey	522
Sharifuzzaman, Md	197, 732
Sharma, Jaibir	491
Sharma, Kirti	421
Sharma, Pallavi	467
Sharma, Sudeep	225

Shaw, Steven W.	613
Shelton, Stefon E.	961
Shen, Gencai	355
Shen, Haixu	53
Shi, Meng	229, 582
Shi, Qin	857
Shi, Runxiao	578
Shi, Zhongyu	263
Shih, Fuchi	566, 574, 791, 1143
Shimokawa, Fusao	213, 775
Shin, Heungjoo	830
Shin, Yoo-Kyum	647
Shiratori, Toshihiro	787
Shokrani, Alborz	697
Siddharth, Anat	149
Sikder, Urmita	483
Sim, Sangjun	724
Simeoni, Pietro	169
Singer, Julian A.	370
Singh, Jujhar	440
Singh, Pushpapraj	347, 570
Snigirev, Viacheslav	149
Soberats, Bartomeu	795
Sobreviela-Falces, Guillermo	554
Song, Cheng	1190
Song, Hye Su	197, 732
Song, Ziliang	405
Soong, Wei-Jen	479
Spector, Sarah O.	653
Stahl-Offergeld, Markus	1099
Steeneken, Peter G.	131
Stemme, Göran	594
Stöckel, Chris	522
Stramm, Till	370
Streque, Jeremy	1139
Su, Rongxuan	1190
Su, Zhaoxi	889
Suetsugu, Ryotaro	534
Sugano, Koji	701
Sugiura, Hirotaka	57, 99
Suh, Bokyung	693
Suh, Seungbeum	91
Sui, Fanping	503, 941
Sun, Baoyun	621
Sun, Chen	945

Sun, Fengpei 1159
Sun, Hongshun 814, 1115
Sun, Jianwen 84
Sun, Ke 351, 881
Sun, Litao 771
Sun, Liuyang 33, 67, 205, 247, 297, 429
Sun, Mingchao 673, 677
Sun, Xiaopeng 845
Sun, Yi 351
Sun, Yiling 997
Sun, Yu 29
Sun, Zhenhuan 965
Sung, Gi-Bong 993
Sung, Wei-Lun 562
Suzuki, Masato 787
Suzuki, Yukio 602, 661

T

Tabrizian, Roozbeh 1179
Tabuchi, Ayumu 301
Tacchini, Riccardo 716
Tai, Yu-Chong 53
Takahashi, Haruna 271
Takahashi, Hidetoshi 221, 499, 610, 760
Takahashi, Tomokazu 787
Takamori, Sho 65, 233, 279, 293, 313
Takao, Hidekuni 213, 775
Takao, Yoshinori 681, 681
Takasato, Minoru 301
Takei, Yusuke 359, 400
Takemura, Hiroki 515
Takeshita, Toshihiro 359, 400
Takeuchi, Shoji 65, 201, 233, 279, 293, 295, 313, 378, 381, 460, 463, 550
Tamayo, Javier 153, 444, 1056
Tan, Jun Ying 396
Tanaka, Shuji 139, 602, 661, 865, 873
Tanaka, Shuma 456
Tanaka, Yo 1013
Tang, Gongbin 631
Tang, Qi 507
Tang, Qiankai 889
Tang, Yuan-Sin 425
Tang, Yue 752
Tao, Chen 205
Tao, Kai 705, 740

Tao, Tiger H. **33, 67, 205, 209, 247, 297, 429, 807**

Tarhan, Mehmet C. **475**

Tas, Niels R. **639**

Tatar, Erdinc **869, 909**

Tatum, Lars P. **483**

Tavakkoli, Hadi **1095**

Tayagui, Ayelen **997**

Teng, Megan **135**

Terao, Kyohei **213, 775**

Teuber, Jeremy **1041**

Thakkar, Pooja **1127, 1131**

Thielen, Brianna **388**

Tian, Hao **149**

Tian, Ji'ao **841**

Tian, Xuedi **177**

Tian, Ye **33**

Tian, Yuxin **285**

Tichy, John A. **728**

Tiggelaar, Roald M. **639**

Toan, Nguyen V. **1103**

Toh, Wei Da **491**

Tohyama, Shugo **305**

Tomizawa, Yasushi **861**

Tong, Xing **728**

Tope, Sayali **657**

Torres Canals, Francesc **1041**

Torres, Francesc **933**

Tottori, Naotomo **271**

Tourrel, Guillaume **452**

Tran, Canh-Dung **366, 1064**

Tran, Dang **366**

Tran, Thomas **1072**

Travnik, Aleš **1107, 1123**

Tsai, Chia-Hsien **1194**

Tsai, Chun-Pu **181, 185, 487, 1155, 1206**

Tsai, Hsiao-En **425**

Tsao, Peggy **957**

Tsao, Pei-Chi **135**

Tsuchiya, Toshiyuki **515, 681**

Tsukamoto, Takashiro **661, 865, 873**

Tu, Kejun **405, 635**

Tu, Liangcheng **837, 841**

Tung, Shao-Siang **1198**

Turan, Bilal **57**

U

Uchida, Kengo 861
Uesugi, Akio 701
Umar, Muhammad 335
Umezawa, Seiji 941, 957
Uranga, Arantxa 933, 937
Uzunoglu, Baha Erim 869

V

van der Heiden, Maurits 153
van Driel, Willem D. 621
van Meer, Berend J. 374
van Zeijl, Henk W. 621
Vazquez, Irma Rocio 467
Verd, Jaume 795
Verma, Satish K. 689
Vescan, Andrei 627
Veske, Tolga 909
Vidal, Frédéric 452
Villanueva, Luis Guillermo 594
Vimercati, Michele 716
Vinayakumar, K.B. 744
Vo, Tuan N.A. 1045
Vollebregt, Sten 72, 131
Vollmann, Morten 925
Voloshin, Andrey 149
Vu, Trung-Hieu 366, 1064

W

Waag, Andreas 1037
Wada, Hiroki 80
Wan, Xiu-Feng 433
Wang, Baosheng 413
Wang, Changhai 598
Wang, Chen 901, 917
Wang, Chih-Hung 479
Wang, Cong 1029
Wang, Daying 103
Wang, Fang 351
Wang, Fei 259, 826, 893, 913
Wang, Han 33
Wang, Huan 323
Wang, Jiachou 881
Wang, Jiachuang 209

Wang, Jie	756
Wang, Kexin	598
Wang, Linlin	917
Wang, Longchun	384, 405
Wang, Luming	281
Wang, Man	578
Wang, Ning	103
Wang, Peng	849
Wang, Qian	841
Wang, Rui N.	149
Wang, Ruoqin	267, 752
Wang, Wei	243, 323
Wang, Wenduo	783
Wang, Xiangyang	949
Wang, Xiaoyi	267, 752
Wang, Xin	1147
Wang, Xiner	67, 205, 247
Wang, Xiong	413
Wang, Xueying	33, 297
Wang, Yang	157, 433, 921
Wang, Yiwei	413
Wang, Yuan	917
Wang, Yudong	263
Wang, Yunong	613
Wang, Yuntong	323
Wang, Yuxi	669
Wang, Zenghui	189, 281
Wang, Zetian	243
Wang, Zhuangzhuang	355
Wang, Ziji	542, 546
Waquier, Louis	153
Wasisto, Hutomo Suryo	973
Wei, Jian	1076
Wei, Shuai	807
Wei, Ting-Chou	123, 127, 685
Wei, Xiaoling	33, 67, 205, 247, 297, 429
Weidner, Michael H.	822
Weig, Eva M.	877
Weng, Wei-Yang	558
Wiegerink, Remco J.	767
Wiendels, Maury	374
Wienecke, Marion	822
Windt, Laura M.	374
Wittemeier, Steffen	522
Wolff, Anders	437

Wong, Man 13, 259, 748
Worsey, Elliott 507
Wu, Cheng-Yen 905
Wu, Feng 263
Wu, Guan-Lin 1194
Wu, Guoqiang 949
Wu, Jinbo 177, 1175
Wu, Junming 546
Wu, Junqiao 483
Wu, Lang 913
Wu, Lixiang 402
Wu, Mingching 685, 720
Wu, Sheng 881, 969
Wu, Shuxian 631
Wu, Tao 177, 413, 669, 969
Wu, Wengang 814, 1115
Wu, Wenjie 837, 841
Wu, Xuezhong 845, 849
Wu, Yi-Xin 1052
Wu, Zhenhua 1005, 1017
Wu, Zhenyu 157
Wu, Zhipeng 115
Wu, Zonglin 631
Wulf, Matthias 1139

X

Xi, Jingqian 917
Xi, Xiang 849
Xi, Ye 384
Xia, Fan 135, 941
Xia, Guoming 857
Xiao, Dingbang 845, 849
Xiao, Yuhao 1159
Xie, Dongcheng 263
Xie, Fei 157
Xie, Huikai 267, 752
Xie, Maosong 281
Xie, Yong 949
Xing, Chong 263
Xing, Xiaoxing 1076
Xing, Yan 606
Xu, Bo 189
Xu, Cheng 818
Xu, Chengjian 33
Xu, Feihong 247

Xu, Feng 631
Xu, Han 243
Xu, Huiping 1190
Xu, Jinghui 1159
Xu, Jiushuai 1037, 1147
Xu, Lei 263, 917
Xu, Liangge 818
Xu, Linbing 673
Xu, Mengfei 384, 405, 635
Xu, Pengcheng 119, 193, 255
Xu, Qingda 384, 405
Xu, Qingmei 323
Xu, Wei 107
Xu, Xiaochao 841
Xue, Ying 323
Xun, Lingxiao 452

Y

Yabuuchi, Kensuke 301
Yadav, Vinit K. 1033
Yalikun, Yaxiaer 1013
Yamada, Genki 213
Yamagata, Chisaki 319
Yamaguchi, Masakazu 1111
Yamanishi, Yoko 271, 417
Yamaoka, Utana 1135
Yamashita, Yu 271
Yanez, Jesús 937
Yang, Bin 384, 405, 635, 712, 783, 929
Yang, Gai 267, 752
Yang, Hao 744
Yang, Heng 351
Yang, Huiran 205, 297, 429
Yang, Lujia 841
Yang, Rui 281
Yang, Shijia 103
Yang, Suhui 1147
Yang, Tung-Lin 425
Yang, Xiaopeng 673
Yang, Zhenchuan 756
Yao, Hulin 177, 1175
Yapici, Murat Kaya 335
Ye, Xiongying 285
Ye, Yifei 33
Yeh, Chun-Chen 1194

Yeh, Hung-Yu 239
Yen, Da-Jen 1143
Yen, Ernest T.-T. 165
Yin, Kuibo 771
Yokokawa, Ryuji 301, 309
Yokota, Takahito 709
Yokoyama, Yoshiyuki 17, 80
Yoo, Dongwoo 327
Yoo, Han Woong 538
Yoo, Seong-Jae 88
Yook, Se-Jin 993
Yoon, Jun-Bo 5, 76
Yoshida, Koki 1084
Yoshida, Shinya 139
You, Minmin 929
Young, Douglas 554
Younkin, Duane 853
Yousuf, Mujeeb 347
Yousuf, S M Enamul Hoque 613, 897
Yu, Haitao 193
Yu, Hongyu 267, 752
Yu, Lei 845
Yu, Ling-Shan 448
Yu, Xiaomei 1202
Yu, Zhenyi 1190
Yuan, Chao 1186
Yuan, Weizheng 740
Yue, Wei 135, 503, 941, 957
Yue, Xiawei 807

Z

Zahed, Md Abu 197, 732
Zaki, Muhammad Faizul 1052
Zamora, Iván 933, 937
Zappa, Dario 251
Zega, Valentina 37, 989
Zeng, Fei 1190
Zhai, Yanfen 402, 1107, 1123
Zhang, Bohan 205, 297, 429
Zhang, Chao 736
Zhang, Chenchen 582
Zhang, Dacheng 885
Zhang, Guoqi 72, 621
Zhang, Haitao 945
Zhang, Haochen 263

Zhang, Haoran	1060
Zhang, Haozhi	119
Zhang, Hemin	621, 865
Zhang, Jian	705, 740, 837
Zhang, Jianfeng	542
Zhang, Jin	157, 542, 546
Zhang, Jinwen	921
Zhang, Jinying	1147
Zhang, Kuikui	33
Zhang, Lejia	413
Zhang, Liping	177, 1175
Zhang, Meixuan	243
Zhang, Menglun	673, 677, 945
Zhang, Pengcheng	281
Zhang, Pingping	209, 807
Zhang, Qifu	285
Zhang, Shibin	177, 1175
Zhang, Shipeng	217
Zhang, Shuai	1190
Zhang, Xin	783
Zhang, Xinyuan	267
Zhang, Yan	263
Zhang, Yao	491
Zhang, Yi	323
Zhang, Yonggui	157
Zhang, Yucheng	402
Zhang, Yulong	84
Zhang, Yuyao	157
Zhang, Ziqi	503
Zhao, Chun	841, 917
Zhao, Jianlong	103, 289, 1005
Zhao, Lurui	965
Zhao, Nan	355
Zhao, Ning	783
Zhao, Qiancheng	833, 885
Zhao, Xiaomeng	177, 1175
Zhao, Xu	1095
Zhao, Yang	857
Zhen, Liyun	712
Zheng, Gang	452
Zheng, Hong-Sen	487, 1155
Zheng, Kevin H.	173
Zheng, Pengcheng	177, 1175
Zheng, Yue	1186
Zhou, Changdong	285

Zhou, Cunkai	33
Zhou, Hang	309
Zhou, Hong	818, 1167
Zhou, Lin	289
Zhou, Min	1175
Zhou, Na	229, 582
Zhou, Xiaoyong	107
Zhou, Xin	845
Zhou, Yangchao	945
Zhou, Yufan	255
Zhou, Zhanxuan	1202
Zhou, Zhitao	33, 67, 205, 247, 297, 429, 807
Zhu, Jiankai	189
Zhu, Linqian	889
Zhu, Xiantao	929
Zhu, Zhezheng	756
Zhu, Ziyi	205
Zhuang, Yi	826, 913
Zimmermann, Sven	522
Zou, Dujuan	205
Zou, Jie	631
Zymelka, Daniel	359, 400

IEEE
445 Hoes Lane
Piscataway, NJ 08854-4141

ISBN 978-1-6654-9309-3